Genetics and Analysis of Quantitative Traits

Genetics and Analysis of Quantitative Traits

Foundations

Bruce Walsh
University of Arizona

Peter M. Visscher
University of Oxford
University of Queensland

Michael Lynch
Arizona State University

OXFORD
UNIVERSITY PRESS

OXFORD
UNIVERSITY PRESS

Great Clarendon Street, Oxford, OX2 6DP,
United Kingdom

Oxford University Press is a department of the University of Oxford. It furthers the University's objective of excellence in research, scholarship, and education by publishing worldwide. Oxford is a registered trade mark of Oxford University Press in the UK and in certain other countries.

First Edition published in 2018
Second Edition published in 2025

Published in the United States of America by Oxford University Press
198 Madison Avenue, New York, NY 10016, United States of America

British Library Cataloguing in Publication Data
Data available

Library of Congress Control Number: 2025933531

ISBN 9780192898180

DOI: 10.1093/oso/9780192898180.001.0001
Printed in the UK by Bell & Bain Ltd., Glasgow

The manufacturer's authorised representative in the EU for product safety is Oxford University Press España S.A., Parque Empresarial San Fernando de Henares, Avenida de Castilla, 2 – 28830 Madrid (www.oup.es/en or product.safety@oup.com). OUP España S.A. also acts as importer into Spain of products made by the manufacturer.

We dedicate this book to current, and future, students of quantitative genetics.
Our task will have been successful if these practitioners build on the basic
concepts presented here to make many of the methods we discuss obsolete.

Contents — Part I (Foundations)

PREFACE

In the late 1980s, we noticed the start of a major phase shift across several subfields of quantitative genetics. The work of Russ Lande and colleagues fundamentally transformed the field of ecological genetics, shifting its focus from single genes to the analysis of biologically relevant traits. The powerful linear mixed models developed by Charles Henderson for animal breeders were starting to leak into other subfields of quantitative genetics, and the growing availability of molecular markers fueled both the hope of actually finding underlying polygenes and the promise of pedigree-free quantitative-genetic analyses. These, and other, developments sparked our desire to provide a modern textbook reviewing both what the dynamic field of quantitative genetics had become and what it was evolving into. The result was *Genetics and Analysis of Quantitative Traits* (Lynch and Walsh 1998). Our timing was fairly auspicious, in that at the date of its publication, linkage-based QTL mapping had reached its zenith, mixed-model and Bayesian usage had greatly expanded, and previously cloistered ideas from various subfields of quantitative genetics were becoming more fully appreciated across the broader enterprise. Twenty years later, we followed this first volume with a second, *Evolution and Selection of Quantitative Traits* (Walsh and Lynch 2018), a major treatment ($\sim$1400 pages) focusing on the population genetics of selection, and selection response in both natural and applied (i.e., breeding) settings.

In the twenty-five years since the publication of our first volume, a second renaissance has occurred, jointly fueled by increased ease in obtaining very densely spaced DNA markers for just about any organism (culminating in whole-genome sequencing) and the accompanying development of high-dimensional statistical tools to handle the resulting tsunami of new data. Linkage-based QTL mapping (using line-cross or pedigrees) was rapidly supplanted by GWAS (LD-driven mapping using very large population samples). Genomic selection and prediction transformed applied breeding and teased medical clinicians with the hope of precision (i.e., individual genome-based) medicine. Finally, the realization that quantitative genetics applies just as well to genomic features (such as mRNA expression or chromatin configuration) as it does to milk production resulted in an explosion of articles with a very quantitative-genetic feel appearing in the leading genomics journals. Hence, a revision of our 1998 book was badly needed. Once again, it appears that our timing is somewhat fortunate, as many of these developments have matured to the point that they are rather stable and highly important.

Throughout, we have attempted to develop central theoretical concepts from first principles. In part, this approach is based on the fact that many recent "advancements" in the field are either natural extensions of existing theory or novel options presented by having access to new types of data. To aid the less statistically sophisticated reader, we have included several chapters and appendices that review essentially all of the statistical tools employed in the book. Wherever possible, we have illustrated theoretical and analytical concepts with empirical examples from diverse settings (e.g., microbes, plants, animals, humans; laboratory, agricultural, domesticated, natural, and clinical populations). As with the first edition, we have tried to structure this book in such a way that encourages its use as a general textbook for various subfields of quantitative genetics (Table 1.1 details different chapter pathways for topic-specific treatments). We also hope that it provides a thorough enough coverage of the literature to prove useful as a standard reference for the field. To aid its use as a general reference, we provide extensive (some would say compulsive-level) indices of both organisms/traits and subjects to help the reader quickly drill down into the relevant sections of interest for any particular topic, and to show how seemly distant topics are often deeply connected.

The massive amount of developments over the past three decades has resulted in the expansion of our 1998 book (which had slightly under 1,000 pages) into a treatment in two parts. Part I (*Foundations*) is a little over 1,200 pages, and covers the foundations of quantitative genetics—Chapters 1–16 in our 1998 treatment—from statistical and genetical basics up thorough GWAS and genetical genomics. Part II (*Estimation and Prediction*, forthcoming), will cover variance component estimation under a variety of classic designs, G × E, maternal effects, threshold and multivariate traits, and genomic selection/prediction (Chapters 17–27 in our 1998 book). It will also add an extensive treatment of multivariate selection and specialized topics (such as line breeding), filling a gap of maternal not covered in our 2018 book.

Essentially every chapter of *Foundations* has been fully rewritten, significantly expanded, and updated. Five new chapters have been added: molecular-based estimation of relatedness (Chapter 8), a new chapter on linear and mixed models (Chapter 10), a new chapter on heterosis and outcrossing (Chapter 13), and two new genomics-based chapters: GWAS (Chapter 20) and quantitative genomics (Chapter 21). We have also expanded and updated the Appendices (from five to nine). Appendix 2 has extensive new material on causation analysis and Mendelian randomization, and we have added appendices on multiple comparisons and meta-analysis (Appendix 6), Bayesian approaches (basics in Appendix 7 and MCMC in Appendix 8), and finally an overview of experimental design (Appendix 9).

Given the vast diversity of this enterprise of modern quantitative genetics, we have tired to be fastidious in tracking down original sources of many of the key concepts, some of which can have rather obscure origins. We have tried to be as faithful as possible in crediting both ideas and presentations to specific authors. However, in many areas our treatment is a gestalt from a large number of very diverse authors (commonly from many different fields), often making attribution of concepts and ideas to specific individuals challenging and problematic. Hence, many treatments have rather extensive literature citations, reflecting the diversity of source material used to shape our presentations.

Acknowledgments

We have been blessed to have wonderful teachers and colleagues to help us navigate the incredibly diverse field of quantitative genetics, ranging from population and evolutionary geneticists, plant and animal breeders, statisticians, and human geneticists. A partial (but not exhaustive) list (we deeply apologize to colleagues we have forgotten) includes Matt Ackerman, Stevan Arnold, Reinhard Burger, Brian and Deborah Charlesworth, Jim Crow, Hong-Wen Deng, Joe Felsenstein, Wilfried Gabriel, Daniel Gianola, Sascha Gimelfarb, Michel Georges, Mike Goddard, Chris Haley, Thomas Hansen, David Houle, Elizabeth Housworth, Alex Kondrashov, Russell Lande, Nick Martin, Bill Muir, Tom Nagylaki, Hans-Peter Piepho, Tim Prout, Kermit Ritland, Chris Schön, Monty Slaktin, Ken Spitze, Robin Thompson, Michael Turelli, Michael Wade, Bruce Weir, John Willis, Jian Yang, and Loïc Yengo. We also thank Andy Sinauer for his considerable encouragement during the writing of our 1998 and 2018 books.

Of special note is the late William Hill, whose contribution to our understanding of quantitative genetics cannot be overstated. Bill was a massive cheerleader and invaluable reviewer for both the 1998 version and our 2018 book, and was still contributing valuable input to the revision until his untimely passing. In many respects, we have tried to create a text to help one think like only Bill could on a vast array of seemly unconnected topics.

Besides these general comments, we also have some author-specific acknowledgments. BW wishes to thank the over 5,000 students from roughly 60 countries who attended his various short courses over the last three decades at locations around the world (25 countries at last count). These international interactions and feedback greatly sharpened his ability to present the material in this book to a truly universal audience (both in terms of academic and cultural backgrounds). While roughly half of these courses were single instructor, he greatly benefited from joint teaching in several long-running programs. These include Bruce Weir's Summer Institute in Statistical Genetics (Trudy Mackay, Marta Wayne, Bill Muir, Guiherme Rosa, Dahlia Nelson, Bruce Weir), the African Plant Breeding Academy (Rita Mumm, Allen van Deynze, Iago Hale), and the Tucson Plant Breeding Institute (Mike Gore,

Lucia Gutiérrez). Finally, my deepest thanks, appreciation, and love to my wonderful wife Lee (Fulmer), who has most generously put up with all my travel and time when working on these various book and teaching projects.

PMV would like to acknowledge the great lifelong mentorship and friendship from the late Bill Hill. Bill's insights, knowledge and memorable phrases ("to which scientific question is this study/paper the answer" and "quantitative genetics isn't dead just yet" come to mind) were well beyond the upper tail of the normal distribution. He also thanks the many colleagues, postdocs and students with whom discussions have broadened his understanding—so many that it is impossible to mention them all. A particular thanks to colleagues who have become dear friends: Mike Goddard, Robin Thompson, Chris Haley, Nick Martin, Josephine Pemberton, Jian Yang, and Loïc Yengo. Last but not least a big thanks to Naomi (Wray)—spouse, mother of their wonderful children, fellow quantitative geneticist and collaborator. Theirs has been an amazingly rewarding joint journey from animal breeding to human genetics, and very satisfying to see that the theory, applications and analysis tools of quantitative genetics are now firmly embedded in studies of human traits and disease.

ML would like to thank the many dozens of graduate student and postdoctoral associates who have passed through his lab, teaching him much more than he likely taught them. While he was at the University of Illinois, John Dudley, Rohan Fernando, Daniel Gianola, Charles Henderson, and Michael Grossman eased his entry into this challenging field; and former department heads Lowell Getz and George Batzli facilitated this transition when ML should have been doing other things that he was hired for. As with BW and PV, Bill Hill also had an indelible impact from an early age.

Finally, we thank people (besides those listed above) who have made comments on one or more chapters: From the University of Queensland (Institute for Molecular Bioscience): Valentin Hivert, Julia Sidorenko, Huanwei Wang, Chloe Yap, Joana Revez, Kathryn Kemper, Yuna Zhang, Beilei Bian, Allan McRae, Mandy Moore, Gunn-Helen Moen, Mulusew Fikere, and David Evans; and Matthew Keller from the University of Colorado (Institute for Behavior Genetics). Again, we apologize to colleagues we have forgotten to list, as this has been essentially a whole-of-community project.

Bruce Walsh, Tucson

Peter Visscher, Brisbane and Oxford

Michael Lynch, Tempe

October 2024

I

Foundations of Quantitative Genetics

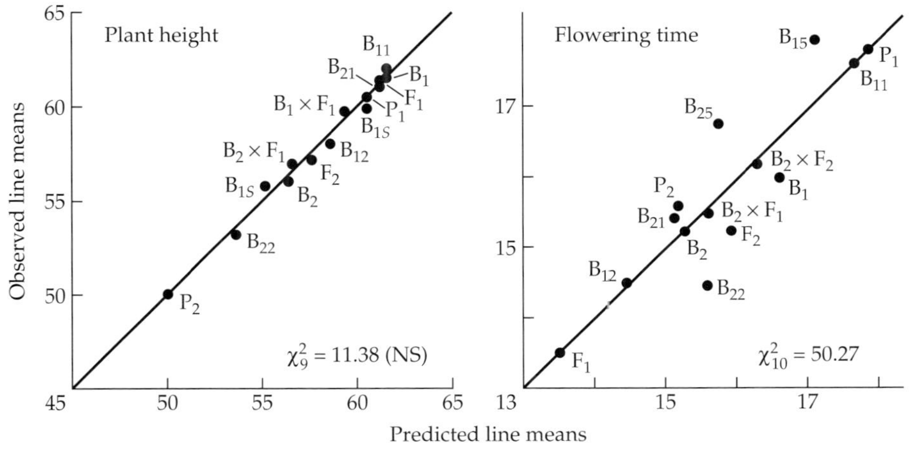

1

An Overview of Quantitative Genetics

The next major explosion is going to be when genetics and computers come together.
Alvin Toffler

Any trait or feature that can be measured and assigned a value can be considered a **quantitative trait**, and the analysis of such traits forms the field of **quantitative genetics** (**QG**). The initial focus of QG was on continuously valued traits such as human height, agricultural production characters (such as grain and milk yield), and morphological features in natural populations of plants and animals. However, QG easily extends to discrete-valued characters, such as binary (presence/absence) traits, most notably human disease risk. As the field continued to mature, QG moved beyond its production and morphological trait roots into much wider categories, such as behavioral and physiological features, as well as from metazoans to single-cell organisms (such as yeasts). Further, QG machinery naturally extends to any measurable *genomic feature*, such as mRNA levels and splicing patterns, methylation status, or the configuration of binding factors on chromatin. As a result, while historically quantitative genetics formed the foundations for plant and animal breeding, evolutionary and ecological genetics, and human genetics, it has now become an integral part of the genomic, and post-genomic, revolutions. Finally, the field is strongly tied to advances in statistical methodology, many of which were motivated by questions arising from the analysis of complex traits (Appendices 2–9).

Most quantitative traits show variation, either within, or between, populations. For **simple** traits, this variation can be largely due to the action of one or two genes, whose genotypic signals overpower most of the environmental (and genetic background) noise. The analysis of such simple traits is a special subset of quantitative genetics. However, most traits are **complex** in the sense that their variability reflects a distribution of expected trait values over a large (and often *very* large) number of different genotypes, overlaid with considerable noise from environmental effects.

For much of its early history, quantitative-genetic machinery (while built upon a Mendelian foundation; Chapters 4–6) was rather agnostic as to the genetics underlying a focal trait, relying just on observations of phenotypes in sets of relatives (Chapter 7). This robust aspect of QG was regarded as a strength, as no genetic details were needed in order to proceed with a wide range of very useful analyses and predictions. This avoidance of any fine genetic details was not a deliberate oversight, but rather an operational decision imposed by the information (or lack thereof) available at the time. However, starting in the early 1980s, the field quickly embraced the additional information emerging from a modest number of molecular markers (such as allozymes, RFLPs, and STRs), then moved on to exploiting abundant SNP information, and finally to using **whole-genome sequencing** (**WGS**, Chapters 8, 17–21). As it became apparent that much of trait variation was not the result of variation in protein sequences, but rather due to subtle variation in regulation, QG fully embraced emerging concepts from functional genomics, to the point where many QG analyses are as focused on upstream molecular features as they are on the downstream final trait values (Chapter 21).

As a result of this fusion of classical quantitative genetics, modern statistics, and genomic/post-genomic data, quantitative genetics today is more important than at any time in its history. QG is the ultimate Phoenix, having risen from (according to its critics) the ashes of irrelevancy numerous times, only to become stronger, more integrated, and more relevant. In this volume, we attempt to not just present the important machinery in the analysis of complex traits but also take the reader on a tour of QG's maturation, from its founding in 1918 by Fisher into a field fully intermeshed with genetics, genomics, evolution, applied breeding, statistics, and bioinformatics.

This book represents the first half of our revision of Lynch and Walsh (1998). Because of the massive growth of the field over the past two decades, and the corresponding increase in material, our complete revision had to be divided into two parts. Here (Part I, consisting of Chapters 1–21 and Appendices 1–9), we cover foundational topics such as basic statistical issues, variance decompositions, resemblance between relatives, line crosses (including inbreeding and heterosis), and the use of marker/sequence information for relatedness estimation, QTL linkage mapping, and GWAS. Part II (starting with Chapter 22) will complete our revision by covering estimation issues, specific designs for using resemblance between relatives, BLUP/REML and genomic prediction, and specialty topics such as G x E, correlated and threshold traits, maternal effects, and so on. We have deliberately written Part I to be self-contained, not requiring any information developed in Part II (although chapters from Part II that further develop specific topics will be referenced here).

We start with a brief overview of the history, and interplay, between classical, quantitative, population, and molecular genetics. We then provide an overview of the rest of the book by discussing the major problems of interest to practitioners of quantitative genetics. The focus of our revision of Lynch and Walsh (1998) is on the genetics and analysis of quantitative traits, while their evolution and selection (for example, various breeding schemes) are examined in considerable detail in Walsh and Lynch (2018; henceforth WL).

A SHORT HISTORY OF QUANTITATIVE GENETICS

From Darwin and Mendel to Fisher 1918

Modern biology rests on the twin pillars of evolution offered by Darwin in his *Origin of Species* (1859) and genetics postulated by Mendel in his *Experiments in Plant Hybridization* (1865). While these two seminal works appeared just six years apart, their impact was felt over very different timescales. Darwin's work had an immediate impact in reshaping biology, but Mendel's work was ignored until being independently discovered by multiple groups 35 years later (DeVries, Correns, and von Tschermak, all in 1900). As a result, the early schism that developed between the Darwinian and Mendelian camps (largely centered around different assumed models of inheritance for discrete versus continuous traits) lasted for almost two decades, with quantitative genetics emerging, in part, as a resolution and unification of these two grand concepts.

In order to explain the approximate intermediacy of progeny phenotypes with respect to those of their parents, Darwin (1859) assumed that continuous characters exhibited **blending inheritance**. This common model of inheritance in Darwin's time postulates that parents mix some kind of biological vital fluid to generate the phenotype of their offspring. Fleeming Jenkin (1867), in what was the first paper in population genetics, pointed out a very serious problem with this model of inheritance. In his review of Darwin's *Origins*, Jenkin (Regius Professor of Engineering at the University of Edinburgh) noted that, under blending inheritance, ***half of any variation is removed each generation***, implying the need of some force to generate sufficient new variation to offset this loss.

Phrased in terms of modern statistics, and recalling (Chapter 3) that the variance of a product of a constant (a) times a random variable (x) is $\sigma^2(ax) = a^2\sigma^2(x)$, his argument was that

$$\sigma^2(z_o) = \sigma^2\left(\frac{z_f + z_m}{2}\right) = \frac{\sigma^2(z_f)}{2^2} + \frac{\sigma^2(z_m)}{2^2} = \frac{\sigma^2(z)}{2}$$

where z_f and z_m are the paternal and maternal trait values (assumed to be uncorrelated), z_o is the offspring value, and we assume that the trait variance is the same in both sexes, so that $\sigma^2(z_f) = \sigma^2(z_m) = \sigma^2(z)$. Thus, any change achieved via natural selection would quickly be diluted away by the assumed nature of inheritance (also see Bulmer 2004). Mendelian genetics provides the solution to this quandary of reduced variation following reproduction. The particulate nature of genes implies that the variation generated by the transmission

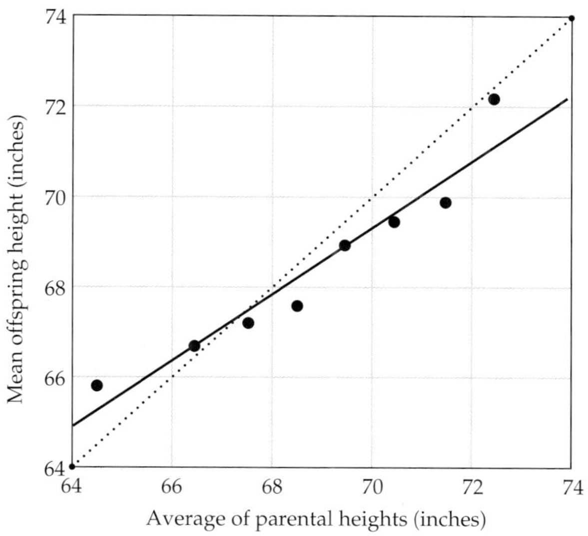

Figure 1.1 The relationship between height of adult children and the average height of their parents (in inches). Circles denote average offspring heights for different 1-inch classes of midparent heights. The best linear fit is given by the solid line, while the dotted line is the expected pattern if mean offspring height were equal to midparent height. (Data from Galton 1886.)

of alternative alleles in the gametes of heterozygous parents (the **segregation variance**) completely restores the genetic variance in each generation (Chapter 8).

Francis Galton, a cousin of Darwin, pointed out another problem. When he plotted the mean heights of offspring (measured at adult age) against the average height of their parents (corrected for differences between the sexes), he obtained a linear relationship (Figure 1.1). However, the slope of the line (being less than one) indicated that offspring were, on average, *less exceptional* than their parents. Parents whose average height was below the mean tended to have offspring taller than themselves but still below the mean. Parents above the mean tended to have offspring shorter than themselves. Galton (1886) called this trend **regression toward mediocrity** (and hence the origin of the term **regression** for such an analysis), further arguing that this trend would erode away any selective progress. He then concluded that evolution must be based upon **sports** (mutations of large effects) rather than on selection acting upon continuous variation. Galton's quantitative approach to the analysis of inheritance marked the founding of the **biometrical school**, from which most of modern statistics can be traced (Stigler 1986; Crow 1993a).

Karl Pearson (1903) pointed out the fault in Galton's logic. Imagine that selection acts in such a way that only the largest parents reproduce, so that the mean height of reproductive adults after selection (μ_s) is greater than the mean height before selection (μ) (Figure 1.2). If μ_o is the average adult height of the offspring of the selected parents, then the **response to selection** across generations is $R = \mu_o - \mu$. Because the slope of the parent–offspring relationship is less than one, the adult offspring are not expected to be as tall (on average) as their selected parents. However, the mean offspring height will still be *greater than the mean height in the parental base population*. Furthermore, the new mean height is stable: if selection on height is stopped, the new height is not expected to decay back to the original value. Thus, the regression toward mediocrity does not pose a serious problem for the theory of evolution by natural selection. Galton apparently failed to realize that selection starts from an *updated mean* each generation and that it is to this new mean value to which the next generation of selection regresses. Although in disagreement with some of Galton's interpretations, Pearson was greatly inspired by Galton's quantitative approach to analysis and went on to develop many methods for the analysis of continuously distributed traits (such as regression and correlation).

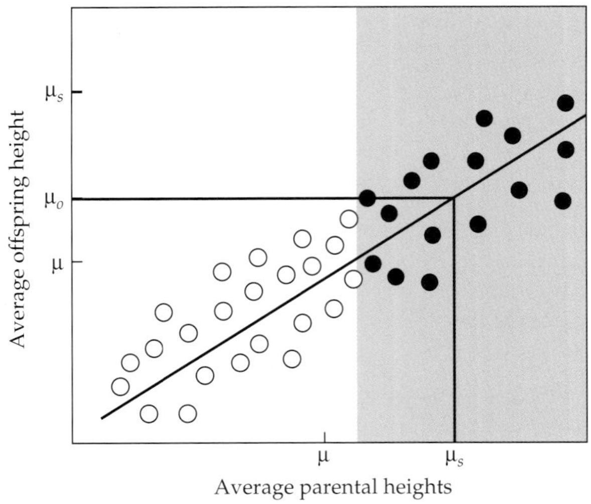

Figure 1.2 Response to selection. The solid line is Galton's regression. The closed circles represent selected parents. In the absence of selection, the mean height for both the parents and offspring is μ. However, if the average height of selected parents is μ_s, the expected height of offspring (μ_o), obtained by reading off the regression, is greater than μ but less than μ_s.

At the time of Mendel's rediscovery in 1900, most of the focus on trait inheritance had been on *continuous* traits, namely those not clearly separable into discrete classes (e.g., Galton 1886, 1889). In contrast, Mendelian genetics centered attention on the inheritance of *discrete* characters such as purple vs. white flower color, smooth vs. wrinkled seeds, and so on. Critically, Mendelian inheritance postulated that genes are *discrete particles*, and hence their heritable components can be recovered in future generations. Imagine crossing red and white parents to create a pink offspring. Under blending inheritance, pure red or pure white individuals could not be recovered for a cross of these offspring as the blended fluid generating a pink offspring would not naturally resolve into its two separate fluid components. In contrast, if a pink offspring is the result of it carrying both a single red and a single white particle (the two alleles), these segregate cleanly in future generations, allowing for the potential of pure red, or pure white (as well as pink) offspring.

Rather than realizing the power of merging the quantitative analysis of inheritance with the particulate nature of transmission, a series of contentious debates instead ensued between the **Mendelians** (led by William Bateson) and the **Biometricians** (led by Pearson and W. F. R. Weldon). The major issue was whether discrete characters had the same hereditary and evolutionary properties as continuously varying characters. The Mendelians held that variation in discrete characters drove evolution through the appearance of new **macromutations** (mutations with large effects, akin to the appearance of sports suggested by Galton). In contrast, the biometricians thought that evolution occurred in very small steps by exploiting continuously distributed variation. In essence, the Mendelians felt that variation in continuous traits was not heritable, while the biometricians felt that it was. This failure to connect the key concepts of inheritance (Mendel) and evolution (Darwin) was largely driven by antagonistic personalities in the different camps (Provine 1971, 2000; Tabery 2004; but see Hogben 1974 for a different perspective), rather than by facts. This animus significantly delayed the fusion of evolution by natural selection with Mendelian inheritance, which would ultimately become known as the **modern synthesis**.

That continuous trait variation could have a discrete (Mendelian) underlying basis was suggested by the British mathematician G. Udny Yule, who gave formal proof for this idea in 1902. Unfortunately for Yule, the only thing that the biometricians and the Mendelians could publicly agree on was the incompatibility of Mendelian genetics and the inheritance of continuous characters. The death of Weldon in 1906, followed by the publication of several key plant breeding experiments from 1908 to 1916, resulted in the rapid emergence of the

multiple-factor hypothesis: continuous variation in a trait is generated, at least in part, by the impact of multiple underlying genes.

George H. Shull, a major figure in American corn breeding (and founder of the journal *Genetics*), noted that self-fertilized corn strains were remarkably uniform in many continuous traits when compared to the outbred populations from which they are derived. His (Shull 1908) explanation (with modern terminology inserted and italicized) was that

> The obvious conclusion to be reached is that an ordinary cornfield is a series of very complex hybrids (*genotypes*) produced by the combination of numerous elementary species (*alleles*). Self-fertilization soon eliminates the hybrid elements (*removes the heterozygosity*) and reduces the strain to its elementary components (*each locus becomes homozygous, resulting in each inbred strain being composed of a single genotype*).

The major implication of Shull's observation was that variation for continuous characters that is lost upon inbreeding must have a Mendelian basis.

Additional support for the multiple-factor hypothesis came from H. Nilsson-Ehle (1909), a Swedish geneticist working with various cereal crops. Many of the characters that he examined yielded 3:1 ratios in the F_2 generation following the cross of two parental strains, consistent with expectations for a single segregating locus with one allele completely dominant over the other. However, there were some striking exceptions. For example, when red-seeded and white-seeded wheat strains were crossed, the F_1 progeny were identical in color and intermediate between the parents, but a diversity of colors, ranging from white to red, was seen in the F_2. Further, in some of the F_2 crosses, a ratio of 63 red to 1 white seeds was observed. Nilsson-Ehle interpreted this to be the result of the segregation of three independent factors, the initial parents being *AABBCC* and *aabbcc*, all members of the F_1 being *AaBbCc* and hence uniform in color, and the F_2 consisting of all possible genotypes, only one of which (*aabbcc*) gives rise to white seed. The probability of obtaining an *aabbcc* offspring from an *AaBbCc* × *AaBbCc* cross is $(1/4)^3 = 1/64$.

From these results, Nilsson-Ehle arrived at two general conclusions. First, ***sexual reproduction can produce a huge diversity of genotypes.*** For example, because a locus with two alleles (*A* and *a*) can produce three genotypes (*AA*, *Aa*, and *aa*), ten diallelic loci can produce $3^{10} \simeq$ 60,000 genotypes. Second, given this huge potential diversity of genotypes, ***apparently new phenotypes appearing within a population may be the result of rare segregants rather than new mutations.*** Nilsson-Ehle, and independently East (1910), offered this as an explanation of **atavisms**, rare individuals that appear to be throwbacks to some previous population. Consider a hybrid population resulting from the cross of two pure (i.e., fully inbred) lines differing at ten loci. The probability of obtaining a specific parental genotype in the F_2 or later generation is $(1/4)^{10}$. Hence, the population size has to be greater than $4^{10} \simeq 10^6$ for there to be much of a chance of observing a parental type. The common observation of **transgressive segregation** (Chapter 18), wherein some F_2 offspring can be more extreme than their parents, is also a natural outgrowth of segregation over multiple loci. If parents are fixed for a mix of plus and minus alleles (increasing, and decreasing, trait values, respectively), then segregation can generate offspring with more plus (or minus) alleles—and hence more extreme phenotypes—than found in either parent.

Subsequent studies quickly confirmed the ideas of Shull and Nilsson-Ehle. For example, East (1911, 1916) and Emerson (1910; Emerson and East 1913) examined quantitative variation in a large number of plants. Typically, strains differing widely in some character were crossed and the variance of the resulting F_1 and F_2 generations recorded. In most of these crosses, especially when the parental populations were formed by repeated self-fertilizations, an **outbreak of variation** was seen in the F_2 (Figure 1.3). Such outbreaks of variation, resulting from the segregation of multiple genotypes from the F_1 heterozygotes, are consistent with the Mendelian model, but completely inconsistent with any blending hypothesis. An extensive historical review of the experimental verification of the multiple-factor hypothesis is given in Chapter 15 of Wright (1968).

The Danish botanist Wilhelm Johannsen (1903, 1909) was among the first to demonstrate that some of the variation in continuous characters was due to environmental rather

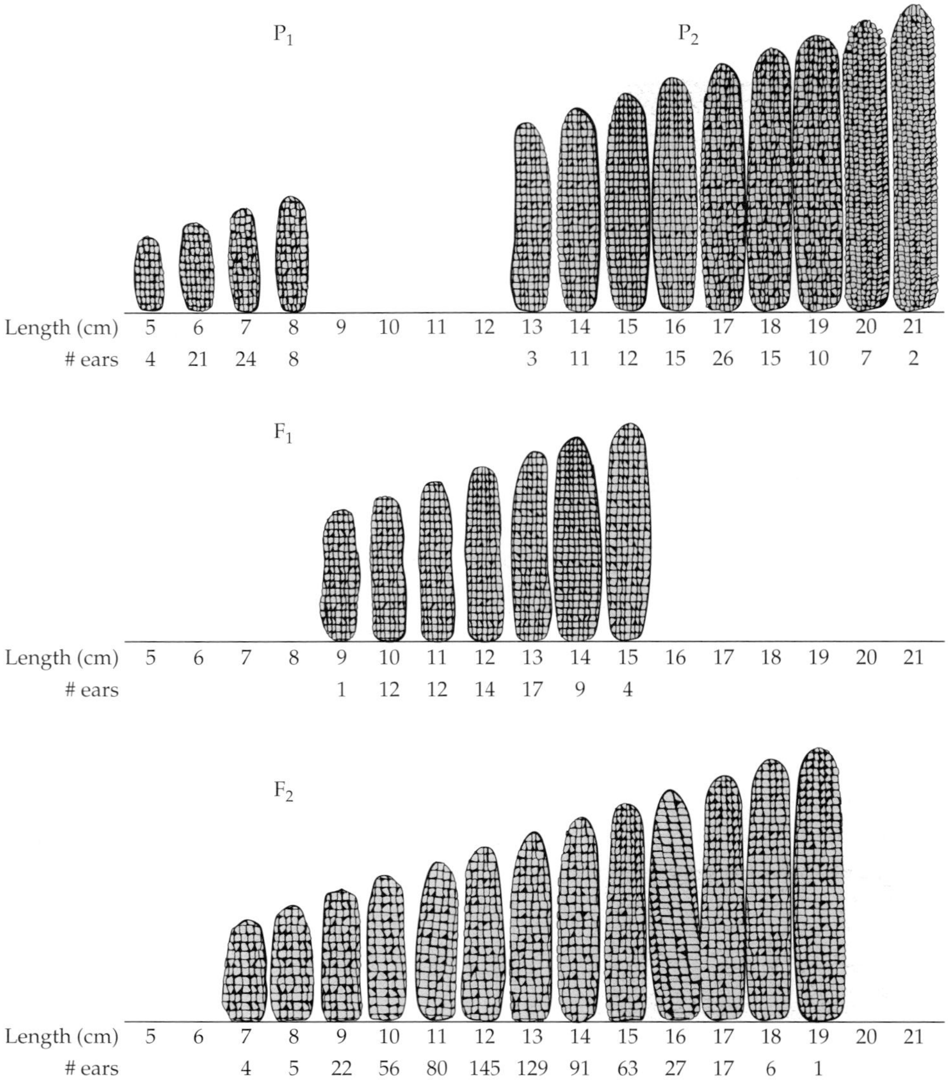

Figure 1.3 The distribution of ear size in the F_1 and F_2 generations formed by crossing two inbred lines of corn differing in ear length. The observed number of ears is given below each size class. The variation seen in the P_1, P_2, and F_1 populations is due entirely to environmental factors, as all individuals in each population have the same genotype, and these three populations show roughly similar amounts of variation. In contrast, the F_2 generation shows considerably more variation, reflecting the diversity of genotypes in this population generated by segregation of genes in the F_1 parents. (Data from East 1911.)

than genetic causes. Starting from a common stock of beans, he produced several inbred lines. For each line, parental seeds of different weights were planted and the mean seed weight in the offspring measured. Johannsen observed that the variation *within a pure line* was not heritable: the mean seed weights of the offspring were essentially independent of the weights of the parental seed (Figure 1.4). In contrast to Galton's data (Figure 1.1), parent–offspring regressions *within lines* were flat (had slopes of zero). The lack of a within-line association in Johannsen's data arose because all parents within a line were genetically identical, in contrast to humans where parents were genetically different. To clarify the

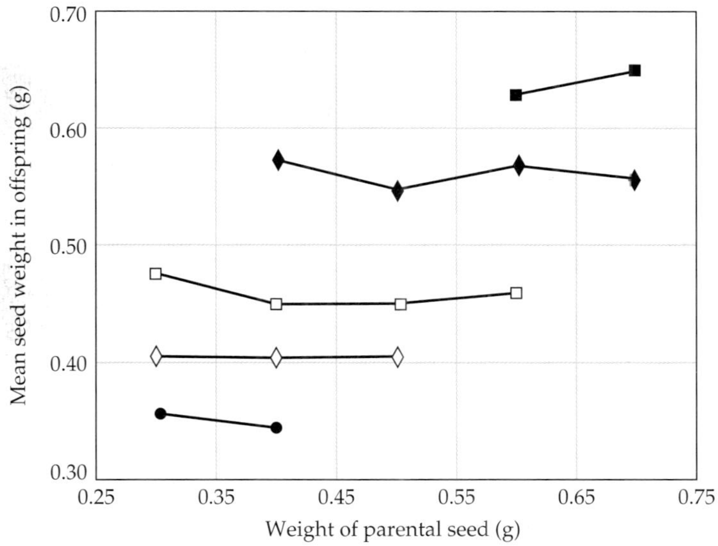

Figure 1.4 Mean offspring seed size as a function of parental seed size for some of Johannsen's pure lines. The data for the different lines are denoted by different symbols. If there is a heritable component to seed weight within a pure line, a line with positive slope is expected, namely, larger parents should yield larger offspring. However, within each line, mean offspring size is essentially independent of the parental phenotype, reflecting the lack of within-line genetic variation. (Data from Johannsen 1903.)

distinction between genetic and environmental effects, Johannsen coined the terms **genotype** (to denote genetically identical members of a pure line) and **phenotype** (the actual observed value for an individual, a compounding of genetic and environmental effects). From this observed lack of within-line selection response, Johannsen concluded that natural selection could never move a character value beyond the level of variation seen in the original population. Like Galton, he felt that macromutations were essential for evolution.

However, Payne (1918) soon demonstrated that selection on *Drosophila* bristle number could result in flies with more extreme phenotypes than those seen in the base population (essentially transgressive segregation). Such a result has been observed in many selection programs involving economically important species of plants and animals: almost always, the range of observed variation underestimates the range of potential variation, often dramatically so (WL Chapters 24–26). By increasing the frequency of favored alleles, selection increases the probability of observing extreme genotypes. Recall from above that if the frequency of the favored allele at each of ten loci is 0.5, then the frequency of the most extreme genotype (a homozygote at all ten loci) is approximately one in a million. However, if selection advances the frequency of the favored allele at each locus to 0.9, the frequency of the most extreme genotype becomes $(0.9^2)^{10} \simeq 0.12$. In other words, about one in eight individuals would exhibit the most extreme genotype.

Shortly after the rediscovery of Mendel's work, a new branch of quantitative biology started to be developed based on the consequences of Mendelian segregation of discrete particles (alleles) in a population. Initially, some geneticists thought that an allele showing dominance would eventually take over a population. Hardy (1908) and Weinberg (1908) showed that this was not the case, marking the formal beginnings of **population genetics**, the science of the behavior of alleles (and genotypes) under various evolutionary forces. Population genetics is highly intertwined with, but still distinct from, quantitative genetics. The former is built around allele and genotype frequencies, while the latter is concerned with the behavior of traits, which depends on *the effects of alleles* (and genotypes) in addition to their frequencies. WL Chapters 2–11 give a comprehensive review of modern population genetics, and detail the merger of population and quantitative genetics.

Fisher 1918: The Formal Foundation of the Quantitative Genetics

The growing consensus by 1916 in favor of the multiple-factor hypothesis set the stage for R. A. Fisher's (1918) brilliant paper, *On the correlation between relatives on the supposition of Mendelian inheritance*. This paper formed the conceptual bridge between Mendelian inheritance with discrete genes and evolution acting on continuous traits (ideas hinted at earlier by Yule 1902, 1906). It was also extremely foundational in the establishment of statistics. The term *variance* was first introduced in this paper, along with the method of **ANOVA** (**the analysis of variance**) that would lead to Fisher's development of the field of **experimental design** (Fisher 1935). Despite its enormous importance, this paper (written in 1916 by Fisher when he was a high school teacher) was initially rejected by the Royal Society of London. As Crow (1972) succinctly stated, "It was apparently too mathematical for the Mendelists and too Mendelian for the biometricians." Norton and Pearson (1976) published the original reviews of Fisher's paper by Karl Pearson and R. C. Punnett, noting that formally Fisher withdrew the paper, rather than having it rejected. It was eventually published by the Royal Society of Edinburgh with the help of Charles Darwin's son, Leonard.

Fisher's paper was built around three key ideas. First, (diploid) parents transmit only *part* of their genotypic value to their offspring, as they only pass along *single alleles*, not *diploid genotypes* (Fisher's decomposition of the genotypic value; Chapter 4). Second, appropriate summary statistics (the **additive genetic** and other **variance components**) largely describe the resemblance between relatives (Chapter 7). Lastly, these variance components can be estimated using phenotypic resemblance between known sets of relatives (Chapter 7, and Chapters 22–33 in Part II).

Fisher used a bit of a simplification for his work, the so-called **infinitesimal model**, which assumes that a large number of genes, each of very small individual effect, underlie most trait genetic variation. His notion of such a genetic model has been often, incorrectly, seized upon as proof that QG is a bit out of touch with reality. As we will see, the machinery of modern QG can easily handle any desired genetic architecture (such as major genes with a polygenic background; Chapter 21).

1920–1950: The Biometrical Age of Quantitative Genetics

Fisher's demonstration that many short-term prediction problems could be completely described using only variance components lead to the rapid expansion of the age of **biometrical genetics**, relying upon only phenotypic information to extract QG information. At nearly the same time as Fisher, Sewall Wright (1921a–1921d), using his method of path analysis (Appendix 2), independently developed many of Fisher's results, and further examined the implications of inbreeding (Chapter 12). During the 1920s and 1930s, Fisher and Wright were joined by J. B. S. Haldane as the primary architects of modern population genetics.

In 1937, Jay Lush published his famous *Animal Breeding Plans*, helped greatly by weekly trips from Iowa State to the University of Chicago to attend Wright's course on quantitative genetics. Using Fisher's (1918) result that the expected slope of the midparent–offspring regression was h^2 (the ratio of the additive genetic variance to the total phenotypic variance), Lush appears to be the first to formally state the **breeders equation**, $R = h^2 S$. This predicts selection response R as the product of heredity h^2 and the selection differential $S = (\mu_s - \mu)$, formalizing (in quantitative-genetic terms) Pearson's idea outlined in Figure 1.2. Lush's book introduced the power of quantitative genetics into the breeding world. Breeders, in turn, went on to develop much of the early machinery of QG in response to needs arising from specific applications.

The biometrical age saw the full fruition of ANOVA designs to estimate a variety of important, and useful, variance components. For example, the various **North Carolina** (**NC**) **designs** (Part II Chapter 25) exploited special features of various plant breeding systems to cleanly estimate dominance variance in an attempt to more fully understand the basis of heterosis in maize. Additional line-cross methods were developed to gain insight into other features, such as the effective number of segregating factors (Chapter 11), detection of nonadditive gene action (Chapter 11), and inbreeding and heterosis (Chapters 12 and 13).

Biometrical methodology was concerned with genetics only in the sense of estimating the appropriate genetic variance components, very much treating any other genetic details as a bit of a black box. This was not really a problem for geneticists at the time, as the gene itself was also a black box. This changed with the rise of molecular biology, starting with the formal demonstration that DNA was the genetic material (Hershey and Chase 1952) and a model for its structure (Watson and Crick 1953).

1950–1980: The Rise of Molecular Genetics and its Divergence from Quantitative Genetics

The interaction between classical and quantitative genetics that played out during the rise of molecular genetics parallels much of the debate between the early Mendelians and biometricians, but with less rancor. During the development of molecular genetics in the 1950s (nicely reviewed by Judson 1979 and Mukherjee 2016), the notion of a gene grew from a fairly nebulous concept into a rather concrete, well understood, object built around a DNA sequence. This resulted in the statistical description of inheritance from quantitative genetics (relying on variances of traits instead of the features of specific genes) becoming far less appealing. Indeed, for many geneticists, quantitative genetics was an anachronism, a crutch that biologists no longer needed. As massive advances were made in the dissection of specific genes with discrete and highly reproducible phenotypes, the result was a diminished interest in continuous traits. Indeed, Fisher's notion of an infinitesimal model—a large number of loci, each with very small effects—was anathema to this molecular way of thinking.

Thus, for three decades, the once harmonious field of genetics bifurcated into molecular and quantitative arms. During this time, biometrical approaches continued to be developed, culminating in Charles Henderson's introduction (1949, 1950, 1963) of linear **mixed models** (BLUP), which elegantly generalized the somewhat rigid ANOVA-based designs for variance component estimation to the complex pedigrees common in animal breeding (Chapters 11, 32, and 33). It is rather fitting that mixed models from breeding would be repurposed many decades later to play key roles in the genomic analysis of complex traits (Chapters 20 and 21).

1980–2000: Molecular Markers, QTL Mapping, and the Age of Semi-major Genes

By the early 1980s, one of the byproducts of the molecular revolution was access to an ever-growing number of molecular-based markers. Geneticists, almost from the time of rediscovery of Mendel, were well versed in using markers to map genes by exploiting the excess of parental gametes seen under linkage (Chapter 5). While a few **quantitative trait loci (QTLs)** were mapped using classical (phenotypic) markers (e.g., Sax's work in 1923 showing linkage in beans between loci for spotting and seed weight), the lack of a sufficient density of markers limited most early attempts at QTL mapping. The bottleneck was technical (lack of markers), not operational (methods for linkage-based mapping methods were well known). In the early 1960s, biochemical approaches for revealing allozymic variation in the protein sequences produced by alternative alleles at a single gene resulted in a modest set (a few dozen) of markers for many species. Starting a decade later, the development of DNA-based markers, such as RFLPs or STRs (Chapter 8), offered up to a few hundred markers. Their availability sparked widespread QTL mapping, especially in crops and laboratory species. As detailed in Chapters 17–19, linkage-based QTL mapping involves the isolation of small chromosomal regions that influence a trait of interest using F_2s or other line-cross progeny. Such studies routinely found that relatively small chromosomal segments (typically on the order of tens of megabases) often account for nontrivial fractions of the trait variation (5% or more; Chapter 18).

The detection of such **semi-major** genes (or, more correctly, *genomic segments*) of modest to large effect was at odds with infinitesimal-like model thinking, and reinforced the view of many molecular geneticists on the importance of single genes (a resurrection of the Mendelian–Biometrician debate). However, the detection of such QTLs also started to open the black box of quantitative genetics, suggesting a road to redemption (at least from the molecular geneticists) whereby one would fairly rapidly be able to isolate a modest number of alleles that generated the bulk of trait variation. Alas, the actual isolation of such candidate genes proved extremely elusive (Chapters 17–21). Eventually, this led to the realization that the large estimated effects of such QTLs were very often substantially inflated by ascertainment bias (so-called **Beavis effects**; Chapter 18) and/or the presence of tightly linked clusters of trait-influencing genes that fractionate upon finer mapping (Chapter 18).

2000–Present: SNPs, GWAS, Whole-genome Sequencing, and the Rise of Quantitative Genomics

The use of linkage-based approaches greatly limited QTL mapping resolution, which was set by the number of recombinants between markers within the population sample of gametes. One or two crossovers per chromosome is expected to be the norm under most linkage designs, resulting in blocks of nonrecombining material typically exceeding tens of megabases. With many genomes on the scale of around 3,000 megabases (such as humans), just 300 well-chosen markers could detect most of these recombination events. Any further increase in marker density quickly becomes redundant, as dense markers tend to be inherited as a block in the gamete sample due to an insufficient number of recombinations.

Thus, the massively increasing number of usable DNA-based markers (such as **single nucleotide polymorphisms, SNPs**), that started to be generated by the late 1990s for model species (such as humans), could not be fully exploited by the limited number of recombinants in a typical linkage mapping population. New approaches would be required, as the mapping roadblock was no longer markers, but rather methodology. This led to the rise of **association mapping** (Chapters 17, 20, and 21), and today **GWAS** (**genome-wide association studies**) are the standard fine-mapping approach in most settings.

The foundation for GWAS to fine-map genes rests on the use of **linkage disequilibrium** (**LD**; Chapter 5)—correlations between alleles at very closely linked loci—in random samples of unrelated individuals from a large population. The reach of LD in a population is typically on the order of less than a kilobase (kb) to a few hundred kb, depending on the population (Chapter 5). LD thus offers much finer resolution than linkage mapping (the latter being on the megabase scale). The GWAS logic, as developed in Chapters 5, 8, and 20, is that a random population sample of individuals contains very small blocks of deep shared ancestry, generating the LD. Such blocks have persisted through hundreds of potential rounds of recombination, randomizing all but the tightest of linked markers.

DNA scoring technology initially used hybridization on **DNA chips** containing up to a few million SNPs, and then moved onto using the scored SNPs to **impute** (predict using a reference population) the genotypes of unscored, but very closely linked, SNPs (Chapter 20). Finally, whole genome sequencing became affordable, initially for a small number of individuals, but now rapidly sequencing thousands of individuals. Indeed, the field is close to the stage where one should consider having the WGS for every phenotyped individual as the norm (ironically, it is now phenotyping, rather than genotyping, that is often the bottleneck). As we detail in Chapter 21, WGS provides very little gain in power (over using a DNA chip) when causal alleles are at least of modest frequency and are in LD with **common SNPs** (a SNP whose minor allele frequency is greater than 1–5%). The power of WGS only shows itself when causal alleles are very rare (Chapter 21), which in turn requires massive sample sizes. Fortunately, GWAS, involving collections of unrelated individuals, is well suited to scale to very large sizes, potentially in the millions of individuals scored.

As detailed in Chapter 21, single GWAS "hits" (significant marker-trait associations) typically account for a very tiny fraction of trait variation. While at first blush this appears to be consistent with Fisher's small-effects model, there is actually a subtle difference. GWAS effect sizes are measured in terms of the amount of *trait variance* explained, and there is an inverse relationship between effect size and allele frequency such that large-effect alleles also tend to be very rare (Chapter 21; WL Chapters 25 and 28). Thus, a small GWAS variance could either reflect a common allele of very small effect or a very rare allele of very large effect. Further, most hits are in noncoding regions, typically rather far from genes (coding regions), making the assignment of a GWAS hit to a particular gene far from trivial (Chapter 21). It appears that most trait variation is generated by regulatory variation, often acting at great distance from the target site. As a result, quantitative geneticists have started to turn their sights to GWAS mapping of genomic features (**quantitative genomics**), such as mRNA expression level, splicing isoform ratios, or methylation/chromatin configurations (Chapter 21). The idea is to leverage this information on upstream effects to obtain greater biological intuition of the processes generating trait variation and to more accurately isolate downstream targets.

In addition to its role in gene mapping, dense marker technology allows classical QG approaches based on the fraction of shared ancestry to be greatly refined. For example, relative-based methods (Chapter 7) assume that two random full sibs each (on average) share one allele at a locus via common ancestry (an allele that is **identical by descent, IBD**). However, due to Mendelian segregation, there is variation about this expected value. At a given locus, full sibs have a 25% chance of sharing no IBD alleles, a 25% chance of sharing two, and a 50% chance of sharing one, yielding to an average IBD sharing of one. Using dense marker information, one can exploit this variation about the expected number of shared alleles, and the actual **realized relationships** can be accurately estimated using marker information (Chapter 8). This use of dense marker information yields much more refined estimates of the fraction of shared genomes between relatives than can be obtained by just using pedigrees.

Thus, we have come full circle, with the genomic (i.e., GWAS) data providing strong support for a version of the infinitesimal-like model (very small marker-effect *variances*) for the genetic architecture of many traits. A quantitative-genetic framework is thus required when using association data for genomic prediction (the framework for genomic selection and genomic-based individualized medicine; Chapters 32 and 33). A quick glance at any of the front-line journals in genomics highlights this growing merger with quantitative genetics, as many functional-genomic features are now routinely examined in a quantitative-genetics framework. Perhaps there is no better exemplar for the journey of QG than human height. It sparked Galton's analysis that started the rise of modern statistics, it motivated Fisher's 1918 paper, which in turn provided the framework for modern breeding. At the genomic level, at the time of this writing, the largest GWAS is a 5.4 million individual meta-analysis (Appendix 6) of human height, which found that over 12,000 sites account for height variation, with a potentially far larger number of much rarer sites waiting to be discovered (Example 21.15).

Quantitative Genetics, the Rise of Statistics, and the Current Statistical Revolution

As mentioned above, the impact of early quantitative-genetic theory extended well beyond the biological sciences. It laid the foundations for modern theoretical and applied statistics. Out of a need for quantitative methods to describe the distributions of continuously distributed characters, Galton provided the empirical motivation for Pearson's formal development of the theory of regression and correlation (Provine 1971; Stigler 1986). Fisher's 1918 paper introduced the concept of variance-component partitioning, upon which the principles of ANOVA are based, and his subsequent contributions had a profound influence on the development of methods for experimental design (Appendix 9) and hypothesis testing (Fisher 1925b, 1935, 1956). Wright's (1921a) method of path analysis (Appendix 2) became broadly embraced by the social and ecological sciences.

As the field of QG continued to mature, more advanced methods arose from interactions between statisticians and quantitative geneticists. Starting with Henderson's work on mixed linear models, the power of random-effects models for many quantitative genetics problems became apparent. As computational advances allowed Fisher's method of maximum likelihood (Appendix 4) to be feasible in more general settings, it quickly gained prominence in quantitative genetics. The ML-based variance estimation method of REML

(Chapter 33) arose from the concern of adjusting the estimated variance for the very large number of fixed effects typically fitted by animal breeders. REML later went on to be one of the workhorse methods in the genomic analysis of complex traits (Chapter 33).

More recently, the complex, high-dimensional nature of modern QG data sets, coupled with further computational and algorithmic advances (Appendix 8), led to the widespread use of Bayesian methods (Appendix 7). As data sets with marker information became both more complex and more numerous, the field embraced evolving statistical concepts such as false discovery rate (FDR) control for multiple comparisons (Appendix 6) and meta-analysis approaches for combining information from multiple data sets (Appendix 6). After the early sojourn of path analysis through the social sciences, it returned with a vengeance in quantitative genetics in the guise of **mediation analysis** to examine the impact of molecular intermediates on potentially downstream traits values (Chapter 21). Path analysis also gave birth to the important field of **Mendelian randomization** for controlling for confounding factors in complex epidemiological settings using quantitative-genetic features (Appendix 2). Advances in QG continue to drive new methods. For example, with whole-genome sequences becoming more routinely used in QG analysis, new statistics on sparse, high-dimensional data will need to be developed. Thus, QG has been both a driver and a grateful recipient of advances in statistical theory and methodology. One final point is worth mentioning. While the impact of the genomics revolution has been obvious, what is less appreciated is that a nearly comparable revolution has occurred in modern statistics. In part, much of this new theory and machinery arose from dealing the size, scale, and complexity of quantitative-genetic data sets.

THE MAJOR GOALS OF QUANTITATIVE GENETICS

Despite the great diversity of fields using quantitative-genetics machinery, most of the questions being examined revolve around a few key issues. Hence, before proceeding, we give a very brief summary of the main issues to be addressed by our revision of Lynch and Walsh (1998).

The Nature of Quantitative-trait Variation

As we have just reviewed, the expression of quantitative characters is typically influenced by both genetic and environmental factors, with patterns of variation that are qualitatively consistent with Mendelian expectations. However, many questions remain. An obvious one, of interest from many perspectives, is *How much of the standing variation in populations is due to genetic causes and how much to environmental ones*? From the standpoint of evolution and applied breeding programs, genetic components of variance are of particular interest because they determine the rates at which characters respond to artificial and natural selection. Environmental variance reduces the efficiency of the response to selection by causing the phenotypes of selected individuals to deviate from their underlying genotypic values. From the perspective of human genetic disorders, the degree to which the expression of undesirable traits is determined by genetic versus environmental causes has broad implications for the development of preventative procedures and genetic counseling strategies.

Many methods exist for partitioning phenotypic variance into its various components (Chapters 22–33), all of which are based on the principle that the phenotypic resemblance between relatives provides information on the degree of genetic differentiation among individuals (Chapter 7). With a sufficiently dense set of molecular markers, we no longer need a *known* collection of relatives, as we can estimate relatedness within a population sample (Chapter 8). Provided this sample has sufficient variance in relatedness, we can apply standard QG methods to estimate variance components (Chapters 8, 32, and 33).

Despite the flexibility of relative-based estimation procedures, virtually all of the methods have some undesirable theoretical and methodological features. For example, the assumed additivity of G and E in the base model, which generates a simple dichotomy between genetic and environmental sources of variation, is often overly simplistic, as genotype × environment interaction may exist (Chapters 27 and 28). Examples include drug interactions with specific genotypes and local adaptation. In the latter case, the performance of a line (or population) is disproportionately better (or worse) relative to other lines in a particular environment (i.e., G and E are no longer additive).

A further complication is that there are several components of both environmental and genetic variation (Chapters 4–6). These different components influence the resemblance between relatives to different degrees, and as a consequence, have substantially different influences on the evolutionary process. They also impact expressions for the joint disease risk occurrence over different sets of relatives. The additive component of the genetic variance (the additive-genetic variance, also known as the variance of breeding values) is of particular interest because it is the primary determinant of the degree to which offspring resemble their parents, which governs the rate of response of a character to selection. When significant nonadditive variance is present, breeders of species that can self (or be cloned) have a variety of options to exploit this other variance (WL Chapters 21–23). Similarly, when dominance is present, disease risk is greater for a pair of full sibs than between parent and offspring.

Ultimately, the pool of genetic variation in a population must be due to a quasi-balance between the forces of selection and random genetic drift, both of which tend to eliminate variation, and the replenishing force of mutation (Chapters 15 and 21; WL Chapters 7 and 28). Recent work has shown that the mutational rate of production of new variation for quantitative traits is remarkably high (Chapter 15), in part due to a very large number of potential mutational target sites (Chapter 21). Further, different traits may have different **genetic architectures** (underlying numbers of loci and the joint distributions of allele effects and frequencies at such loci). These features are functions of both the underlying molecular mechanisms generating trait variation *and* the past evolutionary history of their underlying genetic variants. For example, early-onset diseases likely face different evolutionary pressures than do diseases appearing after reproduction, potentially resulting in rather different genetic architectures. Likewise, a trait may be under no direct selection, but its underlying genetic components may be pleiotropically correlated with effects on other selected traits (Chapters 21 and 26; WL Chapter 28). The machinery from Chapters 18–21 is starting to provide insight into the allele size–frequency joint distribution, which, in turn, informs us somewhat about past evolutionary history.

The Consequences of Inbreeding and Outcrossing

The pattern of change in mean phenotypes in line crosses can yield insight into the mode of gene action, such as the presence of dominance and epistasis (Chapters 11–13). For example, the mean phenotypes of progeny from consanguineous matings commonly differ from those of progeny from random-bred parents. A necessary, but not sufficient, condition for this difference to occur is the presence of nonadditive genetic variance, in particular single-locus dominance (although epistatic terms involving dominance components can also contribute; Chapter 12). Inbreeding effects are almost always deleterious, with the amount of **inbreeding depression** generally increasing linearly with the degree of relatedness between parents (Chapter 12). These observations are consistent with the presence of deleterious recessive alleles segregating at the loci underlying quantitative variation. The mechanisms and consequences of inbreeding have a number of practical implications, e.g., in the design of breeding programs for captive populations of endangered species, the maintenance of inbred lines for biomedical and agricultural research, and the nature of increased disease risk in small, closed human populations. The pattern of fitness decline with inbreeding can also provide insight into the rate and average effects of deleterious mutation (Chapter 12).

While the consequences of crosses between relatives *within* a population can be deleterious, crosses *between* populations can have more complex outcomes. The F_1 hybrids between individuals from *closely related* populations often show enhanced performance relative to their parental populations, a condition called **hybrid vigor** or **heterosis**. As with inbreeding depression, this requires the presence of specific types of nonadditive variance. Like inbreeding depression, single-locus dominance, and dominance epistatic components, can generate heterosis. However, unlike inbreeding depression, additive-by-additive epistatis can also contribute (Chapter 13). One complication is that the amount of heterosis decreases in the F_2 population (the cross of the hybrids). When only dominance variance generates heterosis, this change is straightforward, with only half the initial F_1 heterosis persisting. When epistatic terms contribute, the F_2 decay is more complex (Chapter 13). A more interesting complication arises when the individuals being crossed are from *sufficiently diverged* populations. In this case **outbreeding depression** is often seen, where the hybrid is *less fit* than either parent (in the extreme, being sterile or nonviable). Unlike hybrid vigor, epistasis is required for outbreeding depression, reflecting the breakage in the offspring of favorable combinations of genes built up in the parents. Understanding the genetics of this process is vital to attempts to understand the mechanisms of speciation (Chapter 17).

The Constraints on the Evolutionary Process

Fundamental questions arise in evolutionary biology and in selective breeding programs as to the factors that limit the rate of phenotypic evolution. As noted above, when selection operates on a single trait, the response to selection is roughly proportional to the additive genetic variance for the trait (Chapter 3; WL Chapter 13). However, if the same genes influence the expression of different traits, then an evolutionary change in one trait will necessarily lead to changes in the correlated traits. This can impede the breeding/evolutionary process when there is a conflict between the consequences of selection operating directly on a trait and that operating on correlated traits. Questions of evolutionary tradeoffs have long been the focus of evolutionary ecology, but many of the ideas in that field (e.g., life-history theory) have developed out of simple energetic arguments or comparative surveys on the *phenotypes* of different species. An unambiguous understanding of the constraints on the evolution of systems of complex characters requires information on the magnitude and direction of *genetic* correlations between characters. Quantitative-genetic methodology provides a powerful, but data demanding, means of elucidating these issues. Ultimately, we would like to know the extent to which patterns of genetic correlations within populations are reflected in multivariate patterns of differentiation among species.

A related issue is the unresolved question about the nature of **pleiotropy** (one variant influencing multiple traits), which, along with LD, influences levels of genetic correlations between traits (Chapter 26). How common is pleiotropy, and when it occurs, are its effects limited to a set of traits in within developmental modules, or are they more widespread? Is the degree to which a mutation for a trait influences additional traits a function of its effect size? Resolving the pleiotropic structure of new mutations is critical for a better understanding of the evolutionary forces that maintain quantitative-genetic variation (WL Chapter 28).

The Estimation of Breeding Values and Prediction of Phenotypes

Plant and animal breeding programs are based on choosing the best individuals to serve as parents for the next generation. In the case where offspring are clones of the selected parents, the goal is to choose the parents with the best genotypic (or line) values, as these are exactly what their offspring receive. In the case where offspring are produced by sexually crossing parents, offspring are a mixture of their parental genotypes. In this setting, the goal of breeders is to chose those parents with the best **breeding values**, as this is average amount of their genotypic value that is transmitted to their offspring (Chapter 4). Phenotypes are an imperfect predictor of either of these measures (line/genotypic and breeding values), but their precision can be improved by using information from relatives in a mixed-model framework (BLUP; Chapters 10, 32, and 33). Further, molecular marker information can be used to improve these estimates (**genomic selection**; Chapter 32). A related issue, also examined in Chapter 32, is precision medicine, the use of an individual's marker information to predict their future phenotype (the notion of a **polygenic risk score**).

Statistical and Mathematical Tools for Quantitative Genetics

A number of reviews of helpful statistical and mathematical tools are scattered throughout this book. While we have tried to concentrate these either where they are needed, or summarize them in specific sections (such as the Appendices), we also encourage the reader to use the very extensive subject index when perusing for information about a particular method. Some of the more important methods/tools that we cover are as follows. Chapter 2 introduces basic properties of univariate distributions, such as moments (mean, variance, skew, kurtosis), as well as some less common distributions that arise in certain quantitative-genetic applications (such as the hypergeometric distribution for sampling without replacement and the negative binomial for the waiting time for exactly k events to occur). Bivariate distributions and their metrics (such as correlations, covariances, and regressions) are introduced in Chapter 3, which also examines conditional probability, and, in particular, the powerful theorem of Bayes. Appendix 1 develops the Taylor-series based **delta method** for obtaining approximations for means, variance, and covariances of more complex measures (such as ratios or products). Appendix 2 introduces the method of path analysis for exploring causality, with an extended discussion on Mendelian randomization approaches.

The most useful mathematical machinery in QG is linear (matrix) algebra, which is introduced in Chapter 9, with additional topics (such as generalized inverses) discussed in Appendix 3. Chapter 10 relies heavily on matrix machinery to present both the general linear model (GLM) and the mixed model (an extension of the GLM to include random effects). The latter is one of the main tools in QG, arising in GWAS (Chapter 20), G x E (Chapters 27 and 28), estimation of genetic and breeding values (Chapter 32), genomic selection and prediction (Chapters 32 and 33), meta-analysis (Appendix 6), and experimental design (Appendix 9), to name just a few. The scale on which a trait is measured impacts its analysis, and Chapter 14 discusses various topics in this area. Scales of analysis also serve as an introduction to **generalized linear models**, such as logistic regressions and log-linear models, which more easily handle discrete (i.e., zero/one or count) data than do standard linear models. Such transformations are widely used in linkage-based (Chapter 18) and GWAS (Chapter 20) QTL mapping.

Chapter 16 introduces **mixture distributions** (not to be confused with **mixed models**), weighted sums of simpler distributions (such as a weighted sum of normals). These are commonly used in the search for major genes (Chapter 16) and in QTL mapping (Chapters 17–19). Fisher's method of maximum likelihood (ML)—which provides a unified approach for estimation, assigning standard errors, and hypothesis testing—is reviewed in Appendix 4. Bayesian methods (Appendix 7) can be thought of as an extension of ML methods to more general settings, and generally entail computationally demanding, iterative algorithms, such as Markov Chain Monte Carlo (MCMC) approaches (Appendix 8).

Finally, the calculation of statistical power is highly desirable before embarking on any experiment. Appendix 5 examines power for simple t and z tests, as well as for both fixed and random effects ANOVA, general linear models, and for certain ML settings. Modern quantitative-genetic datasets typically have a large number of tested parameters, raising issues with controlling for multiple testing. Various approaches for control (such as FDR and sequential Bonferroni) are reviewed in Appendix 6. This appendix also examines the related field of meta-analysis: how best to combine signals over a number of studies. Lastly, Appendix 9 reviews basic features in Fisher's theory of experimental design, focusing on analysis under heterogeneous environments.

NAVIGATING THE SCIENCE OF QUANTITATIVE GENETICS

The science of quantitative genetics has been around for over a hundred years, its main principles having been outlined independently by Ronald Fisher (1918) and Sewall Wright (1921a–1921d). It has served as the theoretical basis for most plant and animal breeding programs for almost a century (Lush 1937, 1945; Hanson and Robinson 1963; Turner and

Table 1.1 Suggested topic-specific pathways through Part I.

Topic	Suggested Chapters
Basic quantitative genetics	4–7, 11–13
Analytic tools	2, 3, 9, 10, 14, Appendices 1–9
Agriculture and breeding	11–14, 17–20, Appendix 9
Evolutionary quantitative genetics	12–15, 21, Appendices 4–8
Human genetics	8–10, 12, 16-17, 19–21, Appendices 2, 4–8
Genomic applications	8–10, 16–21, Appendices 2, 4–8

Young 1969; Namkoong 1979; Mayo 1980, 1987; Hallauer and Miranda 1981; Mather and Jinks 1982; Pirchner 1983; Henderson 1984; Wricke and Weber 1986; Ollivier 1988; Hill and Mackay 1989; Gianola and Hammond 1990; Weller 1994; Bos and Caligari 1995; Falconer and Mackay 1996; Kearsey and Pooni 1996; Cameron 1997; Simm 1998; Kinghorn et al. 2000; Gallais 2003; Bernardo 2010, 2020; Hallauer et al. 2010; Schaeffer 2019; Caballero 2020; Xu 2022; Mrode and Pocrnic 2023). It has also played an important role in our understanding of the inheritance of complex human genetic disorders and underpins much of disease genomics (Lander and Schork 1994; Risch and Merikangas 1996; Jansen and Nap 2001; de Koning and Haley 2005; Gibson and Weir 2005; Rockman and Kruglyak 2006; Wellcome Trust Case Control Consortium 2007; Altschuler et al. 2008; Visscher et al. 2012, 2017; Boyle et al. 2017a). Finally, it has also become well entrenched in the field of evolutionary biology (Bulmer 1980; Lande 1988; Barton and Turelli 1989; Boake 1994; Roff 1997; Charmantier et al. 2014; Gienapp et al. 2017; Walsh and Lynch 2018; Arnold 2023).

Hence, this book attempts to reach a very broad audience. As such, the reader might find the detailed level of discussion on some topics to be of less interest than others. How, then, should one navigate this volume? Chapters 4–7 give the foundational concepts that all QG practitioners should know. From here, paths can diverge dramatically given the interests and questions of the reader. Table 1.1 provides some guidance on how to proceed. We have tried to write chapters in such a way that their dependencies on specific previous chapters are clear, while otherwise making them as independent as possible.

2

Properties of Distributions

Science is what we understand well enough to explain to a computer,
Art is all the rest. Donald Knuth

Before delving into the genetics of quantitative variation, it is essential to have a basic understanding of statistics. The foundational statistical concepts and techniques most frequently encountered in quantitative genetics are presented in this and the following chapter. More advanced methods, building on these concepts, are presented in Chapter 10 (general linear and mixed models), while Appendices 1–9 develop additional statistical machinery. For the reader with advanced training in statistical theory, much of these next two chapters will probably be review, and some things may appear to be presented in a nonrigorous manner. Even so, it may still be profitable to skim the following pages to gain familiarity with the notation used throughout the book. In addition, a number of examples will provide some immediate contact with the field of quantitative genetics.

PARAMETERS OF UNIVARIATE DISTRIBUTIONS

Characters studied by biologists are generally of three types: integer-valued, continuous, and categorical (e.g., healthy vs. diseased, color types, etc.). Integer-valued data, such as scale counts in fish or leaf number in plants, are called **meristic characters**. Those measured on a continuous scale are known as **metric characters**. Length, weight, growth rate attributes, and the amount of an mRNA or protein are examples of the latter. Attributes such as survival to a fixed age, disease presence/absence, or whether a specific gene is turned on or off are known as **all-or-none** or **binary characters**, and are a special care of categorical outcomes. Some categorical outcomes can be further specified as **ordinal**, where the outcomes are *ranked,* such as disease severity being scored as mild, moderate, or severe. The idea for ordinal data is that some underlying variable is mapped into increasing higher categorical ranks as it increases in value and passes various threshold values (Chapter 14).

Of course, due to technical limitations, even measures of truly continuously distributed traits must always be artificially placed into discrete categories. Meter sticks, for example, are unable to distinguish between individuals that are 25.2 and 25.3 mm in length. Both would typically be placed in the 25–26 mm category, although the biological reality is that every conceivable length in the 25–26 mm range is possible. Similarly, functional genomic data, such as the number of copies of a specific mRNA or protein, are usually obtained as large integers (counts of molecules generated by the scoring method), which are often then binned into categories.

Suppose that one performs a series of measurements on a collection of individuals. Compilation of the data provides some information on the relative incidence of different trait measures in the broader population. A **univariate distribution** describes the relative frequencies of phenotypes for a single trait, whereas a **bivariate distribution** describes the mutual (or **joint**) distribution of two characters. The joint distribution of more than two traits is referred to as a **multivariate distribution** (Chapter 9). As an example of a bivariate distribution, data for maternal weight and number of offspring are presented for a population of rats in Table 2.1. The data are condensed into the univariate **marginal distributions** of the two traits in the last row and column.

One of the goals of statistics is to fit fairly simple mathematical functions, known as **probability distributions,** to data. If a variable z takes on only discrete values (as with

Table 2.1 The bivariate distribution of mother's weight (in grams) and number of offspring produced for a population of rats. (From Pearson 1910.)

Maternal Weight	Number of Offspring *												
	1	2	3	4	5	6	7	8	9	10	11	12	Totals
50 -	—	—	—	1	3	1	—	—	—	—	—	—	5
60 -	—	—	—	1	6	2	—	—	—	—	—	—	9
70 -	—	—	2	10	17	12	4	—	1	—	—	—	46
80 -	1	1	11	8	18	10	9	3	2	—	—	—	63
90 -	2	5	7	18	30	28	12	5	1	—	—	—	108
100 -	3	5	10	25	37	35	21	7	2	1	—	—	146
110 -	1	4	12	19	38	37	29	6	2	—	—	—	148
120 -	2	6	9	21	36	26	30	14	6	—	1	—	151
130 -	4	4	9	12	35	29	17	17	6	1	1	1	136
140 -	1	4	6	9	12	27	15	6	2	1	—	—	83
150 -	—	3	—	2	13	11	6	6	2	—	—	—	43
160 -	—	2	—	1	11	11	9	3	4	—	—	—	41
170 -	1	—	1	1	2	4	2	2	1	—	1	—	15
180 -	—	—	1	1	—	2	2	2	—	—	—	—	8
190 -	—	—	—	—	—	—	—	—	—	1	—	—	1
Totals	15	34	68	129	258	235	156	71	29	4	3	1	1003

* Each number in the main body of the table refers to the number of observations in a particular bivariate class. For example, 38 animals weighed between 110 and 120 g and produced 5 offspring. The final row and column are the marginal univariate distributions for the two traits.

offspring number), the distribution of z is completely described by giving $\Pr(z = i)$ for each possible outcome i, where Pr stands for probability. For example, for offspring number, $\Pr(z = 1)$ is the proportion of mothers that produce a single offspring, which for the example in Table 2.1 is $15/1003$. Summing over all possible outcomes, $\sum_i \Pr(z = i) = 1$, as the total probability of all possible events is one.

If, on the other hand, z is a continuously distributed variable (as with maternal weight), $\Pr(z = z_i)$ makes no sense because the probability that z takes on any specific *exact value* is infinitesimally small. It is more meaningful to consider the probability that z lies within *a specific range of values* (or *interval*), say z_1 to z_2. This quantity is described by the **probability density function** (or **pdf**), $p(z)$, which satisfies the integral

$$\Pr(z_1 \leq z \leq z_2) = \int_{z_1}^{z_2} p(z)\, dz \tag{2.1a}$$

Namely, the total area under $p(z)$ over the interval of interest. If z_{min} and z_{max} are the upper and lower bounds to z, then $p(z) = 0$ outside of this range, and over the entire range $\int_{z_{min}}^{z_{max}} p(z)\, dz = 1$. Both of these properties are in accord with common sense, as a probability is never negative, and the total probability of all possible outcomes is one. A large number of functions that fulfill these properties have been studied in considerable detail (Kendall and Stuart 1977; Johnson, Kotz, and Balakrishnan 1994, 1995; Johnson, Kemp, and Kotz 2005; Kotz, Balakrishnan, and Johnson 2019). Leemis (1986) provides a useful diagram showing the interconnections between the most common distributions.

A function closely related to the pdf of a distribution is its **cumulative distribution function, cdf(x)**, defined as

$$\text{cdf}(x) = \Pr(z \leq x) = \int_{-\infty}^{x} p(z) dz \tag{2.1b}$$

Equation 2.1b follows the normal convention of integrating from $-\infty$, with the understanding that $p(z)$ takes on a value of zero for $z < z_{min}$. Note that

$$\Pr(z_1 \leq z \leq z_2) = \int_{z_1}^{z_2} p(z)dz = \int_{-\infty}^{z_2} p(z)dz - \int_{-\infty}^{z_1} p(z)dz = \text{cdf}(z_2) - \text{cdf}(z_1) \quad (2.1c)$$

Finally,

$$\Pr(z > x) = \int_x^{\infty} p(z)dz = 1 - \text{cdf}(x) \quad (2.1d)$$

Historically, published tables for statistical distributions were based on cdfs, although now results are directly calculated using statistical packages (such as R).

Example 2.1. Suppose that z is continuously distributed in the range of 0 to ∞ with density function

$$p(z) = \frac{1}{\lambda} e^{-z/\lambda} \qquad \text{for } z \geq 0, \;\; \lambda > 0$$

This is the **negative exponential distribution** (defined by the parameter λ) in which the density has a maximum at $z = 0$ and declines to zero as $z \rightarrow \infty$. Because the integral of $p(z)$ is $-e^{-z/\lambda}$,

$$\int_0^{\infty} p(z)\, dz = -e^{-z/\lambda}\Big|_0^{\infty} = 0 - (-1) = 1$$

showing that $p(z)$ fulfills the two properties of a probability density, as the total area under the pdf is one and $p(z)$ is non-negative. The cdf becomes

$$\text{cdf}(x) = \int_0^x p(z)\, dz = 1 - e^{-x/\lambda}$$

What is the probability that a randomly drawn individual will have z in the range of 1/4 to 1/2?

$$\Pr(1/4 \leq z \leq 1/2) = \int_{1/4}^{1/2} p(z)\, dz = -e^{-z/\lambda}\Big|_{1/4}^{1/2} = e^{-1/(4\lambda)} - e^{-1/(2\lambda)}$$

Note that this equals cdf(1/2) − cdf(1/4). The numerical answer depends on the parameter λ. If, for example, $\lambda = 1/2$, then $\Pr(1/4 \leq z \leq 1/2) = 0.239$.

Before moving on, we emphasize the importance of distinguishing between true **parameters** of distributions and **estimates** of those parameters obtained by sampling. True parameter values can only be obtained if every member of a population is measured with absolute accuracy. We must therefore almost always settle for approximations, the accuracy of which depends on the experimental setting, the measurement scheme, and the sample size. Statisticians often denote parameters of a population with Greek letters and sample estimates with Roman letters. We will adhere to this protocol as much as possible, although there will be some instances where traditional quantitative-genetic notation prevents us from doing so.

The most useful density functions are defined completely by one or two parameters describing the central location and dispersion of the distribution. The most widely used measure of the location is the **arithmetic mean,** μ, also known as the **first moment about**

the origin. If $p(z)$ is the density function of phenotype z, then weighting all values of z by their density leads to

$$\mu = \int_{-\infty}^{+\infty} z\, p(z)\, dz = E(z) \tag{2.2}$$

where $E(z)$ denotes the **expected value** or **expectation** of z. Here, we have arbitrarily put $\pm\infty$ limits on the integral to ensure that the entire range of variation is covered. For discrete characters, $\mu = E(z) = \sum_i z_i \Pr(z = z_i)$. The sample estimate of the mean is generally denoted by $\bar{z}$, and estimated as the average of the n measurements,

$$\bar{z} = \frac{1}{n} \sum_{i=1}^{n} z_i$$

Example 2.2. What is the mean of the distribution discussed in Example 2.1? Because the integral of $(z/\lambda)\, e^{-z/\lambda}$ is $-(z+\lambda)\, e^{-z/\lambda}$,

$$\mu = \int_0^{\infty} z\, p(z)\, dz = \int_0^{\infty} (z/\lambda)\, e^{-z/\lambda}\, dz = -(z+\lambda)\, e^{-z/\lambda} \Big|_0^{\infty} = \lambda$$

Thus, the parameter λ is the mean of the distribution defined by the density function $p(x) = (1/\lambda)\, e^{-z/\lambda}$.

Higher order moments provide measures of the **dispersion** of a frequency distribution. The most familiar and useful such measure is the **variance** (a term introduced in Fisher's 1918 paper). Also known as the **second moment about the mean** (or the **second central moment**), the variance is the expected squared deviation of an observation from the population mean,

$$\sigma^2 = \int_{-\infty}^{+\infty} (z-\mu)^2\, p(z)\, dz = E\left[\, (z-\mu)^2 \,\right] \tag{2.3a}$$

Because $\mu = E(z)$, this quantity can be expressed more simply by expanding $(z-\mu)^2$ to obtain

$$\sigma^2 = E(z^2 - 2z\mu + \mu^2) = E(z^2) - 2\mu E(z) + \mu^2 = E(z^2) - \mu^2 \tag{2.3b}$$

where we have used two useful properties of expectations,

$$E(x+y) = E(x) + E(y) \tag{2.4a}$$

$$E(c\,x) = c\,E(x) \tag{2.4b}$$

for a constant c. Several notations are used for the parametric variance of a distribution. When there is no ambiguity as to the variable being considered, σ^2 suffices. More generally, the variance of z is denoted by σ_z^2 or $\sigma^2(z)$.

A slight complication arises when one wishes to estimate the parameter σ^2 from a random sample of observations from the population. As noted above, the true parameters μ and $E(z^2)$ cannot be known with certainty unless the entire population is sampled. Because the estimated (sample) mean $(\bar{z})$ is a function of the data, ***individual measures tend to be closer to the observed mean than to the true mean***, and as a consequence, observed values of $\overline{z^2} - \bar{z}^2$ (where $\overline{z^2}$ denotes the mean of the squared values of z) ***tend to be slightly less than the parametric value*** $[\, E(z^2) - \mu^2 \,]$. Thus, the estimator $(\, \overline{z^2} - \bar{z}^2 \,)$ is biased in the sense that it tends to *slightly underestimate* the parameter $\sigma^2(z)$ to a degree that decreases with increasing sample size (n).

A major goal of applied statistics is to obtain **unbiased estimators** that account for these kinds of small sample-size limitations. In the case of the variance, the solution is simple, with

$$\text{Var}(z) = \frac{n\left(\overline{z^2} - \bar{z}^2\right)}{n-1} = \frac{1}{n-1}\sum_{i=1}^{n}(z_i - \overline{z})^2 \tag{2.5}$$

providing an unbiased estimate of $\sigma^2(z)$ (for the derivation of this expression, see Example A1.2). Equation 2.5 should be used whenever the true population variance, $\sigma^2(z)$, is being estimated from actual sample data.

The variance is measured in units that are the square of those of the mean, but it is often desirable to describe the dispersion of a frequency distribution on the *same scale* as the mean. The square root of the variance of z is called the **standard deviation** (**SD**) of z. The parametric value is denoted by σ_z, $\sigma(z)$, or just σ, and the statistic by $\text{SD}(z) = \sqrt{\text{Var}(z)}$. The ratio of the SD to the mean, the **coefficient of variation** (**CV**), is frequently used as a relative measure of dispersion. It is known that the statistic $\text{CV}(z) = \text{SD}(z)/\bar{z}$ is a downwardly biased estimator of the parametric index (σ/μ), but the bias is expected to be negligible in most cases (Haldane 1955).

Quantitative geneticists generally rely on the variance as a measure of the dispersion of a distribution. However, additional moments can be informative. For example, the third moment about the mean (μ_3) is a useful measure of the **asymmetry** of a distribution. Also known as the **skewness**, μ_3 is the expected cubic deviation from the mean. As in the case of the variance, it can be expressed in terms of the moments about the origin,

$$\begin{aligned}\mu_3 &= \int_{-\infty}^{+\infty} (z-\mu)^3\, p(z)\, dz = E\left[(z-\mu)^3\right] \qquad (2.6)\\ &= E(z^3) - 3\mu E(z^2) + 3\mu[E(z)]^2 - [E(z)]^3\\ &= E(z^3) - 3\mu E(z^2) + 2\mu^3\end{aligned}$$

Thiele (1889) found that an unbiased sample estimator for μ_3 is

$$\text{Skw}(z) = \frac{n^2\left(\overline{z^3} - 3\,\overline{z^2}\,\bar{z} + 2\,\bar{z}^3\right)}{(n-1)(n-2)} \tag{2.7}$$

where $\overline{z^3}$ denotes the observed mean cubed value of z. The degree of asymmetry can also be described with a dimensionless index, the **coefficient of skewness**, which is estimated by the ratio

$$k_3 = \frac{\text{Skw}(z)}{\text{Var}(z)^{3/2}} \tag{2.8}$$

k_3 is positive when the longer tail of a distribution is to the right, negative when the tail is to the left, and zero for a perfectly symmetrical distribution.

From the above, it follows that

$$\mu_r = \int_{-\infty}^{+\infty} (z-\mu)^r p(z)\, dz \tag{2.9}$$

is a general expression for the rth moment about the mean (the ***r*th central moment**). It also follows that μ_r can always be expressed in terms of moments about the origin $[\,E(z),\, E(z^2),\, \ldots,\, E(z^r)\,]$. As was shown for the variance and the skewness, these terms are obtainable from the binomial expansion of $(z-\mu)^r$.

Finally, we note that when moments are calculated from data that are grouped into classes, as in Table 2.1, a certain amount of bias is introduced because the true measures are assumed to be concentrated at the midpoints of the classes. Provided the total distribution is continuous and falls off smoothly at its extremities, this bias can often be eliminated by application of **Sheppard's correction** (1898). In the case of the variance, the corrected estimate is obtained by subtracting the quantity $\omega^2/12$ from Var(z), where ω is the width of the interval. No correction is required for the third moment about the mean. For details on higher order moments, see Kendall and Stuart (1977, p. 77).

Example 2.3. Utilizing the data for maternal weight (in grams) from Table 2.1, we now summarize the procedures for obtaining estimates of the first three moments.

Weight	z	$n(z)$	$z\,n(z)$	$z^2 n(z)$	$z^3 n(z)$
50 -	55	5	275	15,125	831,875
60 -	65	9	585	38,025	2471,625
70 -	75	46	3450	258,750	19,406,250
80 -	85	63	5355	455,175	38,689,875
90 -	95	108	10,260	974,700	92,596,500
100 -	105	146	15,330	1,609,650	169,013,250
110 -	115	148	17,020	1,957,300	225,089,500
120 -	125	151	18,875	2,359,375	294,921,875
130 -	135	136	18,360	2,478,600	334,611,000
140 -	145	83	12,035	1,745,075	253,035,875
150 -	155	43	6665	1,033,075	160,126,625
160 -	165	41	6765	1,116,225	184,177,125
170 -	175	15	2625	459,375	80,390,625
180 -	185	8	1480	273,800	50,653,000
190 -	195	1	195	38,025	7414,875
Totals		$n =$ 1003	$\sum z\,n(z) =$ 119,255	$\sum z^2\,n(z) =$ 14,812,275	$\sum z^3\,n(z) =$ 1,913,429,875

Note that for each weight category, z is taken arbitrarily to be the midpoint of the measurement interval, so that for the interval 50–60, we take $z = 55$. The frequency of observations in each category, $f(z)$, is equal to $n(z)/n$, where $n(z)$ is the number of observations with phenotype z, and $n = \sum n(z)$ is the total sample size.

The moments about the origin are obtained by dividing the weighted sums in the table by n,

$$\bar{z} = \sum z f(z) = \sum z\,n(z)/n = \frac{119,255}{1003} = 118.90$$

$$\overline{z^2} = \sum z^2 f(z) = \sum z^2 n(z)/n = \frac{14,812,275}{1003} = 14,767.97$$

$$\overline{z^3} = \sum z^3 f(z) = \sum z^3 n(z)/n = \frac{1,913,429,875}{1003} = 1,907,706.75$$

The variance estimated from the pooled data is

$$\text{Var}(z) = \frac{n\,(\,\overline{z^2} - \bar{z}^2\,)}{n-1} = 631.39$$

and application of Sheppard's correction, with $\omega = 10$, reduces this to

$$\text{Var}(z) = 631.39 - \frac{\omega^2}{12} = 623.06$$

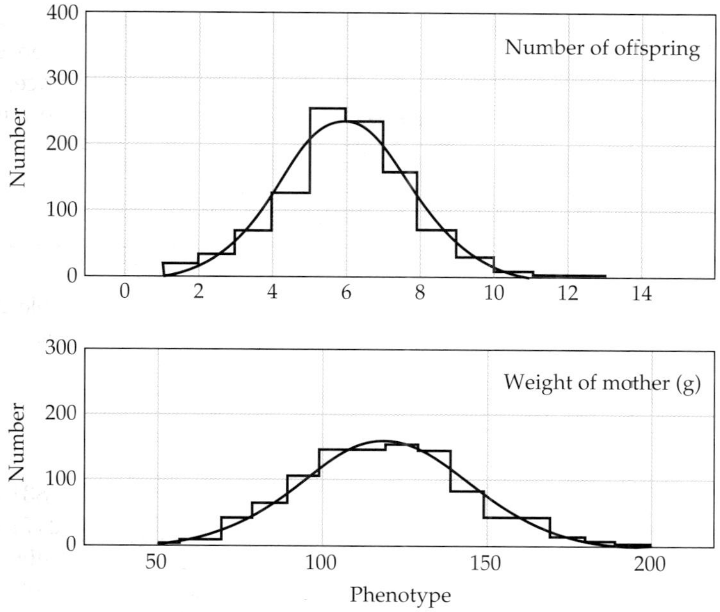

Figure 2.1 Frequency histograms for the two univariate distributions in Table 2.1 and their normal approximations based on the observed means and variances.

The estimated coefficient of variation is then

$$\mathrm{CV}(z) = \frac{[\,\mathrm{Var}(z)\,]^{1/2}}{\bar{z}} = 0.21$$

Finally, from Equations 2.7 and 2.8, the skewness and coefficient of skewness are

$$\mathrm{Skw}(z) = \frac{n^2\,(\,\overline{z^3} - 3\,\overline{z^2}\,\bar{z} + 2\,\bar{z}^3\,)}{(n-1)(n-2)} = 1805.40, \quad k_3 = \frac{\mathrm{Skw}(z)}{[\,\mathrm{Var}(z)\,]^{3/2}} = 0.12$$

THE NORMAL DISTRIBUTION

When large data sets of the type compiled in Table 2.1 are displayed in the form of frequency histograms (Figure 2.1), they often approximate a bell-shaped distribution. Three famous mathematicians, Abraham DeMoivre (1738), Pierre-Simon LaPlace (1778), and Carl Friedrich Gauss (1809), worked out the properties of a very useful description of this form, the **normal distribution**, also referred to as the **Gaussian distribution**. If z is a normally distributed variable, its density function is given by

$$p(z) = (2\pi\sigma^2)^{-1/2} \exp\left[-\frac{(z-\mu)^2}{2\sigma^2}\right] \qquad \text{for } -\infty < z < \infty \tag{2.10}$$

where $\exp(z) = e^z$, with $e \simeq 2.7183$ being the base of natural logarithms, and $\pi \simeq 3.1416$. The normal distribution is a function of only two parameters, the mean (μ) and variance (σ^2). The normal density attains a maximum when $z = \mu$ and declines continuously and symmetrically in both directions as z deviates from μ (Figure 2.1). A normally distributed variable with mean μ and variance σ^2 is often denoted by $z \sim \mathrm{N}(\mu, \sigma^2)$, where $\sim$ means "is distributed as." In future chapters, we often use the notation $\varphi(z, \mu, \sigma^2)$ to denote the probability density (pdf) of a normal, to remind the reader that it is a function of the mean and variance.

The normal distribution plays a central role in statistical theory for two reasons. First, the normal density function has many simple mathematical features that allow the derivation of practical statistical tests. Second, even when actual distributions of phenotypes are inconsistent with the normal density function, after an appropriate scale transformation (Chapter 14), many can be rendered approximately normal.

A general reason why many traits are distributed normally or nearly so is provided by the **central limit theorem**, which states that the sum of a number of independent (or even weakly correlated) random variables approaches normality as the number of variables increases. This is expected to be the case, for example, with a metric character influenced by many environmental factors and a large number of loosely linked genes, each with small additive effects. As a consequence, the normal distribution has been relied upon extensively in quantitative genetics. Whenever an assumption regarding the form of a (continuous) phenotype distribution is necessary, the normal distribution is generally invoked as a first approximation. The normal density function is also often used to define a **Gaussian fitness function** in the theory of stabilizing selection (WL Chapters 16 and 28), the "mean" serving as a measure of the optimum phenotype and the "variance" being inversely related to the intensity of selection (because the fitness function becomes flatter as the width increases).

There are, of course, limitations of the normal density function and of distribution functions in general. For instance, the normal distribution gives small positive values, rather than zero, for negative z, an unrealistic situation for traits such as body size, bone length, or expression level, which cannot take on negative values. Nevertheless, if the mean of a distribution is sufficiently greater than zero, the theoretical incidence of negative values is minuscule, so that truncating the distribution to be non-negative does not induce much bias. It should also be emphasized that the normal distribution is a continuous function, giving positive values for noninteger values of z. It is, therefore, not strictly applicable to meristic traits such as egg number, spine count, or number of transcripts present although it provides a close approximation when the number of classes is large.

It is often convenient to work with a standardized form of Equation 2.10. A **standard normal deviate,** $z' = (z - \mu)/\sigma$, is the deviation of a measure from the population mean in units of standard deviations. Example 2.4 shows that if z is normally distributed with mean μ and variance σ^2, then z' is normal with zero mean and unit variance, i.e.,

$$p(z') = (2\pi)^{-1/2} \exp\left[-\frac{(z')^2}{2}\right] \tag{2.11}$$

Example 2.4. It is known that if y is a function of z, denoted by $f(z)$, then the density function for y is

$$p(y) = \left|\frac{df(z)}{dz}\right|^{-1} p(z)$$

where $|x|$ denotes the absolute value of x. This transformation is valid provided that $df(z)/dz$ exists and is nonzero for all z values for which $p(z) > 0$. This criterion is met by the standard normal deviate. Letting $z' = f(z) = (z - \mu)/\sigma$, then $df(z)/dz = \sigma^{-1}$. Substituting the normal density function for $p(z)$ recovers the **standard normal** or (**unit normal**) **distribution**,

$$p(z') = \left|\frac{1}{\sigma}\right|^{-1} \frac{1}{\sqrt{2\pi\sigma^2}} \exp\left[-\frac{(z-\mu)^2}{2\sigma^2}\right] = (2\pi)^{-1/2} \exp\left[-\frac{(z')^2}{2}\right]$$

Because the normal distribution is symmetrical, the third central moment, $\mu_3 = E[(z - \mu)^3]$, is equal to zero. The fourth central moment, the **kurtosis** $\mu_4 = E[(z - \mu)^4]$, has an

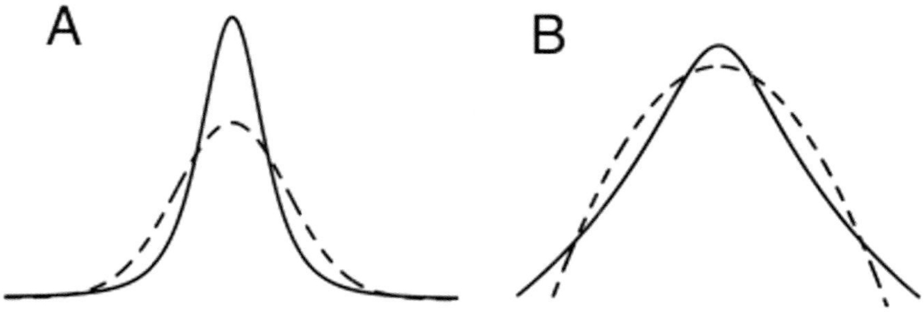

Figure 2.2 An example of a leptokurtic distribution (a t distribution with three degrees of freedom, scaled to have a variance of one; solid line) relative to a unit normal (dashed line). Because kurtosis is defined relative to the variance, one needs to scale both distributions to have the same variance to see the impact of kurtosis. **(A)** A plot of the pdfs for both distributions shows the excessive peakness of the leptokurtic distribution, but its heavier tails are not obvious. **(B)** A plot of log(pdf), which gives an inverted parabola for a normal (dashed line). Note that the leptokurtic distribution (solid line) is both overly peaked and has less mass in intermediate regions around the mean. At the extremes of the distribution, its tails are heavier (the solid line being above the dashed line).

Table 2.2 Types of Kurtosis.

Type	μ_4	k_4	Distribution of Outliers
Leptokurtic	$> 3\sigma^4$	> 0	More extreme than a normal
Mesokurtic	$= 3\sigma^4$	$= 0$	Same as a normal
Platykurtic	$< 3\sigma^4$	< 0	Less extreme than a normal

expected value equal to $3\sigma^4$. Thus, if we let $\text{Kur}(z)$ be the sample estimate of μ_4, the **kurtosis index** is defined as

$$k_4 = \frac{\text{Kur}(z) - 3\,[\,\text{Var}(z)\,]^2}{[\,\text{Var}(z)\,]^2} \tag{2.12a}$$

where

$$\text{Kur}(z) = \frac{n^2(n+1)(\,\overline{z^4} - 4\,\overline{z^3}\,\bar{z} + 6\,\overline{z^2}\,\bar{z}^2 - 3\,\bar{z}^4)}{(n-1)(n-2)(n-3)} \tag{2.12b}$$

For a truly normal distribution, $E[k_4] = 0$. While kurtosis is often presented as a measure of the peakedness of a distribution relative to a normal (i.e., being too broad if $k_4 < 0$, or too narrow if $k_4 > 0$), its importance lies in how it impacts the ***tails of a distribution*** (the extreme high and low values). The tails of a **platykurtic** distribution ($k_4 < 0$; also called **sub-Gaussian**) have less probability mass than those of a normal distribution, resulting in **thinner tails**. Conversely, a **leptokurtic** random variable ($k_4 > 0$; also called **super-Gaussian**) has **fatter** (or **thicker**) **tails**, with more probability mass relative to a normal (Figure 2.2). As we will see, the presence of kurtosis ($k_4 \neq 0$) impacts both p values and the widths of confidence intervals relative to values based on assuming normality. Importantly, a sample skewness significantly different from zero and/or a sample kurtosis significantly different from $3\text{Var}(z)^2$ are indicators for a departure from normality.

The Truncated Normal Distribution

We now consider the properties of a subset of the normal distribution, specifically a tail of the distribution (Figure 2.3). Such a consideration is important in the extreme form of directional selection that is used by most plant and animal breeders and it also underpins

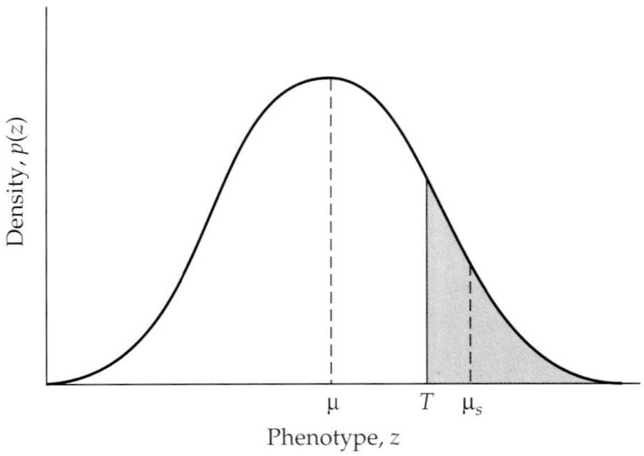

Figure 2.3 A truncated normal distribution (hatched area) with lower limit T and mean μ_s. The total area to the right of T is denoted by $\Phi_T = p(z \geq T)$.

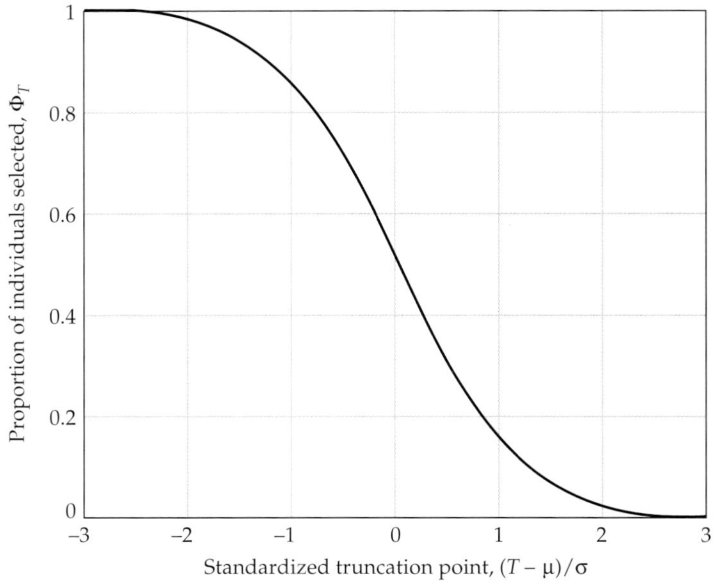

Figure 2.4 The proportion Φ_T of individuals selected from an underlying normal distribution with lower truncation point T, mean μ, and variance σ^2. As the truncation point moves to the right of the mean, i.e., as $(T - \mu)/\sigma$ becomes larger, the fraction of the population to the right of the truncation point approaches zero.

the analysis of threshold traits (Part II Chapter 31). Under **truncation selection**, all individuals below a certain phenotype are culled from the population and hence have zero fitness (WL Chapter 14). The critical phenotype, T, is called the **truncation point**. For a normally distributed phenotype, the mean phenotype of the population above the threshold (i.e., after selection) can be written as

$$\mu_s = E[\, z \mid z \geq T\,] = \int_T^{\infty} z p(z \mid z \geq T)\, dz = \frac{\int_T^{\infty} z p(z)\, dz}{\int_T^{\infty} p(z)\, dz} \tag{2.13}$$

The last step follows in that when computing a mean, the phenotype frequencies must sum to one, and this is accomplished by standardizing the density of phenotype z after selection

$$p(z \mid z \geq T) = \frac{p(z)}{\Pr(z \geq T)} = \frac{p(z)}{\int_T^{\infty} p(z)\, dz}$$

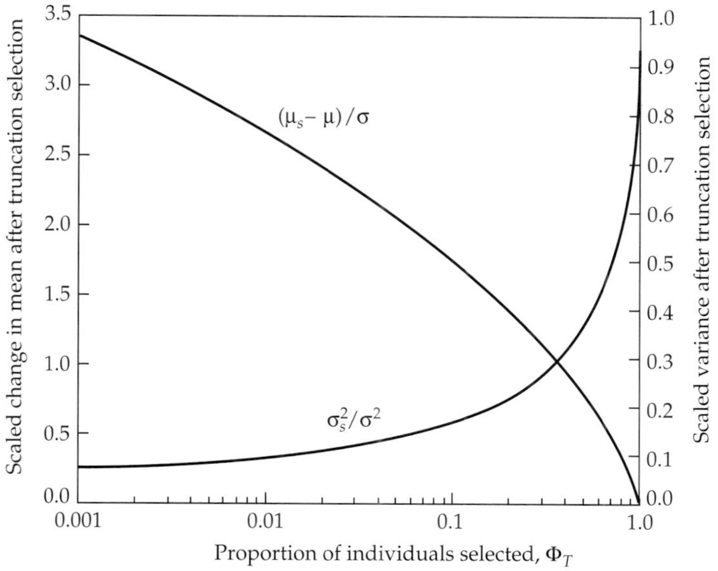

Figure 2.5 The change in the mean phenotype (in units of phenotypic standard deviations) and in the phenotypic variance (relative to the variance prior to selection) as a function of the selection intensity. As the proportion of the population above the threshold (Φ_T) increases, the selection intensity decreases.

Because the denominator, $\int_T^\infty p(z)\,dz$, is the sum of frequencies for phenotypes greater than T (i.e., the fraction of individuals allowed to reproduce), it is a measure of the intensity of selection, and we hereafter denote it as $\Phi_T = \Pr(z > T)$. After integration, the solution to Equation 2.13 is found to be

$$\mu_s = \mu + \sigma \frac{p_T}{\Phi_T} \tag{2.14}$$

where p_T is the height of the standard normal curve at the truncation point, obtained by setting $z' = (T - \mu)/\sigma$ in Equation 2.11,

$$p_T = (2\pi)^{-1/2} \exp\left[-\frac{(T-\mu)^2}{2\,\sigma^2} \right]$$

Values for p_T and ϕ_T can be obtained from standard statistical packages. For example, the `R` command `pnorm(x)` returns the value of $P(U \le x)$, where U denotes a unit normal. Hence, values for Φ_T are obtained using `1 - pnorm(x)`, where x denotes the value of the standardized distance z' of T from μ (see also Figure 2.4). Similarly, the value for p_T can be obtained using `dnorm(x)` using the value of T for `x`.

Equation 2.14 states that if the upper proportion Φ_T of a normally distributed population is selected, the mean will advance $\mu_s - \mu = \sigma p_T/\Phi_T$ units. This change in the mean caused by selection is often denoted by S, the **directional selection differential** (Chapter 3). In units of phenotypic standard deviations, the standardized selection differential is simply $(\mu_s - \mu)/\sigma = p_T/\Phi_T$, which is plotted in Figure 2.5.

In a similar fashion, the trait variance of the selected population can be shown to be

$$\sigma_s^2 = \left[1 + \frac{p_T\, z'}{\Phi_T} - \left(\frac{p_T}{\Phi_T} \right)^2 \right] \sigma^2 \tag{2.15}$$

(Johnson, Kotz, and Balakrishnan 1994), where $\sigma_s^2 < \sigma^2$. The quantity within brackets gives the fraction of phenotypic variance remaining after selection, σ_s^2/σ^2 (Figure 2.5).

Example 2.5. Equation 2.14 can be used to answer another interesting question. What is the average absolute deviation from the mean, $E(\,|\,z-\mu\,|\,)$, for an individual drawn randomly from a normally distributed population? Because the normal distribution is symmetrical, we can set $\Phi_T = 0.5$ to obtain the average deviation to the right of the mean; the average deviation to the left of the mean will be identical in absolute value, but opposite in sign. In this case the truncation point is the mean, $z' = 0$, which when applied to Equation 2.11 yields $p_T = (2\pi)^{-1/2}e^0 = 0.399$. Substituting into Equation 2.14, we obtain $(\mu_s - \mu) = 0.798\,\sigma$. Thus, the average absolute value of individual deviations from the mean is about 80% of the standard deviation. This quantity, known as the **most probable error**, was widely used by statisticians early in the twentieth century, but now the simple standard deviation is usually reported.

CONFIDENCE INTERVALS

Estimates such as $\bar{z}$ and Var(z) vary from one sample to the next because of sampling error, so it is useful to know how far an observed statistic is likely to deviate from the true parameter being estimated. Although the parametric values are unknown, if something is known about the sampling error of the estimate, it is possible to evaluate the probability that the observed value lies within a specific range of the true value. Generally, we do not estimate the sampling error of statistics by sampling populations over and over again, but instead rely on known algebraic expressions that themselves depend on sample statistics. Such a **parametric approach** is heavily dependent on assuming the correct form of the distribution. A more robust **nonparametric** (distribution-free) approach is the **bootstrap method**. This is a computer-intensive approach, wherein a series of **bootstrap samples** is generated by *resampling* from the original population sample *with replacement* to generate a large collection of samples, each of size n. The variation among these bootstrap samples then yields an estimated sampling variance for the statistic of interest.

As an example of the parametric approach, consider an estimate $\bar{z}$ of the mean of a distribution. An important issue here is the probability α that the parameter μ is within a certain range, $\bar{z} \pm \Delta$. By symmetry, this is the same as the probability that $\bar{z}$ lies within the range $\mu \pm \Delta$. Transforming to standardized variables by letting $z' = (\bar{z} - \mu)/\sigma(\bar{z})$, where $\sigma^2(\bar{z})$ is the sampling variance of the mean, then the probability of interest is defined to be

$$1 - \alpha = \Pr[\,(\bar{z} - \Delta) \leq \mu \leq (\bar{z} + \Delta)\,] = \int_{-\Delta/\sigma(\bar{z})}^{+\Delta/\sigma(\bar{z})} p(z')dz' \tag{2.16}$$

The range $\bar{z} \pm \Delta$ defines the **confidence limits** or **confidence interval** (**CI**) for the mean associated with the $(1-\alpha)$ probability level. For example, a value of $\alpha = 0.05$ generates a 95% CI, which implies that 95% of the time such an interval should contain the true value for a random sample from the same population. In applications of Equation 2.16, it is generally assumed that the statistic is **unbiased** (so that the expected value of the statistic equals the true parameter value) and normally distributed. In the case of the mean, this implies that replicate estimates of the mean $(\bar{z})$ should be normally distributed about the parametric value (μ) with sampling variance $\sigma^2(\bar{z})$. Equation 2.16 is then simply an integration over the standardized normal density.

The standardized limits (Δ/σ) relating to α are easily obtained using statistical packages. The quantity Δ/σ, which we denoted by τ, is the range (in standard errors) of values for the deviation between observed statistic and its parametric value that spans a total probability of $1-\alpha$. (Whereas the standard deviation is a measure of the dispersion of *individual measures*, the term **standard error** is usually reserved as a measure of *the dispersion of a* ***statistic***, e.g., the standard deviation, $\sigma(z)$, of z versus the standard error, $\sigma[\,\bar{z}\,]$, of the sample mean, $\overline{z}$.) For any particular probability level, the critical value of τ decreases with increasing sample size, asymptotically approaching a constant. For sample sizes exceeding 40 or so, $\tau \simeq 1.96$ for $\alpha = 0.05$, and $\tau \simeq 2.58$ for $\alpha = 0.01$ (for smaller sample sizes, one uses the t distribution; Example 2.7).

The remaining problem is to obtain an estimate of the sampling variance of the statistic (the square of the standard error of the estimate). In the case of the mean, it is well known that an unbiased estimator of the sampling variance is $\text{Var}(\overline{z}) = \text{Var}(z)/n$, where $\text{Var}(z)$ is the variance of individual measures (Equation 2.5), and n is the number of (uncorrelated) measurements (Appendix 1). Thus, the 95% CI ($\alpha = 0.05$) for the mean is approximately $\overline{z} \pm 1.96\,[\,\text{Var}(z)/n\,]^{1/2}$, namely, the sample mean plus or minus 1.96 times the approximate standard error for the mean. This is approximation as the *exact* standard error, $\sigma(\overline{z}) = \sigma(z)/\sqrt{n}$, requires knowledge of the true variance of z (the Student's t distribution offers a solution; Example 2.7).

Unfortunately, expressions for the sampling variances of other statistics (such as the variance, higher order moments, coefficients of variation, etc.) are usually much more complicated than those for the mean. Appendix 1 outlines procedures that have been used extensively in quantitative genetics to obtain approximate expressions for the sampling variances for such statistics. These expressions are usually referred to as **large-sample variance** estimators because they are functions of observed statistics (e.g., using $\text{Var}(z)$ in place of the unknown $\sigma^2[z]$) whose reliability increases with increasing sample size. A common procedure is to use twice the large-sample standard error as a crude estimate of τ for a 95% confidence limit for a statistic of interest. We emphasize that this assumes that the statistic has a sampling distribution that is close to normal, that the estimator is unbiased, and that the sample size is large enough that the large-sample variance (itself an estimate) is reasonably reliable.

Departures from Normality

When the distribution of a random variable (such as a test statistic of interest) departs from normality, this potentially impacts confidence intervals in two ways. First, *the presence of skew results in asymmetric confidence intervals* (Example 2.6). Second, when the kurtosis index k_4 (Equation 2.12a) departs from zero, the resulting confidence intervals based on Gaussian p values are either too wide ($k_4 < 0$) or too narrow ($k_4 > 0$). To see the impact of kurtosis, first recall that for a standardized normal z', $\Pr(z' \geq 1.645) = 0.05$, so that only 5% of the probability mass for z' exceeds 1.645. Relative to a normal, the probability mass in the tails falls off more quickly for a platykurtic distribution ($k_4 < 0$) and more slowly for a leptokurtic distribution ($k_4 > 0$). Hence, the probability of exceeding 1.645 is greater than 5% in a leptokurtic distribution, and less than 5% in a platykurtic distribution, resulting in confidence intervals (when based on normality assumptions) that are, respectively, too narrow or too wide.

Example 2.6. If the distribution of an estimator displays skew, the confidence intervals become asymmetric. Here, we evaluate the estimate of a variance. Consider a sample, $z_1, \cdots, z_n$, of n values, whose sample variance is

$$\text{Var} = \frac{1}{n-1}\sum_{i=1}^{n}(z_i - \overline{z})^2$$

where $E[\text{Var}] = \sigma^2$. Following the structure of the confidence interval for a mean, one might imagine that the CI has a symmetric form, $\text{Var} \pm C_\alpha$, for σ^2, where C_α varies with α. However, this is not the case. To find the actual form of the CI, suppose that the individual observations are uncorrelated and normally distributed with $z_i \sim N(\mu, \sigma^2)$. For a sample of size n,

$$\text{Var} \sim \left(\frac{\sigma^2}{n-1}\right) \cdot \chi^2_{n-1} \tag{2.17a}$$

namely, $[(n-1)/\sigma^2]$Var is distributed as a chi-square random variable with $n-1$ degrees of freedom (Equation A5.13c), a highly skewed distribution (Figure A5.3).

To construct a confidence interval for σ^2, first note that if $X_{p,n}$ satisfies $\Pr(\chi^2_n \leq X_{p,n}) = p$, then

$$\Pr(X_{\alpha/2,n} \leq \chi^2_n \leq X_{1-\alpha/2,n}) = 1-\alpha \tag{2.17b}$$

The value of $X_{p,n}$ can be obtained using the R command `qchisq(p,n)`. From Equation 2.17a, substituting $(n-1)\text{Var}/\sigma^2$ for χ^2_{n-1}, we have

$$\Pr\left(X_{\alpha/2,n-1} \leq \frac{(n-1)\text{Var}}{\sigma^2} \leq X_{1-\alpha/2,n-1}\right) \tag{2.17c}$$

$$= \Pr\left(\frac{1}{X_{\alpha/2,n-1}} \geq \frac{\sigma^2}{(n-1)\text{Var}} \geq \frac{1}{X_{1-\alpha/2,n-1}}\right) = 1-\alpha \tag{2.17d}$$

yielding

$$\Pr\left[\left(\frac{n-1}{X_{1-\alpha/2,n-1}}\right)\text{Var} \leq \sigma^2 \leq \left(\frac{n-1}{X_{\alpha/2,n-1}}\right)\text{Var}\right] = 1-\alpha \tag{2.17e}$$

This motivates a $(1-\alpha)$% CI for the true variance, σ^2, given the observed sample variance, Var. Suppose we have a sample size of 20 and desire a 95% CI for σ^2. Here $\alpha = 0.05$, $X_{\alpha/2,n-1} = X_{0.025,19}$, and $X_{1-\alpha/2,n-1} = X_{0.975,19}$, which are, respectively, 8.91 and 32.85. From Equation 2.17e, the resulting 95% CI for σ^2 is (19/32.85)Var to (19/8.91)Var, or 0.57·Var to 2.13·Var, which is asymmetric about the sample estimate, Var.

Example 2.7. As mentioned above, using the estimate $\text{Var}(\overline{z})$ in place of the true value $\sigma^2(\overline{z})$ introduces error into the confidence intervals for the mean. This was addressed by Willian Sealy Gosset in 1908, who (for normally disturbed data) obtained the distribution of $\overline{z}/\sqrt{\text{Var}(z)}$. Working for Guinness brewery, Gosset hid his identity by publishing under the name of "Student," resulting in this distribution being called the **Student *t*** (Fisher 1925a; Box 1981; Zabell 2008). This distribution is defined by its degrees of freedom, which is the sample size n (Equation A7.32a). For small n, the kurtosis of the t distribution departs dramatically from a normal, with $k_4 = 6/(n-4)$ (which is defined for $n \geq 5$; the fourth moment is infinite when $n \leq 4$). Hence, the t distribution is leptokurtic ($k_4 > 0$), with thicker tails than a normal (Figure 2.2). The R command `pt(x,n)` returns the value for $\Pr(t_n \leq x)$, where t_n is a t distribution with n degrees of freedom. A unit normal random variable has a 5% probability of exceeding a value of 1.645. The corresponding probabilities for t distributions with 5, 10, and 30 degrees of freedom are, respectively, 8.0%, 6.5%, and 5.5%. The corresponding k_4 values for these distributions are 6, 1, and 0.2. Hence, as n increases, k_4 decreases towards zero and the tail probabilities approach those for a normal.

The R command `qt(p,n)` returns the value for $x_{p,n}$ that satisfies $\Pr(t_n \leq x_{p,n})$. Hence, `qt(0.975,n)` gives the corresponding value for a 95% CI for the mean (corresponding to greater than the lower 2.5% but less than the upper 97.5%). This critical value is 1.96 for a normal, but 2.57, 2.23, and 2.04 for t distributions with, respectively, 5, 10, and 30 degrees of freedom. Hence, the confidence interval based on a Gaussian are too narrow for a t distribution, especially when n is small.

Chebyshev Confidence Intervals

Given the above concerns with constructing confidence intervals when a distribution shows either skew or kurtosis, is there a more robust method for CI construction? One distribution-free approach is given by **Chebyshev's inequality**, which states (for any distribution with finite moments) that

$$P\left(|z-\mu| \leq i\sigma\right) \leq 1 - \frac{1}{i^2} \tag{2.18}$$

For $i = 2, 4$, and 6, this returns values of 0.75, 0.94, and 0.97. Hence, our earlier stated rule of two standard errors does no worse than generate a 75% confidence interval. Noting that the i value (corresponding to the number of standard errors) for a given p satisfies $i_p = \sqrt{1/p}$, the 95% and 99% **Chebyshev confidence intervals** are, respectively, $\bar{z} \pm 4.47\sigma(\bar{z})$ and $\bar{z} \pm 10\sigma(\bar{z})$. Note that these are *substantially* wider than those for a normal distribution, which are $\bar{z} \pm 1.96\sigma(\bar{z})$ and $\bar{z} \pm 2.58\sigma(\bar{z})$, respectively. Hence, while extremely robust, Chebyshev intervals are always too wide. However, if such an interval excludes a hypothesized value (for example, a null of zero), then rejection of that value is on a very sound statistical footing.

DISCRETE DISTRIBUTIONS

We conclude this chapter by introducing a few other important **discrete distributions**, where random variables take on integer (usually non-negative) values. We do so for two reasons. First, many important traits are discrete, such as those that take on **binary values** (e.g., disease presence or absence). While a trait like the number of leaves on a tree is an integer, we usually treat this as a continuous variable, which is a good approximation when the mean trait value is very large. However, other discrete traits, such as number of eggs in a clutch, are often not well approximated as continuous (in part, due to a significant fraction of observations often having values of zero). Second, many features of the genes underlying a trait take on discrete values, such as the number of plus (trait value increasing) alleles that an individual carries, or the number of homozygotes in a sample. These are all examples of **counting distributions**, which consider how many times a particular event (or events) is seen in a sample.

Useful continuous distributions are developed in later chapters. Appendix 5 discusses distributions relating to normals widely used in statistics, such as t, χ^2, and F, while Chapter 9 introduces the multitvariate normal. Appendix 7 examines distributions widely used in modeling, and as priors for a Bayesian analysis, such as the gamma (an extension to the exponential) and beta distributions.

The Binomial and Multinomial Distributions

Suppose that one has n trials or observations (e.g., scoring n individuals), each of which can have only two outcomes (presence/absence or success/failure), and we code each trial with a value of either 0 (failure) or 1 (success). Further, suppose that each trial has the same probability, p, of a success. The **binomial distribution** gives the probability that one sees exactly k successes in n trials, given p, as

$$\Pr(X = k \mid n, p) = \frac{n!}{k!\,(n-k)!}\, p^k\, (1-p)^{n-k} \tag{2.19a}$$

where X is the random variable for the total number of successes, and $k! = k \cdot (k-1) \cdot (k-2) \cdots 2 \cdot 1$ is the factorial function. Equation 2.19a can be computed using the `R` command `dbinom(k, n, p)`. That X follows this distribution is often compactly written as $X \sim \text{Bin}(p, n)$. Two important special cases are

$$\Pr(X = 0) = (1-p)^n \qquad \text{and} \qquad \Pr(X = n) = p^n \tag{2.19b}$$

while the mean and variance of X are

$$\mu_X = np \qquad \text{and} \qquad \sigma^2_X = np(1-p) \tag{2.19c}$$

When $n = 1$ (a single individual), the binomial is also called a **Bernoulli distribution**.

To see the origin of Equation 2.19a, consider the case of $n = 3$ and $k = 1$ (one success, S, and two failures, F). This outcome can arise three different ways, SFF, FSF, and FFS, where the outcomes in each string are ordered from the first to the last trial. Each of these three orders has probability $p(1-p)^2$, yielding a final probability of $3p(1-p)^2$. The general result given by Equation 2.19a follows the same logic by first noting that the combinatory term (the ratio of factorials in Equation 2.19a) computes the different number of ways (orders) that one can achieve exactly k success in n traits ($k = 1, n = 3$ yields a value of 3), with each ordering having success probability of $p^k(1-p)^{n-k}$.

The natural extension of the binomial is to consider situations where the trait value takes on one of m separate outcomes (such as red, green, or blue). Let p_i denote the probability of outcome i, and X_i denote the number of times outcome i appears in the sample. The probability of seeing a specific number, k_i, of each class in a sample of size n is given by the **multinomial distribution**,

$$\Pr(X_1 = k_1, X_2 = k_2, \cdots, X_m = k_m) = \frac{n!}{\prod_{i=1}^m k_i!} \prod_{i=1}^m p_i^{k_i} \tag{2.20a}$$

subject to the constraints

$$\sum_{i=1}^m k_i = n, \quad k_i \geq 0 \qquad \text{and} \qquad \sum_{i=1}^m p_i = 1, \quad p_i \geq 0$$

namely, that outcome counts (k_i) sum to the sample size (n), and that the probabilities of the different outcomes sum to one. The mean and variance associated with each class are

$$\mu_{X_i} = np_i \qquad \text{and} \qquad \sigma^2_{X_i} = np_i(1-p_i) \tag{2.20b}$$

Note that the binomial follows as a special case, with $m = 2$ and $p_2 = 1 - p_1$. Finally, there is a dependency between outcomes (given that they sum to n), which induces a covariance (Chapter 3) between the observations of class i and j,

$$\sigma(X_i, X_j) = -np_ip_j \ \ (\text{for } i \neq j) \tag{2.20c}$$

Example 2.8. Consider a biallelic locus (alleles A and a) where the population frequency of A is 0.4. Assume a sample of size 50 individuals from a population in Hardy–Weinberg equilibrium (e.g., Freq(AA) = Freq(A)^2). What is the probability that the sample contains two or fewer AA homozygotes? We can treat this as a binomial, with $p = 0.4^2 = 0.16$ being the probability that a randomly drawn individual is AA, and let X be the random variable for the number of AA individuals in the sample. From Equation 2.19a,

$$\begin{aligned}
\Pr(X \leq 2) &= \sum_{k=0}^{2} \left(\frac{50!}{k!(50-k)!} \, 0.16^k \, (1-0.16)^{50-k} \right) \\
&= \frac{50!}{0!\,50!} \, 0.16^0 \, 0.84^{50} + \frac{50!}{1!\,49!} \, 0.16^1 \, 0.84^{49} + \frac{50!}{2!\,48!} \, 0.16^2 \, 0.84^{48} \\
&= 0.84^{50} + \left(50 \cdot 0.16^1 \cdot 0.84^{49}\right) + \left[(50 \cdot 49/2) \cdot 0.16^2 \cdot 0.84^{48}\right] \\
&= 0.009
\end{aligned}$$

Alternatively, one can use the R command `pbinom(k, n, p)`, which computes the probability that $\Pr[X \sim \text{Bino}(p,n)] \leq k$. From Equation 2.19c, the expected number of AA is $0.16{\cdot}50 = 8$.

Suppose we want to compute the probability that a sample of size 60 contains exactly 10 AA (X_1), 30 Aa (X_2), and 20 aa (X_3). This follows using the multinomial, with $p_1 = 0.4^2 = 0.16$, $p_2 = 2 \cdot 0.4 \cdot 0.6 = 0.48$, and $p_3 = 0.6^2 = 0.36 = 1 - p_1 - p_2$. Substituting these values into Equation 2.20a yields

$$\Pr(X_1 = 10, X_2 = 30, X_3 = 20) = \frac{60!}{10!\,30!\,20!} 0.16^{10} \, 0.48^{30} \, 0.36^{20} = 0.0143$$

From Equation 2.20b, the expected numbers in the three classes are 0.16·60 = 9.6 (*AA*), 0.48·60 = 28.8 (*Aa*), and *aa* 0.36·60 = 21.6 (*aa*).

The Poisson Distribution

Another common counting distribution is the **Poisson**, which can be used as a stand-alone model, or as an approximation for the binomial. The Poisson assumes that successes arise independently and at a constant rate (the so-called **Poisson process**). Given an *expected number* of successes, λ, in a sample, the Poisson gives the probability that the *observed number* of success X equals k as

$$\Pr(X = k) = \frac{\lambda^k \exp(-\lambda)}{k!}, \quad \text{for } k = 0, 1, \cdots \tag{2.21a}$$

This is easily computed in R using `dpois(k,lam)`. The probability that no events are seen is

$$\Pr(X = 0) = \exp(-\lambda) \tag{2.21b}$$

We denote a Poisson-distributed random variable by $X \sim \text{Pois}(\lambda)$. The Poisson, and extensions thereof, are widely used in quantitative genetics settings where there is a significant probability that a discrete-valued trait takes on a zero value (e.g., Chapter 14; WL Chapter 29).

An interesting feature about the Poisson is that its mean and variance are the same,

$$\mu_X = \lambda \quad \text{and} \quad \sigma_X^2 = \lambda \tag{2.21c}$$

Because of this feature, the Poisson often serves as a benchmark for the amount of variation (relative to the mean) of a distribution. A distribution is called **overdispersed** if its variance is greater than its mean, and **underdispersed** if the variance is less than the mean.

If the number of trials in a sample is known, and each has a constant success rate p, then np is the expected number of successes. Hence, the Poisson is often used to approximate a binomial when the expected number of successes is much less than the number of trials. The reason for the latter restriction is that while the number of successes, X, in a binomial is bounded by the sample size, n, there is no upper bound in the Poisson. The Poisson usually provides an excellent approximation, provided that $\Pr(X > n)$ is negligible.

Example 2.9. Suppose that we are trying to measure the mutation rate for the red colony trait in yeast. Our experiment consists of scoring 5000 colonies, each of which are comprised of exactly 10^5 cells, all starting with a single wildtype (white colony) cell. Suppose that 50 of the 5000 colonies are red, while the rest are white. We can obtain an estimate of the mutation rate using the zero term of a Poisson. Let μ be the unknown per-cell mutation rate, so that $\lambda = 10^5\mu$. The fraction of white colonies (i.e., those showing zero mutations) is 4950/5000 = 0.99. Assuming that mutations arise according to a Poisson process, from Equation 2.21b, we have that $\Pr(X = 0) = 0.99 = \exp(-10^5\mu)$, or that

$$\widehat{\mu} = \frac{-\ln(0.99)}{10^5} = 1 \times 10^{-7}$$

Example 2.10. Let us recalculate the first part of Example 2.8 using the Poisson approximation. The expected number of successes X (homozygotes) in a sample of size 50 is 50· 0.16 = 8. The probability that $X \leq 2$ computed using the Poisson is

$$\Pr(X \leq 2) = e^{-8}\left(\frac{8^0}{0!} + \frac{8^1}{1!} + \frac{8^2}{2!}\right) = e^{-8}\,(1 + 8 + 32) = 0.014$$

Again, this is easily computed in R, as `ppois(k,lam)` returns $\Pr[X \sim \text{Pois}(\texttt{lam}) \leq k]$. While Example 2.8 found the exact answer to be 0.009, the calculation using the Poisson is much simpler.

The Geometric and Negative Binomial Distributions

For a binary trait, the complement to a counting distribution (how many successes are seen in a *fixed sample*) is considering the events unfolding over a *sequence of trials*. This leads to the concept of **waiting times**: how many trials are required to see the first event (geometric) or the first k events (negative binomial). Again, let each trial be independent and have a constant probability, p, of a success. The experiment stops when the required number of successes is first achieved. For example, we screen individuals until we find a recombination between two markers, or (more generally), k such events.

Let T denote a random variable for the number of trials that have taken place when the first event is observed, which is described by the **geometric distribution**, denoted by $T \sim \text{Geo}(p)$, with

$$\Pr(T = k) = (1-p)^{k-1}p, \qquad k = 1, \cdots \tag{2.22a}$$

This is simply obtained as the product of $k-1$ failures $(1-p)^{k-1}$ followed by a success, p. The resulting mean and variance can be shown to be

$$\mu_T = \frac{1}{p} \quad \text{and} \quad \sigma^2_T = \frac{1-p}{p^2} \tag{2.22b}$$

Similarly, the cdf is given by

$$\Pr(T \leq k) = 1 - (1-p)^k \tag{2.22c}$$

The generalization of the geometric is the total number of trials until we first see r successes. The probability of $k+r$ trails (namely, k failures until r successes) is given by the **negative binomial**, $T \sim \text{NB}(r, p)$, with

$$\Pr(T = k + r) = \frac{(k+r-1)!}{k!\,(r-1)!}(1-p)^k p^r, \qquad k = 0, 1, \cdots \tag{2.23a}$$

The combinatorial term follows because the last event in the sequence is always a success (and hence we stop), so the combination is the number of orders for $(r-1)$ events out of a total of $(k+r-1)$. The resulting mean and variance become

$$\mu_T = \frac{r}{p} \quad \text{and} \quad \sigma^2_T = \frac{r(1-p)}{p^2} \tag{2.23b}$$

One of the more important uses of the negative binomial follows from its dispersion (variance to mean ratio), with

$$\frac{\sigma^2_T}{\mu_T} = \frac{(1-p)}{p} = \begin{cases} >1 & \text{for } p < 1/2 \\ 1 & \text{for } p = 1/2 \\ <1 & \text{for } p > 1/2 \end{cases} \tag{2.23c}$$

Hence, the negative binomial can be used to model both overdispersed ($p < 1/2$) and underdispersed ($p > 1/2$) distributions. As such, it is often used in place of the Poisson for modeling discrete data when the variance is different from the mean.

Example 2.11. Suppose that the probability of recombination between two tightly linked markers is $p = 0.001$. From Equation 2.22b, the mean waiting time until a success is $1/p = 1/0.001 = 1000$ trails. Given this result, an investigator scores 1000 individuals. What is the probability of seeing one (or more) recombination events in such a sample? Equation 2.22c yields a probability of

$$1 - (1 - 0.001)^{1000} = 0.632$$

Hence, there is a reasonable chance (37%) that an experiment of this size would *not* contain the desired event, and thus be a failure if the goal is to obtain at least one such event (such as in a breeding program; Chapter 17). How large must the sample size be to have probability 95% than the experiment is a success? From Equation 2.22c, the sample size, $n_{0.95}$, required to have 95% of experiments of this size show at least one success satisfies

$$\Pr(T \leq n_{0.95}) = 1 - (1 - p)^{n_{0.95}} \geq 0.95$$

which rearranges to $(1 - p)^{n_{0.95}} \leq 0.05$ or

$$n_{0.95} = \frac{\ln(0.05)}{\ln(1 - p)} \tag{2.24}$$

This becomes $\ln(0.05)/\ln(0.999) = 2994$, almost three times the initial sample size. This disparity with respect to the expected value (1000) arises because the geometric distribution has a long tail, implying a rare number of events with very large T (time to first success) values.

The Hypergeometric Distribution

Another counting distribution that arises in genomics and population genetics is the **hypergeometric**. This is usually framed as randomly drawing colored balls, without replacement, from an urn. Suppose there are N total balls in the urn, K of which display the desired color (or, more generally, the **desired feature**, such as a gene being a member of a specific gene ontology [GO] category). Now suppose that a sample of $n \leq N$ balls is randomly drawn from the urn (without replacement). What is the probability that there are exactly k balls with the desired feature in the sample? This is given by

$$\Pr(k \mid n, K, N) = \binom{N}{n}^{-1} \binom{K}{k} \binom{N-K}{n-k} \quad \text{for } k = 0, 1, \cdots, \min(n, K) \tag{2.25a}$$

The mean and variance of k are given by

$$\mu = n\,\frac{K}{N} \quad \text{and} \quad \sigma^2 = n\,\frac{(N-K)(N-n)K}{N^2(N-1)} \tag{2.25b}$$

When $N, K \gg n$, we can approximate sampling without replacement by sampling with replacement, and approximate Equation 2.25a using the binomial, with sample size n and success probability $p = K/N$, yielding

$$\Pr(k \mid n, K, N) \simeq \Pr(k \mid n, p = K/N) = \binom{n}{k} p^k \, (1 - p)^{n-k} \tag{2.25c}$$

The hypergeometric distribution is the basis for **Fisher's exact test** for a 2 × 2 contingency table, which is used when the standard large-sample chi-square test is not appropriate (i.e., when the expected value of any cell in the table is small, typically five or less). Chapter 21 examines applications of the hypergeometric distribution when testing for associations using gene sets.

3

Covariance, Regression, and Correlation

Everything should be made as simple as possible, but no simpler.

Albert Einstein

The previous chapter introduced the variance as a measure of the dispersion of a univariate distribution. Additional statistics are required to describe the joint distribution of two or more variables. The **covariance** provides a natural measure of the association between two variables, and appears in the analysis of many problems in quantitative genetics, including the resemblance between relatives, the correlation between characters, the correlation between alleles at either the same locus (inbreeding; Chapter 12) or at different loci (linkage disequilibrium; Chapter 5), and measures of selection (WL Chapters 6, 29, and 30).

As a prelude to the formal theory of covariance and regression, we first provide a brief review of the theory for the distribution of pairs of random variables. We then give a formal definition of the covariance and its properties. Next, we show how the covariance enters naturally into statistical methods for estimating the linear relationship between two variables (least-squares linear regression) and the goodness-of-fit of such linear trends (correlation). Finally, we apply the concept of covariance to several problems in quantitative-genetic theory. More advanced topics associated with multivariate distributions involving three or more variables are taken up in Chapter 9.

JOINTLY DISTRIBUTED RANDOM VARIABLES

The probability of joint occurrence of a pair of random variables (x, y) is specified by their **joint probability density function**, $p(x, y)$, where

$$\Pr(\, y_1 \le y \le y_2,\; x_1 \le x \le x_2 \,) = \int_{y_1}^{y_2} \int_{x_1}^{x_2} p(x, y)\, dx\, dy \tag{3.1a}$$

We often ask questions of the form: What is the distribution of y given that x equals some specified value? For example, we might want to know the probability that parents whose height is 68 centimeters (cm) have offspring with height exceeding 70 cm. To answer such questions, we use $p(y|x)$, the **conditional density** of y given x, where

$$\Pr(\, y_1 \le y \le y_2 \,|\, x \,) = \int_{y_1}^{y_2} p(\, y \,|\, x \,)\, dy \tag{3.1b}$$

Hence, for our height question we wish to compute

$$\Pr(70 \le y < \infty \,|\, x = 68 \,) = \int_{70}^{\infty} p(\, y \,|\, x = 68 \,)\, dy$$

Joint probability density functions, $p(x, y)$, and conditional density functions, $p(y|x)$, are connected by

$$p(x, y) = p(\, y \,|\, x \,)\, p(x) \tag{3.2a}$$

where $p(x) = \int_{-\infty}^{+\infty} p(\, y, x \,)\, dy$ is the **marginal** (univariate) **density** of x.

Two random variables, x and y, are said to be **independent** if $p(x, y)$ can be factored into the product of a function of x only and a function of y only, i.e.,

$$p(x, y) = p(x)\, p(y) \tag{3.2b}$$

If x and y are independent, knowledge of x gives no information about the value of y. From Equations 3.2a and 3.2b, if $p(x, y) = p(x)\, p(y)$, then $p(\, y \mid x\,) = p(y)$.

Conditional Probability and Bayes Theorem

A special case of Equation 3.2a involves the concept of **conditional probability**: if we observe the value of one random variable x (say, the genotype of a marker), how does that inform us about the value of a second random variable y (say, the value of a trait). Rearranging Equation 3.2a gives the **conditional probability of *y* given *x*** as

$$\Pr(y|\, x) = \frac{\Pr(x, y)}{\Pr(x)} \tag{3.3a}$$

where $\Pr(x, y)$ is the **joint probability** of x and y, and $\Pr(x)$ is the **marginal probability** of x.

An important modification of Equation 3.3a is **Bayes theorem**, which serves as the foundation for Bayesian statistics (Appendix 7). To obtain this expression, first rearrange Equation 3.2a (by exchanging the roles of x and y) as

$$\Pr(x, y) = \Pr(x \mid y)\Pr(y)$$

Substituting this result into Equation 3.3a yields Bayes theorem

$$\Pr(y \mid x) = \frac{\Pr(x \mid y)\Pr(y)}{\Pr(x)} \tag{3.3b}$$

Bayes theorem ***flips the conditioning variable***, allowing us to move from $\Pr(x \mid y)$ to $\Pr(y \mid x)$.

Example 3.1. Suppose that the frequency of a homozygous SNP genotype (*MM*) is 0.1, the frequency of a disease is 0.05, and the frequency of individuals with the disease who also have genotype *MM* is 0.04. From Equation 3.3a,

$$\Pr(\text{disease} \mid MM) = \frac{\Pr(\text{disease and } MM)}{\Pr(MM)} = \frac{0.04}{0.10} = 0.40$$

Hence, 40% of all *MM* carriers display the disease. The **relative risk** for the disease for an *MM* carrier is

$$\frac{\Pr(\text{disease} \mid MM)}{\Pr(\text{disease})} = \frac{0.40}{0.05} = 8$$

That is, an individual carrying an *MM* genotype has an eight-fold increase of having the disease relative to a random individual.

As an example of the use of Bayes theorem, we can flip the question around and ask for the probability that an individual with the disease is an *MM* carrier. From Equation 3.3b,

$$\Pr(MM \mid \text{disease}) = \frac{\Pr(\text{disease} \mid MM)\Pr(MM)}{\Pr(\text{disease})} = \frac{0.40 \cdot 0.10}{0.05} = 0.8$$

Hence, 80% of individuals with the disease have an *MM* genotype. Conversely, 20% of disease cases are found in individuals with genotypes other than *MM*.

Example 3.2. As a slightly more involved application of Bayes theorem, consider the impact of a diallelic QTL (a quantitative trait locus underlying a trait). Suppose that the QTL genotypes QQ, Qq, and qq have, respectively, population frequencies of 0.09, 0 42, and 0.49. Further suppose that the distribution of trait values (z) given a QTL genotype are all normal with a common variance of $\sigma^2 = 20$. The mean value of z is a function of the QTL genotype, with values of 75, 70, and 65, respectively, for QQ, Qq, and qq. What is the probability that an individual with a trait value of 75 or larger is a QQ? A Qq? A qq?

To start, first recall our assumptions that the distributions of traits values conditioned on the QTL genotype are normal, with

$$z \,|\, QQ \sim \mathrm{N}(75, 20), \qquad z \,|\, Qq \sim \mathrm{N}(70, 20), \qquad z \,|\, qq \sim \mathrm{N}(65, 20)$$

Consider QQ first. Applying Bayes theorem (Equation 3.3b),

$$\Pr(QQ \,|\, z \geq 75) = \frac{\Pr(z \geq 75 \,|\, QQ) \Pr(QQ)}{\Pr(z \geq 75)}$$

Note that by conditioning on the QTL genotype, we can express $\Pr(z \geq 75)$ as

$$\Pr(z \geq 75 \,|\, QQ) \Pr(QQ) + \Pr(z \geq 75 \,|\, Qq) \Pr(Qq) + \Pr(z \geq 75 \,|\, qq) \Pr(qq)$$

From our assumptions about the conditional distributions, the probability that an individual with QTL genotype QQ exceeds a trait value of 75 is the probability that a normal random variable with mean 75 and variance 20 exceeds 75, namely

$$\Pr(z \geq 75 \,|\, QQ) = \Pr\left[z \geq 75 \,|\, z \sim \mathrm{N}(75, 20)\right]$$

with similar expressions for the other genotypes. Recall that the `R` command `pnorm(x,u,s)` returns $\Pr[z \leq x \,|\, z \sim \mathrm{N}(u, s^2)]$. Hence `1-pnorm(75, 75, sqrt(20))` $= \Pr(z \geq 75 \,|\, QQ)$, which returns a value of 0.5. The `R` command `1-pnorm(75, 70, sqrt(20))`, corresponding to $\Pr(z \geq 75 \,|\, Qq)$, returns a value of 0.132, while `R` finds that $\Pr(z \geq 75 \,|\, qq)$ = 0.013. Thus,

$$\Pr(z \geq 75) = 0.5 \cdot 0.09 + 0.132 \cdot 0.42 + 0.013 \cdot 0.49 = 0.107$$

showing that roughly 11% of the population has z values of 75 or larger. Hence,

$$\Pr(QQ \,|\, z \geq 75) = \frac{\Pr(z \geq 75 \,|\, QQ) \Pr(QQ)}{\Pr(z \geq 75)} = \frac{0.5 \cdot 0.09}{0.107} = 0.422$$

so that 42% of all individuals whose trait value is 75 or larger are QQ. Similarly,

$$\Pr(Qq \,|\, z \geq 75) = \frac{0.132 \cdot 0.42}{0.107} = 0.520 \quad \text{and} \quad \Pr(qq \,|\, z \geq 75) = \frac{0.013 \cdot 0.49}{0.107} = 0.058$$

Even though QQ has a larger mean value, an individual whose trait value exceeds 75 is more likely to be a Qq than a QQ. This somewhat counterintuitive result is simply a consequence of frequency differences, as Qq individuals are roughly five times more common than QQ. This example highlights one of the features of quantitative genetics that distinguishes it from classical genetics: ***the phenotype is a poor predictor of the underlying genotype.***

Expectations of Jointly Distributed Variables

The expectation of a bivariate function, $f(x, y)$, is determined by integration over the joint probability density

$$E[\,f(x, y)\,] = \int_{-\infty}^{+\infty} \int_{-\infty}^{+\infty} f(x, y)\, p(x, y)\, dx\, dy \tag{3.4}$$

Most of this chapter is focused on **conditional expectation,** i.e., the expectation of one variable, given information on another. For example, one may know the value of x (perhaps parental height), and wish to compute the expected value of y (offspring height) given x. In general, conditional expectations are computed by using the conditional density, i.e., for a simple univariate variable

$$E[\,y\,|\,x\,] = \int_{-\infty}^{+\infty} y\,p(\,y\,|\,x\,)\,dy \tag{3.5}$$

If x and y are independent, then $E[y\,|\,x] = E[y]$, the unconditional expectation. Otherwise, $E[\,y\,|\,x\,]$ is a function of the specified x value. For height in humans (Figure 1.1), Galton (1886) observed a linear relationship,

$$E[\,y\,|\,x\,] = \alpha + \beta x \tag{3.6}$$

where α and β are constants. Thus, the conditional expectation of adult height in offspring (y) was linearly related to the average height of the parents (x).

COVARIANCE

Consider a set of paired variables, (x, y). For each pair, subtract the population mean μ_x from the measure of x, and similarly subtract μ_y from y. Finally, for each pair of observations, multiply both of these new measures together to obtain $(x-\mu_x)(y-\mu_y)$, namely the product of their *deviations from their respective means*. The **covariance** of x and y is defined to be the average of this quantity over all pairs of measures in the population,

$$\sigma(x, y) = E[\,(x-\mu_x)\,(y-\mu_y)\,] \tag{3.7}$$

We often denote the covariance by $\sigma_{x,y}$. Because $E[x] = \mu_x$ and $E[y] = \mu_y$, expansion of the product in Equation 3.7 leads to further simplification,

$$\begin{aligned}
\sigma(x, y) &= E[\,(x-\mu_x)\,(y-\mu_y)\,] \\
&= E[\,xy - \mu_y\,x - \mu_x\,y + \mu_x\,\mu_y\,] \\
&= E[x\,y] - \mu_y\,E[x] - \mu_x\,E[y] + \mu_x\,\mu_y \\
&= E[x\,y] - \mu_x\,\mu_y
\end{aligned} \tag{3.8}$$

In words, the covariance is the mean of the pairwise cross-product xy minus the product of the means. The sampling estimator of $\sigma(x, y)$ is similar in form to that for a variance (Equation 2.5),

$$\text{Cov}(x, y) = \frac{n\,(\,\overline{xy} - \overline{x}\cdot\overline{y}\,)}{n-1} \tag{3.9}$$

where n is the number of pairs of observations, and

$$\overline{xy} = \frac{1}{n}\sum_{i=1}^{n} x_i\,y_i$$

The covariance is a measure of association between x and y (Figure 3.1). It is positive if y increases with increasing x, negative if y decreases as x increases, and zero if there is no *linear* tendency for y to change with x. If x and y are independent (and hence $E[xy] = E[x]\,E[y]$), then $\sigma(x, y) = 0$, but *the converse is not true*—a covariance of zero does not necessarily imply independence. (We will return to this shortly; see Figure 3.3.)

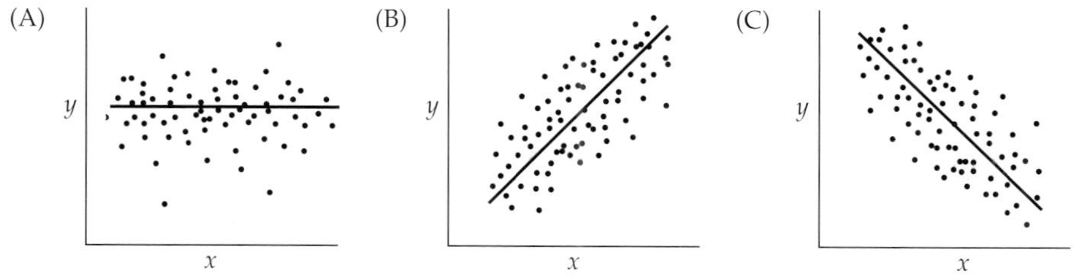

Figure 3.1 Scatterplots for the variables x and y. Each point in the x–y plane corresponds to a single pair of observations (x, y). The line drawn through each scatterplot gives the expected value of y given a specified value of x. **(A)** There is no linear tendency for large x values to be associated with large (or small) y values, so $\sigma(x, y) = 0$. **(B)** As x increases, the conditional expectation of y given x, $E[y \mid x]$, also increases, and $\sigma(x, y) > 0$. **(C)** As x increases, the conditional expectation of y given x decreases, and $\sigma(x, y) < 0$.

Useful Identities for Variances and Covariances

Because $\sigma(x, y) = \sigma(y, x)$, covariances are ***symmetrical***. Furthermore, from the definition of the variance and covariance,

$$\sigma(x, x) = \sigma^2(x) \tag{3.10a}$$

i.e., ***the covariance of a variable with itself is the variance of that variable***. It also follows from Equation 3.8 that, for any constant a,

$$\sigma(a, x) = 0 \tag{3.10b}$$

$$\sigma(a\,x, y) = a\,\sigma(x, y) \tag{3.10c}$$

and if b is also a constant

$$\sigma(a\,x, b\,y) = a\,b\,\sigma(x, y) \tag{3.10d}$$

From Equations 3.10a and 3.10d,

$$\sigma^2(a\,x) = a^2\sigma^2(x) \tag{3.10e}$$

i.e., ***the variance of the transformed variable ax is a^2 times the variance of x***. Likewise, for any constant a,

$$\sigma[\,(a + x), y\,] = \sigma(x, y) \tag{3.10f}$$

so that ***simply adding a constant to a variable does not change its covariance with another variable, nor does it change its variance***.

Finally, the covariance of two sums can be written as a sum of covariances,

$$\sigma[\,(x + y), (w + z)\,] = \sigma(x, w) + \sigma(y, w) + \sigma(x, z) + \sigma(y, z) \tag{3.10g}$$

Similarly, the variance of a sum can be expressed as the sum of all possible variances and covariances. From Equations 3.10a and 3.10g,

$$\sigma^2(x + y) = \sigma^2(x) + \sigma^2(y) + 2\sigma(x, y) \tag{3.11a}$$

which generalizes to

$$\sigma^2\left(\sum_i^n x_i\right) = \sum_i^n \sum_j^n \sigma(x_i, x_j) = \sum_i^n \sigma^2(x_i) + 2\sum_{i<j}^n \sigma(x_i, x_j) \tag{3.11b}$$

As a special case of Equation (3.11b), ***the variance of a sum of uncorrelated variables is just the sum of the variances of each variable***, because the second term in 3.11b is zero for uncorrelated variables.

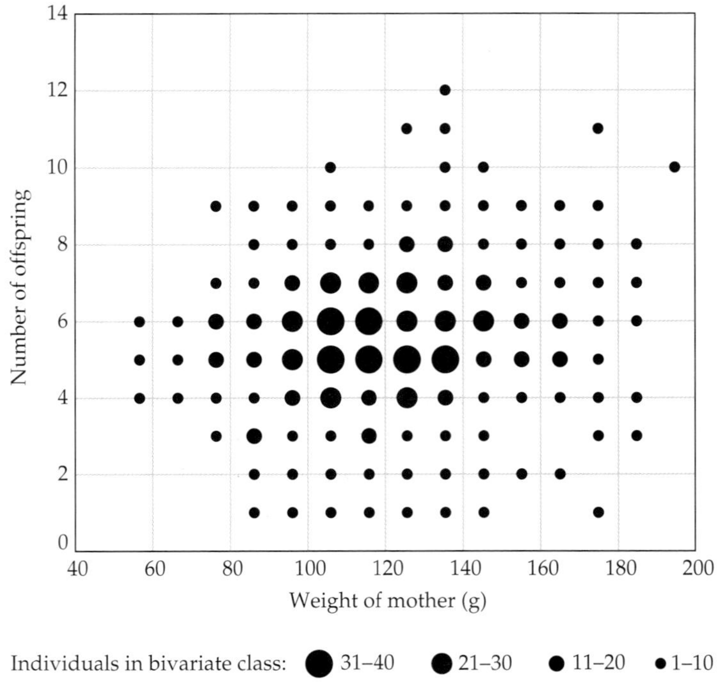

Figure 3.2 A bivariate plot of the relationship between maternal weight and number of offspring for the sample of rats summarized in Table 2.2. Different-sized circles refer to different numbers of individuals in the bivariate classes.

We will make considerable use of the preceding relationships in the remainder of this chapter and in chapters to come. As one application, consider the variance of a *difference*, which is found to be

$$\begin{aligned}\sigma^2(x-y) &= \sigma(x-y, x-y) \\ &= \sigma(x,x) + (-1)\sigma(x,y) + (-1)\sigma(y,x) + (-1)^2\sigma(y,y) \\ &= \sigma^2(x) + \sigma^2(y) - 2\sigma(x,y) \end{aligned} \tag{3.11c}$$

The same logic shows that Equation 3.11c also equals $\sigma^2(y-x)$. Note by comparison to Equation 3.11a, that if x and y are uncorrelated, then the variance of their sum equals the variance of their difference, $\sigma^2(x+y) = \sigma^2(x-y)$. Methods for approximating variances and covariances of more complex functions are outlined in Appendix 1.

REGRESSION

Depending on the causal connections between two variables, x and y, their true relationship may be linear or nonlinear. However, regardless of the true pattern of association, a linear model can always serve as a first approximation. In this case, the analysis is particularly simple,

$$y = \alpha + \beta x + e \tag{3.12a}$$

where α is the y-intercept, i.e., the expected value of y with x = 0 (under the linear model), β is the slope of the line (also known as the **regression coefficient**), and e is the **residual error**. Note that β corresponds to the expected change in y given a unit change in x. Letting

$$\widehat{y} = \alpha + \beta x \tag{3.12b}$$

be the value of y predicted by the model, then the residual error is the deviation between the observed and predicted y value, i.e., $e = y - \widehat{y}$. When information on x is used to predict y, x

is referred to as the **predictor** or **independent variable** and y as the **response** or **dependent variable**.

The objective of linear regression analysis is to estimate the model parameters, α and β, that give the "best fit" for the joint distribution of x and y. The true parameters α and β are only obtainable if the entire population is sampled. With an incomplete sample, α and β are approximated by sample estimators, denoted as a and b. Good approximations of α and β are sometimes obtainable by visual inspection of the data, particularly in the physical sciences, where deviations from a simple relationship are due to errors of measurement rather than biological variability. However, in biology many factors are often beyond the investigator's control. The data in Figure 3.2 provide a good example. While there appears to be a weak positive relationship between maternal weight and offspring number in rats, it is difficult to say anything more precise. An objective definition of "best fit" is required.

Derivation of the Least-Squares Linear Regression

The mathematical method of **least-squares (LS) linear regression**, a method central to much of quantitative genetics, provides one such best-fit solution. Without making any assumptions about the true joint distribution of x and y, least-squares regression minimizes the average value of the squared (vertical) deviations of the observed y values from the regression line predictions. That is, the **least-squares** solution yields the values of a and b that minimize the mean squared residual, $\overline{e^2}$. Other criteria could be used to define "best fit." For example, one might minimize the mean absolute deviations of observed values from predicted values. However, as we will now see, least-squares regression has the unique and very useful property of maximizing the amount of variance in y that can be explained by a linear model.

Consider a sample of n individuals, each of which has been measured for x and y. Recalling the definition of a residual

$$e = y - \widehat{y} = y - a - bx \tag{3.13a}$$

and then adding and subtracting the quantity $(\,\overline{y} + b\,\overline{x}\,)$ on the right side, we obtain

$$e = (\,y - \overline{y}\,) - b\,(\,x - \overline{x}\,) - (\,a + b\,\overline{x} - \overline{y}\,) \tag{3.13b}$$

Squaring both sides leads to

$$\begin{aligned} e^2 = {} & (\,y - \overline{y}\,)^2 - 2\,b\,(\,y - \overline{y}\,)\,(\,x - \overline{x}\,) + b^2(\,x - \overline{x}\,)^2 + (\,a + b\,\overline{x} - \overline{y}\,)^2 \\ & - 2\,(\,y - \overline{y}\,)\,(a + b\,\overline{x} - \overline{y}\,) + 2\,b\,(\,x - \overline{x}\,)(\,a + b\,\overline{x} - \overline{y}\,) \end{aligned} \tag{3.13c}$$

In addition, we consider the average value of e^2 in the sample. The final two terms in Equation 3.13b drop out here because, by definition, the mean values of $(x - \overline{x})$ and $(y - \overline{y})$ are zero, while a, b, $\overline{x}$, and $\overline{y}$ are constants. Finally, by definition, the mean values of the first three terms are directly related to the sample variances and covariance. Thus,

$$\overline{e^2} = \left(\frac{n-1}{n}\right)\left[\,\text{Var}(y) - 2\,b\,\text{Cov}(x,y) + b^2\,\text{Var}(x)\,\right] + (\,a + b\,\overline{x} - \overline{y}\,)^2 \tag{3.13d}$$

The values of a and b that minimize $\overline{e^2}$ are obtained by taking partial derivatives of this function and setting them equal to zero:

$$\frac{\partial\,(\overline{e^2})}{\partial\,a} = 2\,(a + b\,\overline{x} - \overline{y}\,) = 0$$

$$\frac{\partial\,(\overline{e^2})}{\partial\,b} = 2\left[\left(\frac{n-1}{n}\right)[-\text{Cov}(x,y) + b\,\text{Var}(x)\,] + \overline{x}\,(\,a + b\,\overline{x} - \overline{y}\,)\right] = 0$$

The solutions to these two equations are

$$a = \overline{y} - b\,\overline{x} \tag{3.14a}$$

$$b = \frac{\mathrm{Cov}(x,y)}{\mathrm{Var}(x)} \tag{3.14b}$$

Thus, the least-squares estimators for the intercept and slope of a linear regression are simple functions of the observed means, variances, and covariances. From the standpoint of quantitative genetics, this property is exceedingly useful, because such statistics are readily obtainable from phenotypic data. Note that the above analysis also applies when we replace sample estimates ($\overline{x}, \overline{y}, \mathrm{Cov}[x,y], \mathrm{Var}[x]$) by their population values ($\mu_x, \mu_y, \sigma^2[x], \sigma[x,y]$). Making these substitutions into Equation 3.14a and 3.14b recovers expressions for the population values α and β.

The origin of the method of least-squares was the subject of a major priority dispute between two of the early mathematical titans, Gauss and Legendre. Legendre published first, in a 1805 paper on determining the orbit of comets. Although Gauss independently published in 1809, there is evidence that he used this approach as early as 1795 (Plackett 1972; Stigler 1981).

Properties of Least-squares Regressions

Here we summarize some fundamental features and useful properties of the least-squares approach to linear regression analysis:

1. ***The regression line passes through the means of both x and y.*** This relationship should be immediately apparent from Equation 3.14a, which implies $\overline{y} = a + b\,\overline{x}$. Hence, Equation 3.12a can be written to remove a, with

$$(y - \overline{y}\,) = b(x - \overline{x}\,) + e$$

   Namely, a plot of deviations of each variable from its mean yields a plot that goes through the origin.

2. ***The average value of the residual is zero.*** From Equation 3.13a, the mean residual is $\overline{e} = \overline{y} - a - b\,\overline{x}$, which is constrained to be zero by Equation 3.14a. Thus, the least-squares procedure results in a fit to the data such that the sum of (vertical) deviations above and below the regression line are exactly equal, i.e., $\sum e_i = 0$.

3. ***For any set of paired data, the least-squares regression parameters, a and b, define the straight line that maximizes the amount of variation in y that can be explained by a linear regression on x.*** Because $\overline{e} = 0$, it follows that the variance of residual errors about the regression is simply $\overline{e^2}$. As just noted, this variance is the quantity minimized by the least-squares procedure.

4. ***The residual errors around the least-squares regression are uncorrelated with the predictor variable x.*** This statement follows because

$$\begin{aligned}\mathrm{Cov}(x,e) &= \mathrm{Cov}[\,x,(y - a - b\,x)\,] = \mathrm{Cov}(x,y) - \mathrm{Cov}(x,a) - b\,\mathrm{Cov}(x,x)\\ &= \mathrm{Cov}(x,y) - 0 - b\,\mathrm{Var}(x)\\ &= \mathrm{Cov}(x,y) - \frac{\mathrm{Cov}(x,y)}{\mathrm{Var}(x)}\,\mathrm{Var}(x) = 0\end{aligned}$$

   Note, however, that $\mathrm{Cov}(x,e) = 0$ does not guarantee that e and x are *independent*. In Figure 3.3, for example, because of a nonlinear relationship between y and x, the residual errors associated with extreme values of x tend to be negative while those for intermediate values are positive. Thus, if the true regression is nonlinear then $E[\,e\,|\,x\,] \neq 0$ for some x values, and the predictive power of the linear model

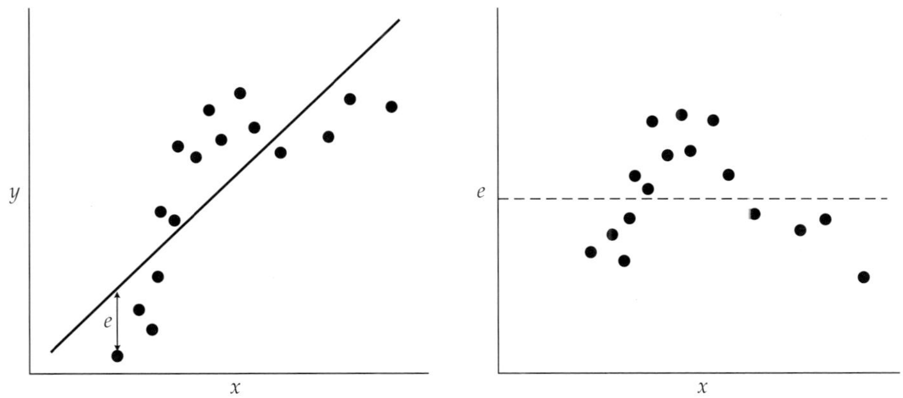

Figure 3.3 A linear least-squares fit to an inherently nonlinear data set. Although there is a systematic relationship between the residual error (e) and the predictor variable (x), e and x are uncorrelated (show no net *linear* trend) when viewed over the entire range of x. The mean residual error ($\overline{e} = 0$) is denoted by the dashed line on the right graph.

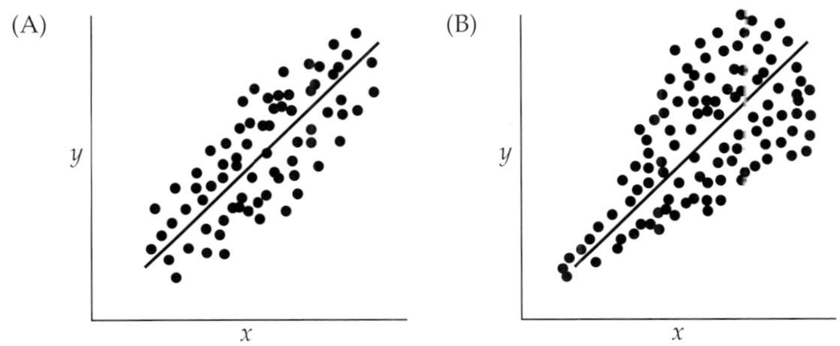

Figure 3.4 The dispersion of residual errors around a regression. **(A)** The regression is homoscedastic: the variance of residuals given x is a constant. **(B)** The regression is heteroscedastic: the variance of residuals increases with x. In this case, higher x values predict y with less certainty.

is compromised. In addition, even if the true regression is linear, the variance of the residual errors may vary with x, in which case the regression is said to display **heteroscedasticity** (Figure 3.4). If the conditional variance of the residual errors given any specified x value, $\sigma^2(\, e \,|\, x\,)$, is a constant (i.e., independent of the value of x), then the regression is said to be **homoscedastic**.

5. There is an important situation in which ***the true regression, the value of $(\,y\,|\,x\,)$, is both linear and homoscedastic, when x are y are bivariate normally distributed.*** The requirements for such a distribution are that the univariate distributions of both x and y are normal and that the conditional distributions of y given x, and x given y, are also normal (Chapter 9). This feature, coupled with a well-developed theory for statistical testing based on Gaussian distributions, makes it generally desirable to work with normally distributed data. For situations in which the raw data are not so distributed, a variety of transformations exist that may render the data close to normality (Chapter 14).

6. It is clear from Equations 3.14a and 3.14b that ***the regression of y on x is different from the regression of x on y unless the means and variances of the two variables are equal.*** This distinction is made by denoting the regression coefficient by $b(y, x)$, $b_{y,x}$, or $b_{y\,|\,x}$ when x is the predictor and y the response variable.

For practical reasons, we have expressed the above properties in terms of the estimators

a, b, Cov(x, y), and Var(x). They also hold when the estimators are replaced by the true parameters α, β, $\sigma(x, y)$, and $\sigma^2(x)$.

Example 3.3. Suppose Cov(x, y) = 10, Var(x) = 10, Var(y) = 15, and $\overline{x} = \overline{y} = 0$. Compute the least-squares regressions of y on x, and of x on y. From Equation 3.14a, $a = 0$ for both regressions. However, $b(y, x) = \text{Cov}(x, y)/\text{Var}(x) = 10/10 = 1$, while $b(x, y) = \text{Cov}(x, y)/\text{Var}(y) = 2/3$. Hence, $\widehat{y} = x$ is the least-squares regression of y on x, while $\widehat{x} = (2/3)y$ is the regression of x on y. As we discuss further in Chapter 10, the estimates $\widehat{y}$ and $\widehat{x}$ based on least-squares are called **BLUEs**, for **best linear unbiased estimates**.

Example 3.4. Our above discussion of regressions assumes that the predictor variables, x_i, are randomly drawn from some population, and our interest is in making inference about the underlying true best linear fit if we could sample the entire population. An alternative use of least-squares regression assumes that the x_i values are a fixed set, rather than being a representative sample from a larger population. In such cases, Equation 3.14a and 3.14b still yield the best (least-squares) linear fit for the observed data.

To see this, we wish to find the values of a and b that minimize $\sum_i e_i^2$, where $e_i = y_i - (a - bx_i)$. As above, we obtain these by finding the values where the partial derivatives of $\sum_i e_i^2$ (with respect to a and b) are zero,

$$\frac{\partial \left[\sum_{i=1}^{n}(y_i - a - bx_i)^2\right]}{\partial a} = 0 \tag{3.15a}$$

$$\frac{\partial \left[\sum_{i=1}^{n}(y_i - a - bx_i)^2\right]}{\partial b} = 0 \tag{3.15b}$$

Recalling from the chain rule that $d[f(z)^2]/dz = 2f(z)d[f(z)]/dz$, Equation 3.15a reduces to

$$-2\sum_{i=1}^{n}(y_i - a - bx_i) = -2\left(n\overline{y} - na - nb\overline{x}\right) = 0$$

which recovers Equation 3.14a

$$a = \overline{y} - b\overline{x} \tag{3.15c}$$

Likewise, after taking partials, Equation 3.15b reduces to

$$\sum_{i=1}^{n}(y_i - a - bx_i)x_i = 0$$

We can simplify this by writing $x_i = x_i - \overline{x} + \overline{x}$ to yield

$$\sum_{i=1}^{n}(y_i - a - bx_i)(x_i - \overline{x}) + \overline{x}\sum_{i=1}^{n}(y_i - a - bx_i) = 0$$

The second summation is zero, as it equals $n\overline{e}$. Next, using Equation 3.15c,

$$\sum_{i=1}^{n}(y_i - \overline{y} + b\overline{x} - bx_i)(x_i - \overline{x}) = 0$$

$$\sum_{i=1}^{n}(y_i - \overline{y})(x_i - \overline{x}) - b\sum_{i=1}^{n}(x_i - \overline{x})^2 = 0$$

yielding

$$b = \frac{\sum_{i=1}^n (y_i - \overline{y})(x_i - \overline{x})}{\sum_{i=1}^n (x_i - \overline{x})^2} = \frac{\sum_{i=1}^n (y_i - \overline{y})(x_i - \overline{x})/(n-1)}{\sum_{i=1}^n (x_i - \overline{x})^2/(n-1)} = \frac{\text{Cov}(x,y)}{\text{Var}(x)} \tag{3.15d}$$

recovering Equation 3.14b.

CORRELATION

For purposes of hypothesis testing, it is often desirable to use a dimensionless measure of association. The most frequently used such measure in bivariate analysis is the **correlation coefficient,**

$$r(x,y) = \frac{\text{Cov}(x,y)}{\sqrt{\text{Var}(x)\,\text{Var}(y)}} \tag{3.16a}$$

Note that $r(x,y)$ is symmetrical, i.e., $r(x,y) = r(y,x)$. Where there is no ambiguity as to the variables being considered, we abbreviate $r(x,y)$ as r. The parametric correlation coefficient is denoted by $\rho(x,y)$, ρ_{xy}, or ρ (when there is no ambiguity) and equals $\sigma(x,y)/[\sigma(x)\sigma(y)]$. The least-squares regression coefficient is related to the correlation coefficient by

$$b(y,x) = r\sqrt{\frac{\text{Var}(y)}{\text{Var}(x)}} \tag{3.16b}$$

An advantage of correlations over covariances is that the former are *scale independent*. This can be seen by noting that if w and c are constants,

$$r(w\,x, c\,y) = \frac{\text{Cov}(w\,x, c\,y)}{\sqrt{\text{Var}(w\,x)\,\text{Var}(c\,y)}} = \frac{w\,c\,\text{Cov}(x,y)}{\sqrt{w^2\,\text{Var}(x)\,c^2\,\text{Var}(y)}} = r(x,y) \tag{3.17a}$$

Thus, scaling x or y by constants does not change the correlation coefficient, although the variances and covariances are affected. Because r is dimensionless with limits of ± 1, it gives a direct measure of the degree of association: if $|r|$ is close to one, x and y are very strongly associated in a linear fashion, while if $|r|$ is close to zero, they are not.

The correlation coefficient has other useful properties. First, ***r* is a standardized regression coefficient** (the regression coefficient resulting from rescaling x and y such that each has unit variance). Letting $x' = x/\sqrt{\text{Var}(x)}$ and $y' = y/\sqrt{\text{Var}(y)}$ gives $\text{Var}(x') = \text{Var}(y') = 1$, implying

$$b(y',x') = b(x',y') = \text{Cov}(x',y') = \frac{\text{Cov}(x,y)}{\sqrt{\text{Var}(x)\,\text{Var}(y)}} = r \tag{3.17b}$$

Thus, when variables are standardized, the regression coefficient is equal to the correlation coefficient regardless of whether x' or y' is chosen as the predictor variable.

Second, ***the squared correlation coefficient measures the proportion of the variance in y that is explained by assuming that E[y|x] is linear.*** The variance of the response variable y has two components: $r^2\,\text{Var}(y)$, the amount of variance accounted for by the linear model (the **regression**, or **model, variance**), and $(1 - r^2)\,\text{Var}(y)$, the remaining variance not accounted for by the regression (the **residual variance**). To obtain this result, we derive the variance of the residual deviation defined in Equation 3.13a,

$$\begin{aligned}
\text{Var}(e) &= \text{Var}(\,y - a - bx\,) = \text{Var}(\,y - bx\,) \\
&= \text{Var}(y) - 2\,b\,\text{Cov}(x,y) + b^2\,\text{Var}(x) \\
&= \text{Var}(y) - \frac{2\,[\,\text{Cov}(x,y)\,]^2}{\text{Var}(x)} + \frac{[\,\text{Cov}(x,y)\,]^2\,\text{Var}(x)}{[\,\text{Var}(x)\,]^2} \\
&= \left(1 - \frac{[\,\text{Cov}(x,y)\,]^2}{\text{Var}(x)\,\text{Var}(y)}\right)\text{Var}(y) = (1 - r^2)\,\text{Var}(y)
\end{aligned} \tag{3.18}$$

Example 3.5. Returning to Table 2.1, the preceding formulae can be used to characterize the relationship between maternal weight and offspring number in rats. Here we take offspring number as the response variable y and maternal weight as the predictor variable x. The mean and variance for maternal weight were found to be $\overline{x} = 118.90$ and $\text{Var}(x) = 623.06$ (Table 2.1). For offspring number, $\overline{y} = 5.49$ and $\text{Var}(y) = 2.94$. To obtain an estimate of the covariance, we first require an estimate of $E[x\,y]$. Taking the xy cross-product of all classes in Table 2.1 (using the midpoint of the interval for the value of x) and weighting them by their frequencies,

$$\overline{xy} = \frac{(1 \cdot 4 \cdot 55) + (3 \cdot 5 \cdot 55) + (1 \cdot 6 \cdot 55) + \cdots + (1 \cdot 10 \cdot 195)}{1003} = 660.14$$

The covariance estimate is then obtained using Equation 3.9,

$$\text{Cov}(x, y) = \frac{1003}{1002}\,[\,660.14 - (118.90 \times 5.49)\,] = 7.39$$

From Equation 3.14b, the slope of the regression is found to be

$$b(y, x) = \frac{7.39}{623.06} = 0.01$$

Thus, the expected increase in the number of offspring per gram increase in maternal weight is about 0.01. How predictable is this change? From Equation 3.16a, the correlation coefficient is estimated to be

$$r = \frac{7.39}{\sqrt{623.06 \times 2.94}} = 0.17$$

Squaring this value, $r^2 = 0.03$. Therefore, only about 3% of the variance in offspring number can be accounted for with a model that assumes a linear relationship with maternal weight.

A TASTE OF QUANTITATIVE-GENETIC THEORY

Directional Selection Differentials and the Robertson–Price Identity

The evolutionary response of a character to selection is a function of the intensity of selection and the fraction of the phenotypic variance attributable to certain genetic effects. As noted at the end of the last chapter, the **directional selection differential**, S, is defined to be the within-generation difference between the mean phenotype μ_s after an episode of selection and the mean before selection μ,

$$S = \mu_s - \mu \tag{3.19}$$

As discussed in WL Chapter 29, with differential reproduction μ_s is given by the fitness-weighted mean.

The degree to which μ_s deviates from μ depends on the survivorship and reproductive rates of individuals with different phenotypes (WL Chapter 29). If all individuals have equal fertility and viability, then $\mu_s = \mu$, $S = 0$, and the population mean phenotype is not expected to change between generations. Now, for simplicity, assume that individuals differ only in the probability of survival to maturity, so that fitness, $W(z)$, is the probability that individuals with phenotype z survive to reproduce. In what follows, no assumptions will be made about the general form of $W(z)$; it may be a continuous or discontinuous function of z, and it may take on values of 0 for some z. If $p(z)$ is the density of phenotype z before selection, then the density after selection is

$$p_s(z) = \frac{W(z)\,p(z)}{\int W(z)\,p(z)\,dz} \tag{3.20}$$

This expression is obtained by noting that $W(z)$ is a weighting factor for phenotype z. The denominator is the **mean individual fitness**, $\overline{W} = E[W]$, the average fitness of a random individual. Letting the **relative fitness** of phenotype z be $w(z) = W(z)/\overline{W}$, Equation 3.20 simplifies to $p_s(z) = w(z)\,p(z)$. It follows that the mean phenotype after selection is

$$\mu_s = \int z\,p_s(z)\,dz = \int z\,w(z)\,p(z)\,dz = E[\,z\,w(z)\,] \tag{3.21}$$

Note also that

$$\overline{w} = \int w(z)\,p(z)\,dz = \frac{1}{\overline{W}}\int W(z)\,p(z)\,dz = \overline{W}/\overline{W} = 1$$

i.e., the mean relative fitness in a population is always equal to one. Because $\mu = E[z]\cdot E[w] = E[z]\cdot 1$, the directional selection differential may be rewritten as

$$S = \mu_s - \mu = E[\,z\,w(z)\,] - E[z]\,E[w] = \sigma[\,z, w(z)\,] \tag{3.22}$$

Thus, the ***directional selection differential is equivalent to the covariance of phenotype and relative fitness***. This relationship, first noted by Robertson (1966), was greatly elaborated on by Price (1970, 1972). We refer to this very useful result as the **Robertson–Price identity**. It applies even when phenotypes vary in reproductive output, provided that the absolute fitnesses, $W(z)$, are weighted accordingly (WL Chapters 6 and 29).

The importance of S can be seen by noting that if the regression of offspring phenotype on that of its average parent is linear with slope β (Figure 1.2), a change in the parental mean phenotype induces an expected change in the mean phenotype across generations equal to

$$R = \Delta\mu = \mu_o - \mu = \beta\,(\mu_s - \mu) = \beta\,S \tag{3.23}$$

where μ_o is the mean phenotype of the offspring of the selected parents. This fundamental relationship, known as the **breeder's equation** (Lush 1937), combines information on the forces of selection (S) with that on inheritance (β) to yield a predictive equation for evolutionary change across generations. A genetic interpretation of the regression coefficient β will be provided in the final section of this chapter. The response to selection is only sporadically mentioned in this volume, with a complete treatment given in Walsh and Lynch (2018).

The Correlation Between Genotypic and Phenotypic Values

Equation 3.23 shows that evolution by natural selection requires heritable variation, as no matter how large S is, the response to selection across generations is zero if $\beta = 0$. Quantification of the correspondence between phenotypic and genotypic values is related to one of the central goals of quantitative genetics: the partitioning of the phenotypic variance into genetic and nongenetic components. The standard approach is to consider the phenotypic value of an individual, z, to be the sum of the total effects of all loci on the trait, G (the **genotypic value**), and an **environmental deviation** E (analogous to the residual error above),

$$z = G + E$$

Using the properties of covariances noted above, the covariance between phenotypic and genotypic values may be written as

$$\sigma_{z,G} = \sigma[\,(G+E), G\,] = \sigma^2_G + \sigma_{G,E} \tag{3.24}$$

The squared correlation coefficient is therefore

$$\rho^2(G,z) = \left(\frac{\sigma_{z,G}}{\sigma_G\,\sigma_z}\right)^2 = \frac{(\sigma^2_G + \sigma_{G,E})^2}{\sigma^2_G\,\sigma^2_z} \tag{3.25a}$$

which simplifies to

$$\rho^2(G, z) = \frac{\sigma_G^2}{\sigma_z^2} \tag{3.25b}$$

if there is no genotype–environment covariance, i.e., if $\sigma_{G,E} = 0$. In this special case, $\rho^2(G, z)$ is simply the proportion of the total phenotypic variance that is genetic. The quantity σ_G^2/σ_z^2 is generally referred to as **heritability in the broad sense** (or **broad–sense heritability**) and abbreviated as H^2.

From Equation 3.24, it can be seen that covariance between genotypic values and environmental deviations ($\sigma_{G,E} \neq 0$) causes the genotype–phenotype covariance to deviate from σ_G^2. Negative covariance between G and E causes a reduction in the correlation between phenotypic and genotypic values, and in extreme cases, can cause $\rho(G, z)$ to become negative. Further details on genotype–environment covariance are covered in Chapter 6 and also in Part II Chapters 27 and 28.

Regression of Offspring Phenotype on Midparent Phenotype

Although the previous example has provided some insight into the genetic basis of phenotypic variation without getting bogged down in genetic complexities, in practice that approach is not very useful. Whereas phenotypic values are easily obtained (they are what we measure), the underlying genetic values are essentially unobservable without an extensive breeding program, and even then, they cannot be determined with complete accuracy.

Fortunately, there are alternative ways to estimate levels of genetic variance of quantitative traits. All such methods are based on the simple fact that related individuals carry copies of many of the same alleles. Consider the resemblance between phenotypes of offspring (z_o) and their midparents (z_{mp}). A midparent value is simply the average phenotype of a mother (z_m) and a father (z_f),

$$z_{mp} = \frac{z_m + z_f}{2}$$

We will confine our attention to a simple genetic situation: a single locus with purely additive gene effects (namely, no dominance or epistasis; Chapter 4), diploidy, random mating, and no selection. Let g_m and g_f, respectively, be the effects of the alleles that an offspring inherits from its mother and father, and g'_m and g'_f be the effects of the alleles that are not transmitted by the parents to this particular offspring. Further letting the environmental effects on the phenotypes of parents and offspring be E_m, E_f, and E_o, the three phenotypes may be expressed as

$$z_m = g_m + g'_m + E_m$$
$$z_f = g_f + g'_f + E_f$$
$$z_o = g_m + g_f + E_o$$

Because the equation for the offspring phenotype contains three terms and that for the midparent phenotype contains six, the complete algebraic expression for the midparent–offspring covariance, $\sigma(z_{mp}, z_o)$, is quite complex. It contains 18 terms. However, provided certain assumptions are met, most of these terms have expected values equal to zero. First, under the assumptions of random mating and no selection, there can be no covariance between the effects of alleles within individuals. Thus, the genes inherited by an offspring have zero covariance with the genes that are not inherited, and the genes in mothers are uncorrelated with those in fathers, i.e., $\sigma(g_m, g'_m) = \sigma(g_m, g_f) = \sigma(g_m, g'_f) = \sigma(g_f, g'_f) = \sigma(g_f, g_m) = \sigma(g_f, g'_m) = 0$. Second, provided there is no genotype–environment covariance, i.e., individuals are not assorted into environments on the basis of their genetic attributes, $\sigma(g_m, E_m) = \sigma(g_m, E_f) = \sigma(g_f, E_m) = \sigma(g_f, E_f) = \sigma(g_m, E_o) = \sigma(g'_m, E_o) = \sigma(g_f, E_o) = \sigma(g'_f, E_o) = 0$. Finally, provided the parents do not transmit their environmental effects to their progeny, i.e., there are no significant maternal or paternal environmental effects, then $\sigma(E_o, E_f) = \sigma(E_o, E_m) = 0$.

Most of these assumptions can be fulfilled in carefully designed experiments (Part II Chapters 22–33). Assuming this is the case, the only potential sources of covariance that exist between midparent and offspring phenotypes are those involving the inherited genes. Thus,

$$\sigma(z_{mp}, z_o) = \sigma\left[\left(\frac{z_m + z_f}{2}\right), z_o\right] = \sigma\left[\left(\frac{g_m + g_f}{2}\right), (g_m + g_f)\right]$$
$$= \frac{\sigma^2(g_m) + \sigma^2(g_f)}{2} \tag{3.26a}$$

Recall that we assumed the genotypic value to be entirely defined by the additive effects of the two alleles at each locus. Thus, under random mating, the total genetic variance in the population is the sum of the variances of maternally and paternally derived genes, $\sigma^2(g_m) + \sigma^2(g_f)$. Because the gene effects are purely additive, this quantity may also be referred to as the **additive genetic variance,** σ_A^2 (or the **additive variance** for short). Thus, provided a number of assumptions are met, the phenotypic covariance between midparent and offspring is equivalent to half the additive genetic variance in the population,

$$\sigma(z_{mp}, z_o) = \frac{\sigma_A^2}{2} \tag{3.26b}$$

In Chapters 5 and 7, it will be shown that this equation holds for any number of loci (with or without dominance) provided they interact additively.

To obtain the expected least-squares regression of offspring on midparent phenotype, the covariance needs to be divided by the variance of midparent phenotypes. Using the properties of variances outlined above, and noting that the phenotypic covariance between parents, $\sigma(z_f, z_m)$, is zero under random mating,

$$\sigma^2(z_{mp}) = \sigma^2\left(\frac{z_m + z_f}{2}\right) = \frac{\sigma^2(z_m) + \sigma^2(z_f)}{4} \tag{3.27a}$$

Thus, provided the phenotypic variance in the two sexes is equal (or has been scaled to be so), the phenotypic variance of midparent values is half the phenotypic variance in the population,

$$\sigma^2(z_{mp}) = \frac{\sigma_z^2}{2} \tag{3.27b}$$

The slope of the least-squares linear regression of offspring phenotype on midparent phenotype is then

$$\beta_{o,mp} = \frac{\sigma(z_o, z_{mp})}{\sigma^2(z_{mp})} = \frac{\sigma_A^2/2}{\sigma_z^2/2} = \frac{\sigma_A^2}{\sigma_z^2} \tag{3.28}$$

Thus, for this special case, the slope of a midparent–offspring regression provides an estimate of the proportion of the phenotypic variance that is attributable to additive genetic factors. Obviously, we have made many assumptions in order to arrive at this expression, and the impact of these will be addressed extensively throughout the rest of the book. The salient issue here—and the historical foundation for quantitative genetics—is that ***inferences concerning the genetic basis of quantitative traits can be extracted from phenotypic measures of the resemblance between relatives*** (Fisher 1918).

In closing, we note that there is an important distinction between the measures of genetic variance that appear in Equations 3.25b and 3.28. In Equation 3.25b, σ_G^2 refers to the total genetic variance, including that due to nonadditive interactions within and among loci, dominance and epistasis, respectively. In Equation 3.28, σ_A^2 refers specifically to the additive component of genetic variation. The ratio σ_A^2/σ_z^2 is known as the **narrow-sense heritability** and is generally abbreviated as h^2. It is possible for the total genetic variance to be entirely additive, but often it is not. This distinction between broad- and narrow-sense heritability and the decomposition of genetic variance into additive and nonadditive components will be covered in detail in Chapters 4, 5, and 7.

The central importance of h^2 can be appreciated by recalling the breeders' equation, Equation 3.23, where β can now be seen to be h^2. Thus, if S is the within-generation change in mean phenotype caused by selection, then the response (the change in offspring mean, R) to selection across generations is (Lush 1937)

$$R = h^2 S \tag{3.29}$$

The narrow-sense heritability can be thought of as the efficiency of the response to selection. If $h^2 = 0$, i.e., if there is no tendency for offspring to resemble their parents, there can be no evolutionary change, regardless of the strength of selection.

4

Properties of Single Loci

Genetics is the science allowing you to blame your parents for everything.

Hetvi

The fact that most principles of quantitative genetics can be expressed without reference to specific genes is precisely why quantitative-genetic analysis is so popular among those who study complex characters. Because this same feature can be cause for suspicion, a primary goal of the next few chapters is to clarify the ways in which quantitative genetics is grounded in fundamental Mendelian concepts. Prior to illustrating the connections between the properties of single genes and the expression and transmission of polygenic traits, we review some very basic and essential vocabulary.

It is well known that the genetic information encoding for characters resides on extremely long strands of deoxyribonucleic acid (DNA) molecules called **chromosomes**. We are still deciphering the function of the vast majority of DNA in organisms, and many believe that a substantial portion of it has no function, at least in multicellular species (reviewed in detail in WL Chapter 8). Historically, the term **gene** referred to a DNA sequence that encodes for a particular product (a protein or RNA), with their chromosomal locations called **loci**. Most organisms have two copies of each of several chromosomes (one copy from each parent), in which case they are said to be **diploid**. Because DNA replication is an imperfect process, mutations arise, and as a consequence the two "copies" of each gene carried by diploid individuals need not be identical. The various forms of a gene are called **alleles**.

This historical definition of a gene, which was largely based on the ability to score functional variants, has been replaced in the genomics era with a much broader definition given that we can now directly score variation in the DNA itself. While it is straightforward to assess whether a sequence codes for a protein or a large RNA, we are still in the early stages of detecting all of the constellation of sequences that can influence gene regulation (Chapter 21). For example, a single base change in a noncoding sequence may influence the expression of a gene on a different chromosome. The former region would be a trait locus in the sense that it can have a functional role and result in trait variation.

Gene loci that exhibit more than one allele are the subject of genetics. Such loci are said to be **polymorphic**, whereas loci at which all gene copies are identical are **monomorphic**. A substantial fraction of the gene pool in many species is polymorphic The possible reasons for this are the subject of a long-standing debate in population genetics and molecular evolution (Kimura 1983; Gillespie 1991; examined in detail in WL Chapter 8). Many mutant alleles are extremely deleterious and are rapidly eliminated by natural selection (WL Chapter 5), while others have only small or no effects at the phenotypic level and remain in the population until they are fixed or lost by chance (WL Chapter 2). Still others are maintained at intermediate levels by a balance between opposing evolutionary forces (WL Chapter 7).

Not all organisms are diploid. Prokaryotes have only a single copy of each gene and are referred to as **haploid**. Many of the protists and basal plants (algae, mosses, and ferns) also have conspicuous haploid stages in their life cycles, as do the fungi and some animals (males of both rotifers and haplo-diploid insects). Organisms with ploidy levels higher than diploid are known as **polyploids**. A **tetraploid** individual contains four sets of homologous chromosomes, whereas a **hexaploid** contains six. Polyploidy is extremely widespread among plants. It is relatively rare among sexual animals, but common among parthenogenetic species.

Even in diploids, some genes are effectively haploid. Such is the case for genes carried in organelles (mitochondria and chloroplasts). Although there may be hundreds of copies of organelle genes per cell, they are generally inherited uniparentally and are essentially all the same. Genes residing on the **sex chromosomes** of organisms with a genetic sex-determination mechanism also have a special ploidy status. In mammals, for example, males carry X and Y chromosomes (and are said to be **heterogametic** because of the mismatching sex chromosomes), whereas females are XX (and are said to be **homogametic**), so that X-linked genes occur only in single "copies" in males. In some organisms, such as birds, moths, and butterflies, the heterogametic (WZ) sex is female, while males are ZZ. In order to distinguish sex chromosomes from the remaining pairs, the latter are referred to as **autosomes**. In this book, unless stated otherwise, we will be dealing with autosomal loci in diploid populations.

The remainder of this chapter is concerned with the quantification of various properties of single loci. We start by reviewing the concepts of allele and genotype frequencies, showing how the two are connected in an ideal situation that is closely approximated in many natural settings. We next show how the phenotypic effects of different alleles can be described in terms of additive and dominance effects. The genotypic frequencies and effects are then incorporated into expressions for the additive and dominance components of genetic variance at the locus. Finally, we show how the additive effects of an individual's genes define its breeding value for any particular trait. While several of the concepts covered in this chapter may seem rather abstract and far removed from the analysis of multilocus traits, they are becoming a bit more tractable as molecular approaches have allowed us to localize **quantitative-trait loci** (**QTLs**). Despite their name, these are not specific loci, but rather are small chromosomal regions showing variation that influence trait values. Under classic QTL mapping using linkage approaches (Chapters 5, 18, and 19), such regions are typically on a megabase (10^6 nucleotides) scale, while finer resolution (on the kilobase scale, 10^4 nucleotides) is offered by **genome-wide association studies** (**GWAS**) which exploit population-level linkage disequilibrium (Chapters 5 and 20). Ideally, the goal is to isolate specific casual sites (often referred to as **quantitative-trait nucleotides**, **QTNs**), but this goal has remained elusive (Chapter 21).

ALLELE AND GENOTYPE FREQUENCIES

The **genotype** at a single locus refers to the pair of alleles that a (diploid) individual carries at that locus. Individuals that have two identical alleles are called **homozygotes**, whereas those that have different alleles are **heterozygotes**. If, for example, we denote the alleles at a particular diallelic locus as B_1 and B_2, there are three possible genotypes: B_1B_1 and B_2B_2 homozygotes, and B_1B_2 heterozygotes. There may, of course, be more than two alleles, and hence more than three genotypes, present at a locus in a population. In particular, with k alleles, there are $k[k+1]/2$ diploid genotypes (k homozygotes and $k(k-1)/2$ heterozygotes).

Allele frequencies are defined uniquely by genotype frequencies. Suppose that P_{11}, P_{12}, and P_{22} represent the proportions of the population that are, respectively, B_1B_1, B_1B_2, and B_2B_2. If these are the only possible genotypes at the locus, then by definition, $P_{11} + P_{12} + P_{22} = 1$. If there are N individuals in the population, then $P_{11}N$ individuals contain two B_1 alleles and $P_{12}N$ individuals contain a single B_1 allele. Because there are a total of $2N$ genes in the population for each autosomal locus, the frequency of the B_1 allele is simply

$$p_1 = \frac{2\,P_{11}\,N + P_{12}\,N}{2\,N} = P_{11} + \frac{1}{2}\,P_{12} \tag{4.1}$$

Thus, the general rule for a diploid autosomal locus is that the frequency of an allele is estimated by the observed frequency of homozygotes plus one-half the observed frequency of all heterozygotes containing that allele.

A direct application of this approach involves **single nucleotide polymorphisms** (**SNPs**), which are usually diallelic, but in theory could segregate as many as four alleles (corresponding to the nucleotides A, G, T, and C). Through either hybridization techniques (such as DNA chips) or high-throughput DNA sequencing, one can score the exact frequencies of the genotypes associated with a particular SNP (Chapter 8). In the absence of sequence data, it is usually impossible to be certain about the genotypic state of any particular locus for complex morphological and behavioral characters. In some cases, however, the majority of the genetic variation for a character depends on a single locus with large effects, allowing for the possibility of directly estimating allele and genotype frequencies. This was the fortuitous case in many of Mendel's classic experiments with peas, and a number of genetic disorders in humans appear to be products of mutant alleles at single loci. Data for a wing-color polymorphism in a British moth are discussed in the following example.

Example 4.1. Fisher and Ford (1947) were able to distinguish three wing-color patterns in the tiger moth *Panaxia dominula,* and, through breeding experiments, the polymorphism was found to result from two alleles segregating at a single locus. WL Examples 9.1 and 9.2 review evidence for selection acting on this specific locus. The following table summarizes the distribution of genotype frequencies observed in a population in 1946

Color Pattern	*dominula*	*medionigra*	*bimacula*	Total
Genotype	B_1B_1	B_1B_2	B_2B_2	
Sample Size (N_{ij})	905	78	3	N = 986
Frequency (P_{ij})	0.918	0.079	0.003	1.000

What are the estimated frequencies of the two alleles? Using Equation 4.1, the frequency of the B_1 allele is found to be

$$p_1 = 0.918 + \frac{0.079}{2} = 0.958$$

and because there are only two alleles, the frequency of B_2 is $p_2 = 1 - p_1 = 0.042$.

THE TRANSMISSION OF GENETIC INFORMATION

The Hardy–Weinberg Principle

From the standpoint of evolutionary analysis, it is crucial to understand how allele and genotype frequencies change from generation to generation (a full treatment of this topic is given in WL Chapters 2–7). Such changes may result from natural selection, mutation, differential migration, inbreeding, or random drift due to gene sampling in finite populations. All of these forces will be considered in due course, but for now we restrict our attention to a highly idealized situation—an autosomal locus uninfluenced by selection and mutation. By assuming the population to be effectively infinite in size and randomly mating, we also eliminate the possibility of inbreeding and random drift. We further assume that generations are discrete and that the population is closed to immigrants.

Although such an idealized situation is never realized perfectly, in many cases (such as over modest time scales) it is close enough to the truth for practical purposes. Under the ideal model, simple and predictable relationships emerge between allele and genotype frequencies, within and between generations. It is therefore an essential point of departure, much like the ideal gas laws in physics.

In sexual populations, individuals do not necessarily produce offspring whose genotypes match their own. Prior to reproduction, sexual individuals produce haploid gametes by a special form of cell division called **meiosis** (Figure 4.1). Thus, with respect to a single

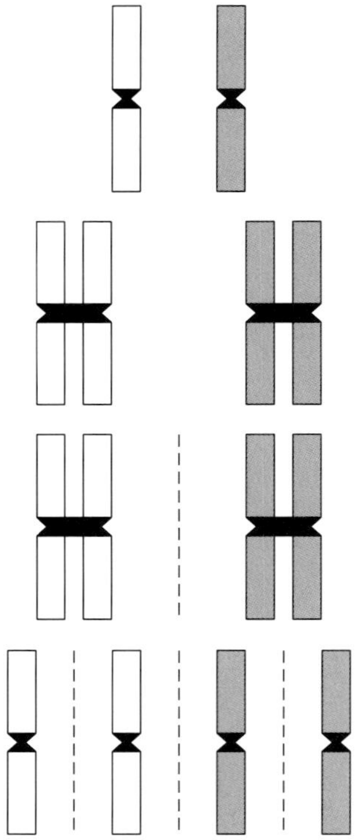

Figure 4.1 Idealized schematic of meiotic production of gametes. Only a single chromosome pair is shown, with the two **homologs** denoted by the different shading. At the onset of meiosis, **sister chromatids** are formed by duplication of each homolog and the homologous pairs come together to form a **tetrad**; although it is not shown, some exchange of material (e.g., **gene conversion** and/or **crossingover**) between homologs may occur at this time. Two meiotic divisions (reductional followed by equational) then produce four haploid products. The maternal and paternal chromosomes migrate to opposite cells during the **reductional division** generally independently for each chromosome pair, and the sister chromatids are isolated into four potential haploid gametes after the **equational division**.

locus, a B_1B_2 heterozygote **segregates** two types of **gametes**—half B_1 and half B_2. The diploid state is restored when gametes from two parents fuse to form a **zygote**. Consequently, at a diallelic locus, a heterozygous parent can potentially produce three types of progeny (B_1B_1, B_1B_2, and B_2B_2), whereas homozygous parents can produce at most two.

Consider a population consisting of separate sexes (**dioecious**) with discrete, nonoverlapping generations. We denote the frequencies of B_1 and B_2 alleles in females in generation 0 by $p_{1f}(0)$ and $p_{2f}(0)$, and those in males by $p_{1m}(0)$ and $p_{2m}(0)$. Under random mating, the expected genotype frequencies in the next generation are obtained from the products of the respective gamete frequencies. For example, because the probability of drawing a B_1 female gamete is $p_{1f}(0)$ and that of drawing a B_1 male gamete is $p_{1m}(0)$, the expected frequency of B_1B_1 zygotes is $p_{1f}(0)\,p_{1m}(0)$. Similarly, the expected frequencies of B_1B_2 and B_2B_2 zygotes are $p_{1f}(0)\,p_{2m}(0) + p_{2f}(0)\,p_{1m}(0)$ and $p_{2f}(0)\,p_{2m}(0)$, respectively. Provided the locus is autosomal, the frequency of the B_1 allele will now be the same in both offspring sexes, because the subpopulations of sons and daughters both acquire half their genes from mothers and half from fathers. Substituting into Equation 4.1, the B_1 allele frequency in generation 1 is

$$\begin{aligned} p_1 &= p_{1f}(0)\,p_{1m}(0) + \frac{p_{1f}(0)\,p_{2m}(0) + p_{1m}(0)\,p_{2f}(0)}{2} \\ &= \frac{p_{1f}(0)\,[\,p_{1m}(0) + p_{2m}(0)\,] + p_{1m}(0)\,[\,p_{1f}(0) + p_{2f}(0)\,]}{2} \\ &= \frac{p_{1f}(0) + p_{1m}(0)}{2} \end{aligned}$$

Likewise, the new frequency for the B_2 allele is $p_2 = 1 - p_1 = [\,p_{2f}(0) + p_{2m}(0)\,]/2$.

Under the conditions of our idealized population, in the next, and in all subsequent, generations, the allele frequencies are both constant and the same in both sexes, with p_1 being the frequency of B_1 and $p_2 = 1 - p_1$ being the frequency of B_2. Further, under random mating (following the logic above), the B_1B_1, B_1B_2, and B_2B_2 genotypes will be found in frequencies p_1^2, $2p_1p_2$, and p_2^2. Such proportions are known as **Hardy–Weinberg (HW) frequencies**, after the two investigators who first pointed out the above relationship (Hardy 1908; Weinberg 1908).

The HW frequencies can also be obtained directly by multiplying out the terms of the binomial expansion, $(p_1 + p_2)^2$. By this means, the Hardy-Weinberg law can be extended to any number of alleles. Suppose, for example, that four alleles (B_1, B_2, B_3, B_4) are present at the locus of interest. The HW frequencies for the various genotypes are obtained by squaring the quantity $(p_1 + p_2 + p_3 + p_4)$. The expected frequency of a genotype homozygous for the B_i allele is p_i^2, while that for a B_iB_j heterozygote is $2p_ip_j$. One application of the HW proportions is a quick quality-control (QC) scan for genomic data, as sequencing and genotyping errors can generate excess departures from HW (relative to the majority of other markers).

Provided that all of the assumptions of the Hardy-Weinberg model are met, we can summarize its implications as follows. First, it takes no more than a single generation to equilibrate and stabilize the gene (allele) frequencies in the two sexes. Second, only one additional generation is required for the stabilization of the (autosomal) genotype frequencies into the predictable HW proportions. These results have obvious implications for the analysis of natural populations. Even if genotype frequencies in a study population are vastly different from HW expectations, for example because of natural selection or population subdivision, they can be rendered close to the idealized proportions by imposing an artificial program of random mating for one or two generations.

Sex-Linked Loci

The preceding results do not extend to sex-linked loci. As noted above, when the male is the heterogametic sex, females are diploid for X-linked loci, but males are haploid. Thus, for every mating pair, there are three X chromosomes, and the frequency of the B_1 allele in the population is $p_1 = [\,p_{1m}(0) + 2p_{1f}(0)]/3$. In the absence of any forces operating differentially on the alleles, this frequency will be maintained indefinitely. However, the gene frequency will not necessarily be p_1 in both of the sexes. Because males only receive an X chromosome from their mother, the male frequency of the B_1 allele in any generation (t) is necessarily equal to the frequency in females in the previous generation $(t - 1)$,

$$p_{1m}(t) = p_{1f}(t-1) \tag{4.2a}$$

On the other hand, fathers and mothers each contribute an X chromosome to their daughters, so the frequency of the B_1 allele in females is equal to the average allele frequency across the two sexes in the previous generation,

$$p_{1f}(t) = \frac{p_{1f}(t-1) + p_{1m}(t-1)}{2} \tag{4.2b}$$

The general solution to these equations, assuming random mating, is

$$p_{1f}(t) - p_1 = \left[-\frac{1}{2}\right]^t \left[\,p_{1f}(0) - p_1\,\right], \qquad p_{1m}(t) = p_{1f}(t-1) \tag{4.2c}$$

Thus, the approach to the equilibrium allele frequency ($p_{1f} = p_{1m} = p_1$) in the two sexes is gradual and oscillatory if the locus is X linked (Figure 4.2). The deviation of the allele

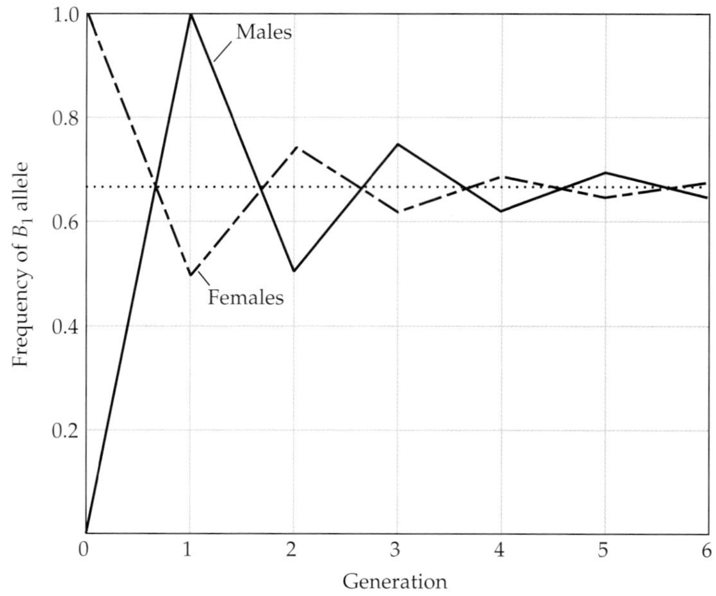

Figure 4.2 The dynamics of gene frequency change for an X-linked gene, B_1, under random mating. An extreme case is illustrated—initially, all females are homozygous for the B_1 gene, $p_{1f}(0) = 1$, while all of the males are haploid for the alternate allele, $p_{1m}(0) = 0$. Consequently, all males contain the B_1 allele in the following generation, while all females are heterozygous. The dotted line represents the population-level gene frequency, $p_1 = [p_{1m}(0) + 2p_{1f}(0)]/3 = 0.67$, towards which both of the sexes converge over time.

frequency from p_1 is halved each generation for both males and females, but the sign changes from generation to generation.

Polyploidy

Another situation in which the Hardy–Weinberg principle is not met exactly arises in polyploid organisms. Because of the high frequency of polyploidy in plants, this case has been examined extensively by Fisher (1947) and Crow (1954), among others. It will only be considered briefly here for a tetraploid species, individuals of which propagate two genes per locus through gametes. The way in which sets of chromosomes assort during meiosis in polyploids depends on the degree of homology between ancestral chromosomes (Marsden et al. 1987). At one extreme are **allopolyploids** that originate by interspecific hybridization. In this case, provided the chromosomes of the parental species are sufficiently different, they will not pair. Meiosis is then identical to that for diploid organisms, except for the doubled number of chromosomes. At the other extreme, **autopolyploids** derive both chromosome sets from the same species. In this latter case, complications for HW can arise because full sets of four, rather than two sets of two, chromosomes can pair during meioses.

For the remainder of our discussion of polyploidy, we will assume that the four sets of chromosomes are sufficiently similar that **tetravalents** (combinations of four homologes), rather than **bivalents**, are formed during meiosis. This condition raises the possibility that some gametes will contain two copies of one of the four genes carried by the parent (i.e., a parent with genotype $B_1B_2B_3B_4$ may produce a B_1B_1 gamete), a result that arises when a **crossover** (a reciprocal exchange of DNA) occurs between replicated arms of two of the four chromosomes during meiosis. The production of such a gamete is referred to as a **double reduction**, and we denote its probability by c. Of the $(1 - c)$ gametes that are not doubly reduced, one-third will contain genes that came from the same parent, and the other two-thirds will contain one paternally derived and one maternally derived gene (Figure 4.3). Here we assume the presence of only two alleles and random assortment of the four homologs. Letting p_i be the frequency of the B_i allele and $p_{ij}(t)$ be the frequency of B_{ij}

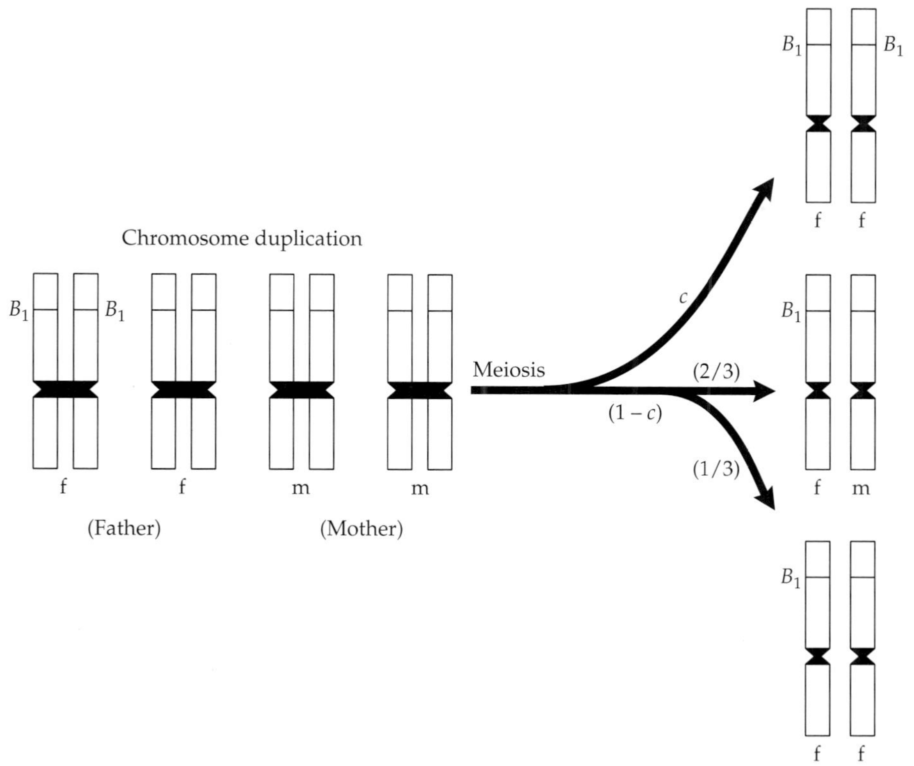

Figure 4.3 The production of three types of (diploid) gametes by a tetraploid individual. This example focuses upon a single paternally derived allele, B_1. The letters f and m refer to chromosomes in an indivdual derived from its father and mother. With four chromosomes (rather than the two of a diploid), the reductional division of meiosis isolates *two* chromosomes at random into each of the resulting two cells. Subsequent equational division generates the gamete types shown at the right. Here c is the probability that the allele of interest will become associated with itself during gametogenesis as a result of a double reduction (i.e., a crossover between the locus and the centromere). If this does not occur (with probability $1 - c$), there is a two-third chance that chromosome B_1 will be associated in a gamete with a maternally derived chromosome and a one-third chance that it will be associated with the other paternally derived chromosome.

gametes in generation t, the following dynamic equations hold:

$$p_{ii}(t) = c\,p_i + \frac{1-c}{3}\left[\,p_{ii}(t-1) + 2\,p_i^2\,\right] \tag{4.3a}$$

$$p_{ij}(t) = \frac{1-c}{3}\left[\,p_{ij}(t-1) + 2p_ip_j\,\right] \tag{4.3b}$$

(Crow and Kimura 1970, pp. 52–53). The equilibrium solution to these equations is obtained by setting $p_{ii}(t) = p_{ii}(t-1)$ and $p_{ij}(t) = p_{ij}(t-1)$, yielding

$$p_{ii} = (1-f)\,p_i^2 + f\,p_i \tag{4.3c}$$

$$p_{ij} = (1-f)\,p_i\,p_j \tag{4.3d}$$

where $f = 3c/(2+c)$. This equilibrium is approached only gradually. The equilibrium genotype frequencies can be obtained as products of the appropriate equilibrium gametic frequencies.

In the absence of crossing-over between homologous pairs of chromosomes, $c = 0$, $f = 0$, and the equilibrium frequency of gamete types is simply equal to the product of the respective allele frequencies. However, if $c > 0$, the equilibrium genotype frequencies are not so simple. Consider the extreme case of free recombination (which is approximately the case when a locus is sufficiently far from its centromere). After chromosomal duplication during gametogenesis, eight chromosomes are assorted, two into each of four gametes. Conditional on one of these being transmitted to a gamete, then of the remaining seven possibilities, one will be identical by descent. Thus, for free recombination, $c = 1/7$, $f = 0.2$, and the equilibrium gamete frequencies are $p_{ii} = 0.2p_i(1+4p_i)$ and $p_{ij} = 0.8p_ip_j$. In essence, if there is any crossing-over, polyploidy results in a sort of "internal inbreeding," reducing the frequency of heterozygous gametes. Wricke and Weber (1986), and especially Gallais (2003), provide a very useful coverage of the many complications that polyploidy introduces in quantitative-genetic formulations.

Age Structure

One final complication with respect to the idealized model is age structure. Up to now we have been assuming a population with discrete, nonoverlapping generations, such as an annual plant with no seed carry-over across years or a univoltine insect. In populations composed of several age classes (the majority of land plants and animals), the generations overlap, and this causes the approach of genotype frequencies towards the HW expectations to be gradual, even in the case of an autosomal locus. This property arises because the genotypes of new recruits are a function of the allele frequencies specific to the reproductive age classes. Juvenile age classes only influence the change in genotype frequencies through mortality, but as they mature they begin to add copies of their genes to the population. The genotype frequencies become stable only after the allele frequencies become homogenized across age classes and sexes.

Of equal significance is the fact that the allele frequencies themselves can be unstable in an age-structured population even in the absence of genotypic differences in age-specific survival and reproduction. Further complexities are introduced by the scheme of mating between the various age classes. All of these subjects are taken up in detail by Charlesworth (1974, 1994) and Gregorius (1976), while Caswell (2001) examines the modeling of age- and stage-structured populations. The critical point is that when newly founded populations have significant age structure, fluctuations in both gene and genotype frequencies may occur for a substantial period of time even in the absence of selection and drift.

Testing for Hardy–Weinberg Proportions

When data are available on genotype frequencies in a population, it is standard practice to cross-check these with their HW expectations. Lack of concordance between the two implies that at least one assumption of the Hardy-Weinberg model is violated, and this often instigates further investigation. Several statistical techniques have been proposed (Weir 2010), the historically most popular of which is the χ^2 (Chi-square) test,

$$X^2 = \sum_{i=1}^{n}\sum_{j\geq i}^{n} \frac{\left(N_{ij} - \widehat{N}_{ij}\right)^2}{\widehat{N}_{ij}} \tag{4.4a}$$

where N_{ij} and $\widehat{N}_{ij}$ are the observed and expected numbers of genotype B_iB_j in a sample.

An alternative, but closely related, approach is the likelihood-ratio based **G-test**,

$$G = -2\sum_{i=1}^{n}\sum_{j\geq i}^{n} N_{ij} \ln\left(\frac{\widehat{N}_{ij}}{N_{ij}}\right) \tag{4.4b}$$

which is becoming more widely used. Such likelihood-based tests have a number of desirable statistical features (Appendix 4). The test statistic G has a sampling distribution very similar to the well-known χ^2 distribution (Appendix 5). That is, if a population in HW equilibrium is sampled many different times and G calculated each time, the frequency distribution of the observed G values will be nearly χ^2 distributed. Thus, the test for WH proportions compares the observed statistic G with the cumulative χ^2 distribution. If G exceeds the level at which there is a 5% chance of obtaining a higher χ^2, then the chance that these data are from a locus showing HW proportions is unlikely.

Regardless of which approach to testing for HW frequencies is taken, it should be kept in mind that some of the conditions underlying the Hardy–Weinberg theorem may be violated without causing detectable departures of observations from expectations. For example, if the product of the survivorships of the two homozygotes is equal to the square of the heterozygote survival, the zygotic frequencies after selection will still be in HW proportions (Lewontin and Cockerham 1959). Thus, a failure to reject the Hardy-Weinberg model should be interpreted with caution. In particular, even a large amount of allele-frequency change over the previous generation still results (under random mating) in the offspring being in HW proportions (albeit with the new allele frequency). Thus Equations 4.4a and 4.4b simply test *one* feature of Hardy–Weinberg—whether genotypic frequencies can be completely described by allele frequencies. This is ensured with random mating when no other evolutionary forces act between the formation of zygotes and the test for HW proportions. The deeper feature of Hardy–Weinberg that is usually untested is the prediction of no allele-frequency change over generations.

Example 4.2. As an example of the application of Equation 4.4b, we return to the data in the table of Example 4.1. The best estimates for the HW expectations are obtained from the observed allele frequencies: $\widehat{N}_{11} = p_1^2 N = 904.9$, $\widehat{N}_{12} = 2\,p_1 p_2 N = 79.3$, and $\widehat{N}_{22} = p_2^2 N = 1.7$. Applying these and the observed values (N_{11}, N_{12}, and N_{22}) from the table,

$$G = -2\,[\,905\,\ln(904.9/905) + 78\,\ln(79.3/78) + 3\,\ln(1.7/3)\,] = 1.0293$$

Under the null hypothesis of HW frequencies, the sampling distribution of G is a function of the number of degrees of freedom, which in the case of the Hardy–Weinberg test is the number of genotypic classes minus the number of allele frequencies that must be estimated from the data minus one. Here, it was necessary to estimate one parameter (p_1) from the data, so there is $3 - 1 - 1 = 1$ degree of freedom. The 95% critical value for a χ_1^2 is 3.841, while the p value for our observed G value is 0.310. Therefore, the observed data are not significantly different from HW expectations. Note that the Chi-square statistic for these data is

$$X = \frac{(905 - 904.9)^2}{904.9} + \frac{(78 - 79.3)^2}{79.3} + \frac{(3 - 1.7)^2}{1.7} = 1.0154$$

which has a p value of 0.314, again showing little departure from HW expectations.

Example 4.3. An interesting application of Hardy–Weinberg is **Snyder's ratio**. The human geneticist Laurence Snyder was interested in testing whether the inability to detect the bitter chemical PTC, a *nontaster*, was a Mendelian recessive. Snyder (1933) noted that normal Mendelian procedures, such as a testcross back to a nontaster (*tt*) to determine whether a taster was homozygous or heterozygous (*TT* or *Tt*), is not an option in human genetics. While all of the offspring of nontaster × nontaster parents were nontasters, consistent with a recessive, Snyder observed 15% nontaster offspring from taster × taster crosses (marriages) and 37% for taster × nontaster crosses. Were these values consistent with his model? Letting q denote the frequency of allele *t*, then under random mating, the frequency of nontaster (*tt*) × taster (TT or Tt) matings is

$$2 \cdot q^2[(1-q)^2 + 2q(1-q)] = 2q^2(1-q)(1-q)$$

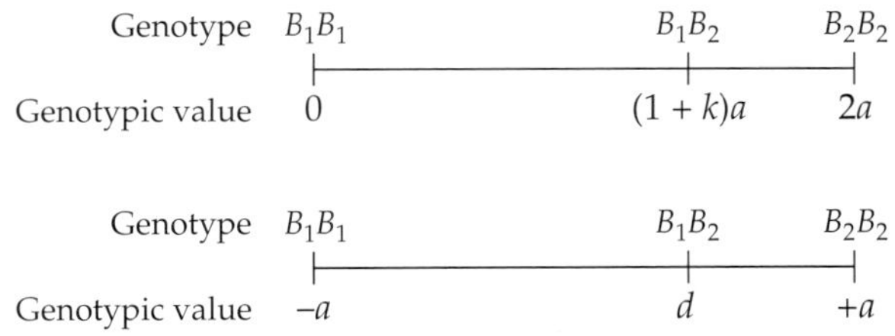

Figure 4.4 Two ways of representing genotypic values for a diallelic locus. A third representation simply adds a constant value, C, to each genotype, so that (for example) the lower representation becomes $C-a, C+d, C+a$. In all representations, $2a$ is the difference between the two homozygotes and $ka = d$ is the departure of the heterozygote from the average value of the two homozygotes.

The leading factor of two arises because the mating could be taster father × nontaster mother, or vise versa. Of these matings, only $Tt \times tt$ produce nontasters, with half of their offspring expected to be nontasters. Hence, the expected fraction of nontaster offspring from taster × nontaster families is

$$\frac{(1/2)2q^2[2q(1-q)]}{2q^2(1-q)(1+q)]} = \frac{q}{1+q} = S$$

Proceeding in the same fashion, the fraction of nontaster offspring from taster × taster matings is S^2. More generally, if we let D denote the (putative) dominant phenotype and R the recessive phenotype, then

Mating	D × D	D × R	R × R
Fraction of R offspring	S^2	S^1	$S^0 = 1$

These expected fractions are Snyder's ratios. With an estimate of q in hand (for example, the square root of the frequency of the R phenotype), one can test for goodness of fit to these ratios. When doing so, Snyder found that nontaster was consistent with a Mendelian recessive. Further discussion of this concept, including extensions to incomplete penetrance and epistasis, can be found in Taylor and Prior (1939), Li (1953), Trankell (1956), Furusho (1960), Doolittle (1968), and Otto et al. (1994).

CHARACTERIZING THE INFLUENCE OF A LOCUS ON THE PHENOTYPE

In Chapter 3, we encountered the concept of partitioning the phenotypic value (z) of an individual into a **genotypic value** (G) and an **environmental deviation** (E),

$$z = G + E$$

where G is the expected phenotype (for a given genotype) resulting from the joint expression of all of the genes underlying the trait. For a multilocus trait, G is a potentially complicated function. For now, however, we are concerned only with the direct contribution of a single autosomal locus, in which case things are quite tractable. We start with the special case in which there are only two alleles. The three genotypic values can then be represented by the scale at the top of Figure 4.4, with $2a$ representing the difference between the mean phenotypes of B_2B_2 and B_1B_1 homozygotes, and k providing a measure of dominance. Alleles B_1 and B_2 behave in a completely additive fashion when $k = 0$, whereas $k = +1$ implies complete dominance of the B_1 allele, and $k = -1$ implies complete dominance of the B_2 allele. If $k > 1$, the phenotypic expression of the heterozygote exceeds that of both homozygotes, and the locus is said to exhibit **overdominance**, whereas $k < -1$ implies **underdominance**.

The fact that we have set the genotypic value of the B_1B_1 homozygote equal to zero may seem troublesome, but it is desirable because it leads to some algebraic simplifications.

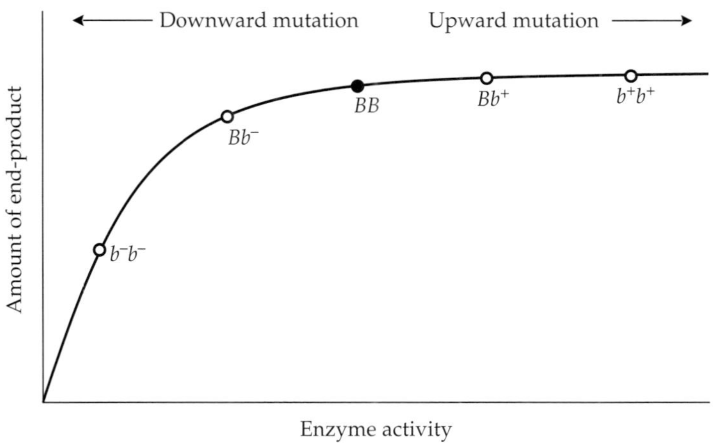

Figure 4.5 The relationship between the activity of a gene product and the **flux** or concentration of an end-product in an enzymatic pathway. *BB* represents the "wild-type" genotype. Upward and downward mutations with the same magnitude of change in enzyme activity are represented as b^+ and b^- alleles. Note that the b^+ allele has only a small effect on the end-product, whereas the downwardly acting mutation b^- has a large homozygous effect, and exhibits pronounced recessiveness relative to *B*.

One can always rescale by the simple addition of some constant, such as in the bottom representation in Figure 4.4. The key parameters are not the actual genotypic values, but rather the difference ($2a$) between homozygotes and the departure ($d = ka$) of the value of the heterozygote from the average of the two homozygotes. In what follows, we use both k and d for measures of dominance, as some expressions are simpler when based on k, while others are when based on d.

Example 4.4. The logic underlying the scaling of genotypic values may be clarified by reference to a particular example—the Booroola (*B*) gene that influences fecundity in the Merino sheep of Australia (Piper and Bindon 1988). Litter size in sheep has a polygenic basis, but in this particular breed, it is determined largely by a single polymorphic locus. The mean litter sizes for the *bb*, *Bb*, and *BB* genotypes based on 685 records are 1.48, 2.17, and 2.66, respectively. Taking these to be estimates of the genotypic values (G_{bb}, G_{Bb}, and G_{BB}), the homozygous effect of the *B* allele is estimated by $a = (2.66 - 1.48)/2 = 0.59$. The dominance coefficient is estimated by taking the difference between *Bb* and *bb* genotypes (i.e., rescaling to give *bb* a value of zero), $2.17 - 1.48 = a(1+k) = 0.69$. Substituting $a = 0.59$, and rearranging to obtain $k = 0.17$ (equivalently, $d = ak = 0.10$). This suggests slight dominance of the Booroola gene, but great confidence cannot be placed on this conclusion. Because the standard errors of the mean genotypic values are approximately 0.09, the midpoint between the two homozygotes, 2.07, is not significantly different from 2.17.

THE BASIS OF DOMINANCE

The presence of dominance complicates many formulations in quantitative genetics, but unfortunately it is a fact of life that cannot be ignored. Since the beginning of the last century, there has been much debate on the genetic and physiological basis of dominance. In the early days, the only genes subject to detailed genetic analysis were those that had a major phenotypic effect. Loci involving such genes are usually characterized by striking levels of dominance. For example, the vast majority of genes with major, deleterious effects on fitness are recessive. Does this then indicate that new mutations are inherently recessive? Fisher

(1928a, 1928b, 1929, 1958) argued that because rare alleles are found almost entirely in the heterozygous state, selection should favor alleles at modifier loci that cause heterozygous carriers of deleterious alleles to resemble the normal homozygote. Implicit in this argument is the assumption that the heterozygote initially encodes for an intermediate phenotype. Using physiological arguments, Wright (1929a, 1929b, 1934a, 1934b) strongly disputed this idea. He also pointed out that although dominance relationships are subject to change, the intensity of selection operating on modifier loci is unlikely to ever be strong enough to be an important evolutionary force. One important caveat to this assertion is that Wright assumed a population in mutation–selection balance, so that the strength of selection against any new modifier allele was on the order of the deleterious mutation rate (WL Chapter 7). There are, however, other settings where the strength of selection on dominance modifiers can be much stronger, such as for an adaptive allele on its way to fixation (Haldane 1956; Mayo and Bürger 1997). The debate between Fisher and Wright was intense and at times bitter, and it scarred their relationship permanently.

Much later, Kacser and Burns (1981) developed a general explanation for dominance based on the biochemical principles of enzyme kinetics. Their **metabolic model** is in good accord with Wright's theory. Most gene products are involved in complex biochemical pathways such that the rate of production of a final end-product (phenotype) is regulated at many steps (Chapter 21; Appendix 2). Consequently, the relationship between gene-product activity (a function of allelic state) and end-product production is often hyperbolic (Figure 4.5). The assumption here is the **safety factor model** (Wright 1929a; Haldane 1930a, 1939; Hurst and Randerson 2000): the concentration of the substrate, not the enzyme, is the rate-limiting step. This safety factor could arise by selection favoring alleles with high expression levels (near the asymptote in Figure 4.5) to buffer environmental perturbations. Under the metabolic model, Kacser and Burns showed that the "wild-type" activity normally lies on or near the plateau of this hyperbolic relationship. This leads to three predictions:

1. ***Mutations with large effects at the phenotypic level will be biased in a downward direction.*** Even if mutations that increase activity occur as frequently as those that decrease it, the former will usually cause imperceptible changes at the phenotypic level. Thus, if a high production rate or end-product concentration is beneficial, we can expect most individually *discernible* mutations to be detrimental. This is a consequence of the nonlinear (hyperbolic) phenotype map.

2. ***The recessivity of downward mutations is an inevitable consequence of the hyperbolic activity–product relationship.*** If we take the heterozygote to be intermediate in activity, the allele producing the homozygote with greater activity will always exhibit dominance on the end-product scale, the degree of dominance diminishing as the heterozygote is placed further out on the plateau.

3. ***The smaller the effect of a mutation, the less pronounced will be the level of dominance.*** Such a result is expected simply because the relationship between the BB, Bb, and bb genotypic values tends towards linearity as the deviations among their activities are reduced. In principle, dominance is much more likely to be a complicating factor for characters whose variation is influenced by one or two genes of large effect than for quantitative characters encoded by numerous loci whose individual effects are indiscernible.

Because the exact form of the relationship in Figure 4.5 can change with a shift in the genetic background, the **Kacser–Burns model** does not rule out the possibility of evolutionary changes in dominance relationships. It does, however, eliminate the necessity of *ad hoc* evolutionary explanations, such as modifier loci, to account for the existence of dominance. A number of authors have noted that the Kacser–Burns framework applies to gene products that are enzymes, but not necessarily to products that are strictly structural or regulatory (Kacser and Burns 1981; Hodgkin 1993; Huber et al. 2018). Indeed, Kondrashov and Koonin (2004) found that for both human and yeast genes, proteins coding for enzymatic functions tended to display more dominance than proteins coding for non-catalytic functions. However, dominance of new mutations is not uncommon among structural genes (Wilkie 1994; Phadnis and Fry 2005; Agrawal and Whitlock 2011). One explanation for this observation is the **poison subunit model**, wherein a single faulty component in a multimeric protein assembly can poison the entire complex.

Two important observations are generally in good accord with the predictions of the Kacser–Burns model. First, in a clever analysis of data on the haploid alga *Chlamydomonas reinhardtii,* Orr (1991) found that when mutations are observed in artificial diploid constructs, they are almost always recessive. Because the heterozygous state almost never exists in a haploid species (except very transiently when zygotes are made), there can be no opportunity for the selection of dominance modifiers; the mutations must be "recessive" at first appearance. Second, as noted by Mukai et al. (1972), Charlesworth (1979), and Crow and Simmons (1983), lethal alleles in *Drosophila* are almost nearly completely recessive, whereas mildly deleterious alleles, whose individual effects are indiscernible, interact in a nearly additive fashion (Chapter 15). Huber et al. (2018) similarly found in *Arabidopsis* that more deleterious mutations were more likely to be recessive than were less deleterious ones. This pattern of larger-effect genes tending to be more recessive was also seen by Phadnis and Fry (2005), who examined a large number of growth-rate mutations in yeast using the single-gene knock-out panel developed by Steinmetz et al. (2002a). A more detailed analysis of the same data by Agrawal and Whitlock (2011) offered a more nuanced view. While they observed the same general pattern as seen by Phadnis and Fry (2005), they also noted that the distribution of dominance coefficients had a large variance, such that some deleterious alleles showed no dominance. They also noted that knock-out mutations in catalytic genes were more recessive than those in structural genes.

Kacser–Burns is not the only pathway-based model with implications for nonadditive genetics effects. See Cornish-Bowden (1987), Savageau (1992), Omholt et al. (2000), Bagheri and Wagner (2004), and Bost and Veitia (2013), and references therein, for details. Additional discussion of the evolution of dominance can be found in the reviews by Mayo and Bürger (1997), Bourguet (1999), Bagheri (2006), Billiard and Castric (2011), and Manna et al. (2011).

FISHER'S DECOMPOSITION OF THE GENOTYPIC VALUE

What Part of G is Passed Onto an Offspring?

One of the foundational pillars of quantitative genetics, the notion of the **additive genetic variance** of a trait, is also one of the field's more nuanced and challenging concepts. Before building the machinery to formally introduce the additive variance, we first consider Fisher's motivational question behind this concept: *What part of the genotype value G of a parent is passed onto its offspring?* The answer is trivial in two settings: offspring produced by cloning (e.g., vegetative propagation or with certain mating systems) and offspring produced by selfing a fully inbred parent. In both cases, the offspring and parental genotypes are identical, and the offspring has the same G value as its parents.

Conversely, the genotypic values for offspring produced by sexual reproduction are rarely the same as either parent. Indeed, offspring from the same parents are expected to show potentially quite dramatic variation in their genotypic values due to **Mendelian segregation** at parental heterozygous loci (e.g, a $B_1B_2C_1C_2$ parent generates four different gametes). These results arise because a sexually produced offspring has two parents, each of which (for diploids) passes along ***single alleles at each locus*** to its offspring, rather than whole genotypes. There is also a ***context-specific aspect to transmission***. Consider a diallelic locus. Whether the resulting offspring from B_1B_1 parent are B_1B_2 or B_1B_1 depends on the probability that the other parent (often a random individual from some target population) donates a B_1 or a B_2 allele. Hence, the ability of a parent to transmit part of its G value to an offspring depends, in part, on contributions from the other parent, and hence is ***population-specific***.

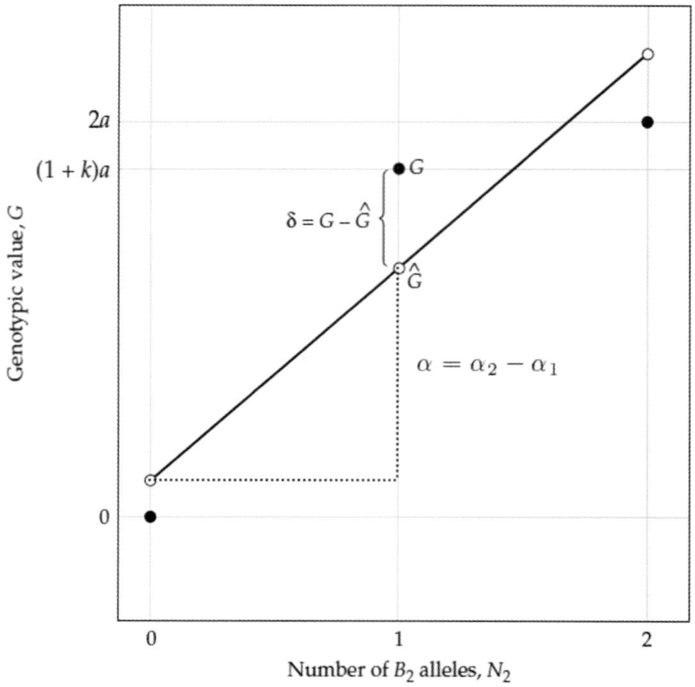

Figure 4.6 Linear least-squares regression (solid line) of the genotypic value G of a single locus on the gene content of allele B_2 (N_2). From left to right, the points represent the B_1B_1, B_1B_2, and B_2B_2 genotypes, corresponding to N_2 values of, respectively, 0, 1, and 2. Solid circles represent the true genotypic values, while open circles are the values, $\widehat{G}$, expected on the basis of average effects alone (Equation 4.7). The deviation (or residual) between G and $\widehat{G}$ for each genotype is δ, the dominance deviation.

Fisher (1918) showed that the basic building block of quantitative-trait transmission in sexual populations is the **additive** (or **average**) **effect of an allele**—the average amount that an individual receiving that allele deviates from the population mean. As we will show, this quantity is not simply an intrinsic feature of a given allele, as it also depends on the population allele frequencies. Hence, the additive effect is a **population-specific** measure that can change with allele frequencies. The sum of additive effects over all loci influencing a trait in an individual is called its **additive genetic value**, or, equivalently, **breeding value**, for that trait. As we will see, the breeding value can be estimated from either designed crosses, knowledge of the trait values and familial relationships among a set of individuals, or marker-trait associations (genomic selection/prediction, polygenic risk scores). The additive genetic variance is simply the ***population variance in breeding values***. If the fraction of trait variance that is additive is small (i.e., the narrow-sense heritability, h^2, is small), then offspring only weakly resemble their parents, which results in (among other things) inefficient selection on a trait (Equation 3.29). To formally develop these concepts, we now consider Fisher's (1918) decomposition of the genetic value.

Fisher's Decomposition

We start by assuming a diallelic locus, whose alleles B_1 and B_2 have frequencies of p_1 and p_2, respectively (where $p_2 = 1 - p_1$). The number of copies of a particular allele (say B_2) in a genotype ($N_2 = 0$, 1, or 2 for diploids) is referred to as the **allele dosage** (or **allele content**). As noted above, unless this allele interacts additively with all other alleles, there will be a nonlinear relationship between the allele content and the genotypic value. It is, nevertheless, useful to consider the best linear approximation to this relationship, as this leads to a partitioning of the genotypic values into their "expected" values based on additivity ($\widehat{G}$) and deviations (residuals) from those expectations resulting from dominance (δ) (Figure 4.6).

The preceding points can be formalized by least-squares regression of genotypic values on the number of B_1 and B_2 alleles in the genotype (N_1 and $N_2 = 2 - N_1$),

$$G_{ij} = \widehat{G}_{ij} + \delta_{ij} = \mu_G + \alpha_1 N_1 + \alpha_2 N_2 + \delta_{ij} \tag{4.5a}$$

The genotypic value (G_{ij}) of genotype $B_i B_j$ is a function of the mean genotypic value in the population (μ_G), the slopes of the regression (α_1 and α_2), the predictor variables (N_1 and N_2), and the residual error (δ_{ij}). Regression theory (Chapters 3 and 10) provides a powerful interpretation of the regression slopes, α_i. Namely, a unit change in the number of i alleles (N_i) results in an expected change of α_i in the genotypic value, so that α_i is the predicted impact on G from an individual carrying a B_i allele. This partitioning of genotypic values into various components is one of several major advances developed in Fisher's 1918 paper. Many of the innovative ideas in this classic paper are presented in his characteristically cursory (some would even say cryptic) manner, but a useful interpretative guide is provided by Moran and Smith (1966).

Unlike the univariate regression discussed in Chapter 3, Equation 4.5a is a **multiple regression**, the properties of which are discussed in Chapter 10. For the two-allele case, however, we can reduce the model to a standard univariate regression by noting that, for any individual, $N_1 = 2 - N_2$, yielding

$$\begin{aligned} G_{ij} &= \mu_G + \alpha_1 (2 - N_2) + \alpha_2 N_2 + \delta_{ij} \\ &= \iota + (\alpha_2 - \alpha_1) N_2 + \delta_{ij} \end{aligned} \tag{4.5b}$$

where $\iota = \mu_G + 2\alpha_1$ is the intercept. We denote the slope of this regression by

$$\alpha = \alpha_2 - \alpha_1 \tag{4.6}$$

and discuss its meaning shortly. Equation 4.5b implies that the genotypic values predicted by the regression are given by

$$\widehat{G}_{ij} = \mu_G + 2\alpha_1 + \alpha N_2 \tag{4.7a}$$

Note that this form also recovers Equation 4.5a,

$$\widehat{G}_{ij} = \mu_G + \alpha_i + \alpha_j = \begin{cases} \mu_G + 2\,\alpha_1 & \text{for } G_{11} \ \ (N_2 = 0) \\ \mu_G + \alpha_1 + \alpha_2 & \text{for } G_{21} \ \ (N_2 = 1) \\ \mu_G + 2\,\alpha_2 & \text{for } G_{22} \ \ (N_2 = 2) \end{cases} \tag{4.7b}$$

We next show that the weighted mean of the coefficients α_1 and α_2 is equal to zero. To accomplish this, return to Equation 4.5a and take expectations, yielding

$$\mu_G = \mu_G + \alpha_1 E(N_1) + \alpha_2 E(N_2) + E(\delta_{ij}) \tag{4.8a}$$

The expected value of the residual δ_{ij} is equal to zero by the properties of least-squares regression, and $E(N_1)/2$ and $E(N_2)/2$ are equivalent, respectively, to p_1 and p_2, the frequencies of the B_1 and B_2 alleles. Thus, Equation 4.8a simplifies to

$$p_1\,\alpha_1 + p_2\,\alpha_2 = 0 \tag{4.8b}$$

showing that the mean value of α_i over all individuals is indeed zero. Finally, from Equations 4.6 and 4.8b and the fact that $p_1 + p_2 = 1$, we obtain

$$\alpha_2 = p_1\,\alpha \qquad \text{and} \qquad \alpha_1 = -p_2\,\alpha \tag{4.9}$$

Table 4.1 Properties of a single segregating diallelic locus under random mating. Alternative versions of these expression follows by noting that $p_1 = 1 - p_2$ and $d = ak$.

Genotype	Gene Content (N)	Genotypic Value (G)	Freq.	$G \cdot N$	N^2	Regression Value ($\widehat{G}$)	Dominance Deviation ($\delta = G - \widehat{G}$)
B_1B_1	0	0	p_1^2	0	0	ι	$-\iota$
B_1B_2	1	$(1+k)a$	$2p_1p_2$	$(1+k)a$	1	$\iota + \alpha$	$(1+k)a - \iota - \alpha$
B_2B_2	2	$2a$	p_2^2	$4a$	4	$\iota + 2\alpha$	$2a - \iota - 2\alpha$

$$\mu_N = 2\,p_1p_2(1) + p_2^2(2) = 2p_2$$

$$\mu_{\widehat{G}} = \iota + 2\,p_1p_2\alpha + 2p_2^2\,\alpha = \iota + 2p_2\alpha$$

$$E(N^2) = 2p_1p_2(1) + p_2^2(4) = 2p_2(1+p_2)$$

$$\mu_\delta = -\iota + 2p_1p_2[(1+k)a - \alpha] + 2p_2^2(a-\alpha) = 0$$

$$\mu_G = 2p_1p_2a(1+k) + 2p_2^2a = 2p_2a(1+p_1k)$$

$$E(\widehat{G}^2) = p_1^2\iota^2 + 2p_1p_2(\iota+\alpha)^2 + p_2^2(\iota+2\alpha)^2 = \iota^2 + 4p_2\alpha\iota + 2p_2\alpha^2(1+p_2)$$

$$E(GN) = 2p_1p_2a(1+k) + 4p_2^2a = 2p_2a[\,2p_2 + p_1(1+k)\,]$$

$$E(\delta^2) = p_1^2\iota^2 + 2p_1p_2[\,(1+k)a - \iota - \alpha\,]^2 + p_2^2(2a - \iota - 2\alpha)^2 = (2p_1p_2ak)^2$$

$$\sigma(G,N) = E(GN) - \mu_G\mu_N = 2p_1p_2a[\,1 + k(p_1 - p_2)\,]$$

$$\sigma_A^2 = E(\widehat{G}^2) - \mu_{\widehat{G}}^2$$

$$\sigma^2(N) = E(N^2) - \mu_N^2 = 2p_1p_2$$

$$\sigma_D^2 = E(\delta^2) - \mu_\delta^2$$

Now recall from Equation 3.14b that the slope of a univariate regression is the covariance between response and predictor variable divided by the variance of the predictor variable. Thus, the slope of the regression in Figure 4.6 is

$$\alpha = \frac{\sigma(G, N_2)}{\sigma^2(N_2)} \tag{4.10a}$$

The terms $\sigma(G, N_2)$ and $\sigma^2(N_2)$ are functions, respectively, of the gene effects (a and k) and allele frequencies (p_1 and p_2). The steps leading up to their computation, under the assumption of random mating, are outlined in Table 4.1. Upon substitution, we obtain

$$\alpha = a\,[\,1 + k\,(\,p_1 - p_2\,)\,] = a\,[\,1 + k\,(\,1 - 2p_2\,)\,] \tag{4.10b}$$

Under the assumption of random mating, α is known as the **average effect of allelic substitution**. It represents the average change in genotypic value that results when a B_2 allele is randomly substituted for a B_1 allele. For the purely additive case ($k = 0$), α is simply equal to a. However, for all other cases, α is also a function of k and of the allele frequencies in the population (Figure 4.7). Such behavior results because, with dominance, the phenotypic effect of a gene substitution depends on the status of the unsubstituted allele. If B_2 is a dominant allele ($k > 0$), then α will be inflated relative to the case of additivity if B_2 is rare ($p_1 > p_2$), but diminished if B_2 is common ($p_1 < p_2$). Thus, except in the case of additivity, the average effect of allelic substitution is not simply a function of the inherent physiological properties of the allele. ***It can only be defined in the context of the population.*** This shows that Fisher's great insight essentially distills to the following key conceptual issue: the intrinsic components of the genotypic value (a and k) alone are not sufficient to describe

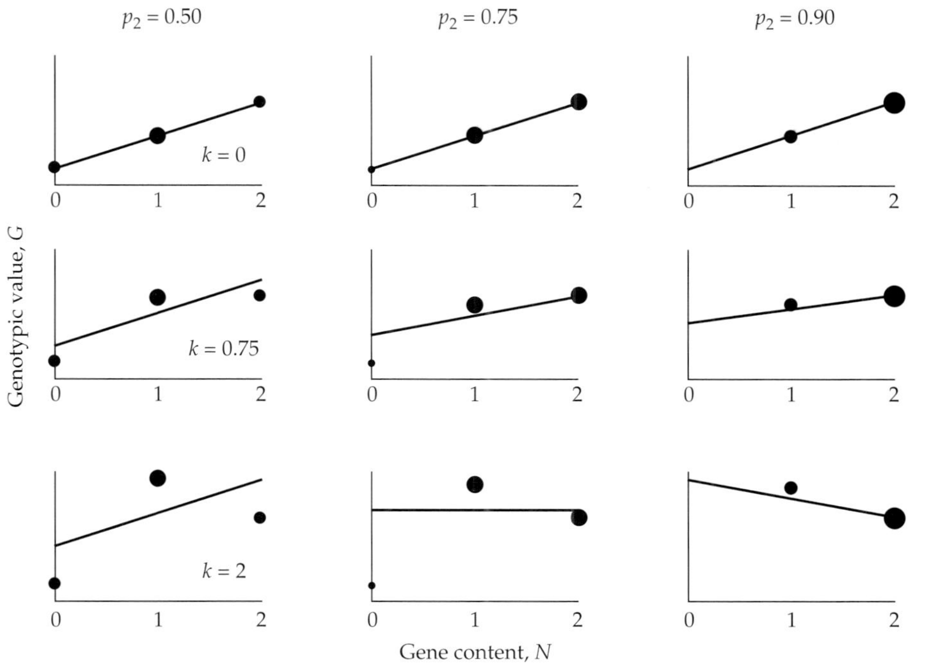

Figure 4.7 The slope α of the linear least-squares regression of genotypic value on gene content as a function of allele frequency, p_2, and degree of dominance, k. The lines denote the regressions, with each of the three points (representing genotypic values) being weighted by their frequency (denoted by the different-sized circles). The columns of graphs give results for different gene frequencies ($p_2 = 0.50$, 0.75, and 0.90), whereas the rows give results for different modes of gene action ($k = 0.00$, additivity; $k = 0.75$, partial dominance; and $k = 2.00$, overdominance). Note that when going across a row (the same model of gene action), the location of the points does not change, but their size (representing their relative abundance) does. Except for the case of complete additivity, the resulting regressions differ with different allele frequencies, as unless all of the points initially fall on a straight line, the different weights of the points change the slope of the optimal linear fit (minimizing of the weighted sum of squared residuals). In the case of overdominance, the slope changes sign as the allele frequency changes; when $p_2 = 0.75$, the slope is zero, i.e., there is no additive genetic variance. Note from Equation 4.10b that the slope $\alpha = a$ when $p_2 = p_1 = 0.5$, regardless of the degree of dominance.

the transmission of the genotypic value from a pair of parents to their offspring. Instead population-weighted metrics of these components (the average effects α_i and dominance deviations δ_{ij}) are required.

PARTITIONING THE GENETIC VARIANCE

Fisher (1918) showed that once the genotypic values have been partitioned in the above manner, it is a relatively simple step to partition the sources of genetic variation at a locus. Recalling the relationship $G = \widehat{G} + \delta$, the total genetic variance may be written as

$$\sigma_G^2 = \sigma^2(\widehat{G} + \delta) = \sigma^2(\widehat{G}) + 2\sigma(\widehat{G}, \delta) + \sigma^2(\delta)$$

From the property of least-squares regression (Chapter 3), the regression prediction ($\widehat{G}$) is uncorrelated with the residual error (in this case, δ). Thus, the total genetic variance attributable to a locus simplifies to the sum of additive and dominance components. Hereafter, we denote these components as σ_A^2 and σ_D^2, such that

$$\sigma_G^2 = \sigma_A^2 + \sigma_D^2 \tag{4.11}$$

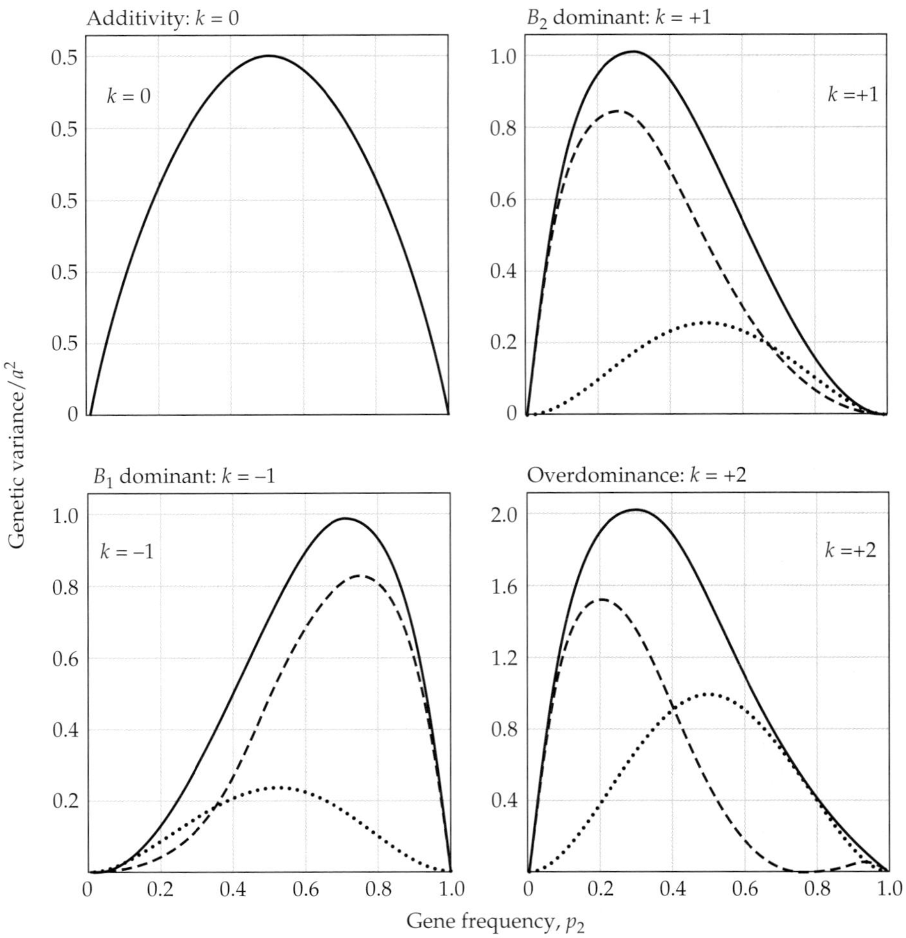

Figure 4.8 The dependence of the components of genetic variance at a locus on the frequency of the B_2 allele. The solid line denotes the total genetic variance, the dashed line the additive genetic variance, and the dotted line the dominance genetic variance. Four cases are illustrated: $k = 0$ (additivity), $k = +1$ (dominance of the B_2 allele), $k = -1$ (dominance of the B_1 allele), and $k = +2$ (a case of overdominance). In the case of additivity, all of the genetic variance is additive. The vertical axes are scaled such that, for any particular case, the actual variances are obtainable by multiplying by a^2, where a is half the difference between homozygous B_1B_1 and B_2B_2 genotypic values.

Statistically speaking, σ_A^2 is the amount of the variance in G that is explained by the regression on N_2 (or equivalently, on N_1), whereas σ_D^2 is the residual variance for the regression. Biologically speaking, σ_A^2 is the genetic variance associated with the average additive effects of alleles (the **additive genetic variance**), and σ_D^2 is the genetic variance associated with dominance deviations (the **dominance genetic variance**).

All of the information necessary to compute these two components of genetic variance for a diallelic locus is contained in Table 4.1, and leads to

$$\begin{aligned}\sigma_A^2 &= 2p_1p_2\alpha^2 \\ &= 2a^2p_1(1-p_1)\left[1+k(1-2p_2)\right]^2 \\ &= 2p_1(1-p_1)\left[a+d(1-2p_2)\right]^2\end{aligned} \tag{4.12a}$$

$$\sigma_D^2 = (2p_1p_2ak)^2 = 4d^2p_1^2(1-p_1)^2 \tag{4.12b}$$

Letting $H = 2p_1(1-p_1)$ denote the heterozygosity, we can also express these variances as

$$\sigma_A^2 = \alpha^2 H \qquad \text{and} \qquad \sigma_D^2 = d^2H^2 \tag{4.12c}$$

Both components of variance depend upon the gene frequencies, the dominance coefficient k, and the homozygous effect a (Figure 4.8). In the case of purely additive allelic effects ($k = 0$ and hence $\alpha = a$), the additive genetic variance reaches a maximum at $p_1 = p_2 = 0.5$, the gene frequency at which heterozygosity H is most pronounced. With dominance, however, the additive genetic variance is maximized at a higher frequency of the recessive allele. This occurs because rare recessive alleles cause little genetic variance, only making a significant impact on trait variance when enough recessive homozygotes appear. Note from Equation 4.12c that σ^2_D is maximized at $p_1 = p_2 = 1/2$ for all values of $k(d)$.

A common misconception is that the relative magnitudes of additive and dominance variance provide information on the additivity of gene action. Equations 4.10b and 4.12a show that this generalization does not hold true. Through its influence on α, dominance contributes to the additive genetic variance, and for certain allele frequencies, can cause σ^2_A to reach much higher levels than in the case of alleles with purely additive effects (Figure 4.8). Even in the case of complete dominance, σ^2_D is unlikely to greatly exceed σ^2_A, and it is often substantially smaller. In the case of overdominance ($k > 1$), probably not a common situation (Chapters 11 and 12), there is always an intermediate gene frequency at which σ^2_A is zero. This occurs when the least-squares regression of G on N_2 has a slope equal to zero (Figure 4.7). At this allele frequency, there is no variance in breeding values, which implies that the expected offspring mean is the same for any parent from this population mated to a random parent.

More formally, from Equation 4.12c, we have

$$\frac{\sigma^2_A}{\sigma^2_D} = \frac{\alpha^2 H}{d^2 H^2} = \frac{a^2\,[\,1 + k\,(\,1 - 2p_2\,)\,]^2}{a^2 k^2 H} = \frac{[\,1 + k\,(\,1 - 2p_2\,)\,]^2}{k^2 H} \tag{4.12d}$$

While this ratio might appear a bit opaque, notice that unless the allele frequencies are intermediate (so that the heterozygosity $H = 2p_1p_2$ is not too small, and hence $1/H$ is not too large), the additive variance is usually larger than the dominance variance. The exception is for a rare recessive allele (e.g., $k = -1, p_2 \ll 1$, in which case $\sigma^2_A/\sigma^2_D \simeq 2p_2 \ll 1$). However, in this case even though $\sigma^2_D > \sigma^2_A$, both variance components are extremely small. Hence, when one allele is somewhat rare, the additive variance (even with large amounts of dominance) is usually larger than the dominance variance (Hill et al. 2008).

Another way to highlight this distinction between genotypic dominance and dominance genetic variance was noted by Cheverud and Routman (1995). They stressed the difference between **physiological dominance**, namely the value of d that is a fixed constant (in our single-locus case), and **statistical dominance**, namely the dominance residual δ_{ij} generated by least-square fit given the allele frequencies. One can have a large value for the former, and yet still have small values of the latter, and hence little dominance variance. We will also note this distinction in Chapter 5 for epistasis, the nonadditive interaction of alleles across loci.

ADDITIVE EFFECTS, AVERAGE EXCESSES, AND BREEDING VALUES

Fisher's partitioning of the genotypic value into additive and dominance components is useful because, in randomly mating diploid species, a parent donates only one allele per locus to each of its offspring. The transmitted allele exhibits its additive effect when randomly combined with a gene from other parents. The dominance deviation of a parent, which is a function of the interaction between the two alleles it carries, is eliminated when gametes (which each carry only one of these alleles) are produced. Thus, one can think of $\widehat{G}$ and δ, respectively, as the heritable and nonheritable components of an individual's genotypic value.

Before clarifying this concept further, however, we need some formal definitions. Two different measures of the effect of an allele were proposed by Fisher (1918, 1941): the average ***excess*** α^*_i and the additive (or average) ***effect*** α_i. As will be shown below, these two measures are equivalent in a randomly mating population, the first having a simple biological interpretation, the second being defined as a least-squares regression parameter.

The **average excess** α_2^* of allele B_2 is the difference between the mean genotypic value of individuals carrying at least one copy of B_2 and the mean genotypic value of a random individual from the entire population,

$$\alpha_2^* = \left(G_{12}\, P_{12\,|2} + G_{22}\, P_{22\,|2} \right) - \mu_G \tag{4.13a}$$

where $P_{ij|i}$ is the conditional probability of a B_iB_j genotype given that one allele is B_i. This is a completely general definition, but initially we will continue to focus on a diallelic locus under random mating, in which case $P_{ij|i} = p_j$, with p_j being the frequency of allele B_j. Under these conditions, Equation 4.13a becomes

$$\alpha_2^* = G_{12}\, p_1 + G_{22}\, p_2 - \mu_G \tag{4.13b}$$

This follows because of all individuals receiving a B_2 allele from one parent, a proportion p_2 (under random mating) receive another B_2 allele from the second parent, while a proportion p_1 receive a B_1 allele from the second parent. Because the genotypic values in these two cases are respectively $2a$ and $a(1+k)$, subtraction of the population mean μ_G (Table 4.1) from the conditional mean yields

$$\begin{aligned} \alpha_2^* &= \{\, p_1[\, a(1+k)\,] + p_2(2a)\,\} - 2\,a\,p_2\,(\,1 + p_1\,k\,) \\ &= p_1\,a\,[\,1 + k\,(\,p_1 - p_2\,)\,] = p_1\,\alpha \end{aligned} \tag{4.14a}$$

namely, a function of the average effect of allelic substitution (Equation 4.10b). In the same manner, the average excess of allele B_1 is found to be

$$\alpha_1^* = -p_2\,a\,[\,1 + k\,(\,p_1 - p_2\,)\,] = -p_2\,\alpha \tag{4.14b}$$

Note that one of the average excesses is positive and the other negative because they are defined as deviations from the population mean genotypic value and hence have expected value zero. In particular,

$$E[\alpha_i^*] = p_1\alpha_1^* + p_2\alpha_2^* = p_1(-p_2\,\alpha) + p_2(p_1\,\alpha) = 0$$

The **additive effects**, α_i, on the other hand, are defined to be the least-squares regression coefficients of genotypic value on gene content. They are obtained by finding the α_1 and α_2 that miminize the mean-squared residual deviation

$$\begin{aligned} M &= E(\,\delta_{ij}^2\,) = E[\,(G_{ij} - \mu_G - \alpha_i - \alpha_j)^2\,] \\ &= (G_{11} - \widehat{G}_{11})^2\,P_{11} + (G_{12} - \widehat{G}_{12})^2\,P_{12} + (G_{22} - \widehat{G}_{22})^2\,P_{22} \end{aligned}$$

where P_{ij} is the frequency of the *ij*th genotype. Again, this is a general definition. For the special case of a randomly mating population (with $P_{11} = p_1^2$, $P_{12} = 2p_1p_2$, and $P_{22} = p_2^2$), setting the partial derivatives of M with respect to α_i equal to zero, and solving gives

$$\alpha_2 = p_1\,a\,[\,1 + k\,(p_1 - p_2)\,] = p_1\,\alpha \tag{4.15a}$$

$$\alpha_1 = -p_2\,a\,[\,1 + k\,(p_1 - p_2)\,] = -p_2\,\alpha \tag{4.15b}$$

Comparing these expressions with Equations 4.14a and 4.14b, we find that additive effects are identical to average excesses in randomly mating populations. The α_i are often referred to as **average effects**, but we use the term *additive effects* to discriminate them from average effects of higher order gene actions (such as dominance).

Table 4.2 Conditional mean genotypic values of progeny under random mating, and their deviations from the mean genotypic value in the population, $\mu_G = 2ap_2(1 + p_1 k)$.

Parental Genotype	Breeding Value	Mean Genotypic Value of Progeny	Deviation of Expected Progeny Mean from μ_G
B_2B_2	$2\alpha_2$	$a[2p_2 + p_1(1 + k)]$	α_2
B_1B_2	$\alpha_1 + \alpha_2$	$a[p_2 + (1 + k)/2]$	$(\alpha_1 + \alpha_2)/2$
B_1B_1	$2\alpha_1$	$ap_2(1 + k)$	α_1

An individual's **breeding value**, hereafter denoted by A, is the sum of the additive effects of its genes. In other words, the breeding value of a B_1B_1 homozygote is simply $2\alpha_1$, that of a heterozygote is $(\alpha_1 + \alpha_2)$, and that of a B_2B_2 individual is $2\alpha_2$. Because breeding values are defined as deviations from the population mean, ***the expected breeding value of a random individual is zero***.

For random-mating populations, an extremely useful relationship emerges from these definitions for additive effects and breeding values. Consider the expected genotypic values of progeny produced by the parental genotypes. In the case of B_2B_2 parents, a proportion p_2 of the offspring will also be B_2B_2, in which case their genotypic value is $2a$, and a proportion p_1 will be B_1B_2 with genotypic value $a(1 + k)$. The average genotypic value of offspring from a B_2B_2 parent is therefore $p_2(2a) + p_1\, a(1 + k) = a\,[\,2p_2 + p_1\,(1 + k)\,]$. When the population mean, μ_G, is subtracted, we obtain (after some simplification) α_2 (Equation 4.15a). Deviations of expected progeny phenotypes from the population mean are given for the other two parental genotypes in Table 4.2. The results in this table show that when mating is random ***the breeding value of a genotype is equivalent to twice the expected deviation of its offspring mean phenotype from the population mean***. The deviation is multiplied by two because only one of the two parental genes is passed on to each offspring.

Thus, we can estimate the breeding value of an individual by mating it to many randomly chosen individuals from the population and taking twice the deviation of its offspring mean from the population mean. This result is a special case of the more general statement that the ***expected offspring deviation from the population mean is given by the average breeding value of its two parents***. Treating one of those parents as random (and hence, a breeding value of zero) recovers our estimate of twice the offspring deviation for the breeding value of a parent. We thus arrive at the answer to Fisher's question on the transmission of G: ***the fraction of G that is, on average, passed from parent to offspring is A/2***. Part II Chapter 32 discusses the estimation of breeding values under very general settings.

Example 4.5. Consider the consequences of the Booroola gene (described in Example 4.4) in two hypothetical random-mating populations with gene frequencies of 0.5 and 0.1. We assume that the phenotypic means within genotypic classes are known without error, so that they are equivalent to the genotypic values. The additive and dominance genetic variances are, respectively, the mean-squared breeding values and the mean-squared dominance deviations because both types of effects have means equal to zero, so that $\sigma^2(x) = E[x^2]$.

	$p_B = 0.5$			$p_B = 0.1$		
	bb	Bb	BB	bb	Bb	BB
Genotypic Value (G_{ij})	1.48	2.17	2.66	1.48	2.17	2.66
Genotype Frequency (P_{ij})	0.25	0.50	0.25	0.81	0.18	0.01
Mean Genotypic Value $\mu_G = P_{bb}G_{bb} + P_{Bb}G_{Bb} + P_{BB}G_{BB}$			2.120			1.616

	$p_B = 0.5$			$p_B = 0.1$		
	bb	Bb	BB	bb	Bb	BB
Additive Effects						
$\alpha_B = p_B G_{BB} + p_b G_{Bb} - \mu_G$			0.295		0.603	
$\alpha_b = p_b G_{bb} + p_B G_{Bb} - \mu_G$			−0.295		−0.067	
Breeding Values						
$A_{ij} = \alpha_i + \alpha_j$	−0.59	0.00	0.59	−0.134	0.536	1.206
$\overline{A} = P_{bb} A_{bb} + P_{Bb} A_{Bb} + P_{BB} A_{BB}$			0.00		0.00	
Dominance Deviations						
$\delta_{ij} = G_{ij} - (\mu_G + \alpha_i + \alpha_j)$	−0.05	0.05	−0.05	−0.002	0.018	−0.162
$\overline{\delta} = P_{bb}\,\delta_{bb} + P_{Bb}\,\delta_{Bb} + P_{BB}\,\delta_{BB}$			0.00		0.00	
Genetic Variance Components						
$\sigma^2_A = P_{bb}\,A^2_{bb} + P_{Bb}\,A^2_{Bb} + P_{BB}\,A^2_{BB}$			0.1740		0.0808	
$\sigma^2_D = P_{bb}\,\delta^2_{bb} + P_{Bb}\,\delta^2_{Bb} + P_{BB}\,\delta^2_{BB}$			0.0012		0.0003	
$\sigma^2_G = \sigma^2_A + \sigma^2_D$			0.1752		0.0811	

The key features from this example are that the breeding values and variance components *depend on the population* (here defined by allele frequency), and that the ratio of additive to genetic variance is > 0.99 in both populations, despite a large *BB* dominance deviation when $p_B = 0.1$.

Although this example is somewhat artificial in that we employed arbitrary gene frequencies, the basic approach has been exploited in the analysis of human genetic disorders. Biochemical and genomic studies are used to identify **candidate loci** that are potential major contributors to the variation of the trait of interest, and the genotypes of random individuals are identified by use of molecular markers or direct sequencing. The average phenotypic values within each genotypic class provide estimates of the genotypic values of candidate genotypes, which can then be used to estimate the fraction of the total phenotypic variance that is associated with the locus. Details on this **measured-genotype approach** are given in Chapter 17.

EXTENSIONS FOR MULTIPLE ALLELES AND NONRANDOM MATING

Although the preceding results were obtained under the assumption of a diallelic locus, they are readily generalized to situations with an arbitrary number of alleles, as well as to nonrandomly mating populations. The algebra necessarily becomes more tedious, but some very useful principles emerge that will be relied upon heavily in subsequent chapters. In addition to presenting this more general treatment, the remainder of the chapter will serve as a review of the concepts introduced earlier in the chapter.

Average Excess

When n alleles are present, the average excess, α_i^*, for any allele B_i is given by

$$\alpha_i^* = \sum_{j=1}^{n} P_{ij|i}\, G_{ij} - \mu_G \tag{4.16a}$$

where $P_{ij|i}$ is the conditional probability of a B_iB_j genotype given that one allele is B_i. Under random mating, $P_{ij|i} = p_j$ and this reduces to

$$\alpha_i^* = \sum_{j=1}^{n} p_j\, G_{ij} - \mu_G \tag{4.16b}$$

where p_j is the frequency of the jth allele.

Example 4.6. Here we show how the average excess α_i^* of an allele i can be related to $\sigma(G, N_i)$, the covariance between genotypic value and the number of copies of that allele. This result will be useful in the following sections. To compute $\sigma(G, N_i) = E(G \cdot N_i) - E(N_i) \cdot E(G)$, we start with the fact that $E(G) = \mu_G$, so we merely require expressions for $E(N_i)$ and $E(G \cdot N_i)$. The mean number of alleles of type i at the locus, $E(N_i)$, is straightforward. Because there are two genes at each locus, and the frequency of allele i is p_i, $E(N_i) = 2p_i$. To obtain $E(G \cdot N_i)$, we use **ordered-genotype** notation, with $P_{[ij]}$ denoting the probability of getting allele i from the mother and allele j from the father. We assume that $P_{[ij]} = P_{[ji]}$. Because the variable N_i takes on only two nonzero values, two and one, the expected cross-product is

$$E(G \cdot N_i) = (G_{ii} \cdot 2) \cdot P_{[ii]} + \sum_{j \neq i}^{n} (G_{ij} \cdot 1) \cdot 2P_{[ij]} = 2p_i \sum_{j=1}^{n} P_{ij|i} G_{ij}$$

where the last step follows from the definition of a conditional genotype probability (Equation 3.3a) as $P_{ij\,|\,i} = P_{[ij]}/p_i$. Putting the above results together, and recalling Equation 4.16a,

$$\sigma(G, N_i) = 2\,p_i \left[\sum_{j=1}^{n} P_{ij|i} G_{ij} - \mu_G \right] = 2\,p_i\,\alpha_i^* \tag{4.17a}$$

Under the assumption of random mating, average excesses are identical to additive effects, and

$$\sigma(G, N_i) = 2\,p_i\,\alpha_i \tag{4.17b}$$

Additive Effects

As in the diallelic case, with n alleles the additive (or average) effects are defined to be the set of α_i that minimizes $E(\delta_{ij}^2)$, obtained from the least-squares solution for the multiple regression

$$G = \mu_G + \sum_{i=1}^{n} \alpha_i\, N_i + \delta \tag{4.18}$$

This expression is the n-allele extension of Equation 4.5a, with N_i being the number of copies of allele i carried by an individual. For example, for the genotype G_{34}, $\sum \alpha_i\, N_i = \alpha_3 + \alpha_4$, and $\delta_{34} = G_{34} - \mu_G - \alpha_3 - \alpha_4$.

Multivariate regressions are covered in detail in Chapter 10, and here we simply cite the basic result: the regression coefficients (i.e., the α_i) are defined by the set of equations

$$\sigma(G, N_i) = \sum_{j=1}^{n} \alpha_j\, \sigma(N_i, N_j) \qquad \text{for } 1 \leq i \leq n \tag{4.19}$$

Expressed in this way, the definitions of the average effects are not immediately transparent, and the general solution to these equations is rather involved (Kempthorne 1957). However, under random mating, the solutions for the α_i are simplified greatly and can be expressed in two ways. First, drawing from Example 4.6

$$\alpha_i = \frac{\sigma(G, N_i)}{2p_i} \tag{4.20a}$$

Second, an equivalent, and even more transparent, solution follows from Equation 4.16b,

$$\alpha_i = \sum_{j=1}^{n} p_j\, G_{ij} - \mu_G \tag{4.20b}$$

i.e., under random mating, the average effects are equal to conditional mean deviations from μ_G.

If mating is nonrandom, but genotype frequencies are given by

$$P_{ii} = (1-f)p_i^2 + f p_i \tag{4.21a}$$

$$P_{ij} = 2(1-f)p_i p_j \tag{4.21b}$$

as occurs under regular inbreeding (Chapter 12), then

$$\alpha_i = \frac{\alpha_i^*}{1+f} \tag{4.22}$$

where f, the inbreeding coefficient, is the fractional reduction of heterozygote frequencies relative to those expected under random mating.

Additive Genetic Variance

To obtain the variance associated with the additive effects, we first need a result from regression theory. Consider the regression $y = \mu + \sum \beta_i\, x_i + e$. Because the total variance of a response variable y equals the variance accounted for by the regression plus the residual variance σ_e^2 (Chapters 3 and 10), it follows that the variance accounted for by the predictor variables is $\sum \beta_i\, \sigma(y, x_i)$. This can be immediately seen by noting

$$\sigma_y^2 = \sigma(y,y) = \sigma\left(y, \mu + \sum \beta_i\, x_i + e\right) = \sum_{i=1}^{n} \beta_i\, \sigma(y, x_i) + \sigma_e^2$$

Drawing the analogy with Equation 4.18, where the additive effects ($\beta_i = \alpha_i$) arise by considering the genotype G as a response variable ($y = G$) and the gene contents N_i as predictor variables ($x = N_i$), the variance associated with the additive effects becomes

$$\sum_{i=1}^{n} \alpha_i\, \sigma(G, N_i)$$

Thus, recalling the result from Example 4.6 that $\sigma(G, N_i) = 2\, p_i\, \alpha_i^*$, the additive genetic variance is

$$\sigma_A^2 = 2 \sum_{i=1}^{n} p_i\, \alpha_i\, \alpha_i^* \tag{4.23a}$$

as noted by Fisher (1941) and Kempthorne (1957). This general definition for the additive genetic variance holds for both randomly and nonrandomly mating populations. In the former case, $\alpha_i = \alpha_i^*$, and Equation 4.23a reduces to

$$\sigma_A^2 = 2 \sum_{i=1}^{n} p_i\, \alpha_i^2 \tag{4.23b}$$

which with $n = 2$ (a diallelic locus) reduces further to Equation 4.12a. Thus, under random mating, σ_A^2 for a locus is simply equal to the mean-squared additive effect, multiplied by two to account for diploidy. More generally, because $E[\,\alpha\,] = 0$, $\sigma_A^2 = E[\,(\alpha_i + \alpha_j)^2\,]$, which can involve covariance terms under nonrandom mating (when alleles at a locus are correlated).

From Equation 4.22, it follows that under regular inbreeding,

$$\sigma_A^2 = 2(1+f)\sum_{i=1}^{n} p_i\,\alpha_i^2 \tag{4.23c}$$

In general, inbreeding inflates the additive genetic variance by causing correlations among the effects of alleles within the same individuals. However, because the α_i themselves are a function of f, inbreeding does not necessarily simply increase the additive genetic variance by the factor $(1+f)$. From Kempthorne (1957, p. 350),

$$\alpha_i = \left(\frac{1-f}{1+f}\right)\alpha_{ir} + \left(\frac{f}{1+f}\right)(G_{ii} - \mu_G) \tag{4.24}$$

where α_{ir} and μ_G, respectively, denote the additive effect of allele *i* (under random mating) and the mean phenotype in the noninbred population. Because $G_{ii} = \mu_G + 2\alpha_i + \delta_{ii}$, if gene action is additive ($\delta_{ii} = 0$ for all i), then $G_{ii} - \mu_G = 2\alpha_{ir}$, so that $\alpha_i = \alpha_{ir} = a$, and the additive genetic variance in an inbred population is, in fact, $(1+f)$ times greater than that under random mating. However, with any level of dominance, $\alpha_i \neq \alpha_{ir}$ under inbreeding, and the change in additive genetic variance with f is not likely to be linear. See WL Chapters 11 and 23 for a full treatment.

Finally, we consider the general definition of the breeding value (A_{ij}) under random mating. Parents with genotype B_iB_j transmit alleles *i* and *j* with equal frequency, while the expected additive effect of the allele contributed by their (random) mates is equal to zero. Thus, the expected deviation of the mean phenotype of offspring of genotype B_iB_j from the population mean is

$$\left(\mu_G + \frac{\alpha_i + \alpha_j}{2}\right) - \mu_G = \frac{A_{ij}}{2} \tag{4.25}$$

which is half the breeding value of the parental genotype. Returning to Equation 4.18,

$$\begin{aligned} G_{ij} &= \mu_G + \alpha_i + \alpha_j + \delta_{ij} \\ &= \mu_G + A_{ij} + \delta_{ij} \end{aligned} \tag{4.26}$$

Thus, the genotypic value at any locus can be decomposed into four quantities: the mean genotypic value for the population, the additive effects of the two genes (whose sum is the breeding value), and a dominance deviation due to the interaction between the genes. Because μ_G is a constant, and A and δ are (by the properties of least-squares regression) uncorrelated, it follows from Equation 4.26 that the genetic variance can be represented as

$$\sigma_G^2 = \sigma^2(\alpha_i + \alpha_j) + \sigma^2(\delta_{ij}) \tag{4.27a}$$

This is a completely general definition, applying even to the case of nonrandom mating (although as noted above, the definitions of the α_i and δ_{ij} change with the degree of inbreeding). For the special case of random mating, α_i and α_j are uncorrelated, and

$$\sigma_G^2 = \sigma_A^2 + \sigma_D^2 \tag{4.27b}$$

Hence, σ_A^2 has a very specific and useful meaning. ***Under random mating, the additive genetic variance is equivalent to the variance of breeding values of individuals in the population.*** Note that if $\sigma_A^2 = 0$, then all breeding values are zero (the mean value). In this case, the expected offspring mean for *any* parent is simply the population mean. Hence, offspring resemble their parents no more than they do a random individual from the population. Likewise, when the additive variance is nearly zero, there is only a very small spread among the breeding values in the populations, so there is correspondingly very little variance in the expected offspring mean for any parent.

Table 4.3 Summary of quantities used to measure genetic effects. Note that *a*, *d*, and *k* are physiological parameters, while the remaining terms in the table are statistical parameters (population-specific functions of the physiological parameters that vary with the allele frequencies).

Homozygous effect, a Dominance coefficient, k, $d = ka$	Intrinsic properties of allelic products. Not functions of allele frequencies, but may vary with genetic background (Example 5.1).
Additive (or average) effect, α_i Average excess, α_i^*	Properties of *alleles* in a particular population. Functions of homozygous effects, dominance coefficients, and genotype frequencies.
Breeding value, A	Property of a particular *individual* in a particular population. Sum of the additive effects of an individual's alleles.
Additive genetic variance, σ_A^2	Property of a particular *population*. Variance of the breeding values of individuals within the population.

SUMMARY

Summing up, the additive effect of an allele, the breeding value of an individual, and the additive genetic variance of a population are hierarchically related measures of genetic effects (Table 4.3). All of this notation can be quite confusing, particularly when the non-subscripted α is used to denote the average effect of allelic substitution. We used the latter quantity in our introduction of the one-locus model for historical reasons and because it provides useful insight into the two-allele situation. However, we will not be using it much in the remainder of the book, nor will we be utilizing the concept of average excess (the latter plays a significant role in considerations of selection response, which is covered in WL Chapters 5 and 6). Unless otherwise noted, we will be dealing with randomly mating populations, so our reference to the additive effect of an allele will be consistent with the conditional mean deviation definition (Equation 4.20b), as well as formally equivalent to a least-squares regression coefficient. Further commentary on the relationship between average excesses and additive effects can be found in Falconer (1985) and Templeton (1987).

5

Sources of Genetic Variation for Multilocus Traits

The opposite of a correct statement is a false statement. But the opposite of a profound truth may well be another profound truth. Niels Bohr

As we proceed from single gene loci to polygenic quantitative characters, it becomes necessary to introduce several new concepts. First, we must consider whether the genotypic values defined for single loci can be combined additively to explain the phenotypic variation associated with multilocus genotypes or whether important nonlinear interactions between loci exist. Second, we must consider whether the inheritance and distribution of genes at one locus are independent of those at other loci. Third, several sources of environmental variance influence the expression of polygenic traits, and this raises questions as to whether trait expression varies with the environmental context and whether specific genotypes are associated with particular environments.

The joint occurrence of all of these sources of variation usually hopelessly obscures our ability to identify the precise breeding value of any single individual. Nevertheless, it is possible to characterize populations with respect to the relative magnitudes of different sources of phenotypic variance. Such a partitioning of variance is useful because different sources of variation have predictable impacts on various evolutionary phenomena. As noted in Chapters 3 and 4, the response to selection is closely associated with the level of additive genetic variance (WL Chapters 13–20). Environmental variance reduces the efficiency of the selection process by obscuring the relationship between genotypes and phenotypes. The response of a population to inbreeding and crossbreeding depends primarily on the level of dominance genetic variance (Chapters 12 and 13).

In the previous chapter, we found that the genetic variance associated with a single locus can be partitioned into additive and dominance components. This approach will be generalized below to account for all of the loci contributing to the expression of a quantitative trait, as well as to allow for variance arising from gene interaction among loci. The sources of variation associated with the environment, alluded to above, are explored in Chapter 6, and the general strategy for the actual measurement of these genetic sources of variance is covered in Chapters 7 and 8. Statistical designs are discussed in Part II Chapters 22–33.

EPISTASIS

In the preceding chapter, the dominance effect at a locus was defined to be the deviation of the observed genotypic value from the expectation based on additive effects. Viewed in this manner, dominance is a measure of nonadditivity of allelic effects *within* loci. **Epistasis** extends this idea to interactions *among* loci. A few comments on the usage of the term epistasis are in order, because it has been used to represent three related, but distinct, concepts (Wade 1992; Phillips 1998, 2008). The term, which translates into *standing upon*, was first coined by Bateson (1909) to denote the condition wherein one gene masks the phenotypic effects of a second, skewing expected segregation ratios of phenotypes. This usage describes the *qualitative effect* of one gene on another, arising from studies of the interaction between major genes. To describe the *quantitative effect* of one gene upon another, Fisher (1918) used the term **epistacy** to denote nonadditivity of individual locus effects, wherein a resulting multilocus genotypic effect was not simply a summation of the component single-locus

Figure 5.1 The three types of floral morphology in a tristylous plant. The central and lateral elongate structures denote, respectively, stigmas and anthers. Floral morphology (stigma length) is a function of two-locus genotypes. There are no *SS* individuals because short morphs are reproductively incompatible with each other.

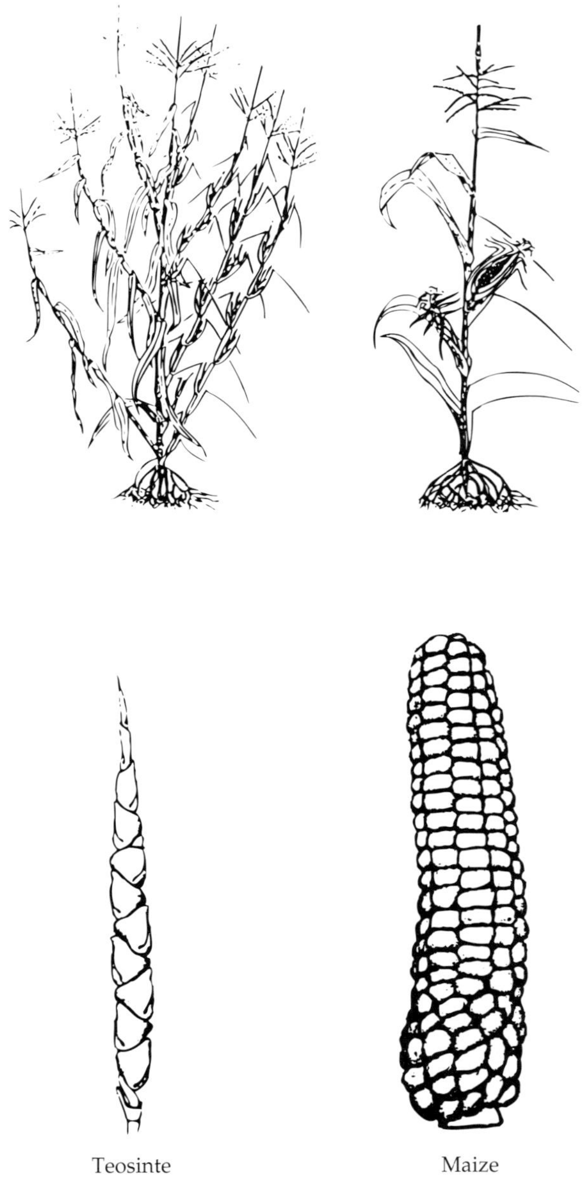

Figure 5.2 Phenotypic differences between domesticated maize *(Zea mays* sp. *mays)* and its progenitor, teosinte *(Zea mays* sp. *parviglumis)*. **Top:** Plant architecture. **Bottom:** Structure of the ear (the female inflorescence). (From Doebley et al. 1990.)

effects. Because this represents the logical extension from major qualitative changes due to Mendelian gene interactions (Bateson's original idea) to changes in measured traits from interactions among the underlying polygenes, Fisher's term was quickly replaced by epistasis in the literature, a trend that we continue here. Finally, molecular biologists often use epistasis to refer to genes whose products interact in a pathway (e.g., Avery and Wasserman 1992). Note that one can have this latter condition occurring without any masking of phenotypes or nonadditive interactions between genes. These different usages for the same term can result in confusion, because observations of nonadditivity among loci (epistasis to a quantitative geneticist) are often framed in terms of interactions on the underlying gene networks (epistasis to a molecular biologist).

A striking example of epistasis among major genes is responsible for a polymorphism in flower morphology in tristylous plants, a group including species of *Eichhornia*, *Lythrum*, *Oxalis*, and *Pontederia* (Weller 1976; Ganders 1979). Long-style morphs have short- and medium-length stamens, mid-style morphs have short and long stamens, and short-style morphs have medium and long stamens (Figure 5.1). Each morph is incompatible with plants of its own type but can freely cross with members of the other types. The polymorphism is determined by two diallelic loci, *short* and *medium*. The presence of an S allele at the first locus results in short-style plants regardless of the genotype at the second locus, whereas ss individuals are mid-styles if they carry an M allele but long-styles if they are mm. Thus, M is dominant to m, but the epistatic interaction of an S allele overrules this relationship.

To provide a more quantitative description of epistasis, consider an individual with alleles A_i and A_j at one locus and B_k and B_l at another. Expanding from Equation 4.26, the genotypic value, G_{ijkl}, can be written as the sum of the effects within loci and a deviation ϵ due to interaction between loci,

$$G_{ijkl} = \mu_G + (\alpha_i + \alpha_j + \delta_{ij}) + (\alpha_k + \alpha_l + \delta_{kl}) + \epsilon_{ijkl} \tag{5.1}$$

When written in this manner, Equation 5.1 glosses over the considerable complexity of epistatic interactions (ϵ_{ijkl}). With only two loci, there are actually three ways in which interactions can arise between loci: additive × additive ($\alpha\alpha$), additive × dominance ($\alpha\delta$), and dominance × dominance ($\delta\delta$). With three loci, there are four additional types of epistasis, ($\alpha\alpha\alpha$), ($\alpha\alpha\delta$), ($\alpha\delta\delta$), and ($\delta\delta\delta$), and the number steadily grows with additional loci.

Example 5.1. As a numerical example of epistasis, consider the genotypic values associated with two diallelic loci in an artificially constructed population of teosinte, the wild progenitor of cultivated maize. The two subspecies are fully interfertile, despite their dramatic differences in plant architecture and inflorescence structure (Figure 5.2). Beadle (1939) hypothesized that substitution of as few as five major genes, plus a number of polygenes, may have been responsible for the teosinte-to-maize transformation, and subsequent studies clearly implicate the importance of a small number of genes with major effects (Doebley and Stec 1993; Dorweiler et al. 1993; Doebley et al. 1995a, 1995b; Hubbard et al 2002; Doebley 2004; Wang et al. 2005b; Studer and Doebley 2011; Whipple et al. 2011; Wills et al. 2013; Wang et al. 2015).

A number of molecular markers can be used to distinguish between genomic segments from maize and teosinte. Through a progressive series of backcrosses and screening of genomic markers, Doebley et al. (1995a) introduced two maize markers, *UMC107* and *BV302*, into a teosinte genetic background. We will denote the maize "alleles" for these two loci as U_M and B_M and the teosinte "alleles" as U_T and B_T. Each of these markers is associated with a tightly linked segment of DNA from its parental origin (Chapter 18). Thus, the U_{TT}/B_{TT} genotype is essentially pure teosinte, whereas the U_{MM}/B_{MM} genotype is homozygous for two small stretches of maize DNA on an otherwise pure teosinte background.

For each of the nine two-locus marker genotypes, Doebley et al. assayed a large number of individuals for several morphological and reproductive traits. The following table gives the results for one such character, the average length of vegetative internodes in the lateral branch (in millimeters). Estimates of the average homozygous-effect coefficients (a) and dominance coefficients (k) (defined in Figure 4.4), conditional on the genotypic states of the alternate locus, are given in the last two columns and rows. These are calculated by setting the three genotypes equal to $C, C + (1 + k)a$, and $C + 2a$, where C is a scaling factor.

	UMC107				
BV302	U_{MM}	U_{MT}	U_{TT}	a	k
B_{MM}	18.0	40.9	61.1	27.0	0.33
B_{MT}	54.6	47.6	66.5	6.0	−2.17
B_{TT}	47.8	83.6	101.7	21.6	0.06
a	14.9	21.4	20.3		
k	1.46	−0.69	−0.73		

Assuming that the observed mean phenotypes of the different marker classes are accurate estimates of the genotypic values, the fact that both a and k vary dramatically with the genetic background provides strong evidence for epistatic interaction between the genes associated with the markers. For example, when the genotype at the *UMC107* locus is U_{MT} or U_{TT}, B_M is quite dominant to B_T $(k \simeq -0.7)$ and the homozygous effect of the B_T allele is $a \simeq 21$. However, on a U_{MM} background, the homozygous effect is reduced to $a \simeq 15$, and the *BV302* locus exhibits overdominance $(k > 1)$. For this and other characters, the authors further demonstrated that the homozygous and dominance effects associated with these markers vary dramatically depending on whether they are expressed in teosinte, maize, or teosinte × maize genetic backgrounds.

Epistasis, Transgenics, and Gene Editing

A common shorthand for epistasis in the literature is G × G, a gene–gene interaction. This notation is commonly used in the context of QTL mapping, wherein the impact of a QTL genotype varies over genetic background (Chapters 18 and 20). Of special interest in commercial agriculture is the presence of T × G interactions, namely does the effect of a **transgene** (the engineering of a gene from another population, often a very distant species, into a genome of interest) vary over background. Setting aside issues with government regulation, transgenes for qualitatively inherited traits—such as insect and herbicide resistance—have been commercially very successful (Linares et al. 2022). In part, such success arises because the major effect of the transgene is rather stable over different genomes (at least for the focal trait), i.e., modest T × G such that the transgene is commercially viable over a number of backgrounds.

However, when moving to more complex traits—such as yield and drought tolerance—T × G interactions can be dramatic (Nuccio et al. 2018; Paul et al. 2018; Simmons et al. 2021). For example, Linares et al. (2022) tested a transgene impacting the ethylene biosynthesis pathway in a number of maize hybrids, by contrasting pairs of hybrids that differed only by the presence/absence of the transgene. Under most hybrid backgrounds, the transgene had a nonsignificant effect on yield, and thus under this screen would have been rejected as a candidate. However, in one particular hybrid background, the transgenic line had a 8% yield advantage over its nontransgenic, but otherwise identical, hybrid. When such significant T × G interactions are present, candidate transgenes could be incorrectly rejected, or advanced, if tested in a limited number of backgrounds (Linares et al. 2022; Khaipho-Burch et al. 2023). These same concerns about strong background epistatic effects also arise with gene editing. However, one advantage of editing is the ability to more rapidly generate the edited constructs over a number of different backgrounds, and hence a more robust testing regime is possible.

A GENERAL LEAST-SQUARES MODEL FOR GENETIC EFFECTS

We now consider a statistical procedure for the definition of epistatic effects, which is a straightforward extension of the one-locus linear model introduced in Chapter 4. We will confine our attention to two loci, as the extension to higher numbers of loci should become obvious (albeit tedious), and we assume random mating throughout.

In the previous chapter, the additive effect of an allele (under random mating) was shown to be equal to the deviation of members of the population with the allele from the population mean phenotype. This definition does not change with the addition of loci. Letting $G_{i...}$ represent the conditional mean phenotype of individuals with allele i at the first locus without regard to the other allele at the locus or to the genotype at the second locus, then

$$\alpha_i = G_{i...} - \mu_G \tag{5.2}$$

Analogous expressions follow for the three other additive effects (α_j, α_k, and α_l). Within each locus, the mean value of the additive effects (weighted by the allele frequencies) is equal to zero.

The dominance effects are defined by considering $G_{ij..}$, the conditional mean phenotype of individuals with alleles i and j at the first locus without regard to genotypic state at the second locus. From Equation 4.26, it follows that

$$\delta_{ij} = G_{ij..} - \mu_G - \alpha_i - \alpha_j \tag{5.3a}$$

$$\delta_{kl} = G_{..kl} - \mu_G - \alpha_k - \alpha_l \tag{5.3b}$$

Here, we have subtracted the mean genotypic value and the additive effects, leaving the dominance effect as the only unexplained portion of the conditional mean at the locus. As in the case of additive effects, the mean dominance deviation at each locus is equal to zero.

The definition of epistatic effects proceeds in a similar fashion. Letting $G_{i.k.}$ be the mean phenotype of individuals with allele i at locus 1 and allele k at locus 2, without regard to the other two alleles, the *ik*th additive × additive effect is

$$(\alpha\alpha)_{ik} = G_{i.k.} - \mu_G - \alpha_i - \alpha_k \tag{5.4a}$$

In other words, $(\alpha\alpha)_{ik}$ is the deviation of the conditional mean $G_{i.k.}$ from the expectation based on the population mean μ_G and the additive effects α_i and α_k. Next, an additive × dominance effect measures the interaction between an *allele* at one locus with a particular *genotype* of another locus. It is defined as the deviation of the conditional mean $G_{i.kl}$ from the expectation based on all lower order effects, in this case the three additive effects, one dominance effect, and two additive × additive effects involving the constituent genes,

$$(\alpha\delta)_{ikl} = G_{i.kl} - \mu_G - \alpha_i - \alpha_k - \alpha_l - \delta_{kl} - (\alpha\alpha)_{ik} - (\alpha\alpha)_{il} \tag{5.4b}$$

Finally, for a dominance × dominance effect,

$$\begin{aligned}(\delta\delta)_{ijkl} = G_{ijkl} &- \mu_G - \alpha_i - \alpha_j - \alpha_k - \alpha_l - \delta_{ij} - \delta_{kl} \\ &- (\alpha\alpha)_{ik} - (\alpha\alpha)_{il} - (\alpha\alpha)_{jk} - (\alpha\alpha)_{jl} \\ &- (\alpha\delta)_{ikl} - (\alpha\delta)_{jkl} - (\alpha\delta)_{ijk} - (\alpha\delta)_{ijl}\end{aligned} \tag{5.4c}$$

This partitioning of the total genotypic value into a series of effects can be summarized as follows. We started with the lowest order effects, the additive effects of alleles. These are defined in a least-squares sense (Chapter 3), in that they account for as much of the variance in genotypic values as possible. We then progressively defined higher order effects, each time accounting for as much of the residual variation as possible. In the two-locus example, $(\delta\delta)_{ijkl}$ describes the final bit of variation not accounted for by additive, dominance, additive × additive, or additive × dominance effects. Summing up terms, we have a complete description of a genotypic value,

$$\begin{aligned}G_{ijkl....} = \mu_G &+ [\alpha_i + \alpha_j + \alpha_k + \alpha_l] + [\delta_{ij} + \delta_{kl}] \\ &+ [(\alpha\alpha)_{ik} + (\alpha\alpha)_{il} + (\alpha\alpha)_{jk} + (\alpha\alpha)_{jl}] \\ &+ [(\alpha\delta)_{ikl} + (\alpha\delta)_{jkl} + (\alpha\delta)_{ijk} + (\alpha\delta)_{ijl}] + (\delta\delta)_{ijkl} + \cdots\end{aligned} \tag{5.5}$$

The open-ended equation merely implies that there are many more terms when more than two loci are involved in the expression of a trait. The parameters in this model depend very much upon the genotype frequencies in the population and can change as allele frequencies change (Example 5.2; Chapter 4). By construction (being deviations from the grand mean), the mean value of each type of effect is always equal to zero.

Equation 5.5 highlights the potential for significant epistasis in the expression of quantitative traits. With n loci underlying a trait, the number of additive effects is of order n, while the number of diallelic and triallelic epistatic effects are potentially of order n^2 and n^3. Thus, even if individual epistatic effects are relatively small, their summed effects may be considerable. This raises a serious issue for quantitative genetics that will become more apparent in later chapters. Although there are many possible types of epistasis, methods for evaluating their quantitative significance have substantial limitations from the standpoint of statistical power and experimental design (Part II Chapters 22–33).

At first glance, Equation 5.5 may seem to be of little practical use. After all, if the genotypic value of an individual cannot be ascertained without error because of environmental contributions to the phenotype, then what hope is there of identifying the individual components? Often there is none. However, a great conceptual advance is achieved when Equation 5.5 is extended to the *population level*. Provided that mating is random and segregation of loci is independent, there is no statistical relationship between the genes found within or among loci. Therefore, the total genetic variance is simply the sum of the variance of the individual effects. Letting $\sigma^2_A = \sigma^2(\alpha_i) + \sigma^2(\alpha_j) + \sigma^2(\alpha_k) + \sigma^2(\alpha_l)$, $\sigma^2_D = \sigma^2(\delta_{ij}) + \sigma^2(\delta_{kl})$, $\sigma^2_{AA} = \sigma^2[(\alpha\alpha)_{ik}] + \sigma^2[(\alpha\alpha)_{il}] + \sigma^2[(\alpha\alpha)_{jk}] + \sigma^2[(\alpha\alpha)_{jl}]$, and so on, the total genetic variance can be expressed as

$$\sigma^2_G = \sigma^2_A + \sigma^2_D + \sigma^2_{AA} + \sigma^2_{AD} + \sigma^2_{DD} + \cdots \tag{5.6}$$

This partitioning of the genetic variance into a series of components, the multilocus analog of Equation 4.11, was developed independently, but with very different approaches, by both Cockerham (1954) and his major professor at the time, Kempthorne (1954).

As we saw with dominance in the previous chapter, one can have very significant amounts of *epistasis*, and yet have rather small *epistatic variances*. Again, as with dominance, this is especially the case when the relevant allele frequencies are not at intermediate values. Recall that by modifying the additive effects of alleles, dominance (a higher order effect than additivity) can inflate or deflate σ^2_A relative to the situation expected under complete additivity (Equation 4.12a; Figure 4.8). Qualitatively similar results apply to epistasis—depending on the frequencies of the genes involved, epistatic interactions (through their statistical influence on the additive and dominance effects of alleles) can greatly inflate the additive and/or dominance components of genetic variance.

Detailed analyses by Hill et al. (2008) and Mäki-Tanila and Hill (2014) found that, under rather general conditions, most of the epistatic effects are loaded onto the additive effect, and hence into the additive variance. For a model with diallelic loci (with effects of 0 and a_i for locus i, where p_i denotes the frequency of the trait-increasing allele) and only additive epistatic interactions ($A \times A$ and $A \times A \times A$), Mäki-Tanila and Hill showed that one can write the additive effect for the trait-increasing allele at locus i as

$$\alpha_i = a_i + 2\sum_{j \neq i} p_j [aa]_{ij} + 4 \sum_{j \neq i} \sum_{k \neq i, k > j} p_j p_k [aaa]_{ijk} \tag{5.7}$$

where a_i, $[aa]_{ij}$, and $[aaa]_{ijk}$, denote, respectively, the fixed constants for the additive value at locus i, the pairwise interactions between alleles at loci i and k, and the three-way interaction between i, j, and k (the derivation is given in Example 5.3). Note that Equation 5.7 is the analog to what we observed in Equation 4.12a with dominance: the additive effect (α_i) is the additive value (a_i) plus a frequency-dependent weight on the dominance contribution (d_i). Summing over loci, the additive variance becomes $\sigma^2_A = \sum H_i \alpha_i^2$, where $H_i = 2p_i(1 - p_i)$ is the heterzygosity at locus i. Just as the dominance variance scales with H^2 (Equation 4.12c), Mäki-Tanila and Hill found that σ^2_{AA} scales with H^2, and σ^2_{AAA} scales with H^3. Hence, the epistatic variance is generally quite small unless the underlying loci all have intermediate allele frequencies. Studies that find a significant amount of epistatic variance usually involve line crosses, where (by construction) all segregating alleles have intermediate frequencies (1/2 or powers of 1/2 for multiple-line crosses; Chapter 18).

Thus, as will be seen in the next example and as cogently pointed out by Jana (1971, 1972), Keightley (1989), and Cheverud and Routman (1995), relatively low magnitudes of epistatic components of *variance* in natural populations are by no means incompatible with the existence of strong epistatic *gene action*. As they did with dominance, Cheverud and Routman (1995) stressed the difference between **physiological epistasis** (the actual values of the genotypes) and **statistical epistasis** (the frequency-dependent best-fit values for $\alpha\alpha$, $\alpha\delta$, and $\delta\delta$). The variances associated with the latter can be small even in the presence of very strong physiological epistasis (Example 5.2). Hence, as with dominance, variance components provide, at best, limited insight into the physiological mode of gene action, as one can have significant interaction, yet have very little *variance* associated with this interaction.

In summary, simple statistical arguments suggest that epistatic interactions are likely to be important in the expression of many quantitative traits as well as in the determinance of levels of additive genetic variance. Wright (1968), for one, argued strongly that epistasis is the rule rather than the exception. For further reviews of the evidence for epistasis, see Wright (1968), Barker (1979), Moreno (1994), Hill et al. (2008), Phillips (2008), Huang et al. (2012), Hemani et al. (2014), Mackay (2014), and Wei et al. (2014).

Example 5.2. To acquire a deeper understanding of the connections between conditional genotypic values, genetic effects, and variance components, we return to the data in Example 5.1, again making the assumption that the entries in the table are accurate estimates of the nine genotypic values associated with the marker loci. Suppose that this population randomly mates, with all alleles having frequency 0.5. What are the genetic effects associated with the different alleles (and their combinations), and what are the expected components of genetic variance associated with the two marker loci? Note that this assumption of immediate allele frequencies is the *most favorable setting* for generating the maximal values for epistatic variances.

Under the assumption that the two loci are unlinked (in actuality, they are on different chromosomes), the frequencies of the two-locus genotypes are defined in the following table. They are simply equal to the products of the Hardy–Weinberg frequencies at the individual loci. Thus, each of the four double-homozygote classes has frequency $(1/4)^2$, while the classes with a single heterozygous locus have frequency $(1/2) \times (1/4)$, and the double-heterozygote has frequency $(1/2)^2$, yielding

	UMC107		
BV302	U_{MM}	U_{MT}	U_{TT}
B_{MM}	1/16	1/8	1/16
B_{MT}	1/8	1/4	1/8
B_{TT}	1/16	1/8	1/16

In the following computations, we let G_{ijkl} denote the genotypic value of individuals with alleles i and j at locus *BV302* and alleles k and l at locus *UMC107*, where i, j, k, and l are either *M* or *T*. Letting P_{ijkl} be the frequency of the *ijkl* genotype, the mean genotypic value in the population is

$$\begin{aligned}\mu_G &= \sum P_{ijkl}\, G_{ijkl} \\ &= [(1/16) \times (18.0 + 61.1 + 47.8 + 101.7)] \\ &\quad + [(1/8) \times (40.9 + 54.6 + 66.5 + 83.6)] + [(1/4) \times 47.6] \\ &= 56.8875\end{aligned}$$

Additive effects. Prior to computing the additive effects, we require the conditional genotypic means associated with each allele. These are obtained by weighting the phenotypic data in the table of Example 5.1 with conditional genotype frequencies. For example, the conditional genotypic mean for individuals containing a B_M allele is

$$G_{B_M\cdots} = \sum_{j,k,l} \Pr(ijkl \mid i = B_M)\, G_{B_M jkl}$$

The conditional probability $\Pr(ijkl \mid i = B_M)$ is the probability that an individual has alleles j, k, and l, given that it is has one B_M allele. This probability is simply the product of the appropriate (Hardy–Weinberg) genotype frequency at the *UMC107* locus and the appropriate allele frequency at locus *BV302*. For example, if $j = B_M$, $k = U_M$, and $l = U_M$, $\Pr(ijkl \mid i = B_M) = 0.5 \times 0.25 = 0.125$. Thus, the genotypic mean conditional on having a B_M allele is

$$\begin{aligned} G_{B_M\cdots} &= (0.5 \times 0.25 \times 18.0) + (0.5 \times 0.5 \times 40.9) \\ &\quad + (0.5 \times 0.25 \times 61.1) + (0.5 \times 0.25 \times 54.6) \\ &\quad + (0.5 \times 0.5 \times 47.6) + (0.5 \times 0.25 \times 66.5) = 47.1500 \end{aligned}$$

The three other conditional means involving single alleles are obtained in a similar manner,

$$G_{B_T\cdots} = 66.6250 \qquad G_{\cdot\cdot U_M \cdot} = 49.3375 \qquad G_{\cdot\cdot U_T \cdot} = 64.4375$$

Recalling Equation 5.2, we obtain the additive effects by subtracting the population mean μ_G from the conditional means,

$$\begin{aligned} \alpha_{B_M} &= -9.7375 \qquad & \alpha_{B_T} &= 9.7375 \\ \alpha_{U_M} &= -7.5500 \qquad & \alpha_{U_T} &= 7.5500 \end{aligned}$$

Noting that each allele has a frequency of 0.5, the mean additive effect of the alleles at each locus is equal to zero.

Dominance effects. To obtain the dominance effects, we require the conditional genotypic means. In this case, for each single-locus genotype, we simply take the average of the three genotypic values associated with the second locus, weighted by their Hardy–Weinberg frequencies. For example, for the B_{MM} genotype,

$$\begin{aligned} G_{B_{MM}\cdot\cdot} &= \sum_{k,l} \Pr(ijkl \mid i, j = B_{MM})\, G_{B_{MM} kl} \\ &= (0.25 \times 18.0) + (0.5 \times 40.9) + (0.25 \times 61.1) = 40.225 \end{aligned}$$

Similarly, the conditional means for the B_{MT} and B_{TT} genotypes are found to be

$$G_{B_{MT}\cdot\cdot} = 54.075 \qquad G_{B_{TT}\cdot\cdot} = 79.175$$

Substituting these values, the population mean, and the appropriate additive effects into Equation 5.3, we obtain the dominance effects as

$$\delta_{B_{MM}} = 2.8125 \qquad \delta_{B_{MT}} = -2.8125 \qquad \delta_{B_{TT}} = 2.8125$$

Weighting these by the Hardy–Weinberg frequencies associated with each genotype, it can be seen that the mean dominance effect at the *BV302* locus is equal to zero. We leave it to the reader to show that for the *UMC107* locus,

$$\delta_{U_{MM}} = 1.9625 \qquad \delta_{U_{MT}} = -1.9625 \qquad \delta_{U_{TT}} = 1.9625$$

Additive $\times$ additive effects. The same logic is used to obtain the additive $\times$ additive effects. Because there are two alleles at each locus, four conditional means (each involving one allele at each locus) must be considered. Each gametic type is found in four two-locus genotypes, in this case with equal frequency 0.25 because all alleles have a frequency of 0.5. Thus, for example, the conditional mean $G_{B_M \cdot U_M \cdot}$ is the average of the genotypic values for the genotypes $B_{MM}\ U_{MM}$, $B_{MT}\ U_{MM}$, $B_{MM}\ U_{MT}$, and $B_{MT}\ U_{MT}$. The four conditional means are

$$G_{B_M \cdot U_M \cdot} = 40.275 \qquad G_{B_M \cdot U_T \cdot} = 54.025$$
$$G_{B_T \cdot U_M \cdot} = 58.400 \qquad G_{B_T \cdot U_T \cdot} = 74.850$$

The additive × additive effects are computed by substituting these values, the population mean, and the appropriate additive effects into Equation 5.4a,

$$(\alpha\alpha)_{B_M \cdot U_M \cdot} = 0.675 \qquad (\alpha\alpha)_{B_M \cdot U_T \cdot} = -0.675$$
$$(\alpha\alpha)_{B_T \cdot U_T \cdot} = 0.675 \qquad (\alpha\alpha)_{B_T \cdot U_M \cdot} = -0.675$$

Again, the mean value of these effects (weighted by their frequencies of occurrence) is zero.

Additive × dominance effects. At this point, we assume that the basic strategy for calculating conditional means is familiar, and we leave it to the reader to verify that the conditional means involving one allele at one locus and two at the other locus are

$$\begin{array}{lll} G_{B_M . U_{MM}} = 36.30 & G_{B_M . U_{MT}} = 44.25 & G_{B_M . U_{TT}} = 63.80 \\ G_{B_T . U_{MM}} = 51.20 & G_{B_T . U_{MT}} = 65.60 & G_{B_T . U_{TT}} = 84.10 \\ G_{B_{MM} U_M \cdot} = 29.45 & G_{B_{MT} U_M \cdot} = 51.10 & G_{B_{TT} U_M \cdot} = 65.70 \\ G_{B_{MM} U_T \cdot} = 51.00 & G_{B_{MT} U_T \cdot} = 57.05 & G_{B_{TT} U_T \cdot} = 92.65 \end{array}$$

The additive × dominance effects follow from substitution of quantities given above into Equation 5.4b,

$$\begin{array}{lll} (\alpha\delta)_{B_M . U_{MM}} = 0.9375 & (\alpha\delta)_{B_M . U_{MT}} = -0.9375 & (\alpha\delta)_{B_M . U_{TT}} = 0.9375 \\ (\alpha\delta)_{B_T . U_{MM}} = -0.9375 & (\alpha\delta)_{B_T . U_{MT}} = 0.9375 & (\alpha\delta)_{B_T . U_{TT}} = -0.9375 \\ (\alpha\delta)_{B_{MM} U_M .} = -4.5750 & (\alpha\delta)_{B_{MT} U_M .} = 4.5750 & (\alpha\delta)_{B_{TT} U_M .} = -4.5750 \\ (\alpha\delta)_{B_{MM} U_T .} = 4.5750 & (\alpha\delta)_{B_{MT} U_T .} = -4.5750 & (\alpha\delta)_{B_{TT} U_T .} = 4.5750 \end{array}$$

Dominance × dominance effects. The nine dominance × dominance effects are computed from the observed genotypic values using Equation 5.4c,

$$\begin{array}{lll} (\delta\delta)_{B_{MM} U_{MM}} = -4.5125 & (\delta\delta)_{B_{MM} U_{MT}} = 4.5125 & (\delta\delta)_{B_{MM} U_{TT}} = -4.5125 \\ (\delta\delta)_{B_{MT} U_{MM}} = 4.5125 & (\delta\delta)_{B_{MT} U_{MT}} = -4.5125 & (\delta\delta)_{B_{MT} U_{TT}} = 4.5125 \\ (\delta\delta)_{B_{TT} U_{MM}} = -4.5125 & (\delta\delta)_{B_{TT} U_{MT}} = 4.5125 & (\delta\delta)_{B_{TT} U_{TT}} = -4.5125 \end{array}$$

Variance components. The final step in this example is to compute the components of genetic variance. Because the average effects are computed in such a way that their means are always equal to zero, the variances are simply equal to the mean-squared effects (weighted by their frequency of occurrence) multiplied by the number of times the effect occurs. For example, for the additive genetic variance, the contribution associated with the *BV302* locus is $2 \times [(0.5 \times (-9.7375)^2) + (0.5 \times 9.7375^2)] = 189.6378$; the mean-squared effect is multiplied by two because there are two additive effects per diploid locus. In the same manner, the additive genetic variance associated with the *UMC107* locus is found to be 114.0050. The total additive genetic variance associated with these two loci is simply the sum of the locus-specific contributions,

$$\sigma_A^2 = 303.6428$$

The dominance genetic variance associated with each locus is equal to the mean-squared dominance effect, $2.8125^2 = 7.9101$ for the *BV302* locus and $1.9625^2 = 3.8514$ for the *UMC107* locus. Thus, the total dominance genetic variance is

$$\sigma_D^2 = 11.7615$$

The additive × additive variance is obtained by multiplying the mean squared additive × additive effect by four, because four such effects contribute to each genotypic value. In this example, all of the additive × additive effects have absolute value 0.675 (as in the case of the additive and dominance effects, the symmetry is a special consequence of all of the gene frequencies being equal to 0.5). Thus, the additive $\times$ additive variance is simply 4×0.675^2,

$$\sigma^2_{AA} = 1.8225$$

For the additive $\times$ dominance variance, there are again four effects contributing to the genotypic value of each individual, two involving *BV302* additive effects and two involving *UMC107* additive effects. The additive $\times$ dominance variance is therefore $(2 \times 0.9375^2) + (2 \times 4.575^2)$,

$$\sigma^2_{AD} = 43.6191$$

Finally, because there is only one dominance $\times$ dominance term per individual, the dominance $\times$ dominance variance is simply the mean-squared effect, 4.5125^2,

$$\sigma^2_{DD} = 20.3627$$

Summing the five components of genetic variance, we obtain the total genetic variance associated with the two markers,

$$\sigma^2_G = 381.2086$$

As a check on the validity of this result, the total genetic variance can be computed directly from the data in the table in Example 5.1 as $\sum P_{ijkl} G^2_{ijkl} - \mu^2_G$, which yields the same answer. Note that roughly 80% of the genetic variance is additive (303.6/381.2).

These results show that despite the existence of strong epistatic interactions between the effects of genes associated with the two markers (Example 5.1), most of the interaction effects are associated statistically with the additive effects of alleles. The three epistatic components of variance account for only 17% of the total genetic variance, and the dominance genetic variance contributes only another 3%. Thus, as emphasized above, the relative magnitude of variance components provides little insight into the actual physiological mode of gene action.

In closing, we note that this example is a somewhat unconventional application of the linear model. Usually, the genes underlying quantitative traits are unknown, in which case it is impossible to estimate their individual effects. However, with the widespread application of molecular markers, this situation is changing (Chapters 17–21). In Chapter 7, we will begin to see that estimation of the components of genetic variance associated with the entire pool of constituent loci underlying a quantitative trait does not require any information about their identity.

Alternative partitioning of epistatic effects have been suggested for a variety of settings, including departures from both Hardy–Weinberg and linkage equilibrium (Van der Veen 1959; Yang, 2004; Zeng et al. 2005; Wang and Zeng 2006, 2009; Alvarez-Castro and Carlborg 2007; Vitezica et al. 2013, 2017; Huang and Mackay 2016).

Example 5.3. This example develops **Kojima's method**, an alternative approach that gives expressions for variance components based on derivatives of the trait mean with respect to allele frequencies (Kojima 1959). Kojima's concern was the analysis of the stability of allele-frequency equilibria in complex traits (WL Chapter 5), which requires expressions involving derivatives of allele frequencies. Kojima noted that this "procedure is expeditious in obtaining the partitions of genotypic variance for many genetic models where the mean can be written as a condensed function of gene frequencies and parameters of the model." At the end of this example, we illustrate this point by using Kojima's method to derive Mäki-Tanila and Hill's (2014) expression for the impact of additive epistasis on the additive variance (Equation 5.7).

Kojima's approach assumes Hardy–Weinberg (single-locus genotypic frequencies are a function of the random union of alleles) and linkage equilibrium (independence of allele frequencies across loci). It starts by expressing the mean of a trait value as the sum of three conditional means for the jth locus,

$$\mu = p_j^2 \, \mu_{Q_jQ_j} + 2p_j(1-p_j) \, \mu_{Q_jq_j} + (1-p_j)^2 \, \mu_{q_jq_j} \tag{5.8a}$$

where, for example, $\mu_{Q_jQ_j}$ is the mean trait value of an individual with j locus genotype Q_jQ_j. Taking the derivative of the trait mean with respect to the frequency (p_j) of Q_j yields

$$\frac{\partial \mu}{\partial p_j} = 2\left[p_j(\mu_{Q_jQ_j} - \mu_{Q_jq_j}) + (1-p_j)(\mu_{Q_jq_j} - \mu_{q_jq_j})\right] = 2\alpha_j \tag{5.8b}$$

namely, twice the average effect of an allelic substitution at locus j (Chapter 4). To see this, writing the mean of the genotypic values at locus j as $2a_j : 1 + a_j(1+k_j) : 0$, the expression in brackets reduces to Equation 4.10b. From Equation 4.12a, we can express the additive genetic variance contributed by this locus as $\sigma^2_A(j) = H_j\alpha_j^2$, where $H_j = 2p_j(1-p_j)$ is the heterozygosity at locus j. Because we have assumed linkage equilibrium, the total additive variance is just the sum of the single locus effects, yielding

$$\sigma^2_A = \sum_{j=1}^{n} \sigma^2_A(j) = \sum_{j=1}^{n} H_j \left(\frac{1}{2}\frac{\partial \mu}{\partial p_j}\right)^2 \tag{5.8c}$$

The expression for the dominance variance follows by taking the second derivative,

$$\frac{\partial^2 \mu}{\partial p_j^2} = 2\left[\mu_{Q_jQ_j} - 2\,\mu_{Q_jq_j} + \mu_{q_jq_j}\right] = 2\delta_j$$

where the dominance variance contributed by locus j becomes $\sigma^2_D(j) = H_j^2\delta_j^2$, which reduces to our previous expression (Equation 4.12b). Summing over loci yields

$$\sigma^2_D = \sum_{j=1}^{n} H_j^2 \left(\frac{1}{2}\frac{\partial^2 \mu}{\partial p_j^2}\right)^2 \tag{5.8d}$$

Kojima showed that his expression for additive $\times$ additive variance,

$$\sigma^2_{AA} = \sum_{j<\ell=1}^{n} H_j H_\ell \left(\frac{1}{4}\frac{\partial^2 \mu}{\partial p_j \partial p_\ell}\right)^2 \tag{5.8e}$$

also reduces to the Cockerham (1954) expressions for σ^2_{AA}. In general, the contribution from a set of $(L+M)$ loci to (L-additive) $\times$ (M-dominance) genetic variance is

$$\sigma^2_{A^L D^M} = \prod_{\ell}^{L}\prod_{j}^{M}\left\{ H_\ell H_j^2 \left[\left(\frac{1}{2^{L+M}}\right)\frac{\partial^{L+M}(\mu)}{\prod_\ell^L \partial q_\ell \prod_j^M \partial q_j^2}\right]^2\right\} \tag{5.8f}$$

where $\prod_\ell^L$ denotes the L-term product (indexed by ℓ). This expression shows the impact of heterozygosity on nonadditive variances. In natural populations, one expects the distribution of minor allele frequencies to be roughly L-shaped (WL Chapter 2), meaning that the heterozygosity values at most loci will be small, diminishing any nonadditive variances. In contrast, when one forms a population by crossing two fully inbred lines, then all segregating loci have a heterozygosity of 1/2 in the F_1 and all subsequent generations, conditions that maximize nonadditive variances.

While the above treatment may seem a bit abstract, these results often allow us to obtain relatively simple expressions for higher order variances. Mäki-Tanila and Hill (2014) examined a model with only additive epistasis with the assumption that genotype q_kq_k contributes zero, Q_kq_k contributes a_k, and Q_kQ_k contributes $2a_k$. Higher order additive terms also appear, but are only nonzero when at least one Q allele is involved, so that (for example), one can write the mean of $Q_1q_1Q_2q_2Q_3q_3$ as $a_1 + a_2 + a_3 + [aa]_{12} + [aa]_{13} + [aa]_{23} + [aaa]_{123}$. Under this model,

$$\mu = 2\sum_i p_i a_i + 4\sum_i\sum_{j>i} p_i p_j [aa]_{ij} + 8\sum_i\sum_{j>i}\sum_{\ell>j>i} p_i p_j p_\ell [aaa]_{ij\ell} + \cdots \tag{5.8g}$$

Applying Equation 5.8b, the average effect of locus i becomes

$$\alpha_i = \frac{1}{2}\frac{\partial \mu}{\partial p_i} = a_i + 2\sum_{j\neq i} p_j [aa]_{ij} + 4\sum_{j\neq i}\sum_{\ell\neq i,\ell>j} p_j p_\ell [aaa]_{ij\ell} + \cdots$$

which recovers Equation 5.7, showing how higher order epistatic interactions load onto the additive effects. Mäki-Tanila and Hill also present expressions for more complex models with dominance.

Extension to Haploids

A large fraction of the more basal organisms, most notably bacteria, many protists, fungi, algae, and mosses, have life cycles that are predominantly haploid. A very brief diploid stage occurs when gametes fuse and chromosomes reassort to form variable haploid progeny. Haploid species have long been used as model systems in genetic studies focused on single loci, but they also offer some interesting possibilities for studies on quantitative-genetic variation (Caten and Jinks 1976; Croft and Jinks 1977; Caten 1979; Steinmetz et al. 2002b; Nogami et al. 2007; Ehrenreich et al. 2010; Liti and Louis 2012; Cubillos et al. 2013). Because haploid organisms have only a single allele at each locus, dominance does not contribute to the genetic variation. Hence, all variance due to nonadditive gene action involves epistasis between additive effects, and Equation 5.6 simplifies to

$$\sigma_G^2 = \sigma_A^2 + \sigma_{AA}^2 + \sigma_{AAA}^2 + \cdots \tag{5.9}$$

Haploid organisms can often be propagated vegetatively (clonally), in which case it is a simple matter to test whether genetic variation can be explained by an additive model. When measures are taken on a large number of replicates of the same genotype, the mean of these measures will be a good estimate of the genotypic value under the experimental setting. If there is no epistatic gene action, the mean phenotype of sexually produced progeny should not be significantly different from the average of the parents.

Figure 5.3 illustrates two analyses of growth rate resulting from crosses between strains of the bread mold, *Aspergillus*. In both cases, there is a wide dispersion of offspring phenotypes due to the reassortment of parental genes. The mean offspring phenotype is not significantly different from that of the midparent in one case, but in the second case there is a substantial depression in offspring growth rate. In this latter case, it appears that different combinations of genes with favorable epistatic effects on growth were sorted into the two parental lines. Sexual reproduction resulted in a breakup of these positive gene interactions. This type of result is observed commonly in crosses between fungal lines (Papa 1970; Merrick 1975), and we will consider it in a broader context in Chapters 11 and 13.

While important morphological and behavioral traits, such as growth rate, can be scored in even simple single-celled organisms, there has been a massive resurgence in the use of unicellular systems to elucidate functional genomic features. For example, it is now straightforward to score the levels of expression of essentially all mRNAs within a genome, creating tens of thousands of traits (Chapter 21). Coupled with a very rapid life cycle and the ability to exploit a haploid genome for gene mapping, yeast has become an extremely powerful model system for quantitative-genetic studies of gene regulation (Steinmetz et al. 2002b; de Koning and Haley 2005; Rockman and Kruglyak 2006; Bloom et al. 2013, 2015; Breunig et al. 2014).

Extension to Polyploids

Partitioning the genetic variance becomes more complex in the case of autopolyploids because of the additional gene effects that need to be taken into account. Consider the effects attributable to single loci. With haploids we need only consider the additive component of variance, while diploid analysis brings in a dominance component to account for diallelic

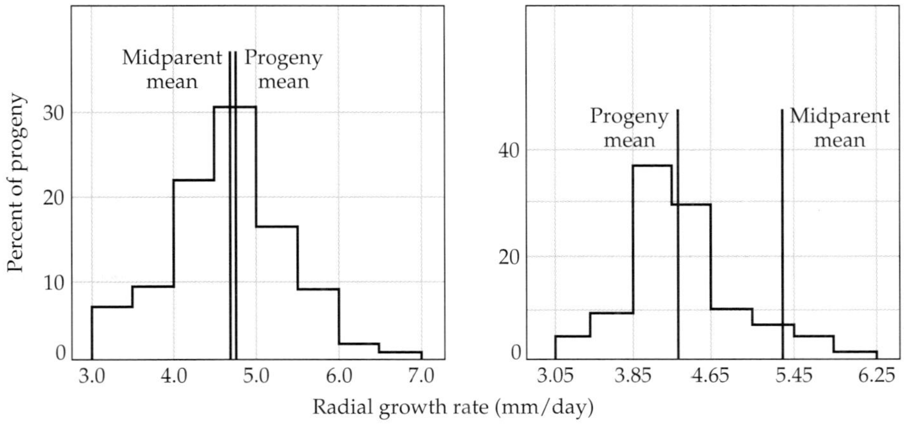

Figure 5.3 Frequency distributions of progeny phenotypes from single crosses in comparison to midparent means. **Left**: *Aspergillus amstelodami* cross 2 × 37. (After Caten and Jinks 1976.) **Right**: *Aspergillus nidulans* cross 7 × 43. (After Jinks et al. 1966.)

interactions. In the case of an autotetraploid with single-locus genotype B_{ijkl}, there are six potential diallelic interactions per locus ($i \times j$, $i \times k$, $i \times l$, $j \times k$, $j \times l$, $k \times l$), as well as four three-allele interactions ($i \times j \times k$, $i \times j \times l$, $i \times k \times l$, $j \times k \times l$), and one four-allele interaction ($i \times j \times k \times l$). Thus, in the absence of interactions between loci, the total genetic variance of a tetraploid can be written as

$$\sigma_G^2 = \sigma_A^2 + \sigma_B^2 + \sigma_T^2 + \sigma_Q^2 \tag{5.10}$$

where σ_B^2, σ_T^2, and σ_Q^2 denote the variances of effects involving bi-, tri-, and quadra-allelic interactions. With epistasis, this expression becomes dramatically more complicated. Even if epistasis only involved pairs of loci, ten epistatic variance terms (σ_{AA}^2, σ_{AB}^2, σ_{AT}^2, σ_{AQ}^2, σ_{BB}^2, etc.) would need to be taken into account, compared with three in the case of diploidy. Further discussion can be found in Kempthorne (1957), Killick (1971), Mather and Jinks (1982), Wricke and Weber (1986), and Gallais (2003).

LINKAGE AND LINKAGE DISEQUILIBRIUM

The use of marker information in quantitative genetics relies on both the coinheritance of alleles at physically close loci and on statistical associations in a population of alleles at different loci. The former is the notion of **linkage**, the latter the concept of **linkage disequilibrium** (**LD**; also called **gametic-phase disequilibrium** because LD can occur between unlinked loci). While both concepts contain the term *linkage*, and both are impacted by how physically close two loci are on a chromosome, they are otherwise rather different. Linkage is a property of the *gametes from an individual*, while LD is a property of a *population of individuals*. As we will show, one can have linkage, but no LD (Figure 5.5), and one can have LD even among unlinked loci (although it usually decays very quickly under random mating).

Linkage

Up to now we have been treating the transmission of alleles at different loci as *independent events*. This follows from Mendel's principle of **independent assortment**, which is generally true for genes on different chromosomes. All seven of the loci that Mendel discussed showed this phenomena, wherein the transmission of an allele at one locus had no impact on the allelic transmission at the second locus. Hence, a double heterozygote ($AbBb$) was expected to produce four equally frequent gametes, AB, Ab, aB, and ab. Mendel was fortunate (or perhaps creative) in his choice of which loci to include, as genes that are sufficiently close on the same chromosome show departures from independent assortment.

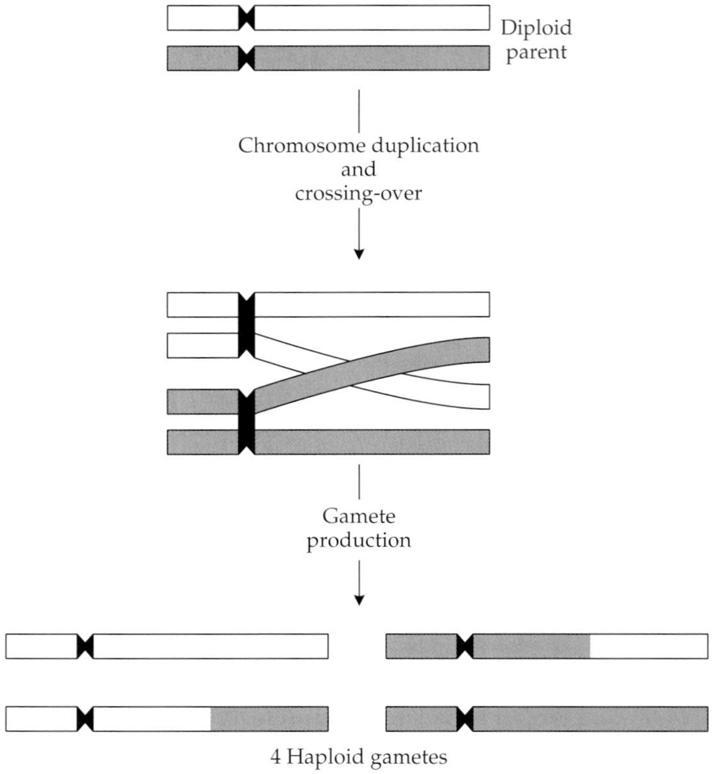

Figure 5.4 A recombination event caused by crossing over during meiosis. Shaded and open bars represent maternally and paternally derived chromosomes; the darkened (and constricted) areas represents the centromere. The two original homologous chromosomes give rise initially to two pairs of sister chromatids, which then recombine and sort independently into four haploid gametes. Gene conversion events—the nonreciprocal exchange of DNA between homologous chromosomes which converts an allele from one homologue into the allele present on the other (WL Chapter 4)—can also occur in the region where potential crossover junctions are initiated.

This departure arises because genes that reside on the same chromosome tend to be inherited as a group, a tendency that declines with increasing distance between the loci. **Recombination** is responsible for this decline. During the meiotic production of gametes, each chromosome replicates to form two identical "**sister**" **chromatids**. Homologous non-sister chromatids then become intimately associated with each other, forming one or more physical connections (recombination intermediates) known as **chiasma** (singular) or **chiasmata** (plural), see Figure 5.4. Each chiasma represents a point of potential transfer of genetic information. A **crossover** occurs when the chromosomal material distal to a chiasma is exchanged completely. However, the molecular process of recombination extends beyond simply generating crossovers. The recombination intermediate (the joining of two distinct DNA molecules) can be resolved in several different ways, some of which produce crossover events, while others do not. On a very local scale around the recombination intermediate (around a few kilobases), heteroduplex DNA (DNA molecules with mismatched [noncomplementary] strands) is generated between the two conjoined DNA molecules. The repair (or segregation by replication) of **heteroduplex DNA** back into homoduplex (exactly complementary) DNA can result in a localized exchange of gene material, without generating crossovers, a process called **gene conversion** (see WL Chapter 4 for more details).

This physical connection of genes at nearby positions on a chromosome is called **linkage**. **Tightly linked** genes are expected to have only a small chance of a recombination, and hence tend to be largely coinherited, while **loosely linked** genes may approach independent assortment even if they residue on the same chromosome. One way to quantify the strength of linkage is through the **recombination frequency**, *c*, between two loci, which is the ***frequency of recombinant gametes*** produced by a doubly heterozygous parent.

Consider an individual with alleles A and B on one chromosome, and a and b on the other. Denoted by AB/ab, this is the **linkage phase** of this double heterozygote. The AB and ab gametes from an AB/ab individual are called **parental gametes**, as they display the same linkage phase as their parent, while Ab and aB are called **recombinant** (or **nonparental**) **gametes**. These latter gametes arise (for linked loci) only when a recombinational event occurs between the loci. Hence, for an AB/ab parent,

Gamete	Frequency	Type
AB	$(1-c)/2$	Parental
ab	$(1-c)/2$	Parental
Ab	$c/2$	Recombinant
aB	$c/2$	Recombinant

For an Ab/aB parent, these gamete classes are reversed, with Ab and aB now being parental, and AB and ab being recombinant. Hence, the presence of linkage is reflected by an **excess of parental gametes** (frequencies greater than 1/4). Note that information on linkage is only provided by double heterozygotes, as single heterozygotes produce only two types of gametes (e.g., $AaBB$ individuals produce equal numbers of AB and aB gametes, regardless of the value of c).

The recombination frequency takes on a minimum value of zero (**complete linkage**) when two loci are so tightly linked that they effectively segregate as a single **supergene.** Free recombination (independent assortment) between loci is denoted by $c = 0.5$. The upper limit is 0.5 rather than 1.0 because crossovers normally involve one rather than two pairs of nonsister chromatids. In principle, if crossovers occurred between loci A and B on both pairs of chromatids, the recombination fraction could exceed 0.5, but this has been observed only rarely (Fisher and Mather 1936).

This excess of parental gametes is the basis of **linkage mapping** approaches, such as classic line-cross **QTL (quantitative trait locus) mapping** (Chapter 18). Consider two linked diallelic loci, a marker locus (with alelles M and m) and a locus influencing trait values (alleles Q and q result in different trait values). In the gametes from a MQ/mq parent, M and Q tend to be coinherited, so that an offspring carrying marker allele M is more likely to contain QTL allele Q than the alternative allele q. This creates a marker-trait association in the immediate descendants of this parent, with M-bearing individuals differing in trait values from those bearing an m allele (the former tending to carry Q, the latter, q).

An immediate consequence of linkage is that one must now use *gamete*, rather than *allele*, frequencies to predict the frequencies of multilocus genotypes, even under random mating. Consider the offspring of a cross between two AB/ab parents. Under independent assortment, the probability of an $AABB$ offspring is simply $\Pr(AA) \cdot \Pr(BB) = (1/4) \cdot (1/4) = 1/16$. However, under linkage,

$$\Pr(AABB) = \Pr(AB \text{ from father}) \cdot \Pr(AB \text{ from mother}) = [(1-c)/2]^2 \tag{5.11a}$$

For complete linkage ($c = 0$), this is 1/4, four times the expected value under independent assortment. Conversely, if the linkage phases of both parents are reversed (Ab/aB), then

$$\Pr(AABB) = \Pr(AB \text{ from father}) \cdot \Pr(AB \text{ from mother}) = c^2/4 \tag{5.11b}$$

which is zero for complete linkage. Finally, if the two parents have different linkage phases ($AB/ab \times Ab/aB$), then $\Pr(AABB) = ([1-c]/2)(c/2) = c(1-c)/4$.

Example 5.4. Linkage was first discovered by Bateson and Killby (1905), who examined a **dihybrid** (i.e., two-trait) **cross** between a **pure-breeding** (i.e., fully inbred) red flower, round pollen pea line and a pure-breeding purple flower, long pollen line. Both of these traits have a simple genetic basis, with the purple (*P*) allele dominant to red (*p*), and long (*L*) pollen shape dominant to round (ℓ). Hence, the cross is $PPLL \times pp\ell\ell$, resulting in an F_1 linkage phase of $PL/p\ell$. Under independent assortment, the expected frequency of a purple, long phenotype in the F_2 is Pr(*P*-)Pr(*L*-) = (3/4)(3/4) = 9/16, where *P*- denotes either *PP* (1/4) or *Pp* (1/2). Similarly, the expected frequency of purple, round is Pr(*P*-)Pr($\ell\ell$) = (3/4)(1/4) = 3/16; red, long is (1/4)(3/4) = 3/16; and red, round is (1/4)(1/4) = 1/16. The values observed by Bateson for 2132 F_2 individuals were as follows:

Phenotype	Genotype	Observed	Expected
Purple, long	*P- L-*	1528	1199
Purple, round	$P\text{-}\ \ell\ell$	106	400
Red, long	*pp L-*	117	400
Red, round	$pp\ \ell\ell$	381	133

While the direct estimation of gamete frequencies was confounded by dominance, it was still clear to Bateson that there was an excess of *PL* and $p\ell$ gametes. We can estimate the recombination frequency between the color and pollen-shape loci from the frequency of red round individuals, which arise as the product of two parental $p\ell$ gametes. Hence, Freq(red round) = 381/2132 = freq($p\ell$)·freq($p\ell$) = $(1-c)^2/4$. Solving yields $1-c=\sqrt{4\cdot 381/2132}=0.845$, or that $c=0.155$. Hence, the expected frequency of *PL* gametes produced by F_1 individuals was $(1-c)/2$ = 42.3%, while $c/2$ = 7.7% were *pL*.

Linkage Disequilibrium: Introduction

While linkage refers to an association between alleles at different loci in the gametes from a *particular parent*, linkage disequilibrium (LD) is an association between alleles at different loci in a *particular population* (Figure 5.5). To formalize this population-level association, consider two diallelic loci, *A* and *B*, and let the population frequencies of the four gamete types be P_{AB}, P_{Ab}, P_{aB}, and P_{ab}, and the frequencies of alleles be p_A, p_a, p_B, and p_b. Observe that allele frequencies directly follow from gamete frequencies, e.g., $p_A = P_{AB} + P_{Ab}$. However, as we saw in Chapter 4 in relating *allele* frequencies to single-locus *genotype* frequencies, the converse is not true—knowledge of *allele frequencies*, by themselves, does not uniquely inform us as to the *gamete frequencies*. As in the single-locus case, to make this transition (from allele to either genotype or gamete frequencies), we need to make some assumptions.

If the allelic state at locus *A* is completely independent of that at locus *B*, then we expect that $P_{AB} = p_A p_B$, $P_{Ab} = p_A p_b$, and so on. Such a setting of population-level independence is called **linkage equilibrium (LE)**. **Multilocus Hardy–Weinberg** is said to occur when individual loci are in Hardy–Weinberg *and* loci are also in LE. In such a setting, the frequencies of multilocus genotypes can be directly computed from allele frequencies, e.g., Freq($AaBB$) = $2p_A p_a p_B^2$. As we showed in Chapter 4, a single generation of random mating is usually sufficient to obtain *single-locus* Hardy–Weinberg equilibrium. Further, under random mating alone, allele frequencies do not change. The situation is rather different when one considers *gamete* frequencies. A single generation of random mating is *not sufficient* to ensure that LE (and hence multilocus Hardy–Weinberg) is reached. Further, although *allele frequencies* are stable over time under random mating, such is *not* the case for *gamete frequencies* that are not in LE.

Many factors—including natural selection, founder effects, migration, and assortative mating—can cause gamete frequencies to depart from LE expectations (WL Chapters 2, 5, 16, and 24). A natural measure of this deviation is the coefficient

$$D_{A_iB_j} = P_{A_iB_j} - p_{A_i} p_{B_j} \tag{5.12a}$$

which was introduced by Lewontin and Kojima (1960). Note that $D_{A_iB_j}$ may be positive

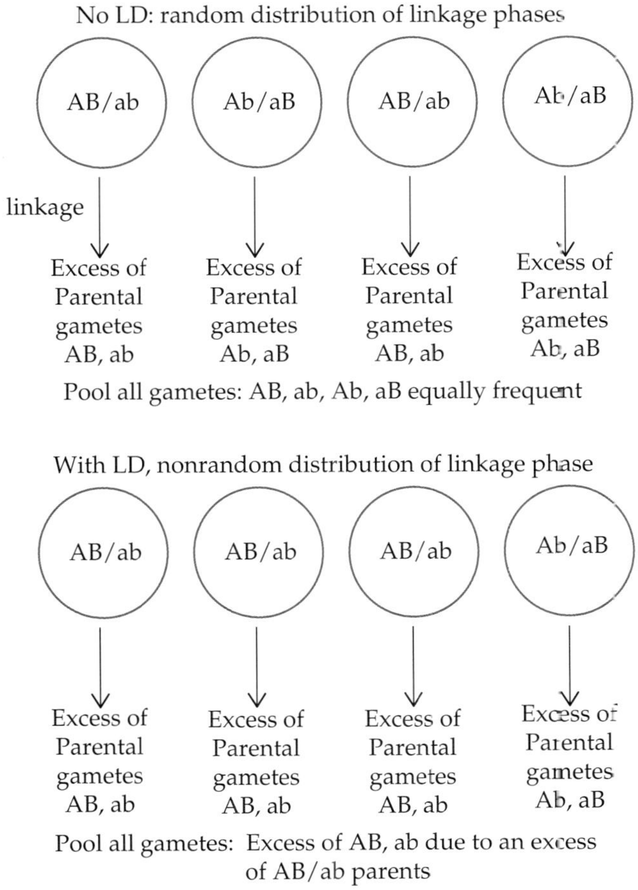

Figure 5.5 Linkage versus linkage disequilibrium. The circles in each panel represent parents, the arrows show the gametes produced by each parent, and the row of parents represents the population. In both panels, the *A* and *B* loci are linked, and hence *each parent* produces an excess of parental gametes. In the upper panel, the population is in linkage equilibrium, in that Freq(*AB*) = 1/4 = freq(*A*)·freq(*B*) = (1/2) · (1/2). The result is that while a correlation among *A* and *B* is seen in the gametes of a *particular parent*, when we *pool* all the gametes, the resulting *population* shows no such correlation. In the bottom panel, there is linkage disequilibrium in this collection of parents as the *AB* haplotype is over-represented (3/8) relative to its linkage equilibrium value (2/8). As a consequence, when we pool all of the gametes from the parents, we see a correlation between the two alleles, in this case, an excess of *AB* and *ab* gametes.

or negative depending on whether A_i and B_j are in **coupling** (A_iB_j gametes are over-represented) or **repulsion** (A_iB_j gametes are underrepresented) disequilibrium. D is often referred to as the **coefficient of linkage disequilibrium,** but because disequilibrium can occur between unlinked genes, many authors prefer to call it the **coefficient of gametic-phase disequilibrium**. Higher order terms for three or more loci are defined in Geiringer (1944), Bennett (1954), and Slatkin (1972).

The index $D_{A_iB_j}$ is equivalent to ***the covariance of the frequencies of A_i and B_j in the same gamete.*** To see this relationship, let $x = 1$ if A_i is present and $x = 0$ otherwise, and let $y = 1$ if B_j is present and $y = 0$ otherwise. Then, $xy = 1$ if both A_i and B_j are present, while $xy = 0$ otherwise, so $\sigma(A_i, B_j) = E[xy] - E[x]E[y] = p_{A_iB_j} - p_{A_i}p_{B_j}$. Note that as a covariance, D is not a standardized measure (Chapter 3). Hence, the value of D, by itself, does not provide a measure of the relative strength of an association. One option, examined in Example 5.5, is to standardize D as a fraction of its maximum absolute value (Lewontin 1964, 1988; Hedrick 1987a; Devlin and Risch 1995). Another, more widely used, standardization is offered by expressing the amount of disequilibrium as a correlation, an idea first proposed by Hill and Robertson (1968). This follows by first noting that

$$\sigma^2(x) = E[x^2] - (E[x])^2 = 1^2 \cdot p_{A_i} - p_{A_i}^2 = p_{A_i}(1 - p_{A_i}) \tag{5.12b}$$

and, similarly, $\sigma^2(y) = p_{B_j}(1 - p_{B_j})$. Hence, we can also express Equation 5.12a as a ***correlation between alleles residing in the same gamete***, with

$$r_{A_iB_j}^2 = \frac{\sigma^2(x,y)}{\sigma^2(x)\,\sigma^2(y)} = \frac{D_{A_iB_j}^2}{\sigma^2(x)\,\sigma^2(y)} = \frac{[P_{A_iB_j} - p_{A_i}p_{B_j}]^2}{p_{A_i}p_{B_j}(1-p_{A_i})(1-p_{B_j})} \tag{5.12c}$$

The use of the correlation coefficient offers a simple test for the presence of LD (Weir 2008), as for a large sample of n gametes,

$$n\, r_{A_iB_j}^2 \sim \chi_1^2 \tag{5.12d}$$

There are many ways in which disequilibrium between loci can originate and many ways by which it can be maintained (Felsenstein 1965, 1974; Hill and Robertson 1966, 1968; Nei 1967; Franklin and Lewontin 1970; Barker 1979; Charlesworth and Charlesworth 1979; Feldman et al. 1980; McVean 2002; Mueller 2004; Slatkin 2008; McVean and Kelleher 2019; WL Chapters 2, 5, 16, 24, and 28). For now, we simply note that, even in the absence of forces that tend to maintain gametic-phase disequilibrium (selection, migration, mutation, and drift), once such disequilibrium is generated, it may persist for many generations.

To see this point, consider the ideal case in which mating occurs randomly in a population that is effectively infinite in size and all systematic forces causing alleles at different loci to become statistically associated are absent. There are two ways to obtain gametes of a particular type: chromosomes of that type may be transmitted intact between generations, or different chromosomes may recombine to form new gene combinations of the appropriate type. If the frequency of gamete type A_iB_j in generation t is $P_{A_iB_j}(t)$, then $(1-c)P_{A_iB_j}(t)$ is the frequency that is passed on without recombination to the next generation. Among the proportion c of new gametes that are recombinants, a fraction $p_{A_i}p_{B_j}$ must contain both A_i and B_j genes, because under random mating, the maternally derived A gene is distributed independently of the paternally derived B gene. Summing terms,

$$P_{A_iB_j}(t+1) = (1-c)P_{A_iB_j}(t) + c\,p_{A_i}p_{B_j}$$

and after subtracting $p_{A_i}p_{B_j}$ from both sides yields

$$D_{A_iB_j}(t+1) = (1-c)D_{A_iB_j}(t) \tag{5.13a}$$

Iterating, this equation generalizes to

$$D_{A_iB_j}(t) = (1-c)^t D_{A_iB_j}(0) \tag{5.13b}$$

The half-life ($t_{1/2}$) of D is obtained by solving $(1-c)^{t_{1/2}} = 0.5$, yielding

$$t_{1/2} = \frac{\ln(0.5)}{\ln(1-c)} \simeq -\frac{\ln(0.5)}{c} = \frac{0.693}{c} \tag{5.13c}$$

Finally, we can also examine the rate of decay of any (squared) correlation, r^2, between alleles. From Equations 5.12c and 5.13c,

$$\begin{aligned} r_{A_iB_j}^2(t) &= \frac{[D_{A_iB_j}(t)]^2}{p_{A_i}p_{B_j}(1-p_{A_i})(1-p_{B_j})} = \frac{(1-c)^{2t}\,[D_{A_iB_j}(0)]^2}{p_{A_i}p_{B_j}(1-p_{A_i})(1-p_{B_j})} \\ &= (1-c)^{2t}\, r_{A_iB_j}^2(0) \end{aligned} \tag{5.13d}$$

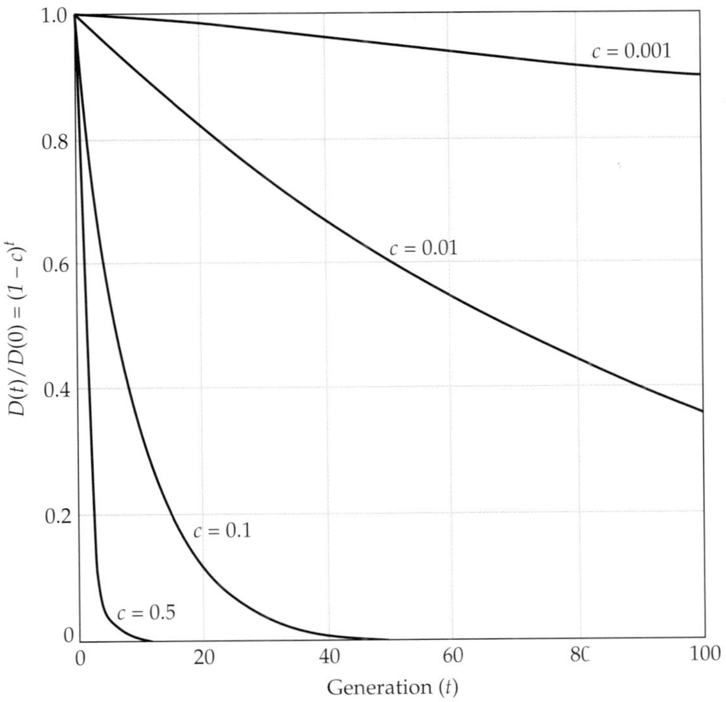

Figure 5.6 The decline, under random mating, of gametic-phase disequilibrium relative to its initial value as a function of the recombination frequency, c.

The half-life for decay in r^2 is just $(t_{1/2})/2$, and hence decays twice as fast as $D(t)$.

The main point of this demonstration is that the disequilibrium between loci decays gradually. Even in the case of unlinked genes ($c = 0.5$), only 50% of the disequilibrium is removed each generation by random mating. With less frequent recombination, the time to attain gametic-phase equilibrium ($D \simeq 0$) can be quite long (Figure 5.6 and Equation 5.13c). An important exception with unlinked loci occurs when we cross (or, equivalently, self) the F_1 offspring (AB/ab) from the cross of two inbred lines ($AABB \times aabb$). Because of independent assortment, the F_2 gametes are in LE. This attainment of LE in just a single generation occurs because this cross is an example of *nonrandom mating*.

The long-term persistence of LD for closely linked sites is the basis for **association mapping** (Chapter 20). When a new mutation ($q \rightarrow Q$) arises at a QTL influencing a trait of interest, this new allele is initially in LD with nearby polymorphic sites. By saturating a chromosome with a sufficient number of markers (typically SNPs), we can capture some of the signal of this causative allele through a marker-trait association generated by this initial LD with nearby markers. For example, suppose Q arose on an M-bearing chromosome. Initially, whenever a chromosome contains a Q, it also contains an M, generating LD. Taking a value of $c = 0.01$ as corresponding to a megabase of DNA (10^6 nucleotides), then if the marker and QTL are, say, 10 kilobases (10^4 nucleotides) apart, $c = 0.01 \cdot 10^4/10^6 = 0.0001$ (see WL Table 4.3). From Equation 5.13c, the expected number of generations for 50% of the initial LD (generated by the appearance of a new mutation) to decay is ~6900. For humans, this is roughly 15,000 years if one assumes a generation time of ~20–25 years. Hence, with sufficiently dense markers, the LD from new mutations with nearby markers should persist for a significant amount of time, and such a correlation between the QTL and marker alleles will generate a marker-trait association in a random sample from the population.

Finally, returning to the behavior of gamete frequencies under random mating, rearranging Equation 5.12a and recalling Equation 5.13b yields

$$P_{A_iB_j}(t) = p_{A_i}p_{B_j} + D_{A_iB_j}(t) \tag{5.14a}$$

$$= p_{A_i}p_{B_j} + (1-c)^t D_{A_iB_j}(0) \tag{5.14b}$$

The first term on the right-hand side of these equations is the LE expectation for the frequency of the A_iB_j haplotype. Because these are based on allele frequencies, they remain unchanged under random mating. However, if initially in LD, gamete frequencies slowly change under random mating as the LD value, D, decays to zero. Gametes with $D(0) < 0$ increase in frequency, eventually reaching their LE value, while gametes with $D(0) > 0$ decrease down to their LE value. Once LE is reached ($D = 0$), then the gamete frequencies remain unchanged under random mating, and we have reached multilocus Hardy–Weinberg equilibrium.

Example 5.5. Because the maximum value of D varies over allele frequencies, Lewontin (1964) proposed a standardized measure

$$D' = \frac{D}{\max|D|}, \qquad \text{where} \quad -1 \le D' \le 1 \tag{5.15a}$$

with

$$\max|D| = \begin{cases} \min[p(1-q), (1-p)q] & \text{for } D \ge 0 \\ \max[-pq, -(1-p)(1-q)] & \text{for } D < 0 \end{cases} \tag{5.15b}$$

To see how this metric differs from r^2, consider the two following situations:

	Case One			Case Two	
	M	m		M	m
Q	20	0	Q	20	0
q	4	6	q	0	10

In case one, $\widehat{p}_M$ = 24/30 = 8/10, $\widehat{p}_Q$ = 20/30 = 2/3, and $\widehat{p}_{MQ}$ = 20/30 = 2/3, yielding

$$\widehat{D}_{MQ} = \widehat{p}_{MQ} - \widehat{p}_M\,\widehat{p}_Q = 2/3 - (8/10)(2/3) = 2/15 = 0.133$$

$$\widehat{r^2}_{MQ} = \frac{(\widehat{D}_{MQ})^2}{\widehat{p}_M(1-\widehat{p}_M)\,\widehat{p}_Q(1-\widehat{p}_Q)} = \frac{(2/15)^2}{(8/10)(2/10)(2/3)(1/3)} = 1/2$$

Because $\widehat{p}_M(1-\widehat{p}_Q) = 4/15$ and $(1-\widehat{p}_M)\,\widehat{p}_Q = 2/15$, max$|D|$ = 2/15 (Equation 5.15b) and

$$D' = \frac{2/15}{2/15} = 1$$

Hence, while D' might suggest complete disequilibrium, its r^2 value is only 0.5. Note that while Q is always found on M-bearing chromosomes, some M-bearing chromosomes instead contain q. Hence, while Q is always associated with M, the converse is not true. The former gives $D' = 1$, while the latter implies that $r^2 < 1$.

For case two, $\widehat{p}_M = \widehat{p}_Q = \widehat{p}_{MQ}$ = 2/3, yielding $D_{MQ} = 2/9 = 0.22$ and $\widehat{r^2}_{MQ} = 1$. Further, $\widehat{p}_M(1-\widehat{p}_Q) = (1-\widehat{p}_M)\,\widehat{p}_Q = 2/9$, giving $D' = 1$. In general, $r^2 = 1$ implies that $|D'| = 1$, but *the converse is not true*. In terms of the 2×2 contingency table, one zero off-diagonal element (one allele is *always* found on the background of the other) implies $|D'| = 1$, while $r^2 = 1$ only when *both* off-diagonal elements are zero.

This example has important implications for how well a marker in LD predicts the effect of a linked causative site (Example 20.1). Suppose that a new mutation (Q) impacting the trait arises on an M-bearing chromosome. In the absence of recombination in our sample of haplotypes, Q is always associated with M, but, conversely, many M-bearing chromosomes instead contain q. While $|D'| = 1$ in this case, the fraction of the trait variance due to Q that is explained by the marker M is only r^2. For example, even though 83% (20/24) of M-bearing chromosomes in case one also contain Q, only 50% of the trait variation attributable to the Q locus is accounted for by the marker locus. Further, if mutations appear at random, the chance that one will arise on the background of a particular marker allele is proportional to the marker frequency at the time of the mutation. Hence, ***more common marker alleles have***

a greater chance of accruing such a linked mutation. Such common marker alelles, however, also have a greater chance of having many of the marker-allele bearing chromosomes in a sample *not* containing the new mutation, and hence a lower value of r^2.

As fully developed by Wray (2005), the key idea is that ***r^2 is large only when the marker and causative alleles have very similar allele frequencies.*** To formally see this, first note that the maximal value of r^2 occurs when a causative *Q* is only found on one marker allele background (*M*). In this setting, let p_M be the frequency of *M*, and $p_Q = \alpha p_M$ be the frequency of *Q*, with $0 < \alpha \leq 1$. Given that the frequency of the *MQ* gamete is $p_Q = \alpha p_M$, then

$$D_{MQ} = p_{MQ} - p_M\, p_Q = p_Q - p_M \cdot p_Q = \alpha p_M(1 - p_M)$$

$$r^2_{MQ} = \frac{[\,\alpha p_M(1 - p_M)]^2}{p_M(1 - p_M)\alpha p_M(1 - p_Q)} = \frac{\alpha(1 - p_M)}{1 - p_Q} \leq \alpha \qquad (5.15c)$$

Example 5.6. The distinction that we have presented between linkage and linkage disequilibrium is simply a function of a *difference in reference populations*. In particular, linkage can be considered as a special case of LD, where the focal population is *the collection of gametes from a single parent*. Consider the LD in the pool of gametes from a AB/ab parent. Here

$$D_{AB} = p_{AB} - p_A\, p_B = (1 - c)/2 - (1/2)(1/2) = (1 - 2c)/4 \qquad (5.16a)$$

$$r^2_{AB} = \frac{D^2_{AB}}{p_A(1 - p_A)\, p_B(1 - p_B)} = \frac{(1 - 2c)^2/4^2}{(1/2)^4} = (1 - 2c)^2 \qquad (5.16b)$$

Finally, note that because $D_{AB}(t) = D_{AB}(0)(1 - c)^t$, the amount of LD remaining after t generations of random mating in the descendants of this population of gametes becomes

$$r^2_{AB}(t) = (1 - 2c)^2(1 - c)^{2t} \qquad (5.16c)$$

Estimation of Gametic-phase Disequilibrium: Phased Data

Accurate estimates of LD values are of importance for several reasons. For example, the LD structure of a genome can provide significant information about the past history of a population (Nordborg and Tavaré 2002; Tenesa et al. 2007; Slatkin 2008; Flury et al. 2010). Ecological factors can induce significant gametic-phase disequilibrium even between unlinked loci. Such an outcome is expected, for example, when a recent hybridization has occurred between two populations or under many forms of selection (WL Chapters 5, 16, and 24). Although genes need not be linked to be in gametic-phase disequilibrium, Equation 5.13c indicates that, all other things being equal, disequilibrium is more likely to persist for linked loci. Hence, extensive LD between loosely (or unlinked) loci implies that such associations were very recently generated. Similarly, the genomic structure of LD within a given population provides an indication of the level of resolution possible under association mapping (Chapter 20). This can be obtained by plotting the LD among pairs of SNPs (typically measured by r^2; Equation 5.12c) as a function of their physical distance (i.e., DNA base pairs) to obtain some indication of the average size of an LD block in a particular population (Chapter 17).

The most straightforward estimates of LD follow when one can directly score gametes or haploid chromosomes in a population, and then substitute estimates of $P_{A_iB_j}$, p_{A_i} and p_{B_j} into either Equation 5.12a or 5.12c. The delicate issue with such an approach is determining linkage phase (generating **phased genotypes**), namely, determining whether an $AaBb$ individual is AB/ab or Ab/aB. Historically, this was accomplished using either **trio data** (sets of both parents and their offspring) to determine the linkage phase (e.g., Gabriel et al. 2002; International HapMap Consortium 2005) or using specific features of the organism to essentially recover haploid products. One application of the latter approach involves *Drosophila*, whose males do not exhibit recombination. Thus, when a homozygous laboratory stock is available, progeny from crosses to wild-caught males can unambiguously yield the gametic states of the previous generation (e.g., Langley et al. 1974). In the genomics era, one can determine the phase for a pair of SNPs directly from sequencing, provided that the distance between the SNPs is less than the average sequencing read length, which with evolving technologies is rapidly approaching the length of a chromosome.

Example 5.7. Males from two wild populations of *Drosophila montana* were crossed with a laboratory stock that was homozygous for unique marker alleles at four esterase loci (Baker and Kaeding 1981). The progeny were then scored electrophoretically for the presence of active or null alleles at the four loci. Only the results of *esterase-3* and *esterase-4*, which exhibited the greatest amount of disequilibrium, are presented here. In the following table A and O denote, respectively, active and null alleles at both loci, so AO, for example, represents chromosomes with an active allele at the first locus (*esterase-3*) and a null at the second locus (*esterase-4*). The entries in the table represent the numbers of observations of each chromosome type, and N denotes the total sample size for each study.

The estimation procedure is from Hill and Robertson (1968) and Hill (1974). The formulae on the right side of the table provide estimates for the active-allele frequencies at the *esterase-3* and *esterase-4* loci, and for the gametic-phase disequilibrium between these loci. The estimates, obtained by substituting the observed incidences of the four chromosomal types, are given in the table.

Chromosome	Utah	Colorado
AA	4	71
AO	50	737
OA	31	457
OO	15	167

Statistic	Utah	Colorado	
N	100	1432	
$\widehat{p}_{A.}$	0.540	0.564	$\widehat{p}_{A.} = (N_{AA} + N_{AO})/N$
$\widehat{p}_{.A}$	0.350	0.368	$\widehat{p}_{.A} = (N_{AA} + N_{OA})/N$
$\widehat{D}_{AA}$	−0.149	−0.158	$\widehat{D}_{AA} = P_{AA} - \widehat{p}_{A.}\widehat{p}_{.A}$
$\widehat{r^2}_{AA}$	0.393	0.436	$\widehat{r^2}_{AA} = \widehat{D}^2_{AA}/[\widehat{p}_{A.}(1-\widehat{p}_{A.})\widehat{p}_{.A}(1-\widehat{p}_{.A})]$

The sampling variance of $\widehat{D}$ is given by

$$\text{Var}(\widehat{D}) = \frac{\widehat{p}_{A.}(1-\widehat{p}_{A.})\widehat{p}_{.A}(1-\widehat{p}_{.A}) + (1-2\widehat{p}_{A.})(1-2\widehat{p}_{.A})\widehat{D} - \widehat{D}^2}{N} \tag{5.17}$$

(Weir 2010), and the standard error of $\widehat{D}$ is simply the square root of Var$(\widehat{D})$. For the two populations, SE$(\widehat{D})$ $= 0.028$ and 0.005. Thus, both populations exhibit significant excess frequencies of repulsion gametes (an active allele at one locus and a null at the other) and have essentially the same level of disequilibrium. Based on the estimated allele frequencies, the number of AO and OA chromosomes expected in the Utah sample under gametic-phase equilibrium is $100 \times [(0.54 \times 0.65) + (0.46 \times 0.35)] \simeq 51$, whereas 81 were observed. For the Colorado population, the equilibrium expectation (740) is also well below the observed value (1194). The existence of disequilibrium between these two loci is not too surprising, as these esterase loci are tightly linked in *Drosophila*.

Estimation of Gametic-phase Disequilibrium: Unphased Data

For most natural populations, the luxury of an inbred marker line, or some other scheme to extract haploid chromosomes, is not an option. Usually, the only available information is the frequency distribution of multilocus genotypes in the study population, which raises some conceptual difficulties. While the gametic composition of most zygotes can be resolved from the genotype (e.g., an *AaBB* individual must have arisen from *AB* and *aB* gametes), the linkage phase of double heterozygotes cannot be resolved definitively (an *AaBb* individual could be either *AB/ab* or *Ab/aB*). Approaches suggested for estimating the amount of LD using such **unphased** data broadly fall into two categories: likelihood approaches and measures of composite disequilibria (associations among single-locus *genotypes*, rather than *gametes*, within an individual).

Likelihood methods (Appendix 4) attempt to infer the gamete frequencies in a population sample, usually under the assumption of random mating (Hill 1974; Clark 1990; Excoffier and Slatkin 1995; Slatkin and Excoffier 1996). These often use the expectation-maximization (EM) method (Appendix 4), an iterative approach where one starts with some initially assumed values for the gamete frequencies, and then updates these values given the data (under the assumption of random union of gametes) until a stable solution is reached. When mating is not random, such an approach can introduce bias, as the updating expectation (E) step assumes random union of gametes. As noted by Osier et al. (1999) and Fallin and Schork (2000), the error is often negligible under inbreeding. Presumably, this is because the increase in homozygosity (relative to random mating) from inbreeding also results in a decrease in ambiguously phased genotypes (double heterozygotes). Conversely, when random mating is violated in such a way as to generate an excess number of heterozygotes relative to Hardy–Weinberg (e.g., admixture), the EM approach can be rather biased (Fallin and Schork 2000).

A different approach is based on the notion of the composite linkage disequilibrium coefficient, a concept introduced by Cockerham and Weir (1977a), who attributed the idea to Burrows. With gametic data, Equation 5.12a gives the *within-gamete* disequilibrium. However, we can also consider the **between-gamete disequilibrium**, namely, the association between A on one chromosome with B on the other chromosome within an individual,

$$D_{A\cdot/\cdot B} = P_{A\cdot/\cdot B} - p_A\, p_B \tag{5.18a}$$

Under random mating in an outbred population, alleles on chromosomes from different parents are unrelated, and $D_{A\cdot/\cdot B} = 0$. However, direct estimation of $D_{A\cdot/\cdot B}$ requires phased data. In particular, both gametes of *AABB*, half the gametes from *AaBB* and *AABb*, and half the gametes from repulsion heterozygotes (*Aa/Bb*) correspond to $A\cdot/\cdot B$, giving

$$P_{A\cdot/\cdot B} = P(AABB) + \frac{1}{2}\left[P(AaBB) + P(AABb) + P(Aa/Bb)\right] \tag{5.18b}$$

Similarly, estimation of $D_{A_iB_j}$ (Equation 5.12a) also requires phased data, as

$$P_{AB} = P(AABB) + \frac{1}{2}\left[P(AaBB) + P(AABb) + P(AB/ab)\right] \tag{5.18c}$$

However, because $P(AB/ab) + P(Ab/AB) = P(AaBb)$, the *sum* of Equations 5.18b and 5.18c *does not* require phased data, as

$$P_{AB} + P_{A\cdot/\cdot B} = 2P(AABB) + P(AaBB) + P(AABb) + \frac{1}{2}P(AaBb) \tag{5.18d}$$

Equation 5.18d motivates **Burrow's composite measure of linkage disequilibrium** (Cockerham and Weir 1977a), the sum of the within- and between-gametic disequilibria,

$$\Delta_{ij} = D_{A_i\cdot/\cdot B_j} + D_{A_iB_j} \tag{5.19a}$$

$$= P_{A_i\cdot/\cdot B_j} + P_{A_iB_j} - 2(p_{A_i}\, p_{B_j}) \tag{5.19b}$$

An unbiased estimator of Δ is given by

$$\widehat{\Delta}_{AB} = \frac{N}{N-1}\left[\frac{4N_{AABB} + 2(N_{AABb} + N_{AaBB}) + N_{AaBb}}{2N} - 2\widehat{p}_A\widehat{p}_B\right] \tag{5.20a}$$

where N is the total sample size, the terms in the numerator are the observed numbers of the four genotypes, and $\widehat{p}_A$ and $\widehat{p}_B$ are gene-frequency estimates (Weir 1979). Under random mating, $D_{A\cdot/\cdot B} = 0$, and hence $\Delta_{AB} = D_{AB}$. Weir and Cockerham (1979) showed that the sampling variance of Equation 5.20a (under random mating) is given by

$$\text{Var}(\widehat{\Delta}) = \frac{\widehat{p}_A\,\widehat{p}_a\,\widehat{p}_B\,\widehat{p}_b}{N-1} + \frac{(2\widehat{p}_A - 1)(2\widehat{p}_B - 1)\widehat{\Delta}}{2N} + \frac{\widehat{\Delta}^2}{N(N-1)} \tag{5.20b}$$

For loci in gametic-phase equilibrium, this reduces to

$$\text{Var}(\widehat{\Delta}) = \frac{\widehat{p}_A\,\widehat{p}_a\,\widehat{p}_B\,\widehat{p}_b}{N-1} \tag{5.20c}$$

Standardized versions of Δ (the analog of D') have been developed, and their properties are reviewed by Hamilton and Cole (2004, 2006, 2008) and Zaykin (2005).

An important extension of features of D to Δ follows by consideration of the covariances and correlations among alleles within a gamete (D) or among genotypes within an individual (Δ). Recall that we defined the gametic (or *genic*) scores x and y for the alleles within a single gamete as being either zero or one (a specified allele is either present or absent). D_{AB} is the covariance between the gametic scores (Equation 5.12a), while r_{AB} is their correlation (Equation 5.12c). This idea extends to Δ, but one instead considers the covariance and correlation between the *genotypic* scores of the two loci within an individual, with $X = 0$, 1, or 2 (for *aa*, *Aa*, or *AA*) and a similar definition for Y for the *B* locus. Weir (2008) and Rogers and Huff (2009) showed that

$$\Delta_{AB} = \frac{1}{2}\sigma(X,Y) \tag{5.21a}$$

Further,

$$r_{AB} \simeq r_{XY} \tag{5.21b}$$

with equality under either random mating, or inbreeding with no recombination. When both inbreeding and recombination are present, the correlation between genotypic scores (r_{XY}) is still a very reasonable estimator for the gametic-phase disequilibrium (r_{AB}). Because of its simplicity in computation (coding SNP genotypes at each locus as 0, 1, or 2, and then computing a correlation among scores) and robustness to nonrandom mating, a number of authors recommend using r_{XY} as the measure of LD with unphased data (Weir 1979, 2008; Schaid 2004a; Laurie et al. 2007; Rogers and Huff 2009; Wellek and Ziegler 2009).

One important testing issue is that while the correlation among gametic scores follows a chi-square distribution under LE (Equation 15.12d), this is not the case for the correlation among genotypic scores (although the two may be close, Weir pers. comm.). One approach to deal with this concern is a randomization test for LD, wherein the genotypes of the two target loci are shuffled over individuals, creating a null distribution of $\widehat{\Delta}$ values under LE (Slatkin and Excoffier 1996). Additional details regarding tests of hypotheses about gametic-phase disequilibrium may be found in Brown (1975), Cockerham and Weir (1977a), Weir (1979, 2008, 2010), Weir and Cockerham (1979), Hedrick (1987a), Lewontin (1988), and Slatkin and Excoffier (1996). Brown (1975) has given considerable attention to the adequacy of sample sizes for detecting disequilibrium, providing useful tables. Unless gene frequencies are intermediate (0.4 to 0.6) at both loci and disequilibrium is strong ($|D| > 0.4$), several hundred to thousands of individuals should be assayed to achieve a reasonable level of statistical power. Estimation procedures for cases in which one or both loci exhibit complete dominance are developed in Hill (1974), and Hill (1975) considers the case when mutliple alleles are present at a locus.

Example 5.8. Genotype frequency data for the M/N and S/s blood group loci for a sample of 1000 Caucasians in England are given in the following table (from Cleghorn 1960). These two loci are known to be very tightly linked; only a few well-established cases of recombination have appeared in thousands of analyses.

	MM	*MN*	*NN*
SS	57	39	3
Ss	140	224	54
ss	101	226	156

Notice that 224 individuals in this data set (24%) have ambiguous phase (are double heterozygotes). Using Equation 4.1, the estimated frequencies of the M and S alleles are

$$\widehat{p}_M = \frac{(57+140+101)+0.5(39+224+226)}{1000} = 0.5425$$

$$\widehat{p}_S = \frac{(57+39+3)+0.5(140+224+54)}{1000} = 0.3080$$

Substitution into Equations 5.20a and 5.20b gives the estimated disequilibrium and its sampling variance,

$$\widehat{\Delta}_{MS} = \frac{N}{N-1}\left[\frac{4N_{MMSS}+2(N_{MMSs}+N_{MNSS})+N_{MNSs}}{2N} - 2\widehat{p}_M\widehat{p}_S\right]$$
$$= 0.071$$

$$\text{Var}(\widehat{\Delta}) = \frac{\widehat{p}_M\,\widehat{p}_N\,\widehat{p}_S\,\widehat{p}_s}{N-1} + \frac{(2\widehat{p}_M-1)(2\widehat{p}_S-1)\widehat{\Delta}}{2N} + \frac{\widehat{\Delta}^2}{N(N-1)} = 0.0000518$$

Finally, substitution of $\widehat{\Delta}_{MS}$ (for $\widehat{\text{D}}_{MS}$), $\widehat{p}_S$, and $\widehat{p}_M$ into Equation 5.12c yields $\widehat{r}^{\,2}_{MS} = 0.095$. The square root of the sampling variance is the standard error of the estimate,

$$\text{SE}(\widehat{\Delta}_{MS}) = 0.0072$$

Because the estimate $\widehat{\Delta}_{MS}$ is 0.071/0.0071 = 9.86 standard errors greater than zero, there appears to be significant disequilibrium between these two loci, with *MS* and *Ns* gametes being overrepresented relative to *Ms* and *NS* gametes. We can refine this statement further by using Chebyshev's inequality (Equation 2.18), which states that the lower bound of being within k standard errors of an estimate is $1 - 1/k^2$, or, in our case, $1 - 1/9.86^2 = 0.990$. Hence, regardless of the form of the distribution of this estimate, there is less than a 1% chance of exceeding the estimated Δ value by chance when the true value is zero.

EFFECT OF DISEQUILIBRIUM ON THE GENETIC VARIANCE

How Widespread is Linkage Disequilibrium?

The genomic distribution of LD block sizes reflects the interplay between local recombination rates and the timing and nature of events generating LD (detailed reviews can be found in WL Chapters 2, 4, and 8). One ubiquitous genome-wide force generating very localized LD is mutation, wherein a new variant initially appears on a single background. Historical demographic events—such as bottlenecks, population expansion, and migration—also create LD. Indeed, the genome-wide pattern of LD has been used to infer the nature and timing of such events (WL Chapter 2). Against this background of genome-wide, small-scale LD effects, selection can generate, at least transiently, much longer blocks of LD. As detailed in WL Chapter 8, favorable new alleles on their way to fixation will drag along a large block of linked sites, generating long haplotypes. Such signatures are a classical approach to search for ongoing selection (WL Chapter 9). Likewise, strong selection on a multilocus epistatic system can also generate LD over large genomic distances (WL Chapter 5), as seen between loci in the mammalian major histocompatibility complex (MHC).

Because LD provides insight into past, and present, evolutionary events influencing a population, a question of much interest is the genomic scales over which LD is present. As with many issues in genetics, initial probes into LD were very much limited by a lack of useful markers. In most cases, LD is expected to be present over only very local scales, as it decays after a modest number of generations for all but the smallest c values (Equation 5.13c). A second issue is that the amount of possible LD is a function of the allele-frequency mismatch between the two alleles being considered. Unless the two alleles have comparable frequencies, their expected LD is small (Equation 5.15c).

One of the first model systems to examine the importance of LD in natural populations was the work started in the 1950s by Dobzhansky and colleagues (summarized in Lewontin et al. 1981), who traced the fate of naturally occurring chromosome inversions in *Drosophila* species, mainly in the Western United States. Inversions tend to suppress recombination, creating very long blocks of LD (often a large fraction of a chromosome, and hence thousands of genes), and Dobzhansky suggested that different inversions were under spatially and temporally varying selection pressure (such as altitudinal gradients).

In an attempt to more finely characterize the LD structure among protein-coding genes, workers in the 1970s turned to **allozyme markers** (e.g., Example 5.7). These are allelic variants coding for different amino-acid sequences that can be detected by running **electrophoresis**. Here, protein extract is placed into slots (lanes) on a slab of gel and their relative movement is measured under an electric current, which can separate sequences (alleles) that differ by a single amino-acid change. The allozyme approach provided the first widespread window into variation at the molecular level (reviewed by Lewontin 1974). Such studies were performed on numerous species, particularly those in the genus *Drosophila*. Most estimates of LD in these early allozyme surveys were nonsignificant, which led a number of authors to conclude that gametic-phase disequilibrium among protein-coding genes was the exception rather than the rule in large, random-mating populations (Langley 1977; Langley et al. 1978; Hedrick et al. 1978). However, the analyses of Brown (1975) suggested that the failure to detect disequilibria in most studies of linked allozyme loci was due to a lack of statistical power resulting from small sample sizes. A meta-analysis (Appendix 6) of results from *Drosophila* by Zapata and Alvarez (1992) found a significant negative relationship between the degree of disequilibrium and the rate of recombination between pairs of protein-coding loci. This pattern is consistent with the idea that gametic-phase disequilibrium is more likely to persist between pairs of closely linked, rather than between weakly linked, or unlinked, loci. Simply put, the scored allozyme loci at the time were not sufficiently linked for significant amounts of LD to persist, as most scored loci were (at least) megabases apart.

What was needed was a much denser set of markers. In the early 1980s, molecular technologies evolved from being able to score a hundred or so protein products (allozymes) to directly probing DNA. Initially, RFLP and SSR markers (Chapter 8) offered a few hundred marker loci, typically scoring variation at sites with little (to no) impact on phenotype, a dramatic departure from allozymes (which score variation in functional products). These first classes of DNA markers soon gave way to the ability to score huge numbers of SNPs and subsequently to our current ability for widespread whole-genome sequencing.

The main finding from using dense DNA markers was that LD between polymorphic sites is the rule in almost every genome, albeit acting usually on a kilobase (kb), as opposed to a megabase, scale. Moreover, the genomic scale over which LD is seen varies dramatically, both over populations and over genomic regions within a population (the latter, in part, reflecting variation in recombination rates and the impact of selection at linked loci; WL Chapters 3 and 8). For example, Buckler and Gore (2007) found that the size of regions influenced by LD is typically less than a kilobase for both wild and landrace populations of maize, but over 100 kb in elite maize lines. They found similar patterns for barley, rice, and soybeans. Numerous studies in humans show a typical LD size of $\sim$50 kb for Europeans, but a much smaller size (typically under 20 kb) in Africans. Further, LD involving common alleles typically extends over shorter distances than LD involving rare alleles (Reich et al. 2001; Wall and Pritchard 2003; Houle and Márquez 2015).

One simple model accounting for such differences in LD structure is differences in the **time to most recent common ancestor** for a particular genomic region in a population. If this time is rather recent, less recombination has occurred, and larger blocks of LD are expected (WL Chapters 2 and 4). Land races have deep ancestry, while elite lines likely have a more recent set of common ancestors. Likewise, under pure drift, common alleles tend to be old alleles, while rare alleles tend to be much more recent (WL Chapter 2). Hence, many more generations of recombination may have occurred between a pair of common alleles than a pair of rare alleles. A confounding factor for rare-allele LD studies was suggested by Houle and Márquez (2015), who also noted this association between rare alleles and long-range disequilibrium in the $\sim$200 inbred lines that comprise the *Drosophila* reference panel (Mackay et al. 2012). They suggested that a random sampling of a large number of rare SNPs over a relatively small number of lines can generate numerous spurious correlations within a sample, a phenomena they called **rarity disequilibrium.** This is not an issue with a modern association studies, given that their sample sizes are typically at least 10^5.

Some insight into this observation of almost universal LD between polymorphic DNA sites over sufficiently small genomic regions follows from the half-life for LD decay (Equation 5.13c). Taking a value of 1% recombination per megabase ($c = 0.01$ per 10^6 nucleotides), one kb of DNA corresponds for a c value of 10^{-4}. (While this is a ball-park average for humans and other animals, as WL Table 4.3 shows, there is great variation in this value.) From Equation 5.13c, for two sites separated by k kilobases, we can write the half-life (in generations) as satisfying $t_{1/2}\, k = 0.639 \cdot 10^4 \simeq 6400$, which rearranges to

$$t_{1/2} \simeq \frac{6400}{k} \tag{5.22}$$

Equation 5.22 shows that the LD generated by an event (such as a new mutation) can persist in the genome because either the time since the event has been short or the distance between sites is small. Sites 10 kb apart whose LD was generated roughly 600 generations ago will still show half the initial D value. LD generated by an event 50 generations ago will have half the initial LD signature extending over a region of k = 6400/50 = 128 kilobases. In regions of reduced rates of recombination, the LD signature extends over a wider region (such as is often seen near centromeres or telomeres).

Finally, we note an important distinction between association of allele *frequencies* at different loci (a population-genetic perspective) and correlations among allelic *effects* for variants that impact traits (a quantitative-genetic perspective). Background processes generating LD, such as mutation or demographic events, should not induce correlations in the signs of allelic effects. Conversely, LD can be generated by the effects of evolutionary forces acting on *specific traits*, which then generates an LD signal in the genome. For example, selection on a trait will result in LD among the underlying trait QTLs, as will trait-based assortative mating (WL Chapters 16 and 24). Such trait-induced evolutionary forces generate LD at the sequence level, but also create *further* associations between *effect sizes* over loci. Even without knowing the causative sites, the latter will generate correlations between the marker-effects for a trait (the marker tagging some signal from the causative site via fine-scale LD). We now (briefly) consider how LD between QTLs can impact the genetic variance components (a topic covered in detail in WL Chapters 5, 16, and 24–26).

Impact of LD on the Genetic Variance

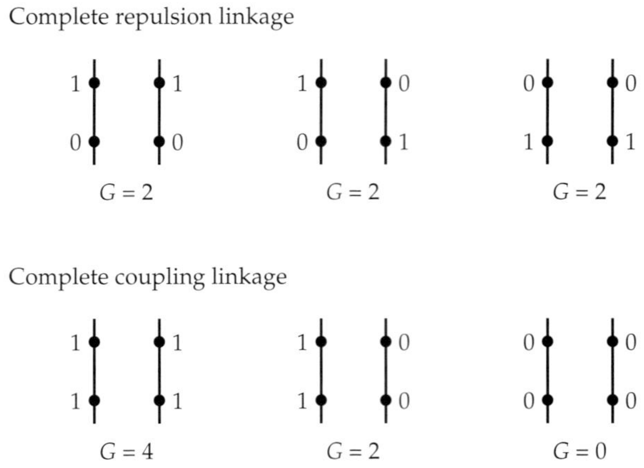

Figure 5.7 Possible two-locus genotypes for extreme cases of complete coupling and complete repulsion disequilibrium. The straight lines represent homologous chromosomes within diploid individuals. The numbers are the additive effects of alleles. The sum of the four effects is the genotypic value, *G*.

While the terms **positive** and **negative disequilibrium** are rather arbitrary for markers—the sign of D can switch when the marker labels are swapped—they take on a specific meaning when used to describe the assocation between the *effects* of alleles at loci underlying a quantitative trait. Positive disequilibrium is said to occur when allelic effects are correlated in gametes. Namely, alleles with a positive influence on a character (**plus alleles**) tend to be associated on some chromosomes (or, more generally, gametes), and those with a negative influence (**minus alleles**) on others. In such cases (also called **coupling disequilibrium**), the observed genetic variation will be inflated relative to the expectation under random assortment. Under negative disequilibrium, allelic effects are negatively correlated, with plus alleles at one locus tending to be associated with minus alleles at another (**repulsion disequilibrium**). If the sign of D is relatively random over the loci underlying a trait, these effects will largely cancel. However, if there is a *directional bias*, the aggregate effects of gametic-phase disequilibrium might be extensive for quantitative traits controlled by large numbers of loci, even if the average level of disequilibrium between pairs of loci is relatively small (as there are n^2 pairwise LD terms). Both selection and assortative mating can generate such a directional bias (WL Chapters 16 and 24).

The impact of coupling versus repulsion disequilibrium is shown in the extreme example given in Figure 5.7, where we assume purely additive allelic effects. In the first case (repulsion disequilibrium), there are only two types of chromosomes: those with genes with additive effect 1 at locus 1 and effect 0 at locus 2, and vice versa. Although there are three types of two-locus genotypes, they all have the same genotypic value (in all cases, the sum of the additive effects is equal to two). Hence, there is no **expressed genetic variance**. The opposite situation is shown in the second example (coupling disequilibrium), where chromosomes have effects of 1 at both loci or effects of 0 at both loci. Assuming random mating and equal frequencies of chromosome types, genotypic values should be in the proportions of 25% 0, 50% 2, and 25% 4. The expressed genetic variance is therefore equal to 2. These extreme situations may be compared to the case in which the two loci are in gametic-phase equilibrium, so that for each of its four genes, an individual has an equal probability of being a 1 or a 0, regardless of the other genic states. From Equation 4.12a, under random mating and with additive gene action, the variance at each locus is $2p(1-p)a^2$, so with $a = 1$ and $p = 0.5$ for both loci, the total genetic variance is equal to one in the absence of disequilibrium.

When dominance effects are present at both loci, gametic-phase disequilibrium can have more complicated effects (Comstock and Robinson 1952; Robinson and Comstock 1955; Avery and Hill 1977, 1979; Weir et al. 1980; Hill and Mäki-Tanila 2015). As in the purely additive case, the additive genetic variance can be either inflated or depressed. The same will be true for the dominance genetic variance, depending on the direction of dominance at pairs of loci. This point can be formalized by considering the multilocus analogs of Equations 4.12a and 4.12b. For n diallelic loci,

$$\sigma_A^2 = 2\sum_{i=1}^{n} \alpha(i)^2 p_i(1-p_i) + 2\sum_{i=1}^{n}\sum_{j\neq i}^{n} \alpha(i)\alpha(j)D_{ij} \tag{5.23a}$$

$$\sigma_D^2 = 4\sum_{i=1}^{n} d_i^2\, p_i^2(1-p_i)^2 + 4\sum_{i=1}^{n}\sum_{j\neq i}^{n} d_i d_j D_{ij}^2 \tag{5.23b}$$

where $\alpha(i)$ is the average effect of allelic substitution at the ith locus (defined in Equation 4.10b) and $d_i = a_i k_i$ is a measure of dominance (Avery and Hill 1977). The equation for the additive genetic variance is obtained by recalling that the variance of a sum is equal to the sum of the variance of all the elements plus twice the sum of the covariances between all pairs of elements (Equation 3.11b). In this case, there are $2n$ variance terms to consider, each with gene-frequency variance $p_i(1-p_i)$, and $2n(n-1)$ covariance terms, each with gene-frequency covariance D_{ij}. (There are $2n(n-1)$ rather than $2n(2n-1)$ covariance terms because $2n^2$ pairs of genes involve contributions from different gametes, whose effects must be uncorrelated under random mating.) Likewise, for σ_D^2, the terms in the double summation are covariances between dominance deviations at loci i and j. Such dominance covariances require associations between both alleles at locus i and both alleles at locus j, and hence the D_{ij}^2 term appears. From Equation 5.12c we can obtain alternative expressions for Equations 5.23a and 5.23b using

$$D_{ij} = r_{ij}\sqrt{p_i p_j (1-p_i)(1-p_j)} \tag{5.23c}$$

A little more insight into these expressions may be acquired by recalling Figure 5.7. In this additive case, $\alpha(1) = \alpha(2) = 1$, $p = 0.5$ for both loci, and each single-locus term in Equation 5.23a is equal to $2 \times 1 \times 0.5 \times 0.5 = 0.5$. In the case of repulsion linkage, there are no 1/1 gametes, so $D = 0 - (0.5)(0.5) = -0.25$, and the two-locus summation term is $2 \times 2 \times 1 \times 1 \times (-0.25) = -1$. Summing up, we find that the observed additive genetic variance is equal to zero, in agreement with the verbal reasoning given above. The previous result for coupling linkage can also be recovered by noting that the contributions from single-locus terms are the same as in the previous example, 0.5 for each locus, whereas the disequilibrium is now $D = 0.5 - (0.5)(0.5) = 0.25$, yielding $\sigma_A^2 = 2$.

When epistatic interactions exist between loci in gametic-phase disequilibrium, the situation becomes extremely complicated. Even when only pairs of loci are considered, the final expression for σ_G^2 requires nearly a printed page (Gallais 1974; Weir and Cockerham 1977), and there is little to be gained by reprinting it here. The main point is that, as in the case of the additive and dominance components of genetic variance, observed levels of epistatic genetic variance can be increased or decreased when gametic-phase disequilibrium is important.

In summary, the components of expressed genetic variance for quantitative traits can be partitioned into expected values under gametic-phase equilibrium (the first summations on the right side of Equations 5.23a and 5.23b) and deviations from these expectations caused by disequilibrium (the double summations). When the **disequilibrium covariance** is negative, we refer to it as **hidden genetic variance** because it is subject to conversion to expressed genetic variance via the breakdown of gametic-phase disequilibria. Indeed, a conceptually simple way to test for the presence of disequilibrium covariance is to expand and randomly mate a population while minimizing selection over several consecutive generations. As gametic-phase equilibrium is approached, the phenotypic variance should converge on the expected equilibrium variance, with the deviation between initial and final levels of expressed genetic variance providing an estimate of the disequilibrium covariance in the

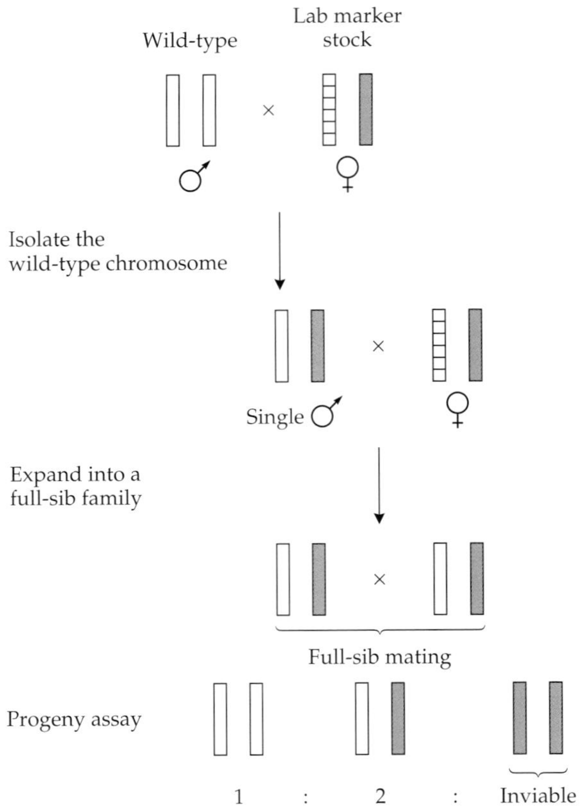

Figure 5.8 The procedure for isolating a drosophilid chromosome and assaying it for viability in the homozygous state. The chromosome is passed on intact between generations because *Drosophila* males do not exhibit recombination and because the marker chromosome suppresses crossing over in females. If the chromosome has normal viability, a 1:2 ratio of wild to marker flies is expected in the progeny generation because the marker chromosome carries recessive lethals.

base population. To our knowledge, such experiments have only been done with hybrid populations of maize (Gardner 1963).

A related prediction, the change in the genetic variance following the cessation of artificial selection, has also (surprisingly) been rarely examined. The only studies we are aware of are two papers by Sorensen and Hill (1982, 1983) on the change in the additive variance in abdominal bristle number following artificial selection in *Drosophila*. Under directional selection to increase the mean number of bristles, selection is expected to generate negative disequilibrium (WL Chapter 16), so that the additive variance should *increase* following the stoppage of selection. Sorensen and Hill (1982) did observe this in one of their two replicate populations. Conversely, disruptive selection (selection for either individuals with many, or few, bristles) is expected to generate positive disequilibrium (WL Chapter 16), with the additive variance declining once selection is stopped. This was indeed seen by Sorensen and Hill (1983).

The Evidence

Hidden genetic variation is expected to be a natural consequence of stabilizing selection, which favors linkage groups for their composite properties without regard to the alleles at individual loci. Returning to Figure 5.7, for example, the three genotypes under complete repulsion disequilibrium are selectively equivalent despite their different gene contents. A most extreme view of this idea was first invoked by Mather (1942, 1943) in his **polygenic balance** model. Mather suggested that the genes for characters under stabilizing selection

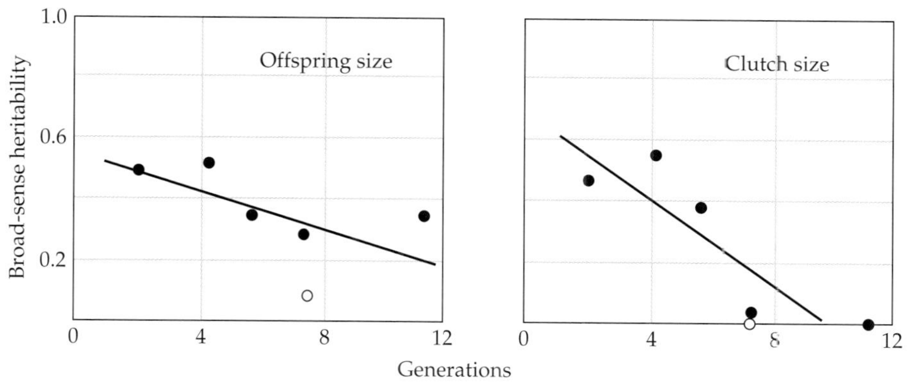

Figure 5.9 Temporal changes in the broad-sense heritability (the fraction of the total phenotypic variance with a genetic basis) for two traits in a *Daphnia* population. Sexual reproduction occurs each year (generation 0), with subsequent generations being asexual. Notice that, for both traits, the expressed genetic variance is nearly zero at the end of the first year (the single open circle), recovers to high levels at the beginning of the next year (the left-most solid circle in each graph), and then declines again towards zero (the remaining solid circles, representing different generations following a sexual reproduction episode). (After Lynch 1984.)

Table 5.1 Viability properties of homozygous intact chromosomes relative to homozygous recombinant chromosomes for five species of *Drosophila*. Only parental chromosomes with near normal viability scores were used.

Species	Chromosome	% Viability Reduction	Reference
D. melanogaster	II	−10	Spiess and Allen 1961
	III	8	
D. persimilis			
South Fork	II	13	Spiess 1959
White Wolf	II	10	
D. prosaltans			
Pirassununga	II	8	Dobzhansky et al. 1959
Rio	II	15	
D. pseudoobscura			
Texas	II	13	Spassky et al. 1958
California	II	26	
D. willistoni	II	3	Krimbas 1961

would lie along chromosomes in such a way that the signs of their contribution to the phenotype would alter. He also suggested that stabilizing selection would lead to the evolution of pairs of chromosomes with complementary features, such that the predominant chromosomal types in a population would be of the forms $+-+-+-\cdots$ and $-+-+-+\cdots$. Although this simple model is an extreme metaphor, subsequent theoretical work (Lewontin 1964; Bulmer 1971, 1974; Chevalet 1988; Gavrilets and Hastings 1993, 1994a, 1995; reviewed in WL Chapters 16, 24, and 28) has upheld the idea that stabilizing selection encourages the development of substantial hidden genetic variance, potentially depressing the level of expressed genetic variance significantly below its LE expectation.

The data are still scant, but several lines of evidence suggest that levels of hidden genetic variance in natural populations may be substantial. First, in *Drosophila*, it is possible to isolate intact chromosomes from natural populations by mating wild males with females from a laboratory strain with a homologous chromosome carrying a dominant visible marker and crossover suppressors (Figure 5.8). The wild chromosome can then be expanded into several full-sibs, each of which carries a marker chromosome. The progeny of crosses within full-sib families provide an assay of viability associated with the homozygous wild chromosome. Because the marker chromosome carries a lethal recessive, a 1:2 ratio of wild-type to marker progeny is expected if the homozygous wild-type is equivalent in viability to the wild/marker heterozygote. Dobzhansky and his colleagues performed this sort of analysis with five species of *Drosophila*, both with wild intact chromosomes and with recombinant wild derivatives. (Recombinant chromosomes are obtained by extracting individuals from the wild and crossing them for one or more generations prior to subjecting males to the crossing scheme in Figure 5.8.) On average, the recombinant chromosomes were associated with lower viabilities than were the intact chromosomes (Table 5.1). This implies that the parental chromosomes extracted from the wild tend to carry (in disequilibrium) groups of genes with favorable epistatic effects on viability. Recombination tends to break up these favorable combinations.

In clonally reproducing organisms, the entire genome is inherited intact, acting as a single linkage group, and natural selection can result in the build-up of extremely high levels of hidden genetic variance (Lynch and Gabriel 1983). This has been demonstrated in populations of the planktonic crustacean *Daphnia*, which periodically punctuate long phases of asexuality with a bout of sexual reproduction (Lynch 1984; Deng and Lynch 1996a). Such populations often exhibit annual cycles of expressed genetic variance for life-history traits (Figure 5.9). When repulsion disequilibrium is favored, clonal selection leads to the rapid erosion of expressed genetic variance and build-up of hidden genetic variance. Following the next episode of sexual reproduction, a fraction of the hidden genetic variance is then reconverted to expressed genetic variance. Qualitatively similar results have been observed when fungal isolates selected for the same phenotype have been crossed (Merrick 1975).

Selection need not always favor the development of hidden genetic variance. If coupling disequilibria are favored, selection will cause the expressed genetic variance to exceed its equilibrium expectations. In this case, the disequilibrium covariance is positive, and recombination would be expected to result in a *reduction* of the expressed genetic variance. Such changes have been observed in a natural population of *Daphnia*, which exhibited an approximately 50% reduction in the levels of expressed genetic variance for life-history traits following a phase of sexual reproduction (Lynch and Deng 1994).

The previous arguments considered how selection can cause accumulation of gametic-phase disequilibria for genes underlying a *single* trait. The same arguments apply to various aspects of multivariate phenotypes. For example, in populations of insects that exploit multiple host plants, one might expect a genetic correlation to evolve such that individuals prefer to feed (or lay eggs) on the plant species upon which they perform best (Rausher 1983; Thompson 1988). Such correlations could result from gametic-phase disequilibria between a set of genes influencing preference and another set influencing performance. (An alternative mechanism, pleiotropy, is that the genes that endow high performance on a particular host plant also encode for behavior that causes preference for that host, but because of the functional differences between these two types of traits, this mechanism seems unlikely.) A number of studies have revealed positive genetic correlations between preference and performance, while others have not (see Chapter 6; WL Chapter 30).

Genetic correlations (Part II Chapter 26) can thus be generated via **pleiotropy**—wherein an allele influences multiple traits—or through the generation of LD, or a combination of both. Pleiotropy is stable over recombination (i.e., pleiotropic-induced correlations do not decay under recombination), whereas genetic correlations generated via LD will decay over time by recombination (unless countered by other evolutionary forces, most notably selection). In populations of garter snakes, a genetic correlation exists such that striped individuals tend to flee directly from predators while spotted individuals tend to use crypsis as a defense (Brodie 1989). Because this correlation involves behavioral and morphological traits (and hence pleoitropy seems unlikely), these genetic correlations might be a result of gametic-phase disequilibrium—a consequence of natural selection favoring individuals that exhibit the right combinations of morphology and behavior.

6

Sources of Environmental Variation

Study hard what interests you in the most undisciplined, irreverent and original manner possible. Richard Feynman

Because all allelic features are subject to molecular stochasticity (Lynch 2024), and metabolic and developmental pathways are influenced to some degree by aspects of the environment, it stands to reason that the expression of most quantitative traits is not completely under genetic control. In some cases, phenotypic responses to changes in the environment are dramatic. For example, in some populations of the tiger salamander (*Ambystoma tigrinum nebulosum*), larvae develop into atypical cannibalistic morphs when raised at high densities (Collins and Cheek 1983). Clones of many planktonic cladocerans and rotifers can be induced to change their external morphology by means of appropriate physical or chemical stimuli (Hutchinson 1967; Havel 1987), and environmentally controlled flight polymorphisms are well documented in locusts (Kennedy 1956) and corixid bugs (Young 1965). However, more often than not, environmental effects are subtle, causing simple amplifications or reductions in sizes of parts, numbers of progeny, rates of growth, physiological performance, gene expression, metabolite levels, and so on.

As with the genetic variance, sources of environmental variance can be partitioned in various ways. It is useful to define two broad categories. **General environmental effects** refer to influential factors that are shared by groups of individuals, the size of the group depending on the context. One example is the impact of the sex of an individual. Other familiar examples include the effects of an experimental treatment (Appendix 9), a patch of habitat, or a growing region. Likewise, through maternal care, mothers will have general effects on their offspring (beyond the direct transmission of genes), usually referred to as **maternal effects**. **Special environmental effects** are residual deviations from the phenotype after accounting for genotype and general environmental effects. These include both measurement error and effects that are unique to individuals (a consequence of microenvironmental variation and random developmental noise). Finally, there is one additional complication with respect to environmental effects. Ideally, the phenotype of an individual can be viewed simply as the sum of its genotypic value and the environmental effects with which it is associated, i.e., $z = G + E$. However, in many cases, different genotypes respond to environmental change in nonparallel ways, a phenomenon known as **genotype $\times$ environment interaction** (**G$\times$E**).

The purposes of this chapter are twofold. First, we illustrate how various classes of environmental effects can be incorporated into the logical framework introduced in the previous chapter for genotypic values. This extension leads to a general linear model for the phenotype, which forms the foundation upon which much of the remainder of the book is built. Second, as background for the remaining chapters, we provide an overview of the biological basis for the various types of environmental effects. Further details will emerge in later chapters. For example, Chapters 27 and 28 (in Part II) are devoted entirely to issues related to genotype $\times$ environment interaction, while Chapter 29 (Part II) examines maternal effects in detail.

EXTENSION OF THE LINEAR MODEL TO PHENOTYPES

Here we let E and e, respectively, denote the contributions of general and specific environmental effects to the phenotypic expression of a character, and let I denote the genotype $\times$ environment interaction effect. The phenotype of the kth individual of the ith genotype exposed to the jth general environmental effect can then be described as a linear function of four components,

$$z_{ijk} = G_i + I_{ij} + E_j + e_{ijk} \tag{6.1a}$$

(As an aside, we note that a common alternative notation for I_{ij} in the literature is GE_{ij}.) Each of these components may have subcomponents. For example, from Equation 5.7, the genotypic value, G_i, is a potentially complicated linear function of the population mean phenotype, the individual's additive-genetic value, and various effects due to dominance and epistasis. As is the case for all genetic effects, the terms I_{ij}, E_j, and e_{ijk} are defined in a least-squares sense as deviations from lower order expectations and, as a consequence, have mean values equal to zero. The population mean phenotype, $\mu_G = \overline{z}_{ijk}$, is the mean phenotype of all genotypes in the population, whereas G_i is the expected phenotype of the particular genotype i averaged over all possible environmental conditions within the experimental setting. The quantity $\mu_G + E_j$ is the mean phenotypic value expected if the *entire population* of genotypes were assayed in the *j*th macroenvironmental setting, whereas $G_i + I_{ij} + E_j$ is the expected phenotype of genotype *i* in that setting. Thus, I_{ij} is the *residual deviation* left after assuming that genotypic and environmental values *act in an additive fashion*. Finally, e_{ijk} is the deviation of an individual's phenotype from the expectation $G_i + I_{ij} + E_j$. As in any least-squares linear model (Chapter 3), the residual deviations e_{ijk} are defined to be uncorrelated with the explanatory variables G, I, and E.

A useful way to think about G×E is to consider the difference between two genotypes, i and ℓ, over a series of environments (such as the differences between two inbred lines assayed over a number of different growing regions). If I is zero, then the expected value of their difference in any growing region is just

$$\text{E}[z_{ij\cdot} - z_{\ell j\cdot}] = G_i + E_j + \text{E}[e_{ij\cdot}] - (G_\ell + E_j + \text{E}[e_{\ell j\cdot}]) = G_i - G_\ell \tag{6.1b}$$

where $z_{ij\cdot}$ is the average value of genotype i in environment j. Hence, the difference in mean performance is *constant* over all environments in the absence of G×E. When G×E is present,

$$\text{E}[z_{ij\cdot} - z_{\ell j\cdot}] = G_i - G_\ell + I_{ij} - I_{\ell j} \tag{6.1c}$$

is an environmentally specific difference.

Example 6.1. As an example of computing genotypic and environmental values, we consider a small data set from Strauss and Karban (1994) on three lines of thrips (*Apterothrix apteris*; an insect in the order Thysanoptera) grown on three clones of their host plant, the seaside daisy (*Erigeron glaucus*). The following table gives the mean performance of thrips, measured as population density after eight generations of growth, in the nine genotype–environment combinations—the thrip lines being denoted as I, II, and III, and the three environments (host plant clones) as 1, 2, and 3. All nine treatments were replicated; the standard errors of the measures were small and are ignored in the following.

	Thrips Line			
Plant	I	II	III	E
1	77	34	47	$-19.22 = E_1$
2	61	159	51	$18.44 = E_2$
3	40	71	107	$0.78 = E_3$
G	59.33	88.00	68.33	

Assuming an equal weight for each cell, and averaging over all nine cells, the mean character value is found to be μ_G = 71.89. The genotypic values of the thrip lines, obtained by averaging the elements within columns, are given in the bottom row of the table. (Note that one could also parameterize the genotypes as $\mu + G$, where now G are deviations from the mean value, in which case the genotypic values respectively become −12.56, 16.11, and −3.56.) The average environmental effects, each defined as the mean performance of all thrip lines in a specific environment (the average within rows) minus the grand mean (μ_G), are given in the final column. For example, E_1 = [(77+34+47)/3] − 71.89. Note that the three values of E average to zero. To obtain the interaction effects, we rearrange Equation 6.1a to

$$I_{ij} = \bar{z}_{ij} - G_i - E_j$$

where i and j, respectively, denote the thrips line and the daisy clone, and $\bar{z}_{ij}$ is the entry in the ith column and jth row of the table. The residual deviation drops out because we have assumed the data to have been obtained without error. Substituting into this equation yields

$$\begin{array}{lll} I_{I,1} = 36.89 & I_{II,1} = -34.78 & I_{III,1} = -2.11 \\ I_{I,2} = -16.78 & I_{II,2} = 52.56 & I_{III,2} = -35.78 \\ I_{I,3} = -20.11 & I_{II,3} = -17.78 & I_{III,3} = 37.89 \end{array}$$

The interaction effects average to zero both within rows and within columns ($\overline{I}_{i\cdot} = \overline{I}_{\cdot j} = 0$). Note also that the magnitudes of the interaction effects tend to be much greater than the magnitudes of the environmental effects, indicating strong genotype × environment interaction. Thus, this study also provides strong evidence of genotype–environment *covariance*. In nature, thrip line I was found living on plant clone 1, line II on clone 2, and line III on clone 3. Thus, the individual clones were associated with the plants on which they best performed.

Because I and e are uncorrelated with the other variables (by construction), using the formula for the variance of a sum (Equation 3.11b), the total phenotypic variance of a population can be written as

$$\sigma_P^2 = \sigma_G^2 + \sigma_I^2 + 2\sigma_{G,E} + \sigma_E^2 + \sigma_e^2 \tag{6.2}$$

The expansion of the total genetic variance, σ_G^2, in terms of additive, dominance, and epistatic components was given in the previous chapter, while σ_E^2 and σ_e^2 are the components of variance due to general and special environmental effects.

The greatest conceptual difficulty with Equation 6.2 is related to the interpretation of σ_I^2 and $\sigma_{G,E}$. The term $\sigma_{G,E}$ refers to **genotype–environment covariance**, which is quite distinct from genotype × environment *interaction*. The latter is concerned with the variation in the phenotypic response of specific genotypes to specific environments, and σ_I^2 is a measure of that variation. Conversely, genotype–environment covariance is a measure of the physical association of particular genotypes with particular environmental general effects. If individuals are randomly distributed with respect to macroenvironments, then $\sigma_{G,E}$ is zero, but σ_I^2 will be nonzero if genotypic values and environmental effects are nonadditive. If, on the other hand, genotypes are nonrandomly distributed, $\sigma_{G,E} \neq 0$. In the absence of a good experimental design (Appendix 9), G–E covariances can occur in many agricultural and human settings. For example, it is not uncommon for a farmer to give more feed to higher milk producers. Attempts to avoid G–E covariances are why blind randomized tests are used in clinical trails, to remove any unconscious bias wherein the healthiest patients (G) are given the new treatment (E) and unhealthier patients given the control.

Even in the most carefully controlled situations, one cannot always rule out the presence of genotype–environment covariance. Nonrandom associations of genotypes and environment can result from limited seed or pollen dispersal in plants and from genetically based dominance hierarchies and other social interactions in animals. Maternal (or paternal) effects can also cause genotype–environment covariance if there is a correlation between parental genotype and ability to provision the young (Chapter 29). This latter source of genotype–environment covariance can occur in an environment that is otherwise completely homogeneous.

Whereas methods exist for the detection of genotype × environment interaction (Chapters 27 and 28), genotype–environment covariance is usually less tangible, contributing an unknown amount to estimates of genetic variance. Take, for example, the simple situation in which genotype × environment interaction is absent and an estimate of the environmental variance is obtained as the phenotypic variance within pairs of monozygotic twins (or among members of a clone), all of which must be genetically identical. Then, from Equation 6.2, it can be seen that the difference between the phenotypic variance and the environmental variance (the among-clone variance) is $\sigma_G^2 + 2\sigma_{G,E}$. Depending on whether $\sigma_{G,E}$ is positive or negative, the true genetic variance will be over- or underestimated.

Example 6.2. The results of an experiment with a clonal plant will help clarify the preceding concepts. Salt marsh cord grass (*Spartina patens*) occurs along much of the Atlantic coast on sand dunes, swale grasslands, and marshes. Silander (1985) removed plants from these three environments and clonally propagated them via rhizomes in the greenhouse. After two years, sufficient material was available to perform a reciprocal transplant experiment in the field. Replicate progeny from each clone were grown at each of three sites: dune, swale, and marsh. A large number of vegetative and reproductive traits were measured, and the data were analyzed by two-way ANOVA with clone and site serving as the main factors.

Trait	Var(G)	Var(E)	Var(I)	Var(e)
Tillers/clone	0.2	34.2	19.9	45.8
Culm height	12.0	56.6	7.0	24.5
Leaves/culm	11.2	19.4	8.4	61.0
Culm diameter	20.6	4.7	0.0	74.7
Longest leaf length	26.7	29.5	6.1	37.7
Longest leaf width	27.5	7.8	3.6	61.3
Third leaf length	25.2	0.0	10.2	64.6
Third leaf width	23.4	3.4	3.6	69.7
Trait averages	18.4	19.5	7.4	54.9

The observed components of variance, given in the above table as percentages of the total phenotypic variance, can be interpreted as follows: the among-clone variance is an estimate of the total genetic variance σ_G^2; the variance among the three sites is an estimate of the general environmental effects variance σ_E^2; the variance within sites (more specifically, among members of the same clone within sites) is an estimate of σ_e^2; and the clone × site variance is an estimate of σ_I^2. Conceivably, some of the observed phenotypic variance may have been caused by maternal effects, but because all members of a clone had the same mother, any variance caused by such effects is compounded with the estimate of σ_G^2. In this experiment, $\sigma_{G,E}$ can be assumed to be zero because individual plants were distributed randomly within treatments. When averages are taken over all of the traits in the study (including those not in the table), the vast majority (55%) of the phenotypic variance is found to be attributable to special environmental effects (σ_e^2). Genotype accounts for an additional 18% of the variance, whereas general environmental effects and genotype × environment interactions account for 19 and 7%, respectively. These results indicate that *Spartina* growth characters are relatively insensitive to what appear to be major changes in habitat.

SPECIAL ENVIRONMENTAL EFFECTS

As noted above, two sources contribute to the special environmental effects variance: internal developmental noise and external microenvironmental heterogeneity. Both sources of variation are generally unpredictable from the standpoint of the individual. Both are unique properties of the population under investigation and can be modified by changing the environmental setting. For unicellular organisms, there is now good information on baseline stochastic variance of mRNA and protein abundance with cells (summarized in Lynch 2024).

Within-individual Variation

In many species (especially animals), it is possible to gain some information about the variance associated with special environmental effects by measuring the same attribute on the right and left sides of bilaterally symmetrical organisms. Differences may arise between such measures because of measurement error, but there is usually a small, but real, random component of asymmetry. This within-individual variation is the finest level at which morphological phenotypic variance can be quantified, and it can sometimes constitute a substantial portion of the total variance of a trait. Leamy (1984, 1992), for example, found that differences between right and left measures account for up to 18% of the phenotypic variance for bone lengths in mice, and similarly high values have been noted for gill-raker and fin measures in rainbow trout (Leary et al. 1992), and for cranial measures in tamarins (a New World monkey; Hutchison and Cheverud 1995).

Following the early suggestions of Mather (1953), Thoday (1953), and Van Valen (1962), numerous investigators have interpreted the variance of right-left measures to be a measure of **developmental noise** (or **developmental instability**). Although it is difficult to actually define the mechanistic basis of such variation, we discuss some correlates below. Soulé (1982) suggested that developmental noise may be a simple consequence of random movement of molecules within developing individuals, while Emlen et al. (1993) and Klingenberg (2003), among others, have considered how variation at the cellular level might translate into differences at the levels of tissues or organs through cellular feedback mechanisms. However, it may be presumptuous to assume that variance in right-left measures is entirely due to internal factors. In no case has the involvement of subtle variation in the external environment from right to left sides been ruled out, although it is known that the extreme asymmetry that develops in lobster claws is a consequence of differential claw use (Govind and Pearce 1986). Thus, it appears prudent to simply denote the right-left variance as within-individual variance σ^2_{ew}, without invoking causality. The total variance resulting from special environmental effects can then be written as the sum of the among-individual and within-individual components,

$$\sigma^2_e = \sigma^2_{ew} + \sigma^2_{ea} \tag{6.3}$$

Van Valen (1962) pointed out the need to distinguish three types of asymmetry. **Directional asymmetry** refers to a consistent bias in one direction such as the tendency of the mammalian heart to be on the left side or of a particular coiling direction (e.g., left = **sinistral**, right = **dextral**) in the shells of snails. For a continuously distributed character, directional asymmetry is detectable as a mean difference between right and left measures that is significantly different from zero. **Antisymmetry** refers to situations in which asymmetry is the rule rather than the exception but is nondirectional. An example is provided by male fiddler crabs, which have equal probabilities of developing right or left signaling claws. **Fluctuating asymmetry (FA)** refers to the common situation in which the difference between the right and left measures is symmetrically (usually normally) distributed around a mean and mode of zero. Studies on developmental homeostasis focus upon the fluctuating form of asymmetry, and we devote the remainder of our attention to it. There is a massive literature on the analysis of FA, and further details of which can be found in reviews by Møller and Swaddle (1997), Polak (2003), Leamy and Klingenberg (2005), Dongen (2006), Graham et al. (2010), and Klingenberg (2015). A nice review of the embryological underpinning left-right asymmetry is given by Levin (2005).

Under the assumption that right and left measures are two of many possible random manifestations of an individual's developmental program, an unbiased estimate of the within-individual variance for a trait is given by

$$\text{Var}(e_w) = \sum_{i=1}^{N} \frac{(r_i - l_i)^2}{2N} - \text{Var}(e_m) \tag{6.4}$$

where N is the number of individuals sampled, r_i and l_i are the right and left measures for the ith individual, and $\text{Var}(e_m)$ is the variance due to measurement error. The

Table 6.1 Summary of studies on the relationship of fluctuating asymmetry to levels of inbreeding and crossbreeding.

Organism	Basis of Comparison	Character(s)	Reference
FA enhanced in inbreds			
Drosophila melanogaster	Inbred line crosses	Bristle number	Mather 1953 Reeve 1960
	Inbreeding variable base population	Wing length	Biémont 1983
Tisbe (copepod)	Inbreeding variable base population	Thoracic leg length	Clarke et al. 1986
Lampsilis, Elliptio (freshwater bivalves)	Natural pops.	Plicae on palps	Kat 1982
Poeciliopsis (fish)	Natural pops.	Fin ray, scale, tooth number	Vrijenhoek and Lerman 1982
Salmo, Salvelinus (Trout species)	Individuals varying in heterozygosity	Fin ray, gill raker, mandibular pores	Leary et al. 1983 1984, 1987
Salmo (Rainbow trout)	Inbreeding variable base population	Fin ray, gill raker, mandibular pores	Leary et al. 1985
Uta (lizard)	Natural pops.	Scale counts	Soulé 1979
Mus (House mouse)	Inbred line cross	Osteometric traits	Leamy 1984, 1992
Saguinus (Tamarin monkey)	Natural pops.	Cranial measures	Hutchison and Cheverud 1995
FA approximately equal in inbreds and outbreds, or variable results			
Drosophila melanogaster	Inbred line cross Inbreeding variable base population	Bristle number	Beardmore 1960 Fowler and Whitlock 1994
Apis (Honeybee)	Inbreeding variable base population	Wing vein length	Brückner 1976 Clarke et al. 1986
Salmo (Rainbow trout)	Hatchery strain crosses	Fin ray, gill raker counts	Ferguson 1986
Lepomis (Bluegill)	Subspecies crosses	Fin ray, scale counts	Felley 1980
Sceloporus (lizard)	Interspecific hybrid zone	Scale counts	Jackson 1973
Mus (House mouse)	Inbred line crosses natural pops.	Molar width	Bader 1965
	Hybrids in natural hybrid zone	Skull traits	Mikula and Macholán 2008
Acinonyx (cheetah)	Species varying in heterozygosity	Craniometric traits Dental dimensions	Wayne et al. 1986 Modi et al. 1987 Kieser and Groeneveld 1991
Drosophila	Populations at various levels of inbreeding	Bristle number	Gilligan et al. 2000
Rivulus (fish)	Natural pops.	Morphological traits	Taylor 2001
FA enhanced in outbreds			
Gasterosteus (Stickleback fish)	Hybrids between natural pops.	Scale counts	Zakharov 1981
Enneacanthus (Banded sunfish)	Interspecific hybrid zone	Fin ray, scale, morph. traits	Graham and Felley 1985

latter quantity is simply the variance among repeated measures of the same trait (on the same side) of the same individual; ideally, Var(e_w) should be estimated in a way that ensures that the investigator has no memory of previous measures. Palmer and Strobeck (1986) recommend making multiple measures on both sides of every individual, treating right and left measures as fixed effects, and performing a two-way analysis of variance. This approach allows one to test for directional asymmetry, as well as to extract Var(e_w) and Var(e_m) directly from the ANOVA mean squares (Chapter 23 in Part II). A number of other metrics of FA have been proposed, and further details on the statistical analysis of FA can be found in Palmer and Strobeck (1986, 1992, 2003), Palmer et al. (1993), Swaddle et al. (1994), Hutchison and Cheverud (1995), Hoffmann and Woods (2003), Graham et al. (2010), and Klingenberg (2015).

Provided that an estimate of the total special environmental effects variance, Var(e), is available, the among-individual component can be estimated as Var(e_a) = Var(e) − Var(e_w). To our knowledge, this has not been done with any organism, but in principle it is straightforward for organisms that can be grown clonally, as the total variance within clones is simply Var(e).

Interest in the within-individual component of variance derives from the idea that relatively asymmetrical individuals are victims of genetic and/or environmental circumstances that enhance the chances of random developmental errors on the two sides of the body. If this were true, Var(e_w) would provide a useful measure of environmental/genetic stress. In an effort to test this idea, many investigators have searched for correlates of fluctuating asymmetry, focusing particularly on the influence of the genetic background. Under the assumption that symmetry is adaptive, and therefore maximized in natural populations, it follows that perturbations to locally adapted gene pools should inflate the level of FA. Such perturbations can be induced artificially in two ways: the average homozygosity can be increased by mating individuals with their close relatives (inbreeding; Chapter 12), or heterozygosity can be enhanced by hybridizing different populations or species (outcrossing; Chapter 13).

A summary of some results indicates that enhanced FA in genetically perturbed individuals is by no means universal (Table 6.1). At least a third of the reported studies indicates no consistent change in asymmetry upon genetic perturbation. However, an objective evaluation of the results is difficult, because many types of crosses have been employed in these studies. Soulé (1982) argued that the reason many inbreds exhibit enhanced FA is that inbreeding results in the production of extreme phenotypes that are more sensitive to developmental perturbations. If this were true, then one would expect the level of FA to be higher in individuals in the tails of phenotype distributions. However, attempts to find such a relationship have been unsuccessful (Soulé and Cuzin-Roudy 1982; Scheiner et al. 1991; Zakharov 1992; Livshits and Smouse 1993; Deng 1997).

There are many plausible explanations for the conflicts in the existing literature. For example, crosses between different strains need not be deleterious and can sometimes have the beneficial effects of masking the expression of deleterious recessives in the parental strains (Chapter 13). Second, many of the existing studies on FA may simply be statistically weak or flawed. The studies summarized in Table 6.1 have used a number of indices other than Equation 6.4. Many investigators have simply employed the average absolute difference between the right and left measures, while others have attempted to correct for scaling that might occur with size changes by dividing the differences by the average measure. Further, very few studies have made any attempt to eliminate the upward bias caused by measurement error. Finally, it is possible that the most adaptive level of FA is not the minimum level. Mather (1953) was able to increase and decrease FA for sternopleural bristle number in *Drosophila melanogaster* by artificial selection. After two generations of relaxed selection, the level of FA in both lines returned to the level in the original base population, suggesting that natural selection favors an intermediate level of FA. Presumably, the complete elimination of asymmetry can only be accomplished at the expense of important fitness characters such as development time.

The effects of *environmental* stress on FA are much more consistent, as σ_{ew}^2 tends to increase in extreme or novel environments (Hoffmann and Parsons 1991; Leung and Forbes 1996; Møller and Swaddle 1997; Hoffmann and Woods 2003). Indeed, FA has been proposed as a general biomarker for detecting environmental stress (Beasley et al. 2013) or assessing animal welfare in agricultural settings (Knierim et al. 2007), but see Kozlov and Zvereva (2015) for the caveat of conformational bias in such studies. FA-stress correlations have been revealed in studies of the exposure of pesticide (Valentine and Soulé 1973; Clarke 1992; McKenzie and Yen 1995). When first exposed to diazinon, the sheep blowfly (*Lucilia cuprina*) initially exhibited an increase in FA, but as the population evolved insecticide resistance, FA returned to normal levels (McKenzie and Clarke 1988). Humans suffering from malnutrition exhibit increases in FA (Bailit et al. 1970), and various types of stress (cold, noise, behavioral modification) inflate FA for tooth morphology and limb bones in mice and rats (Siegel and Doyle 1975a, 1975b, 1975c; Siegel and Mooney 1987). In birds, parasite load is associated with increases in the asymmetry in lengths of tail feathers (Møller 1992). Several studies have reported that individuals with developmental deformities tend to have elevated levels of FA for other traits (Bailit et al. 1970; Woolf and Gianas 1976; Barden 1980; Malina and Buschaung 1984; Leary et al. 1984).

In principle, the logic underlying the use of fluctuating asymmetry as a measure of σ_{ew}^2 can be extended to certain aspects of organisms that are not bilaterally symmetrical. In plants, for example, there are many serially repeated organs such as leaves, and flowers are often radially symmetrical. In cases such as these, more than two measures can be made within each individual, and the within-individual component of variance can be obtained by analysis of variance (Freeman et al. 1993; Møller and Eriksson 1994; Rao et al. 2002). As in the analysis of bilateral symmetry, a critical assumption underlying any such analysis is that the subjects of measurement are products of the same genes. An obvious violation of this assumption would involve a comparison of stem and basal leaves in a herbaceous plant, but the question remains open with more subtle comparisons such as the inner and outer leaves on a branch.

Developmental Homeostasis and Homozygosity

In an influential book that reviewed much of the early literature, Lerner (1954) strongly endorsed the idea that the degree of developmental stability is positively correlated with the overall level of individual heterozygosity. The usual mechanistic explanation for this hypothesis is that heterozygosity acts as a buffer against environmental variation. This effect might occur, e.g., if different allelic products have optimum activities under different environmental conditions.

Some of the logic behind Lerner's ideas were implicit in our discussion of FA, but here we focus on the total effects of heterozygosity on the within- and among-individual components of σ_e^2. Because the main prediction of Lerner's **homeostasis–heterozygosity hypothesis** is that σ_e^2 will be higher for homozygous individuals than for heterozygotes, a conceptually simple test suggests itself. Consider two completely inbred lines and their hybrid (F_1) progeny. All members of the inbred lines will be 100% homozygous, while the F_1 individuals will be uniformly heterozygous at each locus for which the two inbred lines differed. Within each of the three groups, all individuals will be genetically identical, so the observed phenotypic variances within groups can be interpreted as genotype-specific components of environmental variance. Thus, if all three genotypes are equally sensitive to environmental and developmental noise, they should exhibit the same level of phenotypic variance. The survey in Table 6.2 indicates that the phenotypic variance in F_1 hybrids is almost always lower than the average of the parental lines.

Although these kinds of results have been interpreted as strong support for the homeostasis–heterozygosity hypothesis, there are some unresolved issues. In most of the reported studies, the parental lines were not completely homozygous, so the observed phenotypic variances must contain a genetic component. This complicates matters greatly. If there is dominance at a locus, and the frequency of the dominant allele in the F_1 population is

Table 6.2 Survey of studies of phenotypic variation in inbred parental lines and their F_1 hybrids.

Species	Characters	Reference
F_1 less variable		
Drosophila	Bristle number and wing length	Robertson and Reeve 1952
	Viability	Dobzhansky and Spassky 1954
		Dobzhansky and Levene 1955
Mice	Behavioral traits	Hyde 1973
	Skeletal traits	Leamy and Thorpe 1984
	Time to maturity	Yoon 1955
	Weight	Chai 1957
Rats	Weight	Livesay 1930
Cotton	Fiber properties	Meredith et al. 1970
Corn	Vegetative and reproductive traits	Adams and Shank 1959
		Shank and Adams 1960
Tobacco	Vegetative and reproductive traits	Jinks and Mather 1955
Rye	Grain production and plant weight	Pfahler 1966
Sorghum	Grain production	Reich and Atkins 1970
		Jowett 1972
Tomato	Growth rate	Lewis 1954
Equivocal results		
Poeciliopsis (Fish)	Scale and fin ray counts	Angus and Schultz 1983
Arabidopsis	Plant weight	Pederson 1968
Tomato	Reproductive traits	Williams 1960
Drosophila	Wing traits	Whitlock and Fowler 1999
F_1 more variable		
Drosophila	Bristle number	Lewontin 1957
Corn	Grain production	Rowe and Anderson 1964

between 0.5 and 1.0, the average genetic variance in the parental lines will actually exceed that in the F_1 (Lerner 1954). This result is a simple consequence of the expression of the recessive allele being masked in the hybrid line. For a diallelic locus, the variance in the F_1 generation can be found by noting that the frequencies of individuals with genotypic values 0, $(1+k)a$, and $2a$ are q_1q_2, $p_1q_2 + p_2q_1$, and p_1p_2, where p_1 and p_2 are the frequencies of the dominant allele in the two parental populations, and $q_i = 1 - p_i$. Following the procedures introduced in Chapter 4, the total genetic variance within an F_1 population becomes

$$\sigma_G^2 = a^2 \Big\{ \big[(1+k)^2(p_1q_2 + p_2q_1) + 4p_1p_2\big] - \big[(1+k)(p_1q_2 + p_2q_1) + 2p_1p_2\big]^2 \Big\} \qquad (6.5)$$

With certain allele frequencies, the mean genetic variance within the parental lines (obtained from Equations 4.12a and 4.12b) can be more than twice that in the F_1 (Figure 6.1). Thus, the observed patterns in Table 6.2 could be caused in part by **residual heterozygosity** within assumed fully inbred parental lines rather than by differences in environmental sensitivity between inbred and outbred individuals.

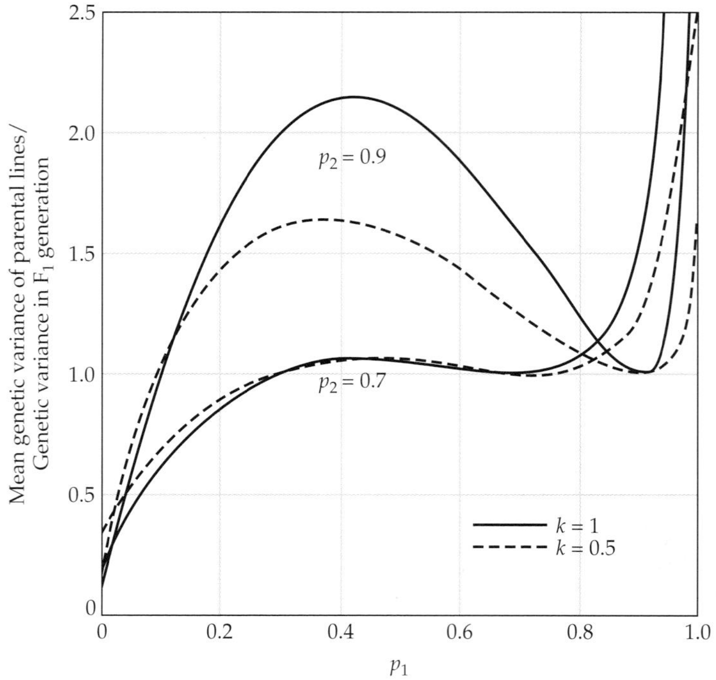

Figure 6.1 The mean single-locus genetic variance within two parental lines relative to that within the F_1 hybrid population. Here p_1 and p_2 are the frequencies of the dominant allele in the two populations. Results are given for complete $(k = 1)$ and partial $(k = 0.5)$ dominance and for two allele frequencies within parental population 2, $p_2 = 0.7$ and 0.9.

Several studies have attempted to test the homeostasis–heterozygosity hypothesis by comparing levels of phenotypic variation within groups of individuals with high versus low heterozygosity for enzyme-coding loci. One limitation of most of these early studies was that genome-wide heterozygosity was assessed by using just a handful of allozyme markers. Data from studies of oysters (Zouros et al. 1980), monarch butterflies (Eanes 1978), *Drosophila melanogaster* (Blanco and Sanchez-Prado 1986), killifish (Mitton 1978), ruffous-collared sparrows (Yezerinac et al. 1992), and humans (Livshits and Kobyliansky 1984) indicate a negative association between morphological variation and heterozygosity. However, studies on pitch pine (Ledig et al. 1983), the snail *Cerion* (Booth et al. 1990), plaice flatfish (McAndrew et al. 1982), herring (King 1984), and brook trout (Hutchings and Ferguson 1992) reveal no relationship, while there appears to be a weak positive relationship in *Daphnia magna* (Yampolsky and Scheiner 1994) and baboons (Bamshad et al. 1994).

One especially interesting observation comes from the ***Drosophila melanogaster*** **Genetic Reference Panel** (**DGRP**) lines (Harbison et al. 2013). The DGRP lines, which are largely (but not fully) inbred show an exception to the inverse relationship between σ_E^2 and levels of heterozygosity. These lines contain varying amounts of residual heterozygosity (which is substantial in several cases) due to segregating chromosomal inversions, with the amount of residual heterozygosity being uncorrelated with σ_E^2.

Finally, a meta-analysis (Appendix 6) by Vøllestad et al. (1999) that considered 41 previous studies (invertebrates, mammals, fish, and one plant) found a very weak negative correlation between enzyme heterozygosity and FA ($r = -0.09$), which implies that only $r^2 \sim 1\%$ of variation in FA is accounted for by the level of enzyme heterozygosity. Interestingly, Vøllestad et al. noted that ectothermic animals showed a significant negative relationship, $r = -0.23 \pm 0.09$ (which is only a $\sim 4\%$ effect), whereas no significant relationship was found for endotherms. One explanation for this finding is based on our previous observation that FA is often more enhanced in stressed individuals, in that ectotherms may be less environmentally buffered, and hence potentially more stressed, than endotherms.

Again, the discordancy between these results may be a statistical artifact. Not only does the problem outlined in Figure 6.1 apply to all of these studies, but there is another serious interpretative difficulty. In most of the analyses cited above, the homozygous groups contain all homozygous individuals regardless of the alleles they are carrying. Thus, for a single diallelic locus contributing to a quantitative trait, the homozygous group would contain *AA* and *aa* individuals with genotypic values 0 and $2a$, while the heterozygous group would only contain individuals with genotypic value $(1 + k)a$. When individuals are classified in this manner, for characters with an additive genetic basis, an inverse relationship is expected between multilocus heterozygosity and the genetic component of phenotypic variation, the most heterozygous group being the least variable (Chakraborty and Ryman 1983; Chakraborty 1987). No special appeal to homeostatic properties of heterozygosity is required to explain the results. On the other hand, when characters have a nonadditive genetic basis, the relationship between heterozygosity and the genetic component of phenotypic variance can be either positive or negative, depending on the allele frequencies and genetic effects (Bishop 1992; Gavrilets and Hastings 1994b).

Interpopulation comparisons in humans (Kobyliansky and Livshits 1983), fox sparrows, and pocket gophers (Zink et al. 1985) have revealed an inverse relationship between *population* estimates of morphological variation and enzyme heterozygosity, consistent with the homeostasis–heterozygosity hypothesis, whereas the relationship appears to be positive in sculpin fishes (Strauss 1991) and ruffous-collared sparrows (Yezerinac et al. 1992). With population-level comparisons, the data are not biased by the artificial grouping of genotypic classes. However, there are still significant interpretative difficulties. The changes in morphological variability with heterozygosity may be due to shifts in the genetic, as well as the environmental, components of variance. If the molecular markers used to estimate heterozygosity levels are linked with the loci contributing to the genetic variance for the characters, then an increase in the latter associated with molecular heterozygosity may completely obscure changes in the environmental component of variance.

In summary, the fundamental problem with most descriptive studies of the relationship between heterozygosity and phenotypic variation is their inability to distinguish between the hypothetical stabilizing effects of heterozygosity on the environmental component of phenotypic variation and the very real effects of heterozygosity on the genetic component of variation. This issue can only be resolved by separating multilocus genotypes into homogeneous groups (based on allelic composition, rather than on total heterozygosity) and by separating the phenotypic component of variance into its genetic and environmental components. A simple way to resolve both of these problems exists for species that can be propagated clonally. All members within clonal groups are necessarily identical with respect to genotype (which can be highly heterozygous), so that the phenotypic variance within a clonal group is entirely due to environmental causes.

Deriving clones from natural populations of *Daphnia*, Deng (1997) used this approach to test the homeostasis–heterozygosity hypothesis. For each parental clone, a selfed offspring genotype was produced (by crowding, *Daphnia* can be induced to produce males, which then fertilize their genetically identical mother or sibs). In populations of *D. pulex* and *D. pulicaria*, the environmental components of variance for life-history characters within inbred clones averaged 75% and 88% higher than within parental clones. These results provide unambiguous support for Lerner's hypothesis.

Several studies with trees have addressed the relationship between annual variation in individual growth rate (a measure of environmental sensitivity) and enzyme heterozygosity. As in Deng's work, these studies are not complicated by the genetic artifacts mentioned above because the units of comparison are the same individuals. Nevertheless, the results are equivocal. Heterozygosity and temporal variation in growth rate are positively correlated in quaking aspen (Mitton and Grant 1980), ponderosa pine (Knowles and Grant 1981), and knobcone pine (Strauss 1987), but negatively correlated in lodgepole pine (Knowles and Mitton 1980), and uncorrelated in pitch pine (Ledig et al. 1983).

We have just summarized an enormous body of work that addresses Lerner's hypothesis. Yet when all of the results are considered, it appears that the acceptance of a general causal relationship between heterozygosity and developmental stability should be postponed until additional adequately designed experiments have been performed.

Genetic Variation in the Environmental Variance

The observation (at least in some cases) of a difference in the environmental variance between homozygotes and heterozygotes shows that *environmental sensitivity can change over genotypes*. That this is not simply a genotype-independent feature (e.g., not just a feature of *heterozygosity per se*), but a property of *specific genotypes*, is supported by several types of observations. First, direct evidence for genetic variation in σ_E^2 is provided by comparing inbred lines. Mackay and Lyman (2005) observed different amounts of environmental variation for bristle number across chromosomal substitution lines of *Drosophila* from a common source population. Similar findings using inbred lines were obtained by Ordas et al. (2008) in maize, by Hall et al. (2007) in *Arabidopsis*, by Ansel et al. (2008) in yeast, and by a number of workers using the *Drosophila melanogaster* DGRP lines (Harbison et al. 2013; Ayroles et al. 2015; Huang et al. 2015; Morgante et al. 2015; Sørensen et al. 2015). DGRP characters showing among-line variation in σ_E^2 included morphological, physiological, and behavioral traits, as well as expression-level traits (i.e., mRNA levels).

Second, at least some major genes are known to impact trait variances. The classic example is the *HSP90* heat shock gene, which codes for a chaperone that ensures proper protein folding. Changes in expression level of this gene can influence the environmental variance for a number of traits (Rutherford and Lindquist 1998; Sangster et al. 2008b; Geiler-Samerotte et al. 2016), but the results are somewhat mixed. When *HSP90* expression is suppressed, it causes increased developmental instability in *Arabidopsis* (Queitsch et al. 2002) but not in *Drosophila* (Milton et al 2003). Variants at the *FTO* (fat mass and obesity-associated protein) locus display different residual variances for body mass in humans (Yang et al. 2012a). However, because *FTO* effects were scored in segregating populations, it is unclear whether its variance effect is due to differences in environmental or genetic background sensitivity. Chapter 14 further examines forces impacting σ_e^2, while Chapter 18 discusses how to map **variance QTLs (vQTLS)**, QTLs that influence σ_E^2. WL Chapter 18 examines selection on σ_e^2 in more detail.

Repeatability

Because the variance among repeated measures on the same individual can only be due to environmental causes (or measurement errors), information on the within-individual component of variance can provide some insight into the possible magnitude of the environmental variance for a trait. For studies with a temporal component, there are two types of within-individual variation. The first, emphasized above, is the variance among measures of homologous characters expressed within individuals at the same time. The second is the variance in expression of a character across time, e.g., temporal variability in the weight of an individual. This second type of variation is not always logically distinct from the first. For example, for animals that are growing, the right and left measures of a bilaterally symmetrical trait may fluctuate on a daily basis. Moreover, one may question whether repeated measures made at wide intervals of time in the same individual should be treated as the measures of the same or different traits (Part II Chapter 26). These issues aside, it follows that an upper-bound estimate of the genetic variance of a trait is provided by

$$\text{Var}(G)_{max} = \text{Var}(z) - \text{Var}(e_w) \tag{6.6}$$

where $\text{Var}(z)$ is an estimate of the total phenotypic variance for the trait, and $\text{Var}(e_w)$ is an estimate of the within-individual component of variance.

Measurement error, solely a function of the investigator, will always inflate estimates of the within-individual component of variance relative to its true value, but because it also contributes to the total phenotypic variance, it cancels out in Equation 6.6. Nevertheless, measurement error is still a problem, because we ordinarily would like to know the fraction of the true phenotypic variance that is accounted for by $\text{Var}(G)_{max}$. Thus, it is desirable to have an estimate of $\text{Var}(z)$ that is free of measurement error. As noted above, for morphological characters that have reached their final stage of development (and are not subject to wear) or for measures made on preserved samples, a simple correction for measurement error can be acquired from repeated measures of the same character. However, the problem is less tractable with behavioral and physiological traits. For such characters, repeated measures may differ because of temporal organismal changes as well as because of measurement error, rendering it essentially impossible to factor out the component of variance due to measurement error.

The expected value of $\text{Var}(G)_{max}$ is necessarily greater than the total genetic variance for the trait because it includes the among-individual component of the special environmental effects variance (σ^2_{ea}) as well as any variance due to general environmental effects (σ^2_E). Thus, letting $\text{Var}(e_m)$ denote the variance associated with measurement error, the **repeatability**

$$r = \frac{\text{Var}(z) - \text{Var}(e_w)}{\text{Var}(z) - \text{Var}(e_m)} \tag{6.7}$$

provides an upper-bound estimate of the broad-sense heritability of a trait (H^2), i.e., of the fraction of the total phenotypic variance that is genetic in basis. The degree to which r exceeds H^2 depends on the magnitude of ($\sigma^2_{ea} + \sigma^2_E$) relative to σ^2_{ew}. In the unlikely event that all of the environmental variance is in the within-individual component, and measurement-error variance has been removed from the denominator, then Equation 6.7 provides an unbiased estimate of H^2. A large value of r offers the *possibility* that a considerable amount of the character variance is genetic, while a small value of r informs us that environmental variance dominates. Further, because $h^2 \leq H^2$, it also follows that $h^2 \leq r$.

Repeatability is often computed as the correlation between two repeated measures (z_1 and z_2) on the same individuals (Falconer and Mackay 1996),

$$r_F = \frac{\text{Cov}(z_1, z_2)}{\text{SD}(z_1)\,\text{SD}(z_2)} \tag{6.8}$$

However, because the variance resulting from measurement error is contained in the denominator of this expression, r_F is downwardly biased, a problem because we are trying to obtain an upper bound on H^2. For a lucid account of the statistical procedures used to estimate repeatability and of common mistakes encountered in the literature, see Lessells and Boag (1987).

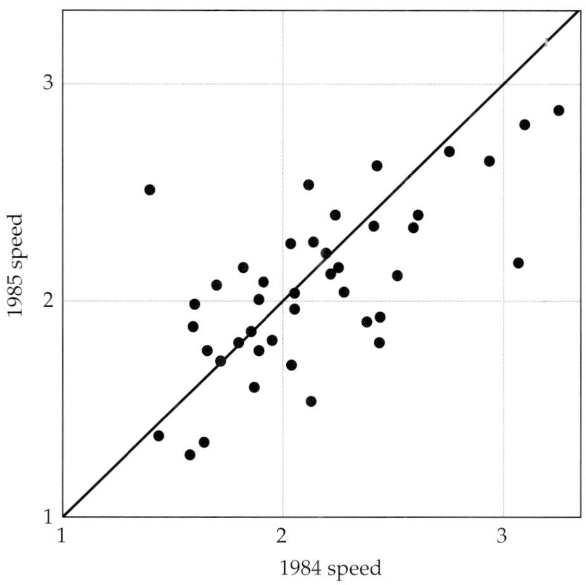

Example 6.3. In an attempt to determine the reliability of single measures of sprint speed as an assessment of performance, Huey and Dunham (1987) looked at the correlation between measures (Falconer's method; Equation 6.8) on wild-caught lizards (*Sceloporus merriami*) in two consecutive years. The repeatability was quite high ($r_F \simeq 0.70$), despite injuries and changes in reproductive condition of the animals between years. Thus, because measurement error could deflate this estimate, 70% or more of the phenotypic variance in running speed in the study population could be a consequence of genetic differences among individuals.

On the other hand, in the sagebrush lizard (*Sceloporus graciosus*), repeatabilities of various aspects of push-up and head-bob displays average only 0.16 (Martins 1991). In this study, ten or more measures were made on each individual over a period of five weeks, and $\text{Var}(G)_{max}$ was estimated by analysis of variance, as the among-individual component of variance. Assuming that measurement error is of minor importance in this study, and noting that the within-individual variance is likely to be greater over longer time spans such that $\text{Var}(z)$ is underestimated, these behavioral traits cannot have very high broad-sense heritabilities.

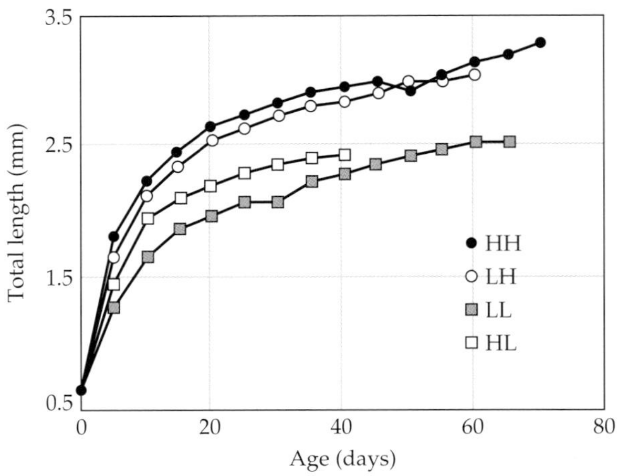

Figure 6.2 Growth trajectories for members of a clone of *Daphnia pulex* raised under various nutritional conditions. The first letter of the key indicates the diet of the mother (H = high, and L = low, food conditions); the second letter refers to the diet of the measured progeny. (From Lynch and Ennis 1983.)

GENERAL ENVIRONMENTAL EFFECTS OF MATERNAL ORIGIN

It is difficult to conceive of a metazoan organism for which some form of maternal effect is not a potential source of variation. The quality of postnatal care in many vertebrates has obvious effects on many aspects of the offspring phenotype. Other types of maternal effect, such as egg or endosperm quantity, are more subtle but nevertheless important (Schaal 1984; Roach and Wulff 1987; Mousseau and Fox 1998).

Because of maternal effects, the phenotypic composition of a population can depend greatly on the ecological setting in the previous generation. Consider the results shown in Figure 6.2 for an experiment in which mothers of a single clone of *Daphnia* were grown on high and low nutritional levels and their progeny were raised at each of these two conditions, to yield four treatments. While the growth trajectory of offspring raised on high food depended only slightly on the maternal environment, low-food progeny whose mothers were well fed grew to substantially larger sizes than those whose mothers were food-limited. The message of this example is that, unless a study population is raised in the environment of interest for at least a generation prior to analysis, one runs the risk that the observed phenotypes may be more a product of the past, than the current, environment.

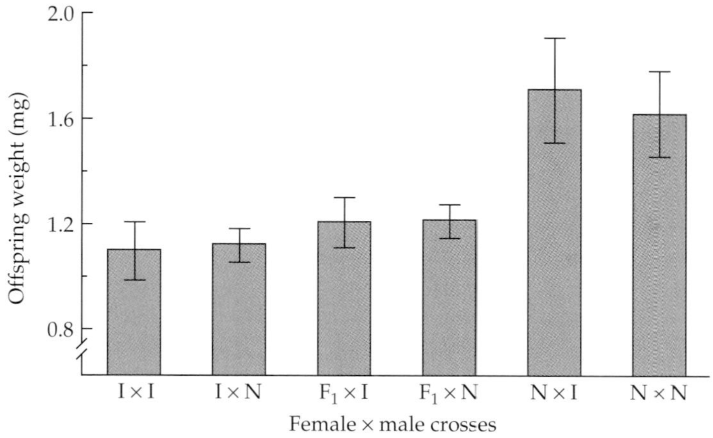

Figure 6.3 Mean offspring sizes at hatching for crosses involving Illinois (I) and North Carolina (N) populations of mosquito fish (*Gambusia affinis*). F_1 refers to a hybrid mother. Vertical lines denote ± 2SE. (From Reznick 1981.)

Although it is useful to think of maternal effects as environmental sources of variation from the standpoint of the recipients (the progeny), it does not follow that the variance of such effects always has an environmental basis. It is well known, for example, that considerable genetic variance exists for milk production in mammals. In plants, the seed coat, which influences dispersal and dormancy in many species, is under the genetic control of the mother but not the father. In the snail, *Lymnaea peregra*, the direction of shell coiling is determined entirely by the maternal genotype, with dextrality being completely dominant (Freeman and Lundelius 1982).

A rather striking example of a genetically based maternal effect is provided by Reznick's (1981, 1982) work on the mosquito fish, *Gambusia affinis*. Because the eggs in this species are fully provisioned prior to fertilization, size at birth depends primarily on the maternal phenotype and hardly at all on the genes inherited through the father. The results of various crosses involving small-egged Illinois fish and large-egged North Carolina fish are shown in Figure 6.3. Note that the mean size at birth of hybrid progeny is purely a function of the maternal parent. This also holds when F_1 females are back-crossed to the parental lines. Here, offspring size is intermediate, as expected, because the F_1 females contain genes from both parental populations.

The relative importance of maternal effects can vary with the age of the mother. Working with highly inbred lines of guinea pigs, Wright (1926) found that the incidence of spotting of the fur increased by approximately 20% with maternal age, while the incidence of polydactyly (extra digits) exhibited a nearly fivefold decline. A dramatic change in the area of daughter fronds accompanies maternal age in duckweed (Ashby and Wangermann 1954). Because the units of comparison in all of these studies are genetically identical, the results can only be explained as a change in environmental sensitivity with maternal age.

Age-specific maternal effects can also influence the genetic properties of populations. For example, in *Drosophila melanogaster*, the genetic variance for bristle number increases with maternal age (Beardmore et al. 1975), and the same is true for caudal-fin ray numbers in guppies (Figure 6.4). However, an investigation in pines revealed little influence of maternal age on the components of variance in the progeny (Lints and Baeten 1981).

In insects, a number of cases are known in which sexually transmitted microorganisms have a pronounced influence on progeny phenotypes and parental reproductive performance (Hoffmann and Turelli 1988; Stouthamer et al. 1993; Moran and Baumann 1994; Wade and Chang 1995; Stevens et al. 2001; Floate et al. 2006; Werren et al. 2008). For example, nearly half of the wild-caught individuals of the parasitic wasp *Nasonia vitripennis* carry one of three sex-ratio distorter organisms (Werren et al. 1981, 1986). One of these,

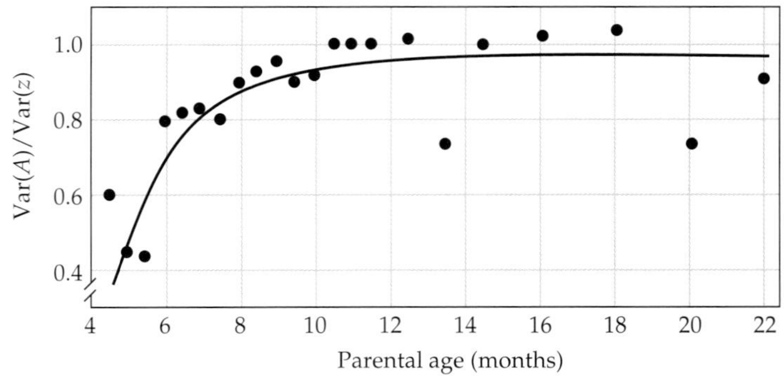

Figure 6.4 Estimated fractions of the phenotypic variance attributable to additive gene action for caudal fin ray number in a laboratory population of guppies. Cohorts of progeny were obtained from groups of parents of various ages. (From Beardmore and Shami 1976.)

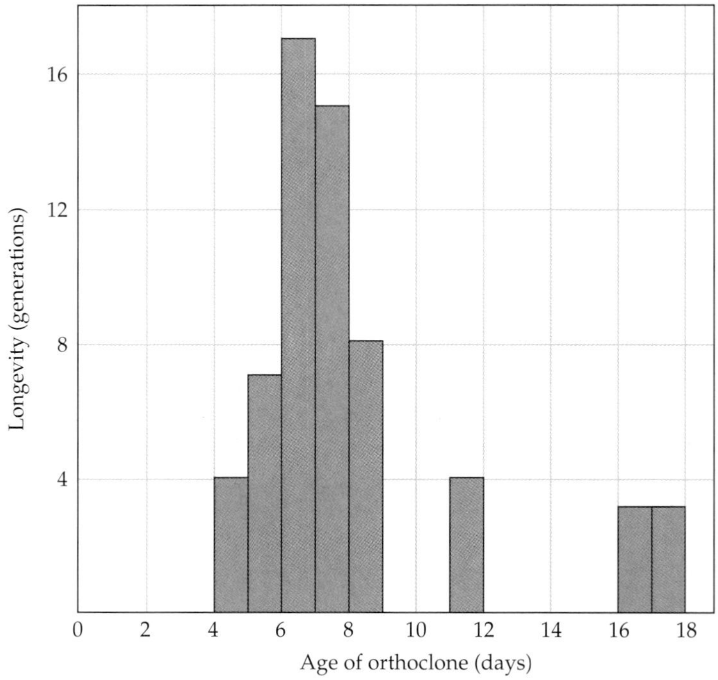

Figure 6.5 The number of generations until extinction as a function of the age at reproduction of orthoclones of the rotifer *Philodina citrina* (Lansing 1947, 1948). The nth orthoclone is a line in which each generation is started from progeny produced by mothers of age n. (After Lints 1978.)

psr (paternal sex ratio), is paternally inherited and causes all-male families, while *msr* (maternal sex ratio) is maternally inherited and results in nearly pure female families. A third, *sk* (son-killer), is maternally and contagiously transmitted and causes death of male eggs.

Once one accepts the significance of maternal effects, the question arises as to whether such effects are transmissible over more than a single generation. A fair amount of work on this subject appears in the gerontological literature. To demonstrate the cumulative effects of maternal age on longevity, Lansing (1947, 1948) worked with the clonal rotifer, *Philodina citrina*. He produced a series of **orthoclones** by propagating successive generations with progeny from mothers of constant age. All of the lines died out eventually, but young and old orthoclones went extinct most rapidly (Figure 6.5). Lansing found that extinction could

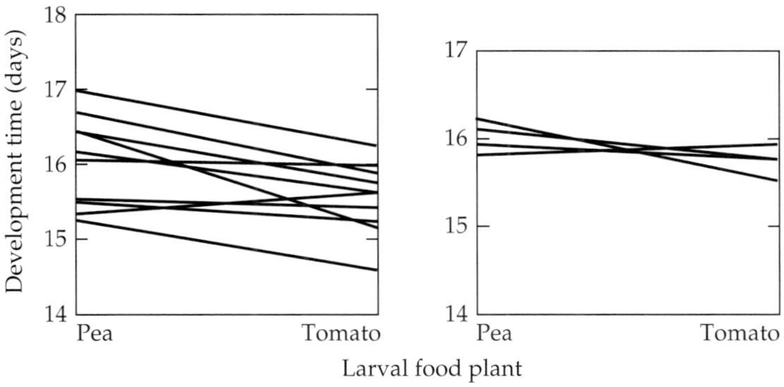

Figure 6.6 The response of development time in the leaf-mining insect *Liriomyza sativae* to a change in larval food plant. In the absence of genotype × environmental interactions, there is a constant difference between genotypes over environments (Equation 6.1b), which result in parallel lines (identical slopes) for the genotypes. The departures from this pattern seen in the figure indicate the presence of G×E. **Left:** Paternal half-sib families of one population. **Right:** Four population means. Differences in the slopes, which result from G×E, are significant for half-sib families, but not for population means. (From Via 1984.)

be avoided in the late-age orthoclones by allowing them to reproduce at younger ages. Thus, the mechanism of the cumulative age effect was most likely cytoplasmic in basis. Cumulative, reversible maternal age effects, now known as **Lansing effects,** have been studied in a number of other organisms, with several characters other than longevity, and with mixed results (Lints 1978; Finch 1990). Aside from these types of studies, however, there is a glaring absence of data on multigenerational transmission of environmental effects. Intuition may suggest that such effects are unlikely to be of significance, but empirical observation would provide a more convincing argument.

GENOTYPE × ENVIRONMENT INTERACTION

The existence of genotype × environment interaction in a population indicates that different genotypes respond to environmental change in different ways. In extreme cases, the ranking of genotypes may be altered by a shift in the environment. These problems are of great concern to breeders of economically important plants and animals because substantial genotype × environment interaction necessitates the development of locally adapted breeds. They are also at the heart of many studies of species adaptations, although few such studies have ever been couched formally in terms of genotype × environment interaction.

A field experiment with the leaf-mining fly, *Liriomyza sativae*, a serious pest of vegetable crops, further illustrates the principal aspects of genotype × environment interaction. Via (1984) obtained animals from adjacent cowpea and tomato fields, produced half-sib families in the lab, and then monitored their performance on both plants Matings were restricted to pairs taken from the same field. Figure 6.6 (left) illustrates the response of the half-sib family means to the larval food plant. In general, the larvae develop more rapidly on tomato, and the variance between families is approximately equal on both hosts. However, the nonparallel responses indicate the existence of genotype × environment interaction. Some families even developed more rapidly on cowpea than on tomato (such a switch in the ranking of genotypes over environments is called a **crossover interaction**). Figure 6.6 (right) illustrates that the mean responses of four *Liriomyza* populations to treatment are not greatly different. Thus, while the within-population analysis indicates that host specialization could evolve in this species, it has not occurred, possibly because of the close proximity of the field sites and the lack of migration barriers between them.

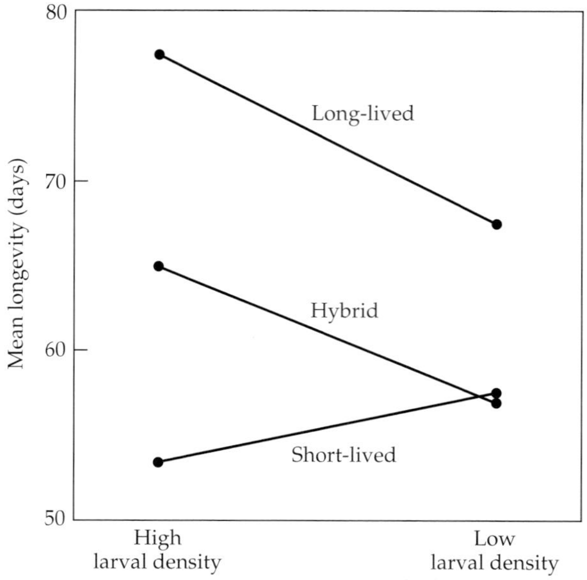

Figure 6.7 The response of adult longevity to density treatment for lines of *Drosophila melanogaster* selected for short and long life spans and for their F_1 hybrids. Genotype × environment interaction is indicated by the nonparallel response of the short-lived line. (From Clare and Luckinbill 1985.)

A second example of genotype × environment interaction involves a selection experiment on longevity in *Drosophila melanogaster* (Clare and Luckinbill 1985). Selection for increased and decreased life span was imposed for about 30 generations, and then the two divergent lines and their F_1 hybrids were assayed for longevity under two conditions—high and low larval density. Under high densities, the same conditions under which selection was practiced, the genes for longevity appear to behave in a perfectly additive manner, the mean phenotype of the hybrids being intermediate to that of the parents (Figure 6.7). However, at low densities, the genes for short life span appear to be completely dominant to those for high life span.

In both of these examples, it was possible to make some inference as to the existence of genotype × environment interaction because members of the same genetic groups were evaluated under well-defined treatments. In the case of field studies, however, individuals cannot usually be assigned to discrete environmental categories. This does not mean that genotype × environment interaction is not important, only that it is often unmeasurable. Unless discrete treatments are employed in an experiment, any genotype × environment interaction will be confounded with the environmental source of variance. Part II Chapters 27 and 28 consider the analysis of G × E in considerable detail.

7

Resemblance Between Relatives

Genetics is when you look like your father. Environment is when you look like your father's best friend. G. Rosa

In the previous two chapters, we found that the phenotypic variance of a trait can theoretically be partitioned into a number of genetic and environmental components. However, the significant practical issue of how these components can be estimated remains. The key to this matter, first pointed out by Fisher (1918) and Wright (1921b), is the fact that various genetic and environmental sources of variance contribute differentially to the resemblance between different types of relatives.

Assuming for the time being an absence of genotype $\times$ environment interaction, and recalling Equation 6.1, let $z_x = G_x + E_x + e_x$ and $z_y = G_y + E_y + e_y$ be the phenotypic values of two members of a particular relationship, such as parent and offspring. As in previous chapters, G, E, and e denote genotypic values, general environmental effects, and special environmental effects, respectively. The phenotypic covariance between relatives x and y thus becomes

$$\begin{aligned}\sigma_z(x, y) &= \sigma[(G_x + E_x + e_x), (G_y + E_y + e_y)] \\ &= \sigma_G(x, y) + \sigma_{G,E}(x, y) + \sigma_{G,E}(y, x) + \sigma_E(x, y)\end{aligned} \tag{7.1a}$$

Because the special environmental effects are random residual deviations, they are uncorrelated among individuals and do not contribute to the resemblance between relatives, i.e., $\sigma_e(x, y) = 0$. The middle two terms, $\sigma_{G,E}(x, y) + \sigma_{G,E}(y, x)$, refer to the covariance of the genotypic value of one member of the pair and the general environmental effect of the other, while the final term, $\sigma_E(x, y)$, is the covariance between general environmental effects. Experiments can often be designed so that all three covariance terms involving E have expected values equal to zero (Part II Chapters 22–25). For now, we ignore the issue of genotype–environment covariance, i.e., we assume $\sigma_{G,E}(x, y) = \sigma_{G,E}(y, x) = 0$. This assumption reduces Equation 7.1a to

$$\sigma_z(x, y) = \sigma_G(x, y) + \sigma_E(x, y) \tag{7.1b}$$

The genetic covariance between relatives, $\sigma_G(x, y)$, merits special attention. Such covariance is a natural consequence of relatives inheriting copies of the same genes. As in the case of the genetic variance, the genetic covariance between relatives can be partitioned into components attributable to additive, dominance, and various epistatic effects. Each term consists of one of the familiar components of genetic variance (Chapter 5) weighted by a coefficient that describes the number of shared genes in pairs of relatives. These coefficients are the initial focus of our attention. Once they are understood, it is a relatively simple step to use the results in Chapter 5 to derive a general expression for the genetic covariance between relatives.

We first consider the ideal situation in which mating is random and loci are unlinked and in gametic-phase equilibrium. The complications that are introduced with linkage, gametic-phase disequilibrium, and assortative mating will then be evaluated. Some of these complications make for difficult reading, but they are realities that cannot be ignored. Next, the environmental causes of the resemblance between relatives will be discussed. This does not fully complete the picture, as still other factors such as sex-linkage, maternal genetic effects, and inbreeding may be of substantial importance in particular cases, but we leave most of our discussion of these to later chapters. Finally, we provide a broad overview of the concept of heritability, a central parameter in many quantitative-genetic formulations.

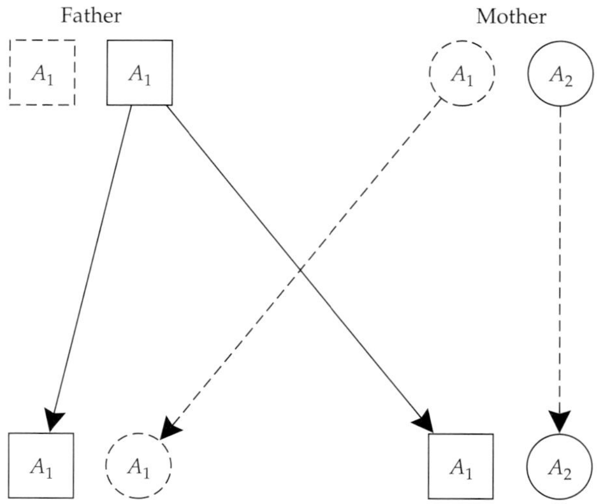

Figure 7.1 The transmittance of genes of two parents to two offspring. All A_1 alleles are identical in *state*. However, of the two A_1 genes carried by the offspring on the left, only the one in the square is identical by *descent* with the A_1 allele carried by the offspring on the right.

MEASURES OF RELATEDNESS

Many relatedness measures have found their way into the population-genetic and sociobiological literature (Wright 1922a; Cotterman 1940; Malécot 1948; Denniston 1974; Jacquard 1974; Orlove and Wood 1978; Michod and Hamilton 1980; Grafen 1985). Not all of these play a central role in quantitative-genetic formulations, but they all share two essential features.

First, relatedness can only be defined with respect to a ***specified frame of reference.*** Technically speaking, all members of a species or population are related to each other to some degree for the simple reason that they contain copies of genes that were present in some remote ancestor in the phylogeny. We avoid this problem by letting the reference population be the base of an observed pedigree. If, for example, no individuals further back than the parental generation have been observed, the usual procedure is to treat that generation as the base and to assume that its members are unrelated. As we detail in Chapter 8, marker information allows this restriction to be significantly relaxed.

Second, all measures of relatedness are based upon the concept of **identity by descent** (**IBD**). Genes that are identical by descent are *direct* descendents of a specific gene carried in some ancestral individual (specifically, the two copies trace back to a single ancestral DNA molecule). The distinction between identity by descent and **identity in state** (also **alike in state, AIS**) is critical. Two genes that have identical nucleotide sequences but have descended from different copies in the reference population are identical in state but not by descent. On the other hand, genes that are identical by descent are necessarily identical in state, barring mutation. The distinction between these two types of identity is clarified in Figure 7.1, where the parental generation is treated as the base population. Although the first offspring contains two A_1 genes, only one of them is identical by descent with the A_1 gene in the second offspring.

Our treatment of relationships is initially based on classical methods for obtaining the *predicted* degree of relationship given a known pedigree. The recent ability to score an essentially endless number of molecular markers in just about any species allows for two important advances, which are examined in Chapter 8. The first is that a known pedigree not needed, as close, to moderately distant, relationships can be reasonably estimated using sufficiently dense marker data. Given that "known" pedigrees can still contain errors (e.g., incorrectly assigned paternities), marker data also provide a check of pedigree data quality. The second is the important distinction between the predicted (or **expected**) value of a

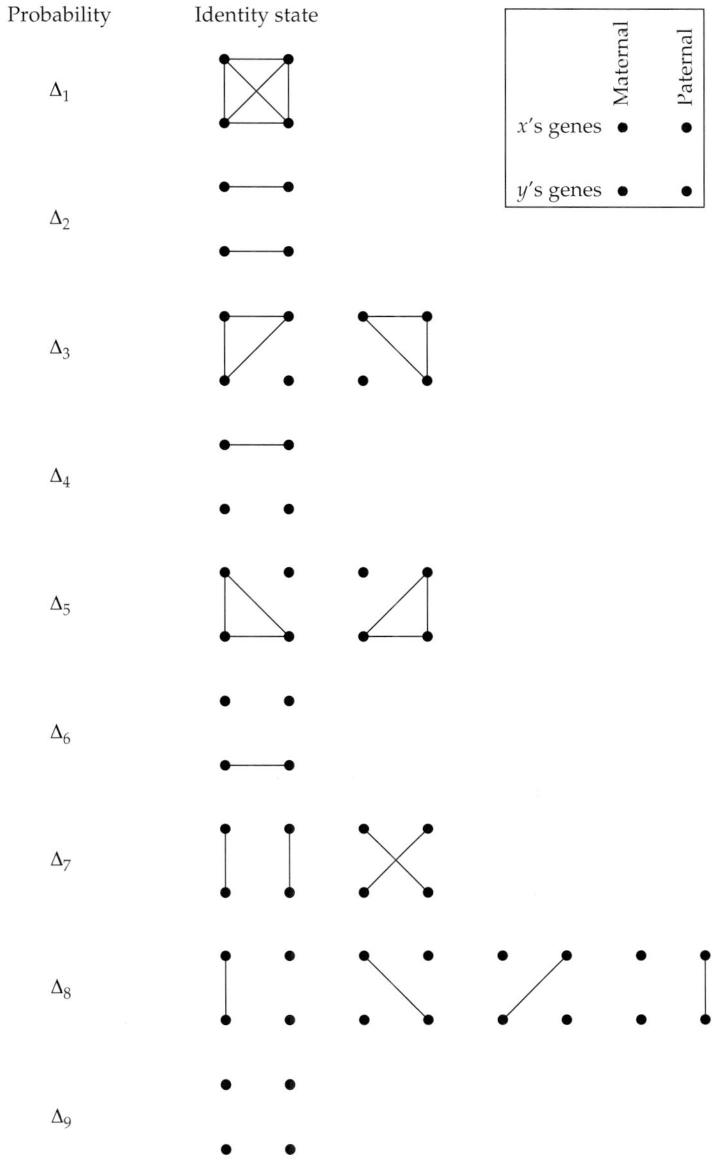

Figure 7.2 The 15 possible states of identity by descent for a locus in individuals x and y, condensed into nine classes, denoted as **identity states**. Alleles that are identical by descent are connected by lines, with horizontal lines implying that the individual is inbred (states 1–6).

relationship for specific types of relatives, and the actual, or **realized**, relationship that occurs for a particular pair of such relatives. A pedigree estimate provides the former, a marker-based estimate the latter. Hence, even with a known and fully accurate pedigree, high-quality marker data provide additional information on the true values of relationships.

Coefficients of Identity

Consider a single locus in two diploid individuals. For the four genes involved, there are 15 possible configurations of IBD due to the fact that identity may exist *within* as well as *between* individuals (Gillois 1964; Figure 7.2). Individuals that contain pairs of alleles that are identical by descent are said to be **inbred**. If we ignore the distinction between maternally and paternally derived genes, the 15 possible configurations reduce to nine **identity states**. These range from state 1 in which the two individuals are inbred and share

Table 7.1 Summary of the properties of each of the nine identity states given in Figure 7.1. Here, Θ_i is the coefficient of coancestry associated with that identity class.

Class	Θ_i	x inbred	y inbred
1	1	Yes	Yes
2	0	Yes	Yes
3	1/2	Yes	No
4	0	Yes	No
5	1/2	No	Yes
6	0	No	Yes
7	1/2	No	No
8	1/4	No	No
9	0	No	No

an allele that is identical by descent (so that all four alleles are identical by descent) to state 9 in which none of the four alleles are identical by descent. Many situations exist in which further simplification can be justified. For example, in large random-mating populations, the probability of the first six identity states is essentially zero.

Associated with each of the nine identity states are probabilities, Δ_1 to Δ_9, which Jacquard (1974) called **condensed coefficients of identity** (Figure 7.2). These coefficients provide a complete description of the probability distribution of IBD between single loci of two individuals. The values that they take on depend on the relationship. For example, suppose that x is a parent and y its offspring, and that neither is inbred. Then, because an individual inherits one and only one gene from its parent, $\Delta_8 = 1$ and all other $\Delta_i = 0$. For noninbred full sibs, there is a 0.5 probability of inheriting the same paternal gene and, independently, a 0.5 probability of inheriting the same maternal gene. Thus, $\Delta_7 = 0.25$ (both pairs of genes are identical by descent), $\Delta_8 = 0.50$ (one pair of genes is identical by descent), and $\Delta_9 = 0.25$ (there is no IBD at the locus), and all other $\Delta_i = 0$.

Coefficients of Coancestry (Θ) and Inbreeding (f)

Suppose now that single alleles are drawn randomly from individuals x and y. The probability that these two alleles are identical by descent, Θ_{xy}, is called the **coefficient of coancestry.** (In the literature, Θ_{xy} is sometimes referred to as the **coefficient of consanguinity, coefficient of kinship,** or **coefficient de parente**). Table 7.1 gives the Θ values associated with each of the nine identity states. For example, either allele from x in a state-3 relationship is IBD to one of the alleles in y, but not the other. Hence, $\Theta = 1/2$ for this class. The same logic holds for state 5, with the roles of x and y reversed. Likewise, a similar argument shows that Θ also equals 1/2 for state 7. Recalling that Δ_i is the probability that a randomly chosen gene pair from relatives x and y corresponds to the ith identity state, Table 7.1 implies that

$$\Theta_{xy} = \Delta_1 + \frac{1}{2}(\Delta_3 + \Delta_5 + \Delta_7) + \frac{1}{4}\Delta_8 \tag{7.2a}$$

Here, we have simply weighted each identity state by its associated Θ value.

An important feature of Θ_{xy} follows by considering a hypothetical offspring (z) of x and y. By the above definition, Θ_{xy} is the probability that the two alleles at a locus in individual z are identical by descent, as the two alleles in z are the result of drawing a single random allele from each of its parents, x and y. The probability that two alleles at a locus in z are IBD is usually called Wright's (1922a) **inbreeding coefficient**, f_z. Thus, ***an individual's inbreeding coefficient is equivalent to its parents' coefficient of coancestry,***

$$f_z = \Theta_{xy} \tag{7.2b}$$

Note that Equation 7.2b dispels a common misunderstanding about inbreeding. If we cross two fully inbred parents (x and y), the resulting offspring are inbred *only* when $\Theta_{xy} > 0$.

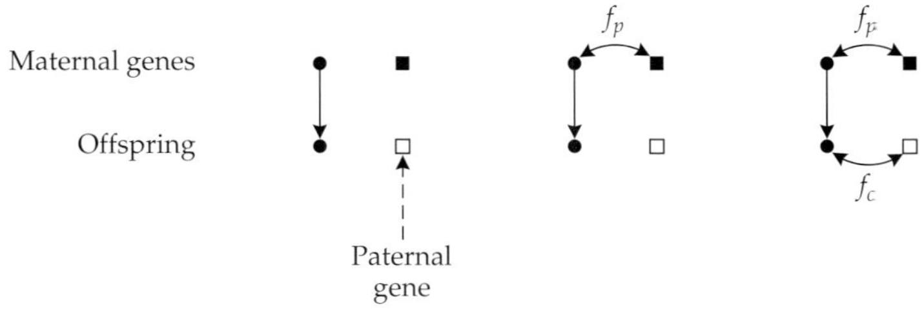

Figure 7.3 The identity of genes by descent for a parent (mother) and her offspring. Circles and squares represent, respectively, maternally and paternally derived genes. **Left:** The mother (denoted by p in the text) is not inbred and her mate (denoted by q) is not a relative (so the offspring is not inbred). **Center:** The mother is inbred ($f_p > 0$) but unrelated to her mate ($\Theta_{pq} = 0$). **Right:** In addition to the mother being inbred, she is related to her mate, so that her offspring is also inbred, with $f_0 = \Theta_{pq}$.

If the two parents are *unrelated*, then their offspring are outbred even when the parents are fully inbred.

We now proceed by example to demonstrate how estimates of Θ_{xy} are derived. The first problem to be tackled is the coefficient of coancestry of an individual with itself, Θ_{xx}. This may seem like a nonsensical task. However, we will soon see that Θ_{xx} is an essential element of all coancestry estimates. Denote the two genes carried by individual x as A_1 and A_2, and then randomly draw a gene from the locus, replace it, and randomly draw another. Θ_{xx} is the probability that the two genes drawn are identical by descent. There are four ways, each with probability $1/4$, in which the genes can be drawn: A_1 both times, A_1 first and A_2 second, A_2 first and A_1 second, and A_2 both times. If two A_1 genes are drawn, they must be identical by descent because they are copies of the same gene. The same applies to a draw of two A_2 genes. Thus, provided that genes A_1 and A_2 are not identical to each other by descent, then Θ_{xx} is simply $(1/4)(1) + (1/4)(1) = 1/2$. We should, however, recognize the possibility that individual x is inbred, in which case the probability that the gene A_1 is identical by descent with the gene A_2 is f_x. Thus, a general expression for the coefficient of coancestry of an individual with itself is

$$\Theta_{xx} = \frac{1}{4}(1 + f_x + f_x + 1) = \frac{1}{2}(1 + f_x) \tag{7.3a}$$

The impact of inbreeding is thus to *increase the coefficient of coancestry*, which will be a common theme throughout. We can rearrange Equation 7.3a to yield the useful identity

$$f_x = 2\Theta_{xx} - 1 \tag{7.3b}$$

Note that if we *self* x, the level of inbreeding in its offspring o is $f_o = \Theta_{xx}$. Hence, from Equation 7.3a we obtain the recursion equation for the effect of selfing on the level of inbreeding,

$$f(t+1) = \frac{1}{2}\,[1 + f(t)] \tag{7.3c}$$

A slightly more complicated situation arises in calculating the coefficient of coancestry between a parent and its offspring. In order to simplify the discussion, we will call the parent (p) of interest the mother, but the same results apply to fathers, provided the locus is autosomal. We first consider the situation in which neither the mother nor her offspring (o) are inbred, i.e., the mother's parents are unrelated, and she is unrelated to her mate (who we denote by q). In that case, of the four ways in which single genes can be drawn from the mother and the child, only one involves a pair that is identical by descent (Figure 7.3, left). Therefore, $\Theta_{po} = 1/4$. Suppose, however, that the mother is inbred (Figure 7.3, center), so that the probability that both of her alleles are identical by descent is f_p. This is the same

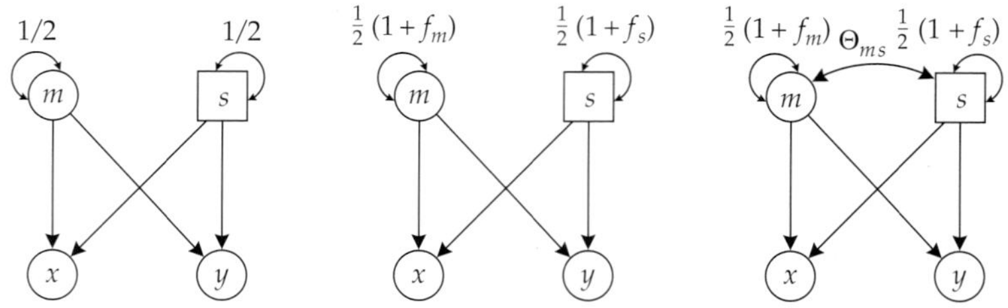

Figure 7.4 Path diagrams (Appendix 2) for analyzing the probability that random genes from two full sibs (x and y) are identical by descent. The path coefficients along single-headed arrows are always equal to $1/2$. **Left:** The parents, m and s, are neither related nor inbred. **Center:** The parents are unrelated, but inbred. **Right:** In addition to being inbred, the parents are related with coefficient of coancestry Θ_{ms}, so that the offspring are inbred to level $f_0 = \Theta_{ms}$.

as the probability that the maternal gene inherited by the offspring is identical by descent with the maternal gene not inherited. The probability of drawing such a gene combination is $1/4$. Therefore, inbreeding in the parent inflates Θ_{po} to $(1+f_p)/4$. With complete inbreeding ($f_p = 1$), both parental alleles are identical by descent, increasing Θ_{po} to $1/2$. Finally, we allow for the possibility that the parents of o are related, so that the offspring is inbred with coefficient $f_o = \Theta_{pq}$ (Figure 7.3, right). It is now necessary to consider the implications of drawing a paternally derived gene from the offspring, the probability of which is $1/2$. Because f_o is equivalent to the probability that maternally and paternally derived genes are identical by descent, the additional parent–offspring identity induced by inbreeding is $f_o/2$. In summary, the most general expression for the coefficient of coancestry for a parent and offspring is

$$\Theta_{po} = \frac{1}{4}(1 + f_p + 2f_o) \tag{7.4a}$$

Again note that the impact of inbreeding is to inflate Θ. Recalling Equations 7.2b and 7.3b, we can alternatively express Equation 7.4a as

$$\Theta_{po} = \frac{1}{4}\left[1 + (2\Theta_{pp} - 1) + 2\Theta_{pq}\right] = \frac{1}{2}\left(\Theta_{pp} + \Theta_{pq}\right) \tag{7.4b}$$

This result foreshadows a simple recursion method for obtaining the Θ values over a very complex pedigree, which will be examined shortly. Often in the literature, Θ_{po} is simply considered to be $1/4$. It should now be clear that this implicitly assumes the absence of inbreeding, which inflates the value of Θ.

We now move on to the coefficient of coancestry of two individuals that share the same father and mother (full sibs). We assume a species with separate sexes so that the mother and father are different individuals, and we again start with the simplest situation, progressively allowing the parents to be inbred and/or related (Figure 7.4). For the analysis of full sibs, as well as for more complicated relationships, the method of path analysis (Appendix 2) provides a useful tool. The elements in Figure 7.4 no longer represent gametes (as in Figure 7.3), but rather individuals.

Let m represent the mother, s the sire (father; we use s in place of f to avoid confusion with use of the latter for inbreeding), and x and y their two offspring. When the parents are neither inbred nor related, there are two paths by which the same gene can be passed to both x and y: $x \leftarrow m \rightarrow y$ and $x \leftarrow s \rightarrow y$. Because both paths have identical consequences, we will simply consider the first of them. First, we note that the probability that both x and y receive the same maternal gene is $1/2$. This is the coefficient of coancestry of the (noninbred) mother with herself, Θ_{mm}, and is represented by the double-headed arrow in the figure. Second, we note that the probability of randomly drawing a maternal gene from individual x is $1/2$, and that the same is true for individual y. Thus, the probability of drawing two maternal genes, identical by descent, one from x and the other from y, is $\Theta_{mm}(1/2)^2 = 1/8$. Adding the same contribution from the paternal path, $x \leftarrow s \rightarrow y$, we obtain the coefficient of coancestry $\Theta_{xy} = 1/4$. Path analysis (Appendix 2) provides a simple way to obtain this result. First, set the path coefficients on all of the single-headed arrows in Figure 7.4 equal to $1/2$. Then, note that the contribution of a path to a correlation between two variables is equal to the product of the path coefficients and the correlation coefficient associated with the common factor (in this case, Θ_{mm} or $\Theta_{ss} = 1/2$).

We now allow for the possibility that the parents are inbred with inbreeding coefficients f_m and f_s, a condition that inflates the coefficient of coancestry of an individual with itself. This is the only necessary change for the path diagram in Figure 7.4 (center). There are still only two paths that lead to genes identical by descent in x and y, and their sum is

$$\Theta_{xy} = \frac{1}{4}\left(\Theta_{mm} + \Theta_{ss}\right) = \frac{1}{4}\left(\frac{1+f_m}{2} + \frac{1+f_s}{2}\right) = \frac{1}{8}\left(2 + f_m + f_s\right) \tag{7.5a}$$

Finally, we allow for the possibility that m and s are related, such that the probability of drawing two genes (one from each of them) that are identical by descent is Θ_{ms}. It is then necessary to consider two additional paths between x and y: $x \leftarrow m \leftrightarrow s \rightarrow y$ and $x \leftarrow s \leftrightarrow m \rightarrow y$ (Figure 7.4, right). Again taking the coefficients on the single-headed arrows to be $1/2$, it can be seen that each of these two new paths makes a contribution $\Theta_{ms}/4$ to Θ_{xy}, where (Equation 7.2b) Θ_{ms} is the level of inbreeding in the offspring, f_o. Adding these to our previous result, we obtain a general expression for the coefficient of coancestry of full sibs,

$$\Theta_{xy} = \frac{1}{8}\left(2 + f_m + f_s + 4f_o\right) \tag{7.5b}$$

which reduces to $\Theta_{xy} = 1/4$ under random mating. Again, we see that inbreeding inflates Θ. Using Equations 7.2b and 7.3b, we can express Equation 7.5b entirely in terms of coefficients of coancestry,

$$\Theta_{xy} = \frac{1}{8}\left[2 + (2\Theta_{mm} - 1) + (2\Theta_{ss} - 1) + 4\Theta_{ms}\right] = \frac{1}{4}\left(\Theta_{mm} + \Theta_{ss} + 2\Theta_{ms}\right) \tag{7.5c}$$

Up to now, we have assumed an autosomal locus. The rules change when the locus of interest is sex-linked. Assuming the male is the heterogametic sex, a male cannot receive an X-linked gene from his father, whereas fathers pass on their X-linked genes to daughters with probability one. Females have two X chromosomes and pass each one on to sons or daughters with the usual probability of one-half. Thus, the protocol for obtaining the coefficient of coancestry for an X-linked locus is similar to that used for autosomal loci except that path coefficients leading from fathers to daughters are replaced by a 1 and those leading from fathers to sons are replaced by a 0. For all paths containing two consecutive males, the probability that X-linked genes are identical by descent is zero.

Pedigree-based Estimates of Θ: Chain-counting

The preceding path technique is readily extended to more distant relationships and more complicated schemes of relatedness (Wright 1921b). The coefficient of coancestry is always the sum of a series of two types of paths between x and y. The first type of path leads from a *single common ancestor* to the two individuals of interest, while the second type passes through *two remote ancestors that are related to each other*. Neither type of path is allowed to pass through the same ancestor more than once. This procedure is summarized by the following equation

$$\Theta_{xy} = \sum_i \Theta_{ii}\left(\frac{1}{2}\right)^{n_i - 1} + \sum_j \sum_{j \neq k} \Theta_{jk}\left(\frac{1}{2}\right)^{n_{jk} - 2} \tag{7.6}$$

where n_i is the number of individuals (including x and y) in the path leading from common ancestor i, and n_{jk} is the number of individuals (including x and y) on the path leading from two different but related ancestors, j and k. Formal proof of this equation can be found in Boucher (1988). Equation 7.6 is often called the method of **chain-counting**.

Example 7.1. One of the first pedigrees to which Wright (1922a) applied his theory of inbreeding was that of Roan Gauntlet, an English shorthorn bull. In the following figure, rectangles and ovals refer to bulls and cows, respectively.

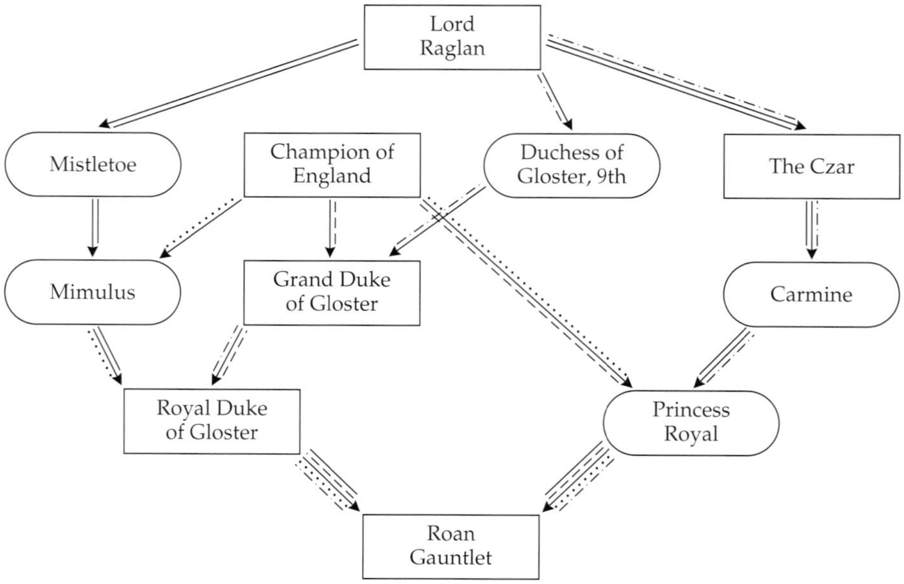

We wish to compute the coefficient of coancestry of the Royal Duke of Gloster and Princess Royal. This is the same as the inbreeding coefficient of their son, Roan Gauntlet. The four possible paths by which alleles identical by descent can be inherited by the Royal Duke and Princess Royal are indicated by the coded lines adjacent to the arrows in the pedigree. Two of these paths (the dotted and the dashed lines) contain four individuals and two (the solid and the dot-and-dash lines) contain seven. Thus, assuming that the remote ancestors, Lord Raglan and Champion of England, are not inbred (so that for both, $\Theta_{ii} = 1/2$) and not related, the coefficient of coancestry of the Royal Duke and Princess Royal is $[2(1/2)^4 + 2(1/2)^7] = 0.141$. This is a slightly closer relationship than that for half sibs (for which $\Theta = 0.125$). Relative to the base population, the alleles at 14% of the autosomal loci in the offspring, Roan Gauntlet, are expected to be identical by descent.

What is the coefficient of coancestry of the Royal Duke of Gloster and Princess Royal with respect to X-linked loci? Of the four paths between these two individuals, only the one traced by the dotted line does not contain consecutive males. Champion of England passes on his X chromosome to each of his daughters, Mimulus and Princess Royal, with probability 1. Mimulus passes that chromosome on to the Royal Duke of Gloster with probability $1/2$. The probability of drawing a specific X chromosome from the Royal Duke is 1 (because he is a male) and from Princess Royal is $1/2$. The coefficient of coancestry for X-linked loci is therefore $1 \cdot 1/2 \cdot 1 \cdot 1/2 = 1/4$. This is substantially greater than the coefficient for autosomal loci over the same path, which is $(1/2)^4 = 1/16$.

Pedigree-based Estimates of Θ: Tabular Method

Chain-counting (Equation 7.6) is a reasonable approach for computing the coefficient of coancestry between a specific pair of relatives in a known pedigree. One caveat is that it is easy to miss paths in complex pedigrees, which results in underestimating Θ. Further, using chain-counting to compute *all* $n[(n-1]/2$ pairwise Θ values in a pedigree of n individuals is very inefficient. An alternative approach that easily accommodates both of these concerns is a simple algorithm called the **tabular** or **recursive** method (Crudent 1949; Emik and Terrill 1949). Here one starts with the **founding members** of a pedigree (individuals with both parents unknown) and then proceeds down the pedigree (i.e., considering all their descendants) in a recursive fashion (Example 7.2). The algorithm is as follows:

(i) ***Founding individuals of the pedigree are assumed to be unrelated and not inbred*** (although this can also be accommodated). These founders are assigned values of $\Theta_{ii} = 1/2$.

(ii) ***Any unknown parent of any future individual is assumed to be unrelated to all others in the pedigree and not inbred,*** and they are also assigned a value of $\Theta_{ii} = 1/2$.

(iii) To compute the self coancestry for individual i, let S_i and D_i denote their sire and dam (father and mother). Applying Equations 7.2b and 7.3a yields

$$\Theta_{ii} = (1 + \Theta_{S_i\,D_i})/2$$

which reduces to 1/2 when $\Theta_{S_i,D_i} = 0$, namely when the parents are unrelated.

(iv) The coancestry between individuals i and j (who are descendants of the founders) is computed as

$$\Theta_{ij} = (\Theta_{i\,S_j} + \Theta_{i\,D_j})/2 = (\Theta_{j\,S_i} + \Theta_{j\,D_i})/2$$

The logic here follows directly from the definition of the coefficient of coancestry, as follows. With probability 1/2, a random allele from j will have come from its sire, and the probability that this allele is IBD with the random allele from i is just $\Theta_{i\,S_j}$, while with probability 1/2 the allele from j is from its mother, giving the IBD probability as $\Theta_{i\,D_j}$. By symmetry, we can reverse the roles and consider random alleles from j and the sire or dam of i. Note that Equation 7.4b (the coefficient of coancestry between a parent p and its offspring o) is just a special case of this expression. Take $i = p$ (the parent) and j as its offspring, with $S_j = p$ and $D_j = q$. Substituting yields $\Theta_{po} = (\Theta_{pp} + \Theta_{pq})/2$.

This recursive approach was extended by Karigl (1981) to compute all nine condensed coefficients of identity (Figure 7.2) for arbitrary pedigrees.

Example 7.2. Again consider Wright's shorthorn pedigree (Example 7.1). Note that Lord Raglan and Champion of England are founders of the pedigree (both parents are unknown, as indicated by the lack of any paths to them from ancestors). Likewise, four other individuals (Mistletoe, Duchess of Gloster, Czar, and Carmine) have an unknown parent (dams in all cases), indicated by a path to them from only one ancestor. We can compactly express all of the pedigree relationships in tabular form, starting with the founders, as follows:

Individual	Number	Sire (S_i)	Dam (D_i)
Lord Raglan	1	unknown	unknown
Champion of England	2	unknown	unknown
Mistletoe	3	1	unknown
Duchess of Gloster, 9th	4	1	unknown
The Czar	5	1	unknown
Mimulus	6	2	3
Grand Duke of Gloster	7	2	4
Carmine	8	5	unknown
Royal Duke of Gloster	9	7	6
Princess Royal	10	2	8
Roan Gauntlet	11	9	10

Applying the recursive rules, for the founders we have $\Theta_{11} = \Theta_{22} = 1/2$, and $\Theta_{12} = 0$. For Mistletoe (i = 3), S_3 = 1 and D_3 = ?, giving $\Theta_{33} = (1 + \Theta_{1?})/2 = 1/2$, where $\Theta_{1?}$ between her sire (1) and (unknown) dam is zero by rule (ii). Likewise,

$$\Theta_{31} = [\Theta_{1S_3} + \Theta_{1D_3}]/2 = (\Theta_{11} + \Theta_{1?})/2 = (1/2)/2 = 1/4$$

$$\Theta_{32} = [\Theta_{2S_3} + \Theta_{2D_3}]/2 = (\Theta_{21} + \Theta_{2?})/2 = 0$$

For Duchess of Gloster (i = 4), S_4 =1 and D_4 = ?, giving $\Theta_{44} = (1 + \Theta_{1?})/2 = 1/2$, while

$$\Theta_{34} = [\Theta_{3\,S_4} + \Theta_{3\,D_4}]/2 = [\Theta_{3\,1} + \Theta_{3\,?}]/2 = (1/4 + 0)/2 = 1/8$$

Proceeding in this fashion down the pedigree, and recalling that $\Theta_{ij} = \Theta_{ji}$, computes all of the other Θ values. Note that this algorithm (with the pedigree coded by associating each pedigree member with the two index numbers for its parents) is easily programmable. Typically, all of the pairwise coancestries for a pedigree are compactly presented as a matrix (Chapter 8). In this case, the result is a (symmetric) matrix with 11 rows and columns, with the element in row i and column j corresponding to Θ_{ij}. The upper left part of this matrix for the first four relatives here is

$$\begin{array}{cccccc} & 1 & 2 & 3 & 4 & \cdots \\ 1 & 1/2 & 0 & 1/4 & 1/4 & \cdots \\ 2 & 0 & 1/2 & 0 & 0 & \cdots \\ 3 & 1/4 & 0 & 1/2 & 1/8 & \cdots \\ 4 & 1/4 & 0 & 1/8 & 1/2 & \cdots \\ \vdots & \vdots & \vdots & \vdots & \vdots & \ddots \end{array}$$

Predicted versus Realized Values of Θ

While Equation 7.2a gives the *expected* value of Θ_{xy} for the pair of relatives x and y, there is a *distribution* of potential Θ values around this mean. Consider two full sibs from noninbred, and unrelated, parents. While the expected value of Θ_{xy} is 1/4, for any particular locus, the actual value is either 1/2, 1/4, or 0, corresponding, respectively, to sharing 2, 1, or 0 IBD alleles (identity states 7, 8, and 9) with associated probabilities of 1/4, 1/2, and 1/4. Hence, random sampling of Θ values over all loci in the sib pair results in a *distribution of values* around the expected value of 1/4 (or, equivalently, a distribution around the expected sharing of 50% IBD alleles). As a result, some full-sib pairs are more similar ($\Theta > 1/4$) and others are less similar ($\Theta < 1/4$) than the expected pedigree-based value. While this concept was fully appreciated in the days of pedigree-based estimates (Fisher 1949; Franklin 1977; Stam 1980), researchers at the time had no ability to actually score the realized amount of IBD sharing. With dense SNP data, we now have this ability (Figure 7.5). As we will see in several later chapters, exploitation of this variance in relationships is an important tool in modern quantitative genetics.

From Table 7.1, the *variance* (at a single locus) in Θ for a specific pair of relatives is given by

$$\sigma^2(\Theta_{xy}) = \sum_{i=1}^{9} \Delta_i(xy)\left(\Theta_i - \overline{\Theta}_{xy}\right)^2 \tag{7.7a}$$

where $\overline{\Theta}_{xy}$ denotes the value given by Equation 7.2a, Θ_i is the value for identity state i, and $\Delta_i(xy)$ is the probability of the relatives x and y being in identity state i. From Table 7.1, there are only four possible Θ_i values: 0 for states 2, 4, 6, and 9; 1/4 for state 8; 1/2 for states 3, 5, and 7; and 1 for state 1. Hence,

$$\begin{aligned}\sigma^2(\Theta_{xy}) =& \overline{\Theta}_{xy}^{\,2}(\Delta_2 + \Delta_4 + \Delta_6 + \Delta_9) + (1/4 - \overline{\Theta}_{xy})^2\Delta_8 \\ &+ (1/2 - \overline{\Theta}_{xy})^2(\Delta_3 + \Delta_5 + \Delta_7) + (1 - \overline{\Theta}_{xy})^2\Delta_1\end{aligned} \tag{7.7b}$$

For only outbred relatives (identity states 7–9), this reduces to

$$\sigma^2(\Theta_{xy}) = \overline{\Theta}_{xy}^{\,2}\Delta_9 + (1/4 - \overline{\Theta}_{xy})^2\Delta_8 + (1/2 - \overline{\Theta}_{xy})^2\Delta_7 \tag{7.7c}$$

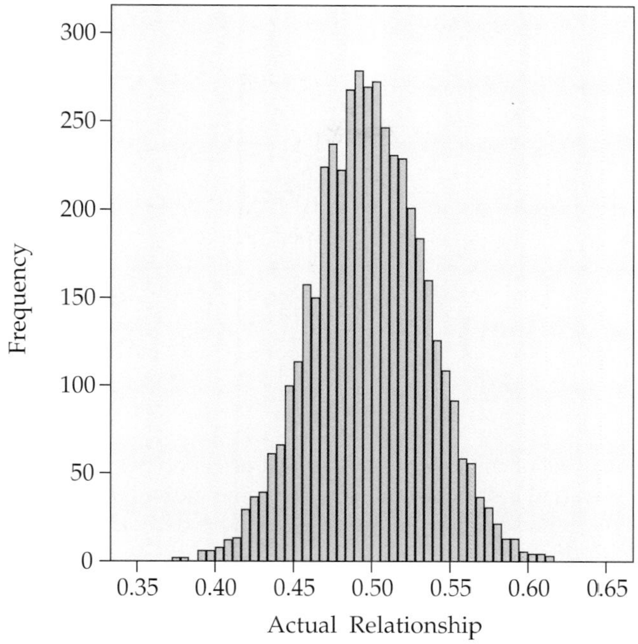

Figure 7.5 Empirical distribution of the actual fraction of shared IBD alleles for 4400 pairs of human full sibs. The realized value of Θ is simply half this value. (After Visscher et al. 2006.)

Hence, there is variation in Θ_{xy} for any pair of relatives, *unless* one of the $\Delta_i = 1$. This only occurs for clones ($\Delta_1 = 1$, $\overline{\Theta}_{xy} = 1$) and, in outbreds, for parent–offspring pairs ($\Delta_8 = 1, \overline{\Theta}_{xy} = 1/4$). The latter occurs because a parent and its offspring (in the absence of inbreeding) always share *exactly* one IBD allele.

An important point is that Equation 7.7b is the variance in Θ associated *with a particular locus*. We then average the locus-specific Θ values over the entire genome. If one has n independently assorting loci, then the genome-wide variance would be $\sigma^2\left(\Theta_{xy}\right)/n$. In reality, the situation is much more complex, as while chromosomes assort independently, sites within a chromosome are correlated due to linkage. We examine this in Chapter 8.

Example 7.3. Consider the (single-locus) variance in Θ for noninbred half and full sibs. For noninbred full sibs, $\Delta_7 = 1/4, \Delta_8 = 1/2, \Delta_9 = 1/4$, and $\overline{\Theta} = 1/4$. Equation 7.7c yields

$$\begin{aligned}\sigma^2\left(\Theta_{xy}\right) &= (1/4)^2(1/4) + (1/4 - 1/4)^2(1/2) + (1/2 - 1/4)^2(1/4)\\ &= 1/32 = 0.03125\end{aligned}$$

for a standard deviation of 0.177 and a coefficient of variation of CV = 0.177/0.25 = 0.707. Turning to half sibs, $\Delta_8 = 1/2, \Delta_9 = 1/2, \overline{\Theta} = 1/8$, and Equation 7.7c yields

$$\sigma^2\left(\Theta_{xy}\right) = (1/8)^2(1/2) + (1/4 - 1/8)^2(1/2) = 1/64 = 0.015625$$

for a standard deviation of 1/8, and a CV of 1. This increase in CV as relatives become more distant is a general phenomenon (Hill and Weir 2011).

The Coefficient of Fraternity, Δ_{xy}

Up to now we have been considering the identity of *single alleles* by descent. Another useful measure is the probability that *single-locus genotypes* of two individuals are identical by descent. The formulation of such a measure, which we denote as Δ_{xy}, is attributable to Cotterman (1954) and was called the **coefficient of fraternity** by Trustrum (1961). The problem is

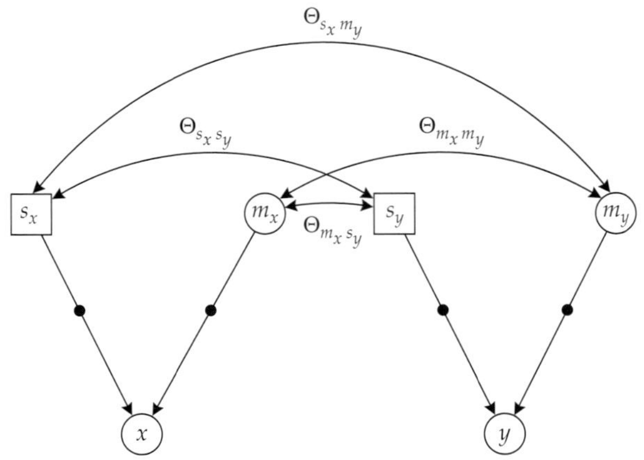

Figure 7.6 The analysis of the identity by descent of genotypes of individuals x and y. s_x and s_y represent sires (fathers) (which may be the same individual) of x and y, respectively, whereas m_x and m_y represent their mothers. Double-headed arrows between two parents represent coefficients of coancestry.

set out in Figure 7.6. Here we denote the mothers of individuals x and y as m_x and m_y, and the fathers (sires) as s_x and s_y. The coefficients of coancestry $\Theta_{m_x m_y}$, $\Theta_{m_x s_y}$, $\Theta_{s_x m_y}$, and $\Theta_{s_x s_y}$ provide measures of the probability of drawing genes identical by descent from all four combinations of parents.

There are two ways by which the genotype of x can be identical by descent with that of y: (i) the gene descending from m_x is identical by descent with that descending from m_y *and* that from s_x is identical by descent with that from s_y, or (ii) the gene from m_x is identical by descent with that from s_y, and that from s_x is identical by descent with that from m_y. Thus, the coefficient of fraternity is obtained as

$$\Delta_{xy} = \Theta_{m_x m_y}\Theta_{s_x s_y} + \Theta_{m_x s_y}\Theta_{s_x m_y} \tag{7.8}$$

In terms of the condensed coefficients of identity, $\Delta_{xy} = \Delta_1 + \Delta_7$, which reduces to Δ_7 in the absence of inbreeding. Ovaskainen et al. (2008) noted that Equation 7.8 is an approximation, requiring that the four probabilities (i.e., the Θ_{ij}) determining Δ_{xy} be independent. This is usually not a serious problem unless the pedigree is highly inbred.

Two examples will suffice to illustrate the use of this equation. First, consider the situation when x and y are full sibs, in which case the mothers are the same individual $(m_x = m_y = m)$, as are the fathers $(s_x = s_y = s)$. Equation 7.8 then reduces to

$$\Delta_{xy} = \Theta_{mm}\Theta_{ss} + \Theta_{ms}^2 \tag{7.9a}$$

If the parents are unrelated, then $\Theta_{ms} = 0$; and if the parents are not inbred, then $\Theta_{mm} = \Theta_{ss} = 1/2$. Substituting these values into the above expression, we obtain $\Delta_{xy} = 1/4$. Expressed in terms of parent (f_s, f_m) and offspring (f_o) levels of inbreeding, Equation 7.9a becomes

$$\Delta_{xy} = \frac{1}{4}\left[(1+f_s)(1+f_m)\right] + f_o^2 \tag{7.9b}$$

showing that any amount of inbreeding inflates the value above $1/4$.

Now consider the case of paternal half sibs, in which case the fathers are the same individual $(s_x = s_y = s)$, but the mothers are different. Now,

$$\Delta_{xy} = \Theta_{m_x m_y}\Theta_{ss} + \Theta_{sm_x}\Theta_{sm_y} \tag{7.9c}$$

Provided that the parents are unrelated and not inbred, then $\Theta_{ss} = 1/2$ and $\Theta_{m_x m_y} = \Theta_{m_x s} = \Theta_{sm_y} = 0$, which yields $\Delta_{xy} = 0$. The genotypes of two individuals cannot be

Table 7.2 Identity coefficients for common relationships under the assumption of no inbreeding, in which case Δ_1 to $\Delta_6 = 0$.

Relationship	Δ_7	Δ_8	Δ_9	Θ_{xy}	Δ_{xy}
Parent–offspring	0	1	0	$\frac{1}{4}$	0
Grandparent–grandchild	0	$\frac{1}{2}$	$\frac{1}{2}$	$\frac{1}{8}$	0
Great grandparent–great grandchild	0	$\frac{1}{4}$	$\frac{3}{4}$	$\frac{1}{16}$	0
Half sibs	0	$\frac{1}{2}$	$\frac{1}{2}$	$\frac{1}{8}$	0
Full sibs, dizygotic twins	$\frac{1}{4}$	$\frac{1}{2}$	$\frac{1}{4}$	$\frac{1}{4}$	$\frac{1}{4}$
Uncle (aunt)–nephew (niece)	0	$\frac{1}{2}$	$\frac{1}{2}$	$\frac{1}{8}$	0
First cousins	0	$\frac{1}{4}$	$\frac{3}{4}$	$\frac{1}{16}$	0
Double first cousins	$\frac{1}{16}$	$\frac{6}{16}$	$\frac{9}{16}$	$\frac{1}{8}$	$\frac{1}{16}$
Second cousins	0	$\frac{1}{16}$	$\frac{15}{16}$	$\frac{1}{64}$	0
Monozygotic twins (clonemates)	1	0	0	$\frac{1}{2}$	1

identical by descent if their maternally (or paternally) derived genes come from unrelated individuals. The identity coefficients for several common relationships are summarized in Table 7.2.

Example 7.4. Returning to the figure in Example 7.1, what is Δ_{xy} for x = Royal Duke of Gloster and y = Princess Royal? Designate the parents as s_x = Grand Duke of Gloster, m_x = Mimulus, s_y = Champion of England, and m_y = Carmine. Noting that Champion of England is the father of Grand Duke of Gloster and Mimulus, $\Theta_{s_x s_y} = \Theta_{m_x s_y} = (1/4)$. Counting the number of individuals in the paths of descent between the remaining two pairs of parents, $\Theta_{m_x m_y} = \Theta_{s_x m_y} = (1/2)^5$. Substituting into Equation 7.8, the probability that x and y have identical genotypes by descent at an arbitrary autosomal locus is

$$\Delta_{xy} = (1/4)(1/2)^5 + (1/2)^5(1/4) = (1/2)^6 \simeq 0.016$$

or about 6% of the (outbred) full-sib value of 0.25.

THE GENETIC COVARIANCE BETWEEN RELATIVES

The bulk of the credit goes to Fisher (1918) and Wright (1921b) for elucidating the connection between the phenotypic resemblance of relatives and the types of genetic variance in populations. However, despite the tremendous advances in these papers, a number of problems remained. Wright did not pursue nonadditive gene action beyond dominance and even then ran into difficulties. Fisher incorporated epistasis but only of the additive × additive type. In 1954, Cockerham and Kempthorne (Cockerham's major professor at the time) independently published papers outlining how the earlier results could be generalized to include any type of gene action. Both authors arrived at the same result by very different routes, but Kempthorne's approach is much simpler and is the one that we will pursue. Cockerham's name will appear many times throughout the book in other contexts.

At the outset, it must be emphasized that the simple results that will emerge below are not obtained without making a number of assumptions: (i) all of the genetic variation is attributable to diploid, autosomal loci; (ii) mating is random; (iii) all loci are unlinked and in gametic-phase equilibrium; (iv) there is no genetic variation for maternal effects; (v) genotype–environment covariance and interaction are unimportant; (vi) there is no sexual dimorphism; and (vii) selection is not operating on the population. In due course, we will relax all of these conditions.

Our first task is to decompose the total genetic covariance between relatives into fundamental components that describe the various types of gene action. We accomplish this by following the logic used in Chapter 5 to partition the genetic variance. Consider a collection of pairs of individuals all of the same type of relationship, and let x and y represent the members of a random pair. From Equation 5.5, taking things out only to the two-locus effects, the genotypic values of the two individuals may be written as

$$\begin{aligned} G_{ijkl..}(x) = \mu_G &+ [\alpha_i^x + \alpha_j^x + \alpha_k^x + \alpha_l^x + \cdots] + [\delta_{ij}^x + \delta_{kl}^x + \cdots] \\ &+ [(\alpha\alpha)_{ik}^x + (\alpha\alpha)_{il}^x + (\alpha\alpha)_{jk}^x + (\alpha\alpha)_{jl}^x + \cdots] \\ &+ [(\alpha\delta)_{ikl}^x + (\alpha\delta)_{jkl}^x + (\alpha\delta)_{kij}^x + (\alpha\delta)_{lij}^x + \cdots] + (\delta\delta)_{ijkl}^x + \cdots \\ G_{ijkl..}(y) = \mu_G &+ [\alpha_i^y + \alpha_j^y + \alpha_k^y + \alpha_l^y + \cdots] + [\delta_{ij}^y + \delta_{kl}^y + \cdots] \\ &+ [(\alpha\alpha)_{ik}^y + (\alpha\alpha)_{il}^y + (\alpha\alpha)_{jk}^y + (\alpha\alpha)_{jl}^y + \cdots] \\ &+ [(\alpha\delta)_{ikl}^y + (\alpha\delta)_{jkl}^y + (\alpha\delta)_{kij}^y + (\alpha\delta)_{lij}^y + \cdots] + (\delta\delta)_{ijkl}^y + \cdots \end{aligned} \tag{7.10a}$$

where i, j and k, l represent alleles at the first and second loci. Fisher (1918) showed that just as the different types of effects are uncorrelated within individuals, they are also uncorrelated between individuals, provided the preceding assumptions are met. Consequently, the genetic covariance between relatives can be expanded into a series of terms, each describing the covariance between the same kinds of effects in two individuals:

$$\sigma_G(x, y) = \sigma_A(x, y) + \sigma_D(x, y) + \sigma_{AA}(x, y) + \sigma_{AD}(x, y) + \sigma_{DD}(x, y) + \cdots \tag{7.10b}$$

Note that if $x = y$, Equation 7.10b reduces to Equation 5.8, the usual expression for the genetic variance.

The remaining task is to express the terms in Equation 7.10b in terms of variance components and coefficients of relationship. We will do this only for the first three terms and then give the general result. First, we evaluate the additive genetic covariance at locus 1. Because the mean value of the effects is zero by definition (Chapter 5), the covariance between x and y caused by the additive effects is equal to the expectation of the cross-product, $E[(\alpha_i^x + \alpha_j^x)(\alpha_i^y + \alpha_j^y)]$. Consider one of the four terms in the expansion, $E[\alpha_i^x \alpha_i^y]$. The two genes of interest may be identical by descent, with probability Θ_{xy}, in which case $E[\alpha_i^x \alpha_i^y] = E[\alpha_i^2]$, which is half the additive genetic variance attributable to locus 1. If the two genes are not identical by descent, then they must be distributed independently so that $E[\alpha_i^x \alpha_i^y] = [E(\alpha_i)] \cdot [E(\alpha_i)] = 0$. These same arguments can be applied to the remaining three terms in $E[(\alpha_i^x + \alpha_j^x)(\alpha_i^y + \alpha_j^y)]$. Thus, the additive genetic covariance is $4\Theta_{xy} E[\alpha_i^2]$, which is twice the additive genetic variance at the locus times the probability that randomly drawn alleles from x and y are identical by descent. Noting that this result applies to all loci and that the distributions of effects at different loci are independent under the assumptions of the model, the additive genetic covariance, obtained by summing over loci, reduces to

$$\sigma_A(x, y) = (2\Theta_{xy})\sigma_A^2 = r_{xy}\sigma_A^2 \tag{7.11a}$$

where $r = 2\Theta$ is **Wright's coefficient of relatedness**.

We now move on to the dominance genetic covariance, which for locus 1 is $E[\delta_{ij}^x \delta_{ij}^y]$. If x and y are identical by descent for both genes at this locus, then $E[\delta_{ij}^x \delta_{ij}^y] = E[\delta_{ij}^2]$, which is the dominance genetic variance attributable to locus 1. The probability of such identity is Δ_{xy}. On the other hand, if x and y do not have identical genotypes by descent, the dominance effects must be distributed independently, and hence $E[\delta_{ij}^x \delta_{ij}^y] = (E[\delta_{ij}]) \cdot (E[\delta_{ij}]) = 0$.

Table 7.3 Coefficients for the components of genetic covariance between different types of diploid relatives under the assumptions of random mating, free recombination, and gametic-phase equilibrium. To obtain the covariance expression for a particular type of relationship, multiply each variance component by its coefficient and sum. For example, the genetic covariance between half sibs is $(\sigma_A^2/4) + (\sigma_{AA}^2/16)$. Blanks indicate values of zero.

Relationship	σ_A^2	σ_D^2	σ_{AA}^2	σ_{AD}^2	σ_{DD}^2
Parent–offspring	$\frac{1}{2}$		$\frac{1}{4}$		
Grandparent–grandchild	$\frac{1}{4}$		$\frac{1}{16}$		
Great grandparent–great grandchild	$\frac{1}{8}$		$\frac{1}{64}$		
Half sibs	$\frac{1}{4}$		$\frac{1}{16}$		
Full sibs, dizygotic twins	$\frac{1}{2}$	$\frac{1}{4}$	$\frac{1}{4}$	$\frac{1}{8}$	$\frac{1}{16}$
Uncle (aunt)–nephew (niece)	$\frac{1}{4}$		$\frac{1}{16}$		
First cousins	$\frac{1}{8}$		$\frac{1}{64}$		
Double first cousins	$\frac{1}{4}$	$\frac{1}{16}$	$\frac{1}{16}$	$\frac{1}{64}$	$\frac{1}{256}$
Second cousins	$\frac{1}{32}$		$\frac{1}{1024}$		
Monozygotic twins (clonemates)	1	1	1	1	1

Again, because these arguments apply to every locus, the dominance genetic covariance, obtained by summing over all loci, is

$$\sigma_D(x, y) = \Delta_{xy}\sigma_D^2 \tag{7.11b}$$

Finally, we consider the epistatic genetic covariance caused by additive × additive effects. For any pair of loci, this involves the 16 cross-product terms in the expectation $E[\{(\alpha\alpha)_{ik}^x + (\alpha\alpha)_{il}^x + (\alpha\alpha)_{jk}^x + (\alpha\alpha)_{jl}^x\}\{(\alpha\alpha)_{ik}^y + (\alpha\alpha)_{il}^y + (\alpha\alpha)_{jk}^y + (\alpha\alpha)_{jl}^y\}]$. Because the results are the same for all terms, it is sufficient to evaluate only one of them. The term $E[(\alpha\alpha)_{ik}^x(\alpha\alpha)_{ik}^y]$ is equivalent to $E[(\alpha\alpha)^2]$ provided that two conditions hold: a random gene drawn from the first locus in x must be identical by descent with one drawn from y, and the same condition must hold at the second locus. If identity by descent does not arise simultaneously for gene pairs drawn from both loci, $E[(\alpha\alpha)_{ik}^x(\alpha\alpha)_{ik}^y] = [E(\alpha\alpha)]^2 = 0$. Now, under the assumption of gametic-phase equilibrium, the probability of identity by descent at locus 1 is independent of that at locus 2. Both probabilities are Θ_{xy} Thus, the probability of joint identity by descent is Θ_{xy}^2, and the covariance caused by additive × additive epistasis between loci 1 and 2 is $16\,\Theta_{xy}^2 E[(\alpha\alpha)^2]$. Noting that $E[(\alpha\alpha)^2]$ is one-fourth the additive × additive genetic variance for a single pair of loci, and summing over all pairs,

$$\sigma_{AA}(x, y) = (2\Theta_{xy})^2\sigma_{AA}^2 = r_{xy}^2\,\sigma_{AA}^2 \tag{7.11c}$$

All three of the evaluated terms from Equation 7.10b reduce to simple functions of identity coefficients (Θ and Δ) and a component of genetic variance, and the same is true for all higher-order terms. Due to the independent distributions of genes at different loci under the assumptions of the model, any covariance due to higher-order epistatic effects is equal to the product of the identity for each component additive effect, the probability of identity for each component dominance effect, and the corresponding variance component. Thus, the covariance attributable to additive × dominance epistasis is

$$\sigma_{AD}(x, y) = (2\Theta_{xy})\Delta_{xy}\sigma_{AD}^2 \tag{7.11d}$$

because there is one additive and one dominance effect involved, while that caused by dominance × dominance epistasis is

$$\sigma_{SD}(x,y) = \Delta_{xy}^2 \sigma_{DD}^2 \tag{7.11e}$$

because there are two dominance but no additive effects involved. Letting n be the number of additive effects and m be the number of dominance effects in a type of gene action, the expression for the covariance between relatives becomes

$$\begin{aligned}\sigma_G(x,y) &= \sum (2\Theta_{xy})^n \Delta_{xy}^m \sigma_{A^n D^m}^2 \\ &= (2\Theta_{xy})\sigma_A^2 + \Delta_{xy}\sigma_D^2 + (2\Theta_{xy})^2 \sigma_{AA}^2 \\ &\quad + (2\Theta_{xy})\Delta_{xy}\sigma_{AD}^2 + \Delta_{xy}^2 \sigma_{DD}^2 + (2\Theta_{xy})^3 \sigma_{AAA}^2 + \cdots \end{aligned} \tag{7.12}$$

Drawing from the coefficients Θ_{xy} and Δ_{xy} given in Table 7.2, explicit expressions for the genetic covariances of common types of relatives are given in Table 7.3. Although these expressions are only expanded to include two-locus epistasis, several things are immediately apparent. First, gene action involving dominance only rarely contributes to the covariance between diploid relatives. It requires that each parent of x be related to a different parent of y (i.e., a nonzero value of Δ_{xy}). Such relationships (full sibs, double first cousins [the offspring from a pair of brothers married to a pair of sisters], and monozygotic twins) are said to be **collateral**. Second, the coefficient for σ_{AA}^2 declines more rapidly with the distance of the relationship than does that for σ_A^2. As noted in Chapters 4 and 5, because the additive genetic variance is a function of all higher order types of gene action, these results should not be misconstrued to imply that the resemblance between relatives is influenced only slightly by dominance and epistatic gene action.

The most useful feature of the expressions in Table 7.3 involves their different coefficients, which permit the estimation of the different variance components from linear combinations of different observed genetic covariances between relatives. For example, ignoring higher order epistasis and environmental sources of covariance, $8 \times$ [parent–offspring covariance $-$ (2 × half-sib covariance)] has an expected value of $8[\,(\sigma_A^2/2+\sigma_{AA}^2/4)-2(\sigma_A^2/4+\sigma_{AA}^2/16)\,] = \sigma_{AA}^2$. Similarly, 2 × [(4 × half-sib covariance) $-$ (parent–offspring covariance)] has an expected value of σ_A^2. Subsequent examples in this chapter will illustrate the utility of these kinds of manipulations. With marker-based estimates of Θ and Δ, an even more refined analysis is possible. For example, Figure 7.5 showed that pairs of full sibs vary in their realized Θ value, and one can use this variation to decouple the effects of shared genes from shared environments in humans (Visscher et al. 2006; Part II Chapter 33).

THE EFFECTS OF LINKAGE AND GAMETIC-PHASE DISEQUILIBRIUM

In deriving the Kempthorne–Cockerham equation (7.12), we assumed that the constituent loci are freely recombining and in gametic-phase equilibrium. We now consider the extent to which the interpretation of observed covariances between relatives needs to be modified in the face of violations of these assumptions. Historically, we have had very little information on either linkage (LD) or (for unlinked loci) gametic-phase disequilibrium (GPD) among the (almost always unknown) loci underlying a particular trait. However, the development of dense-marker scoring systems is starting to offer a few glimpses. We can now survey general genomic patterns of LD using random SNPs (Chapter 5), but this simply informs us as to the patterns of LD among linked markers, but *not* among trait-specific alleles. In theory, one can use marker-tagged casual sites (i.e., GWAS; Chapter 20) for look at LD patterns among *allelic effects*, and this has been done (to some extent) to search for assortative mating (see below). Most of the processes that generate LD between *linked loci* also generate GPD between *unlinked loci* (WL Chapter 16), but the search for such correlations over unlinked loci has a substantial multiple-comparisons burden (facing many of the issues that arise in the search of epistasis among QTLs/SNPs; Chapters 18 and 20). Given that we are still in the exploratory phase of trait-specific LD, we present a few theoretical results to provide some guidance on the impact of LD/GPD on correlations among relatives.

The most complete analyses of this problem were developed by Gallais (1974) and Weir and Cockerham (1977), who allowed for inbreeding as well as linkage and gametic-phase disequilibrium. Neither study went beyond two-locus relationships, as even then the algebraic and notational complexities are enormous. We will take a simpler approach than these authors, first considering the consequences of linkage under the assumption of gametic-phase equilibrium and then evaluating some of the consequences of disequilibrium. In both cases, we will assume that mating is random and that all of the remaining assumptions of the Kempthorne–Cockerham model are fulfilled.

Linkage

Under the assumption of gametic-phase equilibrium, linkage influences only the epistatic components of genetic covariance, which depend upon the multilocus gene *combinations* inherited through gametes. Recall that in Equation 7.12, each component of epistatic variance is weighted by a coefficient of the form $(2\Theta_{xy})^n \Delta_{xy}^m$. The term Θ_{xy}^n is the probability that randomly drawn pairs of genes (one from x and one from y) will be simultaneously identical by descent at n loci, the 2^n coming in because of diploidy. Similarly, Δ_{xy}^m is the probability that the genotypes of x and y are simultaneously identical by descent at m loci. These definitions of the joint probability of events at multiple loci as the products of probabilities at individual loci assume that IBD is *distributed independently* among loci. However, for linked loci, ***IBD is expected to be positively correlated*** because the genes at such loci tend to be inherited together, a feature known as **identity disequilibrium**. Thus, for linked loci, we can anticipate that the multilocus coefficients $(2\Theta_{xy})^n \Delta_{xy}^m$ must be too low, to a degree depending on the recombination frequency.

A general solution to this problem requires the use of **digenic descent coefficients,** which define the probability that two nonalleles (genes from different loci) are copies of genes that were originally contained in the same gamete. Such a condition is known as **equivalence by descent.** The formal theory of digenic descent, developed in great detail by Cockerham and Weir (1968, 1973, 1977a) and Weir and Cockerham (1968, 1969, 1973, 1974, 1977), is algebraically and notationally complex. We will rely upon an approach that is less general but more transparent (Schnell 1961, 1963; Van Aarde 1975). There is no easy entry into this field, but for the adventuresome we suggest the reviews of Weir and Cockerham (1977, 1989) as a starting point.

For our purposes, it will suffice to examine an arbitrary pair of loci, A and B. Consider an individual whose two-locus genotype resulted from the fusion of A_mB_m and A_fB_f gametes (m for mother, f for father). If the recombination fraction for the two loci is c, then this individual will produce gametes in frequencies: $p(A_mB_m) = p(A_fB_f) = (1-c)/2$ and $p(A_mB_f) = p(A_fB_m) = c/2$. Two gametes randomly drawn from this individual can have four possible IBD relationships. Identity exists at both loci with probability $p(A,B) = p^2(A_mB_m) + p^2(A_fB_f) + p^2(A_mB_f) + p^2(A_fB_m)$, at neither locus with probability $p(0,0) = 2p(A_mB_m)p(A_fB_f) + 2p(A_mB_f)p(A_fB_m)$, at only the A locus with probability $p(A,0) = 2p(A_mB_m)p(A_mB_f) + 2p(A_fB_m)p(A_fB_f)$, and at only the B locus with probability $p(0,B) = 2p(A_fB_f)p(A_mB_f) + 2p(A_fB_m)p(A_mB_m)$. Letting $\lambda = (1-2c)^2$ ($\lambda = 1$ for $c = 0$, and 0 when $c = 1/2$), these gametic-identity probabilities simplify to

$$p(A,B) = p(0,0) = \frac{1+\lambda}{4} \tag{7.13a}$$

$$p(A,0) = p(0,B) = \frac{1-\lambda}{4} \tag{7.13b}$$

all of which reduce to $1/4$ with free recombination ($c = 0.5$).

As a specific application of gametic-identity probabilities, we will consider the case of full sibs, first evaluating the covariance caused by additive × additive epistasis. As noted previously, for each pair of loci, the genotypic value of each sib contains four $(\alpha\alpha)$ terms,

Table 7.4 Coefficients for the components of genetic covariance between (noninbred) relatives modified to account for linkage (first five columns) and gametic-phase disequilibrium (last two columns). Here GP represents grandparent, GC represents grandchild, $\lambda = (1-2c)^2$, and t is the generation number for the common ancestors. Blanks denote values of zero.

Relationship	σ_A^2	σ_D^2	σ_{AA}^2	σ_{AD}^2	σ_{DD}^2	$\sigma_{A,A}(0)$	$\sigma_{D,D}(0)$
Parent–offspring	$\frac{1}{2}$		$\frac{1}{4}$			$\frac{(1-c)^t}{2}$	
GP–GC	$\frac{1}{4}$		$\frac{1+\lambda^{1/2}}{16}$			$\frac{(1-c)^t}{4}$	
Great GP–great GC	$\frac{1}{8}$		$\frac{(1+\lambda^{1/2})^2}{64}$			$\frac{(1-c)^t}{8}$	
Half sibs	$\frac{1}{4}$		$\frac{1+\lambda}{16}$			$\frac{(1-c)^t}{4}$	
Full sibs, dizygotic twins	$\frac{1}{2}$	$\frac{1}{4}$	$\frac{2+\lambda}{8}$	$\frac{1+\lambda}{8}$	$\frac{(1+\lambda)^2}{16}$	$\frac{(1-c)^t}{2}$	$\frac{(1-c)^{2t}}{4}$
Uncle (aunt)–nephew (niece)	$\frac{1}{4}$		$\frac{1+\lambda(1+\lambda^{1/2})/2}{16}$			$\frac{(1-c)^t}{4}$	
First cousins	$\frac{1}{8}$		$\frac{1+(1+\lambda)^3}{128}$			$\frac{(1-c)^t}{8}$	
Double first cousins	$\frac{1}{4}$	$\frac{1}{16}$	$\frac{3+(1+\lambda)^3}{64}$	$\frac{1+(1+\lambda)^3}{128}$	$\frac{(1+\lambda)^4}{256}$	$\frac{(1-c)^t}{4}$	$\frac{(1-c)^{2t}}{16}$
Second cousins	$\frac{1}{32}$		$\frac{1+(1+\lambda)^5}{1024}$			$\frac{(1-c)^t}{32}$	
Monozygotic twins	1	1	1	1	1	$(1-c)^t$	$(1-c)^{2t}$

so there are 16 combinations of these terms between individuals. For each of the 16 combinations, the quantity of interest is the joint probability that randomly drawn A genes (one from each sib) are identical by descent and that randomly drawn B genes (one from each sib) are identical by descent. We first draw a random A gene from the two sibs. As noted earlier, these are identical by descent with probability $1/4$. Given that IBD was obtained at the A locus, we now draw the B genes (again, one from each sib). These genes may be identical by descent through two routes: they may both derive from the parent from which the A genes were drawn (with probability $1/4$) or both from the opposite parent (with probability $1/4$). If they come from the same parent as the A gene, the B genes will be identical by descent with conditional probability

$$p(A\,|\,B) = \frac{p(A,B)}{p(A)} = \frac{p(A,B)}{p(A,B)+p(A,0)} = (1+\lambda)/2$$

If they come from the opposite parent, they are identical by descent with probability $1/2$. Summing up, the probability of digenic IBD for two full sibs is $1/4 \cdot 1/4 \cdot [\,(1+\lambda)/2 + 1/2\,] = (2+\lambda)/32$. Given such identity, the contribution to the genetic covariance is $E[(\alpha\alpha)^2]$, which is equivalent to one-fourth of the additive $\times$ additive genetic variance. Thus, after taking account of all 16 pairs of additive $\times$ additive effects, the covariance between full sibs resulting from additive $\times$ additive epistasis is $16 \cdot [(2+\lambda)/32] \cdot [\sigma_{AA}^2/4] = (2+\lambda)\sigma_{AA}^2/8$. With free recombination ($\lambda = 0$), this reduces to $\sigma_{AA}^2/4$, our previous result (Table 7.3). Linkage ($\lambda > 0$) causes the additive $\times$ additive genetic variance to be inflated.

We next consider the additive $\times$ dominance covariance between full sibs. Again, from Equation 7.10a, for any pair of loci, there are 16 combinations of $(\alpha\delta)$ terms in the two sibs. If the two individuals possess an additive $\times$ dominance effect that is identical by descent, the contribution to the genetic covariance will be $\sigma_{AD}^2/4$. To evaluate the probability of such an event, we need to determine the joint probability of drawing single genes that are identical by descent at one locus and genotypes that are identical by descent at the second locus. Such identity is only possible in the eight comparisons for which the additive effects involve the same locus in x and y. As usual with full sibs, the probability of drawing two genes (one from each sib) that are identical by descent is $1/4$. Also as noted above, given that the genes drawn at one locus are identical by descent, the genes for the second locus that descended from

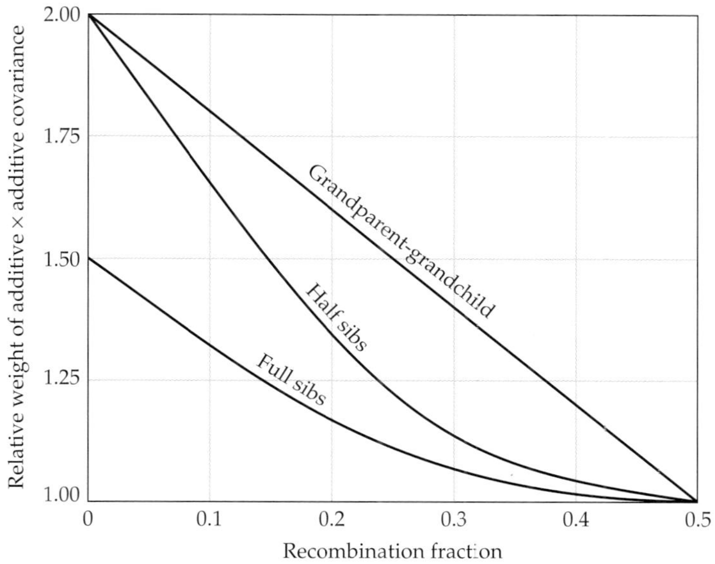

Figure 7.7 Inflation of the additive × additive covariance between relatives caused by linkage, obtained by use of the coefficients in Table 7.4.

gametes from the same parent are identical by descent with probability $(1+\lambda)/2$. The other pair of genes at the second locus are identical by descent with probability $1/2$. Summing up, the additive × dominance covariance becomes $8 \cdot (1/4) \cdot [(1+\lambda)/2] \cdot (1/2) \cdot (\sigma^2_{AD}/4) = (1+\lambda)\sigma^2_{AD}/8$, which reduces to $\sigma^2_{AD}/8$ with free recombination.

Finally, we note that covariance between the single $(\delta\delta)$ term for a pair of loci requires that both sibs inherit maternal gametes that are identical by descent at both loci, and similarly for the paternal gametes. Each of these events occurs with probability $p(A, B)$. Therefore, the dominance × dominance covariance between full sibs is $p^2(A, B)\sigma^2_{DD} = (1+\lambda)^2\sigma^2_{DD}/16$.

The coefficients for full sibs and for other common relationships are summarized in Table 7.4. The main conclusion to be drawn from these results is that *except for parent–offspring and monozygotic twin relationships,* ***linkage inflates the covariance between relatives****, unless there are no epistatic sources of genetic variance.* Linkage has no influence on the parent–offspring covariance because the two individuals always have exactly one gene identical by descent at each locus, and therefore share exactly one of the four additive × additive effects. For grandparent–grandchild and half-sib relations, relative to the situation with free recombination, the additive × additive covariance can be inflated as much as twofold with completely linked loci ($\lambda = 1$), whereas with full sibs the inflation can be no greater than 50% (Figure 7.7). It is of interest to note that whereas the expressions for the covariance of half sibs and for grandparent–grandchild are identical under free recombination, the latter is inflated to a greater degree by linkage unless linkage is complete ($\lambda = 1$). Thus, at least in principle, a comparison of these two types of covariance may shed some light on the presence of linkage for quantitative-trait loci.

Gametic-phase Disequilibrium

Further complications arise when loci are in gametic-phase disequilibrium because this can cause a covariance between the effects of genes carried in the same gamete. The problem has been addressed by Weir et al. (1980), who ignored epistasis and assumed random mating, as we do below. In Chapter 5 (Equations 5.23a and 5.23b), we saw that in the presence of gametic-phase disequilibrium, the total genetic variance can be expressed as

$$\sigma^2_G = \sigma^2_A + \sigma_{A,A} + \sigma^2_D + \sigma_{D,D} \tag{7.14}$$

where $\sigma_{A,A}$ is the contribution due to covariance of additive effects of nonalleles within gametes (the **additive disequilibrium covariance**), and $\sigma_{D,D}$ is the contribution of covariance due to dominance effects of different loci within individuals (the **dominance disequilibrium covariance**).

We now consider the genetic covariance between relatives derived from a base population displaying gametic-phase disequilibrium. The covariance between relatives attributable to equilibrium additive and dominance genetic variance (σ_A^2 and σ_D^2) is the same as given above, so it is only necessary to consider the additional contributions resulting from $\sigma_{A,A}$ and $\sigma_{D,D}$. For situations in which the study population is being maintained in a steady state of disequilibrium from generation to generation (by processes such as natural selection, migration, and/or nonrandom mating; WL Chapters 16 and 24), the modifications to the theory are simple—all of the preceding formulations still apply, except that $(\sigma_A^2 + \sigma_{A,A})$ replaces σ_A^2, and $(\sigma_D^2 + \sigma_{D,D})$ replaces σ_D^2.

The more interesting situation arises when a study population, initially in gametic-phase disequilibrium, is allowed to mate randomly in an environment in which the forces maintaining disequilibria are relaxed. In this case, the covariances between relatives change through time as recombination causes the disequilibrium components of covariance to decay towards zero. In the following, we will consider only the simple (and most common) situation in which the ancestors leading to a relationship are all members of the same generation; more general results can be found in Weir et al. (1980).

We start by considering the additive component of disequilibrium covariance. Returning to Equation 7.10a, we make the distinction that allele i at the first locus and allele k at the second locus are inherited in one gamete, whereas genes j and l are inherited in the other gamete. By using the expression for the variance of a sum, Equation 3.11b, the additive equilibrium genetic variance is defined to be

$$\sigma_A^2 = E[\alpha_i^2] + E[\alpha_j^2] + E[\alpha_k^2] + E[\alpha_l^2] \tag{7.15a}$$

On the other hand, the additive disequilibrium covariance in the base population is

$$\sigma_{A,A}(0) = 2E[\alpha_i \alpha_k] + 2E[\alpha_j \alpha_l] \tag{7.15b}$$

Note that terms involving cross-gamete expectations, e.g., $E[\alpha_i \alpha_j]$ and $E[\alpha_i \alpha_l]$, do not appear in this expression because their expected values are zero under the assumption of random mating (as there is no correlation *between* gametes). In Chapter 5, we found that in the absence of restoring forces, gametic-phase disequilibrium (*within*-gamete correlation) is reduced by a fraction c after each generation of random mating. Thus, the additive disequilibrium covariance remaining after t generations is

$$\sigma_{A,A}(t) = (1-c)^t \sigma_{A,A}(0) \tag{7.15c}$$

To obtain the covariance between relatives associated with the additive equilibrium component of variance, we take expectations of the cross-products of additive effects of genes in two individuals. A similar procedure is followed in obtaining the covariance due to $\sigma_{A,A}$, except that we now focus on pairs of genes at different loci, one in each member of the relationship,

$$\begin{aligned}\sigma_{A,A}(x,y,t) = \big(&E[\alpha_i^x \alpha_k^y] + E[\alpha_i^x \alpha_l^y] + E[\alpha_j^x \alpha_k^y] + E[\alpha_j^x \alpha_l^y] \\ &+ E[\alpha_k^x \alpha_i^y] + E[\alpha_k^x \alpha_j^y] + E[\alpha_l^x \alpha_i^y] + E[\alpha_l^x \alpha_j^y]\big)\end{aligned} \tag{7.15d}$$

For any of the terms in this equation to be nonzero, the genes in them must be equivalent by descent. The key to solving Equation 7.15d is the fact that ***the probability of equivalence by descent for genes at different loci in different gametes is the same as the probability of IBD for alleles at the same locus.*** The reason for this equivalence is that when parents produce a gamete pool, although the two loci (A and B) may be linked, under random mating the gene at the A locus in one parental gamete is independent of the gene at the B

locus in a second gamete. Thus, the probability of any term in Equation 7.15d being nonzero is Θ_{xy}.

From Equation 7.15b, we see that nonzero terms of the form $E[\alpha_i\alpha_k]$ initially have expected values equal to $\sigma_{A,A}(0)/4$. At time t, however, each of the nonzero terms in Equation 7.15d has expected value $\sigma_{A,A}(t)/4$. Because there are eight terms, the covariance between relatives due to additive disequilibrium covariance is $8\,\Theta_{xy}\,\sigma_{A,A}(t)/4$, or from Equation 7.15c,

$$\sigma_{A,A}(x,y,t) = (2\Theta_{xy})(1-c)^t\sigma_{A,A}(0) \tag{7.16}$$

where t denotes the number of generations that the common ancestors are removed from the base population. Thus, for example, the parent–offspring covariance (with $\Theta_{xy} = 1/4$) resulting from additive disequilibrium covariance is $\sigma_{A,A}(0)/2$ if the parents are members of the base population ($t = 0$), $(1-c)\sigma_{A,A}(0)/2$ if the parents are second-generation individuals ($t = 1$), and $(1-c)^2\sigma_{A,A}(0)/2$ if they are third-generation individuals ($t = 3$).

Derivation of the covariance between relatives resulting from dominance disequilibrium covariance follows the same logic just presented. However, in this case, the dominance disequilibrium covariance declines each generation to $(1-c)^2$ of its previous value. This quadratic decline occurs because dominance disequilibria are only maintained if neither of the gametes fusing to form a zygote have undergone recombination. Thus, the dominance disequilibrium covariance is

$$\sigma_{D,D}(t) = (1-c)^{2t}\sigma_{D,D}(0) \tag{7.17a}$$

and the covariance between relatives due to this disequilibrium is

$$\sigma_{D,D}(x,y,t) = \Delta_{xy}(1-c)^{2t}\sigma_{D,D}(0) \tag{7.17b}$$

For example, the covariance between full sibs resulting from dominance disequilibrium covariance is $\sigma_{D,D}(0)/4$, $(1-c)^2\sigma_{D,D}(0)/4$, and $(1-c)^4\sigma_{D,D}(0)/4$, respectively, when the parents are members of the base population, second, and third generations.

The general coefficients for the disequilibrium covariances for common types of relationships, assuming the common ancestor to be a member of generation t (where $t = 0$ denotes the base population), are given in the last two columns of Table 7.4. Unlike all of the equilibrium components of genetic variance, which always cause positive phenotypic covariance between relatives, the components resulting from disequilibrium covariance may be positive or negative depending upon whether loci are in coupling or repulsion disequilibrium.

Strictly speaking, the modified coefficients in Table 7.4 apply to a single pair of loci with recombination frequency c. These expressions could be refined further to account for the influence of linkage and/or gametic-phase disequilibrium of all loci on the covariance between relatives by summing terms over all pairs of loci, weighting each locus by its specific set of variances and covariances. However, without detailed information on the map structure of the constituent loci for a quantitative trait, such refinements would be of little practical value. An alternative approximation can be obtained by assuming that the loci underlying the trait are distributed randomly throughout the genome and using the average recombination frequency $\bar{c}$ in place of c (WL Chapter 24). Estimates of $\bar{c}$ are given for a number of species in Table 11.2.

It may be argued that linkage is of little consequence for the resemblance between relatives in organisms with high chromosome numbers because most pairs of loci will lie on different chromosomes. In this case, $\lambda \simeq 0$, and there is no inflation of the epistatic components of covariance, regardless of the magnitude of gametic-phase disequilibrium. However, regardless of the degree of linkage, gametic-phase disequilibrium is of special concern, as theoretical arguments and some empirical data suggest that it may reach significant levels for characters under selection (Chapter 5; WL Chapters 16 and 24). In this case, even with free recombination ($c = 0.5$), the contribution of disequilibrium covariance to the resemblance between relatives is nonzero. In the following section, we consider a common situation in which the build-up of positive $\sigma_{A,A}$ is inevitable, even with unlinked loci.

Example 7.5. There is one case where the probability of equivalence by descent is unequal to the coefficient of coancestry, which renders Equations 7.17a and 7.17b inappropriate. In monozygotic twins, both members of the pair are products of the same gametes, so the genes inherited in one twin are not independent of those inherited in the other twin through the same parent. Because monozygotic twins have genetic effects at all loci identical by descent, the covariance between monozygotic twins is equivalent to the expressed genetic variance in the population in the present generation. For twins whose parents are members of the base population, the probability that each ancestral gamete contributing to the twin progeny has not experienced a recombination event (between two loci of interest) is $(1 - c)$. Therefore, the covariance between monozygotic twins resulting from additive and dominance effects is

$$\sigma_A(MZ) = \sigma_A^2 + (1 - c)\sigma_{A,A}(0)$$

$$\sigma_D(MZ) = \sigma_D^2 + (1 - c)^2\sigma_{D,D}(0)$$

where the disequilibrium covariances refer to the levels in the parental generation. The sum of these two quantities is the total expressed genetic variance in the population in the twin's generation (ignoring epistasis).

ASSORTATIVE MATING

Although we have assumed a randomly mating population throughout this chapter, it is not unusual for mate choice to be based on aspects of the phenotype. Often, individuals will choose mates whose phenotypes resemble their own (**positive assortative mating**, or **homogamy**). For example, Shine et al. (2001) found positive assortative mating for size in a Canadian population of red-sided garter snakes (*Thamnophis sirtalis*). A meta-analysis by Jiang et al. (2013) of 360 species-trait combinations spanning five animal phyla found that positive assortative mating was not uncommon, with a mean correlation between mates of 0.28. They found that negative assortative mating (**disassortative mating** or **heterogamy**), wherein the phenotype correlation among mates was negative, was rare. In vertebrates, two notable examples of disassortative mating involve body color: wolves in Yellowstone national park (Hedrick et al. 2016) and white-throated sparrows (Tuttle 2003). In plants, assortative mating based on flowering time is common because individuals often produce viable pollen for only short periods of time. Likewise, self-incomparability systems could be regarded as a form of examples of heterogamy. Moura et al. (2021) offered the important caveat that patterns of assortative mating can change over time and space, and that the pooling of data over such different settings can easily result in biased estimates of the nature and strength of assortative mating.

In humans there are significant (and almost always positive) correlations between mates with respect to a number of traits, both morphological and behavioral, such as height, skin color, IQ, social status, religion, level of educational attainment, and conditions such as deafness (Spuhler 1968; Vandenberg 1972; Mare 1995; Schwartz and Mare 2005). Because a phenotypic correlation could be caused by either shared genes or shared environments, a subject of some debate among human biologists is the fraction of these associations that are entirely social (i.e., environmental) in origin. If mate choice is entirely based on social settings, there should be no genetic consequences. Conversely, if mate choice is based on phenotypes, then to the extent that phenotypes are heritable, there should also be a genetic impact for assortative mating, which we examine below.

Genomic data have recently provided some clarity on this debate (at least for some traits). Several studies (e.g., Domingue et al. 2014; Guo et al. 2014; Zou et al. 2015) showed that couples share an excessive number of anonymous SNP alleles relative to some random standard. However, as noted by Abdellaoui et al. (2014), population stratification (Chapter 20) can generate such an excess, and several of these studies did not fully account for its effect. Robinson et al. (2017) further noted that the expected theoretical correlation generated by assortative mating was sufficiently small that most of these studies were under-powered and likely detected some other signal (such as stratification) rather than one created by assortative mating. A more direct test involves using SNPs that are associated with a trait of interest (GWAS; Chapter 20), such as height. Tenesa et al. (2016) found a correlation between the SNP-predicted height of an individual with the phenotypic value of its mate. This was also seen in a more detailed analysis by Robinson et al. (2017), who observed such a correlation for other traits as well, such as BMI (body-mass index) and educational attainment. Hence, at least for some traits in humans, there is a mate-choice component entirely based on phenotype.

Many assortative-mating systems are selective, such that some phenotypes of one or both sexes have a greater ability to attract mates than others. Here, however, we confine our attention to nonselective assortative mating, i.e., we assume that all individuals have an equal opportunity to reproduce, the phenotypic distribution of their available mates being the only limitation. This is the type of assortative mating that Fisher (1918) and Wright (1921d) had in mind when they first attacked the problem from a quantitative-genetic perspective. Almost all of the basic results of the theory were produced by these two pioneers. Fisher's (1918) elaborate treatment of the subject is notoriously difficult, and some uncertainty still exists as to exactly what he meant to say (Moran and Smith 1966; Wilson 1973; Vetta and Smith 1974; Yengo and Visscher 2018).

In some ways, positive assortative mating is like inbreeding, but in other ways it is very different. A simple way to discriminate between the two is to note that inbreeding is choice of mates based on similar *genotypes*, while assortative mating is based on *phenotypes*. If there is genetic variance for the characters that are the targets of mate choice, then positive assortative mating must cause identical alleles to come together more often than in the case of random mating. This will cause an increase in the homozygosity of the population, just as inbreeding does. While prolonged inbreeding can result in a completely homozygous population, positive assortative mating will not generally induce such an extreme genetic structure. It definitely will not if: (i) there are many loci underlying the character upon which mate choice depends; or (ii) there is any variation for the trait due to sources other than additive genetic variance; or (iii) the correlation between mates is less than perfect.

It was shown in Chapter 4 (Equation 4.23c) that inbreeding (in the absence of epistasis) can inflate the genetic variance of a population up to twofold. Strong positive assortative mating has the potential to cause an even greater inflation, while disassortative mating results in a reduction of the genetic variance. The change in variance brought about by assortative mating is *primarily a consequence of a directional build-up of gametic-phase disequilibria*. Positive assortative mating increases the coupling of genes with similar effects, i.e., induces a positive covariance between allelic effects at different loci, while disassortative mating leads to the proliferation of repulsion gametes, in which positive effects at one locus are balanced by negative effects at another. Inbreeding also causes a build-up of gametic-phase disequilibrium, but there is no tendency for extreme gamete types to be formed at the expense of balanced ones, or vice versa.

Example 7.6. Direct observation of positive disequilibrium for several human traits showing assortative mating was seen by Yengo et al. (2018a). These authors started with informative SNP markers detected from previous GWAS analyses for the traits of interest (Chapter 20), and then examined the association between the number of **trait-increasing alleles** (**TIAs**) between even and odd-numbered autosomes. For each individual, the even versus odd chromosome-based trait predictor scores (weighted sums of TIAs) were obtained, and the correlation ϕ among these two scores computed. This contrast nicely controls for shared environmental effects, which should equally impact both the even and odd chromosome sets. The potential of population stratification was controlled by using the first 20 principal components of the marker matrix (this approach is examined in Example 9.13). Using a data set of ~400,000 unrelated individuals of European ancestry, ϕ values of 3.2% and 2.7%, respectively, were found for height and education attainment. These values are consistent with their predicted values under assortative mating (based on trait heritability, fraction of variation accounted for by the predictors, and the phenotypic correlation between mates; see Yengo et al. for details).

A further item of supporting evidence for disequilibrium was seen by comparing the within-individual correlation (ϕ) with the correlations between the SNP-based predictors for a married pair (r_m), the latter computed using an independent data set. Yengo et al. showed that the predicted relationship between these two metrics at assortative mating equilibrium should be $r_m \sim 2\phi$. The slope of a regression for 32 different traits of r_m on ϕ showed a slope of 1.8 (with a standard error of 0.2), consistent with this prediction.

Theory

We now take a more quantitative look at these matters, starting with the consequences of assortative mating for the additive genetic variance and heritability (WL Chapters 16 and 24 provide a more detailed development of the theory). We restrict our attention to the situation in which interactions between loci are additive, because epistasis has not yet been incorporated into the theory. We define ρ_z and ρ_g to be, respectively, the phenotypic and additive genetic correlations between mates and assume that the regression of phenotypes of mates is linear. We also suppose that there are n loci, each contributing equally to the genetic variance of the trait, and define the parameter $\gamma = 1 - [1/(2n)]$.

Starting from a random-mating base population in gametic-phase equilibrium with additive genetic variance σ^2_A, a single generation of assortative mating will shift the additive genetic variance to

$$\sigma^2_A(1) = \left(1 + \frac{\rho_z h^2}{2}\right)\sigma^2_A \tag{7.18}$$

where $h^2 = \sigma^2_A/\sigma^2_z$ is the heritability under random mating (Crow and Kimura 1970). With continued assortative mating, the variance asymptotically approaches the equilibrium

$$\widehat{\sigma}^2_A = \frac{\sigma^2_A}{1 - \gamma\widehat{\rho}_g} \tag{7.19a}$$

where $\widehat{\rho}_g = \rho_z\widehat{\sigma}^2_A/\widehat{\sigma}^2_z$ is the equilibrium genetic correlation between mates. This solution was first obtained by Wright (1921d) for unlinked loci with equivalent effects. Later, Crow and Felsenstein (1968), Bulmer (1980), and Nagylaki (1982) showed that the same result holds with linked loci with arbitrary effects, as long as n (in γ) is replaced by n_e, the effective number of loci. Thus, the general form of Equation 7.19a holds regardless of the map structure of loci. Gimelfarb (1984) proved that the equilibrium is stable. Equation 7.19a can be rewritten as

$$\frac{\widehat{\sigma}^2_A}{\sigma^2_A} = \frac{2 + \left[\sqrt{1 - 4\gamma\rho_z h^2(1-h^2)} - 1\right]/h^2}{2(1-\gamma\rho_z)} \tag{7.19b}$$

which gives the inflation of the equilibrium additive genetic variance relative to that in the random-mating base population (Figure 7.8). The difference $\widehat{\sigma}^2_A - \sigma^2_A$ is the additive disequilibrium covariance, $\sigma_{A,A}$, maintained by assortative mating, which is positive under positive assortative mating and negative under disassortative mating. Upon cessation of nonrandom mating, $\sigma_{A,A}$ decays to zero, and the additive variance returns to its random-mating value, σ^2_A.

Three points can be gleaned from Figure 7.8. First, there is an asymmetry in the response to positive and negative assortative mating. Very strong positive assortative mating can inflate the additive genetic variance to nearly $2n_e\sigma^2_A$ if h^2 is also very high. However, negative

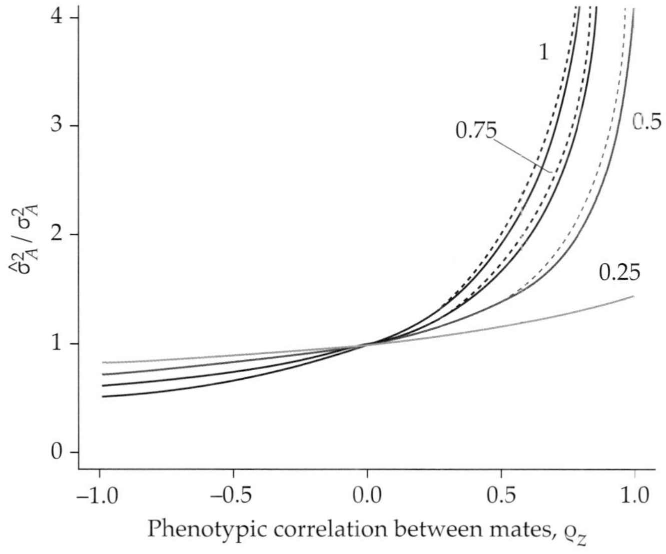

Figure 7.8 Inflation of the additive genetic variance at equilibrium under assortative mating relative to that in an otherwise identical random-mating population for various h^2 values. Solid and dashed lines refer to 10 and 20 effective loci, respectively.

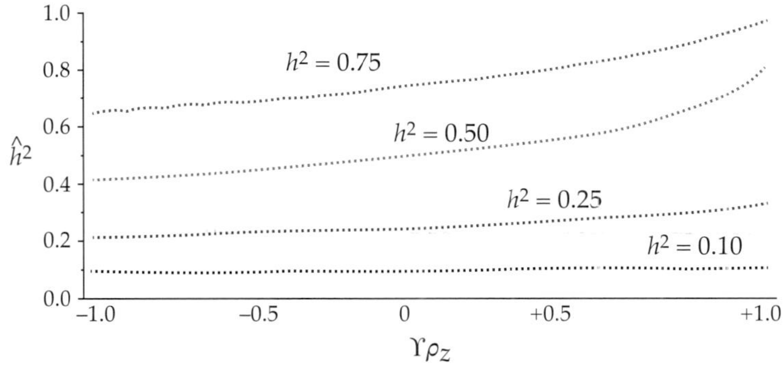

Figure 7.9 The equilibrium heritability, $\widehat{h}^2$, as a function of the heritability under random mating (h^2) and the strength of the assortative mating ($\gamma\rho_z$).

assortative mating can depress the variance to no less than $\sigma_A^2/2$. The little empirical work that has been done in this area is qualitatively consistent with this expectation (Breese 1956; McBride and Robertson 1963; Example 7.7). Second, assortative mating must be fairly strong ($\rho_z^2 \geq 0.2$) and combined with high h^2 to induce much change in the variance. Third, the effective number of loci has a negligible effect unless it is very small.

Finally, consider the equilibrium heritability, $\widehat{h}^2 = \widehat{\sigma}_A^2/\widehat{\sigma}_z^2 = \widehat{\sigma}_A^2/(\widehat{\sigma}_A^2 + \sigma_E^2)$. Following Gomez-Raya and Burnside (1990), Equation 7.19b can be used to show that

$$\widehat{h}^2 = \frac{1 - \sqrt{1 - 4\gamma\rho_z h^2(1 - h^2)}}{2\,\gamma\rho_z\,(1 - h^2\,)} \tag{7.19c}$$

Figure 7.9 plots this as a function of the heritability under random mating (h^2) and the strength of the assortative mating ($\gamma\rho_z$). As expected from their impact on the sign of $\sigma_{A,A}$

positive assortative mating increases the heritability, while disassortative mating decreases it. Again, upon return to random mating, the heritability returns to its random-mating value as $\sigma_{A,A}$ decays to zero. Equation 7.19c implies than an observed heritability seen for a trait under assortative-mating equilibrium is a *biased estimate* of its random-mating value. For example, the observed heritability for human height is around $\widehat{h}^2 = 0.8$. Assuming a phenotypic correlation of $\rho_z = 0.28$ between mates (Example 7.8), one can iteratively solve for h^2 using Equation 7.19c (taking $\gamma \simeq 1$), yielding $h^2 = 0.76$. Hence, the positive association between the effects of height alleles at different loci accounts for about 4% of the observed heritability. We show in Example 7.8 that these same parameters give a much larger change in the additive variation (around 30%). The reason for the smaller change in the heritability is that any change in the additive variance inflates (or deflates if $\rho_z < 0$) *both* the numerator and denominator of the ratio that defines the heritability.

Example 7.7. Gimelfarb (1984) performed an experiment in which lines of *Drosophila melanogaster* were artificially maintained under an absolute regimen of positive and negative assortative mating ($\rho_z \simeq \pm 1.0$). Each generation, 300 male and 300 female flies from each line were scored for abdominal bristle number, and the line was then propagated by performing 300 rank-ordered matings. The change in the phenotypic variance of abdominal bristle number over an eight generation period is shown in the following figure.

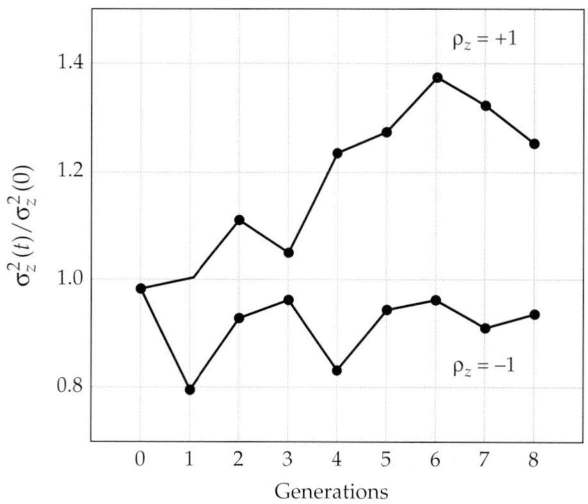

To what extent can these results be reconciled with the theory? Abdominal bristle number in *D. melanogaster* has been the subject of many quantitative-genetic studies. It appears to be nearly completely lacking in nonadditive genetic variance, and random-mating laboratory populations generally exhibit heritabilities of approximately $h^2 = \sigma^2_A/(\sigma^2_A + \sigma^2_e) = 0.5$. Assuming that the effective number of loci underlying the trait is at least five or so, then $\gamma \simeq 1$, and Equation 7.19b predicts that complete negative assortative mating should ultimately reduce the additive genetic variance to approximately 70% of the level in the base population. Thus, scaling the original phenotypic variance to be one so that $\sigma^2_A = \sigma^2_e = 0.50$, we expect complete negative assortative mating to depress the additive genetic variance to approximately $\widehat{\sigma}^2_A = 0.70 \times 0.50 = 0.35$. From the figure, it can be seen that the *phenotypic* variance declined to approximately 90% of the level in the base population. This is close to the theoretical expectation that the phenotypic variance should be 85% of the base-population level, $\widehat{\sigma}^2_z/\sigma^2_z = (\widehat{\sigma}^2_A + \sigma^2_e)/(\sigma^2_A + \sigma^2_e) = (0.35 + 0.50)/(0.50 + 0.50) = 0.85$.

Such an analysis cannot be performed for the line under positive assortative mating because it is unclear that an equilibrium phenotypic variance had been obtained. Nevertheless, a simple expectation can be pointed out for an experiment of this type. With absolute

Table 7.5 Coefficients for the additive and dominance components of the covariance between relatives for an equilibrium population undergoing assortative mating. The equilibrium heritability is $\widehat{h}^2 = \widehat{\sigma}_A^2/\widehat{\sigma}_z^2$. In the absence of assortative mating, $\rho_z = 0$, and all coefficients reduce to those in Table 7.3.

Relationship	$\widehat{\sigma}_A^2$	σ_D^2
Parent–offspring	$\frac{1}{2}(1+\rho_z)$	
Grandparent–grandchild	$\frac{1}{4}(1+\rho_z)(1+\rho_z\widehat{h}^2)$	
Great grandparent–great grandchild	$\frac{1}{8}(1+\rho_z)(1+\rho_z\widehat{h}^2)^2$	
Half sibs	$\frac{1}{4}(1+2\rho_z\widehat{h}^2+\rho_z^2\widehat{h}^2)$	
Full sibs, dizygotic twins	$\frac{1}{2}(1+\rho_z\widehat{h}^2)$	$\frac{1}{4}$
Uncle(aunt)–nephew(niece)	$\frac{1}{4}(1+\rho_z\widehat{h}^2)^2$	$\frac{1}{8}\rho_z\widehat{h}^2$
First cousins	$\frac{1}{8}(1+\rho_z\widehat{h}^2)^3$	$\frac{1}{16}(\rho_z\widehat{h}^2)^2$
Double first cousins	$\frac{1}{4}(1+3\rho_z\widehat{h}^2)$	$\frac{1}{16}$
Second cousins	$\frac{1}{32}(1+\rho_z\widehat{h}^2)^5$	$\frac{1}{64}(\rho_z\widehat{h}^2)^4$
Monozygotic twins	1	1

positive assortative mating, $\rho_z = 1.0$, and a heritability of 0.5, the expected inflation of the additive genetic variance predicted by Equation 7.19b is simply $(2n_e)^{1/2}$. Thus, if this experiment had been carried out to the point at which $\widehat{\sigma}_z^2$ had stabilized, an estimate of the effective number of factors for abdominal bristle number would have been possible.

To this point, we have said nothing about the dominance component of variance. When only two loci contribute to the trait, σ_D^2 can change with assortative mating (Reeve 1961). However, Fisher (1918) argued that if the number of loci is even moderate, the effect will be negligible, and this was later confirmed by Vetta (1976). Thus, for all practical purposes, we can assume that the dominance genetic variance is unaltered by assortative mating.

In addition to creating gametic-phase disequilibrium, assortative mating causes the genotype frequencies for the selected trait to deviate from Hardy–Weinberg expectations. Positive assortative mating leads to a *heterozygote deficit*, while negative assorative mating leads to a *heterozygote excess*. Wright (1921d) and Nagylaki (1982) showed that if the effective number of loci is large, Hardy–Weinberg deviations will be minor unless both the phenotypic correlation between mates and h^2 are very large. This implies that *the vast majority of the change in genetic variance induced by assortative mating is a consequence of gametic-phase disequilibrium* rather than Hardy–Weinberg disequilibrium, i.e., of allelic associations within, rather than between, gametes.

The covariance between relatives in a population undergoing assortative mating has been considered by Fisher (1918), Crow and Felsenstein (1968), Gimelfarb (1981), and Yengo and Visscher (2018), with the last authors also considering the impact of assortative mating on X-chromosome-coded traits. Nagylaki (1978) used path analysis (Appendix 2) to yield a particularly lucid overview for the case of additive gene action. The coefficients of the additive and dominance components of covariance are given for common types of relationships in Table 7.5. Positive assortative mating inflates the additive genetic covariance between all types of relatives, while disassortative mating depresses it. Moreover, for some types of relationships (uncle–nephew, first and second cousins), assortative mating induces dominance genetic covariance where there otherwise would be none. We emphasize that the additive variance appearing in the resemblance between relatives is $\widehat{\sigma}_A^2$, the variance in the current assortatively mating population, not σ_A^2, the variance that would exist in a random-mating population. The latter can always be obtained by rearranging Equation 7.19a. Assuming large n_e ($\gamma \simeq 1$),

$$\sigma_A^2 = \left(1 - \frac{\rho_z \widehat{\sigma}_A^2}{\widehat{\sigma}_z^2}\right) \widehat{\sigma}_A^2 \tag{7.20}$$

Example 7.8. As an example of the application of the preceding theory, we consider an early data set on height in British families (Pearson and Lee 1903). Pearson recruited college students to obtain data from approximately 1,300 families, recording whenever possible the stature of father, mother, and eldest son and daughter (ignoring offspring less than 18 years of age). This was a very large data set for the precomputer era, and it took two years to calculate the statistics by hand. The data are remarkable for their essentially normal distribution and for the linearity of the regressions between relatives.

In the figure below on the left, the histogram gives the observed data for maternal height, while the smooth curve is the fitted normal distribution. On the right, the straight line is the least-squares regression of sons' height on paternal height, the data having been pooled into one-inch size classes for paternal height.

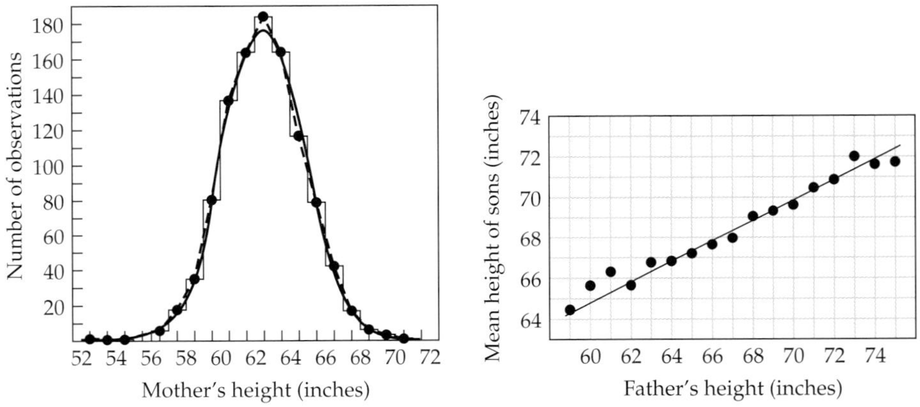

Pearson and Lee do not report the covariances between relatives, but rather the correlations, but this turns out to be quite convenient for the following analysis. The variance of height tends to be larger for males than for females, and this may be expected to influence the regressions depending upon which sexes are involved. However, because a correlation coefficient is equivalent to the covariance between two standardized variables (each with unit variance), we can ignore this problem. The data in the following table indicate that the phenotypic correlations for all four sex-specific parent–offspring combinations are quite consistent, yielding a pooled value of 0.506 ± 0.011. The three full-sib correlations are also consistent with each other and give the pooled value of 0.534 ± 0.019. Finally, we note that there is highly significant assortative mating for height, the estimated value of ρ_z being 0.280 ± 0.028. As a point of reference, a recent meta-analysis by Stulp et al. (2015) of 154 studies over a very wide diversity of cultures obtained a value of $\rho_z = 0.25$ for western populations (and an overall value of 0.23).

Relationship	r	SE
Parent–offspring:		
Father–son	0.514	0.022
Father–daughter	0.510	0.019
Mother–son	0.494	0.024
Mother–daughter	0.507	0.021

Relationship	r	SE
Full sibs:		
Brother–brother	0.511	0.042
Brother–sister	0.553	0.019
Sister–sister	0.537	0.033
Husband–wife	0.280	0.028

We first consider the additive genetic variance in the population. Recall that $\widehat{\sigma}_z^2 = 1$ on the scale of analysis. From Table 7.5, we see that the expected covariance between parent and offspring is $(1+\rho_z)\,\widehat{\sigma}_A^2/2$. Setting this quantity equal to 0.506, substituting 0.280 for ρ_z, and rearranging, we obtain an estimate for $\widehat{\sigma}_A^2$ of 0.791 ± 0.024 (the standard error being obtained by the Taylor expansion delta method; Appendix 1). Thus, unless epistasis is very strong (we have assumed it to equal zero), it appears that approximately 80% of the variance in human height is attributable to additive gene action.

Suppose assortative mating were to be eliminated. What is the expected equilibrium value of the additive genetic variance? Assuming a large effective number of loci ($\gamma \simeq 1$), and substituting the estimates of $\widehat{\sigma}_A^2$ and ρ_z into Equation 7.20, we obtain $\sigma_A^2 \simeq 0.616$. The additive disequilibrium covariance for height is estimated as $0.791 - 0.616 = 0.175$, showing that through the creation of gametic-phase disequilibrium, assortative mating has induced an approximately 28% increase in the additive genetic variance.

We now consider whether the full-sib data are consistent with an additive genetic model. Again from Table 7.5, we see that in the absence of dominance, the covariance between full sibs has an expected value of $\widehat{\sigma}_A^2(1+\rho_z\widehat{h}^2)/2$. Because we are operating on a scale for which $\widehat{\sigma}_z^2 = 1$, this value is also the expected correlation coefficient. Substituting 0.280 for ρ_z and 0.791 for $\widehat{\sigma}_A^2$, we obtain the expectation 0.483 ± 0.024. The difference between the observed and expected value ($0.534 - 0.483 = 0.051$) has a standard error equal to $[(0.024)^2 + (0.019)^2]^{1/2} = 0.031$ and cannot be considered to be significant. Thus, on the basis of the existing data, there are no grounds for rejecting the purely additive model.

In a more recent study, Roberts et al. (1978) performed a census of adult heights in families of a West African population, where assortative mating by height is relatively weak, $\rho_z \simeq 0.10 \pm 0.10$. The correlations between parents and offspring (0.434 ± 0.015), and between full sibs (0.378 ± 0.048), are somewhat lower than those observed in the study of Pearson and Lee. Using the procedures outlined above, these data yield estimates of $\widehat{\sigma}_A^2 \simeq 0.789$ and $\sigma_A^2 = 0.727$. Thus, the expressed additive genetic variances in the two populations are very similar, although a smaller fraction is associated with disequilibrium covariance in the West African population. Roberts et al. also reported a half-sib correlation equal to 0.198 ± 0.059, which can be used to provide a further check on the theory. Substituting the estimate of ρ_z and $\widehat{h}^2 = 0.789$ into the half-sib expression in Table 7.5, the predicted correlation is 0.229, which is not significantly different from the observed value.

There is an interesting historical note regarding the data of Pearson and Lee. Fisher's (1918) demonstration that the data were quite consistent with a Mendelian hypothesis flew in the face of Pearson's notorious non-Mendelian philosophy. Fisher also extended the analyses to other types of relatives, illustrating their consistency with resemblances between first cousins and between grandparents and their grandchildren. However, the literature that Fisher cites for the latter correlations actually involves characters other than height (in one case, eye color)! Thus, at the time, Fisher apparently believed that there was a universal correlation for all characters within a species. This is an interesting twist because Pearson had similar feelings, as amply documented in Pearson and Lee (1903, p. 379): "Thus for most practical purposes we may assume parental heredity for all species and all characters to be approximately represented by a correlation of 0.5." We now know that this is far from the truth.

POLYPLOIDY

Because of the high incidence of polyploidy in plants, some attention needs to be given

Table 7.6 Coefficients for Equation 7.22 for describing the covariance between tetraploid relatives. (After Kempthorne 1957.)

Relationship	Θ_{xy}	Δ_{xy}	φ_{xy}	ϕ_{xy}
Parent–offspring	$\frac{1}{8}$	$\frac{1}{36}$	0	0
Grandparent–grandchild	$\frac{1}{16}$	$\frac{1}{216}$	0	0
Full sibs, dizygotic twins	$\frac{1}{8}$	$\frac{1}{27}$	$\frac{1}{48}$	$\frac{1}{36}$
Half sibs	$\frac{1}{16}$	$\frac{1}{216}$	0	0
Uncle (aunt)–nephew (neice)	$\frac{1}{16}$	$\frac{1}{648}$	0	0
Monozygotic twins	$\frac{1}{4}$	$\frac{1}{6}$	$\frac{1}{4}$	1

to its effects on the covariance between relatives. We will confine our comments to single-locus effects in a randomly mating tetraploid population and assume that the four alleles at each locus assort independently during meiosis. The results for this special case were first worked out by Kempthorne (1953, 1957).

As in the case of diploids, the genotypic values of tetraploids can be partitioned into several independent effects defined (as was the case for diploids) in a least-squares sense. However, because four alleles are present at each locus, in addition to additive (single-allele) and dominance (allele-pair) effects, triallelic and quadra-allelic effects must be considered (Chapter 5). Thus, the single-locus genotypic value of individual x is written as

$$G_{ijkl}(x) = \mu_G + [\alpha_i^x + \alpha_j^x + \alpha_k^x + \alpha_l^x] + [\delta_{ij}^x + \delta_{ik}^x + \delta_{il}^x + \delta_{jk}^x + \delta_{jl}^x + \delta_{kl}^x] \\ + [\gamma_{ijk}^x + \gamma_{ijl}^x + \gamma_{ikl}^x + \gamma_{jkl}^x] + \tau_{ijkl}^x \tag{7.21}$$

As in the case of diploids, the expected value of each of the effects is zero. Thus, the additive and dominance genetic variances for a locus are defined to be $\sigma_A^2 = 4E[\alpha^2]$ and $\sigma_D^2 = 6E[\delta^2]$, while the trigenic and quadragenic variances are $\sigma_T^2 = 4E[\gamma^2]$ and $\sigma_Q^2 = E[\tau^2]$.

The covariances between tetraploid relatives are obtained in the same manner as in the case of diploids. For example, the additive genetic covariance involves 16 cross-products of additive effects (four in individual x times the four in individual y). Each of these cross-products has expectation $E[\alpha^2]$ if the two genes are identical by descent, the probability of which is Θ_{xy}, and expectation zero otherwise. Thus, the expected additive genetic covariance between relatives is $16\Theta_{xy}E[\alpha^2] = 4\Theta_{xy}\sigma_A^2$. In the case of digenic interaction, there are 36 cross-products to consider, each of which has expectation $E[\delta^2]$ if both members of the pair of alleles in x are identical by descent with those in the pair from y (probability Δ_{xy}). Therefore, the digenic covariance is $36\Delta_{xy}E[\delta^2] = 6\Delta_{xy}\sigma_D^2$. Letting φ_{xy} be the probability that all members of random three-gene sets are identical by descent, and ϕ_{xy} be the probabilty that the complete single-locus tetraploid genotypes of x and y are identical by descent, the trigenic and quadragenic covariances between relatives are $4\varphi_{xy}\sigma_T^2$ and $\phi_{xy}\sigma_Q^2$. Thus, assuming random mating, and no epistasis, linkage, or gametic-phase disequilibrium, the covariance between tetraploid relatives can be summarized as

$$\sigma_G(x, y) = 4\Theta_{xy}\sigma_A^2 + 6\Delta_{xy}\sigma_D^2 + 4\varphi_{xy}\sigma_T^2 + \phi_{xy}\sigma_Q^2 \tag{7.22}$$

The coefficients for a number of common relationships are given in Table 7.6.

An interesting consequence of polyploidy is that the covariance between noncollateral relatives (individuals that do not share both parents) can be influenced by dominance. This effect occurs because parents pass two alleles (and hence their interaction) on to their progeny. Thus, the covariance between parent and offspring is $\sigma_A^2/2 + \sigma_D^2/6$, as compared to $\sigma_A^2/2$ in the case of diploidy. Trigenic and quadragenic variance components only appear in expressions of resemblance when collateral relatives are involved.

ENVIRONMENTAL SOURCES OF COVARIANCE BETWEEN RELATIVES

Up to now, we have focused entirely on the genetic causes of resemblance between relatives. There are, however, many circumstances in which the environmental effects on the phenotypes of relatives are correlated. This is an especially pernicious complication in human genetics. The most obvious situation arises when full sibs are raised in a common familial environment, but the resemblance between other types of relatives can also be modified by shared aspects of the environment. For example, in a spatially heterogeneous environment, the resemblance between parents and offspring may be exaggerated by environmental effects if the latter do not randomly disperse following birth. In humans, and probably in other vertebrates, cultural transmission can create a continuity in the behavioral phenotypes of parents and offspring as well as among more distant relatives. In some circumstances, the environment may actually deflate the resemblance between relatives. Consider, for example, trees with limited seed-dispersal abilities. Parent trees that have grown large due to fortuitous circumstances, such as germination in a light gap, may tend to create relatively poor microhabitats for their offspring as a consequence of shading, attraction of herbivores, and so on.

In an effort to deal with the complications of cultural transmission, human quantitative geneticists have gone to great extremes to incorporate various types of environmental covariance into expressions for the resemblance between relatives (Rao et al. 1974; Eaves 1976; Cloninger et al. 1979a, 1979b; Eaves et al. 1988). The result is a family of models that contain large numbers of parameters. Because the number of observable relationships must equal or exceed the number of parameters for there to be any hope of estimating the latter, these models are of little practical utility for most organisms. However, in humans the difficulties are not insurmountable. Due to the frequency of divorce, adoption, and fostering by relatives, and to the availability of records on parentage, many kinds of familial relationships are observable.

Here we consider some of the possibilities, under the assumption of additive gene action, and negligible linkage, gametic-phase disequilibrium, and inbreeding. To illustrate the major points, we will rely on path analysis (Appendix 2), drawing specifically from examples given in Rao et al. (1974). The basic feature of path analysis that we will exploit is its ability to partition the correlation between any two variables into a series of pathways connecting the two variables through causal components. As noted during our proceeding discussion of coefficients of coancestry, the proportional contribution of each pathway to the correlation between relatives is simply equal to the product of various path coefficients and correlation coefficients along the pathway. We start by providing some general definitions, and then show, for several types of relationships, how path analysis can yield expressions for the expected phenotypic correlations in terms of causal genetic and environmental components. We then complete our discussion of environmental effects by providing a worked example involving a large data set on human birth weight.

Before proceeding, we consider how the phenotypic variance can be partitioned in terms of path coefficients. As usual, we consider an individual's phenotype (z) to be the sum of its genotypic value (G), general (shared) environmental effects (E), and special environmental effects (e). We can further partition G into three quantities: the population mean (μ_G), the mean genotypic (breeding) value of the parents ($\bar{A}$), and the deviation of G from $\bar{A}$ caused by segregation of parental genes (S). Thus, an individual's phenotypic value is expressed as

$$z = \mu_G + \bar{A} + S + E + e \tag{7.23}$$

The path coefficients from $\bar{A}$, S, E, and e to the phenotype are denoted by g, s, c, and d, respectively (for expected genotypic value, deviation due to segregation, common environmental effect, and residual deviation). Because S is a random genetic deviation, it is uncorrelated with the other components of Equation 7.23. However, genotype–environment correlation may exist between $\bar{A}$ and E when offspring are raised by their own parents, and we denote it by ρ_{GE}.

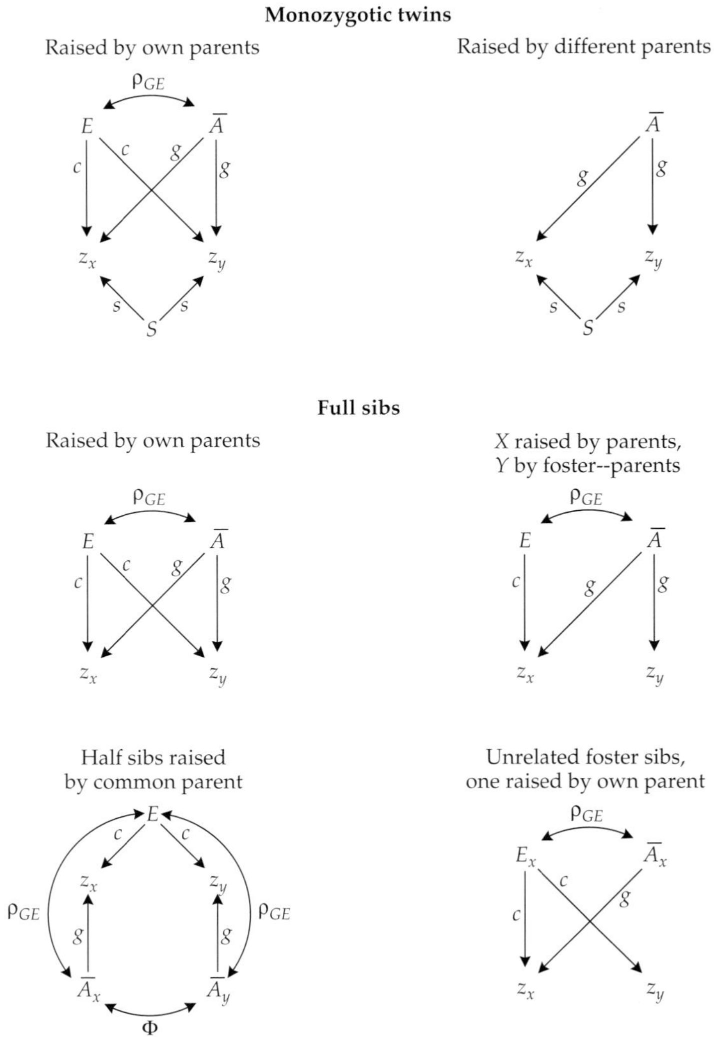

Figure 7.10 Path diagrams for the phenotypic correlation between individuals x and y. All terms are defined in the text.

The phenotypic variance among the pool of individuals raised by biological parents is

$$\sigma_z^2 = \sigma_{\bar{A}}^2 + \sigma_S^2 + \sigma_E^2 + \sigma_e^2 + 2\sigma_{\bar{A},E} \tag{7.24a}$$

Dividing all terms by σ_z^2 provides an expression for the partitioning of the phenotypic variance of offspring raised by their own parents,

$$1 = g^2 + s^2 + c^2 + d^2 + 2g\rho_{GE}c \tag{7.24b}$$

In this expression, g^2, s^2, c^2, and d^2, are, respectively, the fractions of the phenotypic variance attributable to midparent breeding values, segregational, general and special environmental effects, while $2g\rho_{GE}c = 2\sigma(G,E)/\sigma_z^2$ is the fraction resulting from genotype–environment covariance.

Equation 7.24b requires modification in the case of progeny raised in adoptive homes. Under the assumption that there is no correlation between offspring genotype and the environment provided by an adoptive home, then $\rho_{GE} = 0$, and the phenotypic variance of adopted children can differ from that of progeny raised by their own parents. For example, if $\rho_{GE} > 0$ for intact families, the variance among adopted children will be reduced. To account for this, Equation 7.24b needs to be modified to ensure that the sum of the various paths is equal to one. For adopted children, the modification is accomplished by letting

$$1 = \theta^2(g^2 + s^2 + c^2 + d^2) \tag{7.24c}$$

where $\theta^2 = 1/(1 - 2g\rho_{GE}c)$ is the ratio of the phenotypic variances for the two types of offspring. Multiplication by θ^2 has the effect of dividing each of the causal path contributions by the phenotypic variance of progeny raised in adoptive homes rather than by the phenotypic variance of progeny raised by their biological parents.

Finally, we note that it is useful to define

$$h^2 = g^2 + s^2 \tag{7.25}$$

as the heritability of the trait, i.e., the fraction of the total phenotypic variance attributable to additive genetic differences among individuals. With assortative mating,

$$g = h\sqrt{(1 + \rho_g)/2} \tag{7.26a}$$

$$s = h\sqrt{(1 - \rho_g)/2} \tag{7.26b}$$

where, as before, ρ_g is the addtive genetic correlation between mates (i.e., the correlation of their breeding values). Under random mating, $g = s = h/\sqrt{2}$.

To illustrate how these general relationships can be extended to the description of the expected phenotypic correlation between individuals, we now focus on the path diagrams for four specific relationships (Figure 7.10). For simplicity, the diagrams only include those factors that contribute jointly to the phenotypes of both members of a pair of individuals. Thus, the residual environmental deviation (e) never appears, while S (as a random genetic deviation) is only relevant in the case of monozygotic twins.

Because they are genetically identical, monozygotic twins (or **clonemates**) raised by their natural parents share the same general environmental effects, midparent value, and segregation deviation. Allowing for genotype–environment covariance, and summing over all pathways between z_x and z_y, we obtain the phenotypic correlation,

$$\rho(MZ) = s^2 + g^2 + c^2 + 2gc\rho_{GE} = h^2 + c^2 + 2gc\rho_{GE} \tag{7.27}$$

Thus, by itself, the correlation between monozygotic twins raised by their biological parents is not particularly informative, because it is a function of additive genetic variance (h^2), variance due to shared environment (c^2), and genotype–environment correlation $2gc\rho_{GE}$.

Now consider the situation in which monozygotic twins are separated at birth, with each being raised in a different adoptive home. The removal of the common-environment effect eliminates the path $z_x \leftarrow E \rightarrow z_y$ (Figure 7.10), and as noted above, the absence of genotype–environment correlation changes the phenotypic variance in the subpopulation of such twins. Thus, the expected phenotypic correlation between twins raised by different foster parents is $h^2\theta^2$. An estimate of the heritability can be acquired after factoring out θ^2 (the ratio of phenotypic variances, defined above). Expressions for twins living in other combinations of home environments are given in Table 7.7. (Note that when one member of a relationship is living with its parents and the other is living in an adoptive home, the correlation is multiplied by θ, rather than θ^2, because only one member of the pair is from a subpopulation with modified variance.) Part II Chapter 24 treats these issues in considerable detail.

We next consider the correlation between full sibs. The path diagram in this case is identical to that for monozygotic twins, except that the sibs, being products of different gametes, do not share the segregational deviation (S). Thus, the expected correlation between full sibs raised by their biological parents is

$$\rho(FS) = g^2 + c^2 + 2gc\rho_{GE} \tag{7.28}$$

Table 7.7 Expected phenotypic correlations between sibs in terms of path coefficients. All coefficients are defined in the text except b, which is the correlation of environments provided by parents of half sibs. Completely additive gene action is assumed. (After Rao et al. 1974.)

Relationship	Phenotypic correlation
Monozygotic twins:	
Reared by own parents	$h^2 + c^2 + 2gc\rho_{GE}$
One reared by own parents, one by foster parents	$(h^2 + gc\rho_{GE})\theta$
Raised by different foster parents	$h^2\theta^2$
Reared together by foster parents	$(h^2 + c^2)\theta^2$
Full sibs:	
Reared by own parents	$g^2 + c^2 + 2gc\rho_{GE}$
One reared by own parents, one by foster parents	$(g^2 + gc\rho_{GE})\theta$
Reared by different foster parents	$g^2\theta^2$
Half sibs:	
Raised by common parent	$\phi g^2 + c^2 + 2gc\rho_{GE}$
One reared by own parents, one by foster parents	$(\phi g^2 + gc\rho_{GE})\theta$
Reared by different foster parents	$\phi g^2\theta^2$
Reared apart by own parents	$\phi g^2 + c^2 b + gc\rho_{GE}$
Unrelated foster sibs:	
Reared together by same foster parents	$c^2\theta^2$
Reared together by parents of one of them	$(c^2 + gc\rho_{GE})\theta$

As in the case of monozygotic twins, this expression is simplified in situations where one or both sibs are raised in adoptive environments. For example, suppose that one sib is raised by its biological parents, while the other is raised by unrelated foster parents (Figure 7.10). This eliminates the paths $z_x \leftarrow E \rightarrow z_y$ and $z_x \leftarrow \bar{A} \leftrightarrow E \rightarrow z_y$. Moreover, because the second sib comes from a segment of the population without genotype–environment correlation, all of the path coefficients leading to it must be multiplied by θ. Summing over the two paths between x and y, the phenotypic correlation becomes $\theta(g^2 + gc\rho_{GE})$.

A slight complication arises in the case of half sibs. The correlation ρ_g between mid-parent values ($\bar{A}_x$ and $\bar{A}_y$) is no longer 1.0, because there is only one common parent. Rao et al. (1974) showed it to be $\phi = (1 + 3\rho_g)/[2(1 + \rho_g)]$, which reduces to 1/2 under random mating. Summing over all paths between z_x and z_y (Figure 7.10), the correlation between half sibs raised by the parent creating the common environmental effect is

$$\rho(HS) = \phi g^2 + c^2 + 2gc\rho_{GE} \tag{7.29}$$

Again, various simplifications arise when one or both members of the sib pair are raised by foster parents (Table 7.7). For example, when each sib is raised in a different adoptive environment, the only path between sib phenotypes is $z_x \leftarrow \bar{A}_x \leftrightarrow \bar{A}_y \rightarrow z_y$, so the correlation is simply $\phi g^2\theta^2$.

Finally, we note that unrelated individuals fostered by the same set of parents can resemble each other as a consequence of the common environment in the adoptive home. The expected phenotypic correlation between such individuals depends on whether the adoptive parents are the biological parents of one of the foster sibs (Figure 7.10, Table 7.7).

Even more complicated scenarios, for additional types of relatives, have been considered by the authors cited above. However, we assume that the basic principles are clear at this point, and will not pursue these any further here (although we will do so in Chapter 24).

One very notable aspect of the models outlined above is their ability to provide estimates of genotype–environment correlation when data are available on sibs raised in various types of home environments. For example, when covariances are observed for monozygotic twins living in the four types of environmental settings outlined in Table 7.7, joint estimates of h^2, c^2, and ρ_{GE} can be obtained by setting the observed correlations equal to their expected values and solving. Modifications of all of the expressions in Table 7.7 are necessary, however, in the presence of significant sources of nonadditive genetic variance. We close this section with an example of a character whose expression is strongly influenced by shared environmental effects.

Example 7.9. Several large and independent studies have been performed on human birth weight. As can be seen in the following table, the estimated correlations between relatives are quite consistent among studies. For example, the five available full-sib correlations have a narrow range of 0.47 to 0.52. Mi et al. (1986) performed large analyses on several ethnic groups in Hawaii and found only minor differences among them for the correlations between relatives. We will therefore pool the independent estimates where they exist. In keeping with the linear model just outlined, we will assume that dominance and epistatic sources of variance are of negligible importance. In the absence of conflicting data, we will also assume that assortative mating and genotype–environment correlation are negligible, and that general environmental effects are only transmitted through mothers. Under these assumptions, $\rho_G = 0$, $g^2 = h^2/2$, $\rho_{GE} = 0$, and $\theta = 1$.

Relationship	Estimated Correlations	Prediction
Full sibs[2–6]	$0.50, 0.52, 0.47, 0.48, 0.48$	$\frac{h^2}{2} + c^2 = 0.50$
Maternal half sibs[3]	0.58	$\frac{h^2}{4} + c^2 = 0.42$
Paternal half sibs[3]	0.10	$\frac{h^2}{4} = 0.08$
Maternal first cousins[2,6]	$0.14, 0.13$	$\frac{h^2}{8} + \frac{c_G^2}{2} = 0.15$
Paternal first cousins[2,6]	$0.02, 0.06$	$\frac{h^2}{8} = 0.04$
Monozygotic twins[1]	0.67	$h^2 + c^2 = 0.65$
Dizygotic twins[2,3]	$0.59, 0.66$	$\frac{h^2}{2} + c^2 = 0.50$
Half sibs via monozygotic twin parents:		
Maternal[5]	0.31	$\frac{h^2}{4} + c_G^2 = 0.30$
Paternal[5,7]	$-0.03, 0.12$	$\frac{h^2}{4} = 0.08$

References: 1. Penrose (1954a); 2. Robson (1955); 3. Morton (1955a); 4. Billewicz (1972); 5. Nance et al. (1983); 6. Mi et al. (1986); 7. Magnus (1984).

We first consider the additive genetic variance. Inferences about this parameter must be derived from relationships for which shared environmental effects do not influence the covariance. Paternal half sibs and paternal first cousins satisfy these conditions. The expected correlations for these types of relatives are $h^2/4$ and $h^2/8$, respectively, where $h^2 = \sigma_A^2/\sigma_z^2$. Because the observed correlations are 0.10 and 0.04, we obtain independent estimates of h^2 of 4 x 0.10 = 0.40 and 8 x 0.04 = 0.32. Also available is an average correlation of 0.05 for the offspring of monozygotic twin brothers. Such individuals are genetically equivalent to paternal half sibs (the fathers are different individuals, but identical genetically), so this result yields an additional estimate of h^2 = 4 x 0.05 = 0.20. Averaging over all three types of relationship, $h^2 \simeq 0.30$, i. e., additive genetic variance appears to account for approximately 30% of the phenotypic variance.

The data make it very clear that aspects of the maternal environment have a substantial influence on birth weight. For example, the correlation between maternal half sibs is several times greater than that between paternal half sibs, and the same pattern is seen for maternal versus paternal first cousins. The total variation caused by the maternal environment can be obtained from the maternal half-sib correlation, 0.58, whose expectation is $(h^2/4) + c^2$. Subtracting out the additive genetic contribution, c^2 = 0.58 − (0.30/4) = 0.50. Pooling this with an independent estimate of $c^2 = 0.20$ obtained by Magnus (1984), we estimate $c^2 \simeq$ 0.35. Thus, aspects of the mother (in excess of the genes that she contributes to her offspring) account for approximately 35% of the variance in birth weight. (We examine the genetics of maternal effects in detail in Part II Chapter 29.)

There are two ways to partition the maternal effects variance into genetic and environmental components, $c^2 = c_G^2 + c_E^2$. First, monozygotic twin sisters provide the same genetic environment but different home settings for their progeny, which are genetically equivalent to maternal half sibs. The expected correlation between half sibs via this route is therefore $(h^2/4) + c_G^2$. Subtracting $h^2/4$ from the observed correlation, we obtain an estimate of the variance caused by genetic maternal effects, c_G^2 = 0.31 − (0.30/4) = 0.23. Second, the covariance between maternal first cousins is unaffected by common maternal environment, but is influenced by half the genetic maternal variance because the mothers are full sibs; their expected correlation is therefore $(h^2/8) + (c_G^2/2)$. Again equating observed and expected correlations, we obtain a second estimate c_G^2 = 2[0.14 − (0.30/8)] = 0.20. Thus, of the maternal effects variance, approximately two-thirds (0.215/0.35) appears to be caused by the effects of the maternal genotype on the uterine environment that an offspring experiences.

The causes of approximately 35% of the variance remain to be identified. Offspring sex, birth order, and gestation age account for 2, 3, and 4% of the variance, respectively (Penrose 1954a; Morton 1955a; Billewicz 1972; Magnus 1984). Some of these sources of variance presumably fall in the environmental maternal-effects category. Relatively lower weights for first-born offspring account for another 5% of the variation. Morton (1955a) has argued for the existence of dominance genetic variance, but the following argument suggests that this source of variance is negligible. In principle, the correlation between full sibs is $(\sigma_A^2/2 + \sigma_D^2/4 + \sigma_E^2)/\sigma_z^2$. However, from the above $(\sigma_A^2/2 + \sigma_E^2)/\sigma_z^2 \simeq$ (0.30/2) + 0.35 = 0.50, which accounts for the mean observed correlation of 0.49.

The approach that we have taken to analyze these data is not very rigorous from a statistical standpoint, our main objective being to illustrate how correlations derived from several types of relatives can be used to tease out components of variance. Nevertheless, when the estimates of h^2, c^2, and c_G^2 are substituted into the expressions for the expected correlations between relatives, the overall fit to the data is quite good (last column in the preceding table). Thus, variation in human birth weight appears to be largely a function of additive gene action, maternal effects, and special environmental effects, each of which accounts for about a third of the total variance.

THE HERITABILITY CONCEPT

We have now seen, in theory and by example, that the analysis of a series of relationships provides the basis for partitioning the phenotypic variance into its various components. In practice, however, we are often confronted with challenges, aside from the problem of finite resources, that prevent us from ever obtaining exact estimates of variance components. Some of the variance, such as that caused by higher order epistatic interactions, is essentially beyond reach in a statistical sense. Nevertheless, with appropriate experimental designs, most of the fundamental sources of variance (additive and dominance genetic variance, and environmental variance due to common familial environments) can be approximated to a good degree, and levels of confidence attached to them (Part II Chapters 22–25). Most practical applications of quantitative genetics have been concerned with only the additive genetic component of the phenotypic variance, with the remaining components being treated as noise. The ratio σ_A^2/σ_z^2 has come to be known as the **heritability** of a trait (more precisely, the **narrow-sense heritability**).

Table 7.8 Independent estimates of h^2, obtained with Equation 7.30, for temperature tolerance in the marine copepod *Eurytemora affinis*. (From Bradley 1978.)

Relationship	Females	Males
Parent–offspring	0.14 ± 0.44	0.72 ± 0.26
Full sibs	0.20 ± 0.05	0.82 ± 0.04
Paternal half sibs	0.40 ± 0.18	0.84 ± 0.35
Maternal half sibs	0	0.73 ± 0.32

This brings us to an important conceptual issue that has plagued the field of quantitative genetics almost since its inception (Feldman and Lewontin 1975; Bell 1977; Jacquard 1983). The preoccupation with the additive component of genetic variance stems from the desire for a parameter that describes the genetic resemblance between parents and offspring. At the close of Chapter 3 it was shown that the slope of a regression of offspring phenotype on average parental phenotype has expected value σ^2_A/σ^2_z, provided that gene action is purely additive and all of the assumptions underlying the Kempthorne–Cockerham model are met. Moreover, we showed that if selection changes the mean phenotype in the parental generation by S units, the expected evolutionary advance in the offspring generation (relative to that of the parents before selection) is $S\sigma^2_A/\sigma^2_z$. Based on this reasoning, many studies have accepted uncritically the slope of a midparent-offspring regression ($b_{o\bar{p}}$) (or equivalently, twice the slope of mother–offspring or father–offspring regression, $2b_{op}$) as an estimate of σ^2_A/σ^2_z. However, over the last several pages, we have found that the validity of this interpretation requires, among other things, random mating, gametic-phase equilibrium, absence of additive $\times$ additive epistatic genetic variance, and absence of sharded environmental effects. We cannot expect all of these conditions to be fulfilled in many natural populations.

Jacquard (1983) provided a useful discussion of the problems of interpretation of $b_{o\bar{p}}$ and $2b_{op}$ and suggested that these statistics simply be labeled **biometric heritability** without prejudice regarding the mechanisms causing similarity. However, because of the fundamental importance of the ratio σ^2_A/σ^2_z, particularly in selection theory, we will continue to call the latter quantity the heritability, denoting it as h^2 in keeping with Wright's original usage of h as the path coefficient σ_A/σ_z (the correlation between z and A; Appendix 2). Providing an explicit definition eliminates the ambiguity of the usage of h^2 in theoretical contexts, but highlights the practical problems of estimation.

It should now be clear that heritabilities can often be estimated using sets of relatives other than parents and offspring. The general logic behind this approach is that the first term in any genetic covariance expression is $2\Theta_{xy}\sigma^2_A$. Thus, under the assumption that the additive genetic variance is the overwhelming source of phenotypic covariance,

$$h^2 \simeq \frac{\text{Cov}(z_x, z_y)}{2\Theta_{xy}\text{Var}(z)} \tag{7.30}$$

should provide a good approximation to the heritability. Violations of the assumptions of the ideal additive model will usually cause $\text{Cov}(z_x, z_y)/2\Theta_{xy}$ to be an upwardly biased estimator of σ^2_A, overestimating h^2.

The consistency of heritability estimates using different types of relatives provides one approach for assessing bias caused by failures of some of the model assumptions. Clayton et al. (1957) performed such an analysis on abdominal bristle number in a laboratory population of *Drosophila melanogaster*. The estimates of h^2 derived from four types of relatives were consistent with each other: mother–daughter (0.54 ± 0.11), mother–son (0.48 ± 0.11), half sibs (0.48 ± 0.11), and full sibs (0.53 ± 0.07). For this population, the evidence is strong that approximately 50% of the total variance for abdominal bristle number is from additive genetic variance and that the remainder is a function of special environmental effects.

A second example in which heritability estimates are consistent across relationships involves a study of the susceptibility of the marine copepod *Eurytemora affinis* to high

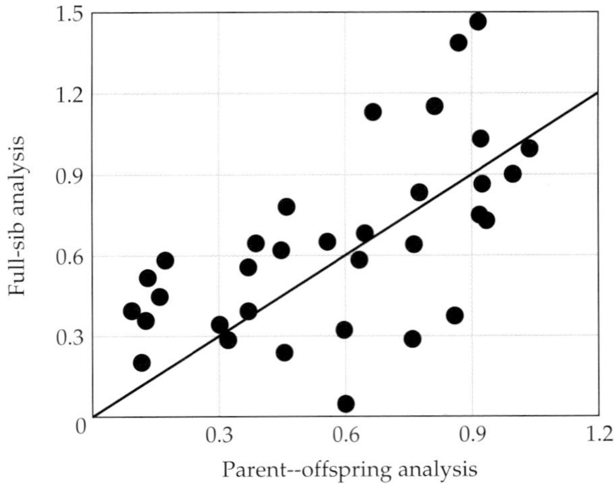

Figure 7.11 The relationship between heritability estimated as 2Cov(PO)/Var(z) from parent–offspring analysis and as 2Cov(FS)/Var(z) from full-sib analysis, for studies in which both estimates are available. Individual data points are for physiological and morphological characters for various natural populations of animals. The straight line gives the expected pattern under a perfect correspondence of the two estimates. (From Mousseau and Roff 1987.)

temperature shock (Table 7.8). Within each sex, four different relationships give fairly consistent results, but there is a clear sexual dimorphism—approximately four times as much variance in males is accounted for by additive genetic variance as in females. The fact that estimates from full sibs and maternal half sibs are not inflated implies that dominance genetic variance and maternal-effects variance are of minor significance. Thus, the data are consistent with the hypothesis that $h^2 \simeq 0.2$ in females and 0.8 in males with the residual variance being attributable to special environmental effects. While this sex-based difference may seem paradoxical, the simple explanation is that males and females perceive temperature tolerance as different (but likely correlated) traits (Part II Chapter 30). If different set of genes influence tolerance in the two sexes, it should not be surprising that it has sex-specific heritabilities.

Such results in which additive genetic variance is the only source of resemblance between relatives are by no means universal in quantitative-genetic analyses. We have already encountered a striking exception with human birth weight, and more will appear in the following chapters. While there is a general tendency for heritability estimates based on parent–offspring and full-sib analyses to be consistent with each other (Figure 7.11), in any particular study, it is incumbent upon the investigator to evaluate whether the inconsistencies between different estimates of h^2 are significant. For example, high levels of dominance genetic variance often exist for fitness-related characters (Crnokrak and Roff 1995; WL Chapter 6).

For practical reasons, the components of variance of natural populations are frequently estimated by assaying a segment of the population in a laboratory setting. Although the goal of such studies is generally to infer the genetic properties of the wild population, laboratory settings often impose a rather substantial change in the environment. It is tempting to speculate that heritability estimates derived from controlled laboratory experiments will be inflated relative to those expressed in the natural environment, where the environmental component of variance might be expected to be magnified by spatial and temporal heterogeneity. However, other outcomes are possible. For example, homeostatic mechanisms, such as habitat selection, which are operable in the field may be rendered inoperable in the laboratory. It is also conceivable that a shift in the environment may induce a change in the additive genetic variance by altering gene expression. We revisit the estimation of h^2 in natural populations using laboratory data in Part II Chapter 22.

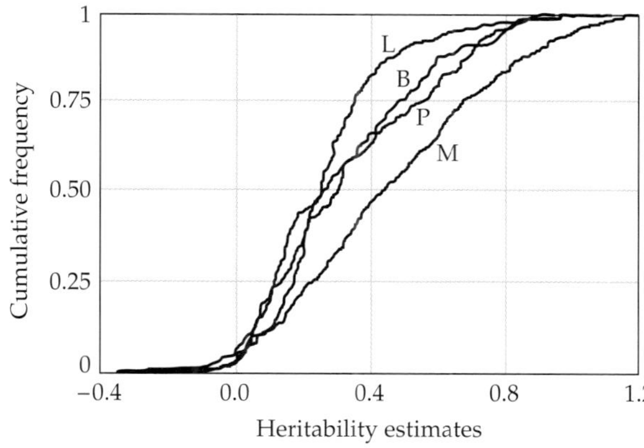

Figure 7.12 Cumulative frequency distributions for heritability estimates derived from numerous wild animal populations. L = life history, B = behavior, P = physiology, M = morphology. These data, from Mousseau and Roff (1987), do not include *Drosophila* studies, which yield a similar pattern (Roff and Mousseau 1987).

The few attempts that have been made to evaluate the sensitivity of heritability estimates to environmental change have yielded a diversity of results. Contrary to expectations, Mackay (1981) found that parent–offspring regressions for sternopleural bristle number and body weight in *Drosophila melanogaster* were *increased* by varying the environment temporally and spatially in the laboratory. For the same species, Coyne and Beecham (1987) found that the parent–offspring regression for abdominal bristle number was not affected by raising the parents in the laboratory (as opposed to the field), while the regression slope for wing length increased by 150%. The change was a consequence of a reduction in the environmental component of variance in the lab-reared parents. Simons and Roff (1994) also observed a general increase in the heritabilities of life-history traits when crickets were raised in the lab, due to a reduction in the environmental component of variance as well as an increase in the genetic component. In a broad survey of the existing data on a diversity of organisms, Weigensberg and Roff (1996) found that there are no *systematic* differences in heritability estimates obtained in the laboratory and in the field; if anything, the latter tend to be slightly higher on average.

In another review, Hoffmann and Parsons (1991) found that heritabilities tend to increase in stressful environments. This observation may be of relevance to the interpretation of some laboratory analyses, in that the laboratory may constitute a form of stress in some cases. However, there are many exceptions to the pattern suggested by Hoffmann and Parsons. In natural populations of birds, for example, heritabilities of bone lengths and body size tend to decline, sometimes to undetectable levels, under poor growth conditions (Gebhardt-Henrich and van Noordwijk 1991; Larsson 1993). There is also human data comparing the heritability of height under different levels of (presumed) environmental stress. Jelenkovi et al. (2016) examined 40 twin cohorts spanning over 100 years (1886 to 1994), finding no support for the hypothesis that the heritability of height was higher in affluent populations versus those with lower living standards, nor did heritability increase as living standards improved. Thus, no strong generalizations emerge from existing studies as to how heritabilities are likely to change in laboratory versus field situations, benign versus harsh environments, novel versus usual conditions, and so forth. The best that can be said is that heritabilities do respond to environmental change, and that substantial care should be taken in extrapolating results beyond the environment in which they are obtained.

If one's sole interest in performing a quantitative-genetic analysis is to demonstrate that the character of interest is heritable, there is probably little point in expending the effort. The outcome is virtually certain. Almost every character in almost every species that has been studied intensely exhibits nonzero heritability. This should come as no surprise, because

Table 7.9 Heritabilities and coefficients of additive genetic (CV_A) and residual (CV_R) variation for representative traits in *Drosophila melanogaster*. Both CV_A and CV_R values are multiplied by 100. Here n is the number of studies, and the median estimates are reported. (After Houle 1992.)

Trait	n	h^2	CV_A	CV_R	CV_R/CV_A
Sternopleural bristles	21	0.44	8.39	7.97	0.95
Wing length	31	0.36	1.56	2.09	1.34
Fecundity	12	0.06	11.90	39.02	3.28
Longevity	7	0.11	9.89	27.73	2.80

mutation brings in a small amount of new genetic variation each generation (Chapter 15; WL Chapter 28). The interesting questions remaining are, How does the magnitude of h^2 differ among characters and species, and why?

One weak generalization that has emerged is that morphological characters tend to have higher heritabilities than life-history traits, with behavioral and physiological characters falling at intermediate levels (Figure 7.12). Although there are numerous exceptions, these results are consistent with the intuitive concept that natural selection will most efficiently reduce the genetic variation for characters closely related to fitness by rapidly driving beneficial genes to fixation and eliminating deleterious ones (Robertson 1955; Fisher 1958; Falconer 1989). However, there are other explanations. As emphasized by Price and Schluter (1991), the relatively low heritabilities of life-history traits may be as much a consequence of relatively high levels of environmental variance as of unusually low levels of genetic variance for such traits. One possible reason for this is that the environmental variance of life-history traits is a function of the variance of all of the other morphological, physiological, and behavioral characters that influence their expression. Alternatively, characters closely related to fitness (life-history traits) may be relatively canalized genetically (Stearns and Kawecki 1994; Chapter 14), such that their expression is relatively insensitive to new mutations. This would result in low levels of genetic variation maintained under selection–mutation balance. Finally, while selection may derive the additive variance to lower values, it may not have a major impact on *nonadditive* genetic variances (WL Chapters 5 and 6).

Evolvability

In comparing the evolutionary potential of different traits/species, it is clearly desirable to use a dimensionless parameter, and one such measure is the heritability (which standardizes the additive variance by comparison to the phenotypic variance). Recall the traditional expression for the rate of evolution of a trait,

$$\Delta\mu = h^2 S \tag{7.31a}$$

This equation neatly separates the forces of selection from the properties of inheritance, such that h^2 can be viewed as the efficiency of the response to selection. The change in the mean in phenotypic standard deviations provides another useful descriptor of evolutionary potential, with

$$\frac{\Delta\mu}{\sigma_z} = h^2 i \tag{7.31b}$$

where $i = S/\sigma_z$ is the standardized selection differential, i.e., the (within-generation) change in the mean caused by selection in units of phenotypic standard deviations.

Houle (1992) has suggested that heritability may not be the best measure of the evolutionary potential of a trait, arguing for the use of the **coefficient of additive genetic variation**, i.e.,

$$CV_A = \frac{\sigma_A}{\mu} \tag{7.31c}$$

where σ_A is the square root of the additive genetic variance and μ is the mean of the trait. In other words, standardizing the additive variance by the *trait mean* (**mean standardization**), rather than the *trait variance* (**variance standardization**). He termed the CV_A metric (originally suggested by Charlesworth 1984) **evolvability**. Because it relies on the mean, this metric is limited to traits that are non-negative and that have a natural value of zero. Height and weight are examples of traits that fit these restrictions, but date number is not, because the zero date on a calendar is arbitrary. Using the definition of CV_A, Equation 7.31a can be rearranged to give the **proportional response** in a trait as

$$\frac{\Delta\mu}{\mu} = \frac{\sigma_A^2}{\sigma_z^2}\frac{S}{\mu} = \left(\frac{\sigma_A^2}{\mu\sigma_z}\right)i = \left(\frac{\sigma_A}{\mu}\right)\left(\frac{S}{\sigma_z}\right)\left(\frac{\sigma_A}{\sigma_z}\right) = CV_A ih \tag{7.31d}$$

Example 7.10. As an example of the difference between the proportional response and the response expressed in standard deviations, consider two traits, both with $h^2 = 0.5$. Trait one has a mean of 10 and a variance of 4, while trait two has a mean of 50 and also a variance of 4. Supposing that $S = 4$, then the change in mean for both traits is $h^2S = 2$, resulting in a response of one standard deviation. Trait one has a new mean of 12, or 120% of its previous mean, while trait two has a new mean of 52, or 104% of its previous mean. Under the change in standard deviations framework, both traits show comparable responses, while as a proportion of the current mean, the change in trait one is much more dramatic.

Houle (1992) and Hansen et al. (2011) observed that h^2 is essentially uncorrelated with evolvability, so that a trait with a lower h^2 could still have high evolvability (i.e., potential for a significant proportional change in the mean), and vice versa. Strikingly, while life-history traits tend to have lower hertabilities, they often display *higher* evolvability values. Table 7.9 shows Houle's (1992) original analysis. Lower values for h^2 for life-history traits appear to arise not from smaller additive variances, but rather from larger values of the residual (environmental plus nonadditive) variance. WL Chapter 6 examines this in more detail.

Finally, consider the special case in which the character is fitness (W). Recall from Equation 3.22 that the selection differential is equivalent to the phenotypic covariance between the character and relative fitness ($w = W/\overline{W}$), i.e., $S = \sigma(z, w) = \sigma(z, W/\overline{W}) = \sigma(z, W)/\overline{W}$, where $\overline{W}$ is mean fitness on the absolute scale. If z is fitness ($z = W$), then $\sigma(z, W) = \sigma(W, W) = \sigma^2(W)$, hence $S = \sigma^2(W)/\overline{W}$, where $\sigma^2(W)$ is the phenotypic variance of fitness. Equation 7.31d then reduces to

$$\frac{\Delta\overline{W}}{\overline{W}} = \frac{\sigma_A^2(W)}{\overline{W}^2} = \sigma_A^2(w) \tag{7.32}$$

where $\sigma_A^2(W)$ and $\sigma_A^2(w)$ are the additive genetic variances of absolute and relative fitness, respectively. Thus, the proportional rate of evolution in mean fitness is equal to the squared coefficient of additive genetic variation of absolute fitness, or equivalently, to the additive genetic variance of relative fitness. This is **Fisher's** (1958) **fundamental theorem of natural selection**, which is examined in detail in WL Chapter 6.

These alternative formulations merely serve to illustrate that there are several ways to measure evolutionary potential, each of which has its own merits in particular contexts (WL Chapters 20, 29, and 30). All of the measures are interchangeable provided that information is available on the phenotypic variance, additive genetic variance, and mean phenotype. As emphasized by Houle (1992), however, many quantitative-genetic studies simply report the heritability of a trait, with no mention of the mean or variance components. This greatly limits the scope of investigation that can be performed with published data. The theory of selection response, including all of the above issues, is discussed in considerable detail in Walsh and Lynch (2018).

8

Marker-based Estimation of Relatedness

Segments of IBD in remotely related individuals are rare but not short. Thompson (2019)

The machinery of quantitative genetics relies heavily on having sets of known relatives. This is not a problem in controlled experiments or when one has access to pedigrees, but these are special settings, and their requirement significantly constrains the universality of these methods. Our recent ability to score tens of thousands of DNA markers rapidly, cheaply, and accurately has opened up essentially any population to a quantitative-genetic analysis, as we can estimate relatedness from this marker information alone.

The idea behind using marker information to estimate the relationship between a pair of individuals is straightforward in concept: relatives should share alleles that are identical by descent (IBD; Chapter 7). Alas, in most settings, we cannot *directly score* IBD alleles, but rather must *infer* the fraction of IBD alleles from information on alike in state alleles (AIS; Chapter 7). IBD alleles are AIS, but the converse is not necessarily true. The amount of IBD sharing can be estimated using either an excess of AIS alleles relative to some neutral expectation or as a correlation between AIS alleles shared by individuals. Implicit in this idea is the notion of some **base population**, from which deviations are measured. Given a randomly mating and unstructured base population, estimated allele frequencies can be used to obtain the expected AIS similarity between two unrelated individuals. The challenging issue with marker data is not simply declaring that two individuals are related, but rather in estimating their *actual degree of relationship*, typically by estimating the coefficient of coancestry, Θ (Chapter 7).

One of the earliest uses of markers involved paternity testing. A famous case from 1943 was the paternity suit of the actress Joan Barry, who claimed that her child was fathered by fellow actor Charlie Chaplin. Chaplin's blood type was *AB*, while both Joan and her child were type *O*. Recall that alleles *A* and *B* are dominant to *i*, with genotype *ii* displaying type *O*. In a tribute to the scientific acumen of juries, Chaplin was declared to be the father. The early crude level of resolution for paternity and forensic issues offered by blood-type testing (**serology**) has since been replaced by highly accurate assessment using DNA markers. In particular, paternity testing and crime-scene analysis (whether a biological sample came from a specified person) is now based (in the US) on just 20 highly polymorphic loci (the CODIS markers), which (roughly speaking) tend to have around ten alleles per locus, all at roughly equal frequencies. Hence the chance of a random individual matching a crime scene DNA profile is on the order of $[2(1/10)^2]^{20} \simeq 10^{-34}$. With a small number of such highly polymorphic markers, we can also assess whether two individuals are very close relatives (parent–offspring, full versus half sibs). Detecting more distant relationships requires significantly more markers, as the expected fraction of IBD alleles scales as $(1/2)^t$, where t is the total number of generations that separate the individuals from their shared ancestor (Chapter 7). A further complication is that there is considerable variation about this expected value, such that many distant relatives will share *no* IBD regions.

Even with a known pedigree, markers are still very useful. First, the veracity of the pedigree can be checked, as **incorrectly ascertained paternities** are not uncommon, even in systems with apparently strong control over matings. For example, Visscher et al. (2002) estimated a sire error rate of $\sim$10% for UK dairy cattle, despite the very widespread use of artificial insemination, while Leroy et al. (2012) found rates of 1–9% for dogs, 1–10% for sheep, and 4% for a French cattle population. Recording errors, as well as the ingenuity of organisms searching for mates, should never be underestimated! Likewise, especially in natural populations, it may be unclear if the offspring from a female are full or half sibs. Such close-relative uncertainties can be easily dealt with using a modest number of highly polymorphic markers. Second, a more subtle issue (first discussed in Chapter 7) is the notion of *expected versus realized relatedness*. Θ values computed from pedigrees represent *expected values*, while the *realized* or *actual* relatedness is a random variable distributed around this expected value. For example, while the expected value of Θ for noninbred full sibs is one-fourth (all pairs of full sibs in a pedigree are assigned this value), the actual relationship for any pair of full sibs can deviate, potentially quite significantly, from 0.25 (Figure 7.5). As we will see in later chapters, this variation around the expected value can be powerfully exploited in a variety of settings. Accurate estimates of such realized relatedness requires *at least* a few thousand markers.

Our presentation is organized as follows. We start with some general comments about marker classes and their associated assays. Next, we examine the use of markers to assign individuals into a few discrete **relationship categories**, such as paternity assessment and whether offspring in a family are half or full sibs. These approaches require only a modest number of markers, and formed the basis for much of the early work on marker-based estimates of relatedness. As marker density greatly increased, so did the ability to detect much more distant relatives. Rather than trying to assign pairs of such relatives into discrete relationship categories, we instead describe their associations using continuous **relatedness** measures, such as Θ_{xy}, f_x, and Δ_{xy}. We first consider a few foundational issues of relatedness measures, such as setting a base population, IBD probability versus allelic correlation frameworks, and the variance in relatedness before considering estimation strategies.

MOLECULAR MARKERS AND GENOMICS

A **marker** is simply a locus with scorable genetic variation. Markers are used in quantitative genetics for *estimating relatedness* (the focus of this chapter), an idea that can be extended to marker-based BLUP and genomic prediction (Part II Chapters 32 and 33). Their other important use is in providing *linkage information* for either linkage-based QTL mapping (such as with a line cross) or association studies (GWAS) that exploit population-level linkage disequilibrium (Chapters 17–21). Our discussion of these different marker systems will also consider their utility for achieving these different objectives.

The ideal class of markers would be those that are: (i) codominant (heterozygotes and homozygotes can be distinguished); (ii) easy and cheap to score; (iii) generally not impacting any trait; (iv) numerous; (v) scattered throughout the genome; and (vi) easily obtainable (a set of markers can be generated for *any* population of interest). As we detail shortly, molecular markers (such as SNPs and STRs) nicely satisfy most of these requirements.

In the classical genetics era, markers were usually **morphological**, impacting the phenotype, such as the recessive *hairy eye* allele and its dominant normal (wild type) allele in *Drosophila melanogaster*. As in this example, such morphological markers are often dominant, complicating their use, but nonetheless, they were employed in very early (and crude) QTL mapping studies in *Drosophila* (Chapter 18). There were also a few early molecular markers, such as blood groups. The first widespread molecular markers used were **allozymes** (also known as **isozymes**), protein variants detected by differences in migration on starch gels in an electric field. The resulting band positions are then revealed using protein-specific stains (Hubby and Lewontin 1966; Lewontin and Hubby 1966). Since the late 1960s, this class of markers has been extensively applied to a variety of population-genetic problems (Lewontin 1974; Charlesworth et al. 2016), and were used in some of the early QTL mapping experiments (Chapter 18). As methods for evaluating variation directly at the DNA level became more widely available during the mid-1980s, DNA-based markers largely replaced allozymes. Because a set of markers tends to be population-specific (or at least largely species-specific), a key feature beyond the *scoring* of markers is in *generating them in the first place*. Given a novel population or species without any existing marker information, can a new set of markers be quickly and cheaply generated? Modern genomics solved this problem.

Properties of DNA Variation

Before briefly reviewing a few legacy systems for scoring markers based on DNA variation (RFLPs, RAPDs, etc.), it is worthwhile to consider the different types of variation that one might see a sample of fully sequenced genomes. The key properties of any such DNA-level variation can be divided into two categories. First, some properties are intrinsic to a class of markers (their stability, genome abundance, levels of polymorphism, and potential restrictions in their genomic coverage, such as coding versus noncoding regions). Second, other properties arise as a result of the assays used to score the markers; most notably, whether heterozygotes can be distinguished from homozygotes (whether the markers are codominant when scored by a particular assay). We group sources of DNA variation as follows:

1. **Single nucleotide polymorphisms** (**SNPs**). These reflect base-pair differences (polymorphisms) at specific single-nucleotide sites. While in theory up to four alleles could be segregating at any single site (representing the four potential DNA bases), most SNPs are biallelic. For these, we code the two alleles as 0 or 1, with the allele coded as 1 denoted as the **reference allele**, often (but not always) taken as the more frequent allele. The **minor-allele frequency**, or **MAF**, denotes the frequency of allele 0. SNP alleles are stable, with mutation rates for nuclear-encoded genes typically on the order of no greater than 10^{-8} per nucleotide per generation, and often much less (WL Table 4.2). SNPs are abundant and widely dispersed throughout the genome, although they tend to be underrepresented in genomic regions of greatly suppressed recombination (such as around many centromeres), likely reflecting the action of both positive and negative selection at linked sites (WL Chapter 8). Their one (minor) deficiency is that a single biallelic SNP is usually not very informative, with a maximum heterozygosity of 1/2 (the maximum value occurs at $p = 1/2$). Further, the minor allele for neutral SNPs is often expected to be rare (WL Equation 2.34), so that most SNPs have a heterozygote frequency far less than 50%. This limitation is easily overcome as one can score a vast number of SNPs, obtaining, in aggregate, a significant amount of polymorphic information. Additionally, there is an ascertainment bias in how SNPs are generally chosen, such that **common SNPs** tend to be greatly overrepresented. Common SNPs are defined as having intermediate allele frequencies: often using MAFs of greater than 1 to 5%, giving heterozygosities of at least 2 to 10%. Finally, most SNPs are regarded as **anonymous DNA markers** (usually—but not always—expected to have no direct impact on a phenotype).

2. **Microsatellites.** A second common category of DNA variation is the tandem repetition of small blocks (**motifs**) of sequence, such as $AGGAGGAGGAGG$. Such **tandem arrays** are often referred to as **satellites** (a holdover from the early days of DNA analysis, when repetitive DNA sequences often formed a distinct band, or *satellite*, when extracted DNA was centrifuged in a salt gradient). The genotype of a tandem array is typically scored as its *number of repeats*, so that a 4,7 heterozyote contains one 4-unit repeat and a 7-unit repeat in the same location on the two homologous chromosomes. The size of the repeat unit forming the array can range from a few bases to thousands of bases, leading to microsatellites, minisatellites, and so on. The most commonly scored are those involving repeating units of just a few bases, and these are called **simple sequence repeats** (**SSRs**) or **simple tandem repeats** (**STRs**). Their major advantage is that any given SSR usually consists of a number of alleles, often of roughly equal frequency. For this reason, the markers used in forensic DNA are well-chosen SSRs that typically contain around 10 alleles of roughly equal frequency, for a heterozygosity of around 90% per SSR (with k equally frequent alleles, the heterozygosity is $[1 - 1/k]$). As a result, just a few SSRs can contain as much polymorphic information as dozens, or even hundreds, of SNPs. The highly polymorphic nature of SSRs arises because of their very high mutation rates (on the order of 10^{-2} to 10^{-4} per generation). This makes ***SSR alleles***

much more unstable than SNP alleles, resulting in signals from distant relationships being partly obscured by mutations changing the allelic state. As a result, SSRs are excellent for close relatives, but become *increasingly problematic* as the most recent common ancestor for a pair of relatives moves further back in time. The same concern arises when using SSRs for linkage information when the common ancestor is several generations back. An association between a particular SSR allele and a nearby QTL can be broken by *mutation* rather than *recombination*. For example, if initally a *5* allele at an SSR and a *Q* allele at a QTL are together in a single linkage block (**haplotype**), an SSR mutation can change this to a $6 - Q$ haplotype. Hence, SSRs may be suitable for linkage mapping (one or a few generations following a cross), but not for association analysis (which uses signals from common relatives that are potentially hundreds of generations back; Chapter 20).

3. **Insertion and deletions** (**indels**). While the most widely used DNA markers in quantitative genetics are either SNPs or SSRs, the genome is a highly dynamic structure, and our final three classes concern such **structural variation**. One common feature are insertions and deletions of various sizes. As with satellites, the scale of such insertions and deletions can vary dramatically, although the larger the indel, the more likely it is to be deleterious (and thus rare in a sample). Indels of just a few bases are rather common (Chapter 15), as are indels generated by mobile-genetic elements, which are abundant in most genomes. Given that such elements can jump into a gene (disrupting its function) and often contain long-range acting *cis* regulatory factors, mobile elements have the potential to influence the expression of nearby QTLs. Mobile elements can also self-excise (jump-out), either cleanly, or leaving some residual sequence behind.

4. **Copy-number variants** (**CNVs**). Potential contributors to quantitative-genetic variation are differences in the number of copies of a particular gene among individuals within a sample. Such copy-number variation is not uncommon, and involves more than just individuals that possess extra copies of the same gene (e.g., Frazer et al. 2009; Zhang et al. 2009; Conrad et al. 2010). More dramatically, CNVs can involve presence/absence differences of *entire genes*, as has been seen between maize lines by Fu and Dooner (2002). Hence, the actual composition of unique genes can vary over individuals. As a result, the notion of a single **reference genome** for a species is now replaced with the notion of a **pan genome**, the aggregate of several sampled genomes to capture all of the distinct genes within a population.

5. **Large-scale variation**. Finally, very large-scale structural variation (at least at the level of moderately sized chromosomal segments)—such as inversions, large duplications, or translocations—can be seen within a sample. Such features are usually over a scale that is not useful for mapping beyond the crude chromosomal assays discussed at the beginning of Chapter 17.

While all of the above classes of DNA variation have been employed for linkage mapping, relationship estimation is almost entirely based on SNPs and SSRs. When possible, ***SNPs are the markers of choice in most settings***, because of their genome-wide abundance, their ease of rapid scoring in very large numbers using modern approaches (e.g., DNA chips and various next-generation sequencing technologies), and their stability.

Legacy Marker Systems

While **whole-genome sequences** (**WGS**) contain all of the available relationship and linkage information in a sample, much of this variation is entirely redundant due the complete correlation of tightly linked markers (LD r^2 values of one; Chapter 5). Hence, for many quantitative-genetic applications, scoring only a subset of genome variation is sufficient. For example, with linkage-based mapping (such as a traditional line-cross QTL analysis;

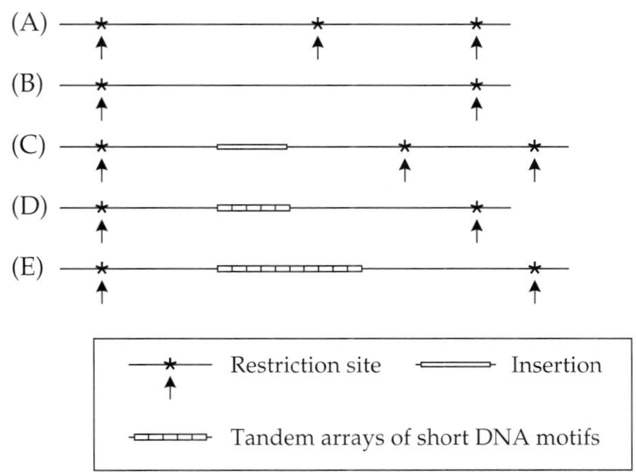

Figure 8.1 A variety of mutational events generate variation in restriction-fragment lengths, scored by the distances between adjacent restriction sites. Single (or multiple) base-pair changes can create or destroy restriction sites, (A) versus (B). Insertions of mobile-genetic elements can create dramatic size differences (C). DNA sequences often exist as tandem arrays of short repeated sequences. Unequal crossing over and/or replication slippage creates variation in the size of such arrays, (D) versus (E).

Chapter 18), the expected block size of nonrecombining regions is large enough that only a modest number of markers are needed to cover the genome. Thus, while we have the ability to score massive numbers of SNPs and STRs, this scale is often not required, and a number of other approaches for scoring DNA variation were used in early marker-based studies. We refer to these older approaches as **legacy marker systems**, as they are occasionally (albeit it diminishingly) still used today by researchers due to the legacy impact of having an existing set of markers on hand and ready to deploy.

One of the simplest approaches is to **digest** DNA with a variety of **restriction enzymes**, each of which cuts the DNA at a specific sequence or **restriction site** (commonly four to six bases). When the digested DNA is run on a gel under an electric current, the fragments separate out according to size. As Figure 8.1 shows, a variety of mutational events can generate fragment length variation. If we attempted to score the entire genome for fragment lengths, the result would be a complete (and uninformative) smear on the gel. Instead, individual bands are isolated from this smear by using labeled **DNA probes** that have base-pair complementarity to particular regions of the genome (often chosen at random). This approach is the basis for assays of **restriction-fragment length polymorphisms** or **RFLPs** (Figure 8.2). Each RFLP probe generally scores a single marker locus with alleles being codominant, as heterozygotes and homozygotes can be distinguished. The number of detectable RFLPs is impressive (Botstein et al. 1980; Doris-Keller et al. 1987; Beckmann and Soller 1983, 1986a, 1986b; Soller and Beckmann 1988).

A particularly interesting RFLP approach involves using tissue-specific cDNA clones as the probes. **cDNAs** are generated from the mRNAs of genes being expressed in that tissue, and probes that detect cDNAs are often called **expressed sequence tags** (**ESTs**). This procedure can allow for enrichment of genes of potential interest in the marker pool, enabling observed differences to be treated as potential candidate loci. For example, Kinzer at al. (1990) used 19 ripening-specific cDNA clones from the tomato (*Lycopersicon esculentum*) to detect polymorphisms in these loci between *L. esculentum* and a wild relative *L. pennellii*. One limitation with using ESTs is that much of quantitative-genetic variation may be regulatory in nature, with causative sites residing outside of (and potentially very far away from) coding regions (Chapter 21). This is less of an issue with linkage mapping due to the large size of the LD blocks generated by a cross, but may be a serious issue in settings where LD blocks are expected to be rather small (such as in a population-level GWAS).

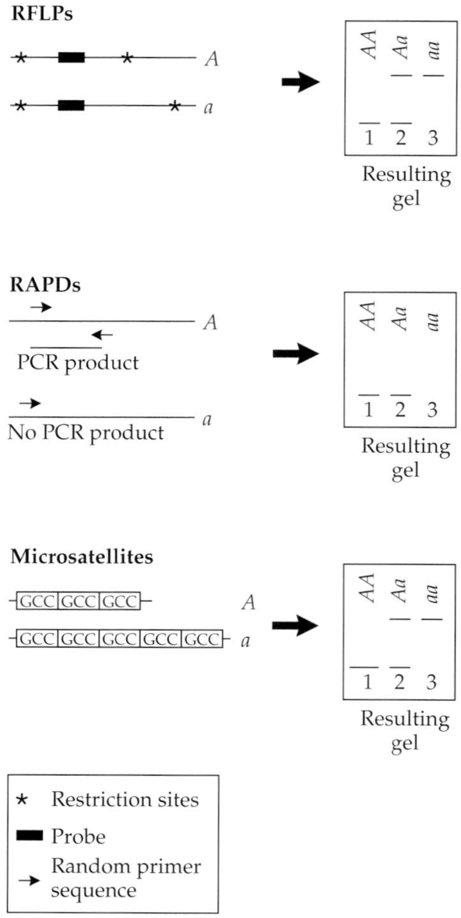

Figure 8.2 Gel-based legacy marker systems. **Top**: RFLPs, restriction-fragment length polymorphisms, variable lengths of restriction fragments in a particular region of the genome, revealed by using a probe for that region. This marker system is codominant (all marker genotypes can be distinguished), as an RFLP heterozygote shows fragments of both lengths. **Middle**: RAPDs, randomly amplified polymorphic DNAs, are revealed by the use of a random primer sequence. A successful PCR reaction requires sequences complementary to the primer in opposite orientation at sufficiently close distance. Running these products out on a gel, both *AA* and *Aa* exhibit a fragment and the site is scored as present, while the fragment does not amplify in *aa* individuals and the site is scored as absent. Hence, RAPDs are dominant markers. **Bottom**: Microsatellite DNAs exhibit variation in the array lengths of short sequences of tandemly repeated DNAs. Fragment length is scored, yielding codominant markers.

A rather different molecular marker assay uses short **primers** for DNA replication via the **polymerase chain reaction** (**PCR**) to delimit fragment sizes. A region flanked (in opposite orientation) by primer binding sequences that are sufficiently close together allows the PCR reaction to replicate this region, generating an amplified fragment. If primer binding sites are missing or are too far apart, the PCR reaction fails and no fragments are generated for that region. This procedure is the basis for **randomly amplified polymorphic DNAs** or **RAPDs** (Williams et al. 1990), sequence polymorphisms detected by using random short sequences as primers (Figure 8.2). RAPDs have an advantage over RFLPs in that a single probe (here, a particular random primer) can reveal several loci at once, each corresponding to different regions of the genome with the appropriate primer sites. However, by their very nature, RAPDs are generally dominant, as they indicate the presence/absence of a particular site, so marker genotypes can be ambiguous (see Figure 8.2).

These basic components for generating markers—scoring the length of various DNA fragments and either amplifying or detecting a specific sequence using DNA hybridization generated by the base-pair complementarity—have been used to generate a number of other legacy marker systems (reviewed in Rafalski and Tingey 1993; Semagn et al. 2006; Grover and Sharma 2016). For example, several mapping studies have used mobile-genetic elements (such as retroviruses) as markers (e.g., Rise et al. 1991; Nuzhdin et al. 1993; Keightley and Bulfield 1993; Ebert et al. 1993; Long et al. 1995). Due to the high rates of movement of some transposable elements, individuals often differ in the presence or absence of elements at particular sites. Scoring their presence/absence by simple hybridization yields dominant markers.

Rapid Scoring of a Large Number of SNPs

As mentioned, the current markers of choice are generally SNPs. One common approach for simultaneously, and inexpensively, scoring a very large number of SNPs is to use **chip technology**, which refers to a number of microarray approaches for scoring thousands to millions of individual SNPs via hybridization technology. Once such a chip is constructed, the same panel of SNPs can be easily scored repetitively over large numbers of individuals. The main limitation is in the set of SNPs chosen to construct the chip, which tend to be biased towards SNPs with MAFs ≥ 0.05 (i.e., common SNPs). A second approach is **genotyping by sequencing (GBS)**, a hybridization and amplification technology that can score a large number of random SNPs in an inexpensive fashion (Elshire et al. 2011; Poland and Rife 2012). Likewise, a variety of next-generation whole-genome sequencing approaches are available (e.g., Day-Williams and Zeggini 2010). Inevitably, quicker and cheaper new technologies will have appeared by the time this book is in print.

Marker information from SNP chip arrays and whole-genome sequences have some fundamental differences (Browning and Browning 2013). The former scores up to a few million SNPs, with a bias toward common SNPs (MAFs $\geq$ 0.01 to 0.05), and relatively modest error-calling rates. The latter scores hundreds of millions of SNPs (in humans), the majority of which are rare (MAF $\leq$ 0.005), and with a much higher error rate, which is especially problematic in that a detected rare variant could be a true SNP or a sequencing error. Common alleles tend to be older mutations, while rare alleles tend to be newer mutations (WL Chapter 2). Hence, the two assay systems, arrays versus sequencing, may be interrogating mutations of different age classes, with common alleles shared by more distant relatives, and rare alleles shared by more recent relationships.

MARKER-BASED ASSIGNMENT OF CATEGORICAL RELATIONSHIPS

Methods for estimating relationships from marker data can be grouped into two approaches: those that are *hypothesis-driven* (e.g., tests for paternity from a pool of candidate males, or determining whether family members are half or full sibs), and those that make *no prior assumptions* about relatedness. These two approaches can be restated as a focus on **categorical relationships** (***assigning*** pairs of individuals into *discrete classes* such as parent–offspring, or full or half sibs) versus continuous measures of **relatedness** (such as ***estimating*** the coefficient of coancestry, Θ). Relationship assignment is typically based on maximum-likelihood methods (Appendix 4) and places individuals into a small number of distinct relationships, such as half sib, full sib, or unrelated (Thompson 1975, 1986a; Thompson and Meagher 1987; Boehnke and Cox 1997; Painter 1997; Broman and Weber 1998; Epstein et al. 2000; Sieberts et al. 2002; Wang 2004; Wang and Santure 2009). These approaches are commonly used in human genetics and forensics. They are also used in studies on the quantitative genetics of natural populations when family units are observed, but with uncertainty about paternity and the related issue of whether sibs are half or full. This approach is typically used to verify connections or fill in missing links in a suspect or incomplete pedigree (Blouin 2003; Pemberton 2008; Huisman 2017). We first consider these assignment methods before turning our attention to estimation of relatedness in the remaining sections.

Paternity Assignment

In the study of natural populations (including humans), one can often assign offspring to a specific mother, forming a natural family unit. Examples include species with extensive maternal care (such as birds, mammals, and some invertebrates) and seeds recovered from an individual plant. When more than one offspring occurs in a family, they could have a single father (full sibs) or multiple fathers (half sibs). A related issue, when applicable, is whether offspring are selfed or outcrossed. A variety of approaches have been proposed for paternity testing, many for the challenging situation where family units may not be obvious. Our focus here is on likelihood-based methods, while variants and other strategies are reviewed by Blouin (2003), Jones and Ardren (2003), Jones et al. (2010), Walling et al. (2010), and Harrison et al. (2013).

We use markers to assess candidate fathers by a two-step process. First, the marker data may **exclude** a candidate. For example, if the mother is M_2M_2 and her offspring is M_2M_3, then any candidate father who does not contain an M_3 allele is immediately excluded (short of a genotyping error). One common approach is to simply use exclusion alone, especially if all but one of the candidates is so excluded (e.g., Dodds et al. 1996; Wang 2007). Second, under **failure to exclude**, we can assign a likelihood ratio of being the father, as opposed to being unrelated to the offspring. We consider four settings for paternity analysis. In increasing order of complexity, these are: (i) known mother and single candidate father; (ii) unknown mother and single candidate father; (iii) known mother and multiple candidate fathers; and (iv) unknown mother and multiple candidate fathers.

We follow Meagher (1986) and Marshall et al. (1998), whose work builds on a large previous literature (see Meagher and Thompson 1986, 1987; Thompson 1986b; Thompson and Meagher 1987; Pena and Chakraborty 1994; and references therein), and uses likelihood ratios (Appendix 4). The basic idea is to contrast the probability (likelihood, L) of the data (D) under two alternative settings—H_1 (the candidate is the father) and H_2 (he is not)—considering the ratio

$$LR(H_1, H_2 \mid D) = \frac{L(D \mid H_1)}{L(D \mid H_2)} \tag{8.1}$$

This ratio quantifies how much more probable the offspring marker genotype is given that the candidate is the father, as opposed to the candidate being unrelated to the offspring. As stressed by Thompson (1986b), this ratio is "*not* a comparison of individuals (fathers and non-fathers) but to two genealogical hypotheses ('father' and 'unrelated') about a given man."

We build the likelihoods for Equation 8.1 as follows. Let the data $D = (g_m, g_o, g_f)$ denote, respectively, the single-locus marker genotypes for the **trio** of mother, offspring, and candidate father (a *pair* of relatives, such as parent and offspring, is called a **dyad**). Recalling from conditional probability (Chapter 3) that $\Pr(x, y) = \Pr(x \mid y)\Pr(y)$, we have

$$L(g_m, g_o, g_f \mid H_1) = T(g_o \mid g_m, g_f)\, P(g_m)\, P(g_f) \tag{8.2a}$$

$$L(g_m, g_o, g_f \mid H_2) = T(g_o \mid g_m)\, P(g_m)\, P(g_f) \tag{8.2b}$$

Note that we have assumed random mating, so that $P(g_m, g_f) = P(g_m)\, P(g_f)$. The key element of Equation 8.2 is the **transmission probability**, T, of obtaining the offspring genotype g_o given the genotype of one (mother, g_m, H_2) or both (mother and candidate father, g_f, H_1) parents. These are the trio transmission probability $T(g_o \mid g_m, g_f)$ under H_1 and the dyad transmission probability $T(g_o \mid g_m)$ under H_2.

To see how these transmission probabilities are computed, consider a biallelic SNP with alleles 0 and 1. For such a marker, homozygotes are denoted by 00 and 11, and heterozygotes by 10. If the mother is 00 and her offspring is 10, then the transmission probability associated with a 11 father is Pr(1 from father)·Pr(0 from mother) + Pr(0 from father)· Pr(1 from mother) = 1·1 + 0·0 = 1. In contrast, the associated transmission probability for a random father becomes p·1 + (1-p)·0 = p, where p denotes the frequency of the 1 allele. Hence,

$$T(g_o = 10 \mid g_m = 00, g_f = 11) = 1$$

$$T(g_o = 10 \mid g_m = 00) = p$$

Note that this approach requires accurate *reference-population* allele frequencies, p. Marshall et al. (1998) present tables of these transmission probabilities for all possible combinations of marker genotypes for the trio, $T(g_o \mid g_m, g_f)$, and for a mother and her offspring, $T(g_o \mid g_m)$.

Substituting Equations 8.2a and 8.2b into Equation 8.1,

$$LR(H_1, H_2 \mid D) = \frac{T(g_o \mid g_m, g_f)\, P(g_m)\, P(g_f)}{T(g_o \mid g_m)\, P(g_m)\, P(g_f)} = \frac{T(g_o \mid g_m, g_f)}{T(g_o \mid g_m)} \tag{8.3a}$$

which yields a value of $1/p$ for our trio data of $g_o = 10$, $g_m = 00$ and $g_f = 11$. Assuming loci assort independently and are in linkage equilibrium, the total likelihood ratio is simply the product of Equation 8.3a over all markers. This resulting product is often called the **paternity index, PI** (Pena and Chakraborty 1994). A common convention (e.g., Slate et al. 2000) is to take an LR value in excess of 20 as strong evidence for paternity (the candidate is 20 times more likely to generate the offspring if he is the father as opposed to being unrelated). This is often stated as an LOD score (natural log of the LR; Chapter 18) in excess of 3, which follows by noting that $\ln(20) \simeq 3$. As a technical aside, the more sophisticated reader might wonder why we do not simply perform a standard LR test, using the large-sample approximation that Equation 8.3a asymptotically follows a χ_1^2 (Appendix 4). The problem is that the null hypothesis (the probability that a specified state is 1.0) lies on the boundary of the parameter space, which compromises the standard χ^2 approximation (Appendix 4).

Using similar logic to that leading to Equation 8.2, when the mother is unknown, the likelihood ratio becomes

$$LR = \frac{T(g_o \mid g_f)\, P(g_f)}{P(g_o)\, P(g_f)} = \frac{T(g_o \mid g_f)}{P(g_o)} \tag{8.3b}$$

Consider the example from above, where the offspring is 10 and the father is 11. Here $P(g_o) = 2p(1-p)$ and $T(g_o \mid g_f) = \Pr(1 \text{ from father}) \cdot \Pr(0 \text{ from mother}) + \Pr(0 \text{ from father}) \cdot \Pr(1 \text{ from mother}) = 1 \cdot (1-p) + 0 \cdot p = 1 - p$, giving LR = $(1-p)/[2p(1-p)] = 1/(2p)$. As these examples illustrate, the known mother setting is more powerful than the unknown mother setting. They also show power is a function of marker allele frequencies, with the signal for IBD increasing as the allele becomes rarer. Hence, ***sharing of rare alleles is a strong signal for IBD sharing***.

While our examples have used SNPs for ease of exposition, paternity testing is often done using SSRs, as the latter are highly polymorphic and consist a constellation of alleles of roughly equal frequency. Meagher (1986) and Meagher and Thompson (1986) noted that the power of a marker locus increases with its number of alleles and the evenness of its allele frequencies (higher power when the frequencies are more equal). Hence, a typical SSR has more power than a typical SNP. Indeed, Städele and Vigilant (2016) estimated that one SSR has roughly the power of six well-chosen SNPs.

Example 8.1. Suppose we have an offspring, its mother, and two potential candidates for the father (male 1 and male 2), which have all been scored at four biallelic SNPs. The resulting genotypes are as follows:

Marker locus	1	2	3	4
Offspring (g_o)	00	10	11	10
Mother (g_m)	00	11	11	10
Male 1 (g_{f1})	11	10	11	10
Male 2 (g_{f2})	00	00	11	10
$T(g_o \mid g_m, g_{f1})$	0	0.5	1	0.5
$T(g_o \mid g_m, g_{f2})$	1	1	1	0.5

The transmission probabilities, T, follow using the logic above. For example, consider marker locus 4 and male 1. The parental genotypes are 10 and 10 and the offspring is also 10. This can occur via two paths: 1 from mom and 0 from dad, each with probability 0.5, or 1 from dad and 0 from mom, giving $T = (0.5 \cdot 0.5) + (0.5 \cdot 0.5) = 0.5$. Note (short of a genotyping error) that marker locus 1 excludes male 1, as this male has a transmission probability of zero.

To compute the LRs, we need the allele frequencies in the reference population, which then yields the components needed to apply Equations 8.3a and 8.3b as

Marker locus	1	2	3	4
p (Freq allele 1)	0.1	0.2	0.5	0.3
$T(g_o \mid g_{f2})$	0.9	0.2	0.5	0.5
$P(g_o)$	0.81	0.32	0.25	0.42
$T(g_o \mid g_m)$	0.9	0.8	0.5	0.5

Applying Equation 8.3a, the LR values for male 2 for the four markers are, respectively, 1/0.9 = 1.11, 1/0.8 = 1.25, 1/0.5 = 2, and 0.5/0.5 = 1. Assuming linkage equilibrium, multiplying these together give the final LR value as 2.78. Now suppose that the mother is unknown. Equation 8.3b gives the LR values for the four markers in male 2 as, respectively, 0.9/0.81 = 1.1, 0.2/0.32 = 0.63, 0.5/0.25 = 2, and 0.5/0.42 = 1.2, giving a total LR of 1.65.

If we are willing to assign a prior probability, π_o, that the candidate is the father, we can use Bayes theorem (Equation 3.3b) to update this probability given the marker data (D). Following Li and Chakravarti (1985) and Thompson (1986b),

$$\Pr(\text{Paternity} \mid D) = \frac{\pi_o \Pr(D \mid H_1)}{\Pr(D)} = \frac{\pi_o \Pr(D \mid H_1)}{\pi_o \Pr(D \mid H_1) + (1-\pi_o)\Pr(D \mid H_2)} = \frac{\pi_o}{\pi_o + (1-\pi_o)/LR} \tag{8.4}$$

where LR is given by Equation 8.3a (known mother) or Equation 8.3b (unknown mother). In the simplest setting, if one has k potential fathers, we could take $\pi_o = 1/k$. More sophisticated approaches can be used that take into account additional information such as time spent with the mother and other behavioral observations (e.g., Neff et al. 2001; Hadfield et al. 2006).

The more general paternity problem is when one has a number of candidate fathers. This commonly arises in natural populations (especially in plants) and has been examined by a number of authors (e.g., Meagher 1986; Meagher and Thompson 1986; Marshall et al. 1998). The first step is excluding candidates to narrow the pool of potential fathers. As mentioned, genotyping errors can result in false exclusions, as can the presence of **null alleles** (denoting such an allele by Ø, 0Ø and 1Ø are erroneously scored as 00 and 11, respectively). Genotype calling errors and null alleles can be built into the likelihood function to accommodate these concerns (e.g., Marshall et al. 1998; Sieberts et al. 2002; Wang 2004; Kalinowski et al. 2006, 2007; Jones and Wang 2016). The next step is to *rank* the nonexcluded males, by computing the LR for each candidate, with the male with the largest LR value taken as the father (the MLE, or maximum-likelihood estimate; Appendix 4). If the mother is unknown, we can follow the same procedure, but now using Equation 8.3b to compute the LR.

While such an approach provides the ML solution for the most likely father among the candidate set, due to sampling, the actual father might not have been included. There is also the problem in deciding if the *most likely* male is *significantly more likely* than the next candidate (Meagher 1986; Marshall et al. 1998; Slate et al. 2000). Letting LR_1 and LR_2 denote the largest and second largest values, Marshall et al. suggested that the statistic $\Delta = \ln(LR_1) - \ln(LR_2)$ be used. In an attempt to account for multiple comparisons (Appendix 6), they evaluate the significance of Δ using simulations based on the number of candidate males and the proportion of males that are sampled. Alternative approaches for this multiple comparison problem have been suggested by Devlin et al. (1988, 1989), Smouse and Meagher (1994), Smouse et al. (1999), Neff et al. (2000a, 2000b), Nielsen et al. (2001), Geber et al. (2003), and Christie (2010).

Note that the above LR assignment approach assumes that a male is either the father or is completely unrelated. In reality, *relatives* of a candidate and/or the offspring (in addition to its mother) may be included in the sample. This is not much of a problem for distant relatives, but complications can arise when close relatives are included in the candidate set. It is especially problematic when sibs of the true father, or undetected sibs of the offspring, are considered as potential fathers. For example, full sibs of an offspring can give *higher* LR values than the true father, a complication that Olsen et al. (2001) referred to as **aunt and uncle (avuncular) effects**. When the mother is known, these concerns only arise with full sibs, but when the mother is unknown, half sibs also cause complications (Thompson and Meagher 1987; Marshall et al. 1998; Olsen et al. 2001; Ford and Williamson 2010). The simplest solution is to estimate Θ directly for each pair (see below), rather than trying to assign individuals into a few discrete relationship categories which may not have sufficient granularity given the genealogy of the population.

Finally, the most challenging situation arises when *both* parents of an offspring are unknown. Meagher and Thompson (1986) suggested that one first use Equation 8.3b to assess paternity and then also use it to assess maternity (using g_f in place of g_m). Among the candidate set of mothers and fathers that this procedure generates, one then tests all pairwise combinations of these high-value (large LR) parents, and the joint pair with the highest LR value is taken as the MLE of the parents. Meagher and Thompson showed this two-step approach is reasonable, as single-parent likelihoods are highly correlated with the joint-parent likelihoods.

Sibship Assignment and Joint Family Likelihoods

There are two rather different settings for sibship assignment. In the simple case, one has a natural family unit (typically a mother and her offspring) and the question is whether these offspring are all full sibs or a collection of full and half sibs (and, if the latter, the secondary question of how many distinct fathers were involved). In the more complex case, one has sampled a natural population of offspring with no obvious mother (or father), raising the question as to how many familial units are present, and then clustering individuals into collections of full sibs potentially nested within larger half-sib families.

While exclusion is possible with marker data from a parent–offspring dyad, such is *not* the case with a putative sib dyad. Here, exclusion would be ruling out the pair being half sibs if the mother was known; or ruling out being either half or full sibs when neither parent is prescribed. Suppose a candidate pair from the sample have genotypes of M_1M_2 and M_3M_4. Such a dyad could be half sibs, full sibs (the parents were M_1M_3 and M_2M_4), or unrelated. While simple exclusion is not possible from sib dyad data alone, we can compute an LR for these different categories (given marker allele frequency estimates). As one adds more individuals into the comparison, exclusion is possible, and especially if more putative sibs in the family are considered. This follows by noting the probability that both possible alleles from a parent are present in the genotypes of n offspring is $1-(1/2)^{n-1}$ (Wang 2007; Wang and Santure 2009).

As this discussion suggests, using multiple-individual (**joint**) likelihoods (units of multiple putative sibs) has significantly more power than pairwise (dyad) LR comparisons. Variants of this approach have been used by a variety of workers (Painter 1997; Almudevar and Field 1999; Thomas and Hill 2000; Thomas et al. 2000; Emery et al. 2001; Smith et al. 2001; Sieberts et al. 2002; Almudevar 2003; Bulter et al. 2004; Wang 2004, 2012; Fernández and Toro 2006; Wang and Santure 2009; Jones and Wang 2010; Almudevar and Anderson 2012; Huisman 2017). The basic logic of such a **full likelihood** approach follows by expanding the transmission probabilities used in simple paternity analysis. The complication is that one must run over all combinations of the possible transmission probabilities for a set of individuals under various assumptions about their relatedness. This is computationally demanding, especially in the more complex setting of no obvious family units. Further, this process must be computed separately for each marker (with the resulting LRs multiplied together). These calculations can be approximated using MCMC-based approaches (such as simulated annealing; Appendix 8), or by cleverly combining information from pairwise likelihoods (Wang 2012).

MARKER-BASED ESTIMATES OF RELATEDNESS: BACKGROUND

Finally, we turn from *assigning* pairs of individuals into *discrete relationship classes* to the *estimation* of *continuous measures* of *relatedness*. The possible single-locus IBD states between a pair of individuals are completely described by Jacquard's (1974) nine condensed coefficients of identity, Δ_i (Figure 7.2). These collapse to just three IBD states in the absence of inbreeding ($\Delta_7, \Delta_8, \Delta_9$; Chapter 7), which are denoted in the human genetics literature by the vector **k** = (k_0, k_1, k_2), corresponding to $(\Delta_9, \Delta_8, \Delta_7)$. Here k_i is the probability that the pair shares i IBD alleles, with $k_2 = 1$ corresponding to monozygotic twins, $k_1 = 1$ to parent–offspring, and $k_0 = 1$ to unrelated individuals. Note that some authors, following Cotterman (1940), denote Δ_8 as $2k_1$, and the k_i are often referred to as **Cotterman coefficients.** The most general setting would be to estimate the likelihoods (or, in a Bayesian setting, the posterior probabilities; Appendix 7) all the Δ_i coefficients for a specific pair (e.g., Example A4.8). However, in the vast majority of cases, our interest is in summary statistics of these identities, namely the coefficient of coancestry (Θ_{xy}; Equation 7.2a), an individual's inbreeding coefficient (f_x), and, more infrequently, the coefficient of fraternity (Δ_{xy}).

The most common reason for estimating relationships is to obtain variance components for quantitative traits (we treat genomic selection/prediction as an extension of variance estimation; Part II Chapters 32 and 33). Recall that $2\Theta_{xy}$ and Δ_{xy} are, respectively, the weights on the additive and dominance variances in the phenotypic covariance between x and y (Chapter 7). More generally, the weight on the k-level additive and ℓ-level dominance variance component, $\sigma^2(A^k D^\ell)$, is $(2\Theta_{xy})^k \Delta_{xy}^\ell$ (Equation 7.12). We use these weights to translate phenotypic covariances into variance estimates (Part II Chapters 22–33). When there is extensive inbreeding, this basic logic still applies, with the added complication that additional variance components arise. This is examined in detail in WL Chapter 11. A second important use of relationship estimates is to ***control for shared ancestry***, such as in a GWAS (Example 9.13; Chapter 20), where the goal is to contrast the effects of a feature (such as a particular SNP) once the impact of cryptic relatedness is removed. A final use is in estimating the level to which a target individual is inbred (estimating its inbreeding coefficient, f). Chapter 12 examines the impact of inbreeding in detail.

Our discussion will first consider a few key conceptual issues, such as the choice of a base population in defining IBD status, IBD probability versus IBD correlational definitions of relatedness, and variance in realized relationships. We then consider a variety of approaches that have been proposed for marker-based estimates of relatedness. While the logic for most methods is relatively independent of the types of markers being used, they often reflect the historical transition from assuming only a few modestly informative markers (alloyzmes, which often contain multiple alleles at modest frequencies) to a few very informative markers (SSRs; e.g., Queller et al. 1993; Blouin et al. 1996) and, finally to a large to huge number of relatively low-information markers (SNPs). There is a very rich (and growing) literature on marker-based relationship estimation. Reviews and commentaries can be found in Ritland (2000), van de Casteele et al. (2001), Blouin (2003), Milligan (2003), Garant and Kruuk (2005), Csilléry et al. (2006), Oliehoek et al. (2006), Weir et al. (2006), Frentiu et al. (2008), Yang et al. (2010, 2011c), Sillanpää (2011a), Toro et al. (2011), Gay et al. (2013), Bérénos et al. (2014), Jensen et al. (2014), Wang (2014, 2016), Conomos et al. (2016), Ackerman et al. (2017), Wang et al. (2017), Weir and Goudet (2017), and Goudet et al. (2018). Huang et al. (2014) provide extensions to polyploids. A masterful review of many of the issues considered here in a population-genetics framework is given by Thompson (2013), while a different perspective on these issues is offered by Speed and Balding (2015).

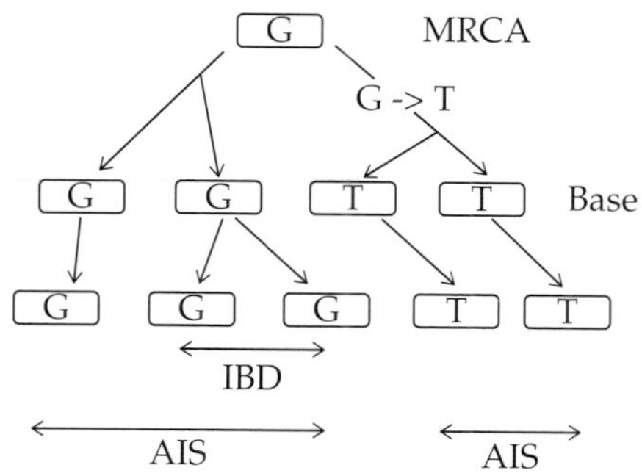

Figure 8.3 The genealogy, or **coalescent structure**, of a sample of five alleles at a single SNP. Arrows denote lines of descent, representing an unspecified number of generations. Note that the sampled alleles all trace back to a single sequence in the most recent common ancestor (MRCA). Note that a mutation (from the **ancestral state** G to the **derived state** T) occurred, so that two of the sampled alleles differ in state from the MRCA. From a genealogical standpoint, all of the G alleles trace back to the MRCA (and hence are all IBD using that individual as the reference). Likewise, both T alleles trace back to a common ancestor (more recent than the MCRA), where the mutation occurred. At some point far more recent than the MRCA, a base (or reference) population is designated. This is usually either taken as the present generation or no more than a modest number of generations back from the present. Conversely, the MRCA is on the order of $4N_e$ generations (N_e being the effective population size; WL Chapters 2 and 3). In the base population, all AIS (alike in state) alleles are assumed to be unrelated (not IBD). By moving the base population forward or backward in time, we change the IBD status of AIS alleles, even though the underlying coalescent structure remains unchanged.

IBD, Base Populations, and the Coalescent

A central organizing concept for treating genetic drift is the **coalescent** of a sample of alleles (variants at a single SNP), namely their genealogy (the *pedigree* of the sampled alleles; Figure 8.3). Under coalescent theory, all existing alleles trace back to some **most recent common ancestor** (**MRCA**) through a series of coalescent events (reviewed in detail in WL Chapter 2). In particular, suppose that k sequences are sampled. At some point in the recent past, two of these sequences had a common ancestor, so that the k genealogies in the sample now coalesce to $k-1$ genealogies. The process continues until the last two remaining genealogies coalesce at the MRCA (Figure 8.3). Under the coalescent framework, all of the alleles are "IBD" to the MRCA, as coalescent theory is *only* concerned with the *genealogy*, and *not* the *allelic states* of its descendants, namely, the contrast between *relatedness* versus *identity*. Mutation is overlaid on this genealogy to generate distinct AIS allelic classes (Figure 8.3).

The strictly genealogical coalescent view of IBD is at odds with the usual notion of IBD status, which further assumes no mutation since the common ancestor, so that IBD alleles *must* also be AIS. As noted by Powell et al. (2010), this apparent conflict is simply a difference in reference points. In particular, we define some **base** (or **reference**) **population** for IBD calculations (Figure 8.3). We assume that all individuals in the base are unrelated and not inbred (although we can modify either of these assumptions; Chapter 7). As shown in Figure 8.3, with respect to the specified base, only two of the sampled alleles are IBD, yet the sample contains multiple AIS alleles (two T and three G). With this same sample of five alleles, we can change the IBD status of these AIS alleles simply by moving the base forward or backward in time. Hence, while the MRCA is at a fixed (but unknown) generation, the generation chosen for the reference population is arbitrary, usually being specified for the convenience of the researcher. The base population is often taken as the current population.

The reason we have focused on a single SNP in the above discussion is due to an important subtlety with the MRCA. It is ***defined for a specific site***, and hence changes over a sequence, in that when recombination occurs, ***the coalescent structures can differ between even closely linked sites***. Thus, there is usually not a common MRCA for all the sites over a gene, much less over a chromosome or genome. A related issue is the notion of an **IBD chromosomal segment**, namely, for the sampled sequences, an unrecombined region tracing back to a single copy in the reference population. In many populations, IBD segments are smaller than a typical gene-coding region. As we will see, segmental IBD is exploited for fine-mapping by GWAS (Chapters 20 and 21).

An important issue that we will not consider in detail here is the estimation of relatedness in highly structured populations, namely, when the larger population consists of a number of subpopulations (demes). This has been examined by Anderson and Weir (2007), Wang (2011b), Thornton et al. (2012), Conomos et al. (2016), and Weir and Goudet (2017), and can be considered a special case of where one sets the reference population. Suppose we have a series of demes that have descended from a common ancestral population, and imagine a rare allele that is common in deme one. Setting the ancestral population as the base, individuals sharing this allele in deme one will be regarded as being closely related. However, if we set deme one as the base (and hence use deme one allele frequencies), the same set of individuals will be regarded, at best, as being weakly related (Anderson and Weir 2007).

There is one final important implication in setting the base population. As we have mentioned, a common use of relationship estimates is to obtain variance components. Estimates of the former (relatedness) vary with the base populations, which, in turn, influences estimates of the latter (variances), such that what we are estimating is the value of these variances in the base population. Hence, ***even when using the same data, as we change the assumed base population, we can change the variance estimates***.

Probabilistic versus Correlational Definitions of Relatedness

The inbreeding coefficient, f, underlies measures of relationship, both within an individual (f) and between individuals (Θ). For inbreeding, the comparison of IBD status is between the two alleles at a locus within an individual. For relationships, Θ_{xy} is based on choosing a random allele from the same locus in individuals x and y, with Θ_{xy} being the average value of f for the four combinations of allelic draws (Chapter 7). Equation 7.2b highlights this connection between inbreeding and coancestry: if z is the offspring of x and y, then $f_z = \Theta_{xy}$. Equation 7.3b provides a further connection in that $2\Theta_{zz} - 1 = f_z$.

Historically, there are two different definitions of f. While Wright (1922a) originally defined f as the ***correlation between gametes***, the popular concept of f being the ***probability of IBD*** is due to Cotterman (1940) and Malécot (1948). Wright (1965), Powell et al. (2010), and Wang (2014) noted that while these concepts are identical in many cases (when the reference population is suitably ancestral to the current one), there are also settings where they differ. While IBD-probability estimates of f and Θ must lie between zero and one, the correlational definition can generate negative values, which in many situations are biologically realistic. A negative estimate of f simply implies that the focal individual is *more heterozygous* than expected by chance in the reference population. As noted by Wang (2014), one example is the F_1 hybrid of two inbred founder lines, which generates a negative value of f (due to the excess heterozygosity in the F_1), regardless of how the parental information is combined to form a reference population. Similarly, if the reference population contains a number of inbred individuals, then an outbred individual can have a negative estimate of f (albeit typically very small), as it is likely to be more heterozygous than a random member of the reference population. Likewise, negative estimates of Θ simply imply that two individuals are *more dissimilar* than two random individuals, again in the context of a reference population. For example, if the reference population consists of a very high fraction of relatives, so that two randomly drawn individuals are likely related, then two unrelated individuals share *less* alleles than the population average, generating a negative θ.

Powell et al. (2010) and Wang (2014) argued that the correlation-based definition of f (and, therefore, Θ) is the most appropriate choice for marker-based estimates that take the current population (and hence the current allele, and genotype, frequencies) as the reference population. For known pedigrees, the obvious choice for a reference population is the collection of founders, but this raises issues of the rather arbitrary way in which founders are chosen. Typically, this is by convenience—known information—rather than from biological motivations. An example of the latter would be an isolated population with known founders (such as a captively bred population). More generally, it is convenient to simply take the current estimated allele frequencies as the reference population, which is assumed in most marker-based studies. Powell et al. (2010) showed the connection between f values associated with different base populations. Let f_1 and f_2 be the values based on two different assumed base populations, and let f_{st} be the correlation between a random gamete drawn from population one and a random gamete drawn from population two. Then

$$f_1 = \frac{f_2 - f_{st}}{1 - f_{st}} \tag{8.5}$$

Additional commentary on these issues is offered by Rousset (2002), Thompson (2013), and Speed and Balding (2015).

Finally, an important point was made by Powell et al. (2010), who argued that ***the objective of most marker-based IBD estimators is to predict the AIS status at the unobserved loci underlying the trait (or traits) of interest***. Obviously, if we knew these causative loci, we could score them directly. If not, we can try to predict their amount of AIS sharing using the marker data as proxies: individuals that share more marker loci IBD are likely to share more AIS causative alleles. A subtle caveat is that the quality of marker-based predictions declines as the distribution of marker allele frequencies (the **allele-frequency spectrum**; WL Chapter 2) departs from the distribution of causal allele frequencies (Chapters 20 and 21). For several reasons, such differences between marker and causal allele frequency spectra are often expected. First, there is typically an ascertainment bias wherein common SNPs (minor-allele frequency at least 1 of 5%) are chosen as markers, skewing the frequency distribution towards higher values. Second, it is thought that many (or most) SNP markers are selectively neutral (or nearly so), while the distribution of causative allele frequencies often reflects past selection (Chapter 21). This includes both direct artificial selection on the trait and prior natural selection. For example, alleles of large effect tend to be at low frequencies in natural populations, likely because they have been under negative natural selection in the past. IBD estimates from pedigree data (which estimate IBD independent of the underlying genetic architecture) are robust to this effect (Powell et al. 2010).

Realized versus Pedigree Kinship

The use of dense marker data highlights the important distinction, introduced in Chapter 7, between **pedigree kinship** and **realized kinship** (Wang et al. 2017). The former is the *expected* value of Θ calculated using a known pedigree (Chapter 7), while the latter is the *actual realization* of Θ for any particular pair of individuals, which can be estimated using dense marker data (e.g., Figure 7.5). This variation about the expected value is generated by **Mendelian sampling** (the segregation of the diploid genotype via meiosis into haploid gametes). Equation 7.7a shows that such variation occurs whenever at least two of the Δ_i probabilities for the pair of relatives are nonzero (which is always true except for clones and parent–offspring; Table 7.2). Dense SNP data can be used to capture this variation in relatedness, giving more accurate weights when using information from relatives (replacing expected values by their actual realizations). This is the basis of the genomic selection method known as **G-BLUP** (**genomic-BLUP**), wherein marker-estimated relatedness values are used in place of pedigree-generated values to improve the BLUP estimates (e.g., VanRaden 2007, 2008; Hayes et al. 2009; Part II Chapters 32 and 33).

Equation 7.7b gives the expected variance in Θ generated by Mendelian sampling at a single locus, $\sigma^2(\Theta)$. With n independently segregating loci, the associated variance for the average of Θ over these loci is simply $\sigma^2(\Theta)/n$. Because chromosomes (generally) assort independently, this result also applies when we sum chromosome-wide estimates. Consider the genome-wide value of Θ for a given pair of individuals (x, y) for a species with k chromosomes,

$$\widehat{\Theta}_{xy} = \sum_{i=1}^{k} w_i \, \widehat{\Theta}_{xy,i} \tag{8.6a}$$

where $\widehat{\Theta}_{xy,i}$ is the estimate for chromosome i, and w_i is the weight placed on this chromosome. Typically, if ℓ_i is the map length of chromosome i, and $\mathcal{L} = \sum^k \ell_i$ is the total genome map length, then $w_i = \ell_i/\mathcal{L}$. Chapter 17 discusses **map lengths—genetic distances** based on recombination, as opposed to **physical distances** based on base pairs—in detail. $\mathcal{L}$ is usually expressed in units of Morgans (the expected total number of crossovers) or centiMorgans, cM (100 cM = 1 Morgan.) The associated genome-wide variance becomes (Risch and Lange 1979)

$$\sigma^2(\widehat{\Theta}_{xy}) = \sum_{i=1}^{k} w_i^2 \, \sigma^2(\widehat{\Theta}_{xy,i}) \tag{8.6b}$$

The delicate issue is that, due to linkage, loci (or markers) on a given chromosome are *not* independent. Hence, if we have n loci on a specific chromosome, the variance associated with that chromosome is *not* $\sigma^2(\Theta)/n$, as there are not n independent sampling events. A number of approximations have been proposed to accommodate this complication from linked, and hence nonindependent, sites (Franklin 1977; Risch and Lange 1979; Suarez et al. 1979; Stam 1980; Stam and Zeven 1981; Donnelly 1983; Hill 1993a, 1993b; Guo 1994a, 1995, 1996; Visscher 1996, 2009; Visscher et al. 2006; Hill and Weir 2011, 2012; Thompson 2013). Most of this work is built around assuming the Haldane mapping function to model the recombination process (Example 8.3). As discussed in detail in Chapter 17, Haldane assumes a Poisson distribution for the number of crossovers within a region and no interference. Note that Equation 8.6b only refers to the **evolutionary variance**, the randomness associated with the Mendelian sampling process alone. There is an additional error, which we ignore here, generated by using a finite number of markers to estimate the Θ value for a given chromosome (or, more widely, genome).

One approach for obtaining the genome-wide variance is to use the single-locus variance (Equation 7.7b) in conjunction with some estimate, n_s, of the number of independently segregating chromosomal segments, taking the variance as $\sigma^2(\Theta)/n_s$. This approach is an approximation, attempting to model the recombinational behavior of a continuous chromosome by using an equivalent number of independently segregating segments (Donnelly 1983). Nonetheless, this is still a useful device for inquiring about some of the underlying factors influencing the variance in Θ. For relatives separated by a single round of meiosis (such as sib pairs), a simple metric for the number of segments is the **recombination** (or **segregation**) **index** (**RI**; Darlington 1939), which is the haploid chromosome number (k) plus the total number of crossovers per gamete (C), with RI = k + C. One can think of this as an approximation for the number of independent blocks of genetic material in a gamete.

As the number of generations back to the common ancestor increases, there is more opportunity for recombination, increasing n_s and thus decreasing the variance. This raises an important point. ***Sets of relatives with the same expected value of*** Θ (such as half sibs versus double first cousins) ***can have different expected standard deviations***. This occurs because although the single-locus variance (Equation 7.7a) is the same in these cases, the expected number of recombinational events differs over the paths separating the compared relatives, resulting in differences in n_s values (Rasmuson 1993; Hill and Weir 2011). An equivalent way of thinking about this concept is that while the single gene IBD probabilities can be the same, the two-locus IBD probabilities differ (Thompson 1988, 2013). A related issue is that the chromosomal-specific variance decreases with the total amount of recombination on that chromosome. Visscher (2006) observed this in humans, and found that the predicted values for $\sigma^2(\widehat{\Theta}_{xy,i})$ based on the map length of the chromosome (Example 8.3) generally correlated well with their empirical values. More generally, the genome-wide variance in Θ increases as chromosome number and/or total map length decreases, both of which have the effect of decreasing n_s.

Hill and Weir (2011) noted that both the skew and the CV increases as relatives become more distant (recall that Example 7.3 showed that the CV of Θ for half sibs was greater than for full sibs). As they succinctly stated: "Although the variance of actual relationships falls as individuals become more distant, its coefficient of variation rises, and so, exacerbated by the skewness, it becomes increasingly difficult to distinguish different pedigree relations from the actual fraction of the genome shared." Hence, ***the concept of discrete relationship classes quickly breaks down as relatives become more distant***, being replaced by continuous measures of relatedness that capture the realized values.

Example 8.2. A number of authors have considered the variance in the fraction of IBD genome sharing between a pair of full sibs. The IBD fraction is given by 2Θ (as Θ concerns drawing single alleles), and hence (Equation 3.10e) this variance is four times the variance of Θ. From Example 7.3, the single-locus variance for noninbred full sibs is 1/32, giving $\sigma^2(2\Theta) = 1/8$, for a standard deviation of 0.354. Using data on the number of crossovers on each human chromosome, Suarez et al (1979) obtained a standard deviation for genome IBD of 0.056 (corresponding to $n_s = 40$) when allowing for recombination, and a value of 0.082 ($n_s = 19$) under the assumption of no recombination (chromosomal segregation only). Using an analytic model of the crossover process, Risch and Lange (1979) obtained a value of 0.04 ($n_s = 78$). Rasmuson (1993) used the estimated segregation index for humans (75 in males and 100 in females) to approximate the standard deviation as $0.354/\sqrt{100} = 0.0354$. Empirical estimates from Visscher et al. (2006) using 4,400 pairs of quasi-independent full-sibs gave a value of 0.036, with realized values of 2Θ ranging from 0.375 to 0.617. Their analytic results (Equation 8.7a) suggested that the effective number of segments per chromosome (n_s) ranged from a high value of 6 for the largest chromosome down to a value of 2 for the smallest, for a total number of 85 effective chromosome segments.

Example 8.3. To see the logic underlying approximations accounting for linkage, we develop expressions for full- and half-sibs variances for the fraction of IBD sharing, 2Θ. We assume no recombinational interference (independence of crossovers) and a Poisson distribution for the number of crossovers along a chromosome (the Haldane map function assumptions; Chapter 17). Consider loci i and j linked together on a chromosome. Recalling Example 7.3, for a single locus in a half sib, $\sigma^2(2\Theta_i) = 4/64 = 1/16$. For two loci,

$$E[2\Theta_i 2\Theta_j] = [2(1-c_{ij})^2 + 2c_{ij}^2]/16$$

where c_{ij} is the recombination fraction between i and j. Hence,

$$\sigma(2\Theta_i, 2\Theta_j) = E[2\Theta_i 2\Theta_j] - E[2\Theta_i]\, E[2\Theta_j] = \frac{(1-2c_{ij})^2}{16} = \frac{\exp(-4m_{ij})}{16}$$

where m_{ij} is the Haldane map distance (this follows from Equation 17.4). Summing over all n loci on a chromosome, the variance in the IBD fraction for that chromosome becomes

$$\sigma^2(2\Theta) = n^{-2}(1/16)\sum_i \sum_j \exp(-4m_{ij}) \simeq \frac{4\ell - 1 + \exp(-4\ell)}{128\,\ell^2} \tag{8.7a}$$

where ℓ is the total map length of the chromosome (in Morgans). Details for the last approximation (which follows for large n) can be found in Guo (1996) and Hill (1993a).

Suppose that there are k chromosomes, the ith of which has length ℓ_i, for a total genome map length of $\mathcal{L} = \sum \ell_i$. Summing over Equation 8.7a yields the variance for the genome fraction of shared IBD alleles in half sibs as

$$\frac{1}{128\mathcal{L}^2}\left[4\mathcal{L} - k + \sum_{i=1}^{k} \exp(-4\ell_i)\right] \tag{8.7b}$$

which can be further approximated as $(4\mathcal{L} - k)/[128\mathcal{L}^2]$. Note that this shows the variance increases with decreasing values of k and/or $\mathcal{L}$. For human autosomes, $k = 22$ and $\mathcal{L} = 35$, giving an approximate variance of

$$(4 \cdot 35 - 22)/[128 \cdot 35^2] = 0.00075$$

for a standard deviation of 0.0274. Visscher (2009) noted that this approximation only differs at the fourth decimal place, 0.0272 versus 0.0274, when compared to the more exact value from Equation 8.7b. The latter can be computed by either using known values of ℓ_i, or by setting $\ell_i = 35/22$ (treating all chromosomes has having the same map length), as (in this case), both returned the same value (0.0272).

Similarly, using the same logic for full sibs yields

$$\frac{1}{64\mathcal{L}^2}\left[4\mathcal{L} - k + \sum_{i=1}^{k} \exp(-4\ell_i)\right] \tag{8.7c}$$

which can be further approximated as $(4\mathcal{L} - k)/[64\mathcal{L}^2]$, yielding a variance of 0.0015 and an associated standard deviation of 0.039. Nice summary tables of the variance for a wide variety of other relatives are given by Guo (1996) and Hill and Weir (2011). The latter also summarize expressions for the skewness.

Local versus Global Relatedness

The combination of Mendelian sampling (generating variation in realized relationships) coupled with the use of dense markers to detect such variation provides a very powerful tool. Consider a collection of full sibs. Because of this sampling, some will be more related, and hence (for a heritable trait) more phenotypically similar, than others. Visscher et al. (2006) used this within-sibship variation to estimate the heritability of human height. An even more granular analysis was proposed by Visscher et al. (2007). Just as Mendelian sampling can generate variation in the realized values of the **global relatedness** (the genome-wide average), it also generates *among-chromosomal differences* in the realized relationship *within* a pair of individuals (**localized relatedness**). Hence, we can estimate the heritability attributed to a *specific chromosome* by contrasting relatedness differences in a sample over just that chromosome. Because longer chromosomes have more recombination, and thus a smaller among-individual variance in fraction of shared IBD, smaller chromosomes tend to show more variance in relatedness, and therefore potentially more power to detect a chromosomal effect. A second implication for these among-chromosomal differences in variance is that smaller chromosomes in distant relatives are more likely to share no IBD segments than are larger chromosomes (Example 8.4).

Example 8.4. The connection between localized relatedness and variation in relatedness is through the notion of IBD chromosome segments. These are continuous stretches of DNA shared by the two focal individuals that have not undergone recombination since the common ancestor (**identity by recombination**). Following Donnelly (1983), Thompson (2013), and Speed and Balding (2015), consider two individuals that last shared a common ancestor g generations in the past. Further suppose there are k chromosomes, a total map length of $\mathcal{L}$, and a total genome size of M megabases (Mb). As relatives become more distant, the probability that they share *no* IBD segments increases. Thompson (2013) showed that one can approximate this by

$$\Pr(\Theta = 0) \simeq \exp\left[-\frac{(2g-1)\mathcal{L}}{2^{2g-1}}\right] \tag{8.8a}$$

Speed and Balding assumed human map lengths of 40.7 Morgans in females and 22.9 Morgans in males, taking $\mathcal{L}$ as their average, 31.8, and a value of $M = 2667$ Mb. For two individuals that last shared a common ancestor five generations in the past,

$$\Pr(\Theta = 0) \simeq \exp\left[-\frac{(2\cdot 5-1)\cdot 31.8}{2^{2\cdot 5-1}}\right] = 0.67$$

Hence, the *majority* of such relatives will share *no* IBD segments. When the common ancestor is three generations back, Equation 8.8a yields 0.007, so that 99.3% of all such pairs will share *at least* one IBD segment.

A closely related issue is the number, n_s, of shared IBD segments. Speed and Balding (2015) approximated this as

$$E[n_s] \simeq \frac{k + 2\mathcal{L}g}{2^{2g-1}} \tag{8.8b}$$

Again taking $g = 5$, $E[n_s] = (22 + 2\cdot 31.8\cdot 5)/(2^9) = 0.66$, while for $g = 3$, $E[n_s] = 6.65$. These values imply that most chromosomes in both of these settings ($g = 3$ and 5) will share *no* IBD segments, generating considerable variance in IBD *among* the chromosomes of a pair of relatives. To further emphasize this point, consider the expected mean size (in megabases) of any such shared segment. Speed and Balding showed that the mean length, *conditional* on at least one shared region being present, is

$$\mu_s \simeq \frac{M}{k + 2\,\mathcal{L}\,g} \tag{8.8c}$$

Even though the majority of individuals with a common relative five generations in the past share no IBD segments, *when* they do, its average length is $2667/(22 + 2\cdot 31.8\cdot 5) = 7.8$ Mb. For $g = 3$, this conditional mean length is 12.5 Mb. This process, wherein fewer and fewer relatives share any IBD segments, but those that do have segments of at least modest size, is what generates the ever increasing skewness in IBD as relatives become more distant (Hill and Weir 2011). Many of the above results appear in the literature as a function of the total number, m, of meioses that separate a pair from their common ancestor, with m replacing $2g$ in the above expressions (Equations 8.8a–8.8c). Chapter 17 further examines the lengths of shared IBD regions.

MARKER-BASED ESTIMATES OF THE COEFFICIENT OF COANCESTRY, Θ

Finally, we turn to the estimation of relatedness statistics from marker data. Again, the issue is how to translate observed AIS (alike in state) sharing at the markers into estimates of IBD sharing at either a global (genome-wide) or local (chromosomal segment) level. This usually requires us to specify a reference population (but see Weir and Goudet 2017; Goudet et al. 2018), with the allele and genotype frequencies in the reference providing us with the expected amount of AIS sharing between random individuals. For a random-mating (i.e., unstructured) population, allele frequencies are sufficient to specify the expected AIS sharing. However, when there is structure in the base population or in the sample (such as inbreeding or when a substantial fraction of the sampled individuals are relatives), allele frequencies alone are not sufficient to adjust AIS status, and we will examine how to accommodate these concerns.

Our discussion largely frames estimators in terms of biallelic SNP markers, but most can easily be extended to multiallelic markers, such as SSRs. Most of these approaches use

single-point (or **SNP-by-SNP**) **estimates**, which treat markers as being independent and then amalgamate these single-locus results into a final composite estimate. Note that single-point estimators do not require any mapping (i.e., positional) information on the markers. As marker density increases, nearby markers become increasingly correlated (linkage disequilibrium; Chapters 5 and 17). This is typically dealt with by **trimming** the number of markers to form a set of largely independent loci (e.g., Purcell et al. 2007; Huisman et al. 2016), but this loses information. More recent estimators exploit information provided by the pattern of linkage disequilibrium, either by weighting adjacent markers by their correlations (e.g., Day-Williams et al. 2011; Speed et al. 2012; Wang et al. 2017) or using the IBD status of *chromosomal segments*, also known as **recombinational IBD** (Example 8.4), as opposed to using IBD *alleles* at individual loci (reviewed by Browning and Browning 2012).

Our focus will be on estimating the three most widely reported relatedness metrics: the coefficient of coancestry (Θ_{xy}; which has the fraction of IBD diploid genome sharing, $2\Theta_{xy}$, as a special case, where we consider both alleles at a locus, rather than a single random one), the inbreeding coefficient of an individual (f_x), and the coefficient of fraternity (Δ_{xy}). We consider these in turn, largely focusing on **method-of-moments** estimators, wherein one generates statistics whose expected value equals the parameter of interest. Maximum likelihood provides a more rigorous statistical framework, but can be computationally quite demanding. Appendix 4 examines ML estimates of continuous relationship metrics.

Single-point IBD Estimators

There are a number of proposed methods that translate data from a SNP into estimates of Θ. The basic approach is as follows. Consider two individuals, x and y. We denote the two alleles in x by a and b (which may be alike in state), and similarly in y by c and d. The **molecular similarity** (or **molecular similarity index**) at locus ℓ between x and y is defined by

$$S_{xy,\ell} = \frac{I_{ac} + I_{ad} + I_{bc} + I_{bd}}{4} \tag{8.9a}$$

where I_{ad} is an indicator function that equals one if a and d are AIS, and otherwise is zero. For biallelic loci (such as most SNPs), $S_{xy,\ell}$ takes on values of 0, 1/2, or 1. A value of zero occurs if x and y are different homozygotes, while $S_{xy,\ell} = 1$ if they are the same homozygote. All other biallelic locus combinations yield a value of 1/2. A value of 1/4 requires at least three distinct alleles, and values of 3/4 do not occur, because if the first three combinations are 1, so is the last (Oliehoek al. 2006). Toro et al. (2002) referred to Equation 8.9a as **molecular coancestry**, as when AIS equals IBD, then $E[S_{xy,\ell}] = \Theta_{xy}$, which follows from the definition of the coefficient of coancestry (Chapter 7). The average over L marker loci yields an estimate of

$$\widehat{\Theta}_{xy} = \frac{1}{L}\sum_{\ell=1}^{L} S_{xy,\ell} \tag{8.9b}$$

The problem with this simple estimator is the equating of AIS with IBD, as unrelated individuals can share AIS alleles. One needs to assign a base (or reference) population and use the expected genotype frequencies in this base to provide a correction for AIS status among randomly drawn individuals. Any such adjustment makes some key assumptions, usually that one has a random sample from the true assumed reference population, and that individuals within the sample are largely unrelated and largely outbred. If the sample contains a sufficient number of inbred individuals and/or relatives, this can bias the correction for alleles that are AIS but not IBD. While this bias may be small, it can potentially be rather large in small populations/samples (Wang 2002; Goudet et al. 2018). A related, but subtle, issue is that the assumed base population *itself* (as opposed to the sample) may contain a significant fraction of relatives and/or inbred individuals (such as might be expected in captive-bred populations). A random pair of individuals from this population has some reasonable chance of being related and/or inbred. IBD estimates in this setting are best thought of as ***differences in IBD status relative to that expected in random draw from the reference population*** (Goudet et al. 2018), and, as such, can take on negative values.

To proceed under the assumption of an appropriate sample from a random-mating reference population, let s_ℓ denote the probability that two randomly drawn alleles in the base are AIS. As shown by Lynch (1988c), the expected value for $S_{xy,\ell}$ is given by

$$E[\,S_{xy,\ell}\,] = \Theta_{xy} + (1 - \Theta_{xy})s_\ell \tag{8.10a}$$

namely, Θ if they are related (IBD, hence AIS), and s_ℓ if they are not related, the latter occurring with probability $1 - \Theta$ (not IBD). Here s_ℓ is the expected AIS sharing among a randomly drawn pair of individuals from the reference population. Underestimating of the amount of AIS sharing between random individuals results in overestimation of relatedness, while overestimation of AIS results in underestimation of relatedness.

For a biallelic locus in Hardy–Weinberg,

$$s_\ell = p_\ell^2 + (1 - p_\ell)^2 = 1 - 2p_\ell(1 - p_\ell) \tag{8.10b}$$

where p_ℓ is the SNP allele frequency in the reference population, a value typically estimated from the sample. When this is done, Queller and Goodnight (1989) recommend first removing the pair under consideration from the sample, and then estimating p_ℓ using this reduced sample. While this is reasonable, the pair may share a rare allele, in the extreme only present in the pair, in which case the reduced sample estimate would be zero. If the sample size is modest to large, including the pair of interest when estimating p_ℓ should introduce little bias, provided the average coancestry in the sample is very close to zero.

Rearranging Equation 8.10a suggests a more general method-of-moments estimator (Lynch 1988c)

$$\widehat{\Theta}_{xy,\ell} = \frac{S_{xy,\ell} - s_\ell}{1 - s_\ell} \tag{8.10c}$$

(note the similarly to Equation 8.5). If we weight all markers equally, then a composite estimate of Θ_{xy} based on L markers (either globally, with the markers spread across the genome, or locally, with the markers restricted to a chromosome or chromosomal segment) is given by

$$\widehat{\Theta}_{xy} = \frac{1}{L}\sum_{\ell=1}^{L} \widehat{\Theta}_{xy,\ell} \tag{8.10d}$$

Lynch and Ritland (1999) suggested that equal weighting is not optimal (in terms of producing the smallest sampling variance for $\widehat{\Theta}_{xy}$), and that each $\widehat{\Theta}_{xy,\ell}$ estimate should be weighted by the inverse of its sampling variance. Eding and Meuwissen (2001) obtained the sample variance of Equation 8.10c, assuming that s_ℓ is known without error, as

$$\sigma^2\left(\widehat{\Theta}_{xy,\ell}\right) = \frac{s_\ell + \Theta_{xy}(1 - 2s_\ell) - \Theta_{xy}^2(1 - s_\ell)}{1 - s_\ell} = w_\ell^{-1} \tag{8.10e}$$

Hence, a more optimal estimator is

$$\widehat{\Theta}_{xy} = \frac{1}{W}\sum_{\ell=1}^{L} w_\ell\,\widehat{\Theta}_{xy,\ell}, \quad \text{with} \quad W = \sum_{\ell=1}^{n} w_\ell = \sum_{\ell=1}^{n} \sigma^{-2}\left(\widehat{\Theta}_{xy,\ell}\right) \tag{8.10f}$$

Equation 8.10e shows that optimal weights are functions of the parameter we are attempting to estimate, but one could substitute an initial estimate and then iterate Equations 8.10c–8.10f until convergence.

Negative estimates of Θ can arise when $S_{xy,\ell} < s_\ell$ over a large number of loci, implying that these individuals are *less related than expected by chance* (relative to random individuals drawn for the reference population). Assuming $s_\ell = 0$ eliminates this problem (the assumption behind Equation 8.9b), but also introduces bias (Speed and Balding 2015; Ackerman et al. 2017). Oliehoek al. (2006) obtained an adjusted value for s_ℓ to ensure that all the $\widehat{\Theta}_{xy}$ are nonnegative, but again this likely introduces some slight bias. Such attempts to force estimates to be nonnegative are largely based on the incorrect interpretation that Θ *must* be positive. As discussed above, negative estimates are both expected and biologically interpretable.

Alternatively, one can base an estimator on the *total number* of shared AIS alleles over L loci (Day-Williams et al. 2011). First, define

$$S_{xy} = \sum_{\ell=1}^{L} S_{xy,\ell} \tag{8.11a}$$

which is the total number of AIS alleles shared between x and y over all scored markers. This has expected value

$$E[\,S_{xy}\,] = \sum_{\ell=1}^{L} E[\,S_{xy,\ell}\,] = \sum_{\ell=1}^{L} [\Theta_{xy} + (1-\Theta_{xy})s_\ell] = L\Theta_{xy} + (1-\Theta_{xy})\sum_{\ell=1}^{L} s_\ell \tag{8.11b}$$

Rearranging yields the **Day-Williams** estimator

$$\widehat{\Theta}_{xy,DW} = \frac{S_{xy} - \sum_{\ell=1}^{L} s_\ell}{L - \sum_{\ell=1}^{L} s_\ell} \tag{8.11c}$$

The distinction between the estimators given by Equations 8.10d and 8.11c is that the former corrects for AIS on a *locus-by-locus* basis, while the latter corrects for the *summed total* of AIS over all loci. Equation 8.10b gives the values of s_ℓ under the assumption of an appropriate sample from a random-mating reference population.

When these assumptions about either the sample, or the base population itself, are incorrect, these AIS corrections can be biased. A simple, but robust, way around this complication is the **Weir–Goudet estimator** (2017). This can be expressed as either the total number (S) or proportion ($M = S/L$) of AIS alleles over L loci,

$$\widehat{\Theta}_{xy,WG} = \frac{S_{xy} - \overline{S}}{L - \overline{S}} = \frac{M_{xy} - \overline{M}}{1 - \overline{M}} \tag{8.12a}$$

where n is the sample size and

$$\overline{S} = \frac{1}{n(n-1)} \sum_{j\neq i}^{n} \sum_{i=1}^{n} S_{ij} \tag{8.12b}$$

is the average sharing over all pairs of individuals in the sample (with $\overline{M} = \overline{S}/L$). The Weir–Goudet estimator measures how much a given pair departs from a *random pair* comparison and, as such, can take on negative values. The latter simply means that this pair is less similar than an average comparison in the sample. This estimator does not attempt to estimate base-population allele frequencies nor does it assume Hardy–Weinberg, rather it takes the *sample as the base population*, which could contain a significant fraction of inbred individuals and/or relatives. Again, for a small sample one could compute $\overline{S}$ by first removing the pair of interest.

Single-point Correlation Estimators

Closely related to IBD probability-based estimators (Equations 8.9–8.12; these are also known as **allele-sharing methods**) are estimators based on the average correlation among alleles. As mentioned, f (and by extension Θ) can be regarded as either an IBD probability or a correlation among alleles (Cotterman 1940; Malécot 1948). This equivalence of Θ with a correlation immediately suggests why some estimates can be negative, with $\Theta < 0$ implying that the two individuals are more *dissimilar* than expected by chance. This provides a complementary viewpoint to negative estimates based on IBD probabilities. Negative estimates of the latter are interpreted as the IBD probability for a specific pair being less than the expected IBD probability for a random pair from the base population.

To proceed, code the two alleles at a given SNP locus (ℓ) as 0 or 1, and let b_ℓ denote the value of a randomly drawn allele from this SNP, where $E[b_\ell] = p_\ell$ is the frequency of the reference allele (1), and $\sigma^2(b_\ell) = p_\ell(1-p_\ell)$. Analogous to Equation 8.10a, the probability that a randomly drawn allele from x and y are both 1 is

$$\Pr(b_{x,\ell} = b_{y,\ell} = 1) = \Theta_{xy} p_\ell + (1 - \Theta_{xy}) p_\ell^2 \tag{8.13a}$$

which rearranges to

$$\begin{aligned}\Theta_{xy,\ell} = \frac{\Pr(b_{x,\ell} = b_{y,\ell} = 1) - p_\ell^2}{p_\ell(1-p_\ell)} &= \frac{E[b_{x,\ell}\, b_{y,\ell}] - (E[b_\ell])^2}{p_\ell(1-p_\ell)} \\ &= \frac{\sigma(b_{x,\ell}, b_{y,\ell})}{\sigma(b_{x,\ell})\sigma(b_{y,\ell})} = \rho(b_{x,\ell}, b_{y,\ell})\end{aligned} \tag{8.13b}$$

namely, the correlation, ρ, between a random allele in x and a random allele in y. Toro et al. (2011) recode Equation 8.13b by using the fraction of reference alleles at locus ℓ in individual x,

$$g_{x,\ell} = \begin{cases} 0 & \text{for } 00 \\ 1/2 & \text{for } 10 \\ 1 & \text{for } 11 \end{cases} \tag{8.13c}$$

to obtain the single-locus estimate of Leutenegger et al. (2003)

$$\widehat{\Theta}_{xy,\ell} = \frac{(g_{x,\ell} - p_\ell)(g_{y,\ell} - p_\ell)}{p_\ell(1-p_\ell)} \tag{8.13d}$$

where p_ℓ is the allele frequency in the base population, which is generally replaced by the average allele frequency, $\widehat{p}_\ell$, in the sample. This use of the sample average (in place of the true value) can introduce significant bias, especially for rare alleles (Yang et al. 2010b).

Similarly, we can consider the correlation in reference allele *copy number* between x and y. Let $T_{x,\ell}$ denote the number of copies of allele 1 (from SNP ℓ) that x carries, where $T_{x,\ell} = 0, 1$, or 2, corresponding, respectively to SNP genotypes of 00, 10, and 11 (note that $T_{x,\ell} = 2g_{x,\ell}$). Hence, $E[T_\ell] = 2p_\ell$, yielding a contribution to the covariance in T_ℓ (the **molecular covariance**; Toro et al. 2011) between two relatives of

$$(T_{x,\ell} - 2p_\ell)\,(T_{y,\ell} - 2p_\ell)$$

The variance in T_ℓ is $\sigma^2(T_\ell) = E[T_\ell^2] - (E[T_\ell])^2$, which under, Hardy–Weinberg, becomes

$$\left[0^2 \cdot (1-p_\ell)^2 + 1^2 \cdot 2 \cdot p_\ell(1-p_\ell) + 2^2 \cdot p_\ell^2\right] - (2p_\ell)^2 = 2p_\ell(1-p_\ell)$$

yielding a correlation in T between x and y at SNP ℓ of

$$\rho(T_{x,\ell}, T_{y,\ell}) = \frac{\sigma(T_{x,\ell}, T_{y,\ell})}{\sigma(T_{x,\ell}) \cdot \sigma(T_{y,\ell})} = \frac{(T_{x,\ell} - 2p_\ell)\,(T_{y,\ell} - 2p_\ell)}{2p_\ell(1-p_\ell)} \tag{8.14a}$$

To relate $\rho(T_{x,\ell}, T_{y,\ell})$ with Θ_{xy}, substituting $T_{x,\ell} = 2g_{x,\ell}$ into Equation 8.14a yields

$$\begin{aligned}\rho(T_{x,\ell}, T_{y,\ell}) &= \frac{(2g_{x,\ell} - 2p_\ell)\,(2g_{x,\ell} - 2p_\ell)}{2p_\ell(1-p_\ell)} \\ &= 2\left[\frac{(g_{x,\ell} - p_\ell)(g_{y,\ell} - p_\ell)}{p_\ell(1-p_\ell)}\right] = 2\widehat{\Theta}_{xy,\ell}\end{aligned} \tag{8.14b}$$

Again, the issue is how to translate the single-locus correlation estimators (Equation 8.14a) into a composite single estimate. Averaging over loci yields

$$\widehat{\Theta}_{xy,V2} = \frac{1}{2L}\sum_{\ell=1}^{L} \rho(T_{x,\ell}, T_{y,\ell}) \tag{8.15a}$$

$$= \frac{1}{2L}\sum_{\ell=1}^{L} \frac{(T_{x,\ell} - 2p_\ell)\,(T_{y,\ell} - 2p_\ell)}{2p_\ell(1-p_\ell)} \tag{8.15b}$$

Widely used (e.g., Hayes et al. 2009; Yang et al. 2010b, 2011c), this estimator appears to be proposed first by VanRaden (2007, 2008) and is often called **VanRaden's second method**. The concern is that SNPs with rare alleles are much more heavily weighted, with weight $\sim 1/q_\ell$ where q is the minor-allele frequency. At first blush, this seems reasonable, as the sharing of rare alleles does indeed provide stronger evidence for IBD than the sharing of common alleles. However, the base-population frequencies of such rare alleles are difficult to estimate with precision, and hence these large weights have very large sampling errors. Yang et al. (2011a) proposed a regression-based correction to adjust Equation 8.15b for using estimated frequencies in place of the true frequencies.

An alternative, and potentially more robust, estimator was also offered by VanRaden (2008), namely **VanRaden's first method**,

$$\widehat{\Theta}_{xy,V1} = \frac{\sum_{\ell=1}^{L} (T_{x,\ell} - 2p_\ell)\,(T_{y,\ell} - 2p_\ell)}{4\sum_{\ell=1}^{L} p_\ell(1-p_\ell)} \tag{8.16a}$$

This estimator weights all loci equally, and can be thought of as the sum of the single-locus covariances in allelic copy number divided by the sum of the variances in allelic copy number. Note that all of these correlation-based estimators (Equations 8.13d, 8.15b, and 8.16a) require unbiased estimates of the base-population allele frequencies and further assume that the base population is an unstructured, randomly mating population. When either of these assumptions fail, correlation-based approaches yield potentially biased estimates. Both of the VanRaden estimators (Equations 8.15b and 8.16a) are special cases of the estimator

$$\widehat{\Theta}_{xy,s} = \frac{\sum_{\ell=1}^{L} (T_{x,\ell} - 2p_\ell)\,(T_{y,\ell} - 2p_\ell)\,[2p_\ell(1-p_\ell)]^{s}}{2\sum_{\ell=1}^{L}[2p_\ell(1-p_\ell)]^{(1+s)}} \tag{8.16b}$$

where $s = 0$ and $s = -1$ correspond, respectively, to VanRaden's first and second methods. The choice of s places different amounts of emphasis on SNPs with rare alleles, with $s = 0$ weighting loci equally, and $s = -1$ assigning more weight to rare alleles.

Example 8.5. Wang et al. (2017) showed that many of the above method-of-moments estimators of Θ (e.g., Equations 8.11c, 8.15b, 8.16b) can be written as special cases of a more general single-locus estimator

$$Z_\ell(a_\ell) = \frac{T_{x,\ell} \cdot T_{y,\ell} - a_\ell(T_{x,\ell} + T_{y,\ell}) + 2a_\ell p_\ell - 4p_\ell^2}{p_\ell(1-p_\ell)} \tag{8.17a}$$

where α_ℓ are user-specificed weights (see below). Specifically, they showed that $E[Z_\ell(a_\ell)] = \Theta$ for all values of a_ℓ (when the true allele frequencies are used). These locus-specific values are then weighted to form the composite estimate

$$\widehat{\Theta}_{xy}(\mathbf{a}, \mathbf{w}) = \sum_{\ell=1}^{L} w_\ell \cdot Z_\ell(a_\ell), \quad \text{where} \quad \sum_{i=1}^{L} w_i = 1 \tag{8.17b}$$

with $\mathbf{a} = (a_1, \cdots, a_L)^T$ the vector of locus-specific a values for Equation 8.17a, and $\mathbf{w} = (w_1, \cdots, w_L)^T$ the vector of weights for the Z_ℓ. From Equation 3.11b,

$$\sigma^2\left[\widehat{\Theta}_{xy}(\mathbf{a},\mathbf{w})\right] = \sum_{i=1}^{L} w_i^2 \sigma^2\left[Z_i(a)\right] + \sum_{j\neq i}^{L}\sum_{i=1}^{L} w_i w_j \sigma\left[Z_i(a), Z_j(a)\right] \tag{8.17c}$$

In the absence of LD, the double sum is zero, as genotypes are uncorrelated over loci. Wang et al. (2017) found that the optimal weights that minimize Equation 8.17c are functions of the true relatedness (e.g., Equation 8.10e). They suggested an improved estimator for Θ by using a two-step process: first, one uses a standard method to provide initial estimates for Θ. These are then substituted into their expressions to obtain the optimal weights in Equation 8.17b to generate an updated estimate.

Usually when estimating Θ, the goal is to obtain all pairwise combinations among a collection of individuals, with the results compactly displayed in matrix form (Chapter 9), namely the **numerator relationship matrix, A**. This matrix is deployed when using BLUP, in genomic prediction, and in correcting for cryptic relationships in a GWAS (Chapters 10 and 20; Part II Chapters 32 and 33). The ijth element of $\mathbf{A}$ is $2\Theta_{ij}$ (the weight on the additive genetic variance; Equation 7.11a), and can be estimated either by using expected values from a pedigree (Chapter 7), or the realized values from marker estimates. In the latter setting, this matrix is usually denoted by $\mathbf{G}$ and is called the **genomic relationship matrix** (VanRaden 2008). Example 9.14 shows how both VanRaden estimators for all pairwise comparisons in a sample can be compactly written in matrix form. Part II Chapter 32 examines methods for combining both pedigree and marker information, generating a **pedigree-genomic relationship matrix, H** (Legarra et al. 2009; Christensen and Lund 2010).

Which Single-point Θ Estimator Should be Used?

Many of the early studies examining the effectiveness of various marker-based estimators of Θ were done in the era when just a few dozen SSRs, or up to a few hundred SNPs, were scored. Generally, these studies found that the correlation between values predicted from known pedigrees and values predicted from markers was modest. Such disparity is is expected in part because we are contrasting an expected value (pedigree) with a specific realization (markers). The latter, provided that we can estimate the realized Θ with precision, is the more accurate measure of relatedness. A second source of error, which does not diminish as one adds more markers, is potentially biased corrections for AIS. Most methods use a Hardy–Weinberg correction for AIS allele sharing in nonrelatives (e.g., Equation 8.10b), which requires accurate estimates of the base-population allele frequencies, and more subtly, that the base population itself is randomly mating. Typically, allele frequencies are estimated from a reference sample, which leads to biased estimates if the sample is not representative of the base or if the sample contains any relatives or inbreds (Wang 2014). These complications can be dealt with by using the reference sample as the base, and then adjusting for the value of a randomly drawn pair from the reference (e.g., the Weir–Goudet estimator; Equation 8.12).

Despite these concerns, more recent studies using significantly more markers found much better performance. For example, good performance has been found using 10^3 SNPs in captively bred populations of zebra finches (Santure et al. 2010), pigs (Lopes et al. 2013), and cattle (Rolf et al. 2010), and in natural populations of soay sheep (Bérénos et al. 2014). Goudet et al. (2018) suggested that 10^4 SNPs will very efficiently estimate realized Θ values in most settings, especially when using the reference sample as the base and an AIS correction based on sample properties (Equation 8.12b rather than Hardy–Weinberg). All of this leads to the current consensus that estimating relatedness using on the order of 10^4 SNPs provides a more accurate picture of shared relatedness than a pedigree constructed without error (Wang 2016). Early contrary views (e.g., Van Horn et al. 2007) were influenced by the poor performance of estimators using only a modest number of markers.

In terms of the relative performance of specific estimators, if the population sample consists mainly of outbred and unrelated individuals, most methods perform equally well, with some methods slightly better for close relatives and others slightly better for more distant relatives. When the sample size, and/or the actual base population, is small, such that both inbreds and relatives are common, then the Weir–Goudet estimator performs better, as it does not assume Hardy–Weinberg and does not rely on estimates of base-population allele frequencies. Their AIS corrector, $\overline{S}$, also takes into account correlations among marker alleles at different loci, so this class of estimators is at least somewhat robust to SNP oversampling (high correlations among SNPs). Methods that are sensitive to rare alleles, namely when single-locus results are weighted differentially (e.g., Equations 8.10f, 8.13d, and 8.15b), are very sensitive to estimates of these frequencies, and can be problematic.

LD-informed Estimators

Estimators leveraging linkage disequilibrium information—correlations between alleles at different markers—fall into two distinct categories. The first approach is a straight-forward extension of combining the single-locus estimators, with the weighting scheme for the site-specific estimates being modified to adjust for correlations among marker alleles at different loci (Day-Williams et al. 2011; Speed et al. 2012; Wang et al. 2017). Equation 8.17c shows their basic variance structure, with the double sum of covariances accounting for the effects of linkage and LD. These estimates return improved (lower sampling variance), and usually more robust, estimates of Θ.

The second approach uses very dense SNP data to locate shared stretches of IBD—**segmental IBD**—in distant relatives (Purcell et al. 2007; Browning 2008; Kong et al. 2008; Albrechtsen et al. 2009; Browning and Browning 2010, 2011, 2012, 2013; Gusev et al. 2009; Manichaikul et al. 2010; Han and Abney 2011; Huff et al. 2011; Moltke et al. 2011; Lawson and Falush 2012; Durand et al. 2014; Li et al. 2014; Glazner and Thompson 2015; Staples et al. 2016; Ramstetter et al. 2017). In Europeans, IBD segments as small as 2 cM (roughly 2 Mb) can routinely be detected using standard SNP arrays (Browning and Browning 2012), which represent individuals with a common ancestor of around 25 generations in the past. Recall from Example 8.4 that most pairs with such distant ancestors likely share *no* IBD segments (Equation 8.8a gives this probability as 99.4%), but *when they do*, the segments are often long enough to be detectable. Applying Equation 8.8c (with the values of M and L used in Example 8.4) gives an expected size of 1.7 Mb for 25 generations, but there is considerable variation about this value, with some segments being substantially longer. The signal used to declare a region as being IBD is an anomalously long run of matching (AIS) SNP alleles across the two individuals. This requires a sufficient number of polymorphic SNPs, with the minimal segment length that can be detected being a function of the amount of polymorphism per unit of recombination. As such, the minimal size can vary dramatically over both populations and species. It can also vary dramatically *within* a genome or chromosome, as regions of low recombination (such as around many centromeres) tend to have greatly reduced levels of polymorphism (WL Chapter 8).

The computational approach for detecting IBD segments is based on **hidden Markov models** (**HMMs**). Full details can be found in most of the above references, but the basic idea is as follows. One starts at a given position on a chromosome with some estimate of the site IBD status based on marker information. The IBD state at an adjacent site is then predicted using a transmission probability (based on map distance and marker information) of remaining in the current state (the adjacent site has the same IBD status as the starting site) or moving to a new state (the new site has a different IBD status). One then moves to the next adjacent site and repeats the process. The resulting output is a probability map for runs of IBD across a chromosomal region. Underpinning this sophisticated mathematics is a rather simple idea: as was the case for rare alleles, ***individuals that share rare*** (highly improbable) ***haplotypes are likely related***.

It is important to recall one of the lessons from Example 8.4, namely, that distant relatives only share short segments of IBD (if they share any at all) that are randomly distributed over their genomes, resulting in a chimeric organization of shared IBD regions over a chromosome. Suppose that individuals x, y, and z all shared the same relative in the distant past. By chance, pairs from this group will share different IBD segments from this ancestor (again, if they share any at all). For example, x and y may share a segment on (say) chromosome 1, but x and z may not, but rather could share a segment on chromosome 3, while y and z share a segment on chromosome 5. Similarly, x and w may share a different common ancestor, so that while part of chromosome 1 in x and y has an IBD segment from the x–y common ancestor, a nearby region in x and new individual w may share a different IBD region from the x–w common ancestor. Hence, when the sample contains a set of individuals with only distant ancestors, ***the genome of a focal individual is a patchwork of IBD segments***, with most of the genome not sharing any IBD blocks with others in the sample, while the rest is a collection of small segments that are IBD with different sets of sampled individuals.

Quantification of the amount of shared IBD segments between a pair of individuals provides a direct measure of the total fraction of the genome shared by the pair (total length of shared IBD segments divided by diploid genome size), and hence a value of 2Θ. While physical distance (e.g., base pairs) are often used, the more appropriate metric is the amount of recombination (e.g., cMs), as the size of IBD regions is a function of their recombination rate. This is the continuous *relatedness* measure for this pair, which is often attempted to be mapped into a discrete *relationship* category, such as fifth cousins. Several of the programs that search for IBD segments return such genealogical estimates, usually in the form of the total number, m, of meioses that separate the two focal individuals. For example, Huff et al. (2011) developed an ML estimator for m, while Purcell et al. (2007) considered the smallest value of m that is consistent with the number and length of IBD segments. Applying Equation 7.6, we can translate an estimate of m into an estimate of Θ, with $\widehat{\Theta} = (1/2)^{\widehat{m}}$. However, as noted earlier, as the common ancestor becomes more distant, both the skew and CV of the amount of IBD sharing increase, creating ever-larger uncertainties about the true nature of the relationship (Hill and Weir 2011). Indeed, Speed and Balding (2015) have argued that the notion of a specific relationship (in discrete categories) has little merit in the era of whole-genome sequencing, as one can directly measure the fraction of two genomes that are shared AIS, a point we also endorse.

MARKER-BASED ESTIMATES OF THE COEFFICIENTS OF INBREEDING (f) AND FRATERNITY (Δ_{xy})

The logic used above for developing estimates for Θ extends to other relatedness metrics. We consider two here: the amount, f_x, of inbreeding in individual x, and the coefficient of fraternity, Δ_{xy}, between x and y. As above, while pedigrees can estimate *expected values* for these quantities (Chapter 7), with sufficiently dense marker data we can estimate their actual *realizations*, which can deviate substantially from their pedigree-expected values. Further, the distribution of recombination events over a genome can also create biases when using pedigree-based f. In particular, for species that have nonuniform recombination rates over chromosomes, even using a modest number of markers can be more informative than using a large pedigree (Knief et al. 2017). As with estimates of θ, correcting f_x and Δ_{xy} estimators for AIS alleles depends on assumptions about the base population.

Single-point Estimators of f

Consider a diallelic locus ℓ with alleles A_ℓ and a_ℓ. With probability f, the two alleles at a randomly chosen locus in an inbred individual are IBD. Hence, the impact of inbreeding on genotype frequencies is straightforward (Chapter 12): (i) decreasing the frequency of heterozygotes relative to Hardy–Weinberg,

$$\text{Freq}(A_\ell a_\ell) = (1-f)2p_\ell(1-p_\ell) \tag{8.18a}$$

and (ii) correspondingly increasing the frequency of homozygotes,

$$\text{Freq}(A_\ell A_\ell) = fp_\ell + (1-f)p_\ell^2 = p_\ell^2 + fp_\ell(1-p_\ell) \tag{8.18b}$$

It is important to stress that an inbred *individual* has a genome consisting of a *mixture* of inbred and outbred *loci* (or, more accurately, *chromosomal segments*), with a fraction f that are inbred (IBD) and a fraction $(1-f)$ that are not. Absent a rare mutation, an inbred locus is never heterozygous, while homozygotes can occur at both inbred (**autozygous**) and outbred (**allozygous**) loci.

The issue in estimating the fraction, f, of autozygosity in an individual's genome is separating autozygous from allozygous homozygotes. This is usually done by comparing the observed homozygosity with the value expected given some reference population. In exactly the same fashion, the deficiency of heterozygotes also provides an estimate of the fraction of the genome that is autozygous. Both these approaches raise, once again, the critical issue of the appropriate reference population. In many settings, the reference population itself may be inbred, so that an estimate of f, as with Θ above, reflects the *relative* difference in f between the focal individual and a random individual from the reference, and thus can be negative.

A number of method-of-moments estimators have been proposed built around the comparison of observed versus expected homozygosity (or heterozygosity) values (reviewed by Caballero et al. 2022). We first consider single-point estimates (f for a specific site), from which a composite estimate for the genome (or chromosomal region) is constructed by some weighting scheme of the point estimates. Their basic structure is as follows. Consider a focal individual (x), and (as above) let $T_{x,\ell}$ denote the copy number of the reference allele—which we arbitrarily take as the major (more common) allele, and denote this allele by 1—at locus ℓ in individual x. $T_{x,\ell}$ takes on values of 0, 1, or 2, corresponding, respectively, for SNP genotypes of 00, 10, and 11. Hence, $T_{x,\ell}(2-T_{x,\ell})$ is only nonzero for a heterozygote (where it has a value of 1). The expected heterozygote frequency when an individual is inbred to level f is given by Equation 8.18a, and we can use this equation to obtain an estimate of f using multilocus data in several ways. First, one could simply average both sides of Equation 8.18a over L markers in the focal individual, yielding the expectation

$$E\left[\frac{1}{L}\sum_{\ell=1}^{L} T_{x,\ell}(2-T_{x,\ell})\right] = (1-f)\frac{1}{L}\sum_{\ell=1}^{L} 2p_\ell(1-p_\ell) \tag{8.19a}$$

which rearranges to give the estimator

$$\widehat{f}_{HOM,1} = 1 - \frac{\sum_{\ell=1}^{L} T_{x,\ell}(2-T_{x,\ell})}{\sum_{\ell=1}^{L} 2p_\ell(1-p_\ell)} = 1 - \frac{O[Het]}{E[Het]} \tag{8.19b}$$

where $O[Het]$ and $E[Het] = \sum_\ell 2p_\ell(1-p_\ell)$ denote, respectively, the numbers of observed and expected marker heterozygotes within an individual. We can equivalently consider Equation 8.19b as an estimate based on the observed excess in homozygotes, rearranging to yield (Purcell et al. 2007)

$$\widehat{f}_{HOM,1} = \frac{O[Hom] - E[Hom]}{L - E[Hom]} \tag{8.19c}$$

which follows from noting that $O[Hom] + O[Het] = E[Hom] + E[Het] = L$. Hence

$$E[Hom] = L - E[Het] = L - \sum_{\ell=1}^{L} 2p_\ell(1-p_\ell) \tag{8.20a}$$

$E[Hom]$ and $E[Het]$ are, respectively, the expected number of homozygous and heterozygous marker loci in a random individual from a randomly mating base population.

These are akin to the AIS corrections in many of the above estimates of Θ, which also critically depended on random mating and known allele frequencies. When inbreeding is present in the base population, these assumptions fail. The approach taken by the Weir–Goudet estimator (Equation 8.12) offers a robust solution to this problem. Instead of calculating the expected number of homozygotes (or heterozygotes) under random mating assumptions, we use the average value of these in the *reference sample*. For example, in a sample of n individuals

$$E_{WG}[Het] = \frac{1}{n}\sum_{i=1}^{n}\left(\sum_{\ell=1}^{L} T_{i,\ell}(2 - T_{i,\ell})\right) \tag{8.20b}$$

with $E_{WG}[Hom] = L - E_{WG}[Het]$. For small samples, one could compute Equation 8.20b with the focal individual excluded. These values are substituted into Equations 8.19b and 8.19c, with the estimated f now being the departure from the average base-population value, which can easily be negative (less inbred than a random individual from the base population).

An alternative estimator for f follows by rearranging Equation 8.18a to give (Li and Horvitz 1953)

$$f = 1 - \frac{\text{freq}(A_\ell a_\ell)}{2p_\ell(1-p_\ell)} \tag{8.21a}$$

which yields

$$\widehat{f}_{HOM,2} = 1 - \frac{1}{L}\sum_{\ell=1}^{L}\frac{T_{x,\ell}(2 - T_{x,\ell})}{2p_\ell(1-p_\ell)} \tag{8.21b}$$

Equation 8.21b differs from Equation 8.19b in that the latter weights all loci equally, while the former places more weight on heterozygotes arising from loci with extreme allele frequencies (where $1/[p_\ell(1-p_\ell)]$ is large). As such, it is a more delicate estimate than Equation 8.19, being overly sensitive to rare alleles (with their corresponding reference-population frequency estimation uncertainties).

The same logic leading to Equation 8.15 can be used to obtain an estimator based on the correlation among uniting gametes (Yang et al. 2010b),

$$\widehat{f}_Y = \frac{1}{L}\sum_{\ell=1}^{L}\gamma_\ell, \quad \text{with} \quad \gamma_\ell = \frac{T_{x,\ell}^2 - (1+2p_\ell)T_{x,\ell} + 2p_\ell^2}{2p_\ell(1-p_\ell)} \tag{8.22a}$$

where the locus-specific estimators simplify to

$$\gamma_\ell = \begin{cases} (1-p_\ell)/p_\ell & \text{for } 1_\ell 1_\ell \;\; (T_{x,\ell}=2) \\ -1 & \text{for } 1_\ell 0_\ell \;\; (T_{x,\ell}=1) \\ p_\ell/(1-p_\ell) & \text{for } 0_\ell 0_\ell \;\; (T_{x,\ell}=0) \end{cases} \tag{8.22b}$$

Because $p_\ell \geq 1/2$ (by construction, as we take 1_ℓ to be the major allele), these weights place more value on the rare homozygote ($0_\ell 0_\ell$) than on the common homozygote ($1_\ell 1_\ell$). Equation 8.22a quantifies how similar the alleles at SNPs are within an individual relative to a randomly mating individual from the base population. As such, it is a biased estimate when the base population itself is inbred, and/or if we have a biased estimate of the base-population allele frequencies. Finally, recall from Equation 7.3b that $f_x = 2\Theta_{xx} - 1$, so that one can repurpose many of the above Θ estimators by simply considering the coancestry of an individual with itself.

Simulations by Caballero et al. (2022) found that when the reference population frequencies are known, all estimators are well behaved and very highly correlated with the true f value. Note that one such setting is when two inbred lines are crossed, and their resulting F_2 taken as the reference, from which inbreeding proceeds (Chapter 12), as here $p = 1/2$ for all segregating F_2 loci. When the reference frequencies are unknown, and relatives and/or inbred individuals are in the sample from which these frequencies are estimated, then Equation 8.19b, using Equation 8.20b to correct for AIS sharing, is the best behaved (Weir and Goudet 2017; Zhang et al. 2022; Caballero et al. 2022).

Several authors have proposed more sophisticated ML-based estimators for f, which have been extended to allow for null alleles and to jointly estimate the allele frequencies along with f (Vogl et al. 2002; Wang 2011a; Hall et al. 2012). These are more computationally feasible than ML estimates of Θ, as one only has to obtain n estimates of f (one for each individual), as opposed to the $n(n-1)/2$ pairwise estimates for Θ.

LD-based Estimators of f: Homozygosity by Descent (HBD) and Runs of Homozygosity (ROHs)

Previously, we used dense SNP data to detect blocks of shared IBD segments *between* two individuals. We extend this approach to detect inbreeding by considering stretches *within* an individual that are IBD between its two homologous chromosomes. This is the extension of the idea of autozygous versus allozygous from loci to *chromosomal segments*, with Browning and Browning (2011) referring to such an autozygous segment as having **homozygosity by descent, HBD**. Hence, an obvious estimator of f would be the fraction of the genome that is HBD. Two different approaches to obtain such estimates have been proposed: **observational** (also known as **rule-based**) and **model-based**. Both approaches start with **runs of homozygosity** (**ROHs**), where a run is defined as a continuous DNA segment that is completely homozygous for all scored markers (Broman and Weber 1999; Chapman and Thompson 2003; Purcell et al. 2007; McQuillan et al. 2008; Howrigan et al. 2011; Keller et al. 2011; Curik et al. 2014; Knief et al. 2015; Zhang et al. 2015; Peripolli et al. 2016; Ceballos et al. 2018). One advantage of using ROH-based estimates is that no reference allele frequencies are required.

The observational estimator for the fraction of the genome that is autozygous is based on simply scoring the observed total length of ROHs within a genome,

$$\widehat{f}_{ROH} = \frac{\text{total length of ROHs}}{\text{genome size}} \tag{8.23a}$$

The key issue with this class of estimators is distinguishing between an allozygous ROH arising by chance in an outbred individual versus an autozygous ROH arising from HBD, with our intuition being that longer ROHs are more likely to be autozygous. Equation 8.23a is a rule-based method, arising by specifying a **size threshold** used to designate a homozygous region as an ROH. Hence, estimates of $\widehat{f}_{ROH}$ for the same individual vary with this assumed threshold. One delicate problem with choosing the size threshold is *linkage disequilibrium*, as a sufficiently short region of DNA persists in the population as essentially a block. Hence, a very short ROH could be consistent with allozygosity, having no more weight than a single homozygous SNP. McQuillan et al. (2008) suggested a minimal threshold size in excess of the 100kb LD block size commonly found in many human populations, whereas later authors suggested much higher threshold values. Obviously, the LD structure of different populations/species results in different threshold levels, see Ceballos et al. (2018) and Meyermans et al. (2022). At a minimum, the ROH threshold should be (at least) several times larger than the average LD block size for the genomic region of interest.

Another issue is that the physical genome size is not the appropriate metric for Equation 8.23a. There is a general trend for genomic regions with low recombination rates to have very reduced variation (WL Chapter 8). Hence, large regions of the genome, such as around many centromeres, can show little variation, inflating the genomic fraction of ROHs (McQuillan et al. 2008). Auton et al. (2009) and Gazal et al. (2014) suggested that a more reasonable estimation approach would be to score both ROHs and total genome size in terms of Morgans, namely *genetic distances* (expected number of crossovers; Chapter 17), rather than *physical distances* (base pairs).

A final issue with ROH-based estimators is more subtle, but much more fundamental. The distribution of HBD lengths (scaled in Morans) is approximately exponential with parameter $1/(2g)$, where g is the number of generations since the common ancestor (Fisher

1954). For example, in a region with an average recombination rate of 1 cM (0.01 Morgans) per Mb of DNA, a threshold size of 5 Mb corresponds to ancestors within the last 10 generations, while a threshold size of 0.5 Mb corresponds to ancestors within the last 100 generations. Hence, ***long ROHs represent very recent inbreeding***, while ***short ROHs may represent ancient inbreeding blocks*** that have yet to be broken up by recombination. Thus, on one hand, the choice of a threshold represents a user-defined tradeoff between the control of false positives (longer ROHs have lower false-positive rates of autozygosity) versus reduced power (ignoring shorter runs that, cumulatively, may be biologically important). More critically, threshold size represents *an assumption* of which time scale is more important: a few more recent, or many more distant, ancestors. Hence, ***the choice of the size threshold is essentially a choice of whether the appropriate reference population is more recent or more ancient***. This lack of a clearly defined threshold generates a certain amount of arbitrariness for ROH-based estimates of f, and, indeed, different estimates for the same individual depending on the chosen threshold value.

A more formal approach for declaring an ROH to be an HBD segment is *model-based*: build models that incorporate SNP-calling and sequencing errors, as well as mutation-rate estimates, to infer ROHs, rather than a strict adherence to exact homozygosity throughout a long sequence. To accommodate these concerns, a number of authors have proposed HMM-based approaches returning posterior probabilities that a given region is HBD (Broman and Weber 1999; Leutenegger et al. 2003; Purcell et al. 2007; Wang et al. 2009; Browning and Browning 2010, 2013; Pemberton et al. 2012; MacLeod et al. 2013; Gazal et al. 2014; Magi et al. 2014; Narasimham et al. 2016; Vieira et al. 2016; Druet and Gautier 2017, 2022; Solé et al. 2017; Szpiech et al. 2017; Bertrand et al. 2018). To distinguish this class of estimators from the observational approach ($\widehat{f}_{ROH}$), we denote their estimates as $\widehat{f}_{HBD}$ (Ceballos et al. 2018; Lavanch and Goudet 2022). Following Bertrand et al. (2018),

$$\widehat{f}_{HBD} = \text{Average genome-wide posterior probability of being HBD} \tag{8.23b}$$

While it might seem that the model-based approach overcomes the issue of choosing a threshold size, this is not really the case. Underlying an HMM is some baseline decay rate, R (in Morgans), of recombination changing in the HBD status between adjacent segments (with e^{-R} being to probability of no switch), which is then weighted by marker information to return posterior probabilities. Here $R = 2g$, where g is the number of generations back to the common ancestor, so that the choice of R (either by default or by using a genome-wide estimator) essentially sets a threshold size. Bertrand et al. (2018) partly addressed this concern by fitting a series of different HMM-based class models, with $R_k = 2g_k$ being the rate for class k, where g_k is the number of generations back to the common ancestor for this class. Hence, the model decay parameter is essentially a probabilistic threshold size. A decay rate of $R = 10$ corresponds to five generations (and associated average HBD length of $1/R_k$ Morgans, or 0.1M = 10 cM), and a rate of $R = 1000$ corresponds to 500 generations (average HBD length of 0.1 cM). Hence, as with rule-based approaches, different choices of R estimate inbreeding contributions from different generations.

An important operational distinction between observational and model-based estimates arises when using **reduced genomic representation** data. While ideally ROH-based estimates use whole-genome sequence (WGS) information, in many cases, one has only some subset of the genome. Observational approaches assume unscored markers within an ROH are HBD, while model-based methods assign these posterior probabilities. As a result, with low to modest marker coverage, model-based estimates perform much better than do rule-based ones (Magi et al. 2014; Bertrandg et al. 2018; Lavanchy and Goudet 2022). When WGS data are used, simulations by Lavanchy and Goudet (2022) found that both classes of estimators work equally well under simple demographies (such as unstructured, and stable, populations), while under more complex demographies, model-based methods outperform rule-based estimates.

One final subtle issue, examined in detail in Chapter 12, is the estimation of f for quantifying the amount of trait-specific inbreeding depression (ID). The concern is that the above estimators for f are *genome-wide* averages, while quantifying ID for a specific trait is based on the average f for *those loci underlying the trait*. While genome-wide estimates of f are good (and potentially exceptional) proxies for average f at the trait QTLs, this not need be the case. The latter may have an excess of low-frequency alleles (relative to the rest of the genome), or may reflect past selection (such as inbred QTL loci being impacted by selection).

Estimators of Δ_{xy}

The last relationship metric we consider is the coefficient of fraternity, Δ_{xy}, the probability that individuals x and y share two alleles IBD (Chapter 7). In the literature, Δ_{xy} is occasionally referred to as the **four-gene relationship coefficient**, with Θ being the **two-gene relationship coefficient**. This metric is of interest in that it is the weight on any dominance variance shared by relatives (Equation 7.11b) and it also arises in studies of inbreeding depression (Chapter 12). Equation 7.8 shows that a nonzero value of Δ_{xy} requires special circumstances, namely that the parents (m_x, s_x and m_y, s_y) of x and y must be related, as

$$\Delta_{xy} = \Theta_{m_x m_y}\Theta_{s_x s_y} + \Theta_{m_x s_y}\Theta_{s_x m_y} \tag{8.24}$$

As a result, nonzero values of Δ_{xy} require that either: (i) *both* the mothers *and* the fathers are related; or (ii) the father of x and the mother of y, *and* the father of y and the mother of x, are related. Such complex relationships in a sample are expected to be much less frequent than sets of relatives that simply share a single IBD allele at a given locus (e.g., Table 7.2). In any sample, the number of pairs of relatives with nonzero Δ_{xy} values is thus expected to be far less than those that have a nonzero Θ_{xy} values, resulting in lower power and less precision when estimating the former relative to the latter. Equation 8.24 also shows that the expected value of Δ scales as the square of Θ. Hence, not only are pairs of relatives with nonzero Δ expected to be much scarcer than relatives with nonzero Θ values, the resulting value of Δ is also expected to be much smaller, and really only of significance between fairly close relatives.

Both method-of-moments (Ritland 1996; Lynch and Ritland 1999; Wang 2002) and likelihood (Milligan 2003; Ackerman et al. 2017) estimators of Δ_{xy} have been proposed. Our focus here is on method-of-moment estimators. The basic structure of likelihood estimators follows using logic similar to that for paternity assessment, see Appendix 4 (in particular, Example A4.8) for details.

For a biallelic marker (such as a SNP), the Lynch and Ritland (1999) and Wang (2002) single-locus estimator based on marker locus ℓ is

$$\widehat{\Delta}_{xy,\ell} = 1 - \frac{(4 - 4I_{xy,2} - 3I_{xy,1})\, h_\ell - 2(1 - I_{xy,2} - I_{xy,1})}{h_\ell^2} \tag{8.25a}$$

where $h_\ell = 2p_\ell(1-p_\ell) = 1 - p_\ell^2 - (1-p_\ell)^2$ is the probability that a random individual is not a homozygote at locus i. The two AIS indicator variables are defined by

$$I_{xy,2} = \begin{cases} 1 & \text{if } x = 00, y = 00 \\ 1 & \text{if } x = 10, y = 10 \\ 1 & \text{if } x = 11, y = 11 \\ 0 & \text{otherwise} \end{cases} \tag{8.25b}$$

Namely, that x and y are AIS for the marker *genotypes*, while I_1 indicates that one individual is a homozygote and the other a heterozygote,

$$I_{xy,1} = \begin{cases} 1 & \text{if } x = 00 \text{ or } 11, y = 10 \\ 1 & \text{if } x = 10, y = 00 \text{ or } 11 \\ 0 & \text{otherwise} \end{cases} \tag{8.25c}$$

Extensions to markers with multiple alleles are given by Lynch and Ritland (1999) and Wang (2002). Again, this estimator assumes known allele frequencies in a random-mating base population, as an unbiased value of h_ℓ depends on these assumptions holding.

As with any single-point estimator, one issue is how one weights individual-locus results to obtain a single composite estimator over all markers in the region of interest (chromosome or entire genome). One could use the Li and Horvitz (1953) scheme of weighting by population heterozygosity, $w_\ell = h_\ell^{-1}$, giving the composite estimator for L independent loci as

$$\widehat{\Delta}_{xy} = \frac{1}{W} \sum_{\ell=1}^{L} w_\ell \, \widehat{\Delta}_{xy,\ell} \tag{8.25d}$$

with $W = \sum w_\ell$. Lynch and Ritland (1999) weight loci using the inverse of the sampling variance (which, in part, depends on the unknown parameters to be estimated), while other weighting schemes are examined by Wang (2002). Improved (smaller variance) composite estimators could be obtained using the approach of Wang et al. (2017), by first estimating the relationship metrics using a standard method, and then substituting these values into the Lynch and Ritland expressions for optimal weights. One could then iterate this approach until the weights stabilize.

An alternative estimator suggested by Zhu et al. (2015) is the dominance counterpart of Equation 8.15b,

$$\widehat{\Delta}_{xy} = \frac{1}{L} \sum_{i=\ell}^{L} \left[\frac{(x_{\ell j} - 2p_\ell^2)\,(x_{\ell k} - 2p_\ell^2)}{4p_\ell^2(1-p_\ell)^2} \right] \tag{8.26a}$$

where the value of $x_{\ell i}$ for locus ℓ in individual i is given by

$$x_{\ell i} = \begin{cases} 0 & \text{genotype is 00} \\ 2p_\ell & \text{genotype is 10} \\ (4p_\ell - 2) & \text{genotype is 11} \end{cases} \tag{8.26b}$$

In general, the sampling variance of $\widehat{\Delta}$ scales as $1/(Ln_a^2)$, where L is the number of markers and n_a the number of alleles per marker (Lynch and Ritland 1999). Using a smaller number of highly polymorphic markers (such as SSRs) can thus be more powerful than using a larger number of less informative markers. In contrast, the sampling variance for estimates of Θ tends to scale as $1/(Ln_a)$ (Lynch and Ritland 1999; Wang 2002), so that adding more markers has the same impact as adding more alleles per marker. The extra sensitivity of estimates of Δ_{xy} to the number of alleles is intuitive, as such estimates are based on the marker *genotypes* of x and y being AIS, and hence the more even allele frequencies are, the greater the power. Given our comments that nontrival values of Δ are generally restricted to rather close relatives, SSRs can have reasonable power in detecting coefficients of fraternity.

9

Introduction to Matrices and Linear Models

Before I came here I was confused about this subject. Having listened to your lecture, I am still confused. But on a higher level. Enrico Fermi

We have already encountered several examples of models in which response variables are linear functions of two or more explanatory (or predictor) variables. For example, we have been routinely expressing an individual's phenotypic value as the sum of genotypic and environmental values. A more complicated example is the use of linear regression to decompose an individual's genotypic value into average effects of individual alleles and residual contributions due to interactions between alleles (Chapters 4 and 5). Such **linear models** not only form the backbone of parameter estimation in quantitative genetics (Part II Chapters 22–33), but are also the basis for incorporating marker (and other genomic) information into modern quantitative-genetic inference (Chapters 20 and 21; Part II Chapters 32 and 33).

This chapter provides the foundational tools for the analysis of linear models, which are developed more fully and extended to the powerful mixed model in Chapter 10. We start by introducing multiple regression, wherein two or more variables are used to make predictions about a response variable. A review of elementary matrix algebra then follows (**binary operations** involving two matrices—e.g., addition or multiplication—and **unary operations** on a single matrix, e.g., its determinant or inverse), starting with matrix notation and building up to matrix multiplication and solutions of simultaneous equations using matrix inversion. Next, we use these results to develop tools for statistical analysis, considering the expectations and covariance matrices of transformed random vectors. We then introduce the multivariate normal distribution, which is by far the most important distribution in quantitative-genetics theory, and conclude with the analysis of the geometry (eigenvalues and eigenvectors) of variance-covariance matrices. Those with strong statistical backgrounds will find little new in this chapter, other than perhaps some immediate contact with quantitative genetics in the examples and familiarization with our notation. Additional advanced topics are covered in Appendix 3 and in WL Appendices 5 and 6.

MULTIPLE REGRESSION

As a point of departure, consider the multiple regression

$$y = \alpha + \beta_1 z_1 + \beta_2 z_2 + \cdots + \beta_n z_n + e \tag{9.1a}$$

where y is the **response variable**, and the z_i are the **predictor** (or **explanatory**) **variables** used to predict the value of the response variable. This multivariate equation is similar to the expression for a simple linear regression (Equation 3.12a) except that y is now a function of n predictor variables, rather than just one. The variables $y, z_1, \ldots, z_n$ represent observed measures, whereas α and $\beta_1, \ldots, \beta_n$ are constants to be estimated using some best-fit criterion. As in the case of simple linear regression, e (the **residual error**) is the deviation between the observed (y) and predicted (or fitted) value ($\widehat{y}$) of the response variable,

$$y = \widehat{y} + e, \quad \text{where} \quad \widehat{y} = \alpha + \sum_{i=1}^{n} \beta_i z_i \tag{9.1b}$$

Recall that the use of a linear model involves no assumptions regarding the true form of relationship between y and $z_1, \ldots, z_n$, nor is any assumption about the residuals being normally distributed required. It simply gives the best linear approximation. Many statistical techniques, including path analysis (Appendix 2) and analysis of variance (Part II Chapter 23), are based on versions of Equation 9.1.

The terms $\beta_1, \ldots, \beta_n$ are known as **partial regression coefficients**. The interpretation of β_i is the expected change in y given a unit change in z_i when all other predictor values are *held constant*. It is important to note that the partial regression coefficient associated with predictor variable z_i often differs from the regression coefficient, β_i', that is obtained in a *univariate* regression based solely on z_i, viz., $y = \alpha + \beta_i' z_i + e$ (Example 9.3). Suppose, for example, that a simple regression of y on z_1 has a slope of zero. This might lead to the suggestion that there is no relationship between z_1 and y. However, it is conceivable that z_1 actually has a strong positive effect on y that is obscured by positive correlations of z_1 with other variables that have negative influences on y. A multiple regression that included the appropriate variables would clarify this situation by yielding a positive value of β_1.

Because it is usually impossible for biologists to evaluate partial regression coefficients by empirically imposing constancy on all extraneous variables, we require a more indirect approach to the problem. From Chapter 3, the covariance of y and a predictor variable is

$$\begin{aligned} \sigma(y, z_i) &= \sigma\left[(\alpha + \beta_1 z_1 + \beta_2 z_2 + \cdots + \beta_n z_n + e), z_i\right] \\ &= \beta_1 \sigma(z_1, z_i) + \beta_2 \sigma(z_2, z_i) + \cdots + \beta_n \sigma(z_n, z_i) + \sigma(e, z_i) \end{aligned} \tag{9.2a}$$

The term $\sigma(\alpha, z_i)$ has dropped out because the covariance of z_i with a constant (α) is zero. By applying Equation 9.2a to each predictor variable, we obtain a set of n equations in n unknowns $(\beta_1, \ldots, \beta_n)$,

$$\begin{array}{llllll} \sigma(y, z_1) = \beta_1 \sigma^2(z_1) & + \beta_2 \sigma(z_1, z_2) + & \cdots + & \beta_n \sigma(z_1, z_n) & + \sigma(z_1, e) \\ \sigma(y, z_2) = \beta_1 \sigma(z_1, z_2) & + \beta_2 \sigma^2(z_2) & + \cdots + & \beta_n \sigma(z_2, z_n) & + \sigma(z_2, e) \\ \quad \vdots \qquad \vdots & \quad \vdots & \ddots & \quad \vdots & \quad \vdots \\ \sigma(y, z_n) = \beta_1 \sigma(z_1, z_n) & + \beta_2 \sigma(z_2, z_n) + & \cdots & + \beta_n \sigma^2(z_n) & + \sigma(z_n, e) \end{array} \tag{9.2b}$$

As in univariate regression, our task is to find the set of constants (α and the partial regression coefficients, β_i) that gives the best linear fit of the conditional expectation of y given $z_1, \cdots, z_n$. Again, the criterion we choose for "best" relies on the **least-squares** approach, which minimizes the squared differences between observed and expected values (i.e., the squared residuals). Thus, our task is to find that set of $\alpha, \beta_1, \cdots, \beta_n$ giving $\widehat{y} = \alpha + \sum \beta_i z_i$ such that $E[(y - \widehat{y})^2 | z_1, \cdots, z_n] = E[\, e^2\,]$ is minimized. Taking derivatives of this expectation with respect to α and the β_i and setting each equal to zero, it can be shown that the set of equations given by Equation 9.2b is, in fact, the least-squares solution to Equation 9.1 (See WL Example A6.4). If the appropriate variances and covariances are known (i.e., we know their true population values), the β_i can be obtained exactly. If these are unknown, as is usually the case, the least-squares estimates b_i are obtained from Equation 9.2b by substituting the observed (estimated) variances and covariances, Var(z_i) and Cov(z_i, z_j), for their (unknown) population values, $\sigma^2(z_i)$ and $\sigma(z_i, z_j)$.

The properties of least-squares multiple regression are analogous to those for simple regression (Chapter 3). First, the procedure yields a solution such that the average deviation of y from its predicted value $\widehat{y}$, $E[e]$, is zero. Hence $E[\,y\,] = E[\,\widehat{y}\,]$, implying

$$\bar{y} = a + b_1 \bar{z}_1 + \cdots + b_n \bar{z}_n \tag{9.3a}$$

Thus, once the fitted values $b_1, \ldots, b_n$ are obtained from Equation 9.2b, the intercept is obtained by $a = \bar{y} - \sum_i^n b_i \bar{z}_i$. Using Equation 9.3a, we can rewrite Equation 9.1a as

$$y - \bar{y} = \sum_{i=1}^{n} \beta_i (z_i - \bar{z}_i) + e \tag{9.3b}$$

implying that

$$y = \bar{y} + \sum_{i=1}^{n} \beta_i (z_i - \bar{z}_i) + e \tag{9.3c}$$

Second, least-squares analysis gives a solution in which the residual errors are uncorrelated with the predictor variables. Thus, the terms $\sigma(e, z_i)$ can be dropped from Equation 9.2b. Third, the partial regression coefficients are entirely defined by variances and covariances. However, unlike simple regression coefficients, which depend on only a single variance and covariance, each partial regression coefficient is a function of the variances and covariances of *all* measured variables. Notice that if $n = 1$, then $\sigma(y, z_1) = \beta_1 \sigma^2(z_1)$, and we recover the univariate solution, $\beta_1 = \sigma(y, z_1)/\sigma^2(z_1)$.

A simple pattern exists in each of the n equations in 9.2b. The ith equation defines the covariance of y and z_i as the sum of two types of quantities: a single term, which is the product of the ith partial regression coefficient and the variance of z_i, and a set of $(n-1)$ terms, each of which is the product of a partial regression coefficient and the covariance of z_i with the corresponding predictor variable. This general pattern suggests an alternative way of writing Equation 9.2b,

$$\begin{pmatrix} \sigma^2(z_1) & \sigma(z_1, z_2) & \dots & \sigma(z_1, z_n) \\ \sigma(z_1, z_2) & \sigma^2(z_2) & \dots & \sigma(z_2, z_n) \\ \vdots & \vdots & \ddots & \vdots \\ \sigma(z_1, z_n) & \sigma(z_2, z_n) & \dots & \sigma^2(z_n) \end{pmatrix} \begin{pmatrix} \beta_1 \\ \beta_2 \\ \vdots \\ \beta_n \end{pmatrix} = \begin{pmatrix} \sigma(y, z_1) \\ \sigma(y, z_2) \\ \vdots \\ \sigma(y, z_n) \end{pmatrix} \tag{9.4a}$$

The table of variances and covariances on the left is referred to as a **matrix**, while the columns of partial regression coefficients and of covariances involving y are called **vectors**. If these matrices and vectors are abbreviated, respectively, as $\mathbf{V}$, $\boldsymbol{\beta}$, and $\mathbf{c}$, then Equation 9.4a can be written even more compactly as

$$\mathbf{V}\boldsymbol{\beta} = \mathbf{c} \tag{9.4b}$$

where $\mathbf{c}$ denotes the vector of covariances between the predictor and response variables.

The standard procedure of denoting matrices as bold capital letters and vectors as bold lowercase letters is adhered to in this book. Notice that $\mathbf{V}$, which is generally called a **covariance matrix**, is symmetrical about the main diagonal. As we shall see shortly, the ith equation in 9.2b can be recovered from Equation 9.4a by multiplying the elements in $\boldsymbol{\beta}$ by the corresponding elements in the ith horizontal row of the matrix $\mathbf{V}$, i.e., $\sum_j \beta_j V_{ij}$. Although a great deal of notational simplicity has been gained by condensing the system of Equations 9.2b to matrix form, this does not alter the fact that the solution of a large system of simultaneous equations is a tedious task if performed by hand. Fortunately, such solutions are rapidly accomplished on computers. Before considering matrix methods in more detail, we present an application of Equation 9.1 to quantitative genetics.

An Application to Multivariate Selection

Karl Pearson developed the technique of multiple regression in 1896, although some of the fundamentals can be traced to his predecessors (Pearson 1920; Stigler 1986). Pearson is perhaps best known as one of the founders of statistical methodology, but his intense interest in evolution may have been the primary motivation for many of his theoretical endeavors. Almost all of his major papers, including the one of 1896, contain rigorous analyses of data gathered by his contemporaries on matters such as resemblance between relatives, natural selection, correlation between characters, and assortative mating (recall Example 7.7). The foresight of these studies is remarkable considering that they were performed prior to the existence of a genetic interpretation for the expression and inheritance of polygenic traits.

Pearson's (1896, 1903) invention of multiple regression developed out of the need for a technique to decompose the observed directional selection on a character into its direct and various indirect components. In Chapter 3 we defined the selection differential S (the within-generation change in the mean phenotype due to selection) as a measure of the total directional selection on a character. However, S cannot be considered to be a measure of the *direct forces* of selection on a character unless that character is uncorrelated with all other selected traits. An unselected character can appear to be under selection if other characters with which it is correlated are under directional selection. Alternatively, a character under strong directional selection may exhibit a negligible selection differential if the indirect effects of selection on correlated traits are sufficiently compensatory.

As a hypothetical example, consider plant fitness measured by the number of visits to a flower by a pollinator. It is observed that larger flowers obtain more pollinators (S for flower size is positive), but also that flower size and nectar volume are positively correlated. Hence, pollinators might be visiting larger flowers simply because the latter have more nectar. A multiple regression of pollinator visit number on both flower size and nectar volume can resolve which trait is the actual target of selection (provided that neither is correlated to other unmeasured targets of selection).

Because he did not employ matrix notation, some of the mathematics in Pearson's papers can be rather difficult to follow. Lande and Arnold (1983) did a great service by extending and rephrasing this work in matrix notation. Suppose that a large number of individuals in a population have been measured for n characters and for fitness. Individual fitness can then be approximated by the linear model

$$w = \alpha + \beta_1 z_1 + \cdots + \beta_n z_n + e \tag{9.5a}$$

where w is the relative fitness (observed fitness divided by the mean fitness in the population, i.e., $w = W/\overline{W}$), and $z_1, \ldots, z_n$ are the phenotypic measures of the n characters. The interpretation of β_i is the expected change in w given a unit change in trait i while all other traits are held constant. Recall from Chapter 3 that the selection differential for the ith trait is defined as the covariance between phenotype and relative fitness, $S_i = \sigma(z_i, w)$. Thus, we have

$$\begin{aligned} S_i = \sigma(z_i, w) &= \sigma(z_i, \alpha + \beta_1 z_1 + \cdots + \beta_n z_n + e) \\ &= \beta_1 \sigma(z_i, z_1) + \cdots + \beta_n \sigma(z_i, z_n) + \sigma(z_i, e) \end{aligned} \tag{9.5b}$$

Note that this expression is of the same form as Equation 9.2b, so that by taking the β_i to be the partial regression coefficients we have $\sigma(z_i, e) = 0$. This expression can also be compactly written as $\mathbf{s} = \mathbf{V}\boldsymbol{\beta}$ (Equation 9.4b), where the vector of covariances ($\mathbf{s}$) has its i element given by S_i. Finally, note that the selection differential of any trait may be partitioned into a component estimating the **direct selection** on the character and the sum of components from **indirect selection** on all correlated characters,

$$S_i = \beta_i \sigma^2(z_i) + \sum_{j \neq i}^{n} \beta_j \sigma(z_i, z_j) \tag{9.5c}$$

It is important to realize that the labels "direct" and "indirect" apply strictly to the specific set of characters included in the analysis; the partial regression coefficients are subject to change if a new analysis includes additional correlated characters that are under selection. Returning to our flower example, suppose that β for size is negative and β for nectar volume positive. This suggests that the pollinators favor smaller flowers with more nectar. The positive correlation between size and nectar volume obscured this relationship, resulting (Equation 9.5c) in a positive value of S for flower size.

Example 9.1. A morphological analysis of the one-spotted stink bug (*Euschistus variolarius*) population performed by Lande and Arnold (1983) provides a good example of the insight that can be gained from a multivariate approach. The bugs were collected along the shore of Lake Michigan after a storm. Of the 94 individuals that were recovered, 39 were alive. All individuals were measured for four characters: head and thorax width, and scutellum and forewing length. The data were then logarithmically transformed to more closely approximate normality (Chapter 14). All surviving bugs were assumed to have equal fitness ($W = 1$), and all dead bugs to have zero fitness ($W = 0$). Hence, mean fitness is the fraction ($p = 39/94 = 0.415$) of individuals that survived, giving **relative fitnesses,** $w = W/\overline{W}$, as

$$w = \begin{cases} 1/p = 2.41 & \text{if the individual survived} \\ 0 & \text{if the individual did not survive} \end{cases}$$

The selection differential for each of the characters is simply the difference between the mean phenotype of the 39 survivors and the mean of the entire sample. These are reported in units of phenotypic standard deviations in the following table, along with the partial regression coefficients of relative fitness on the four morphological characters. Here * and ** indicate significance at the 5% and 1% levels. All of the phenotypic correlations were highly significant.

Character	Selection Differential	Partial Regression Coef. of Fitness	Phenotypic Correlations			
z_i	S_i	β_i	H	T	S	F
Head (H)	–0.11	–0.7	1.00	0.72	0.50	0.60
Thorax (T)	–0.06	11.6**		1.00	0.59	0.71
Scutellum (S)	–0.28*	–2.8			1.00	0.62
Forewing (F)	–0.43**	–16.6**				1.00

The estimates of the partial regression coefficients nicely illustrate two points discussed earlier. First, despite the strong directional selection (β) operating directly on thorax size, the selection differential (S) for thorax size is negligible. This lack of apparent selection results because the positive correlation between thorax width and forewing length is coupled with negative forces of selection on the latter character. Second, there is a significant negative selection differential on scutellum length even though there is no significant direct selection on the character. The negative selection differential in this case is largely an indirect consequence of the strong selection for smaller forewing length. WL Chapters 29 and 30 examine Lande–Arnold fitness estimation in considerable detail. We note in passing that given the 0,1 nature of the response variable, logistic regression (Chapter 14; WL Chapters 14 and 29) is a more appropriate analysis of these data than a linear model.

ELEMENTARY MATRIX ALGEBRA

The solutions of systems of linear equations, such as those introduced above, generally involve the use of matrices and vectors of variables. For those with little familiarity with such constructs and their manipulations, the remainder of the chapter provides an overview of the basic tools of matrix algebra, with a focus on useful results for the analysis of linear models, the workhorse of quantitative genetics.

Basic Notation

A matrix is simply a rectangular array of numbers. Some examples are

$$\mathbf{a} = \begin{pmatrix} 12 \\ 13 \\ 47 \end{pmatrix} \qquad \mathbf{b} = \begin{pmatrix} 2 & 0 & 5 & 21 \end{pmatrix} \qquad \mathbf{C} = \begin{pmatrix} 3 & 1 & 2 \\ 2 & 5 & 4 \\ 1 & 1 & 2 \end{pmatrix} \qquad \mathbf{D} = \begin{pmatrix} 0 & 1 \\ 3 & 4 \\ 2 & 9 \end{pmatrix}$$

A matrix with r rows and c columns is said to have **dimensionality** $r \times c$ (a useful mnemonic for remembering this order is <u>r</u>ailroad <u>c</u>ar). In the examples above, **D** has three rows and two columns, and is thus a 3×2 matrix. An $r \times 1$ matrix, such as **a**, is a **column vector** ($c = 1$), while a $1 \times c$ matrix, such as **b**, is a **row vector** ($r = 1$). A matrix in which the number of rows equals the number of columns, such as **C**, is called a **square matrix**. Numbers are also matrices (of dimensionality 1×1) and are often referred to as **scalars**.

A matrix is completely specified by the **elements** that comprise it, with M_{ij} denoting the element in the ith row and jth column of matrix $\mathbf{M}$. Using the sample matrices above, $C_{23} = 4$ is the element in the second row and third column of $\mathbf{C}$. Likewise, $C_{32} = 1$ is the element in the third row and second column. Two matrices are equal if and only if all of their corresponding elements are equal. Dimensionality is important, as operations on matrices (such as addition or multiplication) are only defined when matrix dimensions agree in the appropriate manner (as is discussed below).

Partitioned Matrices

It is often useful to work with **partitioned matrices** wherein each element in a matrix is itself a matrix. There are several ways to partition a matrix. For example, we could write the matrix $\mathbf{C}$ above as

$$\mathbf{C} = \begin{pmatrix} 3 & 1 & 2 \\ 2 & 5 & 4 \\ 1 & 1 & 2 \end{pmatrix} = \begin{pmatrix} 3 & \vdots & 1 & 2 \\ \cdots & \cdots & \cdots & \cdots \\ 2 & \vdots & 5 & 4 \\ 1 & \vdots & 1 & 2 \end{pmatrix} = \begin{pmatrix} \mathbf{a} & \mathbf{b} \\ \mathbf{d} & \mathbf{B} \end{pmatrix}$$

where

$$\mathbf{a} = (\,3\,), \quad \mathbf{b} = (\,1 \quad 2\,), \quad \mathbf{d} = \begin{pmatrix} 2 \\ 1 \end{pmatrix}, \quad \mathbf{B} = \begin{pmatrix} 5 & 4 \\ 1 & 2 \end{pmatrix}$$

Alternatively, we could partition $\mathbf{C}$ into a single row vector whose elements are themselves column vectors,

$$\mathbf{C} = (\,\mathbf{c}_1 \quad \mathbf{c}_2 \quad \mathbf{c}_3\,) \quad \text{where} \quad \mathbf{c}_1 = \begin{pmatrix} 3 \\ 2 \\ 1 \end{pmatrix}, \quad \mathbf{c}_2 = \begin{pmatrix} 1 \\ 5 \\ 1 \end{pmatrix}, \quad \mathbf{c}_3 = \begin{pmatrix} 2 \\ 4 \\ 2 \end{pmatrix}$$

or as a column vector whose elements are row vectors,

$$\mathbf{C} = \begin{pmatrix} \mathbf{b}_1 \\ \mathbf{b}_2 \\ \mathbf{b}_3 \end{pmatrix} \quad \text{where} \quad \mathbf{b}_1 = (\,3 \quad 1 \quad 2\,), \quad \mathbf{b}_2 = (\,2 \quad 5 \quad 4\,), \quad \mathbf{b}_3 = (\,1 \quad 1 \quad 2\,)$$

As we will shortly see, this partition of a matrix as either a set of row or column vectors forms the basis of matrix multiplication.

Addition and Subtraction

Addition and subtraction of matrices is straightforward. To form a new matrix $\mathbf{A} + \mathbf{B} = \mathbf{C}$, $\mathbf{A}$ and $\mathbf{B}$ must have the same dimensionality ($\mathbf{A}$ and $\mathbf{B}$ have the same number of columns and the same number of rows), so that they have corresponding elements. One then simply adds these corresponding elements, $C_{ij} = A_{ij} + B_{ij}$. Subtraction is defined similarly. For example, if

$$\mathbf{A} = \begin{pmatrix} 3 & 0 \\ 1 & 2 \\ 0 & 1 \end{pmatrix} \quad \text{and} \quad \mathbf{B} = \begin{pmatrix} 1 & 2 \\ 2 & 1 \\ 1 & 0 \end{pmatrix}$$

then

$$\mathbf{C} = \mathbf{A} + \mathbf{B} = \begin{pmatrix} 4 & 2 \\ 3 & 3 \\ 1 & 1 \end{pmatrix} \quad \text{and} \quad \mathbf{D} = \mathbf{A} - \mathbf{B} = \begin{pmatrix} 2 & -2 \\ -1 & 1 \\ -1 & 1 \end{pmatrix}$$

Multiplication

Multiplying a matrix by a **scalar** (a 1×1 matrix) is straightforward. If $\mathbf{M} = a\mathbf{N}$, where a is a scalar, then $M_{ij} = aN_{ij}$. Each element of $\mathbf{N}$ is simply multiplied by the scalar. For example,

$$(-2)\begin{pmatrix} 1 & 0 \\ 3 & 1 \end{pmatrix} = \begin{pmatrix} -2 & 0 \\ -6 & -2 \end{pmatrix}$$

Matrix multiplication is a little more involved. We start by considering the **dot product** of two vectors, as this forms the basic operation of matrix multiplication. Letting $\mathbf{a}$ and $\mathbf{b}$ be two n-dimensional vectors, their dot product $\mathbf{a} \cdot \mathbf{b}$ is a scalar given by

$$\mathbf{a} \cdot \mathbf{b} = \sum_{i=1}^{n} a_i b_i$$

For example, for the two vectors

$$\mathbf{a} = \begin{pmatrix} 1 \\ 2 \\ 3 \\ 4 \end{pmatrix} \quad \text{and} \quad \mathbf{b} = \begin{pmatrix} 4 & 5 & 7 & 9 \end{pmatrix}$$

the dot product is $\mathbf{a} \cdot \mathbf{b} = (1 \times 4) + (2 \times 5) + (3 \times 7) + (4 \times 9) = 71$. The dot product ignores whether the vectors are row, column, or mixed, but is *not defined if the vectors have different lengths*. As we will see, this restriction determines whether the product of two matrices is defined.

The dot-product operator allows us to express systems of equations compactly in matrix form. Consider the following system of three equations and three unknowns,

$$\begin{aligned} x_1 + 2x_2 + x_3 &= 3 \\ 2x_1 - 2x_2 - 4x_3 &= 6 \\ 8x_1 - 4x_2 + 3x_3 &= 9 \end{aligned}$$

This can be written in matrix form as $\mathbf{Ax} = \mathbf{c}$, with

$$\begin{pmatrix} 1 & 2 & 1 \\ 2 & -2 & -4 \\ 8 & -8 & 3 \end{pmatrix} \begin{pmatrix} x_1 \\ x_2 \\ x_3 \end{pmatrix} = \begin{pmatrix} 3 \\ 6 \\ 9 \end{pmatrix}$$

This matrix representation recovers the system of equations using dot products. The dot product of the first row of $\mathbf{A}$ with the column vector $\mathbf{x}$ (which is defined because each vector has three elements) recovers the first equation. The next two equations similarly follow as the dot products of the second and third rows, respectively, of $\mathbf{A}$ on $\mathbf{x}$.

Now consider the matrix $\mathbf{L}_{r \times b} = \mathbf{M}_{r \times c}\mathbf{N}_{c \times b}$ produced by multiplying the $r \times c$ matrix $\mathbf{M}$ by the $c \times b$ matrix $\mathbf{N}$. It is important to note the matching c subscripts, with the number of columns of $\mathbf{M}$ matching the number of rows of $\mathbf{N}$. Partitioning $\mathbf{M}$ as a column vector of r row vectors,

$$\mathbf{M} = \begin{pmatrix} \mathbf{m_1} \\ \mathbf{m_2} \\ \vdots \\ \mathbf{m_r} \end{pmatrix} \quad \text{where} \quad \mathbf{m_i} = \begin{pmatrix} M_{i1} & M_{i2} & \cdots & M_{ic} \end{pmatrix}$$

and $\mathbf{N}$ as a row vector of b column vectors,

$$\mathbf{N} = \begin{pmatrix} \mathbf{n_1} & \mathbf{n_2} & \cdots & \mathbf{n_b} \end{pmatrix} \quad \text{where} \quad \mathbf{n_j} = \begin{pmatrix} N_{1j} \\ N_{2j} \\ \vdots \\ N_{cj} \end{pmatrix}$$

the *ij*th element of **L** is given by the dot product

$$L_{ij} = \mathbf{m_i} \cdot \mathbf{n_j} = \sum_{k=1}^{c} M_{ik} N_{kj} \tag{9.6a}$$

Recall that this dot product is only defined if the vectors $\mathbf{m_i}$ and $\mathbf{n_j}$ have the same number of elements, which requires that the number of columns in **M** must equal the number of rows in **N**. The resulting matrix **L** is of dimension $r \times b$ with

$$\mathbf{L} = \begin{pmatrix} \mathbf{m_1} \cdot \mathbf{n_1} & \mathbf{m_1} \cdot \mathbf{n_2} & \cdots & \mathbf{m_1} \cdot \mathbf{n_b} \\ \mathbf{m_2} \cdot \mathbf{n_1} & \mathbf{m_2} \cdot \mathbf{n_2} & \cdots & \mathbf{m_2} \cdot \mathbf{n_b} \\ \vdots & \vdots & \ddots & \vdots \\ \mathbf{m_r} \cdot \mathbf{n_1} & \mathbf{m_r} \cdot \mathbf{n_2} & \cdots & \mathbf{m_r} \cdot \mathbf{n_b} \end{pmatrix} \tag{9.6b}$$

Note that by using this definition, the matrix product given by Equation 9.4a recovers the set of equations given by Equation 9.2b.

Example 9.2. Compute the product $\mathbf{L} = \mathbf{MN}$ where

$$\mathbf{M} = \begin{pmatrix} 3 & 1 & 2 \\ 2 & 5 & 4 \\ 1 & 1 & 2 \end{pmatrix} \quad \text{and} \quad \mathbf{N} = \begin{pmatrix} 4 & 1 & 0 \\ 1 & 1 & 3 \\ 3 & 2 & 2 \end{pmatrix}$$

Writing $\mathbf{M} = \begin{pmatrix} \mathbf{m_1} \\ \mathbf{m_2} \\ \mathbf{m_3} \end{pmatrix}$ and $\mathbf{N} = (\, \mathbf{n_1} \quad \mathbf{n_2} \quad \mathbf{n_3} \,)$, we have

$$\mathbf{m_1} = (\, 3 \quad 1 \quad 2 \,), \quad \mathbf{m_2} = (\, 2 \quad 5 \quad 4 \,), \quad \mathbf{m_3} = (\, 1 \quad 1 \quad 2 \,)$$

and

$$\mathbf{n_1} = \begin{pmatrix} 4 \\ 1 \\ 3 \end{pmatrix}, \quad \mathbf{n_2} = \begin{pmatrix} 1 \\ 1 \\ 2 \end{pmatrix}, \quad \mathbf{n_3} = \begin{pmatrix} 0 \\ 3 \\ 2 \end{pmatrix}$$

The resulting matrix **L** is 3×3. Applying Equation 9.6b, the element in the first row and first column of **L** is the dot product of the first row vector of **M** with the first column vector of **N**,

$$L_{11} = \mathbf{m_1} \cdot \mathbf{n_1} = (\, 3 \quad 1 \quad 2 \,) \begin{pmatrix} 4 \\ 1 \\ 3 \end{pmatrix} = \sum_{k=1}^{3} M_{1k} N_{k1}$$
$$= M_{11}N_{11} + M_{12}N_{21} + M_{13}N_{31} = (3 \times 4) + (1 \times 1) + (2 \times 3) = 19$$

Computing the other elements yields

$$\mathbf{L} = \begin{pmatrix} \mathbf{m_1} \cdot \mathbf{n_1} & \mathbf{m_1} \cdot \mathbf{n_2} & \mathbf{m_1} \cdot \mathbf{n_3} \\ \mathbf{m_2} \cdot \mathbf{n_1} & \mathbf{m_2} \cdot \mathbf{n_2} & \mathbf{m_2} \cdot \mathbf{n_3} \\ \mathbf{m_3} \cdot \mathbf{n_1} & \mathbf{m_3} \cdot \mathbf{n_2} & \mathbf{m_3} \cdot \mathbf{n_3} \end{pmatrix} = \begin{pmatrix} 19 & 8 & 7 \\ 25 & 15 & 23 \\ 11 & 6 & 7 \end{pmatrix}$$

These straightforward, but tedious, calculations for each element in the new matrix are easily performed on a computer. Indeed, most statistical and math packages have all the matrix operations introduced in this chapter as built-in functions.

As suggested above, certain dimensional properties must be satisfied when two matrices are to be multiplied. Specifically, because the dot product is defined only for vectors of the same length, for the matrix product **MN** to be defined, the number of columns in **M** must equal the number of rows in **N**. Matrices satisfying this row-column restriction are said to **conform.** Thus, while

$$\begin{pmatrix} 3 & 0 \\ 1 & 2 \end{pmatrix}_{2\times 2} \begin{pmatrix} 4 \\ 3 \end{pmatrix}_{2\times 1} = \begin{pmatrix} 12 \\ 10 \end{pmatrix}_{2\times 1}, \quad \text{but} \quad \begin{pmatrix} 4 \\ 3 \end{pmatrix}_{2\times 1} \begin{pmatrix} 3 & 0 \\ 1 & 2 \end{pmatrix}_{2\times 2} \quad \text{is undefined.}$$

Writing

$$\mathbf{M}_{r\times c}\mathbf{N}_{c\times b} = \mathbf{L}_{r\times b}$$

shows that the ***inner indices must match***, while the outer indices (r and b) give the number of rows and columns, respectively, of the resulting matrix. A second key point is that ***the order in which matrices are multiplied is critical***. In general, **AB** is not equal to **BA**. Indeed, even if the matricies conform in one order, they may not in the opposite order. In particular, unless one matrix is $r \times c$ and the other is $c \times r$, the two orders of multiplication will not conform (note that two square matrices, $r = c$, of the same dimension conform in either order).

For example, when the order of the matrices in Example 9.2 is reversed,

$$\mathbf{NM} = \begin{pmatrix} 4 & 1 & 0 \\ 1 & 1 & 3 \\ 3 & 2 & 2 \end{pmatrix} \begin{pmatrix} 3 & 1 & 2 \\ 2 & 5 & 4 \\ 1 & 1 & 2 \end{pmatrix} = \begin{pmatrix} 14 & 9 & 12 \\ 8 & 9 & 12 \\ 15 & 15 & 18 \end{pmatrix}$$

which differs from **MN**. Because order is important in matrix multiplication, it has specific terminology. For the product **AB**, we say that matrix **B** is **premultiplied** by the matrix **A**, or that matrix **A** is **postmultiplied** by the matrix **B**.

Transposition

Another useful matrix operation is **transposition**. The transpose of a matrix **A** is written $\mathbf{A}^T$ (while not used in this book, the notation $\mathbf{A}'$ is also used), and is obtained simply by switching rows and columns of the original matrix, with $A^T_{ij} = A_{ji}$. If **A** is $r \times c$, then $\mathbf{A}^T$ is $c \times r$. As an example,

$$\begin{pmatrix} 3 & 1 & 2 \\ 2 & 5 & 4 \\ 1 & 1 & 2 \end{pmatrix}^T = \begin{pmatrix} 3 & 2 & 1 \\ 1 & 5 & 1 \\ 2 & 4 & 2 \end{pmatrix}$$

$$(7 \quad 4 \quad 5)^T = \begin{pmatrix} 7 \\ 4 \\ 5 \end{pmatrix}$$

A **symmetric matrix** satisfies $\mathbf{A} = \mathbf{A}^T$, and is necessarily square. An important example of a square, symmetric matrix is a covariance matrix (Equation 9.4a). Matrix expressions involving transposes often arise as a result of matrix derivatives (Equations A3.28 and A3.29). For example, the derivative (with respect to a vector, **x**) of **Ax** is the matrix $\mathbf{A}^T$.

A useful identity (when **A** and **B** conform) for transposition is that

$$(\mathbf{AB})^T = \mathbf{B}^T\mathbf{A}^T \tag{9.7a}$$

which holds for any number of conformable matrices, e.g.,

$$(\mathbf{ABC})^T = \mathbf{C}^T\mathbf{B}^T\mathbf{A}^T \tag{9.7b}$$

Vectors in statistics and quantitative genetics are generally written as *column vectors* and we follow this convention by using lowercase bold letters, e.g., **a**, for a column vector and $\mathbf{a}^T$ for the corresponding row vector. With this convention, we distinguish between two vector products, the **inner product** which yields a scalar and the **outer product** which yields a matrix. For the two n-dimensional column vectors $\mathbf{a}$ and $\mathbf{b}$,

$$\mathbf{a} = \begin{pmatrix} a_1 \\ \vdots \\ a_n \end{pmatrix} \qquad \mathbf{b} = \begin{pmatrix} b_1 \\ \vdots \\ b_n \end{pmatrix}$$

their inner product is given by

$$\begin{pmatrix} a_1 & \cdots & a_n \end{pmatrix} \begin{pmatrix} b_1 \\ \vdots \\ b_n \end{pmatrix} = \mathbf{a}^T\mathbf{b} = \sum_{i=1}^{n} a_i b_i = \mathbf{b}^T\mathbf{a} \tag{9.8a}$$

while their outer product yields the $n \times n$ matrix

$$\begin{pmatrix} a_1 \\ \vdots \\ a_n \end{pmatrix} \begin{pmatrix} b_1 & \cdots & b_n \end{pmatrix} = \mathbf{a}\mathbf{b}^T = \begin{pmatrix} a_1b_1 & a_1b_2 & \cdots & a_1b_n \\ a_2b_1 & a_2b_2 & \cdots & a_2b_n \\ \vdots & \vdots & \ddots & \vdots \\ a_nb_1 & a_nb_2 & \cdots & a_nb_n \end{pmatrix} \tag{9.8b}$$

Inner products frequently appear in statistics and quantitative genetics, as they represent *weighted sums*. For example, the regression given by Equation 9.1 can be expressed as

$$y = \alpha + \boldsymbol{\beta}^T\mathbf{z} + e, \quad \text{where} \quad \boldsymbol{\beta} = \begin{pmatrix} \beta_1 \\ \beta_2 \\ \vdots \\ \beta_n \end{pmatrix}, \quad \mathbf{z} = \begin{pmatrix} z_1 \\ z_2 \\ \vdots \\ z_n \end{pmatrix}$$

Another example is a weighted marker score (occasionally called a **polygenic score**), $s = \boldsymbol{\beta}^T\mathbf{z}$, for an individual, where $\mathbf{z}$ is a vector of marker values (e.g., $z_j = 0, 1$, or 2, respectively, for biallelic marker genotypes m_jm_j, M_jm_j, and M_jM_j) for that individual and β_j is the weight assigned to the jth marker (i.e., each copy of M_j adds an amount β_j to the score).

Outer products appear in covariance matrices. Consider an $(n \times 1)$ vector $\mathbf{e}$ of random variables, each with mean zero. The resulting $(n \times n)$ covariance matrix, Cov(e), has its ij element as $E[\, e_ie_j \,] = \sigma_{ij}$, where

$$\text{Cov}(\mathbf{e}) = E[\,\mathbf{e}^T\mathbf{e}\,] = E\left[\begin{pmatrix} e_1 \\ \vdots \\ e_n \end{pmatrix} \begin{pmatrix} e_1 & \cdots & e_n \end{pmatrix}\right]$$
$$= \begin{pmatrix} E[e_1e_1] & E[e_1e_2] & \cdots & E[e_1e_n] \\ E[e_2e_1] & E[e_2e_2] & \cdots & E[e_2e_n] \\ \vdots & \vdots & \ddots & \vdots \\ E[e_ne_1] & E[e_ne_2] & \cdots & E[e_ne_n] \end{pmatrix} = \begin{pmatrix} \sigma_1^2 & \sigma_{12} & \cdots & \sigma_{1n} \\ \sigma_{21} & \sigma_2^2 & \cdots & \sigma_{2n} \\ \vdots & \vdots & \ddots & \vdots \\ \sigma_{n1} & \sigma_{n2} & \cdots & \sigma_n^2 \end{pmatrix}$$

Inverses and Solutions to Systems of Equations

While matrix multiplication provides a compact way of writing systems of equations, we also need a compact notation for expressing the *solutions* of such systems. This is provided by the **inverse** of a matrix, an operation analogous to scalar division. The importance of matrix inversion can be noted by first considering the solution of the simple scalar equation $ax = b$ for x. Multiplying both sides by a^{-1}, we have $(a^{-1}a)x = 1 \cdot x = x = a^{-1}b$. Now consider a square matrix $\mathbf{A}$. The **inverse of A**, denoted $\mathbf{A}^{-1}$, satisfies $\mathbf{A}^{-1}\mathbf{A} = \mathbf{I} = \mathbf{A}\mathbf{A}^{-1}$, where $\mathbf{I}$, the **identity matrix**, is a square matrix with diagonal elements equal to one and all other elements equal to zero. The identity matrix serves the role that 1 plays in scalar multiplication. Just as $1 \times a = a \times 1 = a$ in scalar multiplication, for any matrix $\mathbf{A} = \mathbf{IA} = \mathbf{AI}$. A matrix is called **nonsingular** if its inverse exists. Conditions under which this occurs are discussed in the next section. A useful property of inverses is that if the matrix product $\mathbf{AB}$ is a square matrix (where $\mathbf{A}$ and $\mathbf{B}$ are square and both of their inverses exist), then

$$(\mathbf{AB})^{-1} = \mathbf{B}^{-1}\mathbf{A}^{-1} \tag{9.9}$$

The fundamental relationship between the inverse of a matrix and the solution of systems of linear equations can be seen as follows. For a square nonsingular matrix $\mathbf{A}$, the *unique* solution for $\mathbf{x}$ in the matrix equation $\mathbf{Ax} = \mathbf{c}$ is obtained by premultiplying by $\mathbf{A}^{-1}$,

$$\mathbf{A}^{-1}\mathbf{Ax} = \mathbf{Ix} = \mathbf{x} = \mathbf{A}^{-1}\mathbf{c} \tag{9.10a}$$

If $\mathbf{A}^{-1}$ does not exist ($\mathbf{A}$ is said to be **singular**), then there are either *no solutions* (the set of equations is **inconsistent**), or there are *infinitely many solutions*.

Consider the following two sets of equations,

$$\text{Set one:} \quad \begin{aligned} x_1 + 2x_2 &= 3 \\ 2x_1 + 4x_2 &= 6 \end{aligned} \qquad\qquad \text{Set two:} \quad \begin{aligned} x_1 + 2x_2 &= 3 \\ 2x_1 + 4x_2 &= 3 \end{aligned}$$

For both sets, the left-hand side of the second equation is just twice the left-hand side of the first equation. Set one is **consistent**, with a line of solutions, $x_1 = 3 - 2x_2$. More generally, the solution set of a consistent system could be a plane or hyperplane (whereas it is a point when the coefficient matrix is nonsingular). Set two is inconsistent, as no values of x_1 and x_2 can satisfy both equations.

When $\mathbf{A}$ is either singular or nonsquare, solutions for $\mathbf{x}$ (for a consistent system of equations) can still be obtained using **generalized inverses** (denoted by $\mathbf{A}^{-}$) in place of $\mathbf{A}^{-1}$ (Appendix 3). As we have just seen, the solutions returned in such cases are certain linear combinations of the elements of $\mathbf{x}$, rather than a unique value for $\mathbf{x}$ (see Appendix 3 for details.) Recalling Equation 9.4b, the solution of the multiple regression equation can be expressed as

$$\boldsymbol{\beta} = \mathbf{V}^{-1}\mathbf{c} \tag{9.10b}$$

Likewise, for the Pearson–Lande–Arnold regression giving the best linear predictor of fitness,

$$\boldsymbol{\beta} = \mathbf{P}^{-1}\mathbf{s} \tag{9.10c}$$

where $\mathbf{P}$ is the covariance matrix for phenotypic measures $z_1, \ldots, z_n$, and $\mathbf{s}$ is the vector of selection differentials for the n characters.

Before developing the formal method for inverting a matrix, we consider two extreme (but very useful) cases that lead to simple expressions for the inverse. First, if the matrix is **diagonal** (all off-diagonal elements are zero), then the matrix inverse is also diagonal, with $A_{ii}^{-1} = 1/A_{ii}$. For example,

$$\text{for} \quad \mathbf{A} = \begin{pmatrix} a & 0 & 0 \\ 0 & b & 0 \\ 0 & 0 & c \end{pmatrix}, \qquad \text{then} \qquad \mathbf{A}^{-1} = \begin{pmatrix} a^{-1} & 0 & 0 \\ 0 & b^{-1} & 0 \\ 0 & 0 & c^{-1} \end{pmatrix}$$

Note that if any of the diagonal elements of a diagonal matrix $\mathbf{A}$ are zero, $\mathbf{A}^{-1}$ is not defined, as $1/0$ is undefined. Second, for any 2×2 matrix,

$$\mathbf{A} = \begin{pmatrix} a & b \\ c & d \end{pmatrix}, \qquad \text{then} \qquad \mathbf{A}^{-1} = \frac{1}{ad - bc}\begin{pmatrix} d & -b \\ -c & a \end{pmatrix} \tag{9.11}$$

To check this result, note that

$$\mathbf{A}\mathbf{A}^{-1} = \frac{1}{ad - bc}\begin{pmatrix} a & b \\ c & d \end{pmatrix}\begin{pmatrix} d & -b \\ -c & a \end{pmatrix}$$

$$= \frac{1}{ad - bc}\begin{pmatrix} ad - bc & 0 \\ 0 & ad - bc \end{pmatrix} = \mathbf{I}$$

If $ad = bc$, the inverse does not exist, as division by zero is undefined.

Example 9.3. Consider the multiple regression of y on two predictor variables, z_1 and z_2, so that $y = \alpha + \beta_1 z_1 + \beta_2 z_2 + e$. We solve for the β_i, as the estimate of α follows as $\overline{y} - \beta_1 \overline{z}_1 - \beta_2 \overline{z}_2$. In the notation of Equation 9.4b, we have

$$\mathbf{c} = \begin{pmatrix} \sigma(y, z_1) \\ \sigma(y, z_2) \end{pmatrix} \qquad \mathbf{V} = \begin{pmatrix} \sigma^2(z_1) & \sigma(z_1, z_2) \\ \sigma(z_1, z_2) & \sigma^2(z_2) \end{pmatrix}$$

Recalling that $\sigma(z_1, z_2) = \rho_{12}\,\sigma(z_1)\sigma(z_2)$, Equation 9.11 gives

$$\mathbf{V}^{-1} = \frac{1}{\sigma^2(z_1)\sigma^2(z_2)\,(1 - \rho_{12}^2)}\begin{pmatrix} \sigma^2(z_2) & -\sigma(z_1, z_2) \\ -\sigma(z_1, z_2) & \sigma^2(z_1) \end{pmatrix}$$

The inverse exists provided that both characters have nonzero variance and are not completely correlated ($|\rho_{12}| \neq 1$). Recalling Equation 9.10b, the partial regression coefficients are given by $\boldsymbol{\beta} = \mathbf{V}^{-1}\mathbf{c}$, or

$$\begin{pmatrix} \beta_1 \\ \beta_2 \end{pmatrix} = \frac{1}{\sigma^2(z_1)\sigma^2(z_2)\,(1 - \rho_{12}^2)}\begin{pmatrix} \sigma^2(z_2) & -\sigma(z_1, z_2) \\ -\sigma(z_1, z_2) & \sigma^2(z_1) \end{pmatrix}\begin{pmatrix} \sigma(y, z_1) \\ \sigma(y, z_2) \end{pmatrix}$$

Again using $\sigma(z_1, z_2) = \rho_{12}\,\sigma(z_1)\sigma(z_2)$, this equation reduces to

$$\beta_1 = \frac{1}{1 - \rho_{12}^2}\left[\beta_1' - \rho_{12}\,\frac{\sigma(y, z_2)}{\sigma(z_1)\sigma(z_2)}\right]$$

and

$$\beta_2 = \frac{1}{1 - \rho_{12}^2}\left[\beta_2' - \rho_{12}\,\frac{\sigma(y, z_1)}{\sigma(z_1)\sigma(z_2)}\right]$$

where

$$\beta_1' = \frac{\sigma(y, z_1)}{\sigma^2(z_1)} \quad \text{and} \quad \beta_2' = \frac{\sigma(y, z_2)}{\sigma^2(z_2)}$$

are the *univariate* regression slopes ($y = \alpha' + \beta_i' z_i + e$; Equation 3.14b). Note that only when the predictor variables are uncorrelated ($\rho_{12} = 0$), do the partial regression coefficients β_1 and β_2 reduce to the univariate regression slopes, β_1' and β_2'.

For example, consider our earlier hypothetical example of pollinator visits (y) as a function of flower size (z_1) and nectar volume (z_2). The univariate regression $y = \mu + \beta_1' z_1$ of number of visits as a function of just flower size has a regression slope of β_1', while the regression coefficients on flower size when both size and nectar volume are included in multiple regression ($y = \mu + \beta_1 z_1 + \beta_2 z_2$) is β_1. β_1 and β_1' are only equal when there is no correlation between size and volume.

Determinants and Minors

For a 2×2 matrix, the quantity

$$|\mathbf{A}| = A_{11}A_{22} - A_{12}A_{21} \tag{9.12a}$$

is called the **determinant** of **A**, which more generally is denoted by det(**A**) or |**A**|. As with the two-dimensional case, $\mathbf{A}^{-1}$ exists for a square matrix **A** (of any dimensionality) if and only if det(**A**) $\neq 0$. For square matrices with dimensionality greater than two, the determinant is obtained recursively from the general expression

$$|\mathbf{A}| = \sum_{j=1}^{n} A_{ij}(-1)^{i+j}|\mathbf{A}_{ij}| \tag{9.12b}$$

where i is any fixed row of the matrix **A** and $\mathbf{A}_{ij}$ is the $(n-1) \times (n-1)$ submatrix obtained by deleting the ith row and jth column from **A**. Such a submatrix is known as a **minor.** Each of the n quantities in this equation is the product of three components: the element in the row around which one is working, -1 to the $(i+j)$th power, and the determinant of the ijth minor. In applying Equation 9.12b, one starts with the original $n \times n$ matrix and works down until the minors are reduced to 2×2 matrices whose determinants are scalars of the form $A_{11}A_{22} - A_{12}A_{21}$. A useful result is that the determinant of a diagonal matrix is the product of the diagonal elements of that matrix, so that if

$$A_{ij} = \begin{cases} a_i & i = j \\ 0 & i \neq j \end{cases} \qquad \text{then} \qquad |\,\mathbf{A}\,| = \prod_{i=1}^{n} a_i \tag{9.12c}$$

The next section shows how determinants are used in the computation of a matrix inverse.

Example 9.4. Compute the determinant of

$$\mathbf{A} = \begin{pmatrix} 1 & 1 & 1 \\ 1 & 3 & 2 \\ 1 & 2 & 1 \end{pmatrix}$$

Letting $i = 1$, i.e., using the elements in the first row of **A**,

$$|\mathbf{A}| = 1 \cdot (-1)^{1+1} \begin{vmatrix} 3 & 2 \\ 2 & 1 \end{vmatrix} + 1 \cdot (-1)^{1+2} \begin{vmatrix} 1 & 2 \\ 1 & 1 \end{vmatrix} + 1 \cdot (-1)^{1+3} \begin{vmatrix} 1 & 3 \\ 1 & 2 \end{vmatrix}$$

Using Equation 9.12a to obtain the determinants of the 2×2 matrices, this simplifies to

$$|\mathbf{A}| = [1 \times (3-4)] - [1 \times (1-2)] + [1 \times (2-3)] = -1$$

The same answer is obtained regardless of which row is used, and expanding around a column, instead of a row, produces the same result. Thus, in order to reduce the number of computations required to obtain a determinant, it is useful to expand using the row or column that contains the most zeros. As with all other matrix operations presented in this chapter, the determinant is usually obtained using standard computer packages.

Example 9.5. To see further connections between the determinant and the solution to a set of equations of the form **Ax** = **c**, consider the following two systems of equations:

Set one:
$$\begin{aligned} x_1 + x_2 &= 1 \\ 2x_1 + 2x_2 &= 2 \end{aligned}$$

Set two:
$$\begin{aligned} 0.9999 \cdot x_1 + x_2 &= 1 \\ 2x_1 + 2x_2 &= 2 \end{aligned}$$

The determinant for the coefficient matrix associated with set one is zero, and there is no unique solution, rather a line of solutions, $x_1 = 1 - x_2$. In contrast, the determinant for the matrix associated with set two is nonzero, hence its inverse exists and there is a unique solution. However, the determinant is nearly zero, -0.0002. Such a matrix is said to be **nearly singular**, meaning that although the two sets of equations are distinct, they overlap so closely that there is little additional information from one (or more) of the equations. For this set of equations,

$$\mathbf{A}^{-1} = \begin{pmatrix} -10,000 & 5000 \\ 10,000 & -4999.5 \end{pmatrix}, \quad \mathbf{x} = \begin{pmatrix} 0 \\ 1 \end{pmatrix}$$

While there *technically* is a unique solution, it is *extremely* sensitive to the coefficients in the set of equations, and a very small change (such as through measurement error) can dramatically change the solution. For example, replacing the first equation by $x_1 + 0.9999 \cdot x_2 = 1$, yields the solution of $x_1 = 1, x_2 = 0$.

Computing Inverses

The general solution of a matrix inverse is

$$A_{ji}^{-1} = \frac{(-1)^{i+j}|\mathbf{A}_{ij}|}{|\mathbf{A}|} \tag{9.13}$$

where A_{ji}^{-1} denotes the jith element of $\mathbf{A}^{-1}$, and $\mathbf{A}_{ij}$ denotes the ijth minor of $\mathbf{A}$. The reversed subscripts (ji versus ij) in the left and right expressions arise because the right-hand side computes an element in the transpose of the inverse (see Example 9.6). It can be seen from Equation 9.13 that ***a matrix can only be inverted if it has a nonzero determinant***. Thus, ***a matrix is singular if its determinant is zero***. This occurs whenever a matrix contains a row (or column) that can be written as a weighted sum of the other rows (or columns). In the context of a linear model, this happens if one (or more) of the n equations can be written as a combination of the others, a situation that is equivalent to there being n unknowns but less than n independent equations.

Example 9.6. Compute the inverse of

$$\mathbf{A} = \begin{pmatrix} 3 & 1 & 2 \\ 2 & 5 & 4 \\ 1 & 1 & 2 \end{pmatrix}$$

First, find the determinants of the minors,

$$|\mathbf{A}_{11}| = \begin{vmatrix} 5 & 4 \\ 1 & 2 \end{vmatrix} = 6 \qquad |\mathbf{A}_{23}| = \begin{vmatrix} 3 & 1 \\ 1 & 1 \end{vmatrix} = 2$$

$$|\mathbf{A}_{12}| = \begin{vmatrix} 2 & 4 \\ 1 & 2 \end{vmatrix} = 0 \qquad |\mathbf{A}_{31}| = \begin{vmatrix} 1 & 2 \\ 5 & 4 \end{vmatrix} = -6$$

$$|\mathbf{A}_{13}| = \begin{vmatrix} 2 & 5 \\ 1 & 1 \end{vmatrix} = -3 \qquad |\mathbf{A}_{32}| = \begin{vmatrix} 3 & 2 \\ 2 & 4 \end{vmatrix} = 8$$

$$|\mathbf{A}_{21}| = \begin{vmatrix} 1 & 2 \\ 1 & 2 \end{vmatrix} = 0 \qquad |\mathbf{A}_{33}| = \begin{vmatrix} 3 & 1 \\ 2 & 5 \end{vmatrix} = 13$$

$$|\mathbf{A}_{22}| = \begin{vmatrix} 3 & 2 \\ 1 & 2 \end{vmatrix} = 4$$

Using Equation 9.12b and expanding using the first row of $\mathbf{A}$ gives

$$|\mathbf{A}| = 3|\mathbf{A}_{11}| - |\mathbf{A}_{12}| + 2|\mathbf{A}_{13}| = 12$$

Returning to the matrix defined by Equation 9.13, we obtain

$$\frac{1}{12}\begin{pmatrix} 1\times 6 & -1\times 0 & 1\times -3 \\ -1\times 0 & 1\times 4 & -1\times 2 \\ 1\times -6 & -1\times 8 & 1\times 13 \end{pmatrix} = \frac{1}{12}\begin{pmatrix} 6 & 0 & -3 \\ 0 & 4 & -2 \\ -6 & -8 & 13 \end{pmatrix}$$

and then taking the transpose,

$$\mathbf{A}^{-1} = \frac{1}{12}\begin{pmatrix} 6 & 0 & -6 \\ 0 & 4 & -8 \\ -3 & -2 & 13 \end{pmatrix}$$

To verify that this is indeed the inverse of $\mathbf{A}$, multiply $\mathbf{A}^{-1}$ by $\mathbf{A}$,

$$\frac{1}{12}\begin{pmatrix} 6 & 0 & -6 \\ 0 & 4 & -8 \\ -3 & -2 & 13 \end{pmatrix}\begin{pmatrix} 3 & 1 & 2 \\ 2 & 5 & 4 \\ 1 & 1 & 2 \end{pmatrix} = \frac{1}{12}\begin{pmatrix} 12 & 0 & 0 \\ 0 & 12 & 0 \\ 0 & 0 & 12 \end{pmatrix} = \begin{pmatrix} 1 & 0 & 0 \\ 0 & 1 & 0 \\ 0 & 0 & 1 \end{pmatrix}$$

Again, these tedious calculations are generally performed using matrix inversion routines available in standard packages.

EXPECTATIONS OF RANDOM VECTORS AND MATRICES

Matrix algebra provides a powerful approach for analyzing linear combinations of random variables. Let $\mathbf{x}$ be a column vector containing n random variables, $\mathbf{x} = (x_1, x_2, \cdots, x_n)^T$. We may wish to construct a new univariate (scalar) random variable y by taking some linear combination of the elements of $\mathbf{x}$,

$$y = \sum_{i=1}^{n} a_i x_i = \mathbf{a}^T\mathbf{x}$$

where $\mathbf{a} = (a_1, a_2, \cdots, a_n)^T$ is a column vector of constants. Likewise, we can construct a new k-dimensional vector $\mathbf{y}$ by premultiplying $\mathbf{x}$ by a $k \times n$ matrix $\mathbf{A}$ of constants, $\mathbf{y}_{k\times 1} = \mathbf{A}_{k\times n}\mathbf{x}_{n\times 1}$. Here $\mathbf{y}$ is a vector of k weighted sums, the jth of such sums is

$$y_j = \sum_{i=1}^{n} A_{ji} x_i = \mathbf{a}_j^T\mathbf{x}$$

where $\mathbf{a}_j$ denotes the jth row of $\mathbf{A}$. More generally, an $(n \times k)$ matrix $\mathbf{X}$ of random variables can be transformed into a new $(m \times \ell)$ dimensional matrix, $\mathbf{Y}$, of elements consisting of linear combinations of the elements of $\mathbf{X}$ by

$$\mathbf{Y}_{m\times\ell} = \mathbf{A}_{m\times n}\mathbf{X}_{n\times k}\mathbf{B}_{k\times\ell} \tag{9.14}$$

where the matrices $\mathbf{A}$ and $\mathbf{B}$ are constants with dimensions as subscripted.

If $\mathbf{X}$ is a matrix whose elements are random variables, then the expected value of $\mathbf{X}$ is a matrix $E[\mathbf{X}]$ containing the expected value of each element of $\mathbf{X}$. If $\mathbf{X}$ and $\mathbf{Z}$ are matrices of the same dimension, then

$$E[\mathbf{X} + \mathbf{Z}] = E[\mathbf{X}] + E[\mathbf{Z}] \tag{9.15}$$

This easily follows because the ijth element of $E[\mathbf{X}+\mathbf{Z}]$ is $E[x_{ij}+z_{ij}] = E[x_{ij}] + E[z_{ij}]$. Similarly, the expectation of $\mathbf{Y}$ as defined in Equation 9.14 is

$$E[\mathbf{Y}] = E[\mathbf{AXB}] = \mathbf{A}E[\mathbf{X}]\mathbf{B} \tag{9.16a}$$

For example, for $\mathbf{y} = \mathbf{Xb}$, where $\mathbf{b}$ is an $n \times 1$ column vector,

$$E[\mathbf{y}] = E[\mathbf{Xb}] = E[\mathbf{X}]\mathbf{b} \tag{9.16b}$$

If $\mathbf{X}$ is a matrix of fixed constants, then $E[\mathbf{y}] = \mathbf{Xb}$. Likewise, for $y = \mathbf{a}^T\mathbf{x} = \sum_i^n a_i x_i$,

$$E[y] = E[\mathbf{a}^T\mathbf{x}] = \mathbf{a}^T E[\mathbf{x}] \tag{9.16c}$$

COVARIANCE MATRICES OF TRANSFORMED VECTORS

To develop expressions for variances and covariances of linear combinations of random variables, we first need to introduce the concept of quadratic forms. Consider an $n \times n$ square matrix $\mathbf{A}$ and an $n \times 1$ column vector $\mathbf{x}$. From the rules of matrix multiplication,

$$\mathbf{x}^T\mathbf{A}\mathbf{x} = \sum_{i=1}^n \sum_{j=1}^n A_{ij} x_i x_j \tag{9.17a}$$

Expressions of this form are called **quadratic forms** (or **quadratic products**) as they involve squares (x_i^2) and cross-products ($x_i x_j$), and yield a scalar. A generalization of a quadratic form is the **bilinear form**, $\mathbf{b}^T\mathbf{A}\mathbf{a}$, where $\mathbf{b}$ and $\mathbf{a}$ are, respectively, $n \times 1$ and $m \times 1$ column vectors and $\mathbf{A}$ is an $n \times m$ matrix. Indexing the matrices and vectors in this expression by their dimensions, $\mathbf{b}^T_{1\times n}\mathbf{A}_{n\times m}\mathbf{a}_{m\times 1}$, shows that the resulting matrix product is defined and yields a 1×1 matrix; in other words, a scalar. As scalars, bilinear forms equal their transposes (as the transpose of a scalar simply returns that scalar), giving the useful identity

$$\mathbf{b}^T\mathbf{A}\mathbf{a} = \left(\mathbf{b}^T\mathbf{A}\mathbf{a}\right)^T = \mathbf{a}^T\mathbf{A}^T\mathbf{b} \tag{9.17b}$$

Again let $\mathbf{x}$ be a column vector of n random variables. A compact way to express the n variances and $n(n-1)/2$ covariances associated with the elements of $\mathbf{x}$ is the $n \times n$ matrix $\mathbf{V}$, where $V_{ij} = \sigma(x_i, x_j)$ is the covariance between the random variables x_i and x_j. We will generally refer to $\mathbf{V}$ as a **covariance matrix**, noting that the *diagonal elements represent the variances* and *off-diagonal elements the covariances*. The $\mathbf{V}$ matrix is symmetric ($\mathbf{V} = \mathbf{V}^T$), as

$$V_{ij} = \sigma(x_i, x_j) = \sigma(x_j, x_i) = V_{ji} \tag{9.18a}$$

Note that we can write the covariance matrix as an outer product. Letting $E[\mathbf{x}] = \boldsymbol{\mu}$, then

$$\mathbf{V} = E\left[(\mathbf{x}-\boldsymbol{\mu})(\mathbf{x}-\boldsymbol{\mu})^T\right] \tag{9.18b}$$

This follows as Equation 9.8b gives the ijth element of Equation 9.18b as $E[(x_i - \mu_i)(x_j - \mu_j)] = \sigma(x_i, x_j)$.

Now consider a univariate random variable, $y = \sum c_k x_k$, generated from a linear combination of the elements of $\mathbf{x}$. In matrix notation, $y = \mathbf{c}^T\mathbf{x}$, where $\mathbf{c}$ is a column vector of constants. The variance of y can be expressed as a quadratic form involving the covariance matrix $\mathbf{V}$ for the elements of $\mathbf{x}$,

$$\begin{aligned}
\sigma^2\left(\mathbf{c}^T\mathbf{x}\right) &= \sigma^2\left(\sum_{i=1}^n c_i x_i\right) = \sigma\left(\sum_{i=1}^n c_i\, x_i\, , \sum_{j=1}^n c_j\, x_j\right) \\
&= \sum_{i=1}^n \sum_{j=1}^n \sigma\left(c_i\, x_i, c_j\, x_j\right) = \sum_{i=1}^n \sum_{j=1}^n c_i\, c_j\, \sigma\left(x_i, x_j\right) \\
&= \mathbf{c}^T\mathbf{V}\,\mathbf{c}
\end{aligned} \tag{9.19}$$

Note that if $\mathbf{V}$ is a proper covariance matrix, then $\mathbf{c}^T\mathbf{V}\mathbf{c} > 0$ for all $\mathbf{c}$ (provided $\mathbf{c} \neq \mathbf{0}$), as this quadratic product represents the variance (and hence > 0) of some index of the elements of $\mathbf{x}$. A matrix $\mathbf{A}$ satisfying $\mathbf{c}^T\mathbf{A}\mathbf{c} > 0$ for all $\mathbf{c} \neq \mathbf{0}$ is said to be **positive definite**.

Similarly, the covariance between two univariate random variables created from different linear combinations of $\mathbf{x}$ is given by the bilinear form

$$\sigma(\mathbf{a}^T\mathbf{x}, \mathbf{b}^T\mathbf{x}) = \mathbf{a}^T\mathbf{V}\,\mathbf{b} \tag{9.20}$$

If we transform $\mathbf{x}$ to two new vectors, $\mathbf{y}_{\ell\times 1} = \mathbf{A}_{\ell\times n}\mathbf{x}_{n\times 1}$ and $\mathbf{z}_{m\times 1} = \mathbf{B}_{m\times n}\mathbf{x}_{n\times 1}$, then instead of a single covariance we have an $\ell \times m$ dimensional matrix of covariances, denoted $\boldsymbol{\sigma}(\mathbf{y},\mathbf{z})$, whose ijth element is $\sigma(y_i, z_j)$. Letting $\boldsymbol{\mu}_\mathbf{y} = \mathbf{A}\boldsymbol{\mu}$ and $\boldsymbol{\mu}_\mathbf{z} = \mathbf{B}\boldsymbol{\mu}$, with $E(\mathbf{x}) = \boldsymbol{\mu}$, then $\boldsymbol{\sigma}(\mathbf{y},\mathbf{z})$ can be expressed in terms of $\mathbf{V}$, the covariance matrix of $\mathbf{x}$,

$$\begin{aligned}\boldsymbol{\sigma}(\mathbf{y},\mathbf{z}) &= \boldsymbol{\sigma}(\mathbf{A}\mathbf{x},\mathbf{B}\mathbf{x}) \\ &= E\left[(\mathbf{y}-\boldsymbol{\mu}_\mathbf{y})(\mathbf{z}-\boldsymbol{\mu}_\mathbf{z})^T\right] \\ &= E\left[\mathbf{A}(\mathbf{x}-\boldsymbol{\mu})(\mathbf{x}-\boldsymbol{\mu})^T\mathbf{B}^T\right] \\ &= \mathbf{A}\mathbf{V}\,\mathbf{B}^T \end{aligned} \tag{9.21a}$$

In particular, the covariance matrix for $\mathbf{y} = \mathbf{A}\mathbf{x}$ is

$$\boldsymbol{\sigma}(\mathbf{y},\mathbf{y}) = \mathbf{A}\mathbf{V}\,\mathbf{A}^T \tag{9.21b}$$

so that the covariance between y_i and y_j is given by the ijth element of the matrix product $\mathbf{A}\mathbf{V}\mathbf{A}^T$.

Finally, note that if $\mathbf{x}$ is a vector of random variables with expected value $\boldsymbol{\mu}$, then the expected value of the (scalar) quadratic product $\mathbf{x}^T\mathbf{A}\mathbf{x}$ is

$$E(\mathbf{x}^T\mathbf{A}\mathbf{x}) = \text{tr}(\mathbf{A}\mathbf{V}) + \boldsymbol{\mu}^T\mathbf{A}\boldsymbol{\mu} \tag{9.22}$$

where $\mathbf{V}$ is the covariance matrix for the elements of $\mathbf{x}$, and the **trace** of a square matrix, $\text{tr}(\mathbf{M}) = \sum M_{ii}$, is the sum of its diagonal values (Searle 1971). Note that when x is a scalar and $\mathbf{A} = (1)$, Equation 9.22 collapses to $E[x^2] = \sigma^2 + \mu^2$ (Equation 2.3b). More generally, $E[x_ix_j] = \sigma(x_i, x_j) + \mu_i\mu_j$ (Equation 3.8).

Example 9.7. Consider three traits with the following covariance structure: $\sigma^2(x_1) = 10$, $\sigma^2(x_2) = 20, \sigma^2(x_3) = 30, \sigma(x_1, x_2) = -5, \sigma(x_1, x_3) = 10$, and $\sigma(x_2, x_3) = 0$, yielding the covariance matrix

$$\mathbf{V} = \begin{pmatrix} 10 & -5 & 10 \\ -5 & 20 & 0 \\ 10 & 0 & 30 \end{pmatrix}$$

Further, assume the vector of means for these variables is $\boldsymbol{\mu}^T = (10, 4, -3)$. Consider two new indices constructed from these variables, with $y_1 = 2x_1 - 3x_2 + 4x_3$ and $y_2 = x_2 - 2x_3$. We can express these as inner products, $y_i = \mathbf{c}_i^T\mathbf{x}$, where

$$\mathbf{c}_1 = \begin{pmatrix} 2 \\ -3 \\ 4 \end{pmatrix}, \qquad \mathbf{c}_2 = \begin{pmatrix} 0 \\ 1 \\ -2 \end{pmatrix}$$

From Equation 9.19, the resulting variances become $\sigma^2(y_1) = \mathbf{c}_1{}^T\mathbf{V}\,\mathbf{c}_1 = 920$ and $\sigma^2(y_2) = \mathbf{c}_2{}^T\mathbf{V}\,\mathbf{c}_2 = 140$. Applying Equation 9.20, the covariance between these two indices is $\sigma(y_1, y_2) = \mathbf{c}_1{}^T\mathbf{V}\,\mathbf{c}_2 = -350$, yielding their correlation as

$$\rho(y_1, y_2) = \frac{\sigma(y_1, y_2)}{\sqrt{\sigma^2(y_1)\cdot\sigma^2(y_2)}} = \frac{-350}{\sqrt{920\cdot 140}} = -0.975$$

Finally, consider the function, $y = \mathbf{x}^T\mathbf{A}\mathbf{x}$, with

$$\mathbf{A} = \begin{pmatrix} 1 & -1 & 0 \\ -1 & 3 & 2 \\ 0 & 2 & -4 \end{pmatrix}$$

which yields the quadratic function

$$y = x_1^2 - 2x_1x_2 + 4x_2x_3 + 3x_2^2 - 4x_3^2$$

The expected value of y follows from Equation 9.22, with $E[y] = \boldsymbol{\mu}^T\mathbf{A}\boldsymbol{\mu} + \text{tr}(\mathbf{AV})$, where $\boldsymbol{\mu}^T\mathbf{A}\boldsymbol{\mu} = -16$ and

$$\mathbf{AV} = \begin{pmatrix} 15 & -25 & 10 \\ -5 & 65 & 50 \\ -50 & 40 & -120 \end{pmatrix}$$

which has a trace of $15 + 65 - 120 = -40$, yielding an expected value of $E[y] = -40 + (-16) = -56$. We can also show this directly by taking the expected value of each term in y,

$$E[x_1^2] = \sigma_1^2 + \mu_1^2 = 10 + 10^2 = 110, \quad E[3\,x_2^2] = 3\left(\sigma_2^2 + \mu_2^2\right) = 3(20 + 4^2) = 108,$$

$$E[-4x_3^2] = -4\left(\sigma_3^2 + \mu_3^2\right) = -4(30 + [-3]^2) = -156,$$

$$E[-2\,x_1x_2] = -2\,(\sigma_{1,2} + \mu_1\mu_2) = -2\,(-5 + 10\cdot 4) = -70,$$

$$E[4\,x_2x_3] = 4\,(\sigma_{2,3} + \mu_2\mu_3) = 4\,(0 - 4\cdot 3) = -48$$

which again sums to -56.

THE MULTIVARIATE NORMAL DISTRIBUTION

As we have seen above, matrix notation provides a compact way to express vectors of random variables. We now consider the most commonly assumed distribution for such vectors, the multivariate analog of the normal distribution discussed in Chapter 2. Much of the machinery of quantitative genetics is based on this distribution, which we hereafter denote as the **MVN**, for multivariate normal.

Consider the probability density function for n independent normal random variables, where x_i is normally distributed with mean μ_i and variance σ_i^2, which we denote as $x_i \sim \text{N}(\mu_i, \sigma_i^2)$, with $\mathbf{x} = (x_1.\cdots, x_n)^T$. In this case, because the variables are independent, the joint probability density function for the vector $\mathbf{x}$ is the product of each univariate density,

$$\begin{aligned} p(\mathbf{x}) &= \prod_{i=1}^{n} p(x_i) = \prod_{i=1}^{n} (2\pi)^{-1/2}\sigma_i^{-1} \exp\left(-\frac{(x_i-\mu_i)^2}{2\,\sigma_i^2}\right) \\ &= (2\pi)^{-n/2}\left(\prod_{i=1}^{n}\sigma_i\right)^{-1} \exp\left(-\sum_{i=1}^{n}\frac{(x_i-\mu_i)^2}{2\,\sigma_i^2}\right) \end{aligned} \tag{9.23}$$

We can express this equation more compactly in matrix form by defining the matrices

$$\mathbf{V} = \begin{pmatrix} \sigma_1^2 & 0 & \cdots & 0 \\ 0 & \sigma_2^2 & \cdots & 0 \\ \vdots & \vdots & \ddots & \vdots \\ 0 & \cdots & \cdots & \sigma_n^2 \end{pmatrix} \quad \text{and} \quad \boldsymbol{\mu} = \begin{pmatrix} \mu_1 \\ \mu_2 \\ \vdots \\ \mu_n \end{pmatrix}$$

Because $\mathbf{V}$ is diagonal, its determinant is simply the product of the diagonal elements

$$|\mathbf{V}| = \prod_{i=1}^{n} \sigma_i^2$$

Likewise, $\mathbf{V}^{-1}$ is also diagonal, with ith diagonal element $1/\sigma_i^2$. Hence, using quadratic products, we have

$$\sum_{i=1}^{n} \frac{(x_i - \mu_i)^2}{\sigma_i^2} = (\mathbf{x} - \boldsymbol{\mu})^T \mathbf{V}^{-1} (\mathbf{x} - \boldsymbol{\mu})$$

Substituting in these expressions, Equation 9.23 can be rewritten as

$$p(\mathbf{x}) = (2\pi)^{-n/2} |\mathbf{V}|^{-1/2} \exp\left[-\frac{1}{2}(\mathbf{x} - \boldsymbol{\mu})^T \mathbf{V}^{-1} (\mathbf{x} - \boldsymbol{\mu})\right] \tag{9.24}$$

We will also write this density as $p(\mathbf{x}, \boldsymbol{\mu}, \mathbf{V})$ when we wish to stress that it is a function of the mean vector $\boldsymbol{\mu}$ and the covariance matrix $\mathbf{V}$.

More generally, when the elements of $\mathbf{x}$ are correlated ($\mathbf{V}$ is not diagonal), Equation 9.24 gives the probability density function for a vector of multivariate normally distributed random variables, with mean vector $\boldsymbol{\mu}$ and covariance matrix $\mathbf{V}$. We denote this by

$$\mathbf{x} = \begin{pmatrix} x_1 \\ \vdots \\ x_n \end{pmatrix} \sim \text{MVN}_n(\boldsymbol{\mu}, \mathbf{V})$$

where the subscript indicating the dimensionality of $\mathbf{x}$ is usually omitted. The multivariate normal distribution is also referred to as the **Gaussian distribution**. We restrict our attention to those situations where $\mathbf{V}$ is nonsingular, in which case it is also positive definite (as $\mathbf{c}^T\mathbf{V}\mathbf{c} > 0$ for all vectors $\mathbf{c} \neq 0$; WL Appendix 5). When $\mathbf{V}$ is singular, some of the elements of $\mathbf{x}$ are linear functions of other elements in $\mathbf{x}$, and hence we can construct a reduced vector of variables whose covariance matrix is now nonsingular.

Properties of the MVN

As in the case of its univariate counterpart, the MVN is expected to arise naturally when the quantities of interest result from a large number of underlying variables. Because this condition seems (at least at first glance) to describe many biological systems, the MVN is a natural starting point in biometrical analysis. Further details on the wide variety of applications of the MVN to multivariate statistics can be found in the introductory texts by Morrison (1976) and Johnson and Wichern (2002) and in the more advanced treatment by Anderson (2003). The MVN has a number of useful properties, which we summarize below.

1. *If* $\mathbf{x} \sim$ MVN, *then the distribution of any subset of the variables in* $\mathbf{x}$ *is also MVN.* For example, each x_i is normally distributed and each pair (x_i, x_j) is bivariate normally distributed.

2. *If* $\mathbf{x} \sim$ MVN, *then any linear combination of the elements of* $\mathbf{x}$ *is also MVN.* Specifically, if $\mathbf{x} \sim \text{MVN}_n(\boldsymbol{\mu}, \mathbf{V})$, $\mathbf{a}$ is a vector of constants, and $\mathbf{A}_{m\times n}$ is a matrix of constants, then

$$\text{for} \quad \mathbf{y} = \mathbf{x} + \mathbf{a}, \qquad \mathbf{y} \sim \text{MVN}_n(\boldsymbol{\mu} + \mathbf{a}, \mathbf{V}) \tag{9.25a}$$

$$\text{for} \quad y = \mathbf{a}^T\mathbf{x} = \sum_{k=1}^{n} a_i x_i, \qquad y \sim \text{N}(\mathbf{a}^T\boldsymbol{\mu}, \mathbf{a}^T\mathbf{V}\,\mathbf{a}) \tag{9.25b}$$

$$\text{for} \quad \mathbf{y}_{m\times 1} = \mathbf{A}_{m\times n}\mathbf{x}_{n\times 1}, \qquad \mathbf{y} \sim \text{MVN}_m\left(\mathbf{A}\boldsymbol{\mu}, \mathbf{A}^T\mathbf{V}\mathbf{A}\right) \tag{9.25c}$$

3. *Conditional distributions associated with the MVN are also multivariate normal.* Consider the partitioning of an n-dimensional column vector $\mathbf{x}$ into two components, an $(m \times 1)$ column vector $\mathbf{x_1}$ and an $[(n-m) \times 1]$ column vector $\mathbf{x_2}$ of the remaining variables, e.g.,

$$\mathbf{x} = \begin{pmatrix} \mathbf{x_1} \\ \mathbf{x_2} \end{pmatrix}$$

The mean vector and covariance matrix can be partitioned similarly as

$$\boldsymbol{\mu} = \begin{pmatrix} \boldsymbol{\mu_1} \\ \boldsymbol{\mu_2} \end{pmatrix} \quad \text{and} \quad \mathbf{V} = \begin{pmatrix} \mathbf{V_{x_1x_1}} & \mathbf{V_{x_1x_2}} \\ \mathbf{V}^T_{\mathbf{x_1x_2}} & \mathbf{V_{x_2x_2}} \end{pmatrix} \tag{9.26}$$

where the $m \times m$ and $(n-m) \times (n-m)$ matrices $\mathbf{V_{x_1x_1}}$ and $\mathbf{V_{x_2x_2}}$ are, respectively, the covariance matrices for $\mathbf{x_1}$ and $\mathbf{x_2}$, while the $m \times (n-m)$ matrix $\mathbf{V_{x_1x_2}}$ is the matrix of covariances between the elements of $\mathbf{x_1}$ and $\mathbf{x_2}$. If we condition on $\mathbf{x_2}$, the resulting conditional random variable, $\mathbf{x_1}|\mathbf{x_2}$, is MVN with $(m \times 1)$ mean vector

$$\boldsymbol{\mu}_{\mathbf{x_1}|\mathbf{x_2}} = \boldsymbol{\mu_1} + \mathbf{V_{x_1x_2}}\mathbf{V}^{-1}_{\mathbf{x_2x_2}}(\mathbf{x_2} - \boldsymbol{\mu_2}) \tag{9.27}$$

and $(m \times m)$ covariance matrix

$$\mathbf{V}_{\mathbf{x_1}|\mathbf{x_2}} = \mathbf{V_{x_1x_1}} - \mathbf{V_{x_1x_2}}\mathbf{V}^{-1}_{\mathbf{x_2x_2}}\mathbf{V}^T_{\mathbf{x_1x_2}} \tag{9.28}$$

A proof can be found in most multivariate statistics texts, e.g., Morrison (1976).

4. *If* $\mathbf{x} \sim$ MVN, *the regression of any subset of* $\mathbf{x}$ *on another subset is linear and homoscedastic*. For example, if $\mathbf{x} = (x_1, \cdots, x_n)^T$, then we might be interested in the expected value of the vector $\mathbf{x_1} = (x_1, \cdots x_k)^T$ given the observed values of the vector $\mathbf{x_2} = (x_{k+1}, \cdots x_n)^T$. Rewriting Equation 9.27 in terms of a regression of the predicted value of the vector $\mathbf{x_1}$ given an observed value of the vector $\mathbf{x_2}$, we have

$$\mathbf{x_1} = \widehat{\mathbf{x}}_1 + \mathbf{e}, \quad \text{where} \quad \widehat{\mathbf{x}}_1 = \boldsymbol{\mu_1} + \mathbf{V_{x_1x_2}}\mathbf{V}^{-1}_{\mathbf{x_2x_2}}(\mathbf{x_2} - \boldsymbol{\mu_2}) \tag{9.29a}$$

with

$$\mathbf{e} \sim \text{MVN}_m\left(\mathbf{0}, \mathbf{V}_{\mathbf{x_1}|\mathbf{x_2}}\right) \tag{9.29b}$$

Extending the terminology of a univariate regression (Chapter 3) to this multivariate setting, one can think of $\mathbf{x_1}$ as the response variable (a vector in this case) and $\mathbf{x_2}$ as the predictor variable (also now a vector). Hence, $\mathbf{V_{x_1x_2}}$ is the covariance between the response and predictor variances, while $\mathbf{V_{x_2x_2}}$ is the (co)variance structure of the predictor variables. This generalizes the univariate slope of $\sigma(y,x)\sigma^{-2}(x)$ to an $m \times (n-m)$ matrix of slopes, $\mathbf{V_{x_1x_2}}\mathbf{V}^{-1}_{\mathbf{x_2x_2}}$, in a multivariate setting.

Example 9.8. Consider the regression of the phenotypic value of an offspring (z_o) on that of its parents (z_s and z_d for sire and dam, respectively). Assume that the joint distribution of z_o, z_s, and z_d is multivariate normal. For the simplest case of noninbred and unrelated parents, no epistasis, or genotype–environment correlation, the covariance matrix can be obtained from the theory of correlation between relatives (Chapter 7), giving the joint distribution as

$$\begin{pmatrix} z_o \\ z_s \\ z_d \end{pmatrix} \sim \text{MVN}\left[\begin{pmatrix} \mu_o \\ \mu_s \\ \mu_d \end{pmatrix}, \sigma^2_z \cdot \begin{pmatrix} 1 & h^2/2 & h^2/2 \\ h^2/2 & 1 & 0 \\ h^2/2 & 0 & 1 \end{pmatrix}\right] \tag{9.30a}$$

where the off-diagonal elements, $\sigma^2_A/2 = (h^2/2)\sigma^2_z$, follow from the parent–offspring covariance. Let

$$\mathbf{x_1} = (\, z_o\,), \quad \mathbf{x_2} = \begin{pmatrix} z_s \\ z_d \end{pmatrix}$$

giving

$$\mathbf{V}_{\mathbf{x}_1\mathbf{x}_1} = \sigma_z^2, \quad \mathbf{V}_{\mathbf{x}_1\mathbf{x}_2} = \frac{h^2\sigma_z^2}{2}\begin{pmatrix} 1 & 1 \end{pmatrix}, \quad \mathbf{V}_{\mathbf{x}_2\mathbf{x}_2} = \sigma_z^2 \begin{pmatrix} 1 & 0 \\ 0 & 1 \end{pmatrix}$$

From Equation 9.29a, the regression of offspring value on parental values is linear and homoscedastic with

$$\begin{aligned} z_o &= \mu_o + \frac{h^2\sigma_z^2}{2}\begin{pmatrix} 1 & 1 \end{pmatrix}\sigma_z^{-2}\begin{pmatrix} 1 & 0 \\ 0 & 1 \end{pmatrix}\begin{pmatrix} z_s - \mu_s \\ z_d - \mu_d \end{pmatrix} + e \\ &= \mu_o + \frac{h^2}{2}(z_s - \mu_s) + \frac{h^2}{2}(z_d - \mu_d) + e \end{aligned} \tag{9.30b}$$

where, from Equations 9.28 and 9.29b, the residual error is normally distributed with mean zero and variance

$$\begin{aligned} \sigma_e^2 &= \sigma_z^2 - \frac{h^2\sigma_z^2}{2}\begin{pmatrix} 1 & 1 \end{pmatrix}\sigma_z^{-2}\begin{pmatrix} 1 & 0 \\ 0 & 1 \end{pmatrix}\frac{h^2\sigma_z^2}{2}\begin{pmatrix} 1 \\ 1 \end{pmatrix} \\ &= \sigma_z^2\left(1 - \frac{h^4}{2}\right) \end{aligned} \tag{9.30c}$$

This same approach allows one to consider more complex situations. For example, suppose that the parents assortatively mate (with phenotypic correlation ρ_z). From Table 7.5, the parent–offspring covariance now becomes $\sigma_A^2(1+\rho_z)/2$, and the resulting covariance matrix becomes

$$\mathbf{V} = \sigma_z^2 \cdot \begin{pmatrix} 1 & h^2(1+\rho_z)/2 & h^2(1+\rho_z)/2 \\ h^2(1+\rho_z)/2 & 1 & \rho_z \\ h^2(1+\rho_z)/2 & \rho_z & 1 \end{pmatrix} \tag{9.30d}$$

Similarly, when the parents are inbred and/or related, Equation 7.4b gives the covariances between an offspring and its sire and dam as $(2\sigma_A^2)(\Theta_{ss}+\Theta_{sd})/2$ and $(2\sigma_A^2)(\Theta_{dd}+\Theta_{sd})/2$, respectively, while Equation 7.11a gives the covariance between sire and dam as $\sigma_A(s,d) = 2\Theta_{sd}\sigma_A^2$. Finally, the phenotypic variance of an inbred individual is $\sigma_A^2(1+f) + \sigma_E^2 = \sigma_z^2 + \sigma_A^2 f = \sigma_z^2(1+fh^2)$. Incorporating these expressions, the resulting covariance matrix becomes

$$\mathbf{V} = \sigma_z^2 \cdot \begin{pmatrix} 1+f_o h^2 & h^2(\Theta_{ss}+\Theta_{sd}) & h^2(\Theta_{dd}+\Theta_{sd}) \\ h^2(\Theta_{ss}+\Theta_{sd}) & 1+f_s h^2 & 2h^2\Theta_{sd} \\ h^2(\Theta_{dd}+\Theta_{sd}) & 2h^2\Theta_{sd} & 1+f_d h^2 \end{pmatrix} \tag{9.30e}$$

Recalling Equation 7.3b allows us to entirely write this matrix in terms of the three coefficients of coancestry associated with the parents (Θ_{ss}, Θ_{dd}, and Θ_{sd}), as $f_x = 2\Theta_{xx} - 1$, while $f_o = \Theta_{sd}$. Equation 9.11 easily allows one to invert the 2×2 matrix $\mathbf{V}_{\mathbf{x}_2,\mathbf{x}_2}$ associated with the parents, and using the approach leading to Equationa 9.30b and 9.30c yields the regression and associated residual variance under these more complex cases.

Example 9.9. The previous example dealt with the prediction of the phenotypic value of an offspring given its parental phenotypic values. The same approach can be used to predict an offspring's additive genetic (or breeding) value (A_o) given knowledge of the parental values (A_s, A_d). Again assuming that the joint distribution is multivariate normal and that the parents are unrelated and noninbred, the joint distribution can be written as

$$\begin{pmatrix} A_o \\ A_s \\ A_d \end{pmatrix} \sim \text{MVN}\left[\begin{pmatrix} \mu_o \\ \mu_s \\ \mu_d \end{pmatrix}, \sigma_A^2\begin{pmatrix} 1 & 1/2 & 1/2 \\ 1/2 & 1 & 0 \\ 1/2 & 0 & 1 \end{pmatrix}\right] \tag{9.31a}$$

Proceeding in the same fashion as in Example 9.8, the conditional distribution of offspring additive genetic values, given the parental values, is normal, so that the regression of offspring additive genetic value on parental value is linear and homoscedastic with

$$A_o = \mu_o + \frac{A_s - \mu_s}{2} + \frac{A_d - \mu_d}{2} + e \tag{9.31b}$$

and

$$e \sim \text{N}(0, \sigma_A^2/2) \tag{9.31c}$$

Finally, an important merger of concepts from this and the previous example is the prediction of an offspring's *breeding value* (A_o) given the *phenotypes* (z_s, z_d) of its parents. Assuming multivariate normality,

$$\begin{pmatrix} A_o \\ z_s \\ z_d \end{pmatrix} \sim \text{MVN}\left[\begin{pmatrix} 0 \\ \mu_s \\ \mu_d \end{pmatrix}, \sigma_z^2 \begin{pmatrix} h^2 & h^2/2 & h^2/2 \\ h^2/2 & 1 & 0 \\ h^2/2 & 0 & 1 \end{pmatrix}\right] \tag{9.31d}$$

Elements that differ from Equation 9.30a are that the expected value of A_o is zero (because of breeding values have a mean of zero), and $\sigma^2(A_o) = \sigma_A^2 = h^2\sigma_z^2$. Using the same approach as in Example 9.8 yields

$$A_o = \frac{h^2}{2}(z_s - \mu_s) + \frac{h^2}{2}(z_d - \mu_d) + e, \qquad \sigma_e^2 = \sigma_A^2\left(1 - \frac{h^4}{2}\right) \tag{9.31e}$$

Under more general settings (such as inbred and/or related parents), the covariance matrix in Equation 9.31d is replaced by Equation 9.30e, subject to the minor change that the 1,1 element $(1 + f_0 h^2)$ is replaced by $h^2(1 + f_o)$. Notice that the prediction of breeding values from phenotypic information under very general types of relationships is a function of h^2 and the coefficients of coancestry of the **measured** (i.e., phenotyped) relatives. This serves as a lead-in to the very general method of BLUP for predicting breeding values given a set of phenotypic values on a known group of relatives (Part II Chapter 32).

MATRIX GEOMETRY

Eigenvalues and Eigenvectors

A vector can be thought of as an arrow, corresponding to a direction in space, as it has a *length* and an *orientation*. Similarly, a matrix can be thought of as describing a *vector transformation*, so that when one multiplies a vector by that matrix, it generates a new vector. Such a transformation generally has the effect that the resulting vector is both *rotated* in direction and *scaled* (shrinking or expanding in length) relative to the original vector. Hence, the new vector $\mathbf{b} = \mathbf{Ac}$ usually points in a different direction than $\mathbf{c}$ and has a different length. However, for any square matrix a set of vectors exists that satisfies

$$\mathbf{Ay} = \lambda\mathbf{y} \tag{9.32}$$

Namely, the new vector ($\lambda\mathbf{y}$) is still in the *same direction* (as multiplying a vector by a constant does not change its orientation, except for reflecting it about the origin when $\lambda < 0$), although it is *scaled* so that its new length is an amount λ of the original length. Vectors that satisfy Equation 9.32 are called the **eigenvectors** of that matrix, with λ as the corresponding **eigenvalue** for a given eigenvector. For such vectors, the only action of the matrix transformation is to scale their length by some amount, λ. These vectors thus represent the ***inherent axes associated with the vector transformation given by* $\mathbf{A}$**, and the set of all such vectors, along with their corresponding scalar multipliers, completely describes the geometry of this transformation. Note that if $\mathbf{y}$ is an eigenvector, then so is

$a\mathbf{y}$, as $\mathbf{A}(a\mathbf{y}) = a(\mathbf{A}\mathbf{y}) = \lambda(a\mathbf{y})$, while the associated eigenvalue, λ, remains unchanged. Hence, we typically scale eigenvectors to be of unit length to yield **unit** or **normalized eigenvectors**, which are typically denoted by $\mathbf{e}$.

The resulting collection of eigenvectors and their associated scaling eigenvalues are called the **eigenstructure** of a matrix. This structure provides powerful insight into the geometric aspects of a matrix, such as the major axes of variation in a covariance matrix. Appendix 3 examines the **singular value decomposition** (**SVD**) which generalizes this concept to nonsquare matrices, while WL Appendix 5 examines eigenstructure in more detail.

Notice that if a matrix has an eigenvalue of zero, then from Equation 9.32, there is some vector that satisfies $\mathbf{Ae} = \mathbf{0}$, namely that one can write at least one of the rows of $\mathbf{A}$ as a linear combination of the other rows. Put another way, if one has a set of n equations, then here one or more of these equations are redundant, being a linear combination of other equations. Hence, there is no unique solution to the system of equations. Recall that the determinant of a matrix is zero in such cases, and hence we have an important result that ***the determinant of a matrix is zero when it has one (or more) zero eigenvalues***.

The eigenvalues of an n-dimensional square matrix, $\mathbf{A}$, are solutions of Equation 9.32, which can be written as $(\mathbf{A} - \lambda\,\mathbf{I})\mathbf{y} = \mathbf{0}$. This implies that the determinant of $(\mathbf{A} - \lambda\,\mathbf{I})$ must equal zero, which gives rise to the **characteristic equation** for λ,

$$|\mathbf{A} - \lambda\mathbf{I}| = 0 \tag{9.33a}$$

whose solution yields the eigenvalues of $\mathbf{A}$. This equation can be also be expressed using the **Laplace expansion**,

$$|\mathbf{A} - \lambda\mathbf{I}| = (-\lambda)^n + S_1(-\lambda)^{n-1} + \cdots + S_{n-1}(-\lambda)^1 + S_n = 0 \tag{9.33b}$$

where S_i is the sum of all **principal minors** (minors including diagonal elements of the original matrix) of order i. Finding the eigenvalues thus requires solving a polynomial equation of order n, implying that there are exactly n eigenvalues (some of which may be identical, i.e., **repeated**). In practice, for $n > 2$ this is accomplished numerically, and most analysis packages offer routines to accomplish this task.

Two of these principal minors are easily obtained and provide information on the nature of the eigenvalues. The only principal minor having the same order of the matrix is the full matrix itself, which means that $S_n = |\mathbf{A}|$, the determinant of $\mathbf{A}$. S_1 is also related to an important matrix quantity, the trace,

$$\text{tr}(\mathbf{A}) = \sum_{i=1}^{n} A_{ii} \tag{9.34a}$$

Observe that $S_1 = \text{tr}(\mathbf{A})$, as the only principal minors of order one are the diagonal elements themselves, the sum of which equals the trace. Both the trace and determinant can be expressed as functions of the eigenvalues, with

$$\text{tr}(\mathbf{A}) = \sum_{i=1}^{n} \lambda_i \qquad \text{and} \qquad |\mathbf{A}| = \prod_{i=1}^{n} \lambda_i \tag{9.34b}$$

Hence, $\mathbf{A}$ ***is singular*** ($|\mathbf{A}| = 0$) ***if, and only if, at least one eigenvalue is zero***. As we will see, if $\mathbf{A}$ is a covariance matrix, then its trace (the sum of its eigenvalues) measures its total amount of variation, as the eigenvalues of a covariance matrix are non-negative ($\lambda_i \geq 0$). Another useful result is that ***the diagonal elements in a diagonal matrix correspond to its eigenvalues***.

A final useful concept is the idea of the **rank** of a square matrix, which is its *number of nonzero eigenvalues*. Essentially, this is the amount of information (number of independent equations) that are specified by a matrix. A square $n \times n$ matrix of **full rank** has n nonzero eigenvalues and hence is nonsingular (a unique inverse exists).

Example 9.10. Consider the following matrix

$$\mathbf{A} = \begin{pmatrix} 1 & 2 \\ 2 & 1 \end{pmatrix}. \qquad \text{Hence,} \quad \det(\mathbf{A} - \lambda\mathbf{I}) = \det\begin{pmatrix} 1-\lambda & 2 \\ 2 & 1-\lambda \end{pmatrix}$$

Using Equation 9.12a to compute the determinant and solving Equation 9.33a yields $(1-\lambda)^2 - 4 = 0$, or the quadratic equation $\lambda^2 - 2\lambda - 3 = 0$, which has solutions of 3 and -1. Noting that $S_1 = \text{tr}(\mathbf{A}) = 2$ and $S_2 = \det(\mathbf{A}) = -3$, we also recover this equation using Equation 9.33b, $|\mathbf{A} - \lambda\mathbf{I}| = (-\lambda)^2 + S_1(-\lambda)^1 + S_2 = \lambda^2 - 2\lambda - 3$. Note that the negative eigenvalue implies that $\mathbf{A}$ is *not* a covariance matrix. We can also see this by computing the correlation, with $\rho^2 = 2/\sqrt{1 \cdot 1} > 1$ and thus not a proper correlation. Hence, a symmetric matrix (even with no negative elements) need not be a covariance matrix.

Next, note that the vectors $\mathbf{y}_1^T = (1, 1)$ and $\mathbf{y}_2^T = (1, -1)$ correspond to the eigenvectors of $\mathbf{A}$, as

$$\mathbf{A}\mathbf{y}_1 = \begin{pmatrix} 1 & 2 \\ 2 & 1 \end{pmatrix}\begin{pmatrix} 1 \\ 1 \end{pmatrix} = 3 \cdot \begin{pmatrix} 1 \\ 1 \end{pmatrix}, \quad \text{and} \quad \mathbf{A}\mathbf{y}_2 = \begin{pmatrix} 1 & 2 \\ 2 & 1 \end{pmatrix}\begin{pmatrix} 1 \\ -1 \end{pmatrix} = (-1) \cdot \begin{pmatrix} 1 \\ -1 \end{pmatrix}$$

Here $\mathbf{y}_1$ is an eigenvector associated with $\lambda_1 = 3$, while $\mathbf{y}_2$ is an eigenvector associated with $\lambda_2 = -1$. Recalling that the length of a vector $\mathbf{x}$ is given by $||\mathbf{x}|| = \sqrt{\mathbf{x}^T\mathbf{x}}$ (WL Appendix 5), the lengths of both $\mathbf{y}_1$ and $\mathbf{y}_2$ are $\sqrt{2}$, with the two unit eigenvectors given by $\mathbf{e}_i = \mathbf{y}_i/\sqrt{2}$. Note that $\mathbf{y}_1^T\mathbf{y}_2 = 1 \cdot 1 + (-1) \cdot 1 = 0$, which implies (WL Appendix 5) that these two vectors are **orthogonal** (at right angles) to each other.

Example 9.11. Insight into the role of zero eigenvalues can be gained by considering two example systems of linear equations, where in both cases the corresponding matrix of coefficients is singular. This implies that some of the equations are redundant with each other (i.e., are linear combinations of the others). However, having a determinant of zero is a very coarse measure of the *amount* of redundancy, as it simply indicates that *at least one* equation is redundant. The number of zero eigenvalues goes much further, giving the *number* of redundant equations. Consider the following two sets of equations:

$$\text{Set one:} \quad \begin{aligned} x_1 + x_2 + x_1 &= 1 \\ 2x_1 + 2x_2 + 2x_3 &= 2 \\ 3x_1 + 3x_2 + 3x_3 &= 3 \end{aligned} \qquad \text{Set two:} \quad \begin{aligned} x_1 + x_2 + x_3 &= 1 \\ 2x_1 + 2x_2 + 2x_3 &= 2 \\ x_1 - x_2 - 2x_3 &= 3 \end{aligned}$$

Clearly, set one consists of three redundant equations (all are multiples of each other), while set two contains at least one redundant equation, as the first and second equations are multiples of each other. The corresponding coefficient matrices for these sets become

$$\mathbf{A}_1 = \begin{pmatrix} 1 & 1 & 1 \\ 2 & 2 & 2 \\ 3 & 3 & 3 \end{pmatrix}, \qquad \mathbf{A}_2 = \begin{pmatrix} 1 & 1 & 1 \\ 2 & 2 & 2 \\ 1 & -1 & -2 \end{pmatrix}$$

As expected, both of these matrices have determinants of zero. $\mathbf{A}_1$ has eigenvalues of 6, 0, and 0, while the eigenvalues for $\mathbf{A}_2$ are 2.79, -1.79, and 0. Note in both cases that the trace (6 and 1, respectively) equals the sum of the eigenvalues. Because $\mathbf{A}_1$ contains only one nonzero eigenvalue (it has rank 1), only one relationship can be estimated from set one: the solution is the plane defined as all points satisfying $x_1 = 1 - x_2 - x_3$. Set two, by virtue of having two nonzero eigenvalues (it has rank 2), yields two relationships ($x_2 = x_1 + 5$ and $x_3 = 2x_1 - 4$).

The importance of the number of nonzero eigenvalues will reappear next chapter when we examine the number of fixed effects in a linear model that can be uniquely estimated.

Principal Components of the Variance–covariance Matrix

An important application of eigenstructure is **principal component analysis** (**PCA**), the eigenanalysis of a covariance matrix. Consider a random vector, $\mathbf{x}$, with an associated covariance matrix, $\mathbf{V}$. We are often interested in how the variance of the elements of $\mathbf{x}$ can be decomposed into independent components. For example, even though we may be measuring n variables, only one or two of these may account for the majority of the variation. If this is the case, we may wish to exclude those variables contributing very little variation from further analysis. More generally, if random variables are correlated, then certain linear combinations of the elements of $\mathbf{x}$ may account for most of the variance. PCA extracts these combinations by decomposing the variance of $\mathbf{x}$ into the contributions from a series of orthogonal vectors, the **first principal component** (**PC1**) of which explains the most variation possible for any single vector, the **second principal component** (**PC2**) the next most possible amount, and so on until we account for the entire variance of $\mathbf{x}$. These vectors (or **axes**) correspond to the eigenvectors of $\mathbf{V}$ associated with the largest, second largest, and so on, eigenvalues.

Our starting point for PCA is to note that the eigenvalues of a covariance matrix are never negative, and are all positive if $\mathbf{V}$ is nonsingular. A symmetric matrix, $\mathbf{V}$, with all positive eigenvalues is positive definite, implying that $\mathbf{c}^T\mathbf{V}\mathbf{c} > 0$ for all values of $\mathbf{c} \neq \mathbf{0}$. From Equation 9.19, this quadratic product is non-negative as it corresponds to the variance of the linear combination $\mathbf{c}^T\mathbf{x}$. Because all of the eigenvalues of $\mathbf{V}$ are non-negative, their sum represents the **total variance** implicit in the elements of $\mathbf{x}$. From Equation 9.34b, this sum is simply the trace of $\mathbf{V}$, tr($\mathbf{V}$). A matrix $\mathbf{V}$ is said to be **non-negative definite** if $\mathbf{c}^T\mathbf{V}\mathbf{c} \geq 0$, which happens when all its eigenvalues are non-negative, so that a covariance matrix, even if singular, is always non-negative definite, and a nonsingular covariance matrix is always positive definite.

Suppose that $\mathbf{V}$ is an n-dimensional covariance matrix, and we order its eigenvalues from largest to smallest, $\lambda_1 \geq \lambda_2 \geq \cdots \geq \lambda_n$, with their associated (unit-length) eigenvectors denoted by $\mathbf{e}_1, \mathbf{e}_2, \cdots, \mathbf{e}_n$, respectively. λ_1 is referred to as the **leading eigenvalue**, with $\mathbf{e}_1$ the **leading eigenvector**. It can be shown (WL Appendix 5) that the maximum variance for any linear combination of the elements of $\mathbf{x}$ ($y = \mathbf{c}_1^T\mathbf{x}$, subject to the constraint that $||\mathbf{c}_1|| = 1$), is

$$\max\left[\sigma^2(y)\right] = \max_{||\mathbf{c}_1||=1}\left[\sigma^2\left(\mathbf{c}_1^T\mathbf{x}\right)\right] = \max_{||\mathbf{c}_1||=1}\left[\mathbf{c}_1^T\mathbf{V}\mathbf{c}_1\right] = \lambda_1$$

which occurs when $\mathbf{c}_1 = \mathbf{e}_1$. This vector is the **first principal component** (often abbreviated as **PC1**), and accounts for a fraction $\lambda_1/\text{tr}(\mathbf{V})$ of the total variation in $\mathbf{x}$. We can partition the remaining variance in $\mathbf{x}$ after the removal of PC1 in a similar fashion. For example, the vector $\mathbf{c}_2$, that is orthogonal to PC1 ($\mathbf{c}_2^T\mathbf{e}_1 = 0$) and maximizes the remaining variance can be shown to be $\mathbf{e}_2$, which accounts for a fraction $\lambda_2/\text{tr}(\mathbf{V})$ of the total variation in $\mathbf{x}$. By proceeding in this fashion, we can see that the ith PC is given by $\mathbf{e}_i$, and that the amount of variation it accounts for is

$$\lambda_i \bigg/ \sum_{k=1}^{n} \lambda_k = \frac{\lambda_i}{\text{tr}(\mathbf{V})} \tag{9.35a}$$

Put another way, $\lambda_i/\text{tr}(\mathbf{V})$ is the fraction of that total variance explained by the linear combination $\mathbf{e}_i^T\mathbf{x}$. Hence, the fraction of total variation accounted for by the first k PCs is

$$\frac{\lambda_1 + \cdots + \lambda_k}{\text{tr}(\mathbf{V})} \tag{9.35b}$$

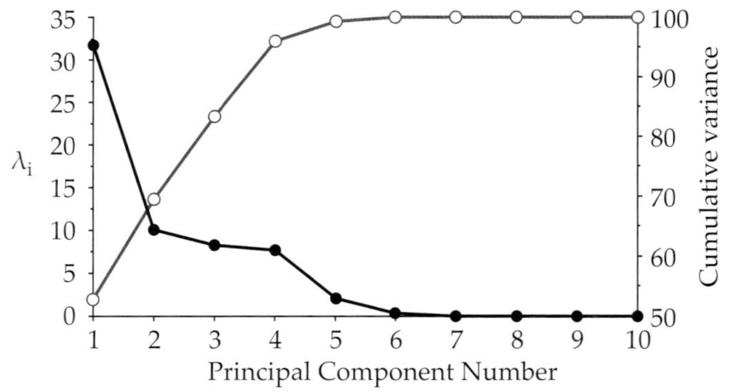

Figure 9.1. A joint scree and cumulative variance plot. The horizontal axis gives the eigenvalue number (λ_i corresponds to PCi), while the vertical axis jointly displays a scree plot (filled circles, the corresponding eigenvalue, λ_i) and the cumulative variance associated with the first i PCs (open circles; Equation 9.35b).

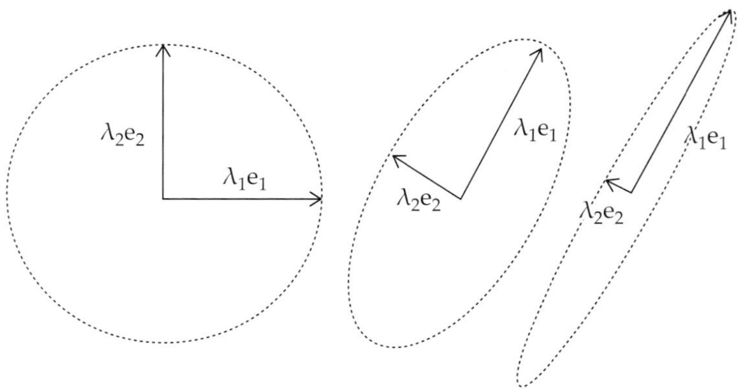

Figure 9.2. The impact of increasing the difference between the major and minor eigenvalues $(\lambda_1 - \lambda_2)$ while their sum is held constant. When the two eigenvalues are equal, the spread of $\mathbf{x}$ values about their mean is spherical, while it becomes increasingly elongated as the difference between the eigenvalues increases. Any tilt in this elongated distribution is generated by correlations between elements of $\mathbf{x}$ (the orientation being given by the direction of the associated eigenvectors, $\mathbf{e}_1$ and $\mathbf{e}_2$).

A graph of Equation 9.35a as a function of k is called a **cumulative variance plot** (Figure 9.1).

Another useful visual display of the eigenstructure is a **scree plot**, which graphs the eigenvalues ranked from largest to smallest (Figure 9.1). The term *scree* refers to the loose pile of rocks that comprise the steep slope of a mountain, as most scree plots display a rapid falloff, akin to what one would see in a scree field. Suppose the eigenvalues of $\mathbf{V}$ are roughly similar in magnitude (a relatively flat scree plot). For three dimensions this implies that the distribution of $\mathbf{x}$ is roughly spherical (i.e., a 3D plot of the elements of $\mathbf{x}$ corresponds to a soccer ball) and hence has little structure. As the eigenvalues become increasingly dissimilar, the scree plots starts to show a dramatic falloff in values, and the distribution of values of $\mathbf{x}$ becomes stretched and elongated, generating some axes with larger, and others with smaller, variances. Figure 9.2 shows the impact in two dimensions when the trace of a matrix (the sum of its eigenvalues) is held constant, while the difference between λ_1 and λ_2 increases.

A nearly flat scree plot indicates very little structure in $\mathbf{V}$, while a typical scree plot (a rapid decline in eigenvalues) indicates that much of the variance is concentrated in a few directions (or **major axes**). Another way to state this is that a small *variance in the eigenvalues*, $\sigma^2(\lambda)$, implies little structure (roughly equal variance in all directions) in the structure of $\mathbf{x}$, while the distribution of $\mathbf{x}$ becomes increasingly concentrated in a smaller number of directions as the variance in the eigenvalues increases.

Example 9.12. Here we perform PCA for the covariance matrix given in Example 9.7,

$$\mathbf{V} = \begin{pmatrix} 10 & -5 & 10 \\ -5 & 20 & 0 \\ 10 & 0 & 30 \end{pmatrix}$$

The eigenvalues and their associated eigenvectors are found to be $\lambda_1 = 34.41, \lambda_2 = 21.12$, and $\lambda_3 = 4.47$, with

$$\mathbf{e}_1 = \begin{pmatrix} 0.400 \\ -0.139 \\ 0.906 \end{pmatrix}, \quad \mathbf{e}_2 = \begin{pmatrix} 0.218 \\ -0.948 \\ -0.238 \end{pmatrix}, \quad \mathbf{e}_3 = \begin{pmatrix} 0.892 \\ 0.287 \\ -0.349 \end{pmatrix}$$

Hence, PC1 accounts for $\lambda_1/\text{tr}(\mathbf{V})$ = 34.41/60 = 57% of the total variation of $\mathbf{V}$.

There are several ways to interpret PC1. First, $\mathbf{e}_1$ is the direction of the maximal axis of variation (e.g., Figure 9.2). Second, $\mathbf{e}_1$ defines a weighted index $y_i = \mathbf{e}_1^T\mathbf{z}$ (a new composite variable),

$$y_1 = 0.400z_1 - 0.139z_2 + 0.906z_3$$

where this new composite variable accounts for 57% of the total variation by itself.

Similarly, PC2 accounts for $\lambda_2/\text{tr}(\mathbf{V})$ = 21.12/60 = 35% of the total variance, and gives the direction of the most variation orthogonal to PC1 ($\mathbf{e}_i^T\mathbf{e}_1 = 0$). The weighted index corresponding to PC2 is

$$y_2 = 0.218z_1 - 0.948z_2 - 0.238z_3$$

Using the two-dimensional vector $\mathbf{y}^T = (y_1, y_2)$ accounts for (34.31+21.12)/60 or 93% of the variation of $\mathbf{x}$. This illustrates one use of PCA, which is for **dimensional reduction**, extracting a set of weighted indices of much lower dimension than the original vector as a proxy for the variation in $\mathbf{x}$.

Finally, another common application of PCs are **PC plots**. Here, each observation is presented by a point in a 2D plot of their PC1 and PC2 scores. Suppose observation i is $\mathbf{z}^T = (1, 0, 1)$. The corresponding y_1 and y_2 values become

$$y_1 = (0.400 \cdot 1) - (0.139 \cdot 0) + (0.906 \cdot 1) = 1.306$$

$$y_2 = (0.218 \cdot 1) - (0.948 \cdot 0) - (0.238 \cdot 1) = -0.02$$

yielding the point $(1.306, -0.02)$ for this observation. Data, especially marker data, are often displayed using such plots to visually look for signals of clustering (population structure).

Example 9.13. As we will see in Chapter 20, an important application of PCA is in controlling for population structure in genome-wide association studies (GWAS). The basic idea of a GWAS is to search for marker-trait associations using densely packed SNPs. For the kth SNP marker (assumed to be biallelic), individuals are grouped into three genotype classes (M_kM_k, M_km_k, and m_km_k), trait means are computed for each class, and a marker effect examined using ANOVA (i.e., an among-group difference in trait means). One simple linear model to test for an effect from SNP k would be the regression

$$z = \mu + \beta_k n_k + e \tag{9.36a}$$

where z is the trait value, μ is the population mean, n_k denotes the number of copies of allele M_k in an indvidual (with values of 0, 1, or 2), and $2\beta_k$ is the difference in average trait value between the two different SNP homozygotes. A significant value of β_k indicates a marker-trait association.

Marker-trait associations can arise from linkage disequilibrium (LD) between the marker and a very closely linked QTL (Chapter 5). However, they can also arise from population structure. Suppose a GWAS sample, unbeknownst to the investigator, consists of two populations, with population one tending to be taller than population two. Further, because of population structure, some marker allele frequencies differ between the populations. A marker that is predictive of group membership (e.g., an allele very common in population one but very rare in population two) will show a marker-trait association even when it is unlinked to any QTLs for height.

This complication from population structure arises when subpopulations in the sample differ in mean trait value (the subpopulation mean $\mu^* \neq \mu$). If one could first adjust for any subpopulation-specific differences, then any remaining marker-trait associations are likely due to LD with nearby QTLs. In a typical GWAS, a very large number of markers are scored, and these provide information on any population structure. To adjust for structure, the investigator first constructs a marker-based $n \times n$ ancestry covariance matrix (Example 9.14) using markers *outside* of those on the chromosome being tested.

A PCA analysis of this covariance matrix would look for the presence of structure by examining either a scree or a cumulative variance plot, and choose the first S PCs by some criteria. The first PC for this matrix is the vector of shared ancestry explaining the most information, where the ith element of PC1 represents the projected ancestry of individual i onto this first axis. More generally, $e_{j,i}$—the ith element of $\mathbf{e}_j$ (PC j)—is the projection of individual i on the jth axis of ancestry. The idea is to predict the mean trait value in a subpopulation (μ^*) by regression on these PCs,

$$z_i = \mu + \sum_{j=1}^{S} v_j e_{j,i} + \beta_k n_k + e \tag{9.36b}$$

$$= \mu^* + \beta_k n_k + e \tag{9.36c}$$

The v_j are the best-fit predictors (partial regression coefficients) of how PC j influences the overall mean (which are fit along with μ and β_k by least-squares regressions). Hence, $\mu^* = \mu + \sum v_j e_{j,i}$ is the predicted mean given the population from which individual i is drawn, leaving $\beta_k n_k$ as any residual effect from marker k (which was *not* used in the population-structure correction).

Example 9.14. A related issue to the population-structure correction in a GWAS is accessing the amount of shared relatedness among all pairwise combinations over a collection of n individuals (Chapter 20). This is done by constructing an $n \times n$ matrix whose ijth element is an estimate of $2\Theta_{ij}$, twice the coefficient of coancestry (Chapters 7 and 8). Note that 2Θ is the expected fraction of the genome shared by two individuals and is also the coefficient on the amount of additive genetic variance that contributes to their phenotypic correlation (Equation 7.11a). When estimated from pedigree data alone, this matrix of relationships is called the **numerator relationship matrix**, and is denoted by $\mathbf{A}$. When estimated solely from marker data (Chapter 8), this is usually denoted by $\mathbf{G}$ and called the **genomic relationship matrix** (Part II Chapters 32 and 33). One delicate issue with marker data is that there is no guarantee the matrix constructed from all pairwise estimators will be non-negative definite (and hence a proper covariance matrix). Ways to both ensure this, and also calculate $\mathbf{G}$ in a single matrix operation, were provided by VanRaden (2007, 2008).

The starting point is the **marker information matrix**, $\mathbf{M}$. For a set of m markers scored over n individuals, $\mathbf{M}$ is $n \times m$, with the ith row corresponding to the m marker genotypes for individual i, while the jth column shows the genotypes for marker j over all n scored individuals. At each SNP, we count the number of copies of the allele (one allele at the SNP is set to value 1, the alternative to value 0; Chapter 8), with M_{ij} denoting the number of reference alleles at locus i in individual j, which takes on values of 0, 1, and 2 for, respectively, SNP genotypes at locus i of 00, 10, and 11 (VanRanden 2007). Alternatively, SNP data are often coded by subtracting one from each category, giving scores of $-1, 0$, and 1 (VanRanden 2008). As a toy example, suppose that four SNPs are scored in each of three individuals, resulting in the 3×4 matrix (using the VanRanden 2008 coding)

$$\mathbf{M} = \begin{pmatrix} 1 & 0 & 1 & 1 \\ 1 & -1 & 1 & -1 \\ -1 & 0 & 0 & 0 \end{pmatrix}$$

This indicates that individual one has genotypes of 11, 10, 11, and 11 at the four markers, individual two has genotypes of 11, 00, 11, and 00, and individual three has genotypes of 00, 10, 10, and 10. The 3×3 and 4×4 matrices, $\mathbf{MM}^T$ and $\mathbf{M}^T\mathbf{M}$, respectively, become

$$\mathbf{MM}^T = \begin{pmatrix} 3 & 1 & -1 \\ 1 & 4 & -1 \\ -1 & -1 & 1 \end{pmatrix} \qquad \mathbf{M}^T\mathbf{M} = \begin{pmatrix} 3 & -1 & 2 & 0 \\ -1 & 1 & -1 & 1 \\ 2 & -1 & 2 & 0 \\ 0 & 1 & 0 & 2 \end{pmatrix}$$

$\mathbf{M}_{n\times m}\mathbf{M}^T_{m\times n}$ corresponds to the $n \times n$ covariance matrix for marker scores (genotypes) among individuals, and as will be shortly demonstrated, can be used to construct $\mathbf{G}$. Diagonal elements correspond to the number of homozygotes for individuals, showing that individuals one, two, and three have, respectively, 3, 4, and 1 homozygotes among their four scored markers. Off-diagonal elements correspond to the similarity (in marker genotypes) of the two individuals. Individuals one and two are more similar across the markers, while one and three, as well as two and three, are more dissimilar (negative values).

Conversely, the $m \times m$ matrix, $\mathbf{M}^T_{m\times n}\mathbf{M}_{n\times m}$, quantifies the covariance over *markers*, with negative values indicating negative associations between alleles at two different markers and positive value indicating positive associations. This is the matrix from which PCs are extracted to correct for population structure (Example 9.13). The dimensionality of this square matrix is the number of markers, m, and usually $m \gg n$, so that this matrix is expected to be singular (the maximum number of positive eigenvalues is the smaller of n and m),

VanRaden (2007, 2008) proposed two methods (Equations 8.15b and 8.16b) for using $\mathbf{M}$ to generate an estimate of $\mathbf{G}$. Define the allele-frequency matrix $\mathbf{P} = (\,\mathbf{p}_1, \mathbf{p}_2, \cdots, \mathbf{p}_m\,)$ whose ith column $\mathbf{p}_i$ is given by $2p_i\mathbf{1}$ (a vector whose values are all $2p_i$, the frequency of the reference allele at locus i) when the VanRanden (2007) marker coding scheme (0, 1, 2) is used. When markers are scored by the VanRanden (2008) coding $(-1, 0, 1)$, then the ith column is given by $2(p_i - 0.5)\mathbf{1}$. Finally, define $\mathbf{Z} = \mathbf{M} - \mathbf{P}$. With these definitions, VanRanden's first method is

$$\widehat{\mathbf{G}}_{VR1} = \frac{\mathbf{ZZ}^T}{2\sum_{i=1}^{m} p_i(1-p_i)} \tag{9.37a}$$

When using the VanRaden (2007) coding, this recovers Equation 8.16a in matrix form. This estimate weights all marker loci equally, as each marker locus estimate is scaled by the same amount (the denominator of Equation 9.37a).

VanRanden's second method weights the information from each locus, by defining the diagonal matrix $\mathbf{D}$ whose iith element is given by

$$D_{ii} = \frac{1}{m\,2\,p_i(1-p_i)} \tag{9.37b}$$

with

$$\widehat{\mathbf{G}}_{VR2} = \mathbf{ZDZ}^T \tag{9.37c}$$

which, when using VanRanden's (2007) coding recovers Equation 8.15b. This estimator places more weight on loci with rare alleles, as in this setting $[p_i(1-p_i)]^{-1}$ is large.

To demonstrate that Equations 9.37a and 9.37c generate non-negative definite matrices, we need to show that $\mathbf{c}^T\mathbf{G}\mathbf{c} \geq 0$ for all vectors $\mathbf{c}$. Ignoring the positive constant in the denominator of Equation 9.37a, we have

$$\mathbf{c}^T\widehat{\mathbf{G}}_{VR1}\,\mathbf{c} = \mathbf{c}^T\mathbf{Z}\mathbf{Z}^T\mathbf{c} = \mathbf{y}^T\mathbf{y} = \sum y_i^2 \geq 0$$

with the vector $\mathbf{y} = \mathbf{Z}^T\mathbf{c}$. Similarly,

$$\mathbf{c}^T\widehat{\mathbf{G}}_{VR2}\,\mathbf{c} = \mathbf{c}^T\mathbf{Z}\mathbf{D}\mathbf{Z}^T\mathbf{c} = \mathbf{z}^T\mathbf{z} = \sum z_i^2 \geq 0$$

where the vector $\mathbf{z} = \mathbf{D}^{1/2}\,\mathbf{Z}^T\mathbf{c}$, with the square-root matrix defined by Equation A3.7a.

10

General Linear and Mixed Models

The first principle is that you must not fool yourself and you are the easiest person to fool. Richard Feynman

Linear models form the backbone of most estimation procedures in quantitative genetics and are extensively used throughout this book. They are generally structured such that a vector, $\mathbf{y}$, of observations of a response variable (y) is modeled as a linear combination of the observed predictor variables (Chapter 9). This chapter introduces some of the basic tools and key concepts underlying the use of such models. We start with models that only involve **fixed effects** (unknown constants to be estimated). Such **general linear models** (**GLMs**) underlie a large number of standard statistical approaches (regression, ANOVA, etc.), all of which have estimators given by simple matrix expressions. We then introduce models which also include **random effects** (effect values are random draws, or **realizations**, from some underlying distribution), which requires estimating the variance of this underlying distribution in order to predict their realized values using the observed value of y. Because they contain both fixed and random effects, these are called **mixed models** (**MMs**), and include the GLM as a special case. Advanced linear model topics are examined in Appendix 3, and specific applications are detailed in Chapter 20 (GWAS), Appendix 6 (meta-analysis), and Appendix 9 (experimental design), and (in Part II) Chapters 27 and 28 (G $\times$ E), 32 (BLUP, Genomic selection/prediction), and 33 (REML).

BASICS OF GENERAL LINEAR MODELS

Equation 9.1a introduced multiple regression, the extension of a univariate regression (Equation 3.12a) for predicting the value of a response variable y as a function of *multiple* sources of information (the predictor variables $x_1, \cdots, x_p$). In quantitative genetics, these predictors could be traits of interest, molecular marker values, or **cofactors** to be removed (adjusted for), such as sex, design blocks, or age effects. Some or all of these predictors could be **indicator variables**, with values of 0 or 1 indicating whether an observation belongs in a particular category or grouping of interest. Models containing only indicator variables are often termed **ANOVA (analysis of variance) models**, while regression usually refers to models in which predictor variables can take on a continuous range of values. Both are special cases of the GLM, wherein the ith observation (y_i) is assumed to be a linear function of p observed and/or indicator variables plus a residual error,

$$y_i = \mu + \sum_{k=1}^{p} \beta_k \, x_{ik} + e_i \tag{10.1}$$

where $x_{i1}, \cdots, x_{ip}$ are the values of the p predictor variables for the ith individual.

Linear Models in Matrix Form

For a vector of n observations, Equation 10.1 can be compactly written in matrix form as

$$\mathbf{y}_{n\times 1} = \mathbf{X}_{n\times k}\, \boldsymbol{\beta}_{k\times 1} + \mathbf{e}_{n\times 1} \tag{10.2}$$

where $\mathbf{y}$ is the $(n \times 1)$ vector of observed y values, $k = 1 + p$, the **design** or **incidence matrix X** is $n \times k$, $\boldsymbol{\beta}$ is a $(k \times 1)$ vector of parameters (usually called **factors** or **effects**) to be estimated, and $\mathbf{e}$ is the vector of **residual errors** (or, simply, **residuals**). Here $\mathbf{y}$ and $\mathbf{X}$

are known (the data), while $\boldsymbol{\beta}$ is estimated, and then used to infer the residuals. For a given value of $\boldsymbol{\beta}$,

$$\widehat{\mathbf{y}} = \mathbf{X}\boldsymbol{\beta} \quad \text{and} \quad \widehat{\mathbf{e}} = \mathbf{y} - \widehat{\mathbf{y}} \tag{10.3}$$

where $\widehat{\mathbf{y}} = \mathbf{X}\boldsymbol{\beta}$ is the $(n \times 1)$ vector of predicted values for a given model.

Before examining how to solve Equarion 10.3 (i.e., estimation of $\boldsymbol{\beta}$), we first present a number of examples to give the reader a feel of how various linear models are expressed in matrix form. This compact representation follows from the definition of matrix multiplication (Equation 9.6a).

Example 10.1. Consider a multiple regression using p predictor variables,

$$y_i = \alpha + \sum_{j=1}^{p} \beta_j x_{ij} + e_i \tag{10.4a}$$

$$= \widehat{y}_i + e_i \tag{10.4b}$$

where x_{ij} is the value of the jth predictor variable in observation i. The vector of information for observation i is $(y_i, \mathbf{x}_i^T)$, where $\mathbf{x}_i^T = (1, x_{i1}, \cdots, x_{ip})$ is the vector of predictor values for that observation. Equation 10.4 decomposes y_i into a **model-predicted value** $\widehat{y}_i$ and a **residual deviation** (or **residual** for short) between the observed and predicted value, $e_i = y_i - \widehat{y}_i$. Given that e_i changes as we change the predictor model, the residuals contain most of the information on model fit and quality.

For n observations, Equation 10.4a can be expressed in GLM matrix form as

$$\mathbf{y} = \begin{pmatrix} y_1 \\ \vdots \\ y_n \end{pmatrix}, \quad \mathbf{X} = \begin{pmatrix} 1 & x_{11} & \cdots & x_{1p} \\ 1 & x_{21} & \cdots & x_{2p} \\ \vdots & \vdots & \ddots & \vdots \\ 1 & x_{n1} & \cdots & x_{np} \end{pmatrix} = \begin{pmatrix} \mathbf{x}_1^T \\ \mathbf{x}_2^T \\ \vdots \\ \mathbf{x}_n^T \end{pmatrix}, \quad \boldsymbol{\beta} = \begin{pmatrix} \alpha \\ \beta_1 \\ \vdots \\ \beta_p \end{pmatrix}, \quad \mathbf{e} = \begin{pmatrix} e_1 \\ \vdots \\ e_n \end{pmatrix}$$

Note from matrix multiplication that $y_i = \mathbf{x}_i^T \boldsymbol{\beta} + e_i$ recovers Equation 10.4a, and

$$\widehat{y}_i = \mathbf{x}_i^T \boldsymbol{\beta} \tag{10.4c}$$

The ith row of $\mathbf{X}$ $(\mathbf{x}_i^T)$ corresponds to the values of the predictor variables for observation i, while the jth column of $\mathbf{X}$ corresponds to the data on parameter j in the experiment. The interpretation of β_i is the change expected in y from a unit change in the predictor variable x_i, while holding all other predictor variables constant.

As a specific example, consider a regression with three predictor variables,

$$y_i = \mu + \beta_1 x_{i1} + \beta_2 x_{i2} + \beta_1 x_{i3} + e_i$$

The predicted value becomes

$$\widehat{y}_i = \mu + \beta_1 x_{i1} + \beta_2 x_{i2} + \beta_1 x_{i3}$$

with $\widehat{e}_i = y_i - \widehat{y}_i$ being the difference between the observed and predicted values. In matrix form, this model becomes

$$\begin{pmatrix} y_1 \\ y_2 \\ \vdots \\ y_n \end{pmatrix} = \begin{pmatrix} 1 & x_{11} & x_{12} & x_{13} \\ 1 & x_{21} & x_{22} & x_{23} \\ \vdots & \vdots & \vdots & \vdots \\ 1 & x_{n1} & x_{n2} & x_{n3} \end{pmatrix} \begin{pmatrix} \mu \\ \beta_1 \\ \beta_2 \\ \beta_3 \end{pmatrix} + \begin{pmatrix} e_1 \\ e_2 \\ \vdots \\ e_n \end{pmatrix}$$

Example 10.2. Consider a half-sib design wherein each of m sires (males) is mated at random to a number of unrelated dams (females) and a single offspring is measured from each cross. The simplest model for this design (i.e., ignoring any other potential cofactors) is

$$y_{ij} = \mu + s_i + e_{ij}$$

where y_{ij} is the phenotype of the jth offspring from sire i, μ is the population mean, s_i is the **sire effect**, and e_{ij} is the residual error (the "noise" remaining in the data after the sire effect is removed). Although this is clearly a linear model, it differs significantly from the regression model described above in that while there are parameters to estimate (the sire effects, s_i), the only measured values are the y_{ij}. Nevertheless, we can express this model in a form that is identical to the standard regression model by using m **indicator variables** to classify the sires of the offspring. The resulting linear model becomes

$$y_{ij} = \mu + \sum_{k=1}^{m} s_k \, x_{ik} + e_{ij}, \quad \text{where} \quad x_{ik} = \begin{cases} 1 & \text{if sire is } k \\ 0 & \text{otherwise} \end{cases}$$

Suppose that three different sires are used in the above half-sib design, with two, one, and three offspring, respectively. In matrix form, $\mathbf{y} = \mathbf{X}\boldsymbol{\beta} + \mathbf{e}$, with

$$\mathbf{y} = \begin{pmatrix} y_{11} \\ y_{12} \\ y_{21} \\ y_{31} \\ y_{32} \\ y_{33} \end{pmatrix}, \quad \mathbf{X} = \begin{pmatrix} 1 & 1 & 0 & 0 \\ 1 & 1 & 0 & 0 \\ 1 & 0 & 1 & 0 \\ 1 & 0 & 0 & 1 \\ 1 & 0 & 0 & 1 \\ 1 & 0 & 0 & 1 \end{pmatrix}, \quad \boldsymbol{\beta} = \begin{pmatrix} \mu \\ s_1 \\ s_2 \\ s_3 \end{pmatrix}, \quad \text{and} \quad \mathbf{e} = \begin{pmatrix} e_{11} \\ e_{12} \\ e_{21} \\ e_{31} \\ e_{32} \\ e_{33} \end{pmatrix}$$

This simple example highlights two issues. First, the model is **overparameterized**, as one could arbitrarily choose a value for μ and adjust each of the sire effects accordingly. We can see this from the column structure of $\mathbf{X}$, as its first column is simply the sum of columns two through four. Hence, $\mathbf{X}$ has only three independent columns (it has rank three) and can only uniquely estimate three parameter combinations. The rank of $\mathbf{X}$ impacts the estimability of model parameters, which we will examine more formally, both below and in Appendix 3. We can easily deal with this issue with the modified model

$$y_{ij} = t_i + e_{ij}$$

where $t_i = \mu + s_i$ is the mean value for sire i, with $\mathbf{X}$ and $\boldsymbol{\beta}$ now becoming

$$\mathbf{X} = \begin{pmatrix} 1 & 0 & 0 \\ 1 & 0 & 0 \\ 0 & 1 & 0 \\ 0 & 0 & 1 \\ 0 & 0 & 1 \\ 0 & 0 & 1 \end{pmatrix}, \quad \boldsymbol{\beta} = \begin{pmatrix} t_1 \\ t_2 \\ t_3 \end{pmatrix}$$

Observe that the difference between two treatments becomes $t_i - t_j = \mu + s_i - \mu - s_j = s_i - s_j$.

The second issue is more subtle, but equally important. Our above treatment assumes that the sire values are **fixed effects** (unknown constants), but in reality they are best treated as **random effects** (realizations of random draws from some unknown distribution). As detailed below, this shift in focus from fixed to random introduces issues in interpretation, but also allows for more efficient use of the data. For example, if the sires are related, then information on a related sire (say sire 3) further informs us as to the value for sire 1. Under a fixed-effect framework, the only information that we can use to estimate the sire-1 effect are *direct observations* on that sire. In this example, those are only the first two observations (only those observations with a 1 in the first column of $\mathbf{X}$, corresponding to data on sire 1). Under a random-effects framework we can also ***borrow information from correlated (indirect) observations***. Hence, if sires 1 and 3 are related, information from the last three observations (direct observations on sire 3) provides additional information on sire 1 that is not accessed in a fixed-effect analysis. We examine fixed versus random effects in detail shortly, with Example 10.10 reframing this example in a random-effects setting.

Example 10.3. Consider a **polynomial regression**,

$$y_i = a + \beta_1 x_i + \beta_2 x_i^2 + \cdots \beta_k x_i^k + e_i \tag{10.5}$$

Although the *data* enter as nonlinear functions, this is still a GLM, as the *linear* in GLM refers to linear with respect to the *parameters being estimated*, not the data. In the case of a polynomial regression, the resulting GLM matrices become

$$\mathbf{X} = \begin{pmatrix} 1 & x_1 & x_1^2 & \cdots & x_1^k \\ 1 & x_2 & x_2^2 & \cdots & x_2^k \\ \vdots & \vdots & \vdots & \cdots & \vdots \\ 1 & x_n & x_n^2 & \cdots & x_n^k \end{pmatrix} \quad \text{and} \quad \boldsymbol{\beta} = \begin{pmatrix} \alpha \\ \beta_1 \\ \vdots \\ \beta_k \end{pmatrix}$$

More generally, any regression of the form

$$y_i = \alpha + \beta_1 \, f(x_i) + \beta_2 \, g(x_i) + \cdots + e \tag{10.6}$$

can be placed in GLM form. Conversely,

$$y_i = \alpha + \exp(-\beta x_i)$$

is *not* a linear model, as it is a nonlinear function of the parameter β. In some cases, a transformation can recover a linear model. For example, the model

$$y_i = \alpha \exp(-\beta x_i)$$

can be written in linear model form as

$$\log(y_i) = \log(\alpha) - \beta x_i$$

Chapter 14 examines general***ized*** linear models, which allow for a certain amount of nonlinearity in the parameters.

Example 10.4. A common addition to many GLMs are **interaction terms**, such as the model

$$y_i = \alpha + \beta_1 x_{i1} + \beta_2 x_{i2} + \beta_3 x_{i2} x_{i1} + e_i \tag{10.7}$$

Here

$$\mathbf{X} = \begin{pmatrix} 1 & x_{11} & x_{12} & x_{11}\,x_{12} \\ 1 & x_{21} & x_{22} & x_{21}\,x_{22} \\ \vdots & \vdots & \vdots & \vdots \\ 1 & x_{n1} & x_{n2} & x_{n1}\,x_{n2} \end{pmatrix} \quad \text{and} \quad \boldsymbol{\beta} = \begin{pmatrix} \alpha \\ \beta_1 \\ \beta_2 \\ \beta_3 \end{pmatrix}$$

The interpretation of an interaction term is that the effect of some of the predictor variables on y are *context-specific*, and hence change as the background values of other predictor variables change. In this example, when x_1 is held constant, a unit change in x_2 changes y by $\beta_2 + \beta_3 x_1$ (i.e., the amount of change in x_2 depends on the current value of x_1). Likewise, a unit change in x_1 changes y by $\beta_1 + \beta_3 x_2$.

Example 10.5. One major strength of GLMs is their *incredible flexibility*, as one can mix and match all of the above modifications (and then some!) to design a model motivated by biological intuition, rather than statistical convenience. As an example, suppose one wants a quadratic regression forced through the origin, with the slope of the quadratic term varying over the sexes. The resulting GLM can be written as

$$y_i = \beta_1 x_i + \beta_2 x_i^2 + \beta_3 s_i x_i^2 + e_i$$

Here s_i is an indicator (0/1) variable for the sex (0 = male, 1 = female). This yields a male quadratic slope of β_2 and a female quadratic slope of $\beta_2 + \beta_3$. The standard linear model hypothesis-testing framework (Appendix 3) can be used to test whether β_3 is significantly different from zero.

Solving Linear Models: Ordinary Least-squares (OLS)

Estimates of the vector $\boldsymbol{\beta}$ of unknown parameters for a GLM are usually obtained by the method of least-squares (**LS**; Chapter 3). This approach uses the observational data given by $\mathbf{y}$ and $\mathbf{X}$, and makes assumptions about the covariance structure of the vector of residual errors, $\mathbf{e}$, leading to either **ordinary least-squares** (**OLS**) or **general least-squares** (**GLS**) solutions. Both approaches estimate $\boldsymbol{\beta}$ by minimizing a sum of squared residuals, with OLS minimizing an unweighted sum, and GLS a weighted sum, as we now detail.

Let $\mathbf{b}$ be an estimate of $\boldsymbol{\beta}$, and denote the vector of y values predicted from this estimate by $\widehat{\mathbf{y}} = \mathbf{Xb}$, so that the resulting vector of residual errors is

$$\widehat{\mathbf{e}} = \mathbf{y} - \widehat{\mathbf{y}} = \mathbf{y} - \mathbf{Xb} \tag{10.8a}$$

The OLS estimate of $\boldsymbol{\beta}$ is the $\mathbf{b}$ vector that minimizes the *unweighted* residual sum of squares,

$$\sum_{i=1}^{n} \widehat{e}_i^{\,2} = \widehat{\mathbf{e}}^T \widehat{\mathbf{e}} = (\mathbf{y} - \mathbf{Xb})^T (\mathbf{y} - \mathbf{Xb}) \tag{10.8b}$$

Taking derivatives, Example A3.6 shows that the OLS estimate satisfies

$$\text{BLUE}_{OLS}(\boldsymbol{\beta}) = \widehat{\boldsymbol{\beta}} = (\mathbf{X}^T\mathbf{X})^{-1}\mathbf{X}^T\mathbf{y} \tag{10.9a}$$

with the OLS estimator of β_i given by the ith element of this column vector. Equation 10.9a is called the **BLUE** (or, more precisely, the **OLS BLUE**), for **best linear unbiased estimator**, of $\boldsymbol{\beta}$.

As an aside

$$\mathbf{H} = \mathbf{X}\left(\mathbf{X}^T\mathbf{X}\right)^{-1}\mathbf{X}^T \tag{10.9b}$$

is referred to as the **hat matrix** (Hoaglin and Welsch 1978), because, for an OLS estimate, the vector of predicted y values is given by

$$\widehat{\mathbf{y}} = \mathbf{X}\widehat{\boldsymbol{\beta}} = \mathbf{X}\left(\mathbf{X}^T\mathbf{X}\right)^{-1}\mathbf{X}^T\mathbf{y} = \mathbf{Hy} \tag{10.9c}$$

$\mathbf{H}$ maps the observed $\mathbf{y}$ values onto their predicted values $\widehat{\mathbf{y}}$. Likewise, define

$$\mathbf{M} = \mathbf{I} - \mathbf{H} = \mathbf{I} - \mathbf{X}\left(\mathbf{X}^T\mathbf{X}\right)^{-1}\mathbf{X}^T \tag{10.9d}$$

as the **absorption matrix** for the fixed effects. Here

$$\mathbf{My} = (\mathbf{I} - \mathbf{H})\mathbf{y} = \mathbf{y} - \widehat{\mathbf{y}} = \mathbf{y} - \mathbf{X}\widehat{\boldsymbol{\beta}} = \widehat{\mathbf{e}} \tag{10.9e}$$

By minimizing the unweighted sum of squared deviations (residuals), OLS assumes that *all residuals contain the same amount of information*. This implies that the *residuals are homoscedastic*, with $\sigma^2(e_i) = \sigma_e^2$ for all i. If some of the residuals have smaller variances, then their corresponding predicted values of y are more precise, and hence their associated residuals should be weighted more heavily than residuals with larger variances. Using an unweighted sum further assumes that *all residuals are uncorrelated*, $\sigma(e_i, e_j) = 0$ for $i \neq j$, as the presence of correlated residuals would change the residual weighting scheme. The OLS assumptions about the residual structure is compactly denoted by $\mathbf{e} \sim (\mathbf{0}, \sigma_e^2\,\mathbf{I})$, namely, that the vector of residuals has a mean vector of zero, and a variance–covariance matrix of

$$\text{Var}(\mathbf{e}) = \sigma_e^2\,\mathbf{I} = \begin{pmatrix} \sigma_e^2 & 0 & \cdots & 0 \\ 0 & \sigma_e^2 & \cdots & 0 \\ \vdots & & \ddots & \vdots \\ 0 & 0 & \cdots & \sigma_e^2 \end{pmatrix} \tag{10.10}$$

Under the OLS assumption, the covariance matrix for the elements of the vector of BLUE estimates is

$$\mathbf{V}_{\widehat{\boldsymbol{\beta}}} = \sigma_e^2\,(\mathbf{X}^T\mathbf{X})^{-1} \tag{10.11a}$$

Hence, the sampling variance for the estimate of β_i is given by the ith diagonal element of the matrix $\mathbf{V}_{\widehat{\boldsymbol{\beta}}}$, while the covariance of this estimator with the BLUE of β_j is the ijth element of $\mathbf{V}_{\widehat{\boldsymbol{\beta}}}$. In applying Equation 10.11a, the residual variance σ_e^2 is usually replaced by the estimated variance,

$$\widehat{\sigma}_e^2 = \frac{1}{n-p}\sum_{i=1}^{n} \widehat{e}_i^{\,2} = \frac{(\mathbf{y} - \mathbf{X}\widehat{\boldsymbol{\beta}})^T(\mathbf{y} - \mathbf{X}\widehat{\boldsymbol{\beta}})}{n-p} \tag{10.11b}$$

where $\widehat{\boldsymbol{\beta}}$ is the BLUE estimate (Equation 10.9a) and p parameters are estimated (p is replaced by the rank of $\mathbf{X}$, when rank[$\mathbf{X}$] $< p$).

It is important to stress that the *only* assumption required to obtain Equations 10.9a and 10.11a is that the covariance matrix for the residuals is $\sigma_e^2\,\mathbf{I}$. While it is often assumed that the residuals must also be normally distributed, this is *not* required for Equations 10.9a and 10.11a to hold. The MVN assumption, however, is required for hypothesis testing and construction of confidence intervals (Appendix 3). If the residuals follow a multivariate normal distribution with $\mathbf{e} \sim \text{MVN}(\mathbf{0}, \sigma_e^2\,\mathbf{I})$, then the OLS estimate is also the maximum-likelihood estimate (Appendix 4).

Finally, we note that the structure of $\mathbf{X}$ informs one as to whether the data contain enough information to uniquely estimate a particular parameter. A parameter is said to be **estimable** if the model returns a unique estimate of its value. If $(\mathbf{X}^T\mathbf{X})^{-1}$ exists (i.e., is nonsingular), then every parameter in a GLM is estimable. What happens when $\mathbf{X}^T\mathbf{X}$ is singular? As discussed in Appendix 3, Equations 10.9a and 10.11a still hold when a **generalized inverse** (or **g-inverse**), denoted $(\mathbf{X}^T\mathbf{X})^-$, is used. In this case, not all of the parameters have unique estimates, rather some can only be uniquely estimated as *linear combinations* (or **contrasts**). For example, a three-parameter model might return an estimate of $\beta_1 = 3$, but only be able to specify that (say) $\beta_2 - 3\beta_3 = 2$. Hence, only two unique *parameter combinations* can be estimated from the data ($\mathbf{X}$ has rank 2).

Lack of estimability often implies a poor experimental design, but can also arise through loss of data from an otherwise well-planned design. Suppose one is examining the effects of height (scored as tall versus short) and sex. If the data (from either poor initial design or through data loss) consist of only tall males and short females, one cannot separate the height and sex effects from each other. Rather, we can only contrast tall males with short females, with the height and sex effects being fully confounded. The number of unique combinations of fixed effects that can be estimated for a given model is given by the rank of $\mathbf{X}$, namely the number of independent columns (columns that cannot be expressed as linear combinations of the other columns; Example 10.2). With n observations and $p < n$ unknowns, $\mathbf{X}$ is an $n \times p$ matrix, so that $\mathbf{X}^T\mathbf{X}$ is $p \times p$. $\mathbf{X}$ is of full rank when its rank is p, and in such cases the inverse of $\mathbf{X}^T\mathbf{X}$ exists. When this inverse does not exist (so that Rank$[\mathbf{X}] < p$), the rank of $\mathbf{X}$ is given by the number of nonzero eigenvalues of $\mathbf{X}^T\mathbf{X}$. When the rank of $\mathbf{X}$ is less than p, a g-inverse must be used. Appendix 3 examines these issues.

Example 10.6. Consider a univariate linear regression where the predictor and response variable both have expected mean zero, so that the regression passes through the origin. The appropriate model becomes

$$y_i = \beta\, x_i + e_i \tag{10.12a}$$

With observations on n individuals, this relationship can be written in GLM form with $\boldsymbol{\beta} = (\beta)$ and design matrix $\mathbf{X} = (x_1,\, x_2,\, \cdots x_n)^T$, implying

$$\mathbf{X}^T\mathbf{X} = \sum_{i=1}^{n} x_i^2 \qquad \text{and} \qquad \mathbf{X}^T\mathbf{y} = \sum_{i=1}^{n} x_i\, y_i$$

Assuming the covariance matrix of $\mathbf{e}$ is $\sigma_e^2\,\mathbf{I}$, then Equation 10.9a gives the OLS estimate of β as

$$\widehat{\beta} = \left(\mathbf{X}^T\mathbf{X}\right)^{-1}\mathbf{X}^T\mathbf{y} = \frac{\sum x_i\, y_i}{\sum x_i^2} \tag{10.12b}$$

while Equations 10.11a and 10.11b give its sample variance as

$$\sigma^2(\,\widehat{\beta}\,) = \left(\mathbf{X}^T\mathbf{X}\right)^{-1}\sigma_e^2 = \frac{\sigma_e^2}{\sum x_i^2} \simeq \frac{\sum_{i=1}^{n}(y_i - \widehat{\beta}x_i)^2}{(n-1)\sum x_i^2} \tag{10.12c}$$

Note that this estimate of β given by Equation 10.12b differs from the standard univariate regression slope (Equation 3.14b) where the intercept value is not assumed to be equal to zero.

Example 10.7. Recall from Equation 9.4b that the vector of partial regression coefficients for a multivariate regression is given by $\mathbf{b} = \mathbf{V}^{-1}\,\mathbf{c}$ (where $\mathbf{V}$ is the estimated covariance matrix and $\mathbf{c}$ is the vector of estimated covariances between $\mathbf{y}$ and $\mathbf{z}$). Here we show that this expression is equivalent to the OLS estimator $\mathbf{b} = (\mathbf{X}^T\mathbf{X})^{-1}\mathbf{X}^T\mathbf{y}$.

For the ith individual, we observe y_i and the values of p predictor variables, $z_{i1}, \cdots, z_{ip}$. Because a regression satisfies $\bar{y} = \alpha + \beta_1\bar{z}_1 + \cdots + \beta_p\bar{z}_p$, subtracting the mean from each observation removes the intercept, with (Equation 9.3b)

$$y_i^* = (y_i - \bar{y}) = \beta_1(z_{i1} - \bar{z}_1) + \cdots + \beta_p(z_{ip} - \bar{z}_p) + e_i$$

For n observations, the resulting linear model $\mathbf{y}^* = \mathbf{X}\boldsymbol{\beta} + \mathbf{e}$ has

$$\mathbf{y}^* = \begin{pmatrix} y_1 - \bar{y} \\ \vdots \\ y_n - \bar{y} \end{pmatrix}, \quad \boldsymbol{\beta} = \begin{pmatrix} \beta_1 \\ \vdots \\ \beta_p \end{pmatrix}, \quad \mathbf{X} = \begin{pmatrix} (z_{11} - \bar{z}_1) & \cdots & (z_{1p} - \bar{z}_p) \\ \vdots & \ddots & \vdots \\ (z_{n1} - \bar{z}_1) & \cdots & (z_{np} - \bar{z}_p) \end{pmatrix}$$

where z_{ij} is the value of character j in the ith individual. Partitioning the design matrix $\mathbf{X}$ into p column vectors corresponding to the n observations on each of the p predictor variables gives

$$\mathbf{X} = (\,\mathbf{x}_1,\ \cdots,\ \mathbf{x}_p\,) \quad \text{where} \quad \mathbf{x}_j = \begin{pmatrix} z_{1j} - \bar{z}_j \\ z_{2j} - \bar{z}_j \\ \vdots \\ z_{nj} - \bar{z}_j \end{pmatrix}$$

giving the jth element of the vector $\mathbf{X}^T\mathbf{y}^*$ as

$$\left(\mathbf{X}^T\mathbf{y}^*\right)_j = \mathbf{x}_j^T\mathbf{y}^* = \sum_{i=1}^{n}(y_i - \bar{y})(z_{ij} - \bar{z}_j) = (n-1)\text{Cov}(y, z_j)$$

showing that $\mathbf{X}^T\mathbf{y}^* = (n-1)\,\mathbf{c}$. Likewise, the jkth element of $\mathbf{X}^T\mathbf{X}$ is

$$\mathbf{x}_j^T\mathbf{x}_k = \sum_{i=1}^{n}(z_{ij} - \bar{z}_j)(z_{ik} - \bar{z}_k) = (n-1)\text{Cov}(z_j, z_k)$$

implying $\mathbf{X}^T\mathbf{X} = (n-1)\mathbf{V}$. Putting these results together gives

$$(\mathbf{X}^T\mathbf{X})^{-1}\mathbf{X}^T\mathbf{y}^* = \mathbf{V}^{-1}\mathbf{c}$$

showing that Equation 9.4b does indeed give the OLS estimates of the partial regression coefficients.

Solving Linear Models: Generalized Least-squares (GLS)

Under OLS, the unweighted sum of squared residuals is minimized, which assumes homoscedastic and uncorrelated residuals. When residual errors are heteroscedastic and/or correlated, OLS estimates of regression parameters and standard errors of these estimates are potentially biased. A more general approach to regression analysis expresses the covariance matrix of the vector of residuals as $\sigma_e^2\,\mathbf{R}$, with $\sigma(e_i, e_j) = R_{ij}\sigma_e^2$. Lack of independence between residuals is indicated by the presence of nonzero off-diagonal elements in $\mathbf{R}$, while heteroscedasticity is indicated by differences in the diagonal elements of $\mathbf{R}$.

Generalized least-squares (GLS) takes these complications into account, and minimizes the weighted sum of squares given by $\mathbf{e}^T\mathbf{R}^{-1}\mathbf{e}$. A special case of GLS is **weighted least-squares** (WLS), which assumes residuals are uncorrelated, but heteroscedastic (such as would occur if the y_i were based on mean values with differing values of n_i; e.g., Example 10.8). As shown in Appendix 3, the use of $\mathbf{e}^T\mathbf{R}^{-1}\mathbf{e}$ follows as the transformation $\mathbf{e}_* = \mathbf{R}^{-1/2}\mathbf{e}$ changes the original vector of residuals ($\mathbf{e}$) into a new vector ($\mathbf{e}_*$) with homoscedastic and uncorrelated residuals, with the GLS solution following from the minimization of the sum $\mathbf{e}_*^T\mathbf{e}_*$. As a result, for the linear model

$$\mathbf{y} = \mathbf{X}\boldsymbol{\beta} + \mathbf{e} \qquad \text{with } \mathbf{e} \sim (0, \sigma_e^2\,\mathbf{R})$$

the GLS estimate of $\boldsymbol{\beta}$ is

$$\text{BLUE}_{GLS}(\boldsymbol{\beta}) = \widehat{\boldsymbol{\beta}} = \left(\mathbf{X}^T\mathbf{R}^{-1}\mathbf{X}\right)^{-1}\mathbf{X}^T\mathbf{R}^{-1}\mathbf{y} \tag{10.13a}$$

(Aitken 1935). The covariance matrix for the GLS estimates of elements of $\boldsymbol{\beta}$ is then

$$\mathbf{V}_{\widehat{\boldsymbol{\beta}}} = \sigma_e^2\left(\mathbf{X}^T\mathbf{R}^{-1}\mathbf{X}\right)^{-1} \tag{10.13b}$$

The estimated residual variance is also modified from Equation 10.11b, with

$$\widehat{\sigma}_e^2 = \frac{(\mathbf{y} - \mathbf{X}\widehat{\boldsymbol{\beta}})^T\mathbf{R}^{-1}(\mathbf{y} - \mathbf{X}\widehat{\boldsymbol{\beta}})}{n - \text{rank}(\mathbf{X})} \tag{10.13c}$$

If residuals are independent and homoscedastic, $\mathbf{R} = \mathbf{I}$, and GLS estimates of $\boldsymbol{\beta}$ reduce to the OLS estimates. If $\mathbf{e} \sim \text{MVN}(\mathbf{0}, \sigma_e^2\,\mathbf{R})$, the GLS estimate of $\boldsymbol{\beta}$ is also the maximum-likelihood estimate.

Example 10.8. A common situation requiring weighted least-squares analysis occurs when residuals are independent but heteroscedastic with $\sigma^2(e_i) = \sigma_e^2/w_i$, where w_i are known positive constants. For example, if each observation y_i is the mean of n_i independent observations (each with uncorrelated residuals with variance σ_e^2), then $\sigma^2(e_i) = \sigma_e^2/n_i$, and hence $w_i = n_i$. Here

$$\mathbf{R} = \text{Diag}(w_1^{-1}, w_2^{-1}, \ldots, w_n^{-1})$$

where Diag denotes a diagonal matrix, giving

$$\mathbf{R}^{-1} = \text{Diag}(w_1, w_2, \ldots, w_n)$$

With this residual variance structure, consider the WLS estimate for the simple univariate regression model $y = \alpha + \beta x + e$. In GLM form,

$$\mathbf{y} = \begin{pmatrix} y_1 \\ \vdots \\ y_n \end{pmatrix}, \qquad \mathbf{X} = \begin{pmatrix} 1 & x_1 \\ \vdots & \vdots \\ 1 & x_n \end{pmatrix}, \qquad \text{and} \qquad \boldsymbol{\beta} = \begin{pmatrix} \alpha \\ \beta \end{pmatrix}$$

Define the following weighted means and cross-products,

$$w = \sum_{i=1}^{n} w_i, \quad \overline{x}_w = \sum_{i=1}^{n} \frac{w_i x_i}{w}, \quad \overline{x^2_w} = \sum_{i=1}^{n} \frac{w_i x_i^2}{w}$$

$$\overline{y}_w = \sum_{i=1}^{n} \frac{w_i y_i}{w}, \quad \overline{xy}_w = \sum_{i=1}^{n} \frac{w_i x_i\, y_i}{w}$$

With these definitions, matrix multiplication and a little simplification yields

$$\mathbf{X}^T\mathbf{R}^{-1}\mathbf{y} = w \begin{pmatrix} \overline{y}_w \\ \overline{xy}_w \end{pmatrix} \qquad \text{and} \qquad \mathbf{X}^T\mathbf{R}^{-1}\mathbf{X} = w \begin{pmatrix} 1 & \overline{x}_w \\ \overline{x}_w & \overline{x^2_w} \end{pmatrix}$$

Applying Equation 10.13a, the GLS estimates of α and β are

$$a = \overline{y}_w - b\overline{x}_w \tag{10.14a}$$

$$b = \frac{\overline{xy}_w - \overline{x}_w\, \overline{y}_w}{\overline{x^2_w} - \overline{x}_w^{\,2}} \tag{10.14b}$$

If all weights are equal ($w_i = c$), these expressions reduce to the standard (OLS) least-squares estimators given by Equation 3.14. Applying Equation 10.13b, the sampling variances and covariance for these estimates become

$$\sigma^2(a) = \frac{\sigma_e^2 \cdot \overline{x^2_w}}{w\,(\overline{x^2_w} - \overline{x}_w^{\,2})}, \qquad \sigma^2(b) = \frac{\sigma_e^2}{w\,(\overline{x^2_w} - \overline{x}_w^{\,2})} \tag{10.14c}$$

$$\sigma(a, b) = \frac{-\sigma_e^2\, \overline{x}_w}{w\,(\overline{x^2_w} - \overline{x}_w^{\,2})} \tag{10.14d}$$

MODEL ASSESSMENT

Overview

Fitting a model is only the first step of any analysis, and should be followed by considerable post-construction assessment. We outline a few key ideas before briefly introducing some useful tests for probing our model assumptions. First, just because a model is *statistically significant* (it has a significantly better fit than a model with no effects beyond a common mean; $y_i = \mu + e_i$) does not mean that it is *biologically relevant*. A good indicator of the latter is the model r^2 value (Equation A3.15a), namely, how much of the variation in the response variable (y) is accounted for by the model. Consider two different models: Model 1 has $p = 10^{-9}$ and $r^2 = 0.01$, while Model 2 has $p = 10^{-4}$ and $r^2 = 0.3$. From a statistical standpoint, Model 1 (with its much more significant p value) might seem superior, but it only accounts for 1% of the variation in y, while Model 2 accounts for 30% and thus is biologically more meaningful than Model 1.

Second, all of the information on how well a model performs is contained in the residuals. As mentioned, normality is *not* required for OLS or GLS estimates, or their standard errors, to be unbiased. It is, however, required for hypothesis testing and the construction of confidence intervals (Appendix A3). Hence, the vector of residuals should be tested for normality, at a minimum by examining their histogram or normal probability plot, or more formally by using standard normality tests (Chapter 14). It is important to stress the difference between the distribution of the dependent variable y and the residual e. The histogram of y values is usually *expected* to be non-normal, as the data likely consist of a number of groups with different means. Even if y values are normally distributed within each group, the distribution of y values in the full sample is a mixture of normals, and hence is likely to show multiple modes, skew, and kurtosis (Chapter 16). In contrast, the distribution of *residuals* reflect errors once we have accounted for the distinction between groups with different means. If one finds that the residuals are not normal, transformation of y to a different scale, such as using $\log(y)$ in place of y, can often normalize the resulting residuals (Chapter 14). More generally, one can use **generalized linear models** (Chapter 14) to handle settings where the residuals are *expected* to be non-normal, such as with binary or discrete data.

Next, **residual plots**, such as $\widehat{e}_i$ versus y_i, or $\widehat{e}_i$ versus $\widehat{y}$, can be highly informative, suggesting outliers and possible biases. While (by construction) the expected value of e is zero under least-squares, they may still show trends despite this restriction. For example, if the spread of residual values increases with y or $\widehat{y}$, this indicates heteroscedasticity (and hence GLS must be used instead of OLS). Modeling using $\log(y)$ in place of y often resolves this issue, which can simply arise as a scaling effect (Chapter 14).

Finally, the distribution of $\widehat{e}_i$ should be independent over both y_i and x_{ij} values if one has an appropriate model. Under the OLS assumption, adjacent residuals (when plotting $\widehat{e}_i$ versus a particular predictor variable, say variable j, x_{ij}) should be uncorrelated. Departures from this pattern can arise if an incorrect model is fitted. Suppose that the true model is a quadratic in variable j (with a maximum in the middle of the data range), but we assumed a linear effect. The latter model would likely still account for some of the variation in y. However, a plot of $\widehat{e}_i$ versus x_{ij} would show blocks (**runs**) of residuals all having the same sign. For example, the first 1/4 of residuals associated with the smallest 25% of x_{ij} values could all be negative, the middle 1/2 positive, and the last 1/4 negative.

A simple check for this type of behavior is the **Wald–Wolfowitz runs test** (Wald and Wolfowitz 1940). Here the test statistic, R_n, for the number of runs (changes in the signs of adjacent residuals) in a sample of size n, is approximately normally distributed with mean $n/2 + 1$ and variance $n/4$. More subtle correlations among (nonadjacent) residuals can arise when the model ignores shared factors, such as observations being for the same family. As will be discussed shortly, such situations are often handled using a mixed model, including these missing shared factors as random effects. We now briefly present a few more formal tests for OLS model assessment, many of which come from econometrics.

The Ramsey RESET Test for Functional Misspecification

Suppose one believes that there might be high-order terms in the model, such as quadratic, cross-product, or even cubic contributions. For a single-variable regression, this is straightforward to test: fit the regression $y_i = \alpha + \beta_1 x_i + \beta_2 x_i^2 + e_i$ and perform a standard t test for whether β_2 is significant. With a multiple regression, the full quadratic-order model becomes

$$y_i = \alpha + \sum_j \beta_j x_{ij} + \sum_j \sum_k \gamma_{jk} x_{ij} x_{ik} + e_i \tag{10.15a}$$

Estimating, and then testing, each of the quadratic terms is rather wasteful in terms of degrees of freedom, and also has rather low power. Ramsey (1969) proposed a clever solution. His **RESET (Regression Equation Specification Error) test** collapses all potential quadratic parameters into a single variable, by fitting the regression

$$y_i = \delta + \sum_j \eta_j x_{ij} + \gamma_1 \widehat{y_i}^2 + e_i \tag{10.15b}$$

where

$$\widehat{y_i} = \alpha + \sum_j \beta_j x_{ij} \tag{10.15c}$$

is the predicted value under the base model. Equation 10.15b uses δ and η_i to stress that these are different from the α and β_i used in Equation 10.15c to obtain $\widehat{y}_i$. Ramsey's idea was that $\widehat{y}_i^2 = (\alpha + \sum_j \beta_i x_{ij})^2$ captures potential quadratic and cross-product terms into a single variable, with a simple t test for the significance of γ_1 determining whether any such effects need to be included. This approach easily extends to higher order effects, e.g.,

$$y_i = \delta + \sum_j^p \eta_j x_{ij} + \gamma_1 \widehat{y_i}^2 + \gamma_2 \widehat{y_i}^3 + e_i \tag{10.15d}$$

where the test for k higher order functional misspecification terms $(\gamma_1, \cdots, \gamma_k)$ follows a $F_{k,n-[p+1+k]}$ test (Appendix 5) for whether any of the γ_i terms are significant.

Tests for Heteroscedasticity

As mentioned, plots of $\widehat{e}_i$ versus either y_i or $\widehat{y}_i$ can suggest departures from the constant-variance assumption of OLS. A number of more formal heteroscedasticity tests have been proposed. One of the simplest is **Goldfeld–Quandt test** (1965), which probes whether $\widehat{e}_i$ is a function of one of the underlying variables (say x_{i1}). The approach is to plot $\widehat{e}_i^2$ versus x_{i1}, and then allocate these residuals into two groups, A and B, with samples sizes of n_A and n_B, respectively. Often, but not always, the lower and upper halves of the x_{i1} comprise the two groups. Using this partition, compute the residual sum of squares for each group, $RSS_x = \sum_{i \in x} \widehat{e}_i^2$, with the test statistic becoming

$$\frac{RRS_B/(n_B)}{RSS_A/(n_A)} \sim F_{n_B, n_A} \tag{10.16a}$$

A significant value indicates evidence for heteroscedasticity.

A related partitioning approach for probing functional misspecification is the **Chow test** (1960). While one could initially fit a single regression to a dataset, a better model might instead fit two regressions, one for values below some threshold in one of the predictor variables, and a second for values above this threshold. An example of such a **structural break** can be seen in Figure 13.6, which examines the genetic gain in maize lines over time. Early breeding methods were much less effective than more recent approaches, resulting in different rates of improvement. The Chow test partitions the data associated with a particular predictor variable x_{ij} into two groups, n_A in the lower group below some proposed threshold and n_B in the upper group. The regression $y_i = \alpha + \sum_{j=1}^{p-1} \beta_j x_{ij}$ is fit for the combined (pooled) group (C) and also separately for each of the two subgroups. The resulting RSS are computed for all three groups, with the Chow test statistic for equality of the two regressions (over groups A and B) given by

$$\frac{(RSS_C - [RSS_A + RSS_B])/p}{(RSS_A + RSS_B)/(n_A + n_B - 2p)} \sim F_{p, n_A + n_B - 2p} \tag{10.16b}$$

This statistic can also be used to estimate the transition value between the two regressions by examining its value over different thresholds for variable j.

A second approach to test for heteroscedastic residuals is to regress some transformation of the residual (square, absolute value, log) as a function of the predictor variables. Focusing on how $\widehat{e}_i$ varies for a single predictor variable (x_i), the **Park test** (1966) models the residuals as

$$\widehat{e}_i^2 = \sigma_0^2 \, x_i^\beta \, \exp(\epsilon_i) \tag{10.16c}$$

Taking logs yields

$$\ln(\widehat{e}_i^2) = \ln(\sigma_0^2) + \beta \ln(x_i) + \epsilon_i \tag{10.16d}$$

Under homoscedasticity, the β term should be zero. The **Glejser test** (Glejser 1969) and its improvements (Im 2000; Machado and Santos Silva 2000) use the same basic logic, but use regressions based on $|\,\widehat{e}_i\,|$, with separate regressions using different functions of the predictor variable (e.g., $x_i, \sqrt{x_i}, 1/x_i, 1/\sqrt{x_i}$). The model with the largest r^2 then used, with a significant β for this best regression (adjusted for multiple comparisons; Appendix 6) indicating heteroscedasticity. One reason for using absolute values over squares is that the former are better behaved for heavy-tailed (leptokurtic; Chapter 2) error distributions (Im 2000).

Moving from considering just a single predictor variable influencing the residuals to the more general setting involving the entire set of predictor variables can be addressed using the **Breusch–Pagan test** (1979). Here, the model fit is

$$\widehat{e}_i^2 = \delta + \sum_j^p \eta_j x_{ij} + \epsilon_i \tag{10.16e}$$

While one could perform a t test to assess the impact of any single predictor variable on the residual, the entire set can be tested at once by asking if Equation 10.16e gives a better fit than the homoscedastic model, $e_i^2 = \delta + \epsilon_i$. From Equation A3.20, the F test for all of the η_j being nonsignificant is given by

$$f_{BP} = \frac{r^2/p}{(1-r^2)/(n-p-1)} \sim F_{p,n-p-1} \tag{10.16f}$$

where r^2 is the coefficient of determination (Equation A3.15a) for the model fit by Equation 10.16e. An alternative test statistic (for large n) is to use $nr^2 \sim \chi_p^2$.

The **White test** (1980) extends the Breusch–Pagan test by allowing heteroscedasticity to be more complex than a simple linear function of the predictor values. Using similar logic to the RESET test, the White test fits the regression

$$\widehat{e}_i^2 = \alpha + \beta_1 \widehat{y}_i + \beta_2 \widehat{y}_i^{\,2} + \epsilon_i \tag{10.16g}$$

Note that the dependency has shifted from individual predictor variables to the predicted *value*. As with Breusch–Pagan, one could test $\beta_1 = \beta_2 = 0$ based on the model-fit r^2 from Equation 12.16g using either an F test (Equation 10.16f with $p = 2$) or $nr^2 \sim \chi_2^2$.

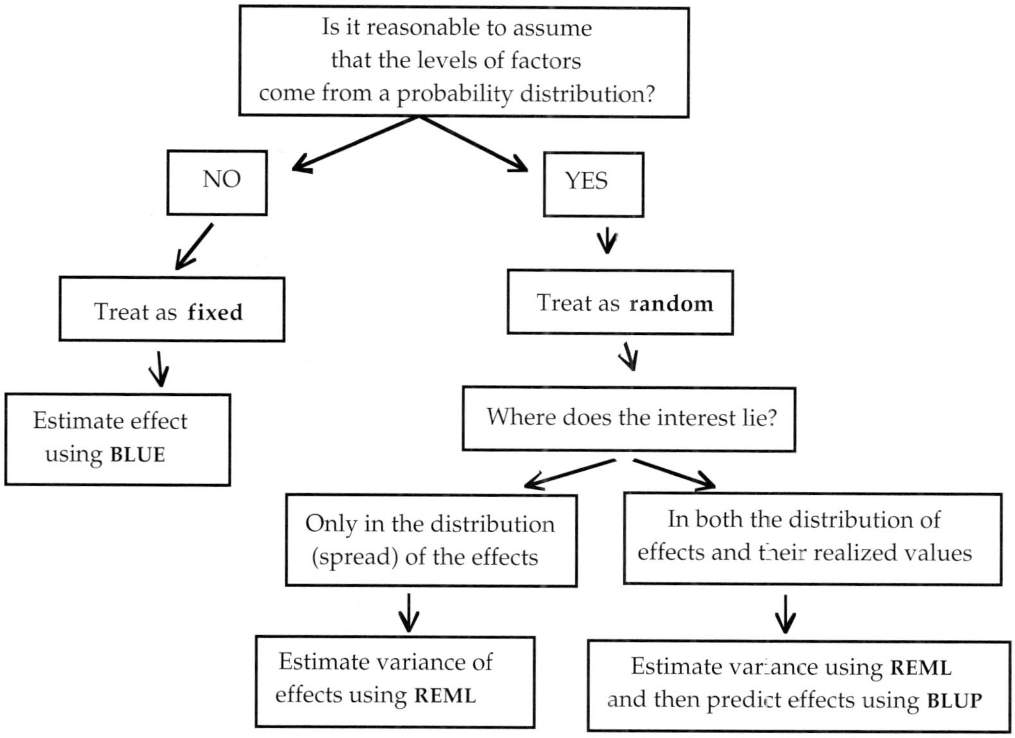

Figure 10.1 A flow chart for considering the roles of fixed and random effects in an analysis. (From Hans-Peter Piepho, reprinted with permission.)

THE GENERAL MIXED MODEL

The mixed model (MM) builds on the GLM, $\mathbf{y} = \mathbf{X}\beta + \mathbf{e}$, by adding one (or more) vectors of random effects, $\mathbf{Zu}$, giving

$$\mathbf{y} = \mathbf{X}\beta + \mathbf{Zu} + \mathbf{e}$$

where $\mathbf{Z}$ (akin to $\mathbf{X}$) is a known matrix and $\mathbf{u}$ is the additional vector of random effects (any linear model always includes the random-effects vector $\mathbf{e}$). As with the residuals of a GLM, one must ***specify the covariance structure*** of $\mathbf{u}$ in order to define the model. Before proceeding further with the analysis of the MM, which is the framework for much of modern quantitative genetics, we disucss the distinction between fixed versus random effects.

Fixed versus Random Effects

Unknown parameters in statistical models are typically classified as either **fixed** or **random** (Figure 10.1). The former are unknown constants to be estimated, whereas the latter are *realizations* drawn from some underlying distribution. While this classification is often treated rather rigidly, the distinction between the two is actually somewhat fluid, and often depends on subtle differences in the assumptions underlying a model. Indeed, for the same data set and essentially the same problem, one investigator might feel justified using a fixed-effect interpretation, while another might choose random effects.

If the effect value is a realization drawn from some underlying distribution, then effects are treated as random. If our interest is only in the values in the sample, then often (but not always; see Example 10.9) they can be treated as fixed effects. Likewise, if one cannot go back after an experiment and resample the same **levels** of a factor, it is best treated as random. For example, one can go back and sample new males and females, or resample from a specific growing location. However, one cannot resample from a specific past year. Hence, one could treat the location as fixed, but the year as random (and hence a year by location interaction is also random).

Factors are often treated as random for convenience (Example 10.9). One setting for this is when a model contains a large number of **nuisance parameters**: parameters that must be estimated, but are not really of any concern to the investigator. Each estimated fixed effect uses a degree of freedom. In contrast, random effects represent realizations (draws) from a particular unknown distribution, typically assumed to be a normal with mean zero (the fixed effects absorb any nonzero mean effect) and variance σ^2. Here, the only parameter to be estimated is this variance, *no matter how many realizations are drawn* from this distribution. Thus, nuisance parameters are often treated as random to save degrees of freedom. Finally, one can always move to a Bayesian framework (Appendices 7 and 8), wherein all parameters are treated as random.

Example 10.9. Suppose that one has a completely randomized design (Appendix 9) of 10 blocks, each of which consists of three replicate plots for each of five different inbred lines, randomly assigned within the block (15 plots per block, for a total of 150 observed plot values). The simplest linear model is

$$y_{ijk} = L_i + B_j + e_{ijk} \tag{10.17a}$$

with y_{ijk} the value of the kth plot in block j of line i (with $1 \leq i \leq 5$, $1 \leq j \leq 10$, $1 \leq k \leq 3$), and we assume the absence of line $\times$ block interactions. Treating all factors as fixed, there are five line effects (L_i) and 10 block effects (B_j), or 15 parameters (degrees of freedom) that must be estimated, namely the column vector of fixed vectors $\boldsymbol{\beta} = (L_1, \cdots, L_5, B_1, \cdots, B_{10})^T$ has 15 terms. Assuming OLS, the matrix form of the model becomes

$$\mathbf{y}_{150\times1} = \mathbf{X}_{150\times15}\boldsymbol{\beta}_{15\times1} + \mathbf{e}_{150\times1}, \qquad \mathbf{e} \sim (0, \sigma_e^2\,\mathbf{I})$$

where any given row of $\mathbf{X}$ is all zeros, except for ones in the columns corresponding to the line and block for that observation.

One does not typically care about the block effects beyond removing any bias they may introduce in the estimate of the L_i. If we could reasonably assume that blocks have independent effects (i.e., blocks that are closer to each other are no more likely to be similar than more distant blocks), then we can treat these as random, with the value (or realization) for a particular block being drawn from some distribution with unknown variance, σ_B^2. The consequence of treating block values as random is that instead of using 10 degrees of freedom for block effects, we *only use one*, that for estimating σ_B^2. Even if a design had 1000 blocks, it would still only use one degree of freedom for blocks under the random-effect model (assuming that all the blocks are uncorrelated).

Under this interpretation, the vector of fixed effect $\boldsymbol{\beta}$ now has five terms (one for each line), while the column vector $\mathbf{b} = (B_1, \cdots, B_{10})^T$ of random effects has ten terms. The model is still given by Equation 10.17a, but we have to further specify the covariance structure for the random block effects (along with the residuals). Assuming independent block effects and OLS residuals, the matrix form of the model is now

$$\mathbf{y}_{150\times1} = \mathbf{X}_{150\times5}\boldsymbol{\beta}_{5\times1} + \mathbf{Z}_{150\times10}\mathbf{b}_{10\times1} + \mathbf{e}_{150\times1}; \quad \mathbf{b} \sim (0, \sigma_B^2\,\mathbf{I}), \mathbf{e} \sim (0, \sigma_e^2\,\mathbf{I}) \tag{10.17b}$$

Any given row of $\mathbf{X}$ or $\mathbf{Z}$ is all zeros, except for ones in the column corresponding to the line ($\mathbf{X}$) or block ($\mathbf{Z}$) for that observation. If one assumes correlation among adjacent blocks, then $\sigma_B^2\,\mathbf{I}$ is replaced by $\sigma_B^2\,\mathbf{R}$, where $\mathbf{R}$ is a nondiagonal matrix whose off-diagonal elements indicate the amount of correlation between each combination of blocks. This requires adding (at least) one additional parameter to estimate, with the random effects now using (at least) two degrees of freedom (see Appendix 9 for examples).

The covariance structures for Equations 10.17a and 10.17b treat the line genotypes as fixed, which implies that we only care about the values of the particular lines in our sample and have no interest in the larger population from which they were drawn. Conversely, if we are indeed interested in the genetic variance, σ_G^2, of line values in the population from which the lines were sampled, we treat the lines as random. Under this random-effects interpretation, we rewrite the model slightly to express the line values as deviations from the overall mean,

$$y_{ijk} = \mu + \ell_i + B_j + e_{ijk} \tag{10.17c}$$

where the realized value for line i becomes $L_i = \mu + \ell_i$. This random-effects interpretation introduces two important features. First, we expect that some lines may be more similar than others due to more recent common ancestry, which can be estimated with either pedigree data (Chapter 7) or molecular markers (Chapter 8). Hence, the covariance matrix need not be diagonal, but rather of the form $\sigma_G^2\mathbf{A}$, where the **relationship matrix A** (Example 9.14) is symmetric. The resulting covariance structure becomes

$$\boldsymbol{\ell} \sim (0, \sigma_G^2\,\mathbf{A}), \quad \mathbf{b} \sim (0, \sigma_B^2\,\mathbf{I}), \qquad \mathbf{e} \sim (0, \sigma_e^2\,\mathbf{I})$$

This is more compactly written as

$$\begin{pmatrix} \boldsymbol{\ell} \\ \mathbf{b} \\ \mathbf{e} \end{pmatrix} \sim \begin{pmatrix} \sigma_G^2\,\mathbf{A} & 0 & 0 \\ 0 & \sigma_B^2\,\mathbf{I} & 0 \\ 0 & 0 & \sigma_e^2\,\mathbf{I} \end{pmatrix} \tag{10.17d}$$

This model has a single fixed effect (the mean, μ) and now $\mathbf{X}$ is simply a column vector of ones. In matrix form, we now have

$$\mathbf{y}_{150\times 1} = \mathbf{X}_{150\times 1}(\mu) + \mathbf{Z}^*_{150\times 5}\boldsymbol{\ell}_{5\times 1} + \mathbf{Z}_{150\times 10}\mathbf{b}_{10\times 1} - \mathbf{e}_{150\times 1}$$

where $\mathbf{Z}$ is as in Equation 10.17b, and any given row of $\mathbf{Z}^*$ is all zeros, except for ones in the column corresponding to the line for that observation.

Second, a critical feature about $\mathbf{A}$ is that random-effects models ***borrow information from correlated observations***. When treating lines as fixed effects (BLUEs), the only information used to estimate a particular line value (say ℓ_3) are those direct observations on line 3 (30 plots in this example). Under a random-effect model (BLUPs), direct observations on line 3 are used, as well as observations on *all* other lines that are related to line 3 (those lines for which the elements A_{3i} and A_{i3} of $\mathbf{A}$ are nonzero). Hence, even if we have no interest in the nature of the population from which the lines were drawn, treating them as random allows us to access this borrowing-of-information feature (provided that some are correlated).

Mixed Models

Consider a column vector $\mathbf{y}$ containing the phenotypic values for a trait measured in n individuals. We assume that these observations are described adequately by a linear model with a $p \times 1$ vector of fixed effects ($\boldsymbol{\beta}$) and a $q \times 1$ vector of random effects ($\mathbf{u}$). The first element of the vector $\boldsymbol{\beta}$ is typically the population mean, and other factors included may be sex, location, year of birth, experimental treatment, and so on. The elements of the vector $\mathbf{u}$ of random effects are often genetic effects such as additive-genetic (breeding) values. In matrix form,

$$\mathbf{y} = \mathbf{X}\boldsymbol{\beta} + \mathbf{Z}\mathbf{u} + \mathbf{e} \tag{10.18}$$

where $\mathbf{X}$ and $\mathbf{Z}$ are, respectively, $n \times p$ and $n \times q$ **incidence matrices** ($\mathbf{X}$ is also called the **design matrix**), and $\mathbf{e}$ is the $n \times 1$ column vector of residual deviations assumed to be distributed independently of the random genetic effects. Often, all of the elements of $\mathbf{Z}$ are 0 or 1, depending upon whether the relevant random effect for a given observation contributes to the value of the response variable, y. Because this model jointly accounts for fixed and random effects, it is generally referred to as a **mixed model** (Eisenhart 1947).

Now consider the means and variances of the component vectors of the MM. By definition, *random effects have mean zero*. Hence, $E(\mathbf{u}) = E(\mathbf{e}) = \mathbf{0}$, implying that $E(\mathbf{y}) = \mathbf{X}\boldsymbol{\beta}$. Denote the $(n \times n)$ covariance matrix for the vector $\mathbf{e}$ of residual errors by $\mathbf{R}$ and the $(q \times q)$ covariance matrix for the vector $\mathbf{u}$ of random genetic effects by $\mathbf{G}$. Excluding the difference among individuals due to fixed effects, from Equation 9.21b and the assumption that $\mathbf{u}$ and $\mathbf{e}$ are uncorrelated, the covariance matrix for the vector of observations $\mathbf{y}$ is

$$\mathbf{V}_{n\times n} = \mathbf{Z}_{n\times q}\mathbf{G}_{q\times q}\mathbf{Z}^T_{q\times n} + \mathbf{R}_{n\times n} \tag{10.19}$$

The first term (which follows from Equation 9.21b) accounts for the contribution from the vector $\mathbf{u}$ of random effects, while the second accounts for the variance due to residual effects. We will generally assume that residual errors have constant variance and are uncorrelated, so that $\mathbf{R}$ is a diagonal matrix, with $\mathbf{R} = \sigma_E^2\,\mathbf{I}$. Hence, $\mathbf{X}\boldsymbol{\beta}$ gives the vector of *mean values* for the observations, while the *spread about this mean* value is influenced by the random effects, $\mathbf{u}$ and $\mathbf{e}$. The ith diagonal element of $\mathbf{V}$ (V_{ii}) gives the variance associated with observation i, while V_{ij} is the covariance between observations i and j.

We are now in a position to contrast the mixed versus general linear models. Under the GLM,

$$\mathbf{y} = \mathbf{X}\boldsymbol{\beta} + \mathbf{e}^* \qquad \text{where} \quad \mathbf{e}^* \sim (\mathbf{0}, \mathbf{V}), \quad \text{implying } \mathbf{y} \sim (\mathbf{X}\boldsymbol{\beta}, \mathbf{V}) \tag{10.20a}$$

where the notation $\sim (\mathbf{a}, \mathbf{V})$ means that the random variable has mean vector $\mathbf{a}$ and covariance matrix $\mathbf{V}$. On the other hand, the MM partitions the vector of residual effects into two components, with $\mathbf{e}^* = \mathbf{Zu} + \mathbf{e}$, giving

$$\mathbf{y} = \mathbf{X}\boldsymbol{\beta} + \mathbf{Zu} + \mathbf{e} \qquad \text{where} \quad \mathbf{u} \sim (\mathbf{0}, \mathbf{G}) \text{ and } \mathbf{e} \sim (\mathbf{0}, \mathbf{R})$$

$$\text{implying } \mathbf{y} \sim (\mathbf{X}\boldsymbol{\beta}, \mathbf{V}) = (\mathbf{X}\boldsymbol{\beta}, \mathbf{ZGZ}^T + \mathbf{R}) \tag{10.20b}$$

When analyzed in the appropriate way, both formulations yield the same estimate of the vector of fixed effects $\boldsymbol{\beta}$, while the MM formulation further allows estimates (more formally, **predictions**) of the vector of random effects $\mathbf{u}$.

Comparing Equations 10.20a with 10.20b shows one important use of adding random effects. In Equation 10.20a, the vector of residuals $\mathbf{e}^*$ is of GLS form, Var($\mathbf{e}^*$) $= \mathbf{V}$, while by adding an appropriate random effect in Equation 10.20b, the vector of residuals $\mathbf{e}$ is often assumed to be of OLS form, Var($\mathbf{e}$) $= \mathbf{R} = \sigma_e^2\,\mathbf{I}$. Shared features (such as individuals experiencing the same family environment) generate correlations among residuals, but these can be removed by adding the appropriate vector (or vectors) of random effects to the model (Part II Chapters 32 and 33).

For the MM, we observe $\mathbf{y}$, $\mathbf{X}$, and $\mathbf{Z}$, while $\boldsymbol{\beta}$, $\mathbf{u}$, $\mathbf{R}$, and $\mathbf{G}$ are generally unknown. Thus, MM analysis involves two complementary estimation issues: (i) *estimation* of the vector of fixed effects ($\boldsymbol{\beta}$) and *prediction* of the vector of random effects ($\mathbf{u}$); and (ii) estimation of the covariance matrices $\mathbf{G}$ and $\mathbf{R}$. These covariance matrices are generally assumed to be functions of a few unknown variance components. Namely, they are usually of the form $\sigma^2\,\mathbf{M}$, where σ^2 is an unknown variance to be estimated and $\mathbf{M}$ is a matrix of known constants. More generally, they can be of the form $\sum_i \sigma_i^2\,\mathbf{M}_i$, where the sum is usually over just a few terms. For the remainder of the chapter, we consider estimators of $\boldsymbol{\beta}$ and $\mathbf{u}$ under the assumption that the variance components associated with $\mathbf{G}$ and $\mathbf{R}$ are known. Part II examines applications of MMs for estimating breeding values (Chapter 32) and variance components (Chapter 33), as well as for genomic selection and prediction (Chapter 32).

Example 10.10. As an expansion of Example 10.2, suppose that three sires are chosen at random from a population, and each crossed to multiple randomly chosen (and unrelated) dams, where each mating yields a single offspring. Two offspring from each sire are evaluated, some in environment 1 and some in environment 2. Let y_{ijk} denote the phenotypic value of the kth offspring of sire i in environment j. The model is then

$$y_{ijk} = \beta_j + u_i + e_{ijk}$$

This model has three random effects (u_1, u_2, u_3), which measure the contribution from each sire, and two fixed effects (β_1, β_2), which describe the influence of the two environments. The model assumes an absence of sire $\times$ environment interaction. As noted above, a total of six offspring were measured. One offspring of sire 1 was assigned to environment 1 and had phenotypic value $y_{1,1,1} = 9$, while the second offspring was assigned to environment 2 and had $y_{1,2,1} = 12$. The two offspring of sire 2 were both assigned to environment 1 and had values of $y_{2,1,1} = 11$ and $y_{2,1,2} = 6$. Finally, one offspring of sire 3 was assigned to environment 1 and had value $y_{3,1,1} = 7$, while the second offspring was assigned to environment 2 and had value $y_{3,2,1} = 14$. The resulting vector of observations can be written as

$$\mathbf{y} = \begin{pmatrix} y_{1,1,1} \\ y_{1,2,1} \\ y_{2,1,1} \\ y_{2,1,2} \\ y_{3,1,1} \\ y_{3,2,1} \end{pmatrix} = \begin{pmatrix} 9 \\ 12 \\ 11 \\ 6 \\ 7 \\ 14 \end{pmatrix}$$

giving the MM as $\mathbf{y} = \mathbf{X}\boldsymbol{\beta} + \mathbf{Z}\mathbf{u} + \mathbf{e}$, where the incidence matrices for fixed and random effects, and the vectors of these effects, are

$$\mathbf{X} = \begin{pmatrix} 1 & 0 \\ 0 & 1 \\ 1 & 0 \\ 1 & 0 \\ 1 & 0 \\ 0 & 1 \end{pmatrix}, \quad \mathbf{Z} = \begin{pmatrix} 1 & 0 & 0 \\ 1 & 0 & 0 \\ 0 & 1 & 0 \\ 0 & 1 & 0 \\ 0 & 0 & 1 \\ 0 & 0 & 1 \end{pmatrix}, \quad \boldsymbol{\beta} = \begin{pmatrix} \beta_1 \\ \beta_2 \end{pmatrix}, \quad \mathbf{u} = \begin{pmatrix} u_1 \\ u_2 \\ u_3 \end{pmatrix}$$

This model is not yet fully specified, as no covariance structures for $\mathbf{u}$ and $\mathbf{e}$ have been proposed. Note that the matrices $\mathbf{X}$ and $\mathbf{Z}$ easily accommodate unbalanced designs.

Estimating Fixed Effects and Predicting Random Effects

While one primary goal of a quantitative-genetic analysis is to estimate variance components, there are also numerous situations in which inferences about fixed effects (such as the effect of a particular location) and/or random effects (such as the breeding value of a particular individual) are the central motivation. As we will see, marker effects can be treated as either fixed or random, depending upon the circumstances (Chapter 20; Part II Chapters 32 and 33). Inferences about fixed effects have come to be called **estimates,** whereas those that concern random effects are known as **predictions.** The most widely used procedures are BLUE and BLUP, referring, respectively, to **best linear unbiased estimator** (of the fixed effects) and **best linear unbiased predictor** (of the random effects). They are *best* in the sense that they minimize the sampling variance, *linear* in the sense that they are linear functions of the observed phenotypes $\mathbf{y}$, and *unbiased* in the sense that $E[\,\text{BLUE}(\boldsymbol{\beta})\,] = \boldsymbol{\beta}$ and $E[\,\text{BLUP}(\mathbf{u})\,] = \mathbf{u}$. Although the method of predicting random effects using BLUP methodology was first discussed by Henderson (1949, 1950), the expression "best linear unbiased predictor" was apparently first used by Goldberger (1962), with the acronym BLUP due to Henderson (1973). BLUP is the method of choice for estimating the breeding values of individuals from field records of large and complex pedigrees, and underpins many marker-based genomic prediction and genomic selection schemes (Chapters 32 and 33).

For the MM given by Equation 10.18, the BLUE of $\boldsymbol{\beta}$ is

$$\widehat{\boldsymbol{\beta}} = \left(\mathbf{X}^T\mathbf{V}^{-1}\mathbf{X}\right)^{-1}\mathbf{X}^T\mathbf{V}^{-1}\mathbf{y} \tag{10.21}$$

with $\mathbf{V}$ as given by Equation 10.19. Notice that this is just the generalized least-squares (GLS) estimator (Equation 10.13a).

We obtain the BLUP for the vector of random effects $\mathbf{u}$ from its regression on $\mathbf{y}$. This follows from Equation 9.29a under the assumption that the joint distribution of $\mathbf{u}$ and $\mathbf{y}$ is multivariate normal. Starting with

$$\mathbf{u} \sim \text{MVN}(\mathbf{0}, \mathbf{G}), \quad \mathbf{y} \sim \text{MVN}(\mathbf{X}\boldsymbol{\beta}, \mathbf{V})$$

and noting (from Equation 9.21a) that the covariance between $\mathbf{u}$ and $\mathbf{y}$ is given by

$$\text{Cov}(\mathbf{u}, \mathbf{y}) = \text{Cov}(\mathbf{u}, \mathbf{X}\boldsymbol{\beta} + \mathbf{Z}\mathbf{u} + \mathbf{e}) = \text{Cov}(\mathbf{u}, \mathbf{u})\,\mathbf{Z}^T = \mathbf{G}\mathbf{Z}^T \tag{10.22a}$$

The joint density of $\mathbf{u}$ and $\mathbf{y}$ thus becomes

$$\begin{pmatrix} \mathbf{u} \\ \mathbf{y} \end{pmatrix} \sim \text{MVN}\left[\begin{pmatrix} \mathbf{0} \\ \mathbf{X}\boldsymbol{\beta} \end{pmatrix}, \begin{pmatrix} \mathbf{G} & \mathbf{G}\mathbf{Z}^T \\ \mathbf{Z}\mathbf{G} & \mathbf{V} \end{pmatrix}\right] \tag{10.22b}$$

In the notation of Equation 9.29a,

$$\mathbf{x_1} = \mathbf{u}, \quad \mathbf{x_2} = \mathbf{y}, \quad \boldsymbol{\mu_1} = \mathbf{0}, \quad \boldsymbol{\mu_2} = \mathbf{X}\boldsymbol{\beta}, \quad \mathbf{V}_{\mathbf{x_1}\mathbf{x_2}} = \mathbf{G}\mathbf{Z}^T, \quad \mathbf{V}_{\mathbf{x_2}\mathbf{x_2}} = \mathbf{V}$$

yielding the BLUP of $\mathbf{u}$ as

$$\begin{aligned} \widehat{\mathbf{u}} &= \boldsymbol{\mu_1} + \mathbf{V}_{\mathbf{x_1}\mathbf{x_2}}\mathbf{V}_{\mathbf{x_2}\mathbf{x_2}}^{-1}(\mathbf{x_2} - \boldsymbol{\mu_2}) \\ &= \mathbf{G}\mathbf{Z}^T\mathbf{V}^{-1}\left(\mathbf{y} - \mathbf{X}\widehat{\boldsymbol{\beta}}\right) \end{aligned} \tag{10.22c}$$

(Henderson 1963). Further, applying Equation 9.28 gives the covariance matrix for $\widehat{\mathbf{u}}$ as

$$\mathbf{G} - \mathbf{G}\mathbf{Z}^T\mathbf{V}^{-1}\mathbf{Z}\mathbf{G} \tag{10.22d}$$

Suppose that we can express $\mathbf{G}$ as $\sigma_G^2\,\mathbf{A}$ and $\mathbf{R}$ as $\sigma_E^2\,\mathbf{B}$. While one might suspect that two variance components (σ_G^2 and σ_E^2) are required to apply Equation 10.22c, in actuality, we only require a single parameter, their ratio, σ_E^2/σ_G^2. To see this, first note that

$$\mathbf{V} = \mathbf{Z}\mathbf{G}\mathbf{Z}^T + \mathbf{R} = \sigma_G^2[\,\mathbf{Z}\mathbf{A}\mathbf{Z}^T + (\sigma_E^2/\sigma_G^2)\,\mathbf{B}\,] \tag{10.23a}$$

Hence,

$$\begin{aligned} \mathbf{G}\mathbf{Z}^T\mathbf{V}^{-1} &= (\sigma_G^2\mathbf{A})\,\mathbf{Z}^T\sigma_G^{-2}[\,\mathbf{Z}\mathbf{A}\mathbf{Z}^T + (\sigma_E^2/\sigma_G^2)\,\mathbf{B}\,]^{-1} \\ &= \mathbf{A}\,\mathbf{Z}^T[\,\mathbf{Z}\mathbf{A}\mathbf{Z}^T + (\sigma_E^2/\sigma_G^2)\,\mathbf{B}\,]^{-1} \end{aligned} \tag{10.23b}$$

In a similar fashion,

$$\begin{aligned} \boldsymbol{\beta} &= \left(\mathbf{X}^T\mathbf{V}^{-1}\mathbf{X}\right)^{-1}\mathbf{X}^T\mathbf{V}^{-1}\mathbf{y} \\ &= \left(\mathbf{X}^T\sigma_G^{-2}[\,\mathbf{Z}\mathbf{A}\mathbf{Z}^T + (\sigma_E^2/\sigma_G^2)\,\mathbf{B}\,]^{-1}\mathbf{X}\right)^{-1}\mathbf{X}^T\sigma_G^{-2}[\,\mathbf{Z}\mathbf{A}\mathbf{Z}^T + (\sigma_E^2/\sigma_G^2)\,\mathbf{B}\,]^{-1}\mathbf{y} \\ &= \left(\mathbf{X}^T[\,\mathbf{Z}\mathbf{A}\mathbf{Z}^T + (\sigma_E^2/\sigma_G^2)\,\mathbf{B}\,]^{-1}\mathbf{X}\right)^{-1}\mathbf{X}^T[\,\mathbf{Z}\mathbf{A}\mathbf{Z}^T + (\sigma_E^2/\sigma_G^2)\,\mathbf{B}\,]^{-1}\mathbf{y} \end{aligned} \tag{10.23c}$$

showing that the BLUE also only depends on the variance ratio (σ_E^2/σ_G^2).

Finally, a few comments on an important feature, **shrinkage**, of BLUPs are in order. Equation 10.22c shows why we speak of *predicting* random effects, as their values are *predicted from a regression*. Consider the ith observation, which we can express as

$$y_i - \widehat{y}_i = u_i + e_i \tag{10.24a}$$

where $\widehat{y}_i$ (the ith element in the vector $\mathbf{X}\widehat{\boldsymbol{\beta}}$, with $\widehat{\boldsymbol{\beta}}$ the BLUE of $\boldsymbol{\beta}$) is the adjustment of observation i for the fixed effects (centering all observations to have an expected value of 0). The central question becomes how much of the deviation between the observed y_i and predicted $\widehat{y}_i$ values is due to the random effect u_i versus the residual e_i. Intuition suggests that if most of the variance is in u, then most of the deviation is due to u_i, while if most of the variation is from residual effects, then very little of the deviation is from u_i. We can more formally illustrate this by supposing only a single observation, so that $\mathbf{Z} = (1)$, $\mathbf{G} = (\sigma_G^2)$, $\mathbf{R} = (\sigma_E^2)$, and $\mathbf{V} = (\sigma_G^2 + \sigma_E^2)$. In this setting, Equation 10.22c reduces to

$$\widehat{u} = \frac{\sigma_G^2}{\sigma_G^2 + \sigma_E^2}\,(y - \widehat{y}\,) = \begin{cases} \simeq 0 & \text{for } \sigma_E^2 \gg \sigma_G^2 \\ \simeq y - \widehat{y} & \text{for } \sigma_G^2 \gg \sigma_E^2 \end{cases} \tag{10.24b}$$

The ratio of σ_G^2 to the total variance $(\sigma_G^2 + \sigma_E^2)$ is the **shrinkage factor** for this random effect. If this ratio is close to one, there is very little shrinkage, with $u_i = y_i - \widehat{y}_i$, which is essentially the estimate obtained by treating u_i as a fixed effect. Conversely, if this ratio is near zero, then almost all of the deviation is due to the residual error, with u_i being assigned a value close to its mean (which, by construction, is zero). Example 10.13 expands on these ideas.

Example 10.11. What are the BLUP values for the sire effects (u_1, u_2, u_3) in Example 10.10? In order to proceed, we require the covariance matrices for the sire effects and the residual errors. We assume that the residual variances within both environments are the same (σ_E^2), so $\mathbf{R} = \sigma_E^2\,\mathbf{I}$, where $\mathbf{I}$ is the 6×6 identity matrix. Assuming that all three sires are unrelated, not inbred, and all are drawn from the same population, $\mathbf{G} = \sigma_S^2\,\mathbf{I}$, where $\mathbf{I}$ is the 3×3 identity matrix and σ_S^2 is the variance of sire effects. Assuming only additive genetic variance, the sire effects (breeding values) are half the sires' additive genetic values (Chapter 7). Because the sires are sampled randomly from an outbred base population, $\sigma_S^2 = \sigma_A^2/4$, where σ_A^2 is the additive genetic variance. Assume that $\sigma_A^2 = 8$ and $\sigma_E^2 = 6$ for a heritability (assuming only additive variances) of $h^2 = \sigma_A^2/(\sigma_A^2 + \sigma_E^2) = 8/(6+8) = 0.57$. With these values, the covariance matrix $\mathbf{V}$ for the vector of observations $\mathbf{y}$ is given by $\mathbf{ZGZ}^T + \mathbf{R}$, or

$$\mathbf{V} = \frac{8}{4}\begin{pmatrix} 1 & 0 & 0 \\ 1 & 0 & 0 \\ 0 & 1 & 0 \\ 0 & 1 & 0 \\ 0 & 0 & 1 \\ 0 & 0 & 1 \end{pmatrix}\begin{pmatrix} 1 & 0 & 0 \\ 0 & 1 & 0 \\ 0 & 0 & 1 \end{pmatrix}\begin{pmatrix} 1 & 1 & 0 & 0 & 0 & 0 \\ 0 & 0 & 1 & 1 & 0 & 0 \\ 0 & 0 & 0 & 0 & 1 & 1 \end{pmatrix} + 6\begin{pmatrix} 1 & 0 & 0 & 0 & 0 & 0 \\ 0 & 1 & 0 & 0 & 0 & 0 \\ 0 & 0 & 1 & 0 & 0 & 0 \\ 0 & 0 & 0 & 1 & 0 & 0 \\ 0 & 0 & 0 & 0 & 1 & 0 \\ 0 & 0 & 0 & 0 & 0 & 1 \end{pmatrix}$$

$$= \begin{pmatrix} 8 & 2 & 0 & 0 & 0 & 0 \\ 2 & 8 & 0 & 0 & 0 & 0 \\ 0 & 0 & 8 & 2 & 0 & 0 \\ 0 & 0 & 2 & 8 & 0 & 0 \\ 0 & 0 & 0 & 0 & 8 & 2 \\ 0 & 0 & 0 & 0 & 2 & 8 \end{pmatrix}, \quad \text{giving} \quad \mathbf{V}^{-1} = \frac{1}{30}\cdot\begin{pmatrix} 4 & -1 & 0 & 0 & 0 & 0 \\ -1 & 4 & 0 & 0 & 0 & 0 \\ 0 & 0 & 4 & -1 & 0 & 0 \\ 0 & 0 & -1 & 4 & 0 & 0 \\ 0 & 0 & 0 & 0 & 4 & -1 \\ 0 & 0 & 0 & 0 & -1 & 4 \end{pmatrix}$$

Using this result, a few simple matrix calculations (Equation 10.21) gives

$$\widehat{\boldsymbol{\beta}} = \begin{pmatrix} \widehat{\beta}_1 \\ \widehat{\beta}_2 \end{pmatrix} = \left(\mathbf{X}^T\mathbf{V}^{-1}\mathbf{X}\right)^{-1}\mathbf{X}^T\mathbf{V}^{-1}\mathbf{y} = \frac{1}{18}\begin{pmatrix} 148 \\ 235 \end{pmatrix}$$

while applying Equation 10.22c yields

$$\widehat{\mathbf{u}} = \begin{pmatrix} \widehat{u}_1 \\ \widehat{u}_2 \\ \widehat{u}_3 \end{pmatrix} = \mathbf{GZ}^T\mathbf{V}^{-1}\left(\mathbf{y} - \mathbf{X}\widehat{\boldsymbol{\beta}}\right) = \frac{1}{18}\begin{pmatrix} -1 \\ 2 \\ -1 \end{pmatrix}$$

Example 10.12. As a prelude to association analysis (Chapters 17, 20, and 21), the effects on a trait from different genotypes at a single candidate gene are often estimated by OLS, using the model

$$y_{ij} = g_i + e_{ij} \tag{10.25a}$$

where y_{ij} is the observed phenotype of the jth individual of genotype i, g_i is the mean genotypic value for the ith genotype at the locus of interest, and e_{ij} is a residual deviation assumed to be independently distributed among individuals. While this model may be reasonable for a random collection of individuals from a large population, when some sampled individuals are relatives, the sharing of alleles at other loci influencing the trait will induce correlations between residuals. If this is the case, OLS analysis can produce biased estimates of the focal QTL effects. When one of the QTL genotypes is very rare, as is often the case, the sampled individuals may be intentionally selected from the same pedigree, so the problem of bias is not trivial.

Use of a MM provides a means of accounting for associations among background QTLs in relatives in a way that eliminates bias in estimates of candidate gene effects. If the relatives in question share only additive effects, then the correlations among residuals are accounted for by the additive genetic relationship matrix $\mathbf{A}$, where A_{ij} is twice the coefficient of coancestry, $2\Theta_{ij}$ (Equation 7.11a). When full sibs are included and dominance is present at background QTLs, both $\mathbf{A}$ and a dominance relationship matrix (Part II Chapter 32) are required. Here we assume that all of the background genetic effects are additive, in which case the simplest MM can be applied,

$$y_{ij} = g_i + e^*_{ij}, \qquad \text{where} \quad e^*_{ij} = a_{ij} + e_{ij} \tag{10.25b}$$

with the contribution from the different single-locus genotypes (g_i) being treated as fixed effects. The additive genetic background effects (a_{ij}) and the residual environmental deviations (e_{ij}) are treated as random effects, both with expected values equal to zero, and with respective variances σ^2_A and σ^2_E. Note that σ^2_A is the background additive genetic variance for the trait in excess of that caused by the focal QTL. In matrix form,

$$\mathbf{y} = \mathbf{Xg} + \mathbf{Za} + \mathbf{e} \tag{10.25c}$$

If there is a single observation for each individual, as we assume below, then $\mathbf{Z} = \mathbf{I}$ and the covariance matrix for the vector of observations $(\mathbf{y})$ becomes

$$\mathbf{V} = \sigma^2_A\,\mathbf{A} + \sigma^2_E\,\mathbf{I} \tag{10.25d}$$

The resulting covariance between the residual errors associated with individuals i and j is $\sigma^2_A A_{ij} = 2\Theta_{ij}\sigma^2_A$. When $\Theta_{ij} \neq 0$,these residuals are correlated, and OLS no longer holds. The error in using OLS to estimate single gene effects is that $\mathbf{A}$ is (incorrectly) assumed to equal an identity matrix, so that $\mathbf{V}$ is incorrectly assumed to be diagonal. From Equation 10.21, the estimates of the QTL means are given by

$$\widehat{\mathbf{g}} = \left(\mathbf{X}^T\mathbf{V}^{-1}\mathbf{X}\right)^{-1}\mathbf{X}^T\mathbf{V}^{-1}\mathbf{y} \tag{10.25e}$$

Kennedy et al. (1992) showed that MM estimates of candidate gene effects are much more reliable than OLS estimates, especially in small selected populations. Building on this approach, several authors (Hoeschele 1988; Hofer and Kennedy 1993; Kinghorn et al. 1993) have proposed BLUP-based segregation analysis (Chapter 16) for estimating the effects of an unknown major gene. Here the elements in the design matrix $\mathbf{X}$ associated with g_i are probabilistic estimates for the major-locus genotypes of each individual. As we develop in Chapter 20, the principles of this example, along with those of Examples 9.13 and 9.14, outline the structure of MMs used in association analysis (GWAS).

Example 10.13. Equations 10.21 and 10.22c provide significant insight into the implications of treating a factor as fixed versus random. Suppose we measure two replicate plots for each of three inbred lines in a single randomized block (Appendix 9), so that the simplest model is

$$y_{ij} = L_i + e_{ij} \tag{10.26a}$$

where y_{ij} is the value of the jth plot of line i. Suppose that the plot values were 6 and 8 for line 1, 10 and 12 for line 2, and 14 and 16 for line 3. Treating the line values as fixed, the resulting GLM, $\mathbf{y} = \mathbf{X}\boldsymbol{\beta} + \mathbf{e}$, has

$$\mathbf{y} = \begin{pmatrix} 6 \\ 8 \\ 10 \\ 12 \\ 14 \\ 16 \end{pmatrix}, \quad \mathbf{X} = \begin{pmatrix} 1 & 0 & 0 \\ 1 & 0 & 0 \\ 0 & 1 & 0 \\ 0 & 1 & 0 \\ 0 & 0 & 1 \\ 0 & 0 & 1 \end{pmatrix}, \quad \boldsymbol{\beta} = \begin{pmatrix} L_1 \\ L_2 \\ L_3 \end{pmatrix} \tag{10.26b}$$

Assuming OLS residuals, Equation 10.9a gives the OLS BLUEs for the line values as

$$\widehat{\beta} = (\mathbf{X}^T\mathbf{X})^{-1}\mathbf{X}^T\mathbf{y} = \begin{pmatrix} 7 \\ 11 \\ 15 \end{pmatrix}$$

Now use this exact same data set, but treat the line values as random. To provide as fair a comparison as possible, assume the lines are unrelated (so that information is not borrowed from correlated observations). The covariance structure for the vector of random line effects is thus $\sigma_G^2\,\mathbf{I}$. The model now becomes

$$y_{ij} = \mu + \ell_i + e_{ij} \tag{10.26c}$$

where μ is the overall mean, and ℓ_i the realization for line i (giving the line value as $L_i = \mu + \ell_i$). Again, assume OLS residuals. The resulting components of the MM, $\mathbf{y} = \mathbf{X}\boldsymbol{\beta} + \mathbf{Z}\mathbf{u} + \mathbf{e}$, become

$$\mathbf{X} = \begin{pmatrix} 1 \\ 1 \\ 1 \\ 1 \\ 1 \\ 1 \end{pmatrix}, \qquad \boldsymbol{\beta} = (\,\mu\,), \qquad \mathbf{Z} = \begin{pmatrix} 1 & 0 & 0 \\ 1 & 0 & 0 \\ 0 & 1 & 0 \\ 0 & 1 & 0 \\ 0 & 0 & 1 \\ 0 & 0 & 1 \end{pmatrix}, \qquad \mathbf{u} = \begin{pmatrix} \ell_1 \\ \ell_2 \\ \ell_3 \end{pmatrix} \tag{10.26d}$$

with a covariance structure of

$$\text{Var}\begin{pmatrix} \mathbf{u} \\ \mathbf{e} \end{pmatrix} = \begin{pmatrix} \sigma_G^2\,\mathbf{I} & \mathbf{0} \\ \mathbf{0} & \sigma_E^2\,\mathbf{I} \end{pmatrix}$$

Using Equation 10.23a, these assumptions ($\mathbf{G} = \sigma_G^2\mathbf{I},\ \ \mathbf{R} = \sigma_E^2\mathbf{I}$) yield

$$\mathbf{V} = \sigma_G^2[\,\mathbf{Z}\mathbf{Z}^T + (\sigma_E^2/\sigma_G^2)\,\mathbf{I}\,]$$

$$= \sigma_G^2\left[\begin{pmatrix} 1 & 1 & 0 & 0 & 0 & 0 \\ 1 & 1 & 0 & 0 & 0 & 0 \\ 0 & 0 & 1 & 1 & 0 & 0 \\ 0 & 0 & 1 & 1 & 0 & 0 \\ 0 & 0 & 0 & 0 & 1 & 1 \\ 0 & 0 & 0 & 0 & 1 & 1 \end{pmatrix} + \sigma_E^2/\sigma_G^2 \begin{pmatrix} 1 & 0 & 0 & 0 & 0 & 0 \\ 0 & 1 & 0 & 0 & 0 & 0 \\ 0 & 0 & 1 & 0 & 0 & 0 \\ 0 & 0 & 0 & 1 & 0 & 0 \\ 0 & 0 & 0 & 0 & 1 & 0 \\ 0 & 0 & 0 & 0 & 0 & 1 \end{pmatrix}\right] \tag{10.26e}$$

One can show for this $\mathbf{V}$ that all values of σ_E^2/σ_G^2 yield

$$\widehat{\beta} = \widehat{\mu} = \left(\mathbf{X}^T\mathbf{V}^{-1}\mathbf{X}\right)^{-1}\mathbf{X}^T\mathbf{V}^{-1}\mathbf{y} = (\,11\,)$$

Equations 10.22c and 10.23b give the solution for the BLUPs of line values as

$$\widehat{\mathbf{u}} = \mathbf{G}\mathbf{Z}^T\mathbf{V}^{-1}\left(\mathbf{y} - \mathbf{X}\widehat{\beta}\right) = \mathbf{Z}^T[\,\mathbf{Z}\mathbf{Z}^T + (\sigma_E^2/\sigma_G^2)\,\mathbf{I}\,]^{-1}\left(\mathbf{y} - \mathbf{X}\widehat{\beta}\right)$$

Consider σ_E^2/σ_G^2 values of 500, 5, 1, 1/5, and $\simeq 0$. Noting that $H^2 = 1/(1 + \sigma_E^2/\sigma_G^2)$, these correspond to broad-sense heritability values of 0.002, 0.17, 0.5, 0.82, and $\simeq 1$, respectively, with corresponding BLUPs of

$$\widehat{\mathbf{u}} = \begin{pmatrix} -0.016 \\ 0 \\ 0.016 \end{pmatrix}, \quad \begin{pmatrix} -1.14 \\ 0 \\ 1.14 \end{pmatrix}, \quad \begin{pmatrix} -2.67 \\ 0 \\ 2.67 \end{pmatrix}, \quad \begin{pmatrix} -3.64 \\ 0 \\ 3.64 \end{pmatrix}, \quad \begin{pmatrix} -4.00 \\ 0 \\ 4.00 \end{pmatrix}$$

These solutions highlight the property of shrinkage that arises with the prediction of random effects. When H^2 is near zero, phenotype is a very poor predictor of the genotypic value. Here most of the deviation of y from μ is due to the residual error e rather than ℓ, and the BLUPs for ℓ are regressed back to near their mean (which is zero for a random effect). Hence, $\widehat{\ell}_i \simeq 0$ and $L_i \simeq \widehat{\mu} = 11$. Conversely, when H^2 is near one, the phenotype is an excellent predictor of genotypic value, so that here most of the deviation of y from μ is due to ℓ. In this setting, there is essentially no shrinkage ($\widehat{\ell}_i \simeq y_i - \mu$), and we recover the fixed-effect solutions, $\widehat{\boldsymbol{\mu}}^T \simeq (\,7 \quad 11 \quad 15\,)$.

There is a tradeoff with using random effects. On one hand, they shrink observed deviations back towards the mean, which is not done by BLUEs. On the other hand, they borrow information from correlated observations (provided $\mathbf{G}$ has nonzero off-diagonal elements) that are not used by BLUEs. Such borrowing of information can result in far less shrinkage than would occur if it was ignored.

Henderson's Mixed-model Equations and MM Standard Errors

Computation of Equations 10.21 and 10.22c requires the inverse of the covariance matrix $\mathbf{V}$. In the above toy examples, $\mathbf{V}^{-1}$ was not difficult to obtain. However, when $\mathbf{y}$ contains many hundreds-of-thousands of observations, as is commonly the case in dairy breeding, the computation of $\mathbf{V}^{-1}$ can be challenging. As a way around this problem, Henderson (1950, 1963, 1973, 1984) offered a more compact method for jointly obtaining $\widehat{\boldsymbol{\beta}}$ and $\widehat{\mathbf{u}}$ in the form of his **mixed-model equations** (**MME**),

$$\begin{pmatrix} \mathbf{X}^T\mathbf{R}^{-1}\mathbf{X} & \mathbf{X}^T\mathbf{R}^{-1}\mathbf{Z} \\ \mathbf{Z}^T\mathbf{R}^{-1}\mathbf{X} & \mathbf{Z}^T\mathbf{R}^{-1}\mathbf{Z}+\mathbf{G}^{-1} \end{pmatrix} \begin{pmatrix} \widehat{\boldsymbol{\beta}} \\ \widehat{\mathbf{u}} \end{pmatrix} = \begin{pmatrix} \mathbf{X}^T\mathbf{R}^{-1}\mathbf{y} \\ \mathbf{Z}^T\mathbf{R}^{-1}\mathbf{y} \end{pmatrix} \tag{10.27}$$

While these expressions may look considerably more complicated than Equations 10.21 and 10.22c, $\mathbf{R}^{-1}$ and $\mathbf{G}^{-1}$ are trivial to obtain if $\mathbf{R}$ and $\mathbf{G}$ are diagonal, and hence the submatrices in Equation 10.27 are much easier to compute than $\mathbf{V}^{-1}$. A second advantage of Equation 10.27 can be seen by considering the dimensionality of the matrix on the left-hand side. Recalling that $\mathbf{X}$ and $\mathbf{Z}$ are $n \times p$ and $n \times q$ respectively, $\mathbf{X}^T\mathbf{R}^{-1}\mathbf{X}$ is $p \times p$, $\mathbf{X}^T\mathbf{R}^{-1}\mathbf{Z}$ is $p \times q$, and $\mathbf{Z}^T\mathbf{R}^{-1}\mathbf{Z}+\mathbf{G}^{-1}$ is $q \times q$. Thus, the matrix that needs to be inverted to obtain the solution for $\widehat{\boldsymbol{\beta}}$ and $\widehat{\mathbf{u}}$ is of order $(p+q) \times (p+q)$, which is often considerably less than the dimensionality of $\mathbf{V}$ (an $n \times n$ matrix). Finally, one can solve Equation 10.27 by Gaussian elimination (Gentle 2007), a computationally efficient method that bypasses direct matrix inversion.

Although there are several ways to derive the MMEs (Robinson 1991; a derivation can be found in WL Example A6.5), Henderson (1950) originally obtained them by assuming that the covariance matrices $\mathbf{G}$ and $\mathbf{R}$ are known and that the densities of the vectors $\mathbf{u}$ and e are each multivariate normal with no correlations between them. In this setting, Equation 10.27 are also the maximum-likelihood estimates of the fixed and random effects. Henderson (1963) later showed that the MMEs do not actually depend on normality, and that $\widehat{\boldsymbol{\beta}}$ and $\widehat{\mathbf{u}}$ are BLUE and BLUP, respectively, under general conditions provided the variances are known.

Example 10.14. Using the values from Examples 10.10 and 10.11, we find that

$$\mathbf{X}^T\mathbf{R}^{-1}\mathbf{X} = \frac{1}{6}\begin{pmatrix} 4 & 0 \\ 0 & 2 \end{pmatrix}, \qquad \mathbf{X}^T\mathbf{R}^{-1}\mathbf{Z} = \left(\mathbf{Z}^T\mathbf{R}^{-1}\mathbf{X}\right)^T = \frac{1}{6}\begin{pmatrix} 1 & 2 & 1 \\ 1 & 0 & 1 \end{pmatrix}$$

$$\mathbf{G}^{-1}+\mathbf{Z}^T\mathbf{R}^{-1}\mathbf{Z} = \frac{5}{6}\begin{pmatrix} 1 & 0 & 0 \\ 0 & 1 & 0 \\ 0 & 0 & 1 \end{pmatrix}, \quad \mathbf{X}^T\mathbf{R}^{-1}\mathbf{y} = \frac{1}{6}\begin{pmatrix} 33 \\ 26 \end{pmatrix}, \quad \mathbf{Z}^T\mathbf{R}^{-1}\mathbf{y} = \frac{1}{6}\begin{pmatrix} 21 \\ 17 \\ 21 \end{pmatrix}$$

After factoring out 1/6 from both sides, the MMEs for these data become

$$\begin{pmatrix} 4 & 0 & 1 & 2 & 1 \\ 0 & 2 & 1 & 0 & 1 \\ 1 & 1 & 5 & 0 & 0 \\ 2 & 0 & 0 & 5 & 0 \\ 1 & 1 & 0 & 0 & 5 \end{pmatrix} \begin{pmatrix} \widehat{\beta}_1 \\ \widehat{\beta}_2 \\ \widehat{u}_1 \\ \widehat{u}_2 \\ \widehat{u}_3 \end{pmatrix} = \begin{pmatrix} 33 \\ 26 \\ 21 \\ 17 \\ 21 \end{pmatrix}$$

Taking the inverse gives the solution

$$\begin{pmatrix} \widehat{\beta}_1 \\ \widehat{\beta}_2 \\ \widehat{u}_1 \\ \widehat{u}_2 \\ \widehat{u}_3 \end{pmatrix} = \frac{1}{270} \begin{pmatrix} 100 & 25 & -25 & -40 & -25 \\ 25 & 175 & -40 & -10 & -40 \\ -25 & -40 & 67 & 10 & 13 \\ -40 & -10 & 10 & 70 & 10 \\ -25 & -40 & 13 & 10 & 67 \end{pmatrix} \begin{pmatrix} 33 \\ 26 \\ 21 \\ 17 \\ 21 \end{pmatrix} = \frac{1}{18} \begin{pmatrix} 148 \\ 235 \\ -1 \\ 2 \\ -1 \end{pmatrix}$$

which is identical to the results obtained in Example 10.11.

A relatively straightforward extension of Henderson's MMEs provides estimates of the standard errors of the fixed and random effects. Partition the inverse of the **left-hand side** (**LHS**) matrix in Equation 10.27 as

$$\begin{pmatrix} \mathbf{X}^T\mathbf{R}^{-1}\mathbf{X} & \mathbf{X}^T\mathbf{R}^{-1}\mathbf{Z} \\ \mathbf{Z}^T\mathbf{R}^{-1}\mathbf{X} & \mathbf{Z}^T\mathbf{R}^{-1}\mathbf{Z} + \mathbf{G}^{-1} \end{pmatrix}^{-1} = \begin{pmatrix} \mathbf{C}_{11} & \mathbf{C}_{12} \\ \mathbf{C}_{12}^T & \mathbf{C}_{22} \end{pmatrix} \tag{10.28}$$

where $\mathbf{C}_{11}$, $\mathbf{C}_{12}$, and $\mathbf{C}_{22}$ are, respectively, $p \times p$, $p \times q$, and $q \times q$ submatrices. Using this notation, Henderson (1975) showed that the sampling covariance matrix for the BLUE of $\boldsymbol{\beta}$ is

$$\boldsymbol{\sigma}(\widehat{\boldsymbol{\beta}}) = \mathbf{C}_{11} \tag{10.29a}$$

while the sampling covariance matrix of the prediction errors $(\widehat{\mathbf{u}} - \mathbf{u})$ is

$$\boldsymbol{\sigma}(\,\widehat{\mathbf{u}} - \mathbf{u}\,) = \mathbf{C}_{22} \tag{10.29b}$$

We consider $\widehat{\mathbf{u}} - \mathbf{u}$ rather than $\widehat{\mathbf{u}}$ as the latter includes variance from both the prediction error and that from the random effects $\mathbf{u}$ themselves. Finally, the sampling covariance of estimated effects and prediction errors is

$$\boldsymbol{\sigma}(\widehat{\boldsymbol{\beta}}, \widehat{\mathbf{u}} - \mathbf{u}\,) = \mathbf{C}_{12} \tag{10.29c}$$

The standard errors of the fixed and random effects are obtained, respectively, as the square roots of the diagonal elements of $\mathbf{C}_{11}$ and $\mathbf{C}_{22}$. Meyer (1989) presented methods for approximating the diagonal elements of the inverse of this matrix (and hence the standard errors) for very large breeding designs where the inverse of the MME matrix may be difficult to compute.

Example 10.15. Consider the MME from Example 10.14. Here for the fixed factors β_1, β_2 and the random effects u_1, u_2, u_3, the inverse of the LHS coefficient matrix is

$$\left(\begin{array}{cc|ccc} 4 & 0 & 1 & 2 & 1 \\ 0 & 2 & 1 & 0 & 1 \\ \hline 1 & 1 & 5 & 0 & 0 \\ 2 & 0 & 0 & 5 & 0 \\ 1 & 1 & 0 & 0 & 5 \end{array}\right)^{-1} = \frac{1}{270} \left(\begin{array}{cc|ccc} 100 & 25 & -25 & -40 & -25 \\ 25 & 175 & -40 & -10 & -40 \\ \hline -25 & -40 & 67 & 10 & 13 \\ -40 & -10 & 10 & 70 & 10 \\ -25 & -40 & 13 & 10 & 67 \end{array}\right)$$

Hence,

$$\mathbf{C}_{11} = \frac{1}{270}\begin{pmatrix} 100 & 25 \\ 25 & 175 \end{pmatrix} \quad \text{and} \quad \mathbf{C}_{22} = \frac{1}{270}\begin{pmatrix} 67 & 10 & 13 \\ 10 & 70 & 10 \\ 13 & 10 & 67 \end{pmatrix}$$

so that, for example,

$$\sigma^2(\widehat{\beta}_1) = \frac{100}{270}, \quad \sigma^2(\widehat{\beta}_2) = \frac{175}{270}, \quad \sigma(\widehat{\beta}_1, \widehat{\beta}_2) = \frac{25}{270}$$

and

$$\sigma^2(\widehat{u}_2 - u_2) = \frac{70}{270}, \quad \sigma(\widehat{u}_1 - u_1, \widehat{u}_3 - u_3) = \frac{13}{270}, \quad \text{and so on.}$$

Finally, we note two other expressions for the BLUP, $\widehat{\mathbf{u}}$, that follow from the MMEs. First, just considering the second row in Equation 10.27, we have

$$\left(\mathbf{Z}^T\mathbf{R}^{-1}\mathbf{X}\right)\widehat{\boldsymbol{\beta}} + \left(\mathbf{Z}^T\mathbf{R}^{-1}\mathbf{Z} + \mathbf{G}^{-1}\right)\widehat{\mathbf{u}} = \mathbf{Z}^T\mathbf{R}^{-1}\mathbf{y} \tag{10.30a}$$

Solving for $\widehat{\mathbf{u}}$ yields

$$\widehat{\mathbf{u}} = \left(\mathbf{Z}^T\mathbf{R}^{-1}\mathbf{Z} + \mathbf{G}^{-1}\right)^{-1}\mathbf{Z}^T\mathbf{R}^{-1}\left(\mathbf{y} - \mathbf{X}\widehat{\boldsymbol{\beta}}\right) \tag{10.30b}$$

which can be shown to be equivalent to Equation 10.22c. Second, we can express the BLUEs and BLUPs in terms of the $\mathbf{C}_{ij}$ submatrices (Equation 10.28) of the inverse of the LHS matrix in the MME, with

$$\begin{pmatrix} \widehat{\boldsymbol{\beta}} \\ \widehat{\mathbf{u}} \end{pmatrix} = \begin{pmatrix} \mathbf{X}^T\mathbf{R}^{-1}\mathbf{X} & \mathbf{X}^T\mathbf{R}^{-1}\mathbf{Z} \\ \mathbf{Z}^T\mathbf{R}^{-1}\mathbf{X} & \mathbf{Z}^T\mathbf{R}^{-1}\mathbf{Z} + \mathbf{G}^{-1} \end{pmatrix}^{-1}\begin{pmatrix} \mathbf{X}^T\mathbf{R}^{-1}\mathbf{y} \\ \mathbf{Z}^T\mathbf{R}^{-1}\mathbf{y} \end{pmatrix} \tag{10.31a}$$

$$= \begin{pmatrix} \mathbf{C}_{11} & \mathbf{C}_{12} \\ \mathbf{C}_{12}^T & \mathbf{C}_{22} \end{pmatrix}\begin{pmatrix} \mathbf{X}^T\mathbf{R}^{-1}\mathbf{y} \\ \mathbf{Z}^T\mathbf{R}^{-1}\mathbf{y} \end{pmatrix} \tag{10.31b}$$

$$= \begin{pmatrix} \left(\mathbf{C}_{11}\mathbf{X}^T + \mathbf{C}_{12}\mathbf{Z}^T\right)\mathbf{R}^{-1}\mathbf{y} \\ \left(\mathbf{C}_{12}^T\mathbf{X}^T + \mathbf{C}_{22}\mathbf{Z}^T\right)\mathbf{R}^{-1}\mathbf{y} \end{pmatrix} \tag{10.31c}$$

REML Estimation of Variance Components

Finally, as noted above, the practical application of Equations 10.21, 10.22c, and 10.27 requires that the variance components be known. Thus, prior to a BLUP analysis, these need to be estimated by ANOVA (Part II Chapter 23) or **restricted maximum likelihood** (**REML**). REML is closely related to BLUP, with (roughly speaking) REML estimates obtained from iterating and updating BLUP estimates until there is suitable convergence (Part II Chapter 33). REML maximizes that part of the likelihood function that is unaffected by fixed effects (Patterson and Thompson 1971). Harville (1977) coined the term **restricted ML**, but Thompson (2008) noted that REML maximizes a *residual* likelihood, and hence preferred the term **residual maximum likelihood**. One advantage of REML estimates (over those obtained by other estimation procedures) is that they are unbiased by the estimates of fixed effects

(Patterson and Thompson 1971). ANOVA estimates of variances under balanced designs are equivalent to REML estimates. Chapter 33 in Part II examines REML in some detail.

Thus, BLUPs are usually obtained by a two-stage approach: variance components are first obtained by REML and then these estimates are used to obtain BLUPs. This two-stage approach is called **empirical BLUP** or **REML/BLUP** (Sorensen and Kennedy 1986; Kennedy and Sorensen 1988; Harville 1990). Kackar and Harville (1981) and Gianola et al. (1986, 1988) showed that using REML estimation does not result in biased values for BLUPs, but that the resulting predictors may not be "best" (there may be other linear predictors with smaller mean-squared errors). While Chapter 33 examines large-sample approximations for the uncertainty in REML variance estimates, these cannot be easily translated into how much additional uncertainty is introduced into BLUPs by using REML *estimates* in place of their *true values*. A more formal procedure is to use a Bayesian approach (Appendices 7 and 8), wherein one draws a value of the variances from their posterior distribution, and then computes the BLUPs given this value. Repeating this procedure thousands of times generates an empirical posterior distribution for BLUPs that fully incorporates any errors introduced by using estimated, instead of true, variances (Appendix 8).

11

Analysis of Line Crosses

When you can do the common things of life in an uncommon way, you will command the attention of the world. George Washington Carver

Distinct populations, such as the isolated demes that comprise some natural populations or local land races and breeds of domesticated species, often exhibit remarkable phenotypic divergence. Such differences are sometimes a simple consequence of environmental influences on phenotypic expression, but genetic differences may arise as a result of local adaptation and random genetic drift. The genetic basis of interdemic differentiation, whether in natural or breeding populations, is of interest for several reasons. With nonadditive gene action, the mean phenotypes of progeny of interdemic crosses (F_1 hybrids from a line cross) will not be intermediate to those of their parents. One manifestation of this phenomenon is **heterosis**, the common observation that hybrids in many domesticated species often have enhanced performance or trait values relative to their parental lines (Chapter 13). Conversely, a problem of some significance in conservation biology is the opposite observation: **outbreeding depression**, wherein hybrids have *reduced* fitness relative to their parental populations (Chapter 13). Thus, depending on circumstances, line crossing can either be very desirable (heterosis) or to be avoided as much as possible (outbreeding depression). For populations that do not normally have an opportunity to interbreed in nature, the nature of their potential hybrids is influenced by local adaptation and drift of the underlying lines, resulting in unpredictable effects on any subsequent hybrids. However, when opportunities for interdemic exchange are common, natural selection may favor specific mating-system properties, including dispersal strategies or reproductive isolation, which enhance or discourage outcrossing (a key to deciphering the mechanisms of speciation; Chapter 17).

The mechanisms of genetic differentiation also have important practical implications beyond prediction of hybrid performance. For example, in artificial-selection programs, the mean phenotypes of selected lines often evolve well beyond the range of variation seen in the base population prior to selection (Chapter 1; WL Chapters 25 and 26). The extent to which such changes are caused by a large number of genes of relatively small effects, as opposed to a few major segregating factors, is an important determinant of whether a search for informative molecular markers is likely to be successful (Chapters 17–20). In addition, the degree to which selection advances made in different lines can be successfully integrated into a single crossbred line is a function of the ways in which genes from the isolated lines interact. Entirely additive gains are easy to stack (or **pyramid**) across lines, while incorporating gains from nonadditive interactions is more problematic.

In this chapter, we show how the judicious choice of line crosses can be used to reveal the relative contributions of additive, dominance, and epistatic effects to population differentiation. A statistical test of the adequacy of alternative genetic models will be presented, and its application to a variety of data sets will be used to show that nonadditive gene action is commonly associated with population differentiation. Several methods for estimating the minimum number of loci responsible for population differentiation will then be discussed. Their application firmly supports the conviction that divergence in most characters of interest to evolutionary biologists, human geneticists, and breeders is influenced by multiple loci. This, however, does not rule out the possibility that a small number of loci are responsible for the majority of the differentiation between species and/or lines within species. Chapter 13 examines both heterosis and outbreeding depression, while Chapter 12 examines the complementary concern of inbreeding depression, the decay in performance upon inbreeding. All three of these phenomena require nonadditive variation.

EXPECTATIONS FOR LINE-CROSS MEANS

We start with two parental populations (P_1 and P_2), each with loci assumed to be in Hardy–Weinberg and gametic-phase equilibrium (note that a fully inbred line is trivially in Hardy–Weinberg, as all loci are fixed for a single allele). For the time being, we also assume that the loci differentiating the two populations are unlinked. An F_1 population is obtained by crossing the P_1 and P_2 lines, and subsequent random mating of F_1 individuals results in an F_2 generation. Because the F_2 population will be in Hardy–Weinberg and gametic-phase equilibrium for unlinked loci, both for genes derived within and between populations, it is logical to treat this as the reference population for the definition of genetic effects.

In previous chapters, we have presented definitions of additive, dominance, and epistatic effects for specific genes and gene combinations within randomly mating populations. The statistical machinery underlying line-cross analysis has parallels with this approach. The concerns here, however, are whether there is a *net difference* between the additive effects of the genes in the P_1 and P_2 populations, whether the genes in P_1 tend on average to be dominant over those in P_2, and whether there are net directional epistatic interactions between P_1 and P_2 genes. Before explicitly defining these **composite effects**, it will be useful to have some indices to describe the **gene content** and the **degree of hybridity** of line-cross derivatives.

For any pair of lines and their derivatives, the genotypes at a locus can be partitioned into three classes: (i) both alleles are from a random sample of P_1 alleles; (ii) both alleles are from a random sample of P_2 alleles; and (iii) one allele is from a random sample of P_1 alleles, while the other is from the P_2 pool of alleles. Let S be the fraction of P_1 genes in a line, and H be the probability that a member of the line has one P_1 and one P_2 gene at a locus. These two indices uniquely specify the expected frequencies of the three classes of genotypes at any autosomal locus:

$$\begin{aligned} S - \frac{H}{2} &= \text{frequency of individuals containing only P}_1 \text{ alleles} \\ H &= \text{frequency of individuals containing one P}_1 \text{ and one P}_2 \text{ allele} \qquad (11.1a) \\ 1 - S - \frac{H}{2} &= \text{frequency of individuals containing only P}_2 \text{ alleles} \end{aligned}$$

Composite effects are formally defined in a least-squares framework, similar to that used for effects at single loci (Chapter 4). Consider first the **composite additive effect**, defined as the *difference* of additive effects of P_1 versus P_2 alleles summed over all loci. A simple expression for this can be obtained by recalling Equation 4.9: for a diallelic locus, the additive effects of the B_1 and B_2 alleles can be written as $-p_2\alpha$ and $p_1\alpha$, where α is the average effect of allelic substitution, and p_1 and p_2 are the frequencies of the B_1 and B_2 alleles. These expressions apply to a randomly mating population, precisely the situation in the F_2 generation. Noting that the F_2 consists of 50% P_1 and 50% P_2 genes, the frequencies of all contrasting alleles (of P_1 versus P_2 origin) are 0.5. Thus, the composite additive effects of P_1 and P_2 genes in the F_2 reference population are equal in absolute value but opposite in sign, and we denote them respectively as $+\alpha^c/2$ and $-\alpha^c/2$, where the superscript c denotes *composite* (as opposed to *single-gene*) effects. The total composite additive effect in the F_2 generation, $[(\alpha^c - \alpha^c)/2]$, is then equal to zero, which is what is desired for a reference population. The total composite additive effect in the F_1 generation is also equal to zero, because every locus contains one P_1 and one P_2 allele. In the P_1 population, however, there are only P_1 alleles, each of which contributes $\alpha^c/2$, giving the total composite additive effect as $(\alpha^c + \alpha^c)/2 = \alpha^c$. In contrast, as each P_2 allele contributes $-\alpha^c/2$, the total composite effect in the P_2 population is $-\alpha^c$. More generally, the contribution of composite additive effects to the mean genotypic value of any line-cross derivative is

$$\left(S - \frac{H}{2}\right)(+\alpha^c) + (H)(0) + \left(1 - S - \frac{H}{2}\right)(-\alpha^c) = (2S-1)\alpha^c = \theta_S\alpha^c \qquad (11.1b)$$

where $\theta_S = 2S-1$ denotes the **source index.** The three terms on the left-most side of Equation 11.1b represent, respectively, the contributions from P_1 homozygotes, P_1P_2 heterozygotes, and P_2 homozygotes. The index θ_S contrasts the expected number of P_1 alleles at a locus in a particular line $(2S)$ with that in the F_2 reference population (1).

The **composite dominance effect** is obtained in a similar manner, again treating the F_2 population as a reference. Returning to Table 4.1, it can be shown that when the frequencies of the two alleles are equal to 0.5, the dominance deviations from the regression on gene content are $-d/2$ for both homozygotes and $d/2$ for the heterozygote. An analogous situation exists for the F_2 of a line cross, which consists of 25% P_1P_1, 50% P_1P_2, and 25% P_2P_2 individuals at each locus. Thus, we denote the composite dominance effects associated with crossbred (P_1P_2) and purebred $(P_1P_1,\ P_2P_2)$ loci as $+\delta^c$ and $-\delta^c$, respectively. More generally, the contribution of composite dominance effects to the genotypic mean of a particular line is

$$\left(S - \frac{H}{2}\right)(-\delta^c) + (H)(+\delta^c) + \left(1 - S - \frac{H}{2}\right)(-\delta^c) = (2H-1)\delta^c = \theta_H \delta^c \tag{11.1c}$$

where $\theta_H = 2H - 1$ is the **hybridity index.** Thus, $\theta_H\delta^c$ yields a value of δ^c for the F_1 population, which consists entirely of hybrids, a value of $-\delta^c$ for the parental lines, and a value of zero for the F_2, maintaining the property that the average composite effects are defined to be zero in the reference population.

There are three important points to note about this approach to interpreting line-cross means. First, the composite effects are denoted as such because they ***summarize the total effects over all loci***. Because some of the effects of individual alleles in population P_1 may be positive and others negative, there is a possibility of considerable cancellation of locus-specific effects. A comparison of the variances within different line-cross derivatives can shed some light on this problem (Mather and Jinks 1982), but we will not take this up here.

Second, provided there is no mating with close relatives, the definition of composite effects does not require that the parental populations be pure (completely homozygous) lines. For any line-cross derivative, the subset of individuals with two P_1 alleles at a particular locus will have the same expected Hardy–Weinberg genotype distribution of P_1 genotypes as the P_1 generation. The same argument applies to loci with two P_2 alleles. Thus, in the absence of inbreeding (which alters the genotypic frequencies within classes; see Chapter 12), differences between the means of various line-cross derivatives cannot be due to a shift in the genotype frequencies *within* the groups P_1P_1, P_1P_2, and P_2P_2. It can only be caused by a shift in the *relative abundances* of these three groups.

Third, the source and hybridity indices are all that are needed to define the contributions of composite epistatic effects to line means. To obtain the general expression, we let $(\alpha_n^c\delta_m^c)$ denote the composite effect involving the interaction of n additive and m dominance effects. (Note that this notation *does not* imply that $(\alpha_n^c\delta_m^c) = \alpha_n^c \cdot \delta_m^c$.) The general expression for the mean genotypic value of a line is then

$$\mu = \mu_0 + \theta_S\alpha_1^c + \theta_H\delta_1^c + \theta_S^2\alpha_2^c + \theta_S\theta_H(\alpha_1^c\delta_1^c) + \theta_H^2\delta_2^c + \cdots \tag{11.1d}$$

where μ_o is the F_2 mean. Note that the coefficient for the composite effect $\alpha_n^c\delta_m^c$ has the form of $(\theta_S)^n\,(\theta_H)^m$. Hill (1982a) provides a formal derivation of Equation 11.1d, and alternative formulations appear in Cockerham (1980), Lynch (1991), and Schnell and Cockerham (1992). Example 11.1 provides a derivation of the result for the additive $\times$ additive component of epistasis. The compositions of the expected line means for common types of crossbreds are given in Table 11.1.

The preceding model applies in the presence of linkage as long as there is no epistasis. Linked genes tend to be inherited as a unit, although they are gradually rendered independent by crossing-over. This process has the effect of altering the likelihood of specific epistatic interactions through progressive rounds of recombination (see Equations 11.2a–11.2c). Provided the parental lines are in gametic-phase equilibrium, the expressions for the P_1, P_2, and F_1 lines still hold, because there is no opportunity for recombination between

Table 11.1 Expected mean phenotypes for various line-cross derivatives in terms of composite additive, dominance, and two-locus epistatic effects, taking the F_2 mean (μ_0) as a point of reference. These expressions assume freely recombining loci. For situations involving self-fertilization, it is further assumed that the parental lines are completely homozygous.

Line	θ_S	θ_H	Expected Mean Phenotype
P_1	1	-1	$\mu_0 + \alpha_1^c - \delta_1^c + \alpha_2^c - \alpha_1^c\delta_1^c + \delta_2^c + \cdots$
P_2	-1	-1	$\mu_0 - \alpha_1^c - \delta_1^c + \alpha_2^c + \alpha_1^c\delta_1^c + \delta_2^c + \cdots$
F_1	0	1	$\mu_0 + \delta_1^c + \delta_2^c + \cdots$
F_2	0	0	μ_0
$B_1 = (P_1 \times F_1)$	$\frac{1}{2}$	0	$\mu_0 + \frac{1}{2}\alpha_1^c + \frac{1}{4}\alpha_2^c + \cdots$
$B_2 = (P_2 \times F_1)$	$-\frac{1}{2}$	0	$\mu_0 - \frac{1}{2}\alpha_1^c + \frac{1}{4}\alpha_2^c + \cdots$
$F_2 \times P_1$	$\frac{1}{2}$	0	$\mu_0 + \frac{1}{2}\alpha_1^c + \frac{1}{4}\alpha_2^c + \cdots$
$F_2 \times P_2$	$-\frac{1}{2}$	0	$\mu_0 - \frac{1}{2}\alpha_1^c + \frac{1}{4}\alpha_2^c + \cdots$
$F_2 \times F_1$	0	0	μ_0
$B_1 \times F_1$	$\frac{1}{4}$	0	$\mu_0 + \frac{1}{4}\alpha_1^c + \frac{1}{16}\alpha_2^c + \cdots$
$B_2 \times F_1$	$-\frac{1}{4}$	0	$\mu_0 - \frac{1}{4}\alpha_1^c + \frac{1}{16}\alpha_2^c + \cdots$
Second backcrosses			
$B_1 \times P_1$	$\frac{3}{4}$	$-\frac{1}{2}$	$\mu_0 + \frac{3}{4}\alpha_1^c - \frac{1}{2}\delta_1^c + \frac{9}{16}\alpha_2^c - \frac{3}{8}\alpha_1^c\delta_1^c + \frac{1}{4}\delta_2^c + \cdots$
$B_1 \times P_2$	$-\frac{1}{4}$	$\frac{1}{2}$	$\mu_0 - \frac{1}{4}\alpha_1^c + \frac{1}{2}\delta_1^c + \frac{1}{16}\alpha_2^c - \frac{1}{8}\alpha_1^c\delta_1^c + \frac{1}{4}\delta_2^c + \cdots$
$B_2 \times P_1$	$\frac{1}{4}$	$\frac{1}{2}$	$\mu_0 + \frac{1}{4}\alpha_1^c + \frac{1}{2}\delta_1^c + \frac{1}{16}\alpha_2^c + \frac{1}{8}\alpha_1^c\delta_1^c + \frac{1}{4}\delta_2^c + \cdots$
$B_2 \times P_2$	$-\frac{3}{4}$	$-\frac{1}{2}$	$\mu_0 - \frac{3}{4}\alpha_1^c - \frac{1}{2}\delta_1^c + \frac{9}{16}\alpha_2^c + \frac{3}{8}\alpha_1^c\delta_1^c + \frac{1}{4}\delta_2^c + \cdots$
Selfed backcrosses			
B_{1s}	$\frac{1}{2}$	$-\frac{1}{4}$	$\mu_0 + \frac{1}{2}\alpha_1^c - \frac{1}{4}\delta_1^c + \frac{1}{4}\alpha_2^c - \frac{1}{8}\alpha_1^c\delta_1^c + \frac{1}{16}\delta_2^c + \cdots$
B_{2s}	$-\frac{1}{2}$	$-\frac{1}{4}$	$\mu_0 - \frac{1}{2}\alpha_1^c - \frac{1}{4}\delta_1^c + \frac{1}{4}\alpha_2^c + \frac{1}{8}\alpha_1^c\delta_1^c + \frac{1}{16}\delta_2^c + \cdots$
Continued selfing from the F_2			
$F_3(S_1)$	0	$-\frac{1}{2}$	$\mu_0 - \frac{1}{2}\delta_1^c + \frac{1}{4}\delta_2^c + \cdots$
$F_4(S_2)$	0	$-\frac{3}{4}$	$\mu_0 - \frac{3}{4}\delta_1^c + \frac{9}{16}\delta_2^c + \cdots$

chromosomes of different parental lines, but those for all subsequent line-cross derivatives will be biased to an extent depending on the map structure of the constituent loci. For a pair of loci with recombination fraction c, the following modifications need to be applied to the expressions for the F_2 and backcross means,

$$\mu(F_2) = \mu_0 + \left(\frac{1-2c}{2}\right)\alpha_2^c + (1-2c)^2\delta_2^c + \cdots \tag{11.2a}$$

$$\mu(B_1) = \mu_0 + \frac{\alpha_1^c}{2} + \left(\frac{1-c}{2}\right)\alpha_2^c + \left(\frac{2c-1}{2}\right)\alpha_1^c\delta_1^c + (1-2c)\delta_2^c + \cdots \tag{11.2b}$$

$$\mu(B_2) = \mu_0 - \frac{\alpha_1^c}{2} + \left(\frac{1-c}{2}\right)\alpha_2^c - \left(\frac{2c-1}{2}\right)\alpha_1^c\delta_1^c + (1-2c)\delta_2^c + \cdots \tag{11.2c}$$

Table 11.2 Genetic map structure and the average recombination frequency ($\bar{c}$) between random pairs of loci throughout the genome. N is the haploid number of chromosomes per genome, and $\mathcal{L}$ is the total map length (in Morgans) in females. For *Drosophila*, we ignore a tiny dot chromosome for which little mapping data are available. In the mosquito *Aedes* and in humans, respectively, recombination in males is approximately 50% and 65% as frequent as that in females; and there is no recombination within chromosomes in male *Drosophila*. For these three taxa, the reported values of $\bar{c}$ are averages for the two sexes. All results are approximations, as genomic maps are continuously being refined with the addition of molecular markers (Chapter 8). All data are from O'Brien (1990), except that for *Phaseolus* (Vallejos et al. 1992), *Pinus* (Plomion et al. 1995), and *Danio* (supplied by J. Postlethwait).

Species	N	$\mathcal{L}$	Lengths of Individual Chromosomes, ℓ_i	$\bar{c}$
Drosophila melanogaster	3	2.77	0.66, 1.03, 1.08	0.365
Drosophila pseudoobscura	4	4.46	0.68, 0.69, 1.01, 2.08	0.386
Aedes aegypti	3	2.28	0.62, 0.80, 0.86	0.380
Caenorhabditis elegans	5	1.61	0.27, 0.31, 0.33, 0.34, 0.36	0.418
Arabidopsis thaliana	5	5.24	0.63, 0.83, 0.98, 1.36, 1.44	0.443
Hordeum vulgare (barley)	7	9.49	0.70, 1.12, 1.15, 1.26, 1.27, 1.59, 2.40	0.465
Neurospora crassa	7	10.02	1.07, 1.13, 1.18, 1.33, 1.49, 1.52, 2.30	0.466
Zea mays (maize)	10	12.10	0.42, 0.78, 0.95, 1.07, 1.12, 1.37, 1.41, 1.55, 1.55, 1.67, 1.76	0.474
Phaseolus vulgaris (bean)	11	8.92	1.05, 1.04, 0.95, 0.92, 0.86, 0.78, 0.74, 0.71, 0.71, 0.60, 0.56	0.471
Pinus pinaster (maritime pine)	12	18.56	1.90, 1.75, 1.69, 1.66, 1.63, 1.61, 1.57, 1.54, 1.54, 1.41, 1.36, 0.90	0.481
Lycopersicon esculentum (tomato)	12	14.91	0.90, 0.92, 0.98, 1.01, 1.04, 1.04, 1.23, 1.34, 1.42, 1.63, 2.11	0.479
Mus musculus (mouse)	20	14.25	0.36, 0.36, 0.49, 0.56, 0.57, 0.57, 0.68, 0.70, 0.71, 0.73, 0.74, 0.78, 0.78, 0.80, 0.81, 0.84, 0.87, 0.89, 1.00, 1.01	0.483
Homo sapiens	23	40.00	0.69, 0.76, 1.12, 1.12, 1.22, 1.24, 1.25, 1.26, 1.38, 1.39, 1.40, 1.47, 1.62, 1.67, 1.67, 1.67, 1.74, 1.75, 1.75, 1.77, 1.92, 2.21, 2.49	0.490
Danio rerio (zebrafish)	25	28.06	0.59, 0.80, 0.84, 0.85, 0.86, 0.87, 0.95, 1.00, 1.02, 1.05, 1.05, 1.09, 1.13, 1.13, 1.14, 1.16, 1.22, 1.22, 1.33, 1.34, 1.39, 1.39, 1.45, 1.53, 1.66	0.489

Example 11.1 presents the derivation of Equations 11.2a–11.2c. The expressions for more advanced crosses (e.g., $B_1 \times F_1$) are complicated because one must account for additional generations of recombination.

Assuming that the epistatic effects between loci are independent of c, these expressions also apply to the total composite effects of all loci when $\bar{c}$, the mean recombination frequency between all pairs of loci, is substituted for c. Because we generally do not know c for any pair of loci, let alone for all of the loci underlying a quantitative trait, the best we can provide is a heuristic guide to the potential significance of linkage. Assuming that the genes are uniformly distributed across all chromosomes, then

$$\bar{c} = 0.5 - \frac{2\mathcal{L} - N + \sum_{i=1}^{N} e^{-2\ell_i}}{4\mathcal{L}^2} \tag{11.3}$$

where N is the number of chromosomes in a haploid set, ℓ_i is the genetic map length of the ith chromosome (in Morgans), and $\mathcal{L} = \sum \ell_i$ is the total map length (Zeng et al. 1990). This expression is based on Haldane's mapping function, $c_{ij} = 0.5(1 - e^{-2m_{ij}})$, which relates the recombination frequency to the genetic map length (m, the expected number of recombinations) between two linked loci, i and j (Chapter 17). Estimates of $\bar{c}$ are given in Table 11.2 for several species for which the genetic maps are reasonably well resolved. For species with a very small number of chromosomes and/or restricted recombination in males, as in *Drosophila* and the mosquito *Aedes*, $\bar{c}$ can be somewhat less than 0.40, but when N exceeds six or so, it tends to be greater than 0.45. For most mammals, N is typically on the order of 15 or more, so most pairs of genes are on different chromosomes, and $\bar{c}$ is very close to 0.5. With $\bar{c} > 0.45$, Equations 11.2a–11.2c are quite close to the expressions in Table 11.1. Thus, unless the genes underlying quantitative traits tend to be aggregated on chromosomes, linkage is unlikely to cause much bias in the interpretation of line-cross means, except perhaps in the case of species such as *Drosophila.*

Example 11.1. All of the composite effects described above are defined in a least-squares sense, and the nice symmetry whereby all effects have the same absolute value but differ in sign is a consequence of a randomly drawn F_2 allele having an equal chance of coming from P_1 or P_2. We now provide a formal derivation of the additive × additive composite effects in the context of a reference population that is both in Hardy–Weinberg and gametic-phase equilibrium. We denote the four gamete types as x_{11}, x_{12}, x_{21}, and x_{22}, where the subscripts refer to the parental sources of alleles (lines 1 or 2) at the first and second locus. All four gamete types have frequencies equal to 0.25 in the F_2 generation. Let the additive × additive effects associated with these gametes be α_{11}, α_{12}, α_{21}, and α_{22}. The effect α_{ij} is defined to be the average residual effect associated with a gamete containing a P_i-derived allele at the first locus and a P_j-derived allele at the second locus, after the additive effects of the two genes have been accounted for (see Equation 5.4a). Under a least-squares framework, the mean residual error is defined to be zero (Chapter 3), which implies

$$\frac{1}{4}\left(\alpha_{11} + \alpha_{12} + \alpha_{21} + \alpha_{22}\right) = 0$$

Furthermore, the mean-squared error is minimized. Noting that the previous expression implies that $\alpha_{22} = -\alpha_{11} - \alpha_{12} - \alpha_{21}$, the function to be minimized is

$$\alpha_{11}^2 + \alpha_{12}^2 + \alpha_{21}^2 + \left(-\alpha_{11} - \alpha_{12} - \alpha_{21}\right)^2$$

Taking the partial derivative with respect to α_{11} and setting it equal to zero, we obtain

$$2\alpha_{11} + \alpha_{12} + \alpha_{21} = 0$$

Subtracting $(\alpha_{11} + \alpha_{12} + \alpha_{21} + \alpha_{22})$, which equals zero, from the above expression, we find that the epistatic effects associated with each of the parental chromosome types are equal, i.e., $\alpha_{11} = \alpha_{22}$. By similar means, it can be shown that $\alpha_{12} = \alpha_{21}$, which when applied to the constraint $(\alpha_{11} + \alpha_{12} + \alpha_{21} + \alpha_{22}) = 0$ implies that

$$\alpha_{11} = \alpha_{22} = -\alpha_{12} = -\alpha_{21}$$

Thus, the additive × additive effects associated with both recombinant chromosome types are equal and opposite in sign to those of the parental chromosomes. Because, in a diploid, there are four combinations of genes at two loci (two within and two between the uniting gametes), we define the effects as

$$\alpha_{11} = \alpha_{22} = +\alpha_2^c/4$$
$$\alpha_{12} = \alpha_{21} = -\alpha_2^c/4$$

In both the P_1 and P_2 populations, all additive $\times$ additive interactions within and between chromosomes are of parental type (α_{ii}), and the composite effect of such interactions is $4(\alpha_2^c/4) = \alpha_2^c$. In the F_1 generation, the two interactions within chromosomes are of parental type, but the pairs of nonalleles between chromosomes are of different parental type, so the composite effect is $2(\alpha_2^c/4) + 2(-\alpha_2^c/4) = 0$.

Now consider the situation in the F_2 generation. Under free recombination, all four gametes are equally frequent, and with random mating, the average additive $\times$ additive epistatic effect within and between uniting gametes is equal to zero, as noted above. Suppose, however, that the two loci are linked, so that a fraction c of gametes are recombinant, and a fraction $(1-c)$ are nonrecombinant. The gamete frequencies are then $p(x_{12}) = p(x_{21}) = c/2$, and $p(x_{11}) = p(x_{22}) = (1-c)/2$. Assuming random mating, the paternal source of an individual's allele at one locus is independent of the maternal source of an allele at a second locus. Thus, the average composite effects associated with additive $\times$ additive interactions *between* uniting gametes is equal to zero. However, because the parental sources of genes *within* gametes will not have been completely randomized, the composite additive $\times$ additive effect within gametes is not zero, but

$$2[c(-\alpha_2^c/4) + (1-c)(\alpha_2^c/4)] = (1-2c)\alpha_2^c/2$$

giving the term in Equation 11.2a.

These results can be extended to the backcross generations. Consider, for example, the situation when F_1 individuals are crossed to members of parental line P_1, creating the B_1 backcross generation. The F_1 parent can contribute each of the four possible gametes, while the P_1 parent contributes only x_{11} gametes. With probability $(1-c)/2$, the offspring genotype is x_{11}/x_{11}, and each of the four possible two-locus interactions involves two P_1 alleles; each such interaction contributes $\alpha_2^c/4$, giving $(1-c)\alpha_2^c/2$. Likewise, with probability $c/2$ the genotype is x_{12}/x_{11}. Here, there are two P_1/P_1 interactions $(2\alpha_2^c/4)$ and two P_1/P_2 interactions $(-2\alpha_2^c/4)$, yielding a total contribution for this genotype of $(c/2)(2\alpha_2^c/4 - 2\alpha_2^c/4) = 0$. In a similar fashion, expectations for the other two genotypes are found to be equal to zero, giving the total contribution in the B_1 backcross as $(1-c)\alpha_2^c/2$.

ESTIMATION OF COMPOSITE EFFECTS

With the preceding model, the expected line means are linear functions of the composite effects. Thus, straightforward procedures can be used to estimate these effects from the observed line means. Generally, when the estimates of k parameters are desired, k types of lines can be identified that allow a solution using simultaneous equations. For example, if the epistatic effects are ignored, the expectations for the P_1, P_2, and F_1 means are simply

$$\mu(P_1) = \mu_0 + \alpha_1^c - \delta_1^c \tag{11.4a}$$
$$\mu(P_2) = \mu_0 - \alpha_1^c - \delta_1^c \tag{11.4b}$$
$$\mu(F_1) = \mu_0 + \delta_1^c \tag{11.4c}$$

Rearranging and substituting observed for expected means (denoted by $\bar{z}$), we obtain the three estimators,

$$\widehat{\mu}_0 = \frac{\bar{z}(P_1) + \bar{z}(P_2) + 2\bar{z}(F_1)}{4} \tag{11.5a}$$

$$\widehat{\alpha}_1^c = \frac{\bar{z}(P_1) - \bar{z}(P_2)}{2} \tag{11.5b}$$

$$\widehat{\delta}_1^c = \frac{2\bar{z}(F_1) - \bar{z}(P_1) - \bar{z}(P_2)}{4} \tag{11.5c}$$

Table 11.3 Coefficients for observed line means in expressions for estimated composite effects using the six-parameter model.

Parameter	P_1	P_2	F_1	F_2	B_1	B_2
μ_0	0	0	0	1	0	0
α_1^c	0	0	0	0	1	–1
δ_1^c	$\frac{1}{4}$	$\frac{1}{4}$	$-\frac{1}{2}$	2	–1	–1
α_2^c	0	0	0	–4	2	2
$\alpha_1^c\delta_1^c$	$\frac{1}{2}$	$-\frac{1}{2}$	0	0	–1	1
δ_2^c	$\frac{1}{4}$	$\frac{1}{4}$	$\frac{1}{2}$	1	–1	–1

With a model that includes all three forms of two-locus epistasis, there are six unknowns, so the mean phenotypes of at least six types of lines need to be evaluated. This is most easily accomplished by assaying both parental lines, their F_1 and F_2 derivatives, and the two backcrosses of F_1 individuals to the parental lines (B_1 and B_2). As in the previous example, the estimated composite effects can then be written as simple linear functions of the observed means (Table 11.3). For example, the composite additive × additive effect is estimated by

$$\widehat{\alpha}_2^c = -4\bar{z}(F_2) + 2\bar{z}(B_1) + 2\bar{z}(B_2)$$

Although we do not consider it in any detail here, it is worth noting that reciprocal crosses between parental lines can be used to estimate maternal effects and the effect of sex chromosomes (Part II Chapters 29 and 30). For example, when males of the P_1 line are crossed to females of the P_2 line, and vice versa, assuming that males are the heterogametic sex, the difference in mean phenotypes of daughters from the two lines provides an estimate of the difference between maternal effects associated with each parental line. In F_1 males, the effect of the X chromosome is superimposed on the maternal-effect difference. However, by making four possible types of crosses between F_1 individuals (two maternal sources of cytoplasm in females × two maternal sources of the X chromosome in males), a clean partitioning of these two sources of composite effects can be obtained (Carson and Lande 1984; Hard et al. 1992; Demuth and Wade 2007a).

Because the estimates of the composite effects are linear functions of the line-cross means, the sampling variances of the estimates can be obtained from the expression for the variance of a sum (Equation 3.11b). Further, as the mean phenotype of each line is estimated independently of the others, there is no covariance between the different mean estimates. Thus, the sampling variance of a composite effect is simply the sum of the sampling variances of the line means used in its estimation, each weighted by the squared coefficient in the estimating equation. For example, for the preceding estimator of $\widehat{\alpha}_2^c$,

$$\text{Var}(\widehat{\alpha}_2^c) = 16\text{Var}[\bar{z}(F_2)] + 4\text{Var}[\bar{z}(B_1)] + 4\text{Var}[\bar{z}(B_2)]$$

where $\text{Var}[\bar{z}(\cdots)]$ is the squared standard error of a line mean.

Finally, we note in passing that one interesting application of line-cross methodology is in the analysis of selection response. In this setting, one grows saved seed from different generations (and often some of their crosses) from a selection experiment (or a breeding program) in a common environment and evaluates their means. This approach, discussed in WL Chapter 18 and often called a **generation-means analysis (GMA)**, allows one to examine the component of selection response (Hammond and Gardner 1974; Smith 1979a, 1979b, 1983; Melchinger and Flachenecker 2006).

Hypothesis Testing

When the number of observations equals the number of parameters to be estimated, the solution of simultaneous equations cannot be used to evaluate the adequacy of the genetic model, as the parameter estimates are constrained to yield the observed line means exactly. This problem is eliminated when the number of observed line means exceeds the number of parameters to be estimated. For example, if data are available for the P_1, P_2, F_1, and F_2 lines, the statistic

$$\Delta = \bar{z}(F_2) - \left(\frac{\bar{z}(P_1) + \bar{z}(P_2)}{4} + \frac{\bar{z}(F_1)}{2} \right) \tag{11.6a}$$

provides a simple test for epistasis. In the absence of epistasis, the expected value of Δ is zero because at every locus, by the Hardy–Weinberg law, the F_2 composition is 25% P_1P_1, 50% P_1P_2, and 25% P_2P_2. The sampling variance of this test statistic is

$$\text{Var}(\Delta) = \text{Var}[\bar{z}(F_2)] + \frac{\text{Var}[\bar{z}(F_1)]}{4} + \frac{\text{Var}[\bar{z}(P_1)] + \text{Var}[\bar{z}(P_2)]}{16} \tag{11.6b}$$

Under the reasonable assumption that the sampling distribution of Δ is approximately normal, the ratio $|\Delta|/\sqrt{\text{Var}(\Delta)}$ provides a simple t test for epistasis. For large samples, if this ratio is greater than 1.96, then the null hypothesis of no epistasis can be rejected with 95% confidence.

A more general approach to hypothesis testing uses general linear models (Chapter 10) to estimate the model parameters and then compares the observed means with the model predictions. With this approach, one can start with a very simple model, evaluate its significance, and gradually add higher order composite effects to the model, until no further significant improvement in the model fit occurs. This procedure, first suggested by Cavalli (1952) and Hayman (1960a), is known as the **joint-scaling test**.

Consider the simple additive model,

$$\bar{z}_i = \mu_0 + \theta_{Si}\alpha_i^c + e_i$$

where the ith line mean ($\bar{z}_i$) has coefficient θ_{Si}, and e_i denotes the deviation of the observed mean from the prediction of the model. In matrix form, letting $\bar{\mathbf{z}}$ be the vector of observed line means, $\mathbf{a}$ be the (2×1) vector of effects μ_0 and α_1^c, and $\mathbf{M}$ be the matrix of coefficients, the linear model becomes

$$\bar{\mathbf{z}} = \mathbf{Ma} + \mathbf{e} \tag{11.7}$$

where $\mathbf{e}$ is the column vector of residual errors, i.e., the vector of deviations between observed and predicted line means. Note that all of the elements in the first column of $\mathbf{M}$ are equal to one, as they are all multipliers for μ_0, whereas the second column contains the coefficients θ_{Si} for the various lines (Table 11.1).

Because the line means may vary with respect to accuracy (reflecting, for example, different sample sizes), they should not be weighted equally in the computation of $\mathbf{a}$ (GLS, not OLS, residuals; Chapter 10). From Equation 10.13a, the GLS solution is

$$\widehat{\mathbf{a}} = (\mathbf{M}^T\mathbf{V}^{-1}\mathbf{M})^{-1}\mathbf{M}^T\mathbf{V}^{-1}\bar{\mathbf{z}} \tag{11.8}$$

where the covariance matrix $\mathbf{V}$ for the residuals is diagonal with diagonal elements equal to the squared standard errors of the means (Example 10.8). The treatment of $\mathbf{V}$ as a diagonal matrix assumes that the measured individuals from the different lines are unrelated, i.e., that there is no sampling covariance between the observed means.

From Equation 10.13b, the sampling variances and covariances of the two parameter estimates, $\widehat{\mu}_0$ and $\widehat{\alpha}_1^c$, are given by the elements of the (2×2) covariance matrix

$$\text{Var}(\widehat{\mathbf{a}}) = \mathbf{C} = (\mathbf{M}^T\mathbf{V}^{-1}\mathbf{M})^{-1} \tag{11.9}$$

The two diagonal elements of $\mathbf{C}$ are $\text{Var}(\widehat{\mu}_0)$ and $\text{Var}(\widehat{\alpha}^c)$, whereas both off-diagonal elements are equal to $\text{Cov}(\widehat{\mu}_0, \widehat{\alpha}^c)$. Sampling covariance arises between estimates of the parameters because they are jointly estimated from a common data set. Letting $\widehat{z}_i = \widehat{\mu}_0 + \theta_{Si}\widehat{\alpha}^c$ be the fitted (predicted) mean phenotype for the ith line, the sampling variance of $\widehat{z}_i$ is simply the variance of a sum,

$$\text{Var}(\widehat{z}_i) = \text{Var}(\widehat{\mu}_0) + 2\theta_{Si}\text{Cov}(\widehat{\mu}_0, \widehat{\alpha}^c) + \theta_{Si}^2 \text{Var}(\widehat{\alpha}^c) \tag{11.10a}$$

Using Equations 9.21b and 11.9, the covariance matrix for the predicted means is

$$\text{Var}(\,\widehat{\mathbf{z}}\,) = \text{Var}(\,\mathbf{M}\,\widehat{\mathbf{a}}\,) = \mathbf{MCM}^T = \mathbf{M}(\mathbf{M}^T\mathbf{V}^{-1}\mathbf{M})^{-1}\mathbf{M}^T \tag{11.10b}$$

Although the least-squares solution is completely general, significance testing requires the assumption of normality. If that assumption is met, the weighted error sum of squares

$$\chi^2 = \sum_{i=1}^{k} \frac{(\bar{z}_i - \widehat{z}_i)^2}{\text{Var}(\bar{z}_i)} \tag{11.11}$$

where k is the number of observed lines, provides a test statistic for the adequacy of the model (Appendix 3). Under the null hypothesis of purely additive gene action, this test statistic will be χ^2 distributed with degrees of freedom (df) equal to the number of lines minus the number of estimated parameters, in this case giving $(k-2)$ df.

If the test statistic is large enough to reject the additive model, the logical next step is to evaluate the additive–dominance model. In this case, the vector $\mathbf{a}$ contains a third element, δ_1^c, and the matrix $\mathbf{M}$ contains a third row consisting of the elements θ_{Hi}. The solution again follows from Equation 11.8, and the new fit is evaluated by use of Equation 11.11, where the χ^2 statistic is now distributed with $k-3$ degrees of freedom, as three parameters are fitted.

Letting χ_A^2 and χ_{AD}^2 denote the test statistics associated with the additive and additive–dominance models, then the difference

$$\Lambda = \chi_A^2 - \chi_{AD}^2 \tag{11.12}$$

is equivalent to a likelihood-ratio test statistic (see Example A4.5). Such statistics are asymptotically χ^2 distributed (for large sample sizes), with degrees of freedom equal to the difference in the number of parameters included in the two models (in this case, one). While the inclusion of dominance in the model will definitely improve the fit, Equation 11.12 provides a test for whether the improvement is significant.

If the additive–dominance model is rejected on the basis of an overly large value for χ_{AD}^2, the next step is to proceed with the analysis of models containing epistatic effects, assuming that enough line means are available for such analysis. At this point, δ_1^c may or may not be dropped from the model depending on its degree of significance in the previous analysis. The significance of the improvement of fit with models containing epistasis can again be evaluated by use of the appropriate likelihood-ratio test, i.e., by the difference in χ^2 test statistics between the modified model and the previous restricted model.

Example 11.2. We now use the joint-scaling test to study the genetic basis of human skin color. The sample consists primarily of residents of Liverpool, England (Harrison and Owen 1964). Pigmentation was measured as the reflectance of the medial aspect of the right upper arm at 545 nm. The P_1 consists of individuals of West African origin and the P_2 of individuals of European descent. The data are as follows

	P_1	P_2	F_1	F_2	B_1	B_2
$\bar{z}_i$	14.4	41.0	28.4	30.3	24.2	34.7
$\text{SE}(\bar{z}_i)$	0.611	0.453	0.581	1.483	1.334	1.122

There is a threefold range of variation in the standard errors of the mean phenotypes, and this translates into a tenfold range in the sampling variances. Clearly, weighted least-squares regression is desirable in this situation.

We start by considering the simplest genetic model, assuming that all gene action is additive within and between loci. Table 11.1 gives the coefficients for the effects μ_0 and α^c as

$$\mathbf{M} = \begin{pmatrix} 1 & 1 \\ 1 & -1 \\ 1 & 0 \\ 1 & 0 \\ 1 & 0.5 \\ 1 & -0.5 \end{pmatrix}$$

The sampling covariance matrix for the line means is diagonal with $V_{ii} = [\text{SE}(\bar{z}_i)]^2$, with

$$\mathbf{V} = \begin{pmatrix} 0.373 & 0.000 & 0.000 & 0.000 & 0.000 & 0.000 \\ 0.000 & 0.205 & 0.000 & 0.000 & 0.000 & 0.000 \\ 0.000 & 0.000 & 0.338 & 0.000 & 0.000 & 0.000 \\ 0.000 & 0.000 & 0.000 & 2.199 & 0.000 & 0.000 \\ 0.000 & 0.000 & 0.000 & 0.000 & 1.780 & 0.000 \\ 0.000 & 0.000 & 0.000 & 0.000 & 0.000 & 1.259 \end{pmatrix}$$

These lead to

$$\mathbf{M}^T\mathbf{V}^{-1}\mathbf{M} = \begin{pmatrix} 12.325 & -2.311 \\ -2.311 & 7.891 \end{pmatrix}$$

which yields the sampling covariance matrix for the parameter estimates (Equation 11.9),

$$\mathbf{C} = (\mathbf{M}^T\mathbf{V}^{-1}\mathbf{M})^{-1} = \begin{pmatrix} 0.086 & 0.025 \\ 0.025 & 0.134 \end{pmatrix}$$

and from Equation 11.8, the parameter estimates become

$$\widehat{\mathbf{a}} = \begin{pmatrix} \widehat{\mu}_0 \\ \widehat{\alpha}^c \end{pmatrix} = \begin{pmatrix} 28.17 \\ -13.07 \end{pmatrix}$$

The standard errors of the parameter estimates are equal to the square roots of the diagonal elements of $\mathbf{C}$,

$$\text{SE}(\widehat{\mu}_0) = (0.086)^{1/2} = 0.29$$
$$\text{SE}(\widehat{\alpha}^c) = (0.134)^{1/2} = 0.37$$

The line means predicted by the model are obtained as $\widehat{\mathbf{z}} = \mathbf{M}\widehat{\mathbf{a}}$, and the sampling variances and covariances of predicted values by Equation 11.10b,

$$\text{Var}(\widehat{\mathbf{z}}) = \mathbf{M}(\mathbf{M}^T\mathbf{V}^{-1}\mathbf{M})^{-1}\mathbf{M}^T$$

The square roots of the diagonal elements of this latter matrix are the estimated standard errors of the *predicted means*. In all cases, the predicted values are very close to the observed means:

	P_1	P_2	F_1	F_2	B_1	B_2
$\widehat{z}$	15.2	41.2	28.2	28.2	21.6	34.7
$\bar{z} - \widehat{z}$	−0.8	−0.2	0.2	2.1	2.6	0.0
$\text{SE}(\widehat{z})$	0.52	0.41	0.29	0.29	0.38	0.31

The test statistic (Equation 11.11), $\chi^2_A = 7.510$, with four degrees of freedom, is not significant as $\Pr(\chi^2_4 \geq 7.510) = 0.11$. Thus, the fitted model with $\widehat{\mu}_0 = 28.17$ and $\widehat{\alpha}^c = -13.07$ appears to adequately explain the data. Reevaluation of the data with the additive–dominance model confirms this conclusion. In this case, the analysis proceeds with

$$\mathbf{M} = \begin{pmatrix} 1 & 1 & -1 \\ 1 & -1 & -1 \\ 1 & 0 & 1 \\ 1 & 0 & 0 \\ 1 & 0.5 & 0 \\ 1 & -0.5 & 0 \end{pmatrix}$$

and

$$\mathbf{a} = \begin{pmatrix} \mu_0 \\ \alpha^c \\ \delta^c \end{pmatrix}$$

yielding the parameter estimates (and associated standard errors):

$$\widehat{\mu}_0 = 28.32\,(0.32), \quad \widehat{\alpha}^c = -13.14\,(0.37), \quad \widehat{\delta}^c = 0.44\,(0.34)$$

Notice that the estimates $\widehat{\mu}_0$ and $\widehat{\alpha}^c$ are very close to those obtained under the purely additive model, and that $\widehat{\delta}^c$ is only slightly greater than its standard error. The test statistic for this analysis is $\chi^2_{AD} = 5.879$. The likelihood-ratio test statistic, $\Lambda = \chi^2_A - \chi^2_{AD} = 1.631$, provides a test of the hypothesis that dominance accounts for a significant proportion of the variance among line means. With one degree of freedom, Λ is not significant, as $\Pr(\chi^2_1 \geq 1.631) = 0.20$.

Line Crosses in *Nicotiana rustica*

Few attempts have been made to estimate more than two-locus epistatic effects using line-cross analysis. However, Mather, Jinks, and their associates did an enormous amount of work with highly inbred lines of tobacco to evaluate the relative importance of various types of higher order gene action (Mather and Jinks 1982). An experiment by Jinks and Perkins (1969) is particularly noteworthy. In addition to the six fundamental line crosses, they created four second-generation backcrosses ($B_1 \times P_1$, $B_1 \times P_2$, $B_2 \times P_1$, $B_2 \times P_2$), the crosses $B_1 \times F_1$ and $B_2 \times F_1$, and the selfed backcrosses (B_{1s} and B_{2s}), and they assayed all of them simultaneously (i.e., in a common garden). With such a large number of lines, there are enough degrees of freedom to test for the significance of three-locus epistatic effects.

In the case of final plant height, a model with fitted parameters (and associated standard errors) $\widehat{\mu}_0 = 57.64 \pm 0.23$, $\widehat{\alpha}^c_1 = 5.32 \pm 0.20$, $\widehat{\delta}^c_1 = 3.55 \pm 0.38$, $\widehat{\alpha}^c_2 = 4.85 \pm 1.28$, and $\widehat{\alpha^c_2\delta^c_1} = 3.73 \pm 1.03$ provided an excellent fit to the data (Figure 11.1). Thus, for this trait, there are significant additive × additive (α^c_2) and additive × additive × dominance ($\alpha^c_2\delta^c_1$) interactions involving genes from different lines, but no significant additive × dominance ($\alpha^c_1\delta^c_1$), dominance × dominance (δ^c_2), or higher order interactions. The most parsimonious three-locus fit for the data on flowering time is obtained with $\widehat{\mu}_0 = 15.91 \pm 0.09$, $\widehat{\alpha}^c_1 = 1.36 \pm 0.15$, $\widehat{\delta}^c_1 = -2.39 \pm 0.22$, and $\widehat{\alpha^c_2\delta^c_1} = 1.81 \pm 0.37$. Here, there are significant differences between observed and predicted line means (Figure 11.1), suggesting the existence of even higher order epistatic interactions. Analyses with other parental lines (Smith 1937b; Hill 1966) led to similar results—final height was usually described adequately by a model incorporating dominance and at least one form of two-locus epistasis, while flowering time was influenced by epistatic interactions between pairs, triplets, and higher numbers of loci. The results from two additional line-cross experiments, involving only the six fundamental generations, indicate that epistasis contributes to line differences in other characters in *Nicotiana* (Table 11.4). In these additional cases, however, effects involving more than pairs of loci need

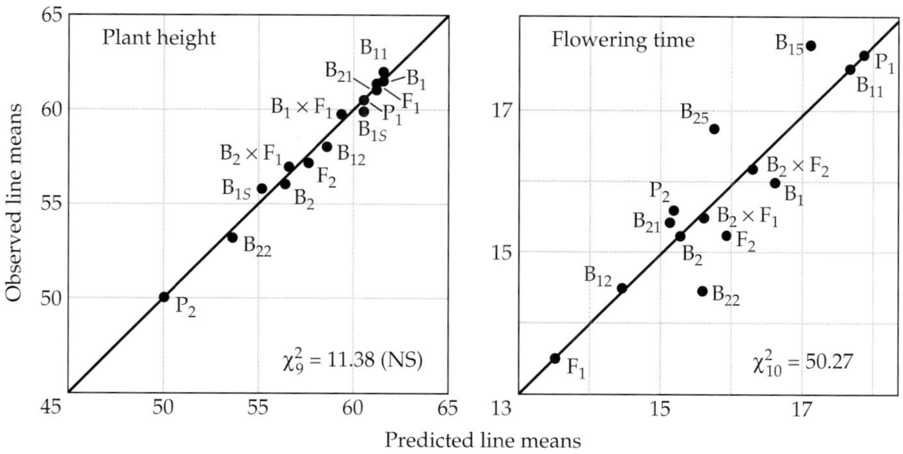

Figure 11.1 Observed versus fitted mean phenotypes for 14 line-cross derivatives from two pure parental lines of tobacco. The diagonal line gives the expected pattern if observed and predicted line means were identical. (After Jinks and Perkins 1969.)

Table 11.4 Composite effects estimated from phenotypic means of two line crosses involving inbred parental stocks of tobacco, $V_1 \times V_5$ (upper rows of data for each trait) and $V_2 \times V_{12}$ (lower rows). Only significant effects are given, and the resulting models provide an adequate fit to the data in all cases. Leaf length was not analyzed in the $V_2 \times V_{12}$ experiment. (Data from Pooni et al. 1985.)

Character	$\widehat{\mu}_0$	$\widehat{\alpha}_1^c$	$\widehat{\delta}_1^c$	$\widehat{\alpha}_2^c$	$\widehat{\alpha_1^c \delta_1^c}$	$\widehat{\delta}_2^c$
Height – 2 wk	5.50	1.19	1.10	1.27		
	4.34	1.97	0.48	0.55		
– 4 wk	15.74	3.53	3.32	3.97		
	15.22	9.96	2.58	3.48		
– 6 wk	49.84	5.02	−2.48	9.95		
	49.14	30.22	11.14	9.21	4.12	
– flowering	73.75	7.99		−5.02		
	83.40	5.84	3.12	−6.21		
– final	126.80	12.57	3.72	−7.58		
	143.44	8.79	21.91			−3.33
Leaf length	19.10	0.35		−1.89		

not be invoked to explain the data, because in no case are the observed means significantly different from the final model predictions.

Additional Data

A summary of results from some other line-cross studies is given in Table 11.5. In most cases, the parental lines are conspecific isolates known at the outset to differ (often substantially) in mean phenotypes. The main message is consistent with the *Nicotiana* results—differentiation of divergent lines almost always involves epistatic effects. We will see in Chapter 14 that epistatic interactions can sometimes be removed by a suitable scale transformation, but it is unlikely that this would be successful for all of the tabulated studies.

VARIANCE OF LINE-CROSS DERIVATIVES

In principle, the joint-scaling test can be used to interpret the variances as well as the means

Table 11.5 A survey of the composite effects estimated in line-cross analyses. The results reported for each analysis describe the most parsimonious genetic model, and all recorded effects are statistically significant. The motivation for the use of logarithmic transformations for some of these traits is discussed in Chapter 14.

Character	$\widehat{\mu}_0$	$\widehat{\alpha}_1^c$	$\widehat{\delta}_1^c$	$\widehat{\alpha}_2^c$	$\widehat{\alpha_1^c\delta_1^c}$	$\widehat{\delta}_2^c$	Reference
Corn							
time to silking	65.19	0.20	−6.16	−4.40	7.33	5.92	Mohamed 1959
time to shed pollen	62.88	−1.57	−4.36	−1.94	4.74	3.69	
Lima beans							
seed size	0.57	−0.25	−0.16	0.04	0.30	0.19	Ryder 1958
Tomatoes							
$\log_{10}$(fruit weight)	0.69	−0.87	0.02	0.12	−0.03		Powers 1951
Pitcher-plant mosquito							
$\log_{10}$(critical photoperiod)	−1.84	−4.45	0.40		1.79	4.18	Hard et al. 1992
Drosophila melanogaster							
ln(longevity)	1.79	−0.06	0.01	0.03			Luckinbill et al. 1988
Drosophila tripunctata							
ovipos. site pref.	0.31	−0.12	0.05	0.18			Jaenike 1987
D. heteroneura × *D. sylvestris*							
ln(head length)	3.00	0.09					Templeton 1977
ln(head width)	3.96	−0.01					
Astyanax (cave fish)							
eye diameter	4.72	−2.43	1.62	1.10	−1.57	−1.24	Wilkens 1971
Mice							
$\log_{10}$(body weight)	1.39	0.18	−0.01	−0.07	−0.01	0.01	Chai 1956
Chickens							
weight	3.49	−0.12	−0.06	−0.08	0.08	0.05	Waters 1931

obtained from line crosses. One of the most important applications of such an analysis is in the interpretation of the outbreaks of variation often seen in an F_2 generation (Figure 1.3). To keep things simple, let us assume that gene action is additive and that the environmental variance (here denoted as σ_E^2) is independent of the genetic background. The expected phenotypic variances for the parental lines and their F_1 offspring are then

$$\sigma^2(P_1) = \sigma_E^2 + \sigma_{A_1}^2 \tag{11.13a}$$

$$\sigma^2(P_2) = \sigma_E^2 + \sigma_{A_2}^2 \tag{11.13b}$$

$$\sigma^2(F_1) = \sigma_E^2 + \frac{1}{2}\sigma_{A_1}^2 + \frac{1}{2}\sigma_{A_2}^2 \tag{11.13c}$$

where $\sigma_{A_1}^2$ and $\sigma_{A_2}^2$ are the additive genetic variances in the P_1 and P_2 lines. The genotypic variance in the F_1 generation follows from the fact that, for each locus, F_1 individuals contain exactly one P_1 allele and one P_2 allele, and that the two haploid sets of alleles contribute variance $\sigma_{A_1}^2/2$ and $\sigma_{A_2}^2/2$.

An additional source of genetic variance will appear in any line-cross derivative for which there is variation among individuals in the proportion of P_1 and P_2 genes. For example, in the F_2 generation there is a 50% probability of being P_1P_2 and 25% probabilities of being P_1P_1 or P_2P_2 at any locus. This variation is in contrast to that in the F_1 generation where *all* individuals are P_1P_2. Letting $-\alpha_i^c/2$ and $+\alpha_i^c/2$ be the mean additive effects of alleles at the ith locus in the P_1 and P_2 lines, then the variance among F_2 individuals attributable to differences between parental lines at this locus is

$$\sigma_s^2(i) = 0.25(-\alpha_i^c)^2 + 0.5\left(\frac{\alpha_i^c}{2} - \frac{\alpha_i^c}{2}\right)^2 + 0.25(\alpha_i^c)^2 = \frac{(\alpha_i^c)^2}{2} \tag{11.14}$$

Summing over all loci, we obtain the **segregational variance,**

$$\sigma_s^2 = \frac{1}{2}\sum_{i=1}^{n}(\alpha_i^c)^2 \tag{11.15}$$

which describes the excess variance that appears in the F_2 generation as a consequence of the segregation of parental-line genes. The expected variance in the F_2 generation is then

$$\sigma^2(F_2) = \sigma_E^2 + \frac{1}{2}\sigma_{A_1}^2 + \frac{1}{2}\sigma_{A_2}^2 + \sigma_s^2 \tag{11.16}$$

To obtain a general expression for the variances within backcross and more advanced generations, we take the following approach, again assuming additive gene action. Recall that if a line has associated parameters H and S, then the proportions of P_1-purebred, P_2-purebred, and crossbred genotypes at any locus are $S - (H/2)$, $1 - S - (H/2)$, and H, respectively. For the ith locus, these classes have genotypic means and variances equal to $(\alpha_i^c,\, \sigma_{A_1i}^2)$, $(-\alpha_i^c,\, \sigma_{A_2i}^2)$, and $(0,\, [\sigma_{A_1i}^2 + \sigma_{A_2i}^2]/2)$, respectively. For any line, the genetic variance associated with that locus can be expressed as

$$\sigma_A^2(i) = E(A_i^2) - \mu_{A_i}^2$$

where A_i denotes the breeding (or additive genetic) value (Chapter 4) of an individual at the locus, and from arguments given above, $\mu_{A_i} = \theta_S\alpha_i^c = (1-2S)\alpha_i^c$. An expression for $E(A_i^2)$ is obtained by averaging over the three possible classes of genotypes,

$$\begin{aligned} E(A_i^2) &= \left(S - \frac{H}{2}\right)[\sigma_{A_1i}^2 + (\alpha_i^c)^2] + H\left(\frac{\sigma_{A_1i}^2 + \sigma_{A_2i}^2}{2}\right) \\ &\qquad + \left(1 - S - \frac{H}{2}\right)[\sigma_{A_2i}^2 + (-\alpha_i^c)^2] \\ &= S\sigma_{A_1i}^2 + (1-S)\sigma_{A_2i}^2 + (1-H)(\alpha_i^c)^2 \end{aligned}$$

Putting these results together, for any derivative line with indices S and H, the variance associated with the ith locus is

$$\sigma_A^2(i) = S\sigma_{A_1i}^2 + (1-S)\sigma_{A_2i}^2 + [4S(1-S) - H](\alpha_i^c)^2 \tag{11.17}$$

Adding the environmental variance, summing over all loci, and recalling Equation 11.15, the expected phenotypic variance for a line with properties S and H is

$$\sigma^2 = \sigma_E^2 + S\sigma_{A_1}^2 + (1-S)\sigma_{A_2}^2 + 2[4S(1-S) - H]\sigma_s^2 \tag{11.18}$$

For example, for the B_1 backcross ($H = 1/2$ and $S = 3/4$) and the B_2 backcross ($H = 1/2$ and $S = 1/4$),

$$\sigma^2(B_1) = \sigma_E^2 + \frac{3}{4}\sigma_{A_1}^2 + \frac{1}{4}\sigma_{A_2}^2 + \frac{1}{2}\sigma_s^2 \tag{11.19a}$$

$$\sigma^2(B_2) = \sigma_E^2 + \frac{1}{4}\sigma_{A_1}^2 + \frac{3}{4}\sigma_{A_2}^2 + \frac{1}{2}\sigma_s^2 \tag{11.19b}$$

We now have all of the statistical machinery necessary to develop a predictive model for line *variances*. Analogous to the model developed above for line *means*, we let $\mathbf{v}$ be the vector of observed phenotypic *variances* for the various lines, $\mathbf{a}$ be the vector of variance components (σ^2_E, σ^2_{A1}, σ^2_{A2}, and σ^2_s), and $\mathbf{M}$ be the matrix of coefficients for the lines, with all elements in the first column being equal to one, and values of S, $(1-S)$, and $2[4S(1-S)-H]$ being entered in the remaining three columns. The linear model for line variances can then be summarized as

$$\mathbf{v} = \mathbf{Ma} + \mathbf{e} \tag{11.20}$$

where $\mathbf{e}$ is the vector of residual errors. The weighted least-squares estimates for the model parameters are given by

$$\widehat{\mathbf{a}} = (\mathbf{M}^T\mathbf{V}^{-1}\mathbf{M})^{-1}\mathbf{M}^T\mathbf{V}^{-1}\mathbf{v} \tag{11.21}$$

where $\mathbf{V}$ is the sampling covariance matrix for the line variances. Provided the individuals of the various lines are unrelated, then all of the elements of $\mathbf{V}$ are zero except those on the diagonal. These are set equal to $2v_j^2/(n_j+2)$, the unbiased estimator of the sampling variance of a variance under the assumption of normality (Equation A1.10c), where n_j is the sample size, and v_j is the observed phenotypic variance of the jth line.

There is one small remaining problem. Unless one is willing to assume that the parental lines are completely homozygous, i.e. $\sigma^2_{A1} = \sigma^2_{A2} = 0$, Equation 11.21 cannot be solved directly. The problem is that $\mathbf{M}$ is singular, because for any line, the coefficient for σ^2_E is equal to the sum of the coefficients for σ^2_{A1} and σ^2_{A2}. This difficulty can be circumvented by deleting the first column from $\mathbf{M}$ and reducing the variance component vector to $\mathbf{a}^T = [\sigma^2(\mathrm{P}_1), \sigma^2(\mathrm{P}_2), \sigma^2_s]$, where $\sigma^2(\mathrm{P}_1) = \sigma^2_E + \sigma^2_{A1}$ and $\sigma^2(\mathrm{P}_2) = \sigma^2_E + \sigma^2_{A2}$ are the phenotypic variances for the two parental lines.

Hayman (1960b) took this analysis a step further in producing a maximum-likelihood procedure (Appendix 4). The diagonal elements of the matrix $\mathbf{V}$ have expectations equal to $2\sigma_j^4/(n_j+2)$, where σ_j^2 represents the expectation of the appropriate entry in $\mathbf{v}$. Under the assumption that the additive model is correct, the projected least-squares values, $\widehat{\mathbf{v}} = \mathbf{M}\widehat{\mathbf{a}}$, should actually be better estimates of the within-line variances than the original elements of $\mathbf{v}$. This implies that the elements of $\widehat{\mathbf{a}}$ should be computed a second time by use of Equation 11.21 after substituting the elements of $\widehat{\mathbf{v}}$ into $\mathbf{V}$. This procedure is then iterated until the estimates of $\widehat{\mathbf{a}}$ stabilize. (Note that during the iterative process, it is only the elements of the covariance matrix $\mathbf{V}$, and not those of the vector of observed variances $\mathbf{v}$, that are modified recursively.)

Because $\mathbf{V}$ is diagonal with $V_{ii} = 2\,\sigma_j^4/(n_j+2)$, Equation A3.11b gives the χ^2 statistic for goodness of fit of the observed variances to the predictions of the additive model as

$$\chi^2 = \sum_{j=1}^{k} \frac{(v_j - \widehat{v}_j)^2}{2\,\widehat{v}_j^{\,2}/(n_j+2)} \tag{11.22}$$

where the $\widehat{v}_j$ are the final estimates of the σ^2_j, and the degrees of freedom associated with the test statistic equal the number of lines (k) minus three (the number of variance components estimated).

Example 11.3. We now apply the joint-scaling test for variances to Harrison and Owen's (1964) data on human skin color. Recall from Example 11.2 that the analysis of means supports the idea that this character has an additive genetic basis. On the other hand, the phenotypic variances, recorded in the following table, appear to be rather inconsistent with the additive model. For example, the F_1 variance is much higher than the average of the P_1 and P_2 variances, and even exceeds that of the F_2. However, because the sampling variance of a variance is quite large (as can be seen in the third row of the table), there is some question as to the significance of these differences.

	P_1	P_2	F_1	F_2	B_1	B_2
v_j	14.918	21.098	31.748	26.382	37.366	37.766
n_j	40	103	94	12	21	30
$2v_j^2/(n_j+2)$	10.597	8.479	20.999	99.430	121.410	89.142

The coefficients for the variance components in the model, $\sigma^2(P_1), \sigma^2(P_2)$ and σ_s^2, are obtained from Equation 11.18, with

$$\mathbf{M} = \begin{pmatrix} 1 & 0 & 0 \\ 0 & 1 & 0 \\ 0.5 & 0.5 & 0 \\ 0.5 & 0.5 & 1 \\ 0.75 & 0.25 & 0.5 \\ 0.25 & 0.75 & 0.5 \end{pmatrix}$$

Inserting the values from the table above, the initial sampling covariance matrix is

$$\mathbf{V} = \begin{pmatrix} 10.597 & 0.000 & 0.000 & 0.000 & 0.000 & 0.000 \\ 0.000 & 8.479 & 0.000 & 0.000 & 0.000 & 0.000 \\ 0.000 & 0.000 & 20.999 & 0.000 & 0.000 & 0.000 \\ 0.000 & 0.000 & 0.000 & 99.430 & 0.000 & 0.000 \\ 0.000 & 0.000 & 0.000 & 0.000 & 121.410 & 0.000 \\ 0.000 & 0.000 & 0.000 & 0.000 & 0.000 & 89.142 \end{pmatrix}$$

Substituting into Equation 11.21, we obtain the initial set of least-squares parameter estimates,

$$\widehat{\mathbf{a}} = (\mathbf{M}^T\mathbf{V}^{-1}\mathbf{M})^{-1}\mathbf{M}^T\mathbf{V}^{-1}\mathbf{v} = \begin{pmatrix} 18.120 \\ 23.681 \\ 14.441 \end{pmatrix}$$

The following table shows how the parameter estimates change over the next several rounds of iterations,

	Iteration						
Estimates	2	3	5	10	15	20	Final SE
$\mathrm{Var}(P_1)$	22.199	22.930	23.446	23.583	23.586	23.586	4.207
$\mathrm{Var}(P_2)$	25.994	25.505	25.185	25.103	25.102	25.102	3.166
$\mathrm{Var}(S)$	17.163	17.038	16.876	16.830	16.829	16.829	10.411
χ^2	14.135	10.432	10.175	10.102	10.101	10.101	

The standard errors are the square roots of the diagonal elements of the final estimate of $(\mathbf{M}^T\mathbf{V}^{-1}\mathbf{M})^{-1}$. Using the final set of parameter estimates, the predicted line variances $(\widehat{v}_j)$, obtained from Equation 11.18, and their standard errors, obtained as the square roots of the diagonal elements of $\mathbf{M}(\mathbf{M}^T\mathbf{V}^{-1}\mathbf{M})^{-1}\mathbf{M}^T$, are

	P_1	P_2	F_1	F_2	B_1	B_2
$\widehat{v}_j$	23.586	25.102	24.344	41.172	32.379	33.137
SE	4.207	3.166	2.306	9.845	5.042	5.187

The final χ^2 value (10.101) is rather large ($\Pr[\chi_3^2 \geq 10.1] = 0.018$), but because of possible non-normality of the data, its statistical interpretation is somewhat questionable. Because the difference between observed and expected line variances is less than two standard errors $(2[\mathrm{SE}(v_j)^2 + \mathrm{SE}(\widehat{v}_j)^2]^{1/2})$ for all lines, there seems to be no strong justification for rejecting the additive model.

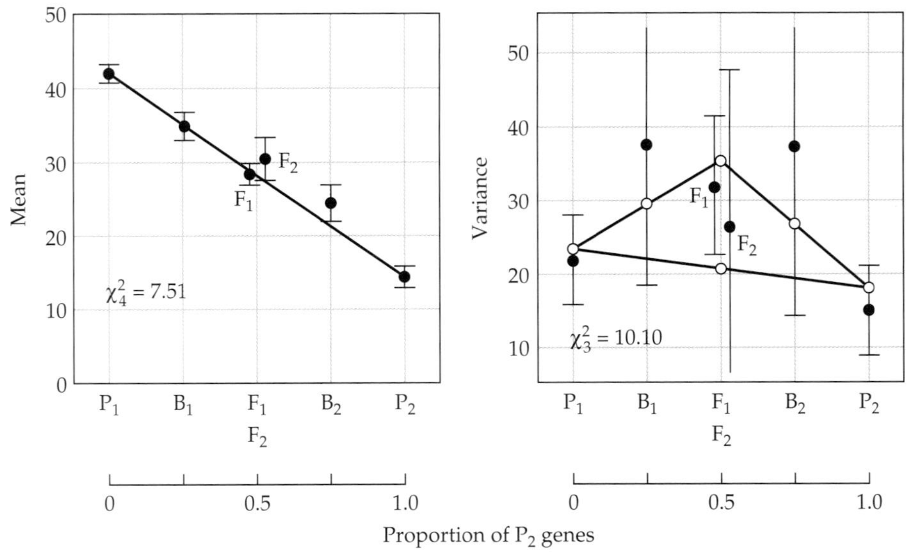

Figure 11.2 Observed means and variances ($\pm$ twice the standard errors) of human skin color in relation to the predictions of the additive model (data from Examples 11.2 and 11.3). The maximum-likelihood predictions are given by the line on the left graph and the open circles on the triangle on the right graph. (From Harrison and Owens 1964.)

A graphical comparison of the observed and predicted means and variances of human skin color (Figure 11.2) serves to illustrate two important points. First, due to the large standard errors of variance estimates, ***scaling tests based on variances are much less powerful than those based on means***. Thus, while the preceding methodology can be generalized to compute the dominance and epistatic components of the segregational variance (Mather and Jinks 1982), the statistical reliability of such analyses is very low, and we will not pursue it further.

Second, the additive model leads to some very simple geometric relationships (Figure 11.2). The expected line means are linear functions of the proportion of genes derived from each parental line. The expected line variances fall on a triangle, the vertices of which represent the expected P_1, P_2, and F_2 variances. The expected F_1 variance lies on the midpoint of the line leading from P_1 to P_2, while the expected B_1 and B_2 variances lie on the midpoints of the lines from P_1 to F_2 and P_2 to F_2, respectively.

BIOMETRICAL APPROACHES TO THE ESTIMATION OF GENE NUMBER

We now turn to a second application of line-cross analysis: estimation of the number of segregating loci responsible for quantitative variation. The subject is of importance for several reasons. First, since the beginning of the last century (Chapter 1), there has been considerable debate as to whether most large evolutionary changes are due to a small number of macromutations or to gradual substitution of minor allelic variants at a large number of loci (Gould 1980; Charlesworth et al. 1982; Gottlieb 1984; Coyne and Lande 1985; Orr and Coyne 1992; Wu and Palopoli 1994; WL Chapter 25). Second, from a more statistical point of view, the desirable properties of normal theory that facilitate quantitative-genetic analysis become violated to a greater degree as the number of segregating loci becomes smaller. The ideal setting for much of quantitative-genetic theory is a very large number of loci, all with small effects, the so-called **infinitesimal model** (Chapter 21; WL Chapter 24). Third, for situations in which most of the genetic variation is a function of one or two genes with major effects, a fine-scale Mendelian analysis (through the direct observation of segregation ratios) will often be possible, provided the environmental component of variance is not of overwhelming importance (Chapter 16).

There are two general approaches to estimating gene number. The **biometrical approach**, which is the subject of this chapter, is based on statistical properties (means and variances) of phenotype distributions, and uses these properties to indirectly infer the number of segregating factors that are likely to be responsible for them. The second approach involves the search for associations between segregating molecular markers and quantitative-genetic variation (Chapters 17–21). While these more direct molecular approaches have largely supplanted biometrical approaches, the latter are both of historical importance and still have utility. For example, Nandamuri et al. (2017) used this approach to estimate the number of genes involved in the regulation of opsin gene expression in cichlid fish.

Although we are ultimately interested in the total number of loci (n) contributing to the variance in trait expression, most estimates of n are actually measures of the **effective number of factors** (n_e) by which the characters in two lines differ. This quantity is equivalent to the number of freely segregating loci with equal effects that would yield the observed pattern of line means and variances. It is important to realize that estimates of n_e do not include the potentially large fraction of loci that do not vary between lines, yet could lead to phenotypic differences given the right mutations. In addition, n_e cannot exceed the number of independently segregating chromosomal segments, which can be approximated by Darlington's (1939) **segregation index**, the number of chromosomes plus the mean number of recombination events per gamete (Chapter 8). In eukaryotes, there are usually one to two recombination events per chromosome. Thus, the maximum possible value of n_e is usually two to three times the haploid chromosome number, although each segregating unit can contain many loci.

In the following section, we will first describe an estimator, $\widehat{n}_e$, for the effective number of factors. We will then describe a more refined estimator, $\widehat{n}$, which provides estimates that are closer to the actual number of loci. Procedures for estimating n_e involve a number of assumptions, the most important of which is additivity of gene action. Thus, prior to analysis, a serious attempt should be made to find a scale on which the observed line means and variances are consistent with the additive model. The joint-scaling tests described on the previous pages provide an approach for testing the effectiveness of various scale transformations (see also Chapter 14).

The Castle–Wright Estimator

The most widely used biometrical method for estimating n_e utilizes information on the phenotypic means and variances of two parental lines and their line-cross derivatives (F_1, F_2, B_1, B_2, etc.). As first developed by Castle (1921) with his graduate student Sewall Wright (1968), the method was intended for use with inbred parental lines. Lande (1981) generalized it for use with genetically variable base populations, and the theory developed below is based on his modifications, and those of Cockerham (1986). In addition to additive gene action, the Castle–Wright estimator assumes unlinked loci and equality of allelic effects, although we relax the latter assumption in the following derivation. It also assumes that all genes with a positive influence on the trait are sorted into one line and all those with negative influences into the other. Hence, such estimates are essentially of the ***difference in the net number*** of plus alleles that the lines differ by, and hence are ***underestimates of the true number of segregating differences***.

Letting α_i^c be the composite additive effect for the *i*th locus, the mean phenotype of the P_1 line can be written as $\mu(P_1) = \mu_0 + \sum_{i=1}^{n} \alpha_i^c = \mu_0 + n\,\overline{\alpha_i^c}$, where $\overline{\alpha_i^c}$ is the average composite additive effect of a locus. That for the P_2 line is $\mu(P_2) = \mu_0 - n\,\overline{\alpha_i^c}$. Thus, the expected difference in mean phenotypes is $\mu(P_1) - \mu(P_2) = 2n\,\overline{\alpha_i^c}$. The segregational variance, defined in Equation 11.15, can also be written as

$$\sigma_s^2 = n[\sigma^2(\alpha_i^c) + (\overline{\alpha_i^c})^2]/2 \qquad (11.23a)$$

where $\sigma^2(\alpha_i^c)$ is the variance of composite effects among loci. The trick to deriving an estimator of n_e is to note that the expected squared difference between parental line means is

$$[\mu(\mathrm{P}_1) - \mu(\mathrm{P}_2)]^2 = E\left[\left(2\sum_{i=1}^{n}\alpha_i^c\right)^2\right] = 4n\left[\overline{(\alpha_i^c)^2} + (n-1)(\overline{\alpha_i^c})^2\right]$$
$$= 4n\left[n(\overline{\alpha_i^c})^2 + \sigma^2(\alpha_i^c)\right] \tag{11.23b}$$

Taking the ratio of these two expressions yields

$$\frac{[\mu(\mathrm{P}_1) - \mu(\mathrm{P}_2)]^2}{\sigma_s^2} = \frac{8\left[n(\overline{\alpha_i^c})^2 + \sigma^2(\alpha_i^c)\right]}{(\overline{\alpha_i^c})^2 + \sigma^2(\alpha_i^c)}$$

which upon rearrangement yields

$$n_e = \frac{[\mu(\mathrm{P}_1) - \mu(\mathrm{P}_2)]^2(1 + C_\alpha)}{8\sigma_s^2} - C_\alpha \tag{11.23c}$$

where $C_\alpha = \sigma^2(\alpha_i^c)/(\overline{\alpha_i^c})^2$ is the squared coefficient of variation of the locus-specific additive effects. In the very unlikely event that all of the assumptions of the model hold, then Equation 11.23c would define the actual number of loci. In recognition of the fact that one or more assumptions are likely to be violated, Equation 11.23c is denoted as a measure of the effective number of factors, n_e. In effect, Equation 11.23c states that the line means and segregational variance are distributed in the same way that would occur if the two populations were differentiated at n_e freely recombining loci with equal and additive effects. Further, it assumes all plus alleles are in one line, so would be an estimate of the *net difference* in plus alleles in the more realistic setting that both the high and low lines contain plus alleles.

In general, C_α is an unobservable quantity, but it must be positive. Ignoring this term for the time being, and substituting observed quantities for expectations, we obtain a biased estimator for the effective number of factors,

$$\widehat{n}_e = \frac{[\bar{z}(\mathrm{P}_1) - \bar{z}(\mathrm{P}_2)]^2 - \mathrm{Var}[\bar{z}(\mathrm{P}_1)] - \mathrm{Var}[\bar{z}(\mathrm{P}_2)]}{8\mathrm{Var}(s)} \tag{11.24a}$$

hereafter referred to as the **Castle–Wright estimator,** where the $\bar{z}(\mathrm{P}_i)$ and $\mathrm{Var}[\bar{z}(\mathrm{P}_i)]$ are, respectively, the observed means and sampling variances of the means for the ith parental line. Estimates of n_e in the literature often ignore the two Var terms in the numerator, but these are required to correct for the sampling error of the estimates of the line means (Cockerham 1986).

When data are available for the backcross generations, the segregational variance estimate, $\mathrm{Var}(s)$, can be obtained by the weighted least-squares procedure described above (Example 11.3). In the absence of backcrosses, it can be computed as a linear function of the observed phenotypic variances within lines, either as

$$\widehat{\mathrm{Var}}(s) = \begin{cases} \mathrm{Var}(\mathrm{F}_2) - \mathrm{Var}(\mathrm{F}_1) & (11.24\text{b}) \\ \text{or} & \\ \mathrm{Var}(\mathrm{F}_2) - \dfrac{2\mathrm{Var}(\mathrm{F}_1) + \mathrm{Var}(\mathrm{P}_1) + \mathrm{Var}(\mathrm{P}_2)}{4} & (11.24\text{c}) \end{cases}$$

The importance of using least-squares estimates whenever possible is seen from Example 11.3, where $\mathrm{Var}(\mathrm{F}_2) - \mathrm{Var}(\mathrm{F}_1)$ is negative, but the iterated least-squares estimate of σ_s^2 is 17.

The large-sample variance of n_e, obtained from the equation for the variance of a ratio under the assumption of normality (Equation A1.19b), is approximately

$$\mathrm{Var}(\widehat{n}_e) \simeq \widehat{n}_e^2\left[\frac{4\{\mathrm{Var}[\bar{z}(\mathrm{P}_1)] + \mathrm{Var}[\bar{z}(\mathrm{P}_2)]\}}{[\bar{z}(\mathrm{P}_1) - \bar{z}(\mathrm{P}_2)]^2} + \frac{\mathrm{Var}[\mathrm{Var}(s)]}{[\mathrm{Var}(s)]^2}\right] \tag{11.25a}$$

If Var(s) is estimated by least-squares analysis, its sampling variance, Var[Var(s)], is obtained directly from the matrix $(\mathbf{M}^T\mathbf{V}^{-1}\mathbf{M})^{-1}$, as described above. Otherwise, it is estimated by the sum of the variances of the variances used to compute Var(s), each weighted by the square of the appropriate coefficient. For example, if Var(s) is estimated by Var$(F_2)-$Var(F_1), then

$$\text{Var}[\text{Var}(s)] = \text{Var}\left[\text{Var}(F_2) - \text{Var}(F_1)\right] = \frac{2[\text{Var}(F_2)]^2}{n_{F_2}+2} + \frac{2[\text{Var}(F_1)]^2}{n_{F_1}+2} \tag{11.25b}$$

where n_{F_2} and n_{F_1} are the sample sizes. This follows because Var(F_1) and Var(F_2) are independent estimates, and because under the assumption of normality, the large-sample variance of a variance is $2\text{Var}^2/(n+2)$ (Equation A1.10c). The variance of the estimator of Var(s) given by Equation 11.24c follows using similar logic.

We noted above that failure to account for the variation in composite effects among segregating loci, i.e., ignoring C_α, will tend to depress $\widehat{n}_e$ below the true number of segregating loci. Additional issues will usually result in a further downward bias. For example, if the genes with positive effects are distributed among both parental lines, the difference between the parental line means will be less than the maximum value because the positive effects of genes at some loci will be canceled by negative effects at others. Consider the case of one line being fixed for a $+$ allele at locus 1 and a $-$ allele at locus 2 and another line being fixed for a $-$ allele at locus 1 and a $+$ allele at locus 2. The number of segregating loci in the F_2 generation is two, but because both parental lines have mean phenotypes equal to zero, the expected value of n_e yielded by Equation 11.24a equals zero. Such problems become immediately apparent when F_2 individuals exhibit phenotypes *outside of the range of variation* in both parental lines, a phenomenon known as **transgressive segregation** (Chapter 18), but the absence of such individuals does not rule out the possibility of transgression. In order to minimize such interpretative difficulties, most investigators utilize parental lines with the maximum range of variation between mean phenotypes (e.g., a cross between a low and a high line for the focal trait). This is often accomplished by artificially selecting lines in the upward and downward direction for several generations prior to crossing.

By inflating the estimated segregational variance, linkage will also cause $\widehat{n}_e$ to be downwardly biased. Letting c_{ij} be the recombination fraction between loci i and j, a more general formula for the segregational variance in the F_2 generation is

$$\sigma_s^2 = \frac{1}{2}\left[\sum_{i=1}^{n}(\alpha_i^c)^2 + \sum_{i=1}^{n}\sum_{j\neq i}^{n}\alpha_i^c\alpha_j^c(1-2c_{ij})\right] \tag{11.26a}$$

where the term on the right is the disequilibrium covariance (generated by linkage creating an excess of parental gametes; Chapter 5). Assuming that the effects of pairs of alleles are uncorrelated with their map distances, then Equation 11.26a simplifies to

$$\sigma_s^2 = \frac{n}{2}\left[\sigma^2(\alpha_i^c) + (\,\overline{\alpha_i^c}\,)^2 + (n-1)(1-2\bar{c})(\,\overline{\alpha_i^c}\,)^2\right] \tag{11.26b}$$

In principle, the disequilibrium contribution to σ_s^2 can be removed by taking the F_2 generation through several additional generations of random mating, because this reduces the disequilibrium covariance by the factor $(1-c)$ each generation (Chapters 5, 18, and 20). A modified estimate of the segregational variance can then be obtained as the difference between the variance in the advanced generation and that in the F_1 line.

An alternative way to deal with the problem of linkage is to use the more general expression for σ_s^2 given by Equation 11.26b, combined with our previous expression for the expected squared difference between line means, to define the relationship between the effective number of factors and the actual number of loci. Substituting Equation 11.26b into 11.23c and rearranging leads to the expression

$$n = \frac{2\,\bar{c}\,n_e + C_\alpha(n_e-1)}{1-n_e(1-2\bar{c})} \tag{11.27}$$

which reduces to Equation 11.23c for the special case in which $\bar{c} = 0.5$. Zeng (1992) suggested that by substituting the estimate $\widehat{n}_e$, obtained by use of Equation 11.24a, into this expression, nearly unbiased estimates of the *actual* number of loci (n) are achievable (under the strong assumption of no transgressive segregation).

In order to take advantage of Zeng's suggestion, we require, at the very least, estimates of $\bar{c}$ and C_α. We have already provided a number of estimates of $\bar{c}$ in Table 11.2, showing how these can be obtained from genetic maps of chromosomes under the assumption of randomly distributed loci. For many species, such detailed information is not available. However, provided the haploid chromosome number M is known, then a downwardly biased estimate of $\bar{c}$ is given by

$$\bar{c} = \frac{M-1}{2M} \tag{11.28}$$

which assumes that recombination only occurs between pairs of genes on different chromosomes (by independent assortment), and that all chromosomes contain equal numbers of genes. Fortunately, estimates of $\bar{c}$ using this approximation are not greatly different than the more refined estimates obtained by use of Equation 11.3. For example, for humans ($M = 23$), maize ($M = 10$), and *Arabidopsis* ($M = 5$), Equation 11.28 yields estimates of $\bar{c} = 0.478$, 0.450, and 0.400, respectively, which contrast with the more exact computations, 0.490, 0.474, and 0.443. Thus, provided the minimum amount of information exists on the cytology of the organism, fairly reasonable estimates of $\bar{c}$ are achievable.

The squared coefficient of variation of effects, C_α, is much more elusive (Chapters 18–20). As outlined in Chapters 18 and 20, estimates of QTL-effect sizes are typically biased upward, especially when detected under low-power settings. Chapter 21 reviews progress on estimating the distribution of effect sizes under GWAS, but these can also be biased by lower power to detect rare alleles of modest effects. Hence, we will use estimates of C_α derived from Keightley's (1994) analysis of data from mutation-accumulation experiments performed on lines of *Drosophila melanogaster* (Chapter 15). For abdominal bristle number, sternopleural bristle number, and viability, C_α is on the order of 6, 24, and 17, respectively. Unfortunately, aside from the fact that these estimates have very large sampling variances, it is unclear how similar the spectrum of effects of spontaneous mutations is to that of the effects of alleles normally segregating in natural populations. To the extent that they are representative, such high values of C_α suggest a very positively skewed (L-shaped) distribution of allelic effects, with a very high density of small effects and a long tail to the right. For comparison, with a half-normal distribution (truncated at the mean) $C_\alpha = 0.57$, with a negative exponential distribution $C_\alpha = 1.0$, and more generally for a gamma distribution with shape and scale parameters α and β (Equations A7.24d and A7.24e)

$$C_\alpha = \frac{\beta}{\alpha(\alpha + \beta + 1)}$$

The modifications suggested above are not trivial, as the magnitude of bias that variation in allelic effects and linkage causes with the Castle–Wright estimator can be quite large. Consider, for example, the situation in which $C_\alpha = 15$, the average of the results reported above, and suppose that the Castle–Wright estimator yields $\widehat{n}_e = 4$. Substituting into Equation 11.27, such an estimate in humans ($\bar{c} = 0.49$) would be compatible with the presence of 53 actual loci, while for *C. elegans* ($\bar{c} = 0.42$), it would imply the presence of 134 loci. Assuming constant allelic effects ($C_\alpha = 0$), the estimate for humans would be essentially unbiased, as $\widehat{n}$ still equals four, but for *C. elegans*, $\widehat{n} = 9$.

By inflating the segregational variance in the F_2, nonadditive gene action is still another factor that has a downward influence on estimates derived by the Castle–Wright estimator. However, provided dominance is the primary source of nonadditivity, then the use of $2\text{Var}(F_2) - \text{Var}(B_1) - \text{Var}(B_2)$ as the estimate of the segregational variance in Equation 11.24a can eliminate most of the problem (Wright 1968; Ollivier and Janss 1993). Because the expectation of this quantity is identical to σ_s^2 defined in Equation 11.26b, Equation 11.27 applies as well.

In summary, we find that violations of the various assumptions of the Castle–Wright model usually conspire to ensure that $\widehat{n}_e$ is an *underestimate* of the actual number of loci contributing to the divergence of lines. Although the bias can be very substantial (Zeng et al. 1990), much of it can be eliminated by making the modifications suggested above, i.e., by first computing $\widehat{n}_e$ by use of Equation 11.24a, then substituting this estimate and estimates of C_α and $\bar{c}$ into Equation 11.27, and solving. An approximate expression for the sampling variance of the improved estimate of the actual number of loci is given by

$$\text{Var}(\widehat{n}) = \frac{4\bar{c}^2(1+C_\alpha)^2\text{Var}(\widehat{n}_e)}{[1-\widehat{n}_e(1-2\bar{c})]^4} \tag{11.29}$$

(Zeng 1992), where $\text{Var}(\widehat{n}_e)$ is defined in Equation 11.25a. Simulations by Zeng (1992) suggest that $\widehat{n}$ provides much more reasonable estimates of the number of loci (n) than does $\widehat{n}_e$. However, the sampling variance of $\widehat{n}$ can be quite large. Even negative estimates are possible, when by chance the estimate of σ_s^2 is negative. (Negativity can occur with the Castle–Wright estimator as well.) Thus, any attempt to estimate n by either approach should be based on large sample sizes (ideally, with hundreds of individuals measured in each line).

Example 11.4. In previous examples involving human skin color, we found that $\bar{z}(\text{P}_1) = 14.4$ and $\bar{z}(\text{P}_2) = 41.0$. Squaring the standard errors of the means yields $\text{Var}[\bar{z}(\text{P}_1)] = 0.373$ and $\text{Var}[\bar{z}(\text{P}_2)] = 0.205$. The estimated segregational variance (Example 11.3) is $\text{Var}(s) = 17.264$, and its sampling variance is obtained by squaring its standard error, $\text{Var}[\text{Var}(s)] = 11.033^2 = 121.724$. Substituting into Equation 11.24a, we obtain $\widehat{n}_e = 5.1$. Substituting into Equation 11.25a, $\text{Var}(\widehat{n}_e) = 10.703$, giving the standard error of $\widehat{n}_e$ as $10.703^{1/2} \simeq 3.3$. Thus, the data suggest the hypothesis that the majority of the genetic difference in skin color between the major races of man is a consequence of a very small number of segregating factors. It should be kept in mind, however, that because of the low degree of accuracy of the estimated segregational variance, $\widehat{n}_e$ is a *highly uncertain* measure of the effective number of factors.

Supposing, for heuristic purposes, that the estimate $\widehat{n}_e$ is accurate, what might the actual number of loci (n) contributing to the character be? From Table 11.2, we know that the mean recombination fraction for randomly distributed genes is extremely high in humans ($\bar{c} = 0.49$). Substituting this and $n_e = 5$ into Equation 11.27, we obtain

$$n = \frac{4.9 + 4C_\alpha}{0.9}$$

Assuming that all loci have equal effects ($C_\alpha = 0$), which seems unlikely, then $n = 5.4$. For $C_\alpha = 1,\ 10,$ and 100, $n = 10,\ 50,$ and 450. Thus, if the squared coefficient of variation of effects is much greater than one, the actual number of loci can greatly exceed the effective number of factors.

A survey of estimates of $\widehat{n}_e$ is given in Table 11.6. It should be emphasized that each estimate only applies to the specific pair of parental lines and that substantial differences would be likely if other parental stocks were used. Furthermore, the data are adequately described by an additive model in only a few cases, so most of the estimates are definitely biased in the downward direction by nonadditive gene action. Despite these limitations, while several of the analyses imply that a dozen or more loci are responsible for the differentiation of characters between parental lines, a number of cases suggest the possibility that a single major factor is involved. The latter conclusion may, of course, be substantially in error due to the approximate and biased nature of the biometrical approach. Nevertheless, the Castle–Wright model serves as a flag for situations in which a leading-factor (major gene) hypothesis (Chapter 16) warrants consideration.

Table 11.6 A sample of estimates (± their standard errors) of the effective number of segregating factors differentiating parental lines, obtained by use of Equation 11.24a. Whenever possible, the segregational variance was obtained by iterated least-squares analysis (Example 11.3). Under the Additive Model column, + and − denote agreement and incompatibility with an additive model.

Species	Character	$\widehat{n}_e$	Additive Model	Reference
Corn	log(% oil + 1.87)	18 ± 2	+	Sprague and Brimhall 1949
	time to silking	1 ± 1	−	Mohamed 1959
	time to shed pollen	1 ± 1	−	
	ln (height)	4 ± 1	−	Emerson and East 1913
	ln (nodes)	5 ± 1	+	
	ln (internode length)	1 ± 1	−	
	ln (ear length)	13 ± 3	−	
	ln (seed weight)	13 ± 3	−	
Lima beans	seed size	17 ± 2	−	Ryder 1958
Red pepper	fruit shape	3 ± 1	−	Khambononda 1950
	fruit weight	13 ± 1	−	
Rice	plant height	1 ± 1	−	Mohamed and Hanna 1964
Goldenrod (*Solidago*)	date of anthesis	6 ± 2	−	Goodwin 1944
Nicotiana langsdorffii × *N. sanderae*	corolla length	13 ± 1	−	Smith 1937b
Mimulus nasutus × *M. guttatus*	ln(flowering time)	1 ± 1	+	Fenster and Ritland 1994
	corolla width	2 ± 1	+	
	stamen level	3 ± 1	+	
Mimulus guttatus × *M. cupriphilus*	flower width	5 ± 2	+	Macnair and Cumbes 1989
	flower height	4 ± 1	+	
	pistil length	18 ± 18	+	
	corolla length	6 ± 3	+	
Tomato	$\log_{10}$ (fruit weight)	12 ± 1	−	Powers 1942
Pearl millet (*Pennisetum*)	height	4 ± 1	−	Burton 1951
	internode length	2 ± 1	−	
	leaves/stem	5 ± 1	−	
		7 ± 1		
Drosophila melanogaster	ln (longevity)	1 ± 1	−	Luckinbill et al. 1988
Drosophila tripunctata	ovipos. site pref.	1 ± 1	−	Jaenike 1987
Drosophila heteroneura × *D. silvestris*	head length	7 ± 4	+	Templeton 1977
	head width	1 ± 1	+	
Cave fish (*Astyanax*)	eye diameter	6 ± 1	−	Wilkens 1971
Chickens	weight	5 ± 1	−	Waters 1931
Mice	$\log_{10}$ (weight)	12 ± 1	−	Chai 1956

Finally, it is important to emphasize that line-cross based estimates of n and n_e are for the number of loci whose (aggregate) alleles have opposite effects on the trait between the two specific reference populations. This is potentially *vastly smaller* than the total number of loci segregating alleles that generate trait variation in a population. Line-cross linkage analysis (Chapters 18 and 19) detects such between-line differences, while population-based association mapping (GWAS; Chapters 20 and 21) attempts to detect *all* segregating variation *within* a *single* population. Hence, the latter likely has a far greater universe of potential loci to detect than the former. Further, the nature of selection shaping the distribution of allelic effects is likely to be different for a pair of line-cross populations (often directly shaped by selection on the focal trait) versus a population sample, where selection on the trait may be indirect, at best (examined in detail in Chapter 21 and WL Chapter 28) .

Effect of the Leading Factor

If the assumptions of additive gene action and unlinked loci hold, then n_e provides some information on the effect of the **leading factor** (the locus accounting for the largest amount of the difference between parental means). Let $\phi_{max} = 2\alpha^c_{max}/[\,\mu(\text{P}_1) - \mu(\text{P}_2)]$ be the proportion of the difference between parental means attributable to the largest factor, and denote its effect by α^c_{max}. Equation 11.26a yields the inequality $\sigma^2_s \geq (\alpha^c_{max})^2/2$, which upon substitution into Equation 11.23c gives

$$\phi_{max} \leq \sqrt{\frac{1 + C_\alpha}{n_e + C_\alpha}} \tag{11.30a}$$

as an estimate of the upper bound on the effect of the leading segregating factor. Because each segregating factor contains one or more loci, ϕ_{max} is also an upper bound on the effect of the leading locus. Lander and Botstein (1989) proposed a simple idea that yields a lower bound estimate for the effect of the leading factor. Under the assumption that one strain contains all "positive" genes and the other all "negative" genes (which might be approximated if the two strains were obtained by intense selection in opposite directions from a common stock), there must be at least one segregating factor with an effect at least as great as $[\,\mu(\text{P}_2) - \mu(\text{P}_1)]/n_e$. Expressed in terms of the proportion of the total difference, this effect is simply

$$\phi_{min} \geq 1/n_e \tag{11.30b}$$

Alternatively, from the standpoint of individual loci, an estimate of the minimum effect of the leading locus is $1/n$. Finally, because (from Equation 11.27) there are at least $2\bar{c}n_e$ loci, the maximum value of the *average* allelic effect is

$$\bar{\phi}_{max} = 1/(2\bar{c}n_e) \tag{11.30c}$$

Estimates of the statistical bounds on the effects of leading factors are of practical importance, because they can potentially provide insight into the likely utility of molecular marker-based searches for loci with major effects (Chapters 17–21). Unfortunately, none of the above-mentioned statistics is particularly informative in this regard. Even when they are reliable, large estimates of the upper bound of the leading factor (Equation 11.30a) do not necessarily imply that any locus actually has a large effect, and although a large ϕ_{min} implies that at least one locus has a major effect, a small ϕ_{min} does not rule out the possibility of several loci with major effects. Further, as we detail in Chapters 20 and 21, a small GWAS effect *does not* imply a small *allelic effect*, as *GWAS-effect* sizes are usually based on *variances*. An allele of large effect that is rare can have a small variance, and an inverse relationship between allele effect size and allele frequency is often seen, with large-effect alleles tending to have very low frequencies (Chapter 21).

Zeng (1992) suggested an alternative approach to predicting the effects of leading factors. Given an estimate of C_α, one first derives the estimate of the number of loci, $\widehat{n}$. This provides an estimate of the mean allelic effect as

$$\widehat{\alpha} = [\bar{z}(\text{P}_1) - \bar{z}(\text{P}_2)]/\widehat{n} \tag{11.31}$$

Together, $\widehat{\alpha}$ and C_α provide the first two moments of the distribution of allelic effects. If the form of the distribution is specified and uniquely defined by the mean and variance, one can then draw $\widehat{n}$ random effects from the distribution, order them, and evaluate the relative contributions of the various factors to the parental line divergence. As noted above, even slightly different values of C_α can lead to very different estimates of n. However, Zeng (1992) found that the number of significant loci (for example, the number that account for 90% of the total divergence) is extremely insensitive to changes in C_α, changing by only five or so over a range in which n changes by hundreds. Thus, for large surveys in which highly reliable estimates of phenotypic means and variances can be acquired, Zeng's approach has promise as a means of estimating the number of major factors.

Example 11.5. Here we present an alternative analytical approach for estimating the number of leading factors and their effects. This approach assumes that something is known about the form of the distribution of allelic effects. Let $p(\alpha)$ be the probability density function of the effects, α_i, and let $F(\alpha)$ be the cumulative frequency distribution, the probability that the effect of a randomly drawn gene is less than α. Because $F(\alpha) = \int_o^\alpha p(x)dx$, then $dF(\alpha)/d\alpha = p(\alpha)$. Suppose now that n genes are randomly drawn from the distribution $p(\alpha)$, and rank ordered in terms of increasing effect, such that α_1 is the smallest effect and α_n is the largest effect (the leading factor). From the perspective of genetic analysis, one would like to know the expected effects of $\alpha_n, \alpha_{n-1}, \alpha_{n-2}$, and so on. The theory of **order statistics** (Harter 1961, 1970a, 1970b; Sarhan and Greenberg 1962; Kendall and Stuart 1977; David 1981) provides a potential solution.

Consider the *r*th smallest value in the set of random draws of n genes. The probability that at least r draws in a sample do not exceed the value α can be modeled as a binomial random variable (Equation 2.19a) with success parameter $F(\alpha)$, with

$$F_r(\alpha) = \sum_{i=r}^{n} \binom{n}{i} [F(\alpha)]^i [1 - F(\alpha)]^{n-i} \tag{11.32a}$$

which leads to the probability density function for the *r*th order statistic,

$$p_r(\alpha) = \frac{dF_r(\alpha)}{d\alpha} = \frac{n!}{(r-1)!(n-r)!} [F(\alpha)]^{r-1} [1 - F(\alpha)]^{n-r} p(\alpha) \tag{11.32b}$$

Thus, the expected value of the *r*th smallest factor is given by

$$E[\alpha_r] = \int_0^\infty \alpha p_r(\alpha) d\alpha \tag{11.32c}$$

which, for the leading factor, reduces to

$$E[\alpha_n] = n \int_0^\infty \alpha [F(\alpha)]^{n-1} p(\alpha) d\alpha \tag{11.32d}$$

An estimate of the proportional contribution of the leading factor to the line differentiation is $2E(\alpha_n)/[\bar{z}(\text{P}_1) - \bar{z}(\text{P}_2)]$.

These expressions only outline a general approach. Their actual implementation requires that one define the form of the probability density function $p(\alpha)$ (for example, a truncated normal or a gamma distribution), and then characterize the function in terms of its parameters (usually the mean and variance of effects). Once an estimate of the number of loci (n) has been acquired (for example, by the use of Equation 11.27), these parameters are specified. The mean effect is estimated by $\bar{\alpha} = [\bar{z}(\text{P}_1) - \bar{z}(\text{P}_2)]/2n$, and the variance is defined by $\bar{\alpha}^2 C_\alpha$.

Extension to Haploids

The Castle–Wright model can be easily extended to the estimation of gene number in haploid organisms (Chovnick and Fox 1953). Because most haploids can be maintained clonally, we assume that the environment is the sole source of variation within the parental lines. The expected means and variances of the derived generations are laid out in Table 11.7. Note that the F_1 generation exhibits the same segregational variance as the F_2 due to the complete segregation of parental genes following fertilization and the production of haploid progeny. This segregational variance is

$$\sigma_s^2 = \sum_{i=1}^{n} 0.5[(\alpha_i^c - 0)^2 + (-\alpha_i^c - 0)^2] = n[\sigma^2(\alpha_i^c) + (\overline{\alpha_i^c})^2] \tag{11.33}$$

Table 11.7 Expected means and variances for line crosses derived from two haploid parental lines, under the assumption of unlinked loci and additive gene action. Composite additive effects are defined as haploid effects, so that parental line divergence is still $2\alpha_1^c$ as in the diploid model.

Line	Mean	Variance
P_1	$\mu_0 - \alpha_1^c$	σ_E^2
P_2	$\mu_0 + \alpha_1^c$	σ_E^2
F_1	μ_0	$\sigma_E^2 + \sigma_s^2$
F_2	μ_0	$\sigma_E^2 + \sigma_s^2$
B_1	$\mu_0 - 0.5\alpha_1^c$	$\sigma_E^2 + \frac{3}{4}\sigma_s^2$
B_2	$\mu_0 + 0.5\alpha_1^c$	$\sigma_E^2 + \frac{3}{4}\sigma_s^2$

which is twice the expectation in the case of diploidy. Thus, the estimation equation for n_e with haploid organisms is the same as in the case of diploidy except that $4\text{Var}(s)$, rather than $8\text{Var}(s)$, appears in the denominator of Equations 11.23c and 11.24a. The equation for the sampling variance for $\widehat{n}_e$ needs to be multiplied by four, but the estimators for gene number, Equations 11.27 and 11.29, still apply provided the segregational variance is estimated in the F_1 generation.

A very similar strategy can be employed with **doubled-haploid (DH) lines**. Such lines, which can be produced by a variety of cytological techniques (Kermicle 1969; Nitzsche and Wenzel 1977; Choo 1981), are homozygous at all loci. If two such parental lines are crossed to produce an F_1 generation, and a random sample of F_1 gametes is used to produce a new series of DHs, the effective number of factors differentiating any two lines can be obtained by computing the segregational variance as half the difference between the F_1 DH variance and the average variance in the P_1 and P_2, and employing Equation 11.24a (Choo and Reinbergs 1982). (The segregational variance is inflated twofold by the enforcement of homozygosity at each locus in DHs.) The most reasonable estimate of n_e obtained with this approach utilizes parental lines with the highest and lowest mean phenotypes in a random sample from the population. However, with even a moderate number of segregating loci, the probability of obtaining the two most extreme lines possible is low unless the number of lines assayed is very large (Example 13.4). Choo and Reinbergs (1982) used this technique to show that at least eight segregating factors contribute to the variation in grain yield, heading date, and plant height in barley. For additional biometrical approaches to gene number estimation in DHs, see Snape et al. (1984). Further extension of the Castle–Wright approach were offered by Comstock and Enfled (1981) who allow for genes to act in a multiplicative (as opposed to additive) fashion, and by Jones (2001) who considered sex linkage and haplodiploidy.

Example 11.6. Croft and Simchen (1965) isolated dikaryotic mycelia from wild populations of the fungus *Collybia velutipes* and from these extracted asexually and sexually derived monokaryotic (haploid) spores. (A dikaryotic mycelium is a filament comprised of fused cells of two different parental origins, where each cell contains two distinct haploid nuclei, one from each parent). The growth rates of germinating spores were then assayed on a laboratory medium. Barring mutations, the growth rate of each asexual propagule is expected to be representative of one of the parental lines, because these propagules contain a single, nonrecombinant nucleus. On the other hand, the sexually derived progeny will exhibit segregational variance. Frequency distributions are given below for both types of offspring for one particular isolate. Note that there is no evidence of transgressive segregation here, as the smallest F_1 is no smaller than the smallest P_1 and the largest F_1 is no larger than the largest P_2, consistent with very few (if any) plus alleles in the smaller line.

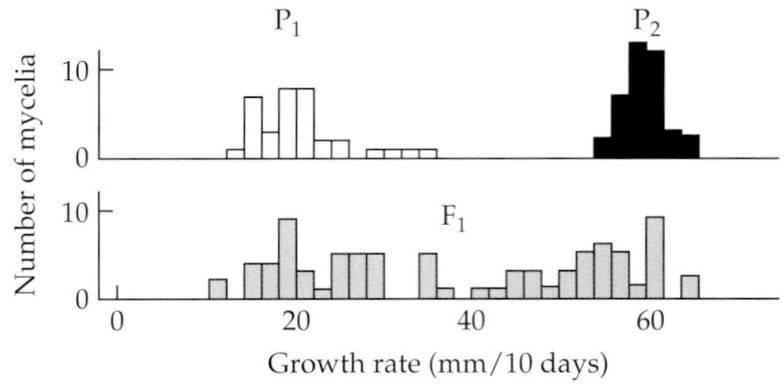

The mean growth rates of the two parental types differ by $\bar{z}(\text{P}_1) - \bar{z}(\text{P}_2) = 39.45$ (mm/10 days), and the sampling variances of the two means are $\text{Var}[\bar{z}(\text{P}_1)] = 2.43$ and $\text{Var}[\bar{z}(\text{P}_2)] = 0.53$. The variance of growth rate among haploid replicates is 26.80, while the excess variance among sexual propagules is 277.80. Taking the latter quantity to be an estimate of the segregational variance, Var(s), substitution into Equation 11.24a (multiplied by 2) yields $\widehat{n}_e = 1.4$. The standard error is approximately 0.5. These results are reasonably consistent with those obtained from four other isolates: $1 \pm 0.2, 5 \pm 0.8, 3 \pm 0.3$, and 1 ± 0.1. Thus, it seems likely that most of the growth rate differences among parental strains may be attributable to one to three loci.

OTHER BIOMETRICAL APPROACHES TO GENE NUMBER ESTIMATION

In outlining the Castle–Wright model, we emphasized several assumptions, violations of which tend to result in underestimation of the actual number of segregating loci. Although two potential problems, gametic-phase disequilibrium and dominance, appear to be reconcilable, two others are less tractable—transgressive segregation (whereby both lines contain plus and minus alleles) and variation among loci for allelic effects. We now consider three approaches that have been developed to circumvent these problems, all involving the use of species that can be self-fertilized.

The Inbred-backcross Technique

Wehrhahn and Allard (1965) developed a useful technique that yields estimates of both the minimum number of genetic factors responsible for the differentiation of two lines and the magnitude of the locus-specific effects. Two pure lines are crossed to form F_1 individuals, each of which is then continually backcrossed to one of the parental populations (the **recurrent parent**; say the P_1) for k generations (i.e., an $\text{F}_1 \times \text{P}_1$ cross, followed by a cross of their progeny to the P_1, the resulting progeny are then crossed to P_1, and so on). The backcross descendants are then inbred for several generations to fix any segregating factors. The rationale for this breeding scheme is that as k becomes large, the probability that any inbred backcross line will retain more than one allele from the **donor parent** (the P_2) becomes small. The effects of individual genes can then be ascertained by comparing the phenotypic means of the recurrent parent (P_1) and the derived lines. A unique advantage of the inbred-backcross technique is that, after single-gene deviant lines have been isolated, lines with pairs of various genes can be constructed to evaluate epistatic effects between specific isolated factors. Likewise, genotype × environment interaction involving individual genes can be examined by growing lines in different environments.

A more formal statement of these arguments is as follows. The probability that a specific gene from the P_2 is incorporated into a specific line after k generations of backcrossing is $p_k = (1/2)^{k+1}$. If n freely segregating factors are responsible for the character difference between the P_1 and P_2, then the probability that a derived line retains just one of the P_2 alleles is

$$p(1) = np_k(1 - p_k)^{n-1} \tag{11.34a}$$

This follows directly from the binomial distribution (Equation 2.19a), because each donor

Table 11.8 The proportion of deviant derived lines that are expected to contain a single gene from the donor parent (last column), for n freely segregating factors and k generations of backcrossing (Data from Wehrhahn and Allard 1965.)

n	k	p_k	$p(1)$	$p(r \geq 1)$	$p(1)/p(r \geq 1)$
4	2	$\frac{1}{8}$	0.335	0.414	0.809
	3	$\frac{1}{16}$	0.206	0.227	0.905
	4	$\frac{1}{32}$	0.114	0.119	0.953
10	2	$\frac{1}{8}$	0.376	0.624	0.510
	3	$\frac{1}{16}$	0.350	0.650	0.735
	4	$\frac{1}{32}$	0.235	0.765	0.863

allele is retained or lost independently with probability p_k. The probability that a derived line contains at least one P_2 gene is

$$p(r \geq 1) = 1 - (1 - p_k)^n \tag{11.34b}$$

The ratio $p(1)/p(r \geq 1)$ is the conditional probability that any deviant derivative line contains only a single P_2 gene. Table 11.8 shows that unless n is very large, this probability is very high after only three or four generations of backcrossing.

If, by statistical comparison of phenotypic means, a fraction $\widehat{p}(r \geq 1)$ of the inbred-backcross lines is observed to differ significantly from the P_1, then a minimum estimate of the number of effective factors can be found by rearrangement of Equation 11.34b,

$$\widehat{n}_{WA} = \frac{\ln[1 - \widehat{p}(r \geq 1)]}{\ln(1 - 0.5^{k+1})} \tag{11.35}$$

(Mulitze and Baker 1985a, 1985b). Unlike the Castle–Wright estimator, Equation 11.35 yields estimates of n_e that are essentially independent of the degree of transgression of gene effects in the P_1 and P_2, i.e., a gene with a low effect in an otherwise high-performing line can be detected when substituted into a low-performing line.

Example 11.7. As shown below, Wehrhahn and Allard (1965) crossed two pure lines of wheat (Ramona and Baart 46) with very different heading dates (flowering times). Two successive backcrosses were made to Ramona (the recurrent parent), followed by three generations of selfing, to produce 69 inbred backcross lines.

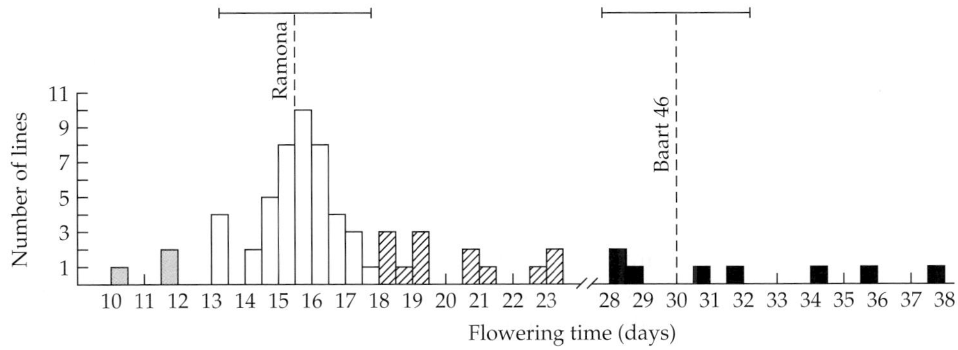

How many lines are expected to contain any specific Baart 46 (donor parent) gene? Because $k = 2$, we have $p_k = (1/2)^3 = 1/8$, and because there are 69 total lines, $69/8 = 8.6$ of these are expected to carry a Baart 46 gene at a specified locus. From the properties of the

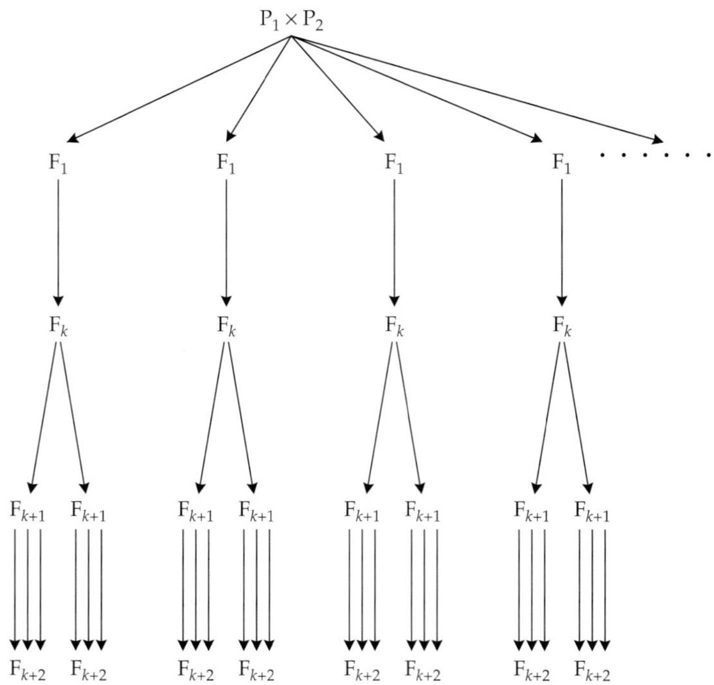

Figure 11.3 The crossing scheme involved in the genotype assay technique of Jinks and Towey (1976). Progeny from all generations beyond the F_1 are obtained by selfing. In generation $k + 2$, multiple progeny are assayed from each of two sublines for each of the isolated selfed lineages.

binomial distribution, the standard error of the estimate is $[69p_k(1 - p_k)]^{1/2} = 2.8$. The distribution of heading date in the derived lines shows three groups of deviants from the Ramona distribution: (i) a group of eight very late lines that appear to contain a factor with major effect (black bars); (ii) a group of 14 lines with slightly late heading dates (striped bars); and (iii) a group of three lines with earlier heading dates (gray bars). The means and 95% confidence limits for the parental lines are given by the vertical and horizontal lines, and the three groups of deviants from the Ramona (recurrent) line are differentially shaded.

By other means, the authors showed rather convincingly that the group of 14 slightly late lines was heterogeneous for two factors. Thus, the difference in heading date between the two parental lines is caused by at least four effective factors, one of which operates in a direction opposite to the others. These four factors accounted for 96% of the line differentiation (80% was due to the leading factor), so if additional loci are involved, their effects must be very small. To see that the observations in the figure are consistent with Equation 11.35, let $\widehat{p}(r \geq 1) = (8 + 14 + 3)/69 = 0.362$. We then obtain $\widehat{n}_{WA} = 3.4$, which rounds up to 4.

Genotype Assay

Jinks and Towey (1976; Towey and Jinks 1977) developed a method for estimating n_e that is similar in philosophy to the inbred-backcross technique. In this case, however, the F_1 progeny of a cross between two pure lines are self-fertilized (instead of backcrossed). Their descendants are also self-fertilized until generation F_k, where k is usually 2 to 5. Two random (selfed) progeny are then raised from each F_k individual and selfed, and their offspring are assayed in a randomized design (Figure 11.3). A comparison of means (t test) and variances (F test) between the two families is used to detect whether one or more loci were segregating in the F_k grandparent. Not all heterozygotes are detected by this

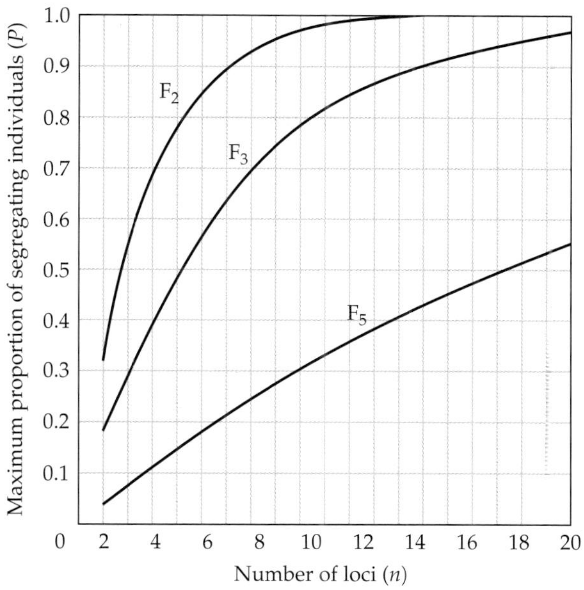

Figure 11.4 The expected fraction of individuals in the F_2, F_3, and F_5 generations detected as segregating by a genotype assay on their grandchildren, under the assumptions leading to Equation 11.36. (From Jinks and Towey 1976.)

approach, because the probability that two randomly chosen offspring of a heterozygous parent will differ in genotype is only 5/8 (the probability that one is BB and the other is Bb is $2 \times (1/4) \times (1/2) = 1/4$, that one is BB and the other is bb is $2 \times (1/4) \times (1/4) = 1/8$, and that one is Bb and the other is bb is $2 \times (1/2) \times (1/4) = 1/4$). However, the observed fraction of segregating grandparents can still be used to make inferences about n.

The logic behind this idea is as follows. Under continuous self-fertilization, heterozygosity is reduced by 50% each generation, so that the probability that an F_k individual is a heterozygote at a particular locus is $(1/2)^{k-1}$. The probability that two assayed progeny from a random F_k individual differ at this locus is then $(5/8) \cdot (1/2)^{k-1} = 5/2^{k+2}$. It follows that the probability that the two descendant sublines differ at least at one segregating locus is

$$P = 1 - \left(1 - \frac{5}{2^{k+2}}\right)^n \tag{11.36}$$

Rearranging and substituting observations for expectations, we obtain another estimator for the effective number of segregating factors,

$$\widehat{n}_{JT} = \frac{\ln(1 - \widehat{P})}{\ln\left(1 - \dfrac{5}{2^{k+2}}\right)} \tag{11.37}$$

There are several potential causes of bias in this estimator, most of which will lead to the usual underestimation of the actual number of loci. First, as in the inbred-backcross technique, the observed fraction of intrapair differences ($\widehat{P}$) is a matter of statistical power, increasing with the size of the assayed families, but decreasing with more stringent criteria for statistical significance. Second, the number of segregating factors will be depressed below the actual number of loci by linkage, the magnitude of this bias decreasing with increasing k (more opportunity for recombination). Hill and Avery (1978) consider this issue in some detail. Third, dominance, epistasis, and gametic-phase disequilibrium can cause the expected phenotypes associated with different genotypes to be the same. Taking

Table 11.9 Minimum number of effective factors responsible for the differentiation of two lines of tobacco (*Nicotiana rustica*) as determined by genotype assay in progressive generations (F_k). Here P is the probability that the two descendant sublines differ at least at one segregating locus. n_{JT} is computed with Equation 11.37 and rounded up to the nearest unit. (Date from Towey and Jinks 1977.)

Grandparent Generation:	F_2	F_3	F_4	F_5	F_6
Flowering time					
$\widehat{P}$	0.200	0.450	0.314	0.233	0.306
$\widehat{n}_{JT}$	1	4	5	7	19
Final height					
$\widehat{P}$	0.400	0.300	0.257	0.200	0.306
$\widehat{n}_{JT}$	2	3	4	6	19

this masking problem into consideration, Jinks and Towey (1976) and Mulitze and Baker (1985a) have derived an alternative to Equation 11.37 that yields an upper (rather than lower) bound to n_e.

The relationship between P and n_{JT} varies rather substantially with the number of generations (k) prior to the genotype assay (Figure 11.4). If there are only a few effective factors, the greatest sensitivity is achieved (i.e., there is a strong response of P to n) when $k = 2$, provided linkage is unimportant. However, if there are more than five segregating loci, an F_2 assay is of little use, whereas an F_5 assay is quite sensitive.

Towey and Jinks (1977) applied the genotype assay to five generations of a cross between two lines of *Nicotiana rustica*. Even after six generations of selfing, there was no obvious decline in the fraction of descendant pairs exhibiting variation, for either flowering time or plant height (Table 11.9). Consequently, the estimates of n_{JT} for both characters increased approximately tenfold throughout the study. The authors interpreted this increase to be a consequence of the gradual elimination of gametic-phase disequilibrium through progressive rounds of recombination. If nothing else, this interpretation justifies our earlier discussion about the bias caused by linkage.

Panse's Technique

This infrequently utilized technique also starts with a cross between two pure lines. The F_1 individuals are self-fertilized to produce an F_2 generation, the members of which are selfed to produce a series of F_3 families. For any segregating locus, half of the F_2 individuals are expected to be heterozygous and the remaining half homozygous. Assuming additive effects, it follows from Chapter 4 that the genetic variance resulting from the *i*th locus in a population in Hardy–Weinberg equilibrium is $2pq\alpha_i^2$. This genetic variance is $\alpha_i^2/2$ for an F_3 family descending from a heterozygote, and zero otherwise. Thus, the average genetic variance within F_3 familes is $\sum_{i=1}^{n} \alpha_i^2/4$, or $\overline{\sigma_A^2} = n\alpha^2/4$ for n loci with equal effects.

Panse (1940) showed how the variance of the genetic variance among F_3 families can be used to estimate n_e. At the *i*th locus, the expected value of this quantity is $(1/2)(\alpha_i^2/2)^2 - (\alpha_i^2/4)^2 = \alpha_i^4/16$. Summing over all loci, and assuming equal effects, this value becomes $\sigma^2(\sigma_A^2) = n\alpha^4/16$. Thus, the ratio of the squared mean genetic variance within F_3 families to the variance of those variances,

$$\widehat{n}_P = \frac{(\overline{\sigma_A^2})^2}{\sigma^2(\sigma_A^2)} \tag{11.38}$$

provides an estimate of the effective number of loci. In addition to its assumption of unlinked loci with additive effects, a major difficulty with this technique is its reliance on genetic variance estimates. It is difficult to determine these with a great deal of precision and even more difficult to procure good estimates of the variance of σ_A^2 among lines.

12

Inbreeding Depression

Nature thus tells us, in the most emphatic manner, that she abhors perpetual self-fertilization.
Darwin (1876)

Inbred individuals are almost always less fit than the progeny of nonrelatives, and, similarly, many traits decline from favorable values under inbreeding. The decline in the mean phenotype with increasing homozygosity within populations, known as **inbreeding depression (ID)**, is often interpreted as heterosis-in-reverse. However, as will be seen below (as well as in Chapter 13), there are some important distinctions between the genetic mechanisms contributing to inbreeding depression in crosses *within* populations and heterosis in crosses *between* populations.

The serious study of the impact of inbreeding begins with Darwin (1868, 1876), although published results go back to at least Knight (1799), and numerous cultures have long traditions against marriage of close relatives (Brown 1991). Darwin's interest was no doubt influenced by the fact that he married his first cousin, Emma Wedgwood. Many of Darwin's ten children had significant health issues, and he was concerned that this was, in part, the result of marrying a close relative. Álvarez et al. (2015) examined Darwin's concern by looking at male fertility in the Darwin–Wedgwood pedigree, finding a significant impact of the husband's inbreeding coefficient on both family size and reproductive duration.

The near universal existence of inbreeding depression bears importantly on many basic issues in evolutionary biology as well as on a number of practical issues in human genetics, agriculture, and conservation biology. For example, the deleterious consequences of self-fertilization are likely to be the leading selective forces responsible for the evolution of various aspects of mating systems in plants (Darwin 1876; Lande and Schemske 1985; Schemske and Lande 1985; Charlesworth and Charlesworth 1987; Uyenoyama 1993; Waller 1993; Carr and Dudash 2003; WL Chapter 23), and of behavioral mechanisms for avoiding mating with close relatives in animals (Shields 1982; Thornhill 1993; Pusey and Wolf 1996; Szulkin et al. 2013). The observations by maize breeders that crosses between inbred lines yield substantially more grain than the inbreds themselves (East 1908; Shull 1908; Jones 1918; Sprague 1983) have given rise to a situation in which most corn farmers are now almost entirely reliant on companies for hybrid seed (Figure 12.1). Finally, the loss of fitness due to the development of inbreeding depression in small populations is a major concern in endangered species management (Templeton and Read 1984; Lacy et al. 1993; Hedrick 1994; Lynch et al. 1995a, 1995b; Lynch 1996; Saccheri et al. 1998; Hedrick and Kalinowski 2000; Richards 2000; Kristensen and Sørensen 2005; O'Grady et al. 2006; Leberg and Firmin 2008; Jamieson and Allendorf 2012; Frankham et al. 2013, 2014; Franklin et al. 2014; Caballero et al. 2017a, 2017b; Pérez-Pereira et al. 2022, 2023).

It is widely appreciated that inbreeding depression is an inevitable consequence of dominance. When gene action is entirely additive (no dominance or epistasis), the average phenotypic effects associated with alleles are independent of the genetic background. Hence, inbreeding depression cannot occur for characters with a purely additive genetic basis (as inbreeding changes *genotype*, not *allele*, frequencies). With dominance, the allelic effect is a function of the genetic background (Equation 4.10b), and thus the average phenotypic effect of an allele changes with a change in *genotype frequencies*, even in the absence of *allele-frequency* change.

Because there are many types of dominance, this simple explanation for inbreeding depression provides only a partial understanding of the full process. We start our discussion by showing how the two competing hypotheses on the genetic mechanism of inbreeding depression—**partially recessive** deleterious alleles versus **overdominance**—lead to some

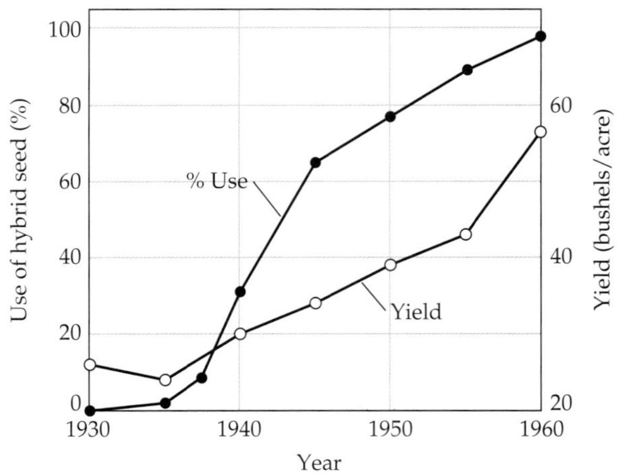

Figure 12.1 Historical change in the reliance on hybrid corn by United States farmers and the increase in mean annual harvest. Duvick (2005) estimated that 50–60% of the yield gain was genetic, with the remainder due to improved agronomic practices. (After Sprague 1983.)

very similar predictions that are in accordance with empirical observations, but also to some major differences that are less easy to resolve empirically (Charlesworth and Charlesworth 1999; Roff 2002; Carr and Dudash 2003; Kristensen and Sørensen 2005; Charlesworth and Willis 2009; Kristensen et al. 2010). Second, we provide a brief outline of the basic statistical issues that arise in the analysis of inbreeding depression. Third, we review the large body of existing data, showing that inbreeding depression exists, at least to some degree, for essentially all characters in all populations of diploid organisms. We close by reviewing how molecular-marker analysis is refining our understanding of some of these issues.

THE GENETIC BASIS OF INBREEDING DEPRESSION

In the absence of allele-frequency change, inbreeding shifts the genotype frequencies in a population in a very simple way. Let f denote the inbreeding coefficient of an individual, i.e., the probability that an individual carries two alleles that are identical by descent (IBD) at a locus (Chapter 7). When zygotes are formed from gametes in a population, the allele in one gamete will be IBD to the allele in the other with probability f. If that allele occurs in the population with frequency p, IBD homozygotes will arise in the population with frequency pf (Table 12.1). With frequency $p^2(1-f)$, homozygotes with two alleles of the same type but not derived from the same common ancestor will arise. By definition, the two alleles in a heterozygote are not IBD, thus heterozygotes will occur in the population at a frequency of $2pq(1-f)$. (Table 12.1). With this information in hand, it is straightforward to derive quantitative expressions for the two mechanistic hypotheses for inbreeding depression.

The **dominance hypothesis** (Davenport 1908; Bruce 1910; Keeble and Pellew 1910; Jones 1917) argues that inbreeding depression is caused by the expression of deleterious recessive genes in homozygous individuals. (We retain the use of the term dominance only for historical reasons; **partial dominance** is a more apt term in that the hypothesis focuses explicitly on dominance due to partially or completely recessive genes). Consider a diallelic locus, where the frequency of the deleterious allele is q, and the fitnesses of the three genotypes are denoted as 1, $1-hs$, and $1-s$ (Table 12.1). Here, s measures the selection against homozygotes for the deleterious allele, and h is a measure of dominance, with $h=0.5$ implying additivity, while $0<h<0.5$ implies that the deleterious allele is partially recessive. Multiplying genotype frequencies (Table 12.1) by their fitnesses, the mean fitness in a population inbred to level f becomes

$$\overline{W}_f = \overline{W}_0 - fpqs(1-2h) \tag{12.1a}$$

Table 12.1 Genotypic frequencies and fitnesses under the two dominance-based hypotheses for inbreeding depression. Two alleles (B, b) are assumed to be present, with respective frequencies p and q.

		Fitness		Phenotype for Arbitrary Character
Genotype	Frequency	Partial Dominance	Overdominance	
BB	$p^2(1-f)+pf$	1	$1-t$	$2a$
Bb	$2pq(1-f)$	$1-hs$	1	$a+d$
bb	$q^2(1-f)+qf$	$1-s$	$1-s$	0

where

$$\overline{W}_0 = 1 - 2pqsh - q^2 s \tag{12.1b}$$

is the mean fitness in the random-mating base population. Provided that $h < 0.5$, then $(1 - 2h)$ is necessarily positive (and hence $\overline{W}_f < \overline{W}_0$). Thus, with partially recessive deleterious alleles, mean fitness is expected to decline linearly with increasing inbreeding coefficient f.

Unless the mutation rate is very high (or the population size is very small), deleterious alleles are expected to be maintained at low frequencies by selection (WL Chapter 7), so it can be assumed in Equation 12.1a that $p = 1-q \simeq 1$, showing that the expected decline in fitness due to complete inbreeding ($f = 1$) at a locus is approximately $qs(1-2h)$. For a randomly mating population in selection–mutation balance, if u is the mutation rate from the beneficial to the deleterious allele and $h \gg \sqrt{\mu/s}$, then $q \simeq u/(hs)$ (Haldane 1927; WL Equation 7.6b). Thus, for large randomly mating populations, the decline in fitness resulting from complete inbreeding at a locus is approximately $[u/(hs)]s(1-2h) = u(1-2h)/h$. Provided that h is independent of s, this result is ***independent of the intensity of selection*** at the locus (s) because of the inverse relationship between the equilibrium frequency of a deleterious allele and its selection coefficient. One subtlety is that one might expect segregating alleles of large deleterious effects (s large) to also be much more recessive (h small), otherwise they would be quickly removed by selection. Such a negative correlation is indeed seen in some cases (e.g., Yang et al. 2017).

A second potential explanation for inbreeding depression is referred to as the **overdominance hypothesis** (East 1908; Shull 1908; Hull 1946). The idea here is that something special about the heterozygous state causes increased vigor relative to both homozygotes. Letting s and t denote the proportional reduction in fitness of the two homozygotes relative to that of the heterozygote (Table 12.1), then

$$\overline{W}_f = \overline{W}_0 - fpq(s+t) \tag{12.2a}$$

where

$$\overline{W}_0 = 1 - p^2 t - q^2 s \tag{12.2b}$$

As in the case of partial recessives, the overdominance hypothesis leads to the prediction that mean fitness will decline linearly with increasing f. However, contrary to the situation with partial recessives, the loss of fitness *increases* with the strength of selection $(s+t)$ maintaining the polymorphism in the random-mating population. In a large randomly mating population, heterozygote superiority leads to a balanced polymorphism with $\widehat{p} = s/(s+t)$ and $\widehat{q} = 1 - \widehat{p} = t/(s+t)$ (Haldane 1927). Hence $f\,[\,\widehat{p}\,\widehat{q}\,(s+t)\,] = f[\,st/(s+t)\,]$. If, for example, $s = t$, the loss of fitness per locus under complete inbreeding is $s/2$.

These alternative hypotheses for inbreeding depression have extremely different evolutionary, and applied, implications. Under the dominance hypothesis, inbreeding depression is an inevitable consequence of recurrent mutation at the genomic level, implying that much of the genetic variation within populations must be associated with the constant influx of deleterious alleles. Although selection *removes* some of these alleles each generation,

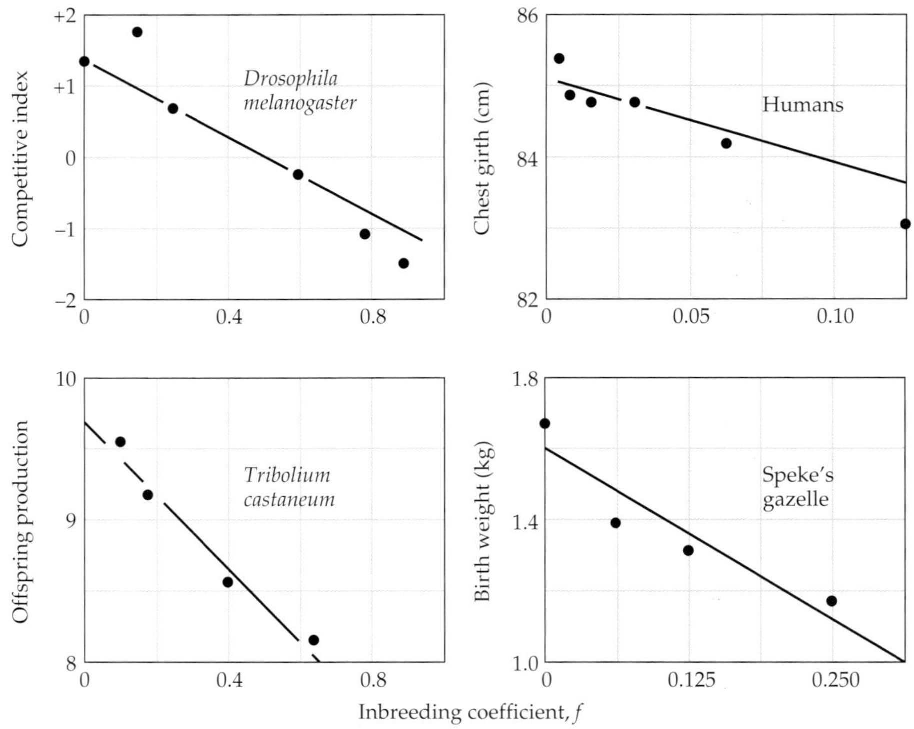

Figure 12.2 Change in mean phenotypes as a function of inbreeding. References: *Drosophila* (Latter and Robertson 1962); humans (Barrai et al. 1964); *Tribolium* (Rich et al. 1984); Speke's gazelle (Templeton and Read 1983).

mutation and/or immigration (especially if from a different locally adapted population) replaces them (WL Chapters 7 and 26). Conversely, under the overdominance hypothesis, variation is *maintained* by selection favoring the heterozygous state at multiple loci. Here, variation is maintained even in the absence of mutational pressure.

Considerable uncertainty exists as to whether overdominance with respect to fitness is a common enough phenomenon to significantly impact ID (Crow 2000; Hedrick 2012; Gao et al. 2014; Charlesworth 2015; Thurman and Barrett 2016). Only rarely has overdominance been suggested by studies with molecular markers, and most of those studies are open to alternative interpretations (discussed below). Nevertheless, as cogently pointed out by Crow (1948, 1952), even if overdominance is quite rare, it warrants serious consideration as a contributing factor in inbreeding depression. To see why, consider the expected reduction in fitness under both hypotheses when gene frequencies are in equilibrium. Under the dominance hypothesis, the maximum inbreeding depression per locus, arising with small values of h, is approximately u/h. Because u is likely to be on the order of 10^{-5} or smaller for most loci, and the evidence suggests that h is usually greater than 0.1 or so (discussed below), the per-locus inbreeding depression resulting from partial dominance is expected to be quite small, $\leq 10^{-4}$. With overdominance, complete inbreeding leads to the loss of the fittest genotype, so the reduction in fitness is potentially quite large. Thus, even if overdominance is a rare phenomenon, only a few such loci need to exist for their contribution to inbreeding depression to rival, or even surpass, that caused by a much larger number of loci displaying partial dominance.

The linear decline in the means of fitness-related traits with an increase in the inbreeding coefficient, observed in many sets of data (Figure 12.2), is consistent with both the partial dominance and overdominance hypotheses. There is, however, a major distinction between the two hypotheses with respect to the expected distribution of mean phenotypes among inbred lines. If overdominance is the major cause of inbreeding depression, *all* inbred

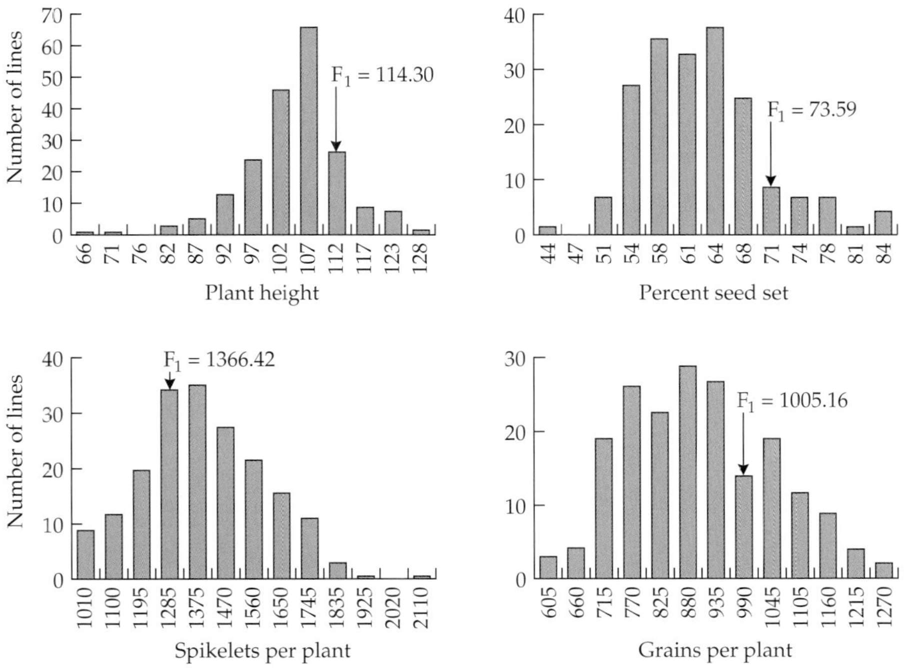

Figure 12.3 Two inbred lines of rice were crossed, from which roughly 200 recombinant inbred lines (RILs; Chapter 17) were extracted. This was accomplished by first crossing the F_1 to obtain an F_2 population, from which 194 individuals were chosen and taken through six rounds of selfing and single-seed descent. The histograms present the distribution of RIL means for four characters. For all of these traits, their F_1 performance (whose values are given by F_1 in the figures) exceeded that of the best parental line (i.e., they display high-parent heterosis; Chapter 13). Note that some of the RILs also exceed the F_1 (and hence outperform the best parent). (From Xiao et al. 1995.)

lines must eventually perform *below* the mean of the randomly mating base population, because a pure line of the best-performing genotype (a heterozygote) cannot be attained. If, on the other hand, partial dominance is the major factor, it should be possible to produce a pure inbred line that performs *at least as well* as the most outstanding member of the base population. If large numbers of loci contribute to the trait of interest, the probability of producing such a line may be quite low. Nevertheless, such lines have been obtained in several studies (Smith 1952; Williams 1959b; Wienhues 1968; Busch et al. 1971; Pooni et al. 1994; Uddin et al. 1994), see Figure 12.3.

A More General Model: Dominance

The preceding discussion has focused on the consequences of inbreeding for *fitness*. A more general account of the change in the mean of an arbitrary character (as opposed to the very special trait of fitness) under inbreeding proceeds as follows. Recalling the genotypic frequencies for an inbred population and then multiplying them by their respective genotypic values (Table 12.1), a general expression for the mean genotypic value for a single diallelic locus is

$$\mu_f = (1-f)[\,p^2(2a) + 2pq(a+d) + q^2(0)\,] + f[\,p(2a) + q(0)\,] = \mu_0 - (2pqd)f \qquad (12.3a)$$

where $\mu_0 = 2p(a+qd)$ is the mean genotypic value in the randomly mating base population. Summing over all loci, the total inbreeding depression is $\mu_f - \mu_0 = -fI$, where

$$I = 2\sum p_i q_i d_i \qquad (12.3b)$$

Recalling (Equation 4.12b) that the dominance genetic variance in a randomly mating population in gametic-phase equilibrium is $\sigma^2_D = \sum(2p_i q_i d_i)^2$, it is clear that ***dominance variance***

***is* necessary, *but not* sufficient**, for inbreeding depression to occur. Note that both I and σ_D^2 are maximized at intermediate allele frequencies ($p_i = q_i = 0.5$, as occurs in the progeny from an inbred line cross). Because the sign of d_i may vary from locus to locus, it is possible for considerable canceling to occur among the effects at different loci, leading to negligible inbreeding depression despite substantial dominance genetic variance. In other words, significant inbreeding depression requires **directional dominance** (mean $d > 0$), namely that *the value of the heterozygote tends to be closer to the value of the larger homozygote*. While the values of I and σ_D^2 can be completely uncoupled (if the signs of the d_i are entirely random), they often tend to be associated. For example, Misztal et al. (1997) examined 14 traits in Holstein cattle, and found that the correlation between values of I and σ_D (taking the square root of the variances so the comparison are on the same scale) was around 80%.

The connection between dominance variance and inbreeding depression when a large number of loci underly a trait deserves further comment. Assuming n equivalent loci (all with the same effect and allele frequency), the dominance variance becomes $\sigma_D^2 = n\,[2dp(1-p)]^2$. Assuming a very large number of underlying loci, then d must scale as $\delta/\sqrt{n}$ in order for the dominance variance to be bounded as n becomes very large (where δ is an arbitrary, but finite, constant). With this scaling,

$$\sigma_D^2 = 4np^2(1-p)^2(\delta/\sqrt{n}\,)^2 = 4\delta^2p^2(1-p)^2 \tag{12.3c}$$

Turning to inbreeding depression, Equation 12.3b becomes

$$I = 2np(1-p)d = 2\sqrt{n}p(1-p)\delta \tag{12.3d}$$

Hence, for this special case of very numerous equivalent loci, $I/\sigma_D \simeq \sqrt{n}$. As noted by Robertson and Hill (1983), as the number of loci showing directional dominance increases, I can become increasingly larger for a fixed value of dominance variance. Biologically this implies that if there is a strong trend for segregating loci to show directional dominance, then when a large number of loci underlies a trait, it can display significant inbreeding depression, even with an apparently small value of σ_D^2.

Hence, while a trait can display very low amounts of ID even when σ_D^2 is large, the converse can also be true, with a trait showing large ID even when σ_D^2 is small. We saw a similar phenomena in Chapter 5, where we noted that a considerable amount of genetic interactions (such as dominance) does not necessarily translate into a significant interaction *variance*. Indeed, for a rare recessive allele ($k = 1, q \simeq 0$), Equation 4.12d implies that $\sigma_A^2/\sigma_D^2 \simeq 2/q$, so that even with complete dominance, essentially all of the genetic variance is additive. Conversely, Equation 12.3b gives the contribution from this locus to I as $\simeq 2qak$, where $2a$ is the difference between homozygotes and $d = ak$ (Chapter 4). Hence, if ID is generated by rare recessive alleles, one expects to find genes involved in ID also being associated with significant additive variance. Indeed, Yengo et al. (2021) found that human genomic regions associated with ID are enriched for sites contributing additive genetic variance for the impacted trait.

There are two (non-exclusive) theories for why directional dominance may be pervasive. First, we previously demonstrated that directional dominance may be a fairly general property of enzymatic systems (Figure 4.5). Second, natural selection provides a filter to decrease the segregating frequency of deleterious alleles that are more dominant. When new deleterious mutations arise, those with heterozygous values that are intermediate between the wildtype and mutant allele are rapidly removed from the population. Conversely, alleles that are recessive tend to persist for longer amounts of time, creating an enriched set of segregating alleles that is biased in favor of directional dominance, even if no such bias exists in the original collection of new mutations. Hence, traits under directional selection are often *expected* to show directional dominance.

Example 12.1. A large empirical study with the flour beetle *Tribolium castaneum* provides some perspective on the role of dominance in inbreeding depression. López-Fanjul and Jódar

(1977) derived 105 lines from a large base population and maintained them by single brother–sister matings for eight generations (to $f = 0.785$). Despite the large sample sizes, the authors could find no evidence that inbreeding caused a shift in the mean rate of egg laying by virgin females at 33 or 28°C. Independent estimates of the heritabilities for these traits, obtained by full-sib correlation and daughter–mother regression, were 0.34 ± 0.02 and 0.33 ± 0.01 at 33°C, and 0.33 ± 0.02 and 0.26 ± 0.02 at 28°C. Recalling from Chapter 7 that heritabilities estimated from full-sib analysis are inflated by dominance genetic variance relative to those obtained by parent–offspring analysis, only for the second temperature was there any evidence of dominance genetic variance, and this is slight (and any additional full-sib variance could be entirely due to maternal effects, rather than dominance; Part II Chapters 23 and 29). Thus, the absence of inbreeding depression for rate of egg laying by virgins is not surprising. However, the study population was not immune to the effects of inbreeding, because two other traits, rate of egg laying by fertilized females and egg viability, exhibited substantial declines with inbreeding.

A More General Model: Epistasis

Because inbreeding depression is a consequence of nonlinear interactions between gene effects, it stands to reason that epistasis may complicate matters. However, provided the base population is in gametic-phase equilibrium, only epistasis involving some *dominance component* (such as AD or DD) contributes to inbreeding depression within populations (Anderson and Kempthorne 1954; Bulmer 1980; Hill 1982a; Lynch 1991). This result arises because although inbreeding causes a change in single-locus *genotypic* frequencies (and hence in terms in Equation 5.6 involving D), it does not (in the absence of selection and gametic-phase disequilibrium) alter the *gametic* frequencies in the population (just as allele frequencies do not change under inbreeding). With this in mind, a general expression for inbreeding depression can be obtained using the composite effects approach (Chapter 11).

Letting $-\delta_1$ be the expected change in the mean caused by single-locus dominance effects (summed over all loci) under complete inbreeding, then $-f\delta_1$ is the expected change at inbreeding level f (δ_1 equals I from Equation 12.3b). The composite additive $\times$ dominance effect may also be altered under inbreeding. It depends only on the inbreeding at single loci and may be represented as $-f(\alpha\delta)$. The composite dominance $\times$ dominance effect depends on whether one ($\delta_2^{[1]}$) or two ($\delta_2^{[2]}$) loci are inbred (have both alleles IBD). Assuming unlinked loci, the probabilities of these two situations are, respectively, $2f(1-f)$ and f^2. Thus, the shift in the mean through the inbreeding-induced alteration of dominance $\times$ dominance effects can be represented by $-2f(1-f)\delta_2^{[1]}$ in the first case and $-f^2\delta_2^{[2]}$ in the second. Summing up terms,

$$\mu_f = \mu_0 - f[\delta_1 + 2\delta_2^{[1]} + (\alpha\delta)] - f^2(\delta_2^{[2]} - 2\delta_2^{[1]}) \tag{12.4a}$$

or more succinctly,

$$\mu_f = \mu_0 - f\Delta_1 - f^2\Delta_2 + \cdots \tag{12.4b}$$

Thus, the expected mean phenotype under inbreeding can be written as a polynomial function of f, with the coefficients Δ_1 and Δ_2 being functions of multiple types of nonadditive gene action. Note that if $\Delta_2 > 0$, the effect of inbreeding depression is *accelerated* as f increases, a condition often called reinforcing, negative, or **synergistic epistasis**. Conversely, if $\Delta_2 < 0$, then the rate of inbreeding depression *decelerates* as f increases, leading to diminishing, positive, or **antagonistic epistasis**.

Equation 12.4b indicates that a net contribution of epistasis to inbreeding depression may sometimes be detected as a nonlinear relationship between the mean phenotype and level of inbreeding. Because the composite inbreeding effects (Δ_i) can be positive or negative, a variety of forms of this relationship is possible. For example, Gulisija et al. (2007) used local regression (LOESS) methods to show nonlinearity in ID with f, finding both synergistic and antagonistic epistasis over their collection of production traits in Jersey cattle.

ID-based evidence of synergistic epistasis was found for pollen size in *Mimulus* (Kelly 2005) and for pupal number in *Drosophila* (Domínguez-García et al. 2019).

Despite these results, the possibility that the epistatic effects involving different loci may cancel each other implies that a lack of nonlinearity cannot be taken as definitive evidence for the absence of important dominance epistatic interactions between loci. Nonetheless, even in the absence of any canceling effect, large departures from linearity are unlikely unless epistasis is very pronounced (Δ_2 large) for the simple reason that f^2 is small relative to f, especially with small f. Moreover, when a nonlinear response is observed, care must also be taken to ensure that it is not simply due to the selective elimination of lines as inbreeding proceeds (e.g., Figure 12.11). Likewise, if selection is occurring on the trait during inbreeding (as might be expected with fitness-related traits), purging (see below) of the more deleterious alleles can occur. This is especially true for deleterious alleles that interact synergistically, as selection would disproportionately remove these, creating a bias against their detection. Hence, loss of lines and/or selection of alleles among surviving lines both result in the value of I in the surviving set of more inbred lines being potentially different—and perhaps *rather* different—from the base-population starting value (Equation 12.3b). Curik et al. (2001) and Domínguez-García et al. (2019) examined other issues involving the regression of the mean on f when epistasis is present.

Finally, we note that although it is often stated that heterosis (the tendency for F_1 phenotypes to exceed the mean phenotype of the two parental lines; Chapter 13) and inbreeding depression are different manifestations of a common phenomenon (directional dominance), this equivalency is not strictly correct. As we will show in Chapter 13, heterosis is genetically equivalent to $(2\delta_1^c - \alpha_2^c)$, where δ_1^c and α_2^c are the composite dominance and composite additive $\times$ additive effects of genes in the two parental lines. On the other hand, under complete inbreeding within populations, the decline in the mean phenotype is defined by the sum $[\delta_1 + (\alpha\delta) + \delta_2^2\,]$, assuming the base population is in gametic-phase equilibrium. Thus, dominance is a factor in both heterosis and inbreeding depression, but it is not a necessary condition for heterosis, which can arise entirely as a function of additive $\times$ additive epistasis (α_2^c). In addition, inbreeding depression, but not F_1 heterosis, is a function of additive $\times$ dominance and dominance $\times$ dominance interactions. These differences in the genetic underpinnings of heterosis and inbreeding depression are a consequence of the extreme degree of gametic-phase disequilibrium that exists in the first generation of a line cross. Further complexities arise when the F_1 progeny of a line cross are subsequently crossed (Lynch 1991). As will be examined in Chapter 13, the presence of additive epistasis (terms involving A, but not D) can generate **outbreeding depression**, the opposite of heterosis, wherein the F_1 and/or the F_2 have *reduced* fitness relative to the parental lines.

A more subtle, but perhaps more important, difference is that ID is a function of *segregating alleles* within the population being inbred, while heterosis depends upon *allele frequency divergence* between the populations being crossed. We develop this point further below and in Chapter 13.

Example 12.2. Some insight into the impact of epistasis on inbreeding depression was offered by the two-locus analysis of Curik et al. (2001). They assumed that the two loci were unlinked and in linkage equilibrium. They further assumed **identity equilibrium**: locus 1 being inbred in an individual does not influence the probability that locus 2 is also inbred—namely, if f is the probability of inbreeding at the first locus, then f^2 is the probability of inbreeding at both loci (more on this concept later in the chapter). They coded the two-locus genotypic values as follows:

	BB	*Bb*	*bb*
CC	$a_1 + a_2 + aa_{12}$	$a_1 + d_2 + ad_{12}$	$a_1 - a_2 - aa_{12}$
Cc	$d_1 + a_2 + ad_{12}$	$d_1 + d_2 + dd_{12}$	$d_1 - a_2 - ad_{12}$
cc	$-a_1 + a_2 - aa_{12}$	$-a_1 + d_2 - ad_{12}$	$-a_1 - a_2 + aa_{12}$

With this coding, the change in the mean under inbreeding becomes

$$\mu_f - \mu_0 = -f\delta_1 - f(\alpha\delta) - f(2-f)\delta_2$$

where p_1 denotes the frequency of allele *B*, p_2 the frequency of allele *C*, and composite terms are

$$\begin{aligned} \delta_1 &= 2\left[p_1(1-p_1)d_1 + p_2(1-p_2)d_2\right] \\ (\alpha\delta) &= 2\left[(2p_2-1)p_1(1-p_1) + (2p_1-1)p_2(1-p_2)\right] ad_{12} \\ \delta_2 &= 4\left[p_1p_2(1-p_1)(1-p_2)\right] dd_{12} \end{aligned}$$

As an example of how these expressions are obtained, consider δ_2, a function of dd_{12}, which only appears when the genotype *BbCc* is present. In inbreds, its appearance requires that *both loci* are noninbred, which occurs with probability $(1-f)(1-f) = 1 - f(2-f)$. Hence, the decrease in the mean of an inbred due to dd_{12} is $f(2-f)[2p_1(1-p_1)]\,[2p_2(1-p_2)]dd_{12}$.

An important item to note is that the additive × dominance term, $(\alpha\delta)$, can ***change sign with allele frequencies***, as $(2p_i - 1)$ can change sign (as p_i moves from below 0.5 to above 0.5). As noted by Curik et al. (2001), one could formally assert that apparent changes in the amount of inbreeding depression over time (purging, see below) may not be due to the purging of deleterious alleles per se, but rather due to changes in the sign of the composite $(\alpha\delta)$ term induced by allele-frequency change. Don Waller (pers comm) suggested this be referred to as **pseudo-purging**.

METHODOLOGICAL CONSIDERATIONS: CONTROLLED CROSSES

A number of difficulties arise in attempts to test for inbreeding depression, some of which are associated with the selective consequences of the inbreeding depression itself. In humans, for example, there is evidence that consanguineous couples, whose early offspring die from the expression of lethal recessives, compensate by reproducing until viable replacements have been born (Schull and Neel 1972; Overall et al. 2002). Some plants may behave in a similar manner by selective abortion of embryos (Willson and Burley 1983). Lack of knowledge of such compensation can lead to underestimates of the deleterious consequences of inbreeding. Keeping these difficulties in mind, we will now consider the statistical aspects of two common approaches to quantifying inbreeding depression using controlled crosses. These matters are taken up in more detail in Lynch (1988a). We examine the analysis of inbreeding depression in natural populations (based on either pedigree, or marker-based, estimates of f, and hence not requiring controlled crosses) later in the chapter.

Before proceeding, a brief introduction to the temporal dynamics of the inbreeding coefficient f under regular systems of (diploid) mating is necessary. The general theory is covered elsewhere (Crow and Kimura 1970; Hartl and Clark 2006; WL Chapters 3 and 23), and because the vast majority of studies on inbreeding depression involve either self-fertilization or full-sib mating, we simply give the results for these special cases. Starting from a random-mating base population at time 0, the average inbreeding coefficient at a locus after t generations of self-fertilization in the absence of selection is

$$f(t) = 1 - \left(\frac{1}{2}\right)^t \tag{12.5a}$$

The quantity $[1 - f(t)] = (1/2)^t$ is equivalent to the fraction of the heterozygosity in the base population that is still present after t generations of selfing. Thus, a single generation of selfing reduces the number of heterozygous loci within individuals by 50%. Thereafter, the heterozygosity declines geometrically towards zero, such that only 1.6% of the original heterozygosity remains after six generations of selfing. With full-sib mated lines, the inbreeding coefficient must be computed with the recurrence equation,

$$f(t) = \frac{f(t-1)}{2} + \frac{f(t-2)}{4} + \frac{1}{4} \tag{12.5b}$$

starting with $f(-1) = f(0) = 0$. Thus, under full-sib mating, the first generation of inbred progeny has $f(1) = 0.25$, i.e., the heterozygosity within individuals is reduced by 25% relative to that in the random-mating base population. The inbreeding coefficient then progressively approaches one, although more slowly than in the case of selfing. More generally, as developed in Chapter 7, the expected level of inbreeding (f_i) for individual i with parents j and k in a pedigree can be expressed in terms of the coefficients of coancestry (θ_{jk}), with

$$f_i = \theta_{jk} \quad \text{and} \quad f_i = 2\theta_{ii} - 1 \tag{12.5c}$$

With autopolyploids, the rate at which homozygosity is obtained is significantly slower than in diploids (Haldane 1930b; Bartlett and Haldane 1934; Wright 1938; Parsons 1959). For example, selfing autotetraploids take about 3.8 times longer to reach a given level of homozygosity as selfing diploids. When double reductions (Figure 4.3) occur, this increases the rate of inbreeding, but even in the most extreme case, this only reduces the time to around 2.9 that of diploids. Hence, for the same number of generations of selfing, autoploids should be less inbred than diploids (also see Equations 13.17a–13.17i; Table13.4).

Single-generation Analysis

A common short-term test of inbreeding depression involves the comparison of the mean phenotypes of offspring from random matings with those from a specific class of consanguineous mating. In any such analysis, both types of individuals should be raised simultaneously in a random design to eliminate the possibility that the differences in means are a product of the environment (Appendix 9). Ideally, the offspring of both types of matings should be derived from several mothers to minimize the importance of maternal effects, and all mothers certainly should be derived from the same base population.

With such an experimental design, an approximate t test can be constructed for the null hypothesis of no inbreeding depression. Here we take the null model to be one of purely additive gene action. Consider the situation in which n progeny are assayed from each of L independent families, both in the control and in the inbred population. Under the null hypothesis of no inbreeding depression, the difference between the observed mean phenotypes of noninbred and inbred offspring ($\Delta\bar{z} = \bar{z}_O - \bar{z}_I$) has expectation zero, and the observed difference must be evaluated against its sampling variance, the sum of the variances of $\bar{z}_O$ and $\bar{z}_I$. Several factors contribute to this variance, as can be seen by referring to the definition of the sample mean

$$\bar{z} = \frac{1}{Ln}\sum_{i=1}^{L}\sum_{j=1}^{n}(A_{i\cdot} + a_{ij} + E_{i\cdot} + e_{ij})$$

where $A_{i\cdot}$ is the mean genotypic value associated with the ith family, a_{ij} is the deviation from that value for the jth member of the family, $E_{i\cdot}$ is the maternal (or, more generally, common-family) effect associated with the ith family, and e_{ij} is the residual environmental effect on the ijth individual.

We start by considering the control. First, the variance in the control mean caused by environmental effects specific to individuals is $\sigma_e^2/(Ln)$ because all such effects are distributed independently among the Ln individuals. Second, the variance caused by maternal (or general environmental, i.e., common-family) effects is σ_E^2/L; it is divided by L because L mothers (families) contribute to the control mean. Third, because the segregational variance within families in the random-mating base population is $\sigma^2(a_{ij}) = \sigma_A^2/2$, and the effects of such residual variation are random with respect to individuals, the contribution of the within-family genetic variance to the variance of the mean is $\sigma_A^2/(2Ln)$. Finally, the among-family variance, $\sigma^2(A_{i\cdot})$, is also $\sigma_A^2/2$, and it contributes $\sigma_A^2/(2L)$ to the sampling variance of the control mean. Summing up terms, the variance of the control (outbred) mean phenotype becomes

$$\sigma^2(\bar{z}_O) = \frac{1}{L}\left[\frac{1}{2}\left(1+\frac{1}{n}\right)\sigma_A^2 + \sigma_E^2 + \frac{\sigma_e^2}{n}\right] \tag{12.6a}$$

Now consider the situation for a sample of progeny derived by selfing L mothers with n progeny sampled per mother. The sampling variance of the mean resulting from general (common-family) and specific environmental effects is exactly the same as in the control. In the first generation of selfing, the *within-family* segregational variance is also identical to that within the control, $\sigma_A^2/2$. However, because of inbreeding, the variance *among families* is σ_A^2, twice that in the control families, where σ_A^2 is still defined as the genetic variance in the base (control) population. Thus, the expected variance of the mean of the sample of selfed progeny is

$$\sigma^2(\bar{z}_S) = \frac{1}{L}\left[\left(1+\frac{1}{2n}\right)\sigma_A^2 + \sigma_E^2 + \frac{\sigma_e^2}{n}\right] \tag{12.6b}$$

The situation for full-sib mating is a little more complicated, but assuming that all progeny within lines are derived from a single brother–sister mating,

$$\sigma^2(\bar{z}_{FS}) = \frac{1}{L}\left[\left(\frac{7}{8}+\frac{3}{8n}\right)\sigma_A^2 + \sigma_E^2 + \frac{\sigma_e^2}{n}\right] \tag{12.6c}$$

The above expressions depart slightly from those in Lynch (1988a) and appear to be more accurate. WL Chapter 23 gives general expressions for different levels of inbreeding under a number of different family structures.

Although these formulae give exact expectations of the variances of control and inbred line means under the additive model, they are difficult to implement unless one has prior information on the components of variance in the base population. However, the basic structures of the formulae yield a very useful result. Note that for the cases of selfing and sib-mating, $\sigma^2(\Delta\bar{z}_S) = \sigma^2(\bar{z}_o) + \sigma^2(\bar{z}_S)$ and $\sigma^2(\Delta\bar{z}_{FS}) = \sigma^2(\bar{z}_o) + \sigma^2(\bar{z}_{FS})$, respectively. In both of these cases, provided the sample size within families (n) is at least two, then $\sigma^2(\Delta\bar{z}) \leq 2\sigma^2(z_O)/L$, where $\sigma^2(z_O)$ is the phenotypic variance within the control line. Thus, a conservative test for inbreeding depression based on a single generation of consanguineous mating employs the test statistic

$$\tau = \frac{|\Delta\bar{z}|}{\text{SD}(z_O)\sqrt{2/L}} \tag{12.7}$$

where $\text{SD}(z_O)$ is the observed phenotypic standard deviation in the random-mating population. Sampling distributions of means are usually approximately normally distributed, so τ may be treated as t-distributed with $L-1$ degrees of freedom.

In the case of self-compatible plants that produce multiple flowers, there is a simple way to further increase the power of a test of inbreeding depression. For any pair of parent plants (A and B), both reciprocal outcrosses (A $\times$ B and B $\times$ A) and two inbreds (A $\times$ A and B $\times$ B) can be produced. Because the two parents contribute equal numbers of genes to both inbred and outbred progeny, variance from general (maternal) environmental effects and parent sampling do not contribute to $\sigma^2(\Delta\bar{z})$ in this case, and for any pair of parent individuals, the test statistic

$$\Delta\bar{z}_{A,B} = \frac{(\bar{z}_{AA}+\bar{z}_{BB})-(\bar{z}_{AB}+\bar{z}_{BA})}{2} \tag{12.8a}$$

has an expected value equal to zero under the null hypothesis of no inbreeding depression. If n replicates are assayed within each of the four groups of progeny, the expected sampling variance of $\Delta\bar{z}_{A,B}$ is

$$\sigma^2(\Delta\bar{z}_{A,B}) = \frac{(\sigma_A^2/2)+\sigma_e^2}{n} \tag{12.8b}$$

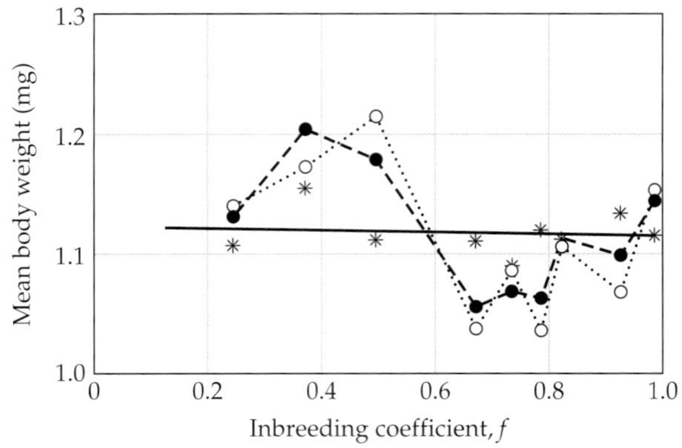

Figure 12.4 The observed trend for mean body weight in inbred (•) and control (○) lines of *Drosophila melanogaster* confounds inbreeding and generational (environmental) trends. The trend from the former can be estimated by use of Equation 12.12, with corrected values given by (∗). Their partial regression is represented by the solid line. (From Kidwell and Kidwell 1966.)

Because the numerator of this expression is less than $\sigma^2(z_O)$, a conservative test for inbreeding depression associated with any pair of parents is provided by

$$\tau = \frac{|\Delta \bar{z}_{A,B}|}{\text{SD}(z_O)/\sqrt{n}} \tag{12.9}$$

where $\text{SD}(z_O)$ is again the phenotypic standard deviation of outcrossed individuals.

Multigenerational Analysis

A common long-term approach to quantifying inbreeding depression on a trait is to regress the mean phenotype on the inbreeding coefficient (Figure 12.2). In studies of this sort, the data usually represent progressively inbred generations derived from the same population, a protocol that introduces a series of statistical issues. First, when the different classes of inbreeding are assayed in different generations (the usual case in animals), the possibility exists that any trend in the mean may be caused by a shift in the environment. An important example of this may be improved husbandry as knowledge about a captive species accrues, resulting in a captive environment that becomes more favorable over time (Hedrick and Kalinowski 2000). Similarly, a newly formed population may show adaptation to captivity, with its fitness increasing as generations of captivity proceed.

Second, the usual assumptions underlying hypothesis testing in ordinary least-squares regression theory are violated in at least two ways: (i) because the means are based upon individuals that are descendants of each other, the data are not independent, and (ii) the sampling variance of the means varies with f because of the loss of genetic variance with inbreeding. The problem of nonindependence of data is a particularly serious one, because it diminishes the effective degrees of freedom in an analysis. For example, once a population of selfers is almost completely inbred, the subsequent generations are no longer free to vary genetically (except by mutation). Nonindependence of data can also cause spurious nonlinearities in the apparent response to inbreeding—if one data point lies above the regression line, the preceding and subsequent ones are likely to do so as well. The result is that regression residuals are likely both correlated and heteroscedastic, requiring a GLS, rather than OLS, regression framework (Chapter 10; WL Chapter 18).

A partial resolution of the nonindependence problem is given below, and the complete solution is outlined in Example 12.4. First, however, some attention needs to be given to the correction of data for environmental shifts between generations. Plant breeders have been able to avoid this issue by storing seed from progressive generations of inbreeding and then growing representatives of all generations simultaneously in a randomized design (Russell et al. 1963; Hallauer and Sears 1973; Cornelius and Dudley 1974). Even here, it is assumed implicitly that seed storage time does not influence performance and that general environmental effects experienced by the parents are not transmitted to the progeny. With most animals, embryo storage is either not currently reliable or not (generally) economically feasible, so there is need for statistical approaches for correcting the data. The question of interest is whether a trend in the inbred-line means is influenced by a temporal shift in the environment. The issue is not trivial as can be seen from the striking parallel directional trend in control and full-sib mated lines of *Drosophila melanogaster* shown in Figure 12.4.

A solution to the environmental trend problem was proposed by Muir (1986a, 1986b), who suggested the use of a **parallel control** as a means of assaying changes in the environment. There are two important considerations in the choice of a control. First, it is essential that the temporal phenotypic changes in the control are entirely attributable to environmental causes. This condition will essentially hold if clones or highly inbred lines are used. A random-bred base population may also serve as an adequate control, provided the character of interest is not modified by selection during the course of the experiment and that the population is large enough that significant genetic drift is unlikely to occur. Second, given a choice of control lines, the one that provides the strongest signal of the environment, i.e., explains a maximum amount of the variance in the inbred line means, is most desirable. Finally, the assumption is no (or very little) genotype $\times$ environment interaction, so that the control and inbred lines equally (additively) react to the environment. WL Chapter 18 examines the use of control populations in more detail.

We start by considering the mean phenotypes of the control ($\bar{z}_C$) and inbred ($\bar{z}_I$) lines at generation t to be functions of general environmental effects common to both of them (E), special environmental effects unique to each of them (e_C and e_I), and genetic change confined to the inbred population, $\Delta\mu_G(t)$,

$$\bar{z}_I(t) = \mu_I(0) + E(t) + e_I(t) + \Delta\mu_G(t) \tag{12.10a}$$

$$\bar{z}_C(t) = \mu_C(0) + E(t) + e_C(t) \tag{12.10b}$$

Because the general environmental effects, $E(t)$, are the only common components of the inbred and control line means, a partial regression of the observed inbred line means, $\bar{z}_I(t)$, on the observed control line means, $\bar{z}_C(t)$, and the inbreeding coefficient, $f(t)$, provides a way of factoring out any general trend of the environment,

$$\bar{z}_I(t) = a + b\bar{z}_C(t) + If(t) + e(t) \tag{12.11}$$

where I is the estimated inbreeding depression (i.e., the expected difference in mean phenotypes of noninbred and completely inbred individuals; Equation 12.3b), and $e(t)$ is the deviation of the mean at generation t from its regression prediction. Applying Equation 12.11, after removing any environmental trend, the corrected means for the inbred lines become

$$\bar{z}_I^*(t) = \bar{z}_I(t) - b[\,\bar{z}_C(t) - \bar{z}_C\,] \tag{12.12}$$

where $\bar{z}_C$ is the mean phenotype of the control lines over all generations. The estimated inbreeding depression (I) is equivalent to the regression of the $\bar{z}_I^*(t)$ on $f(t)$. Figure 12.4 shows a rather striking example of how the application of Muir's approach can overcome a trend obscured by environmental factors. See Rodríguez-Ramilo et al. (2004) for an application of this approach to viabilty under inbreeding in *Drosophila*.

Finally, we return to the problem of hypothesis testing, assuming that the means have been corrected adequately for general environmental trends prior to analysis. Because it ignores the nonindependence of data, OLS regression of $\bar{z}_I^*(t)$ on $f(t)$ leads to downwardly biased estimates of the standard error of I, often by a factor of three or four (Lynch 1988a; WL Chapter 18). An expression for the sampling variance of I, which fully accounts for the correlational structure of the data (under the null hypothesis of a neutral character with an additive genetic basis) is given in Lynch (1988a). A more formal mixed-model solution is outlined in Example 12.4. Lynch's solution, portrayed graphically

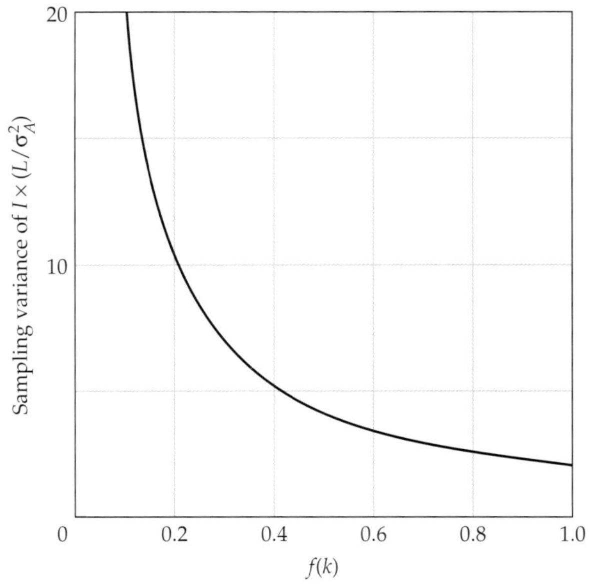

Figure 12.5 The *minimum* sampling variance of the regression coefficient of consecutive line means on their respective inbreeding coefficients, under the assumption of purely additive gene action and ignoring environmental effects. $f(k)$ is the inbreeding coefficient in the final generation. To obtain the actual sampling variance of I, the points on the ordinate must be multiplied by σ_A^2/L, the ratio of the additive genetic variance in the base population to the number of inbred lines; see Example 12.3. (From Lynch 1988a.)

in Figure 12.5, assumes that the regression is performed on a progressive series of inbred lines (e.g., self-fertilization, full-sib mating, or first-cousin mating), starting with $f = 0$ and proceeding for k generations to a final level of inbreeding of $f(k)$. The plotted values are *minimum estimates* of the sampling variance of I, because it is assumed that the variance in the environment makes no contribution to the sampling variance of the means. The sampling variance of I depends primarily on the additive genetic variance in the base population, the number of inbred families, and the level of inbreeding in the final generation. Because the sampling variance of I declines with increasing L (number of lines) and increasing $f(k)$, it is clear that for a fixed amount of resources, the smallest unit of inbreeding (selfing or full-sib mating) should be employed while maximizing the number of lines.

Many of the preceding statistical problems can be avoided when data are available for contemporaneous (and hence presumably experiencing very similar environments) groups of individuals inbred to various degrees. Such is typically the case in the analysis of human and domesticated animal populations where pedigrees are known, and the same can be accomplished in experiments that simultaneously mate various classes of relatives and assay their progeny in a common environment. In both cases, provided the individuals with different levels of f are unrelated, the problem of nonindependent data is eliminated, and provided all individuals are assayed contemporaneously, the need for a temporal control is removed. OLS regression of the group mean phenotypes on f then provides a simple approximation of I and its standard error. Such a regression is only an approximation, as it still has heteroscedastic residuals because the sample variance changes with the level of inbreeding. The delicate issue in natural settings is estimating the level of inbreeding among individuals in a random population sample, as opposed to sampling individuals from a known pedigree. We examine marker-based approaches for estimating f values later in the chapter.

Example 12.3. Consider an experimental design in which the means of $L = 10$ full-sib mated lines are assayed from generation 0 with $f = 0$ to generation 9 with $f(k) = 0.859$ (obtained using Equation 12.5b). Reading off Figure 12.5 at $f(k) = 0.859$, we find the point on the ordinate to be 2.5. The expected sampling variance of the slope I under the null model of no inbreeding depression is obtained by multiplying 2.5 by σ_A^2/L, which gives $\sigma_A^2/4$. If an estimate of σ_A^2 is available, then in this case, two standard errors of the slope are estimated by $\sqrt{\text{Var}(A)}$. Because our treatment ignores environmental sources of variance, it is clear that with this design any regression coefficient whose absolute value is less than the square root of the additive genetic variance in the base population must be considered consistent with the null model of no inbreeding depression.

Example 12.4. One can formally accommodate the complicated residual structure associated with inbreeding experiments by using very minor modifications of the methods for GLS and BLUP analyses of selection experiments developed in WL Chapters 18 and 19. We first sketch the simplest analysis under GLS (Chapter 10). The basic model can be written as

$$R_I(t) = \mu_t - \mu_0 = If(t) + e(t) \tag{12.13a}$$

where $R_I(t)$ denotes the cumulative change in the mean from inbreeding depression. Note that this is akin to the cumulative response to selection that was assumed in WL Chapter 18. To apply a GLS regression framework, the residual error structure (for the means over generations) is required. This is given by the matrix $\mathbf{V}$, whose elements are given by WL Equation 18.15c, namely

$$V_{ij} = \begin{cases} \left(\frac{1}{M_0} + 2f_i\right) h^2\sigma_z^2 + \sigma_z^2/M_i & \text{for } i = j \\ \left(\frac{1}{M_0} + 2f_i\right) h^2\sigma_z^2 & \text{for } i < j \end{cases} \tag{12.13b}$$

where M_i individuals are scored in generation i and σ_z^2 is the initial phenotypic variance. Let $\mathbf{R}$ denote the observed vector of cumulative changes in the mean and $\mathbf{F}$ denote the vector of cumulative levels of inbreeding, whose ith elements are, respectively, $R_I(i)$ and $f(i)$. The resulting GLS estimator for the amount, I, of inbreeding depression is given by

$$\widehat{I}(GLS) = (\mathbf{F}^T\mathbf{V}^{-1}\mathbf{F})^{-1}\mathbf{F}^T\mathbf{V}^{-1}\mathbf{R} \tag{12.13c}$$

whose resulting sampling variance is given by

$$\text{Var}\left[\widehat{I}(\text{GLS})\right] = (\mathbf{F}^T\mathbf{V}^{-1}\mathbf{F})^{-1} \tag{12.13d}$$

WL Example 18.3 provides a worked example (for a selection response experiment), and WL Chapter 18 discusses the incorporation of control populations and sample variances for more complicated designs.

One can test for the presence of epistasis by using a quadratic regression

$$R_I(t) = \mu_t - \mu_0 = If(t) + \beta[f(t)]^2 + e(t) \tag{12.13e}$$

with the same error structure as given by Equation 12.13b. The vector of the GLS estimates for I and β follows from Equation 12.13c with $\mathbf{F}$ now being an $n \times 2$ matrix, whose ith row is given by $[f(i), f^2(i)]$. Epistasis is indicated when β is significantly different from zero.

BLUP machinery (Chapter 10; WL Chapters 19 and 20) can also be used to estimate I. This approach requires significantly more information than a simple GLS estimate, as we need the pedigree (or at least relationship estimates) for every *measured individual* y_i in the sample (rather than just the *generation means* $\overline{z}_i$ that are required for a GLS regression). The only modification required to the approach given in WL Chapter 19 (for the BLUP analysis of selection experiment) is that the observation for individual i now becomes

$$y_i = \mu + A_i + If_i + e_i \tag{12.13f}$$

(WL Equation 23.2c), where the random effect A_i denotes the breeding value for individual i, and f_i is its inbreeding level. Hence, one simply adds an additional fixed effect, I, to the model and then estimates and tests its value (like any standard fixed effect). The relatedness and inbreeding among all the measured relatives (the covariance matrix for the A_i) is accommodated through the relationship matrix, $\mathbf{A}$ (with $A_{ij} = 2\sigma_A^2\theta_{ij}$). Misztal et al. (1997) discusses how to incorporate σ_D^2 into the model. Equation 12.13f also allows for a random population sample, as opposed to series of increasingly inbred generations, to be used to estimate I. The key feature is the incorporation of the covariance structure for breeding values, as the presence of relatives generates correlations among the various A_i, biasing estimates of I (Reid et al. 2008; Becker et al. 2016; Lavanch et al. 2024).

Ritland's Method for Partial Selfers

Most empirical attempts to measure inbreeding depression for *fitness* involve assays of individuals in controlled environments. Because lab conditions often deviate substantially from the situation in nature, one is then left wondering how generalizable the results are to field settings. To eliminate this problem, Ritland (1990a, 1990b) proposed a technique for partially selfing populations of plants that involves essentially no disturbance of individuals in nature and requires no direct estimation of individual fitness. Applying neutral molecular markers (Chapter 8) to progeny arrays, it is possible to estimate the fraction of seed that adults produce by self-fertilization (η), as well as to survey the change in genotype frequencies in a population within and between generations. Genotype frequency changes across generations are a function of the degree of selfing in the parents, while the within-generation changes are a function of genotype-specific fitnesses.

Ritland (1990a, 1990b) suggested several ways in which marker information can be exploited to indirectly infer the fitness consequences of inbreeding. Here, we simply point out the simplest situation, which arises when a population has attained an equilibrium state of inbreeding, i.e., a balance between the production of excess homozygosity by selfing and its loss by selection. The inbreeding coefficient (f) of surviving individuals can be computed using the marker-based approaches discussed in Chapter 8 (and also later in this chapter). The ratio of fitnesses of selfed to outcrossed individuals is then estimated by

$$w = \frac{2(1-\eta)f}{\eta(1-f)} \tag{12.14}$$

Note that when $\eta > 0$ and $f = 0$ (i.e., the adult population is in Hardy–Weinberg equilibrium), $w = 0$, implying that selfed progeny have zero fitness. More general estimators that allow for generational changes in η and f, which appear to be common (Dole and Ritland 1993), are provided in Ritland (1990a, 1990b).

Applications of Equation 12.14 to partially selfing plant populations have generally yielded estimates of w that are slightly lower than those obtained by direct observations of the performance of selfed and outcrossed progeny in experimental populations (Eckert and Barrett 1994; Kohn and Biardi 1995; Schultz and Ganders 1996). Although violations in the assumptions of Ritland's model (such as an absence of biparental inbreeding and an absence of linkage between marker and fitness loci) can lead to biased estimates of w, the bias does not generally appear to be large. Thus, the empirical results tentatively suggest that the inbreeding depression observed in manipulated populations may generally be less severe than that expressed in natural settings. This difference may occur because manipulative studies often fail to fully account for all components of fitness (such as seedling survival) or because the effects of deleterious genes are ameliorated in more benign environments (discussed below).

Epistasis and Inbreeding Depression

An unresolved issue is the extent to which epistasis is involved in inbreeding depression. Recall (Equation 12.4b) that a nonlinear relationship between the mean phenotype and the inbreeding coefficient is an indicator of the presence of epistasis involving dominance effects, which can be tested by using quadratic regressions (Example 12.4). As stressed above, this approach can be biased if lines are lost, or altered, by selection during inbreeding.

A simple way to test for nonlinearity, which avoids the pitfalls of regression, is to compare the change in mean phenotype (per increment in f) between two low levels of f and two high levels of f. Provided the two ranges of f are nonoverlapping, the two observed *changes* are statistically independent, even if the individuals at all four points in time are related. Letting the four observed mean phenotypes, in order of increasing f, be $\bar{z}_1$, $\bar{z}_2$, $\bar{z}_3$, and $\bar{z}_4$, a measure of nonlinearity is then given by

$$\Delta I = \frac{\bar{z}_2 - \bar{z}_1}{\Delta f_L} - \frac{\bar{z}_4 - \bar{z}_3}{\Delta f_H} \tag{12.15a}$$

where $\Delta f_L = f_2 - f_1$, and $\Delta f_H = f_4 - f_3$. Under a linear response, both of the ratios in Equation 12.15a should have the same expected value, but should differ if response is nonlinear. A conservative estimate of the sampling variance of ΔI is given by

$$\text{Var}(\Delta I) = \frac{[\text{SE}(\bar{z}_2)]^2 + [\text{SE}(\bar{z}_1)]^2}{(\Delta f_L)^2} + \frac{[\text{SE}(\bar{z}_4)]^2 + [\text{SE}(\bar{z}_3)]^2}{(\Delta f_H)^2} \tag{12.15b}$$

A test statistic for nonlinearity is then

$$\tau = \frac{|\Delta I|}{\sqrt{\text{Var}(\Delta I)}} \tag{12.15c}$$

which under the null hypothesis of linearity should be t-distributed with degrees of freedom equal to the number of inbred lines in the analysis. To ensure that ΔI is not a function of the differential extinction of lines, only the lines surviving to contribute to $\bar{z}_4$ should be used in such an analysis. Domínquez-García et al. (2019) used this approach to show an acceleration of ID with higher f values for pupal number in *Drosophila*. Willis (1993) used a very similar approach to test for epistasis in life-history traits in the monkey flower (*Mimulus guttatus*). Although he did not correct for line loss, very little evidence for epistasis was found. See Curik et al. (2001) and Domínquez-García et al. (2019) for additional discussion of detection of epistasis by regression on f.

Variance in Inbreeding Depression

Evolutionary biologists interested in the origins of diverse mating systems (e.g., selfing rates), particularly in plants, are quite interested in the potential for *variance* in inbreeding depression among members of the same population (Holsinger 1988; Johnston and Schoen 1994; Uyenoyama et al. 1994; Schultz and Willis 1995). Such variation would seem to be necessary to foster the evolution of alternative forms of mating (WL Chapter 23). To assay whether such variation exists, plant population biologists often use ratios of the fitness of an individual's selfed progeny ($\overline{W}_{Si}$) to their outcrossed progeny ($\overline{W}_{Oi}$) as a measure of its inbreeding depression. This practice raises some statistical problems in that the ratio of expected values is a biased estimate of the expected value of a ratio (Appendix 1). Some of the issues are discussed by Johnston and Schoen (1994), but some of the formulae in their paper are incorrect. Using an improved approximation for the expected value of a ratio (Equation A1.19a), an unbiased estimate of the performance of selfed relative to outcrossed progeny is given by

$$w_i = \frac{\overline{W}_{Si}}{\overline{W}_{Oi}} \left[\frac{1}{1 + [\sigma^2(W_{Oi})/(n_i \overline{W}^{\,2}_{Oi})]} \right] \tag{12.16}$$

where $\overline{W}_{Si}$ and $\overline{W}_{Oi}$ are the observed mean fitnesses of selfed and outcrossed progeny derived from individual i, $\sigma^2(W_{Oi})$ is the variance in fitness of outcrossed progeny, and n_i is the number of outcrossed progeny assayed, all for the ith individual. Using the variance

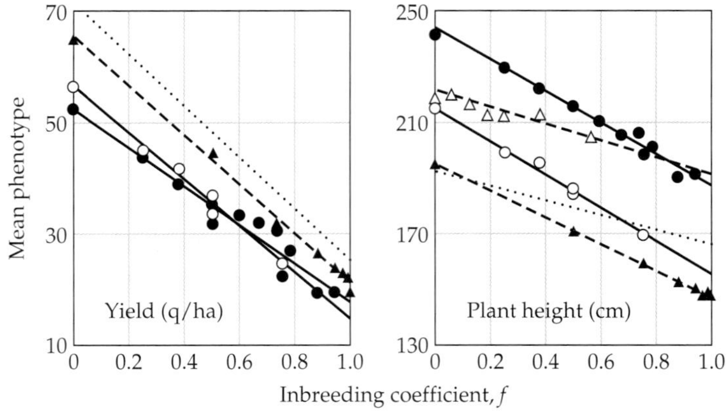

Figure 12.6 The response of mean grain yield and plant height to inbreeding in maize. Data are from: (•, ○) Cornelius and Dudley (1974); (closed triangles) Hallauer and Sears (1973); (Δ) Sing et al. (1967); (· · ·; only the regression line is available) Good and Hallauer (1977). The two studies of Cornelius and Dudley are for the same lines grown in different years. The variation in intercepts is presumably due to differences among base populations as well as among environments in which the experiments were performed.

among the $\widehat{w}_i$ estimates results in an overestimation of $\sigma^2(w_i)$—the variance in inbreeding depression—as the former includes the sampling (error) variance, as $\widehat{w}_i = w_i + e_i$. ANOVA approaches can be used to help separate error variance from the variance in inbreeding depression (Pray and Goodnight 1995), and Moorad and Wade (2005) have suggested specific mating designs to accomplish this goal.

Using measures such as Equation 12.16 to quantify variance in inbreeding depression among individuals raises a number of difficult and unresolved issues. A central problem is that inbreeding depression is not just a property of the *individual*, but of the individual's *prospective mates* as well. It is straightforward enough to estimate an individual's fitness through selfing, but what about the situation with species with separate sexes? An individual's sibs will generally differ with respect to fitness, so the fitness of progeny from full-sib matings will depend on which sibs are employed as mates. The situation is even more extreme when one considers the fitness of individuals produced through outcrossing. Ideally, one would like an estimate of the fitness of outcrossed progeny averaged over all potential mates, but with most species (other than plants), only a small number of matings per individual are possible.

THE EVIDENCE FOR INBREEDING DEPRESSION

Although few of the existing studies of inbreeding depression have fully accounted for all of the difficulties pointed out above, the aggregate of evidence for inbreeding depression is overwhelming. While substantial variation of inbreeding depression exists among species, among traits within species, and even between the sexes (Ebel and Phillips 2016; Clark et al. 2019), almost all organisms exhibit it to some degree. Here we only summarize some of the better-documented cases. An extensive survey of the early literature is available in Wright (1978), and recent reviews are given by Shields (1982), Charlesworth and Charlesworth (1987), Thornhill (1993), Husband and Schemske (1996), Crnokrak and Roff (1999), Keller and Waller (2002), Kristensen and Sørensen (2005), Leroy (2014), and Doekes et al. (2021).

More and better data on the phenotypic consequences of inbreeding are available for maize than for any other organism. Hallauer and Miranda (1981) and Hallauer et al. (2010) review the evidence, which was recognized as early as 1876 by Darwin. There have been some very well-conceived experiments involving prolonged selfing and full-sib mating in lines derived from a genetically diverse base population (Sing et al. 1967; Hallauer and Sears 1973; Cornelius and Dudley 1974; Good and Hallauer 1977; Lamkey and Smith 1987;

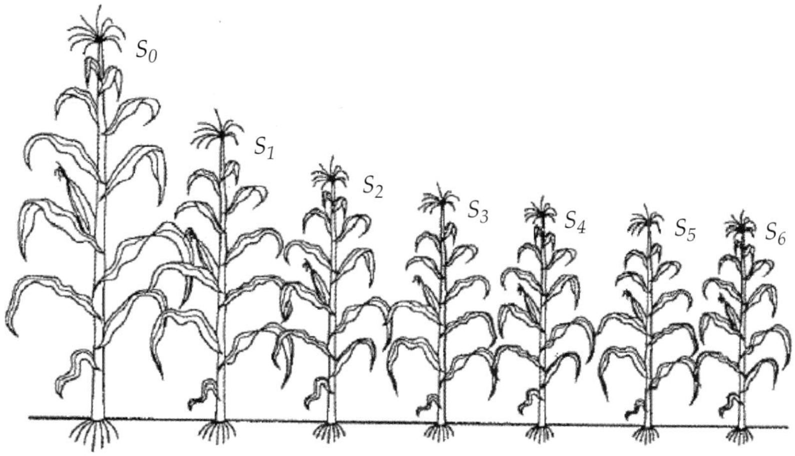

Figure 12.7 Impact of inbreeding on maize height. S_i denotes the number of generations of selfing. S_0 is the hybrid between two inbred lines, and may (depending on the cross) show heterosis, while S_1 is their selfed offspring, in which half the heterotic advantage is removed (Chapter 13). The decline under further generations of selfing reflects inbreeding depression.

Benson and Hallauer 1994). The experiments are very large (involving up to 250 independent lines), and the potential influence of temporal changes in the environment has been minimized by the simultaneous analysis of stored seed. Almost without exception, vegetative, reproductive, and physiological characters exhibit significant shifts in the mean phenotype with inbreeding. Cases have arisen in which the regressions of $\bar{z}$ on f appear to be nonlinear (Hallauer and Sears 1973; Good and Hallauer 1977), but these departures were small. Two characters showing no evidence of nonlinearity are total grain yield and plant height (Figure 12.6). Starting from a genetically diverse base population, complete inbreeding results in an approximately 65% decline in yield and an approximately 25% decline in plant height (Figure 12.7).

Several independent investigations of inbreeding depression have been performed with laboratory stocks of *Drosophila*, a sample of which are listed in Table 12.2. Although substantial variation exists among the results from different studies, a pattern emerges. Primary fitness characters such as viability, fertility, and egg production tend to exhibit very high levels of inbreeding depression (averaging a ~50% reduction in the mean; $I_s = 0.5$), while morphological characters (bristle numbers, body weight and length), which are perhaps more remotely related to fitness, change by only a few percent ($I_s \leq 0.1$), if at all. The latter traits are known to exhibit substantial levels of additive genetic variance, while the fitness characters tend to have lower heritabilities (Mousseau and Roff 1987; WL Chapter 6). Thus, in *Drosophila* there appears to be a major difference in the way the genetic variance for morphological and fitness characters is partitioned: mostly additive for the former, and substantially dominant in the latter. A meta-analysis by DeRose and Roff (1999) of 86 life-history and 15 morphological traits over 54 different species of animals (mostly from wild populations) replicated this pattern of greater inbreeding depression in life-history, as opposed to morphological, traits.

Selection theory helps explain why the additive genetic variance for fitness should be low and why dominance should be directional for fitness-related characters (WL Chapters 5–7, 24–28). Alleles with favorable effects on fitness should move rapidly towards fixation, regardless of their degree of dominance, and dominant alleles with deleterious effects will be eliminated rapidly. However, deleterious recessive alleles will be maintained at low frequencies by mutation pressure. For characters only weakly related to fitness or under stabilizing selection for an intermediate optimum, directional dominance may be less

Table 12.2 A survey of the (scaled) inbreeding depression observed in laboratory populations of *Drosophila*. Here, $I_s = (\bar{z}_O - \bar{z}_1)/\bar{z}_O = 1 - (\bar{z}_1/\bar{z}_O)$, where $\bar{z}_O$ and $\bar{z}_1$ are, respectively, the means of the random-mating base ($f = 0$) and the completely inbred ($f = 1$) population (obtained by linear extrapolation). This scaled change (reduction as a fraction of the original mean) induced by inbreeding allows us to compare its impact over traits. Results marked with an asterisk were obtained from studies involving only one or two chromosomes; in these cases, extrapolation to the entire genome was done by assuming that each major chromosome arm constitutes 20% of the genome, and that the effects are multiplicative across chromosomes. Negative values imply an increase in character value with inbreeding. All data are for *D. melanogaster*, except for *D. subobscura* (Hollingsworth and Maynard Smith 1955), *D. pseudoobscura* (Dobzhansky and Spassky 1963; Dobzhansky et al. 1963; Marinkovic 1967), and *D. willistoni* (Malogolowkin-Cohen et al. 1964).

Character	I_s	Reference
Competitive ability	0.84	Latter et al. 1995
	0.97	Latter and Sved 1994
Egg-to-adult viability	0.57	Garcia et al. 1994
	0.44	Mackay 1985
	0.66*	Malogolowkin-Cohen et al. 1964
	0.48*	Dobzhansky et al. 1963
	0.06	Tantaway and Reeve 1956
Female fertility	0.81	Mackay 1985
	0.18	Tantaway and Reeve 1956
	0.35	Hollingsworth and Maynard Smith 1955
Female rate of reproduction	0.32	Latter et al. 1995
	0.56	Mackay 1985
	0.96	Hollingsworth and Maynard Smith 1955
	0.57	Marinkovic 1967
Male mating ability	0.52*	Hughes 1995
	0.92	Partridge et al. 1985
	0.76	Sharp 1984
Male longevity	0.18*	Hughes 1995
Male fertility	0.00*	Hughes 1995
	0.22*	Dobzhansky and Spassky 1963
Male weight	0.07*	Hughes 1995
	0.10	Mackay 1985
Female weight	−0.10	Kidwell and Kidwell 1966
Abdominal bristle number	0.05	Mackay 1985
	0.06	Kidwell and Kidwell 1966
	0.00	Rasmuson 1952
Sternopleural bristle number	−0.01	Mackay 1985
	0.00	Rasmuson 1952
Wing length	0.03	Tantaway 1957
	0.01	Tantaway and Reeve 1956
Thorax length	0.02	Tantaway 1957

pronounced because mutations that cause a shift in the mean in either direction will be selectively equivalent.

Turning to humans, several large surveys provide firm empirical justification for the incest taboos that exist in many, but not all, societies. Despite this potential for significant inbreeding effects, almost 11% of the global population consists of couples related as at least second cousins ($f \geq 0.0156$) and their progeny (Bittles and Black 2010). Rarely is it possible to obtain data for more extreme situations than first-cousin marriages, but

Table 12.3 The decrease (I) in the mean expected upon complete inbreeding in humans, equal to the mean for noninbred individuals minus the expectation at $f = 1$ (obtained by linear extrapolation). These I values are based on pedigree-based estimates of f. Marker-based estimates of f offer more power (in many settings) and are examined later in the chapter.

Trait	Site	I	Reference
Birth weight (kg)	Japan	5.4*	Morton 1958
	United States	1.7	Slatis and Hoene 1961
Adult height (cm)	Hutterites, U. S.	56	Barrai et al. 1964
	Italy	3	Mange 1964
	Japan	20	Schull 1962
		21	Neel et al. 1970
IQ	Japan	42	Neel et al. 1970
		43	Kudo et al. 1972
		73	Schull and Neel 1965
	United States	42	Slatis and Hoene 1961
Prereproductive survival (%)	Global	70	Bittles and Neel 1994

*This value is obviously too high because it gives a birth weight less than zero.

by linear extrapolation the data are sufficient to demonstrate that more extreme inbreeding would lead to substantial depression in body size and IQ (Table 12.3). With the advent of molecular markers, more refined estimates of f can be generated for otherwise random human samples (see below). For example, in a British population of mainly European extraction, 0.03% (~1/3600) had **extreme inbreeding** (genomes with 10% or greater of their total size being runs of homozygosity; see below).

The consequences of inbreeding for juvenile mortality and the incidence of congenital effects in human are well known and are examined in the next section from a somewhat different perspective (lethal equivalents). One precautionary note is in order. Analyses based on inbreeding depression that rely on natural mating assemblages (as is always true in human studies) run the potential risk that progeny with different levels of f are products of genotypically different groups of parents. If parents that tend to inbreed also tend to be genetically low on the fitness scale, the apparent level of inbreeding depression in the progeny may be substantially exaggerated. As we detail shortly, molecular-marker based estimates of ID for traits weakly correlated with fitness are especially susceptible to **social confounding**, with some confounder (C) influencing both increased levels of inbreeding I and the trait of interest z, creating an association between them ($I \leftarrow C \rightarrow z$).

Because domesticated animals tend to have both higher mean values and greater within-population variance in f, smaller sample sizes are required to detect significant ID effects in livestock relative to humans. Extensive reviews exist on the deleterious consequences of inbreeding in domesticated animals: beef cattle (Dinkel et al. 1968), dairy cattle (Turton 1981; Bezdíček 2017; Howard et al. 2017), dogs (Scott and Fuller 1965), horses (Cothran et al. 1986), poultry (Abplanalp 1990), sheep (Lamberson and Thomas 1984; Wiener et al. 1992a–1992c), swine (Dickerson et al. 1954; Bereskin et al. 1968), and salmonid fishes (Wang et al. 2002). In a meta-analysis of 57 studies covering seven livestock species, Leroy (2014) found that the average scaled change under inbreeding, I_s (Table 12.2; the fractional reduction in terms of the initial mean) was 0.137, while the average when restricted to production traits was 0.351. Leroy argued that the stronger impact of inbreeding on production traits reflected their past history of selection (akin to life-history traits typically showing larger inbreeding effects than other traits). A more expansive analysis of 154 livestock studies by Doekes et al. (2021) found essentially the same results.

A nonlinear response of the mean phenotype to the inbreeding coefficient has been seen in many of these studies. However, because few studies incorporate appropriate controls

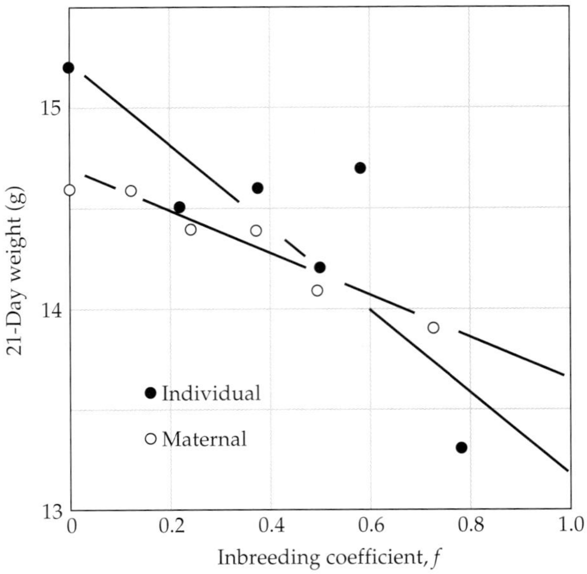

Figure 12.8 The decline in offspring size in laboratory mice as a function of individual (•) and maternal (○) inbreeding. The results of two experiments have been combined after adjusting for mean differences in the control lines. (From White 1972.)

or account for the nonindependence of data in inbred lines, it is difficult to say whether the apparent nonlinearities in the data are simply statistical artifacts, as opposed to real reflections of directional epistasis. In organisms with extensive parental care, still another explanation exists. If **maternal performance** (the ability to raise young) is adversely affected by inbreeding, then an individual's phenotype will be influenced not only by its own level of inbreeding but also by that of its mother. As we will see in the next chapter, the same is also true with heterosis. When males participate in offspring care, inbreeding depression in **paternal performance** can occur, such as detected in song sparrows by Marr et al. (2006).

An elegant experiment performed with laboratory mice (White 1972) illustrates the impact of maternal inbreeding. When progeny with several different levels of inbreeding were crossfostered with mothers inbred to different degrees, maternal inbreeding was found to have approximately half the impact on juvenile weight as individual inbreeding (Figure 12.8). Similarly, Ávila et al. (2013) found that maternal inbreeding accounts for about 50% of the ID effect on *Drosophila* pupal number. Other experiments with mice have verified the effects of maternal inbreeding on progeny performance (Bowman and Falconer 1960; Falconer and Roberts 1960; McCarthy 1967; Nagai et al. 1971), and convincing, but less extensive, data exist for humans (Schull et al. 1970), sheep (Wiener et al. 1992a–1992c), and birds (Sittmann et al. 1966; van Noordwijk and Scharloo 1981). Recent marker-intensive studies have also found an impact of inbreeding on maternal performance in natural populations of Red deer (Huisman et al. 2016) and Soay sheep (Bérénos et al. 2016; Pemberton et al. 2017).

A brief aside is in order here on inbred mouse lines, which are widely used in biomedical research (to provide genetic homogeneity over studies conducted in different labs). It must be stressed that most lines die out during the rapid inbreeding to form the final collection. For example, 95% of all lines from the mouse collaborative cross mapping population (Chapter 18) became extinct, in large part due to male infertility (Shorter et al. 2017). Hence, the combinations of alleles that survived are a very nonrandom sample of alleles in the base population, so that the resulting Δ_i values (Equation 12.4) in the *surviving* inbred lines are likely different from their *base-population* value at the start of inbreeding.

As mentioned earlier, an important interpretational issue with the impact of inbreeding depression is that it can be *environmentally dependent* (Hedrick and Kalinowski 2000; Keller and Waller 2002; Cheptou and Donohue 2011; Pemberton et al. 2017), often with larger effects in more "stressful" environments (Armbruster and Reed 2005; Fox and Reed 2011; Reed et al. 2012). As cogently discussed by Yun and Agrawal (2014), the key issue is not the level of "stress" per se, but rather how the environment translates into increased selection against deleterious alleles. In particular, they found that in *Drosophila*, increasing density (i.e., increased competition) had the strongest impact on inbreeding depression over other perceived environmental stresses.

Hence, results from laboratory settings may downplay the importance of inbreeding depression in nature. Experiments with *Drosophila* support the contention that inbreeding depression is often more severe in more extreme environments (Hoffmann and Parsons 1991; Kondrashov and Houle 1994; Miller 1994), as do those with the flour beetle *Tribolium* (Pray et al. 1994) and with mice (Jiménez et al. 1994). On the other hand, in a very large study comparing 38 human populations, Bittles and Neel (1994) found that the effects of inbreeding on survival to age 10 were independent of the mortality rate of noninbred progeny, which ranged from 3% to 40%. Likewise, while numerous studies with plants have documented increased inbreeding depression under extreme conditions (Antonovics 1968; Schemske 1983; Dudash 1990; Schmitt and Ehrhardt 1990; Wolfe 1993), there were also many cases in which the influence of the environment was negligible (Johnston 1992; Heywood 1993; Charlesworth et al. 1994; Ouborg and Van Treuren 1994; Nason and Ellstrand 1995; Norman et al. 1995). While we know of no cases in which inbreeding depression was more intense in benign laboratory or greenhouse environments than in natural settings, Pemberton et al. (2017) suggest that some caution may be in order in assuming that the laboratory estimates tend to err on the conservative side with respect to the situation in nature. They suggested that artificial settings might sometimes be more challenging than natural populations. In particular, organisms have evolved to handle the stresses in their natural settings, but artificial settings (such as recent captive husbandry of endangered species) may impose stresses they have yet to encounter in nature.

One interesting potential predictor of the direction of environmentally dependent ID is the **phenotypic-variance hypothesis** of Waller et al. (2008). Their basis was Crow's (1958) opportunity for selection measure, which relates the phenotypic variance of a trait under selection with its maximal possible change (see WL Chapter 29 for a detailed discussion). Waller et al. suggested that the impact of an environment on the phenotypic variance is often a better predictor of whether ID increases or decreases, with higher levels of ID often correlating with larger phenotypic variances. Experiments from three different model plant systems—*Brassica rapa* (Waller et al. 2008), *Selene vulgaris* (Sander and Matthies 2016), and *Mimulus guttatus* (Sandner et al. 2021)—offered some support for this idea.

Finally, a number of recent studies have examined changes in gene expression levels under inbreeding. It is usually unclear if such expression changes impact inbreeding depression. One can imagine settings where increased expression at a particular gene decreases fitness, and settings where the converse is true. A very interesting observation on gene expression and fitness ID was offered by García et al. (2012, 2013) in *Drosophila*. Starting from a common base population, a number of inbred lines were extracted, which showed variation in ID. The three lines showing the least amount of ID in fitness (scored by pupal number per vial) were contrasted with the three (surviving) lines showing the largest ID in fitness, and with a control (the starting base population), for gene expression. As expected, significant differences in gene expression were seen over lines. Surprisingly, the three lines showing the *least* amount of ID had the *largest* differences in gene expression relative to the control (the outbred line). The authors suggest that gene regulation was being used to partly buffer the impact of ID on fitness-related traits.

THE NUMBER OF LETHAL EQUIVALENTS

Much of our discussion on inbreeding depression has been framed in terms of fitness. For example, the genetic architecture of ID for various traits is believed to reflect, in part, how changes in those traits impact fitness. There is a very rich literature on the direct impact of inbreeding on fitness, either total lifetime fitness, or its various components. The latter are usually taken to be multiplicative, so that total fitness (number of offspring produced) can be considered as the product of three (or more) components, typically survival, number of mates, and number of offspring per mating. For an exhaustive treatment of fitness estimation and fitness components, see WL Chapter 29.

Lethal Equivalents

The impact of inbreeding depression on fitness is often expressed using the concept of **lethal equivalents**, due to Morton et al. (1956). They showed how the number of lethal equivalents (which can have a very complex underlying genetic architecture) can be estimated using a simple regression technique. While the initial focus of Morton et al. was on summarizing the deleterious consequences of inbreeding for attributes classified by *incidence* (i.e., binary traits)—such as survival (**viability**), disease expression, and so on—the idea of equivalents can easily be extended to any fitness component (see below). Morton et al. defined a lethal equivalent (**detrimental equivalent** for traits other than survival) as any group of genes "that if dispersed in different individuals . . . would cause on average one death." Thus, a lethal equivalent can consist of a single lethal gene or of a large number of mildly deleterious genes. We note in passing that the literature either reports this as **gametic** (**haploid**) or **zygotic** (**diploid**, twice the gametic) **equivalents**. Unless otherwise stated, we use the gametic number of lethal equivalents.

To develop Morton's method, consider the average probability S_f of survivorship for an individual inbred to level f, in contrast to the survival S_0 of an outbred from the same population. Assuming that the environment and all loci act independently (and multiplicatively) in determining survivorship, then

$$S_f = (1 - P_e) \prod_{i=1}^{n} [1 - P_i(f)] \tag{12.17a}$$

where P_e is the genotype-independent probability of dying from environmental causes, and $P_i(f)$ is the probability of dying as a result of deleterious genes at the ith locus when the inbreeding coefficient is equal to f. From Table 12.1,

$$P_i(f) = f q_i s_i + (1 - f)[\, q_i^2 s_i + 2q_i(1 - q_i)(sh)_i \,] \tag{12.17b}$$

where q_i is the frequency of the deleterious allele, and s_i and $(sh)_i$ are the probabilities of mortality for homozygotes and heterozygotes, respectively. If the probability of dying from any single cause is small, then the approximation $(1 - x) \simeq e^{-x}$ implies $\prod_i (1 - x_i) \simeq \exp(-\sum_i x_i)$, yielding

$$S_f \simeq \exp\left[-P_e - \sum_{i=1}^{n} P_i(f) \right] = \exp[-(A + Bf)] \tag{12.18a}$$

where

$$A = P_e + \sum_{i=1}^{n} q_i [q_i s_i + 2(1 - q_i)(sh)_i] \tag{12.18b}$$

is the sum of probabilities of mortality in the random-mating population, and

$$B = \sum_{i=1}^{n} q_i [s_i - q_i s_i - 2(1 - q_i)(sh)_i] \tag{12.18c}$$

is the excess sum of probabilities of mortality that would exist in a completely inbred population. We can also express Equation 12.18a as

$$S_f = S_0 \exp[-Bf], \qquad \text{with} \quad S_0 = \exp(-A) \tag{12.18d}$$

showing that the fractional decline in survival (S_f/S_0) due to inbreeding is $\exp[-Bf]$.

Unfortunately, the mean number of lethal equivalents per gamete, $\sum q_i s_i$, cannot be separated cleanly from other terms in the definitions of A and B. However, because $A+B = P_e + \sum q_i s_i$, the number of lethal equivalents per gamete must be between B and $A+B$, assuming $(sh)_i \geq 0$. As will be seen below, estimates of $A+B$ are usually not much greater than B, so the use of B as an approximation of the effective number of lethals is not greatly troubling.

Estimation of *B*

Logarithmic transformation of Equation 12.18a leads to

$$\ln(S_f) \simeq -A - Bf \tag{12.19a}$$

Thus, the composite measures A and B can be estimated by regressing the natural logarithm of survivorship (or trait incidence) on the inbreeding coefficient (S_f being the fraction of survivors in a cohort of size N_f, all inbred to level f). Morton et al. (1956) recommended the use of weighted least-squares (Example 10.8), weighting the data by $N_f S_f/[1-S_f]$, the inverse of the sampling variance of $\ln(S_f)$, and iterating the regression by substituting expected for observed S_f in the weights. (Recall that a similar procedure was used in the estimation of line-cross variances; Example 11.3).

A problem with this approach arises in cases in which the observed survivorship for certain inbreeding classes is zero as the logarithm of zero is undefined. Templeton and Read (1984) suggested the small sample-size correction

$$S'_f = \frac{1+N'_f}{2+N_f} \tag{12.19b}$$

where N_f and N'_f are the numbers of total and surviving individuals at inbreeding level f. With no observed survivors, this quantity rapidly approaches $\approx 1/N_f$ as the class sample size increases. However, this correction can yield biased estimates of B (Willis and Wiese 1997; Lacy 1997; Kalinowski and Hedrick 1998; Nietlisbach et al. 2019).

To circumvent this concern, Kalinowski and Hedrick (1998) suggested estimators based on maximum likelihood (Appendix 4). Let $n_{i,0}$ and $n_{i,1}$ denote the number of deaths and survivors, respectively, for individuals in inbreeding class i (with value f_i). From Equation 12.18d, the survival probability for class i can be expressed as $S_i = S_0 e^{-Bf_i}$, with the resulting likelihood for m inbreeding classes given by

$$L \propto \prod_{i=1}^{m} (1-S_i)^{n_{i,0}}\, S_i^{n_{i,1}} = \prod_{i=1}^{m} \left(1 - S_0 e^{-Bf_i}\right)^{n_{i,0}} \left(S_0 e^{-Bf_i}\right)^{n_{i,1}} \tag{12.20a}$$

The MLEs for S_0 and B are those values that maximize Equation 12.20a (Appendix 4).

Building on this approach, Glémin et al. (2006), Armstrong and Cassey (2007), and Grueber et al. (2011b) advocated for the use of generalized linear mixed models (GLMMs) for the estimation of B. As detailed in Chapter 14, GLMMs allow for more accurate modeling of residuals when observations are discrete (such as $0/1$ or count data), while expressing the expected value of y_i for a given set of parameter values as a function (g) of a linear model, $E[y_i] = g(\sum \beta_j x_{ij})$. One might initially think that a good GLMM candidate to estimate B would be **logistic regression,**

$$\text{logit}(y_i) = \mathcal{A} + \mathcal{B} f_i + \sum \beta_k x_{ik} \tag{12.20b}$$

where the **logit function** (Equation 14.14b) is the inverse of the logistic function (Equation 14.14a). The GLMM given by Equation 12.20b is said to have a **logit-link**. This link function maps a linear function of the inbreeding coefficient, as well as other cofactors such as sex and age (β_k), into an expected value between zero and one, with resulting binomial residuals (Chapter 14). However, the careful reader will note that we replaced A and B in Equation 12.19a by $\mathcal{A}$ and $\mathcal{B}$ in Equation 12.20b, as A and B arise in $\log(y)$, but *not* in logit(y). Thus, although estimates of $\mathcal{B}$ are occasionally taken as estimates of B (Table 12.4), this is incorrect.

A much better choice for a GLMM is the **log-linear model** (also called **Poisson regression**; Equation 14.17), with a **log-link** function,

$$\ln(y_i) = A + Bf_i + \sum \beta_k x_{ik} \tag{12.20c}$$

with resulting Poisson residuals (although other error structures, such as normal residuals, can be modeled as well). Note that this is the logical extension of Morton's initial regression (Equation 12.19a). Given that solutions to GLMMs are obtained by ML, one can think of the solutions to Equation 12.20a as the simplest log-link GLMM (with no cofactors). The advantages of moving to a full GLMM framework is that it easily handles the addition of cofactors and that most statistical packages have GLMM libraries.

Besides the issue of $\mathcal{B}$ in the logistic regression not formally corresponding to B in Equation 12.19a, Nietlisbach et al. (2019) noted that estimates of $\mathcal{B}$ can change as the value of $\mathcal{A}$ (the background survival rate) changes. This makes comparison of B values over different populations/species/fitness components problematic. The log-link GLMM does not have either of these two problems. While use of a log-link might seem slightly problematic for survival (0/1) data—as the residual structure for a log-link GLMM is Poisson (rather than binomial)—Zou (2004) presented a correction for binary data. Further, when B is computed for more general fitness components (count data such as the number of offspring), Poisson residuals are a more appropriate model.

Practical situations often arise in which one only has data for noninbred individuals ($f = 0$) and a single class of inbred individuals. While a regression is not possible in this case, Equation 12.18d shows that the parameters can be estimated by

$$A = -\ln S_0 \tag{12.21a}$$

$$B = -\frac{\ln(S_f/S_0)}{f} \tag{12.21b}$$

where, for example, f is $1/2$ for self-fertilization and $1/4$ for full-sib mating. Using the delta method (Appendix 1), the large-sample variance for B in this case is found to be

$$\text{Var}(B) \simeq \frac{1}{f^2}\left(\frac{1-S_f}{S_f N_f} + \frac{1-S_0}{S_0 N_0}\right) \tag{12.22}$$

Note that likelihood (Equation 12.20a) also can be applied in this setting (assuming binary data).

Results from Vertebrates

The regression method of Morton et al. (1956) has been extensively applied to humans, with several independent studies indicating that the average number of lethal equivalents per gamete, B, for a number of traits is on the order of one to two (Table 12.4). Results from traits in other vertebrate species are, for the most part, very similar, suggesting on the order of 0.5 to 3 lethal equivalents (for survival) per gamete (Table 12.4). This translates into one to six lethal equivalents per zygote ($2B$ for a diploid), enough to kill the average individual a few times over if fully expressed in the homozygous state.

Of these data, those for the European bison, Holstein cattle, and the Stewart Island robin, which all exhibit no significant lethal load, are the most anomalous. An independent study in Holsteins (using runs of homozygosity; Table 12.8) did indeed find inbreeding depression (for milk yield). Because the bison and robin populations went through strong bottlenecks, it is possible that much of the inbreeding depression was purged (see below) by extensive inbreeding and selection during the population bottleneck. Another explanation is simply pedigree errors, as these three studies did not use molecular markers to verify paternity. Reid et al. (2014) found that paternity errors resulted in significant underestimation

Table 12.4 Estimates of the effective number of lethals per gamete for vertebrates (bounded by B and $A + B$). All estimates were obtained by regression (Equation 12.19a), except those marked by an asterisk (obtained with Equations 12.21a and 12.21b), by † for maximum-likelihood or log-link GLMM-based estimates, and by ‡ for logit-link GLMM-based estimates.

Species	Trait	A	B	Reference
Humans	Survival to maturity			
	France, 1919–1925	0.16	2.87	Morton et al. 1956
	Chicago, 1936–1956	0.18	1.55	Slatis et al. 1958
	Fukuoka, Japan	0.07	0.67	Yamaguchi et al. 1970
	Nagasaki, Hiroshima, and Hirado, 1948–65	0.10	0.67	Schull and Neel 1972
	Survival to age 10	0.20	0.70	Bittles and Neel 1994*
	Conspicuous abnormalities	0.10	1.16	Slatis et al. 1958
	Mental retardation	0.01	0.80	Morton 1978
	Congenital heart disease	0.01	0.32	Gev et al. 1986
Speke's gazelle	1-year viability	0.42	3.75	Templeton and Read 1984
European bison	2-year viability	0.26	0.13	Slatis 1960
Sheep	Survival, 1.5–5 years	0.09	0.39	Wiener et al. 1992c*
Swine	Embryo survival	0.27	1.01	Pisani and Kerr 1961
Cattle	Survival through calving			
Holstein		0.16	0.02	Pisani and Kerr 1961
Jersey		0.18	1.15	
Hereford		0.19	0.64	MacNeil et al. 1989*
Kangaroo rat	Juvenile survival		0.98	Willoughby et al. 2019†
	Lifespan		1.99	
	Number offspring		3.15	
	Lifetime reproductive success		4.66	
Great tit	Survival to fledging	0.36	0.84	van Noordwijk and Scharloo 1981
			1.00	Nietlisbach et al. 2019†
Japanese quail	16-week survival	0.60	1.91	Sittmann et al. 1966
Chicken	18-month survival	0.82	2.10	Pisani and Kerr 1961
Takahē (flightless rail)	Lifetime fitness		8.03	Grueber et al. 2010‡
Collared flycatchers	Juvenile survival		7.47	Kruuk et al. 2002†
Stewart Island robin	Juvenile survival		0.24	Laws and Jamieson 2011‡
North Island robin	Juvenile survival		4.05	Armstrong and Cassey 2007†
Medium ground finch	Juvenile survival		4.47	Keller et al. 2002†
Hawaiian crow	Juvenile survival		6.80	Hoeck et al. 2015‡
Cactus finch	Juvenile survival		4.27	Keller et al. 2002†
Song sparrow	Juvenile survival	0.71	1.32	Keller 1998
			1.40	Nietlisbach et al. 2019 †
Great tit	Juvenile survival		1.30	Szulkin et al. 2007*
Chatham Island robin	Juvenile survival		3.42	Kennedy et al. 2014‡
			1.40	Nietlisbach et al. 2019 †
Atlantic Cod	Hatch survival		0.94	Plough 2016†
Rainbow trout	Fry survival		0.08	Plough 2016†
Steelhead salmon	Three month survival		0.20	Plough 2016†

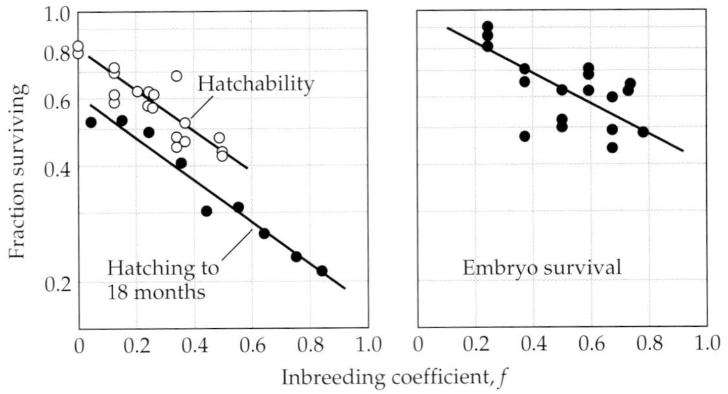

Figure 12.9 Survivorship of white-leghorn chickens (left) and Poland-China pigs (right) as a function of level of inbreeding. (After Pisani and Kerr 1961.)

of inbreeding depression in song sparrows (see Ponzi et al. 2019 for one approach for dealing with pedigree errors). We also note that a study on congenital birth defects, birth weight, and gestational age for people of the Indian state of Tamil Nadu revealed no evidence of inbreeding depression (Rao and Inbaraj 1980). These results, which are quite unusual for humans, may also be related to the decline in the lethal load caused by purging during previous inbreeding. Approximately 40% of the marriages in this population were between second cousins or closer relatives.

An alternative explanation to purging for these low estimates of B follows by noting that inbreeding depression is due to *segregating* deleterious alleles. While purging of these alleles via selection can occur, so can the opposite, wherein deleterious alleles are *fixed* due to small population size. The result is that they are no longer segregating (and hence ID is decreased), but the population is less fit due to these deleterious fixations. We will expand upon this below, as well as in Chapter 13.

A critical assumption underlying lethal-equivalent analysis is that the effects of different loci on survivorship are independent. Directional epistatic interactions between pairs of loci will give rise to nonlinearities in plots of log survival versus inbreeding coefficient, but as noted above, these can be detected only if survivorship estimates are available for several levels of inbreeding ideally over a large range of f values. Other than the data in Figure 12.9, which yielded no evidence of nonlinearity, few data sets are extensive enough to test for this complication.

Results from *Drosophila*

Through the use of **balancer-chromosome** techniques in *Drosophila,* it is possible to obtain a finer picture of the types of deleterious genes that contribute to inbreeding depression. Recall (Figure 5.8) that this procedure enables one to isolate intact chromosomes from natural populations and to assay their homozygous performance with respect to a control chromosome (the **balancer**). By crossing two lines, each one carrying a different chromosome, it is also possible to assay the relative performance of chromosomal heterozygotes. The ratio of the two relative performances provides a measure of the fitness of chromosomal homozygotes ($f = 1$) relative to that of heterozygotes ($f = 0$).

Greenberg and Crow (1960) reasoned that this approach might be exploited to partition the deleterious load in populations into components due to alleles with different magnitudes of effects. The partitioning of fitness classes is arbitrary, but the technique is nevertheless general. The usual procedure has been to classify as **lethals** those chromosomes that, when homozygous, yield less than 10% of the viability observed in random heterozygotes. Chromosomes with relative viabilities greater than 10%, but less than 100%, are referred to as **detrimentals.** Nearly all chromosomes extracted from natural *Drosophila*

Table 12.5 Partitioning of the total number of lethal equivalents (B_T) into the subcomponents resulting from detrimental (B_D) and lethal (B_L) factors. N is the number of studies over which the data are averaged, summarized from Simmons and Crow (1977).

Species	Chromosome	N	B_T	B_D	B_L
Drosophila melanogaster	II	16	0.483	0.236	0.247
	III	3	0.691	0.284	0.407
Drosophila pseudoobscura	II	3	0.450	0.246	0.204
	III	1	0.578	0.352	0.226
Drosophila willistoni	II	1	0.766	0.380	0.386
	III	1	0.690	0.506	0.184

populations have relative viabilities less than one when in the homozygous state, so this categorization encompasses essentially all chromosomes. Denoting the mean viabilities of chromosomal heterozygotes, detrimental chromosomal homozygotes, and all homozygotes as, respectively, S_0, S_D, and S_T, then from Equation 12.21b,

$$B_T = \ln S_0 - \ln S_T \tag{12.23a}$$

$$B_D = \ln S_0 - \ln S_D \tag{12.23b}$$

$$B_L = \ln S_D - \ln S_T \tag{12.23c}$$

The different components of the deleterious load are additive, because they are the summations of effects of individual viability mutations, i.e., $B_T = B_D + B_L$. In other words, B_D (the **deleterious load**) estimates the total number of lethal equivalents resulting from the cumulative effects of all deleterious genes that are individually nonlethal, whereas B_L (the **lethal load**) estimates the additional number of lethal equivalents resulting from recessive lethals being present on a subset of the chromosomes.

Simmons and Crow (1977) summarized a large number of studies employing this approach using the second and third chromosomes in several species of *Drosophila*. The data were remarkably consistent (Table 12.5). The total number of (gametic) lethal equivalents associated with each chromosome, in each species, ranges from 0.5 to 0.8. Noting that chromosomes II and III each comprise approximately 40% of the *Drosophila* genome, these observations suggest that the average drosophilid carries approximately three zygotic lethal equivalents, similar to the situation in vertebrates. Averaging over all of the data, the detrimental and lethal loads per chromosome are, respectively, 0.33 and 0.28. Thus, about half of the total lethal equivalents are associated with lethal recessives, and about one in three chromosomes carries such a gene.

The concern that deleterious alleles do not interact epistatically can be tested with the balancer-chromosome technique. In the absence of average interchromosomal epistasis, the number of lethal equivalents expressed in individuals homozygous for both chromosomes II and III should not be significantly different from the sum of the loads obtained for individuals homozygous for just chromosome II and for just chromosome III. Only a few studies of this nature have been undertaken, and the results are somewhat mixed. The overall picture, albeit a weak one, is that if epistasis exists among the genes on the two chromosomes, it is weak and synergistic (positively reinforcing) (Simmons and Crow 1977).

The other group of invertebrates with a number of estimates (for early survival) are aquaculturally important marine species (summarized by Plough 2016). Estimates of B (mostly based on comparing outbreds with a single inbred group) were 1.06 for scallops, 1.84 for abalone, 1.54 for white shrimp, 6.08 for Eastern oyster, and 3.26 for Pacific oyster. Comparing these results to those from Table 12.4 shows that these B values are on the high end relative to survival estimates for vertebrates.

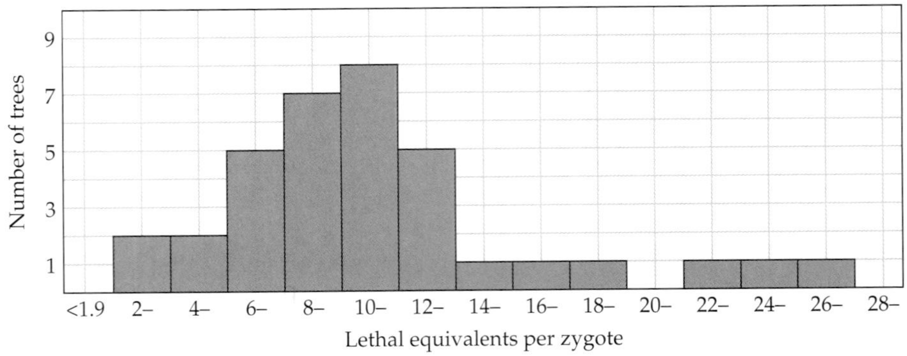

Figure 12.10 Frequency distribution of the mean number of lethal equivalents per zygote ($2B$) for 35 trees in a population of Douglas fir. (After Sorensen 1969.)

Table 12.6 Estimates of lethal equivalents per gamete (B; obtained using Equation 12.21b) affecting early embryonic survival in conifers and herbaceous angiosperms. Estimates for conifers were taken directly from the literature, while those for angiosperms were computed from data on percent germination for outcrossed and selfed seed.

Species	B	Reference
Conifers		
Nobel fir *(Abies procera)*	1.7	Sorensen et al. 1976
Tamarack *(Larix laricina)*	5.4	Park and Fowler 1982
Norway spruce *(Picea abies)*	4.8	Koski 1971
White spruce *(Picea glauca)*	5.0	Fowler and Park 1983
	4.4	Coles and Fowler 1976
Black spruce *(Picea mariana)*	2.4	Park and Fowler 1984
Ponderosa pine *(Pinus ponderosa)*	2.0	Sorensen 1970
Scots pine *(Pinus sylvestris)*	4.4	Koski 1971
	3.6	Savolainen et al. 1992
Loblolly pine *(Pinus taeda)*	4.2	Franklin 1972
	4.8	Bishir and Namkoong 1987
Virginia pine *(Pinus virginiana)*	5.0	Bishir and Namkoong 1987
Douglas fir *(Pseudotsuga menziesii)*	5.0	Sorensen 1969
Short-lived angiosperms		
Begonia hirsuta	0.04	Ågren and Schemske 1993
Begonia semiovata	0.11	Ågren and Schemske 1993
Clarkia tembloriensis	0.07	Holtsford and Ellstrand 1990
Lychnis flos-cuculi	0.39	Hauser and Loeschcke 1994
Mimulus guttatus	0.16	Latta and Ritland 1994
Raphanus sativus	0.01	Nason and Ellstrand 1995
Salvia pratensis	0.67	Ouborg and Van Treuren 1994
Schiedea lydgatei	0.91	Norman et al. 1995

Results from Plants

Equation 12.21b has been used extensively in estimating the number of lethal equivalents in coniferous trees of economic importance. The usual approach has been to compare self-pollinations to outcrosses using a mixture of pollen from several distant trees. Most of the emphasis has been on embryonic mortality, which is easily assayed by counting unfilled seeds. The number of lethal equivalents expressed at this stage is exceptionally high, ranging from one to five per gamete (Table 12.6). Most studies also show a high variance in the number of lethal equivalents per individual ($2B$ for diploids), with few individuals completely free of them and some carrying close to 30 (Figure 12.10). Longer term studies (Park and Fowler 1982, 1984; Fowler and Park 1983; Winn et al. 2011) indicate that most of the lethal equivalents affecting survival are expressed at the embryonic stage, with approximately one to two additional lethal equivalents per gamete influencing subsequent survivorship. It is conceivable that the extraordinarily high mutation load in conifers is a consequence of their long generation time, which may magnify the mutation rate on a per-generation basis (Klekowski 1988). Scofield and Schultz (2006), noting that high-stature plants have many more rounds of mitoses from zygote to gamete relative to small-stature plants, postulated that the per-generation mutation rate is proportional to the number of mitoses. If this is correct, large-statured plants are expected to have higher mutation rates, and stronger ID, relative to small-stature plants. Consistent with this theory, fundamentally different results arise with short-lived herbaceous plants, where B for the probability of germination is consistently less than one (Table 12.6). Such plants do, however, express additional lethal equivalents in the form of survival to maturity and reproductive performance (see Charlesworth and Charlesworth 1987 for a summary).

Finally, Hedrick (1987b) reviewed the extensive literature on genetic load in ferns. The majority of studies have been performed by selfing gametophytes and counting the proportion of spores that germinate, ignoring the load expressed following germination. Again, the mean number of lethal equivalents per gamete, which ranges from 0 to 1.3, appears to be substantially lower than that found in conifers.

Example 12.5. An interesting study on the importance of early selection under inbreeding was presented by Hedrick et al. (2016). Myburg et al. (2014) found that 28 selfed offspring from a single *Eucalyptus* parent had a much higher fraction of heterozygosity in surviving offspring relative to the expectation under Mendelian segregation. The selfed parent was heterozygous at 9590 markers. Heterozygosity at these loci in the 28 offspring averaged 65.5%, a highly significant departure from the expected 50% value. Hedrick et al. noted that the locations of the excesses were widely distributed over the genome. Evidence suggested that this was due to several deleterious alleles, rather than single lethals, within each region. The data further suggested that likely over one hundred genes contributed to inbreeding depression.

As discussed below, this observation of **distorted Mendelian segregation ratios** has been seen in earlier studies, presumably reflecting selection against inbred individuals. For example, studies using Pacific oyster (*Crassostrea gigas*; Launey and Hedgecock 2001; Plough and Hedgecock 2011), flat oysters (*Ostrea edulis*; Bierne et al. 1998), and sugar maples (*Acer saccharum*; Mulcahy 1975) all revealed that initially Mendelian ratios in families became increasingly more skewed toward heterozygotes over time. Hedrick et al. modeled the type of overdominant architecture required to generate such an excess of heterozygosity in the *Eucalyptus* offspring. They found that >20 overdominant loci, each with a selective advantage of 0.6, would be required. Such genes with large effect did not seem to be present.

Under the assumption that the segregation distortion is due to viability genes linked to a marker, QTL mapping can be used to locate and characterize (estimate their selection coefficients) these (Hedrick and Munoa 1990; Launey and Hedgecock 2001; Lui and Xu 2003; Plough and Hedgecock 2011). Such detected locations have been called **vQTL**, for **viability QTLs** (Plough and Hedgecock 2011), although the term vQTL is more commonly used for variance QTLs (Chapter 18). Using this approach, Launey and Hedgecock (2001) and Plough and Hedgecock (2011) estimate B values of between 4 and 6 for early survival in Pacific oyster.

Detrimental Equivalents and Fitness Components

We can generalize Equation 12.18d to an arbitrary trait, such as a fitness component, W. Equation 12.18d becomes

$$W_f = W_0 \exp(-B_W f) \tag{12.24a}$$

The fractional reduction (W_f/W_0), given an amount of inbreeding f, is a function of the detrimental equivalents, B_W, for that fitness component, namely $\exp(-B_W f)$. Taking typical vertebrate survival values (Table 12.4), if $B_W = 1$, then survival in a fully inbred individual is just e^{-1}, or 37%, that of an outbred individual. For $B_W = 2$, this is just 14%. We can also express Equation 12.24a as

$$-\ln\left(\frac{W_f}{W_0}\right) = B_W f \tag{12.24b}$$

Equation 12.24a can also be applied to any arbitrary trait, z. If the trait is continuous, B can be estimated using Equation 12.19a, as residuals are (often) approximately normally distributed. Similarly, if one has only one inbred class, Equation 12.21b can be used. When the trait is an integer-valued fitness component, such as number of mates or offspring per mating, the log-link GLMM is a better choice, as residuals are better fit by a Poisson than by a normal (see WL Chapter 30 for an extensive discussion of residuals in fitness regressions).

One advantage of using B is that it provides a standardized measure for comparing different traits/populations. Table 12.2 introduced another standardized measure for the impact of inbreeding on trait z,

$$\mathrm{I}_s = (\bar{z}_O - \bar{z}_1)/\bar{z}_O = 1 - (\bar{z}_1/\bar{z}_O) \tag{12.24c}$$

This is simply the percentage decrease in trait mean at full inbreeding ($\bar{z}_1$). We can relate B and I_s as follows. First, note that $\bar{z}_1/\bar{z}_O = 1 - \mathrm{I}_s$. For complete inbreeding ($f = 1$), Keller and Waller (2002) noted that Equation 12.24b implies

$$-\ln\left(\frac{\bar{z}_1}{\bar{z}_O}\right) = B_z \tag{12.24d}$$

where we have indexed the detrimental equivalents measure (B_z) to indicate the trait being followed. Hence,

$$B = -\ln(1 - I_s) \tag{12.24e}$$

For example, Brown and Kelly (2019) found that the average fitness (in natural populations) of fully inbred *Mimulus guttatus* individuals was only 3% of their outbred counterparts. Hence $I_s = 0.97$, and Equation 12.24e gives the corresponding number of detrimental equivalents for fitness as $B = -\ln(0.03) = 3.5$.

Using just a single fitness component can give a misleading picture of the true impact of selection (WL Chapter 29). Consider the work of Grueber et al. (2010) on a New Zealand population of the flightless rail, the Takahē (*Porphyrio hochstetteri*). While inbreeding depression was small at most of the individual life-history stages, its cumulative effect was quite substantial. Estimates of the impact of inbreeding depression based on a single fitness component (typically early survival) can thus *significantly underestimate* the total impact of inbreeding depression (O'Grady et al. 2006). Indeed, studies by Glémin et al. (2006) on an endangered mustard (*Brassica insularis*) showed that the effects of inbreeding increased over its lifespan, with an estimate of B for early survival of 0.144, increasing to 0.671 for year-one survival, and to 0.960 for year-two survival. Similar increases in B as more fitness components are considered was seen in the Kangaroo rat by Willoughby et al. (2019), as shown in Table 12.4.

To more formally address the impact of inbreeding over a series of fitness components, consider a set of multiplicative life-history features. Taking total offspring as our measure of fitness (see WL Chapter 29 for other measures, such as accounting for reproductive timing), we can decompose this into various life-history components. One such decomposition is $W = W_j W_m W_{fe}$, namely the total number of offspring W is the product of juvenile survival (W_j), subsequent survival to sexual maturity (W_m), and fecundity (W_{fe}, the number of offspring from a sexually mature individual; WL Chapter 29). Applying Equation 12.24a to each component yields

$$\begin{aligned} W_f &= \left(W_{j,0}e^{-B_j f}\right)\left(W_{m,0}e^{-B_m f}\right)\left(W_{fe,0}e^{-B_{fe} f}\right) \\ &= W_0 \exp(-B_W f) \end{aligned} \tag{12.25}$$

where $B_W = B_j + B_m + B_{fe}$ and $W_0 = W_{j,0}W_{m,0}W_{fe,0}$. The number of detrimental equivalents is thus the sum of the detrimental equivalents for each multiplicative component. Using this approach, O'Grady et al. (2006) noted (via a meta-analysis; Appendix 6) that for vertebrate populations, the average $B \simeq 6$, with average values of $B_j = 1.18$, $B_m = 2.98$, and $B_{fe} = 1.97$. This implies a fully inbred individual has only e^{-6}, or 0.25%, of the fitness of an outbred individual.

Detrimental Equivalents and Minimum Viable Effective Population Size

In small (diploid) populations the amount of inbreeding increases each generation according to

$$f_t = 1 - \left(1 - \frac{1}{2N_e}\right)^t \simeq t/(2N_e), \quad \text{for } t \ll 2N_e \tag{12.26a}$$

where N_e is the effective population size (WL Chapter 3). Managers of small, closed populations (such as an endangered species largely confined to captivity) thus try to maximize N_e (for example, by maximal avoidance of mating between relatives). A critical management issue is the **minimum viable effective population size**, namely the minimal N_e for the population to remain viable over some time period. Early conservation biologists took this value as $N_e \geq 50$, as from Equation 12.26a, this implies no greater than a 1% rate of inbreeding per generation (Soulé 1980; Franklin 1980). However, it was soon realized that the value of B_W is of critical importance, as this measures the *impact* of inbreeding on fitness (Frankham et al. 2014; Pérez-Pereira et al. 2022b). With an estimate of B_W in hand, one can find the critical $N_e(\alpha, \tau)$ such that the fractional reduction in fitness is no more than α after τ generations in captivity. From Equations 12.24a and 12.26a, $N_e(\alpha, \tau)$ satisfies

$$\frac{W_\tau}{W_0} = \exp(-Bf_\tau) = \exp(-B\tau/[2N_e(\alpha,\tau)]) \geq (1-\alpha) \tag{12.26b}$$

Solving gives

$$N_e(\alpha,\tau) \geq \left(\frac{B}{2}\right)\left(\frac{-\tau}{\ln(1-\alpha)}\right) \tag{12.26c}$$

A common target is a reduction of no more that 10% of the starting fitness after five generations, in which case $\tau/\ln(1-\alpha) = -5/\ln(0.9) \simeq 47.5$, yielding $N_e(\alpha,\tau) \geq 24\,B$. Initially, a $B \leq 2$ for survival alone was assumed, giving this critical N_e as ≥ 50 for vertebrates. However, using the more realistic estimate for *total* fitness (the O'Grady estimate of $B_W \simeq 6$) gives the critical N_e as ≥ 150. These estimates ignore the fact that some purging (reduction in B due to selection during inbreeding) can decrease B_W during inbreeding, but Equation 12.27 (below) allows one to accommodate for this effect.

PURGING INBREEDING DEPRESSION

Inbreeding, Extinction, and Purging

Concern has arisen that inbreeding in small captive populations of endangered species has led to a loss in fitness, possibly an irretrievable one (Ralls et al. 1979; Ralls and Ballou 1982a, 1982b; Lacy et al. 1993; Hedrick and Kalinowski 2000; Richards 2000; O'Grady et al. 2006; Leberg and Firmin 2008; Jamieson and Allendorf 2012; Frankham et al. 2013, 2014; Franklin et al. 2014; Theodorou and Couvet 2015; Caballero et al. 2017a, 2017b; Pérez-Pereira et al. 2022a, 2023). Almost every captive animal species for which breeding records have been kept shows an elevated mortality rate in inbred versus outbred progeny. Such reductions in fitness can have a direct impact on extinction rates. Indeed, simulations by O'Grady et al. (2006) found that realistic levels of ID in mammalian and avian species (B for total fitness $\simeq 6$) significantly decreased medium times to extinction.

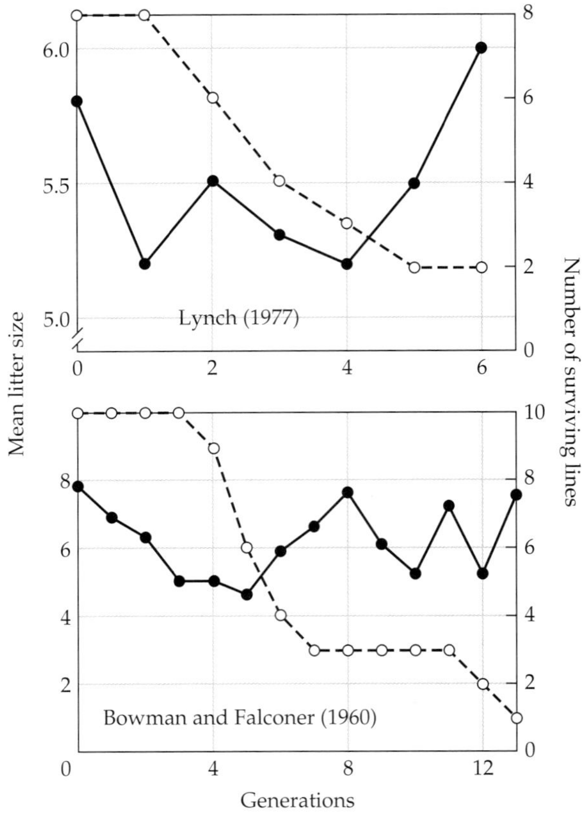

Figure 12.11 Purging of deleterious alleles through inbreeding: the response of mean litter size (●) to inbreeding in the mouse. In both studies, replicate lines were established from the same random-mating base population, and subsequently maintained by single brother–sister matings. Differential extinction of the lines (○) was caused by the failure to produce at least one reproductively competent male and female offspring. In both studies, there initially was a sharp drop in the litter size averaged over all lines, but as the less prolific lines went extinct, the grand mean returned to its initial level. Such results are not expected under the overdominance hypothesis, because with population sizes of only two individuals, most loci are expected to lose all but one allele by random genetic drift within a few generations (WL Chapter 2).

The direct impact of inbreeding on extinction has been seen in natural populations. Saccheri et al. (1998) and Nonaka et al. (2018) found that the local extinction risk significantly increased with inbreeding level in a Finnish metapopulation of the Glanville fritillary butterfly (*Melitaea cinxia*). Turning to wild bird populations, Keller (1998) examined the inbreeding levels in survivors following a crash in a British Columbian population of song sparrows (*Melospiza melodia*). The average inbreeding coefficient of survivors was 0.0065, while the average f was 0.0312 among the 206 birds that perished, with all birds with $f > 0.06$ dying. Finally, Bozzuto et al. (2019) showed that inbreeding has reduced the long-term growth rate in Alpine ibex populations, especially in harsh environments. For these reasons, to limit inbreeding, special efforts are generally made to restrict matings of endangered species to nonrelatives. This option is constrained in many cases because essentially all remaining members of the population are descendants of the same pedigree. For such cases, one strategy is to cross the most distant relatives within the breeding population (e.g., Ballou and Lacy 1995; Fernández and Toro 1999; Fernández and Caballero 2001a, 2001b; Fernández et al. 2004). Similar impacts of inbreeding must eventually arise in closed populations of elite breeds of domesticated species.

As a potential strategy for dealing with such extreme situations, Templeton and Read (1983, 1984) advocated *intentional inbreeding*, combined with rapid population expansion, as a means for exposing deleterious recessive genes, and **purging** them from a remnant population by selection. Obviously, this is a very risky approach for species with low reproductive potential. Implicit in such a management program is the assumption that rare deleterious recessive, rather than overdominant, alleles are the primary agents of inbreeding depression. The observation of purging some of the inbreeding depression under selfing traces back to at least Darwin (1876). He noted that while most inbred lines of the common morning-glory (*Ipomea purpurea*) tended to suffer in vigor and seed set, one of his selfed lines showed surprising vigor. In honor of its performance, he named this line *Hero*.

Purging: Theory

A useful way to think about purging is as extra natural selection against recessive alleles, generated by inbreeding increasing the frequency of deleterious homozygotes (relative to random mating). Consider the simplest case of selection against a fully recessive allele. From WL Equation 5.1b, the expected change in the frequency p of this allele (in an outbred population) is $\Delta_O p \simeq -p^2 s$. However, under inbreeding to level $f\ (\gg p)$, this change is approximately $\Delta_f p \simeq -pfs$, for a ratio of $\Delta_f p/\Delta_O p \simeq f/p$, which can be dramatic for rare alleles, even under weak amounts of inbreeding.

Recall (Equation 12.26a) that, in a finite population, f increases each generation by $\simeq 1/(2N_e)$. Thus, small population size by itself generates inbreeding, which can be further enhanced by crossing more closely related individuals (Glémin 2003; Pérez-Pereira et al. 2023). Purging can thus potentially occur by merely passing a population through a bottleneck. However, as N_e decreases, a countering force arises, as drift becomes increasingly powerful: when $N_e|s| \ll 1$, selection is overpowered by drift (WL Chapter 7). Hence, purging can be important for strongly deleterious alleles (decreasing their contribution to ID), but the resulting lowered N_e allows less deleterious alleles to increase in frequency due via drift, *increasing* their contribution to the load. As a result, any advantage of purging highly deleterious alleles via selection could be more than offset by the advancement of less deleterious ones via higher drift pressure. As we present below, the success of purging depends on the underlying genetic architecture: if the bulk of ID is from rare, highly deleterious alleles, purging can be effective. If, however, the bulk of ID comes from low-frequency mildly deleterious alleles, then purging can be counterproductive. Purging may thus be successful for some traits, but not others, even in the same population.

Because of the impact of drift, attempting purging by simply lowering N_e (as opposed to further incorporating mating between close relatives) will usually increase the frequency at which deleterious alleles are fixed. In full-sib mated lines of mice, purging is only accomplished at the expense of extreme selection (extinction) among replicate lines (Figure 12.11). Another example of the drawbacks of intentional purging was offered by Leberg and Firmin (2008), who examined the impact of bottlenecks on mosquito fish (*Gambusia affinis*). While the passage through serial bottlenecks did not seem to result in any purging, it did result in an increase in the extinction rate and a decline in growth rates. A final point is that because the rate of production of new deleterious alleles by spontaneous mutation is quite high (Chapter 15), any purging of the mutation load from a population is at best a transient phenomenon (García-Dorado 2012).

Theoretical work suggests that, where possible, slow inbreeding over a prolonged period appears to be the most efficient process for purging (Hedrick 1994; Wang et al. 1999; Glémin 2003; Boakes and Wang 2005; García-Dorado 2012; Caballero et al. 2017a). Further, Glémin (2003) stressed that purging by drift alone (the effect of smaller N_e, a largely passive management approach) is less effective than purging by actively mating relatives. However, slow inbreeding and/or forcing nonrandom matings may not be possible in many conservation and breeding settings.

The central point is that any purging strategy represents a delicate balance of inbreeding strength. It must be strong enough to allow selection to efficiently remove deleterious recessives over the time course of the purging, yet still weak enough to avoid fixing mildly deleterious alleles via drift. Fast inbreeding (very small N_e) can quickly purge highly deleterious alleles, either by within-line selection or via among-line selection due to extinction.

However, the cost is the fixation of mildly deleterious alleles, and, more importantly, the potential extinction of lines (and in the extreme, even the entire population) as selection culls less fit individuals (Figure 12.11). Conversely, while slow inbreeding can effectively reduce much of the concealed load from inbreeding depression, it requires much more time, as the weak levels of inbreeding only results in a mild increase in the effective selection coefficient against deleterious alleles. Hence, the failure of many experiments to show significant purging has been attributed to either overly strong inbreeding during the purge phase or a duration under weak inbreeding that was too short to be effective (López-Cortegano et al. 2018). López-Cortegano et al. (2020) suggested that the minimal time scale required to observe purging is on the order of (at least) $\sqrt{N_e}$ generations, or 5, 7, and 10 generations, respectively, for N_e of 25, 50, and 100.

Given the importance of genetic architecture of ID in purging, Willis (1999) examined the role of large- versus small-effect genes on inbreeding depression. He started with 1200 lines from a base population of *Mimulus guttatus* (monkey flower), which were selfed by single-seed descent (SSD) for five generations, resulting in 335 surviving lines. The SSD method allows for the removal of deleterious alleles with very large effects, but the effective population size of an SSD line ($N_e \simeq 1$) is too small to selectively influence genes of even moderate effect (WL Chapter 7). The resulting SSD lines were then intercrossed to form a large "purged" population. The fitness of outcrossed and selfed progeny were examined using parents from both the purged and ancestral populations. In selfed offspring from the purged population, a limited decline in inbreeding depression and a limited increase in inbred fitness was observed, suggesting that while some purging did occur (through the removal of large-effect alleles), the bulk of inbreeding depression was caused by alleles of smaller effect that were not efficiently selected against under SSD. A similar finding that most ID was due to mildly deleterious mutations was made by Baldwin and Schoen (2019), who found that three generations of selection and selfing had no impact on ID for fitness components in populations of the mustard *Leavenworthia alabamica*.

Finally, while conservation biologists tend to be extremely hesitant in applying active purging (especially for species with low reproductive potential), Hedrick and García-Dorado (2016) and López-Cortegano et al. (2021) suggested that commonly applied approaches for maximum avoidance of inbreeding, such as minimum kinship and minimal inbreeding approaches, may be somewhat counterproductive in the long run. Such approaches may tend to minimize the effectiveness of any selection that is occurring, reducing any potential for some modest background-level purging that may improve the population over time. However, minimizing the effectiveness of natural election is a bit of a two-edge sword, as such approaches also reduce the rate of *adaptation to captivity*, a major complication for breeding programs whose goal is to reintroduce populations back in the wild.

Example 12.6 Approaches have been suggested to adjust estimates of the base-population ID values to account for purging. Initial methods were based on fitting regression models that included further coefficients (besides the classical inbreeding coefficient f) to reflect changes in the amount of inbreeding caused by purging (Ballou 1997; Boakes et al. 2007; Gulisija and Crow 2007). A newer approach uses the **inbreeding–purging (IP)** model (García-Dorado 2012; García-Dorado et al. 2016), which uses a **purge-adjusted inbreeding value** g_t (rather than f_t). Using pedigree estimates for f_t, the adjusted inbreeding value g_t is given by the recursive relationship

$$g_t \simeq \left[\frac{1}{2N_e} + \left(1 - \frac{1}{2N_e}\right) g_{t-1}\right] (1 - \gamma f_{t-1}) \tag{12.27a}$$

Here g_t represents the expected f_t value at generation t-weighed by the relative reduction of deleterious recessives via purging, which is quantified by a model-fitted **purging coefficient** γ. When all loci have the same effects, $\gamma = s(1 - 2h)$, reflecting the magnitude of the effect of each deleterious allele that, due to dominance, is only revealed in homozygotes. Under this model, the expected mean in generation t is given by

$$\mu_t \simeq \mu_0 \exp(-B\, g_t) \tag{12.27b}$$

where B, the amount of inbreeding depression in the absence of purging, is also fit by the model. Simulation results (López-Cortegano et al. 2016; Pérez-Pereira et al. 2023) found that the IP approach was generally robust and more powerful than earlier pedigree-based methods.

Example 12.7 López-Cortegano et al. (2021) examined captive populations of four endangered ungulates with extremely low effective population sizes (ranging from 4 to 39) housed (since the 1970s) at the La Hoya experimental station in Almería, Spain. Because of their critically low population sizes, avoidance of inbreeding schemes were used only to choose mates, while there were no constraints on the number of offspring that a female contributes to the next generation (i.e., allowing for selection on offspring number and survival). A significant estimate of γ under the IP model (Equation 12.27a) was taken as evidence for purging. The two populations with intermediate N_e values—dama gazelle (*Nanger dama*) with $N_e \sim 11$ and Cuvier's gazelle (*Gazella cuvieri*) with $N_e \sim 14$—showed substantial purging of ID for early survival (to day 15 after birth). Conversely, the two populations at the extremes—barbary sheep (*Ammotragus lervia*) with $N_e \sim 4$ and dorcas gazelle (*G. dorcas*) with $N_e \sim 39$—showed no such evidence. López-Cortegano suggested that the lack of purging in barbary sheep was a consequence of their effective population size being so low that drift overcomes all but the most deleterious alleles. For dorcas gazelle, they noted that the minimal number of generations to observe purging is on the order of at least $\sim \sqrt{39} = 6$ generations (López-Cortegano et al. 2020), so that time in captivity may not have been sufficiently long to detect its effects. One limitation of these studies is that the effects of *fixation* of deleterious alleles by drift are likely confounded with purging (their *removal* by selection and inbreeding), with fixation of deleterious alleles reducing fitness, while purging increases fitness.

Purging: Observations from Animals

Evidence for purging of deleterious genes has been observed in several laboratory studies with flies (Ehiobu et al. 1989; Bryant et al. 1990; Garcia et al. 1994; Latter et al. 1995; Swindell and Bouzat 2006; Ávila et al. 2010; López-Cortegano et al. 2016; Pérez-Pereira et al. 2021), mice (Bowman and Falconer 1960; Lynch 1977; Connor and Bellucci 1979), *Caenorhabidtis elegans* (Chelo et al. 2018), the bivalve *Argopecten* (Zheng and Zhang 2012), and the gastropod *Physa* (Noël et al. 2016), a fact that provides further support for the dominance hypothesis. Further, experiments can be designed to test aspects of purging theory. For example, López-Cortegano et al. (2016) and Pérez-Pereira et al. (2021) examined the impact of slow inbreeding in *Drosophila*. They found that 100–200 generations of mild inbreeding (modest N_e) produced an eight to ten fold reduction (relative to the base population value) using estimated B values as their measure of ID when the lines were subsequently tested under strong inbreeding. They also noted that the baseline ID was greater under competitive than noncompetitive conditions, but that the amount of purging was also greater under competitive conditions. Bundgaard et al. (2021) also found that slow inbreeding can reduce ID in some, but not all, traits in *Drosophila*.

Turning to small, historically isolated wild animal populations, evidence for natural purging is mixed. Facon et al. (2011) found that invasive populations of ladybird beetles (*Harmonia axyridis*), which presumably passed through bottlenecks, had only a fraction of the inbreeding depression load found in native populations. Visscher et al. (2001) suggested that purging occurred in a small feral population of English Chillingham cattle. Robinson et al. (2018) postulated that purging may account for the long-term persistence and lack of inbreeding depression in island foxes (*Urocyon littoralis*). Populations of this species, distinct from the mainland species, have existed on the Channel Islands off California for over 9,000 years. While these populations display extremely low levels of molecular diversity, morphological data show a lack of congenital defects common to many highly inbred populations, and some of the island populations have strongly rebounded following recent severe bottlenecks.

Evidence that a prolonged bottleneck resulted in at least some purging has also been seen in some populations of birds, namely Stewart Island robins (*Petroica australis rakiura*; Laws and Jamieson 2011), but not for others, such as a flightless rail (the Takahē) *Porphyrio hochstetteri* (Grueber et al. 2010), or the Chatham Island black robin (*Petroica traversi*; Kennedy et al. 2014). Weiser et al. (2016) made an unexpected observation on the last species: while close inbreeding reduced survival when mothers had a modest level of inbreeding, it actually *improved* survival with more inbred mothers. They suggested that this occurs because "a highly inbred chick with highly inbred parents inherits a proven genotype and thus experiences a fitness advantage," a situation potentially akin to Darwin's *Hero* lineage.

Despite the above examples of at least some purging, care needs to be taken in extrapolating the idea of purging as a general conservation tool (Hedrick 1994; Lynch et al. 1995a; Lynch 1996; Hedrick and Kalinowski 2000; Hedrick and García-Dorado 2016). Ballou (1997) found, at best, only very modest support for purging in 25 captive mammalian populations. Similarly, a meta-analysis by Boakes et al. (2007) of 88 species of vertebrates from around 120 zoos inferred some purging, but the authors suggested that its benefits are rarely appreciable. They also suggested (as did Crnokrak and Barrett 2002) that cases where fitness seems to improve during inbreeding may simply reflect adaptation to captivity. One caveat is that the time scale for many of these studies was less than $\sqrt{N_e}$ generations, so that sufficient time may not have passed to see any tangible results from purging.

While almost all vertebrates are outcrossers, a number of nonvertebrate animals are hermaphroditic, and this mating system can impact ID. In particular, Escobar et al. (2011) found a negative correlation between selfing rates and ID in Basommatophoran snails. Just as the mating system can impact the level of inbreeding, so can the genetic system itself (e.g., ploidy). Brucker (1978) and Werren (1993) suggested that inbreeding depression should be reduced for haplodiploid species (such as Hymenoptera and Acarian mites), due to the haploid nature of the male uncovering any deleterious recessives (at least for those traits expressed in males). A meta-analysis (Appendix 6) by Henter (2003) found that while inbreeding depression can be substantial, as a group, haplodiploid taxa tended to experience less inbreeding depression than diploid insects. In a similar fashion, because sex chromosomes (X and W) are haploid in one sex (XY males, WZ females), the expectation is that these should be largely free of deleterious recessives, and thus contribute rather little to ID. Indirect, but strong, evidence for this was seen in *Drosophila* by Schou et al. (2017). They grew populations at difference sizes, following the loss of genetic variation (scored by SNPs). Loss of diversity for autosomes under small effective population sizes was less than predicted by drift alone (suggesting some selection for the retention of some heterozygosity). However, the loss of genetic variation on the X was as predicted from neutral drift. Finally, note that haploid selection should be particularly effective in plants with extensive gametophyte stages (Mulcahy 1976).

Example 12.8 Whole-genome sequencing (WGS) has been deployed to examine the impact of bottlenecks, and potential purging, in wild animal populations (Xue et al. 2015; Grossen et al. 2020; Kleinman-Ruiz et al. 2022). Sequence data can provide (modest) inference on the fitness of variants and fixed sites (Bertorelle et al. 2022). At the crudest level, one can contrast the relative levels of derived (i.e., new mutations) protein-coding missense versus loss of function (LoF) variants (which can often be inferred from sequence data alone, such as very early termination or frameshift variants), with the latter assumed (as a general group) to be more deleterious. On a more sophisticated level, one can use software that predicts the impact of a missense mutations (such as being a conservative replacement, with the derived amino acid being in the same functional class as its ancestor, versus a more radical change, such as having a different charge or other properties). Approaches such as **GERP** (**Genomic Evolutionary**

Rate Profiling; Cooper et al. 2005), based on the relative conservation of a sequence over a phylogeny, can allow noncoding regions to be evaluated as well. The logic is that mutations in more conserved sites are, on average, more deleterious.

Xue et al. (2015) examined WGS data from 44 Gorillas, 13 of which were from the much smaller eastern (and mountain) populations in central Africa. Eastern populations showed a two to three fold reduction in sequence variation, as well as much more extensive LD, relative to their more numerous western lowland counterparts. This is consistent with eastern populations passing through a bottleneck. While eastern gorillas had a larger number of derived missense alleles, they had fewer LoF variants relative to western populations. These observations suggest a purging for highly deleterious (LoF) variants in the eastern population, while their reduced N_e resulted in an increased load of less deleterious mutations.

Grossen et al. (2020) examined WGS data from six ibex species in Europe and North Africa, as well as from the domestic goat (as an outgroup). The European Alpine ibex (*Capra ibex*) survived a strong bottleneck of less than 100 individuals in the 19th century, expanding to a current population of around 50,000. Similar to the findings of Xue et al. (2015), the Alpine ibex carries a reduced load of (predicted) highly deleterious alleles, but an increased load of (predicted) less-deleterious alleles following the bottleneck.

Kleinman-Ruiz et al. (2022) obtained WGS data for 20 Iberian lynx (*Lynx pardinus*) and 28 Eurasian lynx (*Lynx lynx*), the former an endangered species that has passed through multiple bottlenecks, the latter its widespread and abundant sister species. The authors examined the derived deleterious allele count, which includes both segregating and fixed differences, and is thus a measure of *both* the segregating and fixed loads. Similar count totals were seen between the species for putatively neutral, intergenic, and intronic regions. However, relative to its sister species, the Iberian lynx showed ~30% fewer LoF variants and ~11% fewer deleterious missense mutations. Given that the latter two classes are expected (on average) to have larger effects, this observation is consistent with purging in the Iberian lynx. Further, because the derived count considers *both* segregating and fixed differences, purging (of these classes) did not appear to result in an increased fixation load. Further analysis showed that the decrease in deleterious missense variants was seen in high, but not low, recombination regions of Iberian lynx autosomes. This is consistent with theory, as selection is more efficient in regions of higher recombination (WL Chapter 8).

Purging: Observations from Plants

Related to Templeton and Read's (1983, 1984) notion of purging is the idea that obligately self-fertilizing plants should be largely free of inbreeding depression (Wright 1969; Lande and Schemske 1985; Husband and Schemske 1996; reviewed in WL Chapter 23). Here too, some precautionary comments are in order. Byers and Waller (1999) found that purging tended to be a rather "inconsistent" force in plants, with the prediction of reduced inbreeding depression in lines with high rates of selfing showing rather uneven support in the data (e.g., Toczydlowski and Waller 2023). We will examine inconsistency in more detail shortly in the context of the forces driving the evolution of mating systems.

In contrasting segregating versus fixed loads (see below), an important consideration for any analysis of ID is that the effects of close inbreeding on the evolution of inbreeding depression needs to be considered from *two frames of reference*. By eliminating the genetic variation within an isolated population (or within a lineage of obligate selfers), long-term inbreeding can lead to a situation in which any further inbreeding has essentially no genetic consequences, simply because the progeny genotypes are essentially identical to those of their mothers. Once this situation has been reached, then from the standpoint of the *current population*, there is no inbreeding depression. However, as noted above, this need not be the case from the standpoint of the *ancestral population*, if, during the inbreeding process, different deleterious recessive genes have become fixed within different selfing lineages.

Another way to frame this concept is to note that the ancestral load from segregating deleterious alleles is impacted in three ways by inbreeding. First, purging could indeed remove some of the initial deleterious alleles in the daughter (current) population. Second, any reduced N_e can result in the *fixation* in the daughter population of some of these segregating ancestral alleles, creating a **fixation load** relative to the base. Third, any actual ID among the daughter population is a function of their **segregation load**, namely those alleles still segregating in the daughter population. ID is based on comparing inbreds to outcrossers in the daughter, *not* the ancestral, population. One can easily have a reduction in ID in bottlenecked populations relative to their initial values (a reduced segregation load), but the impact of inbreeding via the fixation load could be dramatic. Consistent with this prediction, Angeloni et al. (2011) found that inbreeding depression significantly increased with population size in plants, as did Lohr and Haag (2015), working with microcrustacean *Daphnia magna*. However, this observation of ID decreasing with population size is by no means universal. For example, Spigler et al. (2017) found that ID *increased* with decreasing population sizes in the biennial plant rosepink (*Sabatia angularis*).

The partitioning of deleterious alleles segregating in the ancestral population is nicely illustrated in a study of self-fertilized lines of the normally outcrossing aquatic plant *Eichhornia paniculata* (Barrett and Charlesworth 1991). Inbreeding caused an immediate depression in fitness, but after only two generations of selfing, there was no further decline, suggesting that the vast majority of loci affecting fitness had become homozygous within lines. Nevertheless, despite the absence of inbreeding depression *within* the derived lines, crosses *between* lines exhibited a substantial increase in mean fitness (heterosis; Chapter 13), as expected if the different lines had become fixed for different deleterious recessives. Over time, as new deleterious mutations become sequestered in different selfing lineages, the beneficial effects of outcrossing are expected to increase further (Lynch et al. 1995b). Enhanced fitness in outcrossed progeny of normally self-fertilizing plants has been demonstrated in numerous species (Levin 1989; Charlesworth et al. 1990; Holtsford and Ellstrand 1990; Ågren and Schemske 1993; Van Treuren et al. 1993; Latta and Ritland 1994; Johnston and Schoen 1995). However, there are also numerous examples where the inbred offspring of a parent are more fit than their outbred offspring (e.g., Baskin and Baskin 2017). Clearly, the absence of local inbreeding depression provides no information about the mutation load harbored by a population.

Example 12.9 The various stages in the domestication process of maize provides a model system for purging. The early inbred lines from the 1930s crossed to create the first hybrids had very low fitness, while modern inbred lines have performances that in some cases exceed those of these early commercial hybrids. High ID is not surprising for these early inbred lines, as the ancestor (teosinte) to the land races from which the inbreds were extracted was an outcrosser with a very large effective population size. Domesticated maize landraces first went through a bottleneck, starting from a small subsample of the wild teosinte population. Their descendant lines subsequently underwent purging by breeders attempting to improve these early inbreds. A number of molecular studies have probed the dynamics of the inbreeding load at various stages of this breeding program.

Yang et al. (2017) performed whole-genome sequencing on a panel of 12 elite maize inbred lines and used the GERP approach (mutational impact on fitness being proportional to the conservation of a sequence over a phylogeny) to compute fitness scores on the roughly 350,000 putative deleterious SNPs detected in this panel. SNPs with higher GERP scores (and thus putatively more deleterious) were found at lower frequencies, were enriched in low-recombination regions, and were associated with larger effect sizes on grain yield. The panel of elite lines (as a whole) showed an increase in the fixation of deleterious SNPs (relative to outbred landraces), consistent with fixation of a deleterious load during the bottleneck that formed them. However, they also showed a much smaller fraction of segregating deleterious sites (differential fixation over lines) relative to land races, consistent with purging.

Roessler et al. (2019) found direct evidence for purging by following the dynamics of putative deleterious alleles (predicted, for exomic sequences, by the amount of disruption of a given amino-acid substitution) in a set of 11 maize lines that were selfed for six generations. SNPs predicted to be deleterious were preferentially lost, and were lost at a faster rate in regions of higher recombination (consistent with theory, as selection is more efficient in such regions; WL Chapter 8). Samayoa et al. (2021) contrasted the behavior of 18 traits in a land race of maize with a neighboring population of wild teosinte. Inbreeding depression was greater in maize than in teosinte for 15 of these traits. QTL mapping found that rare large-effect variations were much more common in teosinte than in the maize land race. These observations are consistent with the pattern noted above that a bottleneck can cull large-effect deleterious alleles but can also enhance the fixation of less deleterious alleles, increasing the inbreeding load. This pattern was also seen by Sun et al. (2023), who examined the number and frequency of putatively deleterious mutations (based on GERP scores) using WGS of 350 recent elite maize lines. Again, they found that alleles predicted to be mildly deleterious increased in lines over time, while those predicted to be highly deleterious decreased.

Purging: Complications From Pseudo-overdominance

The process of purging depends on fitness-related loci displaying partial (or complete) dominance. When overdominance (the heterozygote is the most fit) occurs, purging is ineffective, as selection strives to maintain both alleles (WL Chapters 5 and 7). While the general narrative of some purging under inbreeding often holds (providing support for the dominance hypothesis), there are exceptions. As mentioned, the apparent lack of significant purging in many cases could be due to the reduction in ID from inbreeding and selection removing strongly deleterious alleles being countered by drift increasing the frequency of less deleterious alleles. There are, however, more intriguing observations that at first blush might seem inconsistent with the dominance hypothesis.

One is the persistence of **residual heterozygosity** during the formation of highly inbred lines. These are regions that stubbornly hold onto their heterozygosity in the face of strong inbreeding, despite the rest of the genome becoming homozygous (e.g., Gore et al. 2009; McMullen et al. 2009; Brandenburg et al. 2017). A particularly striking example is the work of Smith et al. (2019), who found that about one-third of all genes in a population of honey-bees (*Apis mellifera capensis*) inbred for at least 100 generations displayed heterozygosity. Similar observations for widespread and persistent residual heterozygosity have been seen in presumably highly inbred natural populations of planarian worms (*Schmidtea mediterranea*; Guo et al. 2016) and Scandinavian grey wolves (*Canus lupus*; Kardos et al. 2018b).

One explanation for residual heterozygosity is that genes within these regions display fitness overdominance and are thus able to persist in the face of strong inbreeding. However, an important observation is that these regions are often found in areas of very low recombination. For example, Schou et al. (2017) found that the lowest recombination regions of *Drosophila* showed the slowest loss of heterozygosity during inbreeding. In such low-recombination regions, residual heterozygosity can be explained by the dominance hypothesis, resulting from two (or more) linked loci with deleterious recessives in repulsion-phase disequilibrium (Chapter 5).

To see this, consider two tightly linked loci A and B, where lowercase letters denote recessive deleterious alleles at these loci. To simplify matters, suppose both deleterious alleles are fully recessive ($h = 0$) and are in complete linkage disequilibrium (LD) (we relax these assumptions later in the chapter). If one haplotype is Ab and the other is aB, then Ab/aB heterozygotes will have higher fitness than either Ab/Ab or aB/aB homozygotes, creating the impression of overdominance, with selection maintaining both haplotypes in the population. This dynamic persists in the face of inbreeding until recombination generates the more fit AB gamete. This phenomena, wherein recessive deleterious alleles in repulsion disequilibrium mimics heterozygote advantage, is called **pseudo-overdominance** (Ohta 1971; Sved 1972; Birchler et al. 2006; Waller 2021). In such settings, purging will be ineffective at removing the load from the segregating a and b alleles. If the region of low recombination is large, haplotypes likely contain much larger sets of complementary recessives in repulsion (Waller 2021), so that both the strength of the heterozygous advantage, and the subsequent impact on ID when made homozygous, can be considerable. Such regions resist becoming part of the fixation load, contributing to the segregation load, even under strong inbreeding.

A closely related concept is **associative overdominance** (Frydenberg 1963; Pamilo and Pálsoon 1998), the apparent heterozygote advantage when a *neutral marker* occurs in these complementary haplotyes. Let M_1 and M_2 denote alleles at a marker embedded within repulsion-phase haplotypes, e.g., AM_1b and aM_2B. In such cases, the pseudo-overdominance from selection on A and B creates the appearance of overdominance on the marker (M_1M_2 heterozygotes having higher fitness than M_1M_1 and M_2M_2 homozygotes). While these terms are often used somewhat interchangeably, we restrict pseudo-overdominance to the underlying *selective mechanism* (repulsion-phase linkage of deleterious recessives) and associative overdominance to the *observation* of an associated (neutral) marker displaying apparent overdominance.

Purging: Evolution of Mating Systems

A second potential failure of the strict dominance hypothesis arises from apparent conflicts between theory and observation on the evolution of mating systems (a topic covered in greater detail in WL Chapter 23). Fisher (1941) showed that there is an intrinsic, and strong, fitness advantage for a new mutation that increases the selfing rate. Charlesworth and Charlesworth (1995), and particularly Lande and Schemske (1985), showed that such modifiers will spread unless inbreeding depression imparts at least a 50% reduction in total fitness ($I_s > 0.5$, corresponding to a $B \simeq 0.7$). Further, with high levels of selfing, constant purging should result in little ID. Conversely, under outcrossing, high levels of ID build up, which in turn deters the evolution of selfing. The resulting prediction is that ***mating systems should crystallize into one of two states***. Either to populations that are almost fully selfing with little ID due to purging or to largely outcrossing populations with much higher ID.

How well does this prediction hold? Husband and Schemske (1996) found stronger ID in primarily outcrossing plants (selfing rate $\eta < 0.45$) than in primarily selfing species ($\eta > 0.55$), consistent with the basic theory. However, in a much larger review of impact of ID on plant life-history traits, Winn et al. (2011) found that a more nuanced perspective is needed. While largely selfing species ($\eta > 0.8$) had the smallest ID, the levels of ID were essentially identical for mixed-mating ($0.2 \leq \eta \leq 0.8$) and largely outcrossing ($\eta < 0.2$) species. The basic theory (assuming dominance) predicts that mixed-mating populations are, at best, transient phenomena, and that little ID should occur in them.

The retention of ID under strong selfing is consistent with overdominance being important. It is also consistent under the dominance hypothesis with extensive regions of pseudo-overdominance (Waller 2021; Toczydlowski and Waller 2023). Other explanations have been proposed within the dominance framework that incorporate correlations over loci (such as linkage and identity disequilibrium) greatly reducing the efficiency of purging relative to prediction form single-locus models (e.g., Lande et al. 1994; Roze 2015; Gilbert et al. 2020).

Genetic Rescue

A final strategy for conservation biologists to deal with inbreeding depression is **genetic rescue**: introducing a very limited number of individuals from close by, and related, populations. Because this runs the risk of outbreeding depression (hybrids having lower fitness), we examine this strategy in the next chapter. However, an inbred-based cautionary tale is offered by the work of Hedrick et al. (2019) on an isolated population of grey wolves on Isle Royale in Lake Superior. This population formed around 1950 from mainland grey wolves. Around the late 1990s, a single male immigrated to the population, likely via an ice bridge that formed in 1997. He immediately became the alpha male on the island, with his descendants quickly dominating the ancestry of the population. The initial result of this natural genetic rescue was to likely prevent a collapse of the heavily inbred population during the early 2000s. However, his descendants quickly became the bulk of the population, resulting (by 2019) in a population of only two wolves, with an expected offspring f of almost 0.5. This initially successful rescue was ultimately too successful, with a single individual dominating the pedigree, likely leading to eventual extinction among his descendants.

PARTIAL RECESSIVES VERSUS OVERDOMINANCE

The observation of inbred lines that equal or exceed the average performance of individuals in outcrossed populations is a serious challenge to the contention that overdominance is the primary mechanism of inbreeding depression. Nevertheless, there is still some controversy on the subject, fostered by early puzzling observations with allozyme loci to be considered shortly (heterozygosity–fitness correlations), as well as the above-mentioned observations on the evolution of the selfing rate (mating systems). Further, as just mentioned, pseudo-overdominance can create the appearance of overdominance at linked markers, making it challenging to distinguish between the two competing hypotheses. An independent line of evidence on the relative roles of dominance versus overdominance is thus desired, and fortunately can be obtained from biometrical analyses (inference based on phenotypic summary statistics). None of these biometrical results support the idea that overdominance is a common mode of gene action.

The (*A*+*B*)/*A* Ratio

We previously framed the model for lethal equivalents in terms of partially recessive deleterious alleles. The same logic can be used to redefine the model in terms of overdominant gene action (Table 12.1). This approach again gives rise to Equations 12.18b and 12.18c, but with the definitions of A and B altered to

$$A = P_e + \sum_{i=1}^{n} \left(q_i^2 s_i + p_i^2 t_i \right) \tag{12.28a}$$

$$B = \sum_{i=1}^{n} q_i p_i (s_i + t_i) \tag{12.28b}$$

Morton et al. (1956) noticed a useful feature of this model. Recall (WL Chapter 5) that for a balanced polymorphism maintained by overdominance, the equilibrium allele frequencies are $q_i = t_i/(s_i + t_i)$ and $p_i = s_i/(s_i + t_i)$. Substituting these into the previous expressions,

$$B = \sum_{i=1}^{n} \frac{s_i t_i}{s_i + t_i} \tag{12.29a}$$

$$A = P_e + B \tag{12.29b}$$

Thus, if inbreeding depression is primarily a consequence of overdominance, then the ratio $(A + B)/A = (P_e + 2B)/(P_e + B)$ is constrained to be less than or equal to two in populations that are in a state of balanced polymorphism. Strictly speaking, this result applies to a diallelic locus. With k alleles per locus, the constraint is $(A + B)/A \leq k$ (Crow 1958; Lewontin 1974). Returning to Table 12.4, we see that this ratio is usually on the order of 10 or more. Thus, unless a very large number of alleles are maintained in a delicately balanced polymorphism *at each locus* contributing to inbreeding depression (which seems quite unlikely from theory; WL Chapter 5), the results from lethal-equivalent analyses seem generally inconsistent with the overdominance model.

Estimating the Average Degree of Dominance

Accepting that the linear decline in log fitness in a lethal-equivalent analysis is, in fact, a consequence of multiple partially recessive alleles, then some further inference about the mode of gene action can be made. Using Equations 12.18b and 12.18c, it can be shown that

$$\frac{B}{A+B} \leq 1 - \frac{\sum 2q_i s_i h_i}{\sum q_i s_i} \tag{12.30}$$

Recalling (WL Chapter 7) that under selection–mutation balance, $q_i = u_i/(h_i s_i)$, and upon rearrangement, Equation 12.30 implies that for a population in equilibrium,

$$\tilde{h}_1 = \frac{\sum u_i}{\sum (u_i/h_i)} \leq \frac{A}{2(A+B)} \tag{12.31}$$

The expression on the left is equal to the harmonic mean of the dominance coefficients among *newly arising mutations.* An upwardly biased estimate of this quantity is provided by the ratio $A/[2(A+B)]$, the bias approaching zero as the environmental contribution to mortality becomes negligible ($P_e \to 0$). Application of Equation 12.31 to the data in Table 12.4 shows that, in the vast majority of cases, $\tilde{h}_1$ is in the range of 0.02 to 0.15. It should be kept in mind that $\tilde{h}_1$ will tend to *exceed* the average dominance coefficient of *segregating* deleterious alleles because mutant alleles with stronger effects in heterozygotes (larger h values) are more easily removed by selection. However, because a harmonic mean is always less than the arithmetic mean, these two sources of bias may approximately cancel, leaving $A/[2(A+B)]$ as a reasonable estimator of the arithmetic mean h of segregating alleles. In any event, the data clearly suggest that the majority of deleterious alleles influencing early survival are quite recessive.

When highly inbred lines are available, less biased methods for estimating the average degree of dominance exist, as pointed out by Mukai et al. (1974). Consider a single diallelic locus with the relative fitnesses of the *BB, Bb,* and *bb* genotypes being 1, $(1-hs)$, and $(1-s)$. If a randomly mating base population is inbred (through a series of lines) to complete homozygosity, then (from Table 12.1) a fraction q of the lines will have genotype *bb* and fitness $(1-s)$ at the locus, while the remaining fraction p will have genotype *BB* and fitness 1. Now suppose that the inbred lines are randomly paired and mated. *BB* × *BB* matings will then occur with frequency p^2, giving rise to *BB* progeny with fitness 1. Similarly, *bb* × *bb* matings will occur with frequency q^2, giving rise to *bb* progeny with fitness $(1-s)$, and *BB* × *bb* matings will occur with frequency $2pq$, giving rise to *Bb* progeny with fitness $(1-hs)$. Summing over all loci, the genetic variances for log fitness among inbred lines (G_p), among midparent values (G_{mp}), and among F_1 progeny (G_o) are, respectively,

$$\sigma^2(G_p) = \sum p_i q_i s_i^2 \tag{12.32a}$$

$$\sigma^2(G_{mp}) = \sum p_i q_i s_i^2/2 \tag{12.32b}$$

$$\begin{aligned}\sigma^2(G_o) &= \sum 2p_i q_i s_i^2[\,(1-2p_i q_i)h_i^2 - 2q_i^2 h_i + q_i(1+q_i)/2\,] \\ &\simeq \sum 2p_i q_i s_i^2 h_i^2\end{aligned} \tag{12.32c}$$

The approximation in Equation 12.32c follows from the reasonable assumption that the frequencies of deleterious alleles are kept low by selection, i.e., $q_i \ll 1$. In addition, the covariance among offspring and midparent genotypic values is

$$\begin{aligned}\sigma(G_o, G_{mp}) &= \sum p_i q_i s_i^2[\,h_i(1-2q_i) + q_i] \\ &\simeq \sum p_i q_i s_i^2 h_i\end{aligned} \tag{12.32d}$$

Thus, half the genetic regression of offspring on midparent values has the expected value

$$\frac{b(o,mp)}{2} = \frac{1}{2}\,\frac{\sigma(G_o, G_{mp})}{\sigma^2(G_{mp})} = \frac{\sum p_i q_i s_i^2 h_i}{\sum p_i q_i s_i^2} \tag{12.33a}$$

where the slope follows from Equation 3.14b. Recalling that under selection–mutation balance, $q_i = u_i/(h_i s_i)$ and $p_i \simeq 1$, this expression reduces to

$$\frac{b(o,mp)}{2} = \tilde{h}_2 \simeq \frac{\sum u_i s_i}{\sum u_i s_i/h_i} \tag{12.33b}$$

Like Equation 12.31, this expression is a harmonic mean estimate of the average degree of dominance. In this case, however, each allele is weighted by $(u_i s_i)$, the product of the mutation pressure to the allele and the homozygous mutational effect. If s_i and h_i are uncorrelated, then Equation 12.33b, like 12.31, provides an estimate of the harmonic mean dominance coefficient of new mutations (Watanabe et al. 1976). However, data on newly arisen mutations suggest that such independence of h and s is unlikely (Chapter 15).

Keeping in mind these interpretative limitations, we now consider results for viability in *Drosophila*. Averaging over four studies for which the appropriate data are available (Tantaway 1957; Dobzhansky et al. 1963; Malogolowkin-Cohen et al. 1964; Garcia et al. 1994), we estimate $\tilde{h}_1 \leq 0.14 \pm 0.05$. Thus, consistent with our broader interpretation of the data in Table 12.4, most deleterious segregating genes that influence viability appear to be quite recessive. On the other hand, three studies with *D. melanogaster* (Mukai et al. 1972; Mukai and Yamaguchi 1974; Watanabe et al. 1976) obtained estimates of $\tilde{h}_2$ that average 0.30 ± 0.05. Working with the same species, Hughes (1995) obtained estimates of $\tilde{h}_2$ for additional characters in males: 0.08 for body size, 0.17 for mating ability, 0.14 for mortality rate, and 0.30 for longevity. Finally, data from Wills (1966) and Strickberger (1972) yield estimates of $\tilde{h}_2 = 0.27$ and 0.18 for viability in *D. pseudoobscura*. None of these estimates of $\tilde{h}_2$ are strictly comparable to those for $\tilde{h}_1$, not only because different weights are employed in the two definitions, but also because all of the studies employing Mukai's regression method have excluded chromosomal lines with highly deleterious effects. Thus, to the extent that lethal or semilethal alleles have lower dominance coefficients than do mildly deleterious alleles (as the data clearly suggest; Chapter 15), $\tilde{h}_2$ estimates are expected to be higher than those for $\tilde{h}_1$. Summing up the extensive data for *Drosophila*, most segregating deleterious mutations appear to be recessive, with the average h for all deleterious alleles being on the order of perhaps 0.1, and that for mildly deleterious alleles being on the order of 0.15 to 0.3.

There is an unfortunate void on information on the average degree of dominance in other organisms. However, the regression technique of Mukai et al. (1974) is easily extended to certain species, most notably those that reproduce in nature by obligate self-fertilization. Individuals within such populations are as close to being completely homozygous as one can obtain, and provided that forced outcrosses can be implemented, it is possible to use Equation 12.33b to obtain an estimate of the dominance coefficient averaged over the entire genome. Under obligate selfing, most lethal recessives should be purged, leaving most of the deleterious mutation load in the detrimental class. Johnston and Schoen (1995) used this regression method to obtain estimates of the average degree of dominance in four populations of the annual plant *Amsinckia*. Their average estimates of $\tilde{h}_2$ were: 0.30 for flower number, 0.36 for survivorship from germination to flowering, 0.10 for seed production, and 0.21 for total fitness. These results are close to those for detrimentals in *Drosophila*.

CORRELATION BETWEEN MULTILOCUS HETEROZYGOSITY AND FITNESS

There is a long history of ecological studies finding weak correlations between marker heterozygosity and fitness components (such as growth rate), a phenomena usually denoted in the literature as **heterozygosity–fitness correlations** (**HFCs**). Because this observation has a population-genetic context that has changed over time, some background history is needed in order to provide the proper setting before examining HFCs in detail. As we will see, possible explanations for HFCs reflect many aspects of the overdominance versus partial dominance debate over the cause of inbreeding depression. As with that discussion, the current consensus is that most HFCs simply reflect inbreeding depression arising from dominance (Szulkin et al. 2010). The analysis of HFCs historically marks the first serious marker-based approach for examining the impact of inbreeding. We consider HFCs in some detail before concluding the chapter with more direct marker-based estimates of f and what they tell us about the impact of inbreeding.

The first widespread set of molecular markers were **allozymes** (or **isozymes**), which scored variation in the protein products produced by a given gene (Lewontin and Hubby 1966). Under the method of **electrophoresis**, proteins that vary in total charge travel at

Table 12.7 Possible explanations for heterozygosity–fitness correlations (HFCs).

Direct effects		
	Functional Overdominance Hypothesis	
		Marker locus under direct (and overdominant) selection
Indirect effects		
	Local-effect Hypothesis	
		Marker in LD with selected loci (overdominance or associative overdominance)
		Requires linkage disequilibrium (LD)
		LD block size can be assessed with marker data
	General-effect Hypothesis	
		Inbreeding depression
		Marker heterozygosity predicts genome-wide heterozygosity
		Requires identity disequilibrium (IDE), which in turn requires variance in f
		IDE can be quantified by the g_2 statistic

different speeds on an acrylamide gel under an electric current, resulting in discrete bands corresponding to the different variants when visualized with the appropriate strains (Hubby and Lewontin 1966; Lewontin and Hubby 1966). Allozyme markers were scored over a very wide range of populations and organisms, with one of the early observations being an association in some studies between the amount of allozyme heterozygosity and fitness, with fitness increasing with total heterozygosity (Nevo 1978).

With allozymes, the focus was largely on the direct fitness effects (or lack thereof) of these marker loci. As genomic technology advanced, researchers moved to random DNA markers, most of which were in noncoding regions, and hence were less likely to be directly under selection. HFCs with such markers are thus generated *indirectly* by either selection at linked sites (marker loci in LD with loci under selection; the **local-effect hypothesis**) or by genome-wide effects such as inbreeding, with marker heterozygosity reflecting heterozygosity *across the genome* of an individual (the **general-effect hypothesis**). Table 12.7 summarizes these three competing hypotheses (David et al. 1995; David 1998; Hansson and Westerberg 2002, 2008), which are detailed below.

Under the overdominance hypothesis, one would expect individual fitness to increase with the number of heterozygous allozyme loci per individual, either because the loci themselves are heterotic (direct effects) or because they mark heterotic regions of the genome (the local-effect hypothesis). Because heterozygosity is assumed to be unconditionally advantageous, *this correlation should hold in populations with any level of inbreeding*. In contrast, under the general-effect hypothesis, a correlation between **multilocus heterozygosity** (**MLH**) and fitness should arise *only* if there is a correlation between MLH and the level of individual inbreeding, f. Hence, before examining HFCs in more detail, we first need to consider the nature (if any) of the correlation between MLH and f.

Identity Disequilibrium, Variance in f, and the g_2 Statistic

Up to this point, we have often assumed that all inbred individuals in a sample (often from a controlled cross) have experienced the same value of f. This has mainly been for convenience. In reality, the level of inbreeding varies (often dramatically) over a population sample of individuals, and such variance can generate HFCs.The feature required for heterozygosity–f correlations is **identity disequilibrium** (**IDE**). This refers to situations in which there is a correlation across loci for the probability of alleles being identical by descent, which requires ***variance in the level of inbreeding*** in the sampled population (Weir and Cockerham 1973). It is important to stress the differences between LD and IDE. The former refers to nonrandom association of *alleles* in *gametes* (correlation of alleles within a gamete). As we saw in Chapter 5, the extent of LD can be assessed using markers to quantify LD block size, with the likelihood of the local-effect hypothesis increasing with the average block size (as any given marker is impacted by ever-larger regions of the genome). In contrast, identity disequilibrium refers to nonrandom associations of *diploid genotypes* within *zygotes*: if one locus is IBD homozygous, this provides information on the IBD status of other loci. Linkage is *not* required for IDE, which reflects the genome-wide level of f within an individual. Indeed, for a model with partial selfing, no selection, and no drift, Weir and Cockerham (1973) found that the IDE between two completely linked loci is only twice that for unlinked loci.

As was the case with LD, one can use markers to infer the amount of IDE within a population. A simple test for its presence was suggested by Balloux et al. (2004). Their **heterozygosity–heterozygosity correlation test** randomly partitions marker loci into two groups, and then computes the correlation (over individuals) between the number of heterozygous loci in group one versus group two.

A more formal IDE metric, g_2, was introduced by David et al. (2007). This parameter follows by considering the expectation that *pairs* of loci are jointly heterozygous within an individual, which can be written as

$$E[H_i H_j] = E[H_i]\, E[H_j]\, (1 + g_2) \tag{12.34a}$$

where H_i is a zero-one indicator variable of whether locus i is a heterozygote. If heterozygosity is uncorrelated over loci, then $E[H_i H_j] = E[H_i]\, E[H_j]$ and $g_2 = 0$. If the observation that one locus is heterozygous in an individual inflates the probability that other loci in that individual are also heterozygous, then $g_2 > 0$. David et al. proposed an approximate estimator of g_2, with an improved version offered by Stoffel et al. (2016). This metric is called g_2 because k-way disequilibrium metrics (g_k) were also proposed, with the expectation of joint heterozygosity at k loci within an individual being written as

$$E\left[\prod_{i=1}^{k} H_i\right] = \left(\prod_{i=1}^{k} E\left[H_i\right]\right)(1 + g_k) \tag{12.34b}$$

The connection between g_2 and the variance in inbreeding levels among individuals within the population, $\sigma^2(f)$, was shown by David et al. (2007) to be

$$g_2 = \frac{\sigma^2(f)}{(1 - \overline{f})^2} \tag{12.34c}$$

where $\overline{f}$ is the average level of inbreeding in the population. In the absence of variance in the level of inbreeding, $g_2 = 0$ and there is no IDE. Thus even a highly inbred population can *fail* to show IDE if all individuals are *inbred to the same level* ($\sigma^2(f) = 0$).

A meta-analysis by Miller and Coltman (2014) of estimated g_2 values in vertebrates found that only 26 of 129 estimates were significant. Among all studies, the average value of g_2 was 0.007, while among the studies where g_2 was significant, its average value was 0.025. This implies that, given one locus is heterozygous, the probability that a second is a heterozygote is inflated by just 2.5%. For example, the single-locus expectation for a heterozgyote at locus i (in the absence of information for any other loci) of $(1 - f)2p_i q_i$ increases to $(1 - f)(1 + g_2)2p_i q_i$ when we know that a second locus is a heterozygote. An earlier analysis of 19 vertebrate populations by Grueber et al. (2011a) found a mean value for $\sigma^2(f)$ of 0.0044 and a mean value for $\overline{f}$ of 0.053, suggesting a rough value for g_2 of $0.004/(1{-}0.052)^2 = 0.005$, consistent with the larger summary of Miller and Coltman. An important observation regarding $\sigma^2(f)$ was made by Huisman et al. (2016) in their study on Red deer (*Cervus elaphus*), who found that this variance typically *decreases* with age, presumably reflecting selection against inbred individuals (e.g., Example 12.5).

Given that inbreeding levels are generally modest in most vertebrates, small to modest values of $\sigma^2(f)$, and hence g_2, are expected. However, in organisms with complete selfing, g_2 can be dramatically larger. Consider the situation of a population of fully inbred, but unrelated, individuals. Now suppose that there is a 50 : 50 chance of selfing versus outcrossing.

All of the selfed offspring have $f = 1$, while all of the outcrossed offspring (because parents are unrelated) have $f = 0$. Here $\overline{f} = \overline{f^2} = 0.5$, giving $\sigma^2(f) = \overline{f^2} - \left(\overline{f}\right)^2 = 0.25$. Substituting into Equation 12.34c gives $g_2 = 0.25/(1 - 0.5)^2 = 1$. Under this setting, if one observes a heterozygote in an offspring, that individual must have resulted from an outcross. From Equation 12.34a, for this individual heterozygous at loci i, the probability of a heterozygote at locus j is $(1 - f)(1 + g_2)2p_j(1 - p_j) = (1/2)(1 + 1)2p_j(1 - p_j) = 2p_j(1 - p_j)$, namely its random-mating value. High g_2 values were seen in the work of Toczydlowski and Waller (2023), examining 11 natural populations of the annual jewelweed *Impatiens capensis*. Using 12,500 SNPs, g_2 estimates ranged from 0.036 to 1.068, with an average value of 0.52.

Finally, it is important to note that an estimate of g_2 simply provides information on the amount of IDE within a sample, and this estimate is *trait-independent*. As noted by a number of investigators, one can have a nonsignificant value of g_2 and yet still have a significant HFC, because the former may be estimated with a small set of markers, whereas the latter (when IDE is present) is the cumulative effect over an entire genome (Szulkin et al. 2010; Kardos et al. 2014; Miller and Coltman 2014). Hence, the joint observation of a nonsignificant g_2 and a significant HFC does not rule out the general-effects hypothesis.

Example 12.10 Forstmeier et al. (2012) examined a five-generation pedigree of zebra finches, finding the following distribution of pedigree-expected f values (f_{ped}) among 1210 sampled individuals

Number	Frequency	f_{ped} value
1100	0.909	0
24	0.020	0.01563
6	0.005	0.03125
29	0.024	0.0625
4	0.003	0.125
47	0.039	0.25

For these data, $\overline{f} = 0.01209$ and $\overline{f^2} = 0.00258$. Hence, from Equation 2.5,

$$\text{Var}(f) = \frac{n\left(\overline{f^2} - [\overline{f}\,]^2\right)}{n-1} = \frac{1210\left(0.00258 - 0.01209^2\right)}{1209} = 0.00244$$

As we discuss below, this is a biased underestimate of $\sigma^2(f)$, as it simply uses the *expected value* of f for various sets of relatives (their predigree values), and ignores the variation in their *realizations* of any particular f about this expected value (Chapter 7), which can be directly estimated using marker data (Chapter 8 and below). Substituting into Equation 12.34c gives an estimated g_2 value of

$$\widehat{g_2} = \frac{\text{Var}(f)}{(1 - \overline{f})^2} = \frac{0.00244}{(1 - 0.01209)^2} = 0.00250$$

Generating HFCs Under Indirect Effects: Identity Disequilibrium (IDE)

Two (non-exclusive) mechanisms can indirectly generate HFCs for marker loci not themselves under selection, both of which require correlations between marker loci and loci under selection. The first is LD (the local-effect hypothesis; Table 12.7), which we will examine shortly. The second is IDE, where the number of heterozygous marker loci that an individual carries provides some information on its level of inbreeding (the general-effect hypothesis; Table 12.7). As Equation 12.34c shows, IDE requires not just inbreeding, but more specifically, *variance* in the level of inbreeding over individuals in the population, $\sigma^2(f) > 0$.

Under inbreeding depression, we generally expect a negative correlation between fitness and f, namely, $\rho(w, f) < 0$. How does this translate into a HFC? Szulkin et al. (2010)

noted that the expected correlation between fitness and an estimator, $\widehat{f}$, of f can be expressed as

$$\rho(w, \widehat{f}\,) = \rho(\widehat{f}, f)\, \rho(w, f) \tag{12.35a}$$

Hence, $\rho(w, \widehat{f}\,) < \rho(w, f)$, with the fraction of any true correlation between inbreeding and fitness being captured by HFCs depending on how accurately MLH predicts f. Similarly, the regression slope of fitness (w) on our estimator of f can be written as

$$\beta(w, \widehat{f}\,) = \beta(w, f)\, \beta(f, \widehat{f}\,) \tag{12.35b}$$

Most HFC studies use either H, the total number of heterozygous marker loci in an individual, or the **standardized heterozygosity** (Coltman et al. 1999),

$$H_s = \frac{H}{\overline{H}} \tag{12.36a}$$

namely the number of heterozygous loci standardized by the average number of heterozygous loci in a random individual from the sampled population. Szulkin et al. (2010) showed that the squared correlations between H and H_s with f are

$$\rho^2(f, H) = g_2 \cdot \left(\frac{\overline{H}^2}{\sigma^2(H)} \right) = \sigma^2(f) \left(\frac{\overline{H}^2}{\sigma^2(H)(1-\overline{f})^2} \right) \tag{12.36b}$$

and

$$\rho^2(f, H_s) = \frac{g_2}{\sigma^2(H_s)} = \frac{\sigma^2(f)}{\sigma^2(H_s)(1-\overline{f})^2} \tag{12.36c}$$

where the last steps follows from Equation 12.34c, with $\sigma^2(H)$ and $\sigma^2(H_s)$ being, respectively, the variance among H and H_s values among the individuals in the population (or in our sample).

An important question in the design of HFC studies concerns the required number, and type, of markers needed to detect an HFC generated under the general-effects hypothesis (the presence of *both* IDE in the genome *and* inbreeding depression in the focal trait used as the fitness proxy). This was nicely addressed by Miller et al. (2014), who showed that for m markers, each with an average heterozygosity of $\mathcal{H}$, then

$$\rho^2(f, H_s) = \frac{m\, g_2\, \mathcal{H}}{1 - \mathcal{H} + (m-1) g_2 \mathcal{H}} \tag{12.37a}$$

For example, using the estimated g_2 value ($\simeq$0.003) from Example 12.10, the expected correlation between H_s and f for 1000 SNPs, each with $\mathcal{H} = 0.2$ (which corresponds to a minor-allele frequency of 0.113), is

$$\rho^2(f, H_s) = \frac{m\, g_2\, \mathcal{H}}{1 - \mathcal{H} + (m-1) g_2 \mathcal{H}} = \frac{1000 \cdot 0.003 \cdot 0.2}{1 - 0.2 + 999 \cdot 0.003 \cdot 0.2} = 0.429$$

Hence, this design would pick up just over 40% of the true variance in fitness explained by the level of inbreeding. Note that Equation 12.37a can be rearranged (Miller and Coltman 2014) to yield the required number of markers, m_r, to obtain an expected correlation of r between H_s and f,

$$m_r = \left(\frac{r^2}{1-r^2} \right) \left(\frac{1 - g_2(1+\mathcal{H})}{g_2 \mathcal{H}} \right) \tag{12.37b}$$

Again using the g_2 value from Example 12.10, how many SNPs (with $\mathcal{H} = 0.2$) are required to have H_s account for 90% ($r^2 = 0.9$) of the variance in f? Equation 12.37b yields

$$m_r = \left(\frac{0.9}{1-0.9} \right) \left(\frac{1 - 0.003 \cdot (1+0.2)}{0.003 \cdot 0.2} \right) = 14,946$$

An important caveat for Equations 12.37a and 12.37b is the assumption that markers are *independent*, namely, they are not in LD. The effective population size and linkage map set an upper limit on the number of independent markers (Chapters 5 and 20). Hence, the more proper terminology for Equation 12.37b is not the required number of markers, but rather the **equivalent number of independent markers**, which may be far less than the actual number of markers that were scored.

Equation 12.37a makes the important point that the product of the number of markers and their average heterozygosity ($m\mathcal{H}$) determines the strength of the correlation between f and H_s. Early DNA-based HFC studies typically used no more than a few dozen microsatellites (Chapman et al. 2009). The advantage of this class of markers is that they are highly polymorphic. Roughly speaking, many alleles (length variants; Figure 8.2) at a given microsatellite are at nearly equal frequencies, so that a microsatellite locus with k alleles has an approximate $\mathcal{H}$ value of $(1 - 1/k)$, as there are k different homozygotes, each with approximate frequency $1/k^2$. SNPs, on the other hand, have lower values of $\mathcal{H}$ (as they are typically biallelic with unequal frequencies, so that typically $\mathcal{H} \ll 0.5$), and hence more are required to have the same power as microsatellites. As an example, in a study of two bighorn sheep populations by Miller et al. (2014), microsatellites had $\mathcal{H}$ values of around 0.6, while SNPs had values of around 0.2. Hence, for these values, three times as many SNPs were required to have the same power as microsatellites. Genomic technology has now evolved to the point where it is relatively trivial, and inexpensive, to score thousands (or more) of SNPs quickly. As a result, modern HFC studies are shifting entirely to SNPs.

Miller and Coltman (2014), substituting typical study values in Equation 12.37b, noted that *most HFC studies are significantly underpowered* in terms of the numbers of scored markers. In particular, observe that the correlation between H_s and f increases with the parameter $m\, g_2\, \mathcal{H}$. An investigator can increase this value by scoring more markers (increasing m) or using more polymorphic markers ($\mathcal{H}$). Also, the appropriate sampling design can increase the sample value of g_2 (Miller et al. 2014). This might be accomplished by sampling more individuals (so that the sample includes more inbred individuals), or (more efficiently) by using a hierarchical design that attempts to capture a subset of individuals that are more inbred than random.

In large, randomly mating populations, the association between MLH and f is expected to be negligible because essentially all individuals trace through pedigrees with similar (and very low) levels of inbreeding, rendering the variance in f among individuals, and hence g_2, nearly insignificant. Thus, if the general-effects hypothesis is correct, such populations should not exhibit HFCs. However, even large populations can contain considerable structure, with localized inbreeding and substantial IDE (Vitalis and Couvet 2001), so care must be taken in declaring the general-effect hypothesis to be invalid simply because an HFC is observed in a large population. Further, even in a large stable population, migration can generate IDE, with migrants generally being more outbred relative to residents, generating $\sigma^2(f) > 0$.

Bierne et al. (2000) suggested that one optimal setting to generate HFCs in a large population is during the immediate aftermath of a bottleneck (and subsequent expansion to yield the currently observed large size). A large pre-bottleneck population that is randomly mating is expected to have the potential of considerable inbreeding depression, while a small population might have much of its inbreeding depression purged (see above). Hence, small populations, while potentially having significant IDE, may have little inbreeding depression, while a historically large population has high depression potential but little IDE. A bottleneck generates IDE while keeping the variation needed for inbreeding depression.

Example 12.11. Szulkin et al. (2010) examined an HFC study by Pujol et al. (2005) on an important crop in developing countries, cassava (*Manihot esculenta*). Using eight microsatellites, the mean number of heterozygous loci per individual was $\overline{H} = 3.731$, the sample variance in H was 6.104, and an estimate of g_2 based on these markers was 0.057. Finally, the observed correlation between H and growth rate (taken as a surrogate for fitness) was $\rho^2(w, H) = 0.031$. Equation 12.36b gives an estimate of the correlation between f and H as

$$\rho^2(f, H) = g_2 \cdot \left(\frac{\overline{H}^2}{\sigma^2(H)} \right) = 0.057 \left(\frac{3.731^2}{6.104} \right) = 0.13$$

Hence, H accounts for only 13% of the actual variance in f among individuals. Note that Equation 12.34c implies that $\sigma^2(f) < g_2 = 0.057$. Using the Pujol data, Szulkin et al. (2010) estimated that $\overline{f} = 0.071$, which from Equation 12.34c yields

$$\sigma^2(f) = (1 - \overline{f})^2 g_2 = (1 - 0.071)^2 \cdot 0.057 = 0.049$$

Finally, rearranging Equation 12.35a allows one to estimate the underlying fitness–inbreeding correlation,

$$\rho^2(w, f) = \frac{\rho^2(w, H)}{\rho^2(f, H)} = \frac{0.031}{0.13} = 0.238$$

Thus, while almost 24% of the variation in growth rate was due to differences in inbreeding level, only 13% of this effect was captured by using H, resulting in H accounting for only 3% of the variance in growth rate, as $\rho^2(w, H) = \rho^2(w, f) \cdot \rho^2(f, H) = 0.24 \cdot 0.123 = 0.031$.

Now suppose that this study was carried out with 300 SNPs, each with an average heterozygosity of 0.2. From Equation 12.37a, the expected correlation between H_s and f becomes

$$\rho^2(f, H_s) = \frac{m g_2 \mathcal{H}}{1 - \mathcal{H} + (m-1) g_2 \mathcal{H}} = \frac{300 \cdot 0.057 \cdot 0.2}{1 - 0.2 + 299 \cdot 0.057 \cdot 0.2} = 0.81$$

From Equation 12.37b, the expected number of SNPs (with $\mathcal{H} = 0.2$) in order to have H_s explain 95% of the variation in f is

$$m_r = \left(\frac{r^2}{1 - r^2} \right) \left(\frac{1 - g_2(1 + \mathcal{H})}{g_2 \mathcal{H}} \right) = \left(\frac{0.95}{0.05} \right) \left(\frac{1 - 0.057(1 + 0.2)}{0.057 \cdot 0.2} \right) = 1553$$

Generating HFCs Under Indirect Effects: Linkage Disequilibrium (LD)

The second mechanism for indirectly generating HFCs is LD (the local-effect hypothesis; Table 12.7), which (as shown above) can generate associative overdominance at marker loci linked to deleterious recessives. We can further unpack this phenomenon as follows. As above, let a and b denote deleterious alleles at two loci linked to a neutral marker locus (M). Assuming complete LD, the gametic states associated with the marker locus are AM_1b and aM_2B. Now suppose that each deleterious allele (a, b) reduces fitness by the fraction s in the homozygous state, by hs in the heterozygous state, and that the effects of the two loci are independent and multiplicative. The resulting fitnesses associated with the marker locus become $(1 - s)$ for the two homozygous classes and $(1 - hs)^2$ for the heterozygous class. Under this scenario, the heterozygous marker class will exhibit greater fitness than the homozygous classes provided $h < (1 + s)/2$, showing associative overdominance.

Notice that when fitness is analyzed on the direct scale of measurement, associative overdominance can arise even with additivity (as $h = 0.5 > [1 + s]/2$) or with slight dominance of the deleterious allele. If $h > (1+s)/2$, **associative underdominance** occurs, wherein the marker heterozygotes exhibit *reduced fitness*. These latter two peculiarities disappear if fitness is measured on a logarithmic scale. Associative overdominance or underdominance can still arise in this case, but assuming small s, it depends more simply upon whether h is less than or greater than 0.5, i.e., on whether the deleterious alleles are partially recessive or partially dominant. We highlight this scaling property because many HFC analyses have been performed using data that were not log transformed. The important point is that even for marker loci in Hardy–Weinberg equilibrium (i.e., with no evidence of inbreeding), with no direct effects on fitness, and with no functional overdominance elsewhere in the genome, associative overdominance arises if loci carrying partially recessive deleterious alleles are linked to the marker and in repulsion disequilibrium.

Now consider the opposite situation: coupling disequilibrium, such that the two marked stretches of DNA are aM_1b and AM_2B. The fitnesses associated with the M_1M_1, M_1M_2, and M_2M_2 genotypes are then $(1-s)^2$, $(1-hs)^2$, and 1. Most HFC studies simply consider whether the fitness of heterozygotes exceeds the *average* of the homozygous classes (without respect to homozygous genotype). In this setting, the relevant observation is that associative overdominance with respect to the mean logarithm of fitness for the two homozygous classes will arise if $h < 0.5$, the same conclusion that we arrived at with repulsion disequilibrium. For non-log transformed fitness, the requirement is $h < (2+s)/4$.

The general conclusion is that linked loci carrying partially recessive deleterious alleles in LD, whether in repulsion or in coupling, leads to the ***appearance of overdominance***. The fact that situations fully in accord with the dominance hypothesis can lead to observations fully compatible with the predictions of the overdominance hypothesis is an obvious problem. Because essentially all populations have some degree of structure (either variance in inbreeding and/or gametic-phase disequilibria), it can be difficult to draw rigorous conclusions about the mechanism of inbreeding depression (or about the advantages of heterozygosity at individual loci) from descriptive surveys of the relationship between MLH and fitness-related traits. Fortunately, with the advent of dense marker maps, we can gain some insight into the extent of LD with a genome and of IDE within a sample, both of which can help inform our understanding of HFCs.

We expect that a large (and stable) randomly mating population will have only short blocks of LD, as the time back to the most recent common ancestor will be long, allowing recombination to remove LD for all but the most tightly linked loci (Chapter 5). When LD only extends over very short distances, the chance that a relatively small random set of markers will just happen to capture linked selected loci is expected to be rather unlikely. Hence, as was the case for IDE, we (at first blush) expect that the local-effect hypothesis will be unable to generate any substantial HFCs in a large populations. However, again as with IDE, caution is in order. A currently large population may have experienced bottlenecks and expansions in the past, which can generate significant LD. The use of dense markers can estimate the average LD block size in a population sample, allowing one to probe the likelihood of LD effects.

In summary, both of the indirect hypotheses (local and general effects; Table 12.7) require population structure and demographic features that generate LD and/or IDE. If both types of disequilibria are present, then for local effects to dominate general effects, the cumulative fitness effects of nearby selected loci in LD with a given marker must exceed the cumulative effects from IDE with selected loci in the rest of the genome. As stressed by Szulkin et al. (2010), this is unlikely unless the amount of IDE is extremely small or the LD reach of the scored markers spans a significant fraction of the genome.

HFCs: The Data

Given the above discussion, the most informative MLH surveys come from observations on historically large, random-mating populations, for which gametic-phase and identity disequilibria are likely to be minimized (subject to the caveats mentioned above). Studies of this nature with adequate statistical power raise serious questions about the generality of the overdominance hypothesis when they do not yield a positive correlation between MLH and fitness. What do the data tell us? For most organisms that have been studied, predominantly small, but significant, correlations are often found between heterozygosity and fitness-related characters (Mitton and Grant 1984; Zouros and Foltz 1987), typically accounting for around 1% of the variance (Chapman et al. 2009). There are, however, numerous exceptions in which no such pattern is found.

A substantial body of data for marine bivalves suggests that most populations exhibit a positive correlation between growth rate and *allozyme* MLH (Singh and Zouros 1978; Zouros et al. 1980, 1988; Koehn et al. 1988; Gaffney 1990; Gaffney et al. 1990; David et al. 1995). It should be stressed that these studies often used key enzymes in important metabolic pathways as the markers, enhancing the chance that these markers were under direct selection. Rather strangely, however, samples from these same populations almost always exhibit a *deficiency of heterozygotes*, at least in early-age cohorts. In addition, there appears to be substantial multilocus disequilibrium—an excess of highly homozygous and highly heterozygous individuals. Many studies have documented the elimination of the heterozygote deficiency as cohorts age, which suggests that MLH is positively correlated with survival as well as with growth rate. The presence of disequilibria in new recruits clearly indicates the potential for these results to be a simple consequence of associative overdominance, rather than an intrinsic advantage to allozyme heterozygosity. This conclusion is bolstered by the observation that increases in heterozygosity with cohort aging only appear at loci that initially had heterozygote deficiencies. Given that most marine bivalves are broadcast spawners, it is difficult to envision how a high degree of genetic structure (generating IDE) could arise within populations via restricted mating. However, as Bierne et al. (2000) noted, population bottlenecks can generate HFCs via indirect effects by generating both LD and IDE, and they suggested that this could easily apply to bivalve populations.

An alternative explanation for both the heterozygote deficiencies and the multilocus disequilibria at some loci is the presence of either null alleles or of aneuploidy (missing chromosomes). If either situation is common, as appears to be the case (see Gaffney et al. 1990 for a summary), then a fraction of the individuals that are scored electrophoretically as homozygotes will actually be either active/null heterozygotes or chromosomal haploids. If such individuals have reduced activity for important metabolic and/or developmental functions, as seems likely, their undetected presence will cause a downward bias in the estimated fitnesses of homozygous classes (Foltz 1986). Thus, in addition to yielding apparent heterozygote deficiencies, null alleles and aneuploidy will promote the appearance of heterozygote advantage.

Pogson and Zouros (1994) argued that the failure of random bivalve microsatellite (SSR) markers to show the fitness correlations seen with allozymes is a point in favor of functional overdominance of allozymes. The same observation was also made for rainbow trout (*Oncorhynchus mykiss*) by Thelen and Allendorf (2001). However, the presence of null alleles for allozymes, but not for DNA-based markers, could easily account for this difference. Allozymes score the final protein product, which can fail to be produced by a given gene copy for any number of reasons, resulting in a haploid product from an individual being scored as a homozygote. DNA-based markers generally do not suffer this problem. Hence, DNA markers will not count heterozygotes involving nonfunctional alleles as homozygotes (and likely with reduced fitness). Support for this idea comes from Pogson and Zouros's observation that microsatellite markers, unlike some of the allozymes, showed only small heterozygote deficiencies.

Apparent heterozygote advantages have also been recorded for growth rate in some species of trees (Mitton et al. 1981; Ledig et al. 1983; Strauss 1986; Strauss and Libby 1987). However, other extensive surveys have failed to find any such relationship (Bush and Smouse 1991; Savolainen and Hedrick 1995). In *Pinus radiata,* stands with greater allozyme heterozygote deficiencies exhibit higher correlations between individual heterozygosity and growth rate (Strauss and Libby 1987), results that are reminiscent of those obtained in marine bivalves. If not an artifact of null alleles or aneuploidy, such results may be a simple consequence of multilocus homozygosity acting as a proxy for the variation in degree of inbreeding within individual stands of trees (IDE; the general-effect hypothesis).

Attempts have been made to find an association between MLH and fitness in many other organisms, some with success, others not (reviewed, among others, by Britten 1996; Reed and Frankham 2003; Chapman et al. 2009; Miller and Coltman 2014). Most older (pre-2000) studies were based on small numbers of individuals and a modest number of allozyme loci, and those that found positive correlations are subject to the challenge that the results are an artifact of population structure (generating $\sigma^2(f) > 0$), rather than a consequence of true functional overdominance (Houle 1989a). Three of the largest studies that were performed in this era, *D. melanogaster* by Houle (1989a), fungus beetles by Whitlock (1993), and brook trout by Hutchings and Ferguson (1992), all failed to find an allozyme MLH-fitness association. Conversely, Hoffman et al. (2014) noted that an initially weak association between MLH (based on $\sim$30 microsatellites) and parasite load seen in harbor seals (*Phoca vitulina*) saw the strength of this association increase almost five fold when MLH was based on $\sim$15,000 SNPs. Equation 12.37a predicts such an increase when such a correlation is real.

A number of attempts have been made to develop statistical approaches that could definitively resolve the associative overdominance issue for allozyme data. For example, Smouse (1986) reasoned that a deeper understanding of the mechanisms of inbreeding depression could be obtained by examining the fitnesses of the alternative homozygous classes within loci. Assuming two alleles per locus, under the overdominance hypothesis, the rarer of the two homozygous classes should have the lowest fitness (WL Chapter 5). Smouse's adaptive distance model transiently attracted some followers, until Houle (1994) showed that the model fits cases of associative overdominance as well as cases of functional overdominance.

One proposed experimental approach for determining the importance of local versus general effects in generating HFCs is to examine heterozygosity-fitness correlation in sibs, such as the offspring of selfed individuals (Leary et al. 1987; Fu and Ritland 1994; Bierne et al. 1998; Hansson et al. 2001). The proposed logic is that all members in a sibship have the same value of f, controlling for differences in inbreeding levels. The presence of correlations between MLH and fitness within such families is then taken as evidence *against* the general-effects hypothesis. However, this approach is potentially a bit misleading, as while the *expected* value of f is the same in all of the sibs, the *realized* values show variation about this expectation (Chapters 7 and 8), variation that potentially can be partly captured by variation in heterozygosity among sibs.

Finally, a more promising approach for distinguishing between the alternative hypotheses listed in Table 12.7 is to use dense molecular maps to localize QTLs (Chapters 17–21) involved in inbreeding depression (and heterosis). This is examined in detail in the next chapter in the search for loci underlying heterosis. As hypothesized by Crow (1993c, 2000), for those cases where a QTL involved in heterosis initially appears to show overdominance, finer mapping almost always reveals that it was generated by pseudo-overdominance, with linked deleterious recessive alleles acting as a single gene in the initial linkage block.

Summary: What Generates Most HFCs?

Early allozyme-based HFCs studies were motivated by the search for direct selection via overdominance, although this was only one of three potential explanations (Table 12.7). With the advent of random DNA markers, the focus of HFCs studies has somewhat shifted. In contrast with allozymes, which were often *specifically chosen* because they scored key protein products in important metabolic pathways, *direct* selection on random DNA markers was viewed as unlikely and HFCs involving these markers are likely due to *indirect effects*. These arise via correlations between scored markers and unscored loci under selection, through either LD with nearby selected sites (the local-effect hypothesis) or IDE with selected sites over the entire genome (the general-effect hypothesis). Of these, a stronger impact from general effects is much more likely, as the total fraction of the genome in LD with a modest set of scored markers is likely to be rather small, so that genome-wide effects via IDE would likely swamp any locally based signal. The conclusion is that ***most observed HFCs are simply manifestations of inbreeding depression*** (Szulkin et al. 2010), arising because inbred individuals are less fit, and MLH predicts (or at least ranks) the inbreeding levels of sampled individuals (when IDE is present). Finally, population structure and demography dictate the likelihood of HFCs. Indirect effects are not expected in large, stable, and closed populations, but can be generated by localized mating, bottlenecks, or migration, all of which can be difficult to rule out for any particular population.

MOLECULAR MARKER-BASED ANALYSIS OF INBREEDING DEPRESSION

As detailed above, most information on the impact of inbreeding has historically come from either controlled crosses (regressing trait means on f) or from data on an observed (or inferred) single generation of inbreeding in nature. One could also examine inbreeding depression using pedigree data, with inbreeding estimates, f_{ped}, following from the pedigree coefficients of coancestry (Chapter 7; Equation 12.5c). Unfortunately, these estimates have potential problems beyond the obvious difficulty in obtaining pedigrees for most natural populations. First, they depend on both the accuracy and the depth of the pedigree, as the assumption is that the base of the pedigree (the founders) consists of unrelated and outbred individuals, resulting in f_{ped} tending to underestimate the true value of f. Fortunately, a pedigree going back only three to five generations tends to capture most of the recent inbreeding, so that in many settings, this underestimation is often small (Balloux et al. 2004; Keller et al. 2011; but see Ferenčaković et al. 2017). Second, there is a Mendelian sampling variance associated with f (and other relationships in general; Chapters 7 and 8), such that f_{ped} represents an *expected* value, about which there can be considerable variation for any particular *realization* (Hill and Weir 2011, 2012). Hence, two individuals with the same predicted value of f_{ped} can have rather different realized values, a fact that can be exploited to control for non-genetic factors, such as using within-sib comparisons (e.g., Clark et al. 2019).

As developed in Chapter 8, marker-based estimates of f attempt to address these issues: they can estimate the realized f values with or without a pedigree, and can be applied to essentially any population. Thus, in theory, one could examine the impact of ID from a random collection of individuals from a target population, *provided* that there is *sufficient variation* in the level of inbreeding in the sample.

Runs of Homozygosity (ROHs)

Recall (from Chapter 8) that marker-based estimates of f are either based on genome-wide averaging over single marker (point) estimates (e.g., f_{HOM}, f_Y; Equations 8.19–8.22) or by using dense markers to detect local **runs of homozygosity**, **ROHs** (Broman and Weber 1999; Chapman and Thompson 2003; McQuillan et al. 2008; Keller et al. 2011; Knief et al. 2015; Peripolli et al. 2016; Druet and Gautier 2017; Ceballos et al. 2018; Yengo et al. 2021). The key idea with using ROHs is that they serve as proxies for genomic regions that show **homozygosity by descent** (**HBD**), with the total fraction of the genome that is HBD being the estimate of f. ROHs are commonly found in livestock (Periploii et al. 2016) and in human populations with the potential for some inbreeding. The former is not surprising, as selection is expected to create inbreeding in a population (Morley 1954; WL Chapters 3 and 26), with f increasing as selection proceeds. This was observed by Forutan et al. (2018), who examined North American Holstein dairy cattle, finding that the average number of ROHs (of one MB or larger) increased from 57 in 1990 to 82 by 2016.

As was discussed earlier, there are a few important subtleties with estimates of the fraction of ROHs within a genome (the reader may wish to refresh themselves with the relevant parts of Chapter 8 before proceeding). For **rule-based estimates**, a genomic region is declared to be an ROH when it exceeds some size threshold, based on either physical distance (base pairs) or genetic distance (expected number of recombinants), leading to an estimate of the fraction of the genome that is autozygous (inbred) as

$$\widehat{f}_{ROH} = \frac{\text{total length of ROHs}}{\text{genome size}} \tag{12.38a}$$

The choice of the threshold size for declaring an ROH represents a tradeoff between the control of false positives (longer ROHs have lower false-positive rates of autozygosity) versus reduced power (ignoring shorter runs that, cumulatively, may be biologically important).

Table 12.8 ROH-based estimates of the scaled inbreeding depression, I_s (Equation 12.24c), for cattle and pig traits. Here $*$ and $**$ denote, respectively, significance at the 5% and 1% levels.

Character	I_s	Reference
Holsteins		
205-day milk yield	0.236**	Bjelland et al. 2013
Daily milk yield	0.250**	
Average fat %	0.028	
Average protein %	−0.033	
Iberian pigs: Number of piglets at birth		
ROH $\geq$ 0.5 MB	0.313*	Saura et al. 2015
ROH 0.5 to 5 MB	− 0.760	
ROH $\geq$ 5 MB	0.254*	
Iberian pigs: Number of piglets born alive		
ROH $\geq$ 0.5 MB	0.325**	Saura et al. 2015
ROH 0.5 to 5 MB	− 0.482	
ROH $\geq$ 5 MB	0.256*	

Besides being a statistical issue, threshold-size choice also has more subtle biological implications. ***Long ROHs are consistent with recent inbreeding***, while ***short blocks of ROH are generated by distant inbreeding***, although sufficiently short ROHs could also be allozygous (outbred) regions, due to the inheritance of a set of markers as a single linkage block. The lack of a clearly defined threshold creates some amount of arbitrariness for ROH-based estimates of f, as the smaller the threshold ROH value, the more ancient the assumed base population (Chapter 8). Hence, the choice of a threshold size represents an unstated assumption by the investigator as to whether they feel that more recent, or more ancient, inbreeding events are the main driver of inbreeding depression. Finally, it is often assumed that the genomic distribution of ROH lengths (under inbreeding) should be roughly exponential (Chapter 8). Yengo et al. (2019), however, found that the ROH size distribution in humans was better fit by a mixture of two exponentials, likely reflecting genomic variation in recombination rate.

Another issue with rule-based $\widehat{f}_{ROH}$ estimates arises when only a modest number of markers are scored (the **reduced genomic representation** setting). Unscored markers nested within an ROH based on scored markers are assumed to also be autozygous, which can be problematic under low marker density. In these settings, **model-based estimates** of HBD outperform estimates based on ROHs (see Chapter 8), with

$$\widehat{f}_{HBD} = \text{Average genome-wide posterior probability of being HBD} \tag{12.38b}$$

Even with very dense markers (including whole-genome sequencing), if the population has considerable structure, $\widehat{f}_{HBD}$ can be a better choice than $\widehat{f}_{RHO}$ (Alemu et al. 2021; Lavanch and Goudet 2022), as it assigns posterior probabilities on whether runs are HBD rather simply declaring a particular ROH as being HBD.

Because of the higher frequency of homozygotes in inbred individuals, offspring from the mating of close relatives have higher risks of presenting recessive Mendelian (single major gene) diseases (Vogel and Motulsky 1996). Similarly, several complex diseases (in both humans and livestock) show correlations with ROHs, with a higher fraction of the genome showing ROHs in disease cases as opposed to controls, an observation often called the **homozygosity burden** (reviewed by Ku et al. 2011; Szpiech et al. 2013; Periploii et al. 2016; Ceballos et al. 2018; Pemberton and Szpiech 2018). For example, Keller et al. (2012) examined roughly 22,000 individuals (~9,400 cases, ~12,500 controls) and found that the odds of schizophrenia increased by about 17% for each 1% increase in the fraction of the genome that showed ROHs. This implies a 2.7-fold increase for first-cousin inbreeding, and a 1.29-fold increase for second-cousin inbreeding. Similarly, Howrigan et al. (2016) found an association between ROHs and lower general cognitive ability, with most of the signal from long, rare ROHs (suggesting an impact from very recent inbreeding). However, Johnson et al. (2016), using a sample size almost twice as large as Keller et al. ($\simeq 40,000$), were unable to replicate this correlation between ROHs and schizophrenia risk. This is a strong cautionary tale in that the case-control ascertainment corrections suitable for GWAS (essentially, control for between-sample allele-frequency differences; Chapter 20) may not be sensitive enough to control for more subtle population structures (such as very small differences in recent inbreeding) required when using ROHs. A single generation difference in inbreeding rates between cases and control that would not generate an allele-frequency difference (and hence would not need to be adjusted for in a GWAS) can create a false ROH-trait association.

As with controlled-cross data, one can regress individual trait values on their estimated f_{ROH} values, with (as before) the slope of the regression being the expected inbreeding depression in a fully inbred individual (assuming no epistasis). Table 12.8 presents a few such estimates from domesticated animals. The results of Saura et al. (2015) are especially interesting, as they compared whether most of the inbreeding depression signal was coming from short ROHs (distant ancestors, and hence presumably less deleterious alleles) or long ROHs (more recent ancestors, and hence presumably more deleterious alleles). For the two traits they examined, the estimate of I_s was significant when based on long ROHs (>5 MB), but not when based on intermediate-length ROHs (0.5 to 5.0 MB). However, the effective population size of their pig population was very small, so this could simply be a manifestation of the observation of Keller et al. (2011) that the power for detecting effects from short ROHs decreases with N_e. Conversely, for semen quality traits in cattle, Ferenčaković et al. (2017) found an ROH threshold size of 2 MB gave a better model fit than a threshold of 4 MB, so that much of the ID signal was coming from relatively short ROH regions.

Just as the ROH length associated with the ID signal for a trait provides information about its underlying genetic architecture (e.g., older versus more recent alleles), so too does the point-wise estimator of f that is most strongly associated with the bulk of the ID (as we will now demonstrate). The issue in estimating ID is the ***amount of inbreeding on underlying loci contributing to ID for a focal trait*** (for example, those QTLs showing directional dominance), *not* the *average* genome-wide inbreeding value. While genome-wide estimates of f are good (and potentially exceptional) proxies for average f at the trait QTLs, this need not be the case. Under the partial dominance model, the bulk of ID is due to homozygotes involving low-frequency deleterious alleles that are largely recessive. In such cases, estimates of f that are best able to capture the inbreeding signal from low-frequency alleles will best capture ID. Further, the more deleterious the underlying alleles, the more recent they should be (largely new mutations, else selection would have quickly removed them from the population), and hence would be best captured by long ROHs. Less deleterious alleles contributing to ID would be older and hence best captured by shorter ROHs. Building on this idea of the bulk of ID coming from rare homozygotes, Keller et al (2011) proposed their **homozygous mutation load** (**HML**) metric, focusing on only those homozygotes from rare alleles (i.e., with a population allele frequency below some threshold, such as 1 to 5%). Consistent with this notion, Brown and Kelly (2019) found the number of rare-allele homozygotes was a good predictor of fitness in inbred lines of *Mimulus*. Conversely, if a significant fraction of ID is from moderate-frequency overdominant alleles, the signal from ID would be captured by estimates of f that largely rely on moderate-frequency alleles and short ROHs (as such alleles can be much older).

As noted by Clark et al. (2019), point-based estimates (such as genome-wide averages of excessive marker homozygosity) based on common-allele markers have low power to detect ID effects from nearby rare alleles, as LD is very weak between a common and a rare allele (Example 20.1). As a result, only a small fraction of excess common-marker homozygosity (which provides the estimate of f) is likely to translate into excess causal-site homozygosity (ID for the trait). Conversely, ROHs can detect signals from both rare and common alleles when they are made homozygous by being contained within an ROH. Both

Clark et al. (2019) and Yengo et al. (2021) used this approach to probe the genetic basis of ID in humans. Clark et al. examined 100 traits, 21 of which had sample sizes over 500,000. They found that the signal from the 32 traits with significant ID (involving reproductive success, risky behaviors, cognitive ability, body size, and health) was mainly due to ROHs ($>$1.5 MB), but not from homozygosity tagged by common variants (point-based estimators of f). Hence, the ID was mainly from rare, rather than common, alleles. They noted that this observation was consistent with the partial dominance model (with deleterious recessives kept at low frequency by selection), but not the overdominant model (where causal allele frequencies are predicted to be much higher, and thus accessible to being tagged by common SNPs). Conversely, Yengo et al. (2021) found a more trait-specific pattern. Some of their 11 traits (from a UK Biobank sample of 350,000) had the majority of their ID accounted for by SNP-based point estimates, while others had ROHs being more important.

Homozygosity and ROH Mapping

In 1953, C. A. B. Smith noted that inbred offspring could be used to map recessive disease genes by looking for an excess of homozygotes in a candidate gene region in cases versus controls. However, he quickly dismissed this idea as being wholly impractical given the scarcity of scorable human markers. Bolstered by the ever-increasing availability of such markers, Lander and Botstein (1987) independently proposed this idea and named it **homozygosity mapping**. Morton (1991) rightly noted that this is better called **autozygosity mapping**, as the idea is to search for regions of excess autozygosity in cases versus controls.

A major stumbling block for linkage-based mapping (Chapters 5 and 19) of Mendelian recessive disorders (single recessive genes that have a major impact on disease risk) is that cases are rare in natural populations (on the order of q^2, the squared frequency of the disease allele). As a result, it may be difficult to obtain a sufficient number of the multiple-case families that are needed for traditional linkage analysis (Wong et al. 1986). Most cases are usually found as singletons, and hence contain little traditional linkage information (Chapters 5, 17, and 19; Ott 1991). For example, Lander and Botstein (1987) noted (at the time of their publication) that there are around 100 living cases of Bloom's syndrome, but only nine known families with two or more cases (eight with two and one with three), providing very little traditional linkage information.

However, under homozygosity mapping, even singletons contain useful information. As noted by Lander and Botstein, the fraction of all cases from autozygosity in inbred individuals is $fq/[fq + (1 - f)q^2]$, which is $\simeq 1$ when $q \ll f$. Conversely, the chance that a random location is autozygous is f, for an approximate odds ratio of $1/f$ in support of linkage for each singleton. A singleton case from a first-cousin marriage ($f = 1/16$) thus contains the same amount of linkage information as a family with three affected offspring (Lander and Botstein 1987). In the above-mentioned case of Bloom's syndrome, there were at least 24 cases from marriages of cousins, offering far more information than from the limited number of multiple-case families. Even a modest number (10 to 40) of inbred singletons gives reasonable power to test a particular candidate region (Lander and Botstein 1987). Extensions, and discussions on limitations, of this basic approach are provided by Mueller and Bishop (1993), Houwen et al. (1994), Kruglyak et al. (1995), Miano et al. (2000), Botstein and Risch (2003), Laurier et al. (2006), Leutenegger et al. (2006), Benayoun et al. (2009), and Hildebrandt et al. (2009).

VanRaden et al. (2011) and Sahana et al. (2013) extended this idea to map regions in cattle that contain lethals (or synthetic lethals—linked epistatic loci that combine to generate near lethality). These would appear as haplotypes that are relatively common in the population, but which never appear in ROHs. Similar logic can be applied to crudely localize QTLs for complex traits by using **ROH mapping**. For a binary trait, one searches for chromosomal regions that are enriched for ROHs in cases relative to controls. For continuous traits, one regresses trait value on the presence/absence of ROHs in specific regions or around specific SNPs (Lencz et al. 2007; Kardos et al. 2016; Ferenčaković et al. 2017). The basic idea follows from standard GWAS methods (Chapters 20 and 21). For example, Ferenčaković et al. (2017)

modified the basic GWAS linear model (for the effect of SNP j) as follows

$$z_i = \mu + I\,f_i + \beta_{1j}\,x_{ij} + \beta_{2j}\,\text{ROH}(ij) + e \tag{12.39}$$

Here z_i is the trait value of individual i, f_i its level of genome-wide inbreeding, x_{ij} is the copy number of minor alleles at SNP j in individual i ($x_{ij} = 0, 1, 2$), and ROH(ij) is an indicator variable, equaling one when SNP j in individual i is covered by an ROH. Using this approach, they found 53 SNPs associated with inbreeding depression (β_{2j} significantly less than zero) for cattle semen quality. Similar ROH-mapping approaches were used by Pryce et al. (2014) to localize specific regions of ROHs that resulted in inbreeding depression in days for calving interval and milk yield, and by Saura et al. (2015) for porcine regions influencing fertility (number of piglets).

It is important to stress the difference between the architecture for ID and the architecture for total trait variation. Chapters 17–21 detail approaches for probing the latter, while ROH mapping only focuses on *the gene subset generating an ID signal* (i.e., *segregating sites* displaying *directional dominance*). If rare recessives are a key driver of ID, then there is likely a more stochastic component (over time and populations) to ID gene signals relative to their full trait architecture. For example, Howard et al. (2015) found that different ROH regions impacted ID for Jersey production traits in U.S. versus Australian cattle populations, with the majority of regions at low frequency (as might be expected from sporadic, large effect, and rare, variants). Brown and Kelly (2019) found that clusters of rare-allele homozygotes were good predictors of fitness in inbred lines of *Mimulus*, with different individuals having different clusters. Zhang et al. (2019) observed that large-effect regions involved in ID in potatoes also tended to be line-specific, and that these large-effect regions were typically found in genomic locations with high recombination. In humans, Yengo et al. (2021) similarly found that genomic regions involved in ID were 20-fold enriched in regions of higher recombination. They also found that ID-impacting regions were similarly enriched for conserved sites across mammals (those presumably under strong functional constraints).

Which Marker-based Estimate Should One Use to Detect ID?

What is the best marker-based approach to detect, and quantify, ID in a trait? While the answer initially depends on the number of available markers, the optimal choice becomes a bit more nuanced as the number of markers becomes very large. As noted by Alemu et al. (2021), the choice of the best estimator of f for ID depends "on the frequency, age and effect size of alleles contributing to inbreeding depression and the population demographics." In populations with small N_e, even moderately deleterious alleles can occur at modest frequencies, while when N_e is large, such alleles are almost always very rare. This difference influences which estimator of f is most strongly associated with ID (Alemu et al. 2021; Caballero et al. 2021, 2022; Lavanch et al. 2024).

The least assumption-burdened approach is based on the MLH score, the number of marker loci that are heterozygous in an individual (H), or its standardized value, H_s (Equation 12.36a). Note that an MLH score does not require any knowledge of base-population allele frequencies. As a result, these heterozygosity measures *do not* directly estimate f, but rather attempt to *rank individuals in the sample* from lowest to highest f values (based on H ranks), a procedure that is only appropriate if there is some variance in inbreeding levels, namely that $\sigma^2(f)$, and therefore g_2 (Equation 12.34), are nonzero. As might be expected, MLH is a crude measure of the amount of inbreeding, as MLH and f, while correlated when $g_2 > 0$, do not measure the same quantity. A good example of this is seen in the work of Slate et al. (2004) on Coopworth sheep in New Zealand. They examined ~600 sheep with a 7–10 generation pedigree that were also scored for 138 SSRs markers (roughly equally spaced over the 26 autosomes), finding that MLH was only weakly correlated with pedigree f ($r^2 \simeq 0.03$). Further, pedigree f, but not MLH score, detected evidence of inbreeding depression on several traits.

While methods based on genome-wide averages of single-point estimates of f (Equations 8.19–8.22) are clearly an improvement over MLH approaches—as they directly return estimates of f for a given individual—these have low accuracy when the number of markers is less than a few hundred SSRs (or a few thousand SNPs), even when inbreeding is frequent and severe (Balloux et al. 2004; Slate et al. 2004; Pemberton 2004; Carothers et al. 2006). The situation is even worse for populations experiencing less frequent, and less severe, inbreeding, where $\sigma^2(f)$ and g_2 are expected to be very small. Equations 12.37a and 12.37b provide some insight into the accuracy of single-point methods, given some initial estimate of g_2. Although these equations examine the correlation between f and H_s, they still provide a good benchmark for other single-point approaches.

Given results like these, it was not surprising that pedigree-based estimates, f_{ped}, were regarded until recently as the gold standard for estimates of f (e.g., Pemberton 2004, 2008). Thus, when one has access to only a very modest number of markers, but has a fairly accurate pedigree in hand, use pedigree-based estimates of f, potentially using the markers to check for any obvious pedigree errors (Chapter 8). As the number of markers increases—say at least 10,000 SNPs are scored—in most settings single-point estimators become quite accurate, accounting for upwards of 90% of the true (i.e., realized) f values, and become preferred over pedigree-based estimates (Kardos et al. 2015).

As the marker density increases to where accurate estimates of ROHs (and HBD) become feasible, the natural question for ID estimation is which point estimator of f is best and does this provide a better estimate than using ROHs. There is considerable subtlety in answering these two questions. One addressable concern is that estimates of ROH based on low marker density can poorly predict the actual fraction of HBD regions, so that f_{HBD} should be used in place of f_{ROH} in most settings. The more delicate issue is determining which estimate of f best captures the average f value over *those loci impacting ID* for our focal trait. In an ideal world where the underlying ID QTLs are known, a direct estimate using just these loci would largely resolve this concern, but we usually lack this luxury. This trait-specificity implies that even if the same set of individuals is used to estimate ID for a series of different characters, the optimal estimates of f could vary over traits.

One tradeoff between point and ROH estimates of f is the dependency of the former on a reference population, as most single-point methods require base-population allele frequencies, while ROH does not. However, the chosen size threshold for ROHs (or, equivalently, decay constants for HBD model-based estimates; Chapter 8) is, itself, an assumption about the base population. Not on its allele frequencies, but rather on the time scale that we regard as setting the base. Despite this, it might seem that ROH-based approaches would be preferred because they appear to directly measure f, rather than infer it by the fraction of excessive homozygotes or excessive correlation among alleles at a locus, and they are rather insensitive to allele-frequency estimates. Indeed, several authors have suggested that ROH is favored over single-point methods for quantifying ID (Keller et al. 2011; Gazal et al. 2014; Solé et al. 2017; Kardos et al. 2018a; Nietlisbach et al. 2019).

However, simulations by a number of workers found that the notion that ROH-based estimates of f are optimal is not always the case (Yengo et al. 2017, 2018c; Caballero et al. 2021, 2022; Lavanch et al. 2024). Yengo et al. (2017), using a human population structure, found that modifications of single-point-based estimates of ID can be less biased than ROH-based estimates. Part of this difference arose because ROHs were called using imputed SNPs, which can generate errors (Clark et al. 2019). Yengo et al. (2017) noted a second, and subtle, concern when using single-SNP based methods: the allele frequency and LD structure of marker SNPs might not reflect those of the causative SNPs, which can introduce bias (Chapters 20 and 21). Such differences arise because selection on the causal ID loci can alter both their population frequency and local LD structure relative to the (presumably) neutral markers used for estimating f (WL Chapters 5 and 8). Yengo et al. attempted to resolve this concern by binning (or **stratifying**) marker SNPs into a small number, ℓ, of classes, based on *both* their minor-allele frequencies and their **LD score**. The latter is given by the sum of the pairwise squared correlations between a focal SNP and all scored SNPs within a defined distance (Equation 20.13b), and provides a measure of the LD structure around a given SNP (Yang et al. 2010b; Bulik-Sullivan et al. 2015). They found that such a stratification largely removes any bias for the correlation-based estimator (f_Y; Equation 8.22a), but does not bias-correct the homozygosity-based estimator ($f_{HOM,1}$; Equation 8.19c). Yengo et al. then used f_Y to estimate f within each of the LD minor-allele frequency classes, an approach they refer to as **LD and minor-allele frequency stratified inference** (**LDMS**). A multiple regression is then performed, with the trait value of individual i modeled by

$$z_i = \mu + \sum_{j=1}^{\ell} b_j f_{i,j} + e, \quad \text{with} \quad I = \sum_{j=1}^{\ell} b_j \tag{12.40}$$

where $f_{j,i}$ is the estimate of f (using f_Y) for individual i using SNPs from LDMS category j. An overall estimate of the impact, I, from inbreeding depression (the change in mean under complete inbreeding) is obtained by the sum of the category-specific b_j estimates. They found that ROH estimates of ID were consistently larger (upwardly biased) relative to those based on LDMS-corrected SNP estimators, but that the latter produced more significant examples of ID (due to f_Y having a lower variance than F_{ROH}; Yengo et al. 2018c).

Caballero et al. (2021, 2022) found that among the class of single-point estimators (Equations 8.19–18.22), those giving more weight to rare homozygotes ($f_{HOM,2}$ and f_Y; Equations 8.21b and 8.22a, respectively) were better predictors of ID than were other point-estimators, as these better capture the HML (homozygous mutation load). Their simulations also found that the performance discrepancy between point-based and ROH estimates depended on population demography. In low N_e settings (such as used by Nietlisbach et al. 2019), f_{ROH} gave less biased estimates of ID than single-SNP-based methods. Conversely, with larger N_e (such as used in the simulations of Yengo et al. 2018c), SNP-based methods were better behaved. With small N_e, selection has less of an impact of slightly deleterious alleles, resulting in them occurring at higher frequencies, and thus better captured by certain f estimates. As N_e increases, these alleles become more strongly selected against, shifting the optimal class of f estimators that capture their impact.

Lavanch et al. (2024) considered more complicated settings, in which the sample contains both population and family structure (i.e., sets of relatives). In the absence of structure, most of the above points hold (ROH for low N_e, point estimate for high N_e), but two additional corrections arise under strong structure. First, the regression of trait value on inbreeding should now include a random term (u) for shared polygenic factors,

$$z_i = \mu + I f_i + u + e, \quad \text{where } \text{Cov}(\mathbf{u}) = \sigma_A^2 \mathbf{G} \tag{12.41a}$$

with $\mathbf{G}$ being a marker-based relationship matrix. When population structure is present, they note that using the Weir–Goudet estimator (Equation 8.12a) for the elements of $\mathbf{G}$ (with $G_{ij} = 2\widehat{\Theta}_{ij,WG}$) had the least amount of bias. Second, they found in this setting that a modification (f_L) of f_Y giving a more even weight to all loci,

$$\widehat{f}_L = \frac{\sum_{\ell=1}^{L} \gamma_\ell}{\sum_{\ell=1}^{L} \eta_\ell} \quad \text{with} \quad \gamma_\ell = T_{x,\ell}^2 - (1 + 2p_\ell)T_{x,\ell} + 2p_\ell^2, \quad \eta_\ell = 2p_\ell(1 - p_\ell) \tag{12.41b}$$

was a better estimate of f for ID. Note that f_Y uses the sum of the ratio γ_ℓ/η_ℓ, while f_L considers the ratio of these two sums, resulting in a more even weighting over loci.

Because of these uncertainties, our suggestion is to compute and then report ID estimates using a number of different approaches. This conforms to our general philosophy that when multiple estimators have been proposed, best practices require that (at least) a number of these be used to probe their consistency in the data set of interest.

Example 12.12. Bérénos et al. (2016) examined data from a long-term study of an isolated population of Soay sheep on the Scottish island of St. Kilda, followed since 1985. Information was available on roughly 6,700 individuals, most of which could be placed into a pedigree. Pedigree- and genomic-based estimators of f were compared using ~37,000 SNPs. The table below presents the correlations among various estimators of f: f_{ped} is the pedigree-based estimator (Equation 12.5c); f_Y is Yang's correlation estimator (Equation 8.22a); f_{HOM} is a homozygosity estimator (Equation 8.19c); and f_{ROH} is an ROH estimator (Equation 12.38a) using a 5 MB threshold. Values above the diagonal are correlations based on individuals with four known grandparents, while the values below the diagonal are based on a larger set of individuals where only both parents and one maternal grandparent were known.

	f_{ped}	f_Y	f_{HOM}	f_{ROH}
f_{ped}		0.72	0.67	0.77
f_Y	0.65		0.92	0.87
f_{HOM}	0.60	0.91		0.86
f_{ROH}	0.71	0.86	0.83	

Notice that the ROH estimator had the highest correlation with the pedigree-based estimator (although both point-wise estimators were also highly correlated with f_{ped}), and that all three genomic-based estimators were more correlated with each other than with the pedigree-based estimate. Using the SNP marker data, the estimated value of g_2 was 0.0014 ± 0.0002. Assuming an average heterozygosity of 0.2, Equation 12.37a suggests that the correlation between the point-wise estimate f_H (serving as a surrogate for f_{HOM} or f_Y) and the true f was approximately

$$\rho^2(f, f_H) = \frac{37000 \cdot 0.0014 \cdot 0.2}{1 - 0.2 + 37000 \cdot 0.0014 \cdot 0.2} \simeq 0.93$$

Using these different f estimators, the impact of inbreeding depression was scored on three fitness-related traits (birthweight, hindleg length, and weight at four months). The table below presents estimates of $I_s = (\bar{z}_O - \bar{z}_1)/\bar{z}_O$, the fractional reduction in the mean under complete inbreeding (Equation 12.24c) and its standard error. These values were based on whether the level of inbreeding in the individual, or its mother, was the regression variable, and also partitioned over the method used to estimate f. Significant results at the 5% and 1% levels are, respectively, denoted by $*$ and $**$.

Trait	Focal	f_{ped}	f_{HOM}	f_Y	f_{ROH}
Birthweight	Individual	−0.42 (0.16)	−0.26 (0.29)	−0.06 (0.11)	−0.07 (0.14)
	Maternal	0.01 (0.34)	−1.49 (0.49)**	−0.48 (0.20)**	−0.90 (0.26)**
Hindleg	Individual	−0.18 (0.06)**	−0.30 (0.10)**	−0.09 (0.04)**	−0.12 (0.05)**
	Maternal	−0.07 (0.09)	−0.24 (0.15)	−0.09 (0.06)	−0.11 (0.08)
4 m. weight	Individual	−0.63 (0.17)**	−0.83 (0.26)**	−0.31 (0.11)**	−0.42 (0.14)**
	Maternal	−0.60 (0.27)*	−1.20 (0.42)**	−0.43 (0.17)**	−0.64 (0.22)**

Note that f_{HOM} estimators usually gave the largest values for the amount of inbreeding depression (but also had the largest standard errors), while f_Y estimators gave both the smallest standard errors and smallest estimated values among the genomic-based methods. The genomic-based estimators (f_{HOM}, f_Y, f_{ROH}) were more powerful than the pedigree estimate (f_{ped}) in detecting maternal effects. As mentioned by the authors, part of the increased power under genome-based methods is due to larger sample size, as pedigree-based estimators included only those individuals with pedigree information.

Marker-based Estimates of Inbreeding Depression in Natural Populations

In population surveys on the impact of inbreeding, the amount of variation in f in the sample is critical. If individuals in the sample show too little variation in f, there is very little power (g_2 is too small; Equation 12.34c), requiring a modified sampling design—intentionally enriching for inbreds—to have any power. Examples of the latter approach would be to sample a series of smaller, isolated populations, or from populations known to have higher rates of consanguineous matings.

Keller et al. (2011) found that single-point estimators of f (Equations 8.19–8.22) showed the highest population variation in f, while pedigree-based estimates show the lowest (in large part because f_{ped} is an *expectation*, not a *realization* with its inherent variance about its mean). The population variance in f based on ROHs showed intermediate values. Using ROHs, Keller et al. found that very large samples (10,000 to over 60,000) from a randomly mating human population were needed for sufficient variation in f to have power to detect the impact of inbreeding depression. In a study of 16 health-related traits using roughly 350,000 individuals from 102 cohorts, Joshi et al. (2015) found correlations between height, lung function, cognitive ability, and educational attainment with fraction of the genome in ROHs, implying the presence of directional dominance in these traits. A number of other metabolic traits did not show any such associations. When the expected f value among inbreds is suspected to be small, the ROH threshold should be set lower to capture the effects of distant inbreeding contained in shorter blocks of ROH, although this approach may have poor power if N_e is small (Keller et al. 2011).

Another concern, especially in human populations, is the role of **confounding effects**. Socioeconomic and cultural factors can influence *both* a trait of interest *and* the level of inbreeding within a group (the propensity for mating among relatives). One example of such **social confounding** was seen in a Dutch study by Abdellaoui et al. (2013) that initially found a *negative* correlation between ROH and major depressive disorder (MDD)—apparently, more inbreeding lead to *less* depression. However, when the authors did a more detailed examination of their data, they found that much of the signal was coming from religiosity (R), with individuals scoring high on this trait having increased ROH (due to a higher chance of pairing with like-minded individuals), but also a lower MDD risk. Hence, $MDD \leftarrow R \rightarrow ROH$. In a second study from the same group, Abdellaoui et al. (2015) examined the level of educational attainment. While there was a barely significant ($p = 0.045$) negative association between the genomic fraction of ROHs and the educational attainment of a focal individual, the offspring of more educated parents had fewer ROHs. Abdellaoui et al. were able to show that this was entirely a function of the physical distance between parental birthplaces, with individuals achieving a higher level of education tending to travel further, and, therefore, were less likely to mate with (distant) relatives, compared with individuals who did not migrate. Clark et al. (2019) suggested that one design to control for social confounding is to contrast ROH levels *within sibships*, finding that ID effects estimated from such a design averaged around 20% smaller than estimates based on an uncontrolled design.

13

Heterosis and Outbreeding Depression

With such great benefits, it is no surprise that the breeding of food and future biofuel crops is based on principles that control heterosis, but those principles are still not understood.
Lippman and Zamir (2007)

Inbreeding is the crossing of related individuals from the *same population* (including selfing as a special case), while **crossbreeding** is the mating of unrelated individuals from *different populations*. While the mating of relatives often results in inbreeding depression (Chapter 12), crossbreeding can have two rather different outcomes, depending, in part, on the amount of genetic divergence between the populations being crossed. Agronomists and animal breeders have long known that crossing two distinct lines often has positive effects on fitness-related and agronomic traits in their F_1 progeny (Kölreuter 1766; Knight 1799; Darwin 1876; Lerner 1954; Sheridan 1981; Turton 1981; Sprague 1983), with breeders historically noting that such hybrids showed **luxuriance** for size or growth traits. An F_1 performance that exceeds the average parental performance is generally referred to as **hybrid vigor** or **heterosis** (Shull 1914, 1948). Complicating matters is the frequent observation that when heterosis arises in an F_1 population, much of it is lost in the F_2 and subsequent generations.

Conversely, as the crossed populations become genetically more distinct, their F_1 or F_2 progeny may be significantly *less fit* than the members of the original parental lines, a phenomenon known as **outbreeding depression.** For example, crosses between different species or distantly related populations frequently lead to substantial or complete loss of viability and/or fecundity (Barton and Hewitt 1981; Templeton 1981; Shields 1982; Coyne and Orr 1989a, 1997; Wu and Palopoli 1994). Indeed, outbreeding depression is the genetic basis for post-zygotic species isolation barriers, wherein the offspring of a between-species cross are either inviable or sterile (Chapter 17). Notably, whether one observes heterosis versus outbreeding depression can be *trait specific*, as examples of each can be seen in the *same cross*. Consider the offspring from a female horse (a mare) and a male donkey (a jack). From an agronomic prospective, their F_1 offspring (a mule) is considered more hardy than either of its parents, and hence displays heterosis for these traits. However, it is also sterile, and thus shows outbreeding depression for fitness. These two distinct phenomena resulting from outcrossing are of concern in different applications of quantitative genetics. Breeders are concerned with how to best exploit heterosis for improved agricultural production, while evolutionary and conservation biologists are more interested with outbreeding depression, which (as detailed below) is often assumed to be a byproduct of local adaptation. Heterosis in fitness in crosses between local populations is also of great interest to evolutionary biologists.

Heterosis is often presented (either directly or by inference) as being the complementary process to inbreeding depression (ID). In a sense, this is correct, as both jointly require nonadditive gene interaction (dominance and/or epistasis) in the presence of nonrandom mating. The deficiency of heterozygotes (with respect to random mating) that occurs under inbreeding can result in a decrease in the progeny mean, while the excess of heterozygotes under crossbreeding can cause the progeny mean to increase. When the only nonadditive gene action is dominance, it is often (but not always) the case that a trait showing inbreeding depression will also display heterosis, and vice versa, and hence the two phenomena are coupled. However, when epistasis is present, there can be ***considerable uncoupling of these two phenomena***. The most striking example of this is the observation of joint inbreeding and outbreeding depression for a given trait (Edmands 2006 reviews a few such observations). Even when only dominance is present, uncoupling can also occur because ID is caused by *within-population variation*, whereas heterosis is caused by *allele-frequency diver-*

gence (between-population variation). As a result, *different* sets of segregating alleles may be involved in within- versus between-population crosses, a point we expand on shortly.

Our treatment of crossbreeding is divided into discussions on its genetic underpinnings and on its applications. Our discussion of genetics starts by assuming that only dominance is present, showing that directional dominance is required for heterosis (as was the case for inbreeding depression). We next consider the impact of epistasis, which can result in a larger reduction in F_2 heterosis than predicted from dominance alone, and can also generate outbreeding depression. We then examine results from biometrical and QTL mapping approaches on the nature of both the dominance (partial dominance versus overdominance) and epistatic contribution to heterosis. Our discussion of applications of crossbreeding starts with the agricultural exploitation of heterosis, and contrasts the approaches of plant versus animal breeders. The former typically stop after producing hybrid (F_1) seed (**terminal crosses**), while the latter use much more elaborate schemes, such as rotational crossbreeding (**continual crosses**) and the exploitation of maternal heterosis. We conclude by examining outbreeding depression, which has considerable ramifications in conservation biology and is of great interest to evolutionary biologists.

HETEROSIS AND DIRECTIONAL DOMINANCE

The fundamental distinction between inbreeding depression and heterosis is that the former arises in crosses *within* a population (selfing or mating of relatives), while the latter arises in crosses *between* populations. While the basic concept of heterosis is straightforward, it can be defined in a number of different ways, depending on which reference population is used (Lamkey and Edwards 1999). Under inbreeding, the reference mean is that of an outbred individual (μ_0), while under heterosis, the reference mean is usually the **midparental value**, μ_{MP}. If μ_{P_i} denotes the mean of the ith population, then

$$\mu_{MP} = \frac{\mu_{P_1} + \mu_{P_2}}{2} \tag{13.1a}$$

Note that μ_{MP} is often denoted by $\overline{P}$, the average of the parental lines, and the latter also refers to the average when more than two parental lines are considered (as in the case of synthetic lines; see below). The amount of F_1 heterosis is defined as

$$H_{F_1} = \mu_{F_1} - \mu_{MP} \tag{13.1b}$$

Midparent heterosis is said to occur when the F_1 mean exceeds the midparental value ($H_{F_1} > 0$), while **high-parent heterosis** (occasionally referred to as **heterobeltiosis**) occurs when the F_1 mean exceeds the mean of the best-performing (i.e., the high) parent, with $\mu_{F_1} > \max(\mu_{P_1}, \mu_{P_2})$. Likewise, **low-parent** heterosis, $\mu_{F_1} < \min(\mu_{P_1}, \mu_{P_2})$, can also occur for traits where lower values are better (such as development time or disease susceptibility).

If a trait has only an additive basis, then $\mu_{F_1} = \mu_{MP}$ and there is no heterosis (Chapter 11). Hence, nonadditive gene action is *necessary* for heterosis, but (as with inbreeding depression), it is *not* sufficient. Consider the situation where dominance (but no epistasis) occurs, and let the genotypic values for $q_iq_i : Q_iq_i : Q_iQ_i$ at a locus, i, underlying the trait be $-a_i : d_i : a_i$. Let p_i be the frequency of allele Q_i in population (or line) one and $p_i + \delta_{p_i}$ denote the frequency of Q_i in line two. Assuming the parental populations are in Hardy–Weinberg, then following Willham (1970; Willham and Pollak 1985), these means (for locus i) can be expressed as

$$\begin{aligned} \mu_{P_1} &= (2p_i - 1)a_i + 2p_i(1 - p_i)d_i \\ \mu_{P_2} &= \mu_{P_1} + 2\delta_{p_i}\, a_i - 2\,\delta_{p_i}^2\, d_i \\ \mu_{F_1} &= \mu_{P_1} + \delta_{p_i}\, a_i \end{aligned} \tag{13.2a}$$

While the values of μ_{P_1} and μ_{P_2} follow from assuming Hardy–Weinberg, the mean value of the F_1 is obtained as follows. With probability $p_i \cdot (p_i + \delta_{p_i})$, a Q_i allele is drawn from both populations, yielding a Q_iQ_i genotype with value a_i. Similarly, with probability $(1 - p_i) \cdot (1 - p_i - \delta_{p_i})$, a q_i allele is drawn from both, yielding a q_iq_i genotype with value $-a_i$. Finally, with probability $(1 - p_i) \cdot (p_i + \delta_{p_i})$, a q_i allele is drawn from P_1 and a Q_i from P_2, while with probability $p_i \cdot (1 - p_i - \delta_{p_i})$, a Q_i allele is drawn from P_1 and a q_i from P_2. Both of these situations yield a Q_iq_i and a genotypic value of d_i.

The resulting (midparental) heterosis (for this particular locus) becomes

$$H_{F_1} = \mu_{F_1} - \frac{\mu_{P_1} + \mu_{P_2}}{2} = \delta^2_{p_i} d_i \tag{13.2b}$$

Summing over all loci yields

$$H_{F_1} = \sum_{i=1}^{n} \delta^2_{p_i} d_i \tag{13.2c}$$

Hence, two features are required for heterosis when dominance is present. First, as with inbreeding depression, *directional dominance* is required (i.e., the d_i must tend to be positive). Second, the *strength of heterosis increases with the between-population difference in allele frequencies* (δ_{p_i}) at the underlying trait loci, with no heterosis when the two populations have the same allele frequencies for the trait of interest ($\delta_{p_i} = 0$ for all i). While it is occasionally assumed (or inferred) that heterosis only occurs when fully inbred lines are crossed, Equation 13.2c shows that this is not the case. The importance of crossing inbred lines is that the heterotic effect is largest when the lines are fixed for alternative alleles, as in this setting, $\delta_p^2 = 1$ for any alleles segregating in the F_1, with $H_{F_1} = \sum d_i$, the sum taken over all segregating trait loci. Cress (1966) examined crosses where more than two alleles are segregating, concluding that negative heterosis ($H_{F_1} < 0$) can often occur, even when overdominance is present. Only in the case of complete dominance will the heterosis be non-negative for all potential vectors of allele-frequency differences between the crossed populations.

When dominance (but no epistasis) is present, many, but not *all*, aspects of heterosis and inbreeding depression can be considered as complementary events. The change in means under crossbreeding and inbreeding (to level f) are given by

$$\mu_{F_1} = \mu_{MP} + H_{F_1} \quad \text{and} \quad \mu_f = \mu_0 - fI \tag{13.3a}$$

namely, the sum of a reference mean and an increment, where

$$H_{F_1} = \sum_{i=1}^{n} \delta^2_{p_i} d_i \quad \text{and} \quad I = 2\sum_{i=1}^{n} p_i(1 - p_i)d_i \tag{13.3b}$$

Hence, given the two very important caveats of no epistasis and that the same *segregating alleles* are involved in the cross between lines, and selfing within a line, then inbreeding depression and heterosis are very much complementary events (Filho 1999). If the same set of loci are segregating in the F_1 and in the population being inbred, one does not expect heterosis without also observing inbreeding depression in the trait, and vice versa. Further, in this no epistasis setting, one *cannot* have both inbreeding depression ($I > 0$) *and* outbreeding depression ($H_{F_1} < 0$) for the same trait. When both phenomena occur for the same trait, this implies either a significant role for epistasis and/or that *different* segregating alleles/loci are involved in the inbreeding and crossbreeding situations. An example of the latter would be in a population cross that introduces alleles not segregating within the population being selfed to measure inbreeding depression.

An important feature about heterosis, and one that makes this strategy desirable for commercial seed companies, follows because F_1 genotype frequencies are *not* in Hardy–Weinberg equilibrium, as the mating is nonrandom (an individual from population one is *always* crossed to an individual from population two). The resulting frequency of heterozygotes in the F_1 can be expressed as

$$\Pr(Q_iq_i) = 2\,\overline{p}_i\,(\,1 - \overline{p}_i\,) + \frac{\delta^2_{p_i}}{2} \tag{13.4a}$$

where $\overline{p}_i = [p_i + (p_i + \delta_{p_i})]/2 = p_i + \delta_{p_i}/2$ is the frequency of Q_i in the hybrid population. Hence, the *excess of heterozygotes* in an F_1 relative to a random-mating population with the same allele frequency is $\delta^2_{p_i}/2$. In the resulting F_2, random mating (among the F_1) ensures that genotypes are in Hardy–Weinberg frequencies, with $\Pr(Q_i q_i) = 2\,\overline{p}_i\,(1-\overline{p}_i)$. As a result, half of the initial heterosis generated by dominance disappears, with

$$H_{F_2} = \mu_{F_2} - \frac{\mu_{P_1} + \mu_{P_2}}{2} = \sum_{i=1}^{n} \frac{\delta^2_{p_i} d_i}{2} = \frac{H_{F_1}}{2} \tag{13.4b}$$

A farmer who planted seeds produced by their F_1 plants would see an immediate decline in yield relative to the purchased seed. This **dominance loss** occurs because all segregating loci in the F_1 are of heterozygous origin (the two alleles in any F_1 individual are from different populations), but only half of the F_2 loci are of heterozygous origin (with 1/4 having both alleles from line one and 1/4 with both alleles from line two.) While half the dominance-generated heterosis disappears after one generation of random mating, all subsequent generations show no further changes (as a single generation of random mating generates Hardy–Weinberg proportions at any single locus). As we demonstrate shortly, the presence of epistasis modifies these statements.

Finally, under what conditions does the mean of the F_1 exceed that of both parents (high-parent heterosis)? Summing over all loci, we can rewrite Equation 13.2a as

$$\mu_{F_1} - \mu_{P_1} = \sum_{i=1}^{n} \delta_{p_i} a_i \quad \text{and} \quad \mu_{F_1} - \mu_{P_2} = 2\sum_{i=1}^{n} \delta^2_{p_i} d_i - \sum_{i=1}^{n} \delta_{p_i} a_i$$

Thus, the hybrid exceeds the best parent (high-parent heterosis) when

$$2\sum_{i=1}^{n} \delta^2_{p_i} d_i > \sum_{i=1}^{n} \delta_{p_i} a_i > 0 \tag{13.5a}$$

Note that while overdominance ($d_i > a_i$) facilitates the inequalities in Equation 13.5a, it is not required. For completely inbred lines, it can be shown that Equation 13.5a reduces to

$$2\sum_{i=1}^{n} d_i > \sum_{i=1}^{n} \delta_{p_i} a_i > 0 \tag{13.5b}$$

Note that here $\delta_{p_i} = \pm 1$, depending upon in which parental line the high allele is fixed. The more equal this differential fixation is among lines (i.e., the closer $\Pr(\delta_{p_i} = 1)$ for a random locus is to 0.5), the smaller the right-hand sum, increasing the chance of high-parent heterosis. As we will see in Chapter 18, such a differential fixation of alleles (both lines contain desirable high alleles) is one condition for **transgressive segregation** (*individual* offspring exceeding the value of their parents; Mackay et al. 2021). Melchinger et al. (2007a) gives the conditions required for high-parent heterosis (which they call **better-parent heterosis**) under general epistasis.

Inbreeding versus Panmictic Heterosis

Before proceeding, we need to emphasize a subtle point. While a *pure line* is completely inbred, it consists of only a single genotype, so that randomly drawn genotypes are in Hardy–Weinberg proportions (trivially so, as they are all identical). Conversely, when P_i is a *collection* of inbred lines, this parental population is *not* in Hardy–Weinberg, as it consists of a collection of *different* genotypes (Chapter 12). Equation 13.2c (showing the effect of directional dominance on heterosis) assumes that both parental populations are in Hardy–Weinberg. When one (or both) populations are inbred, the effects of heterosis can be confounded with the effects of inbreeding depression. Under directional dominance, the inbred midparental mean is reduced relative to its random mating value, resulting in an overestimation of the impact of heterosis (Lamkey and Edwards 1999).

To see this, suppose that population P_i is inbred to level f_i. From Equation 13.3a, $\mu_{P_i}^{(f)} = \mu_{P_i}^{(0)} - f_i I_i$, where $\mu_{P_i}^{(0)}$ is the mean under random mating and I_i is the inbreeding depression for the focal trait in population i (Equation 13.3b). The midparental mean $\mu_{MP}^{(f)}$ based on the inbred lines can be related to its random-mating value μ_{MP} as follows,

$$\mu_{MP}^{(f)} = \frac{\mu_{P_1}^{(f)} + \mu_{P_2}^{(f)}}{2} = \frac{\mu_{P_1}^{(0)} - f_1 I_1 + \mu_{P_2}^{(0)} - f_2 I_2}{2} = \mu_{MP} - \frac{f_1 I_1 + f_2 I_2}{2} \tag{13.6a}$$

Namely, the panmictic-midparental mean value minus the average level of inbreeding depression. Note that in both settings (inbred and randomly mating population), the F_1 probability of drawing a Q_i allele from population 1 is p_i and the corresponding probability of drawing Q_i from population 2 is $p_i + \delta_{p_i}$, giving the same value of μ_{F_1},

$$a_i p_i (p_i + \delta_{p_i}) + d_i \left[p_i (1 - [p + \delta_{p_i}]) + (1 - p_i)(p_i + \delta_{p_i}) c - a_i \left[(1 - p_i)(1 - [p_i + \delta_{p_i}]) \right] \right.$$

Hence,

$$H_{F_1}^{(f)} = \mu_{F_1} - \mu_{MP}^{(f)} = H_{F_1} + \frac{f_1 I_1 + f_2 I_2}{2} \tag{13.6b}$$

Lamkey and Edwards (1999) refer to $H_{F_1}^{(f)}$ as the **inbred-midparent heterosis** (the reference midparent is $\mu_{MP}^{(f)}$), while H_{F_1} is the **panmictic-midparent heterosis** (the reference midparent is μ_{MP}). Equation 13.6b shows that there are two components to $H_{F_1}^{(f)}$, a contribution H_{F_1} from the panmictic-midparent heterosis (Equation 13.2c) and a contribution from the removal of inbreeding depression. Lamkey and Edwards proposed the term **baseline heterosis**, H_B, to denote the contribution in the F_1 simply from the recovery of the effects of inbreeding depression reducing the parental means. This is simply the difference between the inbred and panmictic heterosis values,

$$H_B = H_{F_1}^{(f)} - H_{F_1} = \frac{f_1 I_1 + f_2 I_2}{2} \tag{13.6c}$$

Lamkey and Edwards considered the special case of this more general expression, where both populations consists of sets of fully inbreds ($f_1 = f_2 = 1$), so that $H_B = (I_1 + I_2)/2$.

Inbreeding Depression and Heterosis in Subdivided Populations

While both inbreeding depression (ID) and heterosis are impacted by the presence of directional dominance, as we will shortly show, they can be decoupled due to differences in epistasis. In particular, A x A can contribute to heterosis (Equations 13.7a–13.7d), but only epistatic terms involving D (such as A x D or D x D; Equation 12.4a) impact inbreeding depression. As a result, a trait might display heterosis but not ID, or vice versa, simply due to differences in the importance of epistatic components. However, even when only directional dominance is present, there is a more subtle feature of the genetic architecture that can also result in these two phenomena not being complementary.

ID arises as a result of *segregating* alleles within a population (Chapter 12), while heterosis arises from allele-frequency *divergence* between populations (in the extreme, differential fixation). While differential adaptation can drive allele-frequency divergence, so can genetic drift. Divergence under drift is expected to largely influence alleles with (at most) modest effects on fitness, otherwise selection would dominate (Chapter 12; WL Chapter 7). In contrast, inbreeding depression results from segregating variation within a given population, and likely reflects, at least in part, the impact of rare alleles of large effect (Chapter 12). Alleles of smaller effect can also contribute, but these could also be fixed via drift in small populations, and hence only contribute to heterosis (if differentially fixed), but not to ID which requires *segregating* deleterious alleles (Whitlock et al. 2000; Charlesworth 2018).

Much of this theory was discussed in Chapter 12 in the context of the purging of ID. Any initial genetic load of deleterious alleles in the founding population is partitioned into a segregating load within subpopulations, and fixed differences (or at least allele-frequency divergence) between them. As a result, evolutionary biologists are interested in the amount of fitness heterosis generated by crosses between subpopulations, as this serves as a partial proxy for the role of genetic drift in partitioning deleterious alleles over demes (Crow 1948; Lynch 1991; Whitlock et al. 2000; Oakley et al. 2015a, 2015b, 2019).

One prediction under the drift-divergence model is a reduction in heterosis as the effective population sizes of demes (subpopulation) increases (Whitlock et al. 2000; Charlesworth 2018). This has indeed been observed. Lohr and Haag (2015) found that populations of *Daphnia magna* with lower levels of genetic variation (and hence, likely smaller population sizes) showed less inbreeding depression, but more heterosis in crosses. Similar findings of heterosis increasing as N_e decreases were made by Sprigler et al. (2016) for the biennial plant *Sabatia angularis*.

Several studies with natural populations of *Arabidopsis* have also examined the connections between ID within, and heterosis between, demes. Oakley et al. (2019) crossed pairs of natural populations of *A. thaliana* from across Scandinavia and measured lifetime fitness in a common garden. They found that heterosis was common (increased fitness in between-population crosses), but that ID was essentially absent, showing a decoupling of these two features. They suggested this was due to genetic drift differentially fixing mildly deleterious alleles, shifting all of the mutational burden into between-, rather than within-, population load. Willi (2013) and Oakley et al. (2015b) contrasted levels of between-population heterosis for self-compatible (SC) versus self-incompatible (SI) natural populations of *A. lyrata lyrata* in the Great Lakes region of North America. The effective population size in SC populations can be as little as half that for an SI population of similar census size (Pollak 1987; Nordborg 2000). SC populations should thus have experienced more drift (on average) than the SI ones. Willi and Oakley et al. found heterosis for fitness in both SI and SC population crosses, but that SC crosses had a six to seven fold increase in heterosis relative to SI crosses.

ROLE OF EPISTASIS IN HETEROSIS AND OUTBREEDING DEPRESSION

Heterosis can be generated even in the absence of dominance. As first noted by Richey (1942), two completely additive components that act in a multiplicative fashion to generate the final trait value can generate heterosis. Richey's hypothetical example considered plant height as the product of two (assumed independent) components, each of which is additive: number of nodes and internode length. Line one has twice as many nodes, each with half the internode length, as line two. Hence, both lines have the same height, which is scaled to a value of one. The relative number of nodes in the F_1 is (2+1)/2 = 3/2, while the average internode length is (1+1/2)/2 = 3/4, for a height of (3/2)·(3/4) = 1.125, giving H_{F_1} = 0.125 (12.5% midparent heterosis). Richey called this interaction generated by multiplying two additive traits **mock dominance**, although it is now more widely referred to as **multiplicative heterosis** (Dempster 1943; Powers 1944; Williams 1959a; Arunachalam 1977; Minvielle 1987).

As noted by Cockerham (1959) and in more detail by Schnell and Cockerham (1992), the interaction generated by multiplying two additive traits is a form of additive × additive epistasis (Chapter 5). Hence, it should not be surprising that certain types of epistasis can generate heterosis, but what is less clear is when they may do so, and the behavior of any epistatically generated heterosis in the F_2, and subsequent, generations. For example, under what conditions does one see outbreeding depression, instead of heterosis, in hybrids?

The line-cross theory presented in Chapter 11 provides a framework for investigating these questions. As derived in Example 13.1, the heterosis expressed in the F_1 generation is

$$H_{F_1} = \mu_{F_1} - \mu_{MP} = 2\delta_1^c - \alpha_2^c \tag{13.7a}$$

where δ_1^c and α_2^c, are, respectively, the dominance and additive × additive composite effects (see Chapter 11 for details). The value α_2^c is positive when favorable additive epistatic interactions occur between alleles from the same population, while it is negative when the favorable additive × additive combinations occur between alleles from different populations. Similarly (as also shown in Example 13.1), the F_2 heterosis becomes

$$H_{F_2} = \mu_{F_2} - \mu_{MP} = \delta_1^c - \left(\frac{1}{2} + \bar{c}\right)\alpha_2^c - 4\,\bar{c}\,(1-\bar{c})\delta_2^c \tag{13.7b}$$

$$= \frac{H_{F_1}}{2} - \bar{c}\left(\alpha_2^c + 4\,[1-\bar{c}]\delta_2^c\right) \tag{13.7c}$$

where δ_2^c is the dominance × dominance composite effect and $\bar{c}$ is the average recombination rate between epistatic loci that underlie the trait (Willham and Pollak 1985). The value of δ_2^c is positive when favorable D × D epistatic interactions occur between genotypes from the same population. For unlinked loci ($\bar{c} = 0.5$), the F_2 expression reduces to

$$H_{F_2} = \delta_1^c - \alpha_2^c - \delta_2^c = \frac{H_{F_1}}{2} - \left(\frac{\alpha_2^c}{2} + \delta_2^c\right) \tag{13.7d}$$

Hence, the resulting loss of heterosis from the F_1 to the F_2 is

$$\mu_{F_2} - \mu_{F_1} = -\left(\delta_1^c - \alpha_2^c[1/2 - \bar{c}\,] + 4\,\bar{c}\,[1-\bar{c}\,]\delta_2^c\right) \tag{13.8a}$$

$$= -\left(\delta_1^c + \delta_2^c\right) \quad \text{for } \bar{c} = 0.5 \tag{13.8b}$$

Even when linkage is present (provided $c > 0$),

$$\mu_{F_\infty} - \mu_{F_1} = -\left(\delta_1^c + \delta_2^c\right) \tag{13.8c}$$

where μ_{F_∞} denotes the long-term mean (after all LD is removed by recombination). As mentioned above, one component of the change in heterosis from the F_1 to the F_2 is due to dominance loss, as the contribution from single-locus dominance to heterosis declines from $2\delta_1^c$ in the F_1 to δ_1^c in the F_2. The other components of the decline are due to epistasis (α_2^c and δ_2^c).

Several general conclusions can be drawn from these expressions. First, the deviation between the mean phenotype in the F_1 and parental lines is a function of both dominance (δ_1^c) and additive × additive epistasis (α_2^c). Ignoring epistasis, a simple explanation for heterosis is the presence of complementary sets of deleterious recessive genes in both parental lines and the masking of their effects in the F_1 heterozygotes. However, the loss of favorable additive × additive effects (α_2^c) that may exist within populations must also be considered. Although the interactions *within* gametes are preserved in the F_1, those *between* gametes are not. An F_1 line will only exhibit heterosis if the gain in favorable between-population dominance effects ($2\delta_1^c$) exceeds the loss in favorable additive × additive interactions (α_2^c) within populations. If the latter is sufficiently large, outbreeding depression occurs.

Second, half of the heterosis in the F_1 generation is lost by segregation when gametes leading to the F_2 generation are produced ($H_{F_1}/2 = \delta_1^c - \alpha^c/2$), and an *additional* fraction is lost due to the recombination between parental-line genes. This latter quantity, termed **recombination loss** by Dickerson (1969a), is entirely a function of epistatic effects, because recombination does not influence the transmittance of effects associated with single loci. Depending on the signs of α_2^c and δ_2^c, this "loss" may be positive or negative. We examine recombination loss in more detail shortly.

Third, under free recombination, $\mu_{F_2} - \mu_{MP} = \delta_1^c - (\alpha_2^c + \delta_2^c)$, and this is also true in later generations with restricted recombination, provided enough time has passed to eliminate gametic-phase disequilibria. Thus, a hybrid population will ultimately experience outbreeding depression if the net gains due to dominance (δ_1^c) are less than the net losses due to the breakup of favorable additive × additive and dominance × dominance interactions ($\alpha_2^c + \delta_2^c$).

The transition from heterosis to outbreeding depression with increasing genetic distance between parents suggests that there is a fundamental change in predominant gene interactions as mates become more and more distantly related. This observation has led to the suggestion that there must be an optimal degree of outbreeding (Shields 1982; Bateson 1983; Waser and Price 1983; Waser 1993). Dominance is generally believed to be the primary agent of inbreeding depression within most populations (Chapter 12). However, the decline in fitness under outcrossing is usually attributed to a breakup of coadapted gene complexes (favorable epistatic interactions) in the parental lines (Dobzhansky 1948, 1950; Templeton 1986; Lynch 1991; Xiao et al. 1995). Thus, in proceeding from issues of outbreeding enhancement (heterosis) to those of outbreeding depression, at least among theoreticians, there is often a shift in emphasis from interactions *within* loci (dominance) to interactions *among* loci (epistasis).

Example 13.1. Equations 13.7 and 13.8 follow directly from line-cross theory. From Table 11.1,

$$\begin{aligned} \mu_1 &= \mu_0 + \alpha_1^c - \delta_1^c + \alpha_2^c - \alpha_1^c\delta_1^c + \delta_2^c + \cdots \\ \mu_2 &= \mu_0 - \alpha_1^c - \delta_1^c + \alpha_2^c + \alpha_1^c\delta_1^c + \delta_2^c + \cdots \\ \mu_{F_1} &= \mu_0 + \delta_1^c + \delta_2^c + \cdots \end{aligned}$$

Hence, to second-order epistasis, we have

$$\mu_{MP} = \frac{\mu_1 + \mu_2}{2} = \mu_0 - \delta_1^c + \alpha_2^c + \delta_2^c$$

yielding

$$\mu_{F_1} - \mu_{MP} = \mu_0 + \delta_1^c + \delta_2^c - (\mu_0 - \delta_1^c + \alpha_2^c + \delta_2^c) = 2\delta_1^c - \alpha_2^c$$

which recovers Equation 13.7a. Expressions with higher order epistasic terms ($\alpha_4^2, \delta_4^c, \delta_6^c$, etc.) are obtained following similar logic, and can be found in Melchinger et al. (2007a).

Turning to the F_2 mean, $\mu_{F_2} = \mu_0$ for unlinked loci, as the decomposition given in Chapter 11 sets the F_2 population as the base. This generation is used as the reference population because single-locus genotypes are in Hardy–Weinberg and, for unlinked loci, segregation of the F_1 results in linkage equilibrium (Hill 1982a). As an aside, the careful reader might be concerned that, with unlinked loci, only half of the disequilibrium decays each generation under *random mating*. While this is correct, crossing of the F_1 to form the F_2 is an example of *nonrandom mating*, as only heterozygotes (hybrids) are crossed, and independent assortment of unlinked loci generates linkage equilibrium in a single generation. Thus, for unlinked loci

$$\mu_{F_2} - \mu_{MP} = \mu_0 - (\mu_0 - \delta_1^c + \alpha_2^c + \delta_2^c) = \delta_1^c - \alpha_2^c - \delta_2^c \tag{13.9a}$$

which recovers Equation 13.7d.

When loci are linked, Equation 11.2a gives the F_2 mean as

$$\mu_{F_2} = \mu_0 + \left(\frac{1-2c}{2}\right)\alpha_2^c + (1-2c)^2\delta_2^c \tag{13.9b}$$

which reduces to μ_0 when $c = 0.5$. When loci are linked, μ_0 is the mean approached following several generations of random mating to remove any lineage disequilibrium generated by the initial cross (Example 5.6). The departure of the F_2 mean given by Equation 13.9b from this equilibrium value arises because linkage disequilibrium is still present and the population is not in **multilocus Hardy–Weinberg** (Hardy–Weinberg proportions at each locus and linkage equilibrium among loci). Note that the contribution from single-locus dominance effects (δ_1^c) does not enter into μ_{F_2}, as *single loci* are in Hardy–Weinberg in the F_2.

Thus, with linkage,

$$\begin{aligned}\mu_{F_2} - \mu_{MP} &= \left[\mu_0 + \left(\frac{1-2c}{2}\right)\alpha_2^c + (1-2c)^2\delta_2^c\right] - (\mu_0 - \delta_1^c + \alpha_2^c + \delta_2^c) \\ &= \delta_1^c + \left[\left(\frac{1-2c}{2}\right) - 1\right]\alpha_2^c + \left[(1-2c)^2 - 1\right]\delta_2^c \\ &= \delta_1^c - \left(\frac{1}{2} + c\right)\alpha_2^c - 4c(1-c)\delta_2^c \end{aligned} \tag{13.9c}$$

which recovers Equation 13.7b.

Example 13.2. Moll et al. (1965) produced F_1 and F_2 generations from crosses between several lines of maize with varying degrees of genetic divergence "based on ancestral relationships and differences in adaptation." As illustrated in the following figure, when assayed in a common environment, all crosses exhibited heterosis for grain yield in the F_1 and F_2 generations, but this was most pronounced in crosses involving lines with intermediate degrees of divergence (levels 4 and 5). Moreover, at the highest levels of divergence, the performance of the F_1 and F_2 lines converged (in the figure, the data from all crosses have been standardized so that $\overline{z}_{F_2} = 1$).

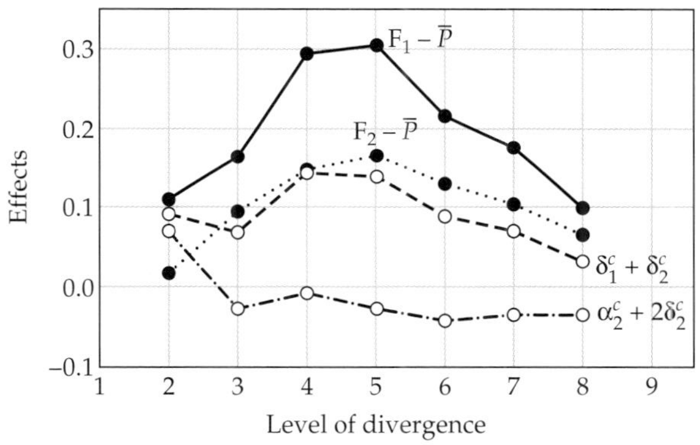

Can any insight into the mode of gene action be inferred from these results? Assuming that $\bar{c} \simeq 0.5$, Equation 13.8b implies that $\overline{z}_{F_1} - \overline{z}_{F_2} \simeq \delta_1^c + \delta_2^c$, the sum of the composite dominance and dominance $\times$ dominance effects. In addition, using the same logic as in Example 13.1 yields

$$\overline{z}_{F_1} + \overline{z}_{MP} - 2\overline{z}_{F_2} \simeq \alpha_2^c + 2\delta_2^c$$

which is the net loss of performance due to segregation and recombination of parental-line gene combinations. As shown in the figure, application of these two formulae suggests that the net effects of dominance between parental lines have a positive influence on grain yield at all levels of divergence, but that the magnitude of this effect is maximized at an intermediate genetic distance. On the other hand, except at the lowest levels of divergence, the estimates of $\alpha_2^c + 2\delta_2^c$ are negative and roughly constant, suggesting favorable epistatic effects *between* genes from *different* sources (contrary to the expectation of individual lines harboring coadapted gene complexes), which increases the amount of heterosis (Equations 13.7a and 13.7c).

Epistasis and Recombination Loss

When only single-locus dominance is present ($\alpha_2^c = \delta_2^c = 0$), the change in the mean in generations subsequent to the F_1 is easily predicted, being $-H_{F_1}/2 = -(\overline{z}_{F_1} - \overline{z}_{MP})/2$ in the F_2 (Equation 13.4b), and then stable thereafter. When epistasis is present (α_2^c and/or δ_2^c nonzero), one can see additional **F_2 breakdown** (Hill 1982), wherein the loss of heterosis is greater than this prediction (Equation 13.7c). Dickerson (1969a, 1972) called this additional decline the recombination loss, as it arises through recombination (or segregation when loci are unlinked) breaking up favorable combinations of alleles from distinct loci that were present in each parental population.

To see this, let A_i denote a random allele at the A locus from population i (1 or 2), with a similar definition for B_i. One can then represent an F_1 individual as A_1B_1/A_2B_2, where each gamete forming the F_1 contains only alleles from one population. Such an individual generates a fraction $(1-c)$ of parental gametes (A_1B_1 and A_2B_2) where both alleles are from the same population, and a fraction c of recombinant (or nonparental) gametes (A_1B_2 and A_1B_2) containing alleles from different populations. The generation of these nonparental gametes further disrupts the transmission of favorable epistatic combinations, leading to additional loss of heterosis. Epistasis depends on pairs of loci, and (for unlinked loci) the F_2 has only half the number of parental gametes (both alleles from the same population) as the F_1. This is akin to the loss of half of the dominance heterosis from the F_1 to the F_2, as the latter has only half of the number of hybrid loci as the former. From Equation 13.7c, the recombination loss in the F_2 is

$$\bar{c}\left(\alpha_2^c + 4\,[1-\bar{c}\,]\delta_2^c\right) \tag{13.10a}$$

which reduces to

$$\frac{\alpha_2^c}{2} + \delta_2^c \tag{13.10b}$$

when loci are unlinked, where the F_2 loss is most pronounced. Kinghorn (1980) suggested that **epistatic loss** is a more appropriate term for this extra decline in heterosis, as it is entirely generated by between-locus interactions.

Dickerson (1972) noted, for unlinked loci, that the quantity

$$R = \alpha_2^c + 2\delta_2^c \tag{13.11a}$$

which is twice the recombination loss in the F_2, provides a general measure for the amount of recombination loss under a variety of crossing schemes used by breeders. As shown in Example 13.2, when loci are unlinked, R can be estimated by

$$\widehat{R} = \overline{z}_{F_1} + \overline{z}_{MP} - 2\,\overline{z}_{F_2} = \alpha_2^c + 2\,\delta_2^c \tag{13.11b}$$

To quantify the expected amount of recombination loss in any particular crossing scheme, Dickerson (1969a, 1972) proposed a coefficient, r_D, to weight R, giving the predicted loss as $r_D R$. While Dickerson called r_D the **coefficient of recombination loss**, we prefer the term **Dickerson coefficient** to avoid confusion between the *actual loss* ($r_D R$), the general metric of loss (R), and the coefficient of loss (r_D). Dickerson defined r_D as the fraction of nonparental gametes that fuse to form the focal cross, so that $r_D = 1/2$ in the F_2 for unlinked loci (as half of the gametes from the F_1 are nonparental). Different paramaterizations for recombination loss have been proposed (Dickerson 1969a, 1972; Kinghorn 1980, 1982, 1983, 1987; Hill 1982; William and Pollak 1985), with Koch et al. (1985) summarizing the connections between them. When loci are linked, the relative weights on α_2^c and δ_2^c, rather than being constant as with R, change depending on the amount of recombination in the crossing scheme (e.g., Equation 13.9b), and the simple expression of $r_D R$ for the loss is only an approximation.

The nature of recombination loss is of considerable interest to animal breeders, who use different crossing schemes to exploit the nonadditive variance that generates heterosis. Unlike the situation in plant breeding, generating a large number of F_1 animals for commercial sale is generally not feasible due to constraints imposed by the reproductive output of most domesticated animals and their associated costs. Rather, two different approaches are used in place of continual production of commercial F_1s, and these differ in their values of r_D.

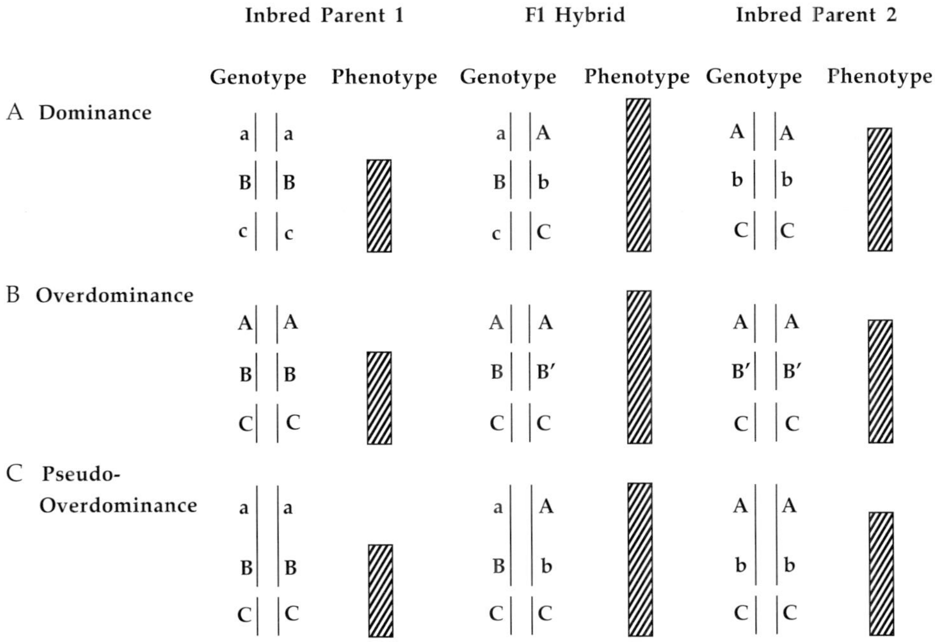

Figure 13.1 Different hypotheses for how dominance can generate heterosis. The figure shows the chromosomal composition of the genotype on the left and the resulting phenotype (indicated by the height of the filled bar) on the right of each genotype. **(A)** Under the dominance (or partial dominance) hypothesis, recessive deleterious alleles are fixed at different loci (*aa* and *cc* in line 1, *bb* in line 2), while all are covered (**dominance complementation**) in the hybrid (*AaBbCc*). **(B)** Under the overdominance hypothesis, the *B* locus shows overdominance, with the heterozygote (*BB′*) having a superior value to either homozygote (*BB*, *B′B′*). **(C)** Under the pseudo-overdominance hypothesis (Chapter 12), the *A* and *B* loci are also fixed for alternative recessive alleles, but now are tightly linked, with parent 1 contributing an *aB* chromosomal segment and parent 2 an *Ab* segment. As with the dominance hypothesis, both deleterious alleles are covered in the hybrid, but the chromosomal segment appears to segregate as a single allele, creating the impression of overdominance. This initial repulsion-phase disequilibrium (association of *a* with *B* and *A* with *b*) decays under recombination. As discussed in Chapter 12, a neutral marker (such as is used for QTL mapping and GWAS; Chapters 18–21) tightly linked to these two repulsion-phase loci would itself show overdominance in a marker-trait association, a condition called associative overdominance. (After Birchler et al. 2003.)

The first is the creation of a **synthetic** (or **composite**) population, such as crossing two lines and then randomly mating the F_1 to form the synthetic population. Because this is just an F_2, the Dickerson coefficient for a two-population synthetic is $r_D = 1/2$. The second class of schemes is **rotational crossbreeding**, the continual crossing of individuals each generation. We detail these schemes below, but the form of a two-breed rotation crossing scheme is $P_1 \times (P_2 \times [P_1 \times \{P_2 \ldots\}])$. The Dickerson coefficient under such a scheme is $2/9$ (Dickerson 1969a), and hence it suffers less recombination loss, making this approach preferable to a synthetic when R is sufficiently large. We detail several complex crossing schemes later in the chapter and examine the importance of maternal heterosis, wherein the use of crossbreed dams (females) often has a significant advantage over using purebreed mothers.

GENETIC NATURE OF HETEROSIS

In our discussion on the genetics of inbreeding depression (Chapter 12), we were mainly concerned with the nature of dominance: is it partial, overdominant, or pseudo-overdominant? While dominance-based epistasis can contribute to inbreeding depression (Equation 12.4a; Example 12.2), its impact is only apparent at very high levels of inbreeding, and thus, in most settings, is usually ignored (Chapter 12). The genetic nature of single-locus dominance in heterosis tracks our interest in the nature of dominance under inbreeding depression, but, in contrast to the situation with inbreeding, epistasis can also be important. In particular, the breakdown of additive $\times$ additive epistasis impacts the F_1 heterosis (Equation 13.7a) and, along with dominance $\times$ dominance epistasis, impacts any recombination loss seen in subsequent generations (Equation 13.7c).

Two different approaches have been used to resolve the genetic nature of heterosis. The classical, or biometrical, approach is statistical, based on line-cross analysis and variance-component estimation (e.g., Jinks 1983; Sprague 1983; Hallauer et al. 2010). More recently, quantitative geneticists have harvested the power of genetic markers and whole-genome sequencing in attempts to isolate ever-smaller chromosomal regions influencing heterosis. These very different approaches are best viewed as complementary, rather than competitive.

Biometrical Approaches: Nature of Dominance

As was the case with inbreeding depression, the question arises as to whether most heterosis is caused by overdominance (Shull 1908; East 1908) or by partial dominance (Davenport 1908; Bruce 1910; Keeble and Pellew 1910). As with inbreeding depression, observe that Equation 13.2c is agnostic with respect to the nature of the required dominance. It simply requires that, in general, the d_i tend to be positive. Overdominance further requires that $d_i > a_i$ (at least for some loci). As with inbreeding depression, **pseudo-overdominance** (recessive alleles at different loci are in repulsion-phase disequilibrium; Chapter 12) can be difficult to distinguish from true overdominance (Figure 13.1). If the former, the evidence for overdominance will disappear over time due to recombination decaying disequilibrium, while if the latter, the signal should persist even after many generations of recombination.

Recall from Chapter 12 that the terms pseudo-overdominance and **associative overdominance** are often used interchangeably in the literature. The former refers to the presence of recessive deleterious alleles in repulsion-phase disequilibrium generating apparent overdominance for a tightly linked chromosomal segment. The latter refers to a neutral marker within this region showing apparent overdominance (i.e., the average trait value for the marker heterozygote exceeds that of both marker homozygotes). Hence, we often discuss associative overdominance in conjunction with QTL/GWAS markers reflecting the manifestation of underlying pseudo-overdominance.

A number of crossing designs have been proposed that allow for direct estimation of the dominance variance, unconfounded by maternal environmental effects (reviewed in detail in Chapter 25 in Part II). In particular, the North Carolina (NC) Designs II and III of Comstock and Robinson (1948, 1952) have been widely used. Under NC Design II (Figure 13.2), the same females (or females from the same inbred line) are mated to the same set of males (or males from the same line), which allows for a direct estimate of the dominance variance (in the absence of epistasis and linkage disequilibrium). The NC Design III (Figure 13.2) backcrosses each F_2 to both parental lines, generating a paired set of line means, $\overline{z}_{1ij}$ and $\overline{z}_{2ij}$, for the two backcrosses from the ith F_2 that are jointly scored in plot j. Let $S_{ij} = \overline{z}_{1ij} + \overline{z}_{2ij}$ and $\Delta_{ij} = \overline{z}_{1ij} - \overline{z}_{2ij}$ represent, respectively, the sum and difference for each such pair. The variances of S and Δ correspond to estimates of the additive and dominance variances (see Part II Chapter 25 for details).

As also discussed in Chapter 25, Comstock and Robinson (1952) noted that one can use estimates of the additive and dominance variance to make inferences on the degree of dominance in the special case where all segregating alleles have frequency 1/2. This condition exists in the F_2 from a cross of two inbred lines. In such settings, the additive variance becomes $\sum_i a_i^2/2$ (Equation 4.12a with $p_i = 0.5$), while the dominance variance (writing the heterozygote as $d_i + a_i = [k_i + 1]a_i$, so that $k_i = d_i/a_i$) is $\sum_i (d_i)^2/4 = \sum_i (k_i a_i)^2/4$ (Equation 4.12b). Hence, twice the ratio of the dominance to additive variance estimates provides a weighted estimate of the dominance coefficient, k. A k value of zero indicates additivity, a value between zero and one ($0 < d \leq a$) suggests partial (k small) to complete

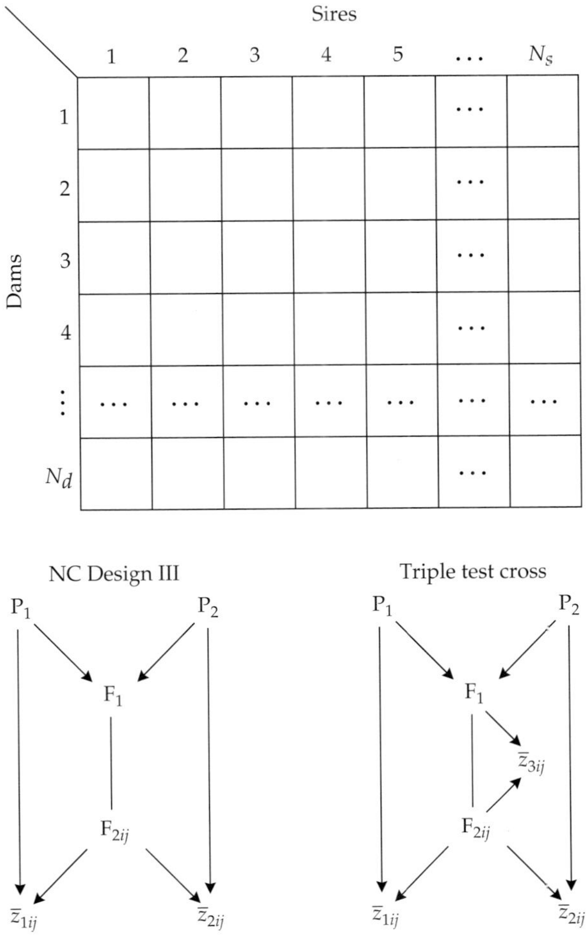

Figure 13.2 Special crossing designs used by biometrical approaches to probe the nature of heterosis. **Top:** Mating scheme for North Carolina Design II. Each cell (representing the offspring of sire i and dam j) contains n observations. **Bottom:** Crossing schemes for North Carolina Design III (left) and the triple test cross (right).

dominance ($k = 1$), and a value greater than one indicates overdominance ($d > a$). One complication noted by Cockerham and Zeng (1996) and Melchinger et al. (2007a) is that additive × additive epistasis biases the dominance estimate (see Equation 13.13).

Hence, the appropriate NC design applied to a cross between inbred lines provides some insight into the average degree of dominance. Pseudo-overdominance (Figure 13.1) can initially generate k values in excess of one, but this value is expected to decline in further generations as random mating erodes any initial linkage disequilibrium (Gardner and Lonquist 1959). Using the Comstock–Robinson approach, Moll et al. (1964) initially obtained estimates of k in excess of one for grain yield in maize, suggesting overdominance. However, this estimate declined to nearly 1 in one line, and to around 3/4 in another following several generations of random mating. Similar findings were seen by Gardner (1963) and Hallauer and Miranda (1981). Thus, while there are few striking examples of true overdominance generating heterosis—*E3* in soybeans (Watanabe et al. 2009), *SFT* in tomatoes (Krieger et al. 2010; see Figure 13.3), *Hd3a* in rice (Huang et al. 2016)—under further scrutiny other examples of apparent overdominance appear to be pseudo-overdominance (Cockerham and Zeng 1996; Dijkhuizen et al. 1996; Graham et al. 1997; Bingham 1999; Garcia et al. 2008; Lariépe et al. 2012). Given that most designs involve very few generations of recombination, such observations are not unexpected and

Figure 13.3 Heterosis in yield caused by overdominance at the *Single Flower Truss* (*SFT*) gene in tomatoes. On the left is the yield for a homozygous wild type parent, on the right is a homozygous mutant parent, and in the middle is their hybrid (heterozygous for the mutant and wild-type alleles). The increased yield results from the suppression of growth termination mediated by *Self Pruning* (*SP*), a gene antagonistic to *SFT*. See Krieger et al. (2010) for details. (Photo courtesy of Zach Lippman, Cold Spring Harbor.)

this ambiguity makes distinguishing between these two explanations often problematic (Luo et al. 2001). For example, Lu et al. (2003) found that 24 of 28 QTLs for grain yield in maize showed overdominance, despite randomly mating the F_2 population for three additional generations to reduce the effects of linkage (the AIC design; Chapter 18). However, Lu et al. noted that such a design would still leave 86% of any initial LD for loci 5 cM apart ($[1-0.05]^3$), and 51% of the LD between loci 20 cM apart. A few generations of random mating is only effective at removing LD under rather loose linkage (Equation 5.13c).

Example 13.3. The NAM (nested association mapping) lines in maize are a series of 5,000 inbred lines generated by crosses of 25 diverse lines to a B73 reference line, generating roughly 200 inbred lines per cross (details in Chapter 18; see Figure 18.12). McMullen et al. (2009) showed that, even after many generations of selfing, a few small regions in some of the NAM lines showed **residual heterozygosity**, wherein an inbred line shows heterozygosity, rather than the expected marker homozygosity (Chapter 12). In these regions, the effect of inbreeding generating homozygosity is countered by selection to maintain heterozygosity. Such an observation is consistent with overdominant loci influencing fitness. However, it is also consistent with pseudo-overdominance, where linked deleterious recessives are in repulsion disequilibrium. McMullen et al. found that residual heterozygosity was higher within 10 cM of the centromeres than along the chromosome arms (4.1% versus 3.2%). Given the recombination suppression around centromeres, they suggested that this observation supports pseudo-overdominance in these regions of reduced recombination. Further analysis by Gore et al. (2009) supported this view, finding that regions of residual heterozygosity were significantly correlated with reduced recombination rates, but not with gene density. Both McMullen et al. and Gore et al. argued that such repulsion-phase configurations arise as a natural consequence of the Hill–Robertson and Bulmer effects (WL Chapters 8 and 16, respectively), which suggests (among other things) that selection tends to drive favorable alleles into repulsion phase, especially in regions of low recombination. Consistent with this hypothesis, Schön et al. (2010) found that many QTL for yield heterosis in maize were in regions encompassing the centromere.

While dominance provides a convenient explanation of heterosis, several objections were initially raised to the strict dominance hypothesis (i.e., the assumption of no over-dominance, $d < a$). The primary objection being that if no overdominance were involved, one would expect to find F_2 individuals that were as good as the F_1. This was generally not seen. However, most studies are unlikely to have sufficiently large sample sizes to find such individuals when the number of loci contributing to heterosis is moderate to large (Example 13.4). When the number of key heterosis genes is modest, selection can quickly recover most of the F_1 performance in a nearly inbred line. Wang et al. (2015) were able to accomplish this for yield and biomass in *Arabidopsis* with just a few generations of recurrent selection. Likewise, if many of the major heterosis genes are known, one could construct a new inbred line that performs almost as well as the F_1. This was done by Hashimoto et al. (2021) with height in sorghum. The F_1 variety *Tentaka* is around four times as tall as either of its parental lines, with five QTLs showing dominance detected for this trait. Pyramiding all five QTL (i.e., five small chromosomal segments) together resulted in an inbred line whose height was close to 85% of the original hybrid. A second example is yield in the elite rice hybrid *Shanyou 63*. Shen et al. (2022) introduced eight heterotic QTL loci into the maternal line (*Zhenshan 97*) for this hybrid, obtaining yield performance that not only exceeded *Zhenshan 97* but was also similar to *Shanyou 63*.

A minor historical objection to the dominance hypothesis was that the F_2 distribution should be skewed (Chapter 2), with the mode greater than the mean, while F_2 populations generally show little skew. Collins (1921) pointed out that lack of skew is also a sample effect, see Figure 13.4. Additionally, Jones (1917) showed that linkage provides another explanation why skew and exceptional F_2 individuals are usually not seen.

An interesting angle on the dominance model was suggested by the work of Fu and Dooner (2002). They examined a 100 kb region around the *bz* gene in two different maize lines, finding that only six of the ten genes in this region were found in both lines. This suggested that even closely related lines may differ in the *actual genes they carry*, allowing their hybrid to have a larger complement of functional genes. We can think of these as truly null alleles, as the entire gene is absent, which is often referred to as **presence/absence variations (PAVs)**. More recent work has shown that variation in the presence/absence of entire genes is a common feature of many species (e.g., Vernikos et al. 2015; Sherman and Salzberg 2020; Qin et al. 2021; Woodhouse et al. 2021; Zhou et al. 2022). These observations led to the terms **core genome** (those genes shared by essentially all individuals), **peripheral genome** (genes whose presence/absence can vary dramatically), and **pan-genome** (the entire collection of genes over the population; Tettelin et al. 2005).

Example 13.4. Just how many F_2 progeny does one have to screen to have a reasonable chance (say 50%) of recovering the largest genotypic value (i.e., one that matches the F_1). We assume complete dominance, so that the presence of a single dominant allele is sufficient to maximize the contribution from a given locus (this is the most favorable setting). With a pure line cross, then with probability 3/4 at least one favorable allele is present at a locus in the F_2. Assuming unlinked loci, the probability of favorable alleles at n loci is $(3/4)^n$. To have a 50% probability that at least one individual has favorable alleles at all n loci requires sampling m individuals, where m satisfies

$$1 - [\,1 - (3/4)^n\,]^m = 0.5$$

This follows as $[1 - (3/4)^n]$ is the probability that a single individual *does not* contain favorable alleles at all loci, and this expression to the mth power is the probability that none of the m

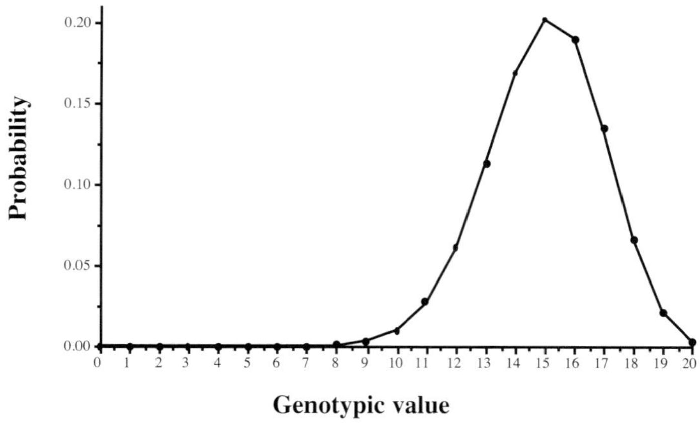

Figure 13.4 The F_2 distribution for 20 completely dominant loci of equal effect ($d = a = 1$). Starting with a cross of pure lines, the probability that a locus contains at least one dominant allele is 3/4, giving a mean value of 20(3/4) = 15, which is also the mode. Note that this mean is below the F_1 mean of 20. Although the distribution is indeed skewed, the investigator is unlikely to observe those individuals with a genotypic value of 8 or less (unless the sample size is enormous). Focusing on the most likely classes of observed individuals (9 to 20), this restricted distribution shows little skew. (After Collins 1921.)

individuals do. Solving yields

$$m_{0.5} = \frac{\ln(0.5)}{\ln[\,1 - (3/4)^n\,]}$$

More generally, to have probability of $100\alpha\%$ requires a sample size of

$$m_\alpha = \frac{\ln(1-\alpha)}{\ln[\,1 - (3/4)^n\,]}$$

For $n = 15, 20, 30$, and 40 loci, $m_{0.5} = 52, 218, 921, 3{,}881$, and 68,924. Hence, while there is a reasonable chance of recovering the best genotype in the F_2 when the number of loci is modest (20 or less), this very quickly becomes impossible with more than 30 loci. Further confounding these numbers is that not only must an optimal genotype be present in the sample, one also must be able to assess its genotypic value with high precision, requiring replication across a number of environments (Appendix 9). One final remark. If there is (on average) partial dominance (i.e., $a > d$), then the heterozygote is not as exceptional as the best possible genotype, and hence F_2 individuals homozygous for favorable alleles at a large number of loci may actually *exceed* the F_1 mean. If all genes have $a > d$, then the above expression for m_α holds, but with 1/4 replacing 3/4.

Biometrical Approaches: Epistasis

Biometrical evidence of epistasis contributing to heterosis can be seen from the observation of an F_2 decline that exceeds the value predicted by dominance ($H_{F_1}/2$), or (more formally) by direct estimates of compositive epistatic effects from line-cross theory (e.g., Equation 13.11b). One of the most widely used line-cross designs for detecting epistasis is the **triple test cross (TTC)**, an extension of NC Design III (Figure 13.2). The latter design crosses F_2 back to both parental lines to generate a series of paired backcrosses for each F_2, with $\overline{z}_{1ij}$ and $\overline{z}_{2ij}$ representing the two backcross means, both measured in plot j, for the ith F_2 individual. The triple test cross adds a third cross, with the F_2 also backcrossed to the

Table 13.1 A few examples of studies detecting significant epistasis using line-cross analysis. The TTC contrast (Equation 13.12) is significant in all examples, indicating epistasis. Joint-scaling tests (Chapter 11) were used to estimate the nature of the epistasis.

Species	Trait	Observation	Reference
Rice	5 yield-related	Significant epistasis in all traits	Singh and Singh 1976
Maize	Ear traits	Significant epistasis in many traits Favorable A x A interactions for several traits	Wolf and Hallauer 1997
		Significant epistasis in all 12 studied traits A x A interactions in half of the traits A x D or D x D interactions in 11 of 12 traits	Paven et al. 2017
Arabidopsis	Biomass traits	Significant epistasis in four traits A x A and D x D significant in four of five traits.	Kusterer et al. 2007a

F_1, yielding a third mean, $\overline{z}_{3ij}$, (also scored in plot j along with the two backcrosses). The contrast

$$\overline{z}_{1ij} + \overline{z}_{2ij} - 2\,\overline{z}_{3ij} \tag{13.12}$$

is nonzero in the presence of epistasis. Given a significant result from the TTC, one can use appropriate combinations of the line-cross means (i.e., joint-scaling tests) to estimate the composite epistatic effects (Chapter 11). Table 13.1 presents a few studies showing that significant epistatic effects are not uncommon, and recombination loss can be of considerable concern in such settings. In such cases, the loss of heterosis from the F_1 to the F_2 (and subsequent) generations can be greater, and perhaps far greater, than predicted from H_{F_1} (Equation 13.7c). Sheridan (1981) found that such excessive F_2 loss is commonplace for traits in domesticated animals. Breeding designs that minimize recombination loss (examined below) should be considered in such settings.

Evidence from QTL Mapping Studies

A second strategy for probing the genetic nature of heterosis is the mapping of factors that influence heterosis to small chromosomal regions. While this approach is called QTL (quantitative trait locus) mapping, in reality, it just maps effects to chromosomal segments that could easily contain a large number of genes, rather than mapping effects to single loci. Mapping of QTLs by linkage approaches (Chapter 18) results in estimation of the composite effects of genes in regions that are usually (at least) tens of megabases in length. The basic idea is very straightforward. Consider a cross between *QQMM* and *qqmm* lines, with *Q* the QTL and *M* the marker allele. If *M* and *Q* are linked, there is an excess of parental gametes produced by their F_1 (Chapter 5), with *MQ* and *mq* gametes being overrepresented. This generates a correlation between the QTL and marker alleles in the progeny, allowing genomic regions to be tagged because markers in those regions show a correlation with trait values. Linkage to a given marker is tested by partitioning F_2 individuals into three groups (the two alternative marker homozygotes and the heterozygote) and using ANOVA to test for among-group differences in trait means.

Two different approaches have been proposed for mapping QTLs underlying heterosis. The first is indirect: perform a standard QTL mapping experiment (such as using F_2 design; Chapter 18), and then infer the impact of detected QTLs on heterosis from the nature of their dominance and epistatic interactions. The second approach requires more specialized crossing designs, and attempts to more directly assess heterosis, such as through the use of **immortalized F_2 populations**, NC Design III, or TTC crossing designs.

The immortalized F_2 population design attempts to address a limitation with a standard F_2 design—lack of replication—as each F_2 individual is genetically unique, with its

Table 13.2 Summary of various QTL mapping studies for heterotic loci in plants. The column labeled Dominance indicates whether the majority of detected loci showed partial to complete dominance (Dom), apparent overdominance (OD), or associative overdominace (AD). Most QTL designs do not have the power to distinguish OD from AO, so that a good number of reported OD loci may instead be due to AD. The Epistasis column indicates whether digenic epistasis (digenic interactions between QTLs) was reported, and its nature (if significant).

Species	Trait	Dominance	Epistasis	Reference
Maize				
	Grain yield	OD	No	Lu et al. 2003
		AD	No	Schön et al. 2010
		OD	No	Frascaroli et al. 2007
		Dom	No	Garcia et al. 2008
		Dom	A x A	Tang et al. 2010
		AD	A x A	Lariépe et al. 2012
		Dom, OD	D x D	Guo et al. 2014
	Height	Dom	No	Lu et al. 2003
		Dom	A x A	Lariépe et al. 2012
	Grain moisture	Dom	No	Lu et al. 2003
		Dom	A x A	Lariépe et al. 2012
	Stalk lodging	Dom	No	Lu et al. 2003
	Yield components	Dom	A x A	Tang et al. 2010
		Dom, OD	No	Liu et al. 2019
		OD	A x A, D x D	Jiang et al. 2015
	Morphological traits	OD, Dom	No	Wei et al. 2015
Rice				
	Yield	AD	No	Xiao et al. 1995
		OD	A x A, D x D	Yu et al. 1997
		OD	A x A	Li et al. 2001
		Dom, OD	D x D	Hua et al. 2003
		Little	A x A	Garcia et al. 2008
		Dom, OD	No	Huang et al. 2016
		Dom	No	Shen et al. 2022
	Yield components	OD	Extensive	Luo et al. 2001
		OD	A x A, D x D	Hua et al. 2002
		Dom	No	Huang et al. 2015
		Dom	No	Li et al. 2016
	Grain weight	OD	D x D	Zhou et al. 2012
Arabidopsis				
	Growth traits	OD	A x A	Melchinger et al. 2007b
	Biomass	OD	A x A, D x D	Kusterer et al. 2007b
		Dom, OD	A x A	Meyer et al. 2010
Cotton (*Gossypium*)				
	Yield	Dom, OD	Extensive	Shang et al. 2016

genotypic value typically estimated using only a single observation on its phenotype. Xing et al. (2002) and Hua et al. (2003) proposed an elegant solution allowing for replication. They accomplished this by first creating 240 RILs (recombination inbred lines; Chapter 18), a set of inbred lines derived by selfing the F_1s from an inbred-line cross. They then randomly divided this collection into two sets, crossing a random member from each set (with no resampling) to make 120 lines. They redid this procedure three times to create a total of 360 such lines, with genetic uniformity within each line, allowing for replication. This collection of lines is essentially equivalent to an F_2 population from the cross of the original inbred lines forming the RILs. Heterotic QTL mapping was then performed by searching for marker-trait associations where the trait value for each cross was its estimated heterotic value, Equation 13.1b (the mean of the cross minus the midparental mean). Liu et al. (2020) review modifications of this design.

Two other popular designs for mapping heterotic QTLs are based on adding marker data to the Design III (Cockerham and Zeng 1996; Melchinger et al. 2007a) and the TTC (Kusterer et al. 2007a; Melchinger et al. 2007b; Reif et al. 2009; He and Zhang 2011) approaches discussed above. One modification of both designs for QTL mapping is that RILs are often used in place of F_2s, allowing for better replication. Melchinger et al. (2007a) found that the estimated QTL dominance effect (under either Design III or the TTC) for a single locus is a composite of both its dominance effect (d) and the additive $\times$ additive interactions (aa) of that locus with all other trait loci. We can see this from Equation 13.7a. This led Melchinger et al. to propose the notion of the **augmented dominance effect** for locus i, defined as

$$d_i^* = d_i - \frac{1}{2}\sum_{j\neq i} aa_{ij} \tag{13.13}$$

with the sum being taken over other loci that have additive $\times$ additive interactions with the focal locus, i. The dominance contrasts in both the Design III and the TTC estimate d_i^*, rather than d_i, and hence estimates of the nature of dominance are biased by the presence of additive $\times$ additive epistasis.

Table 13.2 summarizes the findings from a few of these QTL mapping studies. While examples of apparent overdominant QTLs are often found, given the crude level of resolution, associative overdominance is often hard to rule out. For example, while Stuber et al. (1992) found a major QTL showing overdominance for yield in maize, Graham et al. (1997) showed that, upon finer mapping, the QTL region contained at least two smaller QTLs showing only dominance. Another example is the work of Li et al. (2015), who found two recessive QTLs in tight repulsion-phase disequilibrium influencing height in sorghum hybrids. While these loci were only 3 Mb apart, because of high recombination, their map distance was around 30 cM (Chapter 17), allowing for their effects to be separated.

An interesting comparison is the genetic basis for yield heterosis in the two most important cereals, maize and rice. Yield heterosis in maize appears to be largely from the effects of single-locus dominance, with little impact from epistasis (although Xiao et al. 2021 report large amounts of very small epistasis effects using a different mapping design, akin to NAM). By contrast, heterosis of yield in rice appears to be driven, in large part, by additive x additive effects (Hallauer and Miranda 1981; Yu et al. 1997; Hua et al. 2002, 2003; Garica et al. 2008; Tang et al. 2010; Wei et al. 2015). Jiang et al. (2017) similarly found that epistasis was an important driver for grain yield heterosis in bread wheat (50% was due to A x A, and only 16% due to dominance).

Observations from Autopolyploids Pose Challenges for the Dominance Model

While the dominance hypothesis is compelling, and has significant evidentiary support, there are observations from autopolyploids that are hard to reconcile with the standard dominance complementation model. One is **progressive heterosis**, where heterosis *increases* with the number of *distinct alleles* in a polyploid: *ABCD* tetraploid hybrids display more heterosis than do *AABB* or *CCDD* hybrids (Demarly 1963; Groose et al. 1989; Brichler et al. 2003; Auger et al. 2005; Schnable and Springer 2013; Yao et al. 2013; Fu et al. 2014; Washburn and Birchler 2014; Brichler 2015; Washburn et al. 2019). Phrased another way, single-cross (SC) hybrids ($AAAA \times BBBB$, $CCCC \times DDDD$) show less heterosis than double-cross (DC) hybrids ($AABB \times CCDD$). This is especially striking as the *opposite pattern* is seen in diploids, where SC hybrids usually *outperform* DC hybrids (Sprague and Federer 1951; Rojas and Sprague 1952; Eberhart et al. 1964; Eberhart and Russell 1969).

One explanation for progressive heterosis is the **maximum heterozygosity hypothesis**. While dominance involves the interaction between two alleles at a locus (bigenic effects), higher order interactions can occur in polyploids among different alleles at the same locus (e.g., trigenic and tetragenic effects; Equation 7.21). Demarly (1963) suggested that these higher order interactions could potentially contribute further heterosis, so that the greatest effect would occur in tetragenic (carrying four distinct alleles) individuals (we examine this further in Example 13.9).

Dunbier and Bingham (1975) provided support for the maximum heterozygosity hypothesis for yield in alfalfa through the clever use of haploid-derived autotetraploids (denoted by HD4x). A number of such lines were developed and used to generate single and double crosses. Single crosses of HD4x lines would be expected to carry no more than two alleles, while double crosses could involve up to four distinct alleles. Dunbier and Bingham observed an average performance ranking of DC > SC > HD4x, as expected from the maximum heterozygosity hypothesis. In autotetraploid maize, Sockness and Dudley (1989) similarly found that DC crosses showed more yield heterosis than did SC crosses, as did Washburn et al. (2019) for biomass heterosis. While the latter authors observed progressive heterosis (DC > SC), whole-genome sequencing of the parent and hybrid lines showed results inconsistent with the maximum heterozygosity hypothesis. Under this model, one would expect four distinct alleles at many loci in the DC hybrids, so that while alleles from the A genome might cover alleles in B (and vice versa), as well as for C with D, there would be alleles in the AB pair that would *further* cover alleles in DC, and vice versa. Using bioinformatic approaches to predict deleterious alleles from sequence data (Chapter 12), Washburn et al. found that around 10% of the total mapped genes in their two SC hybrids had complementary genes (one potentially deleterious, the other predicted to be functional). However, in DC hybrids, only 2% of genes showed such potential additional complementation and the authors suggested that this percentage may be an overestimate. This small increase in further complementation was very unlikely to account for the observed 30% to 50% increase in DC heterosis relative to SC values.

Another observation inconsistent with both the dominance and maximum heterozygosity hypotheses comes from work on triploid maize . Yao et al. (2013) found that heterosis for six of nine measured traits differed between AAB versus ABB triploid hybrids. Because A and B were inbred lines, both triploids should show similar complementation in covering recessive deleterious alleles. Reciprocal crosses (switching the roles of pollen and seed parents) showed no evidence of parent-of-origin effects, leading Yao et al. to suggest that heterosis may, in part, be influenced by dosage-sensitive effects (Birchler et al. 2003, 2005, 2015, 2016; Birchler and Veitia 2012).

Gene Regulation, Epigenetics, and Heterosis

A recent development in the study of heterosis is the role of **epigenetics**, in particular, whether the **epigenome** is modified in hybrids (Chen 2010, 2013; Groszmann et al. 2013; He et al. 2013; Geaves et al. 2015; Liu et al. 2022). Epigenetics refers to differences in gene regulation among otherwise identical DNA sequences caused by features such as differential DNA methylation and different chromatin configurations (open versus closed, modifications of histones, etc.). Generally speaking, methylation around a gene suppresses its expression, as does being sequestered within a closed-chromatin region. To date, most of the epigenetic work on heterosis has involved contrasting the **transcriptomes** (the set of all transcribed genes) and/or **methylomes** (the methylation pattern over the genome) between parental lines and their hybrids.

One of the key findings is that ***novel patterns of gene expression are common in hybrids***. First, **extreme-parent heterosis** is often seen: the transcript is either overexpressed (exceeds the high-parent value) or repressed (is less than the low-parent value). As discussed in Chapter 21, regulation of expression is a quantitative trait, and can be influenced by both multiple "genes" (DNA sites) and the environment. Sites impacting a particular transcript could be extremely close to its coding region (presumably reflecting DNA regulatory elements such as promoters, silencers, and enhancers), or could be long-range interactions from sites some distance away (diffusable protein and RNA products, such as transcription factors). The terms *cis* for local effects and *trans* for long-range interactions are often used, but see Chapter 21 for a discussion of some subtleties with this terminology. Heterosis for the expression level of a particular transcript results from interactions between these various elements.

Second, genes in hybrids often show **allele-specific expression** (**ASE**): the gene from one parent is expressed more than the gene from the other parent (Chapter 21). In the extreme, this results in **single-parent expression** (**SPE**), where only the allele from one parent is expressed (which in some cases could be due to PAVs). As a result of SPE, Baldauf et al. (2018, 2002) found that hybrids in maize express around 600–700 more genes than either parent (as a reference, there are around 40,000 core genes in maize). These authors further noted two important features of SPE genes. First, they generally were biased towards **nonsyntenic** genes (those present in maize, but not in its close ancestor, sorghum), while genes present in both (**syntenic**) were underrepresented as SPEs. They also tend to be enriched for transcript factors. Hence, many SPE genes likely represent relatively recent duplication events that have somewhat redundant (and potentially divergent) roles to their core ancestors. Baldauf et al. (2002) showed that the level of (midparent) heterosis correlated with the number of SPEs for the six traits they examined, providing a direct connection between expression of additional genes and heterosis.

Third, changes in ASE may generate overdominance. Shao et al. (2019) found that around 3,330 genes displayed ASE in rice hybrids. Most of these genes had the same parental allele predominately expressed over tested tissues and conditions (temperature and day-length combinations), although its relative advantage might vary. However, around 125 ASE genes showed **direction-shifting**, where the predominate allele changed over tissues/test environments (the allele from one parent was the higher expressed variant in some settings, while the alternative allele was in others). Shao et al. suggested that such direction-shifting variation could allow expression of the two alleles at a given gene to be more optimally tuned over varying conditions, generating overdominance. This idea of overdominance being generated by the different alleles in a heterozygote having different roles over environments was originally suggested in the allozyme era (e.g., Schwartz and Laughner 1969; Lewonton 1974).

In an analogous fashion to direction-shifting with ASE, changes in expression over developmental time can generate heterosis, as elegantly shown by Ni et al. (2009). They examined growth heterosis in allotetraploids formed by crossing *Arabidopsis thaliana* and *A. arenosa*. Excess hybrid growth was driven by temporal expression changes in two circadian clock genes (*CCA1* and *LHY*). These showed negative heterosis (repressed expression) during morning hours, but positive heterosis (increased expression) at dusk, allowing hybrids to produce more starch and chlorophyll than their parents.

Finally, a particularly striking observation is *genes that are expressed in the hybrid, but not in either parent*. Note that this is distinct from SPEs, which are expressed in at least one of the parents. While we have previously discussed the notion that presence-absence variation (PAV) can result in a hybrid expressing more genes than either parent, the observation that the F_1 can express genes not transcribed in *either parent* is somewhat remarkable. Sun et al. (2023) found hundreds of genes that were expressed in rice hybrid *Y900*, but not in either parent. They suggested that transcriptional regulators in one parent influenced the promoters of the other, generating novel expression. Suppose that for a gene G to be expressed it must have both a functional promoter P in *cis* and a *trans*-acting factor T that binds to P. If one parent was $T - pG_1$ and the second was $t - PG_2$ (with the dash indicating that T and P are potentially some distance apart, even unlinked), then neither parent will express G. However, the hybrid would be $T - pG_1/t - PG_2$, expressing allele G_2 (showing SPE; $T \rightarrow P \rightarrow G$). This is an example of additive $\times$ additive epistasis, as expression levels result from the interaction between the T allele at the *trans*-acting site and the P allele at the *cis* site. If the expression of T involves some dominance, then expression of G could have a dominance $\times$ additive epistatic interaction.

Methylomes (the pattern of methylation over a genome) also differ between hybrids and parents. Such differences in methylation can influence gene expression, and can also be *caused* by differences in gene expression. In particular, classes of small RNAs modulate

Table 13.3 Examples of the importance of heterosis for both food production and land conservation. Here % yield advantage is the gain of the hybrid over the best pure-line. While these values are dated, the general picture they suggest is still accurate. Based on Duvick (1999), with modifications for rice.

Crop	% Planted as Hybrids	% Yield Advantage	Annual Added yield (%)	Annual Added Rield (tons/ha)	Annual Rand Savings
Maize	65	15	10	55×10^6	13×10^6 ha
Sorghum	48	40	19	13×10^6	9×10^6 ha
Sunflower	60	50	30	7×10^6	6×10^6 ha
Rice	55	20	11	200×10^6	80×10^6 ha

certain aspects of methylation, and the expression pattern of these RNAs can show heterosis in hybrids, altering methylation patterns (Shen et al. 2012; Ng et al. 2012; Zhang et al. 2014; Shen et al. 2017; Chen et al. 2022).

There are observations from *Arabidopsis* that methylation can play some role in heterosis. First, Shen et al. (2012) showed that heterosis for growth significantly declined in a cross of *Arabidopsis* lines that were demethylated (both parents and their F_1) relative to the same cross when all lines had normal levels of methylation. Second, Lauss et al. (2018) examined heterosis in **epiHybrids**, the F_1 of crosses between genetically identical parents, but with different methylation patterns. Heterosis was seen for some traits, and the authors were able to map **epiQTLs** (Cortijo et al. 2014), differentially methylated regions in the parent lines that impact heterosis in their hybrids. Several of these epiQTL segments were resequenced, and no sequence variation (SNP or structural) was detected, showing that heterotic impact from site was entirely due to methylation differences. In closing, we note that the quantitative genetics of gene expression and the role of molecular pathways underlying trait expression are examined further in Chapter 21 and in Appendix 2.

AGRICULTURAL EXPLOITATION OF HETEROSIS: PLANTS

The exploitation of hybrid vigor in agriculture traces back at least 5,000 years to the Sumerians (Clutton-Brock 1992), who produced mules by crossing horses (*Equus caballus*) with donkeys (*E. asinus*). Modern agriculture has been greatly impacted by hybrids, with the heterosis resulting in significantly increased yield in crops. Such gains directly translate into fewer acres that must be farmed to obtain the same total yield, and these land savings are by no means trivial (Table 13.3). Another critical, yet often overlooked, benefit of hybrids is *uniformity* (all of the F_1 from a cross of two inbred lines are genetically identical), which allows for greater efficiencies in harvesting, such as greatly increased mechanization (Crow 1998; Goldman 1999).

Both animal and plant breeders exploit heterosis, but in rather different ways. Plant breeders have the luxury (in most settings) of very high reproductive rates, long-term storage of genotypes (i.e., seeds), and a lack of maternal effects for most traits (but there are a few examples, e.g., Dapp et al. 2015). In contrast, most species of domesticated animals have low reproductive rates, which generally rules out the plant breeder's approach of generating large numbers of F_1 individuals each year for commercial sale. Further, many animal production traits show strong maternal effects, a trait that itself shows heterosis. Thus, offspring from **crossbred** (hybrid) **dams** have an extra performance boost due to the improved maternal effects (**maternal heterosis**). As a result of these differences, much of the exploitation of heterosis in plants involves generating lines that cross well to produce exceptional hybrids. In contrast, animal breeders tend to use more elaborate crossing schemes, such as rotational crossbreeding. We will first examine exploitation of heterosis in plant systems and then consider specific modifications that are used by animal breeders. Extensive reviews of the application of heterosis in plant breeding can be found in the conference volumes edited by Gowen (1952) and Coors and Pandey (1999).

Heterotic Groups

Lines that cross well have historically been said to display **nicking**. When is this expected? A key concept in the exploitation of heterosis is the notion of **heterotic groups**. Crosses between members *within* a heterotic group display little heterosis, while crosses *between* groups result in significant heterosis. Partitioning lines into sets of two (or more) heterotic groups considerably simplifies a breeder's job, as they can concentrate on crosses between these groups, rather than testing *all* pairwise combinations. While the notion of heterotic groups (or **heterotic patterns**) is the foundation for much of modern plant crossbreeding, it is a surprisingly recent concept. As reviewed by Tracy and Chandler (2006), while the idea was discussed in the early 1970s in some plant-breeding companies (the first appearance of this term in the literature seems to be by Tsotsis 1972), not until the early 1980s did it become mainstream.

Heterotic groups arise from between-group allele-frequency differences at loci underlying the trait of interest. Further, we only care about frequency differences at a *subset of trait loci*—those displaying directional dominance or epistasis—as differences in allele frequencies at purely additive loci have no impact on heterosis. Consider the situation where only dominance is present. From Equation 13.2c, the expected F_1 heterosis is $\sum \delta^2_{p_i} d_i$ (δ being the between line allele-frequency difference), which increases with both the between-group allele frequency differences and with the number of such divergent loci. Inbred lines from different heterotic groups represent the most extreme example, with heterosis equal to $\sum d_i$, with the sum being taken over all differentially fixed loci. The more such fixations, the greater the heterotic effect. Schön et al. (2010) found evidence of such differential fixation for yield QTLs between maize heterotic groups. With epistasis, the general idea is similar, except that now differences in *joint* allele frequencies at two (or more) loci are required. A second important concept is that heterotic groups can *evolve over time*, as groups become more divergent due to selection and drift (Example 13.5). This is akin to our discussion at the start of this chapter on how the divergence between demes in natural populations often generates heterosis for fitness (Whitlock et al. 2000).

Given their importance, how does one detect and/or construct heterotic groups and assign new lines to them? Obviously, the brute-force method of considering all pairwise crosses can be employed (diallels; Part II Chapter 25), but one would like a more efficient approach. If one has a set of lines from *known* heterotic groups, then molecular markers can be used to cluster both known and unknown members, providing a reasonable predictor of the group membership for many unplaced lines (e.g., Barbosa et al. 2003; Reif et al. 2003a; Dhliwayo et al. 2009). However, simply using marker data alone (i.e., without any known reference lines) to predict heterotic groups has been problematic (Charcosset et al. 1991). For example, Sun et al. (2023) performed whole-genome sequencing on the rice lines *R900*, *R9311*, and *Y58S*. The sequence variation between *R900* and *Y58S* was less than that between *R9311* and *Y58S*, yet the yield was much higher for the *R900* × *Y58S* cross (which yields the elite Chinese hybrid *Y900*).

Example 13.2 hinted at one reason why genetic distance is a poor predictor of heterosis, as Moll et al. (1965) observed an intermediate level of divergence for maximal heterosis. As lines became sufficiently divergent, outbreeding depression, rather than heterosis, occurs. Despite this early observation, a number of studies have attempted to use molecular markers to predict heterosis, typically by looking for a correlation between the amount of heterosis and either genetic distance or F_1 heterozygosity (a measure of the genetic differences between the crossed inbred lines). Most of the work has been done on maize yield, with mainly negative results (e.g., Frei et al. 1986; Price et al. 1986; Lamkey et al. 1987; Godshalk et al. 1990; Benchimol et al. 2000; Tracy and Chandler 2006; Dhliwayo et al. 2009; Flint-Garcia et al. 2009), but with a few successes (Lee at al. 1989; Reif et al. 2003a, 2003b). A closer examination suggested that markers can be good predictors for close-line crosses, but poor predictors for more distant crosses (Barbosa et al. 2003; Amorim et al. 2006), in line with the results of Moll et al. (1965). Early marker-based work (using RFLP bands) in chickens found a highly significant negative correlation between band sharing and heterosis in chickens (Gavora et al. 1996; Haberfeld et al. 1996), but this might have occurred because only closely related lines were examined.

The critical point is that genetic divergence, *per se*, is *not* sufficient to define heterotic groups, as the divergence required is *trait specific* (e.g., Flint-Garcia et al. 2009)—typically focusing on yield or yield components—and further restricted to loci with specific nonadditive effects (directional dominance and epistasis). If the allele-frequency divergence at the underlying trait loci is simply a product of genetic drift, then the divergence at random markers may track heterotic potential. However, if trait divergence is partly driven by selection, the pattern of divergence at the underlying trait loci need not track the pattern of neutral divergence. Hence, simply using a metric such as the genetic distance between lines based on random (i.e., presumed neutral) markers is not a fruitful approach for the detection of heterotic groups. Consistent with this observation, Flint-Garcia et al. (2009) examined heterosis for a number of traits in maize, finding that while yield heterosis was poorly predicted using random markers, heterosis for many other traits was fairly well predicted. One might imagine that yield is under considerable selection, but that other traits might be more prone to divergence by drift, in which case neutral divergence may better predict heterosis. (Note that this is yet another example of bias that can arise when the site-frequency spectrum—the distribution of allele frequencies—of the markers differs from the site-frequency spectrum of the causative alleles.) Further, heterotic groups can be trait specific, with group membership potentially shifting as the trait changes. Thus, even when heterosis is found for a focal trait, widespread heterosis in other traits is not guaranteed, nor even expected. In contrast, when inbreeding depression is found, it typically occurs over a number of traits (Chapter 12).

A final important point to stress is that a breeder's goal is exceptional *hybrid* performance, not simply exceptional *heterosis*. As the distinguished maize breeder John Dudley famously said "We sell hybrids, not heterosis," which we will refer to as **Dudley's dictum**. Hybrid performance is the midparental mean plus the heterotic effect. Hence, the goal is not to find the cross with the highest heterosis, which may only yield an average performing hybrid if the parental means are low, but rather the *optimal combination* of parental means plus heterosis that yields the largest overall hybrid value. This suggests the standard strategy of starting the search for exceptional hybrids by using crosses between the highest performing lines from different heterotic groups.

Case Study: Hybrid Corn

Corn (maize) is one of the world's most important cereals, and the leading crop in the United States as measured by acres planted and total yield. Almost all of the U.S. crop is **hybrid corn**, the seed from crosses of elite inbred lines. There are small amounts of **open pollinated** (**OP**) corn (**OPVs**, **open pollinated varieties**), but these are largely restricted to use by small independent farmers. The notion of hybrid corn traces back to Shull (1908, 1909), who noted that, because of the significant high-parent heterosis seen in many corn crosses, the objective of corn breeders should be to find and maintain the best parental lines for hybrids. By maintaining these elite inbred lines and crossing them as needed, genetically identical hybrid seed could be produced year after year. The history of hybrid corn is reviewed by Crabb (1947; 1993), Kiesselback (1951), Anderson and Brown (1952), Crow (1998), Duvick (2001), Tracy and Chandler (2006), and Troyer (2006), while Troyer (1999, 2004) presented a detailed history of most of the founding lines that now comprise the whole of U.S. corn production. Nelson (1993) gives a fascinating discussion of the lives of, and interactions among, Edward East, Rollins Emerson, and Donald Jones, key geneticists behind modern hybrid corn.

The initial problem with the widespread implementation of hybrid seed was that the early inbred lines used by breeders typically had very low seed set. In a (single) cross between two inbred lines, the seed parent is inbred, and although the resulting hybrid seeds (offspring) produce plants with superior yield, each inbred seed *parent* produced only a small number of such seeds. Thus, poor fertility of the initial inbred lines resulted

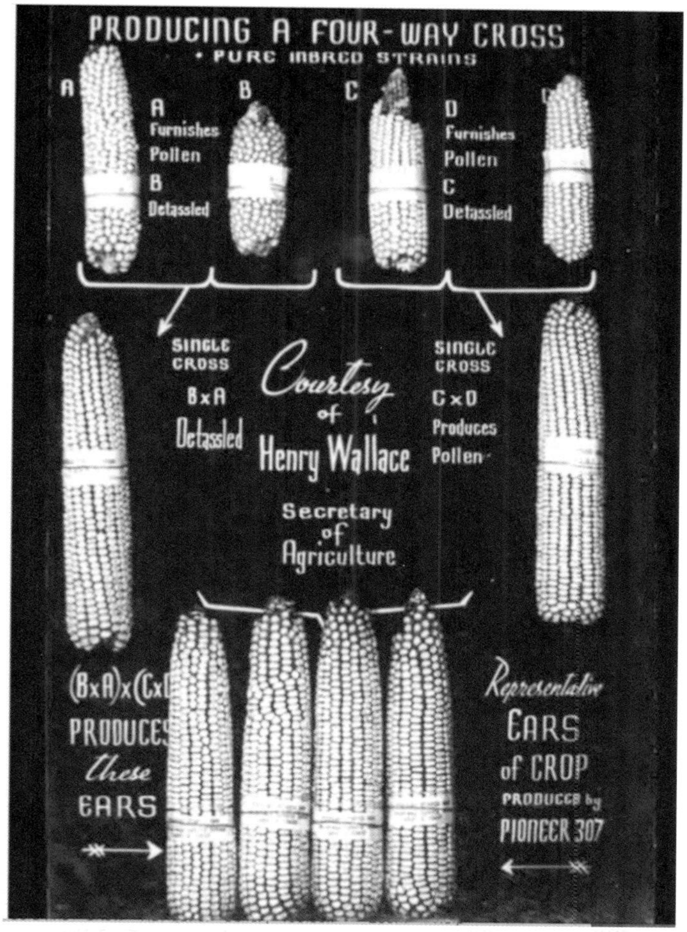

Figure 13.5 A vintage poster from the Pioneer Hi-Bred Company illustrating the use of double crosses in the creation of maize hybrid lines.

in rather few hybrid seeds per plant. Jones (1918, 1922), who was a graduate student of Emerson at the time, suggested that instead of using an inbred line as the seed parent, one instead use a *hybrid* parent, with the hybrid seed for commercial release being produced by a **double cross** instead of a single cross. As shown in Figure 13.5, while single crossses are of the form $A \times B$ or $C \times D$, double (or **four-way**) crosses involve a further cross of two F_1s, $(A \times B) \times (C \times D)$. Because the seed parent is now a hybrid (the progeny from a single cross between two inbred lines, e.g., $A \times B$), it should show heterosis in seed production, resulting in superior seed set. This suggestion directly opened up the vast commercial potential of hybrid corn. A modification was to use a **three-way cross**, where pollen for line A is used to fertilize a $B \times C$ seed parent.

The first sale of hybrid seed was in 1924 by Henry A. Wallace (a farmer who later became Vice-president of the United States), who founded the Pioneer Hi-Bred Corn Company of Iowa the following year. A rapid transition from OPVs to hybrids quickly followed in many midwestern U.S. states. For example, Iowa went from less than 10% hybrids in 1935 to over 90% by 1939. Crow (1998) noted that factors, other than improved yield, likely hastened the adoption of hybrids. First, machine harvesting was being introduced during this time and the greater uniformity of hybrids facilitated mechanized farming. Second, hybrids proved to be more hardy than OPVs during the devastating Dust Bowl drought from 1934 to 1936.

A critical fact that is easily overlooked was that the 1930s also saw the widespread acceptance of **Fisher's principles of experimental design** (Appendix 9), allowing for more

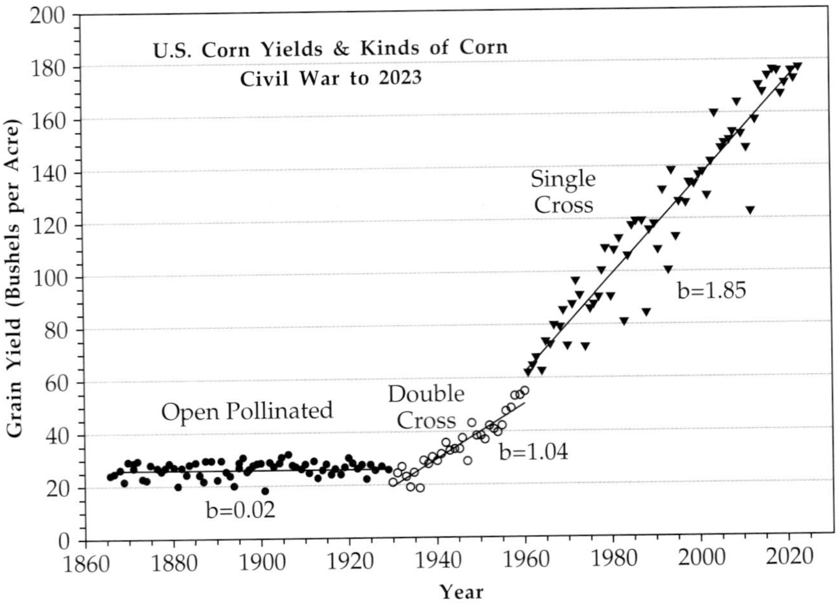

Figure 13.6 Average U.S. corn yields, 1865–2023. Initially, plantings were dominated by open pollinated lines (OPVs), which were largely replaced by double-cross hybrids (four-way crosses of inbred lines; Figure 13.5), and then subsequently by single-cross hybrids. The regression slope (*b*) of yield gain per year are computed for each of these three periods. Figure courtesy of Mark Cooper, using USDA data and building on earlier figures by Troyer (1999, 2007) and Philips (2010).

precise estimation of line effects and greater control over residual noise. Crow (1998) noted that Fisher spend several summers in the early 1930s at Iowa State University, which likely had a great influence on the analysis and testing of potential hybrids. As Crow noted "How much of the increase in agricultural productivity should be credited to Fisher, I don't know. But my guess is considerable." We concur.

From the 1930s to the 1960s, most U.S. hybrid corn was produced by double crosses (Figure 13.5). However, it was soon noted that the seed from single crosses usually outperformed double- and three-way crosses (Sprague and Federer 1951; Rojas and Sprague 1952; Eberhart et al. 1964; Eberhart and Russell 1969). Further, double-cross hybrids were more variable than single-cross (due to segregation in the two single-cross F_1s that are crossed to form the double cross). As breeders were able to select for inbred lines with higher yield, hybrid corn based on single crosses became both commercially feasible and also desirable, given their increased yield and uniformity. Indeed, by the 1960s inbreds had been selected to the point where they often outperformed the initial double-cross hybrids of the 1930s. Ironically, the (initial) inability to create such high performing inbred lines was used by East and Hayes (1912) as evidence against the dominance hypothesis. If heterosis was largely due to overdominance, then no inbred line can match the best hybrid. However, if heterosis is largely due to partial to complete dominance, over time one should be able to stack (or **pyramid**) all of the favorable dominant alleles into a single inbred (e.g., Hashimoto et al. 2021; Shen et al. 2022). Indeed one could, but given the number of underlying yield loci, such selection took time. Since the 1970s, most hybrid corn in the U.S. is the result of single crosses.

The yield gains displayed in Figure 13.6 confound genetic improvement with corresponding dramatic changes in agronomic practices. Indeed, a significant fraction of the gain has come from the development of maize lines that can handle higher levels of stress,

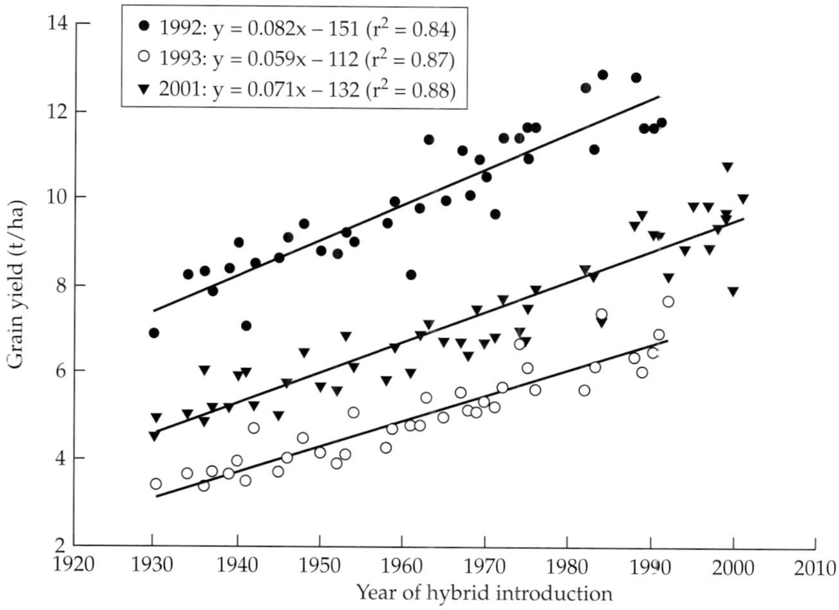

Figure 13.7 Yield (measured in tons per hectare) in maize hybrid lines as a function of year of release. Using remnant seed, all lines were grown in the same set of years, with 1992 being highly favorable, 1993 cool and extremely wet, and 2001 hot and dry. Note that the response is parallel over the three different environments (years), suggesting little genotype × environment interaction (Chapters 27 and 28 in Part II). Such "common-garden" experiments are the cleanest way to separate an observed gain into genetic versus environmental components (WL Chapters 19 and 20 present alternative mixed-model approaches that can also accomplish this goal). This separation is critical, as a yield improvement over time (Figure 13.6) could simply reflect improved agronomic practices, rather than genetic gain. (After Duvick 2005.)

allowing more plants to be planted per acre. Duvick (2005) noted that plant densities have increased by roughly 1.000 plants per hectare (ha) per year in the U.S. corn belt, with densities of around 30,000 plants per ha in 1930, 40,000 plants in the 1960s, 60,000 plants in the 1980s, and 80,000 plants in the 2000s. Figure 13.7 illustrates how much of the yield gain is due to genetics, by growing remnant seeds of the best hybrid from a given historical year under the same conditions in a highly replicated (multiple location) design (Appendix 9). This was done over three different years and shows both the amount of genetic improvement and also that there is little genotype x environment interaction. Hybrid improvement accounted for between 50% and 70% of the total improvement in yield, with the remainder due to improved farming practices (Duvick 2001, 2005).

Recalling Dudley's dictum of "hybrids, not heterosis," hybrid improvement ($\mu_{F_1} = \mu_{MP} + H_{F_1}$) can occur by increasing the performance of the parental inbred lines (μ_{MP}), by increasing the amount of heterosis between crossed lines (H_{F_1}), or both. How much of the improvement in modern U.S. maize hybrids is due to increased heterosis? Very little (Duvick 2005; Troyer and Wellin 2009). Most of the improvement in hybrids arose from the continual improvement of the elite inbred lines that were crossed, improvement that allowed breeders to move from double-cross to single-cross hybrids. Duvick (2005) found that the regression of yield (in tons per ha) on the midparental means of the best inbred lines was 0.048 per year, while the gain in heterosis was only 0.013 per year. Hence, while there was a slight gain in heterosis (H_{F_1}) over time, there was a four fold greater improvement in the inbred lines (μ_{MP}). As noted by Troyer (2006) and Troyer and Wellin (2009), the percentage gain due to heterosis has been *decreasing* over time.

As further developed in Part II Chapter 25, the basic machinery for predicting line-cross means is framed in terms of the **general-combining ability** (**GCA**) of *each line* and the **specific-combining ability** (**SCA**) for *each cross* (Sprague and Tatum 1942). In essence,

GCA is the breeding value of a line (Chapter 4), and (as with breeding values) is a function of population of lines chosen. The predicted mean of a cross is the sum of the maternal (M) and paternal (P) general-combining abilities, GCA(P) + GCA(M). The deviation between the predicted and actual values estimates the specific combining ability for that particular cross. Hence, the expected mean value of the single cross between lines i and j is $\text{GCA}_i + \text{GCA}_j + \text{SCA}_{ij}$. GCA represents the additive genetic contribution, while SCA represents the contribution from a subset of the nonadditive genetic variation in a trait (that which contributes to heterosis). Using this terminology, most of the response in maize hybrids has been through improved GCAs among the elite inbred lines, while the response in SCA, while slightly positive, has a diminishing role in hybrid improvement. Similarly, one can think of heterotic groups as having low within, and high between, group SCAs. Thus, if one knew something about the GCAs of the lines in an initial collection, a quite reasonable starting point is to cross the lines with the highest GCAs from each heterotic group with each other. While the best hybrid could involve a cross between two lines with lower GCAs but a very large SCA, crossing lines with the highest GCAs offers a good starting point.

Example 13.5. The development of heterotic patterns in U.S. maize was examined by Tracey and Chandler (2006). While the lore has been that breeders simply exploited existing geographic variation (e.g., Anderson and Brown 1952), Tracy and Chandler reached a surprisingly different conclusion: "heterotic patterns were created by breeders and are not the result of historical or geographic contingencies," rather "patterns were created by breeders through trial and error from a single race of corn."

A key to the initial division of lines into what eventually became recognized heterotic groups was that maize breeders focused on developing separate exceptional male and female lines. Commercial hybrids are generated by fertilizing elite lines from the female group with pollen from elite male-group lines. Female lines were selected for increased seed yield, to facilitate the transition from double-cross to single-cross hybrids. A second critical component was the ability of female-line members to resist **lodging** (being blown over, or dislodged, during strong storms). Such resistance is critical for mechanical harvesting, which requires upright plants at maturity. As a result of this necessity, the female heterotic pool has a large contribution from members of the **Iowa Stiff Stalk Synthetic** lines (**BSSS**, the *B* being the USDA single-letter code for Iowa, as Illinois got the *I*). Elite male lines focused on (among other things) exceptional pollen production, with lodging resistance being of lesser importance. As the result of these restrictions, the female heterotic group is usually referred to as **Stiff Stalk** (**SS**) and the male group as **Non-Stiff Stalk** (**NSS**).

As noted by Tracey and Chandler (2006), improvement of inbred lines was restricted by breeders to crosses *within* a given group to generate new material, avoiding crosses between groups except to produce commercial hybrids. The issue then became how to organize lines to make such breeding programs more effective. In 1949, the recommendation from the Committee on Grouping of Inbred Lines for Breeding Purposes was "As an arbitrary division, the committee recommends that the lines having an odd entry numbers in the 1948 uniform test of inbreds be tentatively assigned to group A and those having even entry numbers be tentatively assigned to group B" (Anon. 1949). All of the BSSS inbred lines ended up being assigned to group B. The result was crossing only within group A to generate improved A pool elites (which became the current male pool), and crosses only within B to generate improved B pool inbreds (which became the female heterotic group). Decades of breeding under this scheme resulted in the SS female and NSS male heterotic groups that are currently used. Molecular work has supported this view, finding that the initial collection of lines showed no significant clustering, while more recent lines show strong clustering into heterotic groups, due to divergence from this initially rather homogeneous group (Duvick et al. 2004; van Heerwaarden et al. 2012). Reif et al. (2005) examined the nature of heterotic groups from other maize breeding programs and discussed models for their creation, while a nice overview of role of admixture and local adaptation of land races that have contributed to current American and European maize stocks is given by Brandenburg et al. (2017). Finally, Lin et al. (2020) examined how differential selection shaped heterosis in hybrid rice.

Example 13.6. Hybrid corn offers an important cautionary lesson for breeders, namely, the great **Southern Corn Leaf Blight** (**SCLB**) epidemic from 1970 to 1971 (Tatum 1971; Ullstrup 1972). This was one of the most damaging plant epidemics in history: in terms of food energy destroyed it was much larger than the Irish potato blight epidemic in the 1840s that produced widespread famine (Ullstrup 1972). The roots of the epidemic in the United States trace back to what (at the time) appeared to be an elegant genetic solution to a major labor problem in the production of hybrid corn. Corn can self-fertilize, an undesirable feature as it reduces the fraction of hybrid seeds on a plant. To prevent this, corn breeders manually **detassled** plants, removing these pollen-shedding organs, allowing pollen to be only donated from the parents of interest. Typically, one row of pollen parents is planted for every four to six detassled rows of seed parents, so that a pollen parent is within two or three rows of any seed parent (any seed set on the pollen plants is ignored). The discovery of a mitochondrially encoded cytoplasmic factor that produced sterility in males seemed to be a elegant solution around the cost and effort of detassling (Jones and Everett 1949; Jones and Manglesdorf 1951; Rogers and Edwardson 1952). Plants with this T (for Texas, the origin of the line) cytoplasm are denoted by **Tcms** (Texas cytoplasmic male sterity). Seed parents with Tcms cytoplasm produce normal and viable seeds, but no pollen, and thus do not have to be detassled. As a result of this very convenient feature, by 1970, almost 85% of hybrid corn seed in the U.S. contained Tcms. Hence, despite a wide diversity of nuclear genotypes, U.S. corn was close to a clonal monoculture in terms of its mitochondrial (mtDNA) composition.

In 1969, a previous unknown strain of the fungus *Helminthosporium maydis* was detected in a few areas in the midwest corn belt of the central U.S, with Tcms plants being hyper-susceptible to this strain. The result was over a billion dollar loss (in 1970s dollars) and major angst in both the private and public agricultural sectors. Corn is typically a rather disease-free crop (indeed, Ullstrup 1972 noted that in 1970, the U.S. Department of Agriculture considered corn so healthy that it did not employ any full-time corn pathologists in the U.S. corn belt), so this outbreak was even more of a shock and led to a considerable focus on increasing the genetic diversity in crops.

Hybridization in Other Crops

A number of crop breeding programs utilize hybridization for commercial production, mainly for heterosis, but in some cases simply for uniformity (Janick 1999). As would be expected given the diversity of crops, a number of different solutions are used to exploit heterosis. Chapters in the edited volume by Frankel (1983) and the symposium volume edited by Coors and Pandey (1999) provide a nice review of some of these systems: rice (Virmani 1999); sorghum and pearl millet (Axtell et al. 1999); barley (Ramage 1983); rye (Geiger and Miedaner 1999); forage grasses (Kobabe 1983); oilseed crops such as soybean, rapeseed, and sunflower (Miller 1999; Girke et al. 2012); cotton (Meredith 1999); and ornamental (commercial flowers) crops (Reimann-Philipp 1983). While there is an impressive list of vegetable crops utilizing heterosis—carrots, onions, asparagus, spinach, broccoli, cabbage, cauliflower, cucumber, squash, watermelon, tomato, pepper, and eggplant—there are also vegetable crops with features that limit its use (Wehner 1999). The exploitation of heterosis also occurs in trees (Brewbaker and Sun 1999). While conifer breeders typically use same-species crosses, other commercial products are mainly generated by cross-species hybrids (aspens, poplars, and *Eucalyptus*).

Three factors influence the ability to utilize cross-breeding: (i) the existence of heterosis; (ii) the ability to control pollination; and (iii) whether the benefits of (i) offsets the costs of (ii). A good example is wheat (Wilson and Driscoll 1983; Jordaan et al. 1999). Despite an up to 30% yield heterosis, this is not a sufficient gain to be economically viable in the current wheat cropping system. A major reason that hybrid corn has been commercially successful is because of its reproductive structures, with pollen being produced on a long tassel, which allows for relatively easy manual control of pollination. For many species with perfect flowers (containing both male and female structures), manual control of pollination is simply not economically feasible because flowers are too numerous, too small, or both. Male-sterile genes are critical in the creation of commercial hybrids in such species, as is chemical sterilization of pollen. Fu et al. (2014) gives a general overview of some of these solutions, while details for specific crops can be found in the papers cited above.

In contrast to field crops, horticultural crops have much more favorable economic conditions for the commercial exploitation of heterosis. Here, flowers are typically large and easily handled (unlike the grass-family flower heads of many important field crops). Equally important, the yield from each individual plant has significant commercial value (Dobbs 1955), so that individual manipulation of single plants often is economically feasible. Duvick (1999) discusses the required economics and politics for the widespread adoption of hybrid crops.

Retention of Heterosis: Basics of Synthetics

Consider three different agricultural fields. In field one, an equal mixture of P_1 and P_2 are planted, giving a field mean of μ_{MP}, while field two consists of their F_1 hybrids, with mean of $\mu_{MP} + H_{F_1}$. Finally, suppose field three contains the randomly mated offspring of the F_1, for a mean (in the absence of epistasis) of $\mu_{MP} + H_{F_1}/2$. Under continual random mating (again in the absence of epistasis), the F_k for $k > 2$ has the same expected value as the F_2. Thus, while producing hybrids each generation is optimal (in the absence of cost, time, or resource constraints), an open-pollinated variety (the F_2) outperforms the mean of the two parental lines, and (in the absence of allele-frequency change from drift/selection) keeps this improvement ($H_{F_1}/2$) over subsequent generations.

This idea of creating an open-pollinated variety by randomly mating F_1s to maintain some, but not all, of the F_1 heterosis, was first suggested by Hayes and Garber (1919). Such populations are called **synthetic varieties**, and offer one solution for making at least partial use of the heterotic potential available when crossing a series of lines. Synthetics provide the ability to exploit heterosis when pollen control is difficult, such as in species where the flowers are too small and/or too numerous to make controlled crosses practical (e.g., forage grasses, alfalfa, and other legumes). Another important use of synthetics is as reservoirs of desirable germplasm (Sprague and Jenkins 1943) from which future lines can be developed, the Iowa Stiff Stalk Synthetic (BSSS) in maize being one such example.

Constructing a synthetic is straightforward: one starts with a set of n lines (for now, we assumed these are inbred and unrelated, relaxing this shortly), and generates a collection of all $n(n-1)/2$ pairwise crosses (ignoring crosses of a line with itself). The collection of lines resulting from this crossing scheme yields a **synthetic F_1 population**, and we define heterosis for this population in the standard fashion,

$$H_{F_1(syn)} = \mu_{F_1(syn)} - \overline{P} \tag{13.14}$$

where $\overline{P}$ is the trait mean over all parental lines. In the plant breeding literature, this F_1 is also referred to as a **Syn$_1$ population**, with the collection of parents being called the **Syn$_0$**. This F_1 population is then randomly mated to form the **F_2 synthetic population (Syn$_2$)**, which is allowed to randomly mate in all future generations to form the final synthetic population (occasionally referred to as the F_∞). Only a single generation of controlled crossing (the construction of the F_1) is required to generate the final synthetic population, offering a strategy to exploit heterosis without the need for potentially expensive and resource-heavy pollen control each generation.

The resulting single-locus genotype frequencies in the synthetic F_2 are in Hardy–Weinberg, while several additional generations of random mating are required to remove any linkage disequilibrium. What impact does this have on retention of heterosis? Consider first the heterosis generated by single-locus dominance effects. In the F_1, the genotype at any locus is a hybrid, with the two alleles coming from different lines. Conversely, in the F_2 for an n-line synthetic, there is a $1/n$ chance that the two alleles are from the same line, and a $(1-1/n)$ chance that (like the F_1) they are from different lines. Hence, the dominance loss is only $(1 - 1/n)$ as opposed to $1/2$ for the case of $n = 2$ lines, with

$$\mu_{F_2(syn)} = \overline{P} + \left(1 - \frac{1}{n}\right) H_{F_1(syn)} \tag{13.15a}$$

While verbally stated by Wright (1922b) for inbred parental lines, the first formal presentation of Equation 13.15a seems to be Kinman and Sprague (1945), who referred to it as "Wright's formula." We will use the term **Wright's predictor** to avoid confusion with the more classic Wright's formula for the rate of allele-frequency change under selection (WL Equation 5.5b) which is widely used in population genetics.

As we will detail shortly, Equation 13.15a assumes that the founding lines are all unrelated and are diploid. We generally assume the founding lines are fully inbred, but Equation 13.15a also holds for unrelated diploid parental lines with arbitrary levels of inbreeding (Gilmore 1969). If lines are polyploids and/or related, then (as we will detail shortly), adjustments are required for Equation 13.15a. These adjustments express the rate of decay of heterosis in terms of the expected amount of inbreeding at various generations during the formation of the synthetic (Busbice 1969, 1970). Such inbreeding arises from the combining of identical-by-descent alleles in individuals in the Syn_2 (and later) generations due to the synthetic being formed from a fixed number of parents.

Following Powers (1941), a few authors have also considered the situation where gene action is entirely multiplicative. In such cases, Kinman and Sprague (1945) suggested the predictor

$$\widehat{\mu}_{F_2(syn)} = \exp\left(\frac{2}{n^2} \sum_{i<j} \ln(\,\overline{z}_{ij}\,) + \frac{1}{n^2} \sum_{i}^{n} \ln(\,\overline{z}_{i}\,)\right) \tag{13.15b}$$

where $\overline{z}_i$ is the mean of line i and $\overline{z}_{ij}$ the mean of the cross between lines i and j. This expression follows directly from Equation 13.15a and the assumption that gene action is entirely multiplicative, as $x_1 \cdot x_2 \cdots x_n = \exp[\sum_i^n \ln(x_i)]$. Although Equation 13.15b is occasionally credited to Powers, it first appeared in Kinman and Sprague, and is thus best referred to as the **Powers–Kinman–Sprague predictor**.

The creation of synthetic populations seems to be a straightforward method for reducing the decay of heterosis by simply increasing the number of lines (n) when forming the F_1 population. The delicacy with this approach follows from Dudley's dictum of "hybrids, not heterosis." The breeder is interested in the *mean* of the synthetic population, *not* the amount of *heterosis* that is retained. While adding more lines increases the latter (more retention of heterosis), it usually reduces the $\overline{P}$, eventually decreasing the value of $\mu_{F_1} = \overline{P} + H_{F_1}$ as more (and thus lower performing lines) are added. For example, one standard approach is that the two-line synthetic is based on the lines with the two highest GCA values, the three-line on the three highest values, and so on. As more lines are added, the mean GCA, and hence $\overline{P}$, declines. If $\overline{P}$ is reduced sufficiently over the midparental mean using the two highest performing lines, then any extra retention of heterosis may be overwhelmed by a decrease in μ_{F_1} due to $\overline{P}$ starting at a lower value (Example 13.7).

While all of the dominance loss occurs by the F_2 (for a nonselfing diploid), the epistatic (or recombinant) loss takes several generations to reach its limiting value. With $n = 2$, for unlinked loci this occurs in a single generation, but for more than two lines, only 1/2 of the LD decays each generation for unlinked loci. Recall that the recombination loss can be expressed as $r_D R$, where R is given by Equation 13.11a and the Dickerson coefficient, r_D, is the fraction of nonparental gametes in an individual. For an n-breed synthetic where all lines make an equal contribution, then (Dickerson 1972)

$$r_D = 1 - 1/n \tag{13.16a}$$

giving the equilibrium synthetic population mean (following the decay of LD) as

$$\mu_{F_\infty(syn)} = \overline{P} + \left(1 - \frac{1}{n}\right) \left(H_{F_1(syn)} - R\right) \tag{13.16b}$$

Example 13.7. Kinman and Sprague (1945) used data for 10 inbred maize lines and their 45 (non-reciprocal) single crosses to predict the expected yield of various synthetics formed by crossing between 2 and 10 of these lines. They applied both the Wright (Equation 13.15a) and Powers (Equation 13.15b) predictors to see if different assumptions on gene action had a significant impact on the expected F_2 yields. The following graph plots the observed F_1 yields and the expected F_2 values under the different predictors.

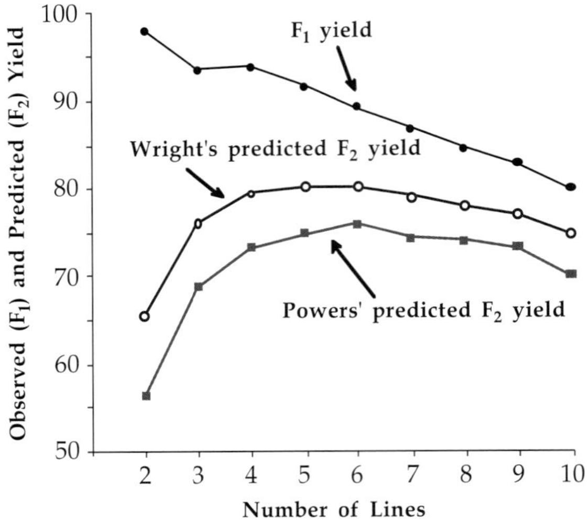

Lines were entered in order of their estimated general combining ability, so for $n = 4$, the top four lines in terms of GCA are used. As expected, as more lines are added, the overall F_1 yield decreases because the best pairs are chosen first, so that $\overline{P}$ decreases with n. Countering this, the expected differential between the F_1 and F_2 means is a decreasing function of n under both Wright's and Powers' expression. Here $n = 6$ is the optimal number of lines under both predictors for this particular trait (yield) in this collection of lines. Approaches for estimating the optimum number of lines can be found in Gallais (1990) and Kutka and Smith (2007).

Retention of Heterosis: Inbreeding in Synthetics

Although later-generation synthetics are random-mating populations, they derive from a set of n parents and hence are, at least to some extent, inbred as individuals can share identical by descent (IBD) alleles from a common parent (Chapter 7). The amount of such inbreeding impacts the retention of heterosis, with smaller amounts of inbreeding resulting in more retention of any initial F_1 heterosis in later generations. Hence, the behavior of the inbreeding coefficient is of critical importance in predicting the mean of a synthetic, which (assuming no epistasis) is a function of the parental allele frequencies and the amount of inbreeding in the equilibrium synthetic population. In the simplest of settings (a nonselfing diploid synthetic population started from n inbred and unrelated parents), the equilibrium level of inbreeding is just $1/n$ and is reached in one generation (the Syn_2).

Busbice (1969) examined the expected inbreeding, and the resulting expected loss of heterosis, under more general settings. For the case of no selfing (i.e., individuals in the synthetic population are self-incompatible), then for a species with ploidy level $2k$ ($k = 1$ for a diploid, 2 for a tetraploid, etc.), the amount of inbreeding f_j in the Syn_j population is

$$f_j = \frac{kr_{j-1} + (k-1)f_{j-1}}{2k-1} \quad \text{for } j \geq 1 \tag{13.17a}$$

where r_j is the probability, in cycle j, that two random alleles from the parents of an individual are IBD. From the definition of the coefficient of coancestry Θ_{ij} (Equation 7.2a), r_0 is simply the average pairwise value of Θ_{ij} among the parents. Similarly, f_0 is the average amount of inbreeding among the parental lines. Note immediately that if the founding parents are unrelated ($r_0 = 0$), then $f_1 = 0$ for a diploid, independent of the level of inbreeding in the parents (as the amount of inbreeding in the offspring is just the coefficient of coancestry of its parents; Equation 7.2b), and hence the Syn$_1$ population is not inbred. This occurs (in a diploid) because two alleles in any Syn$_1$ individual come from different (and hence unrelated) parents. However, this is not the case in the Syn$_2$ and later generations, as there is some chance that the two alleles in an individual trace back to a single copy in one of the founding parents (and hence are IBD).

For a tetraploid, if the parents are unrelated, then $f_1 = f_0/3$, or 1/3 the average inbreeding of the parents, and here the Syn$_1$ is inbred when there is some inbreeding among the founding lines ($f_0 > 0$). Inbreeding values in later generations are obtained by joint iteration of both f_j and r_j. Busbice showed that after the Syn$_1$ generation, r_t essentially reaches it equilibrium value, so that

$$r_e \simeq r_1 = \frac{1}{n}\left(\frac{1+(2k-1)f_0}{2k}\right) + \left(\frac{n-1}{n}\right) r_0 \tag{13.17b}$$

which, when combined with Equation 13.17a, gives the inbreeding in Syn$_t$ generation ($t > 1$) as

$$f_t = r_1\left[1 - \left(\frac{k-1}{2k-1}\right)^{t-1}\right] + f_1\left(\frac{k-1}{2k-1}\right)^{t-1} \tag{13.17c}$$

Hence, the equilibrium level of inbreeding approaches r_1, or

$$f_e = r_e = \frac{1}{n}\left(\frac{1+(2k-1)f_0}{2k}\right) + \left(\frac{n-1}{n}\right) r_0 \tag{13.17d}$$

Note that for a diploid ($k = 1$), f_e is reached after one generation of random mating (i.e., in the Syn$_2$, $f_2 = f_e = r_1$). For a tetraploid, the equilibrium level of inbreeding is approached asymptotically (but is essentially at f_e after four to five generations). Example 13.9 examines the dynamics of f in polyploids in more detail.

In the most common setting, the founding parents are unrelated ($r_0 = 0$), and the equilibrium inbreeding reduces to

$$f_e = \begin{cases} [1+(2k-1)f_0]/(2kn) & \\ 1/(2kn) & \text{for } f_0 = 0 \\ 1/n & \text{for } f_0 = 1 \end{cases} \tag{13.17e}$$

Hence, if the n parents (or lines) are unrelated and completely inbred ($f_0 = 1$), the equilibrium inbreeding is simply one over the number of parents (independent of the ploidy, k), while if the parents are not inbred ($f_0 = 0$) and unrelated ($r_0 = 0$), the equilibrium level of inbreeding, $1/(2kn)$, is a function of both n and ploidy level.

Busbice (1969, 1970) also considered a more general case, where individuals can also self (with constant probability η). This is a reasonable model for many crop species. Busbice found that $r_e \simeq r_1$ is independent of η (this follows because all parents contribute equally to the next generation), and hence is given by Equation 13.17b. The equilibrium inbreeding becomes

$$f_e = r_1 + \frac{\eta(1-r_1)}{\eta + 2k(1-\eta)} \tag{13.17g}$$

This reduces to r_1 when $\eta = 0$ (no selfing) and to 1 when $\eta = 1$ (complete selfing). The expected inbreeding in the Syn$_1$ generation under partial selfing becomes

$$f_1 = \frac{k(1-\eta)r_0 + (\eta/2)(1+(2k-1)f_0) + (k-1)f_0}{2k-1} \tag{13.17h}$$

For a diploid with unrelated parents ($r_0 = 0$), this reduces to $f_1 = (\eta/2)(1+f_0)$. Finally, the expected inbreeding in generation $t > 1$ is

$$f_t = f_e\left[1 - \left(\frac{\eta}{2} + \frac{k-1}{2k-1}\right)^{t-1}\right] + f_1\left(\frac{\eta}{2} + \frac{k-1}{2k-1}\right)^{t-1} \tag{13.17i}$$

The effect of selfing ($\eta > 0$) is to slow the rate at which f_t approaches f_e.

Table 13.4 The equilibrium reduction, $(f_e - f_1)/(f_0 - f_1)$, of the initial heterotic advantage, $H_{F_1(syn)}$, under different assumptions about ploidy $(2k)$, parental inbreeding (f_0), and levels of selfing (η).

Parents completely inbred ($f_0 = 1$), not related ($r_0 = 0$)

$$\frac{f_e - f_1}{f_0 - f_1} = \begin{cases} \dfrac{2 - n\eta}{n(2-\eta)} = \dfrac{1}{n} \quad \text{(for no selfing, } \eta = 0) & \text{Diploids} \\ \dfrac{6 - n(2+3\eta)}{n(4-3\eta)} = \dfrac{3-n}{2n} \quad (\text{for } \eta = 0) & \text{Tetraploids} \\ \dfrac{10 - n(4+5\eta)}{n(6-5\eta)} = \dfrac{5-2n}{3n} \quad (\text{for } \eta = 0) & \text{Hexaploids} \end{cases}$$

No selfing ($\eta = 0$), parents unrelated ($r_0 = 0$), but partly inbred $f_0 > 0$

$$\frac{f_e - f_1}{f_0 - f_1} = \begin{cases} \dfrac{1 + f_0}{2nf_0} & \text{Diploids} \\ \dfrac{3 + f_0(3-2n)}{4nf_0} & \text{Tetraploids} \\ \dfrac{5 + f_0(25-12n)}{18nf_0} & \text{Hexaploids} \end{cases}$$

Busbice (1970) used the above inbreeding expressions to obtain a very general extension of Wright's predictor (Equation 13.15a), allowing for the effect of ploidy ($k \geq 1$), partial inbreeding in the parents ($0 < f_o < 1$), selfing among the Syn progeny ($\eta > 0$), and relatedness ($r_o > 0$) among the founding parents. Busbice started with the assumption that the mean μ_{Syn_t} in the Syn_t generation is a linear function of inbreeding, with

$$\mu_{Syn_t} = A + (1 - f_t)B \tag{13.18a}$$

where f_t is the inbreeding in Syn_t. Busbice (1970) and Sahagún-Castellanos and Villanueva-Verduzco (2007) defined A as the expected mean of genotypes formed by two IBD alleles (with the shared IBD allele randomly drawn from the Syn_0), while H is the mean value of the remaining genotypes, with $B = H - A$. Because Busbice's expression is a simple linear system of equations, if we know the mean values for two generations (and the corresponding levels of inbreeding), we can solve for A and B. Using the means $\overline{P}$ and $\overline{F}_{1(Syn)}$ for the Syn_0 and Syn_1 generations,

$$\overline{P} = A + (1 - f_0)B, \quad \text{and} \quad \overline{F}_{1(Syn)} = A + (1 - f_1)B \tag{13.18b}$$

yielding

$$A = \overline{P} + \frac{f_0 - 1}{f_0 - f_1}\left(\overline{F}_{1(Syn)} - \overline{P}\right), \quad B = \frac{\overline{F}_{1(Syn)} - \overline{P}}{f_0 - f_1}, \quad \text{for } f_0 \neq f_1 \tag{13.18c}$$

Substituting Equation 13.18c into Equation 13.18a yields Busbice's general expression for the expected mean in Syn generation t as

$$\mu_{Syn_t} = \overline{F}_{1(Syn)} - \frac{f_t - f_1}{f_0 - f_1}\left(\overline{F}_{1(Syn)} - \overline{P}\right) \tag{13.18d}$$

Recalling Equation 18.14, we can also express this as

$$\mu_{Syn_t} = \overline{P} + \left(1 - \frac{f_t - f_1}{f_0 - f_1}\right) H_{F_1(syn)} \tag{13.18e}$$

The effects of the number of founders (n), their average amount of inbreeding (f_0), how they are related (r_0), the ploidy level (k), and the partial selfing rate (η) on the mean of the synthetic all enter through the ratio of inbreeding coefficients. In particular, the equilibrium mean of the synthetic population becomes

$$\mu_{Syn_\infty} = \overline{P} + \left(1 - \frac{f_e - f_1}{f_0 - f_1}\right) H_{F_1(syn)} \tag{13.18f}$$

Table 13.4 gives the values of the inbreeding ratios under several common situations, which gives the amount of decay in the initial F_1 heterosis. Finally, note that Equations 13.18c–13.18f are not defined when $f_0 = f_1$. From Equation 13.17h, this only occurs when $f_0 = r_0 = s = 0$, namely that the founding parents are not inbred, not related, and do not self. In this case, $f_e = 1/(2nk)$, and

$$\mu_{Syn_t} = \overline{P} + \left(1 - \frac{1}{2nk}\right) H_{F_1(syn)} \quad \text{for } t \geq 2 \tag{13.18g}$$

While neither the Wright nor Busbice predictors are valid when epistasis is present, Gallais (1975a) provides an expression that allows for additive $\times$ additive epistasis. Wricke and Weber (1986) review the theory of synthetics, while additional treatments are given by Gallais (1976), Gallais and Wright (1980), and Wright (1981).

Example 13.8. The n parents chosen to form a synthetic may themselves be related. A common situation is where n_f unrelated families are chosen, each of which contributes n_o offspring, with $n = n_f\, n_o$. For diploids, if we let the parents of the families be potentially inbred with coefficient f_0, then with probability $1/n_f$ two randomly chosen individuals are from the same family, so that $r_o = \Theta/n_f$, where Θ is the coefficient of coancestry (Chapter 7) for the family in question. In particular,

$$\Theta = \begin{cases} (1+f_0)/8 & \text{half-sibs} \\ (1+f_0)/4 & \text{full-sibs} \\ (1+f_0)/2 & S_1 \text{ sibs} \\ 1 - 2^{-k}(1-f_0) & \text{sibs from an } S_k \text{ line} \\ 1 & \text{fully inbred lines} \end{cases} \tag{13.19}$$

Substitution of the appropriate r_0 value into Equations 13.17a–13.17d allows us to compute the correct inbreeding coefficient and estimate the mean of the synthetic line.

Suppose one has n parents and considers two different ways to form a synthetic. The Syn_0 could consist of the n parents, and we denote the equilibrium level of inbreeding by f_e. One potential drawback with this particular Syn_0 is that the total amount of seed in the Syn_1 may be small. If the species can be selfed, one could instead form the Syn_0 by taking the collection of selfed progeny (S_1 families) from these parents. Their resulting Syn_0 population should be roughly the size of the Syn_1 population using just the parents. Denoting the equilibrium level of inbreeding using S_1 families as f_e', how comparable are equilibrium levels of inbreeding under these two different synthetics? We show here that the levels of inbreeding are identical ($f_e = f_e'$) for the special case of a diploid with no selfing and the initial n parents are unrelated ($k = 1, \eta = 0, r_0 = 0$). Busbice (1970) showed this more generally for arbitrary ploidy, selfing, and allowing the initial parents to potentially be related ($r_0 \neq 0$).

For the synthetic where the Syn_0 consists of the n (unrelated) parents, Equation 13.17d gives

$$f_e = r_1 = \frac{1+f_0}{2n}$$

where f_0 is the (average) level of inbreeding among the founding parents. Now consider the situation where the Syn_0 consists of the S_1 progeny of these n parents. Let f_0' and r_0' denote the initial inbreeding and the degree of relatedness between the Syn_0 individuals. Suppose each selfing produces m offspring, implying that the Syn_0 has $n' = n\,m$ individuals. With probability $1/n$ two random Syn_0 individuals come from the same S_1 family and are related, otherwise they come from different families (and are unrelated). The coefficient of coancestry for sibs from an S_1 family is $(1 + f_p)/2$. Since the average inbreeding of the S_1 parents is $f_p = f_0$, the relationship among two random individuals in the Syn_0 formed from S_1 families becomes

$$r_0' = \frac{1 + f_0}{2n}$$

Note from Equation 13.17d that if n' is large (as would be the case using S_1 families to form the Syn_0), then $r_1' \simeq r_0'$, and hence $f_e' = r_1' \simeq r_0'$. Noting that $r_1 = r_0'$ shows $f_e' = f_e$. Hence, the equilibrium level of inbreeding is the same whether the Syn_0 consists of n parents or the n selfed (S_1) families that descend from them.

Example 13.9. Diploidy ($k = 1$) simplifies the dynamics of inbreeding in a synthetic population relative to its behavior in polyploids ($k > 1$). Provided that selfing does not occur, the equilibrium level of inbreeding for diploids is reached in the Syn_2 population. If epistatic contributions are minimal, the amount of heterosis declines from its maximal value in the Syn_1 to its equilibrium value in the Syn_2. Conversely, when dealing with autopolyploids—autoteraploids such as potatoes and the forage crops alfalfa (*Medicago sativa*) and birdsfoot trefoil (*Lotus corniculatus*)—synthetic populations take many generations to reach their equilibrium inbreeding value. As a result, the mean of the population can *increase* from its value in the Syn_1 generation. This potential increase in heterosis follows by contrasting the inbreeding coefficient in the Syn_1 generation with its equilibrium value. From Equation 13.17h (setting $\eta = 0$) the inbreeding coefficient in the Syn_1 generation is

$$f_1 = \frac{kr_0 + (k-1)f_0}{2k-1} \tag{13.20}$$

which, as noted by Dudley (1964), Dessureaux and Gallais (1969), and Busbice (1969) can be greater than the equilibrium level of inbreeding (Equation 13.17d),

$$f_e = \frac{1}{n}\left(\frac{1 + (2k-1)f_0}{2k}\right) + \left(\frac{n-1}{n}\right) r_0$$

For a tetraploid ($k = 2$) with unrelated, fully inbred parents ($f_0 = 1$, $r_0 = 0$), $f_1 = 1/3$ and $f_e = 1/n$. For $n = 2$, then $f_1 < f_e$ and the equlibrium mean is expected to decline from its Syn_1 value, while for $n \geq 4$, $f_1 > f_e$ and the mean is expected to *increase* each generation as the equilibrium level of inbreeding (which is less than the Syn_1 level) is approached. More generally (as noted by Busbice 1969), when $f_1 > f_e$, f_t approaches f_e from above (i.e., inbreeding decreases down to its equilibrium value), and the amount of initial heterosis retained *increases* over time (until it reaches its equilibrium value). Conversely, f_t approaches f_e from below when $f_1 < f_e$, so that the amount of retained heterosis decreases over several generations.

There is another complication with autopolyploids, namely that the diploid (two-allele) inbreeding coefficient itself can be insufficient to fully describe inbreeding (Busbice and Wilsie 1966; Gallais 1967). This occurs because a diploid locus has only two possible genic states (homozygotes and heterozygotes), while an autopolyploid locus can have up to five different genic states:

State	Representive Genotype	Num. of Alleles
Monogenic	$A_iA_iA_iA_i$	1
Digenic simplex	$A_iA_iA_iA_j$	2
Digenic duplex	$A_iA_iA_jA_j$	2
Trigenic	$A_iA_iA_jA_k$	3
Tetragenic	$A_iA_jA_kA_\ell$	4

Recall that under the maximum heterozygosity hypothesis explanation for progressive heterosis, the greatest effect on heterosis occurs in tetragenic individuals. While autopolyploids have a slower approach to complete homozygosity than do diploids (Haldane 1930b;

Bartlett and Haldane 1934; Wright 1938; Parsons 1959; Demarly 1963; Busbice 1969), they occasionally show the same levels of inbreeding depression for a set number of generations of selfing as do diploids (Dewey 1966; Yao et al. 2020), which might be a reflection of the maximum heterozygosity hypothesis. However, in other studies polyploids show less ID than diploids (as expected from their rate of inbreeding). As summarized in a meta-analysis of Clo and Kolář (2022), there is "no consensus of the consequences of polyploidy on inbreeding depression," and the amount of ID displayed by a polyploid can depend on the time since the polyploidization event.

Dunbier and Bingham (1975) noted that maximum heterozygosity hypothesis has important implications for breeding designs. Suppose one starts with four unrelated monogenic clones (A, B, C, D). The expected frequencies of the four genic classes is different in an equilibrium synthetic versus in a double cross AB × CD, with

State	Freq in Syn	Freq in DC
Monogenic	0.2%	0.0%
Digenic	9.6%	1.2%
Trigenic	49.2%	19.8%
Tetragenic	41.0%	79.0%

If tetragenic interactions are important contributors to heterosis, we would expect a reduction in the mean of the synthetic relative to that for the double cross, which was observed for yield in alfalfa (Dunbier and Bingham 1975).

Retention of Heterosis: Apomixis and Allopolyploids

While synthetics offer some partial relief from the loss of heterosis, the holy grail for the preservation of heterosis across generations is **apomixis**, seed produced by asexual reproduction, effectively immortalizing a hybrid genotype. While uncommon, this breeding system is widespread (Bashaw 1980; Hanna et al. 1999; Abdi et al. 2016; Sailer et al. 2016), although attempts to introduce it into crops by conventional breeding have so far been unsuccessful. Fortunately the cloning of genes involved in some apomixis systems (Khanday et al. 2019) offers the possibility of generating facultative apomixis through transgenics. Indeed, Vernet al. al (2022) were recently able to generate synthetic apomixis with hybrid rice, ushering in an exciting new phase of exploiting of heterosis in crop systems.

As noted by Chen (2010) and Ng et al. (2012), nature offers another approach for immortalizing hybrids, namely **allopolyploids**—hybrids between closely related species—which have been called **doubled interspecific hybrids**. Consider the interspecific cross between species 1 (diploid genome AA) and species 2 (genome BB), with the resulting F_1 hybrid having a genome of AB. As detailed by Chen (2010), many such crosses display exceptional heterosis. However, in their diploid form they are often sterile, as the A and B homoeologous chromosomes are sufficiently divergent that they fail to pair during meiosis, resulting in aneuploid gametes. However, if the genome is doubled, yielding an $AABB$ allotetraploid, pairs of homologous chromosomes can pair during meiosis (A with A, B with B), recovering fertility. Further, their heterosis is stable, as all offspring contain the A and B genes from the original cross (barring eventual divergence via mutation).

One final point about cross-species hybrids deserves comment. While plant biologists have long appreciated the role of hybridization in speciation events (Grant 1981; Goulet et al. 2017), there is a growing appreciation that gene transfer also plays important roles in animal speciation and adaptation (Mallet 2005; Abbott et al. 2013; Payseur and Rieseberg 2016).

AGRICULTURAL EXPLOITATION OF HETEROSIS: ANIMALS

Crossbreeding in Animals: General Concepts

Table 13.5 Estimates of individual, H_I, and maternal heterosis, H_M, in sheep. Results presented as percentage of midparental means, so that the F_1 value for birth weight is 108.1% of the midparents. n_I and n_M indicate the number of estimates used for the reported individual and maternal values. **Prolificacy** is the litter size at birth. (After Nitter 1978.)

Trait	n_I	Mean H_I	n_M	Mean H_M	Total
Birth weight	42	3.2	12	5.1	8.1
Weaning weight	56	5.0	27	6.3	11.3
Fertility	20	2.6	30	8.7	9.3
Prolificacy	20	2.8	31	3.2	6.0
Birth-weaning survival	29	9.8	25	2.7	12.5
Lambs per ewe	20	5.3	25	11.5	16.8
Lambs reared per ewe	20	15.2	25	14.7	29.9
Total weight lambs/ewe	24	17.8	25	18.0	35.8

As the mule illustrates, the importance of between-species hybrids in animal breeding goes back to prehistoric times. The aggressive utilization of crossbreeding to exploit heterosis (as opposed to crosses simply to combine desirable features from different lines) followed Wright's (1922b) extremely influential publication on crossbreeding (and inbreeding) in guinea pigs. General reviews of heterosis in animals are given by Gowen (1952), Sang (1956), and Sheridan (1981).

Animal breeders importantly distinguish between **individual** (or **direct**) and **maternal heterosis**. Individual heterosis is enhanced performance in a *hybrid individual*, while maternal heterosis is enhanced **maternal performance** (such as milk production or higher survival rates of offspring) due to having a *hybrid mother*. Maternal heterosis is often comparable, and can be greater than, individual heterosis (Table 13.5). Maternal and individual heterotic effects can be combined by using **crossbreed dams**. Consider total weight of lambs reared per mated ewe (Table 13.5), which has an 18% individual heterotic advantage in a crossbreed offspring (a single-cross, $A \times B$) and an *additional* 18% advantage (from maternal heterosis) when crossbreed ewes are used in place of purebreed ewes (a three-way cross, $A \times [\, B \times C \,]$). This observation extends beyond sheep. For example, Cundiff et al. (1974a, 1974b) found that maternal heterosis was roughly twice the individual heterosis for several traits in European cattle (*Bos taurus*). This combining of maternal and individual heterotic effects is one reason why three-way crosses are common in animal breeding, generally by crossing a male from line A with a hybrid female (from a $B \times C$ cross). This strategy exploits maternal heterosis in the female, with the sire line usually chosen for its contribution to one or more production traits.

In theory, one could also exploit **paternal heterosis**, increased performance due to paternal effects, but there has been little evidence to date of this effect generally being significant. For example, Bradford et al. (1963) compared purebreed versus crossbreed sire performance, finding no major differences between them except for a slight (4%) elevation of lamb survival, where the hybrid exceeded both parents. If paternal heterosis is important, then a four-way cross—a sire from an $A \times B$ mating crossed to a dam from a $C \times D$ mating—would provide the maximal benefit from all heterotic sources.

A second point that has received substantial attention in the animal-breeding literature is heterosis × environment interaction (Orozco 1976; Barlow 1981; Sheridan 1981; Cunningham 1982). Although the general opinion is that heterosis is more pronounced in suboptimal environments, there are many exceptions to this pattern, and few of the data bearing on the subject come from well-designed experiments.

Case Study: Heterosis in *Bos indicus* × *Bos taurus* Hybrids

Some interesting results on crossbreeding are found in the literature examining crosses between European (*Bos taurus*) and tropical (*Bos indicus*) cattle (reviewed in McDowell 1985). *Bos taurus* breeds have been selected for significant genetic improvement in a number of production traits (such as milk and meat yield), but are adapted to temperate climates. *Bos indicus* breeds are found in tropical countries, and while they show fairly little genetic improvement for production traits (relative to *taurus*), they are well adapted to their local environment. Crosses between *taurus* and *indicus* have been made in the hope of generating higher performing hybrids that are also adapted to tropical environments.

Bos indicus × *B. taurus* hybrids are usually superior to indigenous breeds (i.e., the local breed of *B. indicus*) in milk yield and fitness measures such as calving age and interval. Trail et al. (1985) examined crosses between exotic *B. taurus* breeds (Angus and Red Poll) and indigenous *B. indicus* breeds (Ankole, Boran, and Zebu) in Africa. Crossbred (exotic × indigenous) dams showed superior maternal performance over straightbred indigenous dams. However, while maternal effects were apparently superior, the (single-cross) hybrid progeny of exotic × straightbred (Boran) dams were actually heavier at 24 months than the hybrid progeny of exotic × crossbred dams, suggesting that the 3/4 exotic composition from using crossbreed dams is not as favorable for individual performance as a composition of 1/2 exotic and 1/2 indigenous. Three-breed crosses between an improved-breed sire and a crossbred dam (the offspring of a second improved breed × an indigenous breed) generally tend to do poorer than two-breed crosses. This is contrary to the general superiority of three-breed crosses of *B. taurus* observed in temperate enviroments (McDowell 1985).

Bos indicus × *B. taurus* hybrids show higher levels of heterosis than observed in crosses among breeds within each species. This is not unexpected, as allele frequencies likely have diverged more between species than among the breeds within a species. Heterosis in crosses among *B. indicus* breeds is higher than is observed in crosses among *B. taurus* breeds. For example, Gregory et al. (1985) found that maternal heterosis in crosses within *B. indicus* lines was intermediate between levels produced in *B. indicus* × *B. taurus* hybrids and crosses among *B. taurus* breeds. One explanation is that there is more allele-frequency divergence among the various indigenous *B. indicus* breeds than among the *B. taurus* breeds, due to more local adaptation among *indicus* breeds.

While *indicus* × *taurus* crosses do indeed show significant heterosis, they can also have unanticipated economic disadvantages. Regarding the hybrids, McDowell (1985) noted that

> some farmers have reservations that could influence national breeding programs. The two-breed crossbreed male is not as temperamental as the favored draft breed, e.g., Hariana in India, and does not move as rapidly for plowing or in performing cartage. The smaller hump of the crossbreed is not well suited to handle the traditional wooden yoke. For those reasons the crossed male has a price discrimination against it in the draft market.

Further, crossbreeds usually require supplemental feeding in tropical setting; otherwise they can become nutritionally stressed (McDonnell 1981, 1985).

Hybridization in Other Domesticated Animals

Most production systems for domesticated animals use at least some crossbreeding (Sheridan 1981; Yadav et al. 2018). This includes chicken (Muir and Aggrey 2003), swine (Bidanel 1993; Bittante et al. 1993), cattle (Gregory et al. 1985; Syrstad 1985; Simm 1998), dairy (Swan and Kinghorn 1992; Buckley et al. 2014), goat (Shrestha and Fahmy 2007), and sheep (Jakubec 1977; Simm 1998; Malik and Singh 2006). Aquaculturists have also explored crossbreeding, finding extensive yield heterosis for oysters (Hedgecock and Davis 2007), but little exploitable heterosis for Atlantic salmon (Gjerde and Refstie 1984). Beyond heterosis, one important consideration in choosing which populations to cross is the notion of **breed compatibility**, combining the favorable features from two different lines. Recalling our discussion of general combining ability (GCA), this is the idea of crossing populations which have high GCAs for different, but desirable, traits, to produce a line whose means for these traits are (at least) equal to their midparental values. In a sense, one can think of this as **economic heterosis**, as the combination of these features in a single line is often far more desirable than the economic average value of the two parental lines separately.

Reproductive differences between plants and animals present different opportunities for exploiting heterosis in a planned breeding system. One of the major limitations of cross-

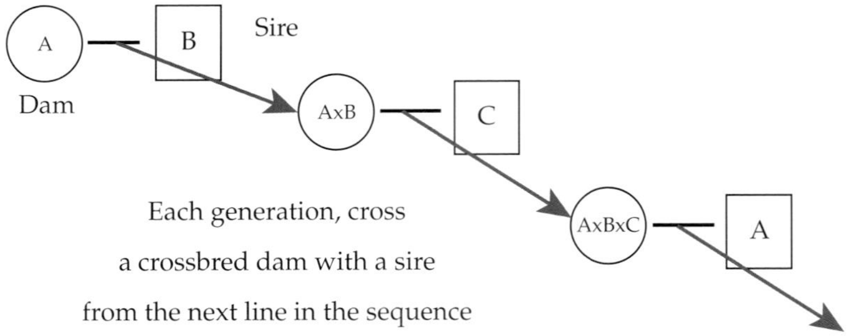

Figure 13.8 A **three-breed rotational** crossing scheme. In each generation, a crossbred dam (circles) is mated (in succession) to a rotation of three different sire (squares) lines. A fraction of the female offspring are used in the next cycle to continue the rotation, while the remainder (all males and some females) are sold commercially. One modification is a **terminal cross rotational** scheme: females that are not advanced within the breeding program are crossed to a sire from a different line, and the resulting offspring are the commercial product. More formally, the figure depicts **sire rotation**, where the sire breed being crossed to the continuing crossbred line of dams is rotated each cycle. A related scheme is **dam rotation** (Dearborn 1969; Dickerson 1972): the same sire breed is used each generation and is crossed to dams generated by rotation from a different set of lines, ideally those showing superior maternal traits.

breeding in plants is pollen control, but sperm control is very easy in animals. Conversely, the generally very low reproductive output of large domesticated animals places severe limitations on the types of commercial crossing schemes that can be used. Chickens have little reproductive limitations, while swine, often with more than a dozen offspring, are more intermediate. Sheep may have two or three offspring per mating, while cattle rarely have more than a single offspring per mating, limiting commercially viable crossing schemes.

Rotational Crossbreeding

In much of plant breeding, **terminal crosses** are used, wherein hybrid individuals are the endpoints and do not reproduce further in the commercial production system. For example, a company sells a farmer F_1 seed, and this seed is generated anew each generation, with the F_1 plants themselves not allowed to reproduce. While such schemes can work in plants with their enormous reproductive potential, they are more difficult in animals. Consider a three-way cross of an A sire to a $B \times C$ dam (with C being the granddam). While this scheme requires only modest numbers of sires from the A and B lines, in large farm animals (i.e., sheep and cows), the number of offspring is on order of the number of dams. Thus, to produce n triple-cross hybrid offspring requires on the order of $2n$ dams (the granddam to produce the $B \times C$ dam, and this dam herself). As an aside, the importance of crossbred dams is such that a potential use of whole-animal cloning would be in the creation of cloned lines from crossbred mothers showing exceptional maternal heterosis. This is the animal equivalence of the plant breeder's search for apomixis for immortalizing a hybrid advantage in the face of Mendelian segregation.

To overcome limitations from low number of offspring per mating, Winters (1952; Winters et al. 1935) suggested a hybridization scheme, **rotational crossbreeding**, that continually recycles hybrid individuals (Figure 13.8). Here, hybrids from the previous generations are crossed (in rotation) to pure lines. For example, a three-breed rotational would use $A \times B$ as the first generation. In generation two, *dams* from the first generation are crossed to line C. In generation three, *dams* from generation two are crossed to sires from line A, and the rotation continues over all three lines in subsequent generations. This approach represents a compromise between trying to maintain maximal heterozygosity within a line (and hence optimize the dominance contribution to heterosis) without having to regenerate the line anew each generation. Further, by restricting the crosses to a pure-line sire and a crossbreed dam, it also fully exploits any maternal heterosis.

The central idea in a rotational scheme is that some of the crossbred dams are used in the next cycle. Male offspring and female offspring not required to keep the rotational population at a desired size can be sold off commercially. There are a number of modifications of this scheme, especially in cattle (Gregory and Cundiff 1980; Gregory et al. 1982; Gosey 1991; Ritchie et al. 1999). One is that female offspring not advanced in the rotation could themselves be crossed to a different sire line that is outside of the rotation (a terminal cross) and the resulting offspring sold. Another variant is that one could use the same sire line, but generate crossbred dams in a cycle among other lines (a rotational cross using dams).

Example 13.10. Consider a three-breed rotational crossbreeding scheme where, as in Figure 13.8, dams from the previous generation are crossed to pure-bred sires in a rotating sequence (line A in one generation, B in the next, C in the third, and then back to A to repeat the cycle). Under this scheme, what fraction of genes from each of the lines are present in any particular generation? The logic is straightforward: in each generation, half of the contribution from the previous dam is passed on, and half the genes are from the sire line. Thus,

		Percentage of Lines		
Generation	Cross	A	B	C
1	$A \times B$	50.0	50.0	0.0
2	$C\times$ gen 1 dam	25.0	25.0	50.0
3	$A\times$ gen 2 dam	62.5	12.5	25.0
4	$B\times$ gen 3 dam	31.3	56.3	12.5
5	$C\times$ gen 4 dam	15.6	28.1	56.3
6	$A\times$ gen 5 dam	57.8	14.1	28.1
7	$B\times$ gen 6 dam	28.9	57.0	14.1
8	$C\times$ gen 7 dam	14.5	28.5	57.0

The asymptotic contributions reached are 57.1% from the current sire line, 28.6% from the sire line used in the previous generation, and 14.3% from the sire line used two generations ago (and the sire line for the next generation).

In a similar fashion, for a four breed (A, B, C, D) rotational scheme,

		Percentage of lines			
Generation	Cross	A	B	C	D
1	$A \times B$	50.0	50.0	0.0	0.0
2	$C\times$ gen 1 dam	25.0	25.0	50.0	0.0
3	$D\times$ gen 2 dam	12.5	12.5	25.0	50.0
4	$A\times$ gen 3 dam	56.3	6.3	12.5	25.0
5	$B\times$ gen 4 dam	28.1	53.1	6.3	12.5
6	$C\times$ gen 5 dam	14.1	26.6	53.1	6.3
7	$D\times$ gen 6 dam	7.0	13.3	26.6	53.1
8	$A\times$ gen 7 dam	53.5	6.6	13.3	26.6

At equilibrium, the line contributions cycle as 53.3%, 26.7%, 13.3%, and 6.7%. As a consequence of the predominant genetic line changing each generation, one consideration of rotational systems is that the cycled breeds be sufficiently compatible such that these compositional changes over cycles do not have commercial liabilities.

One can extend the logic from Example 13.10 to a k-line rotational scheme. The expected fraction of genetic contributions at equilibrium from each of the k lines where crossbred dams are kept and crossed in rotation to purebreed sires are of the form

$$(1/2)\cdot I, \quad (1/2)^2\cdot I, \quad (1/2)^3\cdot I, \quad \cdots, \quad (1/2)^k\cdot I$$

where

$$I = \sum_{i=0}^{\infty}\left(\frac{1}{2^k}\right)^i = 1 + \frac{1}{2^k - 1} \tag{13.21}$$

with the largest fraction ($I/2$) from the line last used as a sire and the smallest fraction ($I/2^k$) for the line to be next used as a sire. The resulting values of I, and the current contributions from the most recent (current) sire (CS) and the next upcoming sire (US), for up to an 8 line crossing scheme are as follows:

k	2	3	4	5	6	7	8
I	1.333	1.143	1.067	1.032	1.016	1.008	1.004
CS	0.667	0.571	0.533	0.516	0.508	0.504	0.502
US	0.333	0.143	0.067	0.032	0.016	0.008	0.004

What are the expected asymptotic average values for a rotational cross and how can one predict this? Expressions allowing for maternal heterosis and recombination loss (i.e., epistasis) are developed below (Table 13.6). Under the assumption of only dominance and no maternal heterosis, some simple predictors were suggested by Carmon et al. (1956). For a two-line (A, B) rotation, the predicted equilibrium mean is

$$\widehat{Rc_2} = \overline{z}_{AB} - \frac{\overline{z}_{AB} - \overline{P}_2}{3}, \quad \text{where} \quad \overline{P}_2 = \frac{\overline{z}_A + \overline{z}_B}{2} \tag{13.22a}$$

Here $\overline{P}_2$ is the average of the two parental lines and $\overline{z}_{AB}$ the mean value of their cross. When heterosis is present ($\overline{z}_{AB} > \overline{P}_2$), the expected mean performance under rotational crossbreeding is less than the single-cross performance ($\overline{z}_{AB}$). Note, however, that the dominance loss in a two-breed rotation is $H/3$, less than the loss of $H/2$ for a two-breed synthetic (Equation 13.15).

For a three-line (ABC) rotational cross, the predicted mean is

$$\widehat{Rc_3} = \overline{SC}_3 - \frac{\overline{SC}_3 - \overline{P}_3}{7}, \quad \text{where} \quad \overline{SC}_3 = \frac{\overline{z}_{AB} + \overline{z}_{AC} + \overline{z}_{BC}}{3} \tag{13.22b}$$

where $\overline{SC}_3$ is the average of the three single crosses between these three lines and $\overline{P}_3$ the average of the three parental lines. The dominance loss is $H/7$ as opposed to $H/3$ under a three-breed synthetic.

For a four-line rotation, the *order of the rotation* matters. Letting the rotation be (A, B, C, D), the predicted long-term performance is

$$\widehat{Rc}_4^{(A,B,C,D)} = \overline{SC}_4 - \frac{\overline{SC}_{na} - \overline{P}_4}{15} \tag{13.22c}$$

where

$$\overline{SC}_4 = \frac{\overline{z}_{AB} + \overline{z}_{AC} + \overline{z}_{AD} + \overline{z}_{BC} + \overline{z}_{BD} + \overline{z}_{CD}}{6} \quad \text{and} \quad \overline{SC}_{na} = \frac{\overline{z}_{AC} + \overline{z}_{BD}}{2}$$

As above, $\overline{P}_4$ is the mean of the original lines, while $\overline{SC}_4$ is the mean of all six possible single crosses between the four lines, and $\overline{SC}_{na}$ is the average of the two single crosses of **nonadjacent lines** in the rotation (and hence varies with the order of rotation used).

Example 13.11. Consider the following data from Kidder et al. (1964) for various crosses of Devon and Brahman cattle. The midparent $\overline{P}$, F_1, two-breed rotational crossbreed Rc_2, the two-breed synthetic S_2 (= F_2), and the backcross (BC, $F_1 \times$ parent) means for several weight-related traits were as follows:

	Means				
Trait	$\overline{P}$	F_1	Rc_2	S_2	BC
Weaning weight	154.2	180.5	178.3	170.1	181.4
12-month weight	210.5	246.8	232.2	212.3	233.6
18-month weight	274.9	315.7	296.6	276.6	295.3
12-18 month weight gain	64.4	68.9	64.4	64.4	61.7

Note that the order of performance for all traits is $F_1 > Rc_2 > S_2$.

How well does Equation 13.22a predict the two-breed rotational crossbreed performance, Rc_2? Here, the F_1 corresponds to $\overline{z}_{AB}$, so that the predicted equilibrium value is

$$\widehat{Rc}_2 = F_1 - \frac{F_1 - \overline{P}_2}{3}$$

For example, for weaning weight

$$\widehat{Rc}_2 = 180.5 - \frac{180.5 - 154.2}{3} = 171.7$$

which is 96% of the observed value (178.3). Similarly, the predicted values (and fraction of the actual values) for 12-month, 18-month, and gain are, respectively, 234.7 (101%), 302.1 (102%), and 67.5 (104%). Hence, for these data, there was a slight tendency to overestimate the true mean, which is to be expected if recombination loss has occurred.

Standard errors for these predicted values can be approximated as follows. We consider the two-breed predictor, but the basic logic easily extends to most of the other predictors developed in this chapter. Rearranging to collect common terms,

$$\widehat{Rc}_2 = \overline{z}_{AB} - \frac{\overline{z}_{AB} - \overline{P}_2}{3} = \left(1 - \frac{1}{3}\right)\overline{z}_{AB} + \left(\frac{1}{6}\right)(\overline{z}_A + \overline{z}_B)$$

Because the estimates of the means are independent and recalling that $\sigma^2(ax) = a^2\sigma^2(x)$ for a constant a (Equation 3.10c), it immediately follows that

$$\sigma^2\left(\widehat{Rc}_2\right) = \left(1 - \frac{1}{3}\right)^2 \sigma^2(\overline{z}_{AB}) + \left(\frac{1}{6}\right)^2 \left(\sigma^2(\overline{z}_A) + \sigma^2(\overline{z}_B)\right)$$

Maternal Heterosis and Recombination Loss

Two complications that were ignored in the previous section are the presence of epistasis (and its associated recombination loss in subsequent generations) and maternal heterosis (which, itself, can also show recombination loss). We examine maternal effects in detail in Part II Chapter 29, but the basic idea is that the environment experienced by an individual can be influenced by its mother (such as milk production and other aspects of maternal care). Maternal effects are often the default assumption for animal traits (particularly for species with maternal care), but rarely assumed for plants. Maternal heterosis is simply the idea that the maternal performance provided by a hybrid mother is superior to that of a purebred mother. The result is that the trait mean of an offspring from a crossbred mother is increased by a maternal heterosis value, H_M. As with any other trait, the maternal heterotic contribution in the F_2 (and subsequent) generations shows a dominance loss, $H_M/2$ and (if certain types of epistasis are present) a further recombination loss (Equations 13.8 and 13.11).

Thus, the maternal heterotic effect from a crossbred mother is expected to decline in the next generation, with her daughters showing less maternal heterosis. The basic approach for incorporating maternal effects was given in a specific case (three-line crosses) by Magee and Hazel (1959), while a more general treatment was developed by Dickerson (1969a, 1972).

As an example of how maternal heterosis and recombination loss impact the means of crosses, consider the offspring from a three-way cross, $A \times (B \times C)$, whose expected mean is given by

$$\mu_{A\times(B\times C)} = \overline{P} + (H_I - R_I/4) + H_M \tag{13.23a}$$

where $\overline{P} = (2\mu_A + \mu_B + \mu_C)/4$ is the weighted parental line average. The various heterotic contributions are obtained as follows. The use of a crossbred mother ($B\times C$) adds a maternal heterotic contribution (H_M), while there are two factors impacting individual heterosis (the two terms in the parentheses). Offspring are fully heterozygous in the sense that all loci have one allele from line A and the other from either B or C, contributing H_I (the average of the individual heterosis from $A \times B$ and $A \times C$ crosses). Secondly, recombination loss appears for individual heterosis, as (for unlinked loci) half of the gametes from the mother are recombinant (of the form B_1C_2 or C_1B_2, as opposed to the parental gametes of B_1B_2 and C_1C_2). Hence, unlike a standard single-cross where all of the gametes that join to form the hybrid are parental in origin (both alleles in a gamete are from the same line), 1/4 of the gametes that made up the three-way hybrid are recombinant (as half of the gametes are from the $B \times C$ parent). Hence, $r_D = 1/4$, for a recombination loss in individual heterosis of $-R_I/4$.

Now suppose that the resulting three-way cross offspring are randomly mated, which we will denote as an F$_2$. Here, the *mother* now has composition $A \times (B \times C)$, and following the logic from above, her maternal heterotic contribution to her offspring is $H_M - R_M/4$. Turning to the composition of individual heterosis, we assume Hardy–Weinberg and linkage equilibrium. Half of the genes in the F$_2$ are from line A, and 1/4 each from lines B and C. Hence, the probability that a locus is homozygous in origin (both alleles from the same population) is $(1/2)^2 + 2(1/4)^2 = 3/8$, so that 5/8 of the time a locus has alleles from different populations, generating a contribution of $(5/8)H_I$. To obtain the recombination loss, we need to compute the Dickinson coefficient, r_D, the probability that the alleles in a *gamete* come from different populations. Under linkage equilibrium, the probability that they come from the same population is also $(1/2)^2 + 2(1/4)^2 = 3/8$, hence $r_D = 1-3/8 = 5/8$, generating a recombination loss in individual heterosis of $-(5/8)R_I$. The resulting F$_2$ mean becomes

$$\mu_{F_2} = \overline{P} + (5/8)(H_I - R_I) + (H_M - R_M/4) \tag{13.23b}$$

Finally, suppose the F$_2$ are randomly mated to form an F$_3$. Because the population is in Hardy–Weinberg and linkage equilibrium, the individual heterosis is the same as the F$_2$, while the maternal heterosis is generated from F$_2$ females, whose heterotic composition now has the form of that for the F$_2$ individual heterosis, yielding

$$\mu_{F_3} = \overline{P} + (5/8)(H_I - R_I) + (5/8)(H_M - R_M) \tag{13.23c}$$

This value is stable in all future generations (assuming random mating). Note that this is a synthetic population, but with unequal, rather than equal, contributions from the three parents. In general, if π_i denotes the contribution for line i (of k total), then the general mean for a synthetic population (at equilibrium) is

$$\mu_{syn} = \overline{P} + \Theta(H_I + H_M) - \Theta(R_I + R_M) \tag{13.24}$$

where

$$\overline{P} = \sum_{i=1}^{k} \mu_i \pi_i, \qquad \text{and} \qquad \Theta = 1 - \sum_{i=1}^{k} \pi_i^2$$

Table 13.6 The impact of heterosis under different crossing systems. The notation used for mating systems has the sire composition to the left of the $\times$ symbol, and the dam composition to the right, so that $(A\times B)\times C$ denotes a crossbred sire $(A\times B)$ mated to a purebred dam (C). For a given mating system, the table gives the coefficients for individual, maternal, and paternal heterosis (H_I, H_M, H_P) and for the associated recombination loss (R_I, R_M, R_P). For example, with a three-breed sire rotational, the average individual heterosis (averaged over cycles) is (6/7) $\overline{H}_I$, where $\overline{H}_I$ is the average individual heterosis for the three crosses. The table (after Dickerson 1972) corrects a minor typo in Dickerson's Table 1, which mistakenly lists the coefficient on H_i for a two-breed sire rotation as 1/2 instead of the correct value of 2/3.

	Heterosis			Recombination Loss		
Mating System	H_I	H_M	H_P	R_I	R_M	R_P
2-breed cross						
$A \times B$	1	0	0	0	0	0
3-breed cross						
$A \times (B \times C)$	1	1	0	1/4	0	0
$(A \times B) \times C$	1	0	1	1/4	0	0
4-breed cross						
$(A \times B) \times (C \times D)$	1	1	1	1/2	0	0
Rotational Crossing (Sire)						
2 sire breeds	2/3	2/3	0	2/9	2/9	0
3 sire breeds	6/7	6/7	0	6/21	6/21	0
4 sire breeds	14/15	14/15	0	14/45	14/45	0
k sire breeds	k_1	k_1	0	$k_1/3$	$k_1/3$	0
Rotational Crossing (Dam)						
2 dam breeds	1	2/3	0	2/9	2/9	0
3 dam breeds	1	6/7	0	6/21	6/21	0
4 dam breeds	1	14/15	0	14/45	14/45	0
k dam breeds	1	k_1	0	$k_1/3$	$k_1/3$	0
Synthetic population						
2 breeds	1/2	1/2	1/2	1/2	1/2	1/2
3 breeds	2/3	2/3	2/3	2/3	2/3	2/3
4 breeds	3/4	3/4	3/4	3/4	3/4	3/4
k breeds	k_2	k_2	k_2	k_2	k_2	k_2

$$\text{where} \quad k_1 = \frac{2^k - 2}{2^k - 1} \quad \text{and} \quad k_2 = \frac{k-1}{k} = 1 - \frac{1}{k}$$

This result (for individual heterosis, but no maternal heterosis or epistasis) is due to Wright (1922b), with the more general version due to Dickerson (1972). Part II Chapter 25 examines the challenging problem of obtaining separate estimates of individual and maternal heterosis, which is complicated in the presence of recombination loss.

Proceeding in this fashion, Dickerson (1969a, 1972) worked out the coefficients for the heterosis and recombination loss terms under a variety of crossing designs, and these are summarized in Table 13.6 (which, for generality, also allow for the potential of paternal heterosis). As an example, consider a three-breed rotation versus a three-breed synthetic (ignoring any paternal heterosis). Table 13.6 gives the heterotic contributions, yielding the expected mean of a three-breed rotational system (averaged over cycles at equilibrium) as

$$\mu_{Rc_3} = \overline{P}_3 + \frac{6}{7}\,(H_I + H_M) - \frac{6}{21}\,(R_I + R_H) \tag{13.25a}$$

while for a three-breed synthetic

$$\mu_{syn-3} = \overline{P}_3 + \frac{2}{3}\,(H_I + H_M) - \frac{2}{3}\,(R_I + R_H) \tag{13.25b}$$

Table 13.7 Different scenarios for when a particular crossbreeding method is optimal. (After Kinghorn 1999.)

Scenario	Recommendation
No cross is better	Purebreed
Individual heterosis is important	Single cross
Both individual and maternal heterosis are important	Three-breed cross
Individual, maternal, and paternal heterosis are important	Four-breed cross
Only two good paternal breeds are available and/or individual heterosis is not important	Backcross
Females are too expensive to buy/produce, or litter size is too small	Rotational cross
Logistical or financial constraints in applying rotational crossbreeding	Synthetic population

where $\overline{P}_3$ is the average of the three parental lines. The resulting expected differences between these two schemes is

$$\mu_{Rc_3} - \mu_{syn-3} = \frac{4}{21}\,(H_I + H_M) + \frac{8}{21}\,(R_I + R_H) \tag{13.25c}$$

Hence, rotational crossbreeding retains more heterosis, and has a smaller recombination loss, than a synthetic. Similarly, for a two-breed rotation versus a two-breed synthetic,

$$\mu_{Rc_2} - \mu_{syn-2} = \frac{1}{6}\,(H_I + H_M) + \frac{5}{18}\,(R_I + R_H) \tag{13.25d}$$

Table 13.6 shows the various tradeoffs for the different crossing systems. Notice, as was found for the two- and three-breed cases, that rotational breeding systems retain more heterosis (both individual and maternal) than do synthetics, and also suffer from less recombination loss. The latter occurs because one parent each cycle contributes half of the gametes, reducing the fraction of nonparental gametes that generates recombination loss in the presence of epistasis. The careful reader might have noted the difference in the coefficient on H_I in sire-based (k_1) versus dam-based (1) rotation schemes (Figure 13.8). In the latter breeding system, the same sire breed (A) is crossed to a rotation of crossbred dams among k lines *not* involving A (Dearborn 1969; Dickerson 1972). As a result, *all* of the heterozygous loci in offspring contain one allele from A and another allele from the crossbred dam lines (which does not contain A). Hence, their offspring retain all of the individual heterosis. Conversely, in sire rotation, the crossbred dams contain a fraction $(1 - k_1)$ of sire-line alleles, so that only k_1 of the loci contain two alleles from different lines.

Sheridan (1981) and Kinghorn (1999) reviewed the conditions that favor particular crossing systems (Table 13.7), balancing constraints in the production system (most notably, number of offspring per mating) with the optimal retention of heterosis and control of recombination loss. While there is significant variation in breeding systems, Kinghorn (1999) noted some general trends. Chickens and pigs with their relatively high reproductive rate tend to use three-way crosses and backcrosses, meat sheep tend to use three-way crosses, temperate beef producers tend to use either rotational systems or composite (synthetic) populations, while tropical beef producers tend to use composites. Wool sheep and dairy mainly use pure lines, largely because exceptional lines for these specific production goals exist in temperate countries.

OUTBREEDING DEPRESSION

When populations are sufficiently divergent, hybrids between them often have *reduced*, not *enhanced*, performance. Recalling Equation 13.7a, $\mu_{F_1} - \mu_{MP} = 2\delta_1^c - \alpha_2^c$, while (Equation

13.7d) for the F_2 (under unlinked loci) $\mu_{F_2} - \mu_{MP} = \delta_1^c - \alpha_2^c - \delta_2^c$. Hence, unless the average dominance deviation is negative ($\delta_1^c < 0$), such outbreeding depression is driven by epistasis, interactions between alleles (additive × additive, α_2^c) or genotypes (dominance × dominance, δ_2^c) at different loci. The idea is that local adaptation favors combinations of alleles that work well together in a particular environment (α_2^c and/or δ_2^c are positive), and that these combinations are randomized in hybrids, decreasing performance. Obviously, this topic is of great interest to evolutionary biologists, as outbreeding depression is often assumed to be a byproduct of adaptation. Further, it is a potential step towards speciation (a topic covered in more detail in Chapter 17). Indeed, the F_1 inviability or sterility in between-species crosses often arises because of between-population incompatibilities between autosomes and sex chromosomes or cytoplasmic background.

Outbreeding depression is also of critical importance in conservation genetics. Suppose that one has a small isolated population with greatly reduced genetic variation, perhaps to the point that it suffers from substantial inbreeding depression. One strategy for its genetic management is to introduce a few individuals from a different, but closely related, population, an idea known as **genetic rescue** (Edmands 2006; Whiteley et al. 2015; Hedrick and Garcia-Dorado 2016; Jørgensen et al. 2021; Pérez-Pereira et al. 2022). If outbreeding depression is important, such a strategy can cause more harm than good. Hence, a deeper understanding of outbreeding depression than we currently possess is of fundamental importance to conservation biologists. Further, outbreeding depression is likely to be population- and/or species-specific.

Frankham et al. (2011) attempted to provide some rough conservation biology guidelines, while Gimond et al. (2013) suggested that mating systems may play a key role in how quickly outbreeding depression evolves. Specifically, they suggest that selfing can foster outbreeding depression, as the effective reduction in recombination during selfing (due to heterozygotes being underrepresented) offers conditions that could allow combinations of favorable genes to quickly evolve. A similar argument could be made for genes within chromosomal inversions, which typically suppress recombination (Example 13.12).

Outbreeding depression is widespread, with a number of empirical studies uncovering striking examples of F_1 and/or F_2 breakdown in crosses among populations of the same species (see Edmands 2006 and Frankham et al. 2011 for reviews). Burton (1987, 1990a, 1990b), Brown (1991), and Edmands (1999) have obtained extensive evidence for the breakdown of physiological competence in crosses between populations of the marine copepod *Tigriopus californicus* inhabiting tide pools separated by only tens of kilometers. Escobar et al. (2008) similarly showed that outbreeding depression in the snail *Physa acuta* exists in crosses between dams as close as a few kilometers apart. Other dramatic evidence of outbreeding depression comes from observations of reduced fitness in crosses of inbred lines of *Drosophila* (Templeton et al. 1976) and plants (Parker 1992) adapted to identical environments. Crosses between outbreeding plants separated by a few to several tens of meters can exhibit substantial reductions in fitness (Waser 1993; Waser and Price 1985, 1994), as can crosses between fish derived from different sites in the same drainage basin (Leberg 1993) and crosses between clones of *Daphnia* from the same pond (Lynch and Deng 1994; Deng and Lynch 1996a). Crosses between French isolates of *Caenorhabditis elegans* showed outbreeding depression over distances as small as 15 meters (Dolgin et al. 2007).

In most of these examples, a decline in performance was observed in the F_1 generation, and data on the F_2 progeny were not obtained. Nonetheless, such results strongly implicate a breakup of favorable additive × additive epistatic effects ($\alpha_2^c > 0$) as a factor contributing to outbreeding depression (Equation 13.7). As Equation 13.8 shows, dominance × dominance epistasis ($\delta_2^c > 0$) can lead to considerable additional decline in the F_2 (beyond the single-locus dominance loss). Indeed, one can have heterosis in the F_1 and outbreeding depression in the F_2 if epistatic dominance is sufficiently large. This has been seen. Fenster and Galloway (2000a; 2000b) observed heterosis in F_1 (and in many F_2) crosses between different populations of the annual legume *Chamaecrista fasciculata*. However, many of the F_3 crosses showed outbreeding depression. Equation 13.7c explains this observation by having moderate linkage between loci that influence dominance × dominance interactions. A second example involves the work by Edmands (1999) on *T. californicus*. F_1 hybrids between populations tend to show heterosis, while most F_2 showed outbreeding depression. Edmands cleverly exploited the lack of crossing over in females of this species to examine the role of recombination. Crosses of both F_1 hybrid males and F_1 hybrid females back to the parental line showed a decline in fitness, with no effect on the mean from the sex of the F_1. Thus, among-chromosome segregation contributed to the recombination loss. However, Edmands also observed that while the means were equal, the recombinational (F_1 male) backcross had a larger variance, suggesting breakup of gene combinations within chromosomes was also important. Studies that report heterosis in the F_1 generation with no evaluation of the F_2 can thus be very misleading. Indeed, of the 39 studies reviewed by Edmands (2006), 44% (17/39) did not show outbreeding depression until at least the F_2.

As these examples illustrate, a particularly difficult issue underlying assessments of the potential for outbreeding depression concerns the generation in which outbreeding depression is revealed. There will always be some loss of heterosis in the F_2 generation, and a breakup of favorable epistatic gene complexes is always implicated when the F_2 performance is not intermediate to that of the F_1 and the mean of the parental lines. However, with low rates of recombination between pairs of epistatically interacting genes, it may take several generations for the negative consequences of mixing coadapted gene complexes to fully emerge.

Given the potential shift from heterosis in the F_1 to outbreeding depression in the F_2, some authors (e.g., Rhode and Cruzan 2005) have suggested that δ_1^c be referred to as the **heteroic component** and $(\alpha_2^c - \delta_2^c)$ as the **outbreeding depression component**. Using the machinery of Chapter 11, these components can be estimated by joint-scaling tests (e.g., Demuth and Wade 2005, 2006, 2007a). While such an approach seems reasonable, an interesting counterexample is given by the work of Oakley et al. (2015a). They examined crosses between *Arabidopsis thalina* for several paired populations from Central Italy with those from Northern Sweden. For two of these Italy × Sweden contrasts, joint-scaling tests found significant, and strong, *negative* δ_1^c values, so that the outbreeding depression was generated more by negative dominance values than epistasis.

A final delicate issue is the measurement of fitness in outbreeding depression studies. A common-garden experiment, while providing a standarized environment, may be misleading, especially if one (or both) of the parental lines is not adaptive to the test environment. Ideally one would test hybrids in the environments of both parents, but logistically this can be very challenging, especially for natural populations. Local adaptation has been repeatedly documented through reciprocal transplant experiments with plants (Clausen et al. 1940, 1945, 1947, 1948), which often exhibit adaptive divergence on spatial scales as small as a few meters (Schemske 1984; Waser and Price 1985; McGraw 1987; Schmitt and Gamble 1990; Galen et al. 1991). Similar results have been obtained with herbivorous insects residing on adjacent, long-lived hosts (Edmunds and Alstad 1978; Karban 1989). In most of these studies, the environmental differences perceived by the organism were not apparent to the investigators, so an absence of obvious ecological differentiation does not provide a compelling argument for ruling out local adaptation.

Example 13.12. One of the classic studies in evolutionary genetics was the early work by Theodosius Dobzhansky on the population genetics of *Drosophila pseudoobscura* chromosome rearrangements in the western United States (his collected works can be found in Lewontin et al. 1981). In the 1950s, this was one of the few systems for examining genetic variation in natural populations. Dobzhansky found that third chromosomes of *pseudoobscura* had three easily recognizable rearrangements (which can be distinguished under a light microscope by their banding patterns). These were referred to as *Standard* (*ST*), *Arrowhead* (*AR*), and *Chiracahua* (*CH*), the last two named for locations where they were collected. Chromosome rearrangements suppress recombination, creating a large region of almost complete linkage disequilibrium. Further, the collection of a given rearrangement can genetically diverge over time as new mutations are differentially accumulated in these nonrecombining regions. Dobzhansky measured the fitness of these rearrangements through competition experiments in population cages, following their frequency changes over time.

Dobzhansky (1950) examined the fitness of rearrangement heterozygotes relative to rearrangement homozygotes (e.g., *ST/CH* versus *ST/ST* and *CH/CH*, and *AR/CH* versus *AR/AR* and *CH/CH*). When the rearrangements came from the same population, heterozygotes showed high-parent heterosis, having higher fitness than either rearrangement homozygote. However, when the rearrangements came from geographically separated populations (California versus Mexico), heterozygotes had lower fitness than either homozygote. Hence, they showed heterosis within a population but outbreeding depression when the rearrangements came from distinct populations.

14

Matters of Scale

We call a thing big or little with reference to what wont it is to be, as we speak of a small elephant or a big rat. D'Arcy Thompson

Generally, the scales on which we take direct measurements are chosen more for their convenience than for their biological relevance or for their amenability to statistical analysis. Common artifacts of scale that complicate the analysis and interpretation of results are the dependence of the variance on the mean, departures from normality, nonadditive interactions, and compression of the data near a natural value of zero. Such complications can often be eliminated by transformation of the raw data to a new scale. A change in scale does not alter the information content of the original data. It simply changes the relationship of character values to one another.

Often, transformations can only be found that satisfy one or two of the desired statistical properties of normality, additivity, and variance independent of the mean. For polygenic traits, it can sometimes be rather difficult to completely eliminate interaction effects because a scale transformation that is successful in eliminating dominance from one locus may create it at another locus, or may create epistasis, and so on. Similarly, a transformation that successfully yields a normal distribution for one population may cause another to deviate substantially from normality. The utility of a particular transformation can also change dramatically with a shift in the environmental background. When these kinds of conflicts arise, the investigator must decide which criteria can be sacrificed in light of the objectives of the analysis.

Scale can have *biological* as well as statistical consequences. For example, a change in body size can result in disproportionate (**allometric**) changes in other correlated characters. Features of development such as **canalization** and **genetic assimilation** can also be direct consequences of scale. Our discussion of scale in this chapter thus considers both statistical and biological issues. Finally, the issue of scale also provides a point of introduction to **generalized** linear models, where on some *underlying scale* variation can be modeled as a standard linear model, which is then transformed into a value on the *observed scale*.

TRANSFORMATIONS TO ACHIEVE NORMALITY

Because many statistical methods, such as hypothesis testing using linear models (e.g., regression analysis and analysis of variance; Chapter 10) are predicated on the assumption that the data (or model residuals) are normally distributed, a standard procedure in most quantitative-genetic investigations is to transform the data to resemble normality as closely as possible prior to analysis. We first consider the **logarithmic transform**, as it is one of the most common, and successful, normalizing transformations.

Log-normal Distributions and the Log Transform

A common departure from normality is positive skewness, such that the tail to the right of the mean is longer than that to the left (as can arise when data are compressed near zero). Such a pattern is often indicative of a **log-normal distribution**, in which case a simple logarithmic transformation gives a normal distribution (Figure 14.1).

The log-normal distribution adequately describes many biological attributes (Wright 1968) and is worth dwelling upon briefly. Suppose that a variable z is log-normally distributed with mean μ_z and variance σ_z^2 on the original scale of measurement. It follows that $y = \ln(z)$ is normally distributed, and that the moments of y are related directly to the

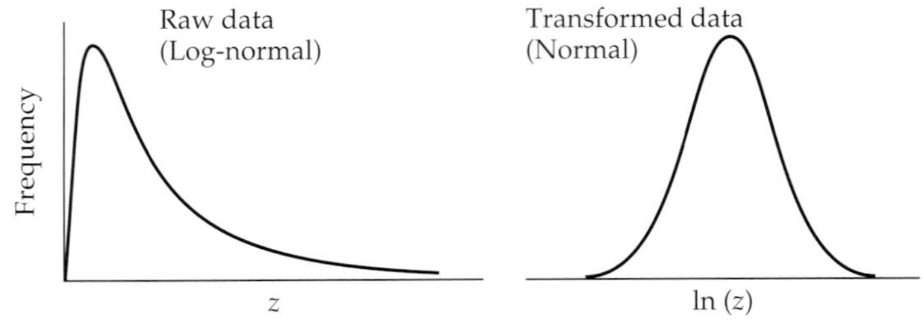

Figure 14.1 Conversion of a log-normal distribution on the original scale to a normal distribution on the log-transformed scale.

moments of z by

$$\mu_y = \ln\, \mu_z - \frac{1}{2} \ln\left(1 + \frac{\sigma_z^2}{\mu_z^2}\right) \tag{14.1a}$$

$$\sigma_y^2 = \ln\left(1 + \frac{\sigma_z^2}{\mu_z^2}\right) \tag{14.1b}$$

(Aitchison and Brown 1966). Because $\ln(1 + x) \simeq x$ for $|x| \ll 1$, when the **coefficient of variation** (σ_z/μ_z) is small (less than ~ 0.3), these expressions are closely approximated by

$$\mu_y = \ln \mu_z - \frac{\sigma_z^2}{2\mu_z^2} \tag{14.2a}$$

$$\sigma_y^2 = \frac{\sigma_z^2}{\mu_z^2} \tag{14.2b}$$

In estimating the moments of a distribution on a log-transformed scale, it is desirable to work directly with the raw data (rather than to simply apply Equations 14.1a and 14.1b) because the original distribution may, in fact, not be log-normal. Occasionally, however, one only has recourse to the sample mean, $\overline{z}$, and sample standard deviation, SD $= \sqrt{\text{Var}(z)}$, on the original scale. In this case, using a Taylor-series approach (Appendix 1), we can still approximate the mean and variance of the log-transformed variables by

$$\overline{y} = \overline{\ln z} \simeq \ln \overline{z} - \frac{1}{2} \ln\left(1 + \text{ CV}^2\right) \simeq \ln \overline{z} - \frac{\text{CV}^2}{2} \tag{14.3a}$$

$$\text{Var}(y) = \text{Var}(\ln z) \simeq \ln\left(1 + \text{CV}^2\right) \simeq \text{ CV}^2 \tag{14.3b}$$

where CV $=$ SD$/\,\overline{z}$ is the estimated coefficient of variation on the original scale. Note that these first-order approximations (identical in form to Equations 14.2a and 14.2b) apply to *any* distribution on the original scale, not just the log-normal. Thus, logarithmic transformation will successfully **stabilize the variance** (removing its dependence on the mean) under a broad range of distributions on the original scale, *provided* the coefficients of variation are roughly constant. On the other hand, logarithmic transformation will only **normalize** the data if the untransformed data are actually log-normally distributed. Methods for testing for normality are discussed shortly.

Galton (1879) first pointed out a possible explanation for the commonness of the log-normal distribution. Suppose the phenotype can be represented as a product of a large number of independent factors $z = x_1 \cdot x_2 \cdot x_3 \cdots x_n$. Upon logarithmic transformation, $y = \ln z = \ln x_1 + \ln x_2 + \ln x_3 + \ldots + \ln x_n$, and under the central limit theorem (Chapter 2), $\ln z$ will tend to normality as n becomes large. Thus, a simple explanation for a log-normal distribution is the existence of ***multiplicative interaction between a large number of factors***.

In situations where data span more than an order of magnitude and there is a ***natural zero value*** (such as percent protein), then the log-transform is often useful, as values get compressed as they approach the natural zero value. A log transform stretches out these out values near zero, countering this natural compression due to scale.

A number of other scale transformations, such as power functions and trigonometric functions, are available if the original data depart from both normality and log-normality (Wright 1968). If, for example, length measures in a population are known to be normally distributed, then it is likely that a cube root transformation would be required to normalize the distribution of weights, because weight is usually proportional to the cube of the length. The **Box–Cox transformation,**

$$y = \frac{z^\lambda - 1}{\lambda} \tag{14.4}$$

provides a fairly general class of transformations for achieving normality, and Box and Cox (1964) developed an approach to find the λ that gives the best fit to normality. This type of transformation usually deals quite adequately with skewness on the original scale (for example, the limit $\lambda \to 0$ is equivalent to the logarithmic transformation), but is less effective with kurtosis ($k_4 \neq 0$; Equation 2.12a). John and Draper (1980) present alternative functions to deal with long tails on symmetrical distributions, and Atkinson (1982) discusses all of these transformations in some detail.

Tests for Normality

There are a variety of methods for determining whether a character is normally distributed, see Chapter 6 of Sokal and Rohlf (1995) for an introduction and Chapter 8 of Wetherill (1986) for more advanced methods. Graphical tests for departures from normality are especially informative. For example, a simple frequency histogram may immediately reveal departures from normality, such as multiple peaks or strong asymmetries. If a character is determined by one (or a few) genes of major effect, a bimodal (or multimodal) distribution can sometimes result (Chapter 16). Such distributions can also result when a character is strongly influenced by a few distinct environmental effects.

With only sample moments in hand, deviations from normality are indicated by significant skewness and/or kurtosis, which have values of 0 and $3\sigma^4$, respectively, under normality (Chapter 2). Bowman and Shenton (1975) proposed a joint test of this, employing the statistic

$$S = \frac{n \cdot k_3^2}{6} + \frac{n \cdot k_4^2}{24} \tag{14.5}$$

where n is the sample size, and from Equations 2.8 and 2.12a, $k_3 = \text{Skw}(z)/\ \text{Var}(z)^{3/2}$ and $k_4 = ([\text{Kur}(z)/\text{Var}(z)^2] - 3)$ are, respectively, the standardized sample skewness and kurtosis, both of which have expected value zero under the assumption of normality. For large sample sizes, S is distributed as a χ^2 with two degrees of freedom. Thus, the hypothesis that a distribution is normal is rejected at the 5% level if $S > 5.99$ and at the 1% level if $S > 9.21$.

A more powerful approach for evaluating normality involves **normal probability plots,** which can be generated by transforming cumulative frequency to a **normal probability** (or **probit**) **scale**. In Chapter 2, we defined Φ_T as the fraction of measures in a distribution that are greater than value T. Here we let $q = (1 - \Phi_T)$ be the cumulative frequency to point T, namely $q = \Pr(z \leq T)$. A given cumulative frequency is rescaled to a normal probability scale by using the **probit transform** prb(q), which is defined as the solution of

$$\Pr[U < \text{prb}(q)] = q \tag{14.6}$$

where U is a standardized (or **unit**) normal random variable with zero mean and unit variance. Given q, prb(q) is obtained from common statistical packages (in `R`, this is `qnorm(q)`). For example, because $\Pr(U < -1) = 0.1587$, the normal probability scale value associated

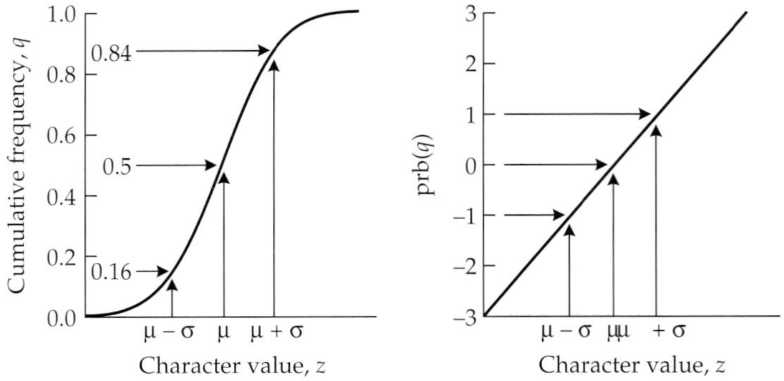

Figure 14.2 The normal probability scale, applied to a normally distributed random variable, z, with mean μ and variance σ^2. **Left**: A plot of cumulative frequency, q, as a function of z gives a sigmoidal curve. **Right**: Rescaling cumulative frequency using the transformation $\text{prb}(q)$ (Equation 14.6) gives a straight line. The scale given by $\text{prb}(q)$ is called a **normal probability scale**. As this figure shows, a character measure with associated $q = 0.50$ (which for a normal distribution corresponds to the mean) returns a normal probability scale value of zero. A character with $q = 0.84$ (which for a normal distribution corresponds to one standard deviation above the mean) returns a value of $\text{prb}(0.84) = +1$, and a character with $q = 0.16$ (which for a normal distribution corresponds to one standard deviation below the mean) returns a value of $\text{prb}(0.16) = -1$.

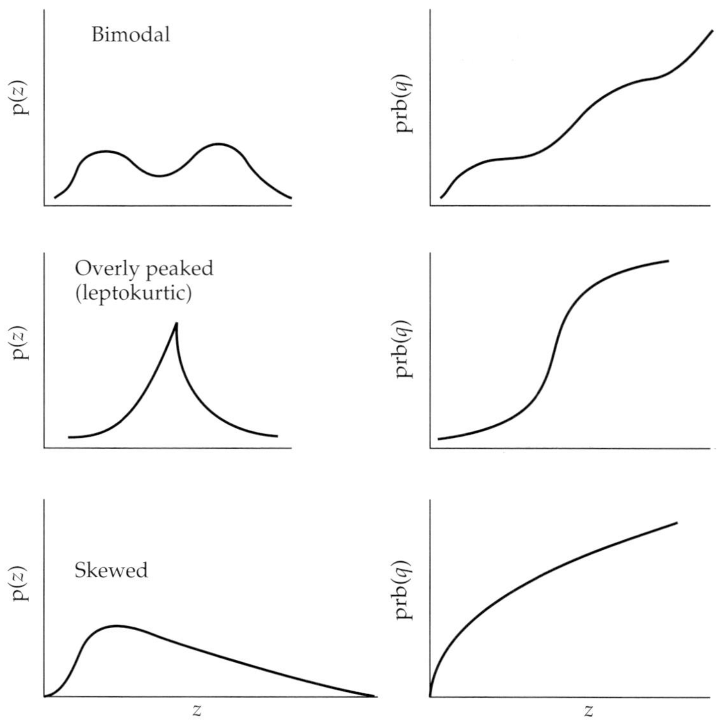

Figure 14.3 Diagnosis of three types of departure from normality using normal probability plots. On the left are probability density functions. On the right are cumulative frequencies plotted on the normal probability scale, $\text{prb}(q)$. **Top**: Bimodal distributions show flat regions when plotted on a normal probability scale. **Middle**: Leptokurtic distributions (overly peaked relative to a normal; Chapter 2) give a sigmoidal curve on a normal probability scale. **Bottom**: Skewed distributions depart from linearity on a normal probability scale by curving downward if the distribution is skewed to the right (as in the figure), or by curving upward if the distribution is skewed to the left.

Table 14.1 Normal probability scale (probit) values, $\text{prb}(0.5+\delta)$, as a function of δ (the deviation for 0.5). Values of $\delta < 0$ follow from the identity $\text{prb}(0.5-\delta) = -\text{prb}(0.5+\delta)$.

δ	0.000	0.099	0.191	0.273	0.341	0.394	0.433	0.460	0.477
$\text{prb}(0.5+\delta)$	0.000	0.250	0.500	0.750	1.000	1.250	1.500	1.750	2.000
δ	0.488	0.494	0.497	0.4987	0.4994	0.4998	0.4999	0.5000	
$\text{prb}(0.5+\delta)$	2.250	2.500	2.750	3.0000	3.2500	3.5000	3.7500	∞	

with a cumulative frequency of $q = 0.1587$ is $\text{prb}(0.1587) = -1$. Because U is symmetrically distributed about zero, it follows that

$$-\text{prb}(0.5-\delta) = \text{prb}(0.5+\delta) \qquad \text{for} \quad 0 \leq \delta \leq 0.5 \tag{14.7}$$

Table 14.1 gives the normal probability scale measures for different values of δ. For example, suppose that only 1.2% of the population is below a given character value, so that $q = 0.012$. Expressing this in the form used by Table 14.1 gives $\text{prb}(0.012) = \text{prb}(0.5 - 0.488) = -\text{prb}(0.5+0.488) = -2.25$. Zero on the probit scale corresponds to the population median ($q = 0.50$), which is also the mean for a normal. For a normally distributed variable, a unit change on the probit scale corresponds to one standard deviation (Figure 14.2).

As shown in Figure 14.2, with a normal distribution, cumulative frequency as a function of z is sigmoidal, but it is linear when transformed to a normal probability scale. Non-normal distributions deviate from linearity on such probability-scale plots, with multimodal, overly peaked, and skewed distributions all exhibiting characteristic departures (Figure 14.3). The normal probability scale forms the basis for the most powerful tests for departures from normality, e.g., the small-sample W test of Shapiro and Wilk (1965) and the large-sample D test of D'Agostino (1971).

Example 14.1. Fisher (1958) gave the following data set (based on unpublished work of Ford and Bull) on the number of vertebrae in herrings. Ignoring the discrete nature of the data, does the normal distribution provide a reasonable fit?

	Number of Vertebrae					
	53	54	55	56	57	58
Population Frequency (%)	0.08	1.06	28.36	61.30	8.91	0.29
Cumulative Frequency, q (%)	0.08	1.14	29.50	90.80	99.71	100.00
$\text{prb}(q)$	–3.14	–1.22	–0.54	1.33	2.76	∞

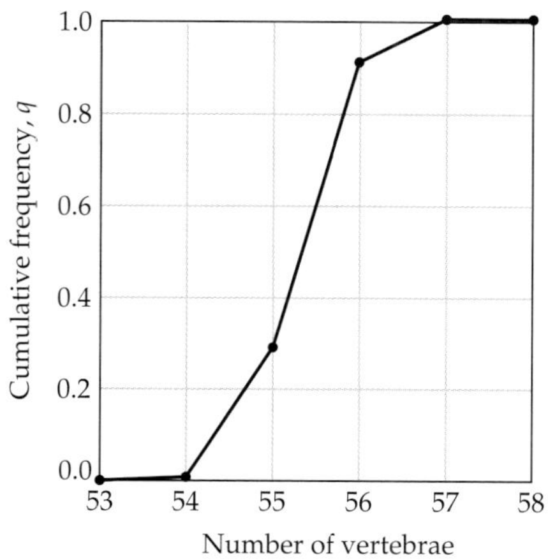

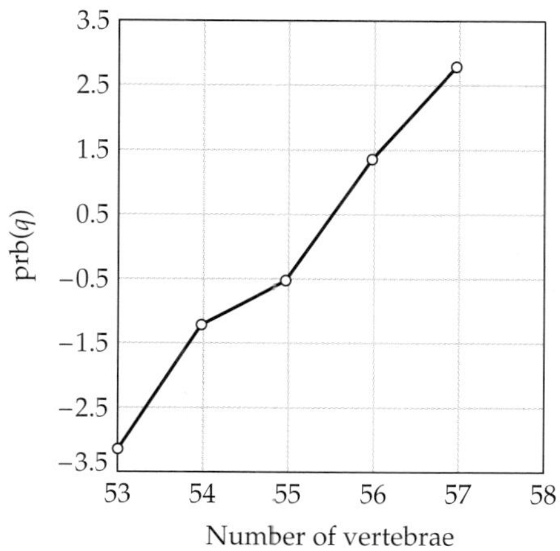

The cumulative frequency associated with a particular character value is the sum of all frequencies up to that point (e.g., the cumulative frequency associated with 55 vertebrae is 0.08 + 1.06 + 28.36 = 29.50%). Plotting cumulative frequency as a function of vertebrae number gives a sigmoidal plot, as shown in the above figure, suggesting a normal distribution. We can also assess the fit to a nomal by transforming the cumulative frequencies to a normal probability scale. For example, the cumulative frequency associated with 55 vertebrae is $q = 0.2950$. The value of prb(q) value associated with 55 vertebrae can be obtained in `R` as `qnorm(0.2950)`, which returns a value of -0.54 (0.54 standard deviations below the mean assuming a normal distribution). Plotting prb(q) versus vertebrae number gives a good linear relationship (see the above figure), suggesting that the normal distribution provides a reasonable description of the data.

STABILIZING THE VARIANCE

For studies concerned with relative levels of variation in different populations, it is useful to know whether the observed differences are simply a consequence of scale. For example, it is not uncommon for the variance to increase as the population mean increases (Figure 14.4). By operating on a scale for which there is no discernible relation between the mean and variance, one can be more confident that any significant differences in observed levels of variance between samples must be attributable to something other than a scaling effect due to differences in the mean. Here, we consider the use of **variance-stabilizing transformations** to render the variance on the transformed scale independent of the mean. Wright (1968, Chapters 10 and 11) gives an excellent discussion of the application of such transformations to distributions of quantitative characters.

Kleckowski's Transformation

We have already encountered the idea that a log transformation can render the variance of different samples independent of the mean when the CV is approximately constant on the underlying scale (Equation 14.3b). For this to be strictly valid, the SD must be directly proportional to the mean, SD(z) $= b\overline{z}$, a line with an intercept of zero. However, a fairly common situation is for the regression of standard deviations on means to have an intercept significantly greater than zero (Figure 14.4). This causes the coefficient of variation to increase dramatically as the mean approaches zero, which in turn implies an inflation of the variance on the logarithmic scale.

Kleckowski (1949) suggested a simple way to eliminate this problem. If the standard deviation on the original scale can be described adequately by the linear regression SD(z) $=$ $a + b\overline{z}$, then it follows by rearrangement that SD(z)$/(\overline{z} + a/b) = b$. Because a/b is a constant, SD(z) $=$ SD$[z + (a/b)]$, so the previous ratio is equivalent to the coefficient of variation of $y = z + a/b$. Thus, the use of the transformed variable $y = z + a/b$ in place of z yields an expected coefficient of variation that is equal to b and thus independent of the mean (also see Example 14.3). This results in the independence of the mean and variance for the log-transformed variable $\ln(y) = \ln(z + a/b)$.

Example 14.2. Consider the *Uta* data in Figure 14.4. Both the intercept and slope of the least-squares regression, SD(z) $= 0.33+0.052\,\overline{z}$, are highly significant (W. C. Kerfoot pers. comm.). Because the relationship between the standard deviation and the mean closely approximates linearity, the transformation

$$y = \ln(z + 0.33/0.052) = \ln(z + 6.35)$$

renders the variance of different populations independent of the mean.

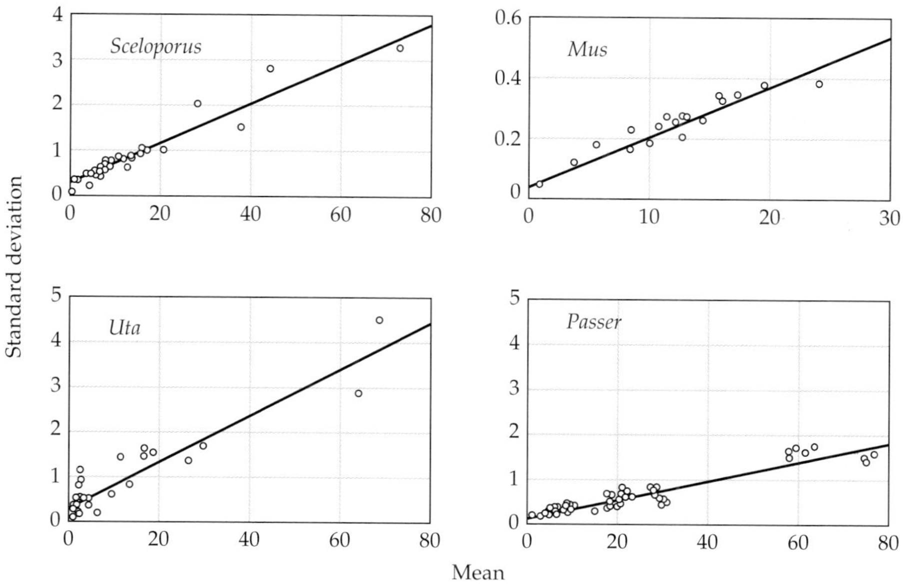

Figure 14.4 Relationship of the standard deviation and the mean for four sets of morphological characters in vertebrates. *Sceloporus* (rough-scaled lizard): meristic traits, several species (Kluge and Kerfoot 1973); *Uta* (side-blotched lizard): scale counts and ratios of morphometric characters (Kluge and Kerfoot 1973); *Mus* (mouse): skeletal and weight characters (several sources given in Soulé 1982); *Passer* (house sparrow): skeletal and feather characters (described in Kluge and Kerfoot 1973).

General Variance-stabilizing Transformations

For more complicated relationships between means and standard deviations, a general formula exists for ascertaining the correct variance-stabilizing transformation. Given the relationship $\sigma_z = f(\mu_z)$, the appropriate transform is given by (Bartlett 1947)

$$y = C \int \frac{dz}{f(z)} \tag{14.8}$$

where C is an arbitrary non-negative constant, generally chosen to set the rescaled variance equal to one. The rescaled character y has a variance that is independent of the mean in any particular population. Thus, the procedure for obtaining a variance-stabilizing transform for any standard deviation–mean relationship, $f(\overline{z})$, is simple. A fit for the function $f(\overline{z})$ is obtained by a least-squares polynomial (or some other nonlinear) regression, and Equation 14.8 is solved using the estimated regression coefficients.

Example 14.3. Suppose $\text{SD}(z) = a + b\,\overline{z}$. Applying Equation 14.8, the variance-stabilizing transform is given by

$$y = C \int \frac{dz}{a + b \cdot z} = \frac{C}{b} \ln\left(z + \frac{a}{b}\right)$$

Note that there is no unique solution for y, as we can multiply y by any constant and still have a variance-stabilizing transform. Letting $C = b$ recovers Kleckowski's correction, $y = \ln(z + a/b)$.

Two other variance-stabilizing transforms commonly appear in the quantitative-genetics literature. First, suppose that the variance (as opposed to the standard deviation) scales with the mean, with $\text{Var}(z) = b\overline{z}$, or that SD = $b\sqrt{\overline{z}}$. Equation 14.8 gives the form of the transformation as

$$y = C \int \frac{dz}{\sqrt{z}} = \sqrt{z}$$

Namely, the **square-root transform** (Bartlett 1936). This transformation is often used for count data, which are often modeled as a Poisson (Equation 2.21a), a distribution whose mean equals its variance. Bartlett (1947) recommends using $\sqrt{z+1/2}$ when the mean of z is small. While the square-root transform has historically been used to make count data closer to a normal distribution, its use is being phased out in favor of more exact analysis, such as log-linear models (Equation 14.17).

Finally, suppose z is binary (and hence can be modeled as a binomial distribution with $n = 1$), so that SD = $\sqrt{p(1-p)}$, where the mean is p, yielding a transformation of

$$y = C \int \frac{dz}{\sqrt{z(1-z)}} = \arcsin(\sqrt{z}\,)$$

This is the **arcsin square-root transform** to standardize frequency data. As with the square-root transform, the use of the arcsin as a normalizing transformation is being phased out in favor of logistic regressions (Equation 14.15b) (Warton and Hui 2011).

The Roginskii–Yablokov Effect

A common departure from a roughly constant value CV with respect to means is an inverse relationship between the CV and mean (CV increases as the mean decreases), which is often seen in functionally related traits. The biological implications of such a relationship have been discussed by many investigators (Pearson and Davin 1924; Roginskii 1959; Yablokov 1974; Rohlf et al. 1983; Kerfoot 1988). We refer to such a relationship as a **Roginskii–Yablokov effect**. Whether such a scaling has important biological underpinnings, as opposed to being a simple statistical artifact, merits some scrutiny.

The most obvious problem in attempting to relate coefficients of variation and means is that the former are inverse functions of the latter, because by definition, $\text{CV}(z) = \text{SD}(z)/\,\overline{z}$. Thus, for the extreme case in which the standard deviation is a constant independent of mean, the expected relationship between the CV and the mean is an inverse hyperbola. On a log–log plot, this is revealed as a linear relationship with a slope of minus one. The data for a number of characters closely approximate this pattern (Figure 14.5), suggesting that in some cases the Roginskii–Yablokov effect is nothing more than a mathematical consequence of the regression of a ratio on its denominator.

A second statistical artifact that can lead to a Roginskii–Yablokov effect is *measurement error* which inflates the variance when the mean is near zero. For characters such as length measurements, which take on only positive values, the SD–mean relationship is constrained to pass through the origin because the standard deviation of measurements must be zero if the mean is zero. However, a linear approximation to the data generally yields a positive y-intercept, suggesting that the SD versus mean relationship is slightly bowed upward near the origin. Such behavior can be a simple consequence of measurement error. Suppose that the SD versus mean relationship is adequately described by the function $\text{SD}(z) = \sqrt{k^2 \cdot \overline{z}^{\,2} + V}$, where k is the CV if we measure without error, and V is the additional variance caused by measurement error (assumed to be independent of the mean). With this relationship, the coefficient of variation approaches k as $\overline{z}$ becomes large, but rapidly increases as $\overline{z}$ becomes smaller than $\sqrt{V}/k$. This explanation is probably more relevant to metric (continuous) characters than meristic (discrete) ones. Presumably, as the mean number of counts per individual declines, so does the sampling variance.

Rohlf et al. (1983) pointed out a third statistical artifact that can lead to a Roginskii–Yablokov effect when the data are log-normally distributed. The discretization of log-normally distributed data into classes (delimited by counts or measurement constraints) leads to a reduction of the mean and an inflation of the variance relative to that expected under a continuous log-normal distribution. This in turn causes an inflation of the CV to a

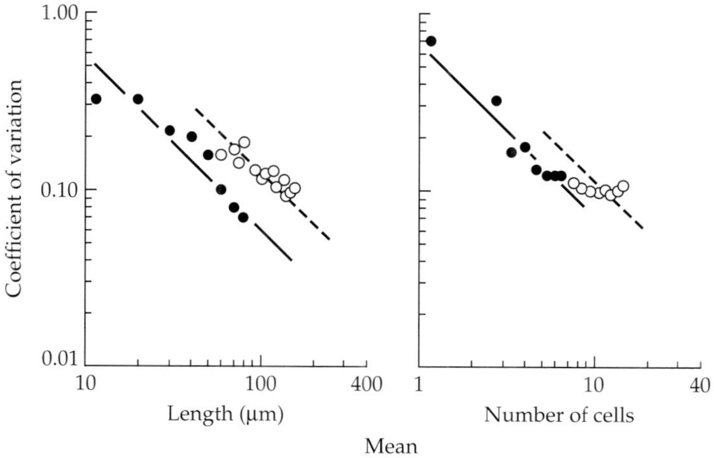

Figure 14.5 Log–log plots of coefficients of variation versus mean phenotypes for lengths and cell counts of tail spines (solid points) and antennules (open points) for samples of the cladoceran crustacean *Bosmina longirostris*. As mentioned in the text, a straight line with a slope of minus one is the expected relationship if the standard deviation is independent of the mean. Only the antennule cell counts deviate greatly from this expectation. (From Kerfoot 1988.)

degree that increases with decreasing means. In principle, the bias caused by discretization can be corrected by applications analogous to Sheppard's correction (Chapter 2) with normally distributed data (Thompson 1951), but this has not been done with most published data.

Despite these statistical artifacts that may be largely (and in some cases, entirely) responsible for observed negative relationships between means and CVs, there are reasons for suspecting that Roginskii–Yablokov effects are sometimes biological in nature. A simple explanation was offered by Pearson and Davin (1924). Consider a character that can be represented as the sum of n developmental units,

$$Z = z_1 + z_2 + \ldots + z_n$$

For example, Z could be composed of a number of smaller bones or a number of specific cells. The variance of Z is then

$$\sigma_Z^2 = \sigma_{z_1}^2 + \sigma_{z_2}^2 + \cdots + \sigma_{z_n}^2 + 2\sigma_{z_1,z_2} + \cdots + 2\sigma_{z_{n-1},z_n}$$

Letting $\overline{\sigma}_{z_i,z_j}$ be the mean of the covariance terms, the squared coefficient of variation is

$$\mathrm{CV}_Z^2 = \frac{\sigma_Z^2}{\mu_Z^2} = \frac{\sum_{i=1}^{n} \sigma_{z_i}^2 + n\,(n-1)\cdot\overline{\sigma}_{z_i,z_j}}{\left(\sum_{i=1}^{n} \mu_i\right)^2} \tag{14.9}$$

Now, for simplicity, let each developmental unit share the same mean (μ_z) and variance (σ_z^2). Then

$$\mathrm{CV}_Z^2 = \mathrm{CV}_z^2 \cdot \left[\frac{1 + (n-1)\,\overline{\rho}}{n}\right] \tag{14.10}$$

where $\mathrm{CV}_z = \sigma_z/\mu_z$ is the coefficient of variation for each developmental unit, and $\overline{\rho}$ is the mean correlation between units. Except in the unlikely event that all of the component parts are perfectly correlated, the fraction on the right is less than one. Thus, the CV of Z

is less than the average CV of the component parts: ***variation in individual components averages out as the sum increases***. Provided that developmental units are similar in size and variance, the Pearson–Davin argument leads to the prediction that ***the CV of size of a structure will decline as the number of component parts increases***, which we denote as the **Pearson–Davin effect**.

The Pearson–Davin argument is supported in a number of cases. Analyzing data of Bader and Hall (1960) on osteometric characters in bats, Lande (1977) found that characters consisting of several bones have CVs of 0.02–0.03, while the individual bones themselves have CVs of around 0.03–0.06. Similarly, Kerfoot (1988) found that while the CV for tail spine length in the cladoceran crustacean *Bosmina* is 0.10, the CV for length of the component cells ranges from 0.28 to 0.36.

A second, and related, biological explanation for the Roginskii–Yablokov effect applies to analyses that compare the same character in different populations. During development, the variance in size of morphological characters often increases initially and then declines as **compensatory growth** focuses most individuals into a narrow range of phenotypes. This pattern of development, also known as **targeted growth** (Riska et al. 1984), would generate a Roginskii–Yablokov effect if the individuals from different samples varied in average age. Those samples containing the oldest, and presumably largest, individuals would be expected to exhibit the lowest CVs.

While much work remains to be done before the underlying determinants of the Roginskii–Yablokov effect can be deciphered, if indeed any generalities are possible, its existence has serious implications for the use of scale transformations. If the CV is not independent of the mean, the routine procedure of log-transforming data prior to analysis will not be successful in stabilizing the variance. Rather, it will cause a negative correlation between the mean and variance on the logarithmic scale. Failure to appreciate the importance of such a scaling effect can lead to interpretative difficulties in long-term studies of natural selection, leading, for example, to the conclusion that directional selection for larger size is accompanied by stabilizing selection about the optimum (Halbach and Jacobs 1971).

The Kluge–Kerfoot Phenomenon

Drawing from morphological data on a number of vertebrates, Kluge and Kerfoot (1973) suggested that traits with high phenotypic variance *within* populations tend to exhibit high levels of divergence *among* populations. They argued that this pattern is due, at least in part, to the variance of a trait within a population being inversely proportional to the intensity of stabilizing selection. However, their conclusion appears to require the assumption that characters that are under strong stabilizing selection within populations have similar optimal phenotypes in different populations. To support their case, Kluge and Kerfoot plotted for various characters the range of means from different populations against the within-population standard deviation, after first dividing both statistics by the overall population mean.

As in the case of the Roginskii–Yablokov effect, the Kluge–Kerfoot phenomenon may be largely a statistical artifact (Sokal 1976; Rohlf et al. 1983). The statistics employed by Kluge and Kerfoot are expected to be intrinsically correlated due to the fact that both contain the same variable (the mean) in the denominator. This problem is exacerbated by the fact that the range of population means is positively related to the within-population variance causing sampling error of the means. In an attempt to eliminate the latter problem, Sokal (1976) employed coefficients of within- and among-population variation extracted by analysis of variance. However, because both CVs still share the same denominator, this procedure has the same problem as Kluge and Kerfoot's analysis. Moreover, as Rohlf et al. (1983) have pointed out, there appears to be a Roginskii–Yablokov effect for the among-population CV just as there is for the within-population CV. Thus, the correlation of within- and among-population variance in many existing studies may be an indirect effect of both measures being correlated with a third (the mean). In a study of morphometric differentiation among house sparrow populations, Baker (1980) showed that the Kluge–Kerfoot

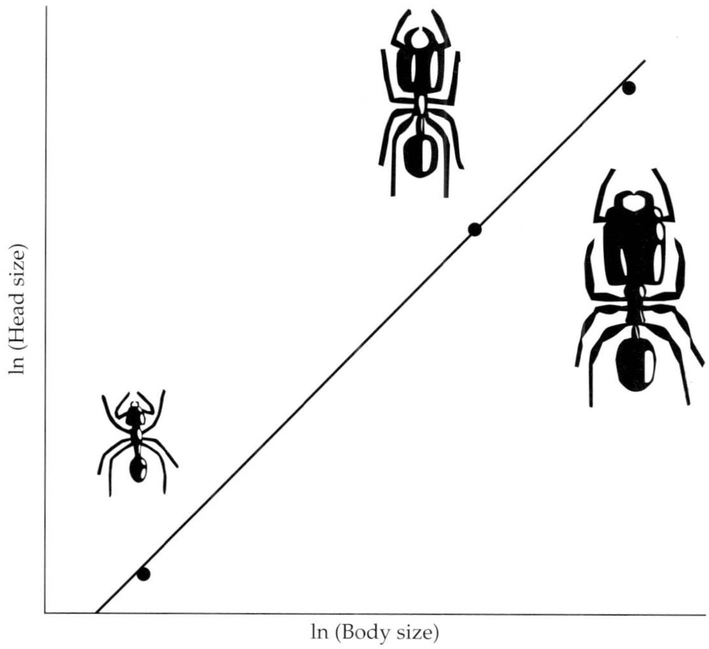

Figure 14.6 Castes of the ant *Pheidole instabilis* display positive allometry for head size. The proportionately larger heads of soldiers can be accounted for simply by changes in size, with head size growing at a faster rate than the rest of the body. (Based on Huxley 1932.)

phenomenon disappeared after the scaling with the mean was eliminated. Until these numerous statistical problems have been accounted for properly with several independent data sets, the biological and evolutionary significance of the Kluge–Kerfoot phenomenon must remain in doubt.

ALLOMETRY: THE SCALING IMPLICATIONS OF BODY SIZE

Many aspects of shape, life histories, behavior, and physiology scale proportionately with body size, both *within* individuals through development and *among* different individuals (Dubois 1897; Thompson 1917, 1943; Huxley 1924, 1932; Gould 1966; Vogel 1981, 1989; Wainwright et al. 1982; Peters 1983; Calder 1984; Schmidt-Nielsen 1984; Fleagle 1985). A simple consequence of such scaling is that the phenotypic variation seen for many characters may be little more than an indirect consequence of variation in body size. Huxley (1924, 1932) noted that the relationship of most characters with body size can be summarized with a particularly simple mathematical expression—a log–log plot of character y versus some measure of body size x (such as weight or length) generally yields a straight line,

$$\ln y = \ln a + b \ln x \qquad (14.11a)$$

This implies a power relationship on the original scale of measurement,

$$y = a\,x^b \qquad (14.11b)$$

Characters that scale with size according to Equation 14.11a are said to display **allometry** (Huxley and Teissier 1936), with the **allometric coefficient** b providing a measure of the scaling of character value with body size. If $b = 1$, **isometric** growth occurs and the ratio of character value to body size is constant. If $b > 1$, **positive allometry** occurs, with the character becoming proportionately larger as size increases. Characters with $b < 1$ display **negative allometry**, becoming proportionately smaller as size increases. Thus, unless

$b = 1$, the shape of an organism changes with size, potentially generating very different morphologies. Figure 14.6 demonstrates allometric scaling for Huxley's classic analysis of the different morphologies of ant castes. The proportionately larger heads associated with soldier ants appear to be a simple scale effect of body size. As nicely summarized by Gayon (2000), there were extensive discussions on relative growth before Huxley's seminal papers (1924, 1932), much of it currently overlooked (likely because it was largely in French).

Three very different types of data can display allometry, and unfortunately these are often confused and/or treated interchangeably. In **ontogenetic** or **growth allometry**, a character is followed through the development of an individual—each data point corresponds to the character at a different developmental stage, ideally in the *same individual*. **Static** or **intraspecific allometry** compares the character in different-sized adults (or individuals of the same age) from the same species or population—each data point represents a *different adult* from a common taxon (e.g., Figure 14.6). **Evolutionary** or **interspecific allometry** examines the character in different species or divergent populations—each data point represents the mean for a *different taxon*. These distinctions are critical, as a character displaying one form of allometry need not display another, e.g., ontogenetic allometry does not necessarily imply evolutionary allometry, and so forth. For further discussions of these differences, see Lande (1979), Cheverud (1982), and Fleagle (1985).

Allometric equations are routinely used to correct for the effects of size by replacing the character value by the deviation from the fitted allometric equation, i.e., by $y - \widehat{y}$, where $\widehat{y}$ is the predicted character value based on Equation 14.11b. Such residuals can then be used to study variation in character values that is independent of the allometric consequences of size variation. Packard and Boardman (1987) give a highly readable account of this approach with numerous examples applied to physiological data.

An important quantity for vertebrate ecologists is the **condition** of an individual, a measure of its well-being or nutritional status (Le Cren 1951). While generally a rather abstract concept, one proposed measure of condition is the residual from an allometric regression of weight on some measure of body size (i.e., size-adjusted weight). The motivation is that individuals in good condition will be heavier than expected given their size, while individuals in poor condition will be underweight. Jakob et al. (1996), Green (2001), and Schulte-Hostedde et al. (2005) discussed the merits of this approach. A related issue is that many behavioral features (e.g., flight and running speed, herding behavior, fighting ability) are impacted by body size. Dial et al. (2008) discuses the impact of allometry on the analysis of such behaviors.

An alternate approach to size correction, especially common in systematics, involves ratios of the character over some measure of body size, e.g., head length/total body length. Unfortunately, there are serious problems with this approach. Unless the character displays isometric growth, ***taking ratios does not remove aspects of body size***, as $y/x = ax^{b-1}$, which is a function of x unless $b = 1$. Further, as noted above, ***ratios introduce spurious correlations:*** the variable y/x is usually negatively correlated with x, even if x and y are themselves independent (Pearson 1897; Tanner 1949; Atchley et al. 1976; Albrecht 1978; Atchley and Anderson 1978; Strauss 1985). Thus, character/size ratios often *confound*, rather than *remove*, size effects and should be avoided. Additional discussion of statistical issues with estimation of allometry coefficients (such as when ordinary least squares is appropriate or whether one should use reduced major axis regression, the latter allowing for measurement error in both the predictor and response variables) can be found in Warton et al. (2006), Packard et al. (2011), and Kilmer and Rodríguez (2016).

REMOVAL OF INTERACTION EFFECTS

Because nonadditive interactions (e.g., genotype $\times$ environment interaction, dominance, and epistasis) can only complicate statistical analysis, additivity is a simplifying assumption underlying much of quantitative-genetic theory. Therefore, in practical applications, it

Table 14.2 Mean weight of offspring (grams at age 33 days) from litter sizes of 1 to 4 for two inbred lines of guinea pigs, for untransformed and transformed $[\log_{10}(x - 80)]$ data. Observe that the untransformed data show a G × E interaction with the difference in strain means being a function of litter size. This dependency is largely eliminated on the transformed scale. (From Wright 1968.)

	Litter Size			
	1	2	3	4
Untransformed data				
Strain 13	299.5	264.0	226.8	203.6
Strain 2	213.4	195.0	171.8	155.4
Difference	86.1	69.0	55.0	48.2
Transformed data				
Strain 13	2.341	2.265	2.167	2.092
Strain 2	2.125	2.061	1.963	1.877
Difference	0.216	0.204	0.204	0.215

is very desirable to find a scale on which genetic and environmental effects are additive. Because there is no compelling reason to expect that a scale removing one aspect of nonadditivity will also eliminate others, nor to expect that the optimal scale for one population will be suitable for all others, it is rare that a wholly satisfactory solution can be found. Nevertheless, experience has shown that considerable simplification is often attainable.

Example 14.4. Suppose the genotypes *BB*, *Bb*, and *bb* have genotypic values of 1, 4, and 9 on the original scale of measurement. Applying the square-root transformation, the genotypic values on this new scale of measurement become 1, 2, and 3. Thus, while *B* is slightly dominant to *b* on the original scale, the gene action is perfectly additive on the transformed scale. The square-root transformation condenses the scale between *Bb* and *bb* relative to that between *BB* and *Bb*, resulting in additivity on the new scale. Provided the genotypic value of *Bb* lies within the range of *BB* and *bb*, a scale will always exist for which *Bb* can be made exactly intermediate to *BB* and *bb*.

In the genetic analysis of populations, nonadditive interactions between alleles (dominance) do not usually cause any insurmountable difficulties. However, because epistasis comes in many forms, all of which are quite difficult to quantify with precision, it is highly desirable to work on a scale for which a model involving no more than additive and dominance effects can be shown to be adequate. In some cases, the appropriateness of a particular transformation in removing nonadditive effects can be examined using joint-scaling tests (Chapter 11).

Genotype × environment interaction can be very difficult to detect in natural populations, because its measurement requires that several distinct genetic groups (e.g., lines or specific QTLs) be scored in a discrete set of environments (Part II Chapters 27 and 28). Such an analysis can often be performed in the laboratory or in a common-garden experiment (Appendix 9), and the results used as a guide to the choice of scale for the field population. However, care should be taken in overextending the results from controlled experiments, as the mode of action of microenvironmental variation in the field may be fundamentally different from that involving the macroenvironmental treatment in the laboratory.

Table 14.2 gives an example of removing a genotype × environment interaction using a transformed scale. The character considered is the mean weight of offspring for two inbred lines of guinea pigs as a function of litter size, the latter being regarded as a component of the offspring's environment. For both strains, there is a substantial decline in offspring weight with increasing litter size, presumably due to competition for maternal care. On the scale of raw measurements, genotype × environment interaction is indicated by the steady decline in the difference between line means with increasing litter size. In the absence of genotype × environment interaction, this difference should be constant across environments (across different litter sizes). Such behavior is seen upon logarithmic transformation. On the other hand, Wright (1968) found that the variance in weight among litter mates was independent of litter size on the original scale but positively correlated with it on the transformed scale. Thus, while logarithmic transformation was able to eliminate genotype × environment interaction of the means, it destabilized the variance.

DEVELOPMENTAL MAPS, CANALIZATION, AND GENETIC ASSIMILATION

We hope that by now the reader has a sense that scale is more than a statistical issue in that by choosing an appropriate scale of analysis, we can often simplify the biological interpretation of characters. For example, allometry shows us that complex changes in character shape can be simple consequences of changes in size. Another potentially important aspect of scale concerns the interpretation of character states as the outcome of the mapping (or rescaling) of some *underlying unobserved* (or **latent**) *variable* (for example, the amount of a particular protein) into an *observable phenotypic space* (such as disease presence or absence). In concluding our discussion of scale, we first examine **developmental maps** for discrete (meristic) traits that transform some underlying continuous variable into an observable discrete phenotype, then explore some of the potential implications for continuously distributed characters, and finally use this mapping to motivate generalized linear models.

Estimating Developmental Maps

Biological systems face a conundrum. On one hand, they must be robust enough to generate a fairly consistent trait value in the face of environmental and genetic noise. On the other hand, they also have to be somewhat flexible, being able to tailor the phenotype to match the environment. Developmental maps provide a useful, albeit imperfect, model for considering some of these issues from a quantitative-genetic perspective.

To motivate the idea behind developmental maps, consider hypertension (high blood pressure) in humans. When blood pressure exceeds some critical (or **threshold**) value, either for genetic or environmental reasons, the individual is said to clinically display hypertension. Thus, classification of individuals as either normal or hypertensive transforms a continuous underlying measure (blood pressure) into a dichotomous character (normal versus hypertensive). Many other diseases appear to result when some underlying (and usually unobserved) physiological variable exceeds a threshold value (Part II Chapter 31).

Letting y be the value of the underlying variable (the **liability**, a value on the **underlying scale**), the observed phenotypic value z can be regarded as a function of the liability, $z = \phi(y)$, where ϕ is the **developmental map** that rescales the underlying (unobserved) liability value y into the phenotypic value z (the **observed scale**). For a simple threshold character, letting y_c be the critical value, and denoting the alternate character states by 0 and 1, then

$$\phi(y) = \begin{cases} 0 & \text{for } \; y < y_c \\ 1 & \text{for } \; y \geq y_c \end{cases}$$

This type of scaling can easily be extended to ordinal characters displaying multiple (ordered) states. For example, a character with three discrete states can be represented as the mapping of some underlying continuous value y by

$$\phi(y) = \begin{cases} \text{character state 1} & \text{for } \; y < y_{c_1} \\ \text{character state 2} & \text{for } \; y_{c_1} \geq y < y_{c_2} \\ \text{character state 3} & \text{for } \; y \geq y_{c_2} \end{cases}$$

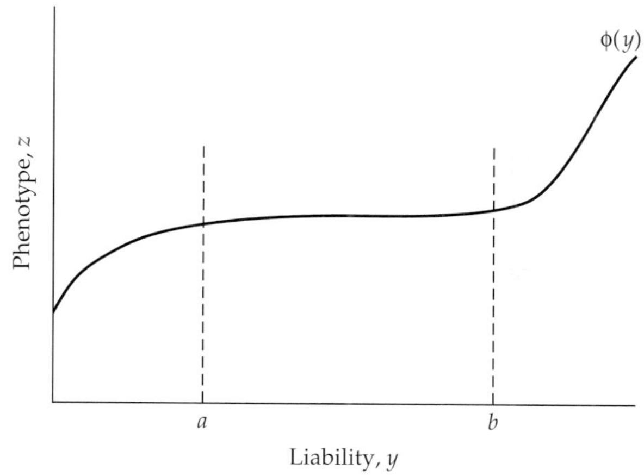

Figure 14.7 A nonlinear developmental map. This map shows a zone of strong canalization for $a \leq y \leq b$, as liability values in this range give essentially the same phenotype.

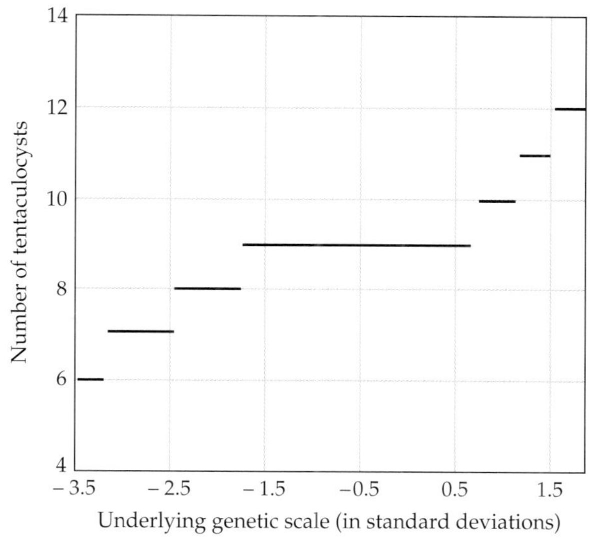

Figure 14.8 Reconstruction of the development map for tentaculocyst number in the jellyfish *Ephyra* using probit analysis under the assumption that liability is normally distributed on some appropriate underlying scale. Strong canalization is seen, with a larger range of liability mapping to the nine-tentaculocyst character state relative to the range mapping to other character states. (Original data of Browne; from Wright 1968.)

Following Waddington (1942), a character (or character state) is said to be **canalized** if the resulting phenotype is fairly insensitive to changes in the underlying environmental and/or genetic factors. Schmalhausen (1949) referred to this as **autoregulation**. While a number of authors (e.g., Gibson and Wagner 2000) make the distinction between robustness (stability of a phenotype over genetic and environmental change) and canalization (the *evolution* of such stability), we will be a bit less formal in the following discussion of these issues.

A continuous character can show a **zone of canalization**—a region of ϕ that is rather flat and within which changes in the underlying genetic and environmental factors result in relatively little (or no) change in the phenotype (Figure 14.7). Assuming that the distribution of the underlying liability values is normal with variance σ^2, the width of the developmental map associated with a particular discrete character state can be estimated in units of σ (Figure 14.8).This procedure is often referred to as **probit analysis** because cumulative frequencies are measured on the probit (normal probability) scale (Equation 14.6).

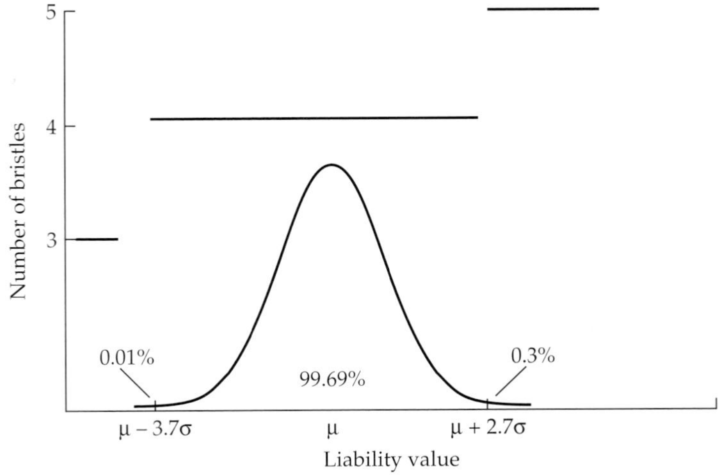

Figure 14.9 Reconstructing a developmental map for *Drosophila* scutellar bristle number. The bulk of the population (99.69%) had four bristles, while 0.01% had three, and 0.3% had five. Assuming a normal distribution of liability on an appropriate underlying scale, the developmental map translating liability into character value (here bristle number) can be estimated via probit analysis. Individuals with a liability value more than 3.7 standard deviations below the mean have three bristles, while individuals whose liability exceeds 2.7 standard deviations above the mean have five bristles. Details are given in Example 14.5.

Suppose the character of interest has k discrete states, which we order from 1 to k. Let $q(m)$ be the cumulative frequency to (and including) class m, and define $x_m = \text{prb}[q(m)]$ to be the probit-scale value associated with $q(m)$, i.e., x_m satisfies $\Pr(U \leq x_m) = q(m)$ where U is a unit normal. The width of liability values that map to class m is estimated by

$$D_m = x_m - x_{m-1} \qquad \text{for } 2 \leq m \leq k-1 \tag{14.12a}$$

where the units of D_m are in terms of number of standard deviations of liability values (Figure 14.9). Because $q(k) = 1$ and $q(0) = 0$, the widths for the first and last classes are not defined, as $\text{prb}[1] = -\text{prb}[0] = \infty$. Using standard results (Kendall and Stuart 1977, p. 254), the sampling variance for the estimator D_m is given by

$$\sigma^2(D_m) = \frac{1}{n}\left\{ \frac{q(m-1)[1-q(m-1)]}{[\varphi(x_m)]^2} + \frac{q(m)[1-q(m)]}{[\varphi(x_{m-1})]^2} - 2\,\frac{q(m-1)[1-q(m)]}{\varphi(x_m)\,\varphi(x_{m-1})} \right\} \tag{14.12b}$$

where $\varphi(x) = \exp(-x^2/2)/\sqrt{2\pi}$ is the unit normal density function.

One reason for using probit analysis to construct developmental maps is that untransformed frequencies can give a misleading picture of the strength of canalization. Consider two populations with identical developmental maps, and with normal distributions of liability values with the same variance but different means. For the first population, suppose that individuals with liability values within one standard deviation of the mean map to the four-bristle class, resulting in 68% of the population having four bristles. In the second population, individuals with liability values between one and three standard deviations above the mean map to the four-bristle class, resulting in only 15.7% of the population having four bristles. In both cases, the width of the four-bristle class is 2σ, but on an untransformed scale, the first population appears to be more canalized for the four-bristle class.

Example 14.5. Sheldon et al. (1964) examined scutellar bristle number in *Drosophila melanogaster* in a number of isogenic lines derived from a common ancestral population. Consider the following data for females from their isogenic line 5, which had 3, 4, or 5 bristles with the following frequencies:

Class	Observed	Frequency	Cumulative Frequency	prb[$q(m)$]
3	1	0.00012	0.00012	−3.70
4	8124	0.99693	0.99705	2.74
5	24	0.00295	1.00000	—

Here n = 8149, x_3 = prb[0.00012] = −3.70, and x_4 = prb[0.99705]= 2.74. Applying Equation 14.12a, the estimated width of the developmental map for class 4 is

$$D_4 = x_4 - x_3 = 2.74 - (-3.70) = 6.44$$

Noting that $q(3)$ = 0.00012, $\varphi(-3.70)$ = 0.0004, $q(4)$ = 0.99705, and $\varphi(2.74)$ = 0.0093, Equation 14.12b gives the sample variance for D_4 as

$$\begin{aligned}\text{Var}(D_4) &= \frac{1}{8149}\left[\frac{0.00012 \times 0.99988}{0.0004^2} + \frac{0.99705 \times 0.00295}{0.0093^2}\right. \\ &\qquad \left. - 2\,\frac{0.00012 \times 0.00295}{0.0004 \times 0.0093}\right] \\ &= 0.096\end{aligned}$$

yielding a standard error of $\sqrt{0.096} = 0.31$ and an estimated width of liability values mapping to the four-bristle class as $6.44\sigma \pm 0.31\sigma$. Sheldon et al. found considerable variation in the strength of canalization across isogenic lines. The mean D_4 of all lines was 5.39, but ranged from 6.44 ± 0.31 down to 3.34 ± 0.10.

Canalization, Selection, and Cryptic Genetic Variation

If the developmental map is highly nonlinear, a simple change in the mean of the underlying liability distribution can result in complex changes in the character distribution. For example, if a population has liability values distributed within a canalized region, as the population is selected away from this region, the character (z) becomes much more variable. As shown in Figure 14.10, this increase in character variance can occur without any change in the variance of the underlying liability distribution. In addition to changing the mean liability, selection can impact the nature of canalization in a trait by changing the liability variable σ^2 and/or the structure of the developmental map ϕ.

As shown by Waddington (1959), there is heritable variation for the strength of canalization, as he was able to select for increased canalization of facet number for bar-eyed mutants of *Drosophila melanogaster*. He did so by splitting families, with one group being reared at 18°C, the other at 25°C. Families showing the smallest difference in facet number between temperatures were chosen to form the next generation. After only five generations of selection, Waddington observed a significant increase in the amount of canalization (Table 14.3).

An interesting example of natural selection for increased canalization was given by Clarke and McKenzie (1987) for Australian sheep blowflies (*Lucilia cuprina*). These flies are a major pest, and in 1955, the insecticide dieldrin was introduced in an attempt to control them. By 1957, most flies in Australia were resistant to dieldrin, and it was replaced by diazinon. Although resistance to diazinon developed in 1976, it is still used. Different

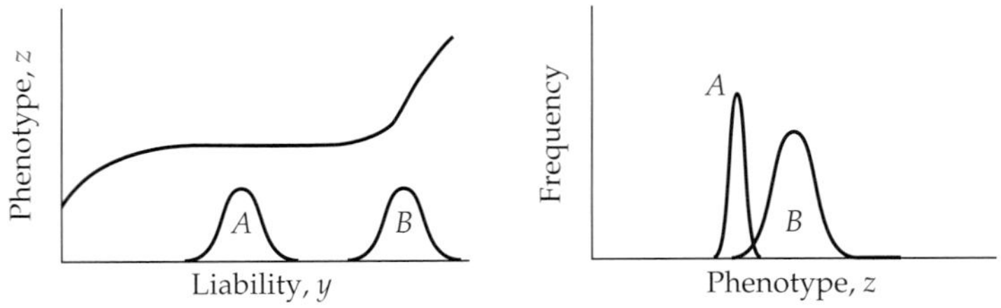

Figure 14.10 When the developmental map is nonlinear, a simple change in the mean of the liability distribution (y, **left**) can result in complex changes in the character distribution (z, **right**). Here, populations A and B have different means, but otherwise identical liability distributions. Besides changing the character value, this simple increase in mean liability also results in an increase in character variance as liability is selected outside the zone of canalization.

Table 14.3 Mean number of eye facets in families (± one standard error) for the control and three lines selected over five generations for increased canalization. (After Waddington 1959.)

Line	Reared at 18°C	Reared at 25°C	Difference
Unselected	156.3 ± 4.8	55.5 ± 4.5	100.8 ± 4.5
Selected A	106.0 ± 1.2	96.5 ± 1.7	14.5 ± 2.3
Selected B	100.7 ± 1.4	95.3 ± 1.7	5.5 ± 2.0
Selected C	111.4 ± 1.7	99.5 ± 2.7	12.9 ± 3.1

loci are responsible for resistance to each insecticide: the *Rop-1* locus for diazinon and the *Rdl* locus for dieldrin. For brevity, we use the locus names to denote their corresponding resistance alleles.

Clarke and McKenzie measured developmental stability for different lines using fluctuating asymmetry (FA; Chapter 6) by comparing the absolute difference in the numbers of three bristle characters on the left versus right sides of flies. *Rdl* alleles cause a significant disruption of developmental canalization, as indicated by a much higher value of fluctuating asymmetry relative to the other strains, while *Rop-1* alleles apparently do not (Table 14.4). However, after 12 generations of continuously backcrossing a *Rop-1*/*Rop-1* resistant strain to the susceptible line M15 while selecting for the retention of *Rop-1*, the amount of fluctuating asymmetry associated with *Rop-1*/*Rop-1* genotypes increased to 5.09 ± 0.08. Clarke and McKenzie interpreted these results as implying that when the *Rop-1* allele arose, it had deleterious pleiotropic effects on development. Strong selection retained the *Rop-1* allele in the population, allowing selection for modifier loci that reduced *Rop-1*'s pleiotropic effects on developmental stability. These modifiers were lost during the continual backcrossing to a strain lacking them (M15 in this case), increasing mean asymmetry. The lack of comparable modifiers for *Rdl* alleles may be due to the fact that dieldrin was extensively used for only a limited period, resulting in insufficient time for the selection of modifiers. More recent work (in blowflies) implicates variation in the *Notch* locus as moderating the asymmetry for the diazinon resistance genotype (Davies et al. 1996; Indrasamy et al. 2000; ffrench-Constant and Bass 2017).

This observation of a major gene disrupting the apparent canalization of a character is fairly common, as revealed in several artificial-selection experiments involving highly canalized characters. For example, scutellar bristle number in *Drosophila* (which is highly canalized at four bristles in almost all *Drosophila* species) could be down-selected to two bristles using the trait variance exposed when the mutant *scute* was introduced (Rendel and Sheldon 1960; Fraser 1963, 1967, 1970; Rendel 1965, 1977, 1979; Rendel et al. 1966). Likewise, **vibrissae number** (whiskers on the nose and fore limbs) in the house mouse, which is normally highly canalized at 19, was able to be considerably down-selected using the variance exposed by the sex-linked gene *Tabby* (Dun and Fraser 1958, 1959; Fraser and Kindred 1960). Still another example is ocellar bristle number in *Drosophila subobscura*,

Table 14.4 Level of fluctuating asymmetry, measured as the absolute value of the differences in three bristle characters on the left and right side for different strains of *Lucilia*. The SWT and M15 lines are susceptible to the insecticides diazinon and dieldrin; the other two lines are fixed for one of the resistance alleles (*Rop-1* for diazinon, *Rdl* for dieldrin) and wild type (susceptible) for the other locus. (After Clarke and McKenzie 1987.)

Strain	Mean Asymmetry
SWT	1.83 ± 0.08
M15	1.81 ± 0.07
Rop-1/Rop-1	1.92 ± 0.07
Rdl/Rdl	3.23 ± 0.08

normally highly canalized at 8, which responded to selection for decreased number when the mutation *ocelliless* was introduced (Sondhi 1961). Again, this change in mean was accompanied by an increase in the variance. In all of these systems, the increase in variance presumably results from the population being moved outside of a zone of canalization (e.g., Figure 14.10). This idea is supported by the fact that, for all three systems, back-selection to return the character values to the original mean value reduced the trait variance. Thus, a return of the mean back to that seen in normal populations also returns the population to the zone of canalization.

While the above examples of changes in canalization could be explained by a simple change in the mean liability value while keeping the developmental map intact, as noted by Waddington (1957, 1959), zones of canalization (the shape of the underlying developmental map, ϕ, itself) could also be subject to selection. Through patient directional selection, Scharloo and colleagues obtained populations of *D. melanogaster* with a mean of eight scutellar bristles per individual, double the normally canalized value of four (reviewed in Scharloo 1988). As in previous examples, this shift in trait mean was accompanied by a substantial increase in genetic variance for the trait as the population moved out of the original zone of canalization. Subsequent artificial stabilizing selection around the new mean eventually doubled the width of the eight-bristle class on the underlying scale. An important caveat on inferring that an outburst of variation following a shift in the mean suggests trait canalization was made by Hermisson and Wagner (2004) and Zhang et al. (2008). They showed that release of hidden variation can occur in the absence of canalization if sufficient epistasis (G $\times$ G) and/or genotype–environment (G $\times$ E) interactions exist and the population experiences a major change in either its environment or genetic background.

One consequence of our developmental-map model of canalization is that **cryptic genetic variation (CGV)** may exist, wherein genetic variation in the underlying liability is not fully expressed as genetic variation in the observed phenotypes. Moving the population outside of a canalization region can result in an increase in genetic variance at the phenotypic level, even without any change in the variance in underlying liability (Figure 14.10). As noted by Paaby and Rockman (2014), "The distinguishing feature of CGV is that the conditions that induce allelic effects are rare or absent in the history of the population, and this rarity limits the opportunities for selection to act on the variation and allows it to accumulate." Unless such normally unexpressed variability has pleiotropic effects on fitness traits, the population-genetic dynamics of underlying loci would follow standard neutral theory (WL Chapter 2), with the level of variation being accrued a function of the effective population size and polygenic mutation rate (Chapter 15; WL Chapters 11 and 28). See Gibson and Dworkin (2004), Paaby and Rockman (2014), and Paaby and Gibson (2016) for additional discussion on the nature, and evolutionary consequences, of CGV.

A final comment. The term CGV can be somewhat confusing to the uninitiated. As developed here, CGV refers to the release of genetic variation uncovered by a dramatic shift in the environment and/or background genotype. However, in other settings, CGV can refer to an entirely different phenomena, namely an increase in genetic variance due to the decay of gametic-phase disequilibrium (Chapter 5). When gametic-phase disequilibrium is present, Equation 5.32a shows that the observed additive genetic variance σ_A^2, is the sum of two components, $\sigma_a^2 + D$, where

$$\sigma_a^2 = 2\sum_{i=1}^{n} \alpha(i)^2 p_i(1-p_i)$$

often called the **genic** variance (WL Chapter 16), is the additive variance expected under linkage equilibrium (LE), while

$$D = 2\sum_{i=1}^{n}\sum_{j\neq i}^{n} \alpha(i)\alpha(j)D_{ij}$$

is the departure due to the presence of disequilibrium, and is negative under many types of selection (WL Chapter 16). Hence, when selection has shaped a trait, often the observed additive variance σ_A^2 can be less than its potential value (σ_a^2) when the population is in LE. Cessation of selection followed by recombination decays D to zero, and the observed additive variance increases.

The existence of canalization has implications beyond evolutionary theory. Gibson (2009; Gibson and Lacek 2020) have suggested that decanalization may be an important factor in the onset of some human diseases, and that this might occur by mutations increasing the variance of the liability distribution, pushing more individuals above a disease threshold. Additional reviews, and perspectives, on canalization can be found in Gibson and Wagner (2000), Sangster et al. (2004), Masel and Siegal (2009), Masel and Trotter (2010), De Visser et al. (2013), Hallgrimsson et al. (2019), Loison (2019), and Takahashi (2019).

Example 14.6 Vigne et al. (2022) were able to isolate a single-nucleotide change that impacts the environmental sensitivity of a trait in *Caenorhabditis elegans*. This species normally lays eggs externally, but under sufficient environmental stress the phenotype of **matricidal hatching** occurs. Here, eggs are retained within the mother, hatching, and eventually killing her. While this can be an adaptive strategy under harsh environments, allowing her offspring to obtain sufficient size to undergo dispause, it obviously is problematic under more benign conditions. The authors isolated a strain, JU751, that displayed high rates of matricidal hatching under benign conditions, showing a lack of sensitivity to the environmental cues relative to the wild type. One (or more) variants in this strain thus canalized the expression of matricidal hatching to the point of making it nearly constitutive. Using QTL mapping with RILs (Chapter 18) enabled the isolation of a 60 kb region containing just three protein-coding genes. A single replacement polymorphism was seen in one of these, KCNL-1, a potassium-channel gene. Using CRISPR-*Cas9* gene editing, they changed this single nucleotide in a wild-type strain to the variant base, generating high rates of matricidal hatching. Reciprocal modification of this nucleotide to the wild-type base in JU751 results in a strain with the normal environmental sensitivity to this trait.

Genetic Assimilation

A potential consequence of moving a population outside of a zone of canalization is **genetic assimilation**, wherein a character state that originally appears to be environmentally determined apparently becomes genetically determined following several generations of selection. This term was coined by Waddington (1953) to account for the behavior of **crossveinless**, a defect in the wing venation pattern of *Drosophila melanogaster*. In the lines studied by Waddington (1952, 1953), no crossveinless flies were found in a base population raised under normal temperatures. However, when temperature-shocked as pupae, some fraction of flies displayed the same phenotype as the crossveinless (*cvl*) mutation.

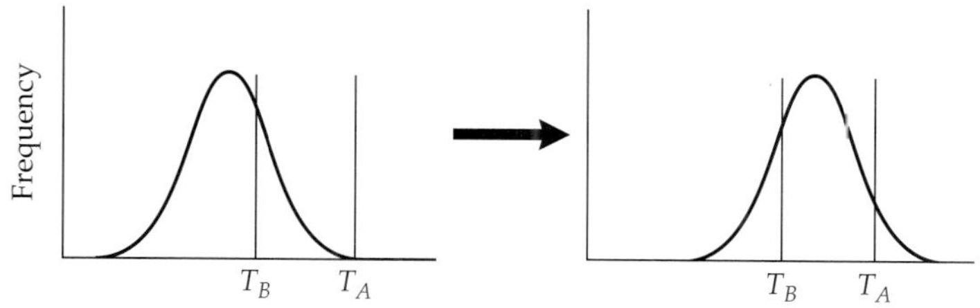

Figure 14.11 A model for the genetic assimilation of the crossveinless character state. Suppose flies display the crossveinless phenotype when the value on some underlying liability scale (horizontal axis) exceeds some threshold value T_A. **Left**: The effect of the temperature shock is to reduce the threshold to $T_B < T_A$, exposing heritable variation in liabilities that are otherwise masked. **Right**: Selection under temperature shock is effective in increasing liability values, eventually shifting the distribution far enough to the right such that crossveinless flies are seen at normal temperatures (liability values exceeding T_A).

Environmentally induced phenotypes that resemble those of known mutations are called **phenocopies** (Goldschmidt 1935), and their presence can cause complications in mapping human disease genes (Chapter 19). Waddington's environmentally induced *cvl* flies were used to form the subsequent generation, with this procedure continually repeated. Not only did the frequency of *cvl* flies induced by temperature shock increase, but *cvl* flies appeared in the untreated populations, reaching frequencies in excess of 95% after 18 generations of selection. Waddington had genetically "assimilated" the crossveinless character state from an apparently environmentally determined character in the base population.

Waddington (1956) presented an even more impressive example of assimilation. In dipterans, the second pair of wings is reduced to a pair of knob-like structures called halteres. The *bithorax* mutant dramatically replaces the halteres with a second set of full wings (a **homeotic** mutation that results in a transformation of body parts). By exposing eggs to ether vapor, a small fraction of bithorax phenotypes was seen. Again, by selecting on these phenocopies, Waddington was able to eventually assimilate bithorax expression in the absence of ether exposure.

Waddington (1942, 1953, 1957) suggested that changes in canalization could account for genetic assimilation, but was rather vague in terms of an explicit model. As Figure 14.11 shows, assimilation could follow as a simple consequence of shifting the liability distribution outside of a zone of canalization. One environment exposes more of the underlying genetic variation in liability than another, making selection in that environment more effective. Introduction of a major mutation can have the same effect.

Example 14.7 Because major genes can disrupt canalization of a specific trait, one might also expect there could be genes whose activity level improves the stabilization of numerous traits. One such example is **HSP90**, which codes a heat-shock protein. Heat-shock proteins are **chaperones**, facilitating the folding of metastable **client** proteins, including correction of improper folds (Lindquist and Craig 1988; Becker and Craig 1994). Among the client proteins of HSP90 are key regulators of growth and development, with HSP90 activity being induced under stress to help stabilize (and properly fold) its clients in the face of environmental challenges (such as temperature shock). Rutherford and Lindquist (1998) noted that individuals with reduced levels of HSP90 often had unusual morphological phenotypes over a multitude of *Drosophila* traits.

Akin to the effects of *scute*, *Tabby*, and *ocelliless*, the "cryptic" variation exposed by HSP90 mutants was often heritable, allowing the authors to select for many phenotypes that eventually became stabilized (assimilated) under normal HSP90 expression. Rutherford and Lindquist (and Rutherford 2000) proposed that HSP90 stabilizes trait expression by **buffering** both genetic and environmental noise. Similar roles for HSP90 (which is widely conserved over all eukaryotes) were seen in *Arabidopsis* by Queitsch et al. (2002). Indeed, Sangster et al. (2008a, 2008b) were able to map *Arabidopsis* QTLs influencing traits that only had an effect under reduced HSP90 activity.

Environmental versus Genetic Canalization

Canalization is the developmental buffering against small perturbations, which may be environmental, genetic, or both. Waddington's initial concern was sensitivity to the environment, with genotypes showing **environmental canalization** (or **environmental robustness**) having lower environmental variances. However, he was also concerned with the fact that a particular genotype may find itself in a variety of *genetic* environments, and that genotypes may also differ in their sensitivities to these different backgrounds (e.g., Table 14.4). **Genetic canalization** (or **genetic robustness**) is the stability of a particular genotype when placed in a variety of different genetic backgrounds. It is a function of the epistasis between a genotype of interest and the universe of genetic backgrounds in which it may find itself (e.g., Example 14.7), partly buffering the impact of new mutational variance (Chapter 15). These two types of sensitivity can be easily confounded, yet they are fundamentally different. Environmental robustness (which requires G $\times$ E) does not necessarily imply genetic robustness (which requires epistasis, G $\times$ G), and vice versa.

Further, it is likely that multiple systems for robustness for different biological processes exist. As mentioned above, one measure of developmental noise is fluctuating asymmetry. *Drosophila* studies have shown that developmental robustness may occur via different mechanisms than environmental buffering. Scharloo (1988) found that lines selected for decreased FA in bristle number showed little differences in between-fly environmental variation. Rutherford (2000) found that while decreased expression of HSP90 impacted between-fly variation, it did not impact FA. Alleles that reduce developmental variance have been called **homeorhetic** (Meiklejohn and Hartl 2002), and in a few cases have been mapped. For example, Leamy et al. (1998) were able to localize QTLs that impacted the amount of FA in mouse skull traits.

Using an appropriate design, the genetic and environmental sensitivities for a particular genotype can be separated. One such design is the use of a series of inbred lines or clones (Fraser and Schadt 2010; Geiler-Samerotte et al. 2013). For a particular genotype of interest (such as a marker locus tagging a quantitative-trait locus), the *among-line* trait variance of a genotype across a series of lines (and hence different genetic backgrounds) is a measure of its genetic sensitivity (G $\times$ G), while the *within-line* variance of the target genotype is a measure of its environmental sensitivity (G $\times$ E).

Example 14.8. Fraser and Schadt (2010) considered mRNA expression levels for thousands of genes over a series of 19 mouse inbred lines. Within each line, roughly 20 individuals were scored for $\sim$160,000 markers. For a given trait (here, the expression level of a specified gene), the within-line variation was contrasted between the two alternative homozygous genotypes at each marker. Because there is essentially no genetic variation within an inbred line, significant differences in the within-line variance over marker genotypes indicate linkage to a QTL influencing environmental robustness (differences in σ_E^2). Conversely, a significant difference in the dispersion of the mean values of the marker genotypes *across* inbred lines (increased among-line variance) indicates that the marker is linked to a QTL influencing genetic robustness. Using this approach, these authors found QTLs for both types of robustness. QTLs for environmental robustness were largely *trans*-acting and sex-specific (different QTLs in the two sexes). In contrast, QTLs for genetic robustness were often *cis*-acting and were not sex-specific. There was no detected overlap between the two classes of QTLs. In reporting their results, the authors used the convention that an eQTL (expression QTL—a QTL impacting the expression level of a gene) mapping close to its target site was regarded as *cis*, while an eQTL mapping further away, or on a different chromosome, was regarded as *trans* (see Chapter 21 for an extended discussion).

One caveat about this study concerns the among-line variance. Given the small number

(19) of genotypic backgrounds (inbred lines), we expect that by chance some markers will be nonrandomly distributed with respect to QTLs that influence the line means. This could result in alternative genotypes at these markers showing different patterns of among-line variance, not because of any epistatic interactions, but rather because they were not sufficiently randomized with respect to background QTLs influencing the mean of the target trait.

Example 14.9. Nagpal et al. (2022) exploited the power of combining genomics with medical databases with extensive environmental records (answers from patient questionnaires such as birth weight, cheese intake, duration of walks, etc.) to search for canalization on human traits. Specifically, they computed polygenic scores (PGS) for ten traits (7 diseases, 3 others) for 400,000 individuals in the UK Biobank. PGS are predictors of trait values based on GWAS markers (Chapter 21; examined in detail in Part II Chapters 32 and 33). Returning to Figure 14.7, one can think of the PGS values as being akin to liability values (y), with the phenotype (z) in this figure being the average penetrance (probability of trait expression) for a given PGS value. For each trait, the authors constructed such a curve for the average penetrance over all environments, and then constructed such maps for 150 specific environments (e.g., low versus high cheese consumption). In some cases, the gap in prevalence increased over specific environments (indicating decanalization), whereas it decreased (showing canalization) in others. Beside the biological interest in such G $\times$ E effects, they can also have profound implications for precision (individualized) medicine.

Is Robustness an Intrinsic, or an Evolved, Property?

While Waddington viewed increased robustness as an evolved feature, the question still persists as to whether robustness is an intrinsic feature of the structure of biological systems, or is an evolved feature. Much of the focus of this debate has been on genetic versus environmental buffering, as the conditions for the evolution of genetic canalization are more restrictive than for environmental canalization (Wagner et al. 1997), in part because the target background under the former is itself continually evolving (reviewed by Flatt 2005 and Hansen 2006). De Visser et al. (2003) added a third hypothesis to this intrinsic-evolved dichotomy, **congruence**. Here selection for environmental and/or developmental robustness generates, as a side effect, some amount of genetic canalization (Wagner et al. 1997; Ancel and Fontana 2000; Meiklejohn and Hartl 2002).

In reality, both intrinsic and evolved features likely underlie robustness, with selection exploiting a system with intrinsic regions of stability by moving a population towards these parameter regions. A good model system for thinking about this issue is the evolution of dominance (De Visser et al. 2003). As Figure 4.5 shows, the kinetic nature of interacting enzyme systems is such that the output (flux) as a function of input (enzyme levels) is usually hyperbolic, with the output reaching a saturation level, above which further input has little effect. Hence, the intrinsic nature of this system has a region of stability—alleles residing on the saturation plateau can have significant perturbations in their underlying input, and yet still have little impact on the flux. However, stability is also an evolved feature, as the alleles were selected towards the region of stability (the saturation level). Similar situations likely exist for many biological networks (Siegal and Bergman 2002; Kitano 2004; Félix and Barkoulas 2015; Osterwalder et al. 2018). For example, small-world, scale-free graphs (the topological representation of many complex biological systems; Appendix 2) tend to be more robust to changes in inputs, and such topologies tend to naturally arise (Chapter 21), leading to a situation for evolution to exploit by moving populations towards these intrinsic stability regions.

Phenotypic Plasticity and Genetic Accommodation

In our discussion of developmental maps, we largely treated the liability value as the evolvable component (as liability has a genetic component), with the shape of the map (such as threshold levels for discrete traits and general curvature for continuous traits) as being largely fixed. Of course, the latter is also evolvable, and we can make a subtle shift in focus whereby now the liability axis is some environmental value, with the mapping giving the **norm of reaction** for a particular genotype (how trait value changes over that environmental variable). Characters often show **phenotypic plasticity**, with trait value changing with the environment, often in an apparently adaptive fashion. West-Eberhard (2003) referred to an adaptive genetic change in the environmental regulation of a trait value as **genetic accommodation**, which can take the form of either increased plasticity or increased robustness (greater or lesser sensitivity). Ehrenreich and Pfennig (2016) noted that genetic assimilation is an extreme example of accommodation, namely a shift such that a formerly environmentally induced phenotype is now fully canalized to display the new value. The evolutionary implication of phenotypic plasticity is a very rich field that will not be further explored here. Reviews, with a strong evolutionary focus, are offered by Schlichting and Pigliucci (1998), West-Eberhard (2003), Badyaev (2005), Pfennig et al. (2010), Moczek et al. (2011), Schlichting and Wund (2014), Ehrenreich and Pfennig (2016), Scheiner et al. (2017), Schneider and Meyer (2017), and Nijhout et al. (2021).

GENERALIZED LINEAR MODELS

Much of classical quantitative genetics is built around **linear models** (Chapter 10), wherein a linear regression, $\widehat{y} = a + \sum_k \beta_k x_k$, is used to predict the value of some outcome ($\widehat{y}$) as a function of a number of predictor variables (the x_k). For example, a linear regression containing marker scores and important environmental features—such as body-mass index (BMI), sex, or cholesterol level—might be used to predict a trait value or a disease risk. The observed trait value can be expressed as its predicted value plus a residual, $y_i = \widehat{y}_i + e_i$, where e_i (the deviation of observed from expected values) is assumed to be normally distributed. Linear models have a number of very desirable features (Chapter 10), and are appropriate in a wide variety of settings (either as exact results or as very good approximations). However, there are also settings where they perform poorly. One limitation occurs with many types of discrete (count) data, where the residual error *deviates quite significantly from the normal distribution* assumed by linear models.

A second limitation is that linear models may predict values that lie *outside of the model parameter space*. Suppose we are trying to predict the probability p that an individual with a given marker genotype and a set of risk factors will get a disease. The resulting predicted probability value (disease risk) should lie between 0 and 1, but one could very easily envisage a linear model predicting a negative value or a value in excess of 1. One can think of this as an ***effect of working on the wrong scale***, and hence we close our discussion of scale effects by showing how a transformation of scale motivates the use of **generalized linear models** (note the **ized**). Such models are based on transforming from *unobserved* value y (or **latent variable**) on some *underlying scale* to an *observed* value z on the **observed scale**. We have already seen an example of this with threshold models, where the latent variable is liability, which maps into a binary (presence/absence) trait z. The two central features of generalized linear models are that: (i) the mean $E[z|y]$ on the observed scale is *not* a simple linear function of the underlying latent variable y, and (ii) the residual error need not be normally distributed.

The basic structure of a generalized linear model can be thought of as an extension of the liability model to more general functions. Let the liability value (y_i) for individual i be modeled as some linear function of predictor variables, x_k (e.g., markers or known environmental factors),

$$y_i = \mu + \sum_k \beta_k x_{k,i} \tag{14.13a}$$

This value on the underlying (latent variable) scale is then mapped by a monotonic function, g, onto the observed scale (z) to give the *expected observed value* of z conditional on the underlying value, y. Specifically, the conditional expectation of z given y_i is

$$E[z \mid y_i] = g(y_i) = g\left(\mu + \sum_k \beta_k x_{k,i}\right) \tag{14.13b}$$

where the function, g, imposes the desired constraint on the mean of z, mapping some unconstrained value y into the desired constraint space, such as $(0, 1)$ or $(0, \infty)$. The inverse g^{-1} is called the **link function**, as it transforms the conditional expectation into a linear model,

$$g^{-1}\left(E[z \mid y_i]\right) = y_i = \mu + \sum_k \beta_k x_{k,i} \tag{14.13c}$$

As with a linear regression, the *actual* observed value is the expected value plus a residual, $z_i = E[z \mid y_i] + e_i$, where e_i no longer need be normal. For example, suppose z is a binary trait and that an underlying y value maps into an expected z value of 0.7, or a 70% chance that such an individual shows the disease (or other trait). Because the actual observation for z is binary—0 (absent) or 1 (present)—the residuals are binomial rather than normal.

We focus here on two particular generalized linear models that commonly appear in quantitative genetics—**logistic regressions** and **log-linear models**—both of which arise in the analysis of count data. The former deals with binary data, the latter with nonnegative integer data assumed to follow a Poisson distribution (Chapter 2). A nice discussion of estimation issues under generalized linear models (beyond the binary and log-linear models presented here) is given by de Villemereuil et al. (2016), while detailed treatments can be found in a number of standard texts, such as McCulloch et al. (2008). Maximum-likelihood (Appendix 4) or Bayesian approaches (Appendix 7) are used for model fitting and analysis.

The Logistic Regression Model for Binary Traits

Characters that display a binary value (presence or absence) are examples of **Bernoulli-distributed** traits (a binomial with $n = 1$, i.e., a single draw). Observed trait values (z) are coded as either 0 (absent) or 1 (present), with the goal being to predict the mean value of z given some information (such as marker genotypes or environmental risk factors) that generates a value y on the liability scale. If $p(y)$ is the probability, for a given y, that z has a value of 1, then $E[z|y] = 1 \cdot p(y) + 0 \cdot [1 - p(y)] = p(y)$. Hence, given a value of y on the latent scale, the observed value z follows a Bernoulli distribution with success parameter $p(y)$.

The function $p(y)$ serves the role of g in Equation 14.13b in mapping from y onto the observed scale (the trait value, z). From a biological standpoint, one imagines that $p(y)$ is a monotonically increasing function of y, approaching 0 for low values (low risk when the liability is sufficiently low) and 1 for high values (almost certain risk when the liability is sufficiently high). One candidate that satisfies these requirements is the **logistic function**,

$$\ell(y) = \frac{\exp(y)}{1 + \exp(y)} = \frac{1}{1 + \exp(-y)} \tag{14.14a}$$

with $\ell(y) \simeq 0$ for $y \ll -1$, $\simeq 1$ for $y \gg 1$, and $\ell(0) = 1/2$. The inverse of the logistic (the link function of Equation 14.13c) is given by the **logit function**, defined as

$$\text{logit}(p) = \ln\left(\frac{p}{1 - p}\right) \tag{14.14b}$$

Namely, the log of the **odds ratio** (probability of the trait being present divided by probability that the trait is absent). If $\ell(y) = p$, then $\text{logit}(p) = y$, so that a logit-transformed p value recovers the liability value,

$$\text{logit}(p|y_i) = y_i \tag{14.14c}$$

For example, suppose that one has 100 identical clones, scored under identical environmental conditions, with 15 of these having a disease. Hence, for this clone, $p = 0.15$, giving its underlying liability value y as

$$y = \text{logit}(0.15) = \ln\left(\frac{0.15}{0.85}\right) = -1.735$$

The approach behind **logistic regression** is to model the value of y on the liability scale as a linear function of the predictor variables of interest,

$$y_i = \mu + \sum \beta_k x_{k,i} \tag{14.15a}$$

To translate this predicted liability value into the observed binary trait value z, we use the logistic function to map y_i into the probability that z_i is one, namely $p(y_i)$, with

$$p(y_i) = \frac{1}{1 + \exp\left[-\left(\mu + \sum \beta_k x_{k,i}\right)\right]} \tag{14.15b}$$

The observed value of z for this value of y follows a Bernoulli distribution with success parameter $p(y)$, which is predicted by Equation 14.15b. Likewise, recalling Equation 14.14c, the predicted value of the logit score (log of the odds ratio) is

$$\text{logit}(p|y_i) = \ln\left(\frac{p(y_i)}{1 - p(y_i)}\right) = \mu + \sum \beta_k x_{k,i} \tag{14.15c}$$

Logistic regressions are a natural choice for the modeling of binary data, as the residual structure of such data is naturally Bernoulli, rather than normally, distributed. Indeed, this is the standard model for association analysis (GWAS) for binary disease traits (Chapter 20).

Log-linear Models for Poisson-distributed Traits

Discrete characters with a large number of possible states, such as numbers of leaves on a tree, can usually be treated as a continuous trait. However, what about a discrete trait with a rather compact distribution? A common example is the number of offspring, such as the clutch size of a bird, which might (say) range from 0 to 6 eggs in an observed sample. This discreteness is of special concern when the trait has a significant probability mass at a particular value (especially zero), as often happens with offspring number (corresponding to failure to reproduce). In such settings, the normal distribution is usually not a good fit, or even a reasonable approximation, for the observed distribution.

A natural way to model such traits is to use the **Poisson distribution** (Chapter 2), where the probability of observing a trait value of k is given by

$$\Pr(z = k) = e^{-\lambda}\,\frac{\lambda^k}{k!} \quad \text{for } k = 0, 1, 2, \cdots \tag{14.16}$$

where $\lambda = E[z]$ is the expected value of the trait, which is constrained to $\lambda > 0$. Motivated by the previous treatment of binary traits, one might imagine that some underlying liability value, y (modeled as a linear regression; Equation 14.15a), is mapped onto a mean value, λ, constrained to be positive (so that it corresponds to a mean of the Poisson). One of the simplest choices that accomplishes this objective is the exponential, $\lambda = \exp(y)$, which maps any y value into $(0, \infty)$, the space of admissible values for the mean of a Poisson.

Hence, the expected value on the observed scale can be expressed as

$$\lambda = \exp(y) = \exp\left(\mu + \sum \beta_k x_k\right) \tag{14.17a}$$

ensuring that $\lambda > 0$ for all y and hence is a proper expectation for a Poisson. In the context of generalized linear models, $g(y) = \exp(y)$, so that the link function $g^{-1}(y)$ is simply $\ln(y)$, as $g^{-1}[g(y)] = \ln[\exp(z)] = y$, yielding

$$\ln(\lambda \mid y) = y = \mu + \sum \beta_k x_{k,i} \tag{14.17b}$$

This is called a **log-linear model**, as the log of the distribution parameter (the Poisson mean λ) is a linear function of the predictor variables. The resulting distribution of an observed trait value z_i, given that its underlying latent value is y_i, is thus a Poisson with mean $\exp(y_i)$.

The inquisitive reader may wonder about the difference between a log transform of data and a log-linear model. If the distribution is centered far away from zero and the probability is spread out over a number of values, then a log transform might be appropriate. The resulting log-transformed data could be treated with a linear model with normally distributed residuals. However, if the distribution is very compact (most of the probability mass resides at just a few values), then a log-linear model may be more appropriate, especially when there is significant probability mass at zero.

However, just because the distribution is compact and has significant probability mass at zero does not ensure that the Poisson distribution will provide a good fit. Modifications of this base model exist, such as a **zero-inflated Poisson**, which has extra probability mass at zero relative to a standard Poisson (e.g., Rodrigues-Motta et al. 2007; see WL Chapter 29), or a **zero-truncated Poisson** for traits without a natural value of zero (WL Chapter 29). Other modifications arise when the trait is overdispersed (variance greater than the means), such as a **generalized** (or **quasi**) **Poisson** (Breslow 1990; Consul and Famoye 1992; Chen and Ahn 1996). A trait could also be underdispersed relative to the Poisson (as seems to consistently be the case for bird clutch size; J. Hadfield pers. comm.). The negative binomial distribution (Equation 2.23a) is often used in place of the Poisson, as its distribution parameter values can be chosen to make it either under- or overdispersed (Equation 2.23c), e.g., ver Hoef and Boveng (2007).

Two final comments. First, while transformations are generally used to make the distribution of residuals closer to normality in order to apply standard linear or mixed models, Fusi et al. (2014) examined **warped linear mixed models**. These jointly estimate the appropriate transform to make that data better behaved for the chosen generalized linear model and then estimate the model parameters. Second, our discussion about generalized linear models has assumed that the underlying variable, y (Equation 14.13a), is entirely a function of *fixed effects*. More generally, this underlying variable could be generated by a mixed model, including both fixed and *random* effects. The resulting class of models are called **generalized linear mixed models**, or **GLMMs**. Their analysis involves a number of subtleties; see Bolker et al (2008), Grueber et al. (2011b), and Bolker (2015) for an introduction and additional references.

II

Quantitative Trait Loci

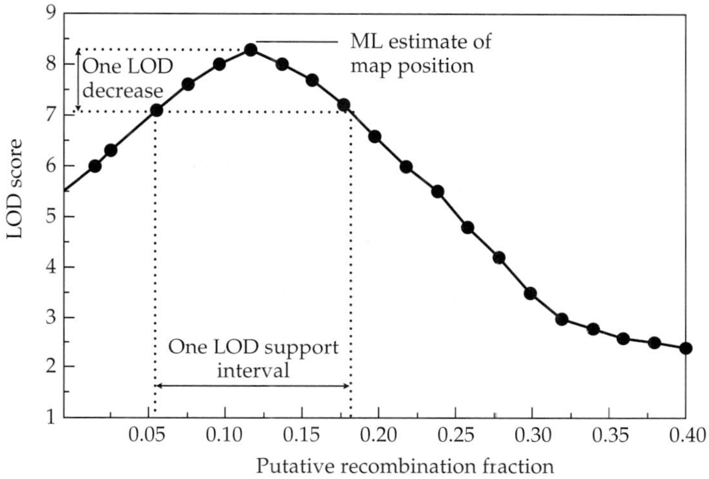

15

Polygenes and Polygenic Mutation

The term polygenic is not intended to imply the existence of any special class of genes called polygenes. Carter (1969)

One of the great strengths of classical quantitative genetics is that a wide variety of questions can be resolved, and outcomes predicted, without detailed knowledge of the underlying **genetic architecture** of a particular trait (Chapters 7, 12–14; Part II Chapters 22–33). Simple phenotypic associations between relatives can provide estimates of genetic variance components (Chapter 7), while line-cross analysis and inbreeding experiments (Chapters 11–13) can provide information on the nature of genetic interactions (dominance and epistasis) and offer some (admittedly crude) insight into the number of loci accounting for the difference in mean character value between lines.

Depending on the question being asked, a deeper understanding of this architecture is often desirable. In particular, knowledge of the joint distribution of the allelic effects and their frequencies at the loci underling a trait is important in a number of applications. For breeders, if genetic variation in a character is mainly due to a few genes with large allelic effects, the response to selection can be considerably different from that predicted from standard models of quantitative genetics that assume a large number of loci with small effects (WL Chapters 24–26). When applied to disease susceptibility genes, the allelic-effects distribution has important consequences for the accuracy of personalized medicine (Part II Chapters 32 and 33). This distribution also provides significant insight into the nature of the evolutionary forces that shaped the causal loci (Chapter 21; WL Chapters 5, 7, and 28). Note that this level of architectural detail (the joint allelic effects-frequency distribution) *does not* require knowing the *actual* underlying variants, much less their molecular mechanisms for generating trait variation. We typically estimate the allelic-effects distribution for a given trait by using proxy markers that are tightly correlated (i.e., in LD) with the actual variants, rather than requiring knowledge of the causal variants themselves (Chapters 18–21).

The highly elusive goal of finding the *actual* genes (or, more formally, **causative sites**) underlying trait variation is one of the holy grails of quantitative genetics. Developmental and systems biologists are interested in the pathway components and interactions that lead to particular trait values. Localizing risk genes can suggest drug targets and possible therapeutics. Finally, knowledge of causal genes can be leveraged to improve breeding and provide evolutionary insight (e.g., searching for signatures of selection on the underlying trait genes; WL Chapters 8–10, 12). Our most accurate resolution is offered by GWAS (Chapter 20), but even here, and even with whole-genome sequences, it has been challenging to assign a highly significant hit to a specific variant within a region (Chapter 21). Further, as will be reviewed in Chapter 21, most causal variants *segregating in populations* appear to generally have small variance effects (consistent with either common alleles of small effect or rare alleles of much larger effect) and the majority of GWAS hits appear in noncoding regions, often some distance from any coding sequences. Hence, much of variation in segregating populations appears to be regulatory in nature. The situation may be somewhat different for *fixed differences between populations*, as many line-cross mapping experiments are performed by choosing parental lines that dramatically differ in a focal trait, often as the result of selection. Any such fixed genetic differences are the result of taking some initial distribution of segregating variation and then strongly filtering it through selection.

In this chapter, we first briefly review potential genetic sources underlying quantitative variation, with a more detailed treatment given in Chapter 21. The basic message is that the ***polygenic mutational target size for a trait is potentially very large***, involving both a significant number of coding regions and an even larger amount of noncoding sequences. The bulk of this chapter concerns the rate at which mutation at these target regions generates new variation for quantitative traits. Polygenic mutation rate estimates have both short- and long-term implications. Over the short term, mutation is a key force in shaping the nature of standing genetic variation, which impacts health, breeding, and adaptation in rapidly changing environments. Over the long term, mutation must ultimately dictate the potential rate of phenotypic evolution. Although the mutational mechanisms responsible for quantitative-trait variation are only starting to be explored, phenotypic data strongly support the idea that mutation is a powerful force at the level of polygenic traits (WL Chapter 27).

POTENTIAL SOURCES OF QUANTITATIVE-GENETIC VARIATION

One of the earliest questions regarding the nature of genes underlying quantitative traits was whether they were fundamentally different from classical **Mendelian genes**. The latter are simply genes with major-effect alleles whose transmission, as scored by phenotype, seems to follow Mendel's basic laws for single-gene inheritance. Such genes could also segregate a number of alleles of lesser (perhaps *far* lesser) effect, and this could account for quantitative-genetic variation. A second possibility is that there are genes that *never* have large-effect mutations (**minor genes**). As such, they would largely escape detection in classical large-effect mutation screens, with allelic variation at these genes generating quantitative variation. Obviously, the most likely scenario is a weighted mixture of these two classes (major and minor genes), raising the question of which accounts for most of the variation for a given trait. Historically, much of this discussion was framed in terms **structural variation**: variation in the final gene product from a given locus, be it slightly different amino-acid (protein), or functional RNA, sequences. (We note that while this is the historical definition of "structural variation" from the evolution literature, genomicists have recently co-opted this term for variation in large-scale genomic features, such as a localized duplication; Chapter 8.) Another contributor is **regulatory variation**: variation in the timing, amount, and celluar location of the gene products (note that this structural versus regulatory distinction is not sharp, as changes in the structure of one gene product—e.g., a transcription factor—can influence the regulation of another). Regulatory sites generating quantitative variation could very easily map at some distance (even on different chromosomes) from the target gene(s) whose structural product(s) they influence. In this case, a genomic search around a site showing a marker-trait association would not yield the target gene whose product influences trait variation. This often (perhaps *very* often) seems to be the case (Chapter 21).

Major Genes and Isoalleles

Is there a real distinction between major- and minor-effects genes or are they just extremes on a continuous distribution? The presence of **isoalleles** at major loci provides some insight. Stern and Schaeffer (1943) initially coined this term to describe alleles whose effects were essentially indistinguishable except under certain environmental settings. They found that three allelic variants of the *cubitus interruptus* locus—which has a major influence on the wing-venation pattern of *Drosophila melanogaster*—all showed the same wild-type phenotypes at 26°C, but exhibited different wing-venation patterns at lower temperatures. Similarly, Muller (1935) and Green (1959) found a large number of isoalleles affecting eye color at the *white* locus of *D. melanogaster*. Because both of these loci were originally detected by displaying alleles of major effect (**major alleles**), the subsequent discovery of isoalleles suggested the possibility of a continuum of allelic effects. Isoalleles have been found at a variety of other loci in a number of organisms, with alleles having a large effect on one character often showing a wide range of **pleiotropic effects** on other traits. Thus, alternate alleles at a locus can display a spectrum of effects, which depends on the genetic and environmental background (Coen et al. 1986; Romeo and McKusick 1994; Botstein and Risch 2003). The *alleles* present at a locus, not the *locus itself*, determine if that locus has a major effect on character variation.

In humans, alleles of small effect are routinely detected at loci that also exhibit rare mutations of large effect. Examples include HDL cholesterol (Cohen et al. 2004; Romeo et al. 2007), blood pressure (Ji et al. 2008), Hirschsprung disease (Emison et al. 2010), and Bardet–Biedl syndrome (Zaghloul et al. 2010). Likewise, loci detected via common alleles with modest effects routinely habor rare alleles of large effects, e.g., Crohn's disease (Rivas et al. 2011) and rheumatoid arthritis (Diogo et al. 2013). These observations presage the debate (examined in Chapters 20 and 21) of whether common diseases are largely caused by common alleles of modest effects or by rare alleles of large effect.

Modifier loci provide a mechanism for "fine tuning" a major allele by epistastically altering its main and/or pleiotropic effects. As noted in Chapter 14, modifiers can sometimes magnify the impact of a major allele, thereby inflating the response to selection. They can also mask the effect of a major allele, producing a zone of canalization. In natural populations of *Drosophila*, extensive genetic variation has been found for loci that modify the effects of the alleles at the *crossveinless* (Milkman 1970), *Beadex-3*, *notchoid*, and *notchoid* loci (Thompson and Spivey 1984), all of which influence wing venation. Modifiers also appear to be involved in the evolution of mimicry in butterflies. Initially, a major allele causes wing patterning to crudely resemble a distasteful model species, with the pattern then subsequently refined by a number of modifier loci (Turner 1977, 1981). The genetics of butterfy mimicry is a very active field, and more recent developments are reviewed by Zhang et al. (2017), Deshmukh et al. (2018), VanKuren et al. (2019), and Van Belleghem et al. (2020). Modifiers can also alter the dominance relationships of major alleles, as demonstrated for the melanic form of the famous black-peppered moth *Biston betularia* (Ford 1975; other examples are discussed in Thompson and Thoday 1972).

Modifiers can have either general or specific effects. An example of the former is given by Thompson (1973, 1974), who found that modifiers selected to alter a specific wing vein in *Drosophila* had similar effects on the phenotypic expression of major alleles at three different loci: *cubitus interruptus*, *short vein*, and *veinlet*. On the other hand, Fraser (1968) found that the phenotypic expression of *Drosophila* bristle loci *scute* and *extravert* was influenced by different sets of modifiers. In this case, the modifiers appear to be more gene-specific than pathway-specific. Through artificial selection experiments, Weber (1992) was able to alter the relative sizes of two very small regions (less than 100 cells across) at the base of the *Drosophila* wing. If this result is general, the control of developmental detail may be modulated at extremely localized scales.

Human disease genetics is perhaps the richest source of information on the nature of the molecular changes underlying major-effects alleles. It should be cautioned that diseases may have rather different features from a randomly chosen trait, so any extrapolation of these results should be treated with at least some skepticism. A nice summary is provided by Botstein and Risch (2003), who found that 59% of (major-effect) disease mutations are single-base missense/nonsense changes (i.e., structural changes in the protein sequence), 22% are deletions (such as whole or partial gene knockouts), 10% are splicing variations, 7% are small insertion/deletion events (including frameshift changes), $<2\%$ are due to complex rearrangements (such as chromosomal translocations), and $<1\%$ are due to other regulatory changes.

Human genetics also offers some insight into the phenotypic effects of structural changes in gene products. Typically, **missense mutations** (changes in the amino-acid sequence that do not involve stop codons) result in less disease severity when they involve **conservative substitutions** (the new amino acid is in the same functional class as the wild type, i.e., in the same charge or hydrophobicity group) as compared with **radical substitutions**, where the new amino-acid residue has significantly different properties from the original (Botstein and Risch 2003). Likewise, larger disruptions tend to have larger effects. For example, mutations generating the severe Duchenne form of muscular dystrophy are caused by frame-shift deletions, while in-frame deletions in the same gene give the milder Becker form (Gillard et al. 1989). Thus, a range of effect sizes can be generated by slightly different protein sequences. An important cautionary example to the view that most major-effect mutations are structural is offered by the *tb1* mutation in maize. This mutation results in a dramatic change in plant architecture (changing single-branched domesticated maize to the multiple-branched architecture of its ancestor, teosinte; Figure 5.2). The causative mutation is the insertion of a *Hopscotch* retroposon roughly 64 kb upstream of the *tb1* coding region, which enhances *tb1* transcription (Studer et al. 2011).

Finally, we are starting to probe the question of whether loci exist at which *all* allelic variation results in only small phenotypic changes. The challenge in addressing this question is ascertainment bias, given the bias towards isolating factors of fairly large effect. As reviewed in Chapter 21, the vast majority of human GWAS hits involve regions not previously associated with large-effect mutations (i.e., previously known trait genes). While this seems to suggest that a large class of genes that can only have small-effect mutations exist, caution is in order. Large-effect mutations may be rare, and it may simply be that not enough alleles have been sampled at a given gene to find these rare larger-effect variants. Genomic insight into this question is offered by systematic, genome-wide single-gene knockout collections that have been generated for yeast (Giaever et al. 2002; Giaever and Nislow 2014) and mice (Skarnes et al. 2011; White et al. 2013). In mice, 42% of the knockouts have discoverable viability effects (Reed et al. 2008), and pleiotropy is widespread. Hence, a mutant allele with a large effect on a target trait might also have such a low fitness that it is never detected. Knockout collections also show that the number of potential underlying loci for a trait is not simply large, but rather can be *vast*. Reed et al. (2008) found that 35% of the roughly 200 single-gene knockout mouse lines surveyed showed changes in weight, with most (92%) of these showing decreased weight. In yeast, while initially 85% of knockouts were deemed "nonessential" in that they showed no obvious effect on growth (Giaever et al. 2002), more careful assays found that around 45% of (haploid) yeast knockouts had (at least) small fitness effects (Breslow et al. 2008; Hillenmeyer et al. 2008).

Constraints on Mapping Resolution

Problems with the actual molecular characterization of loci aside, precise analysis of the phenotypic effects of individual loci underlying quantitative traits remained historically elusive for a purely statistical reason: insufficient number of recombinants in the sample population. As noted in Chapter 11, limitations introduced by a linkage-based analysis restrict the unit of analysis to *effective factors* (linked regions of the genome that may contain multiple loci influencing the trait) rather than to *single genes*, which can lead to biased estimates of the effects of individual loci. For example, if a factor contains linked loci at which alleles influencing character expression are in coupling phase (++ and −−), the effects of individual loci will be overestimated, while if the alleles are in repulsion phase (+− and −+), the effects will be underestimated. The larger the average size of an unrecombined region, the greater the opportunity for bias from lumping the effects of several loci.

Because of this empirical restriction of dealing with *factors* rather than *individual loci*, a reasonable term was needed to denote a genomic region at which segregation contributes to the variance of a quantitative character. Mather (1941) suggested the term **polygene** for a gene of small effect underlying a character. However, for the reasons discussed above, the distinction between major genes (sometimes referred to as **oligogenes**) and polygenes is by no means straightforward (if it even exists at all). A less ambiguous approach is to refer to a *region* underlying quantitative-trait variation as a **QTL**, for **quantitative-trait locus**, as suggested by Geldermann (1975).

Although our technical ability to improve mapping resolution has greatly advanced from the pre-genomics era, this issue of isolating the smallest causative region persists, albeit over a much finer scale (Chapters 20 and 21). The level of resolution is set by the number of meioses in the study design. For linkage-based mapping using line crosses (Chapter 18) or short pedigrees (Chapter 19), only a few rounds of meioses have occurred, resulting in (at best) megabase-level resolution. Hence, a QTL detected via linkage mapping is, in reality, a *rather large region of DNA*, potentially containing tens to hundred of genes. For GWAS mapping based on linkage disequilibrium within a population sample (Chapter 20), hundreds to thousands of historical meioses have occurred (Chapter 8), often resulting in kilobase-level resolution. While the resolution size under a GWAS is often on the scale of one to a few genes, polymorphic sites within this region can still be completely associated (in complete LD), forming a small number of distinct haplotypes. Determining which (if any) of the polymorphic sites within a given haplotype is causative is very problematic (Chapter 21), especially when a phenotypic effect may be the result of an interaction between two (or more) of the sites within that haplotype.

Genetic Pathways to Quantitative Variation

The cartoon view of how genomic variation translates into trait variation is roughly as follows. Some fraction (often rather small) of the genomic DNA codes for RNA transcripts, most of which (mRNAs) are then translated into proteins. However, a growing number of small noncoding RNA transcripts have been shown to be important in gene regulation (He and Hannon 2004; Ambros and Chen 2007; Chen 2009; Storz et al. 2011; Djami-Tchatchou et al. 2017; Tang and Chu 2017). Control over levels of transcription can be influenced by both ***cis*** and ***trans*** factors. *Cis* factors are sequences at which regulatory proteins or RNAs bind to initiate, enhance, or suppress transcription (or facilitate other cellular functions, such as microtubular binding sites for chromosome segregation during cell division). *Cis*-acting factors (by definition) are on the same DNA molecule as the target gene(s) they influence, but could be separated from their target(s) by a considerable distance (e.g., Loots et al. 2000; Studer et al. 2011). DNA is not a rigid rod, but rather a very flexible rope, so that two sites a considerable distance apart on a DNA sequence may be adjacent when the DNA is packaged and attached to the nuclear scaffold. (We note that genome biologists often restrict the use of *cis* to sites that are rather close to the gene of interest; Chapter 21.) *Trans* factors are diffusable products (regulatory proteins or RNAs) that are coded by genes that are not necessarily linked to their target sites. Once a transcript is made, it is usually **processed** (chemically modified). Such **post-transcriptional modifications** can influence translational efficiency and transcript stability. Splicing (in eukaryotes) goes further, often generating several different proteins from the same initial transcript (**alternative splicing**, leading to different mRNA **isoforms**). As detailed in Chapter 21, there is considerable genetic variation at each of these (and many other) regulatory steps.

Control regions act in a modular fashion to evoke a specific regulatory program (e.g., Struhl 1987), and single nucleotide changes within them often result in radical (qualitative) changes in gene expression, sometimes creating major alleles. For example, in *Drosophila,* loss-of-function mutations in the *ultrabithorax* gene result in a duplication of the thorax, implicating regulatory properties of this gene in the development of segmental identity. Waddington (1957) found that bithorax phenocopies can be obtained in otherwise normal flies by exposing early embryos to ether, and that the tendency to express such a phenotype can be advanced by artificial selection (Chapter 14). A dramatic example of genetic assimilation (Chapter 14), such evolutionary change appears to result from allele-frequency change at the *ultrabithorax* locus itself (Gibson and Hogness 1996).

While much of the focus on regulation is at the level of *transcription* (including post-transcriptional RNA processing), a variety of controls can occur at *translation*, such as slight differences in translation efficiency due to sequence variation. Once a protein is made, there are further levels of control in terms of folding, assembly into complexes, transportation around the cell, stability, and so on. Cellular and systems biologists often think in terms of **pathways**, the dynamic, complex networks of interacting components that regulate the production of more downstream products (Chapter 21; Appendix 2). Regulation occurs at this level as well, with the structure of pathways potentially buffering, or enhancing, the impact of perturbations in the nature, timing, and amount of gene-product components in the network. A variant with a large molecular effect, such as a several-fold increase in the amount of transcript or a truncated protein product, may be buffered by a pathway into a very minor effect on the final product (the trait value).

The upshot of this discussion is that there is a **huge role for regulatory variation**, and the strong view of a much larger regulatory mutational target size than for coding sequences (Wright 1968; Wilson 1976; Carroll 2008). The number of bases that mutation can target in all of the relevant coding regions (potential structural variants) for a given trait might be less, and sometimes perhaps *far less*, than the number of mutational targets that impact its regulation. As will be discussed shortly, mutational data in *Drosophila* are consistent with this view, as are emerging GWAS data for human traits such as height (Example 21.15).

Many of the sites impacting regulatory variation could potentially have very indirect effects, seeming to have no obvious biological function. For example, a single-base change at a random, noncoding site might result in that region having a very weak (but nonzero) binding affinity for a small RNA involved in post-translation regulation of genes for our targeted trait. This might generate a very small reduction in the amount of free RNA, resulting in a very small effect on the trait. In contrast, structural changes are rather granular, and hence might be expected to have more dramatic effects, such as a **nonsense mutation** (mutation to premature stop codon) that results in a truncated final product or a frameshift mutation that changes the entire sequence of amino acids past this mutation. Hence, many regulatory variants might be expected to have much smaller effects on a trait than structural variants.

The Molecular Nature of QTL Variation

While variation at "standard" loci (coding sequences and their immediately adjacent control regions) likely accounts for a substantial fraction of trait genetic variance, there are clear situations in which "nonstandard" genetic regions make significant contributions to the variance of a character. This builds on our discussion above that there are a variety of ways in which noncoding sequences can influence gene regulation at both local and very distant sites. For example, Mather (1944) presented evidence that heterochromatic regions (containing highly condensed DNA) can contribute to phenotypic variance. (While such regions were historically thought to be devoid of transcribable genes, this view is changing; Marsano et al. 2019.) Further, whether chromatin is tightly or loosely packaged influences the accessibility of coding regions for transcription and hence is a level of regulation. The classic example is X-chromosome inactivation, wherein only one copy of the X chromosomes in a female mammalian cell remains active. Thus, variation in large-scale genomic features (such as chromosome structure) can potentially generate variation in quantitative traits.

An extreme view of this idea, argued cogently by Cavalier-Smith (1978, 1985, 2005), is that genome size, independent of actual sequence composition, can influence certain aspects of the cellular phenotype (such as cell size and cell-division rate), which have cascading effects on the emergent properties of the phenotype. In plants, for example, there is compelling evidence that genome size variation influences growth rate and morphology (Bachmann et al. 1985; Cullis 1985; Schneeberger and Cullis 1991; Knight and Beaulieu 2008; Lavergne et al. 2010; Balao et al. 2011; Greilhuber and Leitch 2013; Realini et al. 2015). In this sense, variance in genome size translates into a continuum of allelic effects at a single, albeit ill-defined, super-locus, which may or may not be in disequilibrium with the effects of underlying "standard" loci. Diffuse effects like these are very difficult to genetically map.

A more localized example is the impact from small-scale deletions and duplications. That deletions can have an impact is fairly obvious, but the mechanism by which possessing *extra* gene copies can have an effect is less clear. One hypothesis is the concept of maintaining an appropriate **stoichiometry** (relative proportionality) of gene products. For example, when baking a cake, adding too much butter may be worse than not adding butter at all. Developmental systems may be able to buffer the impact of extra products from a small number of duplicates. However, these buffering systems likely become overwhelmed with a sufficiently large number of duplicates (or involving sufficiently important genes, such as rRDA; see below). In humans, the impact of stoichiometry is best demonstrated by the fact that while all nullisomics (possessing only one copy of a specific autosome) are lethal, almost all trisomics (possessing three copies of an autosome) are as well. Because sex chromosomes have evolved systems to maintain stoichiometry of X-linked products between

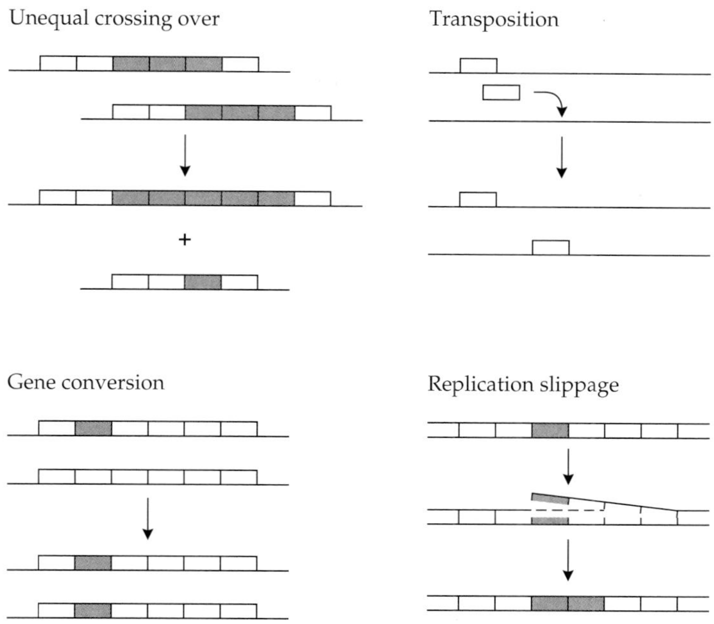

Figure 15.1 Some of the molecular processes that influence genome structure. In **unequal crossing over**, copy-number changes result from recombination between tandemly repeated sequences (usually on sister chromosomes) that are aligned out of register. **Gene conversion**, the nonreciprocal exchange of DNA between two sequences, can change the number of copies of a particular allele that an individual carries. **Transposition** occurs when a mobile genetic element inserts into a new site, usually (but not exclusively) while still retaining a copy in the original position. **Replication slippage** is a rough analog of unequal crossing over, with the single stands of an individual DNA mispairing during replication. These processes all influence the different sources of genomic sequence variation discussed in Chapter 8.

the homogametic and heterogametic sexes (e.g., dosage compensation; Part II Chapter 29), they do not follow this pattern.

A classic study on the impact of changes in gene dosage over small chromosomal scales is the work of Lindsley et al. (1972). These authors used the tricks of *Drosophila* genetics to construct lines that were either haploid or triploid for very small segments, and such **segmental aneuploids** were constructed for a large fraction of the fly genome. They found that a number of such small segments had phenotype effects when haploid (**haplo-insufficient**) and also found a smaller number of regions with impacts when triploid (**triplo-insufficient**). The broad catch-all term of **copy number variation, CNV** (Iafrate et al. 2004; Sebat et al. 2004; Zhang et al. 2009; Conrad et al. 2010,) is used to denote such small-scale duplications and deletions (it is occasionally used to denote larger scale genomic variation as well). We address the impact from CNV on trait variation in Chapters 20 and 21.

Given all of these potential sources of variation, we need to look beyond single-base pair mutations (SNPs), and consider other mutational events that can influence genomic, and possibly trait, variation. Molecular biology has shattered the historical view of a rather quiescent genome (reviewed by Lynch 2007), replacing it with a picture of a genome undergoing a large number of seemingly bizarre processes, including the insertion and excision of mobile gene sequences, changes in gene copy number by unequal crossing over, and gene conversion of large tracts of DNA (Figure 15.1). This emerging view of a dynamic genome blurs the distinctions between mutation, recombination, and DNA repair, suggesting that the mutational origin of quantitative-genetic variation may involve events other than simple point mutations or short insertions/deletions (Mackay 1988, 1989; Frankham 1988).

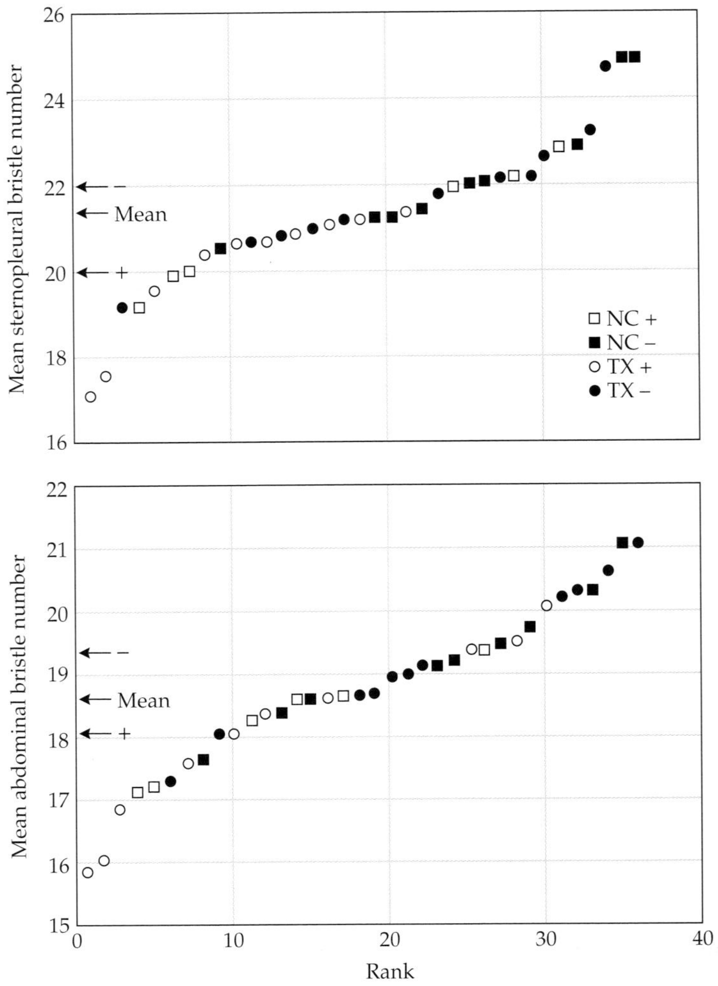

Figure 15.2 Association between chromosomes containing at least one insertional variant in the *achaete-scute* region and sternopleural (**top**) and abdominal (**bottom**) bristle numbers (see Figure 17.1). Squares denote samples from a natural population at Raleigh, NC, and circles denote samples from Texas. Open symbols denote extracted chromosomes containing at least one insertional variant (+), while filled symbols denote chromosomes without insertions (−). The arrows denote the mean bristle number associated with + and − chromosomes and the mean for the entire sample. (From Mackay and Langley 1990.)

In addition to producing loss-of-function alleles by direct insertion into a coding region, insertion of mobile genetic elements (**transposons** and **retrotransposons**) near a gene can alter gene expression, because the elements themselves often contain regulatory sequences. In addition, when transposable elements excise, they frequently leave altered sequences in their wake. Considerable regulatory variation (resulting in a wide range of phenotypic effects) has been found in mutant alleles arising from such excision events, e.g., at the *bronze-1* locus in maize (Schiefelbein et al. 1988), the *pallida* locus in snapdragons (Coen et al. 1986), the *An1*/*An2* anthocyanin synthesis loci in petunias (Gerats et al. 1984), and the *rudimentary* and *rosy* loci in *Drosophila* (Daniels et al. 1985; Tsubota and Schedl 1986). More than 50% of all known major mutations with phenotypic effects in *Drosophila* are associated with transposable element insertions (Finnegan 1992). More recent examples of important mutations arising from mobile element insertion are the above-mentioned *tb1* gene in maize and the classic melanic mutation in the pepper moth *Biston betularia* (van't Hof et al. 2016).

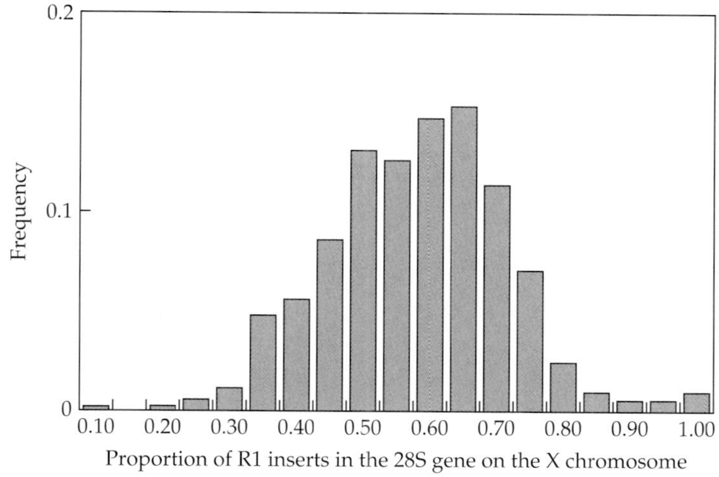

Figure 15.3 The frequency distribution for the fraction of 28S rDNA repeats on the X chromosome that contain an insertion of the mobile element R1 in a population of *Drosophila mercatorum* near Kamuela, Hawaii. A total of 1036 X chromosomes were assayed. (From Hollocher et al. 1992.)

In an investigation of the importance of mobile-element insertions as contributors to quantitative variation in natural populations, Mackay and Langley (1990) examined whether abdominal and sternopleural bristle numbers in *Drosophila* are correlated with insertional variation in the *achaete-scute* region. Because this locus is known to be capable of mutating to major alleles for the presence/absence of sensory hairs, it is a reasonable **candidate locus** for variation in bristle numbers. The authors simply asked whether chromosomes with at least one insertion in the *achaete-scute* region were associated with altered bristle numbers. A strong association was found, with chromosomes bearing inserts having lower numbers of both types of bristles (Figure 15.2). Further analysis suggested that these insertions account for about 5% of the total genetic variation observed in natural populations for both bristle number characters. Variation involving insertions, deletions, and nucleotide substitutions within a 45 kb region surrounding a second neurogenic locus, *scabrous*, accounts for another 10% or so of the genetic variance (Lai et al. 1994).

Multigene families, such as ribosomal (rDNA) gene clusters, are common features of all eukaryotic genomes (Lynch 2007). Copy number variation in such families can manifest into trait variation via two different (but not exclusive) mechanisms: variation in copy *number* and variation in copy *type*. Mutation processes such as gene conversion and unequal crossovers generate both types of variation (reviewed by Lynch 2007). Clear examples of the impact of CNV are seen in ribosomal genes, which typically exhibit copy number variation due to unequal crossing over (Figure 15.1). One of the first examples were the ***bobbed*** mutations in *Drosophila*, which impact a number of features, including bristle morphology. Ritossa et al. (1996) demonstrated that the severity of *bobbed* varied with rDNA copy number. More directly, Frankham and colleagues (reviewed in Frankham 1988) demonstrated that selection for reduced bristle number (one feature of *bobbed*) in *Drosophila* often leads to elevated frequencies of chromosomes with reduced rDNA copy number. In addition, several studies with plant populations selected for grain yield have reported significant associations between rDNA copy number and morphological and life-history traits (Saghai-Maroof et al. 1984; Rocheford et al. 1990; Powell et al. 1992). In flax (*Linum grandiflorum*), CNV in the rDNA, as well as in other regions of the genome, can be induced by a change in growing conditions. Such changes have substantial and heritable effects on plant morphology (Cullis 1981). These results are quite contrary to the general view that while expression of genes within a genome are influenced by the environment, the genomic *sequence* itself is unaffected.

An rDNA example where variation in copy *type* impacts phenotype is the complex syndrome in *Drosophila mercatorum* known as **abnormal abdomen**. This is characterized by reduced age to maturity, reduced longevity, increased early-age fecundity, and retention of juvenile abdominal cuticle in the adult stage (Templeton and Rankin 1978). The syndrome is associated with the insertion of the R1 transposable element into the coding regions of the 28S ribosomal genes, which exist in a tandem array (multiple copies) on the X chromosome (DeSalle et al. 1986; Malik and Eickbush 1999). The expression of abnormal abdomen requires that at least a third of the ribosomal gene copies contain the insert, and the inserted genes must not be underreplicated in the larval fat body (DeSalle and Templeton 1986; Templeton et al. 1989). A population survey indicated a broad distribution of insertion number variants, ranging from chromosomes with almost no insertions to those in which all genes had insertions (Figure 15.3). Moreover, there was an association between numbers of rDNA inserts and the allelic status at a locus determining underreplication, such that both of the components necessary for abnormal abdomen expression tended to cosegregate on the same chromosomes (Hollocher et al. 1992). Thus, the two tightly linked loci act as a multiallelic supergene to determine the expression of the abnormal abdomen syndrome. This particular example may be somewhat generalizable, as R1 elements interrupt the 28S rDNA coding region in a wide variety of insects (Jakubczak et al. 1991; Burke et al. 1998).

THE GENOMIC MUTATION RATE

The new mutational variance for a specific quantitative trait is a product of: (i) the total rate (over all potentially causal sites for that trait) at which new mutations appear, and (ii) the average effects of such new mutations. At first blush, this appears to be a daunting quantity to estimate, as it requires deep knowledge of all potential causal sites at which mutations impacting our trait of interest could appear, the mutation rate over such sites, and the average effect of such mutations. Fortunately, we can decompose this problem into manageable pieces using both genomics and standard quantitative-genetics machinery. Akin to the additive genetic variance of a trait (Chapters 5 and 7), we can estimate the **mutational variance** without knowing its component parts using phenotypic data in an appropriate experimental design. The much more challenging issue is estimating the values of its underlying components. We start with the genome-wide mutation rate, which provides an upper bound on the total mutation rate associated with the causal sites for a particular trait.

Owing to advances in genetics, we have reached the point at which, phylogenetically, few molecular-biological features are as well understood as the genome-wide mutation rate (Lynch et al. 2023). Mutation-rate estimates can be procured by examining sequence divergence over a set of replicate lines from an initially homozygous (or clonally reproducing) ancestral line. While the logic of this approach is clear, one must take care to avoid bias introduced by selection removing, or promoting, specific variants. One would like the raw, not selectively filtered, rate. This can be largely accomplished by regularly passing replicate lines through single progeny (if asexual or self-fertilizing) or single mating pairs. With such small population sizes, random genetic drift largely overwhelms the effects of selection (WL Chapter 7), allowing most new mutations to accumulate in an effectively neutral manner. Mutations arise randomly in the replicate lines, and the average fraction of nucleotide sites containing variants at the end of the study provides an estimate of the mutation rate per nucleotide site. Typically, in unicellular species and many invertebrates, less than one mutation arises per genome per generation, so these kinds of assays need to progress for dozens to hundreds of generations in large numbers of replicate lines to obtain accurate mutation-rate estimates. However, in vertebrate and some land-plant species with large genomes, the rate of mutation can often be ascertained to a good degree of accuracy by simply sequencing the genomes of a few trios of parents and their offspring.

Such experiments have revealed highly significant differences in mutation rates among species. In every major phylogenetic group, there is a negative (nearly inverse) association between the base-substitution (i.e., per-nucleotide) mutation rate and the effective population size of the species (Lynch et al. 2023). This pattern is consistent with the **drift-barrier hypothesis** (Lynch 2008, 2010, 2011), which postulates that, owing to the deleterious nature of most mutations, selection typically operates to reduce the replication-error rate to the lowest possible value, eventually stalling at the point at which the improvement in fitness via further reduction is less than the power of drift ($2N_e|s| \ll 1$ for fitnesses of $1 : 1-hs : 1-s$). All other things being equal, species with larger numbers of functional genomic sites also evolve lower per-base rates, owing to the elevated strength of selection associated with the increased target size for deleterious mutations. The vast majority of mutations involve single nucleotide substitutions, with small insertions/deletions typically arising at $\sim$10% the rate of the former, and larger scale mutations (such as gene-sized insertions/deletions) being rarer still (Lynch et al. 2023).

Across phylogenetic lineages, the range of variation in the per-generation mutation rate per-nucleotide (μ) extends over approximately three orders of magnitude (Lynch et al. 2023). For bacterial species μ ranges from 6×10^{-11} to 3×10^{-8}, with a mean of 2×10^{-9}. For unicellular eukaryotes, the range is 8×10^{-12} to 1×10^{-9}, with a mean of 3×10^{-10}. This reduction in the mutation rate in unicellular eukaryotes is thought to be a consequence of the expansion in genome size, which increases the total fitness consequences of deleterious mutations. However, with a reduction in effective population size, per-nucleotide mutation rates become elevated in multicellular species. In invertebrates, the range is 3×10^{-10} to 9×10^{-9}, with a mean of 4×10^{-9}, with the model organisms *Drosophila* and *Caenorhabditis* both having average estimates of $\sim 4 \times 10^{-9}$. Mammalian rates fall in the range of 2×10^{-9} to 4×10^{-8} with mean 9×10^{-9}, with estimates for human and mouse being 1.3×10^{-8} and 5.6×10^{-9}, respectively. Finally, rates in land plants range from 1×10^{-9} to 6×10^{-8}, with a mean of 2×10^{-8}.

These observations show that total genome mutation rates—(rate per site) $\times$ (genome size)—tend to increase with increasing organism size and complexity, putatively owing to the reduction in effective population size and larger mutational target (the increased number of functional sites overpowering the slight reduction in per-site mutation rate). Provided the numbers of key nucleotide sites are known, such rate measures also provide a basis for estimating and placing upper bounds on the rates of origin of *de novo* mutations for quantitative traits (Kondrashov and Crow 1993).

For example, given the mutation rate cited above, and a haploid genome size of $\sim$180 Mb (megabases) in *D. melanogaster,* $\sim$0.7 new base-substitution mutations are expected to arise per haploid genome per generation. Comparing DNA sequence divergence between fly species and the overall depression of rates of nucleotide substitution relative to the neutral expectation (based on silent sites in protein-coding genes), Halligan and Keightley (2006) surmised that a fraction 0.58 of mutations are purged by purifying selection, which implies a genome-wide deleterious mutation rate of $\sim$0.4 per gamete associated with base substitutions alone. In addition, a survey of 18 transposable element families in this species suggests that the total number of transpositions per gamete is on the order of 0.4 per generation (Nuzhdin and Mackay 1995; Lyman et al. 1996). Assuming the latter are all deleterious, summing these two types of mutation rates and noting that we have ignored other types of mutations (e.g., single nucleotide insertions/deletions, unequal crossing-over, and so on), the total haploid genomic deleterious mutation rate in *Drosophila* appears to be on the order of one per generation. With a haploid genome size of $\sim 3 \times 10^9$ bp, the average human gamete acquires $\sim$40 nucleotide mutations alone (at the zygotic level), with the fraction of these that are deleterious being as high as 0.20 in mammals (Keightley and Gaffney 2003). Obtaining numbers like these for specific quantitative traits is much more difficult, as it requires a knowledge of the genetic loci contributing to trait expression.

Haldane (1935, 1947) and Crow (1992, 1993b) noted that in many animal species the vast majority of mutations appear to arise in males rather than females, perhaps because most mutations arise during periods of cell division. In humans, for example, the mutation rate through males appears to be on the order of four- to ten fold larger than through females.

In the latter, the number of cell divisions between zygote and egg is ~24, regardless of maternal age. In men, however, ~36 cell divisions have occurred in the germ line by the time of puberty (age 13), and thereafter the number increases at a rate of about 23 per year. Thus, for a male of age 20, the number of cell divisions separating zygote and sperm cell is estimated to be 36 + (20−13) · 23 = 200, and by the age of 40, this has increased to 657. However, as reviewed by Crow (1993b), the observed increase in mutation rate with paternal age is even greater than that expected from this simple extrapolation from the increase in number of germline cell divisions (also see Kong et al. 2012).

More recent analyses, based on whole-genome sequencing in a wide range of vertebrates (mammals, birds, and snakes), uphold the elevated mutation rate via males in nearly all species (generally by a factor of 1.5 to 4), but also show that the female mutation rate increases with age (de Manuel et al. 2022). This suggests that sex-biased differences in mutation rates are not a simple consequence of replication errors but also depend on mutagenic processes that are independent of cell division. Many other aspects of mutation-rate variation within and among species are covered in Lynch et al. (2023).

THE MUTATION RATE OF PRODUCTION OF QUANTITATIVE VARIATION

While modern sequencing technology allows one to estimate the genome-wide rate and pattern of DNA *sequence* mutation, translating sequence changes into the rate of mutation input for a particular *character* is a *trait-specific* feature. As such, estimates of the rate of production of new quantitative variation by mutation are normally computed on a per-character, rather than a per-locus or per-nucleotide site, basis. The measure generally sought is the amount of additive genetic variation produced by mutation per generation. This is a composite property of three factors: the genic mutation rate, u; the number of loci underlying the character, n; and the mean squared phenotypic effect of a mutant allele in the heterozygous state.

As in Chapter 4, a convenient scaling for genotypic values for a new mutation at a locus is to set the ancestral homozygous state at 0, the heterozygous state at $a(1+k)$, and the homozygous mutant state at $2a$. Because the frequency of newly arisen mutations at a locus in any generation $(2u_i)$ is very small, the mean trait value for that locus (on this scale) is $\simeq 0$, giving the single-locus variance as $E[\{a(1+k)-0\}^2]$. Summing over all trait loci, the mutational variance becomes

$$\begin{aligned}\sigma_m^2 &= 2\sum_{i=1}^{n} u_i\, E[a_i^2(1+k_i)^2] \\ &= \sigma_{m0}^2 + 2\sigma_{m1}^2 + \sigma_{m2}^2 \end{aligned} \tag{15.1a}$$

where $\sigma_{m0}^2 = U \cdot E[a^2]$, $\sigma_{m1}^2 = U \cdot E[a^2 k]$, and $\sigma_{m2}^2 = U \cdot E[a^2 k^2]$, with $U = 2\sum_{i=1}^{n} u_i$ being the zygotic (diploid) mutation rate for the character (i.e., the expected number of mutations influencing the trait that arise in a newborn each generation), and $E[x]$ denoting the average value of x. The decomposition in Equation 15.1a follows by assuming independence of locus-specific mutation rates and effects (i.e., that u_i, a_i, and k_i are all independent). For mutations with purely additive effects ($k = 0$), the mutational variance simplifies to

$$\sigma_m^2 = \sigma_{m0}^2 = U \cdot E[a^2] \tag{15.1b}$$

A cautionary note in the interpretation of Equation 15.1a is necessary. Strictly speaking, the mutational variance defined in this expression is **nonsegregational mutational variance,** in that it describes the variance produced by a cohort of new mutations that exist *entirely as heterozygotes*, rather than at Hardy–Weinberg frequencies (where the mutations would appear in both heterozygotes and homozygotes). In the case of mutations with additive effects, this definition is not a problem because the variance produced by an allelic effect is independent of the genetic background and uninfluenced by segregation (Chapter 4). With completely recessive mutations ($k = -1$), however, Equation 15.1a defines the mutational variance to be zero, despite the fact that variation will be exposed as the mutant alleles segregate into their homozygous states. For autosomal loci, Hardy–Weinberg frequencies are reached in one generation under monoecy and in two generations under dioecy.

A reasonable way to define the relevant measure of mutational variance for a *segregating population* is to consider the situation in which the only determinants of the dynamics of genetic variance are mutation and random genetic drift. Under this **neutral model** (WL Chapters 2 and 12), considered in detail in Lynch and Hill (1986), populations with constant size eventually reach an equilibrium state, where the input of new variation by mutation is balanced by its loss via random genetic drift (as alleles become fixed). For populations of size one, maintained by self-fertilization and single-seed descent, the equilibrium genetic variance among the progeny is

$$\sigma^2_W = \frac{3}{2}\sigma^2_{m0} + \left(\frac{8\sigma^2_{m1} + 5\sigma^2_{m2}}{4}\right) \tag{15.2a}$$

With full-sib mating, the equilibrium variance is

$$\sigma^2_W = 4\sigma^2_{m0} + 2\left(\frac{6\sigma^2_{m1} + 5\sigma^2_{m2}}{3}\right) \tag{15.2b}$$

and with all larger population sizes (with effective size denoted by N_e), it is very close to

$$\sigma^2_W = 2N_e\left(\sigma^2_{m0} + \frac{2}{3}\left[\sigma^2_{m1} + \sigma^2_{m2}\right]\right) \tag{15.2c}$$

In this latter case, all of the genetic variance is additive, except for $4N_e\sigma^2_{m2}/3$, which defines the dominance component of variance (Lynch and Hill 1986). Note that for recessive mutations ($k < 0$), which are likely to be the rule (Chapters 4 and 12), σ^2_{m1} is negative while σ^2_{m2} is positive, so the contributions from these terms partially cancel (completely canceling in the case of fully recessive mutations). Moreover, the absolute values of both terms are necessarily less than that of σ^2_{m0}. Thus, although not exact in all cases, the equilibrium level of genetic variance in the absence of selection is expected to be close to $2N_e\sigma^2_{m0}$ in most cases. This result justifies the reliance on σ^2_{m0} as a measure of mutational variance, a useful relationship because, as we will soon see, σ^2_{m0} is much easier to estimate than either σ^2_{m1} or σ^2_{m2}.

Estimates of the mutational variance are typically scaled by the environmental variance of the focal trait to yield a **mutational heritability**,

$$h^2_m = \frac{\sigma^2_{m0}}{\sigma^2_E} \tag{15.3}$$

a dimensionless measure suitable for comparing different traits. When mutations have additive effects, h^2_m approximates the heritability of an initially homozygous population following one generation of mutation. (The fact that the denominator in h^2_m is the environmental, rather than the total, variance is not problematical because, as shown below, $\sigma^2_{m0} \ll \sigma^2_E$). An alternative dimensionless measure is the **mutational coefficient of variation**, $\sigma^2_{m0}/\,\overline{z}$, where the denominator denotes the population mean phenotype (Houle et al. 1996).

Estimation from Mutation-accumulation (MA) Experiments

Estimation of the mutational variance is straightforward, but labor intensive. The simplest approach is a **divergence experiment**. Here, replicate lines, extracted from the same ancestral population, are maintained at small enough sizes that the fates of essentially all newly arisen mutations are determined by random genetic drift. With selfed lines maintained by single-progeny descent and full-sib mated lines, all but extremely deleterious alleles will behave in a neutral fashion, so that any census of mutational effects will yield nearly unbiased results. Over time, the mean phenotypes of the replicate lines are expected to drift apart as they randomly accumulate unique mutations. If the mutations have directional effects ($E[a] \neq 0$), a shift in the grand mean phenotype (averaged over all lines) is also expected.

If, prior to expansion, the base population is kept at the same size as the derived lines for a long enough period to be in mutation-drift equilibrium, the expected increment in the among-line component of variance per generation is equal to $2\sigma^2_{m0}$ (Bailey 1959; Lande 1976; Chakraborty and Nei 1982; Lynch and Hill 1986). The simplest way to obtain this result is to note from neutral theory (Kimura 1983; WL Chapters 2 and 12) that the expected number of mutational substitutions at locus i differentiating two populations separated for t generations is $2u_it$. (Letting N denote the census population size, then each generation in each population, an expected $2Nu_i$ new mutations arise. Each of these has an ultimate probability of fixation equal to their initial frequency, $1/[2N]$, so that at equilibrium, the expected number of fixations in a pair of populations each generation is $2 \times 2Nu_i \times 1/(2N) = 2u_i$ at the ith locus.) Upon fixation, each mutation causes an expected increment in the among-population variance of $E[(2a-0)^2]/2 = 2E[a^2]$. Thus, summing over all loci, the expected among-line variance after t generations of divergence is

$$\sigma^2_B(t) = 2\sum_{i=1}^{n}(2u_it)\cdot E[a^2] = 2\sigma^2_{m0}t \tag{15.4a}$$

This is a fairly robust result in that it does not depend on the population size, the degree of dominance of new mutations, or the linkage relationships among loci (Lynch and Hill 1986). Dominance has no influence on the rate of divergence because it has a transient effect, only impacting heterozygotes. Once the mutant allele goes to fixation, a permanent shift in the mean phenotype to $2a_i$ occurs at the locus, regardless of k_i. In principle, the asymptotic rate of divergence might accelerate (or decelerate) when mutations have general epistatic effects (Tachida and Cockerham 1990), but the time scale over which such complications would arise is likely to be quite long (many hundreds to thousands of generations).

To summarize, a conceptually simple way to estimate the mutational variance for a focal trait is to perform a long-term divergence experiment with replicate lines kept at small population sizes, extracting the among-line component of variance (σ^2_B) by analysis of variance (ANOVA; Part II Chapter 23) of phenotypic measures. Division of the estimate of σ^2_B by twice the number of generations of divergence, Var(B)/$(2t)$, provides an estimate of σ^2_{m0}. These results apply strictly to diploid sexual populations, but only slight modifications are required for other forms of reproduction. For example, for haploid populations (sexual or asexual), the preceding expression becomes

$$\sigma^2_B(t) = \sigma^2_{m0}t/2 \tag{15.4b}$$

whereas for diploid asexuals,

$$\sigma^2_B(t) = \sigma^2_m t \tag{15.4c}$$

where σ^2_m is defined as in Equation 15.1a (Lynch 1994). Terms involving dominance arise in the case of diploid asexuals because, in the absence of segregation, mutations are kept permanently in the heterozygous state with effects $a(1+k)$.

There are several potential sources of error in the estimation of the mutational variance with a divergence experiment. First, the expected increase in among-line variance of $2\sigma^2_{m0}$ per generation strictly applies to situations in which the base population has been kept at the same size as the descendant lines for about $6N_e$ or so generations prior to the initiation of a divergence experiment (Lynch and Hill 1986). If the base population is totally homozygous, the rate of divergence among descendant lines will be less than $2\sigma^2_{m0}$ during the early generations of the experiment due to the lack of segregating polymorphisms. On the other hand, if the base population contains excess heterozygosity relative to the equilibrium expectations within the descendant lines, the divergence rate will exceed $2\sigma^2_{m0}$ in the early generations of the experiment owing to segregation of baseline variation. Eventually, under the neutral model, all lines will asymptotically diverge at the rate $2\sigma^2_{m0}$ regardless of the status of the base population. To validate this condition, the among-line variance needs to be compared over multiple time horizons.

A second potentially serious issue in the analysis of a divergence experiment concerns the possibility that environmental effects artificially inflate the among-line component of variance. To guard against such inflation, a set of control lines should be assayed in parallel with the mutation-accumulation lines whenever possible. The control lines should be genetically identical (and maintained in an evolutionarily inert state by, for example, freezing or seed storage) so that any variance among them can only be attributable to environmental causes. Any variance among control lines can then be subtracted from that observed among the mutation-accumulation lines. Maternal effects (Part II Chapters 29 and 30) can be a significant source of inflation of the among-line variance when all of the progeny assayed within a subline are derived from a single mother. Such problems can be eliminated by deriving all assay individuals from different mothers, as this ensures that any variance associated with maternal effects will contribute to the within-line (rather than the among-line) component of variance (Lynch 1985).

Third, **contamination events** can be especially problematic in mutation-accumulation experiments, either deflating estimates of σ^2_{m0} when gene flow accidentally occurs among lines, or inflating it if foreign genotypes immigrate into the experiment. Periodic screening of lines using informative molecular markers can provide an excellent control against contamination (Mackay et al. 1992a; Keightley and Hill 1992). Given that estimates of mutational variance are typically very small, even slight deviations associated with any of these three potential problems can severely bias estimates of mutational variance.

Finally, we note that an alternative approach to estimating mutational variance involves artificial selection on an initially homozygous base population. Recall from Equation 3.29 that the single-generation response to selection is equal to $R = h^2 S$, where S is the selection differential imposed by the investigator. The idea behind the selection approach for estimating the mutational variance is that the standing heritability in a population can be estimated from the ratio of selection response to selection differential, $\widehat{h}^2 = R/S$. As in the case of the neutral model, expressions can be derived for the expected equilibrium heritability of a quantitative trait under truncation selection as a function of the population size and the mutational variance (Hill 1982b, 1982c). Thus, by equating observed and expected heritabilities, an estimate of σ^2_{m0} can be obtained.

Bristle Numbers in *Drosophila*

Much of our knowledge of the properties of polygenic mutation derives from studies on one set of characters in one species: abdominal and sternopleural bristle numbers in *Drosophila melanogaster* (Figure 17.1). These readily quantifiable features have sensory functions associated with the peripheral nervous system, and multiple causal loci have been well characterized at the molecular level. Considering the results from numerous studies (reviewed in Lynch 1988b; Keightley et al. 1993; Houle et al. 1996; Mackay and Lyman 2005; García-Dorado et al. 2000), estimates of the mutational heritabilities for these traits are almost always in the range of 0.002 to 0.005. The effects of individual bristle-number mutations tend to be nonadditive, with mutations with large effects being nearly recessive and sometimes interacting epistatically, and those with small effects being nearly additive on average but with variable levels of dominance (Mackay et al. 1992a; Caballero and Keightley 1994; Fry et al. 1995; Mackay et al. 2005b; Amador et al. 2010).

These observations are notable because quantitative-genetic analyses have frequently led to the suggestion that *Drosophila* bristle numbers have a purely additive genetic basis (Chapter 7). This discrepancy can be reconciled if most mutations for bristle number are deleterious and hence kept at low frequency in natural populations by selection. Because the dominance component of genetic variance is a function of the squared heterozygosity,

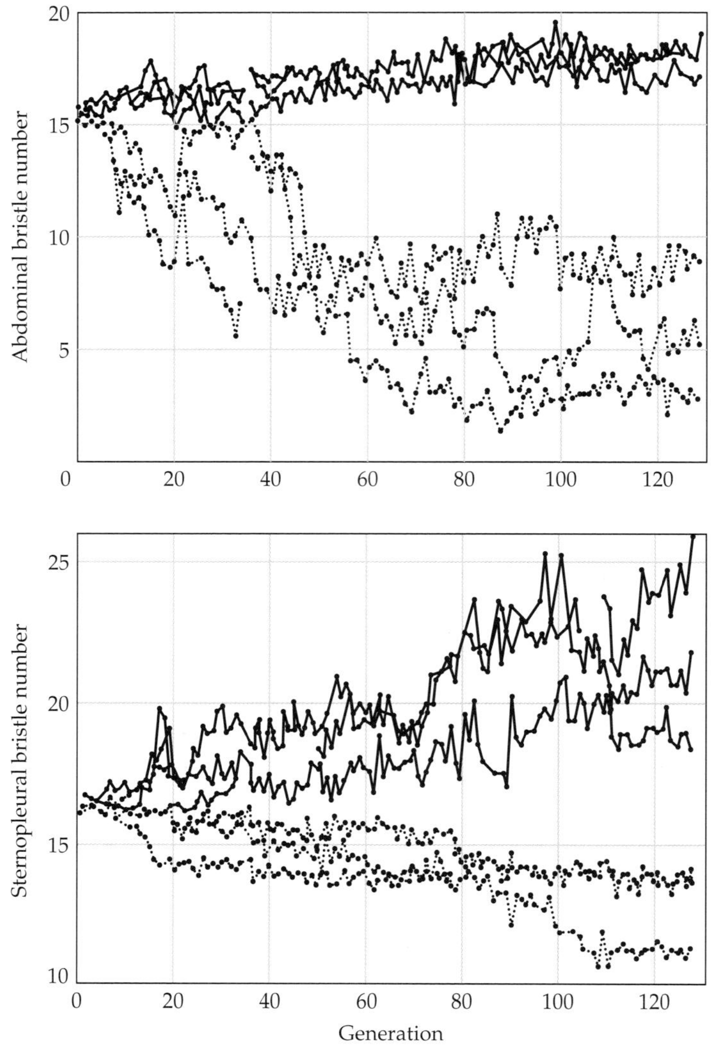

Figure 15.4 Responses to 125 generations of selection for bristle numbers in *Drosophila melanogaster.* For each character, results are given for three upwardly selected and three downwardly selected lines. Up to generation 63, the 10 extreme males and 10 extreme females were selected from a total of 40 measured individuals of each sex; thereafter, 10 pairs were selected from 20 measured individuals of each sex. (From Mackay et al. 1994.)

the vast majority of the genetic variance is additive when recessive alleles are kept at low frequency (Equation 4.12d).

Consistent with this view, most bristle-number mutations with large effects have substantial negative pleiotropic effects on fitness, with some being recessive lethals (Mackay et al. 1992a, 1994; Fry et al. 1995). Evidence for the general negative fitness consequences of such mutations comes from long-term divergence experiments involving initially homozygous lines kept subsequently at average population sizes of 10 individuals (Mackay et al. 1992b). In the absence of any artificial selection, for the first 100 generations or so of isolation, the lines diverge via random genetic drift, but thereafter the rate of dispersion declines dramatically (Mackay et al. 1995). Presumably, this occurs because mutant bristle-number alleles with sufficiently deleterious fitness effects are efficiently purged by natural selection.

Long-term artificial-selection experiments, again starting with homozygous lines, provide further evidence for the negative pleiotropic effects of bristle-number mutations on fitness. Such experiments often culminate in dramatic divergence between high- and low-

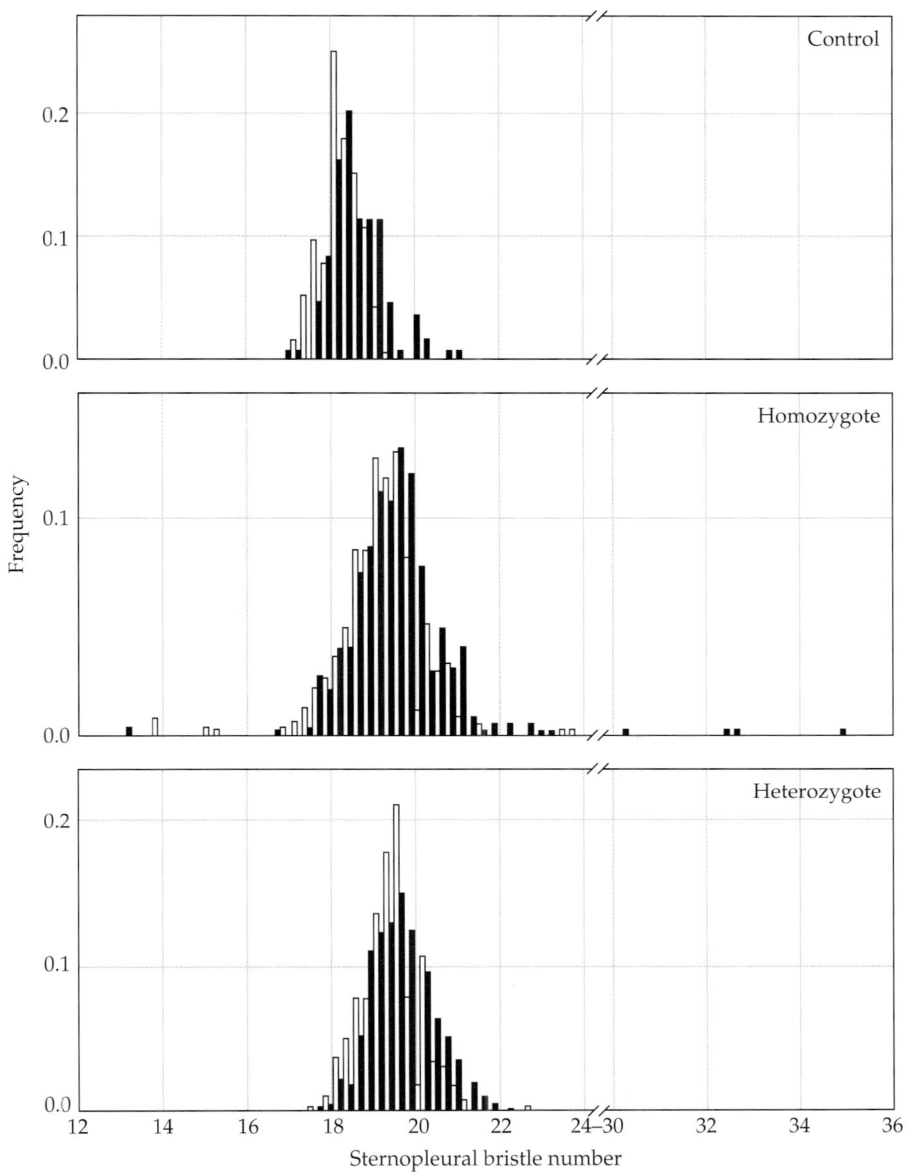

Figure 15.5 The distributions of sternopleural bristle numbers for control, homozygous *P*-element insert lines, and heterozygous *P*-element insert lines, in *Drosophila melanogaster.* Each insert line carries a single *P*-element insertion, as revealed by cytogenetic analysis. The open bars denote second chromosome line means, while the closed bars denote the results for chromosome three. All chromosomes were evaluated on a constant genetic background. (From Lyman et al. 1996.)

selected lines (Clayton and Robertson 1955, 1964; Caballero et al. 1991; Santiago et al. 1992; López and López-Fanjul 1993; Mackay et al. 2005b). For example, after only 125 generations of selection, differences of 12 abdominal bristles and 8 sternopleural bristles (equivalent, respectively, to 7 to 8 phenotypic standard deviations) have been observed (Figure 15.4), and continued selection leads to still greater divergence in the latter character (Mackay et al. 2005b). As in the drift experiments noted above, substantial reductions in fitness are associated with such selective changes (Mackay et al. 1994; Nuzhdin et al. 1995). Moreover, such selection responses, which to a large degree must reflect the distributions of mutational effects (as well as their pleiotropic effects on fitness), are almost always asymmetrical—a downward skew in the case of abdominal bristle number, and an upward skew in the case of sternopleural bristle number (Figure 15.4).

Historically, an attractive feature of *D. melanogaster* as a model system for studying aspects of polygenic mutation has been its amenability to **transposon tagging**. Through the use of appropriate stocks, it is possible to mobilize transposable *P* elements and then subsequently repress their movement (H. Robertson et al. 1988). Hybridization of radiolabeled *P*-element DNA to chromosomal preparations can then be used to ascertain the approximate locations of insertional events. This approach provided a method for categorizing and localizing *P*-element insertions before the era of genome sequencing.

Lyman et al. (1996) used this method to study the effects of single *P*-element insertions on the expression of bristle-number characters. By assaying multiple singly mutated chromosomes on an otherwise constant genetic background, they found that the phenotypic distributions of insertion-associated mutational effects were symmetrical but highly leptokurtic (Chapter 2), with a few inserts having very large effects and the average effect being near zero (Figure 15.5). On average, the mutational effects were partially recessive, with $\bar{k} = -0.37$ for abdominal bristles and $\bar{k} = -0.65$ for sternopleural bristles. There was a weak correlation between viability and the absolute magnitude of effect of a bristle-number mutation, with the viability effects also being highly recessive ($\bar{k} = -1.00$).

The results of Lyman et al. (1996) imply that the mean-squared homozygous effects of individual *P*-element insertions relative to the environmental variance, $E[(2a)^2]/\sigma_E^2$, are approximately 0.28 for abdominal bristles and 0.82 for sternopleural bristles. Based on the mutational heritabilities for these traits, $h_m^2 = U \cdot E[a^2]/\sigma_E^2 = 0.0035$ and 0.0043, respectively, these single-mutation estimates suggest genomic mutation rates (diploid U) of $\sim$0.02 to 0.05 per zygote per generation. Although this extrapolation assumes that *P*-element induced mutations have effects that are comparable to those with other mutational origins, analyses involving the full spectrum of molecular mutations still yield estimates in the range of 0.02 to 0.09 (García-Dorado et al. 1999; Mackay et al. 2005b). Because many of the mutant chromosomes examined in these studies likely had small but undetectable effects, this suggests total gametic mutation rates for bristle numbers as large as 0.1. If this interpretation is correct, then given the total gametic mutation rate of $\sim$1.0 in *D. melanogaster* (noted above) and a total gene number of $\sim$14,000, several hundred loci must influence bristle numbers (far beyond the number known from cell-biological work), either directly or through pleiotropic effects.

Additional Data

Table 15.1 summarizes estimates of the mutational heritabilities for several other organisms and traits. In invertebrates, most of the estimates are in the range of 10^{-3} to 10^{-2}, whereas those for mouse are somewhat higher, ranging up to 0.025. On average, estimates for plants are similar to those for invertebrates, with a slightly broader range perhaps due to sampling error. In highly mutagenic situations, such as when transposable elements are mobilized following hybrid dysgenesis in *Drosophila*, mutational heritabilities can be as high as 0.1 (Mackay 1985b, 1987, 1988). Otherwise, all estimates of h_m^2 within *Drosophila* are within the range of 0.001 to 0.005, except that for viability, which is certainly downwardly biased because chromosomes with highly deleterious mutations were excluded from analyses.

While the procurement of composite measures of mutation such as σ_{m0}^2 is arduous, direct estimation of the actual mutation rate for quantitative characters ($U = 2\sum \mu_i$) is even more difficult. Using the *P*-element tagging approach outlined above for bristle numbers, Clark et al. (1995a) obtained scaled single mutational effects, $E[(2a)^2]/\sigma_E^2$, equal to 0.064 ± 0.005 for body weight and 0.062 ± 0.018 for the activities of 15 enzymes. Comparison of these values with the mutational heritabilities in Table 15.1 yields zygotic mutation rates of 0.29 for body weight and 0.14 for activities of individual enzymes, several times higher than those noted above for bristle numbers. Such high mutation rates cannot be attributable to mutations at the actual enzyme loci, which are expected to occur at the rate of 10^{-6} to 10^{-5} per gene per generation (Mukai and Cockerham 1977). Thus, most of the mutations must

Table 15.1 Estimates of the mutational heritability for a variety of organisms and characters. Where possible the results represent averages over multiple studies, ranging from analyses of drift among initially homozygous lines to artificial selection experiments. Detailed analyses of experiments performed prior to 1996 can be found in Lynch (1988b) and Houle et al. (1996), and a more thorough survey is given in Halligan and Keightley (2009).

Species	Character	h_m^2	Reference
Drosophila melanogaster	Abdominal bristle number	0.0035	See text
	Sternopleural bristle number	0.0043	See text
	Enzyme activities	0.0022	Clark et al. 1995b Harada 1995
	Ethanol resistance	0.0009	Weber and Diggins 1990
	Body weight	0.0047	Clark et al. 1995b
	Wing dimensions	0.0020	Santiago et al. 1992 García-Dorado et al. 2000
	Viability	0.0003	Mukai 1964 Mukai et al. 1972 Cardellino and Mukai 1975 Ohnishi 1977
Caenorhabditis elegans	Body size	0.0039	Azevedo et al. 2002
	Metabolite concentrations	0.0003	Johnson et al. 2020
	Mitotic-spindle properties	0.0004	Farhadifar et al. 2016
	Intrinsic rate of increase	0.0040	Vassilieva and Lynch 1999 Vassilieva et al. 2000
	Survival to maturity	0.0028	Vassilieva and Lynch 1999 Vassilieva et al. 2000
	Longevity	0.0010	Vassilieva and Lynch 1999 Vassilieva et al. 2000
	Total progeny production	0.0014	Keightley and Caballero 1997 Vassilieva and Lynch 1999 Vassilieva et al. 2000
Tribolium castaneum	Pupal weight	0.0091	Goodwill and Enfield 1971
Daphnia pulex	Life-history traits	0.0017	Lynch 1985
Mouse	Lengths of limb bones	0.0234	Bailey 1959
	Mandible measures	0.0231	Festing 1973
	Skull measures	0.0052	Carpenter et al. 1957 Deol et al. 1957 Hoi-Sen 1972
	6-week weight	0.0034	Caballero et al. 1995
Saccharomyces cerevisiae	Life-history traits	0.0005	Hall and Joseph 2010
Arabidopsis thaliana	Life-history traits	0.0059	Schultz et al. 1999 Shaw et al. 2000
Maize	Plant size	0.0112	Russell et al. 1963
	Reproductive traits	0.0073	Russell et al. 1963
Rice	Plant size	0.0030	Oka et al. 1958
	Reproductive traits	0.0028	Sakai and Suzuki 1964

arise in regions of the genome somehow involved in either direct or indirect modifications of enzyme activity, implying that such regions must be quite numerous. This observation is consistent with the notion that much (perhaps most) of quantitative variation is due to regulatory variation (Chapter 21). As in the case of bristle numbers, the distributions of mutational effects in the study of Clark et al. were quite symmetrical, with mutations being equally likely to cause an increase or a decrease in the character, but also highly leptokurtic, with most having pleiotropic effects on multiple traits.

Other estimates, based on simple counting of mutations with major observable effects, are consistent with the view that U for complex traits in multicellular organisms commonly exceeds 0.01. For example, Grewal (1962) examined 27 skeletal traits in inbred strains of lab mice and found that 21 detectable mutations had accumulated in the roughly 72 total generations separating the strains (summing the number of generations over all branches). This gives a rate of $\sim 2 \times [(21/27)/72] = 0.022$ per character per generation per zygote. In a more extensive study using the same traits, Hoi-Sen (1972) obtained essentially the same answer. Studies examining a number of reproductive characters in maize using doubled-haploids (Sprague et al. 1960) and inbred lines (Russell et al. 1963) yield estimates of 0.090 and 0.056 mutations per character per zygote.

These estimates of per-character mutation rates are certainly too low, perhaps substantially so, because mutations of insufficiently large effect are statistically undetectable and hence not counted. As noted above for *Drosophila* bristle number and enzyme activities, numerous lines of evidence suggest that mutations to alleles of small effects arise much more frequently than to alleles of large effect. In a descriptive analysis of highly inbred peanut populations, Gregory (1965a, 1965b) found that deviations of small effect were much more frequent than those of larger effect, although his approach was not statistically rigorous. Mukai (1964) found that the mutation rate over the entire second chromosome of *D. melanogaster* was 20 times higher for mutations with minor viability effects (0.15) than for recessive lethals (0.006). The majority of spontaneous mutations for fitness in bacteria have very small effects, on the order of a 1% decline in fitness per generation, although mutations with larger effects do occur (Kibota 1996; Rana et al. 2021).

If estimates of U in the neighborhood of 0.1 are indeed correct, then typical quantitative traits must be influenced by a large number of loci and/or have unusually high per-locus mutation rates. There is, however, no evidence for more than two to three fold variation in the magnitude of mutation rates across genomic landscapes (Lynch et al. 2023). Consider a typical average length of a protein-coding region, $\sim 10^3$ nucleotides (Lynch 2007, 2024), and a mutation rate per nucleotide site on the high end of what has been observed, 10^{-8} per generation in mammals. In this extreme case, 10^4 loci are required to account for a gametic mutation rate of 0.1. Allowing for another 1 kb of regulatory sequence per gene would only reduce this number by 50%, whereas reducing the mutation rate two fold per generation would increase it by a factor of two.

While one might imagine that thousands of loci can directly impact a character like viability or fitness, such a high number seems unlikely for most other characters, although there has been historically little evidence to support or refute this conjecture. However, recent GWAS results for height (Example 21.15) are hinting that causative sites may indeed number in the thousands, and that most are regulatory. This apparent discrepancy between expected and predicted number of causal loci has been discussed by a number of workers (Grewal 1962; Russell et al. 1963; Hoi-Sen 1972; Grüneberg 1970, Beardmore 1970; Turelli 1984). A possible resolution invokes pervasive pleiotropy and the idea that owing to the integrative nature of metabolic, physiological, and developmental pathways, mutations at the majority of nucleotide sites with functional significance have at least minor indirect effects on nearly all traits (Crow 1992, 1993b; Kondrashov 1995; Boyle et al. 2017a). We will return to this topic in Chapter 21 (also see WL Chapter 28).

Summing up, despite the dozens of MA studies that have been performed in a variety of organisms, it remains unclear whether there are meaningful differences in h_m^2 among species or categories of traits. Conradsen et al. (2022) suggested that much of the observed variation in such measures is a simple consequence of sampling error. It remains to be resolved why h_m^2 (in metazoans) is most commonly on the order of 10^{-3}, regardless of trait and species, i.e., why the recurrent production of mutational variance is typically $\sim$0.1% of the environmental variance for traits.

The possibility also exists that all of the above estimates of the mutational heritability may substantially overestimate the input of genetic variation available to natural selection.

If most spontaneous mutations have negative pleiotropic effects on fitness, as suggested by the aforementioned studies and with further data in the following section, much of the recurrent mutational input of variation may simply represent the price that organisms pay for the privilege of producing rare adaptive mutations.

THE DELETERIOUS EFFECTS OF NEW MUTATIONS

Evolutionary biologists are especially interested in the rate of mutation for fitness and its underlying components such as viability. As first noted by Haldane (1937), large populations are expected to evolve towards a situation where the recurrent input of deleterious mutations is balanced by their selective elimination. Provided the selective disadvantages of mutations exceed the power of random genetic drift, the frequencies of deleterious alleles equilibrate to levels such that the reduction in population mean fitness is solely a function of the mutation rate and independent of the effects of the mutations (Chapter 12; WL Chapter 3). This occurs because the equilibrium frequency of a deleterious allele is inversely proportional to its fitness effect in heterozygotes. Thus, for an understanding of the population-level consequences of deleterious mutation, the genomic deleterious-mutation rate is of central importance.

Further characterization of the properties of deleterious mutations is also important for other issues in evolutionary biology and conservation genetics. In small populations, mildly deleterious mutations do not simply attain equilibrium levels of heterozygosity as defined by mutation–selection balance alone. Rather, mildly deleterious mutant alleles have an appreciable chance of randomly drifting to high frequencies and even fixation. In principle, the accumulation of such mutations can eventually drive a population to extinction when the mutation load exceeds the point at which individuals can replace themselves (Lynch et al. 1993, 1995a, 1995b; Lande 1994). It is also known that pressure from recurrent deleterious mutations can foster the evolution of alternative modes of reproduction (Pamilo et al. 1987; Kondrashov 1988; D. Charlesworth et al. 1992, 1993), with outcrossing populations being much more efficient at eradicating deleterious genes than asexual or obligately selfing populations. The risk of extinction via deleterious-mutation accumulation depends not just on the genomic deleterious-mutation rate, but also on the distribution of mutational effects. Deleterious mutations with selection coefficients on the order of $1/(2N_e)$, where N_e is the effective population size, are of particular concern because such mutations are highly vulnerable to the vagaries of random genetic drift while cumulatively still having significant effects on fitness (Lynch and Gabriel 1990; Gabriel al. 1993; Lande 1994).

To obtain a more than qualitative understanding of these issues, estimates of the genomic deleterious-mutation rate, the average deleterious selective coefficient s, and the dominance coefficient h of new mutations are required. Owing to the practical difficulties noted above, this is a challenging enterprise. Two approaches to the problem have been suggested: (i) the analysis of laboratory MA lines, expanding on the procedures outlined above for the estimation of mutational heritabilities; and (ii) the analysis of the fitness properties of large natural populations under the assumption that they are in mutation–selection balance. A broad overview of the methods and results from MA experiments is presented by Halligan and Keightley (2009).

The Bateman–Mukai Technique

Using estimates of the per-generation changes in the grand mean and the among-line variance for a trait caused by mutation, it is possible to place bounds on the genome-wide rate and the average effect of a new mutation (Bateman 1959; Mukai 1964; Mukai et al. 1972; Crow and Simmons 1983). Although the focus here is primarily on the deleterious consequences of mutations, the statistical approach applies to any character for which mutation causes a directional change in the mean. We will first derive the general results following Lynch (1994), and then apply them specifically to the analysis of fitness characters.

We start with the assumption that the MA lines are kept at a small enough size that essentially all new mutations are effectively neutral, with the vagaries of random genetic drift overwhelming the forces of selection. Roughly speaking, this requires that the selection coefficients of mutant alleles be less than $1/(2N)$ where N is the population size, e.g., $s < 0.5$ for lines maintained by self-fertilization and single-seed descent, and $s < 0.25$ for lines maintained by single brother–sister matings. Letting U denote the diploid genomic mutation rate for the trait, then from arguments given above, the expected number of mutations fixed per generation is simply equal to $U/2$ (the gametic mutation rate). Following the notation used above, conditional on fixation, each mutation causes an expected change in the mean genotypic value equal to $E[2a]$. Thus, assuming there are no substantial epistatic interactions among new mutations, the expected dynamics of change in the mean genotypic value in a MA experiment are given by

$$\bar{g}(t) = \bar{g}(0) + U\bar{a}\,t \tag{15.5}$$

where $\bar{g}(t)$ is the mean genotypic value in generation t. In the following, we let ΔM denote an estimate of the rate of change in the mean, $U\bar{a}$, which may be obtained from a regression of line means on time, or as the total change in the mean $\bar{g}(t) - \bar{g}(0)$ divided by t.

As noted above, the expected rate of increase of the among-line variance in a MA experiment is equal to $2\sigma^2_{m0} = 2UE[a^2]$. Thus, twice the squared rate of change in the mean divided by the rate of change in the variance is equal to $U\,\bar{a}^2/\overline{a^2}$. This quantity reduces to U in the unlikely event that mutations have constant effects. More generally, the ratio is equal to $U/(1 + C_m)$, where

$$C_m = \sigma^2_a/\,\bar{a}^2 \tag{15.6a}$$

is the squared coefficient of variation of mutational effects. Letting ΔV be an estimate of the rate of line divergence ($2\sigma^2_{m0}$) and accounting for the bias produced by sampling error in the estimators ΔM and ΔV, a lower-bound estimate for the diploid genomic mutation rate is

$$\widehat{U}_{min} = \frac{2(\Delta M)^2}{\Delta V(1 + C_{\Delta M})(1 + C_{\Delta V})} \tag{15.6b}$$

where $C_{\Delta M}$ and $C_{\Delta V}$ are squared coefficients of sampling variance (ratios of sampling variance to squared estimates) of ΔM and ΔV, e.g.,

$$C_{\Delta M} = [\text{SE}(\Delta M)/\Delta M]^2 \tag{15.6c}$$

By similar means, it can be shown that an upper-bound estimate of the average mutational effect is given by

$$\widehat{a}_{max} = \frac{\Delta V}{2(\Delta M)(1 + C_{\Delta M})} \tag{15.7}$$

The expected value of $\widehat{a}_{max}$ is actually $(1 + C_m)\,\bar{a}$, so the factor by which it overestimates $\bar{a}$ is identical to the factor by which the mutation-rate estimator underestimates U. Although most attempts to estimate $\widehat{U}_{min}$ and $\widehat{a}_{max}$ have ignored $C_{\Delta M}$ and $C_{\Delta V}$, this causes further bias that is readily avoided.

By crossing lines in which mutations have accumulated uniquely and independently in each line, further insight can be gained on the dominance of mutational effects. For example, starting from a base population in mutation-drift equilibrium, the expected difference between mean phenotypes of F_1 hybrid progeny and that of their contemporary (MA line) parents after t generations of divergence is $UE[ak]t$ (Lynch 1994). Letting ΔD denote an estimate of $UE[ak]$, then

$$\widehat{k} = \frac{\Delta D}{\Delta M(1 + C_{\Delta M})} \tag{15.8}$$

provides an estimate of $\bar{k} \cdot [1 + (\sigma_{a,k}/\bar{a}\bar{k})]$, where $\sigma_{a,k}$ is the covariance of effects a and k in new mutations. Because the signs of all of the terms in this expression can be positive or negative, $\widehat{k}$ can be either an upwardly or downwardly biased estimator of $\bar{k}$, the bias being negligible if $(\sigma_{a,k}/\bar{a}\bar{k}) \ll 1$. Other approaches for estimating the dominance properties of mutations and their influence on the mutational variance are discussed in Mukai (1979), Lynch (1994), and Halligan and Keightley (2009).

The preceding expressions can be interpreted as yielding estimates of the genomic mutation rate and mutational effects that would explain the pattern of change in the mean and variance of a trait in a MA experiment if all mutations had identical effects. The formulae are general, in the sense that they can be applied to any quantitative trait, so long as $\Delta M \neq 0$. However, like all estimators, the Bateman–Mukai estimators (Equations 15.6b and 15.7) are subject to sampling error. Thus, if the estimates ΔM and/or ΔV are not very accurate, then the estimate of U_{min} can easily be substantially above or below its actual value.

Although the preceding derivations apply to diploid sexual lines, they are readily modified for alternative reproductive systems. For example, for haploids (either sexual or asexual), the expected rate of change of the mean is $U\,\overline{a}$, where U is the haploid genomic mutation rate and $\overline{a}$ is the mean effect of a mutational change. As noted above, the rate of divergence of haploid-line means is $\sigma^2_{m0}/2$. Again letting ΔM and ΔV denote estimates of the rates of change of mean and variance, the modified estimators become

$$\widehat{U}_{min} = \frac{(\Delta M)^2}{\Delta V(1 + C_{\Delta M})(1 + C_{\Delta V})}, \tag{15.9a}$$

$$\widehat{a}_{max} = \frac{\Delta V}{\Delta M(1 + C_{\Delta M})} \tag{15.9b}$$

These estimators have also been used in the analysis of *Drosophila* MA experiments involving single chromosomes, although the C_Δ terms involving sampling error (Equation 15.6c) have routinely been ignored. The same estimators apply to diploid asexual populations, except that Equation 15.9b estimates the upper bound on the average heterozygous effect, $E[a(1 + k)]$

Finally, we note that like among-line genetic variances, among-line genetic *covariances* involving *pairs* of traits increase proportionally with generation number. Thus, the pleiotropic effects of mutations on pairs of traits can be quantified by the genetic correlation,

$$r_m(x, y) = \frac{\Delta C_{x,y}}{\sqrt{\Delta V_x\, \Delta V_y}} \tag{15.10}$$

where $\Delta C_{x,y}$ is the rate of change in the among-line covariance between traits x and y (Lynch and Hill 1986; Lynch 1994). For diploid sexual lines and for haploid lines, $r_m(x, y)$ provides an estimate of the correlation between homozygous effects (a_x and a_y), while for diploid asexuals, it estimates the correlation between heterozygous effects ($[a(1 + k)]_x$ and $[a(1 + k)]_y$). In the absence of selection and epistasis, although the among-line variances and covariances continue to grow with the duration of the MA experiment, the genetic correlation is expected to remain constant, although later estimates are expected to be much more accurate as more mutations are sampled.

Maximum-likelihood Procedures

A considerable limitation of the Bateman–Mukai method is its assumption of constant mutational effects, which is certainly not true and likely causes bias in the resultant estimates of mutation rates and effects. The central problem here is that the full distribution of effects is essentially unobservable, owing to the fact that large fractions of mutations typically have effects that are so small as to be indiscernible without astronomical sample sizes. If one had knowledge of the squared coefficient of variation of mutational effects, C_m (Equation 15.6a), correction could be made to the Bateman-Mukai estimator so as to yield unbiased estimates of the mutational parameters. One approach to the problem is to start with an assumed form of the distribution of effects (typically a gamma, a two-parameter distribution; Equation A7.24a), and then to determine the distribution parameters that best fit the observed phenotypic data using a maximum-likelihood (ML) approach (Keightley 1994, 1998) or other model-fitting procedures (García-Dorado 1997; Shaw et al. 2002).

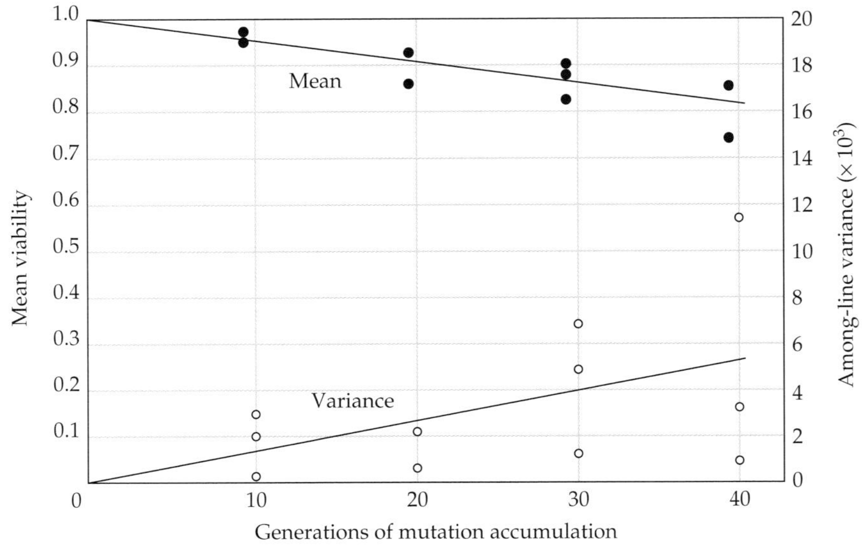

Figure 15.6 Decrease in mean viability and increase in genotypic variance in viability starting from a uniform base population. Second chromosomes of *Drosophila melanogaster* were kept sheltered over a balancer chromosome (see Figure 5.8) and passed through males (which do not exhibit recombination in *Drosophila*). Using this approach, mutations were accumulated on essentially clonally propagated chromosomes for a number of generations before being extracted and made homozygous. The mean viability of these sheltered chromosomes decreased with increasing number of generations of mutation accumulation, while the variance among lines increased. ΔM is estimated from the slope of the regression of mean on numbers of generations of mutation accumulation, while ΔV is estimated from the slope of the variance regression. (From Mukai et al. 1972.)

Typically, a specific form of the distribution of mutational effects is used that can be defined by just two to three parameters, usually accomplished using a gamma distribution of effects and often with separate distributions for mutations with positive versus negative effects weighted by the relative frequency of the two types. One also requires a specific form of the distribution of environmental effects influencing phenotypes, and this is often assumed to be Gaussian in nature with zero mean, and variance to be determined. Finally, the number of mutations/line is assumed to be Poisson-distributed, but the mutation rate itself is an unknown parameter, so taken together at least four to five parameters must be jointly estimated. In principle, if whole genomes have been sequenced at the time of phenotypic characterization, the mutation rate need not be a free parameter, although there would still be a need to evaluate the fraction of mutations with effects on the trait.

Results from Flies, Nematodes, Plants, and Microbes

The preceding approaches have been applied extensively to viability in *Drosophila*. As can be seen in Figure 15.6, when second chromosomes are passed clonally through males (using the techniques outlined in Figure 5.8), where they do not experience recombination, replicate lines exhibit a gradual decline in mean fitness and a gradual increase in among-line variance. Letting hs and $s = 2a$ denote the reductions in viability in mutant heterozygotes and homozygotes (namely, fitnesses of $1 : 1 - hs : 1 - s$), and extrapolating results to the whole genome, the results of Mukai and colleagues (Mukai 1964, 1969, 1979; Mukai and Yamazaki 1968; Mukai et al. 1965, 1972) and of Ohnishi (1977) lead to average Bateman–Mukai estimates of $\widehat{U}_{min} = 0.6$, $s_{max} = 0.06$, and $\overline{h} = 0.36$ (summarized in Simmons and Crow 1977; Crow and Simmons 1983; Lynch et al. 1995b). That is, the data suggest that new mutations for viability arise at a rate approaching one per individual per generation, and on average these mutations are slightly recessive, individually causing a reduction of viability of approximately 6% in homozygotes and approximately 2% ($0.06 \times 0.36 = 0.023$)

in heterozygotes.

These results indicate that the per-generation input of spontaneous mutations is such that, in the absence of effective selection, a *Drosophila* population would experience a $0.3 \times 6\% \simeq 2\%$ reduction in individual viability per generation. There are several reasons why this value might be an underestimate. First, the assays reported above were performed in relatively benign environments. When MA lines of *Drosophila* are assayed in harsher environments, their fitness (relative to controls) can decline much more dramatically, in some cases approaching zero (Kondrashov and Houle 1994; Wang et al. 2014). Future experiments are needed to evaluate whether such genotype $\times$ environment interaction is a consequence of an enhancement of the selective effects of individual mutations (increasing s) and/or of the expression of a larger number of loci (increasing U) in harsh environments. Second, results from Mukai and Yamazaki (1971), Yoshimaru and Mukai (1985), and Houle et al. (1994) suggest that most deleterious mutations have pleiotropic effects on all components of fitness, with mutations for fitness components such as developmental rate, fecundity, longevity, and mating ability having genetic correlations with those for viability near 1.0. This implies that the total impact of deleterious mutations on fitness in *Drosophila* might be greater than that for viability alone. Finally, all of the preceding analyses excluded chromosomes carrying lethal recessive mutations, which arise at a rate of about 0.02 per genome per generation (Crow and Simmons 1983), and cause an average viability reduction of 1% to 5% in heterozygotes, which is very similar to the heterozygous effects of mild detrimentals (Simmons and Crow 1977).

How much bias in the Bateman–Mukai estimates might be caused by unknown variation in mutational effects? Applying an ML technique to the Mukai data and assuming a gamma distribution of effects, Keightley (1994) inferred a highly L-shaped distribution of mutational effects, with most mutations being very close to neutral. However, he also noted that a very broad range of distributions, with squared coefficients of variation of s ranging from $C_m = 0.25$ to 30, are compatible with the data. Such results suggest that the genomic mutation rate to deleterious mutations in *Drosophila* could be as low as 0.8 and possibly as high as 20 (although the latter is far too high to be consistent with the genomic mutation rate known from molecular data). The mean homozygous effect (s) of a new mutation could be as low as 0.002 or as high as 0.05. Although the studies of Mukai and Ohnishi inspired a great deal of thought in evolutionary genetics, there have been a number of concerns with the analyses, including the possibility of line contamination and/or excessive mobile-element activity (Halligan and Keightley 2009). Subsequent work of this sort with other organisms yields a broader perspective, while also opening up additional questions.

Long-term MA lines for the self-fertilizing nematode *Caenorhabditis elegans* yield rather different results than those for flies: an overall rate of fitness decline of only 0.15% per generation, with estimates of $\widehat{U}_{min} \simeq 0.012$ and $s_{max} \simeq 0.20$ for total fitness (Vassilieva and Lynch 1998; Vassilieva et al. 2000; Baer et al. 2005). Using estimates based on fitness components (as opposed to total fitness) does not greatly alter this view, with estimates for total progeny production, viability to maturity, longevity, and body size being $\widehat{U}_{min} \sim 0.024$, 0.008, 0.0005, and 0.0021, and $s_{max} \sim 0.10$, 0.39, 0.47, and -0.54, respectively (Keightley and Caballero 1997; Vassilieva and Lynch 1999; Vassilieva et al. 2000; Azevedo et al. 2002). Application of ML analysis does not qualitatively change these results either, as a model with constant mutational effects is, for the most part, not rejected (Keightley and Caballero 1997; Keightley and Bataillon 2000; Vassilieva et al. 2000). The substantial reduction in U relative to that in *D. melanogaster* cannot be explained by a reduced genome size, as the *C. elegans* genome is only 50% smaller in size but contains ~50% more genes. Moreover, as noted above, both species have very similar per-nucleotide mutation rates.

Data on the quantitative fitness effects of spontaneous mutations are scant for other metazoans. However, Bateman–Mukai analyses of MA lines of the microcrustacean *Daphnia pulex*, which has a genome size similar to that in *Drosophila*, yield estimates of $\widehat{U}_{min}$ for fitness-related traits mostly in the range of 0.01 to 1.0 (mean $\simeq 0.33$) and s_{max} mostly in the

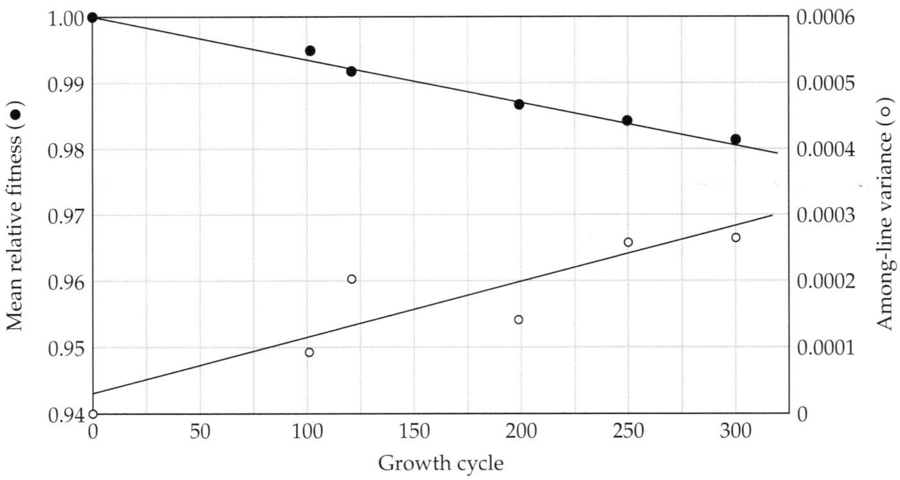

Figure 15.7 Similar plots to Figure 15.6 for 50 lines of the bacterium *E. coli* taken through daily single-cell bottlenecks for 300 days (with approximately 22 cell divisions per day). Here fitness is measured as the exponential rate of clonal expansion in liquid medium. (From Kibota and Lynch 1996.)

range of 0.01 to 0.1 (Lynch et al. 1999; Latta et al. 2013). These results are clearly in closer accord with those for flies than for nematodes.

Direct information on the fitness effects of *de novo* mutations in land plants comes largely from the self-fertilizing mustard *Arabidopsis thaliana.* Starting with a highly inbred lineage, Schultz et al. (1999) maintained 1000 replicate lines by single-seed descent for 10 generations. Compared to the progenitor genotype (stored as seed), total fitness (seed set in a fairly benign environment) exhibited a decline of about 1% per generation. This loss of fitness was a function of cumulative, but smaller, effects on individual fitness components, with values of -0.06%, 0.05%, and 0.4% per generation, respectively, for germination probability, numbers of fruits set, and numbers of seeds per fruit. Application of the Bateman–Mukai technique to the observations on total fitness yielded estimates of $U_{min} = 0.1$ and $\overline{s}_{max} = 0.1$. However, in a separate experiment with this species, starting with a different ancestral genotype, Shaw et al. (2000) observed a symmetrical change in line-specific mean fitnesses and little change in the overall mean, precluding the use of the Bateman–Mukai estimators. In this latter study, ML analysis suggested U in the range of 0.1 to 0.2 (Shaw et al. 2002), although growth of the same lines in a natural environment (as opposed to a more benign greenhouse setting) led to an elevated estimate of $U \simeq 1.0$, with the symmetrical distribution of fitness effects being retained (Rutter et al. 2010).

The differences in the distributions of fitness effects of mutations in these two sets of *A. thaliana* lines remain unexplained, with the Rutter et al. (2010) study suggesting a much larger fraction of beneficial mutations than seen in any other MA experiment. However, it may be telling that a number of mutations in this species have effects that are strongly dependent on the environmental setting (Rutter et al. 2012). The genome size of *Arabidopsis* is $\sim$70% that of *Drosophila,* but contains roughly twice the number of genes, whereas the mutation rate per nucleotide is $\sim$50% higher than in flies (Ossowski et al. 2010; Yang et al. 2015), implying a diploid genomic rate of $\simeq$0.7. Thus, as in the case of *Drosophila,* a large fraction of the total pool of mutations appears to have discernible fitness effects.

Other MA experiments with plants appear to be at least qualitatively consistent with relatively high genome-wide mutation rates to allelic variants with moderately deleterious affects. Schoen (2005) developed MA lines for two annual species of fiddlenecks (*Amsinckia*), and used an ML approach to infer U in the range of 0.1 to 0.4 for flower number and $\overline{s}$ (in the deleterious direction) in the range of 0.02 to 0.18. An experiment with lines involving the wild radish *Raphanus rhaphanistrum* (also an annual) suggested a nearly 2% decline in competitive fitness per generation (Roles and Conner 2008), although this was not significant. As to whether there is a fundamental difference in mutational properties of animals versus plants, there is no evidence of differences in mutagenicity once the proteome and genome sizes are taken into consideration (Lynch et al. 2023).

A few studies have extended the preceding types of analyses to microbes. For example, Kibota and Lynch (1996) performed an MA experiment on the bacterium *Escherichia coli,* taking 50 lines through daily single-cell bottlenecks for 300 days, and periodically assaying the lines for performance. Relative to the initial control line, there was a gradual decline in the mean, combined with an increase in the variance, in growth rate among lines (Figure 15.7). The Bateman–Mukai estimate of $\overline{s}_{max} = 0.012$, and the genomic deleterious-mutation rate estimate, $U_{min} = 1.7 \times 10^{-4}$, is about 17% of the total genomic rate of base-substitution mutation (based on a mutation rate per nucleotide site of 2.3×10^{-10} and a genome size of 4.6 Mb; Lynch et al. 2023). In another bacterium taken through single-cell bottlenecks, *Burkholderia cenocepacia,* the average fitness effect of single mutations was estimated to be ~0.005, with no lines exhibiting an improvement in fitness (Dillon and Cooper 2016; Rana et al. 2021), and the estimated deleterious-mutation rate per haploid genome, 6×10^{-5}, being 6% of the known genome-wide rate of 0.001 (Lynch et al. 2023).

For MA lines of the budding yeast (*Saccaromyces cerevisiae*), ML analysis suggests a deleterious-mutation rate per haploid genome of 5×10^{-5}, approximately half that in *E. coli,* but with much larger fitness effects of mutations, nearly all of which fall in the range of 0.15 to 0.30, with a mean of 0.21 (Zeyl and DeVisser 2001). An independent study by Hall and Joseph (2010) suggests haploid $U_{min} = 1.3 \times 10^{-4}$ and $s_{max} = 0.47$, with these parameter estimates becoming $\sim$ 3 fold less extreme when ML is applied to the data. Results from Wloch et al. (2001) suggest a highly L-shaped distribution of mutations with discernible heterozygous effects, with almost all of these being deleterious, and nearly a third of them being lethal when homozygous (see also Hall and Joseph 2010; Jasmin and Lenormand 2016). Focusing on gene-knockout lines, Agrawal and Whitlock (2011) found an average dominance coefficient of $\overline{h} = 0.2$, with a very wide distribution and a tendency for genes with large homozygous effects (s) to be more recessive. With a base-substitution mutation rate of 3.3×10^{-10} (Lynch et al. 2023) and a genome size of 12.5 Mb, the total haploid genomic mutation rate is $\simeq 0.004$, implying that ~5% of mutations have discernible fitness effects.

Finally, using fully sequenced genomes of MA lines of the unicellular green alga *Chlamydomonas,* Kraemer et al. (2017) estimated average selective effects per mutation of 0.0008 and 0.0011 for competitive fitness and growth rate, respectively, with ML suggesting that 1 to 10% of the total pool of genomic mutations have discernible fitness effects. Further ML analysis of these lines suggested that ~15% of mutations have a positive influence on fitness, with the distributions of both positive- and negative-effect mutations being highly L-shaped (Böndel et al. 2019).

These results from diverse, but select, phylogenetic groups suggest that much work remains to be done before strong generalities can be made with respect to rates of origin and magnitudes of effects of mutations. With knowledge of the mutation rate per nucleotide site from molecular surveys, one can compute the total number of mutations arising per genome per generation, and from comparisons of sequence data in pairs of species, one can compute the fraction of mutations that are removed by purifying selection (Kondrashov and Crow 1993; Halligan and Keightley 2009), providing a compelling approach to estimate the genome-wide deleterious-mutation rate that includes mutations with very small effects. However, a key question from a quantitative-genetics perspective concerns the fraction of the total pool of mutations that has *discernible phenotypic effects* on particular traits. Again, the limits to the phenotypic ascertainment of mutations with very small effects must be acknowledged.

Gratifyingly, the preceding estimates of rates of origin of fitness-altering mutations reside within the bounds dictated by genomic evidence, although the fraction of total mutations with fitness effects falls in a broad range of $\sim$5 to 50%. The average fitness effects of *de novo* mutations are generally in the range of 0.01 to 0.1, although those for *C. elegans* are inexplicably much stronger, and those for *Chlamydomonas* much weaker. Most analyses suggest an L-shaped distribution of mutational effects, with the vast majority of mutations having deleterious effects on fitness, although one set of studies with *Arabidopsis* suggests a nearly symmetrical distribution. Studies in which very small numbers of mutations have arisen per genome frequently reveal a small fraction of lines with improved fitness (Dickinson 2008; Hall and Joseph 2010; Bao et al. 2022), which is not surprising given that we know that adaptive evolution does occur. However, one concern with these types of assessment is that some experimental stocks have been initially somewhat maladapted to the assay environment, magnifying the chances of mutations with beneficial effects. Notwithstanding all of these uncertainties, the data suggest that owing to the combination of mutation pressure and mutational effects, populations with effective sizes smaller than several hundreds of individuals are vulnerable to gradual mutational decay (Lynch and Gabriel 1990; Lande 1994; Lynch et al. 1995a, 1995b).

One final issue is the extent to which spontaneous deleterious mutations interact epistatically. Evolutionary biologists are particularly interested in this issue because the equilibrium fitness load resulting from segregating deleterious mutations depends on the form of the fitness function, i.e., on the relationship between fitness and number of mutations. As noted in Chapter 12, when the effects of mutations at different loci act independently (multiplicative effects on fitness), the equilibrium mean fitness, relative to that of a mutation-free individual, is $\simeq e^{-U}$, where U is the diploid genomic deleterious mutation rate (Haldane 1937). If deleterious mutations interact synergistically, such that fitness declines at an accelerating rate with an increase in mutation number, the equilibrium load is reduced for the simple reason that the strength of selection increases with subsequent mutations (Kimura and Maruyama 1966; Milkman 1978; Crow and Kimura 1979; Kondrashov 1988; B. Charlesworth 1990). On the other hand, with diminishing-returns epistasis, the equilibrium load is increased because the strength of selection (per mutation) weakens as mutations accumulate.

Data on the nature of epistatic interactions among deleterious mutations are scant. If such interactions do exist, they appear to be weak and synergistic (Simmons and Crow 1977; Crow and Simmons 1983). However, for *Drosophila*, almost all of the support for this argument derives from a single observation of Mukai (1969), which suggested that mean fitness in MA lines began to decline at an accelerating rate in later generations. The curvilinearity he observed was not strong, and was primarily a function of two late-generation assays, which were not independent. A related concern associated with such observations is that accelerations in fitness declines may be a consequence of increases in mutation rates rather than epistasis. This might occur if lines with relatively high deleterious-mutation loads experience reduced replication fidelity owing to stress-induced mutagenesis or to the emergence of mutator alleles (Loewe et al. 2003; Ávila et al. 2006; Sharp and Agrawal 2012). The data on these issues are mixed, and for MA lines of yeast, accelerated rates of fitness-decline in later generations (Dickinson 2008) appear not to be due to either epistasis or mutation-rate elevation, but rather to a gradual change in the laboratory environment affecting all isolates regardless of their mutation load (Jasmin and Lenormand 2016).

Analysis of Natural Populations

Mutation-accumulation experiments of the sort described above are extremely labor intensive and require evolutionarily inert controls that are not easily maintained in most species. An alternative approach to characterizing the components of the deleterious-mutation process is to interpret various features of the distribution of fitness in a natural population in light of their expected values under mutation–selection balance. Morton et al. (1956) and Mukai and Yamaguchi (1974) first suggested how the genomic deleterious-mutation rate in a random-mating population might be estimated from the observed inbreeding depression for fitness (Chapter 12), provided an estimate of the average degree of dominance were available. Charlesworth et al. (1990) expanded this approach to obligately selfing populations. Both methods have been generalized by Deng and Lynch (1996b) to yield joint estimators of U, $\overline{h}$, and $\overline{s}$ for mutations affecting total fitness. Their method, which we review below, requires estimates of both the mean and genetic variance for fitness in inbred and outbred generations.

Recall from Chapter 12 (and WL Chapter 7) that the equilibrium frequency of a deleterious allele under mutation–selection balance, assuming $u \ll hs$, is $\sim u/(hs)$, where u is the genic mutation rate, and hs is the fractional loss of fitness in heterozygotes. Using this result, and assuming that mutations are distributed among individuals in a Poisson fashion (Equation 2.21a) and interact multiplicatively to define total fitness (i.e., there is no epistasis), the mean fitness in a random-mating (outcrossing) population can be shown to be

$$\overline{W}_O = W_{max} \exp(-U) \tag{15.11a}$$

with a genetic variance in fitness of

$$\sigma^2_W(O) = \overline{W}^2_O \left(\exp[U \cdot \overline{hs}] - 1\right) \tag{15.11b}$$

Here W_{max} is the expected measure of fitness in a mutation-free individual, and $\overline{hs}$ denotes the mean effect of a mutation in the heterozygous state.

Suppose now that progeny are produced via self-fertilization of a random sample of parental genotypes, and that multiple progeny are assayed per parent. The expected mean fitness and genetic variance in fitness among selfed families are

$$\overline{W}_S = W_{max} \exp\{-(U/4)[2 + (1/\widetilde{h})]\} \tag{15.11c}$$

$$\sigma^2_W(S) = \overline{W}^2_S\{\exp[(U/4)(\overline{s} + (1/4)[\overline{s/h}] + \overline{hs})] - 1\} \tag{15.11d}$$

where $\widetilde{h}$ is the harmonic mean dominance coefficient of new mutations, and $\overline{s/h}$ is the mean value of the ratio s/h for new mutations. These four equations can be rearranged to eliminate the parameter W_{max}, yielding a set of estimators for U, $\overline{s}$, and $\overline{h}$.

First, we define a set of observable measures,

$$\widehat{x} = \ln\left(\frac{\text{Var}(W_O)}{\overline{W}^2_O} + 1\right) \tag{15.12a}$$

$$\widehat{y} = \ln\left(\frac{\overline{W}_S}{\overline{W}_O}\right) \tag{15.12b}$$

$$\widehat{z} = \ln\left(\frac{\text{Var}(W_S)}{\overline{W}^2_S} + 1\right) \tag{15.12c}$$

where Var(W_O) is the observed genetic variance in fitness in the parental generation and Var(W_S) is the observed genetic variance in fitness among selfed families (a variety of methods for estimating genetic variances are covered in Part II Chapters 23–33). Note that $\widehat{y}$, the natural logarithm of the ratio of fitnesses observed in selfed and outcrossed generations, is a measure of inbreeding depression (Chapter 12), whereas $\widehat{x}$ and $\widehat{z}$ are, respectively, functions of the squared coefficients of genetic variation of fitness in the random-mating and selfed populations.

Given information on the means and genetic variances in fitness, $\widehat{x}$, $\widehat{y}$, and $\widehat{z}$ can be computed, allowing estimates of the mutation parameters to be obtained from the solutions of Equations 15.11a–15.11d,

$$\widehat{h} \simeq \frac{1}{8\sqrt{\widehat{z}/\widehat{x}} - 4}\left(C_h + \frac{2 - C^2_{s,h}}{1 + C_{s,h}}\right) \tag{15.13a}$$

$$\widehat{U} \simeq \frac{4\widehat{h}\,\widehat{y}}{2\widehat{h} - 1 - C_h} \tag{15.13b}$$

$$\widehat{s} \simeq \frac{\widehat{x}}{\widehat{U}\,\widehat{h}\,(1 + C_{s,h})} \tag{15.13c}$$

Here, $\widehat{h}$ and $\widehat{s}$ are estimates of the arithmetic mean properties of new mutations. Two new terms appearing in these formulae are functions of the distribution of mutational effects: $C_h = \sigma_h^2/\overline{h}^2$, the squared coefficient of variation of dominance coefficients for new mutations, and $C_{s,h} = \sigma_{s,h}/(\overline{h} \cdot \overline{s})$, the coefficient of covariation between s and h. Rarely will quantitative estimates of these properties be available. However, $C_{s,h}$ is expected to be negative (more deleterious mutations tending to be more recessive), and C_h is necessarily positive, although arguments presented in Deng and Lynch (1996b) suggest that it may be small. In the absence of information on these quantities, Equations 15.13a–15.13c can be still be solved by setting $C_h = C_{s,h} = 0$. The estimators will then be biased—the average degree of dominance and the genomic mutation rate in a downward direction, and the average selection coefficient in an upward direction.

Thus, as in the case of the Bateman–Mukai estimators, the **Deng–Lynch method** yields estimates of one-sided bounds on the mutation parameters. Given these estimates, additional composite measures can also be approximated. For example, the long-term rate of decline in fitness that would result if natural selection were rendered ineffective (as in a very small population) is

$$\Delta M = \frac{\widehat{U}\,\widehat{s}}{2} \tag{15.14}$$

Results in Deng and Lynch (1996b) suggest that Equation 15.14 is an almost unbiased estimator because the downward bias in the estimate $\widehat{U}$ is balanced by the upward bias in the estimate $\widehat{s}$. Moreover, ΔM can be simply estimated by the quantity $\widehat{z}/2$, where $\widehat{z}$ is defined in Equation 15.12c.

Deng and Lynch (1996b) presented a parallel set of equations for the reciprocal situation of outcrossing normally selfing populations. These estimators tend to be less biased by variance in mutational effects, apparently because most of the bias is caused by highly deleterious recessives, which are normally purged from regularly selfing populations (Chapter 12). Further information on the practical application of this technique, its power as a function of sample size, and its sensitivity to violations of the assumptions of the model is provided in Deng and Lynch (1996b).

Although the Deng–Lynch method has not yet been applied to natural populations, a few estimates of U have been obtained from the observed improvement in fitness in outcrossed progeny of normally self-fertilizing plants. Such estimates make use of the estimator of Charlesworth et al. (1990),

$$U = \frac{\widehat{y}}{\widehat{h} - 0.5} \tag{15.15}$$

where $\widehat{y}$ is defined in Equation 15.12b. This method requires an independent estimate of the degree of dominance, $\widehat{h}$. A variety of methods for estimating h were covered in Chapter 12, although it must be stressed that these estimate the average degree of dominance of *segregating mutations*, a selectively filtered subset of the original mutations. As a result, the latter is likely to be somewhat less than the degree of dominance of new mutations (the parameter that should be used in Equation 15.15), because selection more efficiently removes dominant deleterious mutations. In their original survey of the literature, Charlesworth et al. (1990) obtained results suggesting that U must be at least 0.5 for individual fitness components. Johnston and Schoen (1995) obtained estimates of U for total fitness in four populations of fiddlenecks (*Amsinckia*) ranging from 0.24 to 0.87, and Charlesworth et al. (1994) obtained estimates of $U \simeq 0.8$ and 1.5 for two species of *Leavenworthia* (a mustard relative). All of these estimates are in rough accord with the more direct results cited above for MA lines, supporting the idea that the genomic deleterious-mutation rate for total fitness is on the order of 0.2 to 2.0 per individual per generation.

There is one notable limitation with the preceding approach. As there are three observed measures (Equations 15.12a–15.12c), and three parameters to be estimated (Equations 15.13a–15.13c), there are no degrees of freedom to test the validity of the underlying model, which assumes that fitness variance is entirely determined by selection–mutation balance. Kelly (2003) introduced a more elaborate experimental design that allowed for model testing by introducing additional degrees of freedom. Starting with fully inbred lines, single males can be mated to multiple females to create descendant families of full and half sibs. Through subsequent selfing of the derived families, one can then estimate the divergence between inbred and outbred mean phenotypes, the phenotypic covariance among plants within selfed families, the covariance of outbred F_1 individuals and their selfed progeny, and the covariances between half and full sibs. Application of versions of this model to inbred lines of the monkey flower *Mimulus guttatus* led to average estimates of s and h both near 0.04, but ML evaluation also raised issues of the validity of the underlying model. The utility of Kelly's approach, and future extensions of it incorporating additional crosses, is the potential to discriminate the roles that alternative mechanisms might be playing in the maintenance of genetic variation in natural populations.

The Persistence of New Mutations

Provided one has estimates for the standing amount of genetic variance of characters in natural populations, there is another way to use estimates of the mutational variance to gain insight into the strength of selection operating against new mutations. The simplest way to view the problem is to assume that most new mutations are kept rare because of their deleterious nature, and to let hs (assumed to be the average selection against heterozygous mutations) be the fraction of standing genetic variance that is removed each generation by selection. Then, letting $\sigma_G^2(t)$ denote the genetic variance at time t, the dynamics of genetic variance can be represented as $\sigma_G^2(t) = (1 - hs)\,\sigma_G^2(t-1) + \sigma_m^2$. At equilibrium, $\sigma_G^2 = \sigma_m^2/(hs)$ (Barton 1990; Kondrashov and Turelli 1992). Note that this is a variance analog to the equilibrium allele frequency under mutation–selection, $q = \mu/(hs)$. Thus, the ratio σ_m^2/σ_G^2 is a measure of the strength of selection against new mutations, and its reciprocal approximates the mean number of generations that a new mutation persists in a population before being removed by selection (Crow 1992). An alternative and more general definition of σ_G^2/σ_m^2 is the mean number of individuals affected by a mutation before it is eradicated (Li and Nei 1972; García-Dorado et al. 2003). In an infinite population, the mean persistence time and the mean number of affected individuals are equivalent.

Alternative interpretations are required in the extreme case in which mutations are neutral. Under these circumstances, the equilibrium genetic variance is a consequence of a balance between random genetic drift and mutation and is equal to $2N_e\sigma_m^2$, as noted above. The mean persistence time is then $2N_e$ generations.

From estimates of σ_m^2 and σ_G^2, Crow (1992, 1993b) estimated the mean persistence of viability mutations in *Drosophila* to be approximately ~ 40 generations, both for mild detrimentals and recessive lethals. These results are remarkably consistent with estimates of the selection intensity against heterozygous viability mutations cited above. For example, the reciprocal of hs for mildly deleterious mutations is approximately 45. Actual persistence times are a function of the total fitness effects of new mutations, whereas *Drosophila* mutation-accumulation experiments have typically focused only on the viability-to-adulthood component. Thus, the consistency of the two results suggests that the fitness consequences of new mutations may be largely expressed early in life.

A broader review of the data appears in Houle et al. (1996). Summarizing over a diversity of characters and species, all estimates of the mean persistence time were less than 1,000 generations, and more than half were less than 100 generations. The mean persistence times for mutations affecting life-history traits are on the order of 50 generations, again suggesting an average intensity of selection against heterozygotes of about 2%. In *D. melanogaster,* the mean persistence times for fecundity and longevity are approximately 50 and 60 generations, not greatly different from the results for egg-to-adult viability. For morphological traits in this species, mean persistence times tend to be more on the order of 100 generations, much too low to be explained by a neutral model (which would be on the order of $2N_e$ generations). For abdominal and sternopleural bristle numbers, they are on the order of 700 and 400 generations, consistent with the idea that there is weak selection against new bristle-number mutations as a consequence of their pleiotropic effects on fitness (Nuzhdin et al. 1995).

Such low persistence times raise fundamental questions about our understanding of the nature and relevance of standing pools of quantitative-genetic variation. Is such variation available for adaptive change when a population is exposed to a new selective regime? Or is much of it simply a consequence of the recurrent introduction of mutations with unconditionally deleterious side effects on fitness? Under the former view, the ability of a population to respond to selection is directly proportional to the amount of additive genetic variance (Bulmer 1971, 1972, 1980; Lande 1975; Bürger et al. 1989; Houle 1989b). Under the latter view, genetic variation observed within populations is of little relevance to adaptive evolution, as the divergence of mean phenotypes among populations mostly involves rare mutations that have only weak pleiotropic side effects (Robertson 1967; Hill and Keightley 1988; Barton 1990; Keightley and Hill 1990; Kondrashov and Turelli 1992). Because nearly all results from MA experiments support the contention that the majority of new mutations are deleterious, this latter hypothesis must be taken seriously.

We have only provided a brief overview here of the vast literature on the role of mutation–selection balance in the maintenance of quantitative-genetic variation. See WL Chapter 28 for a more general treatment, especially for the various proposed theoretical models (and their deficiencies). We also briefly revisit this topic in Chapter 21, using results from GWAS data.

16

Detecting Major Genes

There are two possible outcomes: if the result confirms the hypothesis, then you've made a measurement. If the result is contrary to the hypothesis, then you've made a discovery.
Enrico Fermi

In some cases, the bulk of character variation (either total or genetic) can be attributed to one or a few **major genes**. For a variety of reasons, it is of great interest to detect such genes. From a biological standpoint, the presence of major genes offers the potential for their isolation and genetic characterization, which in turn may be highly informative as to the underlying biological processes generating trait variation. From a theoretical standpoint, several quantitative-genetic models (especially those dealing with selection response) assume a large number of loci of roughly equal (and small) effects (WL Chapter 24). The validity of these models is severely compromised by the presence of major genes (WL Chapter 25). If one or two major loci account for most of the genetic variation of a trait, essentially any problem of interest can be correctly modeled using the standard machinery of one- and two-locus population genetics (WL Chapter 5).

The observation of a continuous unimodal distribution of phenotypes is often taken as support for a large number of genes of roughly equal effect. It cannot be overstressed that this is an *assumption*. If environmental variation is sufficiently large relative to the effects of any individual gene or if major alleles are at sufficiently low frequency, the effects of segregating major genes can be completely obscured. The most powerful tests for the presence of major genes, which use information from linked molecular-marker loci, are examined in Chapters 17–21. This chapter illustrates how purely phenotypic data can be used to infer the presence or absence of major genes.

Our treatment starts with the simplest form of analysis, **departure-from-normality** tests applied to situations in which no genealogical information is available. We next consider several fairly simple tests that can be used when groups of known sibs are identified. We then introduce mixture models, wherein the phenotypic distribution is assumed to result from a weighted mixture of several underlying distributions (one for each major-locus genotype). Such models form the foundation for much of the discussion over the next several chapters. We conclude by examining **complex segregation analysis** (**CSA**), a collection of likelihood methods that have been historically important in human genetics and have also been applied to other natural populations. The roots of many of the approaches discussed here stem from epidemiological genetics, a subject reviewed by Cavalli-Sforza and Bodmer (1971), Morton et al. (1983), Weiss (1993), Thomas (2004), Ziegler and König (2006), Laird and Lange (2011), and Elson (2017). Additional discussions on the detection of major genes from phenotypic data alone can be found in reviews by Mayo (1989), Hill and Knott (1990), and Le Roy and Elsen (1992).

ELEMENTARY TESTS

While detection of major genes is not an easy task, it is facilitated considerably with experimental populations in which controlled breeding designs can be implemented. For example, a population segregating a major allele initially at low frequency is expected to show a significant increase in heritability after only a few generations of selection (Latter 1965; Frankham and Nurthen 1981; WL Chapter 25), reflecting the increase in additive genetic variation as the rare major allele increases in frequency (Figure 4.8). Heritability can likewise show a sudden decline as a selected major allele approaches fixation.

A related approach for detecting major genes using selection is the **select-and-backcross method** (reviewed by Wright 1952, 1968). The procedure is simple. Two populations

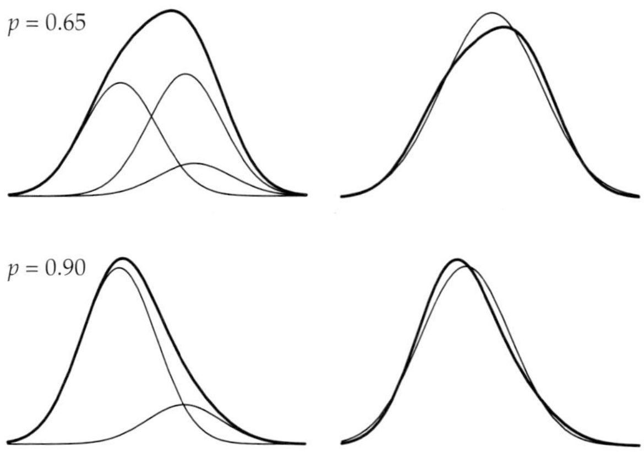

Figure 16.1 Phenotypic distributions resulting from a single diallelic major locus, where each genotype has normally distributed phenotypes with the same variance σ^2. Here the mean genotypic values are $\mu_{qq} = -\sigma$, $\mu_{Qq} = 0.75\sigma$, $\mu_{QQ} = \sigma$, and genotypes are in Hardy–Weinberg proportions, with the frequency of allele q being $p = 0.65$ (**Top**) and $p = 0.9$ (**Bottom**). The left side of each figure shows the three underlying distributions (thin lines) and the resulting mixture distribution (thick line). The right side shows how well the resulting mixture distribution (thick line) fits a normal having the same mean and variance (thin lines).

differing in the character are crossed, with the largest F_1 individuals backcrossed to individuals from the parental population with the smaller mean character value. Continual selection for large character value followed by backcrossing to the smaller line removes genes (from the large line) of minor effect while retaining genes having large effects (Wright's **leading factors**). The ability to maintain a large trait value in the face of continual backcrossing to the smaller line suggests the presence of major genes. Wright (1968) successfully used this approach to isolate leading factors for several traits in guinea pigs, and the Booroola gene in sheep (Examples 4.4 and 4.5) was detected in this manner (Piper and Bindon 1988). Humans and other natural populations that are not amenable to controlled breeding generally preclude such approaches, but as shown below, a random sample of individuals from such a population can sometimes be informative, even in the absence of molecular data.

Departures from Normality

Consider a locus segregating a major allele and assume that the distribution of phenotypes for each of the major-locus genotypes is normal. The resulting distribution, a weighted mixture of normals, is generally *not* normal. When a major gene is segregating, the phenotypic distribution can exhibit multimodality, skewness, and/or kurtosis. Figure 16.1 shows that when major allele frequencies are intermediate, the resulting phenotypic distribution is platykurtic, being more flatly peaked than a normal. The distribution becomes leptokurtic (more peaked than a normal) and skewed when alleles of large effect are at extreme frequencies, i.e., somewhat near zero or one (Hammond and James 1970; O'Donald 1971). When gene frequencies become very close to zero or one, the distribution again more closely approaches normality, as there is effectively only a single genotype in the population.

A variety of tests for normality have been developed, some of which were discussed in Chapter 14, but none of these is particularly powerful for detecting major genes. As a consequence, one must proceed with extreme caution in inferring the *absence* of major genes from a phenotypic distribution that is not significantly different from normal. As Figure 16.1 shows, the departure from normality can be rather small, requiring very large sample sizes for detection. Thoday and Thompson (1976) found that tests based on kurtosis have poor power for detecting major genes. In their simulations, a character with no environmental variation determined by four additive loci (each with two alleles at frequency 0.5 and relative effects of x, $2x$, $3x$, and $4x$) required a sample of roughly 750 individuals to detect non-normality, while with three additive loci of equal effect, sample sizes in excess of 1,000 were required. Conversely, when the distribution significantly departs from normality, one must be equally cautious in inferring the *presence* of a major gene without additional support. For example, a phenotypic distribution can display skew in the absence of major genes due to scale effects (Chapter 14). Similarly, failure to include important cofactors (such as sex or age) can result in a multimodal distribution, which collapses to a single normal when these cofactors are accommodated. As we discuss shortly, mixture models provide a more sophisticated and powerful approach to major-gene detection, assessing whether the fit of the phenotypic distribution is consistent with a mixture of normals.

Tests Based on Sibship Variances

The use of collections of known relatives, rather than random individuals, greatly improves the power of tests for major genes. Several simple tests based on sibship comparisons have been proposed. The presence of major alleles increases the variance within sibships in which the alleles are segregating. For example, if Q and q denote alternative alleles at a major locus, the expected phenotypic variance among sibs from $Qq \times Qq$ parents is greater than for sibs from $qq \times qq$ or $QQ \times QQ$ parents, reflecting segregation within these families, while sibs from $QQ \times Qq$ and $qq \times Qq$ parents display intermediate levels of variance. A number of tests for unequal variances among sibships have been suggested, the simplest being **Bartlett's test** for homogeneity of variances (e.g., Sokal and Rohlf 1995). The difficulty with this approach is that nongenetic causes may underlie differences in sibship variances. Bartlett's test is also rather sensitive to departures from normality. As a way around these problems, Mérat (1968) proposed a test based on the notion that families with significantly elevated variances should also exhibit exceptional platykurtosis and possibly skewness if the inflation of the variance is caused by segregating alleles of large effect.

Fain (1978) proposed an alternative test that utilizes both the means and variances of sibships. The basis for this test is the observation that, for a character determined by many genes of roughly equal effect, there should be no relationship between the variance of the sibship and the phenotypic value of the parents (Pearson 1904; Penrose 1969; Felsenstein 1973; Stark 1976; WL Chapters 16 and 24). If, however, the character is determined by a few genes of large effect, parents with the most extreme phenotypes are likely to be homozygotes, while parents of intermediate phenotypes are more likely to be heterozygotes. This relationship results in a roughly quadratic regression of offspring variance on midparental phenotypic value,

$$\text{Var}(z_i) = a + b_1 \overline{z}_i + b_2 \overline{z}_i^2 \tag{16.1}$$

where $\text{Var}(z_i)$ is the phenotypic variance within the ith sibship, and $\overline{z}_i$ is the midparental value for this sibship. Sibship means replace midparental values when the latter are unknown. A significant value of b_2 is taken as an indication of a major gene. Due to scaling effects (Chapter 14), the variance often increases linearly with the mean, so a significant b_1, by itself, does not necessarily indicate a major gene.

Example 16.1. Bucher et al. (1982) examined a large sample of families that were classified into groups based on cholesterol levels. One particular group (the High group) showed significant heterogeneity of within-sibship variance for cholesterol level ($P < 0.001$ using Bartlett's test of homogeneity of variances), while the other groups did not. Likewise, the regression of the within-sibship variance on the sibship means had a significant quadratic term for the High group, while the quadratic term was not significant in the other groups. These data suggest that a major gene was segregating in at least some of the families forming the High group (i.e., at least one parent was a heterozygote), but was not in the families forming the other groups. Thus, there were additional sources of variance beyond the major gene (such as environmental factors and/or additional polygenes) that contributed to the difference between groups.

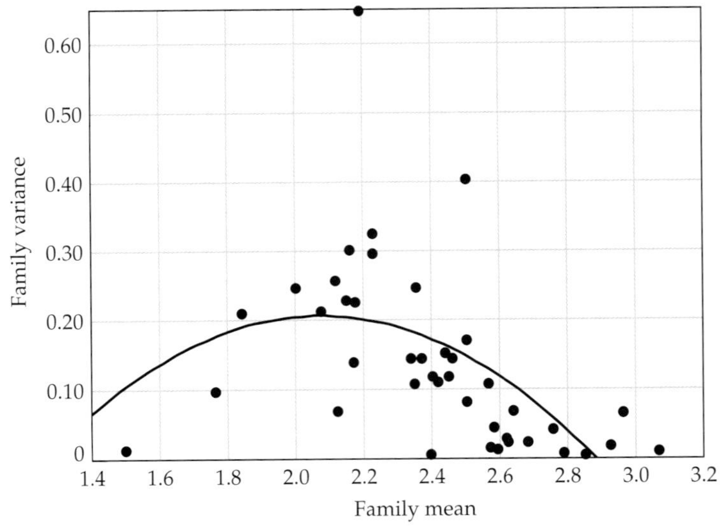

Figure 16.2 An example of Fain's test, plotting family variance as a function of family mean for germination date in *Impatiens capensis*. There is a highly significant quadratic term ($r^2 = 0.32$ versus $r^2 = 0.18$ for a simple linear regression), suggestive of a segregating major gene for germination date in this population. (Data from Mitchell-Olds and Bergelson 1990, kindly provided by Tom Mitchell-Olds.)

A second example of Fain's test is based on Mitchell-Olds and Bergelson's (1990) work on the annual plant *Impatiens capensis*. These authors found for most traits that the regression of within-sibship variance on sibship mean had significant linear (but not quadratic) terms, suggesting that scale effects are common. However, the regression for germination date also showed a significant quadratic term (Figure 16.2), suggesting the presence of a major gene influencing this character.

The power of different sibship variance tests has been considered by several investigators. MacCluer and Kammerer (1984) concluded that both Bartlett's and Fain's test have low power for detecting a major gene even when 100 nuclear families were used. However, they assumed a small number of sibs per family (3 to 5), as their interest was in human populations. On the positive side, they found that these tests are unlikely to give a false indication of a major gene when none is present (under the simplified assumption of normally distributed environmental effects). Mayo et al. (1980) noted that Fain's test is compromised if heterozygotes have lower environmental variances than homozygotes, as this difference partially masks the increased genetic variance in sibships from heterozygous parents.

Example 16.2. Alleles of large effect generally tend to be at low (or very low) frequencies in natural populations (Chapter 21; WL Chapter 25). When such large-effect alleles exist, they give a large signal in the family in which they are segregating, but such families will be rare. Suppose Q denotes the allele of large effect, and let p be its frequency. When $p \ll 1$, there are effectively only two types of **informative families** in which Q is segregating: offspring from $qq \times Qq$ or $Qq \times Qq$ matings. The former are more frequent, while the latter are expected to give a larger signal (a higher within-family variance). Under Hardy–Weinberg, the frequency of $qq \times Qq$ families is $2 \cdot (1-p)^2 \cdot 2\,p(1-p)\,\} = 4\,(1-p)^3 p$, while the frequency of $Qq \times Qq$ families is $4p^2(1-p)^2$, giving the frequency of informative families as $4(-p)^2 p$, of which a fraction $[4p^2(1-p)^2]/[4\,(1-p)^3 p] = p/(1-p)$ are the more informative $Qq \times Qq$ families.

From an experimental design perspective, an investigator would like to know how many random families must be sampled to have a high probability of obtaining at least a few informative ones. For example, how many families must be sampled to have a 95% probability that at least one is informative? Let x denote the frequency of an informative family. The probability that none of n random families are informative thus becomes $(1-x)^n$. Setting $(1-x)^n = 0.05$ and solving for the sample size required to have a 95% chance that at least one family is informative gives $n_{0.95} = \ln(0.05)/\ln(1-x)$. Using this result, and the above values for probability of an informative family, we find that

Major Allele Frequency, p	0.05	0.01	0.005	0.001
$4(1-p^2)p$ = Prob(Informat. fam.)	0.181	0.039	0.020	0.004
% $Qq \times Qq$ families	5	1	0.5	0.1
$n_{0.95}$	15	75	150	749

Major-gene Indices (MGI)

Under polygenic inheritance with additive loci, Karlin et al. (1979) showed that offspring more closely resemble the average of their parents (the midparent) than either individual parent, whereas the converse is true when a major gene is segregating. Exploiting this relationship, they proposed a class of indices for indicating the presence of major genes,

$$\mathrm{MGI}(\gamma) = \frac{E[\,|z_o - (z_f + z_m)/2|^{\gamma}\,]}{E[\,(\,|z_o - z_f|\,|z_o - z_m|)^{\gamma/2}\,]} \tag{16.2}$$

where z_o is the phenotypic value of an offspring whose parents have values z_f and z_m, and γ is a free parameter set by the investigator. Values of $\gamma > 1$ accentuate large deviations between offspring and parents, while values of $\gamma < 1$ accentuate small deviations. If the character is entirely determined by segregation of a major allele with no environmental variation, $\mathrm{MGI}(\gamma) > 1$ for all $\gamma \geq 0$ (independent of the amount of dominance), while if the character is entirely determined by an infinite number of genes of equal effect with a normal distribution of environmental effects, $\mathrm{MGI}(\gamma) < 1$ for all $\gamma \geq 0$. Karlin et al. recommend evaluation of the index at $\gamma = 0.5$, 1, and 2, but this choice of a values was rather *ad hoc*. Improvements and embellishments on this test have been suggested by Carmelli et al. (1979), Karlin and Williams (1981), and Famula (1986), but Le Roy and Elsen (1992) found that both the Bartlett and Fain tests were more powerful than MGI-based tests.

Nonparametric Line-cross Tests

Collins (1967, 1968, 1973), Mode and Gasser (1972), and Birnbaum (1972) independently proposed simple line-cross tests to detect major genes. Suppose a character is completely determined by one locus with two alleles, Q and q. Each genotype has an associated distribution of environmental values, which may differ between genotypes, and the phenotypic distribution in the population is a weighted mixture of the distributions associated with the three underlying genotypes,

$$p(z) = p_{QQ}(z)\Pr(QQ) + p_{Qq}(z)\Pr(Qq) + p_{qq}(z)\Pr(qq) \tag{16.3}$$

where $\Pr(QQ)$ is the probability that a randomly chosen individual is QQ, and $p_{QQ}(z)$ is the distribution of environmental values for a QQ individual, with the other terms defined similarly.

Based on this simple relationship, the test proceeds as follows. Consider two parental populations, P_1 and P_2, fixed for different alleles (Q and q, respectively). If the outbreak of variation in the F_2 is entirely due to segregation at a single locus, then the F_2 distribution is specified exactly by Equation 16.3, with $p_{Qq}(z) = p_{F_1}(z)$, $p_{QQ}(z) = p_{P_1}(z)$, $p_{qq}(z) = p_{P_2}(z)$, $\Pr(QQ) = \Pr(qq) = 1/4$, and $\Pr(Qq) = 1/2$, giving

$$p_{F_2}(z) = \frac{p_{P_1}(z)}{4} + \frac{p_{F_1}(z)}{2} + \frac{p_{P_2}(z)}{4} \tag{16.4}$$

Standard goodness-of-fit criteria (such as a Kolmogorov–Smirnov test; Sokal and Rohlf 1995) are then used to compare the observed and expected distributions. Related tests have also been suggested by Elston (1981b) and Stolk et al. (1984), and these were used by Hagger et al. (1995) to reject the hypothesis of a single dominant gene to account for egg weight differences between two inbred lines of chickens.

This approach can, in theory, be extended to two or more major loci, although tests of more than two loci require an impractical number of crosses. If more than one locus is segregating, knowledge of the parental and F_1 distributions is generally not sufficient to predict the F_2 distribution. Additional information, such as phenotypic distributions of certain backcrosses (e.g., the distribution of B_1, $B_1 \times B_2$, etc.), is required to account for the new F_2 genotypes generated by segregation (Collins 1967, 1968, 1973).

Equation 16.3 provides our first formal introduction to **mixture models**, where the observed distribution is a *weighted mixture of underlying distributions.* The approach taken above is nonparametric in that it does not initially assume any particular form for the underlying distributions. While this is an advantage in some cases, it results in a reduction in power relative to parametric tests that incorporate the correct form of the underlying distributions. We now turn to considerations of parametric mixture models that assume the underlying distributions to be normal. This class of models forms the basis for much of the analysis over the next several chapters.

MIXTURE MODELS

Given that rejection of normality, by itself, is not sufficient to imply a major gene, a more powerful approach is to test whether the phenotypic distribution is consistent with a mixing (or **commingling**) of two or more normals, as would be expected if the distribution of phenotypic values about each major genotype were normal (Tan and Chang 1972; Elston et al. 1974; Boerwinkle et al. 1986; Hoeschele 1988). While this is an improvement over previous methods, such a test is still fraught with a number of potential problems. Even if the phenotypic distribution is consistent with a mixture of normals, there are a variety of explanations other than major genes: for example, the population may be distributed over two or more significantly different environments. Thus, a slightly more powerful approach is to further specify the weights of the underlying normals.

If the distribution is consistent with three underlying normals with Hardy–Weinberg weights p^2, $2p(1-p)$, and $(1-p)^2$, where the allele frequency p is estimated from the data by maximum likelihood (see Example 16.3), then we may have more confidence in a major-locus interpretation. On the other hand, a lack of fit to this model does not necessarily exclude a major gene (e.g., the major gene may not be in Hardy–Weinberg equilibrium or the underlying phenotypic distributions for each major locus genotype may be non-normal). In the absence of information from linked markers (Chapters 17–21), the most powerful mixture-model tests for major genes use information from relatives to specify the weights of the underlying distributions. Before developing this **complex segregation analysis** approach, we first review some general features of mixture models.

The Distribution Under a Mixture Model

Assume the distribution for a trait of interest, z, results from a weighted mixture of several underlying distributions. If there are $i = 1, \cdots, n$ underlying distributions, $p_1(z), \cdots, p_n(z)$, each with frequency $\Pr(i)$, the resulting probability density of an observed variable z is given by a generalization of Equation 16.3,

$$p(z) = \sum_{i=1}^{n} \Pr(i) \cdot p_i(z) \tag{16.5a}$$

It is usually assumed that the underlying distributions are normals, so this becomes

$$p(z) = \sum_{i=1}^{n} \Pr(i) \cdot \varphi(z, \mu_i, \sigma_i^2) \tag{16.5b}$$

where

$$\varphi(z, \mu_i, \sigma_i^2) = \frac{1}{\sqrt{2\pi\sigma_i^2}} \exp\left[-\frac{(z-\mu_i)^2}{2\sigma_i^2}\right] \tag{16.5c}$$

is the probability density function for a normally distributed random variable with mean μ_i and variance σ_i^2. Equation 16.5b has $3n - 1$ parameters to estimate: the $n - 1$ mixing proportions $\Pr(i)$, and the n means and n variances of the underlying distributions. It is usually assumed that all the variances are equal, reducing the number of unknown parameters to $2n$. As detailed shortly, various genetic hypotheses allow us to further specify and evaluate the structure of the mixing proportions.

For the commonly assumed setting of k normals, with mean μ_i, common variance σ^2, and weight p_i, the first four moments of z become

$$E[z] = \sum_{i=1}^{k} p_i\mu_i, \qquad E[z^2] = \sigma^2 + \sum_{i=1}^{k} p_i\mu_i^2$$

$$E[z^3] = \sum_{i=1}^{k} p_i\mu_i\left(\mu_i^2 + 3\sigma^2\right), \qquad E[z^4] = 3\sigma^4 + \sum_{i=1}^{k} p_i\mu_i^2\left(\mu_i^2 + 6\sigma^2\right) \tag{16.5d}$$

The resulting skewness follows by substituting the above results into Equation 2.6

$$\mu_3 = E(z^3) - 3\mu E(z^2) + 2\mu^3 \tag{16.5e}$$

with the scaled skewness, k_3, following from Equation 2.8. Similarly, the kurtosis follows by substituting into Equation 2.9,

$$\mu_4 = E[z^4] - 4E[z^3]\mu_z + 6E[z^2]\mu_z^2 - 3\mu_z^4 \tag{16.5f}$$

with the scaled kurtosis, k_4, following from Equation 2.12a.

Parameter Estimation

Parameters of mixture models are typically estimated by maximum-likelihood procedures (Hasselblad 1966; Day 1969; Everitt and Hand 1981; Redner and Walker 1984; Titterington et al. 1985; McLachlan and Basford 1988; Shoukri and McLachlan 1994), in which case Equation 16.5b gives the likelihood function $\ell(z)$ of the unknown parameters $\Pr(1), \cdots, \Pr(n)$, $\mu_1, \cdots, \mu_n, \sigma_1^2, \cdots, \sigma_n^2$ as a function of the observed value z. As just mentioned, typically one sets most (or all) of the variances equal to each other. One reason for equating variances is that, if they are all free to vary, there can be singularities in the likelihood function.

As an example of how likelihood functions are constructed, consider the situation for a random individual drawn from a population with a single segregating diallelic major locus. Indexing the three genotypes by i where $i = QQ$, Qq, and qq, and assuming that individuals with major-locus genotype i are normally distributed with mean μ_i and common variance σ^2, the resulting likelihood for the jth individual is

$$\ell(z_j) = \Pr(QQ)\, p_{QQ}(z_j) + \Pr(Qq)\, p_{Qq}(z_j) + \Pr(qq)\, p_{qq}(z_j) \tag{16.6a}$$

$$= \Pr(QQ)\, \varphi(z_j, \mu_{QQ}, \sigma^2) + \Pr(Qq)\, \varphi(z_j, \mu_{Qq}, \sigma^2) + \Pr(qq)\, \varphi(z_j, \mu_{qq}, \sigma^2)$$

where z_j is the trait value in the focal individual. This likelihood for an individual is easily generalizable to more complicated genetic models. For example, if there are more than two alleles, or multiple loci, the likelihood has the form of Equation 16.6a with the sum now extending over all n_g multilocus genotypes. For n random (unrelated) individuals, denoting the observed phenotypic values by the vector $\mathbf{z} = (z_1, z_2, \cdots, z_n)$, the overall likelihood is just the product of the n individual likelihoods,

$$\ell(\mathbf{z}) = \ell(z_1, z_2, \cdots, z_n) = \prod_{j=1}^{n} \ell(z_j) \tag{16.6b}$$

Assuming random mating, the Hardy–Weinberg principle describes the frequencies $\Pr(\cdot)$ of the major locus genotypes as a function of p, the frequency of one allele. This leaves five parameters to estimate: p, σ^2, μ_{QQ}, μ_{Qq}, and μ_{qq}.

Appendix 4 reviews the basic features of the maximum-likelihood approach, and the reader may wish to consult this before continuing. Maximum-likelihood estimates (MLEs) are those values of the unknown parameters that maximize the likelihood function when treating the observed data $\mathbf{z} = (z_1, \cdots, z_n)$ as fixed constants. Such estimates can be obtained by numerical maximization of the likelihood function (e.g., Gill et al. 1981; Fletcher 1987) or by other iterative approaches. In particular, **expectation-maximization** (**EM**) methods are both very powerful and very flexible, accommodating missing or incomplete data (Appendix 4). Sample variances and covariances of MLEs can either be obtained directly from the likelihood functions or via approximation methods (e.g., Meyer and Hill 1992). Again, see Appendix 4 for details.

Hypothesis Testing

An important issue in tests for major genes is **model fitting**, i.e., evaluating whether the full model is needed, or if some subset of the model gives essentially the same fit. For example, we might initially assume a mixture of two normals with different means and common variance (as would be the case for a complete recessive), so that the full model has parameters μ_1, μ_2, σ^2, and p. Is the fit using these four parameters significantly better than the fit assuming a single underlying normal with parameters μ and σ^2? For large sample sizes, the **likelihood ratio** (LR) statistic test for whether the full model provides a better fit than a particular subset of the model is

$$\Lambda(\mathbf{z}) = -2\ln\left[\frac{\widehat{\ell}_r(\mathbf{z})}{\widehat{\ell}(\mathbf{z})}\right] = -2\left\{\ln\left[\widehat{\ell}_r(\mathbf{z})\right] - \ln\left[\widehat{\ell}(\mathbf{z})\right]\right\} \tag{16.7}$$

where $\widehat{\ell}(\mathbf{z})$ is the likelihood function evaluated at the MLE for the full model, and $\widehat{\ell}_r(\mathbf{z})$ is the maximum of the likelihood function for the restricted model under which r parameters of the full model are assigned fixed values. Under appropriate conditions, the LR test statistic is approximately distributed as χ^2_r, i.e., as a χ^2 distribution with r degrees of freedom (Wilks 1938); see Appendix 4.

The simplest restricted model assumes no mixture at all, so that the overall distribution is just a single normal distribution with unknown mean and variance. As shown in Appendix 4, the resulting likelihood is just the product of n identical normals with mean μ and variance σ^2. The MLEs in this case are the sample mean $\overline{z}$ and the uncorrected sample variance

$$\text{Var}(z) = \frac{1}{n}\sum_{j=1}^{n}(z_j - \overline{z})^2$$

Note that the ML estimate of the variance is slightly different from the unbiased variance estimator, which divides the sum of squares by $n-1$ (using this latter denominator yields the REML estimate; Part II Chapter 32). Substituting the MLEs of σ^2 and μ gives the maximum value of this restricted likelihood as

$$\widehat{\ell}_r(\mathbf{z}) = \prod_{j=1}^{n}\left\{\frac{1}{\sqrt{2\pi\,\text{Var}(z)}}\exp\left[-\frac{(z_j - \overline{z})^2}{2\,\text{Var}(z)}\right]\right\}$$

Taking logarithms and recalling the definition of Var gives

$$-2\ln\left[\widehat{\ell}_r(\mathbf{z})\right] = n\cdot\left[\ln\text{Var}(z) + \ln 2\pi + 1\right] \tag{16.8}$$

Example 16.3. Consider the likelihood-ratio test statistic for whether a diallelic major gene (in Hardy–Weinberg frequencies, with the phenotypes for each major-locus genotype normally distributed with constant variance) provides a better fit of the data than a single normal distribution. Assume that the data consist of n (unrelated) individuals, randomly chosen from the population. From Equations 16.7 and 16.8, the likelihood-ratio test statistic is given by

$$\Lambda(\mathbf{z}) = -2\left\{\ln\left[\widehat{\ell}_r(\mathbf{z})\right] - \ln\left[\widehat{\ell}(\mathbf{z})\right]\right\}$$
$$= 2\ln\left[\widehat{\ell}(\mathbf{z})\right] + n\cdot[\ln \text{Var}(z) + \ln 2\pi + 1]$$

where

$$\widehat{\ell}(\mathbf{z}) = \max\left[\prod_{j=1}^{n} \ell(z_j)\right]$$

the maximum being taken over all admissible values of p $(0 \leq p \leq 1)$, μ_{QQ}, μ_{Qq}, μ_{qq} $(-\infty < \mu < \infty)$, and σ^2 $(\sigma^2 \geq 0)$, with

$$\ell(z_j) = p^2\cdot\varphi(z_j,\mu_{QQ},\sigma^2) + 2p(1-p)\cdot\varphi(z_j,\mu_{Qq},\sigma^2) + (1-p)^2\cdot\varphi(z_j,\mu_{qq},\sigma^2)$$

Because the full model has five unknown parameters while the reduced model has two (μ, σ^2), the LR test statistic Λ is approximately distributed as χ^2_3. Hence Λ values exceeding 7.82 and 11.4 indicate that a mixture of three normals provides a better fit at the 5% and 1% levels of significance, respectively, than a single normal.

Likelihood-ratio tests require that alternate hypotheses are **nested**, with one model being a subset of the other (i.e., by fixing some parameters of one model we recover the second). If they are not, the large-sample distribution of Λ does not necessarily approach a χ^2. In such cases, various descriptive metrics from **model selection** can be used to compare models (see Burham and Anderson 2002 for a complete treatment), and we focus on two of these, **AIC** and **BIC**. We stress that while both metrics are fully grounded in theoretical principles (Burham and Anderson 2004), comparing values for two different models is largely *ad hoc* in that there is no formal test for significance (i.e., no formal criterion for determining when one model is clearly better than the other).

The idea behind both metrics (as in a likelihood-ratio test) is to *reward goodness of fit* (i.e., smaller values of $-2\ln[L]$ imply a better fit, where L is the model likelihood), but also to *penalize for the number of model parameters*, k. The most widely used model-comparison metric is the **Akaike information criterion** (1974),

$$\text{AIC} = -2\ln(L) + 2k \tag{16.9a}$$

which was adjusted for the sample size, n, by Sugiura (1978),

$$\text{AIC}_c = -2\ln(L) + 2k + \frac{2k(k+1)}{n-k-1} = -2\ln(L) + \frac{2kn}{n-k-1} \tag{16.9b}$$

Burham and Anderson (2004) recommended that AIC_c be used in place of AIC unless $n/k > 40$.

The other widely used metric is the **Bayes information criterion**,

$$\text{BIC} = -2\ln(L) + \ln(n)k \tag{16.9c}$$

which was introduced by Schwarz (1978), and thus is also known as the **Schwarz criterion**. These metrics differ in the penalization of model parameters, with this penalty constant for AIC, but increases with n for BIC. While AIC and BIC are often used somewhat interchangeably, they are actually designed for slightly different purposes. When one of the models being compared is the true one, then BIC picks this model with a probability approaching one in large samples. Conversely, AIC considers the situation where *none* of the candidate models may be correct and then tries to pick among the best of these. Hence, while AIC is typically more widely reported, BIC may be more appropriate for most of the models examined in this chapter. Our recommendation is simply to report both (as they do not always choose the same model). An alternative approach involves the use of resampling methods (Schork and Schork 1989; Churchill and Doerge 1994). Two such methods, permutation tests and bootstrapping, are discussed in Chapter 18.

How does one proceed if the underlying distributions of a mixture model are not normal? MacLean et al. (1976), Elston (1984), and Schork and Schork (1988) suggested that the Box–Cox (1964) power transform (Equation 14.4) can be used to normalize each of the underlying distributions. Recall from Chapter 14 that this transformation is described by a single parameter λ, with $x = (z^\lambda - 1)/\lambda$ for $\lambda \neq 0$ and $x = \ln z$ for $\lambda = 0$. Hence, in the likelihood function, the ith underlying distribution uses $x_i = (z^{\lambda_i} - 1)/\lambda_i$ in place of z, allowing each underlying distribution to have a different transform. If we assume that an observed distribution is generated by k underlying distributions that can be transformed to normality by the appropriate Box–Cox transform, the resulting mixture model can be expressed as a mixture of k normals. The ith normalized underlying distribution has mean μ_i, variance σ_i^2, and transformation parameter λ_i, all of which can be estimated by standard ML methods. This approach offers a test for whether an observed skewed distribution results from: (i) a single naturally skewed distribution, (ii) a mixture of underlying normals, or, (iii) a mixture of underlying distributions that each can be transformed to normality. If the distribution results from a mixture of normals, a standard mixture model should give a significantly better fit than a single transformed distribution (where the common transformation parameter λ is estimated from the data). Whether a mixture model provides a significantly better fit can be evaluated by a likelihood-ratio test. Likewise, the hypothesis that the underlying mixture distributions are themselves skewed can be tested by fitting the λ_i from the data and comparing this to a model with $\lambda_i = 1$ (no transformation).

COMPLEX SEGREGATION ANALYSIS

Starting with Elston and Stewart (1971) and Morton and MacLean (1974), human population geneticists have developed the method of **complex segregation analysis**, or **CSA**, to test between alternative modes of inheritance. Formally, CSA uses a likelihood framework to examine whether the *trait transmission over a pedigree* is consistent with a specific genetic model. This reliance upon pedigree data is one of the major limitations with CSA, as it cannot easily be expanded to a random sample from a large population. As we will see in the next few sections, CSA extends the simple mixture model (Equation 16.6a) by using pedigree information to modify the mixture proportions. Assumptions about the mode of inheritance specify the mixture distribution weights, allowing likelihood-ratio tests for different models of transmission (a single major gene, no major gene but background polygenes, major gene plus background polygenes, and so forth). We illustrate some of these applications below, showing how known relationships among relatives define the transmission probabilities of both major genes and polygenes. The extensive literature on complex segregation analysis is reviewed by Elston and Rao (1978), Boyle and Elston (1979), Elston (1980, 1981a, 1990a), Morton et al. (1983), Jarvik (1998), and Sun (2017).

The use of CSA is certainly on the wane, being replaced by marker-based approaches (Chapters 17–21) that allow for both analysis using large random population samples (as opposed to smaller sample sizes based on pedigrees) and for the localization of any potential major genes, rather then simply suggesting that such genes might be present. However, this does not mean that CSA is not a useful approach, often as a first step before a more detailed analysis is suggested. For example, evidence via CSA of a major locus can be used to justify the further expenditure of resources on a project. Some recent human CSAs include the analysis of esophageal disorders (Sun et al. 2010), musical perfect pitch (Theusch and Gitschier 2011), keratoconus eye disease (Kriszt et al. 2014), autoimmunity disorders (Castiblanco et al. 2015), and facial melasma (Holmo et al. 2018).

While the bulk of the CSA literature deals with human populations, Le Roy et al. (1990) and Knott et al. (1991a, 1991b) reviewed applications in animal breeding. It has also been applied to diseases in domesticated animals, such as bacterial resistance in rainbow trout (Vallejo et al. 2010), obesity in cats (Häring et al. 2011), and diabetes in American Eskimo dogs (Cai et al. 2019). Tourjee et al. (1995) give an application of CSA to plants, examining the genetic basis of flower color in *Gerbera jamesonii*. Many plant (and some animal) applications involve the analysis of line-cross, as opposed to pedigree, data. Likelihood methods related to CSA have been developed for inferring major loci using crosses between inbred lines (Tan and Chang 1972; Elston and Stewart 1973; Tan and D'Angelo 1979; Elston 1984; Janss and Van Der Werf 1992; Loisel et al. 1994; Changjian et al. 1994). Often such analyses use multiple types of crosses: F_2, B_1, B_2, inbred F_2 offspring ($F_{(2:3)}$), etc. Likelihood-based methods for using such data for detecting major genes (often referred to as **joint segregation analysis**) are a combination of Equation 16.4, joint scaling tests (Chapter 11), and mixture models, all placed in a likelihood framework (see Example 16.6).

While complex segregation analysis is the most assumption-burdened test for detecting major genes, it is also the most powerful of the marker-free methods when the assumptions hold (MacCluer et al. 1983; MacCluer and Kammerer 1984). For example, the commingling tests discussed above (e.g., Example 16.3) that simply fit a mixture of normals and use no information from relatives can easily miss major genes that segregation analysis can detect (Kwon et al. 1990). CSA also provides better control over false positives.

Complex segregation analysis assumes normality of the underlying distributions, which greatly simplifies the form of the likelihood function. If this assumption is violated, false detection of a major locus can occur (MacLean et al. 1975; Go et al. 1978; Morton 1984). As mentioned above, if the underlying distributions are suspected to be non-normal, one strategy is to use a likelihood approach that incorporates a transformation parameter (estimated from the data) for each underlying distribution. Alternatively, instead of individually transforming each underlying distribution (through the likelihood function), one could simply apply a single transform to the observed distribution. However, this approach can raise serious issues of interpretation (e.g., Asamoah et al. 1987). While the use of transformations is not a resolved issue, Demenais et al. (1986) suggested that tests that incorporate direct estimates of transmission probabilities (see Example 16.5 below) remove the need to transform the data before performing segregation analysis.

The following sections illustrate some general approaches for constructing likelihood functions for full-sib families under increasingly more general genetic models. We start by assuming only a single diallelic major locus, and then consider separately common-family effects and background segregation of polygenic loci. Likelihoods for more complicated pedigrees or other experimental designs (such as inbred-line crosses) follow using similar arguments (Example 16.6). Before proceeding, we first provide an example to illustrate the types of hypotheses that can be addressed and the types of parameters that need to be incorporated into the likelihood functions underlying complex segregation analysis.

Example 16.4. Morton and MacLean (1974) considered a model with both a segregating diallelic major gene (alleles Q and q) and a completely additive polygenic background. Conditioned on the genotype at the major locus, the distributions of phenotypic values are assumed to be normally distributed with means μ_{QQ}, μ_{Qq}, or μ_{qq}, and common variance $\sigma^2 = \sigma_E^2 + \sigma_A^2$ (the environmental variance plus the additive genetic variance contributed by the background polygenes). Assuming the major-locus genotypes are in Hardy–Weinberg proportions, this model is described by six parameters: p = frequency of Q, the means of each major-locus genotype (μ_{QQ}, μ_{Qq}, μ_{qq}), the environmental variance σ_E^2, and the genetic variance from the polygenic contribution σ_A^2. The resulting likelihood function (see Equations 16.11b and 16.22 below) is complex as it incorporates transmission of both the major alleles and polygenic background from parent to offspring, conditioning over all possible parental genotypes. The amount of support for various genetic hypotheses can be tested using likelihood ratios of appropriate subsets of the full model, as given in the following table:

	Model	Free Parameters	Restricted Parameters
1.	No genetic effects	$\mu,\ \sigma_E^2$	$\mu_{QQ} = \mu_{Qq} = \mu_{qq} = \mu$ $p = 0, \sigma_A^2 = 0$
2.	Major gene, no background polygenes	$\mu_{QQ},\ \mu_{Qq},\ \mu_{qq},$ $p,\ \sigma_E^2$	$\sigma_A^2 = 0$
3.	Background polygenes, no major gene	$\mu,\ \sigma_E^2,\ \sigma_A^2$	$\mu_{QQ} = \mu_{Qa} = \mu_{qq} = \mu$ $p = 0$
4.	Full model: Major gene, background polygenes	$\mu_{QQ},\ \mu_{Qq},\ \mu_{qq},$ $p,\ \sigma_E^2,\ \sigma_A^2$	None

For example, a test of support for a major gene is given by the likelihood ratio using model 1 (a single normal distribution) as the restricted model and model 2 as the full model. The resulting test statistic has $5-2=3$ degrees of freedom, with twice the log of the maximum of the restricted likelihood function given by Equation 16.8. If the major-gene model provides a significant improvement, model 4 (major gene plus polygenic background) can next be tested against the major-gene-only model (2), with the test statistic having $6 - 5 = 1$ degree of freedom. More complicated models are analyzed in a similar fashion.

Likelihood Functions Assuming a Single Major Gene

We start by computing the likelihood for a single individual, then proceed to an entire family, and finally to the collection of all families in our sample. Assume that a single diallelic locus underlies the character and consider the jth offspring from the ith family, o_{ij}, which has father f_i and mother m_i (for notational ease, in the following we use f, m, and o_j, reminding the reader that these, of course, change as we change families). Denote the phenotypic value of this offspring by z_{ij}. Index the major-locus genotypes by g where $g = 1$ for QQ, $g = 2$ for Qq, and $g = 3$ for qq, with g_f, g_m, and g_{o_j} denoting the genotypes of the parents (father and mother) and their jth offspring. Phenotypic values for each major-locus genotype are assumed to be normally distributed with means μ_g and common variance σ^2. Finally, let $\Pr(g_o \,|\, g_f,\, g_m)$ be the probability that an offspring has genotype g_o given that its parents have genotypes g_f and g_m.

Conditioned on the parental genotypes, the likelihood for the ijth offspring is

$$\ell(z_{ij} \,|\, g_f,\, g_m) = \sum_{g_o=1}^{3} \Pr(g_o \,|\, g_f,\, g_m) \cdot \varphi(z_{ij}, \mu_{g_o}, \sigma^2) \tag{16.10a}$$

This conditional likelihood is a mixture model with mixing proportions given by Mendelian segregation. For example, if the father and mother have major-locus genotypes QQ and Qq, then $g_f = 1$ and $g_m = 2$, and

$$\begin{aligned}
\Pr(g_o = 3 \,|\, g_f = 1,\, g_m = 2) &= \Pr(qq \,|\, g_f = QQ,\, g_m = Qq) = 0 \\
\Pr(g_o = 2 \,|\, g_f = 1,\, g_m = 2) &= \Pr(Qq \,|\, g_f = QQ,\, g_m = Qq) = 1/2 \\
\Pr(g_o = 1 \,|\, g_f = 1,\, g_m = 2) &= \Pr(QQ \,|\, g_f = QQ,\, g_m = Qq) = 1/2
\end{aligned} \tag{16.10b}$$

so that with these parents Equation 16.10a reduces to

$$\ell(z_{ij} \,|\, QQ, Qq) = \frac{1}{2} \cdot \varphi(z_{ij}, \mu_{QQ}, \sigma^2) + \frac{1}{2} \cdot \varphi(z_{ij}, \mu_{Qq}, \sigma^2) \tag{16.10c}$$

Conditioned on parental genotype values, each offspring in a family is independent, implying that the likelihood for a full-sib family of n_i offspring is the product of individual likelihoods, giving the conditional likelihood for the ith family as

$$\ell(z_{i\cdot} \mid g_f, g_m) = \prod_{j=1}^{n_i} \ell(z_{ij} \mid g_f, g_m) \tag{16.11a}$$

Because we do not know the QTL genotypes of the parents, the unconditional likelihood for the ith family is obtained by summing over all nine possible pairs of parental genotypes,

$$\ell(z_{i\cdot}) = \sum_{g_f=1}^{3} \sum_{g_m=1}^{3} \ell(z_{i\cdot} \mid g_f, g_m) \Pr(g_f, g_m) \tag{16.11b}$$

Assuming the parents are chosen independently, $\Pr(g_f, g_m) = \Pr(g_f) \cdot \Pr(g_m)$. Further, if genotypes are in Hardy–Weinberg proportions, parental genotype frequencies are completely specified by the frequency p of allele Q, e.g.,

$$\Pr(g_f = 1, g_m = 1) = \Pr(g_f = QQ) \cdot \Pr(g_m = QQ) = p^2 \cdot p^2$$
$$\Pr(g_f = 2, g_m = 1) = \Pr(g_f = Qq) \cdot \Pr(g_m = QQ) = 2p(1-p) \cdot p^2, \text{ etc.}$$

If there are $n_g > 3$ major-locus genotypes (either because of multiple alleles at the major locus or because of several major loci), the appropriate likelihood has sums ranging over the n_g genotypes, and the transmission probabilities are modified to account for the assumed model. Likewise, if the parental phenotypic values (z_f, z_m) are known, these can also be incorporated into the likelihood. Because $\ell(z \mid g) = \varphi(z, \mu_g, \sigma^2)$, the probability that the genotype is g_i given the phenotype is z is

$$\Pr(g_i \mid z) = \frac{\Pr(g_i)\ \varphi(z, \mu_{g_i}, \sigma^2)}{\sum_{j=1}^{n_g} \Pr(g_j)\ \varphi(z, \mu_{g_j}, \sigma^2)} = \frac{\Pr(g_i)\ \varphi(z, \mu_{g_i}, \sigma^2)}{p(z)} \tag{16.12}$$

where $p(z)$ is the phenotypic density function for the entire population (a mixture distribution). Parental phenotypes are then incorporated by replacing $\Pr(g)$ by $\Pr(g \mid z)$. Equation 16.12 follows directly from Bayes theorem (Equations 3.3b and 16.24).

Assuming different families are unrelated, the total likelihood is the product of the individual likelihoods from the n_f families,

$$\ell(\mathbf{z}) = \prod_{i=1}^{n_f} \ell(z_{i\cdot}) \tag{16.13}$$

where $\ell(z_{i\cdot})$ is given by Equation 16.11b. Although there are numerous summation and product indices in this likelihood, there are only five unknown parameters: the three genotypic means, the common variance σ^2, and the QTL allele frequency p.

While the most obvious test for a major gene compares the full model with the restricted model of a single underlying normal, Elston et al. (1975) suggested that a much more robust approach is to treat the **transmission probabilities** $\Pr(g_o \mid g_f, g_m)$ as *unknown parameters* and base hypothesis tests on these. Above, we specified the transmission probabilities based on Mendelian assumptions of inheritance (Equation 16.10b), but we can also treat them as parameters to be estimated. This is most conveniently done by considering the probability, τ_x, that genotype x transmits a Q (and hence transmits q with probability $1-\tau_x$). For a diallelic locus, there are three τ values to estimate, one for each genotype. From the definition of τ, the transmission probabilities can be expressed as

$$\begin{aligned} \Pr(qq \mid g_f, g_m) &= (1-\tau_{g_f})(1-\tau_{g_m}) \\ \Pr(Qq \mid g_f, g_m) &= \tau_{g_f}(1-\tau_{g_m}) + \tau_{g_m}(1-\tau_{g_f}) \\ \Pr(QQ \mid g_f, g_m) &= \tau_{g_f}\,\tau_{g_m} \end{aligned} \tag{16.14}$$

For example, Equations 16.10b become

$$\begin{aligned}\Pr(qq\,|\,g_f = QQ,\ g_m = Qq) &= (1-\tau_{QQ})\,(1-\tau_{Qq})\\ \Pr(Qq\,|\,g_f = QQ,\ g_m = Qq) &= \tau_{QQ}\,(1-\tau_{Qq}) + \tau_{Qq}\,(1-\tau_{QQ})\\ \Pr(QQ\,|\,g_f = QQ,\ g_m = Qq) &= \tau_{QQ}\,\tau_{Qq}\end{aligned} \tag{16.15}$$

so that with these parents, Equation 16.10c becomes

$$\begin{aligned}\ell(z_{ij}\,|\,QQ, Qq) = \tau_{QQ}\,\tau_{Qq} \cdot \varphi(z_{ij}, \mu_{QQ}, \sigma^2)&\\ + \left[\,\tau_{QQ}\,(1-\tau_{Qq}) + \tau_{Qq}\,(1-\tau_{QQ})\,\right] \cdot \varphi(z_{ij}, \mu_{Qq}, \sigma^2)&\\ + (1-\tau_{QQ})\,(1-\tau_{Qq}) \cdot \varphi(z_{ij}, \mu_{qq}, \sigma^2)&\end{aligned}$$

Note that this likelihood reduces to Equation 16.10c using Mendelian segregation transmission probabilities ($\tau_{QQ} = 1$, $\tau_{Qq} = 1/2$, and $\tau_{qq} = 0$).

Elston et al. (1975) suggested that three criteria must be satisfied for acceptance of a major-gene hypothesis: (1) a significantly better overall fit of a mixture model compared with a single normal, (2) failure to reject the hypothesis of Mendelian segregation ($\tau_{QQ} = 1$, $\tau_{Qq} = 1/2$, $\tau_{qq} = 0$), and (3) rejection of the hypothesis of equal transmission for all genotypes ($\tau_{QQ} = \tau_{Qq} = \tau_{qq}$). Criterion (1) reduces false positives due to polygenic background loci, while criteria (2) and (3) offer some robustness against non-normality of the underlying distributions and resemblance due to common environmental effects (Elston 1981a). While incorporation of transmission-probability criteria into likelihood models decreases the possibility of a false positive (Go et al. 1978; Goldin et al. 1981; Demenais et al. 1986), it does so at a cost of decreased power. Loss of power can be significant if the major gene is recessive (Borecki et al. 1995).

The fact that not all families are expected to be segregating the major gene has important consequences for the optimal number and size of families for detecting a major gene. Burns et al. (1984) showed that, for a fixed number of individuals, highest power is generally obtained by examining a moderate number of families of moderate size, as opposed to many small families or a few large families. If a small number of large families is chosen, we run the risk that none of the families are segregating the gene (Example 16.2).

Example 16.5. As a demonstration of the utility of the Elston et al. (1975) criteria for accepting a major-gene hypothesis, consider a study by McGuffin and Huckle (1990), who tested for a genetic basis for attending medical school. The trait here is scored as a binary variable,

$$z = \begin{cases} 1 & \text{attending medical school} \\ 0 & \text{not attending medical school} \end{cases}$$

As discussed below (Equation 16.28), complex segregation analysis can be easily modified to accommodate such binary characters. Of 249 students at the Wales College of Medicine, 15.4% had mothers/fathers who also attended medical school, a 61-fold increase in "risk" (medical school attendance) relative to the general population (0.2%). Taking μ as the population mean, the expected means at an underlying major locus can be modeled by using measures of additivity (a) and dominance (k), and the allele frequency p. General single-locus (a, k, p all estimated) and recessive ($k = 0$, a and p estimated) models were fitted and compared with a null model ($a = k = p = 0$). These single-locus models, which assumed Mendelian transmission probabilities ($\tau_1 = 1, \tau_2 = 1/2, \tau_3 = 0$), were then compared against two alternate transmission models: a generalized model where the three parameters (τ_1, τ_2, τ_3) were estimated from the data, and an equal transmission model ($\tau_1 = \tau_2 = \tau_3 = \tau$). The latter simply fits a mixture model to the data without allowing for Mendelian transmission. The resulting log likelihoods for these models were as follows:

Model	Parameters (in addition to μ) Free	Fixed	Constant + $-2\ln$ (likelihood)
Null	None	$a = k = p = 0$	283.60
Equal transmission	$a, p,$ $\tau_1 = \tau_2 = \tau_3$	$k = 0$	283.60
General single-locus	a, k, p	$\tau_1 = 1, \tau_2 = 1/2,$ $\tau_3 = 0$	120.14
Recessive	a, p	$k = 0, \tau_1 = 1,$ $\tau_2 = 1/2, \tau_3 = 0$	120.14
General transmission	$a, p, \tau_1, \tau_2, \tau_3$	$k = 0$	111.22

The general single-locus model gives a significantly better fit than the null (a single underlying normal) model, with a likelihood-ratio test statistic of 283.60 − 120.14 = 163.46 (three degrees of freedom). Because the recessive model gives the same fit with fewer parameters, it is chosen as the standard for further analysis. The recessive model gives a significantly better fit than the equal transmission model (a mixture distribution not incorporating Mendelian segregation). Thus, criteria (1) and (3) for a major gene hold, because a mixture gives a better fit than a single normal, and the hypothesis of equal transmission ($\tau_1 = \tau_2 = \tau_3$) is rejected. However, the general transmission model (τ_i estimated from the data) gives a significantly better fit than the Mendelian segregation hypothesis ($\tau_1 = 1, \tau_2 = 1/2, \tau_3 = 0$), with a likelihood-ratio test statistic of 120.14 − 111.22 = 8.92 with three degrees of freedom ($P < 0.03$). Thus, these data fail the major-gene criterion (2), as the hypothesis of Mendelian segregation is rejected. Shared environmental effects, rather than major gene effects, likely account for this association between relatives.

Example 16.6. The likelihood-based method of **joint segregation analysis (JSA)** is essentially a fusion of line-cross analysis (Chapter 11) conditioned on major-locus genotypes to yield mixture models. Unlike classical CSA, the mixture proportions are given by the nature of the inbred-line cross (e.g., F_2, B_1, B_2, etc.), as was seen in Equation 16.4. Hence, the often delicate issue of estimating mixture proportions (which can have very flat likelihood surfaces, leading to considerable imprecision of their estimates) is avoided. Conversely, under a JSA one typically estimates many more variances than usually required under most CSA, which can also result in rather flat likelihoods (for the variance estimates).

We illustrate the basic logic of this approach when one has data (trait measures on individuals) from both parental inbred lines, their F_1 and F_2, and both backcrosses, B_1 and B_2. Let P_1 denote the population fixed for the major allele, so that P_1 is QQ, P_2 is qq, and F_1 is Qq, where the major locus genotypic values are $\mu - a$ for qq, $\mu + d$ for Qq, and $\mu + a$ for QQ. Both parental populations or their F_1 show no genetic variation (members from within each of these populations all share the same major-locus and polygenic genotypes). We can decompose the means from each of these populations as

$$\mu_{P_1} = \mu + a + A, \quad \mu_{P_2} = \mu - a - A, \quad \mu_{F_1} = \mu + d + D$$

where A is the composite value from additive polygenic effects, and D the composite value from polygenic dominance effects (Chapter 11). Assuming these lines all have the same environmental variance (σ_E^2),

$$z_{P_1} \sim \mathrm{N}(\mu_{P_1}, \sigma_E^2), \quad z_{P_2} \sim \mathrm{N}(\mu_{P_2}, \sigma_E^2), \quad z_{F_1} \sim \mathrm{N}(\mu_{f_1}, \sigma_E^2)$$

Similarly, conditional on the major-locus genotype, the means for the two backcrosses, $B_1 = P_1 \times F_1$ and $B_2 = P_2 \times F_1$, can be expressed as

$$\mu_{B_1|QQ} = \mu + a + (A + D)/2, \quad \mu_{B_1|Qq} = \mu + d + (A + D)/2$$

and

$$\mu_{B_2|qq} = \mu - a + (D - A)/2, \quad \mu_{B_2|Qq} = \mu + d + (D - A)/2$$

Unlike the P_i and F_1, individuals from backcross generations are a mixture of two possible major-locus genotypes. Because of segregation at the background polygenes, the variance about the major-locus genotypic value is greater than σ_E^2. Further, this background variance may be different in the two backcrosses (as different genotypes are segregating). Denoting these variances as $\sigma_{B_1}^2$ and $\sigma_{B_2}^2$, the distribution of trait values for a random individual from either backcross follows a mixture distribution, with

$$z_{B_1} \sim (1/2)\varphi(z_{B_1}, \mu_{B_1|QQ}, \sigma_{B_1}^2) + (1/2)\varphi(z_{B_1}, \mu_{B_1|Qq}, \sigma_{B_1}^2)$$

and

$$z_{B_2} \sim (1/2)\varphi(z_{B_2}, \mu_{B_2|qq}, \sigma_{B_2}^2) + (1/2)\varphi(z_{B_2}, \mu_{B_2|Qq}, \sigma_{B_2}^2)$$

Finally, we can write the major-locus conditional F_2 means as

$$\mu_{F_2|QQ} = \mu + a + D/2, \quad \mu_{F_2|Qq} = \mu + d + D/2, \quad \mu_{F_2|qq} = \mu - a + D/2$$

and the above logic, along with Equation 16.4, implies that

$$z_{F_2} \sim (1/4)\varphi(z_{F_2}, \mu_{F_2|qq}, \sigma_{F_2}^2) + (1/2)\varphi(z_{F_2}, \mu_{F_2|Qq}, \sigma_{F_2}^2) + (1/4)\varphi(z_{F_2}, \mu_{F_2|QQ}, \sigma_{F_2}^2)$$

The resulting model parameters for likelihood estimation are two major-locus effects (a and d), two composite background polygenic effects (A and D), and four variances ($\sigma_E^2, \sigma_{B_1}^2, \sigma_{B_2}^2$, and $\sigma_{F_2}^2$). Model selection using nested likelihood ratio tests (when appropriate) or the AIC or BIC criteria can be used to not only show support for a major locus (a and/or d significantly different from zero) but also assess where it shows dominance (d significantly different from zero).

JSA has its roots in Tan and Change (1972) and Tan and D'Angelo (1979), and was more formally developed by Elston and Stewart (1973) and Elston (1984), with extensions by Changjian et al. (1994), Jiang et al. (1994), Gai and Wang (1998), Wang et al. (2001), Zhang and Gai (2002), and Zhang et al. (2003). Applications of JSA include beanfly resistance in soybeans (Wang and Gai 2001), flowering time in chickpea (Anbessa et al 2006) and cotton (Hao et al. 2008), and seed number in rapeseed (Zhang et al. 2010a).

Common-family Effects

Members of full-sib families usually share environmental effects, and likelihood functions accounting for these departures have been developed (Morton and MacLean 1974; Knott and Haley 1992a, 1992b). Let the ith family have a common effect c_i, and assume that these effects are normally and independently distributed among families with mean zero and variance σ_c^2. With this modification, the expected phenotypic value of an offspring with genotype g_o from family i is $\mu_{g_o} + c_i$. As before, we assume that the phenotypic values for each genotype (conditional on c_i) are normally distributed with variance σ^2, giving the conditional likelihood for the n_i offspring from this family as

$$\ell(z_{i\cdot} \,|\, g_f, g_m, c_i) = \prod_{j=1}^{n_i} \left[\sum_{g_{o_j}=1}^{3} \Pr(g_{o_j} \,|\, g_f, g_m) \cdot \varphi(z_{ij}, \mu_{g_{o_j}} + c_i, \sigma^2) \right] \tag{16.16}$$

Averaging over all possible values of the common-family effect c_i gives

$$\ell(z_{i\cdot} \,|\, g_f, g_m) = \int_{-\infty}^{\infty} \ell(z_{i\cdot} \,|\, g_f, g_m, c) \cdot \varphi(c, 0, \sigma_c^2)\, dc \tag{16.17}$$

Finally, using the above expression for $\ell(z_{i\cdot} \,|\, g_f, g_m)$, averaging over all possible parental genotypes gives the unconditional likelihood for this family (Equation 16.11b). Assuming the major-locus genotypes are in Hardy–Weinberg proportions, the unconditional likelihood has six unknown parameters: the three genotypic means, the allele frequency p, and the variances σ^2 and σ_c^2. Assuming the n_f families in our pedigree are unrelated, the total likelihood is the product of the individual family likelihoods (Equation 16.13).

The likelihood for the ith family under the restricted model assuming common-family effects, but no major genes, is

$$\ell(z_{i\cdot}) = \int_{-\infty}^{\infty} \ell(z_{i\cdot} \mid c) \cdot \varphi(c, 0, \sigma_c^2)\, dc$$
$$= \int_{-\infty}^{\infty} \left[\prod_{j=1}^{n_i} \varphi\left(z_{ij}, \mu + c, \sigma^2\right) \right] \cdot \varphi\left(c, 0, \sigma_c^2\right)\, dc \tag{16.18}$$

A test for common-family effects but no major gene is given by the likelihood-ratio test using Equation 16.18 versus the likelihood function with $\sigma_c^2 = 0$. The latter is just the likelihood function assuming a single underlying normal, which has its maximum value given by Equation 16.8. Likewise, the likelihood-ratio test for a major gene but no common-family effects compares the full likelihood with a restricted likelihood assuming $\sigma_c^2 = 0$.

Polygenic Background and the Unified Model

Our final modification assumes a background of segregating polygenes in addition to the major gene. The resulting likelihood functions were often unfortunately referred to as **mixed models** in the older human genetics literature, although this is a very different usage of the term from its standard linear-model interpretation (Chapter 10). We prefer using the term **unified model** instead (Lalouel et al. 1983). A variety of such unified-model likelihoods have been proposed for full-sib families (Elston and Stewart 1971; Morton and MacLean 1974; Ott 1979; Lalouel and Morton 1981; Lalouel et al. 1983; Demenais and Bonney 1989; Demenais et al. 1990; Fernando et al. 1994; Stricker et al. 1995b).

We assume the background polygenes to be completely additive, and assume that the background genetic value A is normally distributed with mean 0 and variance σ_A^2. The phenotypic value of an individual with major-locus genotype g and background polygenic value A is assumed to be normally distributed with mean $\mu_g + A$ and variance σ_E^2. Ignoring common-family effects, if the jth sib has background genetic value A_{o_j}, the conditional likelihood for the jth sib in the ith family becomes

$$\ell(z_{ij} \mid g_f,\, g_m,\, A_{o_j}) = \sum_{g_{o_j}=1}^{3} \Pr(g_{o_j} \mid g_f,\, g_m)\, \varphi(z_{ij}, \mu_{g_{o_j}} + A_{o_j}, \sigma_E^2) \tag{16.19}$$

The conditioning on offspring genetic value A_o is removed in two stages. First, A_o is removed by conditioning on parental polygenic values $(A_f,\, A_m)$,

$$\ell(z_{ij} \mid g_f,\, g_m,\, A_f,\, A_m) = \int_{-\infty}^{\infty} \ell(z_{ij} \mid g_f,\, g_m,\, A_{o_j})\, p(A_{o_j} \mid A_f,\, A_m)\, dA_{o_j} \tag{16.20}$$

The additive genetic value of an offspring is assumed to be normally distributed with mean $(A_f + A_m)/2$ and variance $\sigma_A^2/2$, so that the conditional density function is

$$p(A_o \mid A_f,\, A_m) = \varphi\left(A_o, \frac{A_f + A_m}{2}, \frac{\sigma_A^2}{2} \right) \tag{16.21}$$

as developed in Example 9.9. Second, averaging over all possible parental background polygenic values, A_f and A_m, gives a likelihood function for the ith family that is conditioned only on the major-locus genotypes of the parents,

$$\ell(z_{i\cdot} \mid g_f,\, g_m) =$$

$$\int_{-\infty}^{\infty}\int_{-\infty}^{\infty}\left[\prod_{j=1}^{n_i}\ell(z_{ij}\,|\,g_f,\,g_m,\,A_f,\,A_m)\right]\varphi(A_f,0,\sigma_A^2)\,\varphi(A_m,0,\sigma_A^2)\,dA_f\,dA_m \tag{16.22}$$

This expression assumes that parents are drawn at random, are unrelated, and not inbred, but these restrictions can be removed by averaging over an alternative joint distribution of A_f and A_m (Examples 9.8 and 9.9). Finally, Equation 16.22 is substituted into Equation 16.11b to obtain the unconditional likelihood for the entire family. The resulting likelihood has six unknown parameters: the three major-locus means, allele frequency p (assuming genotypes are in Hardy–Weinberg proportions), σ_E^2, and σ_A^2.

Under the restricted model of an additive polygenic background but no major-locus or common-family effects, the likelihood of an individual conditioned on the polygenic values of its parents is

$$\ell(z_{ij}\,|\,A_f,\,A_m)=\int_{-\infty}^{\infty}\varphi(z_{ij},\mu+A,\sigma_E^2)\;\varphi\left(A,\,\frac{A_f+A_m}{2},\,\frac{\sigma_A^2}{2}\right)\,dA \tag{16.23a}$$

giving the unconditional likelihood for the ith family as $\ell(z_{i\cdot})=$

$$\int_{-\infty}^{\infty}\int_{-\infty}^{\infty}\left[\prod_{j=1}^{n_i}\ell(z_{ij}\,|\,A_f,\,A_m)\right]\varphi(A_f,0,\sigma_A^2)\,\varphi(A_m,0,\sigma_A^2)\,dA_f\,dA_m \tag{16.23b}$$

Incorporation of a common-family effect into either Equation 16.22 or 16.23 is straightforward and follows the logic leading to Equation 16.16.

The presence of multiple integrals in the common-family effect and polygenic likelihood functions usually means that considerable computing power is required to obtain the MLEs in even modest pedigrees. Knott et al. (1990, 1991a) provide excellent approximations for these Gaussian integrals decreasing computational requirements. Shoukri and McLachlan (1994) provide details for implementing an EM maximization (Appendix 4). However, because multiple local maxima can occur on the likelihood surface, care must still be taken in numerically computing the global likelihood, as numerical optimization methods may become trapped at local maxima (Demenais et al. 1986; Borecki et al. 1995).

Other Extensions

Extensions allowing for multivariate traits (Blangero and Konigsberg 1991) and genotype $\times$ environment interaction (Eaves 1984; Blangero et al. 1990; Konigsberg et al. 1991; Tiret et al. 1993) have also been developed. Moreover, likelihood functions can be constructed to incorporate any number of fixed effects (e.g., effects due to age, sex, or specific environments) using the general linear mixed model (Chapter 10). Removing such fixed effects prior to analysis increases the power of tests of oligogenic models. An alternative formulation for likelihoods in pedigrees has been suggested by Bonney (Bonney 1984, 1992; Bonney et al. 1989; Demenais and Bonney 1989). These **regressive models** have as their parameters correlations between relatives (i.e., the correlations between parent and offspring, and between full sibs), rather than explicit genetic parameters to express these correlations.

Likelihoods for more general pedigrees follow using the same logic as above: conditioning on all possible genotypes and then averaging over these genotypes to obtain the unconditional likelihood. Several computer packages have been developed for complex segregation analysis in small pedigrees containing on the order of tens of individuals (Elston and Stewart 1971; Hasstedt and Cartwright 1979; Lalouel and Morton 1981; Elston et al. 1986; Lange et al. 1988; Snow and Wijsman 1998), which are compared by MacCluer et al. (1983), Konigsberg et al. (1989), and Snow and Wijsman (1998). However, while small pedigrees can be handled, the computational requirements for complex multigenerational pedigrees (which are common in human genetics) are extremely demanding, making their analysis by classical methods difficult except in special situations. Starting with Elston and Stewart (1971), a number of approaches and approximations have been suggested (Lange and Elston 1975; Cannings et al. 1976, 1978; Ott 1979; Lange and Boehnke 1983; Schork 1991, 1992a; Goradia et al. 1992; Fernando et al. 1993; Hasstedt 1993; Stricker et al. 1995a). One important development is the use of intensive resampling methods, such as the Gibbs sampler (Geman and Geman 1984; Gelfand and Smith 1990) or other MCMC methods (Appendix 8). Here, one randomly samples a large number of the possible genotypes within a pedigree and uses the conditional likelihoods averaged over this set as an estimate of the unconditional likelihood (Thompson and Guo 1991; Guo and Thompson 1992, 1994; Thompson et al. 1993; Janss et al. 1995, 1997; Heath 1997; Heath et al. 1997; Wijsman 2000).

Ascertainment Bias

The preceding likelihood functions assume that families (or more generally, pedigrees) are *chosen at random* from the population as a whole. In many cases *nonrandom* sampling is much more efficient, as even a large random sample can miss pedigrees segregating a rare major gene (Example 16.2). When pedigrees are not chosen at random, the likelihood function must be modified to account for how the observed pedigrees were sampled (**ascertained**). Pedigrees are ascertained through **probands**, individuals who cause a particular pedigree to enter the sample. As first noted by Weinberg (1927), failure to account for how probands are ascertained can bias the analysis. For example, if pedigrees are required to contain at least two affected individuals in order to be included in the analysis, not accounting for this ascertainment scheme can lead to false positives for major loci (Jarvik 1998).

To see the importance of correctly accounting for the sampling scheme, let $\ell(z_{i\cdot} \mid g_f, g_m)$ denote the likelihood function for the ith family, conditioned on the major-locus genotypes of the parents (Equation 16.11a). To obtain the unconditional likelihood, we compute the expectation of this likelihood over all possible parental genotypes giving Equation 16.11b,

$$\ell(z_{i\cdot}) = \sum_{g_f, g_m} \ell(z_{i\cdot} \mid g_f, g_m) \, \Pr(g_f,\, g_m)$$

In our previous treatment, we assumed that parents are chosen at random, so that (for a random-mating population) $\Pr(g_f,\, g_m)$ is entirely determined by p, the frequency of allele Q. However, if parents are *not* chosen at random, $\Pr(g_f,\, g_m)$ is no longer just a function of p, but also of how the parents were sampled.

Unfortunately, using an incorrect model of ascertainment can create as much bias as performing an analysis without considering ascertainment issues. Ascertainment correction is a complicated subject, the details of which we will not pursue further. Some basic concepts are reviewed by Morton (1959), Cavalli-Sforza and Bodmer (1971, see their Appendix II), and Elston (1980, 1981a). Other important papers include Cannings and Thompson (1977), Elston and Sobel (1979), Ewens and Shute (1986), Shute and Ewens (1988a, 1988b), Hodge (1988), and Vieland and Hodge (1995). A nice discussion of some of the subtleties inherent in defining ascertainment schemes is given by Greenberg (1986).

Estimating Individual Genotypic Effects

When a major locus is indicated, the investigator may wish to estimate the genotypes of particular individuals. For example, when the trait is determined by both a major gene and a polygenic background, extreme individuals in some families result from having extreme major-locus genotypes, while in other families they result from having extreme polygenic values. This is a particular problem with studies of human disease where pedigrees are gathered from very wide sampling on the basis of extreme phenotypes. Obviously, one would like to sort out these different causes before proceeding to detailed molecular analyses. For example, in a study of 70 pedigrees displaying high levels of blood cholesterol, Moll et al. (1984) found strong evidence for segregation of a major gene in only three cases, with only polygenic and/or environmental factors influencing cholesterol in the remaining pedigrees. Their strategy was to first fit a unified (major gene plus background polygene) model to the data, and then use the estimated model parameters to predict the major-locus genotype of each parent, given observed parental and offspring phenotypic values.

The key to predicting the major-locus genotype is Bayes theorem (Equation 3.3b), which can be used to estimate the probability of each genotype given the phenotypic values and estimates of the major gene parameters obtained from segregation analysis (such as allele frequencies and genotypic means). The version of Bayes theorem required for this problem is as follows: suppose there are n possible outcomes $(b_1, b_2, \cdots, b_n)$ of a random variable that we cannot observe. Given the observed outcome of a correlated variable X, what is the probability of b_j? From the definition of a conditional probability, $\Pr(b_j \mid X) = \Pr(b_j, X)/\Pr(X)$. We can decompose this further, by noting that $\Pr(b_j, X) = \Pr(b_j)\Pr(X \mid b_j)$ and $\Pr(X) = \sum_i^n \Pr(b_i)\Pr(X \mid b_i)$. Putting these together gives Bayes theorem,

$$\Pr(b_j \mid X) = \frac{\Pr(b_j)\Pr(X \mid b_j)}{\Pr(X)} = \frac{\Pr(b_j)\Pr(X \mid b_j)}{\sum_{i=1}^{n} \Pr(b_i)\Pr(X \mid b_i)} \tag{16.24a}$$

In particular, the probability that an individual with phenotypic value z has genotype j (for $1 \le j \le n$) is

$$\Pr(g_j \mid z) = \frac{\Pr(g_j)\Pr(z \mid g_j)}{\Pr(z)} = \frac{\Pr(g_j)\Pr(z \mid g_j)}{\sum_{i=1}^{n} \Pr(g_i)\Pr(z \mid g_i)} \tag{16.24b}$$

Note that $\Pr(z)$ is simply the distribution of phenotypic values, and the probabilities involving g_i can be computed using the ML estimates of the major gene parameters. More generally, the phenotypes of an individual's offspring and/or additional relatives can be used by replacing the single observation z with a vector $\mathbf{z}$ of phenotypes. Computing $\Pr(\mathbf{z} \mid g_i)$, the probability of the observed vector $\mathbf{z}$ of phenotypes, given that the individual of interest has genotype g_i, is straightforward but can be very tedious (Elston and Stewart 1971; Heuch and Li 1972; van Arendonk et al. 1989; Kinghorn et al. 1993).

Example 16.7. Suppose that a CSA analysis shows strong evidence for a major gene, with estimated means of $\mu_{QQ} = 60$, $\mu_{Qa} = 50$, and $\mu_{qq} = 30$; an estimated frequency of 0.3 for Q; and an estimated variance of $\sigma^2 = 100$. What is the probability for each of the major-locus genotypes for an individual with a trait value of 40? We can solve this using Bayes theorem (Equation 16.24b). Let $\varphi(x, \mu, \sigma^2)$ denote the value of the density function for a normal (with mean μ and variance σ^2) evaluated at x. Assuming Hardy–Weinberg, the values needed for Equation 16.24b for the three genotypes become

	QQ	Qq	qq
$\Pr(g_i)$	0.3^3 0.09	$2 \cdot 0.3 \cdot 0.7$ 0.42	0.7^2 0.49
$\Pr(z \vert g_i)$	$\varphi(40, 60, 100)$ 0.0054	$\varphi(40, 50, 100)$ 0.0242	$\varphi(40, 30, 100)$ 0.0242

Using these values (corresponding to z = 40), we have,

$$p(z) = \sum_{i=1}^{3} \Pr(g_i)\Pr(z \mid g_i) = 0.09 \cdot 0.0054 + 0.42 \cdot 0.0242 + 0.49 \cdot 0.0242$$
$$= 0.0225$$

Applying Equation 16.24b yields

$$\Pr(QQ \mid 40) = \frac{\Pr(QQ)\Pr(40 \mid QQ)}{\Pr(z)} = \frac{0.09 \cdot 0.0054}{0.0225} = 0.021$$

$$\Pr(Qq \mid 40) = \frac{\Pr(Qq)\Pr(40 \mid Qq)}{\Pr(z)} = \frac{0.42 \cdot 0.0242}{0.0225} = 0.452$$

$$\Pr(qq \mid 40) = \frac{\Pr(qq)\Pr(40 \mid qq)}{\Pr(z)} = \frac{0.49 \cdot 0.0242}{0.0225} = 0.527$$

Example 16.8. Consider the value of $\Pr(QQ \mid 40)$ obtained in the previous example. The careful reader might have noticed that the calculation of this conditional probability used the *estimated* (as opposed to the *true*) major-locus parameter values, and hence the value for this conditional probability obtained above is, itself, an estimate. The value of 0.021 obtained in Example 16.6 effectively assumes that the CSA parameters are *known without error*, whereas in fact, parameter estimates have errors and are also correlated. One could obtain large-sample variances and covariances using ML theory (Appendix 4), but how should these be incorporated into some measure of the uncertainty of our estimate of $\Pr(QQ \mid 40)$?

Fortunately, **Bayesian statistics** (Appendix 7) provides a straightforward solution. Under the Bayesian statistical framework, one assumes some prior information, incorporates the likelihood of the information given the data, and returns an updated posterior distribution. Instead of obtaining **point estimates** (e.g., MLE values) of some parameter (or parameters), a Bayesian analysis returns a **posterior *distribution*** of the possible values given the prior and the data. As developed in Appendix 8, a variety of straightforward, but computationally intense, resampling approaches, loosely called MCMC methods, allow for computers to generate draw from the very complex form of most posteriors. In the case of Example 16.6, a single draw would return a vector of the five CSA parameters to account for any correlation structure among these estimates. Using this draw (vector of CSA parameters), one then computes the function of interest using these values (e.g., $\Pr QQ[40]$), recording its value, and repeating this procedure (at a minimum) several thousand times to generate an empirical distribution for the function of interest. The variance of this resulting posterior gives one a measure of uncertainty in the estimate of the function. A number of approaches can be used for this purpose (reviewed in Appendix 7).

ANALYSIS OF DISCRETE CHARACTERS

The likelihood functions developed above can be easily modified to accommodate complex segregation analysis of dichotomous (binary) characters, such as the presence/absence of a disease (e.g., Elston and Rao 1978). Define the **penetrance** θ_g of a genotype g as the probability that a random individual with that genotype displays the trait. Coding the character as

$$y = \begin{cases} 0 & \text{does not display the trait} \\ 1 & \text{displays the trait} \end{cases} \tag{16.25a}$$

gives the likelihood function for an individual with genotype g as

$$\ell(\, y \mid g\,) = (\theta_g)^y\,(1-\theta_g)^{1-y} = \begin{cases} 1-\theta_g & \text{for } y=0 \\ \theta_g & \text{for } y=1 \end{cases} \tag{16.25b}$$

This is the likelihood for a single draw from a binomial (Equation 2.19a with $n = 1, p = \theta_g$), usually referred to as a Bernoulli distribution (Chapter 14).

More generally, if the character has m discrete states, and $\theta_{k,g}$ is the probability that an individual of genotype g has character state k, the likelihood function becomes

$$\ell(\, y \,|\, g\,) = \prod_{k=1}^{m} (\theta_{k,g})^{\delta(y,k)} \qquad \text{where} \qquad \delta(y,k) = \begin{cases} 1 & \text{if } y = k \\ 0 & \text{otherwise} \end{cases} \tag{16.26}$$

This is the generalization of a Bernoulli, with a single draw from a mulitnomial distribution (Equation 2.20a with $n = 1$). Thus, our treatment below of dichotomous characters easily extends to polychotomous traits.

Single-locus Penetrance Model

Assume that a single diallelic locus underlies a dichotomous trait, and denote the penetrances of genotypes QQ, Qq, and qq by θ_1, θ_2, and θ_3, respectively. If allele Q has frequency p, under Hardy–Weinberg expectations, the **population prevalence**, K, of the trait becomes

$$K = p^2 \cdot \theta_1 + 2p(1-p) \cdot \theta_2 + (1-p)^2 \cdot \theta_3 \tag{16.27}$$

As above, likelihood functions are constructed by standard conditioning arguments, using Equation 16.25b. For example, consider a collection of n full sibs from the ith family, with trait values $y_{i1}, \cdots, y_{in}$ (taking values of zero or one). The likelihood for the jth sib from this family, conditioned on it having parental genotypes g_f and g_m, is

$$\ell(\, y_{ij} \,|\, g_f, g_m\,) = \sum_{g_o=1}^{3} \Pr(g_o \,|\, g_f, g_m)\; (\theta_{g_o})^{y_{ij}} \, (1-\theta_{g_o})^{1-y_{ij}} \tag{16.28}$$

Using Equations 16.11a and 16.11b to average over all possible parental genotypes yields the unconditional likelihood for this family. The resulting likelihood has four unknown parameters: allele frequency p, and the gene effects as measured by the three penetrances θ_i. Curtis and Stam (1995) note that this likelihood parameter space can be reduced when an estimate of the population prevalence K is available, as this imposes the restriction given by Equation 16.27 on the four model parameters.

Major Gene Plus a Polygenic Background

The penetrance approach can easily be extended to allow for both a major gene and a polygenic background. For a dichotomous character, the likelihood for an individual with major-locus genotype g and background polygenic value A is

$$\ell(y \,|\, g, A\,) = [\,\psi(g, A)\,]^{y} \; [\,1 - \psi(g, A)\,]^{1-y} \tag{16.29a}$$

where $\psi(g, A)$ is the penetrance for an individual with this genotype. Substituting this likelihood into the previous unified-model likelihoods (Equations 16.19–16.23) allows them to accommodate binary traits. For example, the likelihood $\ell(\, y \,|g_f, g_m, A_f, A_m\,)$ for a particular sib, conditioned on the major-locus genotypes and background polygenic values of its parents is

$$\sum_{i=1}^{3} \int_{-\infty}^{\infty} \ell(\, y \,|\, g_i, A\,) \cdot \Pr(\, g_i \,|\, g_f, g_m\,) \cdot \Pr(\, A \,|\, A_f, A_m\,)\, dA \tag{16.29b}$$

where $\Pr(\, A \,|\, A_f, A_m\,)$, given by Equation 16.21, is a function of the additive genetic variance σ_A^2 of the polygenic values. The conditioning on parental polygenic values is removed by integration, paralleling our development of the unified-model likelihood function (Equations 16.19–16.23). Likewise, common-family environmental effects can be incorporated using an analysis using the same logic leading to Equations 16.16–16.18.

The penetrances $\psi(g, A)$ are usually modeled by assuming an underlying **liability model**. This approach, briefly introduced in Chapter 14 and more fully discussed in Part

II Chapter 31, assumes an underlying normal distribution of liability, z. One trait value is displayed if the liability exceeds some threshold T, while the alternative trait value is displayed if liability lies below the threshold. In this case,

$$\psi(g, A) = \Pr(z > T \mid g, A) = \int_T^\infty \varphi(z, \mu_g + A, 1)\, dz \tag{16.30}$$

As also discussed in Chapter 31, the variance of the liability distribution can always be set equal to one. Defining $\Phi(x) = \Pr(U \le x)$ as the cumulative distribution function for a unit normal U, the integral in Equation 16.30 can be written as

$$\int_{T-(\mu_g+A)}^\infty \varphi(z, 0, 1)\, dz = 1 - \int_{-\infty}^{T-(\mu_g+A)} \varphi(z, 0, 1)\, dz = 1 - \Phi(T - \mu_g - A)$$

Using the useful identity $\Phi(-x) = 1 - \Phi(x)$, Equation 16.30 thus becomes

$$\psi(g, A) = \Phi(\mu_g + A - T) \tag{16.31}$$

Segregation analysis proceeds by substituting this expression into Equation 16.29a. The resulting likelihood for the major-gene-only model has five parameters to estimate: the three major-locus means μ_g, allele frequency q, and the threshold value T. (Alternatively, one can set $T = 1$ and estimate the variance of the liability function.) Examples of this general approach applied to several different genetic models can be found in Thaller et al. (1996).

Finally, we note that an alternative approach for modeling penetrances that avoids having to evaluate the cumulative normal function involves the use of logistic regressions (Chapters 14 and 20). The motivation for this approach is that the logistic function

$$f(x) = \frac{1}{1 + e^{-x}} \tag{16.32}$$

provides a reasonable approximation to the cumulative normal, with

$$\Phi(x) \simeq \frac{1}{1 + \exp(-\psi\, x)} = f(c\,x) \qquad \text{where} \qquad c = \frac{\pi}{\sqrt{3}} \tag{16.33}$$

(Liao 1994). Hence, we can model the penetrance as a logistic function. For example, the penetrance for the major gene and polygenic background becomes

$$\psi(g, A) = f\left(\sum_i \beta_i + \alpha_g + \alpha_A \right) = \frac{1}{1 - \exp\left(-\left[\sum_i \beta_i + \alpha_g + \alpha_A\right]\right)} \tag{16.34}$$

where α_g and α_A are the effects of the major-locus genotype and polygenic background, and the β_i are other factors (such as the effect of sex, specific environments, or specific age groups).

17

Principles of Marker-based Mapping

If we knew what we were doing, it wouldn't be called Research. Albert Einstein.

In a classical Mendelian analysis, scorable genetic variants are used to map and characterize genes. This analysis requires that individual phenotypes are *highly informative* about their underlying genotypes. Quantitative genetics (QG) deals with the opposite setting, wherein the phenotype provides *very little information* on the underlying genotype. Because of this distinction, analysis in the classical QG setting is based on genetic variances rather than on the underlying genes themselves (Chapters 4–7). While this variance-based approach is sufficient in many situations, we ultimately would like to move away from this strictly statistical framework towards direct examination of the effects of the individual **quantitative-trait loci** (**QTLs**) that underlie trait variation.

As discussed below, there are occasions when one can choose suitable **candidate loci** based on biological knowledge of the trait, but in most situations this is not possible. Fortunately, QTLs can be assayed *indirectly* by using linked **marker loci**. This indirect approach had long been recognized (e.g., Payne 1918; Sax 1923), but until the 1980s was regarded as of minor importance because of the lack of a sufficient number of genetic markers to make it a viable method. In the genomics era, this situation changed dramatically, to the point where now a whole genome sequence for any organism can be obtained with not much more effort than was required in the past to score a few dozen molecular markers. As mentioned in Chapter 1, a direct consequence of this new ability to have an essentially endless supply of markers for any organism of interest was a dramatic paradigm shift in the field. Historically, quantitative genetics was largely based on statistical inference of phenotypes of interest by comparing individuals with known degrees of relationship (i.e., the coefficient of coancestry Θ; Equation 7.2a). Modern quantitative genetics uses molecular marker information to augment, or even replace, information on known relationships (Chapter 8).

There are two different, but not exclusive, broad strategies that use marker information. First, as discussed in Chapter 8, markers allow for a quantitative-genetic analysis when relationships among a set of individuals are either ambiguous or even unknown (as is often the case in most natural population settings). Here, marker information is used to estimate the degree of relationships, but otherwise the machinery of quantitative genetics remains the same. Second, and the focus of the next several chapters, is to (largely) ignore relationships and instead search for marker-trait correlations, either using linkage following a cross or within a pedigree, or using linkage disequilibrium (LD) within a population (Chapter 5).

This chapter introduces the basics of marker-based mapping, while the next several chapters consider specific refinements for linkage-based analysis using inbred lines (Chapter 18) and outbred pedigrees (Chapter 19), and linkage-disequilibrium analysis in a population (Chapter 20). Our discussion in this chapter is loosely organized by historical developments of marker-based methods. We start by considering classical approaches for quantifying the effects of entire chromosomes (or chromosomal segments). We then turn to finer genetic resolution, examining a variety of topics concerning markers and genetic maps, including the use of markers to facilitate construction of nearly isogenic lines, construction of specialized mapping populations for linkage-based mapping, and the use of population samples for LD-based mapping, concluding by discussing tests for candidate loci.

CLASSICAL APPROACHES

In some species, the clever use of special biological features (e.g., balancer chromosomes and the lack of male recombination in *Drosophila*) allows one to assay the effects of whole chromosomes on characters of interest. Because such **chromosomal assays** can be done with just a few morphological markers, they were the main method of marker-based trait analysis before the advent of molecular markers. Although restricted to a few species, such assays nonetheless have provided insight into the genetic architecture of a number of traits.

Chromosomal Assays

The idea here is straightforward: using genetic markers, a chromosome (or chromosomal segment) from one line is substituted into an otherwise standard genetic background (usually an inbred line to minimize background variance). A collection of lines in which each chromosome is changed against an otherwise standard background—and hence the ability to screen every individual chromosome—is referred to as a set of **chromosome substitution strains**, or **CSSs** (Nadeau et al. 2000).

Drosophila is highly amenable to such studies given its short generation time, few chromosomes, a well-characterized genetic map, the existence of balancer strains (e.g., Figure 5.8), and the lack of recombination in males. Application of these features allows intact chromosomes from one line to be substituted into another line. Chromosome substitution can also be done in wheat using certain genetic tricks that allow construction of individuals carrying an extra chromosome (e.g., Sears 1953; Law 1966; Snape et al. 1977; Law and Gale 1979; Choo 1983). The effects of large chromosomal segments have also been examined in maize through the use of reciprocal translocations (e.g., D. Robertson 1989) and in mice by using marked chromosomes (e.g., Kluge and Geldermann 1982). Finally, as reviewed by Nadeau et al. (2000) and Shao et al (2008), collections of CSSs have been constructed for probing mouse- and rat-based models of human disease.

In such investigations, given that each chromosome or chromosomal segment typically contains a significant fraction of the total genome, the considerations discussed in Chapter 11 apply, and it is best to speak of ***genetic factors***, rather than *individual genes*. In addition to providing a very coarse estimate of the location of such factors, chromosomal assays provide estimates of the *minimum number* of underlying loci, the degree of dominance measured by the composite (aggregate) effects of loci on the chromosome or chromosomal segment, and the degree of epistasis as measured by interactions between assayed chromosomal units.

With sufficiently large sample sizes, even factors of very small effect can be detected with this approach, and the types of characters that can be analyzed are limited only by the imagination of the investigator. For example, in *D. melanogaster*, Caligari and Mather (1988) used chromosomal assays to localize factors for aggression and competitive response to chromosomes 2 and 3. Sokolowski (1980) examined larval foraging behavior, mapping genetic factors for larval locomotor behavior to chromosome 2 and feeding rate factors to both chromosomes 2 and 3. Luckinbill et al. (1988) localized factors deferring senescence to all major chromosomes, with chromosome 3 accounting for roughly 70% of the observed variation in females. Shao et al (2008) reviewed the results from 90 blood, bone, and metabolic traits in a mouse CSS panel and 54 additional traits in a rat CSS panel, finding that traits tended to be highly polygenic, with strong epistatic interactions between chromosomes.

Estimating the additive, dominance, and epistatic composite effects associated with particular chromosomes is straightforward. For example, let X_B and X_b represent the X chromosomes from two different inbred strains. The composite amount of dominance is estimated by comparing the mean phenotypic values of X_BX_B, X_BX_b, and X_bX_b females in a standardized genetic background. Nonadditive interactions (epistasis) between chromosomes can be detected in a similar manner by considering pairs of chromosomes (Robertson and Reeve 1953; Robertson 1954; Cooke and Mather 1962). Using these approaches, the genetic architectures for a number of characters have been examined in *Drosophila* (Table 17.1). Characters correlated with fitness tend to show epistasis and directional dominance, while those presumably weakly correlated with fitness do not. A serious limitation of these assays is the size of the unit of analysis—a chromosome or large chromosomal segment can contain QTLs having effects in opposite directions, canceling the individual effects and

Table 17.1 The genetic architecture of various characters in *Drosophila* inferred by chromosomal assay. Unless otherwise indicated, all analyses are for *D. melanogaster*. Most of these results were obtained by comparing differences between selected lines. Results for many characters were obtained by both whole and segmental chromosomal assays. Directional dominance occurs when multiple assayed regions show the same direction of dominance (e.g., high values tending to be dominant over low values). Features not examined are denoted by a dash. (Expanded from a shorter table in Kearsey and Kojima 1967.)

	Dominance			
Character	Present	Directional	Epistasis	Reference
Fitness-related Characters				
Viability	Yes	Yes	Yes	Breese and Mather 1960
	—	—	Yes	Seager and Ayala 1982
	No	—	Weak	Ferrari 1987
Male mating activity	—	—	No	Kosuda 1993
Egg hatchability	Yes	Yes	Yes	Kearsey and Kojima 1967
Fecundity	Yes	Yes	—	Keller and Mitchell 1964
Egg-pupal survival	Yes	Yes	—	Keller and Mitchell 1964
Development time	Yes	Yes	—	Keller and Mitchell 1964
	No	—	Weak	Ferrari 1987
Progeny yield	Yes	Yes	Yes	Barnes 1966
Egg production				
D. melanogaster	Yes	Yes	—	Keller and Mitchell 1964
D. pseudoobscura	Yes	—	Yes	Kojima and Kelleher 1963
Physiological Characters				
DDT-resistance	Yes	No	Yes	Dapkus and Merrell 1977
ADH activity	Yes	Yes	Yes	McDonald and Ayala 1978
G6PD activity	—	—	Yes	Miyashita and
6PGD activity	—	—	Yes	Laurie-Ahlberg 1984
Morphological Characters				
Abdominal bristles	Weak	No	Weak	Keller and Mitchell 1962
	Weak	No	Weak	Breese and Mather 1957
Sternopleural bristles	Weak	No	Weak	Breese and Mather 1957
	Weak	No	No	J. Hill 1964
Thorax length	Yes	Yes	Yes	Robertson 1954
	Weak	No	Weak	Keller and Mitchell 1962
Wing length	Yes	Yes	Yes	Robertson 1954
	No	—	No	Keller and Mitchell 1962
Body weight				
D. melanogaster	No	—	Weak	Kearsey and Kojima 1967
D. pseudoobscura	Weak	No	Weak	Frahm and Kojima 1966

leaving a composite effect close to zero. This is especially true of epistatic effects (Hastings 1986).

Natural chromosomal assays become possible when chromosomal rearrangements are segregating within a population. Because chromosomal inversions suppress recombination, alternative inversions can be treated as intact nonrecombining units. This approach was used extensively by Dobzhansky in his studies of the fitnesses of natural inversions in *Drosophila* (Lewontin et al. 1981), and has also been applied to morphological characters. For example, Hasson et al. (1992) found that three inversions on chromosome 2 accounted for at least 25% of the total variance for thorax length in *Drosophila buzzatii* (also see Ruiz et al. 1991). Two of the inversions showed dominance, while a third showed apparent (potentially associative) overdominance. Other studies have detected segregating rearrangements with effects on body size in the grasshopper *Moraba scurra* (White and Andrew 1960), on wing length in the seaweed fly *Coelopa frigida* (Butlin et al. 1982; Wilcockson et al. 1995), and on wing length in *D. melanogaster* (Herández et al. 1993).

Thoday's Method

Thoday (1961, 1979) introduced a considerable refinement for crude chromosomal assays. His method is of substantial interest for purely historical reasons, but it also points out potential limitations of the flanking-marker mapping methods discussed in subsequent chapters. Consider a hypothetical QTL between two linked markers, with allele Q fixed in one inbred line (the **tester line**, also fixed for alleles A and B at the flanking marker loci), and q in another line (fixed for markers a and b). Substituting an aqb chromosome into the tester line gives AQB/aqb heterozygotes, and these individuals are subsequently backcrossed to the tester line. Recombinant aB progeny are then scored for the trait of interest. Because these recombinant chromosomes are either aqB or aQB, a QTL with detectable effects is indicated by the presence of two distinct character classes (corresponding to QQ and Qq individuals). The presence of a linked QTL outside of the interval can also generate distinct classes, but this is not an issue if only one recombination event occurs per chromosome. The main limitation of Thoday's method is the difficulty of isolating single recombinant chromosomes and propagating them intact (without further recombination) in organisms other than *Drosophila*.

Example 17.1. A large sample of Ab recombinant chromosomes was generated from a cross between an ab line with negative character value and an AB tester line with a mean of zero. Each recombinant chromosome was used to generate a line with a recombinant Ab chromosome and an AB tester chromosome in an otherwise genetically homogeneous background. Construction of such lines requires the use of balancer chromosomes, which prevent any further recombination from occurring (see Figure 5.8). The mean character values from the resulting lines formed two distinct clusters, one (comprising 43%) clustering around -1.5, the other clustering around 0. Hence, there appears to be a single factor (at our level of resolution), with the 0 mean class corresponding to Q/Q and the -1.5 class corresponding to Q/q, implying $\mu_{QQ} - \mu_{Qq} = 1.5$. The frequencies of these classes among Ab recombinant chromosomes imply that the A–Q recombination fraction is 43% of the A–B recombination fraction (assuming no interference).

McMillan and Robertson (1974) found that the power of Thoday's method was generally very poor. To achieve high power, both the number of recombinant chromosomes (N) and the number of replicates (n) per recombinant chromosome need to be large. Even if a factor has a large effect, it will remain undetected unless the sample of recombinant chromosomes contains both factor alleles, which requires a sufficiently large N. To see this result, set the distance between the two markers to one, and let f ($0 \leq f \leq 1$) denote the relative distance between the factor and marker A, so that a fraction f of the aB recombinants are aqB and $1-f$ are aQB. The probability that N recombinant (aB) chromosomes all contain the same QTL allele is $f^N + (1-f)^N$. If f is close to zero or one, N has to be considerable for a sample to contain a sufficient number of both QTL genotypes for detection. The power for detection is maximized when the factor is exactly between the two markers (leading to expected equal numbers of aqB and aQB). Power graphs are given by McMillan and Robertson (1974).

Thoday's method can be extended to allow for multiple factors between two markers, but not without complications. As shown in Example 17.2, the estimation of map distances and effects of multiple QTLs between markers *critically* assume that all QTL alleles in the line being examined have *the same directional effect* (McMillan and Robertson 1974). This caveat will be a recurring theme for molecular-marker approaches: ***these methods can be***

compromised if the line being examined contains linked QTL alleles of opposite effects. Data from inbred lines show that such situations are not uncommon (Chapter 18).

Example 17.2. Suppose there are three QTLs between the marker loci A and B, with

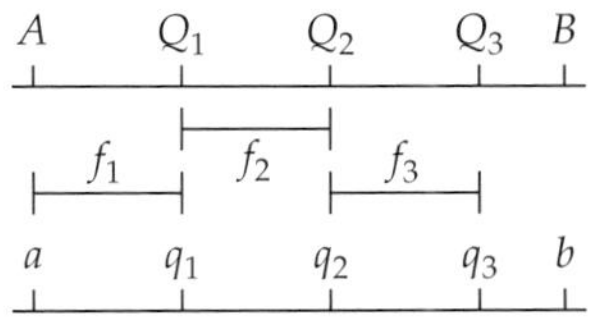

Assuming no double crossovers, let f_1, f_2, f_3, and $1-(f_1+f_2+f_3)$, respectively, be the fractions of recombination between flanking markers that occur between A and the first QTL, between the first and second QTL, the second and third, and the third and B. (As an aside, for detailed power calculations, one can use these frequencies as the success probabilities in a multinomial distribution. Equation 2.20a then returns the probability of obtaining a particular configuration of these three classes given N recombinant chromosomes.) Let Q_i denote alleles from the high line, which have effect $H_i > 0$ relative to alleles q_i from the low line. The expected frequencies and effects of Ab recombinant chromosomes are obtained as follows. As shown in the following figure, a fraction f_1 of Ab chromosomes have no high-line alleles, as the crossover occurred between A and Q_1. Likewise, a fraction f_2 contain high-line allele Q_1 with effect H_1. The last two classes follow similarly. Thoday's method estimates the H_i and f_i by making the *assumption* that all alleles from the high line increase character value (all $H_i > 0$) so that the mean of the largest phenotypic class corresponds to $H_1+H_2+H_3$, the mean of second largest class to H_1+H_2, etc. This ordering allows us to uniquely estimate the H_i and f_i.

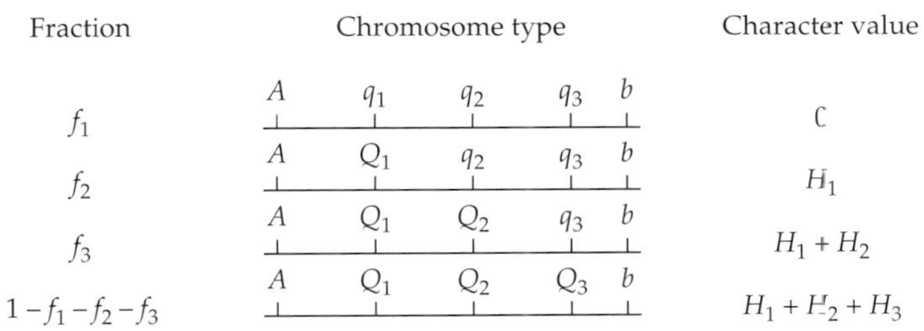

If some of the H_i differ in sign, this ordering breaks down, as illustrated in the following example offered by McMillan and Robertson (1974). Setting the mean of the low line at 0, suppose that 23% of the recombinants have value 0, 7% have value 0.5, 41% have value 1.3, and the remaining 29% have value 3.7. Assuming that the effects of all Q alleles are positive, the second smallest mean corresponds to $H_1 = 0.5$. Likewise, the third smallest mean corresponds to $H_1 + H_2 = 1.3$, and hence $H_2 = 1.3 - 0.5 = 0.8$. Continuing in this fashion gives the estimated values shown below on the left. In actuality, McMillan and Robertson generated these data using a very different configuration than suggested by these estimates, as shown on the right.

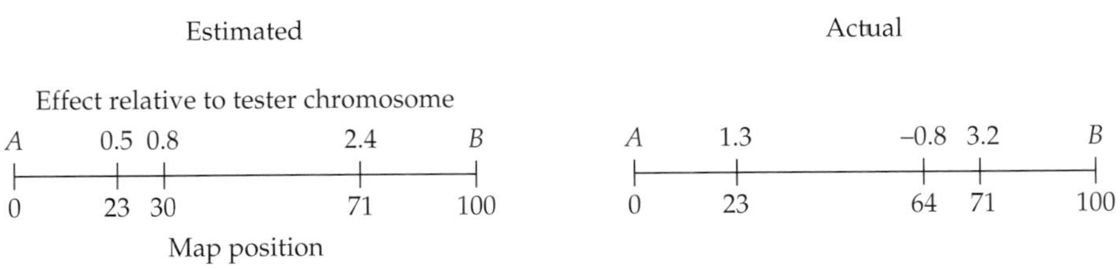

Genetics of *Drosophila* Bristle Number

Drosophila possess a large number of sensory bristles regularly distributed over the adult

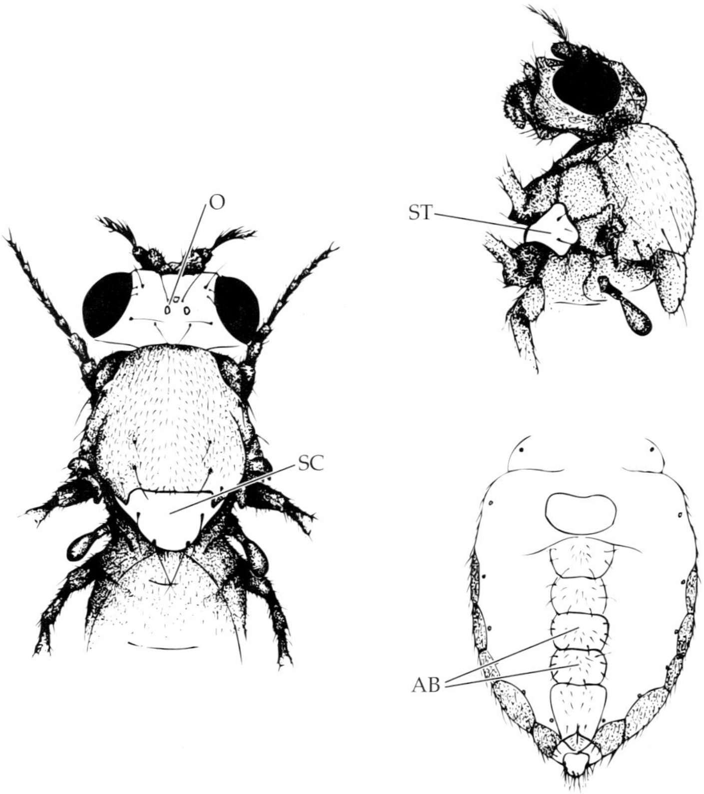

Figure 17.1 Location of head and thoracic bristles in adult *Drosophila*. The most frequently studied bristle systems in quantitative-genetic experiments are sternopleural bristles (ST) on the side of the thorax between the first and second legs, abdominal bristles on the underside of the abdomen (AB), and scutellar bristles (SC) on the upper side of the last thoracic region. Ocellar bristles (O) on the top of the head are also occasionally used.

integument. As discussed in Chapters 14 and 15, a number of bristle systems, most notably sternopleural, abdominal, and scutellar bristles, have been widely used in quantitative-genetic studies (Figure 17.1).

The genetics of sternopleural bristle number have been intensively studied in selected lines using Thoday's method (reviewed in Spickett 1963; Thoday et al. 1964; Spickett and Thoday 1966; Davies and Workman 1971; Davies 1971; Thoday and Thompson 1976; Thoday 1979; Shrimpton and Robertson 1988a, 1988b; Mackay 1995, 1996; Dilda and Mackay 2002; Mackay and Lyman 2005). For example, Spickett and Thoday (1966) examined two selected lines (*vg4* and *vg6*), which showed increases of 15 and 20 bristles, respectively, relative to the standard Oregon R strain. For line *vg6*, 88% of this increase could be accounted for by five factors detectable via Thoday's approach, while three factors (each spanning a distance amounting to less than 2% recombination) accounted for 80% of the increase in line *vg4*. In more detailed analyses restricted to the third chromosome, Davies (1971) and Shrimpton and Robertson (1988a, 1988b) isolated a minimum of 8 and 17 factors, respectively. Extrapolating the latter figure to the rest of the genome yields roughly 60 loci influencing among-line differences in sternopleural bristle number. The distribution of effects for the bristle factors isolated by Shrimpton and Robertson (Figure 17.2) is roughly consistent with results from P-element tagging (Chapter 15), suggesting that sternopleural bristle number is controlled by a few major genes supplemented by numerous genes of smaller effect. Evidence for epistasis was found in most studies.

Spickett (1963) examined the developmental features of the isolated factors from line *vg4* in further detail. One factor increased bristle number over the entire fly by increasing

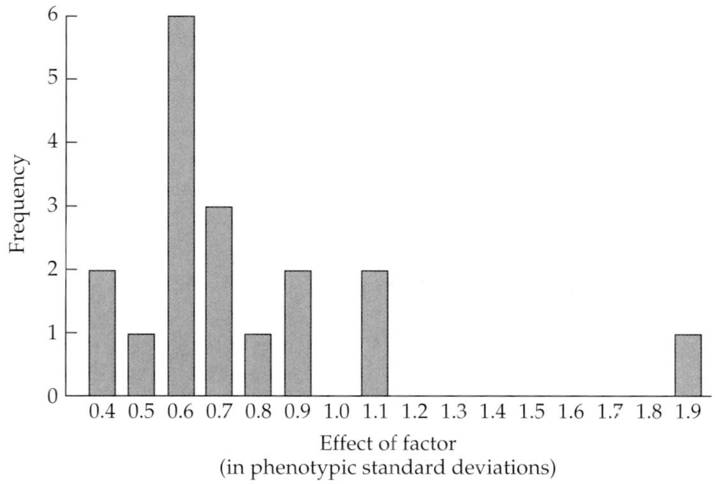

Figure 17.2 Distribution of the effects on sternopleural bristles of *Drosophila melanogaster* third chromosome factors isolated by Shrimpton and Robertson (1988b). This distribution is biased towards large factors, as effects under roughly 0.3 phenotypic standard deviations were not statistically detectable in this study.

total fly cell number. An increase in cell number normally results in a larger fly, but Spickett found a second linked factor that decreased average cell size, resulting in a normal-sized fly with a larger number of cells and more bristles. Two other factors had more local effects. The first was linked to the previous two factors on chromosome 3 and increased the number of bristles in a specific region of the sternopleurite. The second was on chromosome 2 and modified development of a single large bristle into several smaller bristles in this region. Clearly, genetic changes in characters even as superficially simple as bristle number can result from modifications in a variety of developmental pathways.

Genetics of *Drosophila* Speciation

Considerable debate revolves around the genetic basis of speciation. Are such events precipitated by a few genes of major effect (Templeton 1980; Carson and Templeton 1984) or by changes at many loci with individually small effects (Charlesworth et al. 1982; Barton and Charlesworth 1984)? Further, what is the nature of these genetic changes? Are they driven by drift, adaptive selection, or are other forces (such as meiotic drive) involved? Do they involve changes in otherwise relatively normal genes or are unusual genetic features required (e.g., chromosomal rearrangements)? Much of the data on the genetic basis of reproductive isolation via hybrid incompatibility (**postzygotic**, as opposed to pre-mating, or **prezygotic**, barriers) come from crosses between *Drosophila* species that show partial fertility. In discussions of such hybrid incompatibilities, it is important to distinguish between barriers at different stages—**hybrid male sterility** (**HMS**), **hybrid female sterility** (**HFS**), and **hybrid inviability** (**HI**), which can also be sex-specific—as these different outcomes are most likely the result of different genes. A fairly general pattern in *Drosophila* is that HMS, at least initially, tends to be far more common than HFS or HI (Coyne and Orr 1998, 2004). Before examining the impressive advances in speciation genetics from *Drosophila* studies, it is worth noting that reproductive isolation is not speciation, and the generality of lessons from *Drosophila* remains an open issue (Mallet 2006).

Low-resolution mapping from species crosses provides insight into the number and rough chromosomal locations of hybrid incompatibility factors, while more detailed fine-scale mapping offers some insight into potential underlying genes. The observation that started this line of inquiry was that *D. pseudoobscura* × *D. persimilis* hybrids produce normal females but sterile males. Dobzhansky (1936) backcrossed fertile F_1 hybrid females to

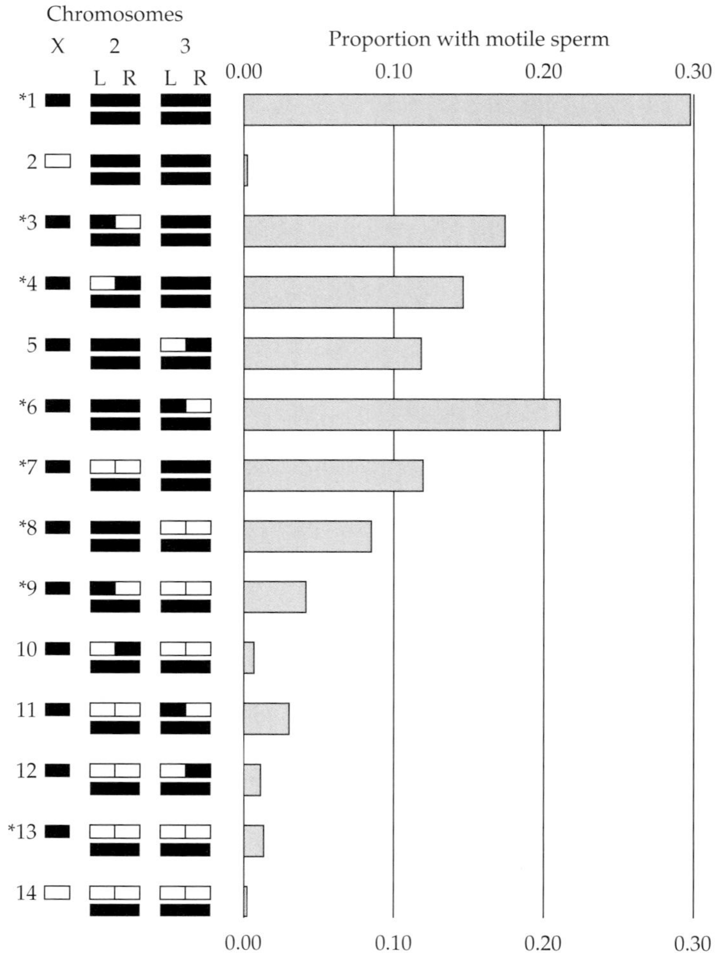

Figure 17.3 Locating male-sterility genes in *D. simulans* × *D. mauritiana* hybrids with chromosomal assays. Fertility is measured by the proportion of males with motile sperm, with chromosome segments from *D. simulans* being shown in black, and those from *D. mauritiana* in white. L and R refer to the left and right arms of chromosomes 2 and 3. All genotypes have a *D. simulans* Y chromosome. Asterisks denote genotypes that produced progeny when back crossed to *D. simulans* females. (From Coyne 1984.)

chromsome-marked males of both species, substituting single chromosomes from one species into an otherwise normal genetic background of the other. He found that testes size (correlated with male fertility) was affected by factors on all five chromosomes, with the strongest effect being on the X chromosome. Orr (1987) found that the reduction in sperm motility in hybrids between these species is largely due to incompatibilities (i.e., epistasis) between factors on the X and Y chromosomes from different species. A finer analysis by Wu and Beckenbach (1983) suggested at least nine factors affecting hybrid male fertility (five X-linked, three autosomal, and one Y-linked).

These studies motivated crosses involving other sibling pairs of *Drosophila* species. For example, Coyne (1984) found that at least five factors are responsible for HMS in *D. simulans* × *D. mauritiana* hybrids, with the main effect again being due to incompatibilities between the X and Y chromosomes of the different species (Figure 17.3). When small sections of the *D. simulans* X chromosome were introgressed into a *D. mauritiana* background by repeated backcrossing, Palopoli and Wu (1994) found four factors influencing HMS, two of which displayed very strong epistasis. Given that they examined less than 20% of the X chromosome, this extrapolates to at least 40 X-linked factors (assuming an equal number of factors were fixed by both species) involved in reproductive isolation between these two species (Davis and Wu 1996). That the genetic units from the above studies are best viewed as *factors*, rather than *single major genes*, was shown by further work from Wu's lab. Introgression of small segments of the X from *D. mauritiana* and *D. sechellia* into *D. simulans* initially suggested that a 500 kilobase region harbored a single major HMS gene (Perez et al. 1993), but a more refined analysis suggested at least two epistatically interacting factors (Perez and Wu 1995). Reed et al. (2008) found that HMS in a cross of *D. arizonae* and *D. mojavensis* was due to multiple epistatic factors mapping to all major chromosomes.

Higher resolution studies were performed by Masly and Presgraves (2007), who introgressed numerous small *D. mauritiana* segments into an otherwise *D. sechellia* background. They made three major conclusions: (i) recessive incompatibilities outnumbered dominant ones; (ii) HMS factors outnumbered all of other types of incompatibilities, and (iii) most X-segment introgressions caused HMS (60%), whereas most autosomal ones did not (18%). Similar observations with other *Drosophila* crosses were made by True et al. (1996a) and Tao et al. (2003).

Fine mapping using very small introgressed segments allowed for the localization of several putative **speciation genes**, major factors from one genetic background that tend to negatively interact with numerous other factors from a different background. The classic example is the ***Odysseus*** (***OdsH***) gene, which causes HMS when a 3kb *OdsH* segment (containing exons 3 and 4) from *D. mauritiana* is introgressed into an otherwise completely *D. simulans* background (Sun et al. 2004). This introgression results in the misexpression of several hundred genes, with autosomal genes being more impacted than X-linked ones (Lu et al. 2012). The ancestor to *OdsH* arose over 60 million years ago from a duplication of the highly conserved *unc-4* homeobox gene (Nunes et al. 2024). *CdsH* shows rapid adaptive evolution in the *D. mauritiana* lineage, with a great excess of replacement over silent substitutions as it evolved from the ancestral embryonic function of *unc-4* into a primarily spermatogenetic function. Such rapid adaptive evolution involving fairly typical genes and a gain (as opposed to loss) of function are common features of most of the handful of isolated genes with large effects on hybrid incompatibility (Orr and Irving 2000; Orr and Presgraves 2000; Orr et al. 2004; Orr 2005).

An important caveat for all of these studies is that they measure the *current* number of factors contributing to reproductive isolation, which can be much greater than the *actual* number involved in the *original* isolation event(s). Some indications of the potential magnitude of this bias can be seen by considering the classic **Dobzhansky–Muller model** (also refered to as the **Bateman–Dobzhansky–Muller model**) of speciation, which assumes that speciation occurs through epistasis, with isolated populations fixing mutually incompatible alleles (Dobzhansky 1936; Muller 1939; Orr 1996; Turelli and Orr 2000; Turelli et al. 2001; Cutter 2012). This model assumes that a base population fixed for *aabb* is split into two isolated populations, with separate mutations, *A* and *B*, arising in each population. In the *bb* background, *A* has no serious negative fitness consequences, so fixation can occur by drift, giving an *AAbb* population. Likewise, the other population fixes allele *B* to yield *aaBB*, where again *B* has no serious fitness effects in the *aa* background. However, *A* and *B*, when together, are incompatible, such that an *AaBb* hybrid has reduced fitness relative to the ancestral population. A number of such incompatibilities are known in species hybrids (see Orr 1995 for a review), and they are related (at least in principle) to synthetic lethals (Chapter 12). Orr (1995) found that under the Dobzhansky–Muller model, the number of fixed incompatible loci separating the two populations should increase with (at least) the square of separation time, and that incompatibilities involving three or more loci evolve more readily than interactions involving pairs of loci (also see Cabot et al. 1994). This accelerated pace of fixing incompatible loci (a phenomenon known as the **snowball effect**) occurs because the fixation of an incompatibility locus facilitates the fixation of additional such loci. Hence, complex interactions among "speciation genes" are expected to be the norm, rather than the exception. This introduces further complications into the already difficult problem of determining how many factors were actually involved in generating the *initial* isolation event(s) that allowed for the accumulation of further incompatibility genes.

Despite these interpretative difficulties, some general conclusions emerge from these studies. First, in *Drosophila* there is a **large-X effect** (also know as **Coyne's rule**; Turelli and Moyle 2007): ***factors with the largest effects on reproductive isolation are located on the X chromosome*** (Coyne and Orr 1989; Coyne 1992). Although originally framed to include all aspects of reproductive isolation, the large-X effect seems to be more general for HMS than for hybrid inviability. The X effect is restricted to reproductive isolation, as no such X chromosome localization is seen for genes influencing morphological differences between *Drosophila* species (e.g., Coyne 1983, 1985; Ford and Aquadro 1996; Liu et al. 1996). While an observed excess of HMS factors is seen on the X relative to the autosomes, essential genes for spermatogenesis are actually *underrepresented* on the X relative to autosomes (Masly and Presgraves 2007; Lu et al. 2010). Hence, the large-X effect for HMS is not simply due to the X containing more male fertility factors than the autosomes. One explanation is dominance (see the discussion below on Haldane's rule). Other explanations are related to the divergence of mechanisms for X chromosome dosage compensation in hybrids (Filatov 2018).

The second (more general) conclusion, first noted by Haldane (1922), is that ***when only one sex in a cross-species hybrid is sterile or inviable, it is usually the heterogametic sex*** (e.g., XY males in most animals, WZ females in birds and butterflies). The *Drosophila* data show that male sterility/inviability arises first (often rather quickly), with female sterility/inviability often developing only after a considerable amount of additional time (Coyne and Orr 1989, 1997). Much has been written about **Haldane's rule** (**HR**) and its possible causes (e.g., Charlesworth et al. 1987; Orr and Coyne 1989; Coyne et al. 1991; Frank 1991a; Read and Nee 1991; Coyne 1992; Coyne and Orr 1993; Orr 1993a, 1993b; Virdee 1993; Wu and Davis 1993; Turelli and Orr 1995, 2000; Orr and Turelli 1996; Sawamura 1996; True et al. 1996a; Zeng 1996; Laurie 1997; Orr 1997; Turelli and Moyle 2007; Koevoets and Beukeboom 2009; Brothers and Delph 2010; Schilthuizen et al. 2011; Demuth et al. 2014; Delph and Demuth 2016; Moran et al. 2017). While most studies of Haldane's rule examine differences between populations, examining within-population variation is also critical (e.g., Wade et al. 1994; Demuth and Wade 2007b; Lachance and True 2010; Cutter 2012). The large-X effect and Haldane's rule are often jointly called the **two rules of speciation** (Coyne and Orr 1989).

While a variety of mechanisms have been proposed to account for HR, Turelli and Orr (1995, 2000; Orr and Turelli 1996) suggested that dominance alone may provide an explanation. If alleles decreasing hybrid fitness are partly recessive, the cumulative effects of sex-linked genes are greater when hemizygous than when heterozygous, resulting in a more pronounced effect in heterogametic hybrids. This effect can be seen by noting that although XX hybrids carry approximately twice the number of deleterious X-linked genes as XY hybrids, if the expression of deleterious effects is less than 50% in heterozygotes, then the overall deleterious effects are smaller in XX hybrids.

While this dominance view for HR appears to be a reasonable explanation for *hybrid inviability*, a more nuanced view is emerging for *sterility*. It has been suggested that reproductive evolution occurs at a faster rate in males than in females, and that such **faster male evolution** can either account for, or augment, HMS in species with male heterogametic sex chromosomes (Wu and Davis 1993; Wu et al. 1996; Kulathinal and Singh 2008). Several studies have examined these competing hypotheses. Presgraves and Orr (1998) contrasted HR in the related mosquito genera *Aedes* and *Anopheles*. While *Anopheles* species have traditional heterogametic males (with a largely inactive Y chromosome), *Aedes* species have a single sex-determining locus, but otherwise their Y chromosomes appear to contain the same genes as the X, removing dominance as a potential source for HR. While both HMS and hybrid inviability follow HR in *Anopheles*, Haldane's rule in *Aedes* holds for sterility (HMS occurs, but not HFS), but not for inviability. In contrast to *Aedes* where both sexes have two active sex chromosomes, in marsupials both sexes are hemizygous, as the paternal *X* chromosome is uniformly silenced. Watson and Demuth (2012) noted that sterility in marsupials generally follows Haldane's rule (with HMS being more common than HFS), consistent with the faster male hypothesis given that dominance does not apply. Conversely, Haldane's rule does not seem to apply to hybrid inviability. While these observations provide support for the faster-male hypothesis for sterility and the dominance hypothesis for inviablity, other studies suggest the opposite conclusion. Butterflies have females (ZW) as the heterogametic sex and males (ZZ) as the homogametic sex. Crosses between *Heliconius melpomene* and *H. cydno* showed sterile females and normal males, with a large "X" (Z) effect, evidence for the dominance as opposed to the faster male hypothesis for sterility (Naisbit et al. 2002).

While mapping studies have provided some insight and clarity into the genetics of speciation, they also raise new questions. In particular, what is the role of **meiotic drive** (**segregation distortion**) factors in speciation? Such factors result in a preferential transmission of an allele from heterozygotes, driving them to fixation in the absence of direct selection (Sandler and Novitski 1957; Jaenike 2001, 2008; Lindholm et al. 2016). Population-genetic models initially suggested a potentially important role for meiotic drive as a speciation process (Frank 1991b; Hurst and Pomiankowski 1991). The idea is that suppressers of meiotic drive factors evolve separately in isolated populations to counter local drive elements. If these suppressers are even slightly recessive, drive is uncovered in hybrids. This hypothesis initially failed to gain traction, as early *Drosophila* crosses did not seem to show the expected drive in hybrids. More recently, however, such segregation distortion has been seen in some hybrids, leading to a reappraisal of its importance (Orr et al. 2006; McDermott and Noor 2010). Tao et al. (2001) localized a factor (which they called *tmy*, for *too much yin*) on chromosome 3 in *D. mauritiana* that impacts both HMS and hybrid male drive in a *D. simulans* background. Similarly, Phadnis and Orr (2009) found a single factor, *Overdrive*, that influences both HMS and segregation distortion in hybrids between the subspecies *D. pseudoobscura pseudoobscura* and *D. p. bogotana*.

GENETIC MAPS

Genetic maps show the ordering of loci along a chromosome and the relative distances between them. As detailed throughout the remainder of this chapter, such maps are essential to the localization of QTLs. The lack of genetic markers historically prevented the construction of detailed maps in all but a few well-studied species with short generation times. However, as predicted by Botstein et al. (1980), molecular markers sparked an explosion of genetic maps in humans, economically important plants and animals, and now for just about any species of interest. The theory for constructing genetic maps is highly refined, and we only introduce a few important issues here. For more detailed reviews, the reader should consult Bailey (1961), Lalouel (1992), and the excellent text by Ott (1991). We note the important distinction between a genetic versus a **physical map**. The latter is based on the actual DNA sequence, and hence distances are in base pairs (typically expressed in units of **kilobases, kb**; **megabases, mb**; or **gigabases, gb**; corresponding to 10^3, 10^6, and 10^9 nucleotides, respectively). In contrast, distances on genetic maps measure the *amount of recombination*. While the gene *order* is the same under either map, the *scaling* of distance between loci is not, as the rate of genetic recombination is not uniform over a genome.

Map Distances versus Recombination Frequencies

Genetic map construction involves both the ordering of loci and the measurement of the recombination-based distance between them. Ideally, distances should be additive so that when new loci are added to the map, previously obtained distances do not need to be radically adjusted. Unfortunately, ***recombination* frequencies *are not additive and hence are inappropriate as distance measures***. To illustrate, suppose that three loci are arranged in the order *A*, *B*, and *C* with recombination frequencies c_{AB}, c_{AC}, and c_{BC}, respectively. Each recombination frequency, c, is the probability that an odd number of crossovers occurs between the markers, while $1 - c$ is the probability of an even number (including zero).

There are two different ways to get an odd number of crossovers over the interval *A*–*C*: (i) an odd number in *A*–*B* and an even number in *B*–*C*, or (ii) an even number in *A*–*B* and an odd number in *B*–*C*. If there is no **interference**, so that the presence of a crossover in one region has no effect on the frequency of crossovers in adjacent regions, these probabilities can be related as

$$c_{AC} = c_{AB}\,(1 - c_{BC}) + (1 - c_{AB})\,c_{BC} = c_{AB} + c_{BC} - 2c_{AB}\,c_{BC} \tag{17.1}$$

This is **Trow's formula** (Trow 1913). More generally, if the presence of a crossover in one region depresses the probability of a crossover in an adjacent region, then

$$c_{AC} = c_{AB} + c_{BC} - 2(1-\delta)c_{AB}\,c_{BC} \tag{17.2}$$

where the **interference parameter** δ ranges from zero if crossovers are independent (no interference, recovering Trow's formula) to one if the presence of a crossover in one region completely suppresses crossovers in adjacent regions (**complete interference**). Note that recombination frequencies are additive only when interference is complete ($\delta = 1$).

Thus, in the absence of very strong interference, observed recombination frequencies can only be considered to be additive if they are small enough that the product $2c_{AB}c_{BC}$ can be ignored. This is not surprising given that the recombination frequency measures only a *part of all recombinant events* (those that result in an odd number of crossovers). A genetic map distance m, on the other hand, attempts to correct for this probability by measuring the ***total number of crossovers*** (both odd and even) between two markers. This is a naturally additive measure, as the number of crossovers between *A* and *C* equals the number of crossovers between *A* and *B* plus the number of crossovers between *B* and *C*.

A number of **mapping functions** attempt to predict the number of crossovers (m) from the observed recombination frequency (c), and, conversely, translate a value of m into a value of c. The simplest, derived by Haldane (1919), assumes that crossovers occur randomly and independently over the entire chromosome, i.e., no interference. Let $p(m, k)$ be the probability of k crossovers between two loci that are m map units apart. Under the assumptions of this model, Haldane showed that $p(m, k)$ follows a Poisson distribution (Equation 2.21a), so that the observed fraction of gametes containing an odd number of crossovers is

$$c = \sum_{k=0}^{\infty} p(m, 2k+1) = e^{-m} \sum_{k=0}^{\infty} \frac{m^{2k+1}}{(2k+1)!} = \frac{1 - e^{-2m}}{2} \tag{17.3}$$

where m is the expected total number of crossovers. Rearranging, we obtain **Haldane's mapping function**, which yields the (Haldane) map distance m as a function of the observed recombination frequency c,

$$m = -\frac{\ln(1-2c)}{2} \tag{17.4}$$

For small c, $m \simeq c$ (as $\ln(1-x) \simeq x$ for $x \ll 1$). For large m, Equation 17.3 shows that c approaches 1/2. Map distance is usually reported in units of **Morgans** (after T. H. Morgan, who first postulated a chromosomal basis for the existence of linkage groups) or as **centiMorgans** (cM), where 100 cM = 1 Morgan. For example, a Haldane map distance of 20 cM ($m = 0.2$) corresponds to a recombination frequency of $c = (1 - e^{-0.2})/2 \simeq 0.16$. Conversely, an observed recombination frequency of $c = 0.3$ corresponds to an expected number of $m = -\ln(1 - 2 \cdot 0.3)/2 = 0.46$ crossovers, or 46 cM (under Haldane's assumption of no interference). Applying Equation 2.21a with $\lambda = 0.49$ expected crossovers gives the probabilities of 0, 1, 2, or 3 crossovers occurring in this region as 0.631, 0.290, 0.039, and 0.006, respectively. Note that 97% (0.29/0.30) of all odd-numbered crossover events (and hence gametes scored as recombinant) are the result of a single crossover, while 2% (0.006/0.3) are the result of a triple crossover.

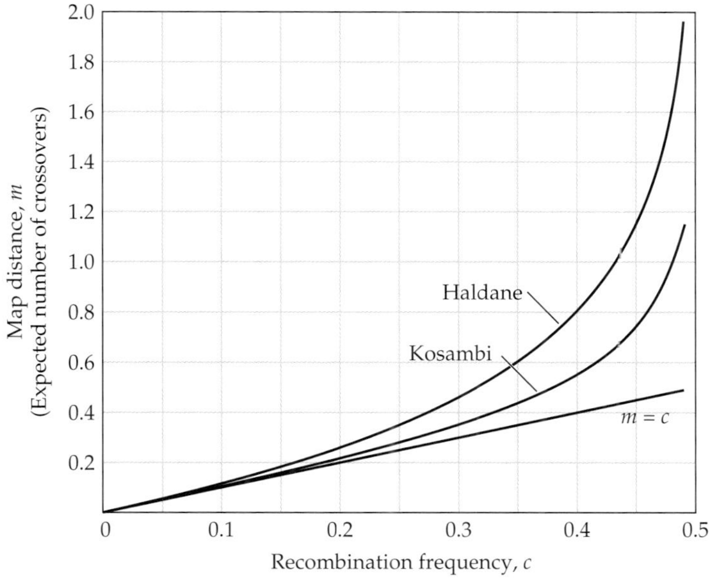

Figure 17.4 Comparison of the Haldane, Kosambi, and simple recombination-frequency ($m = c$) mapping functions, which translate an observed recombination frequency c into an estimate of the expected number of crossovers (the **map distance**, m). The different functions make different assumptions about interference (δ in Equation 17.2). Haldane's function assumes none ($\delta = 0$), Kosambi's assumes a moderate amount, and the simple recombination-frequency model assumes complete interference ($\delta = 1$). For a given c, the Haldane map distance is largest, and the recombination-frequency distance is the smallest. For $c < 0.15$, all three are extremely similar, while the Kosambi and simple function are close for $c < 0.25$.

Although Haldane's mapping function is widely used, several other functions allow for the possibility of crossover interference in adjacent sites (Bailey 1961; Felsenstein 1979; Karlin 1982; Liberman and Karlin 1984; Pascoe and Morton 1987; Ott 1991; Evans et al. 1993; Foss et al. 1993; Goldstein et al. 1995; Zhao and Speed 1996). For example, geneticists often use **Kosambi's** (1944) **mapping function**, which allows for modest interference,

$$m = \frac{1}{4} \ln \left(\frac{1+2c}{1-2c} \right) \tag{17.5}$$

Damodar Dharmananda Kosambi was an Indian mathematician and polymath, and this 1944 paper was his only one on genetics. Figure 17.4 compares the Haldane and Kosambi mapping functions. For small c (< 0.15), both give $m \simeq c$. For large m, both approach $c = 1/2$. The appropriateness of a potential mapping function can be assessed by testing if the computed map distances depart from additivity (e.g., Pascoe and Morton 1987).

There is no universal relationship between map distance and the actual physical distance between loci (Table 17.2). A centiMorgan can correspond to a span of between ten thousand and millions of nucleotide base pairs, depending on the species. Further, even within a genome, there can be rather dramatic differences. For example, crossovers are often suppressed in particular chromosomal regions (increasing the number of base pairs per cM), such as near the centromere and telomeres (chromosome ends), e.g., True et al. (1996b). Further, the rate of recombination is under genetic control with some modifier genes having a general influence throughout the genome and others having a fine-scaled influence on specific chromosomal regions (Brooks 1988). Thus, there may be considerable variation among individuals and among populations in the strength of linkage. The recombination rate can also vary between the sexes. In mammals, for example, females usually have greater map distances than do males. In male *Drosophila*, crossingover is generally entirely suppressed.

Table 17.2 Basic features of the physical and genetic maps of various eukaryotic groups, derived from a large survey of mapping studies involving high-density molecular markers. The grouping "Other unicellular species" includes algae, apicomplexans, ciliates, kinetoplastids, and oomycetes. Numbers in parentheses denote standard errors, and n denotes the number of species surveyed. Map lengths and mean chromosome (Chr.) sizes are in units of Morgans (M). The last column, kilobases per centiMorgan, is simply the average genome size for the group divided by the average genetic map distance and hence is a fairly crude summary statistic. (From Walsh and Lynch 2018).

Group	Total Map Length	Genome Size (Mb)	Haploid Chr. No.	Mean Chr. Size (M)	Kilobases per centiMorgan
Fungi	18.3 (2.2)	36.4 (3.2)	11.9 (1.2)	1.86 (0.36)	20
Other unicellular sps.	10.9 (1.2)	80.9 (23.3)	12.9 (1.2)	0.96 (0.18)	74
Arthropods	18.1 (3.7)	679.6 (172.4)	16.1 (3.4)	1.20 (0.18)	375
Mollusks	9.2 (1.1)	1270.7 (177.2)	13.3 (1.6)	0.71 (0.09)	1313
Nematodes	4.5 (1.2)	97.6 (2.5)	7.3 (1.3)	0.59 (0.05)	217
Fish	16.0 (2.3)	1185.4 (190.5)	25.1 (0.6)	0.63 (0.08)	741
Birds	23.1 (5.4)	1334.0 (48.6)	39.6 (0.4)	0.58 (0.14)	577
Mammals	23.9 (2.5)	3222.0 (108.1)	22.1 (2.2)	1.10 (0.07)	1348
Angiosperms	15.9 (1.6)	2020.3 (434.2)	13.2 (0.9)	1.19 (0.07)	1270

How Many Markers Are Needed?

For many organisms, genetic maps are not yet available, and one must start with randomly obtained markers. As a rough approximation for the number of random polymorphic markers necessary for a desired saturation of the linkage map, assume a circular linkage map of total length L map units. In order to have a fraction p of all loci within m map units of some marker, the required number of randomly distributed markers is

$$n = \frac{\ln(1-p)}{\ln(1-2m/L)} \tag{17.6a}$$

(Lange and Boehnke 1982; Beckmann and Soller 1983). This rearranges to give the proportion of the genome within m map units of a marker as approximately

$$p \simeq 1 - \exp\left(-\frac{2mn}{L}\right) \tag{17.6b}$$

(Jacob et al. 1991). Assuming a circular genome map ignores the effect of chromosome ends, causing Equation 17.6a to underestimate n and Equation 17.6b to overestimate p. A more general expression, which accounts for linear chromosomes, was given by Bishop et al. (1983). Assuming a haploid chromosome number C,

$$p = 1 - \frac{2C\left[(1-x)^{n+1} - (1-2x)^{n+1}\right]}{n+1} - (1-2x\,C)\,(1-2x)^n \tag{17.6c}$$

where $x = m/L$.

An alternative consideration is the average distance between a locus and the nearest random marker. Martin et al. (1991) showed that the expected distance of a gene from the closest of n random markers is

$$E[\,m\,] = \frac{L}{2(n+1)} \tag{17.7a}$$

with the upper 95% confidence interval for this distance given by

$$\frac{L}{2}\left(1 - 0.05^{1/n}\right) \tag{17.7b}$$

It is important to note that the above expressions refer to *randomly chosen* markers. It is obviously much more efficient to use sets of equally spaced markers. If the entire genome is covered by marker loci equally spaced at m map units apart, no locus is more than $m/2$ from any marker, and the average distance of a locus from a marker (assuming a uniform distribution of loci along the chromosome) is $m/4$. Note that evenly spaced markers refer to distances measured on a recombinational, rather than physical distance, scale. Hence, even with a complete reference genome (physical maps) in hand, additional information (localized recombination rates) is required to evenly space the markers.

Example 17.3. The human linkage map is approximately 33 Morgans long with haploid chromosome number $C = 23$. How many random markers are required to achieve a 90% probability that at least one marker is within 10 map units (centiMorgans) of a randomly chosen gene?

Here $L = 3300$ cM, $m = 10$ cM, and $p = 0.9$. The circular-chromosome approximation (Equation 17.6a) gives

$$n = \frac{\ln(1 - .9)}{\ln(1 - 2/330)} \simeq 379$$

Numerically solving Equation 17.6c with $p = 0.9$ gives $n = 404$. Hence, the effect of ignoring chromosome ends underestimates n by about 7%. Suppose that 110 random markers are used. From Equation 17.7a, the expected distance of a particular gene from the nearest of these 110 markers is 16.9 cM, while Equation 17.7b gives the upper 95% confidence interval for this distance as 44.3 cM. In contrast, suppose these 110 markers are not random, but are instead chosen to be equally spaced at 30 cM apart. With this spacing, no locus is more than 15 cM from a marker, and loci are, on average, 7.5 cM from a marker.

DESIGN STRATEGIES FOR DETECTING MARKER-TRAIT ASSOCIATIONS

We now turn our attention to the mapping of individual QTLs (the causative loci that underlie variation in a focal trait). The distinction is often made between **linkage mapping**, exploiting the linkage disequilibrium (LD) generated by an excess of parental gametes *among relatives* (Example 5.6), and **LD mapping** using *a population sample of nonrelatives*. The former allows for QTL mapping even when markers are only loosely linked to causative loci (and hence only a few dozen to a few hundred markers are needed). The latter requires very dense markers (thousands to millions), as population-level LD generally only occurs between very tightly-linked loci (Chapter 5). Given the limitations on the number of available markers, early QTL mapping attempts were linkage-based. More recently, the vast abundance of markers has led to a shift toward population-level QTL mapping. A variety of experimental designs using linkage information in either a cross or a pedigree have been proposed, and we consider these in some detail in Chapters 18 (inbred line crosses) and 19 (outbred crosses and pedigrees), while designs using population-level LD are examined in Chapter 20. Because those chapters focus on methods for estimating both QTL effects and map position, we defer further discussion of most statistical issues until then. The remainder of this chapter is concerned with general design strategies for mapping.

That QTLs, like normal genes, can be mapped (assigned to linkage groups) was first demonstrated by Payne (1918), who found that the X chromosome from selected lines of *Drosophila* contained multiple factors influencing scutellar bristle number. Over the next two decades, a number of workers found associations between Mendelian markers (i.e., distinct phenotypes associated with known single-locus genotypes) and quantitative traits in crosses between inbred lines. For example, Sax (1923) crossed two inbred bean lines (*Phaseolus vulgaris*) differing in seed pigment and weight, with the pigmented parents having heavier seeds than the nonpigmented parents. These crosses demonstrated that seed pigment is determined by a single locus with two alleles, *P* and *p*. Among F_2 segregants from this cross, *PP* and *Pp* seeds were, respectively, 4.3 ± 0.8 and 1.9 ± 0.6 centigrams heavier than *pp* seeds. Hence the *P* allele was linked to a factor (or factors) that act in an additive fashion on seed weight. In a similar manner, Lindstrom (1924) demonstrated linkage between a locus for fruit color and factors for fruit size in tomatoes, while Smith (1937a) observed associations between corolla size differences and several independently segregating flower color genes in tobacco. Similar early studies were reported in maize (Lindstrom 1931), peas (Rasmusson 1927), barley (Wexelesen 1933, 1934), and mice (Green 1931, 1933).

Despite just using single marker loci, these early studies raised the possibility that QTLs could be mapped with some precision, perhaps even allowing for the characterization of individual loci, given a population whose QTLs are in LD with scored markers. Disequilibrium between marker loci and QTLs (correlations between a marker and a QTL allele) is the key requirement, as it creates **marker-trait associations**, with different marker genotypes having different expected values for characters influenced by QTLs linked to, and in LD with, these markers. This simple idea is the foundation for QTL mapping. Creating populations with loci in LD, even when they are only modestly linked, is straightforward: crosses between inbred lines have this property, as do families of sibs and other collections of relatives (Example 5.6). For mapping purposes, crosses between inbred lines have the fewest complications. Consider the cross between $MMQQ$ and $mmqq$ inbred lines, where M and m are scored marker alleles linked to a QTL with alleles Q and q. The F_1 progeny (who are all MQ/mq) from such a cross display maximum disequilibrium and, being genetically identical, are equally informative as parents. As will be examined in detail in Chapter 18, a variety of **mapping populations** (such as F_2s or backcrosses) can be generated by using such F_1 parents.

A number of problems, which are elaborated on in Chapter 19, conspire to make linkage-based QTL mapping considerably more difficult in outbred-line crosses. Not all parents are *guaranteed to be informative* (i.e., are marker-QTL double heterozygotes), as some may be segregating marker alleles, but not linked QTLs ($MmQQ$ or $Mmqq$), or vice versa (e.g., $MMQq$ or $mmQq$). Further, the *marker-QTL linkage phase can differ* between different sets of relatives. For example, marker allele *M* might be associated with QTL allele q in one family, but with *Q* in another. Thus, with outbred populations, marker-trait associations have to be assessed in each set of relatives, rather than (as with inbred-line crosses) averaged over all individuals.

QTL mapping based on the LD among tightly linked loci in a random population sample requires orders of magnitude more markers than linkage-based approaches. However, it also has two significant advantages (Chapter 20). First, mapping based on crosses (linkage mapping) is limited by family size (or pedigree depth), while no such limitation occurs when using population-level LD mapping. Hence, the number of individuals phenotyped is usually on the order of a few hundred in linkage-based studies, but thousands to hundreds-of-thousands in population-level studies. Second, population-level studies have significantly greater mapping resolution, usually on the order of kilobases (i.e., nearly single-gene scale), than linkage-based studies, which typically have only megabase resolution (large chromosomal segment scale). LD block sizes in the former are much smaller as they are shaped by a great many past generations of recombination, as opposed to one or two generations of recombination in most designs for the latter (Equations 5.13d and 5.16c).

Example 17.4. Likelihood functions for QTL mapping follow from standard mixture models (Chapter 16). Consider the simple backcross design, wherein two completely inbred lines (with marker/QTL genotypes *MMQQ* and *mmqq*) are crossed to form an F_1 (*MQ/mq*) which is then backcrossed to the *MMQQ* population. If c denotes the marker-QTL recombination frequency, then a fraction $(1-c)$ of *M*-bearing F_1 gametes contain *Q*, while c contain q. Likewise $(1-c)$ and c of m-bearing gametes contain q and *Q*, respectively. Because the gamete from the parental population is always *MQ*, the conditional probabilities of QTL genotypes given marker genotypes are

$$\Pr(QQ \mid MM) = \Pr(Qq \mid Mm) = 1 - c$$

$$\Pr(Qq \mid MM) = \Pr(QQ \mid Mm) = c$$

Hence, if z_j is the character value for individual j, the likelihood depends on the marker genotype,

$$\ell(z_j) = \begin{cases} (1-c) \cdot \varphi(z_j, \mu_{QQ}, \sigma^2) + c \cdot \varphi(z_j, \mu_{Qq}, \sigma^2), & \text{if marker} = MM \\ c \cdot \varphi(z_j, \mu_{QQ}, \sigma^2) + (1-c) \cdot \varphi(z_j, \mu_{Qq}, \sigma^2), & \text{if marker} = Mm \end{cases}$$

As in Chapter 16, we have assumed that phenotypes are normally distributed with potentially different means for each genotype but a common variance, with $\varphi(z, \mu, \sigma^2)$ denoting the density function for a normal distribution with mean μ and variance σ^2. The total likelihood for n measured backcross individuals is the product of individual likelihoods, $\prod \ell(z_j)$. The maximum-likelihood estimates of the four model parameters (μ_{QQ}, μ_{Qq}, c, σ^2) are obtained by maximizing the likelihood with respect to these variables, treating the observed z_j as fixed constants.

Hypothesis testing follows by standard likelihood-ratio tests (Chapters 16 and 18; Appendix 4). The appropriate test for the presence of a QTL linked to the marker compares the likelihood under the assumed full model with the likelihood under a model of no QTL, in which each individual character value is assumed to be drawn from the same normal with unknown mean μ and variance σ^2. The maximum of this restricted likelihood is given by Equation 16.8. The resulting likelihood-ratio test has $4 - 2 = 2$ degrees of freedom (four free parameters in the full model; μ and σ^2 in the restricted model). Knott and Haley (1992a) showed that this test is not biased by the presence of unlinked QTLs. It is, however, biased if one or more additional QTLs are linked to the marker under consideration.

An observed association between alleles at a polymorphic marker locus and the value of a quantitative trait can result either because of gametic-phase disequilibrium between the marker locus and a QTL or because the marker itself has a pleiotropic effect on the trait. A simple test for pleiotropy versus disequilibrium is to examine a marker-trait association over several generations of random mating. An association due to pleiotropy will not decay over time, while one due to linkage alone will. A caveat is that if linkage is very tight ($c \ll 0.5$), it can take many generations before any substantial decay is observed (Equation 5.16c). If one suspects that the marker itself is the QTL (which, of course, is highly unlikely), a variety of **candidate-locus** approaches (discussed below) can be used to test this hypothesis.

Example 17.5. Following crosses between inbred lines of barley, Powell et al. (1985a, 1985b) found that two loci with recessive dwarfing alleles (*ert* and *denso*) were associated with other quantitative traits: *ert* with reduced seed weight and *denso* with reduced height. By crossing different lines to produce an F_2, it was found that the *ert* locus accounted for roughly 84% of the additive genetic variation in seed weight after both one and three generations of recombination, suggesting that the association between *ert* and seed weight is either due to pleiotropy or very tight linkage. For example, the largest value of c that retains 90% of the originally association satisfies $(1-c)^3 = 0.90$, or $c = 1 - 0.90^{1/3} \simeq 0.03$. In contrast, the *denso* locus accounted for 58% of the additive genetic variance for height in lines undergoing a single round of recombination but only 35% in lines undergoing three rounds. Hence, for the *denso* locus, much of the initial association was due to gametic-phase disequilibrium.

Selective Genotyping and Progeny Testing

Historically, markers were more expensive to score than phenotypes. A strategy proposed to handle this setting when our interest is in a single trait is to first phenotype a number of individuals and then genotype only a *selected subset* of these. Known as **selective genotyping**, this strategy can result in a large increase in power for the simple reason that much of the linkage information resides in individuals with extreme phenotypes (Lebowitz et al. 1987; Lander and Botstein 1989; Carey and Williamson 1991; Darvasi and Soller 1992). However, while selective genotyping offers increased power to *detect* a QTL, it also produces biased estimates of their effects. Unbiased estimates can be obtained by maximum likelihood, either using a likelihood function that directly accounts for the sampling bias (Darvasi and Soller 1992; Muranty and Goffinet 1997; Xu and Vogl 2000) or by treating unscored genotypes as missing values (Lander and Botstein 1989; Johnson et al. 1999). The irony is that marker technology has now advanced to the point where *genotyping is often faster and cheaper than phenotyping*. Indeed, in many settings the bottleneck in marker-trait association studies is the gathering of highly accurate measurements on phenotypes.

This raises the important point that ***marker-trait associations are only as good as the quality of phenotyping***. Usually the reliability of marker-genotype scoring can be assessed with a number of quality control checks (such as departures from Hardy–Weinberg proportions). However, such checks are usually not available for phenotypes. Poor-quality phenotyping results, at a minimum, in a reduction of power. Hence, schemes that improve the precision of phenotyping (such as using good experimental design practices; Appendix 9) improve the power and precision of QTL mapping.

In those organisms where asexual propagation is possible, **replicated progeny** can be used to improve phenotyping precision (Cowen 1988; Simpson 1989; Lander and Botstein 1989; Soller and Beckmann 1990; Knapp and Bridges 1990). The idea here is to reduce the effects of environmental variance by asexually replicating each genotype, using the mean values of these replicated progeny in place of individual values. This is an efficient strategy in that if the heritability of the character is low, even scoring only a few replicated progeny can result in a significant increase in power. Soller and Beckmann (1990) showed that most of the increase in power occurs by measuring 10 or fewer replicated progeny. Besides offering an increase in power, progeny replication allows marker-trait associations to be examined across environments, providing a basis for estimating QTL $\times$ environment interactions (Part II Chapters 27 and 28). A variant approach is to use $F_{2:3}$ (or related) families in place of replicated progeny. Here, one genotypes an F_2 individual, which is then selfed, with the mean of that individual's selfed progeny used as their trait value.

Recombinant Inbred (RILs) and Doubled Haploid (DHLs) Lines

When asexual propagation is not possible, a related method to improve the precision of phenotyping is to use **recombinant inbred lines** (**RILs**). These can be constructed for any organism by taking an F_1 line through multiple rounds of selfing (e.g., Burr and Burr 1991) or multiple generations of brother–sister mating (e.g., Bailey 1981). The resulting lines have essentially no within-line genetic variance (ignoring new mutation and small amounts of residual heterozygosity), whereas the genetic variance among lines is considerable, as each RIL represents a different multilocus genotype. This was the approach used by Mackay and colleagues to study the difficult phenotype of lifespan in *Drosophila* (e.g., Nuzhdin et al. 1997; Ivanove et al. 2015). Longevity is a trait with considerable noise when based on a single measurement (that for a single individual), while the average life span for an RIL of a given genotype can be measured with considerable precision.

A related approach to RILs is the use of **doubled-haploid lines** (**DH** lines or **DHLs**). These are constructed using a variety of reproductive tricks that takes a cell with a single active haploid gamete and then doubles its chromosome number, instantly producing completely homozygous individuals. While DH technology was originally restricted to just a few plant systems (in particular, corn), it has now been widely expanded in plants, with variants of this technology now available for several hundred species (Maluszynski et al.

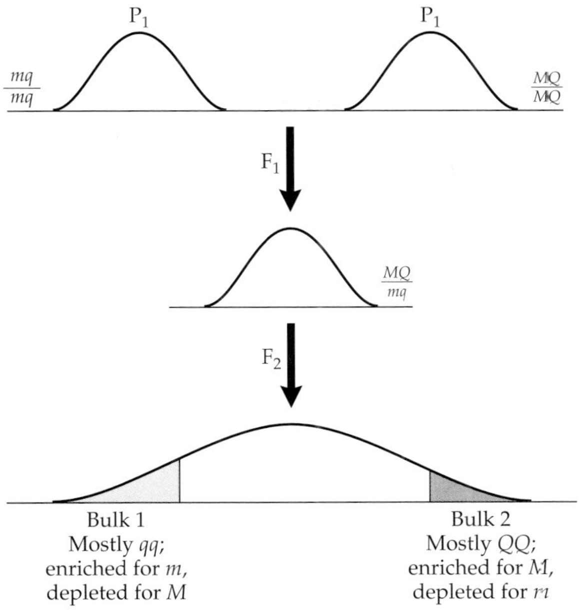

Figure 17.5 Bulked segregant analysis for mapping QTLs. Assume that a major QTL with alleles Q/q is tightly linked to a marker locus with alleles M/m. If we cross two inbred populations ($MMQQ$ and $mmqq$) and then sample the extreme tails of the F_2 phenotype distribution, the lower tail is enriched for qq, and the upper tail is enriched for QQ. Because the marker is closely linked to the QTL, the lower tail is enriched for marker allele m and depleted for M, while the opposite is true in the upper tail.

2003; Forster et al. 2007; Touraev et al. 2009). Following the pioneering work of Streisinger et al. (1981) in zebra fish, DH lines have also been constructed for a number of commercial fish species (Franěk et al. 2020; Hansen et al. 2020). Most cases in fish involve **gynogenetic DHLs**, produced by **gynogenesis** (inactivation of sperm DNA through irradiation or chemical treatment and doubling the haploid egg fertilized with such sperm). However, there are also **androgenetic DHLs**, produced by **androgenesis** (inactivation of the nuclear DNA in the haploid egg and then doubling the sperm DNA of the fertilized egg). Typically, the production of DHLs involves only a single round of recombination, while RILs created by selfing require around two effective rounds of recombination (accounting for the rapid reduction in heterozygotes), while RILs from brother–sister mating require around four effective rounds (Equation 18.5b). Hence, DHLs tend to have poorer mapping resolution than RILs.

Once the considerable work to generate and molecularly characterize a set of RILs or DHLs has been done, any character can be examined, using the previously established marker information to search for marker-trait associations. Only one laboratory needs to characterize the marker genotypes for the set of lines, while subsequent investigators (with these genotypes in hand) can look for marker-trait associations across the set of lines by simply phenotyping traits of interest. Like asexual lineages, RILs and DHLs offer a particularly easy approach for measuring QTL $\times$ environment interactions (Part II Chapters 27 and 28), as the same lines can be raised over different sets of environments. For these reasons, behavioral geneticists often use RILs for mapping behavioral QTLs in mice (e.g., Plomin et al. 1991; Belknap et al. 1993; Crabbe et al. 1994; Flint 2003).

Bulked Segregant Analysis (BSA)

A variant of selective genotyping is **bulked segregant analysis** or **BSA** (Arnheim et al. 1985; Michelmore et al. 1991; Zou et al. 2016). Here individuals are combined (**bulked** or **pooled**) into groups based on trait value (Figure 17.5), with the marker frequencies in each

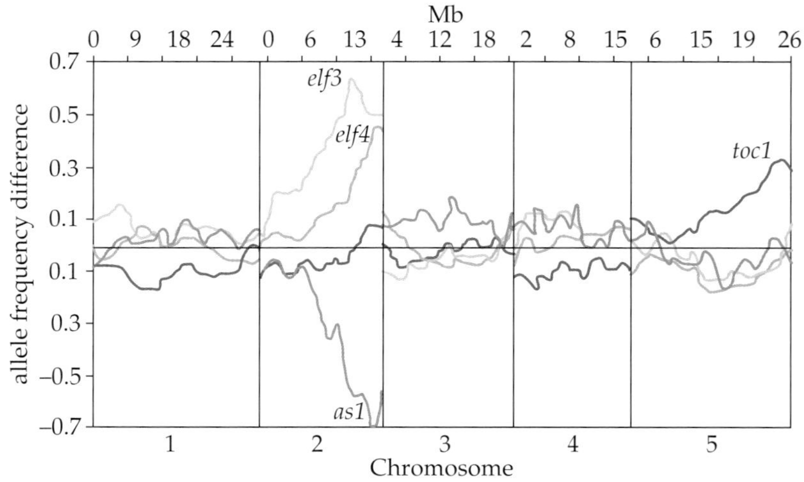

Figure 17.6 Using BSA to map mutations impacting circadian clock and developmental traits in F_2 bulks of *Arabidopsis*. Spline-smoothed curves are plotted of allele-frequency differences between the upper and lower bulks based on hybridization strengths on an Affymetrix gene chip. Four different bulked pairs (represented by the four separate curves) were used to detect mutations in the *early flowering 3* (*efl3*), *early flowering 4* (*efl4*), *time of CAB expression 1* (*toca*), and *asymmetric leaves 1* (*as1*) genes that impacted the bulked trait. (After Hazen et al. 2005.)

bulk then estimated and contrasted. Marker alleles in LD with QTL alleles are expected to have a nonrandom distribution across bulks, in the extreme having a particular marker allele present only in one bulk and the alternative allele present only in the other bulk (Hill 1998b). Unlinked markers (and marker alleles in linkage equilibrium with QTLs) are expected to be randomly distributed across bulks. Figure 17.6 shows an example of this approach, mapping four major genes with a resolution of less than 2 Mb using bulks formed from F_2 populations in *Arabidopsis*.

Early attempts to quantify allele-frequency differences between pools used indirect approaches such as optical density of marker bands (Pacek et al. 1993; Khatib et al 1994; Darvasi and Soller 1994a). The extension of this approach in the genomics era uses quantification via differential hybridization based on either transcriptomics (Borevitz et al. 2003; Brauer et al 2006; Pandit et al 2010) or SNP chips (Hazen et al. 2005; Wenger et al. 2010; Becker et al. 2011). The application of array scoring (RNA or SNPs) in BSA has been called **eXtreme array mapping, XAM** (Woyln et al. 2004). A more direct approach is direct counting of either the number of allelic sequences or the number of transcripts in each pool via very high coverage **next-generation sequencing, NGS,** of the pool (Magwene et al. 2011; Park et al. 2014). NGS data generate a series of counts for each pool, generating a 2×2 contingency table (with rows corresponding to the two alternate marker alleles and columns to the two pools). Magwene et al. (2011) proposed a modification of the G goodness-of-fit statistic (Appendix 4) to test for significance with such NGS data. A power analysis by these authors found that the optimal design was to use rather modest thresholds (around 20%) for bulking phenotypes, as while allele frequencies would be more differentiated using more extreme thresholds, this is countered by fewer members in the pools, compromising power.

Because bulked segregant analysis requires significant enrichment of the alternative QTL alleles in the tails of the trait distribution, it is expected to have poor power for detecting QTLs of small to modest effect (Figure 17.7; Example 17.8). It also requires using either derivatives from a single inbred-line cross (such as F_2s or RILs) or progeny from very large segregating families, as the method fails if groups with the marker and QTL alleles in different linkage phases are lumped. In spite of these limitations, BSA is so straightforward, allowing rapid screening of a very large number of markers, that it is reasonable to

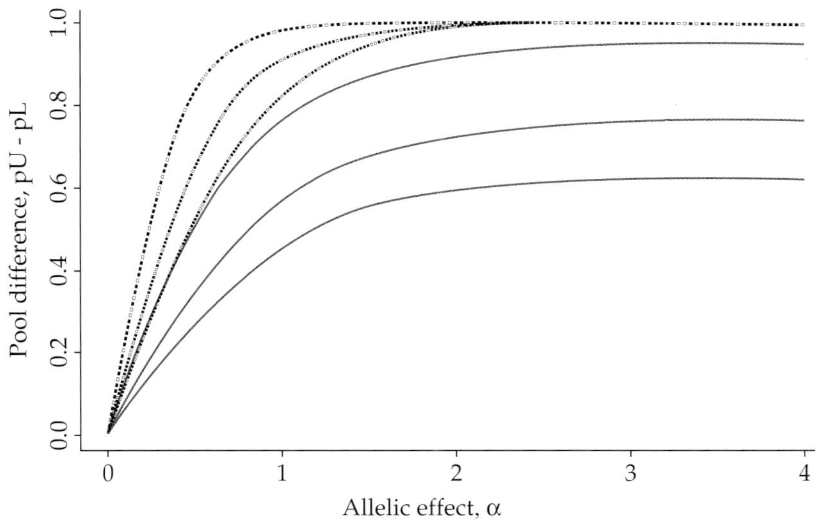

Figure 17.7 The expected QTL allele-frequency difference between the upper and lower bulks of a BSA as a function of the effect (α, in standard deviations) of an additive QTL and the bulk thresholds (δ, which loosely corresponds to the number of standard deviations above and below the mean). The upper three (dotted) lines are the values for RILs and (from upper to lower) corresponds to δ values of 2, 1, and 0.5, while the lower three (solid) lines correspond to an F_2-based BSA with the same ordering for these three δ values. Details in Example 17.8.

consider it in many cases. As shown in Figure 17.7, this approach is most powerful when using RILs. This is due to RIL pools only consisting of QTL homozygotes, while heterozyogtes are present when F_2s are bulked. A further advantage of using RILs is that the replication of these lines provides very precise estimates of their genotypic values (Venuprasad et al. 2009; Vikram et al. 2012). One constraint on using BSA is that a bulk is *trait specific*, so that if one is attempting to map QTLs for multiple traits, independent bulks for each trait must be formed.

Finally, we note that BSA can be used to find *additional markers* linked to a particular region (Giovannoni et al. 1991). Sorting individuals from a cross into separate pools based on alternate alleles at a marker enriches those pools for additional linked markers. Thus, bulking on a marker of interest provides for very efficient screening of a large number of markers. Such an increase in marker density is required for finer mapping of QTL position.

Example 17.6. Yaghoobi et al. (1995) used bulked segregant analysis to map a major gene for root-knot nematode resistance in tomatoes. Forty-eight individuals from an F_1 (resistant $\times$ susceptible) back crossed to the susceptible parent were scored, 25 of which were susceptible, and 23 resistant. This 1:1 segregation ratio is consistent with the hypothesis that a single dominant gene underlies resistance. To map this gene, two bulks of DNA, based on six resistant and six susceptible backcross plants were formed. Each bulk was screened for RAPDs (Chapter 8) using 520 different 10-base primers. Each primer gave on average about eight bands, resulting in 4,160 bands being scored in each pool, about 3% of which varied significantly between pools. These significant primers were then used to probe each of the original 48 backcross plants. One marker was present in 20 of 23 resistant plants and none (out of 25) of the susceptible plants, suggesting linkage to a major resistance gene.

Example 17.7. As a test of BSA, Mansur et al. (1993a) examined four traits—maturity, plant height, lodging (a measure of plant structure), and seed yield—in 284 recombinant inbred lines generated by a cross between two soybean (*Glycine max*) cultivars. The four chosen traits were measured in each of the 284 RILs grown in two distinct environments (Minnesota and Chile) and the 20 highest and lowest performing lines, averaged over both environments, were selected for each trait. DNA was extracted from each of the extreme lines and bulked into a high and low sample for each of the traits. The resulting DNA was tested using radioactive probes for RFLP markers, and the amount of hybridization to each probe was quantified using a phosphoimager. The authors had previously used RFLPs to map a number of QTLs for these and several other traits using maximum-likelihood interval mapping (Chapter 18) in F_2 families. The previous RFLP marker–QTL associations were confirmed, and one marker that showed marginal linkage to maturity and height under interval mapping showed very strong linkage to these traits, as well as to lodging and yield.

Example 17.8. What is the expected QTL allele-frequency difference between the two pools in a BSA? Consider a trait, z, whose mean has been centered at zero. Assume an additive QTL, whose phenotypic distributions for the trait value, z, are normal, with

$$z \mid QQ \sim \mathrm{N}(\alpha, 1), \qquad z \mid Qq \sim \mathrm{N}(0, 1), \qquad z \mid qq \sim \mathrm{N}(-\alpha, 1)$$

Recalling Equation 15.3, the resulting phenotypic distribution for z is a mixture, with a variance of $1 + \sigma^2(\mu_G)$. If (as it often the case), the BSA population is the F_2 between two inbred lines, then freq(Q) = 1/2 and $\sigma^2(\mu_G) = 2(1/4)\alpha^2$, for a trait standard deviation of $\sigma_z = \sqrt{1 + \alpha^2/2} \simeq 1$ for $\alpha < 1$. For RILs, $\sigma^2(\mu_G) = 2(1/2)\alpha^2 = \alpha^2$.

Suppose we contrast the frequency of Q between an upper bulk, U, that consists of all measured individuals whose trait values are at least δ above the mean ($z \geq \delta$), and a lower bulk, L, of individuals whose trait values are less than δ below the mean ($z \leq -\delta$). The expected frequency of Q in the upper bulk is

$$p_U = \Pr(QQ \mid z \geq \delta) + (1/2)\Pr(Qq \mid z \geq \delta)$$

while its expected frequency in the lower bulk is

$$p_L = \Pr(QQ \mid z \leq -\delta) + (1/2)\Pr(Qq \mid z \leq -\delta)$$

As shown in Example 3.2, one can use Bayes theorem to obtain expressions for these various probabilities for QTL genotypes conditioned on trait value. Consider QQ first. Applying Bayes theorem (Equation 3.3b),

$$\Pr(QQ \mid z \geq \delta) = \frac{\Pr(z_{QQ} \geq \delta)\Pr(QQ)}{\Pr(z_{QQ} \geq \delta)\Pr(QQ) + \Pr(z_{Qq} \geq \delta)\Pr(Qq) + \Pr(z_{qq} \geq \delta)\Pr(qq)}$$

Letting $u \sim N(0,1)$ denote a unit normal, note that $z_{QQ} = u + \alpha$, $z_{Qq} = u$, and $z_{qq} = u - \alpha$. The above expression (for an F_2 individual) reduces to

$$\frac{\Pr(u + \alpha \geq \delta)}{\Pr(u + \alpha \geq \delta) + 2\Pr(u \geq \delta) + \Pr(u - \alpha \geq \delta)}$$

$$= \frac{\Pr(u \geq \delta - \alpha)}{\Pr(u \geq \delta - \alpha) + 2\Pr(u \geq \delta) + \Pr(u \geq \delta + \alpha)}$$

Using similar logic to obtain $\Pr(Qq \mid z \geq \delta)$, p_u becomes

$$p_U = \frac{\Pr(u \geq \delta - \alpha) + (1/2)2\Pr(u \geq \delta)}{\Pr(u \geq \delta - \alpha) + 2\Pr(u \geq \delta) + \Pr(u \geq \delta + \alpha)} \tag{17.8a}$$

Similarly,

$$p_L = \frac{\Pr(u \leq -\delta - \alpha) + \Pr(u \leq -\delta)}{\Pr(u \leq -\delta - \alpha) + 2\Pr(u \leq -\delta) + \Pr(u \leq -\delta + \alpha)} \tag{17.8b}$$

For BSA based on RILs from an F_2, the heterozygote terms, $\Pr(u \leq -\delta)$ and $\Pr(u \geq \delta)$, are absent, resulting in higher power (Figure 17.7), with

$$p_U - p_L = \frac{\Pr(u \geq \delta - \alpha)}{\Pr(u \geq \delta - \alpha) + \Pr(u \geq \delta + \alpha)} + \frac{\Pr(u \leq -\delta - \alpha)}{\Pr(u \leq -\delta - \alpha) + \Pr[u \leq -\delta + \alpha)} \tag{17.8c}$$

These expressions can be easily computed using the standard normal cumulative probability function found in any statistical package. For example, if we contrast the upper and lower 20% of the population (which for $\sigma_z^2 \simeq 1$ corresponds to δ = 0.84), the between-bulk difference in allele frequencies, $p_U - p_L$, becomes 0.070, 0.323, 0.532, and 0.683 for α = 0.1, 0.5, 1, and 2, respectively for an F_2 bulk, increasing to values of 0.139, 0.606, 0.890, and 0.995 for a BSA using RILs. Figure 17.7 plots this between-bulk difference for various values of α and δ for both F_2 and RIL designs. A final adjustment is that the above calculations assumed a focus on the frequency of the causative allele, Q, while in fact we are usually following a marker allele, M, linked to the QTL at recombination fraction c. The expected allele-frequency change for the marker follows with α replaced by $\alpha(1 - 2\widetilde{c})$, where

$$\widetilde{c} = \begin{cases} c & \text{for DHLs, F}_2\text{s} \\ 2c/(1+2c) & \text{for RILs formed by selfing} \\ 4c/(1+6c) & \text{for RILs formed by brother–sister mating} \end{cases} \tag{17.8d}$$

QTL Mapping by Marker Allele-frequency Changes in Populations under Selection

Selective genotyping amounts to a single-generation selection experiment, as individuals are chosen from the tails of the trait distribution. As originally suggested by Stuber and Moll (1972) and Stuber et al. (1980), this mapping approach can be generalized by looking for shifts in marker allele frequencies following several generations of selection on a focal trait. Marker loci initially in LD to a QTL for that trait will increase (decrease) in frequency as linked QTL alleles increase (decrease) due to selection. Stuber and colleagues searched for shifts in allozyme frequencies in lines of maize selected for yield. A slightly more powerful design is to divergently select a base population for several generations and then test for significant changes in marker allele frequencies between the up- and down-selected lines (Nuzhdin and Pasyukova 1991; Keightley and Bulfield 1993; Nuzhdin et al. 1993; Ollivier et al. 1997; Parts et al. 2011).

Long-term selection creates two somewhat opposing forces for mapping. While each generation of selection increases the difference in QTL frequencies between divergently selected populations, recombination is expected to decrease the amount of LD between markers and QTLs. Thus, unless the markers and QTLs are tightly linked, fairly rapid QTL allele-frequency changes (i.e., QTLs of modest to large effects) are required for a linked marker to show significant frequency changes (Lebowitz et al. 1987). Keightley and Bulfield (1993) showed that the marker and QTL must be closer than 20 cM for significant marker allele-frequency changes to be likely. One can use selective-sweep theory (WL Chapter 8) to obtain more exact values.

Given the small population sizes of most selection experiments, drift alone is expected to change allele frequencies, motivating the need for tests as to whether observed changes exceed those expected by chance events alone (reviewed in WL Chapter 9). The selection coefficient s on a QTL, which determines how quickly allele frequencies change (WL Equation 5.3), is a function of the QTL effect and the amount of selection on the character (WL Equation 5.21). Thus, s can provide information about the size of QTL effects. A maximum-likelihood approach for estimating s was suggested by Keightley and Bulfield (1993) and Keightley et al. (1996). Here, computer simulations generate the distribution of allele-frequency change between two populations under divergent selection (assuming selection coefficient s) and drift (using an estimate of the effective population size). This procedure is repeated over a range of possible s values, and the value giving the highest probability of the observed change is taken as the ML estimator of s.

Marker alleles associated with body size have been found in several studies of divergently selected lines of mice (Garnett and Falconer 1975; Simpson et al. 1982; Gray and Tait 1993; Keightley and Bulfield 1993; Keightley et al. 1996). For example, after 22 generations of selection, Keightley and Bulfield (1993) found significant changes at four of 16 assorted marker loci (two coat color, two allozyme, 12 proviral inserts). This same set of crosses was further examined using 124 microsatellite loci by Keightley et al. (1996), who found 11 QTLs with estimated effects ranging from 0.17 to 0.28 phenotypic standard deviations. In their pioneering study in maize, Stuber at al. (1980) found that 8 of 20 allozyme markers showed significant differences for yield in divergently selected lines starting from F_2 crosses between inbred lines. To test whether these associations might simply be due to chance, Stuber et al. (1982) independently selected on the markers alone, finding a response for yield in the direction predicted by the original marker-trait associations. Other examples using allele-frequency change to map QTLs in maize include studies by Labate et al. (1999) and Coque and Gallais (2006) for yield, and Wisser et al. (2008) for disease resistance, while De Koeyer et al. (2001) examined yield in oats. Finally, an important special case of using allele-frequency change to map QTLs involves the search for longevity genes (Christensen et al. 2006; Murabito et al. 2012). Here, one contrasts marker–allele frequencies between a control group and a group showing extreme longevity (such as a cohort of humans over 90 years old).

Example 17.9. Nuzhdin et al. (1993) used allele-frequency changes to detect markers associated with fitness in a cross of high and low fitness lines of *Drosophila melanogaster*. These authors used the presence/absence of the mobile genetic elements *mdg1* and *copia* as markers, finding 19 locations on chromosome 2 where the high and low lines differed in the presence/absence of inserted elements. An F_1 was backcrossed to the low line for three generations to generate a base population with high-line alleles at expected frequency $1/32$. This base population was then allowed to reproduce, with marker frequencies sampled after 11, 13, and 17 generations of natural selection. The frequencies of nine high-line markers (all located in a region around the centromere of chromosome 2) showed significantly higher values than expected by drift, suggesting linkage to one (or more) QTLs increasing fitness. The associated selection coefficients estimated by the ML approach of Keightley and Bulfield ranged from 0.3 to 0.7.

MARKER-BASED ANALYSIS USING INTROGRESSION LINES

NILs, HIFs, and STAIRS

The use of **introgression lines, ILs** (Eshed and Zamir 1995), has been proposed for both detection and fine mapping of QTLs. These are the extension of chromosome assays, but instead of whole chromosomes being replaced in an otherwise standardized background, much smaller chromosomal segments are substituted (from a **donor line**) into an otherwise fully homozygous background. Figure 17.8 shows the differences between two common IL population structures, NILs and HIFs. These differ in the uniformity of the homozygous background, either being entirely from one parent (NILs) or a roughly equal mixture from both parents (HIFs).

Nearly isogenic (also **near-isogenic**) **lines, NILs**, are inbred lines containing one or more small regions of DNA from a **donor parent** in an otherwise standard background that is the same over all lines (Muehlbauer et al. 1988; Kaeppler et al. 1993; Monforte and Tanksley 2000). These are also referred to as **congenic strains** (Loosli et al. 1961) or **backcross inbred lines, BILs** (Wehrhahn and Allard 1965; Jeuken and Lindhout 2004; Kooke et al. 2012). NILs are constructed by first crossing a donor parent to an inbred line (the **recurrent parent**) to form an F_1. The resulting offspring are then backcrossed to the recurrent parent for several generations to generate a background consisting of almost entirely recurrent DNA. In species where selfing is allowed, the NIL are formed by at least one generation of

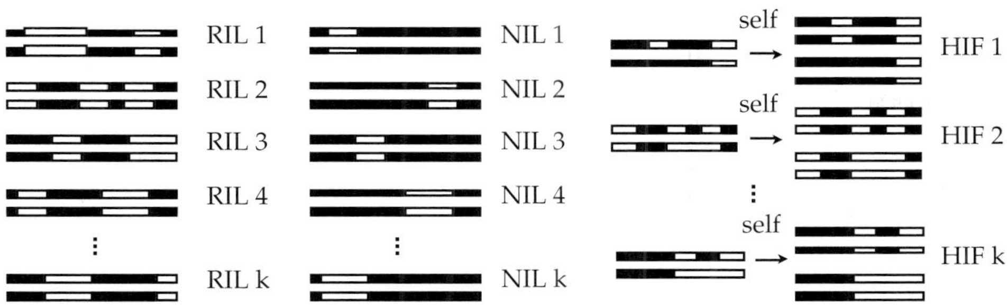

Figure 17.8 Different inbred-line population structures for mapping (RILs) and fine mapping (NILs, HIFs) QTLs. Solid and clear segments represent, respectively, segments of recurrent and donor DNA. **Left:** RILs are a collection of inbred lines that contain roughly equal amounts of random segments from the two parental lines. A contrast of any two RILs, or a RIL with either parental line, compares the cumulative impact from all regions at which the two lines differ. **Center:** NILs provide a much cleaner approach for fine mapping, as comparing an NIL against the recurrent parent contrasts only a single focal region. **Right:** Heterogeneous inbred families (HIFs) represent an intermediate strategy between RILs and NILs. Lines that are not yet fully inbred will still be segregating at a few loci (left side of the panel), with these regions becoming differentially fixed in the resulting family of inbred offspring from that individual (right side of the panel). In such a family, the backgrounds are identical except for the differential fixation of the segregating regions. Hence, within-family comparisons (e.g., two lines from HIF 1) results in an NIL-like contrast of the impact of the differentially fixed region. The key difference between NILs and HIFs is that a region of interest is always contrasted in the same background in NILs (the recurrent parent), while the background for the comparison of HIFs is a roughly equal mixture from both parents. Comparison of a region over a series of different HIFs provides a better measure of epistasis (assayed over multiple genetic backgrounds) than in the NIL setting (assayed over a single background).

selfing following the backcrossing. Where selfing is not possible, after the backcrossing is complete, individuals are repeatedly sib-mated to form the final inbred line.

NILs differ subtly, but significantly, from RILs, see Figure 17.8. While both are inbred and usually start from F_1 parents, RILs are generated by repeated inbreeding of the descendant lines and hence contain about 50% donor DNA, as opposed to the very small fraction of donor DNA expected in NILs due to repeated backcrossing to the recurrent parent. In particular, the expected proportion, $p(b)$, of the donor genome left in an NIL following b generations of backcrossing and a final generation of selfing is

$$p(b) = (1/2)^{b+1} \tag{17.9}$$

Hill (1993b) gives expressions for the variance of $p(b)$, which depends on the number of chromosomes and their map distances. Typically, five to seven generations of backcrossing are performed, giving an expected proportion of donor genome of 1.2% to 0.5%. This repeated backcrossing scheme is the basis for the Wehrhahn–Allard estimator of gene number, discussed in Chapter 11.

Two different approaches of NIL formation have been used to coarse-map and fine-map QTLs: **random NILs** (backcrossing without selection) and **targeted NILs** (selection for either traits or specific markers during backcrossing). Markers greatly facilitate both approaches: (i) by characterizing the size and location of introgressed regions under either random or trait-selected NILs, and (ii) by marker-assisted selection for marker-targeted NILs. In the formation of random NILs, one simply backcrosses to the recurrent parent without any selection. The result is a collection of lines with random introgressed regions from the donor parent in an otherwise recurrent background (Figure 17.8). Because the introgressed segments represent a random sample of the donor genome, one can use a set of random NILs to map QTLs for essentially any trait (Examples 17.10 and 17.11).

The search for QTL regions typically proceeds by testing whether the trait value of a particular NIL is statistically different from that of the recurrent parent. If yes, then one or more QTLs for the trait are located in the regions where the NIL differs from the recurrent parent. The problem of jointly testing multiple NILs is often controlled for by using **Dunnett's test** (Dunnett 1955, 1964) for multiple comparisons against a single control (the recurrent parent). Other methods of analysis of NILs were explored by Falke and Frisch (2011) and Mahone et al. (2012), who found that a linear model approach (treating each introgressed segment as an effect to be estimated), combined with Holm's extension of the Bonferroni correction for multiple testing (Appendix 6), was generally more powerful than Dunnett's test.

While both RILs and random NILs can be used as resource populations for QTL mapping, they represent slightly different strategies. The relative efficiency of these two approaches has been examined (both theoretically and empirically) by several authors (Kaeppler 1997; Keurentjes et al. 2007; Szalma et al. 2007). Generally speaking, RILs have greater power of QTL detection, while NILs offer more precise estimates and the ability to detect QTLs of smaller effect. When epistasis is widespread, RILs offer the advantage over NILs in that a genomic region is tested over several backgrounds versus over a single background in NILs. While using NILs reduces the background genetic noise and thus increases power, it can also provide a misleading picture of the genetic architecture if variation in the genetic background has a significant effect on the target region.

With targeted NILs, one backcrosses to the recurrent parent, but selects for either a specific trait (Example 17.12) or a particular marker (next section), such as that around a segment suspected of containing one, or more, QTLs. While targeted NILs allow for fine mapping, unlike randomly formed lines, these lines are not that useful for mapping QTLs for arbitrary traits, because the introgressed segments do not represent a random sample of donor genome.

Example 17.10. Eshed and Zamir (1995) formed 50 NILs from a cross between the cultivated tomato (*Lycopersicon esculentum*) and a wild relative (*L. pennellii*). Based on analysis with 375 markers, each line contained a single RFLP-defined fragment from *L. pennellii*, averaging around 33 cM (or roughly 3% donor DNA). Comparing the mean value of each NIL against a standard, the authors found 23 QTLs for fruit soluble-solids content and 18 QTLs for fruit mass. By comparison, in an analysis of marker-trait associations in RILs, Goldman et al. (1995) found 7 and 13 QTLs, respectively. Eshed and Zamir subjected two regions to much finer scale mapping, using markers to select for recombinants within these regions. Upon finer analysis, a 55 cM region influencing fruit size was shown to contain at least three separate QTLs. A 37 cM region showing heterosis for soluble-solids yield was shown to result from associative overdominance (Chapters 12 and 13). This latter region was subdivided by recombination into a partially dominant QTL that increased yield and a linked recessive that reduced yield.

Example 17.11. Obando et al. (2008) formed a set of 97 melon (*Cucumis melo*) NILs from a cross between a Spanish (recurrent parent) and a Korean (donor parent) accession. Microsatellite markers were used to characterize the size and locations of the Korean introgressed segments, with the set of random NILs covering roughly 85% of the Korean genome, with an average introgression size of around 41 cM (roughly 3% of the Korean genome). Ten replicates of each NIL, along with 50 replicates of the Spanish line and five replicates of the Korean line, were grown in a completely randomized design (Appendix 9). Roughly thirty morphological and fruit-related traits were scored, and the mean value for each trait for a given NIL replicate was compared against the mean for the control Spanish line using Dunnett's test. Based on this approach, a total of 134 QTLs were detected, 52 for morphological traits, 69 for fruit color traits, and 23 for flavor traits. Every introgressed region showed a QTL for at least one trait, with most having two or more QTLs (an average of 3.5 QTL per genomic region, i.e., each region impacted, on average, 3.5 of the scored traits).

Example 17.12. Targeted NILs are constructed by a select-and-backcross procedure, and hence are not random with regard to the retained donor DNA (Wright 1952; Hill 1998a). "Exotic" sorghum is tall and has a short daylength requirement for flowering, while commercial sorghum cultivars are short (for easier harvesting) and daylength insensitive. Exotic strains are "converted" to commercial cultivars by first crossing to a standard short, daylength-neutral donor parent and continually backcrossing each generation to the exotic parent while selecting for short height and daylength neutrality. The resulting converted strains are expected to contain mostly exotic DNA, but are daylength neutral and short, due to the retention (by selection) of donor QTL alleles influencing these traits. Hence, one approach to mapping QTLs for these traits is to search for regions that are retained in the converted derivatives.

This approach was used as a check of conventional QTL mapping by Lin et al. (1995), who examined 71 markers in nine exotic accessions of *Sorghum bicolor* and their converted derivatives. The authors compared the positions of retained donor regions in these accessions with the positions of QTLs for height and daylength sensitivity mapped in a previous experiment using the F_2 progeny in a cross of *S. bicolor* × *S. propinquum*. Seven of the nine QTLs (six for height, three for daylength) detected using the F_2 cross coincided with regions of donor parent DNA retained in the converted strains despite several generations of backcrossing. One donor region was seen in all nine derivatives, and in the F_2 crosses this region accounted for 55% of the variation in height and 86% of the variation in daylength sensitivity. In contrast, three converted regions (found in at least one line) did not coincide with mapped QTLs. Whether these regions indeed contain QTLs for height or daylength could be examined by crossing the converted strains with their parental exotics and looking for associations between the retained donor markers and trait values.

A second approach for generating ILs is through the use of **heterogeneous inbred families**, or **HIFs** (Haley et al. 1994a; Tuinstra et al. 1997). As shown in Figure 17.8, this approach exploits the fact that while a line undergoing inbreeding is isogenic for much of its genome (having become homozygous), there are regions where it is still segregating alternative segments (residual heterozygosity). Some of the selfed offspring from such parents will be fixed for different segments, generating NILs. For example, after five generations of selfing, roughly 94% of the loci in a line are fixed as homozygotes, while 6% are heterozygotes, resulting in contrasting members of an HIF family being differentially fixed at these segregating regions. The major difference between NILs and HIFs is the nature of the background homozygous genome, which is entirely from the recurrent parent in NILs, but roughly half recurrent and half donor in HIFs (Figure 17.8). The strength of an NIL is that any additional noise from background variation is removed, increasing power. The weakness of NILs is that if epistasis is present, the donor and recurrent regions are only contrasted in the recurrent background, while under HIFs, the contrast is across a variety of different backgrounds.

A final structured inbred population that is occasionally used for QTL fine mapping involves **stepped aligned inbred recombinant strains**, or **STAIRS** (Koumproglou et al. 2002; Perera et al. 2006). This approach represents an intermediate strategy between chromosomal substitution lines and NILs. Here, one generates a series of single recombinants for each chromosome, with donor DNA distant, and recurrent DNA proximal, to the recombination breakpoint. This creates a nested series for each chromosomes that contain successively longer tracts of donor population DNA. Koumproglou et al. generated a series of such lines in *Arabidopis* that allowed for mapping to regions of around 0.5cM by a simple comparison between a few STAIRS lines for the chromosome of interest. As with (random) NILs and HIFs, the construction of a collection of STAIRS requires a considerable amount of time and effort, but the resulting resource makes a powerful tool for mapping any trait showing variation between the chosen recurrent and donor lines.

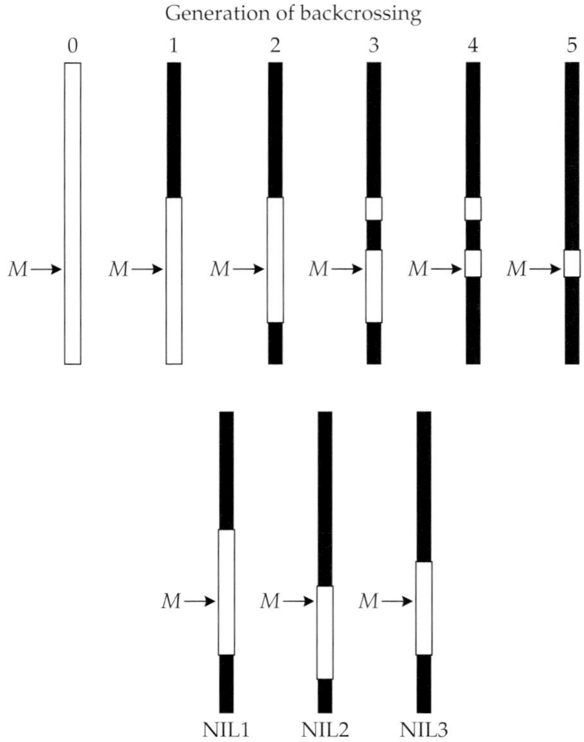

Figure 17.9 **Top:** Construction of a nearly isogenic line (NIL) containing a region of interest, obtained by selecting on a single marker *M* while backcrossing to a standard stock (the recurrent parent). Because the final lines are constructed by selfing, we follow a single (haploid) chromosome, as the NIL has two identical copies of this chromosome. Clear regions are from the donor parent, solid from the recurrent parent. **Bottom:** If several lines are independently selected for the marker, the resulting NILs will contain different retained regions around *M*. With a dense marker map, crosses between such lines allow for finer mapping of the region around *M*.

Marker-based Introgressions

The formation of targeted NILs requires constant backcrossing to the recurrent parent coupled with selection for donor DNA. Historically, selection was phenotypic, retaining backcrossed individuals showing the trait of interest such as disease resistance or a small set of desirable characters (e.g., Example 17.12). In the marker era, one can instead focus on creating NILs around a region of interest that is anchored by one (or more) markers. Such **marker-assisted backcrossing** allows the introgression of specific regions (such as one containing previously detected QTLs) into an otherwise isogenic background (e.g., Young et al. 1988; Tanksley et al. 1989; Paterson et al. 1991; Dudley 1993). Independent NILs formed while selecting on the same marker are expected to differ in the regions retained around the marker (Figure 17.9), and crossing such NILs allows for finer scale mapping. This approach has been used to fine map QTLs (mainly for genes of moderate to large effects, often resistance genes) in maize (Bentolila et al. 1991; Koester et al. 1993; Touzet et al. 1995; Liu et al. 2012), rice (Yu et al. 1991; Ronald et al. 1992; Huang et al. 2013), lettuce (Paran et al. 1991; Den Boer et al. 2013), wheat (Hartl et al. 1993; Schachermayr et al. 1994, 1995; Zheng et al. 2015), barley (Schüller et al. 1992; Hinze et al. 1991; Vu et al. 2010), oats (Penner et al. 1993), soybeans (Kim et al. 2010), and tomatoes (Osborn et al. 1987; Tanksley and Hewitt 1988; Sarfatti et al. 1989; Paterson et al. 1990; Kinkade and Foolad 2013).

As noted by Hospital et al. (1992), marker-assisted selection during the creation of introgression lines has two goals: (i) **foreground selection** to retain a desired donor segment in the face of continual backcrossing to the recurrent line, and (ii) **background selection** to accelerate the removal of donor DNA from the rest of the genome. As the available marker density increases, foreground selection progresses from using single to flanking markers, while the availability of markers outside of the donor region facilitated background selection throughout the rest of the genome. The combination of both marker-assisted foreground and background selection to construct NILs has been called **speed congenics** (Markel et al. 1997; Wakeland et al. 1997; Visscher 1999; Wong 2002).

We start by considering foreground selection using only a single marker in the donor region of interest. Selecting for the retention of a donor marker while backcrossing to the recurrent parent yields an NIL with a block of donor DNA surrounding the marker that can be rather large. Linked undesirable genes can be dragged along with the marker, a phenomenon referred to as **linkage drag** by plant breeders (Brinkman and Frey 1977), and as **Hill–Roberston effects** by evolutionary biologists (Hill and Robertson 1966; WL Chapter 8). Conversely, a single marker may remain associated with several favorable linked QTLs within the introgressed region. The amount of donor DNA present following marker-assisted foreground selection has been examined by several authors (Bartlett and Haldane 1935; Hanson 1959a–1959d; Stam and Zeven 1981; Naveira and Barbadilla 1992; Frisch and Melchinger 2001). The donor DNA remaining in the NILs can either reside on the chromosome under marker selection or on the other (unmarked) chromosomes. We consider these two sources separately.

Suppose there are n chromosomes and a total genome length of $\mathcal{L} = \ell_M + \sum \ell_i$, with ℓ_M being the length of the marker chromosome and ℓ_i being the length of the ith unmarked chromosome (all lengths in Morgans). Consider the DNA retained around the marker, and suppose that the selected marker is in the center of the chromosome. Letting $t = b + 1$, the proportion $p_M(b)$ of donor DNA on this marker chromosome after b generations of backcrossing and a final generation of selfing has expected value

$$E[\,p_M(b)\,] = 2\left(\frac{1 - e^{-t\ell_M/2}}{t\ell_M}\right) \tag{17.10a}$$

(Hanson 1959b), while the variance is

$$\sigma^2[\,p_M(b)\,] = \frac{2}{(t\ell_M)^2}\left[1 - \left(t\ell_M + e^{-t\ell_M/2}\right)e^{-t\ell_M/2}\right] \tag{17.10b}$$

(Naveira and Barbadilla 1992). For $t\ell_M \gg 1$, these are approximately

$$E[\,p_M(b)\,] \simeq \frac{2}{t\ell_M}, \qquad \sigma^2[\,p_M(b)\,] \simeq \frac{2}{(t\ell_M)^2} \tag{17.10c}$$

Because these equations refer to the proportion of this chromosome retained, the corresponding chromosome lengths retained are given by $\ell_M \cdot p_M(b)$, yielding a mean length $\pm$ SD (in Morgans) of $2/t \pm \sqrt{2}/t$ for $t\,\ell_M \gg 1$.

The restriction of the marker being in the center of the chromosome was removed by Stam and Zeven (1981), who showed that using Equations 17.10a–17.10c for a randomly placed marker generally introduces only a small error. Of broader concern is that these equations consider only the *segment around the marker*, while the marked chromosome can also retain blocks of donor DNA elsewhere. Stam and Zeven show that these two effects (noncentrality of the marker and blocks of donor DNA retained outside of the marker) largely cancel, leaving Hanson's original result as a very good approximation.

Turning now to the untagged chromosomes, let $p_{U_i}(b)$ denote the proportion of donor DNA on an untagged chromosome of length ℓ_i. The mean and variance are given by

$$E[\,p_{U_i}(b)\,] = \frac{1}{2^t} \tag{17.11a}$$

$$\sigma^2[\,p_{U_i}(b)\,] = 2\left(\frac{1 - e^{-t\ell_i/2}}{t\,\ell_i\,2^t}\right) - \frac{1}{2^{2t}} \tag{17.11b}$$

as shown by Stam and Zeven (1981). The expected total length of donor DNA retained over all untagged chromosomes thus becomes

$$\sum_{i=1}^{n-1} \ell_i \, E[\, p_{U_i}(b)\,] = \frac{1}{2^t} \sum_{i=1}^{n-1} \ell_i = \frac{\mathcal{L} - \ell_M}{2^t} \tag{17.12a}$$

with variance

$$\sum_{i=1}^{n-1} \ell_i^2 \cdot \sigma^2[\, p_{U_i}(b)\,] \tag{17.12b}$$

Example 17.13. Suppose there are $n = 15$ chromosomes, each of length 75 cM, and that we select for retention of a marker located in the middle of chromosome 1 during five generations of backcrossing followed by a generation of selfing. Applying Equation 17.10a, the expected fraction of donor DNA on chromosome 1 is

$$E[\, p_M(5)\,] = 2\left(\frac{1 - e^{-6\,(0.75/2)}}{0.75 \cdot 6}\right) \simeq 0.40$$

or, in units of chromosome length, $0.4 \cdot 75$ cM = 30 cM. Equation 17.10b gives the standard deviation of the expected proportion retained as 0.23, or 17.25 cM. Likewise, applying Equations 17.11a and 17.11b, the proportion of donor DNA on any untagged chromosome is 0.016 $\pm$ 0.077, or 1.2 $\pm$ 5.8 cM. The fraction of the total expected donor DNA in the NIL contributed by the marked chromosome is $30/[30 + (14 \cdot 1.2)] = 0.64$. Values for other generations are

b	Lengths of Introgressed Segment (cM) Marker Chromosome	All Unmarked	% from Marked Chromosome
2	45.0 $\pm$ 19.1	131.25 $\pm$ 255.9	25.6
7	23.8 $\pm$ 14.8	4.10 $\pm$ 36.7	85.3
10	17.9 $\pm$ 12.0	0.51 $\pm$ 11.3	97.3
12	15.3 $\pm$ 10.5	0.13 $\pm$ 5.2	99.2
15	12.5 $\pm$ 8.7	0.02 $\pm$ 1.7	99.9

Finally, we note that an NIL line created by selecting for retention of a single marker contains ***significantly more donor DNA than an NIL created without any selection***. Here 1,125 cM is the total map length, giving the expected total length of donor DNA in an NIL formed without selection as $1125/2^{b+1}$, or 17.58, 4.39, and 0.55 cM after 5, 7, and 10 generations of backcrossing, respectively.

If donor and recurrent-parent DNAs are sufficiently divergent, recombination between donor and recurrent chromosomes can be greatly reduced, resulting in a much longer segment of donor DNA being retained than expected (e.g., Young and Tanksley 1989a; Paterson et al. 1990). In these cases, the above expressions may give rather serious underestimates. For example, Young and Tanksley (1989a) examined the length of donor DNA introgressed by selecting for a donor marker (resistance to tobacco mosaic virus) in a cross between two species of tomatoes. Even after 11 generations of backcrossing, the length of the introgressed segment in one line was over 51 cM, as compared to the theoretical expectation of 16.7 ± 11.8 cM from Equation 17.10c.

Selection using *multiple*, instead of *single*, markers can greatly accelerate the formation of NILs with very short introgressed segments. First, foreground selection using *flanking markers* around a region of interest can greatly reduce the size of the introgressed region (Young and Tanksley 1989a, 1989b). Second, background selection for markers from the recurrent parent on other chromosomes can accelerate the loss of donor DNA outside of the introgressed region. We consider these two strategies in turn.

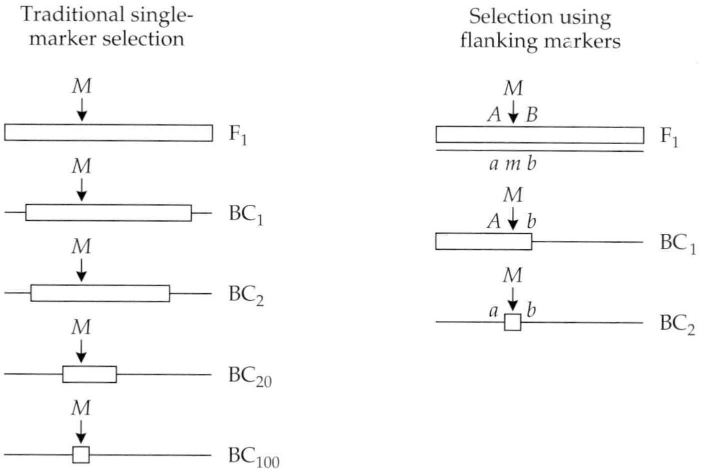

Figure 17.10 Selecting on flanking markers can greatly accelerate the elimination of donor DNA around a selected donor marker *M*. **Left:** Reduction of donor material around a single selected marker leads to an expected value of around 2 cM following 100 generations of selection and backcrossing to the recurrent parent. **Right:** Two generations of selection using a dense marker map can also reduce the introgressed segment to under 2 cM. First, choose two flanking markers (*a* and *b*) from the recurrent parent that are both within 1 cM of *M*, with the donor and recurrent genotypes being *AMB* and *amb*, respectively. In generation one, we select an individual that is either *AMb* or *aMB*. Using this individual as the parent for the second generation, we then select for the recombinant on the opposite side, giving an *aMb* individual with an introgressed region containing *M* that is less than 2 cM. (After Tanksley et al. 1989.)

We consider two-marker foreground selection first. Equation 17.10c shows that 99 backcross generations are required to reduce the average length of a marker-selected introgressed region to 2 cM. However, with a dense marker map, such a region can be obtained in one or two generations by using markers that flank the region of interest (Figure 17.10). If we can screen a very large number of individuals, this can be done in a single generation by looking for a double recombinant. Using flanking markers 1 cM on either side of the region of interest, the probability of a double recombinant (assuming no interference) is 0.01^2, or about one out of every 10,000 progeny. If interference is significant, which is likely for tightly linked markers, far more progeny may need to be scored. This problem can be avoided by screening for two generations, with first-generation individuals showing a single recombinant (on either side of the marker) being used as parents for the second generation.

Example 17.14. Suppose the region of interest is flanked by markers, each at recombination frequency c from the locus of interest. To have a 95% probability of recovering a double recombinant in a single generation (assuming no interference), one must examine roughly $3/c^2$ individuals, or 30,000 individuals for $c = 0.01$ (a 2 cM interval). This follows because the probability of no double recombinants among n sampled chromosomes is $(1-c^2)^n$, giving the probability of at least one double as $1 - (1-c^2)^n$. Setting this equal to 0.95 and solving for n gives $n = -\ln(0.05)/\ln(1-c^2) \simeq 3/c^2$.

How much of a savings can the two-generation method offer? If n_1 first-generation individuals are screened, the probability that at least one of them is a single recombinant is $1-(1-2c)^{n_1}$ (with $2c$ being the probability of recombination on either side of the marker, $1-2c$ is the probability that a chromosome is nonrecombinant with respect to both flanking markers). Using a single recombinant to form the next generation, the probability of a single recombination event on the opposite side of the region is $1-(1-c)^{n_2}$. The probability of obtaining a double recombinant after two generations is the probability of a single recombinant in the first generation times the probability of a single recombinant in the second (this result is not impacted by interference). Thus, to have an overall probability of 95%, we need a probability of 97.5% for each of the two events, as $0.975^2 = 0.95$. The sample size required to have a 97.5% chance of a single recombinant (on either side of the region) in the first generation is obtained by solving $0.975 = 1 - (1-2c)^{n_1}$, yielding $n_1 \simeq 1.85/c$. Likewise, solving $0.975 = 1 - (1-c)^{n_2}$ gives $n_2 \simeq 3.7/c = 2\,n_1$. Using this two-generation approach, to have a 95% probability of recovering a double recombinant, a total of $n_1 + n_2 = 5.55/c$ individuals must be scored. For $c = 0.01$, this is 555 individuals (185 in the first generation, 370 in the second), only 2% of the 30,000 required for the single-generation method. Frisch et al. (1999a) provide a more in-depth analysis of sample size and optimal placement of the flanking markers.

Turning now to background selection, Hillel et al. (1990; 1993) first proposed that selection for recurrent markers (in their case in chickens) could greatly hasten the removal of background donor DNA, although they were overly optimistic about its advantage (Hospital et al. 1992; Groen and Smith 1995; Visscher 1999). In actuality, marker-assisted background selection results in around a two to three generation reduction in the amount of time to reach a desired fraction of recurrent DNA compared to backcrossing without selection (Hospital et al. 1992; Visscher 1996, 1999; Visscher et al. 1996c; Frisch et al. 1999b; Ribaut et al. 2002; Stam 2003).

The optimal number, and placement, of markers for efficient background selection have been examined by a number of authors, with the general conclusion of between two and four markers per Morgan of chromosome length, ideally positioned (for metacentric chromosomes) around 20–30 cM from the telomeres (Hospital et al. 1992; Visscher 1996; Visscher et al. 1996c; Frisch et al. 1999b; Servin and Hospital 2002; Frisch and Melchinger 2005). There is a rapidly diminishing return from adding new markers per chromosome (Herzog and Frisch 2011), as there are simply not enough recombinational events within a sample to exploit this additional fine-mapping information. Hence, the limiting factor is usually the number of recombinations (i.e., number of scored individuals), rather than the ability to finely map their breakpoints (number of markers). This is a recurrent theme in QTL mapping: the number of individuals sampled, rather than the number of markers scored, is generally the limiting factor. ***Far more information is gained in a study by adding more individuals than by increasing the marker density*** (e.g., Frisch and Melchinger 2005). For example, as noted by Kump et al. (2010), the usual fine-mapping strategy of first localization of a QTL using RILs and the subsequent fine mapping by creating NILs for the target region can be significantly improved by combining information from both data sets, as more potential recombinants are scored.

Finally, one can combine reproductive strategies that significantly speed up generation time with marker-assisted selection, an approach termed **supersonic congenics** (Behringer 1997). For many vertebrates, **superovulation and embryo transfer** (hormone treatment that allows for the harvesting of eggs from prepubertal females) dramatically shortens generation times (e.g., Landa et al. 2009). In plants, similar gains can be made using **speed breeding** protocols that use extreme photoperiods and plant hormones to speed development (Ghosh et al. 2018; Watson et al. 2018; Chiurugwi et al. 2019). These rapid advancement-of-generations schemes do not typically allow for the scoring of normal phenotypes, but are suitable to marker–assisted selection.

FINE MAPPING MAJOR GENES USING POPULATION-LEVEL DISEQUILIBRIUM

Limitations of Pedigree-based Fine Mapping

The typical level of mapping resolution for a major gene using the linkage information in a collection of (human) pedigrees is rarely below 1–2 cM (Lange et al. 1985; Bodmer 1986; Boehnke 1994). To a rough approximation, this typically corresponds to (on average) a region of a few megabases in size (Table 17.2), and hence a few hundred potential candidate genes. The limiting factor is the expected number of scoreable recombinants,

$$\lambda = Nn_o c \tag{17.13}$$

in the pedigree. Here N is the number of **informative parents** (individuals that are double heterozygotes for both the marker and the major gene, allowing for recombinants to be detected; Chapter 19), n_0 is the average number of offspring for each such parent, and c the marker-gene recombination frequency. Outside of controlled crosses (Chapter 18), usually only a fraction of parents are informative (Chapter 19), so that N can be considerably smaller than the actual number of individuals in the pedigree. From Equation 17.13, a marker-gene map distance of 1 cM ($c = 0.01$) requires 100 offspring from informative parents ($Nn_o = 100$) to have an expected value of one recombinant ($\lambda = 1$). The chance of not seeing any recombinants is $(1-c)^{Nn_o} \simeq e^{-\lambda}$, or $e^{-1} = 0.37$. More generally, the probability of seeing exactly k recombinants approximately follows for the Poisson distribution, $\lambda^k \exp(-\lambda)/k!$ (Equation 2.21a). Considering the first three terms ($k = 0, 1, 2$) for a set of pedigrees containing 200 total offspring from informative parents ($Nn_o = 200$), the probability of seeing two or fewer recombinants (for $c = 0.1$) is almost 60% (0.586).

Linkage–disequilibrium Mapping (LDM) Using Historical Recombinants

One source of additional linkage information is from **historical recombination**: the descendants of a single ancestral haplotype. Here the expected number of recombinants given a last common ancestor of t generators ago is $2tc$. Hence, two descendant haplotypes with a last common ancestor 100 generations ago have experienced the same expected number of recombination events as 200 offspring from informative parents in a pedigree. This introduces the idea of exploiting population-level LD for gene mapping. Our focus here is for the very special case of mapping a *major gene* (such as an important disease factor), an approach we denote as **linkage-disequilibrium mapping (LDM)**. Chapter 20 examines the much wider use of LD in GWAS approaches for mapping genes of arbitrary effect. This latter approach requires a very dense marker map, but (given sufficient marker density) is applicable to many, if not most, natural populations (Chapter 20).

Here, our focus is a bit different, assuming a lower density of markers with the goal to estimate a map distance between a closely linked marker and the target gene. The basic idea was suggested by Bodmer (1986, cited by Hill and Weir 1994), wherein once the rough (5–10 cM) position of a gene is found,

> . . . it should be possible to saturate the relevant region with further polymorphic markers and look for the ones that have a population association with the trait, rather than increasing the numbers of families analyzed to search for closer linkage. This is the most efficient way of finding closely linked markers, since [human] family data are very inefficient at distinguishing between small recombination fractions such as 0.5% versus 5.0%.

This requires a setting in which the LD around the major gene is sufficiently large to encompass a marker that may be some distance away. Such larger regions of LD can occur in expanding populations, and geneticists have sought out isolated human populations that appear to have recently undergone such expansions for LD mapping of disease genes. The ideal situation occurs when most of the disease alleles descend from a single ancestral mutation, so that all current copies have maintained some of the ancestral marker haplotype. The age of the ancestral mutation from which all current copies derive can neither be too young nor too old. It must be sufficiently old that recombination has reduced the expected size of the retained region to one sufficiently small for fine mapping. However, if the mutation is too old, this region will be too small for analysis (with only modest marker density, no variation may be present around the gene). These conditions are often satisfied in expanding populations that can be traced back to a small number of recent founders. An example is the current Finnish population, which arose from a small founder group about 2,000 years ago (reviewed in de la Chapelle 1993; de la Chapelle and Wright 1998; Peltonen et al. 1999), and other such human populations are discussed by Jorde (1995) and Peltonen et al. (2000). In a rapidly expanding population, all copies of a mutant disease allele often trace back to a single copy present in the founding population or originating shortly after the population expansion began.

Motivated by the Luria–Delbrück theory for estimating mutation rates in exponentially growing populations of bacteria (Luria and Delbrück 1943; Sarkar 1991); Hästbacka et al. (1992) and Lehesjoki et al. (1993) suggested that the same logic can be used to estimate recombination frequencies from population disequilibrium information. The simplest approach proceeds as follows. Suppose a disease allele (of major effect) was either present as a single copy (and hence associated with a single chromosomal haplotype) in the founder population or arose by mutation very shortly after the population was formed. Further assume that there is no **allelic heterogeneity**, so that all disease-causing alleles in the population descend directly from the original mutation, and consider a marker locus tightly linked to the disease locus. The probability that a disease-bearing chromosome has not experienced recombination between the **disease susceptibility** (**DS**) gene and marker locus after t generations is just $(1-c)^t \simeq e^{-ct}$, where c is the marker-DS recombination frequency. Suppose the disease is predominantly associated with a particular haplotype, which presumably represents the ancestral haplotype on which the DS mutant arose. Equating the probability of no recombination to the observed proportion, π, of disease-bearing chromosomes with this predominant haplotype gives $\pi = (1-c)^t$, where t is the age of the mutation or the age of the founding population (whichever is more recent). Hence, one estimate of the recombination frequency is

$$c = 1 - \pi^{1/t} \tag{17.14}$$

Note the key assumption throughout (that greatly limits the generality of this approach): that ***the phenotype unambiguously identifies the presence of a DS allele***. This restriction may be true for some major genes, but not for general genes underlying a complex trait. Hence, this approach is essentially restricted to Mendelizing factors (Kaplan et al. 1995), while with much denser marker maps (and much large population sample sizes), GWAS offers a more general, and more powerful, approach (Chapter 20).

Example 17.15. Hästbacka et al. (1992) examined the gene for diastrophic dysplasis (DTD), an autosomal recessive disease, in Finland. A total of 18 **multiplex** families (showing two or more affected individuals) allowed the gene to be localized to within 1.6 cM from a marker locus (*CSF1R*) using standard pedigree methods. To increase the resolution using pedigree methods requires significantly more multiplex families. Given the excellent public health system in Finland, it was likely that the investigators had already sampled most of the existing families. As a result, the authors turned to LD mapping.

While only multiplex families provide information under standard (i.e., linkage-based) mapping procedures, this is not the case with LD mapping wherein *single affected individuals* (which each contain two DTD-bearing chromosomes) can provide further information. Using LD mapping allowed the sample size to increase by 59 in this study. A number of marker loci were examined, with the *CSF1R* locus showing the most striking correlation with DTD. The investigators were able to unambiguously determine the haplotypes of 152 DTD-bearing chromosomes and 123 normal chromosomes for the sampled individuals. Four alleles (haplotypes) of the *CSF1R* marker gene were detected, with the frequencies for these alleles among normal and DTD chromosomes as follows:

	Chromosome Type			
CSF1R Allele	Normal		DTD	
1-1	4	3.3%	144	94.7%
1-2	28	22.7%	1	0.7%
2-1	7	5.7%	0	0%
2-2	84	68.3%	7	4.6%

Given that the majority of DTD-bearing chromosomes were associated with the rare 1-1 allele (present in only 3.3% of normal chromosomes), the authors suggested that all DTD-bearing chromosomes in the sample descended from a single ancestor carrying allele 1-1. Because 95% of all present DTD-bearing chromosomes contain this allele, $\pi = 0.95$. The current Finnish population traces back to around 2,000 years to a small group of founders, which underwent around $t = 100$ generations of growth. Using these estimates of π and t, Equation 17.14 gives an estimated recombination frequency between the *CSF1R* gene and the DTD gene as $c = 1 - (0.95)^{1/100} \simeq 0.00051$. Thus, the two genes were estimated to be separated by 0.05 cM, or about 50 kb (using the rough rule for humans that 1 cM = 10^6 bp). Subsequent cloning of this gene by Hästbacka et al. (1994) showed it to be 70 kb proximal to the *CSF1R* marker locus. Thus, LD mapping increased precision by about 34-fold over that possible using segregation within pedigrees (0.05 cM versus 1.6 cM).

The DTD example is exceptional in that the disease allele was initially associated with a rare haplotype. Because the majority of current disease-bearing chromosomes have this haplotype, this suggested that the sample of disease alleles was unlikely to contain a significant fraction of new mutants. Lehesjoki et al. (1993) offered a modification of LDM when the predominant marker haplotype among disease-bearing chromosomes is also a common haplotype for normal chromosomes. Let p_{DS} denote the frequency of the most common haplotype of disease-bearing chromosomes and p_n denote the frequency of this haplotype among normal chromosomes. Let π represent the fraction of disease-bearing chromosomes descended from the common ancestor that have not undergone any recombination between the marker and QTL, and α be the proportion of all disease alleles descending from the founder copy (as opposed to being new mutants). We can then divide disease-bearing chromosomes into two classes. First, a fraction $\alpha\,\pi$ of these chromosomes have a disease allele that both traces back to the founder copy (α) and does so through chromosomes that have not undergone recombination between the marker and disease loci (π). Second, a fraction $(1-\alpha\,\pi)$ are either new mutants (which arose on some random haplotype) or a product of recombination between the founder copy and a random haplotype. In either event, the chance that such individuals have the predominant haplotype is just $(1-\alpha\,\pi)\,p_n$. Putting these two results together gives

$$p_{DS} = \alpha\,\pi + (1-\alpha\,\pi)\,p_n \tag{17.15a}$$

Rearranging Equation 17.15a and solving, we find that

$$\delta_p = \alpha\,\pi = \frac{p_{DS} - p_n}{1 - p_n} \tag{17.15b}$$

is a measure of the extent to which disease chromosomes are restricted to the predominant haplotype (Bengtsson and Thomsom 1981). Recalling Equation 17.14,

$$\pi = (1-c)^t = \frac{\delta_p}{\alpha} \tag{17.16}$$

giving an estimate of the recombination frequency as

$$c = 1 - \left(\frac{\delta_p}{\alpha}\right)^{1/t} \tag{17.17}$$

Note that for the DTD data in Example 17.15, $p_n \simeq 0$, so that $\delta_p \simeq p_{DS}$, and we assumed that all disease alleles descended from the founder copy ($\alpha = 1$), so that $c \simeq 1 - (p_{DS})^{1/t}$. Note also that if c is known, one can solve instead for t, the age of the mutation. Risch et al. (1995) used this approach to date the appearance of a mutation causing idiopathic torsion dystonia (ITD) disease in Ashkenazi Jews at about 350 years (WL Chapter 9 discusses other approaches for estimating the age of alleles).

Hästbacka et al. (1992) proposed simple confidence intervals for estimates of c given by Equations 17.14 and 17.17, but Kaplan and Weir (1995) showed that these are downwardly biased, leading to intervals that are too narrow. de la Chapelle and Wright (1998) suggested an improved approximation that incorporates sampling error in the estimation of π. However, maximum-likelihood methods (Appendix 4) generally provide better approximations of the confidence intervals (Kaplan et al. 1995; Terwilliger 1995; Xiong and Guo 1997).

The fraction α of all disease alleles that are direct descendants from the ancestral copy (as opposed to being new mutants) can be obtained as follows. Assuming that selection is weak on carriers (heterozygotes), new mutants that have arisen *after* the initial appearance of the original DS allele in the founding population should comprise a fraction $1-(1-\mu)^t \simeq \mu t$ of all chromosomes, as $(1-\mu)^t$ is the probability that no mutations have occurred. Here μ is the mutation rate at which new disease-causing alleles appear and t is the number of generations since the original mutation arose or since the founding of the population, whichever is more recent. If q is the frequency of disease alleles, then $\mu t/q$ is the expected fraction of disease-carrying chromosomes due to new mutants. Hence, the expected fraction of all disease-carrying chromosomes that descend from the original mutation is

$$\alpha = \left(1 - \frac{\mu t}{q}\right) \tag{17.18}$$

which when applied to Equation 17.17, yields

$$c = 1 - \left(\frac{\delta_p}{1-(\mu t/q)}\right)^{1/t} \tag{17.19}$$

The basic LDM idea of considering the fraction of DS-bearing haplotypes that contain a specific marker allele has been generalized to allow for both multiple alleles and multiple loci (Kaplan et al. 1995; Devlin and Risch 1995; Risch et al. 1995; Terwilliger 1995; Xiong and Guo 1997; Meuwissen and Goddard 2000; Toivonen et al. 2000; Liu et al. 2001; Morris et al. 2002; Schaid 2004b; Zöllner and Pritchard 2005). When multiple markers are used, this leads to tests of the ordering of the markers and the DS gene.

Example 17.16. Sulisalo et al. (1994) examined the major gene for cartilage-hair hypoplasia (CHH), an autosomal recessive disease, in the Finnish population. As in Example 17.15, pedigree information allowed unambiguous determination of the haplotypes associated with most CCH-bearing chromosomes. The authors observed that 85% of these contain a particular allele at marker *D95163*, while only 41% of non-CHH chromosomes carried this allele. Equation 17.15b thus becomes

$$\delta_p = \frac{p_{DS} - p_n}{1 - p_n} = \frac{0.85 - 0.41}{1 - 0.41} = 0.75$$

implying

$$c = 1 - \left(\frac{\delta_p}{\alpha}\right)^{1/t} = 1 - \left(\frac{0.75}{\alpha}\right)^{1/t}$$

To estimate the fraction α of all current CHH alleles that directly trace back to the ancestral copy, first note that the frequency of CHH alleles in Finland is estimated to be 0.0066. Assuming $t = 100$ and $\mu = 1 \times 10^{-5}$,

$$\alpha = \left(1 - \frac{\mu t}{q}\right) = \left(1 - \frac{100 \times 10^{-5}}{0.0066}\right) = 0.85$$

Thus, 85% of all present CHH alleles are estimated to be direct descendants of the founder copy, implying $c = 0.12$ cM. Taking $\mu = 1 \times 10^{-6}$ gives $\alpha = 0.98$ and $c = 0.27$ cM. By contrast, traditional pedigree-based mapping was able to localize the CHH gene to only a 1.7 cM region.

LD mapping has been successfully applied to other disease genes segregating in the Finnish population (Peltonen et al. 1999), such as congenital nephrotic syndrome (Kestilä et al. 1994) and progressive myoclonus epilepsy (Lehesjoki et al. 1993). Kaplan et al. (1995) were able to apply LD mapping to the cystic fibrosis (CF) gene, using much more heterogeneous populations from Europe and elsewhere. The likely reason for success in these cases was that 70% of CF chromosomes worldwide appear to result from a single three-base deletion (Kerem et al. 1989), so that allelic heterogeneity is, at worst, a modest problem. Kaplan et al. (1995) found, however, that LD mapping was not very successful for Huntington's disease (multiple ancestral haplotypes) or Friedreich ataxia (high allelic heterogeneity) using European or North American populations.

Admixture Mapping

One generator of LD that has been exploited in mapping is **admixture**, the recent mixing of two populations that differ in allele frequencies. Suppose a sample of gametes comes from two distinct populations, where m is the fraction from population 1 and $(1-m)$ the fraction from population 2. Even if both populations are in linkage equilibrium, if there are allele-frequency differences (δ) between them at loci A and B ($\delta_A, \delta_B \neq 0$), then the amount of LD between A and B in the sample is

$$D^{(0)} = m(1-m)\delta_A\delta_B \tag{17.20a}$$

Hence, the more even the mixing (m closer to 0.5) and the greater the allele-frequency differences, the larger the initial LD. More generally, if D_i is the A–B disequilibrium in population i, then (Chakraborty and Weiss 1988)

$$D^{(0)} = mD_1 + (1-m)D_2 + m(1-m)\delta_A\delta_B \tag{17.20b}$$

If t generations of random mating follow the initial admixture event, then (Equation 5.13b)

$$D^{(t)} = (1-c)^t D^{(0)} \tag{17.20c}$$

The historical interest in using such **admixture linkage disequilibrium (ALD)** was that when the admixture is recent (t in Equation 17.20c is small), regions that sufficiently differ in allele frequencies retain a LD signal over rather long regions (on the order of 10–30 cM). Hence, a genome-wide scan with a very modest number of markers is feasible. Conversely, in a random sample from most populations, LD exists only between *very tightly* linked markers (those separated by just a few kilobases, e.g., recombination frequencies of $<$ 0.1 cM), a feature that requires (at least) tens-of-thousands of SNPs to be exploited for association mapping (Chapter 20).

In the human genetics setting, Chakraborty and Weiss (1988) proposed that **admixture mapping (AM)** using the ALD generated during the contact between migrating populations could be used to map major genes that differ in allele frequencies between the mixing populations (an idea first suggested by Rife 1954). In particular, family-based tests (such as the TDT; see below) that require sample-level LD are much more powerful when using admixed families, as these have much more widespread LD. Darvasi and Shifman (2005) noted that the admixture mapping strategy falls between the low-marker, poor resolution of linkage mapping and the high-marker, very fine resolution from association studies, with AM very much akin to using AIC lines for linkage-based QTL mapping (Chapter 18). Discussions of sample size and design consideration under AM are given by McKeigue (1997, 1998, 2005), Halder and Shriver (2003), Hoggart et al. (2004), Patterson et al. (2004), and Zhu (2004).

One of the first applications of a genome-wide AM scan was by Zhu et al. (2005), who examined hypertension in an admixture sample of African Americans, representing roughly 3/4 African and 1/4 European ancestry. They used 270 microsatellite markers to scan the genome, with a pooled sample of around 750 admixed cases (high blood pressure) and 570 admixed controls (normal pressure). Based on searching for an excess of African-ancestry markers in the cases versus the controls, they were able to suggest that regions on chromosomes 6 and 21 harbored QTLs that gave higher risk for hypertension. This study highlights one key limitation of AM, namely poor mapping resolution. As our ability to quickly and cheaply score millions of SNP markers evolved, the use of admixture mapping in humans quickly gave way to GWAS (Chapter 20), which offered much higher resolution under much less restrictive settings. However, hybrid zones between populations are common for most species in nature, and AM represents one potential strategy to exploit these settings.

CANDIDATE LOCI

In some cases, there may be sufficient physiological/biochemical information to suspect that certain known loci influence trait expression. In human genetics, it is common practice to directly test for population-level associations between trait value and specific alleles at such **candidate loci** (the **measured-genotype** approach). For example, Boerwinkle and Sing (1987) showed that three common alleles at the human apolipoprotein E locus account for about 8% of the total variation in cholesterol levels. Interestingly, a particular allele of apolipoprotein E also appears to be a major determinant of Alzheimer's disease. The mean age of onset for homozygotes and heterozygotes for this allele are 68.4 ± 1.2 and 75.5 ± 1.0 years, respectively, while the mean age for individuals with no copies of this allele is 84.3 ± 1.3 (Corder et al. 1993). Other candidate loci for Alzheimer's disease are reviewed by Pericak-Vance and Haines (1995) and Selkoe (2001). As Example 17.17 shows, results using the candidate-locus approach can be rather unexpected.

How are candidate loci chosen? One obvious approach is to consider loci known from laboratory studies (such as knockout experiments) to have mutant alleles with major effects on the character of interest, as in natural populations such loci may also be segregating alleles with smaller effects (Chapter 15). Results from several QTL mapping experiments offer some support for this approach (Chapter 18). In maize, for example, Beavis et al. (1991), Edwards et al. (1992), Veldboom et al. (1994), and Berke and Rocheford (1995) found that many QTLs for height map near the locations of known height mutants. Similar findings have been obtained with bristle number in *Drosophila* by Mackay and Langley (1990), Lai et al. (1994), and Long et al. (1995). However, such findings are strongly influenced by conformation bias (focusing more on evidence that fits with prior beliefs), as the true number of QTLs that do not show major knockout effects is unknown (although GWAS studies suggest it may be a considerable fraction, if not the majority; Chapters 20 and 21).

Example 17.17. Winkelman and Hodgetts (1992) examined the growth hormone (GH) gene as a candidate locus for body weight in selected lines of mice. Molecular analysis disclosed the presence of an allele, GH^h, present in all four lines selected for increased weight (being fixed in three of these). An alternative allele, GH^c, was fixed in all five control lines. One of the up-selected lines was crossed to two separate control lines to create two different F_2 populations. The GH^h allele had a significant, but unexpected, effect on body weight in both F_2 populations, as it *decreased* weight. For one population, the genotypes $GH^hGH^h : GH^hGH^c : GH^cGH^c$ had 42-day weights of 29.2 : 30.2 : 31.4, while in the other population these respective weights were 34.6 : 34.9 : 38.8. Thus, the GH^h allele was additive in one of the F_2 backgrounds, but dominant in the other. The GH^h allele behaved rather differently once the F_2 populations were subjected to selection, again increasing in frequency in up-selected lines, and decreasing in frequency in down-selected lines. Winkelman and Hodgetts suggested that the association between the GH^h allele and increased weight was a product of epistatic interactions in mice selected for high weight.

Example 17.18. Kozak et al. (2019) examined time to pupation under diapause-breaking photoperiod and temperature conditions (post-diapause development, PDD, time) in the European corn borer moth (*Ostrinia nubilalis*). Some eastern US populations have responded to shorter winters, evolving a PDD time that is roughly three weeks shorter than the late populations. A cross between an early and a late population found two significant QTLs, both located on the Z chromosome (recall that males in Lepidoptera are the homogametic sex, being ZZ). The first QTL region (QTL1) was roughly 3.1 Mb in size and contained 48 annotated genes, while the second region (QTL2) was 3.7 Mb with 42 annotated genes. Further, the two regions showed a strong epistatic interaction for PDD. Both QTL regions contained a potential candidate gene: the circadian clock gene period (*per*) in QTL1 and pigment-dispersing factor receptor (*Pdfr*) in QTL2.

To further examine these potential candidates, the authors pooled sequence data from five populations, and measured marker allele-frequency divergence among populations differing in PDD time. Across the entire genome, only six SNPs showed exceptional amounts of divergence (Bayes factors in excess of 20; Appendix 7), all of which fell inside these two QTL regions, with three of these within the *per* and *Pdfr* genes. Given the strong interaction between the two QTL regions, the authors expected the target sites of selection to likely be in high LD. In populations of this species, LD roughly spans 2 Mb or less, with sites between 2 and 10 Mb showing much lower amounts (99.9% have values of $r^2 \leq 0.56$). Among SNPs in annotated gene pairs located within or between the two QTL regions (distances of 2–7 Mb), they found 12 outlier pairs, the most extreme of which was between the *per* and *Pdfr* genes. Although roughly 5 Mb apart, they display an LD value of $r^2 = 0.75$. Taken together, these observations provide a strong case that *per* and *Pdfr* genes are major factors in the evolution of PDD time in the early population.

The Transmission/Disequilibrium Test, TDT

When considering dichotomous (presence/absence) traits (such as disease status), the frequency of a particular candidate (or marker) allele in affected (or **case**) individuals is often compared with the frequency of this allele in unaffected (or **control**) individuals. The problem with such **association studies** is that a ***marker-trait association can arise simply as a consequence of population structure***, rather than as a consequence of linkage. Such **population stratification** occurs if the total sample consists of a number of divergent populations (e.g., different ethnic groups) which differ in both candidate-gene frequencies and incidences of the trait. Population structure can severely compromise tests of candidate gene associations, as the next example illustrates. As detailed in Chapter 20, a considerable amount of effort in any GWAS analysis is spent in attempts to control for any effects from population structure.

Example 17.19. Hanson et al. (1995) used segregation analysis (Chapter 16) to find evidence for a major gene for Type 2 diabetes mellitus segregating at high frequency in members of the Pima and Tohono O'odham nations of southern Arizona. In an attempt to map this gene, Knowler et al. (1988) examined how the simple presence/absence of a particular haplotype, Gm^+, was associated with diabetes. Their sample showed the following associations:

Gm^+	Total subjects	% with Diabetes
Present	293	8%
Absent	4,627	29%

The resulting χ^2 value (61.6, 1 df) shows a highly significant negative association between the Gm^+ haplotype and diabetes, making it very tempting to suggest that this haplotype marks a candidate diabetes locus (either directly or by close linkage).

However, the presence/absence of this haplotype is also a very sensitive indicator of admixture with the Caucasian population (who have a much lower diabetes risk). The frequency of Gm^+ was around 67% in Caucasians as compared to $< 1\%$ in full-heritage Pima and Tohono O'odham. When the authors restricted the analysis to such full-heritage adults (over age 35 to correct for age of onset), the association between haplotype and disease disappeared:

Gm^+	Total subjects	% with Diabetes
Present	17	59%
Absent	1,764	60%

Hence, the Gm^+ marker is a predictor of diabetes not because it is linked to genes influencing diabetes but rather because ***it serves as a predictor of whether individuals are from a specific subpopulation with a different disease risk***. Gm^+ individuals usually carry a significant fraction of genes of Caucasian extraction. Because a gene (or genes) increasing the risk of diabetes appears to be present at high frequency in individuals of full-heritage Pima/Tohono O'odham extraction, admixed individuals have a lower chance of carrying this gene (or genes).

The problem of population stratification can be overcome by employing tests that use *family data*, rather than *population data* from unrelated individuals, to provide the case and control samples (Penrose 1939; Woolf 1955; Clarke et al. 1956; Rubinstein et al. 1981; Falk and Rubinstein 1987; Terwilliger and Ott 1992; Spielman et al. 1993; Thomson 1995). One such family-based approach is to consider the transmission (or lack thereof) of a parental marker allele to an affected offspring. Focusing on *transmission within families* controls for associations generated entirely by population stratification and provides a direct test for linkage *provided* that a population-wide association between the marker and disease gene exists (Spielman et al. 1993; Ewens and Spielman 1995). This method can be applied to any family that has at least one affected offspring.

The **transmission/disequilibrium test**, or **TDT** (Spielman et al. 1993), compares the number of times a marker allele is transmitted (T) versus not-transmitted (NT) from a marker heterozygote parent to affected offspring. Under the hypothesis of no linkage, these values should be equal, and the test statistic becomes

$$\chi^2_{td} = \frac{(T - NT)^2}{(T + NT)} \tag{17.21}$$

which follows a χ^2 distribution with one degree of freedom. Equation 17.21 is also known as **McNemar's test of discordance** (Sokal and Rohlf 1995). There is a large literature on the TDT and its extensions (as a group, these methods are often called **family-based association tests**, **FBAT**; or **family-based association analyses**, **FBAA**), with general overviews given by Schaid and Sommer (1994), Spielman and Ewens (1996), Schaid (1998), Rabinowitz and Laird (2000), Zhao (2000), Horvath et al. (2001), Schulze and McMahon (2002), Ewens et al. (2008), Laird and Lange (2006a, 2006b, 2009), Tiwari et al. (2008), Sneller et al. (2009), and Ott et al. (2011).

The signal from a TDT (a significant difference between T and NT) requires *both* linkage (an excess of parental gametes from a parent creating an association in their offspring between a marker allele and a DS allele) *and* population-level LD (gametic phase largely agrees over families, allowing the pooling of values over multiple families). Ewens and Spielman (1995) formalized this by developing expressions for $E[T - NT]$ under different admixture scenarios (parents of affected sibs are from a set of subpopulations, which are either distinct or have started to mix). All of their resulting expressions for this expectation had the form of $2n \cdot (1 - 2c) \cdot \delta$, where n is the number of affected offspring from parents that are marker heterozygotes, c the marker-DS recombination frequency, and δ is a term that is only nonzero when population-level association (LD) occurs. For example, for affected offspring in the third generation after an admixture event,

$$E[T - NT] = 2n\frac{D}{p_{DS}}(1 - 2c) \tag{17.22}$$

where p_{DS} is the frequency of the disease allele, D the disequilibrium between the marker and DS allele, and c their recombination frequency. Hence, TDT has no power if $D = 0$ (no population-level disequilibrium between marker and disease alleles) or $c \simeq 1/2$ (no linkage between the marker and disease loci). As stressed by Ewens and Spielman (1995), the TDT is thus a *joint* test of *both* linkage *and* population-level disequilibrium, leading to its original description as a ***test for linkage in the* presence *of association*** (LD).

A number of papers have examined approximate and exact solutions for the power of a TDT and other FBAT approaches (Sham and Curtis 1995; Risch and Merikangas 1996; Camp 1997, 1999; Spielman and Ewens 1998; Whittaker and Lewis 1998; Knapp 1999b, 1999c; Tu and Whitemore 1999; McGinnis 2000; Chen and Deng 2001; Lange et al. 2002; Diao and Lin 2006; Fisher and Lewis 2008; Glaser and Holmans 2009). Under the null (*either* $c = 1/2$ or $D = 0$), Equation 17.21 is χ^2-distributed. When $c < 1/2$ *and* $D \neq 0$, Equation 17.21 follows a noncentral χ^2 (Appendix 5), with noncentrality parameter (which determines the power of the test; Equation A5.14a) being a function of c, the disequilibrium, and the nature of the disease (dominant, recessive, population frequency). Generally speaking, power is higher for rarer diseases and for recessive diseases. For example, Sham and Curtis (1995), using values for typical diseases, found that the sample sizes (number of trios) required for 90% power (under the assumption that $c \simeq 0$) are around 100 for a rare recessive, 300 for a rare dominant, 400 for a common recessive, and over 1000 for a common dominant.

How are T and NT determined? Consider an M/m parent with three affected offspring. If two of those offspring received this parent's M allele, while the third received m, we score this as two transmitted M, one not-transmitted M. Conversely, if we are following marker m instead, this is scored as one transmitted m, two not-transmitted m. As the following example shows, each marker allele is examined separately under the TDT.

Example 17.20. Copeman et al. (1995) examined 21 microsatellite marker loci in 455 human families with Type 1 diabetes. One marker locus, *D2S152*, had three alleles, with one allele (denoted 228) showing a significant effect under the TDT. Parents heterozygous for this marker transmitted allele 228 to diabetic offspring 81 times, while transmitting alternative alleles to diabetic offspring only 45 times, giving

$$\chi^2 = \frac{(81 - 45)^2}{(81 + 45)} = 10.29$$

which has a corresponding p value of 0.001. As summarized below, the other two alleles (230 and 240) at this marker locus did not show a significant TD effect.

Allele	T	NT	χ^2	p
228	81	45	10.29	0.001
230	59	73	1.48	0.223
240	36	24	2.40	0.121

Hence, this marker appears to be linked to a QTL influencing Type 1 diabetes, with allele 228 in (coupling) LD with an allele that increases the risk for this disease.

Spielman et al. (1993) noted that the TDT can give a false positive if the marker shows **segregation distortion**, wherein heterozygotes preferentially segregate one allele. This distortion can be controlled for by using a standard 2×2 contingency table χ^2 test, considering separately the transmission of a marker allele to affected and unaffected offspring,

	Transmitted	Not Transmitted
Affected Offspring	T_A	NT_A
Unaffected Offspring	T_U	NT_U

with resulting test statistic

$$\chi^2_{td} = \frac{(T_A - NT_A)^2}{(T_A + NT_A)} + \frac{(T_U - NT_U)^2}{(T_U + NT_U)} \tag{17.23}$$

controlling for any segregation distortion. A number of other linkage-based family methods for detecting marker-trait associations that do not require population-wide disequilibrium have been developed, and these are examined in Chapter 19. Chapter 20 examines the GWAS approach of exploiting population-wide disequilibrium between *extremely* close markers that can be detected using very dense marker maps.

The original TDT required **trio data**: genotypes from both parents and an affected offspring. In many cases, especially for diseases with a late age of onset, the genotypes of one (or both) parents may not be available. This issue has been addressed by a number of authors (Curtis and Sham 1995b; Curtis 1997; Laird et al. 1998; Spielman and Ewens 1998; Knapp 1999a, 1999c; Gordon et al. 2004), generally by using marker data from unaffected sibs in lieu of parental data, an approach Spielman and Ewens (1998) called **sib TDT** (or **S-TDT**). A related issue to missing genotypes are genotyping errors, which can impact the TDT. For example, Mitchell et al. (2003) noted that undetected genotyping errors create an apparent overtransmission of common alleles. Gordon et al. (2001, 2004) and Dudbridge (2008) introduced modifications of the TDT that are robust to random genotyping errors.

Other important extensions are to multiple marker loci, i.e., haplotypes (Bickeböller and Clerget-Darpoux 1995; Sham and Curtis 1995b; Clayton 1999; Clayton and Jones 1999; Dudbridge et al. 2000), and to application over pedigrees, as opposed to trio or sib data, leading to **pedigree disequilibrium tests**, **PDTs** (Abescasis et al. 2000; Martin et al. 2000, 2003; Gordon et al. 2004). Finally, while the classic TDT (and its above-mentioned extension) applies to traits with present or absence count data (such as presence or absence of a disease), family-based segregation methods can be extended to quantitative (continuously valued) traits (Allison 1997; Allison et al. 1999a; George et al. 1999; Sun et al. 2000; Zhu and Elston 2001; Malkin et al. 2002; Tiwari et al. 2005; Diao and Lin 2006; Weiner et al. 2017). Such **quantitative transmission/disequilibrium tests**, or **QTDTs**, are reviewed by Ewens et al. (2008).

While the TDT, and its extensions, provide solid control over population structure, they require *rather specialized samples*, namely, family trios (or similar sets of relatives). As such, the total sample size in a TDT tends to be very small, resulting in modest to low power even when testing just a few potential candidates (unless their effect size is substantial). If one wishes to scan a large number of candidates (such as a set of SNPs covering a genome or genomic region), power is essentially non-existent, given the massive multiple-testing corrections required (Chapter 20). Hence, the TDT is best considered as a rather specialized, but gold-standard, test for replication of a signal from one (or a few) markers that are of very high interest (perhaps as a prelude to a rather expensive functional-genomics or gene-editing study). Chapter 21 examines other very-fine mapping methods when very dense SNP data are available.

Estimating Effects of Candidate Loci

While the estimation of genotype means for a candidate locus seems straightforward, there are several potential sources of bias. While it is possible in some settings to distinguish between the direct effects of a candidate locus and its indirect effects due to association with other linked QTLs (e.g., Bovenhuis and Weller 1994), as noted above the presence of LD greatly confounds the interpretation of candidate-locus means (Chapter 21 examines methods for very fine mapping of actual causal sites using dense SNPs). We now consider more subtle sources of bias.

Let z_{ij} denote the phenotypic value of the jth individual with candidate-locus genotype i $(1 \le i \le n_g)$. If μ_i is the mean character value for genotype i, then the simplest model is

$$z_{ij} = \mu_i + \epsilon_{ij} \tag{17.24}$$

If individuals are sampled at random from a large homogeneous population, the residuals are uncorrelated and the μ_i values can be estimated simply by using $\overline{z}_i$. However, when relatives are present in the sample, residuals can be correlated, and one must take the residual covariance matrix into consideration to obtain an unbiased estimator (Example 10.12). The residual error can be decomposed into a genetic component (G) due to segregation at loci other than the target QTL plus the general (E) and specific (e) environmental effects,

$$\epsilon_{ij} = G_{ij} + E_{ij} + e_{ij} \tag{17.25a}$$

Indexing two distinct individuals by j and k,

$$\sigma(\epsilon_j, \epsilon_k) = \sigma(G_j, G_k) + \sigma(E_j, E_k) \tag{17.25b}$$

as special environmental effects are not shared by different individuals (Chapter 6), and G and E are assumed to be uncorrelated. If the relationships among sampled individuals are known, the elements of the covariance matrix can be computed from Equation 17.25b, using the methods of Chapter 7. For example, if i and j are full-sibs, $\sigma(G_j, G_k) = (\sigma^2_A/2) + (\sigma^2_D/4)$, where the variances refer to the contribution of background polygenes. Similarly, full sibs may also share a common maternal environmental effect (m), so that $\sigma(E_j, E_k) = \sigma^2_m$. Estimates of the variance components associated with G and E can be obtained by a variety of methods discussed in detail in Part II (Chapters 22–33). Using these to estimate the residual covariance matrix, genotypic means can be estimated by a GLS regression (Equation 10.25e).

An example of the importance of correcting for common relatives is provided by Bentsen and Klemetsdal (1991), who examined the association between egg productivity and six different MHC haplotypes in chickens. The relative rankings of five of the six haplotype effects differed when comparing uncorrected estimates with those obtained by correcting for shared ancestry. More generally, when additional fixed factors (such as sex- or age-specific effects) are included, mixed-model methods can be used (Boerwinkle et al. 1986; Cowan et al. 1990; Hoeschele and Meinert 1990; Bentsen and Klemetsdal 1991; Blangero et al. 1992; Kennedy et al. 1992). Chapter 32 (in Part II) examines this in some detail.

Many estimates of the amount of variability attributable to a candidate locus are upwardly biased, even for a random population sample. If q_i is the frequency of candidate-locus genotype i, then the variance contributed by this locus is

$$\sigma^2_L = \sum_{i=1}^{n_g} q_i\,(\mu_i - \mu)^2 \qquad \text{where} \qquad \mu = \sum_{i=1}^{n_g} q_i\,\mu_i \tag{17.26}$$

Boerwinkle and Sing (1986) showed that the obvious estimator of the variance

$$s^2 = \sum_{i=1}^{n_g} \widehat{q}_i\,(\,\overline{z}_i - \overline{z}\,)^2 \tag{17.27a}$$

using the estimates $\widehat{q}_i$, $\overline{z}_i$, and $\overline{z}$ in place of their true values is biased because it includes the sampling error of these three estimates. To account for the sampling variances in $\overline{z}_i$

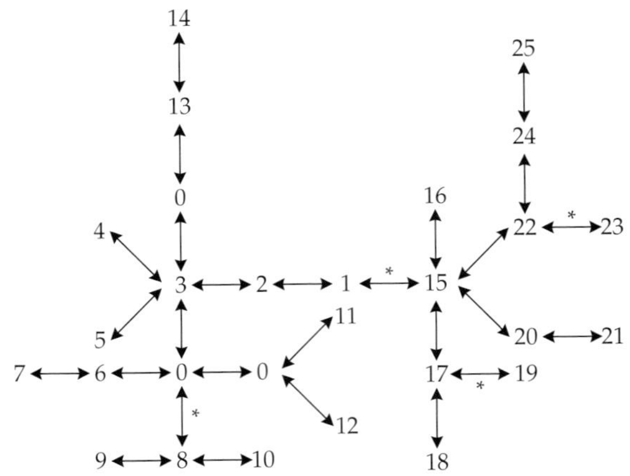

Figure 17.11 Inferred cladogram of haplotypes for a region surrounding the *Drosophila melanogaster* ADH gene. Numbers refer to different haplotypes, while "0" refers to intermediate haplotypes not present in the sample but required to interconnect the existing haplotypes. Distance between two haplotypes on the cladogram provides a measure of relatedness, so that haplotypes 8 and 9 are more closely related than 9 and 6 (one step versus three steps). Using the hierarchical structure implied by this cladogram, a nested ANOVA was performed, with ANOVA levels set by the number of cladogram steps that haplotypes differ by. Four cladogram regions, denoted with asterisks, were found to be associated with significant phenotypic effects. (After Templeton et al. 1987.)

and $\overline{z}$, Boerwinkle and Sing suggest the improved estimator (for a collection of unrelated individuals),

$$s_L^2 = \sum_{i=1}^{n_g} \widehat{q}_i \, (\, \overline{z}_i - \overline{z}\,)^2 - \left(\frac{n_g - 1}{n} \right) \text{Var}(e) \tag{17.27b}$$

where Var(e) is the estimate of the residual variance (within genotypes) and we assume n measured individuals per genotype. While an improvement over Equation 17.27a, this estimator still ignores the sampling variance introduced by using $\widehat{q}_i$ and hence still tends to overestimate the variance accounted for by the candidate locus. Resampling approaches for constructing confidence intervals of both σ_L^2 and its additive and dominance components have been suggested by Kaprio et al. (1991) and Zerba et al. (1996).

A general solution to this reoccurring issue of estimating the sample variance of a composite parameter (σ_L^2) that itself is a function of other parameters that have sampling errors (q_i, μ_i) is to use a full **Bayesian analysis** (Appendices 7 and 8). Under such an analysis, one obtains the posterior distribution of the vector of parameters $(q_1, \cdots, q_{n_g}, \mu_1, \cdots, \mu_{n_g})$ by starting with some prior belief about their values (which is often a flat prior), and then use the observed data values to update this distribution to generate a posterior. As Appendix 8 details, this is usually done by using Markov Chain Monte Carlo methods (MCMC) to generate draws from this posterior distribution. Under such an analysis, a draw from the MCMC sampler returns a vector of values for the q_i and μ_i which are then substituted into Equation 17.26 to generate a corresponding estimate σ_L^2. One repeats this process several thousand times to generate an empirical distribution of σ_L^2 estimates that fully account for the sampling value in both the q_i and μ_i.

Templeton and Sing's Method: Using the Historical Information in Haplotypes

While one can use a single marker to examine whether genetic variation at a candidate locus is associated with character variation, studies typically involve several closely linked markers, often including several polymorphic sites within the gene itself. How should one best extract information from this set of markers? Obviously, it is more powerful to consider **haplotypes** (the multiple-locus genotypes associated with the region of interest) than single markers. One drawback with this approach is that there can be an enormous number of haplotypes, resulting in small sample sizes for each haplotype and a reduction in the power of tests for haplotype-trait associations. What is needed is a logical way to combine information from different haplotypes. In an interesting series of papers, Templeton and Sing (Templeton et al. 1987, 1988, 1992; Templeton 1995) suggest that this can be done by incorporating information on the inferred evolutionary relationships of the sampled haplotypes. We simply outline their basic idea here.

The sequence information in haplotypes can be used to construct a **cladogram** estimating how the different haplotypes are evolutionarily related to each other. With such a cladogram in hand, one can then consider how character value is a function of nested sets of cladogram members, for example by using a nested ANOVA (Chapter 23 in Part II). The motivation behind this approach is that history is the main cause of LD, with associations between loci decaying with time. Hence, the more closely related a set of haplotypes, the more disequilibrium they should display, and the more likely they are to share QTL alleles. Figure 17.11 shows an example of this approach, detecting associations between alcohol dehydrogenase (ADH) activity and haplotypes in a 13-kb region surrounding the ADH gene in *Drosophila*. This approach was used by Klein et al. (2005) in one of the first GWAS studies (Chapter 20) as part of their evidence that a polymorphism in human Complement Factor H was associated with age-related macular degeneration.

18

Linkage Mapping And Characterizing QTLs: Inbred Line Crosses

QTL detection experiments do not map a gene, but rather a genetic effect that might consist of many linked genes within ∼ 30 cM of each other. Mapping loci in inbred crosses can, therefore, disguise the genetic architecture of a QTL. Flint and Mott (2001)

Crosses between completely inbred lines offer an ideal setting for detecting and mapping QTLs by marker-trait associations, as all F_1s are genetically identical and show complete LD for genes differing between lines. A number of crossing designs have been proposed to exploit these features. While usually involving flies and crops, QTL line-cross analysis has also been applied to a number of other plant and animal species, especially mice (e.g., Frankel 1995; Flint and Mackay 2009; Aylor et al. 2011; Atamni et al. 2018). More recently, QTL mapping designs have exploited special biological features of the species being used, such as fungal systems (Liti and Louis 2012). A particularly interesting example is the mapping of QTLs for honey-bees foraging behavior using designs that exploited the haploid nature of the males (Hunt et al. 1995; 2007; Rüppel et al. 2004).

We note that the heyday of linkage-based QTL mapping (i.e., using a set of individuals from a cross or pedigree), that was at its height when our first edition was published, has now likely passed. Thanks to modern genomics, association mapping (using a dense set of markers in a population sample; Chapter 20) is now feasible for just about any species. This allows for a much larger, and (usually) easier to gather, sample than one requiring that all individuals come from a cross or pedigree, coupled with much finer mapping resolution (kilobase, rather than megabase, scales in many cases). However, linkage-based QTL mapping is still viable for many questions and introduces some of the statistical issues for association mapping that we examine more deeply in Chapter 20. Cross-based QTL mapping (this chapter) and association mapping (Chapter 20) also ***examine different genetic architectures***. The former examines the genetic basis of *differentiation* between the parental lines generating the cross, the latter examines the nature of *segregating variation* for the focal trait in a given population. These two architectures can be largely independent, as the former may be shaped by selection for trait divergence, while the later may be shaped (in part) by selection against deleterious alleles (Chapter 15; WL Chapter 28).

While the vast QTL mapping literature may seem complex to the uninitiated, it is based on a few simple ideas. We start by reviewing these foundational building blocks, first considering basic line-cross designs useful for QTL mapping. The key element from which the formal theory of QTL mapping is constructed is the ***conditional probability of a particular QTL genotype given an observed marker genotype***. These probabilities allow a full development of the two principal methods for QTL detection and estimation: least-squares (LS) linear models (which use differences in marker mean) and maximum likelihood (ML, and, by extension, Bayesian methods) which use information from the entire marker-trait distribution. Both approaches are examined in some detail. We conclude our discussion of statistical issues by considering adjustments for multiple comparisons, examine methods that accommodate multiple linked QTLs, develop more complex designs, discuss the bias in effect size estimates introduced by the Beavis effect (or the winner's curse), and consider the sample sizes required to achieve a given power for detecting, and locating, a QTL. Finally, we reward the persevering reader with a review of some interesting case studies that have used inbred lines for QTL mapping.

FOUNDATIONS OF LINE-CROSS MAPPING

The idea behind using marker information to map and characterize QTLs is quite simple: by crossing two inbred lines, the excess of parental over recombination gametes at linked loci (Chapter 5) generates LD in their progeny (Example 5.6). This in turn creates associations between alleles at marker loci and alleles at linked segregating QTLs. A number of diverse experimental designs and statistical methodologies have been developed to exploit this information. Our attempt to make this approach more accessible starts with an overview of some of the basic experimental designs and key tools for the analysis of results from these designs.

Basic Crossing Designs

The large number of possible designs can be categorized by the type of line-cross populations used for generating disequilibrium (e.g., F_2 versus backcross populations), the population in which individuals are genotyped, the population in which phenotypes are scored, and the unit of marker analysis used (e.g., single markers versus interval mapping). We consider these in turn.

Starting with two completely inbred parental lines, P_1 and P_2, a number of line-cross populations derived from their F_1 can be used for QTL mapping. The **F_2**, or **intercross, design** examines marker-trait associations in the progeny from a cross (or selfing) of F_1s, while the **backcross design** (**B** or **BC**) examines marker-trait associations in the progeny formed by backcrossing the F_1 to one (or both) parental lines ($B_i = F_1 \times P_i$). While these are the most widely used designs, other line-cross populations can offer further advantages (and disadvantages). As discussed in Chapter 17, the F_1 can be used to create **recombinant inbred lines** (**RILs**) and **doubled haploid lines** (**DHLs**). While generation of these lines is a laborious process, they allow marker-trait associations to be scored in a completely homozygous background and offer the ability (via replication) of scoring the line phenotype with arbitrary precision and across multiple environments. The F_2 design has an advantage over designs using backcross, RIL, or DHL populations, because it generates three genotypes at each marker locus, which allows the estimation of the degree of dominance associated with detected QTLs. Designs using an F_t population (formed by randomly mating F_1s for $t-1$ generations) allow for even higher resolution of QTL map positions than do F_2s, albeit at the expense of decreased power of QTL detection (due to lower effective marker density). The properties of such **advanced intercross lines** (AILs) are discussed below.

More complex designs can be considered wherein individuals are genotyped in one population, while trait values are scored in a future population derived from the genotyped individuals. Fisch et al. (1996) present a general treatment for such designs. One example is the **$F_{2:3}$ design**, wherein F_2 individuals are genotyped and then selfed to create F_3 families for each individual (Zhang and Xu 2004). The trait value associated with a genotyped individual is estimated by the mean value of their resulting F_3 family. Scoring the phenotype as the mean of several individuals (as opposed to the measurement of a single individual) offers increased power over a standard F_2 design by reducing the sampling variance.

Designs combining information from multiple crosses (Chapter 11) are expected to be more powerful than those involving a single cross, and a number of such schemes have been proposed, in particular, for estimating components associated with dominance and heterosis. One can augment these designs with markers to use them for QTL mapping. Examples include diallel designs whose basic structure is examined in Part II Chapters 23 and 25 (Rebaï and Goffinet 1993; Rebaï et al. 1994a), and Comstock and Robinson's (1952) classic Design III wherein the F_2 from two inbred lines is backcrossed to both parental lines (Cockerham and Zeng 1996). We consider such multiple-line designs later in the chapter. Finally, several investigators have considered designs involving crosses between lines that are not completely inbred, such as a cross of an outbred line to a completely inbred tester (Beckmann and Soller 1988; Dudley 1992; Haley et al. 1994b).

Experimental designs are also classified by the unit of marker analysis chosen by the investigator. Marker-trait associations can be assessed using one-, two-, or multiple-locus marker genotypes. Under a **single-marker analysis**, the distribution of trait values is examined separately for each marker locus. Each marker-trait association test is performed independent of information from all other markers, so that a chromosome with n markers offers n separate single-marker tests. As discussed below, single-marker analysis is generally a good choice when the goal is simple *detection* of a QTL linked to a marker, rather than *estimation* of its position and effects. Under **interval mapping** (or **flanking-marker analysis**), a separate analysis is performed for each *pair* of adjacent marker loci. Here one computes the probability of a putative QTL at a series of discrete locations between the two markers defining the target interval. The use of such two-locus marker genotypes results in $n-1$ separate tests of marker-trait associations along a chromosome with n markers (one for each marker interval). Interval mapping offers a further increase in power of detection (albeit usually a slight one) and more precise estimates of QTL effects and position. As we detail below, both single-marker and interval mapping approaches are biased when multiple QTLs are linked to the marker/interval being considered.

Methods simultaneously using three or more marker loci attempt to reduce or remove such bias. **Composite interval mapping** (Jansen 1993b, 1994b; Zeng 1993, 1994; Jansen and Stam 1994) considers a marker interval plus a few other well-chosen markers in each analysis, so that (as above) $n-1$ tests for interval-trait associations are performed on a chromosome with n markers. **Multiple interval mapping** extends this approach by simultaneously fitting multiple intervals (Kao et al. 1999; Zeng et al. 1999). Finally, **multipoint mapping** considers all of the linked markers on a chromosome simultaneously, resulting in a single analysis for each chromosome (Kearsey and Hyne 1994; Hyne and Kearsey 1995; Wu and Li 1994, 1996a, 1996b; Charmet et al. 1998).

Conditional Probabilities of QTL Genotypes

The basic element upon which the formal theory of QTL mapping is built is the conditional probability that the QTL genotype is Q_k, given the observed (possibly multilocus) marker genotype is M_j. From the definition of a conditional probability (Equation 3.3a), this is

$$\Pr(Q_k \mid M_j) = \frac{\Pr(Q_k M_j)}{\Pr(M_j)} \tag{18.1}$$

The joint and marginal probabilities, $\Pr(Q_k M_j)$ and $\Pr(M_j)$, are functions of the experimental design and the linkage map (the position of the putative QTLs with respect to the marker loci). Computing these probabilities is a relatively simple matter of bookkeeping (see Example 18.1), but can get rather tedious as the number of markers and/or QTLs under consideration increases.

When computing joint probabilities involving more than two loci, one must also account for recombinational interference between loci (Chapter 17). Consider a single QTL flanked by two markers, M_1 and M_2. The gamete frequencies depend on three parameters: the recombination frequency c_{12} between markers, the recombination frequency c_1 between marker M_1 and the QTL, and the recombination frequency c_2 between the QTL and marker M_2. Under the assumption of no interference, $c_{12} = c_1 + c_2 - 2c_1c_2$, while $c_{12} = c_1 + c_2$ under complete interference (Chapter 17). When c_{12} is small, gamete frequencies are essentially identical under either interference assumption. Typically, c_{12} is assumed known, leaving two unknown recombination parameters (c_1 and c_2) under general assumptions about interference. In either case, there is only one parameter to estimate, as assuming complete interference $c_2 = c_{12} - c_1$, and assuming no interference $c_2 = (c_{12} - c_1)/(1 - 2c_1)$. Note that $1/(1-2c_1)$ has values of 1.11, 1.25, 1.67, and 2 for c_1 values of 0.05, 0.1, 0.2, and 0.25, respectively. Hence, for flanking-marker analysis, we restrict attention to the single recombination parameter c_1, the distance from marker locus M_1 to the QTL. When considering analysis of single-marker loci, for notational ease we drop the subscript, using c in place of c_1.

Conditional probabilities involving more than three linked loci are generally dealt with by first assuming an appropriate mapping function on which distances are additive

(Chapter 17), and then translating these distances into recombination frequencies. When a large number of markers is considered, missing marker information can become a problem. Many individuals can be left with incomplete multilocus marker genotypes, potentially excluding them from further analysis. Martínez and Curnow (1992) show how information from linked markers can be used to estimate the genotype at missing or incomplete (i.e., dominant) markers. Predicting the values of missing marker genotypes, **imputation**, is an important concept that arises in association mapping, and will be considered in detail in Chapter 20.

Example 18.1. Consider a single-marker analysis using the F_2 formed by crossing two inbred lines, $MMQQ \times mmqq$. If the recombination frequency between the marker locus and the QTL is c, the expected F_1 gamete frequencies (Chapter 5) are

$$\Pr(MQ) = \Pr(mq) = (1-c)/2, \qquad \Pr(Mq) = \Pr(mQ) = c/2$$

The probability that an F_2 individual is $MMQQ$ is $\Pr(MQ) \cdot \Pr(MQ) = [\,(1-c)/2\,]^2$. Likewise, $2 \cdot \Pr(MQ) \cdot \Pr(mQ) = 2(\,c/2\,)[\,(1-c)/2\,]$ is the probability of an $MmQQ$ individual, and so on. Because the probabilities of the marker genotypes MM, Mm, and mm are $1/4$, $1/2$, and $1/4$, Equation 18.1 gives the F_2 conditional probabilities as

$$\begin{aligned}
&\Pr(QQ \mid MM\,) = (1-c)^2, \quad \Pr(Qq \mid MM\,) = 2c(1-c), \quad \Pr(qq \mid MM\,) = c^2 \\
&\Pr(QQ \mid Mm\,) = c(1-c), \ \Pr(Qq \mid Mm\,) = (1-c)^2 + c^2, \ \Pr(qq \mid Mm\,) = c(1-c) \\
&\Pr(QQ \mid mm\,) = c^2, \quad \Pr(Qq \mid mm\,) = 2c(1-c), \quad \Pr(qq \mid mm\,) = (1-c)^2
\end{aligned}$$

Note that for small c that $\Pr(QQ \mid MM\,) \simeq 1$, $\Pr(Qq \mid Mm\,) \simeq 1$, and $\Pr(qq \mid mmM\,) \simeq 1$.

This same logic extends to multiple marker loci. Suppose the QTL is flanked by two scored markers, and consider the F_2 in a cross of lines fixed for M_1QM_2 and m_1qm_2. What are the conditional probabilities of the QTL genotypes when the marker genotype is $M_1M_1M_2M_2$? Because all F_1s are M_1QM_2/m_1qm_2, under the assumptions of no interference, the frequency of F_1 gametes involving M_1M_2 are

$$\Pr(M_1QM_2) = (1-c_1)(1-c_2)/2, \qquad \Pr(M_1qM_2) = c_1\,c_2/2$$

giving expected frequencies in the F_2 of $M_1M_1M_2M_2$ offspring as

$$\begin{aligned}
\Pr(M_1QM_2/M_1QM_2) &= [\,(1-c_1)(1-c_2)/2\,]^2 \\
\Pr(M_1QM_2/M_1qM_2) &= 2\,[\,(1-c_1)(1-c_2)/2\,]\,[\,c_1\,c_2/2\,] \\
\Pr(M_1qM_2/M_1qM_2) &= (\,c_1\,c_2/2\,)^2
\end{aligned}$$

where $c_2 = (c_{12} - c_1)/(1-2c_1)$. The overall frequency of $M_1M_1M_2M_2$ individuals, $\Pr(M_1M_1M_2M_2)$, is the sum of the three above terms, or $(1-c_{12})^2/4$. Substituting into Equation 18.1 gives

$$\begin{aligned}
\Pr(QQ \mid M_1M_1M_2M_2) &= \frac{(1-c_1)^2(1-c_2)^2}{(1-c_{12})^2} \\
\Pr(Qq \mid M_1M_1M_2M_2) &= \frac{2c_1c_2(1-c_1)(1-c_2)}{(1-c_{12})^2} \qquad (18.2) \\
\Pr(qq \mid M_1M_1M_2M_2) &= \frac{c_1^2\,c_2^2}{(1-c_{12})^2}
\end{aligned}$$

Conditional probabilities for other marker genotypes are computed in a similar fashion. Because $c_1c_2 \leq (c_{12}/2)^2$ is usually very small if c_{12} is moderate to small, essentially all $M_1M_1M_2M_2$ individuals are QQ. For example, assuming $c_1 = c_2 = c_{12}/2$ (the worst case), the conditional probabilities of an $M_1M_1M_2M_2$ individual being QQ are 0.96, 0.98, and 0.99 for $c_1 = c_2 = 0.25, 0.2$, and 0.1.

We now consider the conditional probabilities for other single-marker line cross designs, starting with backcrosses. For a B_1 population, where the F_1 is backcrossed to P_1 (with genotype $MMQQ$), one parental gamete is always MQ. Hence, for a single-marker analysis, there are only two marker genotypes, MM and Mm. Using the frequencies for the four possible gametes (Example 18.1) of the F_1 parent gives the following conditional probabilities

$$\begin{aligned} \Pr(QQ \,|\, MM\,) &= 1-c, \qquad & \Pr(Qq \,|\, MM\,) &= c \\ \Pr(QQ \,|\, Mm\,) &= c, \qquad & \Pr(Qq \,|\, Mm\,) &= 1-c \end{aligned} \tag{18.3a}$$

Likewise, when backcrossing to the P_2 ($mmqq$), the two possible single-locus marker genotypes are Mm and mm, and the conditional probabilities become

$$\begin{aligned} \Pr(qq \,|\, mm\,) &= 1-c, \qquad & \Pr(Qq \,|\, mm\,) &= c \\ \Pr(qq \,|\, Mm\,) &= c, \qquad & \Pr(Qq \,|\, Mm\,) &= 1-c \end{aligned} \tag{18.3b}$$

Historically, backcross designs were often used when only dominant markers (e. g., RAPDs; Chapter 8) were available. However, the backcross design can be driven by the biology of the problem. A good example is genetics of speciation (Chapter 17), wherein backcrosses of a hybrid (F_1) to one species can be viable, while both the reciprocal backcross and the hybrid–hybrid ($F_1 \times F_1$) crosses are not.

For designs involving more than one generation of recombination, the single generation recombination frequency c is simply replaced by a corrected frequency $\widetilde{c}$ that is a function of the particular design. We consider three such designs: advanced intercross lines (AILs), recombinant inbred lines (RILs), and double-haploid lines (DHLs).

AILs (Darvasi and Soller 1995; Liu et al. 1996) are obtained by crossing two inbred lines, but instead of stopping at the F_2, random mating proceeds for t generations, generating an F_t. In this case, unlike the strategy used to create RILs (Chapter 17), inbreeding is avoided by keeping the breeding population size large. As the result of the multiple rounds of recombination, markers in an F_t individual show an expansion of the genetic map relative to an F_2. The expected frequency of a recombinant gamete in the F_t for a pair of loci at recombination fraction c follows from Equation 5.16c,

$$\widetilde{c} = \frac{1-(1-c)^{t-2}(1-2c)}{2} \simeq \frac{t}{2}\,c \tag{18.4}$$

where the approximation holds for $ct \ll 1$ (Darvasi and Soller 1995; Liu et al. 1996). For example, if the marker–QTL recombination frequency is $c = 0.01$, only 1% of the F_2 gametes are recombinant (Mq, mQ), but this increases to 2.5% in an F_5 and 9.1% in an F_{20}. The conditional genotype probabilities for an F_t AIL are given by the F_2-design expressions in Example 18.1, with $\widetilde{c}$ substituted for c.

RILs also involve several generations of recombination, but here genotypes are eventually fixed by inbreeding (Chapter 12). Starting with an MQ/mq F_1 parent, there are only four possible genotypes in the resulting RILs: $MMQQ$, $MMqq$, $mmQQ$, and $mmqq$. The frequency of recombinant gametes (Mq, mQ) in RILs approaches a limiting value of $\widetilde{c} = 2c/(1+2c)$ for selfed lines and $\widetilde{c} = 4c/(1+6c)$ for lines formed by brother–sister mating (Haldane and Waddington 1931). Thus, the expected frequencies of genotypes in RILs are

Line genotype	Frequency
$MMQQ$, $mmqq$	$(1-\widetilde{c})/2$
$MMqq$, $mmQQ$	$\widetilde{c}/2$

While DHLs also have only these four genotypes, they are formed by a single generation of meiosis (if based on using F_1s), so that $\widetilde{c} = c$. If based on F_2s, Equation 18.4 gives $\widetilde{c} = 3c/2 - c^2$. Hence, among either RILs or DHLs, the conditional QTL probabilities are

$$\begin{aligned} \Pr(QQ \,|\, MM) &= 1 - \widetilde{c}, \qquad & \Pr(qq \,|\, MM) &= \widetilde{c} \\ \Pr(QQ \,|\, mm) &= \widetilde{c}, \qquad & \Pr(qq \,|\, mm) &= 1 - \widetilde{c} \end{aligned} \tag{18.5a}$$

where

$$\widetilde{c} = \begin{cases} c, \;\; 1.5c + c^2 & \text{for } F_1 \text{ and } F_2 \text{ DHL, respectively} \\ 2c/(1+2c) & \text{for RILs formed by selfing} \\ 4c/(1+6c) & \text{for RILs formed by brother–sister mating} \end{cases} \tag{18.5b}$$

Values of $\widetilde{c}$ for intermediate generations of inbreeding are given by Teuscher et al. (2006).

Expected Marker-class Means

With these conditional probabilities in hand, the expected trait values for the various marker genotypes follow immediately. Suppose there are N QTL genotypes, $Q_1, \cdots, Q_N$, where the mean of the kth QTL genotype is μ_{Q_k}. The mean value for marker genotype M_j is just

$$\mu_{M_j} = \sum_{k=1}^{N} \mu_{Q_k} \Pr(\, Q_k \,|\, M_j \,) \tag{18.6}$$

The ***QTL effects*** enter through the μ_{Q_k}, while the ***QTL positions*** enter through the conditional probabilities $\Pr(\, Q_k \,|\, M_j \,)$. Equation 18.6 is completely general, allowing for multilocus marker genotypes and multiple QTLs.

Example 18.2. Consider the single-marker F_2 design with a single QTL linked (at recombination frequency c) to the marker. Denote the QTL genotypic values by

$$\mu_{QQ} = \mu + 2a, \quad \mu_{Qq} = \mu + a(1+k), \quad \text{and} \quad \mu_{qq} = \mu$$

where a measures the additive value and k the degree of dominance ($d = ak$; Chapter 4). Applying the conditional probabilities developed in Example 18.1 to Equation 18.6, the mean values for the marker genotypes become

$$\begin{aligned} \mu_{MM} &= \mu + 2a(1-c)^2 + 2c(1-c)(1+k)a \\ \mu_{Mm} &= \mu + 2ac(1-c) + [1 - 2c(1-c)](1+k)a \\ \mu_{mm} &= \mu + 2ac^2 + 2c(1-c)(1+k)a \end{aligned}$$

If the marker and QTL are unlinked ($c = 1/2$), all marker genotypes have the same mean, $\mu + a[1 + (k/2)]$. Rearranging these equations gives

$$(\mu_{MM} - \mu_{mm})/2 = a(1-2c) = a^* \tag{18.7a}$$

$$\frac{\mu_{Mm} - (\mu_{MM} + \mu_{mm})/2}{(\mu_{MM} - \mu_{mm})/2} = k(1-2c) = k^* \tag{18.7b}$$

Hence, one strategy for detecting QTLs is to test for ***significant differences between the mean trait values associated with different marker genotypes***. This is the basis for QTL detection via regression or ANOVA, which we generically refer to as **least-squares (LS) linear model** approaches (Chapter 10).

This example shows that while contrasts of single-marker means can be used to estimate both a^* and k^*, these underestimate the magnitude of a and k by the (unknown) fraction $1 - 2c$. If the marker and QTL are tightly linked, this error is small, but it increases rather dramatically as c approaches 1/2. A small difference between marker–homozygote means is thus compatible with either a tightly linked QTL of small effect or a loosely linked QTL of large effect. As we will shortly show, when multilocus marker genotypes are considered, the use of appropriate combinations of marker means allows for separate estimates of QTL effects (a, k) and position (c).

If there are N QTLs linked to the marker, the ith of which is at recombination frequency c_i from the marker and has associated additive and dominance effects a_i and k_i, then (Edwards et al. 1987),

$$(\mu_{MM} - \mu_{mm})/2 = \sum_{i=1}^{N} a_i^* \tag{18.8a}$$

$$\frac{\mu_{Mm} - (\mu_{MM} + \mu_{mm})/2}{(\mu_{MM} - \mu_{mm})/2} = \sum_{i=1}^{N} a_i^* k_i^* \Big/ \sum_{i=1}^{N} a_i^* \tag{18.8b}$$

where $a_i^* = a_i(1 - 2c_i)$ and $k_i^* = k_i(1 - 2c_i)$. If some of the linked QTLs have effects of opposite sign, some cancellation occurs, reducing the marker-trait association. Moreover, with multiple linked QTLs, the degrees of dominance (k_i) are confounded with the homozygous effects (a_i).

Marker-class means for other designs follow by applying the appropriate conditional probabilities to Equation 18.6. For example, for the B_1 (= $F_1 \times P_1$) design, from Equation 18.3a,

$$\mu_{MM} - \mu_{Mm} = (\mu_{QQ} - \mu_{Qq})(1 - 2c) = a(1 - k)(1 - 2c) \tag{18.9a}$$

Thus, under a backcross design the scaled QTL effects are influenced strongly by the (unknown) degree of dominance k. If Q is completely dominant to q, $k = 1$, and there is no marker–QTL effect. Conversely, if q is dominant to Q, $k = -1$ and the scaled effect becomes $2a(1-2c)$, which is the same as under an F_2 design. Recalling Equation 18.3b, the reciprocal backcross (B_2 = $F_1 \times P_2$) yields a similar expression,

$$\mu_{Mm} - \mu_{mm} = (\mu_{Qq} - \mu_{qq})(1 - 2c) = a(1 + k)(1 - 2c) \tag{18.9b}$$

Note that the ratio of Equation 18.9a to 18.9b gives $(1 - k)/(1 + k)$, so that (provided only a single QTL is linked to the marker) an estimate of k can be obtained if one has access to *both* backcross populations (B_1 and B_2).

The expressions developed in Example 18.2 for F_2 analysis hold for an F_t population, provided $\widetilde{c}$ (given by Equation 18.4) replaces c. For example, $\mu_{MM} - \mu_{mm} = 2a(1-2\,\widetilde{c}\,)$, and so forth. Because $(1-2c) > (1-2\widetilde{c}\,)$, AILs have smaller differences between marker means, and hence reduced power of QTL detection, relative to the F_2 design. Despite this, Darvasi and Soller (1995) advocate the use of AILs for fine-mapping of QTLs, as the expansion of the genetic map (relative to the F_2) offers a higher precision (improved **mapping resolution**) of estimates of QTL position.

For RILs and DHLs, the recombination parameter $\widetilde{c}$ is given by Equation 18.5b, and from Equations 18.5a and 18.6 it follows that

$$\mu_{MM} = \mu_{QQ}\,(1 - \widetilde{c}\,) + \mu_{qq}\,\widetilde{c} \qquad \text{and} \qquad \mu_{mm} = \mu_{QQ}\,\widetilde{c} + \mu_{qq}\,(1 - \widetilde{c}\,) \tag{18.10a}$$

giving

$$\frac{\mu_{MM} - \mu_{mm}}{2} = a\,(1 - 2\,\widetilde{c}\,) = a^* \tag{18.10b}$$

again providing an estimate of a composite parameter of the QTL effect (a) and position (c). Because $\widetilde{c}$ is smallest in DHLs (see Equation 18.5b), the largest marker effect (and greatest power for QTL detection) occurs in this type of line, followed by selfed RILs, and finally by sib-mated RILs. Conversely, the map expansion from recombination during line formation that reduces power increases the precision of QTL position estimates. Hence, confidence intervals are smallest for sib-mated RILs and largest for DHLs.

Finally, note that by considering two-locus (rather than single-locus) marker means, separate estimates of QTL effect and position can be obtained. Taking the genotype at two adjacent marker loci (M_1/m_1 and M_2/m_2) as the unit of analysis, consider the difference between the contrasting double homozygotes in an F_2. If the markers flank a QTL, then under the assumption of no interference, Equation 18.2 (and its analog for $m_1m_1m_2m_2$ probabilities) yields

$$\frac{\mu_{M_1M_1M_2M_2} - \mu_{m_1m_1m_2m_2}}{2} = a\left(\frac{1-c_1-c_2}{1-c_1-c_2+2c_1\,c_2}\right) \simeq a\,(1-2c_1\,c_2) \tag{18.11a}$$

where c_i is the M_i–QTL recombination frequency. Equation 18.11a is essentially equal to a when the distance between flanking markers $c_{12} \leq 0.20$, as here $(1-2c_1\,c_2) \geq 0.98$. Thus, recalling from Equation 18.7a that $\mu_{M_1M_1} - \mu_{m_1m_1} = 2a(1-2c_1)$, we can obtain estimates of the recombination frequencies by substituting Equation 18.11a for a and rearranging to give

$$c_1 = \frac{1}{2}\left(1 - \frac{\mu_{M_1M_1} - \mu_{m_1m_1}}{2a}\right) \simeq \frac{1}{2}\left(1 - \frac{\mu_{M_1M_1} - \mu_{m_1m_1}}{\mu_{M_1M_1M_2M_2} - \mu_{m_1m_1m_2m_2}}\right) \tag{18.11b}$$

Using the same logic (and bookkeeping), estimates for other flanking-marker designs were developed by Knapp et al. (1990), Knapp and Bridges (1990), and Knapp (1991).

Marker Variances and Higher Order Moments

In addition to differing in trait *means*, marker genotypes can also differ in *higher order moments* as well, such as the variance or skew. Such differences can arise in segregating populations (e.g., F_2s) simply as a consequence of changes in the mixture proportions of QTL genotypes among marker genotypes (Equation 16.5d). Such differences are not uncommon. For example, in a cross of tomato species, Weller et al. (1988) found significantly different variances for 28% (40 of 180) of the possible marker-trait associations, while 17% showed significant differences in skewness.

Several workers have suggested the use of these higher order moments for detection of a linked QTL and estimation of its effects (Zhuchenko et al. 1978, 1979; Korol et al. 1981, 1983, 1987; Ginzburg 1983; Asins and Carbonell 1988; Zhang et al. 1992). One difficulty with this approach is that variances (and higher moments) are estimated with far less precision than means, reducing both the power of detection and the accuracy of estimates. Another complication is that not all designs are capable of revealing significant changes in higher moments (e.g., Asins and Carbonell 1988).

RILs and DHLs provide one case where functions of higher-order moments (e.g., a correlation coefficient) can be exploited. Here, single-locus marker information can be used to estimate the recombination frequency c (Hu et al. 1995). As shown in Table 18.1, coding the alternative marker homozygotes as $x = \pm 1$, the expected marker (x) – trait (z) correlation becomes

$$\rho = \frac{\sigma(z,x)}{\sigma(x)\,\sigma(z)} = \frac{a(1-2\widetilde{c})}{\sqrt{a^2+\sigma_e^2}} = \frac{1-2\widetilde{c}}{\sqrt{1+C^2}} \tag{18.12a}$$

Table 18.1 The expected correlation between marker genotype (coded as $x = 1$ for *MM*, $x = -1$ for *mm*) and phenotypic value z can be used to estimate c. We assume that the difference in QTL means is $E(z_{QQ}) - E(z_{qq}) = 2a$ and that the phenotypic distributions conditioned on the QTL genotypes have common (within-line) variance σ_e^2. Because $\sigma(x, z)$ and $\sigma^2(z)$ are unchanged by a change in the mean of z (Chapter 3), we can arbitrarily set $E(z_{QQ}) = -E(z_{qq}) = a$. Coded this way, $E(x) = E(z) = 0$, simplifying calculation of $\sigma(x, z)$ and $\sigma^2(z)$.

Genotype	Freq.	x	$E[z]$
MMQQ	$(1-\widetilde{c})/2$	1	$\mu_{QQ} = a$
MMqq	$\widetilde{c}/2$	1	$\mu_{qq} = -a$
mmQQ	$\widetilde{c}/2$	-1	$\mu_{QQ} = a$
mmqq	$(1-\widetilde{c})/2$	-1	$\mu_{qq} = -a$

$$\sigma(x, z) = E(x \cdot z) = a(1 - 2\widetilde{c}), \qquad \sigma^2(x) = E(x^2) = 1$$

$$\sigma^2(z) = E(z^2) = (1/2)[E(z_{QQ})^2 + \sigma_e^2] + (1/2)[E(z_{qq})^2 + \sigma_e^2] = a^2 + \sigma_e^2$$

where $C = \sigma_e/a$, with a being the QTL effect and σ_e^2 being the within-line variance (see Table 18.1). (The C term was neglected by Hu et al. 1995.) Equation 18.12a assumes no segregation distortion (both RIL homozygotes have equal frequency). Rearranging and letting r be an estimate of ρ suggests the estimator

$$\widehat{\widetilde{c}} = \frac{1 - r\sqrt{1 + C^2}}{2} \leq \frac{1 - r}{2} \tag{18.12b}$$

While the value of C is unknown, by ignoring it one can obtain an upwardly biased estimate of c by first taking $\widetilde{c} = (1 - r)/2$ and then using Equation 18.5b to translate this value of $\widetilde{c}$ into c. Rearranging Equation 18.10b,

$$a = \frac{\mu_{MM} - \mu_{mm}}{2(1 - 2\widetilde{c})} \tag{18.12c}$$

which, upon using $0 \leq \widetilde{c} \leq (1 - r)/2$, gives

$$\frac{\mu_{MM} - \mu_{mm}}{2} \leq a \leq \frac{\mu_{MM} - \mu_{mm}}{2r} \tag{18.12d}$$

Hence, the use of both the observed correlation r and the difference in marker means $(\overline{z}_{MM} - \overline{z}_{mm})$ allows the estimation of upper bounds for both c and a.

Variance QTLs (vQTLs)

While differences in trait variance across marker genotypes could simply reflect differences on the weights on underlying, but equally variable, distributions, they also could have a much more interesting cause: the QTL genotypes themselves may actually have different variances. This is commonly seen among inbred lines (WL Chapter 17), where some marker genotypes are more variable than others. Denoted as **vQTLs (variance QTLs)** by Rönnegård and Valdar (2011) and **veQTLS** (**variance in expression-level traits**) by Huang et al. (2015), such QTLs denote sites where the trait variance differs over marker genotypes. In some settings, vQTLs can be of more interest than QTLs impacting the mean (Weller and Wyler 1992). For example, a reduction in the variance of flowering time shortens the harvesting window, and by reducing costs this may be more significant than changing mean harvesting time per se. While candidate vQTLs could be tagged simply using among-marker differences in variance (e.g., Edwards et al. 1987), a number of more formal approaches to specifically map vQTLs have been proposed, based on modeling the residual variance, e.g., Equations 10.16c, 10.16e, and 10.16g (Ordas et al. 2008; Paré et al. 2010; Struchlain et al. 2010; Visscher and Posthuma 2010; Jimenez-Gomes et al. 2011; Rönnegård and Valdar 2011, 2012; Hothorn et al. 2012; Shen et al. 2012; Cao et al. 2013; Young et al. 2022).

Differences in trait variances among QTL genotypes can result from two different, but not mutually exclusive, causes (WL Chapter 17). They could reflect differences in ***sensitivity to environmental conditions*** ($G \times E$), such that the environmental variance of, say, QQ is 20, while that for qq is 10. Alternatively, they may reflect ***sensitivity to genetic backgrounds*** (epistasis). Indeed, Paré et al. (2010) and Deng and Paré (2011) suggested using variance heterogeneity across markers as a preliminary scan for potentially epistatic loci. Distinguishing between genetic versus environmental sensitivity as causes for vQTLs can be done using panels of RILs (e.g., Fraser and Schadt 2010; see WL Example 17.1). Within a given RIL, one can use replication to estimate the within-line variance very accurately (sensitivity to environmental conditions; G x E). Conversely, one can measure the among-line variance in RILs means for a given marker genotype, which is a direct measure of sensitivity to genetic background (epistasis).

Genome-wide Significance Level with Multiple Tests

A final statistical issue to address is the problem of the proper significance level for an ***entire mapping experiment*** (i.e., the collection of all tests). This is the issue of **multiple comparisons**, which is examined in much greater detail in Appendix 6. The basic concern is as follows. Suppose that 100 markers are tested for departure from the null (no linkage to a QTL), each using an α value of 0.05 (the probability of a false positive: incorrectly declaring linkage when the marker is unlinked to a QTL). If all 100 markers were unlinked to QTLs, we would still (on average) declare $0.05 \cdot 100 = 5$ as being significant (linked to a QTL), resulting in (on avenge) 5 **false positives**. The level of error control, α, for any *single* test is referred to as the **comparison-wise error rate** (**CWER**). In contrast, the **family-wide error rate** (**FWER**), γ, is the probability of one (or more) false positives over the *entire collection of tests*. In the mapping literature, the term **genome-wide error rate** (**GWER**) is often used in place of FWER.

If n independent tests, each with significance level α, are conducted, the probability γ that at least one test shows a false positive (the FWER) is

$$\gamma = 1 - (1 - \alpha)^n \tag{18.13a}$$

For our values of $\alpha = 0.05$ and $n = 100$, $\gamma = 0.9940$. Setting $\alpha = 0.01$ for each individual test, the probability of at least one false positive in 25 tests is 0.22, which increases to 0.633 for 100 tests, and is essentially one for 500 tests. The latter number of tests is not uncommon in QTL mapping studies (n is several orders of magnitude larger in association studies; Chapter 20). Hence, unless we use a very stringent significance value for each test, we run a very high risk of declaring false associations. This tighter stringency for each test results in a higher Type II error rate (failing to declare a linked marker as being significant), resulting in a tradeoff between power of detection and control of the experiment-wide error rate (Appendix 6).

Suppose we wish to achieve an overall significance level γ for the *entire* experiment. With n *independent* tests, the standard **Bonferroni correction** for multiple comparisons, derived by rearranging Equation 18.13a, states that an overall significance level γ requires that each individual test be based on a significance level of

$$\alpha = 1 - (1 - \gamma)^{1/n} \simeq \frac{\gamma}{n} \tag{18.13b}$$

This standard Bonferroni correction is too strict for two reasons. First, sequential Bonferroni corrections (Holm's, Simes-Hochberg, Hommel's; Appendix 6), which follow from procedures based on the ordered p values of the tests (Example A6.4), still control the FWER, but with less stringent control on the CWER. The second issue is that Equation 18.13b assumes *independent* tests. This is appropriate for tests using unlinked markers (such as those on different chromosomes), but is problematic for markers on the same chomosome. Linked markers result in tests generally being positively correlated, which has the effect that the **effective number** of tests, n_e, is *less* than n. Hence, one correction is to adjust for the effective number of tests (usually done separately for each chromosome). Appendix 6 reviews n_e estimates based on the eigenvalues of the correlation matrix of the tests, such as the methods of Cheverud (2000) and Li and Ji (2005) (Equations A6.5a and A6.5b).

A more robust approach for obtaining overall significance levels utilizes resampling procedures such as **permutation tests**, wherein the original analysis is replicated many times on data sets generated by appropriate reshuffling of the original data to simulate draws from the null (Churchill and Doerge 1994; Doerge and Churchill 1996). Let $\mathbf{m}_i$ and z_i denote the vector of marker genotypes and the trait value(s), respectively, for individual i. The original data are of the form $(z_1, \mathbf{m}_1), (z_2, \mathbf{m}_2), \cdots, (z_k, \mathbf{m}_k)$ and one uses some method to search for associations between z_i and one (or more) of the elements of $\mathbf{m}_i$. Under the null, there is no association between z and $\mathbf{m}$. A resampled data set would be (say) of the form $(z_{15}, \mathbf{m}_1), (z_{105}, \mathbf{m}_2), \cdots, (z_{27}, \mathbf{m}_k)$, randomizing z over different draws of $\mathbf{m}$, as would be expected under the null. The test statistic is then computed on this new sample, and this procedure is repeated many times, generating an empirical distribution of the test statistic under the null hypothesis of no marker-trait associations. For example, if the highest test score over all markers in the randomized samples exceeds 4 in only 2.5% of the samples, any marker with a score above 4 is significant at the FWER of $\gamma = 0.025$. Note that one can obtain significance levels on either a per chromosome (using the distribution of randomized scores for that chromosome) or an entire genome basis. Churchill and Doerge suggest that 1,000 resamplings are sufficient for a significance level of 5%, but that 10,000 or more resamplings may be required to generate a stable critical value for the 1% level. By keeping the marker information ($\mathbf{m}_i$) for each individual intact, this approach nicely accounts for missing markers, differences in marker densities, correlations among markers, and any nonrandom segregation of marker alleles. The latter is not uncommon in crosses between widely divergent lines.

A key assumption of permutation tests is that the ***permuted sampling units are exchangeable under the null*** (Welch 1990). Consider the case of selective genotyping (Chapter 17), wherein one phenotypes a sample of individuals, but only genotypes a selected subset of these. In this setting, threshold values must be computed using **stratified permutation tests** (Manichaikul et al. 2007). Here, only the subset of genotypes that have associated phenotypes are shuffled over trait values. Similarly, with family-based QTL mapping (Chapter 19), care must be taken to find the appropriate unit for shuffling.

An alternative strategy for error control under multiple comparisons is the idea of the **false discovery rate** (**FDR**). As detailed in Appendix 6, instead of attempting to limit the *number* of false positives over the *entire set* of tested markers (FWER), one limits the *fraction* of false positives among *the set of tests that are declared to be significant*. For example, setting an FDR control of 5% implies that one expects only 5% of the *significant tests* to be false positives. Conversely, setting an FWER of 5% says that there is only a 5% chance that the *entire* experiment would have one (or more) false positives. The FWER results in more stringent control (and thus less power) than FDR. The FWER setting assumes that essentially none of the tests are true positives, while the FDR framework assumes that some nontrivial fraction of the tests are indeed true positives. Appendix 6 examines the subtle, but quite important, differences between these approaches in detail.

FDR is often used in association studies (Chapter 20), and has been proposed for use in linkage-based QTL mapping (Weller et al. 1998; Lee et al. 2002; Sabatti et al. 2003; Bernardo 2004; Benjamini and Yekutieli 2005). However, Chen and Storey (2006) noted the standard FDR method is not appropriate when a *set of linked markers* are each tested, as signal from the same underlying QTL generates multiple discoveries that are taken as distinct. They instead suggest an FDR-related approach that builds on the Churchill and Doerge resampling approach, essentially looking at the distribution of the *second* highest genome-wide scores (see Appendix 6 for details). We will return to multiple comparison adjustments in Chapter 20, as association mapping has major differences relative to linkage-based mapping. It uses far more markers, many more linked markers, but a smaller *fraction* of correlated markers among these tested (most markers on the same chromosome are uncorrelated in the GWAS setting, as they are essentially in linkage equilibrium; Chapter 5). Further, association mapping can be computationally intensive, limiting the utility of resampling.

Example 18.3. When the marker density on a chromosome is low, Bonferroni corrections are reasonable. Lander and Botstein (1989) called this the **sparse-map** case, where consecutive markers are well-separated (and, as a result, weakly correlated). At the other extreme is the **dense-map** case, where the spacing between consecutive markers approaches zero. Such settings arise when each point on a chromsome generates a test statistic. Lander and Botstein (1989; Lander and Schork 1994; Lander and Kruglyak 1995) approximated this process by using an Orenstein–Uhlenbeck diffusion (LW Appendix 1). Here the value of the test statistics under the null as one moves along a chromosome is approximated by Brownian motion (generating correlations among adjacent tests) with a restoring force that constrains the variance of the process. Their model assumes the Haldane mapping function (Chapter 17) and uniform recombination rates throughout the genome to model the covariance function between marker genotypes (and, hence, between tests based on these genotypes). Using this framework, results from random-walk theory give approximate expressions for the probability that the walk exceeds some value T over some defined duration (here, the genome length).

In particular, if X is the linkage statistics (an LR value, t score, or more general score test), then the FWER, $\gamma_T = \Pr(X > T \text{ somewhere in the genome})$, can be approximately related to the point-test value (CWER), $\alpha_T = \Pr(X > T \text{ at a single site})$, by

$$\gamma_T \simeq [C + 2\rho \cdot G \cdot h(T)]\alpha_T \tag{18.13c}$$

where C is the number of chromosomes, G the genome length (in Morgans), ρ is a measure of the crossover rate in the design (values tabulated in Lander and Kruglyak 1995), and $h(T)$ depends on the test (Lander and Schork 1994). Note by comparison to Equation 18.13b that $[C + 2\rho G h(T)]$ is a measure of the effective number of tests. See Feingold et al. (1993), Dupuis and Siegmund (1999), Piepho (2001a), Zou et al. (2004) and Kao and Ho (2012) for further discussions on dense-map approximations.

Selective Genotyping and Phenotyping

As introduced in Chapter 17, one mapping design is **selective genotyping**: a large sample of individuals are phenotyped, with a subset, usually those showing extreme values, chosen for genotyping. This design is partly a relict of the halcyon days when phenotyping was much cheaper than genotyping. Such a design can produce biased estimates unless suitable corrections are performed (Lander and Botstein 1989; Darvasi and Soller 1992; Muranty and Goffinet 1997; Muranty et al. 1997; Ronin et al. 1998; Johnson et al. 1999; Xu and Vogl 2000; Lee et al. 2014). An especially interesting approach to adjust for bias was suggested by Henshall and Goddard (1999). Typically, QTL detection searches for markers showing an association between trait value as a function of the number of copies of a reference marker allele (trait value conditioned on the marker value). Henshall and Goddard suggested that one instead looks at the expected number of copies of a reference allele as a function of trait value (marker value conditioned on the trait value). In effect, this is a logistic regression (Chapter 14) of allele frequency on trait value. The idea is that the frequency of a focal allele influencing the trait should increase with trait value.

One extension of selective genotyping is to multiple traits. Lin and Ritland (1997) suggested that an individual be genotyped when they display extreme values in at least one of the traits, while Muranty et al. (1997) suggested using a weighted index combining the traits of interest and then selecting individuals with extreme index values. Another variant of selective genotyping is **selective mapping**: the use of selective genotyping to improve map resolution. The idea is to genotype individuals using low marker density (**framework markers**) and then increase the density of genotyping for those individuals showing recombinations within the region of interest (Vision et al. 2000; Ronin et al. 2003; Xu et al. 2005). Finally, one can go full circle with **selective phenotyping**: individuals are chosen to be phenotyped by some marker genotype criteria, such as maximal diversity within target regions (Jin et al. 2004; Sen et al. 2005, 2009; Fu and Jansen 2006).

QTL DETECTION AND ESTIMATION USING LS LINEAR MODELS

We now have all of the necessary machinery in place to consider methods for QTL detection and estimation in greater detail. In a LS linear model framework, detection is very straightforward, while estimation of effects and positions can be more challenging. Equations 18.7–18.11 show the general estimation strategy for linear models: one constructs some appropriate contrast of marker genotype means for the parameter(s) of interest (the **method-of-moments** approach). As we will shortly demonstrate, likelihood (Appendix 4) provides a much better overall framework for *estimation*, so we will focus solely on using linear models for *QTL detection*.

Linear Model Detection of Continuous Traits

As noted above, the simplest test for a marker-trait association involves the comparison of the trait means of alternate marker genotypes. When only two genotypes are compared (such as with single-marker backcross-, RIL-, or DHL-designs), this can be accomplished with a simple *t* test. Many designs, however, involve more than two marker genotypes. For example, the single-marker F_2 design has three marker genotypes: *MM, Mm,* and *mm*. In such cases, all marker genotypic means (or some subset of them) can be compared by using standard LS linear-model approaches, such as ANOVA or regression.

The simplest framework is **gene-dosage regression** (also called the **additive model**), which assumes a single marker locus and only additive effects at a single linked QTL. The phenotypic value z_{ik} of the kth individual with marker genotype i is modeled as a mean value μ plus an (additive) marker effect b and a residual error e_{ik},

$$z_{ik} = \mu + bN_i + e_{ik} \tag{18.14a}$$

where N_i is the number of copies (or **gene dosage**) of the focal marker allele (e.g., values of 0, 1, and 2, for mm, Mm, and MM, respectively). A significant value of b implies one (or more) QTLs linked to the marker. Note that this is a one degree-of-freedom test.

A more general approach that accommodates dominance, and easily extends to multiple markers (i.e., multilocus marker genotypes), is given by the **general genotype model**

$$z_{ik} = \mu + b_i + e_{ik} \tag{18.14b}$$

This is a one-way ANOVA model (Part II Chapter 22), with the presence of a linked QTL being indicated by a significant among-marker variance (i.e., one, or more, of the b_i is significantly different from zero). The associated degrees of freedom is the total number of marker genotypes minus one. Equivalently, we can express this model given by Equation 18.14b as a multiple regression, with the phenotypic value for individual j given by

$$z_j = \mu + \sum_{i=1}^{n} b_i\, x_{ij} + e_j \tag{18.14c}$$

where the x_{ij} are n indicator variables (one for each marker genotype),

$$x_{ij} = \begin{cases} 1 & \text{if individual } j \text{ has marker genotype } i, \\ 0 & \text{otherwise.} \end{cases}$$

The number of marker genotypes (n) in Equations 18.14b and 18.14c depend on both the number of marker loci and the type of design being used. With a single marker, $n = 2$ for a backcross, RIL, or DH design, while $n = 3$ for an F_2 design (using codominant markers). When two or more marker loci are simultaneously considered, b_i corresponds to the effect of a *multilocus* marker genotype, and n is the number of such genotypes considered in the analysis. In the regression framework, evidence of a linked QTL is provided by a significant r^2, the fraction of character variance accounted for by the marker genotypes (Chapter 10). Finally, as mentioned above, the presence of a linked QTL can cause different marker genotypes to have different trait variances. If the difference in variance between marker classes is substantial, the standard ANOVA assumption of variance homogeneity is violated and corrections are required for hypothesis testing (Asins and Carbonell 1988; Xu 1995).

Estimation of dominance requires information on all three genotypes at a marker locus, i.e., an F_2, F_t, or other design (such as *both* backcross populations). In these cases, dominance can be estimated using an appropriate weighted combination of the marker means (e.g., Equations 18.7b and 18.8b). For a single marker locus, a simple test for dominance is whether Equation 18.14b gives a significantly better fit than the strictly additive model (Equation 18.14a).

Epistasis between QTLs can be modeled by including **interaction terms** (Chapter 10). Here, an individual with genotype i at one marker locus and genotype k at a second is modeled as $z = \mu + a_i + b_k + I_{ik} + e$, where a and b denote the single-locus marker effects, and I is the interaction term due to epistasis between QTLs linked to those marker loci. In linear regression form this model becomes

$$z_j = \mu + \sum_i^{n_1} a_i\, x_{ij} + \sum_k^{n_2} b_k\, y_{kj} + \sum_i^{n_1} \sum_k^{n_2} I_{ik}\, x_{ij}\, y_{kj} + e_j \tag{18.14d}$$

where x_{ij} and y_{kj} are indicator variables for two different marker genotypes (with n_1 and n_2 genotypes, respectively). Significant a_i and/or b_k terms indicate significant effects at the individual marker loci (**main** or **marginal effects**), while significant I_{ik} terms indicate epistasis between the effects of the two markers. More formally, we could express these interaction terms using our general epistatic model (Equation 5.5), e.g., Kao and Zeng (2002) and Zeng et al. (2005).

Essentially the same approach can be used to search for QTL $\times$ environment interactions (Part II Chapters 27 and 28) when markers are examined over several environments. The basic model for an individual with marker genotype i measured in the kth environment is $z = \mu + b_i + E_k + I_{ik} + e$, where a significant E_k indicates an environmental effect, while a significant I_{ik} implies a marker $\times$ environment interaction. For example, if the character has significant sex-specific effects, these can be incorporated by using the model $z = \mu + b_i + s_k + I_{ik} + e$ for an individual of marker genotype i and sex k. A significant s_k implies a significant sex effect, while a significant I_{ik} implies a significant marker $\times$ sex interaction. Long et al. (1995) gave an example of the utility of this approach, finding very significant sex-specific effects for bristle number in *Drosophila*. Detecting QTL $\times$ environment interaction is examined more fully in Chapters 27 and 28.

Example 18.4. Edwards et al. (1987) examined two F_2 maize populations. Cross 1 consisted of 1,776 individuals scored for 16 markers, while Cross 2 used a different set of inbred parental lines and consisted of 1,930 individuals scored for 20 markers. As the frequency distribution (below) shows, the detected marker effects (measured by the fraction r^2 of total F_2 phenotypic variance accounted for by each significant marker-trait association) were generally quite small.

A total of 82 vegetative characters were examined, with 60% (Cross 1) and 64% (Cross 2) of all possible marker-trait combinations showing significant effects (at the $\alpha = 0.05$ level) using single-marker ANOVA. On average, each trait showed 10 (Cross 1) and 14 (Cross 2) significant marker associations. Dominance was common, while pairwise epistasis, as tested by incorporating a marker $\times$ marker interaction term into the linear model (Equation 18.14d), was rare. The same two F_2 populations were used by Stuber et al. (1987) to examine 25 yield-related characters, with similar results. In that study, most marker-trait combinations were significant (66% and 72% at the $\alpha = 0.05$ level), and most marker effects were small (over half of the significant associations having r^2 values less than 2%). As a group, yield-related traits displayed more dominance than vegetative traits, but many yield traits were still largely additive. Edwards et al. (1992) examined a subset of the vegetative characters in Cross 2, using a much larger number of markers (114 RFLPs). While only 187 F_2 individuals were scored, 15% of marker-trait associations were significant, and the overall results with respect to the distribution of effects were similar to those for the 1987 experiments

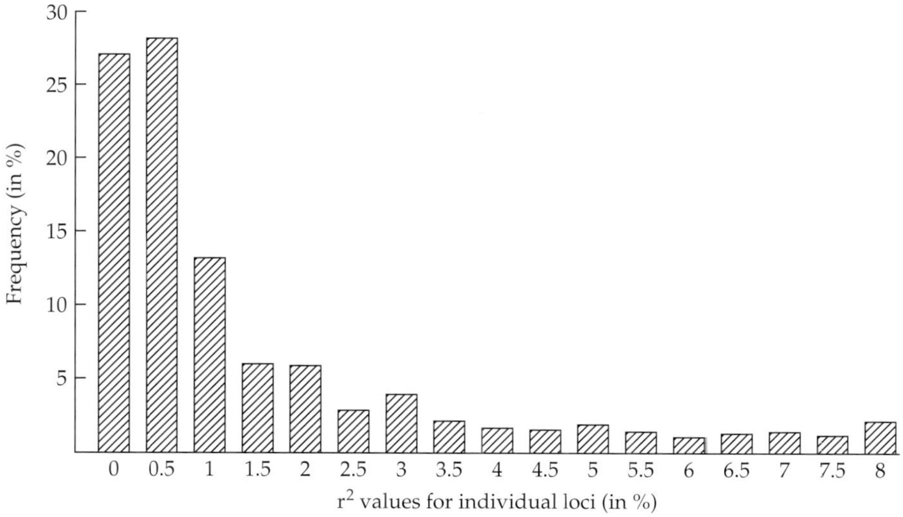

Detecting QTLs for Dichotomous Traits: The Cochran–Armitage Trend Test

Many QTL experiments are concerned with dichotomous (binary) traits, such as disease or pest resistance in crop plants or disease susceptibility in humans. In many cases, one can score quantitative physiological traits contributing to the binary trait, such as blood pressure or number of lesions per leaf, and the above methodology for QTL detection with continuous traits applies. However, such underlying variables are often either unknown or unmeasured, and the data are simply scored as present/absent values (or **cases** versus **controls** in the medical literature). The simplest procedure to detect marker-trait associations in this setting is to test for independence using standard association tables (such as χ^2 or Fisher's exact tests). As shown in the table below, the n total observations are partitioned into counts for each particular class, e.g., n_{P1} is the sample number of Mm individuals showing the trait.

	Marker Genotype			
	mm	Mm	MM	Totals
Present	n_{P0}	n_{P1}	n_{P2}	n_P
Absent	n_{A0}	n_{A1}	n_{A2}	n_A
Totals	n_0	n_1	n_2	n

This same approach easily extended to polychotomous (ordinal) characters. With three marker genotypes and two trait values, the result χ^2 test has two degrees of freedom, with a significant value indicating linkage to one (or more) QTLs.

Note that the contingency table approach is akin to Equation 18.14b, making no assumptions about the trait values (percentage showing the trait) associated with each genotype, hence the two degrees of freedom for the test. A more powerful approach, akin to the gene dosage regression (Equation 18.14a), is the **Cochran–Armitage trend test** (Cochran 1954; Armitage 1955), which uses one degree of freedom to test for a *trend in proportions* as the gene dosage increases. The basic structure of the test is as follows. First one chooses a weight vector $\tau = (\tau_0, \tau_1, \tau_2)$ which is an assumption about the nature of the trend over the three genotypes. The resulting test statistic becomes

$$T = \frac{n_A n_P}{n} \cdot \frac{\left[\sum_{i=0}^{2} \tau_i \left(\frac{n_{Pi}}{n_P} - \frac{n_{Ai}}{n_A} \right) \right]^2}{\sum_{i=0}^{2} \tau_i^2 \frac{n_i}{n} - \left(\sum_{i=0}^{2} \tau_i \frac{n_i}{n} \right)^2} \tag{18.15}$$

where, for large samples, $T \sim \chi_1^2$. The classical Cochran–Armitage test assumes $\tau = (0, 0.5, 1)$, namely an additive trend. Note that one can easily modify Equation 18.15 under the assumption that M is linked to a completely dominant ($\tau = [0, 1, 1]$), or recessive ($\tau = [0, 0, 1]$), QTL allele. Lee (2015) discusses optimal weighting schemes under other scenarios. Note that if we use new weights $k_i = a\tau_i$ (all the weights are scaled by the same constant), the a values in the numerator and denominator cancel, recovering Equation 18.15.

Example 18.5. Consider the following data from Zhang et al. (2005), who examined the association between genotypes at the DNA repair gene *ADPRT* and lung cancer:

	Genotype			
	mm	*Mm*	*MM*	Totals
Present	307 [0.461]	509 [0.502]	184 [0.573]	1000
Absent	359 [0.539]	504 [0.498]	137 [0.427]	1000
Totals	666	1013	321	2000

The number in the square bracket is the fraction of the marker class with (Present) or without (Absent) lung cancer. There appears to be a trend with lung cancer risk increasing with the number of copies of *M*. Is this statistically significant? Performing a standard χ^2 test on this contingency table returns a test statistic value of 10.9664, with an associated p value of $\Pr(\chi_2^2 \geq 10.9664) = 0.0042$. The Fisher-exact test returns (to four decimal places) the same p value.

To compute the Cochran–Armitage statistic (assumping an additive trend), note that

$$\frac{n_A n_P}{n} = \frac{1000 \cdot 1000}{2000} = 500$$

$$\sum_{i=0}^{2} \tau_i \left(\frac{n_{Pi}}{n_P} - \frac{n_{Ai}}{n_A} \right) = 0.5 \left(\frac{509}{1000} - \frac{504}{1000} \right) + 1 \left(\frac{184}{1000} - \frac{137}{1000} \right) = 0.0495$$

$$\sum_{i=0}^{2} \tau_i^2 \frac{n_i}{n} = \frac{1}{4} \cdot \frac{1013}{2000} + 1 \cdot \frac{321}{2000} = 0.2871$$

$$\sum_{i=0}^{2} \tau_i \frac{n_i}{n} = \frac{1}{2} \cdot \frac{1013}{2000} + 1 \cdot \frac{321}{2000} = 0.4138$$

Substituing these values into Equation 18.15 gives

$$T = 500 \cdot \frac{0.0495^2}{0.2871 - 0.4138^2} = 10.5733$$

The associated p value is $\Pr(\chi_1^2 \geq 10.5733) = 0.0011$. Note the nearly four fold smaller p value relative to a standard χ^2 test, which results from the Cochran–Armitage test requiring fewer degrees of freedom.

QTL DETECTION AND ESTIMATION VIA MAXIMUM LIKELIHOOD

Maximum likelihood (ML) methods (Appendix 4) are especially popular in the QTL mapping literature. While LS linear models use only marker means, ML uses the full information from the marker-trait distribution and, as such, is expected to be more powerful. The trade-off is that ML is computationally intensive, requiring rather special programs to solve the likelihood equations, while linear model analysis can be performed with almost any standard statistical package. Further, while modifying the basic model (such as adding extra factors) is rather trivial in the linear model framework, with ML new likelihood functions need to be constructed and solved for each variant of the original model. Although writing down a set of likelihood equations for the model of interest is relatively straightforward (e.g., Example 18.6), obtaining the ML estimates is more difficult. One approach outlined in Appendix 4 is to use specialized algorithms, such as **EM (expectation-maximization) methods** which have been successfully adapted to many of the mixture-model problems in QTL mapping (e.g., Lander and Botstein 1989; Carbonell and Gerig 1991; Carbonell et al. 1992; Luo and Kearsey 1992; van Ooijen 1992; Jansen 1992, 1993a, 1994a, 1996; Luo and Wolliams 1993; Jansen and Stam 1994; Kao and Zeng 1997; Xu et al. 2003; Chen 2005; Xu et al. 2005; Xu and Hu 2010). Alternatively, as we discuss later, a creative use of regressions can often provide excellent approximations to ML solutions. For the remainder of this chapter, we assume that the reader has recently read Chapter 16 and Appendix 4, which introduces much of the ML machinery used here.

Assuming that the distribution of phenotypes for an individual with QTL genotype Q_k is normal with genotypic-specific mean μ_{Q_k} and common variance σ^2, and following the logic of Chapter 16, the likelihood for an individual with phenotypic value z and marker genotype M_j becomes

$$\ell(z \mid M_j) = \sum_{k=1}^{N} \varphi(z, \mu_{Q_k}, \sigma^2) \Pr(Q_k \mid M_j) \tag{18.16}$$

where $\varphi(z, \mu_{Q_k}, \sigma^2)$ denotes the density function for a normal distribution with mean μ_{Q_k} and variance σ^2 (Equation 16.5c), and a total of N QTL genotypes is assumed. This likelihood is a mixture-model distribution (Chapter 16). The mixing proportions, $\Pr(Q_k \mid M_j)$, are functions of the genetic map (the assumed *position*(s) of the QTL(s) with respect to the observed markers) and the experimental design, while the QTL *effects* enter only though the means μ_{Q_k} and variance σ^2 of the underlying distributions.

Example 18.6. Consider the single-marker F_2 design with a single QTL linked to the marker. Making the standard assumption that phenotypes are normally distributed about each QTL genotype, substitution of the F_2 conditional probabilities (Example 18.1) into Equation 18.16 gives the likelihood functions for the three different marker genotypes as

$$\begin{aligned}
\ell(z \mid MM) &= (1-c)^2\varphi(z, \mu_{QQ}, \sigma^2) + 2c(1-c)\,\varphi(z, \mu_{Qq}, \sigma^2) + c^2\varphi(z, \mu_{qq}, \sigma^2) \\
\ell(z \mid Mm) &= c(1-c)\varphi(z, \mu_{QQ}, \sigma^2) + \left[\,(1-c)^2 + c^2\,\right] \varphi(z, \mu_{Qq}, \sigma^2) \\
&\quad + c(1-c)\varphi(z, \mu_{qq}, \sigma^2) \\
\ell(z \mid mm) &= c^2\varphi(z, \mu_{QQ}, \sigma^2) + 2c(1-c)\,\varphi(z, \mu_{Qq}, \sigma^2) + (1-c)^2\varphi(z, \mu_{qq}, \sigma^2)
\end{aligned}$$

as obtained by Weller (1986). For example, if individual one has trait value 10 and marker genotype MM, then its contribution to the likelihood function is

$$\ell(z_1 \mid M_1) = (1-c)^2\varphi(10, \mu_{QQ}, \sigma^2) + 2c(1-c)\,\varphi(10, \mu_{Qq}, \sigma^2) + c^2\varphi(10, \mu_{qq}, \sigma^2)$$

The total likelihood for n F_2 individuals is the product of the individual likelihoods,

$$\ell(\mathbf{z}) = \prod_{i=1}^{n} \ell(z_i \mid M_i)$$

While rather complex, the total likelihood is a function of just five parameters: the QTL position (c), the three QTL means ($\mu_{QQ}, \mu_{Qq}, \mu_{qq}$), and the common variance (σ^2).

The standard likelihood model assumes continuous trait values and normally disturbed residuals, although likelihood models have been developed to map continuous traits with censored observations (Diao et al. 2004). However, a variety of important traits are discrete. Much of human genetics deals with dichotomous (binary) traits, which have binomially distributed residuals (Chapter 2). Plant breeders often score resistance in an ordinal fashion, such as fully resistant, partly resistant, or fully susceptible. Such polychotomous traits have multinomially distributed residuals (Chapter 2). As we saw with segregation analysis (Chapter 16), the likelihood equations can be modified to handle discrete traits by using generalized linear models (Chapters 14 and 20). Dichotomous and polychotomous traits can be modeled through the use of logistic regressions and probit scales (Ghosh et al. 1993; Hackett and Weller 1995; Visscher et al. 1996a; Xu and Atchley 1996; Rao and Xu 1998; Yi and Xu 2000; Deng et al. 2006). While the generalized linear model framework is formally appropriate in such cases, often (as a first pass), the discrete structure of the data is ignored by an investigator, treating them as if they were continuous (e. g., coding alternative binary characters as 0/1) and applying standard ML. When flanking markers are used, this approach gives essentially the same power and precision as methods specifically designed for polychotomous traits (Hackett and Weller 1995; Visscher et al. 1996a), but when single markers are used, this approach can give estimates for QTL position that are rather seriously biased (Hackett and Weller 1995).

A final class of discrete data are those that can take on a value of zero and otherwise have a very limited range of possible values, such as number of eggs laid by a bird. Again, one could treat these discrete values as continuous and use standard ML, but such data are usually poorly approximated by a normal distribution. For example, they often have inflated zero values (an excess of zero values). Such data are best treated as Poisson-distributed (perhaps with extra point mass at zero; Chapter 14, WL Chapter 29). In the generalized linear model framework, such data are modeled using log-linear models (Cui et al. 2006; Cui and Yang 2009). Chen and Liu (2009) and Xu and Hu (2010) present a general EM treatment for using generalized linear models in QTL mapping. An alternative approach for treating non-normally distributed characters is given by Kruglyak and Lander (1995c), who develop a nonparametric interval mapping procedure.

Finally, we note that Bayesian analysis (Appendix 7) is essentially a generalization of likelihood methods, incorporating any prior information to yield a posterior distribution for the unknown parameters. There is a growing literature for Bayesian-based QTL mapping (e.g., Hoeschele and VanRaden 1993a, 1993b; Satagopan et al. 1996; Thaller and Hoeschele 1996a, 1996b; Sillanpää and Arjas 1998; Yi et al. 2003a, 2003b; Zhang et al. 2005; Xu et al. 2009). Sen and Churchill (2001) present a unified treatment of Bayesian QTL mapping, based on conditioning on the unobserved QTL genotype (akin to Equation 18.6), which allows the issues of QTL position and QTL effect to be separated into two independent and manageable parts. Using this partition, they developed an MCMC approach (Appendix 8) to accommodate multiple, interacting QTLs, allowing for non-normally distributed traits, as well as accommodating missing genotype data and genotyping errors. Two powerful uses of Bayesian methods are in estimating the number of QTLs and dealing epistasis, and we will return to these later in the chapter.

Likelihood Maps (Profile Plots)

In the likelihood framework, tests of whether a QTL is linked to the marker(s) under consideration are based on the likelihood-ratio statistic,

$$\text{LR} = -2\ln\left[\frac{\max\ \ell_r(\mathbf{z})}{\max\ \ell(\mathbf{z})}\right]$$

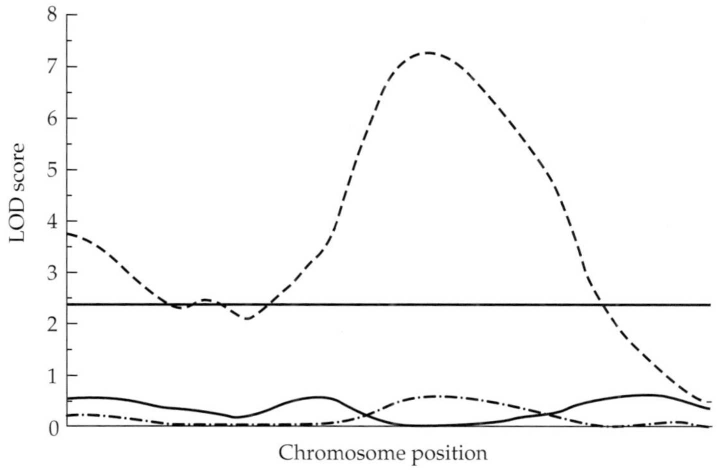

Figure 18.1 Likelihood map (or profile) for QTL positions on chromosome 10 in a cross of two tomato species. Evidence for a QTL is provided when the likelihood function exceeds the significance threshold (indicated by the horizontal line). The upper dashed curve gives the LOD score for fruit pH as a function of map position, showing strong evidence of a QTL near the middle of the chromosome. There is also modest support of a second QTL at the left end of the chromosome. The lower two curves (solid and broken) are for fruit weight and soluble-solid concentration, neither of which shows a significant QTL effect on this chromosome. (After Paterson et al. 1988.)

where $\max\, \ell_r(\mathbf{z})$, given by Equation 16.8, is the maximum of the likelihood function under the null hypothesis of no segregating QTL (i.e., under the assumption that the phenotypic distribution is a single normal). The LR test statistic is approximately χ^2-distributed, with the degrees of freedom given by the extra number of fitted parameters in the full model. For a model assuming a single QTL, most designs have five parameters in the full model (the three QTL means, the common variance, and the QTL position), and two in the reduced model (the mean and variance), giving three degrees of freedom. Certain designs (such as a backcross, RIL, or DHL) involve situations where only two QTL means enter (e.g., QQ and qq for RILs/DHLs, Qq and QQ or qq for a backcross), and here the likelihood-ratio test has two degrees of freedom.

The amount of support for a QTL at a particular map position is often displayed graphically through the use of **likelihood maps** (Figures 18.1 and 18.2), also called **profile plots**, which graph the likelihood-ratio statistic (or a closely related quantity) as a function of map position of the putative QTL. For example, the value of the likelihood map at $c = 0.05$ gives the likelihood-ratio statistic that a QTL is at recombination fraction 0.05 from the marker versus a model assuming no QTL. This approach for displaying the support for a QTL was introduced by Lander and Botstein (1989), who plotted the LOD (**likelihood of odds**) scores (Morton 1955b). The LOD score for a particular value of c is related to the likelihood-ratio test statistic (LR) by

$$\text{LOD}(c) = \log_{10}\left[\frac{\max\, \ell_r(\mathbf{z})}{\max\, \ell(\mathbf{z}, c)}\right] = \frac{\text{LR}(c)}{2\ \ln 10} \simeq \frac{\text{LR}(c)}{4.61} \tag{18.17}$$

showing that the LOD score is simply a constant times the likelihood-ratio statistic. Here $\max\, \ell(\mathbf{z}, c)$ denotes the maximum of the likelihood function given a QTL at recombination frequency c from the marker. Another variant is simply to plot $\max\, \ell(\mathbf{z}, c)$ instead of the likelihood-ratio statistic, as the restricted likelihood, $\max\, \ell_r(\mathbf{z})$, is the same for each value of c. Testing over the entire genome yields a **genomic scan** for QTLs.

The likelihood profile projects the multidimensional likelihood surface (which is a function of the QTL means, the common variance, and the map position) on to a single dimension, that of the map position, c. The ML estimate of c is that which yields the maximum

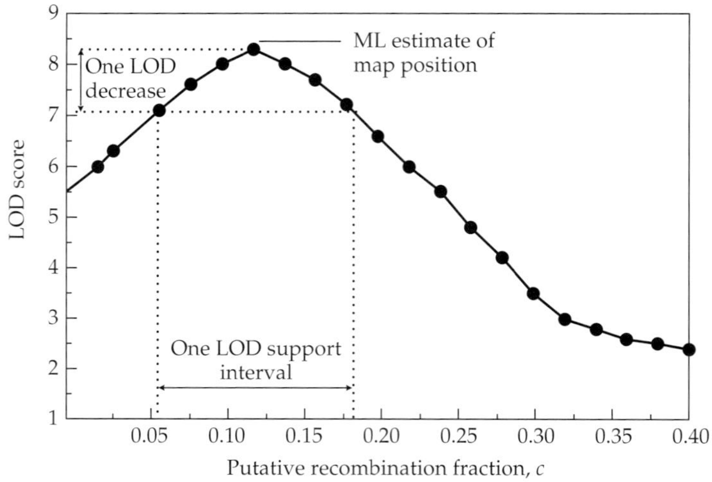

Figure 18.2 Hypothetical likelihood map for the marker–QTL recombination frequency c in a single-marker analysis. Points connected by straight lines are used to remind the reader that likelihood maps are computed by plotting the maximum of the likelihood function for each c value, usually done by considering steps of 0.01 to 0.05. A QTL is indicated if any part of the likelihood map exceeds a critical value. In such cases, the ML estimate for map position is the value of c giving the highest likelihood. Approximate confidence intervals for QTL position (one-LOD support intervals) are often constructed by including the set of all c values giving likelihoods within one LOD score of the maximum value.

value on the likelihood map, and the values for the QTL means and variance that maximize the likelihood given this value of c are the ML estimates for the QTL effects. Thus, in the likelihood framework, *detection* of a linked QTL and *estimation* of its position are coupled: if the likelihood ratio exceeds the critical threshold for that chromosome, it provides evidence for a linked QTL, whose position is estimated by the peak of the likelihood map. If the peak does not exceed this threshold, there is no evidence for a linked QTL.

Precision of ML Estimates of QTL Position

Because ML estimates are approximately normally distributed for large sample sizes, confidence intervals for QTL effects and position can be approximated using the asymptotic sampling variances for the ML estimates (Appendix 4). Approximate confidence intervals are often constructed using the **one-LOD rule** (Figure 18.2), with the confidence interval being defined by all those values falling within one LOD score of the maximum value (Conneally et al. 1985; Lander and Botstein 1989). The motivation for such **one-LOD support intervals** follows from the fact that the large-sample distribution of the LR statistic follows a χ^2 distribution. If only one parameter in the likelihood function is allowed to vary, as when testing whether c equals a particular value (say the observed ML estimate), the LR statistic has one degree of freedom. Because a one-LOD change corresponds to an LR change of 4.61 (Equation 18.17), which for a χ^2 with one degree of freedom corresponds to a significance value of 0.04 (e.g., $\Pr(\chi_1^2 \geq 4.61) = 0.04$), it follows that one-LOD support intervals approximate 95% confidence intervals under the appropriate settings.

While widely used, the one-LOD rule is a *large-sample property*, and, as such, is only an approximation. When the true QTL effect is small, this approach often yields intervals that are *too small*. Mangin et al. (1994a, 1994b) showed that one-LOD confidence intervals have between 60% and 95% probability of actually containing small-effects QTLs, and they developed an improved method for such cases. This observation lead to a series of **LOD-drop rules** using intervals based on decreases in LOD scores in excess of one. For example, simulation studies led van Ooijen (1992) to suggest that support intervals should be based on **two-LOD differences** in order to have a high probability of containing the QTL, while

Dupuis and Siegmund (1999) suggested using a **1.5-LOD support interval**. Very extensive simulations by Manichaikul et al. (2006) showed that the LOD drop value increased with decreasing power (i.e., smaller QTL effects), ranging from around 2 for low power to close to 1.5 to 1 for high power. Visscher and Goddard (2004) showed that the likelihood surface around the MLE is often rather leptokurtic (Chapter 2; also see Figure A4.1), rather than the quadratic required by asymptotic theory for the one-LOD rule to be valid.

A more rigorous approach, although still a large-sample approximation, for obtaining standard errors of both map position and QTL effects is to use the inverse of the Fisher information matrix associated with the likelihood function (Appendix 4). Kao and Zeng (1997) and Chen (2005) showed how to obtain this asymptotic matrix in an EM framework. Bayesian methods (Appendix 7) offer another approach, and tend to have better small-sample properties than likelihood. Example 18.7 outlines a Bayesian credible interval approach, based on areas under the likelihood profile, suggested by Sen and Churchill (2001).

Resampling methods have been suggested as a robust procedure for constructing confidence intervals for QTL position, and Visscher et al. (1996b), Walling et al. (1998, 2002), and Talbot et al. (1999) suggested using a **nonparametric bootstrap** approach (Efron 1979, 1982). Suppose the original data set consists of n individuals. A bootstrap sample is generated by drawing n values, *with replacement*, from the original data set. Such a sample will have some of the original observations present multiple times and others not present at all. A series of N such samples are generated and an estimate (map position in this case) is computed for each, generating a distribution of estimates (the **empirical bootstrap distribution**). The resulting 95% bootstrap confidence interval has as its lower value the estimate corresponding to the 2.5% cumulative frequency point of the empirical bootstrap distribution, while the upper value is that corresponding to the upper 97.5% of the bootstrap distribution. While the bootstrap approach is generally fairly robust, for QTL mapping it appears to be compromised by the bias from ML (and regression) methods tending to place QTLs *exactly* at the markers, as opposed to between them (Walling et al. 2002; Manichaikul et al. 2006). Manichaikul et al. noted that a consequence of this feature is that the accuracy of bootstrap confidence intervals *critically* depends on where the true QTL is relative to the markers, and, as a result, recommend that LOD drop off or Bayesian credible intervals (Example 18.7) be used instead, as these tend to be a bit more robust to true QTL position.

The length of the confidence interval is influenced by the number of individuals sampled, the effect of the QTL in question, and the marker density. Darvasi et al. (1993) showed that precision is not significantly improved by increasing marker density beyond a certain point (around one marker every 5 to 10 cM). Given such a dense map, van Ooijen (1992) found that ML mapping using flanking markers with reasonable sample sizes (200–300 F_2 or backcross individuals) allowed a QTL accounting for 5% of the total variance to be mapped to a 40 cM interval, while one accounting for 10% could be mapped to a 20 cM interval. Unfortunately, these interval sizes are distressingly large for cloning QTLs or even defining their positions to smaller intervals for RIL construction. Equation 18.44b (below) gives a simple approximation for the sample size required for a desired interval length.

One strategy for increasing the precision of mapping is to use lines with expanded genetic maps, such as RILs or AILs. With these designs, estimates of the map position are in terms of the cumulative recombination frequency $\widetilde{c} \simeq \tau c$ (where $\tau > 1$; Equation 18.5), so that the confidence interval for c is reduced by a factor of $1/\tau$. For example, recombinant inbred lines have a two to four fold expansion of the map (Equation 18.5b), and hence reduce the length of the confidence interval for c by 1/2 to 1/4 relative to an F_2. Even more dramatic reductions are possible using advanced intercross lines. A sample size and marker density that yields a 20 cM confidence interval in an F_2 design gives a 3.4 cM confidence interval for the same QTL in an F_{10} design. (This follows from Equation 17.4, which shows that a Haldane distance of 20 cM, corresponding to $c = 0.165$, translates into $\widetilde{c} = c/5 = 0.033$ and a Haldane distance of 3.4 cM with an F_{10} AIL.) Similarly, an F_{20} design would give a 1.7 cM confidence interval.

Example 18.7. Sen and Churchill (2001) proposed an **approximate Bayes credible interval** for QTL position. Bayesian analysis is examined in more detail in Appendix 7, but the basic idea (Equation A7.3d) is that one starts with some initial prior distribution $p(c)$ for QTL position, and then updates this information with data $\mathbf{z}$ via the likelihood $\ell(\mathbf{z}, c)$, to yield posterior distribution for the QTL position,

$$p(c \,|\, \mathbf{z}) = \frac{\ell(\mathbf{z}, c)\, p(c)}{\int \ell(\mathbf{z}, c)\, p(c)\, dc} = \frac{\ell(\mathbf{z}, c)}{\int \ell(\mathbf{z}, c)\, dc}$$

Integration is performed to ensure that the posterior is a proper probability distribution (the area under the posterior is one). The right-hand simplification follows when one assumes a **flat prior**, $p(c) = 1/L$ (L being the length of the chromosome in Morgans), namely we assume, a priori, that all map position values are equally likely QTL locations. Sen and Churchill noted that this distribution can be approximated using profile likelihoods. Using Equation 18.17 to express $\ell(\mathbf{z}, c)$ in terms of LOD scores,

$$p(c \,|\, \mathbf{z}) \simeq \frac{10^{LOD(c)}}{\sum_{i=1}^{L} 10^{LOD(c_i)}}$$

where $LOD(c)$ is the LOD score if the QTL is assumed to be at map position c on a chromosome, and the sum is over the possible c values (moving from one end of a chromosome to the other). Using $p(c \,|\, \mathbf{z})$, a 95% credible interval is simply the smallest interval in c (map position) that contains 95% of the probability mass of c. Manichaikul et al. (2006) found that this approach tended to outperform the bootstrap.

ML Interval Mapping

ML mapping with line crosses usually employs the genotypes of a pair of flanking markers as the unit of analysis. The likelihood functions for such **ML interval mapping (IM)** follow from Equation 18.16 using the appropriate conditional probabilities for QTL genotypes given the two-locus marker genotypes (Jensen 1989; Lander and Botstein 1989; Knapp et al. 1990, Carbonell et al. 1992; van Ooijen 1992; Korol et al. 1996). Example 18.8 shows the basic structure of the resulting likelihood functions. As with single-marker analysis, support for a QTL is evaluated with a likelihood profile over a region (typically a chromosome; Figures 18.1 and 18.5) with the peak of the likelihood profile corresponding to the ML estimate of QTL position within that region and its significance given by a likelihood-ratio test.

Example 18.8. Likelihood functions for interval mapping follow by substituting the appropriate conditional probabilities into Equation 18.16. For example, consider the F_2 formed by crossing two inbred lines. Assuming no interference, from Equation 18.2 the likelihood for marker genotype $M_1M_1M_2M_2$ is

$$\ell(z \,|\, M_1M_1M_2M_2) = \left[\frac{(1-c_1)^2\,(1-c_2)^2}{(1-c_{12})^2}\right] \cdot \varphi(z, \mu_{QQ}, \sigma^2)$$

$$+ \left[\frac{2\,c_1\,c_2\,(1-c_1)\,(1-c_2)}{(1-c_{12})^2}\right] \cdot \varphi(z, \mu_{Qq}, \sigma^2)$$

$$+ \left[\frac{c_1^2\,c_2^2}{(1-c_{12})^2}\right] \cdot \varphi(z, \mu_{qq}, \sigma^2)$$

Likelihoods for the other eight flanking-marker genotypes follow similarly and can be found in Luo and Kearsey (1992), Carbonell et al. (1992), and van Ooijen (1992). Even though these likelihoods involve three recombination parameters (c_{12}, c_1, c_2), the distance between markers (c_{12}) is usually taken as known, and hence $c_2 = (c_{12} - c_1)/(1 - 2\,c_1)$ (assuming no interference) or $c_2 = c_{12} - c_1$ (complete interference). This leaves five parameters to estimate: three QTL means, the common variance σ^2, and the position c_1 of the putative QTL within the interval. Likelihoods for other designs follow using the appropriate conditional probabilities.

One of the first applications of ML interval mapping was performed by Paterson et al. (1988), who examined 237 backcross individuals in a cross between the tomato species *Lycopersicon esculentum* and *L. chmielewskii* for several fruit-related traits (Figure 18.1 gives the chromosome 10 likelihood maps for three traits). By using 68 markers (63 RFLP and 5 isozyme variants), 95% of the genome was within 20 cM of a marker. Six QTLs affecting fruit mass, four affecting concentration of soluble solids, and five affecting fruit pH were detected. A follow-up study (Paterson et al. 1990) using NILs (Chapter 17) detected additional QTLs. However, this finer mapping could not confirm the presence of one putative QTL that showed a highly significant peak on the likelihood map in the 1988 study, suggesting it was a false positive.

With ML interval mapping, the likelihood map for an entire chromosome is constructed by stitching together the likelihood maps for each successive interval. If the order of markers on a particular chromosome is $M_1 - M_2 - M_3 - \cdots - M_n$, the likelihood map for the M_1–M_2 interval is constructed using only marker information from these two loci, the profile plot for the M_2–M_3 interval uses only information from M_2 and M_3, etc.

Given the multiple-test nature of these plots (because each map is actually a set of multiple intervals), the appropriate threshold value for the collection of internal maps that constitutes the likelihood map for a chromosome is debatable. Knott and Haley (1992a) noted that the total number of independent tests is bounded above by the number of intervals examined, but because these intervals are linked, they are not independent tests (Zeng 1993). The lower bound is set by the number of chromosomes examined, as these segregate independently. Hence, we first set a threshold level for each chromosome that ensures a desired genome-wide significance level for the entire collection of chromosomes. If C chromosomes are examined, Equation 18.13b implies that in order to obtain a genome-wide significance level γ, the significance level used to set thresholds for each chromosome is

$$1 - (1 - \gamma)^{1/C} \simeq \gamma/C \tag{18.18}$$

Rebaï et al. (1994b) suggested an improved approach that takes into account differences in chromosome lengths. Turning now to the significance values for intervals on a given chromosome, suppose the chromosome of interest has m intervals and we have set the chromosome-wide significance as γ_i. Simulation studies by Zeng (1994) suggest that if the number of markers is not too large, then, for large sample sizes, the critical value for each interval is approximately given by a χ^2_k value with significance γ_i/m. Here k is the number of free parameters in the likelihood-ratio test. Approximations assuming a dense marker map have been developed (Example 18.3; Lander and Botstein 1989; Feingold et al. 1993; Lander and Schork 1994; Lander and Kruglyak 1995; Piepho 2001a; Kao and Ho 2012), as have those for a sparse marker map (Zeng 1994; Rebaï et al. 1994b). Simulations by Doerge and Rebaï (1996) show that dense marker approximations (assuming a very large number of markers per chromosome) are conservative, with the probability of a test statistic exceeding the α-level threshold being less than α when no QTL is present. For comparison among the intervals on a *specific chromosome*, one can use sequential Bonferroni tests, as well as adjusting for the effective number of tests, both previously discussed above.

Example 18.9. Suppose five chromosomes are used for ML-interval mapping in an F_2 design. Chromosomes 1 through 5 have 10, 5, 20, 30, and 40 markers, respectively. In order to achieve a genome-wide level of significance of $\gamma = 0.10$, what are the approximate critical values for each chromosome? Applying Equation 18.18, the overall level of significance for each chromosome is $1-(1-0.1)^{1/5} = 0.021$.

The critical values for each chromosome vary with the number of markers. For chromosome 1, the significance levels for each test become approximately 0.021/10 = 0.0021. Recall that the degrees of freedom for the test of no QTLs in an F_2 design are 5 − 2 = 3. Because $\Pr(\chi^2_3 > 14.71) = 0.0021$, this implies that the critical values for the likelihood ratios for chromosome 1 is 14.7. Similarly, the critical values for the remaining four chromosomes are 13.2, 16.2, 17.0, and 17.6.

An alternative approach to obtaining critical values is to use permutation tests to set the threshold levels (Churchill and Doerge 1994; Doerge and Churchill 1996). This resampling procedure has the advantage of being robust to the actual distribution of effects. Further, resampling is superior to analytical approximations for data with missing and incomplete marker information, as the permutation test, by keeping genotypes intact during reshuffling, automatically incorporates the special nature of each data set (Doerge and Rebaï 1996). Piepho (2001a) noted that resampling can be computationally intensive, and suggested an approximate threshold approach that is much less demanding. He used results from Davies (1987) for situations where a nuisance parameter is only present in the alternative hypothesis (here, the QTL position, which is absent under the null). Using this theory, Piepho developed an approximation method that uses information from the geometry of the profile plot (the number of times the derivative of the profile plot changes along a chromosome) to obtain a chromosome-wide error rate.

Finally, it should be mentioned that the null hypothesis usually assumed, that of no QTLs, may be misleading. Crossed lines are often chosen *because* they differ in traits of interest, so that there is certainly segregating genetic variance in the F_2 and other line-cross populations. Visscher and Haley (1996) noted that if such background variance is present, it results in a more frequent rejection of the null hypothesis of no QTL than expected. They argued the more appropriate null hypothesis should be that, taking the strain differences into account, the amount of genetic variance explained by a chromosome segment is that expected by chance, and they propose several tests of this hypothesis. Xu (2013) presented general mixed model to account for nonadditive (dominance and epistasis) background variation.

Approximating ML Interval Mapping by Haley–Knott Regressions

One problem with ML estimators is that they can be rather computationally demanding. Among other things, this limits the applicability of resampling methods, which require thousands of ML estimates to be computed per experiment. Fortunately, a simple regression procedure usually gives an excellent approximation of the likelihood map for ML interval mapping (Haley and Knott 1992; Martínez and Curnow 1992; Xu 1995, 1998; Kao 2000). This procedure greatly facilitates resampling, as regressions are quickly and easily computed. Haley and Knott's (1992) idea is to express the regression coefficients as a function of the unknown QTL parameters. Using the Falconer parameterization for genotypic means,

$$\mu_{QQ} = \mu + a, \qquad \mu_{Qq} = \mu + d, \qquad \mu_{qq} = \mu - a \tag{18.19a}$$

this is done by considering the regression

$$z_j = \mu + a \cdot x(M_j) + d \cdot y(M_j) + e_j \tag{18.19b}$$

The variables x and y, which depend on both the flanking-marker genotype of the individual (M) and the assumed map position of the putative QTL, are obtained as follows. Taking the expectation of Equation 18.19b over all individuals with marker genotype M_i gives

$$\mu_{M_i} = \mu + a \cdot x(M_i) + d \cdot y(M_i) \tag{18.20a}$$

From Equation 18.6,

$$\begin{aligned} \mu_{M_i} &= (\mu + a)\,\Pr(QQ \mid M_i) + (\mu + d)\,\Pr(Qq \mid M_i) + (\mu - a)\;\Pr(qq \mid M_i) \\ &= \mu + a \cdot \big[\; \Pr(QQ \mid M_i) - \Pr(qq \mid M_i) \big] + d \cdot \Pr(Qq \mid M_i) \end{aligned} \tag{18.20b}$$

Equating like terms in Equations 18.20a and 18.20b gives the conditional means of x and y, given marker genotype, as

$$x(M_i) = \Pr(QQ \mid M_i\,) - \Pr(qq \mid M_i), \qquad y(M_i) = \Pr(Qq \mid M_i) \tag{18.21}$$

Thus, the x and y values are functions of the conditional QTL probabilities given the flanking-marker genotypes. For the F_2 design with no interference, Equation 18.2 gives

$$x(M_1M_1M_2M_2) = \frac{(1-c_1)^2(1-c_2)^2 - c_1^2\,c_2^2}{(1-c_{12})^2}$$

$$y(M_1M_1M_2M_2) = \frac{2c_1c_2(1-c_1)(1-c_2)}{(1-c_{12})^2}$$

Haley and Knott give expressions for the eight other F_2 marker genotypes, and values for other designs easily follow when the appropriate conditional probabilities are employed. This regression approach was independently suggested by Martínez and Curnow (1992) for the analysis of backcross populations. These authors also detailed how missing marker information can be accommodated (Martínez and Curnow 1994a).

By analogy with likelihood maps, the regression given by Equation 18.19b is computed for each c_1 value within the M_1–M_2 interval, with the value giving the regression with the largest r^2 being taken as the estimate of QTL position. For each c_1 value, Equation 18.21 yields the set of x and y values, allowing μ, a, and d to be estimated by ordinary least-squares regression (Equation 10.9a),

$$\mathbf{b}_{c_1} = \begin{pmatrix} \widehat{\mu} \\ \widehat{a} \\ \widehat{d} \end{pmatrix} = \left(\mathbf{X}_{c_1}^T\,\mathbf{X}_{c_1}\right)^{-1}\mathbf{X}_{c_1}^T\,\mathbf{z} \tag{18.22}$$

where the ith row of the design matrix $\mathbf{X}_{c_1}$ is $(1, x(M_i, c_1), y(M_i, c_1)\,)$, with M_i denoting the marker genotype for individual i.

Haley and Knott show that r^2 plots for this regression are related to likelihood plots. Assuming that phenotypes are normally distributed about each QTL genotype, then if the QTL is completely linked to either marker ($c_1 = 0$ or $c_1 = c$), the residuals for the regression given by Equation 18.19b are normally distributed. In this case, the regression estimates are also ML estimates and the likelihood-ratio test can be expressed as

$$\text{LR} = n\,\ln\left(\frac{\text{SS}_T}{\text{SS}_E}\right) = -n\,\ln(1-r^2) \tag{18.23}$$

where SS_T and SS_E are the total and error (or residual) sums of squares associated with the regression (Equations A3.16a and A3.16c), with the second equality following from Equation A3.15a. If the QTL is not completely linked to either marker, the distribution of residuals follows a *mixture of normals*, as some marker genotype classes will contain different QTL genotypes. ML accommodates this mixture of normals, while LS does not. However, Haley and Knott (1992) and Rebaï et al. (1995) showed that the function given by Equation 18.23 often gives extremely similar values to the true likelihood ratio. Haley and Knott suggested that the number of degrees of freedom appropriate for this test is the number of estimated QTL parameters plus an additional degree of freedom for map position c_1. Xu (1995) noted that this regression approximation tends to overestimate the residual variance, and presented a correction. More generally, if the linear model has additional factors (accounting for, say, differences due to sex and age), the LR test is modified to become

$$\text{LR} = n \ln \left(\frac{\text{SS}_E(reduced)}{\text{SS}_E(full)} \right) \tag{18.24}$$

where the error sums of squares are now for the full model and the reduced model (the latter incorporating all factors but the QTL effects). Improvements building on the Haley–Knott regression approach (such as iteratively reweighted least squares) have been proposed (Xu 1998a, 1998b; Feenstra et al. 2006; Han and Xu 2008).

Example 18.10. Consider the following hypothetical data set: 10 F_2 individuals scored for flanking marker-genotypes M_1/m_1 and M_2/m_2, separated by recombination frequency $c_{12} = 0.30$. The following marker genotypes and their associated character values were observed:

$M_1m_1M_2m_2$	$M_1M_1M_2M_2$	$M_1m_1M_2M_2$	$m_1m_1M_2m_2$	$M_1M_1M_2m_2$
3.9	5.6	3.7	3.9	5.3
$m_1m_1m_2m_2$	$M_1m_1M_2M_2$	$M_1M_1M_2M_2$	$M_1m_1M_2M_2$	$M_1m_1M_2m_2$
1.1	3.6	5.4	3.7	3.3

This yields the observation vector

$$\mathbf{z}^T = (3.9,\ 5.6,\ 3.7,\ 3.9,\ 5.3,\ 1.1,\ 3.6,\ 5.4,\ 3.7,\ 3.3)$$

Assuming no interference, $c_2 = (0.3 - c_1)/(1 - 2c_1)$. For each c_1 value within the M_1–M_2 interval ($0 \leq c_1 \leq 0.3$), a regression is fitted by first using Equation 18.21 to compute the elements of the design matrix for that value of c_1 and then using Equation 18.22 to obtain the regression coefficients. For example, consider three different QTL positions: $c_1 = 0$ (QTL at marker M_1), $c_1 = 0.15$ (QTL in the middle), and $c_1 = 0.3$ (QTL at marker M_2). The resulting regressions for these three c_1 values are

c_1	$\widehat{\mu}$	$\widehat{a}$	$\widehat{d}$	r^2
0.00	3.97	1.47	−0.33	0.730
0.15	3.70	1.89	−0.26	0.732
0.30	2.75	1.65	1.35	0.597

These regressions are obtained using the design matrices

$$\mathbf{X}_0 = \begin{pmatrix} 1 & 0 & 1 \\ 1 & 1 & 0 \\ 1 & 0 & 1 \\ 1 & -1 & 0 \\ 1 & 1 & 0 \\ 1 & -1 & 0 \\ 1 & 0 & 1 \\ 1 & 1 & 0 \\ 1 & 0 & 1 \\ 1 & 0 & 1 \end{pmatrix}, \quad \mathbf{X}_{0.15} = \begin{pmatrix} 1 & 0.00 & 0.85 \\ 1 & 0.91 & 0.09 \\ 1 & 0.35 & 0.60 \\ 1 & -0.56 & 0.40 \\ 1 & 0.56 & 0.40 \\ 1 & -0.91 & 0.09 \\ 1 & 0.35 & 0.60 \\ 1 & 0.91 & 0.09 \\ 1 & 0.35 & 0.60 \\ 1 & 0.00 & 0.85 \end{pmatrix}, \quad \mathbf{X}_{0.3} = \begin{pmatrix} 1 & 0 & 1 \\ 1 & 1 & 0 \\ 1 & 1 & 0 \\ 1 & 0 & 1 \\ 1 & 0 & 1 \\ 1 & -1 & 0 \\ 1 & 1 & 0 \\ 1 & 1 & 0 \\ 1 & 1 & 0 \\ 1 & 0 & 1 \end{pmatrix}$$

To complete the analysis, regressions are computed for the full range of c_1 values over the M_1–M_2 interval, generating the following plot of regression r^2 as a function of c_1.

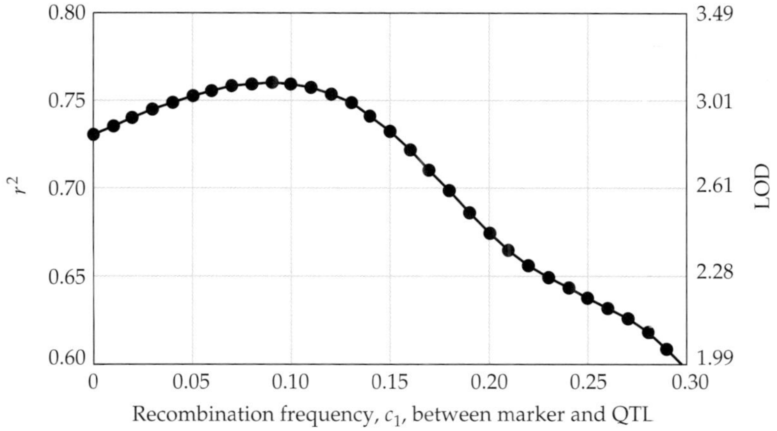

The maximum value of r^2 (0.76) occurs at $c_1 = 0.09$, and the associated regression coefficients are $\widehat{\mu} = 3.90$, $\widehat{a} = 1.76$ and $\widehat{d} = -0.46$. Hence, the data suggest that a QTL lies between these two markers at recombination fraction $c_1 = 0.09$ from marker locus M_1, with estimated genotypic means

$$\widehat{\mu}_{QQ} = \widehat{\mu} + \widehat{a} = 5.66, \quad \widehat{\mu}_{Qq} = \widehat{\mu} + \widehat{d} = 3.44, \quad \widehat{\mu}_{qq} = \widehat{\mu} - \widehat{a} = 2.14$$

Does this example show significant evidence of a QTL? From Equation 18.23, with $n = 10$ and $r^2 = 0.76$, the likelihood ratio (LR) becomes $-10 \cdot \ln(1-0.76)$ = 14.27. Note that only two QTL parameters are fitted (a and d) because the reduced model fits a mean μ. Hence, the critical value for the likelihood ratio is a χ^2 with three degrees of freedom (for a, d, c_1),

$$\Pr[\,\chi^2_3 > -n\ln(1-r^2)\,] = \Pr[\,\chi^2_3 > 14.27\,] = 0.003$$

showing that the QTL effect is indeed significant.

Approximate confidence intervals can be constructed as the set of all those c values giving scores within one LOD of the maximum value. We can translate r^2 values into LOD scores by using LOD = LR/4.61 = $-n\,\ln(1-r^2)/4.61$. The MLE has $r^2 = 0.76$ and $n = 10$, for a LOD score of $-10\,\ln(1{-}0.76^2)/4.61$ = 3.10. Hence any c_1 value with a LOD score of 2.10 or greater is in the one-LOD support interval for QTL position. The resulting interval is c_1 = 0 to 0.28, so that although there is very strong evidence for a QTL, there is extreme uncertainty as to its position within the interval. This is not surprising given the very small sample size.

DEALING WITH MULTIPLE LINKED QTLs

All of the methods discussed so far are best characterized as **one-at-a-time approaches** for mapping QTLs, as they all assume a single QTL linked to the marker(s) of interest. While such methods can detect the presence of multiple QTLs (e.g., finding marker effects on a number of different chromosomes), they cannot discern whether significant effects at several linked markers/intervals are due to a common QTL or due to several linked QTLs. The presence of multiple linked QTLs also introduces serious biases into estimates of QTL effects and positions derived from the one-at-a-time approach. We have already mentioned that several tightly linked QTL, depending on the sign of their effects, can result in either an enhanced (same signs), or reduced (alternating signs), marker signal (Equation 18.8a).

A much more subtle effect is that while the presence of multiple (significant) peaks on a likelihood profile for a given chromosome is generally taken as an indication of multiple QTLs, ***such peaks do not necessarily correspond to the correct QTL positions*** (Martínez

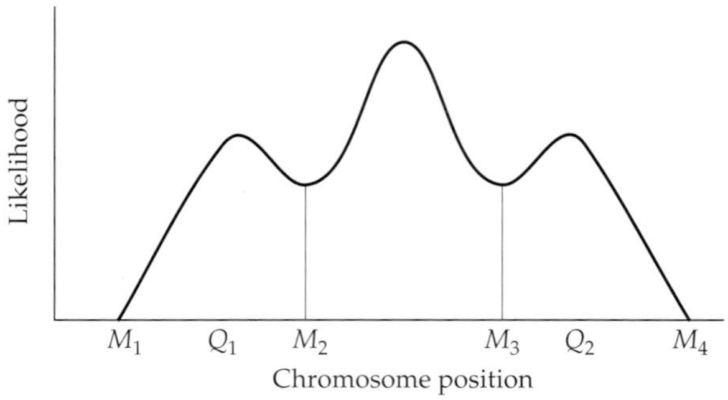

Figure 18.3 A false (or **ghost**) QTL generated by using a single-QTL likelihood function when two linked QTLs are actually present.

and Curnow 1992; Haley and Knott 1992; Wright and Kong 1997). Figure 18.3 gives an example of two linked QTLs embedded within four markers. Using a likelihood function that assumes only a single QTL, interval mapping correctly indicates likelihood peaks in the intervals flanked by M_1–M_2 and M_3–M_4. However, the resulting map also shows a much higher peak between M_2–M_3, incorrectly suggesting the presence of a third QTL in this region. Under a single QTL model, analysis would start with this central peak first.

One approach to mitigate the impact of a detected QTL on the signal from other QTLs is to include the closest marker (or flanking-marker interval) as a cofactor in the analysis. This has the effect of reducing or removing background segregation variance contributed by the detected QTL, reducing the residual variance for the marker/interval under consideration, increasing the power for detection and improving the precision of estimates. Example 18.11 (below) provides a dramatic illustration of this. Adding a cofactor for a detected QTL to Equation 18.14a would yield

$$z_{ik} = \mu + bN_i + hM_i + e_{ik}$$

where M_i is the gene dosage in individuals i from the marker associated with the putative QTL (the cofactor), and N_i the dosage for the current marker being tested. Alternatively, one could first adjust the data to

$$z^*_{ik} = z_{ik} - hM_i$$

and then use these adjusted values to fit additional QTL effects. Dominance effects at the detected QTL can be accommodated by using Equation 18.14c.

While this cofactor approach works well when the detected QTL is unlinked to the region being probed for additional QTLs, it can introduce serious bias when testing for additional QTLs linked to the detected one. In particular, Figure 18.3 shows that using markers associated with the highest peak would force the detected QTL in the wrong location. While specific tests for the presence of linked QTLs in adjacent intervals using sets of three overlapping markers have been suggested (Martínez and Curnow 1992, 1994b; Haley and Knott 1992), these are not without problems (Whittaker et al. 1996).

Most of the single-QTL methods developed above can be extended to multiple QTLs by considering additional marker loci and using conditional probabilities for multilocus genotypes. This approach has been used to develop explicit models for two or three linked QTLs (e.g., Knapp 1991; Haley and Knott 1992; Martínez and Curnow 1992, 1994b; Jansen 1996; Satagopan et al. 1996; Wright and Kong 1997; Goffinet and Mangin 1998). We focus here on three particularly flexible regression-based approaches. The first is **marker-difference regression**, which considers all of the markers on one chromosome in a single analysis by using the regression of differences between the mean values of different genotypes. The second is **composite interval mapping** (**CIM**), which controls for both the effects of linked and unlinked QTLs by using the appropriate marker cofactors. This approach can be extended to **multiple interval mapping** (**MIM**) to simultaneously fit multiple linked intervals (Kao and Zeng 1997; Kao et al. 1999; Kao and Zeng 2002; Chen 2005; Verbyla et al. 2007). Finally, Wright and Mowers (1994) and Whittaker et al. (1996) have shown how positional information for linked QTLs can be extracted from the regression coefficients of a standard multiple regression incorporating several linked markers.

Example 18.11. Lin et al. (1995) examined flowering date through ML interval mapping of 370 F_2 individuals from a cross between cultivated and exotic sorghum (*Sorghum bicolor* × *S. propinquum*). Only a single QTL for flowering date was detected, and this accounted for 85.7% of the total variance. The data were then adjusted to account for the effects of this major gene by using $(z - b_i)$ in place of the trait value z for an individual with genotype i at a marker linked to the major gene, the b_i $(1 \leq i \leq 3)$ being the regression coefficients generated by a standard marker-trait regression using this marker locus. While the uncorrected F_2 phenotypic distribution was clearly bimodal, the adjusted data did not appear to deviate from normality. Using the marker-adjusted data, two additional QTLs for flowering time were found (both unlinked to the original QTL), accounting for an additional 8.3% and 4.2% of the total variance. This example illustrates the potential importance of including additional marker information into the analysis when multiple QTLs are present. In this case, removing the effects of a major unlinked QTL reduced the residual variance sufficiently to enable detection of additional, smaller effect, QTLs.

Marker-Difference Regression

Two groups (Kearsey and Hyne 1994; Hyne and Kearsey 1995; Wu and Li 1994, 1996a, 1996b) proposed a very simple, yet powerful, regression method that simultaneously considers all of the markers on a single chromosome. While the authors refer to this method as **marker regression** or **joint mapping**, we will use the more descriptive term **marker-difference regression**, or **MDR**, to emphasize that this approach is rather different from the regressions that we have considered up to this point. With MDR, each data point in the regression corresponds to a marker *mean value*, rather than to values for *single individuals* (as in our previous regressions). While this data structure results in far fewer points in the regression, the use of means allows the inclusion of individuals missing some marker information and also allows the joint incorporation of information from several experiments. Because the method only considers differences between the marker homozygotes, it is best suited for a collection of RILs or DH lines. It can also be formulated to apply to a BC analysis by contrasting the homozygote and heterozygote of a given marker.

The motivation for MDR follows from Equation 18.7a. We first present the method under the assumption of a single QTL to illustrate the main points before extending it to multiple QTLs. Suppose there are n linked markers along a chromosome containing a single QTL (with alleles Q and q). If the ith marker is at recombination frequency c_i from the QTL, the expected difference between marker homozygote means is

$$y_i = \mu(M_iM_i) - \mu(m_im_i) = 2a(1 - 2c_i)$$

Thus, if we plot the differences y_i versus $(1 - 2c_i)$ for each marker on the chromosome, the resulting n points are expected to fall on a straight line passing through the origin with slope $2a = \mu_{QQ} - \mu_{qq}$. Figure 18.4 illustrates this point, showing two regressions using the same set of marker differences but assuming two different locations for the QTL. The regression computed using the correct position of the QTL is linear, while that assuming the incorrect position is highly nonlinear. As with Haley–Knott regressions, one slides the position of a putative QTL along the chromosome, computing a regression at each point. The regression giving the best fit (i.e., the largest r^2) corresponds to the estimate of QTL position, and the slope of that regression divided by two provides an estimate of the QTL effect, a.

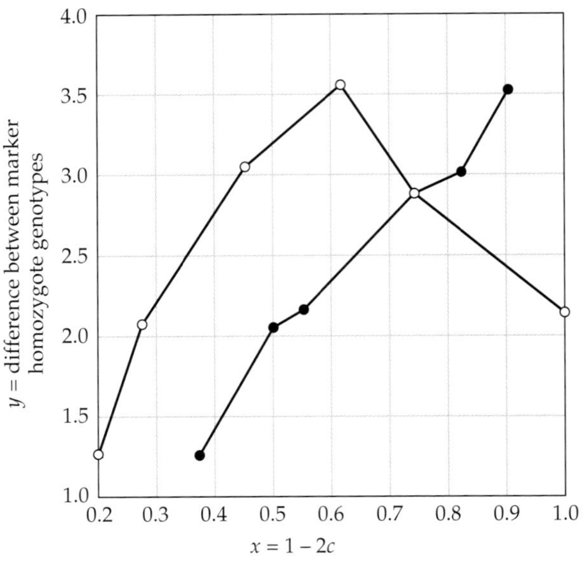

Figure 18.4 Marker-difference regression plot for the data given in Example 18.12. Open circles assume a QTL at map position 90 cM, closed circles a QTL at position 60 cM (the true position). Note that the relationship is linear when the correct position is used, but highly nonlinear under the incorrect position. Further note by comparing the open and solid points that their vertical positions are the same (each corresponding to a difference in marker homozygotes), but their horizontal (location x) values change as we change assumptions as to where the QTL is located (e.g., the solid point at 0.55 corresponds to the open point at 1.0)

To formally develop this approach, suppose that there are n linked markers scored along a single chromosome, and consider the regression

$$y_i = \overline{z}(M_iM_i) - \overline{z}(m_im_i) = \beta\, x_i + e_i \tag{18.25}$$

with the $x_i = 1 - 2c_i$ values obtained by fixing the QTL position and then computing c_i for each marker (the marker-putative QTL distance). Because the residuals are correlated and potentially heteroscedastic, generalized least-squares regression (Chapter 10) must be used, with

$$\widehat{\beta} = \left(\mathbf{X}^T\mathbf{V}^{-1}\mathbf{X}\right)^{-1}\mathbf{X}^T\mathbf{V}^{-1}\mathbf{y} \tag{18.26a}$$

which has sample variance

$$\sigma^2(\widehat{\beta}) = \left(\mathbf{X}^T\mathbf{V}^{-1}\mathbf{X}\right)^{-1} \tag{18.26b}$$

where

$$\mathbf{y} = \begin{pmatrix} y_1 \\ \vdots \\ y_n \end{pmatrix}, \quad \mathbf{X} = \begin{pmatrix} 1-2c_1 \\ \vdots \\ 1-2c_n \end{pmatrix} \tag{18.26c}$$

and

$$\mathrm{V}_{ij} = \begin{cases} \dfrac{\mathrm{Var}(M_iM_i)}{n(M_iM_i)} + \dfrac{\mathrm{Var}(m_im_i)}{n(m_im_i)} & i = j \\[2ex] (1-2c_{ij})\sqrt{\mathrm{V}_{ii}\,\mathrm{V}_{jj}} & i \neq j \end{cases} \tag{18.26d}$$

where $\mathrm{Var}(M_x)$ is the sample variance of $\overline{z}(M_x)$, $n(M_x)$ is the sample size for marker class M_x, and c_{ij} is the distance between markers i and j (Wu and Li 1996a).

Assuming normally distributed residuals, from Equation A3.11a the residual sum of squares,

$$\mathrm{SS}_E = \widehat{\mathbf{e}}^T\mathbf{V}^{-1}\widehat{\mathbf{e}} = (\mathbf{y} - \mathbf{X}\widehat{\beta})^T\mathbf{V}^{-1}(\mathbf{y} - \mathbf{X}\widehat{\beta}) \tag{18.27}$$

follows a χ^2 distribution with $n-2$ degrees of freedom (n data points minus two estimated parameters: the QTL effect $\beta = 2a$ and the assumed position, c). The test for a significant QTL effect compares the SS_E for this regression with that for the regression assuming no marker effect ($y_i = \mu + e_i$). The SS_E for the reduced model is also χ^2-distributed, but with $n-1$ degrees of freedom. Recalling (Equation A5.14c) the additivity property of the chi-square, the difference in residual sums of squares for these two models follows a χ_1^2 distribution under the null hypothesis. Hence, the regression is significant at the α level if SS_E(reduced model) − SS_E(QTL model) exceeds $\chi_1^2(\alpha)$, the α-level cutoff for a χ_1^2. Separate regressions are computed for each chromosome, so that to obtain a genome-wide level of significance γ, each chromosomal regression is tested with significance level $\alpha = 1 - (1-\gamma)^{1/C} \simeq \gamma/C$, where C is the number of chromosomes examined. Piepho (2001b) noted that the χ^2 assumption is only approximate, and suggested an improved significance test for MDR.

Example 18.12. Consider the following hypothetical data (plotted in Figure 18.4) generated by assuming a single QTL with effect $a = 2.0$ at map position 60 cM along a chromosome containing six markers:

Marker Position (cM)	$\overline{z}(M_iM_i) - \overline{z}(m_im_i)$
10	1.26
25	2.06
50	3.04
65	3.54
75	2.90
90	2.15

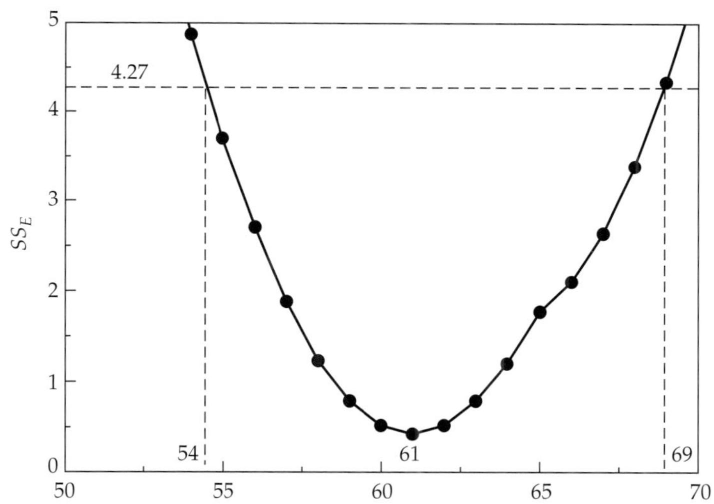

We assume that the variance associated with each marker class is the same with Var(M_x) = 5, and that 50 individuals of each marker class were scored, giving $V_{ii} = 2 \cdot 5/50 = 0.2$. Using this and the c_{ij} values with Equation 18.26d fills out the rest of $\mathbf{V}$. For an MDR analysis, one computes a separate regression for each possible QTL position. Consider the regression for a QTL assumed to be at map position 50 cM. For the first marker, the QTL–marker map distance is 40 cM, which (assuming a Haldane map distance; Equation 17.3) translates into a recombination frequency of

$$c_1 = [1 - e^{(-2 \cdot 0.4)}]/2 \simeq 0.275$$

giving $x_1 = (1 - 2c_1) = 0.45$, and the data point associated with this marker becomes (0.45, 1.26). Computing the remaining data points and applying Equations 18.26 and 18.27 give a regression with SS_E = 11.03. After this procedure is repeated for all positions along the chromosome, the resulting plot of SS_E versus putative QTL position (shown above) exhibits a minimum value (0.43) at map position 61, and hence r^2 is maximized at this position (see Equation A3.15a).

Whether the fit under the single-QTL model is a significant improvement over a model assuming no QTL can be assessed by comparing the error sum of squares of the QTL model (SS_E = 0.43) with the error sum of squares of the reduced (no QTL) model $y_i = \mu + e_i$. Because the QTL model fits an extra parameter, the difference in sums of squares follows a χ^2 distribution with one degree of freedom under the hypothesis of no QTL effect. For the reduced model, SS_E = 19.16, which is obtained by setting $\mathbf{X}$ equal to a vector of ones and applying Equation 18.26a. Hence, the QTL effect is highly significant as

$$\Pr(\chi_1^2 > 19.16 - 0.43) = \Pr(\chi_1^2 > 18.73) = 0.000015$$

The adequacy of the single-QTL model can be assessed by noting that if this model is correct, SS_E follows a χ_4^2 distribution (there are six data points and two fitted parameters, for four degrees of freedom). Because $\Pr(\chi_4^2 > 0.43) = 0.99$, SS_E is not larger than expected by chance, suggesting that there is no need to consider additional QTLs.

Using the estimated map position, the resulting regression has slope 3.84, giving the estimated QTL effect as $\widehat{a} = 3.84/2 = 1.92$. From Equation 18.26b, $\sigma^2(2\,\widehat{a}\,) = \left(\mathbf{X}^T\mathbf{V}^{-1}\mathbf{X}\right)^{-1}$ = 0.16, giving the standard error of $\widehat{a}$ as $\sqrt{0.16}/2 = 0.20$. Because SS_E follows a χ_1^2 distribution, the 95% confidence interval for QTL position contains those values giving regressions with SS_E not exceeding $\chi_1^2(0.05) = 3.84$ of the minimal SS_E value of 0.43 (i.e., SS_E values less than 4.27). This gives the confidence interval for the QTL position as 54 to 69 cM (see figure).

This approach easily extends to multiple QTLs. Recalling Equation 18.8a, if there are N linked QTLs, the jth of which is at recombination frequency c_{ji} from marker i, then (assuming no epistasis),

$$y_i = \mu(M_iM_i) - \mu(m_im_i) = 2a_1(1 - 2c_{1i}) + \cdots + 2a_N(1 - 2c_{Ni}) \tag{18.28a}$$

This immediately suggests the multiple regression

$$y_i = \beta_1 \cdot x_{1i} + \cdots + \beta_N \cdot x_{Ni} + e_i \tag{18.28b}$$

where $x_{ji} = (1 - 2c_{ji})$ and $\beta_j = 2a_j$. The estimates are still given by Equation 18.26a, with $\mathbf{y}$ and $\mathbf{V}$ being defined in the univariate case, and

$$\boldsymbol{\beta} = \begin{pmatrix} \beta_1 \\ \vdots \\ \beta_N \end{pmatrix} \quad \text{and} \quad \mathbf{X} = \begin{pmatrix} 1 - 2c_{11} & \cdots & 1 - 2c_{N1} \\ \vdots & \ddots & \vdots \\ 1 - 2c_{1n} & \cdots & 1 - 2c_{Nn} \end{pmatrix}$$

where N is the number of assumed QTLs, and n is the number of markers. As above, one computes the regression over the set of all possible QTL positions, with the estimates of QTL positions being given by the regression with the smallest SS_E value (or largest r^2). Each additional QTL reduces the degrees of freedom of SS_E by two (one for QTL effect, one for position). The test for whether adding another QTL significantly improves the fit compares the difference in the resulting two error sums of squares (for models assuming N versus $N - 1$ QTLs) with the appropriate critical value for a χ_2^2. Extension of the MDR approach to allow for interacting QTLs was given by Charmet et al. (1998).

Interval Mapping with Marker Cofactors

The careful reader will note that marker-difference regression does not require knowledge of the multilocus marker genotypes of any individual, as all that enters into the analysis are the population means for each separate marker. An alternative approach for dealing

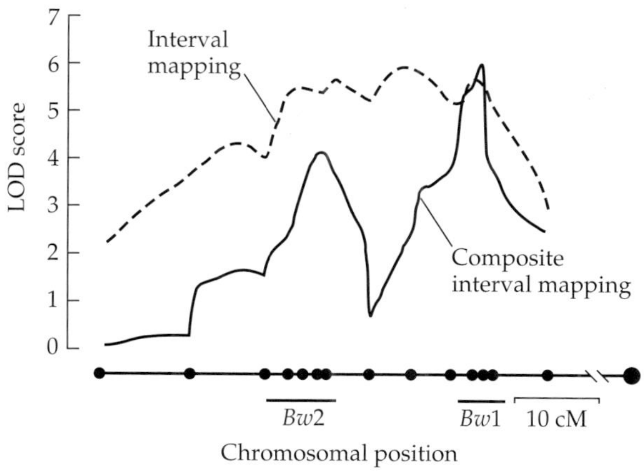

Figure 18.5 Likelihood plots for the X chromosome for QTLs influencing body weight in mice. The likelihood map under standard ML interval mapping (dashed line) shows a single very broad peak. Using the same data, the CIM likelihood map (solid lines) shows two distinct peaks. *Bw1* and *Bw2* denote the two putative body-weight QTLs, and the dots on the chromosome indicate the positions of the marker loci. Note that the ML interval mapping curve shows a ghost QTL, giving the largest peak between the true flanking QTLs (c.f., Figure 18.3). (After Dragani et al. 1995.)

with multiple QTLs that incorporates multilocus marker information from individuals is to modify standard interval mapping to include *additional markers as cofactors* in the analysis. Using appropriate unlinked markers can partly account for the segregation variance generated by unlinked QTLs (Jansen 1992, 1993b; Zeng 1993, 1994), while the effects of linked QTLs can be reduced by including markers linked to the interval of interest (Stam 1991; Zeng 1993, 1994; Rodolphe and Lefort 1993). This general approach of adding marker cofactors to an otherwise standard interval analysis, often referred to as **composite interval mapping (CIM)**, results in substantial increase in the precision of estimates of QTL position (Jansen 1993b, 1994a, 1994b, 1996; Jansen and Stam 1994; Jansen et al. 1995; Zeng 1994; van Ooijen 1994; Utz and Melchinger 1994). Figure 18.5 shows a rather dramatic example of the improvement using CIM over interval analysis.

Suppose the interval of interest is flanked by markers i and $i+1$. One way to incorporate information from additional markers is to consider the sum over some collection of markers outside the interval of interest,

$$\sum_{k \neq i, i+1} b_k \cdot x_{kj} \tag{18.29a}$$

where k denotes a marker locus and j the individual being considered. Letting M_k and m_k denote alternative alleles at the kth marker, the values of the indicator variable x_{kj} depend on the marker genotype of j, with

$$x_{kj} = \begin{cases} 1 & \text{if individual } j \text{ has marker genotype } M_k M_k \\ 0 & \text{if individual } j \text{ has marker genotype } M_k m_k \\ -1 & \text{if individual } j \text{ has marker genotype } m_k m_k \end{cases} \tag{18.29b}$$

This is simply a convenient recoding of the gene-dosage regression (Equation 18.14a) for each marker. Hence, b_k is an estimate of the additive marker effect for locus k. For a backcross or RIL design, each marker has only two genotypes and the indicator variable takes on values 1 and -1. More generally, if there is considerable dominance, the effects of the kth marker locus can be more fully accounted for by adding a dominance term to Equation 18.29a,

$$\sum_{k \neq i, i+1} (b_k \cdot x_{kj} + d_k \cdot w_{kj}) \tag{18.29c}$$

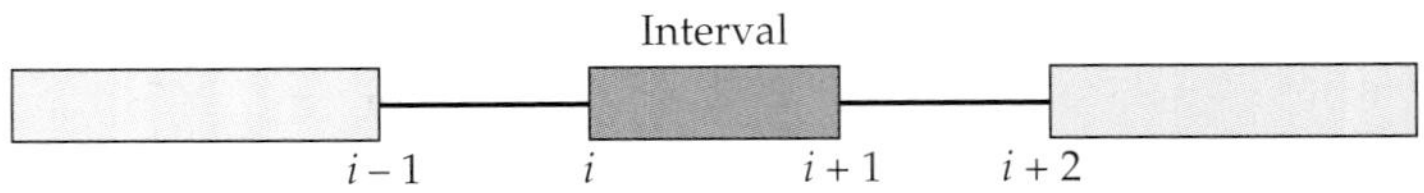

Figure 18.6 Suppose the interval being examined by CIM is between markers i and $i + 1$. Addition of the adjacent markers $i - 1$ and $i + 2$ as cofactors absorbs the effects of any linked QTLs to the left of marker $i - 1$ and to the right of marker $i + 2$. Their inclusion, however, does not remove the effects of QTLs present in the two intervals, $(i - 1, i)$ and $(i + 1, i + 2)$, flanking the interval of interest.

where

$$w_{kj} = \begin{cases} 0 & \text{if individual } j \text{ has marker genotype } M_k M_k \\ 1 & \text{if individual } j \text{ has marker genotype } M_k m_k \\ 0 & \text{if individual } j \text{ has marker genotype } m_k m_k \end{cases} \tag{18.29d}$$

Composite interval mapping proceeds by adding this regression term to the model being considered. For example, upon adding marker cofactors, the Haley–Knott regression focusing on the interval bracketed by markers i and $i + 1$ becomes

$$z_j = \Big[\mu + a \cdot x(M_i) + d \cdot y(M_i) \Big] + \sum_{k \neq i, i+1} b_k \cdot x_{kj} + e_j \tag{18.30}$$

Estimation of the QTL parameters (μ, a, d, c_i) for the interval proceeds as before, e.g., $x(M_i)$ and $y(M_i)$ are given by Equation 18.21 using marker loci i and $i+1$ as the flanking markers, with c_i being the putative QTL- marker i recombination frequency. For each c_i value in the interval, the regression given by Equation 18.30 is fitted (i.e., a, d, and the b_k), and (as before) the c_i value giving the regression with the largest r^2 is taken as the estimate of the QTL position. The significance of the interval can be tested by using Equation 18.24 to compare the full model (Equation 18.30) with the reduced model,

$$z_j = \mu + \sum_{k \neq i, i+1} b_k \cdot x_{kj} + e_j \tag{18.31}$$

which includes the marker cofactors but ignores the interval.

While most QTL mapping papers use some version of ML (or Bayesian) CIM, making it seem a gold standard, there is some debate on just which markers should be added. While there is no single solution, the two markers *directly flanking* the interval being analyzed should always be included. Suppose the interval of interest is delimited by markers i and $i + 1$ (Figure 18.6). Zeng (1994) showed that adding markers $i - 1$ and $i + 2$ as cofactors accounts for all linked QTLs to the left of marker $i - 1$ and to the right of marker $i + 2$. Thus, while these cofactors do not account for the effects of linked QTLs in the intervals immediately adjacent to the one of interest (i.e., the intervals $(i - 1, i)$ and $(i + 1, i + 2)$ in Figure 18.6), they do account for all other linked QTLs.

The number of *unlinked* markers that should be used as cofactors is unclear, as inclusion of too many factors greatly reduces power (Zeng 1994). Jansen and Stam (1994) recommend that the number of cofactors not exceed $2\sqrt{n}$, where n is the number of individuals in the analysis. A first approach would be to include all unlinked markers showing significant marker-trait associations (detected, for example, by standard single-marker regression). If several linked markers from a single chromosome all show significant effects, one might just use the marker having the largest effect. A related strategy, suggested by Jansen (1992, 1993b; Jansen and Stam 1994), is to first perform a multiple regression using all markers and then eliminate those that are not significant. Piepho and Gauch (2001) extend this into a more formal model selection framework.

A multiple-trait extension of composite interval mapping given by Jiang and Zeng (1995) offers improved power for QTL detection and increased precision in estimation (relative to single-trait analysis) by incorporating the correlated error structure among traits (see also Ronin et al. 1995, 1998; Korol et al. 1995; Weller et al. 1997). Jiang and Zeng also develop likelihood-ratio tests for genotype × environment interaction and for tests of pleiotropy versus close linkage (one pleiotropic QTL versus multiple linked QTLs each influencing separate characters).

Hypothesis testing and estimation for CIM follow by simple modifications of the appropriate results for interval mapping. Zeng (1993, 1994) showed that CIM test statistics for linked intervals are only weakly correlated, so that one can approximate each interval as an independent test. Zeng also found that the likelihood ratios within each interval are close to χ^2-distributed, so that an overall significance level of γ for an experiment examining m intervals can be obtained by equating the critical value within each interval to a χ^2 with significance level γ/m.

Resampling methods are easily extended to CIM. Doerge and Churchill (1996) suggested the following permutation test to account for multiple QTLs. A standard permutation test is first used to detect the marker with the greatest marker-trait association. Individuals are then divided (or **stratified**) according to their genotypes at this marker locus, and permutations are performed ***within each stratified group*** to generate new test statistics to find the next most significant QTL. This procedure is repeated until no significant effects are detected. Although permutation and bootstrap approaches are numerically intense, the rapid computation of solutions using Haley–Knott regressions makes these approaches feasible.

Finally, we note that other mapping approaches besides interval mapping can be improved by considering marker cofactors. For example, we can enhance the power of marker-difference regression by including unlinked markers to reduce the residual variance from unlinked QTLs. Because MDR uses the mean values for each marker, the individual data must be adjusted first to remove the effects from unlinked QTLs. Suppose n markers (unlinked to the chromosome of interest) are chosen because they show significant effects. The marker-adjusted value z_j^* of the original trait value of individual j, z_j, is given by

$$z_j^* = z_j - \left(\sum_{k=1}^{n} b_k \cdot x_{kj} \right) \tag{18.32}$$

and an MDR analysis is then performed using these adjusted values.

Detecting Multiple Linked QTLs Using Standard Marker-trait Regressions

Consider the standard gene-dosage multiple regression of trait value on the single-locus genotypes at each of n markers,

$$z_j = \mu + \sum_{k=1}^{n} b_k \cdot x_{kj} + e_j \tag{18.33}$$

where j indexes the individual being considered, and the x_{kj} are given by Equation 18.29b (i.e., only considering additive effects). A rather remarkable finding, due to Wright and Mowers (1994) and Whittaker et al. (1996), is that the regression coefficients b_k for adjacent markers provide information on whether these markers flank a QTL. Further, the b_k can be used in many cases to obtain direct estimates of QTL effect and position.

When a QTL is **isolated**—an interval contains a single QTL and both flanking intervals are free of QTLs—the regression coefficients for the two markers immediately flanking the QTL depend only on this QTL and are not influenced by other linked QTLs (Stam 1991; Zeng 1993). A consequence of this finding is that markers flanking a QTL have regression coefficients of the same sign, while markers not adjacent to a QTL (i.e., there is at least one marker in the regression between the marker of interest and the nearest QTL) have expected regression coefficients of zero. Hence, one can simply scan the regression coefficients to see which intervals show support for a QTL (see Example 18.13).

Whittaker et al. (1996) further showed, for an isolated additive QTL, that the regression coefficients for the flanking markers can be directly used to estimate QTL effect and position.

Suppose markers i and $i+1$ flank an isolated QTL. Whittaker et al. found that for an F_2 population, the estimated distance from marker i to the QTL is

$$c_i = \frac{1}{2}\left[1 - \sqrt{1 - \frac{4\,b_{i+1}\theta_i\,(1-\theta_i)}{b_{i+1} + b_i\,(1-2\theta_i)}}\,\right] \tag{18.34a}$$

where $\theta_i = c_{i,i+1}$ is the distance between the markers. Likewise, an estimate of the QTL's additive effect a, independent of the amount of dominance at this QTL, is given by

$$a^2 = \frac{[\,b_i + (1-2\theta_i)\,b_{i+1}\,]\cdot[\,b_{i+1} + (1-2\theta_i)\,b_i\,]}{1-2\theta_i} \tag{18.34b}$$

where both b_i and b_{i+1} have the same sign as a.

Example 18.13. Whittaker et al. (1996) simulated 2000 F_2 progeny with three chromosomes, each with five markers evenly spaced at 25 cM (implying $c \simeq 0.2$ under Haldane's mapping function). QTLs were placed in the intervals flanked by markers (1, 2), (4, 5), (7, 8), (13, 14), and (14, 15). The multiple regression involving all 15 markers (Equation 18.33) had associated regression coefficients of:

Marker	1	2	3	4	5
b_i	−0.2996	−0.1422	−0.0221	0.2209	0.1956

Marker	6	7	8	9	10
b_i	−0.0189	−0.1922	−0.2404	0.0100	−0.0108

Marker	11	12	13	14	15
b_i	−0.0254	0.0371	0.3019	0.2644	0.3370

Looking for pairs of adjacent regression coefficients that have the same sign and are both significantly different from zero (as judged using standard regression tests, not shown) suggests evidence for QTLs in the intervals (1, 2), (4, 5), (7, 8), (13, 14), and (14, 15). The regression using just these nine markers had essentially the same SS_E as the full regression using all 15 markers, suggesting that none of the omitted markers are adjacent to QTLs (or they are adjacent to multiple linked QTLs whose effects cancel). However, removal of any one of the nine markers results in a regression with a significantly greater error sum of squares, supporting the hypothesis that all of these markers are adjacent to QTLs. Using these nine markers only, the new regression coefficients become

Marker	1	2	4	5	7	8	13	14	15
b_i	−0.2975	−0.1323	0.2296	0.1962	−0.2407	−0.2377	0.3145	0.2640	0.3355

Because the QTLs in intervals (1, 2), (4, 5), and (7, 8) appear to be isolated (no evidence for QTLs in adjacent intervals), Equations 18.34a and 18.34b can be used to estimate their effects and positions. For the QTL in the interval flanked by markers 1 and 2,

$$c_1 = \frac{1}{2}\left[1 - \sqrt{1 - \frac{4\,(-0.1323)\cdot 0.2\,(1-0.2)}{(-0.1323) + (-0.2975)(1-2\cdot 0.2)}}\,\right] = 0.074$$

and the estimate of the squared effect of the QTL is

$$\begin{aligned} a_1^2 &= \frac{[\,(-0.2975) + (1-2\cdot 0.2)(-0.1323)\,]\,[\,(-0.1323) + (1-2\cdot 0.2)(-0.2975)\,]}{1-2\cdot 0.2} \\ &= (0.442)^2 \end{aligned}$$

implying $a_1 = -0.442$ (because the regression coefficients $b_1, b_2 < 0$). Similarly, the estimates for the QTL in the interval (4, 5) are $c_4 = 0.105$ and $a_4 = 0.440$, while for the QTL in (7, 8), we find $c_7 = 0.112$ and $a_7 = -0.494$. The estimated values were rather close to the true values used in the simulations ($a_4 = -a_7 = -a_1 = 0.447$, $c_1 = 0.07$, $c_4 = 0.11$, and $c_7 = 0.11$).

Whittaker et al. (1996) made a final important point that applies to all multiple-QTL methods. Unless a QTL is isolated—it is the only QTL in a particular interval and the flanking intervals lack QTLs—these methods cannot separate out the effects of multiple linked QTLs. In particular, if an interval contains multiple QTLs, we cannot estimate their effects and positions (or even the correct number of QTLs), a point stressed by McMillan and Robertson (1974; Example 17.2). While one obvious solution is simply to increase the marker density to the point where each QTL is indeed isolated, any increase in the marker density must be accompanied by a sufficient increase in sample size to ensure that a *sufficient number of recombination events* have occurred between adjacent markers.

MAPPING INTERACTIONS

Complications with Mapping Epistatic QTLs

When multiple QTLs are present, one must entertain the possibility of epistatic interactions between them. Detecting, and mapping, such interacting loci poses a number of challenges, as we now detail. At first blush this task seems rather doable, as we have already detailed how to modify linear models to search for interactions between two-locus (or higher order) marker genotypes (Equation 18.14d). Just as one could expand the gene-dosage single-marker regression (Equation 18.14a) to allow for dominance (Equation 18.14b), Equation 18.14d can be further decomposed into estimates of specific epistatic interactions (e.g., Kao and Zeng 2002; Zeng et al. 2005; Alvarez-Castro and Carlborg 2007). While detecting epistasis seems straightforward, several critical (and delicate) issues remain. As a result, there is still considerable debate about its importance in QTL mapping (Carlborg and Haley 2004; Malmberg and Mauricio 2005). However, it is also becoming apparent that when epistasis is present, models that incorporate searches for pairwise effects often have higher power for detecting QTL than models that ignore possible interactions (Coffman et al. 2005; Evans et al. 2006; Marchini et al. 2005; Verhoeven et al. 2010).

Most of these concerns revolve around the massive expansion in the number of possible tests. For k markers, there are k potential tests for main effects, where single-locus marker genotypes show significant differences in trait means. However, there are $k(k-1)/2$ additional tests for (two-locus) epistatic interactions, for a total of $k(k+1)/2$. If 100 markers are scored (a modest number for most QTL experiments), there are 4,950 tests for two-marker interactions, for a total of 5,050 tests. Hence, when epistasis is considered, the significance threshold to control the genome-wide error rate (GWER) is considerably more strict than when only single-marker tests are considered. With 100 markers, the standard Bonferroni correction when only single-marker tests are considered to obtain a GWER of 5% would be to test each marker using $\alpha = 0.05/100 = 5 \cdot 10^{-4}$. However, when epistatic interactions are included, to control the GWER all tests must now use $\alpha = 0.05/[100 \cdot 101/2] = 1 \cdot 10^{-5}$, resulting in a significant loss in power. There is also the philosophical issue in that if tests for epistasis are done, the number of *all* tests performed *must* be included in the GWER correction, even if none were significant. Thus, the cost of testing epistasis reduces the power to detect main effects if both classes of tests are performed simultaneously. Conversely, including interaction effects in a model can potentially reduce residual error, increasing the power to detect weak margin effects (Carlborg and Andersson 2002). A second issue is that we usually expect $k(k+1)/2 \gg n$, with potentially far more epistatic parameters to estimate than there are degrees of freedom (a **saturated model**).

The final concern is a more subtle issue in that the *genotypic sample size* for detecting two-locus interactions is less than that for testing single-locus markers. For example, with a design involving n F_2s, the expected number of M_iM_i or m_im_i individuals is $n/4$, while the expected number of $M_iM_iM_jM_j$ or $m_im_im_jm_j$ is $n/16$. Hence, two-locus marker differences are measured with higher standard errors (due to the smaller sample size for each multilocus genotype). This combination of more stringent testing coupled with higher standard errors poses serious power constraints on detection.

Epistatic Mapping Strategies

While little can be done to address the epistatic sample-size issue (short of creating mapping populations to test specific combinations of genotypes), three basic strategies have been proposed to deal with the number of possible tests.

1. **Test only among pairs of markers that both show significant main effects.** If only $s \ll k$ such markers are detected, then the additional number of tests, $s(s-1)/2$ is relatively modest.

2. **Test all interactions between a significant main-effect marker and all other markers.** These first two approaches are called **marginal scans**, requiring at least one marker showing a marginal effect. The rough number of additional tests here is $s(k-1)/2$, which, while far more than the first strategy, is still less than testing all combinations.

3. **Test all combinations, independent of whether they have significant main effects,** a ***full scan.***

Most of the early scans for epistasis relied on strategy 1, making the assumption that if a locus has a strong interaction effect with other loci, at least part of that will also be reflected as a main (marginal) effect. This is often true when the loci involved exhibit somewhat extreme allele frequencies (minor allele frequencies are small). However, by the nature of a two-line cross, all segregating alleles have a minor allele frequency of 1/2, the optimal setting for large interaction effects with potentially small main effects (given the right type of epistasic interaction; e.g., Jana 1971).

What do the data say? Unfortunately, significant epistatic effects between loci with weak (i.e. nonsignificant) main effects can occur in QTL studies. Lark et al. (1995) reported several examples of soybean QTLs where one locus had a significant main effect, but its magnitude was strongly influenced by a second locus (which itself showed no significant main effect). In a mouse QTL experiment on lung cancer susceptibility, Fijneman et al. (1996) found that none of the four *Sluc* QTLs had detectable main effects under interval mapping, while *Sluc1* had a strong interaction with the (unlinked) *Sluc2*, and *Sluc3* interacted with the unlinked *Sluc4*. Similarly, Cubillos et al. (2011) found yeast growth loci only detectable by their interaction effects. A larger study involving 20 yeast traits by Bloom et al. (2015) detected 153 pairwise interactions. The vast majority of these had both loci showing significant marginal effects. However, in 36 cases just one locus was significant, and in 16 cases neither locus was significant. Finally, Holland et al. (1997) found 35 significant QTLs interactions in an oat experiment. Only four of these involved situations where both loci had significant main effects, 28 had only one of the loci with significant main effects, and for three interactions neither of the loci had significant main effects.

These observations suggest that, depending on the nature of the study, strategy 3 must often be employed. Various schemes have been proposed for the challenging task of testing all possible combinations of markers. As summarized by Broman and Speed (2002), this is a **model selection** problem, choosing the best model from a set of candidates. They note that this is a two-step process: (i) choose the model dimensionality (**variable selection**; here number of QTLs and their corresponding associated markers), and then (ii) search within this space for the best (by some criteria) model. A number of approaches have been suggested to accomplish this task.

Kao et al. (1999), Zeng et al. (1999), Siegmund (2004), Storey et al. (2005), Baierl et al. (2006), Stich et al. (2007), Bogdan et al. (2008), and Zou and Zeng (2008) proposed a stepwise selection process wherein one adds components to a regression (marker candidates for direct and/or interaction effects), and then uses some model selection criteria, such as AIC (Equation 16.9a) or BIC (Equation 16.9c), to compare non-nested models. Carlborg et al. (2000) and Carlborg and Andersson (2002) searched for interacting pairs using **genetic algorithms**, a computer science approach for maximization of a complex structure by essentially evolving solutions and having them compete (Holland 1975). Other search strategies have been proposed by Ljungberg et al. (2004), Pattin et al. (2009), and Wei et al. (2010). Jannink and Jansen (2001) proposed a one-dimensional search for loci with strong background interaction effects (which requires more than a two-way cross). Such loci then could be used to seed strategy 2 above. The basic logic of this idea has been independently suggested by a number of workers (Paré et al. 2010; Struchlain et al. 2010; Deng and Paré 2011; Hothorn et al. 2012).

Many of the methods for dealing with epistasis are set within a Bayesian framework (Appendix 7), with a detailed overview of different approaches given by Yi and Shriner (2008). The simulation methodology of MCMC sampler (Appendix 8) can offer an efficient alternative strategy to a **grid search**, wherein QTLs are placed at regular intervals (e.g., 1 cM) across the genome and models are fit given these assumed positions. Such a two-dimensional search examines points on a grid where location (x, y) corresponds to a QTL at genomic position x and a second at genomic position y. The resulting search is on a lower-triangular space ($x + y \leq L$, L being the total map length) and support could be presented as either two-dimensional profile plots (e.g., in the form of a **heat map** over the entire space, where the color of a region shows the strength of support for a pair of QTLs), or by only highlighting significant regions. See Malmberg et al. (2005) for an example of the latter in their search for fitness-related traits in *Arabidopsis*.

Yi et al. (2003a) and Xu (2007) proposed using **stochastic search variable selection (SSVS)**. Introduced by George and McCulloch (1993, 1997), the SSVS approach starts with some initial vector of weights associated with each of the potential model variables and then uses an MCMC sampler to continually update these weights. The resulting variables selected for final model are those that have sufficient posterior weight. Given this set of chosen variables, the optimal model parameters are then estimated, e.g., once the variables for a regression are chosen, the regression is then fit using those variables. Xu (2007) noted that large sample sizes (at least 600) are required for SSVS to be efficient.

A more popular approach is to use a **reversible jump MCMC** (Green 1995). Here, the dimensionality (number of QTL) is treated as a random variable, and one constructs an MCMC with two levels. On one level, the MCMC samples within a fixed number of QTLs. Level two, operating on a longer time scale, allows the sampler to jump up (or down) by one QTL. This generates a posterior distribution for the number of QTL, and the posterior for model parameters (locations and effects) within each QTL number class. Various versions of this method have been proposed for QTL mapping, e.g., Heath (1997), Sillanpää and Arjas (1998), Yi and Xu (2002a), Yi et al. (2003a, 2007), and Yi (2004). A final approach is to use the idea of **shrinkage**, where all of the model parameters are estimated, but estimates are shrunk back towards a mean of zero, so that only those with substantial effects survive the shrinkage. The reader might think that is a version of SSVS, but the latter uses these posterior weights to include or exclude model components which are then fit, whereas the shrinkage approach fits everything, and then shrinks most back to zero. This is done by either treating model effects as random (Chapter 10) or by using penalized regression approaches (Example 20.4). Examples of using these approaches for QTL mapping are given by Wang et al. (2005a), Zhang and Xu (2005), Zhang et al. (2005), Xu (2007, 2010), Yi and Xu (2008), Xu et al. (2009), Cai et al. (2011), and Huang et al. (2015).

Example 18.14. Xu and Jia (2007) examined seven traits in 145 DH barley lines, scored for 127 markers. Each DH line was replicated over an average of 25 plots, with the line means over all plots used for trait values, resulting in very accurate estimates of line genotypic values. Xu and Jia used the method of Xu (2007) to fit 127 main effects and 127·126/2 = 8,001 interaction effects (all pairs of markers). For each effect, REML (Part II Chapter 32) is used to estimate the variance associated with that effect. These variances were then used to shrink the estimate (Equation 10.24b), with those effects with small variances being essentially shrunk to zero.

By construction, the only (pair-wise) epistasis in DH lines is additive-by-additive. Xu and Jia found an average of 11.7 main effects and six interaction effects over the seven traits. The largest main effect accounted for around 18% of the phenotypic variance, while the largest interaction effect was only 2.6%. Taking the sum of all significant effects (judged by 20 reshuffled samples), main effects account for 35% of the variance, while interaction effects for only 6%. Perhaps the most interesting result was that *none* of the significant epistatic effects had *both* of their loci with significant main effects. Very surprisingly, 90% of all interactions had *neither* of the loci with a significant main effect, with 10% having only one locus with a significant main effect.

QTL × E Interactions

Another important potential interaction involves that between the environment and a detected QTL (**QTL × E**). We offer just a few brief remarks on QTL × E, which is examined in detail in Part II Chapter 28. Two different approaches have been used to study this interaction: (i) the consistency of marker-trait associations across environments, and (ii) linear models that incorporate specific terms for marker × environment interactions. The simplest model would be a modification of Equation 18.14d, where b_k is now the effect for enviroment k and I_{ik} the QTL × E interaction term. The consistency measure is a very crude metric of QTL × E, simply asking whether a marker-trait association is detected in all scored environments. If it is, this is generally taken as evidence against G×E, while detection of an association in only some of the environments is often taken as evidence for G×E. However, a QTL can have a significant effect in all environments even in the presence of very significant G×E interaction (Chapter 28). Likewise, low power can result in a QTL being detected in only some of the replicates of an experiment, even when its effects are identical across environments. Consistent with this expectation, Koester et al. (1993) found that QTLs with small effects are less likely to be detected across environments than are QTLs with large effects. By explicitly testing for marker × environment interactions (some of the I_{ik} are significant), linear model methods provide a more sensitive, and accurate, measure of G × E effects (Chapter 28).

SAMPLE SIZES REQUIRED FOR QTL EXPERIMENTS

Before investing the time and expense in a QTL mapping experiment, it is critical to have an understanding of the sample sizes required for the detection of QTLs of specified effects. The probability of detecting a marker-trait association is improved by increasing the difference between means and/or by decreasing the within-marker class (or residual) variance. Increasing the sample size reduces the within-class variance, while changing the experimental design can increase the difference between means. Further, the residual variance can often be decreased by adding explanatory factors to the model, such as sex or age effects.

This section examines a number of issues related to sample size, power, and accuracy. Sen et al. (2007) presented a general tool for aiding in the design of a QTL experiment that considers many of these issues. The important bottom line is that ***most QTL experiments are dramatically underpowered*** (de Koning and Haley 2005). A further point in terms of resource expenditures is that ***adding more individuals to the study is almost always a better strategy than adding more markers***. This is strictly true from a statistical sense. However, this strategy is also true from a practical standpoint, in that it is usually straightforward to store DNA from individuals, allowing for subsequent more dense mapping within focal regions. However, such denser mapping is only appropriate when there are a sufficient number of recombinant events between markers, which is a function of nc. Increasing density lowers c, so that for fixed n (original sample size), at some point not enough recombinants are expected between the dense markers and they become completely redundant.

Power and Sample Size

The following discussion of sample size is restricted to single-marker t tests using the F_2 or backcross designs, where a QTL is indicated if the means for two alternative marker genotypes are significantly different. Using the theory of power calculations (reviewed in Appendix 5), simple expressions can be obtained for these designs. The broad utility of the results developed below is that both theoretical (Simpson 1989, 1992; Haley and Knott 1992; Darvasi et al. 1993; Rebaï et al. 1995; Hu and Xu 2008; Kao and Zeng 2010) and empirical (e.g., Stuber at el. 1992; deVicente and Tanksley 1993; Nodari et al. 1993; Damerval et al. 1994; Champux et al. 1995; Kennard and Harvey 1995; Mayer et al. 2004; Mayer 2005) studies show that t tests and more elaborate flanking-marker methods have very similar power for detection, especially when adjacent markers are no farther than 20 cM apart. Hence, the sample size expressions developed below provide a baseline for most designs.

We start by considering the t test for an F_2 design, where the presence of a linked QTL is indicated when $\overline{z}_{MM} - \overline{z}_{mm}$ is significantly different from zero. Suppose that the marker is completely linked to a single QTL with additive value a, in which case $E(\overline{z}_{MM} - \overline{z}_{mm}) = 2a$. Assuming that the distribution of phenotypes about each QTL genotype has constant variance σ_e^2, then if the numbers of MM and mm individuals measured are n_1 and n_2,

$$\sigma^2\left(\overline{z}_{MM} - \overline{z}_{mm}\right) = \sigma^2\left(\overline{z}_{MM}\right) + \sigma^2\left(\overline{z}_{mm}\right) = \left(\frac{1}{n_1} + \frac{1}{n_2}\right)\sigma_e^2 \tag{18.35}$$

If n total F_2 individuals are scored, we expect only one in four to be a particular marker homozygote, giving $n_1 = n_2 = n/4$ and the expected variance $8\,\sigma_e^2/n$. If $r_{F_2}^2$ denotes the fraction of the total F_2 phenotypic variance $[\sigma_z^2(F_2)]$ due to segregation at the QTL, then $\sigma_e^2 = (1 - r_{F_2}^2)\,\sigma_z^2(F_2)$. Hence, if n is reasonably large, the observed difference in marker means is approximately normally distributed, with

$$\overline{z}_{MM} - \overline{z}_{mm} \sim N\left[2a,\, 8(1 - r_{F_2}^2)\,\sigma_z^2(F_2)/n\right] \tag{18.36a}$$

Under the null hypothesis of no QTL, this difference is distributed as a normal with mean zero and variance $8\sigma_z^2(F_2)/n$.

Using the machinery developed in Appendix 5, the sample size required to have probability $1 - \beta$ of detecting a QTL using a test with an α-level of significance becomes

$$n_{F_2} = \frac{8(1 - r_{F_2}^2)}{\delta_{F_2}^2}\left(\frac{z_{(1-[\alpha/2])}}{\sqrt{1 - r_{F_2}^2}} + z_{(1-\beta)}\right)^2 \tag{18.36b}$$

where $z_{(p)}$ satisfies $\Pr(U \leq z_{(p)}) = p$ with $U \sim N(0,1)$, and

$$\delta_{F_2} = \frac{\mu_{QQ} - \mu_{qq}}{\sigma_z(F_2)} = \frac{2a}{\sigma_z(F_2)} \tag{18.37a}$$

is the difference in QTL means in units of F_2 phenotypic standard deviations. The variance contributed by F_2 segregation at this locus is $\sigma_Q^2(F_2) = a^2(2+k^2)/4$, where k is the dominance coefficient, implying

$$r_{F_2}^2 = \frac{\sigma_Q^2(F_2)}{\sigma_z^2(F_2)} = \frac{a^2(2+k^2)/4}{\sigma_z^2(F_2)} = \frac{\delta_{F_2}^2(2+k^2)}{16} \tag{18.37b}$$

which for a completely additive QTL ($k = 0$) is $r^2_{F_2} = \delta^2_{F_2}/8$. Using Equation 18.37b, we can alternatively express the required sample size in terms of the fraction of variation accounted for by the QTL,

$$n_{F_2} = \left(\frac{1 - r^2_{F_2}}{r^2_{F_2}}\right) \left(\frac{z_{(1-[\alpha/2])}}{\sqrt{1 - r^2_{F_2}}} + z_{(1-\beta)}\right)^2 [1 + (k^2/2)] \tag{18.38}$$

Example 18.15. What sample sizes are required to detect a completely linked QTL using a test with α = 0.05 and β = 0.1 (i.e., a 5% probability of a false positive and a 10% probability of missing a true association)? Note that Pr(U < 1.96) = 0.975 and Pr(U < 1.28) = 0.9, so that $z_{(1-[\alpha/2])} = z_{(0.975)}$ = 1.96 and $z_{(1-\beta)} = z_{(0.9)}$ = 1.28. Substituting these into Equation 18.38 gives the following sample sizes for a completely additive ($k = 0$) and a completely dominant or completely recessive ($k \pm 1$) QTL whose segregation accounts for r^2 of the total F_2 variance:

r^2	0.5	0.3	0.1	0.05	0.01
Additive QTL	16	31	101	206	1046
Dominant QTL	25	46	151	309	1568

Note that the presence of dominance can significantly inflate the required F_2 sample size.

Turning now to the backcross designs, consider $B_1 = F_1 \times P_1$ (i.e., $MQ/mq \times MQ/MQ$). Here $n_1 = n_2 = n/2$, while $\mu_{QQ} - \mu_{Qq} = a(1-k)$, giving

$$\overline{z}_{MM} - \overline{z}_{Mm} \sim N\left[a(1-k),\ 4(1 - r^2_{B_1})\,\sigma^2_z(B_1)/n\right] \tag{18.39a}$$

Using the same logic as above, the required sample size is found to be

$$n_{B_1} = \left(\frac{1 - r^2_{B_1}}{r^2_{B_1}}\right) \left(\frac{z_{(1-[\alpha/2])}}{\sqrt{1 - r^2_{B_1}}} + z_{(1-\beta)}\right)^2 \tag{18.39b}$$

with

$$r^2_{B_1} = \frac{\delta^2_{B_1}}{4}, \quad \text{where} \quad \delta_{B_1} = \frac{a(1-k)}{\sigma_z(B_1)} \tag{18.39c}$$

For the B_2 population, the results are similar, except that the comparison is now $\overline{z}_{Mm} - \overline{z}_{mm}$ and $-k$ replaces k in the above expressions. Comparing the F_2 and the backcross design (for small to modest r^2), the ratio of samples sizes to achieve the same power is approximately

$$\frac{n_{B_1}}{n_{F_2}} \simeq \left[\frac{2}{(1-k)^2}\right] \left[\frac{\sigma_z(B_1)}{\sigma_z(F_2)}\right]^2 \tag{18.40}$$

Thus, if the backcross and F_2 phenotypic variances are the same, the backcross design requires twice as many individuals as an F_2 for a completely additive QTL ($k = 0$). When dominance is present, depending on its direction relative to the backcross population used, the backcross design can require more than twice as many individuals as an F_2 ($k > 0$ for B_1, $k < 0$ for B_2) or fewer individuals than the F_2. (If $k = -1$, the required sample size for the B_1 is only half of that for an F_2 design.) A further complication is that the phenotypic variance is generally rather different in the F_2 and backcross populations due to changes in the variance from segregation of background QTLs. In the F_2 population, all QTL alleles have frequency 1/2, which gives maximum additive variance (provided all QTLs are additive). In a backcross, the minor allele frequency is 1/4, and additive genetic variance is often reduced significantly relative to that in the F_2. Thus, if background QTLs contribute significantly to the character, the backcross can show a reduced variance and more power.

If the QTL is not completely linked to the marker, two corrections are required for the above expressions. First, the difference in means for F_2 homozygous marker genotypes now estimates $2a(1-2c)$. A more subtle correction is that the variance about the marker means increases when $c \neq 0$, as the phenotypic distribution for each marker class is now a *mixture of distributions* with different means (Equation 16.5d). In spite of these complications, to a very good approximation the sample sizes required for a specific power of QTL detection are given by $n_0/(1-2c)^2$, where n_0 is the required sample size under complete linkage (Soller et al. 1976; Soller and Genizi 1978). Thus, the power to detect a linked QTL falls off as $(1-2c)^2$ decreases, being very weak when $c > 0.2$ (25 cM under the Haldane map).

Example 18.16. Suppose we wish to have a 90% chance of detecting (using a test with α = 0.05) a QTL whose segregation accounts for 10% of the total F_2 variance. Further assume that all of the genetic variation at this locus is additive. From Example 18.15, 101 individuals are required to detect this QTL using a completely linked marker. With a marker at recombination frequency c from the QTL, $n = 101/(1-2c)^2$, giving sample sizes of 281, 158, and 125 for c = 0.2, 0.1, 0.05, respectively.

One can increase the power to detect a linked QTL either by increasing the *number of markers* (which decreases c and hence increases the difference between marker means) or by increasing the *number of individuals* genotyped (which decreases the sampling variance). To see the relative importance of each, note from Equation 18.36a that the t statistic has approximate expected value

$$E\left[\frac{\mu_{MM}-\mu_{mm}}{\sigma(\overline{z}_{MM}-\overline{z}_{mm})}\right] \simeq \sqrt{n}\,(1-2c)\left[\frac{a}{\sqrt{2}\,(1-r^2_{F_2})\,\sigma_z(F_2)}\right] \tag{18.41}$$

The term in brackets on the right-hand side is fixed for a given QTL, so that the test statistic scales with the square root of the sample size. Increasing the number of markers results in an increase in the test statistic (by decreasing c), but there is a point of diminishing returns when markers are already closely spaced. For example, for $c = 0.2$ (corresponding to markers spaced 50 cM apart), moving to an infinitely dense map ($c = 0$) requires that only 36% as many individuals be scored to give the same power. However, for markers spaced at $c = 0.1$ and 0.05, these percentages become 81% and 90%.

Darvasi and Soller (1994b) showed, under rather general conditions, that a 20 to 30 cM spacing of markers gives the highest chance of detecting a QTL, given the constraint of scoring a fixed total number of marker genotypes (marker loci $\times$ individuals). Here each QTL is no further than 10 to 15 cM (and on average is within 5 to 7.5 cM) from any marker. Thus, for markers spaced 10 cM or closer, there is really little point in further increasing the marker density when the goal is simple detection of a linked QTL. Increasing marker density does become important if the goal is a highly precise estimate of QTL position or the dissection of a cluster of tightly linked QTLs. However, in order to exploit increased density, we also need to have a large enough sample size to realize a sufficient number of recombinational events between closely linked markers. Without such recombinations, added markers are completely redundant. For example, Mayer et al. (2004) and Mayer (2005) used simulation studies with over 500 F_2 individuals to show that decreasing the marker spacing from 10 to 5 cM significantly improved the precision of QTL estimations.

However, with fewer F_2 individuals, there was not a sufficient number of recombination events to exploit the increased marker density. Thus, it very rarely pays to simply increase marker density without also increasing sample size. One exception is the use of AIL lines, where extra generations (and hence additional meioses) can exploit a denser marker map.

Example 18.17. As mentioned in Example 18.4, Edwards et al. (1987, 1992) examined the same cross of two inbred maize lines with two different designs. Their 1987 design used 1,776 F_2 individuals and 17 markers, while their 1992 design used 187 F_2 individuals and 114 markers. The two designs represent a tradeoff between increased marker density (1992 design) and increased sample size (1987 design), as both examined a somewhat similar number of total marker genotypes (1776 × 17 = 30,200 versus 187 × 114 = 21,300). Comparisons of c values in the two studies is problematic, given that only a fraction of the genome was covered in the 1987 study (about 40% of the genome was within 20 cM of a marker), while under the 1992 design most of the genome was 5 to 10 cM from a marker. Choosing $c = 0.25$ (1987 design) and $c = 0.08$ (1992 design), from Equation 18.41 the expected ratio of t statistics becomes

$$\frac{\sqrt{1776}\,(1 - 2 \cdot 0.25)}{\sqrt{187}\,(1 - 2 \cdot 0.08)} = 1.8$$

showing that (for these c values) the 1987 design had greater power.

ML interval mapping is expected to be somewhat more powerful than the simple single-marker t test, so the above results can be considered as upper bounds for the required sample size, although they are not greatly exaggerated. For example, the power of ML interval mapping to detect QTLs has been examined by several authors (Lander and Botstein 1989; van Ooijen 1992; Carbonell et al. 1993; Darvasi et al. 1993), who concluded that with a reasonable density of markers (one every 20 cM), 250 F_2 individuals are sufficient to detect a QTL whose segregation accounts for at least 5% of the F_2 variation. How does this compare with the required sample size for a t test? Because markers spaced at 20 cM intervals imply that a marker is within 10 cM from the QTL, using the result for $r^2 = 0.05$ from Example 18.15 gives the required sample size for a t test as $206/(1-2\cdot 0.1)^2 = 263$. Hence, the above t-test guidelines are also reasonable for ML interval mapping. Additional power calculations for QTL mapping are presented in Appendix 6.

Power under Selective Genotyping

The idea behind selective genotyping is that, historically, scoring (phenotyping) characters was much less expensive than scoring markers. Under selective genotyping, each choosen individual is genotyped directly, rather than genotyping the bulk mixture of all selected individuals (this latter approach is bulk segregant analysis, discussed in Chapter 17). Hence, if n individuals are scored and genotyped in a normal design, there may be merit in scoring a larger number of individuals $n_z > n$ for the trait value, and then choosing a subset $n_g \leq n$ of these for genotyping. Typically, the uppermost and lowermost fractions (p) of scored individuals are genotyped, giving $n_g = 2p\,n_z \leq n$. Darvasi and Soller (1992) noted that there is typically more linkage information in the tails, rather than the middle, of the trait distribution, so that observations from the tails have more of an impact than observations near the center (Figure 18.7). In particular, they show that selective genotyping by scoring n_z individuals and genotyping $n_g = 2\,p\,n_z$ of these gives the same power as an analysis genotyping all n individuals, when

$$n_z = \frac{n}{2p + 2\,z_{(1-p)}\,\varphi(z_{(1-p)})} \qquad (18.42)$$

Here, $\varphi(z_{(1-p)})$ is the unit normal density function evaluated at $z_{(1-p)}$, where

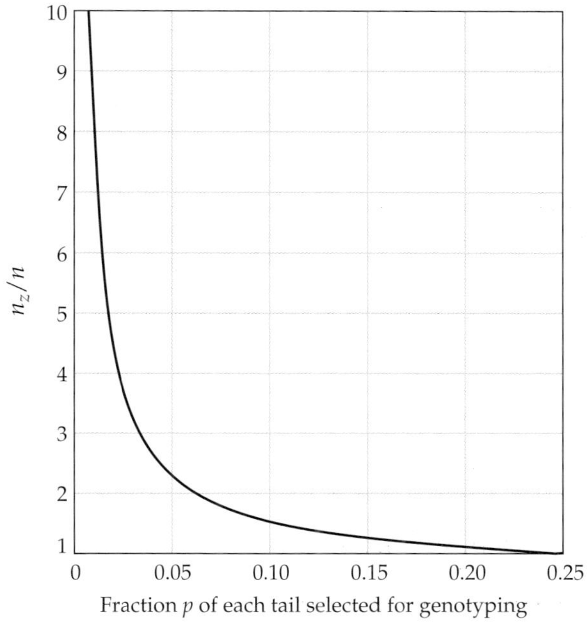

Figure 18.7 Under selective genotyping, n_z individuals are scored for the trait value, with the uppermost and lowermost fraction p of these being genotyped, giving $n_g = 2\,p\,n_z$. Here we plot, as a function of p, the number of individuals scored for the trait (n_z) that yields the same power as scoring and genotyping all individuals in a population of size n.

$\Pr(U > z_{(1-p)}) = p$ with $U \sim N(0,1)$. Figure 18.7 plots the ratio n_z/n as a function of p. For example, selective genotyping using the uppermost and lowermost 10% ($p = 0.1$) of the population requires that $n_z = 1.54\,n$ individuals be phenotyped but only $n_g = 2 \cdot 0.1 \cdot n_z = 0.3\,n$ be genotyped. Because a decrease in p reduces the number of individuals that must be genotyped but increases the number that must be scored for the trait, the optimal p value depends on the relative costs of phenotyping and genotyping each individual (Darvasi and Soller 1992). One of the ironies of the genomics revolution is that *phenotypes* are now often much more expensive to score than genotypes. In theory, one could think about **selective phenotyping**, genotyping a sample of individuals, and then phenotyping those with more extreme values (by some genotype criteria). In essence, this is what genomic selection (Part II Chapter 31) does.

This raises an extremely important issue that is often neglected in mapping studies. There are a variety of metrics (**quality control**, **QC**, approaches) that allow one to judge the quality of marker genotypes long after the experiment was performed (Chapter 20). Conversely, they are generally no robust QC metrics for *phenotypes*. Once the trait data have been collected, there is often no way to examine its quality years later. Outliers, which are very informative if true, might well be measurement or recording errors. Poor trait data have the effect of reducing power. One real-world example is lumping different diseases as the same, confounding linkage information (Chapter 20). There is currently a major push, especially in plant breeding and some model systems (such as yeast and mice), to automate **high-throughput phenotyping**, using robotic systems to measure a number of traits with high accuracy and repeatability (Solberg et al. 2006; Liti et al. 2009, Dhondy et al. 2013; Araus and Cairns 2014; Ohya et al. 2015; Shi 2016; Bazakos et al. 2017). Likewise, human genetics has seen the emergence of a number of national biobanks scoring hundreds of traits (using electronic medical records and patient questionnaires). It is reflection of the power of modern genomics that phenotyping (**phenomics**) (Houle et al. 2010) is now the bottleneck in many mapping experiments, and that high-quality phenotypes (or lack thereof) can make, or break, a very expensive mapping experiment.

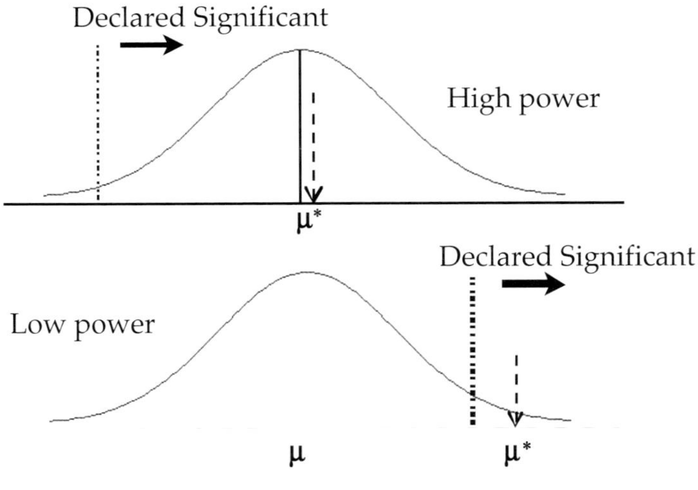

Figure 18.8 The connection between bias and power. Both panels represent the distribution of realizations of a test statistic for a marker effect, with μ the true value. The consequence of choosing only *significant* results is left truncation of this distribution (Chapter 2). In the top panel, the power is very high, as most of the mass of the distribution (i.e., realizations of tests for a marker effect) is to the right of the significance threshold (dotted line). The resulting mean of this left-truncated distribution, μ^*, is close to the true value, and there is little bias. In the bottom panel, the power is low, as most of the distribution is to the left of the threshold, and hence most tests are not declared significant. The resulting mean of the left-truncated distribution (the tiny fraction of realizations declared significant) is substantially greater than the true value, resulting in significant bias. It is immediately apparent that the bias increases as power (area to the right of the threshold) decreases.

The Beavis Effect: Overestimation of Effects Declared to be Significant

The contributions of *detected* (i.e., those *declared to be significant*) QTLs can be substantially (often *very* substantially) overestimated (Lande and Thompson 1990; Beavis 1994; Utz and Melchinger 1994; Melchinger et al. 1998; Göring et al. 2001; Xu 2003). In QTL mapping, this upward bias in estimated effect size is called the **Beavis effect**, based on an influential set of simulations by Beavis (1994) that will be discussed shortly (Figures 18.9 and 18.10). This phenomena is typically called the **winner's curse** in the GWAS literature (e.g., Zöllner and Pritchard 2007).

While the Beavis effect may at first seem a bit perplexing, it is simply a *consequence of sampling from a truncated distribution* (Göring et al. 2001; Xu 2003). As illustrated in Figure 18.8, there is a distribution associated with the realized value of a marker effect, with the true mean effect size, μ, influenced by sampling in any particular experiment. If that realized value exceeds some threshold, it is declared to be significant. The power of the test is the area of the distribution to the right of the significance threshold (Figure A5.1). Under a high power setting (top panel of Figure 18.8), most of the realizations are declared to be significant (most of the probability of the sampling distribution is to the right of the threshold). Given that we *condition* on an effect being significant, the mean value of a *significant result* is the mean, μ^*, of the left-truncated distribution. In the high power setting, this is very close to μ. Conversely, in a lower power setting (bottom panel), only a small fraction of the realizations exceed the threshold (are declared to be significant), so that the mean of the truncated distribution is *much* greater than the true value, $\mu^* \gg \mu$. Further note that this ***bias is expected to increase as power decreases*** (the significance threshold moves further to the right). Xu (2003) formalized this relationship between bias ($\mu^* - \mu$) and the power ($1 - \beta$) of a test (β being the probability of a Type II error, a false negative) by using results for the mean of a truncated normal distribution (Equation 2.14), which yields

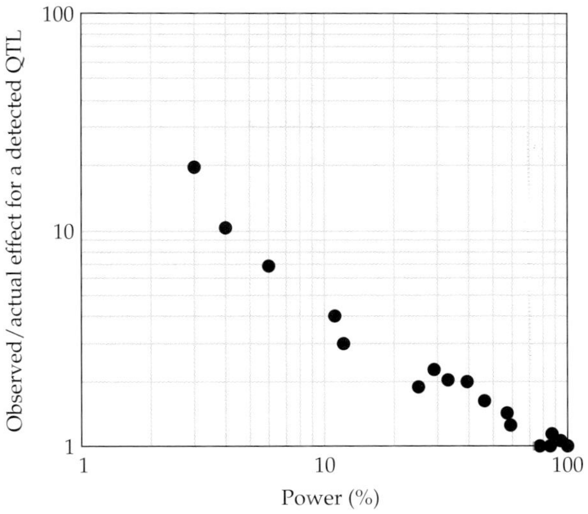

Figure 18.9 Relationship between the probability (power) of detecting a QTL and the amount by which the estimated effect of a *detected* QTL overestimates its actual value. (Based on results from a simulation study of Beavis 1994.)

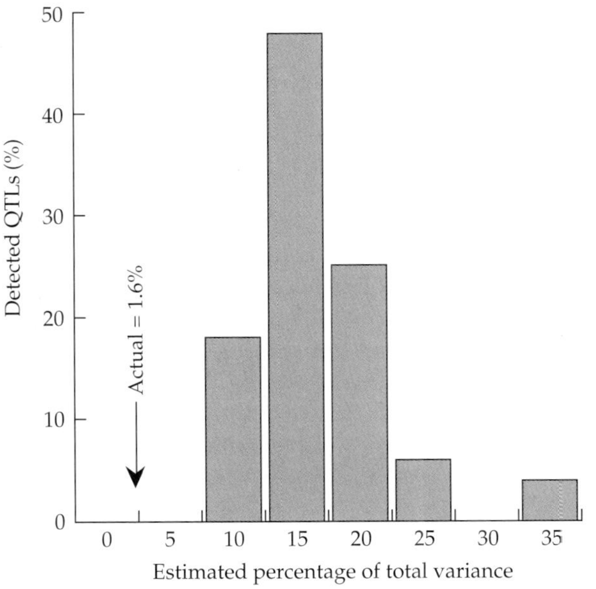

Figure 18.10 Distribution of the estimated effects of *detected* (declared significant) QTLs. Here 40 QTLs, each accounting for 1.6% of the variance, were assumed. Using 100 F_2 individuals, only 4% of such loci were detected. The average estimated fraction of total variation accounted for by each detected QTLs was 16.3%, with the distribution of estimates skewed towards larger values. (From Beavis 1994.)

$$\mu^* - \mu = \sigma \cdot \left(\frac{\varphi(x_{[\beta]})}{1 - \beta} \right) \tag{18.43}$$

Here σ is the sampling variance for the test, $\varphi(z)$ is the value of a unit normal U at value z (Equation 2.11), and $x_{[p]}$ satisfies $\Pr(U \leq x_{[p]}) = p$. The weight on σ has values of 0.80, 1.27, 1.75, 2.07, and 2.67 when the power is 50, 25, 10, 5, and 1%, respectively. For example, for 50% selected, $\mu^* = \mu + 0.8\sigma$. While this might appear to be a small effect, note that σ can often be $\gg \mu$, especially when the latter is small. As an aside, the factor multiplying σ is just the selection intensity associated with saving the upper β of a population (WL Chapter 14).

This inverse relationship between bias and power was seen in a classic set of simulations by Beavis (1994), shown in Figures 18.9 and 18.10. For example, a QTL accounting for 0.75% of the total F_2 variation had only a 3% chance of being detected with 100 F_2 progeny with markers spaced at 20 cM. However, for cases in which such a QTL is detected, the average estimated total variance it accounts for is 15.8% (a 19-fold overestimate of the correct value), highly suggestive of a major gene. With 1,000 F_2 progeny, the probability of detecting such a minor-effect QTL increases to 25%, with each detected QTL accounting for (on average) approximately 1.5% of the total variance (only a twofold overestimate), providing a more correct interpretation of the true genetic architecture. Further, these are the *average* values for the estimates. As shown in Figure 18.10, the distribution of observed effects is skewed, with a few loci having large estimated effects, and the rest small to modest effects. Such an observed distribution of effects, commonplace in QTL mapping studies, have usually been taken as being representative of the true distribution of effects. The simulation studies by Beavis showed that they can be ***spuriously generated by a set of loci with equal effects***.

It is important to stress that the Beavis effect can be much more than a statistical bookkeeping error. In the extreme, it can result in a QTL experiment giving a ***very misleading picture of the genetic architecture.*** Consider the case where all of the QTL effects are small, so that a design has low power ($1 - \beta \simeq 0$). If the number n of QTLs is large, it is likely that at least a few will be detected, $n(1 - \beta) \geq 1$, with their effects significantly overestimated. Such a scenario—wherein we detect a small number of QTLs that appear to account for a significant fraction of the total character variation—can lead to the false conclusion that character variation is largely determined by a few QTLs of major effect (Beavis 1994; Utz and Melchinger 1994). As both Figure 18.8 and Equation 18.43 show, when the power is very small, the bias among *significant* effects is very large.

A variety of approaches have been proposed to adjust for Beavis effects. The most straightforward (**subsampling**, e.g., Example 21.1) is to partition the original sample into detection and estimation subsamples. This is not an efficient use of an often limited sample size. One clever approach to accomplish this partition while still preserving power is based on a modification of bootstrap resampling (Sun and Bull 2005; Sun et al. 2011; Wu et al. 2005, 2006; Poirier et al. 2013; Huang et al. 2018). Recall that a bootstrap sample of size n is generated by drawing (with replacement) n individuals from the original sample. Some individuals will be sampled more than once, while others will not appear in that particular sample. The **out-of-sample** (or **bootstrap resampling bias reduction**, **BR-squared**) method uses the bootstrap sample for detection and the set of individuals *not* drawn (for a given sample) for estimation. This is a nonparametric approach, while a number of authors have suggested parametric methods built around truncated distributions (Figure 18.8), such as truncated normals (Xu 2003; Palmer and Pe'er 2017; Panigraphi et al. 2021), t random variables (Xiao and Boehnke 2011), or noncentral chi-squares (Xie et al. 2021).

Sample Size for a Desired Level of Mapping Resolution

Finally, there is a major difference between *detection* of a marker linked to one (or more) QTLs (the above issues of power), and the *precision* of the QTL location, i.e., the width of the confidence interval (CI) for its position. Darvasi et al. (1993) defined the **resolving power** of a design as the expected width of 95% CI for QTL position assuming a fully saturated map. The resolving power places a lower limit on the effective marker density, as markers closer than the resolving power limit provide no additional information.

Darvasi and Soller (1997) performed extensive computer simulations with a known QTL whose position was estimated using ML mapping under either a BC or F_2 design. They assumed a QTL with effects of $-a : d : a$, where the variance about each genotype, σ_e^2, was set to one. Hence, the genetic variance σ_Q^2 contributed by this QTL is $(a + d)^2/4$ under the BC design and $(2a^2 + d^2)/4$ under the F_2 design, with the fraction r^2 of the *trait* variance accounted for by that QTL being

$$r^2 = \frac{\sigma_Q^2}{\sigma_Q^2 + \sigma_e^2} = \frac{\sigma_Q^2}{\sigma_Q^2 + 1} \tag{18.44a}$$

Their simulations suggested a rather general approximation for the resolving power limit (in cM) of

$$\text{CI} = \frac{3000}{m\,n\,\delta^2} = \frac{530}{n\,r^2} \tag{18.44b}$$

Here the allelic substitution effect $\delta = a + d$ (BC design) and a (F_2, assuming $d = 0$), n is the sample size, and m is the number of **informative meioses** (those involving double heterozygous parents; Chapter 19), which is 1 for the BC design (one F_1 crossed to a homozygous P) and 2 for the F_2 design (two F_1s crossed). While the results of Darvasi and Soller are empirical observations extracted from simulations, Visscher and Goddard (2004) were able to derive them analytically, finding that using values of 3,073 and 499, respectively, in Equation 18.44b is slightly more accurate. They also presented a general expression for other designs based on the noncentrality parameter (Equation A5.14a) of the test statistic.

Example 18.18. Consider a QTL that accounts for 10% of the trait variance in the F_2, where 300 progeny are scored. What is the expected width of the 95% CI for its position? Applying Equation 18.44b with the Visscher–Goddard correction,

$$\text{CI} \simeq \frac{499}{300 \cdot 0.10} = 16.6$$

Hence, even for a major QTL (accounting for 10% of all variation) has a confidence interval length of roughly 17 cM (17 megabases assuming 1 Mb per cM). What sample size is required for a 95% CI of width 1 cM ($\sim$ 1 Mb)? Rearranging Equation 18.44b yields

$$n \simeq \frac{499}{\text{CI} \cdot r^2} = \frac{499}{1 \cdot 0.1} = 4990$$

Conversely, 300 F_2 progeny is a very high-power design for *detecting* a QTL accounting for $r^2 = 0.1$. Applying Equation 18.38 shows that a sample of around 100 has 90% power to detect such a QTL using a test with $\alpha = 0.05$. Hence, if the goal of the QTL experiment is mapping to a reasonably small interval, the typical QTL sample size is far too small. For QTLs accounting for 5% and 2.5%, the expected CIs are, 33.3 and 66.5 cM, respectively, while the corresponding sample size required for a 1 cM interval becomes 9,980 and 19,960.

The Otto–Jones Estimator for the Number of QTLs

As was the case when estimating the number of factors in a line-cross analysis (the Castle–Wright estimator; Chapter 11), a QTL experiment underestimates the actual number of underlying loci generating a between-line divergence. Two factors cause this bias. The first is power, in that sample sizes are usually not sufficiently large to detect QTLs of small effect. In theory, this issue can be addressed by increasing the sample size or by using RILs and sufficient replication. Typically the number of detected QTL increases with the number of progeny in a cross (e.g., Schön et al. 2004 and examples below).

The second factor is the size of an average linkage block in the cross progeny. As was the case for a traditional line-cross experiment, large fractions of any parental chromosome remain intact in most of their progeny due to insufficient recombination. Hence, a single QTL is often a linkage block containing multiple causative loci. Upon attempts to fine-map this QTL it **fractionates**, turning into a series of smaller QTLs. This is a common observation (reviewed by Flint and Mackay 2009; Mackay et al. 2009).

Otto and Jones (2000) attempted to use the detected number (n_d) and the average effect size ($\overline{a}$) of a detected QTL to obtain an estimate of the actual number of QTLs. Their estimate uses some features of the Castle–Wright estimator, and makes two key assumptions. First, it assumes no antagonistic alleles, so that a high line has all + alleles relative to a low line.

Second, it assumes that the QTL effects (a) are drawn from a specified distribution. The logic for their estimator follows along similar lines to our above analysis of the Beavis effect. Assuming an exponential distribution for a, one can take the estimated average value ($\overline{a}$) above the detection threshold (γ) and use these values to estimate the probability remaining to the left of the threshold. There is a conservable body of theory on the distribution of effect sizes fixed by selection during an adaptive walk (WL Chapter 27), which suggests than an exponential distribution of such effects is not unreasonable assumption. As with the Castle–Wright estimator, Otto and Jones start with the observed difference (Δ) in trait mean between the crossed lines. For an exponential distribution of effects they obtain the estimator

$$\widehat{n}_{OJ} = \frac{\Delta}{\overline{a} - \gamma} \tag{18.45a}$$

The threshold value γ is a bit more delicate. It is certainly less than a_{min}, the smallest effect size for the n_d detected QTLs. Otto and Jones suggest the approximate estimate

$$\widehat{\gamma} \simeq \frac{a_{min} n_d - \overline{a}}{n_d - 1} \tag{18.45b}$$

provided that $\overline{a}$ > 25% larger than a_{min} and that $n_d > 3$. Using simulations, they found that Equation 18.45b performed well under these conditions. Finally, they estimated the average effect of an undetected QTL as

$$E[a_{ud}] = \overline{a} \left(1 - \frac{\zeta}{1 - \exp[-\zeta/(1-\zeta)]} \right), \quad \text{with} \quad \zeta = \frac{\gamma}{\overline{a}} \tag{18.45c}$$

Simulations suggested that Equation 18.45a is somewhat robust to the presence of antagonistic alleles, provided that their fraction is modest (less than 15%). A deeper concern would seem to be the Beavis effect, but Otto and Jones note that such effects would inflate *both* the estimates of $\overline{a}$ and γ, while it is the *difference* between these that is used in their estimator. Simulations suggest that unless the number of detected QTLs is small, this is not an unreasonable assumption. Confidence intervals, as well as an extension assuming the more general gamma distribution for effect sizes (of which the exponential is a special case; Appendix 7) are developed in Otto and Jones. One of the more interesting findings of Otto and Jones was that when the detection threshold is small ($\gamma \leq 0.1 \cdot \Delta$), then as the true number of QTLs increases, the number of detected QTLs also initially increases, but this relationship is *not* monotonic. Rather, the expected number of detected QTLs eventually reaches some maximum, and then *declines* afterwards as the number of true QTLs increases.

ADVANCED DESIGNS

Biparental and Multiparental Mapping Populations

We have already introduced the use of a collection (or **panel**) of RILs and DH lines for QTL mapping. One of their advantages is that the line genotypic value for any trait can be measured with arbitrary precision through sufficient replication (Appendix 9). The issue of the optimal number of individuals to score within a line was examined by Belknap (1998) and Crusio (2004). Challenging characters, such as life span, which can have huge variation when based on single individual measurements, become well behaved when one considers the average value of a given line (e.g., Nuzhdin et al. 1997; Mackay et al. 2005a). Similarly, replicated performance of an RIL or DHL over defined environments allows for very accurate estimates of QTL $\times$ environment interactions (e.g., Leips and Mackay 2000). Further, a set of RILs has increased mapping accuracy relative to an F_2 panel due to the additional rounds of recombination during their formation (note that this is *not* true for most sets of DHLs). From Equation 18.5b, a panel of n RILs formed by selfing has roughly the same mapping resolution as a set of $2n$ F_2s, while RILs formed by brother–sister mating are equivalent of roughly $4n$ F_2s.

Table 18.2 Relative strengths and weaknesses of biparental (e.g., F_2s, RILs, etc.), multiparental (e.g., MAGIC, NAM), and GWAS mapping approaches. (After Pascual et al. 2015.)

	Biparental	Multiparental	GWAS
Pop. developmental time	Intermedate	Long	Short
Mapping precision			
Common alleles	+	++	+++
Rare alleles	++	++	+
Total Recombination events	+	++	+++
Sample size	+	++	+++
Detection effect size	Large	Modest	Small
Population structure	No	No	Yes
Strength	Rare alleles	Multiple alleles	High precision
Limitation	Poor precision	Large effort	High LD Pops

A related strategy can be used in yeast, where meiosis generates haploid lines that can be asexually propagated, each instantly forming a clonal lineage (Nogami et al. 2007; Liti and Louis 2012; Cubillos et al. 2013). One can also cross RILs, resulting in panels of **recombinant inbred intercross (RIX) lines** (Threadgill et al. 2002; Zou et al. 2005). In plants, an RIX panel is called an **immortalized F_2 (IMF$_2$)** (Hua et al. 2002; Chapter 13). Futher, one can use an advanced-generation AIL as the starting population for RILs, resulting in **recombinant inbred advanced intercross lines (RIAILs)**, with the notation (Valdar et al. 2006a, 2006b) of RIAIL10 implying RILs starting from an F_{10} AIL base population. RIAILs experience two stages of map expansion, one from the AIL, and the second from the RIL formation. To a good approximation, RIL formation via selfing adds roughly two additional generations of recombination, while brother–sister RILs adds around four. Hence, the final recombination value $\tilde{c}$ for a RIAIL is given by Equation 18.4, using $t+2$ and $t+4$, for selfed and brother–sister RILs. See Winkler et al. (2003) and Teuscher et al. (2005) for a more exact treatment.

Finally, an important mapping resource for examining wild germplasm in an elite germplasm background are **advanced backcross lines** (Tanksley and Nelson 1996). While wild cultivars (such as land races or even wild species) almost certainty contain a rich pool of exploitable genetic diversity, they also underperform current elite cultivars. The advanced backcross design of Tanksley and Nelson uses as the mapping panel a series of inbred lines (either RILs or DHs) constructed from the second or third backcross generation of a donor wild cultivar repeatedly backcrossed to a common parent elite line. For example, (wild × elite) progeny are crossed to elite (BC1), BC1 progeny are backcrossed to elite (BC2), and BC2 progeny are again backcrossed to elite (BC3). The expected fraction of wild alleles in these progeny are, respectively, 1/4, 1/8, and 1/16. Hence, the resulting lines (**AB-RILs** or **AB-DHs**, depending on whether RILs or DHs were used to extract inbred lines) contain small segments of wild material in an otherwise largely elite background. One of the first successful examples of QTL mapping using such panels is the work of Yun et al. (2006) for mapping fungal disease resistance traits in barley. Wang and Chee (2010) reviewed additional applications of this approach in tomatoes, rice, barley, wheat, maize, and cotton.

The advantage of any inbred line panel is that, once genotyped, any trait can be QTL mapped by simply measuring it across the lines and using the existing marker information. At this stage, QTL mapping is solely an issue of accurate phenotyping. Thus, while such panels are expensive and time consuming to produce, once generated they become an important community resource. The major limitation of most panels is that they are usually generated from a **biparent (two-parent) cross**, and thus have limited genetic diversity. Further, such panels are often produced with a specific trait in mind, influencing the set of parents crossed during line formation (e.g., crossing lines with extreme values of a focal trait). While other traits can be analyzed using the resulting panel, phenotypic diversity may be limited. As a result of these diversity limitations, a number of communities have

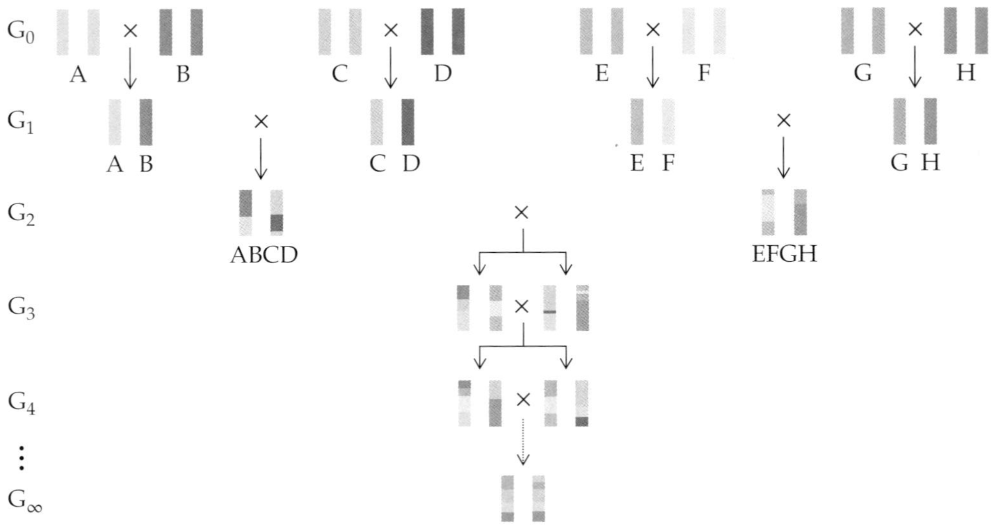

Figure 18.11 An eight-parent (A through H) cross leading to a set of multiparent advanced generation intercross (MAGIC) lines. Following generation 3 (the final wide-population cross), a series of brother–sister crosses are made within sublines until a collection of fully inbred lines are produced. Note the extra rounds for recombination relative to a standard biparental cross. Although there are any number of variations of this basic crossing scheme (such as additional rounds of recombination), the resulting lines are still called MAGICs.

made the expenditure (often as collaborative projects) to develop **multiple-parent crosses**, with the resulting progeny forming a **multiparental population** (**MPP**). A panel of RILs extracted from an MPP shows greater diversity, both genetically and phenotypically, than a standard biparental panel. MPP panels are occasionally referred to as **second generation mapping resources** (Rakshit et al. 2012), and there are two different strategies for creating such panels: MAGIC and NAM lines.

As noted by Pascual et al. (2015), these various mapping strategies—biparental (BP; i.e., classical crosses) and multiparental (MP) linkage mapping, and association mapping (GWAS; Chapter 20)—can be regarded as complementary approaches. Table 18.2 summarizes key strengths and weaknesses of these different approaches. Linkage mapping has limited resolution due to a modest number of recombinations within the sample (with MP greater than BP) and generally much smaller samples than GWAS (again, with MP potentially having larger sample sizes than BP), and hence requiring larger effect sizes for detection. Casual allele diversity is lowest in BP and highest in GWAS. Conversely, GWAS (population-sample) approaches have lower power for rare-allele detection. When such alleles are fortunately included under an BP or MP design, their frequency is either 1/2 (BP) or $(1/k)$ for (a k parent) MP designs, resulting in larger power. By sampling multiple families, MP designs have a larger chance (than BP) of capturing rare alleles.

MAGIC Lines

Multiparent advanced generation intercross (**MAGIC**) **lines** (Mackay and Powell 2006; Cavanagh et al. 2008; Cockram and Mackay 2018)—also called **recombinant inbred heterogeneous stocks, RIHS** (Valdar et al. 2003, 2006a, 2006b)—typically result from four- or eight-way crosses (Figure 18.11). In addition to greater diversity by using multiple founders, the additional rounds of recombination during parental crossing further expands the genetic map relative to a biparental population. Hence, one can think of these as multiple parent AILs, from which RILs are extracted. A few examples of this approach include the **Mouse Collaborative Cross** (Threadgill et al. 2002; Churchill et al. 2004; Aylor et al. 2011; Threadgill and Churchill 2012) involving eight diverse parental lines; the ***Drosophila*** **Syn-**

thetic Population Resource (**DSPR**), a panel of over 1600 RILs formed from 16 diverse parents (King et al. 2012a); the *Arabidopsis* MAGIC lines of Kover et al. (2009) of around 530 RILs from 19 founding parents; the ***Arabidopsis* multiparent RIL population** (**AMPRIL**), a set of over 500 RILs from four founder parents (Huang et al. 2011); a set of 1,091 RILs with eight founding parents in Winter-sown wheat (Mackay et al. 2014), and a panel of haploid lines from a four-parent cross in yeast (Cubillos et al. 2013). Numerous other examples are given in Scott et al. (2020). One indication of the power of the MAGIC design is the result of Valder et al. (2006). They noted that 1,000 MAGIC lines (starting with eight founders) are sufficient to map a QTL accounting for 5% of the phenotypic variance to a region slightly less than one cM, a nearly ten fold reduction relative to an F_2 mapping population (Example 18.18).

The construction of MAGIC lines involves three phases (Valdar et al. 2006a, 2006b), resulting in a large number of variations from the basic crossing structure shown in Figure 18.11.

1. **Mixing**. The first step is the choice of parental lines (by some criteria) followed by a subsequent crossing pattern to generate a **synthetic base population**. The crossing used is often called a **funnel**, distilling the variation from a number of input lines down into a final synthetic product. In addition to the crossing scheme shown in Figure 18.11, other schemes, such as a diallel (Chapter 25 in Part II), namely all pairwise crosses are made and then subsequently intercrossed, have been used (e.g., the AMPRIL lines of Huang et al. 2011). Typically, the crossing pattern during the mixing phase is such that all parental genomes are roughly equally represented in the synthetic base population. However, if differential contributions are desirable, this can be accomplished by modifying the number of times a parent is crossed in this initial phase. Recombination during the mixing phase results in map expansion relative to a biparental cross. Huang et al. (2015) and Scott et al. (2020) discuss founder selection in more detail.

2. **Maintenance.** Once a synthetic base population has been constructed, one has the option of randomly mating this population for several generations to further expand the genetic map. Simulation studies by Yamamoto et al. (2014) found that using less than two cycles of random mating following the formation of the base population had little impact on power. Conversely, using more than six generations significantly improved power under most settings. While random mating should not substantially alter the frequencies of genomic contribution from the founding parents, natural selection can (i.e., selection against deleterious genotypes formed during the mixing phase).

3. **Inbreeding.** Once a sufficient amount of mixing and subsequent recombination has occurred, then either selfing or recurrent brother–sister mating is used to create the final panel of RILs. Note that one could also use doubled haploids constructed from the final stage of the maintenance phase. Selection can be extremely important here, as lines are often lost during inbreeding, especially in species that are largely outbred in nature (Chapter 12). For example, 95% of all attempted RILs in the mouse Collaborative Cross became extinct during inbreeding (Shorter et al. 2017), and only 22% of F_2 plants in the maize NAM panel (Example 18.19) successfully produced lines that survived through five generations of selfing (McMullen et al. 2009).

While QTL detection and mapping in MAGIC lines follow the same basic logic as interval mapping, one complication is that there are more than two potential genotypes at a given site in a RIL. In particular, if k (typically four or eight) inbred lines are crossed to make a MAGIC line, then up to k distinct homozygotes can occur in any locus. At the core of most methods is some marker-based assignment of a given region in a MAGIC line to a founder line. Such assignments are typically probabilistic, e.g., for the marker haploid in a given region there is (say) a 57% chance that it came from parent 1, 20% from parent 2, etc.

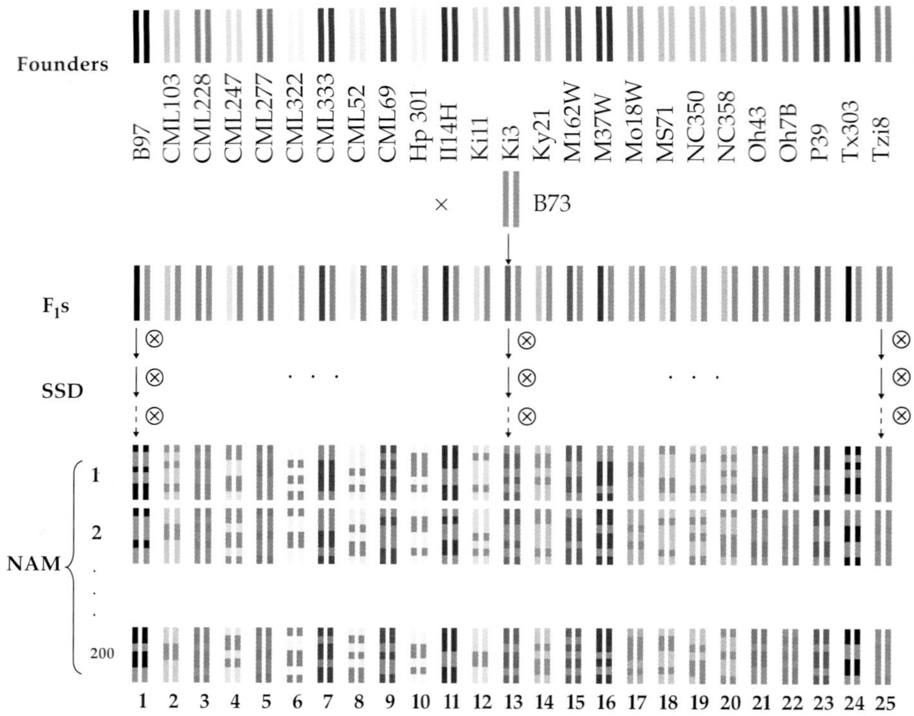

Figure 18.12 The maize NAM lines. Here 25 diverse donor parents were crossed to a common parental line (B73), resulting in 25 F_1 populations. For each cross, single seed descent (SSD) was used to form 200 RILs. The entire collection of 5000 lines forms the maize NAM panel. See Example 18.19 for more details. (From Yu et al. 2008.)

Following these assignments, one uses regression or ANOVA of the trait value on the identity of the founder segment (e.g., Xu 1996; Mott et al. 2000; Valdar et al. 2006a, 2006b; Verbyla et al. 2014; Zhang et al. 2014; Wei and Xu 2016). For example, with k haplotypes in a given region, the additive model (Equation 18.14a) is extended to multiple alleles (haplotypes), yielding

$$z_i = \mu + \sum_{j=1}^{k} b_j N_{ij} + e_{ik}$$

with N_{ij} being the number (0, 1, 2) of haplotype (founder segment) j in individual i (with the sum over the N_{ij} equaling two), and b_j its effect. This formulation raises a number of issues that reoccur throughout the next several chapters, in that each estimated parameter costs a degree of freedom, while many of the b_j are equivalent. For example, we will see this concern in both gene-based GWAS and various tests for rare-allele effects (Chapter 20). Given that many of the haplotypes likely contain the same causal site, the full set of region-specific founder segments in reality consists of $\ell \leq k$ of functionally distinct units impacting the focal trait. The term **allelic series** is often used for the set of functionally distinct haplotypes. Typically ℓ is less than k, although often greater than two (King et al. 2014; Crouse et al. 2020). Power can be improved by amalgamating equivalent haplotypes into the same class (i.e., N_{ij} is now the copy number for haplotype equivalence class j and the sum contains ℓ elements). Approaches for estimating ℓ were explored by Jannink and Wu (2003), Yalcin et al. (2005), Keele et al. (2019), Mosedale et al. (2019), and Crouse et al.

(2020).

There is a growing and increasingly sophisticated literature on the analysis of multiparent populations. Permutation testing and model selection issues are examined by Peirce et al. (2008) and Valdar et al. (2009). Power calculations and related discussions are given by Valdar (2006a), King et al. (2012b), King and Long (2017), and especially Keele et al. (2019). General reviews of MAGIC lines are given by Rakshit et al. (2012), Gatti et al. (2014); Huang et al. (2015), and Scott et al. (2020). Additional, and more technical, discussions are offered by Broman (2005; 2012a, 2012b).

NAM Lines

A second strategy for constructing an advanced mapping panel is a more structured approach, wherein a series of inbred **donor founder parents** (lines) are all crossed to a single **common founder parent**, resulting in a panel of **nested association mapping (NAM) lines** (Figure 18.12). This is loosely analogous to an NC II design wherein full sibs are nested within half sibs (Chapters 23 and 25 in Part II). A NAM panel consists of a series of RILs from each cross (the full-sib analog), with a common parent over all of the crosses (the half sib analog). Originally developed in maize (Yu et al. 2008; McMullen et al. 2009; Giraud et al. 2014) and barley (Maurer et al. 2015), a number of crop species now have NAM panels (Gage et al. 2020; Gireesch et al. 2021). Design issues such as optimal family structure, replication, and crossing schemes have been examined by Yu et al. (2008), Stich (2009), Stich et al. (2010), and Liu et al. (2013).

As stressed by Yu et al. (2008), the power of the NAM design arises because it exploits *two time scales for recombination*. Within any given cross, we are using standard QTL linkage mapping on the associated collection of RILs, and hence relatively crude resolution due to just a few generations of recombination. As a result, the genome of any RIL from that cross is a mosaic of chromosomal segments from the donor and common parent. However, associations between very tightly linked markers within a shared segment *across* donors reflect the impact of recombination over a much deeper time scale. Any such signal is due to population-level LD among the donors, which is expected to persist only between markers that are sufficiently close that any initial LD between them has yet to fully decay. We examine how this signal can be exploited by association mapping in Chapter 20. The second powerful feature of NAMs is that any given QTL allele might be rare (i.e., confined to a single donor line). As such, there is not much of a signal *over* the donor lines. Conversely there is a very strong signal *within* the collection of RILs from its cross to the common parent, as its frequency is expected to be 50%. The NAM approach represents an attempt to maximize the advantages of both linkage and LD mapping while minimizing their limitations. Yu et al. (2008) proposed using a simple regression analysis on marker allele copy number, with a cross-specific mean, for QTL detection, while Buckler et al. (2009) and Li et al. (2011) proposed an extension of CIM to the NAM design.

As with MAGIC populations, there are variations on this basic crossing structure, albeit still within the framework of a single common parent and a diverse collection of donor parents. The notation of **RIL-NAMs** and **DH-NAMs** is occasionally used (e.g., Li et al. 2016) to denote that the final lines were constructed from RILs (typically, using single-seed descent in selfers, where a single seed from each plant is used to form the next selfed generation) or from DHs. Recall that RILs typically have effectively two rounds of recombination relative to DH lines, creating a tradeoff between a slightly more expanded map (RILs) versus potentially much faster (but often more laborious) formation of the final inbred lines (DHs). There are also modifications of NAMs using the advanced backcross approach mentioned above. Here instead of RILs being formed using the F_1s in a series of diverse lines crossed to a common parent, they are constructed using a series of BC2 or BC3 backcrosses to this common (elite) parent, yielding **advanced backcross NAMs, AB–NAM (backcross NAMs, BC–NAM** is also used). For example, in Figure 18.12, instead of the 25 starting crosses for RIL formation being F_1s, they would be (say) BC2s to the common parent. An application of this approach is Nice et al. (2016), where 25 wild barley accessions were twice backcrossed to an elite parent (Rasmusson) before RIL formation proceeds.

Example 18.19. As shown in Figure 18.12, the maize NAM panel consists of 25 very diverse lines crossed to a single parent, B73 (Yu et al. 2008; McMullen et al. 2009). The common parent was chosen based on its widespread use, public physical map, and its available genome sequence. The donor lines consisted of nine temperate lines, more than half tropical lines, two sweet corns, and one popcorn line, capturing a good sample of global maize diversity. While the goal was for 5,000 RILs, the eventual number was 4,700, due (in part) to only modest success in using SSD to create inbred lines (a roughly 80% failure rate). A total of almost 140,000 recombination events occurred over the entire collection, corresponding to around three crossovers per gene (McMullen et al. 2009). Within each nested set of RILs, common-parent-specific (CPS) markers (alleles fixed in B73 that were otherwise rare in the donor parents) could quickly assign regions into having a B73, or a donor parent, origin.

Other Multiple-cross Designs

As highlighted with both MAGIC and NAM lines, one can take any standard line-cross design (Chapter 11) and augment it with markers for QTL mapping. Various line-cross designs are usually employed for more than simply a matter of convenience, such as obtaining sufficient variation (e.g., diallelic crosses; Chapter 25 in Part II), obtaining sufficient control over particular factors (such as maternal effects, e.g., reciprocal diallels; Chapter 25), and (most importantly), to generate desired informative contrasts, such as direct estimates of dominance or epistatic variances (e.g., NC III, triple test cross; Chapter 13). Many of these same concerns arise in QTL mapping (e.g., Cockerham and Zeng 1996; Verhoeven et al. 2006). For example, some MAGIC lines are constructed using diallels to generate maximal diversity in the base population. Similarly, **test cross** designs are often used when heterotic groups are present (Chapter 13), wherein a mapping population (such as a set of F_2, RILs, etc.) are crossed to a common parent from a different heterotic group (e.g., Schön et al. 2004).

Methods for combining QTL mapping information across multiple crosses have been developed by Rebai and Goffinet (1993; 2000), Xu (1998c), Liu and Zeng (2000), Zou et al. (2001), Yi and Xu (2001, 2002b), Li et al. (2005), and Xiao et al. (2016). These build on the fundamental idea of this chapter, conditional QTL probabilities given marker information (Equation 18.1), but involve more than a fair bit of bookkeeping. There are also subtle modeling decisions involved, such as whether to use a fixed- or random-effects analysis, and the appropriate framework (regression, likelihood, or Bayesian). See the aforementioned references for more details.

Segregation Distortion

Finally, many parental populations are chosen because of their wide difference in traits of interest, and marker loci in these inbred-line crosses often show strong segregation distortion (e.g., Vallejos and Tanksley 1983; Edwards et al. 1987; Paterson et al. 1988; Bonierbale et al. 1988; Doebley and Stec 1991; Schön et al. 1993). As discussed in Chapter 17, this is a common feature when there is significant genetic divergence between parental lines. The impact of such distortion on QTL mapping, and designs to exploit some of its features, have been examined by Xu (2008), Xu and Hu (2009), Zhang et al. (2010b), and Cui et al. (2015).

SELECTED APPLICATIONS

The Nature of Transgressive Segregation

QTL mapping experiments provide insight into the nature of **transgressive segregation** (or **transgression**), wherein some F_2 individuals show more extreme trait values than are seen in either parental line. Note that this phenomena is distinct from heterosis (Chapter 13)—the overperformance of the *mean* in the F_1 (which typically declines in the F_2)—rather, it refers to the presence of outlier *individuals* in the F_2. One explanation for such outliers is nonadditive gene action, i.e., epistasis and/or overdominance. Alternatively, transgressive segregation could be caused by the parental lines being fixed for sets of *additive* alleles having opposite effects (**antagonistic alleles**), e.g., one line is fixed for $+-/+-$, the other $-+/-+$, which would generate more extreme genotypes in the F_2 than observed in either parent (e.g., $++/++$ and $--/--$). Such lines are said to contain **complementary QTL alleles.** The antagonistic hypothesis for transgression is the one supported by most QTL studies (Rieseberg et al. 1999, 2003).

For example, Li et al. (1995) observed transgressive segregation for heading date in the cross of *Lemont* and *Teqing* strains of rice (*Oryza sativa*). Using 113 markers (with an average spacing of 19 cM) and 2,418 F_4 lines, three regions that together accounted for 77% of the phenotypic variance in heading date were mapped. While the difference in average heading date between parental strains was just six days, one region from *Lemont* decreased heading date by eight days (relative to its counterpart in *Teqing*), while another from *Teqing* decreased it by seven days. Hence, these lines were fixed for alternative alleles at major loci, resulting in effects that largely canceled.

Transgressive segregation was also observed in 8 of 11 traits measured in a large F_2 population from a cross of *Lycopersicon esculentum* (cultivated tomato) and *L. pennellii*, its wild Peruvian relative (deVicente and Tanksley 1993). Of the 74 QTLs detected for these 11 traits, 36% showed alleles having effects on the character that were antagonistic to parental-line differences (alleles reducing a trait being found in parents from the large line, and vice versa). Pairwise epistasis was ruled out as a major cause for the observed transgressive segregation, as the number of significant epistatic associations did not exceed that expected by chance. However, overdominance (or associative overdominance; Chapter 13) contributed in a few cases, with marker heterozygote means being more extreme than those for marker homozygotes. Likewise, Weller (1987) and Weller et al. (1988) observed that around 25% of the significant marker–QTL relationships in their tomato crosses were opposite in sign from the parental differences. A similar study based on a cross of two phenotypically similar cultivars of soybeans also noted that transgressive segregation due to complementary QTL alleles was quite common (Mansur et al. 1993b).

Transgressive segregation has also been found when lines resistant for certain insect pests or plant pathogens have been crossed to sensitive lines. For example, of seven detected maize QTLs conferring increased resistance to the European corn borer in a resistant × sensitive cross, five came from the resistant parent, while two came from the sensitive parent (Schön et al. 1993). Dirlewanger et al. (1994) similarly found that a sensitive pea line carried a resistance allele for *Ascochyta* fungal blight that was not present in a resistant line.

A more general picture of the ubiquity of transgression was offered in two surveys by Rieseberg and colleagues. Rieseberg et al. (1999) examined 171 studies from a wide range of plant and animal crosses (both natural and domesticated), finding that 91% (155) of these had at least one transgressive trait in the study, and that 44% of the 1,229 total traits showed transgression. Hence, transgression is ***more the rule than the exception***. They noted that transgression tended to be more common in plants than in animals, and was more common in intraspecific, rather than in interspecific, crosses. They also made a series of predictions for when transgression is expected. One seems counterintuitive: *transgression should be more common in crosses between phenotypically similar, rather than phenotypically dissimilar, lines* (e.g., Cubillos et al. 2011). The logic is that more similar lines are likely to be a mixture of different plus and minus alleles, while directional selection might drive down the fraction of antagonistic alleles in more diverged populations. WL Chapter 12 reviewed a number of QTL-based tests for detecting past selection on a trait that exploits this logic, such as an excess of plus alleles (Orr 1998). While Rieseberg et al. (1999) largely focused on the behavior of traits, Rieseberg et al. (2003) examined 3,253 QTLs from 749 traits over 96 studies. Almost two-thirds of the traits had at least one antagonistic allele, meaning that the potential for

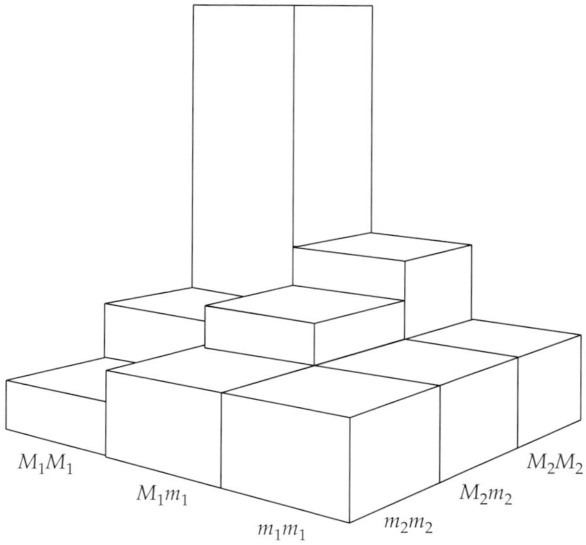

Figure 18.13 An example of epistasis for QTLs influencing protein volume in maize. Height indicates the amount of protein volume for each of the genotypes. Here the $M_1M_1M_2M_2$ marker genotype had the greatest effect on protein volume, while $M_1M_1m_2m_2$ had the smallest. (After Damerval et al. 1994.)

transgression is rather common, as suggested from the trait-only data. Consistent with the trait data, they also found that plants had a significantly higher fraction of antagonistic QTLs than did animals, and that such QTLs were more common in intraspecific, as opposed to interspecific, crosses.

Beyond obvious ramifications for breeding programs, transgressive segregation also has important evolutionary implications. Lewontin and Birch (1966) suggested that inter-species and wide-population hybrids can result in rapid adaptation to new environments (but also recall outbreeding depression; Chapter 13). If transgressive segregation in population crosses is the rule rather than the exception (as seems to be the case), then the (F_2) hybrids from such crosses possess the genetic variability to extend, perhaps considerably, the phenotypic range of a trait relative to either parental population. The results of Rieseberg et al. (1999, 2003) suggested mild caution with this idea in that intraspecific crosses inflate the importance of transgression over what might be expected in interspecific crosses. At a minimum, it is clear that mean phenotypic differences between lines are often very poor predictors of the architecture of their underlying genetic differences.

QTLs Involved in Protein Regulation

Any trait that can be *quantified* (assigned a numerical trait value, even simply as 0 or 1) can be subjected to a quantitative-genetic (QG) analysis. Historically, the traits considered were usually morphological or production phenotypes, such as body weight, height, measures of shapes, or yield. QG analysis then evolved to consider aspects of physiology and behavior. As modern functional genomics blossomed, it was realized that the machinery of QG applies equally to quantifiable molecular traits, such as levels of mRNA transcripts, proteins, and metabolites. QTLs that control the level of expression of a molecular target are denoted as **eQTLs**, for **expression QTLs**, and are a vibrant field of study (Chapter 21). These have also been called **mQTLs**, for **metabolite QTLs**, where referring to control of concentration of metabolites (Breunig et al. 2014).

Here we focus on the seemly modest maize study of Damerval et al. (1994) that started the hunt for eQTLs. They analyzed the spot volumes of 72 anonymous (i.e., unknown) proteins (from a specific seed tissue) separated by high-resolution 2-D polyacrylamide gel

Table 18.3 Number of detected QTLs influencing pollination characters involved in reproductive isolation between *Mimulus cardinalis* and *Mimulus lewisii* and their estimated individual effects (measured by % of variance explained). Due to sampling error, the sum of individual r^2 values exceeds 100% in a few cases. (After Bradshaw et al. 1995.)

	Number of QTLs	% Phenotypic Variance ($r^2 \times 100$)
Pollinator attraction characters		
Petal anthocyanins	2	33.5, 21.5
Petal carotenoids	1	88.3
Corolla width	3	68.7, 33.0, 25.7
Petal width	3	42.4, 41.2, 25.2
Pollinator reward		
Nectar volume	2	53.1, 48.9
Nectar concentration	2	28.5, 23.9
Pollination efficiency		
Stamen length	4	27.7, 27.5, 21.3, 18.7
Pistil length	2	51.9, 43.9

electrophoresis. Genes controlling protein volume are, by definition, **regulatory genes** influencing the amount of that protein. Sixty F_2 individuals were scored with 76 RFLP markers, and both ML-interval mapping and single-marker ANOVA detected a total of 70 QTLs affecting 46 of the 72 proteins. Of these 46 proteins, 25 were influenced by two or more QTLs (up to a maximum of five). Of the 70 detected QTLs, 33 showed strict additivity, while the remaining 37 showed at least some dominance. The amount of variation in protein volume accounted for by a single QTL ranged from 16% (the lower detection limit for this sample size) to 67%, and the cumulative variation accounted for by all detected QTLs for each protein ranged from 37% to 90%. Perhaps the most striking observation was the presence of significant epistasis. Four proteins had QTLs that were only detected through epistasis (their single-locus effects were not significant). In all, 14% of the 72 proteins showed detectable epistasis (Figure 18.13).

QTLs Involved in Reproductive Isolation in *Mimulus*

Bradshaw et al. (1995, 1998) examined the genetic basis of floral differences between sibling species of monkey flower, *Mimulus lewisii* and *M. cardinalis*. Although the ranges of these species overlap and laboratory F_1 hybrids are completely interfertile, hybrid plants are not found in nature. Presumably, this is due to nonoverlap of pollinators. *Mimulus lewisii* shows characters typical of bumblebee pollination: pink flowers with yellow nectar guides, a wide corolla, small volume of highly concentrated nectar, and short anthers and stigma. *Mimulus cardinalis*, on the other hand, shows a typical suite of hummingbird-pollinated characters: red petals lacking nectar guides, a narrow tubular corolla, high nectar volumes, and long anthers and stigma.

Their intial study (denoted as experiment one; Bradshaw et al. 1995) used 93 F_2 plants and 159 markers in a cross between these species, and detected (using ML interval mapping) a number of QTLs for these pollination traits. As shown in Table 18.3, four of the characters appear to each have a QTL accounting for over 50% of the total F_2 variance, while all other characters had a QTL accounting for at least 25% of the total variance. Hence, it appeared that the bulk of the differences in pollination characters (and hence reproductive isolation) could be accounted for by one or two factors (individual loci or a block of tightly linked loci) for each character. However, with these small sample sizes, some caution is in order, given our previous comments about overestimation of QTL effects when power is low (Beavis effects).

To address this concern, a follow-up study (experiment two; Bradshaw et al. 1998) increased the sample size five fold to 465 F_2 plants. The two major QTLs for petal

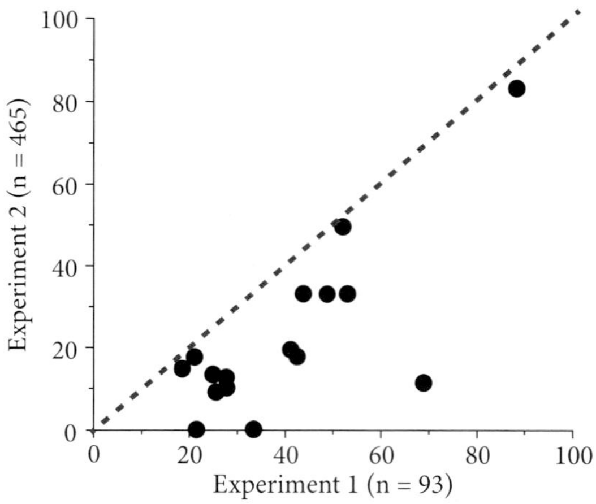

Figure 18.14 Relationship between the effects of detected QTLs for *Mimulus* pollination traits, expressed as percent of F_2 phenotypic variation (r^2), over two crosses with different sample sizes (Bradshaw et al. 1995, 1998). Experiment one measured 96 F_2 plants while experiment two measured 465. Note that all detected QTLs had larger estimated values in the smaller experiment, a clear example of the Beavis effect. The two values of zero from experiment two correspond to the petal anthocyanin QTLs that were not replicated.

anthocyanins from experiment one were not reproduced. However, all of the other QTLs were, albeit (in most cases) with a substantial reduction in the fraction of variation explained (Figure 18.14). Clearly, the Beavis effect inflated the initial importance of regions in experiment one due to its small sample size. Ignoring anthocyanin, of the initial major QTLs ($r^2 \geq 0.25$), less than half retained an $r^2 \geq 0.25$ in experiment two. However, most of the original traits still had large-effect QTLs ($r^2 \geq 0.15$). As might be expected, the larger sample size of experiment two detected 16 additional QTLs not seen in the original cross. In experiment one, the smallest r^2 for a detected QTL was 0.187, while it was 0.033 in experiment two. Further, 7 of the 16 additional QTLs had effects less than the smallest detected value from experiment one.

Following detection of these QTLs, the authors took F_2 hybrids into the field and estimated which traits were the most predictive of bee versus hummingbird attraction (Schemske and Bradshaw 1999; Bradshaw and Schemske 2003). The most predictive traits still had significant single QTLs ($r^2 \geq 0.15$) in the more powerful (and less biased) experiment two. In an especially informative experiment, Bradshaw and Schemske (2003) introgressed alternative alleles at the QTL for yellow carotenoid pigment (which had an $r^2 = 0.83$ in experiment two) into otherwise pure background of the other species. When the yellow allele from *M. lewisii* (bee pollinated) was introduced into an otherwise *cardinalis* background (hummingbird pollinated), the resulting NIL received a 68-fold increase in bee visits relative to the wild type, while a NIL with a *cardinalis* allele in an otherwise *lewisii* background showed a 74-fold increase in hummingbird visits.

QTLs Involved in the Differences Between Maize and Teosinte

Maize and teosinte are dramatically different (see Figure 5.2), to the point that they were originally placed in separate genera. Hybrids, however, are fully interfertile and maize is believed to have resulted from domestication of teosinte (Beadle 1980; Doebley 1992, 2004). In an elegant series of papers, Doebley and colleagues (Doebley et al. 1990, 1994, 1995a, 1995b, 1997; Doebley and Stec 1991, 1993; Dorweiler et al. 1993; Hubbard et al. 2002; Wang et al. 2005b; Studer and Doebley 2011; Whipple et al. 2011; Wills et al. 2013; Wang et al. 2015) characterized many of the genes involved in these dramatic differences.

Maize and teosinte have major differences in plant architecture (Figure 5.2; Table 18.4). Teosinte has multiple long lateral branches, topped with male inflorescences (tassels). In

Table 18.4 Character differences between maize and teosinte (primitive maize race Reventado × *Zea mays parviglumis*). Listed are mean character values (Means), the number of detected QTLs (N), the r^2 value for the largest (Max) and smallest (Min) detected QTLs, and the total r^2 for a model containing all detected QTLs. Locations for the QTLs detected in this cross are plotted as the upper bars in Figure 18.15. (After Doebley and Stec 1993.)

	Means		QTLs			
	Maize	Teos.	N	Max	Min	r^2
Plant Architectural Characters						
Lateral branch internode length (LBIL)	0.7	21.9	5	0.45	0.05	0.63
Number of branches (LIBN)	0.0	5.8	4	0.24	0.04	0.42
% male primary lateral inflorescences (STAM)	0.0	97.0	5	0.23	0.05	0.52
No. secondary ears/lateral branch (PROL)	1.0	8.4	7	0.25	0.04	0.63
Ear Characters						
Cupules along a single rank (CUPR)	37.4	5.3	6	0.25	0.04	0.61
Disarticulation: 1 = none, 10 = full (DISA)	1.0	10.0	6	0.42	0.04	0.60
Glume score: 1 = soft, 10 = hard (GLUM)	1.0	10.0	2	0.41	0.08	0.75
% cupules with only one spikelet (PEDS)	0.0	100.0	5	0.25	0.08	0.69
Number of rows of cupules (RANK)	5.6	2.0	6	0.36	0.05	0.87

maize, these branches are very short and topped with ears (female inflorescences). These differences in plant architecture can be quantified by considering four characters: internode length on lateral branches (small in maize, long in teosinte), the number of branches (none to few in maize, many in teosinte), percentage of lateral branches topped with tassels as opposed to ears (mostly tassels in teosinte, ears in maize), and the number of secondary ears on each lateral branch (few in maize, many in teosinte). Table 18.4 gives the mean values for these characters in maize and teosinte.

Differences in the structure of the female inflorescence (the ear) are even more dramatic (Figure 5.2; Table 18.4). The teosinte ear has 5–10 cupulate fruitcases arranged in pairs. Each of these has a single spikelet that gives rise to a kernel, resulting in 10–20 kernels per teosinte ear. Each mature fruitcase is covered by a hardened outer glume that seals in the kernel, making harvesting very difficult. In contrast, the maize ear is composed of 100 or more cupules (arranged in multiple rows rather than pairs of rows), each cupule containing two spikelets, leading to two kernels per spikelet. These changes result in the maize ear having an order of magnitude more kernels than the teosinte ear. The maize outer glume is soft, so the kernels remain exposed for easy harvesting. Finally, while the teosinte ear easily disarticulates (**shattering** to scatter seeds), kernels on the maize ear stay intact, further facilitating the harvesting of kernels.

In their initial studies, QTLs were mapped in two different crosses, each involving a different primitive maize race and a different subspecies of teosinte (Doebley et al. 1990; Doebley and Stec 1991, 1993). As shown in Table 18.4, a few QTLs of major effect account for most of the differences between characters. These QTLs are mostly in very similar positions in the two crosses (Figure 18.15), with both sets of crosses showing five regions of the maize genome that account for most of the differences. Such results are consistent with Beadle's hypothesis of five major genes accounting for the bulk of the obvious phenotypic differences between maize and teosinte (Beadle 1939). Beadle arrived at this figure by examining 50,000 F_2 maize × teosinte offspring, finding the frequency of all-maize or all-teosinte phenotypes to be $\simeq 1/500$. If n genes are involved, the expected F_2 frequency of either parental genotype is $(1/4)^n + (1/4)^n$. Setting this equal to $1/500$ and solving gives $n = 5$.

Focusing on the five major regions, marker-selected NILs (Chapter 17) were constructed to further characterize the QTLs. Doebley's first target was a QTL on chromosome 4, which accounted for 50% of the variance in glume score. A small maize segment containing this region was introgressed into a teosinte background by three generations of backcrossing and selection for flanking markers (Dorweiler et al. 1993). When NILs with the introgressed

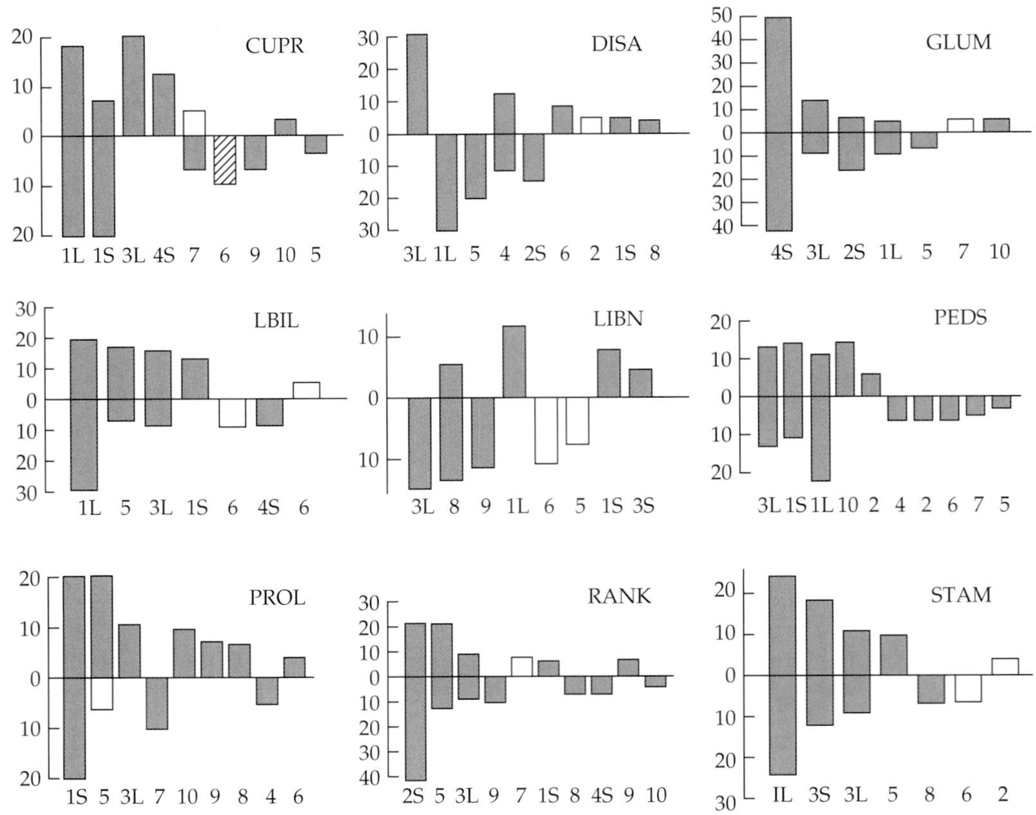

Figure 18.15 QTLs accounting for differences between maize and teosinte in the nine characters listed in Table 18.4. Each bar represents a detected QTL, with bar height indicating its r^2 value. Acronyms for characters are listed in Table 18.4. Chromosome position (chromosome number and short versus long arm) is indicated under each bar. The upper bars refer to QTLs detected in a cross between a primitive maize strain and one subspecies of teosinte (*Z. m. parviglumis*), while the lower bars are for a cross involving a different primitive maize strain and another teosinte subspecies (*Z. m. mexicana*). Aligned bars denote QTLs mapping to very similar positions in both crosses. White bars indicate QTL effects in the opposite direction from parental phenotype (antagonistic effects), while the cross-hatched bar indicates apparent overdominance. (From Doebley and Stec 1993.)

teosinte region were backcrossed to the maize recurrent parent, the resulting F_2 progeny showed two discrete classes for glume score, as would be expected with a single major gene. The putative gene was named *tga1*, for *teosinte glume architecture 1* (and later shown to result from a single amino acid substitution; Wang et al. 2015). A similar analysis of two other regions, QTL-3L and QTL-1L, showed strong epistasis for a number of key traits separating maize and teosinte (Doebley et al. 1995a; see Example 5.1). By using marker-selected introgressed lines, QTL-1L was shown by complementation tests to be the locus *teosinte branched 1* (*tb1*). In maize, mutants at this locus result in teosinte-like features for inflorescence sex (tassels, not ears), and number (many instead of one) and length of lateral branches (long, not short). Doebley found that the joint effects of *tb1* and QTL-3L, by themselves, were sufficient to account for essentially all of the differences in plant architecture between teosinte and maize. Further, these two QTL result in substantial differences in ear architecture. Hence, there is direct evidence that just a few genes can account for a very significant amount of the dramatic differences between teosinte and maize.

Subsequent studies on the *tb1* locus noted an extended region of reduced variation just upstream of this gene in maize, but not teosinte. This is the classical hallmark of a

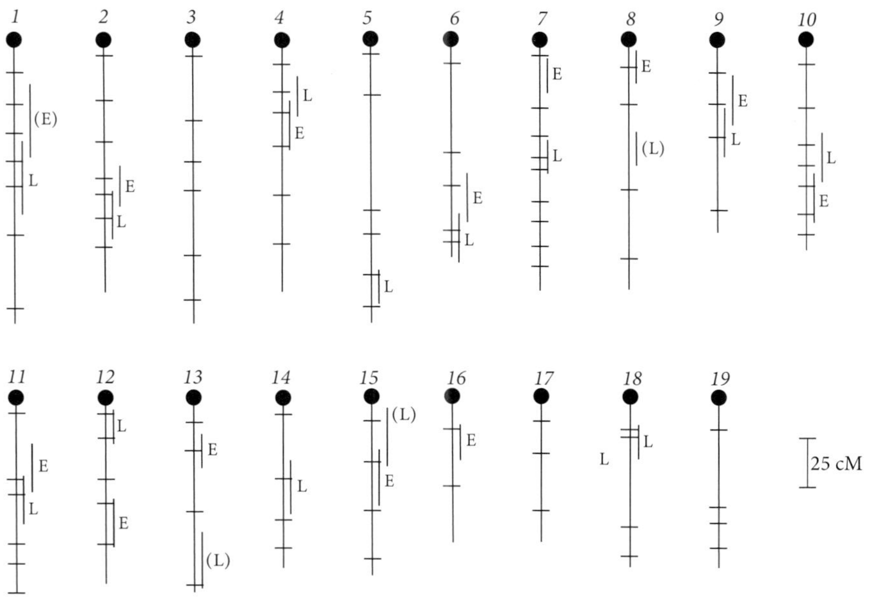

Figure 18.16 Locations for early, E, and late, L, growth QTLs, and early, (E), and late (L) weight QTLs on the 19 mouse autosomes. The marker locations are indicated by hatch marks and QTL confidence intervals by bars (From Vaughn et al. 1999.)

selective sweep (WL Chapter 8), indicating selection on this region in maize. Remarkably, the target site was not the *tb1* coding region, but rather an upstream regulatory region. Using results from sweep theory (WL Chapters 8 and 9), it was estimated that this allele had around a 5% advantage during the domestication phase (Wang et al. 1999; Clark et al. 2004, 2006). This leveraging of QTL information to narrow a search for selection to a small genomic region is routinely done in the search for domestication genes (WL Chapter 9).

A final interesting study using *tb1* addressed the issue of whether this QTL fractionated upon fine mapping (showing that the initial QTL was, in reality, a group of linked loci). Through the use of NILs, Studer and Doebley (2011) found that *tb1* did not fractionate for plant architectural traits (the detected QTL appeared to be a single gene for these traits). However, the effects of *tb1* on ear morphology did appear to fractionate, with *tb1* still having an impact, but its isolated effect was diminished relative to its initial value.

QTLs for Age-specific Growth in Mice

Cheverud et al. (1996) and Vaughn et al. (1999) performed two replicated crosses of the SM/J and LG/J inbred lines (the small and large lines from the Jackson Laboratories). Interval mapping was used to examine age-specific weight (at ages 1 through 10 weeks) and age-specific growth rate. Intercross I (Cheverud et al. 1996) consisted of 535 F_2 mice scored for 75 microsatellite markers, generating 55 intervals that averaged around 28 cM in length. Intercross II (Vaughn et al. 1999) consisted of 510 F_2 mice scored for 96 markers, resulting in 72 intervals averaging around 23cM (Figure 18.16). As Table 18.5 (Intercross I results) shows, considerable numbers of QTLs were found for all characters. All detected QTLs had small effects, with the largest accounting for around 10% of the F_2 phenotypic variance, while the average (detected) effect was around 4% of the F_2 variance. Note also from Table 18.5 that as age increases, so does the number of QTLs for weight and the total r^2 value of these detected loci. Early versus late weight and growth showed different genetic architectures. First,

Table 18.5 QTLs influencing age-specific weight and age-specific growth rates in mice, measured at weekly intervals. Early, Middle, and Late correspond to growth from 1 to 3 weeks, growth from 3 to 6 weeks, and growth from 6 to 10 weeks, respectively, while 6-week refers to growth from 1 to 6 weeks. (After Cheverud et al. 1996.)

	Age-specific Weight (weeks)									
	1	2	3	4	5	6	7	8	9	10
No. of QTLs	7	10	16	13	15	15	14	14	16	17
Total r^2	0.29	0.30	0.56	0.52	0.59	0.63	0.64	0.56	0.67	0.76

	Age-specific Growth			
	Early	Middle	6-week	Late
No. of QTLs	11	12	14	12
Total r^2	0.39	0.51	0.54	0.38

largely distinct sets of QTLs are involved (Figure 18.16). Second, dominance was found to be much more important in early weight and growth than in late weight and growth. The results from Intercross II yielded very similar conclusions, but only 70% of detected QTLs were seen in both studies. As expected, QTLs with larger LOD values in Intercross I were more likely to be replicated in Intercross II. This pair of papers highlights a key feature of many large-sample replicated QTL experiments: coarse characterization of the genetic architecture typically remains unchanged, but replication of specific QTLs is often modest.

QTLs in the Illinois Long-term Selection Lines of Maize

In 1896, C. Hopkins initiated a set of maize lines selected for high- and low-oil and high- and low-protein content (Hopkins 1899; Smith 1908). Selection on these lines continues today, and results from this remarkable study after 76, 90, 100, and 106 generations of selection were summarized by Dudley (1977), Dudley and Lambert (1992, 2004), Moose et al. (2004), and Dudley (2007). The smooth and continuous long-term response in these lines suggested that a large number of genes of relatively small effect underlie the impressive gains that were observed. Crosses of the divergently selected lines have been used in several QTL mapping studies, starting with basic F_2 analysis (Goldman et al. 1993, 1994; Berke and Rocheford 1995) and eventually using RIAILs for finer mapping (Dudley et al. 2004; Laurie et al. 2004; Clark et al. 2006; Willmot et al. 2006; Dudley et al. 2007), as well as backcrosses to one of the founding lines (Wassom et al. 2008a, 2008b).

The first studies were by Goldman et al. (1993, 1994), who crossed the selected (76-generation) high- and low-protein lines and then examined 100 F_3 families (formed by selfing F_2s) using 100 markers spanning the maize genome at an average spacing of about 20 cM. Using single-marker ANOVA, 22 markers on 10 chromosome arms were significantly associated with protein concentration, 19 markers on nine arms were associated with starch concentration, 26 on 13 arms with oil concentration, and 18 on 10 arms with kernel weight. Many of the marker-trait associations extended across clusters of linked markers. In these cases, single-marker ANOVA cannot distinguish between several linked markers all detecting the same linked QTL or multiple QTL.

A multiple regression involving only six (unlinked) markers accounted for 65% of the variation in protein concentration, and this increased to 84% when five significant pairwise epistatic interactions between these markers were incorporated into the regression (using Equation 18.14d). Seven markers accounted for 66% of the variation in starch, increasing to 78% when one significant pairwise epistatic interaction was included. Four marker loci accounted for 43% of the variation in oil concentration, while six markers accounted for 47% of the variation in kernel weight. It seemed very surprising that so few loci could account for such a significant fraction of the differences, especially given the long-term continuous and gradual change in the lines. As we detail shortly, this was almost certainly a Beavis effect, as more recent, and much higher powered, studies found a large number of small-effect QTLs. Berke and Rocheford (1995) examined a cross of two other variant lines from this experiment (high oil with low oil), and found similar results, with six loci accounting for 58% of the genetic variation in oil concentration and seven markers accounting for 56% of the variation in starch concentration.

A biased view of the genetic architecture of a trait from a QTL analysis can arise from two sources: *insufficient recombinant* (lumping the composite effects of several causal loci into single factor) and *low power* (resulting in both an undercounting of the actual number of QTL and overestimation of the effects of detected QTL). Dudley et al. (2004) and Willmot et al. (2006) attempted to address linkage concerns by examining the impact of additional generations of random mating before selfing. Both the first (Generation 70 high × low protein) and second (Generation 70 high x low oil) study compared the results of inbred lines formed from F_2s with those from (random-mating) F_5s. An important limitation is that these two studies used a modest number of markers (64 and 125). Not surprisingly, both studies detected significantly *fewer* QTLs using F_5-derived inbreds relative to using F_2 inbreds. This is expected from a reduction in power due to the expanded map distance generated by additional generations of random mating. Because power declines as $(1-2\tilde{c})^2$, the low marker density was simply not sufficient to adequately cover the expanded map.

Much more powerful designs were used by Laurie et al. (2004), Clark et al. (2006), and Dudley et al. (2007). These authors took generation 70 high versus low crosses, followed by 7–10 generations of random mating before 500 lines were generated by selfing. The resulting lines were then replicated (multiple locations at two years each), and scored for around 500 markers. This improved design attempted to address both the linkage and power limitations of earlier studies. Laurie et al. and Clark et al. used oil crosses (10 generations), while Dudley et al. used protein crosses (seven generations of random mating). Laurie et al. detected around 50 QTL for oil, which accounted for around 50% of the generic variance. Effects were small and largely additive. Surprisingly, around 20% had antagonistic effects (up alleles in the down line, and vice versa). Dudley et al. found similar results for protein.

Limitations of Line-cross QTL Mapping

Despite the impressive statistical machinery developed above for linkage-based QTL mapping using inbred-line crosses, we close on a slightly somber note. Researchers typically perform a QTL mapping experiment to address two, not necessarily independent, questions: (i) what is genetic architecture underpinning the differences between two lines, and (ii) what are the locations of the actual genes underlying these differences. As highlighted in the previous examples, linkage-based QTL mapping is a substantial improvement over classic line-cross analysis (Chapter 11) for probing the genetic architecture of a trait, and can often provide important insight. However, it tends to underestimate the number of QTLs and overestimate the effects of detected QTLs, especially when the sample size is modest (< 500, which is the norm). Hence, it is biased towards presenting a more major-gene view of the architecture (fewer genes of large effect).

Even more problematic is the very poor level of resolution of linkage-based mapping, which tends to be on the tens of megabase scale under the typical sample sizes used. This is an improvement over mapping to specific chromosome arms (Chapter 17), but only moderately helpful for fine mapping. Finally, there is the issue of repeatability, as replicated experiments (using identical line crosses) often fail to find some of the QTLs detected in previous studies. Again, this is most pronounced when sample sizes are small to modest. A compounding issue is that an apparently large QTL detected in an initial study (whose actual value was significantly overestimated by the Beavis effect) might lead the investigator to use a comparably sized population in an attempt to replicate this finding, when in reality a much larger sample size is needed.

A final issue is that an inbred-line cross is probing the nature of *fixed* differences between two lines, typically chosen because they are rather divergent in a focal trait. Hence, many of these genes may have been fixed by directional selection, which can result in large-effect alleles being disproportionately fixed (WL Chapters 25–27). The segregating genetic loci in the progeny of an inbred line cross are thus a rather *unnatural population*, with segregating alleles (or fixed alleles over a final RIL or DHL panel) having frequency one half, and likely enriched for large-effect alleles. In natural populations, segregating large-effect alleles tend to be at very low frequencies (Chapter 21; WL Chapters 24 and 28).

Despite these concerns, there still remain settings, especially in inbred crops, where line-cross mapping may have greater mapping resolution than a GWAS (especially when using second-generation mapping designs, such as MAGIC or NAM panels). The reason is that many elite inbred crops show LD over (at least) hundreds of kilobases. This places a limit on the resolution possible by LD (GWAS) mapping (Myles et al. 2009). Again, however, for linkage-based mapping to have greater resolution, the sample sizes need to be large, allowing more recombination events to be captured. In the words of Cockram and Mackay (2018), "we believe that days of mapping QTL in small populations must come to an end ··· we started with a target of developing 1,000 lines per population: that number now looks to be on the low side." Many practitioners of QTL mapping suffer from what Tversky and Kahneman (1971) called a "*belief in the law of small numbers*," gambling their research hypotheses on small samples without realizing that the odds against them are unreasonably high.

Our next three chapters attempt to use *segregating natural variation*, as opposed to fixed between-line differences, to map QTLs. As such, GWAS are ***very likely probing different genetic architectures than are inbred-line crosses***. Chapter 19 examines linkage mapping using family-based crosses, or more generally, pedigrees. These designs have much less power than inbred-line crosses, and have (at best) similar levels of mapping resolution. Chapter 20 examines association-based mapping, which can use *very* large population samples coupled with very dense marker maps to exploit naturally occurring LD. Such LD often extends over no more than a few tens of kilobase scale (and often much less), as opposed to linkage information (which tends to be on a tens of megabase scale unless the sample size is large), offering substantially improved mapping resolution. Finally, Chapter 21 summarizes the impact of GWAS studies on our current understanding of the genetic architecture of segregating genetic variation.

19

Linkage Mapping And Characterizing QTLs: Outbred Populations of Relatives

Every unhappy family is unhappy in its own way. Tolstoy (Anna Karenina)

For many species, especially humans, the luxury of manipulatable inbred lines is out of the question, in which case one must resort to using outbred individuals for QTL mapping. Often these are *observational studies* as opposed to *controlled crosses*. Outbred populations are also of interest for reasons other than experimental constraints. QTLs detected by inbred-line crosses represent *fixed differences between lines*, or even different species, and the relevance of these results to QTLs *segregating within populations* remains unclear. This fundamental distinction that inbred-line crosses detect QTLs responsible for between-population differences while outbred populations detect QTLs responsible for within-population variation (such as disease risk) makes these complementary, rather than competing, approaches.

There are two general approaches for QTL mapping using outbred individuals: (i) **linkage mapping** using a collection of relatives and a modestly dense set of markers, and (ii) **association mapping** via linkage disequilibrium (LD; Chapter 5), using a random population sample of (ideally) unrelated individuals and a very dense set of markers. Our focus in this chapter is linkage mapping, using the *excess of parental gametes* (Chapter 5) seen in the progeny of a parent to create *marker-trait associations*. This extends the linkage mapping results from the previous chapter to outbred individuals. We do so by considering half- and full-sib families, before turning to more general pedigrees. Hence, to proceed we require *known sets of relatives*. This restriction imposes a significant constraint, and the task of collecting a sufficiently large sample of relatives for adequate power is nontrivial. Note from Example 5.6 that linkage mapping can be considered as a special case of LD mapping, as linkage creates LD in the population of gametes from a parent, which decays by recombination over more distant relatives.

In Chapter 20, we examine population-level (or association) mapping, using the linkage disequilibrium between very closely linked markers (often on a kilobase, or a finer, scale), which generates marker-trait associations only over these very short distances. These associations represent very deep, but short, shared ancestral segments in an otherwise unrelated background genome (Chapter 8). Hence, the presence of relatives is a confounding factor that must be accounted for in an association analysis, and we do so by using some of the machinery developed in this chapter.

Using within-population variation, as opposed to fixed differences between populations, results in *significantly decreased power* for (linkage-based) QTL detection. With inbred crosses, all F_1 parents have identical genotypes (including the same linkage phase), so all individuals are informative, and LD is maximized. Further, QTL effects are expressed as means with inbred crosses (the average value of each QTL genotype), but as genetic variances with outbred crosses (which arises as the metric for QTL effects because of variation in parental genotypes). Because variances are estimated with much less precision than means, estimates from outbred populations are expected to be inherently less precise. All of these complications conspire to make QTL linkage mapping using outbred individuals a far less powerful approach than inbred-line linkage mapping.

As with inbred lines, a variety of designs have been proposed for obtaining samples with the LD required for QTL mapping (i.e., generating marker–QTL correlations). We start our discussion by reviewing the types of outbred parents that are informative for QTL mapping and then consider analyses using collections of relatives, starting with least-squares (LS; ANOVA/regression) and maximum likelihood (ML) methods for family data. We then extend ML methods to pedigrees. We conclude by examining approaches that use comparisons between pairs of relatives—Haseman–Elston regressions (and their extensions) and allele-sharing tests—which, historically, have been widely applied in attempts to map human disease genes.

MEASURES OF INFORMATIVENESS

The major difference between linkage-based QTL analysis using inbred-line crosses versus outbred individuals is that while the parents in the former are *genetically uniform*, parents in the latter are *genetically variable*. This distinction has several consequences. First, only a fraction of the parents from an outbred population are **informative**. For a parent to provide linkage information, it must be heterozygous at both a marker *and* a linked QTL, as only in this situation can a marker–trait association be generated in the progeny (i.e., generating distinguishable parental and recombinant gametes; Chapter 5). Only a fraction of random parents from an outbred population are such double heterozygotes for any particular marker-QTL combination. With inbred lines, F_1s are heterozygous at *all loci that differ between the crossed lines*, so that all (F_1) parents are fully informative. Second, there are only two (equally frequent) alleles segregating at any locus in a (biparental) inbred-line-cross design, while outbred populations can be segregating any number of QTL alleles at any frequency. Finally, in an outbred population, individuals can differ in marker-QTL *linkage phase*, so that an M-bearing gamete might be associated with QTL allele Q in one parent, and with q in another. Thus, with outbred populations, marker–trait associations must be examined *separately for each parent*. With inbred-line crosses, all F_1 parents have identical genotypes (including linkage phase), allowing one to simply average marker-trait associations over all offspring, regardless of their parents.

Before considering specific QTL mapping approaches for outbred populations, some comments on the probability that an outbred family is informative are in order. A parent is **marker-informative** if it is a marker heterozygote, **QTL-informative** if it is a QTL heterozygote, and simply **informative** if it is both. Unless *both* the marker *and* the QTL are highly polymorphic, most parents will not be informative for a given QTL. Given the need to maximize the fraction of marker-informative parents, classes of marker loci successfully used with inbred lines may not be optimal for outbred populations. For example, historically RFLPs, and more recently SNPs, were widely used in inbred lines, but these markers are typically diallelic and hence (at best) have modest polymorphism. Microsatellite marker loci (SSRs/STRs; Chapter 8), on the other hand, are highly polymorphic and hence much more likely to yield marker-informative individuals.

As shown in Table 19.1, there are three types of marker-informative crosses. With a highly polymorphic marker, it may be possible to examine marker-trait associations for both parents. With a **fully informative family** ($M_iM_j \times M_kM_\ell$ or $M_iM_j \times M_iM_k$) all parental alleles can be distinguished (i.e., the parent of origin for a marker alleles in the progeny is unambiguous). Hence, the effects of parents can be examined by comparing the trait values in M_i- versus M_j- offspring and M_k- versus $M_\ell-$ offspring. With a **backcross family** ($M_iM_j \times M_kM_k$), only the heterozygous parent can be examined for marker-trait associations. Finally, with an **intercross family** ($M_iM_j \times M_iM_j$), homozygous offspring (M_iM_i, M_jM_j) are unambiguous as to the origin of parental alleles, while heterozygotes are ambiguous, because allele M_i (M_j) could have come from either parent.

In designing experiments, it is useful to estimate the fraction of families expected to be marker-informative. One measure of this is the **polymorphism information content** (Botstein et al. 1980), or **PIC**, of the marker locus. If p_i is the frequency of allele i, then

$$\text{PIC} = 1 - \sum_{i=1}^{n} p_i^2 - \sum_{i=1}^{n-1} \sum_{j=i+1}^{n} 2\, p_i^2\, p_j^2 \leq \frac{(n-1)^2(n+1)}{n^3} \tag{19.1}$$

Table 19.1 Types of marker-informative matings. Here i, j, k, and ℓ index different marker alleles. These classes of mating can also be classified in terms of the number of marker alleles that are in play (marker alles whose parental origin can be unambiguously determined in offspring).

Fully informative: $M_i M_j \times M_k M_\ell$

Parents are different marker heterozygotes.

All offspring are informative in distinguishing alternative alleles from both parents.

$M_i M_j \times M_i M_k$

Both parents are heterozygotes, but share a common allele.

All offspring are informative in distinguishing alternative parental alleles.

Backcross: $M_i M_j \times M_k M_k$

One parent is a marker heterozygote, the other is a marker homozygote.

All offspring are informative in distinguishing heterozygous parent's alternative alleles.

Intercross: $M_i M_j \times M_i M_j$

Both parents are the same marker heterozygote.

Only homozygous offspring are informative in distinguishing alternative parental alleles.

which is the probability that one parent is a marker heterozygote and its mate has a *different* genotype (i.e., a backcross or fully informative family, but excluding intercross families). In this case, we can distinguish between the alternative marker alleles of the first parent in all offspring from this cross. The upper bound (given by the right-hand side of Equation 19.1) occurs when all marker alleles are equally frequent, $p_i = 1/n$ (which is often roughly correct for SSRs).

Another metric is **proportion of fully informative matings (PFIM)**, where marker alleles from *both* parents can be distinguished in *all offspring*, where

$$\text{PFIM} = \sum_{i=1}^{n-1} \sum_{j=i+1}^{n} 2\,p_i\,p_j \left[\left(\sum_{k=1}^{n-1} \sum_{\ell=k+1}^{n} 2\,p_k\,p_\ell \right) - 2\,p_i\,p_j \right]$$
$$\leq \frac{(n-1)\,(n-2)\,(n+1)}{n^3} \tag{19.2}$$

where i, j, k, and ℓ denote *different* marker alleles (Götz and Ollivier 1992). Again, the upper bound occurs when all alleles are equally frequent. Two comments about PFIM as a measure. First, the restriction that *all* offspring provide information is constraining. Intercross families are *partially* informative, with marker homozygotes, but not heterozgyotes, in the offspring being informative. In such cases, roughly half of the offspring provide linkage information. Second, note that PFIM is zero for any biallelic marker, as both parents have to be heterozygotes, but these would be the same, resulting in an intercross family. Figure 19.1 plots the maximum values of both of these measures as a function of the number of marker alleles. The power of SSRs becomes apparent, given their modest to large number of alleles (usually $n > 5$), coupled with roughly equal allele frequencies.

Of course, a *marker-informative* parent is not necessarily also *QTL-informative*. For a QTL with n_Q alleles, the ith of which has frequency q_i, the probability that a parent is a QTL heterozygote (in a random mating population) is

$$\left(1 - \sum_{i=1}^{n_Q} q_i^2 \right) \leq \left(1 - \frac{1}{n_Q} \right) \tag{19.3}$$

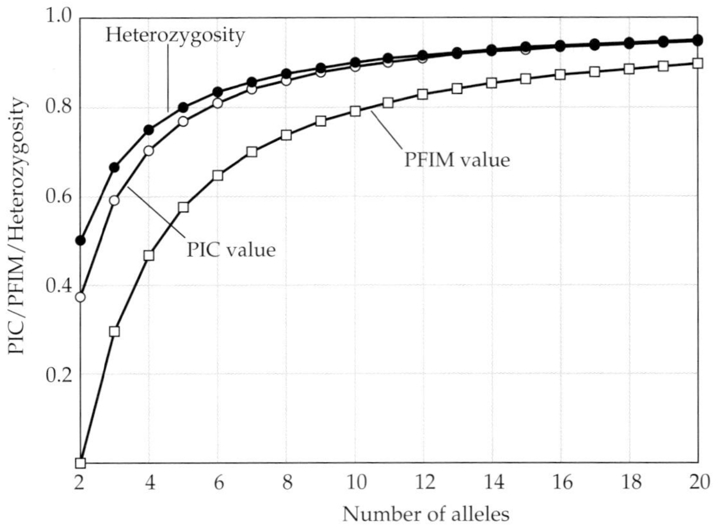

Figure 19.1 The relationship between three measures of marker information: heterozygosity, PIC, and PFIM. Plotted are the maximum values for each measure as a function of the number of marker alleles under the assumption of an even allele-frequency distribution ($p_i = 1/n$).

For a population where parents are in gametic-phase equilibrium, the probability that (at least) one parent is a double (QTL/marker) heterozygote and that alternative marker alleles from this parent can unambiguously be distinguished in all offspring is

$$\Pr(\text{at least one parent fully informative}) = \text{PIC} \times \left(1 - \sum_{i=1}^{n_Q} q_i^2\right) \tag{19.4a}$$

while the probability that both parents satisfy this condition is

$$\Pr(\text{both parents fully informative}) = \text{PFIM} \times \left(1 - \sum_{i=1}^{n_Q} q_i^2\right)^2 \tag{19.4b}$$

Thus, under fairly general settings, we expect most outbred parents to be *uninformative* for a particular QTL–marker combination. For a diallelic QTL in Hardy–Weinberg,

$$\Pr(\text{both parents fully informative}) = \text{PFIM} \times [2q(1-q)]^2 \tag{19.4c}$$

One point to stress, because it can be a source of confusion. Recall Figure 5.5, which shows that while the population of *parents* can be in linkage equilibrium (parental linkage phases are random), their offspring will still show linkage (an excess of parental gametes). It is the latter that is exploited in QTL mapping using relatives.

Example 19.1. Consider a marker locus segregating five alleles in a population. Here,

$$\text{PIC} \leq \frac{4^2 \cdot 6}{5^3} = 0.768 \qquad \text{and} \qquad \text{PFIM} \leq \frac{4 \cdot 3 \cdot 6}{5^3} = 0.576$$

The sample sizes required to see a desired number of informative families follow from the binomial distribution (Equation 2.19), using either PIC or PFIM, respectively, as the success probability, p. From Equation 2.19b, the sample size n to have an average of m successes follows as $n = m/p$. Hence, in order to have (on average) 100 families that are marker-informative for at least one parent, 100/0.768 = 130 randomly drawn families must be examined. If one requires these families to be informative for both parents, then 100/0.576 = 174 families must be examined. However, these numbers are a little misleading. For example, Pr(Bion(0.768, 130) $\geq$ 100—which can be computed in R using the command `1 - pbinom(99, 130, 0.768)`—is only 0.535. This probability increases to 0.903 for n = 138, while 95% power requires a sample size of n = 141. These are the *best-case* scenarios for this marker, occurring if all five alleles have equal frequency. Now suppose that the marker is linked to a QTL with three (functionally distinct) alleles with frequencies 0.5, 0.3, and 0.2. The probability that an individual is a QTL heterozygote is $1 - (0.5^2 + 0.3^2 + 0.2^2) \simeq 0.62$. Thus, to have 100 families with one or both parents fully informative requires sampling (at least) $130/0.62 \simeq 210$ and $174/0.62^2 \simeq 453$ random families, respectively.

SIB ANALYSIS: LEAST-SQUARES LINEAR MODELS

One can use family data to search for QTLs by comparing the trait means between offspring carrying alternative marker alleles from the same parent (i.e., a least-squares, LS, analysis; Chapters 3 and 10). There are a number of ways to accomplish this apparently simple task, and our treatment first considers specific sets of relatives, such as entire half- or full-sib families.

Sibship-based methods for detecting linked QTLs historically developed along two different tracts. Animal and tree breeders typically relied on a *few large half-sib families*, each resulting from a single common parent. This is a sire in animal breeding settings (often using artificial insemination to generate very large familes), but is often a seed parent in tree breeding. Human geneticists, on the other hand, often relied on a *number of small full-sib families*. These different designs not only reflect the biological and sampling limitations present in different species but also mirror concerns about the tradeoff between number of families (increasing the chance of sampling at least one QTL-informative sibship; Example 16.2) and number of sibs per family (increasing the power to detect marker differences, *provided* the family is QTL-informative). Neither design is necessarily optimal, because the highest power (for a fixed sample size) occurs when a single large *informative* sibship is used (A. Hill 1975; Weller et al. 1990; Luo 1993; Kolbehdari et al. 2005; Massault et al. 2008). Unfortunately, although we can gauge if a particular parent is *marker-informative*, we have, at best, only very weak signals that parent is also a *QTL heterozygote* (such as choosing those parents with higher within-family trait variances; e.g., Chatziplis and Haley 2000).

A variety of LS linear model approaches for dealing with sibship data have been proposed. We start by considering the simplest case of a single half-sib family. Care is required when combining information from several half-sibships, and we examine this issue next. Finally, we consider how half-sib analyses are extended to full sibs.

A Single Half-sib Family

Consider the typical animal setting where single sire (male) is crossed to n unrelated dams (females) and a single offspring from each mating is measured. In diary breeding, this is called a **daughter design** (Weller et al. 1990), where only daughter offspring are scored as most traits of interest are related to milk production. In plants, the roles of the sexes in a half-sib family are typically reversed, with the common parent being the seed parent, with pollen donated from a number of unrelated parents. For brevity in our discussion, we will refer to the common parent as a sire.

Ideally, this common parent should be informative at a number of marker loci covering a significant fraction of the genome (as can be accomplished by using a large number of highly informative markers). While we have much less control over the choice of QTL-informative sires, if a major gene is involved, a pilot segregation analysis with a number of candidate sires can be used to estimate the probability that a particular sire is a QTL heterozygote (e.g., Equation 16.24b), and those sires with the highest such probabilities can be used. One might also select those sires showing the greatest within-family trait variance (the idea being that this variance increases with the number of heterozygous QTLs), but offspring variance can be a poor indicator of sire QTL heterozygosity.

Let M_1 and M_2 denote the alternative sire alleles at a heterozygous marker locus, and let z_{ij} denote the phenotype of the jth individual with sire marker allele $i\,(=1,2)$. The simplest linear model for examining whether this marker is linked to a QTL (heterozygous in the sire) is

$$z_{ij} = \mu + m_i + e_{ij} \tag{19.5}$$

where the marker effect m_i is the expected deviation from the mean trait value μ, given that an offspring has sire allele M_i. Because there are only two marker classes, a t test can be used to determine if offspring with M_1 versus M_2 sire alleles have significantly different means,

$$t = \frac{\overline{z}_{M_1} - \overline{z}_{M_2}}{\sqrt{\text{Var}(\overline{z}_{M_1}) + \text{Var}(\overline{z}_{M_2})}} = \frac{\text{MC}}{\text{SE}} \tag{19.6}$$

MC denotes the **marker contrast** between alternative sire alleles, and SE is the standard error of the marker contrast. A significant t value implies that the marker is linked to a QTL that is heterozygous in the sire. Typically, only those offspring whose sire alleles can be determined are included in this analysis. If the marker locus is not very polymorphic, many offspring will be uninformative, being ambiguous as to which sire allele they inherited. Dentine and Cowan (1990) treated these individuals as a missing-data problem, and developed a GLS regression (Chapter 10) that allows them to be included in the analysis.

Finally, consider the expected value of the marker contrast, given the sire genotype. Suppose the sire is M_1Q_i/M_2Q_j and that c is the recombination frequency between the marker and the QTL. The conditional probabilities for alternate QTL alleles in gametes carrying a known sire marker allele are

$$\Pr(\,Q_i\,|\,M_1\,) = \Pr(\,Q_j\,|\,M_2\,) = 1-c, \qquad \Pr(\,Q_j\,|\,M_1\,) = \Pr(\,Q_i\,|\,M_2\,) = c \tag{19.7}$$

Suppose there are n_Q alleles ($Q_1,\cdots,Q_{n_Q}$; with frequencies $q_1,\cdots,q_{n_Q}$) at the focal QTL among the population of dams. Following Chapter 4, decompose the mean value of Q_iQ_j as $\mu_{Q_iQ_j} = \mu + \alpha_i + \alpha_j + \delta_{ij}$, where the deviation from the population mean is determined by the average effects of both alleles ($\alpha_i + \alpha_j$) plus the dominance deviation (δ_{ij}). Assuming random mating, the dam contributes an allele at random, giving the expected value in offspring with sire marker M_1 as

$$\begin{aligned} E\left[\,\mu_{M_1}\,\middle|\,\text{sire} = M_1Q_i/M_2Q_j\,\right] &= (1-c)\sum_{k=1}^{n_Q} \mu_{Q_iQ_k}\,q_k + c\sum_{k=1}^{n_Q} \mu_{Q_jQ_k}\,q_k \\ &= \mu + (1-c)\,\alpha_i + c\,\alpha_j \end{aligned} \tag{19.8}$$

which follows from Equation 4.20b and noting that $\sum \alpha_k\,q_k = \sum \delta_{jk}\,q_k = 0$ (Chapter 4). The expected value for the marker contrast from this sire becomes

$$E\left[\,\mu_{M_1} - \mu_{M_2}\,\middle|\,\text{sire} = M_1Q_i/M_2Q_j\,\right] = (1-2c)\,(\alpha_i - \alpha_j) \tag{19.9}$$

Thus, assuming a single QTL linked to the marker, the expected marker contrast for a given sire is the difference of the average effects of his alleles at this QTL, scaled by the QTL–marker recombination frequency. We note a subtle, but important, point here. Recall from Chapter 4 that in order to define α_i, we need to specify a *reference population*. Here, that population is the one from which the dams were drawn (which specifies the q_i values). If one draws from a different population, the α_i values for this *same sire* might be different (as the q_i values for Equation 19.8 might change).

While a marker contrast may be significantly different from zero *within* a sibship, the expected value of this contrast *across* a random collection of sibships is zero for a population in linkage equilibrium (random linkage phase across parents; see Figure 5.5), as $E(\alpha_x) = 0$. Thus, one cannot simply combine contrasts across families, as these are expected to cancel. On the other hand, *squared* marker-contrasts have expected values that are a function of the additive variance of the linked QTL. To see this relationship, assume that each sire marker allele has sample size $n/2$ and that the within-marker (residual) variance is the same for both marker alleles. Under these assumptions, $\sigma^2(\overline{z}_{M_1}) = \sigma^2(\overline{z}_{M_2}) = \sigma_e^2/(n/2)$, implying that the variance of the observed marker contrast about its expected value within each family is

$$\sigma^2\,(\text{MC}) = E\,(\text{SE}^2) = \sigma^2(\overline{z}_{M_1}) + \sigma^2(\overline{z}_{M_2}) = \frac{4\sigma_e^2}{n} \tag{19.10a}$$

Using $\overline{z}_{M_i} = \mu_{M_i} + \overline{e}_i$, where $E[\,\overline{e}_i] = 0$ and $\sigma^2\,(\overline{e}_i) = \sigma^2(\overline{z}_{M_i})$, and combining these results with Equation 19.9 give the expected value of a squared marker-contrast as

$$\begin{aligned} E\,(\text{MC}^2) &= E\left[\,(\,\overline{z}_{M_1} - \overline{z}_{M_2}\,)^2\,\right] = E\left[\,(\,\mu_{M_1} - \mu_{M_2})^2\,\right] + \sigma^2(\overline{z}_{M_1}) + \sigma^2(\overline{z}_{M_2}) \\ &= (1-2c)^2\,E\left[\,(\alpha_i - \alpha_j)^2\,\right] + \frac{4\,\sigma_e^2}{n} \\ &= (1-2c)^2\sigma_A^2 + \frac{4\,\sigma_e^2}{n} \end{aligned} \tag{19.10b}$$

where the last step follows from $E[\,(\alpha_i - \alpha_j)^2\,] = 2E(\,\alpha_i^2\,) = \sigma_A^2$ (Equation 4.23b). Bovenhuis and Spelman (2000) show that selective genotyping (scoring contrasts using only the more extreme sibs) results in an overestimation of the marker contrast, and present corrections for this bias. Johnson et al. (1999) also addressed this issue.

Example 19.2. An interesting application of QTL mapping methods that is intermediate between true family designs and designs for inbred-line crosses is the **pseudo-testcross** strategy of Grattapaglia et al. (1995). Here, the offspring from a *single cross* were examined as *two separate half-sib families*, by considering each parent separately. Grattapaglia examined 122 F_1 progeny in a cross between two outbreeding tree species *Eucalyptus grandis* (female parent) × *E. urophylla* (male parent). Three growth characters were considered: micropropagation response (MPR), sprouting, and rooting response. Because both parents were outbred, inbred-line-cross methods are not appropriate. The authors instead examined marker-trait associations separately in the gametes from each parent by using RAPD markers that segregate in only one parent. Further, the authors considered pairs of flanking genotypes as the marker genotype, giving rise to interval mapping. Using this approach, a number of QTLs were detected in each parent:

QTLs heterozygous in female parent (*E. grandis*)

MPR	6 QTLs	r^2 range: 0.06 – 0.12	total r^2 = 0.42
Sprouting	4 QTLs	r^2 range: 0.07 – 0.11	total r^2 = 0.23
Rooting	1 QTL	r^2 = 0.09	total r^2 = 0.09

QTLs heterozygous in male parent (*E. urophylla*)

MPR	4 QTLs	r^2 range: 0.07 – 0.11	total r^2 = 0.25
Sprouting	2 QTLs	r^2 range: 0.09 – 0.09	total r^2 = 0.15
Rooting	3 QTLs	r^2 range: 0.07 – 0.21	total r^2 = 0.26

Because only those QTLs that were heterozygous in at least one parent could be detected, there are very likely to be other segregating QTLs in these species.

Several Half-sib Families

Equation 19.10b immediately suggests that one approach for combining information from N (unrelated) half-sib families is to use the sum of the squared t statistics (Equation 19.6) for each (informative) marker contrast,

$$T = \sum_{i=1}^{N} t_i^2 = \sum_{i=1}^{N} \frac{\text{MC}_i^2}{\text{SE}_i^2} \tag{19.11a}$$

where from Equations 19.10a and 19.10b,

$$E(T) \simeq N \frac{E(\text{MC}^2)}{E(\text{SE}^2)} = N\left[1 + (1-2c)^2 \frac{n}{4} \frac{\sigma_A^2}{\sigma_e^2}\right] \tag{19.11b}$$

(We stress that σ_A^2 is the additive genetic variance *due to the focal QTL*, not the trait as a whole.) A value of T significantly greater than N suggests linkage to a QTL segregating in at least one of the sires. If the sample size within each marker class is sufficiently large, so that the observed value of SE^2 can be taken to be very close to the population value of $\sigma^2(\text{MC})$, then the distribution of T is approximately χ^2 with N degrees of freedom.

Rearranging Equation 19.11b suggests the following method-of-moments estimate of $(1-2c)^2\sigma_A^2$,

$$\frac{4\,\text{Var}(e)}{n}\left(\frac{T}{N} - 1\right) \tag{19.12}$$

where $\text{Var}(e)$ is an estimate of σ_e^2 (the within-marker class variance). Aspects of this approach are further examined by Geldermann (1975), Weller et al. (1990), and van der Beek et al. (1995).

A closely related, and more widely used, approach is a nested ANOVA (Lowry and Schultz 1959; Neimann-Sørensen and Robertson 1961), which considers marker effects nested within each sibship. We only briefly consider the properties of nested ANOVA here, with additional comments in Appendix 5 and an extended discussion in Part II Chapter 23. The basic linear model is

$$z_{ijk} = \mu + s_i + m_{ij} + e_{ijk} \tag{19.13}$$

where z_{ijk} denotes the phenotype of the kth individual of marker genotype j from sibship i, s_i is the effect of sire i, m_{ij} is the effect of marker genotype j in sibship i (i.e., the marker effects associated with sire i), and e_{ijk} is the within-marker, within-sibship residual. It is assumed that s, m, and e have means equal to zero, are uncorrelated, and are normally distributed with variances σ_s^2 (the between-sire variance), σ_m^2 (the between-marker, within-sibship variance), and σ_e^2 (the residual, or within-marker within-sibship, variance). A significant marker variance indicates linkage to a segregating QTL, and is tested by using the statistic

$$F = \frac{\text{MS}_m}{\text{MS}_e} \tag{19.14a}$$

Under a balanced design with N sires, each with $n/2$ half-sibs in each marker class, the mean squares are given

$$\text{MS}_m = \frac{n}{2N}\sum_{i=1}^{N}\sum_{j=1}^{2}(\overline{z}_{ij} - \overline{z}_i)^2, \quad \text{MS}_e = \frac{1}{N(n-2)}\sum_{i=1}^{N}\sum_{j=1}^{2}\sum_{k=1}^{n/2}(z_{ijk} - \overline{z}_{ij})^2 \tag{19.14b}$$

Assuming normality, Equation 19.14a follows an *F* distribution (Appendix 5) under the null hypothesis that $\sigma_m^2 = 0$. Assuming a balanced design with N sires, each with $n/2$ half-sibs in each marker class, Equation 19.14a has N and $N(n-2)$ degrees of freedom. Chapter 23 discusses the appropriate degrees of freedom under unbalanced designs. General settings, such as very unbalanced designs, and related sires and/or dams, can be easily handled with the general pedigree methods discussed shortly.

Example 19.3. Three early studies attempted to map QTLs influencing milk production traits in dairy cows using nested ANOVA and allozyme markers (Chapter 8) to examine half-sibships. The pioneering study of Neimann-Sørensen and Robertson (1961) used six markers and six production characters in Red Danish cattle, with 123 sires and 12 to 13 daughters per sire. None of the 36 possible marker-trait associations were significant at the 1% level, while three were significant at the 5% level. Four production characters in Holsteins (14 markers) and Guernseys (15 markers) were examined by Gonyon et al. (1987) and Haenlein et al. (1987), respectively. In Holsteins, the average number of sires was 181, with an average of 5.3 daughters per sire. Of 56 possible marker-character combinations, six were significant at the 1% level, and another three were significant at the 5% level. The sample sizes were smaller in Guernseys, with an average of 88 sires and 5.4 daughters per sire. None of the resulting 64 marker-trait combinations were significant at the 1% level, while three were significant at the 5% level.

Taking into account adjustments for multiple comparisons (Chapter 18; Appendix 6), the Red Danish data did not show convincing evidence of marker effects, as (under the null) the probability of finding ≥ 3 of 36 tests significant at the 5% level is 0.10 (using the binomial, e.g., Example A6.5). The same is true for the Guernsey data, as the probability of ≥ 3 out of 64 tests being significant at the 5% level is 0.40. On the other hand, at least some of the Holstein associations are likely to be significant, even after accounting for multiple comparisons. Given the very low power of this design (especially given the very low marker density), these negative results are not surprising.

The connection between sums of squared marker-contrasts used in t tests (Equation 19.11) and nested ANOVAs can be seen by considering the marker mean squares, MS_m, which can be expressed as a function of the marker contrasts,

$$\text{MS}_m = \frac{1}{N}\sum_{i=1}^{N}\frac{n}{2}\left[\,(\overline{z}_{i1}-\overline{z}_i)^2+(\overline{z}_{i2}-\overline{z}_i)^2\,\right] = \frac{n}{4N}\sum_{i=1}^{N}\text{MC}_i^2 \tag{19.15a}$$

The first equality follows from Equation 19.14b, while the second follows by noting that the mean phenotype of progeny of the ith sire is $\overline{z}_i = (\overline{z}_{i1}+\overline{z}_{i2})/2$, where $\overline{z}_{ij}$ is the mean trait value for marker j in family i. For a balanced design, the mean squares given by Equation 19.14b have expected values of

$$E(\,\text{MS}_m\,) = \sigma_e^2 + (n/2)\sigma_m^2 \qquad \text{and} \qquad E(\,\text{MS}_e\,) = \sigma_e^2 \tag{19.15b}$$

Combining this result with Equations 19.10b and 19.15a gives,

$$\sigma_m^2 = \frac{E(\text{MS}_m) - E(\text{MS}_e)}{n/2} = (1-2c)^2\,\frac{\sigma_A^2}{2} \tag{19.15c}$$

Thus, an estimate of the QTL effect (measured by its additive variance, scaled by the distance between QTL and marker), can be obtained from the observed mean squares.

One immediate drawback of measuring a QTL effect by its variance in an outbred population is that even a completely linked QTL with a large effect can nonetheless have a small σ_m^2. Consider a strictly additive diallelic QTL with allele frequency p, where $\sigma_A^2 = 2a^2p(1-p)$. Even if a is large, the additive genetic variance can still be quite small if the QTL allele frequencies are near zero or one. An alternative way of visualizing this relationship is to note that the probability of a QTL-informative sire is $2p(1-p)$. If this is small, even if a is large, σ_A^2 will be small, as most families will not be informative (i.e., are not QTL heterozygotes). In those rare informative families, however, the between-marker effect is large. Contrast this to the situation with inbred-line crosses, where the QTL effect estimates $2a$, as *all families* are informative, rather than the fraction $2p(1-p)$ seen in an outbred population.

Table 19.2 Sample sizes required to have a 90% chance of detecting a linked QTL using ANOVA with significance level $\alpha = 0.05$. The QTL effect is measured by $\Delta = (1-2c)^2\sigma_A^2/\sigma_e^2$, where c is the marker–QTL recombination frequency. Here σ_e^2 is the residual variance, i.e., the variance within half-sib families not accounted for by the marker effects. N and n denote the number of sires and the number of offspring per sire, respectively, assuming $n/2$ offspring per marker class per sire. $N_T = N\,(2n+1)$ is the total number of genotyped individuals (one for each sire, dam, and offspring). All sires are assumed to be marker informative, and we assume that N is sufficiently large that at least one family is QTL-informative. The number $N(\gamma)$ of random sires required to have probability γ that at least one of them is QTL-informative (for a QTL with n_Q alleles with frequencies $q_1, \cdots, q_{n_Q}$) can be obtained by noting Pr(none of the N sires are informative) $= \left(\sum_{i=1}^{n_Q} q_i^2\right)^{N(\gamma)} = 1-\gamma$. Solving gives $N(\gamma) = \ln(1-\gamma)/\ln\left(\sum_{i=1}^{n_Q} q_i^2\right)$.

	$\Delta = 0.10$		$\Delta = 0.05$		$\Delta = 0.01$	
N	n	N_T	n	N_T	n	N_T
5	150	1,505	290	2,905	1450	14,505
20	40	1,620	110	4,420	540	21,620
100	22	4,500	40	8,100	200	40,100
200	14	5,800	28	11,400	132	53,000

A special modification of the half-sib design is available in fish species that can produce gynogenetic double-haploids (Chapter 17). Here the eggs from a female are fertilized with inactivated sperm. Special tricks are then used to double the chromosomes in the haploid eggs (Franěk et al. 2020), making each individual a double haploid (DH), a homozygote for a gamete from the female. The analysis of this design then proceeds as a typical single, or collection, of half-sibs, but the marker variance is now $\sigma_m^2 = (1-2c)^2\sigma_A^2$, namely double that for a half-sib design (Equation 19.15c). However, countering this larger marker variance is a two fold larger among-individual additive variance in background polygenic effect, inflating the residual variance.

Power of Nested ANOVA Designs

The power of nested ANOVA designs to detect QTLs in a collection of half-sib families has been examined by Soller and Genizi (1978), Weller et al. (1990), Luo (1993), and van der Beek et al. (1995). Details of the power calculations are given in Appendix 5, and some results are presented in Table 19.2. Inspection of this table shows that increasing the number of sibs per family (n) is more efficient than increasing the number of families (N), as the total number of individuals that must be genotyped (N_T) is smaller for small values of N. These power calculations assume that N is sufficiently large so as to include at least one QTL-informative family (Example 16.2). This is a major assumption. A further issue is that our goal is typically to try to sample as many of the segregating QTLs as possible, which necessitates increasing N to capture as many distinct QTL-heterozygous parents as possible.

One approach for increasing power is the **granddaughter design** (Weller et al. 1990), under which each sire produces a number of sons that are genotyped for sire marker alleles (Figure 19.2), and the trait values for each son are taken to be the mean value of the traits in offspring from that son (rather than the direct measures of the son itself). This design was developed for milk-production traits in dairy cows, where the offspring are granddaughters of the original sires. The linear model for this design is

$$z_{ijk\ell} = \mu + g_i + m_{ij} + s_{ijk} + e_{ijk\ell} \tag{19.16}$$

where g_i is the effect of sire i (the grandsire of the measured granddaughters), m_{ij} is the effect of marker allele j ($= 1, 2$) from the ith sire, s_{ijk} is the effect of son k carrying marker allele j from sire i, and $e_{ijk\ell}$ is the residual for the ℓth offspring of this son. Sire marker-allele effects are halved by considering granddaughters (as opposed to daughters), as there is only a 50% chance that the grandsire allele is passed from its son onto its granddaughter.

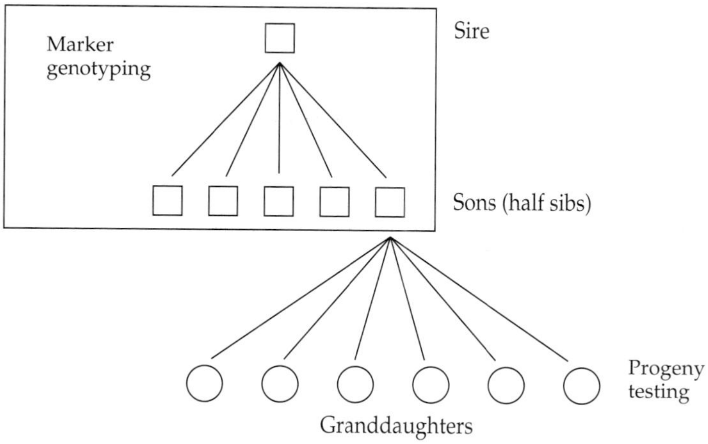

Figure 19.2 The **granddaughter design** of Weller et al. (1990). Here, each sire produces a number of half-sib sons that are scored for the marker genotypes. The character value for each son is determined by **progeny testing**, with the trait value being scored in a large number of daughters (again half sibs) from each son, with his trait value assigned as the mean of his daughters. More general designs allowing for both full- and half-sib families are examined by van der Beek et al. (1995).

However, this reduction in the expected marker contrast is usually more than countered by the smaller standard error associated with each contrast due to the larger number of offspring used to estimate trait values. Van der Beek et al. (1995) examined the power of a number of such three-generation designs, concluding that they generally offer more power than two-generation (parent–offspring) designs.

Finally, Mackinnon and Georges (1992) showed that another factor that can decrease power is selection within the population sample (e.g., using only the highest sires), which has the effect of reducing the between-marker constrasts. Even a small amount of selection acting on the trait of interest can have a large effect on power.

How Many Segregating QTLs?

Recall that any particular sire is unlikely to be a heterozygote for most of the QTLs that are segregating in the broader population (Equation 19.3). Hence, even when a number of sires are tested, this is only examining a fraction of the possible segregating QTLs. With a number of independent sires, one can test each location for a significant marker effect, with a positive result suggesting that the focal sire is indeed a QTL heteroyzygote in a region associated with the test. How can one use these tests for a more general estimate of the number of segregating QTLs in the sample? Chamberlain et al. (2007) suggested that methods for estimating the fraction of true positives in a collection of tests (Appendix 6) provided a useful approach for estimating the actual number of QTLs in the sample. Using a half-sib daughter design and six sire families in dairy cattle, they found that the average sire was heterozygote for around 5 QTL per trait, but also that usually only one sire in the sample was heterozygous for a given QTL.

Half-sib Analysis: Multipoint Regression

As mentioned, variance in marker informativeness (some parents are heterozygous at a focal marker, while others are not) poses a serious limitation for QTL mapping using outbred parents. Multipoint methods introduced in Chapter 18 (exploiting all of the marker information on a given linkage group) provides one solution to this concern (Knott et al. 1994, 1996; Spelman et al. 1996; Vilkki et al. 1997; de Koning et al. 1998, 1999, 2001a, 2001b; Kim et al. 2005). The basic idea is as follows. In a half-sib design, for the common parent (say the sire), one arbitrarily declares one of the chromosomes in a homologous pair as haplotype (gamete) one and the other as haplotype two. Marker information is then used to compute the probability that, at a given position x along this chromosome, the sire contribution is from haplotype one. Let $\pi_{ij}(x)$ denote this probability at chromosome position x for the jth offspring of sire i. At each position x along a chromosomal grid (say at 1 cM increments), one computes the regression of trait value z_{ij} on $\pi_{ij}(x)$,

$$z_{ij} = \mu + s_i + \beta_i(x)\,\pi_{ij}(x) + e_{ij} \tag{19.17}$$

This is the **multipoint regression** modification of Equation 19.13, with the (single) marker effect (m_{ij}) for sire i replaced by $\beta_i(x)$, the between-gamete, within-sire effect at position x. Note that additional fixed effects can easily be added, with μ being replaced by $\mu + \sum \beta_k X_{kij}$. The value of x yielding the regression with the largest r^2 value (where significant) is taken as the QTL location. From Equation 19.9, the expected value of $\beta_i(x)$ is $\alpha_{i,1} - \alpha_{i,2}$, the difference in the average effects of putative QTL alleles at position x in sire i. Note that the $(1-2c)$ from Equation 19.9 does not appear, as it assumed that a QTL lies at exactly position x ($c = 0$). If i is a QTL homozygote at this position, then its expected regression coefficient is zero. If i is a heterozygote, then even with two alleles, its sign can vary over sires (as which QTL allele is assigned to a particular sire haplotype is random).

Combining Line-cross and Half-sib Data

While we have, thus far, framed our discussion of QTL mapping with outbred individuals in terms of family-based analysis, in certain settings we can examine a mapping population using a line-cross analysis. Consider the situation where two lines (or breeds) have either been selected in opposite directions for a target trait or have diverged under domestication in different settings. The resulting lines/breeds would still be outbred (and thus segregating marker alleles), but could vary greatly in the frequency of alleles at a given QTL, in the extreme being differentially fixed. Hence, one could treat a mating between the two lines as a standard line-cross (Beckmann and Soller 1988; Haley et al. 1994b; Knott et al. 1997, 1998; Alfonso and Haley 1998; van Kaam et al. 1998; de Koning 1999, 2000, 2001a, 2001b; Sillanpää and Arjas 1999; Pérez-Enciso and Varona 2000; Maliepaard et al. 2001; Kim et al. 2005). When the two lines/breeds differ dramatically in the frequency of alleles at underlying QTL, this design offers greater power than a family design, with power diminishing as the allele frequency difference shrinks. In particular, the line-cross design is more powerful when the absolute QTL allele frequency difference exceeds 0.6, while the half-sib design is more powerful when this difference is less than 0.4 (Alfonso and Haley 1998; de Koning et al. 2001b; Kim et al. 2005). Note that there can be a bias under this design, in that selection results in larger effect QTLs generally having greater between-line frequency differences (WL Chapters 25 and 26), so that this is an enrichment procedure for larger effect QTLs when trait-selected lines are used.

There are some subtle distinctions between the types of populations used for a line-cross, versus half-sib, design. With any linkage-based design, the goal is to generate *informative parents* (individuals that are double heterozygotes for a QTL and nearby marker) in order to search for marker-trait associations in their offspring. With pure (inbred) lines, the original parents (P_1 and P_2) are fully homozygous and thus completely *uninformative*, and, as a result, no marker-trait associations are expected in their F_1 offspring. These offspring, however, are *highly informative* when used as parents for future generations, with marker–trait associations generated by linkage in their progeny (e.g., F_2, AIL, or other designs; Chapter 18). Conversely, when starting with *outbred* parents, a fraction of these are *already informative* without the need for additional crosses, generating marker-trait associations in their progeny when a parent is a marker–QTL double heterozygote. This is the feature exploited in half-sib designs. However, in order to perform a line-cross analysis, one needs to generate progeny that ***vary in their contribution from the two lines***. While the F_1 from crossing two parents (from different lines) can be informative about QTLs that are heterozygous in one (or both) parents, these offspring have no variation in the fractional contribution from the two lines. All have one allele from line 1 and one from line 2. Variation in line origin does not arise until the F_2, where (in the absence of segregation distortion) 25% have both alleles from line 1, 25% have both from line 2, and 50% have one line 1 and one line 2 allele. Hence, line-cross designs with outbred parents typically span three generations: grandparents are from the two different lines, their F_1 are the parents, and marker-trait associations are scored in their offspring. Conversely, a standard outbred family-based linkage design involves just two generations (parents and offspring).

Clever designs can exploit these two levels of variation (between- versus within-line) by combining information for both line-cross and family-based data. For example, following a cross of two breeds, F_1 individuals are crossed to form a series of half-sib families in the F_2 (e.g., de Koning et al. 2001a). One can then perform both: (i) a line-cross analysis using the entire collection of F_2, and (ii) half-sib analyses among selected subsets of the F_2. To develop these approaches, we first start with the simple F_2 line-cross design: two divergent lines (P_1 and P_2) are crossed with their subsequent F_1 crossed to each other to generate F_2s for mapping. Again, we can use a multipoint approach for QTL mapping (Equation 19.16), but now for position x we compute the probability of which *line* made the contribution. Let $\pi_{1,1,j}(x)$ denote the probability that F_2 individual j is homozygous for contributions from line 1 at position x, with $\pi_{2,2,j}(x)$ defined similarly for line 2. We can also allow for sex-specific contributional differences (Knott et al. 1998), by letting $\pi_{1,2,j}(x)$ denote the probability that j has heterozygous line origins at position x, with a male contribution from line 1 and a female from line 2, with $\pi_{2,1,j}(x)$ similarly defined. QTLs showing such **parent-of-origin effects** (**POE**) (e.g., **imprinting**)—where the expected trait value differs for otherwise identical heterozygotes depending on which sex contributed the allele from a given line—were detected in early pig crosses (de Koning et al. 2000, 2001b, 2002), and have subsequently been found in cattle and sheep (Imumorin et al. 2011). The resulting basic line-cross model becomes (Haley et al. 1994b)

$$z_j = \mu + a \cdot \pi_{Hom,j}(x) + d \cdot \pi_{Het,j}(x) + e_j \tag{19.18a}$$

where

$$\pi_{Hom,j}(x) = [\pi_{2,2,j}(x) - \pi_{1,1,j}(x)] \quad \text{and} \quad \pi_{Het,j}(x) = [\pi_{1,2,j}(x) + \pi_{2,1,j}(x)] \tag{19.18b}$$

As above, the x value generating the model with the largest (significant) r^2 indicates the likely position, with the corresponding values of a and d at this position estimating the QTL effect. Sources of bias in detecting imprinted QTLs have been discussed by de Koning et al. (2002) and Imumorin et al. (2011).

Equation 19.18a only codes a region as arising in a given line, and completely ignores any variation *within* that line. *All* parents from a line are treated equally and assumed to show no variation (in the QTL, while marker variation is allowed). Obviously, there is more information as we can also search for effects when offspring received alternative haplotypes from their parents (the standard half-sib design). A unified treatment combining line-cross and half-sib analysis, developed by Kim et al. (2005), proceeds as follows. Three separate models are fit to the data. The first is the line-cross (**LC**) model for the (multipoint) regression of the jth offspring from sire i,

$$z_{ij} = \mu + s_i + a \cdot \pi_{Hom,ij}(x) + d \cdot \pi_{Het,ij}(x) + e_{ij} \tag{19.18c}$$

which is just Equation 19.18a with a sire effect included and individuals now indexed by ij, the j offspring form sire i. Second, a standard half-sib (**HS**) model (Equation 19.17) is fit, and finally a combined (**CB**) model is fit

$$z_{ij} = \mu + s_i + \beta_i(x)\, \pi_{ij}(x) + a \cdot \pi_{Hom,ij}(x) + d \cdot \pi_{Het,ij}(x) + e_{ij} \tag{19.18d}$$

Based on model fit, Kim et al. defined three classes of QTLs (see their paper for the specific test details). **LC-QTL** are those detected under the LC model (a and/or d significantly different from zero), where the CB model does not give a significantly improved fit. **HS-QTL** are those detected under the HS model (one, or more, of the β_i in Equation 19.17

are significantly different from zero), but do not show a significant effect under LC, and the CB model is not significantly better than the HS model. Finally, **CB-QTL** are those detected under the CB model, with this model giving a significant better fit than either the LC or HS models. The interpretation of these different classes is that LC-QTLs are those that are close to differential fixation between breeds, HS-QTLs are those segregating at similar frequencies within each breed, while CB-QTL are segregating at different frequencies between breeds. Hence, these three classes provide some insight into the within, and between, variation for a set of breeds/lines. Kim et al. applied their approach to 39 growth and quality traits in a cross between Berkshire and Yorkshire pig breeds. A total of 160 QTLs were detected (at the chromosomal 5% level), of which 72 were LC-QTLs, 54 were HS-QTL, and 34 were CB-QTL. Of these, 14 QTL were detected only under the CB model, and were not significant under the LC or HS model.

A Single Full-sib Family

Large half-sib families are not feasible for many species, and one must instead consider full-sib families, which are often much smaller (as is the case in humans). Some domesticated animals such as pigs and poultry can have full-sib families of at least modest size, while some species (e.g., trees, fish, many invertebrates) can have huge full-sib families.

With a single full-sib family, one can again use a simple one-way ANOVA to detect QTLs (Equation 19.5). The only modification required is that the number of offspring marker genotypes ranges from two to four, depending on the marker genotypes of the parents (Table 19.1). With a backcross family ($M_iM_j \times M_kM_k$), the resulting ANOVA has two marker genotypes (M_iM_k and M_jM_k), corresponding to the alternate alleles from the informative parent (and hence the marker contrast $\overline{z}_{M_i} - \overline{z}_{M_j}$). With a fully informative family ($M_iM_j \times M_kM_\ell$), all genotypes are informative as to parental alleles, and the resulting ANOVA considers all four marker genotypes (giving rise to two marker contrasts, $\overline{z}_{M_i} - \overline{z}_{M_j}$ and $\overline{z}_{M_k} - \overline{z}_{M_\ell}$, one for each parent). Finally, while an intercross family ($M_iM_j \times M_iM_j$) has three offspring marker genotypes, heterozygotes provide no information on parental marker alleles and are usually excluded. The ANOVA in this case considers only M_iM_i and M_jM_j offspring (leading to the marker contrast $\overline{z}_{M_iM_i} - \overline{z}_{M_jM_j}$). Obviously, as one considers different markers, a family that is fully informative for one marker may be a backcross family for a second marker, noninformative for a third, and so on.

With a large half-sib family, the effect of a QTL allele from the common parent (typically the sire) is averaged over a large number of random QTL alleles from the other parents, and hence only its additive effect enters into the expected values of marker contrasts. Within a full-sib family, however, each QTL allele from an informative parent is averaged over at most two QTL alleles from the other parent, which results in dominance entering into the expected marker contrasts. To see this result, denote the genotypes of the parents by M_1Q_1/M_2Q_2 and M_3Q_3/M_4Q_4. This notation simply indicates that the marker allele M_i is associated with QTL allele Q_i, and does not indicate the actual state of the allele itself. We allow the possibility that two or more alleles are the same, so that (for example), $M_1 = M_4$ and $Q_1 = Q_2$.

Consider the gametes from the first parent. Among M_1-bearing gametes, $(1-c)$ contain Q_1 while the rest (c) contain Q_2. Likewise, of the M_3-bearing gametes from the second parent, $(1-c)$ contain Q_3 while the rest (c) contain Q_4. The resulting expected genotypic value for M_1M_3 offspring becomes

$$\mu_{M_1M_3} = (1-c)^2\,\mu_{Q_1Q_3} + c(1-c)\,\mu_{Q_1Q_4} + c(1-c)\,\mu_{Q_2Q_3} + c^2\,\mu_{Q_2Q_4} \tag{19.19a}$$

The other three marker means follow similarly. These values can be used to compute the average genotypic value of offspring (within this family) carrying a particular parental marker allele. For example, for M_1,

$$\mu_{M_1} = \frac{\mu_{M_1M_3} + \mu_{M_1M_4}}{2} = (1-c)\left(\frac{\mu_{Q_1Q_3} + \mu_{Q_1Q_4}}{2}\right) + c\left(\frac{\mu_{Q_2Q_3} + \mu_{Q_2Q_4}}{2}\right) \tag{19.19b}$$

Using these results, the expected value of the contrast between offspring bearing alternative marker alleles from the first parent becomes

$$\mu_{M_1} - \mu_{M_2} = (1-2c)\left[\frac{\mu_{Q_1Q_3} + \mu_{Q_1Q_4}}{2} - \frac{\mu_{Q_2Q_3} + \mu_{Q_2Q_4}}{2}\right]$$

$$= (1-2c)\left[(\alpha_1 - \alpha_2) + \left(\frac{\delta_{13} + \delta_{14}}{2} - \frac{\delta_{23} + \delta_{24}}{2}\right)\right] \tag{19.20}$$

where we have again decomposed the QTL genotype as $\mu_{Q_iQ_j} = \mu + \alpha_i + \alpha_j + \delta_{ij}$. As expected, if this parent is a QTL homozygote ($Q_1 = Q_2$), the contrast has expected value zero. Comparing this with the similar contrast in a half-sib design (Equation 19.9) shows that the two are identical in the absence of dominance $\delta_{ij} = 0$. When dominance is present, the contrast in a full-sib design can be larger or smaller than that which would occur in a half-sib design, depending on the dominance relationships between the QTL alleles of the focal individual and those of its mate.

For an intercross family (both parents are the same marker heterozygote, say, M_1M_2), the contrast of interest is between the marker homozygotes, and (noting that $M_1 = M_3$ and $M_2 = M_4$) Equation 19.19a gives

$$\mu_{M_1M_1} - \mu_{M_2M_2} = (1-2c)\,[\,\mu_{Q_1Q_3} - \mu_{Q_2Q_4}\,]$$

$$= (1-2c)\,[\,(\alpha_1 + \alpha_3) - (\alpha_2 + \alpha_4) + (\delta_{13} - \delta_{24})\,] \tag{19.21}$$

Note that even though the parents share marker alleles (so that $M_1 = M_3$ and $M_2 = M_4$), they need not share the corresponding QTL alleles (so that Q_1 and Q_3 are potentially different, as are Q_2 and Q_4).

Several Full-sib Families

As with half-sibs, the contrasts noted above are conditional on the parental genotypes, and have expected value zero when considered over a random collection of parents. Either scaled squared marker-contrasts or nested ANOVA can be used to combine results from a collection of (unrelated) full-sib families into an estimate of the genetic variance associated with the marker alleles (when sires and/or dams are related, pedigree methods, discussed below, can be used). The squared marker-contrasts used in both backcross and fully informative families have expected values (under a balanced design of $n/2$ sibs per marker class) of

$$E(\mathrm{MC}^2) = E[\,(\overline{z}_{M_1} - \overline{z}_{M_2})^2\,]$$

$$= (1-2c)^2\,\left[\,2\,E(\alpha^2) + 4\,(1/2)^2 E(\delta^2)\,\right] + \frac{4\,\sigma_e^2}{n}$$

$$= (1-2c)^2\,(\,\sigma_A^2 + \sigma_D^2\,) + \frac{4\,\sigma_e^2}{n} \tag{19.22}$$

As above, σ_A^2 and σ_D^2 correspond to the contribution from the linked QTL. In the absence of dominance ($\sigma_D^2 = 0$), this equation reduces to the value for half-sibs (Equation 19.10b).

Intercross families use a slightly different marker contrast,

$$\mathrm{MC}_I = \overline{z}_{M_1M_1} - \overline{z}_{M_2M_2}$$

Because the sample sizes for the two marker genotypes are now $n_{M_iM_i} = n/4$, the sampling variance of the marker contrasts is doubled (with $E(\mathrm{SE}_I^2) = 8\sigma_e^2/n$), giving the expected value of the squared marker-contrast as

$$E(\mathrm{MC}_I^2) = (1-2c)^2\,\left[\,4\,E(\alpha^2) + 2E(\delta^2)\,\right] + \frac{8\,\sigma_e^2}{n}$$

$$= 2\left[(1-2c)^2\,(\,\sigma_A^2 + \sigma_D^2\,) + \frac{4\,\sigma_e^2}{n}\right] \tag{19.23}$$

This is twice the expected value for the contrasts for backcross and fully informative families, exactly compensating for the doubling of sampling variance. Hence, for all informative contrasts we have

$$E(\, t^2 \,) \simeq \frac{E(\, \mathrm{MC}^2 \,)}{E(\, \mathrm{SE}^2 \,)} = 1 + \frac{n}{4}\,(1-2c)^2 \left(\frac{\sigma_A^2 + \sigma_D^2}{\sigma_e^2} \right) \tag{19.24}$$

As with Equation 19.11, a test statistic T is obtained as the sum of the t^2 values over all N informative contrasts, with T significantly greater than N being taken as evidence for a linked QTL.

As in the case of half-sib analysis, nested ANOVA is often used in place of sums of squared marker-contrasts. The ANOVA model for full-sib families becomes

$$z_{ijk} = \mu + f_i + m_{ij} + e_{ijk}$$

where f_i, the effect of the ith full-sib family, replaces the half-sib family (or sire) effect s_i in the half-sib design, and the number of marker effects (j) ranges from two to four, depending on the type of family. Again, a significant between-marker within-sibship variance (σ_m^2) indicates linkage to a segregating QTL, and this is tested using the statistic $F = \mathrm{MS}_m/\mathrm{MS}_e$. The use of nested ANOVAs for full-sib families was developed by Jayakar (1970) and A. Hill (1975), while Soller and Genizi (1978), Luo (1993), Knott (1994), van der Beek et al. (1995), and Muranty (1996) examined the power of these designs.

If one ignores the marker heterozygotes in intercross families and considers the marker contrasts for all informative parents, then using the same logic leading to Equation 19.15a shows that the contribution to MS_m is $(n/4)\mathrm{MC}^2$ for each informative contrast in a backcross or fully informative family and $(n/8)\mathrm{MC}_I^2$ for an intercross marker contrast. Recalling Equations 19.22 and 19.23, the expected value of MS_m becomes

$$E(\, \mathrm{MS}_m \,) = (n/2)(1-2c)^2 \left(\frac{\sigma_A^2}{2} + \frac{\sigma_D^2}{2} \right) + \sigma_e^2 \tag{19.25a}$$

so that the variance associated with marker alleles is

$$\sigma_m^2 = (1-2c)^2 \left(\frac{\sigma_A^2}{2} + \frac{\sigma_D^2}{2} \right) \tag{19.25b}$$

This can be estimated by using

$$\mathrm{Var}(m) = \frac{\mathrm{MS}_m - \mathrm{MS}_e}{n/2} \tag{19.25c}$$

Thus, when dominance ($\sigma_D^2 > 0$) is present, full-sib designs are expected to be more powerful for QTL detection than half-sib designs, as σ_m^2 is larger. However, given that σ_D^2 is often $\ll \sigma_A^2$ (Equation 4.12d), any increase in marker-variance from dominance variation is often negligible. Even in the absence of dominance, full-sib designs are generally more powerful, as power depends on the number of *informative marker contrasts* (one for each backcross and intercross family, two for each fully informative family). In the extreme where all N families in a design are fully marker informative, a full-sib design has $2N$ contrasts versus N for a half-sib design with the same number of families.

Adjusting for Multiple Comparisons

As discussed in detail in Chapter 18, a QTL mapping experiment involves multiple hypothesis testing, either at each marker, or more generally, over all positions on a given chromosome. Due to linkage, many of these tests are correlated, so that even with a small to modest number of markers on a given chromosome, Bonferroni corrections are too conservative, as the effective number of tests is less than the number of markers. Chapter 18 discussed corrections for this setting (Cheverud 2000; Li and Ji 2005). However, the most robust procedure is a permutation test (Churchill and Doerge 1994; Doerge and Churchill 1996), where (for line crosses) one keeps the vector of marker information intact for each individual, and then randomizes the trait values over these vectors to generate a draw from the null distribution that is then used to generate the linkage statistic of interest.

A similar permutation approach can be used with family data, ***but with an important caveat***. With inbred lines, the permutation is over the *entire collection of individuals*, as there is no family structure in the population. With family data, the permutation must be *within families*, as the linkage test is whether there is a family-specific parental marker effect. Hence, for half-sibs, the marker data are kept intact, but one randomizes the trait values over a half-sib family, with a separate randomization occurring for each family. Because half-sib families are typically large, randomization is a fairly robust procedure. With full-sib families, the same idea (randomization within each family) still applies, but the concern is that full-sib families are often small, resulting in a very limited number of unique combinations per family, which can compromise permutation tests. More generally, for analysis over a pedigree, standard permutation approach are not appropriate, due to the hierarchical structure of the data (interconnected sets of correlated observations). Lin and Zou (2004) have proposed a Monte Carlo approach for such cases.

Finally, de Koning et al. (1999, 2001b) suggested a modification of the standard Bonferroni adjustment to connect chromosome-wide (π_{chrom}) and genome-wide (π_{genome}) corrections, allowing permutation tests to be run on individual chromosomes and then later adjusted to a genome-wide rate. If the focal chromosome accounts for a fraction γ of the total genome size, then

$$\pi_{genome} = 1 - (1 - \pi_{chrom})^{1/\gamma} \tag{19.26a}$$

and

$$\pi_{chrom} = 1 - (1 - \pi_{genome})^{\gamma} \simeq \gamma\pi_{genome} \tag{19.26b}$$

For example, suppose 15 chromosomes of roughly equal size ($\gamma = 1/15$). From Equation 19.26a, a chromosome-wide significance of 0.01 corresponds to a genome-wide value of 0.14. In order for a genome control of 0.01, Equation 19.26b gives that a chromosome-wide significance threshold of $\pi_{chrom} = 0.00067$ is required.

SIB ANALYSIS: MAXIMUM LIKELIHOOD

ML estimation of QTL position and effects using a set of relatives from an outbred population proceeds as with inbred-line crosses (Chapter 18). One constructs an appropriate likelihood function for the assumed model, obtains the parameter values that maximize this function (given the marker and trait data), and tests alternative hypotheses with likelihood ratios (Appendix 4). One subtle difference for outbred parents is that likelihood maps displaying the amount of support for a putative QTL often show **discontinuities** between adjacent markers (Knott and Haley 1992b); see Figure 19.3. Such discontinuities arise when adjacent loci differ in the fraction of informative individuals in the sample (due to between-locus variance in the marker-allele frequencies, a situation that does not arise with inbred-line crosses). As Figure 19.3 shows, these discontinuities can be avoided by constructing likelihood functions that use all of the marker information on a chromosome (Georges et al. 1995; Kruglyak and Lander 1995b; Kruglyak et al. 1996).

The major complication in applying ML methods to outbred populations is that the resulting likelihood functions are much more complex than for inbred-line crosses due to the need to consider all possible genotypes of measured pedigree members. For example, for an inbred-line cross, all F_1 parents have the same genotype, say MQ/mq. In contrast, in an outbred population, for a diallelic QTL there are 16 possible genotypes for a pair of parents with marker genotype M/m, as each parent could be QM/Qm, QM/qm, qM/Qm, or qM/qm. The situation is even more complex when more than two alleles are segregating at

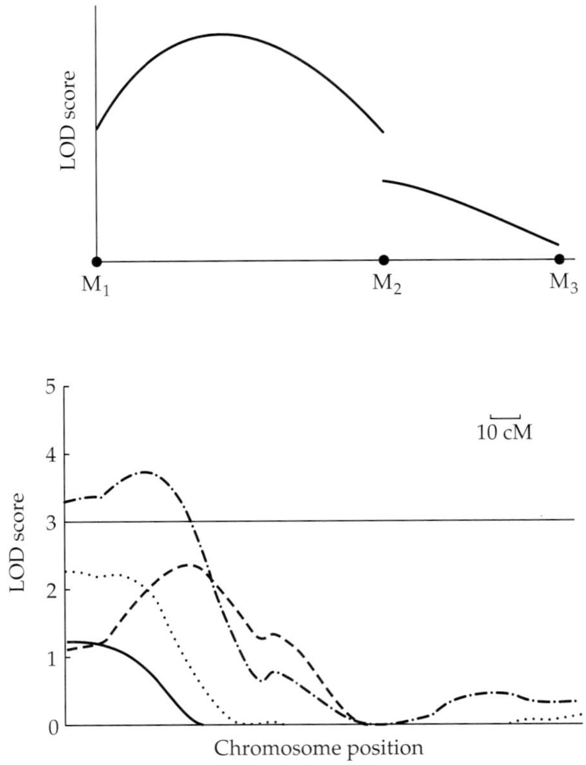

Figure 19.3 **Top:** When adjacent marker loci differ in the amount of heterozygosity, interval mapping in an outbred population gives a likelihood map with discontinuities at some of the marker loci (here, at the second marker, M_2). **Bottom:** Likelihood maps that simultaneously use all markers on a chromosome (instead of two at a time, as with interval mapping) have no such discontinuities. This likelihood map considers bovine chromosome U5(10) and examines four milk production traits in the granddaughters of a single sire. Only fat yield (denoted by – · – ·) showed a significant LOD score (exceeding the critical value of three), indicating a putative QTL on this chromosome. (After Georges et al. 1995.)

the QTL or when one considers two (or more) marker loci simultaneously. For a pedigree with even a modest number of parents, the number of possible genotypes that must be considered is challenging.

LeRoy and Elsen (1995) showed for a single marker that likelihood methods can be considerably more powerful for detecting a QTL in a sib design than LS approaches under loose linkage. However, with multiple markers, Knott et al. (1996) and Baret et al. (1998) found little difference between LS and ML approaches, provided markers were closer than 30 cM. Both studies found that these two approaches have the same power when the marker and QTL are completely linked. One likely explanation for the apparent discrepancy under loose linkage is that while Le Roy and Elsen assumed that only a single QTL was segregating, they did not incorporate a factor into their likelihood function to account for segregation at unlinked sites. Hence, as the marker–QTL recombination frequency approaches 1/2, their likelihood function approaches that for classical segregation analysis (i.e., no marker information; see Chapter 16). Thus, as the marker–QTL linkage becomes looser, the power of the LS linear model decreases to the significance level of the test, while the ML method approaches the power of classical segregation analysis.

Constructing Likelihood Functions

Likelihood functions for detecting QTLs using marker information in sets of sibs follow by suitable modifications of those developed for complex segregation analysis (Chapter 16). Recall that to construct the latter, the likelihood for an individual with trait value z_o is computed by first obtaining the likelihood conditional on the QTL genotypes of its parents (Q_s for the sire and Q_d for the dam),

$$\ell(\, z_o \,|\, Q_s, Q_d) = \sum_{Q_o} \varphi(z_o, \mu_{Q_o}, \sigma^2\,)\, p(Q_o \,|\, Q_s, Q_d)$$

where we make the standard assumption that, conditional on QTL genotype, phenotypes are normally distributed with common variance σ^2. This conditioning is removed by averaging over all possible parental values,

$$\ell(\, z_o\,) = \sum_{Q_s}\sum_{Q_d} \ell(\, z_o \,|\, Q_s, Q_d)\, p(Q_s)\, p(Q_d)$$

When marker data are present, the likelihood for an individual with marker genotype M is obtained by averaging over the probability of each QTL genotype given the marker genotypes,

$$\ell(\, z_o \,|\, M_o, M_s, M_d, Q_s, Q_d) =$$

$$\sum_{Q_o} \varphi(z_o, \mu_{Q_o}, \sigma^2\,)\, p(Q_o \,|\, M_o, M_s, M_d, Q_s, Q_d) \tag{19.27a}$$

where the transmission probabilities $p(Q_o \,|\, M_o, M_s, M_d, Q_s, Q_d)$ are functions of the marker genotypes of the offspring (M_o) and its parents (M_s, M_d). Averaging over all possible parental genotypes gives

$$\ell(\, z_o \,|\, M_o, M_s, M_d) =$$

$$\sum_{Q_s}\sum_{Q_d} \ell(\, z_o \,|\, M_o, M_s, M_d, Q_s, Q_d)\, p(Q_s|\, M_s)\, p(Q_d|\, M_d) \tag{19.27b}$$

In a similar fashion, the likelihood function for any pedigree can (in principle) be computed by suitably conditioning on the genotypes of parents and other relatives.

Likelihood functions for half- and full-sib families have been developed for single markers (Risch 1984; Demenais et al. 1988; Weller 1990; Haley 1991; Amos 1994; Le Roy and Elsen 1995; Mackinnon and Weller 1995; Elsen et al. 1997), interval mapping (Knott and Haley 1992b), and multipoint mapping (Georges et al. 1995; Uimari et al. 1996; Knott et al. 1996). A framework for developing likelihood functions that allows for multiple QTLs is given by Stricker et al. (1995b). Hoeschele and VanRaden (1993a, 1993b) and Uimari et al. (1996) discuss Bayesian modifications of the likelihood functions to allow for the incorporation of prior information (e.g., Smith 1953, 1959).

Example 19.4. Consider Knott and Haley's (1992b) development of a likelihood function for interval mapping in a full-sib family with a segregating diallelic QTL (with alleles *Q* and *q*, with *p* denoting the frequency of *Q*) flanked by two marker loci, *A* and *B*. The transmission probabilities required by Equation 19.27a follow from arguments identical to those used in Chapter 18 and depend on QTL–marker distances and assumptions about interference. For example, if both parents have genotype AQB/aqb, where c_1 is the *A–Q* distance and c_2 the *Q–B* distance, then (assuming no interference) from Example 18.1, for offspring *o* from parents *s* and *d* (for sire and dam, respectively),

$$\Pr\left(o = AQB/AqB \,|\, s, d = AQB/aqb\right) = 2\left[\frac{(1-c_1)(1-c_2)}{2}\right]\left(\frac{c_1\, c_2}{2}\right)$$

The recombination frequency c between these markers is assumed known, in which case c_2 follows as a function of c_1 and c (Chapter 18). For a given pair of parental genotypes, many of these parent–offspring conditional probabilities are zero, as usually only a subset of all possible offspring genotypes can occur for a given pair of parental genotypes (e.g., an offspring cannot be qq if a parent is QQ). The likelihood for all n offspring in a family, conditional on parental genotypes, is

$$\ell(\,\mathbf{z}\,|\,M_\mathbf{o}, M_s, M_d, Q_s, Q_d) = \prod_{j=1}^{n} \ell(\,z_{o_j}\,|\,M_{o_j}, M_s, M_d, Q_s, Q_d) \tag{19.28a}$$

where $M_\mathbf{o}$ denotes the vector of the n offspring marker genotypes. As before, the conditioning on the sire and dam QTL genotype (including phase) is removed by averaging over all possible genotypes, giving the unconditional likelihood for the observed markers as

$$\ell(\,\mathbf{z}\,|\,M_\mathbf{o}, M_s, M_d) =$$

$$\sum_{Q_s}\sum_{Q_d} \ell(\,\mathbf{z}\,|\,M_\mathbf{o}, M_s, M_d, Q_s, Q_d) \cdot \Pr(Q_s\,|\,M_s) \cdot \Pr(Q_d\,|\,M_d) \tag{19.28b}$$

where the sum for each parent is over all QTL genotypes and all QTL-marker phases. In $\Pr(Q_x\,|\,M_x)$, Q_x denotes both the possible QTL genotype and phase with respect to the observed marker M_x. For an individual of known marker genotype, say $A_iA_jB_kB_\ell$, there are two possible marker phases ($A_iB_k\,/\,A_jB_\ell$ and $A_jB_k\,/\,A_iB_\ell$), each of which has four possible QTL genotypes (Q/Q, Q/q, q/Q, q/q), for a total of eight possible marker–QTL phases for each parent. Assuming Hardy–Weinberg proportions hold and that the base population is in linkage equilibrium (so that both marker–QTL phases are equally frequent), the eight possible marker-QTL genotypes for an $A_iB_k\,/\,A_jB_\ell$ individual have expected frequencies as follows:

Marker–QTL Genotype		$\Pr(Q_x\,|\,A_iA_jB_kB_\ell)$
$A_iQ\,B_k\,/\,A_jQ\,B_\ell$,	$A_jQ\,B_k\,/\,A_iQ\,B_\ell$	$p^2/2$
$A_iQ\,B_k\,/\,A_jq\,B_\ell$,	$A_jQ\,B_k\,/\,A_iq\,B_\ell$	$p(1-p)/2$
$A_iq\,B_k\,/\,A_jQ\,B_\ell$,	$A_jq\,B_k\,/\,A_iQ\,B_\ell$	$p(1-p)/2$
$A_iq\,B_k\,/\,A_jq\,B_\ell$,	$A_jq\,B_k\,/\,A_iq\,B_\ell$	$(1-p)^2/2$

where p is the frequency of Q. The likelihood can be suitably modified to account for common-family effects (Knott and Haley 1992b) and background (unlinked) polygenes by using arguments identical to those leading to Equations 16.17 and 16.22. The full likelihood over N unrelated families is just the product of the individual family likelihoods. Equation 19.28b has six parameters to estimate (three QTL means, σ^2, c_1, and p), giving a likelihood-ratio (LR) test with four degrees of freedom when compared to the two degrees of freedom null likelihood (μ, σ^2) of no segregating QTL (Equation 16.8).

The properties of ML estimates using relatives from outbred populations have not been as extensively examined as have those for line-cross populations. Mackinnon and Weller (1995) examined the accuracy of single-marker ML estimation under the half-sib design, finding that while ML estimates of QTL *effects* tend to be reasonably accurate, the same is not the case for estimates of QTL *position*, which have very flat likelihood surfaces. They also observed that estimates of QTL parameters tend to be correlated, so that a poor estimate in one parameter (e.g., map position) can have a detrimental effect on other estimates (e.g., QTL effects). Interval mapping is expected to have a more peaked likelihood surface for QTL position (and a resulting smaller variance), so these concerns may be less important for analyses using multiple markers.

While marker-based likelihood methods have greater power than segregation analysis (Chapter 16) for detecting QTLs (Goldin et al. 1984; Demenais et al. 1988; Knott and Haley 1992b; Le Roy and Elsen 1995), their power is still poor compared to inbred-line-cross designs. Knott and Haley (1992b) showed that the use of flanking markers increases power, as does incorporation of a within-family effect into the likelihood function. Choosing highly polymorphic marker loci further increases power by increasing the frequency of informative sibships. Even with these modifications, however, power remains marginal. With markers at 20 map unit intervals, a QTL whose segregation accounts for 1/3 of the total variance (implying a QTL of massive effect) is barely detectable when 250 full-sib families of four individuals each are used. Further, as with segregation analysis, there is a tradeoff between increasing the number of families versus increasing family size (Chapter 16). For a fixed total number of progeny, one must sample both a sufficient number of families (to ensure that at least one of them is informative) and also a sufficient number of individuals within each family (to have power within an informative family).

MAXIMUM LIKELIHOOD OVER GENERAL PEDIGREES: VARIANCE COMPONENTS

The above likelihood functions explicitly model the transmission of QTL genotypes from parent to offspring. While this method can be extended to multigenerational pedigrees, the number of possible combinations of genotypes for individuals increases exponentially with the number of pedigree members, and solving the resulting likelihood functions becomes increasingly challenging. Further, construction of the likelihood function requires specifying the number of functionally distinct alleles at a focal QTL segregating within the pedigree. This approach is a **fixed-effects** framework, returning separate estimates for each QTL mean and allele frequency, which requires a degree of freedom for each parameter. While such a level of granularity may seem appealing, it is extremely difficult to obtain with any precision. This complexity does not arise with an inbred line cross. Here, only two QTL alleles are segregating, each with frequency 1/2 (in the absence of segregation distortion), restricting the likelihood space to a very manageable dimension.

An alternative approach is to use a **random-effects** framework (Chapters 10; Part II Chapters 32 and 33), constructing likelihood functions using the **variance components** associated with a QTL (or linked group of QTLs) in a genetic region of interest. Here ML estimates a single composite QTL parameter (σ_A^2) instead of attempting to explicitly model all of the underlying genetic details ($\mu_{Q_1Q_1}, \mu_{Q_1Q_2}, \cdots, \mu_{Q_kQ_k}, p_1, \cdots, p_k$). This approach allows for analysis of very general and complex pedigrees, using only a degree of freedom for each estimated variance component. The basic idea is to use marker information to compute the fraction of a genetic region of interest that is *identical by descent* (IBD) between two individuals. Recall that two alleles are IBD if we can trace them back to a single copy in a common ancestor in a specified reference population (Chapter 7), and marker information can be used to estimate the IBD status (Chapter 8).

The random-model approach for QTL mapping was independently developed in animal breeding (Fernando and Grossman 1989) and human genetics (Goldgar 1990). We sketch the basic model of Goldgar first (which is often referred to as the **multipoint IBD method, MIM**), before briefly placing it in a more formal mixed-model framework (Example 19.5). An excellent review of many of the issues arising in variance-component linkage mapping is given by Blangero et al. (2001). Likewise, Part II Chapter 33 gives a more detailed discussion of the estimation of variance components under the likelihood framework.

Consider the simplest case, in which the genetic variance is additive for the QTLs in the region of interest, as well as for background QTLs unlinked to this region. Under Goldgar's model, an individual's phenotypic value is decomposed as

$$z_i = \mu + A_i + A_i^* + e_i \tag{19.29}$$

where μ is the population mean, A is the contribution from the chromosomal interval being examined, A^* is the contribution from QTLs outside this interval, and e is the residual. The random effects A, A^*, and e are assumed to be normally distributed with mean zero and variances σ_A^2, $\sigma_{A^*}^2$, and σ_e^2. Here σ_A^2 and $\sigma_{A^*}^2$ correspond to the additive variances associated with the chromosomal region of interest and background QTLs in the remaining genome, respectively. We assume that none of these background QTLs are linked to the chromosome region of interest so that A and A^* are uncorrelated, and we further assume that the residual e is uncorrelated with A and A^*. Under these assumptions, the phenotypic variance is $\sigma^2_A + \sigma^2_{A^*} + \sigma^2_e$, where the total additive variance, $\sigma^2_A + \sigma^2_{A^*}$, remains constant, but the relative contributions from the two components can change over chromosomal regions.

Assuming no shared environmental effects, then from results in Chapter 7 the phenotypic covariance between two individuals can be decomposed into a contribution from the focal region and a contribution from the background,

$$\sigma(z_i, z_j) = R_{ij}\, \sigma^2_A + 2\Theta_{ij}\, \sigma^2_{A^*} \tag{19.30a}$$

where R_{ij} is the fraction of the chromosomal region shared IBD between individuals i and j, and $2\Theta_{ij}$ is twice Wright's coefficient of coancestry (e.g., $2\Theta_{ij} = 1/2$ for full sibs; Chapter 7). Note that R_{ij} is a random variable whose expected value is $2\Theta_{ij}$. More generally, we could replace the background IBD Θ_{ij} with R^*_{ij}, the marker-based realized IBD relationship in the rest of the genome (Chapter 8), which should be a little bit more powerful.

Goldgar (1990) and Guo (1994a, 1994b) discuss the estimation of R_{ij} from flanking-marker information. For the special case of a nuclear family, Kruglyak and Lander (1995b), building on results of Lander and Green (1987), discussed the estimation of R_{ij} at any point along a chromosome using all marker information (**multipoint mapping**). This has been extended to more general pedigrees by Kruglyak et al. (1996), Sobel and Lange (1996), Almasy and Blangero (1998), Visscher et al. (1999), and Sobel et al. (2001). With a dense marker map, one can estimate the value of R_{ij} for a segment with high precision (Chapter 8).

For a vector $\mathbf{z}$ of observations on n individuals, the associated covariance matrix $\mathbf{V}$ can be expressed as contributions from the region of interest, from the background QTLs, and from the residual effects,

$$\mathbf{V} = \mathbf{R}\,\sigma^2_A + \mathbf{A}\,\sigma^2_{A^*} + \mathbf{I}\,\sigma^2_e \tag{19.30b}$$

where $\mathbf{I}$ is the $n \times n$ identity matrix, and $\mathbf{R}$ and $\mathbf{A}$ are matrices of known constants,

$$\mathbf{R}_{ij} = \begin{cases} 1 & \text{for } i = j \\ \widehat{R}_{ij} & \text{for } i \neq j \end{cases}, \qquad \mathbf{A}_{ij} = \begin{cases} 1 & \text{for } i = j \\ 2\Theta_{ij} & \text{for } i \neq j \end{cases} \tag{19.30c}$$

The elements of $\mathbf{R}$ contain the estimates of IBD status for the region of interest based on marker information, while the elements of $\mathbf{A}$ are given by the pedigree structure, or can be estimated using dense marker information (excluding markers on the chromosome containing the region being tested; Chapter 8).

The resulting likelihood is multivariate normal with mean vector $\boldsymbol{\mu}$ (all of whose elements are μ) and variance–covariance matrix $\mathbf{V}$ (Equation 19.30b). From Equation 9.24,

$$\ell(\,\mathbf{z}\,|\,\mu, \sigma^2_A, \sigma^2_{A^*}, \sigma^2_e\,) = \frac{1}{\sqrt{(2\pi)^n\,|\mathbf{V}|}}\,\exp\left[-\frac{1}{2}(\mathbf{z}-\boldsymbol{\mu})^T\,\mathbf{V}^{-1}\,(\mathbf{z}-\boldsymbol{\mu})\right] \tag{19.31}$$

This likelihood has four unknown parameters (μ, σ^2_A, $\sigma^2_{A^*}$, and σ^2_e). A significant σ^2_A indicates the presence of at least one QTL in the region being considered, while a significant $\sigma^2_{A^*}$ implies background genetic variance contributed from QTLs outside the focal interval. Both of these hypotheses can be tested by LR tests (Appendix 4). The basic approach for a genomic scan using this approach is as follows. One starts with an interval at one end of a chromosome, uses marker data to compute R_{ij} for this interval, and then tests whether σ^2_A is significantly different from zero. One then moves to the adjacent interval, using marker information for this new region to recalculate the elements on $\mathbf{R}$ and estimate its associated σ^2_A, and so on. Note that the elements of $\mathbf{A}$ do not change as one moves over the adjacent intervals (unless $\mathbf{A}$ is estimated from marker data, in which case it may change slightly between chromosomes, due to the exclusion of the focal chromosome markers from its estimation). This correction (via **A**) for shared ancestry (here close relatives) is an important tool in association mapping that is more fully developed in Chapter 20 (also see Example 10.12). While Equation 19.30a handles an arbitrary pedigree, Goldgar's initial focus, and much of the subsequent expansion of this approach, was on unrelated sets of sibs.

Two technical, but important, issues are the choice of ML versus REML variance estimates and the large-sample distribution for the LR test that $\sigma_A^2 = 0$. These issues are more fully developed in Appendix 4 (as well as in Part II Chapter 33), also see Beaty et al. (1985). Briefly, REML is essentially variance estimation once the impact of fixed effects have been removed and, for a variety of reasons, it is generally preferred over standard ML variance estimates (Chapter 33). As for testing, typically an LR test of one parameter would follow a chi-square with one degree of freedom (for large samples). However, this approximation breaks down when the null is on the boundary of the parameter space (Appendix 4). This is the case with testing that a variance is zero. The resulting large-sample distribution of LR test is a mixture of a χ_0^2 (a probability mass at zero) and a χ_1^2, namely $LR \sim (1/2)\chi_0^2 + (1/2)\chi_1^2$ (Chernoff 1954; Self and Liang 1987; Stram and Lee 1994; Baret et al. 1998).

A final issue is that the distribution of residuals in Equation 19.29 may depart from normality, which can inflate the Type I error rate (Allison et al. 1999b; Feingold 2001). Such departures can result if a major locus is segregating outside of the focal region, generating a mixture model for background genetics effects (Chapter 16). Williams and Blangero (1999) noted that the assumption of normality was fairly robust in the absence of extreme outliers and with moderate kurtosis ($k_4 < 2$; Equation 2.12a). Wang et al. (1998) suggested trimming extreme outliers, while other corrections and more robust tests have been offered by Blangero et al. (2000, 2001), Tang and Siegmund (2001), Putter et al. (2002), Wang and Huang (2002), Chen et al. (2004), and Shete et al. (2004).

One important cause of departures from normality is **selective sampling** (i.e., selective genotyping; Chapter 18), when only extreme individuals within a pedigree are genotyped. This results in severely biased estimates of the variance components (Martinez et al. 1998). This bias can be reduced by explicitly working with truncated normal distributions in the likelihood functions (e.g., Sham et al. 2000b), but this can be computationally demanding in a complex pedigree. As we discuss shortly, the classic Haseman–Elston (HE) regression is more robust to departures from normality, but generally less powerful and not as flexible as variance-component methods. However, more recent extension of the HE approach have been shown to have power very close to ML variance components.

The variance-component approach has been extended to distributions other than the normal. Wan et al. (1998) considered traits that follow a log-normal distribution (Chapter 14; Figure 14.1), while binary (dichotomous) traits have been examined by Duggirala et al. (1997), Williams et al. (1999), Yi and Xu (1999), and Williams and Blangero (2004) using a threshold liability model (Chapter 14).

Example 19.5. Equation 19.29 is a strictly random-effects model (the only fixed effect being the common mean). This basic approach can be greatly extended by placing it into a mixed-models framework (Chapter 10), e.g., Fernando and Grossman (1989), Schork (1993), Guo (1994a), van Arendonk et al. (1994), Xu and Atchley (1995), Grignola et al. (1996a, 1996b, 1997), Blangero and Almasy (1997), Hoeschele et al. (1997), and Almasy and Blangero (1998). In particular, suppose that the trait value z_i for observation i is adjusted for m cofactors (trait-influencing factors such as age, sex, or known environmental effects). Incorporation of such additional effects can increase power by reducing the residual variance, σ_e^2, provided that the extra precision overcomes the cost from losing degrees of freedom. The model then becomes

$$z_i = \mu + \sum_{k=1}^{m} \beta_k x_{ik} + A_i + A_i^* + e_i \tag{19.32a}$$

where β_k is the estimated adjustment for cofactor k and x_{ik} is the weight on cofactor k in observation i (see Chapter 10 for examples). As in Goldgar's model, the region-specific effect

(A_i), background polygenic effect (A_i^*), and residual (e_i) are all random, and hence have an associated covariance structure. If desired, dominance can also be incorporated in this framework (e.g., Amos 1994; Gou 1994a; Almasy and Blangero 1998; Williams and Blangero 1999). Zeegers et al. (2004) made the important point that covariate adjustment increases power if the effects of the covariate are independent of the focal QTL, but their inclusion can *decrease* power when the focal QTL interacts with the covariates.

Using the machinery introduced in Chapter 10 (and fully developed in Part II Chapters 32 and 33), we can express Equation 19.32a in mixed-model matrix form as

$$\mathbf{z} = \mathbf{X}\boldsymbol{\beta} + \mathbf{Za} + \mathbf{Zu} + \mathbf{e} \tag{19.32b}$$

where **a** is the vector of region-specific effects and **u** the vector of background polygenic effects. As shown in Chapter 10, the incidence matrix **Z** allows one to account for replicated and missing observations (**Z** = **I** when all observations are singletons, and is nonsquare when observations are missing). The associated covariance structure for the random effects becomes

$$\begin{pmatrix} \mathbf{a} \\ \mathbf{u} \\ \mathbf{e} \end{pmatrix} \sim \begin{pmatrix} 0 \\ 0 \\ 0 \end{pmatrix}, \begin{pmatrix} \sigma_A^2 \mathbf{R} & 0 & 0 \\ 0 & \sigma_{A^*}^2 \mathbf{A} & 0 \\ 0 & 0 & \sigma_e^2 \mathbf{I} \end{pmatrix} \tag{19.32c}$$

If the residuals have some additional structure (e.g., common-family or other shared environmental effects), one can replace $\sigma_e^2 \cdot \mathbf{I}$ with a more general matrix (Chapters 10 and 32). The reader might wonder why we can separately estimate **a** and **u** when they appear to be fully confounded in Equation 19.32a (the model is **Za** + **Zu** = **Z**[**a** + **u**]). The reason (as with a repeatability model; Chapter 32) is that **a** and **u** have *distinct covariance structures*. Note that BLUP machinery (Chapter 10 and 32) can be used to estimate the vector **a**, so that the additive genetic value associated with the focal region *in each individual* can be estimated, offering a much more granular analysis than simply showing that σ_A^2 for the region is significant.

Estimating QTL Position

Goldgar's original approach made no assumptions about how many QTLs were contained in a tested chromosomal segment. As such, it simply provided a test for detecting the presence of one or more QTLs in a given region, along with providing a composite estimate of their effects as measured by the genetic variance attributable to that region. However, if we are willing to assume that only a single QTL resides in the region, Goldgar's method can be modified to estimate QTL position (Amos 1994; Xu and Atchley 1995). The idea is that QTL position enters through estimates of R_{ij}. For example, Amos (1994) showed that for a single marker linked to a QTL at recombination frequency c,

$$\widehat{R}_{ij} = \begin{cases} 1/2 + (1-2c)^2\,(\pi_{ij} - 1/2) & \text{full sibs} \\ 1/4 + (1-2c)^2\,(\pi_{ij} - 1/4) & \text{half sibs} \end{cases} \tag{19.33}$$

where $\widehat{R}_{ij}$ is the estimate of the fraction of QTL alleles IBD between individuals i and j, and π_{ij} is the proportion of IBD marker alleles (the estimation of which is discussed below as well as in Chapter 8). Values for other relatives are given by Amos, allowing most common pedigree structures to be easily incorporated. One then proceeds by treating c as a further parameter to estimate, displaying support for a QTL at any particular position by a standard likelihood profile map. Schork (1993) and Xu and Atchley (1995) considered multiple markers and QTLs, while Kruglyak and Lander (1995b) presented a complete multipoint approach for nuclear families.

Power

Starting with Williams and Blangero (1999), a number of authors have examined the power of the variance-component linkage mapping (e.g., Sham et al. 2000a; Rijsdijk et al. 2001;

Visscher and Hopper 2001; Chen et al. 2004, 2005; Yu et al. 2004; Chen and Abecasis 2006). The underlying machinery used is that the (large-sample) distribution of the likelihood-ratio statistic under the alternative hypothesis follows a **noncentral chi-square distribution** (Appendix 5). A chi-square arises as a sum of squared unit-normals (mean zero, variance one). If the squared normals have nonzero means, then the sum of their squared means is the **noncentrality parameter (NCP)**, λ, and the resulting distribution is a noncentral chi-square ($\lambda = 0$ denotes a **central** chi-square). We denote that v follows a noncentral chi-square distribution with n degrees of freedom and NCP λ by $v \sim \chi^2_{n,\lambda}$, with $\mu_v = n + \lambda$ and $\sigma^2_v = 2(n + 2\lambda)$ (Equation A5.14b). Thus, the impact of the NCP is to shift the distributional mass to the right relative to a central chi-square (Figure A5.3), so that increasing the NCP increases the power (the probability that the test statistic exceeds the null hypothesis critical value, and thus is declared significant).

Under the large-sample properties of likelihood, the NCP for the LR under the alternative (true parameter values are $\boldsymbol{\Theta}_1$) versus the null ($\boldsymbol{\Theta}_0$) is (Equation A5.17a)

$$\lambda = (\boldsymbol{\Theta}_1 - \boldsymbol{\Theta}_0)^T \mathbf{V}^{-1} (\boldsymbol{\Theta}_1 - \boldsymbol{\Theta}_0) \tag{19.34a}$$

where $\mathbf{V}$ is the (large-sample) covariance matrix for the estimates, which is equal to the expected Fisher information matrix $\mathbf{F}$ (Appendix 4). Many pedigrees have a fairly regular structure (such as blocks of otherwise unrelated sibs), and in these cases it is possible to obtain either exact values for $\mathbf{F}$ or to use Taylor-series approximations (Equation A1.4b). An important cautionary note is that this is a *large-sample assumption*, and the sample sizes may not be sufficiently large to approach this limiting distribution (e.g., Yu et al. 2004).

Williams and Blangero (1999) found that the contribution to λ from a single (full) sib-pair becomes

$$\lambda = h_Q^4 \left[\frac{(h^4 + 4)}{2(h^4 - 4)^2} \right] \tag{19.34b}$$

where h_Q^2 is the heritability due to the focal QTL, while h^2 is the total heritability. Hence, a collection of n sib pairs ($2n$ individuals) has an NCP of n times Equation 19.34b. More generally, they found that the contribution of the NCP from any two relatives, x and y, is approximately

$$\lambda_{xy} \simeq h_Q^4 \left[\frac{[h^4(2\theta_{xy})^2 + 1][2\theta_{xy} + 4\Delta_{xy}(1 - 4\theta_{xy})]}{4[h^4(2\theta_{xy})^2 - 1]^2} \right] \tag{19.34c}$$

where Δ_{xy} is the coefficient of fraternity (Equation 7.8). The papers cited above developed slightly better approximations. The key message from these power calculation studies are as follows. First, power (simply for detection) is modest at best (e.g., Example 19.6). More importantly, the noncentrality parameter λ scales as the *square* of the QTL heritability, while under an association analysis power is proportional to the QTL heritability, not its square (Equation 20.8a). Second, for a fixed number of individuals, sibships are more powerful than a nuclear family of the same size (i.e., a sibship of size 4 has more information than two parents and their two offspring). Third, the power increase with larger sibships is greater than linear. In particular, Williams and Blangero (1999; also Rijsdijk et al. 2001) found that the approximate ratio of sample sizes (for a given power) S_m for sibships of size m, for $m > n$, is

$$\frac{S_m}{S_n} \simeq \frac{n(n-1)}{m(m-1)} \tag{19.34d}$$

One way to visualize this result is that a family of k sibs has $k(k-1)/2$ different combinations of pairs, with Equation 19.34d giving the ratio of the number of unique pairs for the two types of families. Relative to $n = 2$, the reduction in number of families is $1/3$, $1/6$, and $1/10$, and for 3, 4, and 5 sib families, respectively. For example, if 1,000 2-sib pairs (2000 individuals) give the desired level of power, then the corresponding numbers become 333 3-sib families (999), 167 4-sib (668), and 100 5-sibs (500). Finally, one should always keep any pedigree information, as this increases power relative to breaking up a pedigree into a series of smaller units. Williams and Blangero presented an example where 16,500 sib pairs (33,000 individuals) were required to obtain sufficient power, which reduced to 3315 individuals if one sampled 69 pedigrees with 48 members each, where the assumed structure for each such pedigree was based on those observed in prior studies.

Example 19.6. Consider the power for detecting a QTL with $h_Q^2 = 0.1$ and $h^2 = 0.4$, namely, the trait has an overall heritability of 40%, while the QTL accounts for 10% of the phenotypic variance. What is the threshold LR value under the null? Recalling that the LR test of a nonzero variance is a mixture distribution, if T_α denotes the threshold for level α, then $\Pr[(1/2)\chi_0^2 + (1/2)\chi_1^2 \leq T_\alpha] = 1 - \alpha$, hence $(1/2) + (1/2)\Pr(\chi_1^2 \leq T_\alpha) = 1 - \alpha$, implying that $\Pr(\chi_1^2 \leq T_\alpha) = 1 - 2\alpha$. For a LOD score of 3 (a standard human-genetics threshold to adjust for multiple comparisons; Ott 1999), which corresponds to $\alpha = 0.0001$, the threshold value is $T_{0.0001} = 13.83$, as $\Pr(\chi_1^2 \leq 13.83) = 0.9998$. Applying Equation 19.34b, the contribution to the NCP from a single sib pair becomes

$$\lambda = \frac{h_Q^4(h^4+4)}{2(h^4-4)^2} = \frac{0.1^2(0.4^2+4)}{2(0.4^2-4)^2} = 0.00141$$

With 10,000 (unrelated) sib pairs, the NCP is 10,000$\times$ 0.00141 = 14.1, with resulting power

$$\Pr(\chi^2_{1,14.1} \geq 13.83) = 0.47$$

or only 47%. For a power of 80%, trial and error finds n = 13,550 sibpairs. From Equation 19.34d, if one moves to sibships of size 4, this reduces to 2,258 families (or 9,032 individuals).

HASEMAN–ELSTON REGRESSION

Starting with Haseman and Elston (1972), human geneticists have developed a number of linkage-based (excess of parental gametes) methods for detecting QTLs using *pairs of relatives* as the unit of analysis. Their idea was to consider the number of alleles IBD between individuals for a given marker. If a QTL is linked to the marker, pairs sharing IBD marker alleles should also tend to share IBD QTL alleles and hence are expected to be *more similar* than pairs not sharing IBD marker alleles. This fairly simple idea is the basis for a number of relative-pair methods (often referred to as **allele sharing** methods), which we now examine in some detail. We first consider Haseman and Elston's original regression method (and its extensions), which deals with continuous characters, and finish by discussing a number of closely related methods that have been developed for dichotomous characters, the latter being motivated by the desire to map human disease genes. Nice reviews of these topics (from a human-genetics perspective) are given by Whittemore (1996), Greenwood and Morgan (2000), Amos and de Andrade (2001), and Feingold (2001, 2002), and (for plants) by Sneller et al. (2009). DeFries (2010) gives an amusing historical anecdote about the initial reaction to HE, while Robertson's (1973) important criticism of the original approach deserves more attention.

Derivation of the Haseman–Elston Regression

If a marker locus is linked to a QTL, the *difference* in character values between two relatives is expected to *decrease* as they share more marker alleles IBD. This is the basis for the HE regression (1972; generalized by Amos and Elston 1989; Amos et al. 1990; Elston 1990b; Barber et al. 2004), which regresses the squared difference in character value between a pair of relatives on the proportion of IBD marker alleles for that pair. Elston et al. (2000) provides a nice review of first-generation HE approaches.

To formally develop the **Haseman-Elston** (**HE**) test, consider a QTL with no dominance. Suppose there are n pairs of relatives (all of the same type, such as all full or all half sibs), and denote the character values for the jth pair by z_{1j} and z_{2j}. The model assumed is

$$z_{ij} = \mu + A_{ij} + e_{ij} \tag{19.35}$$

where A is the QTL effect (with mean zero and additive genetic variance σ_A^2), and e is the residual value. The latter includes all effects not accounted for by the QTL of interest, including environmental effects and contributions from other QTLs unlinked to the marker. The difference in residual values for the jth pair, $e_j = e_{1j} - e_{2j}$, is assumed to have mean zero, variance σ_e^2, and to be uncorrelated with $(A_{1j} - A_{2j})$. Note that the difference between pairs has the especially nice feature of accounting for shared environmental factors, as effects common to both relatives (such as family effects) subtract out. Visscher and Hopper (2001) noted that dizygotic (DZ) twin pairs are likely to provide a more powerful contrast than ordinary sibs, given that the former likely share more environmental features.

The squared difference for each pair, $y_j^D = (z_{1j} - z_{2j})^2$, has expected value

$$\begin{aligned} E(\,y_j^D\,) &= E\left[\,(\,A_{1j} - A_{2j} + e_{1j} - e_{2j})^2\,\right] \\ &= E\left[\,(\,A_{1j} - A_{2j})^2\,\right] + \sigma_e^2 \\ &= 2\,\left[\,\sigma_A^2 - \sigma(\,A_{1j}, A_{2j}\,)\,\right] + \sigma_e^2 \end{aligned} \tag{19.36}$$

From Equation 7.11a, $\sigma(\,A_{1j}, A_{2j}\,) = \sigma_A^2 \cdot \pi_{jt}$, where π_{jt} (= 0, 1/2, or 1) is the proportion of alleles IBD at the QTL. (In what follows, we use the subscripts t and m to distinguish **t**rait versus **m**arker loci.) Hence, the expectation of y^D for pair j, conditional on the proportion of QTL alleles IBD, becomes

$$E(\,y_j^D \,|\, \pi_{jt}\,) = (2\,\sigma_A^2 + \sigma_e^2) - (2\,\sigma_A^2) \cdot \pi_{jt} \tag{19.37}$$

As expected, if $\sigma_A^2 > 0$, the regression of y_j^D on the proportion π_{jt} of *trait* alleles IBD has a negative slope.

While we do not know π_{jt}, we can estimate it using π_{jm}, the proportion of alleles IBD at a *linked marker* locus. (We detail shortly how to estimate π_{jm} using the observed marker information.) Conditioning on the number (i = 0, 1, or 2) of marker alleles IBD gives

$$E(\,y_j^D \,|\, \pi_{jm}\,) = \sum_{i=0}^{2} E\left(\,y_j^D \,|\, \pi_{jt} = \frac{i}{2}\right) \cdot \Pr\left(\,\pi_{jt} = \frac{i}{2} \,\middle|\, \pi_{jm}\right) \tag{19.38}$$

The second term in the summation is the conditional probability for the proportion of QTL alleles IBD given the fraction of marker alleles IBD. These are functions of the type of relatives being considered and the recombination fraction c between the marker and QTL, and are obtained in a similar fashion to the conditional probabilities obtained in Example 18.1. Values of $\Pr(\pi_t \,|\, \pi_m)$ for many common types of relative pairs are given by Bishop and Williamson (1990) and Risch (1990b). Substituting these into Equation 19.38 gives the regression of the squared difference y_j^D on the proportion π_{jm} of *marker* alleles IBD as

$$E(\,y_j^D \,|\, \pi_{jm}\,) = \alpha + \beta \cdot \pi_{jm} \tag{19.39a}$$

where the slope β and intercept α depend on the type of relatives and the marker–QTL recombination fraction c. Zhang et al. (2008) noted that this approach could be extended over p markers to accommodate multiple QTL, with

$$E(\,y_j^D \,|\, \pi_{jm}\,) = \alpha + \sum_{k=1}^{p} \beta_k \cdot \pi_{jm,k} \tag{19.39b}$$

where $\pi_{jm,k}$ is the IBD status at marker k. For full sibs, substituting the appropriate values of $\Pr(\pi_{jt} \mid \pi_{jm})$ into Equation 19.38 gives (Haseman and Elston 1972),

$$\beta = -2\,(1-2c)^2\,\sigma_A^2, \qquad \text{and} \qquad \alpha = 2\,[\,1 - 2(1-c)c\,]\,\sigma_A^2 + \sigma_e^2 \tag{19.39c}$$

A significant negative slope provides evidence of a QTL linked to the marker, with the power of this test scaling with the position $(1-2c)^2$ and QTL effect σ_A^2. The expected slopes for other pairs of relatives are similarly found to be

$$\beta = \begin{cases} -2\,(1-2c)\,\sigma_A^2 & \text{grandparent–grandchild;} \\ -2\,(1-2c)^2\,\sigma_A^2 & \text{half-sibs;} \\ -2\,(1-2c)^2\,(1-c)\,\sigma_A^2 & \text{avuncular (aunt/uncle–nephew/niece).} \end{cases} \tag{19.40}$$

Thus, the HE test is quite simple: for n pairs of the same type of relatives, one regresses the squared difference of each pair on the fraction of alleles IBD at the marker locus. A significant negative slope for the resulting regression indicates linkage to a QTL. This is a one-sided test, as the null (no linkage) is $\beta = 0$ versus the alternative $\beta < 0$.

There are several caveats with this approach. First, different classes of relatives cannot be combined in the standard HE test, requiring separate regressions for each type of relative pair. This can be avoided by modifying the test using an appropriately weighted (GLS; Chapter 10) multiple regression (Olson and Wijsman 1993; Olson 1994). A related issue is that (even when using the same type of relatives), the residuals of the HE regression are *heteroscedastic*, as the variance in phenotype varies with the number of QTL alleles IBD. Using a GLS regression also accounts for this concern, and results in a slight increase in power (Amos et al. 1989). Second, parents and their offspring share *exactly* one allele IBD and hence *cannot* be used to estimate this regression, as there is *no variability in the predictor variable* (Elston 1990b). This same lack of variability holds for clones and monozygotic twins. Finally, QTL position (c) and effect (σ_A^2) are confounded and cannot be separately estimated from the regression slope β. Thus, in its simplest form, the HE method is a *detection* test rather than an *estimation* procedure. This conclusion is not surprising, given that the HE method is closely related to the single-marker linear model. As we discuss shortly, estimation of c and σ_A^2 is possible by extending the HE regression by using IBD status of two (or more) linked marker loci to estimate π_{jt} at arbitrary locations between them (i.e., a multipoint method).

Turning to a QTL with dominance, the resulting model becomes

$$z_{ij} = \mu + A_{ij} + D_{ij} + e_{ij} \tag{19.41}$$

Dominance only causes a departure from the additive model when both QTL alleles are IBD (as can occur in **bilineal relatives**, such as with full sibs; Chapter 7) and hence does not influence most types of relative pairs (**unilineal relatives** that share at most one IBD allele). Define the indicator variables $f_{1,jm}$ and $f_{2,jm}$, where $f_{i,jm} = 1$ if the jth pair share i marker alleles IBD, else it is zero, so that $\pi_{jm} = f_{2,jm} + f_{1,jm}/2$ (Equation 7.2a). For full sibs, Blackwelder and Elston (1982) show that

$$E(\,y_j^D \mid f_{1,jm},\, f_{2,jm}) = \alpha + \beta\cdot\pi_{jm} + \gamma\cdot f_{1,jm} \tag{19.42a}$$

where

$$\beta = -2(1-2c)^2\cdot(\sigma_A^2 + \sigma_D^2) \qquad \text{and} \qquad \gamma = (1-2c)^4\cdot\sigma_D^2 \tag{19.42b}$$

For full sibs with known parental marker genotypes, Amos et al. (1989) showed that ignoring the γ term in Equation 19.42a and simply using $\alpha + \beta\pi_{jm}$ yields an unbiased estimator of β. Hence, if our goal is simply QTL detection, we can ignore the effects of dominance in a full-sib analysis.

Finally, we note a very powerful marriage between modern genomics and the HE approach that is examined in detail in Part II (Chapter 33). Using very dense markers,

Table 19.3 Probability of various pairs of marker genotypes in relatives as a function of the number of IBD alleles between these relatives. Haseman and Elston (1972) noted that there are seven different types of marker pairs. Of these, three share no IBD alleles (Types II, IV, and VII), two share either zero or one alleles (Types III and VI), and two (Types I, V) can share zero, one, or two alleles IBD. Here j, k, ℓ and m index *different* alleles. Here p_i is the frequency of marker allele M_i. (After Elston 1990b.)

Type	Marker Pairs	Pr(M | i = 0)	Pr(M | i = 1)	Pr(M | i = 2)
I	$M_jM_j,\ M_jM_j$	p_j^4	p_j^3	p_j^2
II	$M_jM_j,\ M_kM_k$	$2\,p_j^2\,p_k^2$	0	0
III	$M_jM_j,\ M_jM_k$	$4\,p_j^3\,p_k$	$2\,p_j^2\,p_k$	0
IV	$M_jM_j,\ M_kM_\ell$	$4\,p_j^2\,p_k\,p_\ell$	0	0
V	$M_jM_k,\ M_jM_k$	$4\,p_j^2\,p_k^2$	$p_j\,p_k\,(p_j+p_k)$	$2\,p_j\,p_k$
VI	$M_jM_k,\ M_jM_\ell$	$8\,p_j^2\,p_k\,p_\ell$	$2\,p_j\,p_k\,p_\ell$	0
VII	$M_jM_k,\ M_\ell M_m$	$8\,p_j\,p_k\,p_\ell\,p_m$	0	0

one can accurately estimate the total amount of IBD sharing between a pair of sibs, and hence estimate the additive variance (Visscher et al. 2006). Further, one can look at the IBD sharing for each given chromosome, which can then be used to estimate the additive variance associated with that chromosome (Visscher et al. 2006). In particular, regressing y_i^D for the trait of interest on π_k, the amount of IBD sharing over all of chromosome k, then the estimated additive variance attributable to that chromosome is $\sigma^2_{A_k} = -\beta_k/2$, where β_k is the chromosome-specific HE regression.

Estimating the Number of Marker IBD Alleles

An obvious question is how to determine the number of alleles that are IBD at a given marker. If both parents are *different* marker heterozygotes (i.e., the family is fully informative), IBD status can be determined with certainty in all of their offspring. However, other relative pairs are problematic, as IBD status must be probabilistically assigned rather than determined with certainty. In the absence of pedigree information, Haseman and Elston (1972) and Elston (1990b) suggested the following general estimator, based on Bayes theorem (Equation 3.3b), of π_{jm} for any pair of relatives,

$$p_{jm} = \frac{f_2 \cdot \Pr(M \mid i=2) + (f_1/2) \cdot \Pr(M \mid i=1)}{f_2 \cdot \Pr(M \mid i=2) + f_1 \cdot \Pr(M \mid i=1) + f_0 \cdot \Pr(M \mid i=0)} \tag{19.43}$$

where f_i is the prior probability that the pair of relatives share i genes IBD, and the $\Pr(M \mid i)$ are the probabilities of the observed pair M of marker genotypes given that the pair shares i alleles IBD (Table 19.3). While easy to compute, one must exercise caution in the application of Equation 19.43, as it is rather sensitive to errors in the estimated marker-allele frequencies (Babron et al. 1993) and departures from random mating (Chapter 8). Risch (1990c) developed alternative ML estimators for p_{jm} based on these probabilities, while Hu et al. (1995) showed how to modify Equation 19.43 to accommodate sex-linked markers. Additional robust estimators of relatedness were discussed in Chapter 8.

The probability $\Pr(M \mid i)$ of observing a given pair M of marker genotypes in two relatives given that they share i marker alleles IBD is computed as follows. If the two relatives share no alleles IBD, the frequencies of marker genotype pairs are given by the probability of drawing two such individuals at random. For example, the probability the pair has genotypes M_jM_j and M_jM_k is $2 \cdot p_j^2 \cdot 2\,p_j\,p_k = 4\,p_j^3\,p_k$. If one allele is IBD between the pair of relatives, the shared allele is chosen at random (say M_j with probability p_j) while the remaining allele in each relative is chosen at random. For the previous marker pair, this probability now becomes $p_j \cdot (2p_jp_k) = 2\,p_j^2\,p_k$. If both alleles are IBD, marker genotypes are identical and the probability of observing the pair equals the probability of observing one such genotype, i.e., p_i^2 for $(M_iM_i,\ M_iM_i)$, $2p_i\,p_j$ for $(M_iM_j,\ M_iM_j)$. Using this logic, values for the seven different classes of paired marker genotypes can be computed (Table 19.3). Note that a very similar approach is used in Table A4.1, which forms the basis of likelihood functions for estimating relationships given marker data (see Example A4.8).

Example 19.7. Suppose a marker has four alleles with frequencies $p_1 = 0.5$, $p_2 = 0.25$, $p_3 = 0.15$, and $p_4 = 0.1$. The marker genotypes for three independent pairs of full sibs were found to be: Pair 1 = $(M_1M_1,\ M_1M_1)$; Pair 2 = $(M_3M_4,\ M_3M_4)$; Pair 3 = $(M_4M_4,\ M_4M_4)$. Recall from Chapter 7 that (for full sibs) $f_0 = f_2 = 1/4$, $f_1 = 1/2$. Applying Equation 19.43 and using Table 19.3, the estimated proportions of IBD marker alleles for each of these three pairs are

$$p_{1m} = \frac{(1/4)\,p_1^2 + [(1/2)/2]\,p_1^3}{(1/4)\,p_1^2 + (1/2)\,p_1^3 + (1/4)\,p_1^4} \simeq 0.67$$

$$p_{2m} = \frac{(1/4)\,2\,p_3\,p_4 + [(1/2)/2]\,p_3\,p_4\,(p_3 + p_4)}{(1/4)2\,p_3\,p_4 + (1/2)\,p_3\,p_4\,(p_3 + p_4) + (1/4)\,4\,p_3^2\,p_4^2} \simeq 0.88$$

$$p_{3m} = \frac{(1/4)\,p_4^2 + [(1/2)/2]\,p_4^3}{(1/4)\,p_4^2 + (1/2)\,p_4^3 + (1/4)\,p_4^4} \simeq 0.91$$

These values are substituted in place of π_{jm} when computing the HE regression. By comparing the first and third pairs, it can be seen that rarer alleles are more informative as to IBD status, an observation also seen in Chapter 8 for relatedness estimators. This approach gives a point estimate for the fraction of IBD sharing at a given marker. We will shortly discuss multipoint methods using all of the markers on a chromosome to estimate this value at any location.

Power of the Traditional HE Regression

A number of authors have examined the power of the HE test (Robertson 1973; Blackwelder and Elston 1974, 1982; Amos and Elston 1989; Götz and Ollivier 1992), and large-sample approximations for power have been developed by Blackwelder and Elston (1982) and Amos and Elston (1989). As above, such power calculations involve computing the noncentrality parameter (NCP, λ) of the noncentral chi-square (Equation A5.14a) associated with the alternative hypothesis. The general conclusion from these studies is that the standard HE test has very poor power in most settings, and is best considered to be a method for detecting QTL of large effect. For example, Amos and Elston (1989) found that 320 full-sib pairs are required to have a 90% chance of detecting (at significance level $\alpha = 0.05$) a major gene with a heritability of 0.5 (hence accounting for 50% of the phenotypic variance) when the marker and major gene are completely linked. This number increases to 778 sib pairs if the marker and major gene are separated by recombination fraction $c = 0.1$. The sample size increases as more distant relative pairs are considered, requiring 1,264 grandparent–grandchild pairs; 1,980 half-sib pairs; and 2,445 avuncular pairs. Because these power calculations assume that the fraction π_m of marker alleles IBD is known exactly, in reality, even more pairs are required as marker-uninformative pairs are discarded or π_m is estimated for such pairs.

While these numbers look bleak, the situation is not necessarily as bad with large sibships. In a sibship of size s, there are $s(s-1)/2$ pairwise comparisons, so that single families of 5, 10, and 25 sibs offer 10, 45, and 325 sib pairs. The question remains as to whether all pairwise comparisons within a family are independent. There is some dispute on this point, with Hodge (1984) showing that the number of independent sib pairs approaches $(3/2)s - 2$ for $s > 4$. However, simulation studies have suggested that all pairwise combinations of sibs within a family can often be approximately treated as independent pairs (Blackwelder and Elston 1982, 1985; Amos et al. 1989; Götz and Ollivier 1992; Page et al. 1998; but see Daly and Lander 1996). Wilson and Elston (1993) showed that while all pairs can be used, for small sample sizes the appropriate degrees of freedom for a t test for the regression is given by $(\sum_i [s_i - 1]) - 2$, rather than $(\sum_i \; s_i[s_i - 1]/2\,) - 2$, where s_i is the number of sibs in family i. Thus, the HE test may not be unreasonable for modest sample sizes, *provided* one can obtain a few informative families with a large number of sibs. The irony of the HE approach is that while it was developed for human genetics, the typically small to modest family size of many cultures means that the method is better suited for other species where large family sizes are the norm.

Significant power gains are possible through the use of selective genotyping (Chapter 18), wherein pairs are chosen based on extreme trait values in at least one member (Boehnke and Moll 1989; Carey and Williamson 1991; Cardon and Fulker 1994, Eaves and Meyer 1994; Risch and Zhang 1995, 1996; Gu et al. 1996; Zhang and Risch 1996; Zhao et al. 1997; Dolan and Boomsma 1998; Szatkiewicz et al. 2003; T.Cuenco et al. 2003; Szatkiewicz and Feingold 2004, 2005). As tests with selective genotyping are often based on allele sharing statistics, we discuss them further in the next section. The original logic behind these approaches was the assumption of a far cheaper cost of phenotyping relative to genotyping, which is no longer the case in most settings. Further, these approaches effectively discard much of the phenotyping, which (when these can be genotyped) adds significant power. As a result, selective genotyping is rarely used (at least in human genetics) for continuous traits. However, as we will see below, nonrandom sampling—the case-control (disease-normal) design—is routinely used for binary traits (Example 19.10; Chapter 20), in which a series of cases is extracted from a much larger population sample.

Interval Mapping by a Modified Haseman–Elston Regression

Fulker and Cardon (1994) proposed an interval-mapping modification that allows separate estimates of QTL effect (σ_A^2) and position (c) by considering two linked markers (that flank a single QTL) simultaneously. Recall Equation 19.37,

$$E(\,y_j^D \,|\, \pi_{jt}) = E[\,(z_{1j} - z_{2j})^2 \,|\, \pi_{jt}\,] = (2\sigma_A^2 + \sigma_e^2) - (2\,\sigma_A^2) \cdot \pi_{jt}$$

Fulker and Cardon showed that the proportion π_{jt} of alleles IBD at the trait locus can be estimated by using a regression on $\pi_{jm}^{(1)}$ and $\pi_{jm'}^{(2)}$, the proportion of alleles IBD at two marker loci flanking the QTL, where

$$\widehat{\pi}_{jt} = \frac{1 - \beta_1 - \beta_2}{2} + \beta_1\,\pi_{jm}^{(1)} + \beta_2\,\pi_{jm}^{(2)} \tag{19.44}$$

and

$$\beta_1 = \frac{(1-2c_1)^2 - (1-2c_2)^2(1-2c)^2}{1-(1-2c)^4}, \qquad \beta_2 = \frac{(1-2c_2)^2 - (1-2c_1)^2(1-2c)^2}{1-(1-2c)^4}$$

Here, as elsewhere, c_1 is the recombination frequency between the first marker and the QTL, c_2 between the QTL and the second marker, and c between the two markers (the latter assumed known). As in Chapter 17, c_2 follows directly from c_1 and c under various assumptions about interference (e.g., with complete interference $c_2 = c - c_1$), leaving the β_i as functions of only one unknown parameter, c_1. When the markers are reasonably close ($c < 0.2$), these coefficients simplify to $\beta_{m,1} \simeq c_2/c$ and $\beta_{m,2} \simeq c_1/c$. Substituting into Equation 19.37,

$$E(y_j^D) = \alpha - (2\,\sigma_A^2)\left(\beta_1\,\pi_{jm}^{(1)} + \beta_2\,\pi_{jm}^{(2)}\right) \tag{19.45}$$

where

$$\alpha = \left[\,\sigma_A^2\,(1 + \beta_1 + \beta_2) + \sigma_e^2\,\right]$$

Following the approach of Haley and Knott (1992), the **Fulker-Cardon method** computes this regression for each value of c_1 over the interval $(0 \leq c_1 \leq c)$, with that value giving the regression with the largest r^2 taken to be the estimate of map position, and $2\sigma_A^2$ estimated from the associated regression coefficients (at this optimal value). This method offers increased power for QTL detection relative to the standard HE method (Fulker and Cardon 1994; Cardon and Fulker 1994). This increase in power is small when the single-marker power is either very poor or very high, but can be considerable when single-marker power is modest.

Standard (i.e., two-marker) interval mapping can be extended to **multipoint interval mapping** by considering all linked markers on a chromosome (Fulker et al. 1995; Kruglyak and Lander 1995b). In the approach of Fulker et al., the two-marker regression estimate of the QTL IBD status (Equation 19.44) is replaced by a multiple regression that incorporates all n linked markers on a chromosome,

$$\widehat{\pi}_{jt} = \alpha + \sum_{k=1}^{n} \beta_k \, \pi_{jm}^{(k)} \tag{19.46}$$

where the fraction $\widehat{\pi}_{jt}$ of IBD QTL alleles for the jth pair is predicted using the fractions $\pi_{jm}^{(k)}$ of IBD alleles at each of the n marker loci on the chromosome containing the putative QTL. The β_k are obtained from standard GLS regressions (Equation 10.13), where the vector $\boldsymbol{\beta}^T = (\beta_1, \cdots, \beta_n)$ satisfies $\boldsymbol{\beta} = \mathbf{V}^{-1}\mathbf{C}$, where

$$V_{ik} = \sigma\left(\pi_m^{(i)}, \pi_m^{(k)}\right) = \begin{cases} \text{Var}\left(\pi_m^{(i)}\right) & \text{for } i = k \\ 8\,(1 - 2c_{ik})^2 \; \text{Var}\left(\pi_m^{(i)}\right) \text{Var}\left(\pi_m^{(k)}\right) & \text{for } i \neq k \end{cases} \tag{19.47a}$$

and

$$C_i = \sigma\left(\pi_t, \pi_m^{(i)}\right) = (1 - 2c_i)^2 \; \text{Var}\left(\pi_m^{(i)}\right) \tag{19.47b}$$

Here $\text{Var}(\pi_m^{(i)})$ is the observed sample variance for IBD status at the ith marker (i.e., the sample variance of the $\pi_{jm}^{(i)}$ over all included pairs), c_{ik} is the distance between markers i and k (assumed known), and c_i is the (unknown) distance between the QTL and marker i. Note that this approach is very similar in spirit to the marker-difference regression method (Chapter 18), and has a closely related covariance structure (Equation 18.26d).

Again this regression is computed by first assuming a particular position for the QTL along the chromosome, computing $\widehat{\pi}_{jt}$ using Equation 19.46, and substituting this value into the standard HE regression (Equation 19.37). This process is repeated for all possible positions of the QTL along the chromosome, with the estimate of QTL position taken as that which gives the regression with the largest r^2 value. (This is equivalent to the regression giving the largest t statistic.) Fulker et al. showed that this method offers high precision for mapping and (unlike two-locus interval mapping) accounts for differences in the information content of adjacent markers (as also discussed above for half-sib methods). When only two markers are used, Fulker et al. (1995) showed that QTL position is *biased towards the more informative marker*. Multipoint interval mapping avoids this problem by incorporating all the marker information along the entire chromosome. Because the multipoint regression is easily computed, there is little reason to prefer simple interval mapping (Equation 19.44) over multipoint mapping (Equation 19.46).

An improved multipoint estimator was given by Kruglyak and Lander (1995b). While Equation 19.46 uses marker information to estimate the *mean* value of π_{jt}, Kruglyak and Lander use all the marker information to provide an ML estimate at each point along a chromosome of the actual *distribution* of IBD values (i.e., the probabilities at any particular chromosomal position that the pair shares zero, one, or two IBD alleles). An EM method (Appendix 4) is then applied to compute the regression using this distribution, resulting in greater power than in an analysis based simply on the estimated mean of π_{jt}. As before, regressions are computed at each point along a chromosome, generating a t-statistic plot for the regression significance as a function of putative QTL position.

Example 19.8. Cardon et al. (1994) used interval mapping to examine whether QTLs influencing reading ability were linked to markers on human chromosome 6 (previous work had suggested linkage to this chromosome). For more refined mapping, these authors used a set of five highly informative STR markers (Chapter 8). In the following figure, support for QTL position is given by plotting the regression t statistic as a function of putative QTL position on chromosome 6. The number of alleles for each marker locus ranged from 9 to 13 with associated heterozygosities of 0.6–0.9. Two independent sets of relatives were examined, a collection of 358 siblings from 19 families, and an independent collection of 50 dizygotic twins. Ascertainment of families was done by requiring that one sib in each pair scored very poorly on standardized reading tests (following suitable correction for other environmental factors). Interval mapping on both relative sets shows strong (and independent) support for a QTL near the markers *D6S105* and *TNFB*. Support for a second QTL at 70 cM is suggested from the twin data, but this is not confirmed by the siblings data set. (Figure after Cardon et al. 1994.)

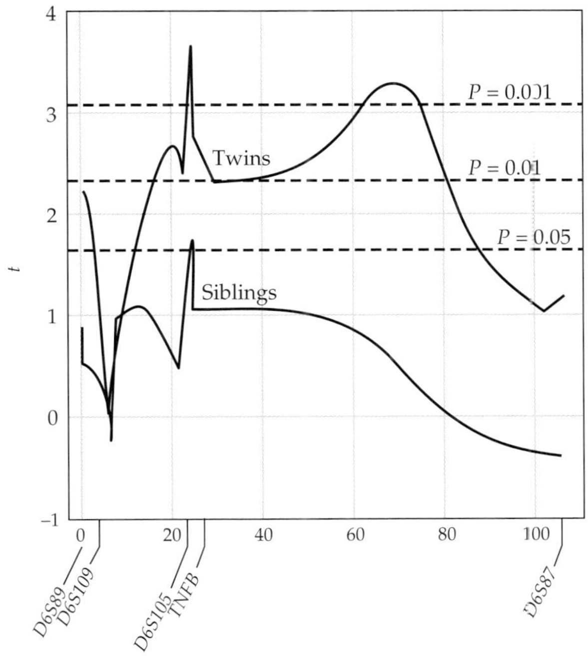

Second-generation Haseman–Elston Regressions

F. Wright (1997) noted that the basic structure of the HE regression (and its extensions discussed above) is less than optimal because information is discarded by just considering the squared difference between pairs. This observation motivated a new class of extensions, which we will call **second-generation HE regressions**. Drigalenko (1998) noted that both the standard HE regression based on the *squared pair-difference*

$$y_i^D = (z_{i,1} - z_{i,2})^2 = [(z_{i,1} - \mu) - (z_{i,2} - \mu)]^2 \tag{19.48a}$$

and regression based on the mean-adjusted *squared pair-sum*

$$y_i^S = [(z_{i,1} - \mu) + (z_{i,2} - \mu)]^2 \tag{19.48b}$$

have the same expected absolute value for the slope. Hence, one could use the (unweighted) average of these two estimates as an improved estimator of β. This weighting is equivalent to considering the product of the mean-adjusted trait values of the pair

$$y_i^P = [(z_{i,1} - \mu)(z_{,i2} - \mu)] \tag{19.48c}$$

(Drigalenko 1998; Elston et al. 2000). Given its similarity to Equation 3.7, Equation 19.48c has been called the **covariance Haseman–Elston**. While in some settings (but not all; Visscher and Hopper 2001) these improvements can be more powerful than a traditional HE regression, they are still underpowered relative to variance components (Forrest 2001; Tang and Siegmund 2001). A number of authors quickly noted that equal weighting of these slope estimates was less than optimal. Rather, one should weight by their sampling variance or some other more optimal scheme, and this can improve power. Further, y_i^D and y_i^S are weakly correlated, and taking this association into account further improves the estimate. This led to a number of second-generation regressions based on differences in weighting, e.g., Xu et al. (2000), Forrest (2001), Sham and Purcell (2001), Visscher and Hopper (2001), Wang et al. (2001), and Franke et al. (2005).

For example, Wang et al. (2001) used the product regression (Equation 19.48c), but instead of correcting all pairs using the grand mean ($\overline{z}$ for μ), they suggested using the family-specific ($\overline{z}_k$ for μ when sibs are from family k). This adjustment assumes that more than two sibs per family are measured, and its motivation is to adjust for family-specific effect (such as shared environmental effects or common background polygenic genotypes). While the original HE regression based on squared differences (y^D) accomplished this goal, this correction is removed when the grand mean is subtracted off in the product term (Equation 19.48c). With only two sibs per family, the Wang method reduces to the standard HE regression.

Sham and Purcell (2001) proposed weighting the slope estimates based on squared pair-differences (D) and squared pair-sums (S) using the correlation r between sibs in the pair,

$$\widehat{\beta}_{SP} = \frac{\widehat{\beta}_D/(1-r)^2 + \widehat{\beta}_S/(1+r)^2}{1/(1-r)^2 + 1/(1+r)^2} \tag{19.48d}$$

In her review of these various modications of the original HE, Feingold (2002) recommended that this **Sham–Purcell estimator** be used whenever estimates of r are available.

Sham et al. (2002) proposed a modification to handle general pedigrees, by swapping the roles of the variables in HE regression, treating the IBD status as the *dependent variable* in a multiple regression on trait value

$$\pi_{jm} = \alpha + \beta_D y_i^D + \beta_D y_j^S + e_j$$

The variances and covariances among the y^D and y^S values are functions of the correlations among the relative pair being examined, so this method requires unbiased estimates of these parameters. However, it is potentially more robust to selective sampling. Wang and Elston (2005) also proposed HE extensions for general pedigrees.

Regression Versus Variance-component Approaches

When the normality assumption is satisfied, ML-based variance-component (ML-vc) approaches should be more powerful than the classic HE regression. However, a concern is the robustness of the ML approach to departures from normality (e.g., by selective trait sampling or intrinsic features of the trait). Conversely, while the original HE regression is generally less powerful, the Type I errors of regressions are much more robust to departures from the assumption of normality of residuals (Miller 1986; Sawilowsky and Blair 1992; Allison et al. 1999b, 2000; Tang and Siegmund 2001). Indeed, the residuals of HE regression are expected to be non-normal, as y is the square of a normal, namely a chi-square. Hence, there appears to be a power versus robustness tradeoff between ML-vc and regression approaches, with ML being more powerful, but regressions being more robust. This led to the search for methods with close to the power of ML, but with the robustness of regressions.

Fortunately, it appears that second-generation HE regressions do just this. By computing the NCP associated with each method, one can compare their relative power (Appendix 5). Sham and Purcell (2001) and Visscher and Hopper (2001) showed that the large-sample NCP for the SP regression estimate (Equation 19.48d) is the same as that for the variance-component estimate (assuming a collection of sib pairs). Hence, these methods have identical power in large samples (a large collection of unrelated sib pairs). For more general pedigrees, Sham et al. (2002) showed that their method has comparable power to ML-vc. These observations are in line with those from Chapter 18, where it was shown the large-sample regression QTL estimators (for line crosses) were close to power to ML methods.

MAPPING DICHOTOMOUS CHARACTERS

Finally, we turn our attention to mapping and characterizing QTLs that underlie dichotomous characters. Many of the above approaches, such as regressions (Kadarmideen and Dekkers 2001; Kadarmideen et al. 2001) and variance components (Duggirala et al. 1997; Williams et al. 1999; Yi and Xu 1999; Williams and Blangero 2004), have been extended to these traits. Our focus here is on a rather different class of approaches developed by human geneticists, whose ultimate goal is the isolation of **disease susceptibility (DS)** genes: QTLs that influence human disease. When a disease can be treated as a continuous character by measuring appropriate underlying continuous variables, such as blood pressure or cholesterol level, the methods discussed in previous sections of this chapter can be applied. However, for many diseases, the appropriate underlying physiological variables are unresolved, and the observed phenotype is naturally dichotomous, with individuals being scored as either normal (unaffected or controls) or affected (or cases). Even when the appropriate underlying variables are known, individuals are often still placed into discrete categories, such as high versus normal blood pressure. We frame our discussion in terms of normal versus affected character states, although our treatment extends to any binary trait.

If the disease has a simple Mendelian basis, wherein a single gene underlies the disease and all individuals of a given genotype have the same character state, standard mapping approaches for simple Mendelian genes can be used (reviewed by Terwilliger and Ott 1994; Ott 1999). However, most human diseases have a more complex genetic basis, with no single gene showing perfect segregation with the disease. Even when genetic causes are largely attributable to a single locus, lack of perfect association can occur for a variety of reasons (Lander and Schork 1994). There can be **incomplete penetrance** (not all susceptible genotypes show the disease), **phenocopies** (individuals display the disease for environmental, rather than genetic, reasons) and **genetic heterogeneity** (multiple alleles at a given locus have different effects on disease susceptibility). Likewise, the disease could have a major gene influenced by polygenic modifiers. Even when individuals contract the disease, the severity, age on onset, etc., is often variable over a cohort of the same major locus genotype (variable **expressivity**), which could reflect both $G \times E$ and $G \times G$ components. All of these factors confound a simple Mendelian analysis, requiring that other methods be used. Amos and de Andrade (2001) noted that some of these confounding factors can often be reduced by focusing on early-onset cases of diseases, as these are less likely to be caused by nongenetic effects, an approach referred to as **enrichment sampling**.

Starting with Penrose (1935, 1954b), human geneticists have developed a variety of methods for linkage-based mapping of DS genes, often using pairs of relatives (typically sibs) as the unit of analysis. These are reviewed by Weiss (1993), Terwilliger and Ott (1994), Lander and Schork (1994), Weeks and Lathrop (1995), McPeek (1999), and Szatkiewicz et al. (2003). One approach is to modify the likelihood functions constructed for segregation analysis of dichotomous characters (Equation 16.29) to incorporate marker information by using Equation 19.27. For a diallelic QTL, the resulting likelihood function has five unknown parameters: QTL position (c), allele frequency (p), and effects (measured by the penetrances ψ_1, ψ_2, ψ_3 of the three QTL genotypes), where

$$\psi_i = \Pr(\text{ Disease} \mid \text{QTL genotype } = i)$$

LR tests for linkage follow by comparing the restricted model assuming $c = 1/2$ (no linkage) against the full model where c is estimated from the data. Instead of attempting to maximize the entire likelihood over a complex pedigree structure, human geneticists usually further reduce the parameter space by assuming p is known, then performing the LR test for linkage under one or more simpler genetic models where at least one of the penetrances is assigned a fixed value. Typically, it is assumed that the DS locus is either completely dominant or fully recessive, so that if D and N denote the disease and normal alleles at the DS locus, the two models generally considered are

$$\begin{array}{lll} \text{Dominance model:} & \psi_{DD} = \psi_{DN} = \psi & \psi_{NN} = \psi_p \\ \text{Recessive model:} & \psi_{DD} = \psi & \psi_{DN} = \psi_{NN} = \psi_p \end{array}$$

where the two parameters to estimate by ML are ψ, the penetrance of a susceptible genotype, and ψ_p, the probability of a phenocopy. Frequently, the model is reduced further by preassigning a value to ψ_p (most often being set to zero). A serious limitation of this approach is that it is *very model-dependent*. Using the wrong genetical model can potentially result in reports of false linkage or (more likely) failure to detect a linked QTL (Clerget-Darpoux et al. 1986; Curtis and Stam 1995a). Further, likelihood methods can be computationally demanding, especially in complex pedigrees. If different fitted genetic models are nested (one is a subset of the other), then LR tests can be used to compare model fit (Appendix 4). If they are not nested, they can still be compared using AIC or BIC model-selection criteria (Equation A4.15).

Relative-pair approaches, on the other hand, are computationally very simple and require no assumptions about the underlying genetic model. Not surprisingly, these methods were historically very popular (before being largely supplanted by genome wide association studies, GWAS) in the early 2000s; Chapter 20), and they form the bulk of our discussion. Such tests are based on the common theme of **allele sharing**—if a marker is linked to a QTL, ***two relatives with the same character state should, on average, share more IBD marker alleles than expected by chance***. This is the same principle underlying the HE regression, again using the linkage signature of an excess of parental gametes (Chapter 5).

Recurrent and Relative Risks for Pairs of Relatives

Before examining relative-pair methods, a brief introduction to measures of risk is in order. Human geneticists have devised a number of measures for the risk that an individual is affected, given that a relative is, and these can be expressed as a function of the genetic parameters of the DS locus. Assigning values to the alternative states of the character of 0 (if unaffected) or 1 (if affected) yields several useful identities. Under this coding, the character follows a Bernoulli distribution (a binomial with $n = 1$) with mean K (the **population prevalence**) and variance $K(1 - K)$. Note that

$$K = \sum_{i=1}^{n} \psi_i \cdot \Pr(\text{QTL genotype } = i)$$

If the character has a genetic basis, *relatives of an affected individual should have a probability greater than K of also being affected*. Letting z_1 and z_2 denote the character states of two relatives, this probability is quantified by the **recurrence risk**,

$$K_R = \Pr(\, z_2 = 1 \mid z_1 = 1\,) \tag{19.49a}$$

which is the conditional probability that a relative of relationship R (say, a full sib, S) of an affected individual is also affected. Because $\Pr(A, B) = \Pr(A \mid B) \cdot \Pr(B)$, it follows that the probability that both relatives are affected is $K_R \cdot K$.

An alternative measure is the **relative risk**, λ_R, the *increase in risk* given that one relative is affected,

$$\lambda_R = \frac{K_R}{K} \tag{19.49b}$$

For example, the population prevalence for type 1 diabetes is 0.4% ($K = 0.004$), while the probability that an individual has diabetes, given that a full sib (S) is diabetic, is 6% ($K_S = 0.06$), giving the relative risk in sibs as $\lambda_S = 0.06/0.004 = 15$ (Davies et al. 1994). Hence, relative to a random individual, your diabetes risk is 15-fold higher if your sib has the disease.

James' identity (1971) relates the recurrence risk (K_R) to the population prevalence (K) and the covariance between relatives,

$$K_R = K + \frac{\text{Cov}(z_1, z_2)}{K} \tag{19.50a}$$

This can be rearranged to yield **Risch's identity** (1990a),

$$\lambda_R - 1 = \frac{\text{Cov}(z_1, z_2)}{K^2} \tag{19.50b}$$

James' identity is completely general and is obtained as follows. Because the z_i are zero-one indicator variables, $z_1 \cdot z_2 = 0$ unless $z_1 = z_2 = 1$, and hence

$$\text{Cov}(z_1, z_2) + K^2 = E(z_1\, z_2) = 1 \cdot \Pr(z_1 = 1,\ z_2 = 1) = K_R \cdot K$$

The first step follows by recalling $E(z_1\, z_2) = \text{Cov}(z_1, z_2) + E(z_1)\, E(z_2)$, where $E(z_i) = K$.

Ignoring epistasis and shared environmental effects, Equation 7.12 yields

$$\text{Cov}(z_1, z_2) = 2\Theta_{12}\, \sigma_A^2 + \Delta_{12}\, \sigma_D^2$$

where σ_A^2 and σ_D^2 are the additive and dominance variances associated with this character (measured on the 0/1 scale), Θ_{12} is the coefficient of coancestry, and Δ_{12} is the coefficient of fraternity for these two individuals (Chapter 7). Hence,

$$K_R = K + \frac{2\Theta_R\, \sigma_A^2 + \Delta_R\, \sigma_D^2}{K} \tag{19.51a}$$

Again note for gene mapping using outbred individuals that QTL effects enter as variances, not means.

Equation 19.51a shows how genetic variances can be estimated using the observed recurrent risks for different sets of relatives (Suarez et al. 1978; Risch 1990a; Olson 1995). For example, assuming only additive and dominance genetic variances are present and rearranging Equation 19.51a gives

$$\sigma_A^2 = 2K(K_{PO} - K) \qquad \text{and} \qquad \sigma_D^2 = 4K(K_S - K_{PO}) \tag{19.51b}$$

where K_{PO} and K_S denote, respectively, the recurrent risk for parent–offspring and full-sib pairs. Likewise, the total genetic variance σ_G^2, even in the presence of epistasis, is

$$\sigma_G^2 = K(K_{MZ} - K) \tag{19.51c}$$

where K_{MZ} is the recurrent risk for monozygotic (identical) twins, as $\text{Cov}(MZ_1, MZ_2) = \sigma_G^2$.

Finally, the rate of decrease of $(\lambda_R - 1)$ in sets of increasingly distant relatives provides a test for epistasis (Risch 1990a). To see this, consider sets of relatives where $\Delta_R = 0$, so that dominance terms can be ignored. In this case, recalling Equation 7.12, Equation 19.50b can be generalized to

$$\lambda_R - 1 = \frac{1}{K^2}\left[\,2\Theta_R\,\sigma_A^2 + (2\Theta_R)^2\sigma_{AA}^2 + (2\Theta_R)^3\sigma_{AAA}^2 + \cdots\right] \tag{19.51d}$$

In the absence of epistasis, $\lambda_R - 1$ decreases by a factor of two when comparing relatives whose coefficient of coancestry also differs by a factor of two (e.g., half sibs versus first cousins; Table 7.3). A greater than twofold decrease in $\lambda_R - 1$ implies epistasis, and, by extension, multiple loci. Risch used this approach to provide evidence for multiple interacting loci for schizophrenia.

Example 19.9. Consider the following data on rheumatoid arthritis. Families were ascertained by having an individual showing the disease, with the presence or absence of this disease in their spouse and in any (full) sibs then scored, yielding

	Rheumatoid Arthritis		
	Present	Absent	Totals
Sibs of Affected	21	475	496
Spouses of Affected	12	661	673
Totals	33	1136	1169

The logic behind using spouses is as a partial control for shared environmental effects. The frequency of this disease in spouses was $12/673 = 0.018$, while the frequency in sibs of affected individuals is $21/496 = 0.042$. Hence the *recurrence risk* (for full sibs) is 4.2%. Taking the spouse value as the prevalence for this population ($K = 0.018$), the *relative risk* becomes

$$\lambda_S = \frac{21/496}{12/673} = \frac{0.042}{0.018} = 2.33$$

Thus, if you have a full sib with arthritis, you are more than twice as likely, compared to a random individual, to have arthritis. Similarly, the probability of two-sib families that are double-affected is expected to be $K^2\lambda_S = 0.018^2 \cdot 2.33 = 0.000755$. Hence, slightly under 0.08% of all two-sib families are doubly affected (both having arthritis). If one simply samples families at random, doubly affected families are rare. Conversely, if one samples families *given* that one sib is observed to have the disease, then the chance that the second sib has it is 4.2%.

Example 19.10. One of the constraints in computing a relative risk is that one must have a good estimate of the disease prevalence for the population from which the relatives were drawn. This can be obtained using a large random population sample, but this is typically a very inefficient approach, as for even a modestly rare disease, only a small fraction of those sampled will display it. An alternative approach is the **case-control** design, wherein one has sampled (by some criteria) a number of cases and then tries to obtain a matching set of controls (disease-free). Under this design, relative risk is not appropriate, as the sample disease frequency is highly skewed relative to its true population frequency. In such settings, the **odds ratio** is typically reported, because it does not depend on the ascertainment (selection) scheme. To see the difference between relative risks and odds ratios, consider the following data examining the association between a disease and a potentially predictive factor (such as a marker genotype):

	Disease		
Factor	Present (Case)	Absent (Control)	Totals
Present	a	b	a+b
Absent	c	d	c+d
Totals	a+c	b+d	

The **odds** of a particular event are defined as the probability of a success divided by the probability of a failure. Hence, if the odds of a particular event are 3:1, a success occurs three times more often than a failure, for a probability of 1/(3+1) = 0.75. The odds associated with a factor (such as a specific genotype or a specific environmental effect, e.g., smoking) are the probability of it being present—in a specific condition (such as being a case)—divided by the probability of it being absent. As a result, the odds for the presence of a factor in a diseased individual are

$$\frac{\Pr(\text{Factor Present} \mid \text{Disease Present})}{\Pr(\text{Factor Absent} \mid \text{Disease Present})} = \frac{a/(a+c)}{c/(a+c)} = \frac{a}{c} \tag{19.52a}$$

Similarly, the odds for the presence of a factor in a normal individual are

$$\frac{\Pr(\text{Factor Present} \mid \text{Disease Absent})}{\Pr(\text{Factor Absent} \mid \text{Disease Absent})} = \frac{b/(b+d)}{d/(b+d)} = \frac{b}{d} \tag{19.52b}$$

The **odds ratio** (**OR**) for the factor under these two different scenarios (odds when disease present versus odds when absent) thus becomes

$$\text{OR} = \frac{a/c}{b/d} = \frac{ad}{bc} \tag{19.52c}$$

Returning to Example 19.9, let the factor here being a sib, giving the odds ratio as

$$\text{OR} = \frac{21 \cdot 661}{12 \cdot 475} = 2.44$$

Hence, a diseased individual is 2.4 times more likely to be a sib (factor present) than a spouse (factor absent).

Assuming that the factor is being a (specific type of) relative (R), with the factor absent being unrelated individuals, the relative risk of R is

$$\lambda_R = \frac{a/(a+b)}{c/(c+d)} = \frac{a}{c} \cdot \frac{c+d}{a+b} \tag{19.52d}$$

$$\simeq \frac{ad}{bc}, \quad \text{when } c \ll d \text{ and } a \ll b \tag{19.52e}$$

Hence, when cases are only a small fraction of the samples with (a) and without (c) the factor, the relative risk is very close to the odds ratio. Case-control designs, however, are often balanced with the number of cases ($a + c$) and the number of controls ($b + d$) being fairly similar, in which case odds ratios and relative risks can be rather different.

Affected Sib-pair (ASP) Tests

When dealing with a dichotomous character, pairs of relatives can be classified into three groups: pairs where both are normal, **singly affected** pairs with one affected and one normal member, and **doubly affected** pairs. The first and last pairs are also called **concordant**, while pairs that differ are called **discordant**. The motivation behind relative-pair tests is that if a marker is linked to a QTL influencing the trait, concordant and discordant pairs should have different distributions for the number of IBD marker alleles.

In addition to being much more robust than ML methods for dichotomous characters, relative-pair tests also have the advantage of selective genotyping in that pairs are usually chosen so that at least one member is affected (e.g., Example 19.9). The pairs of relatives considered are usually full sibs, and a number of variants of these **affected sib pair**, or **ASP**, methods have been proposed (e.g., Day and Simons 1976; de Vries et al. 1976; Green and Woodrow 1977; Fishman et al. 1978; Suarez 1978; Suarez et al. 1978; Badner et al. 1984; Hopper et al. 1984; Suarez and Eerdewegh 1984; Lange 1986a, 1986b; Thompson 1986; Risch 1987, 1990a–1990c; Hauser et al. 1998; Holmans 1998; Faraway 1993; Commenges

1994; Curtis and Stam 1994; Kruglyak and Lander 1995b). Most of these are detection tests, rather than estimation procedures, as they cannot provide independent estimates of QTL effect and position. While our attention focuses on full-sib pairs, this basic approach can easily be applied to any pair of relatives, *provided* (as with HE tests) there is variability in the number of IBD alleles. (This excludes parent–offspring pairs, as these share exactly one allele IBD, and monozygotic twins, which share exactly two alleles IBD.) Most affected sib-pair tests have the basic structure of comparing the observed IBD frequencies (or some statistic based on them) of doubly affected pairs with either their expected values under no linkage or with the corresponding values in singly affected pairs. There are many possible tests based on this idea and most, it seems, have made their way into the literature. We consider three here. Greenwood and Morgan (2000) noted that these tests can be biased if some of the markers show segregation distortion, and show how to correct for this.

Among those n_i pairs with i affected members ($i = 0,\ 1,\ 2$), let p_{ij} denote the frequency of such pairs with j IBD marker alleles ($j = 0,\ 1,\ 2$). From the binomial distribution, the estimator $\widehat{p}_{ij}$ has mean p_{ij} and variance $p_{ij}\,(1 - p_{ij})/n_i$. One ASP test is based on $\widehat{p}_{22}$, the observed frequency of doubly affected pairs that have two marker alleles IBD. Under the assumption of no linkage, $\widehat{p}_{22}$ has mean 1/4 (as full sibs have a 25% chance of sharing both alleles IBD; Chapter 7) and variance $(1/4)(1 - 1/4)/n_2 = 3/(16\,n_2)$, suggesting the test

$$T_2 = \frac{\widehat{p}_{22} - 1/4}{\sqrt{\dfrac{3}{16\,n_2}}} \tag{19.53a}$$

For a large number of doubly affected pairs, T_2 is approximately distributed as a unit normal under the null hypothesis of no linkage. This test is one-sided, as $p_{22} > 1/4$ under linkage.

An alternative approach is to consider statistics that employ the mean number of IBD marker alleles, $p_{i1} + 2\,p_{i2}$. Under the hypothesis of no linkage, this has expected value $1 \cdot (1/2) + 2 \cdot (1/4) = 1$ and variance $[\,1^2 \cdot (1/2) + 2^2 \cdot (1/4)\,] - 1^2 = 1/2$. For doubly affected pairs, the test statistic becomes

$$T_m = \sqrt{2\,n_2}\,\left(\,\widehat{p}_{21} + 2\,\widehat{p}_{22} - 1\,\right) \tag{19.53b}$$

which again for large samples is approximately distributed as a unit normal and is a one-sided test, as $p_{21} + 2\,p_{22} > 1$ under linkage.

Finally, ML-based goodness-of-fit tests (Equation A4.11) can be used (Risch 1990b, 1990c). In keeping with the tradition of human geneticists, ML-based tests usually report LOD (likelihood of odds) scores in place of the closely related LR. (Recall from Equation 18.17 that 1 LR = 4.61 LOD.) Here the data are n_{20}, n_{21}, and n_{22}, the number of doubly affected sibs sharing zero, one, or two marker alleles IBD, with the unknown parameters to estimate being the population frequencies of these classes ($p_{20},\ p_{21},\ p_{22}$). The MLEs for these population frequencies are given by $\widehat{p}_{2i} = n_{2i}/n_2$. The LOD score for the test of no linkage becomes

$$MLS = \log_{10}\left[\,\prod_{i=0}^{2}\left(\frac{\widehat{p}_{2i}}{\pi_{2i}}\right)^{n_{2i}}\,\right] = \sum_{i=0}^{2} n_{2i}\,\log_{10}\left(\frac{\widehat{p}_{2i}}{\pi_{2i}}\right) \tag{19.54}$$

where π_{2i} is the probability that the pair of doubly affected sibs shares i alleles IBD in the absence of linkage to a QTL. (For full sibs, $\pi_{20} = \pi_{22} = 1/4$, $\pi_{21} = 1/2$.) The test statistic given by Equation 19.54 is referred to as the **maximum LOD score**, or **MLS**, with a score exceeding three being taken as significant evidence for linkage (Morton 1955b; Risch 1990b).

Genomic Scans and Multiple Comparisons

Starting in the early 1990s, the increased marker density of the human genetic map initiated the use of **genomic scans** for disease genes: progressively testing all scored markers in a genome for marker-trait associations. While our focus here is using linkage signals over a modest-sized collection of relatives, almost all such scans are now performed by GWAS, using LD signals over a large population sample of nonrelatives (detailed in the next chapter). To perform a linkage-based scan using a set of relatives (such as a collection of sib pairs), the investigator computes an ASP test statistic (such as the MLS value) at each of a number of markers, scanning the genome for markers showing linkage to DS genes. Pioneering studies include scans for type 1 (Davies et al. 1994) and type 2 (Hanis et al. 1996) diabetes and for multiple sclerosis (Ebers et al. 1996).

As discussed earlier in this chapter (and also in Chapter 18), one must adjust any scan for multiple comparisons given that each marker corresponds to a separate (and potentially correlated) test. Further, with multipoint mapping, one is essentially testing each point along a chromosome. While permutation tests provide a robust approach for controlling for false positives (by setting a chromosome-wide threshold to control the error rate), they are difficult to construct for relative-pair data given their very hierarchical structure (but see Wan et al. 1997 for one suggestion). As a result, human geneticists generally use analytic approximations to obtain the genome-wide thresholds to control the false-positive rate over an entire scan. As introduced in Chapter 18, there are two classes of such approximations. The first assumes a **sparse map**, with only a modest number of markers per chromosome being tested and hence adjusted Bonferroni approaches (Appendix 6) can be used. Alternatively, one can assume a **dense map**, wherein the test statistic $t(x)$ is computed at each location x on a chromosome of interest. In this setting, the test statistic is essentially computed at an infinite number of sites, generating a test *curve* for each chromosome (akin to a profile plot; e.g., Figure 19.3). A variety of approximations for this latter setting have been proposed, typically assuming uniform recombination rates along a chromosome of known length (Lander and Bostein 1989; Feingold et al. 1993; Dupuis and Siegmund 1999). For human data, Lander and Kruglyak (1995) suggested using a LOD score of around 4.0 as a signal for **significant linkage**, and a LOD score of 2.0 for **suggestive linkage** (potential locations to follow-up). However, these suggestions have generated much discussion (Witte et al. 1996; Curtis 1996; Lander and Kruglyak 1996; Kruglyak 1996).

Example 19.11. An alternative formulation for the MLS test is to consider each informative parent *separately*, simply scoring whether or not a doubly affected sib pair shares a marker allele from this parent. This approach generates 0 (match, both affected sibs share the allele) or 1 (no match) IBD data. Under the null hypothesis of no linkage, each state (0 or 1) has probability 1/2, and the MLS test statistic becomes

$$MLS = (n - n_1)\log_{10}\left(\frac{1-\widehat{p}_1}{1/2}\right) + n_1 \log_{10}\left(\frac{\widehat{p}_1}{1/2}\right)$$

where n_1 and $\widehat{p}_1 = n_1/n$ are, respectively, the number and frequency of the n sibs sharing the parental allele. This method has the advantage that sibs informative for only one parental marker can still be used. Using this approach, Davies et al. (1994) performed a genome-wide search for markers linked to DS genes influencing human type 1 diabetes. Among doubly affected sibs, one marker on chromosome 6, *D6S273*, had 92 pairs sharing parental alleles and 31 pairs not sharing parental alleles. A second marker on the opposite end of this chromosome, *D6S415*, had 74 pairs sharing parental alleles and 60 not sharing alleles. The MLS scores for these two markers are

$$MLS(\mathit{D6S273}) = 31 \cdot \log_{10}\left(\frac{2 \cdot 31}{123}\right) + 92 \cdot \log_{10}\left(\frac{2 \cdot 92}{123}\right) = 6.87$$

$$MLS(\mathit{D6S415}) = 60 \cdot \log_{10}\left(\frac{2 \cdot 60}{134}\right) + 74 \cdot \log_{10}\left(\frac{2 \cdot 74}{134}\right) = 0.32$$

Thus, the first marker shows significant evidence of linkage, while the second does not. Translating these LOD scores into LR values (the latter being distributed as a χ^2 with one degree of freedom) gives LR = $4.61 \cdot 6.87 = 31.6$ ($p < 0.001$) for *D6S273* and LR = $4.61 \cdot 0.32 = 1.47$ ($p = 0.2$) for *D6S415*.

Power of ASP Tests and Related Issues

The power of different sib-pair tests has been examined by a number of authors (Suarez et al. 1978; Blackwelder and Elston 1985; Goldin and Gershon 1988; Risch 1990a–1990c; Schaid and Nick 1990; Sribney and Swift 1992; Goldin and Weeks 1993; Green and Shah 1993; Shah and Green 1994; Risch and Merikangas 1996.; McGinnis 2000), who concluded that most of the power resides in doubly affected pairs. Blackwelder and Elston (1985) and Schaid and Nick (1990) compared the performance of tests based on the mean number of IBD alleles with those based on a single IBD frequency class (e.g., Equation 19.53b versus 19.53a). If dominance is modest or linkage is loose, T_m is a more powerful test than T_2. However, with strong dominance and tight linkage, the test based on T_2 is slightly more powerful. Sribney and Swift (1992) noted that assortative mating and multiple underlying loci can inflate the number of required sib pairs to achieve a given power. They suggested that the use of **discordant sib trios** (two normal and one affected, or vice versa) can offer higher power in these cases.

The IBD distribution for a marker locus linked to a DS gene was obtained for full sibs by Suarez et al. (1978) and for more general pairs of relatives by Risch (1990b). These expressions are useful for both power calculations (Appendix 5; Figure 19.4) and for showing how the DS gene parameters influence the IBD distribution. Because many tests use only doubly affected full-sib pairs, we restrict attention to this case. It is convenient to express the IBD distribution as the expected frequency under no linkage plus a deviation due to linkage,

$$p_{20} = \frac{1}{4} - d_{20}, \quad p_{21} = \frac{1}{2} - d_{21}, \quad p_{22} = \frac{1}{4} + d_{22}, \tag{19.55a}$$

Here, as above, p_{2i} is the probability that a doubly affected pair shares i IBD marker alleles. If c is the recombination fraction between marker and QTL, these deviations are given by

$$d_{20} = \gamma + 4\,\gamma\,\delta\,c(1-c) \tag{19.55b}$$
$$d_{21} = 2\,\gamma\,\delta\,(1-2c)^2 \tag{19.55c}$$
$$d_{22} = \gamma + 2\,\gamma\,\delta\,[\,(1-c)^2 + c^2] \tag{19.55d}$$

with

$$\gamma = \frac{(1-2c)^2}{4}\left(1 - \frac{1}{\lambda_S}\right) \quad \text{and} \quad \delta = \frac{\lambda_S - \lambda_{PO}}{\lambda_S - 1}$$

where λ_S and λ_{PO} are the relative risks in full-sib and parent–offspring pairs attributable to the DS locus. If there is no dominance at the DS locus ($\lambda_S = \lambda_{PO}$ and hence $\delta = 0$), then $d_{21} = 0$, so this class provides no information on linkage. Further, in this case the magnitude of the deviations for p_{20} and p_{22} are identical (γ), and hence each class provides the same amount of linkage information. When dominance is present, $d_{22} > d_{20}$, implying that the p_{22} class provides the most linkage information. Equation 19.55 assumes that a single linked locus underlies the disease. When multiple loci are involved, the marker IBD distribution is no longer a simple function of the population λ_R values (Risch 1990a, 1990b).

The power using different types of relative pairs was examined by Risch (1990b, 1990c) and Bishop and Williamson (1990). For diseases with large λ_S values, grandparent–grandchild pairs offer the highest power of detection when tightly linked markers are available. For diseases with smaller λ_S values and/or less tightly linked markers, sibs offer the most power. The construction of confidence (likelihood, bootstrap) or credible (Bayesian) intervals for QTL location with ASP data was examined by Papachristou and Lin (2006a, 2006b), Sinha and Luo (2007), and Sinha et al. (2009).

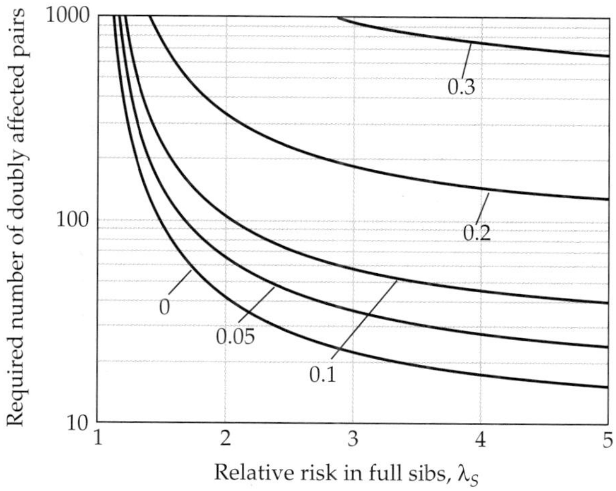

Figure 19.4 Sample size required to detect a QTL using the T_m test (Equation 19.53b). Each curve represents a different value for the recombination fraction c between the QTL and linked marker, and plots the number n_2 of doubly affected sib pairs required to detect (with 75% power at the 5% significance level) a linked QTL with no dominance variance as a function of λ_S for that QTL. If dominance is present, the required n_2 are below the plotted values, so these curves serve as upper bounds on the desirable sample size regardless of the amount of dominance. See Appendix 5 (Equation A5.12c) for these power calculations.

An important issue related to power is the *verification* of previous linkage claims. Such replication was especially problematic in early linkage-based studies claiming associations between markers and certain psychiatric disorders (reviewed in Gershon and Cloninger 1994; Risch and Botstein 1996). An analysis of this problem was presented by Suarez et al. (1994), who found that when the character is determined by several QTLs of roughly similar effects, the number of families required to detect the *first* marker-trait association is much smaller than the number of independent families required to *replicate* the initial claim (also see Bell and Lathrop 1996). In effect, the waiting time (measured as number of sib pairs that must be sampled) required for replication of a claim is far greater than the waiting time to detect an initial association. This result occurs because if the trait is determined by n QTLs of equal effect, and if p is the probability of detecting a given QTL (i.e., the power of the design), then there are n chances for detecting significant marker effects (and hence a power of approximately np), but only one chance of replicating the initial observation (with probability p). This waiting time is especially problematic if a number of rare genes of large effect are involved, as affected individuals in one set of families may arise from one gene, while a second set of families involves another QTL and hence does not replicate the original finding. With the massive sample sizes of modern GWAS, replication is much more common (Wray et al. 2012, 2014; Levinson et al. 2014; Chapter 20).

Multipoint Affected Sib-pair Tests

The evolution of many of the above ASP methods (as was the case for line-cross and half-sib family mapping) was from single marker to single interval to multipoint mapping (simultaneously incorporating information from all linked markers on a chromosome; e.g., Equations 18.26, 19.17, 19.18, and 19.46). This natural progress in statistical methodology tracks improved genomic ability to obtain ever-denser sets of markers. Such an extension for ASP was developed by Kruglyak and Lander (1995b), who used a multipoint method to estimate the IBD distribution at each location along a chromosome. This estimated distribution is then used to compute some ASP statistic for that point, allowing for the generation of a support profile for QTLs as one scans along a chromosome.

For example, for an arbitrary chromosomal map position c, Kruglyak and Lander obtained estimates of $\pi_{20}(c)$, $\pi_{21}(c)$, and $\pi_{22}(c)$—the IBD values at this position in doubly affected sib pairs—which can then be used to compute any ASP statistics, such as MLS. This method allows for a genomic scan of all chromosomal positions, not just those points corresponding to marker loci. As with likelihood maps, estimates of the locations of DS genes are given by local maxima of the test statistic plot, provided they exceed the significance threshold. With the putative DS gene position in hand, estimates of its effect can be obtained by using the estimated IBD distribution at that position.

Recall that the effect of a QTL can be expressed in terms of the relative risks in full-sib and parent–offspring pairs (λ_S and λ_{PO}). First noting that estimates of d_{20} and d_{21} can be obtained from the estimated IBD values, $\widehat{d}_{20} = 1/4 - \widehat{\pi}_{20}$ and $\widehat{d}_{21} = 1/2 - \widehat{\pi}_{21}$, substitution of these into Equations 19.55b and 19.55c (with $c = 0$) and rearranging gives

$$\widehat{\lambda}_S = \frac{1}{1 - 4\,\widehat{d}_{20}} = \frac{1}{4\,\widehat{\pi}_{20}} \quad \text{and} \quad \widehat{\lambda}_{PO} = \frac{1 - 2\,\widehat{d}_{21}}{1 - 4\,\widehat{d}_{20}} = \frac{\widehat{\pi}_{21}}{2\,\widehat{\pi}_{20}} \tag{19.56}$$

If desired, these can be translated into estimates of genetic variances using Equations 19.49b and 19.51b. Kruglyak and Lander (1995a) found that 200 affected pairs were sufficient to localize a gene with effect $\lambda_S = \lambda_{PO} = 5$ to within 1 cM. Localization of a gene with $\lambda_S = 2$ with the same precision requires 700 pairs. This may seem hopeful, given that (from above) diabetes has $\lambda_S = 15$ and rheumatoid arthritis $\lambda_S = 2$. In reality, however, these latter risks are for the *entire set* of DS genes, while the individual gene-specific risks are usually just a small fraction of the total.

Affected Pedigree Member Tests

In many cases, the investigator may have a number of pedigrees, each containing several affected members. While an affected sib-pair analysis could be used, there is potentially more information to be gained by considering *all* affected relatives, not just those that happen to be sibs (or some other group of like relatives). This is the motivation behind an **affected pedigree member** (**APM**) analysis, wherein all pairs of doubly affected relatives within each pedigree are examined in a single analysis. One immediate obstacle to such an analysis is ascertaining the IBD status for marker alleles within each pair of relatives. At a minimum, this requires genotyping all individuals within a pedigree, not just those who are affected. Further, for late-onset diseases (such as Alzheimer's), it is often impossible to genotype the parents, and hence estimates of IBD status of sib pairs can be equivocal, at best. Instead of working with IBD status, Lange (1986a, 1986b) suggested using the **identity by state** (**IBS**) status (Chapter 8) among marker alleles. While IBS status is trivial to ascertain, IBS-based tests are expected to have lower power unless the marker locus is sufficiently polymorphic that IBS status is a strong indicator of IBD status.

While Lange's initial suggestion was to use IBS status for pairs of affected sibs, Weeks and Lange (1988) generalized this approach to consider all pairs of affected members within a pedigree, and a number of extensions have been proposed (Weeks and Lange 1992; Ward 1993; Brown et al. 1994; Schroeder et al. 1994; Whittemore and Halpern 1994; Davis et al. 1996; Song et al. 2002). The APM test of Weeks and Lange is developed as follows. Consider a pair of affected relatives, i and j, which have marker genotypes $M_{i1}M_{i2}$ and $M_{j1}M_{j2}$. If the marker is linked to a DS gene, doubly affected relatives should have more similar marker genotypes than expected by chance alone. The simplest measure of the shared marker alleles between relatives is to consider

$$Z_{ij} = \frac{1}{4}\sum_{k=1}^{2}\sum_{\ell=1}^{2} \delta(M_{ik}, M_{j\ell}) \quad \text{where} \quad \delta(x,y) = \begin{cases} 1 & \text{if } x = y \\ 0 & \text{otherwise} \end{cases} \tag{19.57}$$

which is just the probability that a randomly chosen marker allele from relative i is identical in state to a random marker allele in relative j (the molecular similarity; Chapter 8). Note that Z_{ij} is trivial to compute for *any* pair of affected relatives. Weeks and Lange's test statistic for a pedigree with n (genotyped) affected relatives is

$$T = \frac{Z - E(Z)}{\sigma(Z)} \qquad \text{where} \qquad Z = \sum_{i<j} Z_{ij} \tag{19.58a}$$

the sum being taken over all $n(n-1)/2$ pairs of doubly affected relatives. For sufficiently large sample sizes, T is approximately distributed as a unit normal, where the $E(Z)$ and $\sigma^2(Z)$ values under the null are computed from pedigree information; see Weeks and Lange for details. Improvements were offered by Jung et al. (2006), who suggested using **gene-dropping** simulations (hypothetical alleles are assigned [dropped] to each pedigree founder, with offspring simulated under Mendelian segregation; MacCluer et al. 1986) should be used unless the pedigree is very large.

Because the measure Z_{ij} given by Equation 19.57 weights all genotypes equally, Weeks and Lange suggested that a more reasonable statistic weights matching IBS alleles by some function (f) of their allele frequencies,

$$Z_{ij} = \frac{1}{4} \sum_{k=1}^{2} \sum_{\ell=1}^{2} \delta(M_{ik}, M_{j\ell}) f(p_{M_{ik}}) \tag{19.58b}$$

where $p_{M_{ik}}$ is the frequency of allele M_{ik}. The motivation for this modification is that a match involving a rare allele in distant relatives should be weighted more than a match involving a frequent allele (e.g., Example 19.7). Weeks and Lange suggested using $f = 1/\sqrt{p}$ as a compromise function weighting rare alleles but preserving normality. One criticism of this method is that the weighting functions are rather ad hoc, and it is possible that some weight functions may give significant results while others do not, making the interpretation of linkage somewhat problematic. One solution might be to simply use weights that minimize the sampling variance of the Z_{ij}, which is a standard procedure. A second criticism is that IBS-based tests are sensitive to estimates of the marker-allele frequencies and can give false positives (indications of a linked QTL when none is present) when incorrect frequencies are used (Babron et al. 1993; Van Eerdewegh et al. 1993).

Exclusion Mapping and Information Content Mapping

While the support for a QTL is often presented by likelihood (or related) plots, two other types of plots can provide useful mapping information. **Exclusion mapping** (Edwards 1987; Cook et al. 1980) displays the support for the hypothesis that a QTL is *absent* at a particular position. With a standard LOD score plot, one tests the data against the null hypothesis of no linked QTL. In contrast, under exclusion mapping, LR test statistics are still plotted, but now the null hypothesis is that the position contains a QTL of specified (or greater) effect. An **exclusion LOD score** of -2 or less is taken as sufficient evidence to *exclude* the presence of a gene of this effect or larger segregating within the sample population (Figure 19.5). Using this approach, Ebers et al. (1996) were able to exclude 88% of the genome of their sampled population from containing a DS gene for multiple sclerosis of effect $\lambda_S \geq 3.0$. Similarly, Hyer et al. (1991) and Hanis et al. (1996) were able to exclude regions from containing major genes influencing diabetes. Further details can be found in Hyer et al. (1991), Risch (1993), Kruglyak and Lander (1995b), and Hauser et al. (1998). One major caveat is that even a highly significant exclusion only refers to the sample data. It is possible that a DS gene of major effect is segregating in the population, but has been missed by the sampling scheme.

Another useful type of plot is an **information content map**, which graphs the amount of marker information at each point along a chromosome (Figure 19.6; Kruglyak and Lander 1995b). Here, one plots the r^2 value for a model that predicts the IBD distribution for each point along the chromosome. Regions with high r^2 values imply that the IBD distribution is estimated with little error, so additional markers are not needed to predict IBD status. In contrast, regions with smaller r^2 values would benefit by increasing marker density, as this would improve the ability to predict the IBD status in regions between markers.

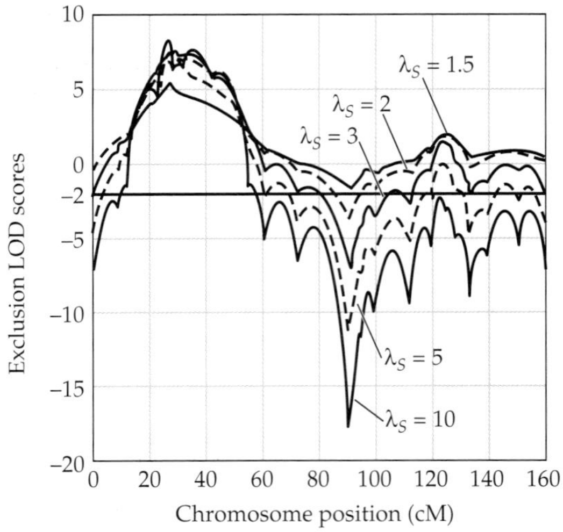

Figure 19.5 LOD scores for exclusion mapping, under the assumption of a DS gene with effect λ_S (and no dominance, so $\lambda_{PO} = \lambda_S$) for a scan of human chromosome 6 for genes influencing diabetes susceptibility (Davies et al. 1994). A LOD score below -2 provides sufficient evidence that a DS gene of specified effect (or greater) is *absent* from this region. For example, a DS gene with $\lambda_S \geq 5$ can be excluded from chromosome positions 70–120 cM and 130–140 cM. (From Kruglyak and Lander 1995b.)

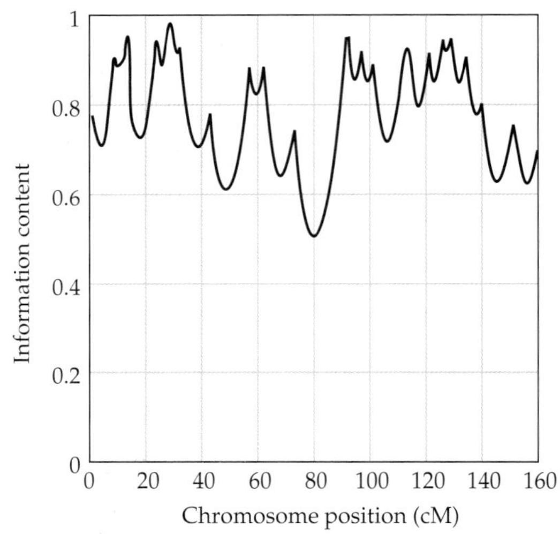

Figure 19.6 Information content mapping of a chromosome 6 diabetes scan (Davies et al. 1994) shows those regions where increasing marker density will provide more linkage information (e.g., 70–90 cM and 140–160 cM). Information content is highest at the markers, falling off in the intervals between markers. (From Kruglyak and Lander 1995b.)

Allele-sharing Tests Using Selected Samples of Continuous Traits

A number of authors have coupled selective sampling of continuous traits with allele-sharing methods to increase power for mapping. The idea—a logical extension of ASP tests—is to use pairs of sibs that are concordant or discordant for *trait values* (i.e., excessively similar or excessively dissimilar) and then conduct some IBD allele-sharing test. As with disease states, concordant pairs should have higher IBD sharing at a trait QTL, while discordant pairs should have lower IBD sharing. Risch and Zhang (1995, 1996) noted that discordant pairs have more power than concordant pairs, in part because the later could result from a common environmental effect inflating trait values. As a result, Eaves and Meyer (1994) and Risch and Zhang (1995, 1996; Zhang and Risch 1996) proposed using **extremely discordant sibling pairs (EDSPs)**, situations where one sib in a pair is in the top 10%, and the other is in the bottom 10%, of trait values. Obviously, such pairs are rare. They would comprise less than 1% of randomly sampled sib pairs when the sib values are independent ($0.1 \cdot 0.1 = 0.01$), and hence a much smaller fraction when sibs are positively correlated (as expected when there is some genetic basis to the trait). Risch and Zhang noted that this strategy reduces the amount of genotyping (to obtain the same power) by 10 to 40-fold. However, they also noted that the smallest detectable effect, for realistic sample sizes, would be one accounting for 10% of the phenotypic variance.

Others have proposed using **moderately discordant sibling pairs (MDSPs)**. Zhao et al. (1997) suggested that a more optimal sampling would be one sib in the upper 10% and the second in the lowest 30%, while Forrest and Feingold (2000) suggested using sibs in the top and bottom 35% of the trait distribution. Gu et al. (1996) proposed a modified EDSP design, **extreme discordant and concordant (EDAC)** sampling, that also includes extremely *concordant* sib pairs, and then tests the difference in pairwise IBD sharing between these two groups (also see Kruse et al. 1997; Knapp 1998; Dolan and Boomsma 1998; Liang et al. 2000; Li and Gastwirth 2001; Szatkiewicz and Feingold 2005). Additional comments and tests are reviewed by Szatkiewicz et al. (2003), T.Cuenco et al. (2003), and Szatkiewicz and Feingold (2004).

Again, the utility of these approaches is now problematic in many settings, in that they assume phenotyping costs are dramatically lower than genotyping cost. In the genomics era, the converse is often true. In reality, one would genotype essentially all phenotyped sib pairs in most settings. Further, the restriction of gathering only sets of relatives is far more onerous than collecting a much larger random population sample needed for a GWAS setting.

LIMITATIONS OF LINKAGE-BASED QTL MAPPING IN OUTBRED POPULATIONS

A number of limitations were raised in the previous chapter for linkage-based QTL mapping using line crosses: poor mapping resolution (megabase scale), low power resulting in overestimation of effects declared to be significant (Beavis effects), and potential fractionation of major QTLs into a number of smaller effect loci as recombination breaks up a linkage block. Most of these issues result from the modest sample sizes (a few hundreds) of typical line-cross designs, something that also plagues most outbred linkage designs. All of these issues, coupled with typically *even lower power*, limits the utility of linkage-based QTL mapping using outbred families.

As a result, our view is that linkage-based mapping (in either a line-cross or outbred setting) is best considered as a method for mapping **oligogenic genes**: QTLs accounting for at least 10% of the phenotypic variance. Genes with smaller effects are very difficult to detect given the standard sampling structure of linkage-based approaches, as the large number of sets of relatives (or large pedigrees) required for sufficient power is generally difficult to obtain. While selective sampling approaches, such as using concordant or discordant sibs, can increase power, one must shift through a much larger number of families to find these. This is the issue of selective genotyping, which is generally not relevant in the post-genomic era of rapid and cheap sequencing. Indeed, for many traits, the resource bottleneck is in the phenotyping, not the genotyping.

As a result of these limitations, coupled with massive advances in genomics, much of the theoretical work on HE regression, variance components, and affected sib methods, reached its heyday in the early 2000s. Following the classic paper of Risch and Merikangas (1996) showing that association studies could have comparable (or greater) power, coupled with a much simpler sampling designs (random individuals, rather than a strict requirement for relatives), work rapidly shifted to GWAS, especially given the exponentially increasing marker density of the human genetic map. Although most high-powered QTL studies now use association mapping (Chapter 20), this method builds on many of the concepts from this chapter.

Finally, there are still situations where linkage mapping may be more feasible than association studies. One setting is where very large families are easy to obtain, such as with trees, fishes, and many invertebrates. Indeed, in many settings obtaining such a large family might be much easier than obtaining a random population sample. Suppose that one wanted to evaluate the genetic architecture of growth rates. A natural starting point would be a set of individuals all at the same developmental stage (such as seeds or eggs). Such a collection would be hard to contemporaneously sample from a population. Conversely, it would be much easier to generate these from a cross, placing the study in a linkage-analysis setting. A second concern, but one that is quickly diminishing as genomics continues to flower, is that a species or population might not yet have the genomic resources in place for dense marker maps needed for GWAS (e.g., Massault et al. 2008). However, the explosive growth of sequencing technologies makes these concerns increasingly irrelevant.

20

Association Mapping

The initial focus of a GWAS should be on the study design, determining the specific trait for investigation, the ascertainment of samples, and the choice of SNPs to be interrogated for association. Without thoughtful consideration of these design issues ... the results of any downstream association will be meaningless. Morris and Cardon (2019)

The original idea behind an association analysis (Chapter 17) was to test a population sample for correlations between different genotypes at a candidate gene and trait values (or disease risk). While the scored genotypes in early studies were often presumed to be causal sites (such as SNPs coding for different amino acids), this was not always the case. More generally, the scored genotypes were simply markers in, or near, a candidate gene, assumed to be in linkage disequilibrium (LD; Chapter 5) with causal sites (QTLs). With the rise of genomic technologies that score hundreds of thousands of markers, this notion of a hypothesis-driven test of the impact of a *specific locus* expanded to an exploratory *scan of the genome* (with no specific candidates in mind), searching for associations over a dense set of markers covering the genome. This strategy is commonly referred to as a **genome-wide association study**, or **GWAS**. In the era of **electronic medical records** (**EMRs**), human geneticists often perform a series of GWAS over thousands of recorded traits, which Denny et al. (2010) and Zhou et al. (2018) referred to as a **phenome-wide association study** (**PheWAS**).

GWAS is now the dominant means of searching for QTLs. While the focus of GWAS studies is typically on humans, equally impressive work has been done in other species, such as with maize using the NAM lines (Chapter 18). As detailed below, a limiting factor to performing a GWAS is the ability to accurately, and cheaply, score vast numbers of markers. Genomics has made this economically feasible in humans, model organisms (e.g., mice, rats, *Drosophila*, *Arabidopsis*), high-value agricultural species (e.g., maize, rice, dairy cattle, chickens), and is becoming increasingly feasible for an ever-growing list of species. While GWAS usually involves either a random sample of individuals from a defined population or a specialized sample of cases with a matching set of controls, plant breeders and investigators using model organisms often use a collection of inbred lines as their GWAS population. Such a mapping population is referred to as an **association panel**.

Much of the basics of a GWAS follow from the marker-trait approaches used in linkage-based mapping (Chapters 18 and 19). This chapter largely focuses on the design and analysis aspects of a GWAS, while Chapter 21 examines the data generated by GWAS (as well as by linkage-based studies) to provide an overview of our current picture of the genetic architecture of traits and diseases. Finally, an important point to stress at the outset of this chapter (and the next) is that even though most methods discussed (at the time of this writing) were often less than a decade old, some are already "legacy" approaches given the speed and scale of development in the field.

MOTIVATION AND HISTORY

LD Assocation Between a Marker and a QTL

The marker signal from a nearby QTL in both linkage-based (Chapters 18 and 19) and association mapping (Chapter 17) is a *marker–trait association*: the different genotypes at the marker have different mean trait values (or differental risks for binary traits). Such signals arise when the structure of the mapping population generates correlations between markers and nearby (linked) QTLs. This naturally occurs in families (due to the excess of parental over recombinant gametes for linked loci; Example 5.6). Such associations also occur in a population due to linkage disequilibrium (LD; Chapters 5 and 17), although spanning a much smaller genomic scale than that generated by linkage following a cross.

To see how naturally occurring LD arises, suppose a QTL allele, q, undergoes a mutation to an new functional allele, Q. If this happens close to a SNP, there is an initial association with the SNP allele on the chromosome (or **background haplotype**) on which the mutation arose. Suppose that an Mq chromosome mutates to MQ. Initially Q is *only* found with marker allele M. At the time of the mutation, $|D'| = 1$ (Example 5.5), namely **complete disequilibrium** (no recombination between the alleles is seen in the sample), as Q-bearing chromosomes *always* contain M. However, it is r^2, the correlation between alleles, that measures how much of the signal at a causative site is imparted onto a nearby marker, with $r^2 = 1$ called **perfect disequilibrium**. As Example 5.5 showed, here $r^2 < 1$, as M-bearing chromosomes can be found *without* Q, so that the $M-Q$ correspondence is not perfect.

Equation 5.13d showed that the intial value of r^2 decays at a rate of $(1-c)^2$ per generation. Thus, if c is very small, the bulk of any initial associations can persist for thousands of generations. Assuming a 1% recombination rate per megabase, then for two loci separated by 10 kilobases, $c = 0.01 \cdot (10/1000) = 0.0001$. The expected time ($t_{1/2}$) for 50% of the initial r^2 value to decay in this setting satisfies $(1-0.0001)^{2t_{1/2}} = 0.5$. Solving gives $t_{1/2} = \ln(0.5)/[2\ln(0.9999)] \simeq 3466$ generations. Assuming a 25 year human generation time, this is roughly 87,000 years. After 10,000 generations, the r^2 value is still 14% of its intial value. If the markers are one kilobase apart, then 98% of the intial r^2 value is present after 1,000 generations (again assuming a c of 0.01 per megabase).

The extent of LD within a population is strongly influenced by its past evolutionary history. The **time to the most recent common ancestor** (**TMRCA**) largely sets the size of the **LD block**, as the longer the TMRCA, the more generations a region has to experience recombination. Because TMRCA scales with the effective population size N_e (WL Chapters 2, 3, and 8), populations with historically large N_e should show smaller average LD-block sizes than historically smaller populations, or than populations that passed through a bottleneck (as is often the case with domesticated lines).

This point is well made by the data on crop LD reviewed by Buckler and Gore (2007). In maize, LD spans less than a kilobase (kb) for wild populations and land races, yet is greater than 100 kb for elite inbreds. In barley, LD spans less than a kb in wild populations, between 80 and 100 kb in land races, and greater than 200 kb in elite lines, while soybeans have LD of around 40–80 kb in wild populations, but greater than 300 kb in elite lines. In humans, Africans show much shorter blocks of LD than do European or Asians. Africa is the ancestral (and hence older) population, while the latter two populations are the result of migrations from this base population that passed through bottlenecks.

Because of population-specific variation in LD block size (which also varies widely *within* the genome of a given population), before performing a GWAS, one should employ a large set of randomly spaced SNPs to examine the extent of LD in the target population or association panel. Such a pilot study informs the investigator of the average marker density needed, and hence the limit of mapping resolution set by the sample LD.

Example 20.1. A common misconception in GWAS studies is the assumption that *markers closer to a causal site will have larger LD values*, and hence larger marker effects (via larger r^2 values). ***Such need not be the case.*** Consider a new causal allele, Q, that arose on an NM marker background (haplotype), where the QTL locus is much closer to M than N. Suppose that recombination is sufficiently rare such that no Q alleles are found on any other haplotypes in the GWAS sample. $D'_{MQ} = D'_{NQ} = 1$ in this case (as Q is *only* found on an N or an M background), but their r^2 values (which determine how much of the actual variance is accounted for by the marker variance) are a function of the marker-allele frequencies.

In particular, Equation 5.15c gives the correlation between M and Q as

$$r_{MQ}^2 = \frac{\alpha_M(1 - p_M)}{1 - p_Q} = \frac{\alpha_M - p_Q}{1 - p_Q} \leq \alpha_M$$

where p_M is the frequency of the marker allele (M) and α_M the fraction of M alleles associated with Q (so that $\alpha_M p_M = p_Q$ is the frequency of Q). Suppose that the frequency of Q is 3%, while the frequencies of M and N are, respectively, 60 and 5%. Here p_M = 0.6 and α_M = 3/60 = 0.05, giving r_{MQ}^2 = (0.05 − 0.03)/(1 − 0.03) = 0.0206, so that the M marker variance accounts for only 2.06% of the causal (actual) variance. Conversely, for the more distant marker, N, p_N = 0.05 and α_N = 3/5 = 0.6, giving r_{NQ}^2 = (0.6 − 0.03)/(1 − 0.03) = 0.588. Thus, the marker variance for the *more distant site* captures almost 60% of the actual (causal) variance, a 30 fold increase over the marker variance for the closer site, M. This is an illustration of the concept from Chapter 5 that r^2 LD values are ***largest when the causal and marker allele frequencies are similar***, and fall off as their absolute frequency difference increases (i.e., as α_M becomes smaller). This impacts the power of a GWAS to detect common versus rare alleles.

Advantages of Association Over Linkage Mapping

The limitations of linkage-based QTL mapping (Chapters 18 and 19) revolve around their need to be *based entirely on sets of relatives* (pairs, trios, families, or more extended pedigrees). This restriction has two consequences. First, it severely constrains the *obtainable sample size* of an experiment, as gathering large sets of relatives is nontrivial. Smaller samples constrain the SNP effect sizes that can be detected, with sample sizes in the thousands needed to reliably detect QTLs of modest effect (less than 5%; Table 19.2). A related issue with linkage-based studies, especially for rare diseases, is that families often show **singletons**, with only one family member being affected, providing no linkage-based information. (Recall from Chapter 17 that we can exploit signals from such singletons when LD is present, e.g., Examples 17.15 and 17.19.) Second, any two random gametes from relatives in a linkage-based design likely have experienced only a very small number of recombination events since their shared ancestor. As a result, among pedigree individuals, LD spans a significant fraction of a chromosome (at least tens of megabases), restricting the level of resolution for fine mapping.

In contrast, it is straightforward (at least conceptually) to obtain a large or very large population sample. Human GWAS studies routinely use tens of thousands of individuals, offering enormous power to *detect* small marker effects (differences in the mean values of genotypes at a given marker). Meta-analysis (Appendix 6) can be used to combine multiple GWAS, generating total sample sizes in the hundreds of thousands to millions. Further, the association signal is generated by very fine-scale LD, often on the order of tens of kilobases (or less). Two individuals in a linkage study have experienced only a few rounds of meioses, while two unrelated individuals from an association sample may be separated by hundreds or thousands of meioses (Chapter 8). As a result of this disparity in recombination history, linkage generates very long-range associations, while LD associations decay very quickly and only persist over very small scales (Chapters 5, 8, and 17). Linkage studies, however, can have an advantage when causal alleles are very rare, as a population-level rare allele can be common within carefully selected pedigrees (such as those showing an unusual number of cases, early-onset cases, or extreme phenotypes).

The History of GWAS

The idea of exploiting population-level LD with nearby markers to fine-map a gene beyond what is possible with linkage-based approaches was suggested by Bodmer (1986), and also by Lande and Thompson (1990), the latter with respect to finding molecular markers for the construction of marker-selection indices. Hästbacka et al. (1992) and Lehesjoki et al. (1993) noted that LD mapping (Chapter 17) allows for fine mapping (resolution of

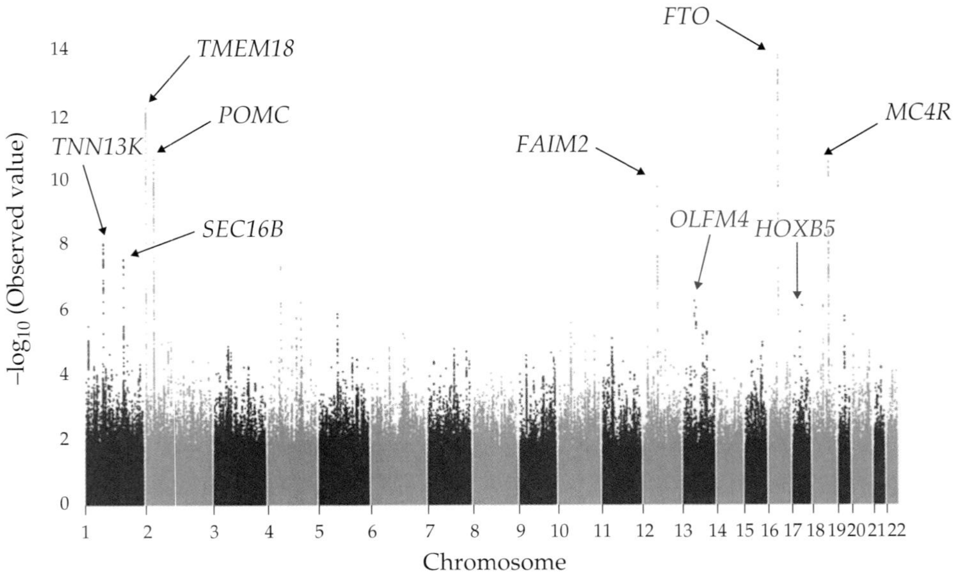

Figure 20.1 A **Manhattan plot**, where each pixel represents the position of a marker in the genome (horizontal axis, here arranged from chromosomes 1 to 22) and its significance, the value of $-\log_{10}(p)$, on the vertical axis. Sufficiently high peaks represent highly significant marker-trait associations, which are usually clustered, as nearby SNPs are typically in LD with each other. For most of the genome, the markers are so dense, and with only modest p values, that the pixels merge into a solid color. As one zooms in on a genomic location, these separate into individual pixels (Figure 20.2). In many plots, clear regions (a horizontal span with no vertical pixels) correspond to the centromeric regions of chromosomes, as these are often marker-poor (have a very low SNP density).

less than one cM; $c \leq 0.01$), even for rare diseases where there are simply not a sufficient number of families to provide the number of meioses required to separate closely linked sites (Example 17.15). This approach *starts with a known major gene* that has been *localized to a relatively small interval via linkage*, which is then fine-mapped using LD. Lyman et al. (1999) were among the first to exploit this idea in nonhuman populations, showing that certain *Drosophila* polymorphisms in the *scabrous* locus had an impact on bristle number. The notion that this same logic could apply in a genomic scan for QTLs was first emphasized in the very influential paper by Risch and Merikangas (1996), who pointed out that association studies have more power than sib-pair (allele-sharing; Chapter 19) approaches for detecting genes of modest to small effects. In contrast to the candidate-gene approach, this is a *hypothesis-free approach* (sites are simply chosen because they collectively span the genome).

Given the costs and logistical efforts involved, it is not surprising that the first large-scale GWAS were for human diseases. One of the first was that of Ozaki et al. (2002), who examined myocardial infarction risk in a Japanese population, using a sample of 1,133 cases and 1,006 controls. Of these, an initial sample of 94 cases and 658 controls were typed for 66,000 SNPs, searching for SNP alleles that were over-represented in the cases. One strong signal was a five-SNP haplotype that spanned 50 kb, which included two SNPs in the lymphotoxon-α, LTA, gene. A follow-up analysis within this region with a set of 1,133 cases and 1,006 controls showed that individuals homozygous for the two LTA SNPs had a significant inflated risk (odds ratio of 1.78) of myocardial infarction.

This pioneering work was followed by two studies on the eye disease age-related macular degeneration (AMD) by Klein et al. (2005) and Dewan et al. (2006). Klein et al. examined 96 cases and 50 controls in Caucasians, scored for 166,000 SNPs. Individuals homozygous for a detected risk marker had a seven fold increase in AMD. This SNP was in LD with an allele of complement factor H with a tyrosine to histidine substitution at residue

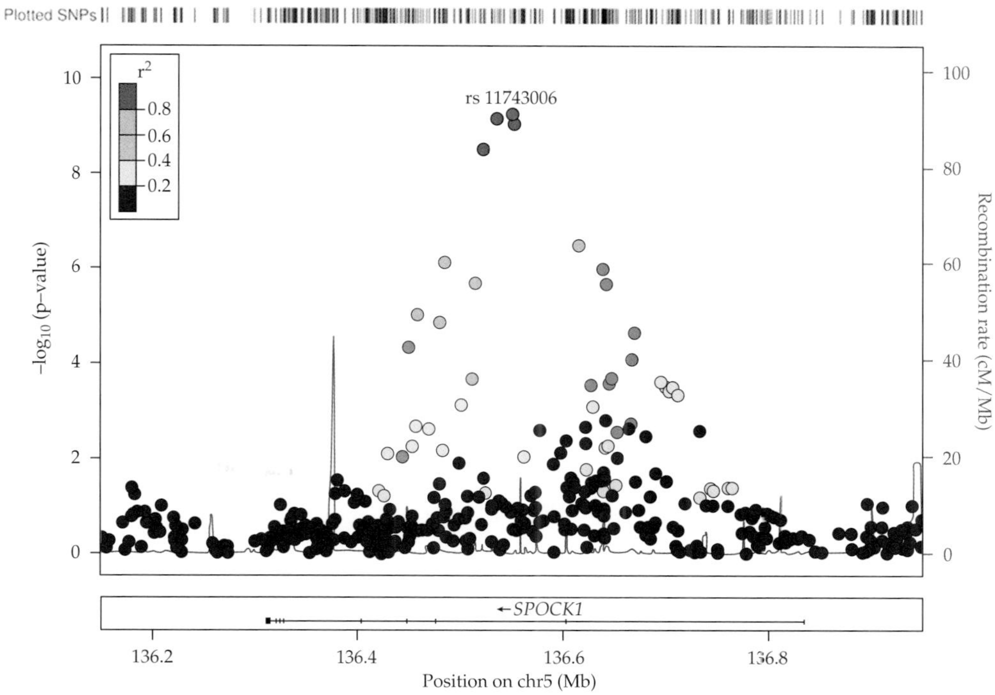

Figure 20.2 A **signal plot** for a cluster of significant SNPs, displaying the fine-scale view of a candidate region, often with functional genomic information. Here, the focus is the region around the *SPOCK1* gene on human chromosome 5, which Chen et al. (2017) found was associated with mathematical ability. The top of the plot shows the location of tested SNPs (tick marks), while the bottom of the plot shows the location on chromosome 5 (136.2 to 136.8 Mb), along with the *SPOCK1* gene and its direction of transcription. The left vertical axis is the $-\log_{10}(p)$ value for a SNP (displayed as a filled circle), while the right vertical axis is the value of localized recombination rate (displayed as a continuous curve; note the recombination hot spot just below 136.4 Mb). Finally, r^2 values for each SNP with respect to the lead SNP (rs11743006) are indicated by the shade of their filled circle. (After Chen et al. 2017.)

402. Dewan et al. examined 96 cases and 130 age-matched controls in a Chinese population, scored for 98,000 SNPs, finding a strong signal for AMD risk in the promoter region of $HTRA1$, a serine protease gene.

While technologically impressive for their time, all three of these studies were woefully underpowered, given the small number of genotyped cases. Hence, they could only detect major genes, and runned the risk of seriously overestimating SNP effects (Beavis effects; Chapter 18). The first truly modern GWAS was the **Wellcome Trust Case Control Consortium, WTCCC**, (2007), which scored 14,000 cases of seven common diseases (roughly 2,000 cases per disease) with 3,000 shared controls, using 500K SNPs genotyped using a Affymetric gene chip array. The WTCCC study established many of the GWAS protocols currently in use today, e.g., stringent quality control (QC) on SNP calling and marker choice, the use of Manhattan (Figure 20.1) and signal (Figure 20.2) plots, adjusting for population structure, use of q–q plots (Figure 20.4) to examine model fit, dealing with multiple comparisons, and stressing that more QTLs of smaller effect would likely be detected in a larger sample size.

Support for a QTL at a particular genomic location is usually displayed by a likelihood profile plot under linkage mapping (e.g., Figures 18.1, 19.3, and 19.5) because the effects of linkage typically span large regions of a chromosome. Conversely, under a GWAS, an LD block typically covers only a very tiny fraction of a chromosome. As a result, each tested marker is associated with a pixel that represents the p value for a test of marker-trait association, such as a Cochran–Armitage trend test (Example 18.5), a chi-square test, or a marker-dosage regression (Equation 18.14a). Note that these are the same single-marker association tests used with linkage studies (Chapters 18 and 19). The resulting plot has marker location on the horizontal axis and (usually) $-\log_{10}(p)$ on the vertical axis, so that higher values indicate greater significance. The result is a smear of pixels (usually so dense that they appear solid), out of which occasionally trickles a few pixels arising above some threshold (Figure 20.1). Such displays are typically called **Manhattan plots** (or, more whimsically, **Dubai plots**), for the skyline of Manhattan with its sky scrapers. An unsuccessful GWAS often generates what some have called a **Manhattan, Kansas plot**, which, like the afore mentioned town, is noted for its flatness and lack of any significant peaks. While a Manhattan plot gives a quick snapshot of the entire genome, **signal plots** (also called **regional association plots**) around significant peaks provide a more focused picture (WTCCC 2007). These depict the fine structure of the SNPs within a small region, and often incorporate additional genomic information, such as known open reading frames and localized recombination rates (e.g., Figure 20.2).

BASICS OF GWAS DESIGNS

The success, or failure, of a planned GWAS often hinges on the amount of care put into the initial study design. The enormous power of a modern GWAS (generated by their massive sample sizes) has a downside in that very small amounts of bias can propagate into false positives with very significant p values. Issues such as careful quality control (QC) when choosing SNPs, care in matching cases and controls, appropriate phenotyping, and population sampling strategies all need meticulous consideration. Even though it was the first truly large-scale GWAS, the WTCCC (2007) study remains the gold standard in dealing with these issues, and anyone contemplating a GWAS should carefully review their paper before proceeding. Members of the WTCCC team have produced a series of protocols and short reviews on basic GWAS design (Zondervan and Cardon 2007), marker selection (Pettersson et al. 2009), QC (Anderson et al. 2010), and basic analysis (Clarke et al. 2011; Morris and Cardon 2019; Wang et al. 2019; Uffelmann et al. 2021).

Gathering the Data

Three basic sampling designs are used in epidemiological studies. A **prospective cohort** study identifies a population (cohort) of individuals and follows them through some defined interval. When done well, these are non-trivial to perform in humans, as they often involve following individuals (and their descendants) over decades. The classic example is the **Framingham Heart study**, started in 1948 in Framingham, Massachusetts with 5,200 adults, and is now on its third generations of participants (Andersson et al. 2019). Another is the **1958 British birth cohort** (Power and Elliott 2006). A limiting factor with these early cohort designs for GWAS is they were often relatively small, so that the number of actual cases of a specific disease they contain may be modest, resulting in low power. However, the recent development of very large **biobanks** that massively scale up these iconic early cohorts by coupling electronic medical records with high-coverage genomic data are removing these early sample-size limitations. The **UK Biobank** is the gold standard. From 2006 to 2010, they recruited over 500,000 volunteers aged 40 to 69, creating a massive cohort wherein even relatively rare diseases have over 5,000 cases. The Biobank incorporates SNP array data, whole exome- and whole genome-sequencing data, plus thousands of phenotypes (EMR codes). A number of such 250,000-plus participant biobanks are being developed in Australia, Asian, Europe, and North America (reviewed in Allen et al. 2024). For example, the direct-to-consumer genetic testing company 23andMe has close to 7 million customers, many of whom self-report phenotypes (such as coffee consumption).

A **cross-sectional** study gathers a single sample over some narrow time window. These are common in natural populations and agricultural settings. As with a cohort study, their limitation for disease mapping is that the resulting number of sampled *cases* may be small. In contrast to these two population sampling schemes, a **case-control** design specifically samples individuals with known cases and tries to match them with appropriate controls (e.g., WTCCC 2007). This is the typical design used for most diseases, and is an example of selective genotyping (Chapters 18 and 19), choosing individuals on the basis of their extreme phenotypes (here, disease presence). This design can be quite powerful, but it is also *critically dependent* on obtaining a control sample that is *well matched with the cases*. For example, the control group (ideally) should be roughly the same constitution as the case group in terms of (among other factors) sex, age, population of origin, and known exposures to environmental risk factors.

When the mapping population consists of an association panel of inbred lines, good experimental design can be used to improve the power of a GWAS (Myles et al. 2009; Appendix 9). In humans, observations are made on single individuals, each (outside of monozygotic twins) representing a unique genotype, typically with their phenotype scored by a single observation. Conversely, when a panel of inbred lines is used, *replication* of each line can significantly reduce the residual variance of line values, significantly increasing power when the individual heritability (h^2) is low (Chapter 17). In these settings, even a modestly sized panel may have power in a GWAS. For example, using only 95 accessions of *Arabidopsis thaliana*, Atwell et al. (2010) were able to locate a number of significant marker effects, in part because they could replicate each line, and thus measure each line value with very high precision and relatively little residual error. Replicative designs for estimating line means are examined in Appendix 9 (and Part II Chapters 24, 27, and 28). One issue with these designs is whether one should do a one-stage or a two-stage analysis. In a two-stage analysis, one first obtains the adjusted entry means (the estimates of the line means following correction for the design, e.g., block, plot, and environmental effects) and then uses these adjusted trait values for the GWAS. In a one-stage analysis (Cullis et al. 1998), these two operations are done simultaneously. Stich et al. (2008) showed that the two-stage analysis performed almost as well as the more exact (and much more computationally demanding) one-stage analysis for a GWAS with replicate lines from an association panel. Experimental design principles can also be used in a collection of outbred individuals. When phenotypes have high measurement error, the use of **repeated measures**—repeated sampling of the trait over time—can result in increased GWAS power (Kang et al. 2008).

Defining the Phenotype

As with linkage-based studies, poor phenotyping can doom a GWAS. At best, inconsistent phenotyping reduces power. This is especially a concern with somewhat ambiguously defined diseases, given that most human disease GWAS are now performed using a consortium of multiple institutions. Hence, very precise criteria, globally reproducible across clinical groups, are essential, such as the McDonald criteria, and their subsequent tweaks, for diagnosing multiple sclerosis (McDonald et al. 2001; Polman et al. 2005, 2011; Thompson et al. 2018). Similarly, careful definition and scoring of diseases in agricultural settings are critical, especially when the study involves a collaboration of multiple groups.

Likewise, for quantitative traits, careful, consistent, and accurate measures that can be applied across multiple investigative groups are essential. While there is considerable focus on QC for *marker data*, a proper GWAS should have *at least* the same level of concern about the *quality of phenotyping*. An important class of quantitative traits in disease studies are **endophenotypes**, intermediate traits presumably upstream of a disease phenotype (Chapter 21). Examples include transcript or protein levels at potential candidate genes, amounts of specific metabolites, or other biomarkers that may indicate disease risk or severity. Because of the impressive buffering of most biological systems, even large perturbations in these upstream actors can map into only small changes in the final phenotype (Chapter 21; Appendix 2). Hence, in some cases, a GWAS on well-chosen endophenotypes may be more powerful, and potentially more informative, than a GWAS directly on the disease itself, as these upstream elements may have larger (relative) effects. Again, the same concept holds in agricultural settings, such as measuring some quantitative feature of a disease (e.g., number of necrotic spots per leaf or parasite load) rather than simply scoring the disease as present or absent. Balancing the potentially increased effect of an upstream endophenotypes is that such traits are often more challenging to measure than the target phenotype. This can result in a much smaller sample size than a GWAS based on the target phenotype, and hence ultimately lower power, even when an upstream effect size is larger.

Many of the above concerns involving both design and phenotyping are undergoing a bit of a paradigm shift in human studies. This is due to the massive sample sizes from biobanks (and other sources, such as 23andMe), exploiting the notion that ***sample sizes trump most other considerations***. As previously mentioned, very large biobanks can still sample rare diseases with sufficient power. Similarly, a noisy proxy phenotype in a very large sample (N_2) can have more power than a well-curated phenotype in a smaller sample (N_1). If ρ_g is the genetic correlation between the precise and noisy phenotype, when $N_2\rho_g^2 > N_1$, the larger but noisier data set wins out. An example of this was the work of Kiefer et al. (2013), who were able to map loci associate with myopia by simply asking 23and Me customers when they first wore glasses.

SNP Selection and QC/QA

We assume here that a DNA chip, or related, technology is used to score a preselected set of markers, with whole-genome sequencing (WGS) examined later in the chapter (and in Chapter 21 as well). When a chip (or some other mass-scoring of a preset collection of markers) is used, five issues can impact a GWAS: (i) coverage; (ii) quality of the marker data; (iii) systematic genotyping bias (such as batch effects); (iv) the appropriateness of the marker set for the population(s) of interest; and (v) the appropriateness of the marker set for the trait(s) of interest. We examine these issues in order.

The **coverage** of a chip (or a set of markers) is the fraction of sites not directly genotyped that are in high LD (and hence somewhat predictable) with genotyped markers. Coverage is usually not an issue with modern chips for humans and high-value agricultural species, but may be a concern in nonmodel species where resources have been more scarce. It is important to stress that coverage is a *joint property of a chip and a target population*. Because the average length of an LD block changes with population history, a chip can have excellent coverage in one population but modest to poor coverage in another. A final issue with coverage is a point raised in the last two chapters: everything else being equal, power is usually increased by *scoring more individuals, rather than more markers*. Spencer et al. (2009), focusing on human chips, reiterated this point, noting that lower coverage chips may result in more power for a GWAS if their use allows for more individuals to be scored.

Quality control, and the related issue of **quality assurance** (**QA**), involve, respectively, procedures to ensure quality as the genomic data is generated, and post-production review of the genotype quality (Laurie et al. 2010). We offer only a few brief comments here, as many QC/QA issues are platform specific. Standard measures of quality are the rate of missing calls (markers that fail to be scored by the genotyping process), departures from Hardy–Weinberg, and mismatches between the stated gender of a sample and the sex chromosome signal. Additional, and more sophisticated, QC metrics are reviewed by Anderson et al. (2010), Laurie et al. (2010), and Morris and Cardon (2019). A critical point, especially with case-control data, is randomizing individuals across genotyping batches. Systematic bias in a batch can impart false signals if the cases are run as one set of batches and controls as another. A similar issue can arise if all controls were genotyped at one center and all cases at another. Even when apparently extracted and scored under identicial conditions, Clayton et al. (2005) noted the presence of scoring biases in some markers between cases and controls, generating spurious associations.

The match of a set of markers to the study population is also critical. Historically, most of the initial human chips were very European-centric and hence could miss important polymorphisms in non-European populations. One might imagine that this issue is equally problematic in many agricultural settings when outlier populations are of interest (such as exotic breeds or land races). Likewise, as mentioned, a chip with good LD coverage in one population may have poor coverage in another.

The final issue with SNP choice is a bit more technical, but its roots follow from Example

20.1. If the allele frequencies of markers do not roughly match the allele frequencies of causal sites, the result is low r^2 values, and hence poor power. The distribution of **minor allele frequencies (MAFs)** over a large set of sites is called the **site frequency spectrum**, and for neutral markers is generally strongly L-shaped, with most markers having rare alleles (the Watterson distribution; WL Equation 2.34). This distribution can be shifted by population history (e.g., bottlenecks or expansions) and past selection (WL Chapters 2, 8, and 9). Conversely, markers chosen for inclusion on a chip are generally biased towards more polymorphic markers (MAFs above 1–5%), namely, a bias toward **common SNPs**. To the extent that the marker frequency spectrum is discordant with the causal frequency spectrum (usually because a shift towards higher frequency values in markers), many causal sites will be missed. We return to this point in some detail below, and examine how WGS impacts GWAS. Lastly, while our focus here is on SNP-based GWAS, other classes of markers (Chapter 8) can also be used. One example are **copy-number variants (CNVs)**, which include duplicated or deleted regions (e.g., Ionita-Laza et al. 2009). Issues in detecting CNVs under different sequencing strategies are reviewed by Teo et al (2012).

Imputation

Even a very dense SNP chip directly scores only a tiny fraction of the existing genetic variation. Fortunately, the presence of strong LD among scored markers allows one to predict (or **impute**) the genotypes at unscored markers, *provided* that one has a scored reference set of haplotypes for these markers. Hence, given a sufficient reference set (representative of the population being sampled), only a modest fraction of these SNPs need actually be scored, allowing one to greatly extend the number of tested SNPs. The idea is similar to the multipoint mapping used for QTL mapping (Chapters 18 and 19), where known recombination rates allowed us to test for a QTL at arbitrary positions between adjacent markers. Imputation (Li et al. 2009), and the closely related issue of haplotype phasing (Li and Stephens 2003; Scheet and Stephens 2006), is a fully mature field, with a highly technical literature, dissecting the virtues of various algorithms to accomplish these goals. Despite this richness, our discussion here will be brief. Applications of imputation in association mapping are reviewed by Servin and Stephens (2007), de Bakker et al. (2008), Guan and Stephens (2008), Marchini and Howie (2010), and Pei et al. (2010).

The basic idea is that one has a set of **genotyped markers** for individual i, a set of **reference markers** (known haplotypes with very dense scored markers), and a set of **target markers** that are present in the reference set, but were not scored in i. More formally, suppose our reference set is N haplotypes consisting of L scored SNPs over a particular genomic region. While the number of haplotypes could be as many of $N = 2^L$, in reality, SNPs are correlated because of LD, and the number of haplotypes is a much more modest number (**strong LD structure**; WL Chapter 9). Let $\mathbf{g}_i^T = (g_{i,1}, g_{i,2}, \cdots, g_{i,L})$ be the vector of SNP genotypes for individual i, where $g_{i,k}$ is the number of reference alleles at marker k. For scored markers, these values are 0, 1, or 2. However, some of these values are *missing* (unscored) and the idea of imputation is to use the reference collection of haplotypes to infer (impute) these missing values. Provided that the level of LD is high, the genotypes at target markers can generally be imputed with fairly high accuracy (e.g., Li and Stephens 2003; Marchini et al. 2007). The idea is that a haplotype with missing marker information is a *mosaic of the reference haplotypes*. Imputation involves estimating the break points between mosaic segments and inferring segment identity for intact regions. This is done using **Hidden Markov models**, which start at the first marker and then progressively move along a chromosome through the L markers. These models use the local recombination rates, along with the observed marker data, to estimate the probability of a break after a marker (change of reference haplotypes) as one moves along the haplotypes for i. If a break occurs, the scored marker information and frequencies of reference haplotypes are used to compute the probability that a specific reference haplotype contributes the next segment. The resulting full set of markers (scored plus imputed) is often referred to as an ***in silico* genotype**. Note that imputation can only make inferences about SNPs that are present in the reference sample.

The accuracy of imputation can be checked by treating a scored marker as unknown and examining how accurately its genotype was predicted. Performing this operation over a large number of markers returns an estimate of the **imputation error rate**. Lin et al. (2010) proposed an **imputation quality score (IQS)**, which adjusts the observed match frequency (P_0) for a called SNP by the probability (P_c) that it matches by chance,

$$IQS = \frac{P_0 - P_c}{1 - P_c}$$

A value of one indicates a perfect performance, while negative values indicate performance worse than expected by chance. More granularity can be obtained by considering the error over different local amounts of LD (e.g., in short- versus long-blocks) or over markers with different MAFs. The imputation error increases as LD and/or MAF decrease. Indeed, rare alleles are difficult to impute with sufficient accuracy. Part of this is due to the size of the reference sample, as imputation error decreases as the reference population size increases (Marchini and Howie 2010). By simulating very large reference sizes, Browning and Browning (2016) and Browning et al. (2018) showed that even rare alleles can be imputed with accuracy if the reference size is sufficiently large. They noted in a simulated reference panel of 200,000 Europeans that markers with a **minor allele copy number (MAC)** of at least nine (for a MAF $\geq 4.5 \times 10^{-5}$) could be imputed with high accuracy ($r^2 \geq 0.8$). Such high imputation accuracy for rare alleles was subsequently shown by Huang et al. (2014) and Hofmeister et al. (2023) using, respectively, 5,000 and 150,000 whole genome sequences from the UK Biobank.

Most imputation methods return a posterior probability distribution of the number of reference alleles at an imputed marker, e.g., $\widehat{g}_{i,k} = (\widehat{g}_{i,k,0}, \widehat{g}_{i,k,1}, \widehat{g}_{i,k,2})$, where $\widehat{g}_{i,k,j}$ is the probability that the imputed copy number for marker k in individual i is j. One can extract a summary statistic from these, such as the **best guess genotype**,

$$\widehat{g}_{i,k,max} = \max_j \left(\widehat{g}_{i,k,j}\right)$$

namely, the genotype with the highest posterior probability, or the mean copy number (**posterior mean** or **allelic dosage**), $\widehat{g}_{i,k,ave} = \widehat{g}_{i,k,1} + 2\,\widehat{g}_{i,k,2}$, and then use these values in an association analysis. Alternatively, one could use the full posterior distribution to compute a weighted association statistic (e.g., Guan and Stephens 2008).

The use of imputation usually results in greater power for a GWAS, even when imputation accuracy is modest (Marchini et al. 2007; Guan and Stephens 2008; Marchini and Howie 2010; Pei et al. 2010). As might be expected from the behavior of the imputation error, power improves with higher LD and with higher MAFs for causal SNPs. A caution, however, is that high *average* imputation accuracy does not improve power in regions of low LD (Pei et al. 2010). The other important use of imputation is in meta-analysis (Appendix 6), wherein the results for a given trait from a number of studies are consolidated into a single analysis. If these studies used different SNPs, imputation can convert them all to a common set (e.g., de Bakker et al. 2008), provided that an appropriate reference sample exists.

BASIC STATISTICAL ANALYSIS OF GWAS DATA

As with single-marker linkage analysis (Chapters 18 and 19), each SNP in a GWAS is tested for a marker-trait association (significant differences in trait means over marker genotypes). We start with metrics for continuous traits and then examine discrete traits. These basic metrics are easily extended by adding cofactors, such as sex or age effects, and effects associated with the design, such as batch and specific-lab effects. A more delicate, but no less important, class of confounding effects are shared ancestry and population structure, and we treat these concerns in the next section.

Continuous Traits

Initially, one usually screens each SNP using the additive (or **gene-dosage**) model (Equation 18.14a; also called a **trend test**). Consider individual i and its genotype at a biallelic SNP marker locus k. We designate one of the SNP alleles as the **reference** (often chosen to be the minor allele) and let $N_{i,k}$ $(= 0, 1, 2)$ be the number of reference alleles at marker k in individual i. These correspond to the nonreference homozygote, the heterozygote, and the reference homozygote, respectively (this could equivalently be coded as $-1, 0, 1$). From Equation 18.14a, the linear model for marker locus k becomes

$$z_i = \mu + b_k N_{i,k} + e_i \tag{20.1a}$$

where z_i denotes the trait value for individual i. For each reference allele added at marker k, the trait mean is changed by b_k. The test statistic for b_k can be written as $T^2 = (\widehat{b_k})^2/\sigma^2(\widehat{b_k})$ which follows an F distribution with 1 and $n-2$ degrees of freedom, which approaches a χ_1^2 distribution for large n.

If the SNP is declared significant, further insight might be gleamed by fitting the general genotype model, with (from Equation 18.29c)

$$z_i = \mu + b_k N_{i,k} + d_k H_{i,k} + e_i \tag{20.1b}$$

where $H_{i,k}$ is an indicator variable for being a heterozygote, so that $H_{i,k} = 1$ if i is a heterozygote at marker k, otherwise it has value zero. This is a two degrees of freedom test, and the significance of dominance ($d_k \neq 0$) can also be tested (one degree of freedom). While the trend test (which uses fewer degrees of freedom) is usually more powerful than the general genotype test, the former has poor power for detecting fully recessive alleles, unless that allele is relatively frequent (Lettre et al. 2007).

Equations 20.1a and 20.1b form the basic framework upon which more complex models are built, usually by adding additional cofactors and random effects. For example, one could incorporate both a general sex effect (a sex-specific mean for the trait), and a sex-specific effect for an allele, using the model

$$z_i = \mu + s_i \mu_s + b_k N_{i,k} + c_k s_i N_{i,k} \tag{20.1c}$$

where s_i is an indicator variable for sex (0 for male, 1 for female). Under this model, μ and $\mu + \mu_s$ are the male and female means for our focal trait. Likewise the impact of each copy of the reference allele on the mean is b_k in males and $b_k + c_k$ in females. Using standard linear model machinery (Chapter 10), the effects of sex on the overall mean ($\mu_s \neq 0$) and/or sex-specific allelic effects ($c_k \neq 0$) can be tested. The general genotype model (Equation 20.1b) can be extended in the same fashion. The interaction with other cofactors is similarly modeled (Morris and Cardon 2019), such as epistasis (e.g., Equation 18.14d).

There are two rather different, but not necessarily exclusive, reasons for including additional factors in a linear model: ***reducing the residual variance*** and ***protecting against false positives***. Including factors that influence the variability of a trait (such as age or sex differences) reduces the residual variance, potentially resulting in increased power to detect SNP effects. We say *potentially* because while their inclusion reduces the residual variance, it does so at the cost of degrees of freedom. Second, as detailed below, variables that introduce trait–SNP correlations in the absence of LD (such as shared relationships and/or population structure) can generate false positives. The inclusion of factors accounting for such **confounding variables** can reduce the false-positive rate.

An important caveat for the choice of covariates was noted by Aschard et al. (2015). Some covariates (e.g., height, body mass) are themselves heritable traits, and their inclusion can bias estimates of SNP effects when there are correlations between the covariate and the focal trait. Aschard et al. recommended either to avoid the inclusion of such covariates, or if one wishes to model them, to do so in a more formal multiple-trait GWAS framework (see below).

Adding SNPs as Cofactors

The power of a GWAS can be improved by incorporating SNPs with significant effects as cofactors when testing additional SNPs (Galarneau et al. 2010; Segura et al. 2012; Yang et al. 2012b). This is an extension of the composite interval mapping (CIM) method used in Chapter 18 for linkage mapping in inbred-line crosses. Recall that one feature of CIM was to include markers with significant effects that reside outside of the focal chromosome being tested. Their inclusion reduces the residual variance, increasing power. Segura et al. (2012) suggested that the same is true for a GWAS, and proposed a stepwise approach for SNP inclusion. Suppose that SNP ℓ is the included cofactor, then when testing SNP k, the model becomes

$$z_i = \mu_i + b_\ell N_{i,\ell} + b_k N_{i,k} + e_i \tag{20.1d}$$

If the SNP cofactor shows dominance, then a $d_\ell H_{i,\ell}$ term can be added to the model. Once the most significant SNP is included in the model as a cofactor, a stepwise approach is used to scan for the next most significant SNP, with this process continuing until some stopping criteria is satisfied. Segura et al. applied this approach to both human and *Arabidopsis* data sets, and were able to identify new associations missed by previous analyses. This method is expected to be most powerful when there are causal sites with (at least) modest effects underlying the trait (as opposed to sites having nearly identical infinitesimal effects). When only summary statistics are available (such as in a meta-analysis setting), the **conditional and joint multiple-SNP analysis of GWAS summary statistics** (**COJO**) approach of Yang et al. (2012b) can be used to condition out cofactors.

Searching for vQTLs

The same logic that searches for correlations between trait mean and marker genotypes can be extended to higher order moments of the trait distribution (Chapter 18). In particular, **variance QTLs** (**vQTLs** or **vSNPs**), introduced in Chapter 18, refer to markers that tag genomic regions that result in differences in the trait *variance*. While of intrinsic interest in their own right, vQTLs are often used as potential candidates in the search for epistatic interactions (and/or G x E), in that an epistatic QTL might have a very weak mean effect (marginal effect), but a strong variance signal (Paré et al. 2010; Deng and Paré 2011).

A very simple approach for detecting vQTLs was suggested by Yang et al. (2012a). The trait value z_i for individual i is first standardized using $y_i = (z_i - \overline{z})/\sigma_z$ and then y_i^2 is fit by a standard regression on marker effects, e.g.,

$$y_i^2 = \mu + b_k N_{i,k} + e_i \tag{20.1e}$$

The advantage with this approach is its simplicity, just using a standard linear-model GWAS with a transformed variable. The drawback is that it can yield negative (i.e., impermissible) estimates of a marker variance. To avoid this concern, variance as a function of marker genotype is typically modeled using log-linear models (Equation 14.17), as this approach ensures non-negative variance estimates. The simplest model for marker k in individual i would be

$$\ln\left[\sigma^2(e_i)\right] = \mu + \beta_k N_{i,k} \tag{20.1f}$$

Hence,

$$\sigma^2(e_i) = \sigma^2 \cdot \exp[\,\beta_k N_{i,k}] \tag{20.1g}$$

Typically, squared residuals, e_i^2, are used in place of the residual variance, as was used in the various tests for heteroscedasticity discussed in Chapter 10, such as the Park test (Equations 10.16c and 10.16d).

Joint-effects models for a given SNP, such as the **double generalized linear model** of Rönnegård and Valdar (2011), fit a standard linear model for mean effects (e.g., Equation 20.1a) and then further model the residual variance as a function of SNP effects (e.g., Equation 20.1e). Using a joint-effects model can be more powerful for detecting a SNP effect than a simple test of mean effects (Cao et al. 2013; Corty and Valdar 2018; Corty et al. 2018;

Young et al. 2022). Further, such a joint approach also controls for scaling effects (Chapter 14). Recall in the latter that the variance can change with the mean, which can be adjusted (variance stabilization) if the mean–standard deviation relationship is known (Equation 14.8). However, this is typically unknown. Further, if the trait distribution is non-normal, mean effects can translate into variance effects. To display which component (mean or variance) provides support for a marker effect, Young et al. introduced the **Manhattan sunset plot**. Here each pixel in the standard Manhattan plot (Figure 20.1) is replaced by a column. The height of the column is partitioned into one color for mean effect and a second color (Young et al. used red, hence the sunset metaphor) indicating the additional support for a variance effect.

Binary Traits: Contingency Table Analysis

Much of the early motivation for human GWAS studies was the search for **disease susceptibility (DS) genes**. In this setting, the case-control design is typically used, where a series of cases are chosen by some criteria, and then matching controls are selected (e.g., WTCCC 2007). The resulting data for a given SNP are in the form of a contingency table, which can be expressed either in terms of genotypes or alleles. For a diallelic SNP, the **genotypic contingency table** is given by

	Marker Genotype			
Trait	mm	Mm	MM	Totals
Present (Case)	n_{P0}	n_{P1}	n_{P2}	n_P
Absent (Control)	n_{A0}	n_{A1}	n_{A2}	n_A
Totals	n_0	n_1	n_2	n

In the contingency table setting, a marker-trait association is indicated by a lack of independence between genotype and trait states (genotype frequencies differ between cases and controls). Thus, one could use a standard chi-square test for independence. This is a two degrees of freedom tests and is the analog of the general genotype model for continuous traits (Equation 20.1b). If one or more of the cell numbers in the table are small, the more precise Fisher's exact test can be used (Chapter 2). Alternatively, one could apply the Cochran–Armitage Trend Test (Equation 18.15), which uses one degree of freedom, and corresponds to the additive model (on the scale of measurement).

Instead of focusing on genotypes, one could instead count the number of reference alleles in each class (two for a homozygote, one for a heterozygote), yielding an **allelic contingency table**

	Marker Allele		
Trait	m	M	Totals
Present (Case)	$2n_{P0} + n_{P1}$	$2n_{P2} + n_{P1}$	$2n_P$
Absent (Control)	$2n_{A0} + n_{A1}$	$2n_{A2} + n_{A1}$	$2n_A$
Totals	n_m	n_M	$2n$

Sasieni (1997) noted that the chi-square test is only appropriate for allelic data when Hardy–Weinberg holds. Under these conditions, the allele test asymptotically approaches the trend test on the genotypic data. There are a number of variants on these basic tests (such as combining results from different tests), which are reviewed by Kuo and Feingold (2010), who concluded that the trend test tends to be the most robust.

Example 20.2. The effects of genotypes on binary traits is usually expressed in terms of odds ratios (Example 19.10). Consider the data (from Example 18.5) of Zhang et al. (2005) on the association between genotypes at the DNA repair gene *ADPRT* and lung cancer:

	Genotype		
	mm	*Mm*	*MM*
Present	307	509	184
Absent	359	504	137
Odds	0.855	1.010	1.343
OR	1.000	1.181	1.571

To see these calcuations, consider the odds for genotype *mm*. From Equation 19.52a,

$$\frac{\text{Pr(Disease Present} \mid mm\,)}{\text{Pr(Disease Absent} \mid mm\,)} = \frac{307/(307+359)}{359/(307+359)} = \frac{307}{359} = 0.855$$

Setting *mm* as the standard, the odds ratio (OR) with respect to the other two genotypes are given in the table. Hence, relative to *mm*, the odds of the disease in *Mm* are increased by 18%, and by 57% for *MM*. Parameterizations for the effects of loci underlying threshold (discrete) traits are examined more detail in Part II Chapter 31.

Example 20.3. The allelic contingency table for the data of Zhang et al. (2005) on the association between *ADPRT* genotypes and lung cancer becomes

	Marker Allele		
	m	*M*	Totals
Present	2·307 + 509 = 1123	2·184 + 509 = 877	2000
Absent	2·359 + 504 = 1222	2·137 + 504 = 778	2000
Totals	2345	1655	4000

The resulting χ_1^2 test statistic for independence is 9.8985, with an associated p value of 0.0017, whereas Example 18.5 computed the Cochran–Armitage statistic as 10.5733 (p = 0.0011).

Binary Traits: Logistic Regression

An alternative strategy for binary traits, which allows for cofactors and other confounders such as relatedness and population structure, is **logistic regression** (Truett et al. 1967; Chapter 14). The idea is that a linear model is constructed on some underlying (or *liability*) scale y (Chapter 14; Part II Chapter 31), which can potentially take on negative values and/or values greater than one. This liability value is then mapped via the logistic function (Equation 14.14a) into a range of (0,1), corresponding to the expected probability, $p(y_i)$, that an individual with a liability score of y_i displays the disease. Hence, we have an **observed value** (the disease state z_i that is either 0 or 1), an **underlying scale** (or **latent value**) y_i, and finally a mapping from the underlying scale y into an *expected value* on the z scale. The actual *observed value* of z_i follows a Bernoulli distribution (a binomial with a single draw; $n = 1$ in Equation 2.19a) with success parameter given $p(y_i)$. From Equation 14.15b, the logistic function mapping from y_i into the expected value of z_i is

$$E[z_i \mid y_i] = p(y_i) = \frac{1}{1+\exp(-y_i)} \tag{20.2a}$$

Recalling Equation 14.14b, y is the predicted value of the **logit score**,

$$\text{logit}(p|y_i) = \ln\left(\frac{p|y_i}{1-p|y_i}\right) = y_i \tag{20.2b}$$

Namely, the liability value y_i corresponds to the **log of the odds** (Examples 19.10 and 20.3), so that e^y corresponds to the expected disease odds for an individual with liability y. Note that Equation 20.2a maps y into p, while Equation 20.2b maps p into y.

Suppose on this underlying y scale, we assume an additive model for the effect of the kth SNP (Equation 20.1a) plus m added cofactors,

$$y_i = \mu + b_k N_{i,k} + \sum_{j=1}^{m} \beta_j x_{i,j} \tag{20.3a}$$

then

$$E[z_i \mid y_i] = p(y_i) = \frac{1}{1 + \exp\left[-\left(\mu + b_k N_{i,k} + \sum_j \beta_j x_{i,j}\right)\right]} \tag{20.3b}$$

and

$$\text{logit}(p|y_i) = \ln\left(\frac{p|y_i}{1 - p|y_i}\right) = \mu + b_k N_{i,k} + \sum_{j=1}^{m} \beta_j x_{i,j} \tag{20.3c}$$

where Equation 20.3c is the predicted value for the log of the odds of the trait being present in individual i.

On the liability scale, the interpretation of the b_k value for a SNP is clear. Under the additive model, each copy of the reference allele changes the log of the odds by b_k, so that e^{b_k} is the change in the odds. Further note that the effects on the frequency scale are multiplicative, as the odds (OD) for y_i can be expressed as

$$\text{OD}(y_i) = \frac{p|y_i}{1 - p|y_i} = \exp(y_i) = \exp\left(\mu + b_k N_{i,k} + \sum_{j=1}^{m} \beta_j x_{i,j}\right) \tag{20.3d}$$

$$= \exp(\mu) \cdot \exp(b_k N_{i,k}) \cdot \exp\left(\sum_{j=1}^{m} \beta_j x_{i,j}\right) \tag{20.3e}$$

$$= \text{OD(average)*OD(marker effect)*OD(Ccfactor effects)} \tag{20.3f}$$

This decomposition shows that the odds for the trait being present in a given individual can be written as the product of the odds for a random individual, the odds for their marker genotype, and the odds associated with any of their additional (known) risk cofactors incorporated into the model.

Logistic regression is often used under a case-control design, and this can have subtle implications for the inclusion of cofactors into the model. In a linear regression setting, or in a logistic regression setting under a *random population sample*, the inclusion of covariates (such as age or sex) usually results in an increase in power. ***Such need not be the case in a case-control sampling design***. Under logistic regression, inclusion of a covariate increases both the estimated marker effect size *and* its sample variance (Robinson and Jewell 1991; Kuo and Feingold 2010; Clayton 2012; Pirinen et al. 2012). This implies that power can actually be *reduced* in some settings by inclusion of a covariate if the decrease in precision is larger than the increase in effect size. Pirinen et al. (2012) examined a simple logistic with and without a single covariate for three low-frequency human diseases, and found that the incursion of the covariate resulted in decreased power. For the case of Ankylosing spondylitis, a disease with a prevalence of 0.0025, the inclusion of a specific HLA risk factor resulted in a logistic regression that needed more than double the case sample size to have the same power as a logistic regression where this factor was excluded. Pirinen et al. found that when the disease is common (prevalence over 20%), covariate inclusion typically increased power, but decreases power when the disease is rare.

What might generate such an effect? Consider a case-control design, with a rare disease impacted by both genetic factors and, independently, by the covariant (e.g., smoking status). In the case population, *both* the rare genetic factor *and* the disease covariate are oversampled relative to their frequencies in the general population, which can create an association between the two in the overall GWAS, impacting power (Mefford and Witte 2012). Zaitlen et al. (2012a, 2012b) developed strategies for the appropriate conditioning on covariates in ascertained case samples that result in improved power.

One final issue that can arise with a logistic regression is the problem of **separation**, in which the likelihood converges, but one of the estimates is infinite. This situation arises when, by chance, one of the covariates perfectly predicts the outcome, and can occur for rare cofactors (such as a rare allele) and/or in small samples or with sparse data. Firth (1993) proposed a penalized likelihood correction for removing ML bias in certain settings, and Heinze and Schemper (2002) noted that the **Firth biased-corrected test** deals with separation (or near separation, where estimates are very unstable). Related approaches are reviewed by Mansournia et al. (2018).

Joint Testing of Multiple Markers: Gene-based GWAS

While above methods test markers (SNPs) one at a time, there are settings where it is advantageous to *simultaneously test a set of markers*. For example, there might be a set of weak signals over a region, that, when combined, result in a much stronger signal. When all of the tested markers reside within a given gene, this is called a **gene-based test for association** (Neale and Sham 2004), and is our focus here. In Chapter 21, we examine more general **gene set analysis** (**GSA**) approaches, where the signals from all of the known genes in a given set or pathway are combined (e.g., Yu et al. 2009; Biernacka et al. 2012; Chen et al. 2019). Changing the unit of analysis to the gene or to a set of genes (or a pathway) can be considered a **secondary analysis** of the original GWAS, and can be accomplished either using the original data, or summary statistics (such as p values, or z or χ^2 scores, for each SNP). We defer the important, and subtle, discussion of what group of SNPs constitutes a gene until Chapter 21.

There are several potential benefits of a gene-based, over a single-SNP based, analysis, but the relative advantage depends on the underlying trait architecture. If a causal gene typically has only a single causal site, then single-SNP analysis is favored. Conversely, if a gene has multiple segregating causal sites, potentially of more modest effect, then a gene-based approach can be more powerful. A gene-based approach can also be ***more robust to genetic heterogeneity***. If there are multiple potential causal sites in the gene whose frequencies vary dramatically over samples, a single-SNP approach can miss these (even in a meta-analysis framework), but a gene-based approach can often capture them. This was seen by Peng et al. (2010), who examined type II diabetes in two different GWAS. They found no replication of *SNPs* (using genome-wide significance), but found seven *genes* that were successfully replicated over the two studies. Finally, the multiple comparisons burden can be somewhat alleviated with a gene-based analysis, as one is testing thousands of genes versus millions of SNPs.

Gene-based GWAS approaches can be broken into two major categories. The first are approaches combining the p values for the scored SNPs in a gene. One advantage of these **combining methods** is that they only need summary-level statistics (and information on the correlation among SNPs for their extensions that do not assume independence). The second approach is more model-based, built around multiple regressions and their extensions (such as using PC scores as the predictor variables, penalized approaches such as ridge regression or the LASSO, or variance-components). While some of these approaches require individual data (which can limit their applicability), others can use just summary statistics. Finally, there is considerable overlap between the multiple-SNP approaches considered here and rare-allele methods discussed later in the chapter.

As reviewed in Appendix 6, there are a number of methods for combining the individual p values for each tested marker into a single global p value for their gene. The foundational approach for such tests is that the distribution of p values under the null follows a uniform distribution. Suppose that one observes ten p values and the lowest 6 are all 0.10 (and hence not individually significant). This is a huge enrichment of small p values (under a uniform, we expect only one in ten to have value ≤ 0.1), suggesting a signal for the *collection* of tests, but one that is too weak to be detected by any *single* test. More complicated distributions—such as using the first k of n ordered p values, or the distribution of p values below some threshold—can be easily computed by simulating draws of a vector of n independent uniform variables (or n correlated uniform variables in more general settings).

One of the most common p-value combining approaches is **Fisher's method** (Equation A6.1a): for k independent tests, where p_i denotes the p value for test i, the sum

$$X^2 = -2 \sum_{i=1}^{k} \ln(p_i) \tag{20.4a}$$

approximately follows a χ^2_{2k} distribution. Loughin (2004) showed that while Fisher's method works well when the signal against the alternative is strong and concentrated in a small fraction of the tests, when a modest signal is spread over a number of tests, **Stouffer's Z score** (Equation A6.2) is more powerful. This method first translates the individual p_i values into unit normal scores, Z_i, and then computes an overall Z score,

$$Z_s = \sum_{i=1}^{k} Z_i / \sqrt{k} \tag{20.4b}$$

Further, as shown by Whitlock (2005), a weighted Z method (Equation A6.2.c) outperforms both Fisher and the unweighted Z when precision varies over tests (as would occur when allele frequencies vary over the tested markers). Bhattacharjee et al. (2012) discussed SNP weighting for a Z-based gene GWAS. The limitation with these approaches is the strong assumption of *independence of tests*, which fails for SNPs within a gene. Fisher's method has been extended to correlated tests by Brown (1975) and Makamb (2003), which requires estimates of the correlation among SNPs (e.g., Moskvina et al. 2011). More generally, one of the above statistics can be computed, and its significance assessed via permutation (see below), but this usually requires access to the original data.

A number of modifications of Fisher's approach have been proposed that may improve its power and performance. Their starting point is to express Fisher's statistic as a product

$$X^2 = -2 \ln \left(\prod_{i=1}^{k} p_i \right) \tag{20.4c}$$

Instead of considering the full product, various *truncated approaches* have been proposed that consider only a subset of the p values, namely those that are sufficiently small. The **truncated product method (TPM)** of Zaykin et al. (2002) considers only those p values below some threshold level (τ),

$$X^2_{trun} = -2 \ln \left(\prod_{i=1}^{k} p_i \, I[\, p_i \leq \tau] \right) \tag{20.4d}$$

where the indicator function $I[\, p_i \leq \tau]$ has value one if $p_i \leq \tau$, otherwise is zero, and the product is taken over nonzero values. The logic is that by discarding modest to large p values, power may be improved, as most of the signal for region-wide significance resides in the smaller p values. A modification is the **rank truncated product method (rank-TPM)** of Dudbridge and Koeleman (2003). Here, one ranks the p_i values from smallest $p_{[1]}$ to largest $p_{[k]}$ and the product is taken over the $K < k$ smallest of these (i.e., using K in place of τ as the threshold). We can extend truncated products to correlated tests, once again using permutation to obtain critical values. Further, these methods can be extended into **adaptive** versions (e.g., Yu et al. 2009; Chen et al. 2013; Yan et al. 2014; Pan et al. 2015), by varying the thresholds (τ or K), and then choosing the threshold yielding the smallest value (with significance again assessed via permutation). Even more generally, Fisher's method is a special case of the **Gamma method** (Zaykin et al. 2007; Biernacka et al. 2012), where the inverses of gamma functions (Table A7.1) are used to place differential weights on p values below some threshold. Zaykin et al. (2007) provides a nice overview of these, and other, combining p-methods.

As mentioned, the significance of these approaches, even when markers are highly correlated, can be assessed using permutations, where the genotypes of an individual are kept intact (preserving their SNP correlation structure), but the phenotypes are randomized (shuffling the phenotypic labels over the genotype vectors). However, this can be computationally intense, and requires access to the original data. Seaman and Müller-Myhsok (2005) developed **direct simulation approaches** (**DSAs**) to sample values from the distribution of truncated products under the null (the p_i follow a uniform distribution, but may be correlated), which is computationally much more efficient, and less restrictive, than a permutation approach.

A related approach to combination methods follows from corrections for multiple comparisons (Appendix 6). Recall that if we wish to have a family-wide error rate (FWER) of γ over the entire collection of tests, the Bonferroni correction over n independent tests uses a significance level for each test of $\alpha = \gamma/n$. One can essentially invert this as follows. Letting $p_{[i]}$ be the ith smallest p value, then the significance of the collection (the probability that none of the SNPs are significant) is approximately $p = np_{[1]}$ (p and $p_{[1]}$ taking the roles of γ and α, respectively). One could also replace n with some measure of the effective number, n_e, of tests (Appendix 6), which can be estimated from the eigenvalues of the correlation matrix of the SNPs (Cheverud 2001; Nyholt 2004; Li and Ji 2005; Patterson et al. 2006; Li et al. 2011). More powerful versions of Bonferroni exists that discount for each significant test (Appendix 6). One such is **Simes–Hochberg correction** (Simes 1986; Hochberg 1988), which (following the logic above) leads to

$$p_{Simes} = \min_j \left(\frac{n\, p_{[j]}}{j} \right) \tag{20.4e}$$

which assumes uncorrelated tests. Li et al. (2011) proposed using an extended version of the Simes test, based on the effective number, n_e, of tests over the entire gene and the effective number, $n_{e[j]}$, based on the first j smallest p values. Their **GATES** (***g**ene-based **a**ssociation **t**est using **e**xtended **S**imes procedure*) test uses

$$p_{GATES} = \min_j \left(\frac{n_e\, p_{[j]}}{n_{e[j]}} \right) \tag{20.4f}$$

Additional discussion on the significance of tests over a set (such as a gene or pathway) is given in Chapter 21 and Appendix 6. A recent interesting addition, which will be discussed when we cover rare-allele tests, is the use of the Cauchy distribution for combining p values in the presence of unknown dependency structures (Pillai and Meng 2016; Liu and Xie 2020).

The second category of gene-based GWAS approaches is based on regressions and their extensions, which often require access to the raw genotype data, although some versions can be extended to just using summary statistics. One approach would be to use **Hotelling's** (1931) $\boldsymbol{T^2}$ **statistic** to jointly test the impact of n markers (Xiong et al. 2002; Chapman et al. 2003; Fan and Knapp 2003). If one considers a multiple regression using all markers in a defined set, $z_i = \mu + \sum \beta_j N_{i,j} + e_i$ (where $N_{i,j}$ is the number of copies of the reference allele at SNP j in individual i), then T^2 is a test that all of the β_j are zero. Recall that T^2 is the generalization of the Student's t test to multiple variables (for a single variable, T^2 reduces to t^2, the square of the t statistic). Again, the idea is that by jointly considering a well-chosen marker set, cumulative small signals might result in a larger significance for the collection as a whole. The tradeoff with a Hotelling-based approach is that inclusion of noncausal markers increases the degrees of freedom, without increasing the noncentrality parameter (the model signal from causal variants), lowering power (Appendix 5). An alternative approach is to use penalized regressions (Example 20.4) that downweight (or remove) model parameters with very low impact.

While Hotelling's test is generally performed using continuous trait data, it can also be modified for binary data. Xiong's T^2-based test for case-control samples is formulated as follows: let $g_{0,i,k}$ denote the genotype for the kth control individual at marker i (using the coding $-1, 0, 1$). For n markers of interest, let $\mathbf{g}_{0,k} = (g_{0,1,k}, g_{0,2,k}, \cdots, g_{0,n,k})^T$ be the vector of marker scores for control individual k, and $\mathbf{g}_{1,j} = (g_{1,1,j}, g_{1,2,j}, \cdots, g_{1,n,j})^T$ be the corresponding genotype vector for case individual j. For n_0 controls and n_1 cases, the Hotelling's T^2 test statistic becomes

$$T^2 = \frac{n_0\, n_1}{n_0 + n_1} \left[(\overline{\mathbf{g}}_0 - \overline{\mathbf{g}}_1)^T \mathbf{S}^{-1} (\overline{\mathbf{g}}_0 - \overline{\mathbf{g}}_1) \right] \tag{20.5a}$$

where $\overline{\mathbf{g}}_i = n_i^{-1} \sum_{k=1}^{n_i} \mathbf{g}_{i,k}$ (for $i = 0, 1$) are the mean vectors for cases and controls and

$$\mathbf{S} = \frac{1}{n_0 + n_1 - 2} \left[\sum_{k=1}^{n_0} (\mathbf{g}_{0,k} - \overline{\mathbf{g}}_0)(\mathbf{g}_{0,k} - \overline{\mathbf{g}}_0)^T + \sum_{j=1}^{n_1} (\mathbf{g}_{1,j} - \overline{\mathbf{g}}_1)(\mathbf{g}_{1,j} - \overline{\mathbf{g}}_1)^T \right] \tag{20.5b}$$

is the pooled sampling covariance matrix. One can think of T^2 as a generalized distance measure between the mean vectors, $\overline{\mathbf{g}}_0$ and $\overline{\mathbf{g}}_1$, of marker frequencies in cases versus controls. Asymptotically,

$$\left(\frac{n_0 + n_1 - n - 1}{n(n_0 + n_1 - 2)} \right) T^2 \sim F_{n, n_0 + n_1 - n - 1} \to \chi^2_n \tag{20.5c}$$

The last step follows because as the denominator degrees of freedom for an F distribution becomes large, it approaches a χ^2 with the numerator degrees of freedom (Appendix 5).

This notion of contrasting the differences in multilocus genotype similarity between some defined set of markers in cases and control has been extended by other authors (e.g., Schaid et al. 2005; Wessel and Schork 2006; Kwee et al. 2008). This approach is often framed using **kernel methods** (Schaid 2010; Larson et al. 2018)—which can be loosely thought of as generalized distance measures—such as the **kernel-based adaptive cluster**, **KBAC**, method of Liu and Leal (2010) and the **kernel-based association test**, **KBAT**, of Mukhopadhyay et al. (2010). Many of these approaches can also be framed as random-effects (RE) models, where the test for a gene effect can be based in terms of variance components (Part II Chapter 33).

Treating SNPs as single markers and then accounting for their correlation structure is one approach for modeling under high levels of LD. The other is to base the unit of analysis on **haplotypes**, the specific configuration of SNP alleles over a small section of a chromosome. Coding the minor allele as one and the major as zero, the haplotype can be encoded as a string of zeros and ones, so that $h_1 = (00001)$ would be a haplotype with major alleles at the first four SNPs and the minor allele at the fifth. By focusing on haplotypes, the unit of analysis moves from a set of diallelic SNPs to a single marker with h alleles (one for each of h defined haplotypes). While conceptually straightforward, this approach raises several issues. The first two are operational: how to define an optimal haplotype length (a sliding window coupled with a cross-validation approach might offer a solution) and how to obtain their phase (extracting the two haplotypes given a diploid multilocus genotype; Chapter 5). The final issue is statistical: haplotype models quickly soak up degrees of freedom. With h haplotypes, the degrees of freedom range from $h - 1$ on the low end (analyzing haplotypes as additive alleles: extending Equation 20.1a and allelic contingency tables to h alleles) to $h(h-1)/2$ on the high end (the full genotype model: extending Equation 20.1b and the genotype contingency tables to all diploid genotypes $h_i h_j$). Further, one would expect a small number of common haplotypes and a larger number of rarer ones. As a result, a haplotype-based analysis often focuses on the relatively common haplotypes, and might lump (collapse) the rest into a single group (similar to rare alleles approaches considered below). As noted above, one approach to deal with such high-dimensional models is to use penalized regressions (Example 20.4)

The basic structure of a haplotype-based analysis treats each haplotype as an allele in an otherwise standard marker analysis, e.g., using the regression $z_i = \mu + \sum_j \beta_j H_{i,j} + e_i$. Here $H_{i,j}$ is either the number of copies of haplotype j in individual i (when they can be scored directly), or can be replaced an expected value, $H_{i,j} = \Pr(\text{haplotype } j \mid \mathbf{g}_i)$, where $\mathbf{g}_i$ is the vector of SNP values for individual i over the region of interest (e.g., Schaid 2002; Schaid et al. 2002).

An intermediate approach that nicely accommodates the challenging issues of defining haplotypes and controlling their dimensionality is to use **principal components** (**PCs**; Chapter 9) of the covariance matrix $\mathbf{C}$ for the markers of interest. Let $\mathbf{e}_i$ be the eigenvector associated with eigenvalue λ_i (ranked from largest to smallest). Recall that the sum of the first k eigenvalues divided by the total sum of eigenvalues gives the amount of variation explained by the first k PCs (Chapter 9). Gauderman et al. (2007), Wang and Abbott (2008), and de Leeuw et al. (2015) suggested that PCs can be used as the predictor variables in a regression, with

$$z_i = \mu + \sum_{j=1}^{k} \beta_j \gamma_{i,j} + e_i, \quad \text{where} \quad \gamma_{i,j} = \mathbf{e}_j^T \mathbf{g}_i \tag{20.5d}$$

Thus, $\gamma_{i,j}$ is the projection of the SNP data from the focal gene ($\mathbf{g}_i$) from individual i onto the jth PC, $\mathbf{e}_j$ (Example 9.12). Gauderman et al. suggested that k should correspond to the number of eigenvalues that account for between 80 and 90% of the total scored SNPs variation in the target region. Gauderman et al. and Ballard et al. (2010) found that PC-based approaches usually outperform haplotype or multiple SNP regressions. One standard concern with any PC method is how to interpret the PC axes (Chapter 9). However, when a causal SNP is found on several different haplotypes (diffusing its haplotype-specific signal), the SNP weighting on the first few PCs could pick out such a signal.

Example 20.4. A common situation that arises in modern quantitative genetics are regressions whose number of parameters p exceeds, often greatly, the sample size n. In a standard least-square regression framework, estimation proceeds using generalized inverses (Appendix 3), resulting in a set of solutions. A more powerful approach is to use **penalized** (or **regularized**) **regressions**. Consider the regression model

$$y_i = \mu + \sum_{j=1}^{p} \beta_j X_{i,j} + e_i$$

In the standard OLS framework, one solves for the β_j that minimizes the sum of squared residuals,

$$\text{RSS} = \sum_{i=1}^{n} e_i^2 = \sum_{i=1}^{n} \left(y_i - \mu - \sum_{j=1}^{p} \beta_j X_{i,j} \right)^2 \tag{20.5e}$$

Penalized regressions start with this framework, and then place constraints on the β_j. Under **ridge regression** (**RR**), one instead minimizes RSS $+\lambda \sum \beta_j^2$ (Hoerl and Kennard 1970), for some **shrinkage parameter** $\lambda > 0$ (Hoerl et al. 1975, Lawless and Wang 1976, and Cule at al. 2011 discuss the choice of λ). An alternative approach is the **least absolute shrinkage and selection operator**, or **LASSO**, which minimizes RSS $+\lambda \sum |\beta_j|$ (Tibshirani 1996). Note that for values of β near zero, β^2 is a much less harsh constraint than $|\beta|$ (as $|\beta| \gg \beta^2$ for $|\beta| \ll 1$), so that the LASSO shrinks most of the β_j to exactly zero (yielding a **sparse estimate**), and hence is often used in **model selection** (choosing the parameters in the final model as those with nonzero βs). The **elastic net** (Zou and Hastie 2005) combines the RR and LASSO approaches, seeking to minimize RSS $+\lambda_1 \sum |\beta_j| + \lambda_2 \sum \beta_j^2$. These approaches can also be extended to generalized linear models, e.g., Le Cressie and van Houwelingen (1992) proposed a logistic ridge regression method.

The use of penalized regressions for marker selection, treating epistatic interactions, and dealing with a set of very highly correlated markers in a GWAS setting has been examined by a number of authors (e.g., Li et al. 2007; Malo et al. 2008; Wu et al. 2009, 2011; Chen et al. 2010; Han and Pan 2010; Zhou et al. 2010; Li et al. 2011; Xu et al. 2012; Gac et al. 2013; Yi et al. 2015; Schaid et al. 2020; Lui et al. 2021).

As with the various p-value combining methods discussed above, given the strong dependency (LD) between SNPs within a genic region, the gold standard for regression-based significance testing is permutation. Here, a gene-based statistic is computed for the observed SNP set, and then trait values are shuffled over individuals, while keeping their marker genotypes intact, generating a test statistic under the null (e.g., the **set-based test** in PLINK; Purcell et al. 2007). When a large number of genes are tested, permutation becomes computationally intense due to the *very* small critical values required to correct for multiple comparisons. To obtain a reasonably stable estimate of the test value corresponding to probability α under the null requires at least $10/\alpha$ permutation samples, which can be the order of 10^7 or greater when a large number of genes are tested. For example, testing 20,000 genes and wanting a genome-wide error rate of $\gamma = 0.05$, α for each gene becomes $0.05/20000 = 2.5 \cdot 10^{-6}$, requiring *at least* 10^7 permutations for a stable estimate.

Given this concern, several creative, and rather diverse, solutions have been suggested to avoid permutation. One interesting approach was suggested by Knijnenburg et al. (2009), drawing on results from extreme value theory (WL Example 27.2). Under rather general assumptions, the extreme tails of a probability distribution converge to a generalized Pareto distribution (WL Equation 27.7), whose parameters can be estimated using ML. This allows a modest number of permutations (say 5,000 to 10,000) to be used to estimate its shape and then extreme values (say the threshold value corresponding to $p = 10^{-8}$) are computed analytically from the resulting distribution.

A second class of solutions is to use simulation-based approaches (drawing test statistic values from their expected null distribution). The underlying logic is that the computational burden for a single simulation draw can be much less than that for a single permutation. Any reduction in burden is important given that the very small α values typically require millions of draws/permutations. Li et al. (2023) presented an overview of many of these approaches when gene-based tests just use SNP summary statistic (such as z, or $z^2 = \chi^2$, values for the test of a significant regression for a given SNP; Equation 20.1f). They do so by framing these summary-statistics-based tests in terms of quadratic products. If $\mathbf{z} \sim \text{MVN}_m(\mathbf{0}, \mathbf{V})$, then the distribution of the quadratic form $Q(\mathbf{z}) = \mathbf{z}^T \mathbf{A} \mathbf{z}$ is

$$Q(\mathbf{z}) \sim \sum_{i=1}^{m} \lambda_i \chi_1^2 \tag{20.5f}$$

namely a weighted sum of χ_1^2 random variable, where the weights λ_i are the eigenvalues of the matrix $\mathbf{VA}$. For many SNP summary-statistics gene-based GWAS tests, z_i is the z score for SNP i (or, more generally, PC_i), with the gene-based test statistics of the form $T = \sum z_i^2$ ($\mathbf{A} = \mathbf{I}$). Here $\mathbf{V}$ is the covariance matrix accounting for correlations among the tests, so that λ_i are the eigenvalues of $\mathbf{V}$ (which is gene-specfic). To obtain the significance (i.e. $\Pr[T > q]$, where q is the observed test statistic value), one samples from the distribution given by Equation 20.5f. Lin (2005) proposed to accomplish this using draws of independent normal random variables which are then adjusted by the correlation structure of the data ($\mathbf{V}$), with the **VEGAS** approach (**versatile gene-based association study**) of Liu et al. (2010) doing something similar. The **fastBAT** approach of Bakshi et al. (2016) uses saddlepoint approximations (e.g., Davies 1973; Kuonen 1999) for the distribution $Q(\mathbf{z})$.

The recent **mBAT** approach of Li et al. (2023) provides a simpler approach for computing p values, while also accounting for the effects of **masking**. Here, the marginal effect of a given SNP (scored by z) is deflated by the presence of nearby SNPs in **negative LD** (alleles with effects of opposite sign are over-represented in haplotypes). More generally, if r is the correlation among two SNPs whose effects are a_1 and a_2, then the observed marginal effect of SNP 1 is a_1+ra_2. If ra_2 is different in sign from a_1, the marginal signal is reduced, resulting in the single SNP test being underpowered. Most forms of selection generate negative LD (WL Chapter 16), so masking could be common. As a result, tests of the sum $\sum z_i^2$ based on the marginal effects of single SNPs could easily miss causal sites.

To account for masking, Li et al. proposed the test

$$T_{mBAT} = \mathbf{z}^T\mathbf{V}^{-1}\mathbf{z} \tag{20.5g}$$

where $\mathbf{V}$ is the correlation matrix among the SNPs being tested within the gene. This is just a multiple regression based on summary statistics. From Equation 20.5f, because $\mathbf{A} = \mathbf{V}^{-1}$, then $\mathbf{VA} = \mathbf{I}$ and $\lambda_i = 1$ for all i. Hence, the distribution of T_{mBAT} is χ^2_m, which is easily computed. Li et al. showed that T_{mBAT} can also be viewed as a test of whether the collection of SNPs account for any trait variance. Given the high correlations among SNPs, the very small eigenvalues of $\mathbf{V}$ may have unstable values, and they suggested using a lower rank approximation of $\mathbf{V}$ (e.g., Equation A3.37 or A3.38). As we flush out later in the chapter when discussing rare alleles, the performance of different tests can vary over genetic architecture (e.g., a few large effects versus many small effects). To provide a more robust test over different architectures, Li et al. constructed their **mBAT-combo** test by Cauchy p-value averaging of their mBAT and fastBAT tests. They found that masking was common in protein-coding genes associated with blood and urine biomarkers, with mBAT-combo and mBAT finding, respectively, 20 and 12% more significant genes than fastBAT alone. For most effects that were only detected under mBAT-combo, a two fold increase in sample size allowed them to be detected using single-SNP approaches.

Detecting Gene–gene (Epistatic) Interactions

As was the case of linkage-based mapping (Chapter 18), it is trivial to extend GWAS to search for epistasis by looking for interactions between marker genotypes. For example, the presence of epistatic interactions between the genotypes at two different SNP markers can be tested as follows. Let $\overline{z}_{ij}$ be the mean value of individuals with two-locus genotype ij, with i indexing the SNP genotypes at locus i (values of 0, 1, 2), and let $\overline{z}_i$ be the mean trait value for single-locus genotype i. A simple test for epistasis is given by

$$T = \sum_{i=1}^{3}\sum_{j=1}^{3} \left(\overline{z}_{ij} - \overline{z}_i - \overline{z}_j + \mu\right)^2$$

Under the null (additive interaction) $E[T] = 0$ and T follows an F distribution with $n - 9$ degrees of freedom (n being the sample size), while $E[T] > 0$ under the alternative (epistasis). In a GWAS setting, this, and other, simple tests can be greatly complicated by three factors: multiple comparisons, power, and the effects of imperfect tagging of causal sites by markers. We consider these in turn. Complications induced by imperfect tagging turn out to have especially subtle effects and can generate false positives in nonobvious settings. There is a growing, and increasingly technical, literature on the search for GWAS epistasis, see Cordell (2009), Wei et al. (2014), and Liu et al. (2021) for more detailed reviews.

The major complication in the search for epistasis under linkage mapping is the massive explosion in the number of comparison as the number of markers increases, basically scaling as m^2 (more exactly, $m[m-1]/2$), a concern that is greatly magnified under a GWAS (given their very large m values). As discussed in Chapter 18, this increase in model space can be handled by either using a hypothesis-free approach (an **exhaustive search** over all such combinations) or by a hypothesis-driven approach (such as testing specific loci, or, more generally, loci first detected by having marginal effects). The latter approach, called a **two-step scan**, can either be based on one, or both, of the markers showing a significant marginal effect (Chapter 18). Marchini et al. (2005) and Evans et al. (2006) examined these different strategies, and found that there were situations where an exhaustive search (despite its far greater multiple-testing burden) had higher power than a two-step approach. This required that the interaction variance is greater, often far greater, than either of the marginal additive variances for the two candidate loci, which in turn, requires that both loci have MAFs near 50% (Chapter 5). In a GWAS setting, satisfying this condition is generally unlikely, even when the pair of loci show strong functional epistasis (differences in genotypic means), as this still generally results in weak statistical epistasis (associated nonadditive variances), the signal detected by a GWAS. Several authors have suggested that, like QTLs showing marginal effects, vQTLs are also candidates in the search for epistastic interactions (as well as G $\times$ E), as the presence of epistasis could generate differences in the variances associated with single-marker genotypes (Paré et al. 2010; Deng and Paré 2011).

Generally speaking, the GWAS signatures for (statistical) epistasis (significant epistatic variances) are weak. As noted by Wei et al. (2014), "*searching for epistasis has contributed rather little to the understanding of complex traits, except for the important observation that large interaction terms are very unlikely to exist between pairwise SNPs.*" However, just as the finding of generally small additive variances *does not* imply the absence of major-effect alleles (they could be just at low frequencies), the observation of (at best) weak epistatic variances does not imply that functional epistasis is generally absent. Rather, even if epistasis occurs, the two-locus genotype frequencies could be such that most effects are loaded onto the additive variance (Chapter 5; Purcell and Sham 2004; Hill et al. 2008; Mäki-Tanila and Hill 2014).

Even under a very marker-dense GWAS, causal QTLs can be imperfectly tagged by marker SNPs (Example 20.1). For a single causal locus, only r^2 of its additive variance is captured by a marker in LD, while only r^4 of its dominance variance is captured by the marker (dominance) variance. With epistasis, even less is captured. If $\sigma^2(A^k D^\ell)$ denotes the true causal epistatic variance, the resulting marker epistatic variance (assuming each causal site is correlated by r to a SNP) becomes $r^{2k} r^{4\ell} \sigma^2(A^k D^\ell)$, so that only fractions r^4, r^6, and r^8, respectively, of the $\sigma^2(AA)$, $\sigma^2(AD)$, and $\sigma^2(DD)$ causal variances are captured by the marker pair (Wei et al. 2014). The impact of imperfect LD on estimating epistatic variances is thus far greater than its impact on detecting single-marker effects (additive variances). Coupled with the least-squares decomposition tending to place most of the variance into additive terms (Chapter 5), it is thus not surprising that estimates of marker epistatic variances, when even detected, tend to be very small. These two features, coupled with the much greater multiple comparison burden, imply that there is generally very low power to detect most epistatic interactions.

A final complication is that even when an epistatic signature is detected, the subtle effects of imperfect tagging of a causal SNP by a marker can result in false positives, resulting in **phantom epistasis**. This can occur over two scales, between linked loci, and, surprisingly, between unlinked loci. As noted by Wood et al. (2014a), Hemani et al. (2014), and de Los Campos et al. (2019), **haplotype effects** can occur when two marker loci are both in imperfect LD with a causal SNP, say $M1 - Q - M2$. When the LD is imperfect, the causal additive variance is not fully accounted for by either marker alone, resulting in their interaction term, capturing part of this residual, being significant. Hence, under a purely additive model, imperfect LD can generate nonadditive signals between markers in LD. Finally, even when marker loci are unlinked *and* in linkage equilibrium with each other, imperfect tagging of causal sites can still generate phantom epistasis. This arises by skewing the distribution of T away from a normal (the effects of a genotypic class now follow a mixture, instead of a normal, distribution; Chapter 16). This can inflate the F statistic over its expected value under the null, generating false positives (see Hemani et al. 2021 for details).

Corrections for Multiple Testing

We previously discussed extracting a single p value for a collection of tests over a focal gene. We now consider the related, but subtly different, task of error control over an entire GWAS, which involves many of the same approaches. The distinction is obtaining a summary p value for a set of tests versus controlling the overall p value over this set (finding the p-

value threshold for each test to ensure a desired FWER, γ). Given the massive numbers of tested markers (either directly or via imputation) in a GWAS—orders of magnitude higher than in a linkage analysis—adjustment for multiple testing is a serious concern. Recall that linkage-based mapping considered two multiple comparison scenarios: the **sparse-map approximation** with roughly independent markers (in which case Bonferroni corrections can provide reasonable control; Appendix 6), and the **dense-map approximation** with correlated markers (essentially testing every position along a chromosome) wherein more elaborate modeling is required (Chapters 18 and 19). The situation under a GWAS falls between these two extremes. Under linkage mapping, the signal from a QTL can propagate over most of a chromosome (albeit with its impact diminishing over distance). In contrast, the LD structure for many populations is such that a chromosome can be considered as a series of very small LD blocks, within which LD is very high (dense map setting of highly correlated tests), but with near independence between blocks (sparse map setting of weakly correlated tests). This difference of long-range correlations under linkage mapping versus only very short-range correlations in a GWAS is a function of the former experiencing only a few meioses per chromosome, whereas the latter may involve thousands (or more) of recombinations since the common ancestor.

Because of LD block structure, the *effective* number of independent tests (n_e) is less than the actual number (n), with the Bonferroni correction testing each marker at $p = \gamma/n_e$ to provide a genome-wide Type I error control of γ (Equations 18.13a and 18.13b). A number of approaches using the eigenstructure of the correlation matrix of tests have been proposed to estimate n_e (e.g., Cheverud 2001; Nyholt 2004, 2005; Ji and Li 2005; Patterson et al. 2006; Appendix 6), most of which are very conservative (n_e is too large) under strong LD structure (Salyakina et al. 2006; Moskvina and Schmidt 2008). However, the **simpleM** method of Gao et al. (2008) seems to perform fairly well in such settings (Gao et al. 2010; Gao 2011; Hendricks et al. 2014; Davis et al. 2016). Ranking the eigenvalues of the test correlation matrix from largest (λ_1) to smallest (λ_m), the **Gao estimator** of n_e (for n tests) is the value that satisfies

$$\sum_{i=1}^{n_e} \lambda_i \Big/ \sum_{i=1}^{n} \lambda_i = C$$

namely the smallest number of eigenvalues that account for a fraction C (typically taken as 0.995) of the total variation (sum of all eigenvalues). One can also use Bayesian approaches to estimate the effective number of tests (Q. Wang et al. 2015; S.-B. Wang et al. 2016b).

An alternative approach is to directly estimate the number of LD blocks and take this as the value of n_e. In their landmark paper promoting association over linkage analysis, Risch and Merikangas (1996) predicted that a modern GWAS might involve testing up to 500,000 sites, for a $p = 1 \times 10^{-7}$ (choosing a γ of 5%). More recent suggestions in humans are around 10^6 LD blocks for European populations yielding $p = 0.05/10^6 = 5 \times 10^{-8}$ (Altshuler et al. 2008; Fadista et al. 2016), and double that number for African populations (Pe'er et al. 2008) with their deeper ancestry, and hence more generations of recombination in a population sample. Estimates of the rough number of independent LD blocks in other species can be much higher (as in outbred maize populations) or much lower (as in a panel of elite inbred lines). In theory, a large set of SNPs with good genomic coverage could be used to estimate the number of blocks (e.g., Example 20.5), or at a minimum, suggest the average spacing of SNPs for adequate power. With a value of n_e in hand, one can further improve on Bonferroni methods by using sequential corrections (Manly et al. 2004; Appendix 6). In the simplest setting (Holm's method), the p values are ordered ($p_{[1]}, \cdots, p_{[n]}$), and if the smallest is rejected (e.g., $p_{[1]} \leq \gamma/n_e$), then the next smallest is tested at $p_{[2]} \leq \gamma/(n_e - 1)$, and so on until failure to reject. More powerful sequential approaches (e.g., Simes–Hochberg and Hommel) are reviewed in Appendix 6.

A second approach to control for multiple comparisons are permutation tests, discussed in both Chapter 18 and in our previous coverage of gene-based GWAS. While (generally) straightforward, with the very large number of tests in a GWAS, one requires estimates of critical test statistics for a given SNP that correspond to extremely small p values (10^{-7}

or smaller). Thus, the standard permutation approach (shuffling a fixed vector of markers over phenotypes) incurs a very heavy computational burden under a GWAS in order to estimate the thresholds associated with such small p values. A further complication is that a permutation requires that one has identified **exchangeable units**, which is not trivial in the presence of population structure. As a result (as discussed above), rather than permuting the original data billions of time, data-free approaches based on sampling from the extreme tails of the distribution of p values (correlated unit random variables) or the test statistics (multivariate normals), have been proposed (Lin 2005; Conneely and Boehnke 2007; Han et al. 2009; Liu et al. 2010; Bakshi et al. 2016), where the correlation structure for these distributions is estimated from the marker data. Hendricks et al. (2014) noted that such applications of **extreme tail theory**, as well as the Gao n_e estimator, were efficient alternatives to permutation. Another option, also discussed above, is that one could perform a modest scale standard permutation and then use the resulting empirical distribution of test statistics under the null to estimate the corresponding Pareto distribution for the extreme values, which then allows critical values for very small p to be obtained analytically (Knijnenburg et al. 2009).

Besides the obvious issue of adjusting for multiple, correlated tests, a much more subtle issues arises in a modern GWAS. Subsequent to the original analysis, imputation and the combination of results from multiple studies are often done, resulting in ***the number of GWAS tests initially performed being rather different from the number of tests that ultimately derive from this data*** (Thomas and Clayton 2004). Consider the setting where one million SNPs are directly scored and tested. Using this group of markers, one could estimate n_e for a Bonferroni-style correction or use a permutation approach based on this set of markers. Now suppose that a reference set of sequences becomes available that allows for the imputation of an addition two million SNPs on the same dataset. This would involve no change in the number of scored individuals, but results in the original data now being used for three million tests. The genome-wide threshold p value obtained via the original permutation is no longer appropriate as additional tests are involved. The very real possibility exists that highly significant markers detected in the initial study would no longer achieve genome-wide significance as these additional markers (and therefore tests) are added. Likewise, estimates of n_e based on correlations among the original tests may also be no longer appropriate.

Such concerns led to the extension of the infinitely dense marker approximation from Chapters 18 and 19 (where all tests on a chromosome are correlated because of insufficient recombination) to an analog of the infinitely dense setting in an LD framework. Here there are a large number of nearly independent islands of SNPs, within which marker correlations are extremely high. Example 20.5 gives one approach for obtaining a limiting threshold value in such **fully saturated** marker settings, effectively adjusting for all current, and possible future, single-marker association tests based on the original data.

Example 20.5. Dudbridge and Gusnanto (2008) proposed an interesting approach for obtaining the fully saturated threshold value via permutation testing using increasing marker density. Their idea was that the threshold value increases with the number of tests, but ultimately approaches some asymptotic value. They took a set of SNPs (from the target population of interest) and randomly assigned them into cases and controls, and then performed a permutation test to obtain the critical p value required for an FWER of γ. They then examined ever-smaller subsets of the SNPs, with $f(k)$ denoting the critical p value for k scored SNPs, which is expected to decrease as k increases. The resulting number of effective tests for a given value of k to obtain a global γ FWER is $n_e(k) = \gamma/f(k)$. The authors suggested fitting a least-squares regression on such data using the **Monod function**

$$n_e(k) = \frac{\mu k}{h+k} + e \tag{20.6a}$$

The estimate of h (the **half-saturation parameter**) is the number of SNPs to achieve half of the threshold value, $f(h) = \mu/2$, and $\widehat{\mu}$ is the limiting value, so that using

$$p = \gamma/\widehat{\mu} \tag{20.6b}$$

provides the fully saturated Type I error control of γ. An alternative estimator for n_e, obtained by fitting a beta distribution (Equation A7.37) for the observed minimum p values over the permutation samples, was developed by Dudbridge and Koeleman (2004), Dudbridge and Gusnanto (2008), and Saffari et al. (2016). See their papers for details.

Dudbridge and Gusnanto applied Equation 20.6a using a set of roughly 360,000 SNPs from a UK Caucasian population. The limiting threshold was around $\widehat{\mu} \simeq 693{,}000$, with Equation 20.6b giving the fully saturated threshold (for $\gamma = 0.05$) as $p = 7.2 \times 10^{-8}$ for this population. They also noted that an eigenvalue-based estimate of n_e (Patterson et al. 2006) based on the 360,000 SNPs returned a value of around 33,000, an order of magnitude too small, while the permutation test for the full set of SNPs returned a value of roughly 228,000, also below the fully saturated value.

Bonferroni, sequential Bonferroni, and resampling are all examples of **family wide error rate (FWER)** approaches (Manly et al. 2004; Rice et al. 2008), where the goal is controlling the experiment-wide (here **genome-wide**) error rate. An alternative approach is the **false discovery rate (FDR)** framework (Appendix 6). The motivation for Bonferroni is the belief that all of the tests are likely under the null, while FDR assumes that some small fraction are likely *not* from the null. In the FDR setting, control is not over *all tests* (as is done with FWER procedures), but rather over ***all tests declared to be significant***. An FDR value of 5% implies that, of the declared significant tests, at most 5% are **false discoveries** (i.e., the true marker effect is zero). As detailed in Appendix 6, the basic underpinning of FDR amounts to adjusting the p-value threshold, π_δ, for declaring a test to be significant until the desired FDR is achieved. Under the assumption that very, very few of the n tests are true positives, the expected number of false discoveries is $n\pi_\delta$, while $n_{S(\delta)}$ is the observed number of tests with $p \leq \pi_\delta$, and hence are declared to be significant. The FDR thus becomes $n\pi_\delta/n_{S(\delta)}$. Decreasing the value of π_δ decreases both $n\pi_\delta$ and $n_{S(\delta)}$, and one adjusts the threshold value until this ratio achieves the desired FDR value (e.g., 0.05); see Appendix 6 for details.

FDR represents an attempt to strike a balance between **sensitivity** (Type II error, the power to detect a difference, e.g., avoiding false negatives) and **specificity** (Type I error, avoiding false positives). The Bonferroni framework, with its extremely conservative p values, has high specificity but poor sensitivity as many markers with small effects do not generate enough of a signal to exceed the extreme p threshold required to be declared significant. The FDR approach lowers this threshold (improving sensitivity), while still providing good specificity (Manly et al. 2004). The goal of an initial GWAS is usually to distill out a small set of candidate regions for future study. In such a setting, an FDR framework is usually better than an FWER approach, as the latter may exclude important regions from future examination. While false positives occur in the FDR candidate set, their number is controlled, and the cost of including such false candidates is often offset by much greater sensitivity, and hence the inclusion of true candidates that might otherwise be excluded. Brzyski et al. (2017) noted that an FDR analysis applied to GWAS is not truly on a SNP-by-SNP basis, as blocks of SNPs in LD will share the same signal from a single causal site. Rather, the FDR unit of analysis is a cluster of SNPs in LD. This is the same issue noted by Chen and Storey (2006) for linkage mapping (Chapter 18 and Appendix 6).

Finally, a rather different approach was suggested by Wacholder et al. (2004), WTCCC (2007), and subsequent authors (Thomas and Clayton 2004; Wakefield 2007, 2008, 2012; Ball 2011), namely using a Bayesian framework. An excellent discussion of Bayesian approaches for multiple GWAS comparisons is given by Stephens and Balding (2009). As suggested by Thomas and Clayton (2004), the basic tenet of this framework is that

> it is not the number of tests performed but rather the prior credibility of the hypotheses that is important in interpreting a set of observed associations. That is, when a hypoth-

esis is unlikely to be true *a priori*, we should require strong evidence to be convinced of its truth.

In this framework, the strength of evidence can be expressed as the **posterior odds ratio** in favor of a true association. Letting T denote the value of the test statistic and τ the significance threshold, then the odds ratio in favor of a true association when the test is deemed significant can be written as

$$\frac{\Pr(H_1 \mid T > \tau)}{\Pr(H_0 \mid T > \tau)} = \frac{\Pr(T > \tau \mid H_1)\Pr(H_1)/\Pr(T > \tau)}{\Pr(T > \tau \mid H_0)\Pr(H_0)/\Pr(T > \tau)}$$
$$= \left[\frac{\Pr(T > \tau \mid H_1)}{\Pr(T > \tau \mid H_0)}\right]\left[\frac{\Pr(H_1)}{\Pr(H_0)}\right] = \frac{1-\beta}{\alpha}\,\frac{\Pr(H_1)}{\Pr(H_0)} \tag{20.7a}$$

where the first step follows from Bayes theorem (Equation 3.3b). Here α is the Type I error rate, β the Type II error rate (for a power of $1-\beta$), and $\Pr(H_1)$ and $\Pr(H_0)$ are the prior probabilities for, and against, an association. To apply Equation 20.7a, one must have some loose idea about the fraction of independent regions that generate associations with the trait, and some details about the effect size (and frequency) in order to specify β. WTCCC (2007) suggested values for $\Pr(H_1)$ in the range of 10^{-4} to 10^{-6}. More recently, Yengo et al. (2022) detected 12,000 (genome-wide) significant SNPs for height out of roughly a million tested, for a $\Pr(H_1)$ value closer to 0.01.

Equation 20.7a is very closely related to Morton's (1955b) **posterior error rate (PER)**—which Wacholder et al. (2004) referred to as the **false positive report probability (FPRP)**—see Appendix 6. Denoting the posterior odds ratio (Equation 20.7a) as PO, an alternative metric is the **posterior probability of association, PPA** (Stephens and Balding 2009), where

$$PPA = \frac{PO}{1+PO} \tag{20.7b}$$

Finally, a more general approach is to replace

$$\frac{\Pr(T > \tau \mid H_1)}{\Pr(T > \tau \mid H_0)} \quad \text{with} \quad \frac{\Pr(T \mid H_1)}{\Pr(T \mid H_0)} \tag{20.7c}$$

Namely, replacing a threshold being exceeded with the actual value, T, of the test statistic for that marker. This ratio of support for the data (T) under the alternative versus the null hypothesis is called a **Bayes factor, BF** (Appendix 7). Computing the BF requires assumptions about the prior distribution of allele effects (and their associated allele frequencies), see Wakefield (2007, 2008, 2012) and Stephens and Balding (2009) for details. Stephens and Balding make the important point that as GWAS analysis moves beyond one-marker-at-a-time considerations to more complex units of interactions, Bayesian approaches can offer much more flexibility than frequentist methods.

Example 20.6. As an application of Equation 20.7a, consider the scenario where ten regions, out of one million LD blocks, influence the trait of interest, and suppose that there is 50% power to detect each effect. To obtain an odds ratio of 10 to 1 in favor of a true effect, rearranging Equation 20.7a gives a required α value of

$$\alpha = \left(\frac{\Pr(H_0 \mid T > \tau)}{\Pr(H_1 \mid T > \tau)}\right)\left([1-\beta]\,\frac{\Pr(H_1)}{\Pr(H_0)}\right)$$
$$= \frac{1}{10} \cdot 0.5 \cdot \frac{10/10^6}{1-10/10^6} = 5 \times 10^{-7}$$

Under these parameters, we expect an average of $0.5 \cdot 10 = 5$ true discoveries and $5 \times 10^{-7} \cdot (10^6 - 10) = 0.5$ false discoveries, for an expected FDR of $0.5/5.5$, or around 9%. The PPA for an odds ratio of 10 (PO = 10) becomes PPA = 10/11 = 0.909. Similarly, for a posterior odds ratio of 20 (with a resulting PPA of 20/21 = 0.952), $\alpha = 2.5 \times 10^{-7}$, which yields an expected 0.25 false discoveries and an expected FDR of 0.25/5.25, or slightly under 5%.

Now suppose a highly polygenic trait, such as height, which may have a large number of regions (say 1,000), but, given their smaller effect sizes, lower power of detection (say $\beta = 0.8$, or a power of 20%). The critical value for a odds ratio of ten becomes

$$\alpha = \frac{1}{10} \cdot 0.2 \cdot \frac{1000/10^6}{1 - 1000/10^6} = 2 \times 10^{-5}$$

In this setting we expected an average of 0.2$\cdot$ 1000 = 200 true discoveries and $2 \times 10^{-5} \cdot (10^6 - 1000) \simeq 20$ false discoveries, for an expected FDR of 20/220, or again around 9%. For a posterior odds ratio of 20:1, $\alpha = 1 \times 10^{-5}$, yielding an expected FDR of 10/210, again slightly under 5%.

Power, Replication, and the Winner's Curse

While it would seem that a modern GWAS, with its very large sample size, has ample power, this ignores two realities. First, most detected GWAS effects are small. This is perhaps not surprising, as the impact of a marker on a GWAS usually appears through the *variance* attributable to the linked causal site. For an additive QTL, this is $2a^2p(1-p)$, where a is the allelic effect, with this signal further attenuated by r^2 due to scoring the effect using a marker in LD with the causal site. In nature, large-effect alleles tend to be at low, to very low, frequencies, most likely as a result of past selection against them (Chapter 21; WL Chapter 28). Further, because common SNPs are typically the markers of choice, the mismatch in frequencies between a marker allele and a rare causal allele results in a small r^2 value (Example 20.1), even under complete disequilibrium ($|D'| = 1$), further reducing the signal from the causal site. Hence, a small GWAS effect *does not* imply a small-effect allele, but rather that the causal site only accounts for a small fraction of the total trait variance. Second, the blessing (and curse) of a modern GWAS is the vast number of markers that are considered, resulting in very small α values being used to test each marker in order to account for multiple comparisons (as just discussed). The cost of such increased specificity is a substantial loss of sensitivity (power).

One consequence of low power, *lack of repeatability*, led to considerable initial skepticism about the validity of GWAS detected sites (often referred to as **GWAS hits**). As noted by Colhoun et al. (2003), poor replication could result from false positives, lack of power, or differences between initial and confirmatory studies (such as using populations segregating different causal alleles or with different LD structures; or variation in how a trait, such as a complex disease, is defined). Much of this initial angst, which was especially common in the psychiatrical genetics community (Levinson et al. 2014), has faded as the sample sizes of GWAS increased, and more care was given to ensure that initial and confirmatory studies used consistent methodologies, resulting in a dramatic improvement in repeatability (Visscher et al. 2012a, 2017; Marigorta et al. 2018; Abdellaoui et al. 2023).

As developed in Chapter 18, one consequence of low power is that the effects of markers *declared to be significant* are overestimated, with the degree of bias increasing as power decreases (Figure 18.8; Equation 18.43). This is called the Beavis effect in QTL linkage mapping, but in GWAS it is commonly referred to as the **winner's curse** (a term from the epidemiological literature, but whose roots trace back to the analysis of competitive bids of oil leases; Capen et al. 1971). The GWAS implications of the winner's curse, and potential corrections, have been extensively discussed in the literature (e.g., Siegmund 2002; Sun and Bull 2005; Wu et al. 2005; Garner 2007; Yu et al 2007; Zöllner and Pritchard 2007; Ghosh et al. 2008; Zhong and Prentice 2008, 2010; Bowden and Dudbridge 2009; Xiao and Boehnke 2009; Faye et al. 2011; Xu et al. 2011; Ferguson et al. 2013; Poirier et al. 2015; Wang et al. 2016; Palmer and Pe'er 2017; Dudbridge and Newcombe 2019; Wang et al. 2020; Xie et al. 2021).

Example 20.7. As we have seen, a number of classic association metrics in a GWAS are either chi-square tests, or can be reformulated as such. In these cases, the test statistic follows a (central) chi-square when there is no association, and a noncentral chi-square (Appendix 5) when there is a marker effect. The key quality for the later distribution is its **noncentrality parameter** (or **NCP**) which can be thought of as the inflation of the test statistic over its value under the null. Specifically, if a test follows a χ^2_k under the null, then its expected value is its degrees of freedom, k, while if the test is from the alternative, its expected value is $k + \lambda$, where λ is the NCP (Equation A5.14a), a function of the sample design and marker effects.

To see how power calculations are performed with noncentral chi-squares, consider the threshold for significance under a chi-square test (with one degree of freedom) when we take $\alpha = 5 \times 10^{-7}$. We can quickly compute the value of this threshold using the `R` command `qchisq(1-5*10^(-7),1)`, which returns a value of 25.26. Hence, any marker whose associated chi-square statistic is less than this value is declared to be nonsignificant. The corresponding critical values for $\alpha = 10^{-6}$ and 10^{-7}, are, respectively, 23.93 and 28.37.

What NCP value is needed to give a high probability that a test of a true effect is declared significant (i.e., high power)? For a test to have power $1-\beta$ (under $\alpha = 5 \times 10^{-7}$), we need to solve for the value of the NCP λ that satisfies $\Pr(\chi^2_{1,\lambda} \geq 25.26) = 1 - \beta$. Recall that (in `R`) $\Pr(\chi^2_{k,\lambda} \leq X)$ is obtained as `pchisq(X,k,λ)`. Suppose that the NCP for a marker under the GWAS design is $\lambda = 15$. What is the power? This is $\Pr(\chi^2_{1,15} \geq 25.26)$, computed as `1- pchisq(25.26,1,15)`, which returns a value of 0.12. Hence, the power of this test is only 12%, meaning it would not be declared significant 88% of the time. What value of λ is required for 80% power? A grid search yields $\lambda = 34.43$. This can also be obtained by using `R` to plot $\Pr(\chi^2_{1,15} \geq 25.26)$ for different values of λ, e.g., `curve(1- pchisq(25.26,1,x), 10,35)` plots the power for λ over (10, 35). Similarly, for this critical value, the corresponding λ values for 50 and 90% power become 25.26 and 39.79. Table 20.1 gives the required NCP values for other situations.

A number of authors have developed expression for power, usually by obtaining expressions for the noncentrality parameter (λ) generated by a specific design (sample size) and marker effects (Schork 2002; Xiong et al. 2002; Chapman et al. 2003; Purcell et al. 2003; Edwards et al. 2005; Visscher and Duffy 2006; Altshuler and Daly 2007; Klein 2007; Yu et al. 2008; Spencer et al. 2009; Yang et al. 2010a; Sham and Purcell 2014; Wang and Xu 2019). As developed by these authors, the basic structure of the NCP is of the form $\lambda = nr^2 g(p, a)$ where n is the sample size, r^2 is the (squared) correlation between marker and causal alleles, and $g(p, a)$ is the causal gene effect (a function of, at least, the allelic effect a and frequency p). We first consider power calculations for a continuous trait under a random sample, and then for a binary trait under a case-control design.

For a continuous trait,

$$\lambda = n\left(\frac{r^2_{MQ}\sigma^2_Q}{1 - r^2_{MQ}\sigma^2_Q}\right) = n\left(\frac{\sigma^2_M}{1-\sigma^2_M}\right) \tag{20.8a}$$

where σ^2_Q is the causal variance and $\sigma^2_M = r^2_{MQ}\sigma^2_Q$ is the corresponding trait variance explained by the marker. We can write the causal variance as $\sigma^2_Q = 2p(1-p)\beta^2$, where β (the slope of a regression of trait value on SNP allele count; Equation 20.1a) is the effect size in phenotypic standard deviations. More generally, if additional cofactors are added to the model, then the standard deviation (for β) refers to the *residual* variance, e.g., Wang and Xu (2019). Because we typically expect $r^2_{MQ}\sigma^2_Q \ll 1$,

$$\lambda \simeq nr^2_{MQ}\sigma^2_Q \tag{20.8b}$$

The power for an imputed site follows by replacing r^2_{MQ} by $r^2_{i\text{-}npQ}$. The required sample size to achieve a desired amount of power follows directly using the critical values in Table

Table 20.1 Critical values for χ^2 tests with one or two degrees of freedom (df) for a given α, and the associated NCP values (λ) required for a desired level of power. See Example 20.7.

			Required λ for a power of		
α	df	Critical	50%	80%	90%
10^{-6}	1	23.93	23.93	32.87	38.11
	2	27.63	26.62	36.11	41.62
$5 \cdot 10^{-7}$	1	25.26	25.26	34.43	39.79
	2	29.02	28.01	37.71	43.35
10^{-7}	1	28.37	28.37	38.05	43.67
	2	32.24	31.23	41.43	47.32
$5 \cdot 10^{-8}$	1	29.72	29.72	39.60	45.33
	2	33.62	32.63	43.02	49.02

20.1. For example, to have 80% power using a significance level of $\alpha = 5 \times 10^{-7}$ requires an NCP value of 34.43. Solving for n

$$n \geq \frac{34.43}{r^2_{MQ}\sigma^2_Q}$$

Suppose the causal site accounts for 0.1% of the total variance and $r^2 = 0.4$, then

$$n = \frac{34.43}{0.4 \cdot 0.001} = 86,075$$

The fraction of this sample size required for 50% power is 25.26/34.43 (the ratio of the NCPs from Table 20.1), or 75% of this sample size (63,150).

Now consider a case-control design, which is a form of selective genotyping (Chapter 18). The power of this design is a function of the difference in marker-allele frequencies between cases and controls (Klein 2007; Yang et al. 2010a; Evans and Purcell 2012). To translate the effect of an underlying locus into the expected case-control allele frequency difference, suppose the three genotypes have (multiplicative) risks of $f : f\gamma : f\gamma^2$, with p the frequency of the risk-increasing allele ($\gamma > 1$), and let K denote the disease incidence (prevalence) in the population. For a design of ηn cases and $(1-\eta)n$ controls,

$$\lambda = nr^2_{MQ}\left(\frac{2p(1-p)(\gamma-1)^2\eta(1-\eta)}{(1-K)^2[1+p(\gamma-1)]}\right) \tag{20.9a}$$

In order to compare the case-control power with that of a continuous trait, Yang et al. (2010a) considered the **threshold-liability model** (Chapter 14; Part II Chapter 31). Under this model, there is an underlying liability value y (assumed to be normally distributed with variance one), with the disease expressed when the liability value exceeds some threshold (T), so that $\Pr(y \geq T) = K$. On this scale, the fraction of the liability variance explained by the causal locus is $\sigma^2_Q \simeq 2p(1-p)(\gamma-1)^2/\iota^2$, where ι is the mean value for the liability above the truncation value, and is given by $\varphi(T)/K$, where $\varphi(x)$ is the unit normal function evaluated at x (the truncated mean of a normal, Equation 2.14). The resulting NCP becomes

$$\lambda \simeq nr^2_{MQ}\sigma^2_Q\left(\frac{\iota^2\eta(1-\eta)}{(1-K)^2}\right) \tag{20.9b}$$

Suppose that a causal locus for a continuous trait (CT) and a causal locus under a case-control (CC) design both have the same variance (σ^2_Q on the observed, and liability, scales, respectively). From Equations 20.8b and 20.9b, the ratio of the two resulting NCP values becomes

$$\frac{\lambda_{CC}}{\lambda_{QT}} = \left(\frac{n_{CC}}{n_{QT}}\right)\left(\frac{\iota^2\eta(1-\eta)}{(1-K)^2}\right) \tag{20.9c}$$

$$\simeq \frac{\iota^2}{4(1-K)^2} \tag{20.9d}$$

with Equation 20.9d applying for a design with equal number of cases and controls ($\eta = 1/2$) and the same total design size ($n_{CC} = n_{QT}$). Consider the relative power for a continuous trait (say height) with that of schizophrenia. For this disease, $K = 0.01$, so that $T = 2.33$ (as for a unit normal U, $\Pr[U \geq 2.33] = 0.01$) and $\iota = \varphi(2.33)/0.01 = 2.64$, giving $\lambda_{CC}/\lambda_{QT} = 2.64^2/[4 \cdot 0.99^2] = 1.78$. Hence, the same power for the case-control design occurs at 1/1.78 = 56% of the sample size of the continuous trait. For example, the power for a design with 120,000 individuals for the continuous traits is the same as for a design with around 33,000 cases and 33,000 controls. When the disease is rare, the power gain from the case-control design more than compensates for the reduction of signal when translating from the liability to the observed (binary) scale.

Multitrait GWAS

As was the case for linkage mapping, marker-trait association machinery can be extended from a single trait to a *vector* of traits. Multivariate traits arise naturally in many settings, such as eQTL mapping (Chapter 21), analyzing correlated traits (Part II Chapter 26), and exploring genotype-by-environment (G $\times$ E) interactions (Part II Chapters 27 and 28). As such, much of our discussion on multiple trait GWAS is spread over these chapters (in Part II), with just a few overview remarks here.

As might be expected given the complexity of the task, a variety of different strategies have been suggested for implementing a multitrait GWAS. These include fully multivariate **direct approaches**, such as estimating the vector $\boldsymbol{\beta}$ of regression coefficients of a focal SNP on each of the traits (e.g., Wisser et al. 2011; Korte et al. 2012; Bolormaa et al. 2014; Zhou and Stephens 2014; Turley et al. 2018; Carlson et al. 2019) or multivariate path analysis (e.g., Grotzinger et al. 2019; Igolkina et al. 2020; Pritikin et al. 2021). For estimation reasons (Chapter 26), direct approaches are generally restricted to just a few traits (five or fewer). With a very large number of traits, **indirect approaches** are often used. These extract some lower dimensional feature(s) from the trait vector, such as using the first PC of the phenotypic covariance matrix, and then map this composite trait in a univariate GWAS (e.g., Aschard et al. 2014; Zhang et al. 2018; Carlson et al. 2019). Approaches have also been suggested for combining information from univariate analyses, such as incorporating the correlation structure between univariate estimates (e.g., Huang et al. 2011; van der Sluis et al. 2013). Finally, while often framed in terms of testing each variant (SNP) separately, one can also use a gene-based (or more generally, set-based) approach as the predictor variables (e.g., Kim et al. 2016). For example, estimating the vector of trait regression coefficient not for a SNP but rather for some composite SNP score over a region/set of interest.

Galesloot et al. (2014), Porter and O'Reilly (2017), and Chung et al. (2019) reviewed a number of multitrait GWAS approaches and found that there is no one method which does best under all settings. However, with correlated traits, all were generally more powerful than performing a series of univariate analyses. A multiple trait GWAS borrows information from correlated traits to improve the estimate of a focal trait, information that is not exploited by a univariate analysis. Even when traits are genetically uncorrelated, they can still be environmentally correlated (Chapters 26)—such as having correlated measurement errors—and incorporating this residual covariance structure improves estimates.

Many of the same general issues about testing and combining information from multiple sources that arose in gene-based GWAS also apply to a multitrait GWAS. For example, in a gene-based setting (one trait, multiple SNPs), one might first test for the significance of the overall gene and then perhaps attempt to detect signals associated with particular variants within such selected genes. Analogously, in a multitrait scan (one SNP, multiple traits), one might first test whether the $\boldsymbol{\beta}$ vector associated with a SNP is significantly different from zero. If so, then one can test particular traits within this vector. Peterson et al. (2016a) proposed a two-step FDR approach in such testing, first controlling the FDR on which variants have a significant nonzero $\boldsymbol{\beta}$ vector and then controlling the FDR for the nonzero elements of $\boldsymbol{\beta}$ within those significant SNPs. Finally, gene-based GWAS can be extended from single to multiple traits, namely the multiple SNPs, multiple traits setting (Deng et al. 2020).

CORRECTING FOR CRYPTIC RELATEDNESS AND POPULATION STRUCTURE

A marker-trait *association* is just that, a *correlation* between the value of a genotype and the value of a trait. While our above discussion has assumed that this correlation arises because a marker locus is in LD with a causal locus (QTL), it can also arise from other factors, such as the presence of undetected relatives in the sample and/or undetected population structure. Given the enormous sensitivity of a modern GWAS to very small influences, these **spurious associations** are potentially very problematic. Chapter 17 discussed the Transmission Disequilibrium Test (TDT), wherein one examines whether a parent heterozygous at a marker passes (transmits), or fails to pass, a specific allele to its offspring. For a random allele, the probability of a transmitted or a nontransmitted event should be equal. However, if we partition individuals by trait value (such as disease presence/absence), this ratio can be skewed for alleles that are linked to causal loci (Example 17.20). By focusing on within-family transmission, population structure is fully controlled when using the TDT, and, given that it specifically uses known relatives, undetected relatives are not a concern. Under a linkage analysis, relatedness is *known, recent, and useful*, while under an association analysis, relatedness is typically *unknown, distant, and a nuisance* (Astle and Balding 2009). Hence, during the early 2000s, the TDT was taken as the gold standard for replication of a proposed candidate gene or association signal. However, this strategy is in conflict with the motivation for transitioning to population-level association analysis, which was to *avoid* the difficulty and resources required to collect a sufficiently large collection of families to have reasonable power. Furthermore, the TDT can be challenging with late-onset diseases, where it may be difficult to sample relatives (parents or sibs) of a case.

As detailed below, in an attempt to move away from family-based studies, a number of approaches have been proposed using marker data to detect relatives and population structure, and then correct for their effects. Examples 9.13 and 9.14 foreshadowed the nature of these corrections, which are typically performed using mixed models (Chapter 10). We will first examine corrections for the presence of relatives before turning to adjustments for population structure. Essentially just one approach has been proposed to adjust for relatives, while multiple approaches have been suggested for population structure. Reviews on a number of the issues discussed below are given by Pritchard et al. (2000b), Pritchard and Donnelly (2001), Hoggart et al. (2003), Voight and Pritchard (2005), Patterson et al. (2006), Tiwari et al. (2008), Astle and Balding (2009), Choi et al. (2009), Price et al. (2010a), Zhang and Deng (2010), Sillanpää (2011b), Sul et al. (2018), and Toosi et al. (2018).

Correcting for Relatives in the Sample

Early GWAS studies typically went out of their way to exclude known relatives (e.g., WTCCC 2007). The reason was simple: relatives share alleles (IBD) and have correlated values for heritable traits (Chapter 7). This induces a correlation between shared alleles and trait values, generating marker-trait associations even at markers unlinked to QTLs. In humans, identification of close relatives within a sample is often (but not always!) straightforward, while detecting more distant relatives can be more problematic. The situation is much more challenging in natural populations, where pedigree relationships are typically not observed, and hence must be inferred from marker data (Chapter 8). The same concern arises when constructing association panels for mapping QTLs using collections of distinct lines (e.g., Atwell et al. 2010). Some of these lines may have recent ancestors, and hence are relatives, which may be unknown to the investigator. The presence of undetected relatives in a sample is referred to as **cryptic relatedness**.

The remedy for these concerns was introduced in Chapter 19 in the form of mixed models (Chapter 10), with a random effect included for relatedness (Equation 19.29; Example 19.5). For example, the additive model for SNP k (Equation 20.1a) now becomes

$$z_i = \mu + b_k N_{i,k} + A_i + e_i \tag{20.10a}$$

where b_k is a fixed effect for the SNP, while A_i is a random effect for the background (additive) polygenic value (the effect of QTLs not in LD with the focal SNP). From Equations 7.11a and 19.30a, these polygenic values are correlated among relatives, with

$$\sigma(A_i, A_j) = 2\Theta_{ij}\,\sigma_A^2 \tag{20.10b}$$

where σ_A^2 is the background additive variance for the trait, and the Θ_{ij} can be estimated from a known pedigree (Chapter 7) or from marker data (Chapter 8). Note that Equation 7.12 generalizes this approach to higher order (nonadditive) shared polygenic effects.

The mixed model given by Equation 20.10a requires specification of the covariance matrix of the A_i random effects. This is given by the $n \times n$ matrix for all of the pairwise Θ_{ij} values among the n individuals in the population sample. The resulting matrix in GWAS studies is often called the **K** (for **kinship**) **matrix** by plant breeders or the **numerator relationship matrix** (**A**) by animal breeders, with the human literature using both. When estimated solely from marker information, this is often called the **genomic relationship matrix** (**GRM**; also the **realized relationship matrix, RRM**). Example 9.14 gives several different marker-based estimates of the GRM. To avoid double-dipping of SNP information (**proximal contamination** and a loss of power; Lippert et al. 2011; Sawcer et al. 2011; Listgarten et al. 2012; Yang et al. 2014), usually the **LOCO** (**leave one chromosome out**) approach is used (Yang et al. 2014). Here, the GRM is computed using marker information for all chromosomes *except* the one on which the tested SNP resides. Hence, the GRM varies slightly over chromosomes. Further, often a subset of the markers is used, trimming out those that are in high LD with each other. Zaitlen et al. (2013) and Jiang et al. (2019) showed that using a **sparse GRM**, truncating small values (e.g., $2\widehat{\Theta}_{ij} < 0.05$ or 0.01) to zero captures essentially as much variation as the full GRM.

Example 20.8. The resulting GWAS mixed model (testing marker k) that accounts for relatives in the sample can be written as

$$\mathbf{z} = \mathbf{X}\boldsymbol{\beta}_k + \mathbf{Z}\mathbf{u} + \mathbf{e}, \qquad \begin{pmatrix} \mathbf{u} \\ \mathbf{e} \end{pmatrix} \sim \begin{pmatrix} \mathbf{0} \\ \mathbf{0} \end{pmatrix}, \begin{pmatrix} \sigma_A^2\mathbf{K} & \mathbf{0} \\ \mathbf{0} & \sigma_e^2\,\mathbf{I} \end{pmatrix} \tag{20.10c}$$

where the elements of GRM **K** are marker-based estimates of $2\Theta_{ij}$ (Example 9.14). The covariance matrix for the residuals (**e**) assumes an OLS structure ($\sigma_e^2\,\mathbf{I}$), but can be replaced by $\sigma_e^2\,\mathbf{R}$ to allow for more general GLS error structures (Chapter 10). Here **u** is the random vector of background polygenic effects (the A_i), the incidence matrix **Z** = **I** when all observations are singletons, and (assuming the only fixed effect beyond an additive SNP effect is the mean, μ) the structure of the vector $\mathbf{X}\boldsymbol{\beta}_k$ is

$$\mathbf{X}\boldsymbol{\beta}_k = \begin{pmatrix} 1 & 2 \\ 1 & 0 \\ 1 & 1 \\ \vdots & \vdots \\ 1 & N_{k,n} \end{pmatrix} \begin{pmatrix} \mu \\ b_k \end{pmatrix} = \begin{pmatrix} \mu + 2\,b_k \\ \mu + 0 \\ \mu + b_k \\ \vdots \\ \mu + N_{k,n}\,b_k \end{pmatrix} \tag{20.10d}$$

In this example, individual one (first row of **X**) is a reference homozygote ($N = 2$) at the focal SNP (marker k), two is a nonreference homozygote ($N = 0$), three is a heterozygote, and so on. Note that under fixed-effect (FE) modeling of SNP effects, this model is *rerun for each tested SNP*. Because σ_A^2 is the polygenic variance accounted for by QTLs *not* in LD with the tested marker, it ***potentially changes over markers***, and hence ***needs to be reestimated for each marker***. The reader might note that the expected change in σ_A^2 over markers should be small, but recall that a large modern GWAS can be influenced by *very* small effects. For all SNPs on the same chromosome, the GRM is unchanged, but the elements in the second column of **X** (corresponding to the genotypes at the focal SNP) varies over markers.

If additional fixed effects (cofactors) are included, these add extra columns to **X** with the corresponding fixed effect added to $\boldsymbol{\beta}_k$ (Chapter 10). For example, if the general genotype model (Equation 20.1b) is fit, then

$$\mathbf{X}\boldsymbol{\beta}_k = \begin{pmatrix} 1 & 2 & 0 \\ 1 & 0 & 0 \\ 1 & 1 & 1 \\ \vdots & \vdots & \vdots \\ 1 & N_{k,n} & H_{k,n} \end{pmatrix} \begin{pmatrix} \mu \\ b_k \\ d_k \end{pmatrix} = \begin{pmatrix} \mu + 2\,b_k \\ \mu + 0 \\ \mu + b_k + d_k \\ \vdots \\ \mu + N_{k,n}\,b_k + H_{k,n}\,d_k \end{pmatrix} \tag{20.10e}$$

Correcting for Population Structure: Genomic Control

Another confounding factor is **population structure** (or **population stratification**), wherein one has unknowingly sampled from several different subpopulations (or **strata**). More generally, there can also be individuals in the sample with different amounts of **admixture** (carrying genes from two, or more, of the subpopulations). Such individuals may not fall into discrete strata, but rather could form a potential continuum of admixture (**continuous admixture populations**). The concern is that if disease risks (or trait means) differ over these strata (or over amounts of admixture), and similarly some SNP frequencies also vary over strata, marker-trait associations can be created (Knowler et al. 1988; Spielman et al. 1993; Lander and Schork 1994; Risch 2000; Cardon and Palmer 2003; Freedman et al. 2004; Marchini et al. 2004; Clayton et al. 2005). For example, suppose allele M (at a particular SNP, unlinked to any QTL) is common in subpopulation 1, but rare in subpopulation 2. Simultaneously, subpopulation 1 has a higher disease risk (or trait mean) than subpopulation 2. The presence of M predicts membership in group 1, which, in turn, predicts a higher disease risk (or larger trait value), generating a marker-trait association. Note that population subdivision may not be obvious to even a careful investigator, leading to the concern of undetected **cryptic stratification** introducing false positives. The original large-scale GWAS (WTCCC 2007) did a careful analysis to ensure that there was little stratification, within both the cases and controls, as well as between them. The latter point is of special concern, in that cases and controls are assumed to be drawn from the same population. Failure to do so creates stratification and the potential for numerous false positives. One potential generator of stratification at the case versus control level is **ascertainment bias**: for example, cases could easily be oversampled from one strata if its members are overrepresented in clinics (Astle and Balding 2009).

Example 17.19 showed a classic example of stratification with diabetes in members of the Pima nation in southern Arizona. Here the focal marker predicted whether an individual had some amount of admixture with Caucasians, a lower risk population. A more recent example is the work of Campbell et al. (2005), examining height in a collection of European Americans. No stratification signal was seen in a panel of roughly 200 SNPs, but a SNP in the *LCT* gene (associated with the persistence of lactase) showed a strong association with height ($p < 10^{-6}$). This allele showed a cline of increasing frequency from southern to northern Europe, roughly matching a similar trend seen in height. When individuals were corrected for ancestry, the signal was greatly reduced overall, and absent in some comparisons.

Increasingly sophisticated population genetics models have examined the impact of stratification and admixture in generating spurious associations (Devlin and Roeder 1999; Pritchard and Rosenberg 1999; Wacholder et al. 2000; Gorroochurn et al. 2004; Rosenberg and Nordborg 2006). A few generalizations emerge from this work. First, spurious associations usually tend to be less severe in admixed populations relative to the corresponding mixture of discrete subpopulations. Second, the largest impact occurs when lower risk subpopulations make up a larger fraction of the sample. Finally, the severity of spurious associations usually decreases as the number of underlying subpopulations increases.

The first general approach to adjust for population structure was the method of **genomic control, GC** (Devlin and Roeder 1999; Bacanu et al. 2000, 2002; Devlin et al. 2001a, 2001b,

2004), using a single correction over all tests (a closely related method was independently proposed by Reich and Goldstein 2001). Suppose a disease GWAS is conducted using a Cochran–Armitage trend test (Equation 18.15). Under the null (no linkage to a QTL), the resulting test statistic at each marker follows a chi-square with one degree of freedom (χ_1^2). Devlin and Roeder (1999) noted that when population structure is present, all of the marker test statistics should be enhanced by a common **genomic inflation factor**, λ. If S_k denotes the test statistic value for marker k, then under the null in the presence of population structure, $S_k \sim \lambda\chi_1^2$, so that $S_k/\lambda \sim \chi_1^2$, suggesting a correction for structure applicable to all tests. Devlin and Roeder showed that the inflation factor (with case-control data) is given by

$$\lambda = 1 + \frac{nF_{ST}\sum_j (D_j - C_j)^2 - 2F_{ST}}{1 + F_{ST}} \tag{20.11a}$$

where each subpopulation has n sampled cases and n sampled controls, D_j and C_j are the fraction of cases (diseased) and controls, respectively, in the jth population, and F_{ST} is a measure of population structure (the fraction of marker variation due to among-population differences; WL Chapter 2). Observe that λ *increases* with the sample size (n), so that larger samples should show *more* bias. For large n and small F_{ST},

$$\lambda \simeq 1 + nF_{ST}\sum_j (D_j - C_j)^2 \tag{20.11b}$$

As first noted by Devlin and Roeder (1999), cryptic relatedness also inflates λ (e.g., Sawcer et al. 2011), and Voight and Pritchard (2005) developed expressions for λ under a variety of relationship scenarios. The latter's conclusion was that deep relationships have a minimum impact of λ, so that a random sample from a large single outbred population should not be impacted. However, sampling bias that favors choosing relatives (such as cases being weakly related) can have a profound impact.

How does one go about estimating λ? Because the expected value of a χ_1^2 is 1 (Equation A5.14b), one might initially consider using the mean value ($\overline{S}$) of test statistics over all markers as an estimate of λ, e.g., the **mean estimator**

$$\widehat{\lambda}_{mean} = \overline{S} \tag{20.12a}$$

The limitation with this approach is that we expect some small fraction of the markers are indeed linked to QTLs, and would generate large S values. Including these when computing the mean value under the null inflates the estimate of λ. Statisticians often deal with such outliers by using the ranks of the data in place of their actual values (e.g., Conover 1999; Sprent and Smeeton 2007), and Devlin and Roeder (1999) used this idea to propose a **median estimator** for λ. Note that $\Pr(\chi_1^2 \geq 0.4549) = 0.50$, so that the expected median value of test statistics under the null is 0.4549 (half of the tests are expected to be above, and half below, this value). By considering ranks, rather than values, extreme outliers do not unduly influence rank-based statistics. Ordering the values of the test statistics over the m markers from smallest ($S_{[1]}$) to largest ($S_{[m]}$), the **Devlin-Roeder estimator** is given by

$$\widehat{\lambda}_{med} = \frac{\text{median}(S_{[1]}, \cdots, S_{[m]})}{0.4549} \tag{20.12b}$$

where median($\cdot$) denotes the value separating the upper half of test scores from the lower half. Other robust estimators for λ were discussed by Wang (2009). Devlin et al. (2004) noted that both Equations 20.12a and 20.12b ignore variation in the S_k, and suggested their **GCF estimator** (genomic control F), where

$$\frac{S_k}{\overline{S}} \sim F_{1,m} \tag{20.12c}$$

so that the GCF-adjusted test statistic follows a scaled F, rather than a scaled χ^2, distribution (for very large m, $F_{1,m}$ approaches a χ_1^2). Equation 20.12c follows because

$$\overline{S} = \frac{1}{m}\sum_{k=1}^{m} S_k \sim \frac{1}{m}\sum^{m} \chi_1^2 \sim \frac{\chi_m^2}{m} \tag{20.12d}$$

with the last step following because the sum of m χ_1^2 is a χ_m^2 random variable (Equation A5.14c). Hence, from the definition of the F distribution (Appendix 5),

$$\frac{S_k}{\overline{S}} \sim \frac{\chi_1^2/1}{\chi_m^2/m} \sim F_{1,m}$$

Dadd et al. (2009) found GCF outperformed GC based on either λ_{mean} (GC$_{mean}$) or λ_{med} (GC$_{med}$). We have framed the above discussion of GC in terms of testing the additive (or trend) model. Zheng et al. (2005) examined the power of using a trend-test based GC when dominance is present, while Zheng et al. (2006) extended GC to the general genotype model.

While conceptually elegant and simple, one immediate flaw with GC is that *all markers are treated equally*. As a result, the *relative rankings of the markers do not change* under genomic control. All marker tests are adjusted by the same amount, when in fact, a more marker-specific approach is required. In Equation 20.11, F_{ST} is the mean value over *all* markers ($\overline{F}_{ST}$), while markers with excessive divergence relative to the average ($F_{ST} \gg \overline{F}_{ST}$) should receive greater adjustments (Price et al. 2006; Kang et al. 2010). Thus we would expect that a proper correction for structure is likely to *change the rankings* of some of the markers. As a result, GC is usually no longer used to correct for population structure. However, as with many ideas in science, GC has been recycled, and is now often regarded as a useful *metric of model fit*. The values of λ before and after a model-based correction are compared to assess how well the model reduced the impact from confounding effects. The rough rule of thumb historically used is that a value of $\lambda \leq 1.05$ implies that the model has largely corrected for any structure (e.g., Price et al. 2010a). However, as we now show, one can observe significant inflation in the *absence of any confounders* if a large number of small-effect QTL underlie a trait (i.e., the trait displays high levels of polygenicity).

Adjusting for Inflation from Polygenicity: LD-Score Regression

GC estimates of λ assume that only a very, very small fraction of tested markers are in LD with QTLs. This assumption fails if a large number of small effect QTLs underpin a trait (the **polygenic model**). Yang et al. (2011a) showed that in this setting, significant inflation can be generated ***even in the absence of any population structure***. In particular, they showed that the impact of polygenes on the genomic inflation factor is

$$\lambda \simeq 1 + n\left(\frac{h^2\,\overline{r^2}\,\overline{s}}{m}\right) \tag{20.13a}$$

Here n is the sample size, h^2 the trait heritability, m the number of tested SNPs, $\overline{s}$ is average number of SNPs in LD with causal variants, and $\overline{r^2}$ is the average r^2 LD between a causal variant and a marker. Notice that, as with inflation from population structure, λ increases with the number of individuals genotyped (n), reflecting the increased power to detect small effects.

Yang et al. (2011a) examined two GWAS studies for human height: the QIMR study of just under 4,000 individuals (Yang et al. 2010b) and the GIANT composite GWAS with a combined sample size of around 184,000 (Lango Allen et al. 2010). Substituting GWAS estimates from the QIMR study into Equation 20.13a suggested a λ median estimate of around 1.03, consistent with the observed value. For the much larger GIANT data set, the observed λ value (based on the median estimator) was 1.55, again consistent with the value predicted by Equation 20.13a. Hence, for a highly polygenic trait, a significant fraction of

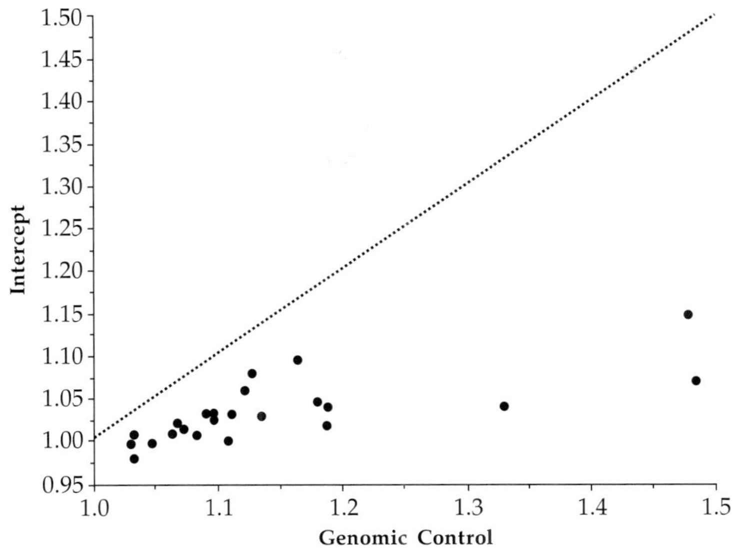

Figure 20.3 Plot of the LD-score regression estimates of λ (the intercept of Equation 20.13c minus one) on the vertical axis versus the λ value estimated from GC on the horizontal axis. The data are from 20 different traits examined in large GWAS studies. Note that many of the GC values are rather large (the two largest, $\simeq 1.5$, are for height and schizophrenia), which might suggest major effects from confounders such as relatedness or population structure. However, the LD-score estimates are all much smaller, suggesting that the vast bulk of the inflation over all these traits is from polygenicity. (Data from Bulik-Sullivan et al. 2015).

the markers will show elevated values, resulting in a genomic inflation factor much larger than one, even in the absence of population structure. Figure 20.3 underscores this point. Further, ***this effect is expected to be more pronounced as the GWAS size increases***.

Bulik-Sullivan et al. (2015) proposed the method of **LD-score regression** to adjust for the inflation due to polygenicity. Their logic was that the impact of population structure should (generally) not scale with the LD reach of a variant, while the impact from polygenicity should (as the marker is impacted by a greater number of causal sites). For a given marker j, they defined its **LD score** as

$$\ell_j = \sum_i r_{ji}^2 \tag{20.13b}$$

where the sum is over scored variants in LD with the focal variant, and r_{ji}^2 the correlation between markers j and i. Under a model where average effect size increases as the MAF decreases (large-effect alleles tend to be rarer; Chapter 21), they showed that the expected test statistic for a noncausal marker j is

$$\mathrm{E}\left[\chi^2 \mid \ell_j\right] = \left[\frac{nh^2}{m}\right] \ell_j + n\alpha + 1 \tag{20.13c}$$

where m is the number of SNPs, h^2 is the trait heritablity, and α denotes the inflation factor from confounders (such as the bracketed term in Equation 20.11b). One then regresses the observed test statistic for marker j on its LD score, ℓ_j (**LD-score regression, LDSC**). Bulik-Sullivan et al. noted that this is most optimally done as a weighted regression (Example 10.8), and discuss various weighting schemes (see their paper for details). Note that the intercept from the fitted regression (which removes the impact of polygenicity) has expected value $1 + n\alpha$, so that the intercept minus one estimates $n\alpha$, the inflation from confounders. Figure 20.3 shows that most of the inflation seen in a number of large GWAS experiments appears to be due to polygenicity.

Correcting for Population Structure: Structured Association Mapping

While GC attempts to *correct* for any population structure, next-generation methods further try to *quantify* the nature of this structure (i.e., how many strata, prediction of strata membership, etc.). These methods are based on the idea that marker data provide information about subpopulation membership. One such signal is that ***differences in allele frequencies between subpopulations generates LD between unlinked markers*** in the combined population sample. To see this, suppose all population 1 individuals are *AABB*, while all population 2 are *aabb*. When the populations are combined into a single sample, assuming equal contributions from both populations, $p_A = p_B = 0.5$, while the LD is

$$D_{AB} = \text{freq}(AB) - \text{freq}(A)\text{freq}(B) = 0.5 - 0.5^2 = 0.25$$

We present this extreme case (differential fixation) to show the basis for this phenomena. In most settings, very small amounts of LD are generated between pairs of unlinked loci that differ in frequency over subpopulations. Such subtle signals require a reasonable number of markers to detect.

This phenomena was exploited by Pritchard and Rosenberg (1999), Pritchard et al. (2000a, 2000b), and Falush et al. (2003, 2007), who assumed that a population sample of interest consisted of k subpopulations, each in Hardy–Weinberg proportions. They then used an MCMC sampler to develop a Bayesian classifier to assign individuals into clusters. Assignment was done in such a way as to minimize the LD between unlinked loci *within* each cluster. Their program, **STRUCTURE**, returns a vector of posterior probabilities of group membership for each sample member. For example, with $k = 3$, this vector might be (0.3, 0.1, 0.6), so that this individual is most likely from group three and least likely from group two. More generally, this could also be interpreted that this individual shows admixture, with 30% of its genes from population 1, 20% from 2, and 60% from three, although Zhu et al. (2002) and Zhang et al. (2003) suggest some caution with this interpretation. Approaches for estimating the number of subpopulations (S) were offered by Pritchard and Rosenberg (1999), Zhu et al. (2002), Evanno et al. (2005), Price et al. (2006), Camus-Kulandaivelu et al. (2007), and Lawson and Falush (2012). The last authors noticed that the presence of highly related individuals in the sample can lead to unreliable group assignment. While SNPs are often the marker of choice, Thornsberry et al. (2001) showed that 141 highly polymorphic SSR markers (Chapter 8) were sufficient to allow STRUCTURE to account for population structure in a set of 92 maize lines.

Pritchard et al. (2000b) proposed using these probabilities of group membership to perform **structured association mapping (SAM)**, essentially testing for association within each subpopulation. Using such latent (underlying, but unmeasured) variables to control for population structure was also suggested by Ripatti et al. (2001), Satten et al. (2001), and Sillanpää et al. (2001). Consider a quantitative (as opposed to a binary) trait. Let the vector $\mathbf{q}_i^T = (q_{i,1}, q_{i,1}, \cdots, q_{i,S-1})$ denote membership probabilities for individual i (because the q_{ij} sum to one, $q_{iS} = 1 - \sum_j^{S-1} q_{ij}$), and let $\mu + v_i$ denote the mean value of the trait from group i. In this setting, the adjusted group-weighted mean for individual i becomes

$$\mu_i = \mu + \sum_{j=1}^{S-1} q_{i,j}\, v_i \tag{20.14a}$$

Note under this coding the μ is the mean for group S. One could equivalently code this so that the v_j are deviations from the overall population mean. The population-structure corrected additive model (Equation 20.1a) becomes

$$\begin{aligned} z_i &= \mu_i + b_k N_{i,k} + e_i \\ &= \left(\mu + \sum_{j=1}^{S-1} q_{i,j}\, v_i \right) + b_k N_{i,k} + e_i \end{aligned} \tag{20.14b}$$

The term in parentheses adjusts for mean trait values varying over groups, removing the effects of structure (to the extent that $\mathbf{q}_i$ captures this). In linear model form, for n individuals and S groups, we can add the vector of population structure corrections as

$$\mathbf{Q}\mathbf{v} = \begin{pmatrix} \mathbf{q}_1^T \\ \mathbf{q}_2^T \\ \vdots \\ \mathbf{q}_n^T \end{pmatrix} \begin{pmatrix} v_1 \\ v_2 \\ \vdots \\ v_{S-1} \end{pmatrix} = \begin{pmatrix} q_{1,1} & q_{1,2} & q_{1,3} & \cdots & q_{1,S-1} \\ q_{2,1} & q_{2,2} & q_{2,3} & \cdots & q_{2,S-1} \\ \vdots & \vdots & \vdots & \cdots & \vdots \\ q_{n,1} & q_{n,2} & q_{n,3} & \cdots & q_{n,S-1} \end{pmatrix} \begin{pmatrix} v_1 \\ v_2 \\ \vdots \\ v_{S-1} \end{pmatrix} \tag{20.14c}$$

where $\mathbf{Q}$ is a known matrix, and the vector $\mathbf{v}$ is estimated (as a fixed effect) from the data. The resulting linear model becomes

$$\mathbf{z} = \mathbf{X}\boldsymbol{\beta} + \mathbf{Q}\mathbf{v} + \mathbf{e} \tag{20.14d}$$

Example 20.8 gives the structure of $\mathbf{X}\boldsymbol{\beta}$ under both the additive and general genotype models.

Correcting for Population Structure: Principal Components

While the STRUCTURE approach is very elegant, it is computationally demanding and its performance in continuous admixture settings is somewhat unclear. An alternative strategy was proposed by Zhu et al. (2002), Zhang et al. (2003), and Price et al. (2006), using the **principal components** (**PCs**) of the covariance matrix $\mathbf{C}$ of marker information (a **PCA**, or **principal component analysis**, also called an **eigenanalysis**). PCs and eigenvectors were discussed in Chapter 9, and the reader might find it useful to review that material before proceeding. Recall that the **first principal component** (**PC1**) is the (unit-length) vector corresponding to the direction of the most variation in the distribution space described by $\mathbf{C}$. This is also the **leading eigenvector** (e_1), namely that associated with the largest eigenvalue (λ_1) of $\mathbf{C}$. The idea is that instead of running a computationally demanding Bayesian classifier, one could take a small number of PCs and use these to capture structure in the marker data, presumably corresponding to distinct subpopulations and/or axes of admixture within the sample. Price et al. (2006) called this approach **EIGENSTRAT** (based on their program name).

The use of principal components to assign individuals into subpopulations traces back to Menozzi et al. (1978). There is sufficient theoretical work to justify this assumption. For example, Patterson et al. (2006) noted that "*natural models of population structure predict that most of the eigenvalues ... will be 'small,' nearly equal, and arise from sampling noise, while just a few eigenvalues will be 'large,' reflecting past demographic events.*" Bryc et al. (2013) showed (for a sufficient number of individuals and markers) that there is a high probability of S large eigenvalues for a population consisting of S well-differentiated subpopulations. Novembre and Stephens (2008) showed that PCs can also pick up signals of spatial differentiation in an otherwise continuous population.

One note of caution: while PCA of population structure is generally powerful, it is not fool-proof. Kimmel et al. (2007) and Zhao et al. (2007) noted that there are situations where PCs do not appropriately correct for population stratification. For example, if cases and controls come from different populations, then a PCA will detect this stratification, but the resulting correction will also remove the QTL association signals. A second complication is that common alleles can show a different stratification signature relative to rare alleles in the same sample (Mathieson and McVean 2012; Zaidi and Mathieson 2020), and PC approaches typically use common alleles. We will return shortly to this point.

Patterson et al. (2006) developed a rather remarkable result, based in theoretical work on the eigenstructure of covariance matrices (Baik, Ben Arous, and Péché 2005). A covariance matrix with just a few expected large eigenvalues shows a **phase-change transition**, where if the amount of population signal is slightly above a threshold value there is a strong eigenvalue signal, while if the amount of signal is slightly below this threshold, no clear eigenvalue signal is present. Patterson called this value the **BBP threshold** (after the authors of the motivating theoretical paper), and noted that with m markers and n individuals, this threshold is

$$\tau_{BBP} = \frac{1}{\sqrt{nm}} \tag{20.15}$$

An F_{ST} value above τ_{BBP} generates a clear eigenvalue signal, while populations with F_{ST} below τ_{BBP} have little signal. For 100,000 markers scored in 1,000 individuals, $\tau_{BBP} = 0.0001$, so that a population structure with $F_{ST} > 0.0001$ should have a strong PC signal.

The PC correction for stratification proceeds as follows. Consider a **marker matrix M** for the scored genotypes at m markers over n individuals, where

$$\mathbf{M}_{n\times m} = \begin{pmatrix} g_{1,1} & g_{1,2} & \cdots & g_{1,m} \\ g_{2,1} & g_{2,2} & \cdots & g_{2,m} \\ \vdots & \vdots & \cdots & \vdots \\ g_{n,1} & g_{n,2} & \cdots & g_{n,m} \end{pmatrix} \tag{20.16a}$$

Here $g_{i,j}$ is the number of copies of the reference (typically the minor) allele at marker j in individual i (taking on values of 0, 1, or 2). Hence, the ith row are the m marker scores for individual i, while the jth column are the values for marker j over the n individuals in the sample. **M** often represents a trimmed subset of the SNPs used in the GWAS, but this need not be the case. Zou et al. (2010) showed that PCs can be influenced by sets of linked SNPs in high LD, falsely suggesting subpopulations. One approach to adjust for this is **thinning**, wherein SNPs are pruned such that adjacent markers in high LD are excluded (e.g., Fellay et al. 2007). However, Zou et al. noted that thinning discards potentially useful information, and instead suggested a weighting scheme to adjust individual SNPs based on their LD scores (Equation 20.13b). Note that the concern here is with *high levels* of LD between *tightly linked* SNPs. The *low levels* of LD expected between *unlinked* SNPs due to the presence of structure are not impacted by these corrections. Finally, the LOCO approach is used, so that each chromosome has a specific **M** matrix, based on all of the used markers *except* for those on the focal chromosome.

Typically, two adjustments are made to the elements of **M** before a PCA. First, the elements are **centered**

$$g_{c,i,k} = g_{i,k} - 2p_k, \qquad \text{where} \qquad p_k = \frac{1}{2n}\sum_{j=1}^{n} g_{j,k} \tag{20.16b}$$

so that the column average is zero. Second, the centered elements are then **normalized** (so that all have the same variance). Note that $g_{i,k}$ is just a binomial random variable (Equation 2.19a; with $n = 2$ and $p = p_k$), and hence (Equation 2.19c) has a variance of $2p_k(1 - p_k)$, leading to

$$g_{n,i,k} = \frac{g_{c,i,k}}{\sqrt{2p_k(1-p_k)}} \tag{20.16c}$$

Denote the centered, normalized marker matrix by $\mathbf{M}^*$. A PCA analysis then proceeds on the covariance between individuals in marker scores. The resulting $n \times n$ covariance matrix is given by

$$\mathbf{C}_{n\times n} = \frac{1}{m}\mathbf{M}^*_{n\times m}(\mathbf{M}^*)^T_{m\times n} \tag{20.16d}$$

where the ijth element of **C** is the inner product of the ith and jth rows of $\mathbf{M}^*$,

$$C_{ij} = \frac{1}{m}\sum_{k=1}^{m} M^*_{ik}M^*_{jk} = \frac{1}{m}\sum_{k=1}^{m}\frac{(g_{i,k}-2p_k)(g_{j,k}-2p_k)}{2p_k(1-p_k)} \tag{20.16e}$$

Note from Equation 8.15b is that this is just the correlation estimate of $2\Theta_{ij}$, which assigns more weight to loci with rare alleles. Hence, if the same set of markers is used to correct for close relatives and to correct for structure, **C** is just the GRM used for kinship.

Let $\mathbf{e}_j$ (length $n \times 1$) denote the jth eigenvector of $\mathbf{C}$, namely that associated with the jth largest eigenvalue. The **projected ancestry**, a_{ij}, of individual i along the jth axis of variation in $\mathbf{C}$ is given by the ith entry in $\mathbf{e}_j$. The correction for differing means over the subgroups follows using the a_{ij}, with the adjusted mean value for individual i given by

$$\mu_i = \mu + \sum_{j=1}^{S} a_{ij}\, v_j \tag{20.17a}$$

where the v_j are fixed effects, estimated by the model. The full model correcting for population structure is given by Equation 20.14b, with Equation 20.14c now using

$$\mathbf{Qv} = \begin{pmatrix} a_{11} & a_{12} & a_{13} & \cdots & a_{1S} \\ a_{21} & a_{22} & a_{23} & \cdots & a_{2S} \\ \vdots & \vdots & \vdots & \cdots & \vdots \\ a_{n1} & a_{n2} & a_{n3} & \cdots & a_{nS} \end{pmatrix} \begin{pmatrix} v_1 \\ v_2 \\ \vdots \\ v_S \end{pmatrix} \tag{20.17b}$$

The number ($S < n$) of nontrivial PCs to include in $\mathbf{Q}$ is a classic PCA problem, with a number of proposed solutions (reviewed by Peres-Neto et al. 2005). One approach is a visual examination of the eigenvalue scree plot (Figure 9.1), given the expectation of a few large, and many small, eigenvalues. However, stopping based on some observed inflection point in the scree plot is ad hoc and often chooses too many PCs. More formal tests have been proposed. For example, Patterson et al. (2006) used the **Tracy–Widom distribution of the largest expected eigenvalue** of a covariance matrix (Tracy and Widom 1994) to estimate the number of significant PCs. Simulations by Shriner (2011, 2012) showed that this approach also systematically overestimated the number of PCs, and that the bias is larger in admixed populations (Patterson et al. also noted overestimate problems under admixture). Shriner suggested that part of this bias may occur because stratified and admixed populations display departures from Hardy–Weinberg (even if the underlying subpopulations are in HW). Shriner found that an alternative stopping rule, **Velicer's minimum average partial test** (Velicer 1976), correctly selected the number of PCs to include. Velicer's criterion stops including PCs when the average of squared corrections (following removal of the included PC effects) starts to increase. Alternatively, Tucker et al. (2014) suggested that as a general rule incorporating five PCs was generally sufficient to correct for stratification in many (real and simulated) settings. A caveat with any stopping rule is that the aim under GWAS is not to include *all* nontrivial PCs, but rather just those that *manage to capture the impact of subpopulation structure and/or admixture.*

Two final comments on the PCA approach. Lee et al. (2011) derived the relationship between PC corrections and genomic control, noting that the latter could be expressed as the sum of squared correlations between a PC and the trait value, each term weighted by the eigenvalue for the corresponding PC. Li and Yu (2008) extended the PC approach by jointly clustering individuals and then also correcting by PC score.

Using Admixed Populations Can Improve the Power of a GWAS

While structured populations are typically treated as a complication that needs to be corrected, starting with Pasaniuc et al. (2011), it was realized that using admixed populations can often *improve* the power of a GWAS relative to a similarly sized unstructured population sample (Zhang and Stam 2014; Atkinson et at. 2021; Hou et al. 2021; Lin et al. 2021). The notion is that differential selection pressures and drift can result in differences in causal allele frequencies between the underlying subpopulations contributing (either directly or via admixture) to the mapping population sample. As a result, causal variants may be at more intermediate frequencies in an admixed population, and hence offer greater power. One measure of such power would be the F_{ST} values of the causal alleles of the focal trait. Lin et al. (2021) estimated that causal-site based F_{ST} values in excess of 0.5 were commonly found for traits in humans. With a strong reference sample (such as can be obtained for some human populations), one can use the haplotypes around a particular SNP allele to infer its ancestry, a proceed called **local ancestry inference** (**LAI**). The **Tractor** approach of Atkinson et al. (2021) leverages such LAI estimates by modifying the basic gene-dosage regression (Equation 20.1a) to

$$z_i = \mu + \sum_{j=1}^{n_p} b_{k,j} N_{i,k,j} + e_i$$

where $N_{i,k,j} (= 0, 1, \text{or } 2)$ is the number of reference alleles (from SNP k) in individual i that originated from ancestral population j. Hence, the modification is to partition the reference allele (at the focal SNP) further into the ancestral population that contributed it to individual i. The gain in power using this approach occurs when the effect size tagged by a marker effect varies substantially over the ancestral populations.

Biobanks and the Resurrection of Family-based Corrections

We began our discussion on corrections for population structure by noting that family-based association tests—such as the TDT and its extensions (e.g., Laird and Lange 2006a, 2006b, 2008; Ott et al. 2011)—fully control for population stratification and cryptic relatedness (assuming sibs from different families are unrelated). Historically, however, *family-based* sampling was very challenging to scale, while *population-based* sampling was relatively trivial to scale to generate massive samples. This shifted most studies to population-based, and hence the need for structure and cryptic relatedness corrections. However, with the advent of very large biobanks (and hence the ability to score thousands of sibs), family-based association studies are starting to regain some traction in the population-based GWAS era. For example, Howe et al. (2022) examined roughly 180,000 sibs (over 19 cohorts) for 25 phenotypes. They found, as did Abdellaoui et al. (2023), that family-based GWAS estimates of effect sizes tended to be smaller than population-based ones (even after controlling for structure), suggesting that the latter are often inflated by factors not fully controlled for outside of family-based comparisons.

Using within-family contrasts requires only a minor modification of the basic GWAS model. Consider the trend test (Equation 20.1a), which regresses phenotype z on the number N of reference alleles at a given marker. Howe et al. (2022) modified the trend test to accommodate sib data by considering z_{ij}, the phenotype of sib j in family i,

$$z_{ij} = \mu + b_{f,k} \overline{N}_{i,k} + b_{w,k} N^*_{ij,k} + e_{ij}$$

where $\overline{N}_{i,k}$ is the average number of reference alleles for marker k in family i and $N^*_{ij,k} = N_{ij,k} - \overline{N}_{i,k}$ is the centered reference allele count for marker k in sib j from family i. For example, for a three-sib family with genotypes 00, 10, and 00 at marker k, then $N_k = 0, 1, 0$ (respectively), $\overline{N}_{i,k} = 1/3$, and $N^*_k = -1/3$, $2/3$, and $-1/3$. In this model, $b_{f,k}$ and $b_{w,k}$ correspond to the among-family and within-family contributions of the marker.

MIXED-MODEL GWAS ANALYSIS

The QK Model

Mixed linear models (Chapter 10) provide an ideal framework for the joint incorporation of corrections for recent (kinship) and distant (population stratification) ancestry. While the above discussion showed the early roots of mixed-model based GWAS analysis (especially when correcting for cryptic ancestry), it was Yu et al. (2006) who jointly incorporated corrections for *both* kinship and stratification. The result was their **QK model**. Their motivation was in the analysis of complex association panels, such as those used by plant breeders. These panels are typically constructed from a very diverse set of inbred lines, leading to a collection of both related lines and sets of lines from very different subpopulations (often representing a global sample of diversity). By their nature, such panels have both recent and deep shared ancestry.

The QK model proceeds by first partitioning fixed effects into three distinct sets of factors ($\boldsymbol{\beta}$, $\mathbf{v}$, b_k,). Two of these sets are common over all markers: the vector of trait covariates $\boldsymbol{\beta}$ (such as sex, age, or environmental effects) and the vector of population structure effects $\mathbf{v}$, while b_k is the specific SNP effect for the marker being tested. This is under an additive model, while under the general genotype model b_k is replaced by the vector (b_k, d_k) of additive and dominance effects (Equation 20.1b). Random effects enter as the vector $\mathbf{u}$ of polygenic values and the vector of residuals ($\mathbf{e}$). The resulting model for testing marker k becomes

$$\mathbf{z} = \mathbf{X}\boldsymbol{\beta} + \mathbf{N}_k b_k + \mathbf{Q}\mathbf{v} + \mathbf{Z}\mathbf{u} + \mathbf{e} \tag{20.18a}$$

The vector $\mathbf{N}_k$ has $N_{i,k}$ as its ith component (the number of copies of the reference allele in individual i), while the vector $\mathbf{Qv}$ adjusts for population structure (either by STRUCTURE or a PCA; Equations 20.14c or 20.17b, respectively).

Because $\mathbf{u}$ and $\mathbf{e}$ are random, and hence $E[\mathbf{u}] = E[\mathbf{e}] = 0$, we can write the vector of adjusted mean values as

$$\boldsymbol{\mu}^* = \mathbf{X}\boldsymbol{\beta} + \mathbf{Q}\mathbf{v} \tag{20.18b}$$

Combining the two random effects as $\mathbf{e}^* = \mathbf{Zu} + \mathbf{e}$, Equation 20.18a can be written as

$$\mathbf{z} = \boldsymbol{\mu}^* + \mathbf{N}_k b_k + \mathbf{e}^* \tag{20.18c}$$

Assuming OLS for $\mathbf{e}$ (Equation 10.10), the variance–covariance matrix for $\mathbf{e}^*$ is

$$\mathbf{V} = \sigma_A^2 \mathbf{ZKZ}^T + \sigma_e^2 \mathbf{I} = \sigma_A^2(\mathbf{ZKZ}^T + \delta^2 \mathbf{I}), \quad \text{where } \delta^2 = \frac{\sigma_e^2}{\sigma_A^2} \tag{20.18d}$$

Here, $\mathbf{K}$ is the genomic relationship (GRM), or kinship, matrix (whose elements are the marker-based estimates of $2\Theta_{ij}$), and these joint corrections for structure ($\mathbf{Q}$) and kinship ($\mathbf{K}$) lead to the term *QK* model. Some authors (e.g., Sun et al. 2010) use P when PCs are used and Q when STRUCTURE is used (e.g., PK versus QK models). Adjusting the data to have a mean of zero (in the absence of a SNP effect) leads to

$$\mathbf{z}^* = \mathbf{z} - \boldsymbol{\mu}^* = \mathbf{N}_k b_k + \mathbf{e}^* \tag{20.18e}$$

With estimates of $\boldsymbol{\mu}^*, \sigma_A^2$, and σ_e^2, this is just a GLS regression (Equation 10.13a), with $\mathbf{V}$ given by Equation 20.18d. One can think of the adjustment for population structure as a fixed effect correction (a change in the mean), while cryptic relatedness enters not through the mean, but rather through correlations among the residuals ($\mathbf{e}^*$ not being of OLS form).

Example 20.9. Because both $\mathbf{Q}$ and $\mathbf{K}$ control for relatedness (distant and more recent, respectively), obvious questions for a QK model are whether both corrections are needed, and if only one correction is applied, which one is better. The answer depends on the (typically unknown) relationship structure of the sample. Yu et al. (2006) considered three traits (flowering time, ear height, ear diameter) in a diverse panel of 277 maize inbred lines. They compared the Q model (Equation 20.18a with no $\mathbf{u}$ term; a structured population analysis), the K model (Equation 20.18a with no $\mathbf{Q}$ term (a single-population model with polygenic control for shared recent ancestry), and the Q+K model (the full version of 20.18a). Over the three traits, all models showed better performance relative to the simple model where both $\mathbf{u}$ and $\mathbf{Q}$ were absent. Among the three corrections for confounding, the performance rankings were Q lowest, K better, and Q+K just slightly better than K alone. For example, for flowering time, roughly 38% of the examined 560 SNPs were significant ($p < 0.05$) under the simple model, while this fraction was reduced (presumably by reducing the fraction of false positives) to 14% for the Q model, 6.1% for the K model, and 6.0% for the Q+K model.

Yu et al. also examined power under different assumed effect sizes. The choice of trait impacted power calculations because the impact of both structure and close relatedness depend on the genetic structure of the particular character. Both the Q and K models had greater power than the simple model. For ear height and flowering time, the Q+K model had the greatest power, while for ear diameter, K had slightly higher power than Q+K.

Zhao et al. (2007) also examined whether the K model alone was sufficient, or if **Q** must also be included in an association panel of 95 *Arabidopsis* accessions. As did Yu et al, they found that while the K model worked well, the Q+K model gave better results. However, when they computed kinship using percentage of shared haplotypes (which they called K*), the K* model alone was essentially as good as the K*+Q model. Conversely, Bradbury et al. (2011) found that the K-only model was superior to other models over a wide range of conditions given the population structure of their collection of Barley lines.

Kinship and Structure: Recent Versus Deep Ancestry

The thoughtful reader might have wondered about our apparently sharp distinction between kinship and population structure (K versus Q), as both are metrics of shared ancestry, the former of recent ancestry, the latter of deeper ancestry. Further, the two corrections often use the same marker information, as the GRM for kinship correction is often used by the PCA to extract axes of population structure. However, this not need be the case, as the GRM for kinship correction and the marker matrix for the PCA correction for structure can contain different sets of markers. Indeed, there are suggestions in the literature that while the structure correction should use a very large set of markers (to capture small signals; e.g., Equation 20.15), kinship may be efficiently captured using a much smaller number of markers (Lippert et al. 2011; Listgarten et al. 2012, 2013; Tucker et al. 2014; Liu et al. 2016; Jiang et al. 2019).

Despite the fact that different set of markers can, in principle, be used for kinship and structure corrections, the issue remains as to why the Q+K model often outperforms either the K or the Q models (Example 20.9). The key is that the *dimensionality* of deep ancestry (i.e., population structure) is different from that for kinship. When structure is present, a random individual shares deep ancestry with a considerable fraction of the population, and this signal is often largely captured by the first few PCs of the marker covariance matrix. In contrast, a random individual (the absence of large extended pedigrees) likely shares recent kinship with only a very, very small fraction of the sampled population. Hence, while recent kinship itself might not be uncommon, it is caused by a large number of very small groups of close relatives. The small signal for each such cluster of close relatives is manifested by very low rank eigenvalues from the GRM. Hence, when the GWAS sample is a mixture of both deep and recent ancestry (such as often occurs with plant association mapping panels), the resulting eigenstructure has a few high-ranking eigenvalues from deep ancestry and a long tail of small eigenvalues from shared recent kinship. While PCs are a poor way to exploit information from these small clusters of recent relatedness, these correlations can be efficiently utilized by the GRM.

Another was to think about independent contributions from Q and K, even when both use the same GRM, is the REML variance estimation framework (Part II Chapter 33). Under a REML analysis, one first removes the impacts from any fixed effects. The REML variance estimate is thus the residual variation that remains after the fixed effects are removed. Because PCAs are entered as fixed effects, the REML estimate for σ_A^2 that forms the basis for the kinship correction extracts the remaining information not captured by the PCs.

Computational Improvements

There has been a considerable amount of theoretical work built around the basic QK structure for a GWAS (Table 20.2), most of it related to making this mixed model computationally more efficient so that it can scale to the size of a modern GWAS ($n > 10^4, m > 10^6$). Using standard methods for solving mixed models (Part II Chapter 33), the run times scale as at least $O(mn^3)$. In the QK model, the vectors $\boldsymbol{\beta}$ and $\mathbf{v}$ of fixed effects, and the polygenic variance σ_A^2 (and its associated vector $\mathbf{u}$ of random polygenic effects) are all **nuisance parameters**, variables that must be included in the model, but that are generally not of much interest by themselves (although we will return to the polygenic variance in Chapter 33). In a GWAS, the sole interest is usually whether b_k is significantly different from zero.

The computationally demanding step in applying the full QK model to each of the m markers is the estimation of the background polygenic additive variance, σ_A^2. In an exact analysis, σ_A^2 must be re-estimated for each different SNP as, in theory, it changes slightly depending on the tested SNP. Indeed, when testing marker k, we could have written this as $\sigma_{A(-k)}^2$ to remind the reader of this fact, but choose not to for ease of presentation. Differences in σ_A^2 among markers are largest when comparing a large-effect marker with a small-effect one. The latter has the large-effect site incorporated into σ_A^2, while the former has it removed. Hence the common assumption of an equal value of the polygenic variance σ_A^2 over all markers (Table 20.2) is not really problematic for markers of very small effects (their removal results in very small marker-specific changes in σ_A^2). With large-effect markers, however, this constant-variance assumption can introduce bias. One simple solution is to incorporate a few of the largest effect SNPs as cofactors (Equation 20.1d), which should result in a more consistent value of σ_A^2 over markers.

Running a full mixed model analysis on the millions of markers (directly scored or imputed) in a modern GWAS is thus computationally very demanding, as standard methods for the estimation of σ_A^2 typically scale as (at least) $O(n^3)$ (Chapter 33). Thus, a GWAS with tens of thousands of individuals and millions of markers is exceptionally challenging if an exact mixed-model analysis is performed on each marker. With σ_A^2 and σ_e^2 estimated, inversion of the $n \times n$ matrix $\mathbf{V}$ (Equation 20.18d)—itself not a trivial operation for a large number of individuals—can be used to obtain exact estimates of the fixed effects (Equation 10.13a). However, as we detail in Chapter 33, estimation of variance is generally an *iterative procedure*, with each step usually involving the inversion of $\mathbf{V}_{[i]} = \sigma_{A[i]}^2(\mathbf{K} + \delta_{[i]}^2\mathbf{I})$, where i indexes the iteration. Because the updated variance components change the value of $\mathbf{V}_{[i]}$, a new inversion is required for each iterative step. Further, in the LOCO setting, the kinship matrix $\mathbf{K}$ is slightly different for each of the tested chromosomes. Given this background, much of the work on mixed-model GWAS analysis starts with some version of Equation 20.18a, and then explores either computational improvements or approximations to make such an analysis feasible for the scale of a modern GWAS. Summaries of the time complexity scaling for the different methods presented below can be found in Svischeva et al. (2012), Zhou and Stephens (2012), Zhou et al. (2018), and Jiang et al. (2019).

The first approach was to develop computational improvements of the mixed-model estimation algorithms (typically by improving the estimation of σ_A^2). One of the first suggestions was the **efficient mixed-model association** (**EMMA**) method of Kang et al. (2008). They showed that using the eigenvalue decomposition (Equation A3.32a) of the GRM $\mathbf{K}$ allowed one to avoid having to invert $\mathbf{V}_{[i]}$ during each iteration of the variance estimation procedure. Computing the initial decomposition has a complexity of order n^3, with subsequent iteration steps having a complexity of order rn (for r iterations). Traditional iterative variance estimation schemes typically have a complexity of order rn^3 (Chapter 33). Under the LOCO setting with markers tested over ℓ chromosomes, there are ℓ separate $\mathbf{K}$ estimates, so that scaling becomes $O(\ell[n^3 + rn])$.

While EMMA reduces the computational complexity for each SNP, a new decomposition most be performed for each marker. Lippert et al. (2011) developed an improvement for setting where the number m_r of SNPs used to estimate the GRM $\mathbf{K}$ is less than n. Their **factored spectrally transformed LMM** (**FaST-LMM**) method requires just a single eigenvalue decomposition that can be used for all tested markers on a given chromosome. Zhou and Stephens (2012) introduced their **genome-wide efficient mixed-model association** (**GEMMA**) which, while also using an eigendecomposition of $\mathbf{K}$, maximizes the likelihood function in a more efficient manner than previous algorithms. As a benchmark,

Table 20.2 Optimization and approximation methods for improving the efficiency of a mixed-model GWAS. Optimization methods attempt to find numerically more efficient methods for solving the exact mixed-model equations. Most approximation approaches start by assuming a constant value for polygeneic variance, rather than the exact model which allows this to vary over markers.

Continuous traits: Optimization of exact solutions		
EMMA	**Efficient mixed-model association**	Kang et al. (2008)
FaST-LMM	**Factored spectrally transformed LMM**	Lippert et al. (2011)
GEMMA	**Genome-wide efficient mixed-model association**	Zhou and Stephens (2012)
Continuous traits: Approximation of exact solutions		
EMMAX	**EMMA expedited**	Kang et al. (2010)
P3D	**Population parameters previously determined**	Zhang et al. (2010)
GRAMMAR	**Genomewide rapid association using mixed model and regression**	Aulchenko et al. (2007)
GRAMMAR-GC		Amin et al. (2007)
GRAMMAR-Gamma		Svishcheva et al. (2012)
BOLT-LMM		Loh et al. (2015a, 2018)
fastGWA		Jiang et al. (2019)
Binary traits: Linear model approximations		
LTMLM	**Liability–threshold mixed linear model**	Hayeck et al. (2015)
LEAP	**Liability estimator as a phenotype**	Weissbrod et al. (2015)
Binary traits: Logistic regression optimization/approximation		
GMMAT	**Generalized linear model association test**	Chen et al. (2016)
SAIGE	**Scalable and accurate implementation of generalized mixed model**	Zhou et al. (2018)
fastGWA-GLMM		Jiang et al. (2021)

Zhou and Stephens noted that the computation times for EMMA, FaST-LMM, and GEMMA for two different human data sets were, respectively, ~ 9 days, 6.8 hours, and 33 minutes for one set and ~ 27 years, 6.2 hours, and 3.3 hours for the other.

This first generation of accelerated methods scaled between $O(mn^2)$ and $O(m^2n)$ (Loh et al. 2015a; 2018). While this was an improvement over the $O(mn^3)$ scaling of classic mixed model analysis, even with these gains, the number of computational steps for a full exact mixed model GWAS for a modern data set is *at least* 10^{14} to 10^{16}. Second-generation methods, **BOLT-LMM** (Loh et al. 2015a; 2018) and **fastGWA** (Jiang et al. 2019), scale even faster, between $O(mn)$ and $O(mn^{3/2})$. Both methods estimate σ_A^2 once (see below) and use clever computational tricks, such as sparse matrix inversion (Jiang et al. 2019).

In addition to more efficient methods for estimating the background polygenic variance associated with each marker, a complementary approach is to approximate the exact model. Several investigators suggested estimating σ_A^2 and σ_e^2 once under the null model of no SNP effect (Equation 20.18a with no $\mathbf{N}$ term), with these estimates subsequently used when testing each marker. This reduces Equation 20.18a to a GLS regression (Equation 10.13a), with $\mathbf{V} = \sigma_A^2\mathbf{K} + \sigma_e^2\mathbf{I}$. This constant polygenic variance approximation was first proposed (for family-based association mapping) by Chen and Abecasis (2007). Kang et al. (2010), using their EMMA variance estimator, called this approach **EMMA eXpedited** (**EMMAX**), while Zhang et al. (2010) called it **population parameters previously determined** (**P3D**). Zhang et al. also suggested that **compression** can be used, wherein individuals with very similar relationship values are clustered as a group, and their mean relatedness used. This has the effect of reducing the size of the kinship matrix, which can speed up calculations. Simulations showed that these approximations generally work well with low heritability traits, but become less precise as heritability increases. They also tended to be less powerful than the exact method. As mentioned above, incorporation of a few of the largest effect SNPs as cofactors should result in a more consistent value of σ_A^2 over markers.

A related approach was motivated by Equation 20.18e, with the idea that one could correct each individual once and then subsequently use the adjusted values in the marker regressions. Aulchenko et al. (2007) called this **genome-wide rapid association using mixed model and regression (GRAMMAR)**. Once again, Equation 20.18a is fit under the null model of no SNP effect (the $\mathbf{N}$ term is not included). With estimates of the vectors of fixed effect (BLUEs for $\widehat{\beta}$ and $\widehat{\mathbf{v}}$) and predictions of the vector of random polygenic effects (BLUPs for $\widehat{\mathbf{u}}$) in hand, the trait value for each individual can be adjusted to remove these effects,

$$\mathbf{z}^* = \mathbf{z} - \widehat{\mathbf{z}}, \quad \text{where} \quad \widehat{\mathbf{z}} = \mathbf{X}\widehat{\beta} + \mathbf{Q}\widehat{\mathbf{v}} + \mathbf{Z}\widehat{\mathbf{u}} \tag{20.19a}$$

The regression for any given SNP is then simply tested using

$$\mathbf{z}^* = \mathbf{N}_k b_k + \mathbf{e} \tag{20.19b}$$

This is just a GLS regression, and estimates for the marker effect, its sampling variance, and its significance are obtained as follows. Let $\mathbf{x}_k$ be the vector of mean-centered counts ($N_{i,k}$) for marker k, whose ith entry is $x_{i,k} = N_{i,k} - 2p_k$. With estimates of σ_A^2 and σ_e^2 in hand, the GLS estimates (Equation 10.13) for the effect from SNP k, its variance, and its test statistic (distributed as χ_1^2), become

$$\widehat{b}_k = \frac{\mathbf{x}_k \mathbf{V}^{-1}\mathbf{z}^*}{\mathbf{x}_k \mathbf{V}^{-1}\mathbf{x}_k}, \qquad \sigma^2\left(\widehat{b}_k\right) = \frac{1}{\mathbf{x}_k \mathbf{V}^{-1}\mathbf{x}_k}, \qquad \chi^2_{[k]} = \frac{(\mathbf{x}_k \mathbf{V}^{-1}\mathbf{z}^*)^2}{\mathbf{x}_k \mathbf{V}^{-1}\mathbf{x}_k} \tag{20.19c}$$

Note that $\mathbf{V} = \sigma_A^2 \mathbf{ZKZ}^T + \sigma_e^2$ is assumed to be the same for all markers (the polygenic variance σ_A^2 is constant over markers).

Aulchenko et al. originally corrected just for relatedness, but their approach easily extends to the full QK model. GRAMMAR turns out to give slightly biased estimates of SNP effects and p values that are conservative (the true p values are slightly less than the reported p values). To account for these effects, subsequent modifications were offered by Amin et al. (2007) (**GRAMMAR-GC**), and Svishcheva et al. (2012) (**GRAMMAR-Gamma**), where SNP effects and test statistics are adjusted by correction factors to reduce bias and make marker-effect tests less conservative.

One final comment on the GRAMMAR approach. At first sight, it appears that Equation 20.19a leads to the adjusted trait values (z_i^*) being free of correlations among relatives, and thus exchangeable. If true, this would allow for permutation tests to easily be applied, randomizing the values of $\mathbf{z}^*$ over the vector of marker information, creating a random draw from the null. Unfortunately, this is only appropriately correct, as BLUPs (predicted values) for the u_i have a correlated error structure (Part II Chapter 32), which compromises the assumption of exchangeability. This concern does not apply to fixed effects, so that the z_i^* are exchangeable with regard to any fixed-effect adjustments.

Mixed Models for Binary Traits

While the above mixed-model adjustments for cryptic relatedness and population structure were presented in the context of a continuous trait, many traits of interest are binary (e.g., disease presence/absence). One approach for such traits is to ignore their dichotomous nature ($z_i = 0, 1$) and simply use the continuous **linear mixed models (LMMs)** developed above (Equation 20.18a), e.g., Sawcer et al. (2011). This is an approximate approach, typically done to reduce the computational burden of an analysis. However, when the number of cases is much less than the number of controls, LMM-based approaches result in excessive false-positives (Loh et at. 2018; Zhou et al. 2018; Jiang et al. 2021). Alternatively, one could perform a more exact—and computationally more demanding—analysis using **generalized linear mixed models (GLMMs)**, such as logistic regression (Chapter 14). We consider these in turn.

The starting point for LMM and GLMM binary trait analysis is to consider a latent variable, which we will call the liability, y_i, associated with individual i. In QK model form, we can express this as

$$y_i = b_k N_{i,k} + \widehat{y}_i \tag{20.20a}$$

where

$$\widehat{y_i} = \mu + \sum_{j=1}^{m} \beta_j x_{i,j} + \sum_j q_{ji} v_j + u_i \tag{20.20b}$$

is the background liability for i, and, as before, our interest is in the significance of b_k, the effect of SNP k. The standard continuous LMM uses $z_i = y_i + e_i$, where the residuals are assumed to be homoscedastic, e.g., $\sigma^2(e_i) = \sigma^2(e)$, namely a constant over all i. This residual structure is invalid for a binary trait, as if $E[y_i] = p_i$, then $\sigma^2(e_i) = p_i(1 - p_i)$, which varies over i. If this variation is small, the LMM approximation may be reasonable. However, Chen et al. (2016) showed that residual heteroscedasticity was rather problematic in an asthma dataset of individuals of Hispanic/Latino heritage. Different subpopulations had different disease risk, and thus had stratification-specific variances. Using a LMM for this data resulted in incorrect p values. A second issue is that many binary traits use a case-control design. If the disease/trait is rare, then cases are a highly nonrandom sample from the population, and failing to accounting for this ascertainment results in a loss of power (Yang et al. 2014).

Hayeck et al. (2015) and Weissbrod et al. (2015) suggested improvements to account for ascertainment using LMM based on the threshold-liability model (Chapter 14; Part II Chapter 31). Under this model, if y_i exceeds some threshold value T, then the disease is present, else it is absent. Hence, $y_i > T$ for cases, while $y_i < T$ for controls. If the disease prevalence and the heritability on the underlying liability scale (Part II Chapter 31) are known, then one can estimate the posterior mean liability for each individuals ($\widehat{y_i}$). This is accomplished using the GRM and the case/control status of all individuals. Then, in the same spirit as Equation 20.19, one performs a standard GWAS regression of $y_i^* = y_i - \widehat{y_i}$ on $N_{i,k}$ for each SNP. This approach was called the **liability–threshold mixed linear model (LTMLM)** by Hayeck et al. (2015) and **liability estimator as a phenotype (LEAP)** by Weissbrod et al. (2015). The improvement in power of these methods over a standard LMM analysis increases with sample size and with the rareness of the disease.

A more exact approach to account for shared relatedness in a binary trait GWAS is to used a generalized linear mixed model, such as a logistic regression (Zhu et al. 2002; Setakis et al. 2006; Zheng et al. 2006; Chen et al. 2016; Banerjee et al. 2018; Shenstone et al. 2018). These formally model the correct (binomial) error structure, allowing for the residual variance to vary over individuals. As we saw above, under a logistic regression framework, the underlying latent value y_i is mapped into the expected value on the observed (binary) trait scale z via a logistic regression,

$$p(y_i) = E[z_i \mid y_i] = \Pr[z_i = 1 \mid y_i] = g(y_i) \tag{20.21a}$$

where the **link function** $g(y)$ is given by the logistic,

$$g(y_i) = \frac{1}{1 + \exp\left[-\left(y_i\right)\right]} \tag{20.21b}$$

where y_i is given by Equation 20.20. This expands on Equation 20.3a by adding fixed-effects terms for population structure (the v_j) and a random effect u_i for the background polygenic value.

Logistic regression introduces a few subtleties typically not present in a LMM analysis. First, the careful reader might recall our above discussion that adding covariates into a logistic regression can actually reduce the power of a GWAS, especially under a case-control design when the disease is fairly rare. This statement still holds for the β_j terms in Equation 20.20b. However, recall that fixed effects are also added to a model to account for confounding factors, reducing the number of false positives. This is the role played by the fixed effects v_j for population structure in Equation 20.20b. Second, the addition of the random effect u_i to control for close relatives makes this a **generalized linear *mixed* model (GLMM)**, whereas our above discussion of logistic regressions (where the elements of y were all fixed) are generalized linear models (Chapter 14). While exact solution to a mixed-model logistic can be obtained using either ML or Bayesian methods, the analysis of GLMM raises a number of computational issues (Breslow and Clayton 1993). As a result, they are even more computationally demanding than an LMM GWAS, facing serious scaling problems to accommodate both the number of individuals n and markers m in a modern GWAS.

One approximate approach, the **generalized linear model association test** (**GMMAT**), was suggested by Chen et al. (2016). Their logic was similar to that used for GRAMMAR (Equation 20.19), in that the mixed-model logistic is fit just once for y_i^* (i.e., the model with no SNP effect), using the mixed-model parameters to predict this value, $\widehat{y}_i^*$, for individual i. One then does the GLS regression

$$z_i - \widehat{y}_i^* = b_k N_{i,k} + e_i \tag{20.22}$$

where the covariance matrix for the e_i can be expressed in terms of parameters estimated in the original mixed logistic regression (see Chen et al. for details). Building on this approach, Zhou et al. (2018) developed a more efficient computational approach for GLMM-based GWAS, **SAIGE** (**Scalable and Accurate Implementation of GEneralized mixed model**). They did so by exploiting optimization/approximation methods (such as saddlepoint approximations; Daniels 1954; Barndorff-Nielsen and Cox 1979; Reid 1998). Finally, Jiang et al. (2021) were able to modify their fastGWA approach to logit scores, resulting in **fastGWA-GLMM**, which runs significantly faster than SAIGE. Both SAIGE and fastGWA allow for unbalanced case/control designs (such as occurs with sampling from biobanks) while controlling the Type I error.

In the broader context of GLMM, there are other candidates, besides the logistic, for the link function g (Equation 20.21b) that maps y into the expected value of z. For example, the threshold-liability model uses the probit function, $g(y) = \Pr(U \leq y)$, where U denotes a unit normal, as its link. The choice of the link function involves assumptions about how additivity on the underlying scale maps into interactions on the observed scale (Clayton 2012). Under a logistic link function, additive terms on the underlying scale become multiplicative odds terms on the observed value (Equation 20.3f).

Assessing Model Fit

In a GWAS, as in any model-fitting endeavor, the investigator wishes to assess model quality and determine whether improvements are required. Because the SNP effect is usually the only parameter not regarded as a nuisance variable, the goal is less about improving overall model fit, and more about improving power and controlling the Type I error rate. WTCCC (2007) suggested the two approaches most widely used to both assess a current GWAS model and to compare it with others: **probability plots** and genomic control (λ) values.

The motivation for probability plots follows from the powerful observation that p values follow a unit-uniform distribution (i.e., a uniform over $[0, 1]$) under the null hypothesis (Appendix 6). For a GWAS, the expectation is that *most* SNP effects are zero (i.e., from the null), but a tiny fraction should be true positives. Hence, there should be a slight excess of very small p values, while the rest of the values should be drawn from a uniform. Probability plots are a useful device to visually inspect if this pattern holds. One of the most common such plots used in GWAS is constructed as follows. With m tested markers, compute $-\ln(p)$ for each and rank these from smallest to largest,

$$-\ln(p)_{[1]}, -\ln(p)_{[2]}, \cdots, -\ln(p)_{[m]}$$

where $-\ln(p)_{[k]}$ is the value of the kth smallest value of $-\ln(p)$. Note that this scaling awards the smallest p values the highest rankings, so that $-\ln(p)_{[1]}$ corresponds to the test with the p value closest to one (so $\ln(p) \simeq 0$), while $-\ln(p)_{[m]}$ corresponds to the test with the smallest p value. This scaling avoids the compression of very small p values near zero. One then plots $-\ln(p)_{[k]}$ against $-\ln(1-[k-1/2]/m)$, the latter representing the expected kth smallest value

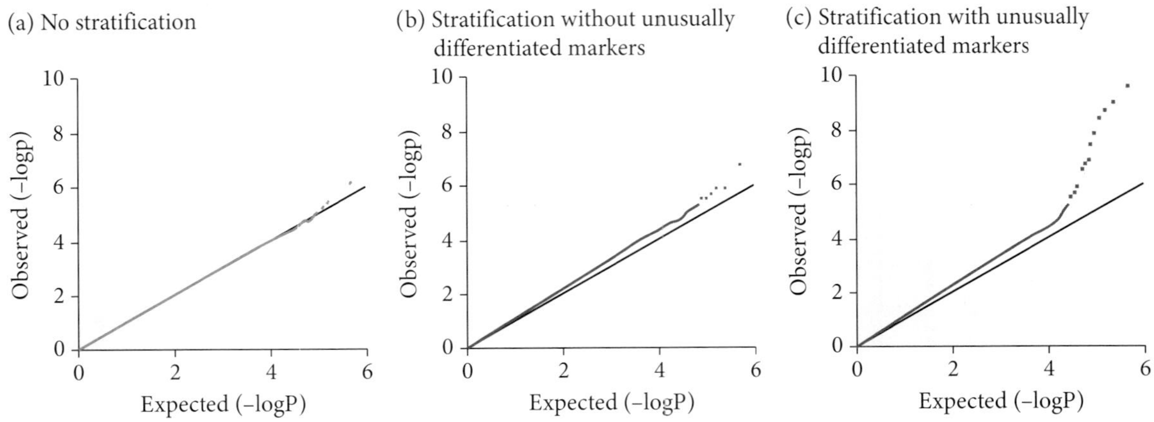

Figure 20.4 Accessing a GWAS model using probability plots. The straight line with a slope of one represents the expected realization between predicted and observed p values under the null (a uniform distribution), while pixels represent p values for individual tests. **Left:** For a population with no stratification, or one corrected for its effects, one might expect a few tests with very small p values (on the far right of the plot) representing true positives, while the rest of the p values follow the expectation under the null. **Center:** When stratification is present, p values are elevated over many of the tests, resulting in a large fraction of the p values exceeding their neutral expectations. **Right:** Stratification coupled with markers showing excessive divergence, resulting in not just departures above the neutral expectation, but in excessive departures. (After Price et al. 2010a.)

under a uniform (the use of $[k-1/2]$ instead of k is a continuity correction). The resulting plot of observed versus expected values should be a straight line with a slope of one, perhaps with a slight, to strong, upturn as k approaches m (representing the small p values of true positives). Deviations from this pattern suggest that the current model needs improvement (Figure 20.4).

There are a number of different approaches for constructing probability plots, which are typically used to test if an empirical distribution matches a theoretical candidate (Wilk and Gnanadesikan 1968; Gerson 1975). The two most common are **p–p** and **q–q** (or **quantile-quantile**) **plots**. If $p_{[k]}$ and $S_{[k]}$ denote the kth smallest p value and test statistic, respectively, then a standard p–p plot graphs $p_{[k]}$ against $(k-1/2)/m$. The common GWAS use of $-\ln(p)$ is a version of a p–p plot, emphasizing the smallest p values, as these same values would be visually indistinguishable from zero if plotted as $p_{[k]}$. Conversely, q–q plots compare the ranked values of the *test statistics* (as opposed to the associated p values of those statistics) with the corresponding expected ranked values from the null distribution. Suppose that the test statistic under the null is a χ^2_1. In a q–q plot, the value of the kth (of m) smallest test statistic is compared with value $f_{[k]}$ that satisfies $\Pr(\chi^2_1 \leq f_{[k]}) = (k-1/2)/m$, generating a point of $(S_{[k]}, f_{[k]})$ in the plot. Again, the expectation under the null is a straight line with a slope of one. WTCCC (2007) used q–q plots, while many other GWAS use the $-\ln(p)$ plots, but both essentially convey the same information. Stirling (1982) discusses the construction of confidence bounds for probability plots.

While an aberrant probability plot is cause for concern, an important cautionary tale was offered by Chen et al. (2016). They observed an apparently well-behaved probability plot for one of their simpler GWAS models. However, a more careful analysis showed this plot resulted from conservative p values in some SNPs balancing out anti-conservative p values from other SNPs. The result was a visually well-behaved probability plot that masked substantial underlying issues.

Typically, a probability plot is presented along with the inflation factor λ for a given model, with the idea that a well-behaved model should also have a small inflation (typically $\lambda < 1.05$). Recall that λ is a general measure of the inflation of the test statistic for a random marker. Models with different complexities (such as Q, K, or Q+K) are often ranked by their

λ values (smaller is better). If a model has a substantial inflation ($\lambda > 1.05$), historically the view was that additional corrections were required. However, as we have seen, this is not strictly correct. If a large number of small effect QTLs underlie the trait, this can inflate λ even after accounting for any confounding effects from structure or cryptic relatedness. Further, this polygenic inflation becomes more pronounced as sample size increases (Equation 20.13a). Indeed, Yengo et al. (2022) found that 60% of all SNPs were significant at the 5% level when tested one-at-a-time (i.e., without a polygenic correction) in a GWAS for height using a sample size of five million people. A more correct genomic-control metric for model comparison is the λ value computed using LD-score regression (Equation 20.13c), which adjusts for polygenicity (Bulik-Sullivan et al. 2015).

GWAS WITH RARE ALLELES

As we discuss in detail in Chapter 21, an ongoing debate is whether a common disease is due to common alleles of small effect or rare alleles of large effect. This is somewhat of a false dichotomy, as numerous studies show that both classes of alleles impact many diseases, and the more appropriate question is their relative contributions. Across a variety of diseases and traits, large-effect alleles generally tend to occur at very low frequencies (Bodmer and Bonilla 2008; WL Chapter 28). One explanation for this observation is that such alleles are likely under negative selection, either by their direct impact on a focal trait, or via pleiotropic effects on other fitness traits (WL Chapter 25). From the theory of mutation–selection balance (WL Chapter 7), we expect such alleles to have frequencies that are inversely proportional to the strength of selection against them. Hence, one view of disease variation is that major-effect alleles (such as amino-acid replacements or knockouts) are under strong selection, and thus tend to be rare, whereas alleles with smaller effects (such a minor regulatory changes) experience weaker selection and thus tend to be at higher frequency. In the extreme where an allele has no impact of fitness, neutral theory predicts (in an idealized equilibrium population) that frequencies follow the Waterson distribution (WL Equation 2.34a), where $\phi(x) = C/x$ (an L-shaped distribution), with x the (derived) allele frequency and ϕ its probability density function. Hence, most neutral alleles are rare. Selection against alleles further shifts this distribution towards even smaller values (as do many demographic features, such as recent population expansions; Kryukov et al. 2009; WL Chapter 9).

A fair assessment of the relative contribution of common versus rare alleles requires unbiased methods that offer equal power of detection of alleles from both classes. The standard one variant at a time GWAS framework discussed above is rather powerful at detecting common (causal) alleles (MAF greater than 1–5%), but is very underpowered for detecting rarer alleles (Li and Leal 2008; Asimit and Zeggini 2010; Bansal et al. 2010). There are two reasons for this. First, for an additive locus, the relative contribution is $\sigma^2_Q = 2a^2p(1-p)$, where a is the allelic effect. For two alleles with the same effect size (a), σ^2_Q increases with the value of the MAF (p). Hence, there is a strong bias in favor of detecting common alleles when the common and rare allele have the same effect size. Given our discussion above, this may be less of a concern, as we might expect a to *increase* as p decreases, so that the average value for σ^2_Q between common and rare alleles might be more similar than initially expected. Even in this case, a second factor, the correlation (r^2) between the marker and causal alleles, results in greatly reduced power for rare alleles. Example 20.1 shows the reason for this: r^2 decreases as the mismatch between the marker and causal allele frequencies increases. By using common SNP alleles as markers, we are biasing detection toward common *causal* alleles, even if common and rare alleles have the same value of σ^2_Q. Because different methodologies from traditional GWAS are often used to search for rare variants, the phrase **rare variant association study** (**RVAS**) has been suggested (Auer and Lettre 2015).

A different approach is thus required to discern the impact of rare alleles. Note that we are now in the realm of **deep resequencing** (high coverage sequencing of a small region), and potentially even whole-genome sequencing (WGS), as rare alleles are typically not scored by standard DNA chips (Asimit and Zeggini 2010), and accurately imputing them requires a very large reference collection of full sequences (for the region, exome, or genome, of interest). The basic logic behind the menagerie of rare-allele approaches (summarized in Table 20.3) is to pick a candidate region (often a single gene), and then test for an excess of rare alleles in this region in the cases over that seen in the controls (or in high versus low trait-valued individuals; Example 20.10). Given that we are amalgamating signals within a gene, it follows that rare-allele test are simply a subclass of gene-based GWAS tests.

As we detail shortly, a variety of approaches and weighting schemes have been proposed to quantify an excess of rare alleles within a candidate region (reviewed by Asimit and Zeggini 2010; Bansal et al. 2010; Basu and Pan 2011; Chen et al. 2011; Pan and Shen 2011; Stitziel et al. 2011; Lee et al. 2012, 2014; Ionita-Laza et al. 2013; Derkach et al. 2014; Moutsianas and Morris 2014; Pan et al. 2014; Moutsianas et al. 2015; Nicolae 2016; Santorico and Hendricks 2016; Povysil et al. 2019). Design issues for detecting rare variants are discussed by Li and Leal (2009), Li et al. (2011), Lee et al. (2014), and Auer and Lettre (2015).

Example 20.10. The basic logic of rare-allele mapping traces back to Cohen et al. (2004). These authors were interested in the impact of three potential human candidate genes (*ABCA1*, *APOA1*, and *LCAT*) on HDL cholesterol (HDL-C) levels. The exons of these genes were sequenced in two different high-low group comparison designs. Variants were classified as either nonsynonymous (NS) mutations that changed the amino-acid sequence of the encoded protein or synonymous (S) variants that did not. The first comparison involved 128 low individuals (HDL-C score less than the fifth percentile) and 128 high individuals (HDL-C are least in the 95th percentile) in the Dallas Heart Study (a diverse collection including both European and African ancestry). The second study was a corresponding group of 155 lows and 108 highs from a more homogeneous Canadian population. Lumping the variants over the three genes, 10 NS and 19 S mutations were found in both the high and low Dallas groups, 15 NS and 7 S mutants restricted to the low group, and 3 NS and 6 S variants were only in the high group. Similar distributions of variants were found in the Canadian study. Hence, NS variants were significantly more common in the low than in the high group. Of the 18 NS variants restricted to one group, all but two were singletons, with these latter two alleles present at four copies each. The PolyPhen program (Ramensky et al 2002; Adzhubei et al. 2010), which attempts to predict the functional impact of a specific amino-acid replacement (benign, possibly damaging, or probably damaging), suggested that these latter two mutations had a benign impact on function. A brief overview of bioinformatical approaches for functional annotation of sequence variation can be found in Auer and Lettre (2015).

A similar study by Ahituv et al. (2007) examined sets of candidate genes in 379 obese and 378 lean individuals. One set consisted of 21 genes with known mutations impacting obesity in mice or humans, and the second set had 37 genes from weight-related pathways. The first group of genes had 46 NS mutations restricted to the obese cohort, but only 26 restricted to the lean group, a significant difference. Further, 19 of NS mutations in the obese cohort were predicted to be deleterious, as opposed to four in the lean group. In contrast, the pathway gene group had 72 NS in the obese group, and 69 in the lean group, which was not significantly different.

These two examples highlight the logic for many rare allele tests: contrast the distribution of rare variants between case/control or high/low (**extreme phenotype**) groups. The above examples focused on **private alleles**, those restricted to just one group, and hence are often called **private allele tests**. More generally, one can also incorporate shared alleles. Further, mutational classes could be differentially weighted (such as a focus on only NS mutations). One could go further, scoring the predicted severity of a NS mutation (or other functional change), and include this information in a weighted comparison statistic.

One important source of bias when using rare alleles in a case-control (or extreme phenotype) design was noted by Li and Leal (2009). If rare variants were discovered by exclusively resequencing only in *cases* and then searching for these same rare variants in controls, the result is a large inflation in false positives. The reason is simple: if an allele is rare, an allele *unrelated to the trait* may by chance be found only in cases. Conversely, a rare allele may only be found in controls. More formally, given a random sample of n (diploid) individuals, for an allele with frequency p,

$$\Pr(\text{at least one allele in sample}) = 1 - (1-p)^{2n} \simeq 1 - e^{-\gamma} \tag{20.23a}$$

where $\gamma = 2np$ is the expected number of allelic copies in the sample, with $e^{-\gamma}$ the probability that no copies are seen. Rearranging Equation 20.23a, the required sample size to ensure a probability π that at least one copy of the rare allele is present is

$$n = \frac{-\ln(1-\pi)}{2p} \tag{20.23b}$$

For example, if $p = 0.001$, the sample size to have a 95% chance of seeing at least one copy is $n = -\ln(0.05)/[0.002] = 1498$. The probability that *no* copies of an allele are found in a sample of n_0 controls while at least one (or more) copies are found in a sample of n_1 cases becomes

$$e^{-\gamma_0}(1 - e^{-\gamma_1}) \tag{20.23c}$$

where $\gamma_i = 2n_i p$. For example, with 3,000 cases and 3,000 controls, and a rare allele ($p = 1/3,000$), the expected number of copies in both samples is 2. The probability that at least one allele is found in the cases is $1 - e^{-2} = 0.8647$, while the probability of at least one copy in the cases and none in the control is $e^{-2}[1 - e^{-2}] = 0.117$. Hence, a number of rare alleles with no impact on the trait would exclusively be found in the cases. More generally, the probability of exactly k copies in the sample follows from the Poisson (Equation 2.21), with $\Pr(k) = \gamma^k \exp(-\gamma)/k!$, for $k = 0, 1, \cdots$.

Burden and Collapsing Tests

The central problem with detecting rare alleles in a standard GWAS framework (such as using a chi-square or regression test) is that the very large sampling variance associated with rare alleles implies very low power, even when their allelic effect is large. A growing number of approaches have addressed this concern building on the basic logic of Example 20.10: contrasting the *total frequency* of rare alleles between comparison groups. These are often called **burden** or **collapsing tests**: *burden* because the presence of rare alleles might imply a fitness burden, and *collapsing* because rare alleles are consolidated into one (or a few) classes, increasing the effective sample size, reducing the sampling variance and thus increasing power. Collapsing tests can be classified as **indicator** (presence/absence of one, or more, rare alleles), **counting** (how many rare alleles are carried by an individual), and **data-adaptive** (scores are weighted by some feature of the data, such as MAF or functional information).

The biological assumption for aggregating rare alleles at a locus into a single class is that **extreme allelic heterogeneity** (**EAH**) is present, where a disease (or extreme trait value) is generated via independent rare mutations. If each mutation has roughly the same effect, then the power of a method that treats them all as a single class is a function of the frequency of that class. For example, Bansal et a1. (2010) considered a rare allele with a frequency of 0.01 in the controls and 0.02 in the cases, where 80% power required 250,000 cases and control. However, if there are five such rare alleles at a candidate gene, then when treated as a single class, the frequencies becomes 0.05 in the controls and 0.10 in the cases, now requiring only 3,000 cases and controls for the same power.

Beyond deciding the appropriate scope of the **testing unit**, or **genomic unit of analysis**, (a single gene or set of genes connected in a pathway), a related issue is whether resequencing in the search for rare alleles uses an **exon-only** or a **whole-gene** approach. The logic for exon-only is that variants in the coding region are more likely to be impactful, and less likely to have no effect (often referred to as **neutral** or **null** variants), with the latter diluting a rare-variant signal (Zawistowski et al. 2010; Pan et al. 2014). Conversely, an exon-only focus assumes that coding mutations are the major source of rare disease/extreme-trait variants, ignoring contributions from regulatory effects in noncoding regions (e.g., Haller et al. 2009). For more general genome-wide scans, one might consider: (i) an **exome scan** over all coding regions in the genome; (ii) using a genomic sliding window; or (iii) scanning all evolutionary conserved regions (which are often noncoding). The problem with such global scans is their relatively modest power within any tested region, coupled with a very high multiple-comparison penalty.

Early versions of burden tests include the **cohort allelic sums test** (**CAST**) of Morgenthaler and Thilly (2007) and the **combined multivariate and collapsing** (**CMC**) method of Li and Leal (2008). CAST simply sums the total number of rare alleles at a target gene (or more generally, region or pathway) and compares the mean number of rare alleles per individual in cases versus controls. Li and Leal (2008) independently proposed a similar collapsing method that compares the fraction of cases carrying rare alleles in a target with the same fraction in controls, which can be tested by using a Fisher's exact test. Li and Leal also suggested an improvement by collapsing variants into a small number of classes based on their allele frequency, and then using a Hotelling's T^2 test (Equation 20.5) over these classes to test significance. This binning approach can accommodate both rare and common alleles, such as collapsing all the rare alleles into one class, and then either considering each common allele separately, or using some collapsing/exclusion scheme on them as well.

While the choice of using only NS mutations (or only private alleles) implicitly assumes a weighting scheme, the next generation of burden tests assigned more formal weights to variants. Madsen and Browning (2009) proposed a **weight sum statistic** (**WSS**) where each variant was weighted by the standard deviation of its frequency,

$$w_j = \sqrt{n_j \widetilde{p}_j (1 - \widetilde{p}_j)}, \quad \text{with} \quad \widetilde{p}_j = \frac{c_j + 1}{2(n_{0,j} + 1)} \tag{21.24a}$$

Here n_j is the total number (cases plus controls) of individuals scored for SNP j, and $\widetilde{p}_j$ corresponds to the **pseudocount** frequency in the controls (adding one to the numerator and denominator to adjust for small-sample issues), where c_j is the number of j alleles seen in the $n_{0,j}$ individuals in the control scored for SNP j. If the allele is absent in the control, then $\widetilde{p}_j = 1/[2(n_j + 1)]$. A **burden score** is then computed for individual i over the m chosen SNPs in region of interest, with

$$S_i = \sum_{j-1}^{m} \frac{I_{ij}}{w_j} \tag{21.24b}$$

where I_{ij} is the indicator function, being one if the minor allele at SNP j is found in i and zero otherwise. One then ranks of the S_i and uses a Wilcoxon test for whether the ranks in cases and controls are nonrandom (such as an excess of higher ranks in cases). Zawistowski et al. (2010) suggested a **cumulative minor-allele test** (**CMAT**) which contrasts weighted minor and major allele counts in cases and controls. Finally, in the **Accumulation of Rare variants Integrated and Extended Locus-specific test** (**ARIEL**) of Asimit et al. (2012), weights on variants are given by a called-sequence quality score (such as a phred score), with rare variants called with higher confidence given more weight.

Using similar logic, burden tests can be constructed for continuous traits. Consider the trend test (Equation 20.1a), but now express the trait value for individual i as a function of gene (or marker region) k as

$$z_i = \mu + b_k \phi_{i,k} + e_i \tag{20.24c}$$

where $\phi_{i,k}$ is some collapsing of the total rare variants at region k in individual i. Morris and Zeggini (2009) suggested using either: (i) the proportion of scored rare alleles in the region for individual i (if $n_{i,k}$ rare SNP sites in region k are scored in individual i and $r_{i,k}$ of these contain the rare allele [a count method], then $\phi_{i,k} = r_{i,k}/n_{i,k}$); or (ii) assigning

ϕ a value of one whenever *any* rare allele in the region is seen in i, else it has value zero (an indicator method). These suggestions are often referred to as the **MZ method** in the literature. As noted by Zawistowski et al. (2010) and Asimit et al. (2012), ϕ can also use **probabilistic genotype calls** (such as the posterior mean allelic dosage discussed above), when rare alleles are either called using low-coverage sequencing (which can be error prone) or by imputation (Mägi et al. 2012).

Significance of any of the above methods can be accessed using standard permutation techniques: keep the genotype information for any individual intact, and then shuffle the phenotypic labels (cases/controls or continuous values). There are two limitations with this approach. First, if population stratification is present, the shuffling has to be done in such a way that stratification information is retained. Second (as discussed previously), when multiple comparisons are performed, the resulting smaller p values per comparison (to control the FWER) requires a substantial increase in the number of permutations per region to obtain sufficiently stable estimates of empirical significance thresholds. If R multiple (independent) regions are being compared, a Bonferroni FWER of γ requires testing each region using γ/R. For testing 10 regions, a 5% FWER requires testing each region at $p = 0.005$, which requires at least 10^4 permutations per region. Hence, scaling this approach to more than a small number of candidate regions is both computationally challenging and results in greatly reduced power. For example, with an exome-wide scan scoring roughly 20,000 genes separately, an FWER of 0.05 requires testing each gene at $\alpha = 2.5 \times 10^{-6}$. Estimating stable empirical thresholds to achieve this level of α requires on the order of 10^7 to 10^8 permutations *per gene*.

Variance Component (Dispersion) Tests

Neale et al. (2011) noted that unweighted collapsing methods (treating all chosen rare alleles as functionally equivalent) can be underpowered. In particular, they noted that in a given gene, one could have neutral rare variants with no impact on risk (with the expectation of being equally represented in cases and controls), variants that increase risk (overrepresented in cases), and **protective variants** that *decrease* risk (underrepresented in cases). Early collapsing methods attempted to account for neutral variants by focusing on variants with apparent functional differences (such as using only nonsynonymous, NS, mutations; Example 20.10). However, by lumping potential protective variants with risk variants, power is decreased. In effect, early collapsing methods required a *mean directionally of effect* for rare variants. In contrast, Neale et al. proposed using *variance*, rather than *mean*, comparisons between cases and controls. Their approach was motivated by observations on the *APOB* gene and triglyceride levels. They examined NS *APOB* mutations in a sample of 96 high and 96 low individuals, finding 18 segregating NS mutations in the combined sample. One variant was present as six copies in the high cohort but absent in the low, while a second variant was absent in the high but with six copies in the low. Hence, the first mutation likely increased risk, while the second was likely protective. If both of these are collapsed into the same category, the result is six copies in both the high and low, and no signal.

To deal with this concern, Neale et al. proposed using the **C-alpha**, **C(α)**, test for **overdispersion** (excess variance) in binomial samples due to an underlying mixture of effects (Neyman and Scott 1966). The idea is that the between-group variance would be larger than expected under binomial sampling (assuming equal chances of an allele being in the cases or controls). More formally, let $p_1 = n_{case}/(n_{control} + n_{case})$ be the fraction of the total sample that are cases ($p_1 = 1/2$ when the cases and controls have the same sample size). Suppose there are n_i total number of copies (in cases plus control) of variant i, with y_i of these in cases. For a variant that has no impact on the trait, $y_i \sim B(p_1, n_i)$, a binomial with sample size n_i and success probability p_1 (Equation 2.19a). In this case, Equation 2.19b gives the expected sample variance as $n_i p_1(1 - p_1)$. However, suppose that the true underlying distribution is a mixture,

$$y_i \sim \pi_r B(n_i, p > p_1) + \pi_p B(n_i, p < p_1) + (1 - \pi_r - \pi_p) B(n_i, p = p_1) \qquad (20.25a)$$

where π_r and π_p are, respectively, the probabilities of a risk or protective allele. Because of the between-class variance (p varying over classes), such a mixture results in a variance larger than the expected binomial variance. The C-alpha test statistic, which tests for such an overdispersion over m groups, is given by

$$T = \sum_{i=1}^{m} \left[(y_i - n_i p_1)^2 - n_i p_1 (1 - p_1) \right] \tag{20.25b}$$

Here $(y_i - n_i p_1)$ is the observed deviation from the neutral expected value, namely a standard burden test. The C-alpha test instead uses the squared deviation and contrasts it with its expected value under neutrality (the binomial variance). For example, for the two *APOB* variants discussed above, $y_1 = 6$ and $y_2 = 0$, their two deviations cancel

$$(y_1 - n_1 p_1) = 6 - 3 = 3, \qquad (y_2 - n_2 p_1) = 0 - 3 = -3$$

Conversely, $n_1 p_1 (1 - p_1) = n_2 p_1 (1 - p_1) = 1.5$, yielding $T = 2[9 - 1.5] = 15$. For large samples $T/\sigma(T) \sim N(0,1)$ under the null, being one-sided test of excessive dispersion (the expression for $\sigma(T)$ can be found in Neale et al.). Note that singletons, by themselves, do not provide information on overdispersion, but the entire collection of singletons can be lumped to form a new category. The *APOB* data found six high group and four low group singletons, giving $y_i = 6, n_i = 10$ in the summation term in Equation 20.25b. Finally, one can use the EM method (Appendix 4) to estimate the mixture proportions (π_r and π_p) in Equation 20.25a, as well as the average value of p for each category (Neale et al. 2011).

A somewhat related approach that also allows for both protective and risk variants is the **kernel-based adaptive cluster** (**KBAC**) test of Liu and Leal (2010). One unusual feature of KBAC is that it uses the frequencies of *multilocus genotypes* as the unit of analysis, and hence allows for the potential of detecting epistatic interactions that could easily be missed in standard burden tests. The basic structure of KBAC is that one considers all of the observed multilocus genotypes in the region of interest that *contain at least one rare allele*. Suppose there are k such combinations, which we label as $G_1, \cdots, G_k$, while G_0 is the collapsed set of all multilocus genotypes *lacking* any rare alleles. KBAC then computes a weighted sum of the *squared* differences between the frequency of G_i in the cases versus controls. The weights are based on the likelihood (under a neutral model) of the observed number of G_i in the cases. For example, if 7 of 10 copies are found in cases, the weight is the probability (under a neutral model) of seeing 7 or fewer copies in the cases. By considering the squared difference, the impact of directional effects is reduced, and Liu and Leal noted that KBAC is also fairly robust to the inclusion of variants with no functional impact.

As noted above, when most of the variants included in the test set are not causal (the **sparse alternative** setting), these methods lose power. Most of the above models arrive at a final summary statistic by taking some weighted sum of individual deviations, or squares of such deviations, e.g., $S = \sum w_i s_i$. Chen et al. (2012) noted an **exponential combination** (**EC**) procedure, using sum of the *exponent* of these individual statistics, e.g., $\sum \exp(w_i s_i / 2)$, is a much more robust approach under the sparse alternative setting.

Regression-based General Framework

The connections between simultaneously testing multiple markers in a normal GWAS (e.g., Equation 20.5), burden tests, and variance component tests—as well as the relative roles of common versus rare alleles—can all be seen by using a multiple regression framework. We do so by extending the trend test (Equation 20.1a) to m markers, whose inclusion is based on some criteria (such as all being within a target gene or in a candidate pathway). Let $N_{i,k}$ denote the copy number for the minor allele at marker k in individual i, and let b_k be the associated regression slope. In its simplest form (assuming only additivity), the multiple-marker trend test becomes

$$z_i = \mu + \sum_{k=1}^{m} b_k N_{i,k} + e_i \tag{20.26a}$$

Note that the multiple regression framework accommodates correlations among markers, and (at this point) makes no distinction between common versus rare alleles. Equation 20.26a can be modified to explicitly model dominance or recessivity by replacing $N_{i,k}$ by $X_{i,k}$, with

$$X_{i,k} = \begin{cases} 0 & \text{homozygous recessive} \\ 1 & \text{otherwise} \end{cases} \quad \text{or} \quad X_{i,k} = \begin{cases} 1 & \text{homozygous recessive} \\ 0 & \text{otherwise} \end{cases} \tag{20.26b}$$

for a dominance or recessive, respectively, assumption.

We can further extend Equation 20.26a by adding cofactors (e.g., Equation 20.18a), and likewise to a logistic regression on binary traits by replacing z_i by $\text{logit}(z_i)$ (Equation 20.2b). The null hypothesis is that the vector $\mathbf{b}^T = (b_1, b_2, \cdots, b_m)$ of regression coefficients is zero. This can be tested using Hotelling's T^2 statistic, resulting in a χ^2_m, and hence a degree of freedom for each marker. If the chosen alleles are common, then there is often sufficient power, *in isolation* (e.g., testing one marker at a time), to detect those with an effect. However, when testing for the significance of a *region* (as opposed to an *individual* maker), inclusion of markers with no impact on the trait reduces power. This occurs by increasing the degrees of freedom associated with the test statistic (χ^2_m versus a χ^2_1) without a corresponding increase in the noncentrality parameter.

Hence, when testing for the significance of a region we would like some criteria to enrich the fraction of markers with true effects. Collapsing methods, with their general focus on rare alleles, represent an attempt to improve power by both increasing the fraction of potentially causal variants and by reframing Equation 20.26a as single degree of freedom statistic. They test the significance of a shared marker effect b_c by rewriting this regression as

$$z_i = \mu + b_c X_i + e_i, \quad \text{with} \quad X_i = \sum_{k=1}^{m} w_k N_{i,k} \tag{20.27}$$

Here X_i is the burden score for individual i, and w_k is the weight for tested marker k. For the set of included markers, one could define X_i in Equation 20.27 as having value one when *any* marker in the testing set has a rare allele, else it has value zero (an indicator approach). One could also set X_i as the total copy number of rare alleles over all scored markers in i (a counting approach). Both these approaches amount to collapsing the multilocus marker genotype into a set of synthetic "super alleles" at a single locus and then performing a standard trend test.

One can view the weights, w_k, as an attempt to increase the signal from potentially causal variants while decreasing the signal for neutral variants. Thus, at one level, the assumed weights provide a simple **masking scheme** for which variants within a region to include ($w_k = 1$) or exclude ($w_k = 0$). As we saw in Example 20.10, one could focus on only private alleles, or on a specific type of mutation (such as NS versus S, motivated by the assumption that NS mutations are more likely to be causal than S mutations). One could go further, and only include NS alleles that are predicted to be possibly or probably damaging, or even just probably damaging. While inclusion of variants is often based on such *a prior* information, Hoffmann et al. (2010) noted that this approach is only optimal when the functional information is very accurate. They proposed their **step-up approach**, akin to stepwise regression, for determining the optimal set of variants within a region to include based on model fit alone (and hence is agnostic to any assumed biological knowledge).

More generally, one could base weights on other features, such as the sampling variance given the marker frequency in controls (Equation 20.24). Several **adaptive burden tests** have been proposed, where the weights are informed by single-marker regression slopes of each variant. The **data-adaptive sums test** (**aSUM**) of Han and Pan (2010) assigns weights of $w_i = -1$ when the univariate marker effect is sufficiently negative and $w_i = 1$ when it is positive. The **estimated regression coefficient** (**EREC**) **test** of Lin and Tang (2011) bases weights on the estimated single-marker regression slopes (adjusted for estimate instabilities when alleles are rare). A number of other adaptive tests have been proposed (e.g., Pan and

Shen 2011; Pan et al. 2014), again using the logic of upweighting variants with causal signatures (such as being NS) and downweighting those without such signatures. Bayesian approaches have also been suggested to filter out noncausal variants (e.g., Quintana et al. 2011; Logsdon et al. 2013). A related approach is the **RareCover** method of Bhatia et al. (2010), which starts with a set of k candidate markers within a region, and extracts the set of these giving the highest correlation with phenotypic value.

A basic issue with the collapsing approach that we have not explicitly considered is what exactly constitutes a rare variant. Ionita-Laza et al. (2013), using results from large sample theory, suggested setting the allele frequency threshold as $\tau = 1/\sqrt{2n}$, with n the sample size. For example, for $n = 500$, the threshold is 0.03, while for $n =$ 10,000, the threshold becomes 0.007. This is a *statistical definition*, while the threshold is usually set for *biological reasons* (enhancing the chance that included markers are causal, or linked to causal sites). Operationally, if the allele frequency threshold is set too high, we may include too many noncausal variants (reducing power), while if set too low we may exclude causal variants. One strategy is to use a **variable threshold** (**VT**) test (Price et al. 2010b), which computes the test statistic over a range of thresholds, and then takes the largest score as the final test value. Again, significance is typically assessed via permutation tests.

An alternative approach for controlling degrees of freedom is to use a random-effects model for the slopes of marker-trait regression (Dandine-Roulland and Perdry 2015). Here the b_k in Equation 20.26a are drawn from a normal distribution with mean zero and variance $w_k\sigma^2$, where the weights are assigned by some criteria. The test for no marker effects reduces to whether the variance component σ^2 is zero, a one degree of freedom test. This is the approach used in the **sequence kernel association test** (**SKAT**) of Wu et al. (2011), which is a generalization of the C-alpha test, and allows for both positive and negative effects. Indeed, it can be shown that C-alpha is a special case of SKAT, but the latter method has greater flexibility in that cofactors are easily incorporated. The **SumSq** (also referred to as the **SSU**) test of Pan (2009), based on $\sum b_k^2$ (the sum of squared estimated slopes), can also be shown to be a special case of SKAT.

Finally, Xu et al. (2012) noted that the machinery of penalized regressions (Example 20.4) offers a powerful approach for incorporating rare variants (also see Zhou et al. 2010). Penalized regression methods are designed to both handle over-parameterized models and models with sparse data (Hastie et al. 2009). Xu et al. showed that both LASSO and RR tended to outperform both burden (VT) and variance tests (SKAT), unless causal alleles are extremely rare (or singletons). Further, both LASSO and RR allow for a joint test of common and rare variants, with Chen et al. (2011) noting that using the LASSO to choose which common variants to include greatly improves power in the setting of many rare, but few common, causal variants.

One closing comment about the regression framework. As stressed above, correction for population structure is often a serious concern in a traditional GWAS, and this is generally accomplished by adding cofactors to Equation 20.26a (e.g., Equations 20.14b and 20.17a). One unresolved issue is that ***the population structure for rare alleles can be rather different from that for common alleles*** (Mathieson and McVean 2012). Common alleles tend to have long evolutionary histories, whereas most rare alleles are rather recent events, potentially leading to different population structures. Further, because of their recent history, rare variants can be spatially clustered (for example, due to different histories of recent expansions, some subpopulations may harbor many more rare variants than others). Hence, using common alleles when constructing cofactor corrections for structure can bias rare-allele tests (but see Listgarten et al. 2013). In theory, one could construct corrections using only low frequency to rare alleles (MAF less than 0.5%), but these can potentially be unstable and may have a flatter eigenvalue distribution (so that more PCs must be included to capture a stratification signal). Zaidi and Mathieson (2020) further examine PC corrections based on only rare alleles. An especially problematic issue was noted by Zhang et al. (2013). They found that PC-based structure correction based on low-frequency variants tended to overadjust in the absence of structure, leading to a substantial loss of power.

Omnibus Tests

A number of authors have examined the power of rare-allele tests (e.g., Basu and Pan 2011; Ladouceur et al. 2012; Lee et al 2012; Moutsianas et al. 2015). When all of the chosen variants have roughly the same effect and act in the same direction, burden tests are the most powerful. Conversely, when the chosen variants are a mixture of neutral, protective, and risk variants, variance component methods (e.g., SKAT) are more powerful. Given that the optimal method depends upon unknown details of the genetic architecture of the trait, several **omnibus tests** have been proposed that combine burden and variance-component tests. The logic is that by combining information from two (or more) tests, there will be more robust power under a random genetic architecture. Conversely, an omnibus test might lose power if one of the extreme architectures is correct (e.g., equal effects in the same direction).

Derkach et al. (2013) proposed using Fisher's method (Equation A6.1a) to combine the p values from a burden test and the SKAT test,

$$T_{Fish} = -2\ln(p_{SKAT}) - 2\ln(p_{burden}) \tag{20.28a}$$

where T_{Fish} follows a χ^2_4 distribution. A related approach is the **Mixed effects Score Test** (**MiST**) of Sun et al. (2013). Their method starts by considering the slope b_k in Equation 20.26a as $b_k = b_t + \delta_k$, where b_t is a fixed effect for the slope when the variant is of type t (such as an NS mutation), assumed to be the same for all type t variants, and δ_k is a random effect unique to variant k. Burden tests assume a constant value for the fixed effect b_t, while variance component tests use the variance of δ_k. Both SKAT and standard burden tests are special cases of this more general slope model, with MiST based on a joint test of the mean effect of a slope (b_t nonzero) and the variance effect of the slope ($\sigma^2[\delta_k] > 0$). They cleverly constructed these two tests to be independent, so that Fisher's test can be used to combine their two probability values (as in Equation 20.28a). They also used **Tippett's method** (1931): for n independent test, their combined p value is given by

$$1 - [1 - p(1)]^n \simeq np(1) \tag{20.28b}$$

where $p(1)$ is the smallest p value over the n tests (Appendix 6). As noted by Westberg (1985), neither Fisher's nor Tippet's method is uniformly superior to the other over all settings. Sun et al. (2013) noted that Fisher's procedure seems to be more powerful when *both* the mean and variance effects of b_k are nonzero, while Tippett's is more powerful when only one component is nonzero. Another common method for combining p values from independent tests is Stouffer's Z score (Equation A6.2). Using the same logic as above, tests could be constructed using Stouffer's method in place of the Fisher or Tippet method. More generally, one can combine the p values from a chosen set of k different tests (e.g., Moutsianas et al. 2015; Liu et al. 2019).

The limitation of standard p-combining methods (Fisher, Tippett, Stouffer) is the requirement that tests are independent. The recently proposed **aggregated Cauchy association test** (**ACAT**) approach of Liu et al. (2019) relaxes this assumption. The key is that they translated p values into **Cauchy random variables**. The Cauchy (the distribution of the ratio of two unit normals) is unusual, in that its density function integrates to one (making it a proper distribution), but none of its moments are finite, because its tails are sufficiently heavy (they do not decay fast enough at large values). As a result of this heavy-tail feature, it is largely insensitive to correlations among p values, especially when the p values are small (Liu et al. 2019; Liu and Xie 2020). The interest in using the Cauchy for combining p values started with a striking finding by Pillai and Meng (2016). Suppose that $\mathbf{y}$ and $\mathbf{x}$ are both MVN with covariance matrix $\mathbf{V}$. Pillai and Meng showed that the weighted sum $\sum w_i\,(y_i/x_i)$ is a standard Cauchy. Noting that y_i/x_i itself is Cauchy, this implies (in least in this setting) that the weighted sum of Cauchy random variables with arbitrary dependency structure is still a Cauchy. Because ACAT upweights small p values, it (like the EC method) is a more robust approach under the sparse alternative setting (many neutral variants in the test set). The ACAT test statistic for combining p values is

$$T_{ACAT} = \sum_{i=1}^{n} w_i \tan[(0.5 - p_i)\pi] \tag{20.29a}$$

with the associated overall p value being

$$p \simeq 0.5 - \frac{\arctan[T_{ACAT}/w]}{\pi} \tag{20.29b}$$

where $w = \sum w_i$. While the expressions in Equation 20.29 might seem a little odd at first, they make more sense by recalling that $\tan(x) \to \infty$ as $x \to \pi/2$ (so that $T_{ACAT} \to \infty$ as $p \to 0$ in Equation 20.29a). Likewise, because $\arctan(x) \to \pi/2$ as $x \to \infty$, in Equation 20.29b, $p \to 0$ as $T_{ACAT} \to \infty$.

Liu's **ACAT-V** test examines each rare variant within the target region, weighting then by a function of their frequency (a modification of Equation 20.24a), and then combines the individual p values using Equation 20.29a. Given the robustness of ACAT under the sparse alternative, ACAT-V easily accommodates regions with a large number of neutral variants. Liu et al. also proposed an omnibus test by combining the p values from six separate tests: ACAT-V, SKAT, and a burden test, where each test is performed assuming equal weight and then recomputed using weights that place more emphasis on rare alleles. The p values from these six test are then combined using ACAT,

$$T_{ACAT-O} = \frac{1}{6} \sum_{i=1}^{6} \tan[(0.5 - p_i)\pi] \tag{20.29c}$$

The idea of using this combination is to construct a test that is relatively robust over different weights, varying directionality of variants, and fraction of neutral variants.

Lee et al. (2012) proposed their **SKAT-optimized** (**SKAT-O**) test, which constructs a linear combination of the a Burden and a SKAT test statistic

$$T_{SKAT-O} = \rho\, T_{SKAT} + (1 - \rho)\, T_{burden} \tag{20.30}$$

where ρ, determined via a grid search over [0,1], is the value that maximizes Equation 20.30. Lee et al. showed how the corresponding p value for this test statistic can be obtained via numerical integration.

As noted by Ionita-Laza et al. (2013), the idea of omnibus tests can be extended to jointly accommodate common and rare alleles. Their **Burden-F** tests compute p values for a burden test using rare alleles and then apply the same test using common alleles, combining the resulting p values using Fisher's method. Similarly, their **SKAT-F** test Fisher-combines the p values from separate SKAT tests based on rare and common alleles. They also constructed **Burden-C** and **SKAT-C** akin to Equation 20.30 by weighted combinations of Burden and SKAT statistics for a test using only common and a test using only rare alleles.

Finally, variations on combining p values have been proposed that weight individual variant $\ln(p_i)$ terms (from a case-control contrast) in Fisher's method by their MAF (using Equation 20.24a). The **σ-MidP** approach of Cheung et al. (2012) used this weighting scheme, but excludes rare alleles with roughly equal counts in cases and controls (to control for the reduction in power from neutral alleles). The **Adaptive combination of P-values for rare variant association testing** (or **ADA**) method of Lin et al. (2014) modifies this rare alleles exclusion criteria. They do so by: (i) using a variable MAF threshold for inclusion of a variant; and (ii) computing separate sums for putative risk alleles (more frequent in cases) and protective alleles (more frequent in controls), with the larger of these two sums being the test statistic for that threshold. This approach is performed over a range of threshold values, with the largest of the test values being the final test statistic. Again, permutation tests are used for significance testing in both of these methods.

Table 20.3 Rare allele mapping approaches. *General modification* procedures can be applied to fine-tune most of the tests. See text for details on specific tests.

Unweighted Burden Tests		
CAST	**Cohort allelic sums test**	Morgenthaler and Thilly (2007)
CMC	**Combined multivariate and collapsing test**	Li and Leal (2008)
MZ		Morris and Zeggini (2009)
Weighted (Adaptive) Burden Tests		
WSS	**Weighted sum statistic**	Madsen and Browning (2009)
CMAT	**Cumulative minor-allele test**	Zawistowski et al. (2010)
aSUM	**Data-adaptive sums test**	Han and Pan (2010)
Step-up		Hoffmann et al (2010)
RareCover		Bhatia et al (2010)
EREC	**Estimated regression coefficient**	Lin and Tang (2011)
ARIEL	**Accumulation of Rare variants Integrated and Extended Locus-specific test**	Asimit et al. (2012)
Variance Component		
SumSq		Pan (2009)
KBAC	**Kernel-based adaptive cluster**	Liu and Leal (2010)
C(α)	**C-alpha**	Neale et al. (2011)
SKAT	**Sequence kernel association test**	Wu et al. (2011)
Omnibus Tests (Burden plus variance component)		
SKAT-O	**SKAT-optimized**	Lee et al. (2012)
	Fisher's method	Derkach et al. (2013)
MiST	**Mixed effects Score Test**	Sun et al. (2013)
ACAT-O	**Aggregating Cauchy Asssociation test**	Liu et al. (2019)
Omnibus Tests (Common plus rare variants)		
Burden-F, Burden-C		Ionita-Laza et al. (2013)
SKAT-F, SKAT-C		
Omnibus Tests (Weighted p values)		
σ-MidP		Cheung et al. (2012)
ADA	**Adaptive combination of *P*-values for rare variant association testing**	Lin et al. (2014)
General Modifications: Variable rare-allele frequency threshold		
VT	**Variable threshold**	Price et al. (2010b)
General Modifications: Sparse alternative setting		
EC	**Exponential combination**	Chen et al. (2012)
ACAT-V	**Aggregating Cauchy Association test**	Liu et al. (2019)

Closing Remarks

As summarized in Table 20.3, a number of rare variant tests have been proposed, built around different assumptions about the distribution of variant effects within the region being considered. As such, in the absence of any knowledge about the trait architecture in the region, there is no uniformly most powerful test. Omnibus tests are a bet-hedging approach, sacrificing a power under specific extreme architectures, while having more power under random architectures. The other issue is that the amount of signal within a given region could easily generate a significant p value, but one not large enough to persist under the much more stringent multiple comparison value required for an exome-wide or genome-wide scan. Unless sample size is massive, rare-variant approaches are best considered as tests over a small to modest number of regions, rather than the whole genome scan of a typical GWAS.

META-ANALYSIS

GWAS projects, especially in humans, are often done under a **consortium model**, wherein k groups perform independent GWAS on a given trait. If one has access to the individual data from each study, then, in theory, one could combine all of the data sets and run a single analysis. Such an approach is often called a **mega-analysis** or a **joint-analysis**. However, this is often not possible. Medical privacy concerns usually prevent members of a consortium from sharing their raw data. Further, for a variety of reasons, the study designs may not be compatible, such as involving different confounding variables or using different markers. The latter issue is typically dealt with by using imputation (de Bakker et al. 2008; Zaitlen and Eskin 2010), creating a set of shared markers over all the studies (this assumes that each study sampled from a population where accurate imputation is possible). Hence, for each shared marker, the consortium typically has only **summary statistics** from each of the k studies (rather than individual data), such as marker p values, estimated effects, and standard errors. A final potential roadblock to performing a full mega-analysis is computational (Panagiotou et al. 2013). While each of the k individual studies may be computationally feasible, a mega-analysis of a full mixed-model may be less so.

The field of **meta-analysis** (the analysis of analyses; Appendix 6) deals with the analysis of such a collection of summary statistics. While the p values for a given marker may be combined over the studies, for a variety of reasons (see Appendix 6) a formal meta-analysis typically deals with the *estimated effects* of a given marker. Lin and Zeng (2009, 2010; also see Olkin and Sampson 1998) showed under rather general conditions that the standard error for a meta-analysis estimate is approximately the same as a mega-analysis estimate using the same data. Hence, in most settings, little efficiency is lost when moving from an analysis based on the individual data to an analysis based on summary statistics from each of the underlying studies. There is a rich, and growing, literature on the application of meta-analysis to GWAS data. de Bakker et al. (2008), Zeggini and Ioannidis (2009), and Evangelou and Ioannidis (2013) offer nice overviews of some of the practical aspects of gathering, and **harmonizing**, the data for a GWAS meta-analysis. General reviews of meta-analysis methodology as applied to GWAS are given by Munafó and Flint (2004), Kavvoura and Ioannidis (2008), Trikalinos et al. (2008), Begum et al. (2012), Panagiotou et al. (2013), and Dudbridge and Newcombe (2019).

Meta-analysis Basics: Fixed Versus Random-effects Analysis

We begin with a quick overview of the machinery of meta-analysis (more fully covered in Appendix 6) before examining specific applications to GWAS. Let T_i be some estimate of the marker effect in study i (such as the odds ratio or regression slope), s_i^2 its sample variance, and θ_i the true value for that study. Because of sampling error,

$$T_i = \theta_i + e_i \tag{20.31a}$$

where we assume that the residuals are independent but heteroscedastic, as $\sigma^2(e_i) = s_i^2$. Under a **fixed-effects (FE) meta-analysis**, we assume that the actual effect size is the *same* over all studies ($\theta_i = \theta$). Recalling generalized least-squares (GLS; Equation 10.13a), the meta-analysis estimate of θ becomes

$$\overline{T} = \frac{\sum_{i=1}^{k} w_i T_i}{\sum_{i=1}^{k} w_i}, \quad \text{where} \quad w_i = \frac{1}{s_i^2} \tag{20.31b}$$

In other words, we use a weighted average, with each study weighted by its precision (studies with smaller standard errors receive larger weights). Assuming similar variances (σ^2) over studies, $E[s_i^2] = \sigma^2/n_i$, so that this scheme places more weight on studies with larger sample sizes. However, variances could easily vary over studies. For example, a smaller study incorporating appropriate cofactor corrections, or with more accurate scoring of the phenotype, may have a smaller standard error than a more poorly designed, but larger, study. When imputation is used, Zaitlen and Eskin (2010) noted that imputation accuracy is likely to vary over studies, and suggested a modification of the weights to reflect this.

The meta-analysis standard error, $s_{\overline{T}}$, for the global (meta) estimate, $\overline{T}$, is

$$s^2_{\overline{T}} = \frac{1}{\sum_{i=1}^k w_i} \tag{20.31c}$$

For the situation where we assume that each individual observation in a given study has the same variance (σ^2), then for k studies with size n_i,

$$\sigma^2(\overline{T}) = \frac{\sigma^2}{\sum_{i=1}^k n_i} = \frac{\sigma^2}{nk}, \quad \text{if } n_i = n \tag{20.31d}$$

The assumption of a single common value for the treatment mean over all studies is often unrealistic, as we might except the true marker effect to vary, at least somewhat, over studies. In this setting, our interest shifts to the variance *among* the actual effects over studies. This leads to the **random-effects (RE) meta-analysis** model

$$T_i = \mu + u_i + e_i \tag{20.32a}$$

where $u_i \sim (0, \sigma^2_u)$. Typically, the effect sizes ($\theta_i = \mu + u_i$) are assumed to be drawn from a normal, $\theta_i \sim \mathrm{N}(\mu, \sigma^2_u)$. In addition to estimating the grand mean (μ), under the RE framework our interest also extends to the **heterogeneity** of the studies, measured by the variation (σ^2_u) among the realized effects. The estimate of μ is also of the form of Equation 20.31b, but with a critical difference. Under an RE model, the weights are now given by

$$w_i = \frac{1}{s^2_i + \widehat{\sigma}^2_u} \tag{20.32b}$$

where $\widehat{\sigma}^2_u$ is the estimate of σ^2_u. Again, the standard error of the meta estimate is given by Equation 20.31c, but now using the RE weights (Equation 20.32b).

This difference in study weighting between FE and RE analyses has several important consequences. First, when $\sigma^2_u = 0$, FE and RE estimates are identical. Second, when $\sigma^2_u > 0$, RE standard errors are larger, and as a result, an RE analysis almost always has lower power than an FE analysis. Third, a subtle feature of RE study-weighting arises when σ^2_u is on the same order as an average value (s^2) of s^2_i (namely, the between-study variance is at least large as the within-study variance). In this setting, the RE w_i tend to ***be more similar over studies***. The result is that, relative to an FE analysis, studies with less precision (for example, due to smaller size) are ***given more weight*** under RE than under FE. In the extreme where $\sigma^2_u \gg s^2_i$, the weights are roughly the same over all studies, independent of their individual precision.

Example 20.11. A modification of the basic RE approach has been suggested by Han and Eskin (2011) and Lee et al. (2017). They noted that one reason for the loss of power under an RE analysis (relative to an FE analysis) is testing against an *incorrect null hypothesis*, and suggested an improved likelihood ratio test (Appendix 4) to account for this. Their argument is as follows. As above, let T_i and s^2_i denote the estimate and its variance. For large sample size, T_i is distributed as $N(\theta_i, s^2_i)$, giving the meta-analysis likelihood as the product of k normals. Under the FE model ($\theta_i = \mu$), the null ($\mu = 0$) and alternative ($\mu \neq 0$) likelihoods become

$$L_{FE,null} = \prod_i^k \frac{1}{\sqrt{2\pi s^2_i}} \exp\left(-\frac{T^2_i}{2s^2_i}\right) \tag{20.33a}$$

$$L_{FE,alt} = \prod_i^k \frac{1}{\sqrt{2\pi s^2_i}} \exp\left(-\frac{(T_i - \mu)^2}{2s^2_i}\right) \tag{20.33b}$$

Substituting the MLE ($\widehat{\theta}$) for θ (which Han and Eskin show is given by Equation 20.31b) into the likelihood for the alternative and computing the likelihood ratio gives a test for the significance of θ (Equation A4.9a).

Under the standard RE model, $T_i \sim N(\mu, s_i^2 + \sigma_u^2)$, giving the classic RE null and alternative likelihoods as

$$L_{RE,null} = \prod_i^k \frac{1}{\sqrt{2\pi(s_i^2 + \sigma_u^2)}} \exp\left(-\frac{T_i^2}{2(s_i^2 + \sigma_u^2)}\right) \tag{20.33c}$$

and

$$L_{RE,alt} = \prod_i^k \frac{1}{\sqrt{2\pi(s_i^2 + \sigma_u^2)}} \exp\left(-\frac{(T_i - \mu)^2}{2(s_i^2 + \sigma_u^2)}\right) \tag{20.33d}$$

Han and Eskin noted that Equation 20.33c is for the null hypothesis of a *mean* zero effect ($\mu = 0$), but still allows for a *variance* in the mean effect ($\sigma_u^2 > 0$). Framed this way, the "null" hypothesis here is a mean effect of zero, but allowing for heterogeneity.

The **RE2 model** of Han and Eskin takes the null as $\mu = 0$ with *no heterogeneity* ($\sigma_u^2 = 0$), resulting in the same null as under the FE model (Equation 20.33a), and the alternative as under the RE model (Equation 20.33d), giving the RE2 likelihoods as

$$L_{RE2,null} = \prod_i^k \frac{1}{\sqrt{2\pi s_i^2}} \exp\left(-\frac{T_i^2}{2s_i^2}\right) \tag{20.33e}$$

$$L_{RE2,alt} = \prod_i^k \frac{1}{\sqrt{2\pi(s_i^2 + \sigma_u^2)}} \exp\left(-\frac{(T_i - \mu)^2}{2(s_i^2 + \sigma_u^2)}\right) \tag{20.33f}$$

As above, the ratio of these two likelihoods (with the MLEs substituted into Equation 20.33f) generates the test statistic, which becomes

$$LR_{HE} = \sum_{i=1}^k \ln\left(\frac{s_i^2}{s_i^2 + \widehat{\sigma_u^2}}\right) + \sum_{i=1}^k \frac{\widehat{\theta}_i^{\,2}}{s_i^2} + \sum_{i=1}^k \frac{(\widehat{\theta}_i - \widehat{\mu})^2}{s_i^2 + \widehat{\sigma_u^2}} \tag{20.33g}$$

which has a large-sample distribution that is a weighted sum of chi-square distributions, namely, $(1/2)(\chi_1^2 + \chi_2^2)$. As discussed in Appendix 4, this mixture arises because one of the null hypothesis values lies on the boundary of the parameter space ($\sigma_u^2 = 0$). Equation 20.33g is partitioned into a heterogeneity component (the first term, testing $\sigma_u^2 = 0$), an FE component (the second term, testing $\mu = 0$), and a final term considering both. Han and Eskin showed that Equation 20.33g is more powerful than the FE test when sufficient heterogeneity is present.

Meta-analysis Basics: Heterogeneity

Despite its lower precision and power, the strength of an RE analysis is in capturing any heterogeneity (variance in true effect sizes) over studies. As we will see below, in many GWAS studies, this is *as important*, indeed in some cases *more important*, than estimating an average effect size. The simplest test for variance in effect sizes is the **Cochran Q** test of heterogeneity,

$$Q = \sum_{i=1}^k \frac{\left(T_i - \overline{T}\right)^2}{s_i^2} \tag{20.34}$$

where (under the null of $\theta_1 = \cdots = \theta_k$, and assuming that the values of T_i are normally distributed) the distribution of Q is χ^2 with $(k-1)$ degrees of freedom. While a standard reported metric in a meta-analysis, Q is *grossly underpowered* when k is small, so that a nonsignificant value *does not* imply a lack of heterogeneity. Indeed, when k is small, often a less strict standard (such as $p \leq 0.1$) is used to declare heterogeneity (e.g., Dudbridge and Newcombe 2019). Conversely, for large k, Q may be *overpowered* in the sense of declaring biologically unimportant heterogeneity as being statistically significant.

While Q is a test statistic for heterogeneity, it can also be used to obtain an estimate of the between-study treatment effect variance, σ_u^2. The **DerSimonian–Laird estimator** is given by

$$\widehat{\sigma}_u^2 = \frac{Q-(k-1)}{S_1 - (S_2/S_1)}, \quad \text{where} \quad S_j = \sum_{i=1}^{k} s_i^{-2j} \quad \text{for} \quad j = 1, 2 \tag{20.35}$$

which is set to zero if negative (DerSimonian and Laird 1986; 2015). Other estimation approaches (e.g., REML; Chapter 10; Part II Chapter 32) have been proposed (Appendix 6).

Higgins and Thompson (2002) noted two problems with quantifying the amount of heterogeneity using either Q or an estimate of σ_u^2. First, the expected value of Q (under the null) is $k-1$, a function of the number of studies. Second, the value of σ_u^2 is dependent on the scale of measurement and test statistic used, and hence not readily comparable over different meta-analyses. Hence, neither Q or σ_u^2 is an optimal measure for general heterogeneity, and they proposed two related metrics for this task. Their ***H*** statistic is given by

$$H = \sqrt{\frac{Q}{k-1}} \tag{20.36a}$$

and measures the excess in Q over its expected value. A modification is the **Mittlbock–Heinzl** (2006) statistic,

$$H_M^2 = \frac{Q-(k-1)}{k-1} = H^2 - 1 \tag{20.36b}$$

Equation 20.36a is the basis for the **Higgins–Thompson index** of heterogeneity

$$I^2 = \frac{H^2-1}{H^2} = \frac{Q-(k-1)}{Q} \tag{20.36c}$$

where

$$E[I^2] = \frac{\sigma_u^2}{\sigma_u^2 + s^2} \tag{20.36d}$$

with

$$s^2 = (k-1)\frac{S_1}{S_1^2 - S_2} \tag{20.36e}$$

measuring the average within-sample variation. The nice feature of I^2 is its natural interpretation as the fraction of the total variation ($\sigma_u^2 + s^2$) due to heterogeneity (σ_u^2). Equation A6.36c can be used to construct approximate confidence intervals for I^2 (as the latter is a function of H). While I^2 is widely used, Nakaoka and Inoue (2009) noted that there are settings where H_M^2 might be a slightly better metric.

Meta-Analysis of a GWAS Collection: Basic Issues

A meta-analysis of a collection of independent GWAS has three aims: (i) the discovery of new marker-trait associations not found in any single study; (ii) replication of any initial associations; and (iii) detection of any variation in marker effects over studies. An early example of the power of a meta-analaysis is from Crohn's disease (Franke et al. 2010). Six previous GWAS studies (comprising a total of 15,000 controls and 6,300 cases) had localized 32 different markers associated with the disease. A meta-analysis of these studies identified 30 additional associations.

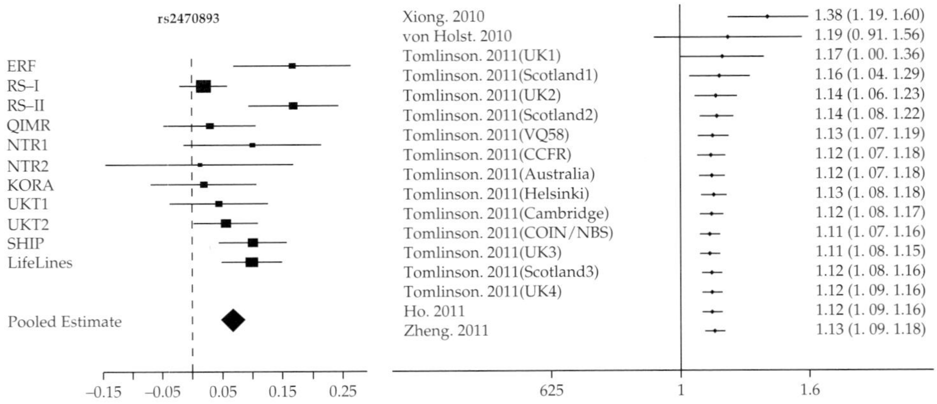

Figure 20.5 **Left:** A meta-analysis **forest plot** for a particular SNP (rs2470893) for a series of GWAS examining coffee consumption. Each row represents a particular study, with middle of the box locating the study estimate, the volume of the box representing the weight given to that study, and the lines on both sides of the box (often called **whiskers**) represent the 95% confidence interval. The vertical dotted line denotes the **line of no effect**. The diamond at the bottom of the plot represents the meta-analysis estimate, with the peak of the diamond denoting the estimate, and the edges denoting the width of the confidence interval. (After Amin et al. 2012.) **Right:** A **cumulative forest plot** for a meta-analysis of SNP rs961253 on colorectal cancer risk. The top row represents the odds ratio estimate in the initial association, and each subsequent row is the meta-analysis of all of the preceeding studies. Note the Proteus phenomena effect, with the initial study having the largest effect. (After Zheng et al. 2012.)

The results of a meta-analysis for a given marker are usually displayed using **forest plots** (Figure 20.5), which provide a visual representation of the individual studies, heterogeneity, and display the final estimates (often both the FE and RE values). These can also be constructed as **cumulative forest plots**, where the top row is the initial study, the second row the meta-analysis using the first two studies, and so on, with the final row giving the full meta-analysis (Figure 20.5). These are usually displayed (from top to bottom) in the chronological order in which the studies appeared. The observation that the first study often has the largest effect has been termed the **Proteus phenomena** by Ioannidis and Trikalinos (2005).

As mentioned, a meta-analysis has multiple objectives. If the goal is simply detecting significant marker effects, then the higher power of an FE analysis is recommended, especially given the very stringent p values required to control the FWER (Begum et al. 2012; Evangelou and Ioannidis 2013; Panagiotou et al. 2013; Dudbridge and Newcombe 2019). Use of the FE model is appropriate because the null hypothesis of no marker effect *jointly* implies $\mu = 0$ (no average effect) *and* $\sigma_u^2 = 0$ (no heterogeneity). In theory, one could have $\mu = 0$ (the *average* marker effect is zero), but still have considerable marker heterogeneity (*individual* studies have significant marker effects). If there is a concern of heterogeneity, then the RE2 model of Han and Eskin (Example 20.11) should be used, as it has greater power than FE when sufficient heterogeneity is present. While an FE analysis is typically used for detection, if heterogeneity is present, the resulting FE confidence intervals for effect size are too narrow, and an RE analysis is more appropriate. Recall that an RE analysis usually downweights larger studies and upweights smaller studies (relative to an FE analysis), which may be of concern in some analyses.

One complication that can arise in a GWAS meta-analysis is when the same subjects overlap in one (or more) studies (such as when shared controls are used). This creates correlations among the study estimates, which is problematic, as standard meta-analysis models assume these are independent. Lin and Sullivan (2009) addressed this concern by developing a GLS estimate that accounts for such correlations. For studies i and j, the expected correlations are

$$r_{ij} \simeq \begin{cases} n_{ij}/\sqrt{n_i n_i} & \text{population sample} \\ \left(n_{ij0}\sqrt{\dfrac{n_{i1}n_{j1}}{n_{i0}n_{j0}}} + n_{ij1}\sqrt{\dfrac{n_{i0}n_{j0}}{n_{i1}n_{j1}}} \right) \Big/ \sqrt{n_i n_j} & \text{case-control} \end{cases} \tag{20.37}$$

where n_{ij}, n_{ij1}, and n_{ij0} are, respectively, the total number of shared subjects, shared cases, and shared controls for studies i and j, while n_i, n_{i1}, and n_{i0} are, respectively, the sample size, number of cases, and number of controls in study i. When studies are independent, the covariance matrix used in the GLS meta-estimate is diagonal, leading to Equation 20.31b. When studies are correlated, the among-study covariance matrix now has off-diagonal elements from these correlations, with estimates following from standard GLS expressions (Equation 10.13a). An analogous situation occurs when a mega-analysis is performed over a series of studies that contain relatives (such as mouse inbred lines). Furlotte et al. (2012) show how to use an estimated relationship matrix to appropriately weight the studies.

While meta-analyses are typically framed in terms of testing the effects of a single marker on a single trait over a collection of studies, their utility is much more general. For example, one could conduct a meta-analysis of a *single marker* over different *traits*, namely a ***search for pleiotropic effects*** of a marker on multiple traits (e.g., Bolormaa et al. 2014). Cotsapas et al. (2011) examined the impact of 107 immune disease-risk SNPs detected for one disease on their impact on other immune-mediated diseases. To do so, they developed a **cross phenotype meta-analysis** (**CPMA**) test. Their concern was, given a detected marker for one disease, how can one test for an association with some, but not necessarily all, of the other diseases. The logic of their test (once again) follows from the assumption of a uniform distribution of p values under the null (Appendix 6). This can be restated as $-\ln(p)$ following an exponential distribution (Appendix 7) with decay rate $\lambda = 1$. Their CPMA test statistic is simply a likelihood ratio of the data under the null ($\lambda = 1$) versus the likelihood under the MLE ($\widehat{\lambda}$) for the fitted λ given the p values for the other diseases. Using this approach, they found that 47 of the 107 SNPs (44%) were associated with some, but not necessarily all, of the other diseases, so that (at least) 44% had pleiotropic effects.

Finally, an efficient meta-analysis heavily relies on stable marker estimates from each individual study, and hence requires that the MAF at the focal marker (in each GWAS) is not too small. We have previously discussed various rare-allele approaches for when this assumption fails (Table 20.3), and these statistics for each study can be used as the entries in a meta-analysis, see Lee et al. (2013) for details. Alternately, a mega-analysis (where possible) might result in a rare allele being sufficiently common in the full dataset to allow for stable estimates of individual effects (Panagiotou et al. 2013). Ma et al. (2013) showed that a joint analysis based on testing single markers can be more powerful than a meta-analysis when the total **minor allele count** (**MAC**) is less than 400. In this setting, they found (with a case-control design) that using logistic regression with a penalized likelihood (the Firth biased-corrected test; Firth 1993; Heinze and Schemper 2002) over the joint data was the most appropriate analysis.

Meta-Analysis of a GWAS Collection: Heterogeneity

As noted by Ioannidis et al. (2007), heterogeneity in GWAS is both very common and *"is a useful aspect of the data, rather than a nuisance, as it can often point to leads that can clarify better the nature of postulated association in the context of meta-analysis."* Heterogeneity can be formally quantified using I^2 (and its confidence interval), and visually displayed with forest plots (Figure 20.5). When significant heterogeneity is present, a **sensitivity analysis** may provide insight into its causal sources. Here, one assesses the effects of removing specific studies on the meta estimate and I^2.

At its simplest level, an understanding of potential sources of heterogeneity, and correcting for them, can result in a more powerful meta-analysis (greater power for detection and tighter confidence intervals). Different trait ascertainment criteria, differential scoring of the trait, and testing the same marker with different genetic models (e.g., fitting a recessive model in some studies and an additive in others) are all potential sources of heterogeneity that can often be easily addressed. More subtle issues involves between-population variation in the structure of LD between markers and causal loci, genetic heterogeneity (causal alleles vary over studies), and differential exposures (variation in environmental and/or genetic backgrounds). A deeper exploration of causes of heterogeneity can result in significant biological insight (e.g., Example 20.12). For example, Zeggini and Ioannidis (2009) found significant heterogeneity in a meta-analysis for the association between Type 2 diabetes (T2B) and the *FTO* locus known to be involved in obesity. In settings where cases and controls were matched for body mass index (BMI), no association between *FTO* and diabetes was found. Hence, it appears that *FTO* predicted obesity, which in turn predicted diabetes, yielding some of the earlier detected associations.

Magosi et al. (2017) noted that heterogeneity is often better thought of as a *study-wide*, rather than a *marker-specific*, effect, with some feature(s) of the study generating a systematic signal over a large set of markers (**systematic heterogeneity**). Hence, there is potentially more power to detect such an effect by jointly considering an appropriate collection of markers. This is the basis for Magosi's **aggregate heterogeneity *M*** statistic, which is computed for each study. M is based on the average of the scaled deviations of observed marker effects from their meta-analytic predicted values. These scaled deviations are referred to as **standardized predicted random effects** (**SPREs**), with the SPRE score for marker i (of m) in study j given by

$$\text{SPRE}_{j,i} = \frac{T_{j,i} - \widehat{\theta}_{j,i}}{\sqrt{\sigma^2_{u,i} + \sigma^2_{j,i} - E^2_{j,i}}} \tag{20.38a}$$

where $T_{j,i}$ is the association statistic for marker i in study j, $\sigma^2_{j,i}$ its sample variance, and $\widehat{\theta}_{j,i}$ its predicted realization. The other two variance terms are the between-study variance for marker i ($\sigma^2_{u,i}$,) and the prediction standard error $E_{j,i}$ (Part II Chapter 32). The resulting M value associated with study j is given by

$$M_j = \frac{1}{m}\sum_{i=1}^{m} \text{SPRE}_{j,i} \tag{20.38b}$$

Extreme values of M_j indicate potential ***outlier studies***, whose systematic effects result in marker values being consistently overpredicted or consistently underpredicted, *relative to the rest of the studies*. Statistical tests follow as M_j is approximately distributed as a unit normal. Magosi et al. applied their approach to a coronary artery disease (CAD) meta-analysis involving 48 different GWAS studies which involved a mixture of ethnicities. Based on M scores, they found that a significant proportion of between-study variation was due to differences in: (i) the amount of early-onset cases; (ii) use of family history for ascertainment; and (iii) inclusion of individuals of East Asian ancestry.

While systematic study effects represent one end of the spectrum of heterogeneity sources, at the opposite end is **marker-specific heterogeneity**. For a given marker, this could be generated by variation in causal alleles over studies, epistatic interactions over different genetic backgrounds, or G $\times$ E interactions caused by major environmental factors varying over studies. Kang et al. (2014) exploited this idea by noting that marker-specific heterogeneity can provide insight into G $\times$ E when study interactions correlate with environmental features. They examined 17 mouse studies (with a total of roughly 5,000 mice), performing separate meta-analyses on 26 markers showing significant effects on HDL cholesterol levels. By considering different mouse lines under different environmental conditions, these studies varied in major environmental features (such as diet), sex, and in genetic backgrounds (the presence/absence of knockout genes impacting other cholesterol pathway genes). For markers with highly significant I^2 values, they attempted to correlate heterogeneity with specific study factors.

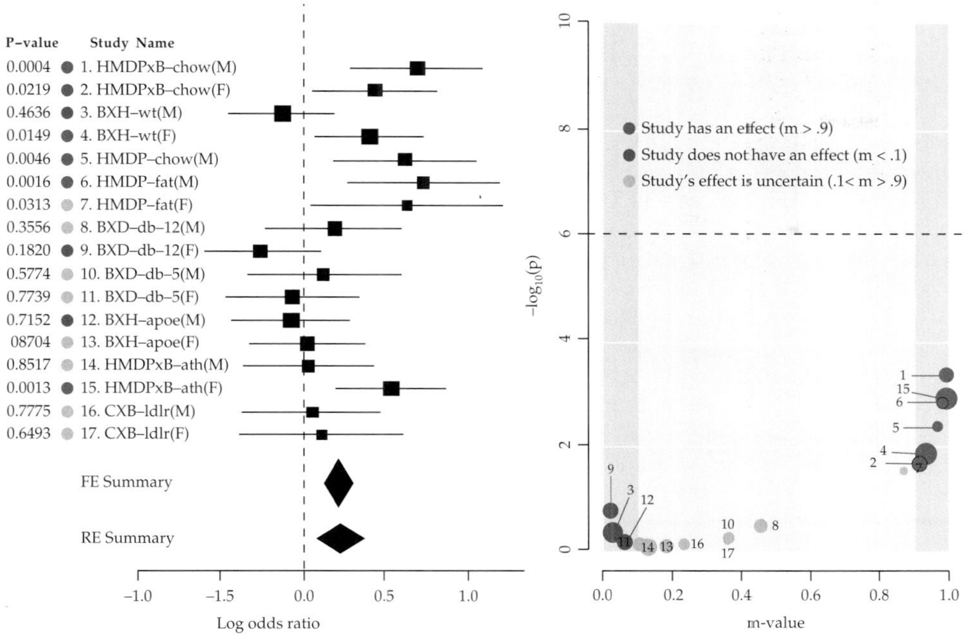

Figure 20.6 **Left:** A forest plot for the *Fabp3* gene in a meta-analysis on mouse HDL levels. Note studies 3 and 4, BXH-wt(M) and BXH-wt(F), which differ only in sex (males versus females). The impact of this marker on females (study 4) was significantly positive (the confidence interval was entirely greater than zero), while the mean impact on weight in males (study 3) was negative, but not significantly so, showing a clear sex effect of this marker in this background (the BXH wild type strain) and a common environment (high fat diet). **Right:** The associated **P–M plot** for these studies. For each study, $-\log_{10}(p)$ is plotted on the vertical axis and m (posterior probability that the effect is present) on the horizontal axis. Note that while the confidence intervals for studies 3 and 4 overlap (left panel)—so that their heterogeneity might simply be the result of sampling noise—their relative locations on the P–M plot offers some clarity (right panel). Study 3 had a low m value (0.03), while study 4 had a high m value (0.93), supporting very little (if any) effect in males, but a strong effect in females. (After Kang et al. 2016.)

At such high-heterogeneity markers, some studies are expected to have interactions while others are not. To determine which particular studies show an effect, Kang et al. used the **m statistic** developed by Han and Eskin (2011). This estimates the posterior probability that an effect exists in a *particular study*, given that the *collection as a whole* is significant. This is an extension of the notion of the posterior probability of association (Equation 20.7b). Han and Eskin suggested that studies with $m < 0.1$ are predicted to have no effect, studies with $m > 0.9$ are predicted to have an effect, and intermediate m values are ambiguous. Han and Eskin (2012) and Kang et al. (2016) proposed that using a **P–M plot** (Figure 20.6) —plotting the p value for a given study and its associated m value—was much more informative as to which studies have effects. Their reasoning was that the m value for a given study borrows cross-study information, while its p value uses only within-study information. For example, they note that markers can have $m > 0.9$, and yet not have very significant p values (e.g., study 2 in Figure 20.6).

Example 20.12. In one of the first human obesity GWAS, Herbert et al. (2006) found a significant signal at the rs7566605 SNP associated with the *INSIG2* gene, with the CC genotype increasing obesity relative to the CG or GG genotypes (a recessive model). This observation proved hard to replicate, occurring in some, but not other, followup studies. Heid et al. (2009) examined whether this lack of repeatability was due to an initial false positive or was a consequence of study heterogeneity. A meta-analysis of 27 studies encompassing 66,000 Caucasians using a case (body mass index, BMI, ≥ 30) versus control (BMI < 30) design found a significant effect (estimated OR of 1.076, $p = 0.023$) under an FE analysis, but not under an RE analysis (OR of 1.051, $p = 0.268$). For the RE analysis, I^2 was 41% with a confidence interval of 6.6% to 62.8%, indicating significant heterogeneity (a p value of 0.015 for Q). The OR and significance values increased as the BMI case threshold increased (using controls with BMI < 25), with OR values of 1.16, 1.18, 1.22, and 1.27 for BMI cutoffs of 32.5, 35, 37.5, and 40.0. They then broke the 27 studies into three sets: 16 general populations (GP), six obese populations (OP), and five healthy populations (HP). In the GP GWAS set, there was a significant effect (OR = 1.092, $p = 0.035$) using the RE model, and a reduction in I^2 (down to 10.9%; confidence interval of 0% to 48.1%). No significant effect was seen in the OP analysis, which showed a high level of heterogeneity (I^2 of 63.2%, Q with a p value of 0.018). Surprisingly, in the HP population, the CC genotype had a significant *protective* effect, with an odds ratio of 0.796 and with no heterogeneity ($I^2 = 0$). The authors suggested that the *INSIG2* gene is associated with *extreme* obesity (and, ironically, decreased BMI in such cases), a signal that can be masked by the study design when an insufficient number of such individuals are sampled.

21

Quantitative Genomics and Probing the Nature of Quantitative Genetic Variation

Whether the goal is discovery of rare variants or common variants, sample sizes are a key limiting factor for furthering our understanding of polygenic diseases, and increasing sample size remains a research priority needed to further ... genetic discoveries. Wray et al. (2018)

The previous chapter examined the powerful GWAS approach for detecting associations between SNPs (and potentially other markers, such as CNVs) and trait values at a target phenotype. In the vast majority of cases, detected markers (GWAS "hits") are not themselves functional, but rather "tag" nearby causal variants via linkage disequilibrium (LD). A GWAS amounts to a very high precision QTL mapping experiment, offering insight into the **genetic architecture** of traits (the number of causal variants and their joint distribution of effects and frequencies). What a GWAS leaves unresolved is the actual molecular nature of causal variants, and how this variation imparts developmental and physiological effects on a target trait or disease. Unfortunately, the high LD that allows one to tag causal variants also impedes their very-fine mapping (i.e., detection of causal **quantitative trait nucleotides, QTNs**), as numerous variants around a causal site may be in almost complete LD.

GWAS was fueled by the rise of high-throughput DNA scoring platforms, culminating in rare allele methods using whole-genome sequence data (Chapter 20). Concurrent with this rise in DNA technologies was the development of other high-throughput functional genomics platforms. These scored intermediate features in the pathway between a DNA sequence and the final trait phenotype, such as **epigenetic modification** of the DNA (changes in chromatin status and configuration) and levels of downstream transcripts, proteins, and metabolites. This chapter examines what has been called the **post-GWAS era** (Zhang et al. 2014; Gallagher and Chen-Plotkin 2018), combining information from GWAS studies and functional genomics to probe the nature of quantitative variation. We attempt to weave together a diverse collection of topics, ranging from a more detailed description of gene regulation to analytic methods for exploiting high-dimensional data sets. We conclude with an overview of our current understanding of the evolutionary forces that have shaped existing genetic variation, how this translates into structuring genetic architectures, and offer some glimpses into the molecular mechanisms underpinning such variation.

We start with high-throughput studies of gene expression (microarrays and RNA-Seq) that provide a snapshot of the **transcriptome** (the genomic set of all transcribed regions). Genome-wide expression studies have had three major impacts on the field of quantitative genetics. First, the amount of transcript for any given gene is a *quantitative trait*, as we can quantify its value, and variation in this value typically has both genetic and environmental components. Thus, we can perform standard quantitative-genetic (QG) tasks such as estimating the heritability of expression levels, search for genetic correlations in expression over different genes, probe its variance components (e.g., how much nonadditive genetic variance is present), test for $G \times E$ (including tissue-, and sex-, specificity), and so on. The thousands of features scored in a single experiment by high-throughput expression platforms thus represent a collection of thousands of quantitative traits, allowing one to explore a distribution of trait architectures. Second, when marker data are jointly gathered with expression data, one can use methods discussed in Chapters 17–20 to map **expression QTLs** (**eQTLs**, or **eSNPs** when tagged by SNPs), markers tagging genetic regions that influence the expression level of a target gene. Finally, if expression levels and trait values are both scored, one can search for **quantitative trait transcripts** (**QTTs**), whose expression levels influence downstream trait values.

While our initial presentation is focused on the transcriptome, it is important to stress that *essentially any functional genomics feature of interest can be treated in a similar manner.* The QG framework used for analysis of the transcriptome easily extends to other genomic features, such as the **proteome** or **metabolome** (the sets, respectively, of all proteins or metabolites within a focal cell or tissue). It also applies to various regulatory intermediates, such as chromatin structure and mRNA processing, leading us to next consider more general **regulatory QTLs** (**regQTLs**) impacting these features, which, in turn, leads to discussions on the method of **transcript-wide association studies** (**TWAS**) and statistical methods to fine-map actual causal sites.

One of the early results from microarray studies was the generation of long lists of **differentially expressed genes** (**DEGs**) between two different cell types (such as normal and diseased). The resulting massive datasets sparked the explosive growth of statistical methods—loosely called **gene-set** and **pathway analyses**—developed to search for patterns among the lists of DEGs, such as enrichment of specific functional classes of genes and/or pathways. We present a modest overview of this vast literature, which offers a number of approaches for extending a GWAS analysis from the SNP or gene level (Chapter 20) to the level of user-defined gene sets (such as known biological pathways or genes in the same functional category). Our increasing ability for high-throughput analyses of genomic features raises issues about how to model and estimate these fine-scale interactions. We briefly introduce the emerging field of **systems biology**, which attempts to model these highly complex systems. As the analysis of gene expression data shows, quantitative genetics and systems biology represent natural starting units for a more holistic fusion of genome biology.

We conclude by examining the current picture of quantitative variation offered by existing GWAS and functional genomics studies. We start with an older debate as to whether the bulk of quantitative genetic variation is due to common alleles of small effect or rare alleles of large effects. We then turn to the observation of "missing heritability," wherein the variance accounted for by using just the significant GWAS hits is only a small fraction of the value seen using phenotypic correlations among relatives (Part II Chapters 22–24). Finally, we consider current hypotheses, such as the **omnigenic model**, that attempt to provide a general framework for the nature of quantitative variation.

As with Chapter 20, much of our focus here is on human studies, because this is where the largest investment has been made in developing genomic resources and tools. The logic and methodologies presented here apply equally well to most other study organisms, and can often be further leveraged by exploiting unique biological features that a focal species might possess. For example, Hymenoptera have haploid males and diploid females, a feature that can be exploited by a clever investigator. Likewise, the ability in some species to generate clones can result in much more accurate measurement of functional genomic features by averaging over a population of clones (Example 21.1), as opposed to using less precise measurements based on single individuals, or more derivative measurements based on tissue cultures developed from these individuals.

PROBING THE TRANSCRIPTOME

Whole-genome Expression Analysis

Paralleling the rise of rapid whole-genome sequencing approaches was the development of a number of **high-throughput profiling technologies** that simultaneously score very large collections of cellular macromolecules, starting with whole-genome transcription analysis (the transcriptome). First-generation transcriptome approaches were based on **microarrays**, which used RNA–DNA or cDNA–DNA hybridization to short gene-specific probes (spotted in an array) to simultaneously score thousands to tens-of-thousands of transcripts (Brown and Botstein 1999; Hedge et al. 2000; Berrar et al. 2003). This was an analog technology, scoring the amount of hybridization by the brightness of a given probe (spot) when made to fluoresce under a laser. This pioneering approach has now been replaced by a digital second-generation technology, **RNA-seq** (**RNA sequencing**). This method uses next-generation sequencing to count the actual number of copies of a given transcript in the sample (Mortazavi et al. 2008; Nagalakshmi et al. 2008; W. Wang et al. 2009). Besides being more accurate, RNA-seq also has more discrimination than a microarray, such as being able to quantify the amounts of alternately spliced products (**isoforms**) of a given primary transcript or detect allele-specific expression (e.g., Battle et al. 2014). The resolution of this technology is such that one can perform **single cell RNA sequencing** (**scRNA-seq**) to generate cell-type specific transcription profiles, which can then be tested for trait predictability (e.g., Watanabe et al. 2019; Võsa et al. 2021). Sources of systematic errors and quality control issues in RNAseq are discussed by Li et al. (2014) and Feng et al. (2015)

Gene expression is expected to be exquisitely tailored to specific tissues, environments, and developmental windows. As such, even when examining a genetically homogenous collection of a unicellular organisms (such as yeast), care must be taken to standardize environmental conditions and development stages (such as rapid growth versus stationary phases). For metazoans, the situation is even more delicate. At a very crude level, one could perform whole-organism RNA extraction, again under the caveat of using consistent environments and developmental stages. However, in metazoans, the focus is usually on a particular tissue or set of tissues deemed relevant to the trait or question under study, such as endosperm in seeds, muscle in cattle, or immortalized cell lines or blood from humans. Accomplishing this standardization limits the tissues/environments/conditions examined, which can skew the biological interpretation of any results by the bias introduced by the choice of tissues and conditions. For example, seed yield may be constrained by root uptake of limiting nutrients, which could be missed by a focus on expression in flower or seed tissues. Even if the transcript change is similar across all tissues, a modest to small change in one tissue may have a more dramatic effect in another.

One of the very surprising results from genomics is that the transcriptome is a much more wild and unruly universe than was perceived just a few decades ago (Johnson et al. 2005; Carninci 2006; Gustincich et al. 2006; Kapranov et al. 2007; Amaral et al. 2008; Kapranov and St Laurent 2012; Deniz and Erman 2017). The historical view was that the vast bulk of the transcriptome was mRNA transcribed from protein-coding genes, along with a few other specialty RNAs (tRNAs, rRNAs) critical for translation. As such, extraction and scoring methods tended to focus on mRNAs, for example by exploiting the presence of a poly-A tail in most processed transcripts. It is now apparent that, at some level, most of the human genome appears to be transcribed. Some sources for these transcripts are obvious, such as those associated with mobile genetic elements and noise from leaky or read-through transcription. However, it is also clear that there are numerous, and very diverse, classes of **noncoding RNAs** (**ncRNAs**; also denoted as **transcripts of unknown function**, or **TUFs**), that play critical roles in gene regulation, roles that are still being resolved. These RNAs are often partitioned by size into either **short ncRNAs** (**sncRNAs**; transcripts less than 200 nucleotides) or **long ncRNAs** (denoted as either **lncRNAs** or **lincRNAs**, for **long intergenetic noncoding RNAs**). Important classes of sncRNAs include **small nuclear RNAs** (**snRNAs**) involved in regulating gene splicing, **small nucleolar RNAs** (**snoRNAs**) involved in modification of functional RNAs (e.g., rRNAs, tRNAs, and snRNAs), and **micro-RNAs** (**miRNAs**) and **small interfering RNAs** (**siRNAs**) involved in post-transcriptional regulation of gene expression. While the roles of lincRNAs are less resolved (Deniz and Erman 2017), given that there are over 15,000 lincRNAs in the human genome their impact is likely nontrivial. Given the unknown roles of many ncRNAs, in a bit of physics envy Johnson et al. (2005) referred to these as **transcriptional dark matter** (or the **dark transcriptome**). It is highly likely that some of this "dark matter" is relevant to quantitative trait variation (St Laurent et al. 2014; Issler and Chen 2015; Li et al. 2016; Gamazon et al. 2018; X. Liu et al. 2019), especially under the view that much of phenotypic variation within a population is due to regulatory variation. Given that many biologists hold the **RNA world** view of the first protocell (with RNA, or its precursors, performing both coding and metabolic functions before being displaced by DNA and proteins), this greatly expanded functional role for RNAs in current cells should really not be that surprising.

Genetical Genomics and eQTLS: Basics

As was discussed in Chapter 18, the first mapping of regulatory QTLs was by Damerval et al. (1994), who detected QTLs influencing the spot volume for anonymous proteins on 2-D gels in maize (Figure 18.13), which was followed by similar studies in mice (Klose et al. 2002; Hartl et al. 2008). Jansen and Nap (2001; de Koning and Haley 2005) coined the term **genetical genomics** for this marriage of QTL mapping and expression of genomic features. Jansen (2003) further suggested that functional analysis in a segregating population can be a more powerful tool than more traditional single-gene perturbations (such as gene modification/knockouts or iRNA silencing). With the development of high throughput RNA expression platforms, the initial focus of genetical genomic studies was on the transcriptome. Most of the early studies used classical linkage-based QTL mapping approaches (Chapters 18 and 19), with the trait value being the amount of transcript from a focal gene, an integrated measure of both its transcription rate and message stability.

The study and mapping of such **eQTLs** (**expression QTLs**; Jansen and Nap 2001; Schadt et al. 2003) that influence the mRNA level of a target gene are of great interest for several reasons. First, as noted by Rockman and Kruglyak (2006), "the road from genotype to phenotype runs through gene expression." Most traditional trait-based GWAS SNPs map to noncoding regions (e.g., Edwards et al. 2013 found that 85% mapped to noncoding regions in humans), and hence likely represent regulatory, as opposed to structural, variants. While regulatory changes can occur at many level —transcriptional, post-transcriptional, translational, post-translational, cell–cell or tissue-tissue interactions, etc —many of these regulatory variants are likely eQTLs. As a result, the location of GWAS SNPs can be very misleading as to the location of causal coding regions (e.g., Example 21.5). Suppose that variation in the amount of transcript in gene Q impacts trait value/disease status. Variation in regulation in transcript abundance might be governed by sites at some distance from the coding region, with the resulting GWAS hits drawing attention away from Q. Conversely, eQTLs can also map close to the coding region for the transcript they influence (see below). In such cases, a correlation between trait value and transcript abundance of a focal gene can provide support for that gene influencing the focal trait. Thus, eQTLs and GWAS SNPs can jointly provide support for causal genes. Finally, the ability to score thousands of transcripts in a single experiment offers a large, relatively unbiased, set of characters upon which to draw influence about the distribution of genetic architectures over this class of traits. Overviews of eQTLs are given by Rockman and Kruglyak (2006), Gilad et al. (2008), Nica and Dermitzakis (2013), Albert and Kruglyak (2015), and Hill et al. (2021).

Early eQTL mapping studies included linkage-based designs using line-crosses (F_2 and RILs; Chapter 18) in yeast (Brem et al. 2002; Yvert et al. 2003; Brem and Kruglyak 2005), mice (Schadt et al. 2003; Bystrykh et al. 2005; Chesler et al. 2005), rats (Hubner et al. 2005; Petretto et al. 2006), maize (Schadt et al. 2003), and *Arabidopsis* (West et al. 2007). Early work in humans was based on outcrossed pedigree designs (Chapter 19), with expression levels being scored using either cell lines or blood extracted from each sampled individual (Schadt et al. 2003; Monks et al. 2004; Morley et al. 2004; Deutsch et al. 2005). As with any linkage-based design, the level of resolution of eQTLs was poor (typically on a megabase scale), especially given the small sample sizes of most of these studies (usually <100 individuals/lines). These initial linkage-based studies were soon followed by LD-based GWAS in humans (Cheung et al. 2005; Stranger et al. 2005, 2007a, 2007; Emilsson et al. 2008; Dimas et al. 2009; Dixon et al. 2009) and maize (Fu et al. 2013; H. Liu et al. 2017), moving from eQTLs to eSNPs. While such **expression-wide association studies** (**EWAS**) offered greater mapping resolution, these early GWAS studies also suffered from very small sample sizes (again, typically less than 100), and thus were dramatically underpowered.

As discussed in Chapter 20, structural DNA variation—such as copy number variation

(CNV), insertions/deletions (indels), and presence/absence of mobile elements—can be used in place of SNPs as markers in a GWAS. Often these structural variants are tagged via LD to nearby SNPs (**tCNVs**, for **tagged copy number variations**; Gamazon and Stranger 2015), and hence their effects are largely captured by a SNP-based GWAS. However, many structural variants are not well tagged by SNPs, and there are reasons to suspect that such variants often have a direct impact on expression levels of nearby transcripts (leading to **eCNVs**). Their impact could simply be due to a duplication resulting in double the amount of transcript, or through more subtle effects such as duplication or deletion of control elements (e.g., enhancers). Consistent with this argument, Gamazon et al. (2011) found that tCNVs were enriched for eQTLs compared to SNPs with matching allele frequencies. Stranger et al. (2007) found that CNVs accounted for around 18% of the total detected variation in gene expression in their study, which is likely an underestimate, as other CNV effects could have been captured by nearby SNPs. Bryois et al. (2014) found that CNVs were more likely to be eQTLs than were SNPs, and that CNVs can act in *trans* as well as *cis*. They also found (as did Chiang et al. 2017 and Uzunovíc et al. 2019) that the effects of structural variants on expression were usually larger than SNP effects. One of the most detailed analysis of the relative contributions from SNPs and structural variants was by Jakubosky et al. (2020), who examined *cis* eQTLs (within 1 MB on either side of the coding region) for 7000 genes showing genetic variation in expression. Over 7 million common variants (SNPs, indels, and other structural variants) were scored, finding 11,000 lead makers impacting expression, with 72% being SNPs, 24% being indels, while the rest were other structure variants.

The major difference between an eQTL/EWAS and a QTL/GWAS experiment is that, in the former, thousands of traits (gene-specific expression levels) are scored at once. Further, many of these traits are highly correlated, reflecting coordinated expression over sets of genes. This trait dimensionality structure introduces two issues. First, it creates a heavy multiple-comparison burden (testing over thousands of individual traits). Coupling this burden of higher stridency levels for each test with the typically small sample sizes of many eQTL experiments made these early studies very underpowered (Gibson and Weir 2005). One consequence of lower power is poor mapping resolution and potentially unstable estimates of eQTL locations (Pérez-Enciso 2004). A second concern is Beavis (winner's curse) effects (Figure 18.8), wherein detected (i.e., declared to be significant) effects are substantially overestimated due to low power (de Koning and Haley 2005).

While many eQTL studies map one transcript at a time (or, conversely, test all transcripts over a given marker, one marker at a time), greater power can be achieved by leveraging information from these correlated values (Kendziorski and Wang 2006). When the number of transcripts greatly exceeds the number of scored samples, methods for mapping low-dimensional correlated traits from Chapter 18 are not applicable. Extensions of standard QTLs methods to handle these high-dimensional, but correlated, traits were developed by Kendziorski et al. (2006), Chen and Kendziorski (2007), Gelfond et al. (2007), Jia and Xu (2007), Chun and Keleş (2009), Zou and Zeng (2009), Wang et al. (2011), Shabalin (2012), Flutre et al. (2013), Davis et al. (2016), and Ongen et al. (2016). An alternative approach has been to use principal components (or, more generally, the singular value decomposition; Appendix 3) to reduce the dimensionality of the transcript data, and then map the resulting PC/SV values as composite traits (Alter et al. 2000; Lan et al. 2003; Biswas et al. 2008). Sul et al. (2013) suggested meta-analytic approaches for combining results over multiple tissues.

Given the potentially complex nature of the transcript correlation structure, permutation tests (Chapter 18) remain the gold standard for significance testing for eQTLs, but the exchangeable (shuffled) unit is no longer single traits. Permutation tests should instead be constructed by *keeping the vector of expression data intact*. If $\mathbf{e}_i$ and $\mathbf{g}_i$ denote the vectors of expression and marker data (respectively) for individual i, then the randomization should be $\mathbf{e}_i$ over $\mathbf{g}_j$, namely holding each vector intact, but shuffling their association.

Example 21.1. Several features about the biology of Baker's yeast (*Saccaromyces cerevisiae*) make it an excellent model system for classical and molecular genetics. Following the cross of two strains, recombinant haploid spores are produced, each of which can be rapidly grown into pure colonies (generating **segregant lines**, the haploid version of RILs). Because each line is a haploid, no dominance is present and any epistasis is additive (e.g., AA, AAA, etc.; Equation 5.9). Finally, yeast has a very high recombination rate, offering increased mapping resolution. Geneticists have long exploited these features, but the lack of any significant morphological variation focused early quantitative-genetic studies on physiological traits such as growth rate. More recently, high throughput systems to phenotype yeast morphological traits have been developed (e.g., Ohya et al 2005), but these still only score a limited number of features. Whole-genome expression analysis introduced thousands of new traits that could be cheaply and rapidly scored, allowing for eQTL mapping. The resulting large sample of eQTLs for different transcripts provides insight into the distribution of genetic architectures underlying expression variation.

An early exploration of yeast expression architectures was done by Brem and Kruglyak (2005), who scored 112 haploid segregant lines for expression level at 5,727 transcripts. These lines were also genotyped using 3,000 markers, with eQTL mapping performed using the standard RIL mapping framework (Chapter 18), modified for haploid lines. While the clonal nature of each segregant allowed phenotypes (expression levels) to be measured with some precision, the power of this design was still modest given the small number of lines. Over 3,500 transcripts showed a significant among-line variation, indicating heritable variation for expression levels. Of these, 2,000 had at least one detectable eQTL, with the authors noting that the other 1,500 heritable transcripts likely had eQTLs with effects too small to be detected. A similar observation was made in mice by Schadt et al. (2003), who detected eQTLs for only about a quarter of the transcripts that showed significant among-line expression differences.

Brem and Kruglyak noted that the difference between the observed among-line variation and the variation attributed to detected eQTLs provides information on the effect size of undetected eQTLs. To remove any Beavis effects (overestimation of effect size when power is weak), they split their 112 lines into a random set of 56 lines for detection, with the remaining set of 56 used to estimate the eQTL effects (this procedure is called a **subsampling approach**). They found that only 3% of highly heritable transcripts were consistent with a single controlling locus, roughly 20% were consistent with one or two locus control, and over half required at least five eQTLs. The latter bound assumed that undetected eQTLs had equal effects, so that the actual number is likely much higher. Transgressive segregation (Chapter 18) was common, where expression levels in some of the lines exceeded the values seen in their parents. Finally, because no dominance is present in haploid lines, departure of the mean of the segregant lines from the midparent value (line-cross analysis; Chapter 11) indicates (additive) epistasis, which was seen in roughly 20% of the highly heritable transcripts. A much more powerful follow-up study by Albert et al. (2018) using over 1,000 segregant lines was able to account for over 70% of the heritability in expression with mapped eQTLs. As predicted by Brem and Kruglyak, control of transcript expression was highly polygenic, with a median number of six eQTLs per transcript, where the majority of variation was accounted for by eQTLs acting at some distance from their target transcript coding region.

Example 21.2. Battle et al. (2014) performed an EWAS using RNA-seq on 922 human samples of whole blood, finding detectable eSNPs for over ten thousand transcripts. Using this large sample of traits, a number of patterns consistent with selection against variants with a large effect on expression emerged. As in Example 21.1, subsampling was used to remove Beavis effects, with the resulting estimated eSNP effect sizes tending to decrease as the variant frequency increased. This same eQTL variant size–frequency pattern was detected in the mustard *Capsella grandiflora* by Josephs et al. (2015) and in rice by Lye et al. (2022). Battle et al. found two other patterns suggesting that transcript level changes were more moderated for genes with potentially large, or important, roles in cellular function. Reduced eQTL variation was seen for both highly conserved genes and for genes whose products were hubs in protein–protein interaction (PPI) networks (Appendix 2). Further, they showed that eQTL effect size was negatively correlated with the number of PPIs for the product of that transcript.

A follow-up study on eQTLs in *C. grandiflora* by Uzunovíc et al. (2019) found that CNVs, in the form of transposable elements (TEs), made a significant contribution to expression variation. They observed that rare TE eQTLs tended to strongly downregulate expression, in contrast to a control group of rare SNPs that showed no net directionality in expression levels. Conversely, common TE eQTLs were more likely to increase expression. These observations suggest TE insertions that downregulate expression are generally selected against, but variants that upregulated expression were often less deleterious.

Genetical Genomics and eQTLS: *Cis* Versus *Trans* Effects

The poor resolution of linkage-based eQTL locations has an important implication in their functional interpretation. In classic genetics, regulatory factors are typically classified as either having ***cis*** or ***trans*** functionality (Haldane 1942; Lewis 1945). A *cis* regulatory site only impacts a gene residing on the same DNA molecule, and typically represents regulatory binding sites such as promoters or regional features such as enhancers, silencers, or insulators. While *cis* sites are usually thought of as being adjacent to the coding sequence they impact, in theory they act at some distance (especially given the tendency of chromatin looping, wherein sites megabases apart may still find themselves in very close proximity in a cell; Rao et al. 2014). *Trans* regulatory factors, on the other hand, are generally envisioned as diffusable products (such as proteins or RNAs) that can impact genes residing on the same, or different, chromosomes (for example, by binding to *cis* sites).

The *cis* terminology (***cis*-eQTL**) was used in the early days of eQTL mapping to refer to an eQTL location that mapped very close to the coding region of the target transcript site, typically within a megabase from the transcription start (**TSS**), or end (**TES**), sites. *Trans* referred to an eQTL that was either at some distance away on the same chromosome, or on an entirely different chromosome, from the target transcript coding region. As noted by Price et al. (2015), the use of these terms was as "a convenient shorthand describing the association" and was not "typically intended as a statement about the underlying biological basis for the association." Given that these are *distance*, as opposed to *functional*, metrics, they have more recently been replaced in the literature with the terms of **local** (or **proximal**) and **distant** (or **distal**) eQTLs (Rockman and Kruglyak 2006; Gilad et al. 2008). Given the poor resolution from linkage mapping, "local" can operationally refer to regions on a megabase scale, which could contain multiple eQTLs.

Note that local eQTLs could functionally act in *trans*. One classic example would be regulatory feedback loops, where the amount of mRNA or protein product from a gene self-regulates its own expression. One can formally distinguish between *cis* versus *trans* functionality using **allele-specific expression**, **ASE** (Wright and Moyer 1966; Knight 2004; Wittkopp et al. 2004, 2008; Battle et al. 2014; Glassberg et al. 2019; Hill et al. 2021). Suppose allele *A* shows a higher level of expression than allele *B* on their original backgrounds. If this is due to a *cis* effect, then the level of expression of *A* should still exceed that of *B* in *A/B* heterozygotes. However, if this is due to a *trans* effect from a factor closely linked to *A*, then both *A* and *B* should show similar expression levels in heterozygotes, as any *trans*-acting factors will operate equally on both. Using this approach, studies in yeast and mice showed that most (but not all) local eQTLs are due to *cis* effects (Doss et al. 2005; Ronald et al. 2005). This pattern was also seen using RNA-seq to score expression in human lymphoblastoid cell lines (Pickrell et al. 2010). Wittkopp et al. (2008) and Emerson et al. (2010) discuss separation of *cis* and *trans* effects when both impact ASE. As noted by Battle et al. (2014), one can take this analysis a step further and map SNPs in LD with causal sites that directly influence ASE, detecting **aseQTLs** (or **aseSNPs**). They did so by treating the ratio of the two transcript types in a heterozygote as a quantitative trait. They used this same logic to map **sQTLs** (or **sSNPs**) that influence the ratio of two different splicing products (**isoforms**) from a transcript. Unexpectedly, some of their detected sSNPs acted in *trans*.

Given that a focal transcript could be impacted by both local and distant factors, rather than provide a profile (or Manhattan) plot (Figure 20.1) for each of the thousands of transcripts, the typical representation of an eQTL experiment is in the form of a **transcriptome map** (Chesler et al. 2005). This is a two-dimensional plot, where each displayed point

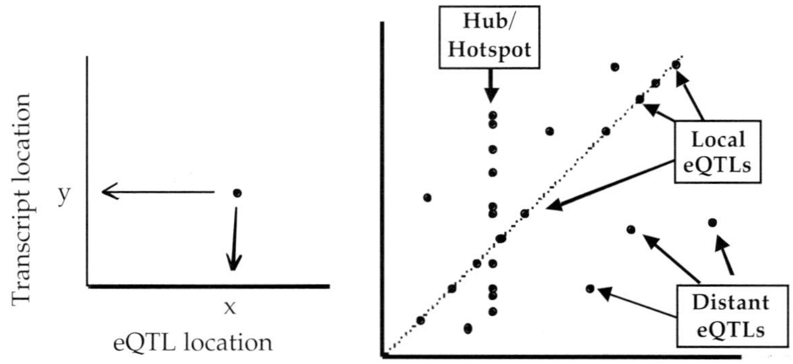

Figure 21.1 A stylized **transcriptome map**, plotting eQTL locations versus the location of the coding region for a transcript. Both axes correspond to genome position, with the horizontal (x) axis denoting a region/marker being tested as an eQTL and the vertical (y) axis the location of the coding region for a transcript (occasionally in the literature these two axes are reversed). A point or pixel at position (x, y) on this map indicates a significant association between a transcript whose coding region is at genomic position y and a marker/region at genomic position x. Points falling on the diagonal correspond to eQTLs that map very close to, or at, the same location as the coding region for the transcript they influence. These have been called ***cis* eQTLs**, but as discussed in the text are better referred to as **local** (**proximal**) **eQTLs**. Points falling off the diagonal correspond to eQTL locations that influence transcripts whose coding regions are at a different location from the eQTL. These have been called ***trans* eQTLs**, but are better referred to as **distant (distal) eQTLs**. A vertical stack of points corresponds to a (small) genomic region that is enriched for eQTLs, and is called a **hotspot** or **hub**, with the eQTLs in that region impacting numerous transcripts.

represents a significant association (which could also be color- or size-coded to represent the strength of the effect; e.g., Jiang et al. 2013). As shown in Figure 21.1, one axis gives the genomic location of the eQTL/eSNP, while the other axis corresponds to the genomic location of the transcript coding region that it impacts. If an eQTL maps close (local) to the coding region of the transcript it impacts, this generates a point on the diagonal of the plot, while if the eQTL maps distal, the point will be off the diagonal (Figure 21.1). When eQTL location is plotted on the horizontal axis (as in Figure 21.1), the presence of **hubs** or **hot spots** (distant eQTLs that impact numerous transcripts) will appear as vertical lines, while they appear as horizontal lines when the roles of axes are reversed (eQTL locations are mapped on the vertical axis). As observed by Lutz et al. (2019; 2022), the genetic architecture of variants at a hub can be rather complicated. In yeast crosses, *trans*-eQTLs mainly occur in hubs, a rather different situation from humans, perhaps reflecting coordinated control of a number of genes when the single-celled yeast experiences an environmental challenge. For example, the protein encoded by the yeast *IRA* gene, a key regulator in the RAS signaling pathway, is a *trans*-eQTL, impacting the level of expression at over a thousand transcripts. The coding region of this gene is over 9,000 bases long, and contains at least seven causal nonsynonymous variants displaying complex epistatic interactions with each other.

Bryois et al. (2014) noted multiple *cis*-eSNPs can impact the same coding region, which can be missed if the lead SNP has a large effect. Akin to removal of the effect of a major allele in standard QTL linkage analysis (Chapter 18), they suggested that the genotype of the first detected *cis*-eSNP be used as a cofactor to search for additional *cis* sites. Using this approach, they found that 20% of the genes in their modest sample of 870 human cell lines had at least two detectable *cis*-eSNPs, a percentage that is certain to rise with larger sample sizes (Gusev et al. 2016). Any collection of such *cis* sites impacting the same transcript should be subsequently tested for epistatic interactions (Chapters 18 and 20).

A common observation (Rockman and Kruglyak 2006; Breitling et al. 2008; Albert and Kruglyak 2015) is that individual *cis* effects tend to be both larger, and more common, than individual *trans* effects. There was considerable debate as to whether this represented reality or was simply a reflection of the low power of most early designs. The biological notion for stronger *cis* effects is that they are likely to be more direct, often focused on just a single transcript, than are *trans* effects, which are diffusive and may only become apparent after interaction with *cis* sites. There is clearly an ascertainment bias in favor of *cis* sites in that many studies only searched for associations with markers immediately adjacent to the transcript coding region. When power is low, the Beavis effect implies that any such sites declared to be significant will likely have an overestimated effect size (de Koning and Haley 2005). Note that tests of only transcript-adjacent (local) markers results in a much lower multiple comparison burden than testing every genomic location (distant) against the focal coding region. If indeed most *trans* sites have weaker effects, their heavier multiple comparison burden will further make them difficult to detect. Note from Example 21.1 that the yeast data suggest eQTLs of modest to weak effect are likely to be the norm, and that most of these are likely *trans*. Conversely, Stamatoyannopoulos (2004) noted a bias in diploids that can result in an *underestimation* of *cis* effect sizes, in that expression level measures incorporate values from *both* alleles (unless allele-specific expression is scored), potentially diluting a strong *cis* effect from one of the alleles.

An alternative approach for decomposing *cis* versus *trans* contributions is to use variance components (Price et al 2011; Lloyd-Jones et al. 2017). While variance-component methods are more formally examined in Part II Chapter 33, the basic approach used by these investigators follows from Equation 19.30a. Here, the expected covariance between two individuals (i and j) given that they shared IBD segments at a focal region (x) is $\sigma(z_i, z_j) = R_{ij,x}\, \sigma^2_A(x)$, where $R_{ij,x}$ is the estimated amount of IBD probability over x for i and j (which can be estimated from marker information; Chapters 8 and 19). By standardizing expression levels to have a variance of one, the associated additive variance for region x now equals the heritability h^2_x associated with genes in that region. To separate the contributions from local versus distal regions on the expression level of a focal gene, Price et al. (2011) used

$$\sigma(z_i, z_j) = R_{ij,x}\, h^2_{g,cis} + R_{ij,x^c}\, h^2_{g,trans}$$

where z_i is the level of the focal transcript (gene g) in individual i, $R_{ij,x}$ is the IBD estimate for region x, and R_{ij,x^c} is the IBD estimate for all regions outside of x (and hence *trans* effects). Here $h^2_{g,cis}$ is the heritability of the expression level of g due to local effects, and $h^2_{g,trans}$ the heritability due to distal effects. Using this approach, Price et al. found that (on average) 37% of the expression heritability in a sample of 19,000 gene transcripts scored over 722 Icelanders were due to *cis* effects for blood-expressed transcripts, while this was 24% for adipose tissue.

Given their weaker power of detection, Kendziorski and Wang (2006) suggested that one approach in a scan for *trans* sites is to test each marker separately, and look for accumulating evidence over the *entire set* of transcripts. Under the null hypothesis of no marker-expression effect, the distribution of p values follows a uniform, so a marker-specific histrogram of the p values for each transcript could be used to estimate the number of significant effects at each marker (e.g., Cotsapas et al. 2011). Appendix 6 discusses a number of these methods, but the complication to applying them to expression data is that we *expect* some transcripts to be highly correlated (Brynedal et al. 2017 suggest adjustments based on the decomposition of the transcript correlation matrix). This also raises concerns about the validity of apparent hubs (Figure 21.1), as Pérez-Enciso (2004) showed via simulation that the correlated structure of transcripts can easily create false positive (or **ghost**) hubs (akin to ghost QTLs; Figure 18.3). A related issue is that subtle environment changes can result in expression shifts of a number of genes, potentially creating ghost hubs if individuals in the sample vary over these conditions (Stamatoyannopoulos 2004).

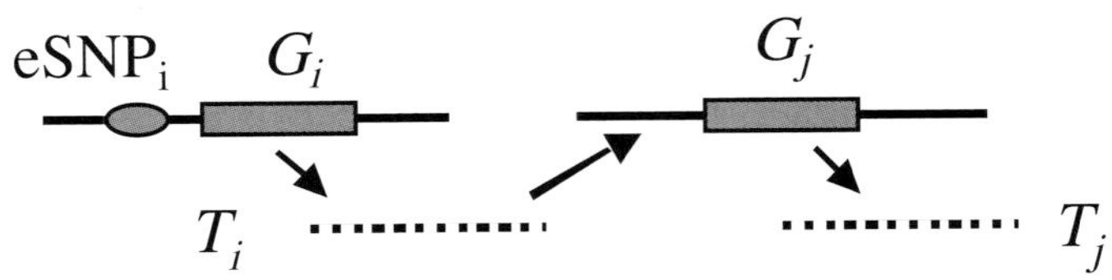

Figure 21.2 The concept of ***cis*-mediation**. The observation is that eSNP i acts at some distance away from a coding region (G_j) to regulate the level of its transcript T_j (eSNP i is a *trans*-eSNP for transcript j). The **mediation hypothesis** is that the impact of eSNP i is through a *cis* effect on the transcript from (local) gene G_i, whose transcript T_i then influences the regulation of transcript T_j of a distant gene G_j. Path analysis methods (Figure 21.3; Appendix 2) allow this idea to be extended over much more complex regulatory networks, as well as providing a framework for estimating direct and indirect effects of any component player (Example 21.3).

As mentioned, there is a huge disparity in testing local versus distal effects. Peterson et al. (2016a) noted that with their human date set of roughly 6.8 million SNPs and 30,100 transcripts, they had 142 million local tests (as an average of 21 genes were within a MB of a typical SNP), and over 200 billion distal tests. As a result, they suggested a hierarchical structure for hypothesis testing, first separating local and distal tests. For local tests, they focused on the entire collection of local SNPs for a given gene (using gene-based GWAS approaches; Chapter 20). Among those genes that showed a significant effect on a local transcript, they then tested each SNP within the gene separately. By first focusing on *sets* of SNPs, the multiple comparisons burden is greatly reduced. Similarly, for distal tests they first tested significance of a given SNP over the entire collection of distal transcripts (considered as a single set; again, modification of gene-based tests can be used), and, if significant, subsequently tested this SNP against each distal transcript.

Recalling the lessons from early GWAS studies (Chapter 20), wherein increasing sample sizes resulted in the discovery of more associated SNPs and greater replication over studies, we expect this same pattern applies to early eQTLs/eSNPs studies given their very small sample sizes. Even more recent studies are still rather modest relative to a modern GWAS, with samples sizes often no greater than a few thousand individuals (e.g., Zeller et al. 2010; Westra et al. 2013). As would be expected for eQTLs of small effect, Bryois et al. (2014) found that eQTL detection improved with increasing sample size. Indeed, Võsa et al. (2021) were able to detect *cis*-eQTLs for roughly 90% of the genes in a study of almost 32,000 individuals. Further, most of these eSNPs replicated over several tissues.

An important breakthrough in the analysis of eQTLs was the hypothesis of ***cis*-mediated *trans* effects**, or ***cis*-mediation** for short (Fehrmann et al. 2011; Jiang et al. 2013; Battle et al. 2014; Bryois et al. 2014; Pierce et al. 2014). As shown in Figure 21.2, this idea postulates that many of the observed *trans* effects are the result of a *cis* effect at the eSNP that impacts the transcript of a local gene, with that transcript then having a *trans* effect on distant transcripts. Formally, suppose that SNP i is observed to have a *trans* effect on the transcript T_j at a distal coding region j. The causality path is eSNP$_i \rightarrow T_i \rightarrow T_j$, in that SNP i *cis*-regulates the level of transcript T_i, which in turn influences transcript T_j. As an aside, the causalty notation $x \rightarrow y$ implies that a change in x while holding all other variables (other than y) constant results in a change in the distribution of y, but a change in y (holding all other variables constant) does not change the distribution of x.

The *cis*-mediation model explains, in part, why *trans* effects appear to be weaker than *cis* effects, as we are scoring a secondary effect. It also suggests strategies to improve *trans* detection. For example, Bryois et al. (2014) focused on only those eSNPs showing a local effect. These were then tested for *trans* effects over the rest of the transcriptome (an idea loosely akin to searching for epistasis by first starting with a QTL having a marginal effect; Chapter 18). This strategy significantly reduces the number of comparisons, resulting in improved power, and allowed Bryois et al. to detect addition *trans*-eSNPs. As outlined in Example 21.3, and discussed in more detail in Appendix 2, multiple regression and path-analytic models can be used to detect, and quantify, mediation effects. In humans, the

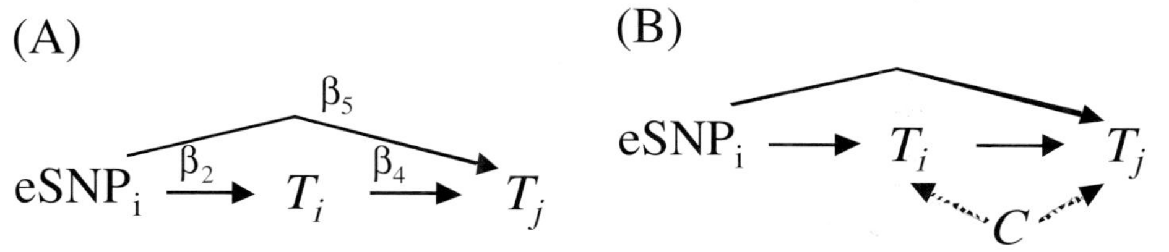

Figure 21.3 Path analysis of a trio (eSNP$_i$, T_i, and T_j) to separate direct and indirect effects. **(A):** The path diagram when only these trio elements are involved (Example 21.3). The direct effect eSNP$_i$ $\rightarrow$ T_j (avoiding T_i) is given by β_5, and the indirect (**mediated**) effect via T_i, eSNP$_i$ $\rightarrow T_i \rightarrow T_j$, is $\beta_2 \cdot \beta_4$. The total effect is given by $\beta_1 = \beta_5 + \beta_2 \cdot \beta_4$. This same logic can be applied to a trio of an eSNP, a transcript it impacts, and a trait value (i.e., replacing T_j with z_k, the value for trait k), or other more complex regulatory pathways (e.g., Example A2.2). **(B):** Mediator confounding occurs, in the simplest case, when an unmeasured factor (a confounding variable C; U, for unmeasured variable, is also used in the literature) impacts both T_i and T_j. In this setting, estimates of the direct and indirect effects can be biased.

observation is that *cis*-mediation is common, but usually is only partial (Example 21.3). One explanation is **mediator confounding** (Figure 21.3), wherein some unmeasured variable (a confounding variable, C) impacts both the *cis* mediator (T_i) and the *trans* transcript (T_j). This can arise when the eSNP used has imperfect LD with the causal SNP, or when measurement error impacts estimates of T_i and/or T_j (Pierce et al. 2014; Yang et al. 2017). Pierce et al. (2014) and Yao et al. (2017) found that many human *trans*-eQTLs are also *cis*-eQTLs for local genes. While partial mediation was common for many of these, complete mediation was rare.

What is the current big-picture view of the QG of RNA expression levels? The bulk of work on eQTLs comes from two very distinct biological systems: multicellular humans and unicellular yeast (recent summaries in Albert and Kruglyak 2015, Albert et al. 2018, and GTEx Consortium 2020). Given the rather modest sample sizes of most studies, a substantial number of additional eQTLs, especially *trans*-eQTLs, are expected to be detected as sample size increases. Despite this limitation, the conclusion is inescapable that eQTLs are ubiquitous, and control of expression for almost all transcripts is polygenic, with *trans* effects generally contributing between two and four fold more variation than *cis* effects (Liu et al. 2019). In both humans and yeast, almost all major transcripts have associated *cis*-eQTLs. In humans, this is greater than 95% for protein-coding regions and 67% for lincRNAs (GTEx Consortium 2020), while most yeast protein-coding genes have detectable *cis*-eQTLs. Further, in humans allelic heterogeneity is common for *cis* sites, with multiple independent *cis* variants being the norm.

The situation for *trans*-eQTLs is a bit more complex. These tend to have much weaker individual effects than *cis*-eQTLs, and are currently vastly undercounted. In humans, roughly one-third of detected *trans*-eQTL show some *cis*-mediation (GTEx Consortium 2020). Further, *cis*-eQTLs tend to be more **tissue-sharing**, while *trans*-eQTLs are more **tissue-specific** (Price et al. 2011; GTEx Consortium 2017; Liu et al. 2017), indicating that even more are awaiting discovery as the tissue pool expands. Finally, *trans*-eQTLs are more enriched for known GWAS hits than are *cis*-eQTLs, which themselves are enriched for GWAS hits relative to random frequency-matched SNPs. Hubs (*trans*-eQTLs that impact numerous transcripts) are seen, but generally impact only a modest number of transcripts.

In contrast to humans, the vast majority (90%) of *trans*-eQTLS effects in yeast map to just around 100 hubs. The median number of transcripts impacted by a yeast hub is 425, with a range of 26 to 4,600 (Albert et al. 2018). Four of these hubs impact over half of the genes showing variation in expression. In contrast to the highly structured *trans* hubs of yeast, the hub structure of humans is far weaker and much more diffuse. Part of our current view of the human hub structure is likely impacted by low power. Indeed, highly structured hubs have been seen in other line-crossed eQTL mapping experiments with model organisms (mice, rats, *Arabidopsis*). In the line-cross setting, segregating alleles have equal frequencies (Chapter 18), as opposed to a GWAS setting where there is lower power to detect effects associated with rare alleles (Chapter 20). Despite these issues, at present it appears that humans and yeast may have rather different *trans* structures.

Example 21.3. As developed by a number of investigators, one can use conditional regressions (path analysis methods; Appendix 2) to both detect, and quantify, the amount of mediation that gene i has on transcript j (Chen et al 2007; Jiang et al. 2013; Pierce et al. 2014; Yang et al. 2017; Yao et al. 2017; Shan et al. 2019; Yao et al. 2020; Wu et al. 2023). This is done using a nested series of regressions to establish causality. Using the notation in Figures 21.2 and 21.3, first consider the association between the dosage of SNP i (the minor allele copy number, N_i = 0, 1, or 2) and the transcript associated with coding region j (T_j),

$$T_j = \alpha_1 + \beta_1 N_i + e_1 \tag{21.1a}$$

One declares SNP i to be a *trans*-eSNP for coding region j when the slope β_1 is significant. This slope measures the **total effect** of SNP i on T_j, the contributions from both direct effects and indirect effects (such as through T_i). Next, we declare SNP i to be a *cis*-eSNP for coding region G_i when the regression

$$T_i = \alpha_2 + \beta_2 N_i + e_2 \tag{21.1b}$$

has a significant slope. Similarly, we declare the T_i has an effect on T_j when β_3 is significant for the regression

$$T_j = \alpha_3 + \beta_3 T_i + e_3 \tag{21.1c}$$

Significant slopes in the above three regressions establish that: (i) SNP i is associated with T_j ($\beta_1 \neq 0$); (ii) SNP i is associated with T_i ($\beta_2 \neq 0$); and (iii) T_i is associated with T_j ($\beta_3 \neq 0$). These univariate regressions, by themselves, do not separate direct from indirect effects. To do so, a multiple regression of T_j is constructed based on both N_i and T_i,

$$T_j = \alpha_4 + \beta_4 T_i + \beta_5 N_i + e_4 \tag{21.1d}$$

If $\beta_5 = 0$, then any effect from SNP i on T_j is simply through its effect on T_i, namely, **full mediation** (the effect of SNP i on T_j is *entirely through* its *cis*-effect on T_i). When both β_4 and β_5 are significant, then **partial mediation** occurs, where *both* T_i and SNP i (the latter through a path *independent* of T_i; Figure 21.3A) impact T_j. Note that this logic need not be restricted to just transcripts, one could measure (say) P_i, the level of protein from gene i, or some other regulatory feature such as methylation, splicing, etc. Modifications of permutation tests to accommodate the correlation structure of mediation analysis are discussed by Jiang et al. (2013) and T. Wang et al. (2020). An excellent overview of mediation analysis is given by Otter et al. (2018).

From the theory of path analysis (Appendix 2), the indirect effect of SNP i on T_j through the path given by T_i, is just the product of the path coefficients, which turns out to be $\beta_2 \cdot \beta_4$ from the above regressions. As shown in Figure 21.3A, the total path effect β_1 assumes the potential of a direct effect β_5 from eSNP$_i$ to T_j ($N_i \rightarrow T_j$) and an indirect of eSNP$_i$ via paths through T_j ($N_i \rightarrow T_i \rightarrow T_j$) with effect $\beta_2 \cdot \beta_4$. Hence, the proportion of the total effect on T_j from eSNP$_i$ mediated via T_i is

$$(\beta_1 - \beta_5)/\beta_1 = \beta_2\beta_4/\beta_1 \tag{21.1e}$$

If there are no unscored correlated factors that impact members of this trio, then the relation $\beta_1 = \beta_5 + \beta_2 \cdot \beta_4$, namely total effect = direct effect plus indirect effect, should hold. If it does not, one is likely missing correlated elements (confounders). Figure 21.3B shows one example. Such confounding could be caused by the focal eSNP$_i$ being in LD with different causal SNPs for the *cis* effect on T_i and the *trans* effect on T_j (Pierce et al. 2014).

Negative values of Equation 21.1e are commonly observed (e.g., Yang et al. 2017), indicating mediator confounding (Figure 21.3B). Even when one has a candidate list of potential confounders to test (e.g., age, sex, environment risk factors, etc.), two variable selection issues arise. First, even assuming that the correct potential confounder variables have been identified, testing all of them in a single regression greatly lowers power. Second, and more problematic, confounder variables can vary over both trios (eSNP$_i$, T_i, T_j) and the tissues in which a trio is tested. Yang et al. (2017) developed **Genomic Mediation analysis with Adaptive Confounding adjustment** (**GMAC**) to address this concern, a trio-tissue specific variable selection approach to choose the appropriate confounder variables (from some candidate list).

One complication with using mediation analysis is that measurement error can result in a false positive for partial mediation when in fact no mediation is present (reviewed by Gastonguay et al. 2023). An alternative approach is to place mediation analysis into a Bayesian model selection framework (Crouse et al. 2022). Consider Figure 21.3A, which has three potential causal paths (β_5, β_2, β_3). A Bayesian analysis would assign posterior probabilities to these various links, such as support for the direct effect (a significant probability that β_5 is nonzero). Complete mediation has this path equal to zero, with the β_2 and β_3 paths being significantly nonzero, while partial mediation has all three paths being nonzero. A Bayesian model selection would assign posterior weights to different models (different set of nonzero paths), and return overall probabilities on the likelihood of various models. This is a more flexible approach and can offer additional biological insight into more problematic cases.

The Epigenome and More General Regulatory QTLs (regQTLs)

The same logic used for mapping QTLs for protein- and transcript-level regulation can be applied to other regulatory features, such as methylation (Gibbs et al. 2010; McRae et al. 2018; Wu et al. 2018), DNAse I sensitivity (Degner et al. 2012), and mRNA splicing (Battle et al. 2014; Qi et al. 2022). Indeed, any genomic or cellular feature that can be quantified (assigned a numerical value, such as a binary 1/0 on-off score) can be treated as a quantitative trait. We refer to QTLs (or SNPs) associated with a scored regulatory feature as a **regulatory QTL** (**regQTL** or **regSNP**). Table 21.1 lists some of the different types of QTLs/SNPs that have been mapped using functional genomics features. Note that these classes are not exclusive, as an eQTL might be the result of regulatory variation at splicing (sQTL), DNAse I sensitivity (dsQTL), methylation (methQTL), or any number of other steps (Example 21.4). regQTLs represent a key step in the merging of QG and functional genomics, a union we refer to as **quantitative genomics**, the natural extension of genetical genomics. The machinery of quantitative genomics provides a powerful analytic framework for extracting signals from the growing tsunami of functional data.

As with eQTLs, regQTLs are conditional, potentially having tissue or developmental-state specificity. For example, even when using the same tissues, one could be scoring a functional feature under steady-state behavior in some settings and under the dynamic response following some environmental perturbation in others. Similarly, ascertaining the direction of causality for regQTLs remains a major problem. A phenotype could *cause* a change in some functional feature, rather than being the *result* of that feature. **Mendelian randomization** (Example 21.10; Appendix 2) offers one approach for assigning causality.

An especially interesting set of regulatory features is the **epigenome** (Susuki and Bird 2008), the structure of the chromatin (and associated binding proteins), as well as other features, that distinguish different tissues between cells with otherwise identical DNA sequences. Examples include methylation of DNA CpG sites, acetylation and methylation of histones, and DNAse I sensitivity regions (Strahl and Allis 2000; G. Wang et al. 2007; Chi et al. 2010; ENCODE Project Consortium 2012, 2020; Romanoski et al. 2015). A growing number of studies highlight the importance, and potential use, of the epigenome. For example, Rakyan et al. (2011b) examined differences in the DNA methylation patterns (the **methylome**) between monozygotic twins (MTZ) that were discordant for childhood-onset type 1 diabetes (T1D). Despite being genetically identical, in discordant twins, one sib displays the disease while the other does not (discordant MTZ are expected to be common unless the disease prevalence is high; Yang et al. 2010c; Visscher et al. 2012b). Rakyan

Table 21.1 A few of the different classes of QTLs (SNPs). The general terminology is to use QTLs generically, especially in a linkage-based analysis to indicate a region, and SNP in a GWAS setting to refer to a SNP showing an association. The QTL/SNP terminology is a bit idiosyncratic, with different versions for some of these abbreviations appearing in the literature.

acQTL/acSNP	Chromatin acetylation QTL/SNP
apaQTL/apaSNP	Alternative polyadenylation QTL/SNP
aseQTL/aseSNP	Allele-specific expression QTL/SNP
caQTL/caSNP	Chromatin accessibility QTL/SNP
cis-xQTL/*cis*-xSNP	*Cis* (local) QTL/SNP for feature x
dsQTL/dsSNP	DNAse I sensitivity QTL/SNP
eQTL/eSNP	RNA expression QTL/SNP
epiQTL/epiSNP	Chromosome epiallele QTL/SNP
hQTL/hSNP	Histone QTL/SNP
haQTL/haSNP	Histone acetylation QTL/SNP
hmQTL/hmSNP	Histone methylation QTL/SNP
meQTL/meSNP	DNA methylation QTL/SNP
methQTL/methSNP	DNA methylation QTL/SNP
miR-QTL/miR-SNP	MicroRNA QTL/SNP
pQTL/pSNP	Protein expression QTL/SNP
pb-xQTL/pb-xSNP	Population-based QTL/SNP for feature x
QTN	Quantitative trait nucleotide
QTT	Quantitative trait transcript
rQTL/rSNP	Ribsome occupancy QTL/SNP
regQTL/regSNP	Regulatory QTL/SNP
sQTL/sSNP	Splicing QTL/SNP
sb-xQTL/sb-xSNP	Sex-based QTL/SNP for feature x
tQTL/tSNP	Trait QTL/SNP
trans-xQTL/*trans*-xSNP	*Trans* (distal) QTL/SNP for feature x
vQTL/vSNP	Variance QTL/SNP

et al. were able to detect 132 CpG sites, differentially methylated between discordant pairs, that correlated with T1D status. Such an association does not imply causality, as cellular perturbations from T1D might *induce* these changes, rather than these changes *causing* T1D. Rakyan et al. then used an independent data set to show that many of these methylation differences could be seen *prior* to the expression of the disease in case individuals.

If environmentally induced methylation events are persistent, then (provided the correct tissue is chosen), one could (in theory) search the methylome for *signatures of past environmental exposures*. Human epigenetic markers can persist over a lifetime (Shah et al. 2014). For example, Heijmans et al. (2008) found that, even after 60 years, Dutch exposed prenatally to famine during a food embargo near the end of World War Two still showed reduced methylation (in peripheral blood) at the insulin-like growth factor II gene relative to appropriate control groups. Breitling et al. (2011) found a single locus that was under-methylated (again, using blood) in smokers (relative to nonsmokers) in an initial sample of 177, and were able to replicate their finding using an independent sample of 316 individuals. Hence, in at least some cases, the epigenome may display persistent signals of past environmental events and exposures. A genome-wide scan for epigenetic features associated with particular traits has been referred to as a **epigenome-wide association study**, or **EpWAS** (EWAS is also widely used in the literature, but we reserve this for expression-wide studies) or a **methylome-wide association study** (**MWAS**). Issues when conducting an EpWAS are reviewed by Rakyan et al. (2011a).

Example 21.4. Two interesting studies attempted to dissect the impact of *cis* variation at various stages in the regulatory pathway (from gene to transcript to protein) using a collection of lymphoblastoid cell lines extracted from 72 Nigerian Yorubas (Battle et al. 2015; Li et al.

2016). The use of cell lines allowed for the detailed measurement of a number of genomic and cellular features, while SNP genotyping of these lines allowed for a GWAS to be conducted for each scored feature. The GWAS was restricted to *cis* (local) QTL by restricting the search to a 20-kb window around the target gene. Battle et al. used RNA-seq to measure standing mRNA levels, ribosome profiling (**ribo-seq**) to measure translation (via occupancy rates, a measure of transitional rates and efficiency; Ingolia et al. 2009), and mass spectroscopy to measure the associated protein levels. Despite the very low power of their design (given their very small sample size), they still found numerous regulatory QTLs at all three steps: 4,400 genes were tested for protein abundance, detecting 300 *cis*-pQTLs; 15,000 genes were scored for ribosome occupancy, yielding 930 *cis*-rQTL; and 16,600 genes were scored for mRNA levels, finding 2,400 *cis*-eQTLs. Among the 4,300 genes scored for all three regulatory phenotypes, 66% of eQTLs had some overlap with both downstream rQTLs (66%) and pQTLs (35%), while 52% of rQTLs overlapped with their corresponding pQTLs. While the eQTL and rQTL effects (when both were present) tended to have a similar effect size, pQTL effects tended to be diminished relative to their upstream eQTL effect. Thus the signal from a change in mRNA expression tends to be at least somewhat buffered when it reaches the protein level. Finally, they detected a class of *cis*-QTLs that influenced protein levels with essentially no effect on mRNA levels (pQTLs that were not eQTLs), suggesting that these might arise via posttranslational regulation.

Li et al. (2016) conducted a more granular analysis using these same lines, adding several new features in addition to those measured by Battle et al. (2015). They examined potential regulatory steps upstream from the transcription start site (TSS) by considering several markers for chromatin modification relating to transcription-factor accessibility. The first was a putative enhancer signal, acetylation of the lysine at position 27 on Histone 3 (H3K27ac) around the TSS, which is associated with enhanced transcriptional activity (Wang et al. 2016a). (Methylation of this lysine, H3K27me3, and of the lysine at position 4, H3K4me1 and H3K4me3, are associated with silencers, enhancers, and promoters, respectively.) They also scored methylation levels and DNAse I sensitivity. While RNA-seq measures the steady-state level of transcript, the actual *rate* of transcription was scored (**4sU-seq**) by using pulse-labeled uridine (s4U). Their RNA-seq analysis included isoform ratios for alternatively spliced products. Regulatory QTLs were detected at each of these steps. They found that around 65% of all eQTLs have effects on chromatin. They also found an uncoupling of expression level and splicing in that sQTLs and eQTLs tend to be independent (only 14 of 275 genes shared the same lead SNP for both). Further, eQTLS are enriched around the TSS, while sQTLs are enriched within the body of the coding sequence (in particular, within the introns that they regulate). While not influencing expression level, 90% of sQTLs generated variation in the final protein sequence. Finally, they noted that GWAS SNPs for four diseases and two traits were enriched for both eQTLS (as noted above) *and* sSNPs, and that the latter appeared to have an effect at least as great as the former. Hence, variation in alternative splicing rates may be an important player underlying quantitative variation.

Deeper insight into the distinction between eSNPs and sSNPs is offered by the analysis of Qi et al. (2022), who examined RNA-seq data from brain cortex samples of roughly 2,500 individuals of European extraction. They noted that sSNPs (changes in the ratio of splicing isoforms, namely alternative transcripts) can occur without changes in overall expression level. Hence, decoupling between sSNPs and eSNPs is possible. How widespread was this? They found 12,800 genes with associated sSNPs (they called these sGenes), while 16,700 genes had associated eSNPs (denoted as eGenes). Of the sGenes, 73% were also eGenes, while among the eGenes, 56% were also sGenes. Hence, simply using eGenes as proxies for sGenes will miss many regulatory regions. On a more granular scale, they found that the actual regSNPs within these genes were more distinct, with 61% of detected sQTLs being distinct from eSNPs, so that simply relying on eSNPs does not capture these associations. While eSNPs were more enriched for TSS, sSNPs were more enriched for RNA-binding protein sites (RBP). Overall, Qi et al. found that the effects of sSNPs were comparable to sSNPs on the heritability of gene expression levels.

Epialleles and epiQTLs

With differential chromatin marking (such as **differentially methylated regions**, **DMRs**) comes the possibility of **epialleles** (Johannes et al. 2008), identical DNA sequences that have functional differences (for a focal trait) based on epigenetic factors. If such marked alleles can be stably transmitted across generations (i.e., they are not reset back to some base state by meiosis or other features of gametogenesis), they can generate cross-generation trait correlations.

One approach for examining the impact of epialleles on quantitative trait variation is the use of **epigenetic recombinant inbred lines** (**epiRILs**), lines that are essentially genetically identical but that differ in methylation patterns. Johannes et al. (2009) constructed such a set of epiRILs in *Arabidopsis thaliana* by crossing otherwise genetically identical lines that differ at the *DDM1* gene. The wild-type allele at this locus is involved in the maintenance of DNA methylation and the silencing of transpose elements (TEs), while mutant *ddm1* plants show a 70% reduction in methylation and some TE movement. The authors found that around 50% of the undermethylated regions in the *ddmq* background regained methylation over two to five generations, and hence most resulting epialleles were not maintained over more than a few generations. However, some sites maintained methylation differences over at least eight generations. Further, there was significant among-line variation in both flowering time and plant height. Some of these differences were likely due to sequences changes from the mobilization of TEs in the *ddm1* background, but others appear entirely to be the result of stable epialleles. A follow-up study by Cortijo et al. (2014) found that one could use standard QTL mapping (using differential methylation states as the "alleles") to detect and map **epiQTLs**, allowing sequencing to rule out genetic changes (i.e., TE insertions/deletions) within these regions. They found that epigenetic changes accounted for 60% to 90% of the among-line differences in their epiRILs for flowering time and root length.

Associations Between regSNPs and GWAS Hits

A common theme throughout Chapter 20 is power, with replicability improving and more tagged SNPs being detected (with better accuracy) as n increases. This concern clearly extends to functional-genomic traits, where historically sample sizes have been modest (typically under a few thousand) relative to more traditional GWAS traits. Indeed, sample-size wise, current functional genomic studies are about two decades behind standard-trait GWAS. Hence, we expect that—as was the case for traditional GWAS—the larger samples sizes of future studies will provide more insight into the roles of, and connections between, the various categories of regSNPs (Table 21.1). For example, Qi et al. (2022) noted that the correlation between eGenes and sGenes (Example 21.4) increased with n, so that weaker associated effects (between splicing and expression levels) become more apparent as sample size increases.

Mostafavi et al. (2023) build on this observation to suggest that, under current sample sizes, regSNPs (functional genomic) and more traditional GWAS (trait-based) hits are "systematically biased toward different types of variants." The general hypothesis is that most GWAS hits are regulatory, and thus these should eventually (in most cases) correspond to hits in various regSNP categories. However, at least for eSNPs, this is currently not the case. Mostafavi et al. noted that only around 40% of GWAS hits colocalize with eSNPs. They also noted differences in the genomic features of the regions harboring GWAS and eSNP hits. The former are enriched for regions with functional annotation (i.e., more GO terms), show signatures of stronger selective constraints, and have complex regulatory patterns over different tissues (more network connections; Appendix 2). The latter (as least for *cis*-eSNPs) are clustered near TSS, are associated with genes showing relaxed selective constraints, tend to have little associated functional annotation, and show simpler regulatory landscapes.

Mostafavi et al. suggest that difference in the strength of selection acting on the *detected variants* might underlie many of these differences. Their basic idea is that alleles of stronger effect are held at lower frequencies by selection, and thus must show larger effects to be detected. Suppose that the strength of association between a genotype and a transcript is β, and the association between the transcript and trait is γ, namely

$$\text{Genotype} \xrightarrow{\beta} \text{expression} \xrightarrow{\gamma} \text{phenotype}$$

Hence, the strength of the association between genotype and trait (a typical GWAS hit) is $\beta\gamma$, while the strength of association with eSNP hits is β. The target of selection is the final trait, for an effect of $\beta\gamma$, with an associated site variance of $2p(1-2)\beta^2\gamma^2$. As we discuss shortly, selection tends to "flatten" (make more uniform) site variance (Simons et al. 2018; O'Connor et al. 2019). This occurs because larger values of $\beta\gamma$ (and hence stronger selection) result in smaller p values, with the increased effect size partly canceling the reduction in p, making the variation per site more uniform. Because the power to detect a SNP effect is a function of its variance (Chapter 20), and because regSNP studies tend to have much smaller sample sizes, only sites with large β values have a sufficiently large variance to be detected (e.g., sites very close to their target). Conversely, with the much larger sample sizes of trait-based GWAS, much smaller variances can be detected, allowing sites with smaller values of $\beta\gamma$ to be detected. Traits with larger values of γ are (generally) under stronger selection, and hence eSNPs at such genes (even with modest β values) would be selected against, resulting in a bias for large β values only at genes with smaller γ values having a detectable signal (Yao et al. 2020). The result is a biased sampling of eSNPs (large β, small γ) relative to GWAS. This bias will decrease as the sample size of regSNP studies approaches that for GWAS.

What is the Nature of GWAS Noncoding SNPs?

One of the surprising observations from GWAS studies was that the vast majority of significant "hits" (SNPs/CNVs statistically associated with trait variation) fall into noncoding regions, and thus are not the result of structural changes (variation in the amino acid sequence of coded proteins). While, by default, these GWAS hits are likely regulatory in nature, can we associate noncoding hits with *specific* regulatory features? One could examine a few model traits in detail for such associations, but these results might be somewhat idiosyncratic, and not that representative of traits in general. A more powerful approach is to use catalogues of noncoding SNP hits assembled over a wide array of traits and diseases to see if these SNPs are enriched for regulatory features (such **enrichment tests** will be discussed shortly). For example, Maurano et al. (2012) showed that a collection of over 5,600 noncoding human SNPs for 200 different diseases and 450 different traits preferentially mapped to DNase I hypersensitive sites (**DHS**) as scored using the ENCODE collection of 85 cell-tissue types. DHS sites are a strong regulatory signal, indicating an open chromatin structure that allows the transcriptional machinery to have greater accessibility to genes. Trynka et al. (2013), Pickrell (2014), Farh et al. (2015), and Gusev et al. (2018) reported similar observations using other human chromatin markers.

An even more direct regulatory connection is the observation in humans that collections of GWAS SNP hits (involving numerous traits and diseases) are significantly enriched for eQTLs. This is especially remarkable in that most of the early studies scored eQTLs using just a single cell line/tissue type, yet the detected eSNPs were enriched for GWAS hits over a wide variety of traits (e.g., Nicolae et al. 2010; Fehrmann et al. 2011; Westra et al. 2013; Battle et al. 2014; Bryois et al. 2014; GTEx Consortium 2017, 2020; Yao et al. 2020; Barbeira et al. 2021). One might imagine that focusing attention on only a single tissue (such as lympholbastoid cell lines) or peripheral blood (a mixture of a limited number of cell types) would only capture a small fraction of eQTLs over the whole organism, and yet even with this restriction, a strong enrichment emerged. Later studies using eQTLs detected over a variety of tissues similarly showing strong enrichment for GWAS SNP hits over a variety of diseases and traits. For example, Gamazon et al. (2018) used a bank of 44 human tissues from 450 individuals, finding that 60% of known trait-associated SNPs were in LD with *cis*-eQTLs detected over one (or more) of the cell lines. They classified detected eSNPs as either **tissue-shared**, or **tissue-specific**, eSNPs, finding that for most (but not all) traits, tissue-shared eSNPs (i.e., detected in more than one tissue) accounted for a greater proportion of trait associations than did tissue-specific eSNPs. Further, at least in humans, local eQTLs tend to be more tissue-shared, while distal eQTLs are often tissue-specific (Price et al. 2011; GTEx Consortium 2017; Liu et al. 2017).

Example 21.5. An important cautionary tale on fine-mapping was offered by Smemo et al. (2014). A set of roughly 90 variants in very high LD that map within a 47 kb region spanning introns 1 and 2 of the *FTO* gene had very strong, and highly reproducible, GWAS hits for human obesity (measured by body mass index, BMI). Individuals homozygous for risk alleles averaged more than 3 kg heavier than individuals homozygous for non-risk alleles. Deletion of *FTO* in mouse models results in leaner mice, while mice overexpressing *FTO* are heavier. Finally, this 47kb region is heavily enriched with *cis*-acting control factors (enhancers, repressors, DNAse I sensitivity sites, TF binding sites). However, *none* of the variants within this region map as eQTLs for *FTO* expression. Smemo et al. found that this region is involved in chromatin looping to a region over a megabase away containing the gene *IRX3*. In a human EWAS using brain tissue, 11 of the *FTO* SNPs associated with BMI were also eSNPs for *IRX3*, but not *FTO*, expression. Further, of the eSNPs associated with *IRX3* expression in either brain or mature adipose tissue, only those expressed in the brain showed highly significant associations with BMI. Hence, the *FTO* GWAS hits appear to be distal eSNPs that impact expression levels of *IRX3* in the brain. The apparent colocalization of *FTO* GWAS hits and mouse knockout effects gave a misleading picture of how these specific causal sites influence human body mass. Further, focusing expression studies solely on one obvious target, mature adipose tissue, would have missed this signal.

An independent study by Claussnitzer et al. (2015), using gene editing in human tissue cultures, offered a rather different finding, highlighting the subtleties of tissue choice. They found strong effects of a particular SNP variant (rs1421085) within this *FTO* region on the expression of *IRX3* and the nearby *IRX5* gene in *precursor* adipocyte cells, resulting in a switch from fat burning to fat storage. This variant disrupted a repressor within this region (*ARID5B*), resulting in the activation of a rather potent early adipocyte enhancer and a doubling of *IRX3* and *IRX5* expression early adipocyte differentiation. Thus, there appear to be potentially several different gene circuits (with different tissue specificity) influencing BMI from genes some distance from the location of the GWAS hits. The different, but not necessarily exclusive, conclusions from these two studies highlight the concern stressed by Barbeira et al. (2018) that researchers need to adopt a more **agnostic scanning** approach when assessing correlations between expression levels and trait values.

A variety of approaches have been proposed to functionally associate a SNP with a particular gene (**S2G**, for **SNP-to-gene**), such as whether the SNP is within an exon, the distance to the TSS, associated fine-mapped *cis*-eQTL, and promoter capture or assessable chromatin information. These approaches are all restricted to detecting relatively closing-acting *cis* regulatory effects. A number of strategies are reviewed by Gazal et al. (2022), who propose an optimal weighting for combining several S2G metrics into a single index.

TRANSCRIPTOME-WIDE ASSOCIATION STUDIES (TWAS)

Quantitative Trait Transcripts, QTT

The logical complement to an eQTL study is to directly search for **quantitative trait transcripts, QTTs** (Passador-Gurgel et al. 2007), whose expression levels are correlated with trait values. At first blush, this idea seems very straightforward. Starting with a sample consisting of both expression and trait data, simple modifications of a variety of GWAS approaches (Chapter 20) can be used to search for QTTs. The resulting analysis is often called a **transcriptome-wide association study** or **TWAS**. We note that there are two flavors of TWAS, based on either using *observed* transcript values (which we denote as an **oTWAS**) or using *predicted* transcript values (given some reference set of eQTLs), a **pTWAS**. A pTWAS tests association based on *genetically predicted* expression levels, while an oTWAS tests association using the *total expression* level (genetic plus environmental contributions). Most TWAS studies, especially in humans, are pTWAS, and indeed the common literature use of TWAS refers to this class of analysis. We examine pTWAS in detail shortly, framing our initial discussion by assuming we have observed transcript values in some appropriate tissue (or set of tissues) given our trait.

Regardless of how transcript values are obtained, a TWAS scores each primary transcript (as opposed to a GWAS scoring each SNP), with the resulting Manhattan plot showing expression level-trait association p values at each transcriptional coding location. One simply changes the predictor variables, replacing the GWAS discrete variable of gene dosage (minor allele copy number, N) with the continuous variable of expression level (T) of a given transcript (which can be extended to the constellation of splicing variants for a given coding region). For example, standard linear regressions can be used for continuous traits, with

$$z_i = \alpha_k + \beta_k \, T_{i,k} + e_i \tag{21.2}$$

where z_i is the phenotype of individual i and $T_{i,k}$ is the transcript amount of gene k in i. As in Chapter 20, the simple intercept α_k can be extended to a set of fixed effects to account for any relevant cofactors. Similarly, with binary traits, one could use logistic regressions (Equation 20.3). The dimensionality of the TWAS test set is on the order of the number of transcripts (in the low tens-of-thousands), as opposed to the number of SNPs (often in the millions), and thus has a much lighter multiple-comparison burden (essentially a gene-based GWAS; Chapter 20). This burden is likely even smaller than the number of transcripts, due to their high correlation structure (e.g., Peterson et al. 2016b).

Unlike a traditional GWAS, a significant transcript effect in a TWAS has immediate biological interpretation. In a SNP-focused GWAS, the sign of a significant SNP effect typically does not convey much of a biological meaning. However, a positive β_k from Equation 21.2 implies that upregulation of transcript T_k increases the trait value (or disease risk), while a negative β_k implies that downregulation of T_k increases the trait value (or risk). As noted by Gamazon et al. (2015), this has immediate therapeutic implications, as targeted downregulation of a gene is usually easier to achieve that targeted upregulation. Hence, genes whose transcripts have positive β_k are better potential targets of therapeutic intervention, as reducing the level of these transcripts would decrease disease risk.

Several important statistical considerations arise when moving from a GWAS to a TWAS. Like SNPs in LD, the predictor variables in a TWAS (transcripts) can be correlated, and one must account for their covariance structure. In the case of SNPs, LD generates these correlations, which do typically not extend outside of a block of tightly linked sites. Trimming SNPs within such a block is often used to reduce the impact from high SNP colinearity. With transcripts, however, there is no genomic locational impact on the correlation structure, as transcripts from unlinked genes can be very highly correlated (coordinately expressed). There are two other, rather subtle, complications when moving from SNPs to transcripts. First, while the SNP genotype of an individual does not change over the environment, *this is not the case with transcripts*. Different environmental exposures can impact which genes are expressed ($E \to T$), and also impact the trait, *independent* of transcript levels ($E \to z$). Such an environmental effect would be a confounder (Figure 21.3B), generating a false associate between transcript level and trait value ($T \leftarrow E \to z$). An appropriate choice of cofactors can often mitigate this effect.

The second (closely related) TWAS complication is that while the direction of GWAS causality between SNP and trait is always unidirectional (SNP $\to$ trait), such that an individual's trait value cannot change their SNP value, *causality can flow in both directions for a trait–transcript pair* ($T_k \to z$ and $z \to T_k$). Flow in the $z \to T_k$ direction is often called **reverse causality**. When the SNP is correlated with both T and z, three base models are possible (Schadt et al. 2005): **causal** (SNP$\to T_k \to z$), **reactive** (SNP$\to z \to T_k$), and **independent** ($z \leftarrow$ SNP $\to T_k$). Consider weight. An environment feature (such as diet) may have a strong impact on trait value, independent of any underlying transcript values. Conversely, as weight increases, the expression levels of genes can change *as a result of the trait value*. The resulting trait–transcript association is driven by the *trait*, not the *transcript* (the reactive model).

The impact of high correlations among predictor variables differs between a GWAS and a TWAS. In a GWAS, a number of significant hits can be generated over a set of SNPs that are all in high LD with the same causal site. Fortunately, the locational clustering of these SNPs usually suggests this possibility. Conversely, a TWAS can have noncausal *transcripts* showing significant associations with a trait because of tight coregulation with a true causal transcript (for the trait). Unlike the case with SNPs, there is no obvious locational clustering to suggest that most are false positives (arising from correlations with a true positive). One potential approach for decoupling the effects of correlated transcripts is to construct the covariance matrix of all n_k transcripts correlated with a focal transcript T_k, and then extract the leading $m_k < n_k$ PCs from this matrix, and use these in a regression,

$$z_i = \alpha_k + \beta_k T_{i,k} + \sum_{j=1}^{m_k} \beta_{k,j} PC_{i,j,k} + e_i \tag{21.3}$$

Here $PC_{i,j,k}$ is the jth PC score for individual i of the covariance matrix for all transcripts associated with the focal transcript k. One could proceed in this manner over each member of the set of correlated transcript (using some threshold value for a minimum absolute correlation) to extract a subset of leading predictors, which could then be simultaneously fit in a multiple regression. Penalized regressions, such as the LASSO (Example 20.4), could also be used for model selection. Given the potential of very high transcript colinearity, these approaches are not fool proof. Further, if there are additional, but unmeasured cofactors (such as important environmental factors that influence both trait and transcript values), false associations can arise.

The above-mentioned methods of path and mediation analysis can be used to probe the causality structure (Appendix 2; Example A2.2), but such an analysis is only as good as the variable selection of possible confounders. As we will see next, one solution to this causality concern is to use *eQTLs as proxies for transcript values*, as changes in the trait value, or the value of other transcripts, will not change the eQTL genotype. This is the instrumental variable approach that Mendelian randomization (Appendix 2) uses to test the impact of some factor (such as high blood pressure) on a disease, free of the effect of confounders or reverse causality (Figure A2.7). If the factor has a direct effect on the trait, genotypes that influence that factor should also influence the downstream trait value.

Example 21.6. Kirst et al. (2004) measured growth rate in clones of 91 backcross individuals from a *Eucalyptus* cross (F_1 offspring of *E. grandus* × *E. globulus* were backcrossed to a *E. grandus* parent). This set of clones was also scored for expression levels at 2,600 genes via a microarray. After adjusting for multiple comparisons, 26 transcripts were significantly correlated with growth, with an additional 11 transcripts added using a slightly less stringent threshold. All of these transcript levels were negatively correlated with growth (lower mRNAs levels were seen in faster growing individuals). A single transcript in the lignin biosynthetic pathway explained 38% of the growth variation, and the majority of the other significant transcripts coded for enzymes in this pathway. An independent analysis found lower amounts of lignin in the faster growing clones. All, save one, of the significant transcripts associated with the lignin pathway were influenced by a single *trans*-acting eQTL on linkage group 9. A second *trans*-eQTL on linkage group 4 also influenced the majority of these transcripts. These two eQTLs colocalized with two QTLs independently mapped for growth (however, confidence intervals for QTL/eQTL locations were very large due to small sample sizes).

TWAS Using SNP-predicted Transcript Values (pTWAS)

The term TWAS as it commonly appears in the human literature refers to a GWAS setting where one has SNP and trait, but no expression, data. To be consistent with this more common use in the literature, in what follows we use TWAS in place of pTWAS. The idea is to leverage the SNP-trait data (without performing any additional genomics) by using the eQTLs previously detected in some **reference transcriptome** population, and then (based on the observed SNP genotypes of an individual) predict the expected values for their unmeasured transcripts (Gamazon et al. 2015; Gusev et al. 2016). In essence, this is a form of *imputation* (Chapter 20). Instead of using the observed SNP genotypes and some larger SNP reference collection to infer *unscored genotypes*, one instead uses the observed SNP genotypes and a reference collection of eQTLs, and then infers the *unscored transcript* values (Schadt et al. 2012 discuss the converse issue of predicting an eSNP genotype from an observed transcript value). For example, in humans one could use eQTLs detected using the **Genotype-Tissue Expression Project** (**GTEx**) set of roughly 50 different tissues from 900 individuals (GTEx Consortium 2017, 2020) or the **Genetic European Variation in Health and Disease** (**GEUVADIS**) set of 460 lymphoblastiod cell lines (Lappalainen et al. 2013). One concern is that eSNP effects are overestimated when power is low, and Huang et al. (2018) suggested using bootstrap approaches (Chapter 18) to obtain less biased estimates. See Wainberg et al. (2019) for a good overview of TWAS advantages, limitations, and best practices.

The massive leverage of a TWAS occurs because once a model for predicting transcript values is built, it can be applied to *any* GWAS study lacking expression data, *provided* two strong assumptions hold. First, that the LD patterns are very similar in the GWAS sample and the transcriptome reference populations. Note that this is the same general restriction as with imputation of missing SNP. Second, that the appropriate tissue type(s) are used as the reference for the trait of interest, or that expression levels in the target tissues have a high level of genetic correlation with the proxy tissue(s) scored. Provided that these assumptions hold, one can take previous (and future) GWAS studies and reanalyze each using TWAS. The resulting shift from testing all the SNPs to testing just the (predicted) transcripts is akin to moving from a SNP-based, to a gene-based, GWAS (Chapter 20), with a huge reduction in the multiple comparisons burden, and, as a result, a potential increase in power. However, any power gain from reduction in the burden of multiple tests could be more than offset by prediction inaccuracies of the expression levels (Huang et al. 2018). It is also worth noting that while we frame TWAS in terms of using local eSNPs to predict the impact of transcript levels on a trait, the exact same approach could be used for other regulatory features. For example, one could use local meSNPs to predict methylation levels, and then examine the impact of the predicted methylation site status on the focal trait.

The starting point for a TWAS is to first map eQTLs using the reference transcriptome data in the desired tissue(s), using SNPs proximal to the coding region in a statistical model to predict its transcript level (i.e., local eSNPs). Let $N_{i,k,j}$ be the number of reference alleles (typically the minor allele count, 0, 1, or 2 at each SNP) for individual i in the reference sample at a proximal SNP $1 \leq j \leq n_k$ for the coding region of transcript k. One then fits the linear model

$$T_{i,k} = \alpha_k + \sum_{j=1}^{n_k} \beta_{k,j} N_{i,k,j} + e_{ik} \tag{21.4a}$$

using a variety of approaches, such as using the **lead** (or **sentinel**) **SNP** (that with the largest effect), or more generally penalized-regression/model-selection approaches such as LASSO (Example 20.4). Distal eSNPs are generally not fitted because of model instability issues given their generally weaker effects coupled with the massive increase in model dimensionality. This statistical justification also has some biological justification, in that local eQTLs tend to be tissue-sharing, while distal eSNPs are often more tissue-specific (GTEx Consortium 2017; Liu et al. 2017; Urbut et al. 2019). Hence, one can use a proxy tissue when the real focal tissue (or tissues) for a trait/disease is unknown, as the proxy may often capture much of the local eQTLs (but could easily miss important distal eQTLs). For example, Qi et al. (2018) examined the correlation of both *cis*-eSNPS and *cis*-meSNPs effects between blood and brain tissue. Blood is far easier to obtain and score for genomic features than brain tissue, and thus much larger sample sizes can be used for the regulatory reference panels (in their study, 500 to 1,000 for brain tissues versus 2,000 to 14,000 for blood). They found a high correlation in effect sizes between the two tissue types, 0.70 for *cis*-eSNPs and 0.78 for *cis*-meSNPs. Hence, the reduction in accuracy in using blood for brain expression is more than offset by far greater reference population sample sizes. Ongen et al. (2017) discuss how to leverage information from an expression panel of different tissues (e.g., GTEx) to determine the causal tissue(s) for a given trait.

Letting n_{k*} denote the number of proximal SNPs retained in the final model for transcript k, the predicted value of transcript k in individual i from the GWAS (with no expression data) becomes

$$\widehat{T}_{i,k} = \alpha_k + \sum_{j=1}^{n_{k*}} \beta_{k,j} N_{i,k,j} \tag{21.4b}$$

with this value then substituted into an oTWAS ($\widehat{T}_{i,k}$ replaces $T_{i,k}$ in Equation 21.3). Gamazon et al. (2015) called this approach **PrediXcan**. More generally, the $\beta_{k,j}$ weights can vary over tissue type. For tissue type h, one can compactly write the vector $\mathbf{t}_i^{(h)}$ of the n_T predicted transcript values for individual i as the product of a tissue-specific $n_s \times n_T$ weight matrix, $\mathbf{W}^{(h)}$, and an n_s-dimensional column vector $\mathbf{g}_i$ of the SNP genotype scores (the $N_{i,k,j}$) for individual i, e.g., $\mathbf{t}_i^{(h)} = \mathbf{W}^{(h)}\mathbf{g}_i$. Because of the use of only local SNPs in the weight matrix, it is very sparse. The power of the TWAS approach is that $\mathbf{W}^{(h)}$ is estimated just once from the transcriptome reference populations, and can then be applied to all relevant GWAS studies (different sets of $\mathbf{g}_i$ vectors). Ideally, a separate TWAS is performed over each of the tissue-specific matrices. Xu et al. (2017) discussed the connections between TWAS and multiple-SNP gene-based GWAS tests (Chapter 20).

There are two shortcomings with this basic TWAS approach that were addressed by subsequent investigators. The first is that any uncertainties in the estimates of $\beta_{k,j}$ are not incorporated into the resulting TWAS (the $\widehat{T}_{i,k}$ values are assumed to be predicted without error). This limitation was addressed by the **collaborative mixed model** (**CoMM**) approach (Yang et al. 2018; Yeung et al. 2019), which *jointly estimates* the effects of a SNP on transcript levels in the transcriptome reference set and its effects on the focal trait in the GWAS. This allows for uncertainty in the predicted transcript levels to be accommodated for in the final TWAS. The cost of this approach is that $\beta_{k,j}$ values must be (jointly) estimated for each new GWAS, as compared to a single estimation that can be used on all future GWAS (using their $\mathbf{g}_i$ values). The second shortcoming is that Equation 21.4b assumes one has access to the full GWAS data set (in particular, the genotypes of each individual). Often one may only have access to the summary statistics, not the individual genotype data. As outlined in Example 21.7, a number of investigators have extended TWAS to the summary data setting.

While using predicted transcript values might appear to be less optimal than using their observed values, as noted by Gamazon et al. (2015), they have two important advantages. By predicting transcription levels solely on the basis of the genotype at the eSNP—what Gamazon et al. (2015) refer to as the **genetically regulated expression** component (**GReX**) of the transcript (eSNP $\rightarrow T_k \rightarrow z$)—any contribution from trait feedback ($z \rightarrow T_k$), which is included in the *observed* transcript value, is not in the TWAS predicted value. Similarly, by conditioning on the eSNP, any environmental contribution is also removed. Note that only common local eSNPs appear in the weights in Equation 21.4b, ignoring any genetic contribution from rare local eSNPs or distal eSNPs. TWAS is thus a test of a nonzero genetic correlation between local control of expression of a target transcript and a focal trait. As mentioned, TWAS is an application of the method of **Mendelian randomization (MR)**, testing for the effect of a factor (transcript T_k) on an outcome (trait value z), where the use of an instrumental variable (*cis*-eSNPs) that predicts T_j controls for bias from reverse causality and confounders (factors influencing both the transcript and trait, but not the genotype). Appendix 2 examines the MR approach in more detail.

Using the TWAS approach, Gusev et al. (2016) leveraged existing GWAS studies to discover 70 new genes associated with obesity-related traits (using expression in blood and adipose tissues as the reference transcriptomes). Their simulation studies showed that when the transcript level was due to multiple causal local eSNPs, TWAS offered greater power than either a standard GWAS or an **eGWAS** (a GWAS with increased power by only testing for associations using significant eSNPs).

Unfortunately, the TWAS approach based upon (predicted) genetically control transcript levels can also generate false positives in many settings (Mancuso et al. 2019; Wainberg et al. 2019). For example, suppose local eSNP$_j$ impacts both transcripts T_i and T_k (**pleiotropy**), but only T_k impacts trait value ($T_i \leftarrow$ eSNP$_j \rightarrow T_k \rightarrow z$). Gene i would still be declared as significant in a TWAS. An alternative setting (**linkage**) occurs when distinct SNPs (j and ℓ) impact the two different transcripts, but these SNPs are in high LD ($T_i \leftarrow$ eSNP$_\ell \leftrightarrow$ eSNP$_j \rightarrow T_k \rightarrow z$). Again, gene i would be declared significant under a TWAS. The **FOCUS** (**fine-mapping of causal gene sets**) method of Mancuso et al. (2019) attempts to remove correlations due to LD among eSNPs from different transcripts. Wainberg et al. (2019) suggested that best practices are to follow-up significant TWAS results by using colocalization methods to control for linkage (to be discussed shortly).

A final (but subtle) concern is that the power of expression studies detecting *cis*-eSNPs is often low, so that only eSNPs of large effects can be detected. If the result is just a single eSNP per transcript, then a TWAS essentially reduces to a candidate SNP association test. When multiple *cis*-eSNPs are found for any given transcript, this is not much of an issue.

Example 21.7. A common setting is that one has access to just the summary statistics from a trait GWAS, rather than the full SNP genotypes of all the study individuals (i.e., we do not know $N_{i,k,j}$). Similarly, the amalgamation of a number of individual GWAS into a single meta-population study usually returns just summary statistics for each SNP (Chapter 20). As shown by a number of investigators, one can still use the weights $\beta_{k,j}$ from Equation 21.4a on the local SNPs around a target transcript T_k to perform a TWAS. For example, the summary statistics extension of PrediXcan (**S-PrediXcan**) by Barbeira et al. (2018) proceeds as follows. As in Equation 21.4a, consider $1 \leq j \leq n_{k*}$ local SNPs around a transcript k. From the GWAS summary statistics, we have the estimated effects for these SNPs on the focal trait, $b_{k,j}$, and their sampling variances, $\sigma^2(b_{k,j})$. From the reference transcriptome, one can estimate the variance of transcript k, $\sigma^2(T_k)$, and either from the GWAS or an equivalent reference population, one can estimate the variance in reference copy number $\sigma^2(N_{k,j})$ for each SNP, which is a function of their population frequencies. An approximate z score for a TWAS for the coding region associated with transcript T_k is then given by

$$z_k \simeq \sum_{j=1}^{n_{k*}} \beta_{k,j} \left(\frac{\sigma(N_{k,j})}{\sigma(T_k)} \right) \left(\frac{b_{k,j}}{\sigma(b_{k,j})} \right) \tag{21.5}$$

where $\beta_{k,j}$ are the regression prediction weights from Equation 21.4a. See Barbeira et al. (2018) for a derivation. A number of such summary statistics-based TWAS approaches have been proposed, such as by Mancuso et al. (2017) testing for a genetic correlation between expression and the trait, the **Summary TWAS** (**S-TWAS**) of Gusev et al. (2016), and the **summary statistic collaborative mixed model** (**CoMM-S^2**) of Yang et al. (2020). The **UTMOST** (**unified test for molecular signatures**) test of Hu et al. (2019) provides a single unified approach for considering all tissues (in the reference set) simultaneously.

STATISTICAL APPROACHES FOR FINE-MAPPING CAUSAL VARIANTS

The high level of LD among blocks of variants that powers a GWAS is also the major impediment to unambiguously declaring a particular variant (or set of variants) to be causal. Depending on the structure of the study population, an LD block could be less than a kb (wild maize), tens of kb (humans, with Europeans having longer blocks than Africans), or over hundreds of kb (association panels of elite cultivars; Buckler and Gore 2007). A shorter LD block does not necessarily mean fewer candidate variants, as the nature of the evolutionary forces generating LD is such that populations with shorter LD blocks also tend to have higher levels of variation (WL Chapters 3 and 8). A further complication is added by recalling Example 20.1: the strength (r^2) of LD is a ***function of the allele-frequency matching between a marker and a causal site***. A nearby marker can actually have a *lower* r^2 value than a more distant marker. When attention is focused to an **association region** tagged with confidence in a GWAS (also called a **risk region** when mapping a disease gene), one is still left with tens to thousands of potential candidate variants. Within this region, p values likely *do not* monotonically become increasingly more significant as one approaches the causal site. The tagged region could also contain multiple causal variants (such as a collection of rare variants), further complicating the use of spatial distribution of p values as an aid for fine mapping of the causal site(s).

Even after sequencing the entire association block (so that *all* variants, including those that are causal, are scored), the **lead**, or **index**, **SNP** (that displays the most significant p value) within that block is ***likely not the causal variant***, especially when power is low and LD is extreme (Ledur et al. 2010; Udler et al. 2010; van de Bunt et al. 2015; Wu et al. 2017; Huang et al. 2018; Schaid et.al. 2018). Simulations by van de Bunt et al. (2015) assuming **whole-genome sequencing** (**WGS**) data still found that the lead SNP corresponded to the causal SNP only 80% of the time when the allele had high frequency and a strong effect ($p = 0.5$; odds ratio, OR, of 1.5), and less than 3% of the time when the allele was less common and of modest effect ($p = 0.05$, OR = 1.1). Hence, even with WGS data and a large population sample, determining the causal variants is far from trivial. The term **QTN** (**quantitative trait nucleotide**) has been used to declare a clear demonstration of a causal SNP (or some other variant, such as a CNV), but this has been very challenging to accomplish in most settings (see Example 21.8 for some exceptions).

Simulations by Wu et al. (2017) highlighted the impact of marker-causal allele frequency mismatch on fine mapping. They examined the distance between the lead SNP and the actual causal site under a variety of assumptions and genotyping schemes (WGS and imputation using different reference sets). Their simulations were based on WGS data of 3,600 individuals with roughly 18 million (retained) genetic variants. Randomly drawn common (minor allele frequency, MAF, > 0.01) and rare (MAF ≤ 0.01) variants were chosen to be causal, with an effect size ranging from 0.2 to 3% of the trait variance. GWAS was then performed using this underlying data and different sets of variants (imputed sets and WGS). When the causal allele was common, 95% of the lead SNP–causal site pairs had an MAF difference of < 0.05. For rare alleles, 95% had an MAF difference of ≤ 0.003. Hence, rare alleles are not tagged by common alleles (the LD, r^2, is just too small). In terms of mapping precision, under WGS, 80% of the lead SNPs for common causal variants were within roughly 10 kb from the causal site. With imputation (instead of WGS; Chapter 20), this distance increased to between 25 and 34 kb (depending on the reference population used). Hence, for common alleles, the mapping resolution using WGS data was only marginally better than using imputed data, and is usually not a cost-effective approach. In contrast, for rare variants, almost 95% of the lead SNPs were with 5 kb of the causal variant using WGS, but only 37% were this close using imputed data. Even assuming WGS data, mapping precision was a function of the causal allele frequency. With very rare causal variants ($p < 0.001$), 98% of the lead SNPs corresponded to causal sites, but this decreased to 30–40% for common causal alleles (MAF > 0.1).

Historically, the hope was that a GWAS would often limit causal variants to a region in which obvious nonsynonymous variants were segregating, with these structural changes in the protein sequences of causal genes driving the majority of trait variation and disease risk. Alas, as we have detailed above, the vast majority of tagged SNPs occur in either noncoding (intergenic) or intronic regions. Regulatory, rather than structural, variation appears to underlie most quantitative trait variation (although protein variation can be generated via the regulatory effects of sSNPs). This has enormous implications for determining a potential set of causal variants within a GWAS region. While we have a fairly good understanding of the impacts of amino acid changes (e.g., Cooper and Shendure 2011), this is not the case for regulatory variation. First, the functional annotation of regulatory sites, especially beyond the TSS and intron-exon boundaries, is still an evolving enterprise. Second, for *trans*-acting regulatory variants, genomic location provides little insight as to the target they influence (e.g., Example 21.5). As noted by Spain and Barrett (2015), the "physical distance of a variant to a gene is not substantive evidence of causality." Indeed, in a schizophrenia GWAS, Gusev et al. (2018) found that the gene closest to the lead SNP was the eventually implicated gene less than a quarter of the time.

Given these concerns, a number of statistical approaches have been developed to prioritize a smaller set of variants from a tagged region for any future functional studies, such as gene editing in model organisms. These fall into two categories: **agnostic approaches** that simply use the correlation (LD) structure among scored variants along with their marginal association statistics, and **annotation-driven** approaches that attempt to leverage additional functional information. The latter could simply be some prior probability of a given class of change having some functional effect (such as a variant in a known promoter box or splicing junction). An important class of annotation-driven approaches searches for colocalizations between a GWAS-detected SNP and a regSNP (such as eQTLs, sQTLs, meQTLs, haQTLs, etc.; Table 21.1). We examine these different approaches in turn. Statistical fine-mapping methods are reviewed by Spain and Barrett (2015), Cannon and Mohlke (2018), Schaid et al. (2018), Sieberts and Schadt (2019), Cano-Gamez and Trynka (2020), and Hutchinson et al. (2020a). While we frame the following discussion on fine mapping in terms of SNPs, the logic applies equally well to other classes of variants (e.g., CNVs).

Selecting Variants and Exploiting Local LD

Suppose a GWAS has been performed, and a number of association regions with genome-wide level significant SNP-trait associations have been detected (e.g., Figure 20.1). The next step is to break these regions down further into blocks of SNPs that are highly correlated (in high LD) with each other, but weakly correlated over blocks. Within each such block, there is a lead SNP (or SNPs) and a collection of tightly linked SNPs in high to very high LD with the lead (e.g., Figure 20.2). The goal is to obtain some minimal **causal set** of SNPs that jointly has a high probability of including the causal SNP (or SNPs), while still containing as few members as possible. In an ideal setting, this set has a membership of one (e.g., Example 21.8).

The simplest analysis, often called the **heuristic approach**, is to consider all SNPs within some (arbitrary) correlation threshold of the lead SNP. Under the heuristic framework, an investigator typically ranks the importance of SNP candidates by their p values, with the largest viewed as the marker most likely to be closest to the causal variant. There are three reasons why this is incorrect. First, as mentioned, a more distal marker can often have a larger r^2 with the causal site than a more proximal marker if the allele frequency match is closer (Example 20.1). Second, p values are a function of the standardized effect, e.g., $\widehat{\beta}/\sigma(\widehat{\beta})$. A marker could have a larger actual effect ($\widehat{\beta}$) but also a larger standard error (e.g., due to having a low-frequency minor allele), resulting in a smaller standardized effect. Finally, there is always statistical noise in the realization of the underlying expected value.

A more formal approach is to use **regression and model selection**: starting with some initial set of SNPs within the association region, a stepwise, multiple, or penalized regression is used to extract the most impactful SNPs after accounting for their LD structure. If the number of SNPs in the region is small, one might fit them all in a multiple regression, but this fails when LD is very high. A more common approach is to use a forward stepwise regression: the best fitting SNP is added to a regression model as a cofactor, and then the next best SNP is selected, and so on until the increase in model fit is no longer significant (e.g., Yang et al. 2012b). Penalized regressions, such as LASSO (Example 20.4), offer a very flexible model-selection approach, as (depending on the penalty function used; Example 20.4), most SNP effects shrink to zero. SNPs remaining in the final model form the reduced set of candidate causal sites. Regression and model-selection approaches provide a formal framework to make full use of the LD structure (the pairwise correlations among all SNPs in the starting set). When using summary data, typically LD estimates from some reference population are used, which can introduce bias if the LD in the reference and study populations have different structures. While regression-based methods are better than heuristic approaches, if SNPs are very highly correlated, even penalized models can be problematic (Schaid et al. 2018). Suppose that two SNPs are almost entirely correlated. A model-selection approach will reject one of the SNPs, which, by chance, could be the causal one (Hormozdiari et al. 2014).

The final, and best, class of methods are Bayesian (also called **probabilistic methods**), based on simple applications of Bayes theorem (Equation 3.3b). They return **posterior inclusion probabilities** (**PIPs**) for each SNP, allowing for the construction of **Bayesian credible sets** (Example 21.8). The rank of SNPs using PIP is often rather different from their p-value ranks (e.g., Maller et al. 2012). Variants not included in the credible sets can be excluded as being causal, allowing the investigator to focus on a smaller set of candidates in more detailed follow-up studies. Bayesian methods can use the full SNP correlation structure and can incorporate additional information in their priors, such as differential weighting for different variants (e.g., promoter mutations given higher weight than variants in random intergenic regions). Example 21.9 outlines their basic structure. Another tool, suggested by Udler et al. (2010), is to use information from two (or more) different populations, an approach called **trans-ethnic mapping** in humans. The logic is that if the same causal variants are segregating in all populations, but under different LD structures, a more precise signal can be generated by a combined analysis (van de Bunt et al 2015; Schaid et al. 2018).

Finally, Wang et al. (2020) noted a problem with just reporting *marker* PIP values. Suppose there are two causal sites within a focal region, sites 1 and 3. Further, suppose that site 2 is in full LD with site 1 ($r^2 = 1$, so that the marker effect sizes using a single-marker regression are equal, $b_2 = b_1$). Likewise, assume noncausal site 4 is in full LD with site 3 ($b_3 = b_4$). Under a model allowing two causal sites, the PIPs for each of these four sites would be 0.5, where in reality a more informative metric would be based on *sets* of markers, e.g.,

$$\Pr[(b_1 \neq 0 \text{ or } b_2 \neq 0)] \quad \text{and} \quad \Pr[(b_3 \neq 0 \text{ or } b_4 \neq 0)]$$

This altered credible set would return the PIPs that at least one *set* member is causative (a set, as opposed to single marker, metric) and provides a more informative view of the biological reality. Wang et al. developed their **sum of single effects** (**SuSiE**) model to accomplish this goal of constructing better credible sets by adding contributions from a regression assuming a single marker effect and then computing a posterior distribution over sets of these effects (see their paper for details).

Example 21.8. Maller et al. (2012) performed a follow-up investigation on 14 associated regions detected in the initial WTCCC study (Chapter 20). They genotyped 5,500 SNPs in 8,000 individuals across 12 genomic regions for three diseases (type 2 diabetes [T2D], coronary artery disease [CAD], and Graves' disease [GD]). Under the (strong) assumption that each association region contained *exactly* one causal variant, they showed that the PIP for SNP i among the candidate set of m SNPs within a region is simply

$$\Pr(\text{SNP}_i \text{ is causal}) = \text{PIP}_i = \text{BF}_i \Big/ \sum_{k=1}^{m} \text{BF}_k \tag{21.6a}$$

where BF_i is the Bayes factor for SNP i (Chapter 20 and Appendix 7). (An alternative expression for Equation 21.6a was given by Udler et al. (2010), replacing the BFs by likelihoods, which assumes a prior that all SNPs have an equal chance of being causal.) Wakefield's (2008) approximate Bayes factors are often used, which are given as follows. Let $\widehat{\beta}_i$ be the (regression slope) estimate of the effect of SNP i, with sample variance σ_i^2. Letting $s_i = \widehat{\beta}_i/\sigma_i$ be the standardized effect for SNP i, with a β prior that is normal with variance σ_b^2, then

$$\text{BF}_i = \frac{\sqrt{1-\gamma_i}}{\exp\left[-\gamma_i s_i^2/2\right]} \qquad \text{where} \quad \gamma_i = \sigma_b^2/(\sigma_i^2 + \sigma_b^2) \tag{21.6b}$$

Using the set of PIP values for a given association region, Udler et al. formed 95% (99%) credible sets by choosing the smallest number of SNPs whose PIP values sum to 0.95 (0.99). For 3 of the 14 regions (two for T2D, one for GD), a single SNP accounted for most of the PIP (>70%), while in four other regions, the number of SNPs in the credible set was small, excluding most of the SNPs from further evaluation. In the remaining seven regions, the credible sets were large (> 70 SNPs). However, in two of these regions, a few SNPs had a PIP > 20%.

A more powerful analysis was offered by Huang et al. (2017), who examined two subtypes of inflammatory bowel (IB) disease: ulcerative colitis and Crohn's disease (we use IB instead of the medical literature abbreviation IBD, reserving the latter to denote identical by descent). Roughly 200 IB loci have been mapped using GWAS, and the authors fine-mapped 94 of these using high-density genotyping in 68,000 individuals (roughly 34,000 cases and controls; with the cases consisting of roughly 19,000 Crohn's and 15,000 ulcerative colitis). With these sample sizes, even very tightly linked markers in high LD could be somewhat decoupled. Several of the original 94 regions could be further broken into two or more independent signals, resulting in a total of 139 independent association regions. Three different Bayesian methods were used to construct credible sets, and only elements present in at least two of these were placed in the final set. The resulting set sizes ranged from 1 to over 400 variants. For 18 of the regions, the 95% credible set had just a single member (**single variant credible set**), 24 others had 2–5 variants in the set, and 27 associations had a SNP with a PIP > 0.5. Hence, with sufficiently large sample sizes, very high confidence of causality can be assigned to a single variant (Udler et al. 2010 presented power calculations for the required sample sizes to exclude a noncausal SNP in high LD). The vast majority of IB signals were associated with both disease subtypes. While several of the regions were enriched for known regulatory signals (e.g., regions with modified histones H3K4me1, H3K4me3, and H3K27ac, indicative of active chromatin), variants with PIP > 0.5 in 21 noncoding regions could not be associated with known regulatory motifs.

Hutchinson et al. (2020b; also Wallace 2013) noted that Beavis effects (inflated effect size estimates for many of the top SNPs) introduced a slight bias on the construction of credible sets, tending to include more members than are needed (in moderate to high power settings). They propose an **adjusted credible set** method to reduce this bias.

Example 21.9. Bayesian methods (Appendix 7) return a posterior probability of inclusion (the probability, given the model assumptions, that a particular SNP is causal). They do so by using Bayes' theorem (Equation 3.3b),

$$\Pr(\text{SNP}_i \text{ is causal}) = \text{PIP}_i = \Pr(c_i = 1 \mid D) = \Pr(D \mid c_i = 1)\Pr(c_i = 1)/\Pr(D)$$

where c_i is a zero-one indicator variable equaling one when SNP i is causal and zero otherwise. The data, D, consist of the vector **s** of standardized marginal association statistics for the k SNPs being considered in the association region and the $k \times k$ correlation matrix **C** between these SNPs. There are numerous way to implement this core idea, depending on assumptions about the number of causal variants within a region, their prior distributions (e.g., using functional information to weight variants), different computational approaches, and so on. Commonly cited methods include **BIMBAM** (Guan and Stephens 2008); **BVSR** (Guan and Stephens 2011); **CAVIAR** (Hormozdiari et al. 2014, 2015); **PAINTOR** (Kichaev et al. 2014, 2016); **SSMR** (Wen 2014); **CAVIARBF** (Chen et al. 2015); **FINEMAP** (Benner et al. 2016); **DAP-G** (Wen et al. 2016); and **SuSIE** (Wang et al. 2020).

To illustrate the basic logic, we outline the **CAusal Variants Identification in Associated Regions** (CAVIAR) method of Hormozdiari et al. (2014, 2015), which allows for $\ell \leq k$ of the SNPs in a focal region to be causal. Let **c** be a k-dimensional vector, whose ith element, c_i, is one if SNP i is causal, and zero otherwise. We make the simplifying assumption that all causal SNPs have the same (standardized) effect β, and that γ is the prior probability that a SNP is causal (both these restrictions can easily be generalized via modifications of the prior). Let s_i denote the standardized marginal association statistic for SNP i. Specifically, the standard gene-dosage regression $z_j = \mu + \beta_i N_{j,i} + e_j$ is fit for SNP i (Equation 20.1a), with $s_i = \widehat{\beta}_i/\sigma(\widehat{\beta}_i)$. Hormozdiari et al. showed that if i is a causal variant in LD with noncausal variant h ($r_{i,h} \neq 0$), then $s_h \sim N(r_{ih}s_i, 1)$ and $\sigma(s_i, s_h) = r_{ih}$. Hence, the vector of association statistics, given the causal SNPs, $\mathbf{s} \mid \mathbf{c}$, is MVN with mean vector $\beta\mathbf{Cc}$ and covariance matrix **C**. As a result, the likelihood of **s**, conditional on **c**, follows from Equation 9.24, with

$$\Pr(\mathbf{s}\,|\,\beta\mathbf{c},\mathbf{C}) \propto \exp\left[-\frac{1}{2}(\mathbf{s}-\beta\mathbf{C}\mathbf{c})^T\,\mathbf{C}^{-1}\,(\mathbf{s}-\beta\mathbf{C}\mathbf{c})\right] \tag{21.7a}$$

Assuming a constant probability γ that any SNP is causal, then the prior on **c** becomes

$$\Pr(\mathbf{c}) \propto \prod \gamma^{c_i}(1-\gamma)^{1-c_i} \tag{21.7b}$$

with the restriction that $\sum c_i = \ell$. The resulting posterior for a particular configuration $\mathbf{c}^*$ of causal sites becomes

$$\Pr(\mathbf{c}^*\,|\,\mathbf{s},\mathbf{C}) = \frac{\Pr(\mathbf{s}\,|\,\beta\mathbf{c}^*,\mathbf{C})\,\Pr(\mathbf{c}^*)}{\sum_{c^*\in C}\Pr(\mathbf{s}\,|\,\beta\mathbf{c}^*,\mathbf{C})\,\Pr(\mathbf{c}^*)} \tag{21.7c}$$

where the set C is all the possible configurations for causal SNPs subject to the constraint that $\sum c_i = \ell$ (ℓ causal variants). For example, assuming a single causal SNP, there are exactly k configurations of one 1 and $(k-1)$ zeros in $\mathbf{c}^*$, with two causal SNPs, there are $k(k-1)$ configurations of two ones and $(k-2)$ zeros, and so forth. For a given choice of ℓ, the posterior value of c_i (the PIP for SNP i) is obtained by summing the c_i values from Equation 21.7c over all of the configurations in C. As an example of how $\mathbf{c}^*$ is constructed with prior information, suppose there are three sites within the focal region, and we assume only one is causal. Further suppose, based on functional annotation (such as a site being an NS variant, near a TSS, etc.), that the three sites show a four fold, no, and a two fold enrichment relative to random variant. The resulting prior weights on these three sites would become (4/7, 1/7, and 2/7).

Colocalization: Exploiting Signals From regSNPs

The TWAS paradigm—a local SNP impacts a transcript, which in turn influences the trait—naturally leads to a deeper discussion of how to leverage functional genomics information to fine-map GWAS hits. With just GWAS summary statistics and SNP LD patterns, one could use functional annotation (when available) to assign weights on the priors of the SNPs in a candidate set when computing their PIPs (e.g., Gagliano et al. 2014; Kichaev et al. 2014; Chen et al. 2016; Wen et al. 2017; Weissbrod et al. 2020). The utility of this approach depends on the annotation accuracy, which is often very poor for many sites with regulatory roles. A biologically more robust approach is an extension of TWAS, namely looking for the **colocalization** between association regions for the trait and association regions for eQTLs influencing a QTT for that trait. More generally, one could use other regulatory QTLs—such as sQTLs, meQTLs, hQTLs, etc.—in place of eQTLs (Table 21.1). There are different levels of granularity associated with a colocalization analysis, depending on both the data set (the amount of recombination in the sample setting the mapping resolution) and the question of interest. Typically, one might use colocalization on a coarse scale to implicate *specific genes* (as in a TWAS), or on a much finer scale to fine-map *specific SNPs* that influence both the trait and the molecular feature of interest.

Early searches of candidate loci, even under the crude resolution of linkage-based mapping, looked for loci in a QTL region whose transcript variation was correlated with trait variation. Wayne and McIntyre (2002) used this approach to identify 34 candidate genes for ovariole number in *Drosophila*. As mentioned above, the problem with simply searching for transcript–trait associations is that the correlation could be generated by the reactive model, where trait variation generates the transcript variation ($z \rightarrow T$), not vice versa ($T \rightarrow z$). Similarly, both z and T could be impacted by a confounder C ($z \leftarrow C \rightarrow T$), creating a correlation between z and T in the absence of $T \rightarrow z$. Hence, simply finding a GWAS hit very near a coding region whose transcript is a QTT is not sufficient evidence that the causal variant acts through the QTT. A more direct connection is provided by

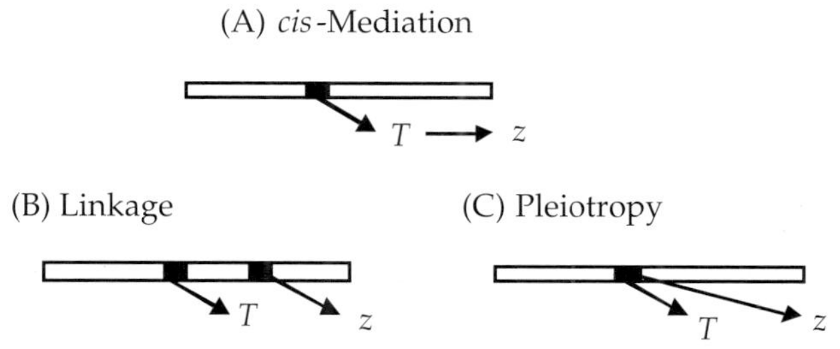

Figure 21.4 An apparent colocalization between a GWAS SNP for trait z (the small black box), transcript (T), and trait (z) could occur through three different pathways. **(A):** Direct *cis*-mediation. The GWAS SNP is an eSNP which directly influences the transcript, which in turn directly influences the trait. **(B):** Linkage. Two tightly linked SNPs are involved. One directly impacts the transcript, the second directly impacts the trait. **(C):** Pleiotropy. The same SNP directly impacts transcript levels and trait values separately, but the transcript level does not impact the trait value. Any combination of these different pathways could be involved, such as a direct *cis*-mediated SNP tightly linked to a separate SNP that only impacts the trait.

demonstrating that the trait GWAS hit is also a regSNP for the QTT that impacts the trait. The initial step in such a demonstration is showing that a GWAS SNP for the trait **colocalizes** with a regSNP.

There are three different levels of a colocalization analysis. The first is simply determining whether there is support for causal SNPs influencing both the focal trait and molecular intermediates (e.g., transcript levels or methylation status) in a given association region. As shown in Figure 21.4, even when a very strong colocalization signal is detected, its interpretation is still unclear. It could be the result of *cis*-mediation, linkage, pleiotropy, or a combination of all three involving multiple, tightly linked, causal variants. Hence, the second step of a colocalization analysis is ruling out *linkage* (Figure 21.4B)—associations generated by tightly linked SNPs with *independent effects* on the two features (T and z). The final step, distinguishing pleiotropy (Figure 21.4B) from *cis*-mediation (Figure 21.4A), requires the type of analysis outlined in Example 21.3. One concern is that multiple, tightly linked *cis*-eQTLs are common (e.g., Zeng et al. 2019), so that allelic heterogeneity (multiple causal variants within the association region) may be the norm (at least for regQTLs).

Early attempts at detecting colocalization between trait and regSNPs were usually visual: looking for alignments of peak SNPs on separate Manhattan plots for the trait and transcript, or using a two-dimensional scatter plot of the trait and transcript $-\log(p)$ values for each SNP (e.g., B. Liu et al. 2019). One immediate issue is that the trait and eSNP p values almost always come from different studies, raising concerns (as in a TWAS) of whether the appropriate tissues (given the focal trait) were used in the eSNP mapping and whether the two different samples had the same LD structure. Another complication is stochasticity. Even with the same underlying causal SNP and with appropriate tissue and population match, the sampled lead SNP for the trait and for the eSNP are likely to map to different locations. For example, if the lead SNP and causal site agree 50% of the time (which is often overly optimistic given our earlier discussion), there is only a 25% chance that the lead GWAS and lead eSNP agree, and there is a good chance that the two lead SNPs will map kilobases apart, even if they have the same underlying causal SNP.

As a result of these concerns, a variety of increasingly more formal approaches have been proposed for testing colocalization. When full (i.e., individual) data are available on both traits, **QTLMatch** tests whether regressions of each character (the focal trait and transcript) against multiple candidate SNPs have proportional slopes, as expected if the SNPs jointly tag a common variant (Plagnol et al. 2009). A variant of this approach was used by Barbeira et al. (2021), who examined the correlation between a SNP's regulatory effect (they looked at both eSNPs and sSNPs) and its effect on the focal trait, finding correlations of 0.18 for eSNPs and 0.25 for sSNPs over a set of almost 75 different human traits/diseases. The

regulatory trait concordance (**RTC**) method of Nica et al. (2010) also requires individual data (at least for the expression sample), and looks at the impact on expression data after including the peak GWAS SNP as a cofactor. The **Sherlock** approach of He et al. (2013) requires only summary statistics, and looks for a concordance between GWAS SNPs and eSNP for the focal transcript. Because there is low SNP filtering, even eSNPs with weak effects (e.g., with nominal, but not genome-wide, significance) are included, and as a result Sherlock can incorporate *trans* signals. Sherlock also has a useful evidence-asymmetry feature in that an eQTL for the focal transcript not corresponding to a GWAS SNP is taken as support against the regSNP-GWAS colocalization, but a GWAS hit that does not correspond to an eQTL for the focal transcript has a neutral impact on the support for colocalization.

These early methods have largely been replaced by Bayesian approaches (e.g., Giambartolomei et al. 2014; Hormozdiari et al. 2016; Wen et al. 2017; Roytman et al. 2018; Wang et al. 2020; Wu et al. 2023). The **coloc** method of Giambartolomei et al. (2014) uses summary statistics (and does not require any LD data), returning the posterior probabilities for five different hypotheses concerning the association of a region with two traits. These are: no association within the region for either trait (PP0); association with one trait, but not the other (PP1 and PP2); association with both traits, but via two independent SNPs (PP3); and association with both traits due to a shared SNP (PP4). A helpful feature of coloc is that it distinguishes evidence for colocalization (high PP4 value) from a lack of power (PP0, PP1, or PP2 have high values). Further, the values of PP3 (linkage) and PP4 (pleiotropy) provide support for distinguishing between these two sources of a colocalization signal. The **multiple-trait-coloc** (**moloc**) method of Giambartolomei et al. (2018) extends this approach to the colocalization of any k features (such as a trait SNP, an eSNP, and a meSNP), but still under the coloc assumption of (at most) a single causal site (in the association region) for each feature. Wallace (2020, 2021) developed approaches to extend coloc to allow for multiple variants (impacting each trait) within an association region. The TWAS **MetaXcan** framework of Barbeira et al. (2018) uses these PP values following an initial TWAS, upweighting inclusion of regions with high PP4 values and removing regions with high PP3 values.

While coloc tests for colocalization *within an association region*, a colocalization fine-mapping approach (*which SNP* within that region), even when multiple causal sites are present, is offered by the **eCAVIAR** (**eQTL and GWAS Causal Variants Identification in Associated Regions**) method of Hormozdiari et al. (2016). This extends their single-trait fine mapping CAVIAR method (outlined in Example 21.9) to jointly consider the causal support for a *pair* of traits. The strength of evidence of colocalization at a specific SNP is given by a **colocalization posterior probability** (**CLPP**). For a vector of M candidate SNPs within an association region, CAVAIR returns PIP_i, the probability that SNP i in this region is causal. Basically, eCAVIAR works by computing PIP_i^z and PIP_i^T, the posterior probability that SNP i is causal for the trait (z) and transcript (T), respectively. The resulting CLPP for SNP i becomes $\text{CLPP}_i = \text{PIP}_i^z \cdot \text{PIP}_i^T$, an approximation that does not account for enrichment.

As noted by Wen et al. (2017) and Hukku et al. (2021), there are connections between fine-mapping, colocalization, and enrichment (next section) that are apparent when one considers the priors for Bayesian methods. Following Example 21.9, let c_i^z and c_i^T be 0/1 indicator variables that SNP i is causal for the trait and transcript, respectively. From Bayes theorem, the resulting expression for the CLPP, given the data D, can be expressed as

$$\Pr(c_i^z = 1, c_i^T = 1 \,|\, D) \propto \Pr(D \,|\, c_i^z = 1, c_i^T = 1) \Pr(c_i^z = 1, c_i^T = 1) \tag{21.8a}$$

Focusing on the prior, we can write this as

$$\Pr(c_i^z = 1, c_i^T = 1) = \Pr(c_i^T = 1 \,|\, c_i^z = 1) \cdot \Pr(c_i^z = 1) \tag{21.8b}$$

$\Pr(c_i^T = 1 \,|\, c_i^z = 1)$—the probability that the eSNP is likely to be causal when the trait SNP is causal — represents the **enrichment** of eSNPs (or any other regSNPs) given a trait GWAS hit. When this conditional probability exceeds $\Pr(c_i^T = 1)$—the chance that a random SNP is an eSNP—then GWAS SNPs are enriched for eSNPs. The **ENLOC** method of Wen et al. (2017) models this enrichment using a logistic regression (Equation 20.2b), with

$$\text{logit}\left[\Pr(c_i^T = 1 \mid c_i^z = 1)\right] = \alpha_0 + \alpha_1 c_i^z \tag{21.8c}$$

where the α_i can be fitted empirically. A value of $\alpha_1 > 0$ measures the log of the odds ratio increase in enrichment when the SNP is causal ($c_i^z = 1$) for the trait. A variety of enrichment tests are reviewed in the next section.

Example 21.10. Another approach for distinguishing linkage from pleiotropy—the latter lumping both traditional pleiotropy (Figure 21.4C) and *cis*-mediated (Figure 21.4A) effects—was suggested by Zhu et al. (2016). Their **SMR** method is based on Mendelian randomization, using GWAS summary statistics. As is more fully developed in Appendix 2, the concept of MR is essentially a subset of path analysis, built around the concept of mediation (Figure 21.3 and Example 21.3). The methodology comes from epidemiology, where one tunes some **instrumental variable** (here, a SNP genotype) that impacts a mediator, that in turn impacts an outcome. Consider a genotype (g, scored by reference allele copy number, N), transcript (T), and trait value (z). Let β_{Nz} and β_{NT} be the regression slopes of the trait and transcript level, respectively, on minor allele copy number for a given SNP (obtained from trait and transcript GWAS). If $g \rightarrow T \rightarrow z$, then assuming no additional path from g to z that is independent of T, $\beta_{Nz} = \beta_{NT} \cdot \beta_{Tz}$, so that $\beta_{Tz} = \beta_{Nz}/\beta_{NT}$ is the mediation effect (Example 21.3). The logic of Zhu et al. is that if the same causal variant underlies both z and T (whether through *cis*-mediation or through the pleiotropy effects of g on both z and T), then the ratio β_{Tz} should have the same expected value for any SNP linked to the causal site (as the correlation between the SNP and causal site appears in both the numerator and denominator of β_{Tz}, and thus cancels). In their **HEIDI** test (**heterogeneity in independent instruments**, the "instrument" being the genotypes at the focal SNP) of pleiotropy versus linkage tests for *heterogeneity* of the β_{Tz} estimates over a set of linked SNPs in the association region. Such heterogeneity is consistent with linkage (different SNPs), but inconsistent with pleiotropy (assuming a single causal variant). Hence, a significant p value (significant heterogeneity) indicates linkage as the source of the colocalization.

An alternative approach to resolving linkage versus pleiotropy when a strong colocalization signal is found was offered by Chun et al. (2017). They reasoned that if the same causal SNP underlies both the trait and expression signal, then this joint evidence should be maximized at markers in the tightest LD with the causal site. Conversely when this association is due to LD between causal sites with independent effects, a different likelihood structure occurs, and they proposed an LR test for this (the same type of linkage versus pleiotropy test has been proposed for linkage-based mapping; Chapter 18).

GENE-SET, PATHWAY, AND NETWORK ANALYSIS

One outcome from the initial wave of whole-genome expression studies was an avalanche of data in the form of lists (often very long lists) of differentially expressed genes between case and control samples. As highlighted above, the development of numerous other high-throughput platforms extends this data to differences in protein levels, methylated sites, splicing isoform ratios, metabolite levels, chromatin modification differences (e.g., the presence of acetylated or methylated histones at a target site), and so on. Hence, while our comments are framed in terms of expression data, the approaches presented here obviously extend over large classes of other genomic and cellular features. In an attempt to extract insight from such data, a number of increasingly sophisticated **knowledge-based approaches** have been developed. These attempt to leverage information from functional genomics databases, such as the functional category of a gene (e.g., a transcription factor, a kinease, etc.), their membership in known pathways, or their interaction partners in networks (e.g., proteins that make physical contact in the cell). Although initially developed for expression data, it was quickly realized that these approaches could be easily modified and applied to GWAS data. We start by briefly considering some of the basic logic of these methods as applied to expression data before focusing on their applications to GWAS. Reviews of expression analysis methods can be found in Drăghici and Krawetz (2003), Allison et al. (2005), Khatri and Drăghici (2005), Curtis et al. (2006), Nam and Kim (2008), Fridley et al. (2010), and Rahmatallah et al. (2016).

Gene Set Analysis of Expression Data: Enrichment Methods

Using a collection of genes as the unit of analysis is loosely called a **gene set analysis (GSA)**, as opposed to an **individual gene analysis (IGA)**, which considers genes one at a time. A GSA is often called a **pathway analysis** (not to be confused with the regression-based method of *path analysis*; Appendix 2) when the gene set members are chosen from a known pathway (e.g., $A \to B \to C$). A GSA is used for two different, but not necessarily independent, reasons. The first is *power*, in that a collection of individually weak IGA signals—such as those showing nominal (e.g., $p < 0.05$), but not genome-wide ($p < 0.05/n$, with n very large), significance—might be boosted under a GSA (akin to gene-based and rare-allele GWAS approaches considered in Chapter 20). The second is that finding significant *sets of genes* (such as members of the same functional class or pathway) may provide greater biological insight than a focus on single genes. If a disease is caused by the *disruption of a pathway*, individuals (and populations) may harbor mutations in different genes in that pathway, resulting in a rather weak *gene-based* signal (Wang et al. 2007; Schadt 2009). A *pathway-based* signal, however, may not only be stronger but could also be more consistent over replicates (e.g., Elbers et al. 2009; Wang et al. 2009).

Example 21.11 shows the initial GSA approach, **enrichment analysis (EA**; also called **over-representation analysis; ORA)**, which asks if there is a modest, but coordinated, shift in expression levels over genes in some known pathway or in some defined category. The basic structure of an enrichment test is that one starts with some bioinformatics database and then examines whether differentially expressed genes are enriched for some feature (Tavazoie et al. 1999; Mootha et al. 2003; Subramanian et al. 2005). In one of the early applications of GSA, Mootha et al. (2003) showed that genes involved in oxidative phosphorylation had coordinately decreased expression levels in the muscle tissue of diabetics. As mentioned in previous sections, enrichment analysis has been widely used to examine whether lead SNPs from a trait GWAS are enriched for SNPs of other features (such as regSNPs) or are overrepresented in specific genomic regions (such as sites with more open chromatin).

Example 21.11. Suppose that one has expression data for 5,000 genes in normal versus cancer cells, 300 of which are declared to show significant differences (using some threshold critical value, say $\alpha = 0.05$). Now suppose we choose (by some criteria) a set of these scored genes (denoted by $\mathcal{G}$; the remainder are in the complement set, $\mathcal{G}^c$), which consists of 100 members, 20 of which show significant differences. The resulting 2×2 contingency table of expression differences versus set membership becomes

	Significant	Not Significant	Totals
Gene set $\mathcal{G}$	20	80	100
Gene set $\mathcal{G}^c$	280	4620	4900
	300	4700	5000

Using a standard chi-square test shows that differentially expressed genes are significantly over-represented ($p = 9 \times 10^{-9}$) in our gene set, $\mathcal{G}$.

With smaller gene sets, the observed value in one (or more) of the entries (cells) in the contingency table can be small, in which case Fisher's exact test is more appropriate. This latter test is based on the **hypergeometric distribution** (Equation 2.25). Under the null hypothesis (no enrichment), the probability of observing at least k genes from a functional category (i.e., $\mathcal{G}$) in a sample of n genes declared to show significant differences is given from the upper tail of the hypergeometric,

$$p = \Pr(X \geq k) = \sum_{i=k}^{n} \Pr(X = i) = \binom{g}{n}^{-1} \left[\sum_{i=k}^{n} \binom{f}{i} \binom{g-f}{n-i} \right] \tag{21.9a}$$

where f is the total number of genes in the functional category and g is the total number of scored genes, where $g \geq f \geq n$ (Drăghici and Krawetz 2003). If n is small relative to f, one can approximate the hypergeometric (sampling without replacement) by a binomial (sampling with replacement) with success parameter $\pi = f/g$ and sample size n (Drăghici and Krawetz 2003), giving

$$p = \Pr(X \geq k) = \sum_{i=k}^{n} \Pr(X = i) = \sum_{i=k}^{n} \binom{n}{i} \pi^i (1-\pi)^{n-i}, \quad \text{with } \pi = f/g \tag{21.9b}$$

There are a variety of enrichment methods based on such 2×2 contingency tables, differing in whether a chi-square, Fisher, hypergeometric, binomial, or other (e.g., normal approximation of a binomial; Doniger et al. 2003) is used to test significance (Khatri and Drăghici 2005; Curtis et al. 2006; Tryputsen et al. 2014).

While Example 21.11 shows the basic logic for enrichment approaches, it leaves open a number of important issues. The first is choosing the genes to include in the tested set ($\mathcal{G}$). This could be *hypothesis-driven*, testing a few candidate gene sets, akin to the candidate gene approach of testing associations (Chapter 17). Alternatively, one could adopt an *exploratory approach*, testing over a wide number of sets, whose gene elements are chosen by some functional criteria. A major tradeoff between these approaches is correcting for multiple comparisons (tests of specific gene sets), which imposes a mild burden in a candidate approach, but could become very large in an exploratory analysis.

One classifier for set construction is to use genes that cluster in the same **Gene Ontology** (**GO**) group (Harris et al. 2004). GO is a hierarchical vocabulary of function (e.g., pigmentation → regulation of pigmentation → regulation of eye pigmentation), with categories becoming larger (and more general) as one moves up the GO hierarchy (toward **parent nodes**, e.g., pigmentation) and narrower and more specific as one moves down the hierarchy (toward **offspring nodes**, e.g., regulation of eye pigmentation). One issue with testing different sets of GO genes is that the hierarchical nature of their labels implies that genes in different groups can be correlated. Zhang et al. (2010) suggested Bayesian approaches to model this dependency structure.

Another criterion is to choose genes from known pathways, for example using the **Kyoto Encyclopedia of Genes and Genomes** (**KEGG**; Kanehisa et al. 2010) database. While deciding which elements constitute a specific pathway might seem straightforward, this is not always the case. A pathway is often treated as a discrete modular item, when, in reality, smaller pathways are often nested within much larger pathways and networks. Hence, the same gene could play key roles in several different user-defined pathways, creating correlations among the analyses. Defining a specific pathway raises issues akin to those of trying to assign SNPs to a particular gene that were discussed in Chapter 20 (in the context of gene-based GWAS). Besides the biological concerns as to what constitutes a reasonable pathway, there are also statistical issues with gene choice (which pathway members to include). As was the case for rare allele methods (Chapter 20), inclusion of too many pathway genes with no effect on a trait in the gene set diffuses any true signal. Conversely, there is a bias towards larger gene sets being significant (Holmans 2010; Wang et al. 2010; Ramanan et al. 2012) unless set size is appropriately controlled (for example, by using permutation for significance testing).

The strength of any GSA depends on the quality of the data used to choose gene set members. If a gene is incorrectly classified (**annotated**) as to its functional group, this lowers the power of an analysis using that gene. Further, a gene can (correctly) be assigned to multiple functional groups. In theory, one could use Bayesian methods to weight set members by some metric of their perceived accuracy of correct assignment (such as was done with imputed SNPs; Chapter 20). The same issue of quality of the bioinformatics data using GO exists with pathways, with a further complication. The nature of a pathway can easily change over cell type, developmental stage, or environment (such a high versus low sugar diet, or well-watered versus drought conditions).

A second issue is that the classification of an expression difference as being significant in an EA is *arbitrary*, depending on the critical-value threshold used. As one changes this threshold, the elements in the 2×2 tables change, potentially changing the significance of the gene set. The **gene set enrichment analysis** (**GSEA**) approach of Mootha et al. (2003), Subramanian et al. (2005), and Efron and Tibshirani (2007) avoids the arbitrariness of thresholds by basing tests on the *ranks* of some expression statistic (such as their p values or score statistics). One then tests for an *enrichment of ranks* in the gene set relative to a control set of scored genes. This results in a running **enrichment score** statistic that changes with each new gene added, yielding a Kolmogorov–Smirnov-like test statistic. At some point the score reaches its maximal value (**maximum enrichment score**, **MES**), after which it starts to decline as further genes are added. The set of genes that yields the MES was called the **leading-edge subset** by Subramanian et al. (2005), and contains the core members of the gene set that provide the greatest signals. While the original GSEA method assumed correlated expressions changes occurred in the same direction (e.g., all up-regulated), Saxena et al. (2006) extended it to an **absolute enrichment** (**AE**) approach that scores absolute (as opposed to signed) changes in expression levels between cases and controls.

A final issue is more subtle, but no less critical: ***What is the appropriate null hypothesis?*** Is it that: (i) there is *no* differential expression for any of the genes in the candidate set ($\mathcal{G}$); or is it that: (ii) the candidate set does not contain *more* differentially expressed genes relative to the rest of the scored genes ($\mathcal{G}^c$). This distinction was first noted by Tian et al. (2005), and formalized by Goeman and Bühlmann (2007), who denoted the first null as **self-contained tests** ($H_{o,sc}$), and the second as **competitive tests** ($H_{o,co}$; such as is performed in a GSEA). Generally speaking, tests against $H_{o,sc}$ are more powerful than tests against $H_{o,co}$, as if $H_{o,sc}$ is true, then so is $H_{o,co}$, while the converse relationship does not hold. As noted by Goeman and Bühlmann, a self-contained test "always has a clear biological meaning. At the same time, it may not always be biologically interesting," as we *expect* some differential expression over the genome. Conversely, a self-contained test does not require information from genes outside of the candidate set.

To examine these different nulls more closely, suppose that the number of differentially expressed genes in $\mathcal{G}^c$ in Example 21.11 was 980 (20%), the same fraction as in $\mathcal{G}$. The resulting 2×2 table now becomes nonsignificant. This is a competitive test ($H_{o,co}$), with the null being that the fraction of differentially expressed genes in the test set is no greater than in the control set. This is a natural comparison in many settings, as the levels of expression may increase over a wide fraction of the genome in (say) diseased versus control individuals. In this situation, the question of interest is whether genes in the control set are *enriched* for differential expression. Conversely, if we restrict our attention to *just* the gene set ($H_{o,sc}$), the expected number of false positives under the null is 5 ($n\alpha = 100 \cdot 0.05$), with the observed value of 20 being highly significant (one can use either Fisher's exact test or a binomial with $\pi = 0.05$ and $n = 100$). Ebrhimpoor et al. (2020) showed that both of these nulls ($H_{o,sc}$, $H_{o,co}$) are special cases of a more general null that they called **simultaneous enrichment**. Here the null is that the fraction of significant differences in $\mathcal{G}$, $\pi(\mathcal{G})$, is $\leq c$. Under the self-contained null, $c = 0$ ($\pi(\mathcal{G}) = 0$); while under competitive null $c = \pi(\mathcal{G}^2)$, namely $\pi(\mathcal{G}) \leq \pi(\mathcal{G}^c)$.

The structure of permutation tests changes under these different nulls (Goeman and Bühlmann 2007). A self-contained test permutes the phenotypic labels while holding the vector of expression data for a single individual constant (often called **subject sampling**). Conversely, a competitive test involves **gene sampling**. Here, the data vector for a given individual consists of phenotypic values, expression data, and labels on which genes (transcripts) are in gene set $\mathcal{G}$ versus set $\mathcal{G}^c$. Permutation now involves keeping the phenotypic and expression data intact, but shuffling the gene set membership labels. For example, if our original set consists of 10 genes, a permutation sample performs the analysis using the same phenotypic and expression data, but the ten genes to be labeled over all individuals as being the case set ($\mathcal{G}$) are randomly chosen from the sample of all genes. As noted by a number of authors, gene-sample permutation assumes independence of expression changes, which is often not the case. Further comments on subtle features of the nature of the competitive null, and their permutation tests, are offered by Efron and Tibshirani (2007), Gatti et al. (2010), Maciejewski (2013), and Debrabant (2017).

Gene Set Analysis of Expression Data: Next-generation Methods

Enrichment analysis (EA) represented the first generation of GSA methods for expression data. As noted by Khatri et al. (2012), these early approaches had a number of limitations. They essentially treated each member in the class as having equal weight, and assumed that each gene was independent of all other genes in the set. For many EA tests, each member is simply a binary data point, counted as either present or absent in a given category.

Second-generation methods, which Khatri et al. refer to as **functional class scoring** (**FCS**), assign weights to each gene in the set, such as their p value or some score statistic based on their amount of differential expression. The problem of how to combine this information, especially when the individual gene metrics are potentially correlated, is one that we have encountered before. In Chapter 20, we examined a variety of methods for aggregating correlated SNP data from a single gene into a gene-specific statistic (gene-based GWAS). All of these approaches (p-value combining approaches such as Fisher, GATES, or rank-truncation; multiple regressions and Hotelling's T; penalized regressions such as LASSO; random-effect variance component models, etc.) have been used to construct FCS tests, where the statistics being combined are based on the expression data for each gene. The discussion from Chapter 20 on hypothesis testing when gene-specific statistics are correlated also applies to FCS methods. Permutation approaches are the gold standard, but simulation and large-sample approximations have been proposed (see Chapter 20 for details). As with first-generation methods, null hypotheses can be framed as either self-contained or competitive tests. Given the number of combinations of different single-gene expression statistics, aggregating statistics over the gene set, and different null hypotheses, it is not surprising that a massive number of gene-set methods have been proposed (Ackermann and Strimmer 2009 examined over 250 different combinations of these elements, while Xie et al. 2021 noted over 100 different published methods/programs).

One of the emerging fruits of the functional genomics revolution is a more detailed understanding of cellular pathways and networks. In particular, we often have detailed information about the **topology** (the structure of connections between elements; Appendix 2) of pathways and networks (such as the web of protein–protein interactions). EA and FCS ignore this information when constructing pathway-summary statistics. **Pathway topology** (**PT**) methods attempt to exploit this additional knowledge, and can be considered as the third generation of GSA approaches (Khatri et al. 2012). For example, Pan (2008) gave higher weights to nearby genes (closer links in the pathway/network topology). Similarly, a network (such as a **protein-protein interaction network**; **PPI**) may not have the directionality of a pathway, but still has a complex topology of interactions (e.g., Figure 21.5). Such information can also be used in GSA. An early such example is Baranzini et al. (2009), who found that certain subsets of the human PPI were enriched in individuals with multiple sclerosis. An overview of these topology-informed methods is given by Mitrea et al. (2013).

We can also consider a fourth generation of methods, **dynamical pathway approaches** (**DPA**), that use information from not only the shape (adjacency and connectivity of elements) but also the **flux** (rate of flow) of products over the pathway, such as the **impact analysis** method of Drăghici et al. (2007). These third- and fourth-generation approaches forge connections between classical QG and functional genomics.

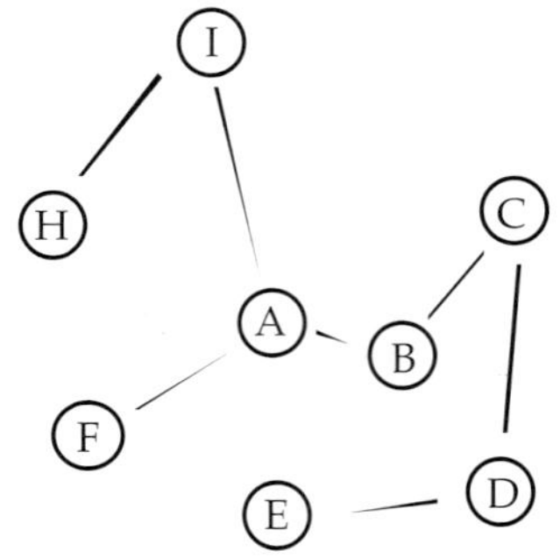

Figure 21.5 A **network** is a collection of **nodes** (the elements A through H), which are connected by **edges**, showing which elements interact with each other. For example, A directly interacts with B, F, and I. In a protein–protein interaction network, this graph implies that protein A contacts proteins B, F, and I in the cell, while B contacts A and C, and so on. Appendix 2 examines the topology (shape and connectiveness) of networks in more detail.

Genome-wide Pathway Analysis (GWPA)

It was quickly realized that expression-based GSA approaches could be applied to GWAS datasets (K. Wang et al. 2007, 2010; Torkamani et al. 2008; Elbers et al. 2009; Hong et al. 2009; Cantor et al. 2010; De la Cruz et al. 2010; Peng et al. 2010; Zhong et al. 2010), leading to the notion of a **genome-wide pathway analysis**, or **GWPA** (Ramanan et al. 2012), also called **GWAS pathway analysis, GWASPA** (Cantor et al. 2010). A GWPA represents the logical extension of the basic unit of a GWAS from a *SNP* to a *gene* to a *gene set*. As with a gene-based GWAS (Chapter 20), motivating factors for a GWPA are the potential of increased power by combining weaker signals, reduced multiple-comparison burden (reduced complexity by reducing the number of tests), greater replication based on pathways (rather than on specific genes), and the hope for greater biological insight. Before proceeding, the reader might find it useful to review the material from Chapter 20 on gene-based GWAS.

Chapter 20 examined two secondary analysis methods to mine hard-won SNP-based GWAS data: gene-based GWAS and meta-analysis. GWPA represents a third class of data-mining approaches, and can be also viewed as a meta-analysis where data (genes) from *within* a single study are distilled into *summary statistics over gene sets*. A gene-based GWAS can similarly be considered a within-study meta-analysis, combining SNPs into gene-based units of analysis. Gene-based GWAS and GWPA from a single study can themselves be placed in a meta-analysis framework to combine results over multiple studies. While a GWPA can be very useful complement to SNP-based or gene-based GWAS, it is best considered as an exploratory analysis to gain additional insight and to guide future research directions.

While the basic logic, and many of the analysis methods, from an expression-based GSA extend to a GWPA, there are two complications, both based on extracting gene-specific scores. Under an expression-based GSA, one has a single variable (mRNA level) for each gene (ignoring alternatively spliced products). What is the corresponding GWPA metric for a gene, which consists of data from a number (perhaps a large number) of SNPs? Specifically, an investigator must decide: (i) which SNPs are assigned to a specific gene; and (ii) how to combine this SNP data into a gene-specific score. There are a number of proposed approaches for the latter issue that were examined in Chapter 20 (gene-based GWAS).

The question of which SNPs comprise a gene is less clear, but is usually accomplished by considering a set of SNPs within some defined window around a coding sequence (akin to considering only proximal SNPs when searching for local eSNPs). This set could be all the SNPs in the window, or a trimmed set, such as removing pairs in very high LD, or by extracting PCs (Chapter 20). This window-based approach can result in a single SNP being assigned to multiple genes in a cluster, which can cause complications. An example of this was highlighted by Sedeño-Cortés and Pavlidis (2014), who noted that Dixson et al. (2014)

assigned a single SNP to multiple members in a gene cluster, each of which had a similar GO annotation, resulting in a GWPA falsepositive. As with rare-allele GWAS methods (Chapter 20), power is improved by excluding SNPs with no apparent functional effect. However, the question of which SNPs are likely functional (especially in noncoding regions) is, at best, poorly resolved (see the discussion above and in Chapter 20). Further, potential control regions for a gene may occur at some distance from the coding sequence, such as long-range *trans*-acting factors. These can potentially be handled in an eSNP framework, using expression data for the target gene (e.g., Zhong et al. 2010).

Because a GWPA involves two levels of data amalgamation—turning collections of SNPs into a *gene score* and turning collections of gene scores into a *set score*—there are two strategies for performing a GWPA: **one-step** versus **two-step** approaches. A one-step approach simultaneously incorporates all of the chosen gene-set SNPs into a single analysis, using the methods from Chapter 20 for combining results for multiple markers. In a two-step analysis, one first generates summary statistics (such as p values) for each gene (e.g., gene-based GWAS), and then combines these scores over the set to obtain the gene-set-level statistic. Which approach is used depends on both data availability and assumptions about the underlying biology. Based on the nature of the gene set, a one-step approach is favored when most of the signal comes from just a few SNPs. In the extreme, a phenotype could be simply the product of one gene and be largely independent of the rest of the pathway. Similarly, when multiple sites within a gene contribute to trait variation, collapsing these to a single gene score can lower power and skew the biological interpretation. Conversely, when there are weak signals over multiple genes, a two-step approach is favored.

In a two-step approach, one can mix and match analysis methods (developed for gene-based GWAS), using one approach within genes and a second for combining the gene-level statistics. For example, the **GRASS** (**g**ene set **r**idge regression in **a**ssociation **s**tudie**s**) method of Chen et al. (2010) uses the LASSO (Example 20.4) to obtain gene-level statistics, and then uses a p value combining method to obtain a gene-set score. The **HYST** (***hy**brid* **s**et-based **t**est) method of Li et al. (2012a) uses GATES (extended Simes; Equation 20.4f) to compute gene-level scores and then combines these using a modification of Fisher's method to allow for correlation among genes. Many two-step methods use the **Min-p approach** for each gene, namely just using the SNP with the smallest p value (i.e., the lead SNP). This has both statistical and biological limitations. Clearly, the more tested SNPs within a gene, the smaller the expected value of Min-p. This gene-size bias can be overcome by using standard permutation methods of keeping the genotypes intact while shuffling phenotypic labels. The biological concern is that if multiple sites in the focal gene are segregating causal alleles, the full impact of the gene is not captured by only using the largest SNP effect.

As with expression-based GSA, an investigator must decide between a self-contained hypothesis (*none* of the gene set elements have an impact on the trait) or a competitive hypothesis (the gene set is not *enriched* for trait-impacting genes relative to the rest of the genome). While one might consider the self-contained hypothesis to be more natural, de Leeuw et al. (2016) strongly argued that the competitive hypothesis is often more appropriate. Their reasoning is that a highly polygenic trait has a large number of underlying genes spread over the genome, so that any random set of genes has some chance of containing causal loci. In this setting, the more natural question becomes whether there is an *overrepresentation* of causal loci in the candidate set.

A final issue is the choice of members in a gene set. One entree into a potential pathway could be through a SNP showing genome-wide significance in a standard GWAS. If such a SNP can be assigned to a known gene, one could then test known associated pathways. Given the ascertainment bias created by choosing a (genome-wide) significant SNP, the entree gene should be removed from the gene set before proceeding. While it could simply be left out of the gene set, due to potential correlated effects among pathway members, a cleaner approach is to include it as a cofactor (as is often done in linkage-based QTL mapping to remove the effect of the leading QTL; Chapter 18). A related issue is that candidate pathways should only be considered when one (or more) of their associated SNPs show a nominal level of significance in the original GWAS (Sedeño-Cortés and Pavlidis 2014), otherwise one could simply search for trait–pathway associations until one is found by chance. Another issue is the size of the tested gene set, given the bias towards larger gene sets being more significant (Holmans 2010; K. Wang et al. 2010; Ramanan et al. 2012). As with many concerns in GSA, this bias can be controlled by using permutation for hypothesis testing. Lastly, beware of false positives generated by a single SNP. The complex systems nature of biological organisms dictates that components/products/intermediates from one pathway can also be involved in several others. Hence, a SNP with genome-wide significance in a GWAS that is involved in several (known) pathways might falsely generate signals in each, when in reality the impact of the gene associated with the SNP on the focal trait is through yet another, undiscovered, pathway.

As would be expected given our above brief comments, there is a massive literature on GWPA methods, reviews of which can be found in Huang et al. (2008), Yu et al. (2009), Holmans (2010), K. Wang et al. (2010), Fridley and Biernacka (2011), Khatri et al. (2012), Ramanan et al. (2012), Schaid et al. (2012), Maciejewski (2013), Mitrea et al. (2013), Tarca et al. (2013), Mooney et al. (2014b), Newton and Wang (2015), de Leeuw et al. (2016), Tamayo et al. (2016), and Xie et al. (2021). Given all of the possible analysis options in a GWPA, there is a real risk of "method-shopping" to find the best p value. Investigators thus must be as transparent as possible about the process of model selection that led to their final result. Some best practices for reporting GWPA results were suggested by Mooney et al. (2014a) and Mooney and Wilmot (2015).

Given all of these options, how should one proceed with a GWPA given a candidate gene set (or a small number of such sets)? A nice case study was given by Gui et al. (2011), looking at Crohn's disease. They examined seven different GWPA approaches, finding that immune-response related pathways tended to be significant, but the number of other detected pathways varied greatly depending on the approach used. Such a variance in outcomes is not unexpected. There is no single omnibus test for trait–pathway associations, as the power of detection is a function of the underlying trait architecture in a given pathway. If some causal pathways mainly influence the trait via a few genes of modest to large effect, while other causal pathways impact trait value through numerous genes of small effect, a method detecting the first class of pathways may have low power for the second class, and vice versa. To address this concern, Gui et al. recommended using several different methods, ideally choosing from self- versus competitive-hypothesis methods, and methods that used information from just a few SNPs in each gene (e.g., Min-p) and methods that incorporate the effects over all of the SNPs (e.g., GRASS).

Describing the Structure of Networks (Systems Biology)

To close our discussion on the ongoing merger of QG and functional genomics, we note that phenotypes are the output of highly complex systems involving a vast number of molecular intermediates that interact with each other and with their environment. The lowest level of organization of such systems are the effects of individual genes on a specified trait (GWAS). At the next level, these genes interact with each other in simple pathways (moving from some starting input to a final product, e.g., a metabolic pathway $A \rightarrow B \rightarrow C \rightarrow D$). Pathway-based GWAS can offer crude glimpses into these interactions by asking if a set of GWAS hits is enriched for targets in a particular pathway (or set of pathways). We can also examine regulatory (as opposed to a metabolic) pathways, and probe some of their features by searching for regQTLs. Individual pathways themselves are embedded into much more complex structures (**networks**), that, at some level, likely encompass every molecular feature within a cell. The newly emerging field of **systems biology**—an eclectic blend of concepts from physics, graph theory, cybernetics, biochemistry, and molecular and cellular biology—attempts to model the organization, evolution, and functional implications of these structures. Brief introductions to the field, mainly focusing on the topology of biological networks, and its implications, are given by Barabási and Oltvai (2004), Vidal et al. (2011), and Hu et al. (2016), while Civelek and Luisis (2014) examines trait variation from a systems biology viewpoint.

As this chapter has highlighted, the machinery of QG applies to any object that shows variation, and this includes network structures. As briefly introduced in Appendix 2, we can describe the static structure of a network with a matrix, $\mathbf{M}$, which describes how objects in a pathway (nodes, such as particular molecular features) are connected by edges (Figures 21.5, A2.9, and A2.10). Edges could be directional (an arrow indicating that one node influences another), or undirected (shown by a line such as in a protein–protein interaction map, indicating which proteins contact each other in a cell). The collection of nodes and their corresponding edges is called the topology of a network, and can obviously change over time. Vidal et al. (2011) have called this structure the **interactome**, and its study **edgetics**.

The next level of resolution beyond topology would be the strength of the edges of a network, and the highest level of resolution would be the dynamics over the network (how the edges change over time, both in terms of connectivity and strength). Very formally, one has a matrix $\mathbf{M}(t)$ describing the network at some time t, and a vector $\mathbf{m}(t)$ whose elements are the concentrations of various molecular features (again, at time t), and we wish to map $[\mathbf{M}(t), \mathbf{m}(t)] \rightarrow [\mathbf{M}(t + \delta t), \mathbf{m}(t + \delta t)]$. The full solution is an almost unthinkably complex problem, but one can chip away at it from two different directions. The first is top-down: do some basic features arise from such systems, largely independent of their component pieces? The answer is yes, and biological robustness (relative insensitivity to perturbations) might be one such **emergent property** of the topology of biological networks (e.g., Barabási and Oltvai 2004; Kitano 2004, 2007; Levy and Siegal 2008; Masel and Siegal 2009; Félix and Barkoulas 2015; Appendix 2). The second direction is bottom-up, in that components of $\mathbf{M}$ (such as the probability at a given node is connected to another specified one, and the strength of this connection) are quantitative traits, and the machinery of QG can be used to probe some of their features, such as the amount of pleiotropy and the nature of any underlying regQTLs. Finally, a growing number of modelers are examining the implications for network structures for generating QG features, such as epistasis and pleiotropy, e.g., Omholt et al. (2000), Ayroles and Zeng (2008), Kliebenstein (2009), and Hu et al. (2011). Balancing the energy and enthusiasm towards applying QG methods to pathway and network variation, Flint and Ideker (2019) caution that "the difficulties in integrating network and genetic data are under appreciated," and run the risk of a number of false positives, reminiscent of those seen in the early days of small-sample GWAS. It is clear that GWAS offers cautionary tales to keep in mind moving forward, the main one being that large sample sizes are critical, and yet most of the current data on network variation are based on very small samples.

WHAT DOES GENOMICS TELL US ABOUT TRAIT ARCHITECTURE?

Finally, we come to the two central questions in QG: (i) what is the nature of molecular processes that translate genetic into phenotypic variation (the **genotype-phenotype map, GP**); and (ii) how has evolution shaped both the nature of standing variation and the systems underlying the GP map. We start by reviewing the debate on whether trait variation is mainly the result of common alleles of small effect or rare alleles of large effect. This naturally leads to the **missing heritability** concern that was expressed in the early days of GWAS, wherein genome-wide significant SNPs accounted for only a small fraction of the heritability estimated using more classical approaches (e.g., Part II Chapters 22–26, 32, and 33). We then briefly examine the architectures of a few classic traits before finishing with models that try to synthesize many of the results from above and from Chapters 17–20 into a framework for how quantitative variation is generated. Our treatment of many of the technical population-genetic issues that arise here will be rather brief. Elsewhere we have reviewed the very rich theory of population genetics (WL Chapters 2–5 and 7–10) and discussed population-genetic models of selection on quantitative traits (WL Chapters 24–28), and we refer the reader to this material for a much deeper discussion.

Common versus Rare Alleles

An ongoing debate in QG is whether the bulk of genetic variance is due to a few alleles of large effect or many alleles of small effect (WL Chapter 28). This argument goes back to the rediscovery of Mendel at the start of the 1900s, and the ensuing debate on whether evolution was driven by mutation creating new alleles of large effect (the Mendelian view) or by selection acting on small-effect alleles over a large number of genes (the biometrician view). As mentioned in Chapter 1, Fisher formed the field of QG by considering Mendelian segregation over a large number of factors. While this framing merged key concepts from the Mendelians (genetics) and biometricians (modern statistics), it did not really answer the question about which effect-size class was evolutionarily more important.

As a result, this debate continues to resurface in many forms (WL Chapters 24–28). In the context of human disease, geneticists quickly separated rare diseases (population frequencies of generally less than one in a thousand) from "common" diseases (those occurring at higher frequencies), such as heart disease, diabetes, obesity, and various cancers. Many rare diseases were found to be caused by rare alleles of large effect, often (but not always) by disrupting protein structure (Botstein and Risch 2003). While it was slowly realized that common diseases likely had an important polygenic component (e.g., Carter 1969), there was divergent opinion on their underlying genetic architecture. The **common disease/common variant (CDCV)** hypothesis (Lander 1996; Cargill et al. 1999; Chakravarti 1999) posited that "common" alleles (frequencies $\geq$ 1– 5%) of small effect underlie most of the disease cases. Conversely, the **common disease/rare variant** hypothesis (**CDRV**)—also called the **heterogeneity hypothesis** (Frayling et al. 1998; Bodmer 1999), **rare variant hypothesis** (Bodmer and Bonilla 2008), or **multiple rare variant hypothesis** (Orozco et al. 2010)—posits that rare alleles underlie common diseases. Under the rare variant model, disease cases represent a massively heterogeneous collection of rare alleles, each of large effect. It is worth stressing that this debate is much deeper than a simple academic exercise, as its resolution has profound implications on the best strategies for isolating causal factors and represents two very different scenarios for the age, and evolution, of causal alleles. Finally, we stress that rare versus common is, of course, a false dichotomy, as both sources contribute variation to complex traits and diseases (Visscher et al. 2012b). The real question is what fraction of the variation is accounted for by a given marker-frequency class (either tagged SNPs, or, more ideally, causal sites). Overviews of many of the surrounding issues are discussed by Bodmer and Bonilla (2008), Gorlov et al. (2008), and Schork et al. (2009).

Practical implications of this debate concern both mapping and functional characterization. Considering the latter first, large-effect alleles often result from disruptions in causal gene products (altered protein or functional RNA sequences) or major disruptions in their regulation. As such, they may offer rather direct signals of causal features. In contrast, smaller effect alleles may represent much more subtle perturbations, such as small changes in regulatory features, and may be rather uninformative about the biology underlying a trait. Indeed, as we have seen, it is challenging to associate a significant SNP effect with a particular gene. Thus, even if most variation is from common alleles, the most important signals for a biologist may reside in rare alleles (Momozawa and Mizukami 2012; Chakravarti and Turner 2016; Faraone 2017; Ferraro et al. 2020; Hyman 2020).

As regards to mapping, the key implication of these two competing hypotheses follows from Example 20.1: the correlation, r^2, between linked alleles rapidly diminishes as their allele frequencies diverge. Hence, the typical marker scored on a SNP chip (usually with a MAF $\geq$ 0.05) does not tag rare causal alleles, but easily tags common causal alleles. Because rare alleles are hard to tag on SNP chips, they are often imputed (Chapter 20). However, imputation accuracy is a function of the underlying LD, and the quality (and especially size) of the reference population (at least ten copies of the allele must be present; Chapter 20). While the imputation quality for rare alleles can be very poor when using reference populations of modest size (e.g., Example 21.15), with the dramatic increases in the number of whole-genome sequences in reference biobanks, alleles with MAF of 0.1%

can be accurately imputed (Browning and Browning 2016; Browning et al. 2018; Huang et al. 2014; Hofmeister et al. 2023). Methods to accommodate rare alleles (Chapter 20) usually proceed by aggregation of variants (e.g., an excess of rare alleles at candidate genes in cases). If the CDCV model is correct, a high-powered GWAS should tag a very significant fraction of the genetic variance of a trait. If the CDRV model is correct, GWAS would only capture a small fraction of the trait variance, and rare allele methods must be used.

Whether causal alleles are mainly rare or mainly common also has profound evolutionary implications, as the frequency of an allele is informative about its history (WL Chapters 2, 8–10). Generally speaking, a *common allele is an old allele*, while a *rare allele is typically young*. This is certainly the case under pure drift (WL Equation 2.12), and also under many forms of selection (WL Chapter 5). Ancient alleles are thus likely no worse than very weakly deleterious, while more deleterious alleles tend to be much younger, and much rarer. While the odd allele or two can quickly rise to moderate frequency by positive selection (such as the sickle cell mutation under a malarial environment), these tend to be exceptions to common alleles being nearly neutral. Hence, the common versus rare allele debate reflects the nature of selection on a random causal allele for a trait (Examples 21.12 and 21.13).

Another interesting difference between loci harboring common versus rare alleles was noted in several population-genetic studies (Hartl and Campbell 1982; Slaktin and Rannala 1997; Reich and Lander 2001). These authors considered the degree of **allelic identity** at a causal locus. This is the probability that two randomly drawn disease alleles from a locus are IBD, the reciprocal of which has been called the effective number of alleles. For a population expansion following a bottleneck (such as modern humans), a locus whose total frequency of disease alleles is very small shows considerable heterogeneity (very low identity between risk alleles and thus a high effective number of such alleles), while a locus where risk alleles are more common shows much less heterogeneity. For example, Pritchard and Cox (2002) noted that a study of 424 Hemophilia B families found that most cases were caused by distinct alleles: there were 167 distinct Hemophilia B mutations, the most frequent of these only accounted for 5% of all mutations. Conversely, the major ΔF508 allele of cystic fibrosis accounts for roughly 2/3 of all European CF-carrying chromosomes. While there is a predominant CF allele, rare alleles also occur, with roughly 270 different mutations seen among 27,000 CF-harboring chromosomes. This heterogeneity difference between loci with common versus rare alleles is a *transient* feature, reflecting past demography, but it is a signal that can persist for thousands of generations until the numerous initial copies of the common allele are slowly replaced by new mutations (Reich and Lander 2001). As noted by McClellan and King (2010), the history of human bottlenecks leads to the seemly paradoxical observation that "most human *variation* is ancient and shared, but most *alleles* are recent and rare." Indeed, Talium et al. (2021) performed WGS on almost 54,000 humans, finding over 400 million variants, 90% of which were SNPs, and the rest indels. Over both variant groups, 97% had frequencies less than 1%, and 46% of each group were singletons (a number that rose to 53% when a subset of 41,000 unrelated individuals was used). The implications of McClellan and King's comment is that common (and older) alleles likely survived migration bottlenecks and are therefore widely dispersed across human populations. Conversely, most alleles are rare variants, often representing variants not present before the migrations out of Africa, and thus are more regional. If most causal variants are common, they are likely widely shared across human populations, but if they are rare, they may be very region-specific.

A hybrid model in the rare versus common debate was offered by Dickson et al. (2010). Their **synthetic association model** (**SA**) postulated that a chromosomal segment tagged by a common SNP might, by chance, have accumulated a number of rare variants, and, in total, these could generate a detectable SNP marker effect. Given the very low expected r^2 between a common and rare allele (Example 20.1), the resulting rare variant effects would have to be rather substantial to generate a detectable maker effect. Orozco et al. (2010) noted that such large-effect variants, if they exist, could be detected within pedigrees (as opposed to a population sample) by standard linkage methods (Chapter 19). They suggested that the

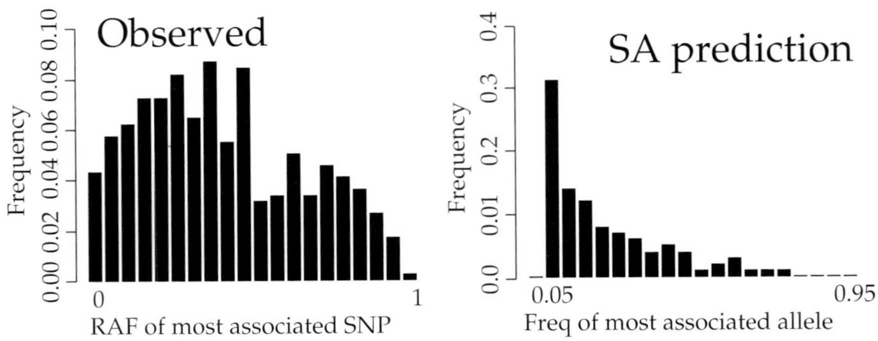

Figure 21.6 **Left:** The observed distribution of frequencies of lead SNPs tagging risk alleles (RAF) for 17 common human diseases. **Right:** This distribution would be highly skewed toward zero (becoming very L-shaped) if synthetic associations between a tag SNP and a collection of rare variants are a common GWAS feature. (After Wray et al. 2011.)

lack of such detectable associations implies that synthetic associations are not a major source of GWAS variance. An even more direct line of evidence against SA was given by Wray et al. (2011). Their simulations found that the frequency distribution for tagged SNP showing the largest marker effect would be highly skewed toward zero if SA were common (Figure 21.6). This skew arises because a haplotype containing a SA of rare variants harbor several "common" SNPs, with the largest signal being assigned to the least common of these (i.e., the common SNP with the lowest population frequency). This is simply a manifestation of the fact that the largest r^2 (and hence largest signal) occurs for the SNP with the least marker-causal SNP frequency mismatch (Example 20.1). As shown in Figure 21.6, a survey of GWAS results for 17 common human diseases instead showed a distribution shifted toward intermediate, rather than low-frequency, SNPs. Finally, simulations by Thornton et al. (2013) showed that tagged markers tend to be in strong LD with only a single deleterious mutation of large effect.

Example 21.12. As mentioned, the rare versus common debate is really about the nature of the **allelic spectrum** for a given trait/disease: the joint distribution of the effect size (a) and frequency (x) for causative alleles. The feature that connects a and x is the strength of selection (s) on a given allele, which generates a population distribution, $\phi(x|s)$, of allele frequencies as a function of the nature of selection. A largely unresolved issue is how the effect size a of a new mutation impacts the nature of selection (s) it experiences. Selection could be generated by a *direct* impact of the focal trait on fitness (such as the difference between early versus late onset alleles) and/or by *indirect* pleiotropic effects on other fitness components. A common assumption (e.g., WL Chapters 25 and 28) is that larger effect mutations are expected to have more deleterious pleiotropic effects, resulting in stronger selection against them, generating an inverse correlation between effect size and allele frequency. This is indeed a fairly common observation across humans, maize, and yeast (e.g., Park et al. 2011; Simmons et al. 2014; Wallace et al. 2014a; Zeng et al. 2018; Bloom et al. 2019; Glassberg et al. 2019; Schoech et al. 2019), but one must correct for Beavis effects that overestimate $|a|$ values for rare allele (e.g., Iles 2008). This inverse relationship has important implications for the impact of rare variants.

To see this, consider the situation where the underlying causal alleles are entirely neutral. In this setting, the effect size should be independent of allele frequency. The simplest model for $\phi(x)$ is the Watterson distribution (WL Equation 2.34b), where the probability (across evolutionary replicates) of a minor allele having frequency x is proportional to $[x(1-x)]^{-1}$. The resulting additive variance contributed by a site with frequency x is thus expected to be

$$\sigma_A^2(x) \cdot \phi(x) \propto 2a^2x(1-x) \cdot [x(1-x)]^{-1} = \text{constant} \qquad (21.10a)$$

The resulting fraction of the total additive variance for a trait under this model from alleles of frequency $x \leq p$ is thus p, implying that rare alleles ($x \leq 0.01$) only account for 1% of the total genetic variation. For rare alleles to have a much greater impact on the total variance, a^2 must increase as x decreases, and/or *more* rare alleles are present than predicted under the Watterson model. Even for strictly neutral alleles, the latter is true in humans, as the Watterson assumption is a long-term stable population size, while populations passing through bottlenecks and subsequent expansion display an excess of rare alleles (WL Chapter 2). Further, selection also inflates the number of rare alleles relative to Watterson. Are these factors sufficient to create a prominent role for rare alleles? At least in humans, models suggest that this is unlikely.

This was nicely illustrated by Zeng et al. (2018) and Schoech et al. (2019), who used a two-component mixture model for the additive effect a of an allele by assuming (for SNP j),

$$p(a_j \mid x) = \delta_0 \cdot \pi_0 + N(0, [2x_j(1-x_j)]^S \sigma_a^2) \cdot (1-\pi_0) \tag{21.10b}$$

where $(1-\pi_0)$ is the **polygenicity**, the fraction of all SNPs that impact a trait, the delta function δ_0 denotes a point mass at zero ($a = 0$), and S is a selection parameter. A value of $S < 0$ implies that average a^2 values increase as x decreases (corresponding to negative selection against alleles), while $S = 0$ corresponds to a neutral assumption of no correlation between a^2 and x. MCMC (Appendix 8) can be used to estimate the model parameters S, σ_a^2, and π_0, an approach Zeng et al. called **BayesS**. Note that by rearranging Equation 2.3b,

$$E[a^2 \mid x, a^2 > 0] = \sigma^2(a \mid a^2 > 0) + (E[a \mid a^2 > 0])^2 = [2x_j(1-x_j)]^S \sigma_a^2 + 0^2 \tag{21.10c}$$

showing (assuming $E[a] = 0$) that the variation associated with SNPs with an MAF of x is

$$([2x_j(1-x_j)]^S \sigma_a^2) \cdot 2x_j(1-x_j) = [2x_j(1-x_j)]^{1+S} \sigma_a^2 \tag{21.10d}$$

Both Zeng et al. (2018) and Schoech et al. (2019) considered over two dozen, largely non-overlapping, traits/diseases from the UK Biobank. Zeng et al. found that all but one of their traits had a negative estimate of S (ranging from -0.609 to 0.012), 24 of which were significantly negative, with a median S of -0.37. The polygenicity $(1 - \pi_0)$ had a median value of 5.4% and ranged from 0.6% to 14.0%. Schoech et al. obtained very similar results for S. Substituting these S values into various population-genetic models for $\phi(x|s)$ showed that no more than 10% of the variance could be due to rare alleles ($x \leq 0.01$).

A more recent study by Zeng et al. (2021) used a summary-statistic version (**SBayesS**) of BayesS, and examined a much larger set of traits, finding almost all had significantly negative estimates of S (with a median value of -0.6). They estimated that, on average across traits, around 1% of the human genome are mutational targets ($1 - \pi_0 \simeq 0.01$) for that trait, with a mean selection coefficient for segregating variants of -0.0007. Relative to other traits, common diseases showed a smaller mutational target size but stronger selection (π_0 smaller, $|S|$ larger). More granularity can be applied to values of S by examining its value over different functional classes of variants (i.e., different annotations), such as nonsynonymous mutations, methylation sites, etc. For example, Gazal et al. (2018), examining 27 traits (from the UK Biobank), found an average value of $S = -1.10$ for nonsynonymous variants and $S = -0.30$ for other variants.

A final complication in the effect size–MAF relationship is that causal alleles in regions of lower LD (relative to the value expected given their MAF) tend to be both more recent and to also have larger effects (Gazal et al. 2017). Again, this is expected if most trait alleles are under weak negative selection. We examine this, and other, complications further in Part II Chapter 33 in a deeper discussion of estimation, and interpretation, of SNP heritability (the fraction of variation accounted for my SNP effects).

Example 21.13. There is a very detailed, and highly technical, literature from evolutionary genetics on the maintenance of quantitative genetic variation (reviewed in WL Chapter 28), and a largely independent corresponding literature that focuses on polygenic risk variants for human diseases (e.g., Pritchard 2001; Pritchard and Cox 2002; Di Rienzo and Hudson 2005;

Peng and Kimmel 2007; Eyre-Walker 2010; Maher et al. 2013; Simons et al. 2014, 2018, 2022; O'Connor et al. 2019). While mathematically impressive, most of these models are *very fragile*, in that apparently trivial changes in assumptions can yield *quantitatively different* predictions. In particular, the choice of modeling how pleiotropy generates correlations between effect size (a) and selection (s) has profound implications. An excellent overview of these issues is given by Johnson and Barton (2005), with a more GWAS-focused review by Sella and Barton (2019), and much more detailed discussion in WL Chapter 28.

One common modeling assumption for maintenance of variation is **stabilizing selection** on our focal trait (fitness declines as one moves away from some optimal value; WL Chapters 5, 16, 29, 30) countered by mutation. If, on a per-locus basis, the strength of selection is weak relative to mutation (the **Gaussian assumption**), the resulting equilibrium distribution of allelic effect sizes is normal, with many small-effect alleles occurring at intermediate frequencies (the common variant model). Conversely, if selection is strong relative to mutation (the **house-of-cards assumption**), we recover the rare variant model, as the resulting distribution is leptokurtic (Chapter 2), with rare alleles of large effects (see WL Table 28.2 for other differences between these models).

As mentioned, when pleiotropy occurs (the variant impacting our focal trait also influences others), the expected outcomes become very model dependent. Simons et al. (2018) developed a model allowing for extensive pleiotropy, wherein a variant impacting a focal trait (with effect a_1) also influences a number of other traits (i.e., generates a vector $(a_1, \cdots, a_n)$ of effects for the n traits it impacts), with the vector of trait values under multivariate stabilizing selection. In this setting, deviation from the optimal value in any direction reduces fitness (WL Chapters 29 and 30), with surfaces of equal fitness satisfying $s = \sum a_i^2/V_s$ (where V_s is the strength of stabilizing selection). Using this assumption, they made a very clever use of Fisher's geometric model for adaptation (WL Chapter 27) to obtain a distribution of effect sizes (a_1) at the focal trait for a mutation with a given selection coefficient, which for a large effective number of traits ($n_e > 10$) is approximately normal, with $a_1 \mid s \sim N(0, sV_s/n_e)$. Applying their model to height and BMI data, they estimated that the fraction of variation arising from alleles of detectable effect sizes is much greater for height (50% of the total variation in height should be detectable using current GWAS sizes) than for BMI (only 15% of the variance should be detectable). Further, they estimated that the total mutation rate for height variants was around five times that for BMI variants.

Distribution of Allelic Effect Sizes

While theoretical models (Examples 21.12 and 21.13) make predictions about the distribution of allelic effect sizes, what do the data say? A number of approaches have been used to estimate the effect-size distribution, starting with either fitting a single exponential or gamma distribution (Otto and Jones 2000; Hayes and Goddard 2001; Weller et al. 2005), and moving to mixture distributions, typically based on weighted sums of normals (Chapter 16). The original use of mixtures traces back to attempts by animal breeders to use marker data to predict breeding value (Meuwissen et al. 2001). The basic data used are typically (but not always) GWAS summary statistics, such as the slope β of the gene-dosage regression (Equation 20.1a) or its logistic regression counterpart (Equation 20.3b), which is essentially an estimate of the allele effect a. One then fits the observed distribution of β values with a mixture model. For example, the **BayesR** method fits a point mass at zero (δ_0) and then three normals with increasingly smaller variances (Erbe et al. 2012), corresponding to allelic classes with increasingly smaller effects,

$$p(a_j) = \pi_0\delta_0 + \pi_1 N(0, \sigma_g^2 \cdot 10^{-2}) + \pi_2 N(0, \sigma_g^2 \cdot 10^{-3}) + \pi_3 N(0, \sigma_g^2 \cdot 10^{-4}) \qquad (21.11a)$$

Under this model, a fraction π_0 of all SNPs have no effect, so that $\pi = (1 - \pi_0)$ is the polygenicity. The π_i and the variance parameter σ_g^2 can be estimated by either ML (Appendix 4) or MCMC (Appendix 8). Holland et al. (2020) coined the term **discoverability** for the effect size variance (essentially, the expected power of detecting variants drawn from that size class), with effects being drawn from the last normal in Equation 21.11a having the lowest discoverability. Note that Equation 21.11a, unlike the models given by Equation 21.10, does not condition the effect size on allele frequency. This is because it effectively assumes the Watterson distribution, so that the expected contribution to the variance from alleles with MAF x is independent of x (Equation 21.10a). We return to this assumption in Part II Chapters 32 and 33.

Moser et al. (2015) applied BayesR to seven of the diseases studied in the WTCCC (Chapter 20): bipolar disorder (BD), coronary artery disease (CAD), Crohn's (CD), hypertension (HT), type 1 and 2 diabetes (T1D, T2D), and rheumatoid arthritis (RA). With the exception of T1D, the other six diseases had an estimate of over 60% of their total variance due to alleles drawn from the smallest effect size (π_4, corresponding to a variance of $\sigma_g^2 \cdot 10^{-4}$), with four of these diseases having over 75% due to this class: CD (94%), HT (88%), CAD (84%), and T2D (78%). The estimates for T1D were roughly 25% of its variance due to alleles drawn from this category, while almost 72% of the variance was attributed to the largest effect group (π_2, variance of $\sigma_g^2 \cdot 10^{-2}$).

Similar mixture models have been examined by Thompson et al. (2015), Zhang et al. (2018), Holland et al. (2020), and O'Connor (2021). All concluded that traits are massively polygenic (normally with at least tens of thousands of estimated causal sites), and that GWAS sample sizes in the hundreds of thousands to millions are required to capture the bulk of the heritability. For example, Zhang et al. used a model with point mass at zero, and then either one or two normals (respectively, their **M2** and **M3** models),

$$a_j \sim \pi_0\delta_0 + \pi_1 N(0,\sigma_1^2) + \pi_2 N(0,\sigma_2^2) \tag{21.11b}$$

with the M2 model not including the last term. M3 generally performed better, and was used to estimate the effect-size distribution for 32 traits. They found that most of the variance was due to small-effect alleles, with the required GWAS sample size to capture 80% of the heritability ranging from a few hundred thousands to several millions. O'Connor (2021) noted that the range of effect size between the 10th and 90th percentiles was around a 100-fold difference, and around 600-fold between the 5th and 95th percentiles. Based on these estimates, genetic architectures are massively polygenic, with a vast range of effect sizes.

Example 21.14. O'Connor et al. (2019) noted that the measure $\pi = (1 - \pi_0)$ for polygenicity is a bit misleading, and a more general metric would be how even effects are distributed over causal loci (essentially, an **effective number of causal sites**). They noted that schizophrenia is underpinned by thousands of small-effect common variants, while Niemi et al. (2018) found that rare severe neurodevelopmental disorders, which are largely expected to be monogenic (caused by rare, highly deleterious mutations), still have around 8% of their variance explained by common alleles of very small effect. Polygenicity as measured by the fraction (π) of SNPs with an impact on the trait would be roughly the same in both these settings, and yet clearly their architectures are very different (polygenetic for schizophrenia; Mendelian, with variable penetrance, for neurodevelopmental disorders). O'Connor suggested that a better polygenicity measure is the *evenness of effect sizes* over the causal sites, which would be high with schizophrenia and very low (dominated by a few loci) with neurodevelopmental disorders.

The contribution of allelic effect to the variance scales as β^2, and the variance of these variance effects scales as β^4. If M SNPs (trimmed for independence) are scored, the expected number that are causal is $M\pi$, where we will now refer to π as the **polygenicity fraction**, while **O'Connor metric** for the effective number of causal SNPs is

$$M_e = M\,\frac{3}{\kappa_4}, \qquad \text{where} \quad \kappa_4 = \frac{E[\beta^4]}{(E[\beta^2])^2} \tag{21.12}$$

where κ_4 is the scaled kurtosis of the β and equals three when β follows a normal distribution (Chapter 2). We denote Equation 21.12 as the **polygenicity kurtosis**. Note that only a fraction π of the β are different from zero, and if these effects follow a normal distribution, then

$3M/\kappa_4 \simeq \pi M$, the same number of causal SNPs predicted using the polygenicity fraction. O'Connor et al. developed an extension of LD regression (Equation 20.13c), **stratified LD fourth moment regression (S-LD4M)**, to estimate M_e.

Applying their estimator to 33 traits (with an average GWAS sample size of 360,000), they found that M_e based on just considering common SNPs (MAF $> 5\%$) had effective number of causal sites ranging from 500 to 30,000 (with a median value around 3000). When low frequency SNPs (MAF from 0.5% to 5%) were used, the estimated M_e values were about 25% of the common M_e values. For example, height had around 3,600 effective causal sites for common alleles and 800 effective sites for low-frequency alleles. Hence, common alleles had a more even distribution of effect sizes than low-frequency alleles.

Given that selection obviously plays a critical role in shaping trait architecture, Simons et al. (2022), building on their 2018 stabilizing selection model (Example 21.12), reached a fairly striking conclusion regarding the distribution of fitness effects for QTL alleles. First, they were able to obtain an approximate expression for the distribution of effect sizes, as a function of the distribution $f(s)$ of fitness effects of new mutations (generated by direct, and pleiotropic, effects) and two scaling parameters: the mutational target size L, and h^2/L, the average per-site heritability. ML estimates of these parameters—including fitting $f(s)$ with a four-knot spine function—was then performed using the estimated β and q values for 95 diverse continuous traits from the UK Biobank. While estimates of h^2, and (especially) L, varied over traits, the data were well-fit by using a single functional form of $f(s)$ over all traits. They concluded that most of variation seen in architectures (at least for this set of human traits) was driven by differences in the scaling parameters H^2 and L. Shockingly, fitness was not a significant driver of *differences* over traits, as most underlying loci for all traits had their fitness values drawn from the *same distribution* of s values. If confirmed, this is indeed a remarkable observation, and we will end the chapter by suggesting some potentially underlying drivers of this observation.

Genetic Architectures and "Missing" Heritability

The decade following the first GWAS was period of both considerable excitement and a good deal of consternation. Despite billions having been spent globally on extensive GWAS studies and the corresponding development of genomic resources, most detected hits were of small effect (each significant SNP individually accounted for only a tiny fraction of the trait heritability) were largely in noncoding regions and cumulatively accounted for only a tiny fraction of the trait heritability. This angst culminated with the highly cited paper of Manolio et al. (2009) on the problem of **missing heritability** not detected by the GWAS studies of the time. For example, human height has a highly repeatable heritability estimate (from resemblance between relatives) of around 0.7–0.8, and despite over 40 significant GWAS hits for height (in 2009), these accounted for only about 5% of the phenotypic variance (a little over 7% of the expected heritability). As Manolio et al. summarized, this was the standard observation over a number of diseases and traits, raising the question of what generated this gap between GWAS-based and relative-based estimates of heritability (the former often called the **SNP heritability**; Part II Chapter 33). More formally, the gap was between the additive variance estimated from SNP effects, $\sum a_i^2 p_i(1-p_i)$ (the sum being taken over all significant SNPs, which can be modified for LD), and value of σ_A^2 estimated from standard relative-based designs (Part II Chapters 22–33).

Proposed explanations for this gap fell into three, not necessarily exclusive, categories. First, this was simply an issue of power with the GWAS designs at the time. Second, there were genetic features not fully captured by a GWAS, such as the effects of CNV, epigenetics, and perhaps other, yet to be discovered, phenomena. The final explanation was that the GWAS results were indeed correct, with the gap arising because relative-based estimates of h^2 were systematically (and dramatically) inflated. A number of different viewpoints were offered on these various resolutions (e.g., Eichler et al. 2010), generating a considerable number of papers, some thoughtful, some speculative, and some rather ignorant of basic features of QG.

As we have hinted throughout this chapter, it now appears that the correct resolution is *lack of power*. Before developing this point, we first discuss overestimation of h^2 by relative-based methods, which has some grain of truth. As detailed in Chapter 7, estimation of the additive variance follows by partitioning the phenotypic covariance for a given pair of relatives in various sources of genetic variation (Equation 7.12) plus for any shared environmental effects. When relatives can only share (at most) single IBD alleles, the shared genetic variance is dominated by σ_A^2, with potentially much smaller contributions for additive epistatic sources (e.g., $\sigma_{AA}^2, \sigma_{AAA}^2$, etc.). For relatives that can share two alleles IBD (e.g., full sibs, monozygotic twins), dominance (and dominance epistatic terms) can also enter into the genetic correlation. Zuk et al. (2012) correctly noted that if these nonadditive terms were sufficiently large, relative-based estimates of h^2 can be significantly inflated.

However, these nonadditive genetic components (i.e., those other than σ_A^2) are generally expected to contribute very little to the resemblance between relatives for two reasons. First, the coefficients on nonadditive variance terms for the contribution for resemblance—$(2\theta)^k\Delta^\ell$ for $\sigma^2(A^kD^\ell)$—are much less than the coefficient on the additive variance (Equation 7.12). Second, and perhaps more fundamentally, in most segregating natural populations, the nonadditive variance components themselves are expected to be much smaller than the additive variance. This is simply a consequence of most genetic variation loading onto the additive component when minor alleles tend to be at low (to very low) frequencies (Hill et al. 2008; Hill 2010; Mäki-Tanila and Hill 2014). As the Watterson distribution shows, this is expected to be the case in natural populations. Consistent with the prediction, GWAS-based estimates of dominance variation in humans are typically very small and cannot account for much of the missing heritability (Zhu et al. 2015; Hivert et al. 2021; Palmer et al. 2023). For example, Hivert et al. considered 70 traits and a sample size of over 250,000, finding an average SNP heritability for additive effects of 0.208 and an average value of 0.001 for dominance. The exception to additive variation overpowering nonadditive variances is in segregating populations formed by inbred line crosses, where segregating allele frequencies (by construction) are 0.5 (Examples 21.16 and 21.17), resulting in more of any underlying nonadditivity mapping into nonadditive *variance* components (Chapter 5).

Relative-based estimates of the additive variance can also be inflated by shared environmental effects among close relatives. Two different approaches can be used to estimate additive variation while minimizing any impact from common environments: use extended pedigrees so that shared environmental effects are likely more diffuse than within a single family (Part II Chapters 32 and 33) or exploit the variation in IBD sharing among the same set of relatives caused by Mendelian segregation (realized versus pedigree kinship; Chapters 7 and 8). A pair of sibs that, by chance, shares more IBD alleles should be more similar than another pair that, also by chance, shares a smaller fraction of IBD alleles, with a regression of their squared difference on the amount of sharing estimating the additive variance. This approach was developed for sib pairs by Visscher et al. (2006) and extended to general sets of relatives by Young et al. (2018). By contrasting IBD differences between sibs (or more general relatives) in the same family (kinship), shared environmental effects are controlled. Both of these approaches have been applied to human traits. Zaitlen et al. (2013), in an analysis of 23 traits in a large Icelandic sample (38,000) with deep pedigrees, found that h^2 estimates based on close relatives (parent–offspring, sibs) were somewhat inflated relative to estimates based on more distant relatives (and hence less likely to share environmental features). For most traits, the inflation was modest (about 15% for height), so that while it does slightly close the gap between GWAS and relative estimates, a very substantial difference still remains. Similar results (relative-based estimates of σ_A^2 were slightly to modestly inflated) were seen when estimates of additive variance based on within family segregation were compared to more traditional estimates (Young et al. 2018; Young 2019; Kemper et al. 2021).

The second explanation for missing heritability—features not captured by a GWAS—is both speculative and increasingly less viable as ever-larger GWAS have closed much of the heritability gap (Example 21.15). However, there is still debate on what is causing the lack of power: common alleles with increasingly smaller effects, or increasingly rarer alleles of large effect. As GWAS sample sizes increases, ever smaller additive variances, $r^2 2p(1-p)a^2$, associated with a marker can be detected, where r^2 is the LD with the tagged marker and p the causal allele frequency. For a rare allele, this is $\sim r^2 2a^2 p$ when the tagging marker is scored directly and $\sim r^2_{imp}(r^2\, 2a^2 p)$ when the marker is first imputed (where r^2_{imp} is the correlation between the actual marker allele and its imputed value). For a common allele, the additive variance tagged by a marker approaches $a^2/4$. This arises as the MAF approaches 0.5, in which case r^2 likely approaches one as most common SNPs are well tagged. For a rare causal allele r^2 (with a common marker SNP) is still expected to be small, being a decreasing function of the causal allele frequency (Example 20.1). It is this expected small value of r^2 that motivated the synthetic association (SA) model, as only a tiny fraction of the true variance of large effect, but rare, alleles would be captured by common SNPs (underestimating their contribution by $1/r^2$). This would be a 10-fold effect for $r^2 = 0.1$ (which would still be a relatively high correlation LD between a common marker allele and a very rare causal allele). However, as we have detailed, the data do not support the SA model (e.g., Figure 21.6).

What impact might whole genome sequencing (WGS) have on the search for missing heritability? The answer depends on whether the heritability gap is due to common alleles of very small effect or very rare alleles of large effect. The impact of WGS is to improve the accuracy of tagging, namely increasing r^2 to one. Common causal alleles are either already captured directly, or tagged with very high accuracy by other common alleles in high LD, so that WGS is unlikely to increase tagging efficiency (Caballero et al. 2015). In this case, only increased sample size (rather than scoring more markers) can generate the increase in power needed to detect common alleles of very small effect.

Rare alleles are a potentially different matter. If rare alleles are accurately captured with high efficiency by imputation from the scored SNPs ($r^2_{imp} \simeq 1$), then WGS, by itself, does not really improve power. However, if very rare alleles are poorly imputed, for example by residing in regions of low LD, then WGS results in a significant increase in power. This was indeed seen for height (Wainschtein et al. 2022; Example 21.15). One caution is that the use of very rare alleles introduces the potential for subtle biases. In particular, as Chapter 20 stressed, corrections for population structure are critical, especially when searching for very small allelic effects in a very large GWAS. Most such corrections are based on common (and thus older) alleles, with the population structure for rare (and thus younger) alleles being potentially rather different. For example, individuals sharing a very rare allele are likely to have shared a recent relative, and thus the potential of sharing a recent common environment (Young 2019).

Human height is one of the best studied traits, and offers considerable insight (and history) into the missing heritability debate. As Example 21.15 highlights, the bulk of the initial missing heritability was due to common alleles of small effect, but very recent whole-genome sequence data find that a substantial proportion is also due to very rare alleles of large effect in low-LD with common SNPs.

Example 21.15. Human height is truly the ultimate quantitative trait. Galton's (1886) work on the resemblance in height between parents and their adult offspring (Figure 1.1) was instrumental to Fisher's 1918 founding of QG. Height has been the subject of a number of ever-larger GWAS projects, and the history of these studies nicely tracks the journey from the beginning of the GWAS era, to navigation through the rough seas of missing heritability, and into the whole genome sequencing era.

While the most widely quoted figure for the heritability of height is 0.8, Yang et al. (2015) suggested that this is a slightly inflated estimate due to shared environmental effects among relatives (typically parent–offspring or sibs). They noted that Hemani et al. (2013) obtained an estimate of $h^2 \simeq 0.7$ based on IBD variance between pairs of sibs. This matches the estimate of 0.7 by Zaitlen et al. (2013) using extended pedigrees. A more recent study by Kemper et al.

(2021) also obtained an IBD-sharing-based estimate of around 0.7 using full sibs, but a larger estimate (0.8) using a more general set of relatives. Hence, in the following discussion we will use the 0.7 value as a *lower bound* for the h^2 that must be explained by markers.

Gudbjartsson et al. (2008) examined 40,000 individuals (mostly of European descent) and discovered 27 genomic regions containing one or more variants associated with height. The estimated effect sizes ranged from 0.3 to 0.6 cm and together explained around 4% of the phenotypic variance in height (6% of h^2, assuming a value of 0.7). By two years later, Lango Allen et al. (2010) had detected almost 200 loci achieving genome-wide significance in a sample of 180,000, which accounted for around 14% of the heritability. One obvious source of missing heritability was the exclusion of markers with nontrivial effects, but which had effect sizes too small to meet the stringent genome-wide significance threshold. Lango Allen examined this issue by considering SNPs that were of nominal, but not genome-wide, significance. When included, the fraction of explained heritability rose to 23%. Four years later, Wood et al. (2014b) had found 700 variants with genome-wide significance in a sample of 253,000, accounting for 20% of the heritability. Removing the restriction to SNPs of genome-wide significance, the most strongly associated 2,000, 3,700, and 9,500 SNPs accounted for 26%, 30%, and 36% of the heritability. Yengo et al. (2018b) performed a meta-analysis with close to 700,000 individuals, finding 3,300 independent marker effects that accounted for roughly 35% of the heritability. Extrapolating from estimates of the effect-size distribution, Zhang et al. (2018) predicted that a sample of over a million would be required to account for 80% of the heritability. How good was this prediction? Yengo et al. (2022) extended their meta-analysis to over 5.4 million individuals, largely European, but with other ancestries as well. While 18% of the roughly one million SNPs examined achieved genome-wide significance, when SNP correlations were removed, this amounted to around 12,000 independent common SNPs. These were located in 7,200 distinct genomic regions, ranging from 70 to 700 kb, covering around 20% of the genome and accounted for 40% of the phenotypic variance ($\simeq$ 60% of h^2), falling far short of the expected value projected from estimates of the effect size distribution. Over 30% of these regions were more than 50 kb away from any known genes.

The clear trend is that the expected number of detected sites increases with sample size, with the roughly 135-fold sample size increase from 2008 to 2022 resulting in a roughly 120-fold increase in genome-wide significant markers, a 270-fold increase in the number of genomic regions, and a 10 fold increase in the amount of heritability assigned to markers. In the words of Yengo et al. (2018b), "increasing GWAS sample sizes continues to deliver." However, Yengo et al. (2022) suggested that GWAS saturation for common alleles may have been reached, as they predict most newly discovered variants will be restricted to the 7,200 genomic regions found in their 5.4 million GWAS meta-analysis. Holland et al. (2020) estimated the number of causal sites as roughly 95,000 (based on estimates of π), with a similar estimate by Boyle et al. (2017a). Because these studies used common SNPs as markers, these effects are, in large part, due to common causal alleles, and thus of relatively modest effects. As noted in Example 21.14, O'Connor et al. (2019) suggested that a better estimate of the polygenicity is the effective number of sites (the more equal the effect size distribution, the larger the effective number). They estimated that height had 3,600 effective causal sites for common alleles (MAF $>$ 0.05) and 800 effective sites for low-frequency alleles (MAF of 0.005 to 0.05). They offered no estimate for very rare alleles (MAF $<$ 0.005).

An alternative to assigning effect sizes to individual markers (fixed-effects modeling) is to treat allelic effects as random and estimate the total amount of variation they explain. Using common markers, Yang et al. (2010b, 2011b, 2015) applied this approach to samples of 4,000 and 44,000, with the marker variance accounting for 62% and 80% of the heritability. Hence, the variance component approach is more powerful for detecting missing heritability, as it includes information from *all* markers, not just those declared significant by some criteria. However, note that the 12,000 variants of Yengo et al. (2022) accounted for over 85% of this common-SNP based estimate of h^2. An especially interesting observation was made by Yang et al. (2011b), who noted that the length of a chromosome was very highly correlated with the amount of variation explained by markers on that chromosome (a similar observation was seen for schizophrenia by Lee et al. 2012). Hence, height-associated markers are, to a first approximation, randomly and uniformly, scattered throughout the genome. However, the more recent analysis of Yengo et al. (2022) found that, while widely scattered, only about 20% of the genome contained regions harboring height variants. Loh et al. (2015b) similarly inferred that over 70% of 1-MB genomic regions in humans harbor at least one causal site for schizophrenia risk.

The variance component approach can also be used with WGS data. Wainschtein et al. (2022) estimated the SNP-associated variance in a sample of 25,500 fully sequenced individuals, considering all SNP alleles present in at least five copies in the sample (a MAF of 0.0001, $\sim$ 900,000 SNPs). Following Yang et al. (2015), they used a variance component model (VC) and estimated a SNP-based heritability of 0.48 (68% if h^2 = 0.7). They then used this same set of SNPs to impute rare alleles, with the heritability estimate using this enlarged set (common plus imputed SNPs) ranging from 0.50 to 0.56 (depending on the imputation model). Thus, including imputed alleles resulted in a slight improvement of the amount of explained variation. They then use WGS for directly score rare alleles, and the resulting heritability estimate improved significantly to 0.7. Most of the improvement came from very rare variants (0.0001 $<$ MAF $<$ 0.001) that were in low LD with common SNPs, and hence with very low imputation accuracy. Thus, while common variants of small effects recover about 2/3 of the relative-based heritability, very rare variants account for the majority of the remaining difference.

Example 21.16. A very interesting, and somewhat different, perspective on missing heritability is offered from the sibship analysis of segregating yeast crosses (which serves as a good representative for any line-cross QTL experiment). Example 21.1 discussed the genetic architecture of yeast eQTLs, while physiological traits (colony growth size under different environmental features) have also been examined (e.g., Ehrenreich et al. 2010; Bloom et al. 2013, 2019) and we focus on the latter. Bloom et al. (2013) examined 46 growth-related traits in just over 1,000 haploid segregant lines from a cross between a laboratory and a vineyard strain. Taking this collection of lines (sibs) as our population, the additive variation within this sibship can be estimated using the variation in IBD sharing among sibs (Chapter 8). Bloom et al. used this approach to estimate narrow-sense heritability values. Broad-sense heritability can be estimated from the between-line variance (Part II Chapter 24), with the difference between the two heritabilities being a measure of the impact of any nonadditive variance (these would only be additive epistatic terms, because the lines are haploid). They then mapped QTLs and examined the fraction of the narrow-sense heritability they recovered, which ranged over traits from 72% to 100%, with a median value of 88%. By considering ever-larger subsets of the lines, they showed that this fraction increased with the number of lines (the analog to accounting for more of the missing heritability by increasing a GWAS sample size). They also found that nonadditive variance was significant, with the difference between narrow- and broad-sense heritability ranging from essentially zero to roughly 50%, depending on the trait. Two-locus epistasis (e.g., $A \times A$) explained only a tiny fraction of this difference.

By construction, there are no rare alleles in the collection of segregating lines, as all have expected frequency 1/2. Bloom et al. (2019) expanded this study by considering multiple crosses, namely 14,000 segregants from all pair-wise crosses between 16 diverse lines (so that the rarest allele frequency is expected to be around 6% in the resulting synthetic population). Again, detected QTLs accounted for the vast majority of the estimated narrow-sense heritability (median value of 68%). However, Bloom et al. noted that rare alleles (those alleles segregating in the lines, but found at frequencies $<$ 0.01 in a 1,000-isolate reference collection) accounted for a disproportionate amount of the additive variance. Such alleles were about 28% of all variants, but their median total contribution to the additive variance was 52%. Note that this does not necessarily provide support for rare alleles having a large impact in a standard GWAS, as their frequencies were artificially high due to the population construction, but does indicate that rare alleles tend to have much larger effects.

Example 21.17. While much of our focus on massive GWAS studies, rightly so, has been on those from human genetics, there is also an impressive list of results from maize (Buckler et al. 2009; Brown et al. 2011; Kump et al. 2011; Poland et al. 2011; Tian et al. 2011; Wallace et al. 2014a, 2014b; Xiao et al. 2016, 2017). Wild maize lines have LD that typically decays in under 2 kb, offering much higher mapping resolution than in humans. This feature, coupled with the NAM line collection (Chapter 18), makes maize a powerful system for fine mapping. The NAM lines are a set of 200 RILs from each cross of 25 maize lines (globally sampled) to a common parent (B73), capturing the variation present over these 26 diverse lines (Figure 18.12). The resulting set of 5,000 inbred lines can be highly replicated, offering a considerable increase in power (estimating the phenotypic value of a genotype by the average over a series of clones versus from a single observation for an outbred individual). A typical NAM GWAS design scored nearly a million plants in four different growing locations over two years (eight environments). Traits examined include flowering time, leaf architecture, pathogen resistance, and inflorescence features (male tassels and female ears). Ignoring inflorescence traits, most traits were dominated by additive alleles of small effect with little detectable epistasis, with most ($>$ 80%) of the additive variance captured by SNPs. G $\times$ E was observed, but the variance it accounted for was much less than the variance from additive effects. Allelic heterogeneity was common, with alleles replicated over some, but not all, families, and different variants segregating at the same genomic location over the collection of families (often with opposite effects with respect on the B73 reference allele). Detected QTL were sometime enriched for known candidate genes, but most QTL did not appear to be associated with any known candidates (loci at which major-effect alleles have been detected). The genetic architecture of inflorescence traits was bit different from the other traits in that they tended to have much larger effects, with ear effects being further larger than tassel effects. Pleiotropic loci were detected that controlled both ear and tassel elongation, again with the ear effects being larger. Brown et al. (2011) suggested that this architectural difference for ears resulted from recent selection on ear morphology during domestication, favoring fixation of large-effect alleles (WL Chapter 27).

Beyond the Infinitesimal Model

The emerging view of quantitative trait variation is that it is massively polygenic, with the average variance attributable to any causal site being very small, and with a range of allelic-effect sizes spanning several orders of magnitude. This pattern is similar to, but subtly different from, Fisher's **infinitesimal model**, in which a very large number of loci, each of very small effect, underlies trait variation (WL Chapter 24). While *allelic effect sizes* do not follow the infinitesimal model, the behavior of the *variance components* associated with the underlying sites is much closer to Fisher's characterization. The small range in the per-site variances (relative to the much larger range of underlying allelic-effect sizes) is a result of the inverse correlation between effect size and allele frequency. This association, best considered as more of a trend than an absolute rule, appears to be at least partly driven by larger effect alleles generally being more deleterious than smaller effect ones.

While we may have some mild clarity on the evolutionary underpinning of trait variation, two apparently perplexing observations pose deep questions for its underlying molecular basis. The first is the massive number of causal sites. For example, estimates of the number of sites impacting height are up to three times the total number of protein-coding genes in the human genome (Example 21.15). Second, almost all variation appears to map to noncoding regions. How can we account for these observations? Clearly, a central issue in the modeling of quantitative variation is the role of gene regulation.

The early days of molecular genetics saw a transition from a focus on simple prokaryotic gene switches (such as the *lac* operon in *E. coli* or the *cro-c1* switch in phage λ) to more elaborate models for eukaryotes where gene regulation was set within a far wider net of regulatory players (e.g., Britten and Davidson 1969). As high throughput functional genomics tools became available, it seemed that most features in a cell were connected to one another. Cellular networks are typically structured as **domains** (or **modules**) of highly interconnected players, separated from other such domains by a few tenuous, but none the less present, connections. More formally, many biological networks appear to be **small world** (Appendix 2), meaning that one can connect any two nodes (or more generally, local hubs organizing subnetworks) via a very small number of steps. Further, as we have seen, there are layers of regulatory elements that can act in much more complex, and far subtler, ways than those envisaged in the early days of simple gene switches with just a few discrete components and sites of action. Thus, it became apparent that trait variation is largely determined by the actions of very complex regulatory networks (e.g., Chakravarti and Turner 2016). One interesting complication to this network view of quantitative variation was noted by He (2017): the structure of biological networks (being small world and scale free; Appendix 2) can generate considerable homeostasis. Hence, networks are both sufficiently robust to perturbations, but also loose enough to allow a certain amount of variation to leak through.

It is against this backdrop that Pritchard and colleagues (Boyle et al. 2017a, 2017b; Liu et al. 2019) proposed their **omnigenic model** (omni = all). Their tenet that "the connected-network aspect of the omnigenic model is a parsimonious model that can potentially explain the major observations" was simply a codification of the widespread belief that one must understand networks in order to understand quantitative variation. Their key innovation was to consider the network generating trait variation as consisting of a limited number of **core genes**, modified (in *trans*) by a much larger set of **peripheral genes** that have very slight individual, but very strong collective, regulatory impacts on the core genes. Under this model, *any* gene that shows regulatory variation in a tissue that impacts the focal trait can potentially have a small effect (hence the term "omni").

While initially a bit nebulous as to what exactly corresponded to a core gene, a more formal definition based on the concept of mediation was proposed by Boyle et al. (2017b) and Liu et al. (2019). Core genes directly impact the trait (for example, by coding for QTTs), while peripheral genes impact the regulation of these traits via mediation. Hence, conditional on core genotypes and expression levels, a trait is not further impacted by the genotypes or transcripts at any peripheral gene. However, as remarked by Liu et al., "it is important to note that our definition of core genes is a simplification of a more complex reality," who further noted that the real operational key to the omnigenic model is that "trait heritability is mainly driven by peripheral genes that *trans*-regulate core genes." Under this model, most large-effect (and therefore rare) variants are from the core genes, being either strong *cis* acting sites (such as eSNPS and sSNPS) or structural variants, while the more numerous (and of smaller effect) common allele sites are almost entirely associated with the peripheral gene set. This view was slightly modified by Liu et al. (2019) who noted that a master regulator peripheral gene (co-regulating a number of the core genes) could have a large effect via the summation of a number of small effects over each impacted gene. A more detailed look at the impact of core genes was offered by Sinnott-Armstrong et al. (2021), who examined the genetic architectures of three traits (urate, IGF-1, and testosterone) whose major pathway components are fairly well defined. While they found that the lead SNPs were highly concentrated in core genes, only 10%–20% of SNP-based h^2 was due to variants in core pathway components. Estimates of the total number of causal sites ranged between 4,000 (testosterone) and 12,000 (urate).

Boyle et al. (2017a) hypothesized that the omnigenic model offered a useful search strategy to better characterize quantitative variation, by focusing on detecting variants in the core genes. They suggested whole exome sequencing in an attempt to detect rare, but functionally informative, mutations in these genes. Such variants, for the most part, would identify at least some of the core genes, which may provide useful biological insight. However, Wray et al. (2018) noted that a key component in GWAS-based functional characterization is always sample size, and that the ever-increasing sample sizes of GWAS can both detect rare alleles while also furthering our understanding of sites with small effects. They suggested a better use of resources is to simply increase GWAS sample sizes in general, rather than amplifying a specific class of sites (e.g., whole exome sequencing).

Finally, the omnigenic model may provide one explanation for the observation of Simons et al. (2022) discussed in Example 21.14—the distribution of selection coefficients at underlying QTLs appears to be largely trait-independent. The proposed web of peripheral genes might generate much of this common distribution. Simons et al. (2022) further noted that their analysis largely ignored rare alleles, exactly those expected to have large selection coefficients, as would likely occur with core genes. Hence, significant fitness differences over traits given their core gene set could occur, but the common background of peripheral genes would drive most of the QTLs of modest effects.

As our discussion about the omnigenic model highlights, the next generation of QG methodology will heavily focus on biological networks as multivariate traits. Some of the machinery offered in this chapter will no doubt be massively extended in the near future to provide biological, computational, and analytic tools to continue the exploration of these structures.

III

Appendices

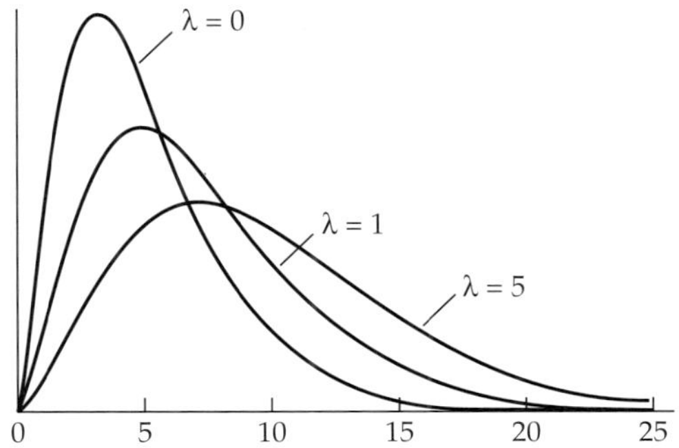

Appendix 1

Expectations, Variances, and Covariances of Compound Variables

Far better an approximate answer to the right question, which is often vague, than the exact answer to the wrong question, which always can be made precise. John Tukey

Many situations arise in quantitative genetics in which a measurement or a statistic is a complicated function of one or more variables. This raises many serious difficulties that are at the heart of statistical analysis. We are interested ultimately in the distribution and accuracy of the observed statistic, but at best we may only know something about the distribution of its component variables, and even that is known with error.

As an example of these problems, consider $\beta_{yx} = \sigma(x,y)/\sigma^2(x)$, the regression coefficient, which is a ratio of a covariance and a variance. With finite sample sizes, the observed statistics Cov(x,y) and Var(x) are only estimates of $\sigma(x,y)$ and $\sigma^2(x)$. Thus, in order to have some confidence in the estimate b_{yx} of β_{yx}, we need to know something about the accuracy of Cov(x,y) and Var(x). Moreover, we need to know how sampling variance of Cov(x,y) and Var(x) translates into sampling variance of b_{yx}, i.e., we need to know the sampling variance of a ratio. The problem is exacerbated because the same individuals are typically used to estimate Cov(x,y) and Var(x), which causes these two estimates to be correlated.

Fortunately, there is a fairly general approximation (often referred to as the **delta method**) for dealing with these issues. This appendix provides a brief overview of the theory underlying the delta method and uses it to derive a number of fundamental results that will prove helpful throughout the book. The delta method is a *distribution-free* approximation, making little assumptions about how the data underlying the statistics are distributed. Bayesian methods (Appendices 7 and 8) provide another, more model-based, approach to these issues. By specifying both some prior distribution for the parameters of interest (often taken to be uninformative, namely, all potential values are equally likely) and a likelihood function, Bayesian methods generate a posterior distribution for the parameters of interest, given any prior information and the current data. MCMC methods (Appendix 8) allow one to draw values from this distribution, and hence by extensive resampling one can generate an empirical probability distribution for any function of interest.

THE DELTA METHOD

Consider an arbitrary expression f, which is a function of x. Performing a Taylor series expansion around an arbitrary constant c,

$$f = f(c) + (x-c)\frac{\partial f(c)}{\partial x} + (x-c)^2\frac{\partial^2 f(c)}{2\,\partial x^2} + (x-c)^3\frac{\partial^3 f(c)}{3\cdot 2\,\partial x^3} + \cdots \tag{A1.1}$$

where $f(c)$ refers to the function evaluated at $x = c$, and the partial derivatives are first evaluated with respect to x, after which c is substituted for x.

Example A1.1. Consider the function $f = \sqrt{x}$. Here, $\partial f/\partial x = 1/(2\sqrt{x})$ and $\partial^2 f/\partial x^2 =$

$-1/(4x^{3/2})$, giving the second-order Taylor series about c as

$$f = \sqrt{x} \simeq \sqrt{c} + \frac{x-c}{2\sqrt{c}} - \frac{(x-c)^2}{8c^{3/2}} + \cdots$$

Expectations of Complex Variables

As our first example of the utility of the Taylor expansion, consider the case where x is a random variable and we wish to determine the expected value of the function f averaged over all x. Generally speaking, ***the mean value of a function is only equal to the function evaluated at the mean of x in the special cases in which the function is linear in x or x is a constant***. For example, a function f is called **convex** when its second derivative is always greater than or equal to zero (e.g., $f = x^2$, $f = e^{-x}$). **Jensen's inequality** states that for any convex function

$$E[f(x)] \geq f\left(E[x]\right) = f(\mu_x) \tag{A1.2}$$

e.g., $E[x^2] \geq \mu_x^2$ and $E[e^{-x}] \geq e^{-\mu_x}$. Hence, we cannot just directly substitute the sample mean when trying to evaluate the mean of some function of the data. However, we can circumvent this problem by expanding f about the mean of x, using Equation A1.1 with $c = \mu_x$, and then taking the expectation,

$$\begin{aligned} E[f] &= E\left[f(\mu_x) + (x-\mu_x)\frac{\partial f(x)}{\partial x}\bigg|_{x=\mu_x} + (x-\mu_x)^2 \frac{\partial^2 f(x)}{2\partial x^2}\bigg|_{x=\mu_x} + \cdots \right] \\ &= f(\mu_x) + E[x-\mu_x]\frac{\partial f(x)}{\partial x}\bigg|_{x=\mu_x} + E[(x-\mu_x)^2]\frac{\partial^2 f(x)}{2\partial x^2}\bigg|_{x=\mu_x} + \cdots \end{aligned} \tag{A1.3a}$$

The last step follows because the derivative terms, evaluated at μ_x, are just constants. Further simplification is possible because, by the definition of a mean, $E[x - \mu_x] = 0$, and $E[(x-\mu_x)^2]$ is the expected variance of x, $\sigma^2(x)$. Thus, ignoring third- and higher order terms,

$$E[f] \simeq f(\mu_x) + \sigma^2(x)\,\frac{\partial^2 f(x)}{2\partial x^2}\bigg|_{x=\mu_x} \tag{A1.3b}$$

This relationship shows that the expected value of f is a function of both the mean and the variance of x if f is a nonlinear function of x. Note for a convex function that the second term is always non-negative, recovering the inequality given by Equation A1.2.

From Example A1.1, if $f = \sqrt{x}$, then $\partial^2 f/\partial x^2 = -1/(4x^{3/2})$, yielding

$$E[\sqrt{x}\,] \simeq \sqrt{\mu_x} - \frac{\sigma^2(x)}{8\,\mu_x^{3/2}} \tag{A1.3c}$$

This same approach can be used to derive expressions for the expectations of functions that depend on more than a single variable. In this case, f is expanded around the means of each of the component variables. With two component variables, e.g., an expansion around μ_x and μ_y leads (after some algebra) to

$$\begin{aligned} E[f] = f(\mu_x,\mu_y) &+ \sigma^2(x)\,\frac{\partial^2 f(x,y)}{2\partial x^2}\bigg|_{x=\mu_x,y=\mu_y} \\ &+ \sigma(x,y)\,\frac{\partial^2 f(x,y)}{\partial x\,\partial y}\bigg|_{x=\mu_x,y=\mu_y} + \sigma^2(y)\,\frac{\partial^2 f(x,y)}{2\partial y^2}\bigg|_{x=\mu_x,y=\mu_y} + \cdots \end{aligned} \tag{A1.4a}$$

When f is a function of n variables, $(x_1, x_2, \ldots, x_n)$, ignoring partials higher than the second order, this generalizes to

$$E[f] \simeq f(\mu_{x_1}, \mu_{x_2}, \cdots, \mu_{x_n}) + \frac{1}{2}\sum_{i=1}^{n} \sigma^2(x_i) \left.\frac{\partial^2 f}{\partial x_i^2}\right|_{x_1=\mu_{x_1},\cdots,x_n=\mu_{x_n}}$$
$$+ \sum_{i=1}^{n}\sum_{j>i}^{n} \sigma(x_i, x_j) \left.\frac{\partial^2 f}{\partial x_i \partial x_j}\right|_{x_1=\mu_{x_1},\cdots,x_n=\mu_{x_n}} \tag{A1.4b}$$

Example A1.2. By definition, the variance, $\sigma^2(x)$, is equal to the average value of $(x_i - \mu_x)^2 = x_i^2 - \mu_x^2$ in a population, where x_i denotes the measure for the ith individual. The population mean, μ_x, is generally not known with certainty, but rather is estimated by $\overline{x}$. Thus, it is tempting to estimate $\sigma^2(x)$ by using the mean value of $x_i^2 - \overline{x}^2$. Here we show that this quantity is a (slightly) downwardly biased estimate of $\sigma^2(x)$, and that there is a simple solution to this problem. Expanding $x_i^2 - \overline{x}^2$, we wish to know whether the expected value of the function

$$f_i = x_i^2 - \left(\frac{\sum_{j=1}^{n} x_j}{n}\right)^2$$

is equal to $\sigma^2(x)$. Under the assumption that the population has been sampled randomly, $\mu_{x_i} = \mu_{x_j} = \mu_x$, $\sigma^2(x_i) = \sigma^2(x_j) = \sigma^2(x)$, and $\sigma(x_i, x_j) = 0$ for all i, j. Equation A1.4b then reduces to

$$E[f_i] = f_i(\mu_x) + \frac{\sigma^2(x)}{2}\left(\frac{\partial^2 f_i}{\partial x_i^2} + \sum_{j \neq i} \frac{\partial^2 f_i}{\partial x_j^2}\right) \tag{A1.5a}$$

The required partial derivatives are

$$\frac{\partial f_i}{\partial x_i} = 2x_i - \frac{2\sum_{j=1}^{n} x_j}{n^2} \qquad \frac{\partial^2 f_i}{\partial x_i^2} = 2\left(1 - \frac{1}{n^2}\right)$$

$$\frac{\partial f_i}{\partial x_j} = -\frac{2\sum_{k=1}^{n} x_k}{n^2} \qquad \frac{\partial^2 f_i}{\partial x_j^2} = -\frac{2}{n^2}$$

Because all higher order partials are equal to zero, Equation A1.5a is *exact* in this case. In addition, substitution of μ_x for x_i and all x_j, shows that $f_i(\mu_x) = 0$. Substituting the partial derivatives into Equation A1.5a gives

$$E\left[x_i^2 - \overline{x}^2\right] = \sigma^2(x) \cdot \left(\frac{n-1}{n}\right)$$

This shows that the mean value of $x_i^2 - \overline{x}^2$ gives a slightly downwardly biased estimate of the population variance, $\sigma^2(x)$. The problem is eliminated by simply multiplying the mean value of $x_i^2 - \overline{x}^2$ by $n/(n-1)$. Thus, an unbiased estimate of $\sigma^2(x)$ is

$$\text{Var}(x) = \left(\frac{n}{n-1}\right)\left(\overline{x^2} - \overline{x}^2\right) \tag{A1.5b}$$

$$= \frac{1}{n-1}\sum_{i=1}^{n}(x_i - \overline{x})^2 \tag{A1.5c}$$

This example foreshadows the REML method for variance estimation, which corrects the sum of squares for the number of fixed effects estimated (Part II Chapter 33). Here only the mean, μ

is estimated. In more complex settings, p fixed effects are estimated, leading to sum of squares term of the form $(x_i - \sum_j^p \beta_{ij})^2$, and the denominator becomes $n-p$. Failure to adjust for the estimation of fixed effects results in biased variance estimates that are too small.

Variances of Complex Variables

The preceding approach can also be used to obtain an expression for the variance of a function. Again expanding around $c = \mu_x$, and substituting for f from Equation A1.1 and $E[f]$ from Equation A1.3a,

$$\begin{aligned}
\sigma_f^2 =& E\left[\{f - E[f]\}^2\right] \\
=& E\left[\left\{\left(f(\mu_x) + (x-\mu_x)\frac{\partial f(\mu_x)}{\partial x} + \cdots\right) - \left(f(\mu_x) + \sigma^2(x)\frac{\partial^2 f(\mu_x)}{2\partial x^2} + \cdots\right)\right\}^2\right] \\
=& E\left[\left\{(x-\mu_x)\frac{\partial f(\mu_x)}{\partial x} + [(x-\mu_x)^2 - \sigma^2(x)]\frac{\partial^2 f(\mu_x)}{2\partial x^2} + \cdots \quad \right\}^2\right]
\end{aligned} \tag{A1.6}$$

Ignoring all but the two lowest order terms, and noting that $E[x-\mu_x] = 0$,

$$\begin{aligned}
\sigma_f^2 \simeq\ & E\left[(x-\mu_x)^2\right]\left[\frac{\partial f(\mu_x)}{\partial x}\right]^2 + 2E\left[(x-\mu_x)^3\right]\left[\frac{\partial f(\mu_x)}{\partial x}\right]\left[\frac{\partial^2 f(\mu_x)}{2\partial x^2}\right] \\
& - 2E[x-\mu_x]\sigma^2(x)\left[\frac{\partial f(\mu_x)}{\partial x}\right]\left[\frac{\partial^2 f(\mu_x)}{2\partial x^2}\right] + E\left[(x-\mu_x)^4\right]\left[\frac{\partial^2 f(\mu_x)}{2\partial x^2}\right]^2 \\
& - 2E\left[(x-\mu_x)^2\right]\sigma^2(x)\left[\frac{\partial^2 f(\mu_x)}{2\partial x^2}\right]^2 + \sigma^4(x)\left[\frac{\partial^2 f(\mu_x)}{2\partial x^2}\right]^2 \\
=\ & \sigma^2(x)\left[\frac{\partial f(\mu_x)}{\partial x}\right]^2 + 2\mu_{3x}\left[\frac{\partial f(\mu_x)}{\partial x}\right]\left[\frac{\partial^2 f(\mu_x)}{2\partial x^2}\right] \\
& + \left[\mu_{4x} - \sigma^4(x)\right]\left[\frac{\partial^2 f(\mu_x)}{2\partial x^2}\right]^2
\end{aligned} \tag{A1.7a}$$

where $\mu_{3x} = E[(x-\mu_x)^3]$ and $\mu_{4x} = E[(x-\mu_x)^4]$ are the third and fourth moments about the mean of x (Chapter 2). When f is a function of two variables, Equation A1.6 expands to

$$\begin{aligned}
\sigma_f^2 = E\Bigg[\Bigg\{ & (x-\mu_x)\frac{\partial f(\mu_x,\mu_y)}{\partial x} + (y-\mu_y)\frac{\partial f(\mu_x,\mu_y)}{\partial y} \\
& + \left[(x-\mu_x)^2 - \sigma^2(x)\right]\frac{\partial^2 f(\mu_x,\mu_y)}{2\partial x^2} + \left[(x-\mu_x)(y-\mu_y) - \sigma(x,y)\right]\frac{\partial^2 f(\mu_x,\mu_y)}{\partial x\,\partial y} \\
& + \left[(y-\mu_y)^2 - \sigma^2(y)\right]\frac{\partial^2 f(\mu_x,\mu_y)}{2\partial y^2} + \cdots \quad \Bigg\}^2\Bigg]
\end{aligned} \tag{A1.7b}$$

where $\sigma(x,y)$ is the covariance of x and y. An approximation, often used in place of Equation A1.7b, is obtained by ignoring all but the first-order terms. Then, if f is a function of n variables,

$$\sigma_f^2 \simeq \sum_{i=1}^{n}\sum_{j=1}^{n}\sigma(x_i,x_j)\left(\frac{\partial f}{\partial x_i}\right)\left(\frac{\partial f}{\partial x_j}\right) \tag{A1.7c}$$

where $\sigma(x_i, x_j)$ is a variance if $i = j$ and a covariance otherwise, and the partial derivatives are evaluated at the expectations for all underlying variables.

Example A1.3. Imagine sampling a population many times independently for n individuals, each time estimating the mean. Can we express the variance of the sample means, $\sigma^2(\overline{x})$, in terms of the variance of individual measures, $\sigma^2(x)$?

Recall that the definition of a sample estimate of the mean is

$$\overline{x} = \frac{x_1 + x_2 + \cdots + x_n}{n}$$

Under the assumption that the population is sampled randomly, there is no covariance between the measures of different individuals, so that Equation A1.7c reduces to

$$\sigma_f^2 \simeq \sum_{i=1}^{n} \sigma^2(x_i) \left(\frac{\partial f}{\partial x_i} \right)^2 \tag{A1.8a}$$

where in this example $f = \overline{x}$. The partial derivative of $\overline{x}$ with respect to each individual measure x_i is simply $1/n$, and assuming a homogeneous population, $\sigma^2(x_i) = \sigma^2(x)$ for all i. Thus, substituting into Equation A1.8a, the sampling variance of a mean is

$$\sigma^2(\overline{x}) = \sum_{i=1}^{n} \frac{\sigma^2(x_i)}{n^2} = \frac{\sigma^2(x)}{n} \tag{A1.8b}$$

i.e., the sampling variance of a mean is equal to the variance of individual measures divided by the sample size. The practical utility of this expression might be questioned because the parameter $\sigma^2(x)$ is something that we can only estimate. However, recall from Example A1.3a that an unbiased estimator of $\sigma^2(x)$ is $\text{Var}(x) = n(\overline{x^2} - \overline{x}^2)/(n-1)$. It follows that

$$\text{Var}(\overline{x}) = \frac{\text{Var}(x)}{n} \tag{A1.8c}$$

is an unbiased estimator of $\sigma^2(\overline{x})$. The square root of $\text{Var}(\overline{x})$ is known as the **standard error of the mean.**

The practice of substituting an observed (and, ideally, unbiased) statistic for a population parameter in sampling-variance equations is widely used to obtain approximate sampling variances of statistics. Because the accuracy of formulations employing such estimates increases with sample size, these formulations are usually referred to as **large-sample variance** expressions, with their square roots yielding **standard errors** of the statistic. It is often possible to use the standard error to construct a **confidence interval** around the estimate, such that the parametric value is encompassed within the confidence limits with a certain probability. However, the construction of confidence intervals for complex functions is generally very difficult not only because of the approximate nature of the variance estimates but also because the forms of distributions of complex functions are usually unknown.

As pointed out in Chapter 2, if it is reasonable to assume that the statistic of interest is approximately normally distributed, then the estimate plus or minus twice the square root of the sampling variance provides an estimate of the 95% confidence interval. Recall that the interpretation of such an interval does not mean that there is a 95% chance that it encloses the unknown parameter. Rather, it implies that if the experiment is replayed numerous times, then 95% of the time the CI will indeed enclose the true value. For variables that are not normally distributed, a more general statement can be made. Recall Chebyshev's theorem (Equation 2.18): if f and μ_f are observed and parametric values of a function, the probability that μ_f is in the range $f \pm k\sigma(f)$ is at least $1-(1/k^2)$. For example, the probability that a parametric value lies within two (true) standard deviations of its estimate is at least 3/4 regardless of the sampling distribution of the estimator. The probability increases to 95% if $k \simeq 4.5$. In other words, ± 4.5 times the standard error of an estimate provides a very conservative estimate of the 95% confidence interval.

Finally, an alternative (but very model-dependent) approach is to use Bayesian methods (Appendix 7). One can use the resulting posterior to generate draws of the random variables of interest (Appendix 8), using these to obtain random draws of the value of any function of interest (i.e., one draws a value of x and then computes $f[x]$). The resulting empirical distribution can be used to obtain most measures of interest on the uncertainty of $f(x)$.

Covariances of Complex Variables

The above procedures can also be used to evaluate the covariance between two compound functions f and g determined by common variables $(x_1, \ldots, x_n)$. Again ignoring all but the first-order terms,

$$\sigma(f, g) = E\left[\{f - E[f]\}\{g - E[g]\}\right] \simeq \sum_{i=1}^{n}\sum_{j=1}^{n} \sigma(x_i, x_j)\left(\frac{\partial f}{\partial x_i}\right)\left(\frac{\partial g}{\partial x_j}\right) \tag{A1.9}$$

Example A1.4. Consider two linear functions of the set of variables $(x_1, \cdots, x_n)$,

$$f = \alpha_1 x_1 + \alpha_2 x_2 + \cdots + \beta_n x_n$$

$$g = \beta_1 x_1 + \beta_2 x_2 + \cdots + \beta_n x_n$$

What is the covariance between f and g? The partial derivatives with respect to the x_i variables are $\partial f/\partial x_i = \alpha_i$ and $\partial g/\partial x_i = \beta_i$. Substituting into Equation A1.9,

$$\sigma(f, g) = \sum_{i=1}^{n} \alpha_i \beta_i \sigma^2(x_i) + \sum_{i=1}^{n}\sum_{j \neq i}^{n} \alpha_i \beta_j \sigma(x_i, x_j)$$

This is an exact result, because here all of the higher order partial derivatives are equal to zero. Equation 9.20 expresses this in matrix form.

SAMPLE VARIANCES OF VARIANCE AND COVARIANCE ESTIMATORS

Because quantitative genetics is often concerned with quantifying components of variance and covariance, it is useful to have expressions for the variance of their estimators. The general rules set forth in the previous sections are readily applied to these calculations, because variances and covariances are linear functions of sums of variables and their squares and cross-products. The rest of this appendix develops some specific applications of these general approximation procedures, largely by example.

Example A1.5. Here we consider, $\sigma^2[\text{Var}(x)]$, the sampling variance of a variance estimate. This quantity can be thought of as the expected variance that would arise among variance estimates obtained from a large number of independent samples from the same population. With n observations $(x_1, x_2, \cdots, x_n)$,

$$f = \text{Var}(x) = \frac{n}{n-1}\left[\frac{\sum_{i=1}^n x_i^2}{n} - \left(\frac{\sum_{i=1}^n x_i}{n}\right)^2\right]$$

Provided that the individual measures of x are obtained from random members of the same population, the covariance between all measures is zero, and the variance associated with each measure is the same, $\sigma^2(x)$. Thus, each of the n observed variables makes an identical contribution to the variance of the estimator Var(x). The partial derivatives with respect to measure x_i, evaluated at its mean, are

$$\frac{\partial f}{\partial x_i} = \frac{n}{n-1}\left(\frac{2x_i}{n} - \frac{2\sum_{j=1}^n x_j}{n^2}\right)\Bigg|_{x_i=\mu_x} = \frac{n}{n-1}\left(\frac{2\mu_x}{n} - \frac{2n\mu_x}{n^2}\right) = 0$$

$$\frac{\partial^2 f}{\partial x_i^2} = \frac{n}{n-1}\left(\frac{2}{n} - \frac{2}{n^2}\right) = \frac{2}{n}$$

Substituting into Equation A1.7a, the variance of Var(x) caused by variation in the ith measure is $[\mu_{4x} - \sigma^4(x)]/n^2$, and summing over all n measures,

$$\sigma^2[\text{Var}(x)] = \frac{\mu_{4x} - \sigma^4(x)}{n} \tag{A1.10a}$$

When x is normally distributed, $\mu_{4x} = 3\sigma^4(x)$ (Kendall and Stuart 1977), giving

$$\sigma^2[\text{Var}(x)] = \frac{2\sigma^4(x)}{n} \tag{A1.10b}$$

Equations A1.10a and A1.10b are exact expressions for the sampling variance of a variance because all partial derivatives higher than second order are equal to zero.

Our final problem is to modify Equation A1.10b in such a way that an unbiased estimate of the sampling variance of $\sigma^2(x)$ can be obtained from the sample statistic Var(x). It is tempting to simply substitute $[\text{Var}(x)]^2$ for $\sigma^4(x)$, but this is not quite correct. Recalling the definition of a variance, $E(z^2) = \sigma^2(z) + \mu_z^2$, and substituting Var$(x)$ for z, and $\sigma^2(x)$ for μ_z, we find that

$$E[\{\text{Var}(x)\}^2] = \sigma^2[\text{Var}(x)] + [\sigma^2(x)]^2$$

Substituting for $\sigma^2[\text{Var}(x)]$ from Equation A1.10b,

$$E[\{\text{Var}(x)\}^2] = \left(1 + \frac{2}{n}\right)\sigma^4(x)$$

which shows that the quantity $n\,[\text{Var}(x)]^2/(n+2)$, rather than $[\text{Var}(x)]^2$, provides (under normality assumptions; Equation A1.10b) an unbiased estimate of $\sigma^4(x)$. Thus, an unbiased estimate of the sampling variance of a variance is given by

$$\text{Var}[\text{Var}(x)] = \frac{2[\text{Var}(x)]^2}{n+2} \tag{A1.10c}$$

General information on the variances and covariances of moments can be found in Chapter 10 of Kendall and Stuart (1977). Letting $m_r = n^{-1}\sum(x_i - \overline{x})^r$ and $\mu_r = E[(x - \mu_x)^r]$

represent observations and expectations for the rth moment about the mean, the following approximations apply to the sampling variances and covariances,

$$\sigma^2(m_r) \simeq \frac{1}{n}\left(\mu_{2r} - \mu_r^2 + r^2\mu_2 \cdot \mu_{r-1}^2 - 2r\mu_{r-1} \cdot \mu_{r+1}\right) \tag{A1.11}$$

$$\sigma(m_r, m_q) \simeq \frac{1}{n}\Big(\mu_{r+q} - \mu_r \cdot \mu_q + r \cdot q \cdot \mu_2 \cdot \mu_{q-1}$$

$$- r \cdot \mu_{r-1} \cdot \mu_{q+1} - q \cdot \mu_{r+1} \cdot \mu_{q-1}\Big) \tag{A1.12}$$

where $\mu_0 = \mu_1 = 0$. Ideally, in the application of any of these formulae, unbiased estimates of the moments (μ_r) should be employed (as illustrated in Examples A1.3 and A1.5). For example, the covariance of an observed mean and a moment about that mean is

$$\sigma(\overline{x}, m_r) = \frac{1}{n}(\mu_{r+1} - r \cdot \mu_2 \cdot \mu_{r-1}) \tag{A1.13a}$$

Thus, an unbiased estimate of the covariance between estimates of the mean and the variance (estimated from the same sample) is

$$\text{Cov}[\,\overline{x}, \text{Var}(x)\,] = \frac{\text{Skw}(x)}{n} \tag{A1.13b}$$

where Skw(x), given as Equation 2.7, is an unbiased estimate of the third moment. For symmetrical distributions, such as the normal, all odd moments have expectations equal to zero, in which case the covariance between mean and variance is zero.

The following results apply to the second-order moments of a bivariate normal distribution (Equation 9.24). The variance of a covariance estimate is

$$\sigma^2\left[\,\text{Cov}(x, y)\,\right] = \frac{\sigma^2(x)\,\sigma^2(y) + \left[\,\sigma(x, y)\,\right]^2}{n} \tag{A1.14}$$

The covariance of the variance and the covariance estimates sharing a common variable is

$$\sigma\left[\,\text{Cov}(x, y), \text{Var}(x)\,\right] = \frac{2\sigma^2(x)\,\sigma(x, y)}{n} \tag{A1.15a}$$

Finally, the covariance of the variance estimates for two variables is

$$\sigma[\,\text{Var}(x), \text{Var}(y)\,] = \frac{2\left[\,\sigma(x, y)\,\right]^2}{n} \tag{A1.15b}$$

All three of these expressions can be used to obtain large-sample variance and covariance estimates by substituting, on the right-hand side of the expressions, observed moments for their expectations. Such expressions are very slightly upwardly biased, by factors of $(n+1)/n$ or $(n+2)/n$, but because the estimates of variances and covariances of higher order moments are highly unreliable unless the sample size is large, this distinction is trivial for most practical purposes.

Expressions for the variances and covariances of moments about the origin (as opposed to about the mean) can be found in Kendall and Stuart (1977, p. 244), and for the variances and covariances of other bivariate moments in Kendall and Stuart (1977, p. 250).

Example A1.6. What is the expected sampling variance of the directional selection differential S? In Chapter 3, it is shown that for any character whose phenotype is denoted by z, S

is equivalent to the covariance between z and relative fitness w, i.e., $S = \sigma(z, w)$. Applying Equation A1.14, we obtain

$$\sigma^2(S) = \frac{\sigma^2(z)\sigma^2(w) + [\,\sigma(z, w)\,]^2}{n} \tag{A1.16a}$$

Note that Equation A1.14 assumes z and w are bivariate normally distributed, implying that we assume that fitness has some optimal value and falls off (roughly quadratically) around this optimum. Some insight into the relative magnitude of the sampling variance of S can be acquired by considering the coefficient of sampling variation,

$$\text{CV}(S) = \frac{\sigma(S)}{E(S)}$$

where $E(S) = \sigma(z, w)$. Letting ρ be the correlation between phenotype and relative fitness, this reduces to

$$\text{CV}(S) = \frac{1}{\rho}\left(\frac{1+\rho^2}{n}\right)^{1/2} \tag{A1.16b}$$

The minimum value of CV(S) arises when the character is the sole determinant of fitness, i.e., $\rho = 1$, in which case CV$(S) = \sqrt{2/n}$. This shows that unless sample sizes are fairly high, the standard error of S relative to its expected value can be quite high: for $n = 50,\ 100,$ and 250, the CVs are, respectively, $\geq$ 0.20, 0.14, and 0.09.

EXPECTATIONS AND VARIANCES OF PRODUCTS

Consider the product $f = uv$, where u and/or v may be variables or functions of variables. Here $\partial f/\partial v = u$, $\partial f/\partial u = v$, $\partial^2 f/\partial u\,\partial v = 1$, and all other partial derivatives are zero. By including all terms involving these three nonzero partials in the application of Equation A1.4a, we obtain the expectation

$$E[uv] = \mu_u\mu_v + \sigma(u, v) \tag{A1.17}$$

where μ_u and μ_v are, respectively, the expected values of u and v. This expression also follows directly from the definition of a covariance, $\sigma(u, v) = E[uv] - \mu_u\mu_v$.

Applying Equation A1.7b,

$$\sigma^2(uv) = E\left[\left\{(u-\mu_u)\mu_v + (v-\mu_v)\mu_u + (v-\mu_v)(u-\mu_u) - \sigma(u,v)\right\}^2\right]$$

After a little algebra, this becomes

$$\sigma^2(uv) = \mu_v^2\sigma^2(u) + \mu_u^2\sigma^2(v) + E[(v-\mu_v)^2(u-\mu_u)^2] - [\,\sigma(u,v)\,]^2$$
$$+\,2\Big(\mu_u\mu_v\sigma(u,v) + \mu_u E[(v-\mu_v)^2(u-\mu_u)]$$
$$+\,\mu_v E[(v-\mu_v)(u-\mu_u)^2]\Big) \tag{A1.18a}$$

If u and v are bivariate normally distributed, then from Kendall and Stuart (1977, p. 85),

$$E[(v-\mu_v)^2(u-\mu_u)^2] = \sigma^2(u)\sigma^2(v) + 2\,[\,\sigma(u,v)\,]^2$$

and

$$E[(v-\mu_v)^2(u-\mu_u)] = E[(v-\mu_v)(u-\mu_u)^2] = 0$$

in which case

$$\sigma^2(uv) = \mu_v^2\sigma^2(u) + \mu_u^2\sigma^2(v) + [\,\sigma(u,v)\,]^2 + 2\mu_u\mu_v\sigma(u,v) + \sigma^2(u)\sigma^2(v) \tag{A1.18b}$$

When u and v are independent, this equation reduces further to

$$\sigma^2(uv) = \mu_v^2\sigma^2(u) + \mu_u^2\sigma^2(v) + \sigma^2(u)\sigma^2(v) \tag{A1.18c}$$

A useful compendium of facts regarding the variances and covariances of products of random variables can be found in Bohrnstedt and Goldberger (1969).

EXPECTATIONS AND VARIANCES OF RATIOS

Many of the statistics utilized in quantitative genetics are ratios of moments. These include the coefficient of variation (Chapter 2), and regression and correlation coefficients (Equations 3.14b and 3.16a). Thus, it is useful to have a general expression for the variance of a ratio, and the delta method yields a good approximation. Letting $f = u/v$, the first-order partials become $\partial f/\partial u = v^{-1}$ and $\partial f/\partial v = -u/v^2$, giving $\partial^2 f/\partial u^2 = 0$, $\partial^2 f/\partial v^2 = 2u/v^3$, and $\partial^2 f/\partial u \partial v = \partial^2 f/\partial v \partial u = -1/v^2$. Evaluating these second-order partials at μ_u and μ_v (again, the expected values of u and v), Equation A1.4a gives

$$E\left[\frac{u}{v}\right] \simeq \frac{\mu_u}{\mu_v}\left(1 + \frac{\sigma^2(v)}{\mu_v^2} - \frac{\sigma(u,v)}{\mu_u\mu_v}\right) \tag{A1.19a}$$

Likewise, from Equation A1.7c,

$$\begin{aligned}\sigma^2(u/v) &\simeq \sigma^2(u)\left(\frac{1}{\mu_v}\right)^2 + \sigma^2(v)\left(\frac{\mu_u}{\mu_v^2}\right)^2 - 2\sigma(u,v)\left(\frac{1}{\mu_v}\right)\left(\frac{\mu_u}{\mu_v^2}\right) \\ &= \left(\frac{\mu_u}{\mu_v}\right)^2\left(\frac{\sigma^2(u)}{\mu_u^2} - \frac{2\sigma(u,v)}{\mu_u\mu_v} + \frac{\sigma^2(v)}{\mu_v^2}\right)\end{aligned} \tag{A1.19b}$$

Both Equations A1.19a and A1.19b are approximations because $\partial f^2/\partial v^2 \neq 0$.

Sampling Variances of Regression and Correlation Coefficients

The least-squares regression coefficient is given by $b = u/v$, where $u = \text{Cov}(x,y)$ and $v = \text{Var}(x)$. To apply Equation A1.19b, we need to know μ_u, μ_v, $\sigma^2(u)$, $\sigma^2(v)$, and $\sigma(u,v)$. Because $\text{Var}(x)$ and $\text{Cov}(x,y)$ are unbiased estimators of the variance and covariance, $\mu_u = \sigma(x,y)$ and $\mu_v = \sigma^2(x)$. Under the assumption that x and y are bivariate normally distributed, we can also use the above results to obtain the variances and covariance of u and v: from Equation A1.14, $\sigma^2(u) = [\sigma^2(x)\sigma^2(y) + \sigma^2(x,y)]/n$; from Equation A1.10b, $\sigma^2(v) = 2\sigma^4(x)/n$; and from Equation A1.15a, $\sigma(u,v) = 2\sigma^2(x)\sigma(x,y)/n$.

Substituting these expressions into Equation A1.19b, we obtain (after some algebra) a result first obtained by Pearson (1896)

$$\sigma^2(b) \simeq \frac{\sigma^2(y)(1-\rho^2)}{n\sigma^2(x)} \tag{A1.20a}$$

where $\rho = \sigma(x,y)/[\sigma(x)\sigma(y)]$ is the correlation coefficient. With much additional algebra, it can also be shown that the sampling variance of a correlation coefficient, $r = \text{Cov}(x,y)/[\text{Var}(x)\text{Var}(y)]^{1/2}$, is

$$\sigma^2(r) \simeq \frac{(1-\rho^2)^2}{n} \tag{A1.20b}$$

Both of these expressions are strictly valid only under the assumption of bivariate normality (Kendall and Stuart 1977; p. 250), and hence are biased by the presence of either a nonzero value for skew or a kurtosis value departing from the Gaussian value of $3\sigma^4$. In practice, large-sample variances for b and r are estimated by substituting observed for expected variances and covariances in Equations A1.20a and A1.20b, and by using $(n-2)$ in place of n in the denominator, although this latter modification is usually of trivial importance.

Example A1.7. Dickerson (1969b) argued that because the numerator (the covariance) of a regression coefficient is estimated with much lower accuracy than the denominator (the variance), the sampling variance of the latter can be safely ignored in computing the standard error of a regression coefficient. Under this assumption, the approximate sampling variance of a regression coefficient is simply $\sigma^2[\text{Cov}(x,y)]/\sigma^4(x)$, which upon substitution for $\sigma^2[\text{Cov}(x,y)]$ becomes

$$\sigma^2(b) \simeq \frac{\sigma^2(y)(1+\rho^2)}{n\sigma^2(x)}$$

Comparing this to Equation A1.20a, it can be seen that ignoring the sampling variance in the denominator leads to a conservative estimate of the standard error of b, i.e., to a standard error that is upwardly biased. The ratio of the standard errors resulting from both expressions is $[(1-\rho^2)/(1+\rho^2)]^{1/2}$. For $\rho = 0.5$, this ratio is 0.77.

Sampling Variance of a Coefficient of Variation

As our final example of finding an approximate (large-sample) sampling variance of a ratio, we consider $\sigma^2[\,\text{CV}(x)\,]$, where the **coefficient of variation,** $\text{CV}(x) = \text{SD}(x)/\overline{x}$. This example shows how the approximations developed above can be successively used for even rather complex functions. As a first step, consider the expectation and variance of SD, the sample estimate of standard deviation. We start with the variance, $\sigma^2[\,\text{SD}(x)\,] = \sigma^2[\sqrt{\text{Var}(x)}\,]$. Applying Equation A1.7c (with $n = 1$ variable), we have

$$\sigma^2(\sqrt{v}\,) \simeq \sigma^2(v)\left(\frac{\partial\sqrt{v}}{\partial v}\right)^2\Bigg|_{\mu_v} = \frac{\sigma^2(v)}{4\,\mu_v} \tag{A1.21}$$

Recalling that $\sigma^2[\,\text{Var}(x)\,] = [(\mu_{4x} - \sigma^4(x)]/n$, and letting $v = \text{Var}(x)$, then $\mu_v = \sigma^2(x)$ and Equation A1.21 gives

$$\sigma^2[\,\text{SD}(x)\,] = \sigma^2[\,\sqrt{\text{Var}(x)}\,] \simeq \frac{\sigma^2[\,\text{Var}(x)\,]}{4\sigma^2(x)} = \frac{\mu_{4x} - \sigma^4(x)}{4n\sigma^2(x)} \tag{A1.22}$$

To compute the approximate expected value of the SD, we apply Equation A1.3b,

$$E[\sqrt{v}\,] \simeq \sqrt{\mu_v} - \sigma^2(v)\,\frac{\mu_v^{-3/2}}{8} \tag{A1.23a}$$

Letting $v = \text{Var}(x)$, $\mu_v = \sigma^2(x)$ and $\sigma^2[\,\text{Var}(x)\,] = [\mu_{4x} - \sigma^4(x)]/n$,

$$\mu_{\text{SD}} = E[\,\text{SD}(x)\,] \simeq \sigma(x) + \frac{\sigma(x)}{8n}\left[1 - \frac{\mu_{4x}}{\sigma^4(x)}\right] \tag{A1.23b}$$

Next, consider $\sigma[\,\overline{x},\ \text{SD}(x)\,]$, the covariance between the standard deviation and the sample mean. Because we know $\sigma[\,\overline{x},\ \text{Var}(x)\,] \simeq \mu_{3x}/n$ (Equation A1.13b), we first use

Equation A1.9 to obtain an approximation for $\sigma[\,\overline{x},\ \mathrm{SD}(x)\,]$ as a function of $\overline{x}$ and $\mathrm{Var}(x)$. Letting $f = u$ and $g = \sqrt{v}$, Equation A1.9 gives

$$\sigma(\,u, \sqrt{v}\,) \simeq \sigma^2(u)\left(\frac{\partial u}{\partial u}\right)\left(\frac{\partial\sqrt{v}}{\partial u}\right) + \sigma^2(v)\left(\frac{\partial\sqrt{v}}{\partial v}\right)\left(\frac{\partial u}{\partial v}\right)$$
$$+\,\sigma(u,v)\left[\left(\frac{\partial u}{\partial u}\right)\left(\frac{\partial\sqrt{v}}{\partial v}\right) + \left(\frac{\partial u}{\partial v}\right)\left(\frac{\partial\sqrt{v}}{\partial u}\right)\right] \tag{A1.24}$$

For our particular case,

$$\frac{\partial u}{\partial v} = \frac{\partial\,\overline{x}}{\partial\,\mathrm{Var}(x)} = 0 \qquad \text{and} \qquad \frac{\partial\,\sqrt{v}}{\partial\,u} = \frac{\partial\,\mathrm{SD}(x)}{\partial\,\overline{x}} = 0$$

so three of the terms in Equation A1.24 are zero, leaving

$$\sigma(\,u, \sqrt{v}\,) = \sigma(u,v)/(2\sqrt{\mu_v})$$

Hence, from Equation A1.13b,

$$\sigma[\,\overline{x},\ \mathrm{SD}(x)\,] \simeq \frac{\sigma[\,\overline{x},\ \mathrm{Var}(x)\,]}{2\sigma(x)} \simeq \frac{\mu_{3x}}{2n\sigma(x)} \tag{A1.25}$$

Everything is now in place for applying Equation A1.19b to obtain $\sigma^2(\mathrm{SD}(x)/\,\overline{x}\,)$. Let $u = \mathrm{SD}(x) = [\mathrm{Var}(x)]^{1/2}$ and $v = \overline{x}$. Hence $\mu_v = \mu_x$, $\sigma^2(v) = \sigma^2(\overline{x}) = \sigma^2(x)/n$ (Example A1.3), and Equations A1.22 and A1.25 provide expressions for $\sigma^2(u)$ and $\sigma(u,v)$. Substituting these into Equation A1.19b gives

$$\sigma^2[\,\mathrm{CV}(x)\,] \simeq \frac{1}{n}\left(\frac{\mu_{\mathrm{SD}}}{\mu_x}\right)^2\left(\frac{\mu_{4x} - \sigma^4(x)}{4\sigma^2(x)\,\mu_{\mathrm{SD}}^2} - \frac{\mu_{3x}}{\mu_x\,\sigma(x)\,\mu_{\mathrm{SD}}} + \frac{\sigma^2(x)}{\mu_x^2}\right) \tag{A1.26a}$$

where μ_{SD} is given by Equation A1.23b. If x is normally distributed, $\mu_{3x} = 0$, $\mu_{4x} = 3\sigma^4(x)$, $\mu_{\mathrm{SD}} = \sigma(x)[\,1 - 1/(\,4n\,)\,]$ and further simplification is possible. Ignoring terms of order n^2 and higher, then

$$\sigma^2[\,\mathrm{CV}(x)\,] \simeq \frac{1}{2n}\left[\frac{\sigma(x)}{\mu_x}\right]^2\left[1 + 2\left(\frac{\sigma(x)}{\mu_x}\right)^2\right] \tag{A1.26b}$$

in which case an estimate for the large-sample variance of the CV is simply

$$\mathrm{Var}[\,\mathrm{CV}(x)\,] \simeq \frac{[\,\mathrm{CV}(x)\,]^2}{2n}\,[1 + 2\{\mathrm{CV}(x)\}^2] \tag{A1.26c}$$

Appendix 2

Path, Causality, and Network Analysis

As humans, we cannot avoid thinking in terms of causality. Asendorpf (2012)

Wright (1921a) developed the method of **path analysis** as a means of interpreting the correlation between two variables in terms of hypothetical **paths of causation** between them. Initially, he was interested in the relative importance of general and specific growth factors for the variation of bone sizes in small mammals (Wright 1918), but he quickly realized the broad utility of his new technique and later applied it to many problems in genetics, agricultural economics, physiology, and ecology. More recently it has been extended and applied widely in systems biology and genomics. Provided the underlying assumptions are kept in mind, path analysis provides an extremely powerful, and conceptually simple, tool. Many of the fundamental principles of quantitative genetics can be derived by its use. It can also be a powerful approach for analyzing multiple levels of molecular data (such as joint data on QTL genotypes, mRNA and protein levels, and trait phenotypes).

The purpose of path analysis is the quantification of the relative contributions of causal sources of variance and covariance ***once a certain network of interrelated variables has been accepted*** (or assumed). While path analysis is not a formal technique proving the actual sources of causality (which can only come from careful experimentation), it does allow one to examine the level of support for different causality structures. In response to periodic abuses and criticism of the technique, Wright (1932, 1934c, 1968, 1983, 1984) repeatedly emphasized these points. Exceptionally lucid accounts of the theory and applications of basic path analysis are given by Li (1975) and Pedhazur (1982). Path analysis has grown from its simple roots and now forms the foundation for the rich fields of **causal analysis** (Spirtes et al. 2000; Pearl 2009; Pearl et al. 2016), **structural equation modeling (SEM)** (Bollen et al. 1989), and **graphical models** (Rosa et al. 2011, 2016; Sinoquet and Mourad 2014; Rohrer 2018), the latter is also widely used in Bayesian modeling (Wu et al. 2010). Chapter 30 in WL examines applications of path analysis in determining the nature of selection on traits in the wild. A nice overview of many of these issues is given by Shipley (2016).

UNIVARIATE ANALYSIS

Through a graphical display, path analysis can greatly facilitate the analysis of a complex problem. Consider, e.g, a system of four measurable variables, one (y) dependent and three (z_1, z_2, and z_3) potential explanatory values. Such a system can be displayed in the form of a **path diagram** (Figure A2.1). Single-headed arrows denote a *direct path* from an explanatory variable to y, implying a cause-and-effect relationship. The connections between the explanatory variables are represented by double-headed arrows. It is assumed that y is a linear function of the z_i, although nonlinear relationships can be incorporated by considering z_i^2. Finally, unless y is known to be completely determined by the observed explanatory variables, an arrow is also drawn from the independent residual term, e.

Figure A2.1 displays only one of several possible path diagrams for our four-variable system. A great deal of information is contained in this diagram. For example, it can easily be seen that z_1 potentially influences y in three ways: directly by the path $z_1 \rightarrow y$ and indirectly by the paths $z_1 \leftrightarrow z_2 \rightarrow y$ and $z_1 \leftrightarrow z_3 \rightarrow y$. Similarly, z_2 influences y through paths $z_2 \rightarrow y$, $z_2 \leftrightarrow z_1 \rightarrow y$, and $z_2 \leftrightarrow z_3 \rightarrow y$, and z_3 influences y through $z_3 \rightarrow y$, $z_3 \leftrightarrow z_2 \rightarrow y$, and $z_3 \leftrightarrow z_1 \rightarrow y$. The independent residual term operates only through path $e \rightarrow y$.

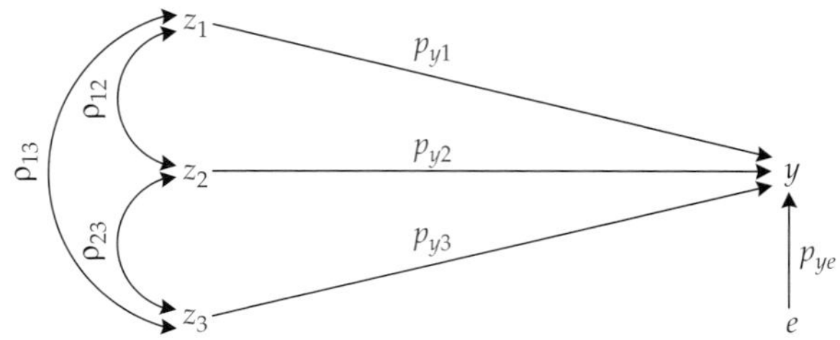

Figure A2.1 Path diagram for the variable y in terms of three explanatory variables (z_1, z_2, and z_3) and a residual error (e). p denotes a path coefficient and ρ a correlation (here between the z_i, standardized to have unit variance).

The labels on the double-headed arrows are the simple correlation coefficients (ρ) between the two denoted variables, while the quantities along single-headed arrows are **path coefficients** (p). The path coefficients for this diagram are standardized partial regression coefficients. If, prior to a multiple regression, each of the variables (y, z_1, z_2, and z_3) is standardized by subtracting the mean and dividing by the standard deviation so that the transformed variables all have zero means and unit variances, the subsequent partial regression coefficients equal the path coefficients.

To see this equivalence, we start with the general linear model

$$y = \alpha + \beta_1 z_1 + \beta_2 z_2 + \cdots + \beta_n z_n + e \tag{A2.1a}$$

where $\beta_1, \beta_2, \cdots, \beta_n$ are partial regression coefficients. Subtracting $\overline{y}$ from the left and its equivalent $(\alpha + \beta_1 \overline{z}_1 + \beta_2 \overline{z}_2 + \cdots + \beta_n \overline{z}_n + \overline{e})$ from the right yields

$$y - \overline{y} = \beta_1(z_1 - \overline{z}_1) + \beta_2(z_2 - \overline{z}_2) + \cdots + \beta_n(z_n - \overline{z}_n) + (e - \overline{e}) \tag{A2.1b}$$

Squaring this expression and taking expectations, we obtain a general expression for the variance of y,

$$\sigma^2(y) = \sum_{i=1}^{n} \beta_i^2 \sigma^2(z_i) + 2 \sum_{i=1}^{n} \sum_{j>i}^{n} \beta_i \beta_j \sigma(z_i, z_j) + \sigma_e^2 \tag{A2.2}$$

The residual variable is uncorrelated with the remaining variables under a least-squares analysis (Chapter 3), so covariance terms involving e do not appear in Equation A2.2. Dividing all terms in Equation A2.2 by $\sigma^2(y)$, recalling that $\sigma(z_i, z_j) = \rho_{ij}\sigma(z_i)\sigma(z_j)$, where ρ_{ij} is the correlation between variables i and j, and defining

$$p_{yi} = \beta_i \left[\frac{\sigma(z_i)}{\sigma(y)} \right] \tag{A2.3a}$$

$$p_{ye} = \frac{\sigma(e)}{\sigma(y)} \tag{A2.3b}$$

we obtain one of the fundamental equations of path analysis,

$$1 = \sum_{i=1}^{n} p_{yi}^2 + 2 \sum_{i=1}^{n} \sum_{j>i}^{n} p_{yi} \rho_{ij} p_{yj} + p_{ye}^2 \tag{A2.4}$$

This expression, known as the **equation of complete determination,** is a simple extension of the multiple regression equation. The p_{yi} are called **path coefficients**, and from Equation A2.3a can be seen to be standardized partial regression coefficients. Thus, the path coefficients are directly obtainable by multiplying partial regression coefficients by ratios of observed standard deviations (Equation A2.3a). Path coefficient p_{yi} may be interpreted as the expected change in y in standard deviations caused by a change in z_i in standard deviations when all other background variables are held constant.

Equation A2.4 greatly expands the utility of multiple regression by *explicitly partitioning the variance* of y into proportional contributions from all of the direct and indirect paths of influence. The contribution from each path is a simple product of the correlation coefficients (for each double-headed arrow) and path coefficients (for each single-headed arrow) along a loop between variables. Thus, the contributions of the direct paths from z_1, z_2, z_3 and e to y are p_{y1}^2, p_{y2}^2, p_{y3}^2, and p_{ye}^2, respectively. The contributions from the indirect paths $y \leftarrow z_1 \leftrightarrow z_2 \rightarrow y$, $y \leftarrow z_2 \leftrightarrow z_3 \rightarrow y$, and $y \leftarrow z_1 \leftrightarrow z_3 \rightarrow y$ are, respectively, $2p_{y1}\rho_{12}p_{y2}$, $2p_{y2}\rho_{23}p_{y3}$, and $2p_{y1}\rho_{13}p_{y3}$. Each of the indirect paths is counted twice because they influence y in both directions (double-headed arrows). For example, the contribution of the path $y \leftarrow z_1 \leftrightarrow z_2 \rightarrow y$ to the variance of y is the same as path $y \leftarrow z_2 \leftrightarrow z_1 \rightarrow y$.

It is important to note that in computing the joint influence of two explanatory variables on y, only the direct correlation between the two variables is considered. Hence, $y \leftarrow z_1 \leftrightarrow z_2 \leftrightarrow z_3 \rightarrow y$ is not a contributing path in Figure A2.1. The entire correlation between z_1 and z_3 is contained in ρ_{13}. Thus, the general rules of path analysis are that: (i) ***there is only one two-headed arrow in any path***, and that (ii) the ***arrows change direction only once in a path***. Unlike correlation coefficients, path coefficients need not have absolute values less than one. Further, the total net contribution over a path can be negative (a negative correlation). The only constraint on Equation A2.4 is that the total contributions sum to one.

Regressions versus Path Analysis

The key distinction between a regression and a path analysis is that regression attempts to ***statistically account*** for the covariance between explanatory variables, while a path analysis further attempts to account for the ***processes generating correlations among the explanatory variables***. Regressions make no assumption about the causality structure generating trait correlations, while Crespi (1990) noted that "path-analytic reasoning assumes that characters are correlated as a result of biological causes that should be used as information rather than adjusted away." A key limitation (and also strength) of a path analysis is that different assumed path diagrams for a set of variables (**causality structures**) can result in rather different biological interpretations (WL Chapter 30).

Example A2.1. As an example of different path diagrams for the same set of variables, consider the following example from Chapter 30 of WL.

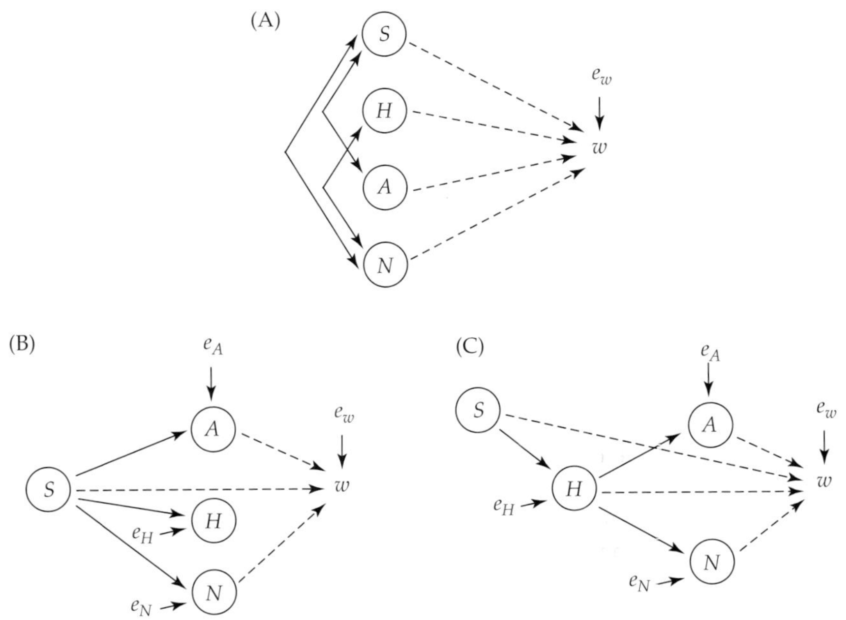

The explanatory variables here are four plant traits: seed weight (S; the weight of the seed that gave rise to the focal individual), plant height (H), leaf area (A), and number of leaves (N). We seek to determine how these influence a measure of fitness, w, namely total seeds from a plant. For emphasis purposes, direct connections to fitness are denoted by dashed arrows. Path (A) is the structure for a standard multiple regression, which makes no assumptions about how the patterns of correlations among the traits arise. Paths (B) and (C) consider how developmental biology impacts the correlation structure among traits. In these structures there are also residuals associated with each of the downstream traits, as the influence of any upstream traits only determines part of their value. In both (B) and (C), the weight of the seed that founded the plant influences the other three traits being considered. Path model (B) assumes no hierarchical developmental structure from seed to the three traits, while path model (C) assumes that seed weight just influences height, and that height then influences the other two traits. Note that the covariance matrix of all three models is identical. Suppose we find that seed weight is correlated with fitness. Path diagrams B and C allow us to determine if S directly impacts fitness, only impacts it indirectly through its impact on downstream traits that in turn directly impact fitness, or has both direct and indirect effects (path analysis can quantify the importance of each).

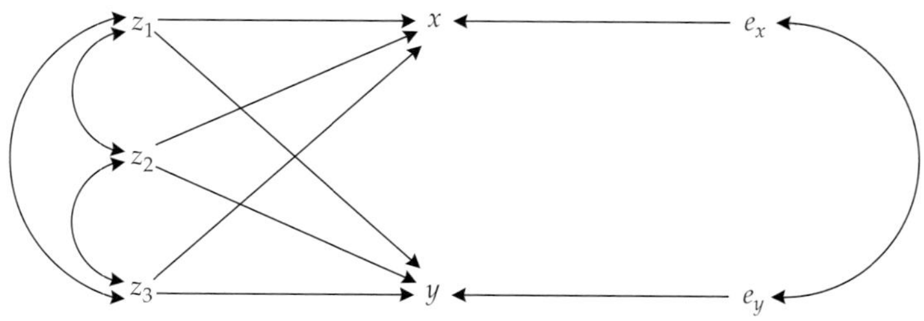

Figure A2.2 Path diagram for a system in which two dependent variables (x and y) are jointly influenced by three explanatory variables (z_1, z_2, and z_3) and residuals (e_x and e_y), which may be correlated. More complicated path diagrams could have one or more of the explanatory variables dependent on upstream explanatory variables, e.g., $z_1 \rightarrow z_2$ and/or $z_1 \rightarrow z_3$ (e.g., Example A2.1). Operationally, when a final path diagram is displayed, many of the proposed connections may not have statistical support, and these are thus often left out. Further, the strength of a path is often indicated by the thickness of the arrow, with thick arrows implying a strong connection and thin arrows a weak connection.

BIVARIATE ANALYSIS

An exceedingly useful property of path analysis is its ability to quantify the degree of association between two variables in terms of one or more mutually shared explanatory variables. Figure A2.2 is identical in form to Figure A2.1 except that z_1, z_2, and z_3 are now causal determinants of two characters, x and y. There are 10 distinct pathways connecting x and y: three direct paths $x \leftarrow z_1 \rightarrow y$, $x \leftarrow z_2 \rightarrow y$, and $x \leftarrow z_3 \rightarrow y$, and seven indirect paths $x \leftarrow z_1 \leftrightarrow z_2 \rightarrow y$, $x \leftarrow z_2 \leftrightarrow z_1 \rightarrow y$, $x \leftarrow z_1 \leftrightarrow z_3 \rightarrow y$, $x \leftarrow z_3 \leftrightarrow z_1 \rightarrow y$, $x \leftarrow z_2 \leftrightarrow z_3 \rightarrow y$, $x \leftarrow z_3 \leftrightarrow z_2 \rightarrow y$, and $x \leftarrow e_x \leftrightarrow e_y \rightarrow y$. The correlation between x and y is simply the sum of the products of path coefficients and correlation coefficients along these paths:

$$\rho_{xy} = p_{x1}p_{y1} + p_{x2}p_{y2} + p_{x3}p_{y3} + p_{x1}\rho_{12}p_{y2} + p_{x2}\rho_{12}p_{y1} + p_{x1}\rho_{13}p_{y3} + p_{x3}\rho_{13}p_{y1} + p_{x2}\rho_{23}p_{y3} + p_{x3}\rho_{23}p_{y2} + p_{xe}e_{xy}p_{ye}$$

where e_{xy} represents the correlation between the residual terms e_x and e_y. This expression

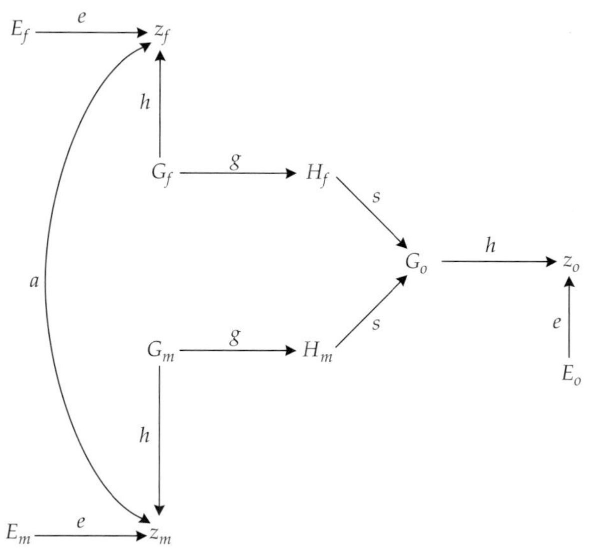

Figure A2.3 Wright's (1921a) path diagram, slightly modified, for the phenotypes of parents and offspring. Variables are defined in the text.

may be generalized to define the correlation between any two variables with n common causal sources of variation,

$$\rho_{xy} = \left(\sum_{i=1}^{n} \sum_{j=1}^{n} p_{xi} \rho_{ij} p_{yj} \right) + p_{xe} e_{xy} p_{ye} \tag{A2.5}$$

For terms in which $i = j$, ρ_{ij} is set equal to one. Just as Equation A2.4 partitions a variance into components, Equation A2.5 partitions a correlation into a series of paths through shared explanatory variables.

APPLICATIONS

Path analysis is very useful in quantitative genetics because explicit statements can often be made about causality. Among the following examples, the first two illustrate how path analysis can be used to derive some fundamental relationships concerning phenotypic correlations.

Phenotypic Correlation Between Parents and Offspring

Much of the methodology of quantitative genetics relies on the comparison of phenotypic measures in close relatives. A common application involves the regression of offspring on parent, which was one of Wright's (1921a) earliest uses of path analysis. Defining an individual's phenotype (z) to be the sum of its genotypic value (G) and an environmental deviation (E), the phenotypes of a father (f) and mother (m) may be written as

$$z_f = G_f + E_f$$
$$z_m = G_m + E_m$$

Provided the parents are not sibs, their environmental deviations may be treated as independent random variables with respect to G. If, however, mates select each other on the basis of phenotypes, a correlation may exist between z_f and z_m. We will denote this correlation by a. Under additive gene action, the genotypic value of an offspring is equal to the sum of the gametic contributions from its father (H_f) and mother (H_m),

$$G_o = H_f + H_m$$

An unambiguous path diagram can be constructed for such a familial structure (Figure A2.3). There are four path coefficients: h from genotypic to phenotypic values, e from environmental effects to phenotypic values, g from genotypic value to gametic value, and s from gametic value to genotypic value. These are assumed to be constant across generations.

We now consider the correlation between the phenotype of a parent and that of its offspring, ρ_{op}. Here we focus on the father–offspring correlation, although in this example, identical results arise for mother–offspring analysis. Regardless of which parent is considered, there are two paths connecting it to its offspring. The first results from the direct gametic contribution that a parent makes to its offspring; i.e., $z_f \leftarrow G_f \rightarrow H_f \rightarrow G_o \rightarrow z_o$. The contribution of this path to ρ_{of} is the product of four path coefficients, $hgsh$. The second path, which only exists under assortative mating, is an indirect route through a mate's gamete, i.e., $z_f \leftrightarrow z_m \leftarrow G_m \rightarrow H_m \rightarrow G_o \rightarrow z_o$. Its contribution to ρ_{of} is $ahgsh$. Summing up,

$$\rho_{of} = h^2 gs(1 + a) \tag{A2.6}$$

A further simplification of this expression, which eliminates the coefficients g and s, is possible. The offspring genotypic value G_o is determined by the two direct paths $H_f \rightarrow G_o$ and $H_m \rightarrow G_o$, each of which contributes a proportion s^2 to the variance of G_o, and by the indirect paths $G_o \leftarrow H_f \leftrightarrow H_m \rightarrow G_o$ and $G_o \leftarrow H_m \leftrightarrow H_f \rightarrow G_o$. The correlation between H_f and H_m is determined by the single path $H_f \leftarrow G_f \leftarrow z_f \leftrightarrow z_m \rightarrow G_m \rightarrow H_m$ and is equal to $h^2 g^2 a$. Therefore, each indirect path makes a proportional contribution of $(hgs)^2 a$ to the variance of G_o. The equation of complete determination for G_o is then

$$1 = 2s^2 + 2(hgs)^2 a$$

which upon rearrangement yields

$$s = [2(1 + h^2 g^2 a)]^{-1/2} \tag{A2.7}$$

Wright (1921a) pointed out that, provided the path coefficients remain constant across generations, the correlation between a genotype and a gamete that *it produces* (G_f and H_f) will be the same as the correlation between a genotype and a gamete that *produced it* (G_o and H_f). It can be seen directly from the path diagram that the first of these correlations is simply g. There are two paths connecting G_o and H_f ($H_f \rightarrow G_o$ and $H_f \leftarrow G_f \rightarrow z_f \leftrightarrow z_m \leftarrow G_m \rightarrow H_m \rightarrow G_o$), however, so their correlation is $s + ghahgs$. Equating this expression to g,

$$g = s(1 + h^2 g^2 a) \tag{A2.8}$$

Multiplying Equations A2.7 and A2.8 together, it can be seen that

$$gs = s^2(1 + h^2 g^2 a) = 0.5$$

Thus, Equation A2.6 simplifies to

$$\rho_{of} = h^2 \left(\frac{1 + a}{2} \right) \tag{A2.9}$$

which in the absence of assortative mating ($a = 0$), reduces to

$$\rho_{of} = \frac{h^2}{2} \tag{A2.10}$$

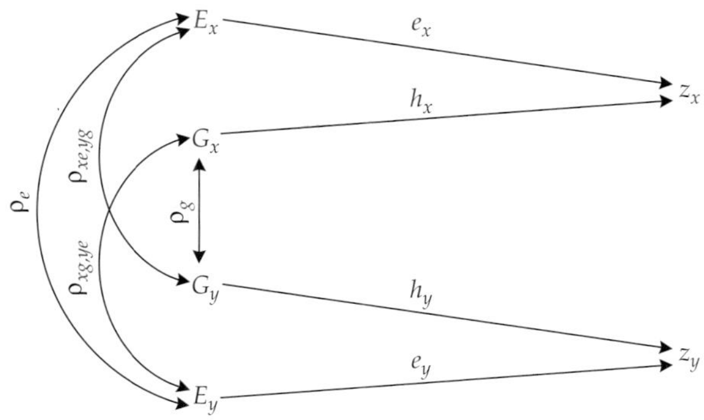

Figure A2.4 Path diagram for the phenotypic correlation between two characters (z_x and z_y) in terms of genetic values (G_x and G_y) and environmental deviations (E_x and E_y).

Mate selection on the basis of phenotypes influences the resemblance between parents and offspring in a particularly simple manner. Perfect disassortative mating ($a = -1$) completely eliminates the correlation between parent and offspring phenotypes, while perfect assortative mating ($a = +1$) doubles it.

Quantitative geneticists have long referred to the fraction of phenotypic variance that is additive genetic in basis as the narrow-sense heritability and abbreviated it as h^2. The use of this notation, particularly the square, may seem puzzling. Returning to Figure A2.3, the origin of h^2 can now be seen to be a historical tribute to Wright's (1921a) path diagram. Under the additive model of gene action, the equation of complete determination for an individual's phenotype is

$$h^2 + e^2 = 1$$

where h^2 is the proportion of the phenotypic variance due to the direct path from the genotypic value.

Correlations Between Characters

Path analysis can also be used to describe the correlation between two different characters, x and y, in the same individual. Here we denote the two phenotypes as

$$z_x = G_x + E_x$$
$$z_y = G_y + E_y$$

where, as usual, E_x has a mean of zero and is independent of G_x, and the same properties apply to E_y and G_y. The path diagram joining the two traits is drawn in its most general form in Figure A2.4. The path $G_x \leftrightarrow G_y$ indicates the possibility of a correlation between genotypic values of the two traits, owing to their expression being mutually determined by shared genes and/or linkage disequilibrium among alleles influencing the two traits (Part II Chapter 26). Correlation between the environmental effects on the two traits is denoted by ρ_e, while those between the genotypic value of one trait and the environmental deviation of the other are indicated by $\rho_{xe,yg}$ and $\rho_{xg,ye}$. The phenotypic correlation between characters x and y, ρ_{xy}, derives from four possible paths: $z_x \leftarrow G_x \leftrightarrow G_y \rightarrow z_y$, $z_x \leftarrow E_x \leftrightarrow E_y \rightarrow z_y$, $z_x \leftarrow E_x \leftrightarrow G_y \rightarrow z_y$, and $z_x \leftarrow G_x \leftrightarrow E_y \rightarrow z_y$. Summing the appropriate products of path and correlation coefficients,

$$\rho_{xy} = h_x \rho_g h_y + e_x \rho_e e_y + e_x \rho_{xe,yg} h_y + e_y \rho_{xg,ye} h_x \qquad \text{(A2.11)}$$

Note that there are only two arrows pointing to z_x and that these come from variables that are uncorrelated (as E_x and G_x are not connected by any paths). The same is true for z_y. Thus, by the equation of complete determination,

$$h_x^2 + e_x^2 = 1$$
$$h_y^2 + e_y^2 = 1$$

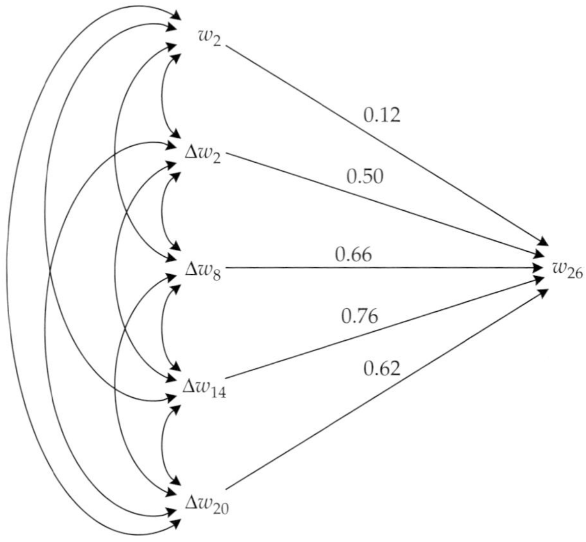

Figure A2.5 Path diagram for weight at day 26 as a function of five growth components for a population of feral pigeons. Numbers on the single-headed arrows are the path coefficients. The correlations between explanatory variables are given in Table A2.1.

Rearranging and substituting $e_x = \sqrt{1 - h_x^2}$ and $e_y = \sqrt{1 - h_y^2}$ into Equation A2.11,

$$\rho_{xy} = h_x h_y \rho_g + \rho_e \sqrt{(1 - h_x^2)(1 - h_y^2)} + \rho_{xe,yg} h_y \sqrt{1 - h_x^2} + \rho_{xg,ye} h_x \sqrt{1 - h_y^2} \tag{A2.12}$$

Thus, the phenotypic correlation between two traits is entirely described in terms of correlations between components of the traits and their heritabilities. Frequently in quantitative-genetic applications, the correlations between E_x and G_y and between E_y and G_x are zero. In this case, Equation A2.12 reduces to

$$\rho_{xy} = h_x h_y \rho_g + \rho_e \sqrt{(1 - h_x^2)(1 - h_y^2)} \tag{A2.13}$$

Growth Analysis

Biological features in which a whole can be considered to be the sum of several individual parts are numerous. For example, the total diet of a predator often consists of several prey species, and the total seed set by a plant can be partitioned into contributions from various flowers. The problem considered here is the size of an individual (or character) at time t, z_t. This can be expressed simply as the sum of the initial size, z_1, and an arbitrary number (n) of subsequent growth increments, z_2 to z_n,

$$z_t = z_1 + z_2 + \cdots + z_n \tag{A2.14}$$

With this type of model, all of the terms on the right necessarily sum to z_t, so there is no residual error term. Moreover, all of the partial regression coefficients (the β_i coefficients on the z_i in the multiple regression equation) are equal to one. Returning to Equations A2.1a and A2.2, it can be seen that the equation of complete determination for z_t reduces to

$$1 = \frac{1}{\sigma^2(z_t)} \left[\sum_{i=1}^{n} \sigma^2(z_i) + 2 \sum_{i=1}^{n} \sum_{j \geq i}^{n} \sigma(z_i, z_j) \right] \tag{A2.15}$$

Table A2.1 Correlations (above diagonal) and path contributions (diagonal and below) of growth components to weight on day 26 for a sample of 100 feral pigeons. Here * and ** denote correlations that are significance at the 5% and 1% levels. The diagonal elements (direct contributions to the variance of w_{26}) are simply the squares of the respective path coefficients given in Figure A2.5. The below-diagonal elements denote the contributions resulting from correlations between characters x and y, and are obtained as $2p_x\rho_{xy}p_y$. (*Source*: D. Droge, unpubl. data.)

	w_2	Δw_2	Δw_8	Δw_{14}	Δw_{20}
w_2	0.014	0.232*	0.100	−0.069	−0.186
Δw_2	0.027	0.255	−0.014	−0.316**	−0.045
Δw_8	0.015	−0.010	0.437	−0.157	−0.096
Δw_{14}	−0.012	−0.244	−0.159	0.584	−0.167
Δw_{20}	−0.027	−0.028	−0.079	−0.158	0.385

Thus, the elements of the variance–covariance matrix for the growth components in Equation A2.14 provide a complete description of the direct and indirect contributions to the variance of size at time t.

As an example of the application of Equation A2.15, the growth dynamics of a population of feral pigeons will be examined. One hundred birds were weighed at regular intervals from shortly after birth to fledging. The path analysis here will consider the weight at day 26 as a function of an initial weight at day 2 plus four subsequent six-day growth increments (days 2–8, 8–14, 14–20, and 20–26). Each growth increment is the difference between adjacent weighings, so letting w_t be the weight on day t, Equation A2.14 becomes

$$\begin{aligned} w_{26} &= w_2 + (w_8 - w_2) + (w_{14} - w_8) + (w_{20} - w_{14}) + (w_{26} - w_{20}) \\ &= w_2 + \Delta w_2 + \Delta w_8 + \Delta w_{14} + \Delta w_{20} \end{aligned}$$

The path diagram (Figure A2.5) illustrates that there are five direct paths and ten indirect paths to w_{26} (each of which must be counted twice). The proportional contributions of these paths are directly obtainable from the variances and covariances of the observed variables and are summarized in Table A2.1. Recalling that for this model all $\beta_i = 1$, the path coefficients can be computed using Equation A2.3a. For example, the contribution of direct path $w_2 \rightarrow w_{26}$ is simply $[p(w_2, w_{26})]^2 = \sigma^2(w_2)/\sigma^2(w_{26})$, while the total contribution from the indirect paths $w_2 \leftrightarrow \Delta w_2 \rightarrow w_{26}$ and $\Delta w_2 \leftrightarrow w_2 \rightarrow w_{26}$ is $2\sigma(w_2, \Delta w_2)/\sigma^2(w_{26})$.

Several aspects of the growth properties of this population are revealed by the path analysis. First, very little of the variation in size at age 26 is accounted for by the size at birth, i.e., $(p_{w_2,w_{26}})^2 = 0.014$. Most of it arises from variation in the post-natal growth rates. Second, all of the indirect paths make negative or negligibly positive contributions to w_{26}. The sum of the direct (diagonal elements of Table A2.1) and indirect (below-diagonal elements) paths are 1.675 and −0.675, respectively. The contribution from the path involving Δw_2 and Δw_{14} ($p = -0.244$) is particularly pronounced because of the highly significant negative correlation between Δw_2 and Δw_{14} ($\rho = -\ 0.316$). On the other hand, while w_2 and Δw_2 are significantly positively correlated ($\rho = 0.232$), their indirect contribution to w_{26} is very small because of the small path coefficient from w_2 to w_{26} ($p_{w_2,w_{26}} = 0.12$; Figure A2.5). The preponderance of negative correlations between growth components is indicative of **compensatory growth**. Individuals that experience early periods of relatively rapid growth generally also experience subsequent periods of slowed growth. More details on this method of growth analysis, as well as estimators for the sampling variance of path coefficients, may be found in Lynch (1988d).

Causality Analysis in Biological Pathways

As introduced in Chapter 21, an important application of path analysis is in disentangling the roles of molecular features (such as QTL genotypes, mRNA and protein levels, chromatin features, methylation marks, etc.) in influencing a trait. The basic idea was introduced

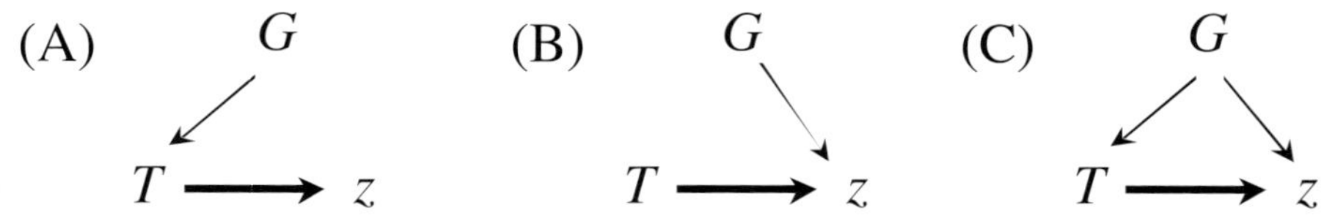

Figure A2.6 Suppose a detailed functional genomic analysis has shown that a focal trait (z) is influenced by the QTL genotype (G) and a quantitative trait transcript (T), potentially from a different locus than the focal QTL. **(A):** The impact of the QTL is as an eQTL that influences the trait only by influencing T levels. **(B):** The QTL has an impact on the trait independent of T levels. **(C):** The QTL acts on both the T and the trait.

by Schadt et al. (2005) and extended by a number of authors (Kulp and Jagalur 2006; Liu et al. 2008; Millstein et al. 2009; Wu et al. 2010). The term **endophenotype** refers to any intermediate molecular phenotype in the mapping of a genotype to a final trait value (such as mRNA, protein, or metabolite levels; histone acetylation/methylation status, etc.). With the rise of functional genomics, a very large number of such intermediate features can be scored (Chapter 21), and critical to our understanding of trait expression is determining the causal structure of how these components interact. Figure A2.6 shows a simple example wherein levels of mRNA (T) from a particular locus influence trait value z (it is **quantitative trait transcript**, QTT, for that trait; Chapter 21). The genotype (G) at a focal QTL also influences the trait phenotype, so that in a simple correlation analysis, both the QTL and QTT appear to influence the trait (both G and T are correlated with z). However, as Figure A2.6 shows, there are three different causal structures (we ignore—for now—the reverse causal structure, QTL $\rightarrow$ trait $\rightarrow$ QQT). Note that these options correspond to three different path diagrams. If (A) is correct, then, conditional on the QTT value, there should be no residual correlation between QTL genotype and trait value. If (B) is correct, conditioning on the QTT value will not change the correlation between QTL genotype and trait, while if (C) is correct, the residual correlation between the QTL and trait will change, but will remain nonzero.

Example A2.2. Consider the situation diagrammed below (left) where a QTL (gene one; represented by a circle) has an impact on both its own transcript level and that at a second gene, both transcipts being represented by triangles in the figure (Mackay et al. 2009). Transcript levels of both genes are correlated with the trait value, as is the genotype at the QTL. How can we determine which factor (or factors) directly impact the trait?

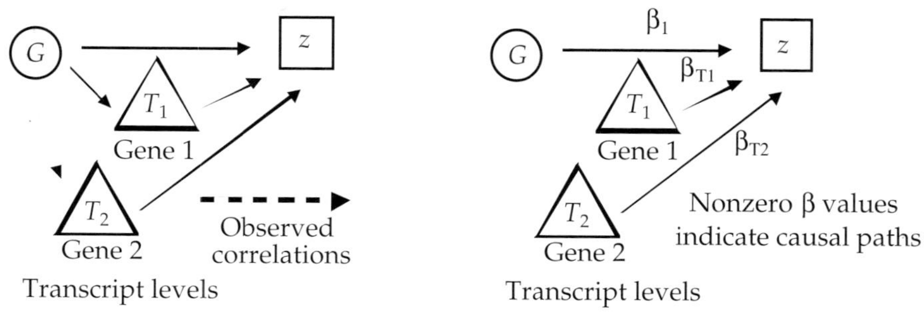

Assuming no other features (confounders) that influence two (or more) of these components (e.g., Figure 21.3B), the general setting is shown on the left in the above figure. The figure on the right shows the direct impacts we wish to test: β_1 for the direct effect of the QTL on the trait, β_{T1} for the direct impact of transcript one level on the trait, and finally β_{T2} for the direct impact of transcript two level on the trait. If we have a set of individuals with data vectors (z, G, T_1, T_2), then these various β can be estimated from the multiple regression

$$z = \mu + \beta_1 G + \beta_{T1} T_1 + \beta_{T2} T_2 + e$$

We can also estimate the strength of the paths from gene one to both transcripts (β_{1Ti}), and thus compute the indirect contribution to the trait, namely $\beta_{1T1}\beta_{T1} + \beta_{1T2}\beta_{T2}$, corresponding to the paths gene one $\rightarrow T_1 \rightarrow$ trait and gene one $\rightarrow T_2 \rightarrow$ trait, respectively. See Example 21.3 (and Figure 21.3) for an application of this approach (often called **mediation analysis**) for determining the impact of one transcription on another, potentially downstream, transcript (*cis*-mediated *trans* effects). As discussed in Example 21.3, while the logic of mediation is straightforward, measurement error can give false signals (Gastonguay et al. 2023), and an alternative approach is to use a Bayesian model selection framework (Crouse et al. 2022).

Finally, some basic terminology for causality. Consider three factors, A, B, and C that have some causal association. The setting where $A \rightarrow B \rightarrow C$ is called a **chain**, and B is said to be a **mediator**. If this is the only path connecting A and C, then conditioning on B results in A and C being uncorrelated. The setting where $A \leftarrow B \rightarrow C$ is called a **fork**, and B is said to be a **confounder**. Conditioning on B again removes the correlation between A and C. Finally, the setting where $A \rightarrow B \leftarrow C$ is called an **inverted fork**, and B is said to be a **collider**. Here conditioning on B can result in a false correlation between A and C (Rohrer 2018).

MENDELIAN RANDOMIZATION

The method of **Mendelian randomization** (**MR**) combines concepts from path analysis, meta-analysis (Appendix 6), and GWAS (Chapter 20) to use genetics as a complement for randomized control trials (RCT; Appendix 9) when examining the causal impact of a potential risk factor on an outcome (e.g., disease). We can illustrate the basic idea as follows. Suppose a large population sample shows a correlation between hair loss and sugar intake (**observational epidemiology**). It could be that: (i) sugar $\rightarrow$ hair loss; (ii) hair loss $\rightarrow$ sugar; (iii) an unmeasured factor (a confounder, U) influences both, hair loss $\leftarrow U \rightarrow$ sugar; or (iv) some combination of these three scenarios. In an ideal setting of an RCT, one would start with a very young cohort (ideally at birth), randomly assign them into high and low sugar diet groups, and follow hair loss over the next fifty years. Of course, such RCTs for assessing the risk of long-term exposure to some factor are generally only possible in laboratories and domesticated species under controlled conditions. Even when such an RCT is feasible, its sample size will almost certainly be too small to detect modest effects (say an odds ratio, OR, < 1.5). Now suppose a GWAS finds a particular allele that naturally generates high sugar levels relative to its alternate allele. One could now look at hair loss as a function of this allele in a random population sample. Further, the massive sample size of a modern GWAS routinely detects very small effects (ORs on the order of 1.05 or less; Chapter 20). The randomization of the treatment (sugar level) occurs via *Mendelian segregation* at conception by randomly assigning alleles to offspring, with any other environmental factors naturally randomized over the two groups. This leads to what Hingorani and Humphries (2005) called "Nature's randomised trials."

Strictly speaking, a truly randomized design would contrast variants *within* a family (e.g., between sibs), while MR is typically performed on a population sample. As with the TDT test (Chapter 17), within-family contrasts remove any complications from population stratification. In a population sample, structure could be present, where both the disease and one of the variants occur in a subpopulation at higher frequencies than the rest of the sample, creating an association between them (Chapter 20). Davies et al. (2019) discuss other sources of bias that can arise in a population-based MR, and how within-family MR accommodates these concerns.

Even in settings where a short-term RCT is at least conceptually possible, MR still has advantages. Difference in cost and effort is an obvious one, using a population sample as opposed to a smaller and more expensive longitudinal RCT. Thus, one can think of an MR as

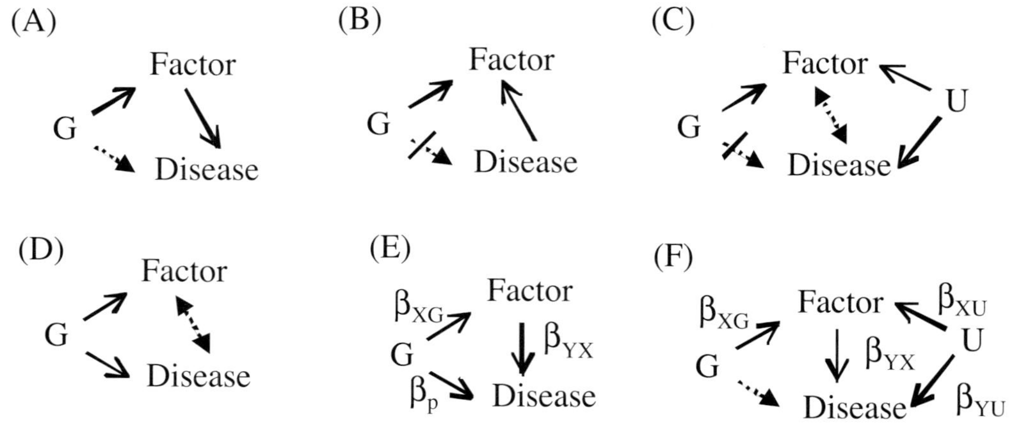

Figure A2.7 Path diagrams for settings where MR will work (A through C), and where it can fail (D), in testing the hypothesis that a factor (X, such as the amount of transcription, where here we assume that the value of X is influenced by genotype G) influences disease risk (Y). Solid arrows indicate causal connections, dashed arrows indicate induced correlations. The **instrumental variable** (or **instrument**), G, is the genotype at a locus influencing the factor being tested. Factor is often denoted as **exposure** in the literature, and disease can be generalized to an **outcome** variable (or state). **(A)** When the SNP influences the factor *and* the factor influences the disease, we see a correlation between the SNP and the disease state. **(B)** MR controls for reverse causality (disease $\rightarrow$ factor). Because the disease cannot alter the genotype, there is no SNP–disease correlation. **(C)** MR controls for a confounder (U) that influences both the factor and disease state. While U results in a correlation between the factor and disease state, there is no correlation between the SNP and disease. **(D)** MR can fail when the SNP has a pleiotropic effect on both the factor and the disease. This induces a correlation between the factor and the disease in the presence of a SNP–disease correlation. Note that pleiotropy could either be strict (the single SNP influences both traits) or LD-induced, where a linked SNP influences a different trait from the target SNP (Chapter 21). More generally, MR can fail when there is a path from G to Y *independent* of X. **(E)** In the presence of a pleiotropic SNP when the factor also influences the trait, the total association β_{YG} between G and Y is the sum of two different paths, one direct (SNP $\rightarrow$ disease, β_p) and one indirect (SNP $\rightarrow$ factor $\rightarrow$ disease; $\beta_{XG}\beta_{YX}$). **(F)** When the factor influences the disease, and both are influenced by a confounder, MR still correctly shows that factor $\rightarrow$ disease, but gives a biased estimate of the true effect, β_{YX}, when G is a weak instrument (Equation A2.17b).

a *screening procedure* for candidate associations for a follow-up RCT (such as in drug discovery settings). While (as discussed below), false positives can arise in MR studies, because of their greater power relative to an RCT, a negative MR result (even if a false negative because of low power) will likely generate a negative RCT result. The caveat to this statement is that MR usually involves a rather small change in the exposure, while an RCT typically imparts a much larger perpetuation, but this may be offset by the fact that MR often represents life-long exposure whereas RCTs only over the course of the experiment. Further, because alleles have been allocated at birth, MR deals with the setting where the critical time of exposure is unknown (i.e., its impact could occur at a younger age than used in the RCT). MR also accommodates cases where withholding of a treatment might be unethical (e.g., Gray and Wheatley 1991) and, conversely, when *applying* a treatment would be unethical, such as examining the health risk from pesticide exposure. If genetic variation exists in the ability to detoxify a suspected agent, this provides an MR solution to test for exposure risk that is controlled for confounders (Davey Smith and Ebrahim 2003). Finally, one can use **intergenerational Mendelian randomization** to test for maternal effects, using the randomization in the mother of alleles influencing a factor suspected to be involved in a maternal risk effect that influences the phenotype of its offspring (Davey Smith and Ebrahim 2004; Lawlor et al. 2017; Evans et al. 2019).

The basic MR concept was proposed by Katan (1986), with Gray and Wheatley (1991) later coining the term Mendelian randomization. Katan noted that low cholesterol levels correlated with increased cancer risk, but it might be that an early (and undetected) tumor decreases cholesterol levels long before the cancer is otherwise displayed (**reverse causality** with cancer → cholesterol). However, certain alleles at the *ApoE* gene naturally have reduced cholesterol levels, starting from birth. Katan reasoned that one could use control of the *ApoE* genotype as a surrogate for control of cholesterol level. If low cholesterol → cancer, then these low-cholesterol alleles should be over-represented in cancer patients. Conversely, if cancer → low cholesterol, then no such association is expected. See Figure A2.7 for the path diagrams for these, and other, settings.

While the logic of MR is straightforward, as with many issues in quantitative genetics, the devil is in the details. As noted by Burgess et al. (2019), MR *"is an important and valuable tool for learning about causal relationships using genetic data, but it is also a fallible one. ··· fundamentally the approach depends on being able to find genetic variants that are plausible proxies for intervention on the risk factor."* One should always perform sensitivity analysis (detailed below) before an MR analysis is considered complete, especially one based on multiple GWAS hits. As noted by Burgess et al. (2017), if no such analysis is done, then any results should *"be viewed as speculative and incomplete"* and *"treated with skepticism."*

The development of MR methodologies is a rapidly growing field that we will only briefly examine. Reviews and discussion are given by Davey Smith and Ebrahim (2003, 2004, 2005), Thomas and Conti (2004), Nitsch et al. (2006), Didelez and Sheehan (2007), Ebrahim and Davey Smith (2008), Lawlor et al. (2008), Sheehan et al. (2008), Brion et al. (2014), Davey Smith and Hemani (2014), VanderWeele et al. (2014), Boef et al. (2015), Burgess and Thompson (2015), Burgess et al. (2015b, 2017a, 2018, 2019), Haycock et al. (2016), Zheng et al. (2017), Hemani et al. (2018), Slob and Burgess (2019), and Zhu (2020). Power calculations are discussed by Brion et al. (2013), Freeman et al. (2013), and Deng et al. (2020). An excellent overview of mediation analysis versus instrumental variables is given by Otter et al. (2018).

Example A2.3. Benn et al. (2017) used MR to examine whether a low level of LDL cholesterol is a risk factor for either Alzheimer's or Parkinson's disease. Cholesterol is a major component of myelin sheaths enclosing neurons in the brain, and observational studies suggest a correlation between lower LDL levels and increased risk for both Alzheimer's and Parkinson's. Indeed, Benn et al. found such a significantly increased risk for Parkinson's in a population sample of 110,000 Danes. This is of serious concern given the widespread use of statins and other cholesterol-lowering drugs to reduce heart disease risk. The authors initially focussed on two genes, *PCSK9* and *HMGCR*, that segregate alleles which naturally lower LDL levels. They first showed that variants at these loci were independent of known confounding risk factors of the LDL cholesterol–disease relationship (age, sex, hypertension, smoking, physical activity, alcohol consumption, educational level, menopausal status). An MR involving 26 variants at these loci found that LDL-lower alleles showed no increased risk for either neurological disorder. The authors then considered 380 variants at other LDL-lowering loci detected from a GWAS. Again, no association between LDL-lowering alleles and increased disease risk was found. Indeed, collectively these alleles showed a significant *reduction* of Alzheimer's risk (OR 0.64, CI 0.52 to 0.79).

Factors probed by MR are not just restricted to presumed molecular intermediates, they can also be modifiable behaviors. One such example is Chen et al. (2008), who used variants at *ALDH2* as surrogates for alcohol consumption. Individuals homozygous for null *ALDH2* alleles experience adverse effects to alcohol, and as a result drink considerably less than other genotypes. Variants at a second locus (*ADH*) similarly influence alcohol consumption, and Howe et al. (2019) used an *ADH* variant (rs1229984) to examine potential causes for the correlation in alcohol consumption among couples. One possibility for this correlation is that the behavior of one partner influences that of the other. A second is mate choice, seeking a partner with a similar alcohol consumption (assortative mating). A third possibility is an unknown confounder influencing both mate choice and alcohol consumption (such as socioeconomic status). Howe et al. found that the rs1229984 status of one partner predicts the phenotypic (consumption) status of the other. Further, it also predicts the rs1229984 status of their partner, suggesting that assortative mating drives this observed relationship.

One final human example is the use of MR to probe the nature of the positive correlation between myopia (near sightedness), M, and years of education, Y (Mountjoy et al. 2002). Does myopia lead to increased education ($M \rightarrow Y$), does years of education (and hence increased reading) lead to increased myopia ($Y \rightarrow M$), or there some confounding factor, C, such as socioeconomic status, with $M \leftarrow C \rightarrow Y$. Mountjoy et al. used an index of 44 markers associated with myopia risk as the instrumental variable (IV), finding no correlation between myopia risk and education. Conversely, using 69 markers for a genetic predisposition for additional education as the IV, they found that individuals predicted to have higher levels of education also had increased amounts of myopia. Hence "exposure" to additional years of education increased the risk of myopia, $Y \rightarrow M$. A number of other interesting human examples can be found in the reviews by Davey Smith and Ebrahim (2004) and Sheehan et al. (2008).

There are obvious applications of MR beyond human studies. For example, in WL Chapter 20, we discussed a hypothetically observed association between seed production (as a proxy of fitness) in plants and alkaloid levels (plant secondary compounds). This might suggest a beneficial role of these compounds, e.g., by reducing insect damage. However, it could also represent a confounding effect, with (say) plants grown in nitrogen-rich soil having both higher fitness and higher alkaloid levels. If a GWAS found SNPs showing variation in alkaloid production, then if alkaloid $\rightarrow$ seeds, higher alkaloid alleles should also have more seeds, while if there is a confounding effect on both, no such association is expected.

The Concept of Instrumental Variables (IVs)

The underpinning of MR is the concept of an **instrumental variable** (**IV**), or **instrument**, an idea originally from economics. This concept first appeared in Appendix B of Philip G. Wright's (1928) *The tariff on animal and vegetable oils*. There was, however, much speculation that Appendix B was actually written by Philip's eldest son, Sewall (yes, the inventor of path analysis). After a detailed analysis, Stock and Trebbi (2003) concluded it was more likely than not that Philip *wrote* the Appendix, but there was strong evidence that the key *idea* came from Sewall. Brief reviews of the utility, and limitations, of IVs in epidemiological studies are given by Greenland (2000) and Hernán and Robins (2006).

To more formally define an IV, suppose the goal is to test whether X is causal to Y, ($X \rightarrow Y$) and (if so) estimate the strength β_{YX} of this effect (how much a unit change in X changes Y). Let U denote the set of all variables that influence *both* X and Y (i.e., all the confounders). Then as shown in Figure A2.8, Z is an IV if it satisfies three conditions: (i) Z is independent of U, (ii) Z is associated with X, and (iii) Z is independent of Y conditional on X and U (this is often called the **exclusion restriction**), namely that there are no paths connecting Z and Y other than the one that runs through X. If any of these conditions fail, Z is said to be an **invalid instrument**. Condition (iii) is the assumption of no (horizontal) pleiotropy when the IV is a genotype, and is often problematic. Condition (i) can fail in the presence of population stratification (Chapter 20) or sampling following selective survival (Nitsch et al. 2006). A more detailed discussion of other routes by which one, or more, of these assumptions can fail is given by VanderWeele et al. (2014), who, along with Glymour et al. (2012), also review validation methods. In what follows, we use Z to denote a general instrumental variable, and G when the instrument is a genetic variant (i.e., the MR setting).

When the IV assumptions hold, then applying the rules of path analysis to Figure A2.7A, we have

$$\beta_{YZ} = \beta_{XZ}\,\beta_{YX}, \quad \text{implying} \quad \beta_{YX} = \frac{\beta_{YZ}}{\beta_{XZ}} \tag{A2.16a}$$

where β_{YZ} and β_{XZ} are the regression slopes of Y and X, respectively, on Z. This leads to the instrumental variable **ratio estimate** (or **Wald estimate**) for β_{YX},

$$\widehat{\beta}_{IV,R} = \frac{\widehat{\beta}_{YZ}}{\widehat{\beta}_{XZ}} \tag{A2.16b}$$

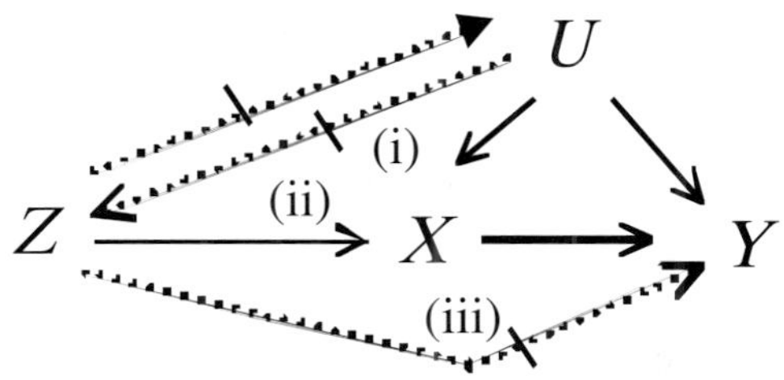

Figure A2.8 The three requirements for Z to be an instrumental variable (or instrument) for the impact of a factor X on an outcome Y. First, any confounder (U) that influences both the factor and the disease (Y) is *not* influenced by Z ($Z \not\to U$). We also assume that the confounder does not influence Z ($U \not\to Z$), which is very reasonable when the instrument is a genotype. Second, Z must directly influence the factor X (path ii is present). Finally, there is *no* direct path from Z to Y (path iii is absent).

Thomas et al. (2007) approximated the standard error using the delta method (Equation A1.19b), whose leading term is

$$\sigma^2(\,\widehat{\beta}_{IV,R}\,) \simeq \frac{\sigma^2(\,\widehat{\beta}_{YZ}\,)}{(\,\widehat{\beta}_{XZ}\,)^2} \tag{A2.16c}$$

Equation A2.16c assumes that β_{XZ} is known without error ($\widehat{\beta}_{XZ} = \beta_{XZ} + e$, where $|e| \ll |\beta_{XZ}|$), commonly referred to in the MR literature as the **NOME** assumption (**NO Measurement Error**; Bowden et al. 2016a). As a result, Equation A2.16c is often called the **NOME weight** for the ratio. Improved approximations were given by Thomas et al. (2016), Thompson et al. (2016), and Bowden et al. (2019), while other approaches for obtaining the SE were discussed by Pierce and Burgess (2013).

When pleiotropy is present (a path from Z to Y without going through X), Equation A2.16b is a biased estimate of β_{YZ}. As outlined in Figure A2.7E, letting β_p be the path coefficient from pleiotropy, the total association between Z and Y becomes

$$\beta_{YZ} = \beta_p + \beta_{XZ}\,\beta_{YX} \tag{A2.17a}$$

yielding

$$\frac{\beta_{YZ}}{\beta_{XZ}} = \beta_{YX} + \frac{\beta_p}{\beta_{XZ}} \tag{A2.17b}$$

The careful reader might notice that even when $\beta_p = 0$, G technically still displays pleiotropy as it impacts both X and (indirectly) Y. The MR literature refers to the effect on Y from G that is fully mediated through X ($\beta_{XZ}\,\beta_{YX}$) as **vertical pleiotropy**, also called **mediated** (Solovieff et al. 2013), **secondary** (Hodgkin 1998), or **spurious pleiotropy** (Gruneberg 1938). Any pleiotropic effect of G on Y that does not pass through X (β_p) is called **horizontal** or **biological pleiotropy** (Solovieff et al. 2013). In our following discussion, any mention of pleiotropy refers to the latter.

Estimating a quantity by the ratio of two estimates can have undesirable statistical properties. An alternative approach that avoids a ratio is to use **two-stage least-squares (2SLS** or **TSLS**) regression. Assume the instrument is a genotype, G. Under 2SLS, one first predicts the value of X given G, $\widehat{X}|G$, and then regresses Y on $\widehat{X}|G$, namely,

$$Y_i = \alpha_{YX} + \beta_{IV,2S}(\widehat{X}_i|G_i) + e_i, \quad \text{where} \quad \widehat{X}_i|G_i = \alpha_{XG} + \beta_{XG}\,G_i, \tag{A2.18}$$

where the estimated slope, $\widehat{\beta}_{IV,2S}$, for the regression of Y on $X|G$ is taken as the estimate for β_{YX}. The 2SLS estimator easily generalizes to multiple instruments (predicting X using a multivariate regression on IVs). This is what a TWAS does, where X is the level of a focal transcript, which is predicted using the (*cis*) eQTL genotypes (G) of an individual (Chapter 21).

A key design concern with any MR is the sampling strategy. In a **single-sample design**, G, X, and Y would be measured on all the members of some sample. However, the intermediate factor X may be more challenging and/or expensive to measure. A **subsample** approach scores X on only some subset of the sample. Pierce and Burgess (2013) showed that, for a strong instrument (defined below), scoring only 20% of the samples for X results in almost the same power as scoring the entire sample. A **two-sample design** can also be used, wherein the association between X and G is measured in one sample and between Y and G in a second. Such a setting allows GWAS summary statistics (Chapters 20 and 21) to be used in place of (often unobtainable) individual-level data. Indeed, these two estimates could be based on a meta-analysis over a number of studies (Appendix 6). Obviously, bias can arise if these different samples have different features (e.g., differences in confounders, population or LD structure, etc.). Minelli et al. (2004), Haycock et al. (2016), and Burgess et al. (2017b) discussed how to use standard sensitivity analysis tools from meta-analysis (heterogeneity tests, forest and funnel plots; Appendix 6) to look for biases and inconsistencies in the data sets. A final issue is Beavis effects, in that a marker detected as being significant (under a low power setting) often has its effect overestimated (Chapters 18 and 20). Richmond et al. (2014) recommended using a subsampling approach in such cases, detecting a G–X association using some part of the sample, and then estimating its effect, β_{XG}, in the remaining part (e.g., Example 21.1).

When using a valid instrument, MR is a rather robust procedure for *demonstrating* that X is causal to Y. However, estimating the actual *strength* of such an association is a bit more challenging. One issue is that even when G is a valid IV, it may still be a **weak instrument**, having a small impact of X relative to the noise in the system. The term weak is slightly misleading, as a very small actual effect (G accounts for only a small fraction of the variance in X) is a weak instrument under small sample size, but becomes strong as the sample size increases. The rough rule in the literature is that the F statistic for the regression of X on G should exceed ten for an instrument to be considered strong (Staiger and Stock 1997; Lawlor et al. 2007). Equation A2.19b shows how this transition from weak to strong occurs as sample size increases.

The ratio estimate of β_{YX} is biased by finite sample size, which is generally only a problem when a weak instrument is used. The directional impact of **weak instrument bias** depends upon the sampling scheme. In a two-sample design, because the estimates of β_{XG} and β_{YG} are from independent data, they are independent and (as we show shortly), the *bias shrinks the estimate of β_{YX} towards zero* (in the direction of the null). Conversely, under a one-sample design, the estimate of β_{YX} is *skewed towards any confounder effects* (Burgess and Thompson 2011; Pierce and Burgess 2013).

To see this distinction, consider Figure A2.7F, and let X_i, Y_i, and U_i be the values of the factor, disease, and confounder in individual i. Then

$$X_i = \beta_{XG}G_i + \beta_{XU}U_i + e_{X_i}, \qquad Y_i = \beta_{YX}X_i + \beta_{YU}U_i + e_{Y_i} \tag{A2.19a}$$

Under a one-sample design, the value of the confounder U_i is shared by X_i and Y_i, while *independent* draws of U influence X and Y in a two-sample design. Under the simplifying assumption of two states of the instrumental variable (1 and 0) and ignoring any effects of e, Burgess and Thompson (2011) showed (for a one-sample design) that the ratio estimator implies

$$\beta_{IV,R} = \beta_{YX} + \beta_{YU}\left(\frac{\Delta\overline{U}}{\beta_{XG} + \beta_{XU}\Delta\overline{U}}\right) \tag{A2.19b}$$

where $\Delta\overline{U} = \overline{U}_1 - \overline{U}_0$ is the random sample differences in the mean confounder values between genotypes. With a weak instrument ($|\beta_{XG}| \ll |\beta_{XU}\Delta\overline{U}|$), then $\beta_{IV,R} \simeq \beta_{YX} + \beta_{YU}/\beta_{XU}$, showing a bias towards the confounder effect (β_{YU}). Assuming n of each of the genotypes in the sample and that $U \sim N(0, \sigma_U^2)$, then $\Delta\overline{U} \sim N(0, 2\sigma_U^2/n)$. Hence, $|\Delta\overline{U}|$

shrinks towards zero as n increases, changing a weak IV into a strong one for sufficiently largely n, with $|\beta_{XG}| \gg |\beta_{XU}\Delta\overline{U}\,|$ and thus $\beta_{IV,R} \simeq \beta_{YX}$.

An issue related to weak instruments is **canalization**, the buffering of fluctuations in intermediate steps in a pathway to limit downstream perturbations of the final end product (Chapter 14). As noted by Davey Smith and Ebrahim (2003), canalization can result in a rather large change in X from a genetic variant imparting only a small change in Y. A similar issue could occur if the variant polymorphism impacts a product that is a few steps removed from the factor we are tying to test. Finally, the basic MR analysis assumes that Z (or G), X, and Y are all *scalars*. We next consider a *vector* **Z** (**G**) of instruments (genotypes) for predicting X. Extensions allowing for a vector **X** (a correlated set of exposure factors) and/or a vector **Y** (a correlated set of outcomes, such as closely related diseases) have been considered by Thompson et al. (2005), Burgess and Thompson (2015), Rees et al. (2016), and Sanderson et al. (2019).

Mendelian Randomization Using Multiple Variants

MR can be a powerful tool if one has a valid strong single instrument (e.g., a variant with a large effect on X). However, most traits are impacted by a large number of loci of modest to small effect (Chapters 20 and 21). As a result, the early days when MR were mostly based on Mendelian single factors (e.g., Davey Smith and Ebrahim 2003) gave way to extensions to exploit polygenic GWAS signals. In this setting there is still a single exposure (X) and outcome (Y) variable, but now MR leverages information from multiple GWAS hits influencing X to test its causality on Y (e.g., Example A2.3).

There are obvious advantages, and pitfalls, to extending an analysis to multiple instruments. Even if each has a fairly weak effect on X, their cumulative effect could be considerable, offering a substantial increase in power. Countering this, as more variants are added, they bring with them the risk that at least some are invalid instruments (such as displaying horizontal pleiotropy). Finally, LD can induce correlations among multiple GWAS hits in the same gene, and one must adjust for these. A variety of solutions for treating these concerns have been suggested, many based on concepts from meta-analysis (Appendix 6).

We start with the strictest assumption that *all* of the k tested variants are *valid instruments*. A second assumption is *functionally consistent effects* over all variants. While each variant may have quantitatively different impacts on X (β_{XG} can vary over G), β_{YX} is the same for all: no matter how a change ΔX is generated, the resulting average change in Y is $\beta_{YX}\Delta X$. An example where this assumption might fail is if one SNP is an eQTL, while another is both an eQLT and an sQTL (Table 21.1). Both SNPs might give the same total level of transcript, but differential splicing may have a greater impact, so that if X is simply total transcript amount, β_{YX} will likely vary over these SNPs.

Because most GWAS-based MRs use a two-sample design, we assume this structure in the rest of our discussion. Let β_{Yj} and β_{Xj} be the estimated regression slopes of Y and X, respectively, on the instrumental variable, variant j. Using Equation A2.16b, one constructs the ratio estimate, $\widehat{\beta}_{IV,R,j}$, for each variant. If the IV assumptions (Figure A2.8) hold, these should all estimate a common parameter (β_{YX}), subject to sampling noise. We can express this as a standard fixed-effects meta-analysis (Equation A6.30a),

$$\widehat{\beta}_{IV,R,j} = \beta_{YX} + e_j, \qquad \text{with} \quad e_j \sim N\left[0, \sigma^2(\widehat{\beta}_{IV,R,j})\right] \tag{A2.20a}$$

This is just a GLM (Chapter 10), with the BLUE estimate of β_{YX} given by the **inverse-variance weighted (IVW) average** (Equation A6.30b),

$$\overline{\beta}_{YX} = \frac{\sum_{j=1}^{k} w_j\, \widehat{\beta}_{IV,R,j}}{\sum_{j=1}^{k} w_j}, \quad \text{where} \quad w_j = \frac{1}{\sigma^2(\widehat{\beta}_{IV,R,j})} \tag{A2.20b}$$

Equation A2.20b assumes the estimates over different variants are uncorrelated, but can be modified to allow for LD (correlations among variants). Using the approximation for the standard error of the ratio estimate given by Equation A2.16c (the NOME assumption),

$$w_j \cdot \widehat{\beta}_{IV,R,j} \simeq \left(\frac{\beta_{Xj}^2}{\sigma^2(\beta_{Yj})} \right) \cdot \left(\frac{\beta_{Yj}}{\beta_{Xj}} \right) = \beta_{Yj}\beta_{Xj}/\sigma^2(\beta_{Yj}) \tag{A2.20c}$$

and Equation A2.20b reduces to the IVW estimator of Burgess et al. (2013),

$$\overline{\beta}_{YX} = \frac{\sum_{j=1}^{k} \beta_{Yj}\beta_{Xj}/\sigma^2(\beta_{Yj})}{\sum_{j=1}^{k} \beta_{Xj}^2/\sigma^2(\beta_{Yj})} \tag{A2.20d}$$

whose standard error is given by Equation A6.30c.

Another approach follows from the concept of **polygenic risk scores** (Part II Chapter 32). The idea is that one takes a candidate set of variants and constructs an index of these to serve as a single instrumental variable. This is the **allele scores** method of Burgess and Thompson (2013). One could either construct an unweighted score (simply counting the number of risk alleles, namely those increasing X) or weight alleles by some scheme (such as by their estimated effects as in a polygenic risk score). The idea is to avoid weak instrument bias by using a single sum as opposed to averaging ratios based on a number of weak instruments. Simulations by Burgess and Thompson showed that this approach can be powerful when all of the variants are valid instruments, but can be severely biased when as few as 10% of the variants are invalid. Further, seriously biased estimates often occur with weighted sums, as the weights are often incorrectly estimated (e.g., Beavis effects). Burgess et al. (2015a) detailed how to use summary data and correlated variants in constructing allele score tests.

A number of more robust estimators allowing some of the instruments (variants) to be invalid (e.g., show pleiotropy) have been proposed. The simplest two were suggested by Bowden et al. (2016b), using the notion that the median is often rather robust to outliers and model departures (e.g., the GC estimator given by Equation 20.12b). Ordering the ratio estimates for the k variants from smallest to largest, the simple **median estimator** takes the median (that value which divides the upper and lower half of the k estimates) as the estimate. Their **weighted median estimator** is a modification of this idea, and proceeds as follows. First, a weight is assigned to each of the estimates (e.g., using Equation A2.20b), which are standardized to sum to one, and the weights ranked from smallest to largest. Again, the median is taken, but now using the cumulative sum of the weights, with the variant whose cumulative sum is 50% has its ratio taken as the group estimate. Their simulations showed that these estimators were consistent even when up to 50% of the information comes from invalid instrumental variables. It should be noted that consistency is a large-sample feature. Nonetheless, median estimates serve as a sensitivity check, in that if the median and IVW estimates are rather similar, one has more confidence in the MR estimate, while if they radically differ, further exploration into the data is likely needed.

A number of regression-based approaches have been proposed to handle the general case of an unknown number of invalid instruments among the variants. The assumption is that any invalid instrument arises entirely from pleiotropy. From Equation A2.17a, the resulting total effect of variant j on Y becomes

$$\beta_{Yj} = \beta_{p,j} + \beta_{YX}\,\beta_{Xj} \tag{A2.21}$$

where the goal is to estimate the shared β_{YX} effect. As noted by Bowden et al. (2017, 2018), one signal of pleiotropy (assuming $\sigma^2(\beta_{p,j}) \neq 0$) is excessive heterogeneity in the ratio estimates obtained from different SNPs, which can be tested using a number of heterogeneity tests from meta-analysis, such as the Cochran's Q (Equation A6.31) or Higgins–Thompson I^2 (Equation A6.37c). Bowden et al. (2019) developed improved weights to yield better performance of the Q test than under the NOME assumption weights (Equation A2.16).

Even in the presence of pleiotropy, IVW gives a consistent estimate of β_{YX} when two conditions are jointly satisfied (Bowden 2017). The first is **balanced pleiotropy**, wherein

$E[\beta_p] = 0$. The second is the **InSide assumption**, for **Instrument Strength independent of direct effects** (Bowden et al. 2015). This is essentially a weaker version of the exclusion restriction assumption (the lack of path (iii) in Figure A2.8), and assumes that the magnitude (and direction) of the pleiotropic effect ($\beta_{p,j}$) is uncorrelated with the direct effect $\beta_{X,j}$ on X. Note that under balanced pleiotropy while there is no *average* effect from pleiotropy, it will result in excess variance of the individual ratio estimates, and hence can be potentially detected by a heterogeneity test.

Several approaches have been suggested to deal with the presence of **directional pleiotropy**, $E[\beta_p] \neq 0$, which attempt to estimate the common β_{YX} in Equation A2.21 under different assumptions on the distribution of $\beta_{p,j}$ (summarized in Bowden et al. 2017; Hemani et al. 2018; Verbanck et al. 2018; Slob and Burgess 2020; Zhu 2020; and Cho et al. 2021). The simplest assumption is that $\beta_{p,j}$ is the same over all variants. In this setting, the β_{YX} estimates will not show excess heterogeneity, but the presence of pleiotropy can be detected by other tests. The most common is the **Mendelian randomization–Egger** (**MR-Egger regression**) approach of Bowden et al. (2015). It is based on the Egger regression test (Equation A6.38) from meta-analysis, where a nonzero intercept indicates the presence of a publication bias (e.g., negative test results are not published). The model becomes

$$\widehat{\beta}_{Yj} = \beta_p + \beta_{YX}\,\widehat{\beta}_{Xj} + e_j, \qquad \text{with} \quad e_j \sim N\left[0, \sigma^2(\widehat{\beta}_{Yj})\right] \tag{A2.22a}$$

The intercept term, β_p, should be zero in the presence of balanced pleiotropy (this is the **MR–Egger intercept test**). This regression accomplishes three goals: testing for the presence of directional pleiotropy ($\beta_p \neq 0$), testing causality of $X \rightarrow Y$ ($\beta_{YX} \neq 0$), and (most problematic) estimating the effect β_{YX}. More generally, Equation A2.22a can still give a consistent estimate of β_{YX} when $\beta_{p,j}$ vary, provided that the InSide assumption holds (Bowden et al. 2017). In this setting, β_p is the average amount of directional pleiotropy.

Turning to estimation, Bowden et al. (2015) showed that the MR–Egger estimate of β_{YX} is slightly biased, with

$$\beta_{YX,MR-Eg} = \beta_{YX}\left(\frac{\sigma^2(\beta_{Xj})}{\sigma^2(\widehat{\beta}_{Xj})}\right) \tag{A2.22b}$$

Because $\sigma^2(\widehat{\beta}_{Xj}) > \sigma^2(\beta_{Xj})$, the estimate of β_{YX} is shrunk towards zero (i.e., back towards the null), a phenomenon known as **regression dilution bias**. Writing $\widehat{\beta}_{Xj} = \beta_{Xj} + e_k$, so that $\sigma^2(\widehat{\beta}_{Xj}) = \sigma^2(\beta_{Xj}) + \sigma^2(e_j)$, we see that if the β_{Xj} are known without error ($\sigma^2(e_j) = 0$, the NOME assumption), then the $\beta_{YX,MR-Eg}$ estimate is consistent. Bowden et al. (2015) noted that this assumption can be tested by quantifying the heterogeneity of the β_{Xj} estimates (the strength of variant j on the exposure factor X) using the Higgins–Thompson I^2 measure of heterogeneity (Equation A6.37c). As discussed in Appendix 6, I^2 is the fraction of variation due to differences in effects (and hence ranges from zero to one). Bowden et al. suggested that if the I^2 statistic for the β_{Xj}, which they denote at I^2_{GX}, in the analysis exceeds 0.9, then the NOME assumption is reasonable, and the shrinkage of the β_{YX} estimate is at most 10%. An I^2_{GX} value close to one implies the variance among the true β_{Xj} values is much larger than their individual sampling errors, $\sigma^2(\beta_{Xj}) \gg \sigma^2(e_j)$. A number of outlier detection approaches have been proposed to fine-tune regression approaches, such as **HEIDI–outlier** (Zhu et al. 2018), **MR–PRESSO** (Verbanck et al. (2018), and **MR–TRYX** (Cho et al. 2020).

As stressed by several authors, both MR–Egger and median estimators should be viewed as *sensitivity analysis methods* for probing assumptions about the MR. Additional statistics have been proposed, such as the Q_R statistic of Bowden et al. (2017). This is just the ratio of the heterogeneity of data around the MR–Egger fitted slope divided by the heterogeneity of data around the IVW slope. A value of Q_R much less than one indicates that the MR–Egger is better and should be given serious consideration. Bowden (2017) stressed the joint use of both Q_R and I^2_{GX} to determine in which case the MR–Egger should be used, while Burgess and Thompson (2017) provide further discussion on how to interpret various MR–Egger findings.

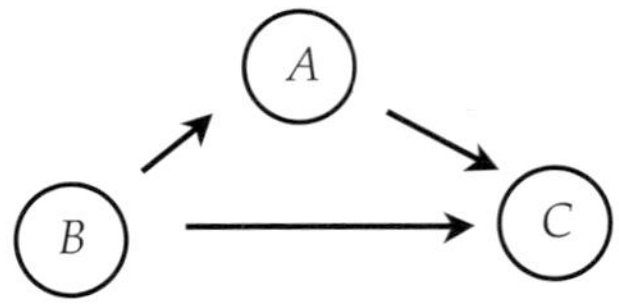

Figure A2.9 A simple directed acyclic graph (DAG). The nodes (A, B, C) indicate elements in a regulatory network, while the arrows (edges) indicate the direction of interaction.

An entirely different approach for dealing with pleiotropy is the use of penalized regressions (Example 20.4). Rees et al. (2019) proposed to allow $\beta_{p,j}$ to vary, but to penalize their total absolute sum. Under the LASSO (which shrinks most $\beta_{p,j}$ to zero, generating variable selection), the estimate is given by

$$\widehat{\beta}_{YX,L\lambda} = \min_{\beta_{p,j},\beta_{YX}} \left(\sum_{j=1}^{k} \frac{(\widehat{\beta}_{Yj} - \beta_{p,j} - \beta_{YX}\widehat{\beta}_{Xj})^2}{\sigma^2(\widehat{\beta}_{Yj})} + \lambda \sum_{j=1}^{k} |\,\beta_{p,j}\,| \right) \tag{A2.23}$$

This returns both the common factor β_{YX} as well as the pleiotropic effects of any retained variants (following model selection). One could also use the LASSO for strict model selection, performing an MR using only those variants whose $\beta_{p,j}$ have been shrunk to zero.

BASIC GRAPH THEORY FOR BIOLOGICAL NETWORKS

One of the rich harvests from functional genomics is elaborate **network structures**, interactions between molecular elements, such as metabolic pathways, regulatory networks, and protein–protein maps (showing which proteins physically interact with each other). Given the rapid accumulation of such information, and its developing importance in quantitative genetics, we conclude by presenting a few basic concepts from graph theory that appear in the network literature. Barabási and Oltvai (2004), Vidal et al. (2011), and Hu et al. (2016) review basic concepts with a focus on biological networks.

Consider Figure A2.9. The items in the network (A, B, and C; for example, proteins) are referred to as **nodes**, and the connections (interactions) between them are called **edges**. An arrow for an edge indicates a directional interaction, while a line indicates an interaction with no assumed directionality. In an **acyclic graph**, no pathways connect a node back to itself through a series of intermediate nodes (e.g., there are no **feedback loops**). Most path diagrams are thus **directed acyclic graphs** (or **DAGs**). We can represent the basic **topology** (shape) of a graph by a matrix **M** containing zeros and ones. In an **undirected graph** (only lines, no arrows), $M_{ij} = 1$ if nodes i and j are connected by an edge, resulting in a square, symmetric matrix (as $M_{ij} = M_{ji}$). In a directed graph, $M_{ij} = 1$ indicates that node i influences node j. In this case, $M_{ij} = M_{ji}$ only if both nodes influence each other, so that the matrix describing the graph is typically not symmetric. For the directed graph in Figure A2.9, the resulting matrix (ordering the columns as A, B, C) becomes

$$\mathbf{M} = \begin{pmatrix} 0 & 0 & 1 \\ 1 & 0 & 1 \\ 0 & 0 & 0 \end{pmatrix}$$

More generally, we can replace the elements of **M** by measures of the strengths (and signs) of interactions between nodal elements. Note that the object **M** is a multi-dimensional quantitative trait, and thus, provided we can measure it in a set of relatives, we can perform standard quantitative-genetic tasks (such as separating genetic from environmental variation). Similarly, with marker data we can also map regQTLs (Table 21.1) that impact various

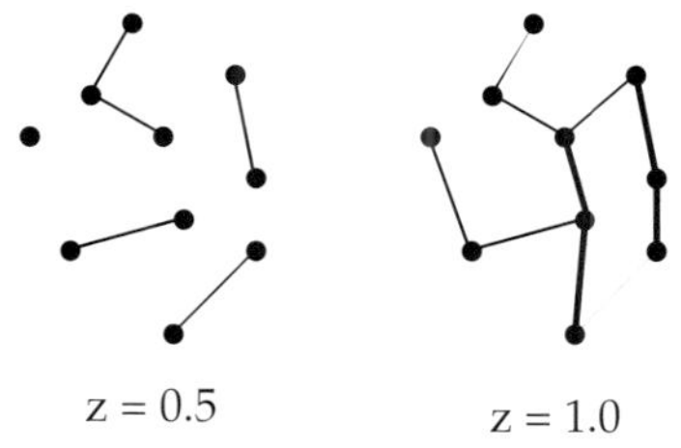

Figure A2.10 The phase transition in ER graphs at $z = 1$, where a giant component forms, in which a large fraction of the nodes in the graph are interconnected.

aspects of **M**, such as its leading eigenvalue (if we wish to extract a univariate trait), or more general features in a multivariate setting.

In order to probe the structure of a biological network, we first need to consider the features of a **random graph**, as this provides one natural null model. Random graphs, where edges are randomly connected to nodes, are often called **Erdõs–Rényi** (1959), **ER, graphs** after the two mathematicians who first considered their basic properties. Formally, we start with n nodes, with p the probability that any two randomly chosen nodes are connected. We consider several metrics for the structure of a graph and compare how the ER values differ from those seen in actual biological networks.

The first descriptor of a graph is its **degree distribution,** $P(k)$, the probability that a randomly chosen node is directly linked to exactly k other nodes. Under ER graphs, $P(k)$ follows a Poisson distribution (Equation 2.21a), with success parameter $z = np$,

$$\Pr(X = k) = \frac{(np)^k \exp(-np)}{k!}, \quad \text{for } k = 0, 1, \cdots \tag{A2.24a}$$

Note that the probability a random node is not connected to any other ($k = 0$) is simply $\exp(-np)$, and that the number of expected connections falls off exponentially.

A second descriptor is the **path length distribution,** d, a measure of how well connected all of the nodes in the graph are to each other. Formally, this is the distribution of smallest number of steps that it takes to go from one node in the graph to any other. For an ER graph, the average path length between any two nodes is roughly given by

$$d_{ER} \sim \ln(n)/\ln(z), \quad z = np \tag{A2.24b}$$

ER graphs display a very striking **phase transition** at $z = 1$ (on average, one connection between any two randomly chosen nodes), forming a **giant component**, wherein a large fraction of the nodes in the graph become interconnected (Figure A2.10). The biological implication of a giant component is that most of the elements in the graph influence (or at least interact with) most other elements. Thus, the network moves from a series of discrete, nonoverlapping modules into a single integrated structure. In a classic paper, Kaufmann (1969) modeled gene circuits as **random Boolean networks**. In Kaufmann's model, genes were randomly connected (formed an ER graph) and simple on/off (Boolean) rules applied (such as if your neighbor is off, you are on). Kaufmann found that such random networks showed giant regulatory components, features that appear to show *complex order* and give all the appearance of being *highly evolved*, yet they were *entirely random*. This raises the key point that ***random graphs can show considerable structure.*** Hence, ***one must be careful in inferring evolution simply because a network shows a complex, and highly integrated, structure.***

A final feature about ER graphs is that they are **small world**, wherein the average path distance between any two nodes is short. The classic example of a small-world graph is the **Kevin Bacon conjecture**. The assertion is that every actor can be traced back to an actor who appeared in a film with Kevin Bacon. The **Bacon number** for an actor is the number

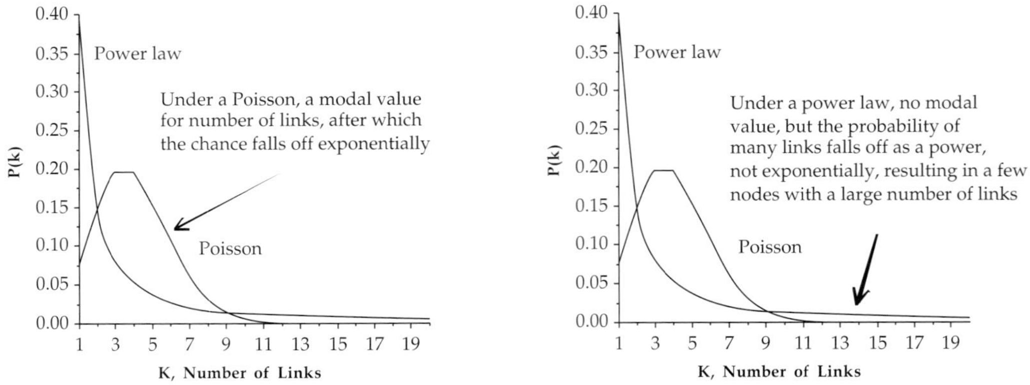

Figure A2.11 Poisson versus power law (scale-free) degree distributions, *P(k)*. Under a Poisson (left; Equation A2.24a), there is some modal number of links, after which the number of links falls off exponentially. In contrast, under a power law distribution (right; Equation A2.24c), there are a few hubs with very many connections.

of actors (nodes) one must go through to find this connection. Mathematicians have a similar concept, an Erdõs number. Erdõs was very prolific, and your **Erdõs number** is how many mathematicians you must go back through to make a connection with someone who wrote a paper with Erdõs. Formally, in a small-world graph, the average path distance is close to that predicted for an ER graph, $d_{ER} \sim \ln(n)/\ln(z)$. The critical feature of small-world graphs is that they ***propagate information very efficiently.*** One can transform a highly regular graph into a small world graph by randomly rewiring a small fraction of the nodes.

How do cellular networks depart from random graphs? Most importantly, their ***degree distribution often follows a power law,***

$$P(k) \sim k^{\gamma} \tag{A2.24c}$$

This is in contrast to the Poisson distribution for k expected under ER graphs (Figure A2.11). Graphs with a power distribution of links are called **scale-free**. In a scale-free graph, a few of the nodes will have ***very many connections***. Such nodes are often called **hubs**, where just a few locations influence a large number of nodes (such as a single eQTL influencing a large number of transcripts; Chapter 21). The important feature of scale-free graphs is that they are ***fairly robust to perturbations***: most randomly chosen nodes can be removed with little effect on the system. While removing a hub has a critical effect, the chance that a randomly removed node is a hub is small. A second feature of scale-free networks is that they are **ultra small**, meaning that their mean path length is actually *shorter* than under a random graph.

Because a scale-free structure gives a network inherent stability, biological homeostasis may just be ***simply a consequence of this structure, rather than a highly evolved feature***. How might such scale-free graphs evolve? The answer turns out to be rather simple: when we add new nodes, the scale-free feature arises when they have a ***slight preference to attach to already established nodes***. This feature is exactly what is thought to happen as gene duplication adds more nodes (elements) to a network (e.g., *Arabidopsis* Interactome Mapping Consortium 2011).

The initial view that most cellular networks appeared to be scale-free has been modified somewhat (Vidal et al. 2011). While metabolic and protein–protein interaction networks tend to be scale-free, genetic regulatory networks often show a mixed behavior. With a directional graph, we can distinguish between the **incoming** and **outgoing** degree distribution, namely the average number of inputs to a node (the number of incoming arrows to a node: how many other nodes directly influence it) and the average of outputs from a node (the number of arrows leaving a node: how many other nodes it directly influences). Regulatory networks often show a scale-free outgoing distribution, meaning that there are a few hubs that can influence numerous nodes. However, the incoming degree distribution is better fitted with an exponential, implying that genes directly regulated by the input from a number of nodes are rare. Han et al. (2004) observed two rather different types of hubs in biological networks: **date hubs**, whose interactions with their partners vary over time, and **party hubs**, which are highly coregulated with their partners. Vidal et al. (2011) suggested that date hubs connect different functional modules of a network with each other, whereas party hubs largely act inside functional modules, but also see Agarwal et al. (2010).

Example A2.4. Apparently well-buffered networks can often be disrupted by single-gene mutations, resulting in a dramatic increase in phenotypic variance (Chapter 14). Rutherford and Lindquist (1998) called genes (where such mutations can arise) **phenotypic capacitors**, with the classic example being heat-shock protein Hsp90, a molecular chaperone (a protein that helps other proteins fold properly). Hsp90 knockouts in both *Drosophila* and *Arabidopsis* show increased phenotypic variance over a number of traits. One hypothesis is that capacitors are often associated with hubs in critical networks. To more broadly search for capacitors, Levy and Siegal (2008) examined phenotypic variances over a suite of morphological features in a collection of 4,700 single-gene knockout lines in yeast (*Saccharomyces cerevisiae*). Roughly 300 ($\sim 5\%$) of these were capacitors, as their knockout lines showed a significant (within-line) increase in the phenotypic variance over a number of traits. Levy and Siegal found that such genes were likely to be at hubs in either networks involving physical interactions between their products (hubs in the protein–protein interaction network) or involving genetic interactions between loci (hubs in the synthetic lethal interaction network). Masel and Siegal (2009) review general issues on robustness, a topic further examined in Chapter 14.

Appendix 3

Further Topics in Matrix Algebra and Linear Models

There's a difference between knowing the path and walking the path. Morpheus

This appendix builds on Chapters 9 and 10, presenting additional results from matrix algebra and linear-model theory. We start by introducing two useful matrix operations, generalized inverses (for solving singular systems of equations) and the square root of a matrix (for obtaining a set of uncorrelated variables). These results are then used for a formal derivation of several properties of generalized least-squares (GLS) estimators. We next examine how linear model sums of squares can be written as quadratic forms and how these sums of squares are used in formal hypothesis testing and in the construction of confidence intervals. We then turn to equivalent linear mixed models (which allow calculations for one model to be performed on a potentially much simpler model), and then conclude with brief introductions to matrix derivatives and matrix decompositions (factorizations).

GENERALIZED INVERSES AND SOLUTIONS TO SINGULAR SYSTEMS OF EQUATIONS

Linear systems of equations are ubiquitous in quantitative genetics and we have presented solutions for such systems by assuming that the appropriate matrices are nonsingular, and hence can be inverted. However, in the real world of large, complex, and unbalanced data sets, the existence of an inverse is by no means guaranteed. Consider the solution of the matrix equation $\mathbf{y} = \mathbf{Ax}$ for the unknown vector $\mathbf{x}$. If $\mathbf{A}$ is square and nonsingular, then $\mathbf{x} = \mathbf{A}^{-1}\mathbf{y}$ is the *unique solution*. However, what happens if $\mathbf{A}$ is singular or is nonsquare? In this case either: (i) the system has no solution and is said to be **inconsistent**; or (ii) there are an *infinite number of solutions* (the solution set is a line, plane, or hyperplane). An example of an inconsistent system is

$$\begin{aligned} x_1 + x_2 &= 1 \\ x_1 + x_2 &= 2 \end{aligned}$$

which cannot be satisfied by any (x_1, x_2). Likewise, a system with an infinite number of solutions is

$$\begin{aligned} x_1 + x_2 &= 1 \\ 2\,x_1 + 2\,x_2 &= 2 \end{aligned}$$

which has a line of solutions of the form $x_2 = 1 - x_1$ for arbitrary x_1. While these two simple systems can be solved by inspection, a more systematic approach is required for arbitrary systems. This is provided by using **generalized inverses**.

Generalized Inverses

Suppose a matrix $\mathbf{A}^-$ exists such that

$$\mathbf{AA}^-\mathbf{A} = \mathbf{A} \tag{A3.1}$$

where $\mathbf{A}$ is $p \times q$ and $\mathbf{A}^-$ is $q \times p$ (note that $\mathbf{A}$ need not be square). Premultiplying both sides of the equation $\mathbf{Ax} = \mathbf{y}$ by $\mathbf{AA}^-$ gives

$$\mathbf{AA}^-\mathbf{Ax} = \mathbf{Ax} = \mathbf{AA}^-\mathbf{y}$$

and hence

$$\mathbf{A}(\mathbf{x} - \mathbf{A}^{-}\mathbf{y}) = \mathbf{0}$$

implying that, if the system is consistent, a solution is

$$\mathbf{x} = \mathbf{A}^{-}\mathbf{y} \tag{A3.2}$$

Given the analogy with the inverse of a nonsingular square matrix, a matrix $\mathbf{A}^{-}$ satisfying Equation A3.1 is called a generalized inverse (**g-inverse**, **conditional inverse**, **pseudo-inverse**) of $\mathbf{A}$. While the notation of using $\mathbf{A}^{-}$ to denote the generalized inverse of $\mathbf{A}$ is widely used in the quantitative genetics, the notation $\mathbf{A}^{+}$ is also used in other literatures.

Unless $\mathbf{A}$ is nonsingular, Equation A3.1 does not define a unique matrix, so we refer to $\mathbf{A}^{-}$ as *a* generalized inverse instead of *the* generalized inverse. A unique generalized inverse, the **Moore–Penrose inverse**, can be obtained by imposing three additional conditions: (i) $\mathbf{A}^{-}\mathbf{A}\mathbf{A}^{-} = \mathbf{A}^{-}$; (ii) $(\mathbf{A}\mathbf{A}^{-})^T = \mathbf{A}\mathbf{A}^{-}$, and (iii) $(\mathbf{A}^{-}\mathbf{A})^T = \mathbf{A}^{-}\mathbf{A}$. However, for our purposes any $\mathbf{A}^{-}$ satisfying Equation A3.1 is sufficient. Equation A3.38 shows how generalized inverses can be computed using the singular value decomposition (Equation A3.36a), with other approaches given by Henderson (1984). More detailed treatment of the properties of generalized inverses is given by Pringle and Rayner (1971), Rao and Mitra (1971), Dhrymes (1978), and Searle (1982), and we summarize some of these results below.

Consistency and Solutions to Consistent Systems

When dealing with linear models for complex designs, it is not immediately clear if the resulting OLS/GLS equations have solutions. Generalized inverses provide a check of consistency, and hence of whether a system of equations has any solutions. A linear system $\mathbf{A}_{m\times q}\,\mathbf{x}_{q\times 1} = \mathbf{y}_{m\times 1}$ is **consistent** if and only if

$$\mathbf{A}\mathbf{A}^{-}\mathbf{y} = \mathbf{y} \tag{A3.3}$$

Given a consistent system, all solutions have the form

$$\mathbf{x} = \mathbf{A}^{-}\mathbf{y} + (\mathbf{I} - \mathbf{A}^{-}\mathbf{A})\mathbf{c} \tag{A3.4}$$

where $\mathbf{c}$ is an arbitrary $q \times 1$ column vector. For example, taking $\mathbf{c} = \mathbf{0}$ recovers Equation A3.2, while if $\mathbf{A}^{-1}$ exists, then $\mathbf{I} - \mathbf{A}^{-1}\mathbf{A} = \mathbf{0}$ and the solution $\mathbf{x} = \mathbf{A}^{-1}\mathbf{y}$ is unique. To see that any expression of the form of Equation A3.4 is a solution, note that

$$\begin{aligned}\mathbf{A}\mathbf{x} &= \mathbf{A}(\mathbf{A}^{-}\mathbf{y} + [\mathbf{I} - \mathbf{A}^{-}\mathbf{A}]\mathbf{c}) \\ &= \mathbf{A}\mathbf{A}^{-}\mathbf{y} + (\mathbf{A} - \mathbf{A}\mathbf{A}^{-}\mathbf{A})\mathbf{c} = \mathbf{y} + (\mathbf{A} - \mathbf{A})\mathbf{c} \\ &= \mathbf{y}\end{aligned}$$

which follows from Equations A3.3 and A3.1, respectively.

Example A3.1. Consider the following system of equations

$$\begin{aligned} x_1 + 2x_2 + 3x_3 &= 5 \\ 2x_1 + x_2 + 2x_3 &= 6 \end{aligned}$$

which can be written in matrix form as $\mathbf{A}\mathbf{x} = \mathbf{y}$, with

$$\mathbf{A} = \begin{pmatrix} 1 & 2 & 3 \\ 2 & 1 & 2 \end{pmatrix}, \qquad \mathbf{x} = \begin{pmatrix} x_1 \\ x_2 \\ x_3 \end{pmatrix}, \qquad \mathbf{y} = \begin{pmatrix} 5 \\ 6 \end{pmatrix}$$

The matrix

$$\mathbf{A}^- = \begin{pmatrix} -11/26 & 9/13 \\ 4/13 & -3/13 \\ 7/26 & -1/13 \end{pmatrix}$$

satisfies $\mathbf{A}\mathbf{A}^-\mathbf{A} = \mathbf{A}$ and thus is a generalized inverse of $\mathbf{A}$. Matrix multiplication shows that $\mathbf{A}\mathbf{A}^- = \mathbf{I}$, implying $\mathbf{A}\mathbf{A}^-\mathbf{y} = \mathbf{y}$. Thus, Equation A3.3 is satisfied and this system of equations is consistent for any $\mathbf{y}$. One solution is $\mathbf{x} = \mathbf{A}^-\mathbf{y}$, or

$$\begin{pmatrix} x_1 \\ x_2 \\ x_3 \end{pmatrix} = \begin{pmatrix} -11/26 & 9/13 \\ 4/13 & -3/13 \\ 7/26 & -1/13 \end{pmatrix} \begin{pmatrix} 5 \\ 6 \end{pmatrix} = \frac{1}{26} \begin{pmatrix} 53 \\ 4 \\ 23 \end{pmatrix}$$

More generally, because

$$\mathbf{I} - \mathbf{A}^-\mathbf{A} = \begin{pmatrix} 1/26 & 2/13 & -3/26 \\ 2/13 & 8/13 & -6/13 \\ -3/26 & -6/13 & 9/26 \end{pmatrix}$$

then from Equation A3.4, any solution to this system of equations has the form

$$\begin{pmatrix} x_1 \\ x_2 \\ x_3 \end{pmatrix} = \frac{1}{26} \begin{pmatrix} 53 \\ 4 \\ 23 \end{pmatrix} + \begin{pmatrix} 1/26 & 2/13 & -3/26 \\ 2/13 & 8/13 & -6/13 \\ -3/26 & -6/13 & 9/26 \end{pmatrix} \begin{pmatrix} c_1 \\ c_2 \\ c_3 \end{pmatrix}$$

which reduces to

$$\begin{pmatrix} x_1 \\ x_2 \\ x_3 \end{pmatrix} = \frac{1}{26} \begin{pmatrix} 53 \\ 4 \\ 23 \end{pmatrix} + c \cdot \begin{pmatrix} 1 \\ 4 \\ -3 \end{pmatrix}$$

where c is an arbitrary constant.

Although an infinite number of solutions exist when $\mathbf{A}$ is singular, particular linear combinations (or **contrasts**) of the elements of $\mathbf{x}$ may have *unique values*. For example, consider the system $x_1 + x_2 = 1$. Here there are an infinite number of solutions for (x_1, x_2), but only a single solution, 1, for the contrast $x_1 + x_2$.

Consider some linear combination $\mathbf{b}^T\mathbf{x} = \sum b_i x_i$. If the vector of constants $\mathbf{b}$ satisfies

$$\mathbf{b}^T\mathbf{A}^-\mathbf{A} = \mathbf{b}^T \tag{A3.5a}$$

then $\mathbf{b}^T\mathbf{x}$ has a unique solution given by

$$\mathbf{b}^T\mathbf{x} = \mathbf{b}^T\mathbf{A}^-\mathbf{y} \tag{A3.5b}$$

To see this, note that Equation A3.4 gives the general solution as

$$\begin{aligned} \mathbf{b}^T\mathbf{x} &= \mathbf{b}^T(\mathbf{A}^-\mathbf{y} + [\mathbf{I} - \mathbf{A}^-\mathbf{A}]\mathbf{c}) \\ &= \mathbf{b}^T\mathbf{A}^-\mathbf{y} + (\mathbf{b}^T\mathbf{I} - \mathbf{b}^T\mathbf{A}^-\mathbf{A})\mathbf{c} \\ &= \mathbf{b}^T\mathbf{A}^-\mathbf{y} + (\mathbf{b}^T - \mathbf{b}^T)\mathbf{c} \\ &= \mathbf{b}^T\mathbf{A}^-\mathbf{y} \end{aligned}$$

which is independent of the arbitrary vector $\mathbf{c}$. Likewise, a vector of contrasts, $\mathbf{Bx}$, has a unique solution $\mathbf{BA}^-\mathbf{y}$, provided $\mathbf{B}$ satisfies $\mathbf{BA}^-\mathbf{A} = \mathbf{B}$.

Example A3.2. Consider the system of equations from Example A3.1. Is there a unique solution for the two linear contrasts $c_1 = x_2 - 4x_1$ and $c_2 = x_3 + 3x_1$? In matrix form,

$$\begin{pmatrix} c_1 \\ c_2 \end{pmatrix} = \begin{pmatrix} x_2 - 4x_1 \\ x_3 + 3x_1 \end{pmatrix} = \mathbf{Bx}$$

where

$$\mathbf{B} = \begin{pmatrix} -4 & 1 & 0 \\ 3 & 0 & 1 \end{pmatrix} \quad \text{and} \quad \mathbf{x} = \begin{pmatrix} x_1 \\ x_2 \\ x_3 \end{pmatrix}$$

Using the generalized inverse for $\mathbf{A}$ from Example A3.1, matrix multiplication shows that

$$\mathbf{BA^-A} = \begin{pmatrix} -4 & 1 & 0 \\ 3 & 0 & 1 \end{pmatrix} = \mathbf{B}$$

Hence, the matrix version of Equation A3.5b gives the unique solution for this vector of contrasts as

$$\begin{pmatrix} c_1 \\ c_2 \end{pmatrix} = \mathbf{BA^-y} = \begin{pmatrix} -4 & 1 & 0 \\ 3 & 0 & 1 \end{pmatrix} \begin{pmatrix} -11/26 & 9/13 \\ 4/13 & -3/13 \\ 7/26 & -1/13 \end{pmatrix} \begin{pmatrix} 5 \\ 6 \end{pmatrix} = \begin{pmatrix} -8 \\ 7 \end{pmatrix}$$

To see that this solution is indeed unique, note that we can rearrange the contrast equations to obtain $x_2 = c_1 + 4x_1$ and $x_3 = c_2 - 3x_1$. Substituting into the original set of equations (Example A3.1),

$$x_1 + 2x_2 + 3x_3 = x_1 + 2(c_1 + 4x_1) + 3(c_2 - 3x_1) = 2c_1 + 3c_2 = 5$$
$$2x_1 + x_2 + 2x_3 = 2x_1 + (c_1 + 4x_1) + 2(c_2 - 3x_1) = c_1 + 2c_2 = 6$$

so that the original set of two equations and three unknowns reduces to a two equation–two unknown system. In matrix form this is

$$\begin{pmatrix} 2 & 3 \\ 1 & 2 \end{pmatrix} \begin{pmatrix} c_1 \\ c_2 \end{pmatrix} = \begin{pmatrix} 5 \\ 6 \end{pmatrix}$$

Since the coefficient matrix is invertible (its determinate is $2 \cdot 2 - 3 \cdot 1 \neq 0$), there is a unique solution for this pair of contrasts ($c_1 = -8$ and $c_2 = 7$).

Estimability of Fixed Effects

The above results have implications for the estimation of fixed effects in the general linear model, $\mathbf{y} = \mathbf{X}\boldsymbol{\beta} + \mathbf{e}$. Recall that the OLS solution for a vector $\boldsymbol{\beta}$ of fixed effects is $\widehat{\boldsymbol{\beta}} = (\mathbf{X}^T\mathbf{X})^{-1}\mathbf{X}^T\mathbf{y}$ (Equation 10.9a). If the design matrix $\mathbf{X}$ has **full column rank** (all columns of $\mathbf{X}$ are independent), $(\mathbf{X}^T\mathbf{X})^{-1}$ exists and the OLS solution for $\boldsymbol{\beta}$ is unique. However, when $(\mathbf{X}^T\mathbf{X})$ is singular (and hence does not have a unique inverse), it is not possible to obtain unique OLS estimates for all the fixed factors in a model. For example, suppose β_1 indicates a sex effect (male versus female) and β_2 indicates the effect of a particular diet. If the design is such that no females used this diet while all males did, we do not have separate information on both sex and diet effects and hence can only estimate $\beta_1 + \beta_2$ rather than being able to estimate both β_1 and β_2 separately.

A linear combination of factors $\mathbf{b}^T\boldsymbol{\beta}$ is said to be **estimable** for a given design matrix $\mathbf{X}$ if there exists some column vector $\mathbf{a}$ that satisfies

$$E[\mathbf{a}^T\mathbf{y}] = \mathbf{b}^T\boldsymbol{\beta} \tag{A3.6a}$$

Estimability thus implies that there is some linear combination $\mathbf{a}^T\mathbf{y}$ of the original data whose expected value equals the desired linear combination of factors. Since $E[\mathbf{y}] = \mathbf{X}\boldsymbol{\beta}$, this definition implies that $\mathbf{b}^T\boldsymbol{\beta}$ is estimable if there exists a column vector $\mathbf{a}$ that satisfies $E[\mathbf{a}^T\mathbf{y}] = \mathbf{a}^T E[\mathbf{y}] = \mathbf{a}^T\mathbf{X}\boldsymbol{\beta} = \mathbf{b}^T\boldsymbol{\beta}$, implying $(\mathbf{a}^T\mathbf{X} - \mathbf{b}^T)\boldsymbol{\beta} = \mathbf{0}$, or that

$$\mathbf{X}^T\mathbf{a} = \mathbf{b} \tag{A3.6b}$$

An alternative (and equivalent) condition is that $\mathbf{b}$ satisfies

$$\mathbf{b}^T(\mathbf{X}^T\mathbf{X})^-(\mathbf{X}^T\mathbf{X}) = \mathbf{b}^T \tag{A3.6c}$$

Henderson (1984) gives other equivalent conditions. Equation A3.5c implies that if $\mathbf{X}^T\mathbf{X}$ is nonsingular, all linear combinations of $\boldsymbol{\beta}$ are estimable. Note that Equation A3.6c is identical to the condition given by Equation A3.5a (taking $\mathbf{A} = \mathbf{X}^T\mathbf{X}$), implying that these solutions are also unique estimates. If estimable, the OLS solution of the vector $\mathbf{b}^T\boldsymbol{\beta}$ given by

$$\text{OLS}(\mathbf{b}^T\boldsymbol{\beta}) = \mathbf{b}^T\left(\mathbf{X}^T\mathbf{X}\right)^-\mathbf{X}^T\mathbf{y} \tag{A3.6d}$$

is unique and is independent of which generalized inverse is used.

Example A3.3. Consider the linear model $\mathbf{y} = \mathbf{X}\boldsymbol{\beta}$, where

$$\boldsymbol{\beta} = \begin{pmatrix} \beta_1 \\ \beta_2 \\ \beta_3 \end{pmatrix} \quad \text{and} \quad \mathbf{X} = \begin{pmatrix} 1 & 1 & 0 \\ 1 & 1 & 0 \\ 0 & 0 & 1 \end{pmatrix}, \quad \text{giving} \quad \mathbf{X}^T\mathbf{X} = \begin{pmatrix} 2 & 2 & 0 \\ 2 & 2 & 0 \\ 0 & 0 & 1 \end{pmatrix}$$

Note that $\mathbf{X}^T\mathbf{X}$ is singular (it has rank 2 as its first two columns are identical), so we cannot obtain unique estimates of all three parameters. For this design matrix, are β_3, $\beta_1 + \beta_2$, and β_1 all estimable? These three combinations correspond to vectors of $\mathbf{b}^T = (0,0,1), (1,1,0)$, and $(1,0,0)$, respectively. For the first two $\mathbf{b}$ vectors, we can find a vector $\mathbf{a}$ that satisfies $\mathbf{X}^T\mathbf{a} = \mathbf{b}$, namely,

$$\mathbf{X}^T\begin{pmatrix} 0 \\ 0 \\ 1 \end{pmatrix} = \begin{pmatrix} 0 \\ 0 \\ 1 \end{pmatrix} \quad \text{and} \quad \mathbf{X}^T\begin{pmatrix} 1/2 \\ 1/2 \\ 0 \end{pmatrix} = \begin{pmatrix} 1 \\ 1 \\ 0 \end{pmatrix}$$

so that, from Equation A3.6b, these two linear combinations, β_3 and $(\beta_1 + \beta_2)$, are estimable. However, because

$$\mathbf{X}^T\begin{pmatrix} a_1 \\ a_2 \\ a_3 \end{pmatrix} = \begin{pmatrix} a_1 + a_2 \\ a_1 + a_2 \\ a_3 \end{pmatrix} \neq \begin{pmatrix} 1 \\ 0 \\ 0 \end{pmatrix}$$

β_1 is not estimable as $a_1 + a_2$ cannot simultaneously equal zero and one, and hence there exists no vector $\mathbf{a}$ that satisfies $\mathbf{X}^T\mathbf{a} = \mathbf{b}$ for this particular combinations of $\mathbf{X}$ and $\mathbf{b}$.

THE SQUARE ROOT OF A MATRIX

The concept of the square root of a symmetric nonsingular matrix provides another useful matrix tool for the analysis of linear models. In particular, using the square root of the covariance matrix transforms a vector of correlated variables into a new vector of variables with covariance matrix **I**, implying that the transformed variables are uncorrelated with unit variance.

Consider a symmetric nonsingular matrix **V** and define $\mathbf{V}^{1/2}$ as the matrix satisfying

$$\mathbf{V}^{1/2}\mathbf{V}^{1/2} = \mathbf{V} \tag{A3.7a}$$

In effect, $\mathbf{V}^{1/2}$ is the square root of a matrix, in that, when squared, we recover **V**. Denoting the inverse of $\mathbf{V}^{1/2}$ as $\mathbf{V}^{-1/2}$, we also have the following properties

$$\mathbf{V}^{-1/2}\mathbf{V}^{1/2} = \mathbf{I}, \quad \mathbf{V}^{-1/2}\mathbf{V}^{-1/2} = \mathbf{V}^{-1}, \quad \text{and} \quad \mathbf{V}^{-1/2}\mathbf{V} = \mathbf{V}^{1/2} \tag{A3.7b}$$

Likewise, both $\mathbf{V}^{1/2}$ and its inverse are symmetric. As shown below, the diagonalization decomposition of **V** leads to expressions for $\mathbf{V}^{1/2}$ (Equation A3.33b) and $\mathbf{V}^{-1/2}$ (Equation A3.33c). Finally, the square root matrix of **I** is simply **I**, with

$$\mathbf{I}^{-1/2}\mathbf{I}^{1/2} = \mathbf{I}, \quad \mathbf{I}^{-1/2}\mathbf{I}^{-1/2} = \mathbf{I}, \quad \text{and} \quad \mathbf{I}^{-1/2}\mathbf{I} = \mathbf{I} \tag{A3.7c}$$

Suppose the random vector **y** has covariance matrix **V** and consider the new vector $\mathbf{z} = \mathbf{V}^{-1/2}\mathbf{y}$. Recalling Equation 9.21b, the resulting covariance matrix for **z** becomes

$$\text{Var}(\mathbf{z}) = \mathbf{V}^{-1/2}\text{Var}(\mathbf{y})\mathbf{V}^{-1/2} = \mathbf{V}^{-1/2}\mathbf{V}\mathbf{V}^{-1/2} = \mathbf{V}^{-1/2}\mathbf{V}^{1/2} = \mathbf{I} \tag{A3.8a}$$

Thus, the transformed variables have unit variance and are uncorrelated. Suppose **y** is an $n \times 1$ column vector with $\mathbf{y} \sim \text{MVN}(\boldsymbol{\mu}, \mathbf{V})$. It follows that

$$\mathbf{z} = \mathbf{V}^{-1/2}(\mathbf{y} - \boldsymbol{\mu}) \sim \text{MVN}(\mathbf{0}, \mathbf{I}) \tag{A3.8b}$$

so that $z_i \sim \text{N}(0, 1)$, and hence the transformed variables are independent unit normals. Thus,

$$\begin{aligned}(\mathbf{y} - \boldsymbol{\mu})^T\mathbf{V}^{-1}(\mathbf{y} - \boldsymbol{\mu}) &= (\mathbf{y} - \boldsymbol{\mu})^T\mathbf{V}^{-1/2}\mathbf{V}^{-1/2}(\mathbf{y} - \boldsymbol{\mu}) \\ &= \mathbf{z}^T\mathbf{z} \\ &= \sum_{i=1}^{n} z_i^2 \sim \chi_n^2 \end{aligned} \tag{A3.9}$$

The last step follows by recalling that the sum of n squared (and independent) unit normal random variables follows a χ^2 distribution with n degrees of freedom (Appendix 5). Thus when **y** is multivariate normal, the quadratic form $(\mathbf{y} - \boldsymbol{\mu})^T\mathbf{V}^{-1}(\mathbf{y} - \boldsymbol{\mu})$ follows a χ^2 distribution. As we will see shortly, Equation A3.9 is the basis for goodness-of-fit tests of linear models. We will return to other matrix decompositions that can be used to remove correlations at the end of this Appendix, where we also examine the relationship between the eigenstructures of **V**, $\mathbf{V}^{1/2}$, and $\mathbf{V}^{-1/2}$.

DERIVATION OF THE GLS ESTIMATORS

One important application of the square root of a matrix is that it allows us to obtain GLS estimators from ordinary least-squares (OLS) estimators. Suppose the linear model is

$$\mathbf{y} = \mathbf{X}\boldsymbol{\beta} + \mathbf{e} \qquad \text{with } \mathbf{e} \sim (0, \sigma_e^2\,\mathbf{R})$$

Premultiplying both sides by $\mathbf{R}^{-1/2}$ gives

$$\mathbf{z} = \mathbf{Z}\boldsymbol{\beta} + \mathbf{f} \qquad \text{with } \mathbf{f} \sim (0, \sigma_e^2\,\mathbf{I})$$

where

$$\mathbf{z} = \mathbf{R}^{-1/2}\mathbf{y}, \qquad \mathbf{Z} = \mathbf{R}^{-1/2}\mathbf{X}, \qquad \mathbf{f} = \mathbf{R}^{-1/2}\mathbf{e}$$

OLS can be applied to this model since the transformed residuals are uncorrelated and homoscedastic. Thus, GLS estimates are obtained from the OLS solution by substituting

$$\mathbf{z} = \mathbf{R}^{-1/2}\mathbf{y} \text{ for } \mathbf{y}, \qquad \mathbf{Z} = \mathbf{R}^{-1/2}\mathbf{X} \text{ for } \mathbf{X}, \qquad \mathbf{f} = \mathbf{R}^{-1/2}\mathbf{e} \text{ for } \mathbf{e} \tag{A3.10}$$

Substituting into the OLS solutions (Equation 10.9a) gives the GLS estimate of β as

$$\begin{aligned}\widehat{\beta} &= \left((\mathbf{X}^T\mathbf{R}^{-1/2})(\mathbf{R}^{-1/2}\mathbf{X}) \right)^{-1} (\mathbf{X}^T\mathbf{R}^{-1/2})(\mathbf{R}^{-1/2}\mathbf{y}) \\ &= \left(\mathbf{X}^T\mathbf{R}^{-1}\mathbf{X}\right)^{-1} \mathbf{X}^T\mathbf{R}^{-1}\mathbf{y}\end{aligned}$$

Likewise, substituting into the OLS covariance expression (Equation 10.11a) gives the resulting covariance matrix for the GLS estimates as

$$\text{Var}(\,\widehat{\boldsymbol{\beta}}\,) = \sigma_e^2 \left(\mathbf{X}^T\mathbf{R}^{-1}\mathbf{X}\right)^{-1}$$

If the residuals follow a multivariate normal distribution, $\mathbf{e} \sim \text{MVN}(\mathbf{0}, \mathbf{V})$, and $\mathbf{y} = \mathbf{X}\beta + \mathbf{e}$ is indeed the correct model, then for $\widehat{\mathbf{y}} = \mathbf{X}\beta$, $\mathbf{y} - \widehat{\mathbf{y}} \sim \text{MVN}(\mathbf{0}, \mathbf{V})$ and it follows from Equation A3.9 that

$$(\mathbf{y} - \widehat{\mathbf{y}})^T \mathbf{V}^{-1} (\mathbf{y} - \widehat{\mathbf{y}}) \sim \chi^2_{n-p} \tag{A3.11a}$$

The degrees of freedom for the χ^2 distribution equal the number of observations minus the number of estimated parameters. Equation A3.11a provides a χ^2 test for the goodness-of-fit of a particular linear model. When $\mathbf{V}$ is diagonal,

$$(\mathbf{y} - \widehat{\mathbf{y}})^T \mathbf{V}^{-1} (\mathbf{y} - \widehat{\mathbf{y}}) = \sum_{i=1}^{n} \frac{(\,y_i - \widehat{y}_i\,)^2}{V_{ii}} \sim \chi^2_{n-p} \tag{A3.11b}$$

QUADRATIC FORMS AND SUMS OF SQUARES

The analysis of linear models relies very heavily on **sums of squares**, which can be expressed in matrix notation as quadratic forms (Equation 9.17a). To introduce the reader to the machinery used to work with sums of squares, we first present expressions for the mean and variance of a quadratic form, and then express linear model sums of squares as quadratic forms.

Moments of Quadratic Forms

When $\mathbf{x}$ is a vector of random variables, the quadratic form $\mathbf{x}^T\mathbf{A}\mathbf{x}$ is a scalar (1×1) random variable. If $\mathbf{x}$ has mean $\boldsymbol{\mu}$ and (nonsingular) covariance matrix $\mathbf{V}$, Equation 9.22 gives the expected value of this quadratic form as

$$E(\mathbf{x}^T\mathbf{A}\mathbf{x}) = \text{tr}(\mathbf{A}\mathbf{V}) + \boldsymbol{\mu}^T\mathbf{A}\boldsymbol{\mu} \tag{A3.12a}$$

where the **trace** of a square matrix, $\text{tr}(\mathbf{M}) = \sum M_{ii}$, is the sum of its diagonal elements. Further, if $\mathbf{x} \sim \text{MVN}(\boldsymbol{\mu}, \mathbf{V})$, then as shown in Searle (1971), the variance of the quadratic form is given by

$$\sigma^2(\mathbf{x}^T\mathbf{A}\mathbf{x}) = 2\,\text{tr}\,(\mathbf{A}\mathbf{V}\mathbf{A}\mathbf{V}) + 4\boldsymbol{\mu}^T\mathbf{A}\mathbf{V}\mathbf{A}\boldsymbol{\mu} \tag{A3.12b}$$

The Sample Variance Expressed as a Quadratic Form

As an introduction to expressing sums of squares as quadratic forms, consider the sample variance for n observations,

$$\text{Var}(x) = \frac{1}{n-1}\sum_{i=1}^{n}(x_i - \overline{x})^2$$

Define the **unit matrix** $\mathbf{J}_{n\times k}$ as an $n \times k$ matrix in which every element is unity, e.g.,

$$\mathbf{J}_{n\times 1} = \begin{pmatrix} 1 \\ \vdots \\ 1 \end{pmatrix} \Big\} n, \qquad \mathbf{J}_{2\times 3} = \begin{pmatrix} 1 & 1 & 1 \\ 1 & 1 & 1 \end{pmatrix}, \qquad \text{etc.}$$

Likewise, define the matrix

$$\mathbf{N} = \frac{1}{n-1}\left(\mathbf{I} - \frac{1}{n}\mathbf{J}\right) = \frac{1}{n-1}\begin{pmatrix} 1-1/n & -1/n & \cdots & -1/n \\ -1/n & 1-1/n & \cdots & -1/n \\ \vdots & \vdots & \ddots & \vdots \\ -1/n & -1/n & \cdots & 1-1/n \end{pmatrix} \tag{A3.13a}$$

where here $\mathbf{J}$ is $n \times n$. Noting that

$$\mathbf{Nx} = \frac{1}{n-1}\left(\mathbf{x} - \frac{1}{n}\mathbf{Jx}\right) = \frac{1}{n-1}\begin{pmatrix} x_1 - \overline{x} \\ \vdots \\ x_n - \overline{x} \end{pmatrix} \tag{A3.13b}$$

it follows that

$$\mathbf{x}^T\mathbf{Nx} = \text{Var}(x) \tag{A3.14a}$$

To see this, observe that

$$\begin{aligned} \mathbf{x}^T\mathbf{Nx} &= \frac{1}{n-1}\begin{pmatrix} x_1 & \cdots & x_n \end{pmatrix}\begin{pmatrix} x_1 - \overline{x} \\ \vdots \\ x_n - \overline{x} \end{pmatrix} \\ &= \frac{1}{n-1}\sum_{i=1}^{n} x_i(\,x_i - \overline{x}\,) = \frac{1}{n-1}\left(\sum_{i=1}^{n} x_i^2 - \overline{x}\sum_{i=1}^{n} x_i\right) \\ &= \frac{1}{n-1}\sum_{i=1}^{n}(x_i - \overline{x}\,)^2 = \text{Var}(x) \end{aligned} \tag{A3.14b}$$

The last step follows by noting that

$$-\overline{x}\sum_{i=1}^{n} x_i = -\overline{x}\,(\,n\,\overline{x}\,) = -2\,\overline{x}\sum_{i=1}^{n} x_i + n\,\overline{x}^2$$

Example A3.4. Because we have expressed $\text{Var}(x)$ as a quadratic form, we can use Equation A3.12a to compute its expected value and Equation A3.12b to compute its sampling variance. If $\mathbf{x} \sim (\boldsymbol{\mu}, \mathbf{V})$, the expected value of $\text{Var}(x)$ is

$$E[\,\text{Var}(x)\,] = E(\mathbf{x}^T\mathbf{Nx}) = \text{tr}(\mathbf{NV}) + \boldsymbol{\mu}^T\mathbf{N}\boldsymbol{\mu}$$

To compute this expression, first note from Equation A3.14b that

$$\boldsymbol{\mu}^T \mathbf{N} \boldsymbol{\mu} = \frac{1}{n-1} \sum_{i=1}^{n} (\mu_i - \overline{\mu})^2$$

Likewise, from Equation A3.13b

$$\mathbf{NV} = \frac{\mathbf{V}}{n-1} - \frac{\mathbf{JV}}{n(n-1)}$$

which has diagonal elements

$$(\mathbf{NV})_{ii} = \frac{1}{n-1} \left(\sigma^2(z_i) - \frac{\sum_j \sigma(z_i, z_j)}{n} \right)$$

After some simplification, we have

$$\text{tr}(\mathbf{NV}) = \sum_{i=1}^{n} (\mathbf{NV})_{ii} = \frac{1}{n} \sum_{i=1}^{n} \sigma^2(z_i) - \frac{2}{n(n-1)} \sum_{i<j} \sigma(z_i, z_j)$$

Putting these results together gives

$$E[\text{Var}(x)] = \frac{1}{n} \sum_{i=1}^{n} \sigma^2(z_i) - \frac{2}{n(n-1)} \sum_{i<j} \sigma(z_i, z_j) + \frac{1}{n-1} \sum_{i=1}^{n} (\mu_i - \overline{\mu})^2$$

where $\overline{\mu} = \sum \mu_i / n$. In the simple situation where all observations have the same mean and variance ($\mu_i = \mu, \sigma^2(z_i) = \sigma^2$) and are uncorrelated, this reduces to

$$E[\,\text{Var}(x)\,] = \sigma^2$$

Turning now to the sample variance of Var(x), if we are again willing to assume that $\mathbf{x}$ is multivariate normal, then from Equation A3.12b,

$$\sigma^2[\text{Var}(x)] = \sigma^2(\mathbf{x}^T \mathbf{N} \mathbf{x}) = 2\,\text{tr}\,[\mathbf{NVNV}] + 4\boldsymbol{\mu}^T \mathbf{NVN} \boldsymbol{\mu}$$

If, for example, $\mathbf{V} = \sigma^2\,\mathbf{I}$ (the x_i are uncorrelated with common variance), then

$$\begin{aligned} \mathbf{NVNV} = \sigma^4\,\mathbf{NN} &= \frac{\sigma^4}{(n-1)^2} \left(\mathbf{I} - \frac{1}{n} \mathbf{J}_{n \times n} \right) \left(\mathbf{I} - \frac{1}{n} \mathbf{J}_{n \times n} \right) \\ &= \frac{\sigma^4}{(n-1)^2} \left(\mathbf{I} - \frac{2}{n} \mathbf{J}_{n \times n} + n^{-2} \mathbf{J}_{n \times n} \mathbf{J}_{n \times n} \right) \end{aligned}$$

The *ij*th element in $\mathbf{J}_{n \times n} \mathbf{J}_{n \times n}$ is n, giving $\mathbf{J}^2_{n \times n} = n\,\mathbf{J}_{n \times n}$. Hence, the *i*th diagonal element of $\mathbf{NVNV}$ is

$$\frac{\sigma^4}{(n-1)^2} \left(1 - \frac{2}{n} + n^{-2} n \right) = \frac{\sigma^4}{n(n-1)}$$

giving tr($\mathbf{NVNV}$) $= \sigma^4/(n-1)$. When all of the means are equal, it follows that $\mathbf{N}\boldsymbol{\mu} = \mathbf{0}$ and the second term in Equation A3.12b vanishes, yielding

$$\sigma^2[\,\text{Var}(x)\,] = \frac{2\sigma^4}{n-1}$$

Sums of Squares Expressed as Quadratic Forms

In the same fashion that we decomposed the total variance into genetic and phenotypic components (Chapters 4–7), we can decompose the total variance of a response vector $\mathbf{y}$ into the variance accounted for by the linear model and the remaining (error or residual) variance. This is typically done by considering three sums of squares, with the **total sum of squares** ($\mathrm{SS_T}$) being the sum of two components, the **error** (or **residual**) **sum of squares** ($\mathrm{SS_E}$) and the **model sum of squares** ($\mathrm{SS_M}$),

$$\mathrm{SS_T} = \mathrm{SS_M} + \mathrm{SS_E}$$

The total sum of squares measures the total variability in the data, while the model sum of squares measures the amount of variation accounted for by the linear model. As noted in our discussions of univariate regression in Chapter 3, the fraction of total variance explained by a linear model is given by the **coefficient of determination**,

$$r^2 = \frac{\mathrm{SS_M}}{\mathrm{SS_T}} = 1 - \frac{\mathrm{SS_E}}{\mathrm{SS_T}} \tag{A3.15a}$$

One issue with r^2 is that goodness-of-fit improves (and hence r^2 increases) as more model parameters are added. As a result, for n observations and p estimated parameters, the **adjusted $\mathbf{r}^2$** (Ezekiel 1930)

$$r^2_{adj} = 1 - \frac{\mathrm{MS_e}}{\mathrm{MS_T}} = 1 - \frac{\mathrm{SS_e}/(n-p)}{\mathrm{SS_T}/(n-1)} \tag{A3.15b}$$

is often reported. This replaces sums of squares (SS) with mean squares (MS), with $r^2_{adj} \leq r^2$, and allows for a more fair comparison of different models by discounting goodness-of-fit as more model parameters are added. The connection between r^2 and r^2_{adj} is

$$r^2_{adj} = 1 - (1 - r^2)\left(\frac{n-1}{n-p}\right) \tag{A3.15c}$$

As summarized in Table A3.1, the sums of squares have different forms under OLS and GLS. Under OLS, the residuals are assumed to be independent with common variance σ^2_e. In this case, each observation/residual is weighted equally, and the total sum of squares is simply

$$\mathrm{SS_T} = \sum_{i=1}^{n} (\, y_i - \overline{y}\,)^2$$

Sums of squares can be expressed as a quadratic form of the vector of observations $\mathbf{y}$, allowing the use of Equations 3A.12a and 3A.12b to obtain their expectations and variances. Recalling Equations A3.14b and A3.13a,

$$\mathrm{SS_T} = \mathbf{y}^T\left(\mathbf{I} - \frac{1}{n}\mathbf{J}\right)\mathbf{y} \tag{A3.16a}$$

where $\mathbf{J}$ is $n \times n$.

Now consider the error sum of squares,

$$\mathrm{SS_E} = \sum_{i=1}^{n} (\, y_i - \widehat{y}_i\,)^2 = \sum_{i=1}^{n} \widehat{e}_i^{\,2}$$

Since $\widehat{\mathbf{e}} = \mathbf{y} - \widehat{\mathbf{y}}$ and $\widehat{\mathbf{y}} = \mathbf{Xb} = \mathbf{X}\left(\mathbf{X}^T\mathbf{X}\right)^{-1}\mathbf{X}^T\mathbf{y}$, we have

$$\mathrm{SS_E} = \widehat{\mathbf{e}}^T\widehat{\mathbf{e}}, \quad \text{where} \quad \widehat{\mathbf{e}} = \left[\mathbf{I} - \mathbf{X}\left(\mathbf{X}^T\mathbf{X}\right)^{-1}\mathbf{X}^T\right]\mathbf{y} \tag{A3.16b}$$

Table A3.1 Summary of useful results for the general linear model, $\mathbf{y} = \mathbf{X}\beta + \mathbf{e}$, under OLS and GLS assumptions for the distribution of residuals. The $\mathbf{Q}$ matrix for the listed sums of squares is the matrix in the quadratic form, $\mathbf{y}^T\mathbf{Q}\mathbf{y}$, while $\mathbf{J}$ is a matrix where each element is 1.

	Ordinary Least Squares, OLS	**Generalized Least Squares, GLS**
Assumed distribution of residuals:	$\mathbf{e} \sim (0, \sigma_e^2\,\mathbf{I})$	$\mathbf{e} \sim (0, \sigma_e^2\,\mathbf{R})$
Least-squares estimator of β:	$\widehat{\beta} = (\mathbf{X}^T\mathbf{X})^{-1}\mathbf{X}^T\mathbf{y}$	$\widehat{\beta} = (\mathbf{X}^T\mathbf{R}^{-1}\mathbf{X})^{-1}\mathbf{X}^T\mathbf{R}^{-1}\mathbf{y}$
Covariance matrix for $\widehat{\beta}$:	$(\mathbf{X}^T\mathbf{X})^{-1}\sigma_e^2$	$(\mathbf{X}^T\mathbf{R}^{-1}\mathbf{X})^{-1}\sigma_e^2$
Estimated residual variance, $\widehat{\sigma_e^2}$	$\dfrac{(\mathbf{y}-\mathbf{X}\mathbf{b})^T(\mathbf{y}-\mathbf{X}\mathbf{b})}{n-\text{rank}(\mathbf{X})}$	$\dfrac{(\mathbf{y}-\mathbf{X}\mathbf{b})^T\mathbf{R}^{-1}(\mathbf{y}-\mathbf{X}\mathbf{b})}{n-\text{rank}(\mathbf{X})}$
Predicted values, $\widehat{\mathbf{y}} = \mathbf{X}\widehat{\beta}$:	$\mathbf{X}(\mathbf{X}^T\mathbf{X})^{-1}\mathbf{X}^T\mathbf{y}$	$\mathbf{X}(\mathbf{X}^T\mathbf{R}^{-1}\mathbf{X})^{-1}\mathbf{X}^T\mathbf{R}^{-1}\mathbf{y}$
Covariance matrix for predicted values, $\widehat{\mathbf{y}}$:	$\mathbf{X}(\mathbf{X}^T\mathbf{X})^{-1}\mathbf{X}^T\sigma_e^2$	$\mathbf{X}(\mathbf{X}^T\mathbf{R}^{-1}\mathbf{X})^{-1}\mathbf{X}^T\sigma_e^2$
SS_T, total sums of squares quadratic form matrix, $\mathbf{Q}$:	$\mathbf{I} - \frac{1}{n}\mathbf{J}$	$\mathbf{R}^{-1} - \frac{1}{n}\mathbf{R}^{-1/2}\mathbf{J}\mathbf{R}^{-1/2}$
SS_M, model sums of squares quadratic form matrix, $\mathbf{Q}$:	$\mathbf{X}\left(\mathbf{X}^T\mathbf{X}\right)^{-1}\mathbf{X}^T - \frac{1}{n}\mathbf{J}$	$\mathbf{R}^{-1}\mathbf{X}\left(\mathbf{X}^T\mathbf{R}^{-1}\mathbf{X}\right)^{-1}\mathbf{X}^T\mathbf{R}^{-1} - \frac{1}{n}\mathbf{R}^{-1/2}\mathbf{J}\mathbf{R}^{-1/2}$
SS_E, error sums of squares quadratic form matrix, $\mathbf{Q}$:	$\mathbf{I} - \mathbf{X}\left(\mathbf{X}^T\mathbf{X}\right)^{-1}\mathbf{X}^T$	$\mathbf{R}^{-1} - \mathbf{R}^{-1}\mathbf{X}\left(\mathbf{X}^T\mathbf{R}^{-1}\mathbf{X}\right)^{-1}\mathbf{X}^T\mathbf{R}^{-1}$
χ^2 goodness-of-fit statistic (assuming residuals are MVN):	$\chi^2 = \dfrac{(\mathbf{y}-\widehat{\mathbf{y}})^T(\mathbf{y}-\widehat{\mathbf{y}})}{\sigma_e^2}$	$\chi^2 = \dfrac{(\mathbf{y}-\widehat{\mathbf{y}})^T\mathbf{R}^{-1}(\mathbf{y}-\widehat{\mathbf{y}})}{\sigma_e^2}$

Expanding this expression and noting that $\mathbf{X}^T\mathbf{X}\left(\mathbf{X}^T\mathbf{X}\right)^{-1} = \mathbf{I}$, this simplifies to

$$SS_E = \mathbf{y}^T\left[\mathbf{I} - \mathbf{X}\left(\mathbf{X}^T\mathbf{X}\right)^{-1}\mathbf{X}^T\right]\mathbf{y} \tag{A3.16c}$$

Finally, the model sum of squares is the difference between the total and error sums of squares,

$$SS_M = SS_T - SS_E = \mathbf{y}^T\left[\mathbf{X}\left(\mathbf{X}^T\mathbf{X}\right)^{-1}\mathbf{X}^T - \frac{1}{n}\mathbf{J}\right]\mathbf{y} \tag{A3.16d}$$

Note that

$$\text{SS}_{\text{M}} = \sum_{i=1}^{n} (\, \widehat{y}_i - \overline{y}\,)^2$$

so that (for OLS) the model sum of squares is the sum of squared deviations of the predicted values from the overall mean.

The sums of squares under GLS are slightly different, as we have to correct for heteroscedasticity and/or the lack of independence among the residuals. Assume that the residuals have covariance matrix $\sigma_e^2\,\mathbf{R}$. From Equation A3.10, $\mathbf{y}$ is replaced by $\mathbf{R}^{-1/2}\mathbf{y}$ and $\mathbf{X}$ is replaced by $\mathbf{R}^{-1/2}\mathbf{X}$ in the above OLS expressions for sums of squares. Hence, the total sum of squares for GLS becomes

$$\begin{aligned} \text{SS}_{\text{T}} &= \mathbf{y}^T\mathbf{R}^{-1/2}\left(\mathbf{I} - \frac{1}{n}\mathbf{J}\right)\mathbf{R}^{-1/2}\mathbf{y} \\ &= \mathbf{y}^T\left[\mathbf{R}^{-1} - \frac{1}{n}\mathbf{R}^{-1/2}\mathbf{J}\mathbf{R}^{-1/2}\right]\mathbf{y} \end{aligned} \tag{A3.17a}$$

Likewise, the error sum of squares becomes

$$\begin{aligned} \text{SS}_{\text{E}} &= \widehat{\mathbf{e}}^T\mathbf{R}^{-1}\widehat{\mathbf{e}} \\ &= \mathbf{y}^T\left[\mathbf{R}^{-1} - \mathbf{R}^{-1}\mathbf{X}\left(\mathbf{X}^T\mathbf{R}^{-1}\mathbf{X}\right)^{-1}\mathbf{X}^T\mathbf{R}^{-1}\right]\mathbf{y} \end{aligned} \tag{A3.17b}$$

and the model sum of squares becomes

$$\text{SS}_{\text{M}} = \mathbf{y}^T\left[\mathbf{R}^{-1}\mathbf{X}\left(\mathbf{X}^T\mathbf{R}^{-1}\mathbf{X}\right)^{-1}\mathbf{X}^T\mathbf{R}^{-1} - \frac{1}{n}\mathbf{R}^{-1/2}\mathbf{J}\mathbf{R}^{-1/2}\right]\mathbf{y} \tag{A3.17c}$$

TESTING HYPOTHESES ABOUT LINEAR MODELS

Because sums of squares are very closely related to the variances accounted for by the various components of a particular linear model, it should not be surprising that these sums form the basis of hypothesis testing. Such tests can be quite involved, especially if we are evaluating the various components of a complex model. Here we consider the simplest case of the nested comparison of a full model with a reduced one.

If the residuals are multivariate-normally distributed with

$$\mathbf{e} \sim \text{MVN}(\mathbf{0}, \sigma_e^2\,\mathbf{I}) \quad \text{for OLS;} \qquad \mathbf{e} \sim \text{MVN}(\mathbf{0}, \sigma_e^2\,\mathbf{R}) \quad \text{for GLS}$$

then (recalling Equations A3.11a and A3.11b), $\text{SS}_{\text{E}}/\sigma_e^2$ is the sum of squared unit normals and hence is χ^2-distributed. In particular, with n observations and p estimated parameters,

$$\frac{\text{SS}_{\text{E}}}{\sigma_e^2} \sim \chi^2_{n-p} \tag{A3.18}$$

as a degree of freedom is lost for each estimated model parameter.

Suppose we have n observations and wish to compare two linear models: a **full model** fitting p parameters and a **reduced model** which uses only a subset ($q < p$) of the parameters in the full model. Do the additional $(p - q)$ fitted parameters provide a significant increase in the amount of variation accounted for by the model? Let SS_{E_f} and SS_{E_r} denote the appropriate (OLS or GLS) error sums of squares for the full and reduced models, respectively. Under the null hypothesis (that the full model provides the same fit as the reduced model), the difference in error sums of squares $(\text{SS}_{\text{E}_r} - \text{SS}_{\text{E}_f})$ is distributed as constant (σ_e^2) times a χ^2_{p-q}. Likewise, from Equation A3.18, $\text{SS}_{\text{E}_f} \sim \sigma_e^2\,\chi^2_{n-p}$. Recalling the definition of the F distribution (the ratio of two scaled chi-square distributions; Appendix 5) it follows that

$$\frac{(\text{SS}_{\text{E}_r} - \text{SS}_{\text{E}_f})\,/\,(p-q)}{\text{SS}_{\text{E}_f}\,/\,(n-p)} = \left(\frac{n-p}{p-q}\right)\left(\frac{\text{SS}_{\text{E}_r}}{\text{SS}_{\text{E}_f}} - 1\right) \tag{A3.19}$$

is distributed as $F_{p-q,n-p}$ under the null hypothesis of no improved fit.

For example, we can ask if a particular linear model accounts for a significant fraction of the variation in y by considering that model versus the simplest reduced model, $y_i = \mu + e_i$ (namely, that none of the model factors have any effect). It is easily seen that the least-squares solution for μ is $\overline{y}$ for OLS and the weighted mean for GLS, giving $\text{SS}_{E_r} = \text{SS}_\text{T}$. Since the number of parameters in the reduced model is $q = 1$, the test for whether a particular linear model accounts for a significant amount of the variation is

$$\left(\frac{n-p}{p-1}\right)\left(\frac{\text{SS}_\text{T}}{\text{SS}_{\text{E}_f}} - 1\right) = \left(\frac{n-p}{p-1}\right)\left(\frac{r^2}{1-r^2}\right) \tag{A3.20}$$

where r^2 is the coefficient of determination for the full model (Equation A3.15). This test statistic follows an $F_{p-1,n-p}$ distribution.

Construction of Confidence Intervals

When residuals are MVN, then OLS/GLM estimates are also MVN, with

$$\widehat{\boldsymbol{\beta}} \sim \text{MNV}\left(\boldsymbol{\beta}, \mathbf{V_b}\right) \tag{A3.21}$$

where $\mathbf{V_b}$ is given by Equation 10.11a for OLS estimates and by Equation 10.13b for GLS estimates. Confidence intervals immediately follow from the MVN property that subsets are also MVN (Chapter 9). Hence, the standard 95% confidence normal interval holds for a given GLM estimate, with

$$\widehat{\beta}_i \pm 1.96\sqrt{V_{ii}} \tag{A3.22a}$$

where V_{ii} is the ith diagonal element in $\mathbf{V_b}$. Similarly, the P value for a two-sided test of the hypothesis that $\beta_i = \beta_{i,o}$ is given by

$$\Pr\left(|U| \geq \left|\frac{\widehat{\beta}_i - \beta_{i,o}}{\sqrt{V_{ii}}}\right|\right) \tag{A3.22b}$$

where U is a unit normal random variable.

Multivariate joint confidence intervals are in the form of *ellipses* (for two dimensions) and *ellipsoids* for higher dimensions. To motivate the form of these intervals, we first consider **Hotelling's T**^{2} statistic. This is essentially the multivariate extension of the classic univariate t test, where for $\overline{z} \sim \text{N}(\mu_0, \sigma^2/n)$,

$$t = \frac{(\overline{z} - \mu_0)^2}{S^2/n}$$

for the null hypothesis that the mean is μ_0 given that the sample variance of z is S^2 (and hence the sample variance for $\overline{z}$ is S^2/n). Squaring both sides, we can express this as

$$t^2 = (\overline{z} - \mu_0)(S^2/n)^{-1}(\overline{z} - \mu_0)$$

Hotelling's T^2 statistic generalizes this to multivariate form, where for $\overline{\mathbf{z}} \sim \text{MNV}(\boldsymbol{\mu}_0, \mathbf{S}/n)$,

$$T^2 = (\overline{\mathbf{z}} - \boldsymbol{\mu}_0)^T(\mathbf{S}/n)^{-1}(\overline{\mathbf{z}} - \boldsymbol{\mu}_0) \tag{A3.23a}$$

Under the null hypothesis ($\boldsymbol{\mu} = \boldsymbol{\mu}_0$),

$$\left(\frac{n-p}{(n-1)p}\right) \cdot T^2 \sim F_{p,n-p} \tag{A3.23b}$$

Confidence ellipsoids follow from these expressions under MVN assumptions. Because $\widehat{\beta} \sim \text{MNV}(\beta, \mathbf{V_b})$, the set of all β values that are consistent (at the $1-\alpha$ percent level) with the observed data satisfies

$$(\widehat{\beta} - \beta)^T \mathbf{V_b}^{-1} (\widehat{\beta} - \beta) \leq \left(\frac{(n-1)p}{n-p} \right) \mathcal{F}_{p,n-p}(1-\alpha) \tag{A3.24}$$

The critical value $\mathcal{F}_{p,n-p}(1-\alpha)$ corresponds to the value of an F distribution with p numerator and $n-p$ denominator degrees of freedom that satisfy

$$\Pr\left[F_{p,n-p} \leq \mathcal{F}_{p,n-p}(1-\alpha)\right] = 1-\alpha$$

The quadratic form given by Equation A3.24 describes an ellipsoid centered at $\widehat{\beta}$ whose axes are given by the eigenvectors of $\mathbf{V_b}$ (WL Appendix 5 examines the analysis of such quadratic forms).

Example A3.5. Suppose that an analysis of a GLM returns the following values

$$\widehat{\beta} = \begin{pmatrix} 1 \\ -2 \end{pmatrix}, \quad \mathbf{V_b} = \begin{pmatrix} 0.11 & -0.02 \\ -0.02 & 0.05 \end{pmatrix}$$

The two univariate 95% confidence intervals become

$$\widehat{\beta_1} \pm 1.96\sqrt{0.11} \ (\pm 0.65), \quad \text{and} \quad \widehat{\beta_2} \pm 1.96\sqrt{0.05} \ (\pm 0.44)$$

or $(0.35, 1.65)$ for β_1 and $(-2.44, -1.66)$ for β_2.

EQUIVALENT LINEAR MIXED MODELS

Two mixed models are said to be **equivalent** if they have the same mean vector $E(\mathbf{y})$ and covariance matrix $\sigma(\mathbf{y}, \mathbf{y})$. The utility of equivalent models is that the parameters of one model can always be expressed as linear combinations of the parameters of any equivalent model. Hence, by choosing an appropriate equivalent model, one can often greatly simplify computations. Chapter 33 in Part II gives examples of this approach, such as the reduced animal model of Quaas and Pollak (1980) and estimating the BLUP values of dominance effects as a function of estimated breeding values. Additional examples from BLUP are given by Henderson (1985). Our purpose here is to briefly introduce the use and construction of equivalent models.

Consider two different linear mixed models, both using the same vector $\mathbf{y}$ of observations but with different assumed vectors of fixed (β vs. β_*) and random ($\mathbf{u}$ and $\mathbf{e}$ vs. $\mathbf{u}_*$ and $\mathbf{e}_*$) effects. Model 1 is

$$\mathbf{y} = \mathbf{X}\beta + \mathbf{Z}\mathbf{u} + \mathbf{e}, \quad \text{where} \quad \mathbf{u} \sim (0, \mathbf{G}) \quad \text{and} \quad \mathbf{e} \sim (0, \mathbf{R})$$

while model 2 is

$$\mathbf{y} = \mathbf{X}_*\beta_* + \mathbf{Z}_*\mathbf{u}_* + \mathbf{e}_*, \quad \text{where} \quad \mathbf{u}_* \sim (0, \mathbf{G}_*) \quad \text{and} \quad \mathbf{e}_* \sim (0, \mathbf{R}_*)$$

Recalling our treatment of general mixed linear models (Chapter 10), Equation 10.19 implies that for model 1,

$$\mathbf{y} \sim (\mathbf{X}\beta, \mathbf{V}), \quad \text{where} \quad \mathbf{V} = \mathbf{Z}\mathbf{G}\mathbf{Z}^T + \mathbf{R}$$

while for model 2,

$$\mathbf{y} \sim (\mathbf{X}_*\boldsymbol{\beta}_*, \mathbf{V}_*), \quad \text{where} \quad \mathbf{V}_* = \mathbf{Z}_*\mathbf{G}_*\mathbf{Z}_*^T + \mathbf{R}_*$$

Thus, these two models are equivalent if

$$\mathbf{X}\boldsymbol{\beta} = \mathbf{X}_*\boldsymbol{\beta}_* \tag{A3.25a}$$

and $\mathbf{V} = \mathbf{V}_*$, or that

$$\mathbf{ZGZ}^T + \mathbf{R} = \mathbf{Z}_*\mathbf{G}_*\mathbf{Z}_*^T + \mathbf{R}_* \tag{A3.25b}$$

Equations A3.25a and A3.25b provide the framework for constructing equivalent models, and hence obtaining models that are potentially easier to analyze. Consider the situation where our interest is in the prediction of random effects and we wish to obtain an equivalent model that considers the same fixed effects but uses a different vector of random effects. (For example, instead of considering a vector of both parental and offspring breeding values, we might simply consider the vector of parental breeding values, using the parental estimates to subsequently estimate the breeding values in their offspring.) If the original model is

$$\mathbf{y} = \mathbf{X}\boldsymbol{\beta} + \mathbf{Zu} + \mathbf{e}, \qquad \text{where} \quad \mathbf{u} \sim (0, \mathbf{G}), \quad \text{and} \quad \mathbf{e} \sim (0, \mathbf{R})$$

and we wish to construct an equivalent model using *any* vector of random effects $\mathbf{u}_* \sim (0, \mathbf{G}_*)$ with

$$\mathbf{y} = \mathbf{X}\boldsymbol{\beta} + \mathbf{Z}_*\mathbf{u}_* + \mathbf{e}_*, \qquad \text{where} \quad \mathbf{u}_* \sim (0, \mathbf{G}_*), \quad \text{and} \quad \mathbf{e}_* \sim (0, \mathbf{R}_*)$$

For these models to be equivalent requires that $\mathbf{V} = \mathbf{V}_*$, hence it immediately follows from Equation A3.25b that the covariance matrix for the vector of new residual values, $\mathbf{e}_*$, is given by

$$\mathbf{R}_* = \mathbf{R} + \mathbf{ZGZ}^T - \mathbf{Z}_*\mathbf{G}_*\mathbf{Z}_*^T \tag{A3.26}$$

Given an estimate of $\mathbf{u}_*$, an estimate of $\mathbf{u}$ can be directly obtained, as parameters of a linear model can always be expressed as linear combinations of the parameters of any equivalent model. In this case, given the BLUP estimate ($\widehat{\mathbf{u}}_*$) of $\mathbf{u}_*$, the BLUP estimate of $\mathbf{u}$ is given by

$$\widehat{\mathbf{u}} = \mathbf{CG}^{-1}\widehat{\mathbf{u}}_* \tag{A3.27}$$

where $\mathbf{C}$ is the covariance matrix between $\mathbf{u}_*$ and $\mathbf{u}$, and $\mathbf{G}$ is the covariance matrix associated with $\mathbf{u}$ (Henderson 1977). This is just the linear regression of $\mathbf{u}_*$ on $\mathbf{u}$ (see Equation 9.27). Note that the vectors $\mathbf{u}_*$ and $\mathbf{u}$ can have different dimensionality, so that if $\mathbf{u}_*$ is $r \times 1$ and $\mathbf{u}$ is $q \times 1$, then $\mathbf{C}$ is an $r \times q$ matrix with $C_{ij} = \sigma(u_{*i}, u_j)$.

DERIVATIVES OF VECTORS AND MATRICES

Our next-to-last topic in matrix algebra concerns the derivatives of vector- and matrix-valued functions. We present a few simple results here, and the reader is referred to Morrison (1976), Graham (1981), and Searle (1982) for more details. Consider first the simplest function of vector $\mathbf{x}$, namely the product of $\mathbf{x}$ and either a vector ($\mathbf{a}$) or matrix ($\mathbf{A}$) of constants. The derivatives of these functions with respect to the vector $\mathbf{x}$ become

$$\frac{\partial\, \mathbf{a}^T\mathbf{x}}{\partial\, \mathbf{x}} = \frac{\partial\, \mathbf{x}^T\mathbf{a}}{\partial\, \mathbf{x}} = \mathbf{a} \tag{A3.28a}$$

$$\frac{\partial\, \mathbf{Ax}}{\partial\, \mathbf{x}} = \mathbf{A}^T \tag{A3.28b}$$

Turning to quadratic forms, if **A** is symmetric, then

$$\frac{\partial\, \mathbf{x}^T\mathbf{A}\mathbf{x}}{\partial\, \mathbf{x}} = 2\mathbf{A}\mathbf{x} \tag{A3.29a}$$

Three useful identities involving quadratic forms follow

$$\frac{\partial\, (\mathbf{a}-\mathbf{x})^T\mathbf{A}(\mathbf{a}-\mathbf{x})}{\partial\, \mathbf{x}} = -2\mathbf{A}(\mathbf{a}-\mathbf{x}) \tag{A3.29b}$$

$$\frac{\partial\, (\mathbf{a}-\mathbf{B}\mathbf{x})^T(\mathbf{a}-\mathbf{B}\mathbf{x})}{\partial\, \mathbf{x}} = -2\mathbf{B}^T(\mathbf{a}-\mathbf{B}\mathbf{x}) \tag{A3.29c}$$

$$\frac{\partial\, (\mathbf{a}-\mathbf{B}\mathbf{x})^T\mathbf{A}(\mathbf{a}-\mathbf{B}\mathbf{x})}{\partial\, \mathbf{x}} = -2\mathbf{B}^T\mathbf{A}(\mathbf{a}-\mathbf{B}\mathbf{x}) \tag{A3.29d}$$

Two final useful identities follow from the chain rule of differentiation,

$$\frac{\partial\, \exp[\, f(\mathbf{x})\,]}{\partial\, \mathbf{x}} = \exp[\, f(\mathbf{x})\,] \cdot \frac{\partial\, f(\mathbf{x})}{\partial\, \mathbf{x}} \tag{A3.30a}$$

$$\frac{\partial\, \ln[\, f(\mathbf{x})\,]}{\partial\, \mathbf{x}} = \frac{1}{f(\mathbf{x})} \cdot \frac{\partial\, f(\mathbf{x})}{\partial\, \mathbf{x}} \tag{A3.30b}$$

Example A3.6. The OLS solution for a linear model is the value of $\boldsymbol{\beta}$ that minimizes the residual sum of squares given **y** and **X**. In matrix form,

$$\sum_{i=1}^{n} e_i^2 = \mathbf{e}^T\mathbf{e} = (\mathbf{y}-\mathbf{X}\boldsymbol{\beta})^T(\mathbf{y}-\mathbf{x}\boldsymbol{\beta})$$

Taking the derivative with respect to $\boldsymbol{\beta}$ and using Equation A3.29c (with $\mathbf{a} = \mathbf{y}$, $\mathbf{B} = \mathbf{X}$, and $\mathbf{x} = \boldsymbol{\beta}$) gives

$$\frac{\partial\, \mathbf{e}^T\mathbf{e}}{\partial\, \boldsymbol{\beta}} = \frac{(\mathbf{y}-\mathbf{X}\boldsymbol{\beta})^T(\mathbf{y}-\mathbf{x}\boldsymbol{\beta})}{\partial\, \boldsymbol{\beta}} = -2\mathbf{X}^T(\mathbf{y}-\mathbf{X}\boldsymbol{\beta})$$

Setting this equal to zero gives $\mathbf{X}^T\mathbf{X}\boldsymbol{\beta} = \mathbf{X}^T\mathbf{y}$, which has solution

$$\boldsymbol{\beta} = \left(\mathbf{X}^T\mathbf{X}\right)^{-1}\mathbf{X}^T\mathbf{y}$$

If $\mathbf{X}^T\mathbf{X}$ is singular, a generalized inverse is used.

Example A3.7. Writing the MVN distribution as

$$\varphi(\mathbf{x}) = a \exp\left(-\frac{1}{2} \cdot (\mathbf{x}-\boldsymbol{\mu})^T\mathbf{V}^{-1}(\mathbf{x}-\boldsymbol{\mu})\right)$$

where $a = \pi^{-n/2}\, |\mathbf{V}|^{-1/2}$, then from Equation A3.30a,

$$\frac{\partial\, \varphi(\mathbf{x})}{\partial\, \mathbf{x}} = \varphi(\mathbf{x}) \cdot \frac{\partial\, \left[\left(-\frac{1}{2}\right) \cdot (\mathbf{x}-\boldsymbol{\mu})^T\mathbf{V}^{-1}(\mathbf{x}-\boldsymbol{\mu})\right]}{\partial\, \mathbf{x}}$$

Applying Equation A3.29b yields

$$\frac{\partial\,\varphi(\mathbf{x})}{\partial\,\mathbf{x}} = -\varphi(\mathbf{x}) \cdot \mathbf{V}^{-1}\,(\mathbf{x} - \boldsymbol{\mu}) \tag{A3.31a}$$

Note that $\varphi(\mathbf{x})$ is a scalar and hence its order of multiplication does not matter, while the order of the other variables (being matrices) is critical. Similarly, we can consider the MVN as a function of the mean vector $\boldsymbol{\mu}$, in which case Equation A3.30a implies

$$\frac{\partial\,\varphi(\mathbf{x},\boldsymbol{\mu})}{\partial\,\boldsymbol{\mu}} = \varphi(\mathbf{x},\boldsymbol{\mu}) \cdot \mathbf{V}^{-1}\,(\mathbf{x} - \boldsymbol{\mu}) \tag{A3.31b}$$

MATRIX DECOMPOSITIONS

Any matrix can be decomposed (or **factored**) into a product of simpler matrices. One important factorization is the singular value decomposition (SVD), which is the foundation for AMMI analysis of genotype × environment interactions (Part II Chapters 27 and 28), and also provides a method for obtaining a generalized inverse (Equation A3.38). With a square symmetric matrix, a special case of the SVD, **diagonalization**, provides significant insight on the eigenstructure of inverses, square roots, and other functions of a diagonalizable matrix. We consider diagonalization first before concluding by briefly examining the SVD.

Diagonalization of a Covariance Matrix

We previously introduced the square-root decomposition of a nonsingular symmetric matrix, $\mathbf{V} = \mathbf{V}^{1/2}\mathbf{V}^{1/2}$ (Equation A3.7a), and showed its utility in generating a new set of uncorrelated random variables (Equation A3.8a). This particular decomposition immediately follows from the more general diagonalization decomposition. In particular, a symmetric matrix $\mathbf{A}$ (such as a covariance matrix) can be **diagonalized** as

$$\mathbf{A} = \mathbf{U}\boldsymbol{\Lambda}\mathbf{U}^T \tag{A3.32a}$$

where $\boldsymbol{\Lambda}$ is a diagonal matrix and $\mathbf{U} = (\,\mathbf{u}_1, \mathbf{u}_2, \cdots, \mathbf{u}_n\,)$ is an **orthonormal matrix**. The columns of such orthonormal matrices have unit length ($\sqrt{\mathbf{u}_i^T\mathbf{u}_i} = ||\mathbf{u}_i|| = 1$) and are orthogonal ($\mathbf{u}_i^T\mathbf{u}_j = 0$ for $i \neq j$). These conditions also imply that $\mathbf{U}$ is a **unitary matrix,** with its inverse being given by its transpose, $\mathbf{U}^{-1} = \mathbf{U}^T$. Unitary matrices generate a rigid rotation (the angle between the vectors $\mathbf{x}$ and $\mathbf{y}$ is the same as the angle between the transformed vectors $\mathbf{Ux}$ and $\mathbf{Uy}$) of the original coordinate system to a new coordinate system given by $\mathbf{U}$ (WL Appendix 5).

If λ_i and $\mathbf{e}_i$ are the ith eigenvalue and its associated unit eigenvector of $\mathbf{A}$, then

$$\boldsymbol{\Lambda} = \text{diag}(\lambda_1, \lambda_2, \cdots, \lambda_n) = \begin{pmatrix} \lambda_1 & 0 & \cdots & 0 \\ 0 & \lambda_2 & \cdots & 0 \\ \vdots & & \ddots & \vdots \\ 0 & \cdots & \cdots & \lambda_n \end{pmatrix} \tag{A3.32b}$$

and

$$\mathbf{U} = (\,\mathbf{e}_1, \mathbf{e}_2, \cdots, \mathbf{e}_n\,) \tag{A3.32c}$$

$\mathbf{U}$ describes a rotation of the original coordinate system to a new coordinate system given by the eigenvectors of $\mathbf{A}$, while the diagonal elements of $\boldsymbol{\Lambda}$ give the amount by which vectors of unit length in the original coordinate system are scaled in the transformed system. If we use the decomposition $\boldsymbol{\Lambda} = \sum_{i=1}^{n} \boldsymbol{\Lambda}_i$, where $\boldsymbol{\Lambda}_i$ is a diagonal matrix whose elements are all zero, except for λ_i, then Equation A3.32a becomes

$$\mathbf{A} = \mathbf{U}\left(\sum_{i=1}^{n} \boldsymbol{\Lambda}_i\right)\mathbf{U}^T = \sum_{i=1}^{n} \mathbf{U}\boldsymbol{\Lambda}_i\mathbf{U}^T = \sum_{i=1}^{n} \lambda_i \mathbf{e}_i\mathbf{e}_i^T \tag{A3.32d}$$

Note that $\mathbf{e}_i\mathbf{e}_i^T$, the outer product of **e** with itself (Equation 9.8b), is an $n \times n$ matrix. Equation A3.32d is called the **spectral decomposition**, **spectral factorization**, or **eigendecomposition** of **A** (WL Appendix 5).

Using Equation A3.32a, it is easy to show that

$$\mathbf{A}^{-1} = \mathbf{U}\boldsymbol{\Lambda}^{-1}\mathbf{U}^T \tag{A3.33a}$$

To see this, note that

$$\mathbf{A}^{-1}\mathbf{A} = \left(\mathbf{U}\boldsymbol{\Lambda}^{-1}\mathbf{U}^T\right)\left(\mathbf{U}\boldsymbol{\Lambda}\mathbf{U}^T\right) = \mathbf{U}\boldsymbol{\Lambda}^{-1}\left(\mathbf{U}^T\mathbf{U}\right)\boldsymbol{\Lambda}\mathbf{U}^T = \mathbf{U}\boldsymbol{\Lambda}^{-1}\boldsymbol{\Lambda}\mathbf{U}^T = \mathbf{U}\mathbf{U}^T = \mathbf{I}$$

Similar logic yields

$$\mathbf{A}^{1/2} = \mathbf{U}\boldsymbol{\Lambda}^{1/2}\mathbf{U}^T \tag{A3.33b}$$

$$\mathbf{A}^{-1/2} = \mathbf{U}\boldsymbol{\Lambda}^{-1/2}\mathbf{U}^T \tag{A3.33c}$$

$$\mathbf{A}^{k} = \mathbf{U}\boldsymbol{\Lambda}^{k}\mathbf{U}^T \quad \text{for any integer } k \tag{A3.33d}$$

Further, the square root decomposition follows as

$$\begin{aligned}\mathbf{A}^{1/2}\mathbf{A}^{1/2} &= \left[\mathbf{U}\boldsymbol{\Lambda}^{1/2}\mathbf{U}^T\right]\left[\mathbf{U}\boldsymbol{\Lambda}^{1/2}\mathbf{U}^T\right] = \mathbf{U}\boldsymbol{\Lambda}^{1/2}\left[\mathbf{U}^T\mathbf{U}\right]\boldsymbol{\Lambda}^{1/2}\mathbf{U}^T \\ &= \mathbf{U}\boldsymbol{\Lambda}^{1/2}\mathbf{I}\boldsymbol{\Lambda}^{1/2}\mathbf{U}^T = \mathbf{U}\boldsymbol{\Lambda}^{1/2}\boldsymbol{\Lambda}^{1/2}\mathbf{U}^T \\ &= \mathbf{U}\boldsymbol{\Lambda}\mathbf{U}^T = \mathbf{A}\end{aligned}$$

Likewise, using Equation A3.32a, we see that premultiplying **A** by $\mathbf{U}^T$ and then postmultiplying by **U** gives a diagonal matrix whose elements are the eigenvalues of **A**

$$\mathbf{U}^T\mathbf{A}\mathbf{U} = \mathbf{U}^T(\mathbf{U}\boldsymbol{\Lambda}\mathbf{U}^T)\mathbf{U} = (\mathbf{U}^T\mathbf{U})\boldsymbol{\Lambda}(\mathbf{U}^T\mathbf{U}) = \boldsymbol{\Lambda} \tag{A3.34}$$

Finally, because $\boldsymbol{\Lambda}$ is diagonal, the ith diagonal elements of $\boldsymbol{\Lambda}^{-1}$, $\boldsymbol{\Lambda}^{1/2}$, $\boldsymbol{\Lambda}^{-1/2}$, and $\boldsymbol{\Lambda}^{k}$ are λ_i^{-1}, $\lambda_i^{1/2}$, $\lambda_i^{-1/2}$, and λ_i^{k}, respectively, implying that if λ_i is an eigenvalue of **A**, then λ_i^{-1}, $\lambda_i^{1/2}$, $\lambda_i^{-1/2}$, and λ_i^{k}, respectively, are eigenvalues of the matrices $\mathbf{A}^{-1}$, $\mathbf{A}^{1/2}$, $\mathbf{A}^{-1/2}$, and $\mathbf{A}^{k}$. Note that Equations A3.33a through A3.33d further imply that the matrices **A**, $\mathbf{A}^{-1}$, $\mathbf{A}^{1/2}$, $\mathbf{A}^{-1/2}$, and $\mathbf{A}^{k}$ all have the same eigenvectors, namely the columns of **U**.

Correlations can be Removed by a Matrix Transformation

As we saw with the square-root transform (Equation A3.8a), a powerful use of diagonalization is that it allows one to extract a set of n uncorrelated variables for any $n \times n$ nonsingular covariance matrix, $\mathbf{V_x}$ (associated with the random vector, **x**). Consider the transformation

$$\mathbf{y} = \mathbf{U}^T\mathbf{x} \tag{A3.35a}$$

where $\mathbf{U} = (\mathbf{e}_1, \mathbf{e}_2, \cdots, \mathbf{e}_n)$ contains the normalized eigenvectors of $\mathbf{V_x}$. Because **U** is an orthonormal matrix, this transformation is a rigid rotation of the axes of the original $(x_1, \cdots, x_n)$ coordinate system to a new system given by $(e_1, \cdots, e_n)$. Applying Equations 9.21b and A3.34, respectively, the covariance matrix for **y** is

$$\mathbf{V_y} = \mathbf{U}^T\mathbf{V_x}\mathbf{U} = \boldsymbol{\Lambda} \tag{A3.35b}$$

where $\boldsymbol{\Lambda}$ is a diagonal matrix whose elements are the eigenvalues of $\mathbf{V_x}$,

$$\sigma(y_i, y_j) = \begin{cases} \lambda_i & \text{if } i = j \\ 0 & \text{if } i \neq j \end{cases}$$

The rigid rotation introduced by $\mathbf{U}$ thus creates a set of n uncorrelated variables, the ith of which is given by

$$y_i = \mathbf{e}_i{}^T \mathbf{x} \tag{A3.35c}$$

This is the length of the projection of $\mathbf{x}$ onto the ith eigenvector of $\mathbf{V_x}$ (WL Appendix 5), so that the axes of the new coordinate system are given by the orthogonal set of eigenvectors of $\mathbf{V_x}$.

Defining the matrix $\mathbf{B}$ as

$$\mathbf{B} = \mathbf{U}\,\boldsymbol{\Lambda}^{-1/2} \tag{A3.35d}$$

the vector $\mathbf{y} = \mathbf{B}^T\mathbf{x}$ has a covariance matrix of $\mathbf{V_y} = \mathbf{I}$, which means that this transformation creates a set of uncorrelated variables, each with unit variance. To see this, note that

$$\begin{aligned} \mathbf{V_y} = \mathbf{B}^T \mathbf{V_x}\, \mathbf{B} &= \left(\mathbf{U}\boldsymbol{\Lambda}^{-1/2}\right)^T \left(\mathbf{U}\boldsymbol{\Lambda}\mathbf{U}^T\right) \left(\mathbf{U}\boldsymbol{\Lambda}^{-1/2}\right) \\ &= \boldsymbol{\Lambda}^{-1/2} \left(\mathbf{U}^T\mathbf{U}\right) \boldsymbol{\Lambda} \left(\mathbf{U}^T\mathbf{U}\right) \boldsymbol{\Lambda}^{-1/2} \\ &= \boldsymbol{\Lambda}^{-1/2}\, \boldsymbol{\Lambda}\, \boldsymbol{\Lambda}^{-1/2} = \mathbf{I} \end{aligned} \tag{A3.35e}$$

Recall that earlier we used another transformation, $\mathbf{y} = \mathbf{V_x}^{-1/2}\mathbf{x}$, where from Equation A3.33c

$$\mathbf{V_x}^{-1/2} = \mathbf{U}\boldsymbol{\Lambda}^{-1/2}\mathbf{U}^T = \mathbf{B}\mathbf{U}^T$$

Because $\mathbf{U}$ is a unitary matrix, the transformations given by $\mathbf{V_x}^{-1/2}$ and $\mathbf{B}$ are simply rigid rotations of each other.

An alternative to both these transformations is the **Cholesky decomposition,** $\mathbf{A} = \mathbf{C}^T\mathbf{C}$, of a square, symmetric matrix $\mathbf{A}$, where $\mathbf{C}$ is a **lower triangular matrix** (all elements above the diagonal are zero). If $\mathbf{C}$ is the Cholesky decomposition for $\mathbf{V_x}$, then $\mathbf{y} = \mathbf{C}^{-1}\mathbf{x}$ also returns a covariance matrix, $\mathbf{V_y}$, of $\mathbf{I}$.

The Singular-Value Decomposition (SVD)

Any $n \times p$ matrix $\mathbf{A}$ can be decomposed as the product of three matrices: an $n \times p$ diagonal matrix $\boldsymbol{\Lambda}$ and two unitary matrices, $\mathbf{U}$ which is $n \times n$ and $\mathbf{V}$ which is $p \times p$. The resulting **singular value decomposition** (**SVD**) is given by

$$\mathbf{A}_{n\times p} = \mathbf{U}_{n\times n}\boldsymbol{\Lambda}_{n\times p}\mathbf{V}^T_{p\times p} \tag{A3.36a}$$

We have indicated the dimensionality of each matrix to allow the reader to verify that each matrix multiplication conforms. The (nonzero) diagonal elements $\lambda_1, \cdots, \lambda_s$ of $\boldsymbol{\Lambda}$ correspond to the **singular values** of $\mathbf{A}$ and are ordered by decreasing magnitude. Returning to the unitary matrices $\mathbf{U}$ and $\mathbf{V}$, we can write each as a row vector of column vectors,

$$\mathbf{U} = (\mathbf{u}_1, \cdots, \mathbf{u}_i, \cdots \mathbf{u}_n), \qquad \mathbf{V} = (\mathbf{v}_1, \cdots, \mathbf{v}_i, \cdots \mathbf{v}_p) \tag{A3.36b}$$

where $\mathbf{u}_i$ and $\mathbf{v}_i$ are n- and p-dimensional column vectors (often called the **left** and **right singular vectors**, respectively). Since both $\mathbf{U}$ and $\mathbf{V}$ are unitary, each column vector has length one and are mutually orthogonal (i.e., if $i \neq j$, $\mathbf{u}_i\mathbf{u}_j^T = \mathbf{v}_i\mathbf{v}_j^T = 0$). Since $\boldsymbol{\Lambda}$ is diagonal, it immediately follows from matrix multiplication that we can write any element in $\mathbf{A}$ as

$$A_{ij} = \sum_{k=1}^{s} \lambda_k\, u_{ik}\, v_{kj} \tag{A3.36c}$$

where λ_k is the kth singular value and $s \leq \min(p, n)$ is the number of nonzero singular values (s is the rank of **A**). Since only the first s diagonal elements of $\boldsymbol{\Lambda}$ are nonzero, the SVD can also be written as

$$\mathbf{A}_{n\times p} = \mathbf{U}_{n\times s}\boldsymbol{\Lambda}_{s\times s}\mathbf{V}^T_{s\times p}, \quad \text{with} \quad \mathbf{U} = (\mathbf{u}_1, \cdots, \mathbf{u}_s), \quad \mathbf{V} = (\mathbf{v}_1, \cdots, \mathbf{v}_s) \tag{A3.36d}$$

This is the **compact SVD**. Since $\boldsymbol{\Lambda}$ is diagonal, we can express Equation A3.36c as

$$A_{ij} = \mathbf{u}_i \boldsymbol{\Lambda} \mathbf{v}_j^T \tag{A3.36e}$$

A final useful identity is that the total variation of **A** (the sum of all its squared values) equals the sum of its squared singular values,

$$\sum_{ij} A_{ij}^2 = \sum_{k=1}^{s} \lambda_k^2 \tag{A3.36f}$$

A critical utility of the SVD arises from the **Eckart–Young theorem** (1936), which relates the best approximation of a matrix by some lower rank (say rank $m < s$) matrix. Define as our measure of goodness of fit between a matrix **A** and a lower rank approximation $\widehat{\mathbf{A}}$ as the sum of squared differences over all elements,

$$\sum_{ij} (A_{ij} - \widehat{A}_{ij})^2$$

Eckart and Young showed that the best fitting approximation $\widehat{\mathbf{A}}$ of rank $m < s$ is given from the first m terms of the singular value decomposition (the **rank-m SVD**),

$$\widehat{A}_{ij} = \sum_{k=1}^{m} \lambda_k \, u_{ik} \, v_{kj} \tag{A3.37a}$$

For example, the best rank-2 approximation is given by

$$A_{ij} \simeq \lambda_1 \, u_{i1} \, v_{j1} + \lambda_2 \, u_{i2} \, v_{j2} \tag{A3.37b}$$

where λ_i is the ith singular value of the **A** matrix, **u** and **v** are the associated singular vectors (see Example A3.8). The fraction of total variation of a matrix accounted for by taking the first m terms in its SVD is

$$\sum_{k=1}^{m} \lambda_k^2 \Big/ \sum_{ij} A_{ij}^2 = \frac{\lambda_1^2 + \cdots + \lambda_m^2}{\lambda_1^2 + \cdots + \lambda_s^2} \tag{A3.37c}$$

Example A3.8. Consider the following performances of three soybean lines in five different environments in New York from a larger dataset given by Gauch (1992). Each entry represents the average of four replications. For environments, the first letter denotes the testing location, while the two numbers denote the year of the test. For ease of presentation, we have rounded entry means and the associated row and column means.

	Genotypes				
Environment	EVAN	WILK	CHIP	Row Mean	$\widehat{E}$
A77	2725	2471	2333	2510	494
V79	1111	578	1278	989	−1027
R81	2038	1386	2350	1925	−91

Environment	EVAN	WILK	CHIP	Row Mean	$\widehat{E}$
I85	1736	1607	1588	1644	−372
G85	3258	2961	2813	3011	995
Column Mean	2174	1801	2072	2016	
$\widehat{G}$	158	−215	56		

Denote the entry means of genotype i in environment j by z_{ij}, the environment and genotype means by $\overline{z}_{\cdot j}$ and $\overline{z}_{i\cdot}$ and the grand mean by $\overline{z}_{\cdot\cdot}$ (= 2016). Under an OLS fixed-effects model, the estimated main effects are given by $\widehat{\mu} = \overline{z}_{\cdot\cdot}$, $\widehat{E}_j = \overline{z}_{\cdot j} - \overline{z}_{\cdot\cdot}$, and $\widehat{G}_i = \overline{z}_{i\cdot} - \overline{z}_{\cdot\cdot}$, and these are given in the above table. The G × E interaction terms (Chapter 6) are estimated by

$$\widehat{GE}_{ij} = z_{ij} - (\widehat{\mu} + \widehat{G}_i + \widehat{E}_j)$$

For example, for EVAN in A77, $\widehat{GE} = 2725 - (2016 + 158 + 494) = 57$. The resulting G × E table of interactions becomes

Environment	Genotypes EVAN	WILK	CHIP
A77	57	176	−233
V79	−36	−196	233
R81	−45	−324	369
I85	−66	178	−112
G85	89	165	−254

Now let's examine SVD approximations (for various ranks) for this table of G × E values, which written in matrix form (using columns for genotypes and rows for environments) as

$$\mathbf{GE} = \begin{pmatrix} 57 & 176 & -233 \\ -36 & -196 & 233 \\ -45 & -324 & 369 \\ -66 & 178 & -112 \\ 89 & 165 & -254 \end{pmatrix}$$

In R, the compact SVD (Equation A3.36d) of a matrix X is given by the command `svd(X)`, returning the compact SVD of **GE** as

$$\begin{pmatrix} 0.40 & 0.21 & 0.18 \\ -0.41 & 0.00 & 0.91 \\ -0.66 & 0.12 & -0.30 \\ 0.26 & -0.83 & 0.11 \\ 0.41 & 0.50 & 0.19 \end{pmatrix} \begin{pmatrix} 746.10 & 0 & 0 \\ 0 & 131.36 & 0 \\ 0 & 0 & 0.53 \end{pmatrix} \begin{pmatrix} 0.12 & 0.64 & -0.76 \\ 0.81 & -0.51 & -0.30 \\ 0.58 & 0.58 & 0.58 \end{pmatrix}$$

The first singular value accounts for $746.10^2/(743.26^2 + 131.36^2 + 0.53^2)$ = 97.0% of the total variation of **GE**, while the second singular value accounts for 3.0%, so that together they account for essentially all of the total variation. The rank-1 SVD approximation of **GE** is given by setting all of the diagonal elements of $\boldsymbol{\Lambda}$ except the first entry to zero,

$$\mathbf{GE}_1 = \begin{pmatrix} 0.40 & 0.21 & 0.18 \\ -0.41 & 0.00 & 0.91 \\ -0.66 & 0.12 & -0.30 \\ 0.26 & -0.83 & 0.11 \\ 0.41 & 0.50 & 0.19 \end{pmatrix} \begin{pmatrix} 746.10 & 0 & 0 \\ 0 & 0 & 0 \\ 0 & 0 & 0 \end{pmatrix} \begin{pmatrix} 0.12 & 0.64 & -0.76 \\ 0.81 & -0.51 & -0.30 \\ 0.58 & 0.58 & 0.58 \end{pmatrix}$$

Similarly, the rank-2 SVD is given by setting all but the first two singular values to zero,

$$\mathbf{GE}_2 = \begin{pmatrix} 0.40 & 0.21 & 0.18 \\ -0.41 & 0.00 & 0.91 \\ -0.66 & 0.12 & -0.30 \\ 0.26 & -0.83 & 0.11 \\ 0.41 & 0.50 & 0.19 \end{pmatrix} \begin{pmatrix} 746.10 & 0 & 0 \\ 0 & 131.36 & 0 \\ 0 & 0 & 0 \end{pmatrix} \begin{pmatrix} 0.12 & 0.64 & -0.76 \\ 0.81 & -0.51 & -0.30 \\ 0.58 & 0.58 & 0.58 \end{pmatrix}$$

The resulting rank-1 and rank-2 approximations for **GE** are

$$\mathbf{GE}_1 = \begin{pmatrix} 34.57 & 189.94 & -224.67 \\ -35.78 & -196.59 & 232.54 \\ -57.51 & -316.03 & 373.81 \\ 22.33 & 122.71 & -145.15 \\ 36.04 & 198.02 & -234.23 \end{pmatrix}, \quad \mathbf{GE}_2 = \begin{pmatrix} 56.95 & 175.95 & -233.05 \\ -36.28 & -196.28 & 232.72 \\ -44.91 & -323.91 & 369.09 \\ -66.03 & 177.97 & -112.03 \\ 88.94 & 164.94 & -254.06 \end{pmatrix}$$

Since there are only three non-zero singular values, the rank-3 SVD returns the matrix **GE**, i.e., $\mathbf{GE}_3 = \mathbf{GE}$. For example, consider the G × E term for EVAN and A77. The rank-1 SVD value for this entry is 34.57, the rank-2 VSD value is 56.95, and the rank-3 (full rank) just the original value of 57.

The SVD immediately allows one to compute a generalized inverse of **A**. Using the compact SVD (Equation A3.36d), it follows that if $\mathbf{A}_{n\times p} = \mathbf{U}_{n\times s}\boldsymbol{\Lambda}_{s\times s}\mathbf{V}^T_{s\times p}$, then

$$\mathbf{A}^-_{p\times n} = \mathbf{V}_{p\times s}\boldsymbol{\Lambda}^{-1}_{s\times s}\mathbf{U}^T_{s\times n} \tag{A3.38}$$

To show this, we need to show that Equation A3.38 satisfies Equation A3.1, $\mathbf{AA}^-\mathbf{A} = \mathbf{A}$. Rearranging and using the fact that **V** and **U** are unity matrices (their inverse is their transpose),

$$\begin{aligned}
\mathbf{AA}^-\mathbf{A} &= \left[\mathbf{U}_{n\times s}\boldsymbol{\Lambda}_{s\times s}\mathbf{V}^T_{s\times p}\right]\left[\mathbf{V}_{p\times s}\boldsymbol{\Lambda}^{-1}_{s\times s}\mathbf{U}^T_{s\times n}\right]\left[\mathbf{U}_{n\times s}\boldsymbol{\Lambda}_{s\times s}\mathbf{V}^T_{s\times p}\right] \\
&= \mathbf{U}_{n\times s}\boldsymbol{\Lambda}_{s\times s}\left[\mathbf{V}^T_{s\times p}\mathbf{V}_{p\times s}\right]\boldsymbol{\Lambda}^{-1}_{s\times s}\left[\mathbf{U}^T_{s\times n}\mathbf{U}_{n\times s}\right]\boldsymbol{\Lambda}_{s\times s}\mathbf{V}^T_{s\times p} \\
&= \mathbf{U}_{n\times s}\boldsymbol{\Lambda}_{s\times s}\mathbf{I}_{s\times s}\boldsymbol{\Lambda}^{-1}_{s\times s}\mathbf{I}_{s\times s}\boldsymbol{\Lambda}_{s\times s}\mathbf{V}^T_{s\times p} \\
&= \mathbf{U}_{n\times s}\boldsymbol{\Lambda}_{s\times s}\boldsymbol{\Lambda}^{-1}_{s\times s}\boldsymbol{\Lambda}_{s\times s}\mathbf{V}^T_{s\times p} \\
&= \mathbf{U}_{n\times s}\boldsymbol{\Lambda}_{s\times s}\mathbf{V}^T_{s\times p} = \mathbf{A}
\end{aligned}$$

Example A3.9. From Equation A3.38, a generalized inverse of **GE** is given by

$$\begin{pmatrix} 0.12 & 0.81 & 0.58 \\ 0.64 & -0.51 & 0.58 \\ -0.76 & -0.30 & 0.58 \end{pmatrix} \begin{pmatrix} \frac{1}{746.10} & 0 & 0 \\ 0 & \frac{1}{131.36} & 0 \\ 0 & 0 & \frac{1}{0.53} \end{pmatrix} \begin{pmatrix} 0.40 & -0.41 & -0.66 & 0.26 & 0.41 \\ 0.21 & 0.00 & 0.12 & -0.83 & 0.50 \\ 0.18 & 0.91 & -0.30 & 0.11 & 0.19 \end{pmatrix}$$

$$= \begin{pmatrix} 0.1989 & 1 & -0.3253 & 0.1170 & 0.2106 \\ 0.1971 & 1 & -0.3271 & 0.1256 & 0.2060 \\ 0.1965 & 1 & -0.3254 & 0.1237 & 0.2058 \end{pmatrix}$$

Denoting this matrix as $\mathbf{GE}^-$, matrix multiplication shows that Equation A3.1 is satisfied, namely that $\mathbf{GE}\,\mathbf{GE}^-\,\mathbf{GE} = \mathbf{GE}$, showing that $\mathbf{GE}^-$ is indeed a generalized inverse of **GE**.

We can also use the SVD to solve much more general sets of linear equations. Until now, we have been assuming n equations and p unknowns, with $n = p$. More generally, we may wish to find solutions for the system

$$\mathbf{A}_{n\times p}\mathbf{x}_{p\times 1} = \mathbf{c}_{n\times 1} \tag{A3.39a}$$

When $n > p$, the system is said to be **overdetermined**, while it is said to be **underdetermined** when $n < p$. We can solve this system using the SVD as follows:

$$\begin{aligned}
\mathbf{A}_{n\times p}\,\mathbf{x}_{p\times 1} &= \mathbf{c}_{n\times 1} \\
\mathbf{U}_{n\times s}\boldsymbol{\Lambda}_{s\times s}\mathbf{V}^T_{s\times p}\,\mathbf{x}_{p\times 1} &= \mathbf{c}_{n\times 1} \\
\mathbf{V}_{p\times s}\boldsymbol{\Lambda}^{-1}_{s\times s}[\mathbf{U}^T_{s\times n}\mathbf{U}_{n\times s}]\boldsymbol{\Lambda}_{s\times s}\mathbf{V}^T_{s\times p}\,\mathbf{x}_{p\times 1} &= \mathbf{V}_{p\times s}\boldsymbol{\Lambda}^{-1}_{s\times s}\mathbf{U}^T_{s\times n}\mathbf{c}_{n\times 1} \\
\mathbf{V}_{p\times s}[\boldsymbol{\Lambda}^{-1}_{s\times s}\boldsymbol{\Lambda}_{s\times s}]\mathbf{V}^T_{s\times p}\,\mathbf{x}_{p\times 1} &= \mathbf{V}_{p\times s}\boldsymbol{\Lambda}^{-1}_{s\times s}\mathbf{U}^T_{s\times n}\mathbf{c}_{n\times 1} \\
[\mathbf{V}_{p\times s}\mathbf{V}^T_{s\times p}]\,\mathbf{x}_{p\times 1} &= \mathbf{V}_{p\times s}\boldsymbol{\Lambda}^{-1}_{s\times s}\mathbf{U}^T_{s\times n}\mathbf{c}_{n\times 1} \\
\mathbf{x}_{p\times 1} &= \mathbf{V}_{p\times s}\boldsymbol{\Lambda}^{-1}_{s\times s}\mathbf{U}^T_{s\times n}\mathbf{c}_{n\times 1} \qquad \text{(A3.39b)}
\end{aligned}$$

where each term above in brackets is an identity matrix. Note from Equation A3.38 that this is just $\mathbf{x}_{p\times 1} = \mathbf{A}^-_{p\times n}\mathbf{c}_{n\times 1}$.

Finally, we can connect the SVD with diagonalization as follows. First suppose that $\mathbf{A}$ is square and symmetric, then

$$\mathbf{A} = \mathbf{U}\boldsymbol{\Lambda}\mathbf{V}^T = \mathbf{A}^T = (\mathbf{U}\boldsymbol{\Lambda}\mathbf{V}^T)^T = \mathbf{V}\boldsymbol{\Lambda}\mathbf{U}^T$$

so that $\mathbf{V} = \mathbf{U}$, recovering Equation A3.32a. Hence, for a symmetric matrix, singular vectors correspond to eigenvectors and singular values to eigenvalues.

Now consider a general matrix, such as an $n \times p$ data matrix, $\mathbf{X}$. Note that the $n \times n$ matrix $\mathbf{X}\mathbf{X}^T$ is symmetric, as $(\mathbf{X}\mathbf{X}^T)^T = (\mathbf{X}^T)^T\mathbf{X}^T = \mathbf{X}\mathbf{X}^T$. Likewise, $\mathbf{X}^T\mathbf{X}$ is a $p \times p$ symmetric matrix. Because $\mathbf{X}^T\mathbf{X}$ and $\mathbf{X}\mathbf{X}^T$ are symmetric, they are also diagonalizable, and we can relate their eigenvalues and eigenvalues to the singular vectors and singular values of $\mathbf{X}$ as follows. Consider $\mathbf{X}^T\mathbf{X}$ first. Using the SVD of $\mathbf{X}$, we have

$$\mathbf{X}^T\mathbf{X} = [\mathbf{U}\boldsymbol{\Lambda}\mathbf{V}^T]^T[\mathbf{U}\boldsymbol{\Lambda}\mathbf{V}^T] = \mathbf{V}\boldsymbol{\Lambda}[\mathbf{U}^T\mathbf{U}]\boldsymbol{\Lambda}\mathbf{V}^T = \mathbf{V}\boldsymbol{\Lambda}\boldsymbol{\Lambda}\mathbf{V}^T = \mathbf{V}\boldsymbol{\Lambda}^2\mathbf{V}^T \qquad \text{(A3.40a)}$$

where we have used $\boldsymbol{\Lambda}^T = \boldsymbol{\Lambda}$, as $\boldsymbol{\Lambda}$ is diagonal. Comparison to Equation A3.23 shows that the elements of $\mathbf{V}$ correspond to the eigenvectors of $\mathbf{X}^T\mathbf{X}$, while the eigenvalues are given by the squares of the singular values of $\mathbf{X}$, $\lambda_i = s_i^2$. Similarly,

$$\mathbf{X}\mathbf{X}^T = \mathbf{U}\boldsymbol{\Lambda}^2\mathbf{U}^T \qquad \text{(A3.40b)}$$

where the elements of $\mathbf{U}$ correspond to the eigenvectors of $\mathbf{X}\mathbf{X}^T$, while its eigenvalues are the same as for $\mathbf{X}^T\mathbf{X}$.

Appendix 4

Maximum Likelihood Estimation and Likelihood-ratio Tests

For the solution of problems of estimation we require a method which for each particular problem will lead us automatically to the statistic by which the criterion of sufficient is satisfied. Such a method is, I believe, provided by the Method of Maximum Likelihood. Fisher (1922)

The method of maximum likelihood (ML), introduced by Fisher (1922), is widely used in human and quantitative genetics and we draw upon this approach throughout the book, especially in Chapters 16–19 (mixture distributions) and Chapter 33 in Part II (variance component estimation). ML provides an *extremely flexible* unified approach for statistical analysis, offering a single framework for (i) parameter estimation; (ii) hypothesis testing; and (iii) quantification of parameter uncertainty (large-sample variances and confidence intervals). Weir (2010) gives a useful introduction with genetic applications, while Cox and Hinkley (1979), Kendall and Stuart (1979), and Edwards (1992) provide more detailed statistical treatments. Sorensen and Gianola (2002) examine applications in quantitative genetics, while Stigler (2007) provides an entertaining overview of the history of ML.

LIKELIHOOD, SUPPORT, AND SCORE FUNCTIONS

The basic idea underlying ML is quite simple. Usually, when specifying a probability density function (pdf), say, a normal with unknown mean μ and unit variance, it is treated as a function of z (the value of the random variable) with the distribution parameters $\boldsymbol{\Theta}$ assumed to be known. In this setting, the model parameters (the vector $\boldsymbol{\Theta}$) are treated as known (fixed), while the outcome (z) is treated as unknown (random). With maximum likelihood estimation, we reverse the roles of the observed values and the distribution parameters by asking: Given a vector of observations **z**, what can we say about $\boldsymbol{\Theta}$? This treats the data **z** as known (fixed), and the parameters $\boldsymbol{\Theta}$ as unknown (random). To specify this alternative interpretation, the density function is denoted as $\ell(\boldsymbol{\Theta} \,|\, \mathbf{z})$, the **likelihood** of $\boldsymbol{\Theta}$ given the observed vector of data **z**. This defines a **likelihood surface**, as $\ell(\boldsymbol{\Theta} \,|\, \mathbf{z})$ assigns a value to each possible point in the $\boldsymbol{\Theta}$-parameter space given the observed data **z**. The **maximum likelihood estimate** (**MLE**) of the unknown parameters, $\widehat{\boldsymbol{\Theta}}$, is the value of $\boldsymbol{\Theta}$ corresponding to the maximum of $\ell(\boldsymbol{\Theta} \,|\, \mathbf{z})$, i.e., the MLE is the value of $\boldsymbol{\Theta}$ that is "most likely" to have produced the data **z**. It is usually easier to find the maximum of a likelihood function by first taking its natural logarithm (ln) and working with the resulting **log-likelihood**

$$L(\boldsymbol{\Theta} \,|\, \mathbf{z}) = \ln \left[\ell(\boldsymbol{\Theta} \,|\, \mathbf{z}) \right] \tag{A4.1}$$

L is also referred to as the **support**. Because the natural log is a monotonic function, $\ell(\boldsymbol{\Theta})$ has the same maxima as $\ln\left[\ell(\boldsymbol{\Theta})\right]$, so that the maximum of L also corresponds to the maximum of the likelihood function. The **score S** of a likelihood function is the first derivative of L with respect to the likelihood parameters, with $S(\theta) = \partial L(\theta)/\partial\theta$ for a single-parameter likelihood function, and

$$\mathbf{S}(\boldsymbol{\Theta}) = \frac{\partial L(\boldsymbol{\Theta})}{\partial \boldsymbol{\Theta}} = \begin{pmatrix} \partial L(\boldsymbol{\Theta})/\partial\theta_1 \\ \vdots \\ \partial L(\boldsymbol{\Theta})/\partial\theta_k \end{pmatrix} \tag{A4.2}$$

for a vector $\boldsymbol{\Theta} = (\theta_1, \cdots, \theta_k)^T$ of k parameters. From elementary calculus it follows that the score evaluated at the MLE (being a maximum) is zero, hence $\mathbf{S}(\widehat{\boldsymbol{\Theta}}) = \mathbf{0}$. This provides one approach for obtaining MLEs.

Example A4.1. Suppose n values, $z_1 \cdots z_n$, are sampled independently from an underlying normal with unknown mean μ and $\sigma^2 = 1$ (unit variance). Letting $\mathbf{z} = (z_1, z_2, \cdots, z_n)^T$, what is the MLE for μ given $\mathbf{z}$? Because the observations are independent, the resulting probability density function for $\mathbf{z}$ is the product of n normal density functions,

$$p(\mathbf{z}, \mu) = \prod_{i=1}^{n} (2\pi)^{-1/2} \exp\left[-(z_i - \mu)^2/2\right]$$
$$= (2\pi)^{-n/2} \exp\left[-\sum_{i=1}^{n}(z_i - \mu)^2/2\right] \tag{A4.3}$$

The log-likelihood (or support) becomes

$$L(\mu \,|\, \mathbf{z}) = \ln\left[\ell(\mu \,|\, \mathbf{z})\right] = -\left(\frac{n}{2}\right)\ln(2\pi) - \frac{1}{2}\sum_{i=1}^{n}(z_i - \mu)^2 \tag{A4.4}$$

which has the score function

$$S(\mu) = \frac{\partial\, L(\mu \,|\, \mathbf{z})}{\partial \mu} = \sum_{i=1}^{n}(\, z_i - \mu\,) = n(\overline{z} - \mu) \tag{A4.5}$$

Setting the score equal to zero and solving gives the MLE as $\widehat{\mu} = \overline{z}$, namely, the sample mean (which is also the method-of-moments estimator).

Large-sample Properties of MLEs

MLEs have several important features when the sample size is *sufficiently large*:

1. **Consistency**: As the sample size increases, the MLE converges to the true parameter value, e.g., $\widehat{\boldsymbol{\Theta}} \to \boldsymbol{\Theta}$.
2. **Invariance**: If $f(\boldsymbol{\Theta})$ is a function of the unknown parameters of the distribution, then the MLE of $f(\boldsymbol{\Theta})$ is $f(\widehat{\boldsymbol{\Theta}})$, i.e., the MLE of a function of the parameters is simply that function evaluated at the MLE. For example, the MLE of $\sqrt{\theta} = (\,\widehat{\theta}\,)^{1/2}$.
3. **Asymptotic normality and efficiency**: As the sample size increases, the sampling distribution of the MLE converges to a normal and (generally) no other estimation procedure has a smaller variance. Hence, for sufficiently large sample sizes, estimates obtained via maximum likelihood typically have the smallest confidence intervals.
4. **Asymptotic (large-sample) Variance**: For sufficiently large sample sizes, the variance of an MLE (assuming a single unknown parameter) is approximately the negative of the reciprocal of the second derivative of the log-likelihood function evaluated at the MLE $\widehat{\theta}$,

$$\sigma^2(\,\widehat{\theta}\,) \simeq -\left(\left.\frac{\partial^2\, L(\theta \,|\, \mathbf{z})}{\partial\, \theta^2}\right|_{\theta=\widehat{\theta}}\right)^{-1} \tag{A4.6}$$

This is just the reciprocal of the curvature of the log-likelihood surface at the MLE. The flatter the likelihood surface around its maximum value (the MLE), the larger the variance; the steeper the surface, the smaller the variance. The minus sign appears because the second derivative is negative (downward curvature) at the maximum of the likelihood function.

Example A4.2. What is the large-sample variance of the MLE for μ from Example A4.1?

$$\frac{\partial^2 L(\mu\,|\,\mathbf{z})}{\partial \mu^2} = \frac{\partial S(\mu\,|\,\mathbf{z})}{\partial \mu} = \frac{\partial \left(\sum_{i=1}^{n} (z_i - \mu) \right)}{\partial \mu} = -n$$

Applying Equation A4.6,

$$\sigma^2\left(\,\widehat{\mu}\,\right) \simeq \frac{1}{n}$$

Using the asymptotic normality of MLEs, the approximate distribution of the MLE is $\widehat{\mu} \sim \mathrm{N}(\mu, n^{-1})$, and the resulting 95% confidence interval for μ is $\widehat{\mu} \pm 1.96/\sqrt{n}$.

The Fisher Information Matrix, F

When estimating a vector of parameters, Equation A4.6 can be generalized by using the **Hessian matrix, H**, the matrix of second partials of the log-likelihood, whose ijth element is given by

$$\mathbf{H}_{ij} = \frac{\partial^2 L(\boldsymbol{\Theta}\,|\,\mathbf{z})}{\partial\, \theta_i\; \partial\, \theta_j} \tag{A4.7a}$$

$\mathbf{H}(\boldsymbol{\Theta}_o)$ refers to the Hessian matrix evaluated at the point $\boldsymbol{\Theta}_o$ and provides a measure of the local curvature of L around that point. The **Fisher information matrix** (**F**), the negative of expected value of the Hessian matrix for L,

$$\mathbf{F}(\boldsymbol{\Theta}) = -E\left[\,\mathbf{H}(\boldsymbol{\Theta})\,\right] \tag{A4.7b}$$

provides a measure of the multidimensional curvature of the log-likelihood surface. Alternately, **F** can be computed as the expected value of the outer product (Equation 9.8b) of the score vector,

$$\mathbf{F}(\boldsymbol{\Theta}) = E\left[\mathbf{S}(\boldsymbol{\Theta})\,\mathbf{S}(\boldsymbol{\Theta})^T\,\right] \tag{A4.7c}$$

The covariance matrix for the MLEs is simply the inverse of the information matrix, with

$$\sigma\left(\widehat{\theta}_i, \widehat{\theta}_j\right) = \left[\mathbf{F}(\boldsymbol{\Theta})^{-1}\right]_{ij} \tag{A4.7d}$$

As in the univariate case, if the likelihood surface is highly curved (very peaked) around the MLE, then the standard errors (being the inverse of the local curvature) are small, while if the likelihood is very flat, the sampling variance is large. A multivariate likelihood surface can have complex geometry, being very steep in some directions and very flat in others, corresponding, respectively, to regions of the parameter space where the data are very informative versus regions where it is rather uninformative. For large sample sizes, **F** is often approximated by the Hessian matrix evaluated at the MLE,

$$\mathbf{F}(\boldsymbol{\Theta}) \simeq -\mathbf{H}(\widehat{\boldsymbol{\Theta}}) \tag{A4.7e}$$

Example A4.3. Suppose n values are sampled independently from a normal with unknown mean and variance. What are the MLEs and their sampling variances? Here $\boldsymbol{\Theta} = (\mu, \sigma)^T$. Noting that $\sum_{i=1}^n (z_i - \mu)^2 = n(\overline{z^2} - 2\,\overline{z}\,\mu + \mu^2)$, the same logic leading to Equation A4.3 shows that the log-likelihood function is

$$L(\mu, \sigma^2 \mid \mathbf{z}) = -\left(\frac{n}{2}\right)\ln(2\pi) - \left(\frac{n}{2}\right)\ln(\sigma^2) - \frac{n\,(\,\overline{z^2} - 2\,\overline{z}\,\mu + \mu^2\,)}{2\,\sigma^2} \tag{A4.8a}$$

Taking derivatives, the score vector becomes

$$\mathbf{S}(\boldsymbol{\Theta}) = \begin{pmatrix} \partial L(\boldsymbol{\Theta})/\partial\mu \\ \partial L(\boldsymbol{\Theta})/\partial\sigma^2 \end{pmatrix} = \left(\frac{n}{\sigma^2}\right)\begin{pmatrix} \overline{z} - \mu \\ \dfrac{\overline{z^2} - 2\,\overline{z}\,\mu + \mu^2}{2\,\sigma^2} - \dfrac{1}{2} \end{pmatrix} \tag{A4.8b}$$

Solving $\mathbf{S}(\widehat{\boldsymbol{\Theta}}) = \mathbf{0}$ gives the MLEs as

$$\widehat{\boldsymbol{\Theta}} = \begin{pmatrix} \widehat{\mu} \\ \widehat{\sigma}^{\,2} \end{pmatrix} = \begin{pmatrix} \overline{z} \\ \overline{z^2} - \overline{z}^{\,2} \end{pmatrix} \tag{A4.8c}$$

As the first step towards computing the Hessian and Fisher matrices, the second partials are found to be

$$\frac{\partial L^2}{(\partial\mu)^2} = -\frac{n}{\sigma^2}, \qquad \frac{\partial L^2}{\partial\mu\,\partial\sigma^2} = -\frac{n(\,\overline{z} - \mu\,)}{\sigma^4} \tag{A4.8d}$$

$$\frac{\partial L^2}{(\partial\sigma^2)^2} = \frac{n}{2\,\sigma^4}\left(1 - \frac{2\,(\,\overline{z^2} - 2\,\overline{z}\,\mu + \mu^2\,)}{\sigma^2}\right) \tag{A4.8e}$$

Because $E(\,\overline{z}\,) = \mu$, the first two derivatives have expected values of $-n/\sigma^2$ and 0, respectively. Likewise, because $E(\,\overline{z^2}\,) = \mu^2 + \sigma^2$, the expected value of Equation A4.8e becomes

$$E\left(\frac{\partial L^2}{(\partial\sigma^2)^2}\right) = \frac{n}{2\,\sigma^4}\left(1 - \frac{2\,(\,\mu^2 + \sigma^2 - 2\,\mu^2 + \mu^2\,)}{\sigma^2}\right) = -\frac{n}{2\,\sigma^4}$$

With the above results, the Fisher matrix becomes

$$\mathbf{F} = -E(\mathbf{H}) = \begin{pmatrix} \dfrac{n}{\sigma^2} & 0 \\ 0 & \dfrac{n}{2\,\sigma^4} \end{pmatrix}$$

Alternatively, evaluating the derivatives at the MLE, $\widehat{\boldsymbol{\Theta}} = (\,\overline{z}, \,\widehat{\sigma}^{\,2})^T$, Equation A4.8d gives values of $-n/\widehat{\sigma}^{\,2}$ and 0, while Equation A4.8e gives $-n/(2\,\widehat{\sigma}^{\,4})$, so that the value of the Hessian matrix evaluated at the MLE becomes

$$\mathbf{H}(\widehat{\boldsymbol{\Theta}}) = -\begin{pmatrix} \dfrac{n}{\widehat{\sigma}^{\,2}} & 0 \\ 0 & \dfrac{n}{2\,\widehat{\sigma}^{\,4}} \end{pmatrix}$$

Applying Equation A4.7d gives the large-sample variances and covariance for the MLEs as

$$\sigma^2(\,\widehat{\mu}\,) = \sigma^2/n \simeq \widehat{\sigma}^{\,2}/n, \quad \sigma^2(\,\widehat{\sigma}^{\,2}) = 2\,\sigma^4/n \simeq 2\,\widehat{\sigma}^{\,4}/n, \quad \sigma(\,\widehat{\mu}, \widehat{\sigma}^{\,2}) = 0$$

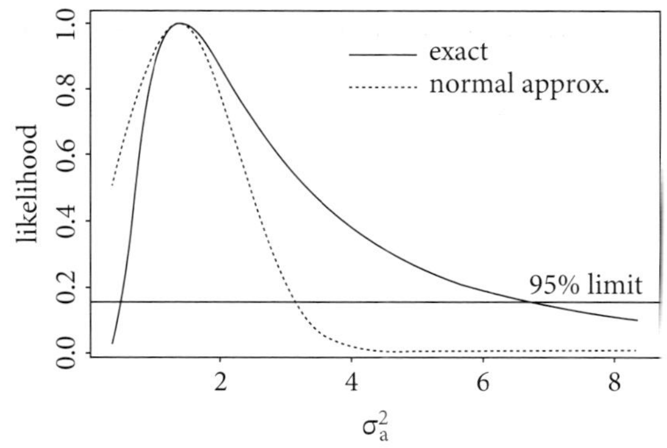

Figure A4.1 A **profile likelihood** (the projection of the likelihood space into a single dimension) for a variance component, showing the 95% confidence space. The likelihood has been rescaled to have a maximal value of one, giving 0.15 as the cutoff for the confidence space (see below). Note that the resulting region is asymmetric about the MLE, while a normality-based interval (dotted line) would be far too narrow. (After Pawitan 2000.)

LIKELIHOOD-RATIO TESTS

Maximum likelihood provides for extremely convenient tests of hypotheses in the form of **likelihood-ratio**, or **LR**, **tests** (reviewed in Chapter 24 of Kendall and Stuart 1979) that examine whether a reduced model provides the same fit as a full model. In particular, we assume a **nested comparison**, where the reduced model is simply the full model with one or more of the parameters assigned set values (such as an effect of zero). The likelihood-ratio test statistic for a nested comparison is given by

$$LR(\widehat{\boldsymbol{\Theta}}_r) = 2\ln\left(\frac{\ell(\widehat{\boldsymbol{\Theta}}\,|\,\mathbf{z})}{\ell(\widehat{\boldsymbol{\Theta}}_r\,|\,\mathbf{z})}\right) = -2\ln\left(\frac{\ell(\widehat{\boldsymbol{\Theta}}_r\,|\,\mathbf{z})}{\ell(\widehat{\boldsymbol{\Theta}}\,|\,\mathbf{z})}\right) = -2\left[L(\widehat{\boldsymbol{\Theta}}_r\,|\,\mathbf{z}) - L(\widehat{\boldsymbol{\Theta}}\,|\,\mathbf{z})\right] \qquad \text{(A4.9a)}$$

where $\ell(\widehat{\boldsymbol{\Theta}}\,|\,\mathbf{z})$ is the likelihood evaluated at the MLE and $\ell(\widehat{\boldsymbol{\Theta}}_r\,|\,\mathbf{z})$ is the maximum of the likelihood function, subject to the restriction that r parameters unconstrained in the full likelihood analysis are assigned fixed values. For sufficiently large sample size, under the null hypothesis, ***the LR test statistic is χ^2_r-distributed***, a χ^2 with r degrees of freedom (Wilks 1938).

Equation A4.9a immediately suggests a method for constructing **LR confidence spaces**. In particular, the $100(1-\alpha)\%$ confidence space for a vector $\boldsymbol{\Theta}_k$ of k parameters is the set of all $(k \times 1)$ vectors $\boldsymbol{\Theta}$ for which $LR(\boldsymbol{\Theta}) \leq X^2_{k,1-\alpha}$, where the critical value satisfies $\Pr(\chi^2_k \leq X^2_{k,1-\alpha}) = 1-\alpha$. Rearranging Equation A4.9a, this condition becomes

$$\ell(\boldsymbol{\Theta}) \geq \exp\left(-X^2_{k,1-\alpha}/2\right) \cdot \ell(\widehat{\boldsymbol{\Theta}}) \qquad \text{(A4.9b)}$$

Namely, the set of all vectors $\boldsymbol{\Theta}$ whose resulting likelihood is at least a fraction

$$\exp(-X^2_{k,1-\alpha}/2) \qquad \text{(A4.9c)}$$

of the likelihood function at its maximum value, $\ell(\widehat{\boldsymbol{\Theta}})$. Equation A4.9b implies that the confidence space can easily be asymmetric about the MLE (Figure A4.1). Further, the space can also be disjoint, as a multimodal likelihood function could have several isolated peaks that satisfy Equation A4.9b. As a result of these features, we refer to a confidence *space* rather than a confidence *interval* (or ellipsoid in multiple dimensions). Because Equation A4.9b relies on the large-sample distribution of the LR test, it is also a large-sample property, and may be inaccurate in small samples and at parameter space boundaries (see below).

For a single variable, the value of Equation A4.9c for a 95% confidence space follows using $X^2_{1,0.95}$ = 3.84, giving $\exp(-3.84/2) = 0.15$, namely all θ whose likelihood function values are at least 15% of the maximal value. In Chapter 18, we introduced the **one LOD confidence interval rule**, which uses a cutoff value of 10% of the MLE value to construct approximate 95% confidence intervals (for a single parameter). Applying Equation A4.9c, $\exp(-X^2/2) = 0.1$, or $X^2 = 4.61$, which for a χ^2_1 corresponds to $\alpha = 0.032$, so that this rule would generate 97% confidence intervals. However, this value is a large-sample property, and (as discussed in Chapter 18) often gives too narrow intervals when the sample size is small.

Example A4.4. Suppose we wish to test the hypothesis that $\mu = 0$ in Example A4.1 (known variance, with $\sigma^2 = 1$). Here the MLE is $\widehat{\mu} = \overline{z}$ and the LR test statistic becomes

$$LR = -2\ln\left(\frac{\ell(0\,|\,\mathbf{z})}{\ell(\widehat{\mu}\,|\,\mathbf{z})}\right) = -2\ln\left(\frac{(2\pi)^{-n/2}\exp(-\sum_{i=1}^{n}(z_i-0)^2/2)}{(2\pi)^{-n/2}\exp(-\sum_{i=1}^{n}(z_i-\overline{z})^2/2)}\right)$$

$$= \sum_{i=1}^{n}\left[\,z_i^2 - (z_i-\overline{z})^2\,\right] = n\,\overline{z}^{\,2}$$

This test statistic is distributed as a χ^2 with one degree of freedom, as one parameter (μ) was assigned a fixed value in the reduced model. Because Prob($\chi^2_1 > 3.84$) = 0.05, the hypothesis $\mu = 0$ is rejected at the 5% level if the LR test statistic ($n\,\overline{z}^{\,2}$) exceeds 3.84.

Now suppose we wish to test this hypothesis under the conditions of Example A4.3, where the variance is also unknown and hence must also be estimated. Here the MLEs for the full model are given by Equation A4.8c. Substituting $\mu = 0$ into Equation A4.8b gives the score function for the restricted model as

$$\left.\frac{\partial L(\sigma^2)}{\partial\sigma^2}\right|_{\mu=0} = \frac{n}{\sigma^2}\left(\frac{\overline{z^2}}{2\,\sigma^2} - \frac{1}{2}\right)$$

giving the MLE for σ^2 under this restriction as $\widehat{\sigma}_r^2 = \overline{z^2}$. Substituting the MLEs into the likelihood functions, and once again using the identity $\sum(z_i-\mu)^2 = n(\,\overline{z^2} - 2\,\overline{z}\,\mu + \mu^2)$ gives the LR test statistic as

$$-2\ln\left(\frac{\ell(\,0,\widehat{\sigma}_r^2\,|\,\mathbf{z}\,)}{\ell(\,\widehat{\mu},\widehat{\sigma}^{\,2}\,|\,\mathbf{z}\,)}\right)$$

$$= -2\ln\left(\frac{(\,\overline{z^2}\,)^{-n/2}\cdot\exp\left[\,-n\,\overline{z^2}\,/\,(\,2\,\overline{z^2}\,)\,\right]}{(\,\overline{z^2} - \overline{z}^{\,2}\,)^{-n/2}\cdot\exp\left[\,-n\,(\,\overline{z^2} - \overline{z}^{\,2})\,/\,2\,(\,\overline{z^2} - \overline{z}^{\,2})\,\right]}\right)$$

$$= n\,\ln\left(\frac{\overline{z^2}}{\overline{z^2} - \overline{z}^{\,2}}\right)$$

Again, for large samples this follows a χ^2_1 distribution as the value of one parameter is assigned a fixed value ($\mu = 0$).

The *G*-test

A common likelihood-ratio based test is the ***G*-test** for goodness of fit, which follows from the multinomial distribution (Equation 2.20a). Consider n observations that have been apportioned into a set of k different categories, denoted by the vector $\mathbf{n} = (n_1, n_2, \cdots, n_k)^T$.

Likewise, let p_i represent the true population frequency of the ith category and let $\mathbf{p} = (p_1, p_2, \cdots, p_k)^T$. From the multinomial distribution, the likelihood of $\mathbf{p}$ given the observations $\mathbf{n}$ is

$$\ell(\mathbf{p} \,|\, \mathbf{n}) = C\, p_1^{n_1}\, p_2^{n_2} \,\cdots\, p_k^{n_k} \tag{A4.10a}$$

where C is the appropriate multinomial coefficient (which is independent of the p_i; Equation 2.20a). It can be shown that the values of p_i that maximize Equation A4.10a (and hence are the MLEs) are $\widehat{p_i} = n_i/n$. This gives the value of the maximum of the likelihood function as

$$\ell(\widehat{\mathbf{p}} \,|\, \mathbf{n}) = C \left(\frac{n_1}{n}\right)^{n_1} \left(\frac{n_2}{n}\right)^{n_2} \cdots \left(\frac{n_k}{n}\right)^{n_k} \tag{A4.10b}$$

In order to test whether the observed data are consistent with a specified vector $\mathbf{q}$ of population frequencies, we need the value of the likelihood function under this constraint. Denoting the expected value for the number of individuals in category i by $\widehat{n}_i = q_i n$, we can write $q_i = \widehat{n}_i/n$. Substitution into Equation A4.10a gives the likelihood under $\mathbf{q}$ as

$$\ell(\mathbf{q} \,|\, \mathbf{n}) = C \left(\frac{\widehat{n}_1}{n}\right)^{n_1} \left(\frac{\widehat{n}_2}{n}\right)^{n_2} \cdots \left(\frac{\widehat{n}_k}{n}\right)^{n_k} \tag{A4.10c}$$

Applying Equation A4.9a yields the likelihood-ratio test (in this case, it is also called the G-test, for goodness of fit) that the observed data are consistent with $\mathbf{q}$,

$$G = -2\ln\left(\frac{\ell(\mathbf{q} \,|\, \mathbf{n})}{\ell(\widehat{\mathbf{p}} \,|\, \mathbf{n})}\right) = -2\sum_{i=1}^{k} n_i \ln\left(\frac{\widehat{n}_i}{n_i}\right) = -2\sum_{i=1}^{k} n_i \ln\left(\frac{q_i}{\widehat{p}_i}\right) \tag{A4.11}$$

Because the k frequencies sum to one, there are $k-1$ unconstrained parameters in the full likelihood, implying that G is asymptotically distributed as a χ^2_{k-1} random variable. Because large sample sizes are required to give the likelihood-ratio test a χ^2 distribution, caution should be exercised in employing this test whenever any expected quantity is less than five (e.g., any $q_i < 5/n$), a problem that can sometimes be avoided by pooling categories.

Likelihood-ratio Tests for the General Linear Model

As a final example of likelihood-ratio tests, consider the general linear model (Chapter 10), $\mathbf{y} = \mathbf{X}\boldsymbol{\beta} + \mathbf{e}$, where we assume that the $n \times 1$ vector of residual errors $\mathbf{e}$ is multivariate normal, with mean vector zero and covariance matrix $\mathbf{V}$, i.e., $\mathbf{e} \sim \text{MVN}(\mathbf{0}, \mathbf{V})$. From Equation 9.24, the density function for $\mathbf{e}$ is

$$(2\pi)^{-n/2} |\,\mathbf{V}\,|^{-1/2} \exp\left(-\frac{1}{2}\mathbf{e}^T\mathbf{V}^{-1}\mathbf{e}\right)$$

Writing the vector of residuals as $\mathbf{e} = \mathbf{y} - \mathbf{X}\boldsymbol{\beta}$ gives the resulting likelihood for $\boldsymbol{\beta}$ and $\mathbf{V}$, conditional on the observed data ($\mathbf{y}$, $\mathbf{X}$), as

$$\ell(\boldsymbol{\beta}, \mathbf{V} \,|\, \mathbf{y}, \mathbf{X}) = (2\pi)^{-n/2} |\,\mathbf{V}\,|^{-1/2} \exp\left(-\frac{1}{2}(\mathbf{y} - \mathbf{X}\boldsymbol{\beta})^T\mathbf{V}^{-1}(\mathbf{y} - \mathbf{X}\boldsymbol{\beta})\right)$$

which has log-likelihood

$$L = \ln \ell = -\frac{n}{2}\ln(2\pi) - \frac{1}{2}\ln |\,\mathbf{V}\,| - \frac{1}{2}(\mathbf{y} - \mathbf{X}\boldsymbol{\beta})^T\mathbf{V}^{-1}(\mathbf{y} - \mathbf{X}\boldsymbol{\beta}) \tag{A4.12}$$

Here $\boldsymbol{\beta}$ is a vector of fixed effects and the matrix $\mathbf{V}$ is a function of k variance components, with $\mathbf{V} = \sum_{i=1}^{k} \mathbf{R}_i\, \sigma_i^2$, where the $\mathbf{R}_i$ are matrices of known constants. Thus, the parameters to be estimated are the vector $\boldsymbol{\beta}$ of fixed effects and the k variances, σ_i^2.

Suppose we wish to compare the relative fit of two models that assume the same covariance structure (i.e., the same **V**), but have different vectors of fixed effects, a vector $\boldsymbol{\beta}_f$ for the full model versus a vector $\boldsymbol{\beta}_r$ for the reduced model that assumes fewer factors (some elements in $\boldsymbol{\beta}_f$ are set to specific values). The likelihood-ratio test statistic becomes

$$LR = -2\left[L(\widehat{\boldsymbol{\beta}}_r \,|\, \mathbf{y}, \mathbf{X}_r) - L(\widehat{\boldsymbol{\beta}}_f \,|\, \mathbf{y}, \mathbf{X}_f) \right]$$
$$= \left[(\mathbf{y} - \widehat{\mathbf{y}}_r)^T \widehat{\mathbf{V}}^{-1} (\mathbf{y} - \widehat{\mathbf{y}}_r) - (\mathbf{y} - \widehat{\mathbf{y}}_f)^T \widehat{\mathbf{V}}^{-1} (\mathbf{y} - \widehat{\mathbf{y}}_f) \right] \qquad \text{(A4.13a)}$$

where $\widehat{\mathbf{y}}_f = \mathbf{X}_f\widehat{\boldsymbol{\beta}}_f$ and $\widehat{\mathbf{y}}_r = \mathbf{X}_r\widehat{\boldsymbol{\beta}}_r$ are the predicted means under the full and reduced models, so that equation A4.13a is just the difference between two weighted sums of squares. For large sample sizes, this test statistic follows a χ^2 distribution with $n_f - n_r$ degrees of freedom, where n_f and n_r are the degrees of freedom for the full and reduced models, respectively. If the reduced model also has a different covariance structure, $\mathbf{V}_r = \sum_{i=1}^{r<k} \mathbf{R}_i\, \sigma_i^2$ (one or more of the σ_i^2 in the full-model **V** set to zero), then $\mathbf{V}_r$ replaces **V** in the left quadratic product in Equation A4.13a, with the degrees of freedom adjusted accordingly.

Example A4.5. Suppose the y_i values are the means of n different populations, e.g., data from a series of populations used in a line-cross analysis (Chapter 11). Assuming the means are independent but with potentially different variances (due to differences in sample sizes, among other things), **V** is a diagonal matrix whose ith element is the variance of the ith mean, $\text{Var}(y_i)$. Recalling Equation A3.11b, the quadratic product in the LR test reduces to

$$(\mathbf{y} - \widehat{\mathbf{y}})^T\, \widehat{\mathbf{V}}^{-1} (\mathbf{y} - \widehat{\mathbf{y}}) = \sum_{i=1}^{n} \frac{(y_i - \widehat{y}_i)^2}{\text{Var}(y_i)}$$

Hence, the likelihood-ratio test statistic for comparing the full model with a reduced model assuming fewer effects is given by

$$\sum_{i=1}^{n} \frac{[\, y_i - \widehat{y}_i(r)\,]^2}{\text{Var}(y_i)} - \sum_{i=1}^{n} \frac{[\, y_i - \widehat{y}_i(f)\,]^2}{\text{Var}(y_i)} \qquad \text{(A4.13b)}$$

which is just the difference in the χ^2 values for the fit of the full and reduced models. This test follows a χ^2 distribution with degrees of freedom given by the difference in degrees of freedom between the full and reduced models.

Wald and Score Tests

In addition to the likelihood ratio (LR) test, two other testing approaches are used in an ML analysis (Figure A4.2). In all cases we are testing whether a vector, $\widehat{\boldsymbol{\Theta}}$, of MLE solutions deviates significantly from some $(k \times 1)$ vector of null hypothesis values, $\boldsymbol{\Theta}_0$. In what follows, the vector of MLE solutions, $\widehat{\boldsymbol{\Theta}}$, is also of length k, and consists of only those parameter values we are testing (parameter values unrestricted under the null are ignored). The **Wald test** (Wald 1943) is given by

$$T_W = (\boldsymbol{\Theta}_0 - \widehat{\boldsymbol{\Theta}})^T \mathbf{F}_{k\times k}^{-1} (\boldsymbol{\Theta}_0 - \widehat{\boldsymbol{\Theta}}) \qquad \text{(A4.14a)}$$

where $\mathbf{F}_{k\times k}^{-1}$ denotes the inverse of the Fisher matrix (Equation A4.7b) evaluated at the null (θ_0). Asymptotically, T_W follows a χ_k^2 distribution. Note by comparison to Equation A3.9 that this test is an approximation based on assuming multivariate normality under the null hypothesis.

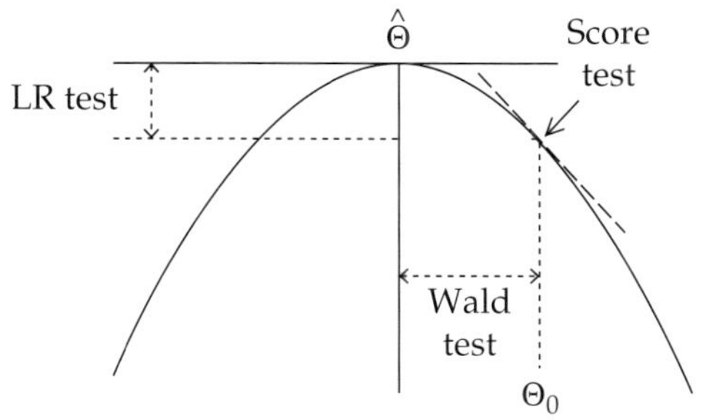

Figure A4.2 A geometric view of the likelihood ratio (LR), Wald, and score tests for comparing the null, Θ_0, against the MLE, $\widehat{\Theta}$. The LR test is the difference in the heights of the likelihood function, the Wald test is the (scaled) distance between the MLE and null, while the score test is the (scaled) derivative of the likelihood function at the null.

For a single parameter, Wald reduces to a standard z test, with

$$\frac{\theta_0 - \widehat{\theta}}{\sigma(\theta)} \sim N(0,1), \quad \text{or equivalently}, \quad T_W = \frac{(\theta_0 - \widehat{\theta})^2}{\sigma^2(\theta)} \sim \chi^2_1, \tag{A4.14b}$$

where $\sigma(\theta)$ follows from the square root of the corresponding element in $\mathbf{H}$. Equation A4.14b follows from A4.14a upon recalling that a squared unit normal follows a χ^2_1 distribution (Appendix 5). Finally, **Wald confidence regions** follow from Equation A4.14a, as the set of all vectors $\boldsymbol{\Theta}$ that satisfy

$$(\boldsymbol{\Theta} - \widehat{\boldsymbol{\Theta}})^T \mathbf{F}^{-1}_{k\times k} (\boldsymbol{\Theta} - \widehat{\boldsymbol{\Theta}}) \leq X^2_{k,1-\alpha}$$

where (as above) the critical value satisfies $\Pr(\chi^2_k \leq X^2_{k,1-\alpha}) = 1 - \alpha$. This quadratic product yields ellipsoidal regions centered around the ML. While Wald intervals are usually easier to compute than LR-based intervals (Equation A4.9b), they can be misleading if the likelihood is not approximately normal around the MLE (Figure A4.1). A nice discussion of Wald versus LR confidence regions is given by Meeker and Escobar (1995), who suggest that, whenever possible, LR be used, especially when sample size is modest. They also recommend examining the profile plots (Figure A4.1) to assess whether the likelihood is roughly normal or shows skew and/or kurtosis, which would compromise the Wald approximation.

The **score test** (Rao 1948), also known as the **Lagrange multiplier test**, is based on the value of the derivative of the likelihood function (the score, S; Equation A4.2) when evaluated at the null parameter value, $\boldsymbol{\Theta}_0$. At $\widehat{\boldsymbol{\Theta}}$ this derivative is zero, so that its value at the null provides a measure of departure from the MLE (Figure A4.2). The resulting test statistic is given by

$$T_S = \mathbf{S}(\boldsymbol{\Theta}_0)^T \mathbf{F}^{-1}_{k\times k} \mathbf{S}(\boldsymbol{\Theta}_0) \tag{A4.14c}$$

where $\mathbf{S}(\boldsymbol{\Theta}_0)$ is the score function evaluated at the null hypothesis value. Asymptotically, T_S also follows a χ^2_k distribution.

Both the score and Wald test assume that the log-likelihood surface L around the MLE is well approximated by a quadratic, which is equivalent to assuming that the likelihood surface ℓ is approximately normally distributed around the MLE, as $\ell \sim \exp[-(\theta - \widehat{\theta})^2/2\sigma^2]$ implies that $L \sim (\theta - \widehat{\theta})^2/2\sigma^2$. The LR test requires evaluation of the likelihood function at both the null and MLE values, while the Wald test requires only the MLE, and the score test does not even require computing the MLE. For sufficiently large sample sizes, these tests are essentially equivalent as the likelihood approaches normality (Cox and Hinkley 1979), while for finite sample size, they are generally ordered as $T_W \geq LR \geq T_S$.

Why these different tests? The short answer is computation. A full LR test (Equation A4.9a) requires obtaining the MLE $\widehat{\Theta}$ and then evaluating two likelihoods (at $\widehat{\Theta}$ and at the null Θ_0). The Wald test (Equation A4.14a) still requires the MLE, but does not require evaluating the likelihood function. Finally, the score test (Equation A4.14b) does not require the MLE, and only evaluates a single likelihood function (through its score, or first derivative) at the null. These three different approaches offer flexibility in developing computationally efficient tests.

Example A4.6. As an example of applying all of the above concepts, suppose we assume a binomial model with success parameter θ. Here

$$\ell(\theta \mid n, k) = \binom{k}{n} \theta^k (1-\theta)^{n-k}$$

$$L(\theta \mid n, k) = \ln(\binom{k}{n}) + k\ln(\theta) + (n-k)\ln(1-\theta)$$

$$S(\theta) = \frac{\partial L(\theta \mid k, n)}{\partial \theta} = \frac{k}{\theta} - \frac{n-k}{1-\theta}, \qquad H(\theta) = \frac{\partial^2 L(\theta \mid k, n)}{\partial \theta^2} = -\frac{k}{\theta^2} - \frac{n-k}{(1-\theta)^2}$$

Solving for $S(\widehat{\theta}) = 0$ yields the MLE as $\widehat{\theta} = k/n$. Substituting into Equation A4.6 gives the large-sample variance as

$$\sigma^2(\widehat{\theta}) = \left.\frac{-1}{H(\theta)}\right|_{\theta=\widehat{\theta}} = -\left(-\frac{k}{(k/n)^2} - \frac{n-k}{(1-k/n)^2}\right)^{-1} = \frac{\widehat{\theta}(1-\widehat{\theta})}{n}$$

Finally, for the Fisher information (Equation A4.7b), because $\mathrm{E}[k] = n\theta$, we have

$$F[\theta] = -E[H(\theta)] = n\left[\frac{1}{\theta} + \frac{1}{1-\theta}\right] = \frac{n}{\theta(1-\theta)}$$

With these results in hand, we now consider asymptotic and LR confidence intervals, and LR, Wald, and score tests.

Suppose that in our sample of $n = 100$, we observed $k = 17$ successes and assume a null of $\theta_0 = 0.4$. Hence, we have $\widehat{\theta} = 0.17$ with standard error of $\sqrt{0.17 \cdot 0.83/100} = 0.0376$. The resulting 95% asymptotic confidence interval (assuming normality) becomes $0.17 \pm 1.965 \cdot 0.0376$, or 0.0962 to 0.2438. From Equation A4.9a, the LR test against $\theta = \theta_0$ becomes

$$\begin{aligned} LR(\theta_0) &= -2\left[\,L(\theta_0 \mid k, n) - L(\widehat{\theta} \mid k, n)\right] \\ &= -2\left\{k\ln(\theta_0) + (n-k)\ln(1-\theta_0) - k\ln\left(\frac{k}{n}\right) - (n-k)\ln\left(\frac{n-k}{n}\right)\right\} \\ &= -2\left\{k\left[\ln(\theta_0) - \ln\left(\frac{k}{n}\right)\right] + (n-k)\left[\ln(1-\theta_0) - \ln\left(\frac{n-k}{n}\right)\right]\right\} \end{aligned}$$

For $k = 17$, $n = 100$, and $\theta_0 = 0.4$, the LR becomes

$$LR(0.4) = -2\left\{17\left[\ln(0.4) - \ln(0.17)\right] + 83\left[\ln(0.6) - \ln(0.83)\right]\right\} = 24.77$$

for a significance of $\Pr(\chi_1^2 > 24.77) = 6.45 \cdot 10^{-7}$. The LR-based 95% confidence interval (CI) is that set of all θ_r such that $LR(\theta_r) \leq 3.841$, as $\Pr(\chi_1^2 > 3.841) = 0.05$. The resulting interval becomes 0.105 to 0.252. Note the slight asymmetry of the CI around the MLE (0.17), which can be seen in the likelihood profile plot,

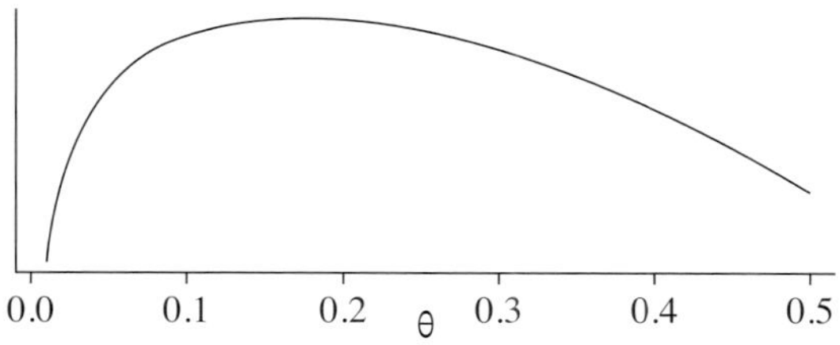

Now consider the Wald test. Equation A4.14b gives this test against the null ($\theta = \theta_0$) as,

$$T_W = n\left[\frac{(0.17-\theta_0)^2}{\theta_0(1-\theta_0)}\right] \sim \chi_1^2$$

For our case of $n = 100$ and $\theta_0 = 0.4$,

$$T_W = 100\left[\frac{(0.17-0.4)^2}{0.4\cdot 0.6}\right] = 22.042$$

for a significance level of $\Pr(\chi_1^2 \geq 22.042) = 2.67 \cdot 10^{-6}$. The 95% Wald confidence interval sets are all those values of θ_r that satisfy

$$100\left(\frac{(\theta_r-0.17)^2}{\theta_r(1-\theta_r)}\right) \leq X^2_{1,0.05} = 3.841$$

yielding the interval as 0.109 to 0.256.

Finally, we turn to the score test. Equation A4.14c gives the score test statistic (against θ_0) as

$$T_S = S(\theta_0)^2 F(\theta_0)^{-1} = \left(\frac{k}{\theta_0} - \frac{n-k}{1-\theta_0}\right)^2\left(\frac{\theta_0(1-\theta_0)}{n}\right)$$

For our setting ($n = 100, k = 14, \theta_0 = 0.4$), this becomes

$$T_S = \left(\frac{17}{0.4} - \frac{83}{0.6}\right)^2\left(\frac{0.4\cdot 0.6}{100}\right) = 21.282$$

for a significance level of $\Pr(\chi_1^2 \geq 21.282) = 3.964 \cdot 10^{-6}$. Computing the 95% CI by those values of θ_r satisfying $S(\theta_r) \leq 3.841$ yields 0.110 to 0.259.

Dealing With Boundaries in the Likelihood Space

When one or more boundaries of the admissible parameter space are finite, the MLE solution may occur at a boundary. For example, the likelihood function for a variance component or a weight in a mixture distribution may take on its maximal value at zero. Likewise, we may wish to test a null value that resides on a boundary, such as a variance being zero. Other examples of boundary nulls arising in quantitative genetics are tests for linkage (where the null is $c = 1/2$) and testing for genotype by environment interactions (where the null is $\rho = 1$, namely a perfect correlation between genotypes over environments; Part II Chapters 27 and 28).

Finite boundaries in the parameter space cause no estimation issues, although the score (first derivative of the log likelihood) can be nonzero at the MLE if the latter is on a boundary. For example, suppose that likelihood for a variance is strictly decreasing away from zero, so that the MLE is at zero, but the derivative of the likelihood function at the MLE is negative. The more subtle issue with a finite space is that adjustments for LR tests are required when the null is assumed to be on the boundary. In this setting, the large-sample distribution of the LR test is no longer χ^2, but rather is a weighted mixture of χ^2 distributions (Chernoff 1954; Moran 1971; Chant 1974; Hopper and Mathews 1982; Self and Liang 1987; Shaw 1987; Stram and Lee 1994; Dominicus et al. 2006; Visscher 2006; Molenberghs and Verbeke 2007; Meyer 2008; Susko 2013).

The simplest setting is testing whether a single variance component is zero, in which case the large-sample distribution of LR test is a mixture of a χ_0^2 (a probability mass at zero) and a χ_1^2, namely $LR \sim (1/2)\chi_0^2 + (1/2)\chi_1^2$ (Chernoff 1954; Self and Liang 1987; Stram and Lee 1994). Because the unconstrained (i.e., the null is assumed to be in the interior of the parameter space) LR test for a single parameter has χ_1^2 distribution, failing to make this adjustment results in tests that are **conservative**: the p values they return are too large. For a test of level $1 - \alpha$, the critical value for this mixture is the $1 - 2\alpha$ critical value for the χ_1^2, so that for tests of α values of 5, 1, 0.5, and 0.1%, the critical values become, respectively, 2.71, 5.41, 6.63, and 9.55 for the mixture distribution, rather than the comparable values of 3.84. 6.63, 7.88, and 10.82 for a χ_1^2.

When conducting a joint test that two of the variance components are zero, the resulting LR distribution is $(1/4)\chi_0^2 + (1/2)\chi_1^2 + (1/4)\chi_2^2$ under many designs (Dominicus et al. 2006; Visscher 2006), but the actual mixture weights can differ. The general theory of computing the mixture weights is given by Self and Liang (1987), but such solutions are often challenging to obtain. The bottom line is that when one is testing a null hypothesis where one or more of the assumed null values are on a boundary, then the p value returned from an unconstrained test is likely too large and thus the corresponding critical values are too large (resulting in lower power). However, some caution is in order with this general statement, as Kopylev and Sinha (2011) noted that there are examples when the resulting distribution is *not* a mixture of χ^2 distribution, and that assuming the unrestricted χ^2 value can be **anticonservative**, resulting in p values that are too small.

Comparing Models That are Not Nested

The likelihood ratio test is ***only valid for comparing nested models***, where the reduced model is a *subset of the full model* obtained by assigning specific values to some of the parameters in the full model. **Non-nested comparisons** can also arise, such as contrasting two different reduced models of the full model. Suppose the full model has three variance components (σ_1^2, σ_2^2, and σ_3^2), and we wish to compare model 1 with only σ_1^2 being nonzero and model 2 with only σ_3^2 nonzero. While both are nested within the *full* model, contrasting these two models with each other is a non-nested comparison.

Such comparisons introduce the broader issue of **model selection** (Burnham and Anderson 2002). While a formal statistical framework occurs for nested comparisons (LR tests), there is an *informal* framework for comparing models that are not nested. Here, various informal statistics are used to compare models, and we focus on two, the AIC and BIC metrics. For both of these metrics, a smaller value means a better model. We stress that while both metrics are fully grounded in theoretical principles (Burnham and Anderson 2004), ***comparing values for two different models is largely ad hoc in that there is no formal test for significance*** (i.e., there is no formal criterion for determining when one model is clearly better than the others). Further, the different metrics can result in different model rankings.

The idea behind both metrics (as in a likelihood-ratio test) is reward for goodness of fit (i.e., smaller values of $-2\ln[L]$ imply better fits), but also to penalize for the number of fitted model parameters, k. This is a natural extension of the adjusted r^2 (Equations A3.15b and A3.15c), which downweights goodness of fit (r^2) by the number of fitted parameters. One of the most widely used model-comparison metrics is the **Akaike information criterion** (Akaike 1973),

$$\text{AIC} = -2\ln(L) + 2k \qquad \text{(A4.15a)}$$

which was adjusted for the sample size, n, by Sugiura (1978),

$$\text{AIC}_c = -2\ln(L) + 2k + \frac{2k(k+1)}{n-k-1} = -2\ln(L) + \frac{2kn}{n-k-1} \tag{A4.15b}$$

AIC_c should be used in place of AIC unless $n/k > 40$ (Burnham and Anderson 2004). Model selection proceeds by computing the AIC values for all of our candidate models, and then choosing the model with the smallest AIC value as the "best" model in the comparison test.

The other widely used metric is the **Bayes information criterion**,

$$\text{BIC} = -2\ln(L) + \ln(n)k \tag{A4.15c}$$

which was introduced by Schwarz (1978), and thus is also known as the **Schwarz criterion**. While AIC and BIC are often used interchangeably, they are actually designed for slightly different purposes. When one of the models being compared is the true model, then BIC picks this model with a probability approaching one in large samples. Conversely, AIC considers the situation where *none* of the candidate models may be correct and then tries to pick among the best of these. Best practice is to typically present both AIC and BIC values, especially if they result in different model rankings. A nice short review of various other model selection criteria can be found in Grueber et al. (2011b).

Quantifying the Variation Explained by a Mixed Model

As first introduced in Chapter 3, coefficient of determination (Equation A3.15)

$$r^2 = \frac{\text{SS}_\text{M}}{\text{SS}_\text{T}} = 1 - \frac{\text{SS}_\text{E}}{\text{SS}_\text{T}}$$

quantifies the amount of total variation in y that is accounted for by a GLM, by contrasting the model residual variance, $\sigma^2(e_i) = \sigma^2(y_i - \widehat{y}_i)$, with the total variance in y (the spread of y around $\overline{y}$). Hence, r^2 quantifies the amount of variation accounted for by the *fixed effects*. However, with a linear *mixed* model, variation around the predicted fixed effects includes *both* residual variation *and* additional variation from other, specified, random effects. In this setting, r^2 does not account for any variation accommodated by the inclusion of specific random effects, and hence does not fully capture the relative fit of a specific model. As reviewed by Sun et al. (2010), a number of extensions for r^2 to mixed models have been proposed. They recommend a likelihood-ratio based extension,

$$r^2_{LR} = 1 - \exp\left(-\frac{2}{n}\left[\ln(L_M) - \ln(L_0)\right]\right) \tag{A4.16}$$

where L_M is the likelihood under the assumed mixed model. Note that this is the *full*, not the REML (Part II Chapter 33), likelihood, as it considers *both* fixed and random effects. L_0 is the likelihood under the simplest model (a single fixed effect, the mean, with residuals as the only random effect). As with the fixed-effects version, Equation A4.16 is the fraction of total variation accounted for by the proposed model, reducing to r^2 in the absence of any random effects (beyond the residual). Note that r^2_{LR} is *not* a model-selection statistic, but rather can provide insight into the quality of the total model fit as well as the importance of a particular model parameter (contrast the values of r^2_{LR} with, and without, this factor).

ITERATIVE METHODS FOR SOLVING ML EQUATIONS

While ML estimation and hypothesis testing with likelihood ratios are conceptually straightforward, in practice it can be quite difficult to accomplish due to the complexities associated with finding the maximum of the likelihood function. Ideally, closed-form solutions for the MLEs can be obtained by deriving the score vector, setting it equal to zero, and solving. However, in many cases this is impractical and numerical approaches must be used. In very simple cases with one or two parameters, a brute force approach relying upon a **grid search** can be used, where one computes a one- or two-dimensional plot of the likelihood surface as a function of the unknown parameters. With more than two variables, this is impractical and a variety of iterative methods have been suggested as alternatives. We discuss two of these here, Newton–Raphson and EM methods (Part II Chapter 33 discusses these methods further in the context of variance-component estimation). Appendix 8 examines another iterative approach, searching via MCMC methods. A potential problem with all iterative methods is that they may not converge to the true MLEs if the likelihood surface contains several local maxima. Further, they can be slow to converge if the likelihood surface is flat. Iterative methods require an initial starting value, and a poor choice can result in the iteration converging to a solution that is a local, but not a global, maximum. Hence, when applying iterative methods, several starting points should be used.

Newton–Raphson Methods

Recall from elementary calculus that one can approximate a function $f(x)$ by expanding it in a power series around a point x_o,

$$f(x) \simeq f(x_0) + (x - x_o)f'(x_o)$$

This suggests one approach for finding roots of the equation $f(x) = 0$. Given some initial guess x_0, an improved value is obtained by solving

$$f(x) = 0 \simeq f(x_0) + (x - x_o)f'(x_o)$$

for x, or

$$x \simeq x_0 - \frac{f(x_0)}{f'(x_0)} \tag{A4.17a}$$

Noting that the score function is zero at the MLE [$S(\widehat{\theta}) = 0$], this suggests one approach for obtaining an iterative solution of the MLE. Applying Equation A4.17a to the score function, so that $f = S$ and $f' = \partial S(\theta)/\partial\theta = \partial L^2(\theta)/\partial^2\theta$, an updated estimate, $\widehat{\theta}^{(k+1)}$, of a current estimate $\widehat{\theta}^{(k)}$ is given by

$$\widehat{\theta}^{(k+1)} = \widehat{\theta}^{(k)} - \left(\left.\frac{\partial L^2(\theta)}{\partial^2\theta}\right|_{\theta=\widehat{\theta}^{(k)}} \right)^{-1} S\big[\widehat{\theta}^{(k)}\big] \tag{A4.17b}$$

which is iterated until $|\,\widehat{\theta}^{(k+1)} - \widehat{\theta}^{(k)}\,|$ is sufficiently small to declare convergence. This is the **Newton–Raphson** method, a member of a class of **quadratic methods**. Such methods involve second partial derivatives of the likelihood function and have a quadratic convergence rate.

The same logic when applied to a multivariate Taylor series (WL Equation A6.7a) implies that a vector $\widehat{\boldsymbol{\Theta}}^{(k)}$ of current estimates is updated by using

$$\widehat{\boldsymbol{\Theta}}^{(k+1)} = \widehat{\boldsymbol{\Theta}}^{(k)} - \mathbf{H}^{-1}\big(\widehat{\boldsymbol{\Theta}}^{(k)}\big)\,\mathbf{S}\big(\widehat{\boldsymbol{\Theta}}^{(k)}\big) \tag{A4.18}$$

where **S** and **H** are the vector of scores and the Hessian matrix, respectively, both evaluated at the current estimate.

One variant of this approach is **Fisher's scoring**, where the Hessian matrix **H** is replaced by its expected value, the negative of Fisher's information matrix **F** (Equation A4.7b),

$$\widehat{\boldsymbol{\Theta}}^{(k+1)} = \widehat{\boldsymbol{\Theta}}^{(k)} + \mathbf{F}^{-1}\big(\widehat{\boldsymbol{\Theta}}^{(k)}\big)\,\mathbf{S}\big(\widehat{\boldsymbol{\Theta}}^{(k)}\big) \tag{A4.19}$$

One advantage of Fisher's scoring is that **F** is usually of a simpler form than **H**, often containing elements equal to zero that are nonzero in **H**. This can make **F** easier to compute and invert (Part II Chapter 33). Further, Fisher's scoring appears to be more robust to poor initial starting choices than the strict Newton–Raphson method (Jennrich and Sampson 1976). In addition to the advantage of quadratic convergence, both Newton–Raphson and Fisher's scoring yield the covariance matrix of MLE estimates from **H** (or **F**) using the final iteration values of $\boldsymbol{\Theta}$ and applying Equation A4.7. Additional quadratic methods are discussed by Kennedy and Gentle (1980).

Expectation-maximization (EM) Methods

Newton–Raphson and related methods require the first and second derivatives of the likelihood function, which can be difficult to obtain and/or computationally demanding (e.g., requiring the repeated inversion of large matrices). An alternative strategy is to use **expectation-maximization (EM) methods**, which were introduced by Dempster et al. (1977) as a very general iterative approach for data sets with missing (or incomplete) data. The idea is that, in many cases, if we had more information about certain observations, MLEs are easily obtained. For example, if observations are drawn from a mixture distribution (Chapter 16), obtaining the MLEs for the means and variances of the underlying distributions is trivial *provided* we know from which distribution each individual observation is drawn. Thus the original data set is treated as incomplete data, missing additional information (e.g., for a mixture model, which distribution a specific observation is drawn from). Using a current estimate of the unknown parameter values, the expected value of the incomplete data is computed (e.g., for a mixture model, the category identity of each individual is estimated). This is the **expectation**, or **E step**. The result is a set of likelihood equations that are considerably easier to solve (the **maximization**, or **M step**) than the full likelihood. The new estimates obtained from the M step are then used to update the expected values, and this approach is iterated until convergence. The EM method refers to a general class of approaches, and there can be several EM versions for solving the same problem.

While EM methods often have fairly simple forms and hence are easy to program, they can be extremely slow to converge to a solution. Balancing this concern, EM methods offer computational advantages over Newton–Raphson methods, as they do not require second derivatives of the likelihood function and they do not directly evaluate the full likelihood function. However, this is a disadvantage in terms of constructing confidence intervals and LR tests, as other approaches must be used to obtain the standard errors of the MLEs and to compute the likelihoods needed for LR tests. Chapter 33 discusses an EM method for computing unknown variance components in linear models, while our focus here is on another broad class of likelihood models used throughout this book, mixture models (introduced in Chapter 16).

EM for Mixture Model Likelihoods

Mixture models naturally arise in a number of quantitative-genetic settings, wherein the observed distribution is a weighted sum of a number of underlying distributions. For example, when a major diallelic locus is segregating in a population, the phenotypic distribution is a weighted sum of the three distributions representing each major locus genotype (Chapter 16). The general likelihood function for a single observation z from the mixture models considered in this book has the form

$$\ell(\boldsymbol{\Theta} \mid z) = \sum_{k=1}^{N} \pi_k \cdot \varphi(z, \mu_k, \sigma^2) \tag{A4.20a}$$

where the distribution is assumed to result from N underlying normals, the kth of which has frequency π_k, mean μ_k, and common variance σ^2, where the normal pdfs (Equation 2.10) are denoted by $\varphi(z, \mu_k, \sigma^2)$. We assume that the number N of underlying distributions is known and wish to estimate the $2N \times 1$ vector $\boldsymbol{\Theta}$ of the N means, the common variance, and the $N - 1$ independent mixing proportions (the π_k). With n individuals independently drawn from this distribution, the full likelihood is

$$\ell(\boldsymbol{\Theta} \mid \mathbf{z}) = \prod_{i=1}^{n} \ell(\boldsymbol{\Theta} \mid z_i) \tag{A4.20b}$$

While appearing rather simple, the full likelihood function is complicated to work with analytically, and numerical approaches are usually employed.

When we observe a particular value, we do not know which underlying distribution (or **category**) it was drawn from. If we knew the **category identity** for each observation, the ML solutions for the mean and variance of the underlying distributions are easily computed. For example, if a single diallelic QTL is segregating, if we could determine whether individuals had QTL genotype *QQ*, *Qq*, or *qq*, then the mean for each genotype and the common variance could easily be estimated. This is the basis of the EM method. We start with some initial guess as to the category identity of each observation, which then allows us to easily compute an ML estimate of the means and variance of the underlying distribution. This guess is in the form of a weight vector for each individual, whose k element, $w(k \mid z)$, is the probability that an individual has the kth QTL genotype given they have trait value z. Using current mean and variance estimates for each category, updated weights can be computed using Bayes' theorem (Equation 3.3b). Because $w(k \mid z) = \Pr(k \mid z)$, applying Bayes' theorem gives

$$w(k \mid z) = \frac{\Pr(k) \cdot \Pr(z \mid k)}{\Pr(z)} = \frac{\pi_k \cdot \varphi(z, \mu_k, \sigma^2)}{\Pr(z)} = \frac{\pi_k \cdot \varphi(z, \mu_k, \sigma^2)}{\sum_{j=1}^{N} \pi_j \cdot \varphi(z, \mu_j, \sigma^2)} \tag{A4.21}$$

These updated weights are then used to obtain new estimates of the category-specific means and the variance, and this procedure is repeated until convergence. Formally, this EM approach proceeds as follows (Aitkin and Wilson 1980):

(1) **Initial step**. Choose initial starting values for the MLEs of the variance $\widehat{\sigma}^{2\,(0)}$ and the vectors of mixture proportions and means,

$$\widehat{\boldsymbol{\pi}}^{(0)} = (\widehat{\pi}_1^{(0)}, \cdots, \widehat{\pi}_N^{(0)}), \qquad \widehat{\boldsymbol{\mu}}^{(0)} = (\widehat{\mu}_1^{(0)}, \cdots, \widehat{\mu}_N^{(0)}) \tag{A4.22}$$

One simple option is to take $\widehat{\pi}_i^{(0)} = 1/N$ and $\widehat{\mu}_i^{(0)} = \overline{z}$, the latter being the population mean.

(2) **E-step**. Define the weight $w^{(1)}(k \mid z_i)$ as the probability that observation z_i is drawn from distribution k given the current estimates of $\widehat{\sigma}^{2\,(0)}$, $\widehat{\boldsymbol{\pi}}^{(0)}$, and $\widehat{\boldsymbol{\mu}}^{(0)}$. From Bayes' theorem, the updated weights become

$$w^{(1)}(k \mid z_i) = \frac{\widehat{\pi}_k^{(0)} \cdot \varphi(z_i, \widehat{\mu}_k^{(0)}, \widehat{\sigma}^{2\,(0)})}{\sum_{j=1}^{N} \widehat{\pi}_j^{(0)} \cdot \varphi(z_i, \widehat{\mu}_j^{(0)}, \widehat{\sigma}^{2\,(0)})} \tag{A4.23}$$

where $\varphi(z_i, \widehat{\mu}_k^{(0)}, \widehat{\sigma}^{2\,(0)})$ is the normal distribution (Equation 2.10) evaluated at the value z_i using mean $\widehat{\mu}_k^{(0)}$ and variance $\widehat{\sigma}^{2\,(0)}$.

(3) **M-step**. Assuming these weights are correct, the updated estimates of the MLEs are obtained as follows:

(a) **Mixing proportions**: Given by the average probability of being in category k,

$$\widehat{\pi}_k^{(1)} = \overline{w}_k^{(1)} = \frac{1}{n} \sum_{i=1}^{n} w^{(1)}(k \mid z_i) \tag{A4.24a}$$

(b) **Means**: Given by the weighted average of the observations,

$$\widehat{\mu}_k^{(1)} = \frac{1}{n} \sum_{i=1}^{n} z_i \left(\frac{w^{(1)}(k \mid z_i)}{\overline{w}_k^{(1)}} \right) \tag{A4.24b}$$

(c) **Variance**: Given by the weighted variance of the observations,

$$\widehat{\sigma}^{2\,(1)} = \frac{1}{n} \sum_{i=1}^{n} \sum_{k=1}^{N} \left(z_i - \widehat{\mu}_k^{(1)} \right)^2 w^{(1)}(k \mid z_i) \tag{A4.24c}$$

These updated estimates are then used to compute new weights, and the whole procedure continues until the iterations converge.

EM Modifications for QTL Mapping

One important application of mixture models involves the use of marker data to map QTLs (Chapters 18 and 19). Here estimates of the category identity are influenced not only by an individual's trait value z but also by their marker genotype m. For example, suppose a single diallelic QTL (with alleles Q and q) is linked to a marker, and an inbred-line cross is used in an attempt to map and characterize this QTL. In this case, the likelihood for an individual with marker genotype m and trait value z is

$$\ell(\boldsymbol{\Theta} \mid z, m) = \sum_{k=1}^{3} \pi_k(m) \cdot \varphi(z, \mu_k, \sigma^2) \tag{A4.25}$$

where

$$\pi_k(m) = \Pr(Q_k \mid m) \qquad \text{and} \qquad \boldsymbol{\Theta} = (\mu_{QQ}, \mu_{Qq}, \mu_{qq}, \sigma^2)$$

where Q_k denotes the kth QTL genotype. The mixing proportions $\pi_k(m)$ are functions of the marker genotype, the QTL position relative to the marker (generally given by the marker–QTL recombination frequency distance c), and the particular design used (Example A4.7).

Support for the presence of a QTL is usually displayed using **likelihood maps**, also called **profile plots** (Chapters 18–19), which plot the maximum value of the likelihood function over all possible values of c, with the c value giving the largest value being taken as the MLE for QTL position. For a given value of c, say c_0, the EM method is used to obtain the MLE for $\boldsymbol{\Theta}$ under the restriction that a QTL is at map position c_0. Here, the weights are again given by Bayes' theorem (Equation A4.23), where the weight for an individual with trait value z and marker genotype m assuming QTL position (c_0) is

$$\Pr(Q_k \mid m, z, c_0) = \frac{\Pr(Q_k \mid m, c_0) \cdot \Pr(z \mid Q_k)}{\Pr(z \mid m, c_0)}$$

$$= \frac{\Pr(Q_k \mid m, c_0) \cdot \varphi(z, \mu_{Q_k}, \sigma^2)}{\sum_{j=1}^{N} \Pr(Q_j \mid m, c_0) \cdot \varphi(z, \mu_{Q_j}, \sigma^2)} \tag{A4.26}$$

Using these weights, updated estimates of the means and variance are obtained as above, and these are substituted back into Equation A4.26 to obtain new weights. This procedure is continued until the iterations converge. Ranging through all possible values of c and plotting the resulting maximum of the likelihood function generates a likelihood map for QTL position. A general EM treatment for QTL mapping is given by Kao and Zeng (1997).

Example A4.7. As an example of accounting for missing marker information, consider QTL mapping in an F_2 design from an inbred-line cross (Chapter 18) using dominant markers (such as RAPDs; Chapter 8). Suppose marker allele M is dominant to allele m, so that the observed marker genotypes are mm and $M-$, the latter consisting of the genotypes MM and Mm. In the F_2, Pr($M-$) = Pr(MM) + Pr(Mm) = 3/4,

$$\Pr(MM \mid M-) = \frac{\Pr(MM)}{\Pr(M-)} = \frac{(1/4)}{(3/4)} = 1/3$$

and likewise $\Pr(Mm \mid M-) = 2/3$. The conditional probability $\Pr(QQ \mid M-)$ that the QTL genotype is QQ given the marker genotype is $M-$ becomes

$$\begin{aligned}\Pr(QQ \mid M-) &= \Pr(QQ \mid MM) \Pr(MM \mid M-) \\ &\quad + \Pr(QQ \mid Mm) \Pr(Mm \mid M-) \\ &= \frac{1}{3} \Pr(QQ \mid MM) + \frac{2}{3} \Pr(QQ \mid Mm)\end{aligned}$$

For the F_2 design (Example 18.1), $\Pr(QQ \mid MM, c) = (1-c)^2$ and $\Pr(QQ \mid Mm, c) = c(1-c)$, giving

$$\Pr(QQ \mid M-, c) = \frac{1}{3}(1-c)^2 + \frac{2}{3} c(1-c) = \frac{1-c^2}{3}$$

Similarly, it can be shown that

$$\Pr(Qq \mid M-, c) = \frac{2(1-c+c^2)}{3} \quad \text{and} \quad \Pr(qq \mid M-, c) = \frac{c(2-c)}{3}$$

Thus, for a given c value, these mixing proportions are fixed constants. Similar logic gives the values for individuals with marker genotype mm. To obtain the weights for the EM method, first index the three QTL genotypes by $k = 1, 2, 3$ for QQ, Qq, and qq, respectively. Given a current estimate of the three QTL means $\widehat{\mu}_k^{(t)}$ and the common variance $\widehat{\sigma}^{2\,(t)}$, the updated weights are obtained from Equation A4.26. For example, the updated weight that an individual is genotype QQ given it has trait value z and marker genotype $M-$ is

$$w^{(t+1)}(1 \mid z, M-, c) = \frac{\Pr(QQ \mid M-, c) \cdot \Pr(z \mid QQ)}{\Pr(z \mid M-, c)}$$

$$\begin{aligned}&= \Pr(QQ \mid M-, c) \cdot \Pr(z \mid QQ) \Big/ \Big[\Pr(QQ \mid M-, c) \cdot \Pr(z \mid QQ) \\ &\quad + \Pr(Qq \mid M-, c) \cdot \Pr(z \mid Qq) + \Pr(qq \mid M-, c) \cdot \Pr(z \mid qq) \Big] \\ &= (1-c)^2 \cdot \varphi(z, \widehat{\mu}_1^{(t)}, \widehat{\sigma}^{2\,(t)}) \Big/ \Big[(1-c)^2 \cdot \varphi(z, \widehat{\mu}_1^{(t)}, \widehat{\sigma}^{2\,(t)}) \\ &\quad + 2(1-c+c^2) \cdot \varphi(z, \widehat{\mu}_2^{(t)}, \widehat{\sigma}^{2\,(t)}) + c(2-c) \cdot \varphi(z, \widehat{\mu}_3^{(t)}, \widehat{\sigma}^{2\,(t)}) \Big]\end{aligned}$$

The probabilities of the two other QTL genotypes follow similarly, as do the weights for individuals with marker genotype mm. Using these updated weights, new estimates of the means and variance are obtained from Equations A4.24b and A4.24c.

ADDITIONAL APPLICATIONS

Estimating Relatedness Metrics Using Marker Data

In Chapter 8, we detailed how ML methods can be used, in conjunction with marker data, to *assign* individuals into prespecified **relationship categories**, such parent–offspring or full–sibs. We also considered a number of method-of-moments approaches to estimate continuous **relatedness metrics**, such as the coefficient of coancestry, Θ. As was briefly mentioned in

Table A4.1 Conditional probabilities for constructing the likelihood function for estimating Jacquard's condensed coefficients of identity (Δ_1 to Δ_9; Chapter 7) from marker data. AIS is the alike-in-state status of the marker locus being contrasted over the two focal individuals. (After a more comprehensive table in Milligan 2003.)

AIS State	Genotypes	$\Pr(S_i \mid \Delta_7)$	$\Pr(S_i \mid \Delta_8)$	$\Pr(S_i \mid \Delta_9)$
S_1	A_iA_i, A_iA_i	p_i^2	p_i^3	p_i^4
S_2	A_iA_i, A_jA_j	0	0	$p_i^2p_j^2$
S_3	A_iA_i, A_iA_j	0	$p_i^2p_j$	$2p_i^3p_j$
S_4	A_iA_i, A_jA_k	0	0	$2p_i^2p_jp_k$
S_5	A_iA_j, A_iA_i	0	$p_i^2p_j$	$2p_i^3p_j$
S_6	A_jA_k, A_iA_i	0	0	$2p_i^2p_jp_p$
S_7	A_iA_j, A_iA_j	$2p_ip_j$	$p_ip_j(p_i+p_j)$	$4p_i^2p_j^2$
S_8	A_iA_j, A_iA_k	0	$p_ip_jp_k$	$4p_i^2p_jp_k$
S_9	A_iA_j, A_kA_ℓ	0	0	$4p_ip_jp_kp_\ell$

Chapter 8, likelihood approaches have also been proposed for estimating these relatedness metrics (Thompson 1975, 1986a; Milligan 2003; Anderson and Weir 2007; Choi et al. 2009; Ackerman et al. 2017). ML allows one to estimate a much broader class of relationships, namely Jacquard's condensed coefficients of identity (Δ_1 to Δ_9; Chapter 7). Using the ML invariance property, substituting the MLEs of the Δ_i into Equation 7.2a gives the MLE for Θ, as well as other composite relationship metrics based on the Δ_i (Ackerman et al. 2017).

We build the likelihood function for $\Delta = (\Delta_1, \cdots, \Delta_9)^T$ as follows (Milligan 2003). Consider a single locus where p_i denotes the frequency of allele A_i, and let S_i correspond to the AIS (marker-alleles alike-in-state) version of Δ_i seen in the pair of individuals being considered. For example, S_1 corresponds to the marker pair (A_iA_i, A_iA_i). Table A4.1 gives the marker genotypes for the other S_i states as well as the conditional probability, $\Pr(S_i \mid \Delta_j)$, of observing AIS state S_i given that the true IBD state is Δ_j. Table 1 of Milligan presents these conditional probabilities for all values of Δ_i, while our table is restricted to Δ_7 to Δ_9, the only potential nonzero values in a random-mating population (Chapter 7).

The resulting likelihood function for the kth marker locus for a pair of individuals that show AIS status S_i is

$$\ell_k(\Delta) = \Pr(S_{k,i} \mid \Delta) = \sum_j \Pr(S_{k,i} \mid \Delta_j)\, \Delta_j \tag{A4.27a}$$

where $S_{k,i}$ denotes that marker locus k has AIS state i. The total likelihood for n independent marker loci (i.e., in linkage equilibrium) is the product of A4.27a over each locus, giving

$$\ell = \prod_{k=1}^{n} \ell_k(\Delta) = \prod_{k=1}^{n} \left(\sum_j \Pr(S_{k,i} \mid \Delta_j)\, \Delta_j \right) \tag{A4.27b}$$

The parameters to maximize in Equation A4.27b are the Δ_j, subject to the simplex constraints $\Delta_j \geq 0$ and $\sum \Delta_j = 1$. One performs this maximization separately for each pair of individuals to obtain MLEs for their Δ_i values. Thus for a sample of n individuals, one performs a series of $[n(n-1)/2]$ ML analyses. For each such analysis, one can also conduct LR ratio tests (e.g., testing full versus half sibs) and obtain approximate (large sample) variances and confidence intervals. As noted in Chapter 8, one interpretation of θ is as deviations from a population mean, in which case it can take on negative values, which (under an ML analysis) is an inadmissible result, as it falls outside of the normal Δ simplex space. Konovalov and Heg (2008) present likelihoods that can take on negative relatedness values.

Example A4.8. As an application of Equation A4.27b, consider six biallelic SNP markers, where we code the alleles as 0 or 1, so that 00 denotes a homozygous 0 genotype. Further, let p_i be the frequency of allele 1 for marker i. Consider the marker genotypes of two individuals given in the table below, which also gives the p_i values for the sampled population.

Marker Locus	1	2	3	4	5	6
Individual 1	00	10	11	10	11	00
Individual 2	00	11	11	10	10	00
AIS State	S_1	S_5	S_1	S_7	S_3	S_1
p (Freq allele 1)	0.1	0.2	0.5	0.3	0.1	0.2

Consider the likelihood for this pair from marker 1. From Equation A4.27a,

$$\begin{aligned}\ell_1 &= \Pr(S_1 \mid \Delta_7)\Delta_7 + \Pr(S_1 \mid \Delta_8)\Delta_8 + \Pr(S_1 \mid \Delta_9)(1 - \Delta_7 - \Delta_8)\\ &= p_1^2\Delta_7 + p_1^3\Delta_8 + p_1^4(1 - \Delta_7 - \Delta_8)\\ &= 0.01\Delta_7 + 0.001\Delta_8 + 0.0001(1 - \Delta_7 - \Delta_8)\end{aligned}$$

where we have replaced Δ_9 by $1 - \Delta_7 - \Delta_8$ (because they must sum to one). The remaining five locus-specific likelihood functions become

$$\begin{aligned}\ell_2 &= 0.2^2 \cdot 0.8\,\Delta_8 + 2 \cdot 0.2^3 \cdot 0.8(1 - \Delta_7 - \Delta_8)\\ \ell_3 &= 0.5^2\Delta_7 + 0.5^3\Delta_8 + 0.5^4(1 - \Delta_7 - \Delta_8)\\ \ell_4 &= 2 \cdot 0.3 \cdot 0.7\Delta_7 + 0.3 \cdot 0.7\Delta_8 + 4 \cdot 0.3^2 \cdot 0.7^2(1 - \Delta_7 - \Delta_8)\\ \ell_5 &= 0.1^2 \cdot 0.9\,\Delta_8 + 2 \cdot 0.1^3 \cdot 0.9(1 - \Delta_7 - \Delta_8)\\ \ell_6 &= 0.2^2\Delta_7 + 0.2^3\Delta_8 + 0.2^4(1 - \Delta_7 - \Delta_8)\end{aligned}$$

where $\ell = \prod_{i=1}^{6} \ell_i$. The MLEs for Δ_7 to Δ_9 for this pair are obtained by numerically solving for the values that maximize ℓ. Applying Equation 7.2a, we can translate the estimates of Δ_i into an estimate of $\widehat{\theta} = \widehat{\Delta}_7/2 + \widehat{\Delta}_8/4$.

Solving The MQM Likelihood Equations

Likelihood-based QTL mapping is extensively discussed in Chapters 18 and 19. While the likelihood functions for models with multiple QTLs can be rather complicated, Jansen (1992, 1993a, 1994a; Jansen and Stam 1994) introduced a very simple regression-based EM approach for maximizing the likelihood equations. Our treatment follows that of Jansen and Stam (1994). Suppose there are n_Q diallelic QTLs influencing the trait of interest and let the genotypes $\mathbf{q}_i\mathbf{q}_i$, $\mathbf{Q}_i\mathbf{q}_i$, and $\mathbf{Q}_i\mathbf{Q}_i$ at the ith QTL have contributions $-a_i$, d_i, and a_i. Assuming no epistasis, the phenotypic value of an individual can be written as the regression

$$z = m + \sum_{i=1}^{n_Q} a_i \cdot x_{ai} + \sum_{i=1}^{n_Q} d_i \cdot x_{di} + e \tag{A4.28a}$$

where e is a random environmental effect and the indicator variables are

$$x_{ai} = \begin{cases} -1 & \text{for } \mathbf{q}_i\mathbf{q}_i \\ 0 & \text{for } \mathbf{Q}_i\mathbf{q}_i \\ 1 & \text{for } \mathbf{Q}_i\mathbf{Q}_i \end{cases} \quad \text{and} \quad x_{di} = \begin{cases} 0 & \text{for } \mathbf{q}_i\mathbf{q}_i \\ 1 & \text{for } \mathbf{Q}_i\mathbf{q}_i \\ 0 & \text{for } \mathbf{Q}_i\mathbf{Q}_i \end{cases} \tag{A4.28b}$$

Thus, if we knew the genotypes at each underlying QTL for the individuals in our sample, we could estimate the a_i and d_i by standard regression methods. Provided that the residual

e is normally distributed with constant variance σ^2 (i.e., the conditional distribution of phenotypic values for each QTL genotype is normal with a common variance σ^2), these regression estimates are also ML estimates. Of course, because we do not know the QTL genotypes for any individual, we cannot directly compute this regression. However, we could treat these QTL genotypes as missing (or incomplete) information that can be inferred from an individual's observed marker genotype and trait value z.

This is accomplished as follows. Instead of using the true values of the indicator variables x_{ai} and x_{di} (which are unknown, as these depend on an individual's QTL genotype), we replace these by the *probability* that an individual has that particular QTL genotype given its observed trait and marker values (these are computed using Equation A4.26). As a simple example, suppose the model assumes only a single QTL, so that in an F_2 design there are three possible QTL genotypes for each marker genotype. Indexing these by $j = 1, 2, 3$ (corresponding to **QQ**, **Qq**, and **qq**, respectively), the single value of (x_{aj}, x_{dj}) for each individual in Equation A4.28b is replaced by three values (corresponding to the three possible genotypes). Each value is weighted by the probability of the corresponding QTL genotype given an individual's observed marker and trait values, so that the weight associated with individual i are

$$(x_{aj}, x_{dj}) = \begin{cases} (1,0) & \text{with probability } W(i,1) = \Pr(\mathbf{QQ}\,|\,\mathbf{m}_i, z_i) \\ (0,1) & \text{with probability } W(i,2) = \Pr(\mathbf{Qq}\,|\,\mathbf{m}_i, z_i) \\ (-1,0) & \text{with probability } W(i,3) = \Pr(\mathbf{qq}\,|\,\mathbf{m}_i, z_i) \end{cases} \tag{A4.29}$$

where $\mathbf{m}$ denotes the vector of observed marker genotypes. Thus the (OLS) regression Equation A4.28a based on each individual's QTL genotype is replaced by a weighted (GLS) regression (Chapter 10) based on an individual's observed marker genotypes and trait value. In the GLS regression each observation is replicated three times (one for each genotype), with each replicate value having the appropriate row in the design matrix $\mathbf{X}$ and weight given by Equation A4.23, yielding

$$\mathbf{z}_{3n\times 1} = \begin{pmatrix} z_1 \\ z_1 \\ z_1 \\ z_2 \\ \vdots \\ z_n \end{pmatrix}, \quad \boldsymbol{\beta} = \begin{pmatrix} m \\ a \\ d \end{pmatrix}, \quad \mathbf{X}_{3n\times 3} = \begin{pmatrix} 1 & 1 & 0 \\ 1 & 0 & 1 \\ 1 & -1 & 0 \\ \vdots & \vdots & \vdots \\ 1 & -1 & 0 \end{pmatrix}$$

with the weight matrix given by

$$\mathbf{W}_{3n\times 3n} = \begin{pmatrix} W(1,1) & 0 & 0 & 0 & \cdots & 0 \\ 0 & W(1,2) & 0 & 0 & \cdots & 0 \\ 0 & 0 & W(1,3) & 0 & \cdots & 0 \\ 0 & 0 & 0 & W(2,1) & \cdots & 0 \\ \vdots & \vdots & \vdots & \vdots & \ddots & \vdots \\ 0 & 0 & 0 & 0 & & W(n,3) \end{pmatrix}$$

If the distribution of phenotypic values, conditioned on QTL genotype, is normal with constant variance σ^2, then provided the weights are correct, the ML estimates of the QTL effects m, a, d, σ^2 are given by the GLS estimators

$$\widehat{\boldsymbol{\beta}} = \left(\mathbf{X}^T\mathbf{W}^{-1}\mathbf{X}\right)^{-1}\mathbf{X}^T\mathbf{W}^{-1}\mathbf{z} \tag{A4.30a}$$

$$\widehat{\sigma}^2 = \frac{1}{n}\left(\mathbf{z} - \mathbf{X}\widehat{\boldsymbol{\beta}}\right)^T \mathbf{W}^{-1}\left(\mathbf{z} - \mathbf{X}\widehat{\boldsymbol{\beta}}\right) \tag{A4.30b}$$

Of course, the initial guess for the weights (**W**) depends on an initial guess of QTL effects ($\boldsymbol{\beta}, \sigma^2$). Updated estimates of effects (the M, or maximization, step) are then used to compute new weights (the E, or expectation, step), which are then used to update **W** to compute a new regression and this procedure is repeated until convergence. As with the Haley–Knott and marker-regressions approaches (Chapter 18), one fixes the QTL position and uses the EM procedure to obtain the estimates of $\boldsymbol{\beta}$ and σ^2 given this position. Each potential QTL position is examined, with the optimal value taken as the ML solution for the position.

More generally, when the model assumes k QTL genotypes, each value is replicated k times (with n_Q biallelic QTLs in an F_2 design, $k = 3^{n_Q}$). With n measured individuals and a model with n_Q QTLs, for a model assuming no epistasis and an F_2 design, there are $2n_Q + 1$ regression parameters (in addition to the variance σ^2), and the dimensionalities of **z**, $\boldsymbol{\beta}$, **X**, and **W** become $n \cdot 3^{n_Q} \times 1$, $(2 \cdot n_Q + 1) \times 1$, $n \cdot 3^{n_Q} \times (2 \cdot n_Q + 1)$, and $n \cdot 3^{n_Q} \times n \cdot 3^{n_Q}$. As above, for each iteration the position of all n_Q QTLs is fixed and Equation A4.30 is used to obtain updated estimates which are then substituted into Equation A4.29 to obtain updated weights for the next iteration. When the solution converges, new values for the QTL positions are tried, with the set of positions giving the maximal value chosen as the ML estimates of position.

Appendix 5

Computing the Power of Statistical Tests

If your experiment needs statistics, you ought to have done a better experiment. Ernest Rutherford

There are two types of errors that one can make when performing a statistical test. A **false positive** (a **type I error**) occurs when the **null hypothesis** is rejected when in fact it is correct. We control for this by setting the **significance level** α of a test (the probability of a false positive) to be small. The other source of error is a **false negative** (a **type II error**), *failing* to reject the null hypothesis when in fact it is false. The **power** of a test is defined to be the probability that the null hypothesis is rejected when it is indeed false. Hence if β is the probability of a false negative, the power is $1 - \beta$ (Figure A5.1).

Before embarking on a potentially very costly experiment, the investigator would like to be certain that the design ensures sufficiently high power given the objectives of the proposed study. Indeed, many experiments that report negative results may in fact have significant biological effects that are swamped out by the high sampling error generated by insufficient sample size. Power depends not only on sample size, experimental design structure (Appendix 9), and the actual values of the unknown distribution parameters being estimated, but also on the assumed level of significance α of the test. Here we consider some of the basics of computing the power of a given design for tests whose statistics are Gaussian (univariate and multivariate normal), χ^2, and F-distributed. These distributions commonly appear in estimation statistics in quantitative genetics. Given that most quantitative geneticists use (at least in some part) the `R` statistical package, we provide the basic `R` commands for many of these power calculations.

POWER OF NORMALLY DISTRIBUTED TEST STATISTICS

Assume the test statistic T is normally distributed under the null hypothesis, with $T \sim N(\mu_0, \sigma_0^2)$. For power calculations, assume the null hypothesis is false and that the test statistic actually has mean μ_1 and variance σ_1^2 (but remains normally distributed), so that $T \sim N(\mu_1, \sigma_1^2)$. To obtain expressions for the required sample size for a given power, it is convenient to first write the sample variance as a function of sample size n, with $\sigma_i^2 = f_i^2/n$ for $i = 0, 1$. (Often $\sigma_0^2 = \sigma_1^2$, in which case $f_0 = f_1$.) Finally, let $z_{(\alpha)}$ satisfy

$$\Pr(U \leq z_{(\alpha)}) = \alpha \tag{A5.1a}$$

where U is a unit normal random variable. Both $z_{(\alpha)}$ given α, and α given z, are easily obtained in `R`, as, respectively, α = `pnorm(z)` and $z_{(\alpha)}$ = `qnorm(`α`)`, for **p**robability and quantile of a (unit) **norm**al. For example, $\Pr(U \leq 1.65) = 0.951$, so that $z_{(0.95)} = 1.65$, and (rounding to two decimal places), `pnorm(1.65)` returns 0.95 and `qnorm(0.95)` returns 1.65.

Two identities involving z_α will prove useful in the following discussion. First, note from Figure A5.2 that

$$\Pr(U > z_{(1-\alpha)}) = \alpha \tag{A5.1b}$$

so that, for example, $\Pr(U > 1.65) = 0.05$. Second, from the symmetry of the normal distribution it can be shown that

$$z_{(\alpha)} = -z_{(1-\alpha)} \tag{A5.1c}$$

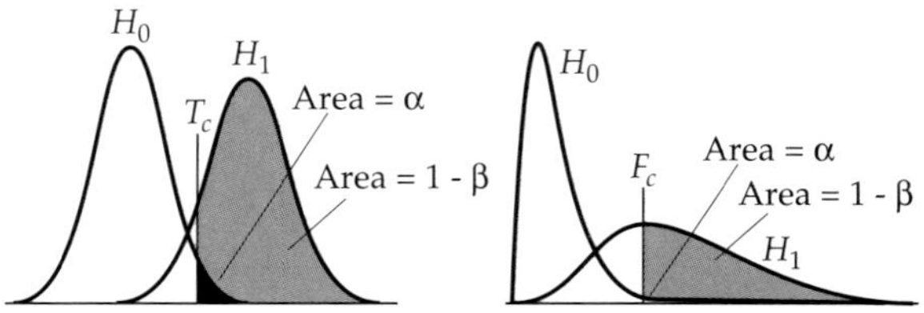

Figure A5.1 Power (1−β) and significance (α) of a normally (**left**) and F-distributed (one-sided) test statistics (**right**). The distributions of the test statistic under the null hypothesis (H_0) and at a particular parameter value under the alternative hypothesis (H_1) are shown. Under the null hypothesis, the probability that the test statistic exceeds a critical value (T_c, F_c) is α, and we define this as the significance of the test (dark shaded area under the H_0 curve). The power of the test for a particular value of the alternative hypothesis is the probability that the test statistic exceeds the critical value (gray shaded area under the H_1 curve), and we denote this by $1-\beta$. Hence, β is the probability that the test statistic is below the critical value (a false negative; a type II error). For these one-sided tests, as the distribution under H_1 shifts further to the right, the power increases (β becomes smaller).

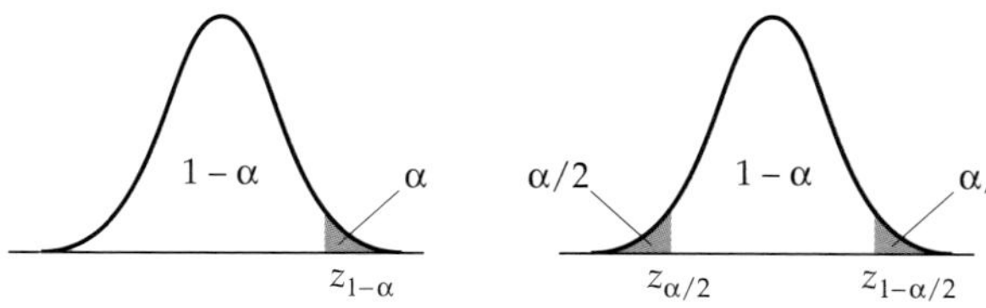

Figure A5.2 Areas under the normal curve for one-sided (left) and two-sided (right) tests of significance at level α.

We now have all the necessary definitions in hand to consider the power of normally distributed tests. Hypothesis tests generally fall into two categories: **one-sided** and **two-sided** (Figure A5.2), and we examine these in turn before considering specific applications.

One-sided Tests

Some hypotheses are naturally one-sided. For example, we may wish to test whether heritability h^2 is significantly different from zero. In this case, true values of $h^2 < 0$ cannot occur, so that the **alternative to the null hypothesis** $h^2 = 0$ is that $h^2 > 0$.

Consider the null hypothesis that the test statistic has mean $\mu = \mu_o$ versus the alternative hypothesis that $\mu = \mu_1$, where $\mu_1 > \mu_0$. Suppose that under the null hypothesis, the test statistic $T \sim N(\mu_o, \sigma_o^2)$, so that $(T - \mu_o)/\sigma_o$ is distributed as a unit normal, implying from Equation A5.1b that

$$\Pr\left(\frac{T-\mu_o}{\sigma_o} > z_{(1-\alpha)}\right) = \Pr\left(T > \mu_o + \sigma_o\, z_{(1-\alpha)}\right) = \alpha \tag{A5.2a}$$

Thus, comparing the observed statistic T with the α-level critical value, $T_c(\alpha)$, where

$$T_c(\alpha) = \mu_o + \sigma_o\, z_{(1-\alpha)} \tag{A5.2b}$$

gives the one-sided test of significance with probability α of a false positive. We **reject the null hypothesis** when $T > T_c(\alpha)$.

Suppose that the **alternative hypothesis** is correct, so that $T \sim N(\mu_1, \sigma_1^2)$. What is the probability that the resulting test statistic is **significant**, namely that it exceeds T_c (and hence a false negative is avoided)? Because it is now $(T - \mu_1)/\sigma_1$ that follows a unit normal, the probability that the test statistic exceeds the critical value $T_c(\alpha)$ given by Equation A5.2b is

$$\Pr\left(T > T_c(\alpha)\right) = \Pr\left(\frac{T-\mu_1}{\sigma_1} > \frac{T_c(\alpha)-\mu_1}{\sigma_1}\right)$$

$$= \Pr\left(U > \frac{\mu_o-\mu_1}{\sigma_1} + z_{(1-\alpha)}\frac{\sigma_o}{\sigma_1}\right) \tag{A5.3a}$$

where U is a unit normal random variable. Noting that $\Pr(U > z_{(\beta)}) = 1-\beta$ (Equation A5.1b), from Equation A5.3a it immediately follows that for the test to have power $1-\beta$ requires

$$\frac{\mu_o-\mu_1}{\sigma_1} + z_{(1-\alpha)}\frac{\sigma_o}{\sigma_1} \le z_{(\beta)} \tag{A5.3b}$$

Using our notation for writing the variance of the test statistic as a function f divided by the sample size n (i.e., $\sigma_i^2 = f_i^2/n$), this becomes

$$\sqrt{n}\,\frac{(\mu_o-\mu_1)}{f_1} + z_{(1-\alpha)}\frac{f_o}{f_1} \le z_{(\beta)} \tag{A5.3c}$$

Because $-z_{(\beta)} = z_{(1-\beta)}$ (Equation A5.1c), this rearranges to give

$$n \ge \frac{f_1^2}{(\mu_1-\mu_o)^2}\left(z_{(1-\alpha)}\frac{f_o}{f_1} + z_{(1-\beta)}\right)^2 \tag{A5.4a}$$

$$= \left(\frac{z_{(1-\beta)}\,f_1 + z_{(1-\alpha)}\,f_0}{\mu_1-\mu_0}\right)^2 \tag{A5.4b}$$

$$\simeq \frac{f_1^2}{(\mu_1-\mu_o)^2}\left(z_{(1-\alpha)} + z_{(1-\beta)}\right)^2 \tag{A5.4c}$$

Equation A5.4c holds when the sample variances under the null hypothesis and true parameter values are approximately equal ($f_0 \simeq f_1$), as is often the case.

Example A5.1. Suppose the null is a normal with mean zero ($\mu_0 = 0$) and variance 4 ($f_o = 2$, hence $\sigma_o^2 = 4/n$). For a 95% one-sided test, Equation A5.2b gives the critical value for sample size n as

$$T_c(0.95) = \mu_o + \sigma_o\,z_{(1-\alpha)} = 0 + 2\cdot 1.65/\sqrt{n} = 3.3/\sqrt{n}$$

for values of 1.04, 0.467, and 0.33 for n of 10, 50, and 100, respectively.

Suppose that the true mean is $\mu_1 = 1$ with variance 4 ($f_1 = 2$). Substituting into Equation A5.3c yields

$$z_{(\beta)} = \sqrt{n}\,\frac{(\mu_o-\mu_1)}{f_1} + z_{(1-\alpha)}\frac{f_o}{f_1} = -\sqrt{n}/2 + 1.65$$

with the power given by the probability that a random draw from H_1 exceeds the critical value, $\Pr(U > z_{(\beta)}) = 1 - \Pr[U \le (-\sqrt{n}/2 + 1.65)] = 1 -$ `pnorm(-sqrt(n)/2 + 1.65)`, which returns values of 0.47, 0.97, and 0.9996, respectively, for n of 10, 50, and 100.

Finally, from Equation A5.4c, the sample size required for 90% power ($\beta = 0.1$, with $z_{(1-\beta)} = z_{0.9} = 1.28$) in a one-sided test with $\alpha = 0.05$ and $\mu_1 = 1$ is

$$n = \frac{f_1^2}{(\mu_1-\mu_o)^2}\left(z_{(1-\alpha)} + z_{(1-\beta)}\right)^2 = 4\cdot(1.65+1.28)^2 = 34.3$$

Two-sided Tests

Other hypotheses are naturally **two-sided**. Here, the null hypothesis is $\mu = \mu_o$ versus the alternative of $\mu \neq \mu_o$. A glance at Figure A5.2 shows that under the null hypothesis,

$$\Pr\left(z_{(\alpha/2)} \leq \frac{T - \mu_o}{\sigma_o} \leq z_{(1-\alpha/2)} \right) = 1 - \alpha \tag{A5.5a}$$

Again recalling Equation A5.1c, this rearranges to become

$$\Pr\left(-\sigma_o\, z_{(1-\alpha/2)} \leq T - \mu_o \leq \sigma_o\, z_{(1-\alpha/2)} \right) = 1 - \alpha$$

giving

$$|\, T - \mu_o \,| > \sigma_o\, z_{(1-\alpha/2)} \tag{A5.5b}$$

as a two-sided test of significance with probability α of a false positive under the null hypothesis. Under the alternative hypothesis that $T \sim N(\mu_1, \sigma_1^2)$, the power of this test is

$$\Pr[\, T - \mu_0 < -\sigma_o\, z_{(1-\alpha/2)} \,] + \Pr[\, T - \mu_0 > \sigma_o\, z_{(1-\alpha/2)} \,]$$

Because $(T - \mu_1)/\sigma_1 \sim U$, this is equivalent to

$$\Pr\left(U < \frac{\mu_o - \mu_1}{\sigma_1} - \frac{\sigma_o}{\sigma_1}\, z_{(1-\alpha/2)} \right) + \Pr\left(U > \frac{\mu_o - \mu_1}{\sigma_1} + \frac{\sigma_o}{\sigma_1}\, z_{(1-\alpha/2)} \right) \tag{A5.6}$$

Using the same logic leading to Equations A5.4a–A5.4c, the sample size n required to give power $1 - \beta$ is found to be

$$n = \frac{f_1^2}{(\mu_o - \mu_1)^2} \left(z_{(1-\alpha/2)}\, \frac{f_o}{f_1} + z_{(1-\beta)} \right)^2 \tag{A5.7a}$$

$$\simeq \frac{f_1^2}{(\mu_o - \mu_1)^2} \left(z_{(1-\alpha/2)} + z_{(1-\beta)} \right)^2 \tag{A5.7b}$$

Thus, the two-sided expressions for n are identical to the one-sided expressions with α replaced by $\alpha/2$.

Example A5.2. One approach for detecting QTLs using a marker locus (with alleles M and m) in inbred-line crosses is to use the test statistic $T = \overline{z}_{MM} - \overline{z}_{mm}$, which compares the mean trait values of alternate marker homozygotes in the F_2 (Chapter 18). This leads to a simple t-test, which for $n > 30$ is essentially a normal test.

How powerful is this method for detecting QTLs? Assuming the marker is completely linked to an additive QTL (no dominance), the mean of T is μ_1 = 2a, the difference between the means of alternate QTL homozygotes. The variance accounted for by the QTL is $(-a)^2/4 + 0^2/2 + a^2/4 = a^2/2$, giving the fraction of the total of the total character variance accounted for by this QTL as $r^2 = (a^2/2)/\sigma_z^2$. As derived in Equation 18.36a, $\sigma^2(T) = \sigma_1^2 = 8(1-r^2)\sigma_z^2/n$, where n is the total number of offspring sampled (only 1/4 of the individuals are a particular homozygote, so that the expected sample size for each homozygote marker class is $n/4$). Under the null hypothesis of no linked QTL, $\mu_0 = 0$ and $\sigma_0^2 = 8\sigma_z^2/n$ (as $r = 0$). This test is two-sided, as a significantly positive, or negative, value of T indicates a linked QTL. Substituting into Equation A5.7a with $f_1^2 = 8(1 - r^2)\sigma_z^2$ and $f_0^2 = 8\sigma_z^2$ gives the sample size required to have power $1 - \beta$ in a test with significant level α as

$$n = \left(\frac{\sqrt{8(1 - r^2)\sigma_z^2}}{0 - 2a} \right)^2 \left(\sqrt{\frac{8\sigma_z^2}{8(1 - r^2)\sigma_z^2}}\, z_{(1-\alpha/2)} + z_{(1-\beta)} \right)^2$$

$$= \left(\frac{1 - r^2}{r^2} \right) \left(\frac{z_{(1-\alpha/2)}}{\sqrt{1 - r^2}} + z_{(1-\beta)} \right)^2 \tag{A5.8}$$

with the last step following by recalling that $2\sigma_z^2/a^2 = 1/r^2$.

Suppose we wish to detect a QTL that accounts for 5% of the total F_2 phenotypic variance ($r^2 = 0.05$). If we set $\alpha = 0.05$ (95% two-sided test; $z_{(1-\alpha/2)} = z_{(0.975)} = 1.96$) and $\beta = 0.1$ (90% power; $z_{(1-\beta)} = z_{0.9} = 1.28$), then from Equation A5.8 the required sample size (number of scored F_2s) becomes

$$n = \left(\frac{1-r^2}{r^2}\right)\left(\frac{z_{(1-\alpha/2)}}{\sqrt{1-r^2}} + z_{(1-\beta)}\right)^2$$

$$= \left(\frac{0.95}{0.05}\right)\left(\frac{1.96}{\sqrt{0.95}} + 1.28\right)^2 = 206$$

More generally, if the QTL–marker recombination frequency is c, then r^2 is replaced by $(1-2c)^2 r^2$. Hence, if $c = 0.10$, then $r^2 = (1-0.2)^2 \cdot 0.05$ for a value of $n = 324$.

One popular approach when considering possible experimental designs is to compute the sample size required to constrain the confidence interval $\mu_1 \pm z_\alpha\, f_1/\sqrt{n}$ to exclude μ_0. Here z_α is the appropriate value to give an α-level one- or two-sided test (depending on the hypothesis being tested). Solving for n,

$$(\mu_1 - \mu_o)^2 \geq \frac{z_\alpha^2\, f_1^2}{n}, \qquad \text{or} \qquad n \geq \frac{z_\alpha^2\, f_1^2}{(\mu_1 - \mu_o)^2}$$

This approach is often used in place of a direct power calculation, but just what is its actual power? From Equations A5.4c and A5.7b, this expression yields the n which gives a test whose power satisfies $z_{(1-\beta)} = 0$ or $\beta = 0.5$ (using this β in Example A5.2 returns a value of $n = 77$). Thus, designs using this sample size have only 50% power, i.e., they are expected to detect a specified effect size only half the time.

Applications: Parent–offspring Regressions

As is detailed in Chapter 7 (as well as in Part II Chapter 22), the slope b_{op} of the parent–offspring regression provides information on genetic variance components. What is the power of such regressions? We first note that N data points are used to compute the regression. For a single-parent regression, N is the number of *parents*, while for a midparent regression, N is the number of *pairs of parents* (we distinguish between these by indexing the regressions by p and $\overline{p}$, respectively). For each parental point, the offspring value could be that for a single offspring ($n = 1$) or the mean value of n offspring (indexed by o and $\overline{o}$, respectively).

Because the increase in power with sample size results from a reduction in the sampling variance, it is again useful to write the sampling variance of b_{op} as some function f divided by the number N of families, $\sigma^2(b_{op}) = f_{op}^2/N$, with

$$f_{op}^2 = (1 - r_{op}^2)\,\frac{\sigma^2(z_o)}{\sigma^2(z_p)} \tag{A5.9a}$$

where r_{op} is the parent–offspring correlation. The sampling variance under the null hypothesis that $b_{op} = 0$ follows by setting $r_{op} = 0$. If the parent-offspring regression has true slope b_{op}, what sample size (measured by number of parents) is required for a test to have power $1-\beta$? Because this is a one-sided test ($b_{op} = 0$ versus $b_{op} > 0$), substituting Equation A5.9a into Equation A5.4c (and hence assuming that the variances under null and alternative hypotheses are essentially equal, i.e., $\sigma_0^2 \simeq \sigma_1^2$) gives

$$N = \left[z_{(1-\alpha)} + z_{(1-\beta)}\right]^2\,\frac{(1-r_{op}^2)}{b_{op}^2}\,\frac{\sigma^2(z_o)}{\sigma^2(z_p)} \tag{A5.9b}$$

$$= \left[z_{(1-\alpha)} + z_{(1-\beta)}\right]^2\,\frac{(1-b_{op}^2)}{b_{op}^2} \tag{A5.9c}$$

where this last step follows by assuming that $\sigma^2(z_o) = \sigma^2(z_p)$ and noting in this case that $r_{op} = b_{op}$ (Equation 3.16b). Equation A5.9b, which assumes single parents, each with a single offspring, can easily be modified to account for midparents and/or multiple offspring per family (Example A5.3). For example, with n measured offspring per family, $\sigma^2(z_{\overline{o}}) = \sigma^2(z_o)[\,t + (1-t)/n\,]$ replaces $\sigma^2(z_o)$, where t is the correlation between sibs (in the absence of dominance and maternal effects, $t = h^2/2$ for full sibs and $h^2/4$ for half sibs; Chapter 7), and

$$r_{\overline{o}p} = \frac{b_{\overline{o},p}}{\sqrt{\gamma(t,n)}}, \quad \text{where} \quad \gamma(t,n) = t + (1-t)/n$$

replaces r_{op}. For a midparent–offspring regression, $\sigma^2(z_{\overline{p}}) = \sigma^2(z_p)/2$ replaces $\sigma^2(z_p)$, and $r_{o\overline{p}} = b_{o\overline{p}}/\sqrt{2}$.

Example A5.3. Suppose 200 (single) parent–offspring pairs are measured (N = 200, n = 1) for a character with h^2 = 0.2 and no epistasis or maternal effects. Assuming a test with α = 0.05, what is the power of this design to detect this regression as being significant? Here $z_{(1-0.05)}$ = 1.65, while from Chapter 7,

$$b_{op} = r_{op} = h^2/2 = 0.1$$

Assuming parent and offspring have equal variance, $\sigma^2(z_o) = \sigma^2(z_p)$, Equation A5.9a gives

$$\sigma^2(b_{op}) = \frac{f^2}{N} = \frac{1 - r_{op}^2}{N} = \frac{1 - 0.1^2}{200} = 0.00495$$

while under the null hypothesis of a slope of zero, $r_{op} = 0$ and hence $\sigma^2(b_{op}) = 1/200 = 0.005$. Equation A5.3 gives the power as

$$\Pr\left(U > \frac{\mu_o - \mu_1}{\sigma_1} + z_{(1-\alpha)}\,\frac{\sigma_o}{\sigma_1}\right) = \Pr\left(U > \frac{(0-0.1)}{\sqrt{0.00495}} + 1.65\,\frac{\sqrt{0.005}}{\sqrt{0.00495}}\right)$$
$$= \Pr(\,U > 0.237\,) = 0.41$$

The probability of a false negative under this design is $\beta = 0.59$, so that the majority of time the observed slope will not be judged to be significantly greater than zero. Measuring only a single offspring per family (n = 1), how many families must be used to have 90% probability that the observed slope is significantly positive (using a test with α = 0.05)? To make use of Equation A5.4, first note that here β = 0.1, with Pr($U \leq$ 1.28) = 0.9, giving $z_{(1-\beta)}$ = 1.28. Hence $\left(z_{(1-\alpha)} + z_{(1-\beta)}\right)^2$ = (1.65 + 1.28)^2 = 8.58. Substituting this result into Equation A5.9c gives the required N as

$$N = 8.58\,\frac{(1 - b_{op}^2)}{b_{op}^2} = 8.58\,\frac{(1 - 0.1^2)}{0.1^2} = 850$$

Now suppose n offspring per parent are scored. Here

$$\frac{\sigma^2(z_{\overline{o}})}{\sigma^2(z_p)} = \frac{\sigma^2(z_o)[\,t + (1-t)/n\,]}{\sigma^2(z_p)} = [\,t + (1-t)/n\,] = \gamma(t,n) \tag{5.10a}$$

Further, $r_{\overline{o},p}^2 = b_{o,p}^2/\gamma(t,n)$ and $b_{\overline{o},p} = b_{o,p}$, so that Equation A5.9b becomes

$$N = \left[z_{(1-\alpha)} + z_{(1-\beta)}\right]^2\,\frac{(1 - r_{\overline{o}p}^2)}{b_{\overline{o}p}^2}\,\frac{\sigma^2(z_{\overline{o}})}{\sigma^2(z_p)}$$
$$= \left[z_{(1-\alpha)} + z_{(1-\beta)}\right]^2\,\frac{[1 - b_{o,p}^2/\gamma(t,n)]}{b_{op}^2}\cdot\gamma(t,n) \tag{A5.10b}$$

Here (assuming full sibs) $t = h^2/2 = 0.1$, giving the required number of parents under different number of sibs (n) as

	1	2	5	10	∞
$\gamma(t,n)$	1	0.55	0.28	0.19	0.10
$r^2_{\overline{o}p}$	0.01	0.02	0.04	0.05	0.10
N	850	463	232	154	77
$N(1+n)$	1700	1389	1392	1694	

where $N(1+n)$ is the total number of measured individuals, namely N parents and Nn sibs. Now turning to midparent–offspring regressions (with a single offspring per parent), $b_{o\overline{p}} = h^2 = 0.2$, $r^2_{o\overline{p}} = b^2_{o\overline{p}}/2 = 0.02$, and

$$\frac{\sigma^2(z_o)}{\sigma^2(z_{\overline{p}})} = \frac{\sigma^2(z_o)}{\sigma^2(z_p)/2} = 2$$

giving the required number of single-offspring, two-parent families as

$$\begin{aligned} N &= \left[z_{(1-\alpha)} + z_{(1-\beta)}\right]^2 \frac{(1-r^2_{o\overline{p}})}{b^2_{o\overline{p}}} \frac{\sigma^2(z_o)}{\sigma^2(z_{\overline{p}})} \\ &= \left[z_{(1-\alpha)} + z_{(1-\beta)}\right]^2 \frac{(1-b^2_{o,\overline{p}}/2)}{b^2_{o\overline{p}}} \cdot 2 \end{aligned} \qquad \text{(A5.10c)}$$

Hence

$$N = 8.58 \cdot \frac{(1-0.02)}{0.2^2} \cdot 2 = 420$$

Finally, suppose that n full-sibs offspring per midparental pair are measured. Here

$$\frac{\sigma^2(z_{\overline{o}})}{\sigma^2(z_{\overline{p}})} = \frac{\sigma^2(z_o)[t+(1-t)/n]}{\sigma^2(z_p)/2} = 2 * \gamma(t,n)$$

Noting that $b_{\overline{o}\,\overline{p}} = b_{o\overline{p}}$, and $r^2_{\overline{o}\,\overline{p}} = b^2_{o\overline{p}}/[2\gamma(t,n)]$, Equation A5.9a becomes

$$\begin{aligned} N &= \left[z_{(1-\alpha)} + z_{(1-\beta)}\right]^2 \frac{(1-r^2_{\overline{o}\,\overline{p}})}{b^2_{o\overline{p}}} \frac{\sigma^2(z_{\overline{o}})}{\sigma^2(z_{\overline{p}})} \\ &= \left[z_{(1-\alpha)} + z_{(1-\beta)}\right]^2 \frac{[1-b^2_{o,\overline{p}}/\{2\gamma(t,n)\}]}{b^2_{o\overline{p}}} \cdot 2\gamma(t,n) \end{aligned} \qquad \text{(A5.10d)}$$

Here (assuming full sibs) $t = h^2/2 = 0.1$, giving the required number of pairs of parents under different number of sibs (n) as

	1	2	5	10	∞
$2\gamma(t,n)$	2.00	1.10	0.56	0.38	0.20
$r^2_{\overline{o}\,\overline{p}}$	0.02	0.04	0.07	0.11	0.20
N	420	227	111	73	34
$N(2+n)$	1260	908	777	876	

where $N(2+n)$ is the number of measured individuals, namely $2N$ parents and Nn sibs.

Applications: QTL Detection Tests Using Doubly Affected Sib Pairs

Chapter 19 examined the use of doubly affected full-sib pairs to detect QTLs influencing a binary character (typically disease presence/absence, so that both sibs in each pair display the disease). If the marker is linked to a QTL influencing the character, we expect the pair members to share more marker alleles than expected by chance. A number of tests based on the number of IBD (identical by descent) marker alleles shared between pair members have been proposed for detecting a linked QTL. Letting p_{2i} and $\widehat{p}_{2i}$ denote the true and estimated fractions of doubly affected pairs sharing $i = 0, 1, 2$ IBD marker alleles (the leading 2 in the subscript indicates we are restricting attention to pairs where both sibs are affected), we consider two such tests here. One test statistic, $T_2 = \widehat{p}_{22}$, is based on the fraction of doubly affected pairs sharing 2 IBD alleles. A second test statistic, $T_m = \widehat{p}_{21}/2 + \widehat{p}_{22}$, corresponds to the mean number of IBD marker alleles shared by a doubly affected pair. The true IBD frequencies can be expressed as deviations from the values expected with no linked QTL,

$$p_{20} = \frac{1}{4} - d_{20}, \qquad p_{21} = \frac{1}{2} - d_{21}, \qquad p_{22} = \frac{1}{4} + d_{22}$$

Under the null hypothesis (the marker is unlinked to a QTL), the d_{2i} are zero, while Equations 19.55b–19.55d give their values as functions of the QTL effects and distance from the marker if a QTL is linked to the marker. Note from Equation 19.55 that $d_{2i} \geq 0$ under linkage.

We examine the power of the test based on $T_2 = \widehat{p}_{22}$ first. This test statistic has mean p_{22} and variance $p_{22}(1 - p_{22})/n$, where n is the number of doubly affected pairs. Expressing the variance of the test statistic as $\sigma^2(T) = f^2/n$, shows that here $f^2 = p_{22}(1 - p_{22})$. Under the null hypothesis of no linked QTL, $p_{22} = 1/4$, giving the mean and variance as $\mu_o = 1/4$ and $f_0^2 = (1/4)(3/4) = 3/16$. When a linked QTL is present, so that $p_{22} = 1/4 + d_{22}$, the mean and variance become

$$E(\widehat{p}_{22}) = \frac{1}{4} + d_{22}, \quad \text{and} \quad f_1^2 = p_{22}(1 - p_{22}) = (1/4 + d_{22})(3/4 - d_{22}) \qquad \text{(A5.11a)}$$

Hence, $(\mu_1 - \mu_0)^2 = d_{22}^2$. The T_2 test is one-sided, as $p_{22} > 1/4$ under linkage to a QTL, and from Equation A5.4b, the required number of doubly affected pairs to have power $1 - \beta$ using a test with significance α is

$$n_{T_2} = \left(\frac{z_{(1-\beta)}\sqrt{(1/4 + d_{22})(3/4 - d_{22})} + z_{(1-\alpha)}\sqrt{3/16}}{d_{22}} \right)^2 \qquad \text{(A5.11b)}$$

In a similar fashion, the expected value for the test statistic T_m is

$$\mu_1 = p_{21} + 2\,p_{22} = \left(\frac{1}{2} - d_{21}\right) + 2\left(\frac{1}{4} + d_{22}\right) = 1 + 2\,d_{22} - d_{21} \qquad \text{(A5.12a)}$$

To compute the variance $\sigma_1^2 = f_1^2/n$, let x denote the fraction of IBD alleles in a randomly chosen individual. Taking expectations gives

$$f_1^2 = E(x^2) - \mu_x^2 = \left[1^2 \cdot \Pr(x = 1) + 2^2 \cdot \Pr(x = 2)\right] - \mu_x^2$$

$$= \left[1^2\left(\frac{1}{2} - d_{21}\right) + 2^2\left(\frac{1}{4} + d_{22}\right)\right] - (1 + 2\,d_{22} - d_{21})^2$$

$$= \frac{1}{2} + d_{21} - (2\,d_{22} - d_{21})^2 \qquad \text{(A5.12b)}$$

Under the null hypothesis of no linked QTLs, $d_{21} = d_{22} = 0$, giving $\mu_0 = 1/2$ and $f_0^2 = 1/2$. Again, this is a one-sided test, as $T_m > 1/2$ for a marker linked to a QTL. Applying Equation A5.4b gives the required number of doubly affected pairs as

$$n_{T_m} = \left(\frac{z_{(1-\beta)}\sqrt{1/2 + d_{21} - (2d_{22} - d_{21})^2} + z_{(1-\alpha)}\sqrt{1/2}}{2d_{22} - d_{21}} \right)^2 \qquad \text{(A5.12c)}$$

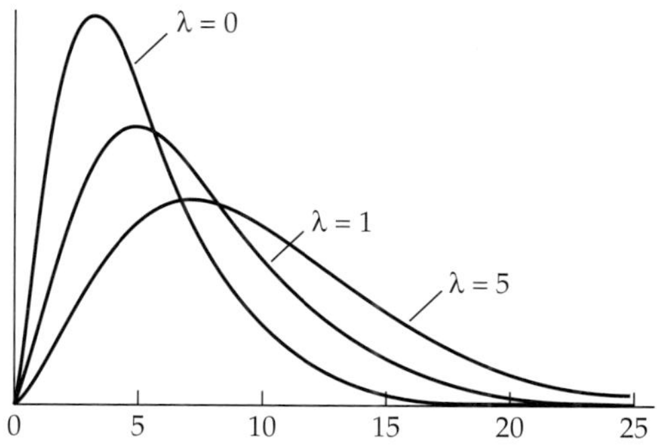

Figure A5.3 The probability distribution function for a noncentral χ^2. As the noncentrality parameter λ increases, the distribution is pulled to the right. Here we plot a χ^2 random variable with n = 5 degrees of freedom and noncentrality parameters λ = 0 (a central χ^2), 1, and 5.

POWER OF CHI-SQUARE TESTS

Central and Noncentral χ^2 Distributions

The χ^2 distribution arises from sums of squared, normally distributed, random variables. If $x_i \sim N(0,1)$, then $u = \sum_{i=1}^n x_i^2 \sim \chi_n^2$, a **central** χ^2 distribution with n degrees of freedom. It follows that the sum of two χ^2 random variables is also χ^2-distributed, so that if $u \sim \chi_n^2$ and $v \sim \chi_m^2$, then

$$u + v \sim \chi^2_{(n+m)} \tag{A5.13a}$$

Two other useful results are that if $x_i \sim N(0, \sigma^2)$, then

$$\sum_{i=1}^{n} x_i^2 \sim \sigma^2 \cdot \chi_n^2 \tag{A5.13b}$$

and for $\overline{x} = n^{-1}\sum_{i=1}^{n} x_i$, then for x_i with a common mean, $x_i \sim N(\mu, \sigma^2)$, then

$$\sum_{i=1}^{n} (x_i - \overline{x})^2 \sim \sigma^2 \cdot \chi^2_{(n-1)} \tag{A5.13c}$$

In this last case, subtraction of the mean causes the loss of one degree of freedom.

A **noncentral** χ^2 arises when the random variables being considered have *nonzero means*. In particular, if $x_i \sim N(\mu_i, 1)$, then $u = \sum_{i=1}^n x_i^2$ follows a noncentral χ^2 distribution with n degrees of freedom and **noncentrality parameter (NCP)**

$$\lambda = \sum_{i=1}^{n} \mu_i^2 \tag{A5.14a}$$

and we write $u \sim \chi^2_{n,\lambda}$. As shown in Figure A5.3, increasing the noncentrality parameter λ shifts the distribution to the right. Hence, the power of a χ^2-based test is a function of the NCP associated with the assumed alternative value. This shift to the right as the NCP increases can also be seen by considering the mean and variance of u,

$$E(u) = n + \lambda \qquad \text{and} \qquad \sigma^2(u) = 2(n + 2\lambda) \tag{A5.14b}$$

It follows directly from the definition that sums of noncentral χ^2 variables also follow a noncentral χ^2 distribution, so that if $u \sim \chi^2_{n,\,\lambda_1}$ and $v \sim \chi^2_{m,\,\lambda_2}$, then

$$(u + v) \sim \chi^2_{(n+m),(\lambda_1+\lambda_2)} \tag{A5.14c}$$

Finally, Equations A5.13b and A5.13c can be generalized to noncentral χ^2 random variables as follows. Suppose $x_i \sim N(\mu_i, \sigma^2)$, then

$$\sum_{i=1}^{n} x_i^2 \sim \sigma^2 \cdot \chi^2_{n,\lambda} \qquad \text{where} \qquad \lambda = \sum_{i=1}^{n} \frac{\mu_i^2}{\sigma^2} \tag{A5.14d}$$

and

$$\sum_{i=1}^{n} (x_i - \overline{x})^2 \sim \sigma^2 \cdot \chi^2_{(n-1),\lambda} \qquad \text{where} \qquad \lambda = \sum_{i=1}^{n} \frac{(\mu_i - \overline{\mu})^2}{\sigma^2} \tag{A5.14e}$$

Sample Variance Confidence Intervals

Provided the individual observations used to estimate a sample variance are uncorrelated and normally distributed with $x_i \sim N(\mu, \sigma^2)$, then Equation A5.13c implies that the sample estimator $\text{Var} = \sum (x_i - \overline{x})^2/(n-1)$ is distributed as

$$\text{Var} \sim \left(\frac{\sigma^2}{n-1} \right) \chi^2_{n-1} \tag{A5.15a}$$

Namely, a scaled chi-square random variable with $n-1$ degrees of freedom. Letting $X_{p,n}$ satisfy $\Pr(\chi^2_n \leq X_{p,n}) = p$, Equation 2.17e gives the confidence intervals for the true variance (σ^2) as

$$\Pr\left[\left(\frac{n-1}{X_{1-\alpha/2,n-1}} \right) \text{Var} \leq \sigma^2 \leq \left(\frac{n-1}{X_{\alpha/2,n-1}} \right) \text{Var} \right] = 1 - \alpha \tag{A5.15b}$$

The value of $X_{p,n}$ can be obtained using the the `R` command `qchisq(p,n)`. Confidence intervals for σ^2 are asymmetrical about the estimated value, Var, and tend to be quite large, even for modest sample sizes.

What is the power of the Var statistic? Specifically, how likely are we to declare the variance to be statistically different from some hypothesized null value, σ_0^2? From Equation A5.15b, the upper and lower critical values for a $100(1-\alpha)\%$ confidence interval for the realized value of Var when the true variance is σ_0^2 are

$$\left(\frac{\sigma_0^2}{n-1} \right) X_{\alpha/2,n-1} \leq \text{Var} \leq \left(\frac{\sigma_0^2}{n-1} \right) X_{1-\alpha/2,n-1} \tag{A5.16a}$$

If Var falls outside this interval, we declare that the true variance is statistically different (at level α) from the assumed value, σ_0^2. The probability β of a false negative is the probability that a sample value of Var falls *within* this interval. We can obtain β by noting that the true distribution is given by $[(n-1)/\sigma_1^2]\,\text{Var} \sim \chi^2_{n-1}$. Multiplying all terms of Equation A5.15a by $(n-1)/\sigma_1^2$ gives the probability β of a sample variance failing to be declared significant as

$$\begin{aligned} \beta &= \Pr\left[\left(\frac{\sigma_0^2}{n-1} \right) \left(\frac{n-1}{\sigma_1^2} \right) X_{\alpha/2,n-1} \leq \chi^2_{n-1} \leq \left(\frac{\sigma_0^2}{n-1} \right) \left(\frac{n-1}{\sigma_1^2} \right) X_{1-\alpha/2,n-1} \right] \\ &= \Pr\left[\left(\frac{\sigma_0^2}{\sigma_1^2} \right) X_{\alpha/2,n-1} \leq \chi^2_{n-1} \leq \left(\frac{\sigma_0^2}{\sigma_1^2} \right) X_{1-\alpha/2,n-1} \right] \end{aligned} \tag{A5.16b}$$

The complement of this event (namely, either less than the lower critical value or greater than the upper critical value) is the power,

$$1 - \beta = \Pr\left[\chi^2_{n-1} \leq \left(\frac{\sigma_0^2}{\sigma_1^2} \right) X_{\alpha/2,n-1} \right] + \Pr\left[\chi^2_{n-1} \geq \left(\frac{\sigma_0^2}{\sigma_1^2} \right) X_{1-\alpha/2,n-1} \right] \tag{A5.16c}$$

Again, there are simple `R` commands to compute these probabilities, with $\Pr(\chi^2_n \leq X)$ given by the command `pchisq(X,n)`, while $\Pr(\chi^2_n \geq X)$ is obtained by `1- pchisq(X,n)`.

Example A5.4. Consider a sample variance estimated from $n = 10$ observations. The lower 2.5% and upper 97.5% critical values are easily obtained in R as `qchisq(0.025,9)` and `qchisq(0.975,9)`, which returns values of 2.700 and 19.023, respectively. Equation A5.15b thus gives the 95% confidence interval (α = 0.05) for the true variance σ^2 as between (9/19.023)Var and (9/2.7)Var, or 0.473·Var to 3.333·Var, implying an uncertainty in σ^2 spanning almost a full order of magnitude.

What observed values of the sample variance are unlikely given an assumed true variance of σ_0^2? From Equation A5.16a, the upper and lower critical values (for a two-sided test with α = 0.05) are $(2.700/9)\sigma_0^2 = 0.3{\cdot}\sigma_0^2$ and $(19.023/9)\sigma_0^2 = 2.11{\cdot}\sigma_0^2$. Finally, what is the power of this design to declare the null model (σ_0^2 is the true variance) false when the correct variance is $\sigma_1^2 = 2\sigma_0^2$? Taking α = 0.05, Equation A5.16c (with $\sigma_0^2/\sigma_1^2 = 1/2$) gives the power as

$$\Pr\left(\chi_9^2 \leq \frac{2.700}{2}\right) + \Pr\left(\chi_9^2 \geq \frac{19.023}{2}\right)$$

which can be calculated in R using `pchisq(1.35,9) + 1- pchisq(9.511,9)`, which returns a value of 0.39, for a type II error rate of 61% when the true variance is twice the assumed variance. A similar calculation assuming $\sigma_1^2 = \sigma_0^2/2$ gives a power of 0.20, or a type II error rate of 80%.

Power of Likelihood-ratio (LR) Tests

Under the null, the likelihood ratio (LR) test statistic (Equation A4.9a) for large samples follows a chi-square distribution whose degrees of freedom are the number of tested parameters (k). As detailed in Appendix 4, when testing against a parameter value on the boundary of the parameter space (such as a variance of zero), the asymptotic LR distribution is now a mixture of chi-squares, so we ignore this complication here, and assume the test is against some vector $\boldsymbol{\Theta}_1$ that is in the interior of the likelihood function space. We can approximate the power of the LR test in these settings (against the null of $\boldsymbol{\Theta}_0$) by noting that when the alternative ($\boldsymbol{\Theta}_1$) is true, the resulting distribution is now a noncentral chi-square with the noncentrality parameter given by the quadratic product

$$\lambda = (\boldsymbol{\Theta}_1 - \boldsymbol{\Theta}_0)^T \mathbf{F}\,(\boldsymbol{\Theta}_1 - \boldsymbol{\Theta}_0) \tag{A5.17a}$$

where $\mathbf{F}$ is the Fisher information matrix (Equation A4.7b: recall for large samples that the covariance matrix $\mathbf{V}$ of the estimates is well approximated by $\mathbf{F}^{-1}$). Hence, under the alternative

$$LR_{H1} \sim \chi^2_{k,\lambda} \tag{A5.17b}$$

with the approximate power for a test of significance at level α given by

$$\Pr\left(\chi^2_{k,\lambda} > X_{1-\alpha,k}\right) \tag{A5.17c}$$

where X is the critical value under the null, namely $\Pr(\chi_k^2 > X_{1-\alpha,k}) = \alpha$. As shown in Example A5.5, the impact of sample size n is through $\mathbf{F}$, which is often of the form $n\mathbf{C}$, where $\mathbf{C}$ is a matrix of constants.

Equation A5.17a follows from the asymptotic distribution of MLE approaching a multivariate normal (Appendix 4). Assume that the vector of means, under both the null and alternative, is MVN with a common covariance matrix, e.g., $\mathbf{x} \sim \text{MVN}(\boldsymbol{\beta}_i, \mathbf{V})$. A common test statistic against the null is given by

$$T^2 = n(\widehat{\boldsymbol{\beta}} - \boldsymbol{\beta}_0)^T \mathbf{V}_{\mathbf{b}}^{-1}(\widehat{\boldsymbol{\beta}} - \boldsymbol{\beta}_0) \tag{A5.17d}$$

where n is the sample size (number of **x** vectors), and $\mathbf{V_b}$ is the sample variance for the estimators. If $\widehat{\boldsymbol{\beta}}$ contains p estimated parameters, then T^2 follows a noncentral F distribution (discussed in the next section) with a noncentrality parameter of

$$\lambda = (\boldsymbol{\beta}_1 - \boldsymbol{\beta}_0)^T \mathbf{V}^{-1} (\boldsymbol{\beta}_1 - \boldsymbol{\beta}_0) \tag{A5.17e}$$

where β_1 is the true value and

$$\left(\frac{n-p}{(n-1)p} \right) \cdot T^2 \sim F\,(p, n-p, \lambda) \tag{A5.17f}$$

Under the asymptotic assumption ($n \to \infty$) leading to Equation A5.17a, the sampling variance matrix approaches $\mathbf{F}^{-1}$, the multiplier on T^2 approaches $1/p$, and $F\,(p, n-p, \lambda) \to [\chi^2(p, \lambda)]/p$, recovering Equation A5.17b.

Example A5.5. Recall from Example A4.3 that the Fisher information matrix for the MLEs for the mean and variance under a normal is

$$\mathbf{F} = n \begin{pmatrix} \dfrac{1}{\sigma^2} & 0 \\ 0 & \dfrac{1}{2\,\sigma^4} \end{pmatrix}$$

The resulting NCP for testing a null of $H_0 = (\mu_0, \sigma_0^2)$ versus the alternative of $H_1 = (\mu_1, \sigma_1^2)$ becomes

$$\lambda = n \left[(\,\mu_1 - \mu_0, \quad \sigma_1^2 - \sigma_0^2\,) \begin{pmatrix} \dfrac{1}{\sigma_1^2} & 0 \\ 0 & \dfrac{1}{2\,\sigma_1^4} \end{pmatrix} \begin{pmatrix} \mu_1 - \mu_0 \\ \sigma_1^2 - \sigma_0^2 \end{pmatrix} \right] = n\gamma$$

where

$$\gamma = \left[\frac{(\mu_1 - \mu_0)^2}{\sigma_1^2} + \frac{(\sigma_1^2 - \sigma_0^2)^2}{2\sigma_1^4} \right]$$

Note that under the alternative, the elements of **F** are based on σ_1^2. Suppose that $\gamma = 0.15$. To have 90% power for a test of $\alpha = 0.05$, we first note that the critical value becomes 5.99. as $\Pr(\chi_2^2 > 5.99) = 0.05$. The required sample size n becomes the solution to $\Pr(\chi^2_{2,n\gamma} > 5.99) = 0.90$, which can be restated as $\Pr(\chi^2_{2,0.15n} \leq 5.99) = 0.10$. Solving gives $n = 84$.

POWER OF *F*-RATIO TESTS

The analysis of variance (ANOVA) is widely used throughout quantitative genetics, e.g., in detecting QTLs via marker–trait associations (Chapters 17–19) and in estimating variance components under balanced experimental designs (Part II Chapters 22–29). ANOVA designs are typically based upon sums of squares (SS_x) and their associated mean squares ($\text{MS}_x = \text{SS}_x/n_x$, with n_x the associated degrees of freedom). ANOVA test statistics are generally given by ratios of mean squares, $\mathcal{F}_{x/y} = \text{MS}_x/\text{MS}_y$. Under normality assumptions, sums of squares are χ^2-distributed, while under the null hypothesis $\mathcal{F}_{x/y}$ is distributed as a (central) F-ratio distribution with n_x (numerator) and n_y (denominator) degrees of freedom, and we denote this as $\mathcal{F}_{x/y} \sim F_{n_x,n_y}$. Thus, a test of level α for the hypothesis of no additional effects on level x (compared to level y) is whether $\mathcal{F}_{x/y} > f_{n_x,n_y,(1-\alpha)}$, where $f_{n_x,n_y,\alpha}$ satisfies

$$\Pr\left[\, F_{n_x,n_y} \leq f_{n_x,n_y,\alpha} \,\right] = \alpha \tag{A5.18a}$$

In R, the command `qf(p,n,m)` returns the value of $f_{n,m,p}$. For example, the upper 95% critical value associated with an $F_{10,30}$ random variable is `qf(0.95.10,30)`, which returns a value of 2.165. Similarly, the probability that an $F_{n,m}$ is less than or equal to x is given by `pf(x,n,m)`, so (for example), the probability that an $F_{3,24}$ exceeds a value of 5 is obtained as `1- pf(5,3,24)`, which returns a value of 0.0078.

Two useful identities are

$$\Pr\left[\, F_{n_x,n_y} > f_{n_x,n_y,(1-\alpha)} \,\right] = \alpha \tag{A5.18b}$$

and

$$f_{n_x,n_y,\alpha} = \frac{1}{f_{n_y,n_x,(1-\alpha)}} \tag{A5.18c}$$

Note that the numerator and denominator degrees of freedom are reversed on the two sides of Equation A5.18c.

The power of the test is the probability that the test statistic $\mathcal{F}_{x/y}$ exceeds this critical value,

$$\Pr\left[\mathcal{F}_{x/y} > f_{n_x,n_y,(1-\alpha)}\,\right] \tag{A5.18d}$$

In order to compute this probability, we require the distribution of the test statistic $\mathcal{F}_{x/y}$ under the alternative hypothesis. Here, the numerator sum of squares follows a noncentral χ^2 distribution, and the resulting $\mathcal{F}_{x/y}$ statistic involving these follows a noncentral F distribution. Hence, before considering power, we follow up on our discussion of central and noncentral χ^2 by describing the properties of central and noncentral F distributions. Proofs of these results can be found in standard texts such as Scheffé (1959), Searle (1971), and Johnson et al. (1995).

Central and Noncentral *F* Distributions

The ratio of two χ^2-distributed variables leads to the F distribution. In particular, if $u \sim \chi^2_n$ and $v \sim \chi^2_m$, then the ratio of these two χ^2 variables divided by their respective degrees of freedom follows a **central *F* distribution** with **numerator** and **denominator degrees of freedom** of n and m (respectively),

$$\frac{u/n}{u/m} = \frac{\chi^2_n/n}{\chi^2_m/m} \sim F_{n,m} \tag{A5.19a}$$

This is occasionally referred to as the **Snedecor *F*** or **Fisher–Snedecor *F*** distribution. Special cases of the F reduce to other distributions. For example, if t_n denotes a Student-t distribution with n degrees of freedom, then

$$t_n^2 \sim F_{1,n} \tag{A5.19b}$$

Further, because

$$\lim_{m\to\infty} \frac{\chi^2_m}{m} \to 1, \quad \text{and thus} \quad \lim_{m\to\infty} F_{n,m} \to \frac{\chi^2_n}{n} \tag{A5.19b}$$

the F distribution can be approximated by a χ^2_n when the denominator degrees of freedom are large.

The **noncentral *F* distribution** results when the numerator χ^2 variable is noncentral. If $u \sim \chi^2_{n,\lambda}$ and $v \sim \chi^2_m$, then $F = (u/n)/(v/m)$ follows a noncentral F distribution with noncentrality parameter λ, and we write $F \sim F_{n,m,\lambda}$. As with the noncentral χ^2, increasing λ shifts the distribution further to the right. Again, this is seen in the mean and variance, with

$$E(F) = \left(\frac{m}{m-2}\right)\left(1+\frac{\lambda}{n}\right) \qquad \text{for } m > 2 \tag{A5.19c}$$

$$\sigma^2(F) = 2\left(\frac{m}{n}\right)^2 \left[\frac{(n+m)^2 + (n+2\lambda)(m-2)}{(m-2)^2(m-4)}\right] \qquad \text{for } m > 4 \tag{A5.19d}$$

While a number of approximations for computing noncentral Fs have been suggested (e.g., Patnaik 1949; Severo and Zelen 1960; Tiku 1965), most quantities of interest can easily be computed in R. For example, `pf(x,n,m,ncp)` returns $\Pr(F_{n,m,ncp} \leq x)$.

Power of Fixed-effects ANOVA Designs

We now have the necessary machinery in hand to perform power calculations for ANOVA designs. We consider fixed-effects designs first, using the simple one-way ANOVA to illustrate the basic approach. Letting y_{ij} denote the value of the jth individual with treatment i, the model is $y_{ij} = \mu + \tau_i + e_{ij}$. We assume N fixed factors (the treatment effects $\tau_1, \cdots, \tau_N$, with $\sum \tau_i = 0$); that the residuals are independent and normally distributed, with $e_{ij} \sim N(0, \sigma_e^2)$; and that the design is balanced with n observations for each treatment. The resulting treatment (or model) and error sums of squares are given by

$$\text{SS}_t = n \sum_{i=1}^{N} (\overline{y}_{i\,\cdot} - \overline{y}_{\cdot\,\cdot})^2, \qquad \text{SS}_e = \sum_{j=1}^{n} \sum_{i=1}^{N} (y_{ij} - \overline{y}_{i\,\cdot})^2 \tag{A5.20a}$$

where $\overline{y}_{i\,\cdot}$ is the sample mean for factor i and $\overline{y}_{\cdot\,\cdot}$ is the overall sample mean. As we will see shortly, these sums of squares have, respectively, $N-1$ and $N(n-1)$ degrees of freedom. The test for a significant treatment effect (at least one $\tau_i \neq 0$) is the $\mathcal{F}_{t/e}$-ratio statistic

$$\mathcal{F}_{t/e} = \frac{\text{MS}_t}{\text{MS}_e} = \frac{\text{SS}_t/(N-1)}{\text{SS}_e/N(n-1)} \tag{A5.20b}$$

The distribution of $\mathcal{F}$ depends on the distribution of the sums of squares. Because $\overline{y}_{i\cdot} \sim N(\tau_i, \sigma_e^2/n)$ and (by construction) $\overline{\tau} = 0$, Equation A5.14e gives

$$\text{SS}_t \sim n \cdot (\sigma_e^2/n) \cdot \chi^2_{N-1,\lambda} \qquad \text{where} \qquad \lambda = \sum_{i=1}^{N} \frac{\tau_i^{\,2}}{\sigma_e^2/n} \tag{A5.21}$$

The noncentrality parameter λ can also be expressed in terms of amount of the total variance attributable to model (treatment) effects, σ_τ^2. Because $\overline{\tau} = 0$, it follows that

$$\sigma_\tau^2 = \frac{1}{N-1} \sum_{i=1}^{N} \tau_i^{\,2} \tag{A5.22a}$$

and hence, from Equation A5.21 it follows that

$$\lambda = n\,(N-1)\,\frac{\sigma_\tau^2}{\sigma_e^2} \tag{A5.22b}$$

Finally, noting that $y_{ij} - \overline{y}_{i\cdot} = (\mu + \tau_i + e_{ij}) - (\mu + \tau_i + \overline{e}_{i\cdot}) = e_{ij} - \overline{e}_{i\cdot}$ and that $e_{ij} \sim N(0, \sigma_e^2)$, we have

$$\text{SS}_e = \sum_{i=1}^{N} \left[\sum_{j=1}^{n} (e_{ij} - \overline{e}_{i\,\cdot})^2 \right] \sim \sum_{i=1}^{N} \sigma_e^2 \cdot \chi^2_{n-1} \sim \sigma_e^2 \cdot \chi^2_{N(n-1)} \tag{A5.23}$$

where the last two identities follow from Equations A5.13c and A5.13a, respectively. Thus the $\mathcal{F}_{t/e}$-ratio test statistic is distributed as

$$\mathcal{F}_{t/e} \sim \left(\frac{\sigma_e^2}{\sigma_e^2} \right) \left[\frac{\chi^2_{N-1,\lambda}/(N-1)}{\chi^2_{N(n-1)}/N(n-1)} \right] \sim F_{N-1,N(n-1),\lambda} \tag{A5.24}$$

where the noncentrality parameter is given by Equation A5.22b.

Under the hypothesis of no treatment effects, $\lambda = 0$ and $\mathcal{F}_{t/e} \sim F_{(N-1),N(n-1)}$. A test with significance level α is thus given by whether $\mathcal{F}_{t/e} > f_{(N-1),N(n-1),(1-\alpha)}$. When at least one treatment effect is nonzero, the test statistic $\mathcal{F}_{t/e}$ follows a noncentral F distribution, and the power of the test is given by

$$\Pr\left[\, F_{N-1,N(n-1),\lambda} > f_{N-1,N(n-1),(1-\alpha)} \,\right] \tag{A5.25}$$

To find the sample size required to give this test power $1-\beta$, we equate Equation A5.25 to $1-\beta$ and solve for n given the fixed values of α, β, N, and $\sum \tau_i^2/\sigma_e^2$ (or, equivalently, σ_τ^2/σ_e^2).

Example A5.6. Consider a fixed-effect design with four factors (N = 4), and suppose that $\sigma_\tau^2/\sigma_e^2 = 1/3$, so that the treatment effects account for $\sigma_\tau^2/(\sigma_\tau^2+\sigma_e^2) = 1/(1+\sigma_e^2/\sigma_\tau^2) = 1/4$ the total variance. If each treatment has a sample size of n = 5, what is the power to detect a significant treatment effect using a test of $\alpha = 0.05$? Here $N(n-1) = 16$, with `qf(0.95,3,16)` returning 3.24 for the critical value ($f_{3,16,0.95}$). Likewise, the noncentrality parameter is

$$\lambda = n\,(N-1)\,(\sigma_\tau^2/\sigma_e^2) = 5 \cdot 3 \cdot (1/3) = 5$$

Using the R command `1 - pf(3.24,3,16,5)` returns the value (the power) of $\Pr(F_{3,16,5} > 3.24)$ as 0.353.

What sample size n is required to give this test 90% power? Here $\alpha = 0.05$, $N = 4$, and $\sigma_\tau^2/\sigma_e^2 = 1/3$ are fixed values. For $n = 15$, we find that $N(n-1) = 56$ and $\lambda = 15$. The critical 95% value for this design becomes $f_{3,56,0.95}$, which is computed by `qf(0.95,3,56)`, which returns a value of 2.769. Substituting these values into Equation A5.24 gives the power as $\Pr(F_{3,56,15} > 2.769)$. This is obtained using `1- pf(2.769,3,56,15)`, which returns a value of 0.896. Increasing n to 16, $N(n-1) = 60$, $\lambda = 16$, and $f_{3,60,0.95} = 2.758$, giving $\Pr(F_{3,60,16} > 2.758) = 0.9167$, for a power of 91.7%.

Application: Power of QTL Mapping in Half-sib Families

As an example of analysis of a more complex design, consider the nested ANOVA design for mapping QTLs in half-sib families (Chapter 19). In the QTL half-sib design, each sire heterozygous for a marker locus is crossed to n dams and a single offspring is scored from each mating. Evidence of a QTL linked to the marker is indicated if the mean trait values for alternative sire alleles at this marker locus are significantly different. Because a different dam is used for each mating, all offspring from the same sire are half sibs. The model is

$$y_{ijk} = \mu + s_i + m_{ij} + e_{ijk} \tag{A5.26}$$

where y_{ijk} denotes the kth offspring of marker genotype j (with $j = 1$ for offspring carrying sire allele 1, $j = 2$ for sire allele 2) from sire i. We assume a completely balanced design, with N sires, each of which have n offspring equally distributed over the two alternative genotypes (the alternative sire marker alleles). The residuals are assumed independent with $e_{ijk} \sim N(0, \sigma_e^2)$ and the test statistic for a significant marker effect is $\mathcal{F}_{m/e} \sim \mathrm{MS}_m/\mathrm{MS}_e$. To compute the power of this test, we need to obtain the distributions of sums of squares. Starting with the error sum of squares,

$$\begin{aligned} \mathrm{SS}_e &= \sum_{i=1}^{N}\sum_{j=1}^{2}\sum_{k=1}^{n/2} (\,y_{ijk} - \overline{y}_{ij\cdot})^2 = \sum_{i=1}^{N}\sum_{j=1}^{2}\sum_{k=1}^{n/2} (e_{ijk} - \overline{e}_{ij\cdot})^2 \\ &\sim \sum_{i=1}^{N}\sum_{j=1}^{2} \sigma_e^2 \cdot \chi^2_{(n/2-1)} = \sigma_e^2 \cdot \chi^2_{2N(n/2-1)} \end{aligned} \tag{A5.27}$$

The last two identities follow from Equations A5.13c and A5.13a, respectively. Hence, there are $N(n-2)$ degrees of freedom associated with SS_e. Turning to the marker sum of squares,

$$\text{SS}_m = \sum_{i=1}^{N} \sum_{j=1}^{2} \sum_{k=1}^{n/2} (\,\overline{y}_{ij\cdot} - \overline{y}_{i\cdot\cdot})^2 = \left(\frac{n}{2}\right) \sum_{i=1}^{N} \sum_{j=1}^{2} (\,\overline{y}_{ij\cdot} - \overline{y}_{i\cdot\cdot})^2 \tag{A5.28a}$$

Because $\sigma^2(\,\overline{e}_{ij\cdot}) = \sigma_e^2/(n/2)$, it follows that $\overline{y}_{ij\cdot} \sim N(m_{ij}, 2\sigma_e^2/n\,)$. Applying Equation A5.14e gives

$$\sum_{j=1}^{2} (\,\overline{y}_{ij\cdot} - \overline{y}_{i\cdot\cdot})^2 = \frac{2\,\sigma_e^2}{n} \cdot \chi^2_{1,\lambda_i} \quad \text{where} \quad \lambda_i = \frac{n}{2\,\sigma_e^2} \cdot (\,m_{i1}^2 + m_{i2}^2) \tag{A5.28b}$$

Substituting this into Equation A5.28a and applying Equation A5.14c gives

$$\text{SS}_m \sim \sigma_e^2 \cdot \chi^2_{N,\lambda} \qquad \text{where} \quad \lambda = \frac{n}{2\,\sigma_e^2} \sum_{i=1}^{N} (\,m_{i1}^2 + m_{i2}^2) \tag{A5.28c}$$

Hence the marker sum of squares follows a noncentral χ^2 distribution with N degrees of freedom and noncentrality parameter λ. It thus remains to compute λ as a function of the unknown QTL parameters.

If the ith sire has marker–QTL genotype $Q_j M_1 / Q_\ell M_2$, then m_{i1} is the expected within-sire deviation for offspring carrying sire marker allele 1 given the sire has this genotype, or

$$m_{i1} = E\,(\,\overline{y}_{i1} - \overline{y}_i \,|\, \text{sire}_i = Q_j M_1 / Q_\ell M_2)$$

Recall from Equation 19.8 that

$$E\,(\,\overline{y}_{i1} \,|\, \text{sire}_i = Q_j M_1 / Q_\ell M_2) = \mu + c\,\alpha_\ell + (1-c)\,\alpha_j$$

while $E\,(\,\overline{y}_i \,|\, \text{sire}_i = Q_j / Q_\ell) = \mu + (\alpha_j + \alpha_\ell)/2$. Putting these together gives

$$m_{i1} = \mu + c\,\alpha_\ell + (1-c)\,\alpha_j - \left[\mu + \left(\frac{\alpha_j + \alpha_\ell}{2}\right)\right] = (0.5 - c)\,(\alpha_j - \alpha_\ell)$$

where α_j is the average effect of allele Q_j and c is the marker–QTL recombination frequency. Because $E(m_i) = 0$, it follows that $m_{i1} = -m_{i2}$ and hence

$$m_{i1}^2 + m_{i2}^2 = 2(0.5 - c)^2 (\alpha_\ell - \alpha_j)^2$$

Taking the expected value over all possible sire QTL genotoypes,

$$E(m_{i1}^2 + m_{i2}^2) = 2(0.5 - c)^2 E\left[\,(\alpha_\ell - \alpha_j)^2\,\right] = 2(0.5 - c)^2 \sigma_A^2$$

where the last step follows from $E[(\alpha_\ell - \alpha_j)^2] = 2E(\alpha^2) = \sigma_A^2$ (Equation 4.23b) being the additive variance attributed to the QTL. The resulting noncentrality parameter for the marker sum of squares becomes

$$\lambda = N \cdot n \left[\frac{(1-2c)^2}{4}\right] \left(\frac{\sigma_A^2}{\sigma_e^2}\right) \tag{A5.28d}$$

Thus, under the hypothesis of no linkage ($c = 0.5$), the test statistic is

$$\mathcal{F}_{m/e} = \frac{\text{MS}_m}{\text{MS}_e} \sim F_{N,N(n-2)}$$

A sire marker effect is declared to be significant at the α-level if $\mathcal{F}_{m/e} > f_{N,N(n-2),(1-\alpha)}$. The power of this test is the probability that the test statistic $\mathcal{F}_{m/e}$ exceeds this critical value,

$$\Pr\left[\, F_{N,N(n-2),\lambda} > f_{N,N(n-2),(1-\alpha)} \,\right] \tag{A5.29}$$

Example A5.7. Suppose we have a completely linked QTL (c = 0) with σ^2_A/σ^2_e = 0.1 (corresponding to a QTL heritability of 1/[1+10] = 0.09). With N = 20 (marker heterozygous) sires and n = 30 sibs/sire, what is the power of detecting this QTL using a significance level of α = 0.05? Here $N(n-2)$ = 560, and `qf(0.95,20,560)` returns 1.589 for $f_{20,560,0.95}$. Because λ = 20·30· 0.1/4 = 15, the power is $\Pr(F_{20,560,15} > 1.589)$, with `1- pf(1.589,20,560,15)` returning a value of 0.596.

How many sibs are required per sire to have 90% power? Solving with increasing n values, we find that for n = 53, $N(n-2)$ = 1020, $f_{20,1020,0.95}$ = 1.581, and λ = 20·36· 0.1/4 = 26.5 for a power of

$$\Pr(F_{20,1020,26.5} > 1.581) = 0.90$$

Power of a Random-effects ANOVA Design

Under a random-effects model, the simple one-way ANOVA model becomes $y_{ij} = \mu + t_i + e_{ij}$, where t is now a ***random variable*** drawn from a normal distribution with variance σ^2_t, so that $t_i \sim N(0, \sigma^2_t)$. Residuals are assumed independent with $e_{ij} \sim N(0, \sigma^2_e)$ as before. Assuming a balanced design with n replicates for each of the N t_i values, the treatment and error sums of squares are also given by Equation A5.20a. As with the fixed-effects model, $\text{SS}_e \sim \sigma^2_e \cdot \chi^2_{N(n-1)}$. However, the distribution of the treatment sum of squares remains a central χ^2, even when $\sigma^2_t > 0$. To see this, note that $\overline{y}_{i\cdot} \sim N(0, \sigma^2)$ with $\sigma^2 = \sigma^2_t + \sigma^2_e/n$. Applying Equation A5.13c, the same logic leading to Equation A5.21 now gives

$$\text{SS}_t \sim n \cdot \sigma^2 \cdot \chi^2_{N-1} = (n\,\sigma^2_t + \sigma^2_e) \cdot \chi^2_{N-1} \tag{A5.30}$$

Hence, the $\mathcal{F}_{t/e}$-ratio statistic is distributed as

$$\mathcal{F}_{t/e} = \frac{\text{SS}_t/(N-1)}{\text{SS}_e/N(n-1)} \sim \frac{(n\,\sigma^2_t + \sigma^2_e)\cdot \chi^2_{N-1}/(N-1)}{\sigma^2_e \cdot \chi^2_{N(n-1)}/N(n-1)}$$

$$\sim \left(1 + n\,\frac{\sigma^2_t}{\sigma^2_e}\right) F_{(N-1),N(n-1)} \tag{A5.31}$$

Note that the term in parentheses is just $E(\text{MS}_t)/E(\text{MS}_e)$. Hence

$$\frac{\mathcal{F}_{t/e}}{1 + n\,(\sigma^2_t/\sigma^2_e)} \sim F_{N-1,N(n-1)} \tag{A5.32a}$$

giving the power of this test as

$$\Pr\left[\, F_{N-1,N(n-1)} > \frac{f_{N-1,N(n-1),(1-\alpha)}}{1 + n\,(\sigma^2_t/\sigma^2_e)} \,\right] \tag{A5.32b}$$

Because, by definition (Equation A5.18b),

$$\Pr\left[\, F_{N-1,N(n-1)} > f_{N-1,N(n-1),\beta} \,\right] = 1 - \beta$$

to have power $1-\beta$ requires that

$$f_{N-1,N(n-1),(1-\alpha)} = \left(1 + n\,\frac{\sigma_t^2}{\sigma_e^2}\right) f_{N-1,N(n-1),\beta} \tag{A5.33}$$

Under more general random-effect models, if the test statistic is $\mathcal{F}_{x/y} = \text{MS}_x/\text{MS}_y$, then

$$\frac{\mathcal{F}_{x/y}}{R_{x/y}} \sim F_{n_x,n_y}, \qquad \text{where} \qquad R_{x/y} = \frac{E(\text{MS}_x)}{E(\text{MS}_y)} \tag{A5.34a}$$

and to obtain power $1-\beta$ requires choosing sample sizes such that

$$f_{n_x,n_y,(1-\alpha)} = \left[\frac{E(\text{MS}_x)}{E(\text{MS}_y)}\right] f_{n_x,n_y,\beta} \tag{A5.34b}$$

Example A5.8. Consider a random-effects model with design parameters similar to Example A5.6, with $\sigma_t^2/\sigma_e^2 = 1/3$, $N = 4$, and $n = 5$. What is the power of this design for a test with $\alpha = 0.05$? Here $f_{3,16,0.95} = 3.24$ and $1 + n\,(\sigma_t^2/\sigma_e^2) = 1 + 5/3 = 8/3$. From Equation A5.32b, the power becomes

$$\Pr\left(F_{3,16} > \frac{3.24}{8/3} = 1.215\right) = 0.336$$

What value of n is required to give 90% power? Trying increasing values of n, we eventually find that for $n = 38$, $N(n-1) = 148$, $1 + n\,(\sigma_t^2/\sigma_e^2) = 13.67$, $f_{3,148,0.95} = 2.67$, where

$$\Pr\left(F_{3,148} > \frac{2.67}{13.67} = 0.195\right) = 0.90$$

Comparing these results to those in Example A5.6 shows the reduced power of a random-effects analysis, relative to a fixed-effects one (where $n = 15$ is needed for 90% power).

Application: Power of the Half-sib Design for Variance Estimation

As an example of computing the power of random-effects ANOVA designs for estimating variance components, consider the standard half-sib design for estimating additive genetic variance. Here, a common parent (typically a sire) is crossed to a number of unrelated dams and a single offspring from each mating is measured. Under this design, the variance in sire effects, s_i, is an estimate of $\sigma_A^2/4$ (see Part II Chapter 23 for a more detailed discussion). Assuming a balanced design, N sires are each mated to n unique and unrelated females. The linear model is $y_{ij} = \mu + s_i + e_{ij}$, where $s \sim (0, \sigma_s^2)$ and $e \sim (0, \sigma_e^2)$. Assume further that s and e are normally distributed. Then, from Equation A5.30, the sire effect sum of squares, SS_s, follows a χ^2 distribution, with

$$\text{SS}_s \sim n \cdot \sigma^2 \cdot \chi^2_{N-1} = (n\,\sigma_s^2 + \sigma_e^2) \cdot \chi^2_{N-1}$$

Likewise, from Equation A5.31, the resulting test statistic $\mathcal{F}_{s/e}$ for a significant sire variance follows an F distribution, with

$$\mathcal{F}_{s/e} = \frac{\text{SS}_s/(N-1)}{\text{SS}_e/N(n-1)} \sim \left(1 + n\,\frac{\sigma_s^2}{\sigma_e^2}\right) F_{(N-1),N(n-1)} \tag{A5.35a}$$

Expressing the sire variance in terms of additive genetic and environmental variance components (ignoring epistasis),

$$\sigma_s^2 = \sigma_A^2/4 = (h^2/4)\sigma_z^2$$

Because $\sigma_z^2 = \sigma_s^2 + \sigma_e^2$, it immediately follows that

$$\sigma_e^2 = \sigma_z^2 - \sigma_s^2 = \sigma_z^2(1 - h^2/4)$$

Substituting these into Equation A5.35a gives

$$\mathcal{F}_{s/e} \sim \left(1 + \frac{n\,h^2}{4 - h^2}\right) F_{(N-1),N(n-1)} \tag{A5.35b}$$

From Equation A5.32b, the power of this test is

$$\Pr\left[F_{N-1,N(n-1)} > \frac{f_{N-1,N(n-1),(1-\alpha)}}{1 + n\,h^2/(4 - h^2)}\right] \tag{A5.35c}$$

Example A5.9. Suppose $h^2 = 0.25$, so that the (scaled) sire variance is $\sigma_s^2/\sigma_z^2 = h^2/4 = 0.0625$. What is the probability of detecting a significant sire variance (with a test of $\alpha = 0.05$) using 10 sires, each mated to 5 dams? Here $N = 10$ and $n = 5$, with $f_{N-1,N(n-1),(1-\alpha)} = f_{9,40,0.95} = 2.12$. Applying Equation A5.35c, the power of this design to detect a sire variance at least this large is

$$\Pr\left[F_{9,40} > \frac{2.12}{1 + 5 \cdot 0.25/(4 - 0.25)} = 1.59\right] = 0.15$$

Hence the power is very poor, with the test indicating a significant sire effect only 15% of the time.

If we keep the number of dams/sire constant at 5, how many sires are required to give this test 90% power? Trial and error shows that with $N = 255$ sires, the critical value becomes $f_{N-1,N(n-1),(1-\alpha)} = f_{254,1020,0.95} = 1.17$, and Equation A5.35c gives

$$\Pr\left[F_{254,1020} > \frac{1.17}{1 + 5 \cdot 0.25/(4 - 0.25)} = 0.88\right] = 0.90$$

which requires producing $255 \cdot 5 = 1275$ offspring.

Note from the form of Equation A5.35c that the number of dams per sire, n, appears to assert a more important role than N. Keeping N constant at 10, how many dams per sire are required to have 90% power? Again trial and error shows for $n = 47$ that $f_{N-1,N(n-1),(1-\alpha)} = f_{9,460,0.95} = 1.90$, and Equation A5.35c gives

$$\Pr\left[F_{9,460} > \frac{1.90}{1 + 47 \cdot 0.25/(4 - 0.25)} = 0.46\right] = 0.90$$

which requires scoring only 470 offspring.

As detailed in Part II Chapter 23, if a total of $T = Nn$ individuals are measured, the optimal number of sires per dam is given by $n = 4/h^2$. Hence $N = T/n = Th^2/4$, and the power under the optimal design for T individuals is

$$\Pr\left[F_{(\{Th^2/4\}-1),T(1-h^2/4)} > \frac{f_{(\{Th^2/4\}-1),T(1-h^2/4),(1-\alpha)}}{1 + 4/(4 - h^2)}\right] \tag{A5.36}$$

Table A5.1 gives the power under the optimal design for various values of T and h^2. Note that with modest, but reasonable, heritabilities ($h^2 = 0.25$), at least 250 individuals must be scored to have a reasonable chance of the half-sib design showing a significant sire variance.

Table A5.1 The power to detect a significant heritability under the optimal sampling scheme for a half-sib design as a function of h^2 and the total number T of individuals measured. For fixed T, the optimal number of dams per sire is $n = 4/h^2$, or 40, 16, 8, 5, and 4 under the heritabilities given below. Power is computed using the design closest to these optimal values, given the constraint that the total number of individuals in the balanced design is T. For example, for $h^2 = 0.1$, $n = 40$ is the optimal value. However, because design must have $N \geq 2$, for $T = 50$, power is computed using $N = 2$ and $n = 25$.

	h^2				
T	0.1	0.25	0.5	0.75	1.0
50	0.12	0.22	0.36	0.47	0.59
100	0.19	0.36	0.58	0.70	0.82
250	0.35	0.65	0.87	0.96	0.99
500	0.57	0.86	0.99	1.00	1.00
750	0.70	0.95	1.00	1.00	1.00
1000	0.80	0.99	1.00	1.00	1.00

For Example A5.9, where $h^2 = 0.25$ and $T = 50$, the optimal number of dams per sire is $n = 4/h^2 = 16$. Rounding up to approach $T = 50$, we take $N = 3$ and $n = 17$ for $T = 51$. Under this design, the power is computed to be 0.234. For $N = 2$ and $n = 25$ ($T = 50$) the power is 0.224, while for $N = 4$ and $n = 12$ ($T = 48$) the power is 0.211. As expected, the optimal design ($n = 17$) does indeed show the largest power.

Appendix 6

Multiple Comparisons: Bonferroni Corrections, False-discovery Rates, and Meta-analysis

Hanlon's razor: *Never attribute to malice that which is adequately explained by stupidity.*

The most entertaining outcome is the most likely—my variant on Occam's razor. Elon Musk

Often one is faced with interpreting a list of p values (the probabilities of the observed outcomes under their null hypotheses), either from a set of independent experiments all testing the *same* hypothesis, or from a single experiment wherein a large number of *different* hypotheses are tested. Both of these are examples of **multiple comparisons**. In the former setting, the problem is how to best combine information from multiple, often rather disparate, studies to make more global statements. In the latter setting, our concern is controlling error over the entire *collection* of tests from a single experiment.

Issues with combining information across studies can be broken into methods for combining p values and more general methods for combining information on *effect size*—even if an effect is *statistically significant*, is it *biologically important*? Given that methods for combining p values and issues related to multiple comparisons within a single experiment share some common features, we examine these first. We then turn to a more detailed look at the field of **meta-analysis** (**MA**), namely, how to most efficiently combine information on effect sizes over a series of studies.

The statistical analysis of a large dataset (or a *collection* of datasets) typically involves not just testing a *single* hypothesis, but rather testing *many* (often *very* many) hypotheses. This is especially true in the genomics era, wherein a single **high-dimensional experiment** may test tens-of-thousands to millions of hypotheses (such as treatment-dependent expression over all of the genes in a genome or marker effects for millions of SNPs). For any particular test, we may assign a preset probability, α, of a **Type I error** (i.e., a **false positive**, declaring a test to be **significant**, namely those tests with $p \leq \alpha$, when in fact the null hypothesis is true; Appendix 5). Under this broad setting, there are two different strategies for controlling error. First, if we expect that the ***vast majority of the hypotheses are truly null***, we are interested in controlling the **experiment-wide error rate**, the probability of a false positive over *all* of the tests (as essentially all will be nulls). The standard approach in this setting is the classic **Bonferroni correction**—obtaining an experiment-wide error rate of π over a set of n comparisons by declaring a test to be significant when $p \leq \alpha = \pi/n$. However, this is usually far too stringent and results in an enormous loss of power. We review sequential-testing schemes to improve this approach, but often the adjustment for multiple comparisons is best accomplished by a ***shift in thinking***.

If we expect some ***reasonable number of the hypotheses to be false***, then trying to avoid *any* false positives is not appropriate. Instead, controlling the *fraction* of false positives among the tests that are declared significant (**discoveries**, $p < \alpha$) is a much better aim. This is especially true in large-scale exploratory experiments whose aim is to discover potential candidates for further study. In this setting, we attempt to control the **false-discovery rate** (**FDR**), namely, the fraction of all tests ***declared to be significant*** that are false positives, as opposed to the Type I error (false-positive) rate. The goal is to find a value, τ, such that the set of tests declared to be significant using $p \leq \tau$ has the desired FDR. As we will show below, the distinction between a false *positive* and a false *discovery* is very subtle but critically important. The former is simply the probability that a test under a true null will be declared significant, while the latter further depends on whether the null is indeed correct. If the null is incorrect, then the probability of a false discovery is zero, while if the null is correct, then the FDR equals the probability of a false positive.

Our treatment of these topics is organized as follows. First, we examine methods for combining p values over independent tests. We then turn to controlling the overall false-positive rate for a collection of tests from a single experiment through the use of Bonferroni corrections and their extensions. Given that the decision to control false positives versus false discoveries hinges to a large extent on the fraction, π_0, of the hypotheses that are truly null, we examine how π_0 can be estimated from the empirical distribution of the p values for a set of tests. Next, we discuss approaches for controlling the FDR (settings where we *expect* some nontrivial fraction of the hypotheses to depart from the null). Finally, we turn to more formal issues in meta-analysis, developing a mixed-model framework for combining estimates across studies. We also examine metrics and graphical tools for assessing possible sources of bias in such an analysis.

COMBINING p VALUES OVER INDEPENDENT TESTS

Hypotheses of interest are often tested in multiple studies, and a key issue is how best to combine the results from a set of such studies into a single global statement (the statistical field of MA—the analysis of analyses). Such studies fall into two categories: repeated tests of the *same hypothesis* (i.e., a SNP marker has no effect on a specific disease over multiple studies or populations), or a collection of tests of *different hypotheses* (i.e., a single SNP marker has no effect on a given *set* of traits/diseases). The most obvious approach (under the first setting of testing the same hypothesis) is simply to pool all of the data and perform a single test, but for a variety of reasons this is often not feasible. For example, different tests of the same hypothesis may involve different methodologies or very different study settings. Further, published papers may not report the full dataset, but rather may just present a few summary statistics. In such settings, one straightforward approach is to consider the list of p values for the collection of experiments that all purport to test the same hypothesis, and try to obtain a single global p value for this entire set. As with normal hypothesis testing, care must be taken in distinguishing between one- and two-tailed hypotheses. For example, we could get a significant p value for a new treatment over a series of tests because it is significantly *worse*, and we would draw an incorrect conclusion if we used combined p values for a two-tailed test when a one-tailed test (the treatment results in an improvement) is more appropriate.

This simple question of how best to combine p values from a set of experiments is potentially fraught with peril for several reasons. First, there is the issue of whether the different studies all really test the same hypothesis (when this is the goal). The investigator must take care to ensure that this is correct before proceeding. Second, there is concern about the so-called **file-drawer effect**, wherein nonsignificant results remain in the file drawer (i.e., are not published), leading to published results being biased toward smaller p values. One general trend seems to be a publication bias in studies with small sample sizes (where nonsignificant results are often not reported), but a reduction in this bias for larger samples (Easterbrook et al. 1991; Dickersin et al. 1992). The presumptive reason for this sample-size effect in publication bias is that smaller studies often lack power, so a nonsignificant result does not necessarily provide strong evidence that the null hypothesis is correct. Conversely, due to the higher power of larger studies, authors may feel more comfortable in publishing negative results. We examine several approaches that attempt to quantify the amount of any such **publication bias** in a MA at the end of this Appendix.

Fisher's χ^2 Method

Fisher (1932b) was among the first to offer a simple approach for combining p values (along with Tippett 1931; see Equation A6.4a), based on the important concept that ***the distribution of realized p values under the null follows a uniform distribution over (0,1)***, namely a **unit**

uniform random variable. Further, if $p \sim$ Uniform(0,1), then $-2\,\ln(p) \sim \chi^2_2$ (Pearson 1938). Hence, under the null, twice the negative natural log of a p value follows a chi-square distribution with two degrees of freedom. If we have k independent tests, then the sum of their log-transformed p values is the sum of k chi-square variables. Such a sum is itself chi-square, with the degrees of freedom given by the sum of the degrees of freedom for the individual chi-squares (Equation A5.13a). These observations form the motivation for **Fisher's combined-probability test**: for k independent tests, where p_i denotes the p value for test i, the sum

$$X^2 = -2\sum_{i=1}^{k} \ln(p_i) \tag{A6.1a}$$

approximately follows a χ^2_{2k} distribution. Fisher's method is a special case of the more general **Gamma method** (Lancaster 1961; Zaykin et al. 2007; Biernacka et al. 2012), based on inverses of gamma functions (Table A7.1). Other approaches, such as results based on the distribution of the sum of n unit uniforms, have also been proposed (e.g., Edgington 1972), see Folkes (1984), Kocak (2017), and Heard and Rubin-Delnchy (2018) for a brief overview.

Example A6.1. Suppose five different groups collected data to test the same hypothesis, and these groups (perhaps using different methods of analysis) reported p values of 0.10, 0.06, 0.15, 0.08, and 0.07. Notice that while none of these individual tests are significant, the trend is clearly that all are "close" to being significant ($\overline{p} = 0.09$). Fisher's statistic becomes

$$X^2 = -2\sum_{i=1}^{5} \ln(p_i) = 24.3921 \qquad \text{with} \qquad \Pr(\chi^2_{10} \geq 24.39) = 0.0066$$

Hence, when taken together, these five tests show a highly significant p value. This is reasonable, as the p values, while not individually significant, still all cluster near small values, while under the null, $\overline{p}$ should be around 0.5 (the average of a unit uniform). In particular, note that two tests each with $p = 0.06$ ($X^2 = 11.25, \Pr[\chi^2_4 \geq 11.25] = 0.024$) offer more support against the null than a single test with $p = 0.05$.

Two common settings for the use of combined p value approaches are gene-based GWAS (whether one, or more, markers within a defined genomic region show a marker effect; Chapter 20), and pathway-based GWAS (one, or more, *genes* within a pathway show an effect; Chapter 21). In the latter setting, there are often two layers of p value combining, first within each gene to obtain gene-wide p values, and then combining these gene-based values to obtain a pathway p value. This can involve using different p value combining methods at these different levels, see Chapter 21 for details. In both gene-, and pathway-, based GWAS, one often expects a number of the tested hypotheses (markers within a gene, genes within a pathway) will show no effects, and their inclusion can result in lower power. One approach for handling settings wherein a fraction of the tests are expected to be true nulls follows by first noting that Equation A6.1a can be expressed as a product,

$$X^2 = -2\,\ln\left(\prod_{i=1}^{k} p_i\right) \tag{A6.1b}$$

One then takes a *truncated* product (Equation A6.1b with $k^* < k$, focusing on the smallest k^* values), which filters out p values from true nulls (which tend to be larger) and then compares this truncated value with that expected under the universal null (all k of the p values are draws from unit uniforms). See Zaykin et al. (2002), Dudbridge and Koeleman (2003), and Sheng and Yang (2013) for details on such **truncated-product** approaches.

Finally, Rice (1990; also see Whitlock 2005) noted that a problem with Fisher's method is that smaller p values are differentially weighted compared to complementary larger p values (e.g., p versus $1 - p$). Equation A6.1a can be rearranged to yield

$$X^2 = -2k\ln(\overline{p}_G) \tag{A6.1c}$$

where $\overline{p}_G$ is the geometric mean of the individual p values, which differentially weights smaller values. Under Fisher's method, an observed p value of (say) 0.001 receives more weight than a complementary value of 0.999, which is as extreme (with $-\ln$ weights of 6.9 versus 0.001). However, under the Z-score transformation—which is obtained by solving $\Pr(U > Z) = p$, where $U \sim \mathrm{N}(0,1)$—the two complementary p values are of equal magnitude (Z scores of -3.09 and 3.09). This motivates the Z-score method, which we now consider.

Stouffer's Z Score

An alternative to Fisher's approach for combining p values was offered by Stouffer et al. (1949; and independently by Liptak 1958), who transformed the individual p values into Z scores (unit normal random variables). The sum of k independent unit normals is itself normal, with a mean of zero and a variance of k. These results lead to **Stouffer's Z score** method: assign a score of Z_i for test i by solving $\Pr(U > Z_i) = p_i$. Let Z_s denote the sum over the transformed p values of k tests, scaled by $k^{-1/2}$ to give it a variance of one, with

$$Z_s = \frac{\sum_{i=1}^k Z_i}{\sqrt{k}} \tag{A6.2a}$$

Because $Z_s \sim N(0,1)$, the overall p value is obtained as

$$p = \Pr(U > Z_s) \tag{A6.2b}$$

As noted by Whitlock (2005), this test was first proposed in a footnote in Stouffer et al.'s sociological study of army life, making it one of the more obscure origins of a statistical method.

Example A6.2. Reconsider the data from Example A6.1. The Z_i values are easily obtained using R, as the command `qnorm(1-p)` returns Z satisfying $\Pr(U \leq Z) = 1 - p$, or (equivalently) $\Pr(U > Z) = p$. For example, Z_1 is calculated by `qnorm(1-0.1)`, or 1.281. Similarly computing the other Z_i values yields

$$\sum_{i=1}^{5} Z_i = 6.754, \qquad \text{hence} \qquad Z_s = \frac{6.754}{\sqrt{5}} = 3.020$$

Because $\Pr(U > 3.020) = 0.00126$, as in Example A6.1, the combined p value is highly significant.

Besides providing symmetric values for large and small p values (i.e., p and $1-p$), a second advantage of the Z-score approach is that one can individually *weight* p values from different tests (Mosteller and Bush 1954; Liptak 1958), as the weighted sum of unit normals is itself a unit normal (while the weighted sum of χ^2 variables—the analog for Fisher's test—is considerably more complex). The resulting weighted version becomes

$$Z_w = \frac{\sum_{i=1}^k w_i Z_i}{\sqrt{\sum_{i=1}^k w_i^2}} \tag{A6.2c}$$

where $Z_w \sim N(0, 1)$. As expected, Z_w (Equation A6.2c) reduces to Z_s (Equation A6.2a) when all the weights are equal. One can either weight by the degrees of freedom in the study or by the reciprocal of the standard error of the estimate. Whitlock (2005) showed that the weighted Z-score method is superior to either X^2 or Z_s when sample size varies over the data. Z_w has higher power and also a higher correlation between its predicted p value and the actual p value obtained if one were able to merge all the samples.

No single p value combining method is uniformly most powerful (Birnbum 1954; Westberg 1985; Loughin 2004; Cheng and Sheng 2017; Kocak 2017; Heard and Rubin-Delanchy 2018). Rather, the choice of method depends on what fraction of the tests are actually false. When most of the tests are true (i.e., from nulls), then Tippett (Equation A6.4a) and Simes (Equation A6.4b) methods discussed below are more powerful, while when most of the tests are false (i.e., most are from the alternative) then Fisher (Equation A6.1a) and Stouffer (Equation A6.2a) are more powerful.

Finally, as detailed later in this Appendix, combining p values is one of the *least powerful* MA approaches, as it leaves much of the information from a collection of studies underutilized. A formal MA requires that studies report standard errors for their estimates. Unfortunately, many studies do not, and simply report p values instead, and in this setting the proceeding methods are the only type of MA available.

BONFERRONI CORRECTIONS AND THEIR EXTENSIONS

We now turn to the complementary problem of determining the significance level, α, for *individual tests* required to control the overall false-positive rate (FPR) over a *collection* of n tests. The typical setting is that a single study or experiment has gathered data and a number of different tests, usually on *different* hypotheses, are performed using these data. Let π denote our desired experiment-wide FPR—the probability of one (or more) false positives *over the entire collection* of n tests being no greater than π. The traditional approach for determining the appropriate α, given n and π, is to use **Bonferroni corrections**.

Standard Bonferroni Corrections

The probability of not making any Type I errors (false positives) over n independent tests, each at level α, is $(1-\alpha)^n$. Hence, the probability, π, of having at least one false positive over the entire collection is simply one minus this value,

$$\pi = 1 - (1-\alpha)^n \tag{A6.3a}$$

If we solve for the α value required for each test,

$$\alpha = 1 - (1-\pi)^{1/n} \tag{A6.3b}$$

This is often called the **Dunn–Ŝidák method**. If we note that $(1-\alpha)^n \simeq 1 - n\alpha$, we obtain the **Bonferroni method** by taking

$$\alpha = \pi/n \tag{A6.3c}$$

Both Equations A6.3b and A6.3c are referred to as **Bonferroni corrections**. In the literature, π is the **family-wide error rate** (**FWER**; also the **genome-wide error rate**, **GWER**), while α is the **comparison-wise error rate** (**CWER**), also referred to as the **point-wise significance level** (**PWSL**).

Example A6.3. Suppose we have $n = 100$ independent tests and wish an overall π value of 0.05. What value of α should be used for each individual test to achieve an experiment-wide FPR of 0.05? The Dunn–Ŝidák correction suggests

$$\alpha = 1 - (1-0.05)^{1/100} = 0.000512$$

while the Bonferroni correction is

$$\alpha = 0.05/100 = 0.0005$$

Note that using such small α values greatly reduces the power for any single test. For example, under a normal distribution, the 95% (two-sided) confidence interval (CI) for the true mean is $\overline{x} \pm 1.96\sqrt{\text{Var}}$, where Var denotes the variance of the sample mean. Moving to an α value of 0.0005 gives the associated CI as $\overline{x} \pm 3.48\sqrt{\text{Var}}$, as $\Pr(|z| \geq 3.48) = 0.0005$ for $z \sim N(0,1)$.

Tippett's Combined p-value Estimator

Rearranging Equation A6.3a yields **Tippett's method** (1931), also called the **Bonferroni-adjusted minimum *p* value**, for combining p values from independent tests. Here π is interpreted as the combined p value of n tests (p_{Tip}), while α is replaced by the smallest p value in the series, $p(1)$, yielding

$$p_{Tip} = 1 - [1 - p(1)]^n \simeq np(1) \tag{A6.4a}$$

Wilkinson (1951) generalized this method to using the jth smallest p value.

Sequential Bonferroni Corrections

Under a strict Bonferroni correction, only those tests whose associated p values are $\leq \pi/n$ are rejected (**declared significant**); all others are **accepted** (or more formally, **fail to be rejected**). This results in a considerable reduction in power if two or more of the hypotheses are actually false. When we reject a hypothesis, one fewer test remains, and the multiple comparison correction should reflect this, resulting in **sequential Bonferroni corrections**. Sequential approaches have increased power compared to standard Bonferroni corrections, as illustrated below in Example A6.4. Shaffer (1995) reviewed these, and other, approaches. The basic structure is that one has a collection of multiple tests, with $H(i)$ denoting the null hypothesis for test i—e.g., the test that marker i has a nonzero effect, in which case $H(i)$ is the null hypothesis of no effect. Rejecting $H(i)$ suggests evidence for a nonzero effect for marker i.

Holm's Method

The simplest of these sequential adjustments is **Holm's method** (Holm 1979). The first step is to order the p values for the n hypotheses being tested from smallest to largest, $p(1) \leq p(2) \leq \cdots \leq p(n)$, and let $H(i)$ be the hypothesis associated with $p(i)$. One proceeds with Holm's method as follows:

- (i) If $p(1) > \pi/n$, accept all n null hypotheses (i.e., none are declared significant).
- (ii) If $p(1) \leq \pi/n$, reject $H(1)$ (i.e., $H(1)$ is declared significant), and consider $H(2)$.
- (iii) If $p(2) > \pi/(n-1)$, accept $H(i)$ (for $i \geq 2$).
- (iv) If $p(2) \leq \pi/(n-1)$, reject $H(2)$ and move onto $H(3)$.
- (v) Proceed with rejecting hypotheses until reaching the first i such that $p(i) > \pi/(n-i+1)$.

We can also apply Holm's method using Equation A6.3a—namely, $\alpha = 1-(1-\pi)^{1/n}$, the Dunn–Ŝidák correction—in place of $\alpha = \pi/n$.

Simes–Hochberg Method

With Holm's method, we stop once we fail to reject a hypothesis. An improvement on this approach is the **Simes–Hochberg correction** (Simes 1986; Hochberg 1988), which effectively *starts backward*, working with the largest p values first.

(i) If $p(n) \leq \pi$, then all hypotheses are rejected.

(ii) If not, $H(n)$ cannot be rejected, and we next examine $H(n-1)$.

(iii) If $p(n-1) \leq \pi/2$, then all $H(i)$ for $i \leq n-1$ are rejected.

(iv) If not, $H(n-1)$ cannot be rejected, and we compare $p(n-2)$ with $\pi/3$.

(v) In general, if $p(n-i) \leq \pi/(n-i+1)$, then all $H(i)$ for $i \leq n-i$ are rejected.

While the Simes–Hochberg approach is more powerful than that of Holm's (see Example A6.4), it is only strictly applicable when the group of tests being jointly considered are independent, whereas Holm's approach does not have this restriction. Hence, the general strategy is to use Holm's method if one is concerned about potential dependencies between tests, and the Simes-Hochberg's method if the tests are independent. Finally, following the logic leading to Equation A6.4a, the Simes correction can also be used to obtain yet another combined p value over a series of n tests, with

$$p_{Simes} = \min_k \left[\frac{n\,p(k)}{k} \right] \tag{A6.4b}$$

Hommel's Method

Hommel's method (1988) is slightly more complicated, but it is more powerful than the Simes–Hochberg correction (Hommel 1989). Under Hommel's method, we reject all hypotheses whose p values are less than or equal to π/k^*, where

$$k^* = \max_i \left(p(n-i+j) > \pi \frac{j}{i} \right) \quad \text{for all } j = 1, \cdots, i$$

Example A6.4 shows how all three of these methods are applied.

Example A6.4. Suppose for $n = 10$ tests, the (ordered) p values are as follows:

i	1	2	3	4	5	6	7	8	9	10
$p(i)$	0.0020	0.0045	0.0060	0.0080	0.0085	0.0090	0.0175	0.0250	0.1055	0.5350
$\frac{\pi}{n-i+1}$	0.0050	0.0056	0.0063	0.0071	0.0083	0.0100	0.0125	0.0167	0.0250	0.0500

For an experiment-wide level of significance of $\pi = 0.05$, the Bonferroni correction is $\alpha = 0.05/10 = 0.005$. Hence, using a strict Bonferroni, we reject hypotheses 1 and 2, and we fail to reject (i.e., we accept) 3 through 10. To apply sequential methods, we use the associated $\pi/(n-i+1)$ values for $\pi = 0.05$, which are given above in the table. Under Holm's method, $p(i) \leq \pi/(n-i+1)$ for $i \leq 3$, and hence we reject H(1) through H(3) and accept the others. Under Simes–Hochberg, we fail to reject H(7) through H(10) [as $p(i) > \pi/(n-i+1)$], but because $p(6) = 0.009 \leq \pi/(n-i+1) = 0.010$, we reject H(6) through H(1).

To apply Hommel's method, we reject all hypotheses whose p values are less than or equal to π/k^*, where

$$k^* = \max_i \left(p(n-i+j) > \pi \frac{j}{i} \right) \quad \text{for all } j = 1, \cdots, i$$

Solving k^* requires an iterative approach, as follows. First, start with $i = 1$. Here, (i = 1, j = 1), $p(10) = 0.5350 > \pi \cdot (1/1) = 0.05$. Now let us try $i = 2$, which yields (for $j = 1, 2$), $p(9) = 0.1055 > \pi(1/2) = 0.025$ and (as above) $p(10) > \pi$. Hommel's condition still holds for $i = 3$, as $p(8) = 0.025 > \pi \cdot (1/3) = 0.0167$, $p(9) > \pi \cdot (2/3) = 0.033$, and $p(10) > \pi$. However, it fails for $i = 4$, as while it holds for $p(7) = 0.175 > \pi \cdot (1/4) = 0.0125$, it fails for ($i = 4, j = 2$) because $p(8) = 0.025 = \pi \cdot (1/2)$. Hence, $k^* = 3$ (because Hommel's condition holds for $k = 3$ but not for $k = 4$), and we reject all hypotheses whose p values are

$\leq 0.05/3 = 0.0167$, which means $H(1)$ through $H(6)$. Note that a strict Bonferroni rejected the fewest null hypotheses and Simes–Hochberg and Hommel's rejected the most null hypotheses (i.e., declared them to be significant), while all methods controlled the experiment-wide false-positive rate at 0.05.

Finally, we can use this toy set of p values to contrast the Simes and Tippett estimates for the combined p value for this sample of tests. For this data set, $p_{Simes} = \min_k [np(k)/k]$, takes on its smallest value at $k = 6$ (Example A6.9), yielding

$$p_{Simes} = \frac{10\, p(6)}{6} = \frac{10 \cdot 0.0090}{6} = 0.015$$

while

$$p_{Tip} = 1 - [1 - p(1)]^{10} = 1 - (1 - 0.002)^{10} = 0.0198$$

Note that $p_{Simes} \leq p_{Tip}$, as p_{Tip} represents the special case of p_{Simes} (the minimum value occurs at $k = 1$).

Dealing with Dependency: Eigenvalue-based Approaches

When different tests share correlated data, this introduces dependency between the p values for these tests. How do we account for this? One approach (Cheverud 2001; Nyholt 2004; Li and Ji 2005; Patterson et al. 2006; Li et al. 2011) is to use the nature of the dependency structure of the data to estimate an **effective number of independent tests**, n_e. This value is then substituted for n in the above methods; e.g., Equation A6.3b becomes $\alpha = 1 - (1 - \pi)^{1/n_e}$. Likewise n_e replaces n in the Tippett (Equation A6.4a) and Simes (Equation A6.4b) estimators. A classic application of this approach is correcting for correlations among tests of marker–trait associations over a set of linked markers in either a QTL mapping experiment or a GWAS (Chapters 18–20).

To proceed, we need to introduce a few facts about the eigenstructure of a *correlation matrix*, $\mathbf{C}$, whose eigenvalues are denoted (from largest to smallest) by $\lambda_1, \cdots, \lambda_n$. First, because $\mathbf{C}$ is a positive-semidefinite matrix, all $\lambda_i \geq 0$ (WL Appendix 5). Second, $\mathbf{C}$ is an $n \times n$ matrix with ones on its diagonal, which makes its trace (the sum of its diagonal elements; Chapter 9) equal to a value of n. The importance of this result is that the trace of a matrix equals the sum of its eigenvalues (Equation 9.34b), which demonstrates that the average eigenvalue of $\mathbf{C}$ is

$$n^{-1} \sum_{i=1}^{n} \lambda_i = n^{-1} n = 1$$

When all of the underlying variables that generate $\mathbf{C}$ are uncorrelated, then $\lambda_1 = \cdots = \lambda_n = 1$, while when all of the observations are completely correlated, then $\lambda_1 = n$ and $\lambda_2 = \cdots = \lambda_n = 0$. These two cases represent the extremes of n independent tests (the former) and one independent test (the latter). As with principal components (Chapter 9), the spread of the eigenvalues tells us about dependency. One metric of this is the variance in the eigenvalues, $\sigma^2(\lambda)$. If all of the eigenvalues are equal, then $\sigma^2(\lambda) = 0$, while if only one eigenvalue is nonzero, then $\sigma^2(\lambda) = n$.

Motivated by the above eigenstructure observations, **Cheverud's method** (2001) computes the effective number of independent tests as

$$n_{e,Cheverud} = n\left(1 - \frac{(n-1)\sigma^2(\lambda)}{n^2}\right), \qquad \text{where} \qquad \sigma^2(\lambda) = \frac{1}{n-1}\sum_{i=1}^{n}(\lambda_i - 1)^2 \quad \text{(A6.5a)}$$

This returns $n_e = n$ when $\sigma^2(\lambda) = 0$ and $n_e = 1$ when $\sigma^2(\lambda) = n$, which matches the expected results from the eigenvalue analysis for these extreme cases. A closely related variance-based estimator was suggested by Patterson et al. (2006).

Li and Ji (2005) noted that Cheverud's method returns an overly large value for n_e (and therefore, less power), especially when used with a large number of moderately correlated tests. While Cheverud's approach considered two extreme cases (n versus one independent test), Li and Ji noted that a set of c identical tests will result in an eigenvalue of $c > 1$, while tests that are only partially correlated with others will generate eigenvalues values <1. Hence, they partitioned an eigenvalue into two parts, its integer value and the remainder, where the integer part implies identical tests (and hence is counted as contributing one independent test), while the remainder represents partial correlations. Formally, the **Li–Ji method** is coded as

$$n_{e,Li-Ji} = \sum_{i=1}^{N} I(\lambda_i \geq 1) + \sum_{i=1}^{N} (\lambda_i - \text{floor}[\lambda_i]) \tag{A6.5b}$$

where the indicator function $I(x \geq 1)$ returns a value of one when $x \geq 1$, and otherwise returns a value of zero. The first sum in Equation A6.5b is the number of eigenvalues of **C** that are ≥ 1. The floor$[x]$ function in the second term corresponds to the largest integer $\leq x$, so the second sum is all of the remainder terms (the effects of partial correlations among tests). Because the sum of the eigenvalues for a correlation matrix is an integer, then the residual sum is also an integer, so that the Li–Ji method always returns an integer value of the effective number of tests. A modification of the Li–Ji approach (**GATES**) was offered by Li et al. (2011). They simply truncated all eigenvalues ≥ 1 to one, and then computed the sum as their estimate of the effective number of tests. Formally,

$$n_{e,GATES} = \sum_{i=1}^{N} I(\lambda_i \geq 1) + \sum_{i=1}^{N} [1 - I(\lambda_i \geq 1)]\ \lambda_i \tag{A6.5c}$$

Note that $n_{e,Li-Ji} > n_{e,GATES}$. Consider the following example from Nyholt (2004), where the eigenvalues of **C** were 7.84, 1.60, 0.24, 0.21, 0.08, 0.03, 0, 0, 0, and 0. Cheverud's method returns $n_{e,Cheverud} = 4.59$, for a critical value (for a 5% FWER) of $\alpha = 0.05/4.59 = 0.0109$. Under Li and Ji, the first two eigenvalues correspond to two independent tests, with the total of their noninteger residuals $0.84 + 0.60 + 0.24 + 0.21 + 0.08 + 0.03 = 2$, for a value of $n_{e,Li-Ji} = 4$ (yielding a critical value of $\alpha = 0.0125$), while $n_{e,GATES} = 1+1+0.24+0.21+0.08+0.03 = 2.56$ (for $\alpha = 0.0195$). Permutations by Nyholt found that a critical value of 0.015 yields a FWER of 0.05, for $n_e = \ln(0.95)/\ln(0.985) = 3.39$. Note (for this example) that while Li–Ji overestimated the effective number of tests, GATES underestimated it. However, their average, $(4+2.56)/2 = 3.28$, is actually very close to the permutation value. Additional corrections for dealing with correlated tests have been proposed by a number of authors (e.g., Zaykin et al. 2002; Owen 2005; Efron 2007; Leek and Storey 2007, 2008; Gao et al. 2008; Moskvina and Schmidt 2008; Galwey 2009; Li et al. 2012b).

Dealing with Dependency: Aggregated Cauchy Approaches

A recent alternative strategy (Pillai and Meng 2016; Liu et al. 2019; Liu and Xie 2020) for dealing with dependency among tests is built around using the **Cauchy distribution**, whose density function is given by

$$\varphi(x) = \frac{1}{\pi\,(1+x^2)} \quad \text{for} \quad -\infty < x < \infty \tag{A6.6a}$$

This is a **standard Cauchy**, akin to a unit normal. (More generally, the Cauchy has both a location x_0 and scale γ parameter, which do not concern us here, with the standard using $x_0 = 0$ and $\gamma = 1$.) The Cauchy arises as the distribution for the ratio of two unit normals (one can also think of it as a t distribution with one degree of freedom). If $\mathcal{C}$ is Cauchy distributed, then its cdf (cumulative distribution function) is given by

$$\Pr(\mathcal{C} \leq x) = \int_{-\infty}^{x} \frac{1}{\pi\,(1+x^2)}\, dx = \frac{\arctan(x)}{\pi} + \frac{1}{2} \tag{A6.6b}$$

and hence

$$\Pr(\mathcal{C} > x) = 1 - \left(\frac{\arctan(x)}{\pi} + \frac{1}{2} \right) = \frac{1}{2} - \frac{\arctan(x)}{\pi} \tag{A6.6c}$$

Recalling that arctan$(x) \to \pi/2$ as $x \to \infty$, shows that $\Pr(\mathcal{C} \leq x) \to 1$ as $x \to \infty$. From Equation A6.6b, it immediately follows that the quantile x_p — the value of x_p satisfying $\Pr[\mathcal{C} \leq x_p] = p$ — is given by solving $p = \arctan(x_p)/\pi + 1/2$, yielding

$$x_p = \tan\left[\pi(p - 1/2)\right] \tag{A6.6d}$$

Because tan$(x) \to \infty$ as $x \to \pi/2$, it follows that $x_p \to \infty$ as $p \to 1$.

For many students, the Cauchy has an unfortunate image, being mainly known as a pathologic distribution (its density function integrates to one, but none of its moments are finite). This feature is due to having heavy tails that do not decline sufficiently fast at large values. A less appreciated feature of the Cauchy is that the weighted sum of standard Cauchys is itself a Cauchy (and, hence, Cauchy random variables do not obey the central limit theorem). Building on this fact, the interest in using the Cauchy for combining p values started with a striking finding by Pillai and Meng (2016). Suppose that $\mathbf{y}$ and $\mathbf{x}$ are both vectors of unit MVNs with correlation matrix $\mathbf{V}$. Pillai and Meng showed that the weighted sum $\sum w_i\,(y_i/x_i)$ is a standard Cauchy. Noting that y_i/x_i itself is Cauchy, this implies (in least in this setting) that the weighted sum of Cauchy random variables with arbitrary dependency structure is still a Cauchy.

Recall that Stouffer's Z score approach translates individual p values into Z values (unit normals), and the weighted sum of these values (which is a normal) is back-transformed into an overall p value. This same approach, but now using the Cauchy, was proposed by Liu et al. (2019) with their **aggregated Cauchy association test** (**ACAT**). As a result of its heavy-tail, the Cauchy is largely insensitive to correlations among p values, especially when the p values are small (Liu et al. 2019; Liu and Xie 2020). The ACAT test statistic for combining p values is

$$T_{ACAT} = \sum_{i=1}^{n} w_i \tan[(0.5 - p_i)\pi] \tag{A6.6e}$$

namely the weighted ($w_i \geq 0$ and $\sum w_i = 1$) Cauchy values associated with the p values of each test (Equation A6.6d), resulting in the weighted sum also being a standard Cauchy. The associated overall p value follows from Equation A6.6c, with

$$p_{ACAT} \simeq 0.5 - \frac{\arctan[T_{ACAT}]}{\pi} \tag{A6.6f}$$

DETECTING AN EXCESS NUMBER OF SIGNIFICANT TESTS

While Bonferroni corrections (and their sequential counterparts) are widely used, when the number of tests is modest to large, their application significantly erodes power, leading to very high **Type II errors** (failing to declare a test significant when the null is false). This tradeoff of Type I versus Type II errors applies to most statistical tests (Appendix 5), and the error of more concern to the investigator determines how to proceed. In many cases, our initial experiment is simply an **enrichment method**: we wish to take a large number of possible hypotheses (e.g., testing each SNP on a chip for marker–trait associations) and extract a subset showing the most support for their alternative hypotheses for further consideration. In such cases, we are often more concerned with Type II errors, as the cost of rejecting a hypotheses that is truly null (i.e., improperly declaring it as significant; a Type I error) may be less than the cost of excluding a hypothesis from the alternative (failure to reject; i.e., a Type II error). In such settings, we would like to calculate the number of hypotheses (n_0; or equivalently, the fraction $\pi_0 = n_0/n$) of true nulls among the n tested hypotheses. The first step toward doing so is to ask if the observed number of significant tests is excessive under the **global null hypotheses** (all hypotheses are truly nulls).

How Many False Positives?

Suppose we perform n independent tests, each with a Type I error rate of α. If all hypotheses are truly null, the number, j, of false positives follows a binomial distribution (Wilkinson 1951), with a "success" probability (a false positive) of α, and n trials (the number of tests), yielding

$$\Pr(j \text{ false positives}) = \frac{n!}{(n-j)!\,j!}(1-\alpha)^{n-j}\,\alpha^j \tag{A6.7a}$$

For n large and α small, this is closely approximated by the Poisson distribution (Equation 2.21a), with Poisson parameter $n\alpha$ (the expected number of false positives), yielding

$$\Pr(j \text{ false positives}) \simeq \frac{(n\alpha)^j e^{-n\alpha}}{j!} \tag{A6.7b}$$

Example A6.5. Suppose 250 independent tests are performed, each with $\alpha = 0.025$ (a 2.5% chance of declaring a result from the null hypothesis to be significant), and 15 tests are declared significant by this criterion. Is this number greater than expected by chance? The expected number of significant tests under the global null hypothesis is $n\alpha = 250 \cdot 0.025 = 6.25$. From Equation A6.7a, the probability of observing 15 (or more) significant tests is

$$\sum_{j=15}^{250} \Pr(j \text{ false positives}) = \sum_{j=15}^{250} \frac{250!}{(250-j)!\,j!}(1-0.025)^{250-j}\,0.025^j$$

We could either sum this series directly or use the cumulative distribution function for a binomial. In R, the probability that a binomial with parameters n and p has a value of i or less is obtained by using the command `pbinom(i,n,p)`. The probability of 15 or greater is one minus the probability of 14 or less, or `1- pbinom(14,250,0.025)`, for which R returns 0.0018. A similar calculation can use the Poisson approximation (Equation A6.7b), with `1-ppois(14,6.25)` returning a value of 0.0021. Given that there is only a 0.2% chance of seeing these many significant tests under the global null, we expect that some of these significant tests are **true discoveries** (those whose associated null hypothesis is incorrect), not false positives. The critical question, of course, is which ones?

Testing for an excessive number of significant tests is a rather crude indicator of the actual number (n_0) of the n tests that are true nulls. It is very possible that $n_0 < n$ and yet we would not detect an excess of significant tests by the above method. Likewise, if an excessive number is detected, what really can we say about n_0 other than $n_0 < n$? For instance, Example A6.5 shows an excess of none significant tests (observed 15, expected 6), but clearly assuming $n_0 = n - 9$ is a bit naive. Finally, the outcome varies with our choice of α. One could easily imagine an excess of significant tests using $\alpha = 0.05$, but not when using $\alpha = 0.01$. Ideally, we would like to have an estimate for n_0 that is independent of the choice of α.

Such estimators follow from the key concept that if the null is correct, random draws of p values *follow a unit uniform distribution*. A more careful examination of the empirical distribution of p values over our sample of tests, rather than simply how many we declare significant, is the key to obtaining estimates of n_0.

Schweder–Spjøtvoll plots

A simple graphical approach using the empirical distribution of p values was suggested by Schweder and Spjøtvoll (1982). If one rank-orders the p values from the smallest $p(1)$ to the

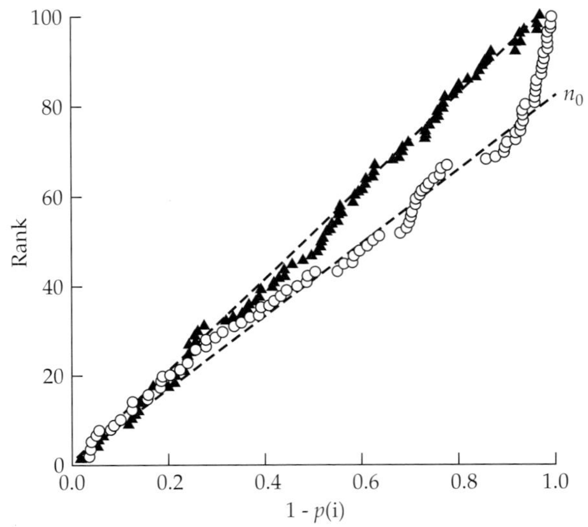

Figure A6.1 A Schweder–Spjøtvoll plot is one approach for detecting departures from a uniform distribution of p values. The p values are ordered from smallest, $p(1)$, to largest, $p(n)$, and one plots the rank of $1 - p(i)$ versus its value. These ranks are reversed from the ranks of $p(i)$, as the rank of $1 - p(n)$, being the smallest value, is 1. Under a uniform, the result is a straight line passing through the origin and the point $(1, n)$. The upper curve (solid triangles), generated by randomly sampling $n = 100$ values from a uniform (0,1), fits this pattern. The lower curve (open circles), generated by simulating p values for 80 true nulls and 20 tests where the alternative was correct, shows an inflation of p values near zero ($1-p$ values near one). This results in a strong departure from linearity near one. Ignoring this upturn and extrapolating the linear fit for the values below this inflection point gives an approximate value of 80 for the value of this projected line when $1 - p = 1$. This yields an estimate of n_0.

largest $p(n)$, a plot of $p(i)$ versus i is a straight line under a uniform. Because our interest is usually in detecting an excessive number of small p values (as would be expected if $n_0 < n$), Schweder and Spjøtvoll suggest plotting $1 - p(i)$ values on the horizontal axis, and then ranks of these values (which are the reverse of the ranks of the $p[i]$) on the vertical axis. For example, the first point is $(1-p[n], 1)$, the second $(1-p[n-1], 2), \cdots$, and the nth $(1-p[1], n)$. If all of the p values are indeed generated from null hypotheses, then these are drawn from a uniform, and the resulting plot will be a straight line (the solid triangles in Figure A6.1). Conversely, if some of the p values are drawn from hypotheses where the null is false, we expect an excess of small p values, and hence an over-abundance of $1 - p$ values near one (the open circles in Figure A6.1).

In addition to providing a quick visual check as to whether the p values follow a uniform, Schweder and Spjøtvoll suggested that these plots can also estimate n_0. One fits the best straight line until the upturn (i.e., inflection point) near one appears, extrapolating this line to obtain the n value for $1-p = 1$ estimates the number of true null hypotheses, n_0. One could use the Chow test (Equation 10.16b) to estimate this inflection point. As shown in Figure A6.1, this gives a value very close to 80 (for the open circles), the correct number of true nulls used to generate this figure.

Estimating n_0: Subsampling From a Uniform Distribution

As suggested by the Schweder–Spjøtvoll plot, the distribution of p values offers insight into the number of truly null hypotheses, n_0. While this plot offers either a simple visual or a more formal regression-based estimator of n_0, it tends to overestimate the number of nulls. Further, it can be difficult to specify exactly where the upturn in the plotted values begins. A number of other estimators have been suggested, again based on a uniform distribution of p values for those tests under the null. Recall that the histogram from a sufficiently large number of draws from a uniform distribution is flat, as all values are equally likely (Figure

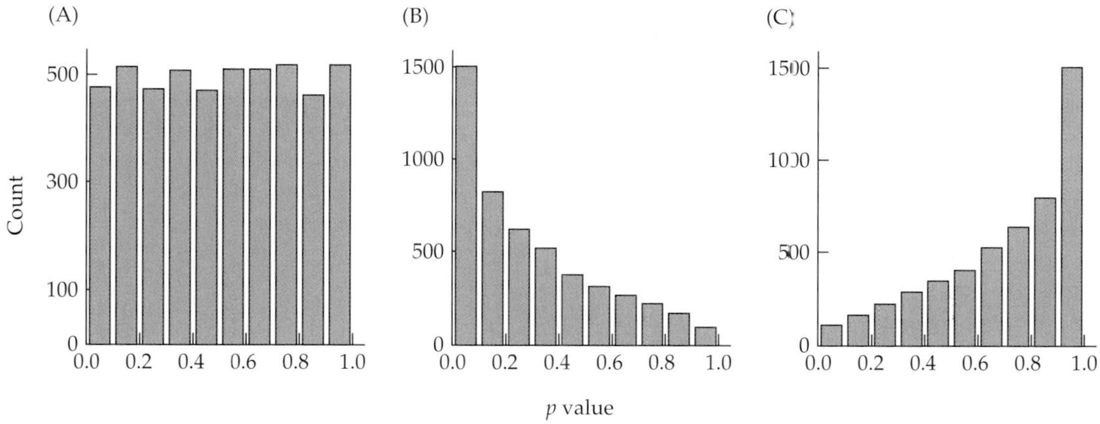

Figure A6.2 Simulated distribution of p values based on 5,000 tests for samples of 25 draws from a normal distribution with a mean of μ and a variance of one. The null hypothesis is $H_0 : \mu \leq 0$. **A:** The distribution of p values when $\mu = 0$ (the null is correct) is uniform. **B:** The distribution when $\mu = 0.2$ is skewed toward an excess of values near zero. **C:** The distribution when $\mu = -0.2$ is skewed toward an excess of values near one.

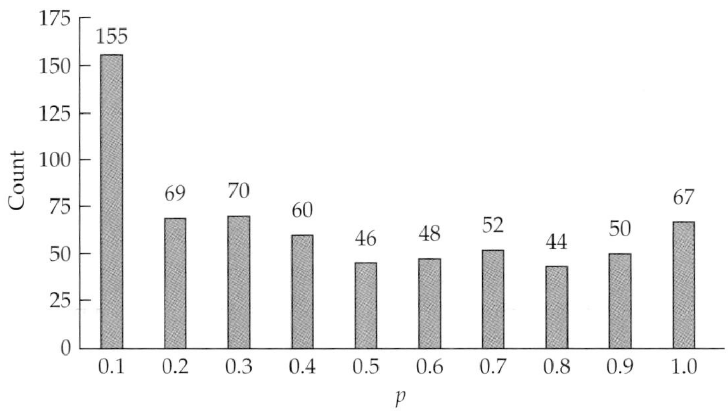

Figure A6.3 An empirical distribution of p values (for n = 644 tests) from Mosig et al. (2001). The number of p values in each of ten bins (of width 0.1) is given above the bars. Note the large excess of values near zero.

A6.2A). However, if the null is false for at least some of the tests, then the distribution of p values is shifted away from uniform, and usually with a skew toward smaller values (Figure A6.2B), but potentially also skewed toward one (e.g., if one-tailed tests are used when a two-tailed test is appropriate; Figure A6.2C).

If the collection of tests contains some alternative hypotheses mixed in with true nulls, we expect the distribution to be a mixture, with fraction, $\pi_0 = n_0/n$, consisting of draws from a uniform and the remaining fraction $(1 - n_0/n)$ from some other distribution. Figure A6.3 plots the empirical distribution of p values from a study by Mosig et al. (2001) on marker–trait associations. While the middle of the distribution appears to be consistent with random sampling around a flat average, there is a large excess of values near zero.

One simple approach for estimating n_0 is to use the average height of the empirical distribution over intermediate p values, excluding values near zero (and potentially one). Recall that the probability density function for a uniform over (0,1) has a very simple form

$$\phi_u(p) = 1 \quad \text{for} \quad 0 \leq p \leq 1 \tag{A6.8a}$$

If there are n_0 truly null tests, then the expected number of p values from these tests that fall within an interval $0 \leq a < b \leq 1$ is simply

$$n_0 \int_a^b \phi_u(p)\, dp = n_0 \int_a^b 1 \cdot dp = n_0(b-a) \tag{A6.8b}$$

Hence

$$\widehat{n}_0(a,b) = \frac{\text{Number of } p(i) \text{ values in } (a,b)}{b-a} \tag{A6.8c}$$

Likewise, an estimate for the fraction $\pi_0 = n_0/n$ of true nulls is

$$\widehat{\pi}_0(a,b) = \frac{\text{Number of } p(i) \text{ values in } (a,b)}{n(b-a)}$$
$$= \frac{\text{Fraction of } p(i) \text{ values in } (a,b)}{b-a} \tag{A4.8d}$$

Example A6.6. What is an estimate of n_o given the data in Figure A6.3? Consider the bins centered around $p = 0.5$. Based on the central three bins (0.4, 0.5, and 0.6), a total of $60 + 46 + 48 = 154$ tests have p values in this interval. From Equation A6.8b, $154 = n_0 \cdot 0.3$ or $n_0 = 154/0.3 = 513$, and hence a fraction, $\pi_0 = n_0/n = 513/644 = 0.80$, of the tests are true nulls. Using the bins from 0.3 to 0.8 yields $n_0 = 322/0.6 = 537$, or $\pi_0 = 537/644 = 0.83$. Hence, it appears that around 80% of the tests are consistent with true nulls. Mosig et al. (2001; also see Nettleton et al. 2006) used an iterative approach (also based on bin counts in the p-value histogram) and arrived at an estimate of $n_0 = 500$ (78%).

Storey and Tibshirani (2003) proposed an estimator of n_o based on the number of p values exceeding some tunable parameter value, λ (taking $a = \lambda$ and $b = 1$ in Equation A6.8b), on the logic that for larger values of λ, most of these draws are from the uniform corresponding to draws from the null. Let $\widehat{\pi}_0(\lambda)$ denote the estimated fraction of truly null hypotheses based on using a tuning value of λ, then

$$\widehat{\pi}_0(\lambda) = \frac{\text{Number of } p(i) \text{ values } > \lambda}{n(1-\lambda)} \tag{A6.9a}$$

and

$$\widehat{n}_0(\lambda) = n \cdot \widehat{\pi}_0(\lambda) = \frac{\text{Number of } p(i) \text{ values } > \lambda}{1-\lambda} \tag{A6.9b}$$

By focusing on the interval $(\lambda, 1)$, the **Storey–Tibshirani estimator** is potentially biased when there is an excess of p values near one. This can happen for a variety of reasons, such as inappropriate assumptions for the test statistic (e.g., the use of one-sided tests when two-sided tests are more appropriate; Figure A6.2C). Both Equation A6.8c and the Storey–Tibshirani estimator (Equation A6.9b) rely on tuning parameters (a and b, and λ, respectively) which define the region of the distribution of p values assumed to be drawn from a uniform (i.e., almost all p values in this interval are assumed to be generated under the null). Nettleton et al. (2006) reviewed these, and other, approaches for estimating n_0 from sampling parts of a presumed uniform and elaborated on their strengths and weaknesses.

One significant concern is that correlated tests can result in either an under- or over-dispersion of p values under the global null hypothesis, resulting in significant departure from a uniform distribution (Efron 2007; Hu et al. 2011; Leek and Storey 2011). This in turn compromises estimates of n_0.

Estimating n_0: Mixture Models

Allison et al. (2002) suggested that π_0 can be estimated by treating the distribution of p

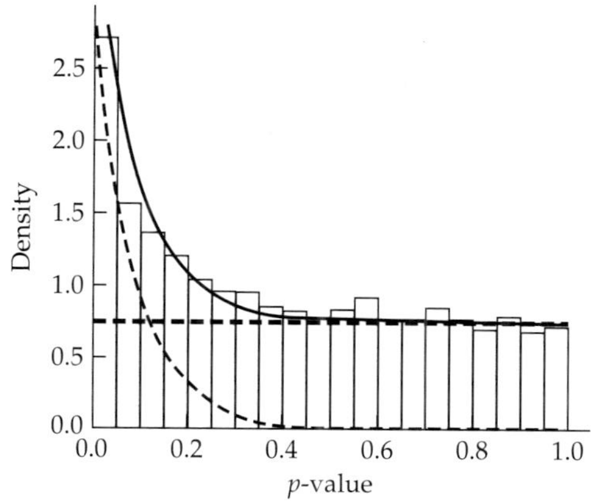

Figure A6.4 The empirical distribution of p values can be treated as a mixture model of a uniform plus a beta distribution (whose shape parameters, a and b, can be estimated via ML), see Equation A6.10b. In this hypothetical example, a weighted mixture of a uniform (horizontal dashed line) and a beta with ($a < 1, b = 1$; dashed curve), yields the mixture distribution (solid curve) that fits the empirical distribution of the p values.

values as a mixture, a fraction π_0 of which comes from a uniform (and hence a uniform distribution function, ϕ_u), while the remainder $(1 - \pi_0)$ are from the distribution, $\phi_A(p)$, of p values when the alternative hypothesis is true (Figure A6.4). While the general form of $\phi_A(p)$ is unknown, a very flexible modeling approach is to assume a beta distribution (Appendix 7; Figure A7.3)

$$\phi_A(p) = \frac{\Gamma(a+b)}{\Gamma(a)\,\Gamma(b)}\, p^{a-1}(1-p)^{b-1} \tag{A6.10a}$$

Under the alternative hypothesis, we expect an increase in p values near zero, which occurs when $a < 1$. Likewise, the beta distribution can easily accommodate an increase in p values near one ($b < 1$). When $a = b = 1$, this simply reduces to a uniform.

Allison et al. (2002) suggested fitting the actual shape by using the data to obtain ML estimates of a and b, as well as our desired parameter, the fraction of true nulls, π_0. The resulting likelihood function for a single p value becomes

$$\ell(p) = (1-\pi_0)\,\phi_A(p) + \pi_0\,\phi_u(p) = (1-\pi_0)\frac{\Gamma(a+b)}{\Gamma(a)\,\Gamma(b)}\, p^{a-1}(1-p)^{b-1} + \pi_0 \tag{A6.10b}$$

with the resulting total likelihood over the n sampled p values (from independent tests) calculated by

$$\ell(\mathbf{p}) = \prod_{i=1}^{n} \ell(p_i) \tag{A6.10c}$$

Standard ML methods (Appendix 4) are used to solve for a, b, and π_0.

While hypothesis testing under a maximum likelihood framework is typically performed using the likelihood ratio (LR) test (Appendix 4), this is not appropriate for tests of the number of components in a mixture, as the LR does not approach a limiting χ^2 distribution, because π_0 is being tested against a boundary value, in this case a value of one (McLachlan 1987). While a modified LR test for mixtures can be constructed that is better behaved (Chen et al. 2001), Allison et al. used a bootstrap approach (McLachlan 1987; Schork 1992b). Here, one first uses the original distribution of p values to compute an LR test statistic for the null of a uniform versus the alternative of a mixture. One then generates **parametric bootstrap samples** by drawing n values of p from the null distribution (here a uniform) and then using this simulated data set to compute an LR test statistic for a mixture. This is done several thousand times to generate an approximate distribution of the LR statistic under the null, which is used to assess significance. For example, if only 0.25% of the bootstrap LR values are equal to (or exceed) the LR value for the original data, the significance is approximately 0.25%. Likewise, approximate standard errors for π_0 can be generated using a **conventional bootstrap** approach. One samples the *original* p values *with replacement* to generate a **bootstrap sample** of size n. This is then used to estimate π_0 (and the other parameters) under a standard ML framework. Several thousand bootstrap samples are generated, and the variation across estimates of π_0 (or any other parameter) over these samples provides an approximate estimate of the sampling variance for $\widehat{\pi}_0$.

Finally, while a beta (or weighted sum of betas) can be used as the functional form for ϕ_A, another approach is to use a nonparametric estimator for this unknown density function. This can be done using a **kernel density estimator**, whereby the form of an unknown density is estimated by using the observed number of counts within a series of bins spanning the distribution in conjunction with an appropriate smoothing function. This approach was used by Robin et al. (2007) and Guedj et al. (2009).

FDR: THE FALSE-DISCOVERY RATE

As mentioned, Bonferroni corrections (and their extensions) are appropriate when we expect that only a very few of the many null hypotheses are false. An alternate setting is where some reasonable fraction of the null hypotheses is *expected* to be false. In such cases, even sequential Bonferroni corrections are likely to be too stringent, resulting in too many false negatives (Type II errors; i.e., a failure to reject a false hypothesis). A different approach is required in these settings, most notably the **false-discovery rate** (**FDR**), introduced by Benjamini and Hochberg (1995), which was motivated by earlier comments by Soriç (1989). Benjamini (2010) reviews its history, as well as recent developments.

The FDR is the fraction of false positives among all the tests that are *declared to be significant*. The motivation for using the FDR is that we may be conducting a very large number of tests, with those that are declared to be significant being subjected to further study. An example would be a search for differential expression over a huge set of genes. The goal of the initial analysis is to distill a large number of candidates down to a reduced set (for further analysis) that is highly enriched for true positives.

In such cases, we are more concerned with making sure all possible true alternatives are included in this reduced set, and we are willing to accept some false positives to accomplish this goal. However, we also do not want to be completely swamped with false positives. The goal is a statistical procedure that results in a significant *enrichment* of true positives (differentially expressed genes in our example), while controlling the fraction of false positives within this enriched set by specifying a value, δ, for the FDR. Choosing an FDR of 5% means that (on average) 5% of the genes that we *declare to be significant* are actually false positives. The flip side is that 95% of those genes (tests) that are declared to be significant do indeed have differential expression. Hence, screening genes with an FDR of 5% results in a significant enrichment of genes that are truly differentially expressed.

To formally motivate the concept of the FDR, suppose a total of n hypotheses are tested (e.g., genes), S of which are judged significant (i.e., the p value for the test is $\leq$ some threshold value, τ). If we had complete knowledge, we would know that n_0 of the hypotheses have the null true and $n_1 = n - n_0$ have the alternative true, and we might find that F of the true nulls were called significant, while T of the alternative trues were called significant, yielding the following table

	Called significant	Called not significant	Total
Null true	F	$n_0 - F$	n_0
Alternative true	T	$n_1 - T$	n_1
Total	S	$n - S$	n

For this experiment, the FDR is the fraction of tests called significant that are actually

Table A6.1 Summary of the multiple comparison parameters used in this Appendix. F denotes the number of false positives, namely, tests under the null that are declared significant.

Parameter	Definition
α	Comparison-wide Type I error (false positive).
β	Type II error (false negative); $1 - \beta$ = power.
π	Family-wide Type I error; $\Pr(F > 0) = \pi$.
δ	False-discovery rate.
π_0	Fraction of all hypotheses that are truly null.
p	Probability of the test statistic under the null.
$p(k)$	kth smallest p value of the n tests.

true nulls, FDR = F/S. (The term **discovery** follows in that a significant result can be considered as a discovery for future work.) As a point of contrast, the normal Type I error (which we can also call the **false-positive rate [FPR]**), which is the fraction of true nulls that are called significant, is F/n_0. Note the critical distinction between these two error rates. While the numerator of each is F, the denominators are considerably different—the total number, S, of *tests called significant* (for FDR), versus the number, n_0, of hypotheses that are *truly null* (FPR). As the threshold value (τ) for significance is changed, so too is the fraction F/S. To obtain an FDR of δ over our experiment, τ is adjusted to find its largest value such that some expectation of F/S is bounded above by δ. Finally, Gadbury et al. (2004) defined the **expected discovery rate** (**EDR**) as T/n_1 (the fraction of all true discoveries declared to be significant), which is the analog of statistical power in this setting.

Another way to see the distinction between the FPR and the FDR is to consider them as probability statements for a single test involving hypothesis i. For the FDR, we condition on the test as being significant,

$$\text{FDR} = \Pr(i \text{ is truly null} \mid i \text{ is declared significant}) = \delta \tag{A6.11a}$$

whereas for the FPR, we condition on the hypothesis being null

$$\text{FPR} = \Pr(i \text{ is declared significant} \mid i \text{ is truly null}) = \alpha \tag{A6.11b}$$

As an introduction to FDR, suppose we are examining 1,000 tests, using a critical value of $\tau = 0.05$ returns 200 of these as significant (those with $p \leq \tau$). Assuming that $n_0 = 1000$, the expected number of false discoveries is 50, for an FDR of 50/200 = 0.25. Now suppose we lower the threshold to $\tau = 0.01$, and 180 tests remain significant, for an FDR of 10/180 = 0.055. If $n_0 = 900$, then FDR becomes FDR of 9/180 = 0.05.

Morton's Posterior Error Rate (PER) and the FDR

Table A6.1 reminds the reader of the various test parameters that arise when multiple comparisons are considered. We now show how these are related. First, the relationship between α, π, and F is as follows. Suppose we have set the FDR (i.e., the Type-I error rate) for an individual test at α. Such a p-value threshold only guarantees that the expected number of false positives is bounded above by $E[F] \leq \alpha \cdot n$. For n independent tests, a π-level experiment-wide false-positive error (setting $\alpha = \pi/n$; namely, the Bonferroni correction) implies that $\Pr(F \geq 1) \leq \pi$, i.e., the probability of *at least one* false positive is no greater than π. To show how α, β, π_0, and δ are related, we first need to introduce the concept of the **posterior error rate** (**PER**).

Fernando et al. (2004) and Manly et al. (2004) both noted that FDR measures are closely related to Morton's (1955b) PER, originally introduced in the context of linkage analysis in humans (this is also referred to as the **false positive report probability [FPRP]**; Wacholder et al. 2004). Morton's PER is simply the probability that a single significant test is a false positive,

$$\text{PER} = \Pr(F = 1 \mid S = n = 1) \tag{A6.12}$$

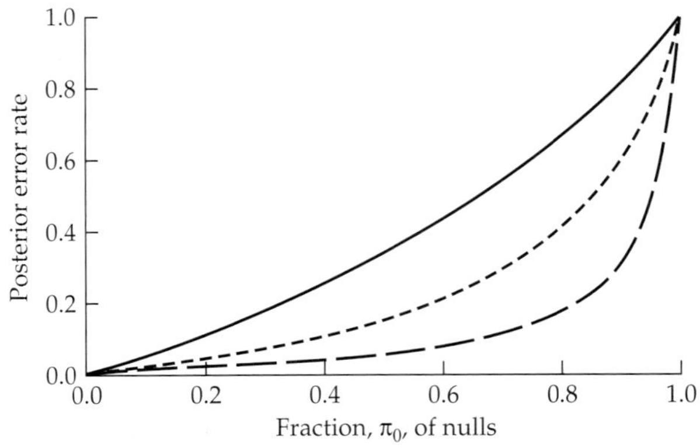

Figure A6.5 Plot of the posterior error rate (Equation A6.15a) for $\alpha = 0.05$, as a function of π_0 (the fraction of cases where the null hypothesis holds) and β (the Type II error, which is one minus the power). The solid curve corresponds to $\beta = 0.9$ (10% power), the short-dashed curve corresponds to $\beta = 0.7$ (30% power), and the long-dashed (lower) curve corresponds to $\beta = 0$ (100% power).

The connection between the FDR and the PER is that if we set the FDR to δ, then the PER for a randomly drawn significant test is also δ.

Framing tests in terms of the PER highlights the **screening paradox**: "Type-I error control may not lead to a suitably low PER" (Manly et al. 2004). For example, we might choose $\alpha = 0.05$, but the PER may be far higher, which means that a test that is *declared to be significant* may have a much larger probability than 5% of being a false positive. The key is that because we are ***conditioning on the test being significant*** (as opposed to conditioning on *the hypothesis being a null*, as occurs with α), S may include either false positives or true positives. The relative fractions of each (and hence the probability of a false positive) is a function of the single test parameters, α and β, and the fraction, π_0, of hypotheses that are truly null. To see this, we apply Bayes' theorem (Equation 3.3b), which yields

$$\text{PER} = \Pr(F = 1 \mid S = n = 1) = \frac{\Pr(\text{false positive} \mid \text{null true}) \cdot \Pr(\text{null})}{\Pr(S = n = 1)} \tag{A6.13}$$

Consider the numerator of Equation A6.13 first. Let $\pi_0 = n_0/n$ be the fraction of all hypotheses that are truly null. The probability that a null is declared significant is simply the Type I error, α, hence

$$\Pr(\text{false positive} \mid \text{null true}) \cdot \Pr(\text{null}) = \alpha \cdot \pi_0 \tag{A6.14a}$$

Turning to the denominator of Equation A6.13, what is the probability that a single (randomly chosen) test will be declared significant, $\Pr(S = n = 1)$? This event can occur because we choose to test a hypothesis that is truly null (π_0) and have a Type I error (α), or because we choose to test a hypothesis that is truly false ($1 - \pi_0$) and avoid a Type II error. For the latter, the power is simply T/n_1, the fraction of all tests under the alternative that is declared to be significant. If we write the power as $1 - \beta$ (β is the Type II error), the resulting probability that a single (randomly drawn) test is significant is

$$\Pr(S = n = 1) = \alpha\pi_0 + (1 - \beta)(1 - \pi_0) \tag{A6.14b}$$

Substituting Equations A6.14a and A6.14b into Equation A6.13 yields

$$\text{PER} = \frac{\alpha \cdot \pi_0}{\alpha \cdot \pi_0 + (1 - \beta) \cdot (1 - \pi_0)} = \left(1 + \frac{(1 - \beta) \cdot (1 - \pi_0)}{\alpha \cdot \pi_0}\right)^{-1} \tag{A6.15a}$$

Figure A6.5 plots Equation A6.15a for various values of π_0 and β.

Sham and Purcell (2014) noted that one can rearrange Equation A6.15a to find the α value to obtain a desired PER value of γ, with

$$\alpha = \left(\frac{\gamma}{1-\gamma}\right)\left(\frac{1-\pi_0}{\pi_0}\right)(1-\beta) \tag{A6.15b}$$

In particular, if there is complete power ($\beta = 0$) and only one of the n tested hypotheses departs from the null ($\pi_0 = 1 - 1/n$), Equation A6.15c reduces to

$$\alpha = \left(\frac{\gamma}{1-\gamma}\right)\left(\frac{1}{n-1}\right) \simeq \frac{\gamma}{n} \tag{A6.15c}$$

which recovers the Bonferroni correction (Equation A6.4).

The Type I error rate, α, of a *random* test, and the PER, γ, for a *significant* test, which are often assumed to be the same, are actually very different. In addition to α, the PER also depends on the power, β, of a test and the fraction, π_0, of tests that are truly null (as these latter parameters influence the probability that a test is declared to be significant). Manly et al. (2004) noted that the PER is acceptably low only if the fraction of alternative hypotheses $(1 - \pi_0)$ is well above α.

Thinking in terms of the PER allows us to consider multiple comparisons in a continuum from Bonferroni-type corrections to using FDR to control the PER. If $\pi_1 = 1 - \pi_0$ is very small, most tests will truly be from the null hypothesis, and we can control the overall false-positive rate with a Bonferroni-type correction. However, if some fraction of the hypotheses is expected to be false (i.e., $1 - \pi_0$ is at least of modest value), then using FDR corrections is a more reasonable approach for controlling the PER.

Example A6.7. In Morton's original application, because there are 23 pairs of human chromosomes, he argued that two randomly chosen genes had a $1/23 \simeq 0.05$ *prior probability of linkage*, namely, $1 - \pi_0 = 0.05$, and thus $\pi_0 = 0.95$. Assuming a Type I error rate of $\alpha = 0.05$ and 80% power to detect linkage ($\beta = 0.20$), applying Equation A6.15a yields a PER of

$$\frac{0.05 \cdot 0.95}{0.05 \cdot 0.95 + 0.80 \cdot 0.05} = 0.54$$

Hence, with a Type I error control of $\alpha = 0.05$, a random test showing a significant result ($p \leq 0.05$) has a 54% chance of being a false positive. This occurs because most of the hypotheses are expected to be null—e.g., if we draw 1,000 random pairs of loci, 950 are expected to be unlinked and we expect $950 \cdot 0.05 = 47.5$ of these to show a false positive. Conversely, only 50 are expected to be linked, and we would declare $50 \cdot 0.80 = 40$ of these to be significant, so that $47.5/87.5 = 0.54$ of the significant results are due to false positives.

What value for α is needed under the above parameters to give a PER of $\gamma = 0.05$? From Equation A6.15b,

$$\alpha = \left(\frac{\gamma}{1-\gamma}\right)\left(\frac{1-\pi_0}{\pi_0}\right)(1-\beta) = \left(\frac{0.05}{1-0.05}\right)\left(\frac{1-0.95}{0.95}\right)(1-0.2) = 0.0022$$

Hence, declaring significance when $p \leq \alpha = 0.0022$ gives a PER of 5%.

Example A6.8. Suppose we set $\alpha = 0.005$ for each test, and assume that the resulting power is essentially 1 (i.e., $\beta \simeq 0$). Consider 5,000 tests under two different settings. First, suppose that the alternative is very rare, with $n_1 = 1$ ($\pi_0 = 0.9998$). Under this setting, we expect $4999 \cdot 0.005 = 24.995$ false positives and one true positive ($1 \cdot [1 - \beta] = 1$), yielding an expected PER of

$$\text{PER} = \frac{24.995}{24.995 + 1} = 0.961$$

Thus, a randomly chosen *significant* test has a 96.1% probability of being a false positive.

Now suppose that the alternative is not especially rare, for example n_1 = 500 (π_0 = 0.9). The expected number of false positives is 4500 · 0.005 = 22.5, while the expected number of true positives is 500, yielding a PER of

$$\text{PER} = \frac{22.5}{522.5} = 0.043$$

The PER is thus rather sensitive to π_0, the fraction of all tests that are truly from the null hypothesis. If π_0 is essentially 1, a PER of δ is obtained using the Bonferroni correction, $\alpha = \delta/n$. However, if π_0 departs even slightly from one (i.e., more than a few of the alternative hypotheses are correct), then the per-test level of α to achieve a desired PER rate is considerably larger (i.e., less stringent) than that given by the Bonferroni correction, namely, $\alpha(\delta) > \delta/n$. For example, for a 0.04 experiment-wide error rate, $\alpha = 0.04/5000 = 8 \cdot 10^{-6}$, which is roughly 625 times smaller than the value of $\alpha = 0.005$ required for a 4% FDR, highlighting the greatly increased power under the FDR framework. This increased power arises because the FDR approach acknowledges that some fraction of the tests are not from the null.

A Technical Aside: Different Definitions of False-discovery Rate

While the FDR for any experiment is simply F/S, there are several subtly different ways to formally define the expectation of this ratio. The original notion of an FDR is due to Benjamini and Hochberg (1995), with modifications suggested by a number of other workers, most notable Storey (2002) and Fernando et al. (2004); see Table A6.2.

While the technical distinction between these different false-discovery rates is important, when actually estimating a FDR from a collection of p values, one is usually left with an expression of the form $E(F)/E(S)$, which consists of the expected number of false positives divided by the expected number of significant tests. Strictly speaking, this is the **proportion of false positives** (**PFP**).

The main distinction between the different FDRs are: (i) the original method of Benjamini and Hochberg (1995), which assumes $n = n_0$ (all hypotheses are nulls); and (ii) all other estimators, which assume n_0 is not necessarily n, and thus attempt to estimate either π_0 or n_0, and then use either to estimate the FDR.

The Benjamini–Hochberg FDR Estimator

The original estimator for the FDR was introduced by Benjamini and Hochberg (1995). Suppose we declare a test to be significant if its p value is at or below some threshold value, $\tau = p(k)$, in which case k of the hypotheses will be declared significant (as $p(k)$ is the kth smallest p value), and $S = k$. If all n of the hypotheses are null, then the expected value of F (the number of false positives) is just $n\,p(k)$. The resulting fraction of all rejected hypotheses that are false discoveries becomes $F/S = n\,p(k)/k$. Hence, the FDR, δ_k, for hypothesis k is bounded by

$$\frac{np(k)}{k} \leq \delta_k \tag{A6.16a}$$

In particular, if we wish to obtain an FDR of δ for the entire experiment, then we reject (i.e., declare as significant) all hypotheses that satisfy

$$p(k) \leq \delta\,\frac{k}{n} \tag{A6.16b}$$

This simple (heuristic) derivation shows why the original Benjamini–Hochberg estimate of the FDR is conservative, as in those settings in which one applies the FDR criteria, the expectation is that some fraction of the hypotheses are not null, and so $n_0 < n$. The correct estimator of the expected number of rejected null hypotheses is $n_0 p(k)$, which leads to

Table A6.2 Measures of false discovery. Here F is the number of false discoveries, S is the number of tests declared significant, and n is the total number of tests. (After Manly et al. 2004.)

	Name	Definition	Reference
FDR	False-discovery rate	$E(\frac{F}{S} \mid S > 0) \Pr(S > 0)$	Benjamini and Hochberg (1995)
pFDR	Positive false-discovery rate	$E(\frac{F}{S} \mid S > 0)$	Storey (2002)
PFP	Proportion of false positives	$E(F)/E(S)$	Fernando et al. (2004)
PER	Posterior error rate	$\Pr(F = 1 \mid S = n = 1)$	Morton (1955b)
FPR	False-positive rate	$\Pr(F > 0)$	

generalized estimate of the FDR, where $\widehat{n}_0$ (e.g., Equations A6.8–A6.10) replaces n. In this case, Equation A6.16a becomes

$$\widehat{\delta}_k = \frac{\widehat{n}_0\, p(k)}{k} \tag{A6.17}$$

Example A6.9. Consider again the 10 ordered p values from Example A6.4. Computing $n\,p(k)/k = 10\,p(k)/k$, where k denotes the test with the kth smallest p value, yields the following table:

k	1	2	3	4	5	6	7	8	9	10
$p(k)$	0.0020	0.0045	0.0060	0.0080	0.0085	0.0090	0.0175	0.0250	0.1055	0.5350
$n\frac{p(k)}{k}$	0.0200	0.0225	0.0200	0.0200	0.0170	0.0150	0.0250	0.0313	0.1172	0.5350

Thus, if we wish an overall FDR value of $\delta = 0.05$, we would reject hypotheses when $n\,p(k)/k \leq \delta = 0.05$, which is satisfied by $H(1)$ through $H(8)$. Notice that this procedure rejects more hypotheses (i.e., returns more discoveries) than any of the sequential Bonferroni methods (Example A6.4).

A (Slightly More) Formal Derivation of the Estimated FDR

Following Storey and Tibshirani (2003), we consider the expected FDR for an experiment where we declare a hypothesis to be significant if its p value is less than or equal to some threshold value, τ. Obviously, as τ becomes smaller, the FDR is smaller (as significant nulls become increasingly less likely). However, if τ is set too small, we lose power (e.g., suppose we set $\tau = \pi/n$; namely, the Bonferroni correction). What we would like to do is to find the expected value of the FDR as a function of the chosen threshold parameter, τ, to allow us to optimally tune this parameter to obtain the desired FDR. If we have a large number of tested hypotheses,

$$E[FDR(\tau)] = E\left[\frac{F(\tau)}{S(\tau)}\right] \simeq \frac{E[F(\tau)]}{E[S(\tau)]} \tag{A6.18}$$

A simple estimate of $E[S(\tau)]$ is given by the observed number of significant tests when the threshold is τ.

To obtain an estimate for $E[F(\tau)]$, we again use the fact that the distribution of p values under the null follows a uniform $(0, 1)$ distribution. Hence,

$$\Pr(p \leq \tau \mid \text{null hypothesis}) = \int_0^\tau \phi_u(p)dp = \tau \tag{A6.19}$$

where $\phi_u(p)$ is the uniform probability density function for p values under the null (Equation A6.8a). Hence, if n_0 of the n tests are truly null, then

$$E[F(\tau)] = n_0 \cdot \Pr(p \leq \tau \,|\, \text{null hypothesis}) = n_0 \cdot \tau \tag{A6.20}$$

Hence,

$$E[FDR(\tau)] = \frac{n_0 \cdot \tau}{S(\tau)} \tag{A6.21}$$

Setting $\tau = p(k)$, then $S(\tau) = k$, and Equation A6.21 becomes $n_0\, p(k)/k$, recovering Equation A6.17. Using the Storey–Tibshirani estimator for n_0 (Equation A6.9b), an estimated value for the FDR using threshold value, τ (and based on the tuning parameter, λ, in the Storey–Tibshirani estimator), becomes

$$\widehat{FDR}(\tau) = n_0 \cdot \frac{\tau}{S(\tau)} = \left(\frac{N[\,p(i)\text{ values } > \lambda\,]}{1-\lambda} \right) \cdot \left(\frac{\tau}{N[\,p(i)\text{ values } \leq \tau\,]} \right) \tag{A6.22}$$

where $N[x]$ is the number of occurrences of event x. Ideally, over a reasonable range of λ values, we expect this estimate to be stable. If λ is set too large, the likelihood that almost all values correspond to draws from a null will be countered by the much smaller sample size (and hence a larger sampling error) from using such a small fraction of the total data.

Under a mixture-model setting (e.g., Equation A6.10), the FDR for a given a significance threshold (τ) is simply the fraction of all true positives that are declared significant divided by the fraction of all tests that are declared significant (i.e., those tests for which $p \leq \tau$). This can be estimated directly from the parameters of the mixture distribution,

$$\text{FDR}(\tau) = \frac{\pi_0\, \text{cdf}_U(\tau)}{\pi_0\, \text{cdf}_U(\tau) + (1-\pi_0)\, \text{cdf}_A(\tau)} = \frac{\pi_0\, \tau}{\pi_0 \tau + (1-\pi_0)\, \text{cdf}_A(\tau)}, \tag{A6.23}$$

where cdf denotes the cumulative distribution function, with

$$\text{cdf}_U(x) = \int_0^x \phi_U(p)dp = x \qquad \text{cdf}_A(x) = \int_0^x \phi_A(p)dp$$

where ϕ_U is the uniform distribution under the null and ϕ_A is the distribution of p values under the alternative (for example, a fitted beta).

Storey's *q* Value

While we can control the FDR for an entire set of experiments, we would also like to have an indication of the FDR for any *particular experiment* (or test) within this family of tests. Intuitively, tests with smaller p values should also have smaller associated FDR values. To address this, Storey (2002; Storey and Tibshirani 2003) introduced the concept of a ***q* value** (as opposed to the p value) for a particular test, where q is the expected FDR rate for tests *within the current experiment* whose p values are at least as extreme as the test of interest. The estimated q value is a function of the p value for that test and the distribution of the entire set of p values from the family of tests being considered, namely,

$$\widehat{q}\,[p(i)] = \min_{\tau \geq p(i)} \widehat{FDR}(\tau) \tag{A6.24}$$

In words, the q value of a test is calculated by using the smallest FDR value over all significance threshold values (τ) such that this threshold is equal to, or greater than, the p value, $p(i)$ of the test.

To see why we used $\min_{\tau \geq p(i)}$ instead of simply setting $q_i = \widehat{FDR}[p(i)]$, recall Example A6.9. This example used the Benjamini–Hochberg estimator for FDR value (which differs from other FDR estimators by a constant, n_0/n). Notice that the smallest FDR occurs for hypothesis 6 (1.5%), and not for hypotheses with smaller p values. This reflects the tradeoff whereby increasing the threshold, τ, for significance results in the declaration of more tests as discoveries, so the ratio $\tau/S(\tau)$ need not monotonically increase as τ increases. As Example A6.9 shows, setting the threshold τ *above* the $p(i)$ value may actually result in a smaller q value, and hence Storey's definition.

Example A6.10. As an example of the interplay between the family-wide error rate (π), and the individual p and q values for a particular test, consider Storey and Tibshirani's (2003) analysis of a microarray data set comparing *BRCA1* and *BRCA2* positive breast cancer tumors. A total of 3,226 genes were examined. Setting a critical p value of $\alpha = 0.001$ detects 51 significant genes (i.e., those with differential expression between the two types of tumors). If we assume that the hypotheses being tested are independent (which is unlikely as expression can be highly correlated across sets of genes; Chapter 21), the probability that there is at least one false positive is $\pi = 1-(1-0.001)^{3226} = 0.96$, while the expected number of false positives is 0.001 · 3226 = 3.2, or 6% (3.2/51) of the declared significant differences. After setting an FDR rate of $\delta = 0.05$, Storey and Tibshirani detected 160 genes that showed significant differences in expression. Of these 160, 8 (5%) are expected to be false positives. Compared to the Bonferroni correction (51 genes, 6% false positives), over three times as many genes were detected, and with a lower FDR rate. Further, Storey and Tibshirani estimated the fraction, π_0, of nulls (genes with no difference in expression) at 67%, which suggests that 33% (or roughly 1,000 of the 3,226 genes) are likely to be differentially expressed between the two tumor types.

This dramatic difference in performance between Bonferroni and FDR control arises because the former enforces very strict control over *any* false positives, resulting in a much smaller set of discoveries. Conversely, FDR is more concerned with the *fraction* of discoveries that are false, and by including more true discoveries, that fraction can be made smaller than under Bonferroni. Because the fraction of null hypotheses that are false in this study is rather substantial, a lower significance threshold includes more of these true discoveries, thus decreasing the FDR.

To contrast the distinction between the p and q values, consider the *MSH2* gene, which has a q value of 0.013 and a p value of $5.50 \cdot 10^{-5}$. This p value implies that the probability of seeing at least this level of difference in expression for a randomly drawn gene from the null hypothesis (no difference in expression) is $5.50 \cdot 10^{-5}$. Conversely, $q = 0.013$ indicates that, *for this experiment*, 1.3% of genes that show differences in expression that are as, or more, extreme (i.e., whose p values are at least as small) as that for *MSH2* are expected to false positives.

Closing Caveats in Using the FDR

While controlling the FDR is a very powerful approach for many multiple-comparison problems, it is not a panacea. One concern is correlations among tests. As mentioned, in this case the null distribution of p values can significantly depart from a uniform, giving biased estimates of π_0 (and thus the FDR). Further, recall that FDR control is accomplished by controlling the *expected* value of the FDR (or some closely related measure, such as the PFP). The *variance* in the FDR across independent experiments can be considerable, especially when the tests are correlated (Owen 2005; Leek and Storey 2011). One approach for treating these concerns is to use Leek and Storey's (2007, 2008) surrogate variable analysis to account for dependencies among the data before the actual p values for individual tests are obtained. A second is Yekutieli's (2008) hierarchical method, wherein one tests nested subsets of the null. A third is to use a **relaxed significant criterion** that is not as strict as Bonferroni, but more stringent than FDR, as we now detail (in the context of QTL mapping).

Chen and Storey's GWER$_k$ Approach for QTL mapping

At first blush, a linkage-based QTL mapping experiment (Chapters 18 and 19), wherein a modest to large number of markers are tested, seems an ideal setting for the use of FDR. For each marker, the null hypothesis is that there is no linkage to a QTL influencing the trait, versus the alternative of linkage to a QTL. Support for linkage is usually displayed in the form of LR profile plots for each chromosome (Figures 18.1 and 18.2), with peaks indicating the genomic locations of likely QTLs. Critical values for the peak heights are obtained using the resampling approach of Churchill and Doerge (1994), wherein one keeps the marker data for each individual intact, but randomizes the trait value over individuals to generate draws from the null (Chapter 18). One then uses the highest peak value that occurs in (say) no more than 5% of the samples as the 5% critical value.

While the use of FDR in QTL mapping was first proposed by Weller et al. (1998) and further explored by a number of authors (e.g., Lee et al. 2002; Sabatti et al. 2003; Benjamini and Yekutieli 2005), Chen and Storey (2006) noted a critical problem. Specifically, the linkage signal from a QTL influences essentially *all the markers on the chromosome* arm on which it resides and so, *as a group*, they all satisfy the *same hypothesis*. Either all are nulls (unlinked to a QTL) or all are failures of the null (linked to a QTL, albeit with differing degrees of a linkage signal). As such, investigators can arbitrarily obtain any FDR level they desire by simply adding or subtracting linked markers. To illustrate this point, Chen and Storey simulated a species with four chromosomes, each with ten markers (scenario A) and a single QTL on chromosome one. With the exact same data, one might wish to fine map by scoring another 40 markers to chromosome one (scenario B). Under scenario A, the critical test statistic value to achieve an FDR of 5% was 3.15, which decreased to 1.52 in scenario B. This greatly reduced stringency arose because of a dramatic reduction in the fraction of true nulls in B ($\pi_0 = 30/80 = 0.375$) relative to A ($\pi_0 = 30/40 = 0.75$), which in turn decreased the posterior error rate of significant tests in B (Figure A6.5).

Chen and Storey's solution to this problem follows from the notion of **suggestive linkage** proposed by Lander and Kruglyak (1995). The latter authors suggested controlling for *no more than* one false positive, as opposed to controlling for *no* false positives under the Bonferroni setting. In linkage settings, the FWER is usually denoted as the **genome-wide error rate (GWER)**. Let GWER_k denote the control for *no more* than k false positives, where GWER_0 is the Bonferroni setting. Chen and Storey suggest using a small value of k, such as one or two. The associated critical value to achieve this level of control can easily be obtained using Churchill and Doerge resampling. Rather than recording the highest peak under draws from the null (corresponding to control over GWER_0), one instead records the value of the kth highest peak. For example, if 5% of all of the simulations have the second highest peak of 3.7 or greater, then using this as the critical value provides for a 5% control over GWER_2.

To a much lesser extent, the same issue occurs in GWAS (Chapter 20) among sets of extremely tightly linked SNPs. However, because the signal in these cases is from the persistence of linkage disequilibrium (LD) over a large number of generations, any common signal is restricted to a set of very tightly linked markers rather than an entire chromosome, and control of the FDR *among* such clusters is appropriate.

FORMAL META-ANALYSIS

Another class of analysis involving multiple comparisons considers comparison *across* studies, rather than trying to adjust for multiple comparisons *within* a single study. Such an analysis of analyses, coined **meta-analysis (MA)** by Glass (1976), is a tool of expanding importance in quantitative genetics. While multiple-comparison corrections involve isolating the significance of *separate* variables from a *single large study* containing many factors, an MM *combines* all information from a *set of studies* in order to increase the power and insight over that from any single study. While the first formal attempt to combine data from multiple studies appears to be Pearson (1904), its formal roots trace back to Fisher (1932b) and Cochran (1937a), following which much of the field was developed in the social sciences. General overviews can be found in Hunter and Schmidt (2005), Cooper et al. (2009), and

Table A6.3 The potential data available for a meta-analysis of k studies. One has reported values for the estimated effect, T_i, for each study. A study may also have m reported **moderator variables** (cofactors such as sex, laboratory, species, etc.). In the simplest setting, the study only reports a p value for whether the effect is significant. In such cases, an overall p value can be obtained from the methods discussed earlier in this Appendix. If the study also reported the standard errors (SE) of the estimates, or (under the assumption of a constant error per observation over all studies) the sample size, then a formal (i.e., model-based) meta-analysis may be performed (as detailed in the text).

	Effect					
Study	Actual	Estimate	Sample size	SE	p value	Moderator variables
1	θ_1	T_1	n_1	s_1	p_1	$M_{11}, M_{12}, \cdots, M_{1m}$
2	θ_2	T_2	n_2	s_2	p_2	$M_{21}, M_{22}, \cdots, M_{2m}$
3	θ_3	T_3	n_3	s_3	p_3	$M_{31}, M_{32}, \cdots, M_{3m}$
$\vdots$	$\vdots$	$\vdots$	$\vdots$	$\vdots$	$\vdots$	$\vdots$
k	θ_k	T_k	n_k	s_k	p_k	$M_{k1}, M_{k2}, \cdots, M_{km}$

Borenstein et al. (2011), and reviews with a specific focus on issues that can arise in biological problems can be found in Harrison (2011), Nakagawa and Santos (2012), Koricheva et al. (2013), and Gurevitch et al. (2018).

Informal, or Narrative, Meta-analysis

Table A6.3 shows the canonical structure of the data for a MA: one has a number of studies, either published or unpublished, dealing with a specific question (such as the association between transcript level and trait phenotype). Study i reports an estimate, T_i, of an effect whose true (and unknown) value is denoted by θ_i. Unfortunately, however, many studies report only T_i and perhaps, p_i, although the latter is often simply reported in binary form (whether or not they are significant at some level) rather than as an actual value.

In settings where studies simply report T_i (and perhaps p_i), only an **informal**, or **narrative**, MA can be performed. Here, one simply presents the statistics on the collection of T_i values, such as a histogram of effects or some metric on their overall distribution (such as their mean, median, or variance). Often these summary statistics are compared over different values of any **moderator variables** (cofactors) associated with the data. For example, a comparison of whether the distribution of QTL effects for a given trait varies over species.

If the p values are only reported in a binary fashion (e.g., significant or nonsignificant), a **vote-counting** (Light and Smith 1971) scheme is used, which details the number of significant differences out of a collection of tests. Again, these results might be contrasted over different moderator variables, such as those over males versus females. If the actual p values are published, then the methods given earlier in this Appendix for combining the (presumably independent) p values can be used. Both the vote-counting and combined p value approaches are examples of the **null hypothesis significance testing** (**NHST**) framework, and represent the *least powerful* form of MA (Hedges and Olkin 1980). Stewart (2010) bemoaned the fact that these approaches have "a long history that lingers beyond its sell-by date." Unfortunately, many published studies in quantitative genetics only present p_i values (either actual values or values coded as significant or not significant), thus limiting the MA to these least-powerful approaches.

Furthermore, most MAs are not simply concerned about whether an effect is significant, as the usual assumption is that it is indeed. Rather, their motivation is in determining the *size of the effect* and whether it varies over cofactors of interest (e.g., whether an association is stronger in males than females). If we focus only on p values, this information will be lost. When a study also reports standard errors for the test statistics (the s_i in Table A6.3), then a **formal MA** can proceed, which allows the meta-analyst to combine estimates of effect sizes, weighted by their precision, over the studies. This exploits the full power of an MA, but critically depends on full transparency in reported studies (at a minimum, reporting values of s_i^2), something that is, surprisingly, often lacking.

Standardizing Effect Sizes

A formal MA proceeds by averaging over the standardized effect sizes for each study. As we detail below, this weights each study by the strength (precision) of the evidence it provides. We will briefly review a few of the common standardizations here, pointing the reader to the general references cited at the start of this section, as well as to Nakagawa and Cuthill (2007), for more details.

The simplest setting is when study i estimates the mean of a quantity of interest, yielding

$$T_i = \overline{z}_i, \qquad \text{with} \qquad s_i^2 = \frac{1}{n_i - 1}\sum_{j=1}^{n_i}(z_{ij} - \overline{z}_i)^2 \tag{A6.25}$$

Another common setting is that in which the study contrasts the means of two different groups, which we will call the treatment (t) versus control (c),

$$T_i = \overline{z}_{i,t} - \overline{z}_{i,c} \tag{A6.26a}$$

Several slightly different standardizations exist in the literature for this case, which are of the form

$$D_i = \frac{T_i}{s_{i,p}} \tag{A6.26b}$$

where $s_{i,p}$ is the pooled standard error for study i. Different standardizations arise from slightly different estimates of this pooled standard error. **Cohen's *d* statistic** uses Equation A6.26b with

$$s_{i,p}^2 = \frac{(n_{i,t} - 1)s_{i,t}^2 + (n_{i,c} - 1)s_{i,c}^2}{n_{i,c} + n_{i,t}} \tag{A6.27}$$

Here, $n_{i,t}$ and $s_{i,t}$ (the latter from Equation A6.25 with $n_i = n_{i,t}$) are the sample size and standard error (respectively) for the treatment group, with similar expressions for the control group, c. This corresponds to the ML estimate of the pooled standard error (and hence is slightly biased; Appendix 4 and Part II Chapter 33).

Another common standardization in the literature is **Hedge's *g***, which uses

$$g_i = \frac{\overline{z}_{i,t} - \overline{z}_{i,c}}{s_{i,p}}, \quad \text{where} \quad s_{i,p}^2 = \frac{(n_{i,t} - 1)s_{i,t}^2 + (n_{i,c} - 1)s_{i,c}^2}{n_{i,c} + n_{i,t} - 2} \tag{A6.28a}$$

which is the unbiased OLS estimate (and hence the REML estimate) of the pooled standard error. One further correction is often applied to g_i, in that the ratio given by Equation A6.28a is a biased estimator of $(\mu_t - \mu_c)/\sigma_p$, which can be adjusted by using

$$g_i\left(1 - \frac{3}{4(n_{i,c} + n_{i,t}) - 9}\right) \tag{A6.28b}$$

Besides comparisons of means, a MA often examines studies measuring the association between two variables, in which case the Pearson correlation coefficient, r_i, is the initial summary statistic. This is a slightly biased estimator of the true correlation coefficient (ρ), but a simple correction returns an unbiased estimate,

$$r_i^* = r_i + \frac{r_i(1 - r_i^2)}{2(n_i - 3)} \tag{A6.29a}$$

Fisher's *z*-transformation is used to both stabilize the variance and remove some of the skew:

$$z_{i,r} = \frac{1}{2} \ln \left(\frac{1 + r_i}{1 - r_i} \right), \qquad \text{with} \qquad s_i^2 = \frac{1}{n_i - 3} \tag{A6.29b}$$

Typically the unbiased estimate (r^*) is used in place of r in the transformed data, and we use $T_i = z_{i,r}$ as the study effect (i.e., the value in Table A6.3). Note that the standard error of the transformed sample correlation is now simply a function of the sample size. Likewise, any estimates based on z can be back-converted to inferences of ρ by using Fisher's z-to-r transform

$$r = \frac{e^{2z_r} - 1}{e^{2z_r} + 1} \tag{A6.29c}$$

Other standardizations exist for other classes of comparisons, such as the log of the odds ratio (Fleiss 1993).

Fixed-effects, Random-effects, and Mixed-model Meta-analysis

Once an appropriate summary statistic, along with its standard error, has been chosen, the next step is to decide whether a fixed, random, or mixed MA should be used. For questions of interest to quantitative geneticists, a mixed-model analysis (Chapter 10) is likely the most appropriate. This is also the most general model, with the fixed-effects and random-effects models following as special cases. However, it will be useful to first consider the structure of these simpler models.

Under a **fixed-effects meta-analysis** (also called the **common-effect model**), we assume that the actual effect size is the *same* over all studies ($\theta_i = \theta$), which yields

$$T_i = \theta + e_i \tag{A6.30a}$$

where we assume that the residuals are independent but heteroscedastic, as $\sigma^2(e_i) = s_i^2$. Under the fixed-effects model, our interest is in combining studies to obtain a better estimate of the common (fixed) effect, θ. This simply involves generalized least-squares (GLS; Equation 10.13a), with the resulting MA global estimate of θ (given the k studies) being

$$\overline{T} = \frac{\sum_{i=1}^k w_i T_i}{\sum_{i=1}^k w_i}, \qquad \text{where} \quad w_i = \frac{1}{s_i^2} \tag{A6.30b}$$

In other words, we use a weighted average, with each study weighted by its precision (studies with smaller standard errors receive larger weights). The MA standard error, $s_{\overline{T}}$, for the global estimate, $\overline{T}$, is

$$s_{\overline{T}}^2 = \frac{1}{\sum_{i=1}^k w_i} \tag{A6.30c}$$

For the situation where we assume that each individual observation in a given study has the same variance, $\sigma^2(T_i) = \sigma^2/n_i$, then for k studies, each of size n,

$$\sigma^2(\overline{T}) = \frac{\sigma^2}{nk} \tag{A6.30d}$$

An obvious next line of inquiry is whether the assumption of a common effect over all studies is reasonable. This can be examined using the **Cochran *Q* test of heterogeneity**,

$$Q = \sum_{i=1}^k \frac{\left(T_i - \overline{T}\right)^2}{s_i^2} \tag{A6.31}$$

where (under the null of $\theta_1 = \cdots = \theta_k$, and assuming that the values of T_i are normally distributed), the distribution of Q is χ^2 with $(k-1)$ degrees of freedom.

One potential reason for a significant Q is that the study consists of different subsets of groups (say, males versus females), with a common effect that is the same in each group but differs among groups. In this case, we can extend the basic model by including a regression on moderator variables,

$$T_i = \theta + \sum_{j=1}^{m} b_j M_{ij} + e_i \tag{A6.32}$$

Often the values of M_{ij} are simply zero-one indicator variables (e.g., 0 for male, 1 for female), but they can be more general regression slopes as well. For example, M_{1j} could be the age of individuals within study j, with a significantly nonzero value of b_1 in Equation A6.32 indicating that the treatment mean varies with age. Again Equation A6.32 is simply a GLS regression, and one can test for moderator-variable effects ($b_j \neq 0$) in the standard regression fashion (Chapter 10).

In most biological settings, the assumption of a single common value for the treatment mean over all studies is unrealistic. For example, in a MA of QTL effect sizes, we *expect* θ_i to vary over studies, and our interest shifts to the variance *among* the actual effects. This leads to the **random-effects meta-analysis** model

$$T_i = \mu + u_i + e_i \tag{A6.33a}$$

where $\mu_i \sim (0, \sigma_u^2)$. Typically, the effect sizes ($\theta_i = \mu + u_i$) are assumed to be drawn from a normal, $\theta_i \sim \mathrm{N}(\mu, \sigma_u^2)$, and are independent of the residuals (which remain heteroscedastic). Under a random-effects analysis, our interest is the variation, σ_u^2, among the realized effects, in addition to their overall grand mean, μ. The estimate for the latter is also of the form of Equation A6.30b, but with a critical difference. Under a random-effects model, the weights are now given by

$$w_i = \frac{1}{s_i^2 + \widehat{\sigma}_u^2} \tag{A6.33b}$$

where $\widehat{\sigma}_u^2$ is the estimate of σ_u^2. One option for obtaining this variance is the **DerSimonian–Laird estimator**, which is based on Cochran's Q value (Equation A6.31),

$$\widehat{\sigma}_u^2 = \frac{Q - (k-1)}{S_1 - (S_2/S_1)}, \quad \text{where} \quad S_j = \sum_{i=1}^{k} s_i^{-2j} \quad \text{for } j = 1, 2 \tag{A6.33c}$$

which is set to zero if it is negative (DerSimonian and Laird 1986; 2015), although other approaches (e.g., REML; Part II Chapter 33) could also be used, and more robust estimates have been suggested by Knapp and Hartung (2003) and Sidik and Jonkman (2005).

This difference in weighting schemes under fixed effects (Equation A6.30b) versus random effects (Equation A6.33b) can have profound implications (a nice discussion is given by Borenstein et al. 2010). In particular, the presence of $\widehat{\sigma}_u^2$ makes the ***random-effect weights more equal over studies***, especially when $\widehat{\sigma}_u^2$ is of the same order as an average value of s_i^2. Under a fixed-effects setting, ***larger studies*** (i.e., those with smaller values of s_i^2) ***are given more weight than smaller studies***. In a random-effects setting, this difference in weights is reduced such that ***larger studies lose influence and smaller studies gain influence***. In the extreme when σ_u^2 is large relative to all of the s_i^2 values, all studies are given roughly equal weight, independent of their sample size.

To see why this occurs, assume we are using the same design that led to Equation A6.30d, namely, k studies all of size n, and all with a common residual variance (σ^2) for each observation. In this case, the variance of the estimate of μ becomes

$$\sigma^2(\overline{T}) = \frac{\sigma^2}{nk} + \frac{\sigma_u^2}{k} \tag{A6.34}$$

showing (unlike the fixed-effects case; Equation A6.30d) that the actual number (k) of studies is at least as important as the total number (nk) of observations over all studies. Under a fixed-effects design, all that matters is nk, as the only error is in estimating the common mean, with the contribution from any particular study being proportional to its sample size. Under a random-effects model, there is an *additional error* from the *among-study variance*, in that the mean realization for each study, $\theta_i = \mu + u_i$, differs, with each new study providing additional information on the among-study variance. This leads to the second important consideration for a random-effects model: power is as much a function of the number of studies (k) as it is of the precision of any particular study (n_i). Indeed, if $\theta_i = \mu + u_i$ is measured without error in any particular study, one still needs a reasonable number of θ_i values to estimate their variance (σ_u^2) with any precision. Hence, not surprisingly, random-effects models have lower power than fixed-effects models.

Given the lower power of a random-effects model, one might be tempted to start with a fixed-effects analysis and only consider random-effects when the fixed-effects model generates a significant Q value (Equation A6.31). Borenstein et al. (2010) strongly caution against this approach. First, if the number of studies is small to moderate, Q can have poor power, so a failure to reject homogeneity among studies could simply be due to low power, and not an absence of among-study variation. Second, although there are costs (lower power) with a random-effects model, this arises because such models make a less stringent assumption: they allow the effect size to vary over studies. This is generally closer to reality, as the assumption of a constant θ value over all studies is unlikely to be correct.

Finally, in many settings, we might expect the grand mean to vary over different categories, such as when a gene's transcript level differs between males and females. Similarly, we may wish to examine whether the distribution of QTL effect size varies between life-history versus morphological traits. The potential of different means over different major categories can be accommodated in a MA model by the use of moderator variables (cofactors). These adjust the mean for a particular class, leading to a **mixed-model meta-analysis.** Suppose that there are $m \ll k$ moderators. The resulting mixed model is

$$T_i = \mu + \sum_{j=1}^{m} b_j M_{ij} + u_i + e_i \tag{A6.35a}$$

where b_j is the (fixed) effect of a moderator, j, which has a value of M_{ij} in study i. Equations A6.32 and A6.33a are special cases of Equation A6.35a, which we can write in general-linear-model form (Chapter 10) as

$$\mathbf{y} = \mathbf{Mb} + \mathbf{u} + \mathbf{e} \tag{A6.35b}$$

where $y_i = T_i$, $\mathbf{b} = (\mu, b_1, \cdots, b_m)^T$, and the ith row of the $k \times (m+1)$ matrix, $\mathbf{M}$, contains the values of the m moderator variables associated with study i (with the first column of $\mathbf{M}$ being all ones for the common μ). The vectors, $\mathbf{u}$ and $\mathbf{e}$, of random effects are assumed uncorrelated, with $\mathbf{e} \sim (\mathbf{0}, \mathbf{R})$ and $\mathbf{u} \sim (\mathbf{0}, \mathbf{G})$, where $\mathbf{R}$ is a known diagonal matrix, diag$(s_1^2, s_2^3, \cdots, s_k^2)$, and $\mathbf{G} = \sigma_u^2 \mathbf{C}$, where $\mathbf{C}$ is a matrix of known constants.

While typically it is assumed that $\mathbf{C} = \mathbf{I}$ (i.e., effects, u_i, are uncorrelated and homoscedastic), we can easily incorporate $\mathbf{C}$ matrices that account for phylogenetic relationships (i.e., correlations) when comparisons are made between species (Hadfield and Nakagawa 2010; Nakagawa and Santos 2012). In such settings, one could use either a Brownian-motion model (WL Equation A1.31) for the elements of $\mathbf{C}$ (divergence under drift) or an Ornstein–Uhlenbeck model (WL Equation A1.33) (divergence under drift and stabilizing selection). The bottom line is that our previous discussions on the great flexibility of mixed models (Chapter 10) also hold when we conduct a MA using a mixed-model framework.

Quantifying Heterogeneity: The Higgins–Thompson I^2 Index

While Cochran's Q provides a test of heterogeneity, it is underpowered when k is small, and overpowered when k is large. Hence, a large amount of heterogeneity can remain undetected when k is small, potentially skewing the analysis towards a fixed-effect model. Conversely, when k is large, even a small amount of biologically unimportant heterogeneity can be detected, skewing the analysis towards a random-effects model. Much akin to the philosophy that it is the size of an effect (not its significance) that is the true target of a MA, we would have to have a more formal quantification of heterogeneity. Higgins and Thompson (2002) propose two related metrics for this task. Their ***H*** statistic is given by

$$H = \sqrt{\frac{Q}{k-1}} \tag{A6.36a}$$

When there is no heterogeneity, the expected value of Q is $k-1$ (the expected value of a χ^2_{k-1}; Equation A5.14b), so that H is the excess in Q scaled by its degrees of freedom (in other settings, H^2 is also known as **Birge's ratio**; Birge 1932). Higgins and Thompson give an approximate 100(1$-\alpha$)% confidence interval for H as

$$\exp\left[\ln(H) \pm z_{1-\alpha/2}\text{SE}\{\ln(H)\}\right] \tag{A6.36b}$$

where z_a satsfies $\Pr(u \le z_a) = a$ for a unit normal U and

$$\text{SE}\{\ln(H)\} = \begin{cases} \dfrac{1}{2}\,\dfrac{\ln(Q)-\ln(k-1)}{\sqrt{2Q}-\sqrt{2k-3}} & \text{if } Q > k \\[2ex] \sqrt{\dfrac{1}{2(k-2)}\left(1-\dfrac{1}{3(k-2)^2}\right)} & \text{if } Q \le k \end{cases} \tag{A6.36c}$$

Recalling our notation from Equation A6.33c, $S_j = \sum_{i=1}^k s_i^{-2j}$ (for $j = 1, 2$), and defining

$$s^2 = (k-1)\frac{S_1}{S_1^2 - S_2} \tag{A6.37a}$$

which can be thought of as the average standard error for a study, yields

$$E[H^2] = \frac{\sigma_u^2 + s^2}{s^2} \tag{A6.37b}$$

where σ_u^2 is the heterogeneity variance (Higgins and Thompson 2002). Rearranging yields the **Higgins–Thompson index**

$$I^2 = \frac{H^2-1}{H^2} \tag{A6.37c}$$

where

$$E[I^2] = \frac{\sigma_u^2}{\sigma_u^2 + s^2} \tag{A6.37d}$$

The nice feature about I^2 is that it has a natural interpretation, namely the fraction of variation due to heterogeneity. Further, Equation A6.36c can be used to construct approximate confidence intervals for I^2 (as the latter is a function of H).

Publication and Other Sources in Bias

The most common concern of a MA is **publication bias**, wherein the chosen sample of studies is nonrandom with respect to all of the actual studies that have been done. Before addressing this, we need to stress that the most serious MA problem is simply ***poor data reporting*** (Gurevitch and Hedges 1999). The lack of inclusion of standard errors in many studies relegates their data to the lower status of a narrative MA. Ideally, *all* of the individual data, and not just summary statistics, should be readily available to the research community. In reality, serious privacy concerns typically constrain the sharing of full data for human studies (such as with pooling GWAS studies; Chapter 20).

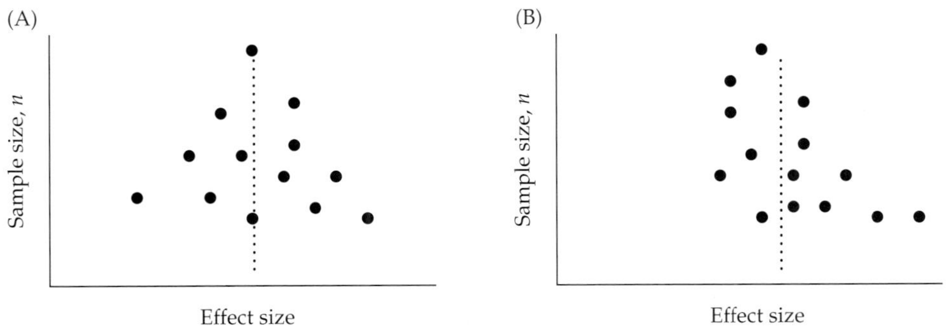

Figure A6.6 A funnel plot contrasts the estimated effect size (on the horizontal axis) of a study versus a measure of its precision (such as the sample size or $1/s_i^2$) on the vertical axis. **A:** An idealized funnel plot showing symmetry about the grand mean (dotted line). **B:** A funnel plot showing **asymmetry**, in this case an excess of larger effect-size estimates in studies with lower precision.

A second concern is **research bias** (Gurevitch and Hedges 1999). For example, in searching for signals of selection on traits, investigators are unlikely to choose a random set of characters. Rather, at least some of the traits will be explicitly chosen because an investigator feels that they are likely to be under selection. This introduces nonrandom sampling, and such a collection of estimates is not an unbiased sample of the nature of selection on a random trait. A related issue is **study bias**, in that human nature is such that investigators will often seek to make their task easier by nonrandomly choosing systems in which it is easier to measure quantities of interest.

These important concerns aside, meta-analysts worry about publication bias, such as the **file-drawer effect**, whereby studies that do not achieve significance are not published but rather are simply left in the file drawer (Rosenthal 1979; Rosenberg 2005). Likewise, studies showing significant effects are more likely to be published in higher profile, and more widely disseminated, journals. Given this concern, a number of heuristic approaches for detecting the potential of publication bias have been proposed (a nice introduction can be found in Møller and Jennions 2001). Perhaps the best way to start any discussion of publication bias, and attempts to adjust for it, is with the comment by Copas and Shi (2000) that "correcting for publication bias is only possible if we are prepared to make unverifiable assumptions."

Methods for treating publication bias consist either of approaches that attempt to *detect it* (funnel plots, rank-correlation tests, Egger regressions) or that attempt to *assess its impact* (fail-safe numbers, trim-and-fill, model selection). Nice reviews of publication bias issues as they relate to evolutionary and ecological studies can be found in Møller and Jennions (2001) and Nakagawa and Santos (2012). While any serious MA should examine publication bias, this is often ignored. In a survey of 100 evolutionary and ecologically related meta-analyses, Nakagawa and Santos found that only 49% attempted to assess publication bias. Among those studies that did, only 45% searched for signals of bias, only 14% attempted to assess the impact of bias, and 41% attempted to do both.

A number of tests for the potential presence of publication bias are constructed around the notion of a **funnel plot** (Light and Pillemer 1984). As shown in Figure A6.6, for each study, one plots the estimated effect size on the horizontal axis and some measure of precision, such as s_i^{-2} or the sample size, on the vertical axis. Under a fixed-effects model, the scatter of points should be broad at the base (reflecting spread about the true mean due to larger standard errors in the estimate), and then narrowing as one moves vertically up in the plot (larger studies). This generates a plot that looks like an inverted funnel, hence the name. Under a fixed-effects model, data high on the vertical axis (i.e., studies with very large sample sizes) should show essentially no spread about the grand mean. Under a random-effects interpretation, even when sampling error is entirely removed (due to a very large sample size), there will still be a spread of values around the grand mean, reflecting the

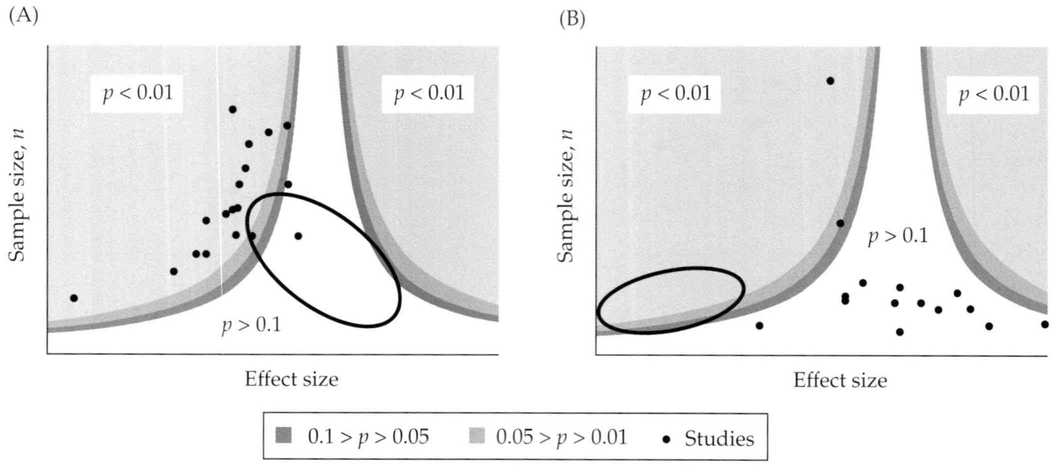

Figure A6.7 Contour-enhanced funnel plots help in the interpretation of asymmetry. Both of the plots show asymmetry, with the open ellipse showing the area of "missing" studies that generates the asymmetry. Such missing values would be added under the trim-and-fill method. The contours (showing regions of statistical significance) show the expected likelihood of study values under the null of no effect. **A:** The "missing" studies are in a region where they are likely to occur by chance ($p > 0.1$), suggesting they are absent due to publication bias. **B:** Here, these studies will all fall into regions unlikely to be seen under the null ($p < 0.05$), and hence may be missing for reasons other than publication bias. (After Peters et al. 2008.)

random effect of sampling the mean for a particular realization. One will still see a funnel, but it will only narrow down (for large n) to a spread given by σ_u^2, namely, the funnel will taper up to a cylinder whose width is a function of σ_u^2.

Figure A6.6A shows an idealized setting, with estimates that are symmetrically distributed about the grand mean. Figure A6.6B shows a situation in which there is asymmetry in the funnel. In this case, there is an excess of large-effect estimates for studies with less precision (i.e., smaller samples). Publication bias can generate such an asymmetry, as studies whose estimated effects are smaller (and therefore either not significant, or only marginally so) are less likely to be published. However, other sources of bias can also generate such an asymmetry, so that its presence does not automatically guarantee that publication bias has occurred. For example, effect sizes could be correlated with their standard errors, as might occur in settings where the realization of an effect size is expected to be small (say, the QTL effect sizes for a trait showing little variation), and so the study collects a larger sample size in order to try to achieve sufficient power (Gurevitch and Hedges 1999).

Funnel asymmetry is an informal visual judgment and, as such, it can be rather misleading (Ioannidis and Trikalinos 2007; Simmonds 2015). However, this basic idea motivated more formal tests. The **Begg–Mazumdar** (1994) **rank correlation test** looks for associations between the ranks of the standard errors (s_i) and study estimates (T_i). The **Egger regression test** (Egger et al. 1997) performs the regression

$$y_i = a + bs_i + e, \quad \text{where} \quad y_i = (T_i - \overline{T}) \tag{A6.38}$$

Under a symmetric funnel plot, the data are distributed symmetrically around zero, and the resulting intercept (a) should be zero. An a value that is significantly different from zero indicates asymmetry. Both of these tests suffer from low power, and they are not recommended when the number of studies is small (ten or fewer). Indeed, Sterne et al. (2011) recommended that these tests "should be used in only a minority of meta-analysis." Further, even a highly significant result does not imply publication bias, and a negative result does not imply that the study is free from such bias (especially if k is relatively small). Another indicator of the potential of publication bias is **time-lag bias**. Here, the initial studies report larger effects than are seen in follow-up studies (this is called the **Proteus phenomena** by

Ioannidis and Trikalinos 2005). This can be tested by simply including the year of publication as a moderator variable in the original analysis (Nakagawa and Santos 2012).

While detecting publication bias can be problematic, *correcting* for it can be even more so. One approach is the **trim-and-fill** method of Duval and Tweedie (2000a, 2000b). Again, this method is based on funnel-plot asymmetry. In step one, smaller (i.e., lower precision) studies are excluded in order to achieve a more symmetric plot, while step two replaces these excluded studies with "missing" studies whose values are imputed from the retained studies. Again, factors other than publication bias can result in asymmetric funnel plots, which compromises this approach.

The use of **contour funnel plots** (Figure A6.7) may provide some additional guidance (Peters et al. 2008). These are funnel plots enhanced by overlaying significance contours, which provides a visual test of whether "missing" studies occur in high- or low-probability regions. The former suggests bias, and the latter suggests simple sampling. Figure A6.7 shows two situations of asymmetry, and the region of the funnel plot with "missing" studies that might be imputed under a trim-and-fill approach. The contour plots show that in Figure A6.7A the missing studies were quite likely to have been seen had there been no publication bias. However, in Figure A6.7B, the missing values are in regions of high significance (and hence could be absent simply by chance, given their low probability under the null), even with a full ascertainment of all studies.

A metric for assessing the *impact* of publication bias is the **fail-safe number** for an analysis (Rosenthal 1979; Rosenberg 2005). This is simply the number of additional studies (i.e., missed studies with no significance) that would have to be added to the analysis to invalidate (i.e., remove the significance of) the current analysis. The rough rule of thumb is that if this number exceeds $5k+10$, then the study is fairly robust (Rosenthal 1979; Rosenberg 2005). Despite often being reported, there are issues with this metric. The most obvious one is that an MA is usually concerned with the average effect size (or the variance in true effect sizes), rather than whether an effect is significant. Hence, even if the fail-safe number indicates that the current study is robust, it is simply robust to overall *significance* and not necessarily to overall *effect size*. Second, as we stressed above, a random effects meta-analysis is generally the most appropriate (which extends to a mixed model if moderator variables are added). Fail-safe numbers in random-effects setting are much smaller than their fixed-effects counterparts and are much more delicate to obtain. Rosenberg (2005) cited an MA based on 71 studies where the fixed-effects fail-safe number (based on slightly different methods) ranged from 7,500 to 8,500, while the fail-safe number under a random-effects model was approximately 9.

The most powerful, and also the most delicate, methods to both detect and adjust for publication bias are model-based approaches. These require a model of the ascertainment scheme. For example, Copas and Shi (2000) proposed an ascertainment model that is very much like the model for a threshold trait (Chapter 14), wherein some underlying latent variable determines whether a study is published or not. In their model, this variable is a function of the standardized effect size, so that studies with smaller standardized effects, and therefore less significance, are less likely to be published. An excellent example of a model-based ascertainment scheme in quantitative genetics was the study by Hersch and Phillips (2004). To obtain the bias in estimates of selection gradients (the strength of natural selection on a particular trait; WL Chapter 30), they simulated each study as a random draw of five gradients from a known distribution, for which the results were only reported when one (or more) of the values was significant. WL Figure 30.7 shows the results of their analysis, which found that such ascertained gradients were overestimated.

Bias When Estimating Magnitudes

Finally, consider a random-effects setting, in which the parameter of interest, θ_i, in a given study (i) is drawn from normal distribution with a mean of μ and a variance of σ_u^2, so $\theta_i \sim N(\mu, \sigma_u^2)$. Suppose we assume the observed effect, T_i, for this study is also normal, but with an additional sampling error, σ_e^2, so that $T_i \sim N(\mu, \sigma_u^2 + \sigma_e^2)$. This additional variance

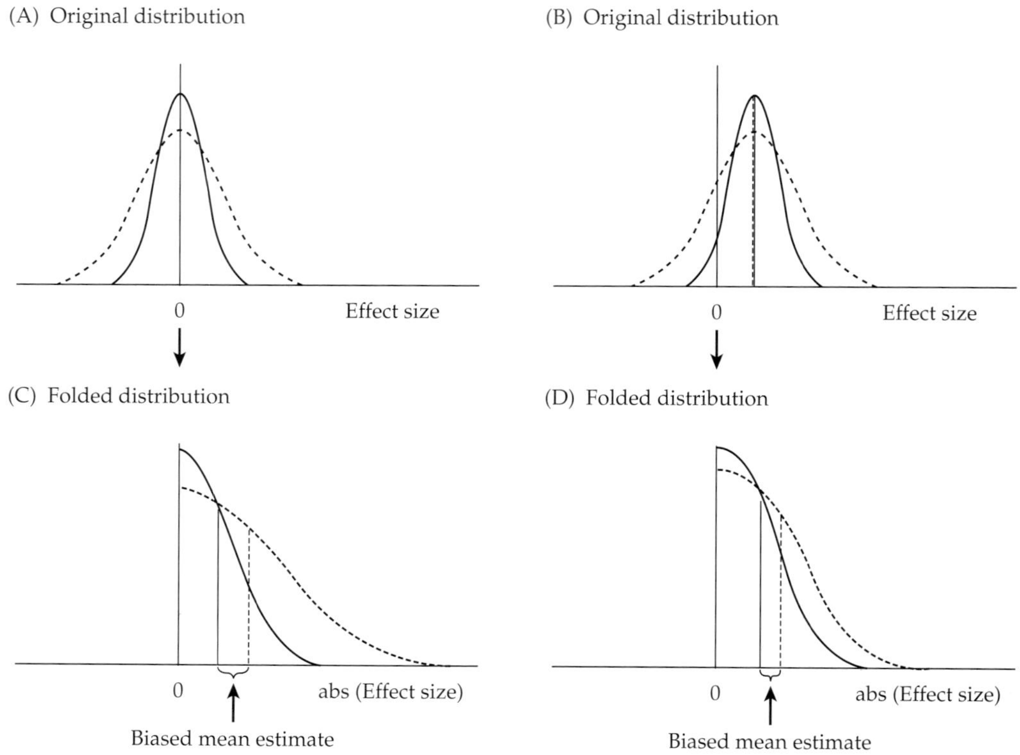

Figure A6.8 The consequences of considering the absolute value, $|\theta|$, of an effect. Here, the true size effect (θ) is normally distributed with a variance of σ^2, whose distribution is indicated by the solid curves. Of course, we do not observe the true value, but rather an estimate, T, which has the same mean as θ, but an additional sampling error, σ_e^2, so that while $\theta \sim N(\mu, \sigma^2)$, $T \sim N(\mu, \sigma^2 + \sigma_e^2)$, which (dashed curves) has a larger total variance. In (**A**) and (**C**), $\mu = 0$, while $\mu > 0$ in (**B**) and (**D**). The distribution of $|\theta|$ is given by **folding** the distribution about zero, with $\Pr(|\theta|) = \Pr(\theta) + \Pr(-\theta)$ for $\theta > 0$, as shown in panels (**C**) and (**D**). While the sampling variance (σ_e^2) does not translate into bias when estimating the mean, as $E[T] = E[\theta]$, it introduces bias when the absolute value of θ is of interest, with $E[|T|] > E[|\theta|]$ when $\sigma_e^2 > 0$, as shown in (**C**) and (**D**). (After Nakagawa and Lagisz 2016.)

has no effect on our estimate of the desired mean, μ, as $E[T_i] = \mu$. However, in many settings in quantitative genetics, our interest may be in the *absolute magnitude*, $|\theta|$, of the effect, rather than its mean value. One example would be the average magnitude of QTL effect size, as the signs of the effects are really not of interest in many settings. As noted by Hereford et al. (2004) and Morrissey (2016), $E[|T|] > E[|\theta|]$ when $\sigma_e^2 > 0$, which will result in *overestimation* of the expected absolute value of the effect. In particular, if $\mu = 0$, then (Example A6.11),

$$E[|\theta|] = \sigma_u \sqrt{2/\pi}, \qquad \text{while} \qquad E[|T|] = \sqrt{\sigma_u^2 + \sigma_e^2} \cdot \sqrt{2/\pi} \tag{A6.39a}$$

which yields a relative error of

$$\frac{E[|T|] - E[|\theta|]}{E[|\theta|]} = \sqrt{\frac{\sigma_e^2}{\sigma_u^2} + 1} - 1 \tag{A6.39b}$$

which can be a significant overestimation of the average magnitude of an effect if the error variance is large relative to the effect variance (Figure A6.8). Morrissey (2016) presented several additional examples where the concern of an MA is in some measure of dispersion about a mean, in which case the error variance associated with using an estimate (T_i) will result in an upwardly biased estimator. Morrissey showed how the use of appropriate mixed models helps to resolve this concern.

Example A6.11. Hereford et al. (2004) and Morrissey (2016) presented general expressions for $E[|x|]$ when $x \sim N(\mu, \sigma^2)$. As shown in Figure A6.8, the distribution of $|x|$ is given by a folded normal distribution, from which it follows that

$$E[|x|] = \sigma\sqrt{\frac{2}{\pi}}\exp\left(-\frac{\mu^2}{2\sigma^2}\right) + |\mu| \cdot \text{erf}\left(\frac{|\mu|}{\sigma\sqrt{2}}\right) \tag{A6.40a}$$

where

$$\text{erf}(x) = \frac{2}{\sqrt{\pi}}\int_0^x \exp(-t^2)\, dt$$

is the error function. Taking $\mu = 0$, Equation A6.40a reduces to Equation A6.39a. Likewise, when $\mu \gg \sigma^2$, Equation A6.40a reduces to

$$E[|x|] \sim |\mu| \tag{A6.40b}$$

which follows because for large values of x, $e^{-x} \to 0$ and $\text{erf}(x) \to 1$.

Appendix 7

Introduction to Bayesian Analysis

Uncertainty is the refuge of hope. Henri Frederic Amiel

The history of statistical methods employed in genetics closely parallels advances in computation. Before the widespread use of computers, **method-of-moments** approaches were common as they are relatively easy to obtain. Here, a summary statistic of the data is computed whose expected value is the parameter of interest (e.g., using the sample mean, $\overline{x}$, as an estimate of the true mean, μ_x, as $E[\,\overline{x}\,] = \mu_x$). In the mid-1970s, maximum-likelihood (ML) methods became much more common-place, as they offer a very flexible platform for statistical analysis (estimation, determining precision, and hypothesis testing), but at the cost of numerically searching an often highly complex multidimensional likelihood surface (Appendix 4). Both these approaches typically return **point estimators** for the variables of interest, along with some measure of their uncertainty. As opposed to these classical (or **frequentist**) approaches, **Bayesian statistics** (which can be viewed as a natural extension of likelihood methods) is concerned with generating the *full distribution* for the parameters, $\boldsymbol{\Theta}$, given the data, $\mathbf{x}$, namely, obtaining the posterior distribution, $p(\boldsymbol{\Theta}\,|\,\mathbf{x})$. As such, Bayesian statistics provides a much more complete picture of the uncertainty in the estimation of the unknown parameters, especially after the confounding effects of nuisance parameters are removed.

Our treatment here is intentionally quite brief. A number of texts have presented excellent treatments of the statistical theory (e.g., Lindley 1965; Berger 1985; Carlin and Louis 2000; Lee 2012; Gelman et al. 2013). Blasco (2017) provided a very lucid introduction to applications in quantitative genetics, while Sorensen and Gianola (2002) offered a more comprehensive treatment. While very deep (and very subtle) differences in philosophy separate hard-core Bayesians from hard-core frequentists (Efron 1986; Glymour 1981), our treatment of Bayesian methods is motivated simply by their use as a powerful statistical tool. This appendix focuses on the basic theory, while the computational approaches that make these methods feasible are briefly examined in Appendix 8.

WHY ARE BAYESIAN METHODS BECOMING MORE POPULAR?

In addition to providing a more formal framework for dealing with parameter uncertainty, two specific features have fueled the rapid growth of Bayesian approaches in genetics and genomics. First, under a Bayesian analysis, all parameters are random effects as opposed to fixed effects (Chapter 10). This has profound implications for degrees of freedom. Consider a gene expression study with 30,000 features (genes of interest), whose mRNA levels are contrasted over a set of 100 normal liver cells versus 100 cancerous ones. If we treat the differential expression level of any particular gene as a fixed effect (an unknown constant to be estimated) we will very quickly use all of the degrees of freedom, given the small sample size. Conversely, if these levels are treated as **random effects**, with the expression difference associated with a particular gene being a random variable drawn from some underlying (and unknown) distribution, then the only degrees of freedom lost will be those used to estimate the associated parameters for this underlying distribution (typically, its variance). Further, prediction of the random realization that corresponds to a particular gene borrows information over all the genes. Thus, a Bayesian analysis can handle high-dimensional experiments in which the number of parameters, p, greatly exceeds the number of observations, n, in a framework that fully manages the uncertainty over all these estimates. Second,

Bayesian methods are ***computationally feasible***, as approaches such as MCMC (Appendix 8) allow high-dimensional datasets to be analyzed in a computationally efficient manner. In settings with a large number of nuisance parameters and/or a high-dimensional dataset, a Bayesian approach not only has considerable appeal, it may be the only approach that is even feasible.

BAYES' THEOREM

The foundation of Bayesian statistics is Bayes' theorem, which was introduced in Chapter 3. From Equation 3.3b,

$$\Pr(\theta \,|\, x) = \frac{\Pr(x \,|\, \theta)\,\Pr(\theta)}{\Pr(x)} \tag{A7.1}$$

In Bayesian statistics, x represents an observable variable (the data), while θ represents a parameter describing the distribution of x. In this setting, $\Pr(\theta)$ is the **prior distribution** of possible parameter values, while $\Pr(\theta \,|\, x)$ is the subsequent **posterior distribution** of θ given the observed data x and the prior. In classical statistics, the unknown parameters are treated as fixed and the data are considered random, whereas under a Bayesian analysis, the data are considered fixed and the unknown parameters that generated the data are considered random.

Equation A7.1 also holds for continuous random variables, with the probability density function (pdf), p, replacing the discrete probability value, Pr. In particular, the continuous multivariate version of Bayes' theorem is

$$p(\boldsymbol{\Theta} \,|\, \mathbf{x}) = \frac{p(\mathbf{x} \,|\, \boldsymbol{\Theta})\, p(\boldsymbol{\Theta})}{p(\mathbf{x})} = \frac{p(\mathbf{x} \,|\, \boldsymbol{\Theta})\, p(\boldsymbol{\Theta})}{\int p(\mathbf{x}, \boldsymbol{\Theta})\, d\boldsymbol{\Theta}} \tag{A7.2}$$

where $\boldsymbol{\Theta} = (\theta_1, \theta_2, \cdots, \theta_n)^T$ is a vector of n (potentially) continuous variables. As with the univariate case, $p(\boldsymbol{\Theta})$ is the assumed prior distribution of the unknown parameters, while $p(\boldsymbol{\Theta} \,|\, \mathbf{x})$ is the posterior distribution given the prior, $p(\boldsymbol{\Theta})$, and the data, $\mathbf{x}$.

The origin of Bayes' theorem has a fascinating history (Stigler 1983). It is named after the Rev. Thomas Bayes, a priest who never published a mathematical paper during his lifetime. The paper in which the theorem appears was posthumously read before the Royal Society by his friend Richard Price in 1764. Stigler suggests it was first discovered by Nicholas Saunderson, a blind mathematician and optician who, at age 29, became Lucasian Professor of Mathematics at Cambridge (the position held earlier by Isaac Newton). This is an example of **Stigler's Law of Eponymy** (Stigler 1980), wherein no discovery or invention is named after its first discoverer (an **eponym**). As is fitting, Stigler's law is self-consistent, as this phenomenon was previously mentioned by Merton (1965).

Example A7.1. Consider a recessive color locus in cattle in which the genotypes *BB* and *Bb* are black, while *bb* is red. Two black-coated parents are crossed, and produce some red offspring, which implies that both parents must be *Bb*. A black-coated son of theirs is crossed to n red dams (*bb*), and all of his offspring are black. What is the posterior probability that he is *BB*?

To solve this problem using Bayes' theorem, we first define the indicator random variable

$$\theta = \begin{cases} 0 & \text{son is } Bb \\ 1 & \text{son is } BB \end{cases}$$

Given that both parents are *Bb*, the expected priors for their offspring are 1/4 for *BB* and 1/2 for *Bb*, resulting in a 3/4 prior for a black-coated offspring. Further, from conditional probability (Equation 3.3a), the prior that a black offspring is *BB* is

$$\Pr(BB \,|\, \text{Black}) = \frac{\Pr(BB, \text{Black})}{\Pr(\text{Black})} = \frac{\Pr(BB)}{\Pr(\text{Black})} = \frac{1/4}{3/4} = 1/3$$

where we used the fact that all *BB* are black, so that Pr(*BB*, Black) = Pr(*BB*). Hence, the prior becomes

$$\Pr(\theta) = \begin{cases} 0 & \text{is} \quad 2/3 \\ 1 & \text{is} \quad 1/3 \end{cases}$$

Further

$$\Pr(\text{all } n \text{ offspring are black} \mid \text{sire is } BB) = 1$$

$$\Pr(\text{all } n \text{ offspring are black} \mid \text{sire is } Bb) = (1/2)^n$$

and

$$\Pr(\text{all } n \text{ black}) = \Pr(\text{all black} \mid BB)^*\Pr(BB) + \Pr(\text{all black} \mid Bb)^*\Pr(Bb)$$

$$= 1 \cdot 1/3 + (1/2)^n \cdot (2/3)$$

If we combine the above values, Bayes' theorem yields

$$\Pr(\theta = 1 \mid n \text{ black offspring}) = \frac{\Pr(n \mid \theta = 1)\Pr(\theta = 1)}{\Pr(n)} = \frac{1 \cdot (1/3)}{1 \cdot (1/3) + (1/2)^n \cdot (2/3)}$$

which returns values of 0.5, 0.67, 0.8, 0.89, 0.94, and 0.998 for $n = 1, 2, 3, 4, 5$, and 10, respectively.

FROM LIKELIHOOD TO BAYESIAN ANALYSIS

The method of maximum likelihood (Appendix 4) and Bayesian analysis are closely related. Suppose $\ell(\boldsymbol{\Theta} \mid \mathbf{x})$ is the assumed likelihood function. Under ML estimation, we would compute the **mode** of the likelihood function (the maximal value of ℓ, as a function of $\boldsymbol{\Theta}$ given the data $\mathbf{x}$), and use the local curvature around the mode to construct confidence intervals. Hypothesis testing follows using likelihood-ratio (LR) statistics. The strengths of ML estimation rely on its *large-sample* properties, namely, that when the sample size is sufficiently large, we can assume both normality of the estimators and that most LR tests follow χ^2 distributions. These features, nice as they are, may not hold for small samples. Conversely, a Bayesian analysis is *exact* for any sample size, given a specified prior.

To transition from a likelihood to a Bayesian analysis, we start with some prior distribution, $p(\boldsymbol{\Theta})$, that captures our initial knowledge (or best guess) about the possible values of the unknown parameters. From Bayes' theorem, the data (likelihood) are combined with the prior to produce a posterior distribution,

$$p(\boldsymbol{\Theta} \mid \mathbf{x}) = \frac{1}{p(\mathbf{x})} \cdot p(\mathbf{x} \mid \boldsymbol{\Theta}) \cdot p(\boldsymbol{\Theta}) \tag{A7.3a}$$

$$= (\text{normalizing constant}) \cdot p(\mathbf{x} \mid \boldsymbol{\Theta}) \cdot p(\boldsymbol{\Theta}) \tag{A7.3b}$$

$$= \text{constant} \cdot \text{likelihood} \cdot \text{prior} \tag{A7.3c}$$

as $p(\mathbf{x} \mid \boldsymbol{\Theta}) = \ell(\boldsymbol{\Theta} \mid \mathbf{x})$ is simply the likelihood function (Appendix 4) and $1/p(\mathbf{x})$ is a constant (with respect to $\boldsymbol{\Theta}$). Consequently, the posterior distribution is often written as

$$p(\boldsymbol{\Theta} \mid \mathbf{x}) \propto \ell(\boldsymbol{\Theta} \mid \mathbf{x})\, p(\boldsymbol{\Theta}) \tag{A7.3d}$$

where the symbol $\propto$ means "proportional to" (equal up to a constant). Note that the constant $p(\mathbf{x})$ normalizes $p(\mathbf{x} \mid \boldsymbol{\Theta}) \cdot p(\boldsymbol{\Theta})$ to one (making the posterior a formal probability distribution), and hence can be obtained by integration,

$$p(\mathbf{x}) = \int_{\boldsymbol{\Theta}} p(\mathbf{x} \mid \boldsymbol{\Theta}) \cdot p(\boldsymbol{\Theta})\, d\boldsymbol{\Theta} \tag{A7.4}$$

The dependence of the posterior on the prior (which can easily be assessed by trying different priors) provides an indication of how much information on the unknown parameter values is contained in the data (the curvature of the likelihood surface). If the posterior is highly dependent on the prior, then the data likely have little signal (a **flat likelihood surface**), while if the posterior is largely unaffected by different priors, then the data are likely highly informative (a sharply peaked likelihood surface). To see this, taking logs on Equation A7.3c yields

$$\log(\text{posterior}) = \log(\text{likelihood}) + \log(\text{prior}) + \text{constant} \tag{A7.5}$$

When the likelihood signal is strong, it largely dominates the prior in the resulting posterior, but when a likelihood is weak, the prior can dominate.

Marginal Posterior Distributions

Often only a subset of the unknown parameters is of concern, with the rest being **nuisance parameters** that are of no interest, but still must be fitted in the model. A strong feature of Bayesian analysis is that one can account for all the uncertainty introduced into the parameters of interest by any uncertainty in the values of nuisance parameters. This is accomplished by integrating the nuisance parameters out of the posterior distribution to generate a **marginal posterior distribution** for the parameters of interest. For example, suppose the mean and variance of data coming from a normal distribution are unknown, but our real interest is only in the variance. Estimating the mean introduces additional uncertainty into our variance estimate, which is not fully captured by standard classical approaches. Under a Bayesian analysis, the marginal posterior distribution for σ^2 is simply

$$p(\,\sigma^2 \,|\, \mathbf{x}) = \int p(\,\mu, \sigma^2 \,|\, \mathbf{x}\,)\, d\mu$$

The resulting marginal posterior for σ^2 captures all of the uncertainty in the estimation of μ that influences the uncertainty in σ^2. This is an especially nice feature when a large number of nuisance parameters must be estimated.

The marginal posterior may involve several parameters (generating **joint marginal posteriors**). Suppose we write the vector of unknown parameters as $\boldsymbol{\Theta} = (\boldsymbol{\Theta}_1, \boldsymbol{\Theta}_{nu})$, where $\boldsymbol{\Theta}_{nu}$ is the vector of nuisance parameters. Integrating over $\boldsymbol{\Theta}_{nu}$ yields the desired marginal for the vector $\boldsymbol{\Theta}_1$ of parameters of interest as

$$p(\boldsymbol{\Theta}_1 \,|\, \mathbf{y}) = \int_{\boldsymbol{\Theta}_{nu}} p(\boldsymbol{\Theta}_1, \boldsymbol{\Theta}_{nu} \,|\, \mathbf{y})\, d\boldsymbol{\Theta}_{nu} \tag{A7.6}$$

While these complex integrals appear quite daunting (and indeed almost always are from an analytic standpoint), generating draws from the marginal distribution is usually very straightforward using MCMC methods (which are examined in Appendix 8).

SUMMARIZING THE POSTERIOR DISTRIBUTION

How do we extract a Bayesian estimator for some unknown parameter, θ? If our mindset is to use some sort of point estimator (as is usually done in classical statistics), then there are a number of candidates. We could follow ML and use the **mode of the posterior distribution** (its maximal value)

$$\widehat{\theta} = \max_{\theta} \left[p(\,\theta \,|\, \mathbf{x}\,) \right] \tag{A7.7a}$$

We could take the **expected value of** θ (its mean) given the posterior

$$\widehat{\theta} = E[\,\theta \,|\, \mathbf{x}\,] = \int \theta\, p(\,\theta \,|\, \mathbf{x}\,) d\theta \tag{A7.7b}$$

Another candidate is the **median of the posterior**, which is more robust than the mean to outliers. Here the estimator satisfies $\Pr(\theta > \widehat{\theta} \mid \mathbf{x}) = \Pr(\theta < \widehat{\theta} \mid \mathbf{x}) = 0.5$, hence

$$\int_{\widehat{\theta}}^{+\infty} p(\,\theta \mid \mathbf{x}\,)d\theta = \int_{-\infty}^{\widehat{\theta}} p(\,\theta \mid \mathbf{x}\,)d\theta = \frac{1}{2} \tag{A7.7c}$$

However, using any of the above estimators, or even all three simultaneously, loses the full power of a Bayesian analysis, as ***the full estimator is the entire posterior density itself***. If we cannot obtain the full form of the posterior distribution, then these estimates of general features of the distribution can be presented. However, as we will see in Appendix 8, we can generally obtain the full posterior by simulation using MCMC sampling, and hence the Bayesian estimate of a parameter is often presented as a frequency histogram (potentially smoothed) of the MCMC-generated samples from the posterior distribution (an **empirical posterior**). Typically, when such a histogram is displayed, it is usually accompanied by one or more of the summary statistics given by Equations A7.7a–A7.7c (as well as other metrics, such as the variance and skewness).

Highest Density Regions (HDRs)

Given the posterior distribution, the construction of uncertainty intervals is straightforward. For example, a $100(1-\alpha)\%$ "confidence interval" is given by any $(L_{\alpha/2}, H_{\alpha/2})$ satisfying

$$\int_{L_{\alpha/2}}^{H_{\alpha/2}} p(\theta \mid \mathbf{x})\, d\theta = 1 - \alpha$$

To reduce the set of possible candidate intervals, one typically uses **highest density regions**, or **HDRs**, where, for a single parameter, the HDR $100(1-\alpha)$ region(s) are the shortest interval(s) giving an area of $(1-\alpha)$. More generally, if multiple parameters are being estimated, the HDR region(s) are those with the smallest *volume* in the parameter space. HDRs are also referred to as **Bayesian confidence intervals** or (better yet) **credible intervals**.

It is critical to note that there is a *profound difference* between a confidence interval (CI) from classical (frequentist) statistics and a Bayesian analysis. The interpretation of a classical CI is that if we were to repeat the experiment a sufficiently large number of times, and construct CIs in the same fashion, the fraction of the resulting collection of CIs that enclose the unknown parameter approaches $(1-\alpha)$. Thus, the frequentist CI is a measure of the *frequency* of occurrences in independent experiments in which the CI encloses the true value (and hence the term frequentist for this type of statistics). In contrast, with a Bayesian HDR, there is a probability of $(1-\alpha)$ that the interval contains the true value of the unknown parameter. While at first blush these two interpretations of CIs appear to be essentially identical, they are not, and indeed they are fundamentally (but subtly) different. Often the CI and Bayesian intervals span essentially the same values, but again the interpretational difference remains. The key point is that the Bayesian prior allows us to make ***direct probability statements*** about θ, while under classical statistics we can only make statements about the behavior of the statistic if we consider ***repeating an experiment a large number of times***. Given the important conceptual difference between classical and Bayesian intervals, Bayesians typically avoid using the term *confidence interval*, using the term *credible interval* instead.

Bayes Factors and Hypothesis Testing

In the classical hypothesis-testing framework, we have two alternatives. The null hypothesis, H_0, that the unknown parameter, θ, belongs to some set or interval, Θ_0 ($\theta \in \Theta_0$), versus the alternative hypothesis, H_1, that θ belongs to the alternative set, Θ_1 ($\theta \in \Theta_1$). Θ_0 and Θ_1 contain no common elements ($\Theta_0 \cap \Theta_1 = \oslash$) and the union of Θ_0 and Θ_1 contains the entire space of values for θ (i.e., $\Theta_0 \cup \Theta_1 = \Theta$).

In the classical frequentist framework, one uses the observed data to test the significance of a particular hypothesis, and (if possible) compute a p value (the probability, p, of observing a value equal to, or more extreme than, that of the test statistic if the null hypothesis is indeed correct). Initially, one would think that the idea of a hypothesis test is trivial in a Bayesian framework, as using the posterior distribution provides the expected p values directly. For example,

$$\Pr(\theta > \theta_0) = \int_{\theta_0}^{\infty} p(\,\theta\,|\,\mathbf{x})\,d\theta \qquad \text{and} \qquad \Pr(\theta_0 < \theta < \theta_1) = \int_{\theta_0}^{\theta_1} p(\,\theta\,|\,\mathbf{x})\,d\theta$$

The fault in this logic under a Bayesian framework is that we also have *prior information* and Bayesian hypothesis testing addresses whether, *given the data*, we are more or less inclined to believe the hypothesis that was suggested from the prior. Hence, the ***prior probabilities influence hypothesis testing***. To formalize this idea, let

$$p_0 = \Pr(\theta \in \Theta_0\,|\,\mathbf{x}) \qquad \text{and} \qquad p_1 = \Pr(\theta \in \Theta_1\,|\,\mathbf{x}) \tag{A7.8a}$$

denote the probabilities, given the observed data, $\mathbf{x}$, that θ is in the null (p_0) and alternative (p_1) hypothesis sets. Note that these are *posterior* probabilities. Because $\Theta_0 \cap \Theta_1 = \oslash$ and $\Theta_0 \cup \Theta_1 = \Theta$, it follows that $p_0 + p_1 = 1$. Likewise, for the *prior* probabilities we have

$$\pi_0 = \Pr(\theta \in \Theta_0) \qquad \text{and} \qquad \pi_1 = \Pr(\theta \in \Theta_1) \tag{A7.8b}$$

Thus the **prior odds** of H_0 versus H_1 are π_0/π_1, while the **posterior odds** are p_0/p_1.

The **Bayes factor**, $\boldsymbol{B_0}$, in favor of H_0 versus H_1 is calculated by the ratio of the posterior odds divided by the prior odds,

$$B_0 = \frac{p_0/p_1}{\pi_0/\pi_1} = \frac{p_0\pi_1}{p_1\pi_0} \tag{A7.9a}$$

The Bayes factor is loosely interpreted as the odds in favor of H_0 over H_1 as given by the data and our prior opinion. Because $\pi_1 = 1 - \pi_0$ and $p_1 = 1 - p_0$, we can also express this as

$$B_0 = \frac{p_0(1-\pi_0)}{\pi_0(1-p_0)} \tag{A7.9b}$$

By symmetry, note that the Bayes factor, B_1, in favor of H_1 versus H_0 is simply $B_1 = 1/B_0$.

Example A7.2. Suppose that the prior distribution of θ is such that $\Pr(\theta > \theta_0) = 0.10$, while for the posterior distribution $\Pr(\theta > \theta_0\,|\,\mathbf{x}) = 0.05$. The latter is significant at the 5% level in a classical hypothesis-testing framework, but the data only double our confidence in the alternative hypothesis relative to our belief based on prior information. If $\Pr(\theta > \theta_0) = 0.50$ for the prior, then a 5% posterior probability would greatly increase our confidence in the alternative hypothesis. Consider the first case in this example, where the prior and posterior probabilities for the null were $\pi_0 = 0.1$ and $p_0 = 0.05$, respectively. The Bayes factor in favor of H_1 versus H_0 is

$$B_1 = \frac{\pi_0(1-p_0)}{p_0(1-\pi_0)} = \frac{0.1 \cdot 0.95}{0.05 \cdot 0.9} = 2.11$$

Similarly, for the second example, where the prior for the null was $\pi_0 = 0.5$,

$$B_1 = \frac{0.5 \cdot 0.95}{0.05 \cdot 0.5} = 19$$

Here, the data showed close to a 20-fold improvement (relative to the prior) in support of H_1. Bayes factors and p values represent fundamentally different approaches to an analysis and are not formally comparable. However, a *loose* interpretation is that a factor of 20 is akin to the level of support of a $p = 0.05$, and a factor of 100 to $p = 0.01$.

When the hypotheses are simple (i.e., single values), say $\Theta_0 = \theta_0$ versus $\Theta_1 = \theta_1$, then

$$p_i \propto p(\theta_i)\, p(\mathbf{x}\,|\,\theta_i) = \pi_i\, p(\mathbf{x}\,|\,\theta_i) \quad \text{for} \quad i = 0, 1$$

Thus

$$\frac{p_0}{p_1} = \frac{\pi_0\, p(\mathbf{x}\,|\,\theta_0)}{\pi_1\, p(\mathbf{x}\,|\,\theta_1)} \tag{A7.10a}$$

and from Equation A7.9a, the Bayes factor (in favor of the null) reduces to

$$B_0 = \frac{p(\mathbf{x}\,|\,\theta_0)}{p(\mathbf{x}\,|\,\theta_1)} \tag{A7.10b}$$

which is simply a *likelihood ratio* (Appendix 4).

When hypotheses are **composite** (containing multiple elements), the situation is slightly more complicated. First, note that the prior distribution of θ conditioned on H_0 or H_1 is

$$p_i(\theta) = p(\theta)/\pi_i \qquad \text{for} \quad i = 0, 1 \tag{A7.11}$$

as the total probability $\theta \in \Theta_i = \pi_i$, so dividing by π_i normalizes the distribution to integrate to one. Thus,

$$\begin{aligned} p_i = \Pr(\theta \in \Theta_i\,|\,\mathbf{x}) &= \int_{\theta \in \Theta_i} p(\theta\,|\,\mathbf{x}) d\theta \\ &= \frac{1}{p(\mathbf{x})} \int_{\theta \in \Theta_i} p(\theta) p(\mathbf{x}\,|\,\theta) d\theta \\ &= \frac{\pi_i}{p(\mathbf{x})} \int_{\theta \in \Theta_i} p_i(\theta) p(\mathbf{x}\,|\,\theta) d\theta \end{aligned} \tag{A7.12}$$

where the second step follows from Bayes' theorem, while the final step follows from Equation A7.11. The Bayes factor in favor of the null hypothesis becomes

$$B_0 = \left(\frac{p_0}{\pi_0}\right)\left(\frac{\pi_1}{p_1}\right) = \frac{\int_{\theta \in \Theta_0} p(\mathbf{x}\,|\,\theta) p_0(\theta) d\theta}{\int_{\theta \in \Theta_1} p(\mathbf{x}\,|\,\theta) p_1(\theta) d\theta} \tag{A7.13}$$

which is a ratio of the weighted likelihoods of Θ_0 and Θ_1.

THE CHOICE OF A PRIOR

Obviously, a critical feature of any Bayesian analysis is the choice of a prior. The key is that when the data have a sufficiently strong signal, even a poor choice of a prior will still not greatly influence the posterior. In a sense, it is an asymptotic (large-sample) property of Bayesian analysis in that all but pathological priors (those with zero probability where the true value lies) can be overcome by sufficient amounts of data. As mentioned above, one can check the impact of the prior by assessing the stability of posterior over a collection of diverse priors. The **location** of a parameter (mean or mode) and its **uncertainty** (variance) of the prior is usually more critical than its actual shape in terms of conveying prior information. The shape (family) of the prior distribution is often chosen to facilitate calculation of the posterior, especially through the use of **conjugate priors** that, for a given likelihood function, return a posterior in the same distribution family as the prior (e.g., a gamma prior returns a gamma posterior when the likelihood is Poisson). We will return to conjugate priors, but first we will discuss other approaches for construction of priors.

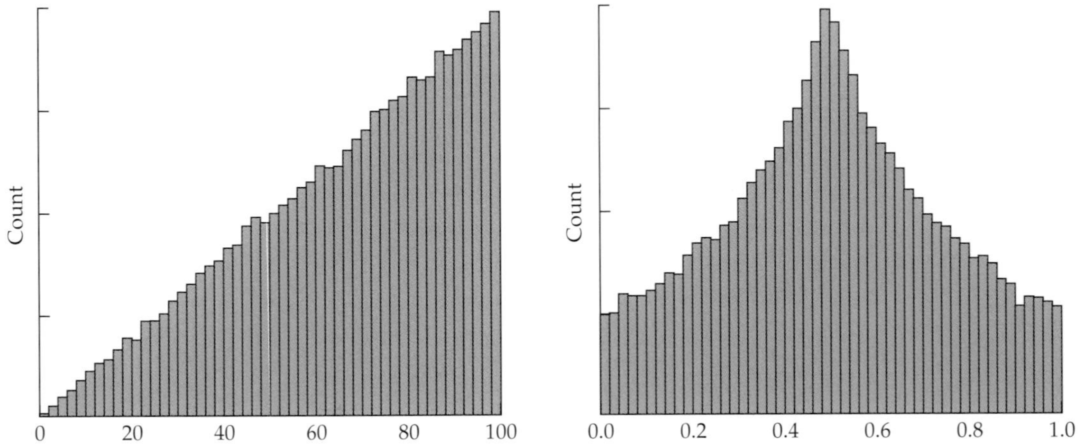

Figure A7.1 A uniform prior on one scale does not result in a flat prior on a transformed scale. Suppose a flat prior on (0,10000) is assumed for both the additive and residual variances. To mimic what happens under MCMC, we display these priors by using the resulting histograms generated from a large number of random draws, with a uniform prior expected to return a flat histogram. **Left:** The resulting prior for the standard deviation of either variance (the square root of a random draw). **Right:** The resulting prior for h^2, the ratio of a random draw for the additive variance divided by this value plus a random draw for the residual variance. Neither of these priors results in a uniform prior (namely, a flat histogram) on the transformed scale.

Diffuse Priors

One of the most commonly used priors is the **flat** or **diffuse** (also called **uninformative** or **naive**) prior, which is simply a constant

$$p(\theta) = \frac{1}{b-a} \qquad \text{for} \qquad a \leq \theta \leq b \tag{A7.14a}$$

This conveys that we have no a priori reason to favor any particular parameter value over another. With a flat prior, the posterior is just a constant C times the likelihood

$$p(\theta \,|\, \mathbf{x}) = C\,\ell(\theta \,|\, \mathbf{x}) \tag{A7.14b}$$

and we typically write that $p(\theta \,|\, \mathbf{x}) \propto \ell(\theta \,|\, \mathbf{x})$. In many cases, classical expressions from frequentist statistics are obtained by Bayesian analysis through assuming a flat prior (e.g., the posterior mode is the MLE).

If the variable (i.e., parameter) of interest ranges over $(0, \infty)$ or $(-\infty, +\infty)$, then, strictly speaking, a flat prior does not exist as, if the constant takes on any nonzero value, the integral does not exist. In such cases a flat prior (i.e., assuming $p[\theta \,|\, \mathbf{x}] \propto \ell[\theta \,|\, \mathbf{x}]$) is referred to as an **improper prior**, and care must be taken to ensure that the product of the prior and the likelihood results in a proper posterior (i.e., $\ell[\theta \,|\, \mathbf{x}]$ has a finite integral over the parameter range). This is by no means certain.

Another complication involved in using a uniform prior arises when the question of interest resides on a *different scale* than that used for the prior. A variable uniform on one scale may be far from uniform on a transformed scale. Figure A7.1 shows two examples based on the assumption that there was a flat prior on the variance. A uniform prior on the variance does *not* result in a uniform prior on the standard deviation (e.g., Van Dongen 2006). Likewise, if one assumes that the additive and residual variances have flat priors, this does not imply a flat prior for h^2, but rather a prior that is sharply peaked at 1/2. When assuming a flat prior, care must be taken that it is truly uninformative on the appropriate scale of biological interest. Otherwise, the choice of what superficially appears as an unbiased prior may instead create a bias that the signal in the data must overcome.

The Jeffreys Prior

Jeffreys (1961) proposed a general prior based on the Fisher information, F, of the likelihood. Recall (Equation A4.7b) that

$$F(\theta \,|\, \mathbf{x}) = -E\left[\frac{\partial^2 \ln \ell(\theta \,|\, \mathbf{x})}{\partial\, \theta^2}\right]$$

The **Jeffreys prior** is as follows:

$$p(\theta) \propto \sqrt{F(\theta \,|\, \mathbf{x})} \tag{A7.15}$$

A full discussion, with derivation, can be found in Lee (2012).

When there are k parameters, **F** is the $k \times k$ Fisher information matrix of the expected second partials, where the elements of **F** are calculated by

$$\mathbf{F}(\,\boldsymbol{\Theta} \,|\, \mathbf{x}\,)_{ij} = -E_x\left[\frac{\partial^2 \ln \ell(\boldsymbol{\Theta} \,|\, \mathbf{x})}{\partial\theta_i\, \partial\theta_j}\right]$$

In this case, the Jeffreys prior becomes

$$p(\boldsymbol{\Theta}) \propto \sqrt{\det[\mathbf{F}(\theta \,|\, \mathbf{x}\,)\,]} \tag{A7.16}$$

Example A7.3. Consider the likelihood of x successes in n independent draws from a binomial with a success parameter of θ,

$$\ell(\theta \,|\, \mathbf{x}) = C\theta^x(1-\theta)^{n-x}$$

where the constant C does not involve θ. Taking logs gives

$$L(\theta \,|\, \mathbf{x}) = \ln\left[\,\ell(\theta \,|\, \mathbf{x})\,\right] = \ln C + x\ln\theta + (n-x)\ln(1-\theta)$$

Thus

$$\frac{\partial L(\theta \,|\, \mathbf{x})}{\partial\theta} = \frac{x}{\theta} - \frac{n-x}{1-\theta}$$

and likewise

$$\frac{\partial^2 L(\theta \,|\, \mathbf{x})}{\partial\theta^2} = -\frac{x}{\theta^2} - (-1)\cdot(-1)\frac{n-x}{(1-\theta)^2} = -\left(\frac{x}{\theta^2} + \frac{n-x}{(1-\theta)^2}\right)$$

Because $E[\,x\,] = n\theta$, then

$$-E\left[\frac{\partial^2 \ln \ell(\theta \,|\, \mathbf{x}\,)}{\partial\,\theta^2}\right] = \frac{n\theta}{\theta^2} + \frac{n(1-\theta)}{(1-\theta)^2} = n\,\theta^{-1}(1-\theta)^{-1}$$

The resulting Jeffreys prior for this likelihood becomes

$$p(\theta) \propto \sqrt{\theta^{-1}(1-\theta)^{-1}} \propto \theta^{-1/2}(1-\theta)^{-1/2}$$

which is a U-shaped beta distribution with parameters $\alpha = \beta = 1/2$ (Equation A7.37a). This prior puts more weight on extreme values relative to assuming a uniform over (0,1), see Figure A7.3.

Example A7.4. Suppose our data consist of n independent draws from a normal distribution with an unknown mean and variance, μ and σ^2. In Example A4.3, we showed that the information matrix in this case is

$$\mathbf{F} = n \begin{pmatrix} \dfrac{1}{\sigma^2} & 0 \\ 0 & \dfrac{1}{2\sigma^4} \end{pmatrix}$$

Because the determinant of a diagonal matrix is the product of the diagonal elements, $\det(\mathbf{F}) \propto \sigma^{-6}$, giving the Jeffreys prior for μ and σ^2 as

$$p(\boldsymbol{\Theta}) \propto \sqrt{\sigma^{-6}} = \sigma^{-3}$$

Because the joint prior does not involve μ, this implies a flat prior for μ (i.e., $p[\mu] = c$). Note here that the prior distributions of μ and σ^2 are independent, as

$$p(\mu, \theta) = c \cdot \sigma^{-3} = p(\mu) \cdot p(\sigma^2)$$

POSTERIOR DISTRIBUTIONS UNDER NORMALITY ASSUMPTIONS

To introduce the basic ideas of Bayesian analysis, as well as treat a common assumption in quantitative genetics, consider the case where data are drawn from a normal (Gaussian) distribution, giving the likelihood function for the ith observation, x_i, as

$$\ell(\mu, \sigma^2 \,|\, x_i) = \frac{1}{\sqrt{2\pi\sigma^2}} \exp\left(-\frac{(x_i - \mu)^2}{2\sigma^2} \right) \tag{A7.17a}$$

If we assume independence, the resulting full likelihood for all n data points (with a sample mean of $\overline{x}$) is

$$\ell(\mu \,|\, \mathbf{x}) = \frac{1}{\sqrt{2\pi\sigma^2}} \exp\left(-\sum_{i=1}^{n} \frac{(x_i - \mu)^2}{2\sigma^2} \right) \tag{A7.17b}$$

$$= \frac{1}{\sqrt{2\pi\sigma^2}} \exp\left[-\frac{1}{2\sigma^2} \left(\sum_{i=1}^{n} x_i^2 - 2\mu n\overline{x} + n\mu^2 \right) \right] \tag{A7.17c}$$

The form of the posteriors given these normal likelihoods is a function of the assumed priors. By using the appropriate conjugate priors, these posteriors follow fairly standard distributions, and hence are easier to work with, as we now demonstrate.

Gaussian Likelihood With Known Variance and Unknown Mean

As a starting point, assume that the variance, σ^2, is known, while the mean, μ, is unknown. For a Bayesian analysis, it remains to specify the prior for μ, $p(\mu)$. Suppose we assume a Gaussian prior, $\mu \sim N(\mu_0, \sigma_0^2)$, with

$$p(\mu) = \frac{1}{\sqrt{2\pi\sigma_0^2}} \exp\left(-\frac{(\mu - \mu_0)^2}{2\sigma_0^2} \right) \tag{A7.18}$$

The mean and variance of the prior, μ_0 and σ_0^2, are referred to as **hyperparameters**. Here, μ_0 specifies a prior location for the parameter (the unknown mean, μ), while σ_0^2 specifies our uncertainty in this prior location—the larger σ_0^2, the greater is our uncertainty. In the limit as $\sigma_0^2 \to \infty$, $p(\mu)$ approaches a flat (and in this case, improper) prior.

A useful device when calculating the posterior distribution is to ignore terms that are constants with respect to the unknown parameters. Suppose $\mathbf{x}$ denotes the data and $\boldsymbol{\Theta}_1$ is a vector of *known* model parameters, while $\boldsymbol{\Theta}_2$ is a vector of unknown parameters. If we can write the posterior as

$$p(\boldsymbol{\Theta}_2 \,|\, \mathbf{x}, \boldsymbol{\Theta}_1) = f(\mathbf{x}, \boldsymbol{\Theta}_1) \cdot g(\mathbf{x}, \boldsymbol{\Theta}_1, \boldsymbol{\Theta}_2) \tag{A7.19a}$$

then

$$p(\boldsymbol{\Theta}_2 \,|\, \mathbf{x}, \boldsymbol{\Theta}_1) \propto g(\mathbf{x}, \boldsymbol{\Theta}_1, \boldsymbol{\Theta}_2) \tag{A7.19b}$$

which follows because $f(\mathbf{x}, \boldsymbol{\Theta}_1)$ is constant with respect to $\boldsymbol{\Theta}_2$.

With the prior given by Equation A7.18, we can express the resulting posterior distribution as

$$\begin{aligned} p(\mu \,|\, \mathbf{x}) &\propto \ell(\mu \,|\, \mathbf{x}) \cdot p(\mu) \\ &\propto \exp\left[-\frac{(\mu - \mu_0)^2}{2\sigma_0^2} - \frac{1}{2\sigma^2} \left(\sum_{i=1}^{n} x_i^2 - 2\mu n \overline{x} + n\mu^2 \right) \right] \end{aligned} \tag{A7.20a}$$

We can factor out additional terms not involving μ to obtain

$$p(\mu \,|\, \mathbf{x}) \propto \exp\left(-\frac{\mu^2}{2\sigma_0^2} + \frac{\mu\,\mu_0}{\sigma_0^2} + \frac{\mu n \overline{x}}{\sigma^2} - \frac{n\mu^2}{2\sigma^2} \right) \tag{A7.20b}$$

Factoring in terms of μ, the term in the exponential becomes

$$-\frac{\mu^2}{2}\left(\frac{1}{\sigma_0^2} + \frac{n}{\sigma^2} \right) + \mu \left(\frac{\mu_0}{\sigma_0^2} + \frac{n\overline{x}}{\sigma^2} \right) = -\frac{\mu^2}{\sigma_*^2} + \frac{2\mu\mu_*}{2\sigma_*^2} \tag{A7.21a}$$

where

$$\sigma_*^2 = \left(\frac{1}{\sigma_0^2} + \frac{n}{\sigma^2} \right)^{-1} \quad \text{and} \quad \mu_* = \sigma_*^2 \left(\frac{\mu_0}{\sigma_0^2} + \frac{n\overline{x}}{\sigma^2} \right) \tag{A7.21b}$$

Finally, by completing the square, we have

$$p(\mu \,|\, \mathbf{x}) \propto \exp\left[-\frac{(\mu - \mu_*)^2}{2\sigma_*^2} + f(\mathbf{x}, \mu_0, \sigma^2, \sigma_0^2) \right] \tag{A7.21c}$$

Recalling Equation A7.19b, we can ignore the second term in the exponential (as it does not involve μ), and the resulting posterior for μ (given the observed data $\mathbf{x}$) becomes

$$p(\mu \,|\, \mathbf{x}) \propto \exp\left[-\frac{(\mu - \mu_*)^2}{2\sigma_*^2} \right] \tag{A7.22a}$$

demonstrating that the posterior density function for μ is a normal with a mean of μ_* and a variance of σ_*^2, namely,

$$\mu \,|\, (\mathbf{x}, \sigma^2) \sim N\left(\mu_*, \sigma_*^2\right) \tag{A7.22b}$$

Notice that the posterior density is in the same form as the prior. This occurred because the prior **conjugated** with the likelihood function—the product of the prior and likelihood returned a distribution in the same family as the prior (but with different distribution parameters). The use of such **conjugate priors** associated with a given family of likelihood functions is a key concept in Bayesian analysis, and we will explore it more fully below.

We are now in a position to inquire about the relative importance of the prior versus the data. Under the assumed prior, the mean (and in this case, the mode as well) of the posterior distribution is

$$\mu_* = \mu_0 \frac{\sigma_*^2}{\sigma_0^2} + \overline{x} \frac{\sigma_*^2}{\sigma^2/n} \tag{A7.23}$$

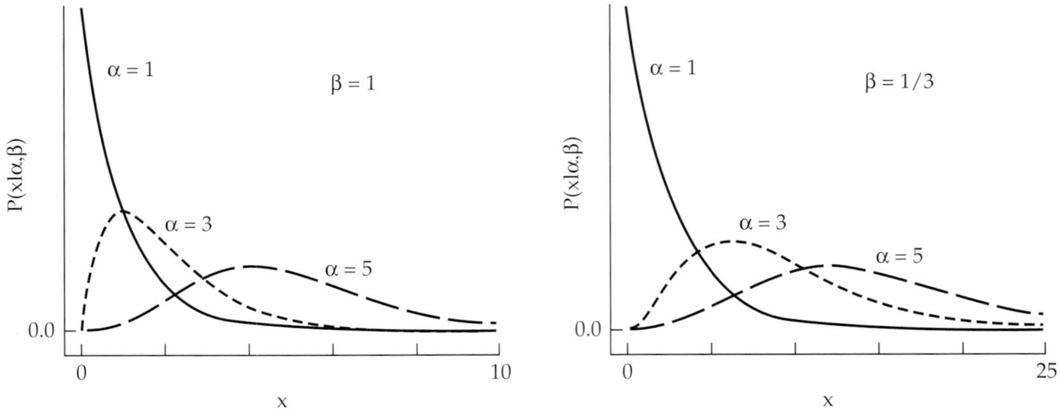

Figure A7.2 The effect of the **shape** (α) and **rate** ($\beta = 1/\lambda$, the inverse of the **scale**) **parameters** on the gamma distribution function. For $\alpha = 1$, the resulting distribution is the simple monotonically decreasing exponential, while for $\alpha > 1$, the distribution is unimodal. The effect of a change in the rate or scale is to keep the general shape but change the scaling with respect to x.

With a very diffuse prior on μ (i.e., $\sigma_0^2 \gg \sigma^2$), $\sigma_*^2 \to \sigma^2/n$ and $\mu_* \to \overline{x}$. Also note from Equation A7.21b that as we collect enough data (i.e., achieve a sufficiently large value of n), $\sigma_*^2 \to \sigma^2/n$ and again $\mu_* \to \overline{x}$, implying that the data, rather than the prior, will be the primarily influence on posterior when the value of n is sufficiently large.

Gamma, χ^2, Inverse-gamma, and χ^{-2} Distributions

Before examining the Gaussian likelihood with unknown variance, a brief aside is needed to develop the **inverse chi-square distribution**, denoted by χ^{-2}. We do this via the gamma and inverse-gamma distributions, as both χ^2 and χ^{-2} are special cases of these distributions.

To motivate the **gamma distribution**, first consider the simple exponential waiting-time distribution, where β is the **rate parameter** (the probability of a success in some small time unit, δ_t, is given by $\beta\,\delta_t$), then the pdf for the exponential is

$$p(x \mid \beta) = \beta e^{-\beta x} \quad \text{for} \quad 0 \leq x < \infty, \quad \beta > 0$$

Because the expected waiting time until a success is $\lambda = 1/\beta$, this can be reparameterized in terms of the **scale parameter** (waiting time) as

$$p(x \mid \beta) = \lambda^{-1} e^{-x/\lambda}$$

The sum of k exponentials with the same rate (or scale) parameter is called an **Erlang distribution**, and it was initially developed for certain problems in telephone queuing theory. Expressed in terms of the rate parameter, the resulting pdf becomes

$$p(x \mid k, \beta) = \frac{\beta^k}{(k-1)!}\, x^{k-1}\, e^{-\beta x} \quad \text{for} \quad 0 \leq x < \infty$$

where the integer k is called the **shape parameter**, with $k = 1$ recovering the exponential.

The gamma distribution follows by allowing the shape parameter to be any positive number, α, with $x \sim \text{Gamma}(\alpha, \beta)$ having its pdf defined by its shape (α) and rate (β) values,

$$p(x \mid \alpha, \beta) = \frac{\beta^\alpha}{\Gamma(\alpha)}\, x^{\alpha-1} e^{-\beta x} \quad \text{for } \alpha,\ \beta,\ x > 0 \tag{A7.24a}$$

Note that the factorial in the Erlang is replaced by the gamma function, $\Gamma(x)$, which is defined by Equation A7.25a. Figure A7.2 shows how changes in these two parameters

Table A7.1 Summary of the functional forms (in terms of x) of various gamma-related distributions. See the text for further details.

Distribution	α	β	$p(x)$/constant
Gamma (α, β)			$x^{\alpha-1}e^{-\beta x}$
Chi-square, χ^2_n	$n/2$	$1/2$	$x^{n/2-1}e^{-x/2}$
Inverse-gamma (α, β)			$x^{-(\alpha+1)}e^{-\beta/x}$
Inverse chi-square, χ_n^{-2}	$n/2$	$1/2$	$x^{-(n/2+1)}e^{-1/(2x)}$
Scaled inverse chi-square, $\chi^{-2}_{(n,\sigma_0^2)}$	$n/2$	$\sigma_0^2/2$	$x^{-(n/2+1)}e^{-\sigma_0^2/(2x)}$

influence the shape of the distribution. Note that, as a function of x

$$p(x \mid \alpha, \beta) \propto x^{\alpha-1}e^{-\beta x} \tag{A7.24b}$$

When expressed in terms of the scale ($\lambda = 1/\beta$) parameter, the pdf becomes

$$p(x \mid \alpha, \lambda) = \frac{\lambda^{-\alpha}}{\Gamma(\alpha)}\, x^{\alpha-1}e^{-x/\lambda}$$

which yields

$$p(x \mid \alpha, \lambda) \propto x^{\alpha-1}e^{-x/\lambda} \tag{A7.24c}$$

Because both the rate and scale versions of the gamma distribution are widely used, take care to know which version your software package is using (e.g., the default in R uses the scale parameter version). We can parameterize a gamma in terms of its mean and variance by noting that

$$\mu_x = \frac{\alpha}{\beta} = \alpha\,\lambda \quad \text{and} \quad \sigma_x^2 = \frac{\alpha}{\beta^2} = \alpha\,\lambda^2 \tag{A7.24d}$$

so that

$$\alpha = \frac{\mu_x^2}{\sigma_x^2} \quad \text{and} \quad \beta = \frac{\mu_x}{\sigma_x^2}, \quad \lambda = \frac{1}{\beta} = \frac{\sigma_x^2}{\mu_x} \tag{A7.24e}$$

$\Gamma(\alpha)$, the **gamma function** evaluated at α (which normalizes the gamma distribution), is defined by

$$\Gamma(\alpha) = \int_0^\infty y^{\alpha-1}e^{-y}dy \qquad \text{for } \alpha > 0 \tag{A7.25a}$$

This is the generalization of the factorial function from the integers to any positive number. If n is an integer, then $(n-1)! = \Gamma(n)$. Using integration by parts, one can show that Γ satisfies the following identities

$$\Gamma(\alpha+1) = \alpha\Gamma(\alpha), \quad \Gamma(1) = 1, \quad \text{and} \quad \Gamma(1/2) = \sqrt{\pi} \tag{A7.25b}$$

The chi-square (χ^2) distribution is a special case of the gamma, as a χ^2 random variable with n degrees of freedom follows a gamma distribution with parameters $\alpha = n/2$ and $\beta = 1/2$ ($\lambda = 2$), namely, $\chi^2_n \sim \text{Gamma}(n/2, 1/2)$, giving the density function as

$$p(x \mid n) = \frac{2^{-n/2}}{\Gamma(n/2)}\, x^{n/2-1}e^{-x/2} \tag{A7.26a}$$

Hence for $x \sim \chi^2_n$,

$$p(x) \propto x^{n/2-1}e^{-x/2} \tag{A7.26b}$$

The **inverse-gamma** distribution will prove useful as a conjugate prior for Gaussian likelihoods with unknown variance. It is defined by the distribution of the random variable $y = x^{-1}$, where $x \sim \text{Gamma}(\alpha, \beta)$. The resulting density function is

$$p(x \mid \alpha, \beta) = \frac{\beta^\alpha}{\Gamma(\alpha)} x^{-(\alpha+1)} e^{-\beta/x} \quad \text{for } \alpha,\ \beta,\ x > 0 \tag{A7.27a}$$

The mean and variance for this distribution are only defined (i.e., finite) if α is sufficiently large, with

$$\mu_x = \frac{\beta}{\alpha - 1} \quad \text{for } \alpha > 1 \quad \text{and} \quad \sigma_x^2 = \frac{\beta^2}{(\alpha-1)^2(\alpha-2)} \quad \text{for } \alpha > 2 \tag{A7.27b}$$

Note for the inverse gamma that

$$p(x \mid \alpha, \beta) \propto x^{-(\alpha+1)} e^{-\beta/x} \tag{A7.27c}$$

If $y \sim \chi_n^2$, then $x = 1/y$ follows an **inverse chi-square distribution**, which is denoted by $x \sim \chi_n^{-2}$. This is a special case of the inverse gamma, with (as for a normal χ^2) $\alpha = n/2$, $\beta = 1/2$. For $n > 4$ (i.e., $\alpha > 2$), the resulting density function is

$$p(x \mid n) = \frac{2^{-n/2}}{\Gamma(n/2)} x^{-(n/2+1)} e^{-1/(2x)} \tag{A7.28a}$$

with a mean and variance of

$$\mu_x = \frac{1}{n-2} \quad \text{and} \quad \sigma_x^2 = \frac{2}{(n-2)^2(n-4)} \tag{A7.28b}$$

The **scaled inverse chi-square distribution** is more typically used in a Bayesian analysis, where the rate parameter, β (which equals 1/2 under a chi-square), is replaced by $\beta = \sigma_0^2/2$, making the resulting pdf

$$p(x \mid n) \propto x^{-(n/2+1)} e^{-\sigma_0^2/(2x)} \tag{A7.29a}$$

where the $1/(2x)$ term in the exponential is replaced by a $\sigma_0^2/(2x)$ term. The scaled inverse chi-square distribution thus involves two parameters (σ_0^2 and n), and is denoted by $\chi^{-2}_{(n,\sigma_0^2)}$ or SI$-\chi^2(n, \sigma_0^2)$. Note that if

$$x \sim \chi^{-2}_{(n,\sigma_0^2)}, \quad \text{then} \quad \sigma_0^2 \cdot x \sim \chi_n^{-2} \tag{A7.29b}$$

which shows that σ_0^2 is a scaling factor on a standard ($\beta = 1/2$) inverse chi-square.

Gaussian Likelihood With Unknown Variance: Scaled Inverse-χ^2 Priors

Suppose data are drawn from a normal distribution with a known mean, μ, but unknown variance, σ^2. The resulting likelihood function can be expressed as

$$\ell(\sigma^2 \mid \mathbf{x}, \mu) \propto (\sigma^2)^{-n/2} \exp\left(-\frac{nS^2}{2\sigma^2}\right) \tag{A7.30a}$$

where

$$S^2 = \frac{1}{n} \sum_{i=1}^{n} (x_i - \mu)^2 \tag{A7.30b}$$

Notice that because we condition on $\mathbf{x}$ and μ (i.e., their values are known), S^2 is a constant. Further observe that, as a function of the unknown variance, σ^2, the likelihood is proportional to a scaled inverse χ^2 distribution (Equation A7.29a). If we take the prior for the unknown variance also as a scaled inverse χ^2 with hyperparameters ν_0 and σ_0^2, the posterior becomes

$$p(\sigma^2 \mid \mathbf{x}, \mu) \propto (\sigma^2)^{-n/2} \exp\left(-\frac{nS^2}{2\sigma^2}\right) (\sigma^2)^{-\nu_0/2-1} \cdot \exp\left(-\frac{\sigma_0^2}{2\sigma^2}\right)$$
$$= (\sigma^2)^{-(n+\nu_0)/2-1} \exp\left(-\frac{nS^2+\sigma_0^2}{2\sigma^2}\right) \tag{A7.31a}$$

Equation A7.29a shows that the resulting posterior is also a scaled inverse χ^2 distribution with parameters $\nu_n = (n+\nu_0)$ and $\sigma_n^2 = (nS^2+\sigma_0^2)$. Hence,

$$\text{the prior } \sigma^2 \sim \chi^{-2}_{\nu_0,\sigma_0^2} \quad \text{yields the posterior} \quad \sigma^2 \mid (\mathbf{x}, \mu) \sim \chi^{-2}_{\nu_n,\sigma_n^2} \tag{A7.31b}$$

Student's *t* Distribution

The final distribution needed for a Bayesian analysis of a Gaussian likelihood is the ***t*** (or **Student's *t***) distribution. Suppose that $x_i \sim N(\mu, \sigma^2)$, so for n independent draws, $\overline{x} \sim N(\mu, \sigma^2/n)$. This implies that $(\overline{x}-\mu)/\sqrt{\sigma^2/n} \sim U$, where $U \sim N(0,1)$ denotes a unit normal. Likewise, the sample variance, $\text{Var}(x)$, follows a scaled chi-square distribution, with $\text{Var}(x) \sim (n-1)\sigma^2\chi^2_{n-1}$ (Equation A5.15a). When the estimated variance, $\text{Var}(x)$, is used in place of the true variance, σ^2, the quantity $(\overline{x}-\mu)/\sqrt{\text{Var}(x)/n}$ follows a t distribution with $n-1$ degrees of freedom, giving rise to the very familiar ***t*-test**. Notice that

$$t_{n-1} = \left(\frac{\overline{x}-\mu}{\sigma/\sqrt{n}}\right)\left(\frac{1}{\sqrt{\text{Var(x)}/\sigma^2}}\right) = \frac{U}{\sqrt{\chi^2_{n-1}/(n-1)}}$$

Thus, a t_ν random variable follows the distribution of a unit normal divided by the square root of a chi-square with ν degrees of freedom,

$$t_\nu = \frac{U}{\sqrt{\chi^2_\nu/\nu}} \tag{A7.32a}$$

Note that $E(\chi^2_\nu) = \nu$, so $E(\chi^2_\nu/\nu) = 1$. Relative to a normal, a t distribution is more peaked and has heavier tails, and this kurtosis becomes more pronounced as ν decreases. Indeed, the tails fall off sufficiently slowly that a t random variable with two degrees of freedom has an infinite variance, while a t with four (or fewer) degrees of freedom has an infinite fourth moment. The coefficient of kurtosis (Equation 2.12a) for a t with $\nu > 4$ degrees of freedom is $k_4 = 6/(\nu-4)$, which approaches the value (zero) for a normal random variable for large values of ν. For $\nu > 30$, the t essentially becomes a unit normal distribution.

As with a unit normal, one can also add scale and location to a standard t_ν-distributed random variable, thus generating a three-parameter family of distributions,

$$t_\nu(\mu,\sigma) = \mu + \sigma \cdot t_\nu \tag{A7.32b}$$

The resulting mean and variance of this distribution are

$$E[t_\nu(\mu,\sigma)] = \mu \quad \text{and} \quad \sigma^2[t_\nu(\mu,\sigma)] = \sigma^2\,\frac{\nu}{\nu-2} \quad \text{for } \nu > 2 \tag{A7.32c}$$

Hence, the choice of μ and σ control, respectively, the location and scale (uncertainty about the location), while ν controls the kurtosis, with heavy tails for values of ν that are small and little kurtosis for $\nu > 20$. The resulting pdf thus becomes

$$p(x \mid \nu, \mu, \sigma) = \frac{\Gamma([\nu+1]/2)}{\Gamma(\nu/2)\sigma\sqrt{\pi\,\nu}}\left[1+\frac{1}{\nu}\left(\frac{x-\mu}{\sigma}\right)^2\right]^{-(\nu+1)/2} \tag{A7.32d}$$

The role of the t distribution in Bayesian statistics is twofold. First, it is often used as a ***more robust prior***, as its heavier tails may better account for outliers. Using a t distribution with low degrees of freedom (often $\nu = 5$) offers a prior that is similar to a normal but allows for more frequent extreme values. The second scenario is that the marginal posterior for μ of a Gaussian likelihood with a normal prior on the mean and an inverse chi-square prior on the variance is a t distribution. This arises after the joint posterior is integrated over all possible σ^2 values (i.e., over an inverse chi-square).

General Gaussian Likelihood: Unknown Mean and Variance

If we put all these pieces together, the posterior density for draws from a normal with the mean and variance both unknown is obtained as follows. First, we write the joint prior by conditioning on the variance,

$$p(\mu, \sigma^2) = p(\mu \,|\, \sigma^2) \cdot p(\sigma^2) \tag{A7.33a}$$

As above, we assume a scaled inverse chi-square distribution for the variance and, conditioned on the variance, a Gaussian prior for the mean with hyperparameters of μ_0 and σ^2/κ_0, namely,

$$\sigma^2 \sim \chi^{-2}_{\nu_0,\sigma_0^2} \quad \text{and} \quad \mu \,|\, \sigma^2 \sim \text{N}\left(\mu_0, \frac{\sigma^2}{\kappa_0}\right) \tag{A7.33b}$$

We write the variance for the conditional mean prior in this way because σ^2 is known (as we condition on it) and we scale σ^2 by the hyperparameter, κ_0. The resulting marginal posterior becomes

$$\sigma^2 \,|\, \mathbf{x} \sim \chi^{-2}_{\nu_n,\sigma_n^2} \quad \text{and} \quad \mu \,|\, \mathbf{x} \sim \text{t}_{\nu_n}\left(\mu_n, \frac{\sigma_n^2}{\kappa_n}\right) \tag{A7.34}$$

where $\text{t}_n(\mu, \sigma^2)$ denotes a t distribution with n degrees of freedom, mean μ, and scale parameter σ^2, and where

$$\nu_n = \nu_0 + n, \qquad \kappa_n = \kappa_0 + n \tag{A7.35a}$$

$$\mu_n = \mu_0 \frac{\kappa_0}{\kappa_n} + \overline{x}\,\frac{n}{\kappa_n} = \mu_0 \frac{\kappa_0}{\kappa_0 + n} + \overline{x}\,\frac{n}{\kappa_0 + n} \tag{A7.35b}$$

$$\sigma_n^2 = \frac{1}{\nu_n}\left(\nu_0\sigma_0^2 + \sum_{i=1}^{n} (x_i - \overline{x})^2 + \frac{\kappa_0 n}{\kappa_n}(\overline{x} - \mu_0)^2\right) \tag{A7.35c}$$

CONJUGATE PRIORS

The above use of prior densities that conjugate the likelihood allowed us to develop analytic expressions of the posterior density. As we will see in Appendix 8, this is critical in developing Gibbs samplers for problems of interest. Table A7.2 summarizes the conjugate priors for several common likelihood functions, with the various families of distributions discussed below.

The Beta and Dirichlet Distributions

With a binomial, each trial (observation) has two possible outcomes and the likelihood is a function of the sample size (number of trials), n, and a single success probability, p (as the two outcomes on any given trial have probabilities of p and $1 - p$). The generalization of this model is the multinomial distribution (Equation 2.20a), where now each trial has k possible outcomes, which requires $k - 1$ success probabilities to describe the likelihood. In particular, for a total of n observations, the probability that n_1 are in category 1, n_2 in category 2, $\cdots$, and n_k in category k is

$$p(n_1, \cdots n_k) = \frac{n!}{n_1!\, n_2! \cdots n_k!}\, p_1^{n_1} \cdots p_k^{n_k} \quad \text{where} \quad \sum_i n_i = n \quad \text{and} \quad \sum_i p_i = 1 \tag{A7.36a}$$

Table A7.2 Conjugate priors for common likelihood functions. If one uses the distribution family of the conjugate prior with its paired likelihood function, then the resulting posterior is in the same distribution family as the prior (albeit, of course, with different parameters).

Likelihood	Conjugate Prior	Equation
Binomial	Beta	A7.37a
Multinomial	Dirichlet	A7.36b
Poisson	Gamma	A7.26a
Normal		
μ unknown, σ^2 known	Normal	A7.17a
μ known, σ^2 unknown	Inverse chi-square	A7.29a
Multivariate normal		
$\boldsymbol{\mu}$ unknown, $\mathbf{V}$ known	Multivariate normal	9.24
$\boldsymbol{\mu}$ known, $\mathbf{V}$ unknown	Inverse-Wishart	A7.40

The conjugate prior for the multinomial likelihood is the **Dirichlet distribution**. If we let $\mathbf{x} = (x_1, x_2, \cdots, x_k)$ denote the k success probabilities, when pdf for $\mathbf{x} \sim$ Dirichlet$(\alpha_1, \cdots, \alpha_k)$ is

$$p(x_1, \cdots x_k \,|\, \alpha_1, \cdots, \alpha_k) = \frac{\Gamma(\alpha_0)}{\Gamma(\alpha_1)\cdots\Gamma(\alpha_k)} x_1^{\alpha_1 - 1} \cdots x_k^{\alpha_k - 1} \tag{A7.36b}$$

where

$$\alpha_0 = \sum_{i=1}^{k} \alpha_i \quad \text{with} \quad \alpha_i > 0, \quad \text{and} \quad \sum_{i=1}^{k} x_i = 1 \quad \text{with} \quad 0 \leq x_i \leq 1 \tag{A7.36c}$$

At first glance, this looks like the multinomial density function (with $\alpha_i - 1 = n_i$). The difference is that the multinomial is calculated over a set of discrete random variables (n_i), thus returning the expected probabilities for any vector of discrete numbers of counts (successes) in each category. Conversely, the Dirichlet treats an equivalent of the vector of outcomes (generalized to non-integers) as fixed and returns the continuous distribution for all possible configurations of the *success parameters* given this data, which means that the data (α_i) are fixed, and the success parameters (x_i) are random. A few key moments of this distribution are

$$\mu_{x_i} = \frac{\alpha_i}{\alpha_0}, \quad \sigma^2(x_i) = \frac{\alpha_i(\alpha_0 - \alpha_i)}{\alpha_0^2(\alpha_0 + 1)}, \quad \text{and} \quad \sigma(x_i, x_j) = -\frac{\alpha_i\,\alpha_j}{\alpha_0^2(\alpha_0 + 1)} \tag{A7.36d}$$

An important special case of the Dirichlet (for $k = 2$ classes) is the **beta distribution**, whose pdf is given by

$$p(x) = \frac{\Gamma(\alpha + \beta)}{\Gamma(\alpha)\Gamma(\beta)} x^{\alpha-1}(1 - x)^{\beta-1} \quad \text{for} \quad 0 \leq x \leq 1, \quad \alpha,\, \beta > 0 \tag{A7.37a}$$

which has a mean and a variance of

$$\mu = \frac{\alpha}{\alpha + \beta} \qquad \text{and} \qquad \sigma^2 = \frac{\alpha\beta}{(\alpha + \beta)^2(\alpha + \beta - 1)} \tag{A7.37b}$$

As Figure A7.3 illustrates, the beta distribution is *extremely flexible*, and can be flat, unimodal, U-, or L-shaped, depending on the choice of α and β.

Wishart and Inverse-Wishart Distributions

The **Wishart distribution** can be thought of as the multivariate extension of the χ^2 distribution. Suppose $\mathbf{x}_1, \cdots, \mathbf{x}_n$ are independent and identically distributed vectors, with

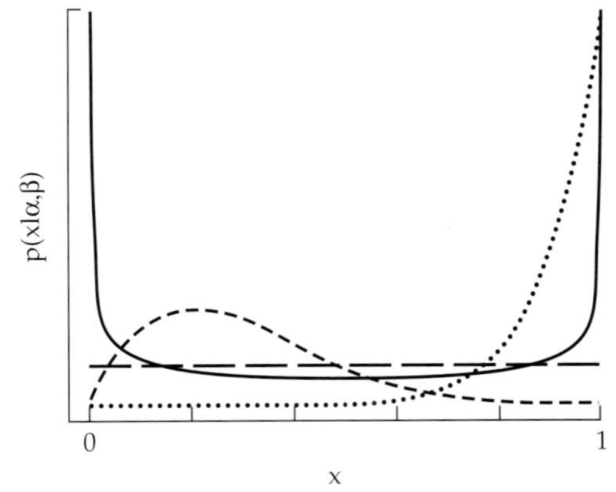

Figure A7.3 For $\alpha = \beta = 1$ (long-dashed curve), the beta distribution is simply the uniform distribution over $(0, 1)$. The pdf for the beta distribution can also be U-shaped ($\alpha = \beta = 0.5$; solid curve), unimodal ($\alpha = 2, \beta = 5$; short-dashed curve), or L-shaped ($\alpha = 10, \beta = 1$; dotted curve). Because the beta distribution is symmetric in α and β, switching their parameter values generates a distribution of the same shape translated about 0.5.

$\mathbf{x}_i \sim \text{MVN}_k(\mathbf{0}, \mathbf{V})$. Using these n draws, and assuming that the mean is known to be zero, the resulting random ($k \times k$ symmetric, positive definite) sample covariance matrix, **W**, is given by

$$\mathbf{W} = \sum_{i=1}^{n} \mathbf{x}_i \, \mathbf{x}_i^T \sim \text{W}_n(\mathbf{V}) \tag{A7.38}$$

This sum is defined as a Wishart distribution with n degrees of freedom and a (matrix) parameter **V**. Recalling that the sum of n squared unit normals follows a χ^2_n distribution, the Wishart is the extension to the multivariate normal. Indeed, for $k = 1$ with $\mathbf{V} = (1)$, the Wishart is simply a χ^2_n distribution, as $\sum x_i^2 \sim \chi^2_n$, because $x_i \sim N(0, 1)$.

The Wishart is the sampling distribution for covariance matrices (just like the χ^2 is associated with the distribution of a sample variance for data drawn from a normal; Equation A5.13c). The pdf of the Wishart distribution is

$$p(\mathbf{W} \,|\, \mathbf{V}) = 2^{-nk/2} \pi^{-k(k-1)/k} \,|\, \mathbf{V} \,|^{-n/2} \,|\, \mathbf{W} \,|^{(n+k+1)/2} \, \frac{\exp\left(-\frac{1}{2}\text{tr}\left[\mathbf{V}^{-1}\mathbf{W}\right]\right)}{\prod_{i=1}^{k} \Gamma\left(\frac{n+1-i}{2}\right)} \tag{A7.39}$$

Recall that the trace (tr) of a matrix is just the sum of its diagonal elements, $\text{tr}(\mathbf{A}) = \sum A_{ii}$ (Chapter 9). See Odell and Feiveson (1966) for an algorithm for generating random draws from the Wishart.

If $\mathbf{Z} \sim \mathbf{W}_n(\mathbf{V})$, then $\mathbf{Z}^{-1} \sim \mathbf{W}_n^{-1}\left(\mathbf{V}^{-1}\right)$, where $\mathbf{W}^{-1}$ denotes the **inverse-Wishart distribution**. The density function for an inverse-Wishart distributed random matrix, **W**, is

$$p(\mathbf{W} \,|\, \mathbf{V}) = 2^{-nk/2} \pi^{-k(k-1)/k} \,|\, \mathbf{V} \,|^{n/2} \,|\, \mathbf{W} \,|^{-(n+k+1)/2} \, \frac{\exp\left(-\frac{1}{2}\text{tr}\left[\mathbf{V}\mathbf{W}^{-1}\right]\right)}{\prod_{i=1}^{k} \Gamma\left(\frac{n+1-i}{2}\right)} \tag{A7.40}$$

which is the distribution of the inverse of the sample covariance matrix.

Appendix 8

Markov Chain Monte Carlo and Gibbs Sampling

God does not play dice with the universe. Albert Einstein

A historical impediment to the more widespread use of Bayesian methods was computational: obtaining the posterior distribution requires the integration of high-dimensional functions. This can be very difficult, but several analytic approximations short of direct integration have been proposed (reviewed by Smith 1991; Evans and Swartz 1995; Tanner 1996). Now, however, the widespread development of **Markov Chain Monte Carlo** (**MCMC**) methods has made obtaining very complex posteriors relatively easy from a conceptual standpoint (although they may still be rather computationally demanding). Indeed, MCMC allows Bayesian approaches to handle much higher dimensional problems than other methods. MCMC approaches are so named because one uses the current sample value to randomly generate the next sample value, thus generating a **Markov chain**.

The realization in the early 1990s (Gelfand and Smith 1990) that one particular MCMC method, the **Gibbs sampler**, is widely applicable to a broad class of Bayesian problems sparked much of the current expansion in the use of Bayesian analysis. MCMC methods have their roots in the **Metropolis algorithm** (Metropolis and Ulam 1949; Metropolis et al. 1953), an attempt by physicists to compute complex integrals by expressing them as expectations of a much simpler probability distribution and then estimating this expectation by drawing samples from that distribution. While the Gibbs sampler had its origins outside of statistics, it is still somewhat surprising that the powerful machinery of MCMC had essentially no impact on the field of statistics until rather recently. General reviews of MCMC approaches can be found in Tanner (1996), Gammerman (1997), Chen et al. (2000), Robert and Casella (2004), and Brooks et al. (2011). Sorensen and Gianola (2002) review MCMC applications in quantitative genetics, which are heavily focused on Gibbs samplers for linear mixed-models (Chapter 10; Part II Chapters 32 and 33). The roundtable discussion on practical issues by Kass et al. (1998) is required reading for anyone contemplating using an MCMC analysis, as is Geyer (2011). Finally, Robert and Casella (2011) present a nice historical overview of the development of MCMC methods.

MONTE CARLO INTEGRATION

To better understand the historical roots of MCMC, we digress for a moment to discuss the original **Monte Carlo** approach, which used random-number generation to approximate integrals. Suppose we wish to compute a complex integral,

$$\int_a^b h(x)\,dx \tag{A8.1a}$$

If we can decompose $h(x)$ into the product of a function, $f(x)$, and a probability density function, $p(x)$, defined over the interval (a,b), then

$$\int_a^b h(x)\,dx = \int_a^b f(x)\,p(x)\,dx = E_{p(x)}[\,f(x)\,] \tag{A8.1b}$$

which means that the integral can be expressed as the expectation of $f(x)$ over the density $p(x)$. After sampling a large number of random draws, $(x_1, \cdots, x_n)$, from the density $p(x)$, we have that

$$\int_a^b h(x)\,dx = E_{p(x)}[\,f(x)\,] \simeq \frac{1}{n}\sum_{i=1}^{n} f(x_i) \tag{A8.1c}$$

This approach is referred to as **Monte Carlo integration**, and it can be used to approximate posterior (or marginal posterior) distributions required for a Bayesian analysis.

Consider the integral $I(y) = \int f(y\,|\,x)\,p(x)\,dx$, which we approximate by

$$\widehat{I}(y) = \frac{1}{n}\sum_{i=1}^{n} f(y\,|\,x_i) \tag{A8.1d}$$

where again x_i are draws from the density $p(x)$. Because we have framed this integral as an expectation, individual random draws have expected value equal to the integral. The spread of those values around their mean allows us to estimate a **Monte Carlo standard error** for our approximation, with

$$\text{SE}^2[\,\widehat{I}(y)\,] = \frac{1}{n}\left[\frac{1}{n-1}\sum_{i=1}^{n}\left(f(y\,|\,x_i) - \widehat{I}(y)\right)^2\right] \tag{A8.1e}$$

where the term in brackets estimates the sampling variance, σ^2. Notice from Equation A8.1e that the standard error scales as $\sigma/\sqrt{n}$, namely, with the square root of the sample size. Thus, a ten fold improvement in precision requires a hundred fold increase in the sample size.

Importance Sampling

Consider any distribution whose probability density function, $p(x)$, has the same support as some target density, $q(x)$ (i.e., is nonzero over the same range of x values). Then,

$$\int f(x)\,q(x)dx = \int f(x)\left(\frac{q(x)}{p(x)}\right)p(x)dx = E_{p(x)}\left[\,f(x)\left(\frac{q(x)}{p(x)}\right)\right] \tag{A8.2a}$$

This forms the basis for the method of **importance sampling**, with

$$\int f(x)\,q(x)dx \simeq \frac{1}{n}\sum_{i=1}^{n} f(x_i)\left(\frac{q(x_i)}{p(x_i)}\right) \tag{A8.2b}$$

where again the x_i values are drawn from the distribution given by $p(x)$. For example, if we are interested in a marginal density as a function of y, $J(y) = \int f(y\,|\,x)\,q(x)dx$, we approximate this by

$$J(y) \simeq \frac{1}{n}\sum_{i=1}^{n} f(y\,|\,x_i)\left(\frac{q(x_i)}{p(x_i)}\right) \tag{A8.3}$$

where x_i are drawn from the approximating density p.

This method is so named in that a judicious choice of $p(x)$ ensures that values for which $f(x)$ is large (important) are appropriately sampled, which can result in a smaller Monte Carlo variance, and hence a more accurate estimate of the integral. Further, it can be faster to carry out in cases where p is more straightforward to sample from than q. Ideally, we would like $p(x)$ to be large when $f(x)q(x)$ is large, namely, $p(x) \propto f(x)q(x)$, although assessing when $q(x)$ is large may be problematic.

Example A8.1. One issue that can arise is when $q(x)$ denotes a posterior distribution whose integration constant is unknown, and hence $q(x)$, as presented, will not be a formal probability distribution if $\int q(x) \neq 1$. If $C^{-1} = \int q(x)$, then $Cq(x)$ is formally a pdf. Fortunately, a simple trick allows us to use importance sampling to obtain C. Because $C \int q(x)dx = 1$, then

$$C^{-1} = \int q(x)dx \simeq \frac{1}{n}\sum_{i=1}^{n} w_i \quad \text{where} \quad w_i = \frac{q(x_i)}{p(x_i)} \tag{A8.4a}$$

with the x_i drawn from $p(x)$. Hence

$$C \cdot \int f(x)\, q(x)dx \simeq \widehat{I} = \sum_{i=1}^{n} w_i f(x_i) \Big/ \sum_{i=1}^{n} w_i \tag{A8.4b}$$

which has an associated Monte Carlo variance of

$$\text{Var}\left(\widehat{I}\right) = \sum_{i=1}^{n} w_i \left[f(x_i) - \widehat{I}\right]^2 \Big/ \sum_{i=1}^{n} w_i \tag{A8.4c}$$

INTRODUCTION TO MARKOV CHAINS

Before introducing two of the most common MCMC methods, the Metropolis–Hastings algorithm and the Gibbs sampler, a few introductory comments on Markov chains are in order. Let X_t denote the value of a random variable at time t, and let the **state space** refer to the range of possible X values. A random variable is said to follow a **Markov process** if the transition probabilities between different values in the state space ***depend only on the random variable's current state***,

$$\Pr(X_{t+1} = s_j \mid X_0 = s_k, \cdots, X_t = s_i) = \Pr(X_{t+1} = s_j \mid X_t = s_i) \tag{A8.5}$$

Thus, the only information about the past needed for a Markov random variable to predict the future is the *current state* of the random variable. Knowledge of the values of earlier states does not influence the transition probability. A **Markov chain** refers to a sequence of random variables, $(X_0, \cdots, X_n)$, generated by a Markov process. A particular chain is defined by its **transition probabilities** (or **transition kernel**), $P(i, j) = P(i \to j)$, which is the probability that a process at state space s_i moves to state s_j in a single step,

$$P(i, j) = P(i \to j) = \Pr(X_{t+1} = s_j \mid X_t = s_i) \tag{A8.6a}$$

We use the notation $P(i \to j)$ to imply a move from i to j, as some texts define $P(i, j) = P(j \to i)$, so we will use the arrow notation to avoid any potential confusion. Let

$$\pi_j(t) = \Pr(X_t = s_j) \tag{A8.6b}$$

denote the probability that the chain is in state j at time t, and let $\boldsymbol{\pi}(t)$ denote the *row* (as opposed to column) vector of the state space probabilities at step (or time) t. We start the chain by specifying a starting row vector, $\boldsymbol{\pi}(0)$. Often all the elements of $\boldsymbol{\pi}(0)$ are zero except for a single element of 1, which corresponds to the process starting in that particular state. As the chain progresses, the probability values become more dispersed over the state space. The probability that the chain has a state value of s_j at time (or step) $t + 1$ is obtained from the **Chapman–Kolomogrov (CK) equation**, which sums over the probability of being in a particular state at the current step value, t, and the transition probability from that state into state s_j

$$\begin{aligned}\pi_j(t+1) &= \Pr(X_{t+1} = s_j) \\ &= \sum_i \Pr(X_{t+1} = s_j \mid X_t = s_i) \cdot \Pr(X_t = s_i) \\ &= \sum_i P(i \to j)\, \pi_i(t) = \sum_i P(i, j)\, \pi_i(t)\end{aligned} \tag{A8.6c}$$

Successive iteration of the CK equations describes the evolution of the chain.

We can more compactly write the CK equations in matrix form as follows: first define the **probability transition matrix, P**, as the matrix whose ijth element is $P(i,j)$, the probability of moving from state i to state j, $P(i \rightarrow j)$. This implies that the rows of $\mathbf{P}$ sum to one, as when considering the ith row, $\sum_j P(i,j) = \sum_j P(i \rightarrow j) = 1$. In matrix form, the Chapman–Kolomogrov equation becomes

$$\boldsymbol{\pi}(t+1) = \boldsymbol{\pi}(t)\mathbf{P} \tag{A8.7a}$$

Hence,

$$\boldsymbol{\pi}(t) = \boldsymbol{\pi}(t-1)\mathbf{P} = \big[\boldsymbol{\pi}(t-2)\mathbf{P}\big]\mathbf{P} = \boldsymbol{\pi}(t-2)\mathbf{P}^2 \tag{A8.7b}$$

Continuing in this fashion yields the probability distribution in generation t as

$$\boldsymbol{\pi}(t) = \boldsymbol{\pi}(0)\mathbf{P}^t \tag{A8.7c}$$

Next define the n-step transition probability, $p_{ij}^{(n)}$, as the probability that the process is in state j, given that it started in state i some n steps ago, namely,

$$p_{ij}^{(n)} = \Pr(X_{t+n} = s_j \mid X_t = s_i) \tag{A8.7d}$$

It immediately follows that $p_{ij}^{(n)}$ is simply the ijth element of $\mathbf{P}^n$, the nth power of the single-step transition matrix. The classic example of a Markov chain in genetics is the Wright–Fisher model of genetic drift (WL Equations 2.1 and 2.2; WL Example 2.1).

Finally, a Markov chain is said to be **irreducible** if there exists a positive integer, n_{ij}, such that $p_{ij}^{(n_{ij})} > 0$ for all ij. That is, all states **communicate** with each other, in that the *process can always move from any one state to any other state* (although this may require multiple steps). Likewise, a chain is said to be **aperiodic** when the number of steps required to move between two states (say x and y) is not always some multiple of a specific integer. Put another way, the chain is not forced into cycles of fixed length between certain states.

Example A8.2. Suppose the state space consists of three possible weather conditions (Rain, Sunny, Cloudy) and that weather patterns follow a Markov process (of course, they do not!). Under this assumption, the probability of tomorrow's weather simply depends on today's weather, and not on any other previous days. If this is the case, the observation that it has rained for three straight days does not alter the probability of tomorrow's weather compared to the situation where (say) it rained today but was sunny for the last week. Suppose the probability transitions given that today is rainy are

P(Rain tomorrow | Rain today) = 0.5
P(Sunny tomorrow | Rain today) = 0.25
P(Cloudy tomorrow | Rain today) = 0.25

This results in the first row of the transition probability matrix (from the current state of rain to the next set of states) being $(0.5, 0.25, 0.25)$. Now suppose that the full transition matrix is

$$\mathbf{P} = \begin{pmatrix} 0.5 & 0.25 & 0.25 \\ 0.5 & 0 & 0.5 \\ 0.25 & 0.25 & 0.5 \end{pmatrix}$$

Note that this Markov chain is irreducible, as all states communicate with each other. Suppose today is sunny. What is the weather expected to be like two days from now? Seven days? Here $\boldsymbol{\pi}(0) = (\,0 \quad 1 \quad 0\,)$, which yields

$$\boldsymbol{\pi}(2) = \boldsymbol{\pi}(0)\mathbf{P}^2 = (\,0.375 \quad 0.25 \quad 0.375\,)$$

and

$$\pi(7) = \pi(0)\mathbf{P}^7 = (\,0.4 \quad 0.2 \quad 0.4\,)$$

Conversely, suppose today is rainy, so $\pi(0) = (\,1 \quad 0 \quad 0\,)$. The expected future weather becomes

$$\pi(2) = (\,0.4375 \quad 0.1875 \quad 0.375\,) \quad \text{and} \quad \pi(7) = (\,0.4 \quad 0.2 \quad 0.4\,)$$

Note that after a sufficient amount of time, the expected weather is *independent of the starting value*. In other words, the chain has reached a stationary distribution, where the state space probability values are independent of the actual starting value.

As the above example illustrates, a Markov chain may reach a **stationary distribution**, π^*, where the vector of probabilities of being in a particular state is *independent of the initial starting distribution*. The stationary distribution satisfies

$$\pi^* = \pi^*\mathbf{P} \tag{A8.8}$$

In other words, π^* is the **left eigenvector** associated with the eigenvalue $\lambda = 1$ of $\mathbf{P}$. (If the row vector $\mathbf{x}$ is a left eigenvector of a square matrix $\mathbf{B}$, then $\mathbf{xB} = \lambda\mathbf{x}$. Taking the transpose of both sides, $\mathbf{B}^T\mathbf{x}^T = \lambda\mathbf{x}^T$, namely $\mathbf{x}^T$ is a standard eigenvector of $\mathbf{B}^T$. Further, the spectral decomposition of $\mathbf{P}$ (Equation A3.32d) implies that the impact of the initial conditions after t steps decays as λ_2^t, where $\lambda_2 < 1$ is the second largest eigenvalue of $\mathbf{P}$. If this eigenvalue is very close to one, the impact of the initial conditions can persist for a substantial amount of time. Specifically, if $\lambda_2 = 1 - \delta$, then $\lambda_2^t = (1-\delta)^t \simeq \exp(-\delta t)$. The transmission matrix from Example A8.2 has $\lambda_2 = 1/4$, so that the impact of the initial conditions declines by $\simeq \exp(-3t/4)$, which is why the stationary distribution is reached so quickly.

One condition for the existence of a stationary distribution is that ***the chain must be irreducible and aperiodic***. When a chain is periodic, it can cycle in a deterministic fashion between states and may never settle down to a stationary distribution (in effect, this cycling *is* the stationary distribution for this chain). Equation A8.33d can be used to show that if $\mathbf{P}$ has no eigenvalues equal to -1, it is aperiodic.

A sufficient condition for a unique stationary distribution is that, for all i and j, the **detailed balance equation** holds, namely,

$$P(j \to i)\,\pi_j^* = P(i \to j)\,\pi_i^* \tag{A8.9}$$

That is, at equilibrium, the amount of probability flux from state j to stage i is exactly matched by the probability flux in the opposite direction (for i to j), so a balance is reached, with no *net* flow of probability over the states. If Equation A8.9 holds for all values of i and j, the Markov chain is said to be **reversible**, and hence Equation A8.9 is also called the **reversibility condition**. Note that this condition implies that $\pi^* = \pi^*\mathbf{P}$, as the jth element of the row vector $\pi^*\mathbf{P}$ is

$$(\pi^*\mathbf{P})_j = \sum_i \pi_i^*\, P(i \to j) = \sum_i \pi_j^*\, P(j \to i) = \pi_j^* \sum_i P(j \to i) = \pi_j^*$$

where the key middle step follows from Equation A8.9, while the last step follows because rows of $\mathbf{P}$ sum to one.

The basic idea of a discrete-state Markov chain can be generalized to a continuous-state Markov process by replacing $\mathbf{P}$ with a probability kernel, $P(x, y)$, that satisfies

$$\int P(x, y)\, dy = 1$$

The continuous extension of the Chapman–Kolomogrov equation becomes

$$\pi_t(y) = \int \pi_{t-1}(x) P(x, y)\, dx \tag{A8.10a}$$

Finally, at equilibrium, the stationary distribution satisfies

$$\pi^*(y) = \int \pi^*(x) P(x, y)\, dx \tag{A8.10b}$$

THE METROPOLIS–HASTINGS ALGORITHM

The roots of MCMC methods trace back to a problem faced by mathematical physicists in applying Monte Carlo integration, namely, generating random samples from some complex distribution in order to apply Equation A8.1. Their solution was the Metropolis–Hastings algorithm (Metropolis and Ulam 1949; Metropolis et al. 1953; Hastings 1970), which is reviewed by Chib and Greenberg (1995).

Suppose our goal is to draw samples from some distribution, $p(\theta) = f(\theta)/K$, where the normalizing constant, K, may be very difficult to compute. Note that Bayesian posteriors are of this form (Equation A7.3b). The **Metropolis algorithm** (Metropolis and Ulam 1949; Metropolis et al. 1953), which generates a sequence of draws from this distribution, is as follows:

1. Start with any initial value, θ_0, satisfying $f(\theta_0) > 0$.

2. Using the current value of θ, sample a **candidate point**, θ^*, from some **jumping distribution**, $q(\theta_1, \theta_2) = \Pr(\theta_1 \rightarrow \theta_2)$, which is the probability of returning a value of θ_2 given a previous value of θ_1. This distribution is also referred to as the **proposal distribution** or **candidate-generating distribution**. The only restriction on the jump density in the Metropolis algorithm is that it is symmetric, with $q(\theta_1, \theta_2) = q(\theta_2, \theta_1)$.

3. Given the candidate point, θ^*, calculate the ratio of the density at the candidate and current, θ_{t-1}, points,

$$\alpha = \frac{p(\theta^*)}{p(\theta_{t-1})} = \frac{f(\theta^*)}{f(\theta_{t-1})}$$

Notice that because we are considering the ratio of $p(x)$ under two different values, the normalizing constant, K, cancels out.

4. If the jump increases the density ($\alpha > 1$), accept the candidate point (set $\theta_t = \theta^*$) and return to step 2. If the jump decreases the density ($\alpha < 1$), then with a probability of α we accept the candidate point, else we reject it and return to step 2. Note that α is a ratio of two likelihoods, so the probability of accepting a move is the ratio of the candidate to current likelihoods.

We can summarize Metropolis sampling as first computing

$$\alpha = \min\left(\frac{f(\theta^*)}{f(\theta_{t-1})}, 1\right) \tag{A8.11}$$

and then accepting a candidate value, θ^*, with probability α (the **probability of a move**). This generates a Markov chain, $(\theta_0, \theta_1, \cdots, \theta_k, \cdots)$, as the transition probabilities from θ_t to θ_{t+1} depend only on θ_t and not on $(\theta_0, \cdots, \theta_{t-1})$. Following a sufficient **burn-in period** (of, say, k steps), the chain approaches its stationary distribution and (as we will demonstrate shortly) samples from the vector $(\theta_{k+1}, \cdots, \theta_{k+n})$ are samples from $p(\theta)$.

Hastings (1970) generalized the Metropolis algorithm by using an arbitrary (as opposed to strictly symmetric) transition probability function, $q(\theta_1, \theta_2) = \Pr(\theta_1 \rightarrow \theta_2)$, and setting the acceptance probability for a candidate point as

$$\alpha = \min\left(\frac{f(\theta^*)\, q(\theta^*, \theta_{t-1})}{f(\theta_{t-1})\, q(\theta_{t-1}, \theta^*)}, 1\right) \tag{A8.12}$$

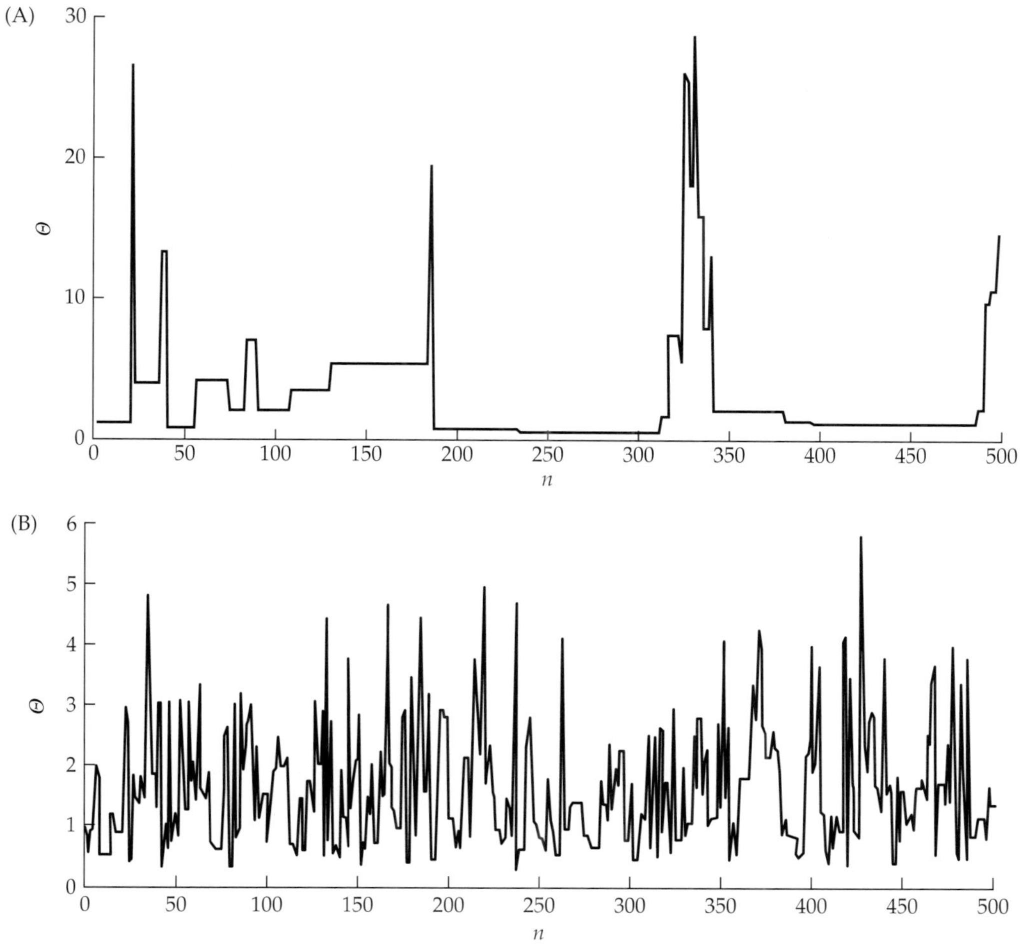

Figure A8.1 Traces for the samplers discussed in Example A8.3. **A:** A sample run when the candidate-generating distribution is a uniform over $(0, 100)$. **B:** A sample run when the candidate-generating distribution is a χ^2_1.

This is the **Metropolis–Hastings algorithm**, and Equation A8.12 is the **Hastings ratio**. When the proposal distribution is symmetric, $q(x, y) = q(y, x)$, we recover the original Metropolis algorithm.

Example A8.3. As a toy example, consider the scaled inverse-χ^2 distribution (Equation A7.29a),

$$p(x) = C \cdot x^{-(n/2+1)} \cdot \exp\left(\frac{-a}{2x}\right)$$

We wish to use the Metropolis algorithm to simulate draws from this distribution with $n = 3$ degrees of freedom and a scaling factor of $a = 4$. Here

$$f(x) = x^{-5/2} \exp[-4/(2x)] = x^{-2.5} \exp[-2/x]$$

Suppose we take as our candidate-generating distribution a uniform over $(0, 100)$. Clearly, there is probability mass above 100 for the scaled inverse-χ^2, but we assume this is sufficiently small that we can ignore it. Now let us run the algorithm. Take $\theta_0 = 1$ as our starting value, and suppose the uniform returns a candidate value of $\theta^* = 39.82$. Computing

α,

$$\alpha = \min\left(\frac{f(\theta^*)}{f(\theta_{t-1})}, 1\right) = \min\left(\frac{(39.82)^{-2.5} \cdot \exp(-2/39.82)}{(1)^{-2.5} \cdot \exp(-2/2 \cdot 1)}, 1\right) = 0.0007$$

Because $\alpha < 1$, the candidate value, θ^*, is accepted with a probability of 0.0007: we randomly draw U from a uniform $(0, 1)$ distribution and accept θ^* if $U \leq \alpha = 0.0007$. If $U > \alpha$, the candidate value is rejected, and we draw another candidate value from the proposal distribution and continue as above. A sample run resulting from first 500 values of θ is plotted in Figure A8.1A. Notice that there are long flat periods (corresponding to all candidate values, θ^*, being rejected). Such a chain is said to be **poorly mixing** (showing long periods between jumps) and is numerically inefficient for sampling the entire probability space defined by the target distribution. Conversely, balancing the poor mixing is the fact that when jumps occur, they tend to be large, thus exploring the extreme values of this distribution.

In contrast, suppose our proposal distribution is a χ_1^2. As this distribution is not symmetric, we must employ Metropolis–Hastings (see Example A8.5 for the details). A resulting Metropolis–Hastings sampling run is shown in Figure A8.1B. Note that the time series looks like **white noise**, and the chain is said to be well mixing. However, note that most jumps are small (< 5), with fewer extreme values sampled than under the uniform proposal distribution.

Example A8.4. This rather technical example shows that the Markov chain generated by the Metropolis–Hastings algorithm has a stationary distribution which corresponds to $p(x)$. Hence, draws from the stationary phase of this chain correspond to random draws from the distribution given by $p(x)$. We do so by showing that, under this chain, $p(x)$ satisfies the detailed balance equation (A8.9), and hence it must be the stationary distribution for the Markov chain.

Under Metropolis–Hastings, we sample from $q(x, y) = \Pr(x \to y \,|\, q)$ and then accept the move with probability $\alpha(x, y)$, so the transition probability kernel is given by

$$\Pr(x \to y) = q(x, y)\,\alpha(x, y) = q(x, y) \cdot \min\left[\frac{p(y)\,q(y, x)}{p(x)\,q(x, y)}, 1\right] \tag{A8.13a}$$

If the Metropolis–Hastings kernel satisfies $P(x \to y)\,p(x) = P(y \to x)\,p(y)$, or

$$q(x, y)\,\alpha(x, y)\,p(x) = q(y, x)\,\alpha(y, x)\,p(y) \quad \text{for all } x, y \tag{A8.13b}$$

then the stationary distribution from this kernel corresponds to draws from the target distribution, $p(x)$.

We show that the balance equation is indeed satisfied with this kernel by considering the three possible cases for any particular (x, y) pair:

1. $q(x, y)\,p(x) = q(y, x)\,p(y)$. Here $\alpha(x, y) = \alpha(y, x) = 1$ implying

$$P(x \to y)\,p(x) = q(x, y)\,p(x) \quad \text{and} \quad P(y \to x)p(y) = q(y, x)\,p(y)$$

Under our assumption that $q(x, y)\,p(x) = q(y, x)\,p(y)$, this reduces to $P(x \to y)\,p(x) = P(y \to x)\,p(y)$, showing that (for this case), the detailed balance equation holds.

2. $q(x, y)\,p(x) > q(y, x)\,p(y)$, in which case

$$\alpha(x, y) = \frac{p(y)\,q(y, x)}{p(x)\,q(x, y)} \quad \text{and} \quad \alpha(y, x) = 1$$

Hence,

$$\begin{aligned} P(x \to y)\,p(x) &= q(x, y)\,\alpha(x, y)\,p(x) \\ &= q(x, y)\,\frac{p(y)\,q(y, x)}{p(x)\,q(x, y)}\,p(x) \\ &= q(y, x)\,p(y) = q(y, x)\,\alpha(y, x)\,p(y) \\ &= P(y \to x)\,p(y) \end{aligned}$$

3. $q(x, y)\, p(x) < q(y, x)\, p(y)$. Here

$$\alpha(x, y) = 1 \quad \text{and} \quad \alpha(y, x) = \frac{q(x, y)\, p(x)}{q(y, x)\, p(y)}$$

Following the same logic just used yields

$$\begin{aligned} P(y \to x)\, p(y) &= q(y, x)\, \alpha(y, x)\, p(y) \\ &= q(y, x) \left(\frac{q(x, y)\, p(x)}{q(y, x)\, p(y)} \right) p(y) \\ &= q(x, y)\, p(x) = q(x, y)\, \alpha(x, y)\, p(x) \\ &= P(x \to y)\, p(x) \end{aligned}$$

which concludes the proof.

Burning-in the Sampler

A key issue in the successful implementation of Metropolis–Hastings, or any other MCMC sampler, is the number of steps (iterations) until the chain approaches stationarity (the length of the **burn-in period**), as initial values are somewhat dependent upon the starting position. Typically the first 1,000 to 50,000 (or more) values of the chain are discarded, and then various convergence tests (see below) are used to assess whether stationarity has indeed been reached.

A poor choice of either a proposal distribution or a starting value can greatly increase the required burn-in time, and an area of current research is whether an optimal starting point and proposal distribution can be found. One suggestion is to initiate the chain as close to the center of the distribution as possible, for example, taking a value close to the distribution's mode (such as using an approximate MLE as the starting value). Geyer (2011), however, offered the simple advice that "any point you don't mind having in a sample is a good starting point."

A chain is said to be **poorly mixing** if it stays in small regions of the parameter space for long periods of time, as opposed to a **well-mixing** chain, which seems to happily explore the entire space (albeit perhaps by very small steps, requiring very many draws to fully survey the entire space). A poorly mixing chain can arise because the target distribution is multimodal and our choice of starting values (or the chance steps that were initially taken) traps the chain near one of the modes (such multimodal posteriors can arise if we have a strong prior that is in conflict with the observed data). Two approaches have been suggested for situations where the target distribution may have multiple peaks. The most straightforward is to use widely dispersed initial values to start several different chains (Gelman and Rubin 1992). A less obvious approach, which we will discuss shortly, is to use **simulated annealing** on a single chain.

Even after an apparently successful burn-in, the real fear for the MCMC practitioner is that the chain is simply showing **pseudo-convergence**. It appears to have reached an equilibrium, when in fact it is simply exploring (albeit perhaps rather slowly) some subset of the entire distribution. If the transition time between different parts of the full distribution is long relative to the sample size, the MCMC user will observe apparent convergence and believe that the simulation is a fair representation of the true posterior. The unfortunate feature is that for *any* MCMC, the skeptic, or a hardened reviewer, can claim (with some justification) that convergence has not been demonstrated. The problem is akin to large-sample results in statistics (such as the large-sample features of ML; Appendix 4): the theory says that for a *sufficiently large* sample, certain desirable properties hold, but usually says rather little about how "large" is large. For a complex posterior, the required chain size might be far, far larger than an investigator envisions. The only real solution is to run the chain for as long as possible. In such a setting, Geyer (2011) argued, a burn-in is not required, and noted that "Burn-in is mostly harmless, which is perhaps why the practice persists. But everyone should understand that it is unnecessary, and those who do not use it are not thereby making an error."

Simulated Annealing

Simulated annealing was developed for finding the maximum of complex functions with multiple peaks where standard hill-climbing approaches may trap the algorithm on a less than optimal peak (Kirkpatrick et al. 1983). The idea is that when we initially start sampling the space, we will accept a reasonable probability of a downhill move in order to explore the entire space. As the process proceeds, we decrease the probability of such downhill moves. The analogy (and hence the term) is the annealing of a crystal as temperature decreases: initially there is a lot of movement, which gets smaller and smaller as the temperature cools. Simulated annealing is very closely related to Metropolis sampling, differing only in that the probability α of a move is given by

$$\alpha_{SA} = \min\left[1, \left(\frac{f(\theta^*)}{f(\theta_{t-1})}\right)^{1/T(t)}\right] = \alpha^{1/T(t)} \tag{A8.14a}$$

where the function $T(t)$ is called the **cooling schedule** (setting $T = 1$ recovers Metropolis sampling), and the particular value of T at any point in the chain is called the **temperature**. For example, suppose $f(\theta^*)/f(\theta_{t-1}) = 0.5$. If $T = 100$, $\alpha = 0.93$, while for $T = 1$, $\alpha = 0.5$, and for $T = 1/10$, $\alpha = 0.0098$. Hence, we start off with a high probability of a jump and then cool down to a very low jump probability. Replacing α with the Hastings ratio (Equation A8.12) allows us to apply simulated annealing to Metropolis–Hastings.

A function with geometric decline is typically used for $T(t)$. For example, to start at T_0 and cool down to a final temperature of T_f over n steps, we can set

$$T(t) = T_0\left(\frac{T_f}{T_0}\right)^{t/n} \tag{A8.14b}$$

More generally, if we wish to cool off to T_f by time n and then keep the temperature constant at T_f for the rest of the run, we can take

$$T(t) = \max\left(T_0\left(\frac{T_f}{T_0}\right)^{t/n}, T_f\right) \tag{A8.14c}$$

In particular, to cool down to Metropolis sampling (by step n), we set $T_f = 1$ and the cooling schedule becomes

$$T(t) = \max\left(T_0^{1-t/n}, 1\right) \tag{A8.14d}$$

Choosing a Jumping (Proposal) Distribution

The Metropolis sampler works with any symmetric proposal (or jumping) distribution, while Metropolis–Hastings allows for more general distributions. So how do we determine our best option for a proposal distribution? There are two general approaches: random walks and independent chain sampling. Under a sampler using a proposal distribution based on a **random-walk chain**, the new value, y, equals the current value, x, plus a random variable, namely, z,

$$y = x + z$$

In this case, $q(x, y) = g(y - x) = g(z)$, the density associated with the random variable z. If $g(z) = g(-z)$, i.e., the density for the random variable z is symmetric (as occurs with a normal or multivariate normal with a mean of zero, or a uniform centered around zero), then we can use Metropolis sampling, as $q(x,y)/q(y,x) = g(z)/g(-z) = 1$. The variance of the proposal distribution can be thought of as a **tuning parameter** that is adjusted to achieve better mixing.

Under a proposal distribution using an **independent chain**, the probability of jumping to point y is *independent* of the current position (x) of the chain, namely, $q(x,y) = g(y)$. The candidate value is simply drawn from a proposal distribution, independent of the current value (see Examples A8.3 and A8.5). Again, any number of standard distributions can be used for $g(y)$. Note that in this case, except for a uniform, the proposal distribution is generally *not symmetric*, as $g(x)$ is generally not equal to $g(y)$ (for $x \neq y$), and Metropolis–Hastings sampling must be used.

As mentioned, we can tune the proposal distribution to adjust the mixing, and in particular the acceptance probability, of the chain. This is generally accomplished by changing the standard deviation (SD) of the proposal distribution. This can be done by tuning the variance for a normal or adjusting the eigenvalues of the covariance matrix for a multivariate normal. Likewise, one can increase or decrease the range, $(-a, a)$, if a uniform is used, or change the degrees of freedom (df) if a χ^2 is used (variance increases with the df; Equation A5.14b). To increase the acceptance probability, one *decreases* the proposal distribution SD. There is a tradeoff in that if the SD is too large, jumps are potentially large (which is good), but are usually not accepted (which is a problem). This leads to high autocorrelation (see below) and very poor mixing, requiring much longer chains. On the other hand, if the proposal SD is too small, moves will generally be accepted (there is a high acceptance probability), but they will also be small, again generating high autocorrelations and poor mixing.

Example A8.5. Suppose we wish to use a χ^2 as our proposal distribution. Recall from Equation A7.26b, that for $x \sim \chi^2_n$,

$$g(x) \propto x^{n/2-1} e^{-x/2}$$

Thus, $q(x,y) = g(y) = C \cdot y^{n/2-1} e^{-y/2}$. Note that $q(x,y)$ is not symmetric, as $q(y,x) = g(x) \neq g(y) = q(x,y)$. Hence, we must use Metropolis–Hastings sampling, with an acceptance probability of

$$\alpha(x,y) = \min\left[\frac{p(y)\, q(y,x)}{p(x)\, q(x,y)}, 1\right] = \min\left[\frac{p(y)\, g(x)}{p(x)\, g(y)}, 1\right] = \min\left[\frac{p(y)\, x^{n/2-1} e^{-x/2}}{p(x)\, y^{n/2-1} e^{-y/2}}, 1\right]$$

If we use the same target distribution as in Example A8.3 (a scaled inverse-χ^2), $p(x) = C \cdot x^{-2.5}\, e^{-2/x}$, the rejection probability becomes

$$\alpha(x,y) = \min\left[\frac{(\, y^{-2.5}\, e^{-2/y}\,)\,(\, x^{n/2-1} e^{-x/2}\,)}{(\, x^{-2.5}\, e^{-2/x}\,)\,(\, y^{n/2-1} e^{-y/2}\,)}, 1\right]$$

Results for a run of the sampler under two different proposal distributions (a χ^2_2 and a χ^2_{10}) are plotted in Figure A8.2. The χ^2_2 has the smaller variance, and thus a higher acceptance probability. Further, it seems to explore more of the distribution space than a χ^2_{10} (compare the number of values above 4 in both traces).

Autocorrelation and Sample Size Inflation

We often expect adjacent members from an MCMC sequence to be positively correlated, and we can quantify the nature of this correlation by using an **autocorrelation function**. Consider a sequence, $(\theta_1, \cdots, \theta_n)$, of length n. Correlations can occur between adjacent members,

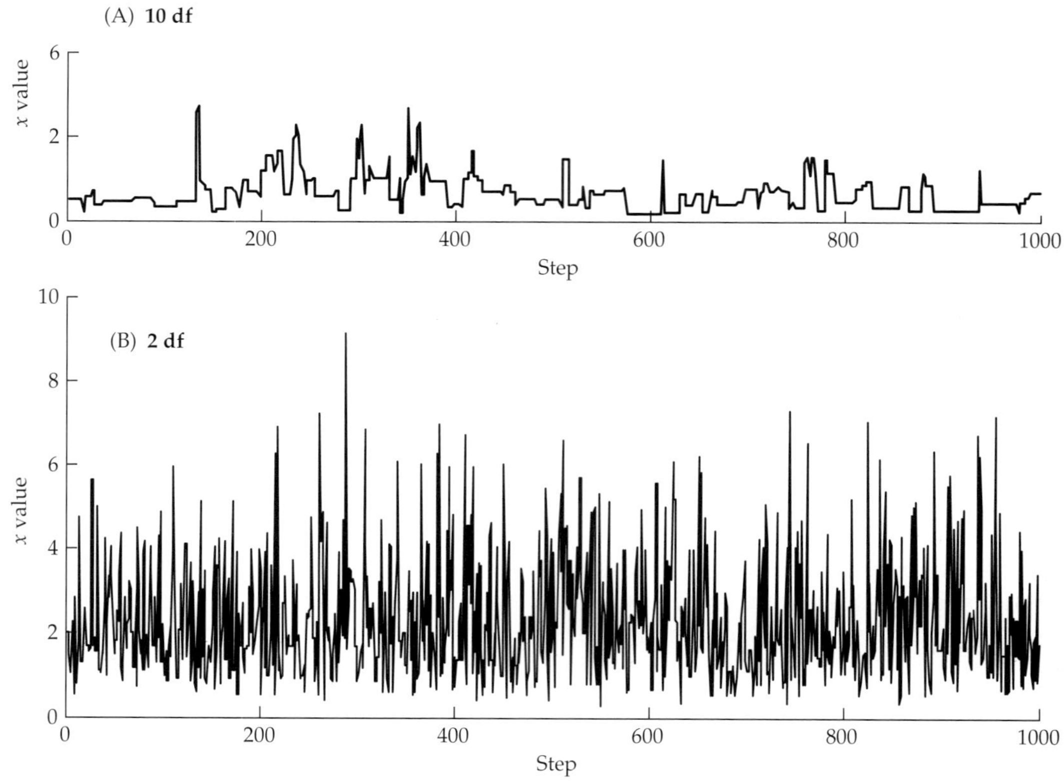

Figure A8.2 Trace plots for the two samplers discussed in Example A8.5. **A:** The proposal distribution is a χ^2_{10}. **B:** The proposal distribution is a χ^2_2.

$\rho(\theta_i, \theta_{i+1}) \neq 0$, and, more generally, between more distant members, $\rho(\theta_i, \theta_{i+k}) \neq 0$. The kth order autocorrelation, ρ_k, can be estimated by

$$\widehat{\rho}_k = \frac{\text{Cov}(\theta_t, \theta_{t+k})}{\text{Var}(\theta_t)} = \frac{\sum\limits_{t=1}^{n-k} \left(\theta_t - \overline{\theta}\right)\left(\theta_{t+k} - \overline{\theta}\right)}{\sum\limits_{t=1}^{n-k} \left(\theta_t - \overline{\theta}\right)^2}, \quad \text{with} \quad \overline{\theta} = \frac{1}{n}\sum_{t=1}^{n} \theta_t \tag{A8.15}$$

An important result from the theory of time-series analysis is that if the values of θ_t are from a stationary (and correlated) process, correlated draws still provide an unbiased picture of the distribution ***provided the sample size is sufficiently large***. The reason for this large-sample stipulation is that, strictly speaking, Equation A8.15 applies only to a stationary process.

Some indication of the required sample size comes from the theory of a **first-order autoregressive process** (AR_1), where

$$\theta_t = \mu + \alpha(\theta_{t-1} - \mu) + \epsilon \tag{A8.16a}$$

where $|\,\alpha\,| < 1$ and ϵ is **white noise**, namely, $\epsilon \sim N(0, \sigma^2)$. Here $\rho_1 = \alpha$ and the kth order autocorrelation is given by $\rho_k = \rho_1^k = \alpha^k$. Under this process, $E(\overline{\theta}) = \mu$ with a standard error of

$$\text{SE}\left(\overline{\theta}\right) = \frac{\sigma}{\sqrt{n}}\sqrt{\frac{1+\rho}{1-\rho}} \tag{A8.16b}$$

The first ratio is the standard error for white noise, while the second ratio, $\sqrt{(1+\rho)/(1-\rho)}$, is the **sample size inflation factor** (**SSIF**), which shows how the autocorrelation inflates the sampling variance relative to independent samples. For $\rho = 0.5, 0.75, 0.9, 0.95$, and 0.99, the associated SSIF values become 3, 7, 19, 39, and 199, respectively. With an autocorrelation of 0.95 (which is not uncommon in a Metropolis–Hastings sequence), roughly 40 times as many iterations are required for the same precision as with an uncorrelated sequence.

One historical strategy for reducing autocorrelation is **thinning** (or **subsampling**) the output, which involves storing only every mth point after the burn-in period. Suppose a Metropolis–Hastings sequence follows an AR_1 model, with $\rho_1 = 0.99$. In this case, sampling every 50, 100, and 500 points gives the correlation between the thinned samples as 0.605 (= 0.99^{50}), 0.366, and 0.007, respectively, returning SSIF values of 2, 1.5, and 1.007. In addition to reducing autocorrelation, thinning the sequence also saves computer memory. However, this throws out a huge amount of the data. If computer memory is not an issue, Geyer (1992) and MacEachern and Berliner (1994) found that it is more efficient (yielding smaller Monte Carlo variances) to keep the entire chain than to use thinning.

The Monte Carlo Variance of a MCMC-Based Estimate

Suppose we are interested in using a burned-in MCMC sequence, $(\theta_1, \cdots, \theta_n)$, to estimate some function, $h(\theta)$, of the target distribution, such as a mean, variance, or specific quantile (namely, a specific cumulative probability value). Because we are drawing random variables, associated with the Monte Carlo estimate, there is a sampling variance of

$$\widehat{h} = \frac{1}{n} \sum_{i=1}^{n} h(\theta_i) \tag{A8.17}$$

A similar issue arose with the error in Monte Carlo integration, where we used the sample variance around $\widehat{h}$ to compute the Monte Carlo variance, dividing this by the number of iterations to yield a standard error (Equations A8.1e and A8.4c). This strategy exploited the fact that draws for Monte Carlo integration are *independent* and therefore uncorrelated. In stark contrast, we expect draws in an MCMC sequence to be highly *dependent*. A standard sample variance estimator ignores these correlations, potentially yielding a highly biased estimate of the Monte Carlo variance (see Equation A8.16b).

One direct approach is to run several chains (or extract subsamples from a very long chain) and use the between-chain variance in $\widehat{h}$ as our variance estimate. Specifically, if $\widehat{h}_j$ denotes the estimate for chain j $(1 \le j \le c)$, where each of the c chains has the same length, then the estimated variance of the Monte Carlo estimate is

$$\text{Var}\left(\widehat{h}\right) = \frac{1}{c-1} \sum_{j=1}^{c} \left(\widehat{h}_j - \widehat{h^*}\right)^2 \quad \text{where} \quad \widehat{h^*} = \frac{1}{c} \sum_{j=1}^{c} \widehat{h}_j \tag{A8.18}$$

Using only a single chain, an alternative approach is to use results from the theory of time series to account for the correlations among members in the chain. To start, we estimate the lag-k autocovariance associated with h by

$$\widehat{\gamma}(k) = \frac{1}{n-1} \sum_{i=1}^{n-k} \left[\left(h(\theta_i) - \widehat{h} \right) \left(h(\theta_{i+k}) - \widehat{h} \right) \right] \tag{A8.19}$$

This is natural generalization of the kth order autocorrelation (Equation A8.15) to the random variable generated by $h(\theta_i)$. Geyer (1992, 2011) obtained the resulting estimate of the Monte Carlo variance as

$$\text{Var}\left(\widehat{h}\right) = \frac{1}{n} \left(\widehat{\gamma}(0) + 2 \sum_{i=1}^{2\delta+1} \widehat{\gamma}(i) \right) \tag{A8.20}$$

Note that the zero-lag term, $\widehat{\gamma}(0)$, is simply the standard variance estimate when observations are uncorrelated, while δ is the largest positive integer satisfying $\widehat{\gamma}(2\delta) + \widehat{\gamma}(2\delta+1) > 0$, (i.e., the higher order, or lag, autocovariances above δ are zero).

One measure of the effects of autocorrelation between elements in the sampler is the **effective chain size**,

$$\widehat{n} = \frac{\widehat{\gamma}(0)}{\text{Var}\left(\widehat{h}\right)} \tag{A8.21}$$

where $\widehat{\gamma}(0)$ is the standard sample variance for uncorrelated observations. In the absence of autocorrelation between members, $\widehat{n} = n$.

CONVERGENCE DIAGNOSTICS

The careful reader will note that we have still not answered the fundamental, and vexing, question of how to determine whether the sampler has reached its stationary distribution. Many tests for stationarity have been proposed, and all can indeed show that stationarity has not yet been reached. However, it is important to stress that ***negative results from these tests are not conclusive evidence that the sampler has achieved stationarity***. Comments from various published reviews on the subject stress this point: "all of the methods can fail to detect the sorts of convergence failure that they were designed to identify" (Cowles and Carlin 1996); "It is clear that the usefulness of convergence assessment methods are limited by their potential unreliability" (Brooks and Roberts 1998); "Diagnostics can only reliably be used to determine a lack of convergence and not detect convergence per se" (Brooks et al. 2003); and "no method works in every case" (El Adlouni et al. 2006). Despite this very serious concern, a variety of methods can certainly show a *failure of convergence* given the burn-in period used for a sampler. The greater question about trust is what we say when convergence diagnostics *do not* indicate a problem. We consider a few basic approaches here, while additional diagnostics were reviewed by Gelman and Rubin (1992), Geyer (1992), Raftery and Lewis (1992b), Cowles and Carlin (1996), Brooks and Roberts (1998), Kass et al. (1998), Mengerson et al. (1999), Robert and Casella (2004), El Adlouni et al. (2006), and Peltonen et al. (2009).

An extra level of complications occurs in models using **revisable jump MCMC**, wherein the chain actually jumps between difference *classes* of models. For example, suppose the number of QTLs, k, is a random variable. While the chain may come to a nearly equilibrium distribution for the parameters associated with a given value of k fairly quickly, one must also sample a sufficient number of jumps *between models* with different values of k to fully explore the model space. One might incorrectly assume that there appears to be convergence when the model quickly equilibrates for a given value of k, but jumps very slowly between different values of k so that only a small sample of the k space has been explored.

Visual Analysis

As shown in Figures A8.1 and A8.2, one should always examine the **time-series trace**, the plot of the random variable being generated versus the number of iterations. In addition to showing evidence for poor mixing, such traces can also suggest a ***minimum*** burn-in period for some starting value. For example, suppose the trace moves very slowly away from the initial value to a rather different value (say after 5,000 iterations), after which it appears to settle down. Clearly, the burn-in period is ***at least*** 5,000 in this case. It must be cautioned that the actual burn-in time ***may be far longer*** than suggested by the trace. Nevertheless, the trace often indicates that more burn-in time is needed. A second approach is to run either several chains or subsample from a much longer chain, and then compare the traces. The expectation under stationarity is that these samples are drawn from the same distribution. Even if visual analysis suggests that two (or more) chains appear to be very similar and are both well-behaved, the concern is that both might be stuck in the same local maximum region of a complex posterior. In such cases, the impression of stationarity and consistency is presented, but significantly longer runs for each chain would show a different outcome.

Correlograms, which plot serial autocorrelations as a function of the time lag, are also very useful in accessing a run from an MCMC sampler. A plot of ρ_k versus k (the kth order autocorrelation versus the lag) should show geometric decay if the sampler series closely follows an AR_1 model. A plot of the **partial autocorrelations** as a function of lag is also useful. The kth partial autocorrelation is the excess correlation not accounted for by a $k-1$ order autoregressive model (AR_{k-1}). Hence, if the first-order model fully fits the data, the second-order partial autocorrelation is zero, as the lagged autocorrelations are accounted for by the AR_1 model (i.e., $\rho_k = \rho_1^k$). Both of these autocorrelation plots can indicate an underlying correlation structure that may not be obvious from the time-series trace.

More Formal Approaches

Gelman and Rubin's (1992) proposal was to first generate m separate chains at initial locations that are widely dispersed in the parameter space, with each returning a burned-in and thinned (such that autocorrelations between estimates are near zero) sequence of length n. Their approach is essentially an ANOVA (Part II Chapter 23), contrasting the estimated among-chain variance with its within-chain value. For the univariate case, let

$$B = \frac{n}{m-1}\sum_{i=1}^{m}(\overline{\theta}_{i\cdot} - \overline{\theta}_{\cdot\cdot})^2 \tag{A8.22a}$$

be the among-chain variance, where $\overline{\theta}_{i\cdot}$ is the mean parameter value for chain i, and $\overline{\theta}_{\cdot\cdot}$ is the grand mean over all chains. Likewise, for the within-chain variance, we define

$$W = \frac{1}{m}\left(\frac{1}{n-1}\sum_{i=1}^{m}\sum_{j=1}^{n}(\overline{\theta}_{ij} - \overline{\theta}_{i\cdot})^2\right) \tag{A8.22b}$$

A combined estimate of the variance is

$$\widehat{\sigma^2} = \frac{n-1}{n}W + \frac{1}{n}B \tag{A8.22c}$$

Gelman and Rubin defined the **potential scale reduction factor** (**PSRF**) as

$$R = \frac{\widehat{\sigma^2}}{W} = 1 + \frac{1}{n}\left(\frac{B}{W} - 1\right) \tag{A8.22d}$$

with values of R near one being consistent with convergence. A plot of R over time should converge to a value near one, if the chains are converging.

Because most MCMCs involve many variables, one could either look at the univariate R trace for each or use the multivariate version suggested by Brooks and Gelman (1998),

$$R = \frac{n-1}{n} + \left(1 + \frac{1}{m}\right)\lambda_1 \tag{A8.23}$$

where λ_1 is the leading eigenvalue of the matrix product $\mathbf{W}^{-1}\mathbf{B}/n$, and where $\mathbf{W}$ and $\mathbf{B}$ are the sample within- and among-covariance matrices for the parameters. Again, a value near 1 is consistent with convergence.

The **Geweke test** (Geweke 1992) splits the sample (after removing a burn-in period) into two parts. If the chain is at stationarity, the means of the two samples should be equal. A modified z-test can be used to compare the two subsamples, and the resulting test statistic is often referred to as a **Geweke z-score**. A value larger than 2 indicates that the mean of the series is still drifting and a longer burn-in is required before monitoring the chain (to extract a sampler) can begin.

A more informative approach is the **Raftery–Lewis test** (Raftery and Lewis 1992a). Here, one specifies a particular quantile, q, of the distribution of interest (typically testing both $q = 2.5\%$ and 97.5%, to give a 95% confidence interval), an accuracy, ϵ, of the quantile, and a power, $1-\beta$, for achieving this accuracy on the specified quantile. With these three parameter values set, the Raftery–Lewis test breaks the chain into a binary (1, 0) sequence, in which an element in the sequence takes on a value of 1 if $\theta_t \leq q$, otherwise it takes on a value of zero. This generates a two-state Markov chain, and the Raftery–Lewis test uses the sequence to estimate the transition probabilities between states. With these probabilities in hand, one can then estimate the number of additional burn-ins (if any) required to approach stationarity, the thinning ratio (how many points should be discarded for each sampled point), and the total chain length required to achieve the preset level of accuracy.

Practical MCMC: How Many Chains and How Long Should They Run?

One can either use a single long chain (Geyer 1992; Raftery and Lewis 1992b) or multiple chains, with each starting from different initial values (Gelman and Rubin 1992). Note that with parallel processing, using multiple chains may be computationally more efficient (given a fixed amount of analysis time) than a single long chain. One suggested approach (Kass et al. 1998) is to use five parallel chains with widely separated starting values, and to start the jth value for parameter θ_i at $\mu_i + (j-2)\sigma_i$, where μ_i and σ_i^2 are the prior values for θ_i. Alternatively, one could randomly sample j from $(-2, 1, 0, 1, 2)$ for each parameter to generate the vector of starting values.

Geyer (1992, 2011), however, forcefully argued that using a single long chain is the best approach. If the chain transition times between different parts of a distribution are long, then a series of shorter chains may entirely miss this problem, even if their starting values are widely dispersed. Theory basically tells us that, *provided* we run a chain for a long enough time, the results will be accurate. The best way to do so is to run a single, long chain. Convergent tests (still recognizing their problems mentioned previously) can be applied to subsections of a long chain as efficiently as they can be applied across shorter, parallel-running chains.

Given these concerns, what is the answer to the critical question about how long to run a chain? **Geyer's dictums** (2011) are that: (i) "the least one can do is to make an overnight run. What better way for your computer to spend its time?"; and (ii) ideally, "one should start a run when the paper is submitted and keep running until the referees' reports arrive."

THE GIBBS SAMPLER

The **Gibbs sampler** (introduced in the context of image processing by Geman and Geman 1984), is a special case of Metropolis–Hastings sampling wherein the random value is *always accepted* (i.e., $\alpha = 1$). The key to the Gibbs sampler is that, at any point in the process, one only considers ***univariate*** conditional distributions—the distribution when all but one of the random variables are assigned fixed values. Such conditional distributions are far easier to simulate than complex joint distributions and usually have simple forms (often they are normals, inverse-χ^2, or other common prior distributions). An important use of conjugate priors (Table A7.2) is that they often result in relatively simple forms for conditional distributions. Thus, one simulates n random variables sequentially from the n univariate conditionals rather than generating a single n-dimensional vector in a single pass using the full joint distribution. Given their great efficiency (every value is accepted), it is often thought that one should strive to generate a Gibbs sampler whenever possible. However, one should consult Geyer (2011) for a different perspective.

To introduce the Gibbs sampler, consider a bivariate random variable (x, y), and suppose we wish to compute one or both of the marginals, $p(x)$ and $p(y)$. The idea behind the sampler is that it is far easier to consider a sequence of conditional distributions, $p(x \mid y)$ and $p(y \mid x)$, than it is to obtain the marginal by integration of the joint density $p(x, y)$, namely, $p(x) = \int p(x, y)dy$. The sampler starts with some initial value, y_0, for y and obtains x_0 by generating a random variable from the conditional distribution $p(x \mid y = y_0)$. The sampler then uses x_0 to generate a new value of y_1, drawing from the conditional distribution based on the value of x_0, $p(y \mid x = x_0)$. The sampler proceeds as follows:

$$x_i \sim p(x \,|\, y = y_{i-1}) \tag{A8.24a}$$

$$y_i \sim p(y \,|\, x = x_i) \tag{A8.24b}$$

Repeating this process k times generates a **Gibbs sequence** of length k, where a subset of points (x_j, y_j) for $1 \leq j \leq m < k$ are taken as our simulated draws from the full joint distribution. One iteration over all the univariate distributions is often called a **scan** of the sampler. To obtain the desired total of m sample values (here each value in the sampler is a vector of realizations of the two random variables), one samples the chain following a sufficient burn-in to remove the effects of the initial starting values, potentially thinning the sequence as well. The Gibbs sequence converges to a stationary (equilibrium) distribution that is independent of the starting values, and by construction this stationary distribution is the target distribution we are trying to simulate (Tierney 1994). As with any MCMC, a guarantee of convergence at some limit is no guarantee that our sample size is sufficiently large to justify the assumption of stationarity for a burned-in (and potentially thinned) sampler.

Example A8.6. Consider the following distribution from Casella and George (1992). Suppose the joint distribution of $x = 0, 1, \cdots, n$ and $0 \leq y \leq 1$ is given by

$$p(x,y) = \frac{n!}{(n-x)!x!}\, y^{\,x+\alpha-1}\, (1-y)^{\,n-x+\beta-1}$$

Note that x is discrete and y continuous. While the joint density is complex, the conditional densities are simple distributions. To see this, first recall that a discrete binomial random variable z has a density (Equation 2.19a) proportional to

$$p(z \,|\, q, n) \propto \frac{q^z(1-q)^{n-z}}{z!(n-z)!} \quad \text{for} \quad 0 \leq z \leq n$$

where $0 < q < 1$ is the success parameter and n the number of traits, which we denote by $z \sim \text{B}(n, p)$. Likewise, recall the density for $z \sim \text{Beta}(a, b)$, a beta distribution (Equation A7.37a) with shape parameters a and b is given by

$$p(z \,|\, a, b) \propto z^{a-1}(1-z)^{b-1} \quad \text{for} \quad 0 \leq z \leq 1$$

Observe that the conditional distribution of x (treating y as a fixed constant) is $x \,|\, y \sim \text{B}(n, y)$, while $y \,|\, x \sim \text{Beta}(x + \alpha, n - x + \beta)$.

The power of the Gibbs sampler is that by computing a sequence of these univariate conditional random variables (a binomial and then a beta), we can compute any feature of either marginal distribution. Suppose $n = 10$ and $\alpha = 1, \beta = 2$. We start the sampler with (say) $y_0 = 1/2$ and then take it through three full iterations.

(i) x_0 is obtained by generating a random $\text{B}(n, y_0) = \text{B}(10, 1/2)$ random variable, returning $x_0 = 5$ in our simulation.

(ii) y_1 is obtained from a $\text{Beta}(x_0+\alpha, n-x_0+\beta) = \text{Beta}(5+1, 10-5+2)$ random variable, returning $y_1 = 0.33$.

(iii) x_1 is a realization of a $\text{B}(n, y_1) = \text{B}(10, 0.33)$ random variable, returning $x_1 = 3$.

(iv) y_2 is obtained from a $\text{Beta}(x_1+\alpha, n-x_1+\beta) = \text{Beta}(3+1, 10-3+2)$ random variable, returning $y_2 = 0.56$.

(v) x_2 is obtained from a $\text{B}(n, y_2) = \text{B}(10, 0.56)$ random variable, returning $x_2 = 7$.

Our particular realization of the Gibbs sequence after three iterations is thus (5, 0.5), (3, 0.33), (7, 0.56). We can continue this process to generate a chain of the desired length. Obviously, the initial values in the chain are dependent upon the value of y_0 chosen to start the chain. This dependence decays as the sequence length increases, so we typically start recording the sequence only after a sufficient number of burn-in iterations have occurred.

When more than two variables are involved, the sampler is extended in the obvious fashion. In particular, the value of the kth variable is drawn from the distribution $p(\theta^{(k)} \mid \boldsymbol{\Theta}^{(-k)})$, where $\boldsymbol{\Theta}^{(-k)}$ denotes a vector containing all of the variables but k. Thus, during the ith iteration of the sample, to obtain the value of $\theta_i^{(k)}$, we draw from the distribution

$$\theta_i^{(k)} \sim p(\theta^{(k)} \mid \theta^{(1)} = \theta_i^{(1)}, \cdots, \theta^{(k-1)} = \theta_i^{(k-1)}, \theta^{(k+1)} = \theta_{i-1}^{(k+1)}, \cdots, \theta^{(n)} = \theta_{i-1}^{(n)})$$

For example, if there are four variables, (w, x, y, z), the sampler becomes

$$\begin{aligned} w_i &\sim p(w \mid x = x_{i-1}, y = y_{i-1}, z = z_{i-1}) \\ x_i &\sim p(x \mid w = w_i, y = y_{i-1}, z = z_{i-1}) \\ y_i &\sim p(y \mid w = w_i, x = x_i, z = z_{i-1}) \\ z_i &\sim p(z \mid w = w_i, x = x_i, y = y_i) \end{aligned}$$

Gelfand and Smith (1990) illustrated the power of the Gibbs sampler to address a wide variety of statistical issues, while Smith and Roberts (1993) showed the natural marriage of the Gibbs sampler with Bayesian statistics for obtaining posterior distributions. A nice introduction to the sampler was presented by Casella and George (1992), while further details can be found in Tanner (1996), Besag et al. (1995), Robert and Casella (2004), and Lee (2013). Sorensen and Gianola (2002) give an especially detailed treatment of the Gibbs samples for the mixed linear model, the workhorse framework for much of quantitative genetics. Note that the Gibbs sampler can be thought of as a stochastic analog to the Expectation-Maximization (EM) approaches (Appendix 4) used to obtain likelihood functions when missing data are present. In the sampler, random sampling replaces the expectation and maximization steps.

Using the Gibbs Sampler to Approximate Marginal Distributions

Any feature of interest for the marginals can be computed from the m realizations of the Gibbs sequence. If $\theta_1, \cdots, \theta_m$ is an appropriately burned-in set of realizations from a Gibbs sampler, the expectation of any function, f, of the random variable θ is approximated by

$$E[f(\theta)]_m = \frac{1}{m} \sum_{i=1}^{m} f(\theta_i) \tag{A8.25a}$$

This is the **Monte Carlo (MC) estimate** of $f(x)$, as $E[f(\theta)]_m \to E[f(\theta)]$ as $m \to \infty$. Likewise, the MC estimate for any function of n variables $(\theta^{(1)}, \cdots, \theta^{(n)})$ is calculated by

$$E[f(\theta^{(1)}, \cdots, \theta^{(n)})]_m = \frac{1}{m} \sum_{i=1}^{m} f(\theta_i^{(1)}, \cdots, \theta_i^{(n)}) \tag{A8.25b}$$

Example A8.7. Although the sequence of length 3 that was computed in Example A8.6 is too short (and too dependent on the starting value) to be a proper Gibbs sequence, for illustrative purposes we can use it to compute MC estimates. The MC estimates of the means of x and y are

$$\overline{x}_3 = \frac{5+3+7}{3} = 5, \quad \overline{y}_3 = \frac{0.5+0.33+0.56}{3} = 0.46$$

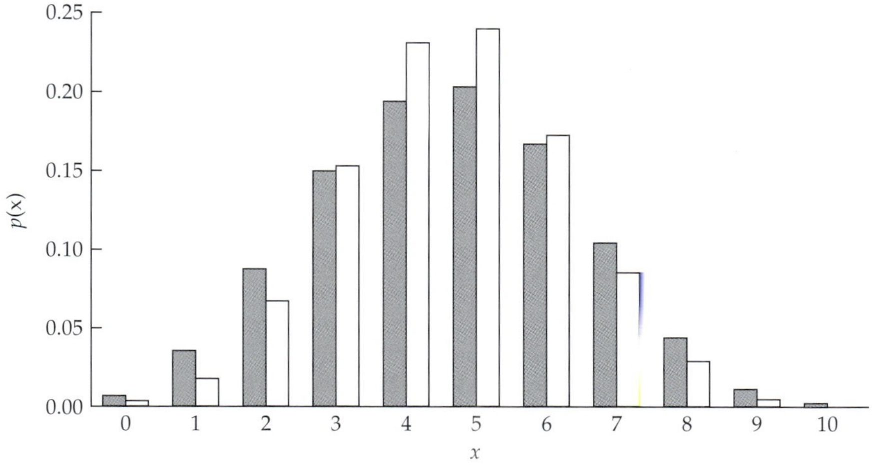

Figure A8.3 The approximation of the marginal posterior distribution for x for the distribution given in Example A8.6 using a Gibbs sequence. The open bars represent the approximation using a binomial with the Gibbs-sequence mean value of y as the success parameter, while the solid bars represent the approximation based on a weighted sum of binomials. See Example A8.8 for details.

We use the subscript 3 throughout to remind the reader of the sample size of this particular Gibbs sampler. Similarly, $(\,\overline{x^2}\,)_3 = 27.67$, $(\,\overline{y^2}\,)_3 = 0.22$, and $(\,\overline{xy}\,)_3 = 2.47$, returning the MC estimates of the variances of x and y as

$$\text{Var}(x)_3 = (\,\overline{x^2}\,)_3 - (\overline{x}_3)^2 = 2.67, \qquad \text{Var}(y)_3 = (\,\overline{y^2}\,)_3 - (\overline{y}_3)^2 = 0.25$$

and their covariance as

$$\text{Cov}(x,y)_3 = (\,\overline{xy}\,)_3 - \overline{x}_3 \cdot \overline{y}_3 = 2.47 - 5 \cdot 0.46 = 0.16$$

One of the most powerful uses of the sampler is to obtain the distribution of complex functions of the sampled random variables. For example, while the mean and variances of $f(x,y) = x/y$ can be approximated using the Delta method (Equations A1.19a and A1.19b), obtaining its sampling distribution is nontrivial. However, it *is trivial* to use the chain to sample values for the distribution of *any* $f(x,y)$. For our toy sampler (Example A8.6), the first three draws of x/y are 5/0.5 = 10, 3/0.33 = 9.09, and 7/0.56 = 12.5. Continuing in this fashion generates an empirical distribution for this function. Obviously, the same strategy works for any vector-valued function of the sampled parameters.

While computing the MC estimate for any *moment* using the sampler is straightforward, computing the actual *shape* of the marginal density is slightly more involved. While one might use the empirical histogram of the Gibbs sequence as a rough approximation of the marginal distribution of x, this turns out to be inefficient, especially for obtaining the tails of the distribution. A better approach is to use the average of the conditional densities $p(x \,|\, y = y_i)$, as the functional form of the conditional density contains more information about the shape of the entire distribution than the sequence of individual realizations, x_i (Gelfand and Smith 1990; Liu et al. 1991). Because

$$p(x) = \int p(x \,|\, y)\, p(y)\, dy = E_y\left[\, p(x \,|\, y)\,\right] \qquad \text{(A8.26a)}$$

one can approximate the marginal density using

$$\widehat{p}_m(x) = \frac{1}{m} \sum_{i=1}^{m} p(x \,|\, y = y_i) \qquad \text{(A8.26b)}$$

Example A8.8. Returning to the Gibbs sequence generated in Example A8.6, recall that the distribution of x given y is binomial, with $x \mid y \sim \mathrm{B}(n, y)$. If we apply Equation A8.26b, the estimate (based on this sequence) of the marginal distribution of x is the weighted sum of three binomials with success parameters of 0.5, 0.33, and 0.56, yielding

$$p_3(x) = \frac{10!}{x!(10-x)!}\left[\frac{0.5^x(1-0.5)^{10-x} + 0.33^x(1-0.33)^{10-x} + 0.56^x(1-0.56)^{10-x}}{3}\right]$$

As Figure A8.3 shows, the resulting distribution (solid bars), although a weighted sum of binomials, departs substantially from the binomial based on the average (here 0.46) of the success parameter (open bars).

Appendix 9

Experimental Design

We often think of experimental design as analogous to recipes in a cookbook. I prefer to think of [it] as a creative series of decisions that are meant to solve one or more problems. Casler (2015)

When dealing with situations where genotypes can be replicated—such as RILs and NAMs for QTL mapping (Chapter 18), line-cross analysis with inbreds (Chapter 11), or probing G $\times$ E (Part II Chapters 27 and 28)—the use of designs that control environmental heterogeneity increases the precision of estimates of features from the replicated unit. There is a massive literature on this subject and we only consider a few of the key concepts here, largely framed in the context of field crop designs that arise in agronomy. Many of these same issues of replication over potentially heterogeneous measurement settings arise in other areas in quantitative genetics, such as in functional genomics experiments (e.g., Yang and Speed 2002). Reviews of general agricultural experimental design can be found in LeClerg et al. (1962), Little and Hillis (1978), Pearce (1983), Gomez and Gomez (1984), Pearce et al. (1988), Mead (1990, 1997), Petersen (1994), John and Williams (1995), Kempton and Fox (1997), Kuehl (2000), Quinn and Koeugh (2002), Piepho et al. (2003), Giesbrecht and Gumpertz (2004), Hinkelman and Kempthorne (2006), Casella (2008), and Mead et al. (2012). A very nice overview of the art of experimental design is given by Casler (2014).

THE NATURE OF PLOTS

The basic **experimental unit** that is replicated when testing lines is a **plot**, a rather vague term that can encompass a multitude of situations, depending upon the biology of the species of interest coupled with logistical and financial concerns. We use the term *plot* for the unit sampled to obtain a mean value for a single replicate of a given line (or genotype). Using a plot average significantly improves the precision of an estimate of the genotypic value of a line. In particular, the error in using the mean value of a plot as an estimate of the true line mean is σ^2_w/n, where σ^2_w is the within-plot variance and n is the number of individuals within a plot. Another motivation for the use of plots is convenience, as it may be difficult to score individual plants, while a plot average is much easier to obtain (this is especially true for cereals and forage crops). In some cases, such as for trees or large shrubs, there may only be a single individual per plot. More generally, a plot could be a row, a set of rows, or some other rectangular configuration of n individuals from a line.

The major tradeoff with any plot design is that the larger the plot size, the greater the chance that individuals within a plot experience significant environmental heterogeneity (and hence an increased value of σ^2_w). Historically, the trend in plant breeding has been a move toward smaller and smaller plots. In the early 1900s, plot sizes for field crops in the United States ranged from 1/40 of an acre to 2 acres, with an average size around 1/10 of an acre (Taylor 1908). The work of Mercer and Hall (1911) led to the use of much smaller plots, while Woods and Stratton (1910) strongly promoted the use of multiple smaller plots in place of a single larger plot. The **rod-row** plot (Norton 1907), which involves rows of rod length (roughly 16 feet or 5 meters) as the basic unit of replication, appears frequently in the classical plant breeding literature. Rod-row and field (i.e., more rectangular) plots generally show good agreement (LeClerg 1966). Plots of even smaller sizes are used when seed or space is limited, such as **short-row** (1.5 meters; St. Martin et al. 1990) or **hill plots** (Bonnett and Bever 1947). The latter (commonly used with cereals) typically involves digging a small hole, which is filled with seed and then covered with soil, although a number of variants on this approach have been used (Hendriksen 1956; Frey 1965). The correlation between trait performance in hill plots versus rod-row settings is mixed (Jellum et al. 1963;

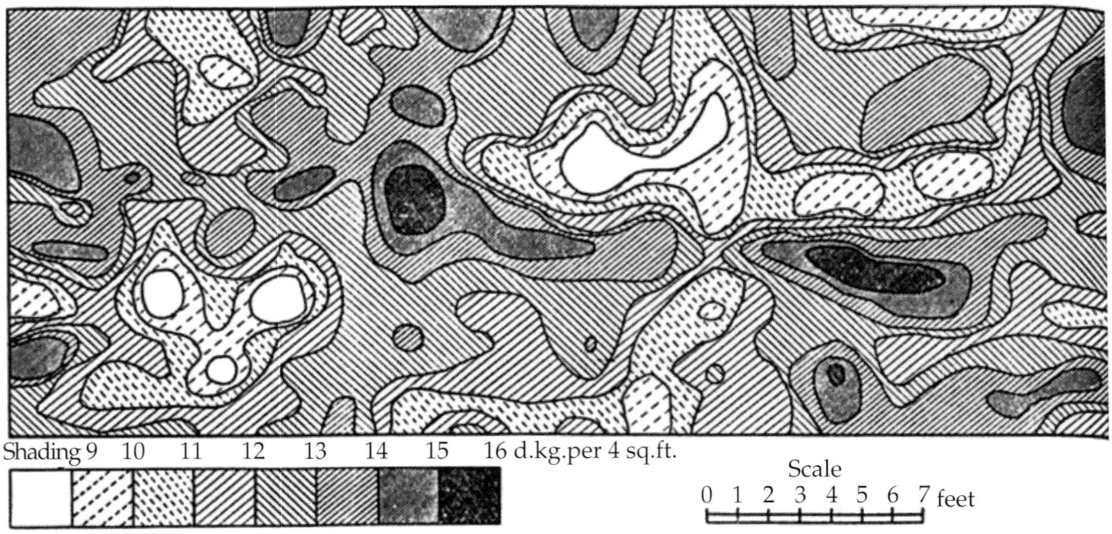

Figure A9.1 A fertility contour map for yield in the wheat variety Waratah grown in a field in Canberra, Australia, in 1934. The map was smoothed using moving averages based on two-foot by two-foot squares. (After Smith 1938.)

Frey 1965; Weber 1984; St Martin 1990; Tragoonrung 1990), leading Frey (1965) to conclude that while hill plots may be useful in preliminary screening (where one is sorting through a large number of lines, often with only a small amount of seed for each), final screening is best performed under conditions as close to agricultural production practices as possible.

One metric for quantifying optimal plot size is based on the **Fairfield–Smith empirical law** (Smith 1938; also see Zhang et al. 1994). This "law" states that the relationship between the variance in yield for a plot of size s, σ_s^2, and the variance of a plot of unit size, σ_1^2, is often of the form

$$\sigma_s^2 = \sigma_1^2 / s^b \tag{A9.1a}$$

or, in log-linear form, which we denote as the **Fairfield–Smith regression**,

$$\ln\left(\sigma_s^2\right) = \ln\left(\sigma_1^2\right) - b\ln(s) \tag{A9.1b}$$

The fitted regression coefficient, b, provides insight into environmental heterogeneity. The smaller the value of b, the higher the correlation between adjacent regions. Conversely, values of b approaching one indicate either a homogeneous environment across the field or *random* heterogeneity between neighboring locations.

There is also the issue of *plot shape*. While the results in the literature are somewhat mixed, the majority point to an advantage of oblong plots over square plots (LeClerg 1966), consistent with the prediction on theoretical grounds that, under patchy soil heterogeneity, oblong plots generally have lower variation than square plots (Dendrinos 1931). This superiority critically depends on using the correct *orientation* of the oblong plots, namely that they are orthogonal to any environmental gradient. Plots of such orientation have higher between-plot, and lower within-plot, variances than do plots whose individual rows align with a gradient (Smith 1938).

For any given experiment, optimal plot shape is thus an empirical issue and can be addressed using either knowledge from previous trials or by performing a **uniformity trial** (Cochran 1937b). Under the latter, the same genotype is grown over a field of interest, and its performance recorded on a very fine scale. Smoothing the local values returns a heat map of field performance (Figure A9.1), which can suggest the optimal plot size (and how it may vary over a field), as well as the nature and direction of any environmental gradients. Such a trial also provides the data to fit a Fairfield–Smith regression. A slightly more formal approach for obtaining the optimal plot configuration is to choose the plot size and shape that has the smallest coefficient of variation in performance. The caveat with using information on prior performance or from a uniformity trial is the potential of G $\times$ E, as environmental gradients might change over the years or seasons, or over different lines, or over different traits. Despite such concerns, a uniformity trial is usually a very sensible first step, especially for plots in an experimental station that are expected to be used over years (or even decades).

There are also concerns about the nature of plots used for genotype testing versus the nature of plots grown for agricultural production. Again, these are G $\times$ E issues: is the testing performance in one environment (say rod-row) a good predictor of the performance in another (the production environment)? For example, a plot consisting of only a single genotype has a very different competitive environment than a plot consisting of a mixture of genotypes. When multiple lines are being simultaneously tested, **guard rows** between pure-line plots are often included. While these are discarded from the analysis, they serve as a barrier to between-line competition, which can alter yield and other traits. Likewise, a plot often consists of 3–5 rows, with the **border rows** being excluded from the analysis. Border and guard rows should contain a mixture of all lines in the treatment to equalize competitive effects (Federer and Basford 1991). One advantage of square plots over oblong ones is that the latter usually have a higher fraction of the material subject to border effects. Portman and Ketata (1997) discuss a number of these issues further.

Finally, just how many plots per genotype (r) are needed? The obvious answer is as many as possible, but, of course, time and resources impose constraints. Letting σ^2 denote the plot-to-plot variance in the mean value for a line, then the standard error for the line mean becomes $\sigma/\sqrt{r}$. In selecting among genotypes, the more relevant question is the standard error associated with the *difference* between two genotypes (or, more generally, treatments). Under the assumption of homoscedasticity (all the genotypes have the same expected plot variance), the sampling variance of any single direct comparison (i.e., both genotypes are in the same block) between two genotypes is $2\sigma^2$ (as each mean has a variance of σ^2). The standard error based on r direct comparisons between two genotypes thus becomes

$$\sigma_d = \sqrt{\frac{2\sigma^2}{r}} \tag{A9.2a}$$

Mead (1997) suggested that to detect a difference of δ between two genotypes requires $3\sigma_d < \delta$. Given a previous estimate, S, of σ_d, the number of required replications becomes

$$r > 18\, S^2/\delta^2 \tag{A9.2b}$$

However, as noted by Mead, formal significance testing in selection decisions is tempered by the expectation that there are true differences between the tested genotypes, and our goal is simply to choose some preset fraction of the best of these. In this setting, Mead suggested choosing r such that $\sigma_d \leq D/6$, where D is the difference between the performance of the *best genotype* in a sample and the *mean performance* in that sample, yielding

$$r > 36 \cdot S^2/D^2 \tag{A9.2c}$$

While the above theory seems straightforward, as Figure A9.1 shows, fields and other experimental settings can be very heterogeneous, even when they superficially appear quite uniform. We now review the basic theory, and a number of specific designs, that attempt to account for the fact that the environment might be structured (i.e., show between-plot correlations). In such settings, the way that genotypes are distributed over structured groups of plots is as important as the number of plots for any given genotype.

FISHER'S PRINCIPLES: REPLICATION, RANDOMIZATION, AND LOCAL CONTROL

Because only an average is measured, *a single plot provides no information on the standard error of the estimated line mean*. Hence, plots themselves must be *replicated*. How such replication should occur is the basis for the massive field of **experimental design**, started (as with many important statistical concepts) by R. A. Fisher. The theoretical foundation of experimental design rests on **Fisher's principles** (1926, 1935; also see Yates 1964 and Box 1980): (i) **replication**; (ii) **randomization**; and (iii) **local control** (or **blocking**). Fisher stressed *replication* of plots (experimental units) to allow for ***estimation of the standard error***. Replication also reduces the error variance through multiple, independent, measurements.

Fisher also stressed *randomization*, the notion that lines should be ***randomly assigned*** within a block to control for any systematic field effects (such as soil moisture or pH gradients). Before Fisher, plots were often laid out in some systematic (as opposed to random) pattern, such that there was often a pattern between replications (e.g., each line was exactly four plots apart). If there are systematic features in the environment, these could potentially interact with the systematic arrangement of replications within a design, confounding the interpretation. With randomization, any systematic features of the environment are now randomized over replicate plots, diminishing their impact.

Local control is the idea that all lines within the same experimental **block *essentially experience the same conditions***, namely, that the block area is regarded as small relative to the scale of field heterogeneity. Hence, the purpose of any particular **blocking structure** is to decompose an experimental setting (such as a field) into a series of blocks that capture much of the heterogeneity in the sense of having a much larger between-block, than within-block, environmental variance. Blocks can also account for sources of variation beyond those from heterogeneity in field conditions; in particular, **operator variation**. If the workers or equipment used for planting the seed, measuring the trait, or plowing the field differ over plots, these sources of error can be controlled by blocking, so that plots in the same block all experience the same operator. Likewise, logistical limitations may prevent performing the same operation on all blocks on the same day, potentially generating day-specific effects. Such operator effects can be rather dramatic, and hence best practices are to ensure that all plots within a block have the same operator, so that (for example), all are harvested on the same day by the same operator. Indeed, even if one assumes a homogeneous field, if it is sufficiently large that not all operations can be done on a single day (or even a single time, such as early morning), then it is best to block the field into sections that are all operated on at the same time. One can always test for block effects, and remove blocks from the model analysis if they are not significant.

COMPLETE BLOCK DESIGNS

Completely Randomized Designs (CRDs)

In an idea setting, field conditions are such that a number of **experimental replicates** (the set of the entire collection of tested lines/treatments) can all occur within a single block. This is the setting for the **completely randomized design** (**CRD**), wherein a set of g lines (genotypes) are each replicated r times over some array of gr plots (such as, but not limited to, a $g \times r$ grid) that is treated as a single block. One allocates genotype plots randomly over the array, and then simply uses a standard one-way ANOVA,

$$y_{ij} = \mu + G_i + e_{ij} \qquad 1 \le i \le g, \quad 1 \le j \le r \tag{A9.3}$$

where the residual errors, e_{ij}, are assumed to be uncorrelated and homoscedastic, namely, $\sigma(e_{ij}, e_{kl}) = \sigma_0^2$ if $ij = kl$, otherwise this covariance is zero.

The use of a CRD in the presence of significant environmental heterogeneity results in higher variances associated with the line means over the entire replicate, as plot-to-plot variation is not accounted for, and thus appears in the line-mean error variance. In particular, if σ_0^2 denotes the variation over plots, and σ^2 ($< \sigma_0^2$) denotes the between-plot variation under more uniform environmental conditions, then $\sigma_0^2 = \sigma_b^2 + \sigma^2$. A standard metric for comparing designs is the variance associated with the difference between two tested genotypes, which (from Equation A9.2a) becomes $2\sigma_0^2/r$ under a CRD.

One potential setting for a CRD is for the screening of a very small number of lines. As the number of lines to be screened increases, so does the area over which they are planted, increasing the risk of field heterogeneity. While appropriate field preparation can reduce some of this variation (e.g., by the use of mechanical seeding devices to ensure uniformity of planting or laser-guided plowing for a field of uniform flatness), other sources of variation (e.g., soil heterogeneity) are generally beyond the direct control of the experimenter. The CRD setting is thus unrealistic beyond even a few genotypes, as environmental heterogeneity is expected to increase with block size.

Randomized Complete Block Designs (RCBDs)

The simplest setting when environmental heterogeneity is present is that there is a strong gradient in one direction. In this setting, with a relatively modest number of genotypes, one possible option is a **randomized complete block design** (**RCBD**; or simply a **complete block design**, **CBD**), a set of r blocks, each of which contains all of the g genotypes (i.e., each block is **complete**, consisting of a full experimental replicate, namely, all of the lines). The model now includes a block effect (b), with the ANOVA becoming

$$y_{ij} = \mu + G_i + b_j + e_{ij} \qquad 1 \leq i \leq g, \quad 1 \leq j \leq r \tag{A9.4}$$

Depending on the assumptions that one is willing to make, blocks can be modeled as either fixed or random effects. If there are just a few blocks, treating them as fixed is best, but with a number of blocks, treating them as random reduces the degrees of freedom used (from r under fixed-effects down to one for the simple model of uncorrelated random block effects).

As above, let σ^2 and σ_0^2 denote, respectively, the among-plot variance within a block and the among-plot variance in the absence of blocking (i.e., the CRD design). When the estimates of these two variance components are very close, one can analyze the experiment as a CRD. This illustrates the tradeoff with any structured design. Adding structure (e.g., blocks) accounts for between-block variance (σ_b^2) that otherwise would be incorporated into the residual variance, but does so at the cost of the additional degrees of freedom required to estimate the block effects. The **relative efficiency**, E^*, of an RCBD with respect to a CRD is defined as the ratio for the expected variances for the difference between the line means of two genotypes, or from Equation A9.2a,

$$E^* = \frac{\sigma_d^2(CRD)}{\sigma_d^2(RCBD)} = \frac{2\sigma_0^2/r}{2\sigma^2/r} = \frac{\sigma_0^2}{\sigma^2} = 1 + \frac{\sigma_b^2}{\sigma^2} \tag{A9.5}$$

When $\sigma_b^2 > 0$, then for the same amount of replication, the standard errors for differences in line means are smaller under an RCBD ($E^* > 1$).

Row–column and Latinized Designs

While the RCBD allows for environmental heterogeneity, it assumes that this variation is largely in one direction, and that the blocks are chosen to capture this heterogeneity (i.e., reduce it within each block). If the environment shows heterogeneity in more than one direction (e.g., Figure A9.1), one option is a **row–column** (**RC**) design. Here, one models row (rb) and column (cb) effects as blocks. For replication j, the model becomes

$$y_{ijk\ell} = \mu + G_i + R_j + rb_{jk} + cb_{j\ell} + e_{ijk\ell} \tag{A9.6}$$

where $y_{ijk\ell}$ denotes the phenotype of genotype i that is in row k and column ℓ of replicate j. As was the case with the block variance for an RCBD, the sizes of the row and column variances indicate whether one can use a less-structured design for analysis. For example, if the row variance is small, then one could employ an RCBD using the column blocks.

A specific RC design is the **Latin square**, a $g \times g$ array of plots where each line appears only once in the same row and only once in the same column, so that the number of

Table A9.1 A few important incomplete block designs. Here g is the number of lines (genotypes), r is the number of replications per genotype, and k ($< g$) is the number of plots per block. For resolvable designs, s is the number of blocks per experimental unit replicate. For balanced designs, λ is the number of direct comparisons between two lines. The values of g, s, r, k, and λ must all be integers. E is the Yates efficiency factor. See the text for full details.

General balanced design		
Not necessarily resolvable	$\lambda(g-1) = r(k-1)$	$E = \dfrac{g(k-1)}{k(g-1)}$
Balanced lattice		
Resolvable, balanced	$g = k^2, s = k, r = k+1$	$E = \dfrac{k}{k+1}$
Simple lattice		
Resolvable, unbalanced	$g = k^2, s = k, r = 2$	$E = \dfrac{k+1}{k+3}$
Triple lattice		
Resolvable, unbalanced	$g = k^2, s = k, r = 3$	$E = \dfrac{2k+2}{2k+5}$
α-lattice		
Resolvable, unbalanced	$g = s \cdot k$	$E \leq \dfrac{(g-1)(r-1)}{(g-1)(r-1) + r(s-1)}$

lines equals the number of replications. The Latin square is the basis of the popular game of Sudoku, in which the same integer may not appear twice in any row or column. Again, because of the expected scaling of environmental heterogeneity with plot size, a Latin square is only appropriate for testing a small number of genotypes. More generally, with **incomplete blocks** (wherein a block does not contain all genotypes), RC designs in which the same line does not occur more than once in a column or a row are said to be **latinized**. One example is a **Youden square**, a Latin design with either rows or columns missing relative to the full design. Piepho et al. (2015) review a number of other latinized designs.

INCOMPLETE BLOCK DESIGNS

Basic Theory

CRD, RCBD, and Latin squares are examples of **complete block designs**, wherein each block contains a full set of the treatments (lines), so that each block corresponds to one complete experimental replicate. These designs may be appropriate after a large number of initial lines are whittled down to a much smaller, more manageable, set of final candidates. However, in the early phases of testing, typically one has a rather large number of lines. In such settings, only a subset of lines can be compared within a block that is assumed to be sufficiently homogeneous (i.e., provides for sufficient local control), and each experimental replicate now consists of a number of blocks. This setting is called an **incomplete block design**, as not all lines appear in all blocks (Yates 1936b). As one might imagine, there are a large number of such designs, with Table A9.1 summarizing those discussed here.

In the CRD setting, one can imagine starting with a deck of gr playing cards (line plots), where there are g different genotypes, each occurring r times. These are shuffled and then randomly placed in the slots of the design grid. Under the RCBD setting, for each of r different grids, one shuffles a set of g cards, and then assigns them at random within each block. Latin squares are a bit more restrictive. Within each replicate, one randomly chooses a genotype, and then randomly assigns g copies in the $g \times g$ array such that no copy appears in either the same row or same column twice. Once the first genotype is assigned, this process continues with the second genotype until the Latin square is complete.

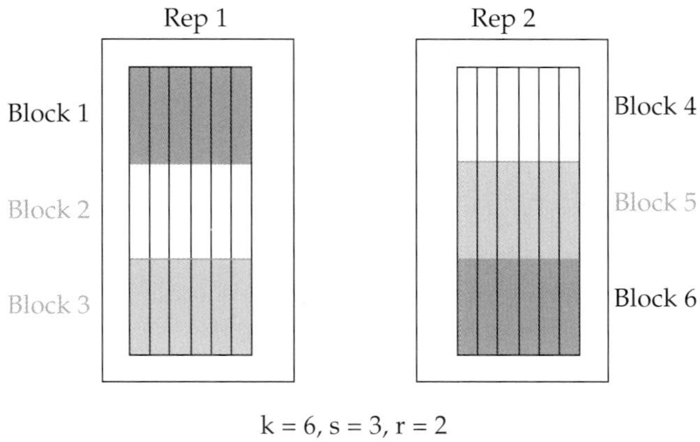

Figure A9.2 An incomplete block design with two replicates ($r = 2$), six plots per block ($k = 6$), and three blocks ($s = 3$) per replicate. For $g = 18$ genotypes, this is a resolvable design, as $sk = g$.

All of these complete block designs impose rather stringent restrictions between number of genotypes (g), the number of replicates per genotype (r), the total number of blocks (n_b), and the number of plots per block (k). Under a CRD, $n_b = 1$ and $k = gr$ (a single block contains all the replicates of all genotypes), while a RCDB takes $n_b = r$ and $k = g$ (r blocks, each of which a contains a complete replicate of all genotypes). In a general incomplete block setting (with each genotype replicated exactly r times), the initial restriction is that

$$gr = n_b k \tag{A9.7a}$$

where $k < g$, namely that the number of plots within a block is less than the number of genotypes being examined. The resulting model (under a resolvable design, see below) for the plot value for line i in the kth plot of replicate j is

$$y_{ijk} = \mu + G_i + R_j + b_{jk} + e_{ijk} \tag{A9.7b}$$

the effects (b_{jk}) of blocks nested within a complete replicate (j) are usually taken as random, while the complete-replicate effects (R_j) are usually taken as fixed.

Because all of the variables in Equation A9.7a must be integers, this imposes limitations on the possible array configurations that one could construct. There is greatly increased flexibility if we allow *unequal replication* (some genotypes are replicated more than others), an issue we return to shortly. As an example of a potential array under an incomplete block design, consider the situation in Figure A9.2, which illustrates an array for $g = 18$, $r = 2$, $k = 6$, and $s = 3$ (three blocks per each complete replicate set of lines).

The construction of incomplete block designs involves two major issues. First, what are the possible array configurations given our choice of g and the total number of plots to score ($n_b\, k$), an issue we return to shortly. Second, how do we best allocate genotypes over such an array. One obvious principle is that of **evenness**, which has two components. First, if possible, equal replication of all genotypes is desired. Second, because the goal of the design is to compare the performance of different genotypes, any genotype should be directly compared against all others (i.e., jointly appear in the same block) in as *even a fashion as possible*.

Consider the following design with $g = 6$ genotypes (A through F), $r = 4$ replicates, $k = 4$ plots per block, $n_b = 6$ blocks (the columns labeled 1 through 6):

1	2	3	4	5	6
A	A	A	A	B	C
B	B	B	C	D	D
C	C	D	E	E	E
D	E	F	F	F	F

In considering comparisons involving line A, A–B and A–C have three direct comparisons, while A–D, A–E, and A–F have two direct comparisons (units together in the same block). Letting σ^2 denote the between-plot variance within each block, from Equation A9.2a the variance associated with the first two pairs of differences is $(2\sigma^2)/3$, while the variance associated with the remaining two pairs is $(2\sigma^2)/2$. Averaging over all pairs, the average variance for direct comparisons is

$$\frac{1}{2}\frac{2\sigma^2}{3} + \frac{1}{2}\frac{2\sigma^2}{2} = \frac{2}{3}\sigma^2$$

This design (wherein any particular genotype is directly compared three times with half of the genotypes and compared twice with the other half), is the most optimal, in that no other design gives a smaller average pairwise comparison variance (Mead 1997). This immediately suggests one metric for comparing different designs, namely the average variance for any pairwise difference. Among the set of designs that are feasible given an investigator's limitations of time and resources, that with ***the smallest average variance of pairwise differences is regarded as being most optimal***.

Indirect Comparisons

An important concept relevant to incomplete block designs is the notion of **indirect comparisons**. In the previous design, each line is compared directly (i.e., within the same block) with every other line, although the number of replications may vary. However, if the number of genotypes is large relative to the number of plots per block, some genotypes may *never* appear together in the same block. Suppose we have two blocks of size three and five genotypes, with the resulting blocks being A B C and C D E. Although there are no *direct* A–D or A–E comparisons, we can indirectly estimate A–D from the difference of A–C in block one and C–D in block two. This follows from our linearity assumption. The variance associated with this indirect estimate is $2(2\sigma^2)$, which is twice as large as that for a direct (i.e., within the same block) comparison. Hence, although we can estimate differences between genotypes that never appear together in the same block, comparing them through intermediates results in a higher sampling variance.

Yates Efficiency Factor, *E*

Returning to the metric of the average variance for all pairwise differences, the **Yates efficiency factor, *E***, (Yates 1936a, 1936b) of an incomplete block design is simply the ratio of its average pairwise difference variance relative to that for an RCBD. Table A9.1 gives E values for a number of classic designs. E values are obtained under the assumption that the within-block variance does not change over block size, and hence simply examine *the cost of the imposed blocking structure*. The **relative efficiency**, E^*, takes into account the expected smaller within-block variances (σ^2) under an incomplete design relative to those under an RCBD (whose baseline value we now denote as σ_0^2),

$$E^* = E \cdot \sigma_0^2/\sigma^2 \tag{A9.8}$$

The cost of blocking (E) is usually more than offset by its smaller within-block variance, so that $E^* > 1$. For example, a design with $E = 0.5$ is more efficient than an RCBD when its within-block variance is less than half that for the block sizes required for as RCBD ($\sigma_0^2 > 2\sigma^2$).

Balance and Resolvablity

We now use the above theory to examine a few incomplete block designs. The basic classification of such designs involves whether they are resolvable and whether they are balanced. In a **resolvable** design, the collection of small incomplete blocks within each complete replication contains all of the treatments. In a **balanced** incomplete block design (**BIBD**), each pair of treatments has the same number of direct comparisons (i.e., the two lines appear together in the same block). In a **partially balanced** incomplete block designs (**PBIBD**), some pairs have c_1 direct comparisons, while the others have c_2 comparisons.

One advantage of using a resolvable design is that if an experimental replicate is lost (for example, each replicate is harvested on a different day, and a weather event compromised the harvest of replicate 3), the design is still valid. With an unresolvable design, the loss of a set of blocks may severely compromise the design. Another advantage of a resolvable design is that one may be examining multiple traits in a given experiment, some of which are only measured in a subset of replicates (due to financial or logistic considerations). In this setting, the design still remains valid, potentially being balanced for some set of traits and unbalanced for others.

One significant constraint with a balanced incomplete block design is that it imposes a relationship between the number of treatments (here, lines, g), the number of replications per treatment (r), the number of plots per block (k), and the number (λ) of times that a specific pair of treatments occurs together in the same block, with Yates (1936b) noting the restriction

$$r(k-1) = \lambda(g-1) \tag{A9.9a}$$

Yates (1936b) showed that the efficiency factor of a design satisfying Equation A9.9a is

$$E = \frac{g(k-1)}{k(g-1)} \tag{A9.9b}$$

For example, with 10 lines ($g = 10$) and four plots per block ($k = 4$), $r = \lambda(9/3) = 3\lambda$. Hence, for a design with $\lambda = 2$ (two within-block comparisons for each pair over the entire design), then $r = 6$, which requires $gr = 60$ total plots. Equation A9.9b gives the efficiency of this design relative to a RCBD as $E = 0.833$. If one instead wants to compare 25 genotypes, with five plots per block ($g = 25, k = 5$) and with $\lambda = 2$, then $r = 2(24/4) = 12$, giving a design with $gr = 300$ total plots, and (from Equation A9.9b) a value of $E = 0.833$. A 250% increase in the number of genotypes results in a 500% required increase in the number of plots.

A further consideration which also limits design flexibility is that the ratio $(g-1)/(k-1)$ *must be an integer*. Hence, the number of genotypes constrains the number of plots per block, $c(k-1) = (g-1)$, where c is some integer. For example, with 25 genotypes, k must satisfy $c(k-1) = 24$, which is only satisfied for blocks of size $k = 2, 3, 4, 7, 9$, and 25.

Lattices: Balanced, Simple, and Triple

One standard BIBD is the **balanced lattice design** (Yates 1936a). For a trial involving nine genotypes (A through I), this has the form

Rep 1			Rep 2			Rep 3			Rep 4		
A	B	C	A	D	G	A	B	C	A	B	C
D	E	F	B	E	H	E	F	H	F	D	E
G	H	I	C	F	I	I	G	D	H	I	G

where the three columns of each replicate correspond to the blocks ($s = 3$). Note that the first column (Block) of Rep. 1 contains genotypes A, D, G, but their actual order within the block is randomized in any actual implementation of this design. Further note that A is contrasted with all other genotypes over the design (A with D and G in Rep 1, A with B and C in Rep 2; A with E and I in Rep 3; and A with F and H in Rep 4). While this design is both balanced and resolvable, to achieve these properties imposes a number of constraints.

First, the number of genotypes must be the square of some integer, $g = k^2$, in which case k becomes the number of plots per block (three in this example). This immediately implies that the number of blocks per replicate, s, also equals k. Finally, in order to achieve balance, the number of replicates becomes $r = k + 1$. Hence, the array structure imposed by the simple lattice design becomes

$$g = k^2, s = k, r = k + 1 \tag{A9.10a}$$

Yates (1936b) showed that efficiency factor for this design is

$$E = \frac{k}{k+1} \tag{A9.10b}$$

For example, if $g = 9$, then $k = 3, s = 3$, and $r = 4$, for a total of 36 plots. Equation A9.10b gives the efficiency factor as $E = 0.75$, so that the variances for any pair of differences are expected to be $1/E$ = 1.33 times larger than those for an RCBD (assuming that the between-plot variances with the blocks are the same under either design). One advantage of a resolvable design is that if the variance of the block effects within a replicate is small, one can revert to an RCBD analysis, treating each experimental replicate as a single block.

The restrictions imposed by Equation A9.10a simply indicate the *structure* of the arrays, not how particular genotypes are *arranged* within this structure. The assignment of genotypes into slots in the design array to achieve balance and resolvability is typically performed using a computer (e.g., the `R` library `agricolae`). The balanced lattice is an example of a **cyclic design**, wherein genotypes are allocated into blocks by using simple shifts from an initial starting block (with the locations of chosen genotypes then randomized within the block). The standard reference for such designs is John and Williams (1995).

Suppose that we again have nine genotypes, but only have the resources for two replicates. If we remove (say) replicates 2 and 4 from the above balanced lattice, it is still resolvable, but now becomes unbalanced. This is an example of a **simple lattice design**, where the value of r in Equation A9.10a is 2, while a **triple lattice design** takes $r = 3$. Table A9.1 gives the efficiency factors for both of these designs. In a review of over 80 lattice experiments, Ma and Harrington (1948) found that the double and triple lattice designs had relative efficiency values (E^*) of 128 and 160% (respectively) over the comparable complete block designs.

Lattices: α-Designs

The balanced, simple, and triple lattice designs all have the restriction that g must be the square of the block size, a significant limitation. In theory, if one has a value of g that is not a perfect square, one could add **checks** (known genotypes for comparisons, such as existing elite lines) into the design, but this is inefficient. Further, when the number of genotypes is modest to large, such as in the first phases of line selection, the number of genotypes to be tested usually makes these designs unfeasible. A much more flexible approach was suggested by Patterson and Williams (1976), which only imposes the constraint that

$$g = sk \tag{A9.11a}$$

Namely, number of genotypes (g) is the product of plots per block (k) and the number of blocks per experimental replicate (s). This family of designs, which is unbalanced but resolvable, is called an **α-lattice**. Patterson and Williams found that the efficiency factor for such a design has an upper bound of

$$E \leq \frac{(g-1)(r-1)}{(g-1)(r-1) + r(s-1)} \tag{A9.11b}$$

While it is an upper bound, Equation A9.11b is usually quite close to the true value.

Given values of g and r (and hence a limitation of gr total plots), Equation A9.11a suggests a number of possible arrays, whose efficiencies can be compared using Equation

A9.11b (Example A9.1). One typically chooses the α-design with the largest value of k (plots per block) that still allows the plots to be relatively homogeneous within each block. While a value of g, s, and k sets the array configuration, the assignment of genotypes into blocks in the array is done by specialized programs that ensure maximal efficiency and resolvability (e.g., the `R` library `agricolae`). All of the details underlying the algorithms to generate such designs under randomization (which often involves a series of cyclic substitutions) can be found in John and Williams (1995). Patterson and Williams (1976) noted that the restriction given by Equation A9.11a can be modified to allow for two block sizes, with $g = s_1 \cdot k + s_2 \cdot (k-1)$, namely s_1 blocks per replicate of k plots, with the remaining s_2 blocks of size $k-1$. These are called **almost equiblock sized α-designs** (**AE α-designs**).

Because α-designs are unbalanced, they are denoted as an **α(0,1)-lattice** when there are either only zero or one direct comparisons among all of the pairs, while an **α(0,1,2)-lattice** design denotes that some pairs have two direct comparisons. Patterson and Williams (1976) stress that $\alpha(0,1)$ designs are preferred, and if one has to use an $\alpha(0,1,2)$ design, then as few as possible of the pairs should be compared twice. They showed that a necessary (but not sufficient) condition for the existence of an $\alpha(0,1)$-lattice is that $k \leq s$ (the number of plots per block $\leq$ the number of blocks per replicate). If an $\alpha(0,1)$-lattice does not exist for a given configuration of g, s, k (i.e., some pairs are directly compared more than once), then an $\alpha(0,1,2)$-lattice usually exists.

Example A9.1 Suppose we have 132 genotypes ($g = 132$) and wish each to be replicated twice ($r = 2$, requiring $gr = 264$ plots). In most cases, the local scale of field variation is expected to be such that a block of 132 plots is likely heterogeneous, so using an RCBD is ruled out. The nearest square is 144, so that using a simple lattice ($r = 2$) requires $k = \sqrt{144} = 12$ plots per block. This block size might still be too large relative to the scale of field heterogeneity. Further, one must add 12 checks at random to complete the square, adding an additional 24 plots. From Table A9.1, the efficiency of this simple lattice design is

$$E = \frac{k+1}{k+3} = \frac{13}{15} = 0.867$$

Alternatively, the possible α-designs for this value of g are given by all those integer combinations of k (the number of plots per block) and s (the number of blocks per complete replication) that satisfy $sk = 132$. A table of these α-designs for ($k \leq 12$) and their associated E values (from Equation A9.11b) is as follows:

k	s	E	$1/E$	k	s	E	$1/E$
1	132	0.33	3.03	6	22	0.76	1.32
2	66	0.52	1.92	11	12	0.86	1.16
3	44	0.60	1.67	12	11	0.87	1.15
4	33	0.67	1.49				

While E increases (and $1/E$ decreases) with k, so does the expected within-block variance. Hence, the desired α-design would be the largest value of k that still gives a relatively homogeneous block. Information from past uniformity trials can assist in this choice. *After* an experiment is performed, the *relative efficiency* (Equation A9.8), E^*, of the design can be judged by comparing the within-block error variance under the design with the error variance generated if the same data were analyzed as an RCBD.

In a summary of 24 cereal variety traits in the UK, Patterson and Hunter (1983) found that the average relative efficiency of lattice designs was around $E^* = 1.3$, so that their use resulted, on average, in a 30% reduction in the between-line differences variance relative to an RCBD. While the effect was greatest for larger trials ($g > 50$), it was still substantial (average $E^* = 1.24$) for smaller traits ($g < 20$). Hoefler et al. (2020) found that the alpha-design was typically the best for medium-sized experiments.

DESIGN ANALYSIS USING GENERAL LINEAR MODELS

All of the above designs can easily be framed in terms of general linear mixed models, of which the general linear model is a special case (when all effects are assumed to be fixed). Because these models are examined in Chapters 10 and 20 (as well as Part II Chapters 27, 28, 32, 33; and WL Chapter 19), we only briefly restate a few results from mixed-model (MM) analysis before expressing some of the above results in MM form. This framework allows us to extend designs to more general settings, such as correlated plot residuals (below) or replication in the presence of G $\times$ E (Part II Chapters 27 and 28). Reviews of the uses of mixed models in field trials are provided by Smith et al. (2005), Cullis et al. (2006), and Smith et al. (2006). An especially nice compact overview is given by Piepho et al. (2003).

The initial analysis of many of the above designs occurs in the ordinary least-squares (OLS) framework, which assumes that residuals are uncorrelated and homoscedastic, namely, the covariance matrix for the residuals becomes $\mathbf{Var}(\mathbf{e}) = \mathbf{R} = \sigma_e^2\,\mathbf{I}$. Under this assumption, the solution to the vector, $\boldsymbol{\beta}$, of fixed effects in the linear model

$$\mathbf{y} = \mathbf{X}\boldsymbol{\beta} + \mathbf{e} \tag{A9.12a}$$

is called the **best linear unbiased estimator**, **BLUE**, and is given by

$$\widehat{\boldsymbol{\beta}} = (\mathbf{X}^T\mathbf{X})^{-}\mathbf{X}^T\mathbf{y} \tag{A9.12b}$$

where $\mathbf{A}^{-}$ denotes a generalized inverse (Appendix 3).

Under more general fixed-effects models, where the residual variances might vary over observations and/or be correlated across observations, the general least-squares (**GLS**) framework is used. Here $\mathbf{Var}(\mathbf{e}) = \mathbf{R}$, which is a symmetric positive-definite matrix (all the eigenvalues of $\mathbf{R}$ are positive; WL Appendix 5). One example where such models arise is when the spatial structure generates correlations between the residuals from adjacent plots within a block (e.g., Equation A9.17). The resulting BLUE is

$$\widehat{\boldsymbol{\beta}} = (\mathbf{X}^T\mathbf{R}^{-1}\mathbf{X})^{-}\mathbf{R}^{-1}\mathbf{X}^T\mathbf{y} \tag{A9.12c}$$

Choosing lines (selection) within the BLUE framework occurs by using BLUE estimates of the genotypic values. Rather than selecting lines based on $\overline{y}_{i\cdot}$ (the mean observed value for line i), one instead chooses lines based on their BLUEs, $\widehat{G}_i$, which are extracted from the appropriate elements of $\widehat{\boldsymbol{\beta}}$ in either A9.12b or A9.12c (Example A9.2). The BLUE approach treats the line values as fixed effects.

An alternative framework, for either genotypic values or block effects, is to consider these as *random effects*. Under the basic mixed linear model, there is a vector ($\boldsymbol{\beta}$) of fixed effects, a vector of random effects ($\mathbf{u}$), covariance structures of $\mathbf{e} \sim (\mathbf{0}, \mathbf{R})$ and $\mathbf{u} \sim (\mathbf{0}, \mathbf{C})$, where $\mathbf{e}$ and $\mathbf{u}$ are uncorrelated. The resulting general mixed model is given by

$$\mathbf{y} = \mathbf{X}\boldsymbol{\beta} + \mathbf{Z}\mathbf{u} + \mathbf{e}, \quad \mathbf{Var}\begin{pmatrix}\mathbf{u}\\ \mathbf{e}\end{pmatrix} = \begin{pmatrix}\mathbf{C} & \mathbf{0}\\ \mathbf{0} & \mathbf{R}\end{pmatrix} \tag{A9.13a}$$

The solutions are given by the mixed-model equations (Equation 10.27),

$$\begin{pmatrix}\mathbf{X}^T\mathbf{R}^{-1}\mathbf{y}\\ \mathbf{Z}^T\mathbf{R}^{-1}\mathbf{y}\end{pmatrix} = \begin{pmatrix}\mathbf{X}^T\mathbf{R}^{-1}\mathbf{X} & \mathbf{X}^T\mathbf{R}^{-1}\mathbf{Z}\\ \mathbf{Z}^T\mathbf{R}^{-1}\mathbf{X} & \mathbf{Z}^T\mathbf{R}^{-1}\mathbf{Z} + \mathbf{C}^{-1}\end{pmatrix}\begin{pmatrix}\widehat{\boldsymbol{\beta}}\\ \widehat{\mathbf{u}}\end{pmatrix} \tag{A9.13b}$$

Example A9.2. Consider a design with four genotypes arrayed in three blocks of three plots each ($g = 4$, $n_b = 3$, and $k = 3$). The genotypes are distributed over the blocks as (1,2,3), (4,1,2), and (2,3,4), with the model

$$y_{ij} = \mu + G_i + b_j + e_{ij} \qquad 1 \le i \le 4, \quad 1 \le j \le 3$$

The assumption is that of OLS, with **Var(e)** $= \mathbf{R} = \sigma_e^2\,\mathbf{I}$, and we consider genotypes and blocks as fixed effects. The linear model becomes $\mathbf{y} = \mathbf{X}\boldsymbol{\beta} + \mathbf{e}$, where

$$\mathbf{y} = \begin{pmatrix} y_{1,1} \\ y_{2,1} \\ y_{3,1} \\ y_{4,2} \\ y_{1,2} \\ y_{2,2} \\ y_{2,3} \\ y_{3,3} \\ y_{4,3} \end{pmatrix}, \quad \mathbf{X} = \begin{pmatrix} 1 & 1 & 0 & 0 & 0 & 1 & 0 & 0 \\ 1 & 0 & 1 & 0 & 0 & 1 & 0 & 0 \\ 1 & 0 & 0 & 1 & 0 & 1 & 0 & 0 \\ 1 & 0 & 0 & 0 & 1 & 0 & 1 & 0 \\ 1 & 1 & 0 & 0 & 0 & 0 & 1 & 0 \\ 1 & 0 & 1 & 0 & 0 & 0 & 1 & 0 \\ 1 & 0 & 1 & 0 & 0 & 0 & 0 & 1 \\ 1 & 0 & 0 & 1 & 0 & 0 & 0 & 1 \\ 1 & 0 & 0 & 0 & 1 & 0 & 0 & 1 \end{pmatrix}, \quad \boldsymbol{\beta} = \begin{pmatrix} \mu \\ G_1 \\ G_2 \\ G_3 \\ G_4 \\ b_1 \\ b_2 \\ b_3 \end{pmatrix}$$

Recall that the ith row of $\mathbf{X}$ corresponds to the ith observation in the $\mathbf{y}$ vector, while the jth column of $\mathbf{X}$ corresponds to the jth fixed effect in $\boldsymbol{\beta}$. For example, going down the columns of $\mathbf{X}$, we see two values of one in columns 2, 4, and 5, and three values of one in column 3, corresponding to two observations on genotypes 1, 3, and 4, and three observations on genotype 2. To see how the basic linear model is recovered, consider the 3rd row of the above set of equations. Matrix multiplication gives $y_{3,1} = \mu + G_3 + b_1 + e_{31}$. Similarly, the 6th row recovers $y_{2,2} = \mu + G_2 + b_2 + e_{22}$.

Now suppose that we have the same data, but treat block effects as random. The model now becomes $\mathbf{y} = \mathbf{X}\boldsymbol{\beta} + \mathbf{Z}\mathbf{b} + \mathbf{e}$, with **Var(b)** $= \mathbf{C} = \sigma_b^2\,\mathbf{I}$ (blocks assumed to be uncorrelated with a common among-block variance, σ_b^2), **Var(e)** $= \mathbf{R} = \sigma_e^2\,\mathbf{I}$, and (with $\mathbf{y}$ as above),

$$\mathbf{X} = \begin{pmatrix} 1 & 1 & 0 & 0 & 0 \\ 1 & 0 & 1 & 0 & 0 \\ 1 & 0 & 0 & 1 & 0 \\ 1 & 0 & 0 & 0 & 1 \\ 1 & 1 & 0 & 0 & 0 \\ 1 & 0 & 1 & 0 & 0 \\ 1 & 0 & 1 & 0 & 0 \\ 1 & 0 & 0 & 1 & 0 \\ 1 & 0 & 0 & 0 & 1 \end{pmatrix}, \quad \boldsymbol{\beta} = \begin{pmatrix} \mu \\ G_1 \\ G_2 \\ G_3 \\ G_4 \end{pmatrix}, \quad \mathbf{Z} = \begin{pmatrix} 1 & 0 & 0 \\ 1 & 0 & 0 \\ 1 & 0 & 0 \\ 0 & 1 & 0 \\ 0 & 1 & 0 \\ 0 & 1 & 0 \\ 0 & 0 & 1 \\ 0 & 0 & 1 \\ 0 & 0 & 1 \end{pmatrix}, \quad \mathbf{b} = \begin{pmatrix} b_1 \\ b_2 \\ b_3 \end{pmatrix}$$

As with $\mathbf{X}$, the ith row of $\mathbf{Z}$ corresponds to observation i, while its jth column corresponds to the jth random effect in $\mathbf{b}$. Again considering the 3rd and 6th row of the above set of equations recovers $y_{3,1} = \mu + G_3 + b_1 + e_{31}$ and $y_{2,2} = \mu + G_2 + b_2 + e_{22}$. The assumed simple covariance structure (independent block effects) requires only one degree of freedom (for estimating σ_b^2). However, as detailed below, the block covariance structure could be more complicated, such as correlations that diminish over blocks, e.g.,

$$\mathbf{C} = \sigma_b^2 \begin{pmatrix} 1 & \rho & \rho^2 \\ \rho & 1 & \rho \\ \rho^2 & \rho & 1 \end{pmatrix}$$

where there are now two degrees of freedom for the block effects (for estimating σ_b^2 and ρ).

In both these designs, we treated genotypes as fixed effects. **BLUE selection** for the best line occurs by choosing the line with the largest estimated G_i value. Note that the best BLUE lines may be different from the best lines chosen based on their average mean values, $\overline{y}_{i\cdot}$. **BLUP selection** is performed by treating genotypes as *random effects*, and choosing those genotypes with the largest BLUP values (Chapter 10 and Part II Chapter 32). If we treat block effects as fixed, then $\mathbf{y} = \mathbf{X}\boldsymbol{\beta} + \mathbf{Z}\mathbf{u} + \mathbf{e}$, where $\mathbf{u}$ is the vector of random genotypic values. One advantage of treating genotypes (lines) as random, is that because they may be correlated (related), we can borrow information from other related lines to improve the estimate of a focal line (Chapter 32). In particular, the covariance matrix for $\mathbf{u}$ is $\sigma_G^2\,\mathbf{K}$, where $\mathbf{K}$ is a kinship matrix corresponding to the correlation among the lines (Chapters 7, 8, and 32). In this setting, (again, with $\mathbf{y}$ as above), then

$$\mathbf{X} = \begin{pmatrix} 1 & 1 & 0 & 0 \\ 1 & 1 & 0 & 0 \\ 1 & 1 & 0 & 0 \\ 1 & 0 & 1 & 0 \\ 1 & 0 & 1 & 0 \\ 1 & 0 & 1 & 0 \\ 1 & 0 & 0 & 1 \\ 1 & 0 & 0 & 1 \\ 1 & 0 & 0 & 1 \end{pmatrix}, \quad \boldsymbol{\beta} = \begin{pmatrix} \mu \\ b_1 \\ b_2 \\ b_3 \end{pmatrix}, \quad \mathbf{Z} = \begin{pmatrix} 1 & 0 & 0 & 0 \\ 0 & 1 & 0 & 0 \\ 0 & 0 & 1 & 0 \\ 0 & 0 & 0 & 1 \\ 1 & 0 & 0 & 0 \\ 0 & 1 & 0 & 0 \\ 0 & 1 & 0 & 0 \\ 0 & 0 & 1 & 0 \\ 0 & 0 & 0 & 1 \end{pmatrix}, \quad \mathbf{u} = \begin{pmatrix} G_1 \\ G_2 \\ G_3 \\ G_4 \end{pmatrix}$$

Note that line mean, BLUE, and BLUP selection approaches might all choose different sets of the best lines. Finally, one could treat *both* blocks and genotype as random, with the model now becoming

$$\mathbf{y} = \mathbf{X}\boldsymbol{\beta} + \mathbf{Z}_1\mathbf{u} + \mathbf{Z}_2\mathbf{b} + \mathbf{e}, \quad \mathbf{Var}\begin{pmatrix} \mathbf{u} \\ \mathbf{b} \\ \mathbf{e} \end{pmatrix} = \begin{pmatrix} \sigma_G^2\mathbf{K} & \mathbf{0} & \mathbf{0} \\ \mathbf{0} & \mathbf{C} & \mathbf{0} \\ \mathbf{0} & \mathbf{0} & \mathbf{R} \end{pmatrix}$$

where $\boldsymbol{\beta} = (\,\mu\,)$ and

$$\mathbf{X} = \begin{pmatrix} 1 \\ 1 \\ 1 \\ 1 \\ 1 \\ 1 \\ 1 \\ 1 \\ 1 \end{pmatrix}, \quad \mathbf{Z}_1 = \begin{pmatrix} 1 & 0 & 0 & 0 \\ 0 & 1 & 0 & 0 \\ 0 & 0 & 1 & 0 \\ 0 & 0 & 0 & 1 \\ 1 & 0 & 0 & 0 \\ 0 & 1 & 0 & 0 \\ 0 & 1 & 0 & 0 \\ 0 & 0 & 1 & 0 \\ 0 & 0 & 0 & 1 \end{pmatrix}, \quad \mathbf{u} = \begin{pmatrix} G_1 \\ G_2 \\ G_3 \\ G_4 \end{pmatrix}, \mathbf{Z}_2 = \begin{pmatrix} 1 & 0 & 0 \\ 1 & 0 & 0 \\ 1 & 0 & 0 \\ 0 & 1 & 0 \\ 0 & 1 & 0 \\ 0 & 1 & 0 \\ 0 & 0 & 1 \\ 0 & 0 & 1 \\ 0 & 0 & 1 \end{pmatrix}, \quad \mathbf{b} = \begin{pmatrix} b_1 \\ b_2 \\ b_3 \end{pmatrix}$$

Blocks: Fixed or Random?

Consider a fairly generic incomplete block design model, with

$$y_{ijk} = \mu + G_i + R_j + b_{jk} + e_{ijk}$$

where R_j is the effect of the set of blocks making up the jth experimental replicate (in a resolvable design) and b_{jk} is the effect of block k in replicate j. With any linear model, the investigator must decide which factors are considered fixed (and hence, require a degree of freedom for each estimated effect) and which factors are considered random (requiring a degree of freedom for each variance or covariance term specified by the distribution of the random effects). The replicate effect, R, is usually treated as fixed. In large part this is because most experiments have only a few experimental replicates, offering very few degrees of freedom with which to estimate a variance for random R values. In such cases, simply treating R as fixed is appropriate, because it is a nuisance parameter, not something whose value we really care about. In settings with more than a few full replicates, treating R as random may be more appropriate, but this is usually in a mutlilocation, multigenerational trial, in which case R is replaced by appropriate G $\times$ E terms (Part II Chapters 27 and 28).

This same logic holds for blocks. If there are only a few blocks, the error in estimating the between-block variance (σ_b^2) can be considerable, while treating blocks as fixed does not use too many degrees of freedom. When there are numerous blocks, treating them as random is quite appropriate. A general rule in the literature is that one has at least 8 to 10 factor levels (here, number of blocks) in order to have a reasonable estimate of the between-block variance, σ_b^2.

A second, more subtle, issue is how to *model* these random block effects (i.e., specify their *covariance structure*). The simplest random-effects model is often used: block effects are independent and homoscedastic. If either assumption is invalid, estimates of random effects when treating them as a simple one degree of freedom random effect (the block variance, σ_b^2) can be highly biased. Spatial effects are likely to cause nearby blocks to be correlated (Example A9.2), and methods developed below, such as trend analysis, can remove this dependency, potentially recovering the simple random-effects model based solely on σ_b^2.

Recovery of Inter-block Information

Consider an incomplete block design in which block one contains genotypes 1 and 2, while block two contains genotypes 2 and 3, with the model $y_{ij} = \mu + G_i + b_j + e_{ij}$. As we have seen, an estimate of the difference $G_1 - G_3$ can be obtained from the difference between $(y_{11} - y_{21})$ in block one and $(y_{32} - y_{22})$ in block two. Note that any block effect cancels in these within-block differences, as, for example,

$$y_{11} - y_{21} = (\mu + G_1 + b_1 + e_{11}) - (\mu + G_2 + b_1 + e_{21}) = G_1 - G_2 + (e_{11} - e_{21})$$

Hence, one estimate of $G_1 - G_3$ is given by

$$\begin{aligned}(y_{11} - y_{21}) - (y_{32} - y_{22}) &= [G_1 - G_2 + (e_{11} - e_{21})] - [G_3 - G_2 + (e_{32} - e_{22})] \\ &= G_1 - G_3 + (e_{11} + e_{22} - e_{21} - e_{32})\end{aligned}$$

Such estimates are based on **intra-block information**. In contrast, one could also consider the difference of the *block sums*, with an estimate for $G_1 - G_3$ being sum of block 1 minus the sum of block 2,

$$([\mu + G_1 + b_1 + e_{11}] + [\mu + G_2 + b_1 + e_{21}]) - ([\mu + G_2 + b_2 + e_{22}] + [\mu + G_3 + b_2 + e_{32}])$$
$$= G_1 - G_3 + b_1 - b_2 + (e_{11} + e_{21} - e_{22} - e_{32})$$

Notice that this **inter-block information** *does* depend on the block effects. The use of this additional information is referred to as the **recovery of inter-block information**. This information can be recovered when blocks are treated as random, but not when they are assumed to be fixed (Yates 1940; Mead et al. 2012; Möhring et al. 2015).

Hence, one motivation for treating blocks as random is the recovery of this information. However, this additional information can be small, and the proper weighting of these two sources of information, intra-block versus inter-block, requires accurate estimates of the variance components, such as σ_b^2. When the number of blocks is small, this variance estimate is very imprecise, and the weighting of the two sources of information can be inaccurate, negating any advantage in using inter-block information. By simulating a large number of different α-lattice and p-rep designs (see below), Möhring et al. (2015) concluded that blocks should always be treated as random when they are numerous (say 10 or more). However, they also found that four or more blocks in an α-lattice design is usually sufficient to support the recovery of inter-block information (provided that the estimate of σ_b^2 is significant, i.e., nonzero). Likewise, for most of the p-rep designs they simulated, treating blocks as random when the estimate of σ_b^2 is significant was also more efficient than treating blocks as fixed (and hence ignoring any inter-block information).

AUGMENTED AND PARTIALLY REPLICATED DESIGNS

Federer Designs

In the early phases of selection from among a set of lines, the collection of candidates is large (often 100 or more) and the breeder typically favors allocating a fixed number of plots to scoring more lines over increased replication of individual lines. As detailed above, one approach for such screening is through the use of incomplete block designs, and, in particular, the rather flexible α-design, wherein individual lines are replicated, but r is usually very small. An alternative approach was suggested by Federer (1956, 1961), which he referred to as an **augmented design**, but is also called a **Federer** or an **unreplicated design.** (Federer originally suggested these be called **hoonuiaku designs**, given they were developed when he was working for the Hawaii Sugar Planters Association—a former student of his stated this translates into "augment to save the boss money.") The idea is simple: a set of checks is used in a standard design into which a set of unreplicated lines that we wish to compare is added (**augmenting** the design). The variance among the replicated checks provides an estimate of the error variance, and differences between unreplicated lines from different blocks are estimated via direct comparisons with the common checks.

As an example, consider the following triple-lattice design with nine replicated checks (A through I) and 28 candidate (and unreplicated) lines (1 through 28). The resulting augmented lattice for the nine checks now has block size $k = 6$ with $s = 3$ blocks per replicate.

Rep 1			Rep 2			Rep 3		
A	B	C	A	D	G	A	B	C
D	E	F	B	E	H	E	F	D
G	H	I	C	F	I	I	G	H
1	2	3	10	11	12	20	21	22
4	5	6	13	14	15	23	24	25
7	8	9	17	18	19	26	27	28

While in practice, the location of genotypes within each block (columns) is randomized, we have placed the checks in the top half of each block for ease of exposition.

Any number of other basic designs can be similarly augmented, with the design based entirely on the checks, followed by unreplicated lines being added to each block. For example, an RCBD with three repeated checks (A, B, C) and 36 unreplicated genotypes becomes

Block 1			Block 2			Block 3		
1	2	3	13	B	14	A	25	26
4	5	B	15	16	17	27	28	29
6	7	8	C	18	19	30	31	32
C	9	10	20	21	A	33	C	34
11	A	12	22	23	24	35	36	B

Here, each set of three columns represents one of the complete blocks (for the checks).

While Federer designs allow one to compare a large series of lines without replicating them, there are costs and limitations. First, the inclusion of genotypes (checks) that we really do not care about reduces the number of testable genotypes, and in the early stages of selection, screening is more concerned with testing more genotypes, not higher replication. Kempton and Gleeson (1997) showed that the inclusion of too many checks retards selection response when the number of plots is fixed. Second, the *choice* of controls is critical. Ideally, one should use at least two different checks with these being closely related to the test lines (e.g., the two parents of a cross generating a series of inbred lines being tested). If the controls do not respond to the environment in the same way as the test lines (i.e., G $\times$ E is present), they can introduce bias, rather than control for environmental effects. Further, competition can occur, which can also compromise the controls. For example, if one has a series of tall and semi-dwarf lines with the checks being a tall parent and a semi-dwarf parent, competition from nearby lines can influence the value of the control. If a semi-dwarf is largely surrounded by tall genotypes it is likely to have a different performance value than when it is surrounded by semi-dwarf genotypes.

Partially Replicated (p-rep) Designs

One limitation of augmented designs is the number of plots allocated to checks. In some settings, this may be appropriate, as checks can represent the current elite lines that one is trying to best. However, an alternative strategy was suggested by Smith et al. (2006) and Cullis et al. (2006). Here, the replicated checks are replaced by a random set of the lines being tested, which they call a **partially replicated** (**p-rep**) **design**. By replacing checks with randomly chosen replicates from the pool of test lines, one can make more efficient use of the number of plots dedicated to the design. Further, when early generation tests are examined over m major environments, replicating a fraction $1/m$ of the lines in each environment allows for all lines to be tested as two replicates in one of the environments. Moehring et al. (2014) showed that this is a very efficient strategy, while Williams et al. (2001, 2014) detailed how to combine features of the α-lattice with p-reps for a more efficient design.

SPATIAL ANALYSIS MODELS

Figure A9.1 showed the task faced by any design incorporating blocks. The idea is to account for field heterogeneity by grouping nearby plots into sufficiently small blocks such that the variance between two plots within a block is much less than the variance between plots from different blocks. This is a rather coarse-grained, and often ad hoc, approach to handling a potentially more fine-grained and nuanced environmental structure (Gusmão 1986; Lin et al 1993). Indeed, a common practice is **post-blocking**, wherein blocks for analysis are assigned *after* the data have been gathered (assuming they still control for other features, such as operator effects).

Consider the CRD design (Equation A9.3), $y_{ij} = \mu + G_i + e_{ij}$, where we assume that the residuals are uncorrelated and have common variance (the OLS assumption). The presence of local heterogeneity can cause residuals from neighboring plots to be correlated, which is one motivation for the use of blocking. However, neighboring plots may still be more similar than more distant plots within the same block, creating a correlation among residuals that decreases with distance. **Spatial models** attempt to control for any such correlations, either by using information from neighbors without regard to the blocking structure, or by starting with an existing blocking structure and then adding a further correction for correlations between residuals (moving from an OLS to a GLS framework). While adjacent plots are usually regarded as being positively correlated, note that competition can induce negative correlations between plots (Dutkowski et al. 2006; Stringer et al. 2011).

While blocking is an attempt at **design-based** control of error, spatial analyses are **model-based** methods that represent an alternative (and sometimes complementary) approach for handling environmental variation. Spatial analysis methods loosely fall into three categories. The first are **nearest neighbor adjustment** methods (**NNA**) that use information from adjacent plots in an attempt to correct for local variation. The second approach is **trend analysis**, wherein one attempts to use the performance data to fit an environment trend, predicting the **yield potential** of a plot given its location. The final approach is to specifically model **correlated residuals** among the plots. Modern spatial analysis models build on these basic ideas by either extending them to more complex models, or combining them, such as a general linear model with trends and correlated residuals. As we detail below, spatial analysis models can easily outperform standard blocked designs. The main issue is that one has a large number of candidate models to choose from, and a reasonable model-selection criteria (such as cross-validation) must be used to pick among them. Because the ranking of genotypes can change depending on the model (e.g., Brownie et al. 1993; Stroup et al. 1994; Singh et al. 2003; Leiser et al 2012), model choice has important consequences and must be given very careful consideration.

An important point to stress is that spatial analysis models all have the ***strong requirement of randomization of genotypes***. While these models attempt to accommodate locally correlated environmental conditions, they all assume that the spatial distribution of genotypes is completely *uncorrelated*. A design wherein family members (or other closely related lines) are grown near to each other compromises spatial models (for example, a small cluster of trees in a forest may be more related to each other than expected by chance). Their genetic similarity would be mapped into environmental similarity, resulting in their estimated G values being regressed back to the mean (zero), underpredicting the value of exceptional lines ($G > 0$) and overpredicting the value of inferior lines ($G < 0$).

The literature on spatial analysis has become increasingly sophisticated and very technical, often requiring very deep diving into time-series analysis. Our aim here is to introduce some of the basic ideas underpinning spatial analysis. Gleeson (1997) gives an excellent nontechnical introduction, while Zimmerman and Harville (1991) provided a nice theoretical overview showing the connections between many approaches. Most of the basic theory behind time-series-based approaches can be found in Gleeson and Cullis (1987), Cullis and Gleeson (1991), Gilmour et al. (1997), and Cullis et al. (1998).

Papadakis and Other NNA Methods

As reviewed by Kempton and Gleeson (1997), the use of information from nearby plots (such as the performance of a check) was an important strategy in the early 1900s before the advent of modern experimental design. Gardner's (1961) notion of stratified mass selection (WL Chapter 13), wherein separate selection decisions are made within each block, is an example of this approach.

One of the first attempts to use of information from neighbors in a more formal way was Papadakis' (1937) nearest neighbor adjustment method, usually called the **Papadakis method** (or, occasionally, **productivity covariate analysis**). The basic idea is as follows. Consider the observed yield, y_{ij}, in a plot at row position i and column position j in a grid. If the experimentalist believes that there is an environmental gradient along a row, then information on the quality of plot (i, j) is provided from plots at positions $(i-1, j)$ and $(i+1, j)$. If there is concern of a gradient along columns, then information is provided from plots $(i, j-1)$ and $(i, j+1)$. A number of variations on the NNA approach follow from whether one considers row only, column only, or both; by how many steps the model employs from the current location (we restrict our discussion to one-step models), and by how to weight this information.

NNA methods can be powerful even when the soil environmental conditions are completely uniform. Suppose that one has a localized outbreak of an insect pest or fungal pathogen in an otherwise uniform field. In setting up a design based on previous trials, a blocking structure might be used that was completely appropriate for the soil heterogeneity, but which could not have predicted these localized effects in the current experiment. NNA methods can correct for these.

NNA methods start by considering the residual associated with a particular location. Suppose that in position ij, the tested genotype is k. The resulting residual becomes

$$\epsilon_{ij} = y_{ij} - \overline{y}_k \tag{A9.14a}$$

where $\overline{y}_k$ is the mean value of genotype k over the entire grid. As one changes grid position, the subtracted mean genotype in Equation A9.14a changes to that for the genotype that occupies the plot of interest. A two-dimensional Papadakis' method assumes the model

$$y_{ij} = \mu + G_{k(ij)} + \beta_r er_{ij} + \beta_c ec_{ij} + e_{ij} \tag{A9.14b}$$

where $G_{k(ij)}$ denotes that the genotype at position ij is k (whose effect is G_k), β_r and β_c are, respectively, row and column cofactors (fitted by the model, akin to regression coefficients), and er and ec are row- and column-residuals, defined by

$$er_{ij} = \frac{\epsilon_{i-1,j} + \epsilon_{i+1,j}}{2}, \qquad ec_{ij} = \frac{\epsilon_{i,j-1} + \epsilon_{i,j+1}}{2} \tag{A9.14c}$$

where $\epsilon_{i,j}$ is given by Equation A9.14a. One could also fit a one-dimensional version, by ignoring either row or column residuals. A model could initially fit both, with one (or both) ignored (and a reduced model refit) if their estimated β coefficient is not significant.

Modifications and extensions of the basic NNA approach were discussed by a number of authors, including Bartlett (1978), Wilkinson et al. (1983), Besag and Kempton (1986), Williams (1986), Pearce (1998), and Piepho et al. (2008). Wilkinson et al. (1983) suggested an iterative approach, wherein one first takes the raw data, then updates Equation A9.14a using the current estimate of G_k from Equation A9.14b in place of $\overline{y}_k$. These adjusted values are then used to rerun the analysis, and this iteration proceeds until stable values are obtained. One can also combine NNA with blocking. For example, Yang et al. (2004) used the model

$$y_{ij} = \mu + G_{k(ij)} + b_i + \beta_c ec_{ij} + e_{ij} \tag{A9.15}$$

where each row is a block (b_i being the block effect for the ith row) and NNA is performed on plots within the same block (i.e., across the columns of a row).

A number of studies have shown that NNA methods can outperform blocked designs by themselves (e.g., Williams 1986; Brownie et al 1993; Lin et al. 1993; Stroup et al. 1994), or can enhance the efficiency when incorporated into a block design (e.g., Ball et al. 1993; Yang et al. 2004).

Trend Analysis

The idea behind NNA methods is to predict the expected **performance potential** of a random genotype that is placed in plot position ij in a specific field. Under Papadakis' method, this is estimated as

$$\mu + \beta_r er_{ij} + \beta_c ec_{ij}$$

where the β are model-fit parameters and the row and column adjustment variables (er and ec) for a specific position are obtained using the residuals from the nearest neighbors (Equation A9.14a). A more general approach, **trend analysis**, was proposed by Kirk et al. (1980). Here one uses information from the *entire grid* of plots (as opposed to just nearest neighbors), and attempts to fit a performance surface over the array, usually by fitting a polynomial for positional effects. For example, a second-order polynomial could be used to predict the yield potential at row i and column j in an array, where

$$\mu + \beta_1\, i + \beta_2\, j + \beta_3\, i^2 + \beta_4\, j^2 + \beta_5\, ij \tag{A9.16a}$$

The trend parameters (the β_i) are fit as standard fixed-effects in a general linear model. For the trend model suggested by Equation A9.16a, the corresponding terms in a GLM would be

$$\mathbf{y} = \begin{pmatrix} y_{11} \\ y_{12} \\ y_{13} \\ \vdots \\ y_{21} \\ y_{22} \\ \vdots \\ y_{ij} \end{pmatrix} = \begin{pmatrix} 1 & 1 & 1 & 1 & 1 & 1 \\ 1 & 1 & 2 & 1 & 4 & 2 \\ 1 & 1 & 3 & 1 & 9 & 3 \\ \vdots & \vdots & \vdots & \vdots & \vdots & \vdots \\ 1 & 2 & 1 & 4 & 1 & 4 \\ 1 & 2 & 2 & 4 & 4 & 4 \\ \vdots & \vdots & \vdots & \vdots & \vdots & \vdots \\ 1 & i & j & i^2 & j^2 & ij \end{pmatrix} \begin{pmatrix} \mu \\ \beta_1 \\ \beta_2 \\ \beta_3 \\ \beta_4 \\ \beta_5 \end{pmatrix} + \mathbf{g} + \mathbf{e} \tag{A9.16b}$$

where **g** and **e** are the vectors for genotypic and residual effects associated with the observed **y** values. While we have framed trend effects as fixed, one could also treat these as random (Gleeson and Cullis 1987; Cullis and Gleeson 1991).

While several empirical studies have found that using a trend analysis can be more efficient than an RCBD (e.g., Tamura et al. 1988; Bowman 1990; Fares et al. 2011), some caution is required when using this approach. The critical issue is the appropriate model. One could use a row-only or column-only model if the assumption is a one-dimensional gradient that aligns with the plot layout. In either case, the usable degrees of freedom is set by the number of plots along a single replicate, offering rather little power and limiting the polynomial degree to no more than second or third order. There is also the choice of which function to use to fit for a trend, and the method used to choose among candidate models. Because models need not be nested (and hence likelihood ratio tests are not appropriate), model choice could be based on AIC or BIC (Singh et al. 2003; Leiser et al 2012), but cross-validation (e.g., Zimmerman and Harville 1991; Grondona et al. 1996) is a more robust choice (especially when the number of fixed effects differs between models). The delicate issue is that one can easily *overfit* (using a polynomial that has too high an order) or *underfit* (too low an order). Extreme caution should be used in fitting any polynomial higher than a quadratic unless one has a very large data set. Tamura et al. (1988) suggested that with between 40 and 70 plots in an array, one should have an upper limit of no more than eight trend parameters.

Correlated Residuals

Modeling a trend is an example of accounting for a **large-scale** (or **global**) effect, while NNA approaches model **small-scale** (or **local**) effects. Another approach for modeling small-scale effects is to incorporate correlations among the residuals into the analysis of a blocked design, namely using a GLS, rather than an OLS, framework. The issue is how to model the correlation structure. Much of the work for spatial analysis of field designs is motivated by the extensive geostatistics literature (Cressie 1991; Gaeton and Guyon 2010).

Consider a simple random-effects model for the yield potential, where we assume no treatment effects (such as in a uniformity trial, where all the genotypes are identical). The observed yield potential for plot i is given by

$$y_i = \mu + t_i + e_i \tag{A9.17a}$$

where t_i is the trend impact on the yield effect and e_i is a measurement error. We allow the trend terms to be correlated—$\sigma(t_i, t_j)$ is not necessarily zero—while the residuals (e_i) are uncorrelated with both each other and with the trend terms.

The expected covariance between two plots, i and j, is assumed to be of the form

$$\sigma(y_i, y_j) = \sigma(t_i, t_j) = \sigma^2 \cdot f\,(d_{ij}) \tag{A9.17b}$$

where d_{ij} is some distance metric between two plots, which could be the difference in number of rows or columns, or the Euclidean distance. The function, f, is assumed to montonically decline to zero. Commonly used candidate covariance functions in the plant breeding literature include

$$f(d) = \begin{cases} \alpha^d & \text{AR1} \quad (\text{for } \alpha \le 1) \\ \exp(-\alpha d^2) & \text{Gaussian} \\ 1 - \alpha d & \text{Linear} \quad (\text{for } d\alpha \le 1) \\ 1 - 1.5(\alpha d) + 0.5(\alpha d)^3 & \text{Spherical} \quad (\text{for } d\alpha \le 1) \end{cases} \tag{A9.17c}$$

Zimmerman and Harville (1991) used an exponential function, $f(d) = \exp(-\gamma d)$, but note that taking $\alpha = \exp(-\gamma)$ shows that this is simply an alternative parameterization of AR1, a **first-order autoregressive model**. Likewise, writing $\alpha = (1-\gamma)$, then for $\gamma \ll 1, \alpha^d \simeq 1-\gamma d$, so that for values of α close to one (high correlations between adjacent plots within a block), the linear variance model is a close approximation to the AR1 model over modest distances. In an analysis of almost 400 German sugar beet and barley trials, Müller et al. (2010) found that the linear and AR1 models usually gave a better fit than either the spherical or Gaussian models.

One graphical approach for examining the covariance structure is the **sample variogram**, also called a **semivariogram**. Consider the half-squared difference of observed yields in a uniformity trial, $\delta_{ij} = (y_i - y_j)^2/2$, which are often referred to as **semivariances**. Taking expected values,

$$E(\delta_{ij}) = \frac{1}{2}E\left[(y_i - y_j)^2\right] = \frac{1}{2}E\left[(\mu + t_i + e_i - \mu - t_j - e_j)^2\right]$$

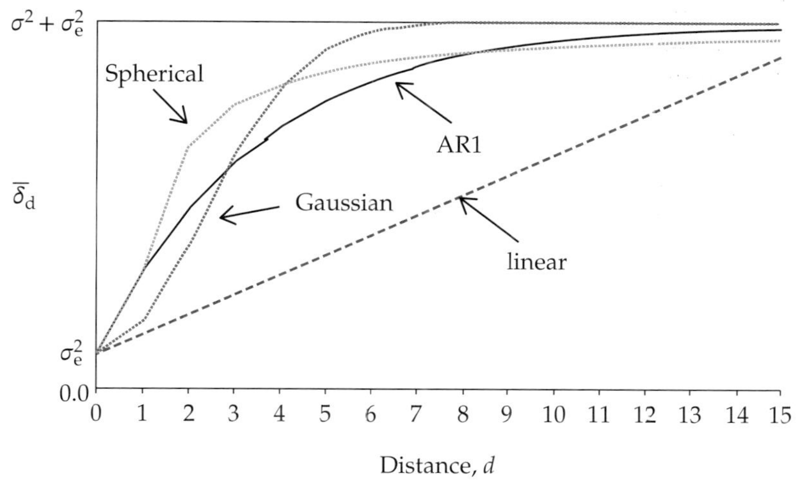

Figure A9.3. The theoretical curves for a variogram for the covariance functions given by Equation A9.17c for predicting $\overline{\delta}_d$, the sample average for the semivariance for sites differing by distance d. The solid line corresponds to an AR1, the long-dashed line to the linear variance function, the dashed and dotted line to the spherical function, and the short-dashed to the Gaussian function.

Expanding and recalling Equation 2.3a, that $E[t_i] = E[e_i] = 0$, Equation 2.3a, and finally that the e_i in Equation A9.17a are uncorrelated, this reduces to

$$E(\delta_{ij}) = \frac{1}{2}E\left[t_i^2 + t_j^2 - 2t_it_j + e_i^2 + e_j^2\right] = \frac{1}{2}\left[\sigma^2(t_i) + \sigma^2(t_j) - 2\sigma(t_i, t_j) + 2\sigma_e^2\right]$$
$$= \sigma^2(1 - f[d_{ij}]) + \sigma_e^2 \qquad \text{(A9.18)}$$

For $i = j$, the expected value is σ_e^2, the measurement error (quaintly referred to as the **nugget** in geostatistics), while the limiting value approaches $\sigma^2 + \sigma_e^2$, which is referred to as the **sill**. The distance until the sill limit is reached is called the **range**. A sample variogram plots the mean values of δ_{ij} as a function of the distance, d_{ij} (Figure A9.3). The intercept corresponds to the value of σ_e^2 (the nugget) and the limiting value to $\sigma^2 + \sigma_e^2$ (the sill). Gilmour et al. (1997) suggested the use of variograms as a graphical tool to help decide which covariance model to use, while Stefanova et al. (2009) presented a few more formal approaches.

Typically, correlated residuals are assumed to be restricted to the plots within a block, with either blocking, or a global trend analysis, accounting for correlations between plots from different blocks. The residual correlation structure is often assumed to be in one direction, in which case d is usually the difference between plot numbers within a row or a column of a block. If one assumes a two-dimensional covariance structure, then one candidate function is the spherical model (Equation A9.17; see WL Example 22.11 for an application). A more common approach, first suggested by Cullis and Gleeson (1991), is to assume a **separable** covariance function, where the covariance between individuals differing by i rows and j columns is given by

$$\sigma^2 f_r(i) f_c(j) \qquad \text{(A9.19a)}$$

The most common version is the **separable AR1 process**, often denoted by AR1 $\times$ AR1. Under this model, the covariance between two plots, separated by i rows and j columns, is

$$\sigma^2 \alpha_r^i \alpha_c^j \qquad \text{(A9.19b)}$$

which allows for row- and column-specific correlations.

Table A9.2 Summary of experimental design options given the issues facing an investigator.

Environment experienced by the experimental unit
Homogeneous: Completely randomized designs (CRDs)
Heterogeneous in one direction: Random complete block designs (RCBDs)
Heterogeneous in two directions:
Latinized and row-column designs
Spatial modeling
Large number of lines
Incomplete Block Designs
α-designs
Unreplicated experiments:
Augmented (Federer) designs
p-rep designs

With an assumed one-dimensional covariance structure, the variogram is a simple graph, while with a two-dimensional covariance structure, the variogram is a three-dimensional response surface (whose x and y axes correspond to distances and whose height is the mean semivariance for a particular pair of distances). With a three-dimensional variogram, one usually looks at its **faces**, the projection onto either row or column distance (Stefanova et al. 2009). A variogram that does not show an asymptotic approach with distance (e.g., showing a peak at intermediate distances followed by a drop-off at longer distances) suggests that either the covariance model is not appropriate or that there is a global trend that has not been captured by the model (Piepho et al. 2008).

Example A9.3. Consider the design from Example A9.2, where the order of the genotypes within the plots is (1,2,3), (4,1,2), and (2,3,4). Assume that genotypes are fixed effects, with blocks and trend effects as random,

$$y_{ij} = \mu + G_{k(ij)} + b_i + t_{ij} + e_{ij} \tag{A9.20a}$$

where the genotype in plot j of block i is line k. We assume that blocks and residuals are uncorrelated (**Var**(**b**) = σ_b^2 **I**, **Var**(**e**) $= \sigma_e^2$ **I**, $\sigma(e,b) = 0$). Finally, we assume that trend terms are correlated within, but not between, blocks.

Under an AR1 model, the covariances among trend effects within the first plot become

$$\sigma(t_{i,1}, t_{i,1}) = \sigma_t^2, \quad \sigma(t_{1,1}, t_{2,1}) = \sigma(t_{2,1}, t_{3,1}) = \alpha\sigma_t^2, \quad \sigma(t_{1,1}, t_{3,1}) = \alpha^2\sigma_t^2$$

Proceeding in a similar fashion over the other blocks yields the 9×9 covariance matrix for the vector of trend effects as

$$\mathbf{Var}(\mathbf{t}) = \sigma_t^2 \begin{pmatrix} 1 & \alpha & \alpha^2 & 0 & 0 & 0 & 0 & 0 & 0 \\ \alpha & 1 & \alpha & 0 & 0 & 0 & 0 & 0 & 0 \\ \alpha^2 & \alpha & 1 & 0 & 0 & 0 & 0 & 0 & 0 \\ 0 & 0 & 0 & 1 & \alpha & \alpha^2 & 0 & 0 & 0 \\ 0 & 0 & 0 & \alpha & 1 & \alpha & 0 & 0 & 0 \\ 0 & 0 & 0 & \alpha^2 & \alpha & 1 & 0 & 0 & 0 \\ 0 & 0 & 0 & 0 & 0 & 0 & 1 & \alpha & \alpha^2 \\ 0 & 0 & 0 & 0 & 0 & 0 & \alpha & 1 & \alpha \\ 0 & 0 & 0 & 0 & 0 & 0 & \alpha^2 & \alpha & 1 \end{pmatrix} \tag{A9.20b}$$

Note the block-diagonal structure of this covariance matrix, where each block corresponds to the correlation of plots within a block, with no correlations between plots from different blocks.

Under the linear variance model,

$$\sigma(t_{i,1}, t_{i,1}) = \sigma_t^2, \quad \sigma(t_{1,1}, t_{2,1}) = \sigma(t_{2,1}, t_{3,1}) = (1-\alpha)\sigma_t^2, \quad \sigma(t_{1,1}, t_{3,1}) = (1-2\alpha)\sigma_t^2$$

and the full covariance variance becomes $\mathbf{Var}(\mathbf{t}) =$

$$\sigma_t^2 \begin{pmatrix} 1 & 1-\alpha & 1-2\alpha & 0 & 0 & 0 & 0 & 0 & 0 \\ 1-\alpha & 1 & 1-\alpha & 0 & 0 & 0 & 0 & 0 & 0 \\ 1-2\alpha & 1-\alpha & 1 & 0 & 0 & 0 & 0 & 0 & 0 \\ 0 & 0 & 0 & 1 & 1-\alpha & 1-2\alpha & 0 & 0 & 0 \\ 0 & 0 & 0 & 1-\alpha & 1 & 1-\alpha & 0 & 0 & 0 \\ 0 & 0 & 0 & 1-2\alpha & 1-\alpha & 1 & 0 & 0 & 0 \\ 0 & 0 & 0 & 0 & 0 & 0 & 1 & 1-\alpha & 1-2\alpha \\ 0 & 0 & 0 & 0 & 0 & 0 & 1-\alpha & 1 & 1-\alpha \\ 0 & 0 & 0 & 0 & 0 & 0 & 1-2\alpha & 1-\alpha & 1 \end{pmatrix} \tag{A9.20c}$$

For Equations A9.20b and A9.20c, the values of σ_t^2 and α are estimated from the data using REML (Part II Chapter 33). To put all of this into a mixed linear model, we combine $t_{ij} + e_{ij}$ into a single residual, ϵ_{ij}, whose covariance structure is given by

$$\mathbf{R} = \mathbf{Var}(\mathbf{t}) + \sigma_e^2\mathbf{I} \tag{A9.20d}$$

and the model becomes

$$\mathbf{y} = \mathbf{X}\boldsymbol{\beta} + \mathbf{Z}\mathbf{b} + \boldsymbol{\varepsilon}, \quad \mathbf{Var}\begin{pmatrix} \mathbf{b} \\ \boldsymbol{\varepsilon} \end{pmatrix} = \begin{pmatrix} \sigma_b^2\mathbf{I} & \mathbf{0} \\ \mathbf{0} & \mathbf{R} \end{pmatrix} \tag{A9.20e}$$

The use of a standard blocked design with an AR1 or AR1 × AR1 covariance structure for the residuals within blocks tends to perform well in both crops (e.g., Singh et al. 2003; Leiser et al. 2012) and forest trees (e.g., Costa e Silva et al. 2001; Saenz-Romero et al. 2001; Dutkowski et al. 2006), but see Hoefler et al. (2020), who argue that the AR1 × AR1 correction often has little benefit.

Finally, as noted by Borges et al. (2018), "*In most situations, models that included spatial correction were better than models with no spatial correlation, but they did not outperform better experimental designs. Therefore, spatial modeling is not a substitute for good experimental design.*"

SUMMARY

Table A9.2 summarizes the various designs options available to an investigator given the issues that they face. Blocking and randomization are critical design components for any experiments, even if the initial design does not survive an experiment (e.g., due to loss of blocks). While environmental control is usually outside of the ability of an investigator, operator control is not, wherein all plots within a block experience the same set of equipment, workers, and management practices. These effects can be substantial, even under an otherwise rather uniform environment. When the blocking design does not fully capture environmental heterogeneity (e.g., blocks are not orthogonal to an environmental gradient), one can still use spatial modeling to adjust for some of these effects.

The basic constraint on any design is the total number of plots scored, gr, where g is the number of genotypes (or lines) tested and r is the average replication per line. Small values of g allow for a high degree of replication, and more classical designs, such as RCBDs or row-column, are often appropriate. Large values of g imply much lower replication, with incomplete block designs or unreplicated experiments as potential options, especially when combined with spatial control (such as an AR1 model for the covariance structure within plots). Federer and Crossa (2001) wisely suggested that one should "design for the experiment, not experiment for the design," an especially apropos point during a cycle of line selection where the design likely changes as an investigator moves from early screening (large g, small r) to later testing (small g, large r).

Literature Cited

References here are arranged alphabetically by author (when involving two or less) and then (for three or more authors) ordered by date as first author et al. (as in the text).

Abbott, R., and 38 others. 2013. Hybridization and speciation. *J. Evol. Biol.* 26: 229–246. [13]

Adbellaoui, A., and 15 others. 2013. Association between autozygosity and major depression: stratification due to religious assortment. *Behav. Genet.* 43: 455–467. [12]

Abdellaoui, A., K. J. H. Verweij, and B. P. Zietsch. 2014. No evidence for genetic assortative mating beyond that due to population stratification. *Proc. Natl. Acad. Sci. USA* 111: E4137. [7]

Abdellaoui, A., and 11 others. 2015. Educational attainment influences levels of homozygosity through migration and assortative mating. *PLoS ONE* 10: e0118935. [12]

Abdellaoui, A., L. Yengo, K. J. H. Verweij, and P. M. Visscher. 2023. 15 years of GWAS discovery: realizing the promise. *Am. J. Hum. Genet.* 110: 1–16. [20]

Abdi, S., Shashi, A. Dwivedi, and V. Bhat. 2016. Harnessing apomixis for heterosis breeding in crop improvement. *In* V. R. Rajpal et al. (eds) *Molecular breeding for sustainable crop improvement*, pp. 79–99. Springer. [13]

Abecasis, G. R., W. O. C. Cookson, and L. R. Cardon. 2000. Pedigree tests of transmission disequilibrium. *Europ. J. Hum. Genet.* 8: 545–551. [17]

Abplanalp, H. 1990. Inbreeding. *In* R. D. Crawford (ed) *Poultry breeding and genetics*, pp. 955–984. Elsevier. Amsterdam, NL [12]

Ackerman, M. S., and 6 others. 2017. Estimating seven coefficients of pairwise relatedness using population genomic data. *Genetics* 206: 105–118. [8, A4]

Ackermann, M., and K. Strimmer. 2009. A general modular framework for gene set enrichment analysis. *BMC Bioinformat.* 10: 47. [21]

Adams, M. W., and D. B. Shank. 1959. The relationship of heterozygosity to homeostasis in maize hybrids. *Genetics* 44: 777–786. [6]

Adzhubei, I. A., and 7 others. 2010. A method and server for predicting damaging missense mutations. *Nat. Methods* 7: 248–249. [20]

Agrawal, A. F., and M. C. Whitlock. 2011. Inferences about the distribution of dominance drawn from yeast gene knockout data. *Genetics* 187: 553–566. [4, 15]

Ågren, J., and D. W. Schemske. 1993. Outcrossing rate and inbreeding depression in two annual monoecious herbs, *Begonia hirsuta* and *B. semiovata*. *Evolution* 47: 125–135. [12]

Ahituv, N., and 18 others. 2007. Medical sequencing at the extreme of human body mass. *Am. J. Hum. Genet.* 80: 779–791. [20]

Aitchison, J., and J. A. C. Brown. 1966. *The log-normal distribution*. Cambridge Univ. Press, Cambridge, UK. [14]

Aitken, A. C. 1935. On least squares and linear combination of observations. *Proc. Royal Soc. Edinburgh A* 55: 42–47. [10]

Aitken, M., and G. T. Wilson. 1980. Mixture models, outliers, and the EM algorithm. *Technometrics* 22: 325–331. [A4]

Akaike, H. 1974. A new look at the statistical model identification. *IEEE Trans. on Automatic Control, AC-19*. 716–723. [16, A4]

Albert, F. W., and L. Kruglyak. 2015. The role of regulatory variation in complex traits and disease. *Nat. Rev. Genet.* 16: 197–212. [21]

Albert, F. W., J. S. Bloom, J. Siegel, L. Day, and L. Kruglyak. 2018. Genetics of *trans*-regulatory variation in gene expression. *eLIFE* 7: e35471. [21]

Albert, R. 2005. Scale-free networks in cell biology. *J. Cell. Sci.* 118: 4947–4957. [A2]

Albrecht, G. H. 1978. Some comments on the use of ratios. *Syst. Zool.* 27: 67–71. [14]

Albrechtsen, A., and 5 others. 2009. Relatedness mapping and tracts of relatedness for genome-wide data in the presence of linkage disequilibrium. *Genet. Epid.* 33: 266–274. [8]

Alemu, S. W., and 6 others. 2021. An evaluation of inbreeding measures using a whole-genome sequenced cattle pedigree. *Heredity* 126: 410–423. [12]

Alfonso, L., and C. S. Haley. 1998. Power of different F_2 schemes for QTL detection in livestock *Anim. Sci.* 66: 1–8. [19]

Allen, N. E., and 24 others 2024. Prospective study design and data analysis in UK Biobank. *Sci. Transl. Med.* 16: eadf4428. [20]

Allison, D. B. 1997. Transmission-disequilibrium tests for quantitative traits. *Am. J. Hum. Genet.* 60: 676–690. [17]

Allison, D. B., M. Heo, N. Kaplan, and E. R. Martin. 1999a. Sibling-based tests of linkage and association for quantitative traits. *Am. J. Hum. Genet.* 64: 1754–1764. [17]

Allison, D. B., and 5 others. 1999b. Testing the robustness of the likelihood-ratio test in a variance-component quantitative-trait loci-mapping procedure. *Am. J. Hum. Genet.* 65: 531–544. [19]

Allison, D. B., J. R. Fernández, M. Heo, and T. M. Beasley. 2000. Testing the robustness of the new Haseman-Elston quantitative-trait loci-mapping procedure. *Am. J. Hum. Genet.* 67: 249–252. [19]

Allison, D. B., and 6 others. 2002. A mixture model approach for the analysis of microarray gene expression data. *Comput. Stat. Data Analysis* 39: 1–20. [A6]

Allison, D. B., X. Cui, G. P. Page, and M. Sabripour. 2005. Microarray data analysis: from disarray to consolidation and consensus. *Nat. Rev. Genet.* 7: 55–65. [21]

Almasy, L., and J. Blangero. 1998. Multipoint quantitative-trait linkage analysis in general pedigrees. *Am. J. Hum. Genet.* 62: 1198–1211. [19]

Almudevar, A. 2003. A simulated annealing algorithm for maximum likelihood pedigree reconstruction. *Theor. Pop. Biol.* 63: 63–75. [8]

Almudevar, A., and E. C. Anderson. 2012. A new version of PRT software for sibling groups reconstruction with comments regarding several issues in the sibling reconstruction problem. *Mol. Ecol. Resources.* 12: 164–178. [8]

Almudevar, A., and C. Field. 1999. Estimation of single-generation sibling relationships based on DNA markers. *J. Argric. Biol. Environ. Stat.* 4: 136–165. [8]

Alter, O., P. O. Brown, and D. Botstein. 2000. Singular value decomposition for genome-wide expression data processing and modeling. *Proc. Natl. Acad. Sci. USA* 97: 10101–10106. [21]

Altshuler, D., and M. Daly. 2007. Guilt beyond a reasonable doubt. *Nat. Genet.* 39: 813–815. [20]

Altschuler, D., M. Daly, and E. S. Lander. 2008. Genetic mapping in human disease. *Science* 322: 881–888. [1, 20]

Alvarez-Castro, J. M., and Ö Carlborg. 2007. A unified model of functional and statistical epistasis and its application in quantitative trait loci analysis. *Genetics* 176: 1151–1167. [5, 18]

Amador, C., A. García–Dorado, D. Bersabe, and C. López–Fanjul. 2010. Regeneration of the variance of metric traits by spontaneous mutation in a *Drosophila* population. *Genetics Res.* 92: 91–102. [15]

Amaral, P. P., M. E. Dinger, T. R. Mercer, and J. S. Mattick. 2008. The eukaryotic genome as an RNA machine. *Science* 319: 1787–1789. [21]

Ambros, V., and X. Chen. 2007. The regulation of genes and genomes by small RNAs. *Development* 134: 1635–1641. [15]

Amin, N., C. M. van Duijn, and Y. S. Aulchenko. 2007. A genomic background based method for association analysis in related individuals. *PLoS One* 12: e1274. [20]

Amin, N., and 52 others. 2012. Genome-wide association analysis of coffee drinking suggest association with *CYP1A1/ CYP1A2* and *NRCAM*. *Mol. Psychiatry* 17: 1116–1129. [20]

Amorim, E. P., and 4 others. 2006. Genetic distance based on SSR and grain yield of inter and intrapopulational maize single cross hybrids. *Maydica* 51: 507–513. [13]

Amos, C. I. 1994. Robust variance-components approach for assessing genetic linkage in pedigrees. *Am. J. Hum. Genet.* 54: 535–543. [19]

Amos, C. I., and M. de Andrade. 2001. Genetic linkage methods for quantitative traits. *Stat. Meth. Med. Res.* 10: 3–25. [19]

Amos, C. I., and R. C. Elston. 1989. Robust methods for the detection of genetic linkage for quantitative data from pedigrees. *Genet. Epidem.* 6: 349–360 (Correction 6: 727). [19]

Amos, C. I., R. C. Elston, A. F. Wilson, and J. E. Bailey-Wilson. 1989. A more powerful robust sib-pair test of linkage for quantitative traits. *Genet. Epidem.* 6: 435–449. [19]

Amos, C. I., and 4 others. 1990. A multivariate method for detecting genetic linkage, with application to a pedigree with adverse lipoprotein phenotype. *Am. J. Hum. Genet.* 47: 247–254. [19]

Anbessa, Y., and T. Warkentin, A. Vandenberg, and R. Ball. 2006. Inheritance of time to flowering in chickpea in a short-season temperate environment. *J. Heredity* 97: 55–61. [16]

Ancel, L. W., and W. Fontana. 2000. Plasticity, evolvability, and modularity in RNA. *J. Exp. Zool.* 288,: 242–283. [14]

Anderson, A. D., and B. S. Weir. 2007. A maximum-likelihood method for the estimation of pariwise relatedness in structured populations. *Genet.* 176: 421–440. [8, A4]

Anderson, C. A., and 5 others. 2010. Data quality control in genetic case-control association studies. *Nat. Protocols* 9: 1564–1583. [20]

Anderson, E., and W. L. Brown. 1952. Origin of corn belt maize and its genetic significance. *In* J. W. Gowan (ed.), *Heterosis*, pp. 124–148. Iowa State College Press, Ames, IA. [13]

Anderson, T. W. 2003. *An introduction to multivariate statistical analysis.* 3rd Ed. John Wiley & Sons, New York, NY. [9]

Anderson, V. L., and O. Kempthorne. 1954. A model for the study of quantitative inheritance. *Genetics* 39: 883–898. [12]

Andersson, C., and 4 others. 2019. 70-year legacy of the Framingham heart study. *Nat. Rev. Cardiology* 16: 687–698. [20]

Angeloni, F., N. J. Ouborg, and R. Leimu. 2011. Meta-analysis on the association of population size and life history with inbreeding depression in plants. *Biol. Conserv.* 144: 35–43. [12]

Angus, R. A., and R. J. Schultz. 1983. Meristic variation in homozygous and heterozygous fish. *Copeia* 1983: 287–299. [6]

Anon. 1949. Minutes of the meeting of the North Central Corn Breeding Technical Committee, USDA, Beltsville, MD. [13]

Ansel, J., and 7 others. 2008. Cell-to-cell stochastic variation in gene expression is a complex genetic trait. *PLoS Genet.* 4: e1000049. [6]

Antonovics, J. 1968. Evolution in closely adjacent populations. V. Evolution of self-fertility. *Heredity* 23: 219–238. [12]

Arabidopsis Interactome Mapping Consortium. 2011. Evidence for network evolution in an *Arabidopsis* interactome map. *Science* 333: 601–601. [A2]

Araus, J. L., and J. E. Cairns. 2014. Field high-throughout phenotyping: the new crop breeding frontier. *Trends Plant Sci.* 19: 52–61. [18]

Armbruster, P., and D. H. Reed. 2005. Inbreeding depression in benign and stressful environments. *Heredity* 95: 235–242. [12]

Armitage, P. 1955. Tests for linear trends in proportions and frequencies. *Biometrics* 11: 375–386. [18]

Armstrong, D. P., and P. Cassey. 2007. Estimating the effect of inbreeding on survival. *Anim. Conserv.* 10: 487–492. [12]

Arnheim, N., C. Strange, and H. Erlich. 1985. Use of pooled DNA samples to detect linkage disequilibrium of polymorphic restriction fragments and human disease: studies of the HLA class II loci. *Proc. Natl. Acad. Sci. USA* 82: 6970–6974. [17]

Arnold, S. J. 2023. *Evolutionary quantitative genetics.* Oxford University Press. Oxford, UK. [1]

Arunachalam, V. 1977. Heterosis for characters governed by two genes. *J. Genet.* 63: 15–24. [13]

Asamoah, A., and 4 others. 1987. Segregation and linkage analysis of dopamine-β-hydrozylase activity in a six-generation pedigree. *Am. J. Med. Genet.* 27: 613–621. [16]

Aschard, H., and 5 others. 2014. Maximizing the power of principal-component analysis of correlated phenotypes in genome-wide association studies. *Am. J. Hum. Genet.* 94: 662–676. [20]

Aschard, H., and 4 others. 2015. Adjusting for heritable covariance can bias effect estimate in genome-wide association studies. *Am. J. Hum. Genet.* 96: 329–339. [20]

Asendorpf, J. B. 2012. Bias due to controlling a collider: A potentially important issue for personality research. *Europ. J. Personality* 26: 391–391. [A2]

Ashby, E., and E. Wangermann. 1954. The effects of meristem aging on the morphology and behavior of fronds in *Lemna minor*. *Ann. N.Y. Acad. Sci.* 57: 476–483. [6]

Asimit, J., and E. Zeggini. 2010. Rare variant association analysis methods for complex traits. *Ann. Rev. Genet.* 44: 293–308. [20]

Asimit, J. L., A. G. Day-Williams, A. P. Morris, and E. Zeggini. 2012. ARIEL and AMELIA: testing for an accumulation of rare variants using next-generation sequencing data. *Hum. Heredity* 73: 84–94. [20]

Asins, M. J., and E. A. Carbonell. 1988. Detection of linkage between restriction fragment length polymorphism markers and quantitative traits. *Theor. Appl. Genet.* 76: 623–626. [18]

Astle, W., and D. J. Balding. 2009. Population structure and cryptic relatedness in genetic association studies. *Stat. Sci.* 24: 451–471. [20]

Atamni, H. A. T., A. Nashel, and F. A. Iraqi. 2018. The collaborative cross mouse model for dissecting genetic susceptibility to infections diseases. *Mamm. Genome* 29: 471–487. [18]

Atchley, W. R., and D. Anderson. 1978. Ratios and the statistical analysis of biological data. *Syst. Zool.* 27: 71–78. [14]

Atchley, W. R., C. T. Gaskins, and D. Anderson. 1976. Statistical properties of ratios. I. Empirical results. *Syst. Zool.* 25: 137–148. [14]

Atkinson, A. C. 1982. Regression diagnostics, transformations and constructed variables. *J. Royal Stat. Soc. B*: 44: 1–36. [14]

Atkinson, E. G., and 13 others. 2021. Tractor uses local ancestry to enable the inclusion of admixed individuals in GWAS and to boost power. *Nat. Genet.* 53: 195–204. [20]

Atwell, S., and 35 others. 2010. Genome-wide association study of 107 phenotypes in *Arabidopsis thaliana* inbred lines. *Nature* 463: 627–631. [20]

Auer, P. L., and G. Lettre. 2015. Rare variant association studies: considerations, challenges and opportunities. *Genome Med.* 7: 16. [20]

Auger, D. L., and 5 others. 2005. Nonadditive gene expression in diploid and triploidhybrids of maize. *Genetics* 169: 389–397. [13]

Aulchenko, Y. S., D.-J. De Koning, and C. Haley. 2007. Genomewide rapid association using mixed model and regression: a fast and simple method for genomewide pedigree-based quantitative trait loci association analysis. *Genetics* 177: 577–585. [20]

Auton, A., and 12 others. 2009. Global distribution of genomic diversity underscores rich complex history of continental human populations. *Genome Res.* 19: 795–803. [8]

Ávarez, G. , F. C. Ceballos, and T. M. Berra. 2015. Darwin was right: inbreeding depression on male fertility in the Darwin family. *Biol. J. Linn. Soc.* 114: 474–483. [12]

Avery, L., and S. Wasserman. 1992. Ordering gene function: the interpretation of epistasis in regulatory hierarchies. *Trends Genet.* 8: 312–316. [5]

Avery, P. J., and W. G. Hill. 1977. Variability in genetic parameters among small populations. *Genet. Res.* 29: 193–213. [5]

Avery, P. J., and W. G. Hill. 1979. Variance in quantitative traits due to linked dominant genes and variance in heterozygosity in small populations. *Genetics* 91: 817–844. [5]

Ávila, V., and 5 others. 2006. Increase of the spontaneous mutation rate in a long–term experiment with *Drosophila melanogaster. Genetics* 173: 267–277. [15]

Ávila, V., C. Amador, and A. García-Dorado. 2010. The purge of genetic load through restricted panmixia in a *Drosophila* experiment. *J. Evol. Biol.* 23: 1937–1946. [12]

Ávila,V., A. Vilas, J. Fernándr, and A. Caballero. 2013. An experimental assessment of artificial within-family selection for fitness in conservation programs. *Conserv. Genet.* 14: 1149–1159. [12]

Axtell, J., and 4 others. 1999. Heterosis in sorghum and pearl millet. *In* J. G. Coors and S. Pandey (eds.), *The genetics and exploitation of heterosis in crops*, pp. 375–386. ASA-CSSA-SSSA Societies, Madison, WI. [13]

Aylor, D. L., and 40 others. 2011. Genetic analysis of complex traits in the emerging collaborative cross. *Gen. Res.* 21: 1213–1222. [18]

Ayroles, J F., and Z.-B. Zeng. 2008. From classical genetics to quantitative genetics to systems biology: modeling epistasis. *PLoS Genet.* 4: e1000029. [21]

Ayroles, J. F., and 7 others. 2015. Behavioral idiosyncrasy reveals genetic control of phenotypic variability. *Proc. Natl. Acad. Sci. USA* 112: 6706–6711. [6]

Azevedo, R. B. R., and 5 others. 2002. Spontaneous mutational variation for body size in *Caenorhabditis elegans. Genetics* 162: 755–765. [15]

Babron, M.-C., M. Martinez, C. Bonaïti-Pellié, and F. Clerget-Darpoux. 1993. Linkage detection by the affected-pedigree-member method: what is really tested? *Genet. Epidem.* 10: 389–394. [19]

Bacanu, S.-A., B. Devlin, and K. Roeder. 2000. The power of genomic control. *Am. J. Hum. Genet.* 66: 1933–1944. [20]

Bacanu, S.-A., B. Devlin, and K. Roeder. 2002. Association studies for quantitative traits in structured populations *Genet. Epid.* 22: 78–93. [20]

Bachmann, K., K. L. Chambers, and H. J. Price. 1985. Genome size and natural selection: observations and experiments in plants. *In* T. Cavalier-Smith (ed.), *The evolution of genome size*, pp. 267–276. John Wiley & Sons, New York, NY. [15]

Bader, R. S. 1965. Fluctuating asymmetry in the dentition of the house mouse. *Growth* 29: 291–300. [6]

Bader, R. S., and J. S. Hall. 1960. Osteometric variation and function in bats. *Evolution* 14: 8–17. [14]

Badner, J. A., A. Chakravarti, and D. K. Wagener. 1984. A test of nonrandom segregation. *Genet. Epidem.* 1: 329–340. [19]

Badyaev, A. V. 2005. Stress-induced variation in evolution: from behavioural plasticity to genetic assimilation. *Proc. Roy. Soc. Lond. B* 272: 877–885. [14]

Baer, C. F., and 10 others. 2005. Comparative evolutionary genetics of spontaneous mutations affecting fitness in rhabditid nematodes. *Proc. Natl. Acad. Sci. USA* 102: 5785–5790. [15]

Bagheri H. C. 2006. Unresolved boundaries of evolutionary theory and the question of how inheritance systems evolve: 75 years of debate on the evolution of dominance. *J. Exp. Zool.* 306B: 329–359. [4]

Bagheri, H. C., and G. P. Wagner. 2004. Evolution of dominance in metabolic pathways. *Genetics* 168: 1713–1735. [4]

Baierl, A., M. Bogdan, F. Frommlet, and A. Futschik. 2006. On locating multiple interacting quantitative trait loci in intercross designs. *Genetics* 173: 1693–1703. [18]

Baik, J., G. Ben Arous, and S. Pécheé. 2005. Phase transition of the largest eigenvalue for nonnull complex sample covariance matrices. *Ann. Prob.* 33: 1643–1697. [20]

Bailey, D. W. 1959. Rates of subline divergence in highly inbred strains of mice. *J. Heredity* 50: 26–30. [15]

Bailey, D. W. 1981. Strategic uses of recombinant inbred and cogenic strains in behavior genetics research. *In* E. S. Gershon et al. (eds.), *Genetic research strategies for psychobiology and psychiatry*, pp. 189–198. Plenum, New York, NY. [17]

Bailey, N. T. J. 1961. *The mathematical theory of genetic linkage.* Clarendon Press, Oxford, UK. [17]

Bailit, H. L., P. L. Workman, J. D. Niswander, and C. J. MacLean. 1970. Dental asymmetry as an indicator of genetic and environmental conditions in human populations. *Hum. Biol.* 42: 626–638. [6]

Baker, A. J. 1980. Morphometric differentiation in New Zealand populations of the house sparrow (*Passer domesticus*). *Evolution* 34: 638–653. [14]

Baker, W. K., and E. A. Kaeding. 1981. Linkage disequilibrium at the alpha esterase loci in a population of *Drosophila melanogaster* from Utah. *Am. Nat.* 117: 804–809. [5]

Bakshi, A., and 6 others. 2016. Fast set-based association analysis using summary data from GWAS identifies novel gene loci for human complex traits. *Sci. Reports* 6: 32894. [20]

Balao, F., J. Herrera, and S. Talavera. 2011. Phenotypic consequences of polypoid and genome size at the microevolutionary scale: a multivariate morphological approach. *New Phyto.* 192: 256–265. [15]

Baldauf, J. A., and 8 others. 2018. Single-parent expression is a general mechanism driving extensive complementation of non-syntenic genes in maize hybrids. *Curr. Biol.* 28: 431–437. [13]

Baldauf, J. A., and 7 others. 2022. Single-parent expression complementation contributes to phenotypic heterosis in maize hybrids. *Plant Physiol.* 189: 1625–1638. [13]

Baldwin, S. J., and D. J. Schoen. 2019. Inbreeding depression is difficult to purge in self-incompatible populations of *Leavenworthia alabamica*. *New Phytol.* 224: 1330–1338. [12]

Ball, R. D. 2011. Experimental design for robust detection of effects in genome-wide case-control studies. *Genetics* 189: 1497–1514. [20]

Ball, S. T., D. J. Mulla, and C. F. Konzak. 1993. Spatial heterogeneity affects variety trials interpretation. *Crop Sci.* 33: 931–935. [A9]

Ballard, D. H., J. Cho, and H. Zhao. 2010. Comparison of multiple-marker association methods to detect association between a candidate region and disease. *Genet. Epidem.* 34: 201–212. [20]

Ballou, J. D. 1997. Ancestral inbreeding only minimally affects inbreeding depressionin mammalian populations. *J. Hered.* 88: 169–178. [12]

Ballou, J. D., and R. C. Lacy. 1995. Identifying genetically important individuals for management of genetic variation in pedigreed populations. *In* J. D. Ballou, M. Gilpin, and T. J. Foose (eds) *Population management for survival and recovery: analytical methods and strategies in small population conservation*, pp. 76–111. Columbia University Press, New York, NY. [12]

Balloux, F., W. Amos, and T. Coulson. 2004. Does heterozygosity estimate inbreeding in real populations? *Mol. Ecol 13* : 3021–3031. [12]

Bamshad, M., M. H. Crawford, D. O'Rourke, and L. B. Jorde. 1994. Biochemical heterozygosity and morphologic variation in a colony of *Papio hamadryas* baboons. *Evolution* 48: 1211–1221. [6]

Banerjee, S., L. Zeng, H. Schunket, and J. Söding. 2018. Bayesian multiple logistic regression for case-control GWAS. *PLoS Gesnet.* 14: e1007856. [20]

Bansal, V., O. Libiger, A. Torkamani, and N. J. Schork. 2010. Statistical analysis strategies for association studies involving rare variants. *Nat. Rev. Genet.* 11: 773–785. [20]

Bao, K., R. H. Melde, and N. P. Sharp. 2022. Are mutations usually deleterious? A perspective on the fitness effects of mutation accumulation. *Evol. Ecol.* 36: 753–766. [15]

Barabási, A.-L., and Z. N. Oltvai. 2004. Network biology: understanding the cell's functional organization. *Nat. Rev. Genet.* 5: 101–113. [21, A2]

Baranzini, S. E., and 15 others. 2009. Pathway and network-based analysis of genome-wide association studies in multiple sclerosis. *Hum. Mol. Genetics* 18: 2078–2090. [21]

Barbeira, A. N., and 14 others. 2018. Exploring the phenotypic consequences of tissue specific gene expression variation inferred from GWAS summary statistics. *Nat. Comm.* 9: 1825. [21]

Barbeira, A. N., and 27 others. 2021. Exploiting the GTEx resources to decipher the mechanisms at GWAS loci. *Genome Biol.* 22: 49. [21]

Barber, M. J., H. J. Cordell, A. J. MacGregor, and T. Andrew. 2004. Gamma regression improves Haseman-Elston and variance components linkage analysis for sib-pairs. *Genet. Epid.* 26: 97–107. [19]

Barbosa, A. M. M., and 5 others. 2003. Relationship of intra-and interpopulation tropical maize single cross hybrid performance and genetic distances computed from AFLP and SSR markers. *Euphytica* 130: 87–99. [13]

Barden, H. S. 1980. Fluctuating dental asymmetry: a measure of developmental asymmetry in Down syndrome. *Am. J. Phys. Anthropol.* 52: 169–173. [6]

Baret, P. V., S. A. Knott, and P. M. Visscher. 1998. On the use of linear regression and maximum likelihood for QTL mapping in half-sib designs. *Genet. Res.* 72: 149–158. [19]

Barker, J. S. F. 1979. Inter-locus interactions: a review of experimental evidence. *Theor. Pop. Biol.* 16: 323–346. [5]

Barlow, R. 1981. Experimental evidence for interaction between heterosis and environment in animals. *Anim. Breed. Abst.* 49: 715–737. [13]

Barndorff-Nielsen, O. E., and D. R. Cox. 1979. Edgeworth and saddlepoint approximations with statistical applications (with discussion). *J. Roy. Statist. Sco. B* 41: 279–312. [20]

Barnes, B. 1966. Environment and selection in *Drosophila melanogaster.* Ph. D. thesis. Birmingham Univ., Birmingham, UK. [17]

Barrai, I., L. L. Cavalli-Sforza, and M. Mainardi. 1964. Testing a model of dominant inheritance for metric traits in man. *Heredity* 19: 651–668. [12]

Barrett, S. C. H., and D. Charlesworth. 1991. Effects of a change in the level of inbreeding on the genetic load. *Nature* 352: 522–524. [12]

Bartlett, M. S. 1936. Square-root transformation in analysis of variation. *J. Roy. Stat. Soc.* 3: 68–78. [14]

Bartlett, M. S. 1947. The use of transformations. *Biometrics* 3: 39–52. [14]

Bartlett, M. S. 1978. Nearest neighbour models in the analysis of field experiments (with discussion). *J. Roy. Stat. Soc. B* 40: 147–174. [A9]

Bartlett, M. S., and J. B. S. Haldane. 1934. The theory of inbreeding in autotetraploids. *J. Genet.* 29: 175–180. [12, 13]

Bartlett, M. S., and J. B. S. Haldane. 1935. The theory of inbreeding with forced heterozygosity. *J. Genet.* 31: 327–340. [17]

Barton, N. H. 1990. Pleiotropic models of quantitative variation. *Genetics* 124: 773–782. [15]

Barton, N. H., and B. Charlesworth. 1984. Genetic revolutions, founder events and speciation. *Ann. Rev. Ecol. Syst.* 15: 133–164. [17]

Barton, N. H., and G. M. Hewitt. 1981. Hybrid zones and speciation. *In* W. R. Atchley and D. S. Woodruff (eds.), *Evolution and speciation: essays in honor of M. J. D. White,* pp. 109–145. Cambridge Univ. Press, Cambridge, UK. [13]

Barton, N. H., and M. Turelli. 1989. Evolutionary quantitative genetics: how little do we know? *Ann. Rev. Genetics* 23: 337–370. [1]

Bashaw, E. C. 1980. Apomixis and its application in crop improvement. *In* W. R. Fehr and H. H. Hadley (eds.), *Hybridization of crop plants*, pp. 45–63. Amer. Soc. Agron. Madison, WI. [13]

Baskin, J. M., and C. C. Baskin. 2017. Seed germination in cleistogamous species: theoretical considerations and a literature survey of experimental results. *Seed Sci. Res.* 27: 84–98. [12]

Basu, S., and W. Pan. 2011. Comparison of statistical test for disease association with rare variants. *Genet. Epid.* 35: 606–619. [20]

Bateman, A. J. 1959. The viability of near-normal irradiated chromosomes. *Internat. J. Rad. Biol.* 1: 170–180. [15]

Bateson, P. 1983. *Mate choice.* Cambridge Univ. Press, Cambridge, UK. [13]

Bateson, W. 1909. *Mendel's principles of heredity.* Cambridge University Press, Cambridge, UK. [5]

Bateson, W., and H. Killby. 1905. Experimental studies in the physiology of heredity: Peas (*Pisum sativum*) *Reports to the Evolution Committee of the Royal Society* 2: 55–80. [5]

Battle, A., and 13 others. 2014. Characterizing the genetic basis of transcriptome diversity through RNA-sequencing of 922 individuals. *Genome Res.* 24: 14–24. [21]

Battle, A., and 6 others. 2015. Impact of regulatory variation from RNA to protein. *Science* 347: 664–667. [21]

Bazakos, C., and 4 others. 2017. New strategies and tools in quantitative genetics: how to go from phenotype to the genotype. *Ann. Rev. Plant Biol.* 68: 435–455. [18]

Beadle, G. W. 1939. Teosinte and the origin of maize. *J. Heredity* 30: 245–247. [5, 18]

Beadle, G. W. 1980. The ancestry of corn. *Sci. Am.* 242: 112–119, 162. [18]

Beardmore, J. A. 1960. Developmental stability in constant and fluctuating environments. *Heredity* 14: 411–422. [6]

Beardmore, J. A. 1970. Viral components in the genetic background? *Nature* 226: 766–767. [15]

Beardmore, J. A., and S. A. Shami. 1976. Parental age, genetic variation and selection. *In* S. Karlin and E. Nevo (eds.), *Population genetics and ecology,* pp. 3–22. Academic Press, New York, NY. [6]

Beardmore, J. A., F. Lints, and A. L. F. Al-Baldawi. 1975. Parental age and heritability of sternopleural chaeta number in *Drosophila melanogaster. Heredity* 34: 71–82. [6]

Beasley, D. A. E., A. Bonisoli-Alquati, and T. A. Mousseau. 2013. The use of fluctuating asymmetry as a measure of environmentally induced developmental instability: A meta-analysis. *Ecol. Indicators* 30: 218–226. [6]

Beaty, T. H., and 5 others. 1985. Use of robust variance components models to analyze triglyceride data in families. *Ann. Hum. Genet.* 49: 315–328. [19]

Beavis, W. D. 1994. The power and deceit of QTL experiments: lessons from comparative QTL studies. *In 49th Annual Corn and Sorghum Research Conference.* pp. 252–268. American Seed Trade Association, Washington, D.C. [18]

Beavis, W. D., D. Grant, M. Albertsen, and R. Fincher. 1991. Quantitative trait loci for plant height in four maize populations and their associations with qualitative genetic loci. *Theor. Appl. Genet.* 83: 141–145. [17]

Becker, A., and 4 others. 2011. Bulk segregant analysis using single nucleotide polymorphism microarrays. *PLoS ONE* 6: e15993. [17]

Becker, J., and E. A. Craig. 1994. Heat-shock proteins as molecular chaperones. *Eur. J Biochem.* 219: 11–23. [14]

Becker, P. J., J. Hegelbach, L. F. Keller, and E. Postma. 2016. Phenotype associated inbreeding biases estimates of inbreeding depression in a wild bird population. *J. Evol. Biol.* 29: 35–46. [12]

Beckmann, J. S., and M. Soller. 1983. Restriction fragment length polymorphisms in genetic improvement: methodologies, mapping and costs. *Theo. Appl. Genet.* 67: 35–43. [8, 17]

Beckmann, J. S., and M. Soller. 1986a. Restriction fragment length polymorphisms and genetic improvement in agricultural species. *Euphytica* 35: 111–124. [8]

Beckmann, J. S., and M. Soller. 1986b. Restriction fragment length polymorphisms in plant genetic improvement. *Oxford Surveys Plant Mol. Cell Biol.* 3: 196–250. [8]

Beckmann, J. S., and M. Soller. 1988. Detection of linkage between marker loci and loci affecting quantitative traits in crosses between segregating populations. *Theor. Appl. Genet.* 76: 228–236. [18, 19]

Begg, C. B. and M. Mazumda. 1994. Operating characteristics of a rank correlation test for publication bias. *Biometrics* 50: 1088–1101. [A6]

Begum, F., D. Ghosh, G. C. Tseng, and E. Feingold. 2012. Comprehensive literature review and statistical considerations for GWAS meta-analysis. *Nucl. Acid. Res.* 40: 3777–3784. [20]

Behringer, R. 1998. Supersonic congenics? *Nat. Genet.* 18: 108. [17]

Belknap, J. K. 1998. Effect of within-strain sample size on QTL detection and mapping using recombinant inbred mouse strains. *Behav. Genet.* 28: 29–38. [18]

Belknap, J. K., and 6 others. 1993. Quantitative trait loci (QTL) applications to substances of abuse: physical dependence studies with nitrous oxide and ethanol in B × D mice. *Behav. Genet.* 23: 213–222. [17]

Bell, A. E. 1977. Heritability in retrospect. *J. Heredity* 68: 297–300. [7]

Bell, J. I., and G. M. Lathrop. 1996. Multiple loci for multiple sclerosis. *Nature Genetics* 13: 377–378. [19]

Benayoun, L., and 9 others. 2009. Genetic heterogeneity in two consanguineous families segregating early onset retinal degeneration: the pitfalls of homozygosity mapping. *Am. J. Med. Genet.* 650–656. [12]

Benchimol, L. L., and 7 others. 2000. Genetic diversity in tropical maize inbred lines: heterotic group assignment and hybrid performance determined by RFLP markers. *Plant Breeding* 119: 491–496. [13]

Bengtsson, B. O., and G. Thomson. 1981. Measuring the strength of associations between HLA antigens and diseases. *Tissue Antigens* 18: 356–363. [17].

Benjamini, Y. 2010. Discovering the false discovery rate. *J. R. Statist. Soc.* B: 72: 405–416. [A6]

Benjamini, Y., and Y. Hochberg. 1995. Controlling the false discovery rate: a practical and powerful approach to multiple testing. *J. Royal Stat. Soc. B* 85: 289–300. [A6]

Benjamini, Y., and D. Yekutieli. 2005. Quantitative trait loci analysis using the false discovery rate. *Genetics* 171: 783–790. [18, A6]

Benn, M., B. G. Nordestgaard, R. Frikke-Schmidt, and A. Tybjaerg-Hansen. 2017. Low LDL cholesterol, *PCSK9* and *HMGCR* genetic variation, and risk of Alzheimer's disease and Parkinson's disease: Mendelian randomization study. *BMJ* 357: j1648. [A2]

Benner, C., and 5 others. 2016. FINEMAP: efficient variable selection using summary data form genome-wide association studies. *Bioinformatics* 32: 1493–1501. [21]

Bennett, J. H. 1954. Panmixia with tetrasomic and hexasomic inheritance. *Genetics* 39: 150–158. [5]

Benson, D. L., and A. R. Hallauer. 1994. Inbreeding depression rates in maize populations before and after recurrent selection. *J. Heredity* 85: 122–128. [12]

Bentolila, S., and 5 others. 1991. Identification of an RFLP marker tightly linked to the *Ht1* gene in maize. *Theor. Appl. Genet.* 82: 393–398. [17]

Bentsen, H. B., and G. Klemetsdal. 1991. The use of fixed-effects models and mixed models to estimate single-gene associated effects on polygenic traits. *Genet. Sel. Evol.* 23: 407–419. [17]

Bérénos, C., P. A. Ellis, J. G. Pilkington, and J. M. Pemberton. 2014. Estimating quantitative genetic parameters in wild populations: a comparison of pedigree and genomic approaches. *Mol. Ecol.* 23: 3434–3451. [8]

Bérénos, C., P. A. Ellis, J. G. Pilkington, and J. M. Pemberton. 2016. Genomic analysis reveals depression due to both individual and maternal inbreeding in a free-living mammal population. *Mol. Ecol.* 25: 3152–3168. [12]

Bereskin, B., and 11 others. 1968. Inbreeding and swine productivity traits. *J. Anim. Sci.* 27: 339–350. [12]

Berger, J. O. 1985. *Statistical decision theory and Bayesian analysis.* Springer-Verlag, New York, NY. [A7]

Berke, T. G., and T. R. Rocheford. 1995. Quantitative trait loci for flowering, plant height, and kernel traits in maize. *Crop Sci.* 35: 1542–1549. [17, 18]

Bernardo, R. 2004. What proportion of declared QTL in plants are false? *Theor. Appl. Genet.* 109: 419–424. [18]

Bernardo, R. 2010. *Breeding for quantitative traits in plants,* 2nd Ed. Stemma Press, Woodbury, MN. [1]

Bernardo, R. 2020. *Breeding for quantitative traits in plants,* 3rd Ed. Stemma Press, Woodbury, MN. [1]

Berrar, D. P., W. Dubitzky, and M. Granzow (eds). 2003. *A practical approach to microarray data analysis.* Kluwer, New York, NY. [21]

Bertorelle, G., and 7 others. 2022. Genetic load: genomic estimates and application in non-model animals. *Nat. Rev. Genet.* 23: 492–503. [12]

Bertrand, A. R., and 4 others, 2018. RZooRoH: An R package to characterized individual genomic autozygosity and identify homozygous-by-descent segments. *Methods Ecol. Evol.* 10: 860–886. [8]

Besag, J., and R. Kempton. 1986. Statistical evaluation of field experiments using neighbouring plots. *Biometrics* 42: 231–251. [A9]

Besag, J., P. J. Green, D. Higdon, and K. L. M. Mengersen. 1995. Bayesian computation and stochastic systems (with discussion). *Statistical Science* 10: 3–66. [A8]

Bezdíček, J. 2017. Effects of inbreeding on milk production, production and type traits in dairy cattle. *Züchtungskundr* 89: 475–491. [12]

Bhatia, G., and 6 others. 2010. A covering method for detecting genetic associations between rare variants and common phenotypes. *PLoS Comput. Biol.* 6: e1000954. [20]

Bhattacharjee, S., and 10 others. 2012. A subset-based approach improves power and interpretation for the combined analysis of genetic association studies of heterogeneous traits. *Am. J. Hum. Genet.* 90: 821–835. [20]

Bickeböller, H., and F. O. Clerget-Darpoux. 1995. Statistical properties of the allelic and genotypic transmission/disequilibrium test for multiallelic markers. *Genet. Epid.* 12: 865–870. [17]

Bidanel, J. P. 1993. Estimation of crossbreeding parameters between Large White and Meishan porcine breeds. III. Dominance and epistatic components of heterosis on reproductive traits. *Genet. Sel. Evol.* 25: 263–281. [13]

Biémont, C. 1983. Homeostasis, enzymatic heterozygosity and inbreeding depression in natural populations of *Drosophila melanogaster. Genetica* 61: 179–189. [6]

Biernacka, J. M., and 4 others. 2012. Use of the gamma method for self-contained gene-set analysis of SNP data. *Eur. J. Hum. Genet.* 20: 565–571. [20]

Bierne, N., S. Launey, Y. Naciri-Graven, and F. Bonhomme. 1998. Early effect of inbreeding as revealed by microsatellite analyses on *Ostrea edulis* larvae. *Genetics* 148: 1983–1906. [12]

Bierne, N., A. Tsitrone, and P. David. 2000. An inbreeding model of associative overdominance during a population bottleneck. *Genetics* 155: 1981–1990. [12]

Billewicz, W. Z. 1972. A note on birth weight correlation in full-sibs. *J. Biosocial Sci.* 4: 455–460. [7]

Billiard, S., and V. Castric. 2011. Evidence for Fisher's dominance theory: how many special cases. *Trends Genet.* 27: 441–445. [4]

Bingham, E. T. 1999. Role of chromosome blocks in heterosis and estimates of dominance and overdominance. *In* J. G. Coors and S. Pandey (eds.), *The genetics and exploitation of heterosis in crops*, pp. 71–87. ASA-CSSA-SSSA Societies, Madison, WI. [13]

Bingham, E. T., R. W. Groose, D. R. Woodfield, and K. K. Kidwell. 1994. Complementary gene interactions in alfalfa are greater in autotetraploids than diploids. *Crop Sci.* 34: 823–829. [13]

Birchler, J. A. 2015. The genetic basis of hybrid vigour. *Nature Plants* 1: 1–2. [13]

Birchler, J. A., and R. A. Veitia. 2012. Gene balance hypothesis: connecting issues of dosage sensitivity across biological disciplines. *Proc. Natl. Acad. Sci. USA* 109: 14746–14753. [13]

Birchler, J. A., D. L. Auger, and N. C. Riddle. 2003. In search of the molecular basis of heterosis. *Plant Cell* 15: 2236–2239. [13]

Birchler, J. A., N. C. Riffle, D. L. Auger, and R. A. Veitia. 2005. Dosage balance in gene regulation: biological implications. *Trends Genet.* 21: 219–226. [13]

Birchler, J. A., H. Yao, and S. Chudalayandi. 2006. Unraveling the genetic basis of hybrid vigor. *Proc. Natl. Acad. Sci. USA* 103: 12957–12958. [12]

Birchler, J. A., A. F. Johnson, and R. A. Veitia. 2016. Kinetics genetics: Incorporating the concept of genomic balance into an understanding of quantitative traits. *Plant Sci.* 245: 128–134. [13]

Birge, R. T., 1932. The calculation of errors by the method of least squares. *Physical Review* 40: 207–227. [A6]

Birnbaum, A. 1954. Combining independent tests of significance. *J. Am. Stat. Assoc.* 49: 122–136. [A6]

Birnbaum, A. 1972. The random phenotype concept, with applications. *Genetics* 72: 739–758. [16]

Bishir, J., and G. Namkoong. 1987. Unsound seeds in conifers: estimation of numbers of lethal alleles and of magnitudes of effects associated with the maternal plant. *Silvae Genetica* 36: 180–185. [12]

Bishop, D. T., and J. A. Williamson. 1990. The power of identity-by-state methods for linkage analysis. *Am. J. Hum. Genet.* 46: 254–265. [19]

Bishop, D. T., C. Cannings, M. Skolnick, and J. A. Williamson. 1983. The number of polymorphic DNA clones required to map the human genome. *In* B. S. Weir (ed), *Statistical analysis of DNA sequence data,* pp. 181– 200. Marcel Dekker, New York, NY. [17].

Bishop, G. R. 1992. Phenotypic variability of polygenic traits. *Acta Zool. Fennica* 191: 133–136. [6]

Biswas, S., J. D. Storey, and J. M. Akey. 2008. Mapping gene expression quantitative trait loci by singular value decomposition and independent component analysis. *BMC Bioinformatics* 9: 244. [21]

Bittante, G., L. Gallo, and P. Montobbio. 1993. Estimated breed additive effects and direct heterosis for growth and carcass traits of heavy pigs. *Livestock Prod. Sci.* 34: 101–114. [13]

Bittles, A. H., and M. L. Black. 2010. Consanguinity, human evolution, and complex diseases. *Proc. Natl. Acad. Sci. USA* 107: 1779–1786. [12]

Bittles, A. H., and J. V. Neel. 1994. The costs of human inbreeding and their implications for variations at the DNA level. *Nat. Genetics* 8: 117–121. [12]

Bjelland, D. W., K. A. Weigel, N. Vukasinovic, and J. D. Nkrumah. 2013. Evaluation of inbreeding depression in Holstein cattle using whole-genome SNP markers and alternative measures of genomic inbreeding. *J. Dairy Sci.* 96: 4697–4706. [12]

Blackwelder, W. C., and R. C. Elston. 1974. Comment on Dr. Roberston's communication. *Behav. Genet.* 4: 97–99. [19]

Blackwelder, W. C., and R. C. Elston. 1982. Power and robustness of sib-pair linkage tests and extension to larger sibships. *Commun. Stat. Theor. Meth.* 11: 449–484. [19]

Blackwelder, W. C., and R. C. Elston. 1985. A comparison of sib-pair linkage tests for disease susceptibility loci. *Genet. Epidem.* 2: 85–97. [19]

Blanco, L. G., and J. A. Sanchez-Prado. 1986. Enzymatic heterozygosity and morphological variance in synthetic populations of *Drosophila melanogaster. Genet. Sel. Evol.* 18: 417–426. [6]

Blangero, J., and L. Almasy. 1997. Multipoint oligogenic linkage analysis in quantitative trait. *Genet. Epid.* 14: 959–964. [19]

Blangero, J., and L. W. Konigsberg. 1991. Multivariate segregation analysis using the mixed model. *Genet. Epidem.* 8: 299–316. [16]

Blangero, J., and 5 others. 1990. Genetic analysis of apolipoprotein A1 in two dietary environments. *Am. J. Hum. Genet.* 47: 414–428. [16]

Blangero, J., S. Williams-Blangero, and J. E. Hixson. 1992. Assessing the effects of candidate genes on quantitative traits in primate populations. *Am. J. Primatol.* 27: 119–132. [17]

Blangero, J., J. T. Williams, and L. Almasy. 2000. Robust LOD scores for variance component-based linkage analysis. *Genet Epid.* 19: S8–S14. [19]

Blangero, J., J. T. Williams, and L. Almasy. 2001. Variance component methods for detecting complex trait loci. *Adv. Genetics* 42: 151–181. [19]

Blasco, A. 2017. *Bayesian data analysis for animal scientists.* Springer, Gewerbestrasse, CH. [A7]

Bloom, J. S., and 4 others. 2013. Finding the sources of missing heritability in a yeast cross. *Nature* 494: 234–237. [5, 21]

Bloom, J. S., and 5 others. 2015. Genetic interactions contribution less than additive effects to quantitative trait variation in yeast. *Nat. Comm.* 6: 8712. [5, 18]

Bloom, J. S., and 4 others. 2019. Rare variants contribute disproportionately to quantitative trait variation in yeast. *eLife* 8: e49212. [21]

Blouin, M. S. 2003. DNA-based methods for pedigree reconstruction and kinship analysis in natural populations. *Trends Ecol. Evol.* 18: 503–511. [8]

Blouin, M. S., M. Parsons, V. Lacaille, and S. Lotz. 1996. Use of microsatellite loci to classify individuals by relatedness. *Mol. Ecol.* 5: 393–401. [8]

Boake, C. R. B. (ed.) 1994. *Quantitative genetic studies of behavioral evolution.* Univ. Chicago Press, Chicago, IL. [1]

Boakes, E. H., and J. Wang. 2005. A simulation study on detecting purging of inbreeding depression in captive populations. *Genet. Res.* 86: 139–148. [12]

Boakes, E. H., J. Wang, and W. Amos. 2007. An investigation of inbreeding depression and purging in captive pedigreed populations. *Heredity* 98: 172–182. [12]

Bodmer, W. F. 1986. Human genetics: the molecular challenge. *Cold Spring Harbor Symp. Quant. Biol.* 51: 1–13. [17, 20]

Bodmer, W. 1999. Familial adenomatous polyposis (FAP) and its gene, *APC. Cytogenet. Cell. Genet.* 86: 99–104. [21]

Bodmer, W., and C. Bonilla. 2008. Common and rare variants in multifactorial susceptibility to common disease. *Nat. Genet.* 40: 695–701. [20, 21]

Boef, A. G. C., O. M. Dekkers, and S. le Cessie. 2015. Mendelian randomization studies: a review of the approaches used and the quality of reporting. *Int. J. Epidem.* 44: 496–511. [A2]

Boehnke, M. 1994. Limits of resolution of genetic linkage studies: implications for the positional cloning of human disease genes. *Am. J. Hum. Genet.* 55: 379–390. [17]

Boehnke, M., and N. J. Cox. 1997. Accurate inference of relationships in sib-pair linkage studies. *Am. J. Hum. Genet.* 61: 423-429. [8]

Boehnke, M., and P. P. Moll. 1989. Identifying pedigrees segregating at a major locus for a quantitative trait: an efficient strategy for linkage analysis. *Am. J. Hum. Genet.* 44: 216–224. [19]

Boerwinkle, E., and C. F. Sing. 1986. Bias of the contribution of single-locus effects to the variance of a quantitative trait. *Am. J. Hum. Genet.* 39: 137–144. [17]

Boerwinkle, E., and C. F. Sing. 1987. The use of measured genotype information in the analysis of quantitative phenotypes in man. III. Simultaneous estimation of the frequencies and effects of apolipoprotein E polymorphism and residual polygenetic effects on cholesterol, betalipoprotein and triglyceride levels. *Ann. Hum. Genet.* 51: 211–226. [17]

Boerwinkle, E., R. Chakraborty, and C. F. Sing. 1986. The use of measured genotype information in the analysis of quantitative phenotypes in man. I. Models and analytical methods. *Ann. Hum. Genet.* 50: 181–194. [16, 17]

Bogdan, M., and 5 others. 2008. Extending the modified Bayesian Information Criterion (mBIC) to dense markers and multiple interval mapping. *Biometrics* 64: 1162–1169. [18]

Bohrnstedt, G. W., and A. S. Goldberger. 1969. On the exact covariance of products of random variables. *J. Am. Stat. Assoc.* 64:1439–1442. [A1]

Bolker, B. M. 2015. Linear and generalized linear mixed models. *In* G. A. Fox (ed.) *Ecological statistics: Contemporary theory and applications*, pp. 306–333. Oxford University Press, Oxford, UK. [14]

Bolker, B. M., and 6 others. 2008. Generalized linear mixed models: a practical guide for ecology and evolution. *Trends Ecol. Evol.* 24: 127–135. [14]

Bollen, K. A. 1989. *Structural equations with latent variables.* Wiley, New York, NY. [A2]

Bolormaa, S., and 9 others. 2014. A multi-trait, meta-analysis for detecting pleiotropic polymorphisms for stature, fatenss and reproduction in breed cattle. *PLoS Genet.* 10: e1004198. [20]

Böndel, K. B., and 7 others. 2019. Inferring the distribution of fitness effects of spontaneous mutations in *Chlamydomonas reinhardtii. PLoS Biol.* 17: e3000192. [15]

Bonierbale, M. W., R. L. Plaisted, and S. D. Tanksley. 1988. RFLP maps based on a common set of clones reveal modes of chromosomal evolution in potato and tomato. *Genetics* 120: 1095–1103. [18]

Bonnet, O. T., and W. B. Bever. 1947. Head-hill method of planting head selections of small grains. *J. Amer. Soc. Agron.* 39: 442–445. [A9]

Bonney, G. E. 1984. On the statistical determination of major gene mechanisms in continuous human traits: regressive models. *Am. J. Med. Genet.* 18: 731–749. [16]

Bonney, G. E. 1992. Compound regression models for family data. *In* J. Ott (ed.), *Models and methods for the genetic analysis of pedigree data,* pp. 28–41. Karger, Basel, CH. [16]

Bonney, G. E., G. M. Dunston, and J. Wilson. 1989. Regressive logistic models for ordered and unordered polychotomous traits: application to affective disorders. *Genet. Epidem.* 6: 211–215. [16]

Booth, C. L., D. S. Woodruff, and S. J. Gould. 1990. Lack of significant associations between allozyme heterozygosity and phenotypic traits in the land snail *Cerion. Evolution* 44: 210–213. [6]

Borecki, I. B., M. A. Province, and D. C. Rao. 1995. Inferring a major gene for quantitative traits by using segregation analysis with tests on transmission probabilities: how often do we miss? *Am. J. Hum. Genet.* 56: 319–326. [16]

Borenstein, M., L. V. Hedges, J. P. T. Higgins, and H. R. Rothstein. 2010. A basic introduction to fixed-effect and random-effects models for meta-analysis. *Res. Syn. Meth.* 1: 97–111. [A6]

Borenstein, M., L. V. Hedges, J. P. T. Higgins, and H. R. Rothstein. 2011. *Introduction to meta-analysis*. Wiley, Chichester, UK. [A6]

Borevitz, J. O., and 8 others. 2003. Large-scale identification of single-feature polymorphisms in complex genomes. *Genome Res.* 13: 513–523. [17]

Borges, A., and 5 others. 2018. Can spatial modeling substitute for experimental design in agricultural experiments? *Crop Sci.* 59: 44–53. [A9]

Bos, I., and P. Caligari. 1995. *Selection methods in plant breeding.* Chapman and Hall, London, UK. [1]

Bost, B., and R. A. Veitia. 2013. Dominance and interlocus interactions in transcriptional activation cascades: models explaining compensatory mutations and inheritance patterns. *Bioessays* 36: 84–92. [4]

Botstein, D., and N. Risch. 2003. Discovering genotypes underlying human phenotypes: past success and mendelian disease, future approaches for complex disease. *Nat. Genet.* 33: 228–237. [12, 15, 21]

Botstein, D., R. L. White, M. Skolnick, and R. W. Davis. 1980. Construction of a genetic linkage map in man using restriction fragment length polymorphisms. *Am. J. Hum. Genet.* 32: 314–331. [8, 17, 19]

Boucher, W. 1988. Calculation of the inbreeding coefficient. *J. Math. Biol.* 26: 57–64. [7]

Bourguet, D. 1999. The evolution of dominance. *Heredity* 83: 1–4. [4]

Bovenhuis, H., and R. J. Spelman. 2000. Selective genotyping to detect quantitative trait loci for multiple traits in outbed populations. *J. Dairy Sci.* 83: 173–180. [19]

Bovenhuis, H., and J. I. Weller. 1994. Mapping and analysis of dairy cattle quantitative trait loci by maximum likelihood methodology using milk protein genes as genetic markers. *Genetics* 137: 267–280. [17]

Bowden, J. 2017. Misconceptions on the use of ME-Egger regression and the evaluation of the InSIDE assumption. *Int. J. Epidem.* 46: 2097–2099. [A2]

Bowden, J., and F. Dudbridge. 2009. Unbiased estimation of odds ratios: combining genomewide association scans with replication studies. *Genet. Epidem.* 33: 406–418. [20]

Bowden, J., G. Davey Smith, and S. Burgess. 2015. Mendelian randomization with invalid instruments: effect estimation and bias detection through Egger regression. *Int. J. Epidem.* 44: 512–525. [A2]

Bowden, J., and 5 others. 2016a. Assessing the suitability of summary data for two-sample Mendelian randomization analyses using MR-Egger regression: the role of the I^2 statistic. *Int. J. Epidem.* 45: 1961–1974. [A2]

Bowden, J., G. Davey Smith, P. C. Haycock, and S. Burgess. 2016b. Consistent estimation in Mendelian randomization with some invalid instruments using a weighted median. *Genet. Epidem.* 40: 304–314. [A2]

Bowden, J., and 5 others. 2017. A framework for the investigation of pleiotropy in two-sample summary data Mendelian randomization. *Stat. Med.* 36: 1783–1802. [A2]

Bowden, J., G. Hemani, and G. Davey Smith. 2018. Invited commentary: detecting individual and global horizontal pleiotropy in Mendelian randomization — a job for the humble heterogeneity statistic? *Am. J. Epidem.* 187: 2681–2685. [A2]

Bowden, J., and 7 others. 2019. Improving the accuracy of two-sample summary-data Mendelian randomization: moving beyond the NOME assumption. *Int. J. Epidem.* 48: 728–742. [A2]

Bowman, D. T. 1990. Trend analysis to improve efficiency of agronomic trials in flue-cured tobacco. *Agron. J.* 82: 499–501. [A9]

Bowman, J. C., and D. S. Falconer. 1960. Inbreeding depression and heterosis of litter size in mice. *Genet. Res.* 1: 262–274. [12]

Bowman, K. O., and L. R. Shenton. 1975. Omnibus test contours for departures from normality based on $\sqrt{b_1}$ and b_2. *Biometrika* 62: 243–250. [14]

Box, G. E. P., and D. R. Cox. 1964. An analysis of transformations. *J. Royal Stat. Soc.* B 26: 211–252. [14, 16]

Box, J. F. 1980. R. A. Fisher and the design of experiments, 1922–1926. *Amer. Stat.* 34: 1–7. [A9]

Box, J. F. 1981. Gosset, Fisher, and the t distribution. *Amer. Stat.* 35: 61–66. [2]

Boyle, E. A., Y. I. Li, and J. K. Pritchard. 2017a. An expanded view of complex traits: from polygenic to omnigenic. *Cell* 169: 1177–1186. [1, 15, 21]

Boyle, E. A., Y. I. Li, and J. K. Pritchard. 2017b. The omnigenic model: response from the authors. *J. Psych. Brain Sci.* 2: S7. [21]

Boyle, C. R., and R. C. Elston. 1979. Multifactorial genetic models for quantitative traits in humans. *Biometrics* 35: 55–68. [16]

Bozzuto, C., and 4 others. 2019. Inbreeding reduces long-term growth of Alpine ibex populations. *Nat. Ecol. Evol.* 3: 1359–1364. [12]

Bradbury, P., T. Parker, M. T. Hamblin, and J.-L. Jannink. 2011. Assessment of power and false discovery rate in genome-wide associations studies using the BarleyCAP germplasm. *Crop Sci.* 51: 52–59. [20]

Bradford, G. E., D. T. Torell, G. M. Spurlock, and W. C. Weir. 1963. Performance and variability of offspring of crossbred and purebred rams. *J. Anim. Sci.* 22: 617–623. [13]

Bradley, B. P. 1978. Genetic and physiological adaptation of the copepod *Eurytemora affinia* to seasonal temperatures. *Genetics* 90: 193–205. [7]

Bradshaw, H. D., Jr., and D. W. Schemske. 2003. Allele substitution at a flower colour locus produces a pollinator shift in monkeyflowers. *Nature* 426: 176–178. [18]

Bradshaw, H. D., Jr., S. M. Wilbert, K. G. Otto, and D. W. Schemske. 1995. Genetic mapping of floral traits associated with reproductive isolation in monkeyflowers (*Mimulus*). *Nature* 376: 762–765. [18]

Bradshaw, H. D., Jr., and 4 others. 1998. Quantitative trait loci affecting differences in floral morphology between two species of monkeyflowers (*Mimulus*). *Genetics* 149: 367–382. [18]

Brandenbrug, J.-T., and 9 others. 2017. Independent introductions and admixtures have contributed to adaptation of European maize and its America counterparts *PLoS Genet.* 13: e1006666. [12, 13]

Brauer, M. J., C. M. Christianson, D. A. Pai, and M. J. Dunham. 2006. Mapping novel traits by array-assisted bulk segregant analysis in *Saccharomyces cerevisiae*. *Genetics* 173: 1813–1816. [17]

Breese, E. L. 1956. The genetical consequences of assortative mating. *Heredity* 10: 323–343. [7]

Breese, E. L., and K. Mather. 1957. The organization of polygenic activity within a chromosome in *Drosophila*. I. Hair characters. *Heredity* 11: 373–395. [17]

Breese, E. L., and K. Mather. 1960. The organization of polygenic activity within a chromosome in *Drosophila*. II. Viability. *Heredity* 14: 375–399. [17]

Breitling, L. P., and 4 others. 2011. Tobacco-smoking-related differential DNA methylation: 27K discovery and replication. *Am. J. Hum. Genet.* 88: 450–457. [21]

Breitling, R., and 10 others. 2008. Genetical genomics: spotlight on QTL hotspots. *PLoS Genet.* 4: e000232. [21]

Brem, R. B., and L. Kruglyak. 2005. The landscape of genetic complexity across 5,700 gene expression traits in yeast. *Proc. Natl. Acad. Sci. USA* 102: 1572–1577. [21]

Brem, R. B., G. Yvert, R. Clinton, and L. Kruglyak. 2002. Genetic dissection of transcriptional regulation in budding yeast. *Science* 296: 752–756. [21]

Breslow, D. K., and 8 others. 2008. A comprehensive strategy enabling high-resolution functional analysis of the yeast genome. *Nat. Methods* 5: 711–718. [15]

Breslow, N. 1990. Tests of hypotheses in overdispersed Poisson regression and other quasi-likelihood models. *J. Am. Stat. Assoc.* 85: 565–571. [14]

Breslow, N. E., and D. G. Clayton. 1993. Approximate inference in generalized linear mixed models. *J Am. Stat. Assoc.* 88: 9–25. [20]

Breunig, J. S., S. R. Hackett, J. D. Rabinowitz, and L. Kruglyak. 2014. Genetic basis of metabolome variation in yeast. *PLoS Genet.* 10: e1004142. [5, 18]

Breusch, T. S., and A. R. Pagan. 1979. A simple test for heteroskedasticity and random coefficient variation. *Econometrica* 47: 1287–1294. [10]

Brewbaker, J. J., and W. G. Sun. 1999. Trees and heterosis. *In* J. G. Coors and S. Pandey (eds.), *The genetics and exploitation of heterosis in crops*, pp. 463–478. ASA-CSSA-SSSA Societies, Madison, WI. [13]

Brinkman, M. A., and K. J. Frey. 1977. Yield component analysis of oat isolines that produce different grain yields. *Crop Sci.* 17: 165–168. [17]

Brion, M.-J. A., K. Shakbazov, and P. M. Visccher. 2013. Calculating statistical power in Mendelian randomization studies. *Int. J. Epidem.* 42: 1497–1501. [A2]

Brion, M.-J. A., B. Benyamin, P. M. Visccher, and G. Davey Smith. 2014. Beyond the single SNP: emerging developments in Mendelian randomization in the "omics" era. *Curr. Epidem. Rep.* 1: 228–236. [A2]

Britten, H. B. 1996. Meta-analysis of the association between multilocus heterozygosity and fitness. *Evolution* 50: 2158–2164. [12]

Britten, R., and E. H. Davidson. 1969. Gene regulation for higher cells: a theory. *Science* 165: 349–357. [21]

Brodie, E. D., III. 1989. Genetic correlation between morphology and antipredator behavior in natural populations of the garter snake *Thamnophis ordinoides. Nature* 342: 542–543. [5]

Broman, K. W. 2005. The genomes of recombinant inbred lines. *Genetics* 169: 1133–1146. [18]

Broman, K. W. 2012a. Genotype probabilities at intermediate generation in the construction of recombinant inbred lines. *Genetics* 190: 403–412. [18]

Broman, K. W. 2012b. Haplotype probabilities in advanced intercross populations. *G3* 2: 199–202. [18]

Broman, K. W., and T. P. Speed. 2002. A model selection approach for the identification of quantitative trait loci in experimental crosses. *J. Royal Stat. Soc. B* 64: 641–656. [18]

Broman, K. W., and J. L. Weber. 1998. Estimation of pairwise relationship in the presence of genotyping errors. *Am. J. Hum. Genet.* 63: 1563–1564. [8]

Broman, K. W., and J. L. Weber. 1999. Long homozygous chromosomal segments in reference families from the centre d'Etude du polymorphisme humain. *Am. J. Hum. Genet.* 65: 1493–1500. [8, 12]

Brooks, L. D. 1988. The evolution of recombination rates. *In* R. E. Michod and B. R. Levin (eds.), *The evolution of sex,* pp. 87–105. Sinauer Assoc., Sunderland, MA. [17]

Brooks, S. P., and G. O. Roberts. 1998. Convergence assessment techniques for Markov chain Monte Carlo. *Stat. Comput.* 8: 319–335. [A8]

Brooks, S. P., P. Giudici, and A. Philippe. 2003. Nonparametric convergence assessment for MCMC model selection. *J. Comput. Graph. Stat.* 12: 1–22. [A8]

Brooks, S., A. Gelman, G. Jones, and X.-L. Meng (eds). 2011. *Handbook of markov chain Monte Carlo.* CRC Press, Boca Raton, FL. [A8]

Brothers, A. N., and L. F. Delph. 2010. Haldane's rule is extended to plants with sex chromosomes. *Evolution* 64: 3643–3648. [17]

Brown, A. F. 1991. Outbreeding depression as a cost of dispersal in the harpacticoid copepod, *Tigriopus californicus. Biol. Bull.* 181: 123–126. [13]

Brown, A. H. D. 1975. Sample sizes required to detect linkage disequilibrium between two or three loci. *Theor. Pop. Biol.* 8: 184–201. [5]

Brown, D. E. 1991. *Human universals.* McGraw-Hill, New York, NY. [12]

Brown, D. L., M. B. Gorin, and D. E. Weeks. 1994. Efficient strategies for genomic searching using the affected-pedigree-member method of linkage analysis. *Am. J. Hum. Genet.* 54: 544–552. [19]

Brown, K. E., and J. K. Kelly. 2020. Severe inbreeding depression is predicted by the "rare allele load" in *Mimulus guttatus. Evolution* 74: 587–596. [12]

Brown, M. B. 1975. A method for combining non-independent, one-sided tests of significance. *Biometrics* 31: 978–992. [20]

Brown, M. D., C. G. Glazner, C. Zheng, and E. A. Thompson. 2012. Inferring co-ancestry in population samples in the presence of linkage disequilibrium. *Genet.* 190: 1447–1460. [8]

Brown, P. J., and 10 others. 2011. Distinct genetic architecture for male and female inflorescence traits of maize. *PLoS Genet.* 7: e1002382. [21]

Brown, P. O., and D. Botstein. 1999. Exploring the new world of the genome with DNA microarrays. *Nat. Genet.* 21: 33–37. [21]

Brownie, C., D. T. Bowman, and J. W. Burton. 1993. Estimating spatial variation in analysis of data from yield traits: a comparison of methods. *Agron. J.* 85: 244–253. [A9]

Browning, B. L., and S. R. Browning. 2011. A fast, powerful method for detecting identity by descent. *Am. J. Hum. Genet.* 88: 173–182. [8]

Browning, B. L., and S. R. Browning. 2013. Detecting identity by descent and estimating genotype error rate in sequence data. *Am. J. Hum. Genet.* 93: 840–851. [8]

Browning, B. L., and S. R. Browning. 2016. Genotype imputation with millions of reference samples. *Am. J. Hum. Genet.* 98: 116–126. [20, 21]

Browning, B. L., Y. Zhou, and S. R. Browning. 2018. A one-penny imputed genome from next-generation reference panels. *Am. J. Hum. Genet.* 103: 338–348. [20, 21]

Browning, S. R. 2008. Estimation of pairwise identity by decent from dense genetic marker data in a population sample of haplotypes. *Genetics* 178: 2123–2132. [8]

Browning, S. R., and B. L. Browning. 2010. High-resolution detection of identity by descent in unrelated individuals. *Am. J. Hum. Genet.* 86: 526–539. [8]

Browning, S. R., and B. L. Browning. 2012. Identity by descent between distant relatives: detection and applications. *Ann. Rev. Hum. Genet.* 46: 617–633. [8]

Bruce, A. B. 1910. The Mendelian theory of heredity and the augmentation of vigor. *Science* 32: 627–628. [12, 13]

Brucker, D. 1978. Why are there inbreeding effects in haplodiploid systems? *Evolution* 32: 456–458. [12]

Brückner, D. 1976. The influence of genetic variability on wing asymmetry in honeybees *(Apis mellifera). Evolution* 30: 100–108. [6]

Bryant, E. H., L. M. Meffert, and S. A. McCommas. 1990. Fitness rebound in serially bottlenecked populations of the house fly. *Am. Nat.* 136: 542–549. [12]

Bryc, K., W. Bryc, and J. W. Silverstein. 2013. Separation of the largest eigenvalues in eigenanalysis of genotype data form discrete subpopulations. *Theor. Pop. Biol.* 89: 34–43. [20]

Bryndeal, B., and 7 others. 2017. Large-scale *trans*-eQTLs affect hundreds of transcripts and mediate patterns of transcriptional co-regulation. *Am. J. Hum. Genet.* 100: 581–591. [21]

Bryois, J., and 11 others. 2014. Cis and trans effects of human genomic variants on gene expression. *PLoS Genet.* 10: e1004461. [21]

Brzyski, D., and 5 others. 2017. Controlling the rate of GWAS false discoveries. *Genetics* 205: 61–75. [20]

Bucher, K. D., H. G. Schrott, W. R. Clarke, and R. M. Lauer. 1982. The Muscatine cholesterol family study: distribution of cholesterol levels within families of probands with high, low and middle cholesterol levels. *J. Chron. Dis.* 35: 385–400. [16]

Buckler, E. S., and M. Gore. 2007. An *Arabidopsis* haplotype map takes root. *Nat. Genet.* 39: 1056–1057. [5, 20, 21]

Buckler, E. S., and 35 others. 2009. The genetic architecture of maize flowering time. *Science* 325: 714–718. [18, 21]

Buckley, F., N. Lopez-Villalobos, and B. J. Heins. 2014. Crossbreeding: implications for dairy cow fertility and survival. *Animal* 8: 122–133. [13]

Bulik-Sullivan, B. K., and 9 others. 2015. LD score regression distinguishes confounding from polygenicity in genome-wide associations studies. *Nat. Genet.* 47: 291–295. [12, 20]

Bulmer, M. G. 1971. The effects of selection on genetic variability. *Am. Nat.* 105: 210–211. [5, 15]

Bulmer, M. G. 1972. The genetic variability of polygenic characters under optimizing selection, mutation, and drift. *Genet. Res.* 19: 17–25. [15]

Bulmer, M. G. 1974. Linkage disequilibrium and genetic variability. *Genet. Res.* 23: 281–289. [5]

Bulmer, M. G. 1980. *The mathematical theory of quantitative genetics.* Oxford Univ. Press, New York, NY. [1, 7, 12, 15]

Bultman, S. J., E. J. Michaud, and R. P. Woychik. 1992. Molecular characterization of the mouse *agouti* locus. *Cell* 71: 1195–1204. [17]

Bundgaard, J., V. Loeschcke, M. F. Schou, and K. Bijlsma. 2021. Detecting purging of inbreeding depression by a slow rate of inbreeding for various traits: the impact of environmental and experimental conditions. *Heredity* 127: 10–20. [12]

Bürger, R., G. P. Wagner, and F. Stettinger. 1989. How much heritable variation can be maintained in finite populations by mutation-selection balance? *Evolution* 43: 1748–1766. [15]

Burgess, S., and S. G. Thompson. 2011. Bias in causal estimates from Mendelian randomization studies with weak instruments. *Stat. Med.* 30: 1312–1323. [A2]

Burgess, S., and S. G. Thompson. 2013. Use of allele scores as instrument variables for Mendelian randomization. *Int. J. Epidem.* 42: 1134–1144. [A2]

Burgess, S., and S. G. Thompson. 2015. *Mendelian randomization: methods for using genetic variants in causal estimation.* Chapman and Hall, Boca Raton, FL. [A2]

Burgess, S., and S. G. Thompson. 2017. Interpreting findings from Mendelian randomization using the MR-Egger method. *Eur. J. Epidem.* 32: 377–389. [A2]

Burgess, S., A. S. Butterworth, and S. G. Thompson. 2013. Mendelian randomization analysis with multiple genetic variants using summarized data. *Genet. Epidem.* 37: 658–665. [A2]

Burgess, S., F. Dudbridge, and S. G. Thompson. 2015a. Combining information on multiple instrumental variables in Mendelian randomization: comparison of allele score and summarize data methods. *Stat. Med.* 35: 1880–1906. [A2]

Burgess, S., N. J. Timpson, S. Ebrahim, and G. Davey Smith. 2015b. Mendelian randomization: where are we now and where are we going. *Int. J. Epidem.* 44: 379–388. [A2]

Burgess, S., D. S. Small, and S. G. Thompson. 2017a. A review of instrumental variable estimators for Mendelian randomization. *Stat. Meth. Med. Res.* 26: 2333–2355. [A2]

Burgess, S., and 4 others. 2017b. Sensitivity analyses for robust causal inference from Mendelian randomization analyses with multiple genetic variants. *Epidem.* 28: 30–42. [A2]

Burgess, S., C. N. Foley, and V. Zuber. 2018. Inferring causal relationships between risk factors and outcomes form genome-wide association study data. *Ann. Rev. Genomics Hum. Genet.* 19: 303–327. [A2]

Burgess, S., C. N. Foley, and V. Zuber. 2019. Inferring causal associations between risk factors and outcomes using genetic variation. *In* D. J. Balding, I. Molke, and J. Marioni (eds.), *Handbook of Statistical Genomics*, 4th Ed., pp. 651–677. Wiley and Sons, New York, NY. [A2]

Burke, W. D., H. S. Malik, C. W. Lathe and T. H. Eickbush. 1998. Are retrotransposons long-term hitchhikers? *Nature* 392: 141–142. [15]

Burnham, K. P., and D. R. Anderson. 2002. *Model selection and multimodel inference: a practical in-formation-theoretic approach,* 2nd Ed. Springer-Verlag, New York, NY. [16, A4]

Burnham, K. P., and D. R. Anderson. 2004. Multimodel inference: understanding AIC and BIC in model selection. *Sociological Methods & Research* 33: 261–304. [16]

Burns, T. L., P. P. Moll, and M. A. Schork. 1984. Comparisons of different sampling designs for the determination of genetic transmission mechanisms in quantitative traits. *Am. J. Hum. Genet.* 36: 1060–1074. [16]

Burr, B., and F. A. Burr. 1991. Recombinant inbred lines for molecular mapping in maize. *Trends Genet.* 7: 55–60. [17]

Burton, G. W. 1951. Quantitative inheritance in pearl millet *(Pennisetum glaucum). Agron. J.* 43: 409–417. [11]

Burton, R. S. 1987. Differentiation and integration of the genome in populations of *Tigriopus californicus. Evolution* 41: 504–513. [13]

Burton, R. S. 1990a. Hybrid breakdown in physiological response: a mechanistic approach. *Evolution* 44: 1806–1813. [13]

Burton, R. S. 1990b. Hybrid breakdown in developmental time in the copepod *Tigriopus californicus. Evolution* 44: 1814–1822. [13]

Busbice, T. H. 1969. Inbreeding in synthetic varieties. *Crop Sci.* 9: 601–604. [13]

Busbice, T. H. 1970. Predicting yield of synthetic varieties. *Crop Sci.* 10: 265–269. [13]

Busbice, T. H., and C. P. Wilsie. 1966. Inbreeding depression and heterosis in an autotetraploids with application to *Medicago sativa* L. *Euphytica* 15: 52–67. [13]

Busch, R. H., K. A. Lucken, and R. C. Frohberg. 1971. F_1 hybrids versus random F_6 line performance and estimates of genetic effects in spring wheat. *Crop Sci.* 11: 357–361. [12]

Bush, R. M., and P. E. Smouse. 1991. The impact of electrophoretic genotype on life history traits in *Pinus taeda. Evolution* 45: 481–498. [12]

Butler, K., C. Field, C. M. Herbinger, and B. R. Smith. 2004. Accuracy, efficiency and robustness of four algorithms allowing full sibship reconstruction from DNA marker data. *Mol. Ecol.* 13: 1589–1600. [8]

Butlin, R. K., I. L. Read, and T. H. Day. 1982. The effects of a chromosomal inversion on adult size and male mating success in the seaweed fly *Coelopa frigida. Heredity* 49: 51–62. [17]

Byers, D. L., and D. M. Waller. 1999. Do plant populations purge their genetic load? Effects of population size and mating history on inbreeding depression. *Ann. Rev. Ecol. Syst.* 30: 479–513. [12]

Bystrykh, L., and 16 others. 2005. Uncovering regulatory pathways that affect hematopoietic stem cell function using 'genetical genomics'. *Nat. Genet.* 37: 225–323. [21]

Caballero, A., 2020. *Quantitative genetics.* Cambridge University Press, Cambridge, UK. [1]

Caballero, A., and P. D. Keightley. 1994. A pleiotropic nonadditive model of variation in quantitative traits.

Genetics 138: 883–900. [15]

Caballero, A., M. A. Toro, and C. López-Fanjul. 1991. The response to artificial selection from new mutations in *Drosophila melanogaster. Genetics* 127: 89–102. [15]

Caballero, A., P. D. Keightley, and W. G. Hill. 1995. Accumulation of mutations affecting body weight in inbred mouse lines. *Genet. Res.* 65: 145–149. [15]

Caballero, A., A. Tenesa, and P. D. Keightley. 2015. The nature of genetic variation for complex traits revealed by GWAS and regional heritability mapping analyses. *Genetics* 201: 1601–1613. [21]

Caballero, A., I. Bravo, and J. Wang. 2017a. Inbreeding load and purging: implications for the short-term survival and the conservation management of small populations. *Heredity* 118: 177–185. [12]

Caballero, A., I. Bravo, and J. Wang. 2017b. The risk of forcing inbreeding in conservation programmes: a reply to Theodorou and Couvet. *Heredity* 119: 51–53. [12]

Caballero, A., B. Villanueva, and T. Druet. 2021. On the estimation of inbreeding depression using different measures of inbreeding from molecular markers. *Evol. Applicat.* 14: 416–428. [12]

Caballero, A., Fernández, B. Villanueva, and M. A. Toro. 2022. A comparison of marker-based estimators of inbreeding and inbreeding depression. *Genet. Sel. Evol.* 54: 82. [8, 12]

Cabot, E. L., A. W. Davis, N. A. Johnson, and C.-I. Wu. 1994. Genetics of reproductive isolation in the *Drosophila simulans* clade: complex epistasis underlying hybrid male sterility. *Genetics* 137: 175–189. [17]

Cai, S. V., T. R. Famula, A. M. Oberbauer, and R. S. Hess. 2019. Heritability and complex segregation analysis of diabetes mellitus in American Eskimo Dogs. *J. Veterinary Internal Med.* 33: 1926–1934. [16]

Cai, X., A. Huang, and S. Xu. 2011. Fast empirical Bayesian LASSO for multiple quantitative trait locus mapping. *BMC Bioinform.* 12: 211. [18]

Calder, W. A. 1984. *Size, function, and life history*. Harvard Univ. Press, Cambridge, MA. [14]

Caligari, P. D. S., and K. Mather. 1988. Competitive interactions in *Drosophila melanogaster*. IV. Chromosome assay. *Heredity* 60: 355–366. [17]

Cameron, N. D. 1997. *Selection indices and prediction of genetic merit in animal breeding*. CAB International, Wallingford, UK. [1]

Camp, N. J. 1997. Genomewide transmission/disequilibrium testing: considerations of the genotypic relative risks at disease loci. *Am. J. Hum. Genet.* 61: 1424–1430. [17]

Camp, N. J. 1999. Genomewide transmission/disequilibrium testing: a correction. *Am. J. Hum. Genet.* 64: 1485–1487. [17]

Campbell, C. D., and 8 others. 2005. Demonstrating stratification in a European American population. *Nat. Genet.* 37: 868–872. [20]

Camus-Kulandaivelu, L., and 4 others. 2007. Evaluating the reliability of *Structure* outputs in case of relatedness between individuals. *Crop Sci.* 47: 887–892. [20]

Cannings, C., and E. A. Thompson. 1977. Ascertainment in the sequential sampling of pedigrees. *Clin. Genet.* 12: 208–212. [16]

Cannings, C., E. A. Thompson, and M. H. Skolnick. 1976. The recursive derivation of likelihoods on complex pedigrees. *Adv. Appl. Prob.* 8: 622–625. [16]

Cannings, C., E. A. Thompson, and M. H. Skolnick. 1978. Probability functions on complex pedigrees. *Adv. Appl. Prob.* 10: 26–61. [16]

Cannon, M. E., and K. L. Mohlke. 2018. Deciphering the emerging complexities of molecular mechanisms at GWAS loci. *Am. J. Hum. Genet.* 103: 637–653. [21]

Cano-Gamez, E., and G. Trynka. 2020. From GWAS to function: using functional genomics to identify the mechanisms underlying complex diseases. *Front. Genet.* 11: A424. [21]

Cantor, R. M., K. Lange, and J. S. Sinsheimer. 2010. Prioritizing GWAS results: a review of statistical methods and recommendations for their application. *Am. J. Hum. Genet.* 86: 6–22. [21]

Cao, Y., and 4 others. 2013. A versatile omnibus test for detecting mean and variance heterogeneity. *Genet. Epidem.* 38: 51–59. [18. 20]

Capen, E. C., R. V. Clapp, and W. M. Campbell. 1971. Competitive bidding in high-risk situations. *J. Petroleum Tech.* 23: 641–653. [20]

Carbonell, E. A., and T. M. Gerig. 1991. A program to detect linkage between genetic markers and nonadditive quantitative trait loci. *J. Heredity* 82: 435. [18]

Carbonell, E. A., T. M. Gerig, E. Balansard, and M. J. Asins. 1992. Interval mapping in the analysis of nonadditive quantitative trait loci. *Biometrics* 48: 305–315. [18]

Carbonell, E. A., and 4 others. 1993. Power studies in the estimation of genetic parameters and the localization of quantitative trait loci for backcross and doubled haploid populations. *Theor. Appl. Genet.* 86: 411–416. [18]

Cardellino, R. A., and T. Mukai. 1975. Mutator factors and genetic variance components of viability in *Drosophila melanogaster*. *Genetics* 80: 567–583. [15]

Cardon, L. R., and D. W. Fulker. 1994. The power of interval mapping of quantitative trait loci, using selected sib pairs. *Am. J. Hum. Genet.* 55: 825–833. [19]

Cardon, L. R., and L. J. Palmer. 2003. Population stratification and spurious allelic association. *Lancet* 361: 598–604. [20]

Cardon, L. R., and 5 others. 1994. Quantitative trait locus for reading disability on chromosome 6. *Science* 266: 276–279. [19]

Carey, G., and J. Williamson. 1991. Linkage analysis of quantitative traits: increased power by using selected samples. *Am. J. Hum. Genet.* 49: 786–796. [17, 19]

Cargill, M., and 16 others. 1999. Characterization of single-nucleotide polymorphisms in coding regions of human genes. *Nat. Genet.* 22: 231–238. [21]

Carlborg, Ö., and L. Andersson. 2002. Use of randomization test to detect multiple epistastic QTLs. *Genet. Res.* 79: 175–184. [18]

Carlborg, Ö., and C. S. Haley. 2004. Epistasis: too often neglected in complex trait studies? *Nat. Genet.* 5: 618–625. [18]

Carlborg, Ö., L. Andersson, and B. Kinghorn. 2000. The use of a genetic algorithm for simultaneous mapping of multiple interaction quantitative trait loci. *Genetics* 155: 2003–2010. [18]

Carlin, B. P., and T. A. Louis. 2000. *Bayes and empirical Bayesmethods for data analysis,* 2nd Ed. Chapman and Hall, New York, NY. [A7]

Carlson, M.O., and 9 others. 2019. Multivariate genomewide association analyses reveal the genetic basis of seed fatty acid composition in oat (*Avena sativa* L.). *G3* 9: 2963–2975. [20]

Carmelli, D., S. Karlin, and R. Williams. 1979. A class of indices to assess major gene versus polygenic inheritance of distributed variables. *In* C. F. Sing and M. Skolnick (eds.), *The genetic analysis of common diseases: applications to predict factors in coronary heart disease,* pp. 259–270. Alan R. Liss, New York, NY. [16]

Carmon, J. L., H. A. Stewart, C. C. Cockerham, and R. E. Comstock. 1956. Prediction equations for rotational crossbreeding. *J. Animl. Sci.* 15: 930–936. [13]

Carninci, P. 2006. Tagging mammalian transcription complexity. *Trends Genet.* 22: 501–510. [21]

Carothers, A. D., and 9 others. 2006. Estimating human inbreeding coefficients: comparison of genealogical and marker heterozygosity approaches. *Ann. Hum. Genet.* 70: 666–676. [12]

Carpenter, J. R., H. Grüneberg, and E. S. Russell. 1957. Genetical differentiation involving morphological characters in an inbred strain of mice. II. American branches of the C57BL and C57BR strains. *J. Morphology* 100: 377–388. [15]

Carr, D. E., and M. R. Dudash. 2003. Recent advances into the genetic basis of inbreeding depression in plants. *Phil. Trans. Roy. Soc. Lond. B* 358: 1071–1084. [12]

Carroll, S. B. 2008. Evo-devo and an expanding evolutionary synthesis: a genetic theory of morphological evolution. *Cell* 134: 25–36. [15]

Carson, H. L., and R. Lande. 1984. Inheritance of a secondary sexual character in *Drosophila silvestris. Proc. Natl. Acad. Sci. USA* 81: 6904–6907. [11]

Carson, H. L., and A. R. Templeton. 1984. Genetic revolutions in relation to speciation phenomena: the founding of new populations. *Ann. Rev. Ecol. Syst.* 15: 97–131. [17]

Carter, C. O. 1969. Genetics of common disorders. *Br. Med. Bull.* 25: 52–57. [12, 21]

Casella, G. 2008. *Statistical design.* Springer, New York, NY. [A9]

Casella, G., and E. I. George. 1992. Explaining the Gibbs sampler. *Am. Stat.* 46: 167–174. [A8]

Casler, M. D. 2015. Fundamentals of experimental design: guidelines for designing successful experiments. *Agron. J.* 107: 692–705. [A9]

Castiblanco, J., and 4 others. 2015. Familial aggregation and segregation analysis in families presenting autoimmunity, polyautoimmunity, and multiple autoimmune syndrome. *J. Immunology Res.* Article ID 572353. [16]

Castle, W. E. 1921. An improved method of estimating the number of genetic factors concerned in cases of blending inheritance. *Proc. Natl. Acad. Sci. USA* 81: 6904–6907. [11]

Caswell, H. 2001. *Matrix population models, 2nd edition* Sinauer, Sunderland, MA. [4]

Caten, C. E. 1979. Quantitative genetic variation in fungi. *In* J. N. Thompson, Jr. and J. M. Thoday (eds.), *Quantitative genetic variation,* pp. 35–60. Academic Press, New York, NY. [5]

Caten, C. E., and J. L. Jinks. 1976. Quantitative genetics. *In* K. D. MacDonald (ed.), *Second international symposium on the genetics of industrial microorganisms,* pp. 93–111. Academic Press, New York, NY. [5]

Cavalier-Smith, T. 1978. Nuclear volume control by nucleoskeletal DNA, selection for cell volume and cell growth rate, and the solution of the DNA C-value paradox. *J. Cell Sci.* 34: 247–268. [15]

Cavalier-Smith, T. (ed.) 1985. *The evolution of genome size.* John Wiley & Son, New York, NY. [15]

Cavalier-Smith, T. 2005. Economy, speed and size matter: evolutionary forces driving nuclear genome miniaturization and expansion. *Ann. Bot.* 95: 147–175. [15]

Cavalli, L. L. 1952. An analysis of linkage in quantitative inheritance. *In* E.C. R. Reeve and C. H. Waddington (eds.), *Quantitative inheritance,* pp. 135–144. His Majesty's Stationary Office, London, UK. [11]

Cavalli-Sforza, L. L., and W. F. Bodmer. 1971. *The genetics of human populations.* W. H. Freeman and Co., San Francisco, CA. [16]

Cavanagh C., M. Morell, I. Mackay, and W. Powell. 2008. From mutations to MAGIC: resources for gene discovery, validation and delivery in crop plants. *Curr. Opin. Plant Biol.* 11: 215–221. [18]

Ceballos, F. C., P. K. Joshi, D. W. Clark, M. Ramsay, and J. F. Wilson. 2018. Runs of homozygosity: windows into population history and trait architecture. *Nat. Rev. Genet.* 19. 220. [8, 12]

Chai, C. K. 1956. Analysis of quantitative inheritance of body size in mice. II. Gene action and segregation. *Genetics* 41: 165–178. [11]

Chai, C K. 1957. Developmental homeostasis of body growth in mice. *Am. Nat.* 85: 49–55. [6]

Chakraborty, R. 1987. Biochemical heterozygosity and phenotypic stability of polygenic traits. *Heredity* 59: 19–28. [6]

Chakraborty, R., and M. Nei. 1982. Genetic differentiation of quantitative characters between populations or species. I. Mutation and random genetic drift. *Genet. Res.* 39: 303–314. [15]

Chakraborty, R., and N. Ryman. 1983. Relationship of mean and variance of genotypic values with heterozygosity per individual in a natural population. *Genetics* 103: 149–152. [6]

Chakraborty, R., and K. M. Weiss. 1988. Admixture as a tool for finding linked genes and detecting that difference from allelic association between loci. *Proc. Natl. Acad. Sci. USA* 85: 9119–9123. [17]

Chakravarti, A. 1999. Population genetics — making sense out of sequence. *Nat. Genet.* 22: 56–60. [21]

Chakravarti, A., and T. N. Turner. 2016. Revealing rate-limiting steps in complex disease biology: the critical importance of studying rare, extreme-phenotype families. *Bioessays* 38: 578–586. [21]

Chamberlain, A. J., H. C. McPartlan, and M. E. Goodard. 2007. The number of loci that affect milk production traits in dairy cattle. *Genetics* 177: 1117–1123. [19]

Champoux, M. C., and 6 others. 1995. Locating genes associated with root morphology and drought avoidance in rice via linkage to molecular markers. *Theor. Appl. Genet.* 90: 969–980. [18]

Changjian, J., P. Xuebiao, and G. Minghong. 1994. The use of mixture models to detect effects of major genes on quantitative characters in plant breeding experiments. *Genetics* 136: 383–394. [16]

Chant, D. 1974. On asymptotic tests of composite hypotheses in nonstandard conditions. *Biometrika* 61: 291–298. [A4]

Chapman, J. M., J. D. Cooper, J. A. Todd, and D. G. Clayton. 2003. Detecting disease associates due to linkage disequilibrium using haplotype tags: a class of tests and determinants of statistical power. *Hum. Hered.* 56: 18–31. [20]

Chapman, J. R., and 4 others. 2009. A quantitative review of heterozygosity-fitness correlations in animal populations. *Mol. Ecol.* 18: 2746–2765. [12]

Chapman, N., and E. Thompson. 2003. A model for the length of tracts of identity by descent in finite random mating populations. *Theor. Popul. Biol.* 64: 141–150. [8, 12]

Charcosset, A., M. Lefort-Buson, and A. Gallais. 1991. Relationship between heterosis and heterozygosity at marker loci: a theoretical computation. *Theor. Appl. Genet.* 81: 571–575. [13]

Charlesworth, B. 1974. The Hardy-Weinberg law with overlapping generations. *Adv. Appl. Prob.* 6: 4–6. [4]

Charlesworth, B. 1979. Evidence against Fisher's theory of dominance. *Nature* 278: 848–489. [4]

Charlesworth, B. 1984. The evolutionary genetics of life histories. *In* B. Shorrocks (ed.), *Evolutionary ecology*, pp. 117–133. Blackwell Scientific, Oxford, UK. [7]

Charlesworth, B. 1990. Mutation-selection balance and the evolutionary advantage of sex and recombination. *Genet. Res.* 55: 199–221. [15]

Charlesworth, B. 1994. *Evolution in age-structured populations.* 2nd Ed. Cambridge Univ. Press, Cambridge, UK. [4]

Charlesworth, B. 2015. Causes of natural variation in fitness: evidence from studies of *Drosophila* populations. *Proc. Natl. Acad. Sci. USA* 112: 1662–1669. [12]

Charlesworth, B. 2018. Mutational load, inbreeding depression and heterosis in subdivided populations. *Mol. Ecol.* 27: 4991–5003. [13]

Charlesworth, B., and D. Charlesworth. 1999. The genetic basis of inbreeding depression. *Genet. Res.* 74: 329–340. [12]

Charlesworth, B., R. Lande, and M. Slatkin. 1982. A neo-Darwinian commentary on macroevolution. *Evolution* 36: 474–498. [11, 17]

Charlesworth, B., J. A. Coyne, and N. H. Barton. 1987. The relative rates of evolution of sex chromosomes and autosomes. *Am. Nat.* 130: 113–146. [17]

Charlesworth, B., D. Charlesworth, and M. T. Morgan. 1990. Genetic loads and estimates of mutation rates in highly inbred plant populations. *Nature* 347: 380–382. [12, 15]

Charlesworth, B., D. Charlesworth, J. A. Coyne, and C. H. Langley. 2016. Hubby and Lewontin on protein variation in natural populations: when molecular genetics came to the rescue of population genetics. *Genetics* 203:1497–1503. [8]

Charlesworth, D., and B. Charlesworth. 1979. Selection on recombination in a multi-locus system. *Genetics* 91: 575–580. [5]

Charlesworth, D., and B. Charlesworth. 1987. Inbreeding depression and its evolutionary consequences. *Ann. Rev. Ecol. Syst.* 18: 237–268. [12]

Charlesworth, D., and B. Charlesworth. 1995. Quantitative genetics in plants: the effect of the breeding system on genetic variability. *Evolution* 49: 911–920. [12]

Charlesworth, D., and J. H. Willis. 2009. The genetics of inbreeding depression. *Nat. Rev. Genet.* 10: 783–796. [12]

Charlesworth, D., M. T. Morgan, and B. Charlesworth. 1992. The effect of linkage and population size on inbreeding depression due to mutational load. *Genet. Res.* 59: 49–61. [15]

Charlesworth, D., M. T. Morgan, and B. Charlesworth. 1993. Mutation accumulation in finite outbreeding and inbreeding populations. *Genet. Res.* 61: 39–56. [15]

Charlesworth, D., E. E. Lyons, and L. B. Litchfield. 1994. Inbreeding depression in two highly inbreeding populations of *Leavenworthia*. *Proc. Royal Soc. Lond. B* 258: 209–214. [12, 15]

Charmantier, A., D. Garant, and L. E. Kruuk. 2014. *Quantitative genetics in the wild*. Oxford University Press, Oxford, UK. [1]

Charmet, G., T. Cadalen, P. Sourdille, and M. Bernard. 1998. An extension of the 'marker regression' method to interactive QTL. *Mol. Breeding* 4: 67–72. [18]

Chatziplis, D. G., and C. S. Haley. 2000. Selective genotyping for QTL detection using sib pair analysis in outbred populations with hierarchical structures. *Genet. Sel. Evol.* 32: 547–560. [19]

Chelo, I. M., and 7 others. 2018. Partial selfing can reduce genetic loads while maintaining diversity during experimental evolution. *G3* 9: 2811–2821. [12]

Chen, H., J. Chen, and J. D. Kalbfleisch. 2001. A modified likelihood ratio test for homogeneity in finite mixture models. *Stat. Meth.* 63: 19–29. [A6]

Chen, H., and 5 others. 2011. Comparison of statistical approaches to rare variant analysis for quantitative traits. *BMC Proceed.* 5: S113. [20]

Chen, H., and 15 others. 2016. Control for population structure and relatedness for binary traits in genetic association studies via logistic mixed models. *Am. J. Hum. Genet.* 98: 653–666. [20]

Chen, H., and 10 others. 2017. A genome-wide association study identifies genetic variants associated with mathematics ability. *Sci. Reports* 7: 40365. [20]

Chen, H., and 45 others. 2019. Efficient variant set mixed model association tests for continuous and binary traits in large-scale whole-genome sequencing studies. *Am. J. Hum. Genet.* 104: 260–274. [20]

Chen, H.-S., R. M. Pfeiffer, and S. Zhang. 2013. A powerful method for combining P-values in genomic studies. *Genet. Epidem.* 37: 814–819. [20]

Chen, J. J., and H. Ahn. 1996. Fitting mixed Poisson regression models using quasi-likelihood methods. *Biom. J.* 38: 81–96. [14]

Chen, L., and J. D. Storey. 2006. Relaxed significance criteria for linkage analysis. *Genetics* 173: 2371– 2381. [18, 20, A6]

Chen, L., G. Davery Smith, R. M. Harbord, and S. J. Lewis. 2008. Alcohol intake and blood pressure: a systematic review implementing a Mendelian randomization approach. *PLoS Med.* 5: e52. [A2]

Chen, L., and 10 others. 2022. Heterosis and differential DNA methylation in soybean hybrids and their parental lines. *Plants* 11: 1136. [13]

Chen, L. S., F. Emmert-Streib, and J. D. Storey. 2007. Harnessing naturally randomized transcription to infer regulatory relationships among genes. *Genome Biol.* 8: 1–13. [21]

Chen, L. S., and 6 others. 2010. Insights into colon cancer etiology via a regularized approach to gene set analysis of GWAS data. *Am. J. Hum. Genet.* 86: 860–871. [20, 21]

Chen, L. S., and 4 others. 2012. An exponential combination procedure for set-based association tests in sequencing studies. *Am. J. Hum. Genet.* 91: 977–986. [20]

Chen, M., and C. Kendziorski. 2007. A statistical framework for expression quantitative trait loci mapping. *Genetics* 177: 761–771. [21]

Chen, M.-H., Q.-M. Shao, and J. G. Ibrahim. 2000. *Monte Carlomethods in Bayesian computation.* Springer-Verlag, New York, NY. [A8]

Chen, W., and 6 others. 2015. Fine mapping causal variants with an approximate Bayesian method using marginal test statistics. *Genetics* 200: 719–736. [21]

Chen, W., and 4 others. 2016. Incorporating functional annotations for fine-mapping causal variants in a Bayesian framework using summary statistics. *Genetics* 204: 933–958. [21]

Chen, W.-M., and G. Abecasis. 2006. Estimating the power of variance component linkage analysis in large pedigrees. *Genet. Epid.* 30: 471–484. [19]

Chen, W.-M., and G. R. Abecasis. 2007. Family-based association test for genomewide association scans. *Am. J. Hum. Genet.* 81: 913–926. [20]

Chen, W.-M., and H.-W. Deng. 2001. A general and accurate approach for computing the statistical power of the transmission disequilibrium test for complex disease genes. *Genet. Epid.* 21: 53–67. [17]

Chen, W.-M., K. W. Broman, and K.-Y. Liang. 2004. Quantitative trait linkage analysis by generalized estimating equations: unification of variance components and Haseman-Elston regression. *Genet. Epid.* 26: 265–272. [19]

Chen, W.-M., K. W. Broman, and K.-Y. Liang. 2005. Power and robustness of linkage tests for quantitative traits in general pedigrees. *Genet. Epid.* 28: 1–23. [19]

Chen, X. 2009. Small RNAs and their roles in plant development. *Ann. Rev. Cell. Dev. Biol.* 35: 21–44. [15]

Chen, Z. 2005. The full EM algorithm for the MLEs of QTL effects and positions and their estimated variances in multiple-interval mapping. *Biometrics* 61: 474–480. [18]

Chen, Z., and J. Liu. 2009. Mixture generalized linear models for multiple interval mapping of quantitative trait loci in experimental crosses. *Biometrics* 65: 470–477. [18]

Chen, Z. J. 2010. Molecular mechanisms of polyploidy and hybrid vigor. *Trends Plant Sci.* 15: 57–71. [13]

Chen, Z. J. 2013. Genomic and epigenetic insights into the molecular bases of heterosis. *Nat. Rev. Genet.* 14: 471–482. [13]

Cheng, L., and X. S. Sheng. 2017. Combinations of "combinations of *p* values". *Empir. Econ.* 53: 3289–350. [A6]

Cheptou, P.-O., and K. Donohue. 2011. Environment-dependent inbreeding depression: its ecological and evolutionary significance. *New Phytologist* 189: 395–407. [12]

Chernoff, H. 1954. On the distribution of the likelihood ratio. *Ann. Math. Stat.* 25: 573–578. [19, A4]

Chesler, E. J., and 12 others. 2005. Complex trait analysis of gene expression uncovers polygenic and pleiotropic networks that modulate nervous system function. *Nat. Genet.* 37: 233–242. [21]

Cheung, V. G., and 5 others. 2005. Mapping determinants of human gene expression by regional and genome-wide associaton. *Nature* 437: 1365–1369. [21]

Cheung, Y. H., G. Wang, S. M. Leal, and S. Wang. 2012. A fast and noise-resilient approach to detect rare-variant associations with deep sequencing data for complex disorders. *Genet. Epidem.* 36: 675–685. [20]

Chevalet, C. 1988. Control of genetic drift in selected populations. *In* B. S. Weir, E. J. Eisen, M. M. Goodman, and G. Namkoong (eds.), *Proceedings of the second international conference on quantitative genetics,* pp. 379–394. Sinauer Assoc., Sunderland, MA. [5]

Cheverud, J. M. 1982. Relationships among ontogenetic, static, and evolutionary allometry. *Am. J. Phys. Anthr.* 59: 139–149. [14]

Cheverud, J. M. 2001. A simple correction for multiple comparisons in interval mapping genome scans. *Heredity* 87: 52–58. [18, 19, 20, A6]

Cheverud, J. M., and E. J. Routman. 1995. Epistasis and its contribution to genetic variance components. *Genetics* 139: 1455–1461. [4, 5]

Cheverud, J. M., and 5 others. 1996. Quantitative trait loci for murine growth. *Genetics* 142: 1305–1319. [18]

Chi, P., C. D. Allis, and G. G. Wang. 2010. Covalent histone modifications — miswritten, misinterpreted and miserased in human cells. *Nat. Rev. Cancer* 10: 457–469. [21]

Chiang, C., and 13 others. 2017. The impact of structural variation human gene expression. *Nat. Genet.* 49: 692–699. [21]

Chib, S., and E. Greenberg. 1995. Understanding the Metropolis-Hastings algorithm. *Am. Stat.* 49: 327–335. [A8]

Chiurugwi, T., S. Kemp, W. Powell, and L. T. Hickey. 2019. Speed breeding orphan crops. *Theor. Appl. Genet.* 132: 607–616. [17]

Choi, Y., E. M. Wijsman, and B. S. Weir. 2009. Case-control association testing in the presence of unknown relationships. *Genet. Epidem.* 33: 668–678. [20]

Cho, Y., and 7 others. 2020. Exploiting horizontal pleiotropy to search for causal pathways within a Mendelian randomization framework. *Nat. Comm.* 11: 1010. [A2]

Choi, Y., E. M. Wijsman, and B. S. Weir. 2009. Case-control association testing in the presence of unknown relationships. *Genet. Epidemiol.* 33: 668–678. [A4]

Choo, T. M. 1981. Doubled haploids for studying the inheritance of quantitative characters. *Genetics* 99: 525–540. [11]

Choo, T. M. 1983. Doubled haploids for locating polygenes. *Can. J. Genet. Cytol.* 25: 425–429. [17]

Choo, T. M., and E. Reinbergs. 1982. Estimation of the number of genes in doubled haploid populations of barley *(Hordeum vulgare). Can. J. Genet. Cytol.* 24: 337–341. [11]

Chovnick, A., and A. S. Fox. 1953. The problem of estimating the number of loci determining quantitative variation in haploid organisms. *Am. Nat.* 87:263–267. [11]

Chow, G. C. 1960. Tests of equality between sets of coefficients in two linear regressions. *Econometrica* 28: 591–605. [10]

Christensen, K., T. E. Johnson, and J. W. Vaupel. 2006. The quest for genetic determinants of human longevity: challenges and insights. *Nat. Rev. Genet.* 7: 436–448. [17]

Christensen, O. F., and M. S. Lund. 2010. Genomic prediction when some animals are not genotyped. *Genet. Sel. Evol.* 42: 2 [8]

Christie, M. R. 2010. Parentage in natural populations: novel methods to detect parent-offspring pairs in large data sets. *Mol. Ecol. Res.* 10: 115–128. [8]

Chun, H., and S. Keleş. 2009. Expression quantitative trait loci mapping with multivariate space partial least squares regression. *Genetics* 182: 79–90. [21]

Chun, S., and 7 others. 2017. Limited statistical evidence for shared genetics effects of eQTLs and autoimmue-disease-associated loci in three major immune-cell types. *Nat. Genet.* 49: 600–605. [21]

Chung, J. G., R. Jun, J. Dupuis, and L. A. Farrer. 2019. Comparison of methods for multivariate gene-based association tests for complex diseases using common variants. *Eur. J. Hum. Genet.* 27: 811–823. [20]

Churchill, G. A., and R. W. Doerge. 1994. Empirical threshold values for quantitative trait mapping. *Genetics* 138: 963–971. [16, 18, 19, A6]

Churchill, G. A., et al. 2004. The collaborative cross, a community resource for the genetic analysis of complex traits. *Nat. Genet.* 36: 1133–1137. [18]

Civelek, M., and A. J. Luisis. 2014. Systems genetics approaches to understand complex traits. *Nat. Rev. Genet.* 15: 34–48. [21]

Clare, H. J., and L. S. Luckinbill. 1985. The effects of gene-environment interaction on the expression of longevity. *Heredity* 55: 19–29. [6]

Clark, A. G. 1990. Inference of haplotypes from PCR-amplified samples of diploid populations. *Mol. Biol. Evol* 7: 111–122. [5]

Clark, A. G., L. Wang, and T. Hulleberg. 1995a. *P*-element-induced variation in metabolic regulation in *Drosophila*. *Genetics* 139: 337–348. [15]

Clark, A. G., L. Wang, and T. Hulleberg. 1995b. Spontaneous mutation rate of modifiers of metabolism in *Drosophila*. *Genetics* 139: 767–779. [15]

Clark, D., J. W. Dudley, T. R. Rocheford, and J. R. LeDeaux. 2006. Genetic analysis of corn kernel chemical composition in the random mated 10 generation of the cross of generations 70 of IHO x ILO. *Crop Sci.* 46: 807–819. [18]

Clark, D. W., et al. 2019. Associations of autozygosity with a broad range of human phenotypes. *Nat. Comm.* 10: 4957. [12]

Clarke, C. A., and 5 others. 1956. ABO Blood groups and secretor character in duodenal ulcer. *Brit. Med. J.* 2: 725–731. [17]

Clarke, G. M. 1992. Fluctuating asymmetry: a technique for measuring developmental stress of genetic and environmental origin. *Acta Zool. Fennica* 191: 31–35. [6]

Clarke, G. M., and 5 others. 2011. Basic statistical analysis in genetic case-control studies. *Nat. Protocols* 6: 121–133. [20]

Clarke, G. M., and J. A. McKenzie. 1987. Developmental stability of insecticide resistant phenotypes in blowfly: a result of canalizing natural selection. *Nature* 325: 345–346. [14]

Clarke, G. M., G. W. Brand, and M. J. Whitten. 1986. Fluctuating asymmetry: a technique for measuring development stress caused by inbreeding. *Aust. J. Biol. Sci.* 39: 145–153. [6]

Clausen, J., D. D. Keck, and W. M. Hiesey. 1940. Experimental studies on the nature of species, volume I: Effect of varied environments on Western North American plants. Publication 520, Carnegie Institute of Washington. Washington, DC. [13]

Clausen, J., D. D. Keck, and W. M. Hiesey. 1945. Experimental studies on the nature of species, volume II: Plant evolution through amphiploidy and autoploidy, with examples from Madiinae. Publication 564, Carnegie Institute of Washington. Washington, DC. [13]

Clausen, J., D. D. Keck, and W. M. Hiesey. 1947. Heredity of geographically and ecologically isolated races. *Am. Nat.* 81: 114–133. [13]

Clausen, J., D. D. Keck, and W. M. Hiesey. 1948. Experimental studies on the nature of species, volume III: Environmental response of climatic races of *Achillea*. Publication 581, Carnegie Institute of Washington. Washington, DC. [13]

Claussnitzer, M., and 18 others. 2015. *FTO* obesity variance circuitry and adipocyte browning in humans. *NEJM* 373: 895–907. [21]

Clayton, D. 1999. A generalization of the transmission/disequilibrium test for uncertain-haplotype transmission. *Am. J. Hum. Genet.* 65: 1170–1177. [17]

Clayton, D. 2012. Link functions in multi-locus genetic models: implications for testing, prediction, and interpretation. *Genet. Epid.* 36: 409–418. [20]

Clayton, D., and H. Jones. 1999. Transmission/disequilibrium tests for extended marker haplotypes. *Am. J. Hum. Genet.* 65: 1161–1169. [17]

Clayton, D. G., and 18 others. 2005. Population structure, differential bias and genomic control in a large-scale, case-control association study. *Nat. Genet.* 37: 1243–1246. [20]

Clayton, G. A., and A. Robertson. 1955. Mutation and quantitative variation. *Am. Nat.* 89: 151–158. [15]

Clayton, G. A., and A. Robertson. 1964. The effects of X-rays on quantitative characters. *Genet. Res.* 5: 410–422. [15]

Clayton, G. A., J. A. Morris, and A. Robertson. 1957. An experimental check on quantitative genetical theory. I. Short-term responses to selection. *J. Genetics* 55: 131–151. [7]

Cleghorn, T. E. 1960. MNSs gene frequencies in English blood donors. *Nature* 187: 701. [5]

Clerget-Darpoux, F., C. Bonaïti-Pellié, and J. Hochez. 1986. Effects of misspecifying genetic parameters in Lod score analysis. *Biometrics* 42: 393–399. [19]

Clo, J., and F. Kolář. 2022. Inbreeding depression in polyploid species: a meta-analysi. *Biol. Lett.* 18: 20220477. [13]

Cloninger, C. R., J. Rice, and T. Reich. 1979a. Multifactorial inheritance with cultural transmission and assortative mating. II. A general model of combined polygenic and cultural inheritance. *Am. J. Hum. Genet.* 31: 176–189. [7]

Cloninger, C. R., J. Rice, and T. Reich. 1979b. Multifactorial inheritance with cultural transmission and assortative mating. III. Family structure and analysis of separation experiments. *Am. J. Hum. Genet.* 31: 366–388. [7]

Clutton-Brock, J. 1992. *Horse power.* Harvard University Press, Cambridge, MA. [13]

Cochran, W. G. 1937a. Problems arising in the analysis of a series of similar experiments. *J. Roy. Stat. Soc.* 4: 102–118. [A6]

Cocharn, W. G. 1937b. A catalogue of uniformity trial data. *J. Roy. Stat. Soc.* 4: 233–255. [A9]

Cochran, W. G. 1954. Some methods for strengthening the common chi-square tests. *Biometrics* 10: 417–451. [18]

Cockerham, C. C. 1954. An extension of the concept of partitioning hereditary variance for analysis of covariances among relatives when epistasis is present. *Genetics* 39: 859–882. [5, 7]

Cockerham, C. C. 1959. Partitions of hereditary variance for various genetic models. *Genetics* 44: 1141–1148. [13]

Cockerham, C. C. 1980. Random and fixed effects in plant genetics. *Theor. Appl. Genet.* 56: 119–131. [11]

Cockerham, C. C. 1986. Modifications in estimating the number of genes for a quantitative character. *Genetics* 114: 659–664. [11]

Cockerham, C. C., and B. S. Weir. 1968. Sib mating with two linked loci. *Genetics* 60: 629–640. [7]

Cockerham, C. C., and B. S. Weir. 1973. Descent measures for two loci with some applications. *Theor. Pop. Biol.* 4: 300–330. [7]

Cockerham, C. C., and B. S. Weir. 1977a. Digenic descent measures for finite populations. *Genet. Res.* 30: 121–147. [5, 7]

Cockerham, C. C., and Z.-B. Zeng. 1996. Design III with marker loci. *Genetics*. 143: 1437–1456. [13, 18]

Cockram, J., and I. Mackay. 2018. Genetic mapping populations for conducting high-resolution trait mapping in plants. *Adv. Biochem. Eng. Biotechnol.* 164: 109–138. [18]

Coen, E. S., R. Carpenter, and C. Martin. 1986. Transposable elements generate novel spatial patterns of gene expression in *Antirrhinum majus*. *Cell* 47: 285–296. [15]

Coffman, C. J., and 6 others. 2005. Model selection in binary trait locus mapping. *Genetics* 170: 1281–1297. [18]

Cohen, J. C., and 5 others. 2004. Multiple rare alleles contribute to low plasma level of HDL cholesterol. *Science* 305: 869–872. [15, 20]

Coles, J. F., and D. P. Fowler. 1976. Inbreeding in neighboring trees in two white spruce plantations. *Silvae Genet.* 25: 29–34. [12]

Colhoun, H. M., P. M. McKeigue, and G. D. Smith. 2003. Problems of reporting genetic associations with complex outcomes. *Lancet* 361: 865–872. [20]

Collins, G. N. 1921. Dominance and the vigor of first generation hybrids. *American Naturalist* 55: 116–133. [13]

Collins, J. P., and J. E. Cheek. 1983. Effect of food and density on development of typical and cannibalistic salamander larvae in *Ambystoma tigrinum*. *Am. Zool.* 23: 77–84. [6]

Collins, R. L. 1967. A general nonparametric theory of genetic analysis. I. Application to the classical cross. *Genetics* 56: 551. [16]

Collins, R. L. 1968. A general nonparametric theory of genetic analysis. II. Digenic models with linkage for the classical cross. *Genetics* 60: 169–170. [16]

Collins, R. L. 1973. Reply to Whitney and Klein. *Genetics* 74: 382–383. [16]

Coltman, D. W. 2005. Testing marker-based estimates of heritability in the wild. *Mol. Ecol.* 14: 2593–2599. [8]

Coltman, D. W., J. G. Pilkington, J. A. Smith, and J. M. Pemberton. 1999. Parasite-mediated selection against inbred Soay sheep in a free-living island populaton. *Evolution* 53: 1259–1267. [12]

Commenges, D. 1994. Robust genetic linkage analysis based on a score test of homogeneity: The weighted rank pairwise correlation statistic. *Genet. Epidem.* 11: 198–200. [19]

Comstock, R. E., and F. D. Enfield. 1981. Gene number estimation when mutliplicative genetic effects are assumed — growth in flour beetles and mice. *Theor. Appl. Genet.* 59: 373–379. [11]

Comstock, R. E., and H. F. Robinson. 1948. The components of genetic variance in populations of biparental progenies and their use in estimating the average degree of dominance. *Biometrics* 4: 254–266. [13]

Comstock, R. E., and H. F. Robinson. 1952. Estimation of average dominance of genes. *In* J. W. Gowen (ed.), *Heterosis*, pp. 494–516. Iowa State College Press, Ames, IA. [5, 13, 18]

Conneally, P. M., and 6 others. 1985. Report of the committee on methods of linkage analysis and reporting. *Cytogen. Cell Genet.* 40: 356–359. [18]

Conneely, K. N., and M. Boehnke. 2007. So many correlated tests, so little time! Rapid adjustment of *P* values for multiple correlated tests. *Am. J. Hum. Genet.* 81: 1158–1168. [20]

Connor, J. L., and M. J. Bellucci. 1979. Natural selection resisting inbreeding depression in captive wild house mice *(Mus musculus)*. *Evolution* 33: 929–940. [12]

Conomos, M. P., A. P. Reiner, B. S. Weir, and T. A. Thornton. 2016. Model-free estimation of recent genetic relatedness. *Amer. J. Hum. Genet.* 98: 127–148. [8]

Conover, W. J. 1999. *Practical nonparametric statistics, 3rd edition*. Wiley and Sons, New York, NY. [20]

Conrad, D. F., and 28 others. 2010. Origins and functional impact of copy number variation in the human genome. *Nature* 464: 704–712. [8, 15]

Conradsen, C., M. W. Blows, and K. McGuigan. 2022. Causes of variability in estimates of mutational variance from mutation accumulation experiments. *Genetics 221*: iyac060. [15]

Consul, P. C., and F. Famoye. 1992. Generalized poisson regression model. *Comm. Stat. Theory Meth.* 21: 89–109. [14]

Cook, P. J. L., and 4 others. 1980. Exclusion mapping illustrated by the MNS blood group. *Ann. Hum. Genet.* 44: 61–73. [19]

Cooke, P., and K. Mather. 1962. Estimating the components of continuous variation. II. Genetics. *Heredity* 17: 211–236. [17]

Cooper, H., L. V. Hedges, and J. C. Valentine. 2009. *The handbook of research synthesis and meta-analysis*, 2nd Ed. Sage Foundation, New York, NY. [A6]

Cooper, G. M., and J. Shendure. 2011. Needles in stacks of needles: finding disease-causal variants in a wealth of genomic data. *Nat. Rev. Genet.* 12: 628–640. [21]

Cooper, G. M., and 5 others. 2005. Distribution and intensity of constraint in mammalian genomic sequences. *Genome Res.* 15: 901–913. [12]

Coors, J. G., and S. Pandey (eds). 1999. *The genetics and exploitation of heterosis in crops*. ASA-CSSA-SSSA Societies, Madison, WI. [13]

Copas, J. and J. Q. Shi. 2000. Meta-analysis, funnel plots and sensitivity analysis. *Biostatistics* 1: 247–262. [A6]

Copeman, J. B., and 15 others. 1995. Linkage disequilibrium mapping of a type 1 diabetes susceptibility gene (*IDDM7*) to chromosomes 2q31-q33. *Nature Genetics* 9: 80–85. [17]

Coque, M., and A. Gallias. 2006. Genomic regions involved in response to grain yield selection at high and low nitrogen fertilization in maize. *Theor. Appl. Genet.* 112: 1205–1220. [17]

Cordell, H. J. 2009. Detecting gene–gene interactions that underlie human diseases. *Nat. Rev. Genet.* 10: 392–404. [20]

Corder, E. H. and 8 others. 1993. Gene dose of apolipoprotein E type 4 allele and the risk of Alzheimer's disease in late onset families. *Science* 261: 921–923. [17]

Cornelius, P. L., and J. W. Dudley. 1974. Effects of inbreeding by selfing and full-sib mating in a maize population. *Crop Sci.* 14: 815–819. [12]

Cornish-Bowden, A. 1987. Dominance is not inevitable. *J. Theor. Biol.* 125: 335–338. [4]

Cortijo, S., and 13 others. 2014. Mapping the epigenetic basis of complex traits. *Science* 343: 1145–1148. [13, 21]

Corty, R. W., and W. Valdar. 2018. QTL mapping on a background of variance heterogeneity. *G3* 8: 3767–3782. [20]

Corty, R. W., and 4 others. 2018. Mean-variance QTL mapping identifies novel QTL for circadian activity and exploratory behavior in mice. *G3* 8: 3783–3790. [20]

Costa e Silva, J., G. W. Dutkowski, and A. R. Gilmour. 2001. Analysis of early tree height in forest genetic trials is enhanced by including a spatially correlated residual. *Can. J. Forest. Res.* 31: 1887–1893. [A9]

Cothran, E. G., J. W. MacCluer, L. R. Weitkamp, and S. A. Guttormsen. 1986. Genetic variability, inbreeding, and reproductive performance in standardbred horses. *Zoo Biol.* 5: 191–201. [12]

Cotsapas, C., and 23 others. 2011. Pervasive sharing of genetic effects in autoimmune disease. *PLoS Genet.* 7: e1002254. [20, 21]

Cotterman, C. W. 1940. A calculus for statistico-genetics. Ph. D. Thesis, Ohio State Univ., Columbus, OH. [7, 8]

Cotterman, C. W. 1954. Estimation of gene frequencies in nonexperimental populations. *In* O. Kempthorne, T. A. Bancroft, J. W. Gowen, and J. L. Lush (eds.), *Statistics and mathematics in biology*, pp. 449–465. Iowa State College Press, Ames, IA. [7]

Cowan, C. M., M. R. Detine, R. L. Ax, and L. A. Schuler. 1990. Structural variation around prolactin gene linked to quantitative traits in an elite Holstein sire family. *Theor. Appl. Genet.* 79: 577–582. [17]

Cowen, N. M. 1988. The use of replicated progenies in marker-based mapping of QTLs. *Theor. Appl. Genet.* 75: 857–862. [17]

Cowles, M. K., and B. P. Carlin. 1996. Markov chain Monte Carlo diagnostics: A comparative review. *J. Am. Stat. Assoc.* 91: 883–904. [A8]

Cox, D. R, and D. R. Hinkley. 1979. *Theoretical statistics*. CRC Press, Boca Raton, FL. [A4]

Coyne, J. A. 1983. Genetic basis of differences in genital morphology among three sibling species of *Drosophila*. *Evolution* 37: 1101–1118. [17]

Coyne, J. A. 1984. Genetic basis of male sterility in hybrids between two closely related species of *Drosophila*. *Proc. Natl. Acad. Sci. USA* 81: 4444–4447. [17]

Coyne, J. A. 1985. Genetic studies of three sibling species of *Drosophila* with relationship to theories of speciation. *Genet. Res.* 46: 169–192. [17]

Coyne, J. A. 1992. Genetics and speciation. *Nature* 355: 511–515. [17]

Coyne, J. A., and E. Beecham. 1987. Heritability of two morphological characters within and among natural populations of *Drosophila melanogaster*. *Genetics* 117: 727–737. [7]

Coyne, J. A., and R. Lande. 1985. The genetic basis of species differences in plants. *Am. Nat.* 126: 141–145. [11]

Coyne, J. A., and H. A. Orr. 1989a. Patterns of speciation in *Drosophila*. *Evolution* 43: 362–381. [13, 17]

Coyne, J. A., and H. A. Orr. 1989b. Two rules of speciation. *In* D. Otte and J. A. Endler (eds.), *Speciation and its consequences*, pp. 180–207. Sinauer Assoc., Sunderland, MA. [13, 17]

Coyne, J. A., and H. A. Orr. 1993. Further evidence against the meiotic drive model of hybrid sterility. *Evolution* 47: 685–687. [17]

Coyne, J. A., and H. A. Orr. 1997. "Patterns of speciation in *Drosophila*" revisited. *Evolution* 51: 295–303. [13, 17]

Coyne, J. A., and H. A. Orr. 1998. The evolutionary genetics of speciation. *Phil. Trans. Roy. Soc. Lond. B* 353: 287–305. [17]

Coyne, J. A., and H. A. Orr. 2004. *Speciation*. Sinauer Associates, Sunderland, MA. [17]

Coyne, J. A., B. Charlesworth, and H. A. Orr. 1991. Haldane's rule revisited. *Evolution* 45: 1710–1714. [17]

Crabb, A. R. 1947. *The hybrid-corn makers: Prophets of plenty*. Rutger Univ. Press, New Brunswick, NJ. [13]

Crabb, A. R. 1993. *The hybrid-corn makers, 2nd ed.* West Chicago Publishing Co., West Chicago, IL. [13]

Crabbe, J. C., J. K. Belknap, and K. J. Buck. 1994. Genetic animal models of alcohol and drug abuse. *Science* 264: 1715–1723. [17]

Crespi, B. J. 1990. Measuring the effect of natural selection on phenotypic interaction systems. *Am. Nat.* 135: 32–47. [A2]

Cress, C. E. 1966. Heterosis of the hybrid related to gene frequency differences between two populations. *Genetics* 53: 269–274. [13]

Cressie, N. A. C. 1991. *Statistics for spatial data*. Wiley, New York, NY. [A9]

Crnokrak, P., and S. C. H. Barrett. 2002. Perspective: purging the genetic load: a review of the experimental evidence. *Evolution* 56: 2347–2358. [12]

Crnokrak, P., and D. A. Roff. 1995. Dominance variance: associations with selection and fitness. *Heredity* 75: 530–540. [7]

Crnokrak, P., and D. A. Roff. 1999. Inbreeding depression in the wild. *Heredity* 83: 260–270. [12]

Croft, J. H., and J. L. Jinks. 1977. Aspects of the population genetics of *Aspergillus nidulans*. *In* J. E. Smith and J. A. Pateman (eds.), *Genetics and physiology of Aspergillus*, pp. 339–360. Academic Press, NY. [5]

Croft, J. H., and G. Simchen. 1965. Natural variation among monokaryons of *Collybia velutipes*. *Am. Nat.* 99: 451–462. [11]

Crouse, W. L., S. N. P. Kelada, and W. Valdar. 2020. Inferring the allelic series at QTL in multiparental populations. *Genet.* 216: 957–983. [18]

Crouse, W. L., and 4 others. 2022. A Bayesian model selection approach to mediation analysis. *PLoS Genet.* 18: e1010184. [21, A2]

Crow, J. F. 1948. Alternative hypotheses of hybrid vigor. *Genetics* 33: 477–487. [12, 13]

Crow, J. F. 1952. Dominance and overdominance. *In* J. E. Gowen (ed.), *Heterosis*, pp. 282–297/. Iowa State College Press, Ames, IA. [12]

Crow, J. F. 1954. Random mating with linkage in polysomics. *Am. Nat.* 88: 431–434. [4]

Crow, J. F. 1958. Some possibilities for measuring selection intensities in man. *Human Biol.* 30: 1–13. [12]

Crow, J. F. 1972. The origins of theoretical population genetics. *Amer. J. Hum. Genet.* 24: 608–609. [1]

Crow, J. F. 1992. Mutation, mean fitness, and genetic load. *Oxford Surv. Evol. Biol.* 9: 3–42. [15]

Crow, J. F. 1993a. Francis Galton: count and measure, measure and count. *Genetics* 135: 1–4. [1]

Crow, J. F. 1993b. How much do we know about spontaneous human mutation rates? *Environ. Mol. Mutagenesis* 21: 122–129. [15]

Crow, J. F. 1993c. Mutation, mean fitness, and genetic load. *Oxford Surveys Evol. Biol.* 9: 3–42. [12]

Crow, J. F. 1998. 90 years ago. The beginning of hybid maize. *Genetics* 148: 923–928. [13]

Crow, J. F. 2000. The rise and fall of overdominance. *Plant Breed. Rev.* 17: 225–257. [12]

Crow, J. F., and J. Felsenstein. 1968. The effect of assortative mating on the genetic composition of a population. *Eugenics Quart.* 15: 85–97. [7]

Crow, J. F., and M. Kimura. 1970. *An introduction to population genetics theory.* Harper & Row, New York, NY. [4, 7, 12]

Crow, J. F., and M. Kimura. 1979. Efficiency of truncation selection. *Proc. Natl. Acad. Sci. USA* 76: 396–399. [15]

Crow, J. F., and M. J. Simmons. 1983. The mutation load in *Drosophila. In* M. Ashburner, H. L. Carson, and J. N. Thompson, Jr. (eds.), *The genetics and biology of Drosophila. Volume 3c,* pp. 1–35. Academic Press, New York, NY. [4, 15]

Crudent, D. 1949. The computation of inbreeding coefficients for closed populations. *J. Heredity* 40: 248–251. [7]

Crusio, W. E. 2004. A note on the effect of within-strain sample sizes on QTL mapping in recombinant inbred strain studies. *Genes Brain Behav.* 3: 249–251. [18]

Csillèry, K., and 7 others. 2006. Performance of marker-based relatedness estimators in natural populations of outbred vertebrates. *Genetics* 173: 2091–2101. [8]

Cubillos, F. A., and 9 others. 2011. Assessing the complex architecture of polygenic traits in diverged yeast populations. *Mol. Ecol.* 20: 1401–1413. [18]

Cubillos, F. A., and 12 others. 2013. High-resolution mapping of complex traits with a four-parent advanced intercross yeast population. *Genetics* 195: 1141–1155. [5, 18]

Cui, Y., and W. Yang. 2009. Zero-inflated generalized Poisson regression mixture model for mapping quantitative trait loci underlying count trait with may zeros. *J. Theor. Biol.* 256: 276–285. [18]

Cui, Y., D. Y. Kim, and J. Zhu. 2006. On the generalized poisson regression mixture model for mapping quantitative trait loci with count data. *Genetics* 174: 2159–2172. [18]

Cui, Y., F. Zhang, J. Xu, Z. Li, and S. Xu. 2015. Mapping quantitative trait loci in selected breeding populations: A segregation distortion approach. *Heredity* 115: 538–546. [18]

Cule, E., P. Vineis, and M. De Iorio. 2011. Significance testing in ridge regression for genetic data. *BMC Bioinfomat.* 12: 372. [20]

Cullis, B. R., and A. C. Gleeson. 1991 Spatial analysis of field experiments — an extension to two dimensions. *Biometrics* 47: 1449–1460. [A9]

Cullis, B., B. Gogel, A. Verbyla, and R. Thompson. 1998. Spatial analysis of multi-environment early generation variety trails. *Biometrics* 54: 1–18. [20, A9]

Cullis, B. R., A. B. Smith, and N. E. Coombes. 2006. On the design of early generation variety trials with correlated data. *J. Agric. Biol. Environ. Stats.* 11: 381–393. [A9]

Cullis, C. A. 1981. Environmental induction of heritable changes in flax: defined environments inducing changes in rDNA and peroxidase isozyme band pattern. *Heredity* 47: 87–94. [15]

Cullis, C. A. 1985. Sequence variation and stress. *In* B. Hohn and E. S. Dennis (eds.), *Genetic flux in plants,* pp. 157–163. Springer-Verlag, New York, NY. [15]

Cundiff, L. V., K. E. Gregory, and R. M. Koch. 1974a. Effects of heterosis on reproduction in Hereford, Angus, and Shorthorn cattle. *J. Anim. Sci.* 38: 711–727. [13]

Cundiff, L. V., K. E. Gregory, F. J. Schwulst, and R. M. Koch. 1974b. Effects of heterosis on maternal performance and milk production in Hereford, Angus, and Shorthorn cattle. *J. Anim. Sci.* 38: 711–727. [13]

Cunningham. E. E. 1982. The genetic basis of heterosis. *Proc. 2nd World Congress Genetic Applied Livestock Production.* Vol 6: 190–205. [13]

Curik, I., J. Sölkner, and N. Stipic. 2001. The influence of selection and epistasis on inbreeding depression estimates. *J. Anim. Breed. Genet*: 247–262. [12]

Curik, I., M. Ferenĉaković, and J. Sölkner. 2014. Inbreeding and run of homozygosity: a possible solution to an old problem. *Live. Sci.* 166: 26–35. [8]

Curtis, D. 1996. Genetic dissection of complex traits. *Nature Genetics* 12: 356–357. [19]

Curtis, D. 1997. Use of siblings as controls in case-control association studies. *Ann. Hum. Genet.* 61: 319–333. [17]

Curtis, D., and P. C. Stam. 1994. Using risk calculation to implement an extended relative pair analysis. *Am. J. Hum. Genet.* 58: 151–162. [19]

Curtis, D., and P. C. Stam. 1995a. Model-free linkage analysis using likelihoods. *Am. J. Hum. Genet.* 57: 703–716. [19]

Curtis, D., and P. C. Stam. 1995b. A note on the application of the transmission disequilibrium test when a parent is missing. *Am. J. Hum. Genet.* 56: 811–812. [17]

Curtis, R. K., M. Orešič, and A. Vidal-Puig. 2006. Pathways to the analysis of microarray data. *Trends Biotech.* 23: 429–435. [21]

Cutter, A. D. 2012. The polymorphic prelude to Bateson–Dobzhansky–Muller incompatibilities. *Trends Ecol. Evol.* 27: 209–218. [17]

Dadd, T., M. E. Weale, and C. M. Lewis. 2009. A critical evaluation of genomic control methods for genetic association studies. *Genet. Epid.* 33: 290–298. [20]

D'Agostino, R. B. 1971. An omnibus test of normality for moderate and large size samples. *Biometrika* 58: 341–348. [14]

Daly, M. J., and E. S. Lander. 1996. The importance of being independent: sib pair analysis in diabetes. *Nature Genetics* 14: 131–132. [19]

Damerval, C., A. Maurice, J. M. Josse, and D. de Vienne. 1994. Quantitative trait loci underlying gene product variation: a novel perspective by analyzing regulation of genome expression. *Genetics* 137: 289–301. [18, 21]

Dandine-Roulland, C., and H. Perdry. 2015. The use of the linear mixed model in human genetics. *Hum. Heredity* 80: 196–206. [20]

Daniels, D. E. 1954. Saddlepoint approximations in statistics. *Ann. Math. Statist.* 25: 631–650. [20]

Daniels, S. B., M. McCarron, C. Love, and A. Chovnick. 1985. Dysgenesis-induced instability of rosy locus transformation in *Drosophila melanogaster*: analysis of excision events and the selective recovery of control element deletions. *Genetics* 109: 95–117. [15]

Dapkus, D., and D. J. Merrell. 1977. Chromosomal analysis of DDT-resistance in a long-term selected population of *Drosophila melanogaster*. *Genetics* 87: 685–697. [17]

Dapp, M., and 7 others. 2015. Heterosis and inbreeding depression of epigenetic *Arabidopsis* hybrids. *Nat. Plants* 1: 15092. [13]

Darlington, C. D. 1939. *The evolution of genetic systems.* Cambridge Univ. Press, Cambridge, UK. [8, 11]

Darvasi, A., and S. Shifman. 2005. The beauty of admixture. *Nat. Genet.* 37: 118–119. [17]

Darvasi, A., and M. Soller. 1992. Selective genotyping for determination of linkage between a marker locus and a quantitative trait locus. *Theor. Appl. Genet.* 85: 353–359. [17, 18]

Darvasi, A., and M. Soller. 1994a. Selective DNA pooling for determination of linkage between a marker locus and a quantitative trait locus. *Genetics* 138: 1365–1373. [17]

Darvasi, A., and M. Soller. 1994b. Optimum spacing of genetic markers for determining linkage between marker loci and quantitative trait loci. *Theor. Appl. Genet.* 89: 351–357. [18]

Darvasi, A., and M. Soller. 1995. Advanced intercross lines, an experimental population for fine genetic mapping. *Genetics* 141: 1199–1207. [18]

Darvasi, A., and M. Soller. 1997. A simple method to calculate resolving power and confidence interval of QTL map location. *Behav. Genet.* 27: 125–132. [18]

Darvasi, A., and 4 others. 1993. Detecting marker-QTL linkage and estimating QTL gene effect and map location using a saturated genetic map. *Genetics* 134: 943–951. [18]

Darwin, C. 1859. *The origin of species by means of natural selection.* Murray, London, UK. [1]

Darwin, C. 1868. *The variation of animals and plants under domestication.* John Murray, London, UK. [12]

Darwin, C. 1876. *The effects of cross and self-fertilization in the vegetable kingdom.* Appleton, New York, NY. [12, 13]

Davenport, C. B. 1908. Degeneration, albinism, and inbreeding. *Science* 28: 454–455. [12, 13]

Davey Smith, G., and S. Ebrahim. 2003. 'Mendelian randomization': can genetic epidemiology contribute to understanding environmental determinants of disease? *Int. J. Epidem.* 32: 1–22. [A2]

Davey Smith, G., and S. Ebrahim. 2004. Mendelian randomization: prospects, potentials, and limitations. *Int. J. Epidem.* 33: 30–42. [A2]

Davey Smith, G., and S. Ebrahim. 2005. What can Mendelian randomization tell us about modifiable behavioural and environmental exposures? *BMJ* 330: 1076–1079. [A2]

Davey Smith, G., and G. Hemani. 2014. Mendelian randomization: genetic anchors for causal inference in epidemiological studies. *Hum. Mol. Genet.* 23: R89–R98. [A2]

David, F. N. 1981. *Order statistics*, 2nd Ed. Wiley, New York, NY. [11]

David, P. 1998. Heterozygosity–fitness correlations: new perspectives on old problems. *Heredity* 80: 531–537. [12]

David, P., B. Delay, P. Berthou, and P. Jarne. 1995. Alternative models for allozyme-associated heterosis in the marine bivalve *Spisula ovalis*. *Genetics* 139: 1719–1726. [12]

David, P., B. Pujol, F. Viard, V. Castella, and J. Goudet. 2007. Reliable selfing rate estimates from imperfect population genetic data. *Mol. Ecol.* 16: 2474–2487. [12]

Davies, A. G., and 7 others. 1996. *Scalloped wings* is the *Lucilia cuprina Notch* homologue and a candidate for the *Modifier* of fitness and asymmetry of Diazinon resistance. *Genetics* 143: 1321–1337. [14]

Davies, J. L., and 15 others. 1994. A genome-wide search for human type 1 diabetes susceptibility genes. *Nature* 371: 130–136. [19]

Davies, N. M., and 5 others. 2019. Within family Mendelian randomization studies. *Hum. Mol. Genet.* 28: R2. [A2]

Davies, R. B. 1973. Numerical inversion of a characteristic function. *Biometrika* 60: 415–416. [20]

Davies, R. B. 1987. Hypothesis testing when a nuisance parameter is present only under the alternative. *Biometrika* 74: 33–43. [18]

Davies, R. W. 1971. The genetic relationship of two quantitative characters in *Drosophila melanogaster*. II. Location of the effects. *Genetics* 69: 363–375. [17]

Davies, R. W., and P. L. Workman. 1971. The genetic relationship of two quantitative characters in *Drosophila melanogaster*. I. Response to selection and whole chromosome analysis. *Genetics* 69: 353–361. [17]

Davis, A. W., and C.-I. Wu. 1996. The broom of the sorcerer's apprentice: the fine structure of a chromosomal region causing reproductive isolation between two sibling species of *Drosophila*. *Genetics* 143: 1287–1298. [17]

Davis, J. R., and 6 others. 2016. An efficient multiple-testing adjustment for eQTL studies that accounts for linkage disequilibrium between variants. *Am. J. Hum. Genet.* 98: 216–224. [20, 21]

Davis, S., M. Schroeder, L. R. Goldin, and D. E. Weeks. 1996. Nonparametric simulation-based statistics in detecting linkage in general pedigrees. *Am. J. Hum. Genet.* 58: 867–880. [19]

Day, N. E. 1969. Estimating the components of a mixture of normal distributions. *Biometrika* 56: 463–474. [16]

Day, N. E., and M. J. Simmons. 1976. Disease susceptibility genes — their identification by multiple case family studies. *Tissue Antigens* 8: 109–119. [19]

Day-Williams, A. G., and E. Zeggini. 2010. The effect of next-generation sequencing technology on complex trait research. *Eur J. Clinical Investi.* 41: 561–567. [8]

Day-Williams, A. G., and 4 others. 2011. Linkage analysis without defined pedigrees. *Genet. Epid.* 35: 360–370. [8]

de Bakker, P. I. W., and 5 others. 2008. Practical aspects of imputation-driven meta-analysis of genome-wide association studies. *Hum. Mol. Genet.* 17: R122–R128. [20]

De Koeyer, D. L., R. L. Phillips, and D. D. Stuthman. 2001. Allelic shifts and quantitative trait loci in a recurrent selection population of oat. *Crop Sci.* 41: 1228–1234. [17]

de Koning, D.-J. and C. S. Haley. 2005. Genetical genomics in humans and model organisms. *Trends Genet.* 21: 377–381. [1, 5, 18, 21]

de Koning, D.-J., P. M. Visscher, S. A. Knott, and C. S. Haley. 1998. A strategy for QTL detection in half-sib populations. *Anim. Sci.* 67: 257–268. [19]

de Koning, D.-J., and 9 others. 1999. Detection of quantitative trait loci for backfat thickness and intramuscular fat content in pigs (*Sus scrofa*). *Genetics* 152: 1679–1690. [19]

de Koning, D.-J., and 5 others. 2000. Genome-wide scan for body composition in pigs reveals important role of imprinting. *Proc. Natl. Acad. Sci. USA* 97: 7947–7950. [19]

de Koning, D.-J., and 6 others. 2001a. Mapping of multiple quantitative trait loci by simple regression in half-sib designs. *J. Anim. Sci.* 79: 616–622. [19]

de Koning, D.-J., and 5 others. 2001b. Detection and characterization of quantitative trait loci for meat quality traits in pigs. *J. Anim. Sci.* 79: 2812–2819. [19]

de Koning, D.-J., H. Bovenhuis, and J. A. M. van Arendonk. 2002. On the detection of imprinted quantitative trait loci in experimental crosses of outbred species. *Genetics* 161: 931–938. [19]

de la Chapelle, A. 1993. Disease gene mapping in isolated human populations: the example of Finland. *J. Med. Genet.* 30: 857–965. [17]

de la Chapelle, A., and F. A. Wright. 1998. Linkage disequilibrium mapping in isolated populations: the example of Finland revisited. *Proc. Natl. Acad. Sci. USA* 95: 12416–12423. [17]

De la Cruz, O., and 4 others. 2010. Gene, region and pathway level analysis in whole-genome studies. *Genet. Epidem.* 34: 222–231. [21]

de Leeuw, C. A., J. M. Mooij, T. Heskes, and D. Posthuma. 2015. MAGMA: Generalized gene-set analysis of GWAS data. *PLoS Comput. Bio.* 10: 1371. [20]

de Leeuw, C. A., B. M. Neale, T. Heskes, and D. Posthuma. 2016. The statistical properties of gene-set analysis. *Nat. Rev. Genet.* 17: 353–364. [21]

de Los Campos, G., D. A. Sorensen, and M. A. Toro. 2019. Imperfect linkage disequilibrium generates phantom epistasis (and perils of big data). *G3*: 1429–1436. [20]

de Manuel, M., F. L. Wu, and M. Przeworski. 2022. A paternal bias in germline mutation is widespread in amniotes and can arise independently of cell division numbers. *eLife* 11: e80008. [15]

de Villemereuil, P, H. Schielzeth, S. Nakagawa, and M. Morrissey. 2016. General methods for evolutionary quantitative genetic inference from generalized mixed models. *Genetics* 204: 1281–1294. [14]

De Visser, J. A. G. M., and 18 others. 2003. Perspective: evolution and detection of genetic robustness. *Evolution*57: 1959–1972. [14]

Dearborn, D. D. 1969. Crossbreeding for beef production. *Nebr. Ext. Circ.* 69-22. [13]

Debrabant, B. 2017. The null hypothesis of GSEA, and a novel statistical model for competitive gene set analysis. *Bioinformat.* 33: 1271–1277. [21]

DeFries, J. C. 2010. Haseman and Elston sib-pair linkage analysis: a brief historical note. *Behav. Genet.* 40: 1–2. [19]

Degner, J. F., and 13 others. 2012. DNAse I sensitivity QTLs are a major determinant of human gene expression. *Nature* 482: 390–394. [21]

Delph, L. F., and J. P. Demuth. 2016. Haldane's rule: genetic bases and their empirical support. *J. Hered.* 107: 383–391. [17]

Demarly, Y. 1963. Genetique des tetaploides et amelioration des plantes. *Ann. Amelio. Plant.* 13: 307–400. [13]

Demenais, F., and G. E. Bonney. 1989. Equivalence of the mixed and regressive models for genetic analysis. Continuous traits. *Genet. Epidem.* 6: 597–617. [16]

Demenais, F., M. Lathrop, and J. M. Lalouel. 1986. Robustness and power of the unified model in the analysis of quantitative measurements. *Am. J. Hum. Genet.* 38: 228–234. [16]

Demenais, F., M. Lathrop, and J. M. Lalouel. 1988. Detection of linkage between a quantitative trait and a marker locus by the lod score method: sample size and sampling considerations. *Am. J. Hum. Genet.* 52: 237–246 [19]

Demenais, F. M., C. Murigande, G. E. Bonney, and D. C. Rao. 1990. Search for faster methods of fitting the regressive models to quantitative traits. *Genetic Epidem.* 7: 319–334. [16]

DeMoivre, A. 1738. *The doctrine of chances* 2nd Ed. (reprinted 1967, Frank Cass, London, UK). [2]

Dempster, A. P., N. M. Laird, and D. B. Rubin. 1977. Maximum likelihood from incomplete data via the EM algorithm. *J. Royal Stat. Soc.* 39: 1–38. [A4]

Dempster, E. R. 1943. Mock dominance. *Science* 97: 464–465. [13]

Demuth, J. P., and M. J. Wade. 2005. On the theoretical and empirical framework for studying genetic interactions within and among species. *Am. Nat.* 165: 524–536. [13]

Demuth, J. P., and M. J. Wade. 2006. Experimental methods for measuring gene interactions. *Annu. Rev. Ecol. Evol. Syst.* 37: 289–316. [13]

Demuth, J. P., and M. J. Wade. 2007a. Population differentiation in the beetle *Tribolium castaneum*. I. Genetics architecture. *Evolution* 61: 494–509. [11, 13]

Demuth, J. P., and M. J. Wade. 2007b. Population differentiation in the beetle *Tribolium castaneum*. II. Haldane's rule and incipient speciation. *Evolution* 61: 694–699. [17]

Demuth, J. P., R. J. Flanagan, and L. F. Delph. 2014. Genetic architecture of isolation between two species of *Silene* with sex chromosomes and Haldane's rule. *Evolution* 68: 332–342. [17]

Den Boer, E., and 5 others. 2013. Fine mapping quantitative resistances to downy mildew in lettuce revealed multiple sub-QTLs with plant stage dependent effects reducing or even promoting the infection. *Theor. Appl. Genet* 126: 2995–3007. [17]

Dendrious, A. D. 1931. The question of shape in field experimentation. *Intl. Assn. Plant Breeders Bull.* 4: 106–113. [A9]

Deng, H.-W. 1997. Decrease of developmental stability upon inbreeding in *Daphnia*. *Heredity* 78: 182–189. [6]

Deng, H.-W., and M. Lynch. 1996a. Change of genetic architecture in response to sex. *Genetics* 143: 203–212. [5, 13]

Deng, H.-W., and M. Lynch. 1996b. Estimation of deleterious-mutation parameters in natural populations. *Genetics* 144: 349–360. [15]

Deng, L., H. Zhang, and K. Yu. 2020. Power calculation for the general two-sample Mendelian randomization analysis. *Genet. Epidem.* 44: 290–299. [A2]

Deng, W., H. Chen, and Z. Li. 2006. A logistic regression mixture model for interval mapping of genetic trait loci affecting binary phenotypes. *Genetics* 172: 1349–1358. [18]

Deng, W. Q., and G. Paré. 2011. A fast algorithm to optimize SNP prioritization for gene-gene and gene-environment interactions. *Genet. Epid.* 35: 729–738. [18, 20]

Deng, Y., and 5 others. 2020. Genome-wide gene-based multi-trait analysis. *Front. Genet.* 11: 437. [20]

Deniz, E., and B. Erman. 2017. Long noncoding RNA (lincRNA), a new paradigm in gene expression control. *Funct. Integr. Genomics* 17: 135–143. [21]

Denniston, C. 1974. An extension of the probability approach to genetic relationships: one locus. *Theor. Pop. Biol.* 6: 58–75. [7]

Denny, J. C., and 9 others. 2010. PheWAS: demonstrating the feasibility of a phenome-wide scan to discover gene-disease associations. *Bioinformatics* 26: 1205–1210. [20]

Dentine, M. R., and C. M. Cowan. 1990. An analytical model for the estimation of chromosome substitution effects in the offspring of individuals heterozygous at a segregating marker locus. *Theor. Appl. Genet.* 79: 775–780. [19]

Deol, M. S., H. Grüneberg, A. G. Searle, and G. M. Truslove. 1957. Genetical differentiation involving morphological characters in an inbred strain of mice. I. A British branch of the C57BL strain. *J. Morphology* 100: 345–375. [15]

Derkach, A., J. F. Lawless, and L. Sun. 2013. Robust and powerful tests for rare variants using Fisher's method to combine evidence of association from two or more complementary tests. *Genet. Epidem.* 37: 110–121. [20]

Derkach, A., J. F. Lawless, and L. Sun. 2014. Pooled association tests for rare genetic variants: a review and some new results. *Stat. Sci.* 29: 302–321. [20]

DeRose, M. A., and D. A. Roff. 1999. A comparison of inbreeding depression in life-history and morphological traits in animals. *Evol.* 53: 1288–1292. [12]

DerSimonian, R., and N Laird. 1986. Meta-analysis in clinical trials. *Controlled Clinical Trails* 7: 177–188. [20, A6]

DerSimonian, R., and N Laird. 2015. Meta-analysis in clinical trials revisited. *Controlled Clinical Trails* 45: 139–145. [20, A6]

DeSalle, R., and A. R. Templeton. 1986. The molecular through ecological genetics of *abnormal abdomen* in *Drosophila mercatorum*. III. Tissue-specific differential replication of ribosomal genes modulates the *abnormal abdomen* phenotype in *Drosophila mercatorum*. *Genetics* 112: 877–886. [15]

DeSalle, R., J. Slightom, and E. Zimmer. 1986. The molecular through ecological genetics of *abnormal abdomen* in *Drosophila mercatorum*. II. Ribosomal DNA polymorphism is associated with the *abnormal abdomen* syndrome in *Drosophila mercatorum*. *Genetics* 112: 861–875. [15]

Deshmukh, R., and 4 others. 2018. Mimicry in butterflies: co-option and a bag of magnificent developmental genetics traits. *WIREs Dev. Biol.* 7: e291. [15]

Dessureaux, L., and A. Gallais. 1969. Inbreeding in heterosis in autotetraploid alfalfa. I. Fertility. *Can. J. Genet. Cytol.* 11: 706–715. [13]

Deutsch, S., and 10 others. 2005. Gene expression variation and expression quantitative trait mapping of human chromosome 21 genes. *Hum. Mol. Genet.* 14: 3741–3749. [21]

deVicente, M. C., and S. D. Tanksley. 1993. QTL analysis of transgressive segregation in an interspecific tomato cross. *Genetics* 134: 585–596. [18]

Devlin, B., and N. Risch. 1995. A comparison of linkage disequilibrium measures for fine-scale mapping. *Genomics* 29: 311-322. [5, 16]

Devlin, B., and K. Roeder. 1999. Genomic control for association studies. *Biometrics* 55: 997–1004. [20]

Devlin, B., K. Roeder, and N. C. Ellstrand. 1988. Fractional paternity assignment: theoretical development and comparison to other methods. *Theor. Appl. Genet.* 76: 369–380. [8]

Devlin, B., K. Roeder, and S.-A. Bacanu. 2001a. Unbiased methods for population-based association studies. *Genet. Epid.* 21: 273–284. [20]

Devlin, B., K. Roeder, and L. Wasserman. 2001b. Genomic control, a new approach to genetic-based association studies. *Theor. Pop. Biol.* 60: 155–166. [20]

Devlin, B., S.-A. Bacanus, and K. Roeder. 2004. Genomic control to the extreme. *Nat. Genet.* 36: 1129–1130. [20]

deVries, R. R. P., and 4 others. 1976. HLA-linked genetic control of host response to *Mycobacterium leprae*. *Lancet* 308: 1328–1330. [19]

Dewan, A., and 12 others. 2006. *HTRA1* promoter polymorphism in wet age-related macular deregulation. *Science* 314: 989–92. [20]

Dewey, D. R. 1966. Inbreeding depression in diploid, tetraploid, and hexaploid crested wheatgrass. *Crop Sci.* 6: 144–147. [13]

Dhliwayo, T., K. Pixley, A. Menkir, and M. Warburton. 2009. Combining ability, genetic distances, and heterosis among elite CIMMYT and IITA tropical maize inbred lines. *Crop Sci.* 49: 1201–1210. [13]

Dhondt, S., N. Wuyts, and D. Inzé. 2013. Cell to whole-plant phenotyping: the best is yet to come. *Trends Plant Sci.* 18: 428–439. [18]

Dhrymes, P. J. 1978. *Mathematics for econometrics.* Springer-Verlag, New York, NY. [A3]

Di Rienzo, A., and R. R. Hudson. 2005. An evolutionary framework for common disease: the ancestral-susceptibility model. *Trends Genet.* 21: 596–601. [21]

Dial, K. P., E. Greene, and D. J. Irschick. 2008. Allometry of behavior. *Trends. Ecol. Evol.* 23: 394–401. [14]

Diao, G., and D. Y. Lin. 2006. Improving the power of association tests for quantitative traits in family studies. *Genet. Epid.* 30: 301–312. [17]

Diao, G., D Y. Lin, and F. Zou. 2004. Mapping quantitative trait loci with censored observations. *Genetics* 168: 1689–1698. [18]

Dickersin, K., Y.-I. Min, and C. L. Meinert. 1992. Factors influencing publication of research results: follow-up of applications submitted to two institutional review boards. *J. Am. Med. Assoc.* 267: 374–378. [A6]

Dickerson, G. E. 1969a. Experimental approaches to utilizing breed resources. *Anim. Breed. Abstracts* 37: 191–202. [13]

Dickerson, G. E. 1969b. Techniques for research in quantitative animal genetics. *In* Amer. Soc. Anim. Sci., *Techniques and procedures in animal science research,* pp. 36–79. Albany, NY. [A1]

Dickerson, G. E. 1972. Inbreeding and heterosis in animals. *Proc. of the Animal Breeding and Genetics Symp. In Honor of Dr. Jay Lush,* pp. 54–77. Amer. Soc. Anim. Sci., Champaign, IL. [13]

Dickerson, G. E., and 9 others. 1954. Evaluation of selection in developing inbred lines of swine. *Mo. Agric. Exp. Sta. Res. Bull.* 551. [12]

Dickinson, W. J. 2008. Synergistic fitness interactions and a high frequency of beneficial changes among mutations accumulated under relaxed selection in *Saccharomyces cerevisiae. Genetics 178*: 1571–1578. [15]

Dickson, S. P., and 4 others. 2010. Rare variants create synthetic genome-wide associations. *PLoS Biol.* 8: e1000294. [21]

Didelez, V., and N. Sheehan. 2007. Mendelian randomization as an instrumental variable approach to casual inference. *Stat. Meth. Med. Res.* 16: 309–330. [A2]

Dijkhuizen, A., J. W. Dudley, and T. R. Rocheford. 1996. Marker-QTL linkages estimated using F_2 and random-mated generations. *Illionis Corn Breeders School* 32: 144–157. [13]

Dilda, C. L., and T. F. C. Mackay. 2002. The genetic architecture of *Drosophila* sensory bristle number. *Genetics* 162: 1655–1674. [17]

Dillon, M. M., and V. S. Cooper. 2016. The fitness effects of spontaneous mutations nearly unseen by selection in a bacterium with multiple chromosomes. *Genetics* 204: 1225–1238. [15]

Dimas, A. S., and 14 others. 2009. Common regulatory variation impact gene expression in a cell type-dependent manner. *Science* 325: 1246–1250. [21]

Dinkel, C. A., D. A. Busch, J. A. Minyard, and W. R. Trevillyan. 1968. Effects of inbreeding on growth and conformation of beef cattle. *J. Anim. Sci.* 27: 313–322. [12]

Diogo, D., and 29 others. 2013. Rare, low-frequency, and common variants in the protein-coding sequence of biological candidates genes from GWASs contribute to risk of rheumatoid arthritis. *Am. J. Hum. Genet.* 92: 15–27. [15]

Dirlewanger, E., and 5 others. 1994. Restriction fragment length polymorphism analysis of loci associated with disease resistance genes and developmental traits in *Pisum sativum* L. *Theor. Appl. Genet.* 88: 17–27. [18]

Dixon, A. L., and 12 others. 2007. A genome-wide associations study of global gene expression. *Nat. Genet.* 39: 1202–1207. [21]

Dixson, L., and 16 others. 2014. Retraction for Dixson et al., Identification of gene ontologies linked to prefrontal-hippocampal functional coupling in the human brain. *Proc. Natl. Acad. Sci. USA* 111: 13582. [21]

Djami-Tchatchou, A. T., N. Sanan-Mishra, K. Ntushelo, and I. A. Dubery. 2017. Functional roles of microRNAs in agronomically important plants — potential as traits for crop improvement and protection. *Front. Plant Sci.* 8: 378. [15]

Dobbs, K. S. 1955. Hybrid vigor in plant breeding. *Proc. Roy. Soc. Lond. B* 144: 185–192. [13]

Dobzhansky, T. 1936. Studies on hybrid sterility. II. Localization of sterility factors in *Drosophila pseudoobscura* hybrids. *Genetics* 21: 113–135. [17]

Dobzhansky, T. 1948. Genetics of natural populations. XVIII. Experiments on chromosomes of *Drosophila pseudoobscura* from different geographical regions. *Genetics* 33: 588–602. [13]

Dobzhansky, T. 1950. Genetics of natural populations. XIX. Origin of heterosis through natural selection in populations of *Drosophila pseudoobscura. Genetics* 35: 288–302. [13]

Dobzhansky, T., and H. Levene. 1955. Developmental homeostasis in natural populations of *Drosophila pseudoobscura. Genetics* 40: 797–808. [6]

Dobzhansky, T., and B. Spassky. 1954. Genetics of natural populations. XXII. A comparison of the concealed variability in *Drosophila prosaltans* with that in other species. *Genetics* 39: 472–487. [6]

Dobzhansky, T., and B. Spassky. 1963. Genetics of natural populations. XXXIV. Adaptive norm, genetic load and genetic elite in *Drosophila pseudoobscura. Genetics* 48: 1467-1485. [12]

Dobzhansky, T., H. Levene, B. Spassky, and N. Spassky. 1959. Release of genetic variability through recombination. III. *Drosophila prosaltans. Genetics* 44: 75–92. [5]

Dobzhansky, T., B. Spassky, and T. Tidwell. 1963. Genetics of natural populations. XXXII. Inbreeding and the mutation and balanced genetic loads in natural populations of *Drosophila pseudoobscura. Genetics* 48: 361–373. [12]

Dodds, K. G., M. L. Tate, J. C. McEwan, and A. M. Crawford. 1996. Exclusion probabilities for pedigree testing farm animals. *Theor. Appl. Genet.* 92: 966–975. [8]

Doebley, J. 1992. Mapping the genes that made maize. *Trends Genet.* 8: 302–307. [18]

Doebley, J. 2004. The genetics of maize evolution. *Ann. Rev. Genet.* 38: 37–59. [5, 18]

Doebley, J., and A. Stec. 1991. Genetic analysis of the morphological differences between maize and teosinte. *Genetics* 129: 285–295. [18]

Doebley, J., and A. Stec. 1993. Inheritance of the morphological differences between maize and teosinte: comparison of results for two F_2 populations. *Genetics* 134: 559–570. [5, 18]

Doebley, J., A. Stec, J. Wendell, and M. Edwards. 1990. Genetic and morphological analysis of maize-teosinte F_2 population: implications for the origin of maize. *Proc. Natl. Acad. Sci. USA* 87: 9888–9892. [5, 18]

Doebley, J., A. Bacigalupo, and A. Stec. 1994. Inheritance of kernel weight in two maize-teosinte hybrid populations: implications for crop evolution. *J. Heredity* 85:191–195. [18]

Doebley, J., A. Stec, and C. Gustus. 1995a. *Teosinte branched1* and the origin of maize: evidence for epistasis and the evolution of dominance. *Genetics* 141: 333–346. [5, 18]

Doebley, J., A. Stec, and B. Kent. 1995b. *Suppressor of sessile spikelets1 (sos1):* a dominant mutant affecting inflorescence development in maize. *Am. J. Bot.* 82: 571–577. [5, 18]

Doebley, J., A. Stec, and L. Hubbard. 1997. The evolution of apical dominance in maize. *Nature* 386: 485–488. [18]

Doekes, H. P., P. Bijma, and J. J. Windig. 2021. How depressing in inbreeding? A meta-analysis of 30 years of research on the effects of inbreeding in livestock. *Genes* 12: 926. [12]

Doerge, R. W., and G. A. Churchill. 1996. Permutation tests for multiple loci affecting a quantitative character. *Genetics* 142: 285–294. [18, 19]

Doerge, R. W., and A. Rebaï. 1996. Significance thresholds for QTL interval mapping tests. *Heredity* 76: 459–464. [18]

Dolan, C. V., and D. I. Boomsma. 1998. Optimal selection of sib pairs from random samples for linkage analysis of a QTL using the EDAC test. *Behav. Genet.* 28: 197–206. [19]

Dole, J., and K. Ritland. 1993. Inbreeding depression in two *Mimulus* taxa measured by multigenerational changes in the inbreeding coefficient. *Evolution* 47: 361–373. [12]

Dolgin, E. S., B. Charlesworth, S. E. Baird, and A. D. Cutter. 2007. Inbreeding and outbreeding depression in *Caenorhabditis* nematodes. *Evolution* 61: 1339–1352. [13]

Domingue, B. W., J. Fletcher, D. Conley, and J. D. Boardman. 2014. Genetic and educational assortative mating among US adults. *Proc. Natl. Acad. Sci. USA* 111: 7996–8000. [7]

Domínguez-García, S., C. García, H. Quesada, and A. Camballero. 2019. Accelerated inbreeding depression suggests synergistic epistasis for deleterious mutations in *Drosophila melanogaster*. *Heredity* 123: 709–722. [12]

Dominicus, A., and 4 others. 2006. Likelihood ratio tests in behavioral genetics: problems and solutions. *Behav. Genet.* 36: 331–340. [A4]

Dongen, S. V. 2006. Fluctuating asymmetry and developmental instability in evolutionary biology: past, present and future. *J. Evol. Biol.* 19: 1727–1743. [6]

Doniger, S. W., and 5 others. 2003. MAPPFinder: using gene ontology and GenMAPP to create a global gene-expression profile form microarray data. *Genome Biol.* 4: R7. [21]

Donnelly, K. P. 1983. The probability that related individuals share some section of the genome identical by descent. *Theor. Pop. Biol.* 23: 34–64. [8]

Doolittle, D. P. 1968. An extension of Snyder's ratios to the case of epistasis. *J. Hered.* 59: 117–119. [4]

Doris-Keller, H., and 32 other authors. 1987. A genetic linkage map of the human genome. *Cell* 51: 319–337. [8]

Dorweiler, J., A. Stec, J. Kermicle, and J. Doebley. 1993. *Teosinte glume architecture1:* a genetic locus controlling a key step in maize evolution. *Science* 262: 233–235. [5, 18]

Doss, S., E. E. Schadt, T. A. Drake, and A. J. Lusis. 2005. *Cis*-acting expression quantitative trait loci in mice. *Genome Res.* 15: 681–691. [21]

Dragani, T. A., and 6 others. 1995. Mapping of body weight loci on mouse chromosome X. *Mammalian Genome* 6: 778–781. [18]

Drăghici, S., and S. A. Krawetz. 2003. Global functional profiling of gene expression data. *In* D. P. Berra, W. Dubitzky, and M. Granzow (eds), *A practical approach to microarray data analysis,* pp. 606–625. Kluwer, New York, NY. [21]

Drăghici, S., and 7 others. 2007. A systems biology approach for pathway level analysis. *Genome Res.* 17: 1537–1545. [21]

Drigalenko, E. 1998. How sib pairs reveal linkage. *Am. J. Hum. Genet.* 63: 1242–1245. [19]

Druet, T., and M. Gautier. 2017. A model-base approach to characterize individual inbreeding at both global and local genomic scales. *Mol. Ecol.* 26: 5820–5841. [8, 12]

Druet, T., and M. Gautlier. 2022. A hidden Markov model to estimate homozygous-by-descent probabilities associated with nested layers of ancestors. *Theor. Pop. Biol.* 145: 38–5. [8]

Dubois, E. 1987. Sur le rapport de l'encéphale avec la grandeur du corps chez les Mammifères. *Bull. Soc. Anthropol. Paris* 4^c séries 8: 337–371. [13]

Dudash, M. R. 1990. Relative fitness of selfed and outcrossed progeny in a self-compatible, protandrous species, *Sabatia angularis* L. *(Gentianaceae):* a comparison in three environments. *Evolution* 44: 1129–1139. [12]

Dudbridge, F. 2008. Likelihood-based association analysis for nuclear families and unrelated subjects with missing genotype data. *Hum. Hered.* 66: 87–98. [17]

Dudbridge, F., and A. Gusnanto. 2008. Estimation of significance threshold for genomewide association scans. *Genet. Epid.* 32: 227–234. [20]

Dudbridge, F., and B. P. C. Koeleman. 2003. Rank truncated product of *P*-values, with applications to genomewide association scans. *Genet. Epidem.* 25: 360–366. [20, A6]

Dudbridge, F,. and B. P. C. Koeleman. 2004. Efficient computation of significance levels for multiple associations in large studies of correlated data, including genomewide association studies. *Am. J. Hum. Geneet.* 75: 424–435. [20]

Dudbridge, F., and P. Newcombe. 2019. Replication and meta-anaysis of genome-wide association studies. *In* D. J. Balding, I. Molke, and J. Marioni (eds.), *Handbook of Statistical Genomics*, 4th Ed., pp. 631–650. Wiley and Sons, New York, NY. [20]

Dudbridge, F., B. P. Koeleman, J. A. Todd, and D. G. Clayton. 2000. Unbiased application of the transmission/disequilibrium test to multilocus haplotypes. *Am. J. Hum. Genet.* 66: 2009–2012. [17]

Dudley, J. W. 1964. A genetic evaluation of methods of utilizing heterozygosis and dominance in autotetraploids. *Crop Sci.* 4: 410–413. [13]

Dudley, J. W. 1977. 76 generations of selection for oil and protein percentage in maize. *In* E. Pollak, O. Kempthorne, and T. B. Bailey, Jr. (eds.), *Proceedings of the international conference on quantitative genetics*, pp. 459–473. Iowa State Univ. Press, Ames, IA. [18]

Dudley, J. W. 1992. Theory for identification of marker locus-QTL associations in population by line crosses. *Theor. Appl. Genet.* 85: 101–104. [18]

Dudley, J. W. 1993. Molecular markers in plant improvement: manipulation of genes affecting quantitative traits. *Crop Sci.* 33: 660–668. [17]

Dudley, J. W. 2007. From means to QTL: The Illinois long-term selection experiment as a case study in quantitative genetics. *Crop Sci.* 47: S20–S31. [18]

Dudley, J. W., and R. J. Lambert. 1992. Ninety generations of selection for oil and protein in maize. *Maydica* 37: 1–7. [18]

Dudley, J. W., and R. J. Lambert. 2004. 100 generations of selection for oil and protein in maize. *Plant Breeding Reviews* 24: 79–110. [18]

Dudley, J. W., and 4 others. 2004. Effects of random-mating on marker-QTL associations in the cross of the Illinois High Protein x Illinois Low Protein maize strains. *Crop Sci.* 44: 1419–1428. [18]

Dudley, J. W., D. Clark, T. R. Rocheford, and J. R. LeDeaux. 2007. Genetic analysis of corn kernel chemical composition in the random mated 7 generation of the cross of generations 70 of IHP x ILP. *Crop Sci.* 47: 45–57. [18]

Duggirala, R., J. T. Williams, S. Williams-Blangero, and J. Blangero. 1997. A variance component approach to dichotomous trait linkage analysis using a threshold model. *Genet. Epid.* 14: 987–992. [19]

Dun, R. B., and A. S. Fraser. 1958. Selection for an invariant character — 'vibrissa number'— in the house mouse. *Nature* 181: 1018–1019. [14]

Dun, R. B., and A. S. Fraser. 1959. Selection for an invariant character, vibrissa number, in the house mouse. *Aust. J. Biol. Sci.* 12: 506–523. [14]

Dunbier, M. W., and E. T. Bingham. 1975. Maximum heterozygosity in alfalfa: Results using haploid-derived autotetraploids. *Crop Sci.* 15: 527 –531. [13]

Dunnett, C. W. 1955. A multiple comparison procedure for comparing several treatments with a control. *J. Am. Stat. Assoc.* 50: 1096–1121. [17]

Dunnett, C. W. 1964. New tables for multiple comparisons with a control. *Biometrics* 20: 482–491. [17]

Dupuis, J., and D. Siegmund. 1999. Statistical methods for mapping quantitative trait loci from a dense set of markers. *Genetics* 151; 373–386. [18, 19]

Durand, E. Y., N. Eriksson, and C. Y. McLean. 2014. Reducing pervasive false-positive identical-by-descent segments detected by large-scale pedigree analysis. *Mol. Biol. Evol.* 31: 2212–2222. [8]

Dutkowski, G. W., and 4 others. 2006. Spatial analysis enhances modelling of a wide variety of traits in forest genetic trials. *Can. J. Forest Res.* 36: 1851–1870. [A9]

Duval, S., and R. Tweedie. 2000a. Trim and fill: a simple funnel-plot–based method of testing and adjusting for publication bias in meta-analysis. *Biometrics* 56: 455–462. [A6]

Duval, S., and R. Tweedie. 2000b. A nonparametric "trim and fill" method of accounting for publication bias in meta-analysis. *J. Am. Stat. Assoc.* 95: 89–98. [A6]

Duvick, D. N. 1999. Heterosis: feeding people and protecting natural resources. *In* J. G. Coors and S. Pandey (eds.), *The genetics and exploitation of heterosis in crops*, pp. 19–29. ASA-CSSA-SSSA Societies, Madison, WI. [13]

Duvick, D. N. 2001. Biotechnology in the 1930s: the development of hybrid maize. *Nat. Rev. Genet.* 2: 69–74. [13]

Duvick, D. N. 2005. Genetic progress in yield of United States maize (*Zea mays* L.). *Maydica* 50: 193–202. [12, 13]

Duvick, D. N., J. S. C. Smith, and M. Cooper. 2004. Long term selection in a commerical hybrid maize program. *Plant. Breed. Rev.* 24: 109–151. [13]

Eanes, W. F. 1978. Morphological variance and enzyme heterozygosity in the monarch butterfly. *Nature* 276: 263–264. [6]

East, E. M. 1908. Inbreeding in corn. *Report Conn. Agric. Exp. Sta.* 1907, pp. 419–428. [12, 13]

East, E. M. 1910. A Mendelian interpretation of variation that is apparently continuous. *Am. Nat.* 44: 65–82. [1]

East, E. M. 1911. The genotype hypothesis and hybridization. *Am. Nat.* 45: 160–174. [1]

East, E. M. 1916. Studies on size inheritance in *Nicotiana*. *Genetics* 1: 164–176. [1]

East, E. M., and H. K. Hayes. 1912. Heterozygosis in evolution and plant breeding. *U. S. Dept. Agric. Bur. Plant Indust. Bull* 243: 1–58. [13]

Easterbrook, P. J., R. Gopalan, J. A. Berlin, and D. R. Matthews. 1991. Publication bias in clinical research. *Lancet* 337: 867–872. [A6]

Eaves, L. 1976. The effect of cultural transmission on continuous variation. *Heredity* 37: 41–57. [7]

Eaves, L. J. 1984. The resolution of genotype × environment interaction in segregation analysis of nuclear families. *Genet. Epidem.* 1: 215–228. [16]

Eaves, L., and J. Meyer. 1994. Locating human quantitative trait loci: Guidelines for the selection of sibling pairs for genotyping. *Behav. Genet.* 24: 443–455. [19]

Eaves, L. J., J. K. Hewitt, and A. C. Heath. 1988. The quantitative genetic study of human developmental change: a model and its limitations. *In* B. S. Weir, E. J. Eisen, M. M. Goodman, and G. Namkoong (eds.), *Proceedings of the second international conference on quantitative genetics*, pp. 297–311. Sinauer Assoc., Sunderland, MA. [7]

Ebel, E. R., and P. C. Phillips. 2016. Intrinsic differences between males and females determine sex-specific consequences of inbreeding. *BMC Evol. Biol.* 16: 36. [12]

Eberhart, S. A., and W. A. Russell. 1969. Yield and stability for a 10-line diallel of single-cross and double-cross maize hybrids. *Crop Sci.* 9: 357 –361. [13]

Eberhart, S. A., W. A. Russell, and L. H. Penny. 1964. Double cross hybrid prediction in maize when epistasis is present. *Crop Sci.* 4: 363 –366. [13]

Ebers, G. C , and 29 others. 1996. A full genome search for multiple sclerosis. *Nature Genetics* 13: 472–476. [19]

Ebert, R. H., and 6 others. 1993. Longevity-determining genes in *Caenorhabditis elegans*: Chromosomal mapping of multiple noninteractive loci. *Genetics* 135: 1003–1010. [8]

Ebrahim, S., and G. Davey Smith. 2008. Mendelian randomization: can genetic epidemiology help redress the failures of observational epidemiology. *Hum. Genet.* 123: 15–33. [A2]

Ebrhimpoor, M., and 4 others. 2020. Simultaneous enrichment analysis of all possible gene-sets: unifying self-contained and competitive methods. *Brief. Bioinfomatics* 21: 1302–1312. [21]

Eckart, C., and G. Young. 1936. The approximation of one matrix by another of lower rank. *Psychometrika* 1: 211–218. [A3]

Eckert, C. G., and S. C. H. Barrett. 1994. Inbreeding depression in partially self-fertilizing *Decodon verticillatus* (Lythraceae): population-genetic and experimental analyses. *Evolution* 48: 952–964. [12]

Edgington, E. S. 1972. A normal curve method of combining probability values from independent experiments. *J. Psychol.* 80: 351–363. [A6]

Eding, H , and T. H. E. Meuwissen. 2001. Marker-based estimates of between and within population kinships for the conservation of genetic diversity. *J. Amin. Breed. Genet.* 118: 141–159. [8]

Edmands, S. 1999. Heterosis and outbreeding depression in interpopulation crosses spanning a wide range of divergence. *Evolution* 53: 1757–1768. [13]

Edmands, S. 2006. Between a rock and a hard place: evaluating the relative risks of inbreeding and outbreeding for conservation and management. *Mol. Ecol.* 15: 463–475. [13]

Edmunds, G. F., and D. N. Alstad. 1978. Coevolution in insect herbivores and conifers. *Science* 199: 941–945. [13]

Edwards, A. W. F. 1992. *Likelihood.* Expanded edition. Johns Hopkins Press, Baltimore, MD. [A4]

Edwards, B, J., and 4 others. 2005. Power and sample size calculations in the presence of phenotype errors for case/control genetic association studies. *BMC Genet.* 6: 18. [20]

Edwards, J. H. 1987. Exclusion mapping. *J. Med. Genet.* 24: 539–543. [19]

Edwards, M. D., C. W. Stuber, and J. F. Wendel. 1987. Molecular-marker-facilitated investigations of quantitative-trait loci in maize. I. Numbers, genomic distributions and types of gene action. *Genetics* 116: 113–125. [18]

Edwards, M. D., T. Helentjaris, S. Wright, and C. W. Stuber. 1992. Molecular-marker-facilitated investigations of quantitative-trait loci in maize. 4. analysis based on genome saturation with isozyme and restriction fragment length polymorphism markers. *Theor. Appl. Genet.* 83: 765–774. [17, 18]

Edwards, S. L., J. Beesley, J. D. French, and A. M. Dunning. 2013. Beyond GWAS: illuminating the dark road from association to function. *Am. J. Hum. Genet.* 93: 779–797. [21]

Efron, B. 1979. Bootstrap methods: another look at the jackknife. *Ann. Stat.* 7: 1–26. [18]

Efron, B. 1982. *The jackknife, the bootstrap and other resampling plans.* SIAM, Philadelphia, PA. [18]

Efron, B. 1986. Why isn't everyone a Bayesian? *Am. Stat.* 40: 1–11. [A7]

Efron, B. 2007. Correlation and large-scale simultaneous significance testing. *J. Amer. Stats Assoc.* 102: 93–103. [A6]

Efron, B., and R. Tibshirani. 2007. On testing the significance of sets of genes. *Ann. Applied Stat.* 1: 107–129. [21]

Egger, M., G. D. Smith, M. Schneider, and C. Minder. 1997. Bias in meta-analysis detected by a simple, graphical test. *BMJ* 315: 629–634. [A6]

Ehiobu, N. G., M. E. Goddard, and J. F. Taylor. 1989. Effect of rate of inbreeding on inbreeding depression in *Drosophila melanogaster. Theor. Appl. Genet.* 77: 123–127. [12]

Ehrenreich, I. A., and 8 others. 2010. Dissection of genetically complex traits with extremely large pools of yeast segregants. *Nature* 464: 1039–1042. [5, 21]

Ehrenreich, I. M., and D. W. Pfennig. 2016. Genetic assimilation: a review of its potential proximate causes and evolutionary consequences. *Ann. Botany* 117: 769–779. [14]

Eichler, E. E., and 6 others. 2010. Missing heritability and strategies for finding the underlying causes of complex disease. *Nat. Rev. Genet.* 11: 446–450. [21]

Eisenhart, C. 1947. The assumptions underlying the analysis of variance. *Biometrics* 3: 1–21. [10]

El Adlouni, S., A.-C. Favre, and B. Bobée. 2006. Comparison of methodologies to assess the convergnce of Markov chain Monte Carlo methods. *Comp. Stat. Data Anal.* 50: 2685–2701. [A8]

Elbers, C. C., and 6 others. 2009. Using genome-wide pathway analysis to unravel the etiology of complex diseases. *Genet. Epidem.* 33: 419–431. [21]

Elsen, J. M., S. Knott, P. Le Roy, and C. S. Haley. 1997. Comparison between some approximate maximum-likelihood methods for quantitative trait locus detection in progeny test designs. *Theor. Appl. Genet/* 95: 236–2456. [19]

Elshire, R. J., and 6 others. 2011. A robust, simple genotyping-by-sequencing (GBS) approach for high diversity species. *PLoS One* 6: e19379. [8]

Elston, R. C. 1980. Segregation analysis. *In* J. H. Mielke and M. H. Crawford (eds.), *Current developments in anthropological genetics. Vol. 1: Theory and methods*, pp. 327–354. Plenum, New York, NY. [16]

Elston, R. C. 1981a. Segregation analysis. *Adv. Hum. Genet.* 11: 63–120. [16]

Elston, R. C. 1981b. Testing one- and two-locus hypotheses for the genetic difference of a quantitative trait between two homozygous lines. *In* E. S. Gershon, S. Matthysse, X. O. Breakefield, and R. D. Ciaranello (eds.), *Genetic research strategies for psychobiology and psychiatry*, pp. 283–293. Plenum, New York, NY. [16]

Elston, R. C. 1984. The genetic analysis of quantitative trait differences between two homozygous lines. *Genetics* 108: 733–744. [16]

Elston, R. C. 1990a. Models for discrimination between alternative modes of inheritance. *In* D. Gianola and K. Hammond (eds.), *Advances in statistical methods for genetic improvement of livestock*, pp. 41–55. Springer-Verlag, Berlin, DE. [16]

Elston, R. C. 1990b. A general linkage method for the detection of major genes. *In* D. Gianola and K. Hammond (eds.), *Advances in statistical methods for genetic improvement of livestock*, pp. 495–506. Springer-Verlag, Berlin, DE. [19]

Elston, R. C. 2017. *Statistical Human Genetics.* Humana Press, New York, NY. [16]

Elston, R. C., and D. C. Rao. 1978. Statistical modeling and analysis in human genetics. *Ann. Rev. Biophys. Bioeng.* 7: 253–286. [16]

Elston, R. C., and E. Sobel. 1979. Sampling considerations in the gathering and analysis of pedigree data. *Am. J. Hum. Genet.* 31: 62–69. [16]

Elston, R. C., and J. Stewart. 1971. A general model for the genetic analysis of pedigree data. *Hum. Hered.* 21: 523–542. [16]

Elston, R. C., and J. Stewart. 1973. The analysis of quantitative traits for simple genetic models from parental, F_1 and backcross data. *Genetics* 73: 695–711. [16]

Elston, R. C., K. K. Nasmboodiri, H. V. Nino, and W. S. Pollitzer. 1974. Studies on blood and urine glucose in Seminole Indians: indications for segregation of a major gene. *Am. J. Hum. Genet.* 26: 13–34. [16]

Elston, R. C., and 5 others. 1975. Studies of the genetic transmission of hypercholesterolemia and hypertriglyceridemia in a 195 member kindred. *Ann. Hum. Genet.* 39: 67–87. [16]

Elston, R. C., and 4 others. 1986. S. A. G. E. — a package of computer programs to perform statistical analysis for genetic epidemiology. Paper presented at the Seventh International Congress of Human Genetics, Berlin, September 22-26, 1986. [16]

Elston, R. C., X. Buxbaum, K. B. Jacobs, and J. M. Olson. 2000. Haseman and Elston revisited. *Genet. Epidem.* 19: 1–17. [19]

Emerson, J. J., and 7 others. 2010. Natural selection on *cis* and *trans* regulation in yeasts. *Genome Res.* 20: 826–836. [21]

Emerson, R. A. 1910. The inheritance of sizes and shapes in plants. *Am. Nat.* 44: 739–746. [1]

Emerson, R. A., and E. M. East. 1913. The inheritance of quantitative characters in maize. *Bull. Agric. Exp. Sta. Neb.* 2. [1, 11]

Emery, A. M., and 4 others. 2001. Assignment of paternity groups without accessto parental genotypes: multiple mating and developmental plasticity in squid *Mol. Ecol.* 10: 1265–1278. [8]

Emik, L. O., and C. E Terrill. 1949. Systematic procedures for calculating inbreeding coefficients. *J. Heredity* 40: 51–55. [7]

Emilsson, V., and 34 others. 2008. Genetics of gene expression and its effect on disease. *Nature* 452: 423–428. [21]

Emison, B. S., and 24 others. 2010. Differential contributions of rare and common, coding and noncoding Ret mutations to multifactorial Hirschsprung disease liability. *Am. J. Hum. Genet.* 87: 60–74. [15]

Emlen, J. M., D. C. Freeman, and J. H. Graham. 1993. Nonlinear growth dynamics and the origin of fluctuating asymmetry. *Genetica* 89: 77–96. [6]

ENCODE Project Consortium. 2012. An integrated encyclopedia of DNA elements in the human genome. *Nature* 489: 57–74. [21]

ENCODE Project Consortium. 2020. Expnaded encyclopedias of DNA elements in the human and mouse genomes. *Nature* 583: 699–710. [21]

Epstein, M. P., W. L. Duren, and M. Boehnke. 2000. Improved inference of relationship for pairs of individuals. *Am. J. Hum. Genet.* 67: 1219–1231. [8]

Erbe, M., and 7 others. 2012. Improving accuracy of genomic predictions within and between dairy cattle breeds with imputed high-density single nucleotide polymorphism panels. *J. Dairy Sci.* 95: 4114–4129. [21]

Erdõs, P., and A. Rényi. 1959. On random graphs. I. *Pub. Mathematicae* 6: 290–297. [A2]

Escobar, J. S., A. Nicot, and P. David. 2008. The different sources of variation in inbreeding depression, heterosis and outbreeding depression in a metapopulation of *Physa acuta*. *Genetics* 180: 1593–1608. [13]

Escobar, J. S., and 9 others. 2011. Patterns of mating-system evolution in hermaphroditic animals: correlations among selfing rate, inbreeding depression and timing of reproduction. *Evolution* 65: 133–1253. [12]

Eshed, Y., and D. Zamir. 1995. An introgression line population of *Lycopersicon pennellii* in the cultivated tomato enables the identification and fine mapping of yield-associated QTL. *Genetics* 141: 1147–1162. [17]

Evangelou, E., and J. P. A. Ioannidis. 2013. Meta-analysis methods for genome-wide associations studies and beyond. *Nat. Rev. Genet.* 14: 379–389. [20]

Evanno, G., S. Regnaut, and J. Goudet. 2005. Detecting the number of clusters of individuals using the software STRUCTURE: a simulation study. *Mol. Ecol.* 14: 2611–2620. [20]

Evans, D. M., and S. Purcell. 2012. Power calculations in genetic studies. *Cold Spring Harbor Protocols* 6: pdb-top069559. [20]

Evans, D. M., J. Marchini, A. P. Morris, and L.R. Cardon. 2006. Two-stage two-locus models in genome-wide association. *PLoS Genet.* 2: e157. [18, 20]

Evans, D. M., and 4 others. 2019. Elucidating the role of maternal environmental exposures on offspring health and disease using two-sample Mendelian randomization. *Int. J. Epidemiol.* 48: 861–875. [A2]

Evans, M., and T. Swartz. 1995. Methods for approximating integrals in statistics with special emphasis on Bayesian integration problems. *Stat. Sci.* 10: 254–272. [A8]

Evans, S. N., M. S. McPeek, and T. P. Speed. 1993. A characterization of crossover models that posses map functions. *Theor. Pop. Biol.* 43: 80–90. [17]

Everitt, B. S., and D. J. Hand. 1981. *Finite mixture distributions*. Chapman and Hall, London, UK. [16]

Ewens, W. J., and N. C. E. Shute. 1986. A resolution of the ascertainment sampling problem. *Theor. Pop. Biol.* 30: 388–412. [16]

Ewens, W. J., and R. S. Spielman. 1995. The transmission/disequilibrium test: history, subdivision, and admixture. *Am. J. Hum. Genet.* 57: 455–464. [17]

Ewens, W. J., M. Li, and R. S. Spielman. 2008. A review of family-based tests for linkage disequilibrium between a quantitative trait and a genetic marker. *PLoS Genet.* 4: e1000180. [17]

Excoffier, L., and M. Slatkin. 1995. Maximum-likelihood estimation of molecular haplotype frequencies in a diploid population. *Mol. Biol. Evol.* 12: 921–927. [5]

Eyre-Walker, A. 2010. Genetic architecture of a complex trait and its implications for fitness and genome-wide association studies. *Proc. Natl. Acad. Sci. USA* 107: 1752–1756. [21]

Ezekiel, M. 1930. *Methods of correlation analysis*. John Wiley and Sons, New York, NY. [A3]

Facon, B., and 8 others. 2011. Inbreeding depression is purged in the invasive insect *Harmonia axyridis*. *Curr. Biol.* 21: 424–427. [12]

Fadista, J., A. K. Manning, J. C. Florez, and L. Groop. 2016. The (in)famous GWAS *P*-value threshold revisited and updated for low-frequency variants. *Eur. J. Hum. Genet.* 24: 1202–1205. [20]

Fain, P. R. 1978. Characteristics of simple sibship variance tests for the detection of major loci and application to height, weight and spatial performance. *Ann. Hum. Genet.* 42: 109–120. [16]

Falconer, D. S. 1985. A note on Fisher's 'average effect' and 'average excess'. *Genet. Res.* 46: 337–347. [4]

Falconer, D. S. 1989. *Introduction to quantitative genetics.* 3rd Ed. Longman Sci. and Tech., Harlow, UK. [7]

Falconer, D. S., and T. F. C. Mackay. 1996. *Introduction to quantitative genetics.* 4th Ed. Longman Sci. and Tech., Harlow, UK. [1, 6]

Falconer, D. S., and R. C. Roberts. 1960. Effects of inbreeding on ovulation rate and foetal mortality in mice. *Genet. Res.* 1: 422–430. [12]

Falk, C. T., and P. Rubinstein. 1987. Haplotype relative risks: an easy reliable way to construct a proper control sample for risk calculations. *Ann. Hum. Genet.* 51: 227–233. [17]

Falke, K. C., and M. Frisch. 2011. Power and false-positive rate in QTL detection with near-isogenic line libraries. *Heredity* 106: 576–584. [17]

Fallin, D., and N. J. Schork. 2000. Accuracy of haplotype frequency estimation for biallelic loci, via the expectation-maximization algorithm for unphased diploid genotype data. *Am. J. Human Genet* 67: 947–959. [5]

Falush, D., M. Stephens, and J. K. Pritchard. 2003. Inference of population structure using multilocus genotype data: linked loci and correlated allele frequencies. *Genetics* 164: 1567–1587. [20]

Falush, D., M. Stephens, and J. K. Pritchard. 2007. Inference of population structure using multilocus genotype data: dominant markers and null alleles. *Mol. Ecol. Notes* 7: 574–578. [20]

Famula, T. R. 1986. Identifying single genes of large effect in quantitative traits using best linear unbiased prediction. *J. Animal Sci.* 63: 68–76. [16]

Fan, R., and M. Knapp. 2003. Genome association studies of complex diseases by case-control designs. *Am. J. Hum. Genet.* 72L 850–868. [20]

Faraone, S. V. 2017. The Omnigenic model: implications for psychiatric genetics. *J. Psych. Brain Sci.* 2: S1. [21]

Faraway, J. J. 1993. Improved sib-pair linkage test for disease susceptibility loci. *Genet. Epidem.* 10: 225–233. [19]

Fares, W. M., H. S. A. Fateh, and A. R. Morsy. 2011. Improving the precision of soybean variety trials using trend analysis models. *Egypt. J. Plant Breed.* 15: 103–116. [A9]

Farh, K. K.-H., and 20 others. 2015. Genetic and epigenetic fine mapping of causal autoimmune disease variants. *Nature* 518: 337–343. [21]

Farhadifar, R., and 4 others. 2016. Mutation is a sufficient and robust predictor of genetic variation for mitotic spindle traits in *Caenorhabditis elegans*. *Genetics* 203: 1859–1870. [15]

Faye, L. L., L. Sun, A. Dimitromanolakis, and S. B. Bull. 2011. A flexible genome-wide bootstrap method that accounts for ranking and threshold-selection bias in GWAS interpretation and replication study designs. *Stats. Med.* 30: 1898–1912. [20]

Federer, W. T. 1956. Augmented (or hoonuiaku) designs. *Hawaiian Planters' Record* LV(2): 191–208. [A9]

Federer, W. T. 1961. Augmented designs with one-way elimination of heterogeneity. *Biometrics* 17: 447–473. [A9]

Federer, W. T., and K. E. Basford. 1991. Competing effects designs and models for two dimensional field arrangements. *Biometrics* 47: 1461–1472. [A9]

Federer, W. T., and J. Crossa. 2001. On the design and analysis of field experiments. *Technical report M-1565*, Dept. of Biometrics, Cornell University. [A9]

Feenstra, B., I. M. Skovgaard, and K. W. Broman. 2006. Mapping quantitative trait loci by an extension of the Haley-Knott regression method using estimating equations. *Genetics* 173: 2269–2282. [18]

Fehrmann, R. S. N., and 27 others. 2011. *Trans*-eQTLs reveal that independent genetic variants associated with a complex phenotype converge on intermediate genes, with a major role for the HLA. *PLoS Genet.* 7: e1002197. [21]

Feingold, E. 2001. Methods for linkage analysis of quantitative trait loci in humans. *Theor. Pop. Biol.* 60: 167–180. [19]

Feingold, E. 2002. Regression-based quantitative-trait-locus mapping in the 21st Century. *Am. J. Hum. Genet.* 71: 217–222. [19]

Feingold, E., P. O. Brown, and D. Siegmund. 1993. Gaussian models for genetic linkage analysis using complete high-resolution maps of identity by descent. *Am. J. Hum. Genet.* 53: 234–251. [18, 19]

Feldman, M. W., and R. C. Lewontin. 1975. The heritability hang-up. *Science* 190: 1163–1168. [7]

Feldman, M. W., F. B. Christiansen, and L. D. Brooks. 1980. Evolution of recombination in a constant environment. *Proc. Natl. Acad. Sci. USA* 77: 4838–4841. [5]

Félix, M.-A., and M. Barkoulas. 2015. Pervasive robustness in biological systems. *Nat. Rev. Genet.* 16: 483–496. [14, A2]

Fellay, J., and 9 others. 2007. A whole-genome association study of major determinants for host control of HIV-1. *Science* 317: 944–947. [20]

Felley, J. 1980. Analysis of morphology and asymmetry in the bluegill sunfish *(Lepomis macrochirus)* in the southeastern United States. *Copeia* 1980: 18–29. [6]

Felsenstein, J. 1965. The effect of linkage on directional selection. *Genetics* 52: 349–363. [5]

Felsenstein, J. 1973. Estimation of number of loci controlling variation in a quantitative character. *Genetics* 74: s78–s79. [16]

Felsenstein, J. 1974. The evolutionary advantage of recombination. *Genetics* 78: 737–756. [5]

Felsenstein, J. 1979. A mathematically tractable family of genetic mapping functions with different amounts of interference. *Genetics* 91: 769–775. [17]

Feng, H., X. Zhang, and C. Zhang. 2015. mRIN for direct assessment of genome-wide and gene-specific mRNA integrity from large-scale RNA-sequencing data. *Nat. Comm.* 6: 7816. [21]

Fenster, C. B., and L. F. Galloway. 2000a. Population differentiation in an annual legume: genetic architecture. *Evolution* 54: 1157–1172. [13]

Fenster, C. B., and L. F. Galloway. 2000b. Inbreeding and outbreeding depression in natural populations of *Chamaecrista fasciculata* (Fabaceae). *Conservation Biol.* 14: 1406–1412. [13]

Fenster, C. B., and K. Ritland. 1994. Quantitative genetics of mating system divergence in the yellow monkeyflower species complex. *Heredity* 73: 422–435. [11]

Ferenčaković, M., J. Sölkner, M. Kapš, and I. Curik. 2017. Genome-wide mapping and estimation of inbreeding depression of semen quality traits in a cattle population. *J. Dairy Sci.* 100: 4721–4730. [12]

Ferguson, J. P., J. H. Cho, C. Yang, and H. Zhao. 2013. Empirical Bayes correction for the winner's curse in genetic association studies. *Genet. Epidem.* 37: 60–68. [20]

Ferguson, M. M. 1986. Developmental stability of rainbow trout hybrids: genomic coadaptation or heterozygosity? *Evolution* 40: 323–330. [6]

Fernandes, S. B., T. M. Casstevens, P. J. Bradbury, and A. E. Lipka. 2022. A multi-trait multi-locus stepwise approach for conducting GWAS on correlated traits. *Plant Genome* 2022: e20200. [20]

Fernández, J., and A. Caballero. 2001a. Accumulation of deleterious mutations and equalization of parental contributions in the conservation of genetic resources. *Heredity* 86: 480–488. [12]

Fernández, J., and A. Caballero. 2001b. A comparison of management strategies for conservation with regard to population fitness. *Conserv. Genet.* 2: 121–131. [12]

Fernández, J., and M. A. Toro. 1999. The use of mathematical programming to control inbreeding in selection schemes. *J. Anim. Breed. Genet.* 116: 447–466. [12]

Fernández, J., and M. A. Toro. 2006. A new method to estimate relatedness from molecular markers. *Mol. Ecol.* 15: 1657–1667. [8]

Fernández, J., M. A. Toro, and A. Caballero. 2004. Managing individuals' contribution to maximize the allelic diversity maintained in small, conserved populations. *Conserv. Biol.* 18: 1358–1367. [12]

Fernando, R. L., C. Stricker, and R. C. Elston. 1993. An efficient algorithm to compute the posterior genotypic distribution for every member of a pedigree without loops. *Theor. Appl. Genet.* 87: 89–93. [16]

Fernando, R. L., C. Stricker, and R. C. Elston. 1994. The finite polygenic mixed model: an alternative formulation for the mixed model of inheritance. *Theor. Appl. Genet.* 88: 573–580. [16]

Fernando, R. L., and 5 others. 2004. Controlling the proportion of false positives in multiple dependent tests. *Genetics* 166: 611-619. [A6]

Ferrari, J. A. 1987. Components of genetic variation associated with second and third chromosome gene arrangements in *Drosophila melanogaster. Genetics* 116: 87–97. [17]

Ferraro, N. M., and 37 others 2020. Transcriptomic signatures across human tissues identify functional rare genetic variation. *Science* 369: eaaz5900. [21]

Festing, M. F. W. 1973. A multivariate analysis of subline divergence in the shape of the mandible in C57BL/Gr mice. *Genet. Res.* 21: 121–132. [15]

ffrench-Constant, R. H., and C. Bass. 2017. Does resistance really carry a fitness cost? *Curr. Opin. Insect Sci.* 21: 39–46. [14]

Fijneman, R. J. A., S. S. de Vries, R. C. Jansen, and P. Dermant. 1996. Complex interactions of new quantitative trait loci, *Sluc1, Sluc2, Sluc3,* and *Sluc4,* that influence the susceptibility to lung cancer in the mouse. *Nat. Genet.* 14: 465–467. [18]

Filatov, D. A. 2018. The two "rules of speciation" in species with young sex chromosomes. *Mol. Ecol.* 27: 3799–3810. [17]

Filho, J. B. M. 1999. Inbreeding depresion and heterosis. *In* J. G. Coors and S. Pandey (eds.), *The genetics and exploitation of heterosis in crops,* pp. 69–80. ASA-CSSA-SSSA Societies, Madison, WI. [13]

Finch, C. E. 1990. *Longevity, senescence, and the genome.* Univ. Chicago Press, Chicago, IL. [6]

Finnegan, D. J. 1992. Transposable elements. *In* D. L. Lindsley, and G. G. Zimm (eds.), *The genome of Drosophila melanogaster,* pp. 1096–1107. Academic Press, New York, NY. [15]

Firth, D. 1993. Bias reduction of maximum likelihood estimates. *Biometrika* 80: 27–38. [20]

Fisch, R. D., M. Ragot, and G. Gay. 1996. A generalization of the mixture model in the mapping of quantitative trait loci for progeny from a biparental cross of inbred lines. *Genetics* 143: 571–577. [18]

Fisher, R. A. 1918. The correlation between relatives on the supposition of Mendelian inheritance. *Trans. Royal Soc. Edinburgh* 52: 399–433. [1, 2, 3, 4, 5, 7]

Fisher, R. A. 1922. On the mathematical foundations of statistics. *Phil. Trans. Royal Soc. Lond. B* 222: 309–368. [A4]

Fisher, R. A. 1925a. Applications of Student's Distribution. *Metron,* 5: 90–104. [2]

Fisher, R. A. 1925b. *Statistical methods for research workers.* Hafner, NY. [1]

Fisher, R. A. 1926. The arrangement of field experiments. *J. Ministry Agricul. Great Britain* 33: 503–513. [A9]

Fisher, R. A. 1928a. The possible modification of the response of the wild type to recurrent mutations. *Am. Nat.* 62: 115–126. [4]

Fisher, R. A. 1928b. Two further notes on the origin of dominance. *Am. Nat.* 62: 571–574. [4]

Fisher, R. A. 1929. The evolution of dominance: a reply to Professor Sewall Wright. *Am. Nat.* 63: 553–556. [4]

Fisher, R. A. 1932b. *Statistical methods for research workers.* Oliver and Boyd, Edinburgh, UK. [A6]

Fisher, R. A. 1935. *The design of experiments.* 8th Ed. Hafner, New York, NY. [1, A9]

Fisher, R. A. 1941. Average excess and average effect of a gene substitution. *Ann. Eugen.* 11: 53–63. [4, 12]

Fisher, R. A. 1947. The theory of linkage in polysomic inheritance. *Phil. Trans. Royal Soc. Lond. B* 233: 55–87. [4]

Fisher, R. A. 1949. *Theory of inbreeding.* Oliver and Boyd, Edinburgh, UK. [7]

Fisher, R. A. 1954. A fuller theory of "junctions" in inbreeding. *Heredity* 8: 187–197. [8]

Fisher, R. A. 1956. *Statistical methods and scientific inference.* 13th Ed. Hafner, New York, NY. [1]

Fisher, R. A. 1958. *The genetical theory of natural selection.* Dover Publ., New York, NY. [4, 7, 14]

Fisher, R. A., and E. B. Ford. 1947. The spread of a gene in natural conditions in a colony of the moth *Panaxia dominula* L. *Heredity* 1: 143–174. [4]

Fisher, R. A., and K. Mather. 1936. Verification in mice of the possibility of more than fifty per cent recombination. *Nature* 137: 362–363. [5]

Fisher, S. A., and C. M. Lewis. 2008. Power of genetic association studies in the presence of linkage disequilibrium and allelic heterogeneity. *Hum. Hered.* 66: 210–222. [17]

Fishman, P., B. K. Suarez, S. E. Hodge, and T. Reich. 1978. A robust method for the detection of linkage in familial diseases. *Am. J. Hum. Genet.* 30: 308–321. [19]

Flatt, T. 2005. The evolutionary genetics of canalization. *Q. Rev. Biol.* 80: 287–316. [14]

Fleagle, J. G. 1985. Size and adaptation in primates. *In* W. L. Jungers (ed.), *Size and scaling in primate biology,* pp. 1–19. Plenum, New York, NY. [14]

Fleiss, J. L. 1993. The statistical basis of meta-analysis. *Stat. Metods Med. Res.* 2: 121–145. [A6]

Fletcher, R. 1987. *Practical methods of optimization.* 2nd Ed. John Wiley, New York, NY. [16]

Flint, J. 2003. Analysis of quantitative trait loci than influence anima behavior. *J. Neurobio.* 54: 46–77. [17]

Flint, J., and T. Ideker. 2019. The great hairball gambit. *PLoS Genet.* 15: e1008519. [21]

Flint, J., and T. F. C. Mackay. 2009. Genetic architecture of quantitative traits in mice, flies, and humans. *Gen. Res.* 19: 723–733. [18]

Flint, J., and R. Mott. 2001. Finding the molecular basis of quantitative traits: successes and pitfalls. *Nat. Rev. Genet.* 2: 437–445. [18]

Flint-Garcia, S. A., and 4 others. 2009. Heterosis is prevalent for multiple traits in diverse maize germplasm. *PLoS One* 4: e7433. [13]

Floate, K. D., G. K. Kyei-Poku, and P. C. Coghlin. 2006. Overview and relevance of *Wolbachia* bacteria in biocontrol research. *Biocontrol Sci. Tech.* 16: 767–788. [6]

Flury, C., and 7 others. 2010. Effective population size of an indigenous Swiss cattle breed estimated from linkage disequilibrium. *J. Anim. Breed. Genet.* 127: 339–337. [5]

Flutre, T., X. Wen, J. Pritchard, and M. Stepends. 2013. A statistical framework for joint eQTL analysis in multiple tissues. *PLoS Genet.* 9: e1003486. [21]

Folks, J. L. 1984. Combination of independent tests. *In* P. R. Krishnaiah and P. K. Send (eds), *Handbook of statistics* 4: 113–120. [A6]

Foltz, D. W. 1986. Null alleles as possible causes of heterozygote deficiencies in the oyster *Crassostrea virginica* and other bivalves. *Evolution* 40: 869–870. [12]

Ford, E. B. 1975. *Ecological genetics,* 4th Ed. John Wiley & Sons, New York, NY. [15]

Ford, M. J., and C. F. Aquadro. 1996. Selection on X-linked genes during speciation in the *Drosophila athabasca* complex. *Genetics* 144: 689–703. [17]

Ford, M. J., and K. S. Williamson. 2010. The aunt and uncle effect revisited–the effect of biased parentage assignment on fitness estimation in a supplemented salmon population. *J. Hered.* 101: 33–41. [8]

Forrest, W. 2001. Weighting improves the "new Haseman-Elston" method. *Hum. Hered.* 52: 47–54. [19]

Forrest, W., and E. Feingold. 2000. Composite statistics for QTL mapping with moderately discordant sibling pairs. *Am. J. Hum. Genet.* 66: 1642–1660. (Correction 66: 2020). [19]

Forster, B. P., E. Herberle-Bors, K. J. Kasha, and A. Touraev. 2007. The resurgence of haploids in higher plants. *Trends Plant Sci.* 12: 368-375. [17]

Forstmeier, W., and 4 others. 2012. Heterozygosity-fitness correlations in zebra finches: microsatellite markers can be better than their reputation. *Mol. Ecol.* 21:3237–3249. [12]

Forutan, M., and 5 others. 2018. Inbreeding and runs of homozygosity before and after genomic selection in North American Holstein cattle. *BMC Genomics* 19: 98. [12]

Foss, E., R. Lande, F. W. Stahl, and C. W. Steinberg. 1993. Chiasma interference as a function of genetic distance. *Genetics* 133: 681–691. [17]

Fowler, D. P., and Y. S. Park. 1983. Population studies of white spruce. I. Effects of self-pollination. *Can. J. For. Res.* 13: 1133–1138. [12]

Fowler, K., and M. C. Whitlock. 1994. Fluctuating asymmetry does not increase with moderate inbreeding in *Drosophila melanogaster. Heredity* 73: 373–376. [6]

Fox, C. W., and D. H. Reed. 2011. Inbreeding depression increases with environmental stress: an experimental study and meta-analysis. *Evolution* 62: 2236–2249. [12]

Frahm, R., and K.-I. Kojima. 1966. Comparison of selection response on body weight under divergent larval density conditions in *Drosophila pseudoobscura. Genetics* 54: 625–637. [17]

Franěk, R., and 6 others. 2020. Isogenic lines in fish — a critical review. *Rev. Aquaculture* 12: 1412–1434. [17, 19]

Frank, S. A. 1991a. Haldane's rule: a defense of molecular drive theory. *Evolution* 45: 1714–1717. [17]

Frank, S. A. 1991b. Divergence of meiotic drive-suppression systems as an explanation for sex-biased hybrid sterility and inviability. *Evolution* 45: 262–267. [17]

Franke, A., et al. 2010. Genome-wide meta-analysis increases to 71 the number of confirmed Crohn's disease susceptibility loci. *Nat. Genet.* 42: 1118–1125. [20]

Franke, D., A. Kleensang, R. C. Elson, and A. Ziegler. 2005. Haseman-Elston weighted by marker informativity. *BMC Genet.* 6: S50. [19]

Frankel, R. (ed.). 1983. *Heterosis: Reappraisal of theory and practice.* Springer-Verlag, Berlin, DE. [13]

Frankel, W. N. 1995. Taking stock of complex trait genetics in mice. *Trends Genet.* 11: 471–477. [18]

Frankham, R. 1988. Exchanges in the rRNA multigene family as a source of genetic variation. *In* B. S. Weir, E. J. Eisen, M. M. Goodman, and G. Namkoong (eds.), *Proceedings of the second international conference on quantitative genetics*, pp. 236–242. Sinauer Assoc., Sunderland, MA. [15]

Frankham, R., and R. K. Nurthen. 1981. Forging links between population and quantitative genetics. *Theor. Appl. Genet.* 59: 251–263. [16]

Frankham, R., and 6 others. 2011. Predicting the probability of outbreeding depression. *Conservation Biol.* 25: 465–475. [13]

Frankham, R., and 4 others. 2013. 50/500 rule and minimum viable populations: response to Jamieson and Allendorf. *Trends Ecol. Evol.* 28: 187–188. [12]

Frankham, R., C. J. A. Bradshaw, and B. W. Brook. 2014. Genetics in conservation management: revised recommendations for the 50/500 rules, Red List criteria and population viability analyses. *Biol. Conserv.* 170: 56–63. [12]

Franklin, E. C. 1972. Genetic load in loblolly pine. *Am. Nat.* 106: 262–265. [12]

Franklin, I., and R. C. Lewontin. 1970. Is the gene the unit of selection? *Genetics* 65: 707–734. [5]

Franklin, I. R. 1977. The distribution of the proportion of the genome which is homozygous by descent in inbred individuals. *Theor. Popul. Biol.* 11: 60–80. [7, 8]

Franklin, I. R. 1980. Evolutionary change in small populations. *In* M. E. Soulé and B. A. Wilcox (eds), *Conservation biology: an evolutionary-ecological perspective*, pp. 135–149. Sinauer, Sunderland, MA. [12]

Franklin, I. R., F. W. Allendorf, and I. G. Jamieson. 2014. The 50/500 rule is still valid–reply to Frankham et al. *Biol. Conserv.* 176: 284–285. [12]

Frascaroli, E., and 7 others. 2007. Classical genetic and quantitative trait loci analyses of heterosis in a maize hybrid between two elite inbred lines. *Genetics* 176: 625–644. [13]

Fraser, A. 1963. Variation of scutellar bristles in *Drosophila*. I. Genetic linkage. *Genetics* 48: 497–514. [14]

Fraser, A. 1967. Variation of scutellar bristles in *Drosophila*. XV. Systems of modifiers. *Genetics* 57: 919–934. [14]

Fraser, A. 1970. Variation of scutellar bristles in *Drosophila*. XVI. major and minor genes. *Genetics* 65: 305–309. [14]

Fraser, A. S. 1968. Specificity of modifiers of *scute* and *extravert* expression. *Genetics* 60: 179. [15]

Fraser, A. S., and B. M. Kindred. 1960. Selection for an invariant character, vibrissa number, in the house mouse. II. Limits to variability *Aust. J. Biol. Sci.* 13: 48–58. [14]

Fraser, H. B., and E. E. Schadt. 2010. The quantitative genetics of phenotypic robustness. *PLoS One* 5: e8625. [14, 18]

Frayling, I., and 12 others. 1998. The APC variants I1307K and E1317Q are associated with colorectal tumors, but not always with a family history. *Proc. Natl. Acad. Sci. USA* 95: 10722–10727. [21]

Frazer, K. A., S. S. Murray, N. J. Schork, and E. J. Topal. 2009. Human genetic variation and its contribution to complex traits. *Nat. Rev. Genet.* 10: 241–251. [8]

Freedman, M. L., and 17 others. 2004. Assessing the impact of population stratification on genetic association studies. *Nat. Genet.* 36: 388–393. [20]

Freeman, D. C., J. H. Graham, and J. M. Emlen. 1993. Developmental stability in plants: symmetries, stress, and epigenesis. *Genetica* 89: 97–119. [6]

Freeman, F., and J. W. Lundelius. 1982. The developmental genetics ofdextrality and sinistrality in the gastropod *Lymnaea peregra. Roux's Arch. Devel. Biol.* 191: 69–83. [6]

Freeman, G., B. J. Cowling, and C. M. Schooling. 2013. Power and sample size calculations for Mendelian randomization studies using one genetic instrument. *Int. J. Epidem.* 42: 1157–1163. [A2]

Frei, O., C. W. Stuber, and M. M. Goodman. 1986. Use of allozymes as genetic markers for predicting performance in maize single cross hybrids. *Crop Sci.* 26: 37–42. [13]

Frenando, R. L., and M. Grossman. 1989. Marker assisted selection using best linear unbiased prediction. *Genet. Sel. Evol.* 21: 467–477. [19]

Frentiu, F. D., and 5 others. 2008. Pedigree-free animal models: the relatedness matrix reloaded. *Proc. Roy. Soc. Lond. B* 275: 639–648. [8]

Frey, K. J. 1965. The utility of hill plots in oat research. *Euphytica* 14: 196–208. [A9]

Fridley, B. L., and J. M. Biernacka. 2011. Gene set analysis of SNP data: benefits, challenges, and future directions. *Europ. J. Hum. Genet.* 19: 837–843. [21]

Fridley, B. L., G. D. Jenkins, and J. M. Biernacka. 2010. Self-contained gene-set analysis of expression data: an evaluation of existing and novel methods. *PLoS ONE* 5: e12693. [21]

Frisch, M., and A. E. Melchinger. 2001. The length of the intact donor chromosome segment around a target gene in marker-assisted backcrossing. *Genetics* 157: 1343–1356. [17]

Frisch, M., and A. E. Melchinger. 2005. Selection theory for marker-assisted backcrossing. *Genetics* 170: 909–917. [17]

Frisch, M., M. Bohn, and A. E. Melchinger. 1999a. Minimum sample size and optimal positioning of flanking markers in marker-assisted backcrossing for transfer of a target gene. *Crop Sci.* 39: 967–975. [17]

Frisch, M., M. Bohn, and A. E. Melchinger. 1999b. Comparison of selection strategies for marker-assisted backcrossing of a gene. *Crop Sci.* 39: 1295–1301. [17]

Fry, J. D., K. A. deRonde, and T. F. C. Mackay. 1995. Polygenic mutation in *Drosophila melanogaster:* genetic analysis of selection lines. *Genetics* 139: 1293–1307. [15]

Frydenberg, O. 1963. Population studies of a lethal mutant in *Drosophila melanogaster*: I. Behaviour in populations with discrete generations. *Hereditas* 50: 89–116. [12]

Fu, D., and 6 others. 2014. Utilization of crop heterosis: a review. *Euphytica* 197: 161–173. [13]

Fu, H., and H. K. Dooner. 2002. Intraspecific violation of genetic colinearity and its implications in maize. *Proc. Natl. Acad. Sci. USA* 99: 9573–9578. [8, 13]

Fu, J., and R. Jansen. 2006. Optimal design and analysis of genetic studies on gene expression. *Genetics* 172: 1993–1999. [19]

Fu, J., and 23 others. 2013. RNA sequencing reveals the complex regulatory network in the maize kernel. *Nat. Commun.* 4: 2832. [21]

Fu, Y.-B., and K. Ritland. 1994. Evidence for the partial dominance of viability genes contributing to inbreeding depression in *Mimulus guttatus. Genetics* 136: 323–331. [12]

Fulker, D. W., and L. R. Cardon. 1994. A sib-pair approach to interval mapping of quantitative trait loci. *Am. J. Hum. Genet.* 54: 1092–1103 (errata 55: 419). [19]

Fulker, D. W., S. S. Cherry, and L. R. Cardon. 1995. Multipoint interval mapping of quantitative trait loci, using sib pairs. *Am. J. Hum. Genet.* 56: 1224–1233. [19]

Furlotte, N. A., and 5 others. 2012. Increasing association mapping power and resolution in mouse genetic studies through the use of meta-analysis for structured populations. *Genetics* 191: 959–967. [20]

Furusho, T. 1960. Estimation of frequency and penetrance of genes responsible for incompletely penetrant, monosomic characters in man. *Jap. J. Hum. Genet.* 5: 11–19. [4]

Fusi, N., C. Lippert, N. D. Lawrence, and O. Stegle. 2014. Warped linear mixed models for the genetic analysis of transformed phenotypes. *Nat. Comm.* 5: 4890. [14]

Gabriel, S.B., and 10 others. 2002. The structure of haplotype blocks in the human genome. *Science* 296: 2225–2229. [5]

Gabriel, W., M. Lynch, and R. Bürger. 1993. Muller's ratchet and mutational meltdowns. *Evolution* 47: 1744–1757. [15]

Gadbury, G. L., and 8 others. 2004. Power and sample size estimation in high dimensional biology. *Stat. Methods Med. Res.* 13: 325–338. [A6]

Gaeton, C., and X. Guyon. 2010. *Spatial statistics and modeling.* Springer, New York, NY. [A9]

Gaffney, P. M. 1990. Enzyme heterozygosity, growth rate, and viability in *Mytilus edulis:* another look. *Evolution* 44: 204–210. [12]

Gaffney, P. M., T. M. Scott, R. K. Koehn, and W. J. Diehl. 1990. Interrelationships of heterozygosity, growth rate and heterozygote deficiencies in the coot clam, *Mulina lateralis. Genetics* 124: 687–699. [12]

Gage, J. L., B. Monier, A. Giri, and E. S. Buckler. 2020. Ten years of the maize nested association mapping population: impact, limitations, and future directions. *Plant Cell* 32: 2083–2093. [18, 21]

Gagliano, S. A., M. R. Barnes, M. E. Weale, and J. Knight. 2014. A Bayesian method to incorporate hundreds of functional characteristics with association evidence to improve variant prioritization. *PLoS One* 9: e98122. [21]

Gai, J.-Y., and J.-K. Wang. 1998. Identification and estimation of a QTL model and its effect. *Theor. Appl. Genet.* 97: 1166–1168. [16]

Galarneau, G., and 5 others. 2010. Fine-mapping at three loci known to affect fetal hemoglobin levels explains additional genetic variation. *Nat. Genet.* 42: 1049–1051. [20]

Galen, C., J. S. Shore, and H. Deyoe. 1991. Ecotypic divergence in alpine *Polemonium viscosum:* genetic structure, quantitative variation, and local adaptation. *Evolution* 45: 1218–1228. [13]

Galesloot, T. E., and 5 others. 2012. A comparison of multivariate genome-wide association methods. *PLoS One* 9: e95923. [20]

Gallagher, M. D., and A. S. Chen-Plotkin. 2018. The post-GWAS era: from association to function. *Am. J. Hum. Genet.* 102: 717–730. [21]

Gallais, A. 1974. Covariances between arbitrary relatives with linkage and epistasis in the case of linkage disequilibrium. *Biometrics* 30: 429–446 (Correction 33: 766). [5, 7]

Gallais, A. 1975a. Prévision de la vigueur et sélection des parents d'une variété synthétique. *Annales de l'Amelioration des Plantes* 25: 233–264. [13]

Gallais, A. 1976. Development and application of prediction formulae for synthetics. *Annales de l'Amelioration des Plantes* 26: 623–628. [13]

Gallais, A. 1990. Theoretical determination of the optimum number of parents for synthetics. *Theor. Appl. Genet.* 79: 417–421. [13]

Gallais, A. 2003. *Quantitative genetics and breeding methods in autopolyploid plants.* INRA, Paris, FR. [1, 4 ,5]

Gallais, A., and A. J. Wright. 1980. A general approach to the concept of variety ability for synthetic varieties. *Theor. Appl. Genet.* 57: 81–87. [13]

Galton, F. 1879. The geometric mean in vital and social statistics. *Proc. Royal Soc. Lond.* 29: 365–367. [14]

Galton, F. 1886. Regression towards mediocrity in hereditary stature. *J. Antropol. Inst. Great Britain and Ireland* 15: 246–263. [1, 3, 21]

Galton, F. 1889. *Natural inheritance.* Macmillan, London, UK. [1]

Galwey, N. W. 2009. A new measure of the effective number of tests, a practical tool for comparing families of non-independent significance test. *Genet. Epidem.* 33: 559–568. [A6]

Gamazon, E. R., and B. E. Stranger. 2015. The impact of human copy number variation on gene expression. *Brief. Funct. Genomics* 14: 352–357. [21]

Gamazon, E. R., D. L. Nicolae, and N. J. Cox. 2011. A study of CNVs as trait-associate polymorphisms and as expression quantitative trait loci. *PLoS Genet.* 7: e100292. [21]

Gamazon, E. R., and 11 others. 2015. A gene-based association method for mapping traits using reference transcriptome data. *Nat. Genet.* 47: 1091–1098. [21]

Gamazon, E. R., and 19 others. 2018. Using an atlas of gene regulation across 44 human tissues to inform complex disease- and trait-associated variation. *Nat. Genet.* 50: 956–967. [21]

Gammerman, D. 1997. *Markov chain Monte Carlo.* Chapman and Hall, New York, NY. [A8]

Ganders, F. R. 1979. The biology of heterostyly. *New Zealand J. Bot.* 17: 607–635. [5]

Gao, H., and 4 others. 2013. Forward LASSO analysis for high-order interactions in genome-wide association study. *Brief. Bioinfromatics* 15: 552–561. [20]

Gao, X. 2011. Multiple testing corrections for imputed SNPs. *Genet. Epidem.* 35: 154–158. [20]

Gao, X., J. Starmer, and E. R. Martin. 2008. A multiple testing correction method for genetic association studies using correlated single nucleotide polymorphism. *Genet. Epidem.* 32: 361–369. [20, A6]

Gao, X., and 4 others. 2010. Avoiding the high Bonferroni penalty in genome-wide association studies. *Genet. Epidem.* 34: 100–105. [20]

Gao, Z., M. Przeworski, and G. Sella. 2014. Footprints of ancient-balanced polymorphisms in genetic variation data from closely related species. *Evolution* 69: 431–446. [12]

Garant, D., and L. E. B. Kruuk. 2005. How to use molecular marker data to measure evolutionary parameters in wild populations. *Mol. Ecol.* 14: 1843–1859. [8]

Garcia, A. F., S. Wang, A. E. Melchinger, and Z.-B Zeng. 2008. Quantitative trait loci mapping and the genetic basis of heterosis in maize and rice. *Genetics* 180: 1707–1724. [13]

García, C., V. Ávila, H. Quesada, and A. Camballero. 2012. Gene-expression changes caused by inbreeding protection again inbreed depression in *Drosophila. Genetics* 192: 161–172. [12]

García, C., V. Ávila, H. Quesada, and A. Camballero. 2013. Are transcriptional responses to inbreeding a functional response to alleviate inbreeding depression? *Fly* 7: 8–12. [12]

Garcia, N., C. López-Fanjul, and A. García-Dorado. 1994. The genetics of viability in *Drosophila melanogaster:* effects of inbreeding and artificial selection. *Evolution* 48: 1277–1285. [12]

García–Dorado, A. 1997. The rate and effects distribution of viability mutation in *Drosophila:* minimum distance estimation. *Evolution* 51: 1130–1139. [15]

García-Dorado, A. 2012. Understanding and prediction the fitness decline of shrunk populations: inbreeding, purging, mutation, and standard selection. *Genetics* 190: 1461–1476. [12]

García–Dorado, A., C. López–Fanjul, and A. Caballero. 1999. Properties of spontaneous mutations affecting quantitative traits. *Genetics Res.* 74: 341–350. [15]

García–Dorado, A., J. Fernández, and C. López–Fanjul. 2000. Temporal uniformity of the spontaneous mutational variance of quantitative traits in *Drosophila melanogaster. Genetics Res.* 75: 47–51. [15]

García–Dorado, A., A. Caballero, and J. F. Crow. 2003. On the persistence and pervasiveness of a new mutation. *Evolution* 57: 2644–2646. [15]

García-Dorado, A., J. Wang, and E. López-Cortegano. 2016. Predictive model and software for inbreeding-purging analysis of pedigreed population. *G3* 6: 3593–3601. [12]

Gardner, C. O. 1961. An evaluation of effects of mass selection and seed irradiation with thermal neutrons on yields of corn. *Crop Sci.* 1: 241–245. [A9]

Gardner, C. O. 1963. Estimates of genetic parameters in cross-fertilizing plants and their implications in plant breeding. *In* W. D. Hanson and H. F. Robinson (eds.), *Statistical genetics and plant breeding*, pp. 225–252. Natl. Acad. Sci., Natl. Res. Council Publ. 982. Washington, D. C. [5, 13]

Gardner, C. O., and J. H. Lonnquist. 1959. Linkage and the degree of dominance of genes controlling quantitative characters in maize. *Agronomy J.* 51: 524–528. [13]

Garner, C. 2007. Upward bias in odds ratio estimates form genome-wide associations studies. *Genet. Epidem.* 31: 288–295. [20]

Garnett, I., and D. S. Falconer. 1975. Protein variation in strains of mice differing in body size. *Genet. Res.* 25: 45–57. [17]

Gastonguay, M. S., G. R. Keele, and G. A. Churchill. 2023. The trouble with triples: examining the impact of measurement error in mediation analysis. *Genet.* 224: iyad045. [21, A2]

Gatti, D. M., and 4 others. 2010. Heading down the wrong pathway: on the influence of correlation within gene sets. *BMC Genomics* 11: 574. [21]

Gatti, D. M., and 12 others. 2014. Quantitative trait locus mapping methods for diversity outbred mice. *G3* 4: 1623–1633. [18]

Gauch, H. G., Jr. 1992. *Statistical analysis of regional yield trials: AMMI analysis of factorial designs.* Elsevier. Amsterdam, NL [A3]

Gauderman, W. J., C. Murcray, F. Gilliland, and D. V. Conti. 2007. Testing association between disease and multiple SNPs in a candidate gene. *Genet. Epidem.* 31: 383–395. [20]

Gauss, C. F. 1809. *Theoria motus corporum coelestium.* (English translation, 1857, Little, Brown and Co., Boston, MA). [2]

Gavora, J. S., R. W. Fairfull, B. F. Benkel, W. J. Cantwell, and J. R. Chambers. 1996. Prediction of heterosis from

DNA fingerprints in chickens. *Genetics* 144: 777–784. [13]

Gavrilets, S., and A. Hastings. 1993. Maintenance of genetic variability under strong stabilizing selection: a two-locus model. *Genetics* 134: 377–386. [5]

Gavrilets, S., and A. Hastings. 1994a. Dynamics of genetic variability in two- locus models of stabilizing selection. *Genetics* 138: 519–532. [5]

Gavrilets, S., and A. Hastings. 1994b. A quantitative-genetic model for selection on developmental noise. *Evolution* 48: 1478–1486. [6]

Gavrilets, S., and A. Hastings. 1995. Dynamics of polygenic variability under stabilizing selection, recombination, and drift. *Genet. Res.* 65: 63–74. [5]

Gay, L., M. Siol, and J. Ronfort. 2013. Pedigree-free estimates of heritability in the wild: promising prospects for selfing populations. *PLoS One* 8: e66983. [8]

Gayon, J. 2000. History of the concept of allometry. *Am. Zool.* 40: 748–758. [14]

Gazal, S., and 5 others. 2014. Inbreeding coefficient estimationwith dense SNP data: Comparison of strategies and application to HapMap III. *Hum. Heredity* 77: 49–62. [8, 12]

Gazal, S., and 10 others. 2017. Linkage disequilibrium-dependent architecture of human complex traits shows action of negative selection. *Nat. Genet.* 49: 1421–1427. [21]

Gazal, S., and 6 others. 2018. Functional architecture of low-frequency variants highlights strength of negative selection across coding and con-coding annotations. *Nat. Genet.* 50: 1600–1607. [21]

Gazal, S., and 12 others. 2022. Combining SNP-to-gene linking strategies to identify disease genes and assess disease omnigenicity. *Nat. Genet.* 54: 827–836. [21]

Gebhardt-Henrich, S. G., and A. J. van Noordwijk. 1991. Nestling growth in the great tit. I. Heritability estimates under different environmental conditions. *J. Evol. Biol.* 3: 341–362. [7]

Geiger, H. H., and T. Miedaner. 1999. Hybrid rye and heterosis. *In* J. G. Coors and S. Pandey (eds.), *The genetics and exploitation of heterosis in crops*, pp. 439–450. ASA-CSSA-SSSA Societies, Madison, WI. [13]

Geiler-Samerotte, K. A., and 5 others. 2013. The details in the distributions: why and how to study phenotypic variability. *Curr. Opin. Biotech.* 24: 752–759. [14]

Geiler-Samerotte, K. A., and 4 others. 2016. Selection transforms the landscape of genetic variation interacting with Hsp90. *PLoS Bio.* 14: e2000465. [6]

Geiringer, H. 1944. On the probability theory of linkage in Mendelian heredity. *Ann. Math. Stat.* 15: 25–57. [5]

Geldermann, H. 1975. Investigations on inheritance of quantitative characters in animals by gene markers. I. Methods. *Theor. Appl. Genet.* 46: 319–330. [15, 19]

Gelfand, A. E., and A. F. M. Smith. 1990. Sampling-based approaches to calculating marginal densities. *J. Am. Stat. Assoc.* 85: 398–409. [16, A8]

Gelfond, J., J. Ibrahim, and F. Zou. 2007. Proximity model for expression quantitative trait loci (eqtl) detection. *Biometrics* 62: 19–27. [21]

Gelman, A., and D. B. Rubin. 1992. Inferences from iterative simulation using multiple sequences (with discussion). *Stat. Sci.* 7: 457 - 511. [A8]

Gelman, A., J. B. Carlon, H. S. Stern, and D. B. Rubin. 2013. *Bayesian data analysis*, 3rd edition. Chapman and Hall, New York, NY. [A7]

Gentle, J. E. 2007. *Matrix algebra: Theory, computations, and applications in statistics*. Springer, New York, NY. [10]

Geman, S., and D. Geman. 1984. Stochastic relaxation, Gibbs distributions and Bayesian restoration of images. *IEEE Transactions on Pattern Analysis and Machine Intelligence* 6: 721–741. [16, A8]

George, E. I., and R. E. McCulloch. 1993. Variable selection via Gibbs sampling. *J. Am. Stat. Assoc.* 88: 881–889. [18]

George, E. I., and R. E. McCulloch. 1997. Approaches for Bayesian variable selection. *Stat. Sin.* 7: 339–373. [18]

George, V., H. K. Tiwari, X. Zhu, and R. C. Elston. 1999. A test of transmission/disequilibrium for quantitative traits in pedigree data, by multiple regression. *Am. J. Hum. Genet.* 65: 236–245. [17]

Georges, M., and 11 others. 1995. Mapping quantitative trait loci controlling milk production in dairy cattle by exploiting progeny testing. *Genetics* 139: 907–920. [19]

Gerats, A. G. M., and 4 others. 1984. Control of anthocyanin synthesis in *Petunia hybrida* by multiple allelic series of the genes *AN1* and *AN2*. *Genetics* 106: 501–508. [15]

Gerber, S., P. Chabrier, and A. Kremer. 2003. FAMOZ: a software for parentage analysis using dominant, codominant and uniparentally inherited markers. *Mol. Ecol. Notes* 3: 479–481. [8]

Gershon, E. S., and C. R. Cloninger. 1994. *Genetic approaches to mental disorders*. American Psychiatric Press, Washington, D. C. [19]

Gerson, M. 1975. The techniques and uses of probability plotting. *The Statistician* 24: 235–257. [20]

Gev, D., N. Roguin, and E. Freundlich. 1986. Consanguinity and congenital heart disease in the rural Arab population in northern Israel. *Hum. Hered.* 36: 213–217. [12]

Geweke, J. 1992. Evaluating the accuracy of sampling-based approaches to the calculation of posterior moments *In* J. M. Bernardo, J. O. Berger, A. P. Dawid, and A. F. M. Smith (eds.), *Bayesian Statistics 4*, pp. 169–193. Oxford University Press, Oxford, UK. [A8]

Geyer, C. J. 1992. Practical Markov chain Monte Carlo (with discussion). *Stat. Sci.* 7: 473–511. [A8]

Geyer, C. J. 2011. Introduction to markov chain Monte Carlo. *In* S. Brooks, A. Gelman, G. Jones, and X.-L. Meng (eds), *Handbook of markov chain Monte Carlo*, pp. 2–48. CRC press, Boca Raton, FL. [A8]

Ghosh, A., F. Zou, and F. A. Wright. 2008 Estimating odds ratio in genome scans: an approximate conditional likelihood approach. *Am. J. Hum. Genet.* 82: 1064–1074. [20]

Ghosh, S., and 11 others. 1993. Polygenic control of autoimmune diabetes in nonobese diabetic mice. *Nat. Genet.* 4: 404–409. [18]

Ghosh, S., and 10 others. 2018. Speed breeding in growth chambers and glasshouses for crop breeding and model plant research. *Nat. Protocols* 13: 2944–2963. [17]

Giaever, G., and C. Nislow. 2014. The yeast deletion collection: a decade of functional genomics. *Genetics* 197: 451–465. [15]

Giaever, G., and 72 others. 2002. Functional profiling of the *Saccharomyces cerevisiae* genome. *Nature* 418: 387–391. [15]

Giambartolomei, C., and 6 others. 2014. Bayesian test for colocalisation between pairs of genetic association studies using summary statistics. *PLoS Genet.* 10: e1004383. [21]

Giambartolomei, C., and 10 others. 2018. A Bayesian framework for multiple trait colocalization from summary association statistics. *Bioinformatics* 34: 2538–2545. [21]

Gianola, D., and K. Hammond (eds.) 1990. *Advances in statistical methods for genetic improvement of livestock.* Springer-Verlag, Berlin, DE. [1]

Gianola, D., J. L. Foulley, and R. L. Fernando. 1986. Prediction of breeding values when variances are not known. *Genet. Sel. Evol.* 18: 485–498. [10]

Gianola, D., S. Im, and R. L. Fernando. 1988. Prediction of breeding values under Henderson's selection model: a revisitation. *J. Dairy Sci.* 71: 2790–2798. [10]

Gibbs, J. R., and 15 others. 2010. Abundant quantitative trait loci exist for DNA methylation and gene expression in human brain. *PLoS Genetics* 6: e00952. [21]

Gibson, G. 2009. Decanalization and the origin of complex disease. *Nat. Rev. Genet.* 10: 134–140. [14]

Gibson, G., and I. Dwokin. 2004. Uncovering cryptic genetic variation. *Nat. Rev. Genet.* 15: 681-690. [14]

Gibson, G., and D. S. Hogness. 1996. Effect of polymorphism in the *Drosophila* regulatory gene *Ultrabithorax* on homeotic stability. *Science* 271: 200–203. [15]

Gibson, G., and K. A. Lacek. 2020. Canalization and robustness in human genetics and disease. *Ann. Rev. Genet.* 54: 189–211. [14]

Gibson, G., and G. Wagner. 2000. Canalization in evolutionary genetics: a stabilizing theory? *BioEssays* 22: 372–380. [14]

Gibson, G., and B. S. Weir. 2005. The quantitative genetics of transcription. *Trend Genet.* 21: 616–623. [1, 21]

Gienapp, P., and 5 others. 2017. Genomic quantitative genetics to study evolution in the wild. *Trends Ecol. Evol.* 32: 897–908. [1]

Giesbrecht, F. G., and M. L. Gumpertz. 2004. *Planning, construction, and statistical analysis of comparative experiments.* Wiley, New York, NY. [A9]

Gilad, Y., S. A. Rifkin, and J. K. Pritchard. 2008. Revealing the architecture of gene regulation: the promise of eQTL studies. *Trends Genet.* 24: 408–415. [21]

Gilbert, K. J., F. Pouyet, L. Excoffier, and S. Peischl. 2020. Transition from background selection to associative overdominance promotes diversity in regions of low recombination. *Curr. Bio.* 30: 101–107. [12]

Gill, P. E., Murray, W. and Wright, M. E. 1981. *Practical optimization.* Academic Press, New York, NY. [16]

Gillard, E. F., and 12 others. 1989. Molecular and phenotypic analysis of patients with deletions within the deletion-rich region of the Duchenne muscular dystrophy (DMD) gene. *Am. J. Hum. Gent.* 45: 507–520. [15]

Gillespie, J. H. 1991. *The causes of molecular evolution.* Oxford Univ. Press, New York, NY. [4]

Gilligan, D. M., and 5 others. 2000. Can fluctuating asymmetry be used to detect inbreeding and loss of genetic diversity in endangered populations? *Anim. Conserv.* 3: 97–104. [6]

Gillois, M. 1964. La relation d'identité génétique. Thesis, Faculté des Sciences, Paris, FR. [7]

Gilmore, E. C. 1969. Effects of inbreeding of paternal lines on predicted yield in synthetics 1. *Crop Sci.* 9: 102–104. [13]

Gilmour, A. R., B. R. Cullis, and A. P. Verbyla. 1997. Accounting for natural and extraneous variation in the analysis of field experiments. *J. Agric. Biol. Environ. Stats.* 2: 269–293. [A9]

Gimelfarb, A. 1981. A general linear model for the genotypic covariance between relatives under assortative mating. *J. Math. Biol.* 13: 209–226. [7]

Gimelfarb, A. 1984. Quantitative characters under assortative mating: gametic model. *Theor. Pop. Biol.* 25: 312–330. [7]

Gimond, C., and 5 others. 2013. Outbreeding depression with low genetic variation in selfing *Caenorhabditis nematodes*. *Evolution* 67: 3087–3101. [13]

Ginzburg, É. K. 1983. Possible localization of genes controlling a quantitative character in self-pollinators. *Genetica* 19: 577–583. [18]

Giovannoni, J. J., R. A. Wing, M. W. Ganal, and S. D. Tanksley. 1991. Isolation of molecular markers from specific chromosomal intervals using DNA pools from existing mapping populations. *Nucl. Acids Res.* 19: 6553–6558. [17]

Giraud, H., and 18 others. 2014. Linkage disequilibrium with linkage analysis of multiline crosses reveals different multiallelic QTL for hybrid performance in the flint and dent heterotic groups of maize. *Genetics* 198: 1717–1734. [18]

Gireesh, C., and 13 others. 2021. Nested association mapping (NAM) populations: present status and future prospects in the genomics era. *Crit. Rev. Plant Sci.* 40: 49–67. [18]

Girke, A., A. Schierholt, and H.C. Becker. 2012 . Extending the rapeseed gene pool with resynthesized *Brassica napus* II: Heterosis. *Theor. Appl. Genet.* 124: 1017–1026. [13]

Gjerde, B., and T. Refstie. 1984. Complete diallel cross between five strains of Atlantic salmon. *Livestock Prod. Sci.* 11: 207–226. [13]

Glaser, B., and P. Holmans. 2009. Comparison of methods for combining case-control and family-based associations studies. *Hum. Heret.* 68: 106–116. [17]

Glass, G. V. 1976. Primary, secondary, and meta-analysis of research. *In* L. Shulman (ed.) *Review of research in education, Vol 5*, pp. 351–379. Peacock, Itasca, IL. [A6]

Glassberg, E. C., and 4 others. 2019. Evidence for weak selective constraints on human gene expression. *Genetics* 211: 575–772. [21]

Glazner, C., and E. Thompson. 2015. Pedigree-free descent-based gene mapping from population samples. *Hum. Hered.* 80: 21–35. [8]

Gleeson, A. C. 1997. Spatial analysis. *In* R. A. Kempton and P. N. Fox (eds), *Statistical methods for plant variety evaluation*, pp. 68–85. Chapman & Hall, London, UK. [A9]

Gleeson, A. C., and B. R. Cullis. 1987. Residual maximum likelihood (REML) estimation of a neighbour model for field experiments. *Biometrics* 43: 277–287. [A9]

Glejser, H. 1969. A new test for heteroskedasticity. *J. Am. Stat. Assoc.*64: 315–323. [10]

Glémin, S. 2003. How are deleterious mutations purged? Drift versus nonrandom mating. *Evolution* 57: 2678–2687. [12]

Glémin, S., L. Vimond, J. R., T. Bataillon, and A. Mignot. 2006. Marker-based investigation of inbreeding depression in the endangered species *Brassica insularis*. *Heredity* 97: 304–311. [12]

Glymour, C. 1981. Why I am not a Bayesian, *In* C. Glymour (ed.), *Theory and evidence*, pp. 63–93. University of Chicago Press, Chicago, IL. [A7]

Glymour, M.. M., E. J. Tchetgen Tchetgen, and J. M. Robins. 2012. Credible Mendelian randomization studies: approaches for evaluating the instrument variable assumptions. *Am. J. Epidem.* 175: 332–339. [A2]

Go, R. C. P., R. C. Elston, and E. B. Kaplan. 1978. Efficiency and robustness of pedigree segregation analysis. *Am. J. Hum. Genet.* 30: 28–37. [16]

Godshalk, E. B., M. Lee, and K. R. Lamkey. 1990. Relationship of restriction fragment length polymorphisms to single-cross hybrid performance of maize. *Theor. Appl. Genet.* 80: 273–280. [13]

Goeman, J. J., and P. Bühlmann. 2007. Analyzing gene expression data in terms of gene sets: methodological issues. *Bioinfomatics* 23: 980–987. [21]

Goffinet, B., and B. Mangin. 1998. Comparing methods to detect more than one QTL on a chromosome. *Theor. Appl. Genet.* 96: 628–633. [18]

Goldberger, A. S. 1962. Best linear unbiased predictors in the generalized linear regression model. *J. Am. Stat. Assoc.* 57: 369–375. [10]

Goldfeld, S. M., and R. E. Quandt. 1965. Some tests for homoscedasticity. *J. Am. Stat. Asso.* 60: 539–547. [10]

Goldgar, D. E. 1990. Multipoint analysis of human quantitative genetic variation. *Am. J. Hum. Genet.* 47: 957–967. [19]

Goldin, L. R., and E. S. Gershon. 1988. Power of the affected-sib-pair method for heterogeneous disorders. *Genet. Epidem.* 5: 35–42. [19]

Goldin, L. R., and D. E. Weeks. 1993. Two-locus models of disease: comparison of likelihood and nonparametric linkage methods. *Am. J. Hum. Genet.* 53: 908–915. [19]

Goldin, L. R., K. K. Kidd, S. Matthysse, and E. S. Gershon. 1981. The power of pedigree segregation analysis for traits with incomplete penetrance. *In* E. S. Gershon, S. Matthysse, X. O. Breakefield, and R. D. Ciaranello (eds.), *Genetic research strategies for psychobiology and psychiatry*, pp. 305–317. Plenum, New York, NY. [16]

Goldin, L. R., and 4 others. 1984. The detection of major loci by segregation and linkage analysis: a simulation study. *Genet. Epidem.* 1: 285–296. [19]

Goldman, I. L. 1999. Inbreeding and outbreeding in the development of a modern heterosis concept. *In* J. G. Coors and S. Pandey (eds.), *The genetics and exploitation of heterosis in crops*, pp. 7–18. ASA-CSSA-SSSA Societies, Madison, WI. [13]

Goldman, I. L., T. R. Rocheford, and J. W. Dudley. 1993. Quantitative trait loci influencing protein and starch concentration in the Illinois long term selection maize strains. *Theor. Appl. Genet.* 87: 217–224. [18]

Goldman, I. L., T. R. Rocheford, and J. W. Dudley. 1994. Molecular markers associated with maize kernel oil concentration in an Illinois high protein × Illinois low protein cross. *Crop Sci.* 34: 908–915. [18]

Goldman, I. L., I. Paran, and D. Zamir. 1995. Quantitative trait locus analysis of a recombinant inbred line population derived from a *Lycopersicon esculentum* × *Lycopersicon cheesmanii* cross. *Theor. Appl. Genet.* 90: 925–932. [17]

Goldschmidt, R. B. 1935. Gen und ausseneigenschaft. *Z. indukt. Abstamm.u. Vereb-Lehre* 69: 38–131. [14]

Goldstein, D. R., H. Zhao, and T. P. Speed. 1995. Relative efficiencies of χ^2 models of recombination for exclusion mapping and gene ordering. *Genomics* 27: 265–273. [17]

Gomez, K. A., and A. A. Gomez. 1984. *Statistical procedures for agricultural research*, 2nd ed. Wiley, New York, NY. [A9]

Gomez-Raya, L., and E. B. Burnside. 1990. The effect of repeated cycles of selection on genetic variance, heritability, and response. *Theor. Appl. Genet.* 79: 568–574. [7]

Gonyon, D. S., and 5 others. 1987. Associations of bovine blood and milk polymorphisms with lactation traits: Holsteins. *Theor. Appl. Genet.* 70: 2585–2598. [19]

Good, R. L., and A. R. Hallauer. 1977. Inbreeding depression in maize by selfing and full-sibbing. *Crop Sci.* 17 935–940. [12]

Goodwill, R., and F. D. Enfield. 1971. Heterozygosity in inbred lines of *Tribolium castaneum*. *Theor. Appl. Genet.* 41: 5–12. [15]

Goodwin, R. H. 1944. The inheritance of flowering time in a short-day species, *Solidago sempervirens* L. *Genetics* 29: 503–519. [11]

Goradia, T. M., K. Lange, P. L. Miller, and P. M. Nadkarni. 1992. Fast computations of genetic likelihoods on human pedigree data. *Hum. Hered.* 42: 42–62. [16]

Gordon, D., S. C. Heath, X. Liu, and J. Ott. 2001. A transmission/disequilibrium test that allows for genotyping errors in the analysis of single-nucleotide polymorphism data. *Am. J. Hum. Genet.* 69: 371–380. [17]

Gordon, D., and 5 others. 2004. A transmission disequilibrium test for general pedigrees that is robust to the presence of random genotyping errors and any number of untyped parents. *Europ. J. Hum. Genet.* 12: 752–761. [17]

Gore, M. A., and 11 others. 2009. A first-generation haplotype map of maize. *Science* 326: 1115–1117. [12, 13]

Göring, H. H. H., J. D. Terwilliger, and J. Blangero. 2001. Large upward bias in estimation of locus-specific effects from genomewide scans. *Am. J. Hum. Genet.* 69: 1357–1369. [18]

Gorlov, I. P., and 4 others. 2008. Shifting paradigm of association studies: value of rare single-nucleotide polymorphisms. *Am. J. Hum Genet.* 82: 100–112. [21]

Gorroochurn, P., S. E. Hodge, H. G. Heiman, and D. A. Greenberg. 2004. Effect of population stratification on case-control association studies. II. False-positive rates and their limiting behavior as number of subpopulations increases. *Hum. Hered.* 58: 40–48. [20]

Gosey, J. A. 1991. Crossbreeding systems and the theory behind composite breeds. *Range Beef Cow Symposium* 23. [13]

Gottlieb, L. D. 1984. Genetics and morphological evolution in plants. *Am. Nat.* 123: 681–709. [11]

Götz, K. U., and L. Ollivier. 1992. Theoretical aspects of applying sib-pair linkage tests to livestock species. *Genet. Sel. Evol.* 24: 29–42. [19]

Goudet, J., T. Kay, and B. S. Weir. 2018. How to estimate kinship. *Mol. Ecol.* 27: 4121–4135. [8]

Gould, S. J. 1966. On the scaling of tooth size in mammals. *Am. Zool.* 15: 351–362. [14]

Gould, S. J. 1980. Is a new and general theory of evolution emerging? *Paleobiol.* 6: 119–130. [11]

Goulet, B. E., F. Roda, and R. Hopkins. 2017. Hybridization in plants: old ideas, new techniques. *Plant Physiol.* 173: 65–78. [13]

Govind, C. K., and J. Pearce. 1986. Differential reflex activity determines claw and closer muscle asymmetry in developing lobsters. *Science* 233: 354–356. [6]

Gowen, J. W. (ed.). 1952. *Heterosis.* Iowa State College Press, Ames, IA. [13]

Grafen, A. 1985. A geometric view of relatedness. *Oxford Surveys Evol. Biol.* 2: 28–89. [7]

Graham, A. 1981. *Kronecker products and matrix calculus with applications.* Halsted Press, New York, NY. [A3]

Graham, G. I., D. W. Wolff, and C. W. Stuber. 1997. Characterization of a yield quantitative trait locus on chromosome five of maize by fine mapping. *Crop Sci.* 37: 601–1610. [13]

Graham, J. H., and J. D. Felley. 1985. Genomic coadaptation and developmental stability within introgressed populations of *Enneacanthus gloriosus* and *E. obesus* (Pisces, Centrarchidae). *Evolution* 39: 104–114. [6]

Graham, J. H., S. Raz, H. Hel-Or, and E. Nevo. 2010. Fluctuating asymmetry: methods, theory, and applications. *Symmetry* 2: 466–540. [6]

Grant, V. 1981. *Plant speciation*, 2nd Ed. Columbia Univ. Pres, New York, NY. [13]

Grattapaglia, D., F. L. Bertolucci, and R. R. Ederoff. 1995. Genetic mapping of QTLs controlling vegetative propagation in *Eucalyptus grandis* and *E. urophylla* using a pseudo-testcross strategy and RAPD markers. *Theor. Appl. Genet.* 90: 933–947. [19]

Gray, A., and A. Tait. 1993. Identification of ornithine decarboxylase as a trait gene for growth in replicated mouse lines divergently selected for lean body mass. *Genet. Res.* 62: 31–37. [17]

Gray, R., and K. Wheatley. 1991. How to avoid bias when comparing bone marrow transplant with chemotherapy. *Bone Marrow Transplant* 7 (Suppl. 3): 9–12. [A2]

Greaves, I. K., and 7 others. 2015. Epigenetic changes in hybrids. *Plant Physiol.* 168: 1197–1205. [13]

Green, A. J. 2001. Mass/length residuals: measures of body condition or generators of spurious results? *Ecology* 82: 1473–1483. [14]

Green, C. V. 1931. Linkage in size inheritance. *Am. Nat.* 65: 502–511. [17]

Green, C. V. 1933. Further evidence of linkage in size inheritance. *Am. Nat.* 67: 377–380. [17]

Green, J. R., and S. Shah. 1993. Power comparison of various sibship tests of association. *Ann. Hum. Genet.* 57: 151–158. [19]

Green, J. R., and J. C. Woodrow. 1977. Sibling method for detecting HLA-linked genes in disease. *Tissue Antigens* 9: 31–35. [19]

Green, M. M. 1959. The discrimination of wild-type isoalleles at the white locus of *Drosophila melanogaster. Proc. Natl. Acad. Sci. USA* 45: 549–553. [15]

Green, P. J. 1995. Reversible jump Markov Chain Monte Carlo computation and Bayesian model determination. *Biometrika* 82: 711–732. [18]

Greenberg, D. A. 1986. The effect of proband designation on segregation analysis. *Am. J. Hum. Genet.* 39: 329–339. [16]

Greenberg, R., and J. F. Crow. 1960. A comparison of the effect of lethal and detrimental chromosomes from *Drosophila* populations. *Genetics* 45: 1153–1168. [12]

Greenland, S. 2000. An introduction to instrumental variables for epidemiologists. *Int. J. Epidem.* 29: 722–729. [A2]

Greenwood, C. M. T., and K. Morgan. 2000. The impact of transmission-ratio distortion on allele sharing in affected sibling pairs. *Am. J. Hum. Genet.* 66: 2001–2004. [19]

Gregorius, H.-R. 1976. Convergence of genetic compositions assuming infinite population size and overlapping generations. *J. Math. Biol.* 3: 179–186. [4]

Gregory, K. E., and L. V. Cundiff. 1980. Crossbreeding in beef cattle: evaluation of systems. *J. Anim. Sci.* 51: 1224–1242. [13]

Gregory, K. E., L. V. Cundiff, and R. M. Koch. 1982. Comparison of crossbreeding systems and breeding stocks used in suckling herds of continental and temperate areas. *Proceed. Second World Congress on Genetics Applied to Livestock Production*, 7, pp. 482–503. [13]

Gregory, K. E., J. C. M. Trail, H. J. S. Marples, and J. Kakonge. 1985. Heterosis and breed effects on maternal and individual traits of BosIndicus breeds of cattle. *J. Anim. Sci.* 60: 1175 –1180. [13]

Gregory, W. C. 1965a. Mutation frequency, magnitude of change, and the probability of improvement in adaptation. *Rad. Botany* 5 (Suppl.): 429–441. [15]

Gregory, W. C. 1965b. Mutation breeding. *In* K. J. Frey (ed.), *Plant breeding*, pp. 189–218. Iowa State Univ. Press, Ames, IA. [15]

Greilhuber, J., and I. A. Leitch. 2013. Genome size and the phenotype. *In* I. A. Leitch, J. Greilhuber, J. Dolezel, and J. F. Wendel, *Plant genome diversity* Volume 2, 323–344. Springer, Vienna, AT. [15]

Grewal, M. S. 1962. The rate of genetic divergence of sublines in the C57BL strain of mice. *Genet. Res.* 3: 226–237. [15]

Grignola, F. E., I. Hoeschele, and B. Tier. 1996a. Mapping quantitative trait loci in outcross populations via residual maximum likelihood. I. Methodology. *Genet. Sel. Evol.* 28: 479–490. [19]

Grignola, F. E., I. Hoeschele, Q. Zhang, and G. Thaller. 1996b. Mapping quantitative trait loci in outcross populations via residual maximum likelihood. II. A simulation study. *Genet. Sel. Evol.* 28: 491–504. [19]

Grignola, F. E., Q. Zhang, and I. Hoeschele. 1997. Mapping linked quantitative trait loci via residual maximum likelihood. *Genet. Sel. Evol.* 29: 529–544. [19]

Groen, A. F., and C. Smith. 1995. A stochastic simulation study of the efficiency of marker-assisted introgression in livestock. *J. Anim. Breed. Genet.* 112: 161–170. [17]

Grondona, M. O., J. Crossa, P. N. Fox, and W. H. Pfeiffer. 1996. Analysis of variety yield trails using two-dimensional separable ARIMA processes. *Biometric* 52: 763–770. [A9]

Groose, R. W., L. E. Talbert, W. P. Kojis, and E. T. Bringham. 1989. Progressive heterosis in autotetraploid alfalfa: studies using two types on inbreds. *Crop Sci.* 29: 1173–1177. [13]

Grossen, C., F. Guillaumne, L. F. Keller, and D. Croll. 2020. Purging of highly deleterious mutations through severe bottlenecks in Alpine ibex. *Nat. Comm.* 11: 1001. [12]

Groszmann, M., and 4 others. 2013. The role of epigenetics in hybrid vigour. *Trends Genet.* 29: 684–690. [13]

Grotzinger, A. D., and 13 others. 2019. Genomic structural equation modelling provides insights into the multivariate genetic architecture of complex traits. *Nat. Hum. Behav.* 3: 513–525. [20]

Grover, A., and P. C. Sharma. 2016. Development and use of molecular markers: past and present. *Crit. Rev. Biotech.* 36: 290–302. [8]

Grueber, C. E., R. J. Laws, S. Nakagawa, and I. G. Jamieson. 2010. Inbreeding depression accumulation across life history stages of the endangered takahe. *Conserv. Biol.* 24: 1617–1625. [12]

Grueber, C. E., J. M. Waters, and I. G. Jamieson. 2011a. The imprecision of heterozygosity-fitness correlations hinders the detection of inbreeding and inbreeding depression in a threatened species. *Mol. Ecol.* 20: 67–79. [12, 14]

Grueber, C. E., S. Nakagawa, R. J. Laws, and I. G. Jamieson. 2011b. Multimodel inference in ecology and evolution: challenges and solutions. *J. Evol. Biol* 24: 699–711. [12, A4]

Gruneberg, H. 1938. An analysis of the pleiotropic effects of a new lethal mutation in the rat (*Mus norvegicus*). *Proc. Roy. Soc. Lond. B* 125: 123–144. [A2]

Grüneberg, H. 1970. Is there a viral component in the genetic background? *Nature* 225: 39–14. [15]

GTEx Consortium. 2017. Genetic effects on gene expression across human tissues. *Nature* 550: 204–213. [21]

GTEx Consortium. 2020. The GTEx Consortium atlas of genetic regulatory effects across human tissues. *Science* 369: 1318–1320. [21]

Gu, C., A. Todorov, and C. D. Rao. 1996. Combining extremely concordant sib pairs with extremely discordant sib pairs provides a cost effective way to linkage analysis of quantitative trait loci. *Genet. Epidem.* 13: 513–533. [19]

Guan, Y., and M. Stephens. 2008. Practical issues in imputation-based association mapping. *PLoS Genet.* 4: e1000279. [20, 21]

Guan, Y. T., and M. Stephens. 2011. Bayesian variable selection regression for genome-wide association studies and other large-scale problems. *Ann. Appl. Stat.* 5: 1780–1815. [21]

Gudbjartsson, D. F., and 36 others. 2008. Many sequence variants affecting diversity of adult human height. *Nat. Genet.* 40: 609–615. [21]

Guedj, M., S. Robin, A. Celisse, and G. Nuel. 2009. Kerfdr: a semi-parametric kernel-based approach to local false discovery rate estimation. *BMC Bioniformatics* 10: 84. [A6]

Gui, H., M. Li, P. C. Sham, and S. S. Cherny. 2011. Comparisons of seven algorithms for pathway analysis using the WTCCC Crohn's disease dataset. *BMC Res. Notes* 4: 386. [21]

Gulisija, D., and J. F. Crow. 2007. Inferring purging from pedigree data. *Evolution* 61: 1043–1051. [12]

Gulisija, D., D. Gianola, and K. A. Weigel. 2007. Nonparametric analysis of the impact of inbreeding on production in Jersey cows. *J. Dairy Sci.* 90: 493–500. [12]

Guo, G., L. Wang, H. Liu, and T. Randall. 2014. Genomic assortative mating in marriages in the United States. *PLOS One* 9: e112322. [7]

Guo, L., and 4 others. 2016. Widespread maintence of genome heterozygosity in *Schmidtea mediterranea*. *Nat. Ecol. Evol.* 1: 0019. [12]

Guo, S. W. 1994a. Computation of identity-by-descent proportions shared by two siblings. *Am. J. Hum. Genet.* 54: 1104–1109. [8, 19]

Guo, S. W. 1994b. Proportion of genes survived in offspring conditional on inheritance of flanking markers. *Genetics* 138: 953–962. [19]

Guo, S.-W. 1995. Proportion of genome shared identical by descent by relatives: concept, computation, and applications. *Am. J. Hum. Genet.* 56: 1468–1476. [8]

Guo, S.-W. 1996. Variation in genetic identity among relatives. *Hum. Hered.* 46: 61–70. [8]

Guo, S. W., and E. A. Thompson. 1992. A monte carlo method for combined segregation and linkage analysis. *Am. J. Hum. Genet.* 51: 1111–1126. [16]

Guo, S. W., and E. A. Thompson. 1994. Monte carlo estimation of mixed models for large complex pedigrees. *Biometrics* 50: 417–432. [16]

Guo, T., and 8 others. 2014. Genetic basis of grain yield heterosis in an "immortalized F_2" maize population. *Theor. Appl. Genet.* 127: 2149–2158. [13]

Gurevitch, J., and L. V. Hedges. 1999. Statistical issues in ecological meta-analyses. *Ecology* 80: 1142–1149. [A6]

Gurevitch, J., and 3 others. 2018. Meta-analysis and the science of research synthesis. *Nature* 55: 175–182. [A6]

Gusev, A., and 7 others. 2009. Whole population, genome-wide mapping of hidden relatedness. *Genom. Res.* 19: 318–326. [8]

Gusev, A., and 24 others. 2016. Integrative approaches for large-scale transcriptome-wide association studies. *Nat. Genet.* 48: 245–252. [21]

Gusev, A., and 19 others. 2018. Transcriptome-wide association study of schizophrenia and chromatin activity yields mechanistic disease insights. *Nat. Genet.* 50: 538–548. [21]

Gusmão. L. 1986. Inadequacy of blocking in cultivar yield trials. *Theor. Appl. Genet.* 72: 98–104. [A9]

Gustincich, S., and 7 others. 2006. The complexity of the mammalian transcriptome. *J. Physiol.* 575: 321–332. [21]

Haberfeld, A., E. A. Dunnington, P. B. Siegel, and J. Hillel. 1996. Heterosis and DNA fingerprinting in chickens. *Poultry Sci.* 75: 951–953. [13]

Hackett, C. A., and J. I. Weller. 1995. Genetic mapping of quantitative trait loci for traits with ordinal distributions. *Biometrics* 51: 1252–1263. [18]

Hadfield, J. D., and S. Nakagawa. 2010. General quantitative genetic methods for comparative biology: phylogenies, taxonomies and multitrait models for continuous and categorical characters. *J. Evol. Biol.* 23: 494–508. [A6]

Hadfield, J. D., D. S. Richardson, and T. Burke. 2006. Towards unbiased parentage assignment: combining genetic, behavioural and spatial data in a Bayesian framework. *Mol. Ecol.* 15: 3715–3730. [8]

Haenlein, G. F. W., D. S. Gonyon, R. E. Mather, and H. C. Hines. 1987. Associations of bovine blood and milk polymorphisms with lactation traits: Guernseys. *Theor. Appl. Genet.* 70: 2599–2609. [19]

Hagger, C., C. Stricker, R. C. Elston, and G. Stranzinger. 1995 Lack of evidence for segregation of a single dominant major gene as the cause of the difference in egg weight between two highly inbred lines of chickens. *Theor. Appl. Genet.* 90: 120–123. [16]

Halbach, U., and J. Jacobs. 1971. Seasonal selection as a factor in rotifer cyclomorphosis. *Naturwissenschaften* 57: 1–2. [14]

Haldane, J. B. S. 1919. The combination of linkage values, and the calculation of distance between the loci of linked factors. *J. Genetics* 8: 299–309. [17]

Haldane, J. B. S. 1922. Sex ratio and unisexual sterility in hybrid animals. *J. Genetics* 12: 101–109. [17]

Haldane, J. B. S. 1927. Mathematical theory of natural and artificial selection. V. Selection and mutation. *Proc. Cambridge Phil. Soc.* 23: 838–844. [12]

Haldane, J. B. S. 1930a. A note on Fisher's theory of the origin of dominance, and on a correlation between dominance and linkage. *Am. Nat.* 64: 87–90. [4]

Haldane, J. B. S. 1930b. Theoretical genetics of autopolyploids. *J. Genet.* 22: 359–372. [12]

Haldane, J. B. S. 1935. The rate of spontaneous mutation of a human gene. *J. Genetics* 33: 317–326. [15]

Haldane, J. B. S. 1937. The effect of variation on fitness. *Am. Nat.* 71: 337–349. [15]

Haldane, J. B. S. 1939. The theory of the evolution of dominance. *J. Genet.* 37: 365–374. [4]

Haldane, J. B. S. 1942. *New paths in genetics.* Harper and Brothers, New York, NY. [21]

Haldane, J. B. S. 1947. The rate of mutation of the gene for hemophilia and its segregation ratios in males and females. *Ann. Eugenics* 13: 262–271. [15]

Haldane, J. B. S. 1955. The measurement of variation. *Evolution* 9: 484–486. [2]

Haldane, J. B. S. 1956. The theory of selection for melanism in Lepidoptera. *Proc. Roy. Soc. Lond. B* 145: 303–306. [4]

Haldane, J. B. S., and C. H. Waddington. 1931. Inbreeding and linkage. Genetics 16: 357–374. [18]

Halder, I., and M. D. Shriver. 2003. Measuring and using admixture to study the genetics of complex disease. *Hum. Genomics* 1: 2–62. [17]

Haley, C. S. 1991. Use of DNA fingerprints for the detection of major genes for quantitative traits in domestic species. *Anim. Genet.* 22: 259–277. [19]

Haley, C. S., and S. A. Knott. 1992. A simple regression method for mapping quantitative trait loci in line crosses using flanking markers. *Heredity* 69: 315–324. [18, 19]

Haley, S. D., and 4 others. 1994a. Heterogeneous inbred populations are useful as sources of near-isogenic lines for RAPD marker localization. *Theor. Appl. Genet.* 88: 337–342. [17]

Haley, C. S., S. A. Knott, and J.-M. Elsen. 1994b. Mapping quantitative trait loci in crosses between outbred lines using least squares. *Genetics* 136: 1995–1207. [18, 19]

Hall, D. W., and S. B. Joseph. 2010. A high frequency of beneficial mutations across multiple fitness components in *Saccharomyces cerevisiae. Genetics* 185: 1397–1409. [15]

Hall, M. C., I. Dworkin, M.C. Ungerer, and M. Purugganan. 2007. Genetics of microenvironmental canalization in *Arabidopsis thaliana. Proc. Natl. Acad. Sci. USA* 104: 13717–13722. [6]

Hall, N., and 4 others. 2012. Maximum likelihood estimation of individual inbreeding coefficients and null allele frequencies. *Genet. Res.* 94: 151–161. [8]

Hallauer, A. R., and J. B. Miranda. 1981. *Quantitative genetics in maize breeding.* Iowa State Univ. Press, Ames, IA. [1, 12, 13]

Hallauer, A. R., and J. H. Sears. 1973. Changes in quantitative traits associated with inbreeding in a synthetic variety of maize. *Crop Sci.* 13: 327–330. [12]

Hallauer, A. R., M. J. Carena, and J. B. Miranda Filho. 2010. *Quantitative genetics in maize breeding*, 3rd Ed. Springer-Verlag, New York, NY. [1, 12, 13]

Haller, G., D. G. Torgerson, C. Ober, and E. E. Thompson. 2009. Sequencing the *IL4* locus in African Americans implicates rare noncoding variants in asthma susceptibly. *J. Allergy Clin. Immunol.* 124: 1204–1209. [20]

Hallgrimsson, B., and 8 others. 2019. The developmental-genetics of canalization. *Semin. Cell. Dev. Biol.* 88: 67–79. [14]

Halligan, D. L., and P. D. Keightley. 2006. Ubiquitous selective constraints in the *Drosophila* genome revealed by a genome–wide interspecies comparison. *Genome Res.* 16: 875–884. [15]

Halligan, D. L., and P. D. Keightley. 2009. Spontaneous mutation accumulation studies in evolutionary genetics. *Ann. Rev. Ecol. Evol. Syst.* 40: 151–172. [15]

Hamilton, D. C., and D. E. C. Cole. 2004. Standardizing a composite measure of linkage disequilibrium. *Ann. Hum. Genet.* 68: 234–239. [5]

Hamilton, D. C., and D. E. C. Cole. 2006. Approximate variance for a standardized composite measure of linkage disequilibrium. *Ann. Hum. Genet.* 70: 535–540. [5]

Hamilton, D. C., and D. E. C. Cole. 2008. Testing for quality of a standardized composite measures of linkage disequilibrium. *Ann. Hum. Genet.* 72: 292–296. [5]

Hammond, J. J., and C. O. Gardner. 1974. Modification of the variety cross diallel model for evaluating cycles of selection. *Crop Sci.* 14: 6–8. [11]

Hammond, K., and J. W. James. 1970. Genes of large effect and the shape of the distribution of a quantitative character. *Aust. J. Biol. Sci.* 23: 867–876. [16]

Han, B., and E. Eskin. 2011. Random-effect model aimed at discovering associations in meta-analysis of genome-wide associations studies. *Am. J. Hum. Genet.* 88: 586–598. [20]

Han, B., and E. Eskin. 2012. Interpreting meta-analyses of genome-wide association studies. *PLoS Genet.* 8: e1002555. [20]

Han, B., H. M. Kang, and E. Eskin. 2009. Rapid and accurate multiple testing correction and power estimation for millions of correlated markers. *PLoS Genet.* 5: e10000456. [20]

Han, F., and W. Pan. 2010. A data-adaptive sum test of disease association with multiple common or rare variants. *Hum. Heredity* 70: 42–54. [20]

Han, J. D., and 10 others. 2004. Evidence for dynamically organized modularity in the yeast protein-protein interaction network. *Nature* 430: 88–93. [A2]

Han, L., and M. Abney. 2011. Identity by descent estimation with dense genome-wide genotype data. *Genet. Epid.* 35: 557–567. [8]

Han, L., and S. Xu. 2008. A Fisher scoring algorithm for the weighted regression method of QTL mapping. *Heredity* 101: 453–464. [18]

Hanis, C. I., and 33 others. 1996. A genome-wide search for human non-insulin-dependent (type 2) diabetes genes reveals a major susceptibility locus on chromosome 2. *Nat. Genet.* 13: 161–166. [19]

Hanna, W., P. Ozias-Akins, and D. Roche. 1999. Apomixis and heterosis. *In* J. G. Coors and S. Pandey (eds.), *The genetics and exploitation of heterosis in crops*, pp. 335–342. ASA-CSSA-SSSA Societies, Madison, WI. [13]

Hansen, T. F. 2006. The evolution of genetic architecture. *Ann. Rev. Eco. Evol. Syst.* 37: 123–157. [14]

Hansen, T. F., C. Pélabon, and D. Houle. 2011. Heritability is not evolvability. *Evol. Biol.* 38: 258–277. [7]

Hansen, T. J., and 7 others. 2020. Production and verification of the first Atlantic salmon (*Salmo salar* L.) clonal lines. *BMC Genet.* 21: 71. [17]

Hanson, L., and 4 others. 1995. Segregation analysis of non-insulin-dependent diabetes mellitus in Pima indians: evidence for a major-gene effect. *Am. J. Hum. Genet.* 57: 160–170. [17]

Hanson, W. D. 1959a. The theoretical distribution of lengths of parental gene blocks in the gametes of an F_1 individual. *Genetics* 44: 197–209. [17]

Hanson, W. D. 1959b. Early generation analysis of lengths of heterozygous chromosome segments around a locus held heterozygous with backcrossing or selfing. *Genetics* 44: 833–837. [17]

Hanson, W. D. 1959c. Theoretical distributions of the initial linkage block lengths intact in the gametes of a population intermated for n generations. *Genetics* 44: 839–846. [17]

Hanson, W. D. 1959d. The breakup of initial linkage blocks under selected mating systems. *Genetics* 44: 857–868. [17]

Hanson, W. D., and H. F. Robinson (eds.) 1963. *Statistical genetics and plant breeding.* Natl. Acad. Sci., Natl. Res. Council Publ. 982. Washington, D. C. [1]

Hansson, B., and L. Westerberg. 2002. On the correlation between heterozygosity and fitness in natural populations. *Mol. Ecol.* 11: 2467–2474. [12]

Hansson, B., and L. Westerberg. 2008. Heterozygosity-fitness correlations within inbreeding classes: local orgenome-wide effects? *Conser. Genetics* 9: 73–83. [12]

Hansson, B. S. Bensch, D. Hasselquist, and M. Akesson. 2001. Microsatellite diversity predicts recruitment in sibling great reed warblers. *Proc. Roy. Soc. Lond. B* 268: 1287–1291. [12]

Hao, J. J., and 4 others. 2008. Inheritance of time of flowering in upland cotton under natural conditions. *Plant Breeding* 127: 383–390. [16]

Harada, K. 1995. A quantitative analysis of modifier mutations which occur in mutation accumulation lines in *Drosophila melanogaster. Heredity* 75: 589–598. [15]

Harbison, S. T., L. J. McCoy, and T. F. C. Mackay. 2013. Genome-wide association study of sleep in *Drosophila melanogaster*. *BMC Genomics* 14: 281. [6]

Hard, J. J., W. E. Bradshaw, and C. M. Holzapfel. 1992. Epistasis and the genetic divergence of photoperiodism between populations of the pitcher-plant mosquito, *Wyeomyia smithii*. *Genetics* 131: 389–396. [11]

Hardy, G. H. 1908. Mendelian proportions in a mixed population. *Science* 28: 49–50. [1, 4]

Häring, T., B. Wichert, G. Dolf, and B. Haase. 2011. Segregation analysis of overweight body condition in an experimental cat population. *J. Heredity* 102 : S28–S31. [16]

Harris, M. A., and 56 others. 2004. The gene ontology (GO) database and informatics resource. *Nucleic Acids Res.* 32: D258–D261. [21]

Harrison, F. 2011. Getting started with meta-analysis. *Methods Ecol. Evol.* 2: 1–10. [A6]

Harrison, G. A., and J. J. T. Owen. 1964. Studies on the inheritance of human skin colour. *Ann. Hum. Genet.* 28: 27–37. [11]

Harrison, H. B., and 4 others. 2013. Relative accuracy of three common methods of parentage analysis in natural populations. *Mol. Ecol.* 22: 1158–1170. [8]

Harter, H. L. 1961. Expected values of normal order statistics. *Biometrika* 48: 151–166. [11]

Harter, H. L. 1970a. *Order statistics and their use in testing andestimation. Volume 1: Tests based on range and studentized range of samples from anormal population.* U. S. Gov. Printing Off., Washington, DC. [11]

Harter, H. L. 1970b. *Order statistics and their use in testing andestimation. Volume 2: Estimates based on order statistics of samples from variouspopulations.* U. S. Gov. Printing Off., Washington, DC. [11]

Hartl, D. L., and R. B. Campbell. 1982. Allelic multiplicity of simple Mendelian disorders. *Am. J. Hum. Genet.* 34: 866–873. [21]

Hartl, D. L., and A. G. Clark. 2006. *Principles of population genetics.* 4th Ed. Sinauer Assoc., Sunderland, MA. [12]

Hartl, L., H. Weiss, F. J. Zeller, and A. Jahoor. 1993. Use of RFLP markers for identification of alleles of the *Pm3* locus conferring powdery mildew resistance in wheat (*Triticum aestivum* L.) *Theor. Appl. Genet.* 86: 959–963. [17]

Hartl, D., and 8 others. 2008. Transcriptome and proteome analysis of early embryonic mouse brain development. *Proteomics* 8: 1257–1265. [21]

Harville, D. A. 1977. Maximum likelihood approaches to variance component estimation and to related problems. *J. Am. Stat. Assoc.* 72: 320–338. [10]

Harville, D. A. 1990. BLUP (Best Linear Unbiased Prediction) and beyond. *In* D. Gianola and K Hammond (eds.), *Advances in statistical methods for genetic improvement of livestock*, pp. 239–276. Springer-Verlag, New York, NY. [10]

Haseman, J. K., and R. C. Elston. 1972. The investigation of linkage between a quantitative trait and a marker locus. *Behav. Genet.* 2: 3–19. [19]

Hashimoto, S., and 11 others. 2021. The dominance model for heterosis explains culm length genetics in a hybrid sorghum variety. *Sci. Report.* 11: 4532. [13]

Hasselblad, V. 1966. Estimating the parameters for a mixture of normal distributions. *Technometrics* 8: 431–444. [16]

Hasson, E., and 5 others. 1992. The evolutionary history of *Drosophila buzzattii.* XXIV. Second chromosome inversions have different average effects on thorax length. *Heredity* 68: 557–563. [17]

Hasstedt, S. J. 1993. Variance components/major locus likelihood approximation for quantitative, polychotomous, and multivariate data. *Genetic Epidem.* 10: 145–158. [16]

Hasstedt, S. J., and P. E. Cartwright. 1979. PAP-pedigree analysis package. Tech. Report 13, Dept. of Medical Biophysics and Computing, Univ. of Utah, Salt Lake City, UT. [16]

Hästbacka, J., and 5 others. 1992. Linkage disequilibrium mapping in isolated founder populations: diastrophic dysplasia in Finland. *Nat. Genet.* 2: 204–211. [17, 20]

Hästbacka, J., and 14 others. 1994. The diastrophic dysplasia gene encodes a novel sulfate transporter: positional cloning by fine-structure linkage disequilibrium mapping. *Cell* 78: 1073–1087. [17]

Hastie, T., R. Tibshirani, and J. H. Friedman. 2009. *The elements of statistical learning: data mining, inference and prediction, 2nd Ed.* Springer-Verlag, New York, NY. [20]

Hastings, A. 1986. Multilocus population genetics with epistasis. II. Equilibrium properties of multilocus models: what is the unit of selection? *Genetics* 112: 157–171. [17]

Hastings, W. K. 1970. Monte Carlo sampling methods using Markov Chains and their applications. *Biometrika* 57: 97–109. [A8]

Hauser, E. R., M. Boehnke, S.-W. Gou, and N. Risch. 1998. Affected-sib-pair interval mapping and exclusion for complex genetic traits: sampling considerations. *Genet. Epidem.* 13: 117–137. [19]

Hauser, T. P., and V. Loeschcke. 1994. Inbreeding depression and mating-distance dependent offspring fitness in large and small populations of *Lychnis flosculi* (Caryophyllaceae). *J. Evol. Biol.* 7: 609–622. [12]

Havel, J. E. 1987. Predator-induced defenses: a review. *In* W. C. Kerfoot and A. Sih (eds.), *Predation: direct and indirect impacts on aquatic communities,* pp. 263–278. Univ. Press New England, Hanover, NH. [6]

Haycock, P. C., and 5 others. 2016. Best (but off-forgotten) practices: the design, analysis, and interpretation of Mendelian randomization studies. *Am. J Clin. Nutr.* 103: 965–978. [A2]

Hayeck, T. J., and 11 others. 2015. Mixed model with correction for case-control ascertainment increases association power. *Am. J. Hum. Genet.* 96: 720–730. [20]

Hayes, B., and M. E. Goddard. 2001. The distribution of the effects of genes affecting quantitative traits in livestock. *Genet. Sel. Evol.* 33: 209–229. [21]

Hayes, B. J., P. M. Visscher, and M. E. Goddard. 2009. Increased accuracy of artificial selection by using the realized relationship matrix. *Genet. Res.* 9: 47–60. [8]

Hayes, H. K., and R. J. Garber. 1919. Breeding small grains in Minnesota. Part II. Barley investigations. *Univ. Minn. Experimental station.* [13]

Hayman, B. I. 1960a. The separation of epistatic from additive and dominance variation in generation means. *Genetica* 31: 371–390. [11]

Hayman, B. I. 1960b. Maximum likelihood estimation of genetic components of variation. *Biometrics* 16: 369–381. [11]

Hazen, S. P., and 8 others. 2005. Rapid array mapping of circadian clock and developmental mutations in *Arabidopsis. Plant Physiology* 138: 990–997. [17]

He, G., H. He, and X. W. Deng. 2013. Epigenetic variations in plant hybrids and their potential roles in heterosis. *J. Genet. Genomics* 40: 205–210. [13]

He, L., and G. J. Hannon. 2004. MicroRNAs: Small RNAs with a big role in gene regulation. *Nat. Rev. Genet.* 5: 522–532. [15]

He, X. 2017. Comment on: An expanded view of complex traits: from polygenic to omnigenic. *J. Psych. Brain Sci.* 2: S2. [21]

He, X., and 6 others. 2013. Sherlock: detecting gene-disease associations by matching patterns of expression QTL and GWAS. *Am. J. Hum. Genet.* 92: 667–680. [21]

He, X.-H., and Y.-M. Zhang. 2011. A complete solution for dissecting pure main and epistatic effects of QTL in triple testcross design. *PLoS One* 6: e24575. [13]

Heard, N. A., and P. Rubin-Delanchy. 2018. Choosing between methods of combining *p* values. *Biometrika* 104: 239–246. [A6]

Heath, S. C. 1997. Markov Chain Monte Carol segregation and linkage analysis for oligogenic models. *Am. J. Hum. Genet.* 61: 748–760. [16, 18]

Heath, S. C., and 4 others. 1997. MCMC segregation and linkage analysis. *Genetic Epid.* 14: 1011–1016. [16]

Hedge, P., and 9 others. 2000. A concise guide to cDNA microarray analysis *Biotechniques* 29: 548–562. [21]

Hedgecock, D., and J. P. Davis. 2007. Heterosis for yield and crossbreeding of the Pacific oyster *Crassostrea gigas. Aquaculture* 272: S17–S29. [13]

Hedges, L. V., and I. Olkin. 1980. Vote-counting methods in research synthesis. *Psycholog. Bull.* 88: 359–369. [A6]

Hedrick, P. W. 1987a. Gametic disequilibrium measures: proceed with caution. *Genetics* 117: 331–341. [5]

Hedrick, P. W. 1987b. Genetic load and the mating system in homosporous ferns. *Evolution* 41: 1282–1289. [12]

Hedrick, P. W. 1994. Purging inbreeding depression and the probability of extinction: full-sib mating. *Heredity* 73: 363–372. [12]

Hedrick, P. W. 2012. What is the evidence for heterozygote advantage selection? *Trends Ecol. Evol.* 27: 698–704. [12]

Hedrick, P. W., and A. García-Dorado. 2016. Understanding inbreeding depression, purging, and genetic rescue. *Trends Ecol. Evol.* 31: 940–952. [12, 13]

Hedrick, P. W., and S. T. Kalinowski. 2000. Inbreeding depression in conservation biology. *Ann. Rev. Ecol Syst.* 31: 39–62. [12]

Hedrick, P. W., and O. Munoa. 1990. Linkage of viability gene to marker loci in selfing organisms. *Heredity* 64: 67–72. [12]

Hedrick, P. W., S. Jain, and L. Holden. 1978. Multilocus systems in evolution. *Evol. Biol.* 11: 101–182. [5]

Hedrick, P. W., D. W. Smith, and D. R. Stahler. 2016. Negative-assortative mating for color in wolves. *Evol.* 70: 757–766. [7]

Hedrick, P. W., J. A. Robinson, R. O. Peterson, and J. A. Vuctich. 2019. Genetics and extinction and the example of Isle Royal wolves. *Anim. Conserv.* 22: 302–309. [12]

Heid, I. M., and 64 others. 2009. Meta-analysi of the *INSIG2* association with obesity including 74,345 individuals: does heterogeneity of estimates relate to study design? *PLoS Genet.* 5: e1000964. [20]

Heijmans, B. T., and 7 others. 2008. Persistent epigenetic differences associated with prenatal exposure to famine in humans. *Proc. Natl. Acad. Sci. USA* 105: 17046–17049. [21]

Heinze, G., and M. Schemper. 2002. A solution to the problem of separation in logistic regression. *Stat. Med.* 21: 2409–2419. [20]

Hemani, G., and 24 others. 2013. Inference of the genetic architecture underlying BMI and height with the use of 20,240 sibling pairs. *Am. J. Hum. Genet.* 93: 865–875. [21]

Hemani, G., and 13 others. 2014. Another explanation for apparent epistasis: Hemani et al. reply Nature 514: E5–E6. [20, 5]

Hemani, G., J. Bowden, and G. Davey Smith. 2018. Evaluating the potential role of pleiotropy in Mendelian randomization studies. *Hum. Mol. Genet.* 27: R2. [A2]

Hemani, G., and 15 others. 2021. Phantom epistasis between unlinked loci. *Nature* 596: E1–E3. [5, 20]

Henderson, C. R. 1949. Estimates of changes in herd environment. *J. Dairy Sci.* 32: 706. [1, 10]

Henderson, C. R. 1950. Estimation of genetic parameters. *Ann. Math. Stat.* 21: 309–310. [1, 10]

Henderson, C. R. 1963. Selection index and the expected genetic advance *In* W. D. Hanson and H. F. Robinson (eds.), *Statistical genetics and plant breeding*, pp. 141–163. Natl. Acad. Sci., Natl. Res. Council Publ. No. 982, Washington, D. C. [1, 10].

Henderson, C. R. 1973. Sire evaluation and genetic trends. *In* Proceedings of the animal breeding and genetics symposium in honor of Dr. J. L. Lush, pp. 10–41. American Society of Animal Science, Champaign, IL. [10]

Henderson, C. R. 1975. Best linear unbiased estimation and prediction under a selection model. *Biometrics* 31: 423–447. [10]

Henderson, C. R. 1977. Best linear unbiased prediction of breeding values not in the model for records. *J. Dairy Sci.* 60: 783–787. [A3]

Henderson, C. R. 1984. *Applications of linear models in animal breeding*. Univ. Guelph, Guelph, Ontario, CA. [1, 10, A3]

Henderson, C. R. 1985. Equivalent linear models to reduce computations. *J. Dairy Sci.* 68: 2367–2377. [A3]

Hendricks, A. E., and 4 others. 2014. Correction for multiple testing in a gene region. *Eur. J. Hum. Genet.* 22: 414–418. [20]

Hendriksen, A. J. T. 1956. The maintenance of winter wheat variety Alba in the Netherlands. *Euphytica* 5: 1–7. [A9]

Henshall, J. M., and M. E. Goddard. 1999. Multiple-trait mapping of quantitative trait loci after selective genotyping using logistic regression. *Genetics* 151: 885–894. [18]

Henter, H. J. 2003. Inbreeding depression and haplodiploidy: experimental measures in a parasitoid and comparisons across diploid and haplodiploid insect taxa. *Evolution* 57: 1793–1803. [12]

Herández, M., J. M. Larruga, A. M. González, and V. M. Cabrera. 1993. Association among quantitative, chromosomal and enzymatic traits in a natural population of *Drosophila melanogaster*. *Genet. Sel. Evol.* 25: 229–248. [17]

Herbert, A., and 23 others. 2006. A common genetic variant is associated with adult and childhood obesity. *Science* 312: 279–283. [20]

Hereford, J., T. F. Hansen, and D. Houle. 2004. Comparing strengths of directional selection; how strong is strong? *Evolution* 58: 2133–2143. [A6]

Hermisson, J., and G. P. Wagner. 2004. The population genetic theory of hidden variation and genetic robustness. *Genetics* 168: 2271–2284. [14]

Hernán, M. A., and J. M. Robins. 2006. Instruments for causal inference: an epidemiologist's dream? *Epidem.* 17: 360–372. [A2]

Hersch, E. I., and P. C. Phillips. 2004. Power and potential bias in field studies of natural selection. *Evolution* 58: 479–485. [A6]

Hershey, A., and M. Chase. 1952. Independent functions of viral protein and nucleic acid in growth of bacteriophage. *J. Gen. Physiol.* 36: 39–56. [1]

Herzog, E., and M. Frisch. 2011. Selection strategies for marker-assisted backcrossing with high-throughput marker systems. *Theor. Appl. Genet.* 123: 251–260. [17]

Heuch, I., and F. H. F. Li. 1972. Pedig — a computer program for calculation of genotype probabilities using phenotype information. *Clin. Genet.* 3: 501–504. [16]

Heywood, J. S. 1993. Biparental inbreeding depression in the self-incompatible annual plant *Gaillardia pulchella* (Asteraceae). *Am. J. Bot.* 80: 545–550. [12]

Higgins, J. P. T., and S. G. Thompson. 2002. Quantifying heterogeneity in a meta-analysis. *Stat. Medicine* 21: 1539–1558. [20, A6]

Hildebrandt, F., and 27 others. 2009. A systematic approach tomapping recessive disease genes in individuals from outbred populations. *PLoS Genet* 5: e1000353. [12]

Hill, A. 1975. Quantitative linkage: a statistical procedure for its detection and estimation. *Ann. Hum. Genet.* 38: 439–449. [19]

Hill, J. 1964. Effects of correlated gene distributions in the analysis of diallel crosses. *Heredity* 19: 27–46. [17]

Hill, J. 1966. Recurrent backcrossing in the study of quantitative inheritance. *Heredity* 21: 85–120. [11]

Hill, M. S., P. V. Zande, and P. J. Wittkopp. 2021. Molecular and evolutionary processes generation variation in gene expression. *Nat. Rev. Genet.* 22: 203–215. [21]

Hill, W. G. 1974. Estimation of linkage disequilibrium in randomly mating populations. *Heredity* 33: 229–239. [5]

Hill, W. G. 1975. Linkage disequilibrium among multiple neutral alleles produced by mutation in finite populations. *Theor. Pop. Biol.* 8: 117–126 [5]

Hill, W. G. 1982a. Dominance and epistasis as components of heterosis. *Z. Tierzüchtg. Züchtgsbiol.* 99: 161–168. [11, 12, 13]

Hill, W. G. 1982b. Rates of change in quantitative traits from fixation of new mutations. *Proc. Natl. Acad. Sci. USA* 79: 142–145. [15]

Hill, W. G. 1982c. Predictions of response to artificial selection from new mutations. *Genet. Res.* 40: 255–278. [15]

Hill, W G. 1993a. Variation in genetic identity within kinships. *Heredity* 71: 652–653. [8]

Hill, W G. 1993b. Variation in genetic composition in backcrossing programs. *J. Hered.* 84: 212–213. [8, 17]

Hill, W. G. 1998a. Selection with recurrent backcrossing to develop congenic lines for quantitative trait loci analysis. *Genetics* 148: 1341–1352. [17]

Hill, W. G. 1998b. A note on the theory of artificial selection in finite populations and application to QTL detection by bulk segregant analysis. *Genet Res.* 72: 55–58. [18]

Hill, W. G. 2010. Understanding and using quantitative genetic variation. *Phil. Trans. Roy. Soc. Lond. B* 365: 73–85. [21]

Hill, W. G., and P. J. Avery. 1978. On estimating number of genes by genotype assay. *Heredity* 40: 397–403. [11]

Hill, W. G., and P. D. Keightley. 1988. Interrelations of mutation, population size, artificial and natural selection. *In* B. S. Weir, E. J. Eisen, M. M. Goodman, and G. Namkoong (eds.), *Proceedings of the second international conference on quantitative genetics*, pp. 57–70. Sinauer Assoc., Sunderland, MA. [15]

Hill, W. G., and S. Knott. 1990. Identification of genes with large effects. *In* D. Gianola and K. Hammond (eds.), *Advances in statistical methods for genetic improvement of livestock*, pp. 477–494. Springer-Verlag, Berlin, DE. [16]

Hill, W. G., and T. F. C. Mackay (eds.) 1989. *Evolution and animal breeding*. CAB International, Wallingford, UK. [1]

Hill, W. G., and A. Mäki-Tanila. 2015. Expected influence of linkage disequilibrium on genetic variance caused by dominance and epistasis on quantitative traits. *J. Anim. Breed. Genet.* 12: 176–186. [5]

Hill, W. G., and A. Robertson. 1966. The effect of linkage on limits to artificial selection. *Genet. Res.* 8: 269–294. [5, 17]

Hill, W. G., and A. Robertson. 1968. Linkage disequilibrium in finite populations. *Theor. Appl. Genet.* 38: 226–231. [5]

Hill, W. G., and B. S. Weir. 1994. Maximum-likelihood estimation of gene location by linkage disequilibrium. *Am. J. Hum. Genet.* 54: 705–714. [17]

Hill, W. G., and B. S. Weir. 2011. Variation in actual relationship as a consequence of Mendelian sampling and linkage. *Genet. Res.* 93: 47–64. [7, 8, 12]

Hill, W. G., and B. S. Weir. 2012. Variation in actual relationship among descendants of inbred individuals. *Genet. Res.* 94: 267–274. [8, 12]

Hill, W. G., M. E. Goddard, and P. M. Visscher. 2008. Data and theory point to mainly additive genetic variance for complex traits. *PLoS Genet.* 4: e1000008. [4, 5, 20, 21]

Hillel, J., and 6 others. 1990. DNA fingerprints applied to gene introgression in breeding programs. *Genetics* 124: 783–789. [17]

Hillel, J., A. M. V. Gibbins, R. J. Etches, and D. M. Shaver. 1993. Strategies for the rapid introgression of a specific gene modification into a commercial poultry flock from a single carrier. *Poultry Sci.* 72: 1197–1211. [17]

Hillenmeyer, M. E., and 13 others. 2008. The chemical genomic portrait of yeast: uncovering a phenotype for all genes. *Science* 320: 362–365. [15]

Hingorani, A., and S. Humphries. 2005. Nature's randomised traits. *Lancet* 336: 1906–1908. [A2]

Hinkelman, K., and O. Kempthorne. 2006. *Design and analysis of experiments. Volumes 1 and 2.* Wiley, New York, NY. [A9]

Hinze, K., and 4 others. 1991. Restriction fragment length polymorphism-mediated targeting of the *ml-o* resistance locus in barley (*Hordeum vulgare*). *Proc. Natl. Acad. Sci. USA* 88: 3691–3695. [17]

Hivert, V., and 7 others. 2021. Estimation of non-additive genetic variance in human complex traits from a large sample of unrelated individuals. *Am. J. Hum. Genet.* 108: 786–798. [21]

Hoaglin, D. C., and R. E. Welsch. 1978. The Hat Matrix in regression and ANOVA. *Am. Stat.* 32: 17–22. [10]

Hochberg, Y. 1988. A sharper Bonferroni procedure for multiple tests of significance. *Biometrika* 75: 800–802. [20, A6]

Hodge, S. E. 1984. The information contained in multiple sibling-pairs. *Genet. Epidem.* 1: 109–122. [19]

Hodge, S. E. 1988. Conditioning on subsets of the data: applications to ascertainment and other genetic problems. *Am. J. Hum. Genet.* 43: 364–373. [16]

Hodgkin, J. 1993. Fluxes, doses and poisons: molecular perspectives on dominance. *Trends Genetics* 9: 1–2. [4]

Hodgkin, J. 1998. Seven types of pleiotropy. *Int. J. Dev. Biol.* 42: 501–505. [A2]

Hoeck, P. E. A, and 4 others. 2015. Effects of inbreeding and parental incubation on captive breeding success in Hawaiian crows. *Biol. Conserv.* 184: 357–364. [12]

Hoefler, R., and 10 others. 2020. Do spatial designs outperform classic experimental designs? *J. Agric. Biol. Environ. Stat.* 25: 523–552. [A9]

Hoerl, A. E., R. W. Kennard, and K. F. Baldwin. 1975. Ridge regression: some simulations. *Commun. Stat. A* 4: 105–123. [20]

Hoeschele, I. 1988. Statistical techniques for detection of major genes in animal breeding data. *Theor. Appl. Genet.* 76: 311–319. [10, 16]

Hoeschele, I., and T. R. Meinert. 1990. Association of genetic defects with yield and type traits: the weaver locus effect on yield. *J. Dairy Sci.* 73: 2503–2515. [17]

Hoeschele, I., and P. M. VanRaden. 1993a. Bayesian analysis of linkage between genetic markers and quantitative trait loci. I. Prior knowledge. *Theor. Appl. Genet.* 85: 953–960. [18, 19]

Hoeschele, I., and P. M. VanRaden. 1993b. Bayesian analysis of linkage between genetic markers and quantitative trait loci. II. Combining prior knowledge with experimental evidence. *Theor. Appl. Genet.* 85: 946–952. [18, 19]

Hoeschele, I., and 4 others. 1997. Advances in statistical methods to map quantitative trait loci in outbred populations. *Genetics* 147: 1445–1457. [19]

Hofer, A., and B. W. Kennedy. 1993. Genetic evaluation for a quantitative trait controlled by polygenes and a major locus with genotypes not or only partly known. *Genet. Sel. Evol.* 25: 537–555. [10]

Hoffman, G. E. 2013. Correcting for population structure and kinship using the linear mixed model: theory and extensions. *PLoS ONE* 8: e75707. [20]

Hoffman, J. I., and 7 others. 2014 . High-throughput sequencing reveals inbreeding depression in a natural population. *Proc. Natl. Acad. Sci. USA* 111: 3775–3780. [12]

Hoffmann, A. A., and P. A. Parsons. 1991. *Evolutionary genetics and environmental stress.* Oxford Univ. Press, New York, NY. [6, 7, 12]

Hoffmann, A. A., and M. Turelli. 1988. Unidirectional incompatibility in *Drosophila simulans:* inheritance, geographic variation and fitness effects. *Genetics* 119: 435–444. [6]

Hoffmann, A. A., and R. E. Woods. 2003. Associating environmental stress with developmental stability: problems and patterns. *In* M. Polak (ed.), *Developmental instability. Causes and consequences*, pp. 387–401. Oxford Univ. Press, New York, NY. [6]

Hoffmann, T. J., N. J. Martin, and J. S. Witte. 2010. Comprehensive approach to analysis rare genetic variants. *PLoS One* 5: e13584. [20]

Hofmeister, R. J., D. M. Riberio, S. Rubinacci, and O. Delaneau. 2023. Accurate rare variants phasing of whole-genome and whole-exome sequencing data in the UK Biobank. *Nat. Genet.* 55: 1243–1249. [20, 21]

Hogben, L. 1974. The origins of theoretical population genetics. *Brit. J. Hist. Sci.* 7: 176–179. [1]

Hoggart, C. J., and 6 others. 2003. Control of confounding of genetic associations in stratified populations. *Am. J. Hum. Genet.* 72: 1492–1504. [20]

Hoggart, C. J., and 4 others. 2004. Design and analysis of admixture mapping studies. *Am. J. Hum. Genet.* 74: 965–978. [17]

Hoi-Sen, Y. 1972. Is subline differentiation a continuing

process in inbred strains of mice? *Genet. Res.* 19: 53–59. [15]

Holland, D., and 9 others. 2020. Beyond SNP heritability: polygenicity and discoverability of phenotypes estimated with a Gaussian mixture model. *PLoS Genet.* 16: e1008612. [21]

Holland, J. 1975. *Adaptation in natural and artificial systems.* Univ. Michigan Press, Ann Arbor, MI. [18]

Holland, J. B., H. S. Moser, L. S. O'Donoughue, and M. Lee. 1997. QTLs and epistasis associated with vernalization response in Oat. *Crop Sci.* 37: 1306–1316. [18]

Hollingsworth, M. J., and J. Maynard Smith. 1955. The effects of inbreeding on rate of development and on fertility in *Drosophila subobscura*. *J. Genetics* 53: 295–314. [12]

Hollocher, H., A. R. Templeton, R. DeSalle, and J. S. Johnston. 1992. The molecular through ecological genetics of *abnormal abdomen* in *Drosophila mercatorum*. III. Components of genetic variation in a natural population of *Drosophila mercatorum*. *Genetics* 130: 355–366. [15]

Holm, S. 1979. A simple sequential rejective multiple test procedure. *Scand. J. Statistics* 6: 65–70. [A6]

Holmans, P. 1998. Affected sib-pair methods for detecting linkage from dichotomous traits: review of the methodology. *Hum. Biol.* 70: 1025–1040. [19]

Holmans, P. 2010. Statistical methods for pathway analysis of genome-wide data for association with complex genetic traits. *Adv. Genet.* 72: 141–179. [21]

Holmo, N. F., and 6 others. 2018. Complex segregation analysis of facial melasma in Brazil: evidence for a genetic susceptibility with a dominant pattern of segregation. *Archives Dermat. Res.* 310: 827–831. [16]

Holsinger, K. E. 1988. Inbreeding depression doesn't matter: the genetic basis of mating-system evolution. *Evolution* 42: 1235–1244. [12]

Holtsford, T. P., and N. C. Ellstrand. 1990. Inbreeding effects in *Clarkia tembloriensis* (Ongraceae) populations with different natural outcrossing rates. *Evolution* 44: 2031–2046. [12]

Hommel, G. 1988. A stagewise rejective multiple test procedure on a modified Bonferroni test. *Biometrika* 75: 383–386. [A6]

Hommel, G. 1989. A comparison of two modified Bonferroni procedures. *Biometrika* 76: 624-625. [A6]

Hong, M.-G., Y. Pawitan, P. K. E. Magnusson, and J. A. Prince. 2009. Strategies and issues in the detection of pathway enrichment in genome-wide association studies. *Hum. Genet.* 126: 289–301. [21]

Hopkins, C. G. 1899. Improvement in the chemical composition of the corn kernel. *Ill. Agric. Exp. Stn. Bull.* 55: 205–240. [18]

Hopper, J. L., and J. D. Mathews. 1982. Extensions to multivariate normal models for pedigree analysis. *Ann. Hum. Genet.* 46: 373–383. [A4]

Hopper, J. L., M. C. Hannah, and J. D. Mathews. 1984. Genetic analysis workshop II: Pedigree analysis of a binary trait without assuming an underlying liability. *Genet. Epidem.* 1: 183–188. [19]

Horel, A. E., and R. Kennard. 1970. Ridge regression: biased estimation for nonorthogonal problems. *Technometrics* 12: 55–67. [20]

Hormozdiari, F., E. Kostem, E. Y. Kang, B. Pasaniuc, and E. Eskin. 2014. Identifying causal variants at loci with multiple signals of association. *Genetics* 198: 497–508. [21]

Hormozdiari, F., G. Kichaev, W.-Y. Yang, B. Pasaniuc, and E. Eskin. 2015. Identification of causal genes for complex traits. *Bioinformatics* 31: i206–i213. [21]

Hormozdiari, F., and 9 others. 2016. Colocalizaton of GWAS and eQTL signals detects target genes. *Am. J. Hum. Genet.* 99: 1245–1260. [21]

Horvath, S., X. Xu, and N. M. Laird. 2001. The family based association test method: strategies for studying general genotype-phenotype associations. *Eur. J. Hum. Genet.* 9: 301–306. [17]

Hospital, F., C. Chevalet, and P. Mulsant. 1992. Using markers in gene introgression breeding programs. *Genetics* 132: 1199–1210. [17]

Hotelling, H. 1931. The generalization of student's ratio. *Ann. Math. Sat.* 2: 360–378. [20]

Hothorn, L. A., O. Libiger, and D. Gerhard. 2012. Model-specific tests on variance heterogeneity for detection of potentially interacting genetic loci. *BMC Genet.* 13: 59. [13]

Hou, K., and 4 others. 2021. On powerful GWAS in admixed populations. *Nat. Genet.* 53: 1631–1633. [20]

Houle, D. 1989a. Allozyme-associated heterosis in *Drosophila melanogaster*. *Genetics* 123: 789–801. [12]

Houle, D. 1989b. The maintenance of polygenic variation in finite populations. *Evolution* 43: 1767–1780. [15]

Houle, D. 1992. Comparing evolvability and variability of quantitative traits. *Genetics* 130: 195–204. [7]

Houle, D. 1994. Adaptive distance and the genetic basis of heterosis. *Evolution* 48: 1410–1417. [12]

Houle, D., and E. J. Márquez. 2015. Linkage disequilibrium and inversion-typing of the *Drosophila melanogaster* genome reference panel. *Genes Genomes Genet.* 5: 1695–1699. [5]

Houle, D., and 6 others. 1994. The effects of spontaneous mutation on quantitative traits. I. Variance and covariance of life history traits. *Genetics* 138: 773–785. [15]

Houle, D., B. Morikawa, and M. Lynch. 1996. Comparing mutational heritabilities. *Genetics* 143: 1467–1483. [15]

Houle, D., D. R. Govindaraju, and S. Omholt. 2010. Phenomics: the next challenge. *Nat. Rev. Genet.* 11: 855–866. [18]

Houwen, R. H. J., and 6 others. 1994. Genome screening by searching for shared segments: mapping a gene for benign recurrent intrahepatic cholestasis. *Nat. Genet.* 8: 380–386. [12]

Howard, J. T., M. Haile-Mariam, J. E. Pryce, and C. Maltecca. 2015. Investigation of regions impacting inbreeding depression and their association with the additive genetic effect for United States and Australia Jersey dairy cattle. *BMC Genomics* 16: 1–13. [12]

Howard, J. T., J. E. Pryce, C. Baes, and C. Maltecca. 2017. Inbreeding in the genomics era: inbreeding, inbreeding depression, and management of genomic variability. *J. Dairy. Sci.* 100: 6009–6024. [12]

Howe, L. J., and 6 others. 2019. Genetic evidence for assortative mating on alcohol consumption in the UK Biobank. *Nat. Comm.* 10: 5039. [A2]

Howe, L. J., et al. 2022. Within-sibship genome-wide association analyses decrease bias in estimates of direct genetic effects. *Nat. Genet.* 54: 581–592. [20]

Howrigan, D. P., and M. A. Simonson, and M. C Keller. 2011. Detecting autozygosity through runs of homozygosity: a comparison of three autozygosity detection algorithms. *BMD Geomics* 12: 460. [8]

Howrigan, D. P., and 45 others. 2016. Genome-wide autozygosity is associated with lower general cognitive ability. *Mol. Psych.* 21: 837–843. [12]

Hu, J. X., C. E. Thomas, and S. Brunak. 2016. Network biology concepts in complex disease comorbidities. *Nat. Rev. Genet.* 17: 615–629. [21]

Hu, S., and 7 others. 1995. Linkage between sexual orientation and chromosome Xq28 in males but not in females. *Nat. Genetics* 11: 248–256. [19]

Hu, T., and 5 others. 2011. Characterizing genetic interactions in human disease association studies using statistical epistasis networks. *BMC Bioinform.* 12: 364. [21]

Hu, X., G. L. Gadbury, Q. Xiang, and D. B. Allison. 2011. Illustrations on using the distribution of a p-value in high dimensional data analysis. *Adv. Appl. Stat. Sci.* 1: 191–213. [A6]

Hu, Y., and 20 others. 2019. A statistical framework for cross-tissue transcriptome-wide association analysis. *Nat. Genet.* 51: 568–576. [21]

Hu, Z., and S. Xu. 2008. A simple method for calculating the statistical power for detecting a QTL located in a marker interval. *Heredity* 101: 48–52. [18]

Hu, Z., and 4 others. 1995. A correlation method for detecting and estimating linkage between a marker locus and a quantitative trait locus using inbred lines. *Theor. Appl. Genet.* 90: 1074–1078. [18]

Hua, J. P., and 5 others. 2002. Genetic dissection of an elite rice hybrid revealed that heterozygotes are not always advantageous for performance. *Genetics* 162: 1885–1895. [13, 18]

Hua, J., and 6 others. 2003. Single-locus heterotic effects and dominance by dominance interactions can adequately explain the genetic basis of heterosis in an elite rice hybrid. *Proc. Natl. Acad. Sci. USA* 100: 2574–2579. [13]

Huang, A., S. Xu, and X. Cai. 2015. Empirical Bayesian elastic net for multiple quantitative trait locus mapping. *Heredity* 11: 107–115. [18]

Huang, B. E., and 8 others. 2015. MAGIC populations in crops: current status and future prospects. *Theor. Appl. Genet.* 128: 999–1017. [18]

Huang, D. W., B. T. Sherman, and R. A. Lempicki. 2008. Bioinformatics enrichment tools: paths toward the comprehensive functional analysis of large gene lists. *Nuc. Acids Res.* 37: 1–13. [21]

Huang, D., and 6 others. 2013. Fine mapping and characterization of BPH27, a brown planthopper resistance gene from wild rice (*Oryza rufipogon* Griff.). *Theor. Appl. Genet* 126: 219–229. [17]

Huang, H., and 45 others. 2017. Fine-mapping inflammatory bowel disease loci to single-variant resolution. *Nature* 547: 173–178. [21]

Huang, J., A. D. Johnson, A.D. and C. J. O' Donnell. 2011. PRIMe: a method for characterization and evaluation of pleiotropic regions from multiple genome- wide association studies. *Bioinformatics* 27: 1201–1206. [20]

Huang, J., and 16 others. 2014. Improved imputation of low-frequency and rare variants using the UK10K haplotype reference panel. *Nat. Comm.* 6: 8811. [20, 21]

Huang, K., K. Ritland, S. Guo, M. Shattuck, and B. Li. 2013. A pairwise relatedness estimator for polyploids. *Mol. Ecol. Res.* 14: 734–744. [8]

Huang, Q. Q., S. C. Ritchie, M. Brozynska, and M. Inouye. 2018. Power, false discovery rate and winner's curse in eQTL studies. *Nucl. Acids Res.* 46: e133. [18, 21]

Huang, W., and T. F. C. Mackay. 2016. The genetic architecture of quantitative trait cannot be inferred from variance component analysis. *PLoS Genet.* 12: e1006421. [5]

Huang, W., and 24 others. 2012. Epistasis dominates the genetic architecture of *Drosophila* quantitative traits. *Proc. Natl. Acad. Sci.* 109: 15553–15559. [5]

Huang, W., and 7 others. 2015. Genetic basis of transcriptome diversity in *Drosophila melanogaster*. *Proc. Natl. Acad. Sci. USA* 112: E6010–E6019. [6]

Huang, X., and 6 others. 2011. Analysis of natural allelic variation in *Arabidopsis* using a multiparent recombinant inbred line population. *Proc. Natl. Acad. Sci. USA* 108: 4488–4493. [18]

Huang, X., and 23 others. 2015. Genomic analysis of hybrid rice varieties reveals numerous superior alleles that contribute to heterosis. *Nat. Comm.* 6: 6258. [13]

Huang, X., and 18 others. 2016. Genomic architecture of heterosis for yield traits in rice. *Nature* 537: 629–633. [13]

Hubbard, L., P. McSteen, J. Doebley, and S. Hake. 2002. Expression patterns and mutant phenotype of *teosinte branched1* correlate with growth suppression in maize and teosinte. *Genetics* 162:1927–1935. [5]

Hubby, J. L., and R. C. Lewontin. 1966. A molecular approach to the study of genic heterozygosity in natural populations. I. The number of alleles at different loci in *Drosophila pseudoobscura*. *Genetics* 54: 577–594. [8, 12]

Huber, C. D., A. Durvauls, A. M. Hancock, and K. E. Lohmueller. 2018. Gene expression drives the evolution of dominance. *Nat. Comm.* 9: 2750. [4]

Hubner, N., and 21 others. 2005. Integrated transcriptional profiling and linkage analysis for identification of genes underlying disease. *Nat. Genet.* 37: 243–253. [21]

Huey, R. B., and A. E. Dunham. 1987. Repeatability of locomotor performance in natural populations of the lizard *Sceloporus merriami*. *Evolution* 41: 1116–1119. [6]

Huff, C. D., and 11 others. 2011. Maximum-likelihood estimation of recent shared ancestry (ERSA). *Genom. Res.* 21: 768–774. [8]

Hughes, K. A. 1995. The inbreeding decline and average dominance of genes affecting male life-history characters in *Drosophila melanogaster*. *Genet. Res.* 65: 41–52. [12]

Huisman, J. 2017. Pedigree reconstruction from SNP data: parentage assignment, sibship clusteringand beyond. *Mol. Ecol. Resources* 17: 1009–1024. [8]

Huisman, J., and 4 others. 2016. Inbreeding depression across the lifespan in a wild mammal population. *Proc. Natl. Acad. Sci. USA* 113: 3585–3590. [8,12]

Hukku, A., and 5 others. 2020. Probabilistic colocalization of genetic variants from complex and molecular traits promise and limitations. *Am. J. Hum. Genet.* 108: 25–34. [21]

Hull, F. H. 1946. Overdominance and corn breeding where hybrid seed is not feasible. *J. Am. Soc. Agron.* 38: 1100–1103. [12]

Hunt, G. J., R. E. Page, Jr., M. K. Fondrk, and C. J. Dullum. 1995. Major quantitative trait loci affecting honey bee foraging behavior. *Genetics* 141: 1537–1545. [18]

Hunt, G. J., and 12 others. 2007. Behavioral genomics of honeybee foraging and next defense. *Naturwissenschaften* 94: 247–267. [18]

Hunter, J. E., and F. L. Schmidt. 2005. *Methods of meta-analysis: Correcting error and bias in research findings*. Sage, Thousand Oaks, CA. [A6]

Hurst, D. L. and J. P. Randerson. 2000. Dosage, deletions and dominance: simple models of the evolution of gene expression. *J. Theor. Biol.* 205: 641–647. [4]

Hurst, L. D., and A. Pomiankowski. 1991. Causes of sex ratio bias may account for unisexual sterility in hybrids: a new explanation of Haldane's rule and related phenomena. *Genetics* 128: 841–858. [17]

Husband, B. C., and D. W. Schemske. 1996. Evolution of the magnitude and timing of inbreeding depression in plants. *Evolution* 50: 54–70. [12]

Hutchings, J. A., and M. M. Ferguson. 1992. The independence of enzyme heterozygosity and life-history traits in natural populations of *Salvelinus fontinalis* (brook trout). *Evolution* 69: 496–502. [6, 12]

Hutchinson, A., J. Asimit, and C. Wallace. 2020a. Fine-mapping genetic associations. *Hum. Mol. Genet.* 19: R82–R89. [21]

Hutchinson, A., H. Watson, and C. Wallace. 2020b. Improving the coverage of credible sets in Bayesian genetic fine-mapping. *PLoS Comput. Biol.* 16: e1007829. [21]

Hutchinson, G. E. 1967. *A treatise on limnology. Vol. II. Introduction to lake biology and the limnoplankton*. John Wiley & Sons, New York, NY. [6]

Hutchison, D. W., and J. M. Cheverud. 1995. Fluctuating asymmetry in tamarin *(Saguinus)* cranial morphology: intra- and interspecific comparisons between taxa with varying levels of genetic heterozygosity. *J. Heredity* 86: 280–288. [6]

Huxley, J. S. 1924. Constant differential growth-ratios and their significance. *Nature* 114: 895–896. [14]

Huxley, J. S. 1932. *Problems of relative growth*. Methuen, London, UK. [14]

Huxley, J. S., and G. Teissier. 1936. Terminology of relative growth. *Nature* 137: 780–781. [14]

Hyde, J. S. 1973. Genetic homeostasis and behavior: analysis, data, and theory. *Behav. Genet.* 3: 233–245. [6]

Hyer, R. N., and 12 others. 1991. High resolution linkage mapping for susceptibility genes in human polgenic diseases: insulin-dependent diabetes mellitus and chromosome 11q. *Am. J. Hum. Genet.* 48: 243–257. [19]

Hyman, S. E. 2020. Wringing biological insight from polygenic signals. *Biol. Psychiatry* 89: 8–10. [21]

Hyne, V., and M. J. Kearsey. 1995. QTL analysis: further uses of 'marker regression'. *Theor. Appl. Genet.* 91: 471–476. [18]

Iafrate, A. J., and 7 others. 2004. Detection of large-scale variation in the human genome. *Nat. Genet.* 36: 949–951. [15]

Igolkina, A. A., and 4 others. 2020. Multi-trait multi-locus SEM model discriminates SNPs of different effects. *BMC Genomics* 21: 490. [20]

Iles, M. M. 2008. What can genome-wide associations studies tell up about the genetics of common disease? *PLoS Genet.* 4: e33. [21]

Im, K. S. 2000. Robustifying Glejser test of heteroskedasticity. *J. Econometrics* 97: 179–188. [10]

Imumorin, I. G., and 7 others. 2011. Genome scan for parent-of-origin QTL effects on bovine growth and carcass traits. *Front. Genet.* 2: 441. [19]

Indrasamy, H., R. E. Woods, J. A. McKenzie, and P. Batterham. 2000. Fluctuating asymmetry for specific bristle characters in *Notch* mutants of *Drosophila melanogaster*. *Genetica* 109: 151–159. [14]

Ingolia, N. T., S. Ghaemmaghami, J. R. S. Newman, and J. S. Weissman. 2009. Genome-wide analysis *in vivo* of translation with nucleotide resolution using ribosome profiling. *Science* 324: 28–223. [21]

International HapMap Consortium. 2005. A haplotype map of the human genome. *Nature* 437: 1299–1320. [5]

Ioannidis, J. P. A., and T. A. Trikalinos. 2005. Early extreme contradictory estimates may appear in published research: the Proteus phenomenon in molecular genetics research and randomized trials. *J. Clin. Epidem.* 58: 543–549. [20, A6]

Ioannidis, J. P. A., and T. A. Trikalinos. 2007. The appropriateness of asymmetry test for publication bias in meta-analysises: a large survey. *CMAJ* 176: 1091–1096. [A6]

Ioannidis, J. P. A., N. A. Patsopoulos, and E. Evangelou. 2007. Heterogeneity in meta-analyses of genome-wide association investigations. *PLoS One* 9: e841. [20]

Ionita-Laza, I., and 4 others. 2009. Genetic association analysis of copy-number variation (CNV) in human disease pathogenesis. *Genomics* 93: 22–26. [20]

Ionita-Laza, I., and 4 others. 2013. Sequence kernel association tests for the combined effect of rare and common variants. *Am. J. Hum. Genet.* 92: 841–853. [20]

Issler, O., and A. Chen. 2015. Determining the role of microRNAs in psychiatric disorders. *Nat Rev. Neurosci.* 16: 201–212. [21]

Ivanov, D. K., and 6 others. 2015. Longevity GWAS using the *Drosophila* genetic reference panel. *J. Gerontolgy A* 70: 1470–1478. [17]

Jackson, J. F. 1973. A search for the population asymmetry parameter. *Syst. Zool.* 22: 166–170. [6]

Jacob, H. J., and 8 others. 1991. Genetic mapping of a gene causing hypertension in the stroke-prone spontaneously hypertensive rat. *Cell* 67: 213–224. [17]

Jacquard, A. 1974. *The genetic structure of populations*. Springer-Verlag, New York, NY. [7, 8]

Jacquard, A. 1983. Heritability: one word, three concepts. *Biometrics* 39: 465–477. [7]

Jaenike, J. 1987. Genetics of oviposition-site preference in *Drosophila tripunctata*. *Heredity* 59: 363–369. [11]

Jaenike, J. 2001. Sex chromosome meiotic drive. *Ann Rev. Ecol. System.* 32: 25–49. [17]

Jaenike, J. 2008. X chromosome drive. *Current Bio.* 18: R508–R511. [17]

Jakob, E. M., S. D. Marshall, and G. W. Uetz. 1996. Estimating fitness: A comparison of body condition indices. *Oikos* 77: 61–67. [14]

Jakubczak, J. L., W. D. Burke, and T. H. Eickbush. 1991. Retroposable elements R1 and R2 interrupt the rRNA genes of most insects. *Proc. Natl. Acad. Sci. USA* 88: 3295–3299. [15]

Jakubec, V. 1977. Productivity of crosses based on prolific breeds of sheep. *Livestock Prod. Sci.* 4: 379–392. [13]

Jakubosky, D., and 13 others. 2020. Properties of structural variants and short tandem repeats associated with gene expression and complex traits. *Nat. Comm.* 11: 2927. [21]

James, J. W. 1971. Frequency in relatives for an all-or-none trait. *Ann. Hum. Genet.* 35: 47–49. [19]

Jamieson, I. G., and F. W. Allendorf. 2012. How does the 50/500 rule apply to MVPs? *Trends Ecol. Evol.* 27: 578–584. [12]

Jana, S. 1971. Simulations of quantitative characters of qualitatively acting genes. I. Nonallelic gene interactions involving two or three loci. *Theor. Appl. Genet.* 41: 216–226. [5, 18]

Jana, S. 1972. Simulations of quantitative characters of qualitatively acting genes. II. Orthogonal subdivision of hereditary variances in two-locus genetics systems. *Theor. Appl. Genet.* 42: 119–124. [5]

Janick, J. 1999. Exploitation of heterosis: uniformity and stability. *In* J. G. Coors and S. Pandey (eds.), *The genetics and exploitation of heterosis in crops*, pp. 335–342. ASA-CSSA-SSSA Societies, Madison, WI. [13]

Jannink, J.-L., and R. Jansen. 2001. Mapping epistatic quantitative trait loci with one-dimensional genome searches. *Genetics* 157: 445–454. [18]

Jannink, J.-L., and X.-L. Wu. 2003. Estimating allelic number and identity in state of QTLs in interconnected families. *Genet. Res.* 81: 133–144. [18]

Jansen, R. C. 1992. A general mixture model for mapping quantitative trait loci by using molecular markers. *Theor. Appl. Genet.* 85: 252–260. [18, A4]

Jansen, R. C. 1993a. Maximum likelihood in a generalized linear finite mixture model by using the EM algorithm. *Biometrics* 49: 227–231. [18, A4]

Jansen, R. C. 1993b. Interval mapping of multiple quantitative trait loci. *Genetics* 135: 205–211. [18]

Jansen, R. C. 1994a. Mapping of quantitative trait loci by using genetic markers: an overview of biometrical models. *In* J. W. van Ooijen and J. Jansen (eds.), *Biometrics in plant breeding: applications of molecular markers*, pp. 116–124. CPRO-DLO, NL. [18, A4]

Jansen, R. C. 1994b. Controlling the type I and type II errors in mapping quantitative trait loci. *Genetics* 138: 871–881. [18]

Jansen, R. C. 1996. A general Monte Carlo method for mapping multiple quantitative trait loci. *Genetics* 142: 305–311. [18]

Jansen, R. C. 2003. Studying complex biological systems using multifactorial perturbation. *Nat. Rev. Genet.* 4: 145–151. [21]

Jansen, R. C., and J.-P. Nap. 2001. Genetical genomics: the added value from segregation. *Trend Genet.* 17: 388–391. [1, 21]

Jansen, R. C. and P. Stam. 1994. High resolution of quantitative trait into multiple loci via interval mapping. *Genetics* 136: 1447–1455. [18, A4]

Jansen, R. C., and 4 others. 1995. Genotype-by-environment interaction in genetic mapping of multiple quantitative trait loci. *Theor. Appl. Genet.* 91: 33–37. [18]

Janss, L. L. G., and J. H. J. Van Der Werf. 1992. Identification of a major gene in F_1 and F_2 data when alleles are assumed fixed in parental lines. *Genet. Sel. Evol.* 24: 511–526. [16]

Janss, L. L. G., R. Thompson, and J. A. M. Van Arendonk. 1995. Applications of Gibbs sampling for inference in a mixed model gene-polygenic inheritance model in animal populations. *Theor. Appl. Genet.* 91: 1137–1147. [16]

Janss, L. L. G., J. A. M. Van Arendonk, and E. W. Brascamp. 1997. Bayesian statistical analyses for presence of single genes affecting meat quality traits in a crossed pig population. *Genetics* 145: 395–408. [16]

Jarvik, G. P. 1998. Complex segregation analyses: uses and limitations. *Am. J. Hum. Genet.* 63: 942–946. [16]

Jasmin, J.–N., and T. Lenormand. 2016. Accelerating mutational load is not due to synergistic epistasis or mutator alleles in mutation accumulation lines of yeast. *Genetics* 202: 751–763. [15]

Jayakar, S. D. 1970. On the detection and estimation of linkage between a locus influencing a quantitative character and a marker locus. *Biometrics* 26: 451–464. [19]

Jeffreys, H. S. 1961. *Theory of probability*, 3rd ed. Oxford University Press, Oxford, UK. [A7]

Jelenkovi, A., and 98 others. 2016. Genetic and environmental influences on adult human height across birth cohorts from 1886 to 1994. *eLIFE* 5: e20320. [7]

Jellum, M. D., C. M. Brown, and R. D. Seif. 1963. Hill and row plot comparison for yield in oats. *Crop. Sci.* 3: 194–196. [A9]

Jenkin, F. 1867. Origins of species. *North British Review* 46: 277–318. [1]

Jensen, H., M. Szulkin, and J. Slate. 2014. Molecular quantitative genetics. *In* A. Charmantier, D. Garant, and L. E. B. Kruuk (eds.), *Quantitative genetics in the wild*, pp. 209–227. Oxford University Press, Oxford, UK. [8]

Jensen, J. 1989. Estimation of recombination parameters between a quantitative trait locus (QTL) and two marker gene loci. *Theor. Appl. Genet.* 78: 613–618. [18]

Jennrich, R. I., and P. F. Sampson. 1976. Newton-Raphson and related algorithms for maximum likelihood variance component estimation. *Technometrics* 18: 11–17. [A4]

Jeong, H., and 4 others. 2000. The large-scale organization of metabolic networks. *Nature* 407: 651–654. [A2]

Jeuken, M. J. W., and P. Lindhout. 2004. The development of lettuce backcross inbred lines (BILs) for exploitation of the *Lactuca saligna* (wild lettuce) germplasm. *Theor. Appl. Genet.* 109: 394–401. [17]

Ji, W., and 9 others. 2008. Rare independent mutations in renal salt handling contributed to blood pressure variation. *Nat. Genet.* 40: 592–599. [15]

Jia, Z., and S. Xu. 2007. Mapping quantitative trait loci for expression abundance. *Genetics* 176: 611–623. [21]

Jiang, C., and Z.-B. Zeng. 1995. Multiple trait analysis of genetic mapping for quantitative trait loci. *Genetics* 140: 1111–1127. [18]

Jiang, C., X. Pan, and M. Gu. 1994. The use of mixture models to detect effects of major genes on quantitative characters in a plant breeding experiment. *Genetics* 136: 383–394. [16]

Jiang, G., and 6 others. 2013. New aQTL SNPs for the CYP2D6 identified by a novel mediation analysis of genome-wide SNP arrays, gene expression arrays, and CYP2D6 activity. *BioMed Res. Inter.* Article 493019. [21]

Jiang, L., M. Ge, H. Zhao, and T. Zhang. 2015. Analysis of heterosis and quantitative trait loci for kernel shape related traits using triple testcross population in maize. *PLoS One* 10: e0124779. [13]

Jiang, L, and 6 others. 2019. A resource-efficient tool for mixed model association analysis of large-scale data.

Nat. Genet. 51: 1749–1755. [20]

Jiang, L., Z. Zheng, H. Fang, and J. Yang. 2021. A generalized linear mixed model asscoiation tool for biobank-scale data. *Nat. Genet.* 53: 1616–1621. [20]

Jiang, Y., and D. I. Bolnick, and M. Kirkpatrick. 2013. Assortative mating in animals. *Am. Nat.* 181: E125–E138. [7]

Jiang, Y., R. H. Schmidt, Y. Zhao, and J. C. Reif. 2017. A quantitative genetic framework highlights the role of epistatic effects for grain-yield heterosis in bread wheat. *Nat. Genet.* 49: 1741–1746. [13]

Jiménez, J. A., K. A. Hughes, G. Alaks, L. Graham, and R. C. Lacy. 1994. An experimental study of inbreeding depression in a natural habitat. *Science* 266: 271–273. [12]

Jimenez-Gomes, J. M., J. A. Corwin, B. Joseph, J. N. Maloff, and D. J. Kliebenstein. 2011. Genomic analysis of QTLs and genes altering natural variation in stochastic noise. *PLoS Genet.* 7: e1002295. [18]

Jin, C., and 4 others. 2004. Selective phenotyping for increased efficiency in genetic mapping studies. *Genetics* 168: 2285–2293. [18]

Jinks, J. L. 1983. Biometrical genetics of heterosis. *In* R. Frankel (ed.) *Heterosis: Reappraisal of theory and practice*, pp. 1–46. Springer-Verlag, Berlin. DE. [13]

Jinks, J. L., and K. Mather. 1955. Stability in development of heterozygotes and homozygotes. *Proc. Royal Soc. Lond. B* 143: 561–578. [6]

Jinks, J. L., and J. M. Perkins. 1969. The detection of linked epistatic genes for a metrical trait. *Heredity* 24: 465–475. [11]

Jinks, J. L., and P. Towey. 1976. Estimating the number of genes in a polygenic system by genotype assay. *Heredity* 37: 69–81. [11]

Jinks, J. L., C. E. Caten, G. Simchen, and H. J. Croft. 1966. Heterokaryon incompatibility and variation in wild populations of *Aspergillus nidulans*. *Heredity* 21: 227–239. [5]

Johannes, F., V. Colot, and R. C. Jansen. 2008. Epigenome dynamics: a quantitative genetics perspective. *Nat. Rev. Genet.* 9: 883–890. [21]

Johannes, F., and 14 others. 2009. Assessing the impact of transgenerational epigenetic variation on complex traits. *PLoS Genet.* 5: e1000530. [21]

Johannsen, W. 1903. *Über Erblichkeit in Populationen und in Reinen Linien.* Gustav Fischer, Jena, DE. [1]

Johannsen, W. 1909. *Elemente der exakten Erblichkeitslehre.* Gustav Fischer, Jena, DE. [1]

John, J. A., and N. R. Draper. 1980. An alternative family of transformations. *Appl. Stat.* 29: 190–197. [14]

John, J. A., and E. R. Williams. 1995. *Cyclic designs and computer-generated designs.* Chapman & Hall, New York, NY. [A9]

Johnson, D. L., R. C. Jansen, and J. A. M. van Arendonk. 1999. Mapping quantitative trait loci in a selectively genotyped outbred populations using a mixture model approach. *Genet Res.* 73: 75–83. [17, 18, 19]

Johnson, E. C., and 33 others. 2016. No reliable association between runs of homozygosity and schizophrenia in a well-powered replication study. *PLoS Genet.* 12: e1006343. [12]

Johnson, J. M., S. Edwards, D. Shoemaker, and E. E. Schadt. 2005. Dark matter in the genome: evidence of widespread transcription detected by microarray tilling experiments. *Trends Genet.* 21: 93–102. [21]

Johnson, L. M., O. J. Smith, D. A. Hahn, and C. F. Baer. 2020. Short–term heritable variation overwhelms 200 generations of mutational variance for metabolic traits in *Caenorhabditis elegans*. *Evolution* 74: 2451–2464. [15]

Johnson, N. L., S. Kotz, and N. Balakrishnan. 1994. *Continuous univariate distributions — 1, 2nd Ed.*. John Wiley & Sons, New York, NY. [2]

Johnson, N. L., S. Kotz, and N. Balakrishnan. 1995. *Continuous univariate distributions — 2.* John Wiley & Sons,New York, NY. [2, A5]

Johnson, N. L., A. W. Kemp, and S. Kotz. 2005. *Univariate discrete distributions, 3rd Ed.*. John Wiley & Sons, New York, NY. [2]

Johnson, R. A., and D. W. Wichern. 2002. *Applied multivariate statistical analysis.* 5th Ed. Prentice-Hall, Upper Saddle River, NJ. [9]

Johnson, T., and N. Barton. 2005. Theoretical models of selection and mutation on quantitative traits. *Phil. Trans Roy. Soc. Lond. B.* 360: 1411–1425. [21]

Johnston, M. O. 1992. Effects of cross and self-fertilization on progeny fitness in *Lobelia cardinalis* and *L. siphilitica*. *Evolution* 46: 688–702. [12]

Johnston, M. O., and D. J. Schoen. 1994. On the measurement of inbreeding depression. *Evolution* 48: 1735–1741. [12]

Johnston, M. O., and D. J. Schoen. 1995. Mutation rates and dominance levels of genes affecting total fitness in two angiosperm species. *Science* 267: 226–229. [12, 15]

Jones, A. G., and W. R. Ardren. 2003. Methods for parentage analysis in natural populations. *Mol. Ecol.* 12: 2511–2523. [8]

Jones, A. G., C. M. Small, K. A. Paczolt, and N. L. Ratterman 2010. A practical guide to methods of parentage analysis. *Mol. Ecol. Resources* 10: 6–30. [8]

Jones, C. D. 2001. Extension of the Castle-Wright effective factor estimator to sex linkage and haplodiploidy. *J. Hered.* 92: 274–276. [11]

Jones, D. F. 1917. Dominance of linked factors as a means of accounting for heterosis. *Genetics* 2: 466–479. [12, 13]

Jones, D. F. 1918. The effects of inbreeding and crossbreeding upon development. *Conn. Agric. Exp. Sta. Bull* 207: 5–100. [12, 13]

Jones, D. F. 1922. The productiveness of single and double first generation corn hybrids. *J. Am. Soc. Agron.* 14: 242–252. [13]

Jones, D. F., and H. L. Everett. 1949. Hybrid field corn. *Conn. Agr. Expt. Sta. Bull* 531: 3–29. [13]

Jones, D. F., and P. C. Manglesdorf. 1951. The production of hybrid corn without detasseling. *Conn. Agr. Exp Sta. Bull.* 550. [13]

Jones, O. R., and J. Wang. 2010. COLONY: a program for parentage and sibship inference from multilocus genotype data. *Mol. Ecol.* 10: 551–555 [8]

Jordaan, J. P., S. A. Engelbrecht, J. H. Malan, H. A. Knobel. 1999. Wheat and heterosis. *In* J. G. Coors and S. Pandey (eds.), *The genetics and exploitation of heterosis in crops*, pp. 411–422 ASA-CSSA-SSSA Societies, Madison, WI. [13]

Jørgensen, D. B., M. Ørsted, and T. N. Kristensen. 2021. Sustained positive consequences of genetic rescues of fitness and behavioral traits in inbred populations of *Drosophila melanogaster*. *J. Evol. Biol.* 35: 868–878. [12]

Jorde, L. B. 1995. Linkage disequilibrium as a gene-mapping tool. *Am. J. Hum. Genet.* 56: 11–14. [17]

Josephs, E. B., Y. W. Lee, J. R. Stichcombe, and S. I. Wright. 2015. Association mapping reveals the role of purifying selection in the maintenance of genome variation in gene expression. *Proc. Natl. Acad. Sci. USA* 112: 15390–15395. [21]

Joshi, P. K., et al. 2015. Directional dominance on stature and cognition in diverse human populations. *Nature* 523: 459–462. [12]

Jowett, D. 1972. Yield stability parameters for sorghum in East Africa. *Crop Sci.* 12: 314–317. [6]

Judson, H. F. 1979. *The eighth day of creation: Makers of the revolution in biology.* Simon and Schuster, New York, NY. [1]

Jung, J., D. E. Weeks, and E. Feingold. 2006. Gene-dropping vs. empirical variance estimation for allele-sharing linkage statistics. *Genet. Epid.* 30: 652–665. [19]

Kackar, R. N., and D. A. Harville. 1981. Unbiasedness of two-stage estimation and prediction for mixed linear models. *Comm. Stat. Theor. Meth.* A10: 1249–1261. [10]

Kacser, H., and J. A. Burns. 1981. The molecular basis of dominance. *Genetics* 97: 639–666. [4]

Kadarmideen, H. N., and J. C. M. Dekkers. 2001. Generalized marker regression and interval QTL mapping methods for binary traits in half-sib family designs. *J. Anim. Breed. Genet.* 118: 297–309. [19]

Kadarmideen, H. N., L. L. G. Janss, and J. C. M. Dekkers. 2001. Power of quantitative trait locus mapping for polygenic binary traits using generalized and regression interval mapping in multi-family half-sib designs. 76: 305–317. *Genet. Res.* [19]

Kaeppler, S. M. 1997. Quantitative trait locus mapping using sets of near-isogenic lines: relative power comparisons and technical considerations. *Theor. Appl. Genet.* 95: 384–392. [17]

Kaeppler, S. M., R. L. Phillips, and T. S. Kim. 1993. Use of near-isogenic lines derived by backcrossing or selfing to map qualitative traits. *Theor. Appl. Genet.* 87: 233–237. [17]

Kalinowski, S. T., and P. W. Hedrck. 1998. An improved method for estimating inbreeding depression in pedigrees. *Zoo Biol.* 17: 481–497. [12]

Kalinowski, S. T., A. P. Wagner, and M. L. Taper. 2006. ML-Relate: a computer program for maximum likelihood estimation of relatednessand relationship. *Mol. Ecol. Notes* 6: 576–579. [8]

Kalinowski, S. T., M. L. Taper, and T. C. Marshall. 2007. Revising how the computer program CERVUS accommodates genotyping error increases success in paternity assignment. *Mol. Ecol.* 16: 1099–1106. [8]

Kanehisa, M., and 4 others. 2010. KEGG for representation and analysis of molecular networks involving diseases and drugs. *Nucleic Acids Res.* 38: D355–D36. [21]

Kang, E. Y., and 7 others. 2014. Meta-analysis identified gene-by- environment interactions as demonstrated in a study of 4,965 mice. *PLoS Genet.* 10: e1004022. [20]

Kang, E. Y., and 5 others. 2016. ForestPMPlot: a flexible tool for visualizing heterogeneity between studies in meta-analysis. *G3* 6: 1793–1798. [20]

Kang, H. M., and 6 others. 2008. Efficient control of population structure in model organism association mapping. *Genetics* 178: 1709–1723. [20]

Kang, H. M., and 7 others. 2010. Variance component model to account for sample structure in genome-wide association studies. *Nat. Genet.* 42: 348–354. [20]

Kao, C. H. 2000. On the differences between maximum-likelihood and regression interval mapping in the analysis of quantitative trait loci. *Genetcs* 156: 855–865. [18]

Kao, C.-H, and H.-A. Ho. 2012. A score-statistic approach for determining threshold values in QTL mapping. *Front. Biosci.* E4: 2670–2682. [18]

Kao, C.-H., and M.-H. Zeng. 2010. An investigation of the power for separating closely linked QTL in experimental populations. *Genet. Res.* 92: 283–294. [18]

Kao, C.-H., and Z.-B. Zeng. 1997. General formulas for obtaining the MLEs and the asymptotic variance-covariance matrix in mapping quantitative trait loci when using the EM algorithm. *Biometrics* 53: 653–665. [18, A4]

Kao, C.-H., and Z.-B. Zeng. 2002. Modeling epistasis of quantitative trait loci using Cockerham's model. *Genetics* 160: 1243–1261. [18]

Kao, C.-H., Z.-B. Zeng, and R. D. Teasdale. 1999. Multiple interval mapping for quantitative trait loci. *Genetics* 152: 1203–1216. [18]

Kaplan, N. L., and B. S. Weir. 1995. Are moment bounds on the recombination fraction between a marker and a disease locus too good to be true? Allelic association mapping revisited for simple genetic diseases in the Finnish population. *Am. J. Hum. Genet.* 57: 1486–1498. [17]

Kaplan, N. L., W. G. Hill, and B. S. Weir. 1995. Likelihood methods for locating disease genes in nonequilibrium populations. *Am. J. Hum. Genet.* 56: 18–32. [17]

Kapranov, P., and G. St. Laurent. 2012. Dark matter RNA: existence, function, and controversy. *Front. Genet.* 3: 60. [21]

Kapranov, P., and 21 others. 2007. RNA maps reveal new RNA classes and a possible function for pervasive transcription. *Science* 316: 1484–1488. [21]

Kaprio, J., R. E. Ferrell, B. A. Kottke, M. I. Kamboh, and C. F. Sing. 1991. Effects of polymorphisms in apolipoproteins E, A-IV, and H on quantitative traits related to risk for cardiovascular disease. *Arterioscler. Thromb.* 11: 1330–1348. [17]

Karban, R. 1989. Fine-scale adaptation of herbivorous thrips to individual host plants. *Nature* 340: 60–61. [13]

Kardos, M., F. W. Allendorf, and G. Luikart. 2014. Evaluating the role of inbreeding depression in heterozygosity-fitness correlations: how useful are tests for identity disequilibrium? *Mol. Ecol. Resources* 14: 519–530. [12]

Kardos, M., G. Luikart, and F. W. Allendorf. 2015. Measuring individual inbreeding in the age of genomics: marker-based measures are better than pedigrees. *Heredity* 115: 63–72. [12]

Kardos, M., H. R. Taylor, H. Ellegren, G. Luikart, and F. W. Allendorf. 2016. Genomics advances the study of inbreeding depression in the wild. *Evol. Applic.* 9: 1205–1218. [12]

Kardos, M., P. Nietlisbach, and P. W. Hedrick. 2018a. How should we compare different genomic estimates of the strength of inbreeding depression? *Proc. Natl. Acad. Sci. USA* 115: E2492–E2493. [12]

Kardos, M., and 9 others. 2018b. Genomic consequences of intensive inbreeding in an isolated wolf population.

Nat. Evol. Evol. 2: 124–131. [12]

Karigl, G. 1981. A recursive algorithm for the calculation of identity coefficients. *Ann. Hum. Genet.* 45: 299–305. [7]

Karlin, S. 1982. Theoretical aspects of genetic map functions in recombination processes. *In* A. Chakravarti (ed.), *Human population genetics: the Pittsburgh symposium*, pp. 209–228. Van Nostrand Reinhold, New York, NY. [17]

Karlin, S., and P. T. Williams. 1981. Structured Exploratory Data Analysis (SEDA) for determining mode of inheritance of quantitative traits. II. Simulation studies on the effect of ascertaining families through high-valued probands. *Am. J. Hum. Genet.* 33: 282–292. [16]

Karlin, S., D. Carmelli, and R. Williams. 1979. Index measures for assessing the mode of inheritance of continuously distributed traits: I. Theory and justifications. *Theor. Pop. Biol.* 16: 81–106. [16]

Kass, R. E., B. P. Carlin, A. Gelman, and R. M. Neal. 1998. Markov chain Monte Carlo in practice: A roundtable discussion. *Am. Stat.* 52: 93–100. [A8]

Kat, P. W. 1982. The relationship between heterozygosity for enzyme loci and developmental homeostasis in peripheral populations of aquatic bivalves (Unionidae). *Am. Nat.* 119: 824–832. [6]

Katan, M. B. 1986. Apolipoprotein E isoforms, serum cholesterol, and cancer. *Lancet* 327: 507–508. [A2]

Kauffman, S. A. 1969. Metabolic stability and epigenesis in randomly constructed genetic nets. *J. Theor. Biol.* 22: 437–467. [A2]

Kavvoura, F. K., and J. P. A. Ioannidis. 2008. Methods for meta-analysis in genetic association studies: A review of their potential and pitfalls. *Hum. Genet.* 123: 1–14. [20]

Kearsey, M. J., and V. Hyne. 1994. QTL analysis: a simple 'marker regression' approach. *Theor. Appl. Genet.* 698–702. [18]

Kearsey, M. J., and K.-I. Kojima. 1967. The genetic architecture of body weight and egg hatchability on *Drosophila melanogaster*. *Genetics* 56: 23–37. [17]

Kearsey, M. J., and H. S. Pooni. 1996. *The genetical analysis of quantitative traits.* Chapman and Hall, London, UK. [1]

Keeble, F., and C. Pellew. 1910. The mode of inheritance of stature and of time of flowering in peas *Pisum sativum*. *J. Genetics* 1: 47–56. [12, 13]

Keele, G. R., W. L. Crouse, S. N. Kelada, and W. Valdar. 2019. Determinants of QTL mapping power in the realized collaborative cross. *G3* 9: 1707–1727. [18]

Keightley, P. D. 1989. Models of quantitative variation in flux in metabolic pathways. *Genetics* 121: 869–876. [5]

Keightley, P. D. 1994. The distribution of mutation effects on viability in *Drosophila melanogaster. Genetics* 138: 1315–1322. [11, 15]

Keightley, P. D. 1998. Inference of genome–wide mutation rates and distributions of mutation effects for fitness traits: a simulation study. *Genetics* 150: 1283–1293. [15]

Keightley, P. D., and G. Bulfield. 1993. Detection of quantitative trait loci from frequency changes of marker alleles under selection. *Genet. Res.* 62:195–203. [8, 17]

Keightley, P. D., and A. Caballero. 1997. Genomic mutation rates for lifetime reproductive output and lifespan in *Caenorhabditis elegans*. *Proc. Natl. Acad. Sci. USA* 94: 3823–3827. [15]

Keightley, P. D., and D. J. Gaffney. 2003. Functional constraints and frequency of deleterious mutations in noncoding DNA of rodents. *Proc. Natl. Acad. Sci. USA* 100: 13402–13406. [15]

Keightley, P. D., and W. G. Hill. 1990. Estimating new mutational variation in growth rate of mice. *In* W. G. Hill, R. Thompson, and J. A. Woolliams (eds.), *Proceedings of the 4th world congress on genetics applied to livestock production.* Edinburgh, UK. [15]

Keightley, P. D., and W. G. Hill. 1992. Quantitative genetic variation in body size of mice from new mutations. *Genetics* 131: 693–700. [15]

Keightley, P. D., T. F. C. Mackay, and A. Caballero. 1993. Accounting for bias in estimates of the rate of polygenic mutation. *Proc. Royal Soc. Lond. B* 253: 291–296. [15]

Keightley, P. D., T. Hardge, L. May, and G. Bulfield. 1996. A genetic map of quantitative trait loci for body weight in mouse. *Genetics* 142: 227–235. [17]

Keller, E. C., Jr., and D. F. Mitchell. 1962. Interchromosomal genotypic interactions in *Drosophila*. I. An analysis of morphological characters. *Genetics* 47: 1557–1571. [17]

Keller, E. C., Jr., and D. F. Mitchell. 1964. Interchromosomal genotypic interactions in *Drosophila*. II. An analysis of viability characters. *Genetics* 49: 293–307. [17]

Keller, L. F. 1998. Inbreeding and its fitness effects in an insular population of song sparrows (*Melospiza melodia*). *Evolution* 52: 240–250. [12]

Keller, L. F., and D. M. Waller. 2002. Inbreeding effects in wild populations. *Trends. Ecol. Evol.* 17: 230–241. [12]

Keller, L. F., P. R. Grant, B. R. Grant, and K. Petern. 2002. Envirnmental conditions affect the magnitude of inbreeding depression in survival of Darwin's finches. *Evolution* 56: 1229–1239. [12]

Keller, M. C., P. M. Visscher, and M. E. Goddard. 2011. Quantification of inbreeding due to distant ancestors and its detection using dense single nucleotide polymorphism data. *Genetics* 189: 237–249. [8, 12]

Keller, M. C., and 10 others. 2012. Runs of homozygosity implicate autozygosity as a schizophrenia risk factor. *PLoS Genetics* 8: e1002656. [12]

Kelly, J. K. 2003. Deleterious mutations and the genetic variance of male fitness components in *Mimulus guttatus*. *Genetics* 164: 1071–1085. [15]

Kelly, J. K. 2005. Epistasis in monkeyflowers. *Genetics* 171: 1917–1931. [12]

Kemper, K. E., and 8 others. 2021. Phenotypic covariances across the entire spectrum of relatedness for 86 billion pairs of individuals. *Nat. Comm.* 12: 1050. [21]

Kempthorne, O. 1953. The correlation between relatives in a simple autotetraploid population. *Genetics* 40: 168–174. [7]

Kempthorne, O. 1954. The correlation between relatives in a random mating population. *Proc. Royal Soc. Lond. B* 143: 103–113. [5, 7]

Kempthorne, O. 1957. *An introduction to genetic statistics.* John Wiley & Sons, NY. [4, 5, 7]

Kempton, R. A., and P. N. Fox. 1997. *Statistical methods for plant variety evaluation*. Chapman & Hall, London, UK. [A9]

Kempton, R. A., and A. C. Gleeson. 1997. Unreplicated trails. *In* R. A. Kempton and P. N. Fox (eds), *Statistical methods for plant variety evaluation*, pp. 86–100. Chapman & Hall, London, UK. [A9]

Kendall, M., and A. Stuart. 1977. *The advanced theory of statistics. Vol. 1. Distribution theory.* 4th Ed. Macmillan, New York, NY. [2, 11, 14, A1]

Kendall, M., and A. Stuart. 1979. *The advanced theory of statistics. Vol. 2. Inference and relationship.* 4th Ed. Macmillan, NY. [A4]

Kendziorski, C. M., and P. Wang. 2006. A review of statistical methods for expression quantitative trait loci mapping. *Mamm. Genome* 17: 509–517. [21]

Kendziorski, C. M., and 4 others. 2006. Statistical methods for expression quantitative trait loci (eQTL) mapping. *Biometrics* 62: 19–27. [21]

Kennard, W. C., and M. J. Harvey. 1995. Quantitative trait analysis of fruit quality in cucumber: QTL detection, confirmation, and comparison with mating-design variation. *Theor. Appl. Genet.* 91: 53–61. [18]

Kennedy, B. W., and D. A. Sorensen. 1988. Properties of mixed-model methods for prediction of genetic merit. *In* B. S. Weir, E. J. Eisen, M. M. Goodman, and G. Namkoong (eds.), *Proceedings of the second international conference on quantitative genetics,* pp. 91–103. Sinauer Assoc., Sunderland, MA. [10]

Kennedy, B. W., M. Quinton, and J. A. M. van Arendonk. 1992. Estimation of effects of single genes on quantitative traits. *J. Anim. Sci.* 70: 2000–2012. [10, 17]

Kennedy, E. S., C. E. Grueber, R. P. Duncan, and I. G. Jamieson. 2014. Severe inbreeding depression and no evidence of purging in an extremely inbred wild species — the Chatham Island black robin. *Evolution* 68: 987–995. [12]

Kennedy, J. S. 1956. Phase transformation in locust biology. *Biol. Rev.* 31: 349–370. [6]

Kennedy, W. J., Jr., and J. E. Gentle. 1980. *Statistical computing.* Marcel Dekker, New York, NY. [A4]

Kerem, B.-S., and 7 others. 1989. Identification of the cystic fibrosis gene: genetic analysis. *Science* 245: 1073–1080. [17]

Kerfoot, W. C. 1988. Defensive spines: inverse relationship between coefficients of variation and size. *Limnol. Oceanogr.* 33: 1412–1429. [14]

Kermicle, J. L. 1969. Androgenesis conditioned by a mutation in maize. *Science* 166: 1422–1424. [11]

Kestilä, M., and 7 others. 1994. Congenital nephrotic syndrome of the Finnish type maps to the long arm of chromosome 19. *Am. J. Hum. Genet.* 54: 757–764. [17]

Keurentjes, J. J. B., and 7 others. 2007. Development of a near-isogenic line population of *Arabidopsis thaliana* and comparison of mapping power with a recombinant inbred line population. *Genetics* 175: 891–905. [17]

Khaipho-Burch, M., and 12 others. 2023. Scale up trials to validate modified crops benefits. *Nature* 621: 471–474. [5]

Khambanonda, I. 1950. Quantitative inheritance of fruit size in red pepper *(Capsicum frutescens* L.). *Genetics* 35: 322–343. [11]

Khanday, I., and 4 others. 2019. A male-expressed rice embryogenic trigger redirect for asexual propagation through seeds. *Nature* 656: 91–96. [13]

Khatib, H., A. Darvasi, Y. Plotski, and M. Soller. 1994. Determining relative microsatellite allele frequencies in pooled DNA samples. *PCR Methods Appl.* 4: 13–19. [17]

Khatri, P., and S. Drăghici. 2005. Ontological analysis of gene expression data: current tools, limitations, and open problems. *Bioinformatics* 21: 3587–3595. [21]

Khatri, P., M. Sirota, and A. J. Butte. 2012. Ten years of pathway analysis: current approaches and outstanding challenges. *PLoS Comput. Biol.* 8: e1002375. [21]

Kibota, T. T. 1996. Spontaneous mutations influencing fitness in *Escherichia coli.* Ph. D. Thesis, Univ. Oregon, Eugene, OR. [15]

Kibota, T. T., and M. Lynch. 1996. Estimate of the genomic mutation rate deleterious to overall fitness in *Escherichia coli. Nature* 381: 694–696. [15]

Kichaev, G., and 7 others. 2014. Integrating functional data to prioritize causal variants in statistical fine-mapping studies. *PLoS Genet.* 10: e1004722. [21]

Kichaev, G., and 6 others. 2016. Improved methods for multi-trait fine mapping of pleiotropic risk loci. *Bioinformatics* 33: 248–255. [21]

Kidder, R. W., M. Koger, J. H. Meade, and J. R. Crockett. 1964. Systems of crossbreeding for beef production in Florida. *Bull. Agric. Experimental Stations, Florida University* No. 673. [13]

Kidwell, J. F., and M. M. Kidwell. 1966. The effects of inbreeding on body weight and abdominal chaeta number in *Drosophila melanogaster. Can. J. Genet. Cytol.* 8: 207–215. [12]

Kiefer, A. K., and 6 others. 2013. Genome-wide analysis points to roles for extracellular matrix remodeling, the visual cycle, and neuronal development in myopia. *PLoS Genet.* 9: e1003299. [20]

Kieser, J. A., and H. T. Groeneveld. 1991. Fluctuating odontometric asymmetry, morphological variability, and genetic monomorphism in the cheetah *Acinonyx jubatus. Evolution* 45: 1175–1183. [6]

Kiesselbach, T. A. 1951. A half-century of corn research. *Am. Scientist* 39: 629–655. [13]

Killick, R. J. 1971. The biometrical genetics of autotetraploids. 1. Generations derived from a cross between two pure lines. *Heredity* 27: 331–346. [5]

Kilmer, J. T., and R. L. Rodríguez. 2016. Ordinary least squares regression is indicated for studies of allometry. *J. Evol. Biol.* 30: 4–12. [14]

Kim, D. H., and 4 others. 2010. Fine mapping of a resistance gene to bacterial leaf pustule in soybean. *Theor. Appl. Genet.* 120: 1443–1450. [17]

Kim, J., Y. Zhang, and W. Pan. 2016. Powerful and adaptive testing for multi-trait and multi-SNP associations with GWAS and sequencing data. *Genetics* 203: 715–731. [20]

Kim, J.-J., and 4 others. 2005. Combined line-cross and half-sib QTL analysis of crosses between outbred lines. *Genet. Res.* 85: 235–248. [19]

Kimmel, G., and 4 others. 2007. A randomization test for controlling population stratification in whole-genome association studies. *Am. J. Hum. Genet.* 81: 895–905. [20]

Kimura, M. 1983. *The neutral theory of molecular evolution.* Cambridge Univ. Press, Cambridge, UK. [4, 15]

Kimura, M., and T. Maruyama. 1966. The mutational load with epistatic gene interactions in fitness. *Genetics* 54: 1337–1351. [15]

King, D. P. F. 1984. Enzyme heterozygosity associated with anatomical character variance and growth in the herring (*Clupea harengus* L.). *Heredity* 54: 289–296. [6]

King, E. G., and A. D. Long. 2017. The Beavis effect in next-generation mapping panels in *Drosophila melanogaster. G3* 7: 1643–1652. [18]

King, E. G., and 7 others. 2012a. Genetic dissection of a model complex trait using the *Drosophila* synthetic population resource. *Genome Res.* 22: 1558–1566. [18]

King, E. G., S. J. Macdonald, and A. D. Long. 2012b. Properties and power of the *Drosophila* Synthetic Population Resource for the routine dissection of complex traits. *Genet.* 191: 935–949. [18]

King, E. G., and 4 others. 2014. Genetic dissection of the *Drosophila melanogaster* female head transcriptome reveals widespread allelic heterogeneity. *PLoS Genet.* 10: e1004322. [18]

Kinghorn, B. 1980. The expression of "recombination loss" in quantitative traits. *Z. Tierzüchtg. Züchtsbiol.* 97: 138–143. [13]

Kinghorn, B. 1982. Genetic effects of crossbreeding. I. Models of merit. *Z. Tierzüchtg. Züchtsbiol.* 99: 59–68. [13]

Kinghorn, B. 1983. Genetic effects of crossbreeding. III. Epistatic loss in crossbred mice. *Z. Tierzüchtg. Züchtsbiol.* 100: 209–222. [13]

Kinghorn, B. 1987. The nature of 2-locus epistatic interactions in animals. Evidence from Sewall Wright's guine pig data. *Theor. Appl. Genet.* 73: 595–604. [13]

Kinghorn, B. 1999. The genetic basis of crossbreeding. *In* B. Kinghorn, B. J. van der Werf, and M. Ryan, (eds.), *Animal breeding: Use of New Technologies*, pp. 35-54. Post Graduate Foundation in Veterinarian Science of the University of Sydney, NSW, AU. [13]

Kinghorn, B. P., B. W. Kennedy, and C. Smith. 1993. A method for screening for genes of major effect. *Genetics* 134: 351–360. [10, 16]

Kinghorn, B. P., J. H .J. Van der Werf, and M. Ryan (Eds.) 2000. *Animal breeding: Use of new technologies.* The Post Graduate Foundation in Veterinarian Science of the University of Sydney, Australia. [1]

Kinkade, M. P., and M. R. Foolad. 2013. Validation and fine mapping of *lyc12.1*, a QTL for increased tomato fruit lycopene content. *Theor. Appl. Genet.* 126: 2163-2175. [17]

Kinzer, S. M., S. J. Schwager, and M. A. Mutschler. 1990. Mapping of ripening-related or -specific cDNA clones of tomato (*Lycopersicon esculentum*). *Theor. Appl. Genet.* 79: 489–496. [8]

Kirk, H. J., J. L. Haynes, and R. J. Monroe. 1980. Applications of trend analysis to horticultural field crops. *J. Am. Soc. Hortic. Sci.* 105: 189–193. [A9]

Kirkpatrick, S., C. D. Gelatt, and M. P. Vecchi. 1983. Optimization by simulated annealing. *Science* 220: 671–680. [A8]

Kirst, M., and 5 others. 2004. Coordinated genetic regulation of growth and lignin revealed by quantitative trait locus analysis of cDNA microarray data in an interspecific backcross of *Eucalyptus*. *Plant Physiol.* 135: 2368–2378. [21]

Kitano, H. 2004. Biological robustness. *Nat. Rev. Genet.* 5: 826–837. [14, 21]

Kitano, H. 2007. Towards a theory of biological robustness. *Mol. Sys. Biol.* 3: 137–143. [21]

Kleczkowski, A. 1949. The transformation of local lesion counts for statistical analysis. *Ann. Appl. Biol.* 36: 139–152. [14]

Klein, R. J. 2007. Power analysis for genome-wide association studies. *BMC Genet.* 8: 58. [20]

Klein, R. J., and 14 others. 2005. Complement factor *H* polymorphism in age-related macular degeneration. *Science* 308: 385–389. [17, 20]

Kleinman-Ruiz, D., and 9 others. 2022. Purging of deleterious burden in the endangered Iberian lynx. *Proc. Natl. Acad. Sci. USA* 199: e2110614119. [12]

Klekowski, E. J., Jr. 1988. Genetic load and its causes in long-lived plants. *Trees* 2: 195–20. [12]

Kliebenstein, D. J. 2009. Advancing genetic theory and applications by metabolic quantitative trait loci analysis. *Plant Cell* 21: 1637–1646. [21]

Klingenberg, C. P. 2003. A developmental perspective on developmental instability: theory, models, and mechanisms. *In* M. Polak (ed.), *Developmental instability. Causes and consequences*, pp. 14–34. Oxford Univ. Press, Oxford, UK. [6]

Klingenberg, C. P. 2015. Analyzing fluctuating asymmetry with geometric morphometrics: concepts, methods, and applications. *Symmetry* 7: 843–934. [6]

Klose, J., and 12 others. 2002. Genetic analysis of the mouse proteome. *Nature* 30: 385–393. [21]

Kluge, A. G., and W. C. Kerfoot. 1973. The predictability and regularity of character divergence. *Am. Nat.* 107: 426–442. [14]

Kluge, R., and H. Geldermann. 1982. Effects of marked chromosome sections on quantitative traits in the mouse. *Theor. Appl. Genet.* 62: 1–4. [17]

Knapp, G., and J. Hartung. 2003. Improved test for a random effects meta-regression with a single covariate. *Stat. Med.* 22: 2693–2710. [A6]

Knapp, M. 1998. Evaluation of a restricted likelihood ratio test for mapping quantitative trait loci with extreme discordant sib pairs. *Ann. Hum. Genet.* 62: 75–87. [19]

Knapp, M. 1999a. The transmission/disequilibrium test and parental-genotype reconstruction: the reconstruction-combined transmission/disequilibrium test. *Am. J. Hum. Genet.* 64: 861–870. [17]

Knapp, M. 1999b. A note on power approximations for the transmission/disequilibrium test. *Am. J. Hum. Genet.* 64: 1177–1185. [17]

Knapp, M. 1999c. Using exact P values to compare the power between the reconstruction-combined transmission/disequilibrium test and the sib transmission/disequilibrium test. *Am. J. Hum. Genet.* 65: 1208–1210. [17]

Knapp, S. J. 1991. Using molecular markers to map multiple quantitative trait loci: models for backcross, recombinant inbred, and doubled-haploid progeny. *Theor. Appl. Genet.* 81: 333–338. [18]

Knapp, S. J., and W. C. Bridges, Jr. 1990. Using molecular markers to estimate quantitative trait locus parameters: power and genetic variances for unreplicated and replicated progeny. *Genetics* 126: 769–777. [17, 18]

Knapp, S. J., W. C. Bridges, Jr., and D. Birkes 1990. Mapping quantitative trait loci using molecular marker linkage maps. *Theor. Appl. Genet.* 79: 583–592. [18]

Knief, U., and 6 others. 2015. Quantifying realized inbreeding in wild and captive animal populations. *Heredity* 114: 397–403. [8, 12]

Knief, U., B. Kempenaers, and W. Forstmeier. 2017. Meiotic recombination shapes precision of pedigree-and marker-based estimates of inbreeding. *Heredity* 118: 239–248. [8]

Knierim, U., and 6 others. 2007. Fluctuating asymmetry as an animal welfare indicator — a review of methodology and validity. *Physiology & Behavior* 92: 398–421. [6]

Knight, C. A., and J. M. Beaulieu. 2008. Genome size scaling through phenotype space. *Ann. Bot.* 101: 759–766. [15]

Knight, J. C. 2004. Allele-specific gene expression uncovered. *Trends Genet.* 20: 113–116. [21]

Knight, T. 1799. An account of some experiments on the fecundation of vegetables. *Philso. Trans. R. Soc. Lond.* 89: 195–204. [12, 13]

Knijnenburg, T. A., L. F. A. Wessels, M. J. T. Reinders, and I. Shmulevich. 2009. Fewer permutations, more accurate P-values. *Bioinform.* 25: i161–i168. [20]

Knott, S. A. 1994. Prediction of the power of detection of marker-quantitative trait locus linkages using analysis of variance. *Theor. Appl. Genet.* 89: 318–322. [19]

Knott, S. A., and C. S. Haley. 1992a. Aspects of maximum likelihood methods for the mapping quantitative trait loci in line crosses. *Genet. Res.* 60: 139–151. [16, 17, 18]

Knott, S. A., and C. S. Haley. 1992b. Maximum likelihood mapping of quantitative trait loci using full-sib families. *Genetics* 132: 1211–1222. [16, 19]

Knott, S. A., C. S. Haley, and R. Thompson. 1990. Approximations to segregation analysis for the detection of major genes. *In* W. G. Hill, R. Thompson, and J. A. Woolliams (eds.), *Proc. 4th World Congr. Genet. Appl. Livestock Prod.*, Vol. 13 pp. 504–507. Edinburgh, UK. [16]

Knott, S. A., C. S. Haley, and R. Thompson. 1991a. Methods of segregation analysis for animal breeding data: parameter estimates. *Heredity* 68: 313–320. [16]

Knott, S. A., C. S. Haley, and R. Thompson. 1991b. Methods of segregation analysis for animal breeding data: a comparison of power. *Heredity* 68: 299–311. [16]

Knott, S. A., J. M. Elsen, and C. S. Haley. 1994. Multiple marker mapping of quantitative trait loci in half sib populations. *Proc. 5th World Congr. Genet. Appl. Livestock. Prod.*, Guelph, Canada 21: 33–36. [19]

Knott, S. A., J. M. Elsen, and C. S. Haley. 1996. Methods for multiple-marker mapping of quantitative trait loci in half-sib populations. *Theor. Appl. Genet.* 93: 71–80. [19]

Knott, S. A., D. B. Neale, M. M. Sewell, and C. S. Haley. 1997. Multiple marker mapping of quantitative trait loci in an outbred pedigree of loblolly pine. *Theor. Appl. Genet.* 94: 810–820. [19]

Knott, S.A., and 11 others. 1998. Multiple marker mapping of quantitative trait loci in a cross between outbred wild boar and large white pigs. *Genetics* 149: 1069–1080. [19]

Knowler, W. C., R. C. Williams, D. J. Pettitt, and A. G. Steinberg. 1988. $Gm^{3;5,13,14}$ and type 2 diabetes mellitus: an association in American indians with genetic admixture. *Am. J. Hum. Genet.* 43: 520–526. [17, 20]

Knowles, P., and M. C. Grant. 1981. Genetic patterns associated with growth variability in ponderosa pine. *Am. J. Bot.* 68: 942–946. [6]

Knowles, P., and J. B. Mitton. 1980. Genetic heterozygosity and radial growth variability in *Pinus contorta. Silvae Genet.* 29: 114–118. [6]

Kobabe, G. 1983. Heterosis and hybrid seed production in fodder grass. *In* R. Frankel (ed.) *Heterosis: Reappraisal of theory and practice*, pp. 124–137. Springer-Verlag, Berlin. DE. [13]

Kobyliansky, E., and G. Livshits. 1983. Relationship between levels of biochemical heterozygosity and morphological variability in human populations. *Ann. Hum. Genet.* 47: 215–223. [6]

Kocak, M. 2017. Meta-analysis of univariate *p*-values. *Comm. Stat. Sim. Comp.* 46: 1257–1265. [A6]

Koch, R. M., G. E. Dickerson, L. V. Cundiff, and K. E. Gregory. 1985. Heterosis retained in advanced generations of crosses among Angus and Hereford cattle. *J. Anim. Sci.* 60 : 1117–1132. [13]

Koehn, R. K., and S. E. Shumway. 1982. A genetic/physiological explanation for differential growth rate among individuals of the American Oyster; *Crassostrea virginica* (Gmelin). *Mar. Biol. Lett.* 3: 35–42. [12]

Koehn, R. K., W. J. Diehl, and T. M. Scott. 1988. The differential contribution by individual enzymes of glycolysis and protein catabolism to the relationship between heterozygosity and growth rate in the coot clam, *Mulina lateralis. Genetics* 118: 121–130. [12]

Koester, R. P., P. H. Sisco, and C. W. Stuber. 1993. Identification of quantitative trait loci controlling days to flowering and plant height in two near isogenic lines of maize. *Crop Sci.* 33: 1209–1216. [17, 18]

Koevoets, T., and L. W. Beukeboom. 2009. Genetics of postzygotic isolation and Haldane's rule in haplodiploids. *Heredity* 102: 16–23. [17]

Kohn, J. R., and J. E. Biardi. 1995. Outcrossing rates and inferred levels of inbreeding depression in gynodioecious *Cucurbita foetidissima* (Cucurbitaceae). *Heredity* 75: 77–83. [12]

Kojima, K. 1959. Role of epistasis and overdominance in stability of equilibria with selection. *Proc. Natl. Acad. Sci. USA* 45: 984–989. [5]

Kojima, K.-I., and T. M. Kelleher. 1963. A comparison of purebred and crossbred selection schemes with two populations of *Drosophila pseudoobscura. Genetics* 48: 57–72. [17]

Kolbehdari, D., G. B. Jansen, L. R. Schaeffer, and B. O. Allen. 2005. Power of QTL detection by either fixed or random models in half-sibs designs. *Genet. Sel. Evol.* 37: 601–614. [19]

Kölreuter, J. G. 1766. Dritte fortsetzung der vorläufigen nachricht von einigen das geschlecht der pflanzen betreffenden versuchen und beobachtungen. Leipzig. [13]

Kondrashov, A. S. 1988. Deleterious mutations and the evolution of sexual reproduction. *Nature* 336: 435–440. [15]

Kondrashov, A. S. 1995. Contamination of the genome by very slightly deleterious mutations: why have we not died 100 times over? *J. Theor. Biol.* 175: 583–594. [15]

Kondrashov, A. S., and J. F. Crow. 1993. A molecular approach to estimating the human deleterious mutation rate. *Human Mutation* 2: 229–234. [15]

Kondrashov, A. S., and D. Houle. 1994. Genotype-environment interactions and the estimation of the genomic mutation rate in *Drosophila melanogaster. Proc. Royal Soc. Lond. B* 258: 221–227. [15]

Kondrashov, A. S., and M. Turelli. 1992. Deleterious mutations, apparent stabilizing selection and the maintenance of quantitative variation. *Genetics* 132: 603–618. [15]

Kondrashov, F. A., and E. V. Koonin. 2004. A common framework for understanding the origin of genetic dominance and evolutionary fates of gene duplications. *Trends Genet.* 20: 287–290. [4]

Kong, A., and 16 others. 2008. Detecting of sharing by descent, long-range phasing and haplotype imputation. *Nat. Genet.* 40: 1068–1075. [8]

Kong, A., and 20 others. 2012. Rate of *de novo* mutations and the importance of father's age to disease risk. *Nature* 488: 471–475. [15]

Konigsberg, L. W., C. M. Kammerer, and J. W. MacCluer. 1989. Segregation analysis of quantitative traits in nuclear families: comparison of three program packages. *Genet. Epidem.* 6: 713–726. [16]

Konigsberg, L. W., J. Blangero, C. M. Kammerer, and G. E. Mott. 1991. Mixed model analysis of LDL-C concentration with genotype-covariate interaction. *Genet. Epidem.* 8: 69–80. [16]

Konovalov, D. A., and D. I. K. Heg. 2008. Technical Advances: A maximum-likelihood relatedness estimator allowing for negative relatedness values. *Mol. Ecol. Res.* 8: 256–263. [A4]

Kooke, R., E. Wijnker, and J. J. B. Keurentjes. 2012. Backcross populations and near isogenic lines. *Meth. Mol. Biol.* 871: 3–16. [17]

Kopylev, L., and B. Sinha. 2011. On the asymptotic distribution of likelihood ratio test when parameters lie on the boundary. *Sankhya* B 73: 20–41. [A4]

Koricheva, J., J. Gurevitch, and K. Mengersen (eds). 2013. *Handbook of meta-analysis in ecology and evolution*. Princeton University Press, Princeton, NJ. [A6]

Korol, A. B., A. A. Zhuchenko, and A. P. Samovol. 1981. Linkage between loci of quantitative characters and marker loci. III. The bias of estimates during disturbance of the original hypothesis. *Genetika* 17: 1234–1247. [18]

Korol, A. B., A. A. Zhuchenko, and I. A. Preigel. 1983. Linkage between loci of quantitative characters and marker loci. IV. Evaluation of parameters by the least-squares method. *Genetika* 19: 594–601. [18]

Korol, A. B., I. A. Preigel, and N. I. Bocharnikova. 1987. Linkage between loci of quantitative characters and marker loci. V. Combined analysis of several markers and quantitative characters. *Genetika* 23: 1421–1431. [18]

Korol, A. B., Y. I. Ronin, and V. M. Kirzhner. 1995. Interval mapping of quantitative trait loci employing correlated trait complexes. *Genetics* 140: 1137–1147. [18]

Korol, A. B., and 5 others. 1996. Estimating variance effects of QTL: an important prospect to increase the resolution power of interval mapping. *Genet. Res.* 67: 187–194. [18]

Korte, A., and 5 others. 2012. A mixed-model approach for genome-wide association studies of correlated traits in structured populations. *Nat. Genet.* 44: 1066–1071. [20]

Kosambi, D. D. 1944. The estimation of map distances from recombination values. *Ann. Eugen.* 12: 172–175. [17]

Koski, V. 1971. Embryonic lethals of *Picea abies* and *Pinus sylvestris*. *Comm. Inst. For. Fenn.* 75:1–30. [12]

Kosuda, K. 1993. A further study of interchomosomal epistatic interaction in male mating activity of *Drosophila melanogaster*. *Heredity* 70: 370– 375. [17]

Kotz, S., N. Balakrishnan and N. L. Johnson. 2019. *Continuous multivariate distributions, 2nd Ed.* John Wiley & Sons, New York, NY. [2]

Koumproglou, R., and 7 others. 2002. STAIRS: a new genetic resources for functional genomic studies in *Arabidopsis*. *Plant J.* 31: 355–364. [17]

Kover, P. X., and 7 others. 2009. A multiparent advanced generation inter-cross to fine-map quantitative traits in *Arabidopsis thaliana PLoS Genet.* 5: e10000551. [18]

Kozak, G. M., and 6 others. 2019. Genomic basis for circannual rhythm in the European corn borer moth. *Curr. Biol.* 29: 3501–3509. [17]

Kozlov, M. V., and E. L. Zvereva. 2015. Confirmation bias in studies of fluctuating asymmetry. *Ecol. Indicators* 57: 293–297. [6]

Kraemer, S. A., and 4 others. 2017. Fitness change in relation to mutation number in spontaneous mutation accumulation lines of *Chlamydomonas reinhardtii*. *Evolution* 71: 2918–2929. [15]

Krieger, U., Z. B. Lippman, and D. Zamir. 2010. The flowering gene *SINGLE FLOWER TRUSS* drives heterosis for yield in tomato. *Nat. Genet.* 42: 459–463. [13]

Krimbas, C. B. 1961. Release of genetic variability through recombination. VI. *Drosophila willistoni*. *Genetics* 46: 323–1334. [5]

Kristensen, T. N., and A. C. Sørensen. 2005. Inbreeding — lessons from animal breeding, evolutionary biology and conservation genetics. *Anim. Sci* 80: 121–133. [12]

Kristensen, T. N., K. S. Pedersen, C. J. Vermeulen, and V. Loeschcke. 2010. Research on inbreeding in the 'omic' era. *Trends Ecol. Evol.* 25: 44–52. [12]

Kriszt, A., and 4 others. 2014. Segregation analysis suggests that keratoconus is a complex non-Mendelian disease. *Acta Ophthalmologica* 92: e562–e568. [16]

Kruglyak, L. 1996. Thresholds and sample sizes. *Nature Genetics* 14: 132–133. [19]

Kruglyak, L., and E. S. Lander. 1995a. High-resolution genetic mapping of complex traits. *Am. J. Hum. Genet.* 56: 1212–1223. [19]

Kruglyak, L., and E. S. Lander. 1995b. Complete multipoint sib-pair analysis of qualitative and quantitative traits. *Am. J. Hum. Genet.* 57: 439–454. [19]

Kruglyak, L., and E. S. Lander. 1995c. A nonparametric approach for mapping quantitative trait loci. *Genetics* 139: 1421–1428. [18]

Kruglyak, L., M. J. Daly, and E. S. Lander. 1995. Rapid multipoint linkage analysis of recessive traits in nuclear families, including homozygosity mapping. *Am. J. Hum. Genet.* 56: 519–527. [12]

Kruglyak, L., M. J. Daly, M. P. Reeve-Daly, and E. S. Lander. 1996. Parametric and nonparametic linkage analysis: A unified multipoint approach. *Am. J. Hum. Genet.* 58: 1347–1363. [19]

Kruse, R., S. A. Seuchter, M. P. Baur, and M. Knapp. 1997. The "possible triangle" test for extreme discordant sib pairs. *Genet. Epidemiol.* 14: 833–838. [19]

Kruuk, L. E. B., B. C. Sheldon, and J. Merila. 2002. Severe inbreeding depression in collared flycatchers (*Ficedula albicollis*). *Proc. Roy. Soc. Lond. B* 269: 1581–1589. [12]

Kryukov, G. V., A. Shpunt, J. A. Stramatoyannopoulos, and S R. Sunyaev. 2009. Power of deep, all-exon resequencing for discovery of human trait genes. *Proc. Natl. Acad. Sci. USA* 106: 3871–3876. [20]

Ku, C. S., N. Naidoo, S. M. Teo, and Y. Pawitan. 2011. Regions of homozygosity and their impact on complex diseases and traits. *Hum. Genet.* 129: 1–15. [12]

Kudo, A., K. Ito, and K. Tanaka. 1972. Genetic studies on inbreeding in some Japanese populations. X. The effects of parental consanguinity on psychometric measurements, school performances and school attendance in Shizuoka school-children. *Jap. J. Hum. Genetics* 17: 231–248. [12]

Kuehl, R. O. 2000. *Design of experiments: Statistical principles of research design and analysis*, 2nd. ed. Duxbury

Press, Pacific Grove, CA. [A9]

Kulathinal, R. J., and R. S. Singh. 2008. The molecular basis of speciation: from patterns to processes, rules to mechanisms. *J. Genet.* 87: 327–338. [17]

Kulp, D. C., and M. Jagalur. 2006. Causal inference of regulatory-target pairs by gene mapping of expression phenotypes. *BMC Genomics* 7: 125. [A2]

Kump, K. L., and 4 others. 2010. Joint analysis of near-isogenic and recombinant inbred line populations yields precise positional estimates for quantitative trait loci. *Plant Genome* 3: 142–153. [17]

Kump, K. L., and 11 others. 2011. Genome-wide association study of quantitative resistance to southern leaf blight in the maize nested association mapping population. *Nat. Genet.* 43: 163–168. [21]

Kuo, C.-L., and E. Feingold. 2010. What's the best statistic for a simple test of genetic association in a case-control study? *Genet. Epid.* 34: 246–253. [20]

Kuonen, D. 1999. Saddlepoint approximations for distributions of quadratic forms in normal variables. *Biometrika* 86: 929–935. [20]

Kusterer, B., and 8 others. 2007a. Analysis of a triple testcross design with recombinant inbred lines reveals a significant role of epistasis in heterosis for biomass-related traits in *Arabidopsis*. *Genetics* 175: 2009–2017. [13]

Kusterer, B., and 7 others. 2007b. Heterosis for biomass-related traits in *Arabidopsis* investigated by quantitative trait loci analysis of the triple testcross design with recombinant inbred lines. *Genetics* 177: 1839–1850. [13]

Kutka, F. J., and M. E. Smith. 2007. How many parents give the highest yield in predicted synthetic and composite populations of maize? *Crop. Sci.* 47: 1905–1913. [13]

Kwee, L. C., and 4 others. 2008. A powerful and flexible multilocus association test for quantitative traits. *Am. J. Hum. Genet.* 82: 386–397. [20]

Kwon, J. M., M. Boehnke, T. L. Burns, and P. P. Moll. 1990. Commingling and segregation analyses: comparisons of results from a simulation study of a quantitative trait. *Genet. Epidem.* 7: 57–68. [16]

Labate, J. A., K. R. Lamkey, M. Lee, and W. L. Woodman. 1999. Temporal changes in allele frequencies in two reciprocally selected maize populations. *Theor. Appl. Genet.* 99: 1166–1178. [17]

Lachance, J., and J. R. True. 2010. X-autosome incompatibilities in *Drosophila melanogaster*: tests of Haldane's rule and geographic patterns within species. *Evolution* 64: 3035–3046. [17]

Lacy, R. C. 1997. Importance of genetic variation to the viability of mammalian populations. *J. Mamm.* 78: 320–335. [12]

Lacy, R. C., A. Petric, and M. Warneke. 1993. Inbreeding and outbreeding in captive populations of wild animal species. *In* N. W. Thornhill (ed.), *The natural history of inbreeding and outbreeding: theoretical and empirical perspectives*, pp. 352–374. Univ. Chicago Press, Chicago, IL. [12]

Ladouceur, M., and 4 others. 2012. The empirical power of rare variant association methods: results from Sanger sequencing in 1,998 individuals. *PLoS Genet.* 8: e1002496. [20]

Lai, C., and 4 others. 1994. Naturally occurring variation in bristle number associated with DNA sequence polymorphisms at the *scabrous* locus of *Drosophila melanogaster*. *Science* 266: 1697–1702. [15, 17]

Laird, N. M., and C. Lange. 2006a. The role of family-based designs in genome-wide associations studies. *Stat. Sci.* 24: 388–397. [17, 20]

Laird, N. M., and C. Lange. 2006b. Family-based designs in the age of large-scale gene-association studies. *Nat. Rev. Genet.* 7: 385–394. [17, 20]

Laird, N. M., and C. Lange. 2008. Family-based methods for linkage and association analysis. *Adv. Genet.* 60: 219–252. [17, 20]

Laird, N. M., and C. Lange. 2011. *The fundamentals of modern statistical genetics.* Springer, New York, NY. [16]

Laird, N. M., D. Blacker, and M. Wilcox. 1998. The sib transmission/disequilibrium test is a Mantel-Haenszel test. *Am. J. Hum. Genet.* 63: 1915. [15]

Lalouel, J. M. 1992. Linkage analysis in human genetics. *In* J. S. Beckmann and T. C. Osborn (eds.), *Plant genomes: methods for genetic and physical mapping*, pp. 167–180. Kluwer Academic, Boston, MA. [17]

Lalouel, J. M., and N. E. Morton. 1981. Complex segregation analysis with pointers. *Hum. Hered.* 31: 312–321. [16]

Lalouel, J. M., D. C. Rao, N. E. Morton, and R. C. Elston. 1983. A unified model for complex segregation analysis. *Am. J. Hum. Genet.* 35: 816–826. [16]

Lamberson, W. R., and D. L. Thomas. 1984. Effects of inbreeding in sheep: a review. *Anim. Breeding Abst.* 52: 287–297. [12]

Lamkey, K. R., and J. W. Edwards. 1999. Quantitative genetics and heterosis. *In* J. G. Coors and S. Pandey (eds). *The genetics and exploitation of heterosis in crops*, pp. 31–48. American Society of Agrnomony, Madison, Wisconsin. [13]

Lamkey, K. R., and O. S. Smith. 1987. Performance and inbreeding depression of populations representing several eras of maize breeding. *Crop Sci.* 27: 695–699. [12]

Lamkey, K. R., A. R. Hallauer, and A. L. Kahler. 1987. Allelic differences at enzyme loci and hybrid performance in maize. *J. Hered.* 78: 231–234. [13]

Lan, H., and 5 others. 2003. Dimension reduction for mapping mRNA abundance as quantitative traits. *Genetics* 164: 1607–1614. [21]

Lancaster, H. O. 1961. The combination of probabilities: an application of orthonormal functions. *Austral. J. Stat.* 3: 20-3. [A6]

Landa, V., V. Zídek, and M. Pravenec. 2010. Generation of rat "supersonic" congenic/conplastic strains using superovulation and embryo transfer. *In* I. Anegon (ed.), *Rat genomcs: methods and protocols*, pp. 267-275. Humana Press, Totowa, NJ. [17]

Lande, R. 1975. The maintenance of genetic variation by mutation in a polygenic character with linked loci. *Genet. Res.* 26: 221–235. [15]

Lande, R. 1976. Natural selection and random genetic drift in phenotypic evolution. *Evolution* 30: 314–334. [15]

Lande, R. 1977. On comparing coefficients of variation. *Syst. Zool.* 26: 214–217. [14]

Lande, R. 1979. Quantitative genetic analysis of multivariate evolution, applied to brain:body allometry. *Evolution* 33: 402–416. [14]

Lande, R. 1981. The minimum number of genes contributing to quantitative variation between and within populations. *Genetics* 99: 541–553. [11]

Lande, R. 1988. Quantitative genetics and evolutionary theory. *In* B. S. Weir, E. J. Eisen, M. M. Goodman, and G. Namkoong (eds.), *Proceedings of the second international conference on quantitative genetics*, pp. 71–84. Sinauer Assoc., Sunderland, MA. [1]

Lande, R. 1994. Risk of population extinction from new deleterious mutations. *Evolution* 48: 1460–1469. [15]

Lande, R., and S. J. Arnold. 1983. The measurement of selection on correlated characters. *Evolution* 37: 1210–1226. [9]

Lande, R., and D. W. Schemske. 1985. The evolution of self-fertilization and inbreeding depression in plants. I. Genetic models. *Evolution* 39: 24–40. [12]

Lande, R., and R. Thompson. 1990. Efficiency of marker-assisted selection in the improvement of quantitative traits. *Genetics* 124: 743–756. [18, 20]

Lande, R., D. W. Schemske, and S. T. Schult. 1994. High inbreeding depression, selective interference among loci, and the threshold selfing rate for purging recessive lethal mutations. *Evoluiton* 48: 965–97. [12]

Lander, E. S. 1996. The new genomics: global views of biology. *Science* 274: 536–539. [21]

Lander, E. S., and D. Botstein. 1987. Homozyogsity mapping: A way to map human recessive traits with the DNA of inbred children. *Science* 236: 1567–1570. [12]

Lander, E. S., and D. Botstein. 1989. Mapping Mendelian factors underlying quantitative traits using RFLP linkage maps. *Genetics* 121: 185–199 (Correction 136: 705). [11, 17, 18, 19]

Lander, E. S., and P. Green. 1987. Construction of multilocus genetic linkage maps in humans. *Proc. Natl. Acad. Sci. USA* 84: 2363–2367. [19]

Lander, E. S., and L. Kruglyak. 1995. Genetic dissection of complex traits: guidelines for interpreting and reporting linkage results. *Nature Genet.* 11: 241–247. [19, A6]

Lander, E. S., and L. Kruglyak. 1996. Genetic dissection of complex traits. *Nat. Genet.* 12: 357–358. [18, 19]

Lander, E. S., and N. J. Schork. 1994. Genetic dissection of complex traits. *Science* 265: 2037–2048. [1, 18, 19, 20]

Lange, C., D. L. DeMeo and N. M. Laird. 2002. Power and design considerations for a general class of family-based association tests: quantitative traits. *Am. J. Hum. Genet.* 71: 1330–1341. [17]

Lange, K. 1986a. The affected sib-pair method using identity by state relations. *Am. J. Hum. Genet.* 39: 148–150. [19]

Lange, K. 1986b. A test statistic for the affected sib-set method. *Ann. Hum. Genet.* 50: 283–290. [19]

Lange, K., and M. Boehnke. 1982. How many polymorphic genes will it take to span the human genome? *Am. J. Hum. Genet.* 34: 842–845. [17]

Lange, K., and M. Boehnke. 1983. Extensions to pedigree analysis. V. Optimal calculation of Mendelian likelihoods. *Hum. Hered.* 33: 291–301. [16]

Lange, K, and R. C. Elston. 1975. Extensions to pedigree analysis. I. Likelihood calculations for simple and complex pedigrees. *Hum. Hered.* 25: 95–105. [16]

Lange, K., L. Kunkel, J. Aldridge, and S. A. Latt. 1985. Accurate and superaccurate gene mapping. *Am. J. Hum. Genet.* 37: 853–867. [17]

Lange, K., D. Weeks, and M. Boehnke. 1988. Programs for pedigree analysis: MENDEL, FISHER, and dGENE. *Genet. Epidem.* 5: 471–472. [16]

Langley, C. H. 1977. Nonrandom associations between allozymes in natural populations of *Drosophila melanogaster*. *In* F. B. Christiansen and T. M. Fenchel (eds.), *Measuring selection in natural populations*, pp. 265–273. Springer-Verlag, Berlin, De. [5]

Langley, C. H., Y. N. Tobari, and K. I. Kojima. 1974. Linkage disequilibrium in natural populations of *Drosophila melanogaster*. *Genetics* 78: 921–936. [5]

Langley, C. H., D. B. Smith, and F. M. Johnson. 1978. Analysis of linkage disequilibria between allozyme loci in natural populations of *Drosophila melanogaster*. *Genet. Res.* 32: 215–230. [5]

Lango Allen, H., et al. 2010. Hundreds of variants clusters in genomic loci and biological pathways affect human height. *Nature* 467: 832–838. [20, 21]

Lansing, A. I. 1947. A transmissible, cumulative and reversible factor in aging. *J. Gerontol.* 2: 228–239. [6]

Lansing, A. I. 1948. Evidence for aging as a consequence of growth cessation. *Proc. Natl. Acad. Sci. USA* 34: 304–310. [6]

LaPlace, P. S. 1778. *Memoire sur les probabilites*. Histoire de l'Academie Royale de Sciences, Annee 1778. pp. 227–332. [2]

Lappalainen, T., and 57 others. 2013. Transcritome and genome sequencing uncovers functional variation in humans. *Nature* 501: 506–511. [21]

Lariépe, A., and 13 others. 2012. The genetic basis of heterosis: multiparental quantitative trait loci mapping reveals contrasted levels of apparent overdominance among traits of agronomical interest in maize (*Zea mays* L.). *Genetics* 190: 795–811. [13]

Lark, K. G., and 4 others. 1995. Interactions between quantitative trait loci in soybean in which trait variation at one locus is conditional upon a specific allele at another. *Proc. Natl. Acad. Sci. USA* 92: 4656–4660. [18]

Larson, N. B., J. Chen, and D. J. Schaid. 2018. A review of kernel methods for genetic association studies. *Genet. Epidem.* 43: 122–136. [20]

Larsson, K. 1993. Inheritance of body size in the barnacle goose under different environmental conditions. *J. Evol. Biol.* 6: 195–208. [7]

Latta IV, L. C., and 5 others. 2013. Genomic background and generation time influence deleterious mutation rates in *Daphnia*. *Genetics* 193: 539–544. [15]

Latta, R., and K. Ritland. 1994. The relationship between inbreeding depression and prior inbreeding among populations of four *Mimulus* taxa. *Evolution* 48: 806–817. [12]

Latter, B. D. H. 1965. The response to artificial selection due to autosomal genes of large effect. I. Changes in gene frequency at an additive locus. *Aust. J. Biol. Sci.* 18: 585–598. [16]

Latter, B. D. H., and A. Robertson. 1962. The effects of inbreeding and artificial selection on reproductive fitness. *Genet. Res.* 3: 110–138. [12]

Latter, B. D. H., and J. A. Sved. 1994. A reevaluation of data from competitive tests shows high levels of heterosis in *Drosophila melanogaster*. *Genetics* 137: 509–511. [12]

Latter, B. D. H., J. C. Mulley, D. Reid, and L. Pascoe. 1995. Reduced genetic load revealed by slow inbreeding in *Drosophila melanogaster*. *Genetics* 139: 287–297. [12]

Launey, S., and D. Hedgecock. 2001. High genetic load in the pacific oyster *Crassostrea gigas*. *Genetics* 159: 255–265. [12]

Laurie, C. C. 1997. The weaker sex is heterogametic: 75 years of Haldane's rule. *Genetics* : 937–951. [17].

Laurie, C. C., and 9 others. 2004. The genetic architecture of response to long-term artificial selection for oil concentration in the maize kernel. *Genetics* 168: 2141–2155. [18]

Laurie, C. C., and 8 others. 2007. Linkage disequilibrium in wild mice. *PLoS Genetics* 3: e144. [5]

Laurie, C. C., and 24 others. 2010. Quality control and quality assurance in genotypic data from genome-wide associations studies. *Genet. Epid.* 34: 591–602. [20]

Laurier, V., and 15 others. 2006. Pitfalls of homozygosity mapping: an extended consanguineous Bardet–Biedl syndrome family with two mutant genes (*BBS 2, BBS 10*), three mutations, but no triallelism. *Europ. J. Hum. Genet.* 14: 1195–1203. [12]

Lauss, K., and 7 others. 2018. Parental DNA methylation states are associated with heterosis in epigenetic hybrids. *Plant Physiol.* 176: 1627–1645. [13]

Lavanch, E., and J. Goudet. 2022. Effect of reduced genomic representation on using runs of homozygosity for inbreeding characterization. *Mol. Ecol. Resour.* 23: 787–802. [8, 12]

Lavanch, E., B. S. Weir, and J. Goudet. 2024. Detecting inbreeding depression in structured populations. *Proc. Natl. Acad. Sci. USA* 121: e2315780121. [12]

Lavergne, S., N. J. Muenke, and J. Molofsky. 2010. Genome size reduction can trigger rapid phenotypic evolution in invasive plants. *Ann. Bot.* 105: 109–116. [15]

Law, C. N. 1966. The location of genetic factors affecting a quantitative character in wheat. *Genetics* 53: 487–498. [17]

Law, C. N., and M. D. Gale. 1979. Cytological markers and quantitative variation in wheat. *In* J. N. Thompson, Jr., and J. M. Thoday (eds.), *Quantitative genetic variation*, pp. 275–293. Academic Press, New York, NY. [17]

Lawless, J., and P. Wang. 1976. A simulation study of ridge and other regression estimators. *Comm. Stat. A* 5: 307–323. [20]

Lawlor, D. A., and 4 others. 2008. Mendelian randomization: using genes as instruments for making causal inferences in epidemiology. *Stat. Med.* 27: 1133–1163. [A2]

Lawlor, D., and 6 others. 2017. Using Mendelian randomization to determine causal effects of maternal pregnancy (intrauterine) exposures on offspring outcomes: Sources of bias and methods for assessing them. *Wellcome Open Res.* 2: 11. [A2]

Laws, R. J., and I. G. Jamieson. 2011. Is lack of evidence of inbreeding depression in a threatened New Zealand robin indicative of reduced genetic load? *Anim. Conserv.* 14: 47–55. [12]

Lawson, D. J., and D. Falush. 2012. Population identification using genetic data. *Ann. Rev. Genomics Hum. Genet.* 13: 337–361. [8, 20]

Le Cren, E. 1951. The length-weight relationship and seasonal cycle in gonad weight and condition in the perch (*Perca fluviatilis*). *J. Anim. Ecol.* 66: 1504–1512. [14]

Le Cressie, S., and J. van Houwelingen. 1992. Ridge estimators in logistic regression. *Appl. Stat.* 41: 191–201. [20]

Le Roy, P., and J. M. Elsen. 1992. Simple test statistics for major gene detection: a numerical comparison. *Theor. Appl. Genet.* 83: 6325–644. [16]

Le Roy, P., and J. M. Elsen. 1995. Numerical comparison between powers of maximum likelihood analysis of variance methods for QTL detection in progeny test designs: the case of monogenic inheritance. *Theor. Appl. Genet.* 90: 65–72. [19]

Le Roy, P., J. Naveau, J. M. Elsen, and P. Sellier. 1990. Evidence for a new major gene influencing meat quality in pigs. *Genet. Res.* 55: 33-40. [16]

Leamy, L. 1984. Morphometric studies in inbred and hybrid house mice. V. Directional and fluctuating asymmetry. *Am. Nat.* 123: 579–593. [6]

Leamy, L. 1992. Morphometric studies in inbred and hybrid house mice. VII. Heterosis in fluctuating asymmetry at different ages. *Acta Zool. Fennica* 191: 111–119. [6]

Leamy, L. J., and C. P. Klingenberg. 2005. The genetics and evolution of fluctuating asymmetry. *Annu. Rev. Ecol. Evol. Syst.* 36: 1–21. [6]

Leamy, L., and R. S. Thorpe. 1984. Morphometric studies in inbred and hybrid house mice. Heterosis, homeostasis and heritability of size and shape. *Biol. J. Linn. Soc.* 22: 233–241. [6]

Leamy, L. J., E. J. Routman, and J. M. Cheverud. 1998. Quantitative trait loci for fluctuating asymmetry of discrete skeletal characters in mice. *Heredity* 80: 509–518. [14]

Leary, R. F., F. W. Allendorf, and K. L. Knudson. 1983. Developmental stability and enzyme heterozygosity in rainbow trout. *Nature* 301: 71–72. [6]

Leary, R. F., F. W. Allendorf, and K. L. Knudson. 1984. Superior developmental stability of heterozygotes of enzyme loci in salmonid fishes. *Am. Nat.* 124: 540–551. [6]

Leary, R. F., F. W. Allendorf, and K. L. Knudson. 1985. Inheritance of meristic variation and the evolution of developmental stability in rainbow trout. *Evolution* 39: 308–314. [6]

Leary, R. F., F. W. Allendorf, and K. L. Knudson. 1987. Differences inbreeding coefficients do not explain the association between heterozygosity at allozyme loci and developmental stability in rainbow trout. *Evolution* 41: 1413–1145. [6, 12]

Leary, R. F., F. W. Allendorf, and K. L. Knudson. 1992. Genetic, environmental, and developmental causes of meristic variation in rainbow trout. *Acta Zool. Fennica* 191: 79–95. [6]

Leberg, P. L. 1993. Strategies for population reintroduction: effects of genetic variability on population growth and size. *Cons. Biol.* 7: 194–199. [13]

Leberg, P. L., and B. D. Firmin. 2008. Role of inbreeding depression and purging in captive breeding and restoration programmes. *Mol. Ecol.* 17: 334–343. [12]

Lebowitz, R. J., M. Soller, and J. S. Beckmann. 1987. Trait-based analyses for the detection of linkage between marker loci and quantitative trait loci in crosses between inbred lines. *Theor. Appl. Genet.* 73: 556–562. [17]

LeClerg, E. L. 1966. Significance of experimental design in plant breeding. *In* K. J. Frey (ed) *Plant breeding*, pp. 243–313. Iowa State University press, Ames, IA. [A9]

LeClerg, E. L., W. H. Leonard, and A. G. Clark. 1962. *Field plot technique*, 2nd ed. Brugess, Minneapolis, MN. [A9]

Ledig, F. T., R. P. Guries, and B. A. Bonefeld. 1983. The relation of growth to heterozygosity in pitch pine. *Evolution* 37: 1227–1238. [6, 12]

Ledur, M. C., N. Navarro, and M. Pérez-Enciso. 2010. Large-sale SNP genotyping in crosses between outbred lines: how useful is it? *Heredity* 105: 173–182. [21]

Lee, C. H., E. Eskin, and B. Han. 2017. Increasing the power of meta-analysis of genome-wide associations studies to detect heterogeneous effects. *Bioinformatics* 33: i379–i388. [20]

Lee, H., and 5 others. 2002. Application of the false discovery rate to quantitative trait loci interval mapping with multiple traits. *Genetics* 161: 905–914. [18, A6]

Lee, H.-I., H.-A. Ho, and C.-H. Kao. 2014. A new simple method for improving QTL mapping under selective genotyping. *Genetics* 198: 1685–1698. [19]

Lee, H. S., and 11 others. 2012. Estimating the proportion of variation in susceptibly to schizophrenia captures by common SNPs. *Nat. Genet.* 3: 247–250. [21]

Lee, M., E. B. Godshalk, K. R. Lamkey, and W. W. Woodman. 1989. Association of restriction fragment length polymorphisms among maize inbreds with agronomic performance of their crosses. *Crop Sci.* 29: 1067–1071. [13]

Lee, P. M. 2013. *Bayesian statistics: An introduction*, 4th ed. Wiley, New York, NY. [A7]

Lee, S., F. A. Wright, and F. Zou. 2011. Control of population stratification by correlation-selected principal components. *Biomet.* 67: 967–974. [20]

Lee, S., M. C. Wu, and X. Lin. 2012. Optimal test for rare variant effects in sequencing association studies. *Biostat.* 13: 762–775. [20]

Lee, S., T. M. Teslovich, M. Boehnke, and X. Lin. 2013. General framework for meta-analysis of rare variants in sequencing association studies. *Am. J. Hum. Genet.* 93: 42–53. [20]

Lee, S., G.R. Abecasis, M. Boehnke, and X. Lin. 2014. Rare-variant association analysis: study designs and statistical tests. *Am. J. Hum. Genet.* 95: 5–23. [20]

Lee, W.-C. 2015. Optimal trend tests for genetic association studies of heterogeneous diseases. *Sci. Reports* 6: 27821. [18]

Leek, J. T., and J. D. Storey. 2007. Capturing heterogeneity in gene expression studies by surrogate variables. *PLoS Genet.* 3: e161. [A6]

Leek, J. T., and J. D. Storey. 2008. A general framework for multiple testing dependence. *Proc. Natl. Acad. Sci. USA* 105: 18718–18723. [A6]

Leek, J. T., and J. D. Storey. 2011. The joint null criterion for multiple hypothesis tests. *Stat. Appl. Genet. Mol. Biol.* 10: Issue 1, Article 28. [A6]

Leemis, L. M. 1986. Relationships among common univariate distributions. *Am. Stat.* 40: 143–146. [2]

Legarra, A., I. Aguilar, and I. Misztal. 2009. A relationship matrix including full pedigree and genomic information. *J. Dairy. Sci.* 92: 4656–4663. [8]

Lehesjoki, A.-E., and 6 others. 1993. Localization of the *EMP1* gene for progressive myoclonus epilepsy on chromosome 21: linkage disequilibrium allows high resolution mapping. *Human Mol. Genet.* 2: 1229–1234. [17, 20]

Leips, J., and T. F. C. Mackay. 2000. Quantitative trait loci for life span in *Drosophila melanogaster*: interactions with genetic background and larval density. *Genetics* 155: 1773–1788. [18]

Leiser, W. L., H. F. Rattunde, H.-P. Piepho, and H. K. Parzies. 2012. Getting the most out of sorghum low-input field trails in West Africa using spatial adjustment. *J. Agron. Crop Sci.* 198: 349–359. [A9]

Lencz, T., and 7 others. 2007. Runs of homozygosity reveal highly penetrant recessive loci in schizophrenia. *Proc. Natl. Acad. Sci. USA* 104: 19942–19947. [12]

Lerner, I. M. 1954. *Genetic homeostasis.* Oliver and Boyd, London, UK. [6, 13]

Leroy, G. 2014. Inbreeding depression in livestock species: review and meta-analysis. *Anim. Genet.* 45: 618–628. [12]

Leroy, G., and 6 others. 2012. An ABC estimate of pedigree error rate: application in dog, sheep and cattle breeds. *Anim. Genet.* 43: 309–314. [8]

Lessells, C. M., and P. T. Boag. 1987. Unrepeatable repeatabilities: a common mistake. *Auk* 104: 116–121. [6]

Lettre, G., C. Lange, and J. N. Hirschhorn. 2007. Genetic model testing and statistical power in population-based association studies of quantitative traits. *Genet. Epid.* 31: 358–362. [20]

Leung B., and M. R. Forbes. 1996. Fluctuating asymmetry in relation to stress and fitness: effects of trait type as revealed by meta-analysis. *Ecoscience* 1: 400–413. [6]

Leutenegger, A.-L., and 6 others. 2003. Estimation of the inbreeding coefficientthrough use of genomic data. *Am. J. Hum. Genet.* 73: 516–523 [8]

Leutenegger, A.-L., and 6 others. 2006. Using genomic inbreeding coefficient estimates for homozygosity mapping of rare recessive traits: application to Taybi-Linder syndrome. *Am. J. Hum. Genet.* 79: 62–66. [12]

Levin, D. A. 1989. Inbreeding depression in partially self-fertilizing *Phlox. Evolution* 43: 1417–1423. [12]

Levin, M. 2005. Left–right asymmetry in embryonic development: a comprehensive review. *Mechanisms Develop.* 122: 3–25. [6]

Levinson, D. F., and 6 others. 2014. Genetic studies of major depressive disorder; why are there no GWAS findings, and what can we do about it? *Biol. Psychiatry* 76: 510–512. [19, 20]

Levy, S. F., and M. L. Siegal. 2008. Network hubs buffer environmental variation in *Sacchaomyces cerevisiae*. *PLoS Biol.* 6: e264. [21, A2]

Lewis, D. 1954. A relationship between dominance, heterosis, phenotypic stability and variability. *Heredity* 8: 333–356. [6]

Lewis, E. B. 1945. The relation of repeats to position effect in *Drosophila melanogaster. Genetics* 30: 137–166. [21]

Lewontin, R. C. 1957. The adaptations of populations to varying environments. *Cold Spring Harbor Symp. Quant. Biol.* 22: 395–408. [6]

Lewontin, R. C. 1964. The interaction of selection and linkage. II. Optimal model. *Genetics* 50: 757–782. [5]

Lewontin, R. C. 1974. *The genetic basis of evolutionary change.* Columbia Univ. Press, New York, NY. [5, 8, 12, 13]

Lewontin, R. C. 1988. On measures of gametic disequilibrium. *Genetics* 120: 849–852. [5]

Lewontin, R. C., and L. C. Birch. 1966. Hybridization as a source of variation for adaptation to new environments. *Evolution* 20: 315–336. [18]

Lewontin, R. C., and C. C. Cockerham. 1959. The goodness of fit test for detecting natural selection in random mating populations. *Evolution* 13: 561–564. [4]

Lewontin, R. C., and J. L. Hubby. 1966. A molecular approach to the study of genic heterozygosity in natural populations. II. Amount of variation and degree of heterozygosity in natural populations of *Drosophila pseudoobscura*. *Genetics* 54: 595–609. [8, 12]

Lewontin, R. C., and K. Kojima. 1960. The evolutionary dynamics of complex polymorphisms. *Evolution* 14: 458–472. [5]

Lewontin, R. C., J. A. Moore, W. B. Provine, and B. Wallace (eds.). 1981. *Dobzhansky's genetics of natural populations I-XLIII*. Columbia University, New York, NY. [5, 13, 17]

Li, A., and J. L. Gastwirth. 2001. A weighted test using both extreme discordant and concordant sib pairs for detecting linkage. *Genet. Epidemiol.* 20: 34–43. [19]

Li, A., and 10 others. 2023. mBAT-combo: a more powerful test to detect gene-trait associations from GWAS data. *Am. J. Hum. Genet.* 110: 30–43. [20]

Li, B., and S. M. Leal. 2008. Methods for detecting associations with rare variants for common diseases: application to analysis of sequence data. *Am. J. Hum. Gent.* 83: 311–321. [20]

Li, B., and S. M. Leal. 2009. Discovery of rare variants via sequencing; implications for the design of complex trait association studies. *PLoS Genet.* 5: e1000481. [20]

Li, C. C. 1953. Some general properties of recessive inheritance. *Am. J. Hum. Genet.* 5: 269–270. [4]

Li, C. C. 1975. *Path analysis — a primer*. Boxwood, Pacific Grove, CA. [A2]

Li, C. C., and A. Chakravarti. 1985. Basic fallacies in the formulation of the paternity index. *Am. J. Hum. Genet.* 37: 809–818. [8]

Li, C. C., and D. G. Horvitz. 1953. Some methods of estimating the inbreeding coefficient. *Am. J. Hum. Genet.* 5: 107–117. [8]

Li, D., and 4 others. 2011. Using extreme phenotype sampling to identify the rare causal variants of quantitative traits in association studies. *Genet. Epidem.* 35: 790–799. [20]

Li, D., and 24 others. 2016. Integrated analysis of phenome, genome, and transcriptome of hybrid rice uncovered multiple heterosis-related loci for yield increase. *Proc. Natl. Acad. Sci. USA* 113: E6026–E6035. [13]

Li, H., and 4 others. 2011. Joint QTL linkage mapping for multiple-cross mating designs sharing one common parent. *PLoS One* 6: e17573. [18]

Li, H., and 12 others. 2014. Relationship estimation from whole-genome sequence data. *PLOS Genet.* 10: e1004144. [8]

Li, J., and L. Ji. 2005. Adjusting multiple testing in multilocus analyses using the eigenvalues of a correlation matrix. *Heredity* 95: 221–227. [18, 19, 20, A6]

Li, J., and 4 others. 2011. The bayesian lasso for genome-wide associations studies. *Bioinformatics* 27: 516–523. [20]

Li, J., A. Bus, V. Spamer, and B. Stich. 2016. Comparison of statistical models for nested association mapping in rapeseed (*Brassica napus* L) through computer simulations. *BMC Plant Biol.* 16: 26. [18]

Li, M.-X., H.-S. Gui, J. S. H. Kwan, and P. C. Sham. 2011. GATES: a rapid and powerful gene-based association test using extended Simes procedure. *Am. J. Hum. Genet.* 88: 283–293. [20, A6]

Li, M.-X., J. S. H. Kwan, and P. C. Sham. 2012a. HYST: a hybrid set-based test for genome-wide association studies, with application to protein-protein interaction-based association analysis. *Am. J. Hum. Genet.* 91: 478–488. [21].

Li, M.-X., J. M. Yeung, S. S. Cherny, and P. C. Sham. 2012b. Evaluating the effective numbers of independent tests and significant p-value thresholds in commercial genotyping arrays and public imputation reference datasets. *Hum. Genet.* 131: 747–756. [A6]

Li, N., and M. Stephens. 2003. Modeling linkage disequilibrium and identifying recombination hotspots using SNP data. *Genetics* 165: 2213–2233. [20]

Li, Q., and K. Yu. 2008. Improved correction for population stratification in genome-wide associations studies by identifying hidden population structure. *Genet. Epid.* 32: 215–226. [20]

Li, R., and 4 others. 2005. Combining data from multiple inbred line crosses improves the power and resolution of quantitative trait loci mapping. *Genetics* 169: 1699–1709. [18]

Li, S., and 13 others. 2014. Detecting and correcting systematic variation in large-scale RNA sequencing data. *Nat. Biotech.* 32: 888–895. [21]

Li, W.-H., and M. Nei. 1972. Total number of individuals affected by a single deleterious mutation in a finite population. *Am. J. Hum. Genet.* 24: 667–679. [15]

Li, X., and 4 others. 2015. Dissecting repulsion linkage in the dwarfing gene *Dw3* region for sorghum plant height provides insight into heterosis. *Proc. Natl. Acad. Sci. US* 112: 11823–11828. [13]

Li, Y., W. K. Sung, and J. J. Liu. 2007. Association mapping via regularized regression analysis of single-nucleotide-polymorphism haplotypes in variable-sized sliding windows. *Am. J. Hum. Genet.* 80: 705–715. [20]

Li, Y., C. Willer, S. Sanna, and G. Abecasis. 2009. Genotype imputation. *Ann. Rev. Genom. Hum. Genet.* 10: 387–406. [20]

Li, Y. I., and 7 others. 2016. RNA splicing is a primary link between genetic variation and disease. *Science* 352: 600–604. [21]

Li, Z., S. R. M. Pinson, J. W. Stansel, and W. D. Park. 1995. Identification of quantitative trait loci (QTLs) for heading date and plant height in cultivated rice (*Oryza sativa* L.). *Theor. Appl. Genet.* 91: 374–381. [18]

Li, Z.-K., and 10 others. 2001. Overdominant epistatic loci are the primary genetic basis of inbreeding depression and heterosis in rice. I. Biomass and grain yield. *Genetics* 158: 1737–1753. [13]

Liang, K.-Y., C.-Y. Huang, and T. H. Beaty. 2000. A unified sampling approach for multipoint analysis of qualitative and quantitative traits in sib pairs. *Am. J. Hum. Genet.* 66: 1631–1641. [19]

Liao, T. F. 1994. *Interpreting probability models: logit, probit, and other generalized linear models*. Sage University Papers Series on Quantitative Applications in the Social Sciences. No. 07-101. Sage, Thousand Oaks, CA. [16]

Liberman, U., and S. Karlin. 1984. Theoretical models of genetic map functions. *Theor. Popul. Biol.* 25: 331–346. [17]

Light, R., and P. Smith. 1971. Accumulating evidence: procedures for resolving contradictions among different research studies. *Harv. Educ. Rev.* 41: 429–471. [A6]

Light, R. J., and D. B. Pillemer. 1984. *Summing up: the science of reviewing research*. Harvard Univ. Press. Cambridge, MA. [A6]

Lin, C. S., M. R. Binns, H. D. Voldeng, and R. Guillemette. 1993. Performance of randomized block designs in field experiments. *Agron. J.* 85: 168–171. [A9]

Lin, D. Y. 2005. An efficient Monte Carlo approach to assessing statistical significance in genome studies. *Bioinformatics* 21: 781–787. [20]

Lin, D. Y., and P. E. Sullivan. 2009. Meta-analysis of genome-wide association studies with overlapping subjects. *Am. J. Hum. Genet.* 85: 862–872. [20]

Lin, D. Y., and Z. Z. Tang. 2011. A general framework for detecting disease associations with rare variants in sequencing studies. *Am. J. Hum. Genet.* 89: 354–367. [20]

Lin, D. Y., and D. Zeng. 2009. Meta-analysis of genome-wide association studies: no efficiency gain in using individual participant data. *Genet. Epidem* 34: 60–66. [20]

Lin, D. Y., and D. Zeng. 2010. On the relative efficiency of using summery statistics versus individual level data in meta-analysis. *Biometrika* 97: 321–332. [20]

Lin, D. Y., and F. Zou. 2004. Assessing genomewide statistical significance in linkage studies. *Genet. Epid.* 27: 202–214. [19]

Lin, J. Z., and K. Ritland. 1997. Quantitative trait loci differentiating the outbreeding *Mimulus guttatus* from the inbreeding *M. platycalyx*. *Genetics* 146: 1115–1121. [18]

Lin, M., and 4 others. 2021. Admixed populations improve power for variant discovery and portability in genome-wide association studies. *Front. Genet.* 12: Article 673167. [20]

Lin, P., and 12 others. 2010. A new statistic to evaluate imputation reliability. *PLoS ONE* 5: e9697. [20]

Lin, W.-Y., X.-Y. Lou, G. Gao, and N. Liu. 2014. Rare variant association testing by adaptive combination of *p*-values. *PLoS ONE* 9: e85728. [20]

Lin, Y.-R., K. F. Schertz, and A. H. Paterson. 1995. Comparative analysis of QTLs affecting plant height and maturity across the Poaceae, in reference to an interspecific sorghum population. *Genetics* 141: 391–411. [17, 18]

Lin, Z., and 14 others. 2020. Divergence selection and genetic introgression shape the genome landscape of heterosis in hybrid rice. *Proc. Natl. Acad. Sci. USA* 117: 4623–4631. [13]

Linares, J. F., and 12 others. 2022. Transgene by germplasm interactions can impact transgene evaluation. *Crop Sci.* 63: 1988–1997. [5]

Lindholm, A. K., and 22 others. 2016. The ecology and evolutionary dynamics of meiotic drive. *Trends Ecol. Evol.* 31: 315–326. [17].

Lindley, D. V. 1965. *Introduction to probability and statistics from a Bayesian viewpoint* (2 Volumes), University Press, Cambridge, UK. [A7]

Lindquist, S., and E. A. Craig. 1988. The heat-shock proteins. *Ann. Rev. Genet.* 22: 631–677. [14]

Lindsley, D. L., and 16 others. 1972. Segmental aneuploidy and the genetic gross structure of the *Drosophila* genome. *Genetics* 71: 157–184. [15]

Lindstrom, E. W. 1924. A genetic linkage between size and color factors in the tomato. *Science* 60: 182–183. [17]

Lindstrom, E. W. 1931. Genetic tests for linkage between row number and certain qualitative genes in maize. *Res. Bull. Iowa State Coll. Agric.* 142: 250–288. [17]

Lints, F. A. 1978. *Genetics and ageing*. S. Karger, Basel, CH. [6]

Lints, F. A., and S. Baeten. 1981. Studies on the descendency of four populations of Koekelaere pines: *Pinus nigra* Arnold, subsp. *laricio* Maire, cv. *Koekelaere*. *Gerontology* 27: 20–31. [6]

Lippert, C., and 5 others. 2011. FaST linear mixed models for genome-wide association studies. *Nat. Method.* 8: 833–835. [20]

Lippman, Z. B., and D. Zamir. 2007. Heterosis revisiting the magic. *Trends Genet.* 23: 60–66. [13]

Liptak, T. 1958. On the combination of independent tests. *Magyar Tudomanyos Akademia Matematikai Kutato Intezetenek Kozlemenyei* 3: 1971–1977. [A6]

Listgarten, J., and 5 others. 2012. Improved linear mixed models for genome-wide association studies. *Nat. Genet.* 9: 525–526. [20]

Listgarten, J., C. Lippert, and D. Heckerman. 2013. FaST-LMM-Select for addressing confounding from spatial structure and rare variants. *Nat. Genet.* 45: 470–471. [20]

Liti, G., and E. J. Louis. 2012. Advances in quantitative trait analysis in yeast. *PLoS Genet.* 8: e1002912. [5, 18]

Liti, G., and 25 others. 2009. Population genomics of domestic and wild yeasts. *Nature* 458: 337–341. [18]

Little, T. M., and F. J. Hillis. 1978. *Agricultural experimentation: Design and analysis*. Wiley, New York, NY. [A9]

Liu, B., A. de la Fuent, and I. Hoeschele. 2008. Gene network inference via structural equation modeling in genetical genomics experiments. *Genetics* 178: 1763–1776. [A2]

Liu, B., and 4 others. 2019. Abundant associations with gene expression complicate GWAS follow-up. *Nat. Genet.* 51: 768–769. [21]

Liu, D. J., and S. M. Leal. 2010. A novel adaptive method for the analysis of next-generation sequencing data to detect complex trait associations with rare variants due to gene main effects and interactions. *PLoS Genet.* 6: e1001156. [20]

Liu, H., and 10 others. 2017. Distant eQTLs and non-coding sequences play critical roles in regulating gene expression and quantitative trait variation in maize. *Molecular Plant* 10: 414–426. [21]

Liu, H., and 25 others. 2019. Genome wide identification and analysis of heterotic loci in three maize hybrids. *Plant Biotech. J.* 18: 185–194. [13]

Liu, J., W. H. Wong, and A. Kong. 1991. Correlation structure and convergence rates of the Gibbs sampler (I): Application to the comparison of estimators and augmentation schemes. Technical Report 299, Dept. Statistics, University of Chicago, Chicago, IL. [A8]

Liu, J., and 5 others. 1996. Genetic analysis of a morphological shape difference in the male genitalia of *Drosophila simulans* and *D. mauritiana*. *Genetics* 142: 1129–1145. [17]

Liu, J. S., and 4 others. 2001. Bayesian analysis of haplotypes for linkage disequilibrium mapping. *Genome Res.* 11: 1716–1724 [17]

Liu, J. Z., and 11 others. 2010. A versatile gene-based test for genome-wide association studies. *Am. J. Hum. Genet.* 87: 139–145. [20]

Liu, J., M. Li, Q. Zhang, and X. Huang. 2020. Exploring the molecular basis of heterosis in plant breeding. *J. Integ. Plant Bio.* 62: 287–298. [13]

Liu, R., and 8 others. 2012. Fine mapping and candidate gene prediction of a pleiotropic quantitative trait locus for yield-related trait in *Zea mays*. *PLoS One* 7: e49836. [17]

Liu, S.-C., and 4 others. 1996. Genome-wide high-resolution mapping by recurrent intermating using *Arabidopsis thaliana* as a model. *Genetics* 142: 247–258. [18]

Liu, W., and 8 others. 2013. Optimum design of family structure and allocation of resources in association mapping with lines from multiple crosses. *Heredity* 110: 71–79. [18]

Liu, W., and 4 others. 2022. From hybrid genomes to heterotic trait output: Challenges and opportunities. *Curr. Opinion Plant Bio.* 66: 102193. [13]

Liu, X., M. Huang, B. Fan, E. S. Buckler, and Z. Zhang. 2016. Iterative usage of fixed and random effect models for powerful and efficient genome-wide association studies. *PLoS Genet.* 12: e1005767. [20]

Liu, X., and 10 others. 2017. Functional architectures of local and distant regulation of gene expression in multiple human tissues. *Am. J. Hum. Genet.* 100: 605–616. [21]

Liu, X., Y. I. Li, and J. K. Pritchard. 2019. *Trans* effects on gene expression can drive omnigenetic inheritance. *Cell* 177: 1022–1034. [21]

Liu, Y, and J. Xie. 2020. Cauchy combination test: a powerful test with analysis *p*-value calculation under arbitrary dependence structures. *J. Am. Statist.. Assoc.* 115: 393–402. [20, A6]

Liu, Y., and Z.-B. Zeng. 2000. A general mixture model approach for mapping quantitative trait loci from diverse cross designs involving multiple inbred lines. *Genet. Res.* 75: 345–355. [18]

Liu, Y., and 5 others. 2019. ACAT: a fast and powerful *p* value combination method for rare-variant analysis in sequencing studies. *Am. J. Hum. Genet.* 104: 410–421. [20, A6]

Liu, Y., and 9 others. 2020. Identification of quantitative trait loci for kernel-related traits and the heterosis for these traits in maize (*Zea mays* L.). *Mol. Genet. Genom.* 95: 121–133. [13]

Liu, Y., and 8 others. 2021. Identifying complex gene-gene interactions: a mixed kernel omnibus testing approach. *Brief. Bioinformat.* 22: 1–15. [20]

Livesay, E. A. 1930. An experimental study of hybrid vigor or heterosis in rats. *Genetics* 15: 17–54. [6]

Livshits, G., and E. Kobyliansky. 1984. Comparative analysis of morphological traits in biochemically homozygous and heterozygous individuals from a single population. *J. Hum. Evol.* 13: 161–171. [6]

Livshits, G., and P. E. Smouse. 1993. Relationship between fluctuating asymmetry, morphological modality and heterozygosity in an elderly Israeli population. *Genetica* 89: 155–166. [6]

Ljungberg, K., S. Holmgren, and Ö. Carlborg. 2004. Simultaneous search for multiple QTL using the global optimization algorithm DIRECT. *Bioinformatics* 20: 1887–1895. [18]

Lloyd-Jones, L. R., and 15 others. 2017. The genetic architecture of gene expression in peripheral blood. *Am. J. Hum. Genet.* 100: 228–237. [21]

Loewe, L., V. Textor, and S. Scherer. 2003. High deleterious genomic mutation rate in stationary phase of *Escherichia coli*. *Science* 302: 1558–1560. [15]

Logsdon, B. A., and 19 others. 2013. A variant Bayes discrete mixture test for rare variant association. *Genet. Epidem.* 38: 21–30. [20]

Loh, P.-R., and 11 others. 2015a. Efficient Bayesian mixed-model analysis increases association power in large cohorts. *Nat. Genet.* 46: 284–290. [20]

Loh, P.-R., and 14 others. 2015b. Contrasting genetic architectures of schizophrenia and other complex diseases using fast variance-components analysis. *Nat. Genet.* 47: 1385–1392. [21]

Loh, P.-R., and 4 others. 2018. Mixed-model association for biobank-scale datasets. *Nat. Genet.* 50: 906–908. [20]

Lohr, J. N., and C. R. Haag. 2015. Genetic load, inbreeding depression, and hybrid vigor covary with population size: an empirical evaluation of theoretical prediction. *Evolution* 69: 3109–3122. [12, 13]

Loisel, P., B. Goffinet, H. Monod, and G. M. De Oca. 1994. Detecting a major gene in an F_2 population. *Biometrics* 50: 512–516. [16]

Loison, L. 2019. Canalization and genetic assimilation: Reassessing the radicality of the Waddingtonian concept of inheritance of acquired characters. *Sem. Cell Devel. Bio.* 88: 4–13. [14]

Long, A. D., and 5 others. 1995. High resolution mapping of genetic factors affecting abdominal bristle number in *Drosophila melanogaster*. *Genetics* 139: 1273–1291. [8, 17, 18]

Loosli, R., E. S. Russell, W. K. Silvers, and J. L. Southard. 1961. Variability of incidence and clinical manifestation of mouse hereditary muscular dystrophy on heterogeneous genetic backgrounds. *Genetics* 46: 347–355. [17]

Loots, G. G., and 6 others. 2000. Identification of a coordinate regulator of interleukins 4, 3, and 5 by cross-species sequence comparisons. *Science* 288: 136–140. [15]

Lopes, M. S., and 6 others. 2013. Improved estimation of inbreeding and kinship in pigs using optimized SNP panels. *BMC Genet.* 14: 92. [8]

López, M. A., and C. López-Fanjul. 1993. Spontaneous mutation for a quantitative trait in *Drosophila melanogaster*. I. Response to artificial selection. *Genet. Res.* 61: 107–116. [15]

López-Cortegano, E., A. Vilas, A. Caballero, and A. García-Dorado. 2016. Estimation of genetic purging under competitive conditions. *Evolution* 70: 1856–1870. [12]

López-Cortegano, E., D. Bersabé, J. Wang, and A. García-Dorado. 2018. Detection of genetic purging and predictive value of purging parameters estimated in pedigreed populations. *Heredity* 121: 38–51. [12]

López-Cortegano, E., E. Moreno, and A. García-Dorado. 2021. Genetic purging in captive endangered ungulates with extremely low effective population sizes. *Heredity* 127: 433–442. [12]

López-Fanjul, C., and B. Jódar. 1977. The genetic properties of egg laying of virgin females of *Tribolium castaneum*. *Heredity* 39: 251–258. [12]

Loughin, T. M. 2004. A systematic comparison of methods for combining *p*-values from independent test. *Compt. Stat. Data Anal.* 47: 467–485. [20, A6]

Lowry, D. C., and F. Shultz. 1959. Testing associations of metric traits and marker genes. *Ann. Hum. Gen.* 23: 83–90. [19]

Lu, H., J. Romero-Severson, and R. Bernardo. 2003. Genetic basis of heterosis explored by simple sequence repeat markers in a random-mated maize population. *Theor. Appl. Genet.* 107: 494–502. [13]

Lu, X., and 12 others. 2010. Genome-wide misexpression of X-linked versus autosomal genes associated with hybrid male sterility. *Genome Res.* 20: 1097–1102. [17]

Luckinbill, L. S., J. L. Graves, A. H. Reed, and S. Koetsawang. 1988. Localizing genes that defer senescence in *Drosophila melanogaster*. *Heredity* 60: 367–374. [11, 17]

Lui, L, and S. Xu. 2003. Mapping viability loci using molecular markers. *Heredity* 90: 459–467. [12]

Luo, L. J., and 9 others. 2001. Overdominant epistatic loci are the primary genetic basis of inbreeding depression and heterosis in rice. II. Grain yield components. *Genetics* 158: 1755–1771. [13]

Luo, Z. W. 1993. The power of two experimental designs for detecting linkage between a marker locus and a locus affecting a quantitative character in a segregating population. *Genet. Sel. Evol.* 25: 249–261. [19]

Luo, Z. W., and M. J. Kearsey. 1992. Interval mapping of quantitative trait loci in an F_2 population. *Heredity* 69: 236–242. [18]

Luo, Z. W., and J. A. Wolliams. 1993. Estimation of genetic parameters using linkage between a marker gene and a locus underlying a quantitative character in F_2 populations. *Heredity* 70: 245–253. [18]

Luria, S. E., and M. Delbrück. 1943. Mutations from bacteria from virus sensitivity to virus resistance. *Genetics* 28: 491–511. [17]

Lush, J. L. 1937. *Animal breeding plans.* Iowa State Univ. Press, Ames, IA. [1, 3]

Lush, J. L. 1945. *Animal breeding plans, 3rd Edition.* Iowa State University Press, Ames, IA. [1]

Lutz, S., C. Brion, M. Kliebhan, and F. W. Albert. 2019. DNA variants affecting the expression of numerous genes in *trans* have diverse mechanism of action and evolutionary histories. *PLoS Genet.* 15: e1008375. [21]

Lutz, S., K. Van Dyke, M. A. Fraru, and F. W. Albert. 2022. Multiple epistatic DNA variants in a single gene affect gene expression in *trans*. *Genetics* 220: iyab208. [21]

Lye, Z., J. Y. Choi, and M. D. Purugganaan. 2022. Deleterious mutations and the rare allele burden on rice gene expression. *Mol. Biol. Evol.* 39: msac 193. [21]

Lyman, R. F., F. Lawrence, S. V. Nuzhdin, and T. F. C. Mackay. 1996. Effects of single *P* element insertions on bristle number and viability in *Drosophila melanogaster. Genetics* 143: 277–292. [15]

Lyman, R. F., C. Lai, and T. F. C. Mackay. 1999. Linkage disequilibrium mapping of molecular polymorphisms at the *scabrous* locus associated with naturally occurring variation in bristle number in *Drosophila melanogaster*. *Genet. Res.* 74: 303–311. [20]

Lynch, C. B. 1977. Inbreeding effects upon animals derived from a wild population of *Mus musculus*. *Evolution* 31: 526–537. [12]

Lynch, M. 1984. The limits to life history evolution in *Daphnia*. *Evolution* 38: 465–482. [5]

Lynch, M. 1985. Spontaneous mutations for life-history characters in an obligate parthenogen. *Evolution* 39: 804–818. [15]

Lynch, M. 1988a. Design and analysis of experiments on random drift and inbreeding depression. *Genetics* 120: 791–807. [12]

Lynch, M. 1988b. The rate of polygenic mutation. *Genet. Res.* 51: 137–148. [15]

Lynch, M. 1988c. Estimation of relatedness by DNA fingerprinting. *Mol. Biol. Evol.* 5: 584–599. [8]

Lynch, M. 1988d. Path analysis of ontogenetic data. *In* L. Persson and B. Ebenman (eds.), *The dynamics of size-structured populations*, pp. 29–46. Springer-Verlag, Berlin, DE. [A2]

Lynch, M. 1991. The genetic interpretation of inbreeding depression and outbreeding depression. *Evolution* 45: 622–629. [11, 12, 13]

Lynch, M. 1994. The neutral theory of phenotypic evolution. *In* L. Real (ed.), *Ecological genetics*, pp 86–108. Princeton Univ. Press, Princeton, NJ. [15]

Lynch, M. 1996. A quantitative-genetic perspective on conservation issues. *In* J. Avise and J. Hamrick (eds.), *Conservation genetics: case histories from nature*, pp. 471–501. Chapman and Hall, New York, NY. [12]

Lynch, M. 2007. *The origins of genomic architecture.* Sinauer Associates, Sunderland, MA. [15]

Lynch, M. 2008. The cellular, developmental, and population–genetic determinants of mutation–rate evolution. *Genetics* 180: 933–943. [15]

Lynch, M. 2010. Evolution of the mutation rate. *Trends Genet.* 26: 345–352. [15]

Lynch, M. 2011. The lower bound to the evolution of mutation rates. *Genome Biol. Evol.* 3: 1107–1118. [15]

Lynch, M. 2024. *Evolutionary cell biology: The origins of cellular architecture.* Oxford Univ. Press, Oxford, UK. [6, 15]

Lynch, M., and H.-W. Deng. 1994. Genetic slippage in response to sex. *Am. Nat.* 144: 242–261. [5, 13]

Lynch, M., and R. Ennis. 1983. Resource availability, maternal effects, and longevity. *Exp. Gerontol.* 18: 147–165. [6]

Lynch, M., and W. Gabriel. 1983. Phenotypic evolution and parthenogenesis. *Am. Nat.* 122: 745–764. [5]

Lynch, M., and W. Gabriel. 1990. Mutation load and the survival of small populations. *Evolution* 44:1725–1737. [15]

Lynch, M., and W. G. Hill. 1986. Phenotypic evolution and neutral mutation. *Evolution* 40: 915–935. [15]

Lynch, M., and K. Ritland. 1999. Estimation of pairwise relatedness with molecular markers. *Genetics* 152: 1753–1766. [8]

Lynch, M., R. Bürger, D. Butcher, and W. Gabriel. 1993. The mutational meltdown in asexual populations. *Heredity* 84: 339–344. [15]

Lynch, M., J. Conery, and R. Bürger. 1995a. Mutation accumulation and the extinction of small populations. *Am. Nat.* 146: 489–518. [12, 15]

Lynch, M., J. Conery, and R. Bürger. 1995b. Mutational meltdowns in sexual populations. *Evolution* 49: 1067–1088. [12, 15]

Lynch, M., and 5 others. 2023. The divergence of mutation rates across the Tree of Life. *EMBO Reports* 24: e57561. [15]

Ma, C., and 4 others. 2013. Recommended joint and meta-analysis strategies for case-control association testing of single low-count variants. *Genet. Epidem.* 37: 539–550. [20]

Ma, R. H., and J. B. Harrington. 1948. The standard errors of different designs of field experiments at the University of Saskatchewan. *Sci. Agr.* 28: 461–474. [A9]

Márquez-Sánchez, F. 1979. An empirical approach for the prediction of maize F_2 synthetics with varying numbers of lines. *Crop Sci.* 19: 439–444. [13]

Márquez-Sánchez, F. 1992. Inbreeding and yield prediction in synthetic maize cultivars made with parental lines: I. Basic methods. *Crop Sci.* 32: 345–349. [13]

Márquez-Sánchez, F. 1993. Inbreeding and yield prediction in synthetic maize cultivars made with parental lines: II. Alternative methods. *Crop Sci.* 33: 1153–1157. [13]

Mäki-Tanila, A., and W. G. Hill. 2014. Influence of gene interaction on complex trait variation with multilocus models. *Genetics* 198: 355–367. [5]

MacCluer, J. W., and C. M. Kammerer. 1984. Power of sibship variance tests to detect major genes. *In* A. Chakravarti (ed.), *Human population genetics: the Pittsburgh symposium*, pp. 125–141. Van Nostrand Reinhold, New York, NY. [16]

MacCluer, J. W., D. K. Wagner, and R. S. Spielman. 1983. Genetic analysis workshop: segregation analysis of simulated data. *Am. J. Hum. Genet.* 35: 784–792. [16]

MacCluer, J. W., J. L. VandeBerg, B. Read, and O. A. Ryder. 1986. Pedigree analysis by computer simulation. *Zoo Biol.* 5: 147–160. [19]

MacEachern, S. N., and L. M. Berliner. 1994. Subsampling the Gibbs sampler. *Amer. Stat.* 48: 188–190. [A8]

Machado, J. A. F., and J. M. C. Santos Silva. 2000. Glejser's test revisited. *J. Econometrics* 97: 189–202. [10]

Maciejewski, H. 2013. Gene set analysis methods: statistical models and methodological differences. *Brief. Bioinformat.* 15: 504–518. [21]

Mackay, I. J., and W. Powell. 2006. Methods for linkage disequilibrium mapping in crops. *Trends Plant Sci.* 12: 57–63. [18]

Mackay, I. J., and 11 others. 2014. An eight-parent multiparent advanced generation inter-cross population for winter-sown wheat: creation, properties, and validation. *G3* 4: 1603–1610. [18]

Mackay, I. J., J. Cockram, P. Howell, and W. Powell. 2020. Understanding the classics: the unifying concepts of transgressive segregation, inbreeding depression and heterosis and their central relevance for crop breeding. *Plant Biotech. J.* 19: 26–34. [12]

Mackay, T. F. C. 1981. Genetic variation in varying environments. *Genet. Res.* 37: 79–93. [7]

Mackay, T. F. C. 1985. A quantitative genetic analysis of fitness and its components in *Drosophila melanogaster. Genet. Res.* 47: 59–70. [12]

Mackay, T. F. C. 1987. Transposable element-induced polygenic mutations in *Drosophila melanogaster*. *Genet. Res.* 49: 225–233. [15]

Mackay, T. F. C. 1988. Transposable element induced quantitative genetic variation in *Drosophila*. *In* B. S. Weir, E. J. Eisen, M. M. Goodman, and G. Namkoong (eds.), *Proceedings of the second international conference on quantitative genetics*, pp. 219–235. Sinauer Assoc., Sunderland, MA. [15]

Mackay, T. F. C. 1989. Mutation and the origin of quantitative variation. *In* W. G. Hill, and T. F. C. Mackay (eds.), *Evolution and animal breeding*, pp. 113–119. CAB International, Wallingford, UK. [15]

Mackay, T. F. C. 1995. The genetic basis of quantitative variation: numbers of sensory bristles of *Drosophila melanogaster* as a model system. *Trends Genet.* 11: 464–470. [17]

Mackay, T. F. C. 1996. The nature of quantitative genetic variation revisited: lessons from *Drosophila* bristles. *BioEssays* 18: 113–121. [17]

Mackay, T. F. C. 2014. Epistasis and quantitative traits: using model organisms to study gene-gene interactions. *Nat. Rev. Genet.* 15: 22–33. [5]

Mackay, T. F. C., and C. H. Langley. 1990. Molecular and phenotypic variation in the *achaete-scute* region of *Drosophila melanogaster*. *Nature* 348: 64–66. [15, 17]

Mackay, T. F. C., and R. F. Lyman. 2005. Drosophila bristles and the nature of quantitative genetic variation. *Phil. Trans. Roy. Soc. B* 360: 1513–1527. [6, 15, 17]

Mackay, T. F. C., R. F. Lyman, and M. S. Jackson. 1992a. Effects of P element insertion on quantitative traits in *Drosophila melanogaster*. *Genetics* 130: 315–332. [15]

Mackay, T. F. C., and 4 others. 1992b. Polygenic mutation in *Drosophila melanogaster*: estimates from divergence among inbred strains. *Evolution* 46: 300–316. [15]

Mackay, T. F. C., J. D. Fry, R. F. Lyman, and S. V. Nuzhdin. 1994. Polygenic mutation in *Drosophila melanogaster:* estimates from response to selection of inbred strains. *Genetics* 136: 937–951. [15]

Mackay, T. F. C., R. F. Lyman, and W. G. Hill. 1995. Polygenic mutation in *Drosophila melanogaster:* nonlinear divergence among unselected strains. *Genetics* 139: 849–859. [15]

Mackay, T. F. C., N. V. Roshina, J. W. Leips, and E. G. Pasyukova. 2005a. Complex genetic architecture of *Drosophila* longevity. *In* E. J. Masoro and S. N. Austand (eds), *Handbook of the biology of aging, 6th Ed.*, pp. 181–216. Academic Press, New York, NY. [18]

Mackay, T. F. C., R. F. Lyman, and F. Lawrence. 2005b. Polygenic mutation in *Drosophila melanogaster:* mapping spontaneous mutations affecting sensory bristle number. *Genetics* 170: 1723–1735. [15]

Mackay, T. F. C, E. A. Stone, and J. F. Ayroles. 2009. The genetics of quantitative traits: challenges and prospects. *Nat. Rev. Genet.* 10: 565–577. [18, A2]

Mackay, T. F. C., and 55 others. 2012. The *Drosophila melanogaster* genetic reference panel. *Nature* 482: 173–178. [5]

Mackinnon, M. J., and M. A. Georges. 1992. The effects of selection on linkage analysis of quantitative traits. *Genetics* 132: 1177–1185. [19]

Mackinnon, M. J., and J. I. Weller. 1995. Methodology and accuracy of estimation of quantitative trait loci parameters in a half-sib design using maximum likelihood. *Genetics* 141: 755–770. [19]

MacLean, C. J., N. E. Morton, and R. Lew. 1975. Analysis of family resemblance. IV. Operational characteristics of segregation analysis. *Am. J. Hum. Genet.* 27: 365–384. [16]

MacLean, C. J., N. E. Morton, R. C. Elston, and S. Yee. 1976. Skewness in commingled distributions. *Biometrics* 32: 695–699. [16]

MacLeod, I. M., and 4 others. 2013. Inferring demography from runs of homozygosity in whole-genome sequence, with correction for sequence errors. *Mol. Biol. Evol.* 30: 2209–2223. [12]

Macnair, M. R., and Q. J. Cumbes. 1989. The genetic architecture of interspecific variation in *Mimulus*. *Genetics* 122: 211–222. [11]

MacNeil, M. D., and 4 others. 1989. Effects of inbreeding and heterosis in Hereford females on fertility, calf survival and preweaning growth. *J. Anim. Sci.* 67: 895–901. [12]

Madsen, B. E., and S. R. Browning. 2009. A groupwise association test for rare mutations using a weight sum statistic. *PLoS Genet.* 5: e10000384. [20]

Magee, W. T., and L. N. Hazel. 1959. General and specific combining ability for 154-day weight among three-line crosses in swine. *J. Anim. Sci.* 18: 790–795. [13]

Magi, A., and 10 others. 2014. H^3 m^2: Detection of runs of homozygosity from whole-exome sequencing data. *Bioinformatics* 30: 2852–2859. [8]

Mägi, R., and 4 others. 2012. Genome-wide association analysis of imputed rare variants: application to seven common complex disease. *Genet. Epidem.* 36: 785–796. [20]

Magnus, P. 1984. Causes of variation in birth weight: a study of offspring of twins. *Clinical Genet.* 25: 15–24. [7]

Magosi, L. E., and 4 others. 2017. Identifying systematic heterogeneity patterns in genetic association meta-analysis studies. *PLoS Genet.* 13: e1006755. [20]

Magwene, P. M., J. H. Willis, and J.K. Kelly. 2011. The statistics of bulk segregant analysis using next generation sequencing. *PLoS Comput. Biol.* 7: e1002255. [17]

Maher, M. C., L. H. Uricchio, D. G. Torgerson, and R. D. Hernandez. 2013. Population genetics of rare variants and complex disease. *Hum. Hered.* 74: 118–128. [21]

Mahone, G. S., D. Borchardt, T. Presterl, and M. Frisch. 2012. A comparison of tests for QTL mapping with introgression libraries containing overlapping and nonoverlapping donor segments. *Crop. Sci.* 52: 2198–2205. [17]

Makambi, K. 2003. Weighted inverse chi-square method for correlated significance test. *J. Appl. Stat.* 30: 225–234. [20]

Mäki-Tanila, A., and W. G. Hill. 2014. Influence of gene interaction on complex trait variation with multilocus models. *Genetics* 198: 355–367. [5, 20, 21]

Malécot, G. 1948. *Les mathématiques de l'hérédité.* Masson, Paris, FR. [7, 8]

Maliepaard, C., and 4 others. 2001. Bayesian versus frequentist analysis of multiple quantitative trait loci with an application to an outbred apple cross. *Theor. Appl. Genet.* 103: 1243–1253. [19]

Malik, B. S., and R. P. Singh. 2006. Evaluation of crossbreeding effects for wool traits in sheep. *Asian-Aust. J. Anim. Sci.* 19: 1536–1540. [13]

Malik, H. S., and T. H. Eickbush. 1999. Retrotransposable elements R1 and R2 in the rDNA units of *Drosophila mercatorum:* abnormal abdomen revisited. *Genetics* 151: 653–665. [15]

Malina, R. M., and P. H. Buschaung. 1984. Anthropometric asymmetry in normal and mentally retarded males. *Ann. Hum. Biol.* 11: 515–531. [6]

Malkin, I., E. Ginsburg, and R. C. Elston. 2002. Increase in power of transmission-disequilibrium tests for quantitative traits. *Genetic Epid.* 23: 234–244. [17]

Maller, J. B., and 35 others. 2012. Bayesian refinement of association signals for 14 loci in 3 common disease. *Nat. Genet.* 44: 1294–1301. [21]

Mallet, J. 2005. Hybridization as an invasion of the genome. *Trends Ecol. Evol.* 20: 229–237. [13]

Mallet, J. 2006. What does *Drosophila* genetics tell us about speciation? *Trend. Ecol. Evol.* 21: 386–393. [17]

Malmberg, R. L., and R. Mauricio. 2005. QTL-based evidence for the role of epistasis in evolution. *Genet. Res.* 86: 89–95. [18]

Malmberg, R. L., S. Held, A. Waits, and R. Mauricio. 2005. Epistasis for fitness-related quantitative traits in *Arabidopsis thaliana* grown in the field and in the greenhouse. *Genetics* 171: 2013–2027. [18]

Malo, N., O. Libiger, and N. J. Schork. 2008. Accommodating linkage disequilibrium in genetic-association analysis via ridge regression. *Am. J. Hum. Genet.* 82: 375–385. [20]

Malogolowkin-Cohen, C., H. Levene, N. P. Dobzhansky, and A. S. Simmons. 1964. Inbreeding and the mutational and balanced loads in natural populations of *Drosophila willistoni. Genetics* 50: 1299–1311. [12]

Maluszynski, K., J. Kasha, B. P. Forster, and I. Szarejok. 2003. *Doubled haploid production in crop plants.* Kluwer, Dordrecht, NL. [17]

Mancuso, N., and 5 others. 2017. Integrating gene expression with summary association statistics to identify genes associated with 30 complex traits. *Am. J. Hum. Genet.* 100: 473–487. [21]

Mancuso, N., and 6 others. 2019. Probabilistic fine-mapping of transcriptome-wide association studies. *Nat. Genet.* 51: 675–682. [21]

Mange, A. P. 1964. Growth and inbreeding of a human isolate. *Human Biol.* 36: 104–133. [12]

Mangin, B., B. Goffient, and A. Rebaï. 1994a. Constructing confidence intervals for QTL location. *Genetics* 138: 1301–1308. [18]

Mangin, B., B. Goffient, and A. Rebaï. 1994b. Constructing confidence intervals for QTL location. *In* J. W. van Ooijen and J. Jansen (eds.), *Biometrics in plant breeding: applications of molecular markers*, pp. 147–152. CPRO-DLO, NL. [18]

Manichaikul, A., J. Dupuis, S. Sen, and K. W. Broman. 2006. Poor performance of bootstrap confidence intervals for the location of a quantitative trait locus. *Genetics* 174: 481–489. [18]

Manichaikul, A., A. A. Palmer, Ś. Sen, and K. W. Broman. 2007. Significant thresholds for quantitative trait locus mapping under selective genotyping. *Genetics* 177: 1963–1966. [18]

Manichaikul, A., and 5 others. 2010. Robust relationships inference in genome-wide association studies. *Bioinformatics* 22: 2867–2873. [8]

Manly, K. F., D. Nettleton, and J. T. G. Hwang. 2004. Genomics, prior probability, and statistical test of multiple hypotheses. *Gen. Res.* 14: 997–1001. [20, A6]

Manna, F., G. Martin, and T. Lenormand. 2011. Fitness landscapes: an alternative theory for the dominance of mutation. *Genetics* 189: 923–937. [4]

Manolio, T. A., and 26 others. 2009. Finding the missing heritability of complex diseases. *Nature* 461: 747–753. [21]

Mansournia, M. A., A. Geroldinger, S. Greenland, and G. Heinze. 2018. Separation in logistic regression: causes, consequences, and control. *Am. J. Epidem.* 187: 864–870. [20]

Mansur, L. M., J. Orf, and K. G. Lark. 1993a. Determining the linkage of quantitative trait loci to RFLP markers using extreme phenotypes of recombinant inbred lines of soybeans (*Glycine max* L. Merr.). *Theor. Appl. Genet.* 86: 914–918. [17]

Mansur, L. M., K. G. Lark, H. Kross, and H. Olliveira. 1993b Interval mapping of quantitative trait loci for reproductive, morphological, and seed traits of soybean (*Glycine max* L.). *Theor. Appl. Genet.* 86: 907–913. [18]

Marchini, J., and B. Howie. 2010. Genotype imputation for genome-wide association studies. *Nat. Rev. Genet.* 11: 499–511. [20]

Marchini, J., L. R. Cardon, M. S. Phillips, and P. Donnelly. 2004. The effects of human population structure on large genetic association studies. *Nat. Genet.* 36: 512–517. [20]

Marchini, J., P. Donnelly, and L. R. Cardon. 2005. Genome-wide strategies for detecting multiple loci that influence complex diseases. *Nat. Genet.* 37: 413–417. [18, 20]

Marchini, J., and 4 others. 2007. A new multipoint method for genome-wide association studies by imputation of genotypes. *Nat. Genet.* 39: 906–913. [20]

Mare, R. D. 1991. Five decades of educational assortative mating. *Am. Sociol. Rev.* 56: 15–32. [7]

Marigorta, U. M., J. A. Rodrigues, G. Gibson, and A. Narvarro. 2018. Replicability and prediction: lessons and challenges from GWAS. *Trend Genet.* 34: 504–517. [20]

Marinkovic, D. 1967. Genetic loads affecting fertility in natural populations of *Drosophila pseudoobscura*. *Genetics* 57: 701–709. [12]

Markel, P., and 6 others. 1997. Theoretical and empirical issues for marker-assisted breeding of congenic mouse strains. *Nat. Genet.* 17: 280–284. [17]

Marr, A. B., and 4 others. 2006. Interactive effects of environmental stress and inbreeding on reproductive traits in a wild bird population. *J. Anim. Ecol.* 75: 1406–1415. [12]

Marsano, R. M., E. Giordano, G. Messina, and P. Dimitri. 2019. A new portrait of constitutive heterochromatin: lessons from *Drosophila melanogaster*. *Trends Genet.* 35: 615–631. [15]

Marsden, J. E., S. J. Schwager, and B. May. 1987. Single-locus inheritance in the tetraploid treefrog *Hyla versicolor* with an analysis of expected progeny ratios in tetraploid organisms. *Genetics* 116: 299–311. [4]

Marshall, T. C., J. Slate, L. E. B. Kruuk, and J. M. Pemberton. 1998. Statistical confidence for likelihood-based paternity inference in natural populations. *Mol. Ecol.* 87: 639–655. [8]

Martínez, O., and R. N. Curnow. 1992. Estimating the locations and the sizes of the effects of quantitative trait loci using flanking markers. *Theor. Appl. Genet.* 85: 480–488. [18]

Martínez, O., and R. N. Curnow. 1994a. Missing markers when estimating quantitative trait loci using regression mapping. *Heredity* 73: 198–206. [18]

Martínez, O., and R. N. Curnow. 1994b. Three marker scanning of chromosomes for QTL in neighboring intervals. *In* J. W. van Ooijen and J. Jansen (eds.), *Biometrics in plant breeding: applications of molecular markers*, pp. 153–162. CPRO-DLO, NL. [18]

Martin, E. R., S. A. Monks, L. L. Warren, and N. L. Kaplan. 2000. A test for linkage and association in general pedigrees: the pedigree disequilibrium test. *Am. J. Hum. Genet.* 67: 146–154. [17]

Martin, E. R., M. P. Bass, E. R. Hauser, and N. L. Kaplan. 2003. Accounting for linkage in family-based tests of association with missing parental genotypes. *Am. J. Hum. Genet.* 73: 1016–1026. [17]

Martin, G. B., J. G. K. Williams, and S. D. Tanksley. 1991. Rapid identification of markers linked to a *Pseudomonas* resistance gene in tomato by using random primers and near-isogenic lines. *Proc. Natl. Acad. Sci. USA* 88: 2336–2340. [17]

Martinez, M. L., N. Vuksinovic, and A. E. Freeman. 1998. Estimating QTL location and QTL variance, in half-sib families under the random model with missing parental genotypes. *J. Anim. Bred. Genet.* 115: 165–180. [19]

Martins, E. 1991. Individual and sex differences in the use of the push-up display by the sagebrush lizard, *Sceloporus graciosus*. *Anim. Behav.* 41: 403–416. [6]

Masel, J., and M. L. Siegal. 2009. Robustness: mechanisms and consequences. *Trends Genet.* 25: 395–403. [14, 21, A2]

Masel, J, and M. V. Trotter. 2010. Robustness and evolvability. *Trends Genet*. 26: 406–414. [14]

Masly, J. P., and D. C. Presgraves. 2007. High-resolution genome-wide dissection of the two rules of speciation in *Drosophila*. *PLoS Bio.* 5: e243. [17]

Massault, C., H. Bovenhuis, C. Haley, and D.-J. de Koning. 2008. QTL mapping designs for aquaculture. *Aquaculture* 285: 23–29. [19]

Mather, K. 1941. Variation and selection of polygenic characters. *J. Genetics* 41: 159–193. [15]

Mather, K. 1942. The balance of polygenic characters. *J. Genetics* 43: 309–336. [5]

Mather, K. 1943. Polygenic inheritance and natural selection. *Biol. Rev.* 18: 32–64. [5]

Mather, K. 1944. The genetic activity of heterochromatin. *Proc. Royal Soc. Lond. B* 132: 308–332. [15]

Mather, K. 1953. Genetical control of stability in development. *Heredity* 7: 297–336. [6]

Mather, K., and J. L. Jinks. 1982. *Biometrical genetics.* 3rd Ed. Chapman and Hall, New York, NY. [1, 5, 11]

Mathieson, I., and G. McVean. 2012. Differential confounding of rare and common variants in spatial structured populations. *Nat. Genet.* 44: 243–246. [20]

Maurano, M. T., and 31 others. 2012. Systematic location of common disease-associated variation in regulatory DNA. *Science* 337: 1190–1195. [21]

Maurer, A., and 8 others. 2015. Modeling the genetic architecture of flowering time control in barley through nested association mapping. *BMC Genomics* 16: 290. [18]

Mayer, M. 2005. A comparison of regression interval mapping and multiple interval mapping for linked QTLs. *Heredity* 94: 599–605. [18]

Mayer, M., Y. Liu, and G. Freyer. 2004. A simulation study on the accuracy of position and effect estimation of linked QT: and their asymptotic standard deviations using multiple interval mapping in an F_2 scheme. *Genet. Sel. Evol.* 36: 455–479. [18]

Mayo, O. 1980. *The theory of plant breeding.* Clarendon Press, Oxford, UK. [1]

Mayo, O. 1987. *The theory of plant breeding, 2nd Edition.* Clarendon, Oxford, UK. [1]

Mayo, O. 1989. Identification of genes which influence quantitative traits. *In* W. G. Hill and T. F. C. Mackay (eds.), *Evolution and animal breeding*, pp. 141–146. CAB International, Wallingford, UK. [16]

Mayo, O., and R. Bürger. 1997. The evolution of dominance: a theory whose time has passed? *Biol. Rev.* 72: 97–110. [4]

Mayo, O., T. W. Hancock, and P. A. Baghurst. 1980. Influence of major genes on variance within sibships for a quantitative trait. *Ann. Hum. Genet.* 43: 419–421. [16]

McAndrew, B. J., R. D. Ward, and J. A. Beardmore. 1982. Lack of relationship between morphological variance and enzyme heterozygosity in the plaice, *Pleuranectus platessa*. *Heredity* 48: 117–125. [6]

McBride, G., and A. Robertson. 1963. Selection using assortative mating in *D. melanogaster*. *Genet. Res.* 4: 356–369. [7]

McCarthy, J. C. 1967. The effects of inbreeding on the components of litter size in mice. *Genet. Res.* 10: 73–80. [12]

McClellan, J., and M.-C. King. 2010. Genetic heterogeneity in human disease. *Cell* 141: 210–217. [21]

McCulloch, C. E., S. R. Searle, and J. M. Neuhaus. 2008. *Generalized, linear, and mixed models*, 2nd Ed. Wiley, Hoboken, NJ. [14]

McDermott, S. R., and M. A. F. Noor. 2010. The role of meiotic drive in hybrid male sterility. *Phil. Trans. Roy. Soc. Lond* B 365: 1265–1272. [17]

McDonald, J. F., and F. J. Ayala. 1978. Genetic and biochemical basis of enzyme activity variation in natural populations. I. Alcohol dehydrogenase in *Drosophila melanogaster*. *Genetics* 89: 371–388. [17]

McDonald, W. I., and 15 others. 2001. Recommended diagnostic criteria for multiple sclerosis: guidelines from the International Panel on the diagnosis of multiplesclerosis. *Ann. Neurol.* 50: 121–127. [20]

McDowell, R. E. 1981. Limitations for dairy production in developing countries. *J. Dairy Sci.* 64: 2463–2475. [13]

McDowell, R. E. 1985. Crossbreeding in tropical areas with emphasis on milk, heath, and fitness. *J. Dairy Sci.* 68: 2418–2435. [13]

McGinnis, R. 2000. General equations for P_t, P_s, and the power of the TDT and the affected-sib-pair test. *Am. J. Hum. Genet.* 67: 1340–1347. [17, 19]

McGraw, J. B. 1987. Experimental ecology of *Dryas octapetala* ecotypes. IV. Fitness response to reciprocal transplanting in ecotypes with differing plasticity. *Oecologia* 73: 465–468. [13]

McGuffin, P., and P. Huckle. 1990. Simulation of Mendelism revisited: the recessive gene for attending medical school. *Am. J. Hum. Genet.* 46: 994–999. [16]

McKeigue, P. M. 1997. Mapping genes underlying ethnic differences in disease risk by linkage disequilibrium in recently admixed populations. *Am. J. Hum. Genet.* 60: 188–196. [17]

McKeigue, P. M. 1998. Mapping genes underlying ethnic differences in disease risk: methods for detecting linkage in admixed populations, by conditioning on parental admixture. *Am. J. Hum. Genet.* 63: 241–251. [17]

McKeigue, P. M. 2005. Prospects for admixture mapping of complex traits. *Am. J. Hum. Genet.* 76: 1–7. [17]

McKenzie, J. A., and G. M. Clarke. 1988. Diazinon resistance, fluctuating asymmetry and fitness in the Australian sheep blowfly, *Lucilia cuprina*. *Genetics* 120: 213–220. [6]

McKenzie, J. A., and J. L. Yen. 1995. Genotype, environment and the asymmetry phenotype. Dieldrin-resistance in *Lucilia cuprina* (the Australian sheep blowfly). *Heredity* 75: 181–187. [6]

McLachlan, G. J. 1987. On bootstapping the likelihood ratio test statistic for the number of components in a normal mixture. *Appl. Stat.* 36: 318–324. [A6]

McLachlan, G. J., and K. E. Basford. 1988. *Mixture models*. Marcel Dekker, New York, NY. [16]

McMillan, I., and A. Robertson. 1974. The power of methods for the detection of major genes affecting quantitative characters. *Heredity* 32: 349–356. [17, 18]

McMullen, M. D., and 30 others. 2009. Genetic properties of the maize nested association mapping population. *Science* 325: 737–740. [12, 13, 18]

McPeek, M. S. 1999. Optimal allele-sharing statistics for genetic mapping using affected relatives. *Genet. Epid.* 16: 225–249. [19]

McQuillan, R., and 20 others. 2008. Runs of homozygosity in European populations. *Am. J. Hum. Genet.* 83: 359–372. [12]

McRae, A. F., and 16 others. 2018. Identification of 55,000 replicated DNA methylation QTL. *Sci. Reports* 8: 17605. [21]

McVean, G. A. T. 2002. A genealogical interpretation of linkage disequilibrium. *Genetics* 162: 987–991. [5]

McVean, G., and J. Kelleher. 2019. Linkage disequilibrium, recombination and haplotype structure. *In* D. J. Balding, I. Molke, and J. Marioni (eds.). *Handbook of Statistical Genomics*, 4th Ed., pp. 55–85. Wiley and Sons, New York, NY. [5]

Mead, R. 1990. *The design of experiments*. Cambridge Univ. Press, Cambridge, UK. [A9]

Mead, R. 1997. Design of plant breeding trials. *In* R. A. Kempton and P. N. Fox (eds), *Statistical methods for plant variety evaluation*, pp. 40–67. Chapman & Hall, London, UK. [A9]

Mead, R., S. G. Gilmour, and A. Mead. 2012. *Statistical principles for the design of experiments*. Cambridge Univ. Press, Cambridge, UK. [A9]

Meagher, T. R. 1986. Analysis of paternity within a natural population of *Chamaelirium luteum*. 1. Identification of most-likely male parents. *Am. Nat.* 128): 199–215. [8]

Meagher, T. R., and E. Thompson. 1986. The relationship between single parent and parent pair genetic likelihoods in genealogy reconstruction. *Theor. Pop. Biol.* 29: 87–106. [8]

Meagher, T. R., and E. Thompson. 1987. Analysis of parentage for naturally established seedlings of *Chamaelirium luteum* (Liliaceae). *Ecology* 68: 803–812. [8]

Meeker, W. Q., and L. A. Escobar. 1995. Teaching about approximate confidence regions based on maximum likelihood estimation. *Am. Stat.* 49: 48–53. [A4]

Mefford, J., and J. S. Witte. 2012. The covariate's dilemma. *PLoS Genet.* 8: e1003096. [20]

Meiklejohn, C. D., and D. L. Hartl. 2002. A single mode of canalization. *Trends Ecol. Evol.* 17: 468–471. [14]

Melchinger, A. E., and C. Flachenecker. 2006. An extension of the Smith model for quantitative genetic analysis of response under recurrent selection. *Plant Breed.* 125: 644–646. [11]

Melchinger, A. E., H. F. Utz, and C. C. Schön. 1998. Quantitative trait locus (QTL) mapping using different testers and independent population samples in maize reveals low power of QTL detection and large bias in estimation of QTL effects. *Genetics* 149: 383–403. [18]

Melchinger, A. E., and 4 others. 2007a. The role of epistasis in the manifestation of heterosis: a systems-oriented approach. *Genetics* 177: 1815–1825. [13]

Melchinger, A. E., and 7 others. 2007b. Genetic basis of heterosis for growth-related traits in *Arabidopsis* investigated by testcross progenies of near-isogenic lines reveals a significant role of epistasis. *Genetics* 177: 1827–1837. [13]

Mendel, G. 1865. Versuche uber pflanzen-hybriden. *Vorgelegt in de Sitzungen.* [1]

Mengerson, K. L., C. P. Robert, and C. Guihenneuc-Jouyaux. 1999. MCMC convergence diagnostics: A review. *In* J. M. Bernardo, J. O. Berger, A. P. Dawid, and A. F. M. Smith (eds.), *Bayesian Statistics 4*, pp. 415–440. Oxford University Press. [A8]

Menozzi, P., A. Piazza, and L. Cavalli-Sforza. 1978. Synthetic maps of human gene frequencies in Europeans. *Science* 201: 786–792. [20]

Mérat, P. 1968. Distributions de frequencies, interpretation du determinisme genetique des characters quantitatifs et recherche de 'genes majeours'. *Biometrics* 24: 277–293. [16]

Mercer, W. B., and A. D. Hall. 1911. Experimental error in field trails. *J. Agr. Sci.* 4: 107–127. [A9]

Meredith, W. R., Jr. 1999. Cotton and heterosis. *In* J. G. Coors and S. Pandey (eds.), *The genetics and exploitation of heterosis in crops*, pp. 451–462. ASA-CSSA-SSSA Societies, Madison, WI. [13]

Meredith, W. R., Jr., R. R. Bridge, and J. F. Chism. 1970. Relative performance of F_1 and F_2 hybrids from doubled haploids and their parent varieties in upland cotton, *Gossypium hirsutum* L. *Crop Sci.* 10: 295–298. [6]

Merrick, M. J. 1975. The inheritance of penicillin titre in crosses between lines of *Aspergillus nidulans* selected for increased productivity. *J. Gen. Microbiol.* 91: 287–294. [5]

Merton, R. K. 1965. *On the shoulders of giants.* Free Press, New York, NY. [A7]

Metropolis, N., and S. Ulam. 1949. The Monte Carlo method. *J. Am. Stat. Assoc.* 44: 335–341. [A8]

Metropolis, N., and 4 others. 1953. Equations of state calculations by fast computing machines. *J. Chem. Phys.* 21: 1087–1091. [A8]

Meuwissen, T. H. E., and M. E. Goddard. 2000. Fine mapping of quantitative trait loci using linkage disequilibria with closely linked marker loci. *Genetics* 155: 421–430. [17]

Meuwissen, T. H. E., B. J. Hayes, and M. E. Goddard. 2001. Prediction of total genetic value using genome-wide dense marker maps. *Genetics* 157: 1819–1829. [21]

Meyer, K. 1989. Approximate accuracy of genetic evaluation under an animal model. *Livestock Prod. Sci.* 21: 87–100. [10]

Meyer, K. 2008. Likelihood calculations to evaluate experimental designs to estimate genetic variances. *Heredity* 101: 212–221. [A4]

Meyer, K., and W. G. Hill. 1992. Approximation of sampling variances and confidence intervals for maximum likelihood estimates of variance components. *J. Anim. Breed. Genet.* 109: 264–280. [16]

Meyer, R. C., and 10 others. 2010. QTL analysis of early stage heterosis for biomass in *Arabidopsis*. *Theor. Appl. Genet.* 120: 227–237. [13]

Meyermans, R., W. Gorssen, N. Buys, and S. Janssens. How to study runs of homozygosity using PLINK? A guide for analyzing medium density SNP data in livestock and pet species. *BMC Genomics* 21: 1–14. [8]

Mi, M. P., M. Earle, and J. Kagawa. 1986. Phenotypic resemblance in birth weight between first cousins. *Ann. Hum. Genet.* 50: 49–62. [7]

Miano, M. G., and 10 others. 2000. Pitfalls in homozygosity mapping. *Am. J. Hum. Genet.* 67: 1348–1351. [12]

Michelmore, R. W., L. Paran, and R. V. Kesseli. 1991. Identification of markers linked to disease-resistance genes by bulked segregant analysis: a rapid method to detect markers in specific genomic regions by using segregating populations. *Proc. Natl. Acad. Sci., USA* 88: 9828–9832. [17]

Michod, R. E., and W. D. Hamilton. 1980. Coefficients of relatedness in sociobiology. *Nature* 288: 694–697. [7]

Mikula, O., and M. Macholán. 2008. There is no heterotic effect upon developmental stability in the ventral side of the skull within the house mouse hybrid zone. *J. Evol. Biol.* 21: 1055–1067. [6]

Milkman, R. D. 1970. The genetic basis of natural variation. X: Recurrence of *cve* polygenes. *Genetics* 65: 289–303. [15]

Milkman, R. D. 1978. Selection differentials and selection coefficients. *Genetics* 88: 391–403. [15]

Miller, J. F. 1999. Oilseeds and heterosis. *In* J. G. Coors and S. Pandey (eds.), *The genetics and exploitation of heterosis in crops*, pp. 399–404. ASA-CSSA-SSSA Societies, Madison, WI. [13]

Miller, J. M., and D. W. Coltman. 2014. Assessment of identity disequilibrium and its relation to empirical heterozygosity fitness correlations: a meta-analysis. *Mol. Ecol.* 23: 1899–1909. [12]

Miller, J. M., and 7 others. 2014. Estimating genome-wide heterozygosity: effects of demographic history and marker type. *Heredity* 112: 240–247 [12]

Miller, P. S. 1994. Is inbreeding depression more severe in a stressful environment? *Zoo Biol.* 13: 195–208. [12]

Miller, R. G. 1986. *Beyond ANOVA: Basics of applied statistics.* Wiley, New York, NY. [19]

Milligan, B. G. 2003. Maximum-likelihood estimation of relatedness. *Genetics* 163: 1153–1167. [8, A4]

Millstein, J., B. Zhang, J. Zhu, and E. E. Schadt. 2009. Disentangling molecular relationships with a causal inference test. *BMC Genet.* 10: 23. [A2]

Milton, C. C., and 4 others. 2003. Quantitative trait symmetry independent of Hsp90 buffering; distinct modes of genetic canalization and developmental stability. *Proc. Natl. Acad. Sci. USA* 100: 13396–13401. [6]

Minelli, C., J. R. Thompson, M. D. Tobin, and K. R. Abrams. 2004. An integrated approach to the meta-analysis of genetic association studies using Mendelian randomization. *Am. J. Epidem.* 160: 445–451. [A2]

Minvielle, F. 1987. Dominance is not necessary for heterosis: a two-locus model. *Genet. Res.* 49: 245–247. [13]

Misztal, I., T. J. Lawlor, and N. Gengler. 1997. Relationships among estimates of inbreeding depression, dominance and additive variance for linear traits in Holsteins. *Genet. Sel. Evol.* 29: 319–326. [12]

Mitchell, A. A., D. J. Cutler, and A. Chakravarti. 2003. Undetected genotyping errors cause apparent overtransmission of common alleles in the transmission/disequilibrium test. *Am. J. Hum. Genet.* 72: 598–610. [17]

Mitchell-Olds, T., and J. Bergelson. 1990. Statistical genetics of an annual plant, *Impatiens capensis*. I. Genetic basis of quantitative variation. *Genetics* 124: 407–415. [16]

Mitrea, C., and 7 others. 2013. Methods and approaches in the topology-based analysis of biological pathways. *Front. Physiology* 4: 278. [21]

Mittlbock, M., and H. Heinzl. 2006. A simulation study comparing properties of heterogeneity measures in meta-analayis. *Stat. Med.* 25: 4321–4333. [20]

Mitton, J. B. 1978. Relationship between heterozygosity for enzyme loci and variation of morphological characters in natural populations. *Nature* 273: 661–662. [6]

Mitton, J. B. 1994. Molecular approaches to population biology. *Ann. Rev. Ecol. System.* 25: 45–69. [12]

Mitton, J. B., and M. C. Grant. 1980. Observations on the ecology and evolution of quaking aspen, *Populus tremuloides,* in the Colorado front range. *Am. J. Bot.* 67: 202–209. [6]

Mitton, J. B., and M. C. Grant. 1984. Associations among protein heterozygosity, growth rate, and developmental homeostasis. *Ann. Rev. Ecol. Syst.* 15: 479–499. [12]

Mitton, J. B., and 4 others. 1981. Associations between heterozygosity and growth rate variables in three western forest trees. *USDA Gen. Tech. Rep. PSW* 48: 27–34. [12]

Miyashita, N., and C. C. Laurie-Ahlberg. 1984. Genetic analysis of chromosomal interaction effects of the activities of the glucose 6-phosphate and 6- phosphogluconate dehydrogenases in *Drosophila melanogaster. Genetics* 106: 655–668. [17]

Moczek, A. P., and 7 others. 2011. The role of developmental plasticity in evolutionary innovation. *Proc. Roy. Soc. Lond. B* 278: 2705–2713. [14]

Mode, C. J., and D. L. Gasser. 1972. A distribution free test for major gene differences in quantitative inheritance. *Math. Biosci.* 14: 143–150. [16]

Modi, W. S., R. K. Wayne, and S. J. O'Brien. 1987. Analysis of fluctuating asymmetry in cheetahs. *Evolution* 41: 227–228. [6]

Moehring, J., E. R. Williams, and H.-P. Piepho. 2014. Efficiency of augmented p-rep designs in multi-environmental trials. *Theor. Appl. Genet.* 127: 1049–1060. [A9]

Mohamed, A. H. 1959. Inheritance of quantitative characters in *Zea mays*. I. Estimation of the number of genes controlling the time of maturity. *Genetics* 44: 713–724. [11]

Mohamed, A. H., and A. S. Hanna. 1964. Inheritance of quantitative characters in rice. I. Estimation of the number of effective factor pairs controlling plant height. *Genetics* 49: 81–93. [11]

Möhring, J., E. Williams, and H.-P. Piepho. 2015. Inter-block information: to recover or not to recover it? *Theor. Appl. Genet.* 128: 1541–1554. [A9]

Molenberghs, G., and G. Verbeke. 2007. Likelihood ratio, score, and Wald tests in a constrained parameter space. *Amer. Stat.* 61: 22–27. [A4]

Moll, P. P., and 5 others. 1984. Detection of genetic heterogeneity among pedigrees through complex segregation analysis: an application to hypercholestererolemia. *Am. J. Hum. Genet.* 36: 197–211. [16]

Moll, R. H., M. F. Lindsey, and H. F. Robinson. 1964. Estimates of genetic variances and level of dominance in maize. *Genetics* 49: 411–423. [13]

Moll, R. H., J. H. Longquist, J. V. Fortuno, and E. C. Johnson. 1965. The relationship of heterosis and genetic divergence in maize. *Genetics* 52: 139–144. [13]

Møller, A. P. 1992. Parasites differentially increase the degree of fluctuating asymmetry in secondary sexual characters. *J. Evol. Biol.* 5: 691–699. [6]

Møller, A. P., and M. Eriksson. 1994. Patterns of fluctuating asymmetry in flowers: implications for sexual selection in plants. *J. Evol. Biol.* 7: 97–113. [6]

Møller, A. P., and M. D. Jennions. 2001. Testing and adjusting for publication bias. *Trends Ecol Evol.* 16: 580–586. [A6]

Møller, A. P., and J. P. Swaddle. 1997. *Asymmetry, developmental stability and evolution*. Oxford Univ. Press, Oxford, UK. [6]

Moltkne, I., and 4 others. 2011. A method for detecting IBD regions simultaneously in multiple individuals — with applications to disease genes. *Genom. Res.* 21: 1168–1180. [8]

Momozawa, Y., and K. Mizukami. 2012. Unique roles of rare variants in the genetic of complex diseases in humans. *J. Hum. Genet.* 66: 11–23. [21]

Monforte, A. J., and S. D. Tanksley. 2000. Development of a set of near isogenic and backcross recombinant inbred lines containing most of the *Lycopersicon hirsutum* genome in a *L. esculentum* genetic background: a tool for gene mapping and gene discovery. *Genome* 43: 803–813. [17]

Monks, S. A., and 8 others. 2004. Genetic inheritance of gene expression in human cell lines. *Am. J. Hum. Genet.* 75: 1094–1105. [21]

Mooney, M. A., and B. Wilmot. 2015. Gene set analysis: a step-by-step guide. *Am. J. Med. Genet.* 168B: 517–527. [21]

Mooney, M. A., J. T. Nigg, S. K. McWeeney, and B. Wilmot. 2014a. Functional and genomic context in pathway analysis of GWAS data. *Trends Genet.* 30: 390–400. [21]

Mooney, M., J. T. Nigg, S. K. McWeeney, and B. Wilmot. 2014b. 'Pitfalls in the application of gene-set analysis to genetics studies': a response. *Trends Genet.* 30: 514–515. [21]

Moorad, J. A., and M. J. Wade. 2005. A genetic interpretation of the variation in inbreeding depression. *Genetics* 170: 1373–1384. [12]

Moose, S. P., J. W. Dudley, and T. R. Rocheford. 2004. Maize selection passes the century mark: a unique resource for 21st century genomics. *Trends Plant Sci.* 9: 358–364. [18]

Mootha, V. K., and 20 others. 2003. PCA-1-α-responsive gene involved in oxidative phosphorylation are coordinately downregulated in human diabetes. *Nat. Genet.* 34: 267–273. [21]

Moran, N., and P. Baumann. 1994. Phylogenetics of cytoplasmically inherited microorganisms of arthropods. *Trends Ecol. Evol.* 9: 15–20. [6]

Moran, P. A., M. G. Ritchie, and N. W. Bailey. 2017. A rare exception to Haldane's rule: Are X chromosomes key to hybrid incompatibilities? *Heredity* 118: 554–566. [17]

Moran, P. A. P. 1971. Maximum-likelihood estimation in non-standard conditions. *Mathematical Proceedings of the Cambridge Philosophical Society* 70: 44–450. [A4]

Moran, P. A. P., and C. A. B. Smith. 1966. Commentary on R. A. Fisher's paper on "The correlation between relatives on the supposition of Mendelian inheritance." *Eugen. Lab. Mem.* 41, Cambridge Univ. Press, Cambridge, UK. [4, 7]

Moreno, G. 1994. Genetic architecture, genetic behavior,

and character evolution. *Ann. Rev. Ecol. Syst.* 25: 31–44. [5]

Morgante, F., and 4 others. 2015. Genetic architecture of micro-environmental plasticity in *Drosophila melanogaster. Sci. Reports* 5: 9785. [6]

Morgenthaler, S., and W. G. Thilly. 2006. A strategy to discover genes that carry multi-allelic or mono-allelic risk for common diseases: A cohort allelic sums test (CAST). *Mut. Res.* 615: 28–56. [20]

Morley, F. H. W. 1954. Selection for economic characters in Australian Merino sheep. IV. The effect of inbreeding. *Aust. J. Agric. Res.* 5: 305–316. [12]

Morley, M., and 6 others. 2004. Genetic analysis of genome-wide variation in human gene expression. *Nature* 430: 743–747. [21]

Morris, A. P., and L. R. Cardon 2019. Genome-wide association studies. *In* D. J. Balding, I. Molke, and J. Marioni (eds.), *Handbook of Statistical Genomics*, 4th Ed., pp. 597–650. Wiley and Sons, New York, NY. [20]

Morris, A. P., and E. Zeggini. 2009. An evaluation of statistical approaches to rare variant analysis in genetic association studies. *Genet. Epidem.* 34: 188–193. [20]

Morris, A. P., J. C. Whittaker, and D. J. Balding. 2002. Fine-scale mapping of disease loci via shattered coalescent modeling of genealogies. *Am. J. Hum. Genet.* 70: 686–707. [17]

Morrison, D. F. 1976. *Multivariate statistical methods.* McGraw-Hill, New York, NY. [9, A3]

Morrissey, M. B. 2016. Meta-analysis of magnitudes, differences and variation in evolutionary parameters. *J. Evol. Biol.* 29: 1882–1904. [A6]

Mortazavi, A., and 4 others. 2008. Mapping and quantifying mammalian transcriptomes by RNA-Seq. *Nat. Methods* 5: 621–628. [21

Morton, N. E. 1955a. The inheritance of human birth weight. *Ann. Hum. Genet.* 20: 125–134. [7]

Morton, N. E. 1955b. Sequential tests for the detection of linkage. *Am. J. Hum. Genet.* 7: 277–318. [18, 19, 20, A6]

Morton, N. E. 1958. Empirical risks in consanguineous marriages: birth weight, gestation time, and measurements of infants. *Am. J. Hum. Genet.* 10: 344–349. [12]

Morton, N. E. 1959. Genetic tests under incomplete ascertainment. *Am. J. Hum. Genet.* 11: 1–16 [16]

Morton, N. E. 1978. Effect of inbreeding on IQ and mental retardation. *Proc. Natl. Acad. Sci. USA* 75: 3906–3908. [12]

Morton, N. E. 1984. Trials of segregation analysis by deterministic and macro simulation. *In* A. Chakravarti (ed.), *Human population genetics: the Pittsburgh symposium*, pp. 83–107. Van Nostrand Reinhold, New York, NY. [16]

Morton, N. E. 1991. Genetic epidemiology of hearing impairment. *Ann. New York Acad. Sci.* 630: 16–31. [12]

Morton, N. E., and C. J. MacLean. 1974. Analysis of family resemblance. III. Complex segregation of quantitative traits. *Am. J. Hum. Genet.* 26: 489–503. [16]

Morton, N. E., J. F. Crow, and H. J. Muller. 1956. An estimate of the mutational damage in man from data on consanguineous matings. *Proc. Natl. Acad. Sci. USA* 42: 855–863. [12, 15]

Morton, N. E., D. C. Rao, and J. M. Lalouel. 1983. *Methods in genetic epidemiology.* S. Karger, Basel, CH. [16]

Mosedale, M., and 6 others. 2019. Identification of candidate risk factor genes for human idelalisib toxicity using a collaborative cross approach. *Toxicol. Sci.* 172: 265–278. [18]

Moser, G., and 5 others. 2015. Simultaneous discovery, estimation and prediction analysis of complex traits using a Bayesian mixture model. *PLoS Genet.* 11: e1004969. [21]

Mosig, M. O., and 5 others. 2001. A whole genome scan for quantitative trait loci affecting milk protein percentage in Israeli-Holstein cattle, by means of selective milk DNA pooling in a daughter design, using an adjusted false discovery rate criterion. *Genetics* 157: 1683–1698. [A6]

Moskvina, V., and K. M. Schmidt. 2008. On multiple-testing correction in genome-wide association studies. *Genet. Epidem.* 32: 567–573. [20, A6]

Moskvina, V., and 5 others. 2011. Evaluation of an approximation method for assessment of overall significance of multiple-dependent tests in genomewide association study. *Genet. Epidem.* 35: 861–866. [20]

Mostafavi, H., J. P. Spence, S. Naqvi, and J. K. Pritchard. 2023. Systematic differences in discovery of genetic effects on gene expression and complex traits. *Nature* 55: 1866–1875. [21]

Mosteller, F., and R. R. Bush. 1954. Selected quantitative techniques. *In*, G. Lindzer (ed), *Handbook of social psychology*, Vol. 1, pp. 289–334. Addison-Wesley, Cambridge, MA. [A6]

Mountjoy, E., and 7 others. 2018. Education and myopia: assessing the direction of causality by mendelian randomisation. *BMJ* 361: k2022. [A2]

Moura, R. Ř., and 4 others. 2021. Assortative mating in space and time: patterns and biases. *Ecol. Lett.* 24: 1089–1102. [7]

Mousseau, T. A., and C. W. Fox (eds). 1998. *Maternal effects as adaptations.* Oxford University Press, Oxford, UK. [6]

Mousseau, T. A., and D. A. Roff. 1987. Natural selection and the heritability of fitness components. *Heredity* 59: 181–197. [7, 12]

Moutsianas, L., and A. P. Morris. 2014. Methodology for the analysis of rare genetic variation in genome-wide association and re-sequencing studies of complex human traits. *Brief. Functional Genomics* 13: 362–370. [20]

Moutsianas, L., and 11 others. 2015. The power of gene-based rare variant methods to detect disease-associated variation and test hypotheses about complex disease. *PLoS Genet.* 11: e1005165. [20]

Mrode, R. A., and I, Pocrnic. 2023. *Linear models for the prediction of the genetic merit of animals.* CABI, Boston, MA. [1]

Muehlbauer, G. J., and 4 others. 1988. Near-isogenic lines — a potential resource in the integration of conventional and molecular marker linkage maps. *Crop Sci.* 28: 729–735. [17]

Mueller, J. C. 2004. Linkage disequilibrium for different scales and applications. *Brief Bioinform.* 5: 355–364. [5]

Mueller, R. F., and D. T. Bishop. 1993. Autozygosity mapping, complex consanguinity, and autosomal recessive disorders. *J. Med. Genet.* 30: 798–799. [12]

Muir, W. M. 1986a. Estimation of response to selection and utilization of control populations for additional information and accuracy. *Biometrics* 42: 381–391. [12]

Muir, W. M. 1986b. Efficient design and analysis of selection experiments. *In* G. E. Dickerson and R. K.

Johnson (eds.), *Proceedings of the 3rd world congress on genetics applied to livestock production.* Agric. Comms., Univ. Nebraska, Lincoln, NE. [12]

Muir, W. M., and S. E. Aggrey. 2003. *Poultry genetics, breeding and biotechnology.* CABI, Wallingford, UK. [13]

Mukai, T. 1964. The genetic structure of natural populations of *Drosophila melanogaster*. I. Spontaneous mutation rate of polygenes controlling viability. *Genetics* 50: 1–19. [15]

Mukai, T. 1969. The genetic structure of natural populations of *Drosophila melanogaster.* VII. Synergistic interaction of spontaneous mutant polygenes controlling viability. *Genetics* 61: 749–761.[15]

Mukai, T. 1979. Polygenic mutation. *In* J. N. Thompson, Jr., and J. M. Thoday (eds.), *Quantitative genetic variation*, pp. 177–196. Academic Press, New York, NY. [15]

Mukai, T., and C. C. Cockerham. 1977. Spontaneous mutation rates at enzyme loci in *Drosophila melanogaster*. *Proc. Natl. Acad. Sci. USA* 74: 2514–2517. [15]

Mukai, T., and O. Yamaguchi. 1974. The genetic structure of natural populations of *Drosophila.* XI. Genetic variability in a local population. *Genetics* 82: 63–82. [12, 15]

Mukai, T., and T. Yamazaki. 1968. The genetic structure of natural populations of *Drosophila melanogaster.* V. Coupling-repulsion effect of spontaneous mutant polygenes controlling viability. *Genetics* 59: 513–535. [15]

Mukai, T., and T. Yamazaki. 1971. The genetic structure of natural populations of *Drosophila melanogaster.* X. Developmental time and viability. *Genetics* 69: 385–398. [15]

Mukai, T., S. I. Chigusa, and I. Yoshikawa. 1965. The genetic structure of natural populations of *Drosophila melanogaster.* III. Dominance effect of spontaneous mutant polygenes controlling viability in heterozygous genetic backgrounds. *Genetics* 52: 493–501. [15]

Mukai, T., S. I. Chigusa, L. E. Mettler, and J. F. Crow. 1972. Mutation rate and dominance of genes affecting viability in *Drosophila melanogaster. Genetics* 72: 335–355. [4, 12, 15]

Mukai, T., R. Cardellino, T. K. Watanabe, and J. F. Crow. 1974. The genetic variance for viability and its components in a local population of *Drosophila melanogaster. Genetics* 78: 1195–1208. [12]

Mukherjee, S. 2016. *The gene: An intimate history*. Simon and Schuster, New York, NY. [1]

Mukhopadhyay, I., E. Feingold, D. E. Weeks, and A. Thalamuthu. 2010. Association tests using kernel-based measures of multi-locus genotype similarity between individuals. *Genet. Epid.* 34: 213–221. [20]

Mulcahy, D.L. 1975 . Differential mortality among cohorts in a population of *Acer saccharum* (Aceraceae) seedlings. *Am. J. Bot.* 62: 422–426. [12]

Mulcahy, D.L. 1976. *Gametic selection in plants and animals.* Elsevier, New York, NY. [12]

Mulitze, D. K., and R. J. Baker. 1985a. Evaluation of biometrical methods for estimating the number of genes. 1. Effect of sample size. *Theor. Appl. Genet.* 69: 553–558. [11]

Mulitze, D. K., and R. J. Baker. 1985b. Evaluation of biometrical methods for estimating the number of genes. 2. Effect of type I and type II statistical errors. *Theor. Appl. Genet.* 69: 559–566. [11]

Müller, B. U., K. Kleinknecht, J. Möhring, and H.-P. Piepho. 2010. Comparison of spatial models for sugar beets and barley trails. *Crop Sci.* 50: 794–802. [A9]

Muller, H. J. 1935. On the incomplete dominance of the normal allelomorphs of white in *Drosophila. J. Genet.* 30: 407–414. [15]

Muller, H. J. 1939. Reversibility in evolution considered from the standpoint of genetics. *Biol. Rev.* 14: 261–280. [17]

Munafó, M. R., and J. Flint. 2004. Meta-analysis of genetic association studies. *Trend Genet.* 20: 439–444. [20]

Murabito, J. M., R. Yuan, and K. L. Lunetta. 2012. The search for longevity and healthy aging genes: insights from epidemiological studies and samples of long-lived individuals. *J. Gerontology A: Biomedical . Medical Sci.* 67: 470–479. [17]

Muranty, H. 1996. Power of tests for quantitative trait loci detection using full-sib families in different schemes. *Heredity* 76: 156–165. [19]

Muranty, H., and B. Goffinet. 1997. Selective genotyping for location and estimation of the effect of a quantitative trait loci. *Biometrics* 53: 629–643. [17, 18]

Muranty, H., B. Goffinet., and F. Santi. 1997. Multitrait and multipopulation QTL search using selective genotyping. *Genet. Res.* 70: 259–265. [18]

Myburg, A. A., and 80 others. 2014. The genome of *Eucalyptus grandis. Nature* 510: 356–362. [12]

Myles, S., and 6 others. 2009. Associating mapping: critical considerations shift from genotyping to experimental design. *Plant Cell* 21: 2194–2202. [18, 20]

Nadeau, J. H., J. B. Singer, A. Matin, and E. S. Lander. 2000. Analysing complex genetic traits with chromosome substitution strains. *Nat. Genet.* 24: 221–225. [17]

Nagai, J., A. J. Lee, and C. G. Hickman. 1971. Preweaning growth of inbred, F_1 hybrid, and random-bred mice as a measure of mother's lactation. *Can. J. Genet. Cytol.* 13: 20–28. [12]

Nagalakshmi, U., and 6 others. 2008. The transcriptional landscape of the yeast genome defined by RNA sequencing. *Science* 320: 1344–1349. [21]

Nagpal, S., R. Tandon, and G. Gibson. 2022. Canalization of the polygenic risk for common diseases and traits in the UK Biobank cohort. *Mol. Bio. Evol.* 39: msac053. [14]

Nagylaki, T. 1978. The correlation between relatives with assortative mating. *Ann. Hum. Genet.* 42: 131–137. [7]

Nagylaki, T. 1982. Assortative mating for a quantitative character. *J. Math. Biol.* 16: 57–74. [7]

Naisbit, R. E., C. D. Jiggins, M. Linares, C. Salazar, and J. Mallet 2002. Hybrid sterility, Haldane's rule and speciation in *Heliconius cydno* and *H. melpomene*. *Genetics* 161: 1517–1526. [17]

Nakagawa, S., and I. C. Cuthill. 2007. Effect size, confidence interval and statistical significance: a practical guide for biologists. *Biol. Rev.* 82: 591–605. [A6]

Nakagawa, S., and E. S. A. Santos. 2012. Methodological issues and advances in biological meta-analysis. *Evol. Ecol.* 26: 1253–1274. [A6]

Nakaoka, H., and I. Inoue. 2009. Meta-analysis of genetic association studies: methodologies, between-study heterogeneity and winner's curse. *J. Hum. Genet.* 54: 615–623. [20]

Nam, D., and S.-Y. Kim. 2008. Gene-set approach for expression pattern analysis. *Brief. Bioinformat.* 9: 189–197. [21]

Namkoong, G. 1979. *Introduction to quantitative genetics in forestry.* Dept. of Agriculture Technical Bulletin No 1588. U. S. Governemnt Printing Office, Washington, DC. [1]

Nance, W. E., and 4 others. 1983. A causal analysis of birth weight in the offspring of monozygotic twins. *Am. J. Hum. Genet.* 35: 1211–1223. [7]

Nandamuri, S. P., B. E. Dalton, and K. L. Carleston. 2017. Determination of the genetic architecture underlying short wavelength sensitivity in Lake Malawi cichilds. *J. Hered.* 108: 379–390. [11]

Narasimhan, V., and 5 others. 2016. BCFtools/RoH: A hidden markov model approach for detecting autozygosity from next-generation sequencing data. *Bioinformatics* 32: 1749–1751. [8]

Nason, J. D., and N. C. Ellstrand. 1995. Lifetime estimates of biparental inbreeding depression in the self-incompatible annual plant *Raphanus sativus*. *Evolution* 49: 307–316. [12]

Naveira, H., and A. Barbadilla. 1992. The theoretical distribution of lengths of intact chromosome segments around a locus held heterozygous with backcrossing in a diploid species. *Genetics* 130: 205–209. [17]

Neale, B. M., and P. C. Sham. 2004. The future of association studies: gene-based analysis and replication. *Am. J. Hum. Genet.* 75: 353–362. [20]

Neale, B. M., and 9 others. 2011. Testing for an unusual distribution of rare variants. *PLoS Genet.* 7: e1001322. [20]

Neel, J. V., and 5 others. 1970. The effects of parental consanguinity and inbreeding in Hirado, Japan. II. Physical development, tapping rate, blood pressure, intelligence quotient, and school performance. *Am. J. Hum. Genet.* 22: 263–286. [12]

Neff, B. D., J. Repka, and M. R. Gross. 2000a. Parentage analysis with incomplete sampling of candidate parents and offspring. *Mol. Ecol.* 9: 515–528. [8]

Neff, B. D., J. Repka, and M. R. Gross. 2000b. Statistical confidence in parentage analysis with incomplete sampling: how many loci and offspring are needed? *Mol. Ecol.* 9: 529–539. [8]

Neff, B. D., J. Repka, and M. R. Gross. 2001. A Bayesian framework for parentage analysis: the value of genetic and other biological data. *Theor. Pop. Biol.* 59: 315–331. [8]

Nei, M. 1967. Modification of linkage intensity by natural selection. *Genetics* 57: 625–641. [5]

Neimann-Sørensen, A., and A. Robertson. 1961. The association between blood groups and several production characteristics in three Danish cattle breeds. *Acta Agric. Scand.* 11: 163–196. [19]

Nelson, O. E. 1993. A notable triumvirate of maize geneticists. *Genetics* 135: 937-941. [13]

Nettleton, D., J. T. G. Hwand, R. A. Caldo, and R. P. Wise. 2006. Estimating the number of true null hypotheses from a histogram of p values. *J. Agr. Biol. Envir. Stats* 11: 337–356. [A6]

Nevo, E. 1978. Genetic variation in natural populations: patterns and theory. *Theor. Popul. Biol.* 13: 121–177.[12]

Newton, M. A., and Z. Wang. 2015. Multiset statistics for gene set analysis. *Ann. Rev. Stat. Appl.* 2: 95–111. [21]

Neyman, J., and E. Scott. 1966. On the use of the $c(\alpha)$ optimal tests of composite hypotheses. *Bill. Intern. Stat. Instit.* 41: 477–497. [20]

Ng, D. W.-K, J. Lu, and Z. J. Chen. 2012. Big roles for small RNAs in polyploidy, hybrid vigor, and hybrid incompatibility. *Curr. Opinion Plant Bio.* 15: 154–161. [13]

Ni, Z., and 7 others. 2009. Altered circadian rhythms regulate growth vigour in hybrids and allopolyploids. *Nature* 457: 327–331. [13]

Nica, A. C., and E. T. Dermitzakis. 2013. Expression quantitative trait loci: present and future. *Phil. Trans. Roy. Soc. B* 368: 20120362. [21]

Nica, A. C., and 6 others. 2010. Candidate causal regulatory effects by integration of expression QTLs with complex trait genetic associations. *PLoS Genet.* 6: e1000985. [21]

Nice, L. M., and 10 others. 2016. Development and genetic characterization of an advanced backcross-nested association mapping (AB-NAM) population of wild × cultivated barley. *Genetics* 203: 1453–1467. [18]

Nicolae, D. L. 2016. Association tests for rare variants. *Ann. Rev. Genom. Hum. Genet.* 17: 117–130. [20]

Nicolae, D. L., and 5 others. 2010. Trait-associated SNPs are more likely to be eQLTs: annotation to enhance discovery from GWAS. *PLoS Genet.* 6: e1000888. [21]

Nielsen, R., D. K. Mattila, P. J. Clapham, and P. J. Palsbøll. 2001. Statistical approaches to paternity analysis in natural populations and applications to the North Atlantic humpback whale. *Genetics* 157: 1673–1682. [8]

Niemi, M. E. K., and 16 others. 2018. Common genetic variants contribute to risk of rare severe neurodevelopmental disorders. *Nature* 562: 268–271. [21]

Nietlisbach, P., and 4 others. 2019. Nonequivalent lethal equivalents: Models and inbreeding metrics for unbiased estimation of inbreeding load. *Evol. Appl.* 12: 266–279. [12]

Nijhout, H. F., A. M. Kidla, and C. C. Hazelwood. 2021. Genetic assimialtion and accommodation: models and mechanism. *Curr. Topic. Develp. Biol.* 141: 337–369. [14]

Nilsson-Ehle, H. 1909. Kreuzungsuntersuchungen an Hafer und Weizen. *Lunds Univ. Årsskrift.*, Series 2, vol. 5, no. 2: 1–122. [1]

Nitsch, D., and 5 others. 2006. Limits to causal inference based on Mendelian randomization: a comparison with randomized controlled traits. *Am. J. Epdiem.* 163: 397–403. [A2]

Nitter, G. 1978. Breeding utilisation for meat production in sheep. *Animal Breeding Abstracts* 46: 131–143. [13]

Nitzsche, W., and G. Wenzel. 1977. *Haploids in plant breeding.* Paul Parey, Hamburg, DE. [11]

Nodari, R. O., and 4 others. 1993. Toward an integrated linkage map of common bean. III. Mapping genetic factors controlling host-bacteria interactions. *Genetics* 134: 341–350. [18]

Nogami, S., Y. Ohya, and G. Yvert. 2007. Genetic complexity and quantitative trait loci mapping of yeast morphological traits. *PLoS Genet.* 3: e31. [5, 18]

Noël, E., and 6 others. 2016. Reduced mate availability leads to evolution of self-fertilization and purging of inbreeding depression in a hermaphrodite. *Evolution* 70: 625–640. [12]

Nonaka, E., and 5 others. 2018. Scaling up the effects of inbreeding depression from individuals to metapopulations. *J. Anim. Ecol.* 88: 1202–1214. [12]

Nordborg, M. 2000. Linkage disequilibrium, gene trees and selfing: an ancestral recombination graph with partial self-fertilization. *Genetics* 154: 923–929. [13]

Nordborg, M. and S. Tavaré. 2002. Linkage disequilibrium: what history has to tell us. *Trends Genet.* 18: 83–90. [5]

Norman, J. K., A. K. Sakai, S. G. Weller, and T. E. Dawson. 1995. Inbreeding depression in morphological and physiological traits of *Schiedea lydgatei* (Caryophyllaceae) in two environments. *Evolution* 49: 297–306. [12]

Norton, B., and E. S. Pearson. 1976. A note on the background to, and refereeing of, R. A. Fisher's 1928 paper 'On the correlation between relatives on the supposition of Mendelian inheritance'. *Notes Records Roy. Soc. Lond* 31: 151–162. [1]

Norton, J. B. 1907. Notes on breeding oats. *Am. Breeders Assoc. Rep.* 4: 280–285. [A9]

Novembre, J., and M. Stephens. 2008. Interpreting principal component analysis of spatial population genetic variation. *Nat. Genet.* 40: 646–649. [20]

Nuccio, M. L., M. Paul, N. J. Bate, J. Cohn, and S. R. Cutler. 2018. Where are the drought tolerant crops? An assessment of more than two decades of plant biotechnology effort in crop improvement. *Plant Sci.* 273: 110–119. [5]

Nunes, W. V. B., and 11 others. 2024. A comprehensive evolutionary scenario for the origin and neofunctionalization of the *Drosophila* speciation gene *Odysseus* (*OdsH*). *G3* 14: jkad 299. [17]

Nuzhdin, S. V., and T. F. C. Mackay. 1995. The genomic rate of transposable element movement in *Drosophila melanogaster. Mol. Biol. Evol.* 12: 180–181. [15]

Nuzhdin, S. V., and E. G. Pasyukova. 1991. New approach to polygene mapping. Location of genes controlling fitness level in related *Drosophila melanogaster* stocks. *Genetika* (Russ) 27: 849–859. [17]

Nuzhdin, S. V., P. D. Keightley, and E. G. Pasyukova. 1993. The use of retrotransposons as markers for mapping genes responsible for fitness differences between related *Drosophila melanogaster* strains. *Genet. Res.* 62: 125–131. [8, 17]

Nuzhdin, S. V., J. D. Fry, and T. F. C. Mackay. 1995. Polygenic mutation in *Drosophila melanogaster:* the causal relationship of bristle number to fitness. *Genetics* 139: 861–872. [15]

Nuzhdin, S. V., and 4 others. 1997. Sex-specific quantitative trait loci affecting longevity in *Drosophila melanogaster*. *Proc. Natl. Acad. Sci. USA* 94: 9734–9739. [17, 18]

Nyholt, D. R. 2004. A simple correction for multiple testing for single-nucleotide polymorphisms in linkage disequilibrium with each other. *Am. J. Hum. Genet.* 74: 765–760. [20, A6]

Nyholt, D. R. 2005. Evaluation of Nyholt's procedure for multiple testing correction — Author's reply. *Hum. Hered.* 60: 61–62 [20]

Oakley, C. G., J. Ågren, and D. W. Schemske. 2015a. Heterosis and outbreeding depression in crosses between natural populations of *Arabidopsis thaliana. Heredity* 115: 73–82. [13]

Oakley, C. G., J. P. Spoelhof, and D. W. Schemske. 2015b. Increased heterosis in selfing populations of a perennial forb. *AoB Plants* 7: plv122: [13]

Oakley, C. G., S.Lundemo, J. Ågren, and D. W. Schemske. 2019. Heterosis is common and inbreeding depression absent in natural populations of *Arabidopsis thaliana. J. Evol. Biol.* 32: 592–603. [13]

Obando, J., and 6 others. 2008. Identification of melon fruit quality quantitative trait loci using near-isogenic lines. *J. Am. Soc. Horticultural Sci.* 133: 139–151. [17]

O'Brien, S. J. (ed.) 1990. *Genetic maps*, 5th Ed. Cold Spring Harbor Press, Cold Spring Harbor, NY. [11]

O'Connor, L. J. 2021. The distribution of common-variant effect sizes. *Nat. Genet.* 53: 1243–1249. [21]

O'Connor, L. J., and 5 others. 2019. Extreme polygenicity of complex traits is explained by negative selection. *Am. J. Hum. Genet.* 105: 456–476. [21]

Odell, P. L., and A. H. Feiveson. 1966. A numerical procedure to generate a sample covariance matrix. *Am. Stat. Assoc. Journ.* 61: 198–203. [A7]

O'Donald, P. 1971. The distribution of genotypes produced by alleles segregating at a number of loci. *Heredity* 26: 233–241. [16]

O'Grady, J. J., and 5 others. 2006. Realistic levels of inbreeding depression strongly affect extinction risk in wild populations. *Biol. Conserv.* 133: 42–51. [12]

Ohnishi, O. 1977. Spontaneous and ethyl methanesulfonate induced mutations controlling viability in *Drosophila melanogaster*. II. Homozygous effect of polygenic mutations. *Genetics* 87: 529–545 [15]

Ohta, T. 1971. Associative overdominance caused by linked detrimental mutations. *Genet. Res.* 18: 356–362. [12]

Ohya, Y., and 24 others. 2005. High-dimensional and large-scale phenotyping of yeast mutations. *Proc. Natl. Acad. Sci. USA* 102: 19015–19020. [21]

Ohya, Y., Y. Kimori, H. Okada, and S. Ohnuki. 2015. Single-cell phenomics in budding yeast. *Mol. Biol. Cell* 26: 3920–3925. [18]

Oka, H. I., J. Hayashi, and I. Shiojiri. 1958. Induced mutations of polygenes for quantitative characters in rice. *J. Heredity* 49: 11–14. [15]

Oliehoek, P. A., J. J. Windig, J. A. N. van Arendonk, and P. Bijma. 2006. Estimating relatedness between individuals in general populations with a focus on their use in conservation programs. *Genetics* 173: 483–496. [8]

Olkin, I., and A. Sampson. 1998. Comparison of meta-analysis versus analysis of variance of individual patient data. *Biometrics* 54: 317–322. [20]

Ollivier, L. 1988. Current principles and future prospects in selection of farm animals. *In* B. S. Weir, E. J. Eisen, M. M. Goodman, and G. Namkoong (eds.), *Proceedings of the second international conference on quantitative genetics*, pp. 438–450. Sinauer Assocs., Sunderland, MA. [1]

Ollivier, L., and L. L. G. Janns. 1993. A note on the estimation of the effective number of additive and dominant loci contributing to quantitative variation. *Genetics* 135: 907–909. [11]

Ollivier, L., L. A. Messer, M. F. Rothschild, and C. Legault. 1997. The use of selection experiments for detecting quantitative trait loci. *Genet. Res.* 69: 227–232. [17]

Olsen, J. B., C. Busack, J. Britt, and Paul Bentzen. 2001. The aunt and uncle effect: an empirical evaluation of the confounding influence of full sibs of parents on pedigree reconstruction. *J. Hered.* 92: 243–247. [8]

Olson, J. M. 1994. Some empirical properties of an all-relative-pairs linkage test. *Genet. Epidem.* 10: 87–102. [19]

Olson, J. M. 1995. Multipoint linkage analysis using sib pairs: an interval mapping approach for dichotomous outcomes. *Am. J. Hum. Genet.* 56: 788–798. [19]

Olson, J. M., and E. M. Wijsman. 1993. Linkage between quantitative trait and marker loci: methods using all relative pairs. *Genet. Epidem.* 10: 87–102. [19]

Omholt, S. W., E. Plante, L. Øyehuang, and K. Xiang. 2000. Gene regulatory networks generating the phenomena of additivity, dominance and epistasis. *Genetics* 155: 969–980. [4, 21]

Ongen, H., and 4 others. 2016. Fast and efficient QTL mapper for thousands of molecular phenotypes. *Bioinformatics* 32: 1479–1485 [21]

Ongen, H., and 6 others. 2017. Estimating the causal tissue for complex traits and disease. *Nat. Genet.* 49: 1676–683. [21]

Ordas, B., R. A. Malvar, and W. G. Hill. 2008. Genetic variation and quantitative trait loci associated with developmental stability and the environmental correlation between traits in maize. *Genet. Res.* 90: 385–395. [6, 18]

O'Reilly, P. F., and 6 others. 2012. MultiPhen: joint model of multiple phenotypes can increase discovery in GWAS. *PLoS One* 7: e34861. [20]

Orlove, M. J., and C. L. Wood. 1978. Coefficients of relationship and coefficients of relatedness in kin selection: a covariance form for the RHO formula. *J. Theor. Biol.* 73: 679–686. [7]

Orozco, F. 1976. Heterosis and genotype-environment interaction: theoretical and experimental aspects. *Bulletin Technique du Departement de Genetique Animale* 24. [13]

Orozco, G., J. C. Barrett, and E. Zeggini. 2010. Synthetic association in the context of genome-wide association scan signals. *Hum. Mol. Genet.* 19: R137–R144. [21]

Orr, H. A. 1987. Genetics of male and female sterility in hybrids of *Drosophila pseudoobscura* and *D. persimilis*. *Genetics* 116: 555–563. [17]

Orr, H. A. 1991. A test of Fisher's theory of dominance. *Proc. Natl. Acad. Sci. USA* 88: 11413–11415. [4]

Orr, H. A. 1993a. A mathematical rule of Haldane's rule. *Evolution* 47: 1606–1611. [17]

Orr, H. A. 1993b. Haldane's rule has multiple genetic causes. *Nature* 361: 532–533. [17]

Orr, H. A. 1995. The population genetics of speciation: the evolution of hybrid incompatibilities. *Genetics* 139: 1805–1813. [17]

Orr, H. A. 1996. Dobzhansky, Bateson, and the genetics of speciation. *Genetics* 144: 133–135. [17]

Orr, H. A. 1997. Haldane's rule. *Ann. Rev. Ecol. System.* 28: 195–218. [17]

Orr, H. A. 1998. Testing natural selection vs. genetic drift in phenotypic evolution using quantitative trait locus data. *Genetics* 149: 2099-2104. [18]

Orr, H. A. 2005. The genetic theory of adaptation: A brief history. *Nat. Rev. Genet.* 6: 119–127. [17]

Orr, H. A., and J. A. Coyne. 1989. The genetics of postzygotic isolation in the *Drosophila virilis* group. *Genetics* 121: 527–537. [17]

Orr, H. A., and J. A. Coyne. 1992. The genetics of adaptation: a reassessment. *Am. Nat.* 140: 725–742. [11]

Orr, H. A., and S. Irving. 2000. Genetic analysis of the Hybrid male rescue locus of *Drosophila. Genetics* 155: 225–231. [17]

Orr, H. A., and D. C. Presgraves. 2000. Speciation by postzygotic isolation: forces, genes and molecules. *BioEssays* 22: 1085–1094. [17]

Orr, H. A., and M. Turelli. 1996. Dominance and Haldane's rule. *Genetics* 143: 613–616. [17]

Orr, H. A., J. P. Masly, and D. C. Presgraves. 2004. Speciation genes. *Curr. Opin Genet. Develop.* 14: 675–679. [17]

Orr, H. A., J. P. Masly, and N. Phadnis. 2006. Speciation in Drosophila: from phenotypes to molecules. *J. Heredity* 98: 103–110. [17]

Osborn, T. C., D. C. Alexander, and J. F. Fobes. 1987. Identification of restriction fragment length polymorphisms linked to genes controlling soluble solids content in tomato fruit. *Theor. Appl. Genet.* 73: 350–356. [17]

Osier, M., and 8 others. 1999. Linkage disequilibrium at the ADH2 and ADH3 loci and risk of alcoholism. *Am. J. Human Genet.* 64: 1147–1157. [5]

Ossowski, S., and 7 others 2010. The rate and molecular spectrum of spontaneous mutations in *Arabidopsis thaliana*. *Science* 327: 92–94. [15]

Osterwalder, M., and 19 others. 2018. Enhancer redundancy provides phenotypic robustness in mammalian development. *Nature* 554: 239–243. [14]

Ott, J. 1979. Maximum likelihood estimation by counting methods under polygenic and mixed models in human pedigrees. *Am. J. Hum. Genet.* 31: 161–175. [16]

Ott, J. 1991. *Analysis of human genetic linkage*. Revised Edition. Johns Hopkins, Baltimore, MD. [12, 17, 18]

Ott, J., Y. Kamatani, and M. Lathrop. 2011. Family-based designs for genome-wide association studies. *Nat. Rev. Genet.* 12: 465–474. [17, 20]

Otter, T., M. J. Pachali, S. Mayer, and J. R. Landwher. 2018. Causal inference using mediation analysis or instrumental variables — full mediation in the absence of condition independence. *Marketing* 40: 41–57. [21, A2]

Otto, P. A., O. Frota-Pessoa, and S. P. Polcan. 1994. Snyder's ratios with incomplete penetrance. *J. Hered.* 85: 331–335. [4]

Otto, S. P., and C. D. Jones. 2000. Detecting the undetected: estimates the total number of loci underlying a quantitative trait. *Genetics* 156: 2093–2107. [18, 21]

Ouborg, N. J., and R. Van Treuren. 1994. The significance of genetic erosion in the process of extinction. IV. Inbreeding load and heterosis in relation to population size in the mint *Salvia pratensis*. *Evolution* 48: 996–1008. [12]

Ovaskainen, O., J. M. Cano, and J. Merilä. 2008. A Bayesian framework for comparative quantitative genetics. *Proc. Roy .Soc. B* 275: 669–678. [7, A8]

Overall, A. D. J., M. Ahmad, and R. A. Nichols. 2002. The effect of reproductive compensation on recessive disorders within consanguineous human populations. *Heredity* 88: 474–479. [12]

Owen, A. B. 2005. Variance of the number of false discoveries. *J. Roy. Stat. Soc. B* 67: 411–426. [A6]

Ozaki, K., and 10 others. 2002. Functional SNPs in the lymphotoxon-α gene that are associated with susceptibility to myocardial infraction. *Nat. Genet.* 32: 650–654 (correction 33: 107). [20]

Paaby, A. B., and G. Gibson. 2016. Cryptic genetic variation in evolution's developmental genetics. *Biology* 5: 28. [14]

Paaby, A. B., and M. V. Rockman. 2014. Cryptic genetic variation: evolution's hidden substrate. *Nat.*

Rev. Genet. 15: 247–258. [14]

Pacek, P., A. Sajantila, and A.-C. Syvänen. 1993. Determination of allele frequencies at loci with length polymorphism by quantitative analysis of DNA amplified from pooled samples. *PCR Methods Appl.* 2: 313–317. [17]

Packard, G. C., and T. J. Boardman. 1987. The misuse of ratios to scale physiological data that vary allometrically with body size. *In* M. E. Feder, A. F. Bennett, W. W. Burggren, and R. B. Huey (eds.), *New directions in ecological physiology*, pp. 216–239. Cambridge Univ. Press, Cambridge, UK. [14]

Packard, G. C., G. F. Birchard, and T. J. Boardman. 2011. Fitting statistical models in bivariate allometry. *Biol. Rev.* 86: 549–563. [14]

Page, G. P., C. I. Amos, and E. Boerwinkle. 1998. The quantitative LOD score: test statistic and sample size for exclusion and linkage of quantitative traits in human sibships. *Am. J. Hum. Genet.* 62: 962–968. [19]

Painter, J. 1997. Sibship reconstruction without parental information. *J. Agric. Biol. Envron. Stat.* 2: 212–229. [8]

Palmer, A. R., and C. Strobeck. 1986. Fluctuating asymmetry: measurement, analysis, patterns. *Ann. Rev. Ecol. Syst.* 17: 391–421. [6]

Palmer, A. R., and C. Strobeck. 1992. Fluctuating asymmetry as a measure of developmental stability: implications of non-normal distributions and power of statistical tests. *Acta Zool. Fennica* 191: 57–72. [6]

Palmer, A. R., and D. Strobeck. 2003. Fluctuating asymmetry analysis revisited. *In* M. Polak (ed.), *Developmental instability. Causes and consequences*, pp. 279–319. Oxford Univ. Press, Oxford, UK. [6]

Palmer, A. R., C. Strobeck, and A. K. Chippindale. 1993. Bilateral variation and the evolutionary origin of macroscopic asymmetries. *Genetica* 89: 201–218. [6]

Palmer, C., and I. Pe'er. 2017. Statistical correction of the Winner's Curse explains replication variability in quantitative trait genome-wide association studies. *PLoS Genet.* 13: e1006916. [18]

Palopoli, M. F., and C.-I. Wu. 1994. Genetics of hybrid male sterility between *Drosophila* sibling species: a complex web of epistasis is revealed in interspecific studies. *Genetics* 138: 329–341. [17]

Pamilo, P., and S. Pálsoon. 1998. Associative overdominance, heterozygosity and fitness. *Heredity* 81: 381–389. [12]

Pamilo, P., M. Nei, and W.-H. Li. 1987. Accumulation of mutations in sexual and asexual populations. *Genet. Res.* 49: 135–146. [15]

Pamler, C., and I. Pe'er. 2017. Statistical correction of the winner's curse explains replication variability in quantitative trait genome-wide association studies. *PLoS Gemet.* 13: e1006916. [20]

Palmer, D. S., and 11 others. 2023. Analysis of genetic dominance in the UK biobank. *Science* 379: 1341–1349. [21]

Pan, W. 2008. Network-based model weighting to detect multiple loci influencing complex diseases. *Hum. Genet.* 124: 225–23. [21]

Pan, W. 2009. Asymptotic tests of association with multiple SNPs in linkage disequilibrium. *Genet. Epidem.* 33: 497–507. [20]

Pan, W., and X. Shen. 2011. Adaptive tests for association analysis of rare variants. *Genet. Epidem.* 35: 381–388. [20]

Pan, W., and 4 others. 2014. A powerful and adaptive associated test for rare variants. *Genetics* 197: 1081–1095. [20]

Pan, W., I.-Y. Kwak, and P. Wei. 2015. A powerful pathway-based adaptive test for genetic association with common or rare variants. *Am. J. Hum. Genet.* 97: 86–98. [20]

Panagiotou, O. A., C. J. Willer, J. N. Hirschhorn, and J. P. A. Ioannidis. 2013. The power of meta-analysis in genome-wide association studies. *Ann. Rev. Genom. Hum. Genet.* 14: 441–465. [20]

Pandit, A., and 13 others. 2010. Combining QTL mapping and transcriptome profiling of bulked RILs for identification of functional polymorphism for salt tolerance genes in rice (*Oryza sativa* L.). *Mol. Genet. Genomics* 284 : 121–136. [17]

Panigrahi, S., J. Zhu, and C. Sabatti. 2021. Selection-adjusted inference: an application to confidence intervals for cis-eQTL effect sizes. *Biostat.* 22: 181–197. [18]

Panse, V. G. 1940. A statistical study of quantitative inheritance. *Ann. Eugenics* 10: 76–105. [11]

Papa, K. E. 1970. Inheritance of growth rate in *Neurospora crassa:* crosses between previously selected lines. *Can. J. Genet. Cytol.* 12: 1–9. [5]

Papachristou, C., and S. Lin. 2006a. A comparison of methods for intermediate fine mapping. *Genet. Epid.* 30: 677–689. [19]

Papachristou, C., and S. Lin. 2006b. A two-step procedure for constructing convenience interval of trait loci with application to a rheumatoid arthritis dataset. *Genet. Epid.* 30: 18–29. [19]

Papadakis, J. 1937. Méthode statistique pour des expériences sur champ. *Inst. Amel. Plantes a Salonique, Bulletin 23.* [A9]

Paré, G., N. R. Cook, P. M. Ridker, and D. I. Chaseman. 2010. On the use of variance per genotype as a tool to identify quantitative trait interaction effects: a report from the women's genome health study. *PLoS Genet.* 6: e1000981. [18, 20]

Paran, I., R. Kesseli, and R. Michelmore. 1991. Identification of restriction fragment length polymorphism and random amplified polymorphic DNA markers linked to downy mildew resistance genes in lettuce, using near-isogenic lines. *Genomes* 34: 1021–1027. [17]

Park, J.-H., and 8 others. 2011. Distribution of allele frequencies and effect sizes and their interrelationships for common genetic susceptibility variants. *Proc. Natl. Acad. Sci. USA* 108: 18026–18031. [21]

Park, R. E. 1966. Estimation with heteroscedastic error terms. *Econometrica* 34:: 888. [10]

Park, Y., and 10 others. 2014. ABCC transporters mediate insect resistance to multiple Bt toxins revealed by bulk segregant analysis. *BMC Bio.* 12: 1–15. [17]

Park, Y. S., and D. P. Fowler. 1982. Effects of inbreeding and genetic variances in a natural population of tamarack (*Larix laricina* (Du Roi) K. Koch) in eastern Canada. *Silvae Genetica* 31: 21–26. [12]

Park, Y. S., and D. P. Fowler. 1984. Inbreeding in black spruce (*Picea mariana* (Mill.) B. S. P.): self-fertility, genetic load and performance. *Can. J For. Res.* 14: 17–21. [12]

Parker, M. A. 1992. Outbreeding depression in a selfing annual. *Evolution* 46: 837–841. [13]

Parsons, P. A. 1959. Some problems in inbreeding and

random mating in tetrasomics. *Agronomy J.* 51: 465–467. [12, 13]

Partridge, L., T. F. C. Mackay, and S. Aitken. 1985. Male mating success and fertility in *Drosophila melanogaster. Genet. Res.* 46: 279–285. [12]

Parts, L., and 13 others. 2011. Revealing the genetic structure of a trait by sequencing a population under selection. *Genome Res.* 21: 1131–1138. [17]

Pasaniuc, B., and 42 others. 2011. Enhanced statistical tests for GWAS in admixed populations: assessment using African Americans from care and a breast cancer consortium. *PLoS Genet.* 7: e1001371. [20]

Pascoe, L., and N. E. Morton. 1987. The use of map functions in multipoint mapping. *Am. J. Hum. Genet.* 40: 174–183. [17]

Pascual, L., and 13 others. 2016. Dissecting quantitative trait variation in the resequencing era: complementarity of bi-parental, multi-parental and association panels. *Plant Sci.* 242: 120–130. [18]

Passador-Gurgel, G., and 4 others. 2007. Quantitative trait transcripts for nicotine resistance in *Drosophila melanogaster. Nat. Genet.* 39: 264–268. [21]

Paterson, A. H., and 5 others. 1988. Resolution of quantitative traits into Mendelian factors by using a complete RFLP linkage map. *Nature* 335: 721–726. [18]

Paterson, A. H., J. W. DeVerna, B. Lanini, and S. D. Tanksley. 1990. Fine mapping of quantitative trait loci using selected overlapping recombinant chromosomes, in an interspecies cross of tomato. *Genetics* 124: 735–742. [17, 18]

Paterson, A. H., S. D. Tanksley, and M. E. Sorrells. 1991. DNA markers in plant improvement. *Adv. Agron.* 46: 39–90. [17]

Patnaik, P. B. 1949. The noncentral χ^2 and F distributions and their approximations. *Biometrika* 36: 202–232. [A5]

Patterson, H. D., and A. E. Hunter. 1983. The efficiency of incomplete block designs in National List and Recommended List cereal variety trails. *J. Agr. Sci. Camb.* 101: 427–433. [A9]

Patterson, H. D., and R. Thompson. 1971. Recovery of interblock information when block sizes are unequal. *Biometrika* 58: 545–554. [10]

Patterson, H. D., and E. R. Williams. 1976. A new class of resolvable incomplete block designs. *Biometrika* 63: 83–92. [A9]

Patterson, N., and 11 others. 2004. Methods for high-density admixture mapping of disease genes. *Am. J. Hum. Genet.* 74: 979–1000. [17]

Patterson, N., A. L. Price, and D. Reich. 2006. Population structure and eigenanalysis. *PLoS Genet.* 2: e190. [20, A6]

Patti, K. A., and 7 others. 2009. A computationally efficient hypothesis testing method for epistasis analysis using multifactor dimensionality reduction. *Genet. Epidemi.* 31: 87–94. [18]

Paul, M. J., M. L. Nuccio, and S. S. Basu. 2018. Are GM crops for yield and resilience possible? *Trends Plant Sci.* 23: 10–16. [5]

Pavan, R., and 4 others. 2017. Detection of epistasis through triple test cross (TTC) analysis in maize (*Zea mays* L.). *J. Appl. Nat. Sci.* 9: 2496–2501. [13]

Pawitan, Y. 2000. A reminder of the fallibility of the Wald statistic: likelihood explanation. *Am. Stat.* 54: 54–56. [A4]

Payne, F. 1918. The effect of artificial selection on bristle number in *Drosophila ampelophila* and its interpretation. *Proc. Natl. Acad. Sci. USA* 4: 55–58. [1, 17]

Payseur, B. A. and L. H. Rieseberg. 2016. A genomic perspective on hybridization and speciation. *Mol. Ecol.* 25: 2337–2360. [13]

Pearce, S. C. 1983. *The agricultural field experiment: A statistical examination of theory and practice.* Wiley, New York, NY. [A9]

Pearce, S. C. 1998. Field experimentation on rough land: the method of Papadakis reconsidered. *J. Agric. Sci.* 131: 1–11. [A9]

Pearce, S. C., G. M. Clarke, G. V. Dyke, and R. E. Kempson. 1988. *A manual of crop experimentation.* Oxford University Press, Oxford, UK. [A9]

Pearl, J. 2009. *Causality: models, reasoning, and inference,* 2nd Ed. Cambridge University press, Cambridge, UK. [A2]

Pearl, J., M. Glymour, and N. P. Jewell. 2016. *Causal inference in statistics: A primer.* John Wiley and Sons, Chichester, UK. [A2]

Pearson, E. S. 1938. The probability transformation for testing goodness of fit and combining independent tests of significance. *Biometrika* 30: 134–148. [A6]

Pearson, K. 1896. Contributions to the mathematical theory of evolution. III. Regression, heredity and panmixia. *Phil. Trans. Royal Soc. Lond. A* 187: 253–318. [9, A1]

Pearson, K. 1897. Mathematical contributions to the theory of evolution — on a form of spurious correlation which may arise when indices are used in measurement of organs. *Proc. Royal Soc. Lond.* 60: 489–498. [14]

Pearson, K. 1903. Mathematical contributions to the theory of evolution. XI. On the influence of natural selection on the variability and correlation of organs. *Phil. Trans. Royal Soc. Lond.* A 200: 1–66. [1, 9]

Pearson, K. 1904. Mathematical contributions to the theory of evolution. XII. On a generalized theory of alternative inheritance, with special reference to Mendel's laws. *Phil. Trans. Royal Soc. Lond. A* 203: 53–86. [16]

Pearson, K. 1910. Darwinism, biometry, and some recent biology. *Biometrika* 7: 368–385. [2]

Pearson, K. 1920. Notes on the history of correlation. *Biometrika* 13: 25–45. [9]

Pearson, K., and A. G. Davin. 1924. On the biometric constraints of the human skull. *Biometrica* 16: 328–363. [14]

Pearson, K., and A. Lee. 1903. On the laws of inheritance in man. I. Inheritance of physical characters. *Biometrika* 2: 357–462. [7]

Pederson, D. G. 1968. Environmental stress, heterozygote advantage and genotype-environment interaction in *Arabidopsis. Heredity* 23: 127–138. [6]

Pedhazur, E. J. 1982. *Multiple regression in behavioral research.* 2nd Ed. Holt, Rinehart and Winston, Fort Worth, TX. [A2]

Pe'er, I., R. Yelensky, D. Altshuler, and M. J. Daly. 2008. Estimation of the multiple testing burden for genomewide association studies of nearly all common variants. *Genet. Epid.* 32: 381–385. [20]

Pei, Y.-F., L. Zhang, J. Li, and H.-W. Deng. 2010. Analyses and comparison of imputation-based association methods. *PLoS One* 5: e10827. [20]

Peirce, J. L., and 7 others. 2008. Genome reshuffling for advanced intercross permutation (GRAIP): simulation and permutation for advanced intercross population analysis. *PLoS ONE* 3: e1977. [18]

Peltonen, L., A. Jalanko, and T. Varilo. 1999. Molecular genetics the Finnish disease heritage. *Hum. Mol. Genet.* 8: 1913–1923. [17]

Peltonen, L., A. Palotie, and K. Lange. 2000. Use of population isolates for mapping complex traits. *Nat. Rev. Genet.* 1: 182–190. [17]

Peltonen, J., J. Venna, and S. Kaski. 2009. Visualizations for assessing convergence and mixing of Markov chain Monte Carlo simulations. *Compt. Stat. Data Anal.* 53: 4453–4470. [A8]

Pemberton, J. 2004. Measuring inbreeding depression in the wild: the old ways are the best. *Trends Ecol. Evol.* 19: 613–615. [12]

Pemberton, J. 2008. Wild pedigrees: the way forward. *Proc. Roy. Soc. Lond. B* 275: 613–621. [8, 12]

Pemberton, J. M., P. E. Ellis, J. G. Pilkington, and C. Bérénos. 2017. Inbreeding depression by environment interactions in a free-living mammal population. *Heredity* 118: 64–77. [12]

Pemberton, T. J., and Z. A. Szpiech. 2018. Relationship between deleterious variation, genome autozygosity, and disease risk: Insights from the 1000 genome project. *Am. J. Hum. Genet.* 102: 658–675. [12]

Pemberton, T. J., and 5 others. 2012. Genomic patterns of homozygosity in worldwide human populations. *Am. J. Hum. Genet.* 91: 275–292. [8]

Pena, S. D. J., and R. Chakraborty. 1994. Paternity testing in the DNA era. *Trends Genet.* 10: 204–209. [8]

Peng, B., and M. Kimmel. 2007. Simulation provide support for the common disease–common variant hypothesis. *Genetics* 175: 763–776. [21]

Peng, G., and 11 others. 2010. Gene and pathway-based second-wave analysis of genome-wide association studies. *Europ. J. Hum. Genet.* 18: 111–117. [20, 21]

Penner, G. A., and 4 others. 1993. Identification of a RAPD marker linked to the oat stem rust gene *Pg3*. *Theor. Appl. Genet.* 85: 702–705. [17]

Penrose, L. S. 1935. The detection of autosomal linkage in data which consist of pairs of brothers and sisters of unspecified parentage. *Ann. Eugen.* 6: 133–138. [19]

Penrose, L. S. 1939. Some practical considerations in testing for genetic linkage in sib data. *Ohio J. Sci.* 39: 291–296. [17]

Penrose, L. S. 1954a. Some recent trends in human genetics. *Carylogia* 6 (suppl.): 521–530. [7]

Penrose, L. S. 1954b. The general purpose sib-pair linkage test. *Ann. Eugen.* 18: 120–124. [19]

Penrose, L. S. 1969. Effects of additive genes at many loci compared with those of a set of alleles at one locus in parent-child and sib correlations. *Ann. Hum. Genet.* 35: 15–21. [16]

Pereira, M. G., and M. Lee. 1995. Identification of genomic regions affecting plant height in sorghum and maize. *Theor. Appl. Genet.* 90: 380–388. [17]

Perera, S. A. C. N., T. M. Wilkes, and M. J. Kearsey. 2006. More precise QLT mapping using STAIRS. *Tropical Agricultural Research* 18: a23. [17]

Peres-Neto, P. R., D. A. Jackson, and K. M. Somers. 2005. How many principal components? Stopping rules for determining the number of non-trivial axes revisited. *Comput. Stat. Data. Anal.* 49: 974–997. [20]

Perez, D. E., and C.-I. Wu. 1995. Further characterization of the *Odysseus* locus of hybrid sterility in *Drosophila simulans*: one gene is not enough. *Genetics* 140: 201–206. [17]

Perez, D E., C.-I. Wu, N. A. Johnson, and M.-L. Wu. 1993. Genetics of reproductive isolation in the *Drosophila simulans* clade: DNA marker-assisted mapping and characterization of a hybrid-male sterility gene, *Odysseus (Ods)*. *Genetics* 133: 261–275. [17]

Pérez-Enciso, M. 2004. *In silico* study of transcriptome genetic variation in outbred populations. *Genetics* 166: 547–554. [21]

Pérez-Enciso, M., and L. Varona. 2000. Quantitative trait loci mapping in F_2 crosses between outbred lines. *Genetics* 155: 391–405. [19]

Pérez-Pereira, A., and 7 others. 2021. Long-term exhaustion of the inbreeding load in *Drosophila melanogaster*. *Heredity* 127: 373–383. [12]

Pérez-Pereira, A., A. Caballero, and A. García-Dorado. 2022a. Reviewing the consequences of genetic purging on the success of rescue programs. *Conserv. Genet.* 23: 1–17. [12]

Pérez-Pereira, N., Jinliang W., H. Quesada, and A. Caballero. 2022b. Prediction of the minimum effective size of a population viable in the long term. *Biodiversity & Conservation* 31: 2763–2780. [12]

Pérez-Pereira, A., E. López-Cortegano, A. García-Dorado, and A. Caballero. 2023. Prediction of fitness under different breeding designs in conservation programs. *Anim. Conserv.* 26: 86–102. [12]

Pericak-Vance, M. A., and J. L. Haines. 1995. Genetic susceptibility to Alzheimer disease. *Trends Genet.* 11: 504–508. [17]

Peripoli, E., and 5 others. 2016. Runs of homozygosity: current knowledge and applications in livestock. *Anim Genet.* 48: 255–271. [8, 12]

Peters, J. L., and 4 others. 2008. Contour-enhanced meta-analysis funnel plots help distinguish publication bias from other causes of asymmetry. *J. Clinc. Epid.* 61: 991–996. [A6]

Peters, R. H. 1983. *The ecological implications of body size.* Cambridge University Press, Cambridge, UK. [14]

Peterson, C. B., M. Bogomolov, Y. Benjamini, and C. Sabatti. 2016a. Many phenotypes without many false discoveries: error controlling strategies for multitrait association studies. *Genet. Epidem.* 40: 45–56. [20]

Peterson, C. B., M. Bogomolov, Y. Benjamini, and C. Sabatti. 2016b. TreeQTL: hierarchical error control for eQTL findings. *Bioinfomratics* 32: 2556–2558. [21]

Petersen, R. G. 1994. *Agricultural field experiments: Design and analysis.* Marcel Dekker, New York, NY. [A9]

Petretto, E., and 4 others. 2006. Integrated gene expression profiling and linkage analysis in the rat. *Mamm. Genome* 17: 480–489. [21]

Pettersson, F. H., and 6 others. 2009. Marker selection for genetic case-control association studies. *Nat. Protocols* 4: 743–753. [20]

Pfahler, P. L. 1966. Heterosis and homeostasis in rye (*Secale cereale* L.). I. Individual plant production of varieties and intervarietal crosses. *Crop Sci.* 6: 397–401. [6]

Pfennig, D. W., and 5 others. 2010. Phenotypic plasticity's impacts on diversification and speciation. *Trend. Ecol. Evol.* 25: 459–467. [14]

Phadnis, N. and J. D. Fry. 2005. Widespread correlations between dominance and homozygous effects of mutations: implications for theories of dominance. *Genetics* 171: 385–392. [4]

Phadnis, N., and H. A. Orr. 2009. A single gene causes both male sterility and segregation distortion in *Drosophila* hybrids. *Science* 323: 376–379. [17]

Phillips, P. C. 1998. The language of gene interaction. *Genetics* 149: 1167–1171. [5]

Phillips, P. C. 2008. Epistasis — the essential role of gene interactions in the structure and evolution of genetic systems. *Nat. Rev. Genet.* 9: 855–867. [5]

Phillips, R. L. 2010. Mobilizing science to break yield barriers. *Crop Sci.* 50: S99–S108. [13]

Pickrell, J. K. 2014. Joint analysis of functional genomic data and genome-wide association studies of 18 human traits. *Am. J. Hum. Genet.* 94: 559–573. [21]

Pickrell, J., and 9 others. 2010. Understanding mechanisms underlying human gene expression variation with RNA sequencing. *Nature* 464: 768–772. [21]

Piepho, H.-P. 2001a. A quick method for computing approximate thresholds for quantitative trait loci detection. *Genetics* 157: 425–432. [18]

Piepho, H.-P. 2001b. Significance testing for QTL mapping by marker difference regression. *Theor. Appl. Genet.* 102: 1099–1102. [18]

Piepho, H. P., and H. G. Gauch Jr. 2001. Marker pair selection for mapping quantitative trait loci. *Genetics* 157: 433–444. [18]

Piepho, H. P., A. Büchse, and K. Emrich. 2003. A hitchhiker's guide to mixed models for randomized experiments. *J. Agon. Crop. Sci.* 189: 310–322. [A9]

Piepho, H.-P., C. Richter, and E. Williams. 2008. Nearest neigbhor adjustment and linear variance models in plant breeding trails. *Biometerical J.* 50: 164–189. [A9]

Piepho, H.-P., E. R. Williams, and V. Michel. 2015. Beyond Latin squares: a brief tour of row-column designs. *Agron. J.* 107: 2263–2270. [A9]

Pierce, B. L., and S. Burgess. 2013. Efficient design for Mendelian randomization studies: subsample and 2-sample instrumental variable estimators. *Am. J. Epidem.* 178: 1177–1184. [A2]

Pierce, B. L., and 16 others. 2014. Mediation analysis demonstrates that *trans*-eQTL are often explained by *cis*-mediation: a genome-wide analysis among 1,800 south Asians. *PLoS Genet.* 10: e1004818. [21]

Pillai, N. S., and X.-L. Meng. 2016. An unexpected encounter with the Cauchy and Lévy. *Ann. Statist.* 44: 2089–2097. [20 A6]

Piper, L. R., and B. M. Bindon. 1988. The genetics and endrocrinology of the Booroola sheep *F* gene. *In* B. S. Weir, E. J. Eisen, M. M. Goodman, and G. Namkoong (eds.), *Proceedings of the second international conference on quantitative genetics*, pp. 270–280. Sinauer Assoc., Sunderland, MA. [4]

Pirchner, F. 1983. *Population genetics in animal breeding*. 2nd Ed. Plenum, New York, NY. [1]

Pirinen, M., P. Donnelly, and C. C. A. Spencer. 2012. Including known covariance can reduce power to detect genetic effects in case-control studies. *Nat. Genet.* 44: 848–851. [20]

Pisani, J. F., and W. E. Kerr. 1961. Lethal equivalents in domestic animals. *Genetics* 46: 773–786. [12]

Plackett, R. L. 1972. The discovery of the method of least squares. *Biometrika* 59: 239–251. [3]

Plagnol, V., D. J. Smyth, J. A. Todd, and D. G. Clayton. 2009. Statistical independence of the colocalized association signals from type 1 diabetes and *RPS26* gene expression on chromosome 12q13. *Biostastistics* 10: 327–334. [21]

Plomin, R., G. E. McClearn, G. Gora-Maslak, and J. M. Neiderhiser. 1991. Use of recombinant inbred strains to detect quantitative trait loci associated with behavior. *Behav. Genet.* 21: 99–116. [17]

Plomion, C., N. Bahrman, C.-E. Durel, and D. M. O'Malley. 1995. Genomic mapping in *Pinus pinaster* (maritime pine) using RAPD and protein markers. *Heredity* 74: 661–668. [11]

Plough, L. V. 2016. Genetic load in marine animals: a review. *Cur. Zoo.* 62: 567–579. [12]

Plough, L. V., and D. Hedgecock. 2011. Quantitative trait locus analysis of stage-specific inbreeding depression in the Pacific oyster *Crassostrea gigas*. *Genetics* 189: 1473–1486. [12]

Pogson, G. H., and E. Zouros. 1994. Allozyme and RFLP heterozygosities as correlates of growth rate in the scallop *Placopecten magellanicus*: a test of the associative overdominance hypothesis. *Genetics* 137: 221–231. [12]

Poirier, J. G., and 5 others. 2015. Resampling to address the winner's curse in genetic association analysis of time to event. *Genet. Epidem.* 39: 518–528. [18, 20]

Polak, M. (ed.) 2003. *Developmental instability. Causes and consequences*. Oxford Univ. Press, Oxford, UK. [6]

Poland, J. A., and T. W. Rife. 2012. Genotyping-by-sequencing for plant breeding and genetics. *Plant Genome* 5:92–102. [8]

Poland, J. A., P. J. Bradbury, E. S. Buckler, and R. J. Nelson. 2011. Genome-wide nested association mapping of a quantitative resistance to northern leaf blight in maize. *Proc. Natl. Acad. Sci. USA* 108: 6893–6898. [21]

Pollak, E. 1987. On the theory of partially inbreeding finite populations. I. Partial selfing. *Genetics* 117: 353–360. [13]

Polman, C. H., and 13 others. 2005. Diagnostic criteria for multiple sclerosis: 2005 revisions to the "McDonald criteria". *Ann. Neurol.* 58: 840–846. [20]

Polman, C. H., and 17 others. 2011. Diagnostic criteria for multiple sclerosis: 2010 revisions to the McDonald criteria. *Ann. Neurol.* 69: 292–302. [20]

Ponzi, E.. L. F. Keller, and S. Muff. 2019. The simulation extrapolation technique meets ecology and evolution: a general and intuitive method to account for measurement error. *Methods Ecol. Evol.* 10: 1734–1748. [20]

Pooni, H. S., J. L. Jinks, and J. F. F. de Toledo. 1985. Predicting and observing the properties of second cycle hybrids using basic generations and inbred line $\times$ F_1 crosses. *Heredity* 54: 121–129. [11]

Pooni, H. S., P. S. Virk, D. T. Coombs, and M. K. U. Chowdhury. 1994. The genetical basis of hybrid vigour in a highly heterotic cross of *Nicotiana tabacum*. *Theor. Appl. Genet.* 89: 1027–1031. [12]

Portmann, P., and H. Ketata. 1997. Field plot technique. *In* R. A. Kempton and P. N. Fox (eds), *Statistical methods for plant variety evaluation*, pp. 9-18. Chapman & Hall, London, UK. [A9]

Porter, H. F., and P. F. O'Reilly. 2017. Multivariate simulation framework reveals performance of multi-trait GWAS methods. *Sci. Reports* 7: 38837. [20]

Povysil, G., and 5 others. 2019. Rare-variant collapsing analyses for complex traits: guidelines and applications. *Nat. Rev. Genet.* 20: 747–759. [20]

Powell, J. E., P. M. Visscher, and M. E. Goddard. 2010. Reconciling the analysis of IBD and IBS in complex trait studies. *Nat. Rev. Genet.* 11: 800–805. [8, 12]

Powell, W., W. T. B. Thomas, P. D. S. Caligari, and J. L. Jinks. 1985a. The effects of major genes on quantitatively varying characters in barley. 1. The GP *ert* locus. *Heredity* 54: 343–348. [17]

Powell, W., P. D. S. Caligari, W. T. B. Thomas, and J. L. Jinks. 1985b. The effects of major genes on quantitatively varying characters in barley. 2. The *denso* and daylength response loci. *Heredity* 54: 349–352. [17]

Powell, W., and 4 others. 1992. Association between rDNA alleles and quantitative traits in doubled haploid populations of barley. *Genetics* 130: 187–194. [15]

Power, C., and J. Elliott. 2006. Cohort profile: 1958 British birth cohort (national child development study). *Int. J. Epidem.* 35: 34–41. [20]

Powers, L. 1941. Inheritance of quantitative charaacters in crosses of two species of *Lycopersicon*. *J. Agr. Res.* 63: 149–174. [13]

Powers, L. 1942. The nature of the series of environmental variances and the estimation of the genetic variances and the geometric means in crosses involving species of *Lycopersicon*. *Genetics* 27: 561–575. [11]

Powers, L. 1944. An expansion of Jones's theory for the explanation of heterosis. *Am. Nat.* 78: 275–280. [13]

Powers, L. 1951. Gene analysis by the partitioning method when interactions of genes are involved. *Bot. Gaz.* 113: 1–23. [11]

Pray, L. A., and C. J. Goodnight. 1995. Genetic variation in inbreeding depression in the red flour beetle *Tribolium castaneum*. *Evolution* 49: 176–188. [12]

Pray, L. A., J. M. Schwartz, C. J. Goodnight, and L. Stevens. 1994. Environmental dependency of inbreeding depression: implications for conservation biology. *Cons. Biol.* 8: 562–568. [12]

Presgraves, D. C., and H. A. Orr. 1998. Haldane's rule in taxa lacking a hemizygous X. *Science* 282: 952–954. [17]

Price, A. L., and 5 others. 2006. Principal components analysis corrects for stratification in genome-wide association studies. *Nat. Genet.* 38: 904–909. [20]

Price, A. L., N. A. Zaitlen, D. Reich, and N. Patterson. 2010a. New approaches to population stratification in genome-wide association studies. *Nat. Rev. Genet.* 11: 459–463. [20]

Price, A. L., and 6 others. 2010b. Pooled association test for rare variants in exon-resequencing studies. *Am. J. Hum. Genet.* 86: 832–838. [20]

Price, A. L., and 5 others. 2011. Single-tissue and cross-tissue heritability of gene expression via identity-by-descent in related or unrelated individuals. *PLoS Genet.* 7: e1001317. [21]

Price, A. L., C. C. A. Spencer, and P. Donnelly. 2015. Progress and promise in understanding the genetic basis of common diseases. *Proc. Roy. Soc. B* 282: 20151684. [20, 21]

Price, G. R. 1970. Selection and covariance. *Nature* 227: 520–521. [3]

Price, G. R. 1972. Extension of covariance selection mathematics. *Ann. Hum. Genet.* 35: 485–490. [3]

Price, S. C., and 4 others. 1986. Relationships between performance and multilocus heterozygosity at enzyme loci in single-cross hybrids of maize. *J. Hered.* 77: 341–344. [13]

Price, T., and D. Schluter. 1991. On the low heritability of life-history traits. *Evolution* 45: 853–861. [7]

Pringle, R. M., and A. A. Rayner. 1971. *Generalized inverse matrices with applications to statistics.* Griffin, London, UK. [A3]

Pritchard, J. K. 2001. Are rare variants responsible for susceptibility to complex diseases? *Am. J. Hum. Genet.* 69: 124–137. [21]

Pritchard, J. K., and N. J. Cox. 2002. The allelic architecture of human disease genes: common disease–common variants ... or not? *Hum. Mol. Genet.* 11: 2417–2423. [21]

Pritchard, J. K., and P. Donnelly. 2001. Case-control studies of association in structured or admixed populations. *Theor. Pop. Biol.* 60: 227-237. [20]

Pritchard, J. K., and N. A. Rosenberg. 1999. Use of unlinked genetic markers to detect population stratification in association studies. *Am. J. Hum. Genet.* 65: 220–228. [20]

Pritchard, J. K., M. Stephens, and P. Donnelly. 2000a. Inference of population structure using multilocus genotype data. *Genetics* 155: 945–959. [20]

Pritchard, J. K., M. Stephens, N. A. Rosenberg, and P. Donnelly. 2000b. Association mapping in structured populations. *Am. J. Hum. Genet.* 67: 170–181. [20]

Pritikin, J. N., and 4 others. 2021. GW-SEM 2.0: Efficient, flexible, and accessible multivariate GWAS. *Beh. Genet* 51: 343–357. [20]

Provine, W. B. 1971. *The origins of theoretical population genetics.* Univ. Chicago Press, Chicago, IL. [1]

Provine, W. B. 2001. *The origins of theoretical population genetics: with a new afterword.* University of Chicago Press, Chicago, IL. [1]

Pryce, J. E., M. Haile-Mariam, M. E. Goddard, and B. J. Hayes. 2014. Identification of genomic regions associated with inbreeding depression in Holstein and Jersey dairy cattle. *Genet. Sel. Evol.* 46: 71. [12]

Pujol, B., P. David, and D. McKey. 2005. Microevolution in agricultural environments: how a traditional Amerindian farming practice favors heterozyogsity in cassava (*Manihot esculeta* Crantz, Euphorbiaceae.) *Evol. Lett.* 8: 138–137. [12]

Purcell, S., and P. C. Sham. 2004. Epistasis in quantitative trait locus linkage analysis: interaction or main effect? *Behav. Genet.* 34: 143–152. [20]

Purcell, S., S. S. Cherny, and P. C. Sham. 2003. Genetic power calculator: design of linkage and association genetic mapping studies of complex traits. *Bioinformatics* 19: 149–150. [20]

Purcell, S., and 10 others. 2007. PLINK: a tool set for whole-genome association and population-based linkage analysis. *Am. J. Hum. Genet.* 81 559–575. [8, 20]

Pusey, A., and M. Wolf. 1996. Inbreeding avoidance in animals. *Trends Ecol. Evol.* 11: 201–206. [12]

Putter, H., L. A. Sandkuijl, and J. C. van Houwelingen. 2002. Score test for detecting linkage to quantitative traits. *Genet. Epid.* 22: 345–355. [19]

Qi, T., and 16 others. 2018. Identifying gene targets from brain-related traits using transctriptomic and methylomic data from blood. *Nat. Comm.* 9: 2282. [21]

Qi, T., and 6 others. 2022. Genetic control of RNA splicing and its distinct role in complex trait variation. *Nat. Genet.* 54: 1355–1363. [21]

Qin, P., and 28 others. 2021. Pan-genome analysis of 33 genetically diverse rice accessions reveals hidden genomic variations. *Cell* 184: 3542–3553. [13]

Quaas, R. L., and E. J. Pollak. 1980. Mixed model methodology for farm and ranch beef cattle testing programs. *J. Anim. Sci.* 51: 1277–1287. [A3]

Queitsch, C., T. A. Sangster, and S. Lindquist. 2002. Hsp90 as a capacitor of phenotypic variation. *Nature* 417: 618–624. [6, 14]

Queller, D. C., and K. F. Goodnight. 1989. Estimating relatedness using genetic markers. *Evolution* 43: 258–275. [8]

Queller, D. C., J. E. Strassmann, and C. R. Hughes. 1993. Microsatellites and kinship. *Trend Ecol. Evol.* 8: 285–288. [8]

Quinn, G. P., and M. K. Koeugh. 2002. *Experimental design and data analysis for biologists.* Cambridge Univ. Press, Cambridge, UK. [A9]

Quintana, M. A., J. L. Berstein, D. C. Thomas, and D. V. Conti. 2011. Incorporating model uncertainty in detecting rare variants: the Bayesian risk index. *Genet. Epidem.* 35: 638–649. [20]

Rabinowitz, D., and N. Laird. 2000. A unified approach to adjusted association tests for population admixture with arbitrary pedigree structure and arbitrary missing marker information. *Hum. Hered.* 50: 211–223. [17]

Rafalski, J. A., and S. V. Tingey. 1993. Genetic diagnostics in plant breeding: RAPDs, microsatellites and machines. *Trends Genet.* 9: 275–280. [8]

Raftery, A. E., and S. Lewis. 1992a. How many iterations in the Gibbs sampler? *In* J. M. Bernardo, J. O. Berger, A. P. Dawid, and A. F. M. Smith (eds.), *Bayesian Statistics 4*, pp. 763–773. Oxford University Press. [A8]

Raftery, A. E., and S. Lewis. 1992b. Comment: One long run with diagnostics: Implementation strategies for Markov Chain Monte Carlo. *Stat. Sci.* 7: 493–497. [A8]

Rahmatallah, Y., F. Emmert-Streib, and G. Glazko. 2016. Gene set analysis approaches for RNA-seq data: performance evaluation and application guideline. *Brief. Bioinformat.* 17: 393–407. [21]

Rakshit, S., A. Rakshit, and J. V. Patil. 2012. Multiparent intercross populations in analysis of quantitative traits. *J. Genet.* 91: 111–117. [18]

Ralls, K., and J. Ballou. 1982a. Effect of inbreeding on juvenile mortality in some small mammal species. *Lab. Anim.* 16: 159–166. [12]

Ralls, K., and J. Ballou. 1982b. Effects of inbreeding on infant mortality in captive primates. *J. Primatol.* 3: 491–505. [12]

Ralls, K., K. Brugger, and J. Ballou. 1979. Inbreeding and juvenile mortality in small populations of ungulates. *Science* 206: 1101–1103. [12]

Ramage, R. T. 1983. Heterosis and hybrid seed in barely. *In* R. Frankel (ed.) *Heterosis: Reappraisal of theory and practice*, pp. 71–93. Springer-Verlag, Berlin. DE. [13]

Ramanan, V. K., L. Shen, J. H. Morre, and A. J. Saykin. 2012. Pathway analysis of genomic data: concepts, methods, and prospects for future development. *Trends Genet.* 28: 323–332. [21]

Ramensky, V., P. Bork, and S. Sunyaev. 2002. Human nonsynonymous SNPs: server and survey. *Nucleic Acids Res.* 30: 3894–3900. [20]

Ramsey, J. B. 1969. Tests for specification errors in classical linear least squares regression analysis. *J. Roy. Stat. Soc. B.* 31: 350–371. [10]

Ramstetter, M. D., and 7 others. 2017. Benchmarking relatedness inference methods with genome-wide data from thousands of relatives. *Genetics* 207: 75–82. [8]

Rana, A., and 5 others. 2021. Precise measurement of the fitness effects of spontaneous mutations by droplet digital PCR in *Burkholderia cenocepacia. Genetics* 219: iyab117. [15]

Rao, C. R. 1948. Large sample tests of statistical hypotheses concerning several parameters with applications to problems of estimation. *Math. Proc. Cambridge Phil. Soc.* 44: 50–57. [A4]

Rao, C. R., and S. K. Mitra. 1971. *Generalized inverse of matrices and its applications.* John Wiley & Sons, New York, NY. [A3]

Rao, D. C., N. E. Morton, and S. Yee. 1974. Analysis of family resemblance. II. A linear model for familial correlation. *Am. J. Hum. Genet.* 26: 331–359. [7]

Rao, G. Y., S. Andersson, and B Widén. 2002. Flower and cotyledon asymmetry in *Brassica cretica*: genetic variation and relationships with fitness. *Evolution* 56: 690–698. [6]

Rao, P. S. S., and S. G. Inbaraj. 1980. Inbreeding effects on fetal growth and development. *J. Med. Genet.* 17: 27–33. [12]

Rao, S. Q., and S. Xu. 1998. Mapping quantitative trait loci for categorical traits in four-way crosses. *Heredity* 81: 214–224. [18]

Rao, S. S. P., and 10 others. 2014. A 3D map of the human genome at kilobase resolution reveals principles of chromatin looping. *Cell* 159: 1665–1680. [21]

Rasmuson, M. 1952. Variation in bristle number of *Drosophila melanogaster. Acta Zoologica* 33: 1–31. [12]

Rasmuson, M. 1993. Variation in genetic identity within kinships. *Hered.* 70: 266–268. [8]

Rasmusson, J. 1927. Genetically changed linkage values in *Pisum. Hereditas* 10: 1–152. [17]

Rausher, M. D. 1983. Variability for host preference in insect populations: mechanistic and evolutionary models. *J. Insect Physiol.* 31: 873–889. [5]

Raykyan, V. K., T. A. Down, D. J. Balding, and S. Beck. 2011a. Epigenome-wide association studies for common human diseases. *Nat. Rev. Genet.* 12: 529–541. [21]

Raykyan, V. K., and 15 others. 2011b. Identification of Type I diabetes-associated DNA methylation variable positions that precede disease diagnosis. *PLoS Genet.* 7: e1002300. [21]

Read, A., and S. Nee. 1991. Is Haldane's rule significant? *Evolution* 45: 1707–1709. [17]

Realini, M. F., L. Poggio, J. Cámara-Hernández, and G. E. González. 2015. Intra-specific variation in genome size in maize: cytological and phenotypic correlates. *AoB PLANTS* 8: plv138. [15]

Rebaï, A., and B. Goffinet. 1993. Power of tests for QTL detection using replicated progenies derived from a diallel cross. *Theor. Appl. Genet.* 86: 1014–1022. (Correction 92: 128–129). [18]

Rebaï, A., and B. Goffinet. 2000. More about quantitative trait locus mapping with diallel designs. *Genet. Res.* 74: 243–247. [18]

Rebaï, A., B. Goffinet, B. Mangin, and D. Perret. 1994a. Detecting QTLs with diallel schemes. *In* J. W. van Ooijen and J. Jansen (eds.), *Biometrics in plant breeding: applications of molecular markers*, pp. 170–177. CPRO-DLO, NL. [18]

Rebaï, A., B. Goffinet, and B. Mangin. 1994b. Approximate thresholds of interval mapping tests for QTL detection. *Genetics* 138: 235–240. [18]

Rebaï, A., B. Goffinet, and B. Mangin. 1995. Comparing power of different methods of QTL detection. *Biometrics* 51: 87–99. [18]

Redner, R. A., and H. F. Walker. 1984. Mixture densities, maximum likelihood and the EM algorithm. *SIAM Review* 26: 195–239. [16]

Reed, D. H., and R. Frankham. 2003. Correlation between fitness and genetic diversity. *Conserv. Biol.* 17: 230–237. [12]

Reed, D. H., C. W. Fox, L. S. Enders, and T. N. Kristensen. 2012. Inbreeding-stress interactions. Evolutionary and conservation consequences. *Ann. N.Y. Acad. Sci.* 1256: 33–48. [12]

Reed, D. R., M. P. Lawler, and M.G.Tordoff. 2008. Reduced body weight is a common effect of gene knockout in mice. *BMC Genetics* 9: 1–6. [15]

Reed, L. K., B. A. LaFlamme, and T. A. Markow. 2008. Genetic architecture of hybrid male sterility in *Drosophila*: analysis of intraspecies variation for interspecies isolation. *PLoS One* 3: e3076. [17]

Rees, J. M. B., A. M. Wood, F. Dudbridge, and S. Burgess. 2016. Extending the MR-Egger method for multivariable Mendelian randomization to correct for both measured and unmeasured pleiotropy. *Stat. Med.* 36: 4705–4718. [A2]

Rees, J. M. B., A. M. Wood, and S. Burgess. 2019. Robust methods in Mendelian randomization via penalization of heterogeneous causal estimates. *PLoS One* 14: e0222362. [A2]

Reeve, E. C. R. 1960. Some genetic tests on asymmetry of sternopleural chaeta number in *Drosophila*. *Genet. Res.* 1: 151–172. [6]

Reeve, E. C. R. 1961. A note on non-random mating in progeny tests. *Genet. Res.* 2: 195–203. [7]

Reich, D. E., and D. B. Goldstein. 2001. Detecting association in a case-control study while correcting for population stratification. *Genet. Epid.* 20: 4–16. [20]

Reich, D. E., and E. S. Lander. 2001. On the allelic spectrum of human disease. *Trends Genet.* 17: 502–510. [21]

Reich, D. E., and 10 others. 2001. Linkage disequilibrium in the human genome. *Nature* 411: 199–204. [5]

Reich, V. H., and R. E. Atkins. 1970. Yield stability of four population types of grain sorghum, *Sorghum bicolor* (L.) Moench, in different environments. *Crop Sci.* 10: 511–517. [6]

Reid, J. M., P. Arcese, and L. F. Keller. 2008. Individual phenotype, kinship, and the occurrence of inbreeding in song sparrows. *Evolution* 62: 887–899. [12]

Reid, J. M., and 5 others. 2013. Pedigree error due to extra-pair reproduction substantially biases estimates of inbreeding depression. *Evol.* 68: 802–815. [12]

Reid, N. 1998. Saddlepoint methods and statistical inference. *Stat. Sci.* 3: 213–238. [20]

Reif, J. C., and 8 others. 2003a. Genetic distance based on simple sequence repeats and heterosis in tropical maize populations. *Crop Sci.* 43: 1275–1282. [13]

Reif, J. C., and 8 others. 2003b. Use of SSRs for establishing heterotic groups in subtropical maize. *Theor. Appl. Genet.* 107: 947–957. [13]

Reif, J. C., A. R. Hallauer, and A. E. Melchinger. 2005. Heterosis and heterotic patterns in maize. *Maydica* 50: 215–223. [13]

Reif, J. C., and 6 others. 2009. Unraveling epistasis with triple testcross progenies of near-isogenic lines. *Genetics* 181: 247–257. [13]

Reimann-Philipp, R. 1983. Heterosis in ornamentals. *In* R. Frankel (ed.) *Heterosis: Reappraisal of theory and practice*, pp. 234–259. Springer-Verlag, Berlin. DE. [13]

Relton, C. L., and G. Davey Smith. 2012. Two-step epigenetic Mendelian randomization: a strategy for establishing the causal role of epigenetic processes in pathways to disease. *Int. J. Epidem.* 41: 161–176. [A2]

Rendel, J. M. 1965. Bristle pattern in *scute* stocks of *Drosophila melanogaster*. *Am. Nat.* 99: 25–32. [14]

Rendel, J. M. 1977. Canalisation in quantitative genetics. *In* E. Pollak, O. Kempthorne, and T. B. Bailey, Jr. (eds.), *Proceedings of the international conference on quantitative genetics*, pp. 23–28. Iowa State Univ. Press, Ames, IA. [14]

Rendel, J. M. 1979. Canalization and selection. *In* J. N. Thompson, Jr., and J. M. Thoday (eds.), *Quantitative genetic variation*, pp. 139–156. Academic Press, New York, NY. [14]

Rendel, J. M., and B. L. Sheldon. 1960 Selection for canalization of the scute phenotype in *Drosophila melanogaster*. *Aust. J. Biol. Sci.* 13: 36–47. [14]

Rendel, J. M., B, L. Sheldon, and D. E. Finlay. 1966. Selection for canalization of the scute phenotype. II. *Am. Nat.* 100: 13–31. [14]

Reznick, D. 1981. "Grandfather effects": the genetics of interpopulation differences in offspring size in the mosquito fish. *Evolution* 35: 941–953. [6]

Reznick, D. 1982. Genetic determination of offspring size in the guppy *(Poecilia reticulata)*. *Am. Nat.* 120: 181–188. [6]

Rhode, J. M., and M. B. Cruzan. 2005. Contributions of heterosis and epistasis to hybrid fitness. *Am. Nat.* 166: E124–E139. [13]

Ribaut, J. M., C. Jiang, and D. Hoisington. 2002. Simulation experiments on efficiencies of gene introgression by backcrossing. *Crop Sci.* 42: 557–565. [17]

Rice, T. K., N. J. Schork, and D. C. Rao. 2008. Methods for handling multiple testing. *Adv. Genet.* 60: 293–308. [20]

Rice, W. H. 1990. A consensus combined *p*-value test and the family-wide significance of component tests. *Biometrics* 46: 303–308. [A6]

Rich, S. S., A. E. Bell, D. A. Miles, and S. P. Wilson. 1984. An experimental study of genetic drift for two quantitative characters in *Tribolium*. *J. Hered.* 75: 191–195. [12]

Richards, C. M. 2000. Inbreeding depression and genetic rescue in a plant metapopulation. *Am. Nat.* 155: 383–394. [12]

Richey, F. D. 1942. Mock-dominance and hybrid vigor. *Science* 96: 280–281. [13]

Richmond, R. C., and 5 others. 2014. Assessing causality in the association between child adiposity and physical activity levels. a Mendelian randomization analysis. *PLoS Med.* 11: e1001618. [A2]

Rieseberg, L. H., M. A. Archer, and R. K. Wayne. 1999. Transgressive segregation, adaptation and speciation. *Heredity* 83: 363–372. [18]

Rieseberg, L. H., A. Widmer, A. M. Arntz, and J. M. Burke. 2003. The genetic architecture necessary for transgressive segregation is common in both natural and domesticated populations. *Phil. Trans. Roy. Soc. Lond. B* 358: 1141–1147. [18]

Rife, C. D. 1954. Populations of hybrid origin as source material for the detection of linkage. *Am. J. Hum. Genet.* 6: 26–33. [17]

Rijsdijk, F. V., J. K. Hewitt, and P. C. Sham. 2001. Analytic power calculation for QTL linkage analysis of small pedigrees. *Eur. J. Hum. Genet.* 9: 335–340. [19]

Ripatti, S., J. Pitkäniemi, and M. Sillanpää. 2001. Joint modeling of genetic association and population stratification using latent class models. *Genet. Epid.* 21: S409–S414. [20]

Risch, N. 1984. Segregation analysis incorporating linkage markers. I. Single-locus models with an application to type I diabetes. *Am. J. Hum. Genet.* 36: 363–386. [19]

Risch, N. 1987. Assessing the role of HLA-linked and unlinked determinants of disease. *Am. J. Hum. Genet.* 40: 1–14. [19]

Risch, N. 1990a. Linkage strategies for genetically complex traits. I. Multilocus models. *Am. J. Hum. Genet.* 46: 229–241. [19]

Risch, N. 1990b. Linkage strategies for genetically complex traits. II. The power of affected relative pairs. *Am. J. Hum. Genet.* 46: 229–241. [19]

Risch, N. 1990c. Linkage strategies for genetically complex traits. III. The effect of marker polymorphism on analysis of affected relative pairs. *Am. J. Hum. Genet.* 46: 242–253 (Correction 51: 673–675). [19]

Risch, N. 1993. Exclusion mapping of complex diseases. *Am. J. Hum. Genet.* 53: A185. [19]

Risch, N. 2000. Searching for genetic determinants in the new millennium. *Nature* 405: 847–856. [20]

Risch, N., and D. Botstein. 1996. A manic depressive history. *Nat. Genet.* 12: 351–353. [19]

Risch, N., and K. Lange. 1979. Application of a recombination model in calculating the variance of sib pair genetic identity. *Ann. Hum. Genet.* 43: 177–186. [8]

Risch, N., and K. Merikangas. 1996. The future of genetic studies of complex human diseases. *Science* 273: 1516–1517. [10, 17, 19, 20]

Risch, N. J., and H. Zhang. 1995. Extreme discordant sib pairs for mapping quantitative trait loci in humans. *Science* 268: 1584–1589. [19]

Risch, N. J., and H. Zhang. 1996. Mapping quantitative trait loci with extreme discordant sib pairs: Sampling considerations. *Am. J. Hum. Genet.* 58: 836–843. [19]

Risch, N., and 8 others. 1995. Genetic analysis of idiopathic torsion dystonia in Ashkenazi Jews and their recent descent from a small founder population. *Nat. Genet.* 9: 152–159. [17]

Rise, M. L., W. N. Frankel, J. M. Coffin, and T. N. Seyfired. 1991. Genes for epilepsy mapped in the mouse. *Science* 253: 669–673. [8]

Riska, B., W. R. Atchley, and J. J. Rutledge. 1984. A genetic analysis of targeted growth in mice. *Genetics* 107: 79–101. [14]

Ritchie, H., D. Banks, D. Buskirk, J. Cowley, and D. R. Hawkins. 1999. Crossbreeding systems for beef cattle. *Michigan State University Extension Bulletin E-2701.* [13]

Ritland, K. 1990a. Gene identity and the genetic demography of plant populations. *In* A. H. D. Brown, M. T. Clegg, A. L. Kahler, and B. S. Weir (eds.), *Plant population genetics, breeding, and genetic resources,*pp. 181–199. Sinauer Assoc., Sunderland, MA. [12]

Ritland, K. 1990b. Inferences about inbreeding depression based on changes of the inbreeding coefficient. *Evolution* 44: 1230–1241. [12]

Ritland, K. 1996. A marker-based method for inferences about quantitative inheritance in natural populations. *Evolution* 50: 1062–1073. [8]

Ritland, K. 2000. Marker-inferred relatedness as a tool for detecting heritability in nature. *Mol. Ecol.* 9:1195–1204. [8]

Ritossa, F. M., K. C. Atwood, and S. Spiegelman. 1996. A molecular explanation of the bobbed mutants of *Drosophila* as partial deficiencies of "ribosomal" DNA. *Genetics* 54: 819–834. [15]

Rivas, M. A., and 41 others. 2011. Deep resequencing of GWAS loci identifies independent rate variant associated with inflammatory bowl disease. *Nat. Genet.* 343: 1066–1073. [15]

Roach, D. A., and R. D. Wulff. 1987. Maternal effects in plants. *Ann. Rev. Ecol. Syst.* 18: 209–236. [6]

Robert, C. P., and G. Casella. 2004. *Monte Carlo statistical methods,* 2nd Ed Springer Verlag. [A8]

Robert, C., and G. Casella. 2011. A short history of Markov chain Monte Carlo: Subjective recollections from incomplete data. *In* S. Brooks, A. Gelman, G. Jones, and X.-L. Meng (eds), *Handbook of markov chain Monte Carlo,* pp. 102-115. CRC press, Boca Raton, FL. [A8]

Roberts, D. F., W. Z. Billewicz, and I. A. McGregor. 1978. Heritability of stature in a west African population. *Ann. Hum. Genet.* 42: 15–24. [7]

Robertson, A. 1955. Selection in animals: synthesis. *Cold Spring Harbor Symp. Quant. Biol.* 20: 225–229. [7]

Robertson, A. 1966. A mathematical model of the culling process in dairy cattle. *Anim. Prod.* 8: 95–108. [3]

Robertson, A. 1967. The nature of quantitative genetic variation. *In* R.A. Brink and E. D. Styles (eds.), *Heritage from Mendel*, pp. 265–280. Univ. Wisconsin Press, Madison, WI. [15]

Robertson, A. 1973. Linkage between marker loci and those affecting a quantitative trait. *Behav. Genet.* 3: 389–391. [19]

Robertson, A., and W. G. Hill. 1983. Population and quantitative genetics of many linked loci in finite populations. *Proc. Roy. Soc. Lond. B* 219: 252–264. [12]

Robertson, D. 1989. Understanding the relationship between qualitative and quantitative genetics. *In* T. Helentjaris and B. Burr (eds.), *Development and application of molecular markers to problems in plant genetics*, pp. 81–87. Cold Spring Harbor Press, Cold Spring Harbor, NY. [17]

Robertson, F. W. 1954. Studies in quantitative inheritance V. Chromosome analyses of crosses between selected and unselected lines of different body sizes in *Drosophila melanogaster. J. Genetics* 52: 494–520. [17]

Robertson, F. W., and E. C. R. Reeve. 1952. Heterozygosity, environmental variation and heterosis. *Nature* 170: 286. [6]

Robertson, F. W., and E. C. Reeve. 1953. Studies in quantitative inheritance. IV. The effects of substituting chromosomes from selected lines into different genetic backgrounds. *J. Genetics* 51: 586–610. [17]

Robertson, H. M., and 5 others. 1988. A stable genomic source of *P* element transposase in *Drosophila melanogaster. Genetics* 118: 461–470. [15]

Robin, S., A. Bar-Hen, J. J. Daudin, and L. Pierre. 2007. A semi-parametic approach for mixture models: application to local false discovery rate estimation. *Comput. Stat. Data Analysis* 51: 5483–5493. [A6]

Robinson, G. K. 1991. That BLUP is a good thing: the estimation of random effects. *Stat. Sci.* 6: 15–51. [10]

Robinson, H. F., and R. E. Comstock. 1955. Analysis of genetic variability in corn with reference to probable effects of selection. *Cold Spring Harbor Symp. Quant. Biol.* 20: 127–136. [5]

Robinson, J. A., and 4 others. 2018. Purging of strongly deleterious mutations explains long-term persistence and absence of inbreeding depression in island foxes. *Curr. Biol.* 28: 3487–3494. [12]

Robinson, L. D., and N. P. Jewell. 1991. Some surprising results about covariate adjustment in logistic-regression models. *Int. Stat. Rev.* 59: 227–240. [20]

Robinson, M. R., and 21 others. 2017. Genetic evidence of assortative mating in humans. *Nat. Hum. Behav.* 1: 0016. [7]

Robson, E. B. 1955. Birth weight in cousins. *Ann. Hum. Genet.* 19: 262–268. [7]

Rocheford, T. R., J. C. Osterman, and C. O. Gardner. 1990. Variation in the ribosomal DNA intergenic spacer of a maize population mass-selected for high grain yield. *Theor. Appl. Genet.* 79: 793–800. [15]

Rockman, M. V., and L. Kruglyak. 2006. Genetics of global gene expression. *Nature* 7: 862–872. [1, 5, 21]

Rodolphe, F., and M. Lefort. 1993. A multiple-marker model for detecting chromosomal segments displaying QTL activity. *Genetics* 134: 1277–1288. [18]

Rodrigues-Motta, M., and 4 others. 2007. A zero-inflated Poisson model for genetic analysis of number of mastitis cases in Norwegian Red cows. *J. Dairy Sci.* 90: 5306–5315. [14]

Rodríguez-Ramilo, S. T., and 4 others. 2004. Mutation-selection balance accounting for genetic variation for viability in *Drosophila melanogaster* as deduced from an inbreeding and artificial selection experiment. *J. Evol. Bio.* 16: 528–541. [12]

Roessler, K., and 9 others. 2019. The genome-wide dynamics of purging during selfing in maize. *Nat. Plants* 5: 980–990. [12]

Roff, D. A. 1997. *Evolutionary quantitative genetics*. Chapman and Hall, New York, NY. [1]

Roff, D. 2002. Inbreeding depression: tests of the overdominance and partial dominance hypotheses. *Evolution* 56: 768–775. [12]

Roff, D. A., and T. A. Mousseau. 1987. Quantitative genetics and fitness: lessons from *Drosophila*. *Heredity* 58: 103–118. [7]

Rogers, A. R., and C. Huff. 2009. Linkage disequilibrium between loci with unknown phase. *Genetics* 182: 839–844. [5]

Rogers, J. S., and J. R. Edwardson. 1952. The utilization of cytoplasmic male-sterile hybrids in the production of corn hybrids. *Agronomy J.* 44: 8-13. [13]

Roginskii, Y. Y. 1959. Some results of using the quantitative method to study morphological variability. (In Russian). *Arkhiv. Anat. Gistol. Embriol.* 36: 83–89. [14]

Rohlf, F. J., A. J. Gilmartin, and G. Hart. 1983. The Kluge-Kerfoot phenomenon — a statistical artifact. *Evolution* 37: 180–202. [14]

Rohrer, J. M. 2018. Thinking clearly about correlations and causation: graphical causal models for observational data. *Adv. Methods Practives Psychol. Sci.* 1: 27–42. [A2]

Rojas, B. A., and G. F. Sprague. 1952. A comparison of variance components in yield trails. III. General and specific combining ability and their interaction with location and years. *Agron. J.* 44: 462–466. [13]

Roles, A. J., and J. K. Conner. 2008. Fitness effects of mutation accumulation in a natural outbred population of wild radish (*Raphanus raphanistrum*): comparison of field and greenhouse environments. *Evolution* 62: 1066–1075. [15]

Rolf, M. M., and 7 others. 2010. Impact of reduced marker set estimation of genomic relationship matrices on genomic selection for feed efficiency in Angus cattle. *BMC Genet.* 11: 24. [8]

Romanoski, C. E., and 4 others. 2015. Roadmap for regulation. *Nature* 518: 314–316. [21]

Romeo, G., and V. A. McKusick. 1994. Phenotypic diversity, allelic series and modifier genes. *Nat. Genet.* 7: 451–453. [15]

Romeo, S., and 6 others. 2007. Population-based resequencing of *ANGPTL4* uncovers variations that reduce triglycerides and increase HDL. *Nat. Genet.* 39: 513–516. [15]

Ronald, J., R. B. Brem, J. Whittle, and L. Kruglyak. 2005. Local regulatory variation in *Saccharomyces cerevisiae*. *PLoS Genet.* 1: e25. [21]

Ronald, P. C., and 6 others. 1992. Genetic and physical analysis of the rice bacterial blight disease resistance locus, *Xa21*. *Mol. Gen. Genet.* 236: 113–120. [17]

Ronin, Y. I, V. M. Kirzhner, and A. B. Korol. 1995. Linkage between loci for quantitative traits and marker loci: multi-trait analysis with a single marker. *Theor. Appl. Genet.* 90: 776–786. [18]

Ronin, Y. I., A. B. Korol, and J. I. Weller. 1998. Selective genotyping to detect quantitative trait loci affecting multiple traits: interval mapping analysis. *Theor. Appl. Genet.* 97: 1169–1178. [18]

Ronin, Y., and 4 others. 2003. High-resolution mapping of quantitative trait loci by selective recombinant genotyping. *Genetics* 164: 1657–1666. [18]

Rönnegård, L., and W. Valdar. 2011. Detecting major genetic loci controlling phenotypic variability in experimental crosses. *Genetics* 188: 435–447. [18, 20]

Rönnegård, L., and W. Valdar. 2012. Recent developments in statistical methods for detecting genetic loci affecting phenotypic variability. *BMC Genet.* 13: 630. [18]

Rosa, G. J. M., and 5 others. 2011. Inferring causal phenotype networks using structural equation models. *Genet. Sel. Evol.* 43: 6. [A2]

Rosa, G. J. M., V. P. S. Felipe, and F. Penagaricano. 2016. Applications of graphical models in quantitative genetics and genomics. *In* H. K. Kadaermideen (ed.), *System biology in animal production and health, Vol 1*, pp. 95–116. Springer, CH. [A2]

Rosenberg, M. S. 2005. The file-drawer problem revisted: a general weighted method for calculating fail-safe numbers in meta-analysis. *Evolution* 59: 464–468. [A6]

Rosenberg, N. A., and M. Nordborg. 2006 A general population-genetic model for the production by population structure of spurious genotype-phenotype associations in discrete, admixed or spatially distributed populations. *Genetics* 173: 1665–1678. [20]

Rosenthal, R. 1979. The "file drawer problem" and tolerance for null results. *Physchol. Bull.* 86: 638–641. [A6]

Rousset, F. 2002. Inbreeding and relatedness coefficients: what do they measure? *Heredity* 88: 371–380. [8]

Rowe, P. R., and R. H. Anderson. 1964. Phenotypic stability for a systematic series of corn genotypes. *Crop Sci.* 6: 563–566. [6]

Roytman, M., G. Kichaev, A. Gusev, and B. Pasaniuc. 2018. Methods for fine-mapping with chromatin and expression data. *PLoS Genet.* 14: e1007240. [21]

Roze, D. 2015. Effects of interference between selected loci on the mutation load, inbreeding depression, and heterosis. *Genetics* 20: 745–757. [12]

Rubinstein, P., and 6 others. 1981. Genetics of HLA disease associations: the use of the haplotype relative risk (HRR) and the "haplo-delta" (Dh) estimates in juvenile diabetes from three racial groups. *Hum. Immun.* 3: 384. [17]

Ruiz, A., and 5 others. 1991. Genetic variance for body size in a natural population of *Drosophila buzzatii*. *Genetics* 128: 739–750. [17]

Rüppel, O., T. Pankiw, and R. E. Page, Jr. 2004. Pleiotropy, epistasis and new QTL: the genetic architecture of honey bee foraging behavior. *J. Hered.* 95: 481–491. [18]

Russell, W. A., G. F. Sprague, and L. H. Penny. 1963. Mutations affecting quantitative characters in long-term inbred lines of maize. *Crop Sci.* 3: 175–178. [12, 15]

Rutherford, S. L. 2000. From genotype to phenotype: buffering mechanisms and the storage of genetic information. *Bioessays* 22: 1095–1105. [14]

Rutherford, S. L., and S. Lindquist. 1998. Hsp90 as a capacitor for morphological evolution. *Nature* 396: 336–342. [6, 14, A2]

Rutter, M. T., F. H. Shaw, and C. B. Fenster. 2010. Spontaneous mutation parameters for *Arabidopsis thaliana* measured in the wild. *Evolution* 64: 1825–1835. [15]

Rutter, M. T., and 8 others. 2012. Fitness of *Arabidopsis thaliana* mutation accumulation lines whose spontaneous mutations are known. *Evolution* 66: 2335–2339. [15]

Ryder, E. J. 1958. The effects of complementary epistasis on the inheritance of a quantitative character, seed size in lima beans. *Agron. J.* 50: 298–301. [11]

Sabatti, C., S. Service, and N. Freimer. 2003. False discovery rates in linkage and association genome screens for complex disorders. *Genetics* 164: 829–833. [18, A6]

Saccheri, I., and 5 others. 1998. Inbreeding and extinction in a butterfly metapopulation. *Nature* 392: 491–494. [12]

Saenz-Romero, C., E. V. Nordheim, R. P. Guries, and P. M. Crump. 2001. A case study of a provenance/progeny test using trend analysis with correlated errors and SAS PROC MIXED. *Silvae Genetica* 50: 127–134. [A9]

Saffari, A., and 6 others. 2016. Estimation of a significant threshold for epigenome-wide association studies. *Genet. Epidem.* 42: 20–33. [20]

Saghai-Maroof, M. A., K. M. Soliman, R. A. Jorgensen, and R. W. Allard. 1984. Ribosomal DNA spacer length polymorphism in barley: Mendelian inheritance chromosomal location and population dynamics. *Proc. Nat. Acad. Sci. USA* 81: 8014–8018. [15]

Sahagún-Castellanos, J. 1998. Efficiency of three methods for prediction of performance of synthetic varieties. *J. Genet. Breed.* 52: 143–150. [13]

Sahagún-Castellanos, J., and C. Villanueva-Verduzco. 2007. On the meaning of Busbice's prediction formula. *Maydica* 52: 145–150. [13]

Sahagún-Castellanos, J., A. Peña-Lomelí, D. Arellano-Suarez, and J. E. Rodríguez-Pérez. 2023. Inbreeding and genetic erosion from a finite model of a synthetic formed with single crosses. *Plants* 12: 541. [13]

Sahana, G., and 4 others. 2013. Novel harmful recessive haplotypes identified for fertility traits in Nordic Holstein cattle. *PLoS One*: e82909. [12]

Sailer, C., B. Schmid, and U. Grossniklaus. 2016. Apomixis allows the transgenerational fixation of phenotypes in hybrid plants. *Curr. Biol.* 26: 331–337. [13]

Salyakina, D., and 4 others. 2005. Evaluation of Nyholt's procedure for multiple testing correction. *Hum. Hered.* 60: 19–25. [20]

Samayoa, L.F., and 10 others. 2021. Domestication reshaped the genetic basis of inbreeding depression in a maize landrace compared to its wild relative, teosinte. *PLoS Genet.* 17: pgen. 1009797. [12]

Sanderson, E., G. Davery Smith, F. Windmeijer, and J. Bowden. 2019. An examination of multivariable two-sample in the single-sample and two-sample summary data settings. *Int. J. Epidem.* 48: 713–727. [A2]

Sandler, L., and E . Novitski. 1957. Meiotic drive as an evolutionary force. *Am. Nat.* 91: 105–110. [17]

Sandner, T. M., and D. Matthies. 2016. The effects of stress intensity and stress type on inbreeding depression in *Silene vulgaris*. *Evolut.* 70: 1225–1238. [12]

Sandner, T. M., D. Matthies, and D. M. Waller. 2021. Stresses affect inbreeding depression in complex ways: Disentangling stress-specific genetic effects from effects of initial size in plants. *Heredity* 127: 347–356. [12]

Sang, J. H. 1956. Hybrid vigour and animal production. *Animal Breed. Abstracts* 24: 1–6. [13]

Sangster, T. A., S. Lindquist, and C. Queitsch. 2004. Under cover: causes, effects and implications of Hsp90 mediated genetic capacitance. *Bioessays* 26: 348–362. [14]

Sangster, T. A., and 6 others. 2008a. HSP90 affects the expression of genetic variation and developmental stability in quantitative traits. *Proc. Natl. Acad. Sci. USA* 105: 2963–2968. [6, 14]

Sangster, T. A., and 9 others. 2008b. HSP90-buffered genetic variation is common in *Arabidopsis thaliana*. *Proc. Natl. Acad. Sci. USA*.105: 2969–2974. [14]

Santiago, E., and 4 others. 1992. The distribution of spontaneous mutations on quantitative traits and fitness in *Drosophila melanogaster*. *Genetics* 132: 771–781. [15]

Santorico, S. A., and A. E. Hendricks. 2016. Progress in methods for rare variant association. *BMC Genet.* 17: 6. [20]

Santure, A. W., and 5 others. 2010. On the use of large marker panels to estimate inbreeding and relatedness: empirical and simulation studies of a pedigreed zebra finch population typed at 771 SNPs. *Mol. Ecol.* 19: 1439–1451. [8]

Sarfatti, M., J. Katan, R. Fluhr, and D. Zamir. 1989. An RFLP marker linked to the *Fusarium oxysporum* resistance gene *I2*. *Theor. Appl. Genet.* 78: 755–759. [17]

Sarhan, A. E. and B. G. Greenberg. 1962. *Contributions to order statistics*. Wiley, New York, NY. [11]

Sarkar, S. 1991. Haldane's solution of the Luria-Delbrück distribution. *Genetics* 127: 257–261. [17]

Sasieni, P. D. 1997. From genotypes to genes: doubling the sample size. *Biometrics* 53: 1253–1261. [20]

Satagopan, J. M., B. S. Yandell, M. A. Newton, and T. C. Osborn. 1996. A Bayesian approach to detect quantitative trait loci using Markov chain monte carlo. *Genetics* 144: 805–816. [18]

Satten, G. A., W. D. Flanders, and Q. Yang. 2001. Accounting for unmeasured population structure in case-control studies of genetic association using a novel latent-class model. *Am. J. Hum. Genet.* 68: 466–477. [20]

Saura, M., and 6 others. 2015. Detecting inbreeding depression for reproductive traits in Iberian pigs using genome-wide data. *Genet. Sel. Evol.* 47: 1. [12]

Savageau, M. A. 1992. Dominance according to metabolic control analysis: major achievement or house of cards? *J. Theor. Biol.* 154: 131–136. [4]

Savolainen, O., and P. Hedrick. 1995. Heterozygosity and fitness: no association in Scots pine. *Genetics* 140: 75–766. [12]

Savolainen, O., K. Kärkkäinen, and H. Kuittinen. 1992. Estimating numbers of embryonic lethals in conifers. *Heredity* 69: 308–314. [12]

Sawamura, K. 1996. Maternal effects as a cause of exceptions for Haldane's rule. *Genetics* 143: 609–611. [17]

Sawcer, S., et al. 2011. Genetic risk and a primary role for cell-mediation immune mechanisms in multiple sclerosis. *Nature* 476: 214–219. [20]

Sawilowsky, S. S., and R. C. Blair. 1992. A more realistic look at the robustness of Type II error properties of the *t* test to departures from population normality. *Psych. Bull.* 111: 352–360. [19]

Sax, K. 1923. The association of size differences with seed-coat pattern and pigmentation in *Phaseolus vulgaris*. *Genetics* 8: 552–560. [1, 17]

Saxena, V., D. Orgill, and I. Kohane. 2006. Absolute enrichment: gene set enrichment analysis for homeostatic systems. *Nucl. Acids Res.* 34: e151. [21]

Schaal, B. A. 1984. Life-history variation, natural selection, and maternal effects in plant populations. *In* R. Dirzo and J. Sarukhán (eds.), *Perspectives on plant population ecology*, pp. 188–206. Sinauer Assoc., Sunderland, MA. [6]

Schachermayr, G., and 5 others. 1994. Identification and localization of molecular markers linked to the *Lr9* leaf rust resistance gene of wheat. *Theor. Appl. Genet.* 88: 110–115. [17]

Schachermayr, G., and 5 others. 1995. Identification and localization of molecular markers linked to the *Agropyron elongatum*-derived leaf rust resistance gene *Lr24* in wheat. *Theor. Appl. Genet.* 90: 982–990. [17]

Schadt, E. E. 2009. Molecular networks as sensors and drivers of common human disease. *Nature* 461: 218–223. [21]

Schadt, E. E., and 13 others. 2003. Genetics of gene expression surveyed in maize, mouse and man. *Nature* 422: 297–302. [21]

Schadt, E. E., and 20 others. 2005. An integrative genomics approach to infer causal associations between gene expression and disease. *Nat. Genet.* 37: 710–717. [21, A2]

Schadt, E. E., S. Woo, and K. Hao. 2012. Bayesian method to predict individual SNP genotypes from gene expression data. *Nat. Genet.* 44: 603–608. [18, 21]

Schaeffer, L. R. 2019. *Animal models.* Privately published, available on Amazon. [1]

Schaid, D. J. 1998. Transmission disequilibrium, family controls, and great expectations. *Am. J. Hum. Genet.* 63: 935–941. [17]

Schaid, D. J. 2002. Relative efficiency of ambiguous vs. directly measured haplotype frequencies. *Genet. Epidmen.* 23: 426–443. [20]

Schaid, D. J. 2004a. Linkage disequilibrium testing when linkage phase is unknown. *Genetics* 166: 505–512. [5]

Schaid, D. J. 2004b. Evaluating associations of haplotypes with traits. *Genet. Epid.* 27: 348–364. [17]

Schaid, D. J. 2010. Genome similarly and kernel methods. I: Advancements by building on mathematical and statistical foundations. *Hum. Hered.* 70: 109–131. [20]

Schaid, D. J., and T. G. Nick. 1990. Sib-pair linkage tests for disease susceptibility loci: common tests vs. the asymptotically most powerful test. *Genet. Epidem.* 7: 359–370. [19]

Schaid, D. J., and S. S. Sommer. 1994. Comparison of statistics for candidate-gene association studies using cases and parents. *Am. J. Hum. Genet.* 55: 402–409. [17]

Schaid, D. J., and 4 others. 2002. Score tests for association between traits and haplotypes when linkage phase is ambiguous. *Am. J. Hum. Genet.* 70: 425–434. [20]

Schaid, D. J., and 4 others. 2005. Nonparametric test of association of multiple genes with human disease. *Am. J. Hum. Genet.* 76: 780–793. [20]

Schaid, D. J., and 9 others. 2012. Using the gene ontology to scan multilevel gene sets for associations in genome wide association studies. *Genet. Epidem.* 36: 3–16. [21]

Schaid, D. J., W. Chen, and N. B. Larson. 2018. From genome-wide associations to candidate causal variants by statistical fine-mapping. *Nat. Rev. Genet.* 19: 491–504.

Schaid, D. J., J. P. Sinnwell, N. B. Larson, and J. Chen. 2020. Penalized variance components for association of multiple genes with traits. *Genet. Epidem.* 44: 665–675. [20]

Scharloo, W. 1988. Selection on morphological patterns. *In* G. de Jong (ed.), *Population genetics and evolution*, pp. 230–250. Springer-Verlag, New York, NY. [14]

Scheet, P., and M. Stephens. 2006. A fast and flexible statistical model for large-scale population genotype data: applications to inferring missing genotypes and haplotypic phase. *Am. J. Hum. Genet.* 78: 629–644. [20]

Scheffé, H. 1959. *The analysis of variance.* John Wiley & Sons, New York, NY. [A5]

Scheiner, S. M., R. L. Caplan, and R. F. Lyman. 1991. The genetics of phenotypic plasticity. III. Genetic correlations and fluctuating asymmetries. *J. Evol. Biol.* 4: 51–68. [6]

Scheiner, S. M., M. Barfield, and R. D. Holt. 2017. The genetics of phenotypic plasticity. XV. Genetic assimilation, the Baldwin effect, and evolutionary rescue. *Ecol. Evol.* 7: 8788–8803. [14]

Schemske, D. W. 1983. Breeding system and habitat effects in three neotropical *Costus* (Zingiberaceae). *Evolution* 37: 523–539. [12]

Schemske, D. W. 1984. Population structure and local selection in *Impatiens* (Balsaminaceae), a selfing annual. *Evolution* 37: 523–539. [13]

Schemske, D. W., and H. D. Bradshaw, Jr. 1999. Pollinator preference and the evolution of floral traits in monkeyflowers (*Mimulus*). *Proc. Natl. Acad. Sci. USA* 96: 11910–11915. [18]

Schemske, D. W., and R. Lande. 1985. The evolution of self-fertilization and inbreeding depression in plants. II. Empirical observations. *Evolution* 39: 41–52. [12]

Schiefelbein, J. W., D. B. Furtek, H. K. Dooner, and O. E. Nelson, Jr. 1988. Two mutations in a maize *bronze-1* allele caused by transposable elements of the *Ac-Ds* family alter the quantity and quality of gene product. *Genetics* 120: 767–777. [15]

Schilthuizen, M., M. C. W. G. Giesbers, and L. W. Beukeboom. 2011. Haldane's rule in the 21st century. *Heredity* 107: 95–102. [17]

Schlichting, C., and M. Pigliucci 1998. *Phenotypic evolution: a reaction norm perspective.* Sinauer Assoc. Sunderland, MA. [14]

Schlichting, C. D., and M. A. Wund. 2014. Phenotypic plasticity and epigenetic marking: an assessment of evidence for genetic accommodation. *Evolution* 68: 656–672. [14]

Schmalhausen, I. I. 1949. *Factors of evolution.* Blakiston, Philadelphia, PA. [14]

Schmidt-Nielsen, K. 1984. *Scaling: Why is animal size so important*? Cambridge Univ. Press, Cambridge, UK. [14]

Schmitt, J., and D. W. Ehrhardt. 1990. Enhancement of inbreeding depression by dominance and suppression in *Impatiens capensis*. *Evolution* 44: 269–278. [12]

Schmitt, J., and S. E. Gamble. 1990. The effect of distance from the parental site on offspring performance and inbreeding depression in *Impatiens capensis:* a test of the local adaptation hypothesis. *Evolution* 44: 2022–2030. [13]

Schnable, P. S., and N. M. Springer. 2013. Progress toward understanding heterosis in crop plants. *Ann. Review Plant Biol.* 64: 71–88. [13]

Schneeberger, R. G., and C. A. Cullis. 1991. Specific alterations associated with the environmental induction of heritable changes in flax. *Genetics* 128: 619–630. [15]

Schneider, R. F., and A. Meyer. 2017. How plasticity, genetic assimilation and cryptic genetic variation may contribute to adaptive radiations. *Mol. Ecol.* 26: 330–350. [14]

Schnell, F. W. 1961. Some general formulations of linkage effects in inbreeding. *Genetics* 46: 947–957. [7]

Schnell, F. W. 1963. The covariance between relatives in the presence of linkage. *In* W. D. Hanson and H. F. Robinson (eds.), *Statistical genetics and plant breeding,* pp. 468–483. Natl. Acad. Sci. Res. Council Publ. 982, Washington, D.C. [7]

Schnell, F. W., and C. C. Cockerham. 1992. Multiplicative vs. arbitrary gene action in heterosis. *Genetics* 131: 461–469. [11, 13]

Schoech, A. P., and 9 others. 2019. Quantification of frequency-dependent genetic architectures in 25 UK Biobank traits reveal action of negative selection. *Nat. Comm.* 10: 790. [21]

Schoen, D. J. 2005. Deleterious mutation in related species of the plant genus *Amsinckia* with contrasting mating systems. *Evolution* 59: 2370–2377. [15]

Schön, C. C., and 4 others. 1993. Mapping and characterization of quantitative trait loci affecting resistance against second-generation European corn borer in maize with the aid of RFLPs. *Heredity* 70: 648–659. [18]

Schön, C. C., and 5 others. 2004. Quantitative trait locus mapping based on resampling in a vast maize testcross experiment and its relevance to quantitative genetics for complex traits. *Genetics* 167: 485–498. [18]

Schön, C. C., B. S. Dhillon, H. F. Utz, and A. E. Melchinger. 2010. High congruency of QTL positions for heterosis of grain yield in three crosses of maize. *Theor. Appl. Genet.* 120: 321–332. [13]

Schork, N. J. 1991. Efficient computation of patterned covariance matrix mixed models in quantitative segregation analysis. *Genetic Epidem.* 8: 29–46. [16]

Schork, N. J. 1992a. Extended pedigree patterned covariance matrix mixed models for quantitative phenotype analysis. *Genetic Epidem.* 9: 73–86. [16]

Schork, N. 1992b. Bootstrapping likelihood ratios in quantitative genetics. *In* R. Lepage and L. Billard (eds.), *Exploring the limits of the bootstrap*, pp. 389–393. Wiley, New York. [A6]

Schork, N. J. 1993. Extended multipoint identity-by-descent analysis of human quantitative traits: efficiency, power, and modeling considerations. *Am. J. Hum. Genet.* 53: 1306–1319. [19]

Schork, N. J. 2002. Power calculations for genetic association studies using estimated probability distributions. *Am. J. Hum. Genet.* 70: 1480–1489. [20]

Schork, N. J., and M. A. Schork. 1988. Skewness and mixtures of normal distributions. *Comm. Stat. Theor. Meth.* 17: 3951–3969. [16]

Schork, N. J., and M. A. Schork. 1989. Testing separate families of segregation hypotheses: bootstrap methods. *Am. J. Hum. Genet.* 45: 803–813. [16]

Schork, N. J., S. S. Murray, K. A. Frazer, and E. J. Topol. 2009. Common vs. rare allele hypotheses for complex diseases. *Curr. Opion. Genet. Develop.* 19: 212–219. [21]

Schou, M. F., and 4 others. 2017. Unexpected high genetic diversity in small populations suggest maintenance by associative overdominance. *Mol. Ecol.* 26: 6510–6523. [12]

Schroeder, M., D. L. Brown, and D. E. Weeks. 1994. Improved programs for the affected-pedigree-member method of linkage analysis. *Genet. Epidem.* 11: 68–74. [19]

Schüller, C., G. Backes, G. Fischbeck, and A. Jahoor. 1992. RFLP markers to identify the alleles on the *Mla* locus conferring powdery mildew resistance in barley. *Theor. Appl. Genet.* 84: 330–338. [17]

Schull, W. J. 1962. Inbreeding and maternal effects in the Japanese. *Eugen. Quart.* 9: 14–22. [12]

Schull, W. J., and J. V. Neel. 1965. *The effects of inbreeding on Japanese children.* Harper & Row, New York, NY. [12]

Schull, W. J., and J. V. Neel. 1972. The effects of parental consanguinity and inbreeding in Hirado, Japan. V. Summary and interpretation. *Am. J. Hum. Genet.* 24: 425–453. [12]

Schull, W. J., H. Nagano, M. Yamamoto, and I. Komatsu. 1970. The effects of parental consanguinity and inbreeding in Hirado, Japan. I. Still births and prereproductive mortality. *Am. J. Hum. Genet.* 22: 239–262. [12]

Schulte-Hostedde, A. I., B. Zinner, J. S. Millar, and G. J. Hickling. 2005. Restitution of mass-size residuals: validating body condition indices. *Ecology* 86: 155–163. [14]

Schultz, S. T., and F. R. Ganders. 1996. Evolution of unisexuality in the Hawaiian Islands: a test of microevolutionary theory. *Evolution* 50: 842–855. [12]

Schultz, S. T., and J. H. Willis. 1995. Individual variation in inbreeding depression: the roles of inbreeding history and mutation. *Genetics* 141: 1209–1223. [12]

Schultz, S. T., M. Lynch, and J. H. Willis. 1999. Spontaneous deleterious mutation in *Arabidopsis. Proc. Natl. Acad. Sci. USA* 96: 11393–11398. [15]

Schuze, T. G., and F. J. McMahon. 2002. Genetic association mapping at the crossroads: Which test and why? Overview and practical guidelines. *Am. J. Med. Genet.* 114: 1–11. [17]

Schweder, T., and E. Spjøtvoll. 1982. Plots of *p*-values to evaluate many tests simultaneously. *Biometrika* 69: 493–502. [A6]

Schwartz, C. R., and R. D. Mare. 2005. Trends in educational assortative mating from 1940 to 2003. *Demography* 42: 621–646. [7]

Schwartz, D., and W. J. Laughner. 1969. A molecular basis for heterosis. *Science* 166: 626–627. [13]

Schwarz, G. 1978. Estimating the dimension of a model. *Annals. Stat.* 6: 461–464. [16, A4]

Scofield, D. G., and S. T. Schulz. 2006. Mitosis, stature and evolution of plant mating systems: low-Φ and high-Φ plants. *Proc. Roy. Soc. Lond. B* 273: 275–282. [12]

Scott, J. P., and J. L. Fuller. 1965. *Genetics and social behavior of the dog.* Univ. Chicago Press, Chicago, IL. [12]

Scott, M. F., and 22 others. 2020. Multi-parent populations in crops: a toolbox integrating genomics and genetic mapping with breeding. *Heredity* 125: 396–416. [18]

Seager, R. D., and F. J. Ayala. 1982. Chromosome interactions in *Drosophila melanogaster*. I. Viability studies. *Genetics* 102: 467–483. [17]

Seaman, S. E., and B. Müller-Myhsok. 2005. Rapid simulation of *P* values for product methods and multiple-testing adjustment in association studies. *Am. J. Hum. Genet.* 76: 399–408. [20]

Searle, S. R. 1971. *Linear models*. John Wiley & Sons, New York, NY. [9, A3, A5]

Searle, S. R. 1982. *Matrix algebra useful for statistics.* John Wiley and Sons, New York, NY. [A3]

Sears, E. R. 1953. Nullisomic analysis in common wheat. *Am. Nat.* 87: 245–252. [17]

Sebat, J., and 20 others. 2004. Large-scale copy number polymorphisms in the human genome. *Science* 305: 525–528. [15]

Sedeño-Cortés, A. E., and P. Pavlidis. 2014. Pitfalls in the application of gene-set analysis to genetics studies. *Trends Genet.* 30: 513–514. [21]

Segura, V., and 6 others. 2012. An efficient multi-locus mixed-model approach for genome-wide association studies in structure populations. *Nat. Genet.* 44: 825–830. [20]

Self, S. G., and K.-Y. Liang. 1987. Asymptotic properties of maximum likelihood estimators and likelihood ratio tests under nonstandard conditions. *J. Amer. Stat. Assoc.* 82: 605–610. [19, A4]

Selkoe, D. J. 2001. Alzheimer's disease: genes, proteins, and therapy. *Physiological Rev.* 81: 741–766.

Sella, G., and N. H. Barton. 2019. Thinking about the evolution of complex traits in the era of genome-wide association studies. *Ann. Rev. Genom. Hum. Gene.* 20: 461–493. [21]

Semagn, K., Å. Bjørnstad, and M. N. Ndjiondjop. 2006. An overview of molecular marker methods for plants. *African J. Biotech.* 5: 2540–2568. [8]

Sen, Ś., F. Johannes, and K. W. Borman. 2009. Selective genotyping and phenotyping strategies in a complex trait context. *Genetics* 181: 1613–1626. [18]

Sen, Ś., J. M. Satagopan, and G. A. Churchill. 2005. Quantitative trait locus study designs from an information perspective. *Genetics* 170: 447–464. [18]

Sen, S., and G. A. Churchill. 2001. A statistical framework for quantitative trait mapping. *Genetics* 159: 371–387. [18]

Sen, S., J. M. Sataopan, K. W. Broman, and G. A. Churchill. 2007. R/qtlDesign: inbred line cross experimental design. *Mamm. Genome* 18: 87–93. [18]

Servin, B., and F. Hospital. 2002. Optimal positioning of markers to control genetic background in marker-assisted backcrossing. *J. Hered.* 93: 214–217. [17]

Servin, B., and M. Stephens. 2007. Imputation-based analysis of association studies: candidate regions and quantitative traits. *PLoS Geet.* 3: e114. [20]

Setakis, E., H. Stirnadel, and D. J. Balding. 2006. Logistic regression protects again population structure in genetic association studies. *Gen. Res.* 16: 290–296. [20]

Severo, N. C., and M. Zelen. 1960. Normal approximation to the chi-square and noncentral *F* probability distributions. *Biometrika* 47: 411–416. [A5]

Shabalin, A. A. 2012. Matrix eQTL: ultra fast eQTL analysis via large matrix operations. *Bioinformatics* 28: 1353–1358. [21]

Shaffer, J. F. 1995. Multiple hypothesis testing. *Ann. Rev. Psychol.* 46: 561–584. [A6]

Shah, S., and J. R. Green. 1994. The distribution of IQ: a powerful sibship test of association. *Am. J. Hum. Genet.* 58: 163–173. [19]

Sham, P. C., and D. Curtis. 1995. An extended transmission/disequilibrium test (TDT) for multi-allele marker loci. *Ann. Hum. Genet.* 59: 323–336. [17]

Sham, P. C., and S. Purcell. 2001. Equivalence between Haseman-Elston and variance-components linkage analysis for sib pairs. *Am. J. Hum. Genet.* 71: 238–253. [19]

Sham, P. C., and S. M. Purcell. 2014. Statistical power and significance testing in large-scale genetic studies. *Nat. Rev. Genet.* 15: 335–346. [20, A6]

Sham, P. C., S. S. Cherny, S. Purcell, and J. K. Hewitt. 2000a. Power of linkage versus association analysis of quantitative traits, by the use of variance-component models, for sibship data. *Am. J. Hum. Genet.* 66: 1616–1630. [19]

Sham, P. C., J. H. Zhao, S. Cherny, and J. Hewitt. 2000b. Variance-component QTL linkage analysis of selection and non-normal samples: conditioning on trait values. *Genet. Epid.* 19: S22–S28. [19]

Sham, P. C., S. Purcell, S. S. Cherny, and G. R. Abecasis. 2002. Powerful regression-based quantitative-trait linkage analysis of general pedigrees. *Am. J. Hum. Genet.* 66: 1616–1630. [19]

Shah, S., and 16 others. 2014. Genetic and environmental exposures constrain epigenetic drift over the human life course. *Genome Res.* 24: 1725–1733. [21

Shan, N., Z. Wang, and L. Hou. 2019. Identification of trans-eQTLS using mediation analysis with multiple mediators. *BMC Bioinformatics* 20: 126. [21]

Shang, L., and 5 others. 2016. Epistasis together with partial dominance, over-dominance and QTL by environment interactions contribute to yield heterosis in upland cotton. *Theor. Appl. Genet.* 129: 1429–1446. [13]

Shank, D. B., and M. W. Adams. 1960. Environmental variability within inbred lines and single crosses of maize. *J. Genetics* 57: 119–126. [6]

Shao, H., and 13 others. 2008. Genetic architecture of complex traits: large phenotypic effects and pervasive epistasis. *Proc. Natl. Acad. Sci. USA* 105: 19910–19914. [17]

Shao, L., and 11 others. 2019. Patterns of genome-wide allele-specific expression in hybrid rice and the implications on the genetic basis of heterosis. *Proc. Natl. Acad. Sci. USA* 116: 5653–5658. [13]

Shapiro, S. S., and M. B. Wilk. 1965. An analysis of variance tests for normality (complete samples). *Biometrika* 52: 591–611. [14]

Sharp, N. P., and A. F. Agrawal. 2012. Evidence for elevated mutation rates in low–quality genotypes. *Proc. Nat. Acad. Sci. USA* 109: 6142–6146. [15]

Sharp, P. M. 1984. The effect of inbreeding on competitive male mating ability in *Drosophila melanogaster*. *Genetics* 106: 601–612. [12]

Shaw, F. H., C. J. Geyer, and R. G. Shaw. 2002. A comprehensive model of mutations affecting fitness and inferences for *Arabidopsis thaliana*. *Evolution* 563: 453–463. [15]

Shaw, R. G. 1987. Maximum-likelihood approaches to quantitative genetics of natural populations. *Evolution* 41: 812–826. [A4]

Shaw, R. G., D. L. Byers, and E. Darmo. 2000. Spontaneous mutational effects on reproductive traits of *Arabidopsis thaliana*. *Genetics* 155: 369–378. [15]

Sheehan, N. A., V. Didelez, P. R. Burton, and M. D. Tobin. 2008. Mendelian randomisation and causal inference in observational epidemiology. *PLoS Medicine* 5: e177. [A2]

Sheldon, B. L., J. M. Rendel, and D. E. Finlay. 1964. The effect of homozygosity on developmental stability. *Genetics* 49: 471–484. [14]

Shen, G., and 7 others. 2022. Assembly of yield heterosis of an elite rice hybrid is promising by manipulating dominant quantitative trait loci. *J. Integ. Plant Bio.* 64: 688–701. [13]

Shen, H., and 11 others. 2012. Genome-wide analysis of DNA methylation and gene expression changes in two *Arabidopsis* ecotypes and their reciprocal hybrids. *Plant Cell* 24 : 875–892. [13]

Shen, X., M. Pettersson, L. Rönnegård, and Ö. Carlborg. 2012. Inheritance beyond plan heritability: variance-controlling genes in *Arabidopsis thaliana*. *PLoS Genet.* 8: e1002839. [18]

Shen, Y., and 8 others. 2017. Analysis of transcriptional and epigenetic changes in hybrid vigor of allopolyploid *Brassica napus* uncovers key roles for small RNAs. *Plant J.* 91: 874–893. [13]

Sheng, X., and J. Yang. 2013. An adaptive truncated product method for combining dependent p values. *Econ. Lett.* 119: 180–182. [A6]

Shenstone, E., and 5 others. 2018. An assessment of the performance of the logistic mixed model for analysis binary traits in maize and sorghum diversity panels. *PLoS One* 13: e0207752. [20]

Sheppard, W. E. 1898. On the calculation of the most probable values of frequency constants from data arranged according to equidistant divisions of a scale. *Proc. Lond. Math. Soc.* 29: 353–380. [2]

Sheridan, A. K. 1981. Crossbreeding and heterosis. *Anim. Breed. Abst.* 49: 131–144. [13]

Sherman, R. M., and S. L. Salzberg. 2020. Pan-genomics in the human genome era. *Nat. Rev. Genet.* 21: 243–254. [13]

Shete, S., and 5 others. 2004. Effects of winsorization on power and type I error of variance components and related methods of QTL detection. *Behav. Genet.* 34: 153–159. [19]

Shi, Y., and 27 others. 2016. Unmanned aerial vehicles for high-throughput phenotyping and agronomic research. *PLoS ONE* 11: e0159781. [18]

Shields, W. M. 1982. *Philopatry, inbreeding, and the evolution of sex.* SUNY Press, Albany, New York, NY. [12, 13]

Shine, R., D. O'Connor, M. P. LeMaster, and R. T. Mason. 2001. Pick on someone your own size: ontogenetic shifts in mate choice by male garter snakes result in size-assortative mating. *Anim. Behav.* 61: 1133–1141. [7]

Shipley, B. 2016. *Cause and correlation in biology*, 2nd Ed. Cambridge University Press, Cambridge, UK. [A2]

Shorter, J. R., and 18 others. 2017. Male infertility is responsible for nearly half of the extinction observed in the mouse collaborative cross. *Genetics* 206: 557–572. [12, 18]

Shoukri, M. M., and G. J. McLachlan. 1994. Parametric estimation in a genetic mixture model with application to nuclear family data. *Biometrics* 50: 128–139. [16]

Shrestha, J. N. B., and M. H. Fahmy. 2007. Breeding goats for meat production: 2. Crossbreeding and formation of composite population. *Small Ruminant Res.* 67: 93–112. [13]

Shrimpton, A. E., and A. Robertson. 1988a. The isolation of polygenic factors controlling bristle score in *Drosophila melanogaster*. I: Allocation of third chromosome sterno-pleural bristle effects to chromosome sections. *Genetics* 118: 437–443. [17]

Shrimpton, A. E., and A. Robertson. 1988b. The isolation of polygenic factors controlling bristle score in *Drosophila melanogaster*. II: Distribution of third chromosome bristle effects within chromosome sections. *Genetics* 118: 445–459. [17]

Shriner, D. 2011. Investigating population stratification and admixture using eigenanalysis of dense genotypes. *Hered.* 107: 413–420. [20]

Shriner, D. 2012. Improved eigenanalysis of discrete subpopulations and admixture using the minimum average partial test. *Hum. Hered.* 73: 73–83. [20]

Shull, G. H. 1908. The composition of a field of maize. *Rpt. Am. Breed. Assoc.* 4: 296–301. [1, 12, 13]

Shull, G. H. 1909. A pure-line method in corn breeding. *Rpt. Am. Breed. Assoc.* 5: 51-59. [13]

Shull, G. H. 1914. Duplicate genes for capsule-form in *Bursa bursa-pastoris*. *Z. Ind. Abstr. Ver.* 12: 97–149. [13]

Shull, G. H. 1948. What is heterosis? *Genetics* 33: 439–446. [13]

Shute, N. C. E., and W. J. Ewens. 1988a. A resolution of the ascertainment sampling problem. II. Generalizations and numerical results. *Am. J. Hum. Genet.* 43: 374–386. [16]

Shute, N. C. E., and W. J. Ewens. 1988b. A resolution of the ascertainment sampling problem. III. Pedigrees. *Am. J. Hum. Genet.* 43: 387–395. [16]

Sidik, K., and J. N. Jonkman. 2005. A note on variance estimation in random effects meta-regression. *J. Biopharmaceutical Stat.* 15: 823–838. [A6]

Sieberts, S. K., and E. E. Schadt. 2019. Inferring causal associations between genes and disease via the mapping of expression quantitative trait loci. *In* D. J. Balding, I. Molke, and J. Marioni (eds.), *Handbook of Statistical Genomics*, 4th Ed., pp. 696–733. Wiley and Sons, New York, NY. [20]

Sieberts, S. K., E. M. Wusman, and E. A. Thompson. 2002. Relationship inference from trios of individuals, in the presence of typing errors. *Am. J. Hum. Genet.* 70: 170–180. [8]

Siegal, M. L., and A. Bergman. 2002. Waddington's canalization revisited: developmental stability and evolution. *Proc. Natl. Acad. Sci. USA* 99: 10528–10532. [14]

Siegel, M. I., and W. J. Doyle. 1975a. Stress and fluctuating limb asymmetry in various species of rodents. *Growth* 39: 363–369. [6]

Siegel, M. I., and W. J. Doyle. 1975b. The differential effects of prenatal and postnatal audiogenic stress on fluctuating dental asymmetry. *J. Exp. Zool.* 191: 211–214. [6]

Siegel, M. I., and W. J. Doyle. 1975c. The effects of cold stress on fluctuating asymmetry in the dentition of the mouse. *J. Exp. Zool.* 191: 211–214. [6]

Siegel, M. I., and M. P. Mooney. 1987. Perinatal stress and increased fluctuating asymmetry of dental calcium in the laboratory rat. *Am. J. Phys. Anthropol.* 73: 267–270. [6]

Siegmund, D. 2002. Upward bias in estimation of genetic effects. *Am. J. Hum. Genet.* 71: 1183–1188. [20]

Siegmund, D. 2004. Model selection in irregular problems: Applications to mapping quantitative trait loci. *Biometrika* 91: 785–800. [18]

Silander, J. A., Jr. 1985. The genetic basis of the ecological amplitude of *Spartina patens*. II. Variance and correlation analysis. *Evolution* 39: 1034–1052. [6]

Sillanpää, M. J. 2011a. On statistical methods for estimating heritability in wild populations. *Mol. Ecol.* 20: 1324–1332. [8]

Sillanpää, M. J. 2011b. Overview of techniques to account for confounding due to population stratification and cryptic relatedness in genomic data association analyses. *Hereidty* 106: 51–519. [20]

Sillanpää, M. J., and E. Arjas. 1998. Bayesian mapping of multiple quantitative trait loci from incomplete inbred line cross data. *Genetics* 148: 1373–1388. [18]

Sillanpää, M. J., and E. Arjas. 1999. Bayesian mapping of multiple quantitative trait loci from incomplete outbred offspring data. *Genetics* 151: 1605–1619. [19]

Sillanpää, M. J., and 4 others. 2001. Bayesian association mapping for quantitative traits in a mixture of two populations. *Genet. Epid.* 21: S692–S699. [20]

Simes, J. R. 1986. An improved Bonferroni procedure for multiple tests of significance. *Biometrika* 73: 75–754. [20, A6]

Simm, G. 1998. *Genetic improvement of cattle and sheep*. Farming Press, Suffolk, UK. [1, 13]

Simmonds, M. 2015. Quantifying the risk of error when interpreting funnel plots. *System. Rev.* 4: 24. [A6]

Simmons, C. R., and 7 others. 2021. Successes and insights of an industry biotech program to enhance maize agronomic traits. *Plant Sci.* 307: 110899. [5]

Simmons, M. J., and J. F. Crow. 1977. Mutations affecting fitness in *Drosophila* populations. *Ann. Rev. Genet.* 11: 49–78. [12, 15]

Simons, A. M., and D. A. Roff. 1994. The effect of environmental variability on the heritabilities of traits of a field cricket. *Evolution* 48: 1637–1649. [7]

Simons, Y. B., M. C. Turchin, J. K. Pritchard, and G. Sella. 2014. The deleterious mutation load is insensitive to recent population history. *Nat. Genet.* 46: 220–224. [21]

Simons, Y. B., A. Bullaughey, R. R. Hudson, and G. Sella. 2018. A population genetic interpretation of GWAS findings for human quantitative traits. *PLoS Biol.* 16: e2002985. [21]

Simons, Y. B., and 4 others. 2022. Simple scaling laws control the genetic architectures of human complex traits. *bioRxiv.* 10.04.509926. [21]

Simpson, E., and 5 others. 1982. *H-2* associated differences in replicated strains of mice divergently selected for body weight. *Immunogenetics* 15: 63–70. [17]

Simpson, S. P. 1989. Detection of linkage between quantitative trait loci and restriction fragment length polymorphism using inbred lines. *Theor. Appl. Genet.* 77: 815–819. [17, 18]

Simpson, S. P. 1992. Correction: detection of linkage between quantitative trait loci and restriction fragment length polymorphisms using inbred lines. *Theor. Appl. Genet.* 85: 110-111. [18]

Sing, C. F., R. H. Moll, and W. D. Hanson. 1967. Inbreeding in two populations of *Zea mays* L. *Crop Sci.* 7: 631–636. [12]

Singh, M., and 5 others. 2003. Spatial variability models to improve dryland field trials. *Exper. Agric.* 39: 151–160. [A9]

Singh, S., and R. B. Singh. 1976. Triple test cross analysis in two wheat crosses. *Heredity* 37: 173–177. [13]

Singh, S. M., and E. Zouros. 1978. Genetic variation associated with growth rate in the American oyster (*Crassostrea virginica*). *Evolution* 32: 342–353. [12]

Sinha, R., and Y. Luo. 2007. Efficient intermediate fin mapping: confidence set inference with likelihood ratio test statistic. *Genet. Epid.* 8: 922–936. [19]

Sinha, R., and 4 others. 2009. Bayesian intervals for linkage locations. *Genet. Epid.* 33: 604–616. [19]

Sinnott-Armstrong, N., S. Naqvi, M. Rivas, and J. K. Pritchard. 2021. GWAS of three molecular traits highlights core genes and pathways alongside a highly polygenic background. *eLife* 10: e58615. [21]

Sinoquest, C., and R. Mourad (eds). 2014. *Probabilistic graphical models for genetics, genomics, and postgenomics.* Oxford University Press, Oxford, UK. [A2]

Sittmann, K., H. Abplanalp, and R. A. Fraser. 1966. Inbreeding depression in Japanese quail. *Genetics* 54: 371–379. [12]

Skarnes, W. C., and 17 others. 2011. A conditional knockout resource for the genome-wide study of mouse gene function. *Nature* 474: 337–342. [15]

Slate, J., T. Marshall, and J. Pemberton. 2000. A retrospective assessment of the accuracy of the paternity inference program CERVUS. *Mol. Ecol.* 9: 801–808. [8]

Slate, J., and 6 others. 2004. Understanding the relationship between the inbreeding coefficient and multilocus heterozygosity: theoretical expectations and empirical data. *Heredity* 93: 255–265. [12]

Slatis, H. M. 1960. An analysis of inbreeding in the European bison. *Genetics* 45: 275–287. [12]

Slatis, H. M., and R. E. Hoene. 1961. The effects of consanguinity on the distribution of continuously variable characters. *Am. J. Hum. Genet.* 13: 28–31. [12]

Slatis, H. M., R. H. Reis, and R. E. Hoene. 1958. Consanguineous marriages in the Chicago region. *Am. J. Hum. Genet.* 10: 446–464. [12]

Slatkin, M. 1972. On treating the chromosome as the unit of selection. *Genetics* 72: 157–168. [5]

Slatkin, M. 2008. Linkage disequilibrium — understanding the evolutionary past and mapping the medical future. *Nat. Rev. Genet.* 9: 477–485. [5]

Slatkin, M., and L Excoffier. 1996. Testing for linkage disequilibrium in genotypic data using the Expectation-Maximization algorithm. *Heredity* 76: 377–383. [5]

Slatkin, M., and B. Rannala. 1997. The sampling distribution of disease-associated alleles. *Genetics* 147: 1855–1861. [21]

Slob, E. A. W., and S. Burgess. 2019. A comparison of robust Mendelian randomization methods using summary data. *Genet. Epidem.* 44: 313–329. [A2]

Smemo, S., and 24 others. 2014. Obesity-associated variants within *FTO* form long-range functional connections with *IRX3*. *Nature* 507: 371–375. [21]

Smith, A., B. Cullis, and R. Thompson. 2001. Analyzing variety by environment data using multiplicative mixed models and adjustments for spatial field trend. *Biometrics* 57: 1138–1147. [A9]

Smith, A., B. R. Cullis, and R. Thompson. 2005. The analysis of crop cultivar breeding and evaluation trials: an overview of current mixed model approaches. *J. Agricul. Sci.* 143: 449–462. [A9]

Smith, A. B., P. Lim, and B. R. Cullis. 2006. The design and analysis of multi-phase plant breeding experiments. *J. Agricul./ Sci.* 144: 393–409. [A9]

Smith, A. F. M. 1991. Bayesian computational methods. *Phil. Trans. R. Soc. Lond. A* 337: 369–386. [A8]

Smith, A. F. M., and G. O. Roberts. 1993. Bayesian computation via the Gibbs sampler and related Markov chain Monte-Carlo methods (with discussion). *J. Roy. Stat. Soc. Series B* 55: 3-23. [A8]

Smith, B. R., C. M. Herbinger, and H. R. Merry. 2001. Accurate partition of individuals into full-sib families from genetic data without parental information. *Genetics* 158: 1329–1338. [8]

Smith, C. A. B. 1953. The detection of linkage in human genetics. *J. Royal Stat. Soc. Ser. B* 15: 153–184. [12, 19]

Smith, C. A. B. 1959. Some comments on the statistical methods used in linkage investigation. *Am. J. Hum. Genet.* 11: 289–304. [19]

Smith, H. F. 1938. An empirical law describing heterogeneity in the yields of agricultural crops. *J. Agric. Sci.* 28: 1–23. [A9]

Smith, H. H. 1937a. Inheritance of corolla color in the cross *Nicotiana Langsdorfii* by *N . Sanderae*. The relation between genes affecting size and color in certain species of *Nicotiana*. *Genetics* 22: 347–360. [17]

Smith, H. H. 1937b. The relation between genes affecting size and color incertain species of *Nicotiana*. *Genetics* 22: 361–375. [11]

Smith, H. H. 1952. Fixing transgressive vigour in *Nicotiana rustica*. *In* J. W. Gowen (ed.), *Heterosis,* pp. 161–164. Iowa State College Press, Ames, IA. [12]

Smith, L. H. 1908. Ten generations of corn breeding. *Illinois Agri. Exp. Stn. Bull.* 218. [18]

Smith, N. M. A., and 9 others. 2019. Strikingly high levels of heterozygosity despite 20 years of inbreeding in a clonal honey bee. *J. Evol. Biol.* 32: 144–152. [12]

Smith, O. S. 1979a. A model for evaluating progress from recurrent selection. *Crop Sci.* 19: 223–226. [11]

Smith, O. S. 1979b. Application of a modified diallel analysis to evaluate recurrent selection for grain yield in maize. *Crop Sci.* 19: 819–822. [11]

Smith, O. S. 1983. Evaluation of recurrent selection in BSSS, BSCB1, and BS13 maize populations. *Crop Sci.* 23: 35–40. [11]

Smouse, P. E. 1986. The fitness consequences of multiple-locus heterozygosity under the multiplicative overdominance and inbreeding depression models. *Evolution* 40: 946–958. [12]

Smouse, P. E., and T. R. Meagher. 1994. Genetic analysis of male reproductive contributions in *Chamaelirium luteum* (L.) Gray (Liliaceae). *Genetics* 136: 313–322. [8]

Smouse, P. E., T. R. Meagher, and C. J. Kobak. 1999. Parentage analysis in *Chamaeliriumluteum* (L.) Gray (Liliaceae): why do some males have higher reproductive contributions? *J. Evol. Biol.* 12: 1069–1077. [8]

Snape, J. W., C. N. Law, and A. J. Worland. 1977. Whole chromosome analysis of height in wheat. *Heredity* 38: 25–26. [17]

Snape, J. W., A. J. Wright, and E. Simpson. 1984. Methods for estimating gene numbers for quantitative characters using doubled haploid lines. *Theor. Appl. Genet.* 67: 143–148. [11]

Sneller, C. H., D. E. Mather, and S. Crepieux. 2009. Analytical approaches and population types for finding and utilizing QTL in complex plant populations. *Crop Sci.* 49: 363–380. [17, 19]

Snow, G. L., and E. M. Wijsman. 1998. Pedigree analysis package (PAP) vs. MORGAN: model selection and hypothesis testing on a large pedigree. *Genetic Epidem.* 15: 355–369. [16]

Snyder, L. H. 1932. The inherence of taste deficiency in man. *Ohio J. Sci.* 32: 436–440. [4]

Snyder, L. H. 1934. Studies in human inheritance. X. A table to determine the proportion of recessives to be expected in various mating involving a unit character. *Genetics* 19: 1–17. [4]

Sobel, E., and K. Lange. 1996. Descent graphs in pedigree analysis: applications to haplotyping, location scores, and marker-sharing statistics. *Am. J. Hum. Genet.* 58: 1323–1337. [19]

Sobel, E., H. Sengul, and D. E. Weeks. 2001. Multipoint estimation of identity-by-descent probabilities at arbitrary positions among marker loci on general pedigrees. *Hum. Hered.* 52: 121–131. [19]

Sockness, B. A., and J. W. Dudley. 1989. Performance of single and double cross autotetraploid maize hybrids with different levels of inbreeding. *Crop Sci.* 29: 875–879. [13]

Sokal, R. R. 1976. The Kluge-Kerfoot phenomenon reexamined. *Am. Nat.* 110: 1077–1091. [14]

Sokal, R. R., and F. J. Rohlf. 1995. *Biometry.* 2nd Ed. W. H. Freeman and Co., New York, NY. [14, 16, 17]

Sokolowski, M. B. 1980. Foraging strategies of *Drosophila melanogaster*: a chromosomal analysis. *Behav. Genet.* 10: 291–302. [17]

Solberg, L. C., and 19 others. 2006. A protocol for high-throughput phenotyping, suitable for quantitative trait analysis in mice. *Mamm. Genome* 17: 129–146. [18]

Solé, M., and 6 others. 2017. Age-based partitioning of individual genomic inbreeding levels in Belgian Blue cattle. *Genet. Sel. Evol.* 49: 92. [8, 12]

Soller, M., and J. S. Beckmann. 1987. Cloning quantitative trait loci by insertional mutagensis. *Theor. Appl. Genet.* 74: 369–378. [17]

Soller, M., and J. S. Beckmann. 1988. Genomic genetics and the utilization for breeding purposes of genetic variation between populations. *In* B. S. Weir, E. J. Eisen, M. M. Goodman, and G. Namkoong (eds.), *Proceedings of the second international conference on quan-*

titative genetics, pp. 161–188. Sinauer Assoc., Sunderland, MA. [8]

Soller, M., and J. S. Beckmann. 1990. Marker-based mapping of quantitative trait loci using replicated progenies. *Theor. Appl. Genet.*. 80: 205–208. [17]

Soller, M., and A. Genizi. 1978. The efficiency of experimental designs for the detection of linkage between a marker locus and a locus affecting a quantitative trait in segregating populations. *Biometrics* 34: 47–55. [18, 19]

Soller, M., T. Brody, and A. Genizi. 1976. On the power of experimental designs for the detection of linkage between marker loci and quantitative loci in crosses between inbred lines. *Theor. Appl. Genet.* 47: 35–39. [18]

Solovieff, N., and 4 others. 2013. Pleiotropy in complex traits: challenges and strategies. *Nat. Rev. Genet.* 14: 483–495. [A2]

Sondhi, K. C. 1961. Selection for a character with a bounded distribution of phenotypes in *Drosophila subobscura*. *J. Genetics* 57: 193–221. [14]

Song, K. K., E. Feingold, and D. E. Weeks. 2002. Statistics for nonparametric linkage analysis of X-linked traits in general pedigrees. *Am. J. Hum. Genet.* 70: 181–191. [19]

Sorensen, D. and D. Gianola. 2002. *Likelihood, Bayesian and MCMC methods in quantitative genetics.* Springer, New York, NY. [A4, A7, A8]

Sorensen, D. A., and W. G. Hill. 1982. Effect of short term directional selection on genetic variability: experiments with *Drosophila melanogaster*. *Heredity* 48: 27–33. [5]

Sorensen, D. A., and W. G. Hill. 1983. Effects of disruptive selection on genetic variance. *Theor. Appl. Genet.* 65: 173–180. [5]

Sorensen, D. A., and B. W. Kennedy. 1986. Analysis of selection experiments using mixed model methodology. *J. Anim. Sci.* 63: 245–258. [10]

Sorensen, F. 1969. Embryonic genetic load in coastal Douglas fir *Pseudotsuga menziesii* var. *menziesii*. *Am. Nat.* 103: 389–398. [12]

Sorensen, F. 1970. Self-fertility of a central Oregon source of ponderosa pine. *USDA Forest Serv. Res. Pap. PNW-109*. Pac. Northwest Forest and Range Exp. Stn., Portland, OR. [12]

Sorensen, F. C., J. F. Franklin, and R. Woolard. 1976. Self-pollination effects on seed and seedling traits in noble fir. *Forest. Sci.* 22: 155–159. [12]

Sørensen, P., and 4 others. 2015. Genetic control of environmental variation of two quantitative traits of *Drosophila melanogaster* revealed by whole-genome sequencing. *Genetics* 201: 487–497. [6]

Soriç, B. 1989. Statistical "discoveries" and effect size estimation. *J. Am. Statist. Ass.* 84: 608–610. [A6]

Soulé, M. E. 1979. Heterozygosity and developmental stability: another look. *Evolution* 33: 396–401. [6]

Soulé, M. E. 1980. Thresholds for survival: maintaining fitness and evolutionary potential. *In* M. E. Soulé and B. A. Wilcox (eds), *Conservation biology: an evolutionary-ecological perspective*, pp. 151–169. Sinauer, Sunderland, MA. [12]

Soulé, M. E. 1982. Allomeric variation. 1. The theory and some consequences. *Am. Nat.* 120: 751–764. [6, 14]

Soulé , M. E., and J. Cuzin-Roudy. 1982. Allomeric variation. 2. Developmental instability of extreme phenotypes. *Am. Nat.* 120: 765–786. [6]

Spain, S. L., and J. C. Barrett. 2015. Strategies for fine-mapping complex traits. *Hum. Mol. Genet.* 2: R111–R119. [21]

Spassky, B., N. Spassky, H. Levene, and T. Dobzhansky. 1958. Release of genetic variability through recombination. I. *Drosophila pseudoobscura*. *Genetics* 43: 844–867. [5]

Speed, D., and D. J. Balding. 2015. Relatedness in the post-genomic era: is it still useful? *Nat. Rev. Genet.* 16:33–44. [8]

Speed, D., G. Hermani, M. R. Johnson, and D. J. Balding. 2012. Improved heritability estimation from genome-wide SNPs. *Am. J. Hum. Genet.* 91: 1011–1021. [8]

Spelman, R. J., and 4 others. 1996. Quantitative trait loci analysis for five milk production traits on chromosome six in the Dutch Holstein-Friesian population. *Genetics* 144: 1799–1807. [19]

Spencer, C. C. A., Z. Su, P. Donnelly, and J. Marchini. 2009. Designing genome-wide association studies: sample size, power, imputation, and the choice of genotyping chip. *PLoS Genet.* 5: e1000477. [20]

Spickett, S. G. 1963. Genetic and developmental studies of a quantitative character. *Nature* 199: 870–873. [17]

Spickett, S. G., and J. M. Thoday. 1966. Regular response to selection. 3. Interaction between located polygenes. *Genet. Res.* 7: 96–121. [17]

Spielman, R. S., and W. J. Ewens. 1996. The TDT and other family-based tests for linkage disequilibrium and association. *Am. J. Hum. Genet.* 59: 983–989. [6]

Spielman, R. S., and W. J. Ewens. 1998. A sibship test for linkage in the presence of association: the sib transmission/disequilibrium test. *Am. J. Hum. Genet.* 62: 450–458. [17]

Spielman, R. S., R. E. McGinnis, and W. J. Ewens. 1993. Transmission test for linkage disequilibrium: the insulin gene region and insulin-dependent diabetes mellitus (IDDM). *Am. J. Hum. Genet.* 52: 506–516. [17, 20]

Spiess, E. B. 1959. Release of genetic variability through recombination. II. *Drosophila persimilis*. *Genetics* 44: 43–58. [5]

Spiess, E. B., and A. C. Allen. 1961. Release of genetic variability through recombination. VII. Second and third chromosomes of *Drosophila melano-gaster*. *Genetics* 46: 1531–1553. [5]

Spigler, R. B., K. Theodorou, and S.-M. Change. 2017. Inbreeding depression and drift load in small populations at demographic disequilibrium. *Evolution* 71: 81–94. [12, 13]

Spirtes, P., C. Glymour, and R. Scheines. 2000. *Causation, prediction and search*, 2nd Ed. MIT Press, Cambridge, MA. [A2]

Sprague, G. F. 1983. Heterosis in maize: Theory and practice. *In* R. Frankel (ed.), *Monographs on theoretical and applied genetics, Vol. 6. Heterosis*, pp. 47–70. Springer-Verlag, Berlin, DE. [12, 13]

Sprague, G. F., and B. Brimhall. 1949. Quantitative inheritance of oil in the cornkernel. *Agron. J.* 41: 30–33. [11]

Sprague, G. F., and W. Federer. 1951. A comparison of variance components in corn yield traits. II. Error, year × variety, location × variety and variety components. *Agron. J.* 43: 535–541. [13]

Sprague, G. F., and M. T. Jenkins. 1943. A comparison of synthetic varieties, multiple crosses, and double crosses in corn. *J. Agron.* 35: 137–147. [13]

Sprague, G. F., and L. A. Tatum. 1942. General vs. specific combining ability in single crosses of corn. *J. Am. Soc. Agron.* 34: 923–932. [13]

Sprague, G. F., W. A. Russell, and L. H. Penny. 1960. Mutations affecting quantitative traits in the selfed progeny of double monoploid maize stocks. *Genetics* 45: 855–866. [15]

Sprent, P., and N. C. Smeeton. 2007. *Applied nonparametric statistical methods, 4th Ed.* Chapman and Hall, New York, NY. [20]

Spuhler, J. N. 1968. Assortative mating with respect to physical characteristics. *Eugen. Quart.* 15: 128–140. [7]

Sribney, W. M., and M. Swift. 1992. Power of sib-pair and sib-trio linkage analysis with assortative mating and multiple disease loci. *Am. J. Hum. Genet.* 51: 773–784. [19]

St. Laurent, G., Y. Vyatkin, and P. Kapranov. 2014. Dark matter RNA illuminates the puzzle of genome-wide association studies. *BMC Med.* 12: 97. [21]

St. Martin, S. K., B. W. Dye, and B. A. McBalin. 1990. Use of hill and short-two plots for selection of soybean genotypes. *Crop Sci.* 30: 74–79. [A9]

Städele, V., and L. Vigilant. 2016. Strategies for determining kinship in wild populations using genetic data. *Ecol. Evol.* 6: 6107–6120. [8]

Staiger, D., and J. Stock. 1997. Instrumental variables regression with weak instruments. *Econometrica* 65: 557–586. [A2]

Stam, P. 1980. The distribution of the fraction of the genome identical-by-descent in finite random mating populations. *Genet. Res.* 35: 131–155. [7, 8]

Stam, P. 1991. Some aspects of QTL analysis. *Proceedings of the Eighth Meeting of the Eucarpia Section Biometrics on Plant Breeding*, Brno, Czechoslovakia. pp. 24–32. [18]

Stam, P. 2003. Marker-assisted introgression: speed at any cost. *In* T. J. L. van Hintum, A. Lebeda, D. Pink, and J. W. Schut (eds.), *Eucarpia Leafy Vegetables*, pp. 117-124. [17]

Stam, P., and A. C. Zeven. 1981. The theoretical proportion of the donor genome in near-isogenic lines of self-fertilizers based on backcrossing. *Euphytica* 30: 227–238. [8, 17]

Stamatoyannopoulos, J. A. 2004. The genomics of gene expression. *Genomics* 84: 449–457. [21]

Staples, J., and 6 others. 2016. PADRE: Pedigree-aware distant-relationship estimation. *Am. J. Hum. Genet.* 99: 154–162. [8]

Stark, A. E. 1976. On the method of Penrose of estimating the number of effective factors contributing to a character. *Ann. Hum. Genet.* 39: 465–470. [16]

Stearns, S. C., and T. J. Kawecki. 1994. Fitness sensitivity and the canalization of life-history traits. *Evolution* 48: 1438–1450. [7]

Stefanova, K. T., A. B. Smith, and B. R. Cullis. 2009. Enhanced diagnostics for the spatial analysis of field trials. *J. Agric. Biol. Environ. Stats.* 4: 392–410 [A9]

Steinmetz, L. M., and 10 others. 2002a. Systematic screen for human disease genes in yeast. *Nat. Genet.* 31: 400–404. [4]

Steinmetz, L. M., and 6 others. 2002b. Dissecting the architecture of a quantitative trait locus in yeast. *Nature* 416: 326–330. [5]

Stephens, J. C., D. Briscoe, and S. J. O'Brien. 1994. Mapping by admixture linkage disequilibrium in human populations: limits and guidelines. *Am. J. Hum. Genet.* 55: 809–824. [17]

Stephens, M., and D. J. Balding. 2009. Bayesian statistical methods for genetic association studies. *Nat. Rev. Genet.* 10: 681–690. [20]

Stern, C., and E. W. Schaeffer. 1943. On wild-type isoalleles in *Drosophila melanogaster*. *Proc. Natl. Acad. Sci. USA* 29: 361–367. [15]

Sterne, J. A. C., and 18 others. 2011. Recommendations for examining and interpreting funnel plot asymmetry in meta-analyses of randomised controlled trials. *MBJ* 343: d4002 [A6]

Stevens, L., R. Giordano, and R. F. Fialho. 2001. Male-killing, nematode infections, bacteriophage infection, and virulence of cytoplasmic bacteria in the genus Wolbachia. *Ann. Rev. Ecol. System.* 32: 519–545. [6]

Stewart, G. 2010. Meta-analysis in applied ecology. *Biol. Lett.* 6: 78–81. [A6]

Stich, B. 2009. Comparison of mating designs for establishing nested association mapping populations in maize and *Arabidopsis thaliana*. *Genetics* 183: 1525–1534. [18]

Stich, B., and 6 others. 2007. Power to detect higher-order epistatic interactions in a metabolic pathway using a new mapping strategy. *Genetics* 176: 563–570. [18]

Stich, B., and 5 others. 2008. Comparison of mixed-model approaches for association mapping. *Genetics* 178: 1745–1754. [20]

Stich, B., and 4 others. 2010. Optimal allocation of resources for QTL detection using nested association mapping strategy in maize. *Theor. Appl. Genet.* 120: 553–561. [18]

Stigler, S. M. 1980. Stigler's law of eponymy. *Trans. NY Acad. Sci. Series II* 39: 147–158. [A7]

Stigler, S. M. 1981. Gauss and the invention of least squares. *Ann. Stat.* 9: 465–474. [3]

Stigler, S. M. 1983. Who discovered Bayes's theorem? *Am. Stat.* 37: 290–296. [A7]

Stigler, S. M. 1986. *The history of statistics.* Harvard Univ. Press, Cambridge, MA. [1, 9]

Stigler, S. M. 2007. The epic story of maximum likelihood. *Stat. Sci.* 22: 598–620. [A4]

Stirling, W. D. 1982. Enhancements to aid interpretation of probability plots. *J. Roy. Stat. Soc. D* 31: 211–220. [20]

Stitziel, N. O., A. Kiezun, and S. Sunyaev. 2011. Computation and statistical approaches to analysis variant identified by exome sequencing. *Genome Biol.* 12: 227. [20]

Stock, J. H., and F. Trebbi. 2003. Who invented instrumental variable regression? *J. Econom. Perspectives* 17: 177–194. [A2]

Stoffel, M. A., and 6 others. 2016. inbreedR: an R package for the analysis of inbreeding based on genetic markers. *Method. Ecol. Evol.* 7: 1331–1339. [12]

Stolk, J. M., and 6 others. 1984. Inheritance of adrenal phenylethanolamine *N*-methyltransferase activity in the rat. *Genetics* 108: 633–649. [16]

Storey, J. D. 2002. A direct approach to false discovery rates. *J. Roy. Stat. Soc. B:* 64: 479–498. [A6]

Storey, J. D., and R. Tibshirani. 2003. Statistical significance for genomewide studies. *Proc. Natl. Acad. Sci. USA* 100: 9440–9445. [A6]

Storey, J. D., J. M. Akey, and L. Kruglyak. 2005. Multiple locus linkage analysis of genomewide expression in yeast. *PLoS Bio.* 3: e267. [18]

Storz, G., J. Vogel, and K. M. Wassarman. 2011. Regulation by small RNAs in bacteria: expanding frontiers. *Mol. Cell* 43: 880–891. [15]

Stouffer, S. A., and 4 others. 1949. *The American solder, Vol. 1: Adjustment during army life.* Princeton University Press, Princeton, NJ. [A6]

Stouthamer, R., J. A. J. Breeuwer, R. F. Luck, and J. H. Werren. 1993. Molecular identification of microorganisms associated with parthenogenesis. *Nature* 361: 66–68. [6]

Strahl, B. D., and C. D. Allis. 2000. The language of covalent histone modifications. *Nature* 403: 41–45. [21]

Stram, D. O., and J. W. Lee. 1994. Variance components testing in the longitudinal mixed effects model. *Biometrics* 50: 1171–1177. Correction *Biometrics* 51: 1196. [19, A4]

Stranger, B. E., and 11 others. 2005. Genome-wide association of gene expression variation in humans. *PLoS Genet.* 378. [21]

Stranger, B. E., and 13 others. 2007a. Population genomics of human gene expression. *Nat. Genet.* 39: 1217–1224. [21]

Stranger, B. E., and 16 others. 2007b. Relative impact of nucleotide and copy number variation on gene expression phenotypes. *Science* 315: 848–853. [21]

Strauss, R. E. 1985. Evolutionary allometry and variation in body forms in the South American catfish genus *Corydoras* (Callichthyidae). *Syst. Zool.* 34: 381–396. [14]

Strauss, R. E. 1991. Correlations between heterozygosity and phenotypic variability in *Cottus* (Teleostei: Cottidae): character components. *Evolution* 45: 1950–1956. [6]

Strauss, S. H. 1986. Heterosis at allozyme loci under inbreeding and crossbreeding in *Pinus attenuata. Genetics* 113:115–134. [12]

Strauss, S. H. 1987. Heterozygosity and developmental stability under inbreeding and crossbreeding in *Pinus attenuata. Evolution* 41: 331–339. [6]

Strauss, S. H., and W. J. Libby. 1987. Allozyme heterosis in radiata pine is poorly explained by overdominance. *Am. Nat.* 130: 879–890. [12]

Strauss, S. Y., and R. Karban. 1994. The significance of outcrossing in an intimate plant-herbivore relationship. I. Does outcrossing provide an escape from herbivores adapted to the parent plant? *Evolution* 48: 454–464. [6]

Streisinger, G., and 4 others. 1981. Production of clones of homozygous diploid zebra fish (*Brachydanio rerio*). *Nature* 291: 293-296. [17]

Strickberger, M. W. 1972. Viabilities of third chromosomes of *Drosophila pseudoobscura* differing in relative competitive fitness. *Genetics* 72: 679–689. [12]

Stricker, C., R. L. Fernando, and R. C. Elston. 1995a. An algorithm to approximate the likelihood for pedigree data with loops by cutting. *Theor. Appl. Genet.* 91: 1054–1063. [16]

Stricker, C., R. L. Fernando, and R. C. Elston. 1995b. Linkage analysis with an alternative formulation for the mixed model of inheritance: the finite polygenic mixed model. *Genetics* 141: 1651–1656. [16, 19]

Stroup, W. W., P. S. Baenizger, and D. K. Mulitze. 1994. Removing spatial variation from wheat yield trails: a comparison of methods. *Crop Sci.* 86: 62–66. [A9]

Struchlain, M. V., and 4 others. 2010. Variance heterogeneity analysis for detection of potentially interacting genetic loci: method and its limitations. *BMC Genet.* 11: 92. [18]

Struhl, K. 1987. Promoters, activator proteins, and the mechanism of transcriptional initiation in yeast. *Cell* 49: 295–297. [15]

Stuber, C. W, and R. H. Moll. 1972. Frequency changes of isozyme alleles in a selection experiment for grain yield in maize (*Zea maize* L.). *Crop Sci.* 12: 337–340. [17]

Stuber, C. W., and 4 others. 1980. Allozyme frequency changes associated with selection for increased grain yield in maize. *Genetics* 95: 225–236. [17]

Stuber, C. W., M. M. Goodman, and R. H. Moll. 1982. Improvement of yield and ear number resulting from selection at allozyme loci in a maize population. *Crop Sci.* 22: 737–740. [17]

Stuber, C. W., M. D. Edwards, and J. F. Wendel. 1987. Molecular-marker-facilitated investigations of quantitative-trait loci in maize. II. Factors influencing yield and its component traits. *Crop Sci.* 27: 639–648. [18]

Stuber, C. W., and 4 others. 1992. Identification of genetic factors contributing to heterosis in a hybrid from two elite inbred lines using molecular markers. *Genetics* 132: 823–839. [13, 18]

Student. 1908. The probable error of a mean. *Biometrika* 6: 1–25. [2]

Studer, A. J., and J. F. Doebley. 2011. Do large effect QTLs fractionate? A case study at the maize domestication QTL *teosinte branched 1. Genetics* 188: 673–681. [5, 18]

Studer, A., Q. Zhao, J. Ross-Ibarra, and J. Doebley. 2011. Identification of a functional transposon insertion in the maize domestication gene *tb1*. *Nat. Genet.* 43: 1160–1165. [15]

Stulp, G., M. J. P. Simons, S. Grasman, and T. V. Pollet. 2015. Assortative mating for human height: A meta-analysis. *Am. J. Hum. Biol.* 29: e22917. [7]

Suarez, B. K. 1978. The affected sib pair IBD distribution for HLA disease susceptibility genes. *Tissue Antigens* 12: 87–93. [19]

Suarez, B. K., and P. V. Eerdewegh. 1984. A comparison of three affected-sib-pair scoring methods to detect HLA-linked disease susceptibility genes. *Am. J. Med. Genet.* 18: 135–146. [19]

Suarez, B. K., J. Rice, and T. Reich. 1978. The generalized sib pair IBD distribution: its use in the detection of linkage. *Ann. Hum. Genet.* 42: 87–94. [19]

Suarez, B. K., T. Reich, and P. M. Fishman. 1979. Variability in sib pair genetic identity. *Hum. Hered.* 29: 37–41. [8]

Suarez, B. K., C. L. Hampe, and P. Van Eerdewegh. 1994. Problems of replicating linkage claims in psychiatry. *In* E. S. Gershon and C. R. Cloninger (eds.), *Genetic approaches to mental disorders*, pp. 23–46. American Psychiatric Press, Washington, D. C. [19]

Subramanian, A., and 10 others. 2005. Gene set enrichment analysis: a knowledge-based approach for interpreting genome-wide expression profiles. *Proc. Natl. Acad. Sci. USA* 102: 15545–15550. [21]

Sugiura, N. 1978. Further analysis of the data by Akaike's information criterion and the finite corrections. *Comm. Stat. Theor. Meth.* A7: 13–26. [16, A4]

Sul, J. H., and 4 others. 2013. Effectively identifying eQTLs from multiple tissues by combining mixed model and meta-analytic approaches. *PLoS Genet.* 9: e1003491. [21]

Sul, J. H., L. S. Martin, and E. Eskin. 2018. Population structure in genetic studies: confounding factors and mixed models. *PLoS Genet.* 14: e1007309. [20]

Sulisalo, T., and 6 others. 1994. High-resolution linkage-disequilibrium mapping of the cartilage-hair hypoplasia gene. *Am. J. Hum. Genet.* 55: 937–945. [17]

Sun, F. Z., W. D. Flanders, Q. H. Yang, and H. Y. Zhao. 2000. Transmission/disequilibrium tests for quantitative traits. *Ann. Hum. Genet.* 64: 555–565. [17]

Sun, G., and 6 others. 2010. Variation explained in mixed-model association mapping. *Heredity* 105: 333–340. [20, A4]

Sun, J., Y. Zheng, and L. Hsu. 2013. A unified mixed-effect model for rare-variant association in sequencing studies. *Genet. Epidem.* 37: 334–344. [20]

Sun, L., and S. Bull. 2005. Reduction of selection bias in genomewide studies by resampling. *Genet. Epidem.* 28: 352–367. [18, 20]

Sun, L., and 5 others. 2011. BR-squared: a practical solution to the winner's curse in genome-wide scans. *Hum. Genet.* 129: 545–552. [18, 20]

Sun, S., C.-T. Ting, and C.-I. Wu. 2004. The normal function of a speciation gene, *Odysseus*, and its hybrid sterility effect. *Science* 305: 81–83. [17]

Sun, S., and 8 others. 2023. Unraveling prevalence and effects of deleterious mutations in maize elite lines across decades of modern breeding. *Mol. Bio. Evol.* 40: msad170. [12]

Sun, X., 2017. Segregation analysis using the unified model. *In* R. C. Elston (ed), *Statistical human genetics*, p. 233–256. Humana Press, New York, NY. [16]

Sun, X., and 10 others. 2010. Segregation analysis of Barrett's esophagus and associated adenocarcinomas. *Cancer Epidemiology Prevention Biomarkers* 19: 666–674. [16]

Sun, Z., and 22 others. 2023. Dissecting the genetic basis of heterosis in elite super-hybrid rice. *Plant Physiol.* 192: 307–325. [13]

Susko, E. 2013. Likelihood ratio tests with boundary constraints using data-dependent degrees of freedom. *Biometrika* 100: 1019–1023. [A4]

Suzuki, M. M., and A. Bird. 2008. DNA methylation landscapes: provocative insights from epigenomics. *Nat. Rev. Genet.* 9: 465–476. [21]

Sved, J. A. 1972. Heterosis at the level of the chromosome and at the level of the gene. *Theor. Pop. Biol.* 3: 491–506. [12]

Svishcheva, G. R., and 4 others. 2012. Rapid variance components-based method for whole genome association analysis. *Nat. Genet.* 44: 1166–1170. [20]

Swaddle, J. P., M. Witter, and I. C. Cuthill. 1994. The analysis of fluctuating asymmetry. *Anim. Behav.* 48: 986–989. [6]

Swan, A. A., and B. P. Kinghorn. 1992. Symposium: Dairy crossbreeding: Evaluating and exploitation of crossbreeding in dairy cattle. *J. Dairy Sci.* 5: 624–639. [13]

Swindell, W. R., and J. L. Bouzat. 2006. Reduced inbreeding depression due to historical inbreeding in *Drosophila melanogaster*: evidence for purging. *J. Evol. Biol.* 19: 1257–1264. [12]

Syrstad, O. 1985. Heterosis in *Bos taurus–Bos indicus* crosses. *Livestock Prod. Sci.* 12: 299–307. [13]

Szalma, S. J., and 4 others. 2007. QTL mapping with near-isogenic lines in maize. *Theor. Appl. Genet.* 114: 121–1228. [17]

Szatkiewicz, J. P., and E. Feingold. 2004. A powerful and robust new linkage statistic for discordant sibling pairs. *Am. J. Hum. Genet.* 75: 906–109. [19]

Szatkiewicz, J. P., and E. Feingold. 2005. QTL mapping with discordant and concordant sibling pairs: new statistics and new design strategies. *Genet. Epid.* 28: 326–340. [19]

Szatkiewicz, J. P., K. T. Cuenco, and E. Feingold. 2003. Recent advances in human quantitative-trait-locus mapping: comparison of methods for discordant sibling pairs. *Am. J. Hum. Genet.* 73: 874–885. [19]

Szpiech, Z. A., and 6 others. 2013. Long runs of homozygosity are enriched for deleterious variation. *Am. J. Hum. Genet.* 93: 90–102. [12]

Szpiech, Z. A., A. Blant, and T. J. Pemberton. 2017. *GARLIC*: Genome Autozygosity regions likelihood-basedinference and classification. *Bioinformatics* 33: 2059–2062. [12]

Szulkin, M., D. Garant, R. H. McCleery, and V. C. Sheldon. 2007. Inbreeding depression along a life-history continuum in the great tit. *J. Evol. Biol.* 20: 1531–1543. [12]

Szulkin, M., N. Bierne, and P. David. 2010. Heterozygosity-fitness correlations: a time for reappraisal. *Evolution* 64: 1202–1217. [12]

Szulkin, M., and 3 others. 2013. Inbreeding avoidance, tolerance, or preference in animals? *Trends Ecol. Evol.* 28: 205–211. [12]

T.Cuenco, K., J. P. Szatkiewicz, and E. Feingold. 2003. Recent advances in human quantitative-trait-locus mapping: comparison of methods for selected sibling pairs. *Am. J. Hum. Genet.* 73: 863–873. [19]

Tabery, J. G. 2004. The "Evolutionary Synthesis" of George Udny Yule. *J. History Biol.* 37: 73–101. [1]

Tachida, H., and C. C. Cockerham. 1990. Evolution of neutral quantitative characters with gene interaction and mutation. *In* N. Takahata and J. F. Crow (eds.), *Population biology of genes and molecules*, pp. 233–249. Baifukan, Tokyo, JP. [15]

Takahashi, K. H. 2019. Multiple modes of canalization: links between genetic, environmental canalization and developmental stability, and their trait-specificity. *Semin. Cell. Dev. Biol.* 88: 14–20. [14]

Talbot, C. J., and 5 others. 1999. High-resolution mapping of quantitative trait loci in outbred mice. *Nat. Genet.* 21: 305–308. [18]

Talium, D., et al. 2021. Sequencing of 53,831 diver genomes form the NHLBI TOPMed program. *Nature* 590: 290–297. [21]

Tamayo, P., G. Steinhardt, A. Liberzon, and J. P. Mesirov. 2016. The limitation of simple gene set enrichment analysis assuming gene independence. *Stat. Method. Med. Res.* 25: 472–487. [21]

Tamura, R. N., L. A. Nelson, and G. C. Naderman. 1998. An investigation of the validity and usefulness of trend analysis for field plot data. *Agron. J.* 80: 712–718. [A9]

Tan, W. Y., and W. C. Chang. 1972. Convolution approach to genetic analysis of quantitative characters of self-fertilized populations. *Biometrics* 28: 1073–1090. [16]

Tan, W. Y., and H. D'Angelo. 1979. Statistical analysis

of joint effects of major genes and polygenes in quantitative genetics. *Biom. J.* 21: 179–192. [16]

Tang, H. K., and D. Siegmund. 2001. Mapping quantitative trait loci on oligogenic models. *Biostatistics* 2: 147–162. [19]

Tang, J., and C. Chu. 2017. MicroRNAs in crop improvement fine-tuners for complex traits. *Nat. Plants* 3: 1-11. [15]

Tang, J., and 8 others. 2010. Dissection of the genetic basis of heterosis in an elite maize hybrid by QTL mapping in an immortalized F_2 population. *Theor. Appl. Genet.* 120: 333–340. [13]

Tanksley, S. D., and J. Hewitt. 1988. Use of molecular markers in breeding for soluble solids content in tomato — a reexamination. *Theor. Appl. Genet.* 75: 811–823. [17]

Tanksley, S. D., and J. C. Nelson. 1996. Advanced backcross QTL analysis: a method for the simultaneous discover and transfer of valuable QTLs from unadapted germplasm into elite breeding lines. *Theor. Appl. Genet.* 92: 191–203. [18]

Tanksley, S. D., N. D. Young, A. H. Paterson, and M. W. Bonierbale. 1989. RFLP mapping in plant breeding: new tools for an old science. *Biotechnol.* 7: 257–264. [17]

Tanksley, S. D., M. W. Ganal, and G. B. Martin. 1995. Chromosome landing: a paradigm for map-based gene cloning in plants with large genomes. *Trends Genet.* 11: 63– 68. [17]

Tanner, J. M. 1949. Fallacy of per-weight and per-surface area standards and their relation to spurious correlation. *J. Appl. Physiology* 2: 1–15. [14]

Tanner, M. A. 1996. *Tools for statistical inference*, 3rd ed. Springer-Verlag, New York, NY. [A8]

Tantawy, A. O. 1957. Genetic variance of random-inbred lines of *Drosophila melanogaster* in relation to coefficients of inbreeding. *Genetics* 42: 121–136. [12]

Tantawy, A. O., and E. C. R. Reeve. 1956. Studies in quantitative inheritance. IX. The effects of inbreeding at different rates in *Drosophila melanogaster. Z. Indukt. Abst. Ver. bungslehre* 87: 648–667. [12]

Tao, Y., D. L. Hartl, and C. C. Laurie. 2001. Sex-ratio segregation distortion associated with reproductive isolation in *Drosophila. Proc. Natl. Acad. Sci. USA* 98: 13183–13188. [17]

Tao, Y., S. Chen, D. L. Hartl, and C. C. Lauries. 2003. Genetic dissection of hybrid incompatibilities between *Drosophila simulans* and *D. mauritiana*. I. Differential accumulation of hybrid male sterility effects on the X and the autosomes. *Genetics* 164: 1383–1397. [17]

Tarca, A. L., G. Bhatti, and R. Romero. 2013. A comparison of gene set analysis methods in terms of sensitivity, prioritization and specificity. *PLoS ONE* 8: e79217. [21]

Tatum, L. A. 1971. The Southern Corn Leaf Blight epidemic. *Science* 171: 1113–1116. [13]

Tavazoie, S., and 4 others. 1999. Systematic determination of genetic network architecture. *Nat. Genet.* 22: 281–285. [21]

Taylor, D. S. 2001. Physical variability and fluctuating asymmetry in heterozygous and homozygous populations on *Rivulus marmoratus. Can. J. Zool.* 79: 766–778. [6]

Taylor, F. W. 1909. The size of experimental plots for field crops. *Proc. Amer. Soc. Agron.* 1: 56–58. [A9]

Taylor, G. L., and A. M. Prior. 1939. Blood groups in England. III. Discussion of the family material. *Ann. Eugen.* 9: 18–44. [4]

Templeton, A. R. 1977. Analysis of head shape differences between two interfertile species of Hawaiian *Drosophila*. *Evolution* 31: 330–341. [11]

Templeton, A. R. 1980. The theory of speciation by the founder principle. *Genetics* 92: 1011–1038. [17]

Templeton, A. R. 1981. Mechanisms of speciation — a population genetic approach. *Ann. Rev. Ecol. Syst.* 12: 23–48. [13]

Templeton, A. R. 1986. Coadaptation and outbreeding depression. *In* M. E. Soulé (ed.), *Conservation biology: the science of scarcity and diversity*, pp. 105–116. Sinauer Assoc., Sunderland, MA. [13]

Templeton, A. R. 1987. The general relationship between average effect and average excess. *Genet. Res.* 49: 69–70. [4]

Templeton, A. R. 1995. A cladistic analysis of phenotypic associations with haplotypes inferred from restriction endonuclease mapping. V. Analysis of case/control sampling designs. Alzheimer's disease and the apoprotein E locus. *Genetics* 140: 403–409. [17]

Templeton, A. R., and M. A. Rankin. 1978. Genetic revolutions and control of insect populations. *In* R. H. Richardson (ed.), *The screwworm problem*, pp. 83–112. Univ. Texas Press, Austin, TX. [15]

Templeton, A. R., and B. Read. 1983. The elimination of inbreeding depression in a captive herd of Speke's gazelle. *In* C. M. Schonewald-Cox, S. M. Chambers, F. MacBryde, and L. Thomas (eds.), *Genetics and conservation: a reference for managing wild animal and plant populations*, pp. 241–261. Benjamin/Cummings, Menlo Park, CA. [12]

Templeton, A. R., and B. Read. 1984. Factors eliminating inbreeding depression in a captive herd of Speke's gazelle (*Gazella spekei*). *Zoo Biol.* 3: 177–199. [12]

Templeton, A. R., C. F. Sing, and B. Brokaw. 1976. The unit of selection in *Drosophila mercatorum*. I. The interaction of selection and meiosis in parthenogenetic strains. *Genetics* 82: 349–376. [13]

Templeton, A. R., E. Boerwinkle, and C. F. Sing. 1987. A cladistic analysis of phenotypic associations with haplotypes inferred from restriction endonuclease mapping. I Basic theory and an analysis of alcohol dehydrogenase activity in *Drosophila*. *Genetics* 117: 343–351. [17]

Templeton, A. R., C. F. Sing, A. Kessling, and S. Humphries. 1988. A cladistic analysis of phenotypic associations with haplotypes inferred from restriction endonuclease mapping. II. The analysis of natural populations. *Genetics* 120: 1145–1154. [17]

Templeton, A. R., H. Hollocher, S. Lawler, and J. S. Johnston. 1989. Natural selection and ribosomal DNA in *Drosophila*. *Genetics* 31: 296–303. [15]

Templeton, A. R., K. A. Crandall, and C. F. Sing. 1992. A cladistic analysis of phenotypic associations with haplotypes inferred from restriction endonuclease mapping. III. Cladogram estimation. *Genetics* 132: 619–633. [17]

Tenesa, A., and 6 others. 2007. Recent human effective population size estimated from linkage disequilibrium. *Genome Res.* 17: 520–526. [5]

Tenesa, A., K. Rawlik, P. Navarro, and O. Canela-Xandri. 2016. Genetic determination of height-mediated mate choice *Genome Biol.* 16: 269. [7]

Teo, S. M., and 4 others. 2012. Statistical challenges associated with detecting copy number variations with next-generation sequencing. *Bioinformatics* 28: 2711–2718. [20]

Terwilliger, J. D. 1995. A powerful likelihood method for the analysis of linkage disequilibrium between trait loci and one or more polymorphic marker loci. *Am. J. Hum. Genet.* 56: 777–787. [17]

Terwilliger, J. D., and J. Ott. 1992. A haplotype-based haplotype relative risk statistic. *Hum. Hered.* 42: 337–346. [17]

Terwilliger, J. D., and J. Ott. 1994. *Handbook of human genetic linkage*. Johns Hopkins Univ. Press, Baltimore, MD. [19]

Tettelin, H., and 45 others. 2005. Genome analysis of multiple pathogenic isolates of *Streptococcus agalactiae*: implications for the microbial "pan-genome". *Proc. Natl. Acad. Sci. USA* 102: 13950–13955. [13]

Teuscher, F., V. Guiard, P. E. Udolph, and G. A. Brockmann. 2005. The map expansion obtained with recombinant inbred strains and intermated recombinant inbred populations for finite generation designs. *Genetics* 170: 875–879. [18]

Teutonico, R. A., and T. C. Osborn. 1994. Mapping of RFLP and qualitative trait loci in *Brassica rapa* and comparison to the linkage maps of *B. napus*, *B. oleracea*, and *Arabidopsis thaliana*. *Theor. Appl. Genet.* 89: 885–894. [17]

Thaller, G., and I. Hoeschele. 1996a. A Monte Carlo method for Bayesian analysis of linkage between single markers and quantitative trait loci. I. Methodology. *Theor. Appl. Genet.* 93: 1161–1166. [18]

Thaller, G., and I. Hoeschele. 1996b. A Monte Carlo method for Bayesian analysis of linkage between single markers and quantitative trait loci. II. A simulation study. *Theor. Appl. Genet.* 93: 1167–1174. [18]

Thaller, G., L. Dempfle, and I. Hoeschele. 1996. Maximum likelihood analysis of rare binary traits under different modes of inheritance. *Genetics* 143: 1819–1829. [16]

Thelen, G. C., and F. W. Allendorf. 2001. Heterozygosity-fitness correlations in rainbow trout: effects of allozyme loci or associative overdominance? *Evolution* 55: 1180–1187. [12]

Theodorou, K., and D. Couvet. 2015. The efficiency of close inbreeding to reduce genetic adaptation to captivity. *Heredity* 114: 38–47. [12]

Theusch, E., and J. Gitschier. 2011. Absolute pitch twin study and segregation analysis. *Twin Res. Hum. Genet.* 14: 173–178. [16]

Thoday, J. M. 1953. Components of fitness. *Symp. Soc. Exp. Biol.* 7: 96–113. [6]

Thoday, J. M. 1961. Location of polygenes. *Nature* 191: 368–370. [17]

Thoday, J. M. 1979. Polygene mapping: uses and limitations. *In* J. N. Thompson, Jr., and J. M. Thoday (eds.), *Quantitative genetic variation*, pp. 219–233. Academic Press, New York, NY. [17]

Thoday, J. M., and J. N. Thompson, Jr. 1976. The number of segregating genes implied by continuous variation. *Genetica* 46: 335–344. [16, 17]

Thoday, J. M., J. B. Gibson, and S. G. Spickett. 1964. Regular responses to selection. II. Recombination and accelerated response. *Genet. Res.* 5: 1–19. [17]

Thomas, D. C. 2004. *Statstical methods in genetic epidemiology*. Oxford University Press, New York, NY. [16]

Thomas, D. C., and D. G. Clayton. 2004. Betting odds and genetic associations. *J. Nat. Cancer Inst.* 96: 421–423. [20]

Thomas, D. C., and D. V. Conti. 2004. Commentary: The concept of 'Mendelian randomization'. *Int. J. Epidem.* 33: 21–25. [A2]

Thomas, D. C., D. A. Lawlor, and J. R. Thompson. 2007. Re: estimation of bias in nongenetic observations studies using "Mendelian triangulation" by Bautista et al. *Ann. Epidem.* 17: 511–513. [A2]

Thomas, S. C., and W. G. Hill. 2000. Estimating quantitative genetic parameters using sibships reconstructed from marker data. *Genetics* 155: 1961–1972. [8]

Thomas, S. C., J. M. Pemberton, and W. G. Hill. 2000. Estimating variance components in natural populations using inferred relationships. *Heredity* 84: 427–436. [8]

Thompson, A., J., and 29 others. 2018. Diagnostic criteria for multiple sclerosis: 2017 revisions to the McDonald criteria. *Lancet Neurol.* 17: 162–173. [20]

Thompson, D'A. W. 1917. *On growth and form*. Cambridge Univ. Press, Cambridge, UK. [14]

Thompson, D'A. W. 1943. *On growth and form* (revised ed.) Cambridge Univ. Press, Cambridge, UK. [14]

Thompson, E. A. 1975. The estimation of pairwise relationships. *Ann. Hum. Genet.* 39: 173–188. [8, [A4]

Thompson, E. A. 1986a. *Pedigree analysis in human genetics*. Johns Hopkins Univ. Press, Baltimore, MD. [8, A4]

Thompson, E. A. 1986b. Likelihood inference of paternity. *Am. J. Hum. Genet.* 39: 285–287. [8]

Thompson, E. A. 1988. Two-locus and three-locus gene identify by descent in pedigrees. *IMA J. Math. Appl. Med. Biol.* 5: 261–280. [8]

Thompson, E. A. 2013. Identity by descent: variation in meiosis, across genomes, and in populations. *Genetics* 194: 301–326. [8]

Thompson, E. A. 2019. Descent-based gene mapping in pedigrees and populations. *In* D. J. Balding, I. Molke, and J. Marioni (eds.), *Handbook of Statistical Genomics*, 4th Ed., pp. 573–596. Wiley and Sons, New York, NY. [8]

Thompson, E. A., and S. W. Guo. 1991. Evaluation of likelihood ratios for complex genetic models. *IMA J. Math. Appl. Med. Biol.* 8: 149–169. [16]

Thompson, E. A., and T. R. Meagher. 1987. Parental and sib likelihoods in genealogy reconstruction. *Biometrics* 43 : 585–600. [8]

Thompson, E. A., S. Lin, A. B. Olshen, and E. M. Wijsman. 1993. Monte carlo analysis on a large pedigree. *Genet. Epidem.* 10: 677–682. [16]

Thompson, G. 1986. Determining the mode of inheritance of RFLP-associated diseases using the affected sib-pair method. *Am. J. Hum. Genet.* 39: 207–221. [19]

Thompson, H. R. 1951. Truncated lognormal distributions. *Biometrika* 38: 414–422. [14]

Thompson, J., C. Minelli, and M. F. Del Greco. 2016. Mendelian randomization using public data from genetic consortia. *Int. J. Biostat.* 12: 2. [ ]

Thompson, J. N. 1988. Evolutionary ecology of the relationship between oviposition preference and performance of offspring in phytophagous insects. *Entomol. Exp. Appl.* 47: 3–14. [5]

Thompson, J. N., Jr. 1973. General and specific effects of modifiers of mutant expression. *Genet. Res.* 22: 211–215. [15]

Thompson, J. N., Jr. 1974. Studies of the nature and function of polygenic loci in *Drosophila*. I. Comparison of genomes from selected lines. *Heredity* 33: 373–387. [15]

Thompson, J. N., Jr., and W. E. Spivey. 1984. Organization of polygenic system: cell death modifiers from natural populations of *Drosophila melanogaster*. *Genet. Res.* 44: 261–269. [15]

Thompson, J. N., Jr., and J. M. Thoday. 1972. Modification of dominance byselection in the homozygote. *Heredity* 29: 285–292. [15]

Thompson, J. R., and 4 others. 2005. Meta-analysis of genetic studies using Mendelian randomization — a multivariate approach. *Stat. Med.* 224: 2241–2254. [A2]

Thompson, R. 2008. Estimation of quantitative genetic parameters. *Proc. Roy. Soc. Lond. B* 275: 679–686. [10]

Thompson, W. K., and 10 others. 2015. An empirical Bayes mixture model for effect size distributions in genome-wide association studies. *PLoS Genet.* 11: e1005717. [21]

Thomson, G. 1995. Mapping disease genes: family-based association studies. *Am. J. Hum. Genet.* 57: 487–498. [17]

Thornhill, N. W. (ed.) 1993. *The natural history of inbreeding and outbreeding.* Univ. Chicago Press, Chicago, IL. [12]

Thornsberry, J. M., and 5 others. 2001. *Dwarf8* polymorphisms associate with variation in flowering time. *Nat. Genet.* 28: 286–289. [20]

Thornton, K., A. Foran, and A. Long. 2013. Properties and modeling of GWAS when complex disease risk is due to non-complementing deleterious mutations in genes of large effect. *PLoS Genet.* 9: e1003258. [21]

Thornton, T., and 5 others. 2012. Estimating kinship in admixed populations. *Am. J Hum. Genet.* 91: 122–138. [8]

Threadgill, D. W., and G. A. Churchill. 2012. Ten years of the collaborative cross. *G3* 2: 153–156. [18]

Threadgill, D. W., K. W. Hunter, and R. W. Williams. 2002. Genetic dissection of complex and quantitative traits: from fantasy to reality via a community effort. *Mamm. Genome* 13: 175–178. [18]

Thurman, T. J., and R. D. Barrett. 2016. The genetic consequence of selection in natural populations. *Mol. Ecol.* 25: 1429–1448. [12]

Tian, F., and 9 others. 2011. Genome-wide association study of leaf architecture in the maize nested association mapping population. *Nat. Genet.* 43: 159–162. [21]

Tian, L., and 5 others. 2005. Discovering statistically significant pathways in expression profiling studies. *Proc. Natl. Acad. Sci. USA* 102: 13544–12549. [21]

Tibshirani, R. 1996. Regression shrinkage and selection via the lasso. *J. Roy. Stat. Soc.* B 58: 267–288. [20]

Tierney, L. 1994. Markov chains for exploring posterior distributions (with discussion). *Ann. Stat.* 22: 1701–1762. [A8]

Tiku, M. L. 1965. Laguerre series forms for noncentral χ^2 and F distributions. *Biometrika* 52: 415–428. [A5]

Tippett, L. M. C. 1931. *The methods of statistics.* Williams and Norgate, London, UK. [20, A6]

Tiret, L., L. Abel, and R. Rakotovao. 1993. Effect of ignoring genotype-environment interaction on segregation analysis of quantitative traits. *Genet. Epidem.* 10: 581–586. [16]

Titterington, D. M., A. F. M. Smith, and U. E. Makov. 1985. *Statistical analysis of finite mixture distributions.* John Wiley & Sons, New York, NY. [16]

Tiwari, H. K., and 4 others. 2005. New joint covariance and marginal-based tests for association and linkage for quantitative traits for random and nonrandom sampling. *Genet. Epid.* 12: 865–870. [17]

Tiwari, H. K., and 5 others. 2008. Review and evaluation of methods correcting for population stratification with a focus on underlying statistical principles. *Hum. Hered.* 66: 67–86. [17, 20]

Toczydlowski, R. H., and D. M. Waller. 2023. Failure to purge: population and individual inbreeding effects on fitness across generations of wild *Impatiens capensis*. *Evolution* 77: 1315–1329. [12]

Toivonen, H. T. T., and 7 others. 2000. Data mining applied to linkage disequilibrium mapping. *Am. J. Hum. Genet.* 67: 133–145. [17]

Toosi, A., R. L. Fernando, and J. C. M. Dekkers. 2018. Genome-wide mapping of quantitative trait loci in admixed populations using mixed linear model and Bayesian multiple regression analysis. *Genet. Sel. Evol.* 50: 32. [20]

Torkamani, A., E. J. Topel, and N. J. Schork. 2008. Pathway analysis of seven common disease assessed by genome-wide association. *Genomics* 92: 265–272. [21]

Toro, M., and 5 others. 2002. Estimation of coancestry in Iberian pigs using molecular markers. *Conserv. Genet.* 3: 309–320. [8]

Toro, M., L. A. García-Cortés, and A. Legarra. 2011. A note on the rationale for estimating genealogical coancestry from molecular markers. *Genet. Sel. Evol.* 43: 27. [8]

Touraev, A., B. P. Forster, and S. M. Jain. 2009. *Advances in haploid production in higher plants.* Springer, Berlin, DE. [17]

Tourjee, K. R., J. Harding, and T. G. Byrne. 1995. Complex segregation analysis of *Gerbera* flower color. *Heredity* 74: 303–310. [16]

Touzet, P., R. G. Winkler, and T. Helentjaris. 1995. Combined genetic and physiological analysis of a locus contributing to quantitative variation. *Theor. Appl. Genet.* 91: 200–205. [17]

Towey, P., and J. L. Jinks. 1977. Alternative ways of estimating the number of genes in a polygenic system by genotype assay. *Heredity* 39: 399–410. [11]

Tracy, C., and H. Widom. 1994. Level-spacing distributions and the Airy kernel. *Comm. Math. Phys.* 159: 151–174. [20]

Tracy, W. F., and M. A. Chandler. 2006. The historical and biological basis of the concept of heterotic patterns in corn belt dent maize. *In* K. R. Lamkey and M. Lee (eds.) *Plant breeding: The Arnel R. Hallauer International Symposium*, pp. 219–233. Ames, IA. Wiley-Blackwell. [13]

Tragoonrung, S., P. M. Hayes, and B. L. Jones. 1990. Comparison of hill and row plots for agronomic and quality traits in spring malting barley (*Hordeum vulgare* L.) *Can. J. Plant Sci.* 70: 61–69. [A9]

Trail, J. C. M., K. E. Gregory, H. J. S. Marples, and J. Kakong. 1985. Comparison of *Bos tarus-Bos indicus* breed croses with straightbred *Bos indicus* breeds of cattle for maternal and individual traits. *J. Anim. Sci.* 60: 1181–1187. [13]

Trankell, A. 1956. Penetrance calculus in population genetics. *Am. J. Hum. Genet.* 8: 43–48. [4]

Trikalinos, T. A., G. Salanti, E. Zintzaras, and J. P. A. Ioannidis. 2008. Meta-analysis methods. *Adv. Genet.* 60: 311–334. [20]

Trow, A. H. 1913. Forms of reduplication: primary and secondary. *J. Genetics* 2: 313–324. [17]

Troyer, A. F. 1999. Background of US hybrid corn. *Crop Sci.* 39: 601–626. [13]

Troyer, A. F. 2004. Background of US hybrid corn II. *Crop Sci.* 44: 370–380. [13]

Troyer, A. F. 2006. Adaptedness and heterosis in corn and mule hybrids. *Crop Sci.* 46: 528–543. [13]

Troyer, A. F., and E. J. Wellin. 2009. Heterosis decreasing in hybrids: yield test inbreds. *Crop Sci.* 49: 1969–1976. [13]

True, J. R., B. S. Weir, and C. C. Laurie. 1996a. A genome-wide survey of hybrid incompatibility factors by the introgression of marked segments of *Drosophila mauritiana* chromosomes into *Drosophila simulans*. *Genetics* 142: 819–837. [17]

True, J. R., J. M. Mercer, and C. C. Laurie. 1996b. Differences in crossover frequency and distribution among three sibling species of *Drosophila*. *Genetics* 142: 507–523. [17]

Truett, J., J. Cornfield, and W. Kannel. 1967. A multivariate analysis of the risk of coronary heart disease in Framingham. *J. Chron. Dis.* 20: 511–524. [29]

Trustrum, G. B. 1961. The correlations between relatives in a random mating diploid population. *Proc. Camb. Phil. Soc.* 57: 315–320. [7]

Trynka, G., and 6 others. 2013. Chromatin marks identify critical cell types for fine mapping complex trait variants. *Nat. Genet.* 45: 124–130. [21]

Tryputsen, V., J. Cabrera, A. De Bondt, and D. Amaratunga. 2014. Using Fisher's method to identify enriched gene sets. *Stat. Biopharm. Res.* 6: 154–162. [21]

Tsotis, B. 1972. Objectives of industry breeders to make efficient and significant advances in the future. *In* D. Wilkinson (ed.), *Proceeding 27th Ann. Corn and Sorghum Research Conference*. American Seed Trade Association, Washington, DC. [13]

Tsubota, S., and P. Schedl. 1986. Hybrid dysgenesis-induced revertants of insertions at the 5′ end of the *rudimentary* gene in *Drosophila melanogaster*: Transposon-induced control mutations. *Genetics* 114: 165–182. [15]

Tu, I.-P., and A. S. Whitemore. 1999. Power and association and linkage test when the disease alleles are unobserved. *Am. J. Hum. Genet.* 64: 461–649. [17]

Tucker, G., A. L. Price, and B. Berger. 2014. Improving the power of GWAS and avoiding confounding from population stratification with PC-Select. *Genetics* 197: 1045–1049. [20]

Tuinstra, M. R., G. Ejeta, and P. B. Goldsbrough. 1997. Heterogeneous inbred family (HIF) analysis: a method for developing near-isogenic lines that differ at quantitative trait loci. *Theor. Appl. Genet.* 95: 1005–1011. [17]

Turelli, M. 1984. Heritable genetic variation via mutation-selection balance: Lerch's zeta meets the abdominal bristle. *Theor. Pop. Biol.* 25: 138–193. [15]

Turelli, M., and L. C. Moyle. 2007. Asymmetric postmating isolation: Darwin's corollary to Haldane's rule. *Genetics* 176: 1059–1088. [17]

Turelli, M., and H. A. Orr. 1995. The dominance theory of Haldane's rule. *Genetics* 140: 389–402. [17]

Turelli, M., and H. A. Orr. 2000. Dominance, epistasis and the genetics of postzygotic isolation. *Genetics* 154: 1663–1679. [17]

Turelli, M., N. H. Barton, and J. A. Coyne. 2001. Theory and speciation. *Trends Ecol. Evol.* 16: 330–343. [17]

Turley, P., and 19 others. 2018. Multi-trait analysis of genome-wide association summary statistics using MTAG. *Nat. Genet.* 50: 229–237. [20

Turner, H. N., and S. S. Y. Young. 1969. *Quantitative genetics in sheep breeding*. Cornell University Press, Ithaca, NY. [1]

Turner, J. R. G. 1977. Butterfly mimicry: the general evolution of adaptation. *Evol. Biol.* 10: 163–206. [15]

Turner, J. R. G. 1981. Adaptation and evolution in *Heliconius*: a defense of neo-Darwinism. *Ann. Rev. Ecol. Syst.* 12: 99–121. [15]

Turton, J. D. 1981. Crossbreeding of dairy cattle — a selective review. *Anim. Breed. Abst.* 49: 293–300. [12, 13]

Tuttle, E. M. 2003. Alternative reproductive strategies in the white-throated sparrow: behavioral and genetic evidence. *Behav. Ecol.* 14: 425–432. [7]

Tversky, A., and D. Kahneman. 1971. Belief in the law of small numbers. *Psychol. Bull.* 76: 105–110. [18]

Uddin, M. N., F. W. Ellison, L. O'Brien, and B. D. H. Latter. 1994. The performance of pure lines derived from heterotic bread wheat hybrids. *Aust. J. Agric. Res.* 45: 591–600. [12]

Udler, M. S., J. Tyrer, and D. E. Easton. 2010. Evaluating the power to discriminate between highly correlated SNPs in genetic association studies. *Genet. Epidem.* 34: 463–468. [21]

Uffelmann, E., and 8 others. 2021. Genome-wide association studies. *Nat. Rev. Methods Primers* 1: 1–21. [20]

Uimari, P., G. Thaller, and I. Hoeschele. 1996. The use of multiple markers in a Bayesian method for mapping quantitative trait loci. *Genetics* 143: 1831–1842. [19]

Ullstrup, A. J. 1972. The impacts of the southern corn leaf blight epidemics of 1970-71. *Ann. Rev. Phytopathology* 10: 37–50. [13]

Urbut, S. M., G. Wang, P. Carbonetto, and M. Stephens. 2019. Flexible statistical methods for estimating and testing effects in genomics studies with multiple conditions. *Nat. Genet.* 51: 187–195. [21]

Utz, H. F., and A. E Melchinger. 1994. Comparison of different approaches to interval mapping of quantitative trait loci. *In* J. W. van Ooijen and J. Jansen (eds.), *Biometrics in plant breeding: applications of molecular markers*, pp. 195–204. CPRO-DLO, NL. [18]

Uyenoyama, M. K. 1993. Genetic incompatibility as a eugenic mechanism. *In* N. W. Thornhill (ed.), *The natural history of inbreeding and outbreeding: theoretical and empirical perspectives*, pp. 60–73. Univ. Chicago Press, Chicago, IL. [12]

Uyenoyama, M. K., K. E. Holsinger, and D. M. Waller. 1994. Ecological and genetic factors directing the evolution of self-fertilization. *Oxford Surv. Evol. Biol.* 9: 327–381. [12]

Uzunovíc, J., E. B. Josephs, J. R. Stinchcombe, and S. I. Wright. 2019. Transposable elements are important contributors to standing variation in gene expression in *Capsella grandifora*. *Mol. Biol. Evol.* 36: 1734–1745. [21]

Valdar, W., J. Flint, and R. Mott. 2003. QTL fine-mapping with recombinant-inbred heterogeneous stocks and in vitro heterogeneous stocks. *Mamm. Genome* 14: 830–838. [18]

Valdar, W., J. Flint, and R. Mott. 2006a. Simulating the collaborative cross: power of quantitative trait loci detection and mapping resolution in large sets of recombinant inbred strains of mice. *Genetics* 72: 1783–1797. [18]

Valdar, W., and 9 others. 2006b. Genome-wide genetic association of complex traits in heterogeneous stock mice. *Nat. Genet.* 8: 879–887. [18]

Valdar, W., C. C. Holmes, R. Mott, and J. Flint. 2009. Mapping in structured populations by resample model averaging. *Genetics* 182: 1263–1277. [18]

Valentine, D. W., and M. E. Soulé. 1973. Effects of $p \cdot p'$-*DDT* on developmental stability of pectoral fin rays in the grunnion, *Leuresthestenuis*. *Fish. Bull. U.S.* 71: 921–926. [6]

Vallejo, R. L., and 7 others. 2010. Evidence of major genes affecting resistance to bacterial cold water disease in rainbow trout using Bayesian methods of segregation analysis. *J. Anim. Sci.* 88: 3814–3832. [16]

Vallejos, C. E., and S. D. Tanksley. 1983. Segregation of isozyme markers and cold tolerance of an interspecific backcross in tomato. *Theor. Appl. Genet.* 66: 241–247. [18]

Vallejos, C. E., N. S. Sakiyama, and C. D. Chase. 1992. A molecular marker-based linkage map of *Phaseolus vulgaris* L. *Genetics* 131: 733–740. [11]

Van Aarde, I. M. R. 1975. The covariance of relatives derived from a random mating population. *Theor. Pop. Biol.* 8: 166–183. [7]

van Arendonk, J. A. M., C. Smith, and B. W. Kennedy. 1989. Method to estimate genotype probabilities at individual loci in farm livestock. *Theor. Appl. Genet.* 78: 735–740. [16]

van Arendonk, J. A. M., B. Teir, and B. P. Kinghorn. 1994. Use of multiple genetic markers in prediction of breeding values. *Genetics* 137: 319–329. [19]

Van Belleghem, S. M., and 4 others. 2020. Perfect mimicry between *Heliconius* butterflies is constrained by genetics and development. *Proc. Roy. Soc. B* 287: 20201267. [15]

van de Bunt, M., and 5 others. 2015. Evaluating the performance of fine-mapping strategies at common variant GWAS loci. *PLoS Genet.* 11: e1005535. [21]

van de Casteele, T., P. Galbusera, and E. Matthysen. 2001. A comparison of microsatellite-based pairwise relatedness estimators. *Mol. Ecol.* 10: 1539–1549. [8]

van der Beck, S., J. A. M. van Arendonk, and A. F. Groen. 1995. Power of two- and three-generation QTL mapping experiments in an outbred populations containing full-sib or half-sib families. *Theor. Appl. Genet.* 91: 1115– 1124. [19]

van der Sluis, S., D. Posthuma and C. V. Dolan, C.V. 2013. TATES: efficient multivariate genotype-phenotype analysis for genome-wide association studies. *PLoS Genet* 9: e1003235. [20]

Van der Veen, J. H. 1959. Tests of non-allelic interaction and linkage for quantitative characters in generations derived form two diploid pure lines. *Genetics* 30: 201–232. [5]

Van Dongen, S. 2006. Prior specification in Bayesian statistics: three cautionary tales. *J. Theor. Biol.* 242: 90–100. [A7]

Van Eerdewegh, P., C. L. Hampe, B. K. Suarez, and T. Reich. 1993. Alzheimer's disease: a piscatorial trek. *Genet. Epidem.* 10: 395–400. [19]

van Heerwaarden, J., M. B. Hufford, and J. Ross-Ibarra. 2012. Historical genomics of North American maize. *Proc. Natl. Acad. Sci. USA* 109: 12420–12425. [13]

Van Horn, R. C., J. Altmann, and S. C. Alberts. 2008. Can't get there from here: inferring kinship from pairwise genetic relatedness. *Am. Behav.* 75: 1173–1180. [8]

van Kaam, J. B. C. H. M, and 7 others. 1998. Whole genome scan for quantitative trait loci affecting body weigh in chickens using a three generation design. *Livestock Prod. Sci.* 54: 133–150. [19]

van Noordwijk, A. J., and W. Scharloo. 1981. Inbreeding in an island population of the great tit. *Evolution* 35: 674–688. [12]

Van Noort, V., B. Snel, and M. A. Huynen. 2004. The yeast coexpression network has a small-world, scale-free architecture and can be explained by a simple model. *EMBO Reports* 5: 280–284. [A2]

van Ooijen, J. W. 1992. Accuracy of mapping quantitative trait loci in autogamous species. *Theor. Appl. Genet.* 84: 803–811. [18]

van Ooijen, J. W. 1994. Comparison of a single-QTL model with an approximate multiple-QTL model for QTL mapping. *In*, J. W. van Ooijen and J. Jansen (eds.), *Biometrics in plant breeding: applications of molecular markers*, pp. 205–212. CPRO-DLO, NL. [18]

Van Treuren, R., R. Bijlsma, N. J. Ouborg, and W. Van Delden. 1993. The effect of population size and plant density on outcrossing rates in *Salvia pratensis*. *Evolution* 47: 1094–1104. [12]

Van Valen, L. 1962. A study of fluctuating asymmetry. *Evolution* 16: 125–142. [6]

Vandenberg, S. G. 1972. Assortative mating, or who marries whom? *Behav. Genet/* 2: 127–157. [7]

VanderWeele, T. J., E. J. Tchetgen Tchetgen, M. Cornelis, and P. Kraft. 2014. Methodological challenges in Mendelian randomization. *Epidem.* 25: 427–435. [A2]

VanKuren, N. W., D. Massardo, S. Nallu, and M. R. Kronforst. 2019. Butterfly mimicry polymorphisms highlight phylogenetic limits of gene reuse in the evolution of diverse adaptations. *Mol. Biol. Evol.* 36: 2842–2853. [15]

VanRaden, P. M. 2007. Genomic measures of relationship and inbreeding. *INTERBULL Bull.* 37: 33–36. [8, 9]

VanRaden, P. M. 2008. Efficient methods to compute genomic predictions. *J. Dairy Sci.* 91: 4414–4423. [8, 9]

VanRaden, P. M., K. M. Olson, D. J. Null, and J. L. Hutchison. 2011. Harmful recessive effects on fertility detected by absence of homozygous haplotypes. *J. Dairy Sci.* 94: 6153–6161. [12]

van't Hof, A. E., and 8 others. 2016. The industrial melanism mutation in British peppered moths is a transposable element. *Nature* 534: 102–105. [15]

Vassilieva, L., and M. Lynch. 1999. Accumulation of spontaneous mutations in *Caenorhabditis elegans*. *Genetics* 151: 119–129. [15]

Vassilieva, L., A. M. Hook, and M. Lynch. 2000. The fitness effects of spontaneous mutations in *Caenorhabditis elegans*. *Evolution* 54: 1234–1246. [15]

Vaughn, T. T., and 6 others. 1999. Mapping quantitative trait loci for murine growth: a closer look at genetic architecture. *Genet. Res.* 74: 313–322. [18]

Veldboom, L. R., M. Lee, and W. L. Woodman. 1994. Molecular marker- facilitated studies in an elite maize population. I. Linkage analysis and determination of QTL for morphological traits. *Theor. Appl. Genet.* 88: 7–16. [17]

Velicer, W. F. 1976. Determining the number of components from the matrix of partial correlations. *Psychometrika* 41: 321–327. [20]

Venuprasad, R., and 7 others. 2009. Identification and characterization of large-effect quantitative trait loci for grain yield under lowland drought stress in rice using bulk-segregant analysis. *Theor. Appl. Genet.* 120: 177–190. [17]

Ver Hoef, J. M., and P. L. Boveng. 2007. Quasi-Poisson vs. negative binomial regression: how should we model overdispersed count data. *Ecology* 88: 2766–2772. [14]

Verbanck, M., C.-Y. Chen, B. Neale, and R. Do. 2018. Detection of widespread horizontal pleiotropy in causal relationship inferred from Mendelian randomization between complex traits and disease. *Nat. Genet.* 50: 693–698. [A2]

Verbyla, A. P., B. R. Cullis, and R. Thompson. 2007. The analysis of QTL by simultaneous use of the full linkage map. *Theor. Appl. Genet.* 116: 95–111. [18]

Verhoeven, K. J. F., J.-L. Jannink, and L. M. McIntyre. 2006. Using mating designs to uncover QTL and the genetic architecture of complex traits. *Heredity* 96: 139–149. [18]

Verhoeven, K. J. F., G. Casella, and L. M. McIntyre. 2010. Epistasis: obstacle or advantage for mapping complex traits? *PLoS One* 5: e12264. [18]

Vernet, A., and 6 others. 2022. High-frequency synthetic apomixis in hybrid rice. *Nat. Comm.* 13: 7963. [13]

Vernikos, G., D. Medini, D. R. Riley, and H. Tettelin. 2015. Ten years of pan-genome analyses. *Curr. Opin. Microbio.* 23: 148–154. [13]

Vetta, A. 1976. Dominance variance in Fisher's model of assortative mating. *Ann. Hum. Genet.* 39: 447–453. [7]

Vetta, A., and C. A. B. Smith. 1974. Comments on Fisher's theory of assortative mating. *Ann. Hum. Genet.* 38: 243–248. [7]

Via, S. 1984. The quantitative genetics of polyphagy in an insect herbivore. I. Genotype-environment interaction in larval performance on different host plant species. *Evolution* 38: 881–895. [6]

Vidal, M., M. E. Cusick, and A.-L. Barabási. 2011. Interactome networks and human disease. *Cell* 144: 986–998. [21, A2]

Vieira, F. G., M. Fumagalli, A. Albrechsten, and R. Nielsen. 2013. Estimating inbreeding coefficient from NGS data: Impact of genotype calling and allele frequency estimation. *Genome Res.* 23: 1852–1861. [8]

Vieira, F. G., A. Albreschtsen, and R. Nielsen. 2016. Estimating IBD tracts from low coverage NGS data. *Bioinformatics* 32: 2096–2102. [8]

Vieland, V. J., and S. E. Hodge. 1995. Inherent intractability of the ascertainment problem for pedigree data: a general likelihood framework. *Am. J. Hum. Genet.* 56: 33–43. [16]

Vigne, P., and 11 others. 2021. A single-nucleotide change underlies the genetic assimilation of a plastic trait. *Sci. Adv.* 7: eabd9941. [14]

Vikram, P., and 7 others. 2012. Bulk segregant analysis: An effective approach for mapping consistent-effect drought grain yield QTLs in rice. *Field Crops Res.* 134: 185–192. [17]

Vilkki, H. J., and 4 others. 1997. Multiple marker mapping of quantitative trait loci of Finnish dairy cattle by regression. *J. Dairy Sci.* 80: 198–204. [19]

Villanueva, B., and 7 others. 2021. The value of genomic relationship matrices to estimate levels of inbreeding. *Genet. Sel. Evol.* 53: 1–17. [12]

Virdee, S. 1993. Unraveling Haldane's rule. *Trends Ecol. Evol.* 8: 185–187. [17]

Virmani, S. S. 1999. Exploitation of heterosis for shifting the yield frontier in rice. *In* J. G. Coors and S. Pandey (eds.), *The genetics and exploitation of heterosis in crops*, pp. 423–438. ASA-CSSA-SSSA Societies, Madison, WI. [13]

Vision, T. J., and 4 others. 2000. Selective mapping; a strategy for optimizing the construction of high-density linkage maps. *Genetics* 155: 407–420. [18]

Visscher, P. M. 1996. Proportion of the variation in genetic composition in backcrossing programs explained by genetic markers. *J. Heredity* 87: 136–138. [8, 17]

Visscher, P. M. 1999. Speed congenics: accelerated genome recovery using genetic markers. *Genet. Res.* 74: 81–85. [17]

Visscher, P. M. 2006. A note on the asymtotic distribution of likelihood ratio tests to test variance components. *Twin Res. Hum. Genet.* 9: 490–495. [A4]

Visscher, P. M. 2009. Whole genome approaches to quantitative genetics. *Genetica* 136: 351–359. [8]

Visscher, P. M., and D. L. Duffy. 2006. The value of relatives with phenotype but missing genotypes in associations studies for quantitative traits. *Genet. Epidem.* 30: 30–36. [20]

Visscher, P. M., and M. E. Goddard. 2004. Prediction of the confidence interval of quantitative trait loci location. *Behav. Genet.* 34: 477–482. [18]

Visscher, P. M., and C. S. Haley. 1996. Detection of putative quantitative trait loci in line crosses under infinitesimal genetic models. *Theor. Appl. Genet.* 93: 691–702. [18]

Visscher, P. M., and J. L. Hopper. 2001. Power of regression and maximum likelihood methods to map QTL from sib-pair and DZ twin data. *Ann. Hum. Genet.* 65: 583–601. [19]

Visscher, P. M., and D. Posthuma. 2010. Statistical power to detect genetic loci affecting environmental sensitivity. *Behav. Genet.* 40: 728–733. [18]

Visscher, P. M., C. S. Haley, and S. A. Knott. 1996a. Mapping QTLs for binary traits in backcross and F_2 populations. *Genet. Res.* 68: 55–63. [18]

Visscher, P. M., R. Thompson, and C. S. Haley. 1996b. Confidence intervals in QTL mapping by bootstrapping. *Genetics* 143: 1013–1020. [18]

Visscher, P. M., C. S. Haley, and R. Thompson. 1996c. Marker-assisted introgression in backcross breeding programs. *Genetics* 144: 1923–1932. [17]

Visscher, P. M., and 4 others. 1999. Detecting QTLs for uni- and bipolar disorder using a variance component method. *Psychiatric Genet.* 9: 75–84. [19]

Visscher, P. M., D. Smith, S. J. G. Hall, and J. A. Williams. 2001. A viable herd of genetically uniform cattle. *Nature* 409: 303. [12]

Visscher, P. M., J. A. Woolliams, D. Smith, and J. L. Williams. 2002. Estimation of pedigree errors in the UK dairy population using microsatellite markers and the impact on selection. *J. Dairy Sci.* 85: 2368–2375. [8]

Visscher, P. M., and 7 others. 2006. Assumption-free estimation of heritability from genome-wide identity-by-descent sharing between full siblings. *PLoS Genet.* 2: e41. [7, 8, 19, 21]

Visscher, P. M., and 13 others. 2007. Genome partitioning of genetic variation for height from 11,214 sibling pairs. *Am. J. Hum. Genet.* 81: 1104–1110. [8]

Visscher, P, M. A. Brown, M. I. McCarthy, and J. Yang. 2012a. Five years of GWAS discovery. *Am. J. Hum. Genet.* 90: 7–24. [1, 20]

Visscher, P. M., M. E. Goddard, E. M. Derks, and N. R. Wray. 2012b. Evidence based psychiatric genetics, AKA the false dichotomy between common and rare variant hypotheses. *Mol. Psychiatry* 17: 474–485. [21]

Visscher, P. M., and 6 others. 2017. 10 years of GWAS discovery: biology, function, and translation. *Am. J. Hum. Genet.* 101: 5–22. [1, 20]

Vitalis, R. and D. Couvet. 2001. Estimation of effective population size and migration rate from one- and two-locus identity measures. *Genetics* 157: 911–925. [12]

Vitezica, Z. G., L. Varona. and A. Legarra. 2013. On the additive and dominant variance and covariances of individuals within the genome selection scope. *Genetics* 195: 1223–1230. [5]

Vitezica, Z. G., A. Legarra, M. A. Toro, and L. Varona. 2017. Orthogonal estimates of variance for additive dominance, and epistatic effects in populations. *Genetics* 206: 1297–1307. [5]

Vogel, F., and G. W. Motulsky. 1996. *Human Genetics. Problems and approaches*, 3rd Ed. Springer, Berlin, DE. [12]

Vogel, S. 1981. *Life in moving fluids: the physical biology of flow.* Princeton Univ. Press, Princeton, NJ. [14]

Vogel, S. 1989. *Life's devices: the physical world of animals and plants.* Princeton Univ. Press, Princeton, NJ. [14]

Vogl, C., A. Karhu, G. Moran, and O. Savolainn. 2002. High resolution analysis of mating systems: inbreeding in natural populations of *Pinus radiata*. *J. Evol. Biol.* 15: 433–439. [8]

Voight, B. F., and J. K. Pritchard. 2005. Confounding from cryptic relatedness in case-control association studies. *PLoS Genet.* 1: e32. [20]

Vøllestad, L. A., K. Hindar, and A. P. Møller. 1999. A meta-analysis of fluctuating asymmetry in relation to heterozygosity. *Heredity* 83: 206-218. [6]

Võsa, U., et al. 2021. Large-scale *cis*- and *trans*-eQTL analysis identify thousands of genetic loci and polygenic scores that regulate blood gene expression. *Nat. Genet.* 53: 1300–1310. [21]

Vrijenhoek, R. C., and S. Lerman. 1982. Heterozygosity and developmental stability under sexual and asexual breeding systems. *Evolution* 36: 768–776. [6]

Vu, G. T. H., and 6 others. 2010. Fine mapping and syntenic integration of the semi-dwarfing gene *sdw3* of barley. *Funct. Integ. Genomics* 10: 509–521. [17]

Wacholder, S., N. Rothman, and N. Caporaso. 2000. Population stratification in epidemiologic studies of common genetic variants and cancer: quantification of bias. *J. Natl. Cancer Inst.* 92: 1151–1158. [20]

Wacholder, S., and 4 others. 2004. Assessing the probability that a positive report is false: an approach for molecular epidemiology studies. *J. Natl. Cancer Instit.* 6: 434–442. [A6]

Waddington, C. H. 1942. Canalization of development and the inheritance of acquired characters. *Nature* 150: 563–565. [14]

Waddington, C. H. 1952. Selection of the genetic basis for an acquired character. *Nature* 169: 278. [14]

Waddington, C. H. 1953. Genetic assimilation of an acquired character. *Evolution* 7: 118–126. [14]

Waddington, C. H. 1956. Genetic assimilation of the bithorax phenotype. *Evolution* 7: 118–126. [14]

Waddington, C. H. 1957. *The strategy of the genes.* Macmillan, New York, NY. [14, 15]

Waddington, C. H. 1959. Canalization of development and genetic assimilation of acquired characters. *Nature* 183: 1654–1655. [14]

Wade, M. J. 1992. Epistasis. *In* E. F. Keller and E. A. Lloyd (eds), *Keywords in evolutionary biology*. pp. 87–91. Harvard Univ. Press, Cambridge, MA. [5]

Wade, M. J., and N. W. Chang. 1995. Increased male fertility in *Tribolium confusum* beetles after infection with the intracellular parasite *Wolbachia*. *Nature* 373: 72–74. [6]

Wade, M. J., N. A. Johnson, and G. Wardle. 1994. Analysis of autosomal polygenic variation for the expression of Haldane's rule in flour beetles. *Genetics* 138: 791–799. [17]

Wagner, G. P., G. Booth, and H. Bagheri-Chaichian. 1997. A population genetic theory of canalization. *Evolution* 51: 329–347. [14]

Wainberg, M., and 14 others. 2019. Transcriptome-wide association studies: opportunities and challenges. *Nat. Genet.* 51: 592–599. [21]

Wainschtein, P., and 51 others. 2022. Assessing the contribution of rare variants to complex trait heritability from whole-genome sequence data. *Nat. Genet.* 54: 263–273. [21]

Wainwright, S. A., W. D. Biggs, J. D. Currey, and J. M. Gosline. 1982. *Mechanical design in organisms.* Princeton Univ. Press, Princeton, NJ. [14]

Wakefield, J. 2007. A Bayesian measure of the probability of false discovery in genetic epidemiology studies. *Am. J. Hum. Genet.* 81: 208–227. [20]

Wakefield, J. 2008. Bayes factors for genome-wide association studies: comparison with *p*-values. *Genet. Epid.* 33: 79–86. [20, 21]

Wakefield, J. 2012. Commentary: genome-wide significance thresholds via Bayes factors. *Int. J. Epidem.* 41: 286–291. [20]

Wakeland, E., and 4 others. 1997. Speed congenics: a classic technique in the fast lane (relatively speaking). *Immunology Today* 18: 472–477. [17]

Wald, A. 1943. Tests of statistical hypotheses concerning several parameters when the number of observations is large. *Trans. Am. Math. Soc.* 54: 426–482. [A4]

Wald, A., and J. Wolfowitz. 1940. On a test whether two samples are from the same population. *Ann. Math. Stat.* 11: 147–162. [10]

Wall, J. D., and J. K. Pritchard. 2003. Haplotype blocks and linkage disequilibrium in the human genome. *Nat. Rev. Genet.* 4: 587–597. [5]

Wallace, C. 2013. Statistical testing of shared genetic control for potentially related traits. *Genet. Epidem.* 37: 802–813. [21]

Wallace, C. 2020. Eliciting priors and relaxing the single causal variant assumption in colocalisation analyses. *PLoS Genet.* 16: e1008720. [21]

Wallace, C. 2021. A more accurate method for colocalisation analysis allowing for multiple causal variants. *PLoS Genet.* 17: e1009400. [21]

Wallace, J. G., S. J. Larsson, and E. S. Buckler. 2014a. Entering the second century of maize quantitative genetics. *Heredity* 112: 30–38. [21]

Wallace, J. G., and 5 others. 2014b. Association mapping across numerous traits reveals patterns of functional variation in maize. *PLoS Genet.* 10: e1004845. [21]

Waller, D. M. 1993. The statics and dynamics of mating system evolution. *In* N. W. Thornhill (ed.), *The natural history of inbreeding and outbreeding: theoretical and empirical perspectives*, pp. 97–117. Univ. Chicago Press, Chicago, IL. [12]

Waller, D. M. 2021. Addressing Darwin's dilemma: can pseudo-overdominance explain persisting inbreed depression and load? *Evolution* 75: 779–793. [12]

Waller, D. M., J. Dole, and A. J. Bersch. 2008. Effects of stress and phenotypic variation on inbreeding depression in *Brassica rapa*. *Evolution* 62: 917–931. [12]

Walling, C. A., J. M. Pemberton, J. D. Hadfield, and L. E. B. Kruuk. 2010. Comparing parentage inference software: reanalysis of a red deer pedigree. *Mol. Ecol.* 19: 1914–1928. [8]

Walling, G. A., P. M. Visscher, and C. S. Haley. 1998. A comparison of bootstrap methods to construct confidence interval in QTL mapping. *Genet. Res.* 71: 171–180. [18]

Walling, G. A., and 4 others. 2002. On the mapping of quantitative trait loci at marker and non-marker locations. *Genet. Res.* 79: 97–106. [18]

Walsh, B., and M. Lynch. 2018. *Evolution and selection of quantitative traits*. Oxford University Press, Oxford, UK. [Most chapters]

Wan, Y., J. Cohen, and R. Guerra. 1997. A permutation test for the robust sib-pair linkage method. *Ann. Hum. Genet.* 61: 79–87. [19]

Wan, Y., and 4 others. 1998. Genetic linkage analysis using lognormal variance components. *Ann. Hum. Genet.* 62: 521–530. [19]

Wang, A. D., N. P. Sharp, and A.l F. Agrawal. 2014. Sensitivity of the distribution of mutational fitness effects to environment, genetic background, and adaptedness: a case study with *Drosophila*. *Evolution* 68: 840–853. [15]

Wang, B., and P. W. Chee. 2010. Application of advanced backcross quantitative trait locus (QTL) analysis in crop improvement. *J. Plant Breeding Crop Sci.* 2: 221–232. [18]

Wang, B., S. Sverdlov, and E. Thompson. 2017. Efficient estimation of realized kinship from single nucleotide polymorphism genotypes. *Genetics* 205: 1063–1078. [8]

Wang, D., and 4 others. 2001. Transformation of sib-pair values for the Haseman-Elston method. *Am. J. Hum. Genet.* 68: 1238–1249. [19]

Wang, G. G., C. D. Allis, and P. Chi. 2007. Chromatin remodeling and cancer, part I: covalent histone modifications. *Trends Mol. Med.* 13: 363–372. [21]

Wang, G., A. Sarkar, P. Carbonetto, and M. Stephens. 2020. A simple new approach to variable selection in regression, with application to genetic fine mapping. *J. Roy. Stat. Soc. B* 82: 1273–1300. [21]

Wang, G., and S. Xu. 2019. Statistical power in genome-wide association studies and quantitative trait locus mapping. *Heredity* 123: 287–306. [20]

Wang, H., and 6 others. 2005a. Bayesian shrinkage estimation of quantitative trait loci parameters. *Genetics* 170: 465–480. [18]

Wang, H., and 8 others. 2005b. The origin of the naked grains of maize. *Nature* 436: 714–719. [5, 18]

Wang, H., and 4 others. 2015. Evidence that the origin of naked kernels during maize domestication was caused by a single amino acid substitution in *tga1*. *Genetics* 200: 965–974. [5, 18]

Wang, J. 2002. An estimator for pairwise relatedness using molecular markers. *Genetics* 160: 1203–1215. [8]

Wang, J. 2004. Sibship reconstruction from genetic data with typing errors. *Genetics* 166: 1963–1979. [8]

Wang, J. 2007. Parentage and sibship exclusions: higher statistical power with more family members. *Heredity* 99: 205–217. [8]

Wang, J. 2011a. A new likelihood estimator and its comparison with moment estimators of individual genome-wide diversity. *Heredity* 107: 433–443. [8]

Wang, J. 2011b. Unbiased relatedness estimation in structured populations. *Genetics* 187: 887–901. [8]

Wang, J. 2012. Computationally efficient sibship and parentage assignment from multilocus marker data. *Genetics* 191: 183–194. [8]

Wang, J. 2014. Marker-based estimates of relatedness and inbreeding coefficients: an assessment of current methods. *J. Evol. Biol.* 27: 518–530. [8]

Wang, J. 2016. Pedigrees or markers: Which are better in estimating relatedness and inbreeding coefficient? *Theor. Pop. Biol.* 107: 4–13. [8, 12]

Wang, J., and J. Gai. 2001. Mixed inheritance model for resistance to agromyzid beanfly (*Melanagromyza sojae* Zehntener) in soybeans. *Euphytica* 122: 9–18. [16]

Wang, J., and A. W. Santure. 2009. Parentage and sibship inference from multilocus genotype data under polygamy. *Genetics* 181: 1579–1594. [8]

Wang, J., R. Guerra, and J. Cohen. 1998. A statistically robust variance component approach for quantitative trait linkage analysis. *Ann. Hum. Genet.* 62: 349–359. [19]

Wang, J., W. G. Hill, D. Charlesworth, and B. Charlesworth. 1999. Dynamics of inbreeding depression due to deleterious mutations in small populations: mutation parameters and inbreeding rate. *Genet. Res.* 74: 165–178. [12]

Wang, J., D. W. Podlich, M. Cooper, and I. H. DeLacy. 2001. Power of the joint segregation analysis method for testing mixed major-gene and polygene inheritance models of quantitative traits. *Theor. Appl. Genet.* 103: 804–816. [16]

Wang, J., and 4 others. 2019. Inferring gene-disease association by an integrative analysis of eQTL: genome-wide associations study and protein-protein interaction data. *Hum. Hered.* 83: 117–129. [21]

Wang, K. 2009. Testing for genetic association in the presence of population stratification in genome-wide association studies. *Genet. Epid.* 33: 637–645. [20]

Wang, K., and D. Abbott. 2008. A principal component regression approach to multilocus genetic association studies. *Genet. Epidem.* 32: 108–118. [20]

Wang, K., and J. Huang. 2002. A score-statistic approach for the mapping of quantitative-trait loci with sibships of arbitary size. *Am. J. Hum. Genet.* 70: 412–424. [19]

Wang, K., M. Li, and M. Bucan. 2007. Pathway-based approaches for analysis of genomewide association studies. *Am. J. Hum. Genet.* 81: 1278–1283. [21]

Wang, K., and 27 others. 2009. Diverse genome-wide association studies associate the IL12/IL23 pathway with Crohn disease. *Am. J. Hum. Genet.* 84: 399–405. [21]

Wang, K., M. Li, and H. Hakonarson. 2010. Analysing biological pathways in genome-wide association studies. *Nat. Rev. Genet.* 11: 843–854. [21]

Wang, L., and 5 others. 2015. Hybrid mimics and hybrid vigor in *Arabidopsis*. *Proc. Natl. Acad. Sci. USA* 112: E4959–E4967. [13]

Wang, M., R. Li, and S. Xu. 2020. Deshrinking ridge regression for genome-wide association studies. *Bioinfomratics* 36: 4154–4162. [20]

Wang, M. H., H. J. Cordell, and K. van Steen. 2019. Statistical methods for genome-wide association studies. *Seminars in Cancer Biol.* 55: 53–60. [20]

Wang, P., and 9 others. 2011. A model selection approach for expression quantitative trait loci (eQTL) mapping. *Genetics* 187: 611–621. [21]

Wang, Q., J. Wei, and S. Xu. 2015. An efficient empirical Bayes method for genomewide association studies. *J. Anim. Breed. Genet.* 133: 253–263. [20]

Wang, R.-L., and 4 others. 1999. The limits of selection during maize domestication. *Nature* 398: 236–239. [18]

Wang, S., J. J. Hard, and F. Utter. 2002. Salmonid inbreeding: a review. *Rev. Fish Biol. & Fisheries* 11: 301–319. [12]

Wang, S., C. Haynes, F. Barany, and J. Ott. 2009. Genome-wide autozygosity mapping in human populations. *Genet. Epid.* 33: 172–180. [8]

Wang, S., and 12 others. 2016a. Modeling *cis*-regulation with a compendium of genome-wide histone H3K27ac profiles. *Genome Res.* 26: 1417–1429. [21]

Wang, S.-B., and 9 others. 2016b. Improving power and accuracy of genome-wide association studies via a multi-locus mixed linear model methodology. *Sci. Reports* 6: 19444. [20]

Wang, T., and R. C. Elston. 2005. Two-level Haseman-Elston regression for general pedigree data analysis. *Genet. Epid.* 29: 12-22. [19]

Wang, T., and Z.-B. Zeng. 2006. Models and partition of variance for quantitative trait loci with epistasis and linkage disequilibrium. *BMC Genet.* 7: 9. [5]

Wang, T., and Z.-B. Zeng. 2009. Contribution of genetic effects to genetic variance components with epistasis and linkage disequilibrium. *BMC Genet.* 10: 52. [5]

Wang, T., and 6 others. 2020. eQTLMAPT: fast and accurate eQTL mediation analysis with efficient permutation testing approaches. *Frontiers Genet.* 10: 1309. [21]

Wang, W., M. Gerstein, and M. Synder. 2009. RNA-Seq: a revolutionary tool for transcriptomics. *Nat. Rev. Genet.* 10: 57–63. [21]

Ward, P. J. 1993. Some developments on the affected-pedigree-member method of linkage analysis. *Am. J. Hum. Genet.* 52: 1200-1215. [19]

Warton, D. I., and F. K. C. Hui. 2011. The arcsine is asinine: the analysis of proportions in ecology. *Ecology* 92: 3–10. [14]

Warton, D. I., I. J. Wright, D. S. Falster, and M. Westoby. 2006. Bivariate line-fitting methods for allometry. *Biol. Re.* 81: 259–291. [14]

Waser, N. M. 1993. Population structure, optimal outcrossing, and assortative mating in angiosperms. *In* N. W. Thornhill (ed.), *The natural history of inbreeding and outbreeding: theoretical and empirical perspectives,* pp. 1–13. Univ. Chicago Press, Chicago. IL. [13]

Waser, N. M., and M. V. Price. 1983. Optimal and actual outcrossing in plants, and the nature of plant-pollinator interaction. *In* C. E. Jones, and R. J. Little (eds.), *Handbook of experimental pollination biology,* pp. 341–359. Van Nostrand Reinhold, New York, NY. [13]

Waser, N. M., and M. V. Price. 1985. Reciprocal transplant experiments with *Delphinium nelsonii* (Ranunculaceae). Evidence for local adaptation. *Am. J. Bot.* 72: 1726–1732. [13]

Waser, N. M., and M. V. Price. 1994. Crossing-distance effects in *Delphinium nelsonii:* outbreeding and inbreeding depression in progeny fitness. *Evolution* 48: 842–852. [13]

Washburn, J. D., and J. A. Birchler. 2014. Polyploids as a "model system" for the study of heterosis. *Plant. Reprod.* 27: 1–5. [13]

Washburn, J. D., M. J. McElfresh, and J. A. Brichler. 2019. Progressive heterosis in a genetically defined tetraploid maize. *J. Genet. Genomics* 46: 389–396. [13]

Wassom, J. J., V. Mikkelineni, M. O. Bohn, and T. R. Rocheford. 2008a. QTL for fatty acid composition of maize kernel oil in Illinois High Oil x B73 backcross-derived lines. *Crop Sci.* 48: 69–78. [18]

Wassom, J. J., and 8 others. 2008b. QTL associated with maize kernel oil, protein, and starch concentrations; kernel mass; and grain yield in Illinois High Oil x B73 backcross-derived lines. *Crop Sci.* 48: 243–252. [18]

Watanabe, K., and 4 others. 2019. Genetic mapping of cell type specificity of complex traits. *Nat Comm.* 10: 3222. [21]

Watanabe, S., and 11 others. 2009. Map-based cloning of the gene associated with soybean maturity locus *E3*. *Genetics* 182: 1251–1262. [13]

Watanabe, T. K., O. Yamaguchi, and T. Mukai. 1976. The genetic variability of third chromosomes in a local population of *Drosophila melanogaster. Genetics* 82: 63–82. [12]

Waters, N. F. 1931. Inheritance of body weight in domestic fowls. *Rhode Island Agric. Exp. Stn. Bull.* 228: 7–103. [11]

Watson, A., and 10 others. 2018. Speed breeding is a powerful tool to accelerate crop research and breeding. *Nat. Plants* 4: 23–29. [17]

Watson, E. T., and J. P. Demuth. 2012. Haldane's rule in marsupials: what happens when both sexes are functionally hemizygous? *J. Heredity* 103: 453–458. [17]

Watson, J. D., and F. H. Crick. 1953. Molecular structure of nucleic acids: a structure for deoxyribose nucleic acid. *Nature* 171: 737–738. [1]

Wayne, M. L., and L. M. McIntyre. 2002. Combining mapping and arraying: an approach to candidate gene identification. *Proc. Natl. Acad. Sci. USA* 99: 14903–14906. [21]

Wayne, R. K., W. S. Modi, and S. J. O'Brien. 1986. Morphological variability and asymmetry in the cheetah *(Acinonyx jubatus),* a genetically uniform species. *Evolution* 40: 78–85. [6]

Weber, K. E. 1992. How small are the smallest selectable domains of form? *Genetics* 130: 345–353. [15]

Weber, K. E., and L. T. Diggins. 1990. Increased selection response in larger populations. II. Selection for ethanol vapor resistance in *Drosophila melanogaster* at two population sizes. *Genetics* 125: 585–597. [15]

Weber, W. E. 1984. Selection in early generations. *In* W. Lange, A. C. Zeven, and N. G. Hogenboom (eds.), *Efficiency in plant breeding: Proceedings 10th congress of the European assocation for research in plant breeding, EUCARPIA*, pp. 72–81. Wageningen, NL. [A9]

Weeks, D. E., and K. Lange. 1988. The affected-pedigree-member method of linkage analysis. *Am. J. Hum. Genet.* 42: 315–326. [19]

Weeks, D. E., and K. Lange. 1992. A multilocus extension of the affected-pedigree-member method of linkage analysis. *Am. J. Hum. Genet.* 50: 859–868. [19]

Weeks, D. E., and G. M. Lathrop. 1995. Polygenic disease: methods for mapping complex disease traits. *Trends Genet.* 11: 513–519. [19]

Wehner, T. C. 1999. Heterosis in vegetable crops. *In* J. G. Coors and S. Pandey (eds.), *The genetics and exploitation of heterosis in crops*, pp. 287–398. ASA-CSSA-SSSA Societies, Madison, WI. [13]

Wehrhahn, C., and R. W. Allard. 1965/ The detection and measurement of the effects of individual genes involved in the inheritance of a quantitative character. *Genetics* 51: 109–119. [11, 17]

Wei, J., and S. Xu. 2016. A random-model approach to QTL mapping in multiparent advanced generation intercross (MAGIC) populations. *Genetics* 202: 471–486. [18]

Wei, W.-H., S. Knott, C. S. Haley, and D.-J. de Koning. 2010. Controlling false positives in the mapping of epistatic QTL. *Heredity* 104: 401–409. [18]

Wei, W.-H., G. Hemani, and C. S. Haley. 2014. Detecting epistasis in human complex traits. *Nat. Rev. Genet.* 15: 722–733. [5, 20]

Wei, X., and 8 others. 2015. Heterotic loci for various morphological traits of maize detected using a single segment substitution lines test-cross population. *Mol. Breed.* 35: 94. [13]

Weigensberg, I., and D. A. Roff. 1996. Natural heritabilities: can they be reliably estimated in the laboratory? *Evolution* 50: 2149–2157. [7]

Weinberg, W. 1908. Ueber den Nachweis der Vererbung beim Menschen. *Jh. Ver. Vaterl. Naturk. Wurttemb.* 64: 368–382. [1, 4]

Weinberg, W. 1927. Mathematische grundlagen der probandenmethode. *Z. Induktive Abstammungs Vererbungslehre* 48: 179–228. [16]

Weiner, D. J., et al. 2017. Polygenic transmission disequilibrium confirms that common and rare variation act additively to create risk for autism spectrum disorders. *Nat. Genet.* 49: 978–985. [17]

Weir, B. S. 1979. Inferences about linkage disequilibrium. *Biometrics* 35: 235–254. [5]

Weir, B. S. 2008. Linkage disequilibrium and association mapping. *Annu. Rev. Genomics Hum. Genet.* 9: 129-142. [5]

Weir, B. S. 2010. *Genetic data analysis, 3rd Ed.* Sinauer Assoc., Sunderland, MA. [4, 5, A4]

Weir, B. S., and C. C. Cockerham. 1968. Pedigree mating with two linked loci. *Genetics* 61: 923–940. [7]

Weir, B. S., and C. C. Cockerham. 1969. Group inbreeding with two linked loci. *Genetics* 63: 711–742. [7]

Weir, B. S., and C. C. Cockerham. 1973. Mixed self and random mating at two loci. *Genet. Res.* 21: 247–262. [7, 12]

Weir, B. S., and C. C. Cockerham. 1974. Behavior of pairs of loci in finite monoecious populations. *Theor. Pop. Biol.* 6: 323–354. [7]

Weir, B. S., and C. C. Cockerham. 1977. Two-locus theory in quantitative genetics. *In* E. Pollak, O. Kempthorne, and T. B. Bailey, Jr. (eds.), *Proceedings of the international conference on quantitative genetics*, pp. 247–269. Iowa State Univ. Press, Ames, IA. [5, 7]

Weir, B. S., and C. C. Cockerham. 1979. Estimation of linkage disequilibrium in randomly mating populations. *Heredity* 42: 105–111. [5]

Weir, B. S., and C. C. Cockerham. 1989. Complete characterization of disequilibrium at two loci. *In* M. W. Feldman (ed.), *Mathematical evolutionary theory*, pp. 86–110. Princeton Univ. Press, Princeton, NJ. [7]

Weir, B. S., and J. Goudet. 2017. A unified characterization of population structure and relatedness. *Genetics* 206: 2085–2103. [8]

Weir, B. S., C. C. Cockerham, and J. Reynolds. 1980. The effects of linkage and linkage disequilibrium on the covariances of noninbred relatives. *Heredity* 45: 351–359. [5, 7]

Weir, B. S., A. D. Anderson, and A. B. Hepler. 2006. Genetic relatedness analysis: modern data and new challenges. *Nat. Rev. Genet.* 7: 771–780. [8]

Weiser, E. L., C. E. Grueber, E. S. Kennedy, and I. G. Jamieson. 2016. Unexpected positive and negative effects of continuing inbreeding in one of the world's most inbred wild animals. *Evolution* 70: 154–166. [12]

Weiss, K. M. 1993. *Genetic variation and human disease: principles and evolutionary approaches.* Cambridge Univ. Press, Cambridge, UK. [16, 19]

Weissbrod, O., C. Lippert, D. Geiger, and D. Heckerman. 2015. Accurate liability estimation improves power in ascertained case-control studies. *Nat. Methods* 12: 332–334. [20]

Weissbrod, O., and 13 others. 2020. Functionally informed fine-mapping and polygenic localization of complex trait heritability. *Nat. Genet.* 52: 1355–1363. [21]

Welch, W. J. 1990. Construction of permutation test. *J. Amer. Statist.* 85: 693–698. [18]

Wellcome Trust Case Control Consortium (WTCCC). 2007. Genome-wide association study of 14,000 cases of seven common diseases and 3,000 shared controls. *Nature* 447: 661–676. [1, 20]

Wellek, S., and A. Ziegler. 2009. A genotype-based approach to assessing the association between single nucleotide polymorphisms. *Hum. Hered.* 67: 128–139. [5]

Weller, J. I. 1986. Maximum likelihood techniques for the mapping and analysis of quantitative trait loci with the aid of genetic markers. *Biometrics* 42: 627–640. [18]

Weller, J. I. 1987. Mapping and analysis of quantitative trait loci in *Lycopersicon* (tomato) with the aid of genetic markers using approximate maximum likelihood methods. *Heredity* 59: 413–421. [18]

Weller, J. I. 1990. Experimental designs for mapping quantitative trait loci in segregating populations. *In* W. G. Hill, R. Thompson, and J. A. Woolliams (eds.), *Proc. 4th World Congr. Genet. Appl. Livestock Prod.*, Vol. 13, pp. 113–116. Edinburgh, UK. [19]

Weller, J. I. 1994. *Economic aspects of animal breeding.* Chapman and Hall, London, UK. [1]

Weller, J. I., and A. Wyler. 1992. Power of different sampling strategies to detect quantitative trait loci variance effects. *Theor. Appl. Genet.* 83: 582–588. [18]

Weller, J. I., M. Soller, and T. Bordy. 1988. Linkage analysis of quantitative traits in an interspecific cross of tomato (*Lycopersicon esculentum* × *Lycopersicon pimpinellifolium*) by means of genetic markers. *Genetics* 118: 329–339. [18]

Weller, J. I., Y. Kashi, and M. Soller. 1990. Power of daughter and granddaughter designs for determining linkage between marker loci and quantitative trait loci in dairy cattle. *J. Dairy Sci.* 73: 2525–2537. [19]

Weller, J. I., J. Z. Song, Y. I. Ronin, and A. B. Korol. 1997. Designs and solutions to multiple trait comparisons. *Anim. Biotech.* 8: 107–122. [18]

Weller, J. I., and 4 others. 1998. A new approach to the problem of multiple comparisons in the genetic dissection of complex traits. *Genetics* 150: 1699–1706. [18, A6]

Weller, J. I., M. Shlezinger, and M. Ron. 2005. Correcting for bias in estimation of quantitative trait loci effects. *Genet. Sel. Evol.* 37: 501–522. [21]

Weller, S. G. 1976. The genetic control of tristyly in *Oxalis* section *Ionoxalis*. *Heredity* 37: 387–393. [5]

Wen, X. 2014. Bayesian model selection in complex linear systems, as illustrated in genetic association studies. *Biometrics* 70: 73–83. [21]

Wen, X., Y. Lee, F. Luca, and R. Pique-Regi. 2016. Efficient integrative multi-SNP association analysis via deterministic approximation of posteriors. *Am. J. Hum. Genet.* 98: 1114–1129. [21]

Wen, X., R. Pique-Regi, and F. Luca. 2017. Integrating molecular QTL data into genome-wide genetic association analysis: probabilistic assessment of enrichment and colocalization. *PLoS Genet.* 13: e1006646. [21]

Wenger, J. W., K. Schwartz, and G. Sherlock. 2010. Bulk segregant analysis by high-throughput sequencing reveals a novel xylose utilization gene from *Saccharomyces cerevisiae*. *PLoS Genet.* 6: e1000942. [17]

Werren, J. H. 1993. The evolution of inbreeding in a haplodiploid organisms. *In* N. W. Thornhill (ed.),*The natural history of inbreeding and outbreeding*, pp. 42–94. Univ. Chicago Press, Chicago, IL. [12]

Werren, J. H., S. W. Skinner, and E. L. Charnov. 1981. Paternal inheritance of a daughterless sex ratio factor. *Nature* 293: 467–468. [6]

Werren, J. H., S. W. Skinner, and A. M. Huger. 1986. Male-killing bacteria in a parasitic wasp. *Science* 231: 990–992. [6]

Werren, J. H., L. Baldo, and M. E. Clark. 2008. *Wolbachia*: master manipulators of invertebrate biology. *Nat. Rev. Microbiol.* 6: 741–751. [6]

Wessel, J., and N. J. Schork. 2006. Generalized genome distance-based regression methodology for multilocus association analysis. *Am. J. Hum. Genet.* 79: 792–806. [20]

West, M. A. L., and 6 others. 2007. Global eQTL mapping reveals the complex genetic architecture of transcript-level variation in *Arabidopsis*. *Genetics* 175: 1441–1450. [21]

West-Eberhard, M. J. 2003. *Developmental plasticity and evolution*. Oxford University Press. Oxford, UK. [14]

Westberg, M. 1985. Combining independent statistical tests. *J. Roy. Stat. Soc. D* 34: 287–296. [20, A6]

Westra, H.-J., and 55 others. 2013. Systematic identification of *trans* eQTL as putative drivers of known disease associations. *Nat. Genet.* 45: 1238–1242. [21]

Wetherill, G. B. 1986. *Regression analysis with applications*. Chapman and Hall, London, UK. [14]

Wexelsen, H. 1933. Linkage between quantitative and qualitative characters in barley. *Hereditas* 17: 323–341. [17]

Wexelsen, H. 1934. Quantitative inheritance and linkage in barley. *Hereditas* 18: 307–348. [17]

Whipple, C. J., and 9 others. 2011. *Grassy tillers1* promotes apical dominance in maize and responds to shade signals in the grasses. *Proc. Natl. Acad. Sci. USA* 108: E506–E512. [5, 18]

White, H. 1980. A heteroskedasticity-consistent covariance matrix estimator and a direct test for heteroskedasticity. *Econometrica* 48: 817–838. [10]

White, J. K., and 15 others. 2013. Genome-wide generation and systematic phenotyping of knockout mice reveals new roles for many genes. *Cell* 154: 452–464. [15]

White, J. M. 1972. Inbreeding effects upon growth and maternal ability in laboratory mice. *Genetics* 70: 307–317. [12]

White, M. J. D., and L. E. Andrew. 1960. Cytogenetics of the grasshopper *Moraba scurra*. V. Biometric effects of chromosomal inversions. *Evolution* 14: 284–291. [17]

Whiteley, A. R., S. W. Fitzpatrick, W. C. Funk, and D. A. Tallmon. 2015. Genetic rescue to the rescue. *Trends Ecol. Evol.* 30: 42–49. [13]

Whitlock, M. 1993. Lack of correlation between heterozygosity and fitness in forked fungus beetles. *Heredity* 70: 574–581. [12]

Whitlock, M. C. 2005. Combining probability from independent tests: the weighted Z-method is superior to Fisher's approach. *J. Evol. Biol.* 18: 1368–1373. [20, A6]

Whitlock, M. C., and K. Fowler. 1999. The changes in genetic and environmental variance with inbreeding in *Drosophila melanogaster*. *Genetics* 152: 345–353. [6]

Whitlock, M. C., P. K. Ingvarsson, and T. Hatfield. 2000. Local drift load and the heterosis of interconnected populations. *Heredity* 84: 452–457. [13]

Whittaker, J. C., and C. M. Lewis. 1998. The effect of family structure on linkage test using allelic association. *Am. J. Hum. Genet.* 63: 889–897. [17]

Whittaker, J. C., R. Thompson, and P. M. Visscher. 1996. On the mapping of QTL by regression of phenotypes on marker type. *Heredity* 77: 23–32. [18]

Whittemore, A. S. 1996. Genome scanning for linkage: an overview. *Am. J. Hum. Genet.* 59: 704–716. [19]

Whittemore, A. S., and J. Halpern. 1994. A class of tests for linkage using affected pedigree members. *Biometrics* 50: 118–127. [19]

Wiener, G., G. J. Lee, and J. A. Woolliams. 1992a. Effects of rapid inbreeding and of crossing of inbred lines on the body weight growth of sheep. *Anim. Prod.* 55: 89–99. [12]

Wiener, G., G. J. Lee, and J. A. Woolliams. 1992b. Effects of rapid inbreeding and of crossing inbred lines on the growth of linear body dimensions of sheep. *Anim. Prod.* 55: 101–114. [12]

Wiener, G., G. J. Lee, and J. A. Woolliams. 1992c. Effects of rapid inbreeding and of crossing of inbred lines on conception rate, prolificacy and ewe survival of sheep. *Anim. Prod.* 55: 115–121. [12]

Wienhues, F. 1968. Long-term yield analyses of heterosis in wheat and barley: variability of heterosis, fixation of heterosis. *Euphytica* 17: 49–62. [12]

Wijsman, E. M. 2000. Monte Carlo Markov chain methods and model selection in genetic epidemiology. *Computational Stat. Data Analysis* 32: 349–360. [16]

Wilcockson, R. W., C. S. Crean, and T. H. Day. 1995. Heritability of a sexually selected character expressed in both sexes. *Nature* 374: 158-159. [17]

Wilk, M. B., and R. Gnanadesikan. 1968 Probability plotting methods for the analysis of data. *Biometrika* 55: 1–17. [20]

Wilkens, H. 1971. Genetic interpretation of regressive evolution processes: studies on hybrid eyes of two

Astyanax cave populations (Characidae, Pisces). *Evolution* 25: 530–544. [11]

Wilkie, A. O. 1994. The molecular basis of genetic dominance. *J. Med. Genet.* 31: 89–98. [4]

Wilkinson, B. 1951. A statistical consideration in psychological research. *Psychol. Bull.* 48: 156–158. [A6]

Wilkinson, G. N., S. R. Eckert, T. W. Hanock, and O. Mayo. 1983. Nearest neighbour (NN) analysis of field experiments (with discussion). *J. Roy. Stat. Soc. B* 45: 151–211. [A9]

Wilks, S. S. 1938. The large sample distribution of the likelihood ratio for testing composite hypotheses. *Ann. Math. Statist.* 9: 60–62. [16, A4]

Willham, R. L. 1970. Genetic consequences of crossbreeding. *J. Anim. Sci.* 30: 690 –693. [13]

Willi, Y. 2013. Mutational meltdown in selfng *Arabidopsis lyrata*. *Evolution* 67: 806–815. [13]

William, R. L., and E. Pollak. 1985. Theory of heterosis. *J. Dairy Sci.* 68: 2411–2417. [13]

Williams, E. R. 1986. A neighour model for field experiments. *Biometrika* 73: 279–287. [A9]

Williams, E., H.-P. Piepho, and D. Whitaker. 2011. Augmented p-rep designs. *Biomet. J.* 53: 19–27. [A9]

Williams, E. R., J. A. John, and D. Whitaker. 2014. Construction of more flexible and efficient p–rep designs. *Austral. New Zealand J. Stat.* 56: 89–96. [A9]

Williams, J. G. K., and 4 others. 1990. DNA polymorphisms amplified by arbitrary primers are useful asgenetic markers. *Nucl. Acids Res.* 18: 6531–6535. [8]

Williams, J. T., and J. Blangero. 1999. Power of variance component linkage analysis to detect quantitative trait loci. *Ann. Hum. Genet.* 63: 545–563. [19]

Williams, J. T., and J. Blangero. 2004. Power of variance component linkage analysis — II. Discrete traits. *Ann. Hum. Genet.* 68: 620–632. [19]

Williams, J. T., P. van Eerdewegh, L. Almasy, and J. Blangero. 1999. Joint multipoint linkage analysis of multivariate qualitative and quantitative traits. I. likelihood formulation and simulation results. *Am. J. Hum. Genet.* 65: 1134–1147. [19]

Williams, W. 1959a. Heterosis and the genetics of complex characters. *Heredity* 15: 327–328. [13]

Williams, W. 1959b. The isolation of pure lines from F_1 hybrids of tomato and the problem of heterosis in inbreeding crop species. *J. Agric. Sci.* 53: 347–353. [12]

Williams, W. 1960. Relative variability of inbred lines and F_1 hybrids in *Lycopersicon esculentum*. *Genetics* 45: 1457–1465. [6]

Willis, J. H. 1993. Effects of different levels of inbreeding on fitness components in *Mimulus guttatus*. *Evolution* 47: 864–876. [12]

Willis, J. H. 1999. The role of genes of large effect on inbreeding depression in *Mimulus guttatus*. *Evolution* 53: 1678–1691. [12]

Willis, K., and R. J. Wiese. 1997. Elimination of inbreeding depression from captive populations: Speke's gazelle revisited. *Zoo. Bio.* 16: 9–16. [12]

Willmot, D. B., J. W. Dudley, T. R. Rocheford, and A. Bari. 2006. Effects of random mating on marker-QTL associations for grain quality traits in the cross of Illinois High Oil x Illinois Low Oil. *Maydica* 51: 187–199. [18]

Willoughby, J. R., P. M. Waser, A. Brüniche-Olsen, and M. R. Christie. 2019. Inbreeding load and inbreeding depression estimated from lifetime reproductive success in a small, dispersal-limited population. *Heredity* 123: 192–201. [12]

Wills, C. 1966. The mutational load in two natural populations of *Drosophila pseudoobscura*. *Genetics* 53: 281–294. [12]

Wills, D. M., and 6 others. 2013. From many, one: genetic control of prolificacy during maize domestication. *PLoS Genet.* 9: e1003604. [5, 18]

Willson, M. F., and N. Burley. 1983. *Mate choice in plants: tactics, mechanisms, and consequences*. Monog. Pop. Biol. 19. Princeton Univ. Press, Princeton, NJ. [12]

Wilson, A. C. 1976. Gene regulation in evolution. *In* F. J. Ayala (ed.), *Molecular evolution*, pp. 225–236. Sinauer Assoc., Sunderland, MA. [15]

Wilson, A. F., and R. C. Elston. 1993. Statistical validity of the Haseman-Elston sib-pair test in small samples. *Genet. Epidem.* 10: 593–598. [19]

Wilson, P., and C. J. Driscoll. 1983. Hybrid wheat. *In* R. Frankel (ed.) *Heterosis: Reappraisal of theory and practice*, pp. 94-123. Springer-Verlag, Berlin. DE. [13]

Wilson, S. R. 1973. The correlation between relatives under the multifactorial model with assortative mating. *Ann. Hum. Genet.* 37: 189–215. [7]

Winkelman, D. C., and R. B. Hodgetts. 1992. RFLPs for somatotropic genes identify quantitative trait loci for growth in mice. *Genetics* 131: 929–937. [17]

Winkler, C. R., and 4 others. 2003. On the determination of recombination rates in intermated recombinant inbred populations. *Genetics* 164: 741–745. [18]

Winn, A. A., and 10 others. 2011. Analysis of inbreeding depression in mixed-mating plants provides evidence for selective interference and stable mixed mating. *Evolution* 65: 3339–3359. [12]

Winters, L. M. 1952. Rotational crossbreeding and heterosis. *In* J. W. Gowen (ed.), *Heterosis*, pp. 371–377. Iowa State College Press, Ames. IA. [13]

Winters, L. M., O. M. Kiser, P. S. Jordan, and W. H. Peters. 1935. A six years study of crossbreeding swine. *Minn. Agr. Exp. Stat. Bull.* 320. [13]

Wisser, R. J.,and 4 others. 2008. Selection mapping of loci of quantitative disease resistance in a diverse maize population. *Genetics* 180: 583–599. [17]

Wisser, R. J., and 6 others. 2011. Multivariate analysis of maize disease resistances suggests a pleiotropic genetic basis and implicates a *GST* gene. *Proc. Natl. Acad. Sci. USA* 108: 7339–7344. [20]

Witte, J. S., R. C. Elston, and N. J. Schork. 1996. Genetic dissection of complex traits. *Nat. Genet.* 12: 355–356. [19]

Wittkopp, P. J., B. K. Haerum, and A. G. Clark. 2004. Evolutionary changes in *cis* and *trans* gene regulation. *Nature* 430: 85–88. [21]

Wittkopp, P. J., B. K. Haerum, and A. G. Clark. 2008. Regulatory changes underlying expression differences within and between *Drosophila* species. *Nat. Genet.* 40: 346–350. [21]

WL. See Walsh, B., and M. Lynch. 2018.

Wloch, D. M., K. Szafraniec, R. H. Borts, and R. Korona. 2001. Direct estimate of the mutation rate and the distribution of fitness effects in the yeast *Saccharomyces cerevisiae*. *Genetics* 159: 441–452. [15]

Wocholder, S., and 4 others. 2004. Assessing the probability that a positive report is false: An approach for molecular epidemiology studies. *J. Nat. Cancer Inst.* 96: 434–442. [20]

Wolf, D. P., and R. Hallauer. 1997. Triple testcross analysis to detect epistasis in maize. *Crop Sci.* 37: 763–770. [13]

Wolfe, L. M. 1993. Inbreeding depression in *Hydrophyllum appendiculatum:* role of maternal effects, crowding, and parental mating history. *Evolution* 47: 374–386. [12]

Wolyn, D. J., and 7 others. 2004. Light-response quantitative trait loci identified with composite interval and eXtreme array mapping in *Arabidopsis thaliana. Genetics* 167: 907–917. [17]

Wong, F. L., R. M. Cantor, and J. I. Rotter. 1986. Sample-size considerations and strategies for linkage analysis in autosomal recessive disorders. *Am. J. Hum Genet.* 39: 25–37. [12]

Wong, G. T. 2002. Speed congenics: applications for transgenic and knock-out mouse strains. *Neuropeptides* 36: 230–236. [17]

Wood, A. R., and 9 others. 2014a. Another explanation for apparent epistasis. *Nature* 514: E3–E5. [20]

Wood, A. R., et al. 2014b. Defining the role of common variation in the genomic and biological architecture of adult human height. *Nat. Genet.* 46: 1173–1186. [21]

Woodhouse, M. R., and 6 others. 2021. A pan-genomic approach to genome databases using maize as a model system. *BMC Plant Bio.* 21: 1–10. [13]

Woods, T. B., and F. J. Stratton. 1910. Interpretation of experimental results. *J. Agr. Sci.* 3: 417–559. [A9]

Woolf, B. 1955. On estimating the relation between blood group and disease. *Ann. Hum. Genet.* 19: 251–253. [17]

Woolf, C. M., and A. D. Gianas. 1976. Congenital cleft lip and fluctuating dermatoglyphic asymmetry. *Am. J. Hum. Genet.* 28: 400–403. [6]

Wray, N. R. 2005. Allele frequencies and the r^2 measure of linkage disequilibrium: Impact on design and interpretation of association studies. *Twin Res. Hum. Genet.* 8: 87–94. [5]

Wray, N. R., S. M. Purcell, and P. M. Visccher. 2011. Synthetic associations created by rare variants do not explain most GWAS results. *PLoS Biol.* 9: e1000579. [21]

Wray, N. R., and 31 others. 2012. Genome-wide association study of major depressive disorder: new results, meta-analysis, and lessons learned. *Mol. Psych.* 17: 36–48. [19]

Wray, N. R., and 5 others. 2014. Research review: polygenic methods and their application to psychiatric traits. *J. Child Psychology Psychiatry* 55: 1068–1087. [19]

Wray, N. R., and 4 others. 2018. Common disease is more complex than implied by the core gene omnigenic model. *Cell* 173: 1573–1580. [21]

Wricke, G., and W. E. Weber. 1986. *Quantitative genetics and selection in plant breeding.* Walter de Gruyter and Co., New York, NY. [1, 4, 5, 13]

Wright, A. J. 1981. Quantitative genetics of diploid synthetic varieties. *In* A. Gallais (ed.), *Quantiative genetics and breeding methods*, pp. 137–157. INRA, Lusignan, FR. [13]

Wright, A. J., and R. P. Mowers. 1994. Multiple regression for molecular-marker, quantitative trait data from large F_2 populations. *Theor. Appl. Genet.* 89: 305–312. [18]

Wright, D. A., and F. H. Moyer. 1996. Parental influences on lactate dehydrogenase in the early development of hybrid frogs in the genus *Rana. J. Exper. Zoo.* 163: 215–229. [21]

Wright, F. A. 1997. The phenotypic difference discards sib-pair QTL linkage information. *Am. J. Hum. Genet.* 60: 740–742. [19]

Wright, F. A., and A. Kong. 1997. Linkage mapping in experimental crosses: The robustness of single-gene models. *Genetics* 146: 417–425. [18]

Wright, S. 1918. On the nature of size factors. *Genetics* 3: 367–374. [A2]

Wright, S. 1921a. Correlation and causation. *J. Agric. Res.* 20: 557–585. [1, A2]

Wright, S. 1921b. Systems of mating. I. The biometric relations between parents and offspring. *Genetics* 6: 111–123. [1, 7]

Wright, S. 1921c. Systems of mating. II. The effects of inbreeding on the genetic composition of a population. *Genetics* 6: 111–123. [1]

Wright, S. 1921d. Systems of mating. III. Assortative mating based on somatic resemblance. *Genetics* 6: 144–151. [1, 7]

Wright, S. 1922a. Coefficients of inbreeding and relationship. *Am. Nat.* 56: 330–339. [7, 8]

Wright, S. 1922b. The effects of inbreeding and crossbreeding on guinea pigs. *US Dept. Ag. Bulletin* 1090. [13]

Wright, S. 1926. Effects of age of parents on characteristics of the offspring. *Am. Nat.* 60: 552–559. [6]

Wright, S. 1929a. Fisher's theory of dominance. *Am. Nat.* 63: 274–279. [4]

Wright, S. 1929b. The evolution of dominance: comment on Dr. Fisher's reply. *Am. Nat.* 63: 556–561. [4]

Wright, S. 1932. General, group and special size factors. *Genetics* 17: 603–619. [A2]

Wright, S. 1934a. Physiological and evolutionary theories of dominance. *Am. Nat.* 68: 25–53. [4]

Wright, S. 1934b. Professor Fisher on the theory of dominance. *Am. Nat.* 68: 562–565. [4]

Wright, S. 1934c. The method of path coefficients. *Ann. Math Stat.* 5: 161–215. [A2]

Wright, S. 1938. The distribution of gene frequencies in populations of polyploids. *Proc. Natl. Acad. Sci. USA.* 24: 372–377. [12, 13]

Wright, S. 1952. The genetics of quantitative variability, *in* E. C. R. Reeve and C. H. Waddington (eds.), *Quantitative inheritance*, pp. 5–4. Her Majesty's Stationary Office, London, UK. [16, 17]

Wright, S. 1965. The interpretation of population structure by *F*-statistics with special regard to systems of mating. *Evolution* 19: 395–420. [8]

Wright, S. 1968. *Evolution and the genetics of populations. I. Genetic and biometric foundations.* Univ. Chicago Press, Chicago, IL. [1, 5, 11, 14, 15, 16, A2]

Wright, S. 1969. *Evolution and the genetics of populations. II. The theory of gene frequencies.* Univ. Chicago Press, Chicago, IL. [12]

Wright, S. 1978. *Evolution and the genetics of populations. III. Experimental results and evolutionary deductions.* Univ. Chicago Press, Chicago, IL. [12]

Wright, S. 1983. On "Path analysis in genetic epidemiology: a critique." *Am. J. Hum. Genet.* 35: 757–768. [A2]

Wright, S. 1984. Diverse uses of path analysis. *In* A. Chakravarti (ed.), *Human population genetics: the Pittsburgh symposium*, pp. 1–34. Van Nostrand Reinhold, New York, NY. [A2]

WTCCC. See Wellcome Trust Case Control Consortium

Wu, C.-I., and A. T. Beckenbach. 1983. Evidence for extensive genetic differentiation between the sex-ratio and the standard arrangement of *Drosophila pseudoobscura* and *D. persimilis* and identification of hybrid sterility factors. *Genetics* 105: 71–86. [17]

Wu, C.-I., and A. W. Davis. 1993. Evolution of postmating reproductive isolation: the composite nature of Haldane's rule and its genetic bases. *Am. Nat.* 142: 187–212. [17]

Wu, C.-I., and M. F. Palopoli. 1994. Genetics of postmating reproductive isolation in animals. *Ann. Rev. Ecol. Syst.* 27: 283–308. [11, 13]

Wu, C. I., N. A .Johnson, and M. F. Palopoli. 1996. Haldane's rule and its legacy: Why are there so many sterile males? *Trends Ecol. Evo.* 11: 281–284. [17]

Wu, L. Y., and 4 others. 2005. Resampling methods to reduce the selection bias in genetic effect estimation in genome-wide scans. *BMC Genet.* 6: 1–5. [18, 20]

Wu, L. Y., L. Sun, and S. B. Bull. 2006. Locus-specific heritability estimation via the bootstrap in linkage scans for quantitative trait loci. *Hum. Hered.* 62: 84–96. [18]

Wu, M. C., and 5 others. 2011. Rare-variant association testing for sequencing data with sequence kernel association test. *Am. J. Hum. Genet.* 89: 82–93. [20]

Wu, T. T., and 4 others. 2009. Genome-wide association analysis by lasso penalized logistic regression. *Bioinformatics* 25: 714–721. [20]

Wu, W. R., and W.-M. Li. 1994. A new approach for mapping quantitative trait loci using complete genetic marker linkage maps. *Theor. Appl. Genet.* 89: 535–539. [18]

Wu, W. R., and W.-M. Li. 1996a. Model fitting and model testing in the method of joint mapping of quantitative trait loci. *Theor. Appl. Genet.* 92: 477–482. [18]

Wu, W. R., and W.-M. Li. 1996b. Joint mapping of quantitative trait locus using F_2 populations. *Theor. Appl. Genet.* 93: 1156–1160. [18]

Wu, X.-L., B. Heringstad, and D. Gianola. 2010. Bayesian structural equation models for inferring relationships between phenotypes: a review of methodology, identifiability, and applications. *J. Anim. Breed. Genet.* 127: 3–15. [A2]

Wu, Y., Z. Zhang, P. M. Visscher, and J. Yang. 2017. Quantifying the mapping precision of genome-wide association studies using whole-genome sequencing data. *Genome Biol.* 18: 1–10. [21]

Wu, Y., and 14 others. 2018. Integrative analysis of omics summary data reveals putative mechanisms underlying complex traits. *Nat. Comm.* 9: 918. [21]

Wu, Y., and 5 others. 2023. Joint analysis of GWAS and multi-omics QTL summary statistics reveals a large fraction of GWAS signals shared with molecular phenotypes. *Cell Genomics* 3: 100344. [21]

Wuchty, S. 2000. Scale-free behavior in protein domain networks. *Mol. Biol. Evol.* 18: 1694–1702. [A2]

Xiao, J., J. Li, L. Yuan, and S. D. Tanksley. 1995. Dominance is the major genetic basis of heterosis in rice as revealed by QTL analysis using molecular markers. *Genetics* 140: 745–754. [12, 13]

Xiao, R., and M. Boehnke. 2009. Quantifying and correcting for the winner's curse in genetic association studies. *Genet. Epidem.* 33: 453–462. [20]

Xiao, Y., and 27 others. 2016. Genome-wide dissection of the maize ear genetic architecture using multiple populations. *New Phyt.* 210: 1095–1106. [18, 21]

Xiao, Y., and 4 others. 2017. Genome-wide association studies in maize: praise and stargaze. *Molecular Plant* 10: 359–374. [21]

Xiao, Y., and 17 others. 2021. The genetic mechanism of heterosis utilization in maize improvement. *Genome Bio.* 22: 148. [13]

Xie, C., S. Jauhari, and A. Mora. 2021. Popularity and performance of bioinformatics software: the case of gene set analysis. *BMC Bioinformat.* 22: 191. [21]

Xie, F., S. Wang, W. D. Beavis, and S. Xu. 2021. Estimation of genetic variance contributed by a quantitative trait locus: correcting the bias associated with significance test. *Genetics* 219: iyab115. [20]

Xing, Y. Z., and 5 others. 2002. Characterization of the main effects, epistatic effects and their environmental interactions of QTLs on the genetic basis of yield traits in rice. *Theor. Appl. Genet.* 105: 248–257. [13]

Xiong, M., and S.-W. Guo. 1997. Fine-scale genetic mapping based on linkage disequilibrium: theory and applications. *Am. J. Hum. Genet.* 60: 1513–1531. [17]

Xiong, M., J. Zhao and E. Boerwinkle. 2002. Generalized T^2 test for genome association studies. *Am. J. Hum. Genet.* 70: 1257–1268. [20]

Xu, C., Y.-M. Zhang, and S. Zu. 2005. An EM algorithm for mapping quantitative resistance loci. *Heredity* 94: 119–128. [18]

Xu, C., X. Wang, Z. Li, and S. Xu. 2009. Mapping QTL for multiple traits using Bayesian statistics. *Genet. Res.* 91: 23–37. [18]

Xu, C., and 5 others. 2012. Multiple regression methods show great potential for rate variant association tests. *PLoS One* 7: e41694. [20]

Xu, L. Z., R. V. Craiu, and L. Sun. 2011. Bayesian methods to overcome the winner's curse in genetic studies. *Ann. Appl. Stat.* 5: 201–231. [20]

Xu, S. 1995. A comment on the simple regression method for interval mapping. *Genetics* 141: 1657–1659. [18]

Xu, S. 1996. Mapping quantitative trait loci using four-way crosses. *Genet. Res.* 68: 175–181. [18]

Xu, S. 1998a. Further investigation on the regression method of mapping quantitative trait loci. *Heredity* 80: 364–373. [18]

Xu, S. 1998b. Iteratively reweighted least squares mapping of quantitative trait loci. *Behav. Genet.* 28: 341–355. [18]

Xu, S. 1998c. Mapping quantitative tai loci using multiple families of line crosses. *Genetics* 148: 517–524. [18]

Xu, S. 2003. Theoretical basis of the Beavis effect. *Genetics* 165: 2259–2268. [18]

Xu, S. 2007. An empirical Bayes method for estimating epistatic effects of quantitative trait loci. *Biometrics* 63: 513–521. [18]

Xu, S. 2008. Quantitative trait locus mapping can benefit from segregation distortion. *Genetics* 180: 2201–2208. [18]

Xu, S. 2010. An expectation-maximization algorithm for the Lasso estimation of quantitative trait locus effects. *Heredity* 105: 483–494. [18]

Xu, S. 2013. Mapping quantitative trait loci by controlling polygenic background effects. *Genetics* 195: 1209–1222. [18]

Xu, S. 2022. *Quantitative genetics*. Springer, Cham, CH. [1]

Xu, S., and W. R. Atchley. 1995. A random model approach to interval mapping of quantitative trait loci.

Genetics 141: 1189–1197. [19]

Xu, S., and W. R. Atchley. 1996. Mapping quantitative trait loci for complex binary diseases using line crosses. *Genetics* 143: 1417–1424. [18]

Xu, S., and Z. Hu. 2009. Mapping quantitative trait loci using distorted markers. *Intern. J. Plant Genomics* 410825. [18]

Xu, S., and Z. Hu. 2010. Generalized linear model for interval mapping of quantitative trait loci. *Theor. Appl. Genet.* 121: 47–63. [18]

Xu, S., and Z. Jia. 2007. Genomewide analysis of epistatic effects for quantitative traits in barley. *Genetics* 175: 1955–1963. [18]

Xu, S., and C. Vogl. 2000. Maximum likelihood analysis of quantitative trait loci under selective genotyping. *Heredity* 84: 525–537. [17]

Xu, S., and 4 others. 2003. An EM algorithm for mapping binary disease loci: application to fibrosarcoma in a four-way cross mouse family. *Genet. Res.* 82: 127–138. [18]

Xu, X., S. Weiss, X. Xu, and L. J. Wei. 2000. A unified Haseman-Elston method for testing linkage with quantitative traits. *Am. J. Hum. Genet.* 67: 1025–1028. [19]

Xu, Z., C. Wu, P. Wei, and W. Pan. 2017. A powerful framework for integrating eQTL and GWAS summary data. *Genetics* 207: 893–902. [21]

Xue, Y., and 22 others. 2015. Mountain gorilla genomes revel the impact of long-term population decline and inbreeding. *Science* 348: 242–245. [12]

Yablokov, A. V. 1974. *Variability of mammals.* Amerind, New Delhi, IN. [14]

Yadav, V., and 6 others. 2018. Crossbreeding systems of livestock. *Pharma Innov. J.* 7: 8–13. [13]

Yaghoobi, J., I. Kaloshian, Y. Wen, and V. M. Williamson. 1995. Mapping a new nematode resistance locus in *Lycopersicon peruvianum. Theor. Appl. Genet.* 91: 457–464. [17]

Yalcin, B., J. Flint, and R. Mott. 2005. Using progenitor strain information to identify quantitative trait nucleotides in outbred mice. *Genetics* 171: 673–681. [18]

Yamaguchi, M., T. Yanase, H. Nagano, and N. Nakamoto. 1970. Effects of inbreeding on mortality in Fukuoka population. *Am. J. Hum. Genet.* 22: 145–155. [12]

Yamamoto, E., and 6 others. 2014. Effect of advanced intercrossing on genome structure and on the power to detect linked quantitative trait loci in a multi-parental population: a simulation study in rice. *BMC Genet.* 15: 50. [18]

Yampolsky, L. Y., and S. M. Scheiner. 1994. Developmental noise, phenotypic plasticity, and allozyme heterozygosity in *Daphnia. Evolution* 48: 1715–1722. [6]

Yan, Q., and 7 others. 2014. Kernel-machine testing coupled with a rank-truncated method for genetic pathway analysis. *Genet. Epidem.* 38: 447–456. [20]

Yang, C., and 5 others. 2018. CoMM: a collaborative mixed model to dissecting genetic contributions to complex traits by leveraging regulatory information. *Bioinformatics* 35: 1644–1652. [21]

Yang, F., and 4 others. 2017. Identifying *cis*-mediators for *trans*-eQTLs across many human tissues using genomic mediation analysis. *Genome Res.* 27: 1859–1871. [21]

Yang, J., N. R. Wray, and P. M Visscher. 2010a. Comparing apples and oranges: equating the power of case-control and quantitative trait association studies. *Genet. Epid.* 34: 254–257. [20]

Yang, J., and 11 others. 2010b. Common SNPs explain a large proportion of the heritability for human height. *Nat. Genet.* 42: 565–569. [8, 12, 20, 21]

Yang, J., P. M. Visscher, and N. R. Wray. 2010c. Sporadic cases are the norm for complex disease. *Eur. J. Hum. Genet.* 18: 1039–1044. [21]

Yang, J., and 21 others. 2011a. Genomic inflation factors under polygenic inheritance. *Eur. J. Hum. Genet.* 19: 807–812. [20]

Yang, J., and 23 others. 2011b. Genome partitioning of genetic variation for complex traits using common SNPs. *Nat. Genet.* 43: 519–525. [21]

Yang, J., S. H. Lee, M. E. Goddard, and P. M. Visscher. 2011c. GCTA: a tool for genome-wide complex trait analysis. *Am. J. Hum. Genet.* 88: 76–82. [8]

Yang, J., and 10 others. 2012a. FTO genotype is associated with phenotypic variability of body mass index. *Nature* 490: 267–272. [6, 20]

Yang, J., and 18 others. 2012b. Conditional and joint multiple-SNP analysis of GWAS summary statistics identified additional variants influencing complex traits. *Nat. Genet.* 44: 369–375. [20, 21]

Yang, J., and 4 others. 2014. Advantages and pitfalls in the application of mixed-model association methods. *Nat. Genet.* 46: 100–106. [20]

Yang, J., and 24 others. 2015. Genetic variance estimation with imputed variants find negligible missing heritability for human height and body mass index. *Nat. Genet.* 47: 1114–1120. [21].

Yang, J., and 7 others. 2017. Incomplete dominance of deleterious alleles contributes substantially to trait variation and heterosis in maize. *PLOS Genet.* 13: e1007019. [12]

Yang, R.-C. 2004. Epistasis of quantitative trait loci under different gene action models. *Genetics* 167: 1493–1505. [5]

Yang, R.-C., T. Z. Ye, S. F. Blade, and M. Bandara. 2004. Efficiency of spatial analysis of field pea variety trails. *Crop Sci.* 44: 49–55. [A9]

Yang, S., and 7 others. 2015. Parent–progeny sequencing indicates higher mutation rates in heterozygotes. *Nature* 523: 463–467. [15]

Yang, Y. H., and T. Speed. 2002. Design issues for cDNA microarray experiments. *Nat. Rev. Genet.* 3: 579–588. [A9]

Yang, Y., and 9 others. 2020. CoMM-S^2: a collaborative mixed model using summary statistics in transcriptome-wide association studies. *Bioinformatics* 36: 2009–2016. [21]

Yates, F. 1936a. A new method of arranging variety trails involving a larger number of varieties. *J. Agric. Sci.* 26: 424–455. [A9]

Yates, F. 1936b. Incomplete randomized blocks. *Ann. Eugen.* 7: 122–140. [A9]

Yates, F. 1940. The recovery of inter-block information in balanced incomplete block designs. *Ann. Eugen.* 10: 317–325. [A9]

Yates, F. 1964. Sir Ronald Fisher and the design of experiments. *Biometrics* 20: 307–321. [A9]

Yao, C., and 9 others. 2017. Dynamic role of *trans* regulation of gene expression in relation to complex traits. *Am. J. Hum. Genet.* 100: 571–580. [21]

Yao, D. W., L. J. O'Connor, A. L. Price, and A. Gusev. 2020. Quantifying genetic effects on diseases mediated by assayed gene expression levels. *Nat. Genet.* 52: 626–633. [21]

Yao, H., A. D. Gray, D. L. Auger, and J. A. Birchler. 2013. Genomic dosage effects on heterosis in triploid maize. *Proc. Natl. Acad. Sci. USA* 110: 2665–2669. [13]

Yao, H., and 5 others. 2020. Inbreeding depression in genotypically matched diploid and tetraploid maize. *Front. Genet.* 11: 564928. [13]

Yengo, L., and P. M. Visscher. 2018. Assortative mating on complex traits revisited: double first cousins and the X-chromosome. *Theor. Pop. Biol.* 124: 51–60. [7]

Yengo, L., and 6 others. 2017. Detection and quantification of inbreeding depression for complex traits from SNP data. *Proc. Natl. Acad. Sci. USA* 114: 8602–8607. [12]

Yengo, L., and 13 others. 2018a. Imprint of assortative mating on the human genome. *Nat. Hum. Behav.* 2: 948–954. [7]

Yengo, L., and 10 others. 2018b. Meta-analysis of genome-wide association studies for height and body mass index in $\sim$ 700,000 individuals of European ancestry. *Hum. Mol. Genet.* 27: 3641–3649. [21]

Yengo, L., and 6 others. 2018c. Estimation of inbreeding depression from SNP data. *Proc. Natl. Acad. Sci. USA* 115: E2494–E2495. [12]

Yengo, L., N. R. Wray, and P. M. Visccher. 2019. Extreme inbreeding in a European ancestry sample from the contemporary UK population. *Nat. Comm.* 10: 3719. [12]

Yengo, L., and 5 others. 2021. Genomic partitioning of inbreeding depression in humans. *Am. J. Hum. Genet.* 108: 1488–1501. [12]

Yengo, L., et al. 2022. A saturated map of common genetic variants associated with human height. *Nature* 610: 704–712. [21]

Yeung, K.-F., Y. Yang, C. Yang, and J. Liu. 2019. CoMM: A collaborative mixed model that integrates GWAS and eQTL data sets to investigate the genetic architecture of complex traits. *Bioinformatins Biol.* 13: 1–6. [21]

Yezerinac, S. M., S. C. Lougheed, and P. Handford. 1992. Morphological variability and enzyme heterozygosity: individual and population level correlations. *Evolution* 46: 1959–1964. [6]

Yi, N. 2004. A unified Markov Chain Monte Carlo framework for mapping multiple quantitative trait loci. *Genetics* 167: 967–975. [18]

Yi, N., and D. Shriner. 2008. Advances in Bayesian multiple quantitative trait loci mapping in experimental crosses. *Heredity* 100: 240–252. [18]

Yi, N., and S. Xu. 1999. A random model approach to mapping quantitative trait loci for complex binary traits in outbred populations. *Genetics* 153: 1029–1040. [19]

Yi, N., and S. Xu. 2000. Bayesian mapping of quantitative trait loci for complex binary traits. *Genetics* 155: 1391–1403. [18]

Yi, N., and S. Xu. 2001. Bayesian mapping of quantitative trait loci under complicating mating designs. *Genetics* 157: 1759–1771. [18]

Yi, N., and S. Xu. 2002a. Mapping quantitative trait loci with epistatic effects. *Genet. Res.* 79: 185–198. [18]

Yi, N., and S. Xu. 2002b. Linkage analysis of quantitative trait loci in multiple line crosses. *Genetica* 114: 217–230. [18]

Yi, N., and S. Xu. 2008. Bayesian LASSO for quantitative trait loci mapping. *Genetics* 179: 1045–1055. [18]

Yi, N., S. Xu, and D. B. Allison. 2003a. Bayesian model choice and search strategies for mapping interacting quantitative trait loci. *Genetics* 165: 867–883. [18]

Yi, N., V. George, and D. B. Allison. 2003b. Stochastic search variable selection for identifying multiple quantitative trait loci. *Genetics* 164: 1129–1138. [18]

Yi, N., and 5 others. 2007. An efficient Bayesian model selection approach for interacting quantitative trait loci models with many effects. *Genetics* 176: 1865–1877. [18]

Yi, N., and 4 others. 2015. Penalized multimarker vs. single -marker regression methods for genome-wide association studies for quantitative traits. *Genetics* 199: 205–220. [20]

Yoon, C. H. 1955. Homeostasis associated with heterozygosity in the genetics of time of vaginal opening in the house mouse. *Genetics* 40: 297–309. [6]

Yoshimaru, H., and T. Mukai. 1985. Relationships between the polygenes affecting the rate of development and viability in *Drosophila melanogaster. Jap. J. Genet.* 60: 307–334. [15]

Young, A. I. 2019. Solving the missing heritability problem. *PLoS Genet.* 15: e1008222. [21]

Young, A. I., and 9 others. 2018. Relatedness disequilibrium regression estimates heritability without environmental bias. *Nat. Genet.* 50: 1304–1310. [21]

Young, A. I., F. L. Wauthier, and P. Donnelly. 2022. Identifying loci affecting trait variability and detecting interactions in genome-wide association studies. *Nat. Genet.* 50: 1608–1614. [18 , 20]

Young, E. C. 1965. General development in British Corixidae. *Proc. Royal Entomol. Soc. Lond. A* 40: 159–168. [6]

Young, N. D., and S. D. Tanksley. 1989a. RFLP analysis of the size of chromosomal segments retained around the *Tm-2* locus of tomato during backcross breeding. *Theor. Appl. Genet.* 77: 353–359. [17]

Young, N. D., and S. D. Tanksley. 1989b. Restriction fragment length polymorphism maps and the concept of graphical genotypes. *Theor. Appl. Genet.* 77: 95–101. [17]

Young, N. D., D. Zamir, M. W. Ganal, and S. D. Tanksley. 1988. Use of isogenic lines and simultaneous probing to identify DNA markers tightly linked to the *Tm-2a* gene in tomato. *Genetics* 120: 579–585. [17]

Yu, J., and 11 others. 2006. A unified mixed-model method for association mapping that accounts for multiple levels of relatedness. *Nat. Genet.* 38: 203–208. [20]

Yu, J., J. B. Holland, M. D. McMullen, and E. S. Buckler. 2008. Genetic design and statistical power of nested association mapping in maize. *Genet.* 178: 539–551. [18, 20]

Yu, K., and 6 others. 2007. Flexible design for following up positive findings. *Am. J. Hum. Genet.* 81: 540–551. [20]

Yu, K., and 7 others. 2009. Pathway analysis by adaptive combination of p-values. *Genet. Epidem.* 33: 700–709. [20, 21]

Yu, S. B., and 7 others. 1997. Importance of epistasis as the genetic basis of heterosis in an elite rice hybrid. *Proc. Natl. Acad. Sci. USA* 94: 9226–9231. [13]

Yu, X., S. A. Knott, and P. M. Visscher. 2004. Theoretical and empirical power of regression and maximum-likelihood methods to map quantitative trait loci in general pedigrees. *Am. J. Hum. Genet.* 75: 17–26. [19]

Yu, Z. H., D. J. Mackill, J. M. Bonman, and S. D. Tanksley. 1991. Tagging genes for blast resistance in rice via linkage to RFLP markers. *Theor. Appl. Genet.* 81: 471–476. [17]

Yule, G. U. 1902. Mendel's laws and their probable relations to intra-racial heredity. *New Phytologist* 1: 193–207, 222–238. [1]

Yule, G. U. 1906. On the theory of inheritance of quantitative compound characters on the basis of Mendel's laws — a preliminary note. *Report of the Conference on Genetics* 140–142. [1]

Yun, L, and A. F. Agrawal. 2014. Variation in the strength of inbreeding depression across environments: effects of stress and density dependence. *Evol.* 68: 3599–3606. [12]

Yun, S., and 8 others. 2006. Validation of quantitative trait loci for multiple diseases resistance in barley using advanced backcross lines developed with a wild barley. *Crop Sci.* 46: 1179–1186. [18]

Yvert, G., and 7 others. 2003. *Trans*-acting regulatory variation in *Saccharomyces cerevisiae* and the role of transcription factors. *Nat. Genet.* 35: 57–64. [21]

Zabell, S. L. 2008. On Student's 1908 article "the probable error of a mean". *J. Amer. Stat. Assoc.* 103: 1–7. [2]

Zaghloul, N. A., and 11 others. 2010. Functional analyses of variants reveal a significant role for dominant negative and common alleles in oligogenic Bardet-Biedl syndrome. *Proc. Natl. Acad. Sci. USA* 107: 10602–10607. [15]

Zaidi, A. A., and I. Mathieson. 2020. Demographic history mediates the effect of stratification in polygenic scores. *eLife* 9: e61548. [20]

Zaitlen, N., and E. Eskin. 2010. Imputation aware meta-analysis of genome-wide association studies. *Genet. Epidem.* 4: 537–542. [20]

Zaitlen, N., and 47 others. 2012a. Informed conditioning on clinical covariates increases power in case-control association studies. *PLoS Genet.* 8: e1003032. [20]

Zaitlen, N., and 15 others. 2012b. Analysis of case-control association studies with known risk variants. *Bioinform.* 28: 1729–1737. [20]

Zaitlen, N., and 6 others. 2013. Using extended genealogy to estimate components of heritability for 23 quantitative and dichotomous traits. *PLoS Genet.* 9: e1003520. [20, 21]

Zakharov, V. M. 1981. Fluctuating asymmetry as an index of developmental homeostasis. *Genetica (Pol.)* 13: 241–256. [6]

Zakharov, V. M. 1992. Population phenogenetics: analysis of developmental stability in natural populations. *Acta Zool. Fennica* 191: 7–30. [6]

Zapata, C., and G. Alvarez. 1992. The detection of gametic disequilibrium between allozyme loci in natural populations of *Drosophila*. *Evolution* 46: 1900–1917. [5]

Zawistowski, M., and 5 others. 2010. Extending rare-variant resting strategies: analysis of noncoding sequence and impute genotypes. *Am. J. Hum. Genet.* 87: 604–617. [20]

Zaykin, D. V. 2004. Bounds and normalization of the composite linkage disequilibrium coefficient. *Genet. Epdiem.* 27: 252–257. [5]

Zaykin, D. V., L. A. Zhivotovsky, P. H. Westfall, and B. S. Weir. 2002. Truncated product method for combining *P*-values. *Genet. Epidem.* 22: 170–185. [20, A6]

Zaykin, D. V., and 4 others. 2007. Combining *p*-values in large-scale genomics experiments. *Pharmaceutical Stat.* 6: 217–226. [20, A6]

Zeegers, M., F. Rijsdijk, and P. Sham. 2004 Adjusting for covariates in variance components QTL linkage analysis. *Behav. Genet.* 34: 127–133. [19]

Zeggini, E., and J. P. A. Ioannidis. 2009. Meta-analysis in genome-wide association studies. *Parmacogenomics* 10: 191–201. [20]

Zeller, T., and 26 others. 2010. Genetics and beyond — the transcriptome of human monocytes and disease susceptibility. *PLoS One* 5: e10693. [21]

Zeng, B., and 14 others. 2019. Comprehensive multiple eQTL detection and its application to GWAS interpretation. *Genetics* 212: 905–918. [21]

Zeng, J., and 17 others. 2018. Signatures of negative selection in the genetic architecture of human complex traits. *Nat. Genet.* 50: 746–753. [21]

Zeng, J., and 12 others. 2021. Widespread signatures of natural selection across human complex traits and functional genomic categories. *Nat. Comm.* 12: 1164. [21]

Zeng, L.-W. 1996. Resurrecting Muller's theory of Haldane's rule. *Genetics* 143: 603–607. [17]

Zeng, Z.-B. 1992. Correcting the bias of Wright's estimates of the number ofgenes affecting a quantitative character: a further improved method. *Genetics* 131: 987–1001. [11]

Zeng, Z.-B. 1993. Theoretical basis for separation of multiple linked gene effects in mapping of quantitative trait loci. *Proc. Natl. Acad. Sci. USA* 90: 10972–10976. [18]

Zeng, Z.-B. 1994. Precision mapping of quantitative trait loci. *Genetics* 136: 1457–1468. [18]

Zeng, Z.-B., D. Houle, and C. C. Cockerham. 1990. How informative is Wright's estimator of the number of genes affecting a quantitative character? *Genetics* 126: 235–247. [11]

Zeng, Z.-B., C.-H. Kao, and C. J. Basten. 1999. Estimating the genetic architecture of quantitative traits. *Genet Res.* 74: 279–289. [18]

Zeng, Z.-B., T. Wang, and W. Zou. 2005. Modeling quantitative trait loci and interpretation of models. *Genetics* 169: 1711–1725. [5, 18]

Zerba, K., R. E. Ferrell, and C. F. Sing. 1996. Genotype-environment interaction: apolipoprotein E (*ApoE*) gene effects and age as an index of time and spatial context in the human. *Genetics* 143: 463–478. [17]

Zeyl, C., and J. A. G. M. DeVisser. 2001. Estimates of the rate and distribution of fitness effects of spontaneous mutation in *Saccharomyces cerevisiae*. *Genetics* 157: 53–61. [15]

Zhang, C., and 9 others. 2019. The genetic basis of inbreeding depression in potato. *Nat. Genet.* 51: 374–378. [12]

Zhang, F., and H.-W. Deng. 2010. Correcting form cryptic relatedness in population-based association studies of continuous traits. *Hum. Hered.* 69: 38–33. [20]

Zhang, F., W. Gu, M. E. Hurles, and J. R. Lupski. 2009. Copy number variation in human health, disease, and evolution. *Ann. Rev. Genom. Hum. Genet.* 10: 451–481. [8, 15]

Zhang, H., and N. Risch. 1996. Mapping quantitative-trait loci in humans by use of extreme concordant sib pairs: selected sampling by parental phenotypes. *Am. J. Hum. Genet.* 59: 951–957. [19]

Zhang, J., and D. O. Stam. 2014. The role of local ancestry adjustment in association studies using admixed populations. *Genet. Epidem.* 38: 502–515. [20]

Zhang, L., and 6 others. 2010a. Inheritance of seeds per silique in *Brassica napus* L. using joint segregation analysis. *Field Crops Res.*116: 58–67. [16]

Zhang, L., and 8 others. 2010b. Effects of missing marker and segregation distortion on QTL mapping in F_2 populations. *Theor. Appl. Genet.* 121: 1071–1082. [18]

Zhang, L., and 7 others. 2014. Small RNAs as important regulators for the hybrid vigour of super-hybrid rice. *J. Experimental Botany* 65: 5989–6002.

Zhang, M., and 4 others. 2005. Mapping multiple quantitative trait loci by Bayesian classification. *Genetics* 169: 2305–2318. [18]

Zhang, Q., and 4 others. 2015. Estimation of inbreeding using pedigree, 50k SNP chip genotypes and full sequence data in three cattle breeds. *BMC Genetics* 16: 88. [8]

Zhang, Q. S., J. Goudet, and B. S. Weir. 2022. Rank-invariant estimation of inbreeding coefficients. *Heredity* 128: 1–10. [8]

Zhang, R., A. W. Warrick, and D. E. Myers. 1994. Heterogeneity, plot shape effect and optimum plot size. *Geoderma* 62: 183–197. [A9]

Zhang, S., Z. Zhu, and H. Zhao. 2003. On a semiparametic test to detect associations between quantitative traits and candidate genes using unrelated individuals. *Genet. Epid.* 24: 44–56. [20]

Zhang, S., J. Cao, Y. M. Kong, and R. H. Scheuermann. 2010. GO-Bayes: a gene ontology-based overresentation analysis using a Bayesian approach. *Bioinformatics* 26: 905–911. [21]

Zhang, W., and 4 others, 2017. Tracing the origin and evolution of supergene mimicry in butterflies *Nat. Comm.* 8: 1-11. [15]

Zhang, W., and 9 others. 2018. PCA-based multiple-trait GWAS analysis: a powerful model for exploring pleiotropy. *Animals* 8: 239. [20]

Zhang, X., and 10 others. 2005. Polymorphisms in DNA base excision repair genes *ADPRT* and *XRCC1* and risk of lung cancer. *Cancer Res.* 65: 722–726. [18, 20]

Zhang, X., S. D. Bailey, and M. Lupien. 2014. Laying a solid foundation of Manhattan — 'setting the foundational basis for the post-GWAS era'. *Trends Genet.* 30: 140–149. [21]

Zhang, X.-S. 2008. increase in quantitative variation after exposure to environmental stress and/or introduction of a major mutation: $G \times E$ interaction and epistasis or canalization? *Genetics* 180: 687–695. [14]

Zhang, X. F., J. A. Mosjidis, and Z. L. Hu. 1992. Methods for detection and estimation of linkage between a marker locus and quantitative trait loci. *Plant Breeding* 109: 35–39. [18]

Zhang, Y., and J. Gai. 2002. An expansion of the joint segregation analysis of P_1, F_1, P_2, F_2, and $F_{(2:3)}$ generations in the mixed major gene plus polygenes inheritance analysis of quantitative traits. *J. Biomath.* 17: 363–368. [16]

Zhang, Y., W. Guan, and W. Pan. 2013. Adjustment for population stratification via principal components in association analysis of rare variants. *Genet Epidem.* 37: 99–109. [20]

Zhang, Y., G. Qui, J.-H. Park, and N. Chatterjee. 2018. Estimation of complex effect-size distribution using summary-level statistics from genome-wide association studies across 32 complex traits. *Nat. Genet.* 50: 1318–1326. [21]

Zhang, Y.-M., and S. Xu. 2004. Mapping quantitative trait loci in F_2 incorporating phenotypes of F_3 progeny. *Genetics* 166: 1981–1993. [18]

Zhang, Y.-M., and S. Xu. 2005. A penalized maximum likelihood method for estimating epistatic effects of QTL. *Heredity* 95: 96–104. [18]

Zhang, Y.-M., J.-Y. Gai, and Y.-H. Yang. 2003. The EIM algorithm in the joint segregation analysis of quantitative traits. *Genetics Res.* 81: 157–163. [16]

Zhang, Y.-M., and 5 others. 2005. Mapping quantitative trait loci using naturally occurring genetic variance among commercial inbred lines of maize (*Zea mays* L.) *Genetics* 169: 2267–2275. [18]

Zhang, Y.-M., H.-Y. Lü, and L.-L. Yao. 2008. Multiple quantitative trait loci Haseman-Elston regression using all markers on the entire genome. *Theor. Appl. Genet.* 117: 683–690. [19]

Zhang, Z., and 10 others. 2010. Mixed linear model approach adapted for genome-wide association studies. *Nat. Genet.* 42: 355–360. [20]

Zhang, Z., W. Wang, and W. Valdar. 2014. Bayesian modeling of haplotype effects in multiparent populations. *Genet.* 198: 139–156. [18]

Zhao, H. 2000. Family-based association studies. *Stat. Meth. Med. Res.* 9: 563–587. [17]

Zhao, H., and T. P. Speed. 1996. On genetic map functions. *Genetics* 142: 1369–1377. [17]

Zhao, H., H. Zhang, and J. I. Rotter. 1997. Cost-effective sib-pair designs in the mapping of quantitative-trait loci. *Am. J. Hum. Genet.* 60: 1211–1221. [19]

Zhao, K., and 10 others. 2007. An *Arabidopsis* example of association mapping in structured samples. *PLoS Genet.* 3: e4. [20]

Zheng, A., and 9 others. 2012. The SNP rs961253 in 20p12.3 is associated with colorectal cancer risk: a case-control study and a meta-analysis of the published literature. *PLoS One* 7: e34625. [20]

Zheng, G., B. Freidlin, Z. Li, and J. L. Gastwirth. 2005. Genomic control for associations studies under various genetic models. *Biometrics* 61: 186–192. [20]

Zheng, G., B. Freidlin, and J. L. Gastwirth. 2006. Robust genomic control for associations studies. *Am. J. Hum. Genet.* 78: 350–356. [20]

Zheng, H., L. Li, and G. Zhang. 2012. Inbreeding depression for fitness-related traits and purging the genetic load inthe hermaphroditic bay scallop *Argopecten irradians irradians* (Mollusca: Bivalvia). *Aquaculture* 366: 27–33. [12]

Zheng, J., and 7 others. 2017. Recent developments in Mendelian randomization studies. *Curr. Epidem. Rep.* 4: 330–345. [A2]

Zheng, Z., and 10 others. 2015. Fine mapping of a large-effect QTL conferring Fusarium crown rot resistance on the long arm of chromosome 3B in hexaploid wheat. *BMC Genomics* 16: 850. [17]

Zhong, H., and R. L. Prentice. 2008. Bias-reduced estimators and confidence intervals for odds ratio in genome-wide association studies. *Biostats.* 9: 621–634. [20]

Zhong, H., and R. L. Prentice. 2010. Correcting "winner's curse" in odds ratios from genomewide association finding for major complex human diseases. *Genet. Epidem.* 34: 78–91. [20]

Zhong, H., and 4 others. 2010. Integrating pathway analysis and genetics of gene expression for genome-wide association studies. *Am. J. Hum. Genet.* 86: 581–591. [21]

Zhou, G., and 8 others. 2012. Genetic composition of yield heterosis in an elite rice hybrid. *Proc. Natl. Acad. Sci. USA* 109: 15847–15852. [13]

Zhou, H., M. E. Sehl, J. S. Sinsheimer, and L. Lange. 2010. Association screening of common and rare genetic variants by penalized regression. *Bioinformatics* 26: 2375–2383. [20]

Zhou, W., and 18 others. 2018. Efficiently controlling for case-control imbalance and sample relatedness in large-scale genetic association studies. *Nat. Genet.* 50: 1335–1341. [20]

Zhou, X., and M. Stephens. 2012. Genome-wide efficient mixed-model analysis for association studies. *Nat. Genet.* 44: 821–824. [20]

Zhou, X., and M. Stephens. 2014. Efficient multivariate linear mixed model algorithms for genome-wise association studies. *Nat. Methods* 11: 407–409. [20]

Zhou, Y., and 16 others. 2022. Assembly of a pangenome for global cattle reveals missing sequences and novel structural variation, providing new insights into their diversity of evolutionary history. *Genome Res.* 32: 1585–1601. [13]

Zhu, X. 2020. Mendelian randomization and pleiotropy analysis. *Quant. Biol.* 1–11. [A2]

Zhu, X., and R. C. Elston. 2001. Transmission/disequilibrium tests for quantitative traits. *Genetic Epid.* 20: 57–74. [17]

Zhu, X., S. L. Zhang, H. Zhao, and R. S. Cooper. 2002. Association mapping, using a mixture model for complex traits. *Genet. Epid.* 23: 181–196. [20]

Zhu, X., R. S. Cooper, and R. C. Elston. 2004. Linkage analysis of a complex disease through the use of admixed populations. *Am. J. Hum. Genet.* 74: 1136–1153. [17]

Zhu, X., and 10 others. 2005. Admixture mapping for hypertension loci with genome-scan markers. *Nat. Genet.* 37: 177–181. [17]

Zhu, Z., and 17 others. 2015. Dominance genetic variation contributes little to the missing heritability for human complex traits. *Am. J. Hum. Genet.* 96: 377–385. [8, 21]

Zhu, Z., and 10 others. 2016. Integration of summary data from GWAS and eQTL studies predicts complex trait gene targets. *Nat. Genet.* 48: 481–487. [21]

Zhu, Z., and 10 others. 2018. Causal associations between risk factors and common diseases inferred from GWAS summary data. *Nat. Comm.* 19: 224. [A2]

Zhuchenko, A. A., A. P. Samovol, A. B. Korol, and V. K. Andryushchenko. 1978. Linkage between loci of quantitative characters and marker loci. I. Model. *Genetika* 14: 771–778. [18]

Zhuchenko, A. A., A. P. Samovol, A. B. Korol, and V. K. Andryushchenko. 1979. Linkage between loci of quantitative characters and marker loci. II. Influence of three tomato chromosomes on variability of five quantitative characters in backcross progeny. *Genetika* 15: 672–683. [18]

Ziegler, A., and I. R. König. 2006. *A statistical approach to genetic epidemiology*. Wiley-VCH. Weinheim, DE. [16]

Zimmerman, D. I., and D. A. Harville. 1991. A random field approach to the analysis of field-plot experiments and other spatial experiments. *Biometrics* 47: 223–239. [A9]

Zink, R. M., M. F. Smith, and J. L. Patten. 1985. Association between heterozygosity and morphological variance. *J. Heredity* 76: 415–420. [6]

Zöllner, S., and J. K. Pritchard. 2005. Coalescent-based association mapping and fine mapping of complex trait loci. *Genetics* 169: 1071–1092. [17]

Zöllner, S., and J. Pritchard. 2007. Overcoming the winners curse: estimating penetrance parameters for case-control data. *Am. J. Hum. Genet.* 80: 605–615. [18, 20]

Zondervan, K. T., and L. R. Cardon. 2007. Designing candidate gene and genome-wide case-control association studies. *Nat. Protocols* 2: 2492–2500. [20]

Zou, C., P. Wang, and Y. Xu. 2016. Bulked sample analysis in genetics, genomics and crop improvement. *Plant Biotech. J.* 14: 1941–1955. [17]

Zou, F., B. S. Yandell, and J. P. Fine. 2001. Statistical issues in the analysis of quantitative traits in combined crosses. *Genetics* 158: 1339–1346. [18]

Zou, F., F. Fine, J. Hu, and D. Lin. 2004. An efficient resampling method for assessing genome-wide statistical significance in mapping quantitative trait loci. *Genetics* 168: 2307–2316. [18]

Zou, F., and 6 others. 2005. Quantitative trait locus analysis using recombinant inbred intercrosses: theoretical and empirical considerations. *Genetics* 170: 1299–1311. [18]

Zou, F., S. Lee, M. R. Knowles, and F. A. Wright. 2010. Quantification of population structure using correlated SNPs by shrinkage principal components. *Hum. Hered.* 70: 9–22. [20]

Zou, G. A. 2004. modified Poisson regression approach to prospective studies with binary data. *Am. J. Epidem.* 159: 702–706. [12]

Zou, H., and T. Hastie. 2005. Regularization and variable selection via the elastic net. *J. Roy. Stat. Soc. B* 67: 301–320. [20]

Zou, J. Y., and 8 others. 2015. Genetic and socioeconomic study of mate choice in Latinos reveals novel assortment patterns. *Proc. Natl. Acad. Sci. USA* 112: 13621–13626. [7]

Zou, W., and Z.-B. Zeng. 2008. Statistical methods for mapping multiple QTL. *Int. J. Plant Genet.* 2008: 286561. [18]

Zou, W., and Z.-B. Zeng. 2009. Multiple interval mapping for gene expression QTL analysis. *Genetica* 137: 125–134. [21]

Zouros, E., and D. W. Foltz. 1987. The use of allelic isozyme variation for the study of heterosis. *Isozymes* 13: 1–59. [12]

Zouros, E., S. M. Singh, and H. E. Miles. 1980. Growth rate in oysters: an overdominant phenotype and its possible explanations. *Evolution* 34: 856–867. [6, 12]

Zouros, E., M. Romero-Dorey, and A. L. Mallet. 1988. Heterozygosity and growth in marine bivalves: further data and possible explanations. *Evolution* 42: 1332–1341. [12]

Zuk, O., E. Hechter, S. R. Sunyaev, and E. S. Lander. 2012. The mystery of missing heritability: genetic interactions create phantom heritability. *Proc. Natl. Acad. Sci. USA* 109: 1193–1198. [21]

Author Index

Organism and Trait Index

Subject Index